MW00632724

Neuroimmune Pharmacology

Neuroimmune Pharmacology

Edited by

Tsuneya Ikezu, M.D., Ph.D.
University of Nebraska Medical Center
Omaha, NE

and

Howard E. Gendelman, M.D.
University of Nebraska Medical Center
Omaha, NE

Topic Editors

Serge Przedborski, M.D., Ph.D.
Columbia University
New York, NY

Kalipada Pahan, Ph.D.
Rush University
Chicago, IL

Alexander V. Kabanov, Ph.D.
University of Nebraska Medical Center
Omaha, NE

Jonathan Kipnis, Ph.D.
University of Virginia
Charlottesville, VA

Managing Editor

Robin Taylor
University of Nebraska Medical Center
Omaha, NE

Tsuneya Ikezu, M.D., Ph.D.
Department of Pharmacology and Experimental Neuroscience
University of Nebraska Medical Center
Omaha, NE, 68198-5880
Email: tikezu@unmc.edu

Howard E. Gendelman, M.D.
Department of Pharmacology and Experimental Neuroscience
University of Nebraska Medical Center
Omaha, NE, 68198-5880
Email: hegendel@unmc.edu

ISBN-13: 978-0-387-72572-7 e-ISBN-13: 978-0-387-72573-4

Library of Congress Control Number: 2007940389

© 2008 Springer Science+Business Media, LLC
All rights reserved. This work may not be translated or copied in whole or in part without the written permission of the publisher (Springer Science+Business Media, LLC, 233 Spring Street, New York, NY-10013, USA), except for brief excerpts in connection with reviews or scholarly analysis. Use in connection with any form of information storage and retrieval, electronic adaptation, computer software, or by similar or dissimilar methodology now known or hereafter developed is forbidden.
The use in this publication of trade names, trademarks, service marks, and similar terms, even if they are not identified as such, is not to be taken as an expression of opinion as to whether or not they are subject to proprietary rights.

Printed on acid-free paper.

9 8 7 6 5 4 3 2 1

springer.com

Acknowledgments

This interdisciplinary work would not have seen fruition without the tireless, dedicated, and selfless support of one very important person, Robin Taylor. Robin is a driver of details. Her undaunted enthusiasm, drive for excellence and tirelessness inspires all of those who engage and know her. The links between editors, authors, administrators, and publishers have flown seamlessly under her watchful eye. As a managing editor, her skills remain without parallel.

Andrea Macaluso and all staff at Springer USA who are inspired leaders in publishing and medical biocommunication. Their leadership directed the need for such a comprehensive and interdisciplinary work but made themselves always available at even a moment's notice. We are proud to have linked with such talented individuals and ones so dedicated to the pursuit of excellence in all they do.

To our leaders and visionaries, John Gollan, Harold Maurer, Thomas Rosenquist, Clarence Ueda, and Carol Swarts, a simple thank you for all your support, through good times and bad, seems too simple. The gift of your ears and your thoughtful responses and unbridled determination to bring out only the best in us is appreciated beyond words.

To our chapter contributors and reviewers whose expertise and dedications were invaluable for completing the tasks at hand.

To Seiko Ikezu and Bonnie Bloch our partners in life journeys and best friends.

To Yohei and Michiko Ikezu and Soffia Gendelman, our parents, navigators, and visionary role models.

To Yumiko Aoyama for tireless support of family and research and Anthony Johnson for life lessons and mentorship.

To our children and sources of life joys, Clark and David Ikezu and Sierra and Adam Gendelman and Lesley Gendelman-Ehrenkranz and Miles Ehrenkranz.

We salute the *Journal of Neuroimmune Pharmacology* that provided the inspiration for this work. To Opendra (Bill) Narayan who passed away suddenly during the final completion of this work. Your mentorship, science, and personal contributions to us and to the field of neuroimmune pharmacology will be a lasting monument.

Tsuneya Ikezu
Howard E. Gendelman

Preface

In the past two decades, enormous strides have been made in our understanding of the relationships between inflammation, innate immune responses, adaptive immune responses, and degenerative human diseases. The developing information has mostly appeared in specialty journals that have dealt only with isolated aspects of these tightly related fields. As a result, contemporary scientists have had a difficult time finding sources, even in review articles, that provide an integrated picture. This volume, by assembling chapters that demonstrate the relationship between these historically separated fields, overcomes that difficulty. There are sections on immunology of the nervous system, diseases that result from immunological dysfunction, current therapeutic approaches, and prospects for the future. Overall, it integrates cutting-edge neuroscience, immunology, pharmacology, neurogenetics, neurogenesis, gene therapy, adjuvant therapy, proteomics, and magnetic resonance imaging. It is a rich harvest and readers will gain a perspective that has not previously been so readily available. Exposure to such a wealth of ideas is bound to inspire readers to undertake new and productive research initiatives.

The modern era of research into neuroinflammation and its relationship to neurodegenerative diseases began in the 1960's with the elaboration by Ralph van Furth of the monocyte phagocytic system. He injected labeled monocytes into animals and followed their migration and maturation into resident phagocytes in all body tissues. This provided closure between Metchnikoff's 1882 discovery of mesodermal attack cells in starfish larvae, which he named phagocytes, and del Rio Hortega's 1919 discovery of phagocytic mesodermal cells entering the brain, which he named microglia. Hortega's results had always been questioned and for more than two decades the controversy continued as to whether microglia were truly phagocytes of mesodermal origin or were merely typical brain cells of epidermal origin. Were they truly the effecter cells in brain inflammation or were they merely housekeeping cells with an as yet undefined role? Resolving the controversy required development of the techniques of immunohistochemistry and monoclonal antibody production. These tools for exploring brain biochemistry at the cellular level opened new vistas for understanding brain functioning and the pathogenesis of human disease. Using these tools, our laboratory and that of Joseph Rogers in Sun City demonstrated that HLA-DR was strongly expressed on activated microglia. The identification of HLA-DR, a well-known leukocyte marker displayed by antigen presenting cells, on these cells vindicated both Hortega and van Furth. The way was paved for many productive investigations exploring the properties of microglial cells and their relationship to inflammation and immune responses. This example of a conjunction between a fundamental concept and technical advances to establish its validity has been repeated many times since, as the chapters in this volume illustrate.

For a time, the concept that the brain is immunologically privileged held sway amongst neuroscientists. This was based on a narrow view that only the invasion of brain by lymphocytes could be taken as evidence of an inflammatory response. Since lymphocytes and antibodies are relatively restricted in their ability to cross the blood brain barrier, the brain was supposedly isolated from self-attack. But immunochemistry, coupled with newly developed molecular biological techniques, revealed that a spectrum of inflammatory mediators, including many known to cause tissue damage, were produced within the brain by resident brain cells. These discoveries required entirely new interpretations as to the nature of neuroinflammation and its relationship to immune responses. The innate immune system, operating at the local level in the brain, has clearly proved to be the first line of defense. Indeed, the basic discoveries from studying the response of brain in a variety of neurological diseases, is causing a reevaluation of a number of peripheral degenerative disorders where innate immune responses, which had previously been ignored, have been shown to play a critical role in their pathogenesis. In other words, those studying the brain are providing immunologists with revolutionary new concepts regarding classical peripheral diseases. The insights of Part 1 of this volume need to be interpreted in this broader context.

Part 2 moves from the general to the specific. Individual neurological disorders and details of their pathogenesis are presented. They involve disorders where innate immune responses predominate, as in Alzheimer's disease, to others such as multiple sclerosis, where adaptive immune responses predominate, and still others, which seem to involve both. We have suggested that diseases involving self-damage generated by innate immune responses be defined as autotoxic to differentiate them from classical autoimmune diseases where self damage is generated by adaptive immune responses. The common theme, however, is the involvement of microglia as the effecter cells.

Part 2 also deals with neurogenetics. No field of neuroscience is moving so rapidly. The methodology for linking familial disease to DNA mutations commenced in the late 70's through identification of restriction fragment linked polymorphisms. By 1983, when James Gusella and his colleagues demonstrated a linkage of the G8 fragment to Huntington disease only about 18 markers were known. Now over 500,000 single nucleotide polymorphisms have been localized so that every centimorgan of the human genome can be explored. This advance has been coupled with rapid methods for sequencing DNA. The section on Genetics must be regarded as the tiny tip of a giant iceberg where much below the surface will soon be revealed.

The ultimate objective of neuroscientists studying human disease is to find more effective treatments. Part 3 covers the pharmacology of existing drugs, as well as describing approaches now in clinical evaluation, and those still at the bench level. Some of these include concepts that depart from established therapeutic approaches giving the reader much food for thought.

To complete the picture, there is a final chapter on imaging of the brain. The brain is inaccessible, and imaging provides a view of how it functions in vivo and how it is affected by neurological disease. Most importantly, it now provides methods for objectively measuring the effects of therapeutic agents in diseases where progressive brain degeneration occurs.

In summary, this is a volume not to be put on the shelf as a reference text, but to be read cover to cover by aspiring neuroscientists.

Dr. Patrick L. McGeer
Professor Emeritus
Kinsmen Laboratory of Neurological Research
University of British Columbia, Canada

Contents

Part 2: Nervous System Diseases and Immunity: Clinical Descriptions

Contributors

Jinsy A. Andrews, M.D.
Clinical Fellow
Eleanor and Lou Gehrig MDA/
ALS Research Center
Department of Neurology
Columbia University
New York, NY, USA

Stanley H. Appel, M.D.
Professor and Chair
Department of Neurology
Methodist Neurological Institute
Houston, TX, USA

Alison E. Baird, FRACP, Ph.D.
Stroke Neuroscience Unit
National Institute of Neurological Disease and Stroke
National Institutes of Health
Bethesda, MD, USA

William A. Banks, M.D.
Professor
Department of Internal Medicine
Veterans Affair Medical Center-St. Louis
Saint Louis University School of Medicine
St. Louis, MO, USA

Elena V. Batrakova, Ph.D.
Research Assistant Professor
Department of Pharmaceutical Sciences
Center for Drug Delivery and Nanomedicine
College of Pharmacy
University of Nebraska Medical Center
Omaha, NE, USA

Richard A. Bessen, Ph.D.
Associate Professor
Department of Veterinary Molecular Biology
Montana State University
Bozeman, MT, USA

Keshore R. Bidasee, Ph.D.
Associate Professor
Department of Pharmacology and Experimental Neuroscience
University of Nebraska Medical Center
Omaha, NE, USA

Michael D. Boska, Ph.D.
Professor
Department of Radiology
University of Nebraska Medical Center
Omaha, NE, USA

William J. Bowers, Ph.D.
Associate Professor
Department of Neurology
Center for Aging and Developmental Biology
University of Rochester School of Medicine and Dentistry
Rochester, NY, USA

David B. Bylund, Ph.D.
Professor
Department of Pharmacology and Experimental Neuroscience
University of Nebraska Medical Center
Omaha, NE, USA

Carl B. Camras, M.D.
Professor and Chair
Department of Ophthalmology and Visual Sciences
University of Nebraska Medical Center
Omaha, NE, USA

Pawel S. Ciborowski, Ph.D.
Assistant Professor
Department Pharmacology and Experimental Neuroscience
University of Nebraska Medical Center
Omaha, NE, USA

Subhajit Dasgupta, Ph.D.
Research Associate
Department of Oral Biology
College of Dentistry
University of Nebraska Medical Center
Lincoln, NE, USA

Jean D. Deupree, Ph.D.
Professor
Department of Pharmacology and Experimental Neuroscience
University of Nebraska Medical Center
Omaha, NE, USA

Toby K. Eisenstein, Ph.D.
Professor and Co-Director
Center for Substance Abuse Research
Department of Microbiology and Immunology
Temple University School of Medicine
Philadelphia, PA, USA

Nathan Erdmann, B.S.
Graduate Student
Department of Pharmacology and Experimental Neuroscience
University of Nebraska Medical Center
Omaha, NE, USA

Howard J. Federoff, M.D., Ph.D.
Professor
Departments of Neurology and
Microbiology and Immunology
Center for Aging and Developmental Biology
University of Rochester School of Medicine and Dentistry
Rochester, NY, USA

Mary L. Filipi, M.S.N., Ph.D.
Assistant Professor
Department of Neurological Sciences
University of Nebraska Medical Center
Omaha, NE, USA

Kazunori Fugo, M.D. Ph.D.
Visiting Fellow
Viral Immunology Section
Neuroimmunology Branch
National Institute of Neurological Disorders and Strokes
National Institutes of Health
Bethesda, MD, USA

Howard E. Gendelman, M.D.
Professor and Chair
Department of Pharmcology and Experimental Neuroscience
University of Nebraska Medical Center
Omaha, NE, USA

Seymour Gendelman, M.D.
Clinical Professor
Department of Neurology
Mount Sinai Medical Center
New York, NY, USA

Anuja Ghorpade, Ph.D.
Professor
Vice Chair Cell Biology and Genetics
Director Brain Bank
University of North Texas Health Science Center
Fortworth, TX, USA

Ralf Gold, M.D.
Professor and Chair
Department of Neurology
St. Josef Hospital
Ruhr University
Bochum, Germany

Marcia N. Gordon, Ph.D.
Professor
Department of Molecular Pharmacology and Physiology
University of South Florida
Tampa, FL, USA

Paul H. Gordon, M.D.
Assistant Professor
Eleanor and Lou Gehrig MDA/ALS Research Center
Department of Neurology
Columbia University
New York, NY, USA

Christian W. Grant, Ph.D.
Fellow
Viral Immunology Section
Neuroimmunology Branch
National Institute of Neurological Disorders and Stroke
National Institutes of Health
Bethesda, MD, USA

G. Jean Harry, Ph.D.
Head, Neurotoxicology Group
Laboratory of Neurobiology
National Institute of Environmental Health Sciences
National Institutes of Health
Research Triangle Park, NC, USA

Kathleen M. Healey, M.S.N., Ph.D.
Nurse Practitioner
Department of Neurological Sciences
University of Nebraska Medical Center
Omaha, NE, USA

Terry D. Hexum, Ph.D.
Professor
Department of Pharmacology and Experimental Neuroscience
University of Nebraska Medical Center
Omaha, NE, USA

Yunlong Huang, Ph.D.
Post-Doctoral Research Associate
Department of Pharmacology and Experimental Neuroscience
University of Nebraska Medical Center
Omaha, NE, USA

Tsuneya Ikezu, M.D., Ph.D.
Associate Professor and Vice Chair for Research
Department of Pharmacology and Experimental Neuroscience
University of Nebraska Medical Center
Omaha, NE, USA

Steven Jacobson, Ph.D.
Chief
Viral Immunology Section,
Neuroimmunology Branch
National Institute of Neurological Disorders and Stroke
National Institutes of Health
Bethesda, MD, USA

Malabendu Jana, Ph.D.
Research Associate
Department of Oral Biology
College of Dentistry
University of Nebraska Medical Center
Lincoln, NE, USA

Michelle C. Janelsins, M.S.
Graduate Student
Department of Microbiology and Immunology
Center for Aging and Developmental Biology
University of Rochester School of Medicine and Dentistry
Rochester, NY, USA

Clinton Jones, Ph.D.
Professor
Department of Veterinary-Biomedical Sciences
University of Nebraska-Lincoln
Lincoln, NE, USA

Alexander V. Kabanov, Ph.D.
Professor
Department of Pharmaceutical Sciences
Center for Drug Delivery and Nanomedicine
University of Nebraska Medical Center
Omaha, NE, USA

David E. Kaminsky, B.S.
Center for Substance Abuse Research
Department of Microbiology and Immunology
Temple University School of Medicine
Philadelphia, PA, USA

Jonathan Kipnis, Ph.D.
Assistant Professor
Department of Neuroscience
University of Virginia
Charlottesville, VA, USA

Dennis L. Kolson, M.D., Ph.D.
Associate Professor
Department of Neurology
University of Pennsylvania School of Medicine
Philadelphia, PA, USA

Qingzhong Kong, Ph.D.
Assistant Professor
Department of Pathology
Case Western Reserve University
Cleveland, OH, USA

Leila Kump, M.D.
Chief Resident
Department of Ophthalmology
University of Nebraska Medical Center
Omaha, NE, USA

Helmar C. Lehmann, M.D.
Resident
Department of Neurology
Heinrich-Heine University
Duesseldorf, Germany

M. Patricia Leuschen, Ph.D.
Associate Professor
Departments of Genetics, Cell Biology and Anatomy; Neurological Sciences
University of Nebraska Medical Center
Omaha, NE, USA

Miguel G. Madariaga, M.D.
Assistant Professor
Department of Internal Medicine
University of Nebraska Medical Center
Omaha, NE, USA

Eyal Margalit, M.D., Ph.D.
Associate Professor
Chief of Retina Service
Department of Ophthalmology
University of Nebraska Medical Center
Omaha, NE, USA

Clyde E. Markowitz, M.D.
Assistant Professor
Department of Neurology
Hospital of the University of Pennsylvania
Philadelphia, PA, USA

Eliezer Masliah, M.D.
Professor
Department of Neurosciences and Pathology
University of California San Diego
San Diego, CA, USA

Richard J. Miller, Ph.D.
Professor
Department of Molecular Pharmacology
and Biological Chemistry
Northwestern University School of Medicine
Chicago, IL, USA

Dave Morgan, Ph.D.
Professor
Department of Molecular Pharmacology and Physiology
Director of Basic Neuroscience Research
Director, Alzheimer Research Laboratory
University of South Florida
Tampa, FL, USA

R. Lee Mosley, Ph. D.
Assistant Professor
Department of Pharmacology and Experimental Neuroscience
University of Nebraska Medical Center
Omaha, NE, USA

Norbert Müller, M.D.
Professor
Hospital for Psychiatry and Psychotherapy
Ludwig-Maximilians-Universität
München, Germany

L. Charles Murrin, Ph.D.
Professor
Department of Pharmacology and Experimental Neuroscience
University of Nebraska Medical Center
Omaha, NE, USA

Prasad R. Padala, M.D.
Assistant Professor
Department of Psychiatry
University of Nebraska Medical Center
VA Medical Center
Omaha, NE, USA

Kalipada Pahan, Ph.D.
Professor
Department of Neurological Sciences
Rush University Medical Center
Chicago, IL, USA

Hui Peng, M.D.
Instructor
Department of Pharmacology and Experimental Neuroscience
University of Nebraska Medical Center
Omaha, NE, USA

Yuri Persidsky, M.D., Ph.D.
Professor
Departments of Pathology and Microbiology/
Pharmacology and Experimental Neuroscience
University of Nebraska Medical Center
Omaha, NE, USA

Frederick Petty, M.D., Ph.D.
Professor
Department of Psychiatry
Creighton University
VA Medical Center
Omaha, NE, USA

Larisa Y. Poluektova, M.D., Ph.D.
Assistant Professor
Department of Pharmacology and Experimental Neuroscience
University of Nebraska Medical Center
Omaha, NE, USA

Murali Prakriya, Ph.D.
Assistant Professor
Department of Molecular Pharmacology and
Biological Chemistry
Northwestern University
Chicago, IL, USA

Serge Przedborski, M.D., Ph.D.
Professor and Chair
Departments of Neurology, Pathology, and Cell Biology
Center for Neurobiology and Behavior
Center for Motor Neuron Biology and Diseases
Columbia University
New York, NY, USA

Rahil T. Rahim, Ph.D.
Post Doctoral Fellow
Fels Institutes for Cancer Research and Molecular Biology
Center for Substance Abuse Research
Temple University School of Medicine
Philadelphia, PA, USA

Sriram Ramaswamy, M.D.
Assistant Professor
Department of Psychiatry
Creighton University
Omaha, NE, USA

Ashley D. Reynolds, B.S.
Graduate Student
Department of Pharmacology and Experimental Neuroscience
University of Nebraska Medical Center
Omaha, NE, USA

Thomas J. Rogers, Ph.D.
Professor
Center for Substance Abuse Research
Department of Pharmacology
Fels Institutes for Cancer Research and Molecular Biology
Temple University School of Medicine
Philadelphia, PA, USA

Wojciech Rozek, Ph.D.
Post Doctoral Research Associate
Departments of Biochemistry and Molecular Biology; and Virology
National Veterinary Research Institute
Pulawy, Poland

Ramendra N. Saha, Ph.D.
Department of Oral Biology
College of Dentistry
University of Nebraska Medical Center
Lincoln, NE, USA

Sam Sanderson, Ph.D.
Associate Professor
School of Allied Health Professions
University of Nebraska Medical Center
Omaha, NE, USA

Eric Scholar, Ph.D.
Professor
Department of Pharmacology and Experimental Neuroscience
University of Nebraska Medical Center
Omaha, NE, USA

Michal Schwartz, Ph.D.
Professor
Department of Neurobiology
Weizmann Institute of Science
Rehovot, Israel

Markus J. Schwarz, M.D.
Privatdozent
Hospital for Psychiatry and Psychotherapy Ludwig-Maximilians-Universität
München, Germany

Kazim A. Sheikh, M.D.
Associate Professor
Department of Neurology
Johns Hopkins University School of Medicine
Baltimore, MD, USA

Ericka P. Simpson, M.D.
Assistant Professor
Department of Neurology
Methodist Neurological Institute
Houston, TX, USA

Michael Slifer, M.D.
Assistant Professor
Miami Institute for Human Genomics
University of Miami
Miami, FL, USA

Samantha S. Soldan, Ph.D.
Department of Neurology
University of Pennsylvania School of Medicine
Philadelphia, PA, USA

David K. Stone, M.S.
Department of Pharmacology and Experimental Neuroscience
University of Nebraska Medical Center
Omaha, NE, USA

Susan Swindells, M.B.B.S.
Professor
Department of Internal Medicine
University of Nebraska Medical Center
Omaha, NE, USA

Kang Tang, M.D.
Research Associate
Department of Pharmacology and Experimental Neuroscience
University of Nebraska Medical Center
Omaha, NE, USA

Mark P. Thomas, Ph.D.
Assistant Professor
School of Biological Sciences
University of Northern Colorado
Greeley, CO, USA

Wallace B. Thoreson, Ph.D.
Professor
Department of Ophthalmology and Visual Sciences
University of Nebraska Medical Center
Omaha, NE, USA

Myron Toews, Ph.D.
Professor
Department of Pharmacology and Experimental Neuroscience
University of Nebraska Medical Center
Omaha, NE, USA

Klaus V. Toyka, FRCP/FAAN
Professor and Chair
Unit for Multiple Sclerosis / Neuroimmunology
Department of Neurology
Julius Maximilians University
Würzburg, Germany

Deven Tuli, M.D.
Glaucoma Fellow
Department of Ophthalmology
and Visual Sciences
University of Nebraska Medical Center
Omaha, NE, USA

Jeffery M. Vance, M.D., Ph.D.
Professor and Chair
Division of Human Genetics
Director of Center of Genomic Medicine
Miami Institute for Human Genomics
University of Miami
Miami, FL, USA

Matthew L. White, M.D.
Associate Professor
Department of Radiology
University of Nebraska Medical Center
Omaha, NE, USA

Nicholas Whitney, B.S.
Department of Pharmacology and Experimental Neuroscience
University of Nebraska Medical Center
Omaha, NE, USA

Gregory F. Wu, M.D., Ph.D.
Department of Neurology
University of Pennsylvania School of Medicine
Philadelphia, PA, USA

Huangui Xiong, M.D., Ph.D.
Associate Professor
Department of Pharmacology and Experimental Neuroscience
University of Nebraska Medical Center
Omaha, NE, USA

Albert A. Yen, M.D.
Physician
Department of Neurology
Methodist Neurological Institute
Houston, TX, USA

Jialin Zheng, M.D.
Associate Professor
Department of Pharmacology and Experimental Neuroscience
University of Nebraska Medical Center
Omaha, NE, USA

Abbreviations

Neuroimmune Pharmacology Abbreviations

^{1}H	proton
^{1}H-MRSI	proton magnetic resonance spectroscopic imaging
2D SDS-PAGE	two-dimensional polyacrylamide gel electrophoresis
3'UTR	3'-untranslated region
3HK	3-hydroxykynurenine
5-ASA	5-aminosalicyclic acid
5-HIAA	5-hydroxyindole acetic acid
5-HT	5-hydroxy tryptophan
5-HT	5-hydroxytryptamine
6-OHDA	6-hydoxydopamine
8-OHdG	8-hydroxy-2'-deoxyguanosine
γc	common γ-chain
AAV	adeno-associated virus
Ab	antibody
Aβ	amyloid-β
ABP	actin-binding protein
AC	anterior chamber
ACAID	anterior chamber associated immune deviation
αCamKII	α-calcium/calmodulin-dependent protein kinase II
ACh	acetylcholine
AChE	acetylcholinesterase
ACTH	adrenocorticotrophic hormone (pro-opiomelanocortin; POMC)
AD	Alzheimer's disease
ADAM	a disintegrin and metalloprotease
ADCC	antibody-dependent cellular cytotoxicity
ADEM	acute disseminated encephalomyelitis
AF	activation factor
Ag	antigens
AGE	advanced glycation endproducts
AGM	aorta-gonad mesonephros
AHSCT	autologous hematopoietic stem cell transplantation
AICA	anterior inferior cerebellar artery
AID	activation-induced (cytidine) deaminase
AIDP	acute inflammatory demyelinating polyradiculoneuropathy
AIDS	acquired immunodeficiency syndrome
AIF	apoptosis inducing factor
AIRE	autoimmune regulator
AIS	anterior chamber associated immune deviation (ACAID)-inducing signal
ALS	amyotrophic lateral sclerosis
AMAN	acute motor axonal neuropathy
AMN	adrenomyeloneuropathy
AMPA	α-amino-3-hydroxy-5-methylisoxazole-4-propionic acid
AMSAN	acute motor-sensory axonal neuropathy
ANS	autonomic nervous system
AP-1	activating protein-1
APAF 1	apoptotic peptidase activating factor 1
APC	antigen presenting cells
aPKC	atypical protein kinase C
aPL	antiphospholipid antibody
APO1	apoptosis antigen 1 (Fas/CD95)
apoE	apolipoprotein E
APP	β-amyloid precursor protein
APPs	acute phase proteins
ARE	AU-rich response element
ARMD	age-related macular degeneration
ART	antiretroviral therapy
ASL	arterial spin-labeled
ASTIN	acute stroke therapy by inhibition of neutrophils
AT	adoptive transfer
ATL	adult T cell leukemia
AVP	arginine vasopressin
AZT	3'-azidothymidine/zidovudine

BACE	beta-site amyloid precursor protein cleaving enzyme
BBB	blood-brain barrier
BCR	B-cell receptor
BCRP	breast cancer resistance protein
BDNF	brain-derived neurotrophic factor
bFGF	basic fibroblast growth factor (FGF2)
bHLH	basic helix-loop-helix
BHV-1	bovine herpesvirus-1
BLIMP-1	B lymphocyte-induced maturation protein-1
BLV	bovine leukemia virus
BM	bone marrow
BMP	bone morphogenetic protein
BMVEC	brain microvascular endothelial cell
BrdU	bromodeoxyuridine
BRM's	biological response modifiers
BSE	bovine spongiform encephalopathy
BTLA	B and T lymphocyte attenuator
C/EBP	CCAAT box/ enhancer binding protein
C3d	complement C3 fragment d
C4d	complement C4 fragment d
CA	cornu ammonis
CaMKII	calcium/calmodulin-dependent protein kinase II
CAMs	cell adhesion molecules
CARD15	NOD2/caspase recruitment domain 15
CB	cannabinoid
CB1	cannabinoid receptor 1
CB2	cannabinoid receptor 2
CBA	cytokine bead arrays
CBF	cerebral blood flow
CBF1	C promoter binding factor 1
CBP	cAMP-response element binding protein (CREB)-binding protein
CBV	cerebral blood volume
CCT	central corneal thickness
CD	cluster of differentiation
CD11b	complement component 3 receptor 3 subunit (integrin alpha M; ITGAM)
CD40L	CD40 ligand (TNFSF5)
CDR	complementarity-determining region
CDV	canine distemper virus
CFT	2-β-carbomethoxy-3β-(4-fluorophenyl) tropane
cGMP	cyclic guanosine 5'-monophospate
CGRP	calcitonin-gene related peptide
CHAT	choline acetyltransferase
CHN	congenital hypomyelinating neuropathy
Cho	choline
CIDP	chronic inflammatory demyelinating polyradiculoneuropathy
CINC1	cytokine-induced neutrophil chemoattractant-1
CJD	Creutzfeldt-Jakob disease
CLA	cutaneous leukocyte antigen
CLP	common lymphoid progenitor
CMAP	compound motor action potential
CMC	critical micelle concentration
CMP	common myeloid precursor
CMT	Charcot-Marie-Tooth (disease)
CMV	cytomegalovirus
CNG	cGMP-gated
CNPase	cyclic nucleotide 3' phospohydrolase
CNS	central nervous system
CNTF	ciliary neurotrophic factor
CO	cytochrome oxidase
COMT	catechol-O-methyltransferase
Con A	concanavalin A
COP-1	copolymer-1
COX	cyclooxygenase (prostaglandin-endoperoxide synthase; PTGS)
CR	complement receptor
CRalBP	cellular retinal binding protein (retinaldehyde binding protein 1; RLBP1)
CRE	cAMP-responsive element
Cre	creatine
CREB	cAMP-response element binding protein
CRH	corticotrophin-releasing hormone
CRP	C-reactive protein
CRVO	central retinal vein occlusion
CSF	cerebrospinal fluid
CSF-1R	colony stimulating factor receptor
CSPG	chondroitin sulfate proteoglycans
CT	computed tomography
CTA	computed tomographic angiography
cTEC	cortical thymic epithelial cell (TEC)
CTL	cytotoxic T lymphocyte
CTLA-4	cytotoxic T lymphocyte antigen-4
CVO	circumventricular organ
Cx	connexin
cyto c	cytochrome c
D1R	type-1 family of dopamine receptors
D^9-THC	D^9-tetrahydrocannabinol
DA	Daniel's strain of Theiler's virus
DA	dopamine
DAF	decay-accelerating factor
DAG	diacylglycerol
DAT	dopamine transporter (solute carrier family 6A3; SLC6A3)
DβH	dopamine-β-hydroxylase
DC	dendritic cells
DCX	doublecortin
ddC	dideoxycytidine
ddI	dideoxyinosine
DG	dentate gyrus
dGTP	deoxyguanosine triphosphate

DHA	docosahexaenoic acid
Dhh	desert hedgehog
DIGE	DIfference Gel Electrophoresis
DISC	death-inducing signaling complex
DN	double/dominant negative
Doc2	double C2 protein
DOR	delta-opioid receptor
Dox	doxorubicin
Dox	doxycycline
DRPLA	dentatorubral-pallidoluysian atrophy
DSI	depolarization induced suppression of inhibition
DSPN	distal sensory peripheral neuropathy
DSS	Dejerine-Sottas syndrome
DTH	delayed-type hypersensitivity
DTI	diffusion tensor imaging
DWI	diffusion weighted imaging
E2F	early–region-2 transcription factor
EAE	experimental allergic/autoimmune encephalomyelitis
EAN	experimental allergic/autoimmune neuritis
EAU	experimental autoimmune uveitis
EBV	Epstein-Barr virus
ECT	electroconvulsive therapy
EDSS	expanded disability status scale
EEG	electroencephalography
EGC	embryonic germ cell
EGF	epidermal growth factor
EGFR	epidermal growth factor receptor
ELAVL4	embryonic lethal, abnormal vision, Drosophila-like 4
eLTP	early long-term potentiation
EMG	electromyography
EndoG	endonuclease G
EOAD	early-onset Alzheimer's disease
EPI	echo-planar imaging
EPSP	excitatory postsynaptic potential
ER	endoplasmic reticulum
ERK2	extracellular signal-regulated kinase 2
ESC	embryonic stem cell
ESI-MS/MS	electrospray ionization-mass spectrometry/mass spectrometry
ET	endothelin
ETP	early T lineage progenitor
FA	fractional anisotropy
FAD	familial Alzheimer's disease
fALS	familial amyotrophic lateral sclerosis
FasL	Fas ligand (FASLG)
FcgR-1	Fc receptor, IgG, high affinity-1
fCJD	familial Creutzfeldt-Jakob disease
FcR	Fc receptor
FDC	follicular dendritic cell
FDG	fluorodeoxyglucose
FDOPA	6-[(18)F]fluoro-L-dopa
FFI	fatal familial insomnia
FGF	fibroblast growth factor
FID	free induction decay
FIV	feline immunodeficiency virus
FKN	fractalkine (CX3CL1)
fMRI	functional magnetic resonance imaging
Foxp3	forkhead box P3 transcription factor
FRC	fibroblastic reticular cell
FRS2	fibroblast growth factor receptor substrate 2
FS	Fisher syndrome
FTD	frontotemporal dementia
FT-ICR	fourier transformed ion cyclotron resonance mass spectrometry
Fz/PCP	Ffizzled/planar cell polarity
GA	glatiramer acetate
GABA	gamma-aminobutyric acid
GAD	glutamate decarboxylase
GalC	galactocerebroside
GALT	gut-associated lymphoid tissue
GAP	GTPase activating protein
GAPDH	glyceraldehyde 3 phosphate dehydrogenase
GBM	glioblastoma multiforme
GBS	Guillain-Barré syndrome
GC	germinal center
GCDC	germinal center dendritic cell
GCL	ganglion cell layer
GL	granular cell layer
GC-MS	gas chromatography combined with mass spectrometry
G-CSF	granulocyte-colony stimulating factor
Gd	gadolinium
GDF	growth and differentiation factor
GDNF	glial-derived neurotrophic factor
GEF	GDP-GTP exchange factor
GFAP	glial fibrillary acidic protein
GFP	green fluorescent protein
GKAP	guanylate kinase-associated protein
GLC1A	chromosome 1 open-angle glaucoma gene
Gln	glutamine
GM-CSF	granulocyte macrophage-colony stimulating factor
GMP	granulocyte monocyte precursor
GnRH	gonadotropin-releasing hormone
GPCR	G-protein-coupled receptor
GPI	glycosylphosphatidylinositol
GR	glucocorticoid receptor
GRIP	glucocorticoid receptor-interacting protein
GRO-α	growth-related oncogene alpha
GSH	glutathione

GSS	Gerstmann-Straussler-Scheinker disease
GSTO1	glutathione s-transferase omega-1
G^T	G-protein transducin
GUCY	guanylate cyclase
HAD	HIV-associated dementia
HAM/TSP	HTLV-I associated myelopathy/tropical spastic paraparesis
HAT	histone acetyltransferase
HBsAg	hepatitis B surface antigen
HCMV	human cytomegalovirus
HD	Huntington's disease
HDAC	histone deacetylase
HES	hairy and enhancer of split homolog
HEV	high endothelial venule
HFS	high frequency stimulation
Hh	hedgehog
HHH	hypervolemic-hemodilution and hypertensive
HHV-6	human herpes virus-6
HIV-1	human immunodeficiency virus type 1
HIVE	human immunodeficiency virus encephalitis
HLA	human leukocyte antigen
HLH	helix-loop-helix
HMG-CoA	3-hydroxy-3-methylglutaryl coenzyme A
HNE	4-Hydroxy-2-nonenal
hnRNP-A1	heterogeneous nuclear ribonuclear protein-A1
HPA	hypothalamic-pituitary-adrenal
HSC	hematopoietic stem cell
Hsp	heat-shock protein
HSV	herpes simplex virus
HSVE	herpes simplex virus-mediated encephalitis
HTLV	human T-cell lymphotrophic virus type
HTRA2	high temperature requirement serine protease 2
Htt	huntingtin
HveA	herpesvirus entry mediator A
I-1	regulatory protein inhibitor-1
IBS	inflammatory bowel syndrome
ICAM	intracellular adhesion molecule
ICAT	isotope coded affinity tags
ICH	intracerebral hemorrhage
iCJD	iatrogenic Creutzfeldt-Jakob disease
ICOS	inducible co-stimulatory molecule
Id	inhibitor of differentiation
IDE	insulin degrading enzyme
IDO	indoleamine 2,3-dioxygenase
IE	immediate early
IF	intermediate filament
IFN	interferon
Ig	immunoglobulin
IGF	insulin-like growth factor
IgG	immunoglobulin G
IGIV	Immunoglobulin intravenous therapy
Ihh	Indian hedgehog
IIDD	idiopathic inflammatory demyelinating disease
IκB	inhibitory kappa B
IKK	IκB kinase
IL	interleukin
IL1RA	IL-1 receptor antagonist
ILBD	incidental Lewy body disease
ILM	inner limiting membrane
IM	intramuscular
IMPDH	inosine monophosphate dehydrogenase
iNKR	inhibitory natural killer cell receptor
INL	inner nuclear layer
iNOS	inducible nitric oxide synthase (NOS2A)
IOP	intraocular pressure
IP	interferon-inducible protein
InsP3R	inositol 1,4,5-triphosphate receptor
IPL	inner plexiform layer
IRAK	IL-1 receptor associated protein kinase
IRBP	interphotoreceptor retinoid-binding protein
IRF	interferon regulatory factor
ISCOM	immunostimulating complexes
ISH	in situ hybridization
IT15	interesting transcript 15
ITAM	immunoreceptor tyrosine-based activation motif
ITGAM	integrin, alpha M
ITR	inverted terminal repeat
IVDU	intravenous drug use
IVIg	intravenous immunoglobulin
JAK	Janus kinase
JEV	Japanese encephalitis virus
JNK	c-Jun N-terminal kinase
KLH	keyhole limpet hemocyanin
KOR	kappa opioid peptide receptor
KYN	kynurenine
KYNA	kynurenic acid
LAK	lymphokine-activated killer (cell)
L-AP4	L-2-amino-4-phosponobutyric acid
LAT	latency associated transcript
LB	Lewy bodies
LC	locus coeruleus
LCA	leukocyte common antigen
LC-FTICR MS	liquid chromatography fourier transform ion cyclotron resonance mass
LC-MS	liquid chromatography combined with mass spectrometry

LC-UV-SPE-NMR liquid chromatography, UV detection, solid phase extraction, and nuclear
LD linkage disequilibrium
LDL low density lipoprotein
LFA-1 leukocyte function-associated antigen-1 (integrin beta 2; ITGB2)
LGN lateral geniculate nucleus
LIF leukemia inhibitory factor
Lingo-1 Leucine-rich-repeat and Ig domain containing Nogo-receptor interacting protein-1
LMN lower motor neuron
LOAD late-onset Alzheimer's disease
LOS lipooligosaccharide
LPA lysophosphatidic acid
LPS lipopolysaccharide
LRR leucine-rich repeat
LRRK2 leucine-rich repeat kinase 2
LT lymphotoxins
LTD long-term depression
LTP long-term potentiation
LTR long-terminal repeat
LT-βR lymphotoxin beta receptor

MAC membrane attack complex
MAdCAM-1 mucosal addressin cell adhesion molecule-1
MAG myelin associated glycoprotein
MAML mammalian mastermind-like
MAO monoamino-oxidase
MAP microtubule-associated protein
MAPK mitogen-activated protein kinases
Mash1 mammalian achaete-scute homologue 1
MBGI myelin-based growth inhibitor
MBP myelin basic protein
MC-1R melanocortin-1 receptor
MCMD minor cognitive motor disorder
MCP membrane cofactor protein (CD46)
MCP-1 monocyte chemoattractant protein-1 (CCL2)
M-CSF macrophage-colony stimulating factor (CSF1)
MD major depression
MDD major depressive disorder
MDM monocyte-derived macrophages
MDP muramyl-dipeptide
MDP monocytes and dendritic cell progenitor
MDR multidrug resistant
MEG magnetoencephalography
MEPP miniature end-plate potential
MFS Miller Fisher syndrome
MHC major histocompatibility complex
MHC-II class II major histocompatibility complex
MHPG methoxy-hydroxy-phenylethanolamine glycol
MHV mouse hepatitis virus
mI myoinositol
MIF migration inhibitory factor
Mint1 Munc-18 interacting protein 1
MIP macrophage inflammatory protein
MJO Machado-Joseph disease
MME membrane metallo-endopeptidase (neprilysin)
MMP matrix metalloprotinease
MMSE Mini-Mental State Examination
MNGC multi-nucleated giant cell
Mn-SOD manganese superoxide dismutase
MOAT multi-specific organic anion transporter
MOBP myelin associated/oligodendrocyte basic protein
MOG myelin oligodendrocyte glycoprotein
MOI multiplicity of infection
MOR mu opioid receptor
MOSP myelin/oligodendrocyte specific protein
MP mononuclear phagocytes
MPA mycophenolic acid
MPL monophosphoryl lipid A
MPO myeloperoxidase
MPP^+ 1-methyl-4-phenylperydinium
MPTP 1-methyl-4-phenyl-1,2,3,6-tetrahydropyridine
MR mineralocorticoid receptors
MRA magnetic resonance angiography
MRI magnetic resonance imaging
MRP multidrug resistance protein
MRS magnetic resonance spectroscopy
MRSI magnetic resonance spectroscopic imaging
MS multiple sclerosis
MSA multisystem atrophy
MSCs myelinating Schwann cells
MSN medium spiny neuron
mSOD1 mutant Cu2+/Zn2+ superoxide dismutase 1
MSRV multiple sclerosis retrovirus
MT magnetization transfer
mTEC medullary thymic epithelial cell
mTOR mammalian target of rapamycin
MTR magnetization transfer ratio
MUC1 mucin type 1 glycoprotein
MuLV murine leukemia virus
Munc-18 mammalian homologue of unc-18
MVE Murray Valley encephalitis virus
MZ marginal zone

NAA N-acetyl-aspartate
NAC N-acetyl cysteine
NADPH nicotinamide adenine dinucleotide phosphate
NCAM neural cell adhesion molecule
NE norepinephrine

NET	norepinephrine transporter
NeuN	neuronal nuclei
NF	neurofilament
NF-κB	nuclear factor-κ-B
NFAT	nuclear factor of activated T lymphocytes
NFL	nerve fiber layer
NFT	neurofibrillary tangles
Ng-CAM	neuronal-glial cell adhesion molecule (L1/NILE)
NGF	nerve growth factor
NgR	Nogo-66 receptor
NICD	Notch intracellular domain
NK	natural killer (cells)
NKT	natural killer T (cells)
NMDA	N-methyl-D-aspartate
NMDAR	N-methyl-D-aspartate receptors
NMJ	neuromuscular junction
NMO	neuromyelitis optica
NMR	nuclear magnetic resonance
NMSCs	nonmyelinating Schwann cells
nNOS	neuronal nitric oxide synthase
NNRTIs	nonnucleoside analogue reverse transcriptase inhibitors
NO	nitric oxide
NOS	nitric oxide synthase
NOT	nucleus of the optic tract
NPC	neural progenitor cell
NPY	neuropeptide Y
NPZ-8	neuropsychological Z score for 8 tests
NR	nuclear receptor
NRG	neuregulin
NRL	nuclear receptor ligand
NRTI	nucleoside analogue reverse transcriptase inhibitors
NSAID	non-steroidal anti-inflammatory drug
NSC	neural stem cell
NSE	neuron specific enolase
NSF	N-ethylmaleimide sensitive factor
NT	3-nitrotyrosine
NTF	neurotrophin
NVU	neurovascular unit
OB	olfactory bulb
OCB	oligoclonal band
OE	olfactory epithelium
OHT	ocular hypertension
OL	oligodendrocyte
OLM	outer limiting membrane
OMgp	oligodendrocyte-myelin glycoprotein
OMP	olfactory marker protein
ONH	optic nerve head
ONL	outer nuclear layer
OP	oligodendrocyte progenitors
OPC	oligodendrocyte progenitor cell
OPCA	olivopontocerebellar atrophy
OPL	outer plexiform layer
ORF	open reading frame
ORN	olfactory response neuron
OSP	oligodendrocyte-specific protein
OVA	ovalbumin
P_0	myelin protein zero
$p75^{NTR}$	p75 neurotrophin receptor (nerve growth factor receptor; NGFR)
PACAP	pituitary adenylate cyclase activating polypeptide
PAF	platelet activating factor
PAG	periaqueductal gray
PAMP	pathogen-associated molecular pattern
PARP	poly(ADP-ribose) polymerase
PASAT	Paced Auditory Serial Addition Test
PBBS	peripheral benzodiazepine binding sites
PBL	peripheral blood lymphocyte
PBMC	peripheral blood mononuclear cell
PBR	peripheral benzodiazepine receptor
PCP	phencyclidine
PD	Parkinson's disease
PD1	program death-1
PDGF	platelet-derived growth factor
PDTC	pyrrolidine dithiocarbamate
PE	plasma exchange
PEG	polyethylene glycol
PEI	polyethyleneimine
PENK	proenkephalin
PERG	pattern electroretinogram
PET	positron emission tomography
PG	prostaglandin
Pgp	P-glycoprotein
PHF	paired helical filament
PI	phospatidylinositol
PI3K	phosphatidylisositol-3-kinase
PICA	posterior inferior cerebellar artery
PICK1	protein interacting with C kinase 1
PKA	cAMP-dependent protein kinase
PKG	protein kinase G
PLGA	poly(D,L-lactide-coglycolide)
PLP	proteolipid protein
PMCA	plasma membrane bound Ca2+-ATPase
PMD	Pelizaeus-Merzbacher disease
PMN	polymorphonuclear (leukocyte)
PMP22	peripheral myelin protein 22
PNS	peripheral nervous system
POAG	primary open-angle glaucoma
polyQ	polyglutamine
POMC	pro-opiomelanocortin
POU3F2	POU class 3 homeobox 2
PP	protein phosphatase
PPAR	peroxisome proliferator activated receptor

PPF paired-pulse facilitation
PPG poly(propylene glycol)
PP-MS primary progressive multiple sclerosis
PR photoreceptor
PrPc cellular prion protein
PrP^{res} protease resistance prion
PrPsc disease-associated prion protein
PRR pattern recognition receptor
PS presenilin (PSEN)
PSA-NCAM poly-sialylated form of the neural cell adhesion molecule
PSCs perisynaptic Schwann cells
PSD postsynaptic density
PSP progressive supranuclear palsy
PSW periodic sharp wave
Ptc patched, a hedgehog receptor
PTP post-tetanic potentiation
PVL periventricular leukomalacia
PVN paraventricular nucleus

RA rheumatoid arthritis
Rag recombination-activating gene
RAGE receptor for advanced glycation end product
RANTES regulated upon activation normal T-cell expressed and secreted (CCL5)
RAS renin-angiotensin system
Rb retinoblastoma
REM rapid eye movement
RER rough endoplasmic reticulum
RF radiofrequency
RFLPs restriction fragment length polymorphisms
RIM Rab3-interacting molecule
RIP receptor interacting protein
RMS rostral migratory stream
RNAi RNA interference
RNS reactive nitrogen species
ROCK Rho kinase
ROI reactive oxygen intermediate
RORγ retinoic-acid-receptor-related orphan receptor-γ
ROS reactive oxygen species
RPE retinal pigment epithelial (cells)
RR-MS relapsing and remitting multiple sclerosis
RTK receptor tyrosine kinase
rt-PA recombinant tissue plasminogen activator
RT-PCR reverse transcription polymerase chain reaction
RyR ryanodine receptor

sALS sporadic amyotrophic lateral sclerosis
SAP synapse-associated protein
SAPAP SAP-associated protein (discs, large homolog-associated protein-1; DLGAP1)
SAPK stress-activated protein kinase (JNK, MAPK8)
sAPP secreted β-amyloid precursor protein
SBMA spinobulbar muscular atrophy
SCs Schwann cells
SC superior colliculus
SCA-3 spinocerebellar ataxia-3
scFv single chain Fv antibodies
SCID severe combined immunodeficiency
sCJD sporadic Creutzfeldt-Jakob disease
SCPs Schwann cell precursor
sCrry soluble complement receptor-related protein y
SDF-1 stromal cell-derived factor 1 (CXCL12)
SEC sinus endothelial cell
SELDI-TOF surface enhanced laser desorption ionization time-of-flight
SER smooth endoplasmic reticulum
SERCA sarco(endo)plasmic reticulum Ca2+-ATPase
SERT serotonin transporter
sFI sporadic fatal insomnia
SGLPG sulfated glucuronyl lactosaminyl paragloboside
SGZ subgranular zone
Shh sonic hedgehog
sIg surface immunoglobulin
sIL-2R soluble IL-2 receptor
SITA Swedish interactive thresholding algorithm
SIV simian immunodeficiency virus
SLE systemic lupus erythematosus
SMAC second mitochondrial-derived activator of caspase
SMase sphingomyelinase
SMN survival motor neuron gene
SN substantia nigra
SNAP sensory nerve action potential
SNAP-25 synaptosome-associated protein of 25,000 daltons
SNARE NSF attachment receptor
SNpc substantia nigra pars compacta
SNPs single nucleotide polymorphisms
SNS sympathetic nervous system
SOCS suppressors of cytokine signaling
SOD1 superoxide dismutase 1
SP1 specificity protein 1
SPECT single photon emission computed tomography
SPG-II spastic paraplegia type II
SR-A scavenger receptor type A
SRBCs sheep red blood cells

SREBP	sterol regulatory element binding protein
SRF	serum-response factor
SSRI	selective serotonin reuptake inhibitors
STAT	signal transducers and activators of transcription
STP	short-term potentiation
SV5	simian virus 5
SVZ	subventricular zone
SWAP	short wavelength automated perimetry
T	tesla
TBE	tick-borne encephalitis virus
TBP	TATA-binding protein
TCA	tricarboxylic acid
TCR	T-cell receptor
TCV	T cell vaccination
TDO	tryptophan 2,3-dioxygenase
TE	echo time
TEC	thymic epithelial cell
Teff	T effector cells
TF	transcription factor
TG	trigeminal ganglia
TGF	transforming growth factor
Th1	T helper type 1 cell
Th2	T helper type 2 cell
TIA	transient ischemic attack
TIR	Toll/IL-1 receptor
TJ	tight junction
TLR	toll-like receptor
TMEV	Theiler's mouse encephalomyelitis virus
TMT	trimethyltin
TN	terminal nuclei
TNF	tumor necrosis factor
TNFR	tumor-necrosis factor receptor
TOR1	target of rapamycin 1
TRAF	TNF-receptor mediated factor
TRAIL	TNF-related apoptosis inducing ligand (TNFSF10)
TRANCE	TNF-related activation-induced cytokine (TNFSF11)
TRANCER	TRANCE receptor (TNFRSF11A)
TRE	tax responsive element
Treg	T regulatory cells
Trk	receptor tyrosine kinase
TRP	transient receptor potential
TSA	tissue-specific antigen
Tyk2	protein tyrosine kinase 2
UACA	uveal autoantigen with coiled domains and ankyrin repeats
UMN	upper motor neuron
V1	primary visual cortex
VAMP	vesicle-associated membrane protein
VCAM	vascular cell adhesion molecule
VEGF	vascular endothelial growth factor
VEP	visual evoked potential
VIP	vasoactive intestinal peptide
VMAT-2	vesicular monoamine transporter-2 (solute carrier family 18; SLC18A2)
VZV	varicella-zoster virus
WKAH	Wistar-King-Aptekman-Hokudai
WNV	West Nile virus
X-ALD	X-adrenoleukodystrophy
XIAP	X-linked inhibitor of apoptosis protein
ZO-1	zonula occludens-1 (TJP1)

References

HUGO Gene Nomenclature Committee: http://www.genenames.org/index.html
"Abbreviations and Symbols for Chemical Names": www.blackwell-symergy.com/doi/pdf/10.1111/j.1432.1033.1967.tb00070.x

1
Introducing Neuroimmune Pharmacology

Howard E. Gendelman and Tsuneya Ikezu

Keywords Neuroscience; Immunology; Pharmacology

Neuroscience, Immunology, and Pharmacology are broad disciplines that, without argument, impact upon a large component of what we come to know as biomedical science (Elenkov et al., 2000; Gendelman, 2002; McGeer and McGeer, 2004). Each, by themselves and even more so when put together, is multidisciplinary and require, for the student, both a broad knowledge and deep understanding of molecular and cellular biology. The linking of the disciplines is ever more challenging when they are placed together. The reasons are that each must first be understood as a single entity. The bridges between disciplines are what we now call multidisciplinary science and require another level of insight. When combined they form the basis of self, our engagement with the environment, as well as disease. Indeed, drugs that influence organ function, aging, and tissue homeostasis and repair can improve clinical outcomes.

A special feature of "human kind" among other species is the presence of an extraordinary complex immune system that can be used to protect against a plethora of harmful microbial pathogens, including viruses, bacteria, and parasites, as well as abnormal cells and proteins (Petranyi, 2002). This underlies the complexity of the human genome which encodes expansive immune-related genes not found in lower species (Hughes, 2002). When the immune system is compromised, disease occurs and often does with ferocity; a range of clinical manifestations ensue that follows as a consequence of neurodegenerative, psychiatric, cancer, and infectious diseases, or those elicited by the immune system's attack on itself. The latter is commonly referred to as "destructive" autoimmunity (Christen and von Herrath, 2004). Interestingly enough, the immune system may sometimes be an impediment to therapy. Indeed, modulating its function is required for long term and successful organ transplantation (Samaniego et al., 2006). On balance, modulation of the immune system can affect "neuroprotective" responses for certain diseases (Schwartz, 2001).

Like the immune system, the nervous system contains surveillance functions and also possesses a number of functional roles that include mentation, movement, reasoning, sensation, vision, hearing, learning, breathing, and most behaviors. The nervous system contains defined tissue structures such as the brain, spinal cord, and peripheral nerves. On the cellular level, it includes networks of nerve cells with a variety of functional activities, complex networks and communications, supportive and regulatory cells, called glia, and a protective barrier that precludes the entry of a variety of macromolecules, cells, and proteins. It also possesses connections throughout the body that permits it function. Neuroscience is the discipline used to explore each of the nervous system regions and cells that include their networks and modes of communication in health and disease. As humans we have 100 billion neurons that are each functional units contained within the nervous system. Molecular and biochemical studies along with cell and animal systems were each used alone and together to explore and define neural biology. The tasks of neurosciences are to better understand the brain's function in the context of ontogeny, organism development, and aberrations during disease. Neuroscience is an interdisciplinary field and evolved as such during the past quarter of century. It includes neurobiology, neurochemistry, neurophysiology, mathematics, psychology, computer neuroscience, and learning and behavior. Even more recently, it has included the field of immunology (Sehgal and Berger, 2000). In a historical context, the brain, for a long time, was considered an immune privileged organ, meaning that its protective shield or barrier, commonly termed the blood–brain barrier, served to protect, defend, and as a consequence exclude ingress of toxins, cells, and pathogens (Streilein, 1993; Becher et al., 2000). Nonetheless, this is balanced by the fact that resident inflammatory cells do exist inside the brain and are capable of producing robust immune responses. In recent years, we have come to accept that it is the mononuclear phagocytes (MP; perivascular macrophage and microglia) that are disease perpetrators, while the astrocyte serves as "supportive" and "homeostatic" in nurturing neurons and in protection against the ravages of disease (Gendelman, 2002; Simard and Nedergaard, 2004; Trendelenburg and Dirnagl, 2005). The neuron in this neuroimmune model is the passive recipient of the battles that rage between the MP and the astrocyte. Findings that have emerged over the past half-decade have challenged this model. We now know that dependent upon environmental cues

T. Ikezu and H.E. Gendelman (eds.), *Neuroimmune Pharmacology*.
© Springer 2008

and disease microglia, astrocytes, and other neural cell elements including endothelial cells and oligodendrocytes possess immunoregulatory functions. We also know that microglia and astrocytes dependent upon the environment and stimuli can be supportive, destructive or both. Even more importantly neurons can secrete immunoregulatory factors and engage directly into cell-cell-environmental cue stimulations. To make the system perhaps even more complex, local neuroimmune processes can result in the recruitment of T cells and enticement of the adaptive immune response, significantly affecting disease outcomes (Schwartz and Kipnis, 2004). All in all things appear more complex than once thought even in the past decade.

These three disciplines and the complexities inherent in each academic field is perhaps the most multidisciplinary serving to bring scientists and clinicians together with knowledge of neurobiology, immunology, pharmacology, biochemistry, cellular and molecular biology, virology, genetics, gene therapy, medicinal chemistry, nanomedicine, proteomics, pathology, and physiology. Even more than immunology and neuroscience, pharmacology integrates a broad knowledge in scientific disciplines enabling the pharmacologist a unique perspective to tackle drug-, hormone-, immune-, and chemical-related pathways as they affect human health and behavior. Drug actions and therapeutic developments form the basis of such discoveries but the central understanding of how they act provide vision for further research to improve human well-being and health.

This textbook is unique in scope by serving to investigate the intersection of this new discipline. Neuropharmacologists study drug actions including neurochemical disorders underlying a broad range of diseases such as schizophrenia, depression, and neurodegenerative diseases (such as Alzheimer's and Parkinson's diseases). Drugs can also be used to examine neurophysiological or neurobiochemical changes as they affect brain, behavior, movement, and mental status. Immunopharmacology seeks to control the immune response in the treatment and prevention of disease. Research does include immunosuppressant agents used in organ transplant as well as developing agents that affect bone marrow function and cell differentiation in cancer therapies.

What then defines the field of *Neuroimmune Pharmacology*? Is it simply a field that intersects the three disciplines of neuroscience, immunology, and pharmacology in seeking to better define the epidemiology, prevention, and treatment of immune disorders of the nervous systems (Figure 1.1). Are these disorders limited in their scope in affecting behavior, cognition, motor and sensory symptoms, or do they also involve developmental and degenerative disorders? It is clear that the immune system is linked, in whole or part, to diseases that develop as a consequence of genetic abnormalities and a

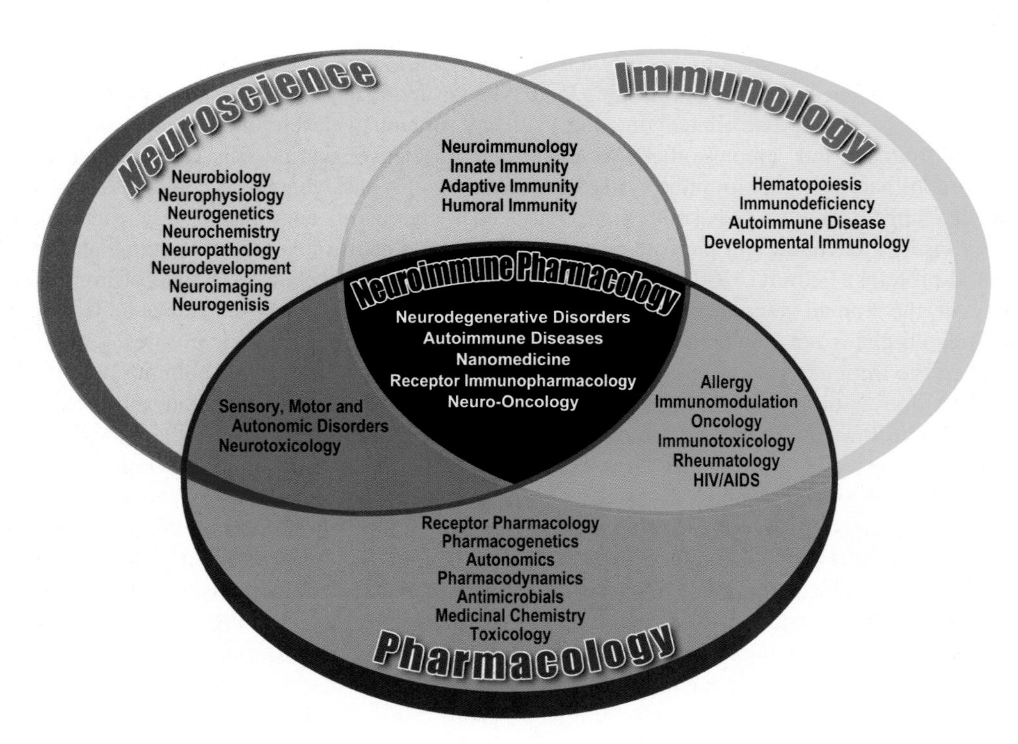

FIGURE 1.1. Neuroimmune Pharmacology. The intersection of the specific disciplines of neuroscience, immunology, and pharmacology are illustrated in this VENN diagram.

broad range of environmental cues (including microbial infections, and abused drugs), and toxins. So, where does Neuroimmune Pharmacology find its niche? These can occur, in part, as a consequence of neuropeptides, neurotransmitters, cytokines, chemokines, and abused drugs. Like much in science, we are left with more questions than answers. In the end, we seek avenues for translational research and better understanding of disease mechanisms. Diseases are together linked to microbial agents, by inflammatory processes, by emergence of cancerous cells or tumors, by stress, by environmental cues, and by genetic disturbances. No matter the cause harnessing the immune processes for pharmacological benefit will, at days end, provide "real" solutions to positively affect some of the most significant and feared disorders of our century.

What do we seek to accomplish by this textbook? First, we would be remiss in not acknowledging the pivotal discoveries made by others when research fields intersect (Rock and Peterson, 2006). These include the discovery and characterization of the guanosine triphosphate (GTP) binding and prion proteins (Gilman, 1995; Rodbell, 1995; Prusiner, 1998, 2001), neurotransmission and memory functions (Carlsson, 2001; Greengard, 2001; Kandel, 2001), and odorant receptors (Buck, 2000; Axel, 2005). We posit that new discoveries can and will be made through the intersections of Neuroscience, Immunology, and Pharmacology and as such sought to define it for the student. The notion that inflammation contributes in significant manner to neurodegeneration and significantly beyond autoimmune diseases is brought front and center and demonstrated without ambiguity for multiple sclerosis, peripheral neuropathies, Alzheimer's and Parkinson's disease, amyotrophic lateral sclerosis as well as for microglial infections of the nervous system including NeuroAIDS where microglial activation is central to disease processes (Appel et al., 1995; Toyka and Gold, 2003; McGeer and McGeer, 2004; Ercolini and Miller, 2005; Gendelman, 2002). Perhaps most importantly, we have laid the groundwork for how the immune system can be harnessed either through its modulation, through altering blood–brain barrier integrity and function, and/or by drug-delivery strategies that target the brain.

No doubt this textbooks is an expansive read for the student and scholar alike. To this end, we are humbled by its realization. These chapters lay only the beginnings to what we believe will be a large future footprint into the integration between neuroscience, immunology, and pharmacology.

References

Appel SH, Smith RG, Alexianu M, Siklos L, Engelhardt J, Colom LV, Stefani E (1995). Increased intracellular calcium triggered by immune mechanisms in amyotrophic lateral sclerosis. Clin Neurosci 3:368–374.

Axel R (2005). Scents and sensibility: A molecular logic of olfactory perception (Nobel lecture). Angew Chem Int Ed Engl 44:6110–6127.

Becher B, Prat A, Antel JP (2000). Brain-immune connection: Immunoregulatory properties of CNS-resident cells. Glia 29:293–304.

Buck LB (2000). The molecular architecture of odor and pheromone sensing in mammals. Cell 100:611–618.

Carlsson A (2001). A half-century of neurotransmitter research: Impact on neurology and psychiatry (Nobel lecturey). Biosci Rep 21:691–710.

Christen U, von Herrath MG (2004). Initiation of autoimmunity. Curr Opin Immunol 16:759–767.

Elenkov IJ, Wilder RL, Chrousos GP, Vizi ES (2000). The sympathetic nerve—an integrative interface between two supersystems: The brain and the immune system. Pharmacol Rev 52:595–638.

Ercolini AM, Miller SD (2005). Role of immunologic cross-reactivity in neurological diseases. Neurol Res 27:726–733.

Gendelman HE (2002). Neural immunity: Friend or foe? J Neurovirol 8:474–479.

Gendelman HE, Persidsky Y (2005). Infections of the nervous system. Lancet Neurol 4:12–13.

Gilman AG (1995). Nobel lecture G proteins and regulation of adenylyl cyclase. Biosci Rep 15:65–97.

Greengard P (2001). The neurobiology of dopamine signaling. Biosci Rep 21:247–269.

Hughes AL (2002). Natural selection and the diversification of vertebrate immune effectors. Immunol Rev 190:161–168.

Kandel ER (2001). The molecular biology of memory storage: A dialogue between genes and synapses. Science 294:1030–1038.

McGeer PL, McGeer EG (2004). Inflammation and the degenerative diseases of aging. Ann NY Acad Sci 1035:104–116.

Petranyi GG (2002). The complexity of immune and alloimmune response. Transpl Immunol 10:91–100.

Prusiner SB (1998). Prions. Proc Natl Acad Sci USA 95:13363–13383.

Prusiner SB (2001). Shattuck lecture—neurodegenerative diseases and prions. N Engl J Med 344:1516–1526.

Rock RB and Peterson PK (2006). Microglia as a pharmacological target in infectious and inflammatory diseases of the brain. J Neuroimm Pharm 1:117–117.

Rodbell M (1995). Nobel lecture signal transduction: Evolution of an idea. Biosci Rep 15:117–133.

Samaniego M, Becker BN, Djamali A (2006). Drug insight: Maintenance immunosuppression in kidney transplant recipients. Nat Clin Pract Nephrol 2:688–699.

Schwartz M (2001). Harnessing the immune system for neuroprotection: Therapeutic vaccines for acute and chronic neurodegenerative disorders. Cell Mol Neurobiol 21:617–627.

Schwartz M, Kipnis J (2004). A common vaccine for fighting neurodegenerative disorders: Recharging immunity for homeostasis. Trends Pharmacol Sci 25:407–412.

Sehgal A, Berger MS (2000). Basic concepts of immunology and neuroimmunology. Neurosurg Focus 9:e1.

Simard M, Nedergaard M (2004). The neurobiology of glia in the context of water and ion homeostasis. Neuroscience 129:877–896.

Streilein JW (1993). Immune privilege as the result of local tissue barriers and immunosuppressive microenvironments. Curr Opin Immunol 5:428–432.

Toyka KV, Gold R (2003). The pathogenesis of CIDP: Rationale for treatment with immunomodulatory agents. Neurology 60:S2–S7.

Trendelenburg G, Dirnagl U (2005). Neuroprotective role of astrocytes in cerebral ischemia: Focus on ischemic preconditioning. Glia 50:307–320.

Part 1
Immunology of the Nervous System

2
Innate and Adaptive Immunity in Health and Disease

Howard E. Gendelman and Eliezer Masliah

Keywords Neuroinflammation; Innate and adaptive immunity; Neurodegenation, Synapse loss

Neuroinflammatory processes play a significant role in health and disease of the nervous system. These regulate development, maintenance, and sustenance of brain cells and their connections. Linked to aging, epidemiologic, animal, human, and therapeutic studies all support the presence of a neuroinflammatory cascade in disease. This is highlighted by the neurotoxic potential of microglia. In steady state, microglia serve to protect the nervous system by acting as debris scavengers, killers of microbial pathogens, and regulators of innate and adaptive immune responses. In neurodegenerative diseases, activated microglia affect neuronal injury and death through production of glutamate, proinflammatory factors, reactive oxygen species, quinolinic acid amongst others and by mobilization of adaptive immune responses and cell chemotaxis leading to transendothelial migration of immunocytes across the blood–brain barrier and perpetuation of neural damage. As disease progresses, inflammatory secretions engage neighboring glial cells, including astrocytes and endothelial cells, resulting in a vicious cycle of autocrine and paracrine amplification of inflammation perpetuating tissue injury. Such pathogenic processes contribute to neurodegeneration. Research from others and our own laboratories seek to harness such inflammatory processes with the singular goal of developing therapeutic interventions that positively affect the tempo and progression of human disease (Crutcher et al., 2006).

As the life expectancy of the human population continues to increase, the possibility of developing neuro inflammatory and neurodegenerative diseases have increased considerably during the past 50 years. Of the neurodegenerative disorders, Alzheimer's disease continued to be the leading cause of dementia in the aging population. Traditionally, neurodegenerative disorders have been define as conditions where there is selective loss of neurons within specific region of the brain accompanied by astrogliosis. However, in the past 20 years, we have learned that the pathological process leading to the disfunction of selected circuitries in the brain initiates with damage to the synapses rather than with the loss of neurons. In fact, neuronal loss is a late event that is probably preceded by damage to axons and dendrites followed by shrinkage of the neuronal cell body and abnormal accumulation of filamentous proteins.

Therefore, the revised concept of neurodegeneration suggest that neuronal injury initiates at the synaptic junction and propagates throughout selected circuitries leading to neuronal dysfunction which resolves in the classical clinical symptoms characteristic to each of the neurodegenerative disorders (Hashimoto and Masliah, 2003). So, for example, in Alzheimer's disease early damage to the synapses between the entorhinal cortex and the molecular layer of the dentate gyrus (perforant pathway) resolves in the short-term memory deficits characteristic of this dementing disorder. Later on disconnection of the cortico-cortico fibers in the frontal, parietal, and temporal cortex resolved in more severe memory deficits, and alterations in executive functions, and abstraction. Degeneration of connections between the nucleus basalis of Meynert and the neocortex resolves in attention and memory deficits that usually associated with loss of cholinergic neurons. Other circuitries and neuronal populations are also affected in Alzheimer's disease illustrating the complexity of these disorders and the fact that the concept of single population is affected is limited. That is the case with several other disorders including Parkinson's disease where degeneration is not limited to the dopaminergic system, but also involves the limbic system, the raphe nucleus, the insula, and other systems.

In response to the injury neurons produce adhesions molecules and trophic factors that recruit astroglial and microglial cells to participate in the process of repair of the damage. In addition the microvasculature and other glial systems might also participate in the process. Thus, neurodegeneration is accompanied by astrogliosis, microgliosis, and microvascular remodeling. While astroglial cells initially produce trophic factors and cytokines that aid in the tissue repair, eventually these factors could amplify the inflammatory response, increase vascular permeability and result in microglial activation, which in turn might lead to the production of more proinflammatory cytokines and chemokines. A critical balance between

T. Ikezu and H.E. Gendelman (eds.), *Neuroimmune Pharmacology*.
© Springer 2008

the repair and the proinflammatory factors often determines the future rate and progression of the degenerative process.

The understanding of the mechanisms of neurodegeneration and inflammatory response in these neurological conditions has seen a tremendous progress in the past 10 years. It is now recognized that probably small soluble misfolded protein aggregates denominated oligomers are responsible for the injury. So, for example, in Alzheimer's disease A beta protein oligomers might damage the synapses in the limbic system while in Parkinson's disease a-synuclein oligomers damage the axons in the striatum and cortical regions. While significant progress has been made in understanding the fundamental mechanisms for the neuronal injury, less is known about the reasons for the selective neuronal vulnerability characteristic to these neurological conditions.

Reference

Crutcher KA, Gendelman HE, Kipnis J, Perez-Polo JR, Perry VH, Popovich PG, Weaver LC (2006) Debate: "Is increasing neuroinflammation beneficial for neural repair?" J Neuroimm Pharmacol 3:195–211.

Hashimoto M, Masliah E (2003). Cycles of aberrant synaptic sprouting and neurodegeneration in Alzheimer's and dementia with Lewy bodies. Neurochem Res 28:1743–1756.

3 Anatomical Networks: Structure and Function of the Nervous System

Eliezer Masliah

Keywords Connectivity; Trafficking; Microcirculation; Synapses; Neuroanatomy; Cerebrum; Cerebellum; Brainstem

3.1. Introduction

The understanding of the gross and fine structure of the brain is of fundamental importance to elucidate how the nervous system works, how it interacts with other peripheral organs and tissues and how it responds to external stimuli. The functions and activity of the nervous system are not only regulated by its intrinsic wiring but also by interactions with cellular components of peripheral tissues. Of significant importance are the interactions between the nervous system with the immune and endocrine systems. While for the former, interactions involve trafficking of immune and hematopoietic cells into the brain; the latter interactions are primarily of a chemical nature, mediated by hormones and growth factors that reach the nervous system via the circulation.

In this context, the main objective of this chapter will be to provide an overview to the structure and organization of the nervous system that is relevant to the understanding of the unique interactions between the nervous and the immune system and to elucidate the pathogenesis of Alzheimer's disease (AD), Parkinson's disease (PD), HIV encephalitis (HIVE), and other neurodegenerative and neuroinflammatory disorders (Ringheim and Conant, 2004; Gendelman, 2002; Ho et al., 2005).

The central nervous system (CNS) has been traditionally considered an immunologically privileged site; however, it should be viewed as an immunological specialized region. In fact, cells from the immune system such as lymphocytes and macrophages constantly circulate through the nervous system under physiological conditions. Interactions between these trafficking immune cells and neuronal and glial components of the nervous system are critical in maintaining the stability and functional activity of selected neuronal circuitries involved in memory formation, sleep, and regulation of hormonal production (Avital et al., 2003; Opp and Toh, 2003). For example, a recent study showed that trafficking of T-cells into the adult hippocampus contributes to maintenance of neurogenesis and learning (Ziv et al., 2006). Remarkably, recent evidence shows that not only does the immune system regulate neuronal function but also conversely neuronal activity regulates the immune response. For example, recent studies have shown that cholinergic neurotransmission is capable of inhibiting pro-inflammatory cytokine release and protects against systemic inflammation (Pavlov and Tracey, 2005).

Therefore, immunological reactions in the nervous system are not exclusively related to pathological conditions such as viral infections, autoimmune diseases, and inflammatory disorders (Owens et al., 2005; Eskandari et al., 2003) but also may be involved in regulation of neural homeostasis under physiological conditions and stress (Buller, 2003). In fact immune reactions that occur in the nervous system take a unique character which is probably determined by the very specialized local anatomy. Important anatomical characteristics of the nervous system that determine the unique nature of the neuro-immune reactions include the relative lack of lymphatic drainage, the lack of endogenous antigen presenting cells, and the selective permeability of the blood–brain barrier (BBB).

The distribution and patterns of migration of neuroimmune cells are regulated by genetically encoded programs, patterns of connectivity in anatomical regions and blood flow. Moreover, and as it will be described later, there are endogenously derived neuroimmune cells such as the astrocytes and microglia as well as cells derived from the peripheral circulation such as lymphocytes and macrophages. There are three routes for leukocyte entry into the nervous system: circulation through the subarachnoid space which is mediated by P-selectin, migration across the choroid plexus which is mediated by PECAM, and extravasation from postcapillary venules mediated by P-selectin, lymphocyte adhesion molecule (LFA-1) and intercellular adhesion molecule (ICAM) (Engelhardt and Ransohoff, 2005).

In summary, the nervous system has a unique circulatory organization based on the flow of the cerebrospinal fluid

T. Ikezu and H.E. Gendelman (eds.), *Neuroimmune Pharmacology*.
© Springer 2008

(CSF), the microvasculature, the BBB and the regulation of the blood flow by the neural activity that provides a series of mechanisms to selectively regulate interactions between the nervous system and trafficking cells from peripheral tissues. This chapter provides the structural framework for the better understanding of these interactions under physiological conditions and in neurological diseases.

3.2. Gross Anatomical Structure of the Brain

3.2.1. General Organization of the Central Nervous System

3.2.1.1. Introduction

The nervous system is divided into the CNS and the peripheral nervous system (PNS). The CNS is composed of the brain and spinal cord and a triple membranous covering denominated meninges. The outer membrane is the dura, the intermediate arachnoid, and the innermost pial membrane. The complexity of the CNS, with its millions of neurons and connections can be summarized to six major divisions: (1) cerebral hemispheres, (2) diencephalon, (3) midbrain, (4) pons and cerebellum, (5) medulla, and (6) spinal cord (Martin, 1989) (Figure 3.1). The PNS consists of nerves connected to the brain and spinal cord (cranial and spinal nerves) and their branches within the body (Crossman and Neary, 2005). Spinal nerves serving the upper or lower limbs join to form the brachial or lumbar plexus, respectively, within which fibers are distributed into named peripheral nerves. The PNS also includes some groups of peripherally located nerve cell bodies that are located within ganglia (e.g. dorsal root ganglia). Neurons that detect changes in, and control the activity of, the internal organs are denominated autonomic nervous system (ANS). Its components are present in both the central and peripheral nervous systems. The ANS is divided into two anatomically and functionally distinct components called the sympathetic and parasympathetic divisions, which generally have antagonistic effects on the structure that they innervate. The ANS innervates smooth muscle, myocardium and secretory glands and it is an important

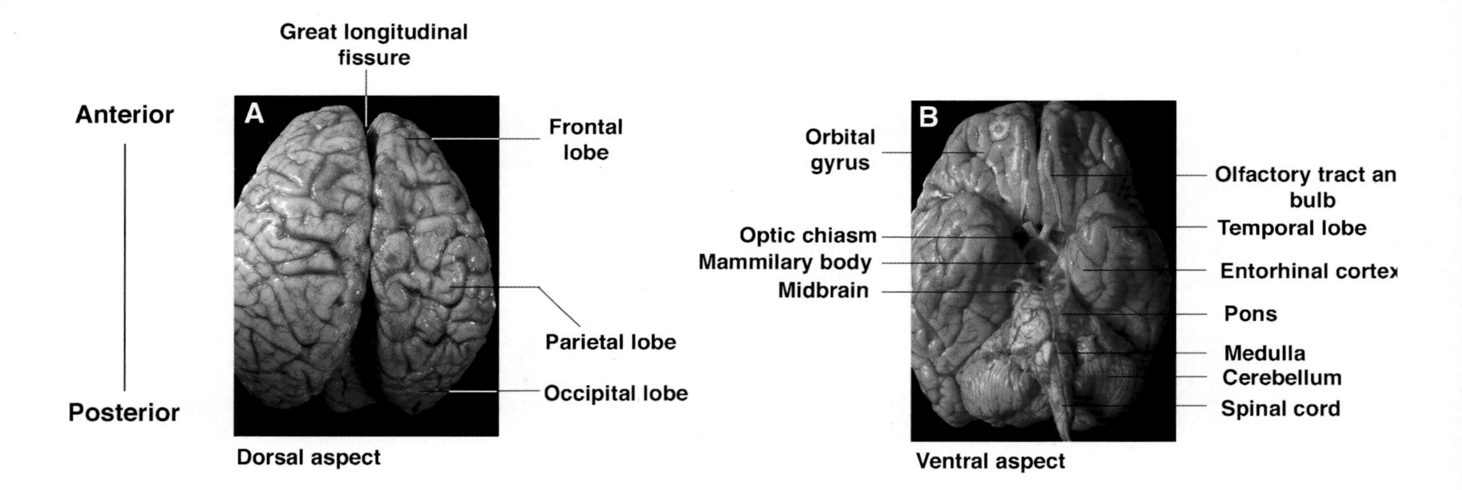

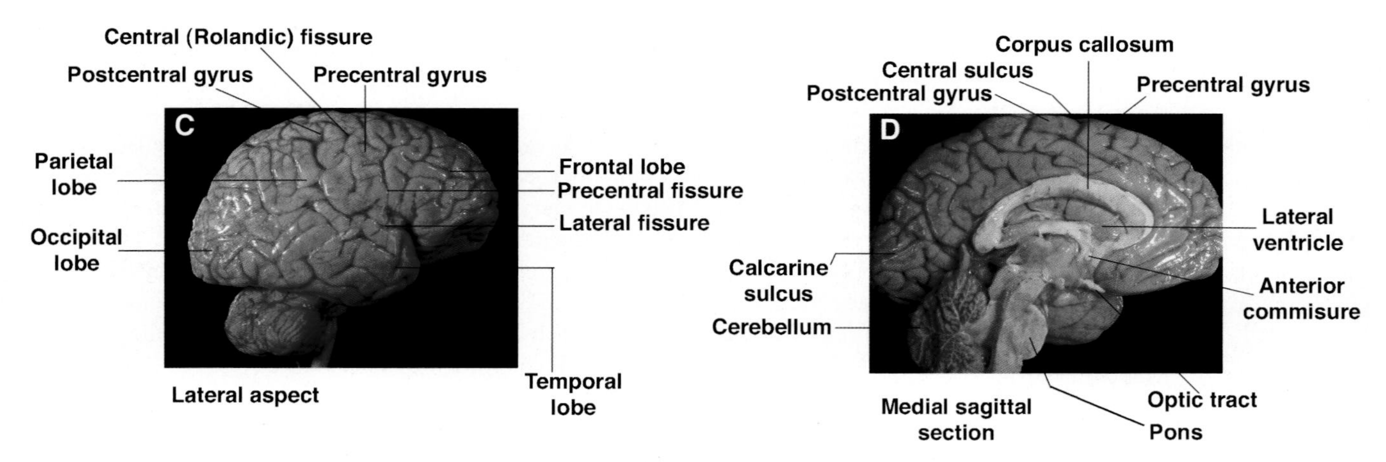

FIGURE 3.1. Gross external structures of the brain.

player of the homeostatic mechanisms that regulate the internal environment of the body.

3.2.1.2. Cerebral Hemispheres

The cerebral hemispheres are divided into two parts by the interhemispheric (sagittal) fissure. Interconnecting the two hemispheres is the corpus callosum (Martin, 1989) (Figure 3.1). The cerebral hemispheres have four major parts, which are the cerebral cortex, basal ganglia (or striatum), hippocampal formation, and amygdala (Figure 3.2). The human cerebral cortex is a highly convoluted structure. The elevated convolutions are known as gyri and the folds that separate the gyri are called sulci. The cerebral cortex includes the frontal, parietal, temporal, and occipital regions (Figure 3.1). In these regions the cerebral cortex usually has six layers. The frontal and parietal lobes are separated by the central sulcus (Rolandic sulcus) which separates two functional regions of the cortex namely the primary motor cortex which is located in the precentral gyrus and the primary somatosensory cortex which is distributed in the postcentral gyrus (Figure 3.1). The central sulcus extends from the longitudinal fissure along the midline ventrally almost into the lateral cerebral sulcus (Sylvian sulcus) (Figure 3.1). The frontal lobe, the largest of the cerebral lobes, extends from the central sulcus to the frontal pole. The function of the precentral gyrus is to integrate motor functions from different brain regions. The neurons are organized in a somatotopic manner, which means that different parts of the precentral gyrus are associated with distinct parts of the body both anatomically and functionally. Immediately anterior to the premotor cortex there are three parallel gyri, the superior middle and inferior frontal gyri. These areas involve processing of executive functions such as abstraction, thinking, cognition, language and emotion. In patients with AD and frontotemporal dementia (FTD), these areas are heavily damaged resulting in the characteristic profile of cognitive impairment typical of these patients. Part of the inferior frontal cortex (left side) includes Broca's motor speech area, an area important in the formulation of motor components of speech. The parietal lobes include the somatosensory cortex within the postcentral gyri (Figure 3.1). The remainder of the parietal lobe is divided into a superior and inferior lobes separated by the interparietal sulcus. The supramarginal gyrus and the angular gyrus divide the inferior parietal lobe, these regions receive input from the auditory and visual cortex and are involved in integration and discrimination of perception. Ventral to these gyri and extending into the temporal cortex is Wernicke's area; this structure is involved in language comprehension. While injury to Broca's area results in broken language, damage to Wernicke region results in difficulty understanding language.

The temporal cortex is situated inferior to the lateral sulcus and is divided into superior middle and inferior temporal gyri (Figure 3.2). On the inner aspect of the superior temporal cortex are the gyri of Heschl, which is the primary auditory region. The inferior portion of the temporal lobe is involved in vision function and parts of the medial temporal lobe are involved in olfactory functions. The temporal lobe also includes the hippocampus and parahippocampal cortex. The hippocampus and amygdala are the second and third components of the cerebral hemispheres and are located under the cortical surface (Figure 3.2). The hippocampus is involved in memory formation while the amygdala modulates the action of the autonomic nervous system, hormone release and emotions. These two structures are part of the limbic system which includes the cingulate cortex as well as part of the diencephalon and midbrain. Together the limbic system plays a central role in the regulation of cognitive functions and mood. The occipital cortex or striate cortex is involved in integration of visual information.

The second major component of the cerebral hemisphere is the basal ganglia, which includes the caudate, putamen and globus pallidus (Figure 3.2). A white matter tract, denominated anterior commissure, divides the inferior aspect of the putamen and globus pallidus and separates the major cholinergic center in the brain also known as nucleus Basalis of Meynert or substantia innominata (Figure 3.2). The loss of cholinergic input in patients with AD is related to degeneration of this region, which is dependent on NGF for survival. The caudate and putamen are divided by the anterior capsule. The basal ganglia participate in control of movement but also have a role in behavior. In patients with Huntington's disease, this brain structure is severely affected.

3.2.1.3. The Diencephalon

Rostral to the brainstem lies the forebrain, consisting of the diencephalon and cerebral hemispheres (Crossman and Neary, 2005). The two sides of the diencephalon are separated by the lumen of the third ventricle, whose lateral walls they constitute (Figure 3.2). The diencephalon consists of four main subdivisions in a dorsoventral direction: the epithalamus, thalamus, subthalamus, and hypothalamus (Figure 3.2). The epithalamus is small and its most recognizable component is the pineal gland, which lies in the midline immediately rostral to the superior colliculi of the midbrain. The thalamus is by far the largest component of the diencephalon and it forms much of the lateral wall of the third ventricle. The thalamus plays an important role in sensory, motor and cognitive functions and has extensive connections with layers IV and V of the cerebral cortex. The thalamus is a central structure for relaying information to the cerebral hemispheres (Figure 3.2). The thalamus includes a large complex of nuclei that includes three principal nuclear masses (anterior, medial, and lateral) divided by the internal medullary lamina. Within the internal medullary lamina lie the intralaminar nuclei. On the lateral aspect of the thalamus, the reticular nucleus can be found. The hypothalamus forms the lower part of the walls and floor of the third ventricle (Figure 3.2). It is a highly complex and important region because of its involvement in many systems, most notably the neuro-endocrine system, the limbic system, and the ANS. From the ventral aspect of the hypothalamus in the midline emerges the infundibulum or pituitary stalk,

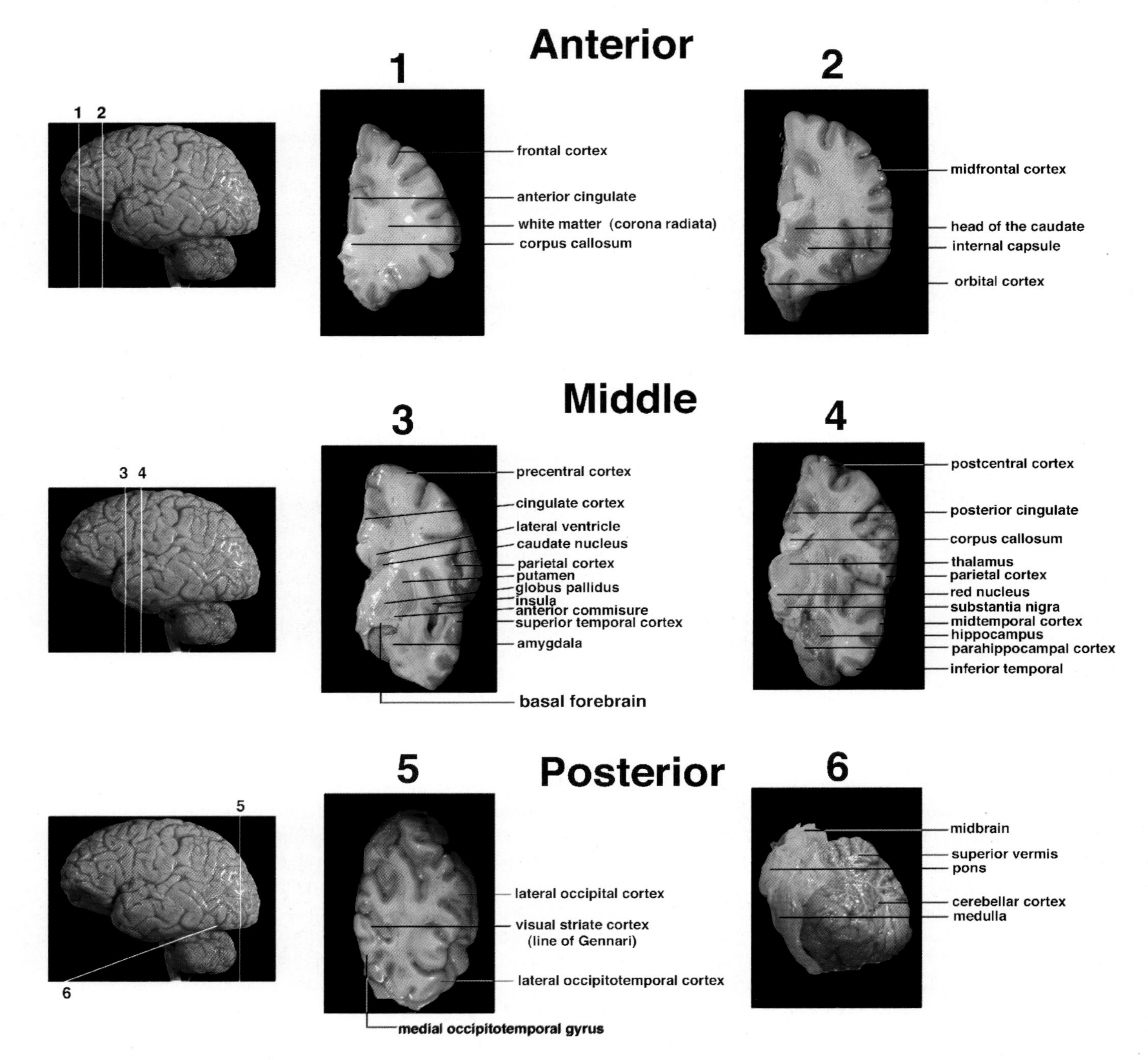

FIGURE 3.2. Gross internal structures of the central nervous system.

to which is attached the pituitary gland. The hypothalamus integrates the functions of the ANS and endocrine hormone release from the pituitary gland. The subthalamus is located under the thalamus and dorsal lateral to the hypothalamus it contains the subthalamic nucleus and the zona incerta. The subthalamic nucleus is connected to the globus pallidus and substantia nigra and is important in the control of movement. This circuitry is often affected in patients with PD and is amenable to surgical manipulation.

3.2.1.4. *The Brainstem*

The midbrain, pons, cerebellum, and medulla constitute the brainstem (Figure 3.2). The brainstem has three general functions. The first is to receive sensory information from cranial structures and to control muscles of the head. This function of the brainstem is similar to that of the spinal cord. The cranial nerves are constituents of the PNS that provide the sensory and motor innervation of the head and therefore are analogous to the spinal nerves. The second is related to the fact that the brainstem contains neural circuits that transmit information from the spinal cord to higher brain regions and back. Finally, the integrated actions of the medulla, pons, and midbrain regulate the levels of awareness and arousal. This function is mediated by a diffuse collection of structures in the brainstem, denominated the reticular formation. In addition to these general functions, the various divisions of the brainstem provide specific sensory and motor functions. The medulla and the pons play a role in the vital regulation of the blood pressure and respiratory functions. The midbrain, also known as the mesencephalon, is divided into dorsal and ventral parts at the level of the aqueduct (Figure 3.2). The dorsal part is known as the tectum and includes the colliculi. The most ventral part of the midbrain tegmentum contains the substantia nigra, which consists of pars compacta and pars reticularis. The substantia nigra contains melanin pigmented neurons that produce dopamine. Degeneration of the substantia nigra is characteristic of patients with PD and results in disconnection between the midbrain and the caudo-putamen region. Other important structures contained in the midbrain are the oculomotor nucleus, the red nucleus, the periaqueductal grey and the medial lemniscus. Furthermore, the branches of the third cranial nerve emerge from the midbrain.

The pons (Figure 3.2) is further divided into ventral and dorsal sections. The ventral section contains the pontocerebellar fibers and the pontine nucleus. Corticospinal fibers run longitudinally. Other important structures in the pons include the locus ceruleus, which is main source of adrenergic fibers in the CNS. The neurons in the locus ceruleus are pigmented and produce epinephrine and norepinephrine. This brainstem structure is often affected in patients with AD, PD and depression. The medulla is divided into a caudal, mid, and rostral portions. The caudal portion includes the nucleus of the spinal tract of the trigeminal nerve. The mid portion of the medulla includes the nucleus gracilis and cuneatus as well as the decussation of the pyramids. The rostral medullary portion includes the inferior olivary nucleus which connects with the cerebellum. Damage to the pons and medulla is almost always life threatening.

The cerebellum regulates body movements, and may do so by controlling the timing of skeletal muscle contractions. The cerebellum and pons are considered together because they develop from the same portion of the embryonic brain. The cerebellum is attached to the brainstem by a large mass of nerve fibers that lie lateral to the fourth ventricle on either side (Figure 3.2). The cerebellum is divided into three sections by the inferior, middle, and superior cerebellar peduncles. These contain nerve fibers between the medulla, pons, and midbrain, respectively, and the cerebellum. The largest and most prominent is the middle cerebellar peduncle. The cerebellum consists of an outer layer of grey matter, the cerebellar cortex, surrounding a core of white matter. The cortical surface is highly convoluted to form a regular pattern of narrow, parallel folds or folia. The cerebellar cortex contains three layers the outer or molecular, the Purkinje cells, and the inner granular layer. The cerebellar white matter consists of nerve fibers running to and from the cerebellar cortex. The white matter has a characteristic branching, tree-like arrangement in section, as its ramifications reach towards the surface. The cerebellum is involved with the coordination of movement.

3.2.1.5. The Spinal Cord

The spinal cord has the simplest organization of all six major divisions. It participates in the control of limb and trunk musculature, in visceral functions, and in the processing of sensory information from these structures. Also, it is a conduit for the flow of information to and from the brain. The spinal cord is the only portion of the central nervous system that has a clear external segmental organization, reminiscent of its embryonic and phylogenetic origins. The spinal cord is divided into the central thoracic lumbar and sacral segments. While in the cerebrum the gray matter is in the cortex and the white in the core, in the spinal cord the gray matter is central and the white is peripheral.

At cross section, the gray matter is the cord from the anterior and posterior columns, which contains the motoneurons and sensory neurons, respectively. An important feature of each spinal cord segment is the presence of a pair of roots (or associated branches or rootlets) called the dorsal and ventral roots. The dorsal roots contain sensory axons whereas the ventral roots contain motor axons. These sensory and motor axons, which are part of the peripheral nervous system, become mixed in the spinal nerves en route to their peripheral targets. The spinal nerves, which are also components of the peripheral nervous system, transmit sensory information to the spinal cord and motor commands to the muscles and viscera.

3.2.2. Internal Organization of the Central Nervous System

3.2.2.1. Projection and Connections in the Brain

The cerebral hemisphere and diencephalon have a more complex organization than that of the brainstem and spinal cord (Martin, 1989). The thalamus relays information from subcortical structures to the cerebral cortex via two different functional classes of nuclei; namely, those that are for relay and those that are for diffuse projection. Three of the four anatomical divisions of the thalamus serve relay functions (anterior, medial, and lateral nuclei) and one is a diffuse projection nuclei (intralaminar). Thalamic neurons send the axons to the cerebral cortex via the internal capsule, as do cortical neurons that project to subcortical sites. There are two major somatosensory pathways: the dorsal column of the medial lemniscal system, which mediates tactile, and vibration, and the anterolateral system, which mediates pain and temperature sense.

There are three other major somatosensory cortical areas, which include the primary, secondary, and tertiary somatosensory cortical areas and are somatotopically organized. The secondary somatosensory cortex and the posterior parietal cortex receive their projections from the primary somatosensory cortex and the posterior insular cortex receives input from the secondary somatosensory cortex. Corticortical projections as well as callasol connections are made by neurons of layers two and three. Descending projections to the striatum, brainstem, and spinal cord originate from neurons in layer five, while projections to the thalamus originate from neurons in layer six.

Another important source of connectivity is that of the limbic system which includes the cingulate, hippocampus, and amygdala. The hippocampal formation is integrated by an infolding of the inferomedial part of the temporal lobe into the lateral ventricle along the choroid fissure. The dentate gyrus is distributed in between the parahippocampal gyrus and the hippocampus and contains a layer of granular neurons and the molecular layer, which receives connection from the entorhinal cortex and the parahippocampal gyrus. The hippocampal formation receives projections from the inferior temporal cortex via the entorhinal cortex and from contralateral fibers via the fornix. Efferent fibers merge on the ventricular surface of the hippocampus as the fimbria. The limbic system includes intrinsic as well as extrinsic connectivity. Connections between the entorhinal cortex and the molecular layer of the dentate gyrus (perforant pathway), Mossy fibers, CA1-CA4, and subiculum integrate the intrinsical connections. These structures are often affected in AD and are responsible for the short-term memory loss in this condition. The extrinsic limbic connections are between the cingulate gyrus, the hippocampal formation, the amygdala, and the septum, which in turn connect with the hypothalamus. Overall the limbic system network should be considered as functional units that include the prefrontal cortex, cingulate cortex, amygdaloid nucleus, limbic thalamus, nucleus accumbens, anterior hypothalamus, and raphe nucleus (Morgane et al., 2005).

3.2.2.2. Laminar Organization of the Cerebral Cortex

The cerebral cortex is the structure to which the dorsal column-medial lemniscal system projects and the origin of the corticospinal tract. It has a characteristic structure with neurons that are organized into layers. Different cortical regions contain characteristically different numbers of cell layers (Figure 3.3). Most of the cerebral cortex contains at least six cell layers, and this cortex is termed the isocortex (Figure 3.3). Because the isocortex dominates the cerebral cortex of phylogenetically higher vertebrates, it is also termed neocortex. In contrast to the isocortex, the allocortex contains fewer than six layers. Although present in higher vertebrates, the allocortex dominates the cortex of phylogenetically more primitive vertebrates. The phylogenetically oldest type of allocortex, the archicortex, constitutes the hippocampal formation and contains three cell layers, which are the molecular, granular, and pyramidal. The paleocortex, thought to be a more advanced allocortex, is associated with areas that mediate olfactory function. The neocortex comprises the major sensory, motor, and association areas. Regions of neocortex that serve different functions have a different microscopic anatomy. Areas that regulate sensation have a well-developed layer IV. This is the layer to which most thalamic neurons from the sensory relay nuclei project. The primary visual cortex has this morphology. In contrast, the primary motor cortex has a thin layer IV and a thick layer V. Layer V contains the pyramidal neurons that project to the spinal cord, via the corticospinal tract. Association areas of the cerebral cortex, such as prefrontal and parietal association cortex have a morphology that is intermediate between those of sensory cortex and motor cortex. In summary, based primarily on differences in the thickness of cortical layers and on the sizes and shapes of neurons there are two types of cortex. The neocortex (or isocortex) has six layers; the allocortex has fewer than six layers and includes the archicortex of the hippocampus and the paleocortex of the olfactory regions.

3.2.2.3. Neuronal Subtypes and Patterns of Interconnectivity

The activity of the CNS depends on the complex patterns of connectivity among neurons and associated glial cells (Figure 3.4). The neurons are composed of a neuronal cell body, axons and dendrites (Figure 3.4A). The axons have a terminal end that constitutes the presynaptic site of the synapse. The dentrites have an apical site and multiple branches and spines. The connections among neurons are denominated synapses. Synapses occur between axons and dendrites, axons and cell bodies, and axons and axons. Neurons are excitatory and inhibitory. Excitatory neurons produce glutamate while inhibitory neurons produce GABA. Other neurotransmitters include acetylcholine as well as neuropeptides. Excitatory neurons are usually pyramidal (Figure 3.4A) and multipolar. Inhibitory neurons, also known as interneurons, are bipolar or pseudo unipolar and contain calcium binding proteins such as calbindin (Figure 3.4B), parvalbumin and calretinin. Types of interneurons include the Martinotti cells, chandelier cells, double bouquet cells, giant basket cells, Cajal Retzius cells and bipolar cells. These sensory neurons receive information through dendrites and transmit that information to the lCNS via axon terminals (Siegel and Sapru, 2006). Retinal bipolar cells, sensory cells of the cochlear, and vestibular ganglia are included in this category.

Pseudo-unipolar neurons have a single process that arises from the cell body and divides into two branches. One of these branches projects to the periphery, while the other projects to the CNS. Each branch has the structural and functional characteristics of an axon. Information collected from the terminals of the peripheral branch is transmitted to the CNS via the terminals of the other branch. Unipolar neurons are relatively

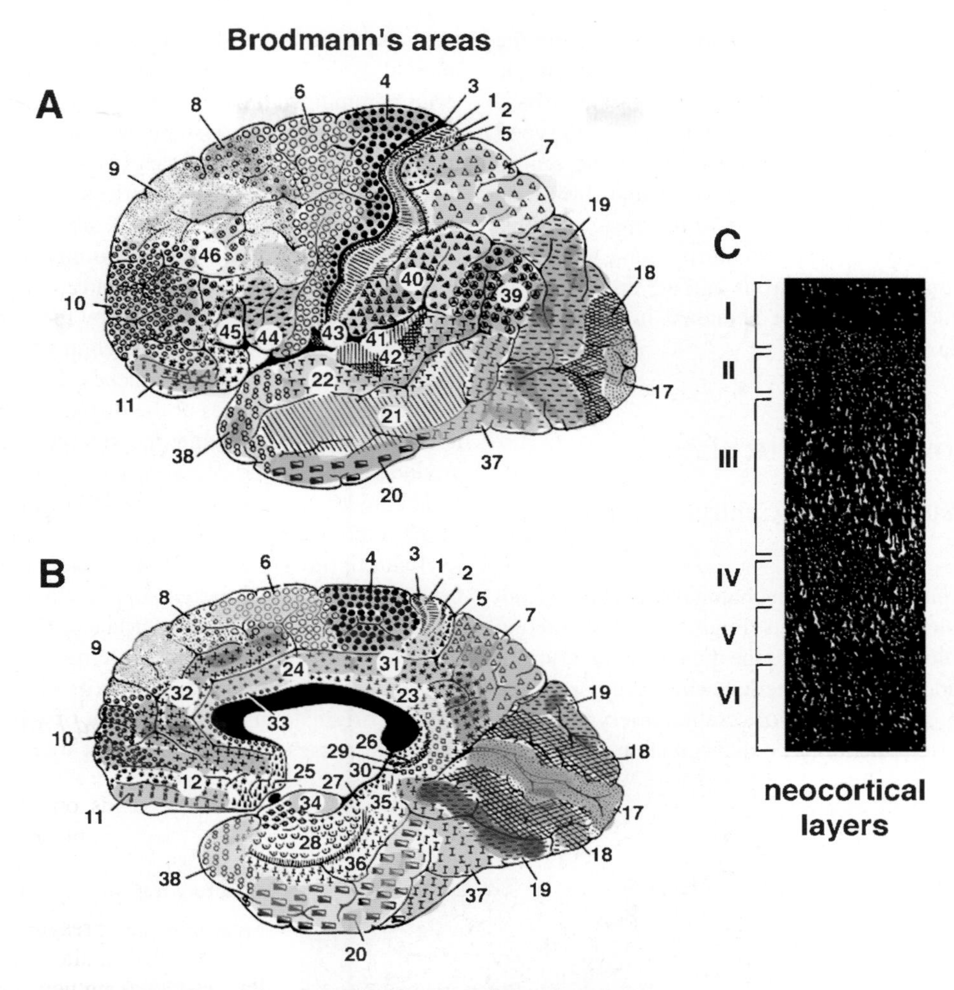

FIGURE 3.3. Diagrammatic representation of the neocortical layers and Broadman areas.

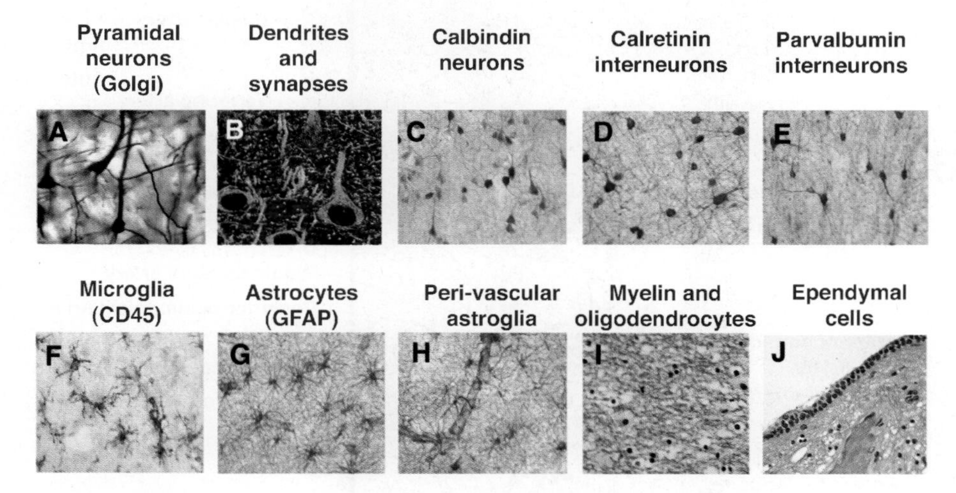

FIGURE 3.4. Cellular components of the central nervous system. (A) Neurons impregnated with Golgi silver satin, (C) Calbindin immunoreactive interneurons in the neocortex, (G) astrocytes immunoreactive with an antibody against GFAP, (H) peri-vascular astrocytes components of the BBB, (I) oligodendrocytes and white matter tracts stained with luxol fast blue, (F) ependimal cells around the periventricular zone.

rare in vertebrates. In these neurons, dendrites arise from the apical part of the cell body and axons form the base where the dendrites are located. Neurons can also be divided into principal or projecting neurons also known as type I or Golgi type I neurons. Principal neurons (e.g., motor neurons in the ventral horn of the spinal cord) have very long axons and form long fiber tracts in the brain and the spinal cord. Intrinsic neurons, also known as type II or Golgi type II neurons, have very short axons. These neurons are interneurons and are considered to have inhibitory function. They are abundant in the cerebral and cerebellar cortex.

3.3. Cerebrovascular Circulation

3.3.1. Blood Supply to the Central Nervous System

The cerebral hemispheres and diencephalon receive blood from the anterior and posterior circulations (Figure 3.5). The cerebral cortex receives its blood supply from the three cerebral arteries: the anterior and middle cerebral arteries, which are part of the anterior circulation, and the posterior cerebral artery, which is part of the posterior circulation. The diencephalon, basal ganglia, and internal capsule are supplied from branches of the internal carotid artery, the three cerebral arteries, and the posterior communicating artery (Crossman and Neary, 2005). The anterior and posterior systems are interconnected by two networks of arteries: (1) the circle of Willis, which is formed by the three cerebral arteries, the posterior communicating artery, and the anterior communicating artery, and (2) terminal branches of the cerebral arteries, which anastomose on the superior convexity of the cerebral cortex (Figure 3.5). The arterial supply of the cerebral cortex is provided by the distal branches of the anterior, middle, and posterior cerebral arteries. These branches are often termed "cortical" branches (Lee, 1995). The anterior cerebral artery originates at the division of the internal carotid artery, and courses within the interhemispheric fissure and around the rostral and dorsal surfaces of the corpus callosum.

The middle cerebral artery, which originates at the division of the internal carotid artery, passes through the lateral sulcus (Sylvian fissure) en route to the lateral convexity of the cerebral hemisphere, to which it supplies blood. The middle cerebral artery travels along the surface of the insular cortex, over the inner surface of the frontal, temporal, and parietal lobes, and appears on the lateral convexity. The posterior cerebral arteries originate at the bifurcation of the basilar artery, and each one passes around the lateral margin of the midbrain.

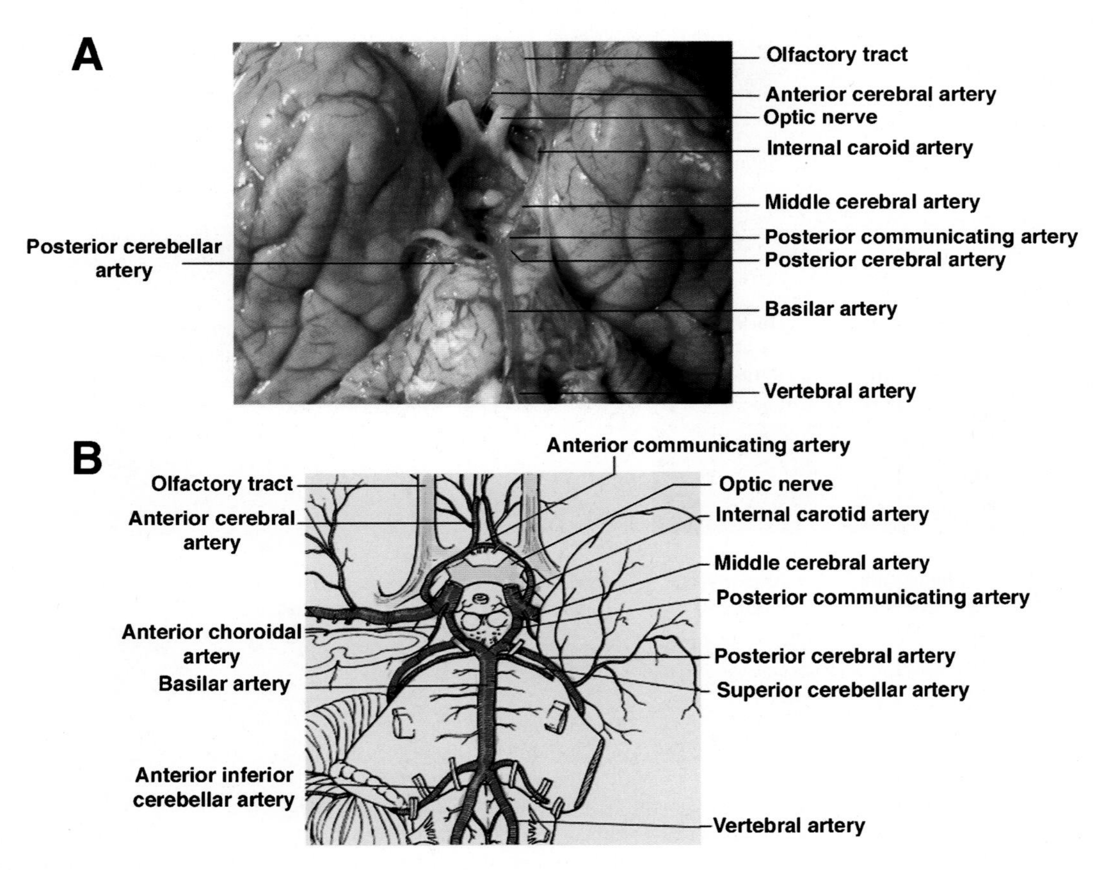

FIGURE 3.5. Gross external appearance of the circle of Willis and subarachanoid vessels.

The posterior cerebral artery supplies the occipital lobe and portions of the medial and inferior temporal lobe. The arterial supply of the spinal cord is derived from the vertebral arteries and the radicular arteries. The brain is supplied by the internal carotid arteries (the anterior circulation) and the vertebral arteries, which join at the pontomedullary junction to form the basilar artery (collectively termed the posterior circulation). The brainstem is supplied by the posterior system. The medulla receives blood from branches of the vertebral arteries as well as from the spinal arteries and the posterior inferior cerebellar artery (PICA). The pons is supplied by paramedian and short circumferential branches of the basilar artery. Two major long circumferential branches are the anterior inferior cerebellar artery (AICA) and the superior cerebellar artery. The midbrain receives its arterial supply primarily from the posterior cerebral artery as well as from the basilar artery. The venous drainage of the spinal cord drains directly to the systemic circulation. By contrast, veins draining the cerebral hemispheres and brain stem drain into the dural sinuses. Cerebrospinal fluid also drains into the dural sinuses through unidirectional valves termed arachnoid villi.

3.3.2. Immune Cell Trafficking Through the Cerebral Vascular Network

The subarachnoid vessels penetrate the cortical structures and branch to produce the microvascular network which is in turn surrounded by the cellular components of the BBB through which immune cells may need to traffic (Figure 3.5B). The capillary endothelium of the BBB contains tight junctions that control the movement of leukocytes. In addition these cells are surrounded by a basement membrane and the processes of astroglial cells. The routes for trafficking of leukocytes into the nervous system are through the choroid plexus, subarachnoid space, and perivascular space (Ransohoff et al., 2003). In the first route, leukocytes migrate from the blood to CSF across the choroid plexus. In this pathway of migration, leukocytes travel across the fenestrated endothelium of the choroid-plexus, migrate through the stromal core villi, interact with epithelial cells of the chorioid plexus and enter the CSF at its site of formation. This pathway is likely to be one route by which immune cells enter the CSF under physiological conditions using P-selectin and PECAM.

For the second route, leukocytes reach the CNS from blood to subarachnoid space. In this pathway, leukocytes travel from the internal carotid artery, across postcapillary venules at the pial surface of the brain into the subarachnoid space and the Virchow-Robin perivascular spaces. There, they might encounter cells of the monocyte/myeloid lineage that are competent for antigen presentation. The perivascular regions, where there is direct communication with the CSF compartment, are considered probable sites of lymphocyte-APC interaction and therefore, of immune surveillance of CNS. This pathway is also dependent on P-selectin.

The third pathway involves leukocyte immigration from blood to parenchymal perivascular space. In this third pathway, immune cells can enter the parenchyma directly, passing from internal carotids through the branching vascular tree of arterioles and capillaries and finally extravasating through postcapillary venules. In this case, leukocytes are required to cross the BBB and the endothelial basal lamina. Trafficking of activated lymphocytes across a resting cerebrovascular endothelium is a low-efficient event. This process is dependent on P-selectin, LFA-1 and ICAM (Greenwood et al., 2002).

3.4. Glial Cell Types

The supporting cells located in the CNS are called neuroglia or glial cells. They are relatively nonexcitable and more abundant. Neuroglia has been classified into the following groups: astrocytes, oligodendrocytes, microglia, and ependymal cells (Siegel and Sapru, 2006) (Figure 3.4). Astrocytes are the largest and have a stellate (star-shaped) appearance because their processes extend in all directions (Figure 3.4G). Their nuclei are ovoid and centrally located. The astrocytes provide support for the neurons, a barrier against the spread of transmitters from synapses, and insulation to prevent the electrical activity of one neuron from affecting the activity of neighboring neurons. Some transmitters (for example, glutamate and y-aminobutyric acid [GABA]), when released from nerve terminals in the CNS, are taken up by astrocytes, thus terminating their action. The neurotransmitters taken up by astrocytes are processed for recycling. When extracellular K^+ increases in the brain due to local neural activity, astrocytes take up K^+ via membrane channels and help to dissipate K^+ over a large area because they have an extensive network of processes. Astrocytes are further divided into the following subgroups: protoplasmic astrocytes, fibrous astrocytes, and Muller cells. Protoplasmic astrocytes are cells present in the gray matter in close association with neurons. Because of their close association with the neurons, they are considered satellite cells and serve as metabolic intermediaries for neurons. They give out thicker and shorter processes, which branch profusely. Several of their processes terminate in expansion called end-feet. The neuronal cell bodies, dendrites, and some exons are covered with end-feet joined together to form a limiting membrane on the inner surface of the pia mater (glial limiting membrane) and outer surface of blood vessels (called perivascular lining membrane). The perivascular end-feet may serve as passage for the transfer of nutrients from the blood vessels to the neurons across the BBB (Figure 3.4H). Abutting of processes of protoplasmic astrocytes on the capillaries as perivascular end-feet is one of the anatomical features of the blood–brain barrier. Fibrous astrocytes are found primarily in the white matter between nerve fibers. Several thin, long, and smooth processes arise from the cell body; these processes show little branching. Fibrous astrocytes function to repair damaged tissue, which may result in scar formation. Muller Cells are modified astrocytes present in the retina.

The oligodendrocytes are involved in the myelination process (Figure 3.4I). The oligodendrocytes present in the gray matter are called perineural oligodendrocytes. Oligodendrocytes are smaller than astrocytes and have fewer and shorter branches. Their cytoplasm contains the usual organelles (e.g., ribosomes, mitochondria, and microtubules), but they do not contain neurofilaments. In the white matter, oligodendrocytes are located in rows along myelinated fibers and are known as interfascicular oligodendrocytes.

The microglia are the smallest of the glial cells and are involved in phagocytosis and neuroinflammatory response in the CNS (Figure 3.4F). These cells are probably derived from monocytes from the bone marrow. They usually have a few short branching processes with thorn-like endings. These processes arising from the cell body give off numerous spine-like projections. They are scattered throughout the nervous system. When the CNS is injured, the microglia become enlarged, mobile, and phagocytic. Ependymal cells consist of three types of cells: ependymocytes, tanycytes, and choroidal epithelial cells. Ependymocytes are cuboidal or columnar cells (Figure 3.4G) that form a single layer of lining in the brain ventricles and the central canal of the spinal cord. They possess microvilli and cilia. The presence of microvilli indicates that these cells may have some absorptive function. The movement of their cilia facilitates the flow of the cerebrospinal fluid (CSF). Tanycytes are specialized ependymal cells that are found in the floor of the third ventricle, and their processes extend into the brain tissue where they are juxtaposed to blood vessels and neurons. Tanycytes have been implicated in the transport of hormones from the CSF to capillaries of the portal system and from hypothalamic neurons to the CSF. Choroidal epithelial cells are modified ependymal cells. They are present in the choroid plexus and are involved in the production and secretion CSF. They have tight junctions that prevent the CSF from spreading to the adjacent tissue.

3.5. Brain Regions Linked to Neurodegeneration and Other Neurological Diseases

As the life expectancy of the human population continues to increase, the possibility of developing neuroinflammatory and neurodegenerative diseases has increased considerably during the past 50 years. Of the neurodegenerative disorders, Alzheimer's Disease continues to be the leading cause of dementia in the aging population. Traditionally, neurodegenerative disorders have been defined as conditions in which there is selective loss of neurons within specific regions of the brain accompanied by astrogliosis. However, in the past 20 years, we have learned that the pathological process leading to the disfunction of selected circuitries in the brain initiates with damage to the synapses rather than with the loss of neurons. In fact, neuronal loss is a late event that is probably preceded by damage to axons and dendrites followed by shrinkage of the neuronal cell body and abnormal accumulation of filamentous proteins.

Therefore, the revised concept of neurodegeneration suggests that neuronal injury initiates at the synaptic junction and propagates throughout selected circuitries leading to neuronal dysfunction, which resolves in the classical clinical symptoms characteristic to each of the neurodegenerative disorders (Hashimoto and Masliah, 2003). For example, in Alzheimer's Disease, early damage to the synapses between the entorhinal cortex and the molecular layer of the dentate gyrus (perforant pathway) results in the short term memory deficits characteristic of this dementing disorder. Later on disconnection of the cortico-cortico fibers in the frontal, parietal, and temporal cortex results in more severe memory deficits, alterations in executive functions, and abstraction. Degeneration of connections between the nucleus basalis of Meynert and the neocortex results in attention and memory deficits usually associated with loss of cholinergic neurons. Other circuitries and neuronal populations are also affected in Alzheimer's Disease, illustrating the complexity of these disorders and the fact that the concept of single population is affected needs to be expanded to multiple populations. This is the case with several other disorders including Parkinson's Disease, where degeneration is not limited to the dopaminergic system but also involves the limbic system, the raphe nucleus, the insula, and other systems.

In response to injury, neurons produce adhesion molecules and trophic factors that recruit astroglial and microglial cells to participate in the process of repair. In addition, the microvasculature and other glial systems might also participate in the process. Thus, neurodegeneration is accompanied by astrogliosis, microgliosis, and microvascular remodeling. While astroglial cells initially produce trophic factors and cytokines that aid in tissue repair, eventually these factors could amplify the inflammatory response by increasing vascular permeability resulting in microglial activation, which in turn might lead to the production of more proinflammatory cytokines and chemokines. A critical balance between the repair and proinflammatory factors often determines the future rate and progression of the degenerative process.

The understanding of the mechanisms of neurodegeneration and inflammatory response in these neurological conditions has undergone a tremendous progress in the past 10 years. It is now generally accepted that small soluble misfolded protein aggregates denominated oligomers are responsible for the injury. So for example, in Alzheimer's Disease, A-beta protein oligomers might damage the synapses in the limbic system while in Parkinson's Disease, a-synuclein oligomers damage the axons in the striatum and cortical regions. While significant progress has been made in understanding the fundamental mechanisms for the neuronal injury, less is known about the reasons for the selective neuronal vulnerability characteristic to these neurological conditions.

Summary

The understanding of the gross and fine structure of the brain is of fundamental importance to elucidate how the nervous system works, how it interacts with other peripheral organs and tissues and how it responds to external stimuli. The main objective of this chapter will be to provide an overview to the structure and organization of the nervous system that is relevant to the understanding of the unique interactions between the nervous and the immune system and to elucidate the pathogenesis of AD, PD, HIVE and other neurodegenerative and neuroinflammatory disorders,†. The nervous system has a unique circulatory organization based on the flow of CSF, the microvasculature the BBB and the regulation of the blood flow by the neural activity that provides a series of mechanisms to selectively regulate interactions between the nervous system and trafficking cells from peripheral tissues. This chapter provides the structural framework for the better understanding of these interactions under physiological conditions and in neurological diseases and includes sections of gross anatomical structure of the brain, projections, cellular variety, and cerebrovascular circulation.

Review Questions/Problems

1. Interconnecting the two hemispheres is the

a. corpus callosum
b. anterior commissure
c. Rolandic fissure
d. Sylvia fissue
e. Precentral gyrus

2. The central fissure divides the

a. visual cortex
b. sensory and motor cortex
c. limbic system
d. brain stem
e. cerebellum

3. Correlate the brain regions with the neurotransmitter

a. acetyl choline _____
Neocortex, hippocampus
b. glutamate _____
Locus ceruleus
c. norepinephrine _____
S. Nigra
d. dopamine _____
Basal forebrain

4. Correlate the brain region with the disease

a. Parkinson's disease _____
Neocortex, hippocampus, basal forebrain
b. Alzheimer's disease _____
Basal Ganglia
c. Huntington's disease _____
S. Nigra
d. Aphasia _____
Broca's Area

5. The primary auditory cortex or gyri of Heschl is located in the

a. frontal lobe
b. parietal lobe
c. occipital lobe
d. temporal lobe
e. brainstem

6. The middle cerebral artery irrigates the

a. fronto-temporal cortex
b. visual cortex
c. brainstem
d. cerebellum

7. Neurons that produce GABA are also known as

a. pyramidal cells
b. inhibitory interneurons
c. microglia
d. astrocytes

8. The perforant pathway connects the

a. entorhinal cortex with the forebrain
b. hippocampus with the hypothalamus
c. entorhinal cortex with hippocampus
d. cerebellum and brainstem
e. hippocampus and midbrain

Acknowledgments. The author would like to thanks Ms. Maria Alonso for invaluable help preparing this chapter. The author is supported by NIH grants AG5131, AG18440, MH6512 and MH62962.

References

Avital A, Goshen I, Kamsler A, Segal M, Iverfeldt K, Richter-Levin G, Yirmiya R (2003) Impaired interleukin-1 signaling is associated with deficits in hippocampal memory processes and neural plasticity. Hippocampus 13:826–34.

Buller KM (2003) Neuroimmune stress responses: Reciprocal connections between the hypothalamus and the brainstem. Stress. 6(1):11–17.

Crossman AR, Neary D (2005) Neuroanatomy: An illustrated color text. Churchill Livingstone, London, UK.

Engelhardt B, Ransohoff RM (2005) The ins and outs of T-lymphocyte trafficking to the CNS: Anatomical sites and molecular mechanisms. Trends Immunol 26:485–495.

Eskandari F, Webster J, Sternberg EM (2003) Neural immune pathways and their connection to inflammatory diseases. Arthritis Res Ther 5:251–265.

Gendelman HE (2002) Neural immunity: Friend or foe? J Neurovirol 8:474–9.

Greenwood J, Etienne-Manneville S, Adamson P, Couraud PO (2002) Lymphocyte migration into the central nervous system: Implication of ICAM-1 signaling at the blood-brain barrier. Vascular Pharmacol 38:315–22.

Hashimoto M, Masliah E (2003) Cycles of aberrant synaptic sprouting and neurodegeneration in Alzheimer's and dementia with Lewy bodies. Neurochem Res 28:1743–56.

Ho GJ, Drego R, Hakimian E, Masliah E (2005) Mechanisms of cell signaling and inflammation in Alzheimer's disease. Curr Drug Targets Inflamm Allergy 4:247–56.

Lee RMKW (1995) Morphology of cerebral arteries. Parmac Ther 66:149–173.

Martin JH (1989) Neuroanatomy: Text and Atlas. Appleton & Lange, Norwalk Connecticut.

Morgane PJ, Galler JR, Mokler DJ (2005) A review of systems and networks of the limbic forebrain/limb midbrain. Prog Neurobiol 75:143–160.

Opp M, Toh LA (2003) Neural-immune interactions in the regulation of sleep. Front Biosci 8:768–779.

Owens T, Babcock AA, Millward JM, Toft-Hansen H (2005) Cytokine and chemokine inter-regulation in the inflamed or injured CNS. Brain Res Rev 48:178–184.

Pavlov VA, Tracey KJ (2005) The cholinergic anti-inflammatory pathway. Brain Behav Immun 19:493–9.

Ransohoff RM, Kivisaekk P, Kidd G (2003) Three or more routes for leukocyte migration into the central nervous system. Nature 3:569–581.

Ringheim GE, Conant K (2004) Neurodegenerative disease and the neuroimmune axis (Alzheimer's and Parkinson disease, and viral infections) J Neuroimmunol 147:43–49.

Siegel A, Sapru HN (2006) Essential Neuroscience. Baltimore Maryland, Philadelphia Pennsylvania, USA: Lippincott Williams & Wilkins.

Ziv Y, Ron N, Butovsky O, Landa G, Sudai E, Greenberg N, Cohen H, Kipnis J, Schwartz M (2006) Immune cells contribute to the maintenance of neurogenesis and spatial learning abilities in adulthood. Nat Neurosci 9:268–275.

4
The Blood Brain Barrier

William A. Banks

Keywords Active transport; Adsorptive endocytosis and transcytosis; Brain endothelial cell; Cytokine; Endothelin; Facilitated diffusion; Immune cell; Neurovascular unit; Transmembrane diffusion

4.1. Introduction

The blood-brain barrier (BBB) is intimately involved in regulation of the neuroimmune axis. It first prevents the unrestricted mixing of the fluids of the central nervous system with the blood. If then selectively regulates the exchange of cells, cytokines, and other solutes between the central nervous system and the blood. This regulatory aspect of the BBB is accomplished through various mechanisms and is itself affected by neuroimmune physiologica and pathophysiologic phenomena. This chapter will explore the intimate connects between the BBB and other components of the neuroimmune axis.

4.2. Development and Structure of the Blood-Brain Barrier

Evidence for an interface between the circulation and the central nervous system (CNS) dates back to the end of the nineteenth century (Bradbury, 1979). The best known of those early studies were done by a young Paul Erlich who found that some dyes did not stain the brain after their peripheral injection. Erlich concluded erroneously that the lack of staining was because these dyes did not bind to brain tissue. Several decades later, Goldmann, a student of Erlich's, found that these dyes could stain the brain when injected intravenously. Thus, these dye studies were reinterpreted as evidence in favor of some sort of barrier between the CNS and blood.

The location and nature of that barrier was controversial through much of the twentieth century. Elegant studies by Davson and colleagues identified the barrier at the vascular level. However, alternative opinions were held until classic studies were conducted with the electron microscope by Reese and co-workers in the late 1960s (Reese and Karnovsky, 1967; Brightman and Reese, 1969). Previous work had shown no difference between vascular beds of peripheral tissues and the CNS when studied grossly or at the light microscope level. However, Reese and co-workers found numerous ultrastructural differences. These included a much-reduced rate of pinocytosis and an absence of intracellular fenestrations. The most widely discussed finding, however, is the presence of tight junctions between adjacent endothelial cells. The tight junctions, low rate of pinocytosis, and low number of intracellular fenestrations effectively eliminate capillary gaps and pores. This, in turn, essentially eliminates the production of a plasma-derived ultrafiltrate and hence the leakage of serum proteins into the brain.

From this single change, the lack of a production of an ultrafiltrate, evolves a large number of consequences for CNS function. Obviously, it is the basis of the restriction of protein access, which first defined the BBB in late nineteenth century. The need for an efficient lymphatic system is eliminated, but the lack of a lymphatic system means that the CNS needs other methods to rid itself of the free water and wastes produced by metabolism and the secretions of the choroid plexus. Without production of an ultrafiltrate, the CNS depends on other methods to extract nourishment from the blood. The BBB addresses this need with a large number of selective transporters for substances from electrolytes to regulatory proteins (Davson and Segal, 1996e; Davson and Segal, 1996d). Because the CNS is not equipped to handle an ultrafiltrate, its reintroduction, as with hypertensive crisis, can result in increased intracranial pressure and encephalopathy (Al-Sarraf and Phillip, 2003; Johansson, 1989; Mayhan and Heistad, 1985).

4.2.1. Components of the BBB

The BBB is not a single barrier, but several barriers, which are in parallel. This contrasts with the testis-blood barrier, which consists of several barriers in series. The most studied of these barriers is the vascular barrier and perhaps the least studied are the barriers, that interface between the circumventricular organs (CVO) and the rest of the CNS.

T. Ikezu and H.E. Gendelman (eds.), *Neuroimmune Pharmacology*.
© Springer 2008

4.2.1.1. *Vascular BBB*

The vascular BBB occurs because of the modifications noted by Reese and co-workers in the endothelial cells, which comprise the capillary bed and line the venules and arterioles of the CNS. It is likely that these three regions are highly specialized. For example, immune cells primarily cross at the venules and most of the classic transporters are located at the capillaries (Engelhardt and Wolburg, 2004). No CNS cell is more than about 40 μm from a capillary. This means that a substance, that can cross the vascular BBB can immediately access the entire CNS. Substances that cross the vascular BBB can be either flow-dependent or not dependent on flow rate. A flow-dependent substance is one in which the BBB extracts from the blood nearly the maximal amount possible (Kety, 1987). The only way to increase the amount of the substance entering the brain is to increase the flow rate to the brain. Glucose is an example of a flow-dependent substance (Rapoport et al., 1981). A brain region that is particularly active has its increased demand for glucose met by an increase in regional blood flow. In contrast, transport of a cytokine such as tumor necrosis factor (TNF) is not flow dependent. Only a small percent of the TNF in blood is extracted by brain via the saturable transporter for TNF located at the BBB (Banks et al., 1991). Alterations of blood flow within physiological limits do not alter the uptake of TNF from blood by brain. However, extreme changes in the rate of blood flow or capillary tortuosity can result in rheological changes, such as the loss of laminar flow. Such alterations likely occur in stroke, AIDS, and Alzheimer's disease (de la Torre and Mussivand, 1993; Nelson et al., 1999). This may result in impaired permeation of flow-dependent and non-flow dependent substances.

4.2.1.2. *Choroid Plexus*

The choroid plexus are bags composed of epithelial cells that project into the ventricles and contain a capillary plexus (Johanson, 1988). The capillaries do not have barrier function and so produce an ultrafiltrate, which fills the bag. The epithelial cells have tight junctions and so prevent the ultrafiltrate from entering the ventricular space. Unlike the capillaries, the epithelial cells of the choroid plexus have a high rate of vesicular turnover, which is responsible for the production of the cerebrospinal fluid (CSF). However, the CSF is not an ultrafiltrate, but a secreted substance. The choroid plexus also has many selective transport systems, some of which are specific to it or are enriched in comparison to the vascular BBB.

4.2.1.3. *Tanycytic Barrier*

The CNS of mammals contains seven regions of the brain where the vasculature does not fully participate in a BBB (Gross and Weindl, 1987). These regions have at least one side that faces a ventricle and so are termed circumventricular organs (CVOs). Together, they comprise about 0.5% of the brain by weight. Their capillaries allow the production of an ultrafiltrate and so their cells are in more intimate contact with the circulation. They are known to play vital roles as sensing organs for critical peripheral events; for example, they act as emetic centers and are important in blood pressure modulation (Johnson and Gross, 1993; Ferguson, 1991). They can relay their signals to the rest of the brain by neurons, which project from them to distant brain regions or project to them from other brain regions. However, the mixing of their interstitial fluids with that of adjacent brain tissue has been shown to be limited in most studies (Peruzzo et al., 2000; Plotkin et al., 1996; Rethelyi, 1984). Diffusion through brain tissue is poor and this alone would tend to produce a limit to mixing within a few hundred microns of the CVO (Cserr and Berman, 1978; de Lange et al., 1995). However, most studies also find a physical barrier to diffusion. The epithelial cells, which line the ventricles form tight junctions when they are over CVOs, thus limiting CVO-to-CSF diffusion. A functional barrier also exists for the diffusion of substances from the CVO to the adjacent brain region (Peruzzo et al., 2000; Plotkin et al., 1996; Rethelyi, 1984). Recent work has shown that bands of tanycytes limit diffusion out of the median eminence to the adjacent arcuate nucleus.

4.2.2. Perinatal Development and Special Characteristics of the Neonatal BBB

The idea that the rodent perinatal BBB is not developed has been extensively revised over the last few decades (Davson and Segal, 1996b). Many of the differences in the BBB of developing and adult animals, which were ascribed to an immature brain, are now known to be adaptions to the altered demands of the CNS. For example, some amino acids enter the CNS of neonates more rapidly than the CNS of adults not because the BBB is defective, but because BBB amino acid transporters altered to favor their transport. The BBB is slave to the CNS and so the reason the BBB transports them more avidly is because they are in greater demand by a developing CNS. Most of these adaptations involve transporter mechanisms. In comparison, non-saturable passage is not altered with development (Cornford et al., 1982).

This theme of the BBB adjusting its transporter capabilities to serve the CNS is repeated throughout development, probably also with aging, and even in disease conditions. One of the most dramatic examples is the developmental loss of the mannose 6-phosphate receptor, which transports the enzyme β-glucuronidase across the BBB. Transport function is very robust in neonates but is absent in adult animals (Urayama et al., 2004). Transgenic mice that do not produce β-glucuronidase develop a muccopolysachharidosis, which recapitulates Sly's disease (Sands et al., 1994). Because of the developmental differences in the BBB expression of the mannose 6-phosphate receptor, the brain disease of neonates but not of adults responds to the peripheral administration of glucuronidase (Vogler et al., 1999).

Work with marsupials show that their BBB is largely intact in what is essentially the *in utero* period (Dziegielewska et al., 1988). As reviewed elsewhere (Urayama et al., 2004), work by several other laboratories has repeatedly shown that not only is the perinatal rodent BBB not leaky to albumin, but that the vascular space is smaller per gram of brain. Some evidence suggests, however, that the barrier may be more leaky to much smaller molecules such as sucrose.

There are other aspects of barrier function that also change with development. The tanycytic barrier between the median eminence and the arcuate nucleus develops after birth in the rodent (Peruzzo et al., 2000). This means the arcuate nucleus is very vulnerable to circulating neurotoxins during the neonatal period. For example, monosodium glutamate destroys the arcuate nucleus with resulting obesity when given intravenously to a neonate, but not an adult. The epithelial cells, which line the ventricles of the brain, have tight junctions even over non-CVO sites in neonates, but not adults. Thus, neonates have a CSF-brain barrier, which limits the diffusion of substances between brain tissue and CSF.

4.2.3. Concept of the Neurovascular Unit and Comparison to Peripheral Vascular Beds

The endothelial cell is an anatomical location of barrier function and of the various saturable transporters (Figure 4.1). Capillary beds from peripheral tissues have numerous intracellular and intercellular pores and fenestrations and high rates of pinocytosis that account for their leakiness. The brain endothelial cell engages in comparatively little pinocytosis, has few intracellular pores or fenestrations, and intercellular pores or gaps are eliminated because of tight junctions.

However, the brain endothelial cell does not function in isolation. The abluminal (brain side) of the capillary is encased in a basement membrane 40–80 nm thick. This membrane does not act as a barrier to molecules but may restrict viral-sized particles (Muldoon et al., 1999). It also holds pericytes in close approximation to the endothelial cell (Balabanov and Dore-Duffy, 1998). The pericyte may be a pluripotent cell and may also act in opposition to astrocytes (Deli et al., 2005). Astrocytes project endfeet that surround the capillary in what looks at the ultrastructural level like a mesh or netting. Astrocytes secrete substances that tend to encourage tight junction formation. Pericytes tend to oppose this action of astrocytes. All three of these cell types (pericytes, astrocytes, and endothelial cells) secrete a variety of substances, including cytokines, into their local environment (Fabry et al., 1993). This greater BBB complex is in further communication with other cell types in the CNS, most notably microglia and neurons. The microglia may, in turn, be at equilibrium with circulating macrophages (Williams and Hickey, 1995). Other immune cells also enter and exit the CNS at unknown rates and frequencies as influenced by yet to be determined factors. Clearly, secretions of prostaglandins, nitric oxide, and cytokines from each of these cells are important for intercellular communication and can influence endothelial cell permeability (Chao et al., 1994; Nath et al., 1999; Shafer and Murphy, 1997).

The concept of the neurovascular unit (NVU) emphasizes the interactive role that cells and events within the CNS and in the circulation play on BBB permeability, as well as the role that the consequences of BBB permeability play on them. The concept of the NVU includes other factors long known to influence the penetration of substances across the BBB, such as degradation, sequestration, and serum protein binding. The encompassing concept of the NVU is particularly useful when considering the next section, the mechanisms of transport across the BBB.

4.3. Mechanisms of Transport Across the BBB

Substances can enter or exit the CNS by a variety of mechanisms. Some of these mechanisms are operational in both the blood-to-brain (influx) or the brain-to-blood (efflux) directions, whereas others are unidirectional.

4.3.1. Blood to CNS

Saturable and non-saturable modes predominate influx. Within each of these categories are a diverse number of mechanisms. These different mechanisms tend to favor certain groups or types of substances.

4.3.1.1. Non-Saturable Passage

The hallmark of non-saturable passage is that the percent of material crossing into the CNS is not affected by the amount of material available for transport. The two main mechanisms of non-saturable passage are transmembrane diffusion and the extracellular pathways. The former is much better studied and its principles are widely applied by industry for the development of CNS drugs; the latter has received much less attention.

4.3.1.1.1. Transmembrane Diffusion

The major non-saturable mechanism by which small molecules cross the BBB is by membrane or transmembrane diffusion (Rapoport, 1976). The major determinant of passage is the degree to which the substance is lipid soluble. A substance that is too lipid soluble will be unable to repartition into the brain's interstitial fluid and so become trapped in the cell membranes of the BBB. A ratio of about 10:1 in favor of lipid vs aqueous solubility is near ideal for maximal passage across the BBB. The second most important determinant is molecular weight with passage being favored for smaller molecules. Other physicochemical determinants, such as charge, can occasionally become dominant for specific compounds. Many exogenous substances, including many drugs with CNS activity, enter the brain predominantly by way of transmembrane diffusion. Morphine and ethanol are prime examples of common substances, which cross the BBB by this mechanism (Oldendorf, 1974).

There seems to be no absolute molecular weight cut-off for transmembrane diffusion. A previous study, which had thought to define such an absolute limit had discovered, in retrospect, early evidence for an efflux system (Levin, 1980). The largest substance to date noted to have a measurable uptake by brain

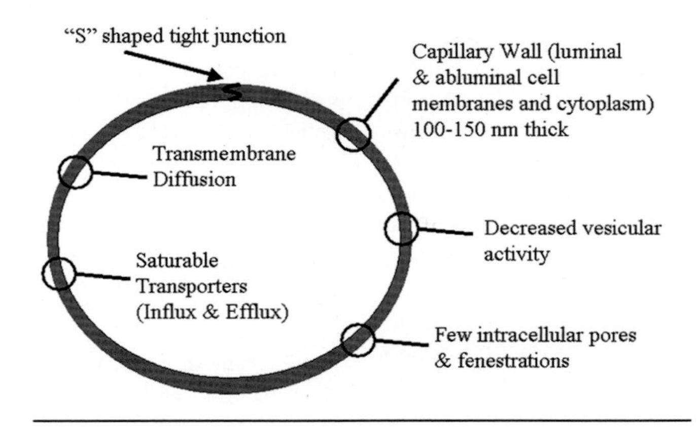

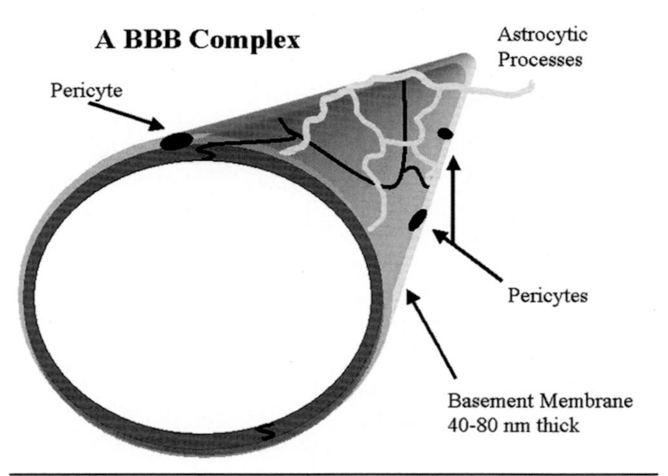

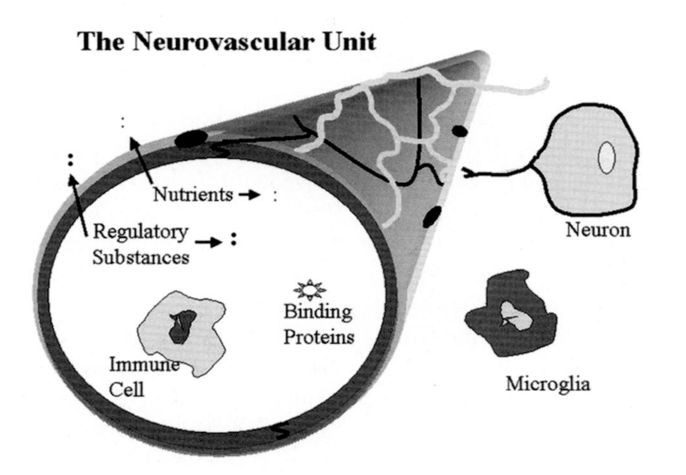

FIGURE 4.1. The vascular blood-brain barrier: three levels of complexity. The upper panel illustrates the brain endothelial cell. This is the functional and anatomical site of both barrier function and of saturable and non-saturable mechanisms of passage. The major modifications allowing both barrier function and selective penetration of substances are indicated. The middle panel illustrates other cell types and structures important in BBB function. Pericytes are embedded in a basement membrane and astrocytes form a net-like structure over the capillary bed. Both cell types are in paracellular communication with the brain endothelial cells. Pericytes and astrocytes to some extent oppose each others effects on BBB functions. The lower panel illustrates the neurovascular unit, a concept, that emphasizes integration of peripheral, BBB, and central interactions.

by way of transmembrane diffusion is cytokine-induced neutrophil chemoattractant-1 (CINC1), with a MW of about 7.8 kDa (Pan and Kastin, 2001a). A surprisingly large number of small, lipid soluble compounds cross the BBB at a rate which is considerably greater or lesser than that predicted by their physicochemical characteristics (Oldendorf, 1971; Oldendorf, 1974). Binding to serum proteins and efflux systems are the major factors, that decrease influx and the presence of a saturable transporter is the major factor that increases influx.

4.3.1.1.2. Extracellular Pathways

Some protein is present in the CSF, showing that the BBB is not absolute. The amount of protein in CSF, however, is very small, being about 0.5%, or 1/200th, of that in plasma. The CSF is not an ultrafiltrate, but a secreted fluid. This means that the relative and absolute concentrations of proteins, electrolytes, minerals, and other substances can differ tremendously to that of plasma. The extracelluar pathways are another avenue by which substances can enter the CNS (Balin et al., 1986; Broadwell and Sofroniew, 1993). These pathways represent what have sometimes been termed functional leaks at discreet areas of the brain, including the large vessels of the pial surface and subarachnoid space, the circumventricular organs, the nasal epithelium, the sensory ganglia of spinal and cranial nerves, and some deep brain regions, such as the nucleus tractus solitarius (Broadwell and Banks, 1993).

The amount of a substance that can enter the brain by the extracellular pathways is small. However, this route may be therapeutically relevant for compounds that have favorable peripheral pharmacokinetics, such as a long serum half-life and a small volume of distribution (Banks, 2004). Therapeutic antibodies and erythropoietin can access the brain by way of the extracellular pathways (Banks et al., 2004a; Banks et al., 2002; Banks et al., 2005a; Kozlowski et al., 1992) and this may underlie their therapeutic benefits (Alafaci et al., 2000; Ehrenreich et al., 2002; Erbyraktar et al., 2003; Morgan et al., 2000; Janus et al., 2000; DeMattos et al., 2002; Farr et al., 2003; Hock et al., 2003).

4.3.1.2. Receptor-Mediated and Saturable Transporters

Saturable processes represent a diverse group of mechanisms. Included in this group for purposes of discussion are two processes, which share some characteristics with the saturable systems: diapedesis and adsorptive endocytosis/transcytosis.

4.3.1.2.1. Active Transport vs Facilitated Diffusion

Saturable transporters (Yeagle, 1987) can be divided into those that require energy (active transport) and those that do not (facilitated diffusion). Both are dependent on a protein that acts as the transporter, may have co-factors, or be modulated by disease processes. Energy requiring systems can be unidirectional; that is, they may have only an influx or efflux component. Non-energy requiring saturable transport (facilitated diffusion) is bidirectional; that is, it transports substances in both directions with net flux being from the side of higher concentration to the side of lower concentration.

Most known saturable transporters at the BBB are facilitated diffusion systems (Davson and Segal, 1996c). For example, GLUT-1, the transporter for glucose, is a facilitated diffusion transporter. If the level of glucose is artificially raised above that of serum (or if radioactive glucose is introduced into the CNS, but not the serum), efflux of glucose can be shown.

4.3.1.2.2. Transcytotic vs Transmembrane Transport

Saturable transporters can also be categorized based on whether they use pores or vesicles to transport their ligands across the BBB. In the pore system, the molecule crosses from one side of the cell membrane to the other by passing through a cavity in the transporter protein. The substance is thus transported either into or out of the cytoplasm of the BBB cell; a second set of transporters on the opposing cell membrane completes the transfer across the BBB or the substance can rely on transmembrane diffusion. With vesicular transport, the transported substance binds to a receptor, invagination produces a vesicle, which is then routed to the opposite membrane, and the contents of the vesicle are released from the cell surface. Most small molecules, such as glucose and amino acids, use pores. Pore systems may be either active or facilitated diffusion systems. Vesicular transporters, on the other hand, are energy requiring and so are characterized by unidirectional transport. It is reasonable to assume that very large molecules would be required to use vesicles rather than pores to cross, but the molecular weight at which vesicles would be requisite is not known. It has been proposed that interleukin-2 (IL-2) is transported (Drach et al., 1996) by p-glycoprotein (P-gp). As P-gp is a pore system (Begley, 2004), IL-2 would be the largest substance currently known to be transported by a pore system. Peptides much smaller than IL-2 are known to cross by vesicular dependent pathways (Shimura et al., 1991; Terasaki et al., 1992). It is clear, then, that the size of the ligand alone does not dictate the need for vesicular transport.

4.3.1.2.3. Diapedesis of Immune Cells

A major shift in thinking about the relation of immune cells to the CNS and BBB has occurred over the last few decades. The CNS was once viewed as separate from the immune system and sterile in terms of immune cell occupancy except under conditions of brain infection. It is now clear that immune cells patrol the normal CNS, although the rate at which they enter and exit is not known. A major type of brain cell, the microglia, is known to be derived from peripheral macrophages, although the extent to which the pools of peripheral macrophages and microglia mix in the normal postnatal condition is unknown.

4.3.1.2.4. Adsorptive Endo- and Transcytosis

Adsorptive endocytosis occurs when a glycoprotein on the brain's endothelial surface binds another glycoprotein in ligand like fashion (Broadwell et al., 1988; Broadwell, 1989). This second glycoprotein (the ligand) may be free or attached to the surface of a virus or immune cell (Mellman et al., 1986). The binding can initiate endocytosis with the subsequent vesicle having several potential fates (Banks and Broadwell,

1994). In some cases, the vesicle is routed to lysosomes, the glycoprotein destroyed, and the vesicle rerouted to the endothelial cell surface for discharge of contents. In other cases, the vesicle can be routed to the Golgi complex and endoplasmic reticulum. In other cases still, the vesicle can be discharged at the endothelial cell surface opposite to that of uptake. In this case, the vesicle has crossed the width of the endothelial cell, and hence crossed the BBB, in a transcytotic event. What determines the fate of these vesicles is largely unknown, but at least some vesicles can engage in more than one fate (Broadwell, 1993). It may be that binding of a large amount of glycoprotein to the endothelial cell can overwhelm the lysosomal pathway and result in the vesicles being routed to the trancytotic or Golgi complex pathways.

Several principles of adsorptive endocytosis and transcytosis are clear. Many of the glycoprotein ligands are toxic and endocytosis may represent a mechanism to rejuvenate or repair the membrane (Raub and Audus, 1990; Vorbrodt and Trowbridge, 1991; Westergren and Johansson, 1993). Many viruses coopt adsorptive endocytosis mechanisms to invade and infect brain endothelial cells and adsorptive transcytosis to invade the brain (Marsh, 1984; Chou and Dix, 1989; Schweighardt and Atwood, 2001). These processes may also be related to diapedesis as many of the events of immune cell passage across the BBB resemble these endocytic mechanisms. For example, both LFA-1 and ICAM, important to immune cell passage across the BBB, are glycoproteins. Although adsorptive endocytosis is in some sense saturable because of a finite amount of any single glycoprotein on a cell surface, it is not easy to demonstrate classical saturable kinetics for this process. In fact, excess glycoprotein can sometimes further stimulate endocytosis and so lead to a paradoxic increase, rather than decrease, in the rate of passage across the BBB (Banks et al., 1997). Glycoprotein distribution on brain endothelial cells is polarized; that is, a glycoprotein may be enriched on either the luminal or abluminal membranes (Vorbrodt, 1994; Zambenedetti et al., 1996). The tight junctions act as a fence to keep the glycoproteins confined to their respective sides of the endothelial cell (Deli et al., 2005). This means that the movement of a glycoprotein molecule (or a virus whose coat displays that glycoprotein) can be unidirectional as its transcytosis can only be initiated from the side of the brain endothelial cell, which contains the ligand's complementary glycoprotein (Villegas and Broadwell, 1993; Broadwell, 1989). The possession and distribution of glycoproteins similarly dictate which viruses can invade the brain; neurovirulent viruses, which invade the brain as free virus (as opposed to entering in Trojan horse fashion inside an infected immune cell) can do so because they possess a glycoprotein ligand capable of binding to the BBB.

4.3.2. CNS to Blood

Traditionally, passage in the brain-to-blood direction (efflux) has been neglected. However, efflux often accounts for the inability of otherwise effective drugs to accumulate in the CNS. Pharmacogenomic studies have suggested that the individual variation in efflux mechanisms may explain why some individuals are less sensitive to the CNS effects of drugs or more sensitive to their toxicities (LÜscher and Potschka, 2002; Fellay et al., 2002). Efflux mechanisms are important to the homeostasis of the CNS, ridding the brain of toxins (Taylor, 2002). The rate of efflux can be, in addition to synthesis and degradation, an important determinant of the level of a substance produced within the CNS (Chen et al., 1997; Chen and Reichlin, 1998; Maness et al., 1998).

4.3.2.1. Non-Saturable

Efflux, like influx, has both saturable and non-saturable mechanisms. Transmembrane diffusion occurs for both influx and efflux. Other mechanisms, such as bulk flow, are unique for efflux.

4.3.2.1.1. Transmembrane Diffusion

Many of the principles that govern influx by transmembrane diffusion are also important in efflux. The dramatic role that efflux by transmembrane diffusion can play can be illustrated by comparing the fate of small, lipid soluble molecules to that of a protein after intrathecal administration. Intrathecal application of small, lipid soluble molecules, such as anesthetics, can have a local effect on spinal cord function but have little or no effect on the brain (Bernards, 1999). These substances readily cross the brain endothelial cell by transmembrane diffusion and do this as easily in the brain-to-blood direction as in the blood-to-brain direction. Therefore, they are cleared from the CSF before they are able to reach the brain (McQuay et al., 1989). In contrast, proteins such as leptin are too large and water soluble to undergo much transmembrane diffusion (LeBel, 1999; McCarthy et al., 2002). Leptin can reach the brain after intrathecal administration in amounts sufficient to produce effects on feeding through the hypothalamus (McCarthy et al., 2002; Shyng et al., 1993).

Efflux by transmembrane diffusion can also contribute to the poor diffusion of substances within brain parenchyma. Diffusion within the interstitial space of the brain is dependent on Brownian motion and the production of metabolic free water as driving forces and so is very slow (Cserr, 1984; Cserr and Berman, 1978). However, efflux by non-saturable (and saturable) mechanisms can further reduce the distance a substance will ultimately diffuse. For example, the less lipid soluble drug atenolol can diffuse about 3 times further into brain tissue than can the more lipid soluble drug acetaminophen (de Lange et al., 1993).

4.3.2.1.2. Bulk Flow and CSF Lymphatic Drainage

Bulk flow refers to the reabsoption of CSF into the blood at the arachnoid villi (Davson and Segal, 1996a). Any substance dissolved in CSF will enter the blood by this mechanism (Jones and Robinson, 1982; Pollay and Davson, 1963). In some cases, the levels of a substance in blood achieved after injection into the CSF can be sustained longer and at higher levels than after an intravenous bolus (Maness et al., 1998; Chen and Reichlin, 1998; Chen et al., 1997). This is because the central injection acts similarly to an intravenous infusion, slowly delivering drug to the blood. CSF drains from the brain at the level of the cribriform plate into the cervical lymphatic system (Widner et al., 1987; Yamada et al., 1991). This may be the dominant route for CSF drainage at normal CSF pressures (Boulton et al., 1999). This can provide a direct route from the CNS to the cervical lymphatics (Oehmichen et al., 1979), as has been illustrated for gp120, the glycoprotein of the human immunodeficiency virus, HIV-1 (Cashion et al., 1999). This route to the lymphatics may explain why substances injected into the brain can produce a different immune response than when the substance is injected peripherally (Cserr and Knopf, 1992; Knopf et al., 1995).

4.3.2.2. *Saturable Transport*

The last decade has seen a huge increase in the interest of efflux by saturable mechanisms. Just as efflux by transmembrane diffusion can limit diffusion of a substance within the CNS, so can the presence of a saturable efflux transporter (Blasberg, 1977). Much of this interest centers around the multi-drug efflux transport systems (Begley, 2004), most notably P-gp. However, other efflux transporters for peptides, proteins, endogenous substances, and drugs are known to play important roles in physiology and disease (Drion et al., 1996; Martins et al., 1997; Mealey et al., 2001). For example, peptide transport system-1 is a major regulator of brain levels of methionine enkephalin, an endogenous opiate, that suppresses voluntary ethanol drinking (Plotkin et al., 1998). Depression and recovery of peptide transport systems-1 with ethanol drinking may relate to alcohol withdrawal seizures (Banks and Kastin, 1994; Banks and Kastin, 1989). IL-2 is currently the only cytokine known to be transported by a saturable efflux system; some have postulated this transporter may be P-gp. Poor accumulation of protease inhibitors, antibiotics, AZT, anti-cancer drugs and many other substances occurs because of efflux systems (Fellay et al., 2002; Glynn and Yazdanian, 1998; King et al., 2001; Lee et al., 1998; LÜscher and Potschka, 2002; Masereeuw et al., 1994; Spector and Lorenzo, 1974). Impaired efflux of amyloid β protein, the protein believed to cause Alzheimer's disease, develops with aging in mice which overexpress amyloid precursor protein, thus promoting further accumulation within brain of amyloid β protein (Ghersi-Egea et al., 1996; Banks et al., 2003; Deane et al., 2004). Evidence suggests that impaired transport develops in humans as well and so may be a major mechanism for induction of Alzheimer's disease (Tanzi et al., 2004; Shibata et al., 2000).

4.4. Neuroimmune Interactions

The above discussion of BBB fundamentals is tailored towards understanding the role of the BBB in neuroimmune interactions. Below are specific examples of how the BBB is involved in neuroimmune interactions.

4.4.1. Receptors that are Expressed on BBB for Receptor: Ligand Interactions

An important distinction for understanding the function of the BBB is that of receptors vs transporters. The term receptor has undergone a transformation of its usage since its introduction in the late nineteenth century when it was first used to denote some physiological function. Eventually, the term receptor was used to denote a physical binding site through which a drug or hormone could exert its effects on a cell. In the 1980s, a distinction was made between receptor and binding sites, the former being coupled to intracellular machinery that translated its binding into a cellular effect. Binding sites on the brain endothelial cell can represent transporters, but they can also represent traditional receptors, that is, binding sites coupled to intracellular machinery. For example, brain endothelial cells have both insulin receptors and transporters. As a result, insulin is transported across the BBB to exert effects inside the CNS, but insulin also alters a number of functions of the brain endothelial cell. As examples of the latter, insulin alters the BBB transport of AZT (Ayre et al., 1989), tryptophan (Cangiano et al., 1983) and leptin (Kastin and Akerstrom, 2001) and alters brain endothelial cell alkaline phosphatase activity (Catalan et al., 1988). Many in vitro BBB studies have assumed that a binding site represents transporter function and many in vivo studies are so designed as to not consider whether receptors as well as transporters may exist at the BBB. However, a great deal of indirect evidence and some direct evidence indicates that the vascular BBB and the choroid plexus probably possess a large variety of receptors, which can alter BBB functions. Besides insulin, substances which bind to and alter the function of brain endothelial cells include mu opiate receptor ligands (Baba et al., 1988; Vidal et al., 1998; Chang et al., 2001), cytokines (Ban et al., 1991; Cunningham et al., 1992; van Dam et al., 1996; Vidal et al., 1998; Moser et al., 2004; Khan et al., 2003), leptin (Kastin et al., 2000; Bjorbaek et al., 1998), acetylcholine (Grammas and Caspers, 1991), adrenergics (Walsh et al., 1987; Kalaria and Harik, 1989), glutamate (Koenig et al., 1992; Krizbai et al., 1998), and chemokines (Sanders et al., 1998).

4.4.2. Permeability to Cytokines and Related Substances

The BBB is known to transport several cytokines in the blood-to-brain direction. For example, the BBB transports the IL-1's, IL-6, and TNF by three separate transport systems. Additionally, nerve growth factor, brain derived neurotrophic factor, interferons, neurotrophins, and leukemia inhibitory factor (Poduslo and Curran, 1996; Pan et al., 1997b; Pan et al., 1998b; Pan et al., 1998a) are also transported across the BBB. In some cases, the same gene which gives rise to a cytokine's receptor also produces the cytokine's transporter, whereas in other cases the receptor and transporter are immunologically distinct proteins (Banks and Kastin, 1992; Pan and Kastin,

2002). In general, transporters occur throughout the CNS, including the spinal cord, although the transport rate can vary greatly among CNS regions (Pan et al., 1998b; Pan et al., 2002; Banks et al., 1994; McLay et al., 1997). Enough cytokine is transported into the brain to affect CNS function. For example, IL-1α crosses the BBB at the posterior division of the septum where it mediates cognitive impairments (Banks et al., 2001).

The cytokine transporters are not static but respond to physiological and pathological events. The transport rates of IL-1 and TNF each show diurnal variations (Pan et al., 2002; Banks et al., 1998b). The transport rate of TNF is altered in animals with experimental allergic/autoimmune encephalomyelitis (EAE), spinal cord injury, or blunt trauma to the brain (Pan et al., 1996; Pan et al., 1997a; Pan and Kastin, 2001b; Pan et al., 2003b).

4.4.3. Permeability to Other Neuroimmune Substances

Other substances with neuroimmune actions are handled by the BBB in a variety of ways. Monoamines are largely excluded by the BBB (Hardebo and Owman, 1990; Kalaria et al., 1987) and opiates and opiate peptides as a rule enter the brain by transmembrane diffusion but are transported by saturable systems in the brain-to-blood direction (King et al., 2001; Elferink and Zadina, 2001; Banks and Kastin, 1990). Pituitary adenylate cyclase activating peptide, a member of the VIP/secretin/PACAP family, has immune functions (Arimura, 1992). Transport of its two major forms across the BBB is complex, involving both brain-to-blood and blood-to-brain components (Banks et al., 1993). Its blood-to-brain transport is altered with injury (Somogyvari-Vigh et al., 2000). Some of the other immune active substances whose passage across the BBB has been investigated are melanocyte stimulating hormone (Wilson et al., 1984), corticotrophin releasing hormone (Martins et al., 1996), and enkephalins (Banks et al., 1986; Elferink and Zadina, 2001).

4.4.4. Permeability to Immune Cells

As discussed above, immune cells cross the BBB by the highly regulated process of diapedesis. The mechanism by which immune cells cross the BBB has also been greatly clarified by recent work. Two major assumptions about how immune cells would enter the CNS have not withstood investigation. The first assumption was that immune cells would enter the CNS by leaking across a disrupted BBB. However, disruptions to the BBB are usually mediated by increased vesicular activity in the endothelial cells (Vorbrodt et al., 1995; Lossinsky et al., 1983; Mayhan and Heistad, 1985). These vesicles of 100 nm or so could not accommodate the passage of an immune cell 10,000 nm in diameter. Even in diseases where there is both increased immune cell trafficking into the CNS and a disrupted BBB, there is often a mismatch between the site of immune cell entry and BBB disruption (Engelhardt and Wolburg, 2004).

The second major assumption is that immune cells would cross between opposing endothelial cells taking the paracellular route. However, some cells favor a transcellular route. These immune cells tunnel through venular endothelial cells leaving the intercellular tight junctions intact (Wolburg et al., 2005). This tunneling process is complex and is initiated when LFA-1 on an immune cell binds to ICAM on the brain endothelial cell. Other paracellular messengers, which likely include cytokines, are then released (Male, 1995; Persidsky et al., 1997). Protrusions and invaginations of the endothelial cell and protrusions of the immune cell occur, with the immune cell possibly using the tight junction as an initial anchoring site (Lossinsky et al., 1991). Other ligands which have been postulated to play a role in this transcytotic process include PECAM, VE-cadherin, members of the JAM family and CD99 (Engelhardt and Wolburg, 2004). Some plasma inevitably accompanies the passage of the immune cell, which can give the appearance of a disrupted BBB (Greenwood et al., 1995; Avison et al., 2004; Persidsky et al., 2000).

4.4.5. Permeability to Viruses

Whether a virus is neurovirulent or not depends largely on its ability to cross the BBB (Chou and Dix, 1989). This should not necessarily be assumed, as viruses could induce neurotoxicity without themselves crossing the BBB by several mechanisms. For example, shed viral proteins might cross the BBB as could circulating cytokines whose release from peripheral sources was induced by the virus. However, most neurovirulent viruses seem to do their major damage directly after entering and replicating within the CNS. Some viruses can replicate within brain endothelial cells and are subsequently shed into the CNS (Cosby and Brankin, 1995). Other viruses invade the CNS by crossing the BBB (Nakaoke et al., 2005). Initial uptake by either route involves events reminiscent of adsorptive endocytosis as discussed above. Viral glycoproteins bind to brain endothelial cell (or choroid plexus) glycoproteins to initiate endocytosis. As with adsorptive endocytosis, the virus-containing vesicle is subsequently routed to various membrane systems, which can include discharge to the original side of uptake or transcytosis. Sialic acid and heparan sulfate are common components of the glycoproteins involved in viral uptake by the BBB (Schweighardt and Atwood, 2001; Banks et al., 2004c). In some cases, the functional glycoprotein is known. For example, rabies can bind to acetylcholine and nerve growth factor receptors (Schweighardt and Atwood, 2001). Without an appropriate luminal or basal glycoprotein with which to bind to the BBB cell, the virus is largely excluded from the CNS.

4.4.6. Secretion of Neuroimmune-Active Substances

The brain endothelial cells and the epithelial cells of the choroid plexus are capable of secreting a large number of neuroimmune active substances. These include ILs (Fabry et al., 1993; Hofman et al., 1999; Reyes et al., 1999), TNF (Lee et al., 2001), nerve growth factor, (Moser et al., 2004), endothelin (Didier et al., 2002), monocyte chemoattractant peptide (Chen et al., 2001), nitric oxide (Mandi et al., 1998), RANTES (Simpson et al., 1998), and prostacyclin (Faraci and Heistad, 1998). Some of these substances are secreted spontaneously, and many of them can be stimulated with immunoactive substances such as lipopolysaccharide (LPS), bacteria, or viral proteins (Reyes et al., 1999; Vadeboncoeur et al., 2003; Hofman et al., 1999; Didier et al., 2002). The unique architecture of the BBB allows it to receive input from one of its surfaces and to secrete substances into the other. For example, LPS applied to the abluminal surface of brain endothelial cells in monolayer cultures will enhance release of IL-6 from the luminal surface (Verma et al., 2005).

4.4.7. Modulation of BBB Function by Neuroimmune Substances

Traditionally, neuroimmune modulation has been thought of in terms of disruption of the BBB. However, as the review above indicates, transporter functions are also vulnerable to manipulation by neuroimmune elements. Alterations in transport function are likely to be a more common event than disruption, as the latter is likely seen only with extreme pathology, whereas the former is likely a physiological, as well as a pathological, aspect of neuroimmune regulation. Other functions of the BBB, such as brain endothelial cell secretions, are also clearly affected by neuroimmune events.

4.4.7.1. Agents that Increase Permeability Through the BBB

Disruption was the first BBB function noted to be perturbed in neuroimmune disease. However, the review above makes it clear that an increase in the BBB permeability can be induced for specific agents by increasing their blood-to-brain transport rate or inhibiting their brain-to-blood efflux rate.

4.4.7.2. Regulation of BBB Integrity and Tight Junction Function

The classic example of BBB disruption is that seen in multiple sclerosis and the animal model of that disease, EAE (Pozzilli et al., 1988; Butter et al., 1991; Juhler et al., 1984). LPS and treatment with cytokines such as TNF have also been shown to induce BBB disruption (Megyeri et al., 1992). As discussed above, paracellular (through tight junctions) and transcellular mechanisms of transport exist. Although either can underlie BBB disruption, classic studies have shown that the major cause of increased protein leakage across the BBB for almost every kind of insult to the BBB or to the CNS is mediated by transcytotic mechanisms (Lossinsky et al., 1983; Vorbrodt et al., 1995). Nevertheless, recent advances in understanding tight junction assembly and regulation have encouraged many to investigate paracellular mechanisms. Both tight junction function and transcytosis are regulated events, although it is unclear to what extent protein leakage into the brain may be altered under physiological conditions. To some extent, paracellular and transcytotic routes likely involve some of the same cellular machinery, such as the cytoskeleton. TNF is known to induce rearrangements in cytoskeletal architecture. Additionally, cerebral ischemia, diabetes mellitus, and even intense pain are associated with alterations in the expression and cellular distribution of tight junction proteins and opening of the BBB (Brown and Davis, 2002; Chehade et al., 2002; Huber et al., 2002). The importance of regulatory processes in BBB disruption is vividly illustrated by the paradoxic finding that maximal disruption does not occur at the time of the CNS injury, even when the event is traumatic, but hours or days later (Baldwin et al., 1996). It is thought that it is the peripheral and central responses, such as cytokine release, to CNS injury rather than the CNS injury itself, which results in BBB disruption.

4.4.7.3. Regulation of Saturable Transporters

Regulation of both influx and efflux transporters are influenced by neuroimmune events. Additionally, cytokine transporters are also affected by various CNS events.

4.4.7.3.1. LPS and Blood-to-Brain Transporters

LPS increases the transport of cisplatin, insulin, and the HIV-1 viral coat glycoprotein gp120, but not of TNF or pituitary adenylate cyclase activating polypeptide (Banks et al., 1999; Minami et al., 1998; Nonaka et al., 2005; Osburg et al., 2002; Xaio et al., 2001). LPS affects leptin transport (Nonaka et al., 2004) through peripheral mechanisms and increases pituitary adenylate cylcase activating polypeptide binding to receptors on the BBB but does not alter transport. CNS injuries such as ischemia or trauma to the spinal cord induce a cascade of events, which can affect the transport of neuroimmune substances across the BBB as discussed below.

4.4.7.3.2. Cytokines and P-gp

Efflux systems are also altered by immune modulators. In vitro studies suggest that TNF and interferon gamma regulate P-gp (Theron et al., 2003; Stein et al., 1996). Interestingly, IL-2 appears to be both a substrate for P-gp and a modulator of its activity (Bonhomme-Faivre et al., 2002; Castagne et al., 2004; Drach et al., 1996). Because P-gp regulates the brain concentration of so many drugs and endogenous substances, immunomodulation could affect many other responses. For example, brain levels of exogenous opiate drugs such as morphine (King et al., 2001), endogenous opiates such β-endorphin (Kastin et al., 2002), and neurotoxins such as cyclosporine (Sakata et al., 1994) would all be expected to be increased in patients given IL-2. Immunomodulation of efflux systems, therefore, could have a major effect on CNS metabolism and the response to drugs.

4.5. Role of BBB in Neuroimmune Diseases

The above review has emphasized BBB/neuroimmune interactions under normal physiological conditions. However, the BBB is intimately involved in neuroimmune diseases as well. The BBB can be a target of such disease, its functions may be adaptive to disease, or it can be a contributor to the disease process. Below are some examples of the ways in which the BBB is altered in diseases with neuroimmune processes.

4.5.1. TNF Transport and EAE

TNF has a biphasic effect on many neuroimmune processes, with too little or too much producing harmful effects (Pan et al., 1997c). TNF mediates many of its pathological effects through its central receptors and transport of circulating TNF is one source of CNS TNF (Gutierrez et al., 1993; Osburg et al., 2002; Pan and Kastin, 2002). Induction of EAE is partially dependent on TNF and IL-1 (Schiffenbauer et al., 2000). Immune cell invasion in general and during EAE in particular is dependent on TNF-modulated expression of ICAM and VCAM on brain endothelial cells and of LFA-1 on immunocytes (Male, 1995; Barten and Ruddle, 1994). Finally, the saturable transport across the BBB of TNF itself is greatly increased in EAE (Pan et al., 1996).

4.5.2. CNS Injuries and Cytokine Transport

As noted above, CNS injuries can produce a disruption of the BBB, but this disruption is temporally dissociated from the injury (Pan et al., 1997a; Baldwin et al., 1996; Banks et al., 1998a). This dissociation is because the disruption is the consequence of the reactions to injury rather than to injury itself. Not surprisingly, then, CNS injuries can also produce complex alterations in the BBB transport of cytokines. Besides the example of TNF in EAE given above, TNF transport is also increased in spinal cord injury (Pan et al., 1997a; Pan et al., 2003b; Pan and Kastin 2001b). This increase is neither confined to the site of injury, homogeneous through out the CNS, nor related to the disruption pattern of the BBB (Pan et al., 2005; Pan et al., 1997a; Pan and Kastin, 2001b). It is also temporally and regionally independent of the changes in BBB transport rates of other cytokines and immunoactive substances whose transport rates are also altered with CNS injury (Banks et al., 1998a; Pan et al., 1998b). The pattern also is dependent on the type of CNS injury (Pan and Kastin, 2001b).

4.5.3. Antiretrovirals and the BBB

A major problem in treating viruses that can invade the CNS, such as HIV-1, is that antiretrovirals often cross the BBB poorly (Thomas, 2004). The major problem, however, is not that these substances are especially limited by their rate of transmembrane diffusion, but that they are nearly all substrates for efflux systems. For example, AZT is 16 times more lipid soluble than sucrose and so should cross much more rapidly, but actually crosses at the same rate (Wu et al., 1998). AZT is a ligand for at least two efflux systems (Masereeuw et al., 1994; Takasawa et al., 1997; Wang and Sawchuck, 1995) and the protease inhibitors are all substrates for P-gp (Lee et al., 1998). P-gp is expressed by immune and other cells as well with three major phenotypic clusters in humans. Those with higher expression of P-gp, and therefore less able to accumulate protease inhibitors in tissues, are more resistant to treatment for HIV-1 (Fellay et al., 2002).

4.5.4. Immune Cell Invasion

Immune cell trafficking into the CNS is important in mediating neuroimmune diseases. Immune cell invasion is an early event in multiple sclerosis and EAE (Wolburg et al., 2005). Infected immune cells are a mechanism by which HIV-1 (Koyanagi et al., 1997; Nottet et al., 1996) and perhaps prions (Klein et al., 1997) invade the CNS.

Immune cell passage across the BBB is, in turn, affected by immune modulators. LPS and the HIV-1 immunoactive protein Tat increase expression by brain endothelial cells of ICAM and VCAM (Pu et al., 2003; Nottet et al., 1996) and monocytes treated with LPS have an increased rate of passage across the BBB (Persidsky et al., 1997). In vitro studies suggest that these events may be mediated through IL-1β and IL-6 (De Vries et al., 1994).

4.5.5. Efflux of NeuroAIDS-related Proteins and Cytokines

Because the BBB prevents the effective accumulation of many of the antivirals, the CNS can act as a reservoir of virus. This reservoir could potentially reinfect the peripheral tissues. The CNS-to-blood movement of HIV-1 has not been investigated, but movement of two of its proteins has been. The coat glycoprotein gp120 is cleared by non-saturable mechanisms (Cashion et al., 1999). However, it has a propensity to be reabsorbed predominately by nasal drainage. As a result, it drains by way of lymphatic vessels directly to the cervical lymphatic nodes. If whole virus also takes this route, then that means that lymph nodes could be directly reinfected without the virus having to enter the circulation where it could be exposed to antiviral agents.

A CNS reservoir of virus could affect the peripheral immune system by a mechanism, which does not involve reinfection of peripheral tissues. Tat, like gp120, is also reabsorbed with the CSF into the blood by a non-saturable mechanism (Banks et al., 2005b). Proteins that are enzymatically resistant in blood, such as Tat, gp120, and cytokines, can achieve high levels in blood even when their only source is the CSF. Production of these proteins within the CNS with subsequent reabsorption with the CSF into blood could be a way in which CNS virus produces toxic effects at peripheral tissues.

To date, IL-2 is the only cytokine known to be transported from the brain to the blood by a saturable transporter (Banks et al., 2004b). This transporter, along with binding to plasma proteins and robust degradation by the BBB or CNS, effectively prevents much IL-2 from entering the brain. Evidence suggests that this transporter is likely P-gp. P-gp activity is decreased with HIV encephalitis (Persidsky et al., 2000) and this could lead to blood-borne IL-2 entering the CNS. Chronic IL-2 administration induces stereotypic behaviors and is used in an animal model of schizophrenia (Zalcman, 2001; Zalcman, 2002). Therefore, an enhanced entry of IL-2 is one mechanism by which HIV-1 could induce behavioral changes.

Summary

The BBB intimately interacts with cells and their secretions that are in both the CNS and periphery. Some neuroimmune substances, exemplified by cytokines, can cross the BBB directly and also have direct effects on the BBB. The BBB is itself a source of neuroimmune substances and can receive signals from one side, for example the brain side, and release substances in response to that signal from its other side. The passage of immune cells across the BBB is a highly regulated event as is the passage of viruses and viral particles. Overall, the BBB is an important component of the neuroimmune axis and the only component, which is simultaneously physically in both the peripheral and central compartments of the neuroimmune system.

Review Questions/Problems

1. The major cell type comprising the vascular blood-brain barrier is:

a. epithelial
b. endothelial
c. muscle
d. adipose
e. a and c

2. Characteristics of the vascular blood-brain barrier are:

a. tight junctions
b. increased vesicular activity in the blood-to-brain direction
c. a reduced number of saturable transport systems
d. a and b
e. a and c
f. b and c

3. The neurovascular unit:

a. would not include circulating immune cells or their products
b. would include circulating immune cells and their products
c. would exclude pericytes
d. is primarily concerned with vagal control of cerebral blood flow
e. is not relevant to mammalian physiology

4. Current evidence best supports a model by which immune cells cross the BBB by:

a. crossing a blood-brain barrier that is already disrupted
b. tunneling between brain endothelial cells (paracellular pathway)
c. tunneling through brain endothelial cells (transcytotic pathway)
d. occurs only in disease states
e. a and d
f. b and d
g. c and d

5. Passage of substances from the CNS to the blood can occur by which pathway(s)

a. transmembrane diffusion
b. endocytosis by vascular epithelial cells
c. with reabsorption of the cerebrospinal fluid
d. a and b
e. a and c
f. b and c

6. Which statement is false about cytokines and the blood-brain barrier:

a. a blood-to-brain saturable transport system has been described for IL-6
b. cytokines can disrupt the blood-brain barrier
c. cytokines are secreted by brain endothelial cells
d. a hallmark of cytokine transporters is their lack of change with CNS diseases
e. cytokines can cross the blood-brain barrier in amounts that can affect CNS function

7. Viruses can cross the blood-brain barrier

a. by infecting immune cells which then cross the blood-brain barrier
b. by crossing as free virus
c. by first attaching to brain endothelial cell glycoproteins
d. by first attaching to receptors on brain endothelial cells
e. all of the above

8. P-gp is

a. a blood-to-brain saturable transport system
b. is a vesicular-mediated mechanism of crossing the brain endothelial cell
c. has protease inhibitors as substrates
d. does not transport opiate peptides
e. is not modulated by IL-2
f. all of the above

9. The major reason that the antiviral AZT has limited uptake by brain is:

a. it is too lipid insoluble to cross by transmembrane diffusion
b. it is a P-gp substrate

c. it is too unstable in blood
d. it is transported out of the brain
e. it binds too strongly to immune cells

10. From the list below, the one largely excluded from the CNS by the BBB is:

a. monoamines
b. cytokines
c. immune cells
d. opiates
e. viruses

References

Al-Sarraf H, Phillip L (2003) Effect of hypertension on the integrity of blood brain and blood CSF barriers, cerebral blood flow and CSF secretion in the rat. Brain Res 975:179–188.

Alafaci C, Salpietro G, Grasso G, Sfacteria A, Passalacqua M, Morabito A, Tripodo E, Calapai G, Buemi M, Tomasello F (2000) Effect of recombinant human erythropoietin on cerebral ischemia following experimental subarachnoid hemorrhage. Eur J Pharmacol 406:219–225.

Arimura A (1992) Pituitary adenylate cyclase activating polypeptide (PACAP): Discovery and current status of research. Regul Pept 37:287–303.

Avison MJ, Nath A, Greene-Avison R, Schmitt FA, Bales RA, Ethisham A, Greenberg RN, Berger JR (2004) Inflammatory changes and breakdown of microvascular integrity in early human immunodeficiency virus dementia. J Neurovirol 10:223–232.

Ayre SG, Skaletski B, Mosnaim AD (1989) Blood-brain barrier passage of azidothymidine in rats: Effect of insulin. Res Comm Chem Path Pharmacol 63:45–52.

Baba M, Oishi R, Saeki K (1988) Enhancement of blood-brain barrier permeability to sodium fluorescein by stimulation of mu opioid receptors in mice. Naunyn-Schmied Arch Pharmacol 37:423–428.

Balabanov R, Dore-Duffy P (1998) Role of the CNS microvascular pericyte in the blood-brain barrier. J Neurosci Res 53:637–644.

Baldwin SA, Fugaccia I, Brown DR, Brown LV, Scheff SW (1996) Blood-brain barrier breach following cortical contusion in the rat. J Neurosurg 85:476–481.

Balin BJ, Broadwell RD, Salcman M, el-Kalliny M (1986) Avenues for entry of peripherally administered protein to the central nervous system in mouse, rat, and squirrel monkey. J Comp Neurol 251:260–280.

Ban E, Milon G, Prudhomme N, Fillion G, Haour F (1991) Receptors for interleukin-1 (α and β) in mouse brain: Mapping and neuronal localization in hippocampus. Neuroscience 43:21–30.

Banks WA (2004) Are the extracellular pathways a conduit for the delivery of therapeutics to the brain? Curr Pharm Des 10:1365–1370.

Banks WA, Broadwell RD (1994) Blood to brain and brain to blood passage of native horseradish peroxidase, wheat germ agglutinin and albumin: Pharmacokinetic and morphological assessments. J Neurochem 62:2404–2419.

Banks WA, Farr SA, La Scola ME, Morley JE (2001) Intravenous human interleukin-1α impairs memory processing in mice: Dependence on blood-brain barrier transport into posterior division of the septum. J Pharmacol Exp Ther 299:536–541.

Banks WA, Jumbe NL, Farrell CL, Niehoff ML, Heatherington A (2004a) Passage of erythropoietic agents across the blood-brain barrier: A comparison of human and murine erythropoietin and the analog Darbopoetin alpha. Eur J Pharmacol 505:93–101.

Banks WA, Kastin AJ (1989) Inhibition of the brain to blood transport system for enkephalins and Tyr-MIF-1 in mice addicted or genetically predisposed to drinking ethanol. Alcohol 6:53–57.

Banks WA, Kastin AJ (1990) Editorial Review: Peptide transport systems for opiates across the blood-brain barrier. Am J Physiol 259:E1–E10.

Banks WA, Kastin AJ (1992) The interleukins -1α, -1 β, and -2 do not disrupt the murine blood-brain barrier. Int J Immunopharmac 14:629–636.

Banks WA, Kastin AJ (1994) Brain-to-blood transport of peptides and the alcohol withdrawal syndrome. In: Models of Neuropeptide Action (Strand FL, Beckwith B, Chronwall B, Sandman CA, eds), pp 108–118. New York, NY: New York Academy of Sciences.

Banks WA, Kastin AJ, Akerstrom V (1997) HIV-1 protein gp120 crosses the blood-brain barrier: Role of adsorptive endocytosis. Life Sci 61:L119–L125.

Banks WA, Kastin AJ, Arimura A (1998a) Effect of spinal cord injury on the permeability of the blood-brain and blood-spinal cord barriers to the neurotropin PACAP. Exp Neurol 151:116–123.

Banks WA, Kastin AJ, Brennan JM, Vallance KL (1999) Adsorptive endocytosis of HIV- 1gp120 by blood-brain barrier is enhanced by lipopolysaccharide. Exp Neurol 156:165–171.

Banks WA, Kastin AJ, Ehrensing CA (1994) Transport of blood-borne interleukin-1α across the endothelial blood-spinal cord barrier of mice. J Physiol (London) 479:257–264.

Banks WA, Kastin AJ, Ehrensing CA (1998b) Diurnal uptake of circulating interleukin-1α by brain, spinal cord, testis and muscle. Nim 5:36–41.

Banks WA, Kastin AJ, Fischman AJ, Coy DH, Strauss SL (1986) Carrier-mediated transport of enkephalins and N-Tyr-MIF-1 across blood-brain barrier. Am J Physiol 251:E477–E482.

Banks WA, Kastin AJ, Komaki G, Arimura A (1993) Passage of pituitary adenylate cyclase activating polypeptide 1–27 and pituitary adenylate cyclase activating polypeptide 1–38 across the blood-brain barrier. J Pharmacol Exp Ther 267:690–696.

Banks WA, Niehoff ML, Zalcman S (2004b) Permeability of the mouse blood-brain barrier to murine interleukin-2: Predominance of a saturable efflux system. Brain Behav and Immun 18:434–442.

Banks WA, Ortiz L, Plotkin SR, Kastin AJ (1991) Human interleukin (IL) 1α, murine IL-1α and murine IL-1β are transported from blood to brain in the mouse by a shared saturable mechanism. J Pharmacol Exp Ther 259:988–996.

Banks WA, Pagliari P, Nakaoke R, Morley JE (2005a) Effects of a behaviorally active antibody on the brain uptake and clearance of amyloid beta proteins. Peptides 26:287–294.

Banks WA, Robinson SM, Nath A (2005b) Permeability of the blood-brain barrier to HIV-1 Tat. Exp Neurol 193:218–227.

Banks WA, Robinson SM, Verma S, Morley JE (2003) Efflux of human and mouse amyloid β proteins 1–40 and 1–42 from brain: Impairment in a mouse model of Alzheimer's disease. Neuroscience 121:487–492.

Banks WA, Robinson SM, Wolf KM, Bess JW, Jr., Arthur LO (2004c) Binding, internalization, and membrane incorporation of human immunodeficiency virus-1 at the blood-brain barrier is differentially regulated. Neuroscience 128:143–153.

Banks WA, Terrell B, Farr SA, Robinson SM, Nonaka N, Morley JE (2002) Transport of amyloid β protein antibody across the blood-brain barrier in an animal model of Alzheimer's disease. Peptides 23:2223–2226.

Barten DM, Ruddle NH (1994) Vascular cell adhesion molecule-1 modulation by tumor necrosis factor in experimental allergic encephalomyelitis. J Neuroimmunol 51:123–133.

Begley DJ (2004) ABC transporters and the blood-brain barrier. Curr Pharm Des 10:1295–1312.

Bernards CM (1999) Epidural and intrathecal drug movement. In: Spinal Drug Delivery (Yaksh TL, ed), pp 239–252. New York: Elsevier.

Bjorbaek C, Elmquist JK, Michl P, Ahima RS, van Beuren A, McCall AL, Flier JS (1998) Expression of leptin receptor isoforms in rat brain microvessels. Endocrinology 139:3485–3491.

Blasberg RG (1977) Methotrexate, cytosine arabinoside, and BCNU concentration in brain after ventriculocisternal perfusion. Cancer Treatment Reports 61:625–631.

Bonhomme-Faivre L, Pelloquin A, Tardivel S, Urien S, Mathieu MC, Castagne V, Lacour B, Farinotti R (2002) Recombinant interleukin-2 treatment decreases P-glycoprotein activity and paclitaxel metabolism in mice. Anit-Cancer Drugs 13:51–57.

Boulton M, Flessner M, Armstrong D, Mohamed R, Hay J, Johnston M (1999) Contribution of extracranial lymphatics and arachnoid villi to the clearance of a CSF tracer in the rat. Am J Physiology 276: R818–R823.

Bradbury M (1979) The Concept of a Blood-Brain Barrier. New York: John Wiley and Sons Ltd.

Brightman MW, Reese TS (1969) Junctions between intimately apposed cell membranes in the vertebrate brain. J Cell Biol 40:648–677.

Broadwell RD (1989) Transcytosis of macromolecules through the blood-brain barrier: A cell biological perspective and critical appraisal. Acta Neuropathol (Berl) 79:117–128.

Broadwell RD (1993) Endothelial cell biology and the enigma of transcytosis through the blood-brain barrier. Adv Exp Med Biol 331:137–141.

Broadwell RD, Balin BJ, Salcman M (1988) Transcytotic pathway for blood-borne protein through the blood-brain barrier. Proc Natl Acad Sci USA 85:632–636.

Broadwell RD, Banks WA (1993) Cell biological perspective for the transcytosis of peptides and proteins through the mammalian blood-brain fluid barriers. In: The Blood-Brain Barrier (Pardridge WM, ed), pp 165–199. New York: Raven Press, Ltd.

Broadwell RD, Sofroniew MV (1993) Serum proteins bypass the blood-brain barrier for extracellular entry to the central nervous system. Exp Neurol 120:245–263.

Brown RC, Davis TP (2002) Calcium modulation of adherens tight junction function: A potential mechanism for blood-brain barrier disruption after stroke. Stroke 33:1706–1711.

Butter C, Baker D, O'Neill JK, Turk JL (1991) Mononuclear cell trafficking and plasma protein extravasation into the CNS during chronic relapsing experimental allergic encephalomyelitis in Biozzi AB/H mice. J Neurol Sci 104:9–12.

Cangiano C, Cardelli-Cangiano P, Cascino A, Patrizi MA, Barberini F, Rossi F, Capocaccia L, Strom R (1983) On the stimulation by insulin of tryptophan transport across the blood-brain barrier. Biochem Int 7:617–627.

Cashion MF, Banks WA, Bost KL, Kastin AJ (1999) Transmission routes of HIV-1 gp120 from brain to lymphoid tissues. Brain Res 822:26–33.

Castagne V, Bonhomme-Faivre L, Urien S, Reguiga MD, Soursac M, Gimenez F, Farinotti R (2004) Effect of recombinant interleukin-2 pretreatment on oral and intravenous digoxin pharmacokinetics and P-glycoprotein activity in mice. Drug Metab Dispos 32:168–171.

Catalan RE, Martinez AM, Aragones MD, Miguel BG, Robles A (1988) Insulin action on brain microvessels; effect on alkaline phosphatase. Biochem Biophys Res Commun 150:583–590.

Chang SL, Felix B, Jiang Y, Fiala M (2001) Actions of endotoxin and morphine. Adv Exp Med Biol 493:187–196.

Chao CC, Gekker G, Sheng WS, Hu S, Tsang M, Peterson PK (1994) Priming effect of morphine on the production of tumor necrosis factor-alpha by microglia: Implications in respiratory burst activity and human immunodeficiency virus-1 expression. J Pharmacol Exp Ther 269:198–203.

Chehade JM, Hass MJ, Mooradian AD (2002) Diabetes-related changes in rat cerebral occlusin and zonula occludens-1 (ZO-1) expression. Neurochem Res 27:249–252.

Chen G, Castro WL, Chow HH, Reichlin S (1997) Clearance of ^{125}I-labelled interleukin-6 from brain into blood following intracerebroventricular injection in rats. Endocrinology 138:4830–4836.

Chen G, Reichlin S (1998) Clearance of [^{125}I]-tumor necrosis factor-α from the brain into the blood after intracerebroventricular injection into rats. NeuroImmunoModulation 5:261–269.

Chen P, Shibata M, Zidovetzki R, Fisher M, Zlokovic BV, Hofman FM (2001) Endothelin-1 and monocyte chemoattractant protein-1 modulation in ischemia and human brain-derived endothelial cell cultures. J Neuroimmunol 116:62–73.

Chou S, Dix RD (1989) Viral infections and the blood-brain barrier. In: Implications of the Blood-Brain Barrier and Its Manipulation, Volume 2: Clinical Aspects (Neuwelt EA, ed), pp 449–468. New York: Plenum Publishing Corporation.

Cornford EM, Braun LD, Oldendorf WH, Hill MA (1982) Comparison of lipid-mediated blood-brain-barrier penetrability in neonates and adults. Am J Physiol 243:C161–C168.

Cosby SL, Brankin B (1995) Measles virus infection of cerebral endothelial cells and effect on their adhesive properties. Vet Microbiol 44:135–139.

Cserr HF (1984) Convection of brain interstitial fluid. In: Hydrocephalus (Shapiro K, Marmarou A, Portnoy H, eds), pp 59–68. New York: Raven Press.

Cserr HF, Berman BJ (1978) Iodide and thiocyanate efflux from brain following injection into rat caudate nucleus. Am J Physiol 4:F331–F337.

Cserr HF, Knopf PM (1992) Cervical lymphatics, the blood-brain barrier and the immunoreactivity of the brain: A new view. Immunol Today 13:507–512

Cunningham ET, Jr., Wada E, Carter DB, Tracey DE, Battey JF, De Souza EB (1992) In situ histochemical localization of type I interleukin-1 receptor messenger RNA in the central nervous system, pituitary, and adrenal gland of the mouse. J Neurosci 12:1101–1114.

Davson H, Segal MB (1996a) Blood-brain-CSF relations. Physiology of the CSF and Blood-Brain Barriers, pp 257–302. Boca Raton: CRC Press.

Davson H, Segal MB (1996b) Ontogenetic aspects of the cerebrospinal system. Physiology of the CSF and Blood-Brain Barriers, pp 607–662. Boca Rataon: CRC Press, Inc.

Davson H, Segal MB (1996c) Special aspects of the blood-brain barrier. Physiology of the CSF and Blood-Brain Barriers, pp 303–485. Boca Raton: CRC Press.

Davson H, Segal MB (1996d) Special aspects of the blood-brain barrier. Physiology of the CSF and Blood-Brain Barriers, pp 303–485. Boca Raton: CRC Press.

Davson H, Segal MB (1996e) The proteins and other macromolecules of the CSF. Physiology of the CSF and the Blood-Brain Barrier, pp 573–606. Boca Raton: CRC Press.

de la Torre JC, Mussivand T (1993) Can disturbed brain microcirculation cause Alzheimer's disease? Neurological Research 15:146–153.

de Lange EC, Bouw MR, Mandema JW, Danhof M, De Boer AG, Breimer DD (1995) Application of intracerebral microdialysis to study regional distribution kinetics of drug in rat brain. Br J Pharmacol 116:2538–2544.

de Lange ECM, Bouw MR, Danhof M, De Boer AG, Breimer DD (1993) Application of intracerebral microdialysis to study regional distribution kinetics of atenolol and acetaminophen in rat brain. The Use of Intracerebral Microdialysis to Study the Blood- Brain Barrier Transport Characteristics of Drugs (thesis, Leiden/Amsterdam Center for Drug Research) Sinteur, Leiden, pp. 93–106.

De Vries HE, Moor AC, Blom-Roosemalen MC, De Boer AG, Breimer DD, van Berkel TJ, Kuiper J (1994) Lymphocyte adhesion to brain capillary endothelial cells in vitro. J Neuroimmunol 52:1–8.

Deane R, Wu Z, Sagare A, Davis J, Du Yan S, Hamm K, Xu F, Parisi M, LaRue B, Hu HW, Spijkers P, Guo H, Song X, Lenting PJ, Van Nostrand WE, Zlokovic BV (2004) LRP/amyloid beta-peptide interaction mediates differential brain efflux of Abeta isoforms. Neuron 43:333–344.

Deli MA, Abraham CR, Kataoka Y, Niwa M (2005) Permeability studies on in vitro blood-brain barrier models: Physiology, pathology, and pharmacology. Cell Mol Neurobiol 25:59–127.

DeMattos RB, Bales KR, Cummins DJ, Paul SM, Holtzman DM (2002) Brain to plasma amyloid-β efflux: A measure of brain amyloid burden in a mouse model of Alzheimer's disease. Science 295:2264.

Didier N, Banks WA, Creminon C, Dereuddre-Bosquet N, Mabondzo A (2002) Endothelin-1 production at the in-vitro blood-brain barrier during HIV infection. Neuroreport 13:1179–1183.

Drach J, Gsur A, Hamilton G, Zhao S, Angerler J, Fiegl M, Zojer N, Raderer M, Haberl I, Andreeff M, Huber H (1996) Involvement of P-glycoprotein in the transmembrane transport of interleukin-2 (IL-2), IL-4, and interferon-gamma in normal human T lymphocytes. Blood 88:1747–1754.

Drion N, Lemaire M, Lefauconnier JM, Scherrmann JM (1996) Role of p-glycoprotein in the blood-brain transport of colchicine and vinblastine. J Neurochem 67:1688–1693.

Dziegielewska KM, Hinds LA, Mollgard K, Reynolds ML, Saunders NR (1988) Blood-brain, blood-cerebrospinal fluid and cerebrospinal fluid-brain barriers in a marsupial (*MACROPUS EUGENII*) during development. J Physiol 403:307–388.

Ehrenreich H, Hasselblatt M, Dembowski C, Depek L, Lewczuk P, Stiefel M, Rustenbeck H- H, Breiter N, Jacob S, Knerlich F, Bohn M, Poser W, Rither E, Kochen M, Gefeller O, Gleiter C, Wessel TC, De Ryck M, Itri L, Prange H, Cerami A, Brines M, Siren A-L (2002) Erythropoietin therapy for acute stroke is both safe and beneficial. Molecular Medicine 8:495–505.

Elferink RPJO, Zadina JE (2001) MDR1 P-glycoprotein transports endogenous opioid peptides. Peptides 22:2015–2020.

Engelhardt B, Wolburg H (2004) Minireview: Transendothelial migration of leukocytes: Through the front door or around the side of the house? Eur J Pharmacol 34:2955–2963.

Erbyraktar S, Grasso G, Sfacteria A, Xie QW, Coleman T, Kreilgaard M, Torup L, Sager T, Erbayraktar Z, Gokmen N, Yilmaz O, Ghezzi P, Villa P, Fratelli M, Casagrande S, Leist M, Helboe L, Gerwein J, Christensen S, Geist MA, Pedersen LO, Cerami-Hand C, Wuerth JP, Cerami A, Brines M (2003) Asialoerythropoietin is a nonerythropoiectic cytokine with broad neuroprotective activity in vivo. Proc Natl Acad Sci USA 100:6741–6746.

Fabry Z, Fitzsimmons KM, Herlein JA, Moninger TO, Dobbs MB, Hart MN (1993) Production of the cytokines interleukin 1 and 6 by murine brain microvessel endothelium and smooth muscle pericytes. J Neuroimmunol 47:23–34.

Faraci FM, Heistad DD (1998) Regulation of the cerebral circulation: Role of endothelium and potassium channels. Physiol Rev 78:53–97.

Farr SA, Banks WA, Uezu K, Sano A, Gaskin FS, Morley JE (2003) Antibody to beta-amyloid protein increases acetylcholine in the hippocampus of 12 month SAMP8 male mice. Life Sci 73:555–562.

Fellay J, Marzolini C, Meaden ER, Black DJ, Buclin T, Chave J-P, Decosterd LA, Furrer H, Opravil M, Pantaleo G, Retelska D, Ruiz L, Schinkel AH, Vernazza P, Eap CB, Telenti A (2002) Response of antiretrovial treatment in HIV-1 infected individuals with allelic variants of the multidrug resistance transporter 1: A pharmacogenetic study. Lancet 359:30–36.

Ferguson AV (1991) The area postrema: A cardiovascular control centre at the blood-brain interface? Can J PhysiolPharmacol 69:1026–1034.

Ghersi-Egea JF, Gorevic PD, Ghiso J, Frangione B, Patlak CS, Fenstermacher JD (1996) Fate of cerebrospinal fluid-borne amyloid β-peptide: Rapid clearance into blood and appreciable accumulation by cerebral arteries. J Neurochem 67:880–883.

Glynn SL, Yazdanian M (1998) In vitro blood-brain barrier permeability of nevirapine compared to other HIV antiretroviral agents. J Pharm Sci 87:306–310.

Grammas P, Caspers ML (1991) The effect of aluminum on muscarinic receptors in isolated cerebral microvessels. Res Comm Chem Path Pharmacol 72:69–79.

Greenwood J, Bamforth S, Wang Y, Devine L (1995) The blood-retinal barrier in immune- mediated diseases of the retina. In: New Concepts of a Blood-Brain Barrier (Greenwood J, Begley DJ, Segal MB, eds), pp 315–326. New York: Plenum Press.

Gross PM, Weindl A (1987) Peering through the windows of the brain. J Cereb Blood Flow Metab 7:663–672.

Gutierrez EG, Banks WA, Kastin AJ (1993) Murine tumor necrosis factor alpha is transported from blood to brain in the mouse. J Neuroimmunol 47:169–176.

Hardebo JE, Owman C (1990) Enzymatic barrier mechanisms for neurotransmitter monoamines and their precursors at the blood-brain barrier. In: Pathophysiology of the Blood-Brain Barrier (Johansson BB, Owman C, Widner H, eds), pp. 41–55.Amsterdam: Elsevier.

Hock C, Konietzko U, Streffer JR, Tracy J, Signorell A, Muller-Tillmanns B, Lemke U, Henke K, Moritz E, Garcia E, Wollmer MA, Umbricht D, de Quervain DJ, Hofmann M, Maddalena A, Papassotiropoulos A, Nitsch RM (2003) Antibodies against beta-amyloid slow cognitive decline in Alzheimer's disease. Neuron 38:547–554.

Hofman F, Chen P, Incardona F, Zidovetzki R, Hinton DR (1999) HIV-tat protein induces the production of interleukin-8 by human brain-derived endothelial cells. J Neuroimmunol 94:28–39.

Huber JD, Hau VS, Borg L, Campos CR, Egleton RD, Davis TP (2002) Blood-brain barrier tight junctions are altered during a 72-h exposure to lamba-carrageenan-induced inflammatory pain. Am J Physiol 283:H1531–H1537.

Janus C, Pearson J, McLaurin J, Mathews PM, Jiang Y, Schmidt SD, Chishti MA, Horne P, Heslin D, French J, Mount HTJ, Nixon RA, Mercken M, Bergeron C, Fraser PE, George-Hyslop P, Westaway D (2000) A β peptide immunization reduces behavioral impairment and plaques in a model of Alzheimer's disease. Nature 408:979–982.

Johanson CE (1988) The choroid plexus-arachnoid membrane-cerebrospinal fluid system. In: Neuromethods; The Neuronal Microenvironment (Boulton AA, Baker GB, Walz W, eds), pp 33–104. Clifton, NJ: The Humana Press.

Johansson BB (1989) Hypertension and the blood-brain barrier. In: Implications of the Blood-Brain Barrier and its Manipulation, Volume 2: Clinical Aspects (Neuwelt EA, ed), pp 389–410. New York: Plenum Publishing Co.

Johnson AK, Gross PM (1993) Sensory circumventricular organs and brain homeostatic pathways. FASEB J 7:678–686.

Jones PM, Robinson ICAF (1982) Differential clearance of neurophysin and neurohypophysial peptides from the cerebrospinal fluid in conscious guinea pigs. Neuroendocrinology 34:297–302.

Juhler M, Barry DI, Offner H, Konat G, Klinken L, Paulson OB (1984) Blood-brain and blood-spinal cord barrier permeability during the course of experimental allergic encephalomyelitis in the rat. Brain Res 302:347–355.

Kalaria RN, Harik SI (1989) Increased alpha 2- and beta 2-adrenergic receptors in cerebral microvessels in Alzheimer disease. Neurosci Lett 106:233–238.

Kalaria RN, Mitchell MJ, Harik SI (1987) Correlation of 1-methyl-4-phenyl-1,2,3,6- tetrahydropyridine neurotoxicity with blood-brain barrier monoamine oxidase activity. Proc Natl Acad Sci USA 84:3521–3525.

Kastin AJ, Akerstrom V (2001) Glucose and insulin increase the transport of leptin through the blood-brain barrier in normal mice but not in streptozotocin-diabetic mice. Neuroendocrinology 73:237–242.

Kastin AJ, Akerstrom V, Pan W (2000) Activation of urocortin transport into brain by leptin. Peptides 21:1811–1817.

Kastin AJ, Fasold MB, Zadina JE (2002) Endomorphins, Met-enkephalin, Tyr-MIF-1 and the P-glycoprotein efflux system. Drug Metab Dispos 30:231–234.

Kety SS (1987) Cerebral circulation and its measurement by inert diffusible tracers. In: Encyclopedia of Neuroscience, Volume I (Adelman G, ed), pp 206–208. Boston: Birkhuser.

Khan NA, DiCello F, Nath A, Kim KS (2003) Human immunodeficiency virus type 1 tat-mediated cytotoxicity of human brain microvascular endothelial cells. J Neurovirology 9:584–593.

King M, Su W, Chang A, Zuckerman A, Pasternak GW (2001) Transport of opioids from the brain to the periphery by P-glycoprotein: Peripheral actions of central drugs. Nature Neuroscience 4:221–222.

Klein MA, Frigg R, Flechsig E, Raeber AJ, Kalinke U, Bluethmann H, Bootz F, Suter M, Zinkernagel RM, Aguzzi A (1997) A crucial role for B cells in neuroinvasive scrapie. Nature 390:687–690.

Knopf PM, Cserr HF, Nolan SC, Wu TY, Harling-Berg CJ (1995) Physiology and immunology of lymphatic drainage of interstitial and cerebrospinal fluid from the brain. Neuropathol Appl Neurobiol 21:175–18

Koenig H, Trout JJ, Goldstone AD, Lu CY (1992) Capillary NMDA receptors regulate blood-brain barrier function and breakdown. Brain Res 588:297–303.

Koyanagi Y, Tanaka Y, Kira J, Ito M, Hioki K, Misawa N, Kawano Y, Yamasaki K, Tanaka R, Suzuki Y, Ueyama Y, Terada E, Tanaka T, Myasaka M, Kobayashi T, Kumazawa Y, Yamamoto N (1997) Primary human immunodeficiency virus type 1 viremia and central nervous system invasion in a novel hu-PBL-immunodeficient mouse strain. J Virol 71:2417–2424.

Kozlowski GP, Sterzl I, Nilaver G (1992) Localization patterns for immunoglobulins and albumins in the brain suggest diverse mechanisms for their transport across the blood-brain barrier (BBB) In: Progress in Brain Research (Ermisch A, Landgraf R, Rühle HJ, eds), pp 149–154. Amsterdam: Elsevier.

Krizbai IA, Deli MA, Pestenacz A, Siklose L, Szabo CA, Andras I, Joo F (1998) Expression of glutamate receptors on cultured cerebral endothelial cells. J Neurosci Res 54:814–819.

LeBel CP (1999) Spinal Delivery of neurotrophins and related molecules. In: Spinal Drug Delivery (Yaksh TL, ed), pp 543–554. Amsterdam: Elsevier.

Lee CGL, Gottesman MM, Cardarelli CO, Ramachandra M, Jeang KT, Ambudkar SV, Pastan I, Dey S (1998) HIV-1 protease inhibitors are substrates for the MDR1 multidrug transporter. Biochemistry 37:3594–3601.

Lee YW, Hennig B, Fiala M, Kim KS, Toborek M (2001) Cocaine activates redox-regulated transcription factors and induces TNF-alpha expression in human brain endothelial cells. Brain Res 920:125–133.

Levin VA (1980) Relationship of octanol/water partition coefficient and molecular weight to rat brain capillary permeability. J Med Chem 23:682–684.

Lossinsky AS, Pluta R, Song MJ, Badmajew V, Moretz RC, Wisniewski HM (1991) Mechanisms of inflammatory cell attachment in chronic relapsing experimental allergic encephalomyelitis: A scanning and high-voltage electron microscopic study of the injured mouse blood-brain barrier. Microvasc Res 41:299–310.

Lossinsky AS, Vorbrodt AW, Wisniewski HM (1983) Ultracytochemical studies of vesicular and canalicular transport structures in the injured mammalian blood-brain barrier. Acta Neuropathol (Berl) 61:239–245.

LÜscher W, Potschka H (2002) Role of multidrug transporters in pharmacoresistance to antiepileptic drugs. J Pharmacol Exp Ther 30:7–14.

Male D (1995) The blood-brain barrier—No barrier to a determined lymphocyte. In: New Concepts of a Blood-Brain Barrier (Greenwood J, Begley DJ, Segal MB, eds), pp 311–314. New York: Plenum Press.

Mandi Y, Ocsovszki I, Szabo D, Nagy Z, Nelson J, Molnar J (1998) Nitric oxide production and MDR expression by human brain endothelial cells. Anticancer Res 18:3049–3052.

Maness LM, Kastin AJ, Farrell CL, Banks WA (1998) Fate of leptin after intracerebroventricular injection into the mouse brain. Endocrinology 139:4556–4562.

Marsh M (1984) The entry of enveloped viruses into cells by endocytosis. Biochem J 218:1–10.

Martins JM, Banks WA, Kastin AJ (1997) Acute modulation of the active carrier-mediated brain to blood transport of corticotropin-releasing hormone. Am J Physiol 272:E312–E319.

Martins JM, Kastin AJ, Banks WA (1996) Unidirectional specific and modulated brain to blood transport of corticotropin-releasing hormone. Neuroendocrinology 63:338–348.

Masereeuw R, Jaehde U, Langemeijer MWE, De Boer AG, Breimer DD (1994) *In vivo* and *in vitro* transport of zidovudine (AZT) across the blood-brain barrier and the effects of transport inhibitors. Pharm Res 11:324–330.

Mayhan WG, Heistad DD (1985) Permeability of blood-brain barrier to various sized molecules. Am J Physiology 248:H712–H718.

McCarthy TJ, Banks WA, Farrell CL, Adamu S, Derdeyn CP, Snyder AZ, LaForest R, Litzinger DC, Martin D, LeBel CP, Welch MJ (2002) Positron emission tomography shows that intrathecal leptin reaches the hypothalamus in baboons. J Pharmacol Exp Ther 307:878–883.

McLay RN, Kimura M, Banks WA, Kastin AJ (1997) Granulocyte-macrophage colony-stimulating factor crosses the blood-brain and blood-spinal cord barriers. Brain 120:2083–2091.

McQuay HJ, Sullivan AF, Smallman K, Dickenson AH (1989) Intrathecal opioids, potency and lipophilicity. Pain 36:111–115.

Mealey KL, Bentjen SA, Gay JM, Cantor GH (2001) Ivermectin sensitivity is associated with a deletion mutation of the mdr1 gene. Pharmacogenetics 11:727–733.

Megyeri P, Abraham CS, Temesvari P, Kovacs J, Vas T, Speer CP (1992) Recombinant human tumor necrosis factor alpha constricts pial arterioles and increases blood-brain barrier permeability in newborn piglets. Neurosci Lett 148:137–140.

Mellman I, Fuchs R, Helenius A (1986) Acidification of the endocytic and exocytic pathways. Ann Rev Biochem 55:663–700.

Minami T, Okazaki J, Kawabata A, Kuroda R, Okazaki Y (1998) Penetration of cisplatin into mouse brain by lipopolysccharide. Toxicology 130:107–113.

Morgan D, Diamond DM, Gottschall PE, Ugen KE, Dickey C, Hardy J, Duff K, Jantzen P, DiCarlo G, Wilcock D, Connor K, Hatcher J, Hope C, Gordon M, Arendash GW (2000) A β peptide vaccination prevents memory loss in an animal model of Alzheimer's disease. Nature 408:982–985.

Moser KV, Reindl M, Blasig I, Humpel C (2004) Brain capillary endothelial cells proliferate in response to NGF, express NGF receptors and secrete NGF after inflammatiion. Brain Res 1017:53–60.

Muldoon LL, Pagel MA, Kroll RA, Roman-Goldstein S, Jones RS, Neuwelt EA (1999) A physiological barrier distal to the anatomical blood-brain barrier in a model of transvascular delivery. Am J Neuroradiol 20:217–222.

Nakaoke R, Ryerse JS, Niwa M, Banks WA (2005) Human immunodeficiency virus type 1 transport across the in vitro mouse brain endothelial cell monolayer. Exp Neurol 193:101–109.

Nath A, Conant K, Chen P, Scott C, Major EO (1999) Transient exposure to HIV-1 Tat protein results in cytokine production in macrophages and astrocytes: A hit and run phenomenon. J Biol Chem 274:17098–17102.

Nelson PK, Masters LT, Zagzag D, Kelly PJ (1999) Angiographic abnormalities in progressive multifocal leukoencephalopathy: An explanation based on neuropathologic findings. Am J Neuroradiol 20:487–494.

Nonaka N, Hileman SM, Shioda S, Vo P, Banks WA (2004) Effect of lipopolysaccharide on leptin transport across the blood-brain barrier. Brain Res 1016:58–65.

Nonaka N, Shioda S, Banks WA (2005) Effect of lipopolysaccharide on the transport of pituitary adenylate cyclase activating polypeptide across the blood-brain barrier. Experimental Neurology 191:137–144.

Nottet HS, Persidsky Y, Sasseville VG, Nukuna AN, Bock P, Zhai QH, Sharer LR, McComb RD, Swindells S, Soderland C, Gendelman HE (1996) Mechanisms for the transendothelial migration of HIV-1-infected monocytes into brain. J Immunol 156:1284–1295.

Oehmichen M, Gruninger H, Wietholter H, Gencic M (1979) Lymphatic efflux of intracerebrally injected cells. Acta Neuropathol 45:61–65.

Oldendorf WH (1971) Brain uptake of radio-labelled amino acids, amines and hexoses after arterial injection. Am J Physiol 221:1629–1639.

Oldendorf WH (1974) Lipid solubility and drug penetration of the blood-brain barrier. Proc Soc Exp Biol Med 147:813–816.

Osburg B, Peiser C, Domling D, Schomburg L, Ko YT, Voight K, Bickel U (2002) Effect of endotoxin on expression of TNF receptors and transport of TNF-alpha at the blood-brain barrier of the rat. Am J Physiol 283:E899-E908.

Pan W, Banks WA, Fasold MB, Bluth J, Kastin AJ (1998a) Transport of brain-derived neurotrophic factor across the blood-brain barrier. Neuropharmacology 37:1553–1561.

Pan W, Banks WA, Kastin AJ (1997a) BBB permeability to ebiratide and TNF in acute spinal cord injury. Exp Neurol 146:367–373.

Pan W, Banks WA, Kastin AJ (1997b) Permeability of the blood-brain barrier and blood- spinal cord barriers to interferons. J Neuroimmunol 76:105–111.

Pan W, Banks WA, Kastin AJ (1998b) Permeability of the blood-brain barrier to neurotrophins. Brain Res 788:87–94.

Pan W, Banks WA, Kennedy MK, Gutierrez EG, Kastin AJ (1996) Differential permeability of the BBB in acute EAE: Enhanced transport of TNF-α. Am J Physiol 271:E636–E642.

Pan W, Cornelissen G, Halberg F, Kastin AJ (2002) Selected contributions: Circadian rhythm of tumor necrosis factor-alpha uptake into mouse spinal cord. J Appl Physiol 92:1357–1362.

Pan W, Kastin AJ (2001a) Changing the chemokine gradient: CINC1 crosses the blood-brain barrier. J Neuroimmunol 115:64–70.

Pan W, Kastin AJ (2001b) Increase in TNF alpha transport after SCI is specific for time, region, and type of lesion. Exp Neurol 170:357–363.

Pan W, Kastin AJ (2002) TNF α transport across the blood-brain barrier is abolished in receptor knockout mice. Exp Neurol 174:193–200.

Pan W, Kastin AJ, Pick CG (2005) The staircase test in mice after spinal cord injury. Int J Neuroprotect Neuroregen 1:32–37.

Pan W, Kastin AJ, Rigai T, McLay R, Pick CG (2003a) Increased hippocampal uptake of tumor necrosis factor alpha and behavioral changes in mice. Exp Brain Res 149:195–199.

Pan W, Zadina JE, Harlan RE, Weber JT, Banks WA, Kastin AJ (1997c) Tumor necrosis factor-α: A neuromodulator in the CNS. Neurosci Biobehav Rev 21:603–613.

Pan W, Zhang L, Liao J, Csernus B, Kastin AJ (2003b). Selective increase in TNF alpha permeation across the blood-spinal cord barrier after SCI. J Neuroimmunol 134:111–117.

Persidsky Y, Stins M, Way D, Witte MH, Weinand M, Kim KS, Bock P, Gendelman HE, Fiala M (1997) A model for monocyte migration through the blood-brain barrier during HIV-1 encephalitis. J Immunol 158:3499–3510.

Persidsky Y, Zheng J, Miller D, Gendelman HE (2000) Mononuclear phagocytes mediate blood-brain barrier compromise and neuronal injury during HIV-1-associated dementia. J Leukoc Biol 68:413–422.

Peruzzo B, Pastor FE, Blazquez JL, Schobitz K, Pelaez B, Amat P, Rodriguez EM (2000) A second look at the barriers of the medial basal hypothalamus. Exp Brain Res 132:10–26.

Plotkin SR, Banks WA, Kastin AJ (1996) Comparison of saturable transport and extracellular pathways in the passage of interleukin-1α across the blood-brain barrier. J Neuroimmunol 67:41–47.

Plotkin SR, Banks WA, Kastin AJ (1998) Enkephalin, PPE, mRNA, and PTS-1 in alcohol withdrawal seizure-prone and -resistant mice. Alcohol 15:25–31.

Poduslo JF, Curran GL (1996) Permeability at the blood-brain barrier and blood-nerve barriers of the neurotrophic factors: NGF, CNTF, NT-3, BDNF. Mol Brain Res 36:280–286.

Pollay M, Davson H (1963) The passage of certain substances out of the cerebrospinal fluid. Brain 86:137–150.

Pozzilli C, Bernardi S, Mansi L, Picozzi P, Iannotti F, Alfano B, Bozzao L, Lenzi GL, Salvatore M, Conforti P, Fieschi C (1988) Quantitative assessment of the blood-brain barrier permeability in multiple sclerosis using 68-Ga-EDTA and positron emission tomography. J Neurol Neurosurg Psych 51:1058–1062.

Pu H, Tian J, Flora G, Lee YW, Nath A, Hennig B, Toborek M (2003) HIV-1 Tat protein upregulates inflammatory mediators and induces monocyte invasion into the brain. Mol Cell Neurosci 24:224–237.

Rapoport SI (1976) Blood Brain Barrier in Physiology and Medicine. Raven Press, New York.

Rapoport SI, Ohata M, London ED (1981) Cerebral blood flow and glucose utilization following opening of the blood-brain barrier and during maturation of the rat brain. Fed Proc 40:2322–2325.

Raub TJ, Audus KL (1990) Adsorptive endocytosis and membrane recycling by cultured primary bovine brain microvessel endothelial cell monolayers. J Cell Sci 97:127–138.

Reese TS, Karnovsky MJ (1967) Fine structural localization of a blood-brain barrier to exogenous peroxidase. J Cell Biol 34: 207–217.

Rethelyi M (1984) Diffusional barrier around the hypothalamic arcuate nucleus in the rat. Brain Res 307:355–358.

Reyes TM, Fabry Z, Coe CL (1999) Brain endothelial cell production of a neuroprotective cytokine, interleukin-6, in response to noxious stimuli. Brain Res 851:215–220.

Sakata A, Tamai I, Kawazu K, Deguchi Y, Ohnishi T, Saheki A, Tsuji A (1994) *In vivo* evidence for ATP-dependent and p-glycoprotein-mediated transport of cyclosporin A at the blood-brain barrier. Biochem Pharmacol 48:1989–1992.

Sanders VJ, Pittman CA, White MG, Wang G, Wiley CA, Achim CL (1998) Chemokines and receptors in HIV encephalitis. AIDS 12:1021–1026.

Sands MS, Vogler C, Kyle JW, Grubb JH, Levy B, Galvin N, Sly WS, Birkenmeier EH (1994) Enzyme replacement therapy for murine mucopolysaccharidosis type VII. J Clin Invest 93:2324–2331.

Schiffenbauer J, Streit WJ, Butfiloski E, LaBow M, Edward C, Moldawer 3rd LL (2000) The induction of EAE is only partially dependent on TNF receptor signaling but requires the IL-1 type 1 receptor. Clin Immunol 95:117–125.

Schweighardt B, Atwood WJ (2001) Virus receptors in the human central nervous system. J Neurovirol 7:187–195.

Shafer RA, Murphy S (1997) Activated astrocytes induce nitric oxide synthase-2 in cerebral endothelium via tumor necrosis factor alpha. GLIA 21:370–379.

Shibata M, Yamada S, Kumar SR, Calero M, Bading J, Frangione B, Holtzman DM, Miller CA, Strickland DK, Ghiso J, Zlokovic BV (2000) Clearance of Alzheimer's amyloid-β 1–40 peptide from brain by LDL receptor-related protein-1 at the blood-brain barrier. J Clin Invest 106:1489–1499.

Shimura T, Tabata S, Ohnishi T, Terasaki T, Tsuji A (1991) Transport mechanism of a new behaviorally highly potent adrenocorticotropic hormone (ACTH) analog, ebiratide, through the blood-brain barrier. J Pharmacol Exp Ther 258:459–465.

Shyng SL, Huber MT, Harris EC (1993) A prion protein cycles between the cell surface and an endocytic compartment in cultured neuroblastome cells. Biol Chem 268:15922–15928.

Simpson JE, Newcombe J, Cuzner ML, Woodrofe MN (1998) Expression of monocyte chemoattractant protein-1 and other beta-chemokines by resident glia and inflammatory cells in multiple sclerosis. J Neuroimmunol 84:238–249.

Somogyvari-Vigh A, Pan W, Reglodi D, Kastin AJ, Arimura A (2000) Effect of middle cerebral artery occulsion on the passage of pituitary adenylate cyclase activating polypeptide across the blood-brain barrier in the rat. Regul Pept 91:89–95.

Spector R, Lorenzo AV (1974) The effects of salicylate and probenecid on the cerebrospinal fluid transport of penicillin, aminosalicylic acid and iodide. J Pharmacol Exp Ther 188:55–65.

Stein U, Walther W, Shoemaker RH (1996) Modulation of mdr1 expression by cytokines in human colon carcinoma cells: An approach for reversal of multidrug resistance. British Journal of Cancer 74:1384–1391.

Takasawa M, Terasaki T, Suzuki H, Sugiyama Y (1997) In vivo evidence for carrier-mediated efflux transport of 3'azido-3'deoxythymidine and 2',3'-dideoxyinosine across the blood- brain barrier via a probenecid-sensitive transport system. J Pharmacol Exp Ther 281:369–375.

Tanzi RE, Moir RD, Wagner SL (2004) Clearance of Alzheimer's Abeta peptide: The many roads to perdition. Neuron 43:608.

Taylor EM (2002) The impact of efflux transporters in the brain on the development of drugs for CNS disorders. Clinical Pharmacokinetics 41:81–92.

Terasaki T, Takakuwa S, Saheki A, Moritani S, Shimura T, Tabata S, Tsuji A (1992) Absorptive-mediated endocytosis of an adrenocorticotropic hormone (ACTH) analogue, ebiratide, into the blood-brain barrier: Studies with monolayers of primary cultured bovine brain capillary endothelial cells. Pharm Res 9:529–534.

Theron D, de Lagerie SB, Tardivel S, Pelerin H, Demeuse P, Mercier C, Mabondzo A, Farinotti R, Lacour B, Roux F, Gimenez F (2003) Influence of tumor necrosis factor-alpha on the expression and function of P-glycoprotein in an immortalized rat brain capillary endothelial cell line, GPNT. Biochem Pharmacol 66:579–587.

Thomas SA (2004) Anti-HIV drug distribution to the central nervous system. Current Pharmaceutical Design 10:1313–1324.

Urayama A, Grubb JH, Sly WS, Banks WA (2004) Developmentally regulated mannose 6- phosphate receptor-mediated transport of a lysosomal enzyme across the blood-brain barrier. Proc Natl Acad Sci USA 101:12663.

Vadeboncoeur N, Segura M, Al-Numani D, Vanier G, Gottschalk M (2003) Proinflammatory cytokine and chemokine release by human brain microvascular endothelial cells stimulated by Streptococcus suis serotype 2. FEMS Immunology and Medical Microbiology 35:49–58.

van Dam AM, De Vries HE, Kuiper J, Zijlstra FJ, De Boer AG, Tilders FJH, Berkenbosch F (1996) Interleukin-1 receptors on rat brain endothelial cells: A role in neuroimmune interaction? FASEB J 10:351–356.

Verma S, Nakaoke R, Dohgu S, Banks WA (2006) Release of cytokines by brain endothelial cells: A polarized response to lipopolysaccharide. Brain Behav Immun. 20:449–545.

Vidal EL, Patel NA, Wu G, Fiala M, Chang SL (1998) Interleukin-1 induces the expression of mu opioid receptors in endothelial cells. Immunopharmacology 38:261–266.

Villegas JC, Broadwell RD (1993) Transcytosis of protein through the mammalian cerebral epithelium and endothelium: II. Adsorptive transcytosis of WGA-HRP and the blood-brain and brain-blood barriers. J Neurocytol 22:67–80.

Vogler C, Levy B, Galvin NJ, Thorpe C, Sands MS, Barker JE, Baty J, Birkenmeier EH, Sly WS (1999) Enzyme replacement in murine mucopolysaccharidosis type VII: Neuronal and glial response to beta-glucuronidase requires early initiation of enzyme replacement therapy. Pediatr Res 45:838–844.

Vorbrodt AW (1994) Glycoconjugates and anionic sites in the blood-brain barrier. In: Glycobiology and the Brain (Nicolini M, Zatta PF, eds), pp 37–62. Oxford: Pergamon Press.

Vorbrodt AW, Dobrogowska DH, Ueno M, Lossinsky AS (1995) Immunocytochemical studies of protamine-induced blood-brain barrier opening to endogenous albumin. Acta Neuropathol (Berl) 89:491–499.

Vorbrodt AW, Trowbridge RS (1991) Ultrastructural study of transcellular transport of native and cationized albumin in cultured sheep brain microvascular endothelium. J Neurocytol 20: 998–1006.

Walsh RJ, Slaby FJ, Posner BI (1987) A receptor-mediated mechanism for the transport of prolactin from blood to cerebrospinal fluid. Endocrinology 120:1846–1850.

Wang Y, Sawchuck RJ (1995) Zidovudine transport in the rabbit brain during intravenous and intracerebroventricular infusion. J Pharm Sci 7:871–876.

Westergren I, Johansson BB (1993) Altering the blood-brain barrier in the rat by intracarotid infusion of polycations: A comparison between protamine, poly-L-lysine and poly-L-arginine. Acta Physiol Scand 149:99–104.

Widner H, Jonsson BA, Hallstadius L, Wingardh K, Strand SE, Johansson BB (1987) Scintigraphic method to quantify the passage from brain parenchyma to the deep cervical lymph nodes in the rat. Eur J Nucl Med 13:456–461.

Williams KC, Hickey WF (1995) Traffic of hematogenous cells through the central nervous system. Curr Top Microbiol Immunol 202:221–245.

Wilson JF, Anderson S, Snook G, Llewellyn KD (1984) Quantification of the permeability of the blood-CSF barrier to α-MSH in the rat. Peptides 5:681–685.

Wolburg H, Wolburg-Buchholz K, Engelhardt B (2005) Diapedesis of mononuclear cells across cerebral venules during experimental autoimmune encephalomyelitis leaves tight junctions intact. Acta Neuropathol 109:181–190.

Wu D, Clement JG, Pardridge WM (1998) Low blood-brain barrier permeability to azidothymidine (AZT), 3TC, and thymidine in the rat. Brain Res 791:313–316.

Xaio H, Banks WA, Niehoff ML, Morley JE (2001) Effect of LPS on the permeability of the blood-brain barrier to insulin. Brain Res 896:36–42.

Yamada S, DePasquale M, Patlak CS, Cserr HF (1991) Albumin outflow into deep cervical lymph from different regions of rabbit brain. Am J Physiol 261:H1197–H1204.

Yeagle P (1987) Transport. The Membranes of Cells. Academic Press, Inc., Orlando, pp. 191–215.

Zalcman SS (2001) Interleukin-2 potentiates novelty—and GBR 12909-induced exploratory activity. Brain Res 899:1–9.

Zalcman SS (2002) Interleukin-2-induced increases in climbing behavior: Inhibition by dopamine D-1 and D-2 receptor antagonists. Brain Res 944:157–164.

Zambenedetti P, Giordano R, Zatta P (1996) Indentification of lectin binding sites in the rat brain. Glycoconjugate Journal 13:341–346.

5
Anterior Chamber and Retina

Leila Kump and Eyal Margalit

Keywords ACAID; Autoimmunity; Immune deviation; Ocular immunology; Ocular inflammation

5.1. Introduction

The goal of this chapter is to introduce the immune response processes and mechanisms that occur in different compartments of the eye. This is done to explain the relationship between ocular anatomy and immunogenic inflammation, and to understand the consequences of ocular immune system's malfunction.

5.2. Anatomy and Physiology

The eye is an organ that transmits light from the surrounding environment onto the retina. The cornea, iris, and crystalline lens cooperate to form a focused image on the retina. The neurons of the retina capture the image and encode it for accurate transmission via the optic nerve to the visual centers in the brain. Figure 5.1 shows a horizontal cross-section of the human eye. Light entering the eye is refracted by the cornea and the crystalline lens, and is aperture-limited by the iris. The lens is suspended by a ring of thin fibers (zonules) that are attached to a membrane encapsulating the lens. Its refractive power can be adjusted by contraction or relaxation of a muscle ring called the ciliary muscle. This allows one to focus on an image as the object's distance from the observer varies.

The lens separates the anterior and posterior chambers. Thus, the eye contains 3 compartments: anterior chamber (AC), posterior chamber and vitreous cavity. The AC is a space between the iris and the cornea, which is filled with aqueous humor. On average, it is 3 mm deep, with a volume of about 250 µL. The posterior chamber is also filled with aqueous humor; its location is posterior to the iris and ciliary body and anterior to the lens and the vitreous face. The largest compartment of the eye is the vitreous cavity, which is filled with vitreous gel. Its average volume is 6 mL. This compartmental division is a practical one. It is used extensively in clinical practice of ophthalmology. Differential diagnosis of a large number of infectious and inflammatory disorders is based on the anatomic location of the process. It is also closely related to the main purpose of this chapter to describe the immune processes taking place in different compartments of the eye.

5.3. Anterior Chamber

5.3.1. Anatomy and Physiology

The cornea serves as the front part of the AC of the eye. Its exterior is covered by the precorneal tear film, which lubricates, nourishes and protects the corneal surface. The iris and the pupil represent the posterior border of the AC. The AC angle is an important structure which is comprised of: Schwalbe's line, Schlemm's canal and trabecular meshwork, scleral spur, anterior border of the ciliary body and the iris (Figure 5.2). Aqueous humor that fills the AC is produced by the ciliary epithelium located in the posterior chamber. The fluid flows through the pupil and is drained by the trabecular meshwork into Schlemm's canal and subsequently into the episcleral vessels. This passage is named *the conventional pathway*. Aqueous humor is also drained by *a uveoscleral pathway* across the ciliary body into the supraciliary space.

5.3.2. Anterior Chamber Associated Immune Deviation (ACAID)

The phenomenon of immune privilege was initially described by Dooremal in 1873 (van Dooremal, 1873). He noticed that tumor cells injected into the AC of the eye formed growing tumors unlike tumor cells injected into skin or other organs. Medawar discovered that transplants between genetically different individuals usually were destroyed due to the ability of the immune system to detect alien molecules on the graft. However, skin grafts placed in the AC of the eye and in the brain of rabbits survived much

T. Ikezu and H.E. Gendelman (eds.), *Neuroimmune Pharmacology*.
© Springer 2008

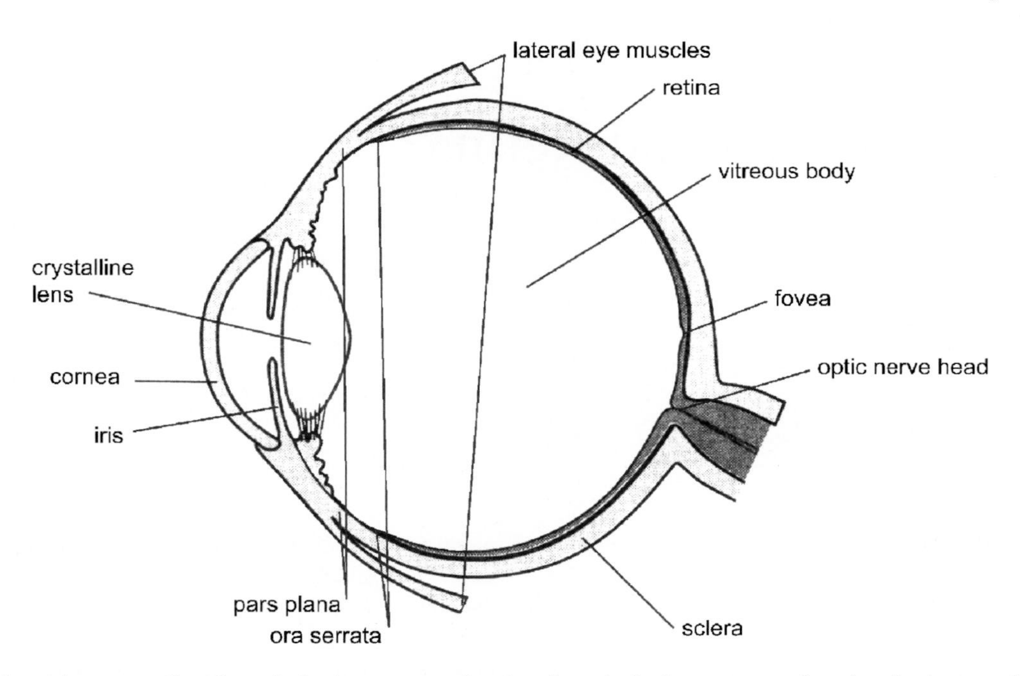

FIGURE 5.1. Horizontal cross-section through the human eye, showing the principal structures referred to in the text. (Illustration from "Ocular Anatomy, Embryology and Teratology" 1982, used with permission from Williams and Wilkins, Philadelphia, PA.)

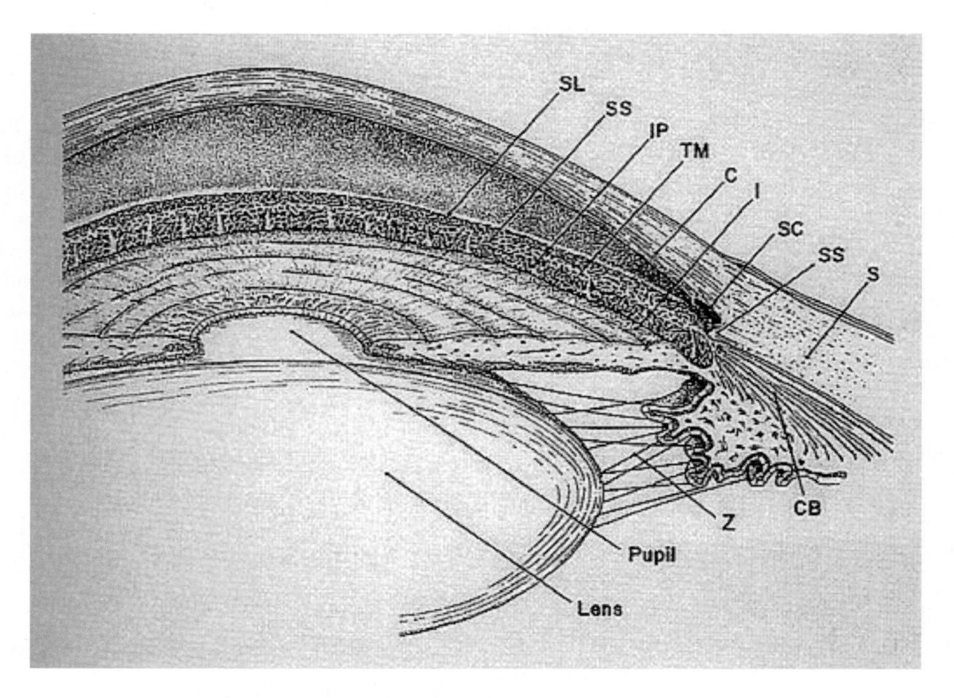

FIGURE 5.2. Drawing of the structures of the angle of the AC and ciliary body. SL, Schwalbe's line; SS, scleral spur; IP, iris process; TM, trabecular mashwork; C, cornea; I, iris; SC, Schlemm's canal; S, sclera; CB, ciliary body; Z, zonular fibers. (Illustration from "Ocular Anatomy, Embryology and Teratology" 1982, used with permission from Williams and Wilkins, Philadelphia, PA.)

longer. Medawar called the AC of the eye and the brain "immune privileged sites." In 1948 Medawar's fundamental work explained that immune privilege was a special circumstance where the laws of transplantation immunology did not apply (Medawar, 1948). Immune privilege was defined as a prolonged, sometimes indefinite, survival of organ or tissue grafts at special body/organ sites (Barker and Billingham, 1977). Additional examples of such immune privileged sites would be the cornea, lens, vitreous cavity, and subretinal space; testis, ovary, adrenal cortex, pregnant uterus, and certain tumors.

Anatomic structures were considered key elements of immune privilege in earlier studies. The existence of a blood-tissue barrier and the absence of afferent lymphatic drainage pathways led to the belief that antigens of the grafts placed in immune privileged sites remained physically sequestered from the immune environment. As such, the immune system never became aware of them. However, later studies demonstrated that the maintenance of immune privilege is actually a dynamic process combining immunoregulatory forces combined with anatomic structure to allow the survival of grafts in privileged sites (Niederkorn et al., 1980; Niederkorn 1990; Ksander and Streilein, 1993; Tompsett et al., 1990; Streilein, 1995). In other words, the blood-brain and blood-ocular barriers exist, but their creation and existence are actively maintained. The list of factors responsible for the ocular immune privilege is presented in Table 5.1.

It is now recognized that both active and passive factors of immune privileged sites contribute to their status. It has been known for over 100 years that the AC of the eye possessed qualities allowing a long-term survival of tissue and tumor grafts (van Dooremal, 1873). In the late 1970s Kaplan and Streilein discovered that antigenic cells placed into the AC were not only detected by the immune system, but also elicited a downregulation of alloimmune responses (Kaplan et al., 1975; Kaplan and Streilein, 1977).

It became clear that antigens are not physically sequestered from immune recognition in the AC, but that the consequence of their detection is a deviation from the expected immune response. Although AC injection of allogeneic lymphoid cells aroused the humoral arm of the immune response resulting in the production of antibodies against donor histocompatibility antigens, the cell-mediated component was impaired. Studies in the early 1980s led to the knowledge that a wide variety of antigens (alloantigens, tumor-specific antigens, soluble protein antigens, haptens, viral-encoded antigens) placed in the AC of mice produced a similar deviant pattern of systemic immune responses. Niederkorn, Streilein, and Shadduck coined the term Anterior Chamber Associated Immune Deviation (ACAID) to designate this unusual response (Niederkorn et al., 1980). From multiple studies it became clear that ACAID is an antigen-specific systemic immune response to eye-derived antigens, a response that represents a selective deficiency of T cell functions that mediate delayed hypersensitivity to the antigen (Th1 cells), and B cells that secrete complement-fixing antibodies. However, this systemic immune response retains primed, clonally expanding cytotoxic T cell precursors and B cells that secrete IgG1, a noncomplement-fixing antibody. Lymphoid tissues of mice with ACAID contain three populations of regulatory T cells: $CD8^+$ T cells suppressing expression of delayed hypersensitivity; $CD4^+$ T cells suppressing the induction of Th1 cell responses; and $CD8^+$ T cells that inhibit B cells from switching to IgG isotypes that fix complement.

In late 1980s experiments with samples of AC's aqueous humor showed that it possessed some immunomodulatory properties, but these properties were not global, i.e., aqueous humor does not inhibit all immune reactions (Kaiser et al., 1989). Suppression of the following phenomena by aqueous fluid was found:

1. Activated macrophage production of nitric oxide and reactive oxygen intermediate (Taylor et al., 1998).
2. Neutrophil-mediated lysis of target cells (Miyamoto et al., 1996).
3. Lysis of target cells by natural killer (NK) cells (Apte and Niederkorn, 1996).
4. Conventional antigen-presenting cell (APC) activation of Th1-type cells in-vitro (Takeuchi et al., 1999).
5. C1q from binding to antibody-coated erythrocytes.
6. C3 cleavage to C3b via the classical or the alternative complement pathways (Goslings et al., 1998).

As we learn more about the ocular immunosuppressive environment, the list of immuneomodulatory molecules and factors within aqueous humor continues to grow.

Streilein and Kaplan have discovered that the spleen plays a crucial role in the immune deviation phenomena in rats (Streilein and Kaplan, 1979). This was confirmed in later experiments in mice (Streilein and Niederkorn, 1981; Wilbanks et al., 1991). Following an injection of the heterologous protein antigen ovalbumin (OVA) into the AC, an ACAID-inducing signal (AIS) is released to the blood stream. For example, when naïve mice received intravenous injection of

TABLE 5.1. Factors responsible for ocular immune privilege.

Passive	Active: soluble factors	Active: cell-surface factors
Blood-tissue barrier	Transforming growth factor-β2	CD95 ligand
Tissue fluids drain IV	Alpha-melanocyte stimulating hormone	CD55, CD59, CD 46
Deficient lymphatics	Vasoactive intestinal peptide	
Reduced expression of major histocompatibility class I and II molecules	Calcitonin gene-related peptide	
Reduced antigen presenting cells with altered function	Somatostatin	
	Thrombospondin	
	Macrophage migration inhibitory factor	
	IL-1 receptor antagonist	
	Free cortisol	

blood from mice that had been injected with OVA into the AC 48 h prior, the naïve mice acquired OVA-specific ACAID. It was postulated that eye-derived antigen-bearing antigen presenting cells (APCs) escaped through the trabecular meshwork, and cells migrated via the bloodstream to the spleen. In the spleen the regulatory cells that mediate ACAID eventually emerged. These findings indicate that the eye contains nascent APCs that pick up the antigen and deliver it to the spleen via the blood. The spleen induces ACAID, without which, immune privilege in the AC cannot be sustained.

Further understanding of the ACAID phenomenon is attributed to Hara et al. who developed an in-vitro generated ACAID-inducing signal (Hara et al., 1992). These investigators incubated conventional APC (from peritoneal exudate) overnight in the presence of acid-treated aqueous humor and then pulsed with OVA. When injected into naïve mice these APCs induced ACAID. Since then, it became possible to perform experiments with ACAID-inducing APCs that could be grown in vitro. Takeuchi and Streilein conducted series of experiments that led them to conclude that ACAID-inducing APCs were formed by pulsing active-TGFβ-2-pretreated-peritoneal-macrophages with OVA. The principal action of TGF-beta is to inhibit activation and proliferation of lymphocytes. It is a family of molecules named TGFβ-1, TGFβ-2 and TGFβ-3. Cells of the immune system mostly produce TGFβ-1. This family inhibits immunologic, inflammatory responses and has immunosuppressive effect. The above-mentioned APCs were capable of activating T-cell receptors (TCR) transgenic OVA-specific CD4 and CD8 T cells in vitro (Takeuchi et al., 1998; Takeuchi et al., 1999). It was found that a universal property of ACAID-inducing APCs created in vitro is the capacity to deviate naïve CD4+ and CD8+ T cells away from their typical differentiation pathway toward a pathway that might enable them to be regulators. Normal aqueous humor contains large amounts of thrombospondin (a glycoprotein involved in the clotting cascade). The gene controlling the formation of this protein is constantly active in ocular parenchymal cells. Thrombospondin and TGFβ-2 may be the key players in creating immune deviation when antigens are introduced into immune-privileged sites.

Unique qualities of ACAID-inducing APCs have been analyzed by many researchers in order to understand their ability to alter the function of responding T cells (Takeuchi et al., 1998; Faunce et al., 2001; Masli et al., 2002). TGFβ-2 treated peritoneal macrophages usually activate Th1 cells, and express normal levels of class I and II major histocompatibility proteins. However, TGFβ-2 exposed APCs fail to upregulate CD40, and they secrete only small amounts of IL-12. IL-12 serves as a mediator and an inducer of innate immune responses to intracellular microbes. It also activates NK cells, promotes their cytolytic activity and development of Th1 cells. CD40 plays various roles in macrophage, dendritic cell, and endothelial cell activation. Expression of IL-12 and CD40 is not upregulated even in the presence of responding T cells. Interestingly, the supernatants of TGFβ-2-treated APCs contain active TGFβ-1, which is usually produced by bone-marrow-derived cells.

In addition, treatment with active TGFβ-2 has a profound effect on the genetic arrangement of APCs, as discovered by a study of macrophage hybridoma #59, which is a laboratory–created cell line for immunologic studies (Wetzig et al., 1982). It is apparent that the thrombospondin gene is an early target of TGFβ-2 and is upregulated. Thrombospondin has a number of actions that promote the ACAID-inducing properties of TGFβ-treated APCs: 1) it binds to CD36 on APCs and tethers latent TGFβ to the APC surface; 2) it binds to CD47 on APCs and on responding T cells, potentially forming a cellular bridge that stabilizes the APC/T cell interaction; 3) with CD47 on T cells, thrombospondin alters signaling through the TCR in a manner that deviates the cell's functional program; and 4) it promotes conversion of latent TGF-β to its active form in the environment between APC and T cells.

As mentioned above, the spleens of mice that receive OVA into the AC acquire three populations of regulatory T cells. These cells are produced within seven days (Wetzig et al., 1982; Niederkorn and Streilein, 1983; Waldrep and Kaplan, 1983; Wilbanks and Streilein, 1990). One population is CD4+. When it is adoptively transferred into naïve recipients, it suppresses the induction of OVA-specific delayed hypersensitivity. A second population of regulatory cells is CD8+, which inhibits the expression of delayed hypersensitivity. So, generation of ACAID correlates with regulatory T cells that suppress the induction and expression of delayed hypersensitivity. A third population of regulatory CD8+ cells arises following AC injection of OVA. This population prevents OVA-specific B cells from class-switching their immunoglobulin isotype to the isotypes that fix complement: IgG2a, IgG2b, and IgG3. The second population of regulatory cells, efferent CD8+ T cells that suppress delayed hypersensitivity expression, has gained a lot of attention because of its uniqueness to ACAID. These cells are relatively easy to detect and evaluate. They can be evaluated systemically by transferring large numbers of spleen cells intravenously into mice previously sensitized to OVA and challenged for delayed hypersensitivity a few hours later. The regulator cells can be also evaluated locally in a "local adoptive transfer reaction" in which putative regulator cells are mixed with OVA-specific T cells and antigen, and the mixture is injected intradermally into the ear pinnae of naïve mice. Thus, researchers have been able to elucidate a mechanism of ACAID by demonstrating the production of T regulatory cells that inhibit the local expression of delayed hypersensitivity.

In order to induce immune deviation only two cells need to interact: the "tolerogenic" APC and the T cell. These two cells interact under the influence of innate immune cells within the lymphoid organs. It has been found that ACAID does not occur in the absence of gamma/delta T cells, (Skelsey et al., 2001; Xu and Kapp, 2001), natural killer T (NKT) cells (Sonoda et al., 1999), or B cells (D'Orazio et al., 1998a). The innate cells that have been most exclusively studied are the APC and NKT cells. The NKT cell is important in peripheral tolerance induction in association with the AC (Sonoda et al., 1999).

The antigen-transporting cells may direct the multicellular organization of the innate cells (specifically, NKT cell) with the appropriate adaptive immune cell precursors. The soluble factors secreted by the antigen-transporting cell recruit specific cell types and factors to create an immunosuppressive environment favorable to the induction of tolerance. The antigen that is injected into the AC is carried from the eye by the indigenous APCs. These transporting APCs are unique, possibly because they are bathed in eye-specific immunosuppressive factors that induce their special phenotype. It has been shown that the eye-derived macrophages selectively increased the chemokine MIP-2 (Faunce et al., 2001). While MIP-2 was previously thought to be a neutrophil chemoattractant, it was found that that MIP-2 was a strong chemoattractant for NKT cells as well. NKT cells are absolutely required for the development of the efferent T regulator cell in ACAID. When NKT cells were removed by use of antibodies in vivo, or when NKT deficient mice were used, the efferent T regulator cells were not generated (D'Orazio and Niederkorn, 1998b).

5.3.3. ACAID and Other Forms of Immune Regulation and Tolerance

Immunologic tolerance is defined as unresponsiveness to an antigen and it is induced by prior exposure to that antigen. When specific lymphocytes encounter antigens, three possible outcomes may follow: the lymphocytes are activated, leading to immune responses; the lymphocytes are inactivated or eliminated, leading to tolerance; or the antigen is ignored. Different forms of the same antigen may induce an immune response or tolerance or may elicit no response. Tolerance comes in many forms. Central tolerance ensures that the repertoire of mature lymphocytes cannot recognize ubiquitous self antigens, which are antigens most likely to be present in the degenerative lymphoid organs. This mechanism is mainly responsible for the elimination of self-reactive lymphocytes from the mature repertoire, and thus for self/nonself discrimination and therefore suppression of autoimmune disease. Peripheral tolerance, also called "active tolerance," is induced by recognition of antigens without adequate levels of the costimulators that are required for lymphocyte activation. T cell activation requires not only the recognition of antigen, but also the recognition of costimulators. These are also called "signal 2." If costimulators are not present while the antigen is being recognized, T cell anergy occurs. This also occurs in response to persistent and repeated stimulation by self-antigens in the peripheral tissues.

ACAID is a special form of peripheral tolerance. About 98% of the antigen that is inoculated into the eye is carried into the bloodstream within 6h of injection (Wilbanks and Streilein, 1989). It was suggested that ACAID is just a manifestation of intravenously induced tolerance. However, ACAID was demonstrated to be distinctly different from intravenously induced tolerance. Intravenous tolerance is mediated by a CD8+ afferent T suppressor/regulator cell, and in intravenous tolerance no regulatory efferent T cell is generated (Heuer et al., 1982). In ACAID, the afferent regulatory T cell is CD4+ and the efferent regulator is CD8+. Further, ACAID does not occur in mice that are deficient in NKT cells. Intravenously induced tolerance to the same antigens can be induced readily in NKT cell knockout (KO) mice (Sonoda et al., 1999).

Both ACAID and oral tolerance produce efferent CD8+ T regulator cells that are dependent on intact NKT cells. While low dose oral tolerance and ACAID correlate with the presence of efferent CD8+ regulatory cells, the cells are not functionally identical. The efferent T cells of oral tolerance respond to antigen by secreting IL-10 and TGF-β, whereas the efferent T suppressor cells of ACAID secrete TGF-β only (Weiner, 1997). However, NKT cells are needed for both forms of CD8+ T regulatory cell-mediated tolerance.

Can we learn whether ACAID is central or peripheral tolerance by examining the T helper cells that are involved? T helper cells were found to comprise two functionally distinct types (type 1 and 2) (Mosmann and Coffman, 1989; Mosmann, 1992). Th1 cells responded to antigen stimulation by secreting IL-1 and IFNγ. Th1 cells mediate delayed hypersensitivity reactions, and promote the switch of B cells to Ig isotypes that fix complement (IgG2a/b, IgG3 in mice). Th2 cells respond to antigen stimulation by secreting IL-4, IL-5, IL-6, IL-10, and IL-13, but they fail to secrete IFNγ. Th2 cells promote humoral responses rich in noncomplement–fixing IgG1, IgA, and IgE antibodies. Also, Th1 and Th2 cells are able to cross-regulate each other: Th1 cells suppressing Th2 cells and vice versa. Cross-regulation of Th1 and Th2 cells has been called "immune deviation." It was speculated that ACAID is a Th2 response evoked by an unusual route of antigen delivery (the eye). It has been reported that mice genetically deficient in IL-10 production fail to acquire ACAID (D'Orazio and Niederkorn, 1998). Some experiments dispute the idea that ACAID is a Th-2 mediated response (Streilein et al., 2001). It is currently thought that ACAID represents a unique form of tissue-dependent regulation of systemic immunity that resembles, but is mechanistically different from tolerance induced by some other routes and procedures. ACAID is a form of peripheral tolerance, unlike central tolerance, where clonal deletion and/or anergy occur within the thymus. Once induced, ACAID persists for a very long period of time. ACAID is an actively acquired and actively maintained manifestation of ocular immune privilege.

5.3.4. ACAID, Ocular Immune Diseases, and Implications for Therapy

ACAID may have either beneficial or deleterious effects on ocular inflammatory disease. Ocular inflammation that is elicited by Tcell-mediated delayed hypersensitivity and by compliment fixing antibodies that impair or destroy vision. ACAID tries to protect the eye from such injury by selectively suppressing innate (production of phagocytic cells and

NK cells, blood proteins and cytokines) and adaptive (lymphocyte and antibody production) immune responses of the intense proinflammatory type (when large amounts of proinflammatory cytokines are present and are ready to excite inflammation). For certain infectious pathogens the entire array of immune defense mechanisms is mobilized in order to eliminate the exciting agent. If immune privilege and ACAID prevail, resistance to infectious agents cannot be mediated by delayed hypersensitivity T cells and by complement-fixing antibodies because these components are not active in such an environment. In this situation the eye faces a dilemma: whether to sustain immune privilege and prevent intraocular inflammation, thus risking the infection to ruin the eye, or withdraw immune privilege and kill the pathogen, but risk the eye destruction from uncontrolled inflammation.

Thus, immune privilege and ACAID could have both positive and negative effects in the eye. So what is the role of such a mechanism? ACAID might play a role in protecting the eye against autoimmune attack directed at strong ocular autoantigens such as arrestin, interphotoreceptor retinol binding protein, and rhodopsin.

Some speculate that ACAID-based immunotherapy may be beneficial in immune-mediated diseases of the eye and a variety of other organs. This is due to the fact that the immune deviation of ACAID produces T regulatory cells that are effective in inhibiting both Th1 and Th2 responses (both primary and secondary responses). Cd1d-reactive NKT cell-dependent tolerance or ACAID induced by inoculation of antigen into the eye may contribute to self-tolerance and prevention of autoimmune responses in organs and tissues in general.

There are reports of correlations of defective or deficient NKT cells in a number of autoimmune diseases in mice and humans (Sumida et al., 1995; Baxter et al., 1997, Wilson et al., 1998; Illes et al., 2000; Mieza et al., 1996; Zeng et al., 2000; Shi et al., 2001; Nagane et al., 2001). In humans, diabetes, systemic sclerosis, myasthenia gravis, and multiple sclerosis and in mice, a lupus model, are associated with NKT cell deficiencies. It was demonstrated that the adoptive transfer of NKT cells prevented diabetes in NOD mice. However, it is not clear whether the effectiveness of this treatment was actually due to the establishment of immune deviation of Tr cell-mediated active tolerance.

It is known that ACAID contributes to ocular tumor survival and long-term survival of orthoptic corneal allografts. Some experiments showed that removal of NKT cells prevented the continued acceptance of ocular tumors and caused the elimination of the immune privilege of the eye. Without functioning NKT cells, mice were not able to accept orthoptic corneal allografts for prolonged periods of time (Sonoda et al., 2002). This data demonstrates the essential role of NKT cells in the generation of ACAID. New information about ACAID may lead to application of ACAID mechanisms in prevention and treatment of immune-mediated inflammatory diseases in humans.

5.4. Retina

5.4.1. Anatomy

A more complete coverage of this topic appears in Chap. 11 of this book. In short, the retina is a thin transparent layered structure lining the posterior eye wall. The main cell types include photoreceptors (rods and cones), which capture the light; bipolar and ganglion cells, which pass the visual signal on toward the optic nerve, and horizontal and amacrine cells, which provide lateral interactions among cells in neighboring locations. Figure 5.3 demonstrates a diagram of cross-sectioned retinal layers (**A**), a preparation of normal donor retina (**B**), and tissue from a donor who suffered from a retinal degeneration, causing the disappearance of the photoreceptor layer (**C**).

The layers of normal cross-sectioned retina (from outer to inner retina) are:

- Retinal pigment epithelium (RPE) and its basal lamina
- Rod and cone inner and outer segments
- Outer nuclear layer
- Outer plexiform layer
- Inner nuclear layer
- Inner plexiform layer
- Ganglion cell layer
- Nerve fiber layer
- Internal limiting membrane

A single layer of RPE cells fulfills the role of sustaining the metabolism of the photoreceptors: The metabolic level of the photoreceptor outer segments is among the highest in the human body. RPE cells supply nutrients and oxygen, regenerate phototransduction products, and digest debris shed by the photoreceptors.

Photoreceptors, the cells capturing the incoming light, come in two main classes: rods, whose high internal gain allows vision at very low light levels, and cones, in short, medium, and long wavelength types to allow color perception. In both classes of cells the actual light capture and conversion takes place in the outer segment—visible as a band with long/thin elements (rods) toward the left edge and shorter/stubbier elements (cones) in the center of Figure 5.3B. The cell's inner segment, located in the outer nuclear layer (ONL), provides the transduction to secondary neurons and regulates cell function.

The outer plexiform layer (OPL) contains the contacts between photoreceptors, horizontal and bipolar cells, allowing the first stage in retinal image processing to take place. This processing provides coupling between neighboring rods and/or cones, allowing for the smooth transition between rod and cone function during light/dark adaptation and for dealing with differences, and rapid changes, in local illumination. The inner nuclear layer (INL) contains the cone and rod bipolar cells. Cone bipolar cells form the first stage of image processing in the visual system, combining output from different cone types for color processing. In the periphery, they gather information from multiple cones, and save image transmission bandwidth by reducing resolution. All rod bipolar cells receive input from multiple rods, allowing reliable signal processing at very low light levels. Also located in the INL are the cell bodies of horizontal and amacrine cells.

In the inner plexiform layer (IPL), extensive interactions take place through synaptic contacts: between bipolar cell axons and retinal ganglion cell dendrites; between rod and cone on-bipolar cells through rod amacrine cells (thus merging the rod output signals into the cone pathway); between amacrine, ganglion, and bipolar cells (providing inhibitory feedback to strengthen ganglion cell properties); and between bipolar cells and interneurons (performing a feedback function between neighboring bipolar cells).

The ganglion cell layer (GCL) contains the cell bodies of retinal ganglion cells, with their axons running across the retinal surface (nerve fiber layer) toward the optic nerve head, and on through the optic nerve to the lateral geniculate nucleus in the mid-brain. The inner retinal blood supply (outside the foveal avascular zone), the nerve fiber layer, and a thin membrane (the inner limiting membrane) form the most superficial retinal structures.

In addition to the neuronal cell types responsible for visual signal transmission the retina has several types of supporting cells, similar in function to the RPE cells supporting the photoreceptor outer segments. The most important of these are the Mueller cells, whose principal role appears to be to buffer and balance electrolyte concentrations in the extracellular space, in response to the activity of the retinal neurons, especially photoreceptors.

Figure 5.3C shows a cross-section of the retina, but this tissue came from a donor who had lost useful vision due to a retinal degeneration. There is a complete loss of photoreceptor outer segments, the almost complete absence of cells in the ONL (i.e., photoreceptor inner segments). However, the inner retinal cells are preserved in substantial numbers.

5.4.2. Anatomy and Physiology: Retino-Cortical Pathway

The eye and the central nervous system are connected through the fibers of the optic nerve. These fibers, with a diameter of about 1 μm, are actually the axons of retinal ganglion cells. Inside the eye, the fibers run along the retinal surface toward

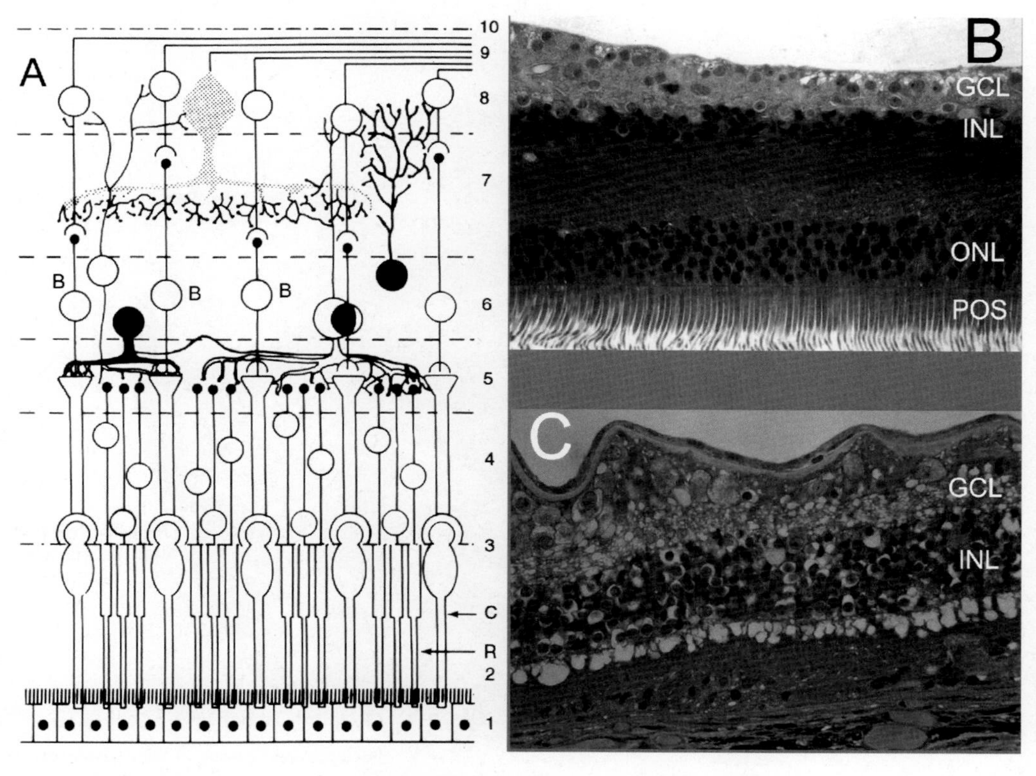

FIGURE 5.3. A: Schematic representation of the cell layers in the retina. Light entering the retina passes through the nerve fiber (9), ganglion cell (8), inner nuclear (6)—containing bipolar cells (B), and outer nuclear (4) layers, before being absorbed in the outer segments of the rods (R) and cones (C). B: Cross-section through the retina near the fovea, showing healthy photoreceptor outer segments (POS), multiple layers of photoreceptor cell nuclei in the outer nuclear layer (ONL; labeled black), bipolar cell nuclei in the inner nuclear layer INL) and ganglion cell bodies in the ganglion cell layer (GCL). C: Retina of a patient with a long history of retinal degeneration, and bare light perception in the last years of life, showing lack of photoreceptor outer segments and cell bodies in comparison with B. (Illustration from "Neuroprosthetics: Theory and Practice," Chapter 17, 2004. Used with permission from World Scientific Publishing Co.)

the optic nerve head in a characteristic pattern, such that fibers of the upper and lower retinal halves remain separated, and fibers close to the horizontal meridian, but far from the nerve head, arc away from this line to allow room for fibers originating closer to the nerve head. This orderly arrangement causes the fibers from the foveal area (which form 15–20% of all nerve fibers) to be located in the temporal quadrant of the optic nerve, at least for the initial portion of its trajectory.

After the axons enter the optic nerve, each fiber is encapsulated with a myelin sheath, formed by a class of cells called oligodendrocytes; this sheath decreases the membrane conductance of the axons, increasing the conduction velocity and the length over which impulses can be conducted without severe attenuation. Only at the so-called Ranvier nodes is the myelin sheath interrupted, allowing the impulses to be reinforced by virtue of the gating properties of the local membrane.

A cross-section through the human visual pathways can be seen in Figure 5.4. One may note that the predominant pathway leads from the eye to the lateral geniculate nucleus (LGN) of the thalamus, and from there to the occipital part of the cortex, while less important numbers of fibers branch off to a tectal area, the superior colliculus (SC), and to a number of pre-tectal nuclei. We will briefly discuss these subcortical pathways below.

The LGN and cortical areas exist in duplicate in the two halves of the brain. Each deals with one half of the visual world: the optic nerves from the two eyes meet in a structure called the optic chiasm, where fibers from the two nasal retinas cross over to combine with those from the temporal retina of the fellow eye; consequently each LGN and cortical hemisphere receive visual information from two corresponding retinal halves on their own side, and thus from the contralateral half of the visual field.

The LGN has a layered structure, with layers devoted to nerve fibers coming from each eye, and subsequent pairs of layers receiving axons of different ganglion cell types. Significant interaction between layers does not occur; therefore, binocular interactions such as those required for stereopsis do not take place until the level of the visual cortex. The gateway function of the LGN, which in other mammals such as the cat appears to play a crucial role in adaptation and attention, and through which signals from the two eyes can mutually inhibit each other, is thought to be less prominent in primates, including humans. Yet anatomical feedback connections from a number of subcortical nuclei onto the LGN are as extensive in monkey as in cat, and gating functions related to circadian rhythms and other systemic conditions is therefore plausible in primates as well.

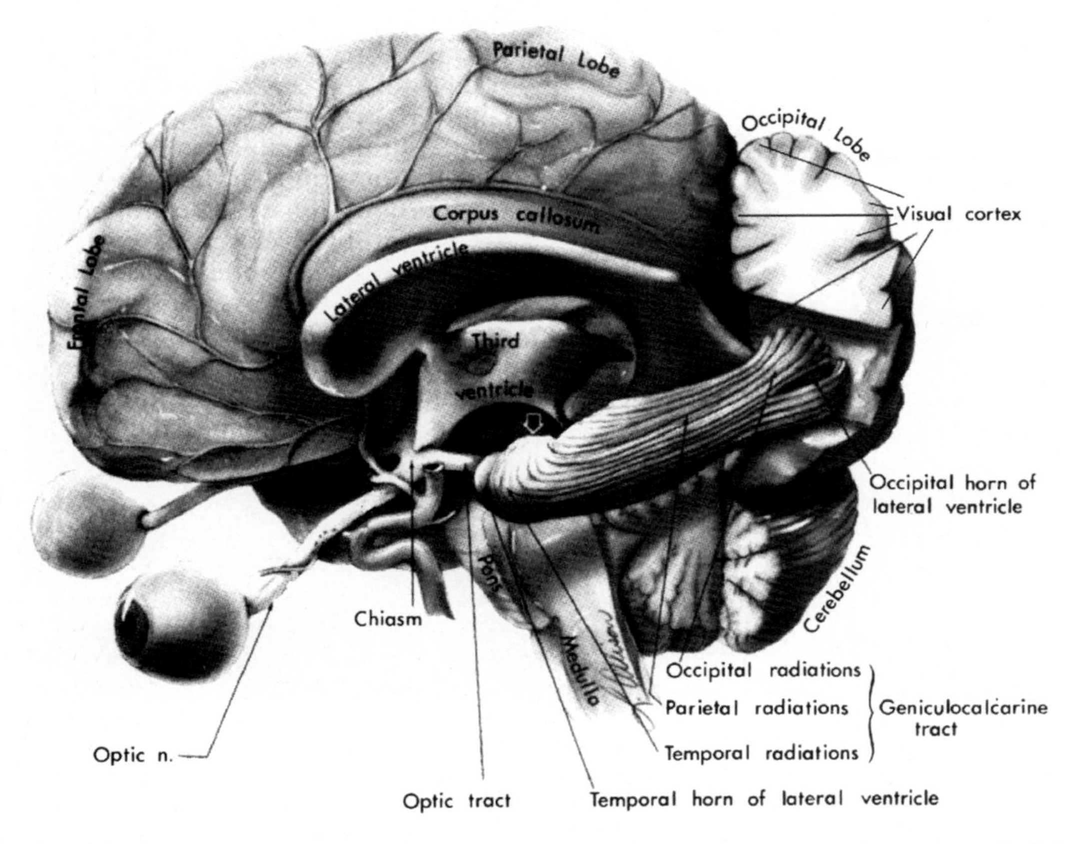

FIGURE 5.4. Structure and location of the human primary visual pathways, in relation to other major brain structures. The left cerebral hemisphere, with the exception of the occipital cortex, has been removed; the left LGN is hidden by the optic radiations (arrow). (Illustration from "Neuroprosthetics: Theory and Practice," Chapter 17, 2004. Used with permission from World Scientific Publishing Co.)

Forward pathways from the LGN lead to the primary visual cortex (V1, also called striate cortex; these fibers form the optic radiation), but also to higher visual cortical areas and to subcortical areas such as the superior colliculus (SC). The role of the extrastriate cortical pathways is still a topic of extensive investigation; from clinical cases it is evident, however, that patients with lesions to the striate cortex retain little or no useful vision. The roles of the tectal pathways, including mutual connections between cortical areas, the SC and the pulvinar, are also far from completely understood. It is thought, for example, that cortical connections with midbrain areas are essential for maintaining and shifting attention, rather than for processing detailed visual information.

5.4.3. Anatomy and Physiology: Visual Cortex

The visual cortex occupies the occipital and parts of the parietal and temporal lobes of the cerebral cortex. Like the entire cortex, it forms a highly folded structure, with a thickness of approximately 1.5 mm. It is surrounded by the cerebrospinal fluid, several layers of meninges- pia mater, arachnoid, dura, and the skull.

Like the retina, the visual cortex is a layered structure, in which different cell groups perform different tasks. Along its two-dimensional surface, one finds an orderly mapped representation of the outside world. Unlike the retina, the cortex consists of multiple areas, hierarchically organized, each of which performs a partial processing task in the analysis of the scene around us. At the present time, over 30 visual cortical areas per hemisphere are recognized in monkey, and a smaller number of more expanded areas are thought to exist in humans.

The first cortical representation, in the striate cortex, is shown schematically in Figure 5.5. It presents a straightforward map of the visual world, but contains four major transformations.

The projection from the LGN (and thus retinal ganglion cells) onto V1 input cells has approximately constant density, which means that the central visual field is highly overrepresented in the visual cortex: Roughly 20% of V1 represents the retinal fovea, and thus the central 1°–2° of the visual field, with rapid drop-off of the density toward the periphery. This nonhomogeneous map is conveniently expressed by the cortical magnification factor, M('), i.e., the number of mm of cortex devoted to 1° of retina, as a function of eccentricity.

The folding of the human cortex, prompted by the evolutionary expansion of higher (cognitive) processing has resulted in an arrangement where most of the peripheral visual field is represented in portions of V1 that are buried in the medial walls and sulci of the cortical hemispheres. Only the foveal representation, situated along the border of V1 and the adjacent area V2, is exposed at the surface of the occipital cortex. More peripheral visual field areas are represented along the medial walls of the cerebral hemispheres, and in deep sulci embedded within these areas.

V1 has an intricate columnar structure on a scale of approximately 1 mm^2, in which cells are arranged in intersecting patterns with different receptive field characteristics: One direction contains bands of cells receiving alternating projections from the two eyes, and roughly perpendicular to these ocular dominance bands, but curving into a radial "pinwheel" arrangement centered around singularities within these bands, one finds cell populations with a gradual shift in preferred stimulus orientation in superficial (2 and 3) and deep (5) cortical layers. Independently one finds cells with more straightforward properties: So-called "simple" cells in layers 4a/b/c receiving input directly from the LGN, and cytochrome oxidase (CO)-rich blobs in layers 3 and 5 in which cells show clear color specificity. The widely held notion that the striate cortex in primates contains a regular arrangement of repeating "hypercolumns" has been challenged by a more plausible arrangement in which ocular dominance and orientation specificity self-organize independently around CO-rich blobs in the course of visual development. Whichever model may be closest to reality, it is clear that the striate cortex does not provide homogeneity and isotropy, even on a local scale. Homogeneity and anisotropy may be even more pronounced in higher cortical areas (Freund, 1973).

As explained previously for ganglion cell and LGN characteristics, even the signals arriving in the cortex do not carry simple point-to-point representations of the visual field. From the complex processing network in V1 hypercolumns emerge feed-forward signals to higher cortical areas. In turn, feedback signals from these higher areas further influence local signal processing. Moreover, subcortical signals modulate the activity level in response to state changes such as sleep and arousal.

5.4.4. Anatomy and Physiology: Subcortical Pathways

While the visual pathway to the LGN and striate cortex receives the great majority of retinal ganglion cell axons, there exist subcortical pathways as well, formed by optic nerve fibers projecting to pretectal nuclei and to the pulvinar. In primates, the projections to the pregeniculate nucleus and pulvinar are thought to be of minor importance, and may be thought of as anatomical evolutionary remnants. For example, in lower mammals, ablation of striate cortex at birth allows these projections to greatly increase in density, leading to the development of crude functional vision, but similar experiments in newborn monkeys show neither the proliferation of projections nor appreciable acquisition of visual function (Cowey et al., 1994; Cowey et al., 2001).

However, other projections in primates, in particular those to the pretectal nucleus of the optic tract (NOT) and the terminal nuclei (TN) of the accessory optic system, have been demonstrated to play an important role in the rapid control of eye position through vestibulo-ocular reflex, saccades, and sustained fixation.

Detailed studies of anatomy and physiology of the primate eye movement system over the last several decades in awake, trained animal models have shown that the NOT receives information on "retinal slip," i.e., generalized displacement of the

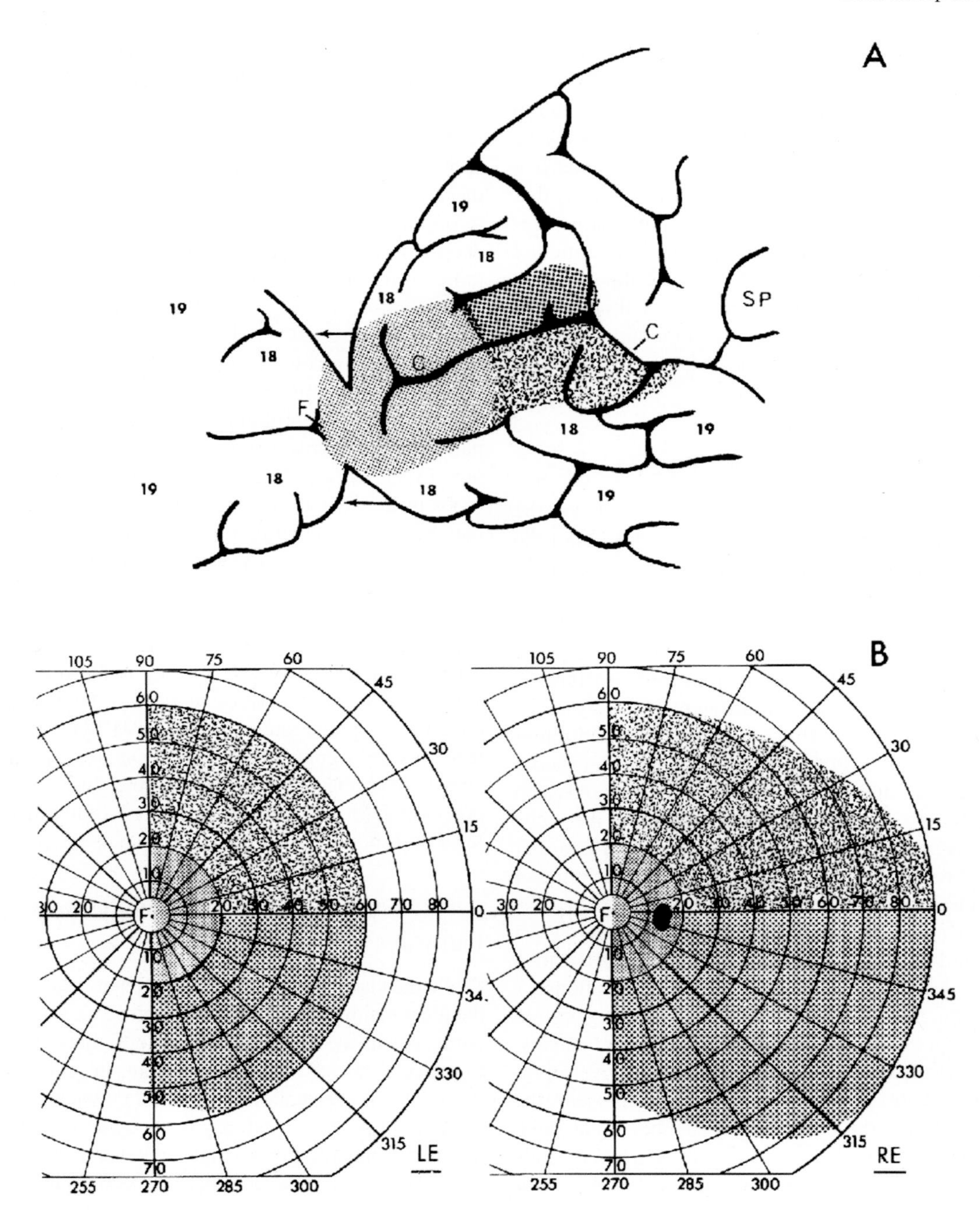

FIGURE 5.5. V1 projection of the visual field. The medial wall and part of the occipital surface of the left cerebral hemisphere (A) and the corresponding visual fields for the two eyes (B) are shown. Note that the projection of the fovea (F) and a narrow surrounding hemicircle of the visual field project onto the occipital cortex, with the projection of the vertical meridian adjacent to area V2, whereas more peripheral areas—including most of the macula—projects to the medial wall of the cortex, with much of the projection buried in the calcarine fissure (C). Also note that the left hemisphere receives information from the right visual hemifield, that the superior visual field projects to the inferior part of V1—i.e., gross localization is preserved from retina to V1, and that corresponding retinal locations in the two eyes project to the same cortical location. No such locations exist for the far nasal segment of the right retina (60°–90°), as the bridge of the nose blocks the corresponding area in the left eye. (Illustration from "Neuroprosthetics: Theory and Practice," Chapter 17, 2004. Used with permission from World Scientific Publishing Co.)

retinal image (Mustari and Fuchs, 1989). This retinal slip signal is encoded as a velocity signal, and serves as input to the neural integrator in the nucleus prepositus hypoglossi (Mustari and Fuchs, 1989). Pathways between the NOT and primary visual cortex (as well as multiple similar projections between cortical and subcortical structures) are also known to exist, and have been shown to compensate in part for lesions to the NOT or its retinal input (Mustari and Fuchs, 1989; Hoffmann, 1996).

5.5. Retinal Immunology

Distinctive immunologic characteristics of the eye were well documented and led to the concept of immunologic privilege. The success of corneal transplantation established the eye as a site for tissue transplantation (von Hippel, 1888). Later, the harvesting and transplanting photoreceptors or retinal pigment epithelial (RPE) cells into subretinal space further demonstrated this characteristic (Silverman et al., 1992, Silverman and Hughes, 1989; Kaplan et al., 1997). This is influenced by the existence of immunologic privilege in the subretinal space and immunogenicity of the transplanted tissue.

As mentioned above, immune privilege in the eye is thought to be the result of a combination of anatomic features with a special molecular environment. An absence of lymphatic drainage, blood-ocular barrier, the presentation of antigen to the host via the circulation (camero-splenic axis), and the reduced expression of MHC class I and II molecules on resident cells in the AC are the anatomic features that contribute to immune privilege (Streilein et al., 1997). In a similar manner, the subretinal space is protected by a blood-ocular barrier, via tight junctions between RPE cells, as well as the vascular endothelium of the retinal circulation. The subretinal space has no lymphatic drainage and has reduced expression of MHC class I and II molecules on parenchymal cells of the neurosensory retina (Wang et al., 1987).

Some of the molecular mechanisms necessary to achieve and maintain immune privilege in the AC also exist in subretinal space.

1. Complement inhibition: There have been several complementary regulatory proteins identified immunohistologically in the eye. Their role is to protect intraocular tissue from complement-mediated destruction. There is a selective expression of CD59 (inhibiting formation of complement) in the neurosensory retina, but not membrane cofactor protein (MCP or CD 46) or decay-accelerating factor (DAF or CD55), which are expressed within the AC, and both of which participate in the regulation of complement activation (Bora et al., 1993).
2. Fas ligand-mediated apoptosis: Fas-ligand is a membrane protein that is a member of the TNF family of proteins expressed on activated T cells. Fas ligand binds to Fas, thus stimulating a signaling pathway leading to apoptotic cell death of the Fas-expressing cell. Fas ligand is expressed in various ocular tissues, including the RPE and neurosensory retina. Experiments in vivo and in vitro suggest that constitutive expression of Fas ligand in a tissue leads to a deletion of Fas+ T cells that enter the tissue (Griffith et al., 1997; Jorgenson et al., 1998).
3. Immunosuppressive molecules: Aqueous humor contains several immunomodulatory substances, including TGF-β_2, free cortisol, IL-1 receptor antagonist, substance P and vasoactive intestinal peptide. It has been shown that RPE secretes TGF-β2 as well. This immunosuppressant is found in the vitreous gel of the eye.

Thus, the existence of a limited immune privilege in the subretinal space is the result of antigen-specific inhibition of cellular and humoral responses. After the introduction of an antigen into the subretinal space, a suppression of delayed-type hypersensitivity to soluble-protein antigens and a delayed rejection of allogeneic RPE cells are seen. These phenomena are the result of the inhibition of cell-mediated immunity. In the AC, delayed-type hypersensitivity is inhibited while the cytotoxic antibody response is exaggerated. In the subretinal space, both antibody production and delayed-type hypersensitivity are inhibited. This is an important difference between the immunologic privilege process within the AC and the subretinal space.

Most of the experimental studies involving placement of alloantigens in the subretinal space have been performed in rodents, and much about the nature of the immune privilege in the subretinal space remains unknown. Also, the immunogenetic disparity between donor and host is not known, so naturally, a question arises, whether immunosuppression is necessary at the time of allogeneic transplant. The survival of the allogeneic transplant is dependent on the presence of immune privilege at that site, as well as the immunogenicity of the transplanted tissue. There are studies demonstrating that RPE cells express MHC class I but not class II molecules. They can express class II molecules when stimulated with interferon-γ (Liversidge et al., 1988; Liversidge et al., 1998). The RPE also express ICAM-1, a molecule necessary for T-cell activation (Liversidge et al., 1990). The expression of MHC class II and ICAM-1 on the apical plasma membrane of cultured RPE suggests that these molecules may play an important role in the presentation of antigen to the host (Percupo et al., 1990; Osusky et al., 1997).

There are many retinal diseases for which currently no effective treatment exists. The potential usefulness of retinal transplantation as a treatment option can be explored once the knowledge of the parameters determining the immunologic rejection of allogeneic retinal transplant and biologic mechanisms of cells transplanted into subretinal space are better known.

5.5.1. Retinal Antigens and Autoimmunity

In 1968 Wacker & Lipton developed an excellent animal model of experimental autoimmune uveitis (EAU) (Wacker and Lipton, 1968). The EAU model has been used for several decades in eliciting immune mechanisms, the identification of pathogenic epitopes of autoantigens in the eye in animals, and the evaluation of therapeutic strategies.

Retinal antigens, such as S-antigen (arrestin), interphotoreceptor retinoid-binding protein (IRBP), rhodopsin, recoverin, and phosduscin, appear to hold uveitogenic properties. Immunization with these antigens or their fragments can induce ocular inflammation in susceptible strains of laboratory animals. New eye autoantigens have been discovered, such as uveal autoantigen with coiled domains and ankyrin repeats (UACA) in patients with Vogt-Koyanagi-Harada disease (Yamada et al., 2001).

EAU models using S-antigen and IRBP (peptides derived from these proteins) contributed mostly to our knowledge of retinal autoimmunity. S-antigen, a 48-kDa protein (arrestin), is one of the antigens used to induce EAU (Singh et al., 1988). It is the first retinal autoantigen that has been implicated in the pathogenesis of uveitis (Hirose et al., 1989; Merryman et al., 1991). The main action of arrestin is blocking the interaction of rhodopsin with the G-protein transducin in the phototransduction cascade. Immunization of susceptible animals with S-antigen induces a predominantly CD4+ T-cell-mediated inflammatory response in the retina, uveal tract, and the pineal gland. S-antigen has been implicated in the pathogenesis of uveitis through molecular mimicry. It has been demonstrated that several exogenous antigens, such as baker's yeast, *Escherichia coli*, hepatitis B virus, streptococcal M5 protein, Moloney murine sarcoma virus, and baboon endogenous virus, and several endogenous antigens, such as human leukocyte antigen B-derived peptide, tropomyosin antigens share sequence homology with uveitogenic peptide M of S-antigen (Shinohara et al., 1990). Antistreptococcal monoclonal antibodies were found to recognize several uveitogenic peptides of S-antigen, thus suggesting that immunological mimicry between self and exogenous antigens from an infectious agent may be a potential mechanism in the pathogenesis of uveitis in humans (Lerner et al., 1995).

IRBP is a major protein (1264 amino acid residues) of the interphotoreceptor matrix. It functions as a transporter of retinoids between the retina and RPE. It is found in both the eye and the pineal gland. A spectrum of disease ranging from hyperacute to chronic relapsing disease could be induced with variable doses of this antigen. The inflammation is located at the photoreceptor layer, producing histopathology similar to that seen in uveitis, retinal vasculitis, granuloma, focal serous detachments, loss of photoreceptors, and formation of sub-RPE infiltrates resembling to Dalen-Fuchs nodules (Caspi et al., 1988). The relatively long duration of disease activity in the murine IRBP EAU model makes it a good model for evaluation of therapeutic strategies in established disease (Chan et al., 1990).

Retinal autoimmunity is a complex mechanism that has previously been thought to be always pathological. However, based on recent studies, retinal autoimmunity now appears to have certain neuroprotective qualities. One such study showed that vaccination with peptides derived from IRBP resulted in protection of retinal ganglion cells from glutamate-induced death or death as a consequence of optic nerve injury (Mizrahi et al., 2002). It is evident that the immune system not only protects the body against invading pathogens but also protects it from toxic substances released by the body's own tissues during stress and trauma. So, the autoreactive cells that induce neuroprotection and those that induce autoimmune disease may share the same qualities, indicating their potential to be protective and destructive at the same time (Kipnis et al., 2002). Lymphocytes reactive to retinal antigens have been found in healthy individuals (Mizrahi et al., 2002; Nussenblatt et al., 1980). The presence of circulating autoreactive cells in healthy humans suggests that immunoregulatory mechanisms are probably in place to prevent retinal autoimmunity. It appears that the ability to protect the eye from inflammation and injury does not purely depend on mechanisms of immune privilege but instead on a precise regulation of autoimmunity.

Summary

The eye has been recognized as an immune-privileged site for more than 100 years. Medawar demonstrated this by showing a prolonged, often indefinite, survival of organs or tissue grafts in the anterior chamber of the eye (Medawar, 1948). Immune privilege is a dynamic process in which immunoregulatory mechanisms combined with anatomical factors maintain the vitality of grafts in privileged sites. ACAID is the best-studied immune-privilege phenomenon in the eye. Although it is an anterior-chamber phenomenon, there is enough evidence to show that there are some of the same mechanisms at work in the vitreous and subretinal space (Jiang et al., 1993). Therefore understanding ACAID could possibly further our knowledge of retinal autoimmunity.

Historically retinal autoimmunity has been considered pathogenic. So, active suppression of retinal immunity was thought to be necessary for the health of the eye. However, later studies have demonstrated the presence of retinal autoantibodies in normal controls (Yamamoto et al., 1993). Animal optic nerve injury studies suggested possible beneficial roles of retinal autoimmunity in controlling collateral damage to the retinal ganglion cells (Kipnis et al., 2002). Thus, retinal autoimmunity can be viewed as both protective and destructive phenomenon.

With the growing sophistication of molecular immunology techniques, there is more hope for elucidation of complex mechanisms of ocular inflammation.

Review Questions/Problems

1. How many compartments the eye contains?

a. Three: cornea, anterior chamber and posterior chamber
b. Two: anterior and posterior chamber
c. Four: cornea, anterior chamber, posterior chamber and vitreous cavity
d. Three: anterior chamber, posterior chamber and vitreous cavity

2. What is the volume of vitreous cavity on average?

a. 2 mL
b. 4 mL

c. 6mL
d. 8mL
e. 10mL

3. Anterior chamber angle includes all of the following structures except:

a. Schwalbe's line
b. Sclera
c. Ciliary body
d. Lens
e. Iris
f. Schlemm's canal
g. Trabecular meshwork

4. Aqueous is produced by:

a. ciliary epithelium
b. corneal endothelium
c. Schlemm's canal
d. trabecular meshwork
e. iris roots

5. All of the following are immune privileged sites except:

a. cornea
b. conjunctiva
c. lens
d. vitreous cavity and subretinal space
e. testis and ovary

6. Which statement is not true regarding immune privilege?

a. The phenomenon of immune privilege was initially described in the nineteenth century.
b. Immune privilege is defined as a prolonged, sometimes indefinite, survival of organ or tissue grafts at special body/organ sites.
c. Immune privilege is a dynamic process maintained by anatomic structures and immunoregulatory forces.
d. Grafts placed in immune privileged sites remain sequestered from the immune environment.
e. Both active and passive factors of immune privileged sites and tissues contribute to the privileged status.

7. All the following is true about anterior chamber immune associated deviation (ACAID) except:

a. ACAID is an antigen-specific systemic immune response to eye-derived antigens.
b. ACAID is an actively acquired and actively maintained manifestation of ocular immune privilege.
c. In ACAID aqueous humor possesses immunomodulatory properties and inhibits all immune reactions.
d. ACAID is a form of peripheral tolerance, unlike central tolerance, where clonal deletion and/or anergy occur within the thymus.
e. Once induced, ACAID persists for a very long period of time.

8. Visual cortex occupies the following structures:

a. occipital and parts of the parietal and temporal lobes of the cerebral cortex
b. parietal and parts of occipital and temporal lobes of the cerebral cortex
c. lateral geniculate body, occipital and frontal lobes of the cerebral cortex
d. frontal and parts of occipital and temporal lobes of the cerebral cortex
e. occipital and parts of frontal and parietal lobes of the cerebral cortex

9. Which substance found in the eye is not uveitogenic?

a. S-antigen (arrestin)
b. TGF-β
c. interphotoreceptor retinoid-binding protein (IRBP)
d. recoverin
e. phosduscin

10. Which molecules do not possess immunosuppressive properties?

a. free cortisol
b. IL-1 receptor antagonist
c. substance P
d. rhodopsin
e. vasoactive intestinal peptide

References

Apte RS, Niederkorn JY (1996) Isolation and characterization of a unique natural killer cell inhibitory factor present in the anterior chamber of the eye. J Immunol 156:2667–2673.

Barker CF, Billingham RE (1977) Immunologically privileged sites. Adv Immunol 25:1–54.

Baxter AG, Kinder SJ, Hammond KJ, Scollay R, Godfrey DI (1997) Association between alphabetaTCR+CD4-CD8- T-cell deficiency and IDDM in NOD/Lt mice. Diabetes 46:572–582.

Bora NS, Gobleman CL, Atkinson JP, Pepose JS, Kaplan HJ (1993) Differential expression of the complement regulatory proteins in the human eye. Invest Ophthalmol Vis Sci 34:3579–3584.

Caspi RR, Roberge FG, Chan CC, Wiggert B, Chader GJ, Rozenszain LA, Lando Z, Nussenblatt RB (1988) A new model of autoimmune disease. Experimental autoimmune uveoretinitis induced in mice with two different retinal antigens. J Immunol 140:1490–1495.

Chan CC, Caspi RR, Ni M et al. (1990) Pathology of experimental autoimmune uveoretinitis in mice. J Autoimmun 3:247–255.

Cowey A, Stoerig P, Bannister M (1994) Retinal ganglion cells labeled from the pulvinar nucleus in macaque monkeys. Neuroscience 61:691–705.

Cowey A, Johnson H, Stoerig P (2001) The retinal projection to the pregeniculate nucleus in normal and destriate monkeys. Eur J Neurosci 13:279–290.

D'Orazio TJ, Niederkorn JY (1998a) Splenic B cells are required for tolerogenic antigen presentation in the induction of anterior chamber-associated immune deviation (ACAID). Immunology 95:47–55.

D'Orazio TJ, Niederkorn JY (1998b) A novel role for TGF-beta and IL-10 in the induction of immune privilege. J Immunol 160:2089–2098.

Faunce DE, Sonoda KH, Streilein JW (2001) MIP-2 recruits NKT cells to the spleen during tolerance induction. J Immunol 166:313–321.

Freund J-H (1973) Neuronal mechanisms of the lateral geniculate body. Vol VII/3B pp. 177–246 in Jung R (ed). Handbook of sensory physiology Springer Berlin.

Goslings WRO, Prodeus AP, Streilein JW, Carrol MC, Jager MS, Taylor AW (1998) A small molecular weight factor in aqueous humor acts on C1q to prevent antibody dependent complement activation. Invest Ophthalmol Vis Sci 39:989–995.

Griffith TS, Brunner T, Fletcher SM, Gren DR, Ferguson TA (1997) Fas ligand-induced apoptosis as a mechanism of immune privilege. Science 270:1189–1192.

Hara Y, Caspi RR, Wiggert B, Dorf M, Streilein JW (1992) Analysis of an in vitro generated signal that induced systemic immune deivation similar to that elicited by antigen injected into the anterior chamber of the eye. J Immunol 149:1531–1538.

Heuer J, Bruner K, Opalka B, Kolsch E (1982) A cloned T-cell line from a tolerant mouse represents a novel antigen specific suppressor cell type. Nature 296:456–458.

Hirose S, Singh VK, Donoso LA Shinohara T, Kotake S, Tanaka T, Kuwabara T, Yamaki K, Geri I, Nussenblatt RB (1989) An 18-mer peptide derived from the retinal S antigen induces uveitis and pinealitis in primates. Clin Exp Immunol 77:106–111.

Hoffmann KP (1996) Comparative neurobiology of the optokinetic reflex in mammals. Rev Bras Biol 56S1 2:303–314.

Illes Z, Kondo T, Newcombe J, Oka N, Tabita T, Yamamura T (2000) Differential expression of NK T cell V alpha 24J alpha Q invariant TCR chain in the lesions of multiple sclerosis and chronic inflammatory demyelinating polyneuropathy. J Immun 164:4375–4381.

Jiang LQ, Jorquera M, Streilein JW (1993) Subretinal space and vitreous cavity as immunologically privileged sites for retinal allografts. Invest Ophthalmol Vis Sci 34:3347–3354.

Jorgenson A, Wiencke AK, la Cour M, Koestel CG, Madsen HO, Hamann S, Liu GM, Scerfig E, Prause JU, Svejgaard A, Odum N, Nissen MH, Roepke C (1998) Human RPE cell-induced apoptosis in activated T cells. Invest Ophthalmol Vis Sci 39:1590–1599.

Kaiser CJ, Ksander BR, Streilein JW (1989) Inhibition of lymphocyte proliferation by aqueous humor. Regul Immunol 2:42–49.

Kaplan HJ, Streilein JW (1977) Immune response to immunization via the anterior chamber of the eye I. F1-lymphocyte–induced immune deviation. J Immunol 118:809–814.

Kaplan HJ, Streilein JW, Stevens TR (1975) Transplantation immunology of the anterior chamber of the eye II. Immune response to allogeneic cells. J Immunol 115:805–810.

Kaplan HJ, Tezel TH, Berger AS, Wolf ML, Del Priore LV (1997) Human photoreceptor transplantation in retinitis pigmentosa. A safety study. Arch Ophthalmol 115:1168–1172.

Ksander BR, Streilein JW (1993) Regulation of the immune response within privileged sites: In Mechanisms of Regulation of Immunity Chemical Immunology. (Granstein R, ed) Basel, Karger pp 117–145.

Kipnis J, Mizrahi T, Yoles E, Ben-Nun A, Schwartz M (2002) Myelin specific Th1 cells are necessary for post-traumatic protective autoimmunity. J Neuroimmunol 130:78–85

Lerner MP, Donoso LA, Nordquist RE, Cunningham MW (1995) Immunological mimicry between retinal S-antigen and group A streptococcal M proteins. Autoimmunity 22:95–106.

Liversidge JM, Sewell HF, Forrester JV (1988) Human RPE cells differentially express MHC class II (HLA DP, DR and DQ) antigen in response to in vitro stimulation with lymphokine or purified IFN-γ. Clin Exp Immunol 73:489–494.

Liversidge JM, Sewell HF, Forrester JV (1990) Interaction between Lymphocytes and cells of the blood-retina barrier: Mechanisms of T lympohocyte adhesion to human retinal capillary endothelial cells and RPE cells in vitro. Immunology 71:390–396.

Liversidge JM, Sewell HF, Thomson AW, Forrester JV (1998) Lymphokine-induced MHC class II antigen expression on cultured RPE cells and the influence of cyclosporine A. Immunology 63:313–317.

Masli S, Turpie B, Hecker KH, Streilein JW (2002) Expression of thrombospondin in TGF beta-treated APCs and its relevance to their immune deviation-promoting properties. J Immunol 168(5):2264–2273.

Medawar P (1948) Immunity to homologous grafted skin III The fate of skin homografts transplanted to the brain, to the subcutaneous tissue and to the anterior chamber of the eye. Br J Exp Pathol 29:58–69.

Merryman CF, Donoso LA, Zhang XM, Heber-Katz E, Gregerson DS (1991) Characterization of a new, potent, immunopathogenic epitope in S-antigen that elicits T cells expressing V beta 8 and V alpha 2-like genes. J Immunol 146:75–80.

Mieza MA, Itoh T, Cui JQ, Makino Y, Kawano T, Tsushida K, Koike T, Shirai T, Yogita H, Matsuzawa A, Koseki H, Taniguchi M (1996) Selective reduction of V alpha 14+NK T cells associated with disease development in autoimmune-prone mice. J Immunol 156:4035–4040.

Miyamoto K, Ogura Y, Hamada M, Nishiwaki H, Hiroshiba N, Honda Y (1996) In vivo quantification of leukocyte behavior in the retina during endotoxin-induced uveitis. Invest Ophthalmol Vis Sci 37:2708–2715.

Mizrahi T, Hauben E, Schwartz M (2002) The tissue-specific self-pathogen is the protective self-antigen: The case of uveitis. J Immunol 169:5971–5977.

Mosmann TR (1992) T lymphocyte subsets, cytokines, and effector functions. Ann NY Acad Sci 664:89–92.

Mosmann TR, Coffman RI (1989) TH1 and TH2 cells: Different patterns of lymphokine secretion lead to different functional properties. Annu Rev Immunol 7:145–173.

Mustari MJ, Fuchs AF (1989) Response properties of single units in the lateral terminal nucleus of the accessory optic system in the behaving primate. J Neurophysiol 61:1207–1220.

Nagane Y, Utsugisawa Obara D, Tohgo H (2001) NKT-associated markers and perforin in hyperplastic thymuses from patients with Myasthenia gravis. Muscle Nerve 24:1359–1364.

Niederkorn J, Streilein JW, Shadduck JA (1980) Deviant immune responses to allogeneic tumors injected intracamerally and subcutaneously in mice. Invest Ophthalmol Vis Sci 20:355–363.

Niederkorn JY (1990) Immune privilege and immune regulation in the eye. Adv Immunol 48:191–226.

Niederkorn JY, Streilein JW (1983) Intracamerally induced concomitant immunity: Mice harboring progressively growing intraocular tumors are immune to spontaneous metastases and secondary tumor challenge. J Immunol 131:2670–2674.

Nussenblatt RB, Gery I, Ballintine EJ, Wacker WB (1980) Cellular immune responsiveness of uveitis patients to retinal S-antigen. Am J Ophthalmol 89:173–179.

Osusky R, Dorio RJ, Arora Y, Ryan SJ, Walker SM (1997) MHC class II positive RPE cells can function as antigen-presenting cells for microbial superantigen. Ocul Immunol Inflamm 5:43–50.

Percupo CM, Hooks JJ, Shinohara T, Caspi R (1990) Detrick B Cytokine-mediated activation of a neuronal retinal resident cell provokes antigen presentation. J Immunol 145:4101–4107.

Shi F, Ljunggren HG, Sarvetnick N (2001) Innate immunity and autoimmunity: From self-protection to self-destruction. Trends immunol 22:97–101.

Shinohara T, Singh VK, Tsuda M, Yamaki K, Abe T, Suzuki S (1990) S-antigen: From gene to autoimmune uveitis. Exp Eye Res 50:751–757.

Silverman MS, Hughes SE (1989) Transplantation of photoreceptors to light-damaged retina. Invest Ophthalmol Vis Sci 30:1684–1690.

Silverman MS, Hughes SE, Valentino TL, Liu Y (1992) Photoreceptor transplantation: Anatomic electrophysiologic, and behavioral evidence for the functional reconstruction of retinas lacking photoreceptors. Exp Neurol 115:87–94.

Skelsey ME, Mellon J, Niederkorn JY (2001) Gamma delta T cells are needed for ocular immune privilege and corneal graft survival. J Immunol 166:4327–4333.

Singh VK, Nussenblatt RB, Donoso LA, Yamaki K, Chan CC, Shinohara T (1988) Identification of a uveitopathogenic and lymphocyte proliferation site in bovine S-antigen. Cell Immunol 115:413–419.

Sonoda KH, Exley M, Snapper S, Balk S, Stein-Streilein J (1999) CD1-reactive natural killer T cells are required for development of systemic tolerance through an immune-privileged site. J Exp Med. 190:1215–1226.

Sonoda KH, Taniguchi H, Stein-Streilein J (2002) Long-term survival of corneal allografts is dependent on intact CD1d-reactive NKT cells. J Immunol 168:2028–2034.

Streilein JW (1995) Unraveling immune privilege. Science 270:1158–1159.

Streilein JW, Kaplan HJ (1979) In Immunology and Immunopathology of the eye. Paris Masson et Cie pp.174–180.

Streilein JW, Niederkorn JY (1981) Induction of anterior chamber-associated immune deviation requires an intact, functional spleen. J Exp Med 153:1058–1067.

Streilein JW, Takeuchi M, Taylor AW (1997) Immune privilege, T-cell tolerance and tissue-restricted autoimmunity. Hum Immunol 52:138–143.

Streilein JW, Katagiri K, Zhang-Hoover J, Mo JS, Stein-Streilein J (2001) Invest OphthalmolVis Sci 42:S523. Is ACAID aTh2-mediated response?

Sumida T, Sakamoto A, Murata H, Makino Y, Takahashi H, Yoshida S, Nishioka K, Iwamoto I, Taniguchi M (1995) Selective reduction of T cells bearing invariant V alpha 24J alpha Q antigen receptor in patients with systemic sclerosis. J Exp Med 182:1163–1168.

Takeuchi M, Alard P, Streilein JW (1998) TGF-β promotes immune deviation by altering accessory signals of antigen-presenting cells. J Immunol 160:1589–1597.

Takeuchi M, Alard P, Verbik D, Ksander B, Streilein JW (1999) Anterior chamber-associated immune deviation-inducing cells activate T cells, and rescue them from antigen-induced apoptosis. Immunology 98:576–583.

Taylor AW, Yee DG, Streilein JW (1998) Suppression of nitric oxide generated by inflammatory macrophages by calcitonin gene-related peptide in aqueous humor. Invest Ophthalmol Vis Sci 39:1372–1378.

Tompsett E, Abi-Hanna D, Wakefield D (1990) Immunological privilege in the eye: A review. Curr Eye Res 9:1141–1150.

Van Dooremal JC (1873) Die Entwicklung der in fremden Grund versetten lebenden Gewebe. Albrecht Van Graefes Arch Ophthalmol 19:358–373.

Von Hippel A (1888) Eine neue methode der Hornhauttransplantation. Arch Ophthalmol Leipz 34:108.

Wacker WB, Lipton MM (1968) Experimental allergic uveitis. II. Serologic and hypersensitive responses of the guinea pig following immunization with homologous retina. J Immunol 101:157–165.

Waldrep JC, Kaplan H (1983) Anterior chamber associated immune deviation induced by TNP-splenocytes (TNP-ACAID). Systemic tolerance mediated by suppressor T-cells.Invest Ophthalmol Vis Sci. Aug 24(8):1086–1092.

Wang HM, Kaplan HJ, Chan WC, Johnson M (1987) The distribution and ontogeny of MHC antigens in murine ocular tissue. Invest Ophthalmol Vis Sci 28:1383–1389.

Weiner HL (1997) Oral tolerance: Immune mechanisms and treatment of autoimmune diseases. Immunol Today 18:335–343.

Wetzig R, Foster C, Greene M (1982) Ocular immune responses I. Priming of A/J mice in the anterior chamber with azobenzenearsonate-derivatized cells induces second-order-like suppressor T cells. J Immunol 128:1753–1762.

Wilbanks GA, Streilein JW (1989) The differing patterns of antigen release and local retention following anterior chamber and intravenous inoculation of soluble antigen. Evidence that the eye acts as an antigen depot. Regional Immunol 2:390–398.

Wilbanks GA, Streilein JW (1990) Distinctive humoral repsonses following anterior chamber and intravenous administration of soluble antigen. Evidence for active suppression of IgG2α-secreting B-cells. Immunology 71:566–572.

Wilbanks GA, Mammolenti M, Sreilein JW (1991) Studies on the induction of anterior chamber associated immune deviation (ACAID) II. Eye-derived cells participate in generating blood-borne signals that induce ACAID. J Immunol 146:3018–3024.

Wilson SB, Kent SC, Patton KT, Orban T, Jackson RA, Exley M, Porcelli S, Schatz DA, Atkinson MA, Balk SP, Strominger JL, Hafler DA (1998) Extreme Th1 bias of invariant Valpha24JalphaQ T cells in type 1 diabetes. Nature 391:177–181.

Xu Y, Kapp JA (2001) γδ T cells are critical for the induction of anterior chamber-associated immune deviation. Immunology 104:142–148.

Yamamoto JH, Minami M, Inaba G, Masuda K, Mochizuki M (1993) Cellular autoimmunity to retinal specific antigens in patients with Behçet's disease. Br J Ophthalmol 77:584–589.

Yamada K, Senju S, Nakatsura T, Murata Y, Ishihara M, Nakamura S, Ohno S, Negi A, Nishimura Y (2001) Identification of a novel autoantigen UACA in patients with panuveitis. Biochem Biophys Res Commun 280:1169–1176.

Zeng D, Lee MK, Tung J, Brendolan A, Strober S (2000) Cutting edge: A role for CD1 in the pathogenesis of lupus in NZB/NZW mice. J Immunol 164:5000–5004.

6
Hippocampus and Spatial Memory

Huangui Xiong

Keywords Amygdala; Cytokine; Dentate gyrus; EPSCs; EPSPs; Hippocampus; Limbic system; Long-term potentiation; Neurotransmitter; NMDA receptor; Spatial memory; Synapse; Temporal lobe

6.1. Introduction

The hippocampus is a brain structure located inside the temporal lobe. It forms a part of the limbic system and plays an important role in the formation, consolidation and retrieval of episodic memories. It has been shown that repetitive activation of excitatory synapses in the hippocampus causes an increase in synaptic strength that last for hours or days designated as long-term potentiation (LTP). It is widely believed that LTP provides an important key to understand the cellular and molecular mechanisms by which memories and formed and stored. Sensory information enters the hippocampus mainly through the perforant pathway consisting of the axons of neurons in layer II and III of the entorhinal cortex. The perforant path axons terminate on the dendrites of the dentate gryus granular cells. Then information flow through the hippocampus from the dentate gyrus to the CA3, the CA1 and the subiculum, forming hippocampal intrinsic trisynaptic circuit. In addition to its "traditional" role in learning and memory, the hippocampus is also involved in neuroimmunity. Lesion of hippocampus alters immunity and neuronal functions in the hippocampus are modulated by a variety of immune active molecules.

6.2. Anatomy of the Hippocampus

The hippocampus, so named because its shape vaguely resembles that of a seahorse, is a symmetrical structure located inside the medial temporal lobe on both sides of the human brain. It is a curved sheet of cortex folded into the medial surface of the temporal lobe. In transverse sections from rodent brain, the hippocampus has the appearance of two interlocking Cs with three distinct sub-fields, the dentate gyrus, the hippocampus proper (cornu ammonis, CA) and the subiculum. Although there is a lack of consensus to terms describing the hippocampus and the adjacent cortex, the term hippocampal formation generally applies to the dentate gyrus, fields CA1—CA3 (and CA4 is frequently called the hilus and considered part of the dentate gyrus) and the subiculum. The CA1-CA3 fields make up the hippocampus proper.

In animals, the hippocampus is among the phylogenetically oldest parts of the brain. It occupies most of the ventroposterior and ventrolateral walls of the cerebral cortex in rodents. However, the hippocampus occupies less of the telencephalon in proportion to cerebral cortex in primates, especially in humans. The significant development of hippocampus volume in primates correlates with overall increase of brain mass and neocortical development.

The anatomy of the hippocampus has been studied intensively in rodents since Cajal (1911) published his famous drawing illustrating the main cells, connections and flow of impulse traffic in the hippocampal formation (Figure 6.1). Subsequent studies using physiological, biochemical and axonal tracing techniques have added details in the cytoarchitecture and connections of the hippocampal formation, making it one of the most studied and best known structures in the brain (Raisman et al., 1966; Swanson and Cowan, 1975; Swanson et al., 1981; Frotscher, 1985).

One striking feature of hippocampal circuitry is the pattern of afferent termination. Major hippocampal afferents originating from entorhinal cortex and ipsilateral and contralateral hippocampal subfields synapse on the dendrites of the principal cells in a laminated pattern. For instance, hippocampal commissural and associational fibers synapse within the proximal one-third of the granule cell dendritic field, which is close to the cell body layer. The massive perforant path fibers terminate topographically in the outer two-thirds of the dendritic field. Afferents also have a laminar organization in the hippocampal proper.

Another important feature of the hippocampus is its intrinsic circuitry. Information flow through the hippocampus proceeds from dentate gyrus to CA3 to CA1 to the subiculum, forming the principal intrinsic trisynaptic circuit (Figure 6.1). CA2 represents only a very small portion of the hippocampus and its presence is often ignored in accounts of hippocampal function, though it is notable that this small region seems

T. Ikezu and H.E. Gendelman (eds.), *Neuroimmune Pharmacology*.
© Springer 2008

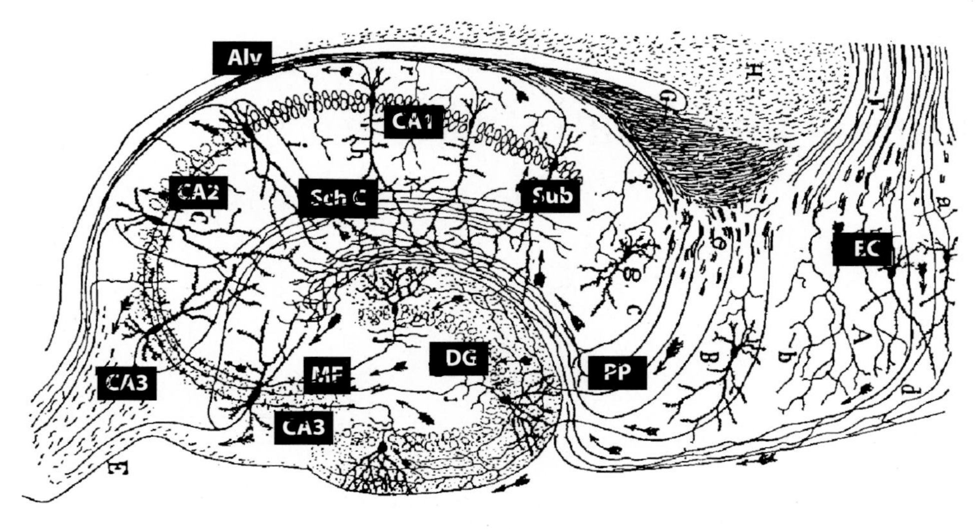

FIGURE 6.1. Schematic drawing by Ramon y Cajal (1911) of the main cells, connections and flow of impulse traffic in the hippocampus. Alv: alvus, CA: Cornu Ammonis, DG: dentate gyrus, EC: entorhinal cortex, MF: mossy fibers, PP: perforant path, Sch C: ffer collaterals, Sub: subiculum

unusually resistant to conditions that usually cause large amounts of cellular damage, such as epilepsy.

6.2.1. Dentate Gyrus

6.2.1.1. Cytoarchitecture

The dentate gyrus is a sharply folded trilayered cortex that forms a cap over the free edge of Ammon's horn (C or V shape in rodents). Its cell layer, the granule layer, contains densely packed granule cells, which are the principal neurons in the dentate gyrus. These granule cells have small (about 10 μm in diameter), spherical cell bodies, which are stacked approximately 4–10 cells thick in the granule layer. The molecular layer contains the dendrites of the granule cells, which typically show a conically shaped dendrite field. Because the dendrites emerge only from the top or apical portion of the cell body, granule cells are considered to be monopolar neurons. The granule cell dendrites extend perpendicularly to the granule cell layer, into the overlying molecular layer where they receive synaptic connections from several sources. The axons of granule cells are called mossy fibers because of the peculiar appearance of their synaptic terminals. They originate from the basal portion of the cell body and extend into the polymorphic cell layer in the hilus, a transition area between the dentate gyrus and hippocampus proper. The mossy fibers synapse onto some of the neurons in hilar area, such as mossy cells in the polymorphic cell layer. However, most mossy fibers project to the stratum lucidum of the CA3 region where they terminate onto proximal apical dendrites of the pyramidal CA3 cells. In the dentate gyrus, the most prominent class of interneurons is called the pyramidal basket cell, of which there are at least five types. The cell bodies of these neurons are typically located at the border between the granule cell layer and the polymorphic cell layer, and their axons innervate the cell body of the granule cells. There are also interneurons in the molecular layer, the most interesting of which is an axo-axonic cell that terminates on the initial axon segments of the granule cells. Excitatory interneurons, the mossy cells, can also be found in the molecular layer. Their axons only project to the molecular layer of the ipsilateral and contralateral side.

Interestingly, the dentate gyrus is one of the few brain regions where neurogenesis takes place. Neurogenesis is thought to play a role in the formation of new memories. It has also been found to be increased in response to both antidepressants and physical exercise, implying that neurogenesis may improve symptoms of depression.

6.2.1.2. Fiberarchitecture

6.2.1.2.1. Afferents to the Dentate Gyrus—The Perforant Path

As emphasized by Cajal (1911) and later corroborated by Lorente de No (1934), the main input to the dentate gyrus is from the entorhinal cortex (but also perirhinal cortex, among others) by way of a fiber system called the perforant path. It is the major input to the hippocampus. The axons of the perforant path arise principally in layers II and III of the entorhinal cortex, with minor contributions from the deeper layers IV and V. Axons from layers II/IV project to the granule cells of the dentate gyrus and pyramidal cells of the CA3 region, while those from layers III/V project to the pyramidal cells of the CA1 and the subiculum. The perforant fibers are laminated afferents to the dentate molecular layer and terminate in the outer two thirds of the dentate molecular layer. Thus, the perforant path represents a massive input to the dentate gyrus.

Perforant path input from the entorhinal cortex layer II enters the dentate gyrus and is relayed to region CA3. Region CA3 combines this input with signals from the entorhinal cortex layer II, makes extensive connections within the region, and sends connections to region CA1 through a set of fibers called the Schaffer collaterals. Region CA1 receives input from region CA3 and the entorhinal cortex layer III, then projects to the subiculum in addition to sending information along the output paths of the hippocampus.

The perforant path is composed of the medial perforant path and the lateral perforant path generated respectively at the medial and lateral portions of the entorhinal cortex. It was in the perforant pathway that long-term potentiation (LTP) was first discovered (Bliss and Gardner-Medwin, 1973). The medial perforant path synapses onto the inner one-third dendritic area of the granule cells, while the lateral perforant path does it onto the outer two-thirds dendrites of these same cells. Approximately 60–70% of the total dendritic fields of individual granule cells are taken up with the perforant input (Desmond and Levy, 1982). Other afferents to the dentate gyrus originate from brain stem raphe neuclei (Moore and Halaris, 1975), the locus cerules (Room et al., 1981) and hypothalamic supramammillary nucleus (Wyss et al., 1979).

6.2.1.2.2. Efferents from the Dentate Gyrus—The Mossy Fibers

The dentate gyrus does not project to other brain regions. Within the hippocampal formation, it only projects to CA3 via the mossy fibers (Figure 6.2). The mossy fibers are the axons of dentate gyrus granule cells. They extend from the dendate gyrus to CA3 pyramidal cells, forming their major output. Mossy fiber synapses on CA3 neurons display distinctive terminal boutons, with multiple transmitter release sites and post-synaptic densities. The mossy fiber terminals are large (3–6 μm) and make asymmetric synaptic contact on the dendritic shaft and dendritic spines (thorny excrescences) of CA3 pyramidal cells. Multiple granule cells can synapse onto a single CA3 pyramidal cell. This pathway is studied extensively as a model for the functional roles of kainate receptors in synaptic plasticity. For instance, LTP is *N*-methyl-*D*-aspartate (NMDA) receptor-independent in this pathway, but it appears to involve pre-synaptic kainate receptors. The mossy fibers are glutamatergic, however they have also been shown to be immunoreactive for GABA and opiate peptides (dynorphin and enkephalin).

There are two targets for mossy fibers: the ipslateral CA3 fields of hippocampal proper and the mossy cells scattered throughout the hilus. The mossy fibers descend to the hilus and ramify. Some of these collaterals branch within the hilus, while a long branch projects to CA3 area. After traveling through the hilus, the long branches group into two bundles. The infrapyramidal bundle contacts the thorny spines on proximal basal dendrites in CA3, while the suprapyramidal bundle forms the lucidum layer in CA3. Electron microscopy has revealed that varicosities in the granule cell axons contain enormous numbers of synaptic vesicles (Blackstad and Kjaerheim, 1961).

6.2.2. The Hippocampus Proper (CA1-CA3 Fields)

6.2.2.1. Cytoarchitecture

Hippocampus proper is a U-shaped fold of cortex composed of CA1—CA3 fields (Figure 6.1). The narrow layer of pyramidal neurons extends from subiculum to the hilus of the dentate gyrus. Adjacent to the subiculum is a field of tightly packed medium-sized cells, which Lorente de No (1934) named the CA1. Next to CA1 are fields CA2 and CA3 containing large, less densely packed cells. CA2 pyramidal cells can be distinguished from CA3 pyramidal cells in Golgi preparations by the absence of characteristic thorny spines on the proximal apical dendrites, but not by Nissl-staining.

In Golgi preparations, the pyramidal cells in hippocampal proper are characterized by two groups of dendrites, the apical dendrite and the basal dendrites. The apical dendrite is a single pole extending from the pyramidal apex through the stratum radiatum into the molecular layer, giving off a few branches in the CA1 field of the stratum radiatum. In the molecular layer many secondary and tertiary branches of the main dendrite produce a dense tuft. The basal dendrites originate from the base of pyramidal cells and branch in the dendritic layer between the pyramidal cell layer and the white matter surrounding the hippocampus, forming another dense tuft. The axon originates from the proximal segment of basal dendrite and travels to the alveus where it may bifurcate, especially in the CA1 field. The bifurcated axons in the CA1 field travel with one branch toward the subicular area and with the other toward the fimbria. Within the oriens layer, the axons of the CA3 pyramidal neurons give off recurrent branches, constituting the Schaffer collaterals. The Schaffer collaterals make synaptic connection to the CA1 neurons.

6.2.2.2. Fiberarchitecture

The CA3 pyramidal cells give rise to highly collateralized axons, which project both to within the hippocampus (CA3, CA2 and CA1) and also to the same fields in the contralateral hippocampus via the commissural fibers. Some of the CA3 axons project to the lateral septal nucleus. All of the CA3 and CA2 pyramidal cells give rise to highly divergent projections to all subfields of the hippocampus. The main afferents to the CA1 pyramidal cells come from the Schaffer collateral/

commissural pathway. This pathway is derived from axons that project from the CA3 region of the hippocampus to the CA1 region. The axons either come from CA3 neurons in the same hippocampus (ipsilateral) or from an equivalent structure in the opposite hemisphere (contralateral). These latter fibers are termed commissural fibers, as they cross from one hemisphere of the brain to the other. While Schaffer collaterals are often illustrated as extending only through the stratum radiatum, it should be emphasized that both the stratum radiatum and the stratum oriens of CA1 are heavily innervated by CA3 axons. This pathway is extensively utilized in the study NMDA receptor-dependent LTP and LTD.

Unlike the CA3 field, pyramidal cells in CA1 do not give rise to a major set of collaterals that distribute within CA1, i.e. they have few associational connections. The axon can give rise to collaterals that terminate on basal dendrites of other CA1 cells. While the axon can give rise to collaterals that terminate on basal dendrites of other CA1 cells, it is clear the massive associational network that is so apparent in the CA3 is largely missing in the CA1. The CA1 also receives a fairly substantial input from the amygdaloid complex and is the first hippocampal field that originates a return projection to the entorhinal cortex. As with the CA3, however, the CA1 field receives light noradrenergic and serotonergic projections.

The septal nucleus is the major extrahippocampal target for the pyramidal cells from all CA fields. Axons from both CA1 and CA3 terminate in the septal nucleus. The fibers from dorsal hippocampal project to dorsomedial areas of the lateral septal nucleus, while progressively more ventral parts of the hippocampus terminate in correspondingly more lateroventral bands in the lateral septal nucleus (Meibach and Siegel, 1977b). Some branches of the CA1 axons terminate sparsely in olfactory bulb (de Olmos et al., 1978) and prefrontal cortex (Swanson, 1981).

6.2.3. The Subicular Complex

The subiculum curves anteriorly and laterally to wrap around the posterior extension of the dentate gyrus. It borders the medial entorhinal cortex and field of CA1 (Figure 6.1). The subiculum can be divided into three distinct cytoarchitectural areas. The parasubiculum borders the medial entorhinal cortex and contains moderately packed medium-sized cells. Next to the parasubiculum is presubiculum, which is characterized by a superficial lamina of densely packed small cells. The subiculum proper is the third distinct area. It borders anterolaterally to the field of CA1 and has a loosely packed pyramidal cell layer and a wide molecular layer.

Hippocampal intrinsic afferents to the subiculum originate from other hippocampal regions. The CA1 neurons send a dense projection to the subiculum and the CA3 pyramidal cells project to all parts of the subicular complex through the Schaffer collateral pathway (Swanson et al., 1978). The extrahippocampal projections to the subiculum mainly originate from the raphe nuclei (Conrad et al., 1974), locus coeruleus (Jones and Moore, 1977; Haring and Davis, 1985), amygdaloid nuclei (Krettek and Price, 1977), septal nucleus (Meibach and Siegel, 1977a) and peririhnal cortex (Kosel et al., 1983).

The subiculum is an important area for hippocampal efferents. The brain target regions for subicular projections include, but are not limited to, thalamic nuclei (Aggleton et al., 2005), reunions nucleus(Herkenham, 1978), mammillary body of the hypothalamus (Aggleton et al., 2005) and amygdala (Kishi et al., 2006). The para-subiculum and pre-subiculum project heavily to the anterior thalamic nuclei (Sikes et al., 1977; Cohen and Eichenbaum, 1993).

6.3. Role of the Hippocampus in Learning and Memory

6.3.1. Memory Functions of the Hippocampus

The role of the hippocampus in learning and memory has been the focus of neuroscience studies since Scoville and Milner reported the case of HM who had severe anterograde amnesia following bilateral medial temporal lobe resection (Scoville and Milner, 1957). The pattern of impaired and spared memory functions displayed by HM prompted a now commonly acknowledged characterization of the temporal lobe amnesic syndrome. It is widely believed that damage to the hippocampus disrupts declarative memory processes and more specifically episodic memory functions (Tulving and Markowitsch, 1998; Tulving, 2001). Declarative memories have been further divided into episodic and semantic memories. Episodic memory is concerned with conscious recall of specific episodes, and semantic memory with the storage of factual information. The memory functions that are spared in temporal lobe lesions have been classified as nondeclarative or procedural memories (Squire, 1992; Cohen and Eichenbaum, 1993). Nondeclarative or procedural memory processes are thought to operate automatically and do not include information about where or when learning experience took place. The pattern of memory deficits in humans with temporal lobe damage prompted the question, what role did the hippocampus play in memory?

The precise role of the hippocampus in learning memory remains to be determined. Accumulating evidence suggest that hippocampus has an essential role in the formation of new memories about experienced events. Damage to the hippocampus usually results in profound difficulties in forming new memories (anterograde amnesia), and normally also affects access to memories prior to the damage (retrograde amnesia). Although the retrograde effect normally extends some years prior to the brain damage, in some cases older memories remain. This sparing of older memories leads to the idea that consolidation over time involves the transfer of memories out of the hippocampus to other parts of the brain. However, it is difficult to test the sparing of older memories experimentally. Also, in some cases of retrograde amnesia, the sparing appears to affect memories formed decades before the damage to the hippocampus occurred, so its role in maintaining these older memories remains controversial.

Animal studies indicate that the hippocampus plays a role in storing and processing spatial information. In rodents, the firing rate of hippocampal neurons was found to correlate to the location of the animal in a test environment and these cells are referred to as place cells (O'Keefe and Dostrovsky, 1971; O'Keefe and Conway, 1978). There are many thousands of different place cells, which can be activated in response to a location in a particular environment. Different place cells have different place fields, which are not fixed in absolute space and are relative to spatial cues. The discovery of place cells led to the idea that the hippocampus might act as a cognitive map — a neural representation of the layout of the environment (O'Keefe and Nadel, 1978). There has been considerable support for the cognitive mapping theory from lesion and unit recording studies. It has been reported that rats with hippocampal lesions exhibit impairment in learning as detected in radial arm (Jarrard, 1983), T-maze (Bannerman et al., 2001) and Morris water maze, where lesioned rats had poor performance in finding the hidden platform (Morris et al., 1982; Morris et al., 1986). Indeed, neuroimaging studies in humans revealed that the hippocampus becomes active during spatial navigation, suggesting that the hippocampus in humans contributes to the encoding and retrieval of spatial information. This finding corresponds well with the results in animal studies, which show a significant deficit in spatial navigation resulting from hippocampal damage.

6.3.2. Synaptic Mechanisms of Memory

The nature of the physiological basis of learning and memory remains an enigma in neurobiology. The assumption that information is stored in the brain as changes in synaptic efficacy was initially proposed by Ramon y Cajal and later refined by Hebb (1949). This assumption was not tested until the early 1970s when Timothy Bliss and Terje LØmo made an important discovery that brief high frequency electrical stimulation of an excitatory pathway to the hippocampus produced a long-lasting enhancement in the strength of the stimulated synapses. This effect is now known as long-term potentiation (LTP).

LTP is expressed as a persistent increase in the size of the evoked synaptic response recorded from single cells or group of cells. It can be induced by high frequency stimulation (HFS, typically 100 Hz) or other types of stimulation, such as theta-burst stimulation. These stimulus paradigms resemble the synchronized firing patterns and frequencies that occur in the hippocampus during learning (Otto et al., 1991), making them useful experimental means to generate "learning activities" in the hippocampus. Over the past 30 years, LTP has been intensively studied because it is the leading experimental model for the synaptic changes that may underlie learning and memory.

6.3.2.1. Basic Properties of LTP

Although LTP was first demonstrated at the perforant path synapses on the granule cells in the dentate gyrus (Bliss and Gardner-Medwin, 1973), the majority of experiments on understanding the mechanisms of LTP have been performed on the Schaffer collateral/commissural synapses on the CA1 pyramidal cells in the hippocampus.

LTP in the hippocampus has three basic properties: cooperativity, associativity and input-specificity (Bliss and Collingridge, 1993). Cooperativity refers to the fact that long-lasting synaptic enhancement following HFS increases with the number of stimulated afferents (McNaughton et al., 1978). Threshold stimulus intensity during HFS is required for synaptic enhancement. 'Weak' HFS, which activates relatively few fibers, does not produce LTP, whereas strong stimulation at the same frequency and for the same duration produces LTP. Associativity means that a weak input (small number of stimulated afferents) can be potentiated if it is active at the same time as a strong tetanus (large number of stimulated afferents) is applied to a separate but convergent input (Bliss and Collingridge, 1993). This associativity has often been viewed as a cellular analog of associative or classic conditioning (Malenka and Nicoll, 1999). Another basic property of LTP is its input-specificity. When LTP is induced by repetitive stimulation the increase in synaptic strength usually does not occur in other synapses (on the same cell) that are not active at the time of repetitive stimulation (Bliss and Collingridge, 1993). This property increases the storage capacity of individual neurons (Malenka and Nicoll, 1999).

One remarkable feature of LTP is that it can be induced by a brief HFS, lasting less than or equal to a second and consisting of stimulation frequencies well within the range of normal axon discharging. Longevity is an additional feature of LTP. Once induced, LTP can persist for many hours in brain slices *in vitro* or in an anaesthetized animal, and for days or weeks (possibly even a lifetime) in a freely moving animal.

6.3.2.2. Mechanisms of Hippocampal LTP

It is well accepted that activation of postsynaptic NMDA receptors, a subtype of glutamate receptors, is required for the induction of LTP in the hippocampus. The key role that the NMDA receptors play in LTP induction relies on the voltage-dependent block of its channel by Mg^{2+} (Ascher and Nowak, 1988). In this way the NMDA receptor channel complex behaves as a molecular detector for LTP induction. To trigger the induction of LTP, two events must occur simultaneously: the cell membrane must be sufficiently depolarized to expel Mg^{2+} from NMDA channels at the same time L-glutamate binds to NMDA receptors and promotes the opening of these receptor-ligand-gated ion channels. The membrane depolarization can be achieved by repetitive tetanic stimulation of synapses or by directly depolarizing the cell while continuing low frequency stimulation of synapses (Gustafsson et al.,

1987). At Schaffer collateral—CA1 pyramidal cell synapses, Na^+ ions passing through the AMPA receptors are responsible for this membrane depolarization. When the membrane depolarization is sufficient and reaches a certain level, it expels Mg^{2+} from the NMDA receptor channel, allowing Ca^{2+} as well as Na^+ to enter the cell. The influx of Ca^{2+} through NMDA receptor channel raises intracellular Ca^{2+} and triggers the induction of LTP.

Considerable evidence now links this rise in postsynaptic Ca^{2+} concentration to the induction of LTP. The most compelling evidence in support of this model comes from experimental results wherein LTP induction can be blocked by pharmacological inhibition of NMDA receptors (Collingridge et al., 1983), or prevented by the injection of a Ca^{2+} chelator into the postsynaptic neuron (Lynch et al., 1983; Yang et al., 1999). Ca^{2+} imaging studies have demonstrated that tetanic stimulation directly increases Ca^{2+} within dendrites and spines as a result of NMDA receptor activation (Regehr and Tank, 1990; Perkel et al., 1993; Yuste and Denk, 1995). Although NMDA receptors are the primary source of Ca^{2+} entry into the dendrites and spines, activation of dendritic voltage-gated Ca^{2+} channels and Ca^{2+} release from intracellular stores also elevate Ca^{2+} levels and contribute to the induction of LTP. However, the mechanisms underlying the Ca^{2+} channel-dependent LTP may differ from NMDA receptor-dependent LTP (Malenka and Nicoll, 1999).

6.3.2.3. Expression of LTP

It has long been a challenge for neurobiologists to identify whether the increase in synaptic strength is mediated primarily through a pre- or post-synaptic mechanism. Available experimental results indicate that the increase in the synaptic strength could be either pre- or post-synaptic or both, or through an extrasynaptic mechanism, such as reduction in uptake of glutamate by glial cells resulting in an elevated concentration of glutamate at synaptic cleft. Evidence for pre-synaptic mechanisms comes from experiments measuring the overflow of radiolabeled or endogeneous L-glutamate in hippocampus before and after the induction of LTP (Bliss et al., 1986). In addition, quantal analysis of synaptic transmission reveals the proportion of synaptic failures decreases after the induction of LTP (Kullmann and Siegelbaum, 1995; Malenka and Nicoll, 1999). As synaptic failures represent the failure of neurotransmitter release or silent synapses, it was concluded that the induction of LTP is the consequence of an increase in the probability of neurotransmitter release. Evidence supporting this conclusion came from the finding that the variation around the mean of EPSCs decreased during LTP (Kullmann and Siegelbaum, 1995; Malenka and Nicoll, 1999). If the probability of neurotransmitter release increases during LTP, then the quantal content will, on average, increase and the coefficient of variation will decrease, because the coefficient of variation (SD/mean) is inversely proportional to the quantal content (Malenka and Nicoll, 1999).

Experimental results from other studies have shown a postsynaptic mechanism for LTP induction. First, paired-pulse facilitation (PPF) is not altered after the induction of LTP and this has been interpreted as evidence for a post-synaptic facilitation of LTP. PPF occurs when two pre-synaptic stimuli are delivered with a short interval (50–200 ms) and is thought to result from residual Ca^{2+} in the pre-synaptic terminal following the first stimulus, enhancing release during the second stimulus (Manabe et al., 1993). Various manipulations known to increase transmitter release do cause a decrease in the facilitation ratio, because release is already enhanced to near saturation during the first stimulus (Manabe et al., 1993). The failure in the alteration of PPF ratio suggests a post-synaptic locus of LTP expression.

A number of studies have shown that during LTP the AMPA receptor component of the EPSC is selectively enhanced, with little or no change in the component of NMDA receptors (Larkman and Jack, 1995), though both receptors are frequently co-localized at individual synapses. Pharmacological modulation of transmitter release affects AMPA and NMDA components equally, arguing for a selective postsynaptic alteration in either the density or properties of AMPA receptors (Kullmann and Siegelbaum, 1995).

6.4. Neuroimmunomodulation via Hippocampus

One of the important recent advances in understanding the biological basis of neurodegenerative disorders is the recognition that there is extensive communication between the central nervous system (CNS) and the immune system. Initial evidence that the immune system may communicate with the CNS was provided by Besedovsky et al. (1977), who observed that activation of the immune system was accompanied by changes in hypothalamic, autonomic and endocrine processes. The existence of neural-immune interactions is now supported by abundant evidence showing that the immune system communicates with the CNS through immunotransmitters (primarily cytokines) leading to direct CNS activation (Berkenbosch et al., 1987; Sapolsky et al., 1987) or by release of CNS-derived cytokines, and that the CNS regulates the immune system via neurotransmitters, hormones and neuropeptides. It has been shown that cells of immune system can synthesize and release several immunomodulatory hormones, neuropeptides and catecholamines (Blalock, 1989, 1994). For example, lymphocytes and macrophages produce the endogenous opioid peptides, norepinephrine, and epinephrine (Lolait et al., 1984; Harbour et al., 1987; Engler et al., 2005). A recent study (Rivest, 2003) shows that the CNS responds to systemic bacterial infection with innate immune reaction without pathogen's direct access to the brain. Whether caused by a microbe, trauma, toxic metabolite, autoimmunity, or as part of a broader degenerative process, activation of the immune system results in changes in the activity of discrete populations of brain neurons, including hippocampal neurons. Accumulating evidence indicates that these mechanisms are relevant for the course of infectious, inflammatory, autoimmune and neoplastic diseases.

6.4.1. Lesion of Hippocampus Affects Immunity

It has been shown that during immune challenge the hippocampus exhibits time-dependent changes in neurotransmitter levels and that an intact hippocampus is essential for the normal humoral immunity for the primary immune response in rats (Devi et al., 2004). Lesions of the dorsal hippocampus were found to produce a transient increase in splenocytes and thymocytes, as well as increased T-cell mitogen responses (Brooks et al., 1982). Lesions of the hippocampus were also found to cause differential effects on humoral immunity depending on the lesions of different subfields of the hippocampus (Pan and Long, 1993). Axotomy of afferent fibers within the molecular layer of the dentate gyrus caused activation of neural-immune elements in the slice cultures (Coltman and Ide, 1996). In addition, electrical stimulation of hippocampus increased the number of neutrophils and phagocytic index while also decreasing the number of lymphocytes and plasma corticosterone level in rats (Devi et al., 1993). Lesions in hippocampus induced by kainic acid resulted in elevated antibody production including IgM and IgG (Nance et al., 1987). Taken together, these studies show that lesions (or stimulation) in the hippocampus affect immune functions.

6.4.2. Immunomodulation of Neuronal Functions in Hippocampus

Hippocampal neuronal activities have been examined thoroughly in studies on neuroendocrine, autonomic and cognitive function, as well as psychomotor behavior. Circumstantial evidence indicates that hippocampal physiology can be modulated by the immune system (Jankowsky and Patterson, 1999; Jankowsky et al., 2000). This modulation could be achieved through proinflammatory cytokines including, but not limited to, interleukin-1β (IL-1β), IL-2, IL-6, tumor necrosis factor alpha (TNF-α) and interferon gamma (INF-γ) (Wrona, 2006). Indeed, a number of cytokines, including the aforementioned ones, are expressed in the hippocampus and alteration of their expression levels can affect hippocampal functions. Studies have shown that inflammatory cytokines, released in response to the detection of foreign substances (antigens), influence ion channel activities, intracellular Ca^{2+} homeostasis, membrane potentials, and suppress or enhance the induction of LTP in the hippocampus (Koller et al., 1997). Extensive experimental results have implicated IL-1 as the most likely candidate for key immunotransmitter, communicating immunological activation to the brain including the hippocampus (Besedovsky et al., 1975; Besedovsky et al., 1986). It is worth to point out that cytokines rarely work in isolation. For instance, the release of IL-1β is usually associated with the release of the other proinflammatory cytokines, such as TNF-α and IL-6, which are indeed expressed in the hippocampus.

IL-1β and its receptors are expressed in the hippocampus (Farrar et al., 1987; Ban et al., 1991; Cunningham and De Souza, 1993). High density of binding sites for IL-1β has been detected in the hippocampus, with highest density in the dentate gyrus (Takao et al., 1990). It has been shown that peripheral immune activation by lipopolysaccharide (LPS) up-regulates IL-1β mRNA expression and increases IL-1β protein in the hippocampus (Laye et al., 1994; Nguyen et al., 1998). This suggests that immune activation may modulate hippocampal function via release of immune active molecules, such as IL-1β. As hippocampus is a brain region involved in learning and memory, the immune associated upregulation of IL-1β mRNA and protein expression may interrupt hippocampal functions such as learning and memory. Indeed, IL-1β suppresses the induction of LTP in the CA1 and CA3 areas of the hippocampus as well as in the dentate gyrus (Katsuki et al., 1990; Bellinger et al., 1993; Cunningham et al., 1996; O'Connor and Coogan, 1999; Xiong et al., 2000), while having no significant effects on excitatory postsynaptic potential (EPSP) evoked by low frequency stimulation. The IL-1β-mediated suppression of LTP is antagonized by an IL-1β receptor antagonist, suggesting that IL-1β inhibits LTP through IL-1β receptors. In addition to IL-1, IL-2 had similar effects on LTP in hippocampus. Application of recombinant IL-2 inhibited the induction of both short-term potentiation (STP) and LTP. It also inhibited post-tetanic potentiation (PTP) and LTP maintenance without affecting basal synaptic transmission (Tancredi et al., 1990). Moreover, IL-2 deficiency results in altered hippocampal cytoarchitecture (Beck et al., 2005).

LTP was also suppressed by TNF-α in both the CA1 region (Tancredi et al., 1992) and the dentate gyrus (Cunningham et al., 1996). In the CA1, the induction of LTP was inhibited by TNF-α if the tetanic stimulation was given at least 50 min after TNF-α application. In contrast to IL-1 and IL-2, TNF-α increased basal synaptic transmission in the CA1 region of the slices acutely exposed to TNF-α (Cunningham et al., 1996) but not in the dentate gyrus. The underlying mechanisms, by which TNF-α increased basal synaptic transmission, have not been determined. Brief treatment of hippocampal slices with TNF-α did not influence LTP, while long-lasting application (>50 min) of TNF-α inhibited LTP.

Summary

The hippocampus is a symmetrical structure located inside the medial temporal lobe on both sides of the human brain. In cross-sections, the hippocampus consists of two interlocking sheets of cortex with three distinct sub-regions: the dentate gyrus, the hippocampus proper (CA1—CA3) and the subiculum. The hippocampus has a highly defined laminar structure with visible layers of pyramidal cells arranged in rows. A striking feature of hippocampus is its connection circuitry.

The connections within the hippocampus generally follow this laminar format and are largely unidirectional. They form well-characterized closed loops that originate mainly in the adjacent entorhinal cortex. Thus information flow through the hippocampus proceeds from the dentate gyrus to the CA3 to the CA1 to the subiculum, forming the principal trisynaptic circuit. Together with the adjacent amygdyla and entorhinal cortex, the hippocampus forms the central axis of the limbic system and plays an important role in spatial learning and awareness, navigation, episodic/event memory, and neuroimmunomodulation.

Acknowledgments. The author thanks Dr. Daniel Monaghan and Mr. James Keblesh for critical reading of the manuscript. Supported by NIH grant NS041862.

Review Questions/Problems

1. **Where is the hippocampus located in the human brain?**
2. **What are the three subfields in the hippocampus?**
3. **What are the principal cells in the dentate gyrus and the CA1 field?**
4. **CA1 region belongs to which subfield in hippocampus?**
5. **The main input to the dentate gyrus originates from which part of brain and via which fiber path?**
6. **The medial and lateral perforant paths project to which part of the dendrite tuft of the granule cells?**
7. **Mossy fibers originate from which part of the hippocampus and synapse onto the neurons of which field(s)?**
8. **How many groups of dendrites do a CA1 neurons have and what is the name for each group? From which part of a CA1 neuron does the axon originate?**
9. **What is the main afferent pathway to the CA1 pyramidal cells and from where does this pathway originate?**
10. **Briefly describe information flow through the hippocampus**
11. **Damage to the hippocampus disrupts which type of memory?**
12. **What are the three basic properties of LTP? Define each of the three basic properties.**
13. **What is the key role that NMDA receptors play in LTP induction and what two simultaneously occurring events are needed for LTP induction?**

References

Aggleton JP, Vann SD, Saunders RC (2005) Projections from the hippocampal region to the mammillary bodies in macaque monkeys. Eur J Neurosci 22:2519–2530.

Ascher P, Nowak L (1988) The role of divalent cations in the N-methyl-D-aspartate responses of mouse central neurones in culture. J Physiol 399:247–266.

Ban E, Milon G, Prudhomme N, Fillion G, Haour F (1991) Receptors for interleukin-1 (alpha and beta) in mouse brain: Mapping and neuronal localization in hippocampus. Neuroscience 43:21–30.

Bannerman DM, Yee BK, Lemaire M, Wilbrecht L, Jarrard L, Iversen SD, Rawlins JN, Good MA (2001) The role of the entorhinal cortex in two forms of spatial learning and memory. Exp Brain Res 141:281–303.

Beck RD, Jr., King MA, Ha GK, Cushman JD, Huang Z, Petitto JM (2005) IL-2 deficiency results in altered septal and hippocampal cytoarchitecture: Relation to development and neurotrophins. J Neuroimmunol 160:146–153.

Bellinger FP, Madamba S, Siggins GR (1993) Interleukin-1 beta inhibits synaptic strength and long-term potentiation in the rat CA1 hippocampus. Brain Res 628:227–234.

Berkenbosch F, van Oers J, del Rey A, Tilders F, Besedovsky H (1987) Corticotropin-releasing factor-producing neurons in the rat activated by interleukin-1. Science 238:524–526.

Besedovsky H, Sorkin E, Keller M, Muller J (1975) Changes in blood hormone levels during the immune response. Proc Soc Exp Biol Med 150:466–470.

Besedovsky H, Sorkin E, Felix D, Haas H (1977) Hypothalamic changes during the immune response. Eur J Immunol 7:323–325.

Besedovsky H, del Rey A, Sorkin E, Dinarello CA (1986) Immunoregulatory feedback between interleukin-1 and glucocorticoid hormones. Science 233:652–654.

Blackstad TW, Kjaerheim A (1961) Special axo-dendritic synapses in the hippocampal cortex: Electron and light microscopic studies on the layer of mossy fibers. J Comp Neurol 117:133–159.

Blalock JE (1989) A molecular basis for bidirectional communication between the immune and neuroendocrine systems. Physiol Rev 69:1–32.

Blalock JE (1994) Shared ligands and receptors as a molecular mechanism for communication between the immune and neuroendocrine systems. Ann N Y Acad Sci 741:292–298.

Bliss TV, Gardner-Medwin AR (1973) Long-lasting potentiation of synaptic transmission in the dentate area of the unanaestetized rabbit following stimulation of the perforant path. J Physiol 232:357–374.

Bliss TV, Collingridge GL (1993) A synaptic model of memory: Long-term potentiation in the hippocampus. Nature 361:31–39.

Bliss TV, Douglas RM, Errington ML, Lynch MA (1986) Correlation between long-term potentiation and release of endogenous amino acids from dentate gyrus of anaesthetized rats. J Physiol 377:391–408.

Brooks WH, Cross RJ, Roszman TL, Markesbery WR (1982) Neuroimmunomodulation: Neural anatomical basis for impairment and facilitation. Ann Neurol 12:56–61.

Cajal R (1911) Histologie du systeme nerveux de l'homme et des vertebres Vol. 2. Instituto Ramon y Cajal: Madrid.

Cohen NJ, Eichenbaum H (1993) Memory, amnesia and the hippocampus system. MIT Press: Cambridge.

Collingridge GL, Kehl SJ, McLennan H (1983) The antagonism of amino acid-induced excitations of rat hippocampal CA1 neurones in vitro. J Physiol 334:19–31.

Coltman BW, Ide CF (1996) Temporal characterization of microglia, IL-1 beta-like immunoreactivity and astrocytes in the dentate gyrus of hippocampal organotypic slice cultures. Int J Dev Neurosci 14:707–719.

Conrad LC, Leonard CM, Pfaff DW (1974) Connections of the median and dorsal raphe nuclei in the rat: An autoradiographic and degeneration study. J Comp Neurol 156:179–205.

Cunningham AJ, Murray CA, O'Neill LA, Lynch MA, O'Connor JJ (1996) Interleukin-1 beta (IL-1 beta) and tumour necrosis factor (TNF) inhibit long-term potentiation in the rat dentate gyrus in vitro. Neurosci Lett 203:17–20.

Cunningham ET, Jr., De Souza EB (1993) Interleukin 1 receptors in the brain and endocrine tissues. Immunol Today 14:171–176.

de Olmos J, Hardy H, Heimer L (1978) The afferent connections of the main and the accessory olfactory bulb formations in the rat: An experimental HRP-study. J Comp Neurol 181:213–244.

Desmond NL, Levy WB (1982) A quantitative anatomical study of the granule cell dendritic fields of the rat dentate gyrus using a novel probabilistic method. J Comp Neurol 212:131–145.

Devi RS, Namasivayam A, Prabhakaran K (1993) Modulation of non-specific immunity by hippocampal stimulation. J Neuroimmunol 42:193–197.

Devi RS, Sivaprakash RM, Namasivayam A (2004) Rat hippocampus and primary immune response. Indian J Physiol Pharmacol 48:329–336.

Engler KL, Rudd ML, Ryan JJ, Stewart JK, Fischer-Stenger K (2005) Autocrine actions of macrophage-derived catecholamines on interleukin-1 beta. J Neuroimmunol 160:87–91.

Farrar WL, Kilian PL, Ruff MR, Hill JM, Pert CB (1987) Visualization and characterization of interleukin 1 receptors in brain. J Immunol 139:459–463.

Frotscher M (1985) Mossy fibres form synapses with identified pyramidal basket cells in the CA3 region of the guinea-pig hippocampus: A combined Golgi-electron microscope study. J Neurocytol 14:245–259.

Gustafsson B, Wigstrom H, Abraham WC, Huang YY (1987) Long-term potentiation in the hippocampus using depolarizing current pulses as the conditioning stimulus to single volley synaptic potentials. J Neurosci 7:774–780.

Harbour DV, Smith EM, Blalock JE (1987) Splenic lymphocyte production of an endorphin during endotoxic shock. Brain Behav Immun 1:123–133.

Haring JH, Davis JN (1985) Retrograde labeling of locus coeruleus neurons after lesion-induced sprouting of the coeruleohippocampal projection. Brain Res 360:384–388.

Hebb D (1949) The organization of behavior. New York: John Wiley and Sons.

Herkenham M (1978) The connections of the nucleus reuniens thalami: Evidence for a direct thalamo-hippocampal pathway in the rat. J Comp Neurol 177:589–610.

Jankowsky JL, Patterson PH (1999) Cytokine and growth factor involvement in long-term potentiation. Mol Cell Neurosci 14:273–286.

Jankowsky JL, Derrick BE, Patterson PH (2000) Cytokine responses to LTP induction in the rat hippocampus: A comparison of in vitro and in vivo techniques. Learn Mem 7:400–412.

Jarrard LE (1983) Selective hippocampal lesions and behavior: Effects of kainic acid lesions on performance of place and cue tasks. Behav Neurosci 97:873–889.

Jones BE, Moore RY (1977) Ascending projections of the locus coeruleus in the rat II. Autoradiographic study. Brain Res 127:25–53.

Katsuki H, Nakai S, Hirai Y, Akaji K, Kiso Y, Satoh M (1990) Interleukin-1 beta inhibits long-term potentiation in the CA3 region of mouse hippocampal slices. Eur J Pharmacol 181:323–326.

Kishi T, Tsumori T, Yokota S, Yasui Y (2006) Topographical projection from the hippocampal formation to the amygdala: A combined anterograde and retrograde tracing study in the rat. J Comp Neurol 496:349–368.

Koller H, Siebler M, Hartung HP (1997) Immunologically induced electrophysiological dysfunction: Implications for inflammatory diseases of the CNS and PNS. Prog Neurobiol 52:1–26.

Kosel KC, Van Hoesen GW, Rosene DL (1983) A direct projection from the perirhinal cortex (area 35) to the subiculum in the rat. Brain Res 269:347–351.

Krettek JE, Price JL (1977) Projections from the amygdaloid complex and adjacent olfactory structures to the entorhinal cortex and to the subiculum in the rat and cat. J Comp Neurol 172:723–752.

Kullmann DM, Siegelbaum SA (1995) The site of expression of NMDA receptor-dependent LTP: New fuel for an old fire. Neuron 15:997–1002.

Larkman AU, Jack JJ (1995) Synaptic plasticity: Hippocampal LTP. Curr Opin Neurobiol 5:324–334.

Laye S, Parnet P, Goujon E, Dantzer R (1994) Peripheral administration of lipopolysaccharide induces the expression of cytokine transcripts in the brain and pituitary of mice. Brain Res Mol Brain Res 27:157–162.

Lolait SJ, Lim AT, Toh BH, Funder JW (1984) Immunoreactive beta-endorphin in a subpopulation of mouse spleen macrophages. J Clin Invest 73:277–280.

Lorente de No R (1934) Studies on the structure of the cerebral cortex. II Continuation of the study of ammonic system. J Psychol Neurol 46:113–177.

Lynch G, Larson J, Kelso S, Barrionuevo G, Schottler F (1983) Intracellular injections of EGTA block induction of hippocampal long-term potentiation. Nature 305:719–721.

Malenka RC, Nicoll RA (1999) Long-term potentiation—a decade of progress? Science 285:1870–1874.

Manabe T, Wyllie DJ, Perkel DJ, Nicoll RA (1993) Modulation of synaptic transmission and long-term potentiation: Effects on paired pulse facilitation and EPSC variance in the CA1 region of the hippocampus. J Neurophysiol 70:1451–1459.

McNaughton BL, Douglas RM, Goddard GV (1978) Synaptic enhancement in fascia dentata: Cooperativity among coactive afferents. Brain Res 157:277–293.

Meibach RC, Siegel A (1977a) Efferent connections of the septal area in the rat: An analysis utilizing retrograde and anterograde transport methods. Brain Res 119:1–20.

Meibach RC, Siegel A (1977b) Efferent connections of the hippocampal formation in the rat. Brain Res 124:197–224.

Moore RY, Halaris AE (1975) Hippocampal innervation by serotonin neurons of the midbrain raphe in the rat. J Comp Neurol 164:171–183.

Morris RG, Garrud P, Rawlins JN, O'Keefe J (1982) Place navigation impaired in rats with hippocampal lesions. Nature 297:681–683.

Morris RG, Anderson E, Lynch GS, Baudry M (1986) Selective impairment of learning and blockade of long-term potentiation by an N-methyl-D-aspartate receptor antagonist, AP5. Nature 319:774–776.

Nance DM, Rayson D, Carr RI (1987) The effects of lesions in the lateral septal and hippocampal areas on the humoral immune response of adult female rats. Brain Behav Immun 1:292–305.

Nguyen KT, Deak T, Owens SM, Kohno T, Fleshner M, Watkins LR, Maier SF (1998) Exposure to acute stress induces brain interleukin-1beta protein in the rat. J Neurosci 18:2239–2246.

O'Connor JJ, Coogan AN (1999) Actions of the pro-inflammatory cytokine IL-1 beta on central synaptic transmission. Exp Physiol 84:601–614.

O'Keefe J, Dostrovsky J (1971) The hippocampus as a spatial map. Preliminary evidence from unit activity in the freely-moving rat. Brain Res 34:171–175.

O'Keefe J, Nadel L (1978) The Hippocampus as a Cognitive Map. Oxford: Oxford University Press.

O'Keefe J, Conway DH (1978) Hippocampal place units in the freely moving rat: Why they fire where they fire. Exp Brain Res 31:573–590.

Otto T, Eichenbaum H, Wiener SI, Wible CG (1991) Learning-related patterns of CA1 spike trains parallel stimulation parameters optimal for inducing hippocampal long-term potentiation. Hippocampus 1:181–192.

Pan Q, Long J (1993) Lesions of the hippocampus enhance or depress humoral immunity in rats. Neuroreport 4:864–866.

Perkel DJ, Petrozzino JJ, Nicoll RA, Connor JA (1993) The role of Ca2+ entry via synaptically activated NMDA receptors in the induction of long-term potentiation. Neuron 11:817–823.

Raisman G, Cowan WM, Powell TP (1966) An experimental analysis of the efferent projection of the hippocampus. Brain 89:83–108.

Regehr WG, Tank DW (1990) Postsynaptic NMDA receptor-mediated calcium accumulation in hippocampal CA1 pyramidal cell dendrites. Nature 345:807–810.

Rivest S (2003) Molecular insights on the cerebral innate immune system. Brain Behav Immun 17:13–19.

Room P, Postema F, Korf J (1981) Divergent axon collaterals of rat locus coeruleus neurons: Demonstration by a fluorescent double labeling technique. Brain Res 221:219–230.

Sapolsky R, Rivier C, Yamamoto G, Plotsky P, Vale W (1987) Interleukin-1 stimulates the secretion of hypothalamic corticotropin-releasing factor. Science 238:522–524.

Scoville WB, Milner B (1957) Loss of recent memory after bilateral hippocampal lesions. J Neurol Neurosurg Psychiatry 20:11–21.

Sikes RW, Chronister RB, White LE, Jr. (1977) Origin of the direct hippocampus—anterior thalamic bundle in the rat: A combined horseradish peroxidase—Golgi analysis. Exp Neurol 57:379–395.

Squire LR (1992) Memory and the hippocampus: A synthesis from findings with rats, monkeys, and humans. Psychol Rev 99:195–231.

Swanson LW (1981) A direct projection from Ammon's horn to prefrontal cortex in the rat. Brain Res 217:150–154.

Swanson LW, Cowan WM (1975) Hippocampo-hypothalamic connections: Origin in subicular cortex, not ammon's horn. Science 189:303–304.

Swanson LW, Wyss JM, Cowan WM (1978) An autoradiographic study of the organization of intrahippocampal association pathways in the rat. J Comp Neurol 181:681–715.

Swanson LW, Sawchenko PE, Cowan WM (1981) Evidence for collateral projections by neurons in Ammon's horn, the dentate gyrus, and the subiculum: A multiple retrograde labeling study in the rat. J Neurosci 1:548–559.

Takao T, Tracey DE, Mitchell WM, De Souza EB (1990) Interleukin-1 receptors in mouse brain: Characterization and neuronal localization. Endocrinology 127:3070–3078.

Tancredi V, Zona C, Velotti F, Eusebi F, Santoni A (1990) Interleukin-2 suppresses established long-term potentiation and inhibits its induction in the rat hippocampus. Brain Res 525:149–151.

Tancredi V, D'Arcangelo G, Grassi F, Tarroni P, Palmieri G, Santoni A, Eusebi F (1992) Tumor necrosis factor alters synaptic transmission in rat hippocampal slices. Neurosci Lett 146:176–178.

Tulving E (2001) Episodic memory and common sense: How far apart? Philos Trans R Soc Lond B Biol Sci 356:1505–1515.

Tulving E, Markowitsch HJ (1998) Episodic and declarative memory: Role of the hippocampus. Hippocampus 8:198–204.

Wrona D (2006) Neural-immune interactions: An integrative view of the bidirectional relationship between the brain and immune systems. J Neuroimmunol 172:38–58.

Wyss JM, Swanson LW, Cowan WM (1979) Evidence for an input to the molecular layer and the stratum granulosum of the dentate gyrus from the supramammillary region of the hypothalamus. Anat Embryol (Berl) 156:165–176.

Xiong H, Zeng YC, Lewis T, Zheng J, Persidsky Y, Gendelman HE (2000) HIV-1 infected mononuclear phagocyte secretory products affect neuronal physiology leading to cellular demise: Relevance for HIV-1-associated dementia. J Neurovirol 6(Suppl 1):S14–S23.

Yang SN, Tang YG, Zucker RS (1999) Selective induction of LTP and LTD by postsynaptic [Ca2+]i elevation. J Neurophysiol 81:781–787.

Yuste R, Denk W (1995) Dendritic spines as basic functional units of neuronal integration. Nature 375:682–684.

7
Glial and Neuronal Cellular Compostion, Biology, and Physiology

Kalipada Pahan

The brain has the unique ability to affect so many things from movement to emotion to cognitive abilities. It can, together with life events, transform a person into an artist, a priest, a scientist, or a teacher. To achieve any goal, all of our body systems need to collaborate with one other. It is the brain that controls and co-ordinates the activity of all in response to environmental cues and demands with the assistance of all of the nervous system. Each and every demand is detected by our senses and messages are judged by merit, which in turn, directs particular responses. Taken together, our nervous system is concerned about sensory input and motor output. Sensory nerves collect information about the body's internal and external environment and convey it to the central nervous system (CNS). Motor nerves carry instructions on what to do. Let me describe it with an example. When we feel hungry, our internal environment generates a sensory input that provides us the awareness of hunger. Then sensory input from the external environment provides us the information on how to obtain food. Consequently, motor output is generated in the external environment to get and swallow food. Then motor activity in the internal environment assists us with the food intake to the extent until we get the nod from our sensory input from the internal environment that enough food has been consumed.

This input and output business is simply dictated by a group of cells forming the basis of the supercomputing system of the brain—called neurons. This particular group of cells forces us to revere brain as an elite organ. It is the "neuron doctrine" of Professor Cajal that contributed to the basic understanding of the organization of the CNS. As nobody in this world is unable to survive alone, neurons are also not an exception from this universal rule. Therefore, there are other cells in the CNS, called glial cells (Figure 7.1). Glial cells have diverse functions that are necessary for the proper development and function of the complex nervous system. The growing number of links between glial malfunction and human disease has generated great interest in glial cell biology.

Axons are surrounded by while matter coating called myelin that consists of a layer of proteins packed between two layers of lipids. This myelin coating enables axons to conduct impulses between the brain and other parts of the body. Myelin is synthesized by specialized cells—oligodendrocytes in the CNS and Schwann cells in the peripheral nervous system (PNS). Although the composition of CNS and PNS myelins are not same, they are assigned for the same function—to promote efficient transmission of a nerve impulse along the axon. Schwann cells have an intimate association with axons and each Schwann cell forms myelin around a single axon, and lines up along the axon to define a single internode. On the other hand, one oligodendrocyte extends several processes and can myelinate upto 1–40 axons with distinct internodes. In addition to myelination, Schwann cells are able to migrate and phagocytose debris from the PNS. However, oligodendrocytes do not have such activity.

The major cells in the CNS are astrocytes that are believed to support the entire structure of the microenvironment (Liedtke et al., 1996) together with endothelial cell lining. In addition to structural support, astrocytes have many other important functions, such as food supply, water balance, ion homeostasis, regulation of neurotransmitters, detoxification of ammonia, organizing information network, and release of neuropeptides and neurotrophins. Because neurotrophins support the growth of neurons and astrocytes are the major producer of neurotrophins in the CNS, these cells also play an important role in neurogenesis. The other important cell type is microglia, the resident macrophages in the CNS. As happens in other organs, cells in the CNS also undergo natural cell death. Then microglia keep the CNS microenvironment clean by scavenging these dead cell bodies. In addition, when immune responses are generated within the CNS or from outside the CNS, microglia, being the primary CNS immune cells, receive and pass on that response to other cells (Carson, 2002; Rock et al., 2004). Under physiological condition, the immune response usually ends up with a logical conclusion leading to the development of a better neuroimmune system. Another less characterized glial cell type is Bergmann glia that are composed of unipolar protoplasmic astrocytes in the

T. Ikezu and H.E. Gendelman (eds.), *Neuroimmune Pharmacology*.
© Springer 2008

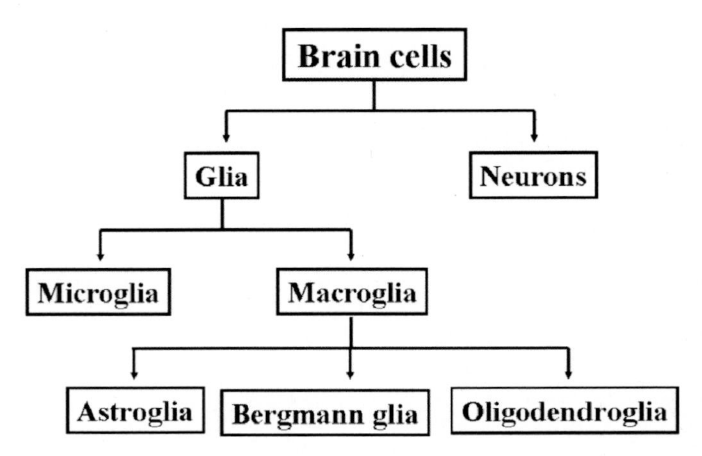

FIGURE 7.1. Classification of brain cells.

cerebellar cortex. These are associated with granule cells in the developing cerebellum and with Purkinje cells in the adult cerebellum.

As part of our life, we experience stress, trauma, infection, injury etc. and come in contact with various toxic substances. Although brain is separated from the rest of our body by a well-defined blood-brain barrier, brain perceives and faces each of these challenges (Figure 7.2). Therefore, after these insults, the injured brain cell usually tries to either die or survive. When an injured cell dies, the death process usually becomes associated with increased production of proinflammatory molecules, decreased production of anti-inflammatory molecules, increase in T-helper 1 (Th1) response, decreased production of growth factors, increased expression of death genes (e.g. bad, bax), decreased expression of survival genes (e.g. bcl_2, X-IAP, survivin), and over-activation of tumor suppressor genes (e.g. p53). On the other hand, during the survival of an injured cell, opposite phenomena are observed (Figure 7.2). If the survival process is accompanied by a very high Th2 response, huge production of growth factors, abnormal cell growth, and mutation of tumor suppressor genes, the injured cell may like to live as a cancerous cell. Although everybody is receiving some kind of insults or injuries, everybody does not get the disease because in healthy human beings, there is a proper balance between cell death and cell survival. When this invaluable balance is lost, we see the disease (Figure 7.2). Although there are four major cell types in the CNS, under neuroinflammatory and neurodegenerative stress conditions, only neurons and oligodendroglia succumb to cell death. On the other hand, astroglia and microglia do not die but undergo activation and gliosis under the same condition. Although glial activation is not always bad, when activated glia are forced to amplify the stress response to such an extent that it goes out of control, it helps in neuroimmune and neurodegenerative pathologies (Gonzalez-Scarano and Baltuch, 1999; Eng et al., 1992; Reier, 1986).

Another aspect that also plays a vital role in unsettling the balance between physiology and pathophysiology is dose or amount of a biomolecule. Each and every biomolecule is important for our cells. However, any good thing either in excess or in less is not good any more. For example, nitric oxide (NO), an important CNS signaling molecule, exerts profound effects depending on its dose. In neuronal nitric oxide synthase (–/–) mice, when neurons do not produce much NO, male mice become very aggresive and fight with each other (Nelson et al., 1995) much like terrorists do all the time! When this molecule is present in physiological amount, neurons feel happy and relaxed due to the activation of guanylate cyclase (GC)-cyclic GMP (cGMP) pathway and proper neurotransmission (Hawkins et al., 1998). On the other hand, when NO is produced within the CNS in excess, a bitter mood prevails over this happy mood as excess NO starts a "Hurricane Katrina" damaging cellular powerhouse mito-

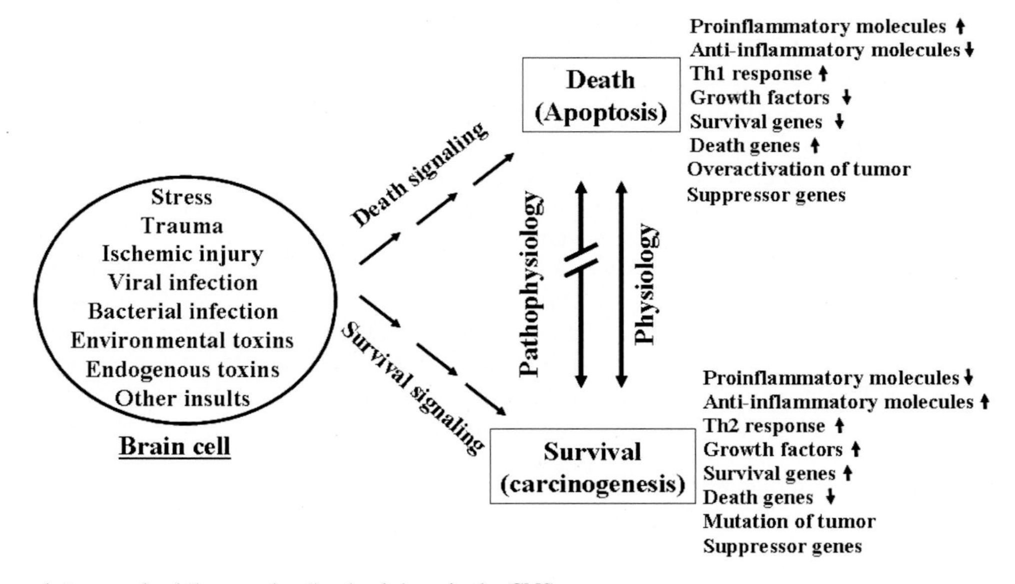

FIGURE 7.2. Balance between physiology and pathophysiology in the CNS.

chondria, inactivating essential enzymes (e.g. anti-oxidant enzymes), castrating multifunctional transcription factors (e.g. NF-κB), oxidizing lipids, and destroying many other biomolecules (Mitrovic et al., 1994; Radi et al., 1991).

Once Sir Charles Sherrington, 1932 Nobel Laureate of Medicine, described brain as "an enchanted loom where millions of flashing shuttles weave a dissolving pattern, always a meaningful pattern though never an abiding one". Even after more than a hundred years of research, enchanting discoveries are still coming out about this organ. We are still more or less in the dark about how to control these flashing shuttles and how to direct them to follow a certain meaningful pattern. In the following chapters, we have made an honest attempt to cover most of the known aspects of CNS physiology. A one-year old boy has about 100 billion neurons. As we grow older, neurons are lost and not replaced. However, contrary to this century-old dogma, recently it has been demonstrated that humans are able to generate new nerve cells throughout their life. Therefore, we also provide some cues for future scientists to develop the wit of transforming old and depressed minds into perpetually youthful and active minds.

Acknowledgements. This study was supported by grants from National Institutes of Health (NS39940 and NS48923), National Multiple Sclerosis Society (RG3422A1/1) and Michael J. Fox Foundation for Parkinson Research.

References

Carson MJ (2002) Microglia as liaisons between the immune and central nervous systems: Functional implications for multiple sclerosis. Glia 40:218–231.

Eng LF, Yu AC, Lee YL (1992) Astrocytic response to injury. Prog Brain Res 94:353–365.

Gonzalez-Scarano F, Baltuch G (1999) Microglia as mediators of inflammatory and degenerative diseases. Annu Rev Neurosci 22:219–240.

Hawkins RD, Son H, Arancio O (1998) Nitric oxide as a retrograde messenger during long-term potentiation in hippocampus. Prog Brain Res 118:155–172.

Liedtke W, Edelmann W, Bieri PL, Chiu FC, Cowan NJ, Kucherlapati R, Raine CS (1996) GFAP is necessary for the integrity of CNS white matter architecture and long-term maintenance of myelination. Neuron 17:607–615.

Mitrovic B, Ignarro LJ, Montestruque S, Smoll A, Merrill JE (1994) Nitric oxide as a potential pathological mechanism in demyelination: Its differential effects on primary glial cells in vitro. Neuroscience 61:575–585.

Nelson RJ, Demas GE, Huang PL, Fishman MC, Dawson VL, Dawson TM, Snyder SH (1995) Behavioural abnormalities in male mice lacking neuronal nitric oxide synthase. Nature 378:383–386.

Radi R, Beckman JS, Bush KM, Freeman BA (1991) Peroxynitrite oxidation of sulfhydryls. The cytotoxic potential of superoxide and nitric oxide. J Biol Chem 266:4244–4250.

Reier PJ (1986) Gliosis following CNS injury: The anatomy of astrocytic scars and their influences on axonal elongation. Astrocytes 3:263–324.

Rock RB, Gekker G, Hu S, Sheng WS, Cheeran M, Lokensgard JR, Peterson PK (2004) Role of microglia in central nervous system infections. Clin Microbiol Rev 17:942–964.

8
Astrocytes, Oligodendrocytes, and Schwann Cells

Malabendu Jana, Subhajit Dasgupta, Anuja Ghorpade, and Kalipada Pahan

Keywords Astrocytes; Glial precursors; Calcium excitability; Astrogliosis; Oligodendrocytes; CNS myelination; Oligodendroglial death; Oligodendroglial regeneration; Schwann cells; PNS myelination

8.1. Introduction

Central nervous system (CNS) is composed of two major cell types: neuron and glia. Astrocytes and oligodendrocytes belong to the latter category. Astrocytes, through an intricate network surrounding blood vessels, play an important role in supplying food, water and ions from periphery to the CNS and maintain CNS homeostasis. Astrocytes also play an active role in neurogenesis. However, under inflammatory or neurodegenerative conditions, astrocytes produce pro-inflammatory mediators and take active part in the ongoing events. Neurons in the CNS are covered by myelin sheath that maintains conduction of nerve impulse. Consistently, the CNS houses oligodendrocytes for myelin synthesis. On the other hand, Schwann cells are the myelinating cells in the peripheral nervous system (PNS). Balanced expression of several genes and activation of transcription factors critically regulate the entire complicated functional network of astrocytes, oligodendrocytes and Schwann cells. Keeping a birds' eye view, this chapter delineates genesis and functional aspects of astrocytes, oligodendrocytes and Schwann cells.

8.2. Historical View

For decades, astrocytes and oligodendrocytes were considered as silent partners of neurons in the CNS. It was known that astrocytes, like neurons, were unable to transmit messages as they did not possess voltage and ion gated channels. With the advancement of science, it is now well accepted that astrocytes possess ion channels as well as G-protein coupled receptors necessary to sense and respond to neuronal activities. Recent advancements also reveal that oligodendrocytes, apart from myelinating neurons in the CNS, secrete some growth factors to help neuronal growth and development. On the other hand, under disease conditions, astroglia undergo proliferation and gliosis. Activated astroglia also secrete neurotoxic molecules that may be involved in the loss of neurons in neurodegenerative disorders and the damage of oligodendroglia in neuroinflammatory demyelinating disorders.

The present chapter focuses on biology and functional aspects of astrocytes and oligodendrocytes ranging from their genesis to their enormous role in maintaining CNS homeostasis along with their role in CNS pathology. The biology and function of PNS myelinating Schwann cells has been discussed later as a separate section (Section 6).

8.3. Development of Astrocytes and Oligodendrocytes in the CNS

The vertebrate nervous system including neurons, astrocytes, oligodendrocytes, and other cells originates from a flat sheet of neuroepithelial cells, constituent of the inner lining of neural plate along the dorsal surface of embryo (Fujita, 2003). These neuroepithelial cells are the earliest precursors in the developing CNS.

8.3.1. Generation of Glial Precursor Cells

During neurogenesis, neuroblasts are first derived from stem cells and then migrate peripherally to the mantle and marginal layers in the developing brain. After that, DNA synthesis in neurons is completely ceased and the progenitor cells enter into the phase of gliogenesis in the neural tube. These glioblasts are functionally different but morphologically indistinguishable from the multipotent stem cells and eventually differentiate first into functional astrocytes and then

T. Ikezu and H.E. Gendelman (eds.), *Neuroimmune Pharmacology.*
© Springer 2008

oligodendrocytes. The quiescent form of the glioblasts called microglia comes after these events.

Differentiation of cortical progenitor cells is being controlled by some transcription factors having basic-helix-loop-helix (bHLH) motifs. These are NeuroD, Neurogenin, Mash, Olig, Id, and Hes families of protein. The restricted and time-dependent binding of these transcription factors with corresponding DNA sequences present in the promoter of different developmental genes determines the outcome of final cell types. Recent developments (Gotz and Barde, 2005; Alvarez-Buylla et al., 2001) show that neuron and glia are generated from same progenitors/precursors.

8.3.2. Signaling Events Driving the Precursors to Functional Cells: Astrocytes and Oligodendrocytes

The signaling events like hedgehog and notch regulate genesis of functionally distinguished glia and neurons from multipotent stem cells. Hedgehog (Hh) family of signaling molecules are the key organizers of tissue patterning during embryogenesis (Altaba et al., 2002). In mouse, three Hh genes have been identified. These are Desert hedgehog (Dhh), Indian hedgehog (Ihh) and Sonic hedgehog (Shh). The Shh plays a vital role in the development of CNS. In mammals and birds, Shh is the only hedgehog family member that is reported to be expressed in normal CNS.

The oligodendrocyte progenitors (OPs) in caudal as well as ventral neural tube originate under the influence of Shh protein secreted from ventral midline. At this initial stage, Shh patterns the ventral neuroepithelium by controlling the expression of a set of transcription factors PAX6, NKX 2.2, high mobility group protein SOX10, and basic helix-loop-helix proteins Olig1 and Olig2. These Olig genes and SOX10 are co-expressed in cells before the appearance of PDGF-α on OPs. These PDGF-positive OPs then proliferate and migrate away from the ventricular surface to all parts of the CNS before differentiating into functional myelin forming mature cells.

Notch signaling (Yoon and Gaiano, 2005) first specifies glial progenitors and then functions in those cells to promote astrocytes versus oligodendrocytes fate. The ligands of the Notch signaling pathway are expressed in differentiating neurons. The receptors Notch are transmembrane proteins and are found on neural stem cells. Upon ligand binding, intracellular domain (NICD) of Notch is cleaved by γ-secretase which then enters into the nucleus to form a complex with C promoter binding factor (CBF1) and mastermind-like (MAML). Then the complex (NICD:CBF1:MAML) binds to promoter regions of target genes Hes and Herp and up-regulates corresponding HES/HERP proteins. These proteins are bHLH transcriptional regulators that antagonize proneural genes like Mash 1 and neurogenins. As a result, it blocks neuronal differentiation.

8.4. Astrocytes: Biology and Function

In the middle of the nineteenth century, German anatomist and pathologist Rudolph Virchow was wondering about the group of cells in the brain that surround the neurons and fill the spaces between them. Dr. Virchow named these cells as "neuroglia" means "neural glue." He used the term "glue" to represent the gluing function of these cells to hold the neurons in place. Nowadays "neuroglia" is collectively used for all glial cells in the CNS. Later on, due to "star-shaped" appearance, the major neuroglial cells were named as "Astrocytes" (Astra: star; cyte: cells).

8.4.1. Morphology and Markers

Morphologically, astrocytes can be classified into two types: fibrous astrocytes and protoplasmic astrocytes (Brightman and Cheng, 1988). Fibrous astrocytes are located predominantly in white matter and possess fewer but longer processes. These processes form cytoplasmic bundles of intermediate filaments (IFs). The major constituent of these filaments are glial fibrillary acidic protein (GFAP). Under light microscope, the fibrous astrocytes look like a star-shaped cell body with finer processes. These processes are extended for long distances and contain abundant IFs.

The protoplasmic astrocytes, on the other hand, have more complex morphology. They contain highly branched processes that form membranous sheets surrounding the neuronal processes, cell bodies and end-feet on capillaries. In contrast to fibrous astrocytes, these cells have fewer IFs and a greater density of organelles.

Apart from the ultrastructural study, astrocytes can also be identified on the basis of marker proteins (Table 8.1).

8.4.2. Heterogeneous Population of Astrocytes in the CNS

In the CNS, many cells share some characteristics with astrocytes. These "*astrocytes-like*" cells are pituicytes, tanycytes,

TABLE 8.1. Markers of astrocytes.

Marker	Function	Cellular localization	Molecular weight
GFAP	Major constituent of intermediate filament found mostly in adult astrocytes	Cytoplasm	50 kDa (predicted)
EAAT1	Transport of amino acids	Cytoplasm	59.5 kDa
Glutamine synthase	In CNS, the enzyme found only in astrocytes; it catalyzes conversion of glutamate to glutamine	Cytoplasm	43 kDa
S-100	Ca-binding proteins	Cytoplasm/ Nucleus	21–24 kDa

GFAP, glial fibrillary acidic protein; EAAT1, excitatory amino acid transporter ependymal cells, and Müller glia (Brightman and Cheng, 1988). "Bergmann glia" or Golgi epithelial cell, one of the astrocyte subtypes, is found mainly in cortical region of the cerebellum. The soma of such cell type is located in the Purkinje cell layer; they extend very long processes that end at the glia limitans of pia mater and large blood vessels. Also, there are astrocytes in white matter with more protoplasmic topology and with mixed fibrous and protoplasmic features.

On the basis of morphology and antigenicity, astrocytes from optic nerve cultures were designated as type 1 and type 2 (Raff et al., 1984). The type 1 astrocytes were originally defined as flat, polygonal cells that expressed GFAP but did not bind anti-ganglioside monoclonal antibodies A2B5 or R24 and LB1 except rat neural antigen 2 (Ran 2). These type 1 astrocytes proliferate well in the presence of epidermal growth factor and are found during gliogenesis in early developmental stage. On the other hand, type 2 astrocytes are found as $GFAP^{+}A2B5^{+}$ cells in rat optic nerve culture.

However, it is yet to know whether these morphologically distinct heterogeneous populations of astrocytes are also different in their function or such morphological differences are merely intrinsic.

8.4.3. Physiological Role of Astrocytes in the CNS

8.4.3.1. Maintaining CNS Homeostasis

8.4.3.1.1. Providing Structural Support

Astrocytes have long been considered as structural support cells of the brain. The anatomy of brain microvasculature shows that astrocytic end feet constitute an envelope around blood vessels (Kacem et al., 1998). Astrocytic processes are positioned beneath the pial membrane and the ependymal surface and thereby segregate the CNS parenchyma from external environment (Figure 8.1). The cytoplasmic processes of astrocytes form a close network around the synaptic complex and maintain synaptic integrity (Newman, 2003).

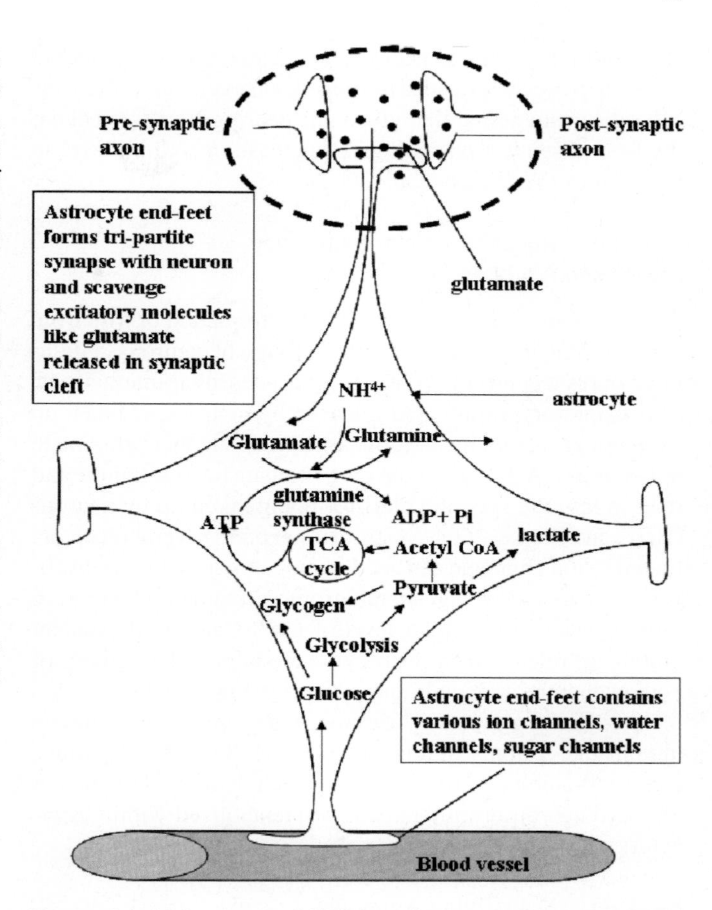

FIGURE 8.1. Maintenance of CNS integrity by astrocytes. Astrocytes endfeet forms a network on blood capillary and regulate transfer of water, ions and sugars. The pre-synaptic position of astrocytes is critical for the uptake of excitotoxic glutamate from neuronal synapse.

8.4.3.1.2. Maintaining Water Balance

Water is essential in the CNS for formation and maintenance of cerebrospinal fluid. Water enters into the CNS either through diffusion due to difference in osmotic pressure or through some specified channels. Astrocytes, through membrane-bound transporter system, maintain water and ionic homeostasis in the brain. The co-transporter system like "sodium-glutamate co-transporter" (Na^{+}-glutamate, EAAT1) and sodium-potassium-chloride ion co-transporter (Na^{+}-K^{+}-Cl^{-} co-transporter, NKCC) are located on astroglial membrane and regulate astrocytic water transport into the CNS (Figure 8.1).

Apart from the co-transporters, astrocytic perivascular system in the brain involves membrane-bound water channels, called aquaporins (Nagelhus et al., 2004). These water channels specifically mediate water fluxes within the brain. Among several members of the aquaporin family, aquaporin 4 and 9 are expressed in astrocytes. The activity of aquaporins is regulated by transmembrane G-protein coupled receptor (GPCR) family of hormonal receptors.

8.4.3.1.3. Maintaining IonHomeostasis

Astrocytes are responsible in maintaining extracellular K^{+} ion concentration at a level compatible with neuronal function. Astrocytes form a syncytium through which it efficiently redistributes K^{+} from perineuronal to perivascular space. Such redistribution of K^{+} is mediated by inwardly rectifying K^{+} ion channels. One such K^{+} ion channel Kir 4.1 is expressed in astrocytes surrounding neuronal synapses as well as blood vessels in the brain (Nagelhus et al., 2004).

In addition to having K^{+} channels, astrocytes bear plenty of other ion channels; function of many of them is still under research. Astrocyte cell surface bears an atypical sodium channel Nax that is assumed to be under voltage-gated sodium channel family. Nax is exclusively localized to perineuronal lamellate processes extended from ependymal cells and astrocytes. It has been suggested that glial cells bearing Nax channel are the first to sense a physiological increase in sodium level in body fluids (Watanabe et al., 2002).

8.4.3.1.4. Regulating Neurotransmitter and Amino Acid Levels

Astrocytes are active participants in the formation of tri-partite synapse and modulate synaptic activity of neurons. Glutamate plays a central role in astrocytic-neuronal interactions. This excitatory amino acid released by neurons, is taken up by astrocytes from the neuronal synapses via their glutamate transporters. Astrocytes convert glutamate into glutamine and release into the synaptic cleft for being taken up by neurons (Hertz and Zielke, 2004). Astrocytes express several receptors linked to ion channels and second messenger pathways. Activation of receptors e.g. metabotropic glutamate receptor, in turn, elevates intracellular level of Ca^{2+}. Calcium-dependent glutamate release from astrocytes modulates the activity of both excitatory and inhibitory synapses (Figure 8.1).

Apart from glutamate, astrocytes also uptake neurotransmitters like gamma amino butyric acid (GABA), aspartate, taurine, β-alanine, serotonin, and catecholamines. The fate of all these neurotransmitters is to be metabolized within astrocytes.

8.4.3.1.5. Detoxifying Ammonia

Ammonia is toxic for the CNS. Ammonia toxicity may result in neurological abnormalities leading to seizures, mental retardation, brain edema, convulsion, and coma. One of the most important enzymes that catalyze the formation of ammonia in the brain is glutamate dehydrogenase. This enzyme catalyzes reversible oxidative deamination of glutamate and produces ammonia particularly in astrocytes, thereby provides a mechanism for the removal of excess nitrogen from certain amino acids. Brain lacks carbamoyl phosphate synthase 1 and ornithine transcarbamylase, essential enzymes for the urea cycle, and thereby unable to remove accumulated ammonia (Cooper and Plum, 1987). However, astrocytes convert excess ammonia to glutamine via glutamine synthase (Figure 8.1). The excreted glutamine from astrocytes is taken up by neurons. In fact, either in physiological condition or even in hyperammonemic condition, rapid conversion of ammonia to glutamine in astrocytes is the predominant detoxification event in the CNS (Bak et al., 2006).

8.4.3.2. Supplying Energy

Glucose and ketone bodies are the primary source of energy in mammalian brain under normal physiological conditions. In comparison to its weight, which is only 2–3% of total body weight, brain consumes up to one fourth of body's total glucose supply.

8.4.3.2.1. Glycolysis

Astrocytes are the major food depot in the CNS. The food is stored in the forms of glycogen. Several studies suggest that glycogen phosphorylase and synthase are predominantly localized in astrocytes. Glucose is utilized in astrocytes mainly via glycolysis (Wiesinger et al., 1997). Deprivation of glucose in cultured astrocytes results in reduction in ATP/ADP ratio and membrane depolarization. Sugars enter into the metabolic pathway through phosphorylation which is considered as rate determining step. Astrocytes express hexokinase 1, the primary isoform of hexokinase in the CNS. This enzyme is mostly localized in mitochondria, only about 30% of it is found in cytosol.

8.4.3.2.2. Oxidative Metabolism

In order to generate energy in the form of ATP, sugars are bound to enter into oxidative metabolic pathway, the tri carboxylic acid cycle (TCA cycle) (Figure 8.1). Glycolysis generates 2 molecules of ATP and the TCA cycle generates 30 more ATP molecules from one molecule of glucose. The formation of energy in astrocytes is either through utilization of glucose under normal physiological condition or from reserve food storage glycogen via gluconeogenesis.

8.4.3.3. Organizing the Information Network in the CNS

As has been discussed earlier, astrocytes outnumber neurons by about ten to one in the CNS; and yet, historically they were considered to be a sort of glue (γλια) or connective tissues of the CNS. In the recent years though, it is increasingly clear that astrocytes form an integral and active component of the information network in the CNS and have received the "stardom" reflecting their morphology (Haydon, 2001; Nedergaard et al., 2003; Ransom et al., 2003). Astrocytes thus are a critical participant of "the tripartite synapse" (Araque et al., 1999; Perea and Araque, 2002). Indeed, not too long ago an entire book dedicated to "The Tripartite Synapse" that discussed in excellent detail, the anatomical and functional basis of neuroglial interactions, astrocyte calcium excitability, and the role of astrocyte in regulation of synaptic function (Volterra et al., 2002). Any discussion on astrocyte biology is thus incomplete without a consideration of their role in information transfer and intercellular communications in the brain. The details of the molecular mechanisms of this process via specific signaling events in health and disease will be discussed later in this book (see chapter by Pahan and Bidasee). This section highlights the current information on the glial communication networks of metabolite transport through gap junctions and the importance of calcium as a hallmark regulator of glial function.

8.4.3.3.1. Role of GAP Junctions

In the brain, astrocytes form a syncytium, or a network of integrated cells (Scemes, 2000). Such a syncytium consists of astrocytes that have cytoplasmic continuity in adjacent cells through gap junctions (Bennett et al., 2003). Gap junctions serve as a conduit between two astrocytes and consist of two hemichannels, also called connexones. These hemichannels or connexones are contributed by the juxtaposed cells and together form the gap junction (Figure 8.2). Hemichannels assembled in the endoplasmic reticulum consist of junctional proteins belonging to the family of connexins (Contreras et al., 2003; Saez et al., 2003). Although a variety of connexins are expressed by astrocytes, connexin (Cx) 43 is the predominant astrocyte connexin (Theis et al., 2005). These gap junctions form channels or pores about 1–1.5 nm in diameter and permit transfer of small metabolites including, but not limited to, NAD, ATP, glutamate and Ca2+ (Saez et al., 2005). Gap junctions formed with Cx43 are also permeable to dyes, such as Lucifer yellow or propidium, which serve as experimental tools for studies on gap junction function. The presence of gap junctions between astrocytes and the tripartite synapse consisting of the conventional synapse ensheathed by astrocyte processes together serve as models for neuroglial interactions. Indeed, evidence from astrocyte-neuron co-cultures has demonstrated that the presence of astrocytes in neuronal cultures increases the number of synapses and their efficiency. On the other hand, gap junctional communication and function can be regulated by neurons (Rouach et al., 2004). It is now generally accepted that astrocytes are likely involved in a variety of neurodegenerative diseases; however, alterations in the gap junction communication in pathological conditions, regulation of hemichannels and connexin expression and function is largely unresolved (Nakase and Naus, 2004).

8.4.3.3.2. Role of Calcium

While neurons are most prominently identified with their electrical excitability and astrocytes lack such electrical impulses, calcium waves that propagate through gap junctions have emerged as the parallel mechanism to that of the transfer of electrical impulse from one neuron to another. This of course by no means suggests that calcium communication is unique to astrocytes; indeed, such signals are commonly used by a variety of cells and neurons are no exception to this. In neurons, calcium signals lead to an instant integrated elecrical and chemical communication in synaptic cells. In both cultured astrocytes and astrocytes in intact brain slices, excitation of one cell can form a calcium wave transferred to several neighboring cells in multiple directions. This involves elevated calcium in a single cell followed by elevated intra-

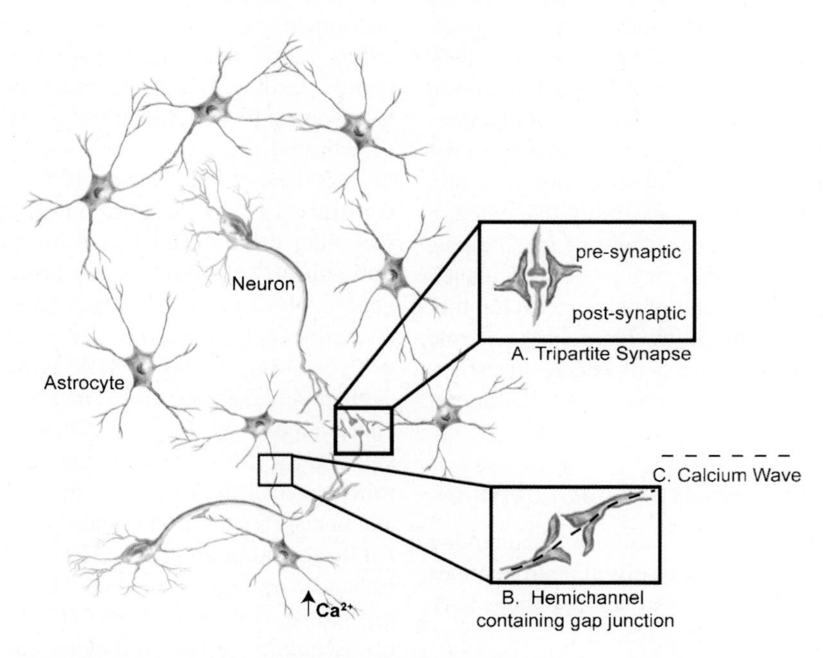

FIGURE 8.2. The astrocyte nexus: The astrocyte processes engulf the synapse forming what is now known as the tripartite synapse, consisting of pre- and post-synaptic elements as the two components along with the astrocyte ensheathment as the third. The brain astrocytes communicate with each other through intercellular connections forming a large network on interconnected cells. These cells join through gap junctions that form a channel through which small molecules can pass from one cell to the next. Calcium, one of the molecules that can pass through these gap junctions can lead to exponential transfers within the astrocyte nexus forming a calcium wave. Such connexin 43 containing functional gap junctions are primarily observed in the astrocytes in the brain.

cellular calcium in other cells. Such transfer of the calcium wave has been related to the Cx43 gap junction coupling of astrocytes both in vitro and in situ (Schipke and Kettenmann, 2004). Mobilization of intracellular calcium is also widely used by astrocytes as a prominent cell signaling mechanism in response to a variety of stimuli both in physiological and pathological conditions (Verkhratsky and Kettenmann, 1996). Calcium ions form an important cellular messenger and can provide an exquisitely sensitive mechanism for signaling, depending on their amount, distribution, amplitudes, and time course.

Even more exciting is that neuronal activity can stimulate such calcium communication in astrocytes and vice versa (Perea and Araque, 2005a). Thus, the role of calcium in the function of the tripartite synapse has received significant recent attention (Araque and Perea, 2004; Hirrlinger et al., 2004; Perea and Araque, 2005b). Ca^{2+} excitability of glia is observed in response to a variety of stimuli (Verkhratsky and Kettenmann, 1996). Such an elevation in intracellular calcium in a single astrocyte thus leads to the elevated calcium in the neighboring cells or the calcium wave described above. The complexity of calcium regulation in astrocytes is even greater, as has been revealed by recent studies showing that the calcium oscillations in a single cell may not even encompass the entire cell volume, but remain restricted to certain microdomains or certain processes within the astrocyte. Thus, we see the beauty of the autonomous functioning of a specific part of the cell that may have encompassed a syncytium of other cells or may ensheathe certain synapses, and provide at the same time the possibility of chemical coupling of the entire cell when given the appropriate stimulus (Carmignoto and Pozzan, 2002; Kettenmann and Filippov, 2002; Nett and McCarthy, 2002). The question as to what such calcium excitability of glia does in the neuroglial interactions is evolving. Certainly, from the example set by neurons where calcium signaling is tightly integrated with chemical release, the possibility that calcium excitability of astrocytes leads to the release of metabolites, which may in turn act in an autocrine or a paracrine manner, has also been investigated. The arena of calcium excitability of glia, their functional syncytium in the brain and their role in the tripartite synapse, all are subjects of intense investigation and will prove valuable in the complete understanding of neuroglial function.

8.4.3.4. *Releasing Neuropeptides and Neurotrophins*

Apart from its function as "support" cells by maintaining integrity of the CNS, astrocytes release several neuropeptides and neurotrophins. So far, four families of neuropeptides have been demonstrated to be expressed in astrocytes:

A. *Renin-angiotensin family*: The major function of renin-angiotensin system (RAS) in periphery is to maintain body-fluid homeostasis and regulate blood pressure.

B. *Endothelins*: Endothelins (ETs), a group of vasoactive peptides, acts as growth factor, exerting different functions like induction of proliferation, protein synthesis or changes in morphology. The ET1 also increases the rate of glucose 6-phosphate utilization via pentose phosphate pathway.

C. *Enkephalins*: The pentapeptide enkephalins are the ligand of orphan receptors in brain and are present mostly as precursor form in astrocytes.

D. *Neurotrophins*: Astrocytes are capable of releasing several growth factors and neurotrophic factors including epidermal growth factor (EGF), transforming growth factor (TGF), insulin like growth factor-1 (IGF-1), fibroblast growth factor-2 (FGF-2), brain-derived neurotrophic factor (BDNF), glial-derived neurotrophic factor (GDNF), and neurotrophin-3 (NT-3) (Wu et al., 2004).

8.4.3.5. *Facilitating Neurogenesis*

One of the key advances in the field of neurobiology is the discovery that astroglial cells can generate neurons not only during development, but also throughout adult life and potentially even after brain lesion. It has been shown that in neurogenic regions of adult brain (ventricular zone, hippocampus, and olfactory subependyma), astrocytes secrete factors like FGF-2, IGF-1, Shh, BDNF, GDNF, and NT-3 that induce neurogenesis (Altaba et al., 2002) and support the growth of neurons and neural progenitor cells (Wu et al., 2004).

8.4.4. Role of Astrocytes in CNS Disorders

8.4.4.1. *Activation of Astrocytes and Gliosis*

Recent evidence suggests that astrocytes might act as immunocompetent cells within the brain (Shrikant and Benveniste, 1996). Astrocytes react to various neurodegenerative insults rapidly, leading to vigorous astrogliosis. This reactive gliosis is associated with alteration in morphology and structure of activated astrocytes along with its functional characteristics (Eddleston and Mucke, 1993). The astrocytic processes construct a bushy network surrounding the injury site, thus secluding the affected part from the rest of the CNS area. Subsequently, astrogliosis has been implicated in the pathogenesis of a variety of neurodegenerative diseases, including Alzheimer's disease (AD), Parkinson's disease, inflammatory demyelinating diseases, HIV-associated dementia (HAD), acute traumatic brain injury, and prion-associated spongiform encephalopathies (Eng and Ghirnikar, 1994). Although activated astrocytes secret different neurotrophic factors for neuronal survival, it is believed that rapid and severe activation augments/initiates inflammatory response leading to neuronal death and brain injury (Tani et al., 1996; Yu et al., 1993). Enhanced up-regulation of GFAP is considered as a marker for astrogliosis (Eng et al., 1994). GFAP increases at the periphery of ischemic lesion following neurodegenerative insults (Chen et al., 1993). Senile plaques, a pathologic hallmark of Alzheimer's disease, are associated with GFAP-positive activated astrocytes (Nagele et al., 2004). It is reported that in various neuroinflammatory diseases, the increased GFAP expression corresponds to the severity of astroglial activation (Eng et al., 1992; Eng and Ghirnikar, 1994).

Recently our lab showed that various neurotoxins increase the expression of GFAP in astrocytes via nitric oxide (NO) (Brahmachari et al., 2006) suggesting that scavenging of NO may be an important mechanism in attenuating astrogliosis. Although the activation of NF-kB is involved in neurotoxin-induced production of NO in astrocytes, once NO is produced, it does not require the activation of NF-κB to induce the expression of GFAP (Figure 8.3). However, NO induces/increases the expression of GFAP in astrocytes via guanylate cyclase (GUCY)—cyclic GMP (cGMP)—protein kinase G (PKG) pathway (Figure 8.3).

8.4.4.2. Release of Pro-inflammatory Molecules

Upon severe activation in response to various neurodegenerative and neuroinflammatory challenges, astrocytes secrete various pro-inflammatory molecules including pro-inflammatory cytokines (TNF-α, IL-1α, IL-1β, IL-6, and lymphotoxin), chemokines, reactive oxygen species, reactive nitrogen species, and eicosanoids (Brosnan et al., 1994; Gendelman et al., 1994; Meeuwsen et al., 2003; Van Wagoner et al., 1999). These secreted pro-inflammatory molecules play an important role in the pathogenesis of various neurological disorders (Heales et al., 2004). In cultured murine astrocytes, bacterial lipopolysaccharides (LPS) act as a prototype inducer of various inflammatory responses. LPS is capable of inducing the expression of pro-inflammatory cytokines and inducible nitric oxide synthase (iNOS) in rat primary astrocytes (Pahan et al., 1997) but unable to induce the expression of iNOS in human astrocytes (Jana et al., 2005).

Among different pro-inflammatory cytokines (IL-1β, TNF-α, and IFN-γ) tested, only IL-1β alone is capable of inducing iNOS in human primary astrocytes (Jana et al., 2005). Similarly, among different cytokine combinations, the combinations involving only IL-1β as a partner are capable of inducing iNOS in human astrocytes (Figure 8.4). The combination of IL-1β and IFN-γ induces the expression of iNOS at the highest level in human astrocytes. Different pro-inflammatory transcription factors are involved in the transcription of iNOS in various cell types including astrocytes (Kristof et al., 2001; Liu et al., 2002; Pahan et al., 2002; Xie et al., 1994). All the three cytokines independently induce the activation of AP-1 while IL-1β and TNF-α but not IFN-γ induces the activation of NF-κB. However, among three cytokines, only IL-1β is capable of inducing the activation of CCAAT box/enhancer-binding proteinβ (C/EBPβ) (Figure 8.4) suggesting an essential role of C/EBPβ in the expression of iNOS in human primary astrocytes (Jana et al., 2005). In addition to pro-inflammatory cytokines, viral double-stranded RNA (Auch et al., 2004) and HIV-1 Tat also induce the expression of iNOS and the production of NO in human astrocytes (Liu et al., 2002).

8.4.4.3. Do astrocytes Present Antigen Under Autoimmune Response?

The CNS has long been known as "immunological privileged site" as it is secluded by BBB from peripheral immune system. However, this hypothesis is gradually becoming wrong. Microglia are capable of functioning as antigen-presenting cells (APC) as they express MHC I and II molecules (Carpentier et al.,

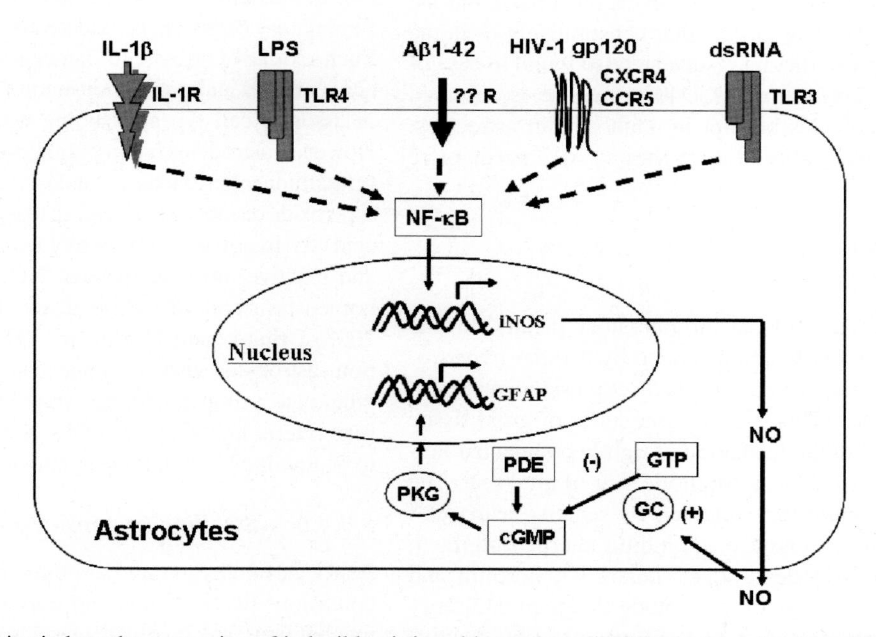

FIGURE 8.3. Various neurotoxins induce the expression of inducible nitric oxide synthase (iNOS) via the activation of NF-κB. Nitric oxide produced from iNOS then induces the activation of guanylate cyclase (GUCY) that catalyzes the production of cGMP. Inhibition of phosphodiesterase may also increase the level of cGMP. Cyclic GMP utilizes protein kinase G (PKG) to increase the expression of GFAP. IL-1R, IL-1 receptor; TLR4, toll-like receptor4; GPCR, G protein-coupled receptor; TLR3, toll-like receptor 3.

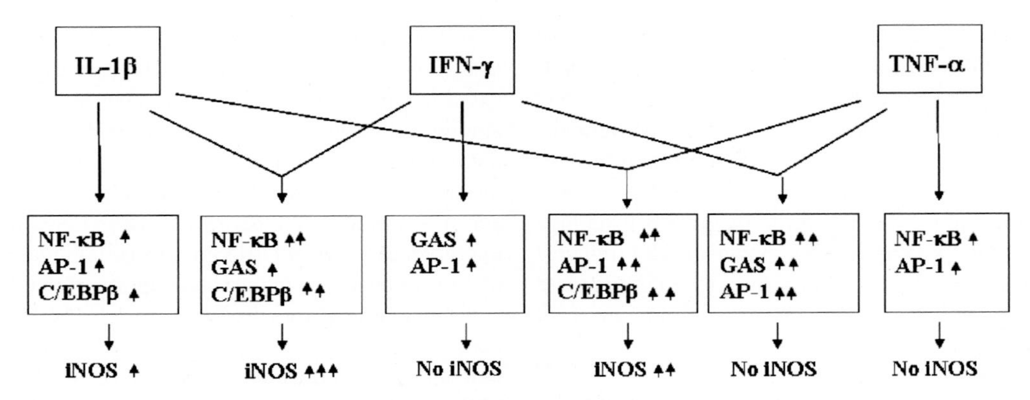

FIGURE 8.4. Expression of iNOS by various pro-inflammatory cytokines in human primary astrocytes. TNF-&bdotalpha;and IFN-γ alone or in combination are unable to induce the expression of iNOS. On the other hand, IL-1B alone or in combination with other cytokines induce iNOS in human astrocytes. Activation of AP-1 and GAS together by IFN-γ is not sufficient for the expression of iNOS. Activation of AP-1 and NF-κB together by TNF-α is also not sufficient for iNOS expression. Activation of AP-1, NF-κB and GAS together even at a higher level by the combination of TNF-α and IFN-γ compared to that induced by individual cytokines is also not sufficient for the expression of iNOS. However, Il-1β capable of activating C/EBPβ, AP-1 and NF-κB induced iNOS in human astrocytes suggesting an important role of C/EBPβ in the expression of iNOS in human astrocytes.

2005; Dong and Benveniste, 2001). In addition, microglia also expresses co-stimulatory molecules B7.1 and B7.2 molecules which play a role during antigen presentation. Another possible candidate as CNS APC is astrocyte. Expression of MHC II in astrocyte upon stimulation with IFN-γ or viruses has been demonstrated both *in vivo* and *in vitro*. However, capability of astrocytes as APC is still a controversial point. Examination of CNS tissues in MS, shows expression of B7-1 or B7-2 co-stimulatory molecules on macrophages and microglia but not on astrocytes. Human astrocytes also do not express co-stimulatory molecules B7-1 or B7-2. On the other hand, murine astrocytes express B7-1 or B7-2 either constitutively or in the presence of IFN-γ. Conflicting results are also found in case of CD40 expression. For example, CD40 expression is observed in fetal human astrocytes but not in adult human astrocytes. Therefore, functional ability of astrocytes as APC needs more research.

8.4.4.4. Formation of Glial Scar: A Double-Edged Sword

Astrocytes play a dual role in inflammatory insults. In one hand, activated astrocytes, characterized by cellular hypertrophy, proliferation and increased GFAP expression represent anisomorphic gliosis. This is the consequence of gross tissue damage and results in the formation of tightly compacted limiting glial margin termed as astrogliotic scar or glial scar. The pro-inflammatory molecules released by reactive astrocytes in the scar cause tissue damage and inhibit neurite outgrowth as well as induce oligodendrocytes death. Chondroitin and keratin sulphate proteoglycans are among the main inhibitory extracellular matrix molecules that are produced by reactive astrocytes in the glial scar and are believed to play a crucial part in failure to regeneration. On the other hand, isomorphic gliosis, formed in response to insult, results in improved recovery and regeneration of the damaged tissue (Eddleston and Mucke, 1993; Silver and Miller, 2004). At the sites distant from injury, activated astrocytes get transformed to a more pronounced stellate shape with increased production of anti-oxidants and soluble growth factors that coordinate tissue remodeling in enhancing the survival of adjacent neurons and glia.

8.4.4.5. Trying to Defend Neurons Against Oxidative Stress and Excitotoxic Damage

One of the hallmarks of various neurodegenerative and neuroinflammatory disorders is oxidative stress-induced CNS damage. Such oxidative stress can damage lipids, proteins and nucleic acids of cells and power-house mitochondria causing cell death in assorted cell types including neurons and oligodendroglia. However, astrocytes having high levels of anti-oxidant enzymes (glutathione peroxidase, catalase, glutathione reductase, and superoxide dismutase) and anti-oxidants (gluthathione and ascorbic acid) try to absorb reactive oxygen species ($O_2^{=}$, O_2^{-}, and OH·) and reactive nitrogen species (NO, $ONOO^-$), maintain redox homeostasis and defend the insulted CNS (Chen and Swanson 2003; Dringen and Hirrlinger, 2003; Wilson, 1997). In addition, astrocytes also scavenge detrimental molecules such as glutamate, produced during synaptic transmission through neurons (Hertz and Zielke, 2004). Astrocytes convert glutamate to glutamine by glutamine synthetase.

8.4.4.6. Swelling of Astrocytes

Astrocytes undergo rapid swelling in certain acute pathological conditions like ischemia and traumatic brain injury. Different mechanisms are involved in such swelling process of astrocytes. Some of these are, decreasing extracellular fluid osmolarity, intracellular acidosis, formation of ammonia, increase in Na^+, K^+, $2Cl^-$ co-transporter system, and due to drastically elevated levels of arachidonic acid and its metabolites. Alteration in glutamate metabolism and accumulation of glutamine and its transamination product, alanine is another possible cause of astrocytes swelling. In ischemic condition or in acute brain trauma, proton accumulation in cytoplasm cause astroglial cell swelling predominantly via activation of Na^+/H^+ and Cl^-/HCO_3^- exchangers.

8.4.4.7. *Undergoing Apoptosis Under Acute Insults*

Although astrocytes usually undergo proliferation and gliosis in various neurodegenerative disorders, under acute insults, astrocytes may undergo apoptosis. *Ex vivo* cell culture studies demonstrate that tumor suppressor protein p53 plays a role in neuronal as well as astrocyte apoptosis (Bonni et al., 2004). HIV-1 infection of the central nervous system (CNS) frequently causes dementia and other neurological disorders in which apoptosis of astrocytes along with neuron has been found. HIV-1 infection of primary brain cultures induces the receptor tyrosine kinase c-kit and causes apoptosis of brain cells including astrocytes (He et al., 1997). The importance of c-Kit in apoptosis of astrocytes has further been confirmed by overexpressing c-Kit in an astrocyte-derived cell line in the absence of HIV-1. The mechanism of c-kit induction by HIV-1 involves transactivation of the c-kit promoter by the HIV-1 Nef protein.

8.5. Oligodendrocytes: Biology and Function

Camillo Golgi was the first to give a good description of glia. A few years later, Cajal, student of Rio Hortega (1921) showed that there are two quite distinct cell types of neuroglia besides astrocytes using silver carbonate impregnation technique which he named oligodendrocyte (OL) and microglia. OLs are specifically the myelin-forming cells of the CNS.

8.5.1. Markers and Morphological Characteristics of Various Developmental Stages of Oligodendrocytes

Among different brain cells, the development of OL has been well characterized. During differentiation, oligodendrocyte lineage cells (early oligodendrocyte progenitors, oligodendrocyte progenitors, pro-oligodendrocytes, immature oligodendrocytes, and mature oligodendrocytes) (Figure 8.5) express stage-specific components that serve as markers of lineage progression (Table 8.2). Morphological characteristics of various developmental stages are shown in Figure 8.5.

8.5.2. Biological Role of Oligodendrocytes in the CNS

The major biological role of OL is myelination. However, OL may also promote neuronal survival, axonal growth and process formation. Neuronal function is also influenced by OL-derived soluble factors that induce sodium channel-clustering along axons. Neurotrophins (NGF, BDNF, and NT3) produced from OL may provide the trophic support for both OL and local neurons.

PSA-NCAM, polysialylated form of neural cell adhesion molecule; PDGFR-α, platelet-derived growth factor receptor α; MBP, myelin basic proteins; PLP, proteolipid protein; DM20, isoform of PLP; MOG, myelin/oligodendrocyte glycoprotein; MAG, myelin-associated glycoprotein; CNPase, 2′,3′-cyclic nucleotide 3′-phosphodihydrolase (Baumann and Pham-Dinh, 2001; Deng and Poretz, 2003).

8.5.2.1. *Myelinating CNS Neurons*

Myelination is a sequential multi-step process in which a myelinating cell recognizes and adheres to an axon, then ensheathes, wraps and ultimately excludes its cytoplasm from the spiraling process to form compact myelin. An OL is able to myelinate upto 40 axons depending on its localization. Myelin is composed of lipids and proteins, most of which are

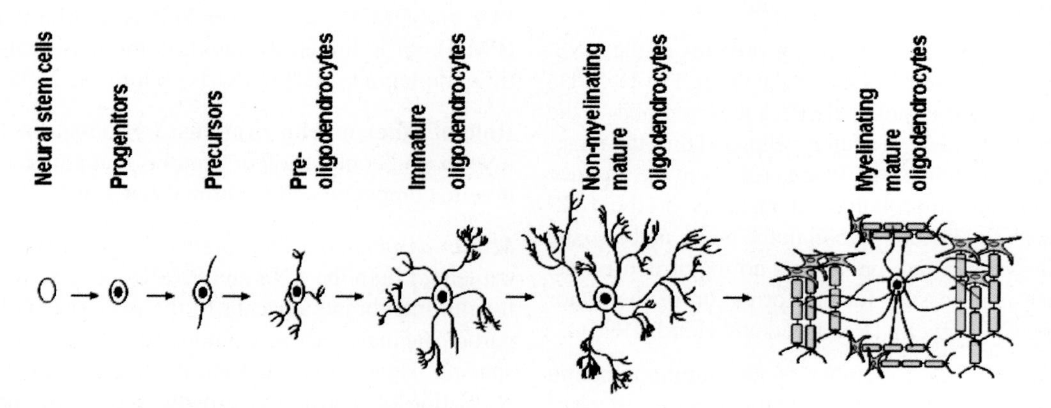

FIGURE 8.5. Different stages of oligodendroglial development.

TABLE 8.2. Stage-specific markers of oligodendrocytes.

Developmental stages	Markers	Detection	Characterization
Precursor	PSA-NCAM, Nestin, PDGFR-α,	Anti-PDGFR-α antibody	PSA-NCAM$^+$ /Nestin$^+$/
A2B5$^-$			
	DM-20		
Oligodendrocyte Progenitor cells(OPCs)	NG2/AN2+ proteoglycan, PDGFR-α protein or mRNA, GD3-related gangliosides, DM-20, CNPase	Anti-NG2 antibody, Anti-PDGFR-α antibody, A2B5 antibody	A2B5$^+$ /O4$^-$
Pro/pre-oligodendrocyte	PDGFRα, O4, GD3, NG2/AN2+, PLP/DM20, CNPase	Anti-NG2 antibody, Anti-PDGFR-α antibody, O4 antibody	A2B5$^+$ /O4$^+$
Immature oligodendrocytes	GalC, O4, CNPase, PLP/DM20	O4, O1, CNPase	A2B5-/MBP-/ R-mAb+
Mature oligodendrocytes	GalC, O4, CNPase, MBP, PLP, MAG, MOG	MBP, MOG, PLP, MAG	A2B5-/MBP+

specific for the myelin sheath. The major proteins are MBP, PLP, CNPase, and MAG. In the CNS, axonal factors play a critical role in the myelination process and thickness of the myelin, and that myelination in the CNS depends on a balance between positive and negative axonal signal (Sherman and Brophy, 2005).

8.5.2.1.1. Role of Proteins

Role of CNPase: CNPase was first identified in CNS myelin and it represents about 4% of the total myelin protein in the CNS. It possesses enzymatic activity that catalyses the hydrolysis of 2′, 3′-cyclic nucleotides into their corresponding 2′-neucleotides. However, to date any substrates of this enzyme has not been detected in the brain. Therefore, precise role of this enzyme in brain is unknown. However, it is one of the earliest myelination-specific polypeptides, synthesized by oligodendrocytes prior to the appearance of the myelin structural proteins (MBP and PLP) and its synthesis persists into the adulthood, suggesting a role in the synthesis and maintains of the myelin sheath. Over-expression of CNPase in transgenic mice perturbs myelin formation and creates aberrant OL membrane expansion. Recently, CNP knockout mouse study shows that CNP protein is required for maintaining the integrity of para-nodes and disruption of the axo-glial signaling at this site causes the progressive axonal degeneration (Rasband et al., 2005).

Role of MBP: MBP is one of the major proteins of the CNS, and it constitute about 25–30% of the total protein. The 18.5 kDa isoform of myelin basic protein (MBP) is the exemplar of the family, being most abundant in adult myelin, and thus the most-studied. Shiverer mutation of MBP gene results in the absence of MBP proteins and morphological analysis of the CNS reveals an almost total lack of myelin in the brain, and also the existing myelin is abnormal, presenting no major dense line. Therefore, MBP is necessary for the formation of the major dense line in the CNS myelin (Readhead and Hood, 1990).

Role of MOG: MOG is a member of the immunoglobulin super family, preferentially localized on the outside surface of myelin sheath and on the surface of OL process. Immunocytochemical studies demonstrate that the expression of MOG is late in OL differentiation compared with other major myelin proteins. It is used as a surface marker of oligodendrocyte maturation. This specific CNS protein is a minor component of myelin, constituting 0.01–0.05% of total myelin proteins. It is a 26–28 kDa integral glycoprotein and like other myelin proteins it may exist in multiple forms. Since the location of this protein in outermost surface of the myelin, it is easily accessible to a humoral immune response. MOG not only binds C1q but also may be the protein in myelin responsible for complement activation (Johns and Bernard, 1999).

Role of PLP/DM20: PLP is the most abundant intra-membrane protein and represents about 50–60% of the total protein in the CNS. It is localized predominantly in compact myelin. DM-20 and PLP arise from alternative splicing of a genomic transcript and differ by a hydrophilic peptide segment of 35-amino acids long, the presence of which generates the PLP product. PLP is necessary for normal myelin compaction, but the molecular mechanism for the adhesive function of these proteins is not known. In addition, it has been shown that PLP/DM20 play a metabolic role in maintaining axonal metabolism (Knapp, 1996) and also play an important role in the formation of intraperiod line and in maintaining axonal integrity. In human, mutation of PLP and DM20 gene causes Pelizaeus-Merzbacher disease (PMD), an X-linked dysmyelinating neuropathy, and spastic paraplegia type II (SPG-II) (Duncan, 2005).

Role of other myelin proteins: Besides these four proteins, myelin also contains other proteins that play a critical role in myelin compaction and neuronal function.

Myelin-associated glycoprotein (MAG): MAG is a minor constituent of both the CNS and PNS myelin. MAG found on the myelin membrane adjacent to the axon. MAG is believed to participate in axonal recognition and adhesion, inter-membrane spacing, signal transduction during glial cell differentiation, regulation of neurite out growth, and in the maintenance of myelin integrity.

Myelin associated/oligodendrocyte basic protein (MOBP): MOBP is abundantly expressed in the CNS myelin and shares several characteristics with MBP. MOBP is synthesized by mature OL and localized at the major dense line, suggesting a role in the myelin compaction process.

P2: P2, a basic protein with a molecular weight about 13.5 kDa, is located on the cytoplasmic side of the compact myelin membranes. It may serve as lipid carrier and thus could be involved in the assembly, remodeling and maintenance of myelin (Garbay et al., 2000).

Oligodendrocyte-specific protein (OSP/claudin-11): OSP/claudin-11 and PLP are both tetraspan proteins concentrated in CNS myelin. OSP represents about 7% of total myelin proteins. They possibly play an important role in myelin formation and maintenance due to their localization and concentration in membrane sheaths. Individual OSP/claudin-11- and PLP-null mice have relatively normal-appearing myelin and mild neurological deficits due to their compensatory role. However, when both OSP/claudin-11 and PLP genes are knocked out, mice show severe neurological deficits, markedly abnormal myelin compaction, and smaller axon diameters (Chow et al., 2005).

Cx32: Cx32, an integral membrane protein, is structurally related to PMP22 with four hydrophobic transmembrane domains. Recent studies show that it is also expressed on some areas of the CNS myelin and corresponding myelinating OL. Cx32 is preferentially expressed in oligodendrocytes in the CNS and in Schwann cells in the PNS. In addition to forming gap junctional channels, Cx32 also forms functional hemichannels. Mutation in Cx32 causes a common peripheral demyelinating neuropathy, X-linked Charcot-Marie-Tooth disease (Gomez-Hernandez et al., 2003).

Oligodendrocyte-myelin glycoprotein (OMgp): It is a glycosylated protein with molecular mass 120 kDa. It is located in the para-nodal areas of myelin. During injury, it inhibits the axonal growth by interacting with Nogo-66 receptor (NgR) complex.

Myelin/oligodendrocyte specific protein (MOSP): MOSP is a novel surface protein which is exclusively expressed in CNS myelin. It also plays an important role in membrane/cytoskeleton interactions during the formation and maintenance of CNS myelin.

8.5.2.1.2. Role of Lipids

In the CNS, lipids play an important role in myelin formation along with various protein molecules. One of the major biochemical characteristics that distinguish myelin from other biological membranes is its' high lipid-to-protein ratio. About 70–80% of the dry weight of myelin is comprised of lipid components and 20–30% protein. In every mammalian species, myelin contains cholesterol, phospholipids and glycolipids in molar ratios ranging from 4:3:2 to 4:4:2. In mature brain, cholesterol is the major lipid in myelin (about 20–25%) but generally normal myelin does not contain any cholesterol ester. Cholesterol helps to increase membrane thickness and fluidity as well as ion leakage through membranes which may be relevant to its property of electrical insulation.

Other abundant lipids in myelin are galactosylcerebrosides (Gal-C) and their sulfated derivatives (sulfatides). GalC represent 20% lipid dry weight in mature myelin. Immunological and chemical perturbation studies indicate that these lipids are involved in oligodendrocyte differentiation, myelin formation and myelin stability. These galactolipid-deficient animals exhibit severe tremor, hindlimb paralysis and display electrophysiological deficits in both CNS and PNS (Baumann and Pham-Dinh, 2001).

8.5.2.1.3. Molecules Involved in Positive and Negative Regulation of Myelination

The formation and maintenance of the myelin sheath require the coordination of a number of gene products. While some gene products facilitate myelination, some others try to suppress myelin formation. In the following lines, we describe such positive and negative regulatory mechanisms.

Molecules involved in positive regulation of myelination: OLIG1 and OLIG2 are closely related basic helix-loop-helix transcription factors that are expressed in myelinating OL and their progenitor cells in the developing CNS. Both OLIG1 and OLIG2 are positive regulators of myelination. Specifically OLIG1 has an essential role in oligodendrocyte differentiation and myelination, as it regulates the transcription of major myelin-specific genes MBP, PLP and MAG. On the other hand, OLIG2 is required for the initiation of oligodendrogliogenesis but its role in myelination is controversial (Xin et al., 2005).

Another important molecule that stimulates myelination is GPI-linked neural cell recognition molecule F3/contactin. This is a physiological ligand of Notch that signals via DTX1 to promote the development of OL (Hu et al., 2003; Popko, 2003). Additionally, F3 also transduces signals to glial intracellular Fyn, which then interacts with Tau protein to mediate myelination. As expected, different neurotrophins also favor myelination through the maintenance of oligodendroglial cell health and viability. For example, neurotrophin-3 (NT3) is known to induce both survival and proliferation of oligodendrocyts. NT3 interacts with TrkC to activate CREB that plays a critical role in proliferation and maturation of OPCs, and in the expression of myelin genes (MBP, P2, P0 MOG, PLP, and MAG) and anti-apoptotic gene Bcl-2. In addition, recent studies have identified many other molecules (e.g. PAX3, PPAR-δ, MyT1, SOX, GTX, Sp1, SCIP/Oct6/Tst-1) that may function as positive regulators of myelination (Wegner, 2000).

Molecules involved in negative regulation of myelination: Bone morphogenic proteins (BMP4s) should have a role in regulating bone density! Yes, they do have and in addition, these important molecules also regulate oligodendrocyte development. At early stage, BMPs regulate cell lineage decision and at later stage, they inhibit cell specialization in OL. For example, BMP4 signaling inhibits the generation of OL and enhances the generation of astrocytes from neural progenitor cells both *in vitro* and *in vivo*. BMP4 induces the expression of all four members of the inhibitor of differentiation (ID) family of helix-loop-helix transcriptional inhibitors and blocks oligodendrocyte lineage commitment through the interaction with OLIG1 and OLIG2 (Samanta and Kessler, 2004).

LINGO-1, a transmembrane protein containing a leucine-rich repeat (LRR) and immunoglobin domain, functions as a component of the NgR1/p75 and NgR1/Taj (Troy) signaling complexes. Recent studies show that LINGO1 is also expressed in OL where it negatively regulates oligodendrocyte differentiation and axonal myelination by down-regulating the function of Fyn kinase and up-regulating the activity of RhoA-GTPase. Lack of LINGO-1 expression promotes more axonal myelination due to increased expression of myelin gene such as MBP, CNPase and MOG in OL. LINGO-1 knockout mice also show earlier onset of CNS myelinatin (Mi et al., 2005).

8.5.3. Fate of OL in CNS Pathology

Death of OL and subsequent myelin loss has been reported in a variety of myelin disorders including, multiple sclerosis (MS), X-adrenoleukodystrophy (X-ALD), adrenomyeloneuropathy (AMN), vascular dementia, periventricular leukomalacia (PVL), hypoxia, and ischemia. Several factors that might be associated with OL death in these pathophysiological conditions are discussed below.

8.5.3.1. Role of Autoimmune Trigger in the Death of OL

MS and experimental allergic encephalomyelitis (EAE), an animal model of MS, are autoimmune diseases of the CNS mediated by T cells recognizing self-myelin proteins including MBP, MOG, and PLP. T cells are activated in the periphery by unknown antigens of both myelin and non-myelin origins. After activation, T cells cross the blood-brain barrier and invade into the brain where they accumulate and proliferate in response to antigen re-stimulation. These activated T cells secrete different pro-inflammatory molecules which stimulate not only the resident glial cells (microglia and astroglia) but also other infiltrating cells. $CD4^+$ and $\gamma\delta$ T cells express Fas-L which is found to be associated with oligodendroglial death. Furthermore, infiltrating $CD8^+$ T cells interact with MHC class 1 surface receptor of OL and in turn cause oligodendroglial lysis. T cell-derived perforin may also be responsible for oligodendroglial death (Scolding et al., 1990).

8.5.3.2. Role of Cytokines in the Death of OL

Cytokines are important mediators in the inflammatory demyelination observed in MS, EAE, X-adrenoleukodystrophy (X-ALD), and Theiler's virus infection. In these pathologies, pro-inflammatory cytokines and others factors released by endogenous glial cells and/or infiltrated macrophages and $CD4^+$ Th1 cells, accumulate and exert pleiotropic effects on OL. At lower concentrations, these cytokine may be involved in normal development of the nervous system while following brain trauma or inflammatory insults, the overproduction of these cytokines may result in a homeostatic imbalance and may contribute to the outcome of the pathological event. Various cytokines can directly kill OL or it may also affect other signaling pathways that could be involved in the susceptibility of OL. For example, IFN-γ produced by T cells may induce oligodendroglial apoptosis and cell death via JAK-STAT pathway. Another pro-inflammatory cytokine TNF-α induces oligodendroglial death via death signaling pathways (e.g. death inducing signaling complex (DISC), ceramide signaling pathway and stress-activated protein kinase pathways (SAPK) (Buntinx et al., 2004). IL-1 is a strong stimulus for TNF-α release from astrocytes and microglia. Both IL-1 and TNF-α are capable of inhibiting the expression of myelin genes via redox-sensitive mechanism (Jana and Pahan, 2005).

8.5.3.3. Role of Nitric Oxide in the Death of OL

Nitric oxide (NO), a short-lived and highly reactive free radical, is an important physiological messenger in the CNS. However, high level of NO in the CNS has been associated with different type of neurodegenerative diseases such as MS and EAE. During CNS inflammation, activated microglia, astrocytes and infiltrating cells express inducible nitric oxide synthase (iNOS) producing excessive amount of NO. OL at different stages of differentiation are differentially sensitive to NO. For example, OPCs and immature oligodendroctyes are more susceptible than mature OL to NO. However underlying mechanisms are poorly understood. It has been shown that NO reacts with superoxide generated by NADPH oxidase from activated glial cells and infiltrating cells to form peroxynitrite, the most reactive NO derivative. This peroxynitrite plays a critical role in the death of OL (Li et al., 2005).

8.5.3.4. Role of Oxidative Stress in the Death of OL

Reactive oxygen species (ROS) and reactive nitrogen species (RNS) leading to oxidative stress have been implicated as mediators of demyelination and axonal damage in both MS and EAE. Oxidative stress can damage lipids, proteins and nucleic acids of cells and mitochondria potentially causing OL cell death. During oxidative stress-induced oligodendroglial apoptosis, cytochrome c is released from damaged mitochondria, which in turn leads to the activation of the death-related caspases 3 and 9. Another study by Vollgraf et al. (1999) shows that mature OL exposed to oxidative stress undergo chromatin segmentation, condensation via mechanisms involving transcriptional activation of the immediate early stress genes (c-fos and c-jun). An induction of Bax protein has also been reported under oxidative stress condition in OL (Mronga et al., 2004).

8.5.3.5. Role of Ceramide in the Death of OL

Ceramide, the lipid second messenger and a hydrolyzed product of sphingomyelin, is involved in apoptosis of OL. In pathological conditions, pro-inflammatory cytokines (TNF-α and IL-1β) released specifically from activated microglia and astroglia leads to the activation of sphingomyelinases and production of ceramide in OL via the redox-sensitive mechanism (Singh et al., 1998). Furthermore, a direct role of oxidative stress-induced ceramide production via activation of neutral sphingomyelinase has been recently demonstrated in OL. During Alzheimer's disease, amyloid-β is aggregated in the plaque region that also causes OL death by activating the NSMase-ceramide cascade via redox-sensitive pathways (Lee et al., 2004). In addition, ceramide can inhibit inwardly rectifying K^+ currents and cause depolarization in OL (Hida et al., 1998).

8.5.4. Regeneration of OL

Recent evidences indicate that OPs remain intact both in normal white matter and in demyelinating CNS of patients with MS. OPs can give rise to new OL in experimental conditions and have the ability to repopulate areas from where they are missing (Franklin, 2002).

8.5.4.1. Molecules Involved in the Regeneration of OL

Recent identification of several genes associated with regeneration of OL has become helpful to understand the mechanism of remyelination and formulate a strategy for possible therapeutic intervention in demyelinating disorders. It has been shown that during demyelination, the expression of several genes such as, Nkx2.2, Olig1, Olig2, BMP4, and Fyn are increased in OPs (Lubetzki et al., 2005). Although functions of each of these genes are not clearly understood, some of these genes may lead to differentiation of the quiescent OPs to mature remyelinating OL and help oligodendrocyte regeneration in demyelinated areas (Zhao et al., 2005).

8.5.4.2. Role of Schwann Cells in the Regeneration of OL

During demyelination, after macrophages/microglia remove myelin debris and glial scar, SCs enter into the CNS where they remyelinate axons in the absence of reactive astrocytes. During this period, survival of SCs requires the axon-derived trophic factors (Zhao et al., 2005). Schwann cell can also produce some growth factors like IGF1, FGF2 and PDGF, which promote the migration of OPs and maturation into myelinating OL.

8.5.4.3. Role of Thyroid Hormone in the Regeneration of OL

Thyroid hormone (TH) plays an important role by regulating several stages of oligodendrocyte development and maturation. Oligodendrocytes express TH receptors and during demyelination, TH increases the expression of NGF. This NGF, in turn may lead to an increase in maturation of OPs and remyelination through the activation of Notch-Jagged signaling pathway. TH hormone also up-regulates the expression of PDGFR-α, MBP and CNPase in CNS tissues of animals with MS (Calza et al., 2005).

8.6. Schwann Cells (SCs): Peripheral Glia

The peripheral nervous system contains a number of distinct glial cells, each of which is intimately associated with different parts of the neurons or with specific neuronal cell types. Earlier, these cells were known as the supporting cells of the PNS but recent studies delineate their multifunctional role. These cells are of two types: satellite cells and Schwann cells (SCs). Satellite cells surround the neuronal cell bodies in dorsal root sensory ganglia and in sympathetic and parasympathetic ganglia. These cells help to maintain a controlled microenvironment around the nerve cell body, providing electrical insulation and a pathway for metabolic exchange. The other cells named after German physiologist Schwann are flattened cells with an elongated nucleus oriented longitudinally along the nerve fiber. Surface of all axons in peripheral nerves are ensheathed by non-myelinating or myelinating SCs.

8.6.1. Classification of Schwann Cells (SCs)

In the mature nervous system, SCs can be divided into three classes based on their morphology, biochemistry and function: myelinating Schwann cells (MSCs), non-myelinating Schwann cells (NMSCs) and perisynaptic Schwann cells (PSCs) (Figure 8.6).

MSCs are well characterized and they wrap around axons with a diameter of 1 μm or greater, including all motor neurons and some sensory neurons. This is a mystery why they wrap a specific diameter of the axons. Smaller diameter axons including many sensory and all post-ganglionic sympathetic neurons are myelinated by NMSCs. The NMSCs provide the metabolic and mechanical support to the axon. The NMSCs appear latter than MSCs. They express higher levels of GFAP, p75NTR and cell adhesion molecule L1 compare to MSC. The PSCs located at the neuromuscular junction incompletely wrap around the pre-synaptic terminal of motor axons. They help to maintain a stability of the neuromuscular junction and regulate synaptic transmission (Corfas et al., 2004).

8.6.2. Schwann Cell Development

SCs originate from the neural crest cells, a transient population of cells migrating away from the dorsal part of the neural tube. The signaling pathway and their detailed migratory route are not clearly known. Neural Crest cells are multipotent cells that differentiate to form neurons and glia of the PNS, and also additional cell and tissue types such as melanocytes and connective tissue of the head. Several molecules (e.g. ErbB3, transcription factor SOX10, AP2α and Ets1, the N-Cadherin 6, the low affinity receptor for nerve growth factor p75NTR) have been shown to play important roles during the detachment of neural crest from neural tube (Jessen and Mirsky, 2005).

Markers of lineage progression: Characterization of a number of specific biochemical markers has increased our knowledge on the stages of SC maturation, both *in vivo* and *in vitro*. Some of the biochemical markers have been shown to overlap partially (Table 8.3).

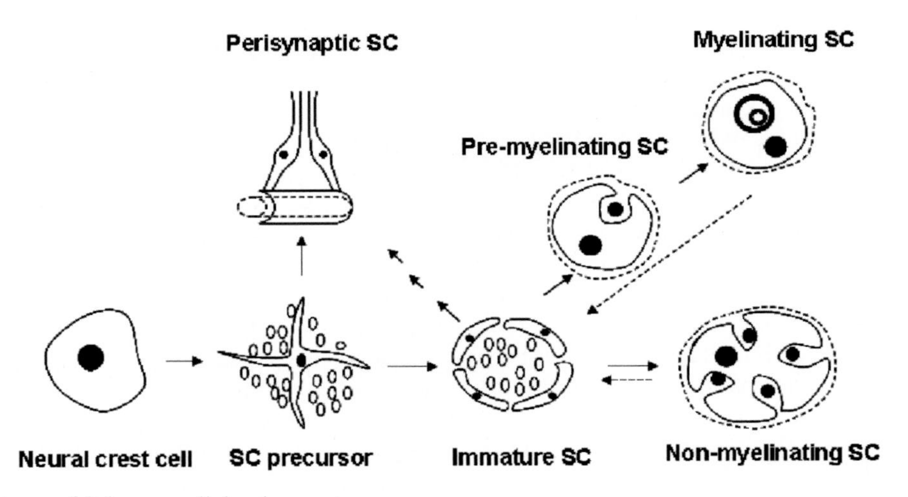

FIGURE 8.6. Different stages of Schwann cell development.

TABLE 8.3. Stage-specific markers of Schwann cells.

Stages	Markers
Neural crest cells	SOX10, AP2α
Schwann cell precursor (SCPs)	Cadherin19, AP2α, low level P0, GAP43, F-spondin, SOX10, BFABP, DHH
Immature Schwann cells	S100β, GFAP, low level of P0, SOX10, O4 antigen, BFABP, DHH
Myelinating Schwann cells	P0, PMP-22, MBP
Non-myelinating Schwann cells	NCAM, GFAP

SOX10, SRY (sex determining region Y) box 10; AP2α, activator protein 2α; DHH, desert hedgehog; GAP43, growth associated protein 43; P0, protein zero; O4, lipid antigen; BFABP, brain fatty acid-binding protein; S100, calcium-binding protein; PMP22, peripheral 22 kDa myelin protein, MBP, myelin basic protein; GFAP, glial fibrillary acidic protein; NCAM, neuronal cell-adhesion molecule.

SOX10, SRY (sex determining region Y) box 10; AP2α, activator protein 2α; DHH, desert hedgehog; GAP43, growth associated protein 43; P0, protein zero; O4, lipid antigen; BFABP, brain fatty acid-binding protein; S100, calcium-binding protein; PMP22, peripheral 22 kDa myelin protein, MBP, myelin basic protein; GFAP, glial fibrillary acidic protein; NCAM, neuronal cell-adhesion molecule.

8.6.3. Signaling Pathways Involved in Survival, Migration and Death of SCs

8.6.3.1. Survival

The survival of immature SCs in late embryonic and pre-natal nerves is probably controlled by a balance between factors that support survival and factors that cause death. Axon-derived neuregulin family (NRG-1, NRG-2, and NRG-3) have been implicated in the biological processes of SCs including fate specification, proliferation, survival, migration, regulating the extent of myelination, and triggering demyelination. It is believed that the interaction between several NRG ligands with different ERB receptors (ErbB2, ErbB3, and ErbB4) on SCs plays a critical role in regulating these steps (Garratt et al., 2000; Michailov et al., 2004). SCs can support their own survival by producing a number of growth factors such as IGF2, NT3, PDGF-β, LIF, and lysophosphatidic acid (LPA) in an autocrine fashion. The autocrine survival circuits are probably important in maintaining the survival of SCs in injured nerves. However, SCPs may need some signal for their survival from neurons.

8.6.3.2. Migration

During development of the PNS, neural crest cells migrate along the outgrowing axons and proliferate in order to produce sufficient number of cells for myelination of axons. Various factors or signaling molecules present on the neighboring cells effect Schwann cell migration in cell culture and it is possible that these signals lead to SCs movements during radial sorting *in vivo*. Integrins, a subgroup of adhesion receptors mediate interaction between cytoplasm and the extracellular environment. This interaction influences the migration of neural crest cells, axonal out growth and SCs differentiation. Integrins can interact with different growth factors, cell adhesion molecules (NCAM and F3) and intracellular cytoskeleton or adaptors proteins. This interaction is crucial for conformational changes and movement of SCs. Several growth factors that regulate the migration of SCs include NRG1, BDNF, GDNF, NT3, and IGF-1. The majority of these molecules are also expressed by SC itself (Yamauchi et al., 2005; Iwase et al., 2005).

8.6.3.3. Death

NGF acting via the p75 neutrophin receptor promotes cell death during the SC injury or infection via activation of c-jun-N-terminal kinase (JNK). It has been found that neonatal p75 neurotrophin receptor mutant mice are less prone to cell death after nerve transaction. Here it is to be noted that the same neurotrophin signaling pathway may also promote survival of

SCs via activation of NF-κB. Although mechanisms behind p75-mediated death of SCs are poorly understood, Yeiser et al. (2004) have shown that NGF signaling through the p75 receptor is deficient in TRAF-6 (–/–) mice and that NGF is unable to kill TRAF-6 (–/–) SCs. In addition, TGFβ is also known to cause apoptosis of SCs via JNK in culture (Jessen and Mirsky, 2005).

8.6.4. Differences Between OL and SC

Although both OL and SC share the common task of synthesizing myelin, there are some differences between the two cell types. See Table 8.4.

8.6.5. Biological Roles

8.6.5.1. Myelinating Peripheral Neurons

SCs cover most part of the PNS neurons by myelin sheath. Although the PNS myelin is mainly formed by the differentiation of the plasma membrane of SCs, myelination of mammalian PNS is a very complex developmental process. It requires intricate timing of several gene expression and cellular interactions between the axon and differentiated SCs (Michailov et al., 2004). In the PNS, mature SCs express Dhh, a family member of the Hh signaling proteins that is involved in the formation of peripheral nerve sheath and is also responsible for the formation of nerve-tissue barrier. Therefore, it has been found that Dhh mutant mice are defective in nerve barrier formation and unable to protect themselves against inflammatory responses. The activity of Dhh is regulated by several molecules such as Notch1, Hes5, MASH-1, and others (Parmantier et al., 1999). Another protein NDRG1 that is abundantly expressed in the cytoplasm of SCs rather than myelin sheath is also essential for maintenance of the myelin sheaths in peripheral nerves.

In addition, some well-known transcription factors such as, KROX20, NF-κB, SOX10, OCT-6, and POU class 3 homeobox 2 (POU3F2) also play an important role in PNS myelination. KROX-20, a master regulator for myelinating SCs, appears to be fundamental in controlling SC differentiation, regulating the expression of a number of genes including periaxin, P0, MBP, and PMP22 by interacting with NAB (NGF1-A binding) proteins. Mutation of this transcription factor Krox-20 is associated with lethal human neuropathy such as congenital hypomyelinating neuropathy (CHN), Dejerine-Sottas syndrome (DSS) and the Charcot-Marie-Tooth (CMT) disease. OCT-6 and POU3F2 have been implicated in the expression of Krox-20 and may therefore positively regulate myelination (Ghislain and Charnay, 2006; Mattson, 2003).

8.6.5.2. Tissue Repair/Regeneration

The SCs play a pivotal role during the event of mechanical damage such as spinal or peripheral nerve injury due to their regenerative properties. SCs in the distal stumps of adult animals can survive for several months in the absence of axons due to injury/insult and these SCs provide both trophic factors and adhesive substrates that promote axonal regeneration and restore the original function. After nerve injury, SCs can transform their phenotype from differentiated myelinating state to the de-differentiating state. During this process, there is also up-regulation of regeneration-associated genes such as the neurotrophin receptor p75 NTR, neuregulin and their receptors (erbB2, erbB3, erbB4), and GAP-43. They also produce different trophic factors (GDNF, TGF-β, IGF-2, NT3, PDGF-β, and LIF), adhesion molecules (L1, NCAM), extra-cellular

TABLE 8.4. Differences between oligodendrocytes (OL) and Schwann cells (SC).

OL	SC
1. OLs are present only in the CNS.	1. SCs are the major glial cells in the PNS.
2. The sub ventricular zone (SVZ), which is present in late gestational and early post-natal mammalian brain, is a major source of OL.	2. SCs are originated from the neural crest.
3. Oligodendroglial developmental steps are irreversible.	3. Fully differentiated SCs retain an unusual plasticity throughout the life and can readily de-differentiate to form cells similar to immature SCs.
4. One oligodendrocyte extends several processes and can myelinate upto 1 to 40 axons with distinct internodes. node.	4. One SC has an intimate association with axon and each SC forms myelin around a single axon, and lines up along the axon to define a single inter-
5. GM4, one of the most abundant lipids of the CNS, is present in OL.	5. Some glycolipids such as sulfated glucuronyl paragloboside and its derivatives are specific to SC.
6. CNS myelin contains more choline and plasmalogens than PNS myelin (Garbay et al., 2000).	6. PNS myelin contains more ethanol phosphoglycerides than CNS myelin.
7. Basic proteins (MBP and PLP) are major constituents of CNS myelin (about 80% of the total protein) (Baumann and Pham-Dinh, 2001).	7. Glycoproteins (P0 and PMP22) are major constituents of PNS myelin.
8. OLs have phagocytic activity.	8. SCs do not have phagocytic activity.
9. OLs migrate slower and divide and remyelinate at a slower rate than SCs.	9. SCs migrate faster and divide and remyelinate at a faster rate than OLs.
10. OLs are less resistant than SCs to injury.	10. SCs are more resistant than OLs to injury.

matrix molecules (laminin, tenascin), proteoglycans, and collagen type IV in an autocrine/paracrine manner, thereby providing a favorable environment for axonal re-growth and their own survival (Jessen and Mirsky, 1999).

Summary

Astrocytes, OL and SC are not silent partners of others anymore as thought a couple of decades earlier. Recent works have put these cell types in the forefront of neuroscience research. Although astrocytes being the major cell type in the CNS get more attention than the other two cell types, both OL and SC play an equally important role in human health and disease through myelination of neurons in the CNS and PNS respectively. As a result, thousands of cutting-edge research articles are coming out each year describing biological and functional aspects of astrocytes, OL and SC. Therefore, now it is an uphill task to compile everything about these three important cell types in a single chapter. However, here we have made an honest attempt to briefly delineate major biological and functional aspects of these cell types. Although there are vast body of evidence that implicate dysfunction and dysregulation of astrocytes, OL and SC in a number of human neurological diseases, we are still more or less in the dark to draw an unifying picture from these data. An improved understanding of their genesis and function in both healthy and diseased conditions is necessary for better preservation of brain in physiological conditions and for better repairing of this organ in pathophysiological situations.

Acknowledgements. This study was supported by grants from National Institutes of Health (NS39940 and NS48923), National Multiple Sclerosis Society (RG3422A1/1) and Michael J. Fox Foundation for Parkinson Research.

Review Questions/Problems

1. During developmental stage which cell types come last:

a. microglia
b. neuron
c. astrocyte
d. oligodendrocyte

2. If there is formation of abnormal Shh protein during embryogenesis:

a. genesis of astrocyte will be affected
b. genesis of oligodendrocytes will be affected
c. genesis of neurons will be affected
d. brain development will be impaired
e. all are true
f. none are true

3. Enzyme glutamine synthase is not found in

a. astrocytes
b. oligodendrocytes
c. none of the above is true
d. all are true

4. Tripartite synapse is formed by

a. neurons
b. microglia
c. astrocytes
d. microglia and astrocytes
e. neuron and astrocyte

5. In the central nervous system, major role of astrocyte is to

a. scavange glutamine
b. scavange cell debris
c. produce ATP
d. all the above
e. none the above

6. In the CNS, glycogen is found only in

a. oligodendrocyte
b. microglia
c. astrocyte
d. neuron
e. all the above

7. In astrogliosis, astrocytes form

a. cluster all around the CNS
b. bushy network surrounding the injury site
c. all are true

8. Schwann cells but not oligodendrocytes have phagocytic activity.

a. True
b. False

9. Unmyelinated axons generally have a smaller diameter than myelinated axons.

a. True
b. False

10. A single Schwann cell forms myelin around one and only one axon while a single oligodendrocyte forms myelin around several separate axons.

a. True
b. False

11. Oligodendrocytes progenitors are identified by A2B5 antibody whereas pre-oligodendrocytes are identified by O4 antibody.

a. True
b. False

12. Which of the sequential stage is correct for oligodendroglial development?

a. Neural stem cells, progenitors, pre-oligodendrocytes, precursors and mature oligodendrocytes.
b. Progenitors, precursors, pre-oligodendrocytes, mature oligodendrocytes and neural stem cells.
c. Neural stem cells, precursors, progenitors, pre-oligodendrocytes and mature oligodendrocytes.
d. Pre-oligodendrocytes, precursors, mature oligodendrocytes, progenitors and neural stem cells.
e. None of the above.

13. Action potentials are conducted rapidly through (choose one)

a. myelinated axons
b. unmyelinated axons
c. large diameter axons
d. small diameter axon
e. both a and c

14. Which of the following molecule is not a part of the peripheral nervous system?

a. LINGO-1
b. BDNF
c. CNTF
d. PDGF
e. EGF

15. Which of the following functions in the nervous system is not provided by the oligodendrocytes?

a. Ensheath axons
b. Supply neurotrophic factors
c. Form the node of Ranvier
d. Phagocytic properties to remove debris.

16. Non-myelinating Schwann cell is characterized by the following properties EXCEPT

a. Wraps axons greater than 1 μm.
b. Appear later than myelinating Schwann cells.
c. Myelinate all post-ganglionic sympathetic neurons.
d. Produce more p75NTR and GFAP.
e. Provide mechanical and metabolic support to the neuron.

17. Gliogenesis and Neurogenesis during development

a. occur simultaneously during development
b. follow this sequence i.e. gliogenesis followed by neurogenesis
c. occur one after the other after the first is completed
d. occur sequentially with overlapping periods

References

Altaba AR, Palma V, Dahmane N (2002) Hedgehog-Gli signaling and the growth of the brain. Nat Rev Neurosci 3:24–33.

Alvarez-Buylla A, Garcia-Verdugo JM, Tramontin AD (2001) A unified hypothesis on the lineage of neural stem cells. Nat Rev Neurosci 2:287–293.

Araque A, Perea G (2004) Glial modulation of synaptic transmission in culture. Glia 47:241–248.

Araque A, Parpura V, Sanzgiri RP, Haydon PG (1999) Tripartite synapses: Glia, the unacknowledged partner. Trends Neurosci 22:208–215.

Auch CJ, Saha RN, Sheikh FG, Liu X, Jacobs BL, Pahan K (2004) Role of protein kinase R in double-stranded RNA-induced expression of nitric oxide synthase in human astroglia. FEBS Lett 563:223–228.

Bak LK, Schousboe A, Waagepeters HS (2006) The glutamate/GABA-glutamine cycle: Aspects of transport, neurotransmitter homeostasis and ammonia transfer. J Neurochem 98:641–653.

Baumann N, Pham-Dinh D (2001) Biology of oligodendrocytes and myelin in the mammalian central nervous system. Physiol Rev 81:871–927.

Bennett MV, Contreras JE, Bukauskas FF, Saez JC (2003) New roles for astrocytes: Gap junction hemichannels have something to communicate. Trends Neurosci 26:610–617.

Bonni P, Cicconi S, Cardinate A, Vitale C, Serafino AL, Ciotti MT, Marlier LN (2004) Oxidative stress induces p 53-mediated apoptosis in glia: P53 transcription independent way to die. J Nerosci Res 75:83–95.

Brahmachari S, Fung YK and Pahan K (2006) Induction of glial fibrillary acidic protein expression in astrocytes by nitric oxide. J Neurosci 26:4930–4939.

Brightman MW, Cheng JHT (1988) Cell membrane interactions between astrocytes and brain endothelium. In: The Biochemical Pathology of Astrocytes. (Norenberg MD, Hertz L, Schousboe A, eds.), pp. 21–39. New York, NY: Alan R. Liss.

Brosnan CF, Battistini L, Raine CS, Dickson DW, Casadevall A, Lee SC (1994) Reactive nitrogen intermediates in human neuropathology: An overview. Dev Neurosci 16:152–161

Buntinx M, Gielen E, Hummelen PA, Raus J, Ameloot M, Steels P, Stinissen P (2004) Cytokine-induced cell death in human oligodendroglial cell lines II: Alterations in gene expression induced by interferon-γ and tumor necrosis factor-α. J Neurosci Res 76:846–861.

Calza L, Fernandez M, Giuliani A, D'Intino S, Pirondi S, Sivilia S, Paradisi M, Desordi N, Giardino L (2005) Thyroid hormone and remyelination in adult central nervous system: A lesson from an inflammatory-demyelinating disease. Brain Res Brain Res Rev 48:339–346.

Carmignoto G, Pozzan T (2002) Calcium oscillations as a signaling system that mediates the bi-directional communication between neurones and astrocytes. In: The Tripartite Synapse: Glia in Synaptic Transmission(Volterra A, Magistretti PJ, Haydon PG, eds), pp 151–163. New York: Oxford University Press.

Carpentier PA, Begolka WS, Olson JE, Elhofy A, Karpus WJ, Miller SD (2005) Differential activation of astrocytes by innate and adaptive immune stimuli. Glia 49:360–374.

Chen H, Cabon F, Sun P, Parmantier E, Dupouey P, Jacque C, Zalc B (1993) Regional and developmental variations of GFAP and actin mRNA levels in the CNS of jimpy and shiverer mutant mice. J Mol Neurosci 4:89–96.

Chen Y, Swanson RA (2003) Astrocytes and brain injury. J Cereb Flow Metab 23:137–149.

Chow E, Mottahedeh J, Prins M, Ridder W, Nusinowitz S, Bronstein JM (2005) Disrupted compaction of CNS myelin in an OSP/Claudin-11 and PLP/DM20 double knockout mouse. Mol Cell Neurosci 29:405–413.

Contreras JE, Saez JC, Bukauskas FF, Bennett MV (2003) Gating and regulation of connexin 43 (Cx43) hemichannels. Proc Natl Acad Sci USA 100:11388–11393.

Cooper AJL, Plum P (1987) Biochemistry and Physiology of brain ammonia. Physiol Rev 67:40–519.

Corfas G, Velardez MO, Ko CP, Ratner N, Peles E (2004) Mechanisms and roles of Axon-Schwann cell interactions. J Neurosci 24:9250–9260.

Deng W, Poretz RD (2003) Oligodendroglia in developmental neurotoxicity. Neurotoxicol 24:161–178.

Dong Y, Benveniste EN (2001) Immune function of astrocytes. Glia 36:180–190.

Dringen R, Hirrlinger J (2003) Glutathione pathways in the brain. Biol Chem 384:505–516.

Duncan ID (2005) The PLP mutants from mouse to man. J Neurol Sci 228:204–205.

Eddleston M and Mucke L (1993) Molecular profile of reactive astrocytes—implications for their role in neurologic disease. Neurosci 54:15–36.

Eng LF and Ghirnikar RS (1994) GFAP and astrogliosis. Brain Pathol 4:229–237.

Eng LF, Yu AC, Lee YL (1992) Astrocytic response to injury. Prog Brain Res 94:353–365.

Franklin JM (2002) Why does remyelination fail in multiple sclerosis? Nat Rev Neurosci 3:705–714.

Fujita S (2003) The discovery of the Matrix cell, the identification of the multipotent neural stem cell and the development of the central nervous system. Cell Struct Funct 28:205–228.

Garbay B, Heape AM, Sargueil F, Cassagne C (2000) Myelin synthesis in the peripheral nervous system. Prog Neurobiol 61:267–304.

Garratt AN, Britsch S and Birchmeier C (2000) Neuregulin, a factor with many functions in the life of a Schwann cell. Bioessays 22:987–996.

Gendelman HE, Genis P, Jett M, Zhai OH, Nottet HS (1994) An experimental model system for HIV-1 induced brain injury. Adv Neuroimmunol 4:189–193.

Ghislain J, Charnay P (2006) Control of myelination in Schwann cells: A Krox20 Cis-regulatory element integrates Oct6, Brn2 and Sox10 activities. EMBO Rep 7:52–58.

Gomez-Hernandez JM, Miguel MD, Larrosa B, Gonzalez D, Barrio L (2003) Molecular basis of calcium regulation in connexin-32 hemichannels. Proc Natl Acad Sci USA 100:16030–16035.

Gotz M, Barde Y (2005) Radial Glial cells: Defined and major intermediates between embryonic stem cells and CNS neurons. Neuron 46:369–372.

Haydon PG (2001) GLIA: Listening and talking to the synapse. Nat Rev Neurosci 2:185–193.

He J, DeCastro CM, Vandenbark GR, Busoglio J,Gabuzda D (1997) Astrocyte apoptosis induced by HIV-1 transactivation of the c-kit proto oncogene. Proc Natl Acad Sci USA 94:3954–3959.

Heales SJ, Lam AA, Duncan AJ, Land JM (2004) Neurodegeneration or neuroprotection: The pivotal role of astrocytes. Neurochem Res 29:513–519.

Hertz L and Zielke HR (2004) Astrocyte control of glutamatergic activity: Astrocytes as stars of the show. Trends Neurosci 27:735–743.

Hida H, Takeda M, Soliven B (1998) Ceramide inhibits inwardly rectifying K^+ currents via a Ras- and Raf-1-dependent pathway in cultured oligodendrocytes. J Neurosci 18:8712–8719.

Hirrlinger J, Hulsmann S, Kirchhoff F (2004) Astroglial processes show spontaneous motility at active synaptic terminals in situ. Eur J Neurosci 20:2235–2239.

Hu QD, Ang BT, Karsak M, Hu WP, Cui XY, Duka T, Takeda Y, Chia W, Sankar N, Ng YK, Ling EA, Maciag T, Small D, Trifonova R, Kopan R, Okono H, Nakafuku M, Chiba S, Hirai H, Aster JC, Schachner M, Pallen CJ, Watanabe K, Xiao ZC (2003) F3/contactin acts as a functional ligand for Notch during oligodendrocyte maturation. Cell 115:163–175.

Iwase T, Jung CG, Bae H, Zhang M, Soliven B (2005) Glial cell line-derived neurotrophic factor-induced signaling in Schwann cells. J Neurochem 94:1488–1499.

Jana M, Pahan K (2005) Redox regulation of cytokine-mediated inhibition of myelin gene expression in human primary oligodendrocytes. Free Radic Biol Med 39:823–831.

Jana M, Anderson JA, Saha RN, Liu X, Pahan K (2005) Regulation of inducible nitric oxide synthase in proinflammatory cytokine-stimulated human primary astrocytes. Free Radic Biol Med 38:655–664.

Jessen KR, Mirsky J (1999) Why do Shwann cells survive in the absence of axons? Ann N Y Acad Sci 883:109–115.

Jessen KR, Mirsky R (2005) The origin and development of glial cells in pheripheral nerves. Nat Rev Neurosci 6:671–682.

Johns TG, Bernard CC (1999) The structure and function of myelin oligodendrocytes glycoprotein. J Neurochem 72:1–9.

Kacem K, Lacombe P, Seylaz J, Bonvento G (1998) Structural organization of the perivascular astrocytes endfeet and their relationship with the endothelial glucose transporter: A confocal microscopy study. Glia 23:1–10.

Kettenmann H, Filippov V (2002) Signaling between neurones and Bergmann glial cells. In: The Tripartite Synapse: Glia in Synaptic Transmission (Volterra A, Magistretti PJ, Haydon PG, eds). New York: Oxford University Press, pp 139–150.

Knapp PE (1996) Proteolipid protein: Is it more than just a structural component of myelin? Dev Neurosci 18:297–308.

Kristof AS, Marks-Konczalik J, Moss J (2001) Mitogen-activated protein kinases mediate activator protein-1-dependent human inducible nitric oxide synthase promoter activation. J Biol Chem 276:8445–8452.

Lee JT, Xu J, Lee JM, Ku G, Han X, Yang DI, Chen S, Hsu CY (2004) Amyloid-beta peptide induces oligodendrocyte death by activating the neutral sphingomyelinase-ceramide pathway. J Cell Biol 164:123–131.

Li J, Baud O, Vartanian T, Volpe JJ, Rosenberg PA (2005) Peroxinitrite generated by inducible nitric oxide synthase and NADPH oxidase mediates microglial toxicity to oligodendrocytes. Proc Natl Acad Sci USA 102:9937–9941.

Liu X, Jana M, Dasgupta S, Koka S, He J, Wood C, Pahan K (2002) Human immunodeficiency virus type 1 (HIV-1) Tat induces nitric oxide synthase in human astroglia. J Biol Chem 277:39312–39319.

Lubetzki C, Williams A, Stankoff B (2005) Promoting repair in multiple sclerosis: Problems and prospects. Curr Opin Neurol 18:237–244.

Mattson MP (2003) Insulating axons via NF-kB. Nat Neurosci 6:105–106.

Meeuwsen S, Persoon-Deen C, Bsibsi M, Ravid JM, Noort JM (2003) Cytokine, chemokine and growth factor gene profiling of cultured human astrocytes after exposure to proinflammatory stimuli. Glia 43:243–253.

Mi S, Miller RH, Lee X, Scott ML, Shulag-Morskaya S, Shao Z, Chang J, Thill G, Levesque M, Zhang M, Hession C, Sah D, Trappy B, He Z, Jung V, McCoy JM, Pepinesky RB (2005) Lingo-1 negatively regulates myelination by oligodendrocytes. Nat Neurosci 8:747–750.

Michailov GV, Sereda MW, Brinkmann BG, Fischer TM, Haug T, Birchmeier C, Role L, Lai C, Schwab MH, Nave KA (2004) Axonal neuregulin-1 regulates myelin sheath thickness. Science 304:700–703.

Mronga T, Stahnke T, Goldbaum O, Richter-Landsberg C (2004) Mitochondrial pathway is involved in hydrogen-peroxide-induced apoptotic cell death of oligodendrocytes. Glia 46:446–455.

Nagele RG, Wegiel J, Venkataraman V, Imaki H, Wang KC, Wegiel J (2004) Contribution of glial cells to the development of amyloid plaques in Alzheimer's disease. Neurobiol Aging 25:663–674.

Nagelhus EA, Mathiisen TM, Ottersen OP (2004) Aquaporin-4 in the central nervous system: Cellular and subcellular distribution and coexpression with Kir 4.1. Neuroscience 129:905–913.

Nakase T, Naus CC (2004) Gap junctions and neurological disorders of the central nervous system. Biochim Biophys Acta 1662: 149–158.

Nedergaard M, Ransom B, Goldman SA (2003) New roles for astrocytes: Redefining the functional architecture of the brain. Trends Neurosci 26:523–530.

Nett W, McCarthy KD (2002) Hippocampal astrocytes exhibit both spontaneous and receptor-activated Ca^{2+} oscillations. In: The Tripartite Synapse: Glia in synaptic transmission (Volterra A, Magistretti PJ, Haydon PG, eds), pp 127–138. New York: Oxford University Press.

Newman EA (2003) New roles for astrocytes: Regulation of synaptic transmission. Trends Neurosci 26:536–542.

Pahan K, Jana M, Liu X, Taylor BS, Wood C, Fischer SM (2002) Gemfibrozil, a lipid-lowering drug, inhibits the induction of nitric-oxide synthase in human astrocytes. J Biol Chem 277:45984–45991.

Pahan K, Sheikh GF, Namboodiri AMS, Singh I (1997) Lovastatin and phenylacetate inhibit the induction of nitric oxide synthase and cytokines in rat primary astrocytes, microglia and macrophages. J Clin Invest 100:2671–2679.

Parmantier E, Lynn B, Lawson D, Turmaine M, Namini SS, Chakrabarti L, McMahon AP, Jessen KR and Mirsky R (1999) Schwann cell-derived Desert hedgehog controls the development of peripheral nerve sheaths. Neuron 23:713–724.

Perea G, Araque A (2002) Communication between astrocytes and neurons: A complex language. J Physiol Paris 96:199–207.

Perea G, Araque A (2005a) Synaptic regulation of the astrocyte calcium signal. J Neural Transm 112:127–135.

Perea G, Araque A (2005b) Glial calcium signaling and neuron-glia communication. Cell Calcium 38:375–382.

Popko B (2003) Notch signaling: A rheostat regulating oligodendrocyte differentiation? Dev Cell 5:668–669.

Raff MC, Abney ER, Millar RH (1984) Two glial lineages diverge prenatally in rat optic nerve. Dev Biol 106:53–60.

Ransom B, Behar T, Nedergaard M (2003) New roles for astrocytes (stars at last). Trends Neurosci 26:520–522.

Rasband MN, Tayler J, Kaga Y, Yang Y, Lappe-Siefke C, Nave KA, Bansal R (2005) CNP is required for maintenance of axon-glia interactions at nodes of Ranvier in the CNS. Glia 50:86–90.

Readhead C, Hood L (1990) The dysmyelinating mouse mutations shiverer (shi) and myelin deficient (shimld). Behav Genet 20:213–234.

Rio Hortega DP (1921) Histogenesis y evolucion normal; exodo y distribucion regional de la microglia. Memor Real Soc Esp Hist Nat 11:213–268.

Rouach N, Koulakoff A, Giaume C (2004) Neurons set the tone of gap junctional communication in astrocytic networks. Neurochem Int 45:265–272.

Saez JC, Contreras JE, Bukauskas FF, Retamal MA, Bennett MV (2003) Gap junction hemichannels in astrocytes of the CNS. Acta Physiol Scand 179:9–22.

Saez JC, Retamal MA, Basilio D, Bukauskas FF, Bennett MV (2005) Connexin-based gap junction hemichannels: Gating mechanisms. Biochim Biophys Acta 1711:215–224.

Samanta J, Kessler JA (2004) Interactions between ID and OLIG proteins mediate the inhibitory effects of BMP4 on oligodendroglial differentiation. Development 131:4131–4142.

Scemes E (2000) Components of astrocytic intercellular calcium signaling. Mol Neurobiol 22:167–179.

Schipke CG, Kettenmann H (2004) Astrocyte responses to neuronal activity. Glia 47:226–232.

Scolding NJ, Jones J, Compston DA, Morgan BP (1990) Oligodendrocyte susceptibility to injury by T-cell perforin. Immunology 70:6–10.

Sherman DL, Brophy PJ (2005) Mechanisms of axon ensheathment and myelin growth. Nat Neurosci Rev 6:683–690.

Shrikant P and Benveniste EN (1996) The central nervous system as an immunocompetent organ: Role of glial cells in antigen presentation. J Immunol 157:1819–1822.

Silver J and Miller JH (2004) Regeneration beyond the glial scar. Nat Neurosci 5:146–156.

Singh I, Pahan K, Khan M, Singh AK (1998) Cytokine-mediated ceramide production is redox sensitive: Implications to proinflammatory cytokine-mediated apoptosis in demyelinating diseases. J Biol Chem 273:20354–20362.

Spiryda LB (1998) Myelin protein zero and membrane adhesion. J Neurosci Res 15:137–146.

Tani M, Glabinski AR, Tuohy VK, Stoler MH, Estes ML, Ransohoff RM (1996) In situ hybridization analysis of glial fibrillary acidic protein mRNA reveals evidence of biphasic astrocyte activation during acute experimental autoimmune encephalomyelitis. Am J Pathol 148:889–896.

Theis M, Sohl G, Eiberger J, Willecke K (2005) Emerging complexities in identity and function of glial connexins. Trends Neurosci 28:188–195.

Van Wagoner NJ, Oh JW, Repovic P, Benveniste EN (1999) Interleukin-6 (IL-6) production by astrocytes: Autocrine regulation by IL-6 and the soluble IL-6 receptor. J Neurosci 19:5236–5244.

Verkhratsky A, Kettenmann H (1996) Calcium signalling in glial cells. Trends Neurosci 19:346–352.

Vollgraf U, Wegner M, Richter-Landsberg C (1999) Activation of AP-1 and nuclear factor-kappaB transcription factors is involved in hydrogen peroxide-induced apoptotic cell death of oligodendrocytes. J Neurochem 73:2501–2509.

Volterra A, Magistretti PJ, Haydon PG (eds) (2002) The Tripartite Synapse: Glia in Synaptic Transmission. New York: Oxford University Press.

Watanabe E, Hiyama TY, Kodama R, Noda M (2002) NaX sodium channel is expressed in non-myelinating Schwann cells and alveolar type II cells in mice. Neurosci Lett 330:109–113.

Wegner M (2000) Transcriptonal control in myelinating. Glia 31: 1–14.

Wiesinger H, Hamprecht B, Dringen R (1997) Metabolic pathways for glucose in astrocytes. Glia 21:22–34.

Wilson JX (1997) Antioxidant defense of the brain: Role for astrocytes. Can J Physiol Pharmacol 75:1149–1163.

Wu H, Friedman WJ, Dreyfus CF (2004) Differential regulation of neurotrophin expression in basal forebrain astrocytes by neuronal signals. J Neurosci Res 76:76–85.

Xie QW, Kashiwabara Y, Nathan C (1994) Role of transcription factor NF-kappa B/Rel in induction of nitric oxide synthase. J Biol Chem 269:4705–4708.

Xin M, Yue T, Ma Z, Wu FF, Gow A, Lu QR (2005) Myelinogenesis and axonal recognition by oligodendrocytes in brain are uncoupled in Olig1-null mice. J Neurosci 25:1354–1365.

Yamauchi J, Miyamoto Y, Tanoue A, Shooter EM and Chan JR (2005) Ras activation of a Rac1 exchange factor, Tiam1, mediates neurotrophin-3-induced Schwann cell migration. Proc Natl Acad Sci USA 102:14889–14894.

Yeiser EC, Rutkoski NJ, Naito A, Inoue J, Carter BD (2004) Neurotrophin signaling through the p75 receptor is deficient in traf6–/– mice. J Neurosci 24:10521–10529.

Yoon J, Gaiano N (2005) Notch signaling in the mammalian central nervous system: Insights from mouse mutants. Nat Neurosci 8:709–715.

Yu AC, Lee YL, Eng EL (1993) Astrogliosis in culture: I. The model and the effect of antisense oligonucleotides on glial fibrillary acidic protein synthesis. J Neurosci Res 34:295–303.

Zhao C, Fancy SP, Magy L, Urwin JE, Franklin RJM (2005) Stem cells, progenitors and myelin repair. J Anat 207:251–258.

9
Macrophages, Microglia, and Dendritic Cells

Anuja Ghorpade, Howard E. Gendelman, and Jonathan Kipnis

Keywords Antigen presenting cells; Microglia; Mononuclear phagocytes; Oligodendrocytes glycoprotein; Neural progenitor cells; Neurodegeneration; Neuroprotection; Reactive oxygen species; Toll-like receptors

9.1. Introduction

Celebrated as the "decade of the brain," the ten years between 1990 and 2000 gave rise to an increased awareness of brain function in health and disease. At the cellular level, none other than the microglia emerged as stars of the decade despite being only 10% of the total brain cell population. The profiles of microglia overpower, in part, neurons (Streit et al., 1999). In this chapter, we provide an overview of the diverse functions of microglia including their ontogeny, immune functions, and the spectrum of their responses during neurodegeneration. Microglia are the brain's mononuclear phagocytes (MP) and are influenced by their microenvironment (Streit, 2006). Phagocytosis, killing, secretion of bioactive factors, and antigen presentation are function shared amongst all MP including microglia. Such roles in both health and disease are developed and discussed, in some detail, in this chapter.

9.2. MP Ontogeny

MP includes a family of terminally differentiated cells with shared innate immunological function such as monocytes, macrophages, dendritic cells (DC), and microglia (Flaris et al., 1993). Originally, macrophages were considered part of the reticuloendothelial system, which was in 1969 replaced by the current MP designation. Macrophages in Greek, means "big eaters" principally referring to their phagocytic function in nonspecific or innate defense [reviewed by Fujiwara (Fujiwara and Kobayashi, 2005)]. Cells originate from circulating monocytes to ingest and destroy cellular debris and pathogens as well as stimulating lymphocytes to respond to antigens. The term macrophage, was used nearly a century ago by Elie Mechnikoff in 1882, when he discovered phagocytosis and explained its function, the first experimentally based theory in immunology. The concept was accepted in 1908 when Metchnikoff won the Nobel Prize and proved to be the first official recognition for the existence of immunology (Frolov, 1985). Nonetheless, it was nearly 12 years later when the contributions of W. Ford Robertson, Santiago Ramón y Cajal, Pio del Rio-Hortega, and Wilder Penfield developed the insights and discrimating histology that lead to the term microglia that was inevitably coined by del Rio-Hortega in 1919 (Kitamura, 1973). The word "mesoglia" was made for microglia in the beginning of the 20th century to dictate their unique origins as compared to neurons and macroglia. del Rio-Hortega was the one who conducted the first systematic studies on microglia and is considered the "father"" of these cell types although F. Nissl and F. Robertson made earlier observations of the cell's morphology. Many of Hortega's observations are valid even today (Hortega, 1919, 1932) including the fact that resting microglia show characteristic elongated, almost bipolar cell bodies with spine-like processes.

The debate about the nature and identity of microglial cells has continued over several decades. In the 1980s, new immunohistochemical markers and lectins were discovered that provided miroglia with improved histological characterization (Mannoji et al., 1986). It became clear with the advent of newer antigenic markers that microglia shared phenotypic characteristics and lineage-related properties with bone marrow-derived monocytes and macrophages of the MP-lineage system (Lee et al., 1992).

9.2.1. MP Differentiation and Development

Macrophages are renewed on an ongoing basis by the influx of monocytes from the circulation and by local division of immature MP that have arised from precursors in the bone marrow. The cells develop from monoblasts to promonocytes then to circulating monocytes. During microbial infections and an inflammatory response monocyte production increases

T. Ikezu and H.E. Gendelman (eds.), *Neuroimmune Pharmacology*.
© Springer 2008

from the bone marrow, a process regulated by colony stimulating factors produced from both macrophages and lymphocytes. The ontogeny of the macrophage are highly primitive and cells with MP properties can be found in the yolk sac (Moore and Metcalf, 1970). These cells express CD11b, the mannose receptor, and share other macrophage-like functions. In the bone marrow, the common myeloid precursor for macrophages and neutrophils is called colony forming unit-granulocyte macrophage (CFU-GM). This leads to the formation of monoblasts, the most immature cells of the MP system, which further develop into promonocytes and then monocytes. The newly formed monocytes remain in the bone marrow for about 24 h and then enter the peripheral blood. Bone marrow progenitors give rise to monocytes, macrophages, and DC (van Furth, 1982). There is an established lineage relationship between macrophages and DC. These two cell types have in common essential roles in the development, differentiation, and maintenance of tissue homeostasis. Perhaps even more important is their shared roles in tolerance, and regulation of inflammation and immunological responses. Circulating monocytes can also give rise to several tissue-macrophages, Among the cells now recognized as being derived from a monocyte pool are histiocytes (in connective tissues), Kupffer cells (in the liver), osteoclasts (in the bone), microglial cells (in the brain), synovial type A cells, interdigitating cells, alveolar cells (in the lung), osteoclasts (in bone), and Langerhans cells (of the skin) (Nibbering et al., 1987; Sluiter et al., 1987).

9.2.1.1. Histological and Immunohistochemical Characterization of Monocytes and Macrophages

Macrophages are generally large, irregularly shaped cells and contain a kidney-shaped or round nucleus (Figure 9.1). Monocytes usually have ample cytoplasm in contrast to lymphocytes. The surface of a macrophage is not smooth and presents a ruffled appearance. Differentiated cells usually bear microvilli that can be visualized with localization of vimentin, an MP-specific intermediate filament protein (Figure 9.1). Microglia, the resident central nervous system (CNS) macrophages, present a ramified morphology in the resting stages. Upon activation or injury, these cells can revert to an activated state and appear amoeboid in nature (Figure 9.1).

9.2.1.2. Microglia

Shown by lectin immunohistochemistry, microglia are present in the embryonic brain by 13 weeks of gestation (Billiards et al., 2006). Blood-borne monocytes migrate to the fetal brain and after the formation of blood–brain barrier, remain to form the resident population. There is functional and morphological variability in the microglia observed from 13 to 18 weeks of gestation indicating differential derivation (Chan et al., 2007). Ramified cells are found more in the cortex and amoeboid microglia, also known as brain macrophages, are found more in the germinal matrix. Moreover, in the first few weeks of the postnatal period, amoeboid microglia reduce in number giving

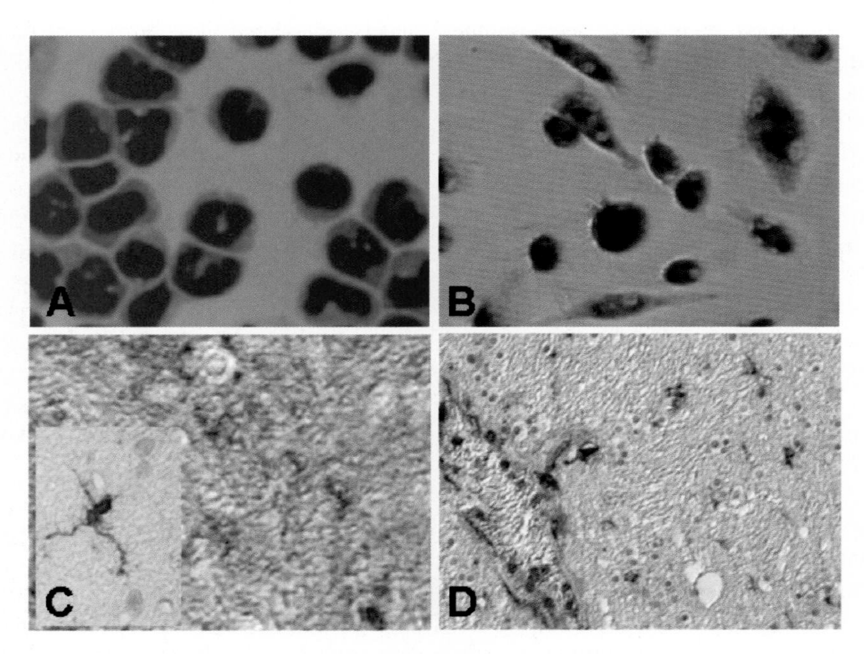

FIGURE 9.1. Morphological and immunohistochemical characteristics of monocytes and macrophages. Panel A demonstrates Wright's stain demonstrating the classical monocyte kidney-shaped nucleus. Panel B demonstrates the fine processes on microglial cells called the fimbriae. In the brain, microglial cells demonstrate a typical ramified phenotype as shown in panel C. In neuroinflammatory diseases, such as HAD, cells from the periphery infiltrate the brain as shown in panel D. A blood vessel in brain parenchyma is surrounded by perivascular macrophages.

rise to cells bearing longer, thinner, and branched processes (Monier et al., 2006). In addition to their capacity to proliferate in situ in response to injury, morphologically microglia appear as a distinct class of MP. The normal phagocytic functions of blood macrophages are considerably downregulated in ramified microglia; hence they are occasionally referred to as resting microglia. In response to injury or upon activation, they show a typical graded response and differentiate into immunologically active cells with functional and morphologic plasticity. Chimeric animal models have been extensively used to determine the myeloid origin of CNS microglial (Hickey et al., 1992; Lassmann et al., 1993; Hickey and Williams, 1996).

9.2.1.3. DC

DC are unique antigen presenting cells (APC) in that these cells can initiate the primary immune response. Both lymphoid and myeloid precursors can give rise to DCs. Evidence for the myeloid ontogeny of DCs was obtained when myeloid-committed precursors gave rise to mature DCs in vivo under the influence of granulocyte monocyte-colony stimulating factor (GM-CSF). However, transfer of lymphoid precursor cells in an irradiated mouse can give rise to DCs in addition to other cells of the lymphoid lineage, indicating their differentiation through the lymphoid lineage (Ardavin et al., 2001; Ardavin, 2003). In vitro, both human monocyte subsets can form DC when exposed to interleukin (IL)-4 and GM-CSF (Wu et al., 2001). Recently, a third population has been identified on the basis of CD64 expression. These cells resemble DCs and also have monocyte features (Tacke and Randolph, 2006).

Antigen presentation is essential for the maintenance of antigen-specific T cell responses in the nervous system (Groux et al., 2004; Karman et al., 2004). How this occurs remains incompletely defined. Although it is now well accepted that DC are the most potent APC type responsible for any primary immune response and occur through a strong and sustained ability to activate naive T cells the role of these cells in CNS immunity is hotly contested. The reasons are numerous. First, the differential phenotypes of DCs make their functional characterization in the nervous system difficult. Second, DCs are both capable of initiating the immune responses and interestingly, of peripheral tolerance. Third, DCs constitutively express major histocompatibility complex (MHC) class II molecules and although upregulation of these molecules in response to inflammatory stimuli differentiation of their role in health and disease is difficult. Fourth, DCs actively participate in the humoral immune response. CD11c as well as CD205 and MHC class II can tract DC but are not always exclusive. DCs serve as nature's adjuvants. DCs express the pattern-recognition receptors such as toll-like receptors (TLRs) and can strongly influence the quality and the robustness of the immune response. DC-based therapeutic vaccines and TLRs present attractive ways to harness the innate and adaptive immune response in combating neurodenerative disease.

9.2.2. MP-Specific Markers and Cellular Heterogeneity

Macrophages express a variety of surface markers that can be broadly classified into families of receptors including adhesion molecules, chemokines, lectins, advanced glycation endproducts (AGE), lipoprotein enzymes, cytokines, and Fc and complement. MP show well-developed enzymes including, but not limited to, nonspecific esterase expressed on the plasma membrane, acid phosphatase, alpha-glucuronidase, lysozyme, and peroxidase. Many of these are used for localization, identification, and isolation of MP. Table 9.1 provides a comprehensive list of these molecules. Several of the markers that are used for identification and localization of macrophages are denoted with an asterisk. Among the receptor families, scavenger receptors, glycosylphosphatidylinositol (GPI)-anchored receptors such as CD14, integrins (complement receptor), immunoglobulin superfamily (FcR), chemokine receptors, NK-like-C-type lectin (Dectin-1), C-type lectin (DC-SIGN), and the TLRs are important for immune recognition by macrophages (Gordon and Taylor, 2005).

Peripheral blood monocytes also exhibit heterogeneity in size, granularity, and shape of the nucleus. In addition, two distinct subpopulations of monocytes exist in circulation on

TABLE 9.1. Surface markers for macrophages/microglial.

Chemokine Receptors
CCR1, CCR2, CCR3, CCR5, CCR8, CCR9, CXCR1, CXCR2, CXCR4, CX3CR1
Complement Receptors
C3b, C3d, C3bi, C1q
Cytokine Receptors
Interleukins 1, 2, 3, 4, 6, 7, 10, 13, 16, 17
Interferon α, β, γ
Colony stimulating factors
Fc Receptors
IgG2a, IgG2b, IgG1, IgG3, IgA, IgE
Fibronectin & Laminin Receptors
Mannose, fucose, galactose
Integrins (CFA-1, CR3, CR4, VLA-4)
Toll-like receptors (Pattern Recognition)
CD4, CD14, CD16, CD45, CD68, CD163, HLA-DR

the basis of CD14 expression. CD14, a part of the lipopolysaccharide (LPS) receptor, is a surface marker used commonly for identification of monocytes. Differential expression of CD14 and CD16-the FcγRIII can segregate peripheral monocytes into $CD14^{hi}CD16^{-}$ and CD14+ CD16+ types of monocytes. The classic monocytes, $CD14^{hi}CD16^{-}$ cells, express higher amounts of MHC Class II and FcγRII molecules. CD14+ CD16+ monocytes express the chemokine receptor CCR5, a highly relevant receptor in human immunodeficiency virus (HIV)-1-infection of macrophages, whereas the $CD14^{hi}CD16^{-}$ express CCR2 (Gordon and Taylor, 2005).

Although most of the tissue macrophages arise from circulating monocytes, it has been shown that local proliferation of monocytes does contribute to this pool. DCs differ in the surface marker expression depending on their lineage. In mice, the two subsets express high levels of CD11c, MHC Class II, and the costimulatory molecules CD86 and CD40. The most commonly used marker to differentiate between the two DC populations is CD8α expressed only on the lymphoid lineage and not on the myeloid lineage. In humans, CD14+ and CD11c+ monocyes and lineage-negative $CD11c^{-}$ IL3Ra are the two precursor pools of DCs. The immunobiology of DCs, ontogeny and the different outcomes of the immune response as influenced by DCs are extensively reviewed by Banchereau et al. (2000).

9.2.3. Heterogeneity of CNS Macrophages and Microglia

The heterogeneity that is archetypal to the MP member prevails within the localized CNS environment. Microglia, perivascular macrophages, meningeal macrophages, and choroids plexus macrophages are some examples of the MP-originated cells in the brain. The classical microglial cells (Figure 9.1) are bone-marrow-derived, reside in the brain parenchyma, and exhibit an arborized or ramified morphology in the resting stage. Typically, microglia have a small soma and ramified processes. Although these cells harbor the ability to perform all macrophage functions, in homeostasis, these cells have severely downregulated most of these functions and appear to be resting brain macrophages. Perivascular macrophages (Figure 9.1) are those cells that line the blood vessels and do not migrate deep into the brain parenchyma. Meningeal macrophages and choroids-plexus macrophages describe the locations of these cells. Meninges, the membranes that surround the brain or choroids-plexus (the location of the blood and cerebrospinal fluid interface), have macrophages that comprise the CNS MP pool. Depending on the type of the microglial cell, the kinetics of its repopulation may be varied. For example, meningeal macrophages may be replaced by bone-marrow-derived monocytes rather rapidly, whereas perivascular and choroids-plexus microglia replacement is relatively slower.

Among the subpopulations of monocytes, CCR2+ Ly6C+ monocytes are defined as the inflammatory monocytes, which are recruited to the site of inflammatory injury. The responses of these "inflammatory" monocytes to different stimuli are intriguing. Depending on the stimulus, these monocytes may assume phenotypes that appear to be distinct in vitro. However, the existence of such phenotype plasticity in vivo, its significance, and whether these phenotypes are overlapping or mutually exclusive in vivo is currently unclear.

9.2.4. Markers to Distinguish Microglia from Other Brain Macrophages

Indeed, one of the challenges in developing transgenic mouse models with microglia-specific expression is the lack of a unique promoter that will express only in the microglia population. Thus, most of the strategies that attempt to distinguish microglia from other brain macrophages are based on quantitative differences rather than qualitative parameters (Flaris et al., 1993). The most commonly used method for isolation of microglial cells has been to enrich the $CD45^{low}$ population (parenchymal micrgoglia) from the $CD45^{hi}$ population of peripheral and other macrophages (Dick et al., 1995). For rat cells, ED2 is used as a unique marker for perivascular macrophages. Recently, the scavenger receptor, CD163 is proposed to be the human homolog of this protein however, additional evidence is needed to confirm this. In humans, monkeys, and mice, CD163 is expressed at high levels on all circulating monocytes and perivascular macrophages rather than parenchymal microglia in humans, monkeys, and mice (Kim et al., 2006). Specifically, the CD14+ CD16+ monocytes express high levels of CD163.

9.3. Macrophages and Microglia: Biology and Function

According to the dominant viewpoint of the scientific community, microglial cells first become present in CNS tissue during the embryonic and early postnatal phases of development (Barron, 1995; Kaur et al., 2001). During this time when the brain is dynamic and undergoing constant remodeling, microglia are believed to assist in the clearance of undesirable cells through apoptosis (Thomas, 1992; Polazzi and Contestabile, 2002). In the mature brain under physiological conditions, resting microglia adopt a ramified morphology and serve as resident immune cells with very slow turnover (Wierzba-Bobrowicz et al., 1995; Tanaka et al., 1998). Thus, microglia could be viewed as the first line of the innate immune mechanism in the brain. Macrophages and microglia, both MP, arise from hematopoietic stem cells in the bone marrow as described above. After passing through the monoblast and promonocyte states of the monocyte stage, they enter the blood. Then, they enter tissues and increase in size, phagocytic activity, and lysosomal enzyme content and become macrophages. The morphology of macrophages varies among different tissues and between normal and pathological states, and not all macrophages can be identified by morphology alone. However, most macrophages are large cells with a round or indented nucleus, a well-developed Golgi apparatus, abundant endocytotic vacuoles, lysosomes, phagolysosomes, and a plasma membrane covered with ruffles or microvilli. The functions of macrophages include nonspecific phagocytosis and pinocytosis, specific phagocytosis of opsonized microorganisms mediated by Fc receptors and complement receptors, killing of ingested microorganisms, digestion and presentation of antigens to T and B lymphocytes, and secretion of a large number of diverse products, including many enzymes (lysozyme, collagenases, elastase, and acid hydrolases), several complement components and coagulation factors, some prostaglandins and leukotrienes, and several regulatory molecules. In this section, we will review the basic functions of the macrophage as a part of the immune surveillance mechanism.

9.3.1. The Four R's of Immune Response

As the immune responses involving MP participate in host defense mechanisms both against invading organisms and against injury, they involve four major components that include: Recognition, Recruitment, Removal, and Resolution/Repair. Both innate and adaptive immunity is involved in the process of recognition. The pattern recognition concept for innate immunity has evolved into a well accepted scheme of how immune recognition by APC is carried out using so-called pathogen-associated molecular patterns (PAMPs) by a limited number of germline encoded APC pattern recognition receptors (PRRs). In contrast, the adapative arm of immune recognition involves a large repertoire of specific T and B cell receptors against specific peptides, generated by somatic recombination. The recruitment process of the immune response is largely mediated by the process of chemotaxis via chemokines generated at the site of injury. The recruitment of immune cells at the site of injury involves slowing of the cells in circulation, rolling against the vessel wall, binding to the wall close to the site of injury, followed by transmigration into the damaged tissues. This process plays a critical role in infiltration of peripheral cells into the brain during neurodegenerative diseases. Once at the site of injury, MP undertake the process of removal, which involves phagocytosis and intracellular clearance of the foreign bodies, apoptotic cells, and pathogens. These are discussed in some detail below.

9.3.2. Phagocytosis and Intracellular Killing

The phrase "to phagocytose" literally means "to eat." This is an important function of MP that is shared with neutrophils. Phagocytosis is a critical defense mechanism operational not only against a variety of invading microorganisms including bacteria, viruses, fungi, and protozoa but also in cleanup of debris in tissue microenvironment. The process of phagocytosis involves amoeboid movements of cellular pseudopodia from the approaching MP. Three predominant pathways including the Fc-receptor-mediated phagocytosis, complement receptor-mediated phagocytosis, and the mannose receptor-mediated phagocytosis exist (Aderem and Underhill, 1999). Ligation of these receptors to their specific ligands leads to cytoskeletal rearrangements in the MP that facilitate internalization of the particles. The presence of opsonins, which are recognition molecules that bind to both the invading particle and the MP, facilitate this process. Most commonly, immunoglobulin G (IgG) molecules and complement fragments serve as opsonins. This microglial function is deficient in the brain parenchyma in neurodegenerative diseases such as Alzheimer's Disease (AD), where deposits of amyloid beta (Aβ) are inefficiently cleared by the brain MP leading to senile plaque burden. Once the particles are surrounded and engulfed inside the MP, the process of intracellular clearance begins. The production of intracellular reactive oxygen intermediates (ROIs) is an important contributor to clearance as well as other enzymes. Fc receptor-mediated phagocytosis results in the release of large amounts of oxygen and arachidonic acid metabolites. Another mechanism is considered to involve charge compensation across the phagocytic vacuole membrane. The pumping of electrons into the phagocytic vacuole creates a charge that needs to be compensated. During this charge replenishment, the environment inside the vacuole favors enzymatic activity and microbial clearance (Segal, 2005). Additionally, this process of clearance does not occur until the phagosome with the engulfed particles fuses with the lysosomes to form the secondary lysosome also referred to as the phagolysosome. As mentioned above, MP in tissues are relatively long-lived and do not turnover rapidly. The long life span of most tissue macrophages allows for sustained phagocytic and clearance abilities in the repopulated microenvironments.

9.3.3. Antigen Processing and Presentation

MP are adorned with a remarkable ability to ingest a foreign moiety, digest it partially and present the broken down peptides in the context of the MHC proteins to the T cells. This whole process of antigen processing and presentation has been extensively reviewed in several articles (Frei and Fontana, 1997; Watts, 1997; Perry, 1998; Banchereau et al., 2000). Both MHC Class I and II molecules are thus presenting the protein environment of the MP for T cell surveillance. A majority of the peptides derived from foreign proteins are usually presented in the context of MHC Class II molecules rather than MHC Class I. Antigen processing is in part affected by factors intrinsic to the endosomal compartment, including pH, protease levels, and reducing potential. Other factors, such as the presence and vicinity of invariant chains for MHC Class II binding, interactions of the antigen with other proteins and modulation of lysosomal enzymatic activities may also influence antigen presentation. In essence, effective antigen presentation is achieved with a good balance between release of epitopes and their destruction. In addition, MHC Class II expression is necessary but not sufficient for presentation to T cells as additional costimulatory molecules are required for successful antigen presentation. The readout system for antigen presentation process is a functional one. The success of antigen presentation is usually measured as a T cell proliferation response. However, not all antigen presentation phenomena necessarily lead to proliferative responses and may be restricted to cytokine production without proliferation.

The brain presents a special case for MHC Class II-related antigen presentation since microglial cells, the largest of the brain MP population, do not express high levels of Class II. However, Class II expression on microglia is readily enhanced upon activation. Indeed, upregulation of MHC Class II or human leukocyte antigen (HLA)-DR is a sign of microglial activation and is commonly associated with CNS pathologies and with neural injury. Thus, undoubtedly for antigen presentation in the brain at homeostasis, the perivascular macrophages are more important as they constitutively express MHC Class II. A new paradigm for the role of microglial cells in early immune responses in the brain has emerged in the form of TLR activation and this will be discussed in the current opinion section below (Gordon, 2002).

9.3.4. Secretion of Immune Factors

The innate pro-inflammatory response of these cells is activated upon exposure to LPS, prostaglandin or other TLR ligands, leading to production of classical proinflammatory cytokines including tumor necrosis factor (TNF)-α. On the other hand, classical activation by interferon (IFN)-γ and LPS leads to production of TNF-α and also increased secretion of reactive oxygen species (ROS) and inducible nitric oxide synthase (iNOS).

Macrophages and all MP in general, are known to produce a plethora of immune factors that are secreted into the environment and can act in a paracrine and/or autocrine manner. The microenvironment affects timing, dosage, and the receptiveness of the target cells. Moreover, there are very few of these immune factors that are solely attributed to MP, and multiple other cell types express immune mediators. Nonetheless, several of these factors may work in concert with others and it is difficult to delineate individual versus combined effects. Thus, the varied biological effects of these factors are not mutually exclusive and thus the net effect of the response in the microenvironment may be unpredictable. Table 9.2 provides an overview of the major secretory families of MP. Among the proinflammatory cytokine family, TNF-α and IL-1 have the most pleitropic effects.

9.4. Microglia and Neurodegenerative Diseases

9.4.1. Overview

Microglia are very sensitive in detecting changes in their microenvironment and react promptly to any alteration. These cells become readily activated following injury, during a neurodegenerative process, or upon interacting with misfolded proteins or invading pathogens. (Nakajima and Kohsaka, 1998; Streit et al., 1999; Mrak and Griffin, 2005; Vilhardt, 2005). Upon activation, microglia quickly proliferate, become hypertrophic, and upregulate a variety of surface receptors, including CD11b and those involved in antigen presentation to T cells, that is, MHC Class I and II and costimulatory molecules. Activation of microglia also induces changes in their phenotype as resting ramified cells become amoeboid and retract their processes upon activation (Thomas, 1992; Gehrmann, 1996). These cells can interact with T cells, activate T cells, and produce cytokines for further boost of an immune response. Besides morphological changes and upregulation of surface receptors, activated microglia also produce a plethora of secreted factors such as growth factors, pro- and anti-inflammatory cytokines, ROS, nitric oxide (NO), and glutamate. As would be expected from the variety of factors produced, some can potentially promote neuronal survival whereas others exacerbate neuronal degeneration (Table 9.3). Microglial activation is incited not only by bacterial antigens and pro-inflammatory cytokines, but also by self-compounds that exceed physiological concentration or acquire an altered structure (e.g., aggregation). For example, Aβ peptides associated with senile plaques in AD induce activation of microglia (Sriram and Rodriguez, 1997; Rangon et al., 2003). A similar effect is obtained with a protein found in lewy bodies associated with Parkinson's disease (PD), α-synuclein (Iseki et al., 2000;

TABLE 9.2. Microglial/macrophage secretory factors.

Complement
C1, C2, C3, C4, C5
Cytokines
TNFα
IL1, IL6, IL-10, IL-12, IL-15
IFN-α, IFN-γ
TGF-β
M-CSF, GM-CSF
CXC Chemokines (IL-8, GROα, IP-10)
FGF, PDGF, BDNF
Arachidonic Acid Metabolites
Prostaglandin E2, leukotrienes
Platelet activating factors
Reactive Oxygen Species
Nitric oxide, peroxynitrite, superoxide
Hydrogen peroxide

TABLE 9.3. Microglial factors that affect neuronal protection/destruction.

Molecule	Effect on neuronal survival
NGF	Neuroprotective
BDNF	Neuroprotective
NT-3	Neuroprotective
NT-4	Neuroprotective
TGF-β	Neuroprotective
IL-10	Neuroprotective
IGF-1	Neuroprotective
VEGF	Neuroprotective
NO	Neurodestructive
H_2O_2	Neurodestructive
Glutamate	Neurodestructive
Superoxide	Neurodestructive
Hydroxyl radicals	Neurodestructive
Peroxynitrite	Neurodestructive

Kitamura et al., 2001; Zhang et al., 2005). Due to the rapid response of microglia to environmental changes and upregulation of its activation markers following injury and during most neurodegenerative conditions, activated microglia are often found in conjunction with damaged sites in postmortem human tissue and in animal models. Indeed, activated microglia were colocalized with neuritic plaques in the brain regions of AD patients (Benveniste et al., 2001; Hoozemans et al., 2001; Meda et al., 2001).

As microglial cells are of the hematopoietic lineage, they respond to bacterial antigens and acquire a phagocytic phenotype in vitro. Under such stimuli, microglia produce NO and glutamate. Thus, conditioned media obtained from activated microglia leads to some extent of neurotoxicity in cultured neurons. Based on these observations, activated microglia were suggested to be neurotoxic and their proximity to damaged neurons led to a theory for microglia-mediated neurodegeneration (Liu et al., 2002; Liu and Hong, 2003; Kadiu et al., 2005). However, while microglia can contribute to neurodegenerative processes through the release of a variety of neurotoxic compounds, it has been shown that microglia can also protect neurons from degeneration by secreting other compounds (Grunblatt et al., 2000; Liu et al., 2002; Morale et al., 2006). Therefore, it should be remembered that the inflammatory process leading to the injured or diseased nerve is not a process mediated by a single type of cell, but it is rather a complex process of cellular interactions. As such, a discovery of activated microglia in conjunction with injured tissue does not necessarily imply that these cells are a part of the pathology as they could also be a part of the physiological healing process.

9.4.2. Acute Injury

Acute injury initiates a process of secondary degeneration, where neurons that were undamaged by primary injury but located in close proximity to damaged neurons degenerate due to high levels of neurotoxic compounds in the environment (Yoles and Schwartz, 1998). Why does the microglial response following this type of injury attract so much scientific attention? The answer to this question is simple—microglia are the major class of cells that populate the site of injury. Astrocytes, the main nonneuronal cells in the CNS, leave injured tissue and create the glial scar, thus delineating the site of injury (Moalem et al., 1999; Butovsky et al., 2001). Therefore, in this model it appears that the cells that can provide trophic support to injured neurons are mainly microglia.

The microglial response to injury is very rapid and within several hours activated ED1 positive microglial cells can be clearly found in the injured sites (Moalem et al., 1999; Hauben et al., 2000; Butovsky et al., 2001). Activated microglia demonstrate features of specialized macrophages and acquire rod-like or "bushy" morphology as seen in the white matter or gray matter, respectively (Sorensen et al., 1996). Activated microglia engage in the same activity as most tissue macrophages including the killing of microorganisms, stimulation of inflammation, presentation of antigens, and liquefying and phagocytosing nonviable tissue. In injury activated cells express higher levels of MHC Class I and II molecules and show upregulation of complement type 3 receptor (CR3) and lectin binding proteins (Nelson et al., 2002).

Since activated microglia are implicated in exacerbating brain diseases, the biochemical markers and pathways that are stimulated by microglia in injury have been extensively studied. Several recent evidences clearly demonstrate that microglial phenotype after injury could be either destructive or protective (Butovsky et al., 2001; Nelson et al., 2002; Butovsky et al., 2005). In response to various stimuli, microglial cells will produce different sets of dominant compounds that differentially affect the fate of neural tissue. The partial list of common compounds can be found in Table 9.3.

One of the factors that controls microglial phenotype and determines the effect on neuronal survival is the adaptive immune response evoked by injury, namely the CD4+ T cells. T cells control not only the phenotype but also the microglial population size. In immune deficient rodents, microglial activation following injury is attenuated. Similar results are obtained in rodents with malfunctioning T cells. Moreover, a controlled T cell response changes the profile of molecules predominantly released by microglial cells (Shaked et al., 2004). Microglia activated following injury or in vitro by bacteria-derived antigens, such as LPS express high levels of pro-inflammatory mediators and possess a phagocytic phenotype. Following interaction with T cells, fewer neurotoxic factors are produced by microglia whereas production of neuroprotective growth factors and anti-inflammatory molecules is increased. Moreover, activation of iNOS in injury-activated microglia is high but can be down regulated upon interaction with T cells. Likewise, release of glutamate, one of the major neurotoxic agents produced by microglia, is altered as a result of interaction with T cells. Microglia that were activated by T cells reverse the glutamate transporters, which results in buffering rather than a release of glutamate (Butovsky et al., 2005, 2006a; Shaked et al., 2005). In normal brain tissue, glutamate buffering is mediated by astrocytes. Due to the lack of astrocytes at the site of injury, the role that microglia take in glutamate clearance presumably affects neuronal survival significantly. Microglia in the injured CNS tissue can be viewed as local guardians. If these guardians fail to correctly read incoming stress signals, then they will not develop the phenotype needed to fight off the threat, or alternatively, in fighting it off, the price they pay (in terms of death of neighboring neurons) is likely to outweigh the benefit (Schwartz et al., 2003, 2006).

9.4.3. Multiple Sclerosis

The role of activated microglia in neuronal survival during inflammatory conditions has been extensively studied in the models of experimental allergic/autoimmune encephalomyelitis (EAE) and in the animal model of Multiple Sclerosis (MS) (Antel and Owens, 1999; Becher et al., 2000; Biernacki et al., 2005; Jack et al., 2005). A rapid recruitment of blood-borne monocytes, an activation of resident microglia and perivascular macrophages and the recruitment of autoimmune T cells are among the most consistent changes observed in MS and EAE. The initial and subsequently developed neural lesions in MS are characterized by a T lymphocyte-dominated inflammatory response, although monocytes, macrophages, and B cells are also present. The large variety of symptoms commonly attributed to MS reflects the heterogeneity of the disease (Lassmann et al., 2001). Two main kinds of demyelinating processes, each with its specific pathology, may be seen (alone or in various combinations) in patients with MS. Macrophage-mediated demyelination demonstrates radial expansion of the lesions, which are composed of inflammatory infiltrates (T cells and macrophages). Activated macrophages and microglia are associated with myelin destruction, attributable in part to macrophage cytotoxicity caused, for example, by TNF-α or ROS (Probert et al., 2000). The second type of demyelination is antibody-mediated. The lesions are similar to those seen in macrophage-mediated demyelination, with evidence of immunoglobulins and activated complement at active sites of myelin destruction. The inflammation is T cell-mediated and is activated by macrophages/microglia, with complement-mediated lysis of antibody-targeted myelin. Similar lesions are found in animal models of EAE induced by active or passive immunization with epitopes of myelin oligodendrocyte glycoprotein (MOG). Additionally, T cells in combination with demyelinating anti-MOG antibodies (Linington et al., 1988) induce this type of demyelination.

As cells of the innate immune system, microglia serve as sensors of events occurring within the inflamed CNS. APC properties of microglia allow infiltrating immune cells to accumulate in the CNS (Nelson et al., 2002). The presence of antigen specific T cells at the sites of inflammation allows for bidirectional feedback in the CNS; thus, microglial-T cell regulation will play a crucial role in the propagation and resolution of inflammatory lesions. Monocytes infiltrating along with the T cells represent an important source of tissue macrophages that, once present in the CNS parenchyma, can fulfill parallel functions to microglia. Soluble products derived from autoimmune Th1 cells upregulate expression of MHC Class II, costimulatory molecules (CD80/86), CD40, and the adhesion molecule CD54 (ICAM-1) on microglia. Coculture of naive T cells with macrophages/microglia modulated by activated Th1 cells has been shown to promote a Th1-type phenotype in the responding T cells. Th1 soluble products also enhance the production of chemokines and cytokines (CXCL10/IP10, TNFα, and IL-6) by microglia (Sorensen et al., 1999; Flugel et al., 2001).

The acute CNS inflammatory response in MS and in EAE is self-limited, which is a contributing factor to the functional recovery that follows most clinical relapses (Antel and Owens, 1999; Jack et al., 2005; Prat and Antel, 2005). Histopathologic analyses indicate that a significant proportion of the infiltrating T cells (more than 70%) undergo programmed cell death and are subject to phagocytosis by microglia and macrophages. The rate of phagocytosis of apoptotic lymphocytes by microglia significantly exceeds that of nonapoptotic cells. Such phagocytosis down regulates the immune functions of microglia, as measured by the production of proinflammatory cytokines and T-cell priming/proliferation. The down regulation of APC functions in microglia after phagocytosis of cells programmed to die may represent an active mechanism to limit the inflammatory response that characterizes the acute MS lesion (Williams et al., 1994; Prat et al., 2001).

Microglia/macrophages are central components of the MS lesion throughout all phases of the disease course and they are armed with a battery of molecules capable of damaging cells. The acute MS lesion has been classically defined based on the presence of phagocytes ingesting myelin and of infiltrating lymphocytes. The chronic MS lesion, on the other hand, which is characteristic of the progressive phase of the disease, is dominated by the presence of activated microglia/macrophages without a prominent lymphocyte component. Myelin itself has been shown to bind and activate complement. Cross-linking of FcRs on microglia can induce an oxidative burst in these cells (Iribarren et al., 2002). The production of ROS after a phagocytic burst could represent an important source of toxic microglia-derived mediators in the CNS, particularly in combination with reactive nitrogen species, present on MS plaques due to upregulation of iNOS (Brosnan et al., 1994; Mitrovic et al., 1996; Gebicke-Haerter, 2001).

Recent studies attribute a novel role to microglia as both supporters and blockers of oligodendrocyte renewal from the endogenous neural progenitor cells (NPC) pool in the adult CNS (Butovsky et al., 2006a, b). The in vitro findings showed that microglia activated with IL-4, in part via production of Insulin-like growth factor (IGF)-I and down regulation of TNF-α, were remarkably potent in counteracting the impediment to oligodendrogenesis induced by high-dose IFN-γ. In vivo, IL-4-activated microglia supported oligodendrogenesis and clinical recovery in rodents in which severe inflammatory conditions are known to evoke clinical symptoms of transient or chronic EAE. Injection of IL-4 treated microglia into the CSF of animals with acute or chronic EAE caused an increase in the number of newly formed microglia. Most of the new microglia expressed MHC Class II and IGF-I (Butovsky et al., 2006a, b). Recent evidence supports the active participation of IGF-I in maintenance of the integrity and homeostasis of the CNS. This growth factor was shown, for example, to play an important role in the differentiation and survival of oligodendrocytes and to be beneficial in the treatment of EAE. It seems reasonable to assume that the IGF-I produced by IL-4-activated microglia is responsible, at least in part, for the shift to a Th2 phenotype and thus for the increased number of MHC Class II+ microglia expressing IGF-I. The increased oligodendrogenesis in mice correlated with a higher incidence of newly formed MHC Class II+ microglia (Butovsky et al., 2006a; Schwartz et al., 2006).

9.4.4. Alzheimer's Disease

Alzheimer's disease is accompanied by widespread loss of neurons and synapses in the brain. AD brains also show disruption of the blood–brain barrier and general metabolic deficit. The neuropathology of AD is mediated, at least in part, by amyloid plaques and neurofibrillary tangles (Durany et al., 1999).

Microglia are key players in the pathology and repair of AD (Eikelenboom et al., 2002; Rogers et al., 2002; Streit, 2004; Kim and de Vellis, 2005). Several laboratories have provided evidence that microglia demonstrate an apparently increased state of activation with age, thus providing partial explanation for the fact that aged brains are more vulnerable to developing lesions, which are composed of activated microglia. AD plaques are composed of a basic unit—Aβ peptide. In animals models of AD, microglia are a major component of amyloid plaques (Aisen, 1996; Marx et al., 1998; Marx et al., 1999; Bornemann and Staufenbiel, 2000; Emmerling et al., 2000). However, the fact that plaques can form in different ways may explain why some plaque types (dense-cored) are associated with microglial cells while others (diffuse) are mostly not. Beside Aβ, amyloid plaques contain many other substances such as lysosomal enzymes, cellular DNA, AGE etc., which are known to be sufficient to activate microglia. The hypothesis that microglia potentiate the toxicity of Aβ peptide is most directly substantiated by the physical proximity between amyloid fibrils and microglia (Streit, 2004; Streit et al., 2005).

According to the neuroinflammation theory of AD, the key pathological mechanism of AD is activation of the microglial cell. The neuroinflammation theory has originated from those studies that showed clustering of microglial cells within amyloid deposition in the human brains. These studies have been reinforced by numerous publications showing immunological activity of microglia. Proinflammatory molecules such as cytokines, complement components and MHC Class II receptors were detected in the AD brain in association with microglia. Studies on cultured microglia demonstrate that these cells can produce, in response to Aβ, a variety of neurotoxins (such as proteolytic enzymes, cytokines, complement proteins, ROS, N-methyl-D-aspartate (NMDA)-like toxins and reactive nitrogen intermediates). Several in vitro works proposed that Aβ may bind receptors for advanced glycation endproducts (RAGE) on microglia and thus stimulate intracellular signals leading to activation of transcription nuclear factor (NF)-κB. Early evidence from epidemiological studies supported the neuroinflammatory hypothesis, suggesting a beneficial effect of the prolonged use of nonsteroidal anti-inflammatory drugs (NSAIDs) in reducing the risk of developing AD (Scali et al., 2000; Hoozemans et al., 2001; Eikelenboom and van Gool, 2004; Gasparini et al., 2005).

Recent results showed that exposure of microglia to Aβ, reminiscent of exposure to LPS, resulted in a cytotoxic microglial phenotype. Cytotoxicity was associated with upregulation of TNF-α. The cytotoxicity induced by LPS or by Aβ could be partly overcome by pretreatment of the microglia with IL-4 *in-vitro* (Butovsky et al., 2005). Other studies have shown that the induction of TNF-α expression in microglial cells by Aβ is dependent on the fibrillary state of the peptide; thus, for example, Aβ (1–40) (which is nonfibrillar) does not induce TNF-α expression (Goodwin et al., 1995; Campbell et al., 2004; Rosenberg, 2005). Moreover, a recent study showed that the LPS receptor CD14, a key receptor for innate immunity, interacts with fibrils of Alzheimer's amyloid peptide. Neutralization of the receptor with anti-CD14 antibodies, or genetic deficiency of CD14, significantly reduces the microglial activation induced by these fibrils and attenuates the neurotoxic effect.

9.4.5. HIV-1-Associated Dementia

HIV-1 infects CD4+ cells of the immune system. Thus, in the brain, the primary targets of HIV-1 replication are the cells of the MP lineage. This includes microglia, perivascular macrophages, multinucleated giant cells formed post-viral infection and blood-monocyte-derived macrophages that infiltrate the brain during disease. The importance of activated microglia in the neuropathogenesis of HIV-1-associated dementia (HAD) cannot be understated. Indeed, virus-infected and/or immune activated MP secrete a variety of neurotoxic factors that affect neural function in HAD. Although monocytes and microglia remain the major cellular targets of HIV-1 in the brain, HIV-1 infection appears to be necessary, but not sufficient, to produce neurological disease (Nottet et al., 1996; Gendelman et al., 1997; Ghorpade, 1997). Thus, it is theorized that infected brain MP elicit disease through their secretory factors which induce both autocrine and paracrine cytokine production. These factors affect neural function through a complex circuitry of intercellular interactions between MP and various neural elements such as astrocytes, endothelial cells and neurons. The monocyte subset that expresses CD14 and CD16 is important in HIV-CNS disease (Fischer-Smith et al., 2001). This monocyte subset expands in the disease state and higher percentages of the CD14+ CD16+ monocytes correlate with HIV-1 disease progression. Similar to numbers of immune competent MP, activated microglia are known to produce a variety of cytokines that are upregulated in HAD. Pro-inflammatory cytokines include IL-1β, TNF-α, and IL-6 among others.

9.4.6. Parkinson's Disease

Parkinson's disease is a common neurodegenerative disorder characterized clinically by resting tremor, slowness of movements, rigidity, and postural instability. The pathology of PD is associated with a dramatic loss of dopamine containing neurons in the substantia nigra pars compacta (SNpc) and their termini within the striatum, resulting in subsequent loss of striatal dopamine (Vila et al., 2001; Barcia et al., 2003). The detection of elevated levels of pro-inflammatory cytokines and evidence of oxidative stress-mediated damage in postmortem PD brains led to the notion of microglial involvement in the degenerative process. Indeed, a consistent pathological feature of the disease is the chronic activation of microglial cells around degenerating neurons. With the exception of rare familial forms, the majority of PD cases are sporadic. Some epidemiological studies indicate a correlation between early-life brain injury and late development of PD, suggesting that inflammation in the brain, and specifically microglial activation plays a crucial role in the pathogenesis of the disorder, at least in its early stage. Additional evidence to support the notion of inflammatory processes in the brain as a pre-conditioning factor for development of neurodegenerative disorders, such as PD, is that people exposed to certain viruses or other infectious agents have an increased probability to develop PD or Parkinsonian-like syndromes (McGeer and McGeer, 1997; Vila et al., 2001; Teismann and Schulz, 2004; Hald and Lotharius, 2005; Nagatsu and Sawada, 2005).

Interestingly, the SN is an area extremely rich in microglia. As discussed above, microglia may serve as a source of ROS. Dopaminergic neurons in the SN are known to have a reduced antioxidant capacity, evidenced by a reduced level of the intracellular thiols, rendering them uniquely vulnerable to a variety of oxidative insults (Mirza et al., 2000; Koutsilieri et al., 2002; Hald and Lotharius, 2005). Therefore, it is reasonable to suggest that activation of microglia as a result of brain injury or infectious agent may serve as a risk factor and may trigger a cascade of events culminating in a massive death of dopaminergic neurons. There are several compounds that selectively damage dopaminergic neurons through microglial activation, such as rotenone (Gao et al., 2003), diesel exhaust particles (Block et al., 2004), and Aβ (Qin et al., 2002). Most of these toxins have a dual mode: they damage dopaminergic neurons via microglial activation at low concentrations, whereas at high concentrations, they induce a direct neuronal death. LPS, on the other hand, is toxic to dopaminergic neurons only via activation of microglia, and not directly. It should be kept in mind that the phagocytic phenotype of microglia induced by infectious agents and other compounds results in production of ROS, among other neurotoxic factors (listed in Table 9.3); thus, their neurodestructive effect is not restricted to dopaminergic neurons. Due only to the unique susceptibility of dopaminergic neurons to oxidative stress is the microglial neurotoxic effect so robust on these neurons.

Alpha-synuclein is a major component of lewy bodies, the morphological hallmark of PD. However, the pathophysiological role of this protein in degeneration of dopaminergic neurons is not understood. It was thought that alpha-synuclein exerts a direct damage on dopaminergic neurons; however, recent works demonstrated that microglia are capable of potentiating alpha-synuclein-induced neuronal degeneration. Alpha-synuclein leads to production of extra-cellular superoxide, increased production of ROS and induces morphological changes in microglia (Eells, 2003; Hishikawa et al., 2003). It should be mentioned, however, that microglia can also produce glial-derived neurotrophic factor (GDNF), a major factor shown to beneficially affect dopaminergic neurons (Connor, 2001; Chiang et al., 2005).

9.4.7. Glaucoma and Ocular Disorders

The origin of retinal microglia is controversial. Initial studies suggested neuroepithelial cells as an origin of microglia in the retina, whereas later studies provide evidence supporting a hematopoietic origin. Microglial precursors enter the developing retina via the ciliary margin prior to the organization of retinal vasculature and these cells differentiate into ramified parenchymal microglia. These cells express DC45, and MHC Class I and II molecules. In the developing rodent retina, most microglia are found at the vitreal surface at 12 days of gestation, and by 16 days of gestation, most microglial cells appear in a regular array across the retina. Microglia are largely confined to the ganglion cell layer and inner plexiform layer (Chen et al., 2002). While resting microglia are spread in different parts of the retina forming a network of cells with a potential immune effector function, their role in the normal retina is under-investigated. Microglia are considered to play a role in host defense against invading microorganisms upon injury and breakdown of retinal–blood barrier (Chen et al., 2002). Neurodegeneration activates microglial cells and facilitates their phagocytotic activity. Thus, these cells clean up cell debris from the damaged retina. However, an abnormal accumulation of activated microglia could also result in retinal dystrophy. Under inflammatory conditions in the eye, microglia serve as APCs and thus allow activation and proliferation of autoimmune T cells, which augment local inflammatory responses with potential destructive effects to adjacent neurons. Presentation of self-antigens by retinal microglial cells serves as a trigger for experimental autoimmune uveitis. Neuroprotective T cells, on the other hand, also require the presence of microglia at the site of injury to facilitate a neuroprotective response (Hauben et al., 2000; Butovsky et al., 2001; Kipnis et al., 2002). In glaucoma, for example, microglia function by clearing the debris and the deleterious breakdown products from degenerating retinal ganglion cells and their axons (Chen et al., 2002; Tezel and Wax, 2004). Boosting the T cell immune response further facilitates this beneficial effect and also leads to production of growth factors from microglia, which further facilitate neuroprotection (Schori et al., 2001; Bakalash et al., 2005). However, microglial cells also produce neurotoxic compounds and overactivation of microglia in the glaucomatous optic nerve head may be associated with immune-mediated neurodegeneration. In the retina, as with previously described neurodegenerative conditions, microglia serve a dual role in neuroprotection/neurotoxicity, which is determined by environmental factors and an adaptive immune response.

Retinal detachment associated with vitreoretinopathy is a complication of ocular trauma and vitreoretinal surgery. The mechanism leading to the detachment is not completely understood (Weller et al., 1991; McGillem and Dacheux, 1998; Valeria Canto Soler et al., 2002). Several studies suggest a role of MP, however; their origin was not detected. Recently a distinct population of proliferating cells, presumably of microglial origin, was identified. These studies suggested that several soluble factors [e.g., transforming growth factor (TGF)-β] produced by invading macrophages and local activated microglia might underlie the mechanism for development of vitreoretinal pathology (Weller et al., 1990). Microglia may also be involved in the pathogenesis of age-related macular degeneration, which is associated with hypertrophy of retinal microglia as well as with diabetic retinopathy, since elevated numbers of microglia are also present (Chen et al., 2002; Caicedo et al., 2005; McGeer et al., 2005).

9.4.8. Neurogenesis

In the adult brain, the subgranular zone of the hippocampus and subventricular zone of lateral ventricles are the two neurogenic active zones that throughout adult life give rise to new neurons and glia cells from NPC. It was found that irradiation blocks differentiation of neural stem cells in vivo due to an inflammatory response caused by activated microglia. Blocking inflammation, elicited by either irradiation or injection of bacterial LPS in the rat brain using nonsteroid anti-inflammatory drugs prevented impairment of hippocampal neurogenesis (Monje et al., 2003). Similar impairment of hippocampal neurogenesis by injection of LPS was prevented by minocycline, a drug that blocks microglial activation (Ekdahl et al., 2003a; Ekdahl et al., 2003b). Thus, inflammatory mediators/cytokines released by microglia during an immune response to injury or disease strongly influence neurogenesis and neuronal function (Kempermann et al., 2003; Kempermann and Kronenberg, 2003). Proinflammatory cytokine IL-6 is the key factor involved in inhibition of neurogenesis in vitro; administration of an antibody to IL-6 selectively blocked the IL-6 effect and restored hippocampal neurogenesis in vivo (Monje et al., 2003).

Other studies have associated microglial activation with improved neurogenesis (Butovsky et al., 2006a; Ziv et al., 2006). Indeed, an enriched environment was shown to facilitate endogenous neurogenesis (Kempermann et al., 1998a, b). Examination of brains from animals kept in an enriched environment demonstrated activated microglial cells located in close proximity to hippocampal NPC. These microglia expressed MHC Class II molecules and also showed immunoreactivity to IGF-1 (Ziv et al., 2006), a major factor facilitating neurogenesis. Interestingly, these microglial cells were shown to interact with T cells (of yet unknown antigenic specificity). Depletion of T cells (using T cell deficient mice) abolished activation of microglial cells, which led to abrogation of the enriched-environment-induced facilitation of neurogenesis (Ziv et al., 2006). Similar results were obtained with minocycline, which blocks microglial activation. Similar to neurodegenerative conditions, microglia may play a dual role in neurogenesis, which is dictated by their phenotype determined in turn by environmental clues.

9.5. Future Perspectives

9.5.1. The Neuroimmune Synapse

Gap junctions in the brain have been predominantly studied in astrocytes. Gap junctions provide continuity in the cytoplasm of neighboring cells. Such gap junctions provide astrocytes with cellular transfer mechanisms for small molecules and secondary messengers. Although classically microglia were not considered to be gap-junction coupled, recent evidence suggests that under certain conditions, such as in injury, connexin (Cx) proteins are upregulated in microglial cells and may provide cell-to-cell continuity and communication in the areas of injury. Gap junctions consist of hemichannels contributed by cells juxtaposed to each other. The Cx family of proteins is involved in the formation of gap junctions. Microglia have been shown to express both Cx43 and Cx36, members of the Cx family of proteins and these observations, have implications in the neural cell network in inflammation and injury (Eugenin et al., 2001; Takeuchi et al., 2006).

9.5.2. TLRs Signaling in Macrophages and Microglia

TLRs are some of the recently recognized PRRs important in the early responses to pathogens. The pattern recognition concept was originally proposed by Janeway (Medzhitov and Janeway, 2000). Indeed, in microglial cells, TLRs are presumed to be key sensors of infection and initiate the innate immune responses. TLR members are identified on the basis of the molecular patterns and ligands they recognize. Lipid structures are recognized by TLR2 whereas LPS is recognized by TLR4. Viral and bacterial nucleic acids are recognized by TLR3, 7, 8, and 9, whereas dsRNA is recognized by TLR3. The signal transduction of activated-TLRs is classified into two major pathways. One pathway involves the activation of NF-κB, which is a master switch for production of a variety of cytokines and effector molecules. The alternate activation pathway involves activation of microtubule-associated protein (MAP) kinase p38 and cJun NH2 terminal kinase (JNK). Most TLR (except for TLR3) signaling involves activation of a novel protein kinase IL-1 receptor associated protein kinase (IRAK) -1 and -4. The current discussion in the literature pertains to the mechanisms of TLR signaling during microglial responses to injury, and the potential of microglial TLRs to serve as rapid sensors and initiate innate immunity in the brain (Lee and Lee, 2002; O'Neill, 2006).

9.5.3. DC-Based Vaccines

The immune system thus applies both the innate and the adaptive immune responses as combined host defense mechanisms that operate in the periphery and in tissues including the brain. The main feature of adaptive immunity is its elasticity in generating diverse and yet specific immune responses.

However, the immune system is almost blind to all antigens that are provided in a soluble form; therefore, the use of adjuvants has a dramatic immune-stimulating effect. DC, the most powerful antigen-presenting cells, are capable of capturing, processing, and presenting antigens to T- and B-cells. This permits DC the capability for breaking self-tolerance. DC can stimulate cytotoxic T-lymphocytes to clear tumor or virus-infected cells. Vaccination with DCs can also affect clearance of tumors without eliciting autoimmunity (Zhang et al. 2007). They may also induce neuroprotective adaptive immune activities for models of neurodegenerative diseases (Schwartz et al., 2006).

Summary

Macrophages, microglia, and DC are the primary cell elements involved in the clearance and inactivation of microbial pathogens, in immune surveillance, and in homeostatic functions in all body tissues including the nervous system. Paradoxically, during disease these cells compromise are principal mediators of brain injury mediated through secretion of a variety of inflammatory neurotoxic factors and through precipitating adaptive immune responses. Modulation of macrophage and dendritic cell function mediated through environmental cues, microbial pathogens, or disordered protein structure or regulation leads to a wide range of pathological outcomes. The net result of a complex series of autocrine and paracrine amplification of immunoregulatory factors leads to intracellular events in neighboring neurons and glia that affect disease. Importantly, such events can now be controlled by immune intervention strategies that can prevent or slow neural tissue damage. Taming the function outcomes of macrophage and microglial activation is a frontier for research for neuroimmune pharmacology.

Review Questions/Problems

True/False

1. **Microglia are solely neuroprotective cell.**
2. **Microglia are solely neurodestructive cells.**
3. **Microglial activation is always associated with impaired neurogenesis.**
4. **Microglia are not found in the retina but only in the optic nerve.**
5. **Retinal detachment associated with vitreoretinopathy as a result of complication of ocular trauma is mediated solely by microglia.**
6. **Microglia could be a source of IGF-1 in the brain.**
7. **Microglia can clear amyloid plaques, but not very efficiently.**
8. **Th1 soluble products enhance the production of chemokines and cytokines (CXCL10/IP10, TNF, and IL-6) by microglia.**
9. **T cells in combination with anti-MOG antibodies induce demyelination.**
10. **Astrocytes are the major cells of interest in the mechanically injured CNS.**
11. **Which of the following describe the shared functions of mononuclear phagocytes?**
    a. Phagocytosis
    b. Intracellular killing of microbes and elimination of tissue debris
    c. Secretion of bioactive factors
    d. Antigen presentation
    e. All of the above
12. **Mononuclear phagocytes are a family of terminally differentiated cells with shared innate immune functions and include all of the following with the exception of?**
    a. Monocytes
    b. Macrophages
    c. Dendritic Cells
    d. Histiocytes
    e. CD4+ T lymphocytes
    f. Microglia
13. **The ontogeny of tissue macrophages includes the following in order of development?**
    a. Monoblasts to promonocytes to monocytes to tissue macrophages
    b. Promonocytes to monoblasts to monocytes to tissue macrophages
    c. Monoblasts to monocytes to histiocytes to Kupffer cells
    d. Monocytes to promonocytes to tissue macrophages
14. **All are histological features of monocytes and macrophages except?**
    a. Large, irregularly shaped cells with a kidney-shaped or round nucleus.
    b. Ample cytoplasm in contrast to lymphocytes
    c. Smooth surface
    d. Differentiated cells commonly bear microvilli that are visualized with localization of vimentin
    e. After activation cells revert to an activated state and appear amoeboid in nature
15. **Macrophages express a number of surface markers that can be classified broadly into family of receptors that include all of the following except?**
    a. Adhesion molecules
    b. Chemokines

c. Lectins
d. Advanced glycation endproducts
e. TCR

16. Describe the neuroinflammatory theory of Alzheimer's disease

17. How can microglial responses be modulated during neurodegenerative conditions?

18. Is there a dual role of microglia in health and disease? How does this occur?

References

Aderem A, Underhill DM (1999) Mechanisms of phagocytosis in macrophages. Annu Rev Immunol 17:593–623.

Aisen PS (1996) Inflammation and Alzheimer disease. Mol Chem Neuropathol 28:83–88.

Antel JP, Owens T (1999) Immune regulation and CNS autoimmune disease. J Neuroimmunol 100:181–189.

Ardavin C (2003) Origin, precursors and differentiation of mouse dendritic cells. Nat Rev Immunol 3:582–590.

Ardavin C, Martinez del Hoyo G, Martin P, Anjuere F, Arias CF, Marin AR, Ruiz S, Parrillas V, Hernandez H (2001) Origin and differentiation of dendritic cells. Trends Immunol 22:691–700.

Bakalash S, Shlomo GB, Aloni E, Shaked I, Wheeler L, Ofri R, Schwartz M (2005) T-cell-based vaccination for morphological and functional neuroprotection in a rat model of chronically elevated intraocular pressure. J Mol Med 83:904–916.

Banchereau J, Briere F, Caux C, Davoust J, Lebecque S, Liu YJ, Pulendran B, Palucka K (2000) Immunobiology of dendritic cells. Annu Rev Immunol 18:767–811.

Barcia C, Fernandez Barreiro A, Poza M, Herrero MT (2003) Parkinson's disease and inflammatory changes. Neurotox Res 5:411–418.

Barron KD (1995) The microglial cell. A historical review. J Neurol Sci 134(Suppl):57–68.

Becher B, Prat A, Antel JP (2000) Brain-immune connection: Immuno-regulatory properties of CNS-resident cells. Glia 29:293–304.

Benveniste EN, Nguyen VT, O'Keefe GM (2001) Immunological aspects of microglia: Relevance to Alzheimer's disease. Neurochem Int 39:381–391.

Biernacki K, Antel JP, Blain M, Narayanan S, Arnold DL, Prat A (2005) Interferon beta promotes nerve growth factor secretion early in the course of multiple sclerosis. Arch Neurol 62:563–568.

Billiards SS, Haynes RL, Folkerth RD, Trachtenberg FL, Liu LG, Volpe JJ, Kinney HC (2006) Development of microglia in the cerebral white matter of the human fetus and infant. J Comp Neurol 497:199–208.

Block ML, Wu X, Pei Z, Li G, Wang T, Qin L, Wilson B, Yang J, Hong JS, Veronesi B (2004) Nanometer size diesel exhaust particles are selectively toxic to dopaminergic neurons: The role of microglia, phagocytosis, and NADPH oxidase. FASEB J 18:1618–1620.

Bornemann KD, Staufenbiel M (2000) Transgenic mouse models of Alzheimer's disease. Ann NY Acad Sci 908:260–266.

Brosnan CF, Battistini L, Raine CS, Dickson DW, Casadevall A, Lee SC (1994) Reactive nitrogen intermediates in human neuropathology: An overview. Dev Neurosci 16:152–161.

Butovsky O, Hauben E, Schwartz M (2001) Morphological aspects of spinal cord autoimmune neuroprotection: Colocalization of T cells with B7—2 (CD86) and prevention of cyst formation. FASEB J 15:1065–1067.

Butovsky O, Talpalar AE, Ben-Yaakov K, Schwartz M (2005) Activation of microglia by aggregated beta-amyloid or lipopolysaccharide impairs MHC-II expression and renders them cytotoxic whereas IFN-gamma and IL-4 render them protective. Mol Cell Neurosci 29:381–393.

Butovsky O, Ziv Y, Schwartz A, Landa G, Talpalar AE, Pluchino S, Martino G, Schwartz M (2006a) Microglia activated by IL-4 or IFN-gamma differentially induce neurogenesis and oligodendrogenesis from adult stem/progenitor cells. Mol Cell Neurosci 31:149–160.

Butovsky O, Landa G, Kunis G, Ziv Y, Avidan H, Greenberg N, Schwartz A, Smirnov I, Pollack A, Jung S, Schwartz M (2006b) Induction and blockage of oligodendrogenesis by differently activated microglia in an animal model of multiple sclerosis. J Clin Invest 116:905–915.

Caicedo A, Espinosa-Heidmann DG, Pina Y, Hernandez EP, Cousins SW (2005) Blood-derived macrophages infiltrate the retina and activate Muller glial cells under experimental choroidal neovascularization. Exp Eye Res 81:38–47.

Campbell A, Becaria A, Lahiri DK, Sharman K, Bondy SC (2004) Chronic exposure to aluminum in drinking water increases inflammatory parameters selectively in the brain. J Neurosci Res 75:565–572.

Chan WY, Kohsaka S, Rezaie P (2007) The origin and cell lineage of microglia—New concepts. Brain Res Rev 53:344–354.

Chen L, Yang P, Kijlstra A (2002) Distribution, markers, and functions of retinal microglia. Ocul Immunol Inflamm 10:27–39.

Chiang YH, Borlongan CV, Zhou FC, Hoffer BJ, Wang Y (2005) Transplantation of fetal kidney cells: Neuroprotection and neuroregeneration. Cell Transplant 14:1–9.

Connor B (2001) Adenoviral vector-mediated delivery of glial cell line-derived neurotrophic factor provides neuroprotection in the aged parkinsonian rat. Clin Exp Pharmacol Physiol 28:896–900.

Dick AD, Ford AL, Forrester JV, Sedgwick JD (1995) Flow cytometric identification of a minority population of MHC class II positive cells in the normal rat retina distinct from CD45lowCD11b/c+ CD4low parenchymal microglia. Br J Ophthalmol 79:834–840.

Durany N, Munch G, Michel T, Riederer P (1999) Investigations on oxidative stress and therapeutical implications in dementia. Eur Arch Psychiatry Clin Neurosci 249(Suppl 3):68–73.

Eells JB (2003) The control of dopamine neuron development, function and survival: Insights from transgenic mice and the relevance to human disease. Curr Med Chem 10:857–870.

Eikelenboom P, Bate C, Van Gool WA, Hoozemans JJ, Rozemuller JM, Veerhuis R, Williams A (2002) Neuroinflammation in Alzheimer's disease and prion disease. Glia 40:232–239.

Eikelenboom P, van Gool WA (2004) Neuroinflammatory perspectives on the two faces of Alzheimer's disease. J Neural Transm 111:281–294.

Ekdahl CT, Claasen JH, Bonde S, Kokaia Z, Lindvall O (2003a) Inflammation is detrimental for neurogenesis in adult brain. Proc Natl Acad Sci USA 100:13632–13637.

Ekdahl CT, Zhu C, Bonde S, Bahr BA, Blomgren K, Lindvall O (2003b) Death mechanisms in status epilepticus-generated neurons and effects of additional seizures on their survival. Neurobiol Dis 14:513–523.

Emmerling MR, Watson MD, Raby CA, Spiegel K (2000) The role of complement in Alzheimer's disease pathology. Biochim Biophys Acta 1502:158–171.

Eugenin EA, Eckardt D, Theis M, Willecke K, Bennett MV, Saez JC (2001) Microglia at brain stab wounds express connexin 43 and in vitro form functional gap junctions after treatment with interferon-gamma and tumor necrosis factor-alpha. Proc Natl Acad Sci USA 98:4190–4195.

Fischer-Smith T, Croul S, Sverstiuk AE, Capini C, L'Heureux D, Regulier EG, Richardson MW, Amini S, Morgello S, Khalili K, Rappaport J (2001) CNS invasion by CD14+/CD16+ peripheral blood-derived monocytes in HIV dementia: Perivascular accumulation and reservoir of HIV infection. J Neurovirol 7:528–541.

Flaris NA, Densmore TL, Molleston MC, Hickey WF (1993) Characterization of microglia and macrophages in the central nervous system of rats: Definition of the differential expression of molecules using standard and novel monoclonal antibodies in normal CNS and in four models of parenchymal reaction. Glia 7:34–40.

Flugel A, Hager G, Horvat A, Spitzer C, Singer GM, Graeber MB, Kreutzberg GW, Schwaiger FW (2001) Neuronal MCP-1 expression in response to remote nerve injury. J Cereb Blood Flow Metab 21:69–76.

Frei K, Fontana A (1997) Antigen presentation in the CNS. Mol Psychiatry 2:96–98.

Frolov VA (1985) I Mechnikoff's contribution to immunology. J Hyg Epidemiol Microbiol Immunol 29:185–191.

Fujiwara N, Kobayashi K (2005) Macrophages in inflammation. Curr Drug Targets Inflamm Allergy 4:281–286.

Gao HM, Hong JS, Zhang W, Liu B (2003) Synergistic dopaminergic neurotoxicity of the pesticide rotenone and inflammogen lipopolysaccharide: Relevance to the etiology of Parkinson's disease. J Neurosci 23:1228–1236.

Gasparini L, Ongini E, Wilcock D, Morgan D (2005) Activity of flurbiprofen and chemically related anti-inflammatory drugs in models of Alzheimer's disease. Brain Res Rev 48:400–408.

Gebicke-Haerter PJ (2001) Microglia in neurodegeneration: Molecular aspects. Microsc Res Tech 54:47–58.

Gehrmann J (1996) Microglia: A sensor to threats in the nervous system? Res Virol 147:79–88.

Gendelman HE, Ghorpade A, Persidsky Y (1997) The neuropathogenesis of HIV-1 associated dementia. In: Peterson, Remington (eds), In Defense of the Brain pp 290–304. Blackwell Sco.

Ghorpade A (1997) Laboratory models for microglia-HIV interactions. In: Gendelman HE, Lipton SA, Epstein L, Swindells S (eds), The Neurology of AIDS pp 86–96. New York.

Goodwin JL, Uemura E, Cunnick JE (1995) Microglial release of nitric oxide by the synergistic action of beta-amyloid and IFN-gamma. Brain Res 692:207–214.

Gordon S (2002) Pattern recognition receptors: Doubling up for the innate immune response. Cell 111:927–930.

Gordon S, Taylor PR (2005) Monocyte and macrophage heterogeneity. Nat Rev Immunol 5:953–964.

Groux H, Fournier N, Cottrez F (2004) Role of dendritic cells in the generation of regulatory T cells. Semin Immunol 16:99–106.

Grunblatt E, Mandel S, Youdim MB (2000) Neuroprotective strategies in Parkinson's disease using the models of 6-hydroxydopamine and MPTP. Ann NY Acad Sci 899:262–273.

Hald A, Lotharius J (2005) Oxidative stress and inflammation in Parkinson's disease: Is there a causal link? Exp Neurol 193:279–290.

Hauben E, Butovsky O, Nevo U, Yoles E, Moalem G, Agranov E, Mor F, Leibowitz-Amit R, Pevsner E, Akselrod S, Neeman M, Cohen IR, Schwartz M (2000) Passive or active immunization with myelin basic protein promotes recovery from spinal cord contusion. J Neurosci 20:6421–6430.

Hickey W, Williams KC (1996) Mononuclear phagocyte heterogeneity and the blood-brain barrier: A model for Neuropathogenesis. In: Gendelman HE, Lipton, SA, Epstein, L, Swindells, S (eds), The Neurology of AIDS pp 61–72. New York: Chapman and Hall.

Hickey WF, Vass K, Lassman H (1992) Bone marrow derived elements: An immunohistochemical and ultrastructural survey of rat chimeras. J Neuropathol Exp Neurol 51:246–256.

Hishikawa N, Niwa J, Doyu M, Ito T, Ishigaki S, Hashizume Y, Sobue G (2003) Dorfin localizes to the ubiquitylated inclusions in Parkinson's disease, dementia with Lewy bodies, multiple system atrophy, and amyotrophic lateral sclerosis. Am J Pathol 163: 609–619.

Hoozemans JJ, Rozemuller AJ, Veerhuis R, Eikelenboom P (2001) Immunological aspects of alzheimer's disease: Therapeutic implications. BioDrugs 15:325–337.

Hortega PR (1919) El tercer elemento de los centros nerviosos. Poder fagocitario y movilidad de la microglia. Bol Soc Esp Biol Ano ix:154–166.

Hortega PR (1932) Microglia. Cytology and Cellular Pathology of the Nervous System. In: Penfield, W (ed), pp 481–584. New York: Hoeber.

Iribarren P, Cui YH, Le Y, Wang JM (2002) The role of dendritic cells in neurodegenerative diseases. Arch Immunol Ther Exp (Warsz) 50:187–196.

Iseki E, Marui W, Akiyama H, Ueda K, Kosaka K (2000) Degeneration process of Lewy bodies in the brains of patients with dementia with Lewy bodies using alpha-synuclein-immunohistochemistry. Neurosci Lett 286:69–73.

Jack C, Ruffini F, Bar-Or A, Antel JP (2005) Microglia and multiple sclerosis. J Neurosci Res 81:363–373.

Kadiu I, Glanzer JG, Kipnis J, Gendelman HE, Thomas MP (2005) Mononuclear phagocytes in the pathogenesis of neurodegenerative diseases. Neurotox Res 8:25–50.

Karman J, Ling C, Sandor M, Fabry Z (2004) Dendritic cells in the initiation of immune responses against central nervous system-derived antigens. Immunol Lett 92:107–115.

Kaur C, Hao AJ, Wu CH, Ling EA (2001) Origin of microglia. Microsc Res Tech 54:2–9.

Kempermann G, Kronenberg G (2003) Depressed new neurons—adult hippocampal neurogenesis and a cellular plasticity hypothesis of major depression. Biol Psychiatry 54:499–503.

Kempermann G, Kuhn HG, Gage FH (1998a) Experience-induced neurogenesis in the senescent dentate gyrus. J Neurosci 18:3206–3212.

Kempermann G, Brandon EP, Gage FH (1998b) Environmental stimulation of 129/SvJ mice causes increased cell proliferation and neurogenesis in the adult dentate gyrus. Curr Biol 8:939–942.

Kempermann G, Gast D, Kronenberg G, Yamaguchi M, Gage FH (2003) Early determination and long-term persistence of adult-generated new neurons in the hippocampus of mice. Development 130:391–399.

Kim SU, de Vellis J (2005) Microglia in health and disease. J Neurosci Res 81:302–313.

Kim WK, Alvarez X, Fisher J, Bronfin B, Westmoreland S, McLaurin J, Williams K (2006) CD163 identifies perivascular macrophages in normal and viral encephalitic brains and potential precursors to perivascular macrophages in blood. Am J Pathol 168:822–834.

Kipnis J, Mizrahi T, Yoles E, Ben-Nun A, Schwartz M (2002) Myelin specific Th1 cells are necessary for post-traumatic protective autoimmunity. J Neuroimmunol 130:78–85.

Kitamura T (1973) The origin of brain macrophages—some considerations on the microglia theory of Del Rio-Hortega. Acta Pathol Jpn 23:11–26.

Kitamura Y, Ishida Y, Takata K, Kakimura J, Mizutani H, Shimohama S, Akaike A, Taniguchi T (2001) Alpha-synuclein protein is not scavenged in neuronal loss induced by kainic acid or focal ischemia. Brain Res 898:181–185.

Koutsilieri E, Scheller C, Grunblatt E, Nara K, Li J, Riederer P (2002) Free radicals in Parkinson's disease. J Neurol 249(Suppl 2):II1–II5.

Lassmann H, Bruck W, Lucchinetti C (2001) Heterogeneity of multiple sclerosis pathogenesis: Implications for diagnosis and therapy. Trends Mol Med 7:115–121.

Lassmann H, Schmied M, Vass K, Hickey WF (1993) Bone marrow derived elements and resident microglia in brain inflammation. Glia 7:19–24.

Lee SC, Liu W, Brosnan CF, Dickson DW (1992) Characterization of primary human fetal dissociated central nervous system cultures with an emphasis on microglia. Lab Invest 67:465–476.

Lee SJ, Lee S (2002) Toll-like receptors and inflammation in the CNS. Curr Drug Targets Inflamm Allergy 1:181–191.

Linington C, Bradl M, Lassmann H, Brunner C, Vass K (1988) Augmentation of demyelination in rat acute allergic encephalomyelitis by circulating mouse monoclonal antibodies directed against a myelin/oligodendrocyte glycoprotein. Am J Pathol 130:443–454.

Liu B, Hong JS (2003) Role of microglia in inflammation-mediated neurodegenerative diseases: Mechanisms and strategies for therapeutic intervention. J Pharmacol Exp Ther 304:1–7.

Liu B, Gao HM, Wang JY, Jeohn GH, Cooper CL, Hong JS (2002) Role of nitric oxide in inflammation-mediated neurodegeneration. Ann NY Acad Sci 962:318–331.

Mannoji H, Yeger H, Becker LE (1986) A specific histochemical marker (lectin Ricinus communis agglutinin-1) for normal human microglia, and application to routine histopathology. Acta Neuropathol (Berl) 71:341–343.

Marx F, Blasko I, Grubeck-Loebenstein B (1999) Mechanisms of immune regulation in Alzheimer's disease: A viewpoint. Arch Immunol Ther Exp (Warsz) 47:205–209.

Marx F, Blasko I, Pavelka M, Grubeck-Loebenstein B (1998) The possible role of the immune system in Alzheimer's disease. Exp Gerontol 33:871–881.

McGeer EG, McGeer PL (1997) The role of the immune system in neurodegenerative disorders. Mov Disord 12:855–858.

McGeer EG, Klegeris A, McGeer PL (2005) Inflammation, the complement system and the diseases of aging. Neurobiol Aging 26 Suppl 1:94–97.

McGillem GS, Dacheux RF (1998) Migration of retinal microglia in experimental proliferative vitreoretinopathy. Exp Eye Res 67:371–375.

Meda L, Baron P, Scarlato G (2001) Glial activation in Alzheimer's disease: The role of Abeta and its associated proteins. Neurobiol Aging 22:885–893.

Medzhitov R, Janeway C, Jr. (2000) Innate immune recognition: Mechanisms and pathways. Immunol Rev 173:89–97.

Mirza B, Hadberg H, Thomsen P, Moos T (2000) The absence of reactive astrocytosis is indicative of a unique inflammatory process in Parkinson's disease. Neuroscience 95:425–432.

Mitrovic B, Parkinson J, Merrill JE (1996) An in vitro model of oligodendrocyte destruction by nitric oxide and its relevance to multiple sclerosis. Methods 10:501–513.

Moalem G, Monsonego A, Shani Y, Cohen IR, Schwartz M (1999) Differential T cell response in central and peripheral nerve injury: Connection with immune privilege. Faseb J 13:1207–1217.

Monier A, Evrard P, Gressens P, Verney C (2006) Distribution and differentiation of microglia in the human encephalon during the first two trimesters of gestation. J Comp Neurol 499:565–582.

Monje ML, Toda H, Palmer TD (2003) Inflammatory blockade restores adult hippocampal neurogenesis. Science 302:1760–1765.

Moore MA, Metcalf D (1970) Ontogeny of the haemopoietic system: Yolk sac origin of in vivo and in vitro colony forming cells in the developing mouse embryo. Br J Haematol 18:279–296.

Morale MC, Serra PA, L'Episcopo F, Tirolo C, Caniglia S, Testa N, Gennuso F, Giaquinta G, Rocchitta G, Desole MS, Miele E, Marchetti B (2006) Estrogen, neuroinflammation and neuroprotection in Parkinson's disease: Glia dictates resistance versus vulnerability to neurodegeneration. Neuroscience 138:869–878.

Mrak RE, Griffin WS (2005) Glia and their cytokines in progression of neurodegeneration. Neurobiol Aging 26:349–354.

Nagatsu T, Sawada M (2005) Inflammatory process in Parkinson's disease: Role for cytokines. Curr Pharm Des 11:999–1016.

Nakajima K, Kohsaka S (1998) Functional roles of microglia in the central nervous system. Hum Cell 11:141–155.

Nelson PT, Soma LA, Lavi E (2002) Microglia in diseases of the central nervous system. Ann Med 34:491–500.

Nibbering PH, Leijh PC, van Furth R (1987) Quantitative immunocytochemical characterization of mononuclear phagocytes I. Monoblasts, promonocytes, monocytes, and peritoneal and alveolar macrophages. Cell Immunol 105:374–385.

Nottet HSLM, Persidsky Y, Sasseville VG, Nukuna AN, Bock P, Zhai QH, Sharer LR, McComb R, Swindells S, Soderland C, Rizzino A, Gendelman HE (1996) Mechanisms for the transendothelial migration of HIV-1-infected monocytes into brain. J Immunol 156:1284–1295.

O'Neill LA (2006) How Toll-like receptors signal: What we know and what we don't know. Curr Opin Immunol 18:3–9.

Perry VH (1998) A revised view of the central nervous system microenvironment and major histocompatibility complex class II antigen presentation. J Neuroimmunol 90:113–121.

Polazzi E, Contestabile A (2002) Reciprocal interactions between microglia and neurons: From survival to neuropathology. Rev Neurosci 13:221–242.

Prat A, Antel J (2005) Pathogenesis of multiple sclerosis. Curr Opin Neurol 18:225–230.

Prat A, Biernacki K, Wosik K, Antel JP (2001) Glial cell influence on the human blood-brain barrier. Glia 36:145–155.

Probert L, Eugster HP, Akassoglou K, Bauer J, Frei K, Lassmann H, Fontana A (2000) TNFR1 signalling is critical for the development of demyelination and the limitation of T-cell responses during immune-mediated CNS disease. Brain 123 (Pt 10):2005–2019.

Qin L, Liu Y, Cooper C, Liu B, Wilson B, Hong JS (2002) Microglia enhance beta-amyloid peptide-induced toxicity in cortical and mesencephalic neurons by producing reactive oxygen species. J Neurochem 83:973–983.

Rangon CM, Haik S, Faucheux BA, Metz-Boutigue MH, Fierville F, Fuchs JP, Hauw JJ, Aunis D (2003) Different chromogranin immunoreactivity between prion and a-beta amyloid plaque. Neuroreport 14:755–758.

Rogers J, Strohmeyer R, Kovelowski CJ, Li R (2002) Microglia and inflammatory mechanisms in the clearance of amyloid beta peptide. Glia 40:260–269.

Rosenberg PB (2005) Clinical aspects of inflammation in Alzheimer's disease. Int Rev Psychiatry 17:503–514.

Scali C, Prosperi C, Vannucchi MG, Pepeu G, Casamenti F (2000) Brain inflammatory reaction in an animal model of neuronal degeneration and its modulation by an anti-inflammatory drug: Implication in Alzheimer's disease. Eur J Neurosci 12:1900–1912.

Schori H, Kipnis J, Yoles E, WoldeMussie E, Ruiz G, Wheeler LA, Schwartz M (2001) Vaccination for protection of retinal ganglion cells against death from glutamate cytotoxicity and ocular hypertension: Implications for glaucoma. Proc Natl Acad Sci USA 98:3398–3403.

Schwartz M, Butovsky O, Bruck W, Hanisch UK (2006) Microglial phenotype: Is the commitment reversible? Trends Neurosci 29:68–74.

Schwartz M, Shaked I, Fisher J, Mizrahi T, Schori H (2003) Protective autoimmunity against the enemy within: Fighting glutamate toxicity. Trends Neurosci 26:297–302.

Segal AW (2005) How neutrophils kill microbes. Annu Rev Immunol 23:197–223.

Shaked I, Porat Z, Gersner R, Kipnis J, Schwartz M (2004) Early activation of microglia as antigen-presenting cells correlates with T cell-mediated protection and repair of the injured central nervous system. J Neuroimmunol 146:84–93.

Shaked I, Tchoresh D, Gersner R, Meiri G, Mordechai S, Xiao X, Hart RP, Schwartz M (2005) Protective autoimmunity: Interferon-gamma enables microglia to remove glutamate without evoking inflammatory mediators. J Neurochem 92:997–1009.

Sluiter W, Hulsing-Hesselink E, Elzenga-Claasen I, van Hemsbergen-Oomens LW, van der Voort van der Kleij-van Andel A, van Furth R (1987) Macrophages as origin of factor increasing monocytopoiesis. J Exp Med 166:909–922.

Sorensen JC, Dalmau I, Zimmer J, Finsen B (1996) Microglial reactions to retrograde degeneration of tracer-identified thalamic neurons after frontal sensorimotor cortex lesions in adult rats. Exp Brain Res 112:203–212.

Sorensen TL, Tani M, Jensen J, Pierce V, Lucchinetti C, Folcik VA, Qin S, Rottman J, Sellebjerg F, Strieter RM, Frederiksen JL, Ransohoff RM (1999) Expression of specific chemokines and chemokine receptors in the central nervous system of multiple sclerosis patients. J Clin Invest 103:807–815.

Sriram S, Rodriguez M (1997) Indictment of the microglia as the villain in multiple sclerosis. Neurology 48:464–470.

Streit WJ (2004) Microglia and Alzheimer's disease pathogenesis. J Neurosci Res 77:1–8.

Streit WJ (2006) Microglial senescence: Does the brain's immune system have an expiration date? Trends Neurosci 29:506–510.

Streit WJ, Walter SA, Pennell NA (1999) Reactive microgliosis. Prog Neurobiol 57:563–581.

Streit WJ, Conde JR, Fendrick SE, Flanary BE, Mariani CL (2005) Role of microglia in the central nervous system's immune response. Neurol Res 27:685–691.

Tacke F, Randolph GJ (2006) Migratory fate and differentiation of blood monocyte subsets. Immunobiology 211:609–618.

Takeuchi H, Jin S, Wang J, Zhang G, Kawanokuchi J, Kuno R, Sonobe Y, Mizuno T, Suzumura A (2006) Tumor necrosis factor-alpha induces neurotoxicity via glutamate release from hemichannels of activated microglia in an autocrine manner. J Biol Chem 281:21362–21368.

Tanaka S, Suzuki K, Watanabe M, Matsuda A, Tone S, Koike T (1998) Upregulation of a new microglial gene, mrf-1, in response to programmed neuronal cell death and degeneration. J Neurosci 18:6358–6369.

Taylor PR, Martinez-Pomares L, Stacey M, Lin HH, Brown GD, Gordon S (2005) Macrophage receptors and immune recognition. Annu Rev Immunol 23:901–944.

Teismann P, Schulz JB (2004) Cellular pathology of Parkinson's disease: Astrocytes, microglia and inflammation. Cell Tissue Res 318:149–161.

Tezel G, Wax MB (2004) The immune system and glaucoma. Curr Opin Ophthalmol 15:80–84.

Thomas WE (1992) Brain macrophages: Evaluation of microglia and their functions. Brain Res Rev 17:61–74.

Valeria Canto Soler M, Gallo JE, Dodds RA, Suburo AM (2002) A mouse model of proliferative vitreoretinopathy induced by dispase. Exp Eye Res 75:491–504.

van Furth R (1982) Current view on the mononuclear phagocyte system. Immunobiology 161:178–185.

Vila M, Jackson-Lewis V, Guegan C, Wu DC, Teismann P, Choi DK, Tieu K, Przedborski S (2001) The role of glial cells in Parkinson's disease. Curr Opin Neurol 14:483–489.

Vilhardt F (2005) Microglia: Phagocyte and glia cell. Int J Biochem Cell Biol 37:17–21.

Watts C (1997) Capture and processing of exogenous antigens for presentation on MHC molecules. Annu Rev Immunol 15:821–850.

Weller M, Heimann K, Wiedemann P (1990) The pathogenesis of vitreoretinal proliferation and traction: A working hypothesis. Med Hypotheses 31:157–159.

Weller M, Esser P, Heimann K, Wiedemann P (1991) Retinal microglia: A new cell in idiopathic proliferative vitreoretinopathy? Exp Eye Res 53:275–281.

Wierzba-Bobrowicz T, Gwiazda E, Poszwinska Z (1995) Morphological study of microglia in human mesencephalon during the development and aging. Folia Neuropathol 33:77–83.

Williams K, Ulvestad E, Antel J (1994) Immune regulatory and effector properties of human adult microglia studies in vitro and in situ. Adv Neuroimmunol 4:273–281.

Wu L, Vandenabeele S, Georgopoulos K (2001) Derivation of dendritic cells from myeloid and lymphoid precursors. Int Rev Immunol 20:117–135.

Yoles E, Schwartz M (1998) Degeneration of spared axons following partial white matter lesion: Implications for optic nerve neuropathies. Exp Neurol 153:1–7.

Zhang S, Wang Q, Miao B (2007) Review: dendritic cell-based vaccine in the treatment of patients with advanced melanoma. Cancer Biother Radiopharm 22:501–507.

Zhang W, Wang T, Pei Z, Miller DS, Wu X, Block ML, Wilson B, Zhou Y, Hong JS, Zhang J (2005) Aggregated alpha-synuclein activates microglia: A process leading to disease progression in Parkinson's disease. FASEB J 19:533–542.

Ziv Y, Ron N, Butovsky O, Landa G, Sudai E, Greenberg N, Cohen H, Kipnis J, Schwartz M (2006) Immune cells contribute to the maintenance of neurogenesis and spatial learning abilities in adulthood. Nat Neurosci 9:268–275.

10
Neuronal and Glial Signaling

Murali Prakriya and Richard J. Miller

Keywords Action potential; Calcium; Glia; Ion channel; LTP; Neuron; Neurotransmitter; Potassium; Resting membrane potential; Sodium; Structure; Synapse

10.1. Introduction

The major function of the nervous system is to rapidly transfer and integrate information with a view to organizing the diverse functions of multicellular organisms. Our present understanding as to how nerve cells communicate originates with studies in the nineteenth century on the anatomy of the nervous system. Work conducted by Camillo Golgi and Santiago Ramon y Cajal defined the fine structure of neurons for the first time. Golgi favored the idea that all the nerves in the nervous system existed as a sort of reticular net, rather than as separate entities. Cajal, in contrast, concluded that neurons were independent entities and that a minute gap existed between the ends of nerve fibers and other nerve cells or the muscles that they innervate. This gap was ultimately named a "synapse" by the great neurologist Sir Charles Sherrington in 1897, from the Greek meaning "to clasp". Cajal also argued that neurotransmission was basically a unidirectional process, information being received by dendrites and being transmitted unidirectionally along axons. The actual existence of the synapse as a structure was not confirmed until the development of the electron microscope in the 1950s. Around this time a fierce debate took place between two opposing sets of scientists (Valenstein, 2005). One set, primarily electrophysiologists, held the view that information was transmitted between nerves and between nerves and muscles by purely electrical processes. The other group, primarily pharmacologists, suggested that chemical messenger molecules were released by the presynaptic nerve and carried the information across the synapse. The nature of the chemicals that constitute neurotransmitters was gradually revealed by the work of several important investigators who characterized the effects of different substances that mimicked or blocked the actions of neurotransmitters on fast skeletal muscles and in the autonomic nervous system. This finally culminated in the demonstration by Otto Loewi in 1921 that the vagus nerve secreted a chemical ("Vagusstoff") that mediated the slowing effect of vagal stimulation on the heart. This substance turned out to be acetylcholine.

The term glia, or rather neuroglia (nerve glue), was first introduced by Virchow in 1859 who conceived of these cells as being neutral elements concerned in holding the nervous system together and playing a supportive role. Further studies by Ramon y Cajal and his colleagues divided glia into astrocytes, oligodendrocytes and microglia. We now know that glial cells are far from inactive support systems and are engaged in diverse dynamic interactions with nerve cells. Nissl suggested that microglia had the capacity for migration and phagocytosis and this has indeed proved to be the case. As opposed to other major cell types in the brain, the microglia are of mesenchymal origin and can generally be thought of as resident cells with macrophage like properties. Signaling between neurons and glia is involved in dynamically integrating the functions of the nervous system in health and disease. In general, signaling in the nervous system requires us to understand the properties of the different types of neurotransmitters, how they engage receptors and how this impacts the electrical and other properties of neurons and glial cells.

10.2. Electrogenesis and the Action Potential

10.2.1. The Resting Membrane Potential

The most common molecules in the body are water and inorganic molecules such as sodium, potassium and chloride ions. A feature that is common among all living cells is that the concentrations of these ions are different in the extracellular and intracellular compartments. The extracellular fluid is high in sodium (Na^+) and chloride (Cl^-) ions, but low in potassium (K^+) ions (Figure 10.1). In contrast, the intracellular solution is

T. Ikezu and H.E. Gendelman (eds.), *Neuroimmune Pharmacology*.
© Springer 2008

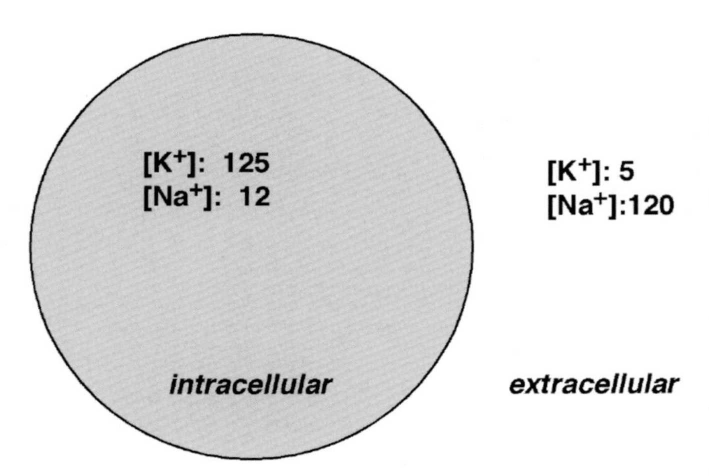

FIGURE 10.1. Composition of major ions in the intracellular and extracellular compartments.

low in Na^+ and Cl^-, but high in K^+. This difference is maintained and regulated by control mechanisms residing in the plasma membrane of the cell, a phospholipid bilayer with protein molecules inserted into it. The plasma membrane acts as a selectively permeable barrier permitting some molecules to cross while excluding others. When a pathway or "channel" for the movement of these charged molecules opens up across the plasma membrane, a phenomenon known as "gating", the net movement of ions is governed not only by their concentration gradients, but also by the electrical potential difference across the plasma membrane. As we shall discuss below, neurons express proteins in their plasma membranes that act as channels allowing the selective permeability of different ions.

How does an electrical potential come into this picture? The movement of ions through selective pores in the membrane gives rise to a charge separation across the phospholipid bilayer, essentially giving rise to a "capacitor" and a potential difference across the membrane. It takes the movement of only a very small number of charges to give rise to a substantial membrane potential difference across the membrane. As a result, ions that are propelled down their concentration gradient encounter the electrical force that opposes their movement down this concentration gradient. Equilibrium for ion movement across the membrane is reached when the electrical force exactly balances the diffusional force arising from the concentration gradient. A relationship that describes the value of the electrical potential reaches at this equilibrium condition is given by the Nernst equation:

$$V = \frac{RT}{zF} \ln\left(\frac{[X]_{out}}{[X]_{in}}\right)$$

where V is the membrane voltage at equilibrium, $[X]_{out}$ and $[X]_{in}$ are the extracellular and intracellular concentrations of the ion being examined, R is the gas constant, T is the absolute temperature, z is the valence of the ion, and F is Faraday's constant. It is important to remember that the Nernst equation applies only to one ion at a time and only to ions that can cross the plasma membrane. For an ion in question, with a certain extracellular and intracellular concentration, the value of the electrical potential given by the Nernst equation is called the equilibrium potential.

For the typical ionic concentrations in mammalian cells (Figure 10.1), the equilibrium potentials for K^+ and Na^+ can be calculated from the Nernst equation to be −80 mV and +57 mV, respectively. If the plasma membrane is permeable only to K^+ and no other ion, the membrane potential would be determined solely by the equilibrium potential for K^+ and would be −80 mV. Real cells, however, are permeable to more than one ion at a time, and as a consequence, their resting membrane potential is influenced not only by the movements of K^+, but also other ions, primarily Na^+. For a typical cell that has a large K^+ permeability at rest, if we increase the Na^+ permeability to the membrane very slightly, the net effect would be to depolarize the membrane potential away from the equilibrium potential of K^+ and toward the equilibrium potential of Na^+. There is a struggle between Na^+ on the one hand, tending to make the V_m equal to +57 mV, and K^+ and Cl^- on the other hand, which push to make V_m equal to −80 mV. An equation that quantitatively relates these factors is the Goldman-Hodgkin-Katz equation (also referred to as the constant field equation because of the assumption made that the membrane field between the intra and extra-cellular compartment varies at a constant rate with distance). For a cell that is permeable to Na^+, K^+ and Cl^-, this equation can be written as:

$$V_m = \frac{RT}{zF} \ln\left(\frac{p_K[K^+]_o + p_{Na}[Na^+]_o + p_{Cl}[Cl^-]_i}{p_K[K^+]_i + p_{Na}[Na^+]_i + p_{Cl}[Cl^-]_o}\right)$$

This equation is similar to the Nernst equation except that it simultaneously takes into account the contributions of all three permeant ions. It indicates that the membrane potential is governed by two factors: (1) the ionic concentrations, which determine the equilibrium potentials for the ions, and, (2) their relative permeabilities, which determine the relative importance of a particular ion in governing where V_m lies. For many cells, including most neurons and immune cells, this equation can be simplified: the chloride term can be dropped altogether because the contribution of chloride to the resting membrane potential is insignificant. In this case, the Goldman equation becomes:

$$V_m = \frac{RT}{zF} \ln\left(\frac{p_K[K^+]_o + p_{Na}[Na^+]_o}{p_K[K^+]_i + p_{Na}[Na^+]_i}\right)$$

Because it is easier to measure relative ion permeabilities than the absolute permeabilities, this equation can be rewritten in a slightly different form:

$$V_m = 58\log\left(\frac{a[\mathrm{K}^+]_o + [\mathrm{Na}^+]_o}{a[\mathrm{K}^+]_i + [\mathrm{Na}^+]_i}\right)$$

where term $a = p_K/p_{Na}$ is the permeability of K^+ relative to Na^+, and the term RT/F has been evaluated at room temperature and converted to log. In most cells at rest, the ratio a is about 50, resulting in a membrane potential of −71 mV for a cell with ionic composition as shown in Figure 10.1.

Because the steady-state membrane potential lies between the equilibrium potentials for Na^+ and K^+, there is a constant movement of K^+ out from the cell and Na^+ into the cell. To ensure that this does not lead to a progressive decline in the concentration gradients across the membrane, all cells have a Na–K pump, which uses the hydrolyses of ATP to simultaneously pump K^+ into the cell and push Na^+ out. The constant fluxes of K^+ and Na^+ constitute electrical currents across the cell membrane, and at steady-state, these currents cancel each other out so that the net membrane current is zero.

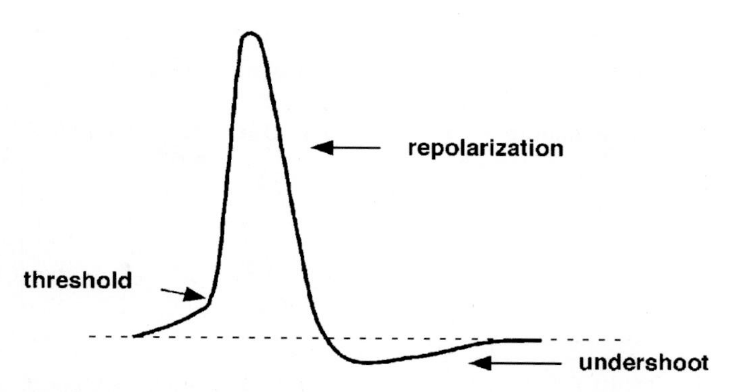

FIGURE 10.2. Various phases of the action potential. Once a depolarization reaches a certain threshold, the membrane potential moves rapidly in a regenerative manner toward E_{Na}. The opening of K^+ channels coupled with the inactivation of Na^+ channels causes the membrane potential to repolarize. The prolonged hyperpolarization (undershoot) results from slow closing of open K^+ channels following rapid repolarization.

10.2.2. The Action Potential

As noted above, the membrane potential is governed by the relative permeabilities for K^+ and Na^+. If the K^+ permeability is greater than the Na^+ permeability, the membrane potential is closer to E_K. Conversely, if the Na^+ permeability far exceeds the K^+ permeability, the membrane potential should be closer to E_{Na} than E_K. In excitable cells such as neurons, ionic permeabilities are not fixed, but can be varied resulting in the occurrence of transient, dramatic changes in the membrane potential. It is these transient changes in the membrane potential brought about by changes in ionic permeabilities that underlies the action potential, a fundamental basic signal that sub-serves communication between all brain cells.

The typical profile of membrane potential changes during an action potential is illustrated in Figure 10.2. Following a small depolarization to a "threshold" value, there is a sudden, large jump in the membrane potential during which the potential transiently moves in the positive direction (referred to as a depolarization) and actually reverses in sign for a brief period. After peaking at a positive voltage, the membrane potential begins an equally rapid return toward its resting value, and transiently becomes more negative than its normal resting value. The last part of the action potential is a slow return to its resting value that often lasts several milliseconds. This is called the undershoot of the action potential. This basic electrical signal pattern is fundamental to neurons and is the basis of information transfer between neurons of the brain.

The key to understanding the origin of the action potential lies in the factors that influence the membrane potential of the cell as exemplified by the Goldman relationship. Recall that the membrane potential of the cell lies somewhere between E_K and E_{Na}. At rest, because the relative permeability of the membrane is much higher for K^+ than Na^+, the point at which the membrane potential lies is closer to E_K. If the Na^+ permeability suddenly increases dramatically, then the membrane potential would correspondingly shift toward E_{Na}. For example, if p_K/p_{Na} changed from 50 at rest to 0.02, then the membrane potential would swing from −71 mV to 53 mV. After a brief delay, if the p_K/p_{Na} changed back to 50, the membrane potential would be expected to return to −71 mV. It is these transient changes in Na^+ permeability that are responsible for the swings in membrane potential from near E_K toward E_{Na} and back during the action potential.

10.2.3. The Sequence of Activation and Inactivation of Na⁺ and K⁺ Channels During an Action Potential

A dramatic increase in Na^+ permeability requires a dramatic increase in the number of channels that allow Na^+ to enter the cell. Thus, the resting p_{Na} is only a small fraction of what it could be because most membrane sodium channels are closed at rest. What stimulus causes the hidden Na^+ channels to reveal themselves? It turns out that the activation of these Na^+ channels is triggered by membrane depolarization. When V_m is at its usual resting levels around −70 mV, these Na^+ channels are closed and p_{Na} is low. However, depolarization causes the channels to open. Because the voltage-activated Na^+ channels respond to depolarization, the response of the membrane to depolarization is regenerative, and thus explosive (Figure 10.3). A small depolarization of the membrane opens Na^+ channels, which causes influx of Na^+ into the cell and additional depolarization, which in turn opens more Na^+ channels. This explains the all-or-none nature of the action potential: once it is triggered, it runs to completion.

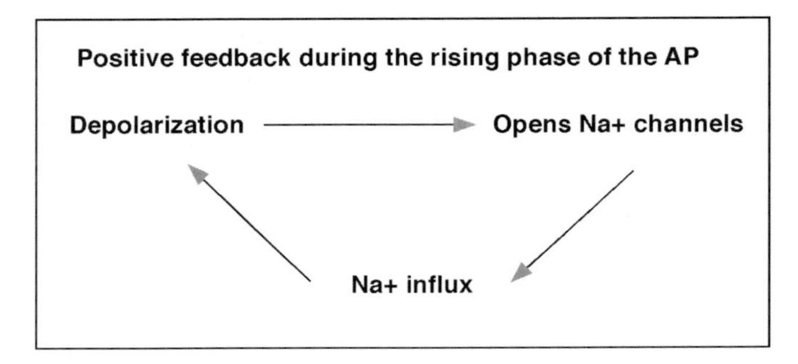

FIGURE 10.3. Behavior of Na^+ channels and resulting changes in membrane potential during the rising phase of the action potential.

What causes V_m to return to its resting potential following the regenerative depolarization during the action potential? Two processes cause this: (1) the time-dependent inactivation of the depolarization induced increase in p_{Na}, and (2) a delayed increase in the p_K initiated by depolarization. A few milliseconds following the opening of voltage-activated Na^+ channels by depolarization, an "inactivation gate" is triggered which plugs the ion permeation pathway of Na^+ channels and shuts down channel activity, causing a dramatic decrease in p_{Na}. Almost at the same time as this is happening, depolarization-activated K^+ channels begin to open up, causing the p_K to dramatically increase. The closure of Na^+ channels and the opening of K^+ channels cause a dramatic increase in the p_K/p_{Na} ratio, thereby shifting the balance strongly in favor of E_K. Therefore, the repolarizing phase of the action potential arises from the simultaneous decline in p_{Na} and an increase in p_K. In fact, the increase in p_K lasts for several milliseconds after the Na^+ channels close, causing the p_K/p_{Na} ratio to be higher than at rest. This causes the V_m to be pushed closer to E_K, explaining the undershoot that follows fast repolarization.

How does the action potential propagate along the nerve fiber? This is easily explained by the basis for the generation of the action potential. As we have just seen, the stimulus for an action potential is a depolarization of sufficient strength to open large numbers of Na^+ channels to cause a regenerative membrane depolarization. Once such a depolarization occurs in one part of the cell, it brings neighboring regions above the threshold, setting up an action potential in that region. This in turn causes other neighboring regions to reach threshold and triggers the same gating schemes of Na^+ and K^+ channels to produce action potentials in those regions. The "inactivation" of Na^+ channels ensures that once an action potential has occurred in one region, it cannot immediately occur again in that region until Na^+ channels have recovered from inactivation. This causes the unidirectional movement of the action potential, away from the region that triggered the initial rise in membrane potential, and typically down the nerve fiber. This is the basis of the fundamental long-distance signal of the nervous system.

10.2.4. Transmission of Signals Between Neurons: Voltage-Activated Ca^{2+} Channels Mediate Neurotransmitter Release

Propagation of action potentials typically occurs from the somatic regions of neurons, through the axon and into the "terminals"—the tiny axon branches that terminate in synapses on neighboring neurons or end organs. As described below, invasion of the action potential into nerve terminals causes fusion of pre-packaged neurotransmitter-filled vesicles with the plasma membrane, releasing neurotransmitter across the synapse, and resulting in the activation of receptors on neighboring cells. What process couples the electrical action potential signal in the nerve terminal to the release of neurotransmitter? This is mediated by Ca^{2+} influx from the extracellular space into the nerve terminal through voltage-activated Ca^{2+} channels. Compared to Na^+ and K^+, Ca^{2+} is present in much lower amounts in the extracellular space (1–2 mM) and was therefore ignored in the previous discussions of resting membrane potential and action potentials. However, as a chemical messenger inside the cell, Ca^{2+} mediates a variety of critical signaling functions. Nature has evolved such that the intracellular concentration of Ca^{2+} is of the order of only 100 nM. With the Ca^{2+} concentration in the extracellular space of 1–2 mM, this creates a 10,000-fold concentration gradient across the membrane. In addition, with these ionic concentrations, the Nernst equation indicates that the equilibrium potential for Ca^{2+} is also very positive. Therefore, near the resting potential, both the concentration and electrical gradients promote the movement of Ca^{2+} into the cell. When a conduit for Ca^{2+} entry, such as a voltage-activated Ca^{2+} channel opens up, Ca^{2+} rushes into the presynaptic intracellular space, elevating the local Ca^{2+} concentration, and resulting the fusion of neurotransmitter-filled vesicles through series of very rapid signal transduction events.

Voltage-activated Ca^{2+} channels are activated by membrane depolarization and represent a large family of related channels with a wide tissue distribution. They are found ubiquitously in neurons, muscles, and endocrine cells as well as in many epithelial and endothelial cells. In addition to neurotransmitter release, they mediate a variety of essential functions in the body including muscle contraction, insulin secretion, gene expression, modulation of signal transduction events and in excess can cause cell death.

10.2.5. Membrane Properties of Glial Cells

Intermingled with the neurons in the brain are a variety of other cell types. The most common of these "satellite" cells are glial cells. These make up virtually about one half of the total volume of the brain and exist in several forms such as astrocytes, oligodendrocytes, and Schwann cells. Membrane properties of glial cells exhibit fundamental differences from neurons, the chief difference being their passive nature. Unlike neurons, most glial cells are not excitable and do not fire action potentials. Membrane potential measurements of glia indicate a relatively negative resting membrane potential—around −90 mV in contrast to −70 mV for most neuronal cells. This arises from the fact that, in contrast to neurons, the membranes of glia such as astrocytes and oligodendrocytes are permeable almost exclusively to K^+.

Glial cells play several essential roles in neuronal function. Schwann cells form the well-studied myelin sheaths around large axons of peripheral nerves, enabling faster propagation of action potentials along nerve fibers by effectively increasing the membrane resistance of the fibers. The end-feet of astrocytic cells helps form the blood-brain barrier, which limits what substances cross over from the vasculature into the brain. Glial cells are also involved in guiding axons to their targets during neuronal development and during regeneration of nerve fibers after injury.

One particularly intriguing function of glial cells related to their unique membrane properties is the uptake of excess extracellular K^+ by astrocytes. When neurons fire repeatedly, K^+ accumulates in the extracellular space. Pumps and transporters in the neighboring astrocytes take up the excess K^+ and store it to protect neurons from the depolarization that could result from the increase in extracellular K^+ concentration. What happens next to the excess K^+ taken up? Astrocytes are connected to each other by electrical synapses—essentially cytoplasmic bridges between neighboring cells, forming sheets of physically connected cells. As a result, the K^+ taken up by astrocytes in one area is shuttled to neighboring astrocytes through the cytoplasmic bridges to draw it away from areas of high extracellular K^+. It has been discovered that astrocytes lining membranes around the blood vessels have significantly higher K^+ channel density than the other cells of this network. The K^+ taken up by the astrocytic network is eventually extruded by so called "end-feet" specializations of high K^+-channel density directly into the blood vessels. By this mechanism, the high K^+ permeability of astrocytes protects neurons from excess depolarization that could result from K^+ efflux into the extracellular space.

10.2.6. The Structure of Channel Proteins

A remarkable feature of ion channels is that once open, they promote the diffusion of ions down their concentration gradient often with extremely high selectivity and at extremely high rates (tens of millions of ions per second). What essential common structural elements confer ion channels with these properties? As might be expected, a first requirement is the existence of a pore region for passage of ions. The phospholipid bilayer is a hostile, low dielectric barrier to the passage of hydrophilic and charged ions. The amino acids of ion channel proteins provide a comfortable conducting hydrophilic pathway across the hydrophobic interior of the membrane. As a result, ion channels are necessarily transmembrane proteins with domains that span the membrane. Ion channels are usually constructed from the assembly of several subunits. A second requirement is the existence of a gating mechanism that regulates the transport of ions across the pore. Channel gating is controlled by external factors (voltage, for voltage-activated channels, ligands for ligand-gated channels). Gating arises from the movements of protein domains that open or occlude the ion permeation pathway. Finally, the channel pores are often ion selective, discriminating between ions of varying sizes and charges to enable the passage of only a single ionic species.

How nature has evolved to solve these issues can be understood by exploring the structure of the K^+ channel. The crystal structure of a non-voltage gated, two transmembrane spanning K-channel from the bacteria *Streptomyces lividans* (KcsA K^+ channel) has been solved to 2 Å resolution (Doyle et al., 1998; Zhou et al., 2001). Like its mammalian voltage-gated K^+ channel counterparts, the KcsA channel is a tetramer. Each subunit of this tetramer has only two membrane-spanning segments (instead of six for the mammalian channels), but it closely resembles the fifth and sixth transmembrane regions (S5 and S6) of the mammalian voltage-gated K^+ channels in its amino acid sequence. The two transmembrane segments in each subunit are α helices, with a peripheral and an inner helix (Figure 10.4D) that run almost in parallel through the membrane. The inner helix, which corresponds to S6 in the well-characterized *Shaker* potassium channel of *Drosophila* forms the lining of the inner part of the pore. The four helix pairs are like the support poles for an inverted teepee (with the top inside), widely separated near the outer membrane surface and converging toward a narrow zone near the inner cytoplasmic surface. The inner helices are tilted with respect to the membrane normal by about 25° and are slightly kinked with the wider part facing the outside of the cell allowing the structure to form the pore region near the extracellular surface of the membrane. This region contains the K^+ channel signature sequence, forming the selectivity filter, which discriminates between K^+ and Na^+ ions. Within the selectivity filter, the orientation of the amino acid side chains preclude their participation in ion coordination, leaving this function to the oxygen atoms of the main chain carbonyls. They form an oxygen ring coordinating dehydrated K^+ ions. Roughly in the middle of the membrane is a water-filled cavity, lined by hydrophobic amino acids, and at the bottom of the cavity is another constriction that is likely involved in channel gating (Yellen, 2002).

How does this structure explain the key issues of high ion selectivity, high ion throughput, and gating of K^+ channels? Selectivity for K^+ over Na^+ is explained by the binding of the carbonyl oxygens of the selectivity filter to K^+ ions. The spatial geometry of the selectivity filter and its energetics for ion binding is perfectly tuned for K^+, the channel's natural ligand, but not for Na^+ and other ions (Yellen, 2001; Zhou et al., 2001). As a consequence, hydrated K^+ ions moving into the selectivity filter seamlessly lose their attached water molecules to form bonds with the selectivity filter. For Na^+, the pore does not adjust to form good bonding, resulting in poor dehydration and migration of the water-attached Na^+

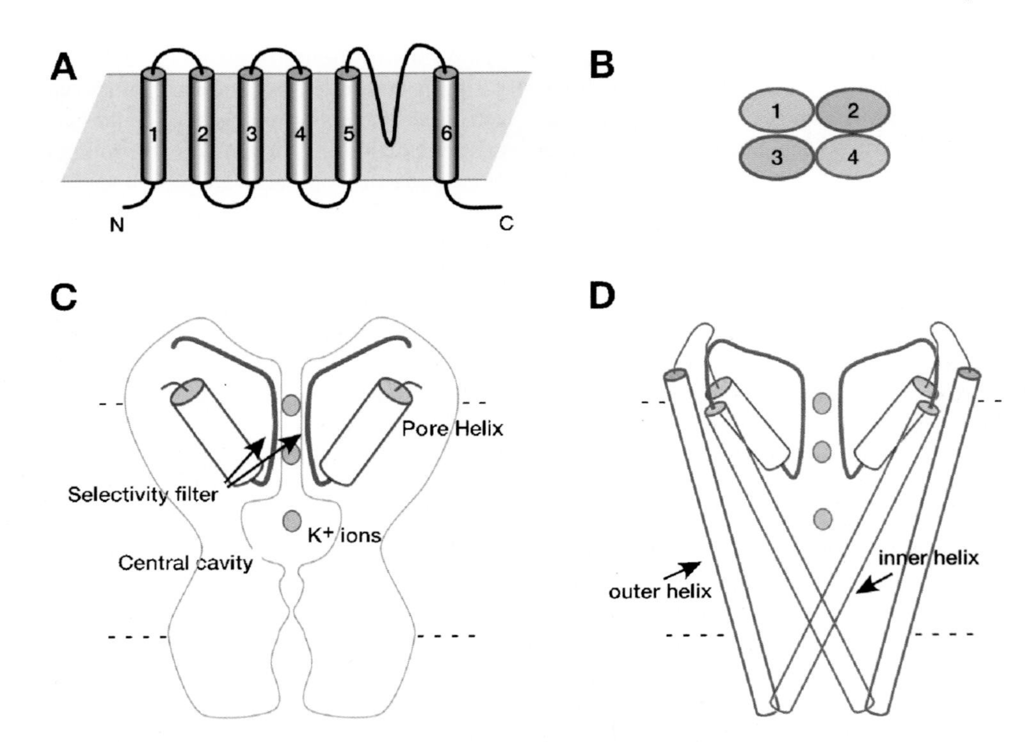

FIGURE 10.4 **(A)** Topology of a single subunit of a voltage-gated K^+ channels. Hydropathy studies predict the presence of 6 transmembrane α-helices. The sequence that spans the pore region is present between the S5 and S6 transmembrane helices. **(B)** Four identical copies of the K^+ channel subunit shown in A assemble together to form the walls of the K^+ channel. **(C)** Cross-section of an open K^+ channel, based on the crystal structure of the bacterial K^+ channel, KcsA. The selectivity filter, the wide intracellular cavity, and pore helix dipoles are highlighted. **(D)** A diagram derived from the high-resolution structure of the KcsA channel, showing the cross-over of the inner helices, corresponding to the S6 transmembrane segment of the mammalian K^+ channels. The four inner helices produce a narrow opening that provides access to the water-filled cavity in the middle of the membrane protein. (adapted from Yellen, 2002).

into the selectivity filter. The tuning of the selectivity filter for K^+ bonding also allows for its high throughput. New K^+ ions entering the selectivity filter from the water-filled cavity expel one or more of the resident ions to the opposite side on the extracellular side and produce net transport. Because the water-filled cavity is hydrophobic, K^+ ions in the cavity are less happy and impatient to get into the comfort zone of the selectivity filter, eliciting high throughput rate of ion movement. The structural basis for the phenomenon of gating is less well understood and currently the topic of intense discussion. It has been shown that the narrow constriction near the intracellular side created by the crossover of the two outer helices and lined by hostile hydrophobic residues, can close the ion permeation pathway and terminate the supply of K^+ ions to the selectivity filter (del Camino and Yellen, 2001). This has been proposed to be the primary closing process. Additionally, by adjusting the tuning of K^+-bonding and hence the rate of ion throughput, the selectivity filter itself has been proposed to underlie the opening/closing of the channel (Yellen, 2002).

The elucidation of the K^+ channel pore structure has provided us with real insight and confirmation of abstract ideas on pores, filters, gates, and ion binding sites. Structures of other channels such as Na^+ and Ca^{2+} channels await elucidation. Given the differences in ion selectivity and regulation between K^+ channels and these other channels, many differences are to be expected. However, fundamental similarities between these channels such as their pseudotetrameric subunit composition, voltage-dependent gating, and selectivity for specific ions suggest that the lessons gained from examining K^+ channel structure are also likely to extend to these and other ion channel families (Hille, 2001).

10.3. Neurotransmitters and Neurotransmission

10.3.1. Classical Neurotransmitters

What types of molecules act as neurotransmitters and how does the process of neurotransmission proceed? Following the identification of acetylcholine (ACh) as a chemical neurotransmitter by Leowi, noradrenaline (or norepinephrine, NE) was identified through the work of Dale, Cannon and others as the neurotransmitter at many sympathetic neuroeffector junctions. Although ACh and NE are quite different from each other from the chemical point of view, they do share certain key features in terms of the way they act as neurotransmitters. These features are also shared by numerous other substances that have subsequently been demonstrated to act as neurotransmitters, including other biogenic amines (e.g. dopamine and serotonin), amino acids (e.g. glutamate, GABA, and glycine) as well as many peptides (e.g. Substance P, NPY, CGRP, and the endorphins). Because of these shared features an informal consensus has gradually been reached setting out the "rules" for demonstrating that a substance acted as a neurotransmitter at a particular synapse. These rules are something like this:

(1) The potential neurotransmitter substance should be localized in the presynaptic neuron together with the enzymatic machinery for its biosynthesis.
(2) The substance should be released by stimulation of the presynaptic nerve. Because transmitter release has been shown to be dependent on the influx of Ca^{2+} ions through voltage dependent Ca^{2+} channels located in the nerve terminal (see above), evoked transmitter release should be Ca^{2+} dependent.
(3) Drugs that block synaptic transmission at a particular synapse should also block the effects of the substance when it is directly applied to the postsynaptic cell.
(4) A mechanism should exist (in addition to free diffusion) for terminating the action of the proposed neurotransmitter. Inhibition of this mechanism should prolong the time course of action of the proposed neurotransmitter.

It is clear that these criteria are fulfilled in the cases of ACh and NE acting as neurotransmitters at different synapses. Let us consider, for example, the effects of ACh at the synapse made by a motor neuron with fast skeletal muscle ("neuromuscular junction").

(1) ACh is found to be stored within the terminals of motor neurons. Detailed analysis has demonstrated that ACh is stored within small packages called synaptic vesicles that are concentrated around "active zones" on the presynaptic membrane. These active zones have been identified as specialized sites for neurotransmitter containing vesicle release. The enzyme for synthesizing ACh from choline and acetyl-CoA, choline acetyltransferase, is also found within the presynaptic terminal. Choline acetyltransferase is found in the cytoplasm. When ACh is synthesized it is pumped into synaptic vesicles by means of a specific carrier molecule located in the vesicle membrane. Once released, ACh subsequently diffuses across the synapse and activates nicotinic ACh receptors localized on the plasma membrane of the postsynaptic muscle cell producing depolarization of the muscle (see below).
(2) Stimulation of the presynaptic nerve results in the release of ACh in a Ca dependent manner. Release of ACh can be demonstrated using a bioassay of the type originally employed by Loewi or more commonly nowadays by a direct chemical method.
(3) Neurotransmission at these synapses can be inhibited by the drug d-tubocurarine, an antagonist of nicotinic ACh receptors. Direct application of ACh mimics the effects of nerve stimulation, e.g. depolarization of the muscle and muscle contraction. D-tubocurarine also inhibits both of these effects.
(4) The enzyme acetylcholinesterase (AChE) is localized at the synapse and degrades ACh released by the presynaptic nerve, thus limiting its actions. Inhibition of the effects of AChE (e.g. with a cholinesterase inhibitor such as physostigmine) prolongs the time course of action of ACh or of stimulation of the presynaptic nerve.

These observations are clearly consistent with the view that ACh acts as a chemical neurotransmitter at these synapses, indeed, they furnish the necessary "proof" of this proposed hypothesis.

Similarly, if we consider the effects of NE at noradrenergic synapses (Figure 10.5).

(1) As with ACh, NE can be shown to be stored within vesicles localized to the presynaptic terminal. The enzymes responsible for NE biosynthesis are also found within the presynaptic terminal. The synthesis of NE is more complicated than the single step involved in ACh synthesis. In the case of NE an entire biosynthetic pathway exists with the initial step catalyzed by the enzyme tyrosine hydroxylase (TH) being rate limiting (Figure 10.5). TH activity can be precisely regulated at several levels (see legend to Figure 10.5). For example, it is subject to feedback inhibition by its products such as the catecholamines dopamine and NE. Following its biosynthesis, NE is pumped into synaptic vesicles using a specific pump localized in the vesicle membrane.
(2) Stimulation of NE containing nerves is associated with the Ca^{2+} dependent release of NE.
(3) Neurotransmission resulting from NE release can be blocked by a variety of drugs that block adrenergic receptors. These may be α or β blockers depending on the situation. Application of NE mimics the effects of nerve stimulation and these effects can also be inhibited by the same drugs. For example, vasodilation produced by stimulation of sympathetic innervation of blood vessels can be blocked by a blocker of α adrenergic receptors such as phentolamine.
(4) Rather than being metabolized directly like ACh, the synaptic terminals of noradrenergic nerves express a high affinity uptake system for NE. Following its postsynaptic actions, NE is retaken up into the nerve terminals from which it was released via this high affinity uptake system and can be repackaged into synaptic vesicles. Drugs that block the high affinity presynaptic uptake system (e.g. cocaine and the tricyclic antidepressant amitryptaline) enhance the effects of presynaptic nerve stimulation or of exogenously applied NE.
(5) The properties of classical neurotransmitters, as described for ACh and NE, can be applied with some variations to numerous other substances which have subsequently been determined to act as neurotransmitters. For example, in the case of amino acid transmitters such as GABA or glutamate, we would note that much of the released neurotransmitter is cleared from synapses by high affinity uptake into glial cells (astrocytes), in addition to nerve terminals. Peptide neurotransmitters, such as one of the endor-

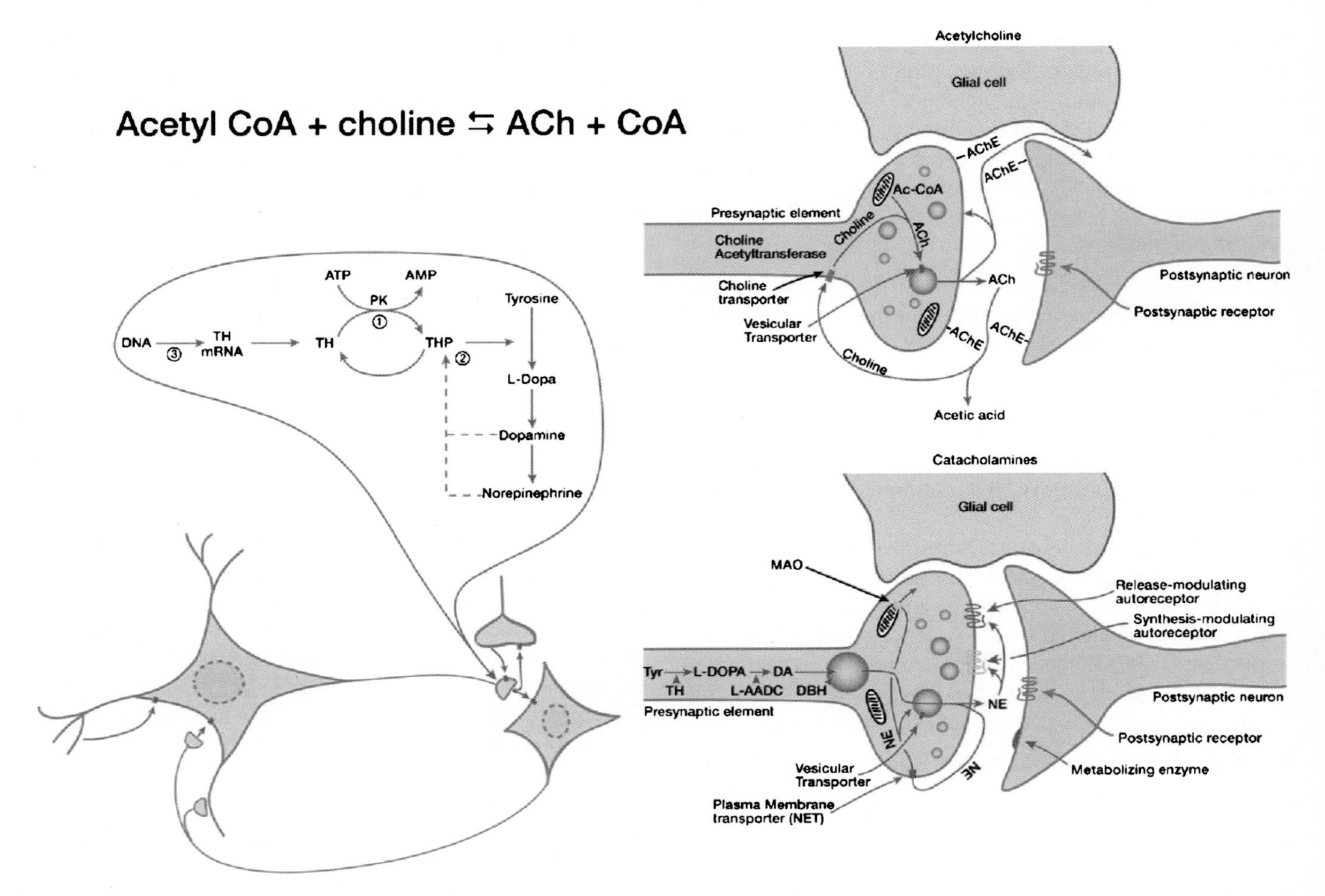

FIGURE 10.5. (a) The enzyme choline acetyltransferase catalyzes the synthesis of acetylcholine (ACh) from Acetyl-CoA and choline. (b) Diagram illustrates the different phases of ACh synthesis, release and degradation by a cholinergic neuron. (c) Biosynthetic pathway for catecholamines such as norepinephrine (NE) in a noradrenergic neuron. Note that the rate of biosynthesis of a catecholamine transmitter can be regulated at three levels. First, feedback inhibition of the rate limiting enzyme tyrosine hydroxylase (TH) by NE, adrenaline (E) or dopamine (DA). Second, phosphorylation of TH by second messenger regulated kinases and finally, at the protein level of TH by transcriptional control of mRNA transcription. (d) Diagram illustrates the different phases of NE synthesis, release and uptake in a typical noradrenergic neuron.

phins, are synthesized in the soma of the neuron as inactive precursor molecules after which they are packaged into synaptic vesicles. These are then transported down the axon to the nerve terminal. The active peptide is cleaved from its precursor by proteases copackaged in the vesicle during its transport from the soma to the terminal. In response to nerve stimulation and Ca^{2+} influx the contents of the vesicle, including the peptide neurotransmitter are released.

As will be appreciated, in all of these cases neurotransmitters are ultimately packaged into synaptic vesicles. These synaptic vesicles are released from the terminal in response to a rise in intracellular Ca^{2+}. In most instances the source of this Ca^{2+} is the extracellular medium, and Ca^{2+} moves into the nerve terminal through voltage sensitive Ca^{2+} channels. A complex of proteins holds the vesicle in a "primed" state at presynaptic release sites, the active zones discussed above. Some of the proteins involved in this complex are provided by the vesicle (v-SNAREs) and some by the presynaptic active zone (t-SNAREs). One or more proteins, in this complex senses the rise in Ca^{2+} and vesicle fusion and transmitter release ensues. The fact that neurotransmitters are stored within synaptic vesicles provides an anatomical basis for the properties of neurotransmitter release when recorded electrophysiologically. In particular, in electrophysiological recordings, transmitters are observed to be released in discrete packages or "quanta". The evoked release of these quanta is dependent on Ca^{2+} influx via voltage dependent Ca^{2+} channels situated in close juxtaposition to the active zones, exactly where transmitter-containing vesicles are docked.

10.3.2. Novel Neurotransmitters

Although these traditional views of neurotransmitter function cover the properties of many neurotransmitters, it is now clear that numerous substances that act as neurotransmitters do not fit easily into this description. Consider, for example, the endocannabinoids (Figure 10.6). Neurotransmitters in this class (e.g. anandamide and 2-arachidonoylglycerol, 2-AG) are lipid molecules derived from arachidonic acid which act as endogenous activators of the same receptors as the drug cannabis, the CB1 and CB2 cannabinoid receptors (Piomelli 2003). These transmitters show several features that differ from the above model. First, endocannabinoids are usually synthesized and released from the postsynaptic rather than the presynaptic cell. Secondly, although their synthesis is Ca^{2+} dependent, endocannabinoids are not stored in synaptic vesicles.

Following their synthesis endocannabinoids leave their cell of origin by a diffusion driven mechanism, rather than as discrete quanta. Endocannabinoids are frequently synthesized postsynaptically and then diffuse across the synapse to produce effects on the presynaptic neuron. Thus, in this case neurotransmission may be viewed as occurring in the reverse direction! Furthermore, consider a molecule like NO, which is a gas and so also diffuses very easily. Here again its synthesis is dependent on an increase in Ca^{2+}. Once synthesized, however, NO can leave the cell by diffusion and enter target cells the same way. As in the case of endocannabinoids the information carried by NO may well travel backwards across the synapse, influencing presynaptic functions following its postsynaptic release. Both the endocannabinoids and NO are clearly neurotransmitters. They are responsible for the activity dependent transfer of information across synapses. However, their *modus operandi* is not at all traditional. Indeed, we could now list of whole host of molecules that can be released from neurons following electrical stimulation and which can then transfer information across synapses in the forward or back-

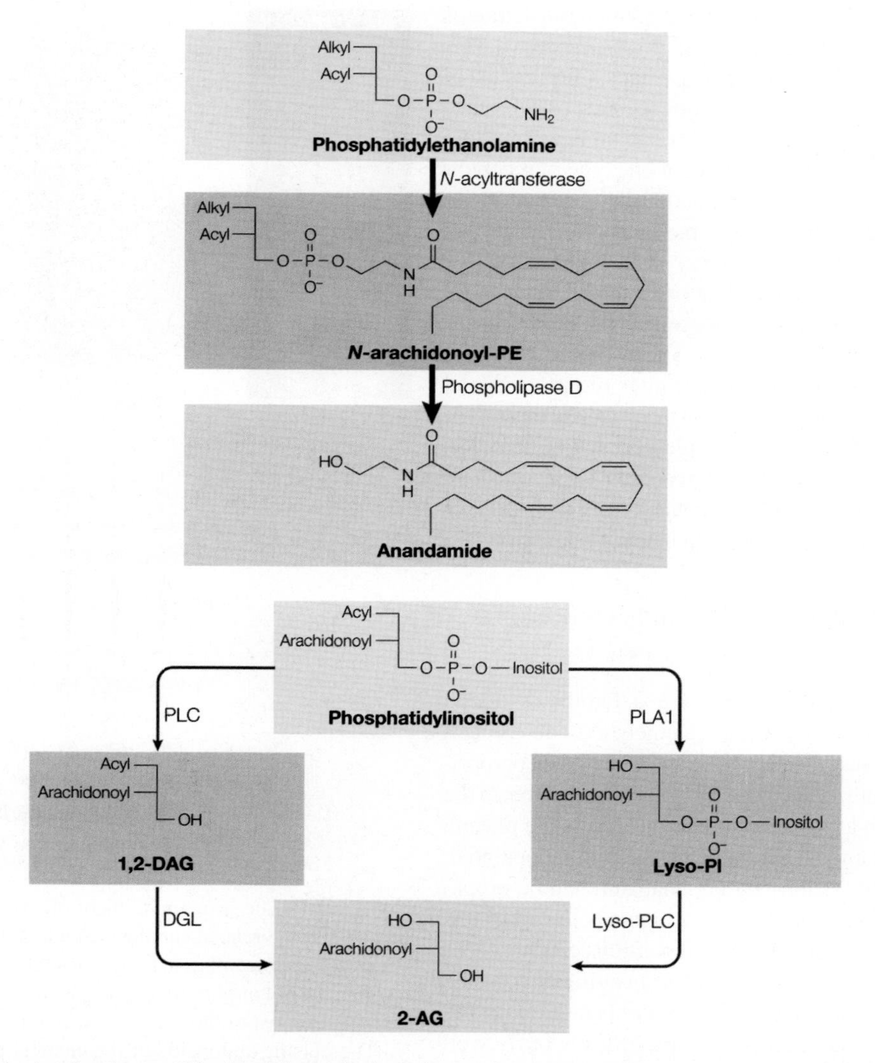

FIGURE 10.6. The biosynthesis of endocannabinoids. Biosynthetic pathways for the biosynthesis of the major endocannabinoids anandamide and 2-arachidonoylglycerol (2-AG). Note that the enzymes N-acyltransferase in anandamide biosynthesis. The biosynthesis of 2-AG can proceed via two different routes and is also dependent on an increase in Ca^{2+}. Figures reprinted from Piomelli (2003) with permission from Nature Publishing Group.

ward direction. These include a variety of growth factors and cytokines. Given the wide range of molecules that can apparently act as neurotransmitters, it is also true that information transfer across synapses can vary over a vast time scale ranging from very rapid electrical events taking a few milliseconds to much longer events taking hours/days and which involve changes in protein synthesis or gene transcription. As we shall now discuss, the type of information involved will depend on the receptor mechanism employed to decode the action of the neurotransmitter. Neurotransmitter receptors are expressed by neurons, glia and microglia and will be involved in decoding information transfer between all of these cell types.

10.4. Neurotransmitter Receptors

Once a neurotransmitter is released it will diffuse across the synaptic gap and interact with its target cell. The type of information "transmitted" to the target cell will depend both on the nature of the neurotransmitter and its receptor. As discussed above, this information transfer can be very rapid (perhaps a few milliseconds) or relatively slow (hundreds of milliseconds or even longer). Sometimes a single neurotransmitter can produce multiple types of signals if it can activate more than one type of receptor. Indeed, more than one receptor for a particular transmitter can be expressed simultaneously by the same target cell. If we our consider our archetypal neurotransmitter ACh for example, its actions can be very rapid due to activation of the nicotinic class of ACh receptors, or its actions can be somewhat slower, due to activation of muscarinic ACh receptors. The effects of ACh typify the two major classes of neurotransmitter receptors-ligand gated ion channels (ionotropic receptors) that mediate rapid synaptic transmission, and G-protein coupled receptors (GPCRs, sometimes also called metabotropic receptors) that mediate slower synaptic transmission. Let us consider the structures of these receptors and how they function in neurotransmission.

10.4.1. Ligand-Gated Ion Channels (Ionotropic Receptors)

Ionotropic neurotransmitter receptors are a family of ligand gated ion channels. Normally, these channels exist in the cell membrane in a closed state. However, upon binding the appropriate neurotransmitter they open transiently and ions permeate the channel. This results in a redistribution of ions across the plasma membrane of the cell and a change in the membrane potential. This change in potential is the "signal" that can then be propagated by the target cell. A change in membrane potential of this type is a very rapid event taking only a few milliseconds.

In the case of the nicotinic ACh receptors at the neuromuscular junction, the structure of the receptor has been extensively studied and the structural basis for its properties is fairly well established (Figure 10.7). Nicotinic receptors are made up of a pentamer of subunits, which surround a central channel region that spans the membrane. X-ray crystallographic and electron microscopic studies have revealed the overall structure of the receptor. The majority of its mass exists extracellularly in the form of a "vestibule" in which ions congregate. The receptor then enters the cell membrane as a channel or "pore" region, which narrows to highly restricted area in the middle of the membrane. Less of the mass of the receptor is found on the

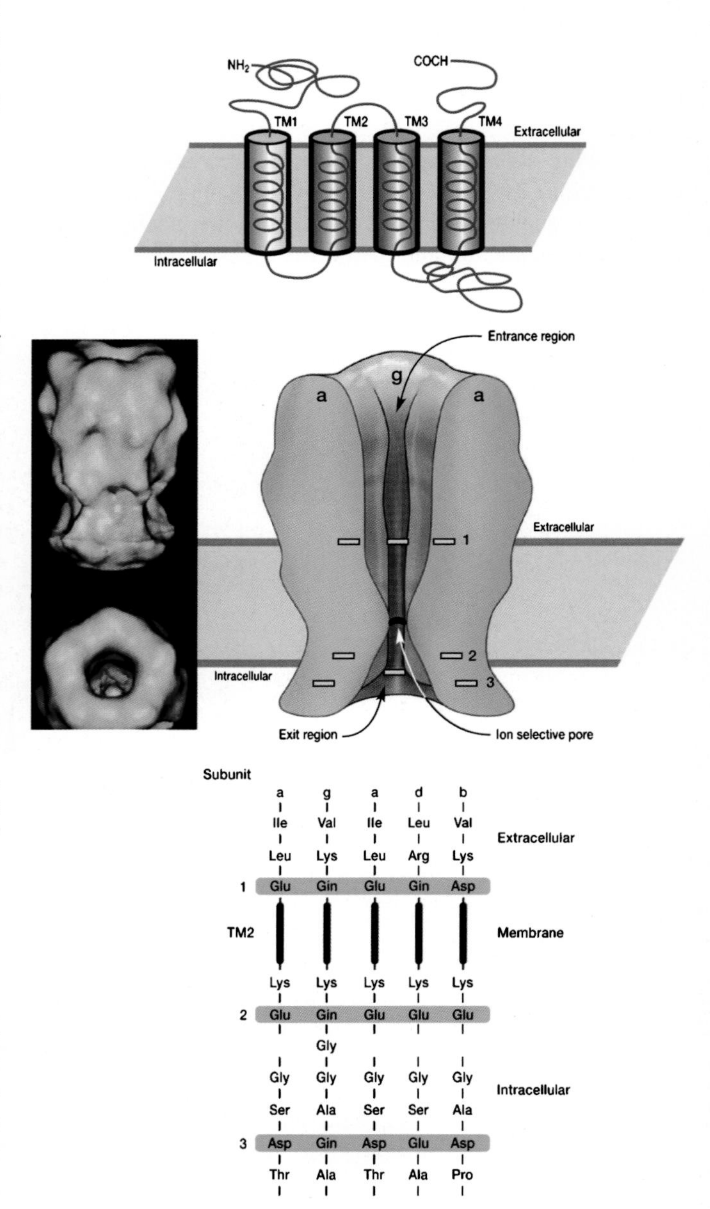

FIGURE 10.7. The structure of nicotinic ACh receptors. Receptors are pentamers of 5 related subunits (a) which form a barrel like array in which the transmembrane region of each subunit (TM2) surrounds a central channel or pore region (b) electron micrographs have demonstrated that most of the mass of the channel is located extracellularly (c) The outside and inside of the channel pore region is surrounded by rings of negatively charged amino acids. These rings of negative charges will repel anions and attract cations resulting is a channel with cationic selectivity.

intracellular (cytoplasmic) side of the membrane. Intracellularly, receptors such as nicotinic receptors are often seen to be associated with cytoskeletal proteins that serve to localize them to the appropriate portion of the cell membrane, particularly opposite active vesicle release zones in the presynaptic terminal. The 5 subunits that make up the nicotinic receptor are all related transmembrane glycoproteins with molecular weights of around 25 KDa. Each of the subunits is a transmembrane protein that crosses the membrane 4 times (transmembrane regions 1–4; TM1-4), so that their N and C termini are both extracellular. The five subunits are arranged like a barrel in which the central ion channel or pore is created by one TM2 region supplied by each subunit. The five subunits that make up the receptor consist of two α subunits, one β subunit, one γ subunit and either a δ or ε subunit. Although these are all proteins of closely related sequence, only the α subunits have binding sites for ACh. ACh molecules bind cooperatively to the receptor producing a conformational change and increasing the probability of channel opening or "gating." When two ACh molecules bind there is a high probability of channel opening. Once open, the channel is permeable to cations, which permeate the pore region moving down their electrochemical gradients. As discussed above, at normal resting potentials Na moves into the cell and K moves out. The net result of this ionic flux is depolarization of the cell as the membrane potential moves toward the reversal potential of the nicotinic receptor channel-i.e. 0 mV. Sufficient depolarization of the postsynaptic cell will trigger an action potential. If receptors are exposed to ACh for longer periods of time they change conformation once again to a form a "desensitized" state, in which ACh remains bound to the receptor but the channel is closed. The selectivity of nicotinic receptors for cations over anions is engendered by an interesting structural motif. As can be seen in Figure 10.7, both the extra and intracellular regions of the receptor just outside the cell membrane are surrounded by a ring of amino acids with a net negative charge. Hence any cation will be attracted into the channel whereas any anion will be repelled. Nicotinic receptors in other parts of the nervous system, including the sympathetic ganglia and the brain, are also thought to be composed of pentameric arrays of subunits. These subunits are members of an extended gene family of nicotinic receptor subunits that can associate in multiple different combinations of α and β subunits (Gotti and Clementi, 2004). For example, in the brain many nicotinic receptors may exist as pentamers of the $\alpha4\beta2$ combination. Indeed, some receptors e.g. $\alpha7$ exist as a homomeric array of the same subunits. The basic properties of all of these nicotinic receptors are the same but some of the details differ. For example, in addition to being permeable to Na^+ and K^+, some receptors (e.g. $\alpha7$ pentamers) are also highly permeable to Ca^{2+}. Given the importance of Ca^{2+} as an intracellular second messenger, this may have important signaling consequences.

The pentameric arrangement of subunits used to construct nicotinic receptors is also utilized to make up other types of ligand gated ion channels including the GABA-A, glycine and 5-HT3 receptors. Here again the subunits used are members of the same extended gene family of proteins as those that make up nicotinic receptors, and the basic structure of the receptor as well as its mechanism of action are similar. Interestingly, the GABA-A and glycine receptors are anion (chloride and bicarbonate) rather than cation permeable channels, a property based on the same structural motif that makes the nicotinic receptors cation selective. However, in this case the channels possess rings of positive rather than negative amino acids. GABA-A receptors are highly important as they mediate the rapid effects of GABA at most of the synapses in the brain. Furthermore, these receptors are also the targets for many pharmacologically significant substances including anxiolytic and hypnotic drugs. As in the case of nicotinic receptors the precise subunit composition and properties of GABA-A receptors vary throughout the neuraxis depending on the precise combination of subunits utilized. The ability to target drugs to these various subclasses or receptors allows for the production of agents with selective anxiolytic or hypnotic properties, for example.

Another important family of ligand gated ion channels is the receptors for the excitatory amino acid neurotransmitters such as glutamate and aspartate (Figure 10.8). These act as neurotransmitters at the vast majority of excitatory synapses in the brain. These receptors are also multisubunit ion channel arrays, in this case consisting of tetramers of related subunits (Wollmuth and Sobolevsky, 2004). The proteins that make up these subunits are all related in structure but are quite different from those that make up the nicotinic and GABA-A receptor family.

In the case of glutamate receptors the basic structure of each subunit is a protein that crosses the membrane 3 times. The TM2 or P region, which again forms the pore of the channel, does not cross the membrane entirely and forms a loop like structure that folds back into the membrane. Indeed, the basic structure of this pore region bears similarities to that described above for K^+ channels. The overall structure of each subunit consists of a large N-terminal extracellular region attached to a transmembrane region that forms the channel. It is interesting to note that the large extracellular region is related in structure to ancient bacterial periplasmic amino acid binding proteins. Similar extracellular motifs also make up the N-terminal of metabotropic glutamate receptors, although in this case they are attached to a 7 transmembrane core rather than an ion channel motif (Jingami et al., 2003 Figure 10.8). In the case of the ionotropic glutamate receptors, binding of glutamate to the "clam shell" region in the extracellular portion of the receptor closes the clamshell, initiating a conformational change that eventually results in channel gating. There are basically 3 types of glutamate receptors, named for archetypal agonists that activate each class—these being AMPA, kainate and NMDA receptors. The subunits that make up these receptors (GluR subunits) are all related and form an extended gene family. Although their basic structures

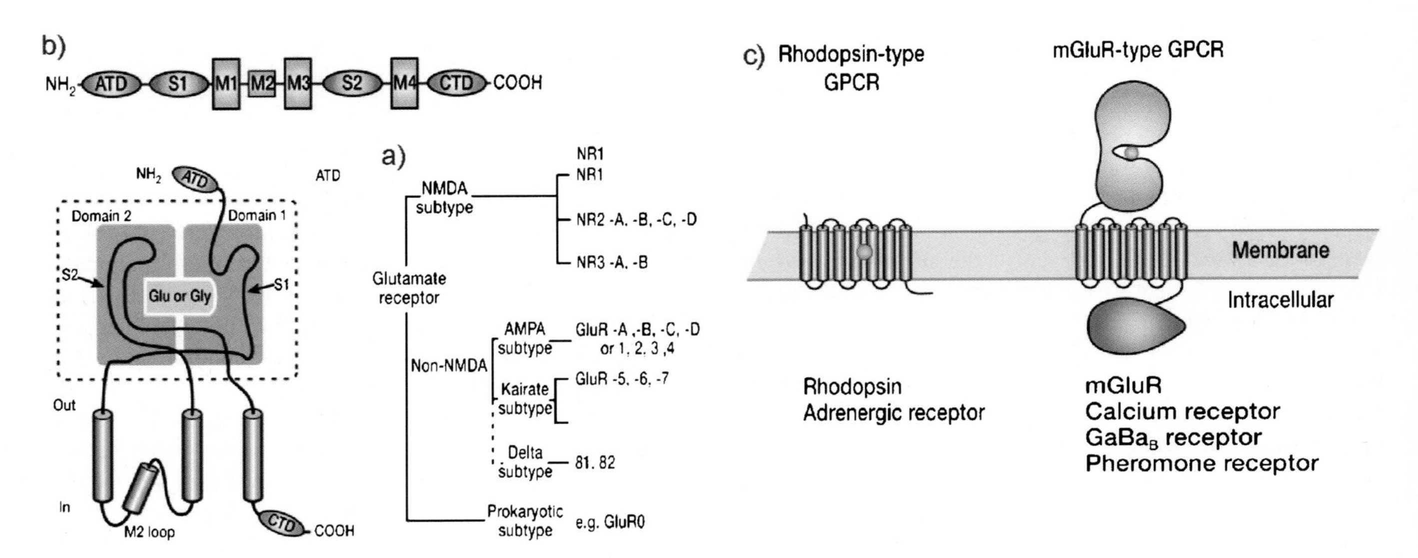

FIGURE 10.8. The structures of ionotropic and metabotropic glutamate receptors. (a) Family tree of related protein subunits that constitute different types of ionotropic glutamate receptors. (b) Typical domain structure of a glutamate receptor subunit. The extracellular regions show homologies to bacterial periplasmic binding proteins. (c) Structure of a metabotropic glutamate receptor. In this case the extracellular region also bears homology to a bacterial periplasmic binding protein, but this time it is attached to a 7-transmembrane G protein coupled receptor motif. Metabotropic glutamate receptors probably function as dimmers or other higher order arrays.

and gating mechanisms are similar, there are some interesting details that distinguish each type. The AMPA receptors are made up of tetrameric arrays of the GluR 1–4 (or A-D) subunits. These receptors are basically cation selective channels that are relatively impermeable to Ca. However, it was demonstrated that the structure of the channel as coded in the genome was slightly different. The originally coded GluR2 subunit possessed a single glutamine residue in the TM2/P loop region. This is the region of the channel that makes up the cation permeable channel pore. In contrast to the sequence coded in the genome, the subunit normally found to exist in functional AMPA receptors expressed in cells has an arginine residue in exactly the same position. The triplet code for Gln is CAG. It was shown that neurons express a highly specific adenine deaminase that "edits" this triplet to produce C-Inosine-G, which is then read by the protein synthetic machinery as CGG or arginine. It was observed that Gln or Arg containing channels differed in their Ca^{2+} permeabilities, the extra positive charge associated with Arg repelling Ca^{2+} ions and making the channel relatively Ca^{2+} impermeable. It is possible, however, that in some cases editing of GluR2 may be incomplete or that some tetrameric AMPA receptors may not include an edited GluR2 subunit and are therefore Ca^{2+} permeable. Thus, a single base change produced by RNA editing can completely change an important property of these AMPA receptors, i.e their permeability to Ca^{2+}.

In the case of NMDA receptors Ca^{2+} permeability is also an important issue and is a key to understanding certain forms of glutamate mediated synaptic plasticity (see below). These properties of NMDA receptors can be readily observed through a consideration of their electrophysiological properties (Figure 10.9). Let us compare the properties of AMPA and NMDA receptors. When recorded in physiological solutions, currents carried via AMPA receptors display current/voltage (I/V) relationships that strictly follow Ohm's law. On the other hand the I/V relationships for NMDA receptors exhibit a region of "negative slope conductance" similar to that displayed by voltage dependent channels, such as voltage sensitive Na^+ or Ca^{2+} channels (see above). An important discovery was that this behavior was not due to the intrinsic gating properties of the NMDA receptor protein itself. It was observed that if Mg^{2+} was removed from the physiological bathing solution in which the measurements were made, currents carried by NMDA receptors behaved in a strictly Ohmic manner just like the related AMPA receptors. Further investigations revealed that NMDA receptors can be blocked by Mg^{2+} at hyperpolarized membrane potentials. As the cell depolarizes the block by Mg^{2+} is relieved. This behavior means that glutamate will only activate an NMDA mediated current when the cell is relatively depolarized. This has important implications for the mechanisms underlying some types of synaptic plasticity (see below).

10.4.2. G-Protein-Coupled Receptors/GPCRs

In addition to ligand gated ion channels neurotransmitters often activate GPCRs. These receptors are all based on a structural motif in which the receptor protein crosses the membrane 7 times in a serpentine like manner so that that N-terminal is extracellular and the C-terminal is intracellular. This arrange-

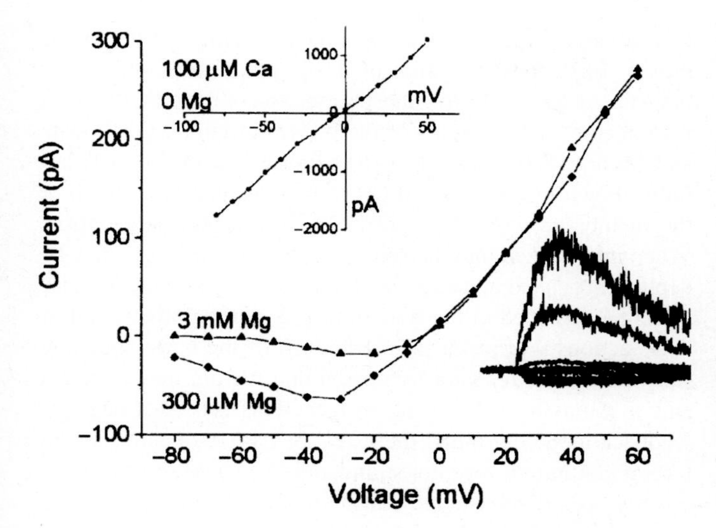

FIGURE 10.9. Biophysical properties of the N-methyl-D-aspartic acid receptor. The current voltage (i/v) relationship measured in physiological solutions shows a region of "negative slope conductance" at hyperpolarized potentials. Little inward current is observed until the cell is substantially depolarized. However, when Mg^{2+} is removed from the bathing medium the i/v relationship follows Ohm's law and is represented by a straight line. The reason for this behavior is that Mg^{2+} block NMDA receptors in a voltage dependent manner (see main text).

ment results in the formation of three intracellular loops between the different transmembrane regions of the protein. These loops can be used to transduce signals to the inside of the cell and, in particular, to heterotrimeric G-proteins. GPCRs constitute an extremely large gene family and include many variations on the same basic structural motif (Figure 10.10). However, in virtually every instance the mechanism of action is basically the same. Binding of the agonist produces a conformational change in the receptor that leads to activation of the G-protein. G-protein activation results in its dissociation into α or βγ subunits-either of which can then mediate signal transduction within the cell. In the classical description of GPCR function the downstream signal is typically produced through the regulation of an enzyme like adenylate cyclase or phospholipase C. Activation of these enzymes produce second messenger molecule such as cAMP, diacylglycerol (DAG) or Ca^{2+} that then activate protein kinases or other effectors. For example, phosphorylation of ion channels by activated kinases changes their gating properties and so transduces the message to the level of electrical signaling. However, this classical model is only one of many possible alternatives. The subunits of the G-protein may bypass the kinase activation step and interact directly with the channel protein, resulting in direct channel gating. For example, direct interaction of G-protein βγ subunits with K^+ channels or N-type Ca^{2+} channels can

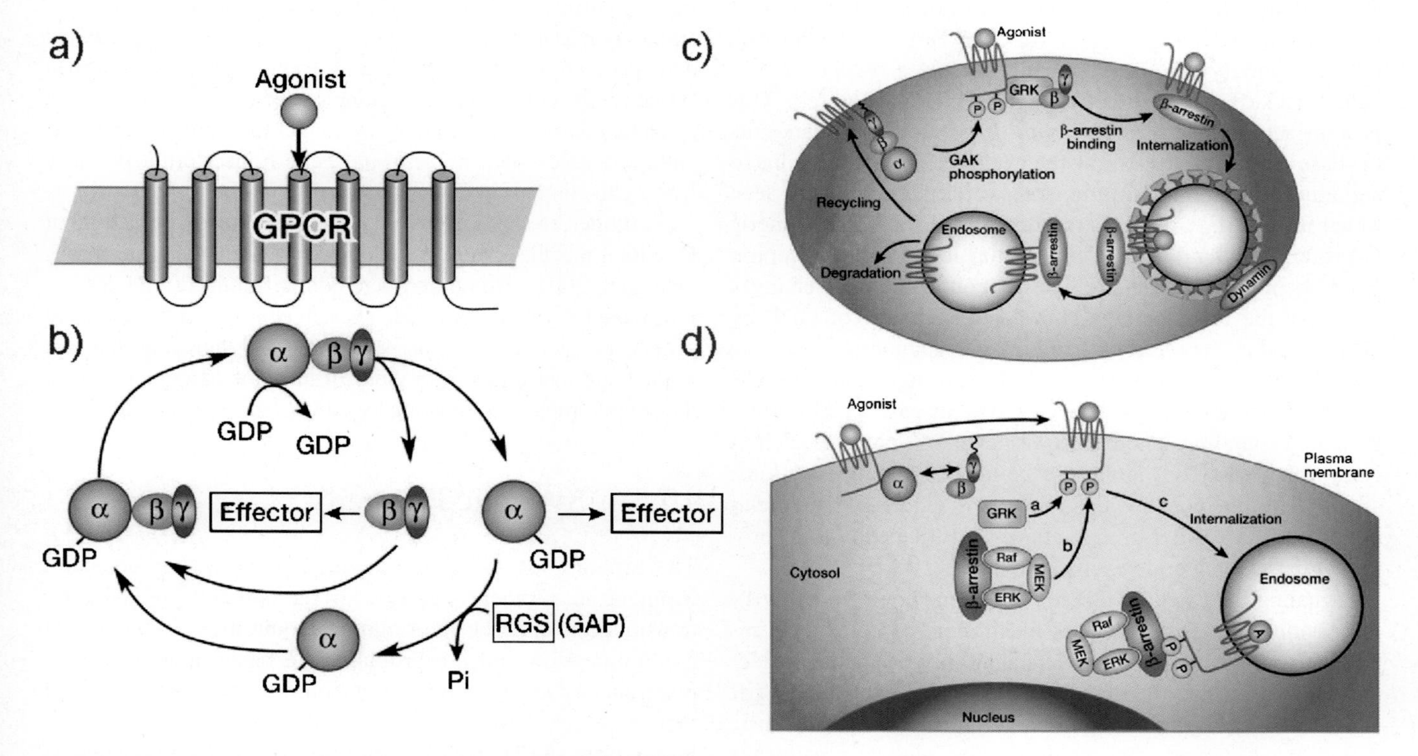

FIGURE 10.10. G-protein coupled receptor (GPCR) structure and function. (a) The basic 7-transmembrane structure of a GPCR is common to all members of the family. (b) The cycle of heterotrimeric G-protein activation. The cycle is initiated by stimulation of GTP/GDP exchange on the G-protein alpha subunit produced by an agonist induced conformational change in the GPCR. (c) Activation of a GPCR also produces interactions with proteins of the β-arrestin family that mediate uncoupling of the GPCR from its G-protein and receptor internalization. (d) Arrestins can also act as scaffold proteins that bring together members of the MAP kinase pathway and so activate MAP kinase signaling.

directly modulate channel function by activating K^+ channels or inhibiting N-channels. Such effects can directly modulate neurotransmitter release at nerve terminals. Although such a model is generally applicable to the activation of GPCRs, in the case of certain GPCRs that are important in the nervous system the basis for signal transduction is not entirely clear and may not even involve a traditional G-protein. Thus, for GPCRs of the frizzled or smoothened families, which are involved in the important Wnt and hedgehog signaling pathways, the precise mechanisms of signal transduction are not known and the role of G-proteins is unclear. Thus, as we shall discuss below, activation of a heterotrimeric G-protein may be only one way in which a GPCR can ultimately transduce information.

How does a GPCR normally work from the molecular point of view? In order to answer this question we should first consider how G-proteins produce their effects (Figure 10.10). For heterotrimeric G proteins the α-subunit has intrinsic GTPase activity. Normally GDP is bound to this subunit. In order to initiate a cycle of G-protein mediated signaling this GDP moiety must be replaced by a GTP molecule. Binding of GTP results in dissociation of the G-protein into α and β subunits and signaling ensues. Once the GTP has been hydrolyzed the heterotrimer reforms and signaling ceases. In order for these events to be carried out two proteins are important in addition to the G-protein. The first of these is a GEF or "guanyl nucleotide exchange factor". This protein is responsible for stimulating the initial exchange of GTP for GDP. Because the intrinsic GTPase activity of most α-subunits is low a second protein called a GAP (GTPase activating protein) is required. This protein acts upon the GTP bound α–subunit to enhance its GTPase activity and so allow the cycle of G-protein mediated signaling to terminate. In this context the GPCR can be seen to act as a GEF. Under resting conditions the heterotrimeric G-protein is bound to the intracellular loops of the receptor. Binding of the agonist produces a conformational changes such that the receptor now stimulates exchange of GDP by GTP bound to the α-subunit and initiates signaling. Thus, the GPCR is an agonist activated GEF. The arrangement of GPCR and heterotrimeric G-protein is a special case of G-protein mediated signaling. In other cases "small" G-proteins such as Ras, which carry out numerous cellular signaling functions, are not heterotrimers but also function in GTP regulated manner. In this case other proteins have GEF like activity to promote guanyl nucleotide exchange.

Signaling mediated by GPCRs has been shown to be a very information rich event which is much more complex than initially supposed. In order to understand this we should consider the entire sequence of events that takes place when a GPCR is activated. Here again there are many variations on the basic theme, but it appears that there are other signaling pathways that are activated in addition to G-proteins. Binding of the agonist to the receptor initiates conformational changes in the intracellular C terminal region. This allows diverse residues in this region to be phosphorylated by kinases of the GRK (GPCR kinase) family. Phosphorylation of these residues has numerous effects. First, the interaction of the receptor and G-protein is interrupted. Secondly, the phosphorylation of the GPCR allows it to interact with proteins of the β-arrestin family (Lefkowitz and Shenoy, 2005 and Figure 10.10). This interaction was initially shown to allow the GPCR to be relocated to a region of the membrane (coated pit) resulting in receptor endocytosis. When internalized into the cell in this way the receptor may be dephosphorylated and recycled back to the cell surface or else it can be degraded in the lysozomes. As will be obvious all of these actions will result in interference of the basic signaling functions of the receptor by uncoupling it from the G-protein and/or removing it from the surface. This acts as a negative feedback loop controlling the extent of GPCR signaling in the face of continuous receptor stimulation. However, it has subsequently been demonstrated that β-arrestin like molecules have a very large number of additional receptor related functions. For example, β-arrestins can act as scaffolding proteins for the intermediates of the MAP kinase signaling cascade. Thus, once activated by a GPCRs arrestins might bind MAP kinases allowing for their mutual phosphorylation and activation. The overall effect of this is to inhibit the initial signaling of the GPCR and to redirect it down the MAP kinase pathway. Indeed, it has been frequently observed that activation of a GPCR can produce activation of ERK or other MAP kinases (Figure 10.10). Other studies have linked GPCR/β-arrestin interactions to effects on the ubiquitination/proteasome pathway. Thus, the diversity of signaling initiated by a GPCR can potentially be very great and can operate over a wide time course. Rapid signaling events can influence the activity of ion channels, electrical excitability and synaptic communication, whereas longer term effects can influence processes such as neuronal gene transcription possibly leading to changes in neuronal structure or phenotype.

A further recent insight into the mechanism of action of GPCRs is that they often act as dimers or other higher order arrays (Milligan, 2004). Homo and heterodimerization of GPCRs has now been frequently reported. These receptor dimers may have properties, including agonist selectivity and signaling, that are unique and that differ from those of receptor monomers. Thus, the overall impact of signaling by GPCRs may be very diverse.

10.5. Synaptic Plasticity

The strength of synaptic communication is not constant. Moment to moment changes in the strength of synaptic transmission underlie the ongoing requirements of neuronal communication and are probably the basis of long lasting phenomena such as learning and memory. We now understand that there are numerous forms of this synaptic "plasticity". Some forms of plasticity last for brief periods of time whereas others last for many hours or days-perhaps encompassing the lifetime of the organism. In each case changes in the amount of transmitter released, postsynaptic sensitivity to the transmitter or combinations of these processes are involved in producing these effects. In order to give a general idea as to what these important and widespread neuronal signaling processes involve, we shall discuss two different examples of synaptic plasticity.

10.5.1. Long-Term Potentiation (LTP)

Long-term potentiation and its flip side, long-term depression (LTD), are probably the most intensively studied of all forms of synaptic plasticity (Collingridge et al., 2004 and Figure 10.11). LTP was first observed *in vivo* using rabbits by Bliss and Lomo who demonstrated that a brief high frequency stimulation of the perforant path input to the dentate gyrus of the hippocampus from the entorhinal cortex produced a long lasting (many hours) potentiation of the extracellularly recorded field potential. This phenomenon, or something akin to it, has now been shown to exist at other hippocampal synapses and at many other synapses throughout the brain. Subsequent investigations both *in vivo* and using hippocampal slice preparations revealed LTP to have some interesting properties. For example, if a neuron receives both a weak and a strong input, in which the weak input is not sufficient to produce potentiation, the weak input can be potentiated if it is paired in time with a tetanic stimulation of the strong input. This property of "associativity" was seen to be a neurophysiological correlate of the proposal made by Canadian psychologist Donald Hebb that coincident activity between two synaptically coupled neurons would cause increases in the synaptic strength between them-a postulate that was made to explain how long-term phenomena such as memory could be represented synaptically. From the point of view of the present discussion the mechanism underlying LTP at many excitatory synapses has been shown to depend on the biophysical properties of NMDA receptors as discussed above. Indeed, the induction of LTP at numerous synapses has been shown to depend both on NMDA receptor activation and the associated influx of Ca^{2+} into the postsynaptic neuron. It will be recalled that the function of NMDA receptors is critically dependent on block by physiological concentrations of Mg^{2+}. Accordingly, when synaptic activity is low and the cell maintains a relatively high resting potential NMDA receptors are blocked. At low rates of presynaptic stimulation the synaptic release of glutamate will result in activation of postsynaptic AMPA receptors. Although AMPA receptor activation will result in some postsynaptic depolarization, this will not be sufficient to relieve the Mg^{2+} block of postsynap tic NMDA receptors, and so no NMDA receptor associated

a)

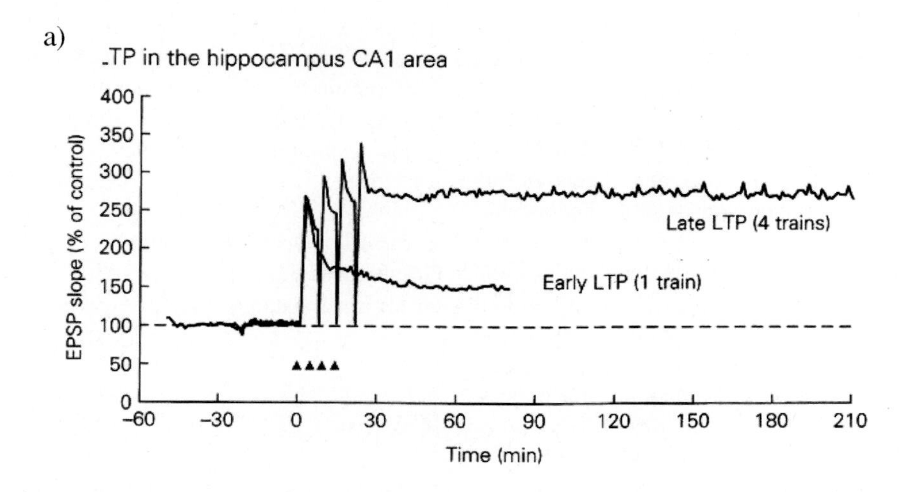

b)

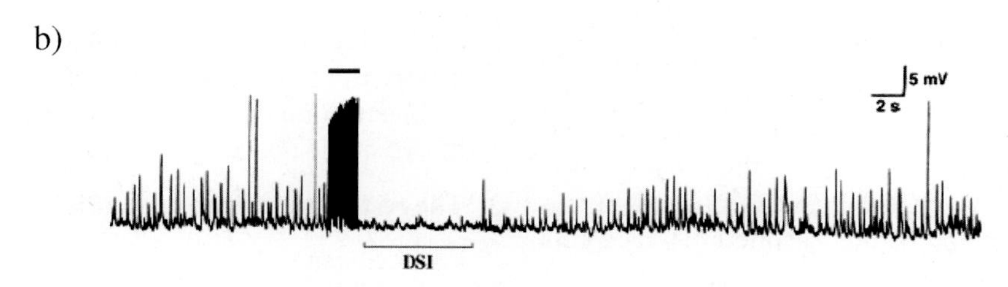

FIGURE 10.11. Examples of receptor mediated synaptic plasticity. (a) Long-term potentiation. (LTP). Synaptic transmission at hippocampal synapses (in this example the CA3/CA1 synapse) is potentiated following application of a tetanus (period of rapid stimulation) to the presynaptic nerve. This phenomenon is dependent on the activation of synaptic AMPA and NMDA receptors (see main text). (b) Depolarization suppression of inhibition (DSI). Diagram shows how endocannabinoids produce DSI in the hippocampus. Synaptic potentials recorded as hippocampal neurons are depressed following a tetanus.

current will result. Now consider the situation when a tetanus is applied to the presynaptic nerve resulting in increased glutamate release. The degree of postsynaptic depolarization produced by AMPA receptor stimulation may now sufficient to relieve the block of NMDA receptors by Mg^{2+}. Once Mg^{2+} leaves the receptor and the receptor is unblocked, its activation by glutamate will not only result in the influx of Na^+ but also that of Ca^{2+}. Subsequent studies have shown that it is this influx of Ca^{2+} that acts as a second messenger stimulating the changes responsible for maintained increases in synaptic function, primarily the insertion of new AMPA receptor subunits into the postsynaptic cell membrane and hence an increase in the strength of glutamate mediated signaling. There are numerous variations on this basic model including observations that, depending on the precise pattern of pairing of synaptic inputs, resulting synaptic transmission may be decreased rather than increased, a phenomenon known as Long-Term Depression (LTD). Thus, synaptic strength at central glutamate synapses can be tuned up or down depending on the overall requirements of the synapse in question, a process dependent on glutamate receptor mediated signaling.

10.5.2. Depolarization-Induced Suppression of Inhibition (DSI)

It has been shown that when the pyramidal neurons of the CA1 field of the hippocampus are depolarized, the inhibitory GABA mediated input to these cells is transiently suppressed (Figure 10.11). This phenomenon is known as "depolarization induced suppression of inhibition" (DS1, Diana and Marty, 2004). As with LTP/LTD described above, phenomena of this type occur widely throughout the nervous system. The mechanism of this effect and its dependence on receptor signaling is interesting. It has been shown that DSI is dependent on the influx of Ca^{2+} into the postsynaptic cell (e.g. the pyramidal neuron in this example). However, this results in a reduction of GABA release presynaptically. This means the signal that is responsible for this effect must be transmitted backwards across the synapse. It will be recognized that this is something that applies to endocannabinoid signaling in the brain. Indeed, it has been shown that DS1 can be inhibited by blockers of CB1 cannabinoid receptors (Piomelli, 2003). Thus, the mechanism involved appears to be as follows. Influx of Ca^{2+} into the postsynaptic cell as a result of postsynaptic depolarization, activates endocannabinoid synthesis. The endocannabinoid molecules then leave the cell and diffuse back across the synapse where they bind to CB1 receptors situated on presynaptic terminals. Activation of presynaptic CB1 receptors produces activation of heterotrimeric G-proteins. Binding of the G-protein $\beta\gamma$ subunits to voltage dependent Ca^{2+} channels in the nerve terminal, inhibits these channels. Thus, less Ca^{2+} enters the terminal in response to an action potential and less transmitter is released. The metabolism of the endocannabinoid results in the transient nature of the phenomenon.

Summary

Neurons communicate with each other, as well as with glial cells, through the propagation of action potentials and the release of chemical neurotransmitters across synapses. In this chapter we discuss the molecular processes that are responsible for the electrical excitability of neurons that allows the generation of action potentials as well as the electrical properties of glia cells. The nature of the resting membrane potential in cells and the structure and function of voltage dependent Na, K and Ca channels are discussed. Many different types of substances can act as neurotransmitters. The first neurotransmitter to be discovered was acetylcholine, but now many other substances including biogenic amines and peptides are also known to be neurotransmitters. Conventional neurotransmitters carry information across synapses in an anterograde manner, although some recently defined neurotransmitters such as the endocannabinoids carry information in a retrograde fashion. Neurotransmitters act on different types of receptors. The main classes of receptors are ligand gated ion channels, such as the nicotinic acetylcholine receptor and G-protein coupled receptors, which include the majority of receptors for biogenic amines. Activation of ligand gated ion channels results in a rapid change in the membrane potential, whereas activation of a G-protein coupled receptor can result in numerous changes in ion gradients and second messenger systems. Neurotransmission is not a unvarying process and the strength of synaptic communication can vary on a moment to moment basis – a process known as synaptic plasticity. Different types of synaptic plasticity exist and may involve the participation of receptors for glutamate or for endocannabinoids for example.

Acknowledgment. The authors thank William Wassom, Graphic Designer at the University of Nebraska Medical Center for his assistance with the figures.

Review Questions/Problems

1. The undershoot of an action potential occurs from

a. Prolonged opening of voltage-activated Na^+ channels
b. Opening of voltage-activated K^+ channels
c. Inactivation of Na^+ channels
d. All of the above

2. The resting membrane potential is largely influenced by

a. K^+ channels
b. Na^+ channels
c. none of the above
d. all of the above

3. What is a "selectivity filter" of an ion channel? Do all channels contain a selectivity filter?

4. How many pore-forming K^+ channel subunits assemble to form a functional channel?

a. One
b. Two
c. Four
d. None of the above

5. Voltage-gated Ca^{2+} channels mediate

a. Fusion of neurotransmitter-filled vesicles in the presynaptic terminals
b. Repolarization of the action potential
c. Generation of the resting membrane potential
d. All of the above

6. Name four properties that you would expect a chemical substance to possess if it functions as a neurotransmitter at a synapse you are investigating.

7. What is an "endocannabinoid"? What evidence exists that these substances act as neurotransmitters?

8. How many protein subunits does a nicotinic acetylcholine receptor possess?

a. More than one
b. Lots
c. Five
d. Not sure, but definitely around 5 (…..or perhaps 6)

9. How do ionotropic glutamate receptors regulate the movement of Na and Ca ions into neurons? Why is Ca influx through glutamate receptor channels important?

10. What is the basic structure of a G-protein coupled receptor? How does agonist binding to the receptor lead to the activation of heterotrimeric G-proteins?

11. Which of the following apply to the arrestin class of proteins?

a. They are scaffold proteins that can bind members of the MAP kinase pathway.
b. They can sometimes act as receptors for opioid peptides.
c. They are involved in down regulating the effects of activation of GPCRs
d. They act as blockers of ionotropic glutamate receptors, or something like that.

References

Collingridge G, Isaac JT, Wang WT (2004) Receptor trafficking and synaptic plasticity. Nat Rev Neurosci 5:952–962.

del Camino D, Yellen G (2001) Tight steric closure at the intracellular activation gate of a voltage-gated K+ channel. Neuron 32: 649–656.

Diana MA, Marty A (2004) Endocannabinoid-mediated short-term plasticity. Br J Pharmacol 142:9–19.

Doyle DA, Morais Cabral J, Pfuetzner RA, Kuo A, Gulbis JM, Cohen SL, Chait BT, MacKinnon R (1998) The structure of the potassium channel: Molecular basis of K^+ conduction and selectivity. Science 280:69–77.

Gotti C, Clementi F (2004) Neuronal nicotinic receptors: From structure to pathology. Prog Neurobiol 74:363–396.

Hille B (2001) Ion Channels of Excitable Membranes. 3rd edition. Sinauer Associates, Inc.

Jingami H, Nakanishi S, Morikawa K (2003) Structure of the metabotropic glutamate receptor. Curr Opin Neurobiol 13:271–278.

Lefkowitz RK, Shenoy SK (2005) Transduction of receptor signals by β-arrestins. Science 308:512–517.

Milligan G (2004) G-protein coupled receptor dimerization: Function and ligand pharmacology. Mol Pharmacol 66:1–7.

Piomelli D (2003) The molecular logic of endocannabinoid signaling. Nat Rev Neurosci 4:873–884.

Valenstein ES (2005) The War of the Soups and the Sparks. Columbia University Press, NY.

Wollmuth LP, Sobolevsky AI (2004) Structure and gating of the glutamate receptor ion channel. Trends Neurosci 27:321–328.

Yellen G (2001) Keeping K^+ completely comfortable. Nat Struct Biol 8:1011–1013.

Yellen G (2002) The voltage-gated potassium channels and their relatives. Nature 419:35–42.

Zhou Y, Morais-Cabral JH, Kaufman A, MacKinnon R (2001) Chemistry of ion coordination and hydration revealed by a K^+ channel-Fab complex at 2.0 A resolution. Nature 414:43–48.

11 The Vertebrate Retina

Wallace B. Thoreson

Keywords Amacrine cell; Bipolar cell; Cone; Fovea; Horizontal cell; Müller cells; Photoreceptor; Retina; Retinal ganglion cell; Rod

11.1. Introduction

The retina, like other parts of the CNS, derives embryologically from the neural tube. Retinal neurons and glia therefore have many properties in common with other CNS tissue, but they also exhibit specialized response properties and proteins that have evolved to serve the retina's function in transducing light energy into nerve signals and analyzing the visual image. For example, to encode small changes in light intensity, many retinal neurons respond to light with graded changes in membrane potential and do not exhibit sodium-dependent action potentials. In addition to possessing the specialized proteins needed for phototransduction (e.g., rhodopsin and transducin), the retina contains a number of proteins specialized for the transmission and processing of visual information such as the metabotropic glutamate receptor, mGluR6, in ON type bipolar cells; the $GABA_c$ receptor (Lukasiewicz et al., 2004); and the alpha 1F calcium channel subtype in rod photoreceptors. This chapter summarizes some of the special features of retina that help it to transduce and process light.

11.2. Anatomy

The retina is a thin sheet of neural tissue (150–400 μm thick) at the back of the eye upon which the visual image is focused by the cornea and lens (Figure 11.1A). Behind the neurosensory retina is a layer of pigmented epithelial cells known as the retinal pigment epithelium (RPE, Figure 11.1) and beyond the RPE is a dense bed of capillaries within the choroid. Surrounding the choroid is the sclera, made up of densely woven collagen fibers that help to encase and protect the globe.

The retina has a highly organized laminar structure that is similar in all vertebrate species (Figure 11.1B; Dowling, 1987; Rodieck, 1998). It is oriented so that photoreceptors lie at the back, adjacent to the RPE. Thus, to reach the photoreceptors light must pass through overlying layers of the retina. The photoreceptor (PR) layer consists of the outer and inner segments of rod and cone photoreceptors. Separating the inner segments and photoreceptor cell bodies is the outer limiting membrane (OLM) formed from apical processes of glial Müller cells. The outer nuclear layer (ONL) contains the cell bodies of photoreceptors, which make synaptic contact with horizontal and bipolar cells in the outer plexiform layer (OPL). The inner nuclear layer (INL) contains cell bodies of horizontal, bipolar, and amacrine cells. Synaptic contacts among bipolar, amacrine and retinal ganglion cells are made in the inner plexiform layer (IPL). Anterior to the IPL, closer to the front surface of the retina, is the ganglion cell layer (GCL), which contains cell bodies of retinal ganglion cells. Axons from retinal ganglion cells create a nerve fiber layer (NFL) at the inner surface of the retina. The axons join together as they exit the eye to form the optic nerve, which projects to higher visual centers. The absence of retina and photoreceptors at the optic nerve head creates a small blind spot (Figure 11.1A). The vitreal surface of the retina is bounded by an inner limiting membrane (ILM) formed by Müller cell endfeet.

The retina contains ~60 types of retinal neurons grouped into five major classes: photoreceptors (3–4 types), horizontal cells (2 types), bipolar cells (10–12 types), amacrine cells (~30 types), and ganglion cells (10–15 types) (Masland, 2001; Wässle, 2004). Photoreceptors release L-glutamate to stimulate bipolar cells, which in turn release L-glutamate onto ganglion cells forming a throughput pathway for the transmission of light signals through the retina (Figure 11.2). Horizontal and amacrine cells are predominantly inhibitory interneurons with processes extending laterally in the outer and inner plexiform layers, respectively. As light signals are transmitted from photoreceptors to bipolar cells to ganglion cells, they are modified by inhibitory synaptic feedback from horizontal cells onto photoreceptors and bipolar cells in the outer retina and from amacrine cells onto bipolar and ganglion cells in the inner retina. Photoreceptors and second order retinal neurons (bipolar and horizontal cells) respond to light with graded changes in their membrane potential (Dowling, 1987; but see Kawai et al., 2001). Action potentials are typically observed only in third order neurons of the retina (amacrine and ganglion cells). The use of graded responses by cells early in the visual

T. Ikezu and H.E. Gendelman (eds.), *Neuroimmune Pharmacology.*
© Springer 2008

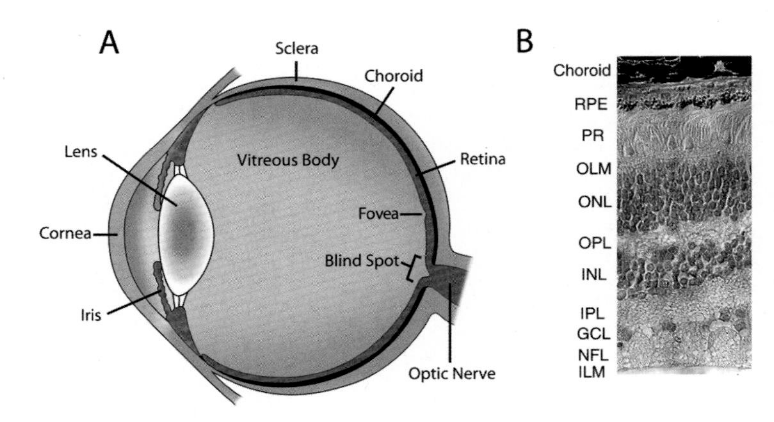

FIGURE 11.1. (A) Schematic diagram of a primate eye. (B) Cross section of primate retina showing the different layers.

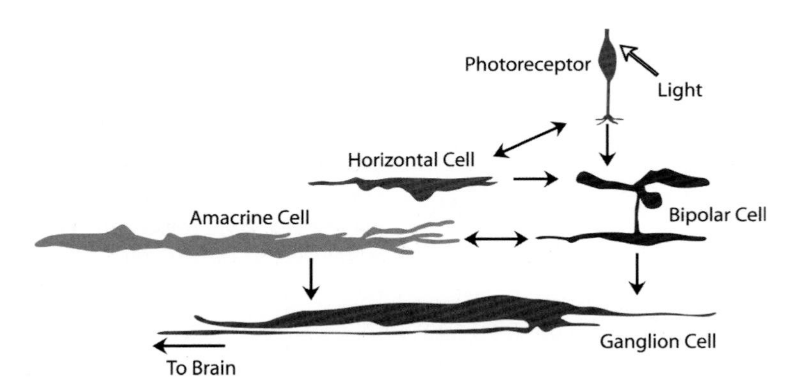

FIGURE 11.2. The five major retinal cell types (photoreceptors, bipolar cells, horizontal cells, amacrine cells, and ganglion cells) and their synaptic connections.

pathway is likely related to the ability of graded responses to transmit more information than a spike code (Laughlin, 2001). By contrast, the use of action potentials by ganglion cells is necessary to propagate information over the greater distances needed to reach higher visual centers.

11.3. Cell Types

11.3.1. Rod and Cone Photoreceptor Cells

11.3.1.1. Outer Segments and Phototransduction

There are two major types of photoreceptors cells: rods and cones (Ebrey and Koutalos, 2001). The structure and phototransduction proteins of rods are specialized to allow them to respond to dim lights. Cones are less sensitive to light, but provide high acuity and color vision. Rods and cones are named for their rod- and cone-shaped outer segments, respectively. The outer segments of both cell types contain the proteins necessary for phototransduction, packaged into disc-shaped organelles. Cone discs are formed by invaginations of the outer segment plasma membrane whereas rod discs are completely sequestered within the outer segment.

The principal job of photoreceptors is to transduce light into an electrical signal. Phototransduction involves a cascade of enzymes in the outer segment (Arshavsky et al., 2002) (Figure 11.3). It is initiated by absorption of a photon by rhodopsin (or cone-specific opsins). Rhodopsin is a G-protein coupled receptor with homology to other G-protein coupled receptors (GPCRs) (e.g., beta-adrenergic, muscarinic, etc.). However, unlike GPCRs that are activated by the binding of a neurotransmitter ligand, the activating ligand of opsin is a light-sensitive chromophore molecule, vitamin A aldehyde (or retinal), bound within a pocket of the opsin protein. Absorption of a photon by this chromophore initiates a conformational transition of 11-*cis*-retinal, which is bent around the 11-*cis* carbon position, into all-*trans* retinal, which has a straight chain configuration (Figure 11.3). The conformational change in retinal produces conformational changes in the seven *trans*-membrane domains of rhodopsin, causing it to assume an active configuration known as metarhodopsin. The activated GPCR, metarhodopsin, stimulates exchange of GTP for GDP on the associated G-protein, transducin (G_T). Activated alpha subunit of transducin (α_T) stimulates the activity of a cGMP-specific phosphodiesterase, which hydrolyzes cGMP. The plasma membrane of outer segments contains nonselective cation channels permeable to Na^+, K^+, and Ca^{2+} that are opened

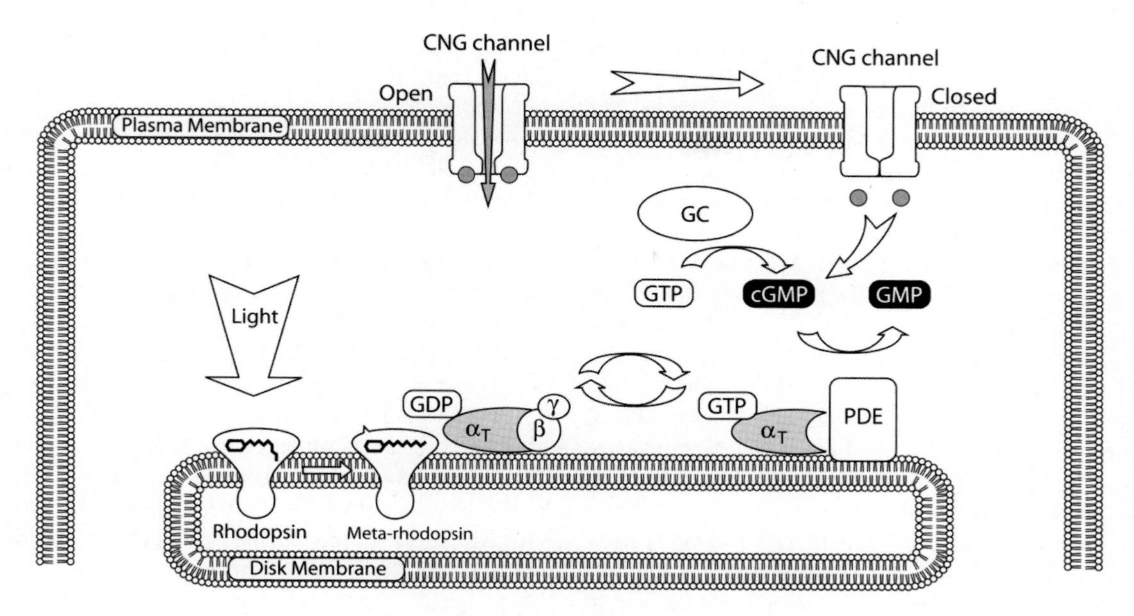

FIGURE 11.3. The enzymatic cascade responsible for phototransduction.

by the binding of cGMP to their intracellular face. Light-induced reduction in cytoplasmic [cGMP] causes some cGMP-gated cation (CNG) channels to close. When open in darkness, Na^+ and Ca^{2+} influx through CNG channels causes photoreceptors to depolarize. Conversely, light-induced closure of these channels causes photoreceptors to hyperpolarize. Because brighter lights cause a larger number of CNG channels to close, the amplitude of membrane hyperpolarization is graded with intensity.

The enzymatic transduction cascade initiated by photon absorption is terminated by two mechanisms: (1) Light-activated meta-rhodopsin is inactivated in a process that begins with its phosphorylation by rhodopsin kinase. Following phosphorylation of rhodopsin at three sites by rhodopsin kinase, the protein arrestin binds to rhodopsin and arrests further activity. (The actions of rhodopsin kinase are analogous to the regulation of beta-adrenergic receptors by beta-adrenergic receptor kinase.) (2) Light activated transducin is shut down by its intrinsic GTPase activity, which converts GTP into GDP and thereby deactivates the G-protein. The intrinsic GTPase activity of transducin is accelerated by binding of an accessory protein, RGS/Gβ5, to the transducin alpha subunit.

Tremendous amplification of the light signal by phototransduction in rods allows them to detect absorption of a single photon (Field et al., 2005). To accomplish this feat, each activated metarhodopsin molecule catalyzes hundred to thousands of transducin molecules and although each transducin activates only a single PDE molecule, every PDE molecule hydrolyzes thousands of cGMP molecules. The net result is that activation of a single rhodopsin molecule by a single photon of light causes the degradation of 10^5–10^6 cGMP molecules causing many cGMP-gated cation channels to close and producing a small but detectable change in membrane potential.

Mutations in phototransduction proteins are a major cause of retinitis pigmentosa and other photoreceptor degenerations (Kennan et al., 2005). Mutations in rhodopsin, phosphodiesterase and cGMP-gated cation channels can all produce retinitis pigmentosa. Mutations in rod-specific transducin produce night blindness whereas mutations in cone-specific transducin produce achromatopsia (rod monochromacy). Similarly, by preventing cones from responding to light, mutations in cone-specific cGMP-gated cation channels also lead to achromatopsia. For an updated list of genes involved in retinal diseases, consult the "Retinal Information Network" (http://www.sph.uth.tmc.edu/Retnet/).

11.3.1.2. Light Adaptation

Photoreceptors can respond to only a limited range of intensities before their responses saturate. To maintain responsiveness over the large range of intensities encountered in the world, the phototransduction apparatus adapts its sensitivity to increased light levels to maintain a constant relative response to increments in illumination ($\Delta I/I$ = constant response) (Burkhardt, 2001; Fain et al., 2001). This process of light adaptation is largely due to calcium-sensitive adjustments of the phototransduction enzyme cascade. As mentioned above, cGMP-gated cation channels are permeable to Ca^{2+} allowing Ca^{2+} to enter the outer segments when these channels are open in darkness. This steady Ca^{2+} influx is countered by the extrusion of Ca^{2+} from outer segment by Na^+/Ca^{2+} exchangers. The closing of cGMP-gated channels in response to light diminishes the influx of Ca^{2+} but its efflux via the Na^+/Ca^{2+} exchanger continues. The concentration of Ca^{2+} in the outer segment therefore diminishes in light. This decrease in $[Ca^{2+}]_i$ has two main effects on phototransduction: (1) Decreased $[Ca^{2+}]$ enhances the activity of guanylyl cyclase stimulating the production of cGMP. The resulting increase in cGMP opens cGMP-gated channels leading to depolarization of the photoreceptor membrane potential. (2) Decreased [Ca^{2+}] increases the affinity of cGMP for cGMP-gated cation channels to further promote their opening. The membrane depolarization that results from reopening of cGMP-gated cation channels restores the operating range of the photoreceptor cell, allowing it to hyperpolarize in response to another flash of light.

While calcium-dependent adaptation is sufficient to maintain the sensitivity of cones over a large range of intensities, at extremely high intensities there is also a contribution from photochemical adaptation in which a significant fraction of unbleached chromophore (11-*cis*-retinal) is bleached to all-*trans*-retinal (Burkhardt, 1994). At most light levels, there is a sufficient reservoir of 11-*cis*-retinal in the outer segments so that the bleaching of chromophore molecules does not appreciably limit the sensitivity of the opsin molecule, as long as levels of 11-*cis*-retinal are soon restored. (As presented later, restoration of chromophore levels after bleaching requires participation of the RPE.) However, at extremely high light levels, the rate and amount of bleaching is sufficient to limit the availability of chromophore and thereby limit the sensitivity of opsin to light.

The combined effects of calcium-dependent and photochemical adaptation allow cones to maintain constant incremental sensitivity to light over 10^7-fold changes in intensity (Burkhardt, 1994). Rods, sensitive to dimmer lights, maintain relatively constant sensitivity over a thousand-fold change in intensity. Rod and cone systems together therefore allow the visual system to perform the impressive feat of maintaining relatively constant incremental sensitivity to light over a ten billion-fold range of intensities. For comparison, pupillary constriction or dilation contributes a 16-fold change in sensitivity, although it has the advantage of being relatively rapid, occurring within a couple hundred milliseconds compared to minutes for full light adaptation.

11.3.1.3. Photoreceptor Inner Segment, Soma, and Synaptic Terminal

The inner segment of photoreceptors is packed with mitochondria that fuel its tremendous metabolic demands. Sodium ions continuously entering the outer segments through cGMP-gated cation channels in darkness are extruded by Na/K-ATPases in the inner segments. The continuous consumption of ATP by these pumps makes photoreceptor cells among the most metabolically active in the body.

Below the inner segment is the soma and nucleus, which connects at its base to the axon and synaptic terminal. Photoreceptors release glutamate at ribbon synapses (Heidelberger et al., 2005). Synaptic ribbons are specialized for sustained release of neurotransmitter and are also found in the terminals of retinal bipolar cells, as well as vestibular and cochlear hair cells. Synaptic ribbons receive their name because of their planar structure in photoreceptor terminals, although bipolar and hair cell ribbons are more spherical in shape.

The ribbon is composed mainly of the structural protein, Ribeye, but also includes a kinesin motor protein, KIF3A, and Rab3-interacting protein, RIM. Ribbons are attached to the synaptic active zone by bassoon, and its structural relative, piccolo. Although the ribbon appears to anchor a readily releasable pool of vesicles, molecular motors do not appear to be involved in vesicle movements near the active zone. RIM protein mutations have been implicated in an autosomal dominant rod-cone dystrophy (Johnson et al., 2003).

Glutamate release from synaptic terminals of photoreceptor and bipolar cells is regulated by calcium influx through L-type calcium channels (Heidelberger et al., 2005). The use of L-type channels at ribbon synapses contrasts with the reliance on N, P, and Q type channels for neurotransmission at conventional synapses of spiking neurons. A retina-specific L-type channel, alpha 1F (CaV1.4), is localized to rod terminals. Mutations in this channel produce a congenital stationary night blindness (Bech-Hansen et al., 1998).

11.3.1.4. The Fovea

In humans, the center of the visual image is focused on the fovea in the macula lutea, the region of highest acuity in the retina (Kolb et al., 2005). Although primates are the only mammals with a fovea, many birds and lizards also possess a fovea. Unlike more peripheral retina where light must pass through overlying retinal layers to reach the photoreceptors, overlying neurons are displaced at the fovea to diminish light scattering. At the center of the resulting foveal pit, the only structures between the outer segments and vitreal surface are cone axons. Cone axons contain xanthophyll pigments that give the macula lutea its characteristic yellow color. The fovea contains only cones and, at its very center, is even free of blue-sensitive cones. Visual acuity, the ability to resolve fine spatial details, is limited by the spacing between cones in the fovea and the density of cones parallels visual acuity in the retina. Acuity can be as high as 20/10 at the foveal center, but falls off rapidly toward the retinal periphery. Although we are not typically conscious of these movements, the eye is constantly in motion, making small microsaccades to allow the high acuity fovea to scan various points and thus construct a high- resolution image at higher visual centers. Loss of the macular region (e.g., in macular degeneration) thus leads not only to a loss of central vision but also to a general loss of high acuity vision.

11.3.1.5. Cones and Cone Opsins

The retinas of Old World primates have three cone subtypes with different spectral sensitivities: short-wavelength (S or blue-sensitive) cones, middle-wavelength (M or green-sensitive) cones, and long-wavelength (L or red-sensitive) cones. Trichromacy evolved in Old World primates 40 million years ago with duplication of a single M/L ancestral pigment gene followed by divergence into separate M and L pigments. Differences in spectral sensitivity among different cone types arise from the presence of different cone opsins. Cone opsins are 40% homologous to rhodopsin, S cone opsins are 40% homologous with M and L cone opsins, but M and L cone opsins are 97% homologous with one another. The differences in spectral absorbance among different opsins result from differences in a small number of amino acid residues that alter the position of hydroxyl groups close to 11-*cis*-retinal. The 30 nm difference in spectral absorbance between primate M and L cones is determined primarily by three amino acids: alanine vs. serine at position 180 (~4 nm), phenylalanine vs. tyrosine at position 277 (~10 nm) and alanine vs. threonine at position 285 (~16 nm) (Deeb, 2005).

Different photopigments are encoded by genes located on different chromosomes (Deeb, 2005). The gene for rhodopsin is found on chromosome 3, the S cone pigment gene on chromosome 7, and M and L cone pigment genes on the X chromosome. M and L cone pigments are found in a tandem array on the X chromosome. Recombination between these genes on adjacent X chromosomes is the most frequent cause of color vision anomalies. For this reason, most color vision defects involve red-green color vision and are X-linked recessive. Males with anomalous M or L cone pigments are most common (anomalous trichromats), with a frequency of ~6% among European males, while 2% are entirely lacking in one pigment or the other (dichromats).

11.3.2. Horizontal Cells

Horizontal cells are laterally arborizing interneurons in the outer retina that receive excitatory synaptic inputs from photoreceptors and make inhibitory synapses onto cones and bipolar cells. The neurotransmitter released from horizontal cells is predominantly GABA. A light-evoked reduction of glutamate release from photoreceptors causes horizontal cells to hyperpolarize by reducing the activation of AMPA-type glutamate receptors. With the exception of certain fish retina, there appear to be no NMDA receptors in horizontal cells of adult retina (Thoreson and Witkovsky, 1999). Rats and mice have only one type of horizontal cell, but most other mammals have two types (Masland, 2001). Because both types hyperpolarize to light of all visible wavelengths, they are sometimes referred to as luminosity-type horizontal cells. Many nonmammalian vertebrates have additional chromaticity or color-opponent horizontal cell types that depolarize to certain wavelengths and hyperpolarize to other wavelengths.

Horizontal cells in most vertebrates have large receptive fields due to extensive gap junction coupling among cells. Large receptive fields allow horizontal cells to measure illumination from a wide area. Inhibitory feedback from horizontal cells to cones and bipolar cells subtracts the mean luminance level measured over a wide area from signals transmitted to the inner retina about local luminance changes (Kamermans and Spekreijse, 1999). As discussed later, this negative feedback contributes to formation of the center-surround arrangement of visual receptive fields, important for the detection of edges.

11.3.3. Bipolar Cells

Bipolar cells transmit signals from photoreceptors to ganglion cells. They receive glutamatergic input from photoreceptors, inhibitory influences from horizontal cell contacts at their dendrites in the outer retina, and additional inhibitory inputs from amacrine cell contacts at their terminals in the inner retina. Bipolar cells release L-glutamate at ribbon synapses that contact amacrine and ganglion cells in the inner plexiform layer. There are 9–11 types of cone bipolar cell and a single type of rod bipolar cell in mammalian retina (Masland, 2001; Wässle, 2004). The different cone bipolar cells can be grouped into two major physiological subtypes: cone ON bipolar cells that depolarize to light and cone OFF bipolar cells that hyperpolarize to light. All rod bipolar cells in the mammalian retina are of the ON, depolarizing type. In lower vertebrates, many ON and OFF bipolar cells receive mixed rod and cone inputs.

ON and OFF responses of bipolar cells result from the presence of different glutamate receptors in the two cell types. OFF bipolar cells possess KA and AMPA-type ionotropic glutamate receptors, but not NMDA receptors (Thoreson and Witkovsky, 1999). Thus, like horizontal cells, the synapse from cones to OFF bipolar cells is sign-conserving, that is, light-evoked hyperpolarization of the cone reduces the depolarizing influence of AMPA/KA receptors thereby causing the OFF bipolar cell to hyperpolarize.

ON bipolar cells do not possess ionotropic glutamate receptors but are instead activated by a metabotropic glutamate receptor, mGluR6 (Slaughter and Awatramani, 2002). mGluR6 is a G-protein coupled receptor that acts via the G-protein, G_o, to close nonselective cation channels. Thus, light-evoked cessation of glutamate release from photoreceptors causes these cation channels in ON bipolar cells to open and thereby depolarizes the cell. By contrast with OFF bipolar and horizontal cells, the synapse from photoreceptors to ON bipolar cells is therefore sign-inverting.

ON and OFF bipolar cells excite ON- and OFF-type ganglion and amacrine cells, respectively. ON and OFF pathways remain segregated into the lateral geniculate nucleus (LGN) of the thalamus suggesting this segregation has functional significance. Saturating mGluR6 with the selective agonist, L-2-amino-4-phosphonobutyric acid (L-AP4), produces an acute deficit in the perception of positive contrast (i.e., bright spots on a dark background) (Schiller et al., 1986). It has therefore been suggested that ON bipolar cells preferentially encode information about positive contrast and OFF bipolar cells preferentially encode information about negative contrast. However, ON and OFF bipolar cells can respond equally well to positive and negative contrast steps (Burkhardt, 2001). Furthermore, although mutations in mGlu6 leading to a loss of rod ON bipolar cell function produce night blindness, mGlu6 mutations in ON bipolar cell function do not produce obvious deficits in contrast perception at higher light levels (Dryja et al., 2005). Thus, the role played in contrast perception by the segregation of different ON and OFF pathways remains unclear.

11.3.4. Amacrine Cells

Amacrine cells are laterally interconnecting interneurons and most contain the inhibitory neurotransmitters GABA, glycine or both. Amacrine cells are excited by glutamate released from bipolar cells. This glutamate acts principally on KA and AMPA receptors although many amacrine cells also possess NMDA receptors (Thoreson and Witkovsky, 1999). By anatomical and neurochemical criteria, amacrine cells can be classified into 29 different types (Masland, 2001; Wässle, 2004). With the exception of glutamate, almost every type of neurotransmitter is present in at least one type of amacrine cell. Physiological responses of amacrine cells include transient and sustained depolarizations at light onset (ON cells), light offset (OFF cells), or both (ON/OFF cells). By contrast with the graded light responses of bipolar, horizontal, and photoreceptor cells, many amacrine cells exhibit sodium-dependent action potentials making their responses much more transient.

Many amacrine cells have relatively dedicated functions. Some examples include:

1. Large, radially symmetric starburst amacrine cells help create directional selectivity in directionally selective ganglion cells (Taylor and Vaney, 2003).
2. Dopaminergic amacrine cells with widespread dendritic arborizations increase release of dopamine in response to increases in global illumination. Dopamine diffuses throughout the retina to influence cells as far away as the RPE. The increased release of dopamine by light modifies cell function to optimize retinal responses in daylight (Witkovsky, 2004).
3. AII amacrine cells transfer rod signals from rod bipolar cells to ganglion cells (Bloomfield and Dacheux, 2001).

11.3.5. Ganglion Cells

Retinal ganglion cells are the output cells of the retina. Their axons course along the vitreal surface of the retina and bundle together to exit the eye as the optic nerve. Ganglion cells are excited by glutamate released from bipolar cells acting on both NMDA and non-NMDA (KA- and AMPA-type) glutamate receptors (Thoreson and Witkovsky, 1999).

There are 10–15 types of ganglion cells in mammalian retina (Masland, 2001; Wässle, 2004). The two most common types in primate retina are M (magnocellular) cells and P (parvocellular) cells. The various types of ganglion cells remain generally segregated in their projections to the LGN: M ganglion cells project to M cell layers of the LGN, P cells to the P cell layers, and bistratified cells project predominantly to the koniocellular (interlaminar) regions. Primate M and P cells are analogous to Y and X cell types in cat retina.

M cells have large cell bodies and large dendritic arborizations resulting in large receptive fields (Rodieck, 1998). M cells are classified anatomically as parasol ganglion cells. The large receptive fields of M cells limit their contribution to fine feature analysis. Instead, their output is primarily related to motion and other changes in illumination.

In contrast to M cells, P cells have small cell bodies with small dendritic arborizations resulting in small receptive fields. P cells are also wavelength-selective. P cell output thus contributes to fine feature analysis and color vision. P cells are classified anatomically as midget ganglion cells.

S cones do not provide direct inputs into M and P-type ganglion cells, which receive inputs only from M and L cones. S cones instead provide inputs into two different types of ganglion cells: blue OFF cells and small bistratified blue ON cells.

Recent studies have revealed a population of ganglion cells that are intrinsically light sensitive, that is, they do not require photoreceptor inputs in order to respond to light (Fu et al., 2005). The intrinsic light-sensitivity of these cells is conferred by the presence of the photopigment, melanopsin. Intrinsic light responses of these cells are slow and exhibit minimal adaptation. Although the mechanism of phototransduction employed by melanopsin remains under investigation, early results indicates that it may involve transient receptor potential (TRP) channels similar to those used for invertebrate phototransduction. Thus, intrinsically photosensitive ganglion cells may utilize a more primitive phototransduction mechanism than photoreceptor cells. Melanopsin-containing ganglion cells are very large but few in numbers. They project to the suprachiasmatic nucleus where their tonic responses to light provide important signals for setting circadian rhythms.

11.4. Circuitry

In addition to transducing the incoming light, the retina plays a number of important roles in the initial process of analyzing visual information. In the following section, we consider the retinal circuitry employed for analysis of edges, color, directional selectivity, and scotopic vision.

11.4.1. Edge Detection and Center-Surround Receptive Fields

The detection of contrast edges is enhanced by the center-surround arrangement of receptive fields in cones, bipolar cells, and ganglion cells (Kuffler 1953; Baylor et al 1971). How does this center-surround organization improve edge detection? Consider an ON type ganglion cell (Figure 11.4) in which the circular center of the receptive field is excited by light. Its center-surround arrangement is imparted by the presence of an annular inhibitory region flanking this central excitatory region. A small spot of light illuminating only the excitatory center strongly excites this cell but an annulus of light falling only on the inhibitory surround strongly inhibits it. Full field illumination, which stimulates both the excitatory center and the inhibitory surround, thus produces smaller

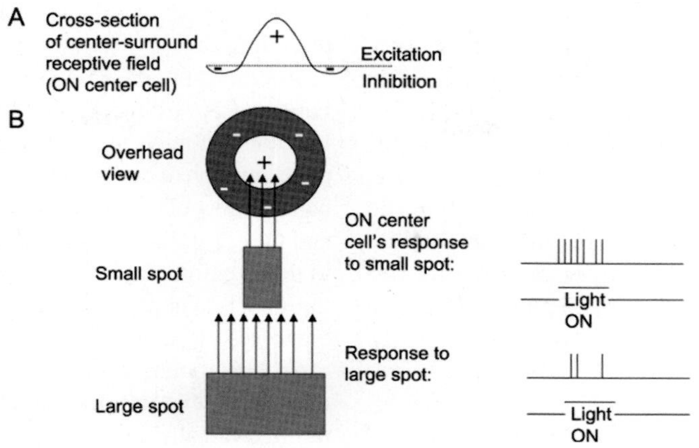

FIGURE 11.4. Center-surround arrangement of receptive fields in the retina enhances responses to small spots, edges, and bars of light. (A) Cross section of the center/surround receptive field of an ON center cell. Light falling on the center of the receptive field excites the cell whereas light falling in the periphery produces inhibition. (B) Overhead view of the center/surround arrangement of a receptive field of an ON center ganglion cell illustrating differences in the trains of action potentials evoked by a small spot or full field illumination.

changes in ganglion cell output. A bar or edge of light illuminating the entire excitatory center along with small portions of the inhibitory surround evokes a stronger response than full field illumination, but not as strong as a spot of light illuminating just the excitatory center. The net result of this center-surround receptive field arrangement is that cells respond more strongly to spots, bars and edges than to full field illumination. These same kinds of considerations can be extended to OFF type cells that are excited by light decrements in the center and inhibited by light decrements in the surround. Another way to think about center-surround inhibition is that inhibitory feedback from the broad receptive fields of horizontal and amacrine cells subtracts the mean luminance level from signals transmitted to ganglion cells about local luminance changes. In machine vision, the mathematical equivalent of the center-surround receptive field is implemented to spatially differentiate the image and thus enhance the detection of edges. This is useful in robotics and as a component of artificial visual stimulating systems for visually deprived humans.

11.4.2. Color

The retina initiates the process of analyzing color in the world. The presence of three spectrally distinct types of cones in the primate retina (L, M, and S) provides the physiological basis for trichromatic vision (R, G, and B). The responses of an individual cone do not vary with wavelength, but only with the number of photons absorbed. Thus, one can obtain an identical response from a green-sensitive cone with either green or red light as long as one adjusts the intensities of the two lights to provide equivalent photon capture by the green-sensitive cone. For this reason, color discrimination requires comparisons between inputs from different classes of cones. Synaptic comparisons between different types of cones produce responses that are color opponent, that is, cells that respond to one wavelength by depolarizing but to another wavelength by hyperpolarizing. In nonmammalian vertebrates, color opponency is evident in horizontal and bipolar cells (Twig et al., 2003). In mammalian retinas, color opponency is first detected in ganglion cells (Dacey and Packer, 2003).

There are red/green and blue/yellow opponent cell types in the retina. These two classes of opponent neurons contribute to the perception of color opponent after-images (e.g., the illusory appearance of red produced after gazing steadily at a field of green). In primate retina, the red/green hue axis of color vision utilizes separate retinal circuits and ganglion cell types from the blue/yellow axis. Red/green opponency in primate P-type ganglion cells arises from the selective segregation to center and surround of M and L cone inputs via distinct bipolar cells (Dacey and Packer, 2003). Via specialized S cone bipolar cells, S cones provide the input for responses to blue light in the receptive field center of blue ON bistratified and blue OFF-type ganglion cells. Responses to yellow light in the receptive field surround are generated from a sum of L and M cone signals.

11.4.3. Directional Selectivity

Some magnocellular ganglion cells are excited by stimuli moving along one axis of the receptive field (e.g., upward) but show little response to stimuli moving in the opposite direction (e.g., downward). Surprisingly, directionally selective ganglion cells with directional asymmetries in their receptive field receive their main synaptic inputs from symmetrically radiating starburst amacrine cells (Taylor and Vaney, 2003). Starburst amacrine cells provide both excitatory (cholinergic) and inhibitory (GABAergic) inputs into ganglion cells. The directional selectivity in postsynaptic ganglion cells results from the fact that cholinergic excitation from starburst amacrine cells occurs earlier than GABAergic inhibition when visual stimuli move in the preferred direction, but GABAergic inhibition precedes excitation when stimuli move in the opposite, nonpreferred direction. Thus, GABAergic inhibition more strongly dampens the response of the ganglion cell to stimuli moved in the nonpreferred direction compared to the preferred direction. The mechanisms responsible for this asymmetric synaptic output of acetycholine and GABA from starburst amacrine cells remain under investigation.

11.4.4. Rod Pathways

Scotopic or low-light vision is mediated by rod photoreceptors. The mammalian retina has a duplex organization in which rods communicate with ganglion cells using a largely separate circuit from cones. The primary rod circuit begins with rods communicating with ON-type rod bipolar cells that, unlike cone bipolar cells, do not contact ganglion cells directly but instead contact AII amacrine cells. AII amacrine cells form gap junctions with cone ON bipolar cell terminals and inhibitory, sign-inverting glycinergic synapses with cone OFF bipolar cells. Output from AII amacrine cells, therefore, feeds rod signals into cone bipolar cells driving both ON and OFF center ganglion cells. AII amacrine cells provide the primary pathway for rod signals at very low light levels, but at higher light levels there is also a contribution from direct rod inputs into OFF bipolar cells and the transmission of rod signals through gap junctions into neighboring cones (Bloomfield and Dacheux, 2001).

There is tremendous synaptic convergence in the rod pathway with as many as 75,000 rods converging onto a single ganglion cell. By contrast, cones show much less convergence, with a 1:1 connection of cones to midget bipolar cells to midget ganglion cells in the foveal center. This difference accounts for the high visual acuity mediated by cone circuits. On the other hand, the large convergence of signals in the rod pathway facilitates the perception of very dim flashes of light. The ability of individual rods to respond to a single photon of light combined with convergence allows us to perceive the absorption of as few as a dozen photons within 100 ms (Field et al., 2005).

11.5. Glia

The predominant retinal glial cell is the Müller cell. Müller cells are radial glia that span the retina from the OLM to ILM and ensheathe virtually every cell in the retina. They play a number of important roles in maintaining homeostasis (Newman and Reichenbach, 1996). For example, Müller cells are a primary storage depot for glycogen in the retina that can provide metabolites (e.g., lactic acid) to neurons during times of metabolic stress. Müller cells also help to remove and redistribute metabolic waste products. They exhibit high levels of glutathione that can help protect the retina from oxidative stress. Müller cells also possess Na/HCO_3 cotransport mechanisms and carbonic anhydrase to stabilize pH levels in the retina.

Another important role of Müller cells is the spatial redistribution of K^+ from regions of high concentration to regions of low concentration in order to maintain a stable extracellular K^+ concentration of about 3 mM (Kofuji and Newman, 2004). When neuronal depolarization causes K^+ levels to increase in the IPL and OPL, excess K^+ ions enter Müller cells processes. K^+ influx in the plexiform layers is accompanied by a simultaneous efflux through K^+ channels clustered at the vitreal surface and along blood vessels. Conversely, a reduction in K^+ accompanying neuronal hyperpolarization is accompanied by an efflux of K^+ out of Müller cells into the plexiform layers. The current flowing through radially oriented Müller cells as a result of this spatial buffering of K^+ produces measurable *trans*-retinal potentials, such as slow PIII of the electroretinogram. It was once believed that Müller cell K^+ currents were the predominant mechanism responsible for the ERG b-wave, but more recent studies suggest that the b-wave primarily reflects ON bipolar cell responses.

Müller cells are major sites for the uptake and removal of neurotransmitters, most notably glutamate and GABA. Glutamate transport into neurons is smaller and slower than transport into Müller cells and thus uptake into Müller cells is the principal mechanism responsible for the initial removal of extracellular glutamate following synaptic activation (Pow, 2001). In addition to neurotransmitter transporters, Müller cells possess neurotransmitter receptors and can release neuroactive substances (e.g., ATP) (Newman, 2004). Thus, activity of retinal neurons can influence Müller cells and Müller cell activity can in turn influence adjacent neurons.

The retina also contains three other types of glia: astrocytes, oligodendrocytes and microglia. Retinal astrocytes are located primarily in the NFL and oligodendrocytes form the myelin sheath of axons in the optic nerve. Hematopoetically derived microglia are small stellate cells that, when quiescent, associate with inner retinal blood vessels.

Infection or damage to the retina stimulates Müller cell gliosis (Garcia and Vecino, 2003) and migration of microglia to the injured area to assist in phagocytosing debris from dying cells. Microglia and Müller cells both release cytokines in response to injury but the release of proinflammatory cytokines can sometimes exacerbate cell damage during retinal disease. Cytokines released by Müller cells include vascular endothelial growth factor and transforming growth factor beta, which promote neovascularization, as well as basic fibroblastic growth factor. At least in some species, Müller cells respond to injury and cytokines by dedifferentiating into progenitor cells that can give rise to neurons (Fischer and Reh, 2003).

11.6. Retinal Pigment Epithelial Cells

The RPE is a monocellular epithelium of hexagonal cells whose name reflects the facts that it contains melanin pigment granules and forms the outermost layer of the retina. The apical processes of RPE cells ensheathe outer segments of rods and cones and each RPE cell contacts 20–30 rods (Marmor and Wolfensberger, 1998; Strauss, 2005). Melanin granules are concentrated in the apical processes and cytoplasm of RPE cells but are nearly absent from basal cytoplasm. By absorbing stray photons of light that have passed through photoreceptor outer segments, melanin granules improve the optical isolation of individual photoreceptors.

Neighboring RPE cells are connected by tight junctions that help to create the blood/retinal barrier separating the neurosensory retina from fenestrated capillaries in the choroid. The basement membrane of RPE, together with the adjacent basement membrane of the choroid, forms a structure known as Bruch's membrane. RPE cells possess a number of organic and ion transporters to help move polar molecules across the blood retinal barrier. These include transporters for amino acids, folate, ascorbic acid, myo-inositol, organic anions, glucose and lactate.

Photoreceptors continually synthesize new phototransduction proteins for incorporation into newly formed outer segment discs. As new discs are formed at the base of the outer segment, older discs are phagocytosed by RPE cells at the tip of the outer segment. Each outer segment disc is formed and shed in ~2 weeks. The phagocytosis of discs occurs in circadian bursts (rods at the end of night, cones at the end of day).

The bleached, all-*trans* form of the chromophore retinal cannot be converted back into photosensitive 11-*cis*-retinal within photoreceptor cells but requires participation of the RPE (Lamb and Pugh, 2004). After the conversion of 11-*cis*-retinal to all-*trans*-retinal by light, all-*trans*-retinal is rapidly converted to all-*trans*-retinol (all-*trans*-vitamin A) in the photoreceptor outer segment. All-*trans*-retinol is then transported out of the photoreceptor cell, through the interstitial space separating the photoreceptor outer segments and RPE, and into the RPE via a process involving interstitial retinoid binding proteins (IRBP). Upon entering the RPE, retinol is bound to cellular retinol binding protein. All-*trans*-retinol is converted to 11-*cis*-retinol by retinol isomerase (RPE65) and 11-*cis*-retinol is converted back to 11-*cis*-retinal (11-*cis*-vitamin A aldehyde) by 11-*cis*-retinol dehydrogenase. 11-*cis*-retinal is transferred to cellular retinal binding protein (CRalBP) for transport to the RPE cell surface. Finally, regenerated 11-*cis*-retinal is transported back to photoreceptors via IRBP. By depriving photoreceptors of chromophore, mutations of proteins in the visual cycle (e.g., retinol isomerase [RPE65], 11-*cis*-retinol dehydrogenase, CRalBP, and IRBP) can lead to photoreceptor degeneration.

The interstitial space between RPE and photoreceptors contains a sticky interphotoreceptor matrix consisting of glycoproteins, proteoglycans, and hyaluronic acid that helps the retina adhere to the back of the eye. Retinal adhesion is also promoted by extrusion of water from the RPE to choroid. Basolateral Cl channels and apical Na/K/2Cl transporters are particularly important for water transport out of the RPE. Because photoreceptors are not physically bound to the RPE, the retina can detach from the RPE with a strong blow to the eye, fluid build-up behind the retina (rhegmatogenous detachment), or traction from overlying cells in proliferative vitreoretinopathy. Detachment of photoreceptor cells from the adjacent RPE prevents recycling of the photopigment which blocks phototransduction by depriving opsin of sufficient chromophore. Unless repaired, retinal detachment therefore results in blindness.

One by-product of photoisomerization in photoreceptor outer segments is A2E (*N*-retinylidene-*N*-retinylethanolamine). A2E is ingested by RPE cells when they phagocytose outer segments. However, A2E cannot be enzymatically degraded by RPE cells and thus accumulates in these cells, where it becomes a major component of lipofuscin granules (Sparrow and Boulton, 2005). Stargardt's macular degeneration involves a defect in the ABCR transporter that increases accumulation of A2E suggesting it may play a role in macular degeneration. The damage to the RPE associated with A2E accumulation may result from stimulation of apoptosis in A2E-laden RPE cells produced by exposure to blue light.

A major risk factor for age-related macular degeneration (ARMD) is the presence of drusen deposits (Zarbin, 2004). White drusen spots visible with an ophthalmoscope are formed by deposits between the RPE and Bruch's membrane. Drusen formation involves inflammatory reactions and RPE cells overlying drusen often show signs of impending cell death. Studies on genetic factors contributing to ARMD have singled out mutations in complement factor H as a major risk factor (Edwards et al., 2005; Haines et al., 2005; Klein et al., 2005). This has led to the suggestion that mutations in complement factor H promote an over-active inflammatory response that contributes to drusen formation leading to RPE cell damage and ARMD.

11.7. Blood Supply

Blood is supplied to the retina by the central retinal artery and choroidal blood vessels (Oyster, 1999). The central retinal artery arises from the ophthalmic artery, which in turn branches off the internal carotid artery. Upon entering the retina, the central retinal artery branches into deep capillary beds in the INL and superficial capillary beds in the GCL. Endothelial cells of retinal capillaries are joined by tight junctions, contributing to the blood/retinal barrier. There is little or no autonomic regulation of the retinal circulation; blood flow through these capillaries is instead primarily controlled by autoregulation (Wangsa-Wirawan and Linsenmeier, 2003). Retinal capillaries drain into the central retinal vein.

Choroidal vessels derive from posterior and short posterior ciliary arteries that, like the central retinal artery, branch off from the ophthalmic artery. The choroidal circulation forms a dense bed of fenestrated capillaries, known as the choriocapillaris, adjacent to the basolateral surface of the RPE. The flow rate through the choriocapillaris is among the highest in the body and the arterio-venous drop in PO_2 is minimal. This high flow rate supplies the energetically demanding photoreceptors with large amounts of oxygen and maintains the retina at a constant temperature despite changing levels of radiant energy focused onto the back of the eye. Oxygen for the photoreceptors comes primarily from the choroid. The high level of oxygen consumption by photoreceptor cells in darkness produces a PO_2 of zero at the level of the inner segments in mammalian retina (Linsenmeier, 1986). Under these conditions, there is no oxygen available for the remainder of the retina from the choroid. Other retinal neurons instead receive their oxygen from the retinal circulation. (By contrast with mammals, many cold-blood vertebrates lack retinal capillaries and rely on the choroid to supply the entire retina with oxygen and other nutrition.) While capable of autoregulation (Kiel and Shepherd, 1992), blood flow in the choroicapillaris is also regulated by autonomic inputs. Choroidal capillaries drain into four vortex veins, one from each quadrant of the eye.

Summary

The retina is an outpost of the CNS with neuronal structures and proteins specialized for the transduction of light signals into a neural code that the brain can interpret. Mutations of proteins involved in phototransduction or synaptic transmission through the retina produce visual deficits ranging from subtle color vision defects to complete blindness. Although normally isolated from blood-mediated immune responses by the blood-retinal barrier, inflammatory responses, including gliosis and those mediated by the complement system, are important contributors to retinal degeneration, particularly ARMD. Studying immune responses in the retina and their pharmacological manipulation therefore presents a promising avenue for the treatment and prevention of retinal degeneration.

Review Questions/Problems

1. **Briefly summarize the major steps in phototransduction in rods.**
2. **What are the calcium-dependent steps in light adaptation?**
3. **What are the major physiological features of the two most common types of retinal ganglion cells in the primate retina?**
4. **In mammalian retina, which capillary beds supply oxygen to the photoreceptors and which to the remaining parts of the retina?**
5. **For an ON-type ganglion cell, describe how the center-surround organization of visual receptive fields improves edge detection.**
6. **Describe the mechanisms responsible for directional selectivity in retinal ganglion cells.**
7. **Describe the rod pathway used in scotopic vision.**
8. **Describe the key enzymatic steps in the visual cycle converting bleached all-*trans*-retinal back into light-sensitive 11-*cis*-retinal.**
9. **Describe the spatial redistribution of K^+ by Müller cells.**
10. **Summarize the enzymatic mechanisms by which phototransduction is terminated after photostimulation.**
11. **Which of the following is the predominant glial cell type in the retina?**
    a. Astrocyte
    b. Müller cell
    c. Microglia
    d. Schwann cell
    e. RPE cell
12. **In which layer of the retina do bipolar cell terminals contact ganglion cell dendrites?**
    a. Outer nuclear layer
    b. Outer plexiform layer
    c. Inner nuclear layer
    d. Inner plexiform layer
    e. Ganglion cell layer
13. **Which of the following statements is true about the fovea?**
    a. The fovea contains both rods and cones
    b. The fovea is the region of the retina where ganglion cell axons exit the eye.
    c. The fovea is the region of the retina responsible for the highest acuity vision in humans and primates.
    d. The location of the fovea on the retina is not a fixed anatomical feature but varies with focus.
    e. At the center of the fovea, light must first pass through ganglion, amacrine, horizontal and bipolar cells before reaching photoreceptor outer segments.
14. **Upon which chromosome are genes for M cone pigments located?**
    a. X chromosome
    b. Y chromosome
    c. Chromosome 3
    d. Chromosome 7
    e. Chromosome 10
15. **Which of the following statements is true about ON and OFF type retinal bipolar cells?**
    a. ON bipolar cells possess KA/AMPA receptors and OFF bipolar cells possess NMDA receptors.
    b. ON bipolar cells possess NMDA receptors and OFF bipolar cells possess mGluR6.
    c. ON bipolar cells possess mGluR6 and OFF bipolar cells possess NMDA receptors.

d. ON bipolar cells possess mGluR6 and OFF bipolar cells possess KA/AMPA receptors.
e. ON bipolar cells possess KA/AMPA receptors and OFF bipolar cells possess mGluR6.

16. Which of the following is NOT considered to be a major role of Müller cells?

a. Spatial redistribution of potassium
b. Neurotransmitter uptake and removal
c. Homeostatic maintenance of retinal pH levels
d. Glycogen storage
e. Promoting retinal adhesion to the back of the eye

17. Which of the following is NOT considered to be a major role of RPE cells?

a. Phagocytosis of injured cells
b. Formation of blood-retinal barrier
c. Absorption of stray light
d. Photopigment recycling
e. Phagocytosis of outer segment discs

18. Which of the following proteins is a major constituent of synaptic ribbons?

a. Retinol dehydrogenase
b. Melanin
c. Ribeye
d. RGS/Gβ5
e. Bestrophin

19. Which of the following is the main reason that retinal detachment leads to blindness?

a. The retina dies from oxygen deprivation after separating the retina from the choroidal blood supply.
b. Separation of the RPE and photoreceptor outer segments prevents visual pigment recycling.
c. Retinal detachment shears off photoreceptor outer segments because they are covalently bound to the RPE.
d. Photoreceptors die because of the inability of the RPE to phagocytose photoreceptor outer segments.
e. Retinal detachment moves the retina out of the optical plane of focus.

20. What is the light-sensitive molecule found in intrinsically photosensitive retinal ganglion cells of mammalian retina?

a. Melanopsin
b. Transducin
c. Rhodopsin
d. Cryptochrome
e. Peropsin

References

Arshavsky VY, Lamb TD, Pugh Jr EN (2002) G-proteins and phototransduction. Annu Rev Physiol 64:153–187.

Baylor DA, Fuortes MG, O'Bryan PM (1971) Receptive fields of cones in the retina of the turtle. J Physiol 214:265–294.

Bech-Hansen NT, Naylor MJ, Maybaum TA, Pearce WG, Koop B, Fishman GA, Mets M, Musarella MA, Boycott KM (1998) Loss-of-function mutations in a calcium-channel alpha1-subunit gene in Xp11.23 cause incomplete X-linked congenital stationary night blindness. Nat Genet 19:264–267.

Bloomfield SA, Dacheux RF (2001) Rod vision: Pathways and processing in the mammalian retina. Prog Retin Eye Res 20:351–384.

Burkhardt DA (1994) Light adaptation and photopigment bleaching in cone photoreceptors in situ in the retina of the turtle. J Neurosci 14:1091–1105.

Burkhardt DA (2001) Light adaptation and contrast in the outer retina. Prog Brain Res 131:407–418.

Dacey DM, Packer OS (2003) Colour coding in the primate retina: Diverse cell types and cone-specific circuitry. Curr Opin Neurobiol 13:421–427.

Deeb SS (2005) The molecular basis of variation in human color vision. Clin Genet 67:369–377.

Dowling JE (1987) The Retina: An Approachable Part of the Brain. Cambridge, Mass: Harvard University Press.

Dryja TP, McGee TL, Berson EL, Fishman GA, Sandberg MA, Alexander KR, Derlacki DJ, Rajagopalan AS (2005) Night blindness and abnormal cone electroretinogram ON responses in patients with mutations in the GRM6 gene encoding mGluR6. Proc Natl Acad Sci USA 102:4884–4889.

Ebrey T, Koutalos Y (2001) Vertebrate photoreceptors. Prog Retin Eye Res 20:49–94.

Edwards AO, Ritter R 3rd, Abel KJ, Manning A, Panhuysen C, Farrer LA (2005) Complement factor H polymorphism and age-related macular degeneration. Science 308:421–424.

Fain GL, Matthews HR, Cornwall MC, Koutalos Y (2001) Adaptation in vertebrate photoreceptors. Physiol Rev 81:117–151.

Field GD, Sampath AP, Rieke F (2005) Retinal processing near absolute threshold: From behavior to mechanism. Annu Rev Physiol 67:491–514.

Fischer AJ, Reh TA (2003) Potential of Muller glia to become neurogenic retinal progenitor cells. Glia 43:70–76.

Fu Y, Liao HW, Do MT, Yau KW (2005) Non-image-forming ocular photoreception in vertebrates. Curr Opin Neurobiol 15:415–422.

Garcia M, Vecino E (2003) Role of Muller glia in neuroprotection and regeneration in the retina. Histol Histopathol 18:1205–1218.

Haines JL, Hauser MA, Schmidt S, Scott WK, Olson LM, Gallins P, Spencer KL, Kwan SY, Noureddine M, Gilbert JR, Schnetz-Boutaud N, Agarwal A, Postel EA, Pericak-Vance MA (2005) Complement factor H variant increases the risk of age-related macular degeneration. Science 308:419–421.

Heidelberger R, Thoreson WB, Witkovsky P (2005) Synaptic transmission at retinal ribbon synapses. Prog Ret Eye Res 24:682–720.

Johnson S, Halford S, Morris AG, Patel RJ, Wilkie SE, Hardcastle AJ, Moore AT, Zhang K, Hunt DM (2003) Genomic organisation and alternative splicing of human RIM1, a gene implicated in autosomal dominant cone-rod dystrophy (CORD7). Genomics 81:304–314.

Kamermans M, Spekreijse H (1999) The feedback pathway from horizontal cells to cones. A mini review with a look ahead. Vision Res 39:2449–2468.

Kawai F, Horiguchi M, Suzuki H, Miyachi E (2001) Na(+) action potentials in human photoreceptors. Neuron 30:451–458.

Kennan A, Aherne A, Humphries P (2005) Light in retinitis pigmentosa. Trends Genet 21:103–110.

Kiel JW, Shepherd AP (1992) Autoregulation of choroidal blood flow in the rabbit. Invest Ophthalmol Vis Sci 33:2399–2410.

Klein RJ, Zeiss C, Chew EY, Tsai JY, Sackler RS, Haynes C, Henning AK, SanGiovanni JP, Mane SM, Mayne ST, Bracken MB, Ferris FL, Ott J, Barnstable C, Hoh J (2005) Complement factor H polymorphism in age-related macular degeneration. Science 308:385–389.

Kofuji P, Newman EA (2004) Potassium buffering in the central nervous system. Neuroscience 129:1045–1056.

Kolb H, Fernandez E, Nelson R (2005) Webvision: The organization of the retina and visual system. http://webvision.med.utah.edu/

Kuffler SW (1953) Discharge patterns and functional organization of mammalian retina. J Neurophysiol 16:37–68.

Lamb TD, Pugh Jr EN (2004) Dark adaptation and the retinoid cycle of vision. Prog Retin Eye Res 23:307–380.

Laughlin SB (2001) Efficiency and complexity in neural coding. Novartis Found Symp 239:177–192.

Linsenmeier RA (1986) Effects of light and darkness on oxygen distribution and consumption in the cat retina. J Gen Physiol 88:521–542.

Lukasiewicz PD, Eggers ED, Sagdullaev BT, McCall MA (2004) GABAC receptor-mediated inhibition in the retina. Vision Res 44:3289–3296.

Marmor MF, Wolfensberger TJ (1998) The Retinal Pigment Epithelium: Function and Disease. New York: Oxford University Press.

Masland RH (2001) The fundamental plan of the retina. Nat Neurosci 4:877–886.

Newman EA (2004) Glial modulation of synaptic transmission in the retina. Glia 47:268–274.

Newman E, Reichenbach A (1996) The Muller cell: A functional element of the retina. Trends Neurosci 19:307–312.

Oyster CW (1999) The Human Eye: Structure and Function. Sunderland, Mass: Sinauer Associates, Inc.

Pow DV (2001) Amino acids and their transporters in the retina. Neurochem Int 38:463–484.

Rodieck RW (1998) The First Steps in Seeing. Sunderland, Mass: Sinauer Associates, Inc.

Schiller PH, Sandell JH, Maunsell JH (1986) Functions of the ON and OFF channels of the visual system. Nature 322:824–825.

Slaughter MM, Awatramani GB (2002) On bipolar cells: Following in the footsteps of phototransduction. Adv Exp Med Biol 514:477–492.

Sparrow JR, Boulton M (2005) RPE lipofuscin and its role in retinal pathobiology. Exp Eye Res 80:595–606.

Strauss O (2005) The retinal pigment epithelium in visual function. Physiol Rev 85:845–881.

Taylor WR, Vaney DI (2003) New directions in retinal research. Trends Neurosci 26:379–385.

Thoreson WB, Witkovsky P (1999) Glutamate receptors and circuits in the vertebrate retina. Prog Retin Eye Res 18:765–810.

Twig G, Levy H, Perlman I (2003) Color opponency in horizontal cells of the vertebrate retina. Prog Retin Eye Res 22:31–68.

Wangsa-Wirawan ND, Linsenmeier RA (2003) Retinal oxygen: Fundamental and clinical aspects. Arch Ophthalmol 121:547–557.

Wässle H (2004) Parallel processing in the mammalian retina. Nat Rev Neurosci 5:747–757.

Witkovsky P (2004) Dopamine and retinal function. Doc Ophthalmol 108:17–40.

Zarbin MA (2004) Current concepts in the pathogenesis of age-related macular degeneration. Arch Ophthalmol 122:598–614.

12
Lymphocytes and the Nervous System

Larisa Y. Poluektova

Keywords Antibodies; B cells; Effector memory T cells; Follicular dendritic cells (FDCs); Immune privileged; Regulatory T cells; Stat4; Stat6; T-Bet

12.1. Introduction

The immune system maintains the integrity of the human body by recognition and elimination of new antigens coming from outside and inside generated modified substances. The control of immune reactions to infections, tumors and degeneration in the brain is different than in other tissues. The differences in immune control in the brain became clear with the development of transplantology. In 1873 Dutch ophtalmologist van Dooremaal observed prolonged mouse skin graft survival in the dog eye, and later in 1948 Sir Peter Medawar, while trying to explain success of corneal homograft transplantation, performed classic transplantation of skin to the brain and anterior chamber of the eye of non-immunized or immunized rabbits (previously transplanted with the same donor's skin sample to the chest). He highlighted two mechanisms involved in the immunologic control in brain: existing systemic immunity to donor's skin and the status of vasculature. Tissue grafted in the brain or eye survived longer than in any other place in the body without sensitized lymphocyte infiltration via new vessels of the graft. A few more places in the human body obtained definition of "privileged" sites (testis and a placenta) based upon immune tolerance to the graft. Only direct transplantation/injection allows backward introduction of "unknown" antigens. This process in brain parenchyma (not meninges and choroids plexes) is also different than in non-privileged sites due to the absence of endogenous professional antigen-presenting cells capable to trigger immune responses (Hart and Fabre, 1981), and because of the specific properties of non-professional antigen-presenting microglial cells in parenchyma (Matyszak and Perry, 1996; Matyszak, 1998). However, *in situ* inflammation induced by pathogens or abnormal proteins with complementary peripheral adaptive immune responses will induce brain damage and appearance of tertiary lymphoid structures (Aloisi and Pujol-Borrell, 2006). The role of professional antigen-presenting cells, microglial cells and regulatory T cells in these events will be reviewed in other chapters (4, 9, 18–24, 37, 42, 46).

As of today, the control of immune reactions in the brain attributed mostly to the specific cell membrane-bound molecules, soluble factors and brain-associated immune deviation (BRAID) reviewed by (Niederkorn, 2006).

In this chapter we will review the development and function of lymphocytes, which include T cells (thymus derived), B cells (associated with antibody production) and natural killer (NK) cells (effectors of innate immunity).

The specialized functions of lymphocyte subsets are reflected in the differing patterns of molecules and genes they express. These molecules often underlie the unique function of the subsets they mark. For instance, CD3 defines T cells because it is an essential component of the TCR complex; CD4 is associated with MHC class II recognition, mostly in "helper" (Th) responses, whereas CD8 is associated with MHC class I recognition and cytotoxic responses, and B cells are defined by their immune globulin (Ig) expression or production. Immunological research during much of the 1980–1990 era was directed at identifying and characterizing leukocyte markers, first through mAb production, and then through gene sequencing. This led to the discovery of most of the CD molecules that are used today to mark leukocyte subsets and define subset function (http://www.pathologyoutlines.com/ cd100300.html).

The final stage of lymphocyte differentiation is antigen-specific memory cells. Only these cells are capable of surveillance in the brain. The mechanisms of survilance of perivascular and parenchymal sites are reviewed in the following articles (Ransohoff et al., 2003; Engelhardt and Ransohoff, 2005).

It is important to discuss lymphopoiesis and lymphocyte function before discussing lymphocyte-CNS interactions during diseases of the nervous system in this textbook.

T. Ikezu and H.E. Gendelman (eds.), *Neuroimmune Pharmacology*.
© Springer 2008

12.2. Overview of Embryonic Lymphopoiesis

Today, the common concept of lymphoid tissue ontogeny results from the understanding that circulating stem/progenitor cell populations migrate from the yolk sack into the aorta-gonad mesonephros (AGM), then into organ anlages, to initiate hematopoiesis in fetal liver and bone marrow, bursa of Fabricius, spleen and T cells lymphopoiesis in the thymus. The generation of lymphocytes from adult hematopoietic stem cells (HSCs) is different from embryonic development and will be discussed separately. Our knowledge about ontogeny of the immune system developed mainly from studies on mice, where the performing of in vivo experiments to test precursors activities of certain cell populations was possible. By contrast, most information about human lymphopoiesis came from *in vitro* studies. The principles outlined in mouse models seem to be generally consistent with some underlining mechanisms of human lymphoid system development, but the cell surface phenotypes of human transitional cell populations are often different from those in the mouse.

12.2.1. T Cells

The thymus is a primary lymphoid organ involved in the development of T cells. T cells are primary effectors of adaptive immune responses. The human thymus develops from the endoderm at approximately embryonic day 35. The thymic epithelium and surrounding mesenchyme derived from the cephalic region of the neural crest play an important role in T cell development. The epithelium undergoes morphological changes, beginning to express high levels of major histocompatibility class (MHC) II complex and of epidermal growth factor-like transmembrane protein Delta 1. This protein interacts with the Notch receptors from the family of epidermal growth factor-like transmembrane proteins that control different aspects of tissue development and homeostasis, and plays a critical role in T cell development.

At the embryonic stage, lymphomyeloid commitment precedes thymic development in $CD45^+CD117^+CD34^+$ hematopoietic clusters of AGM region. In addition, multi-lineage progenitors restricted to T, NK and macrophage lineage (pTM) or just T lineage (pT), or B and B plus macrophage (pBM) were identified in the AGM region. At the time of entry into the thymic epithelium, immigrant cells express $CD45^+$, $CD117^+$, $CD44^+$, $CD34^+$, and $\alpha4$ integrin, but are negative for CD62L, CD25, and Thy-1.2 in mice. An important cytokine that regulates survival of human lymphoid progenitors is interleukin (IL)-7 (see Chapter 15). Before exposure to epithelial inductive influence by thymic epithelial stromal cells (TEC), lymphomyeloid progenitors express the alpha chain of the IL-7 receptor). Upon differentiation into functionally mature lymphocytes (CD4 and CD8), IL-7Rα appears to be downregulated and cells lose thymic homing capacity. The transmembrane receptor Notch is essential during early T lineage development. In the absence of Notch1-mediated signals, T cell development is arrested at an early stage and B cells accumulate intrathymically.

12.2.2. B Cells

B-lymphocytes are generated during a lifetime by differentiation from hematopoietic stem progenitors. At the mature stage, they possess a system that can sense the presence of microorganisms and contribute to their destruction by secreting immunoglobulins (Igs). The precise site of B cell commitment in ontogenesis is not known. The B cell precursors derive from the intraembryonic para-aortic region. B cell commitment might take place in the omentum and fetal liver. These progenitors express $CD117^+AA4.1^+CD34^+CD45^+CD19^+Sca\text{-}1^-$. Fetal liver has long been considered the initial site of B cell commitment, and from then on B progenitors expand in a synchronous wave-like pattern reaching a peak in the perinatal stage. Even after birth B cells may develop at sites other than the bone marrow. The myeloid suppressing transcription factor Pax-5 plays a critical role in B cell commitment, while the chemokine receptor CXCR4 and its ligand SDF-1α play a pivotal role of B cell development in the fetal liver (see Chapter 13.)

12.2.3. NK Cells

Natural killer (NK) cells are one component of the innate immune system and have the ability to both lyse target cells and provide an early source of immunoregulatory cytokines. As with B cells, the precise site of NK cell commitment in ontogenesis is not known. The population of fetal liver $CD34^+CD38^+$ cells contains committed NK precursors, which are unable to develop into T cells. The fetal thymus also contains an immature $CD34^-CD5^-CD56^-$ population with a clonogenic NK cell potential, as well as T/NK precursors with bipotentiality. The general consensus is that the bone marrow is the site for NK development in adults.

12.2.4. Cells Derived from Hematopoietic Precursors and Involved in Immune Responses in the Brain

Brain tissue contains only one type of specialized resident cell of macrophage lineage called microglia. Several waves of population in the brain by cells of macrophage lineage have been proposed: primitive macrophages from yolk sac with high proliferation capacity, myeloid precursors from fetal liver with ability to proliferate, and bone marrow derived non-dividing (or very limited division) precursors in adult. Two types of resident brain macrophages, microglia and perivascular macrophages, exist with non-selective, but differently expressed human markers. Both types are $CD45^{lo}CD11c^+CD11b^+(Mac\text{-}1^{+/-})CD64^{+/-}CD68^{+/-}RCA\text{-}1^{+/-}MHC\text{-}classII^{+/-}Ham56^+$ and function in the immune system as effectors of innate immunity. Microglia and perivascular macrophages are major participants in the establishment of adaptive immune reaction and immune privilege *in situ* in the parenchyma and the meninges. The nature of microglia and other brain residents from monocyte lineage will be explored in Chapter 9.

12.3. Postnatal Development of Lymphocytes

12.3.1. Hematopoietic Stem Cells

The generation of lymphocytes from adult HSCs is different from embryonic development. HSCs are defined by their unique capacity to self-renew as well as their ability to differentiate into all blood cell lineages. Therefore, transplanted HSCs are crucial for reconstitution of hematopoiesis in patients following bone marrow (BM) ablation. Although representing approximately only 0.05–0.1% of total BM cells, virtually all HSCs activity has been shown to be contained within the lineage negative or low ($Lin^{-/lo}$) $Sca1^{+}$c-$Kit(CD117)^{lo/hi}$ (LSK) $CD34^{-/lo}$ HSC compartment. The LSK HSC pool can be further subdivided for short-term repopulating cells using additional markers. As HSCs begin to express high levels of CD34, CD27 and another tyrosine kinase type receptor, Flk-2/Flt3, they lose long-term repopulating activity.

To date it is still unclear whether the four lymphoid lineages—T, B, NK, and DCs develop from common lymphoid progenitors (CLPs, Figure 12.1), or whether they are derived from either lymphoid-restricted stem cells or multipotent progenitors. However, at least in the adult murine bone marrow, a population of Lin-IL-$7R^{+}$Thy-1^{-}Sca-$1^{lo}$$CD117^{lo}$ cells has been shown to contain CLPs. This has been determined by their capacity to develop into lymphoid cells while being unable to support myeloid differentiation. It is possible that between common myeloid precursors (CMP) and granulocyte monocyte precursors (GMP), exist a population of progenitors for monocytes and dendritic cells (MDP), that are positive for CX_3CR1 (fractalkine receptor) (Fogg et al., 2006). These novel progenitors give rise to monocytes, to several subsets of macrophages, and to steady-state $CD11c^{+}$ $CD8^{+}$ and $CD11c^{+}$ $CD8^{-}$ dendritic cells *in vivo*. They are devoid of lymphoid, erythroid, and megakaryocytic potential and, also lack granulocyte differentiation potential.

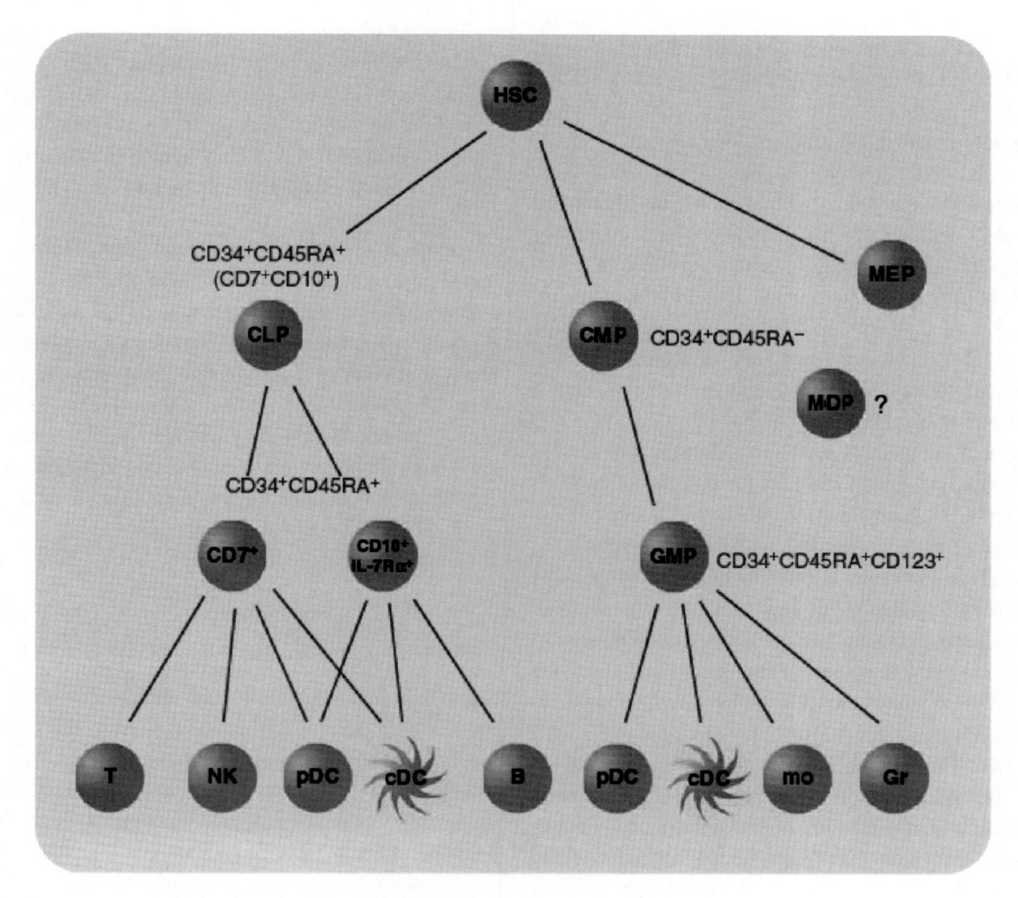

FIGURE 12.1. Model for lymphocytes development. Modified from Blom and Spits (2006).

12.3.2. T Cells Development, TCR, MHC, and CD4-CD8 Commitment

Postnatal lymphomyeloid progenitors are present in bone marrow and circulate in the blood. They do not express IL-7Rα. Expression of this receptor reappears on CLP in adult bone marrow before entry into the thymus and final commitment to T cells. The thymus does not provide an environment for self-renewal. Seeding of the thymus with bone marrow progenitors is required to maintain thymopoiesis throughout adult life. $CD25^-$ (IL-2 receptor alpha chain) and $CD117^+$ expression have been used to characterize progenitor migration from blood into the thymus. Intrathymic development of T cells is based on three major sets of genes: promiscuous tissue-specific antigen (TSA) expression on thymic epithelial cells (TEC), MHC class I and II molecules expression on antigen-presenting cells (APC) such as cortical TEC, medullary TEC, thymic dendritic cells (DCs) and macrophages, and T cell receptor (TCR) molecule expression on mature T cells.

12.3.2.1. The Thymic Epithelial Cells

The nature of TEC precursors and pathways of maturation is still under debate. The most accepted concept today is that TEC come from the self-renewing precursors of endoderm origin, positive for expression of keratin 5 and 8. TEC precursors cells undergo a maturation pathway beginning as cortical TEC (cTEC) without expression of a transcriptional regulator known as an autoimmune regulator (AIRE), advancing to an immature medullary TEC (mTEC) with low levels of AIRE expression, then maturing into mTEC AIRE-positive cells, and finally into Hassall's corpuscles. TEC express tissue-specific antigens representing all parenchymal organs, and including up to 5–10% of all currently known genes, including fetal stage-, tumor- and sex-associated genes. TSA expression increases with TEC maturation. Brain-, intestine- and liver-related TSA have the highest presentation on mTEC. TECs do not form a static scaffold, but similar to stratified epithelia of the skin and gut have a steady state turnover with a half-life of 3–4 weeks. Together with a turnover rate of two weeks for thymic DCs, this makes the APC population in the medulla a highly dynamic compartment (Derbinski and Kyewski, 2005). Those, mTEC and thymic DCs play a central role for the establishment of central tolerance (Kyewski and Klein, 2006).

Interaction with TEC is the major step in T cell education in the thymus and in establishment of self-tolerance. The cross-talk between derived early T lineage progenitors (ETPs) with MHC II-positive cTEC gives rise to $CD4^-CD8^-$ double/dominant negative (DN) thymocytes. DN thymocytes undergo proliferative expansion, lose B- and NK potential and commit to T cell lineage with the onset of TCR β-chain rearrangement. Following T cell lineage commitment, positive selection occurs through the interaction of now $CD4^+CD8^+$ double positive cells with cTEC bearing MHC I and II. Cross-talk between mTEC and DCs expressing TSA leads to deletion (or induction of anergy) of self-reacting T cells, a process known as negative selection. Negative selection results in the generation of CD4 or CD8 positive T cells. Maturation of ETPs into CD4 or CD8 T cells take place in corticomedullary junction and medulla. Following education and establishment of self tolerance, single positive CD4 or CD8 cells leave the thymus (Bommhardt et al., 2004). However, the ectopic expression of antigens on mTEC is very low and several are not expressed at all. Indeed, it seems unbelievable that a tiny population of mTEC could realistically reproduce the tremendous antigenic diversity created by post-translational modifications, alternative mRNA splicing and differential peptide processing. Moreover, intrathymic gene expression cannot induce tolerance to the broad range of innocent external antigens derived from flora or food. It was proposed that thymus-tropic DCs that patrol the periphery and then return to the blood might constitute a mechanism to prevent misguided T cell responses toward at least some of these unpredictable antigens. Normal blood contains small numbers of circulating differentiated DC, which can capture blood-borne microorganisms. These DCs use a classical multistep adhesion cascade for homing to the thymus through microvessels in the corticomedullary junction (Bonasio et al., 2006). They can induce antigen-specific clonal deletion and participate in maintaining central tolerance to brain-born antigens throughout human life as well as participate in the induction of regulatory T cells for establishment of acquired peripheral tolerance (Goldschneider and Cone, 2003). Intrathymal expression of brain antigens such as putative multiple sclerosis (MS) autoantigens B-crystallin, S100, proteolipid protein (PLP) spliced variant, but not myelin oligodendrocyte glycoprotein (MOG)-α and MOG-β isoforms were found in the medullary compartment in humans (Bruno et al., 2002). During ontogeny, oligodendrocytes first express the fetal form termed golli-MBP, which postnatally is replaced by the adult form, termed classic MBP. The switch from golli-MBP to classic MBP does not occur in mTECs. Instead, golli-MBP remains the predominant transcript throughout adulthood, possibly favoring self-tolerance to epitopes present in this isoform (Kyewski and Klein, 2006). Thus, T cells specific for epitopes encoded only by the full-length form of PLP or classic MBP, respectively, may escape intrathymal selection and contribute to the initiation and maintenance of the early phase of multiple sclerosis (Chapter 18, 21).

12.3.2.2. T Cells Receptor and CD4/CD8 Commitment

Signaling through the T cell receptor controls key events in the life of T cells: their development in the thymus from common lymphoid progenitors, the survival of naive T cells following their exit from the thymus, and the differentiation of these cells into effector populations with discrete functional profiles. Mature T cells express either as αβ- or γδ-T-cell receptors that recognize specific antigens. These receptors are generated by a rearrangement of gene segments that result in the formation of genes encoding the α-,β-,γ-, and δ-chains of the receptors.

The choice of γδ *vs.* αβ chains of TCR expression happens at the stage of pre-TCR by selection of cells with productive TCRβ rearrangements irrespective of the Vβ gene segment used. Rearrangement of TCR generates receptors with different affinity and avidity to an unlimited amount of peptides (derived from self and non-self proteins) in the context of MHC molecules expressed on thymic APC/TEC.

Thymocyte fate and selection are related by the degree of TCR downregulation and internalization after TCR/peptide/MHC engagement. Maximal TCR downregulation is correlated with negative selection (highest avidity, agonists), and suboptimal TCR downregulation is correlated with positive selection (lowest avidity, antagonists). Intermediate stages in development of early thymocytes from triple negative ($CD3^-CD4^-CD8^-$) into double positive ($CD3^+CD4^+CD8^+$) are presented in Table 12.1.

Further lineage commitment is directed by cellular interactions with MHC molecules. Cellular interaction with MHC class I antigens generates CD8 cells. Interaction with MHC class II antigens generates CD4 cells (reviewed in Kappes et al., 2005). The Lck protein tyrosine kinase activity upon "strong" TCR signaling favors development of $CD4^+$ cells, whereas "weak" TCR signaling and reduction of Lsk activity favors development of $CD8^+$ cells. TEC and thymic hematopoietic DCs are able to induce the proliferation and differentiation of $CD4^+CD8^-CD25^-$T cells into $CD4^+CD25^+FOXP3^+$ (forkhead box P3) regulatory T cells (Treg). This induction depends on peptide/MHC class II interactions, and the presence of IL-2 (Watanabe et al., 2005).

TABLE 12.1. Markers for intermediate stages in T cell development.

	Thymus						
Blood	TSP	ETP	Pre-T	Small CD4 ISP	Large CD4 ISP	icTCRβ+/ ic TCRβ- EDP	CD3εlo DP
CD34	+lo	+	+	+	–	–	–
CD1α	–	–	+	+	+	+	+
icTCRβ	–	–	–	–	+	+/–	+
icCD3ε	?	+/–	+	+	+	+–	+
CD2	–	+/–	+	+	+	+–	+
CD5	–	+/–	+	+	+	+–	+
CD7	+lo	+	+	+	+	+–	+
CD4	–	–	–	+	+	+	+
CD8α	–	–	–	–	–	+	+
CD8β	–	–	–	–	–	–	+
CD45RA	+lo	+	–	–	–	–	–
CD45RO	–	–	–	–	+	+/–	+
IL-7Rα	–	+	+	+	+	+	+
DJβ	–	+	+	+	+	+	+
V-D-Jβ	–	–	–	+	+	+	+
V-Jα	–	–	–	–	–	–	+

Abbreviation: TSP, thymus seeding progenitors; ETP, early thymic progenitor; ISP, immature single-positive progenitor, icTCR, intracytoplasmic TCR; EDP, early double-positive

These mature, but naive T cells exit peripheral blood and seed lymphoid organs in T-cell specific zones to be acquired by adaptive immune responses for elimination of infected or tumor cells, support for humoral immune responses, formation of immunologic memory, prevention of excessive tissue damage, and facilitation of tissue regeneration (see Chapter 12).

12.3.2.3. NKT-Cell Development

One lineage choice that occurs rather late in T-cell ontogeny is the development of NKT cells. NKT cells express markers of both NK cells and T lymphocytes. These include Ly49 family receptors NK1.1 and TCR in mice. Mature NKT cells are $CD44^{hi}CD69^+$, a phenotype consistent with an activated cell. A hallmark of NKT cells is copious IL-4 and IFN-γ secretion promptly upon TCR activation. Because of this property, NKT cells most likely possess regulatory functions. They suppress multiple autoimmune phenomena, and in some cases, inhibit tumor metastases. Unlike conventional αβ T cells that are selected by classical MHC class I and II, the vast majority of NKT cells are selected by the non-polymorphic class I-like molecule, CD1d.

NKT-cell development is dependent on the protease cathepsin L. As NKT cells react to lipids, cathepsin L may process a lipid carrier protein that is required for loading lipids onto CD1d. A strong candidate for the lipid carrier protein is prosaposin, a precursor that is cleaved into a series of lipid transfer proteins. As the structure of lipids capable of binding CD1d is not as diverse as peptides, it is perhaps not surprising that NKT cells exhibit a limited TCR repertoire. The CD1d-restricted NKT cells also differ from mainstream T cells in co-receptor expression. They are $CD4^+$ or double negative, and never $CD8^+$.

12.3.2.4. T Cells and Brain

The long living neurons and synapses that provide long-term memory have to be strongly protected from damage by inflammatory/autoreactive $CD4^+$ and cytolytic/cytotoxic $CD8^+$ T cells. Brain parenchyma does not support migration or survival of naïve T cells (Hickey, 2001). The highly specialized nature of the cerebral vasculature endothelial cell lining with tight junctions provides a diffusion barrier for hydrophilic molecules (see Chapter 38) and forms a physical barrier to lymphocytes. However, lymphocytes capable to enter perivascular spaces, without disruption of this barrier via emperipolesis (Engelhardt, 2006). Moreover, a small amount of memory T cells predominantly consisted of $CD4^+/CDRA^-/CD27^+/CD69^+$ activated central memory T cells expressing high levels of CCR7 and L-selectine are present in cerebrospinal fluid of patients with noninflamed CNS (Engelhardt and Ransohoff, 2005). Choroid plexus stroma always contains $CD3^+$ T cells (Kivisakk et al., 2003). These cells represent a circulating pool of long living lymphocytes and express other tissue specific homing determinants, such as skin homing molecules cutaneous leukocyte antigen (CLA) and CCR4 chemokine receptor and gut tissue committed determinants integrin α4β7 and CCR9 chemokine receptor (Kivisakk et al., 2006).

For a short period of time, T cells can be found in perivascular spaces and make parenchymal foci as an inflammatory reaction to acute viral or bacterial infection, during autoimmune and degenerative diseases (see Chapters 18–24). However, such physiologic responses in the brain parenchyma are quickly terminated by neuron-produced gangliosides via the downregulation of MHC class I and II co-stimulatory molecular expression on glial cells (Massa, 1993). Several soluble factors can suppress activation and proliferation of T cells in response to local antigens presented by microglia. For example, vasoactive intestinal peptide (VIP), which is a neurotransmitter in the suprachiasmatic nucleus, modulates T cells differentiation, and affects the profile of cytokine secretion by cells of macrophage lineage (Delgado et al., 2002). During inflammation in the brain a wide spectrum of anti-inflammatory and immunosuppressive substances were observed: IL-10, TGF-β, indoleamine dioxygenase, prostaglandins etc. T cells could be eliminated from brain parenchyma by apoptosis mediated by Fas-Fas ligand (CD95/CD95L) interaction. Fas ligand is constitutively expressed throughout the CNS (neurons, astrocytes, oligodendrocytes, microglia, and vascular endothelium), and its expression increases in response to inflammation. Antigen-priming and activation of lymphocytes also stimulates expression of CD95 (the receptor for Fas ligand). Through Fas-Fas-L interaction, endothelial cells as well as astrocytes, prevent access of activated T cells (Choi and Benveniste, 2004).

Lymphocyte behavior in brain tissue depends upon functional status of microglial cells, parenchymal and perivascular macrophages.

Activation of brain microglial cells, DCs and macrophages may also be stopped by factors directly produced by neurons, such as CD200 molecules. This protein from the Ig superfamily plays an important role in the downregulation of myeloid cell function, which expresses a receptor to this ligand. Another soluble factor chemokine, fractalkine, produced by neurons and astrocytes, downregulates inflammatory activity of microglia/macrophages via its receptor CX3CR1 (Cardona et al., 2006; Charo and Ransohoff, 2006).

Persistence of T cells in the brain is a rare event and is associated with chronic pathology forming tertiary lymphoid structure that provides support and activation signals by DCs and macrophages. These structures are present in perivascular spaces, meninges and choroid plexus, and were studied mostly in multiple sclerosis (see Chapter 18).

12.3.3. B Cells

12.3.3.1. Development in Bone Marrow

In adults, B cells are generated from B-lineage committed precursors in the bone marrow (BM). Newly formed B cells express antigen-specific surface antibodies (sIgM$^+$ and sIgD$^+$). In the BM, B cells are subjected to multiple selective pressures that purge autoreactive B cells and guide differentiation of the remaining cells into functionally distinct peripheral B cell compartments (Table 12.2). Naive B cells then enter the circulation and migrate to the spleen and lymph nodes.

TABLE 12.2. Markers for early stages in B cell development

	Bone marrow						
	CLP	Early B	Pro-B	Pre-BI	Large pre-BII	Small pre-BII	Immature B
CD34	+	+	+	−	−	−	−
CD10	+	+	+	+	+	+	+
IL-7Rα	+	+	+	+−	−	−	−
CD19	−	−	+	+	+	+/−	+
CD79α	−	+	+	+	+	+	+
TdT	−	−	+	−	−	−	−
RAG	−	−	+	+	−	+	+
Vpre-B	−	+	+	+	+	−	−
μH	−	−	+/−	+	+	+	+
pre-BCR	−	−	−	−	+	−	−
IgH	GL	DJ_H	V_HDJ_H	V_HDJ_H	V_HDJ_H	V_HDJ_H	V_HDJ_H
κL	GL	GL	GL	GL	GL	V_LJ_L	V_LJ_L
cycling	−	−	−	+	+	−	−
Pax-5	−	−	+	+	+	+	+
sIgM	−	−	−	−	−	−	+

The generation of B-lymphocytes from HSCs is controlled by two cytokine receptors (Flk2/Flt3 and IL-7R) and six transcription factors (PU.1, Ikaros, E2A, Bcl11a, EBF, and Pax-5). Ikaros and PU.1 act in parallel pathways to control the development of lymphoid progenitors, in part, by regulating the expression of essential signaling receptors Flt3, CD117, and IL-7Rα. Generation of the earliest B cell progenitors depends on transcriptional factors E2A and EBF that coordinately activate the B cell gene expression program and immunoglobulin heavy-chain gene rearrangements at the onset of B-lymphopoiesis. Pax5 restricts developmental options of lymphoid progenitors to the B cell lineage by repressing transcription of lineage-inappropriate genes and by simultaneously activating expression of B-lymphoid signaling molecules. Two other transcriptional factors LEF1 and Sox4 contribute to the survival and proliferation of pro-B cells in response to extracellular signals. Finally, IRF4 and IRF8 together control the termination of pre-B cell receptor signaling, and thus, promote differentiation to small pre-B cells undergoing light-chain gene rearrangements (Singh et al., 2005). Productive variable-region and joining-region rearrangements (V_LJ_L rearrangements) of the immunoglobulin light chain are completed in pre-B cells. Expression of the B-cell receptor (BCR) drives pre-B cells to the immature B-cell stage. Newly formed B cells are exported to peripheral lymphoid tissues as functionally immature or transitional intermediates cells.

12.3.3.2. Distribution of B Cells and Function

Selection of newly formed B cells to transitional, follicular and marginal-zone (MZ) B cells depends on integrated signals from several classes of surface receptors such as, the BCR and co-signals, the tumor-necrosis factor receptor (TNFR) family, and the G-protein-coupled receptors (GPCRs).

Follicular and MZ long-lived compartments participate differentially in B-cell effector functions such as the germinal-centre (GC) reaction, long-term memory, antigen presentation, and antibody and plasma-cell (PC) generation. Different strengths of BCR signaling are required for the development of the three mature B-cell subsets: peritoneal B cells require the strongest BCR signal, follicular B cells require an intermediate BCR signal, and marginal-zone B cells require a weaker BCR signal.

During B lymphocyte development, antibodies are assembled by random gene segment re-assortment to produce a vast number of specificities. A potential disadvantage of this process is that some of the antibodies produced are self-reactive, and in humans, the majority (55–75%) of all antibodies expressed by early immature B cells display self-reactivity, including polyreactive and anti-nuclear specificities. Most auto-antibodies are removed from the population at discrete checkpoints during B cell development. Inefficient checkpoint regulation would lead to substantial increases in circulating auto-antibodies. So, an important role of the BCR at the immature B-cell stage is to induce efficient elimination of these potentially harmful cells. This can be achieved in three ways: immature B cells are eliminated through negative selection (BCR-induced cell death), immature B cells are inactivated (anergy), or immature B cells revise the specificity of their BCRs (receptor editing).

12.3.3.3. Brain and B Cells

In contrast to T cells, in different immunopathologies, the brain provides a fostering environment to B cells. Primary central nervous system (CNS) lymphomas are usually of B cell origin. The cerebrospinal fluid (CSF) of patients with chronic infections and autoimmune diseases of the CNS typically contains remarkably stable oligoclonal Ig bands. In the CNS of multiple sclerosis patients, clonally expanded B cells and plasma cells persist. Ectopic B cell follicles develop in the meninges of patients with secondary progressive MS, and B cell differentiation may be recapitulated in the CNS of MS patients (Krumbholz et al., 2006) (see Chapters 18–24).

12.3.4. Adult NK Cells

Human NK cells comprise ~15% of all lymphocytes and are defined phenotypically by their expression of CD56 and lack of CD3 expression. Two distinct populations of human NK cells can be identified based upon their cell-surface density of CD56. The majority (~90%) of human NK cells have low-density expression of CD56 (CD56^{dim}) and express high levels of Fcγ receptor III (FcγRIII, CD16), whereas ~10% of NK cells are CD56^{bright}CD16^{dim} or CD56^{bright}CD16^{-} (Table 12.3).

Over the past decade a number of phenotypic and functional properties of human NK cells have been characterized. There is now good evidence to suggest that distinct immunoregulatory roles can be assigned to the CD56^{bright} and CD56^{dim} NK-cell subsets. Peripheral NK cells express, in addition to CD16 and CD56, CD161 (NKR-P1A). Low-affinity FcgRIII on the surface of NK cells binds to Ab-coated (opsonized) targets and signals through associated subunits containing an immunoreceptor tyrosine-based activation motif (ITAM) and directs antibody-dependent cellular cytotoxicity (ADCC).

TABLE 12.3. Markers for NK cell development.

	Commitment		Maturation		
	CLP	T/NK precursor	NKp	NKi	NK
CD34	+	+	+lo	–	–
CD45RA	+	+	+	+	+
CD7	–	+	+lo	–	–
CD10	+	–	–		
CD122	–	+lo	+	+	+
α4β7	–	–	+	–	–
CD16	–	–	–	+/–	+/–
CD56	–	–	–	–	+hi
CD161	–	–	–	+	+

Of significance, the CD56^{bright} NK-cell subset expresses the high-affinity IL-2R constitutively and can produce interferon gamma (IFN-γ) in response to picomolar concentrations of IL-2. Since IL-2 is produced only by T cells and not NK cells, the expression of high-affinity IL-2Rαβγ on CD56^{bright} NK cells is necessary to promote cytokine cross-talk between NK cells and T cells in secondary lymphoid organs. NK cells constitutively express several receptors for monocyte-derived cytokines (monokines), including IL-1, IL-10, IL-12, IL-15, and IL-18, and most likely receive some of their earliest activation signals from monocytes during the innate immune response.

12.4. Organization of the Secondary Lymphoid Tissues

12.4.1. Embryonic Development of Lymphoid Tissues

Development of the lymphatic system begins with the invagination of endothelial cells from veins and the formation of lymphoid sacs. At the location of the lymph sacs, connective tissue protrudes into these lymph sacs, forming the very first anlagen of the lymph nodes. At this moment, differentiation of mesenchymal cells into specialized cells, known as lymphoid organizer cells, initiates the formation of lymph nodes.

Platelet-derived growth factor (PDGF), fibroblast growth factor, as well as TGF superfamily of growth factors, all of which are crucial for the differentiation of mesenchymal cells, are likely to be involved in the earliest phases of lymphoid organ formation. However, no molecules have been identified that direct the early specification of mesenchymal cells into specific lymph node organizer cells.

Mesenchymal specification of lymph node organizer cells is expected to be independent from lymphotoxin beta receptor (LT-βR) signaling. Stromal organizer cells mediate attraction and retention of hemopoietic cells (lymphoid tissue inducers, LTi), resulting in accumulation and clustering of cells (Mebius, 2003; Cupedo and Mebius, 2005).

For the generation of LTi cells from IL-7R-expressing hematopoietic precursors, expression of retinoic-acid-receptor-related orphan receptor-γ (RORγ) by LTi cells is mandatory. LTi cells trigger the LT-βR on stromal cells through expression of $LT\alpha_1\beta_2$. $LT\alpha_1\beta_2$ is induced by ligation of either the TNF-related activation-induced cytokine (TRANCE) that signals through TRANCE receptor (TRANCER) and/or IL-7R. Functional LTi cells need to express CXCR5 as well as CCR7 to respond to chemokines involved in lymph node formation such as CXCL13, CCL19, and CCL21.

During the generation of functional lymphoid tissue organizers, LT-βR signaling leads to expression of mucosal addressin cell adhesion molecule (MAdCAM-1), intercellular adhesion molecule 1 (ICAM-1), and vascular cell-adhesion molecule 1 (VCAM-1) on endothelial cells, as well as the production of the chemokines CXCL13, CCL19, and CCL21 by stromal cells and HEVs (Carlsen et al., 2005).

A few days before birth, lymphocytes and dendritic cells (DCs) are recruited to the lymph-node and Peyer's-patch anlagen. Because T cells and B cells themselves express LT-$\alpha_1\beta_2$, recruitment of lymphocytes is amplified by a second positive-feedback loop and culminates in the formation of a mature lymph node or Peyer's patch consisting of segregated T-cell zones and B-cell follicles, HEVs and additional specialized mesenchymal cells.

IL-7R signaling has been reported to be essential for the development of Peyer's patches, but this pathway is also involved in the generation of some lymph nodes. In the absence of a functional components of the IL-7R signaling pathway such as IL-2R common γ-chain (γc) and Janus kinase 3 (Jak3) or in the absence of IL-7, peripheral lymph nodes do not develop. Although lymphopenia has been suggested as an explanation for the inability to detect lymph nodes in IL-7R-deficient and γc-deficient mice, lymph nodes can easily be found in severe combined immunodeficient (SCID), recombination-activating gene (Rag) 1 product and Rag2-deficient mice, which also lack mature lymphocytes.

Inducer, precursor $CD45^+CD4^+CD3^-$ cells also differentiate into antigen-presenting cells, as well as natural killer cells, but not to B or T cells. Detailed phenotyping of adult LTi (or accessory) $CD45^+CD4^+CD3^-$ cells has shown that they are similar to a $CD4^+$ cell population.

12.4.2. Lymph Node Architecture

Lymph nodes have two components: the stroma with its fixed cells and the parenchyma with its migrating cell populations. In contrast to non-lymphoid parenchymal organs, the stromal network has an indispensable functional role.

Fibroblastic reticular cells (FRCs) constitute the main stromal population and form its internal three-dimensional network. FRCs express surface molecules and produce "homing" chemokines for T cells, B cells, and DCs. FRCs provide anatomical arrangements that influence the traffic patterns of lymphocytes facilitating lymphocyte "crawling" along preformed "corridors". Moreover, FRCs typically wrap reticular fibers (mainly made up of type III and IV collagen, elastin and laminin) to form a set of interconnected channels called the FRC conduit system.

This system allows for efficient and rapid transfer of soluble molecules, such as antigens, chemokines and immune complexes from the subcapsular sinus to the deeper cortex, the high HEVs, and perhaps the follicular compartment.

Lymph nodes can undergo rapid and profound hypertrophy during an immune response, and FRCs are deeply involved in matrix reorganization and the remodeling of lymph node architecture.

The conduit system and sinuses are lined by sinus endothelial cells (SECs). SECs are flattened cells that do not form a continuous layer, especially in the wall of the medullary sinus, but contain intercellular gaps or pores. Gaps have also been demonstrated in the floor of the subcapsular sinus. FRCs and SECs might be ontogenically related and serve to filtrate lymph fluid collected from all peripheral tissues, including brain interstitial tissue and cerebrospinal fluid.

In the subcapsular sinus, macrophages and DCs sample the lymph and remove microorganisms and debris contained there. FRCs and SECs cells also transport and/or process antigenic material for presentation to B and T cells. However, soluble lymph-borne material can restrict access to the lymphocyte compartment and the cortex.

Migrating naive T and B cells enter lymph node via HEVs, migrate through the net of dendritic cells and exit via single (one) for each nodule efferent lymph vessel back to circulation. Soluble antigen and circulating DCs enter via afferent lymphatic vessels at multiple sites along the capsule. All secondary lymphoid tissues have some structural similarity where the major component is a lymphoid follicle with germinal center, follicular mantle and marginal zone (Figure 12.2).

12.4.3. Antigen-Presenting Cells in Lymph Node

The lymph node harbors a composite population of dendritic cells (DCs)—residents and emigrants—capable of selection, stimulation and elimination of effector lymphocytes. These cells trigger adaptive immunity and/or tolerance. In the human lymph node T-cell compartment, two primary subsets of DCs, compared to six in mice, have been identified: monocytoid $CD11b^-CD11c^+$ and plasmacytoid interferon (IFN)-α producing $CD123^+CD11b^-CD11c^-$. The last subset clusters with T cells around HEVs.

$CD11c^+$ DCs are present as stellate cells whose delicate processes or "veils" extend in many directions, these are called interdigitating DCs.

In the B cell compartment, follicular DCs (FDCs) orchestrate B cell function. Whether FDCs originate from hematopoietic progenitors or from stromal elements is still a controversy. New evidence suggests the presence of two types of FDCs within the human germinal centers: the classic FDCs

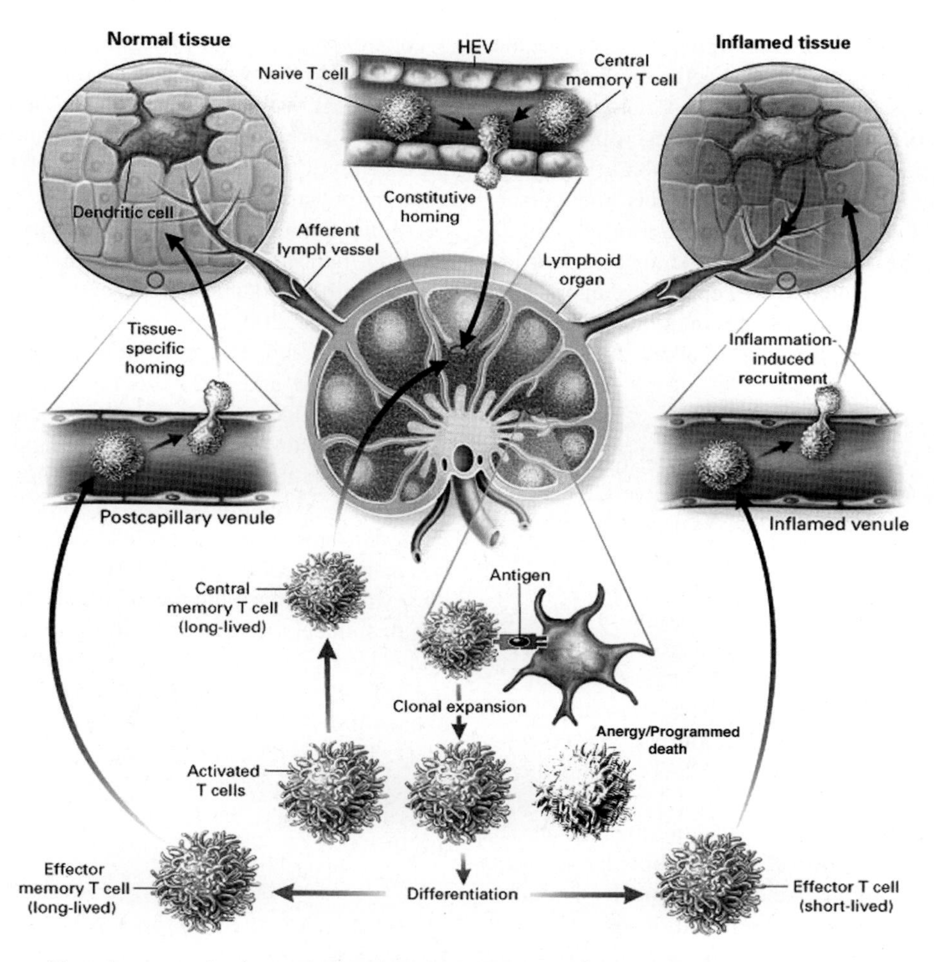

FIGURE 12.2. Pathways of lymphocytes selection and migration. Modified from Von Andrian and Mackay (2000).

that express CD21 and fibroblast-related markers represent stromal cells and the CD3⁻CD4⁻CD11c⁻ germinal center dendritic cells (GCDC) that represent hematopoietic cells and may be analogous to the antigen-transporting cells described in mice.

Under specific conditions *in vitro*, human monocytes can downregulate expression of the hematopoietic marker CD45 and function as FDCs (Heinemann and Peters, 2005). For example, FDCs are generally considered to be the major source of CXCL13 (also called B cell-attracting chemokine-1) both in normal and aberrant lymphoid tissue. The most CXCL13-expressing cells in rheumatoid arthritis and ulcerative colitis are of monocyte/macrophage lineage. They are located not only in irregular lymphoid aggregates within an FDC network but also within and near smaller collections of B cells in diseased tissue where no FDCs are detected.

The same type of macrophage-derived FDCs could support B cells in multiple sclerosis patients (Krumbholz et al., 2006).

These cells occupy the outer layers of primary and secondary lymph node follicles—the light zone and the follicular mantle. FDCs constitute a three-dimensional sponge-like network formed by interdigitation of adjacent cells. As shown by ultrastructural studies, these cells present filiform or beaded dendrites with characteristic "ball-of-yarn" convolutions.

FDCs play a pivotal role in promoting B-cell proliferation and differentiation in germinal centers. They are very efficient in trapping and retaining antigen–antibody complexes through CD21–CD35 and FcγRIIB receptors for long periods and presenting them as iccosomes (immune-complex-coated bodies) to memory B cells. More details of lymph node functional organization were described in several reviews (Von Andrian and Mempel, 2003; Crivellato et al., 2004).

Spleen also contains lymph node-like organized structures, however this organ is multifunctional. For detailed information read Cupedo and Mebius's (2005) review.

12.4.4. Lymph Node as the Home for Primary Adaptive Immune Responses and Peripheral Tolerance

12.4.4.1. Common Rules of Adaptive Immune Responses

The major function of the immune system is the control of our biologic individuality. The strength and quality of immune responses is based on highly polymorphic MHC molecules and various recombinations of T and B cell receptors.

Cytokine/chemokine milieu induced by external, self and self-modified proteins, loaded into DCs/macrophages and delivered from affected tissues (including the brain) by lymph into regional draining lymph nodes controls the balance between induction of immunity or tolerance. Both the induction of antigen-specific immunity and tolerance rely on the direct interaction of DCs with naive T cells. These interactions occur in the T cell zone of lymph nodes in the vicinity of high endothelial venules (Von Andrian and Mackay, 2000).

12.4.4.2. CD4⁺ Cell Polarization

Depending on the co-stimulatory and cytokine signals that are provided by DCs at the time of priming, naive $CD4^{+}CD25^{-}$ T cells differentiate into the different types of helper cells (Th1, Th2, and T follicular helper). Activation of naive $CD4^{+}$ T cells through the TCR elicits low but detectable transcription of both genes IL-4 and IFN-γ. Expression of the T-box transcription factor, T-bet, plays a critical role in Th1 development and supports transcriptional competence at the IFN-γ locus and selective responsiveness to IL-12. As stimulation progresses in the presence of signaling through the essential IL-12 receptor (via transcription factor STAT4), primed cells express IL-2 and become Th1 helper cells.

Another transcriptional factor GATA3 is involved in Th2 differentiation and contributes to chromatin remodeling events that favor IL-4 gene stable demethylation and IL-4 production. Signaling through IL-4 receptor, and downstream with STAT6, favors stable production of IL-4 (as well as IL-5, IL-13) and differentiation in Th2 cells (Ansel et al., 2003; Kalies et al., 2006). It is unclear whether Tfh cells differentiate as a separate lineage at the time of T-cell priming or whether they emerge at a later stage from non-polarized, primed T cells or even from polarized Th2 or Th1 (less possible) cells.

12.4.4.3. CD4⁺ Cells Interaction with B Cells

A proportion of newly primed T cells with upregulated expression of CXCR5 migrate toward the T-cell-zone-follicle boundary where they might provide cognate help to antigen-specific B cells. In this location they can interact with B cells and also with $CD4^{+}CD3^{-}$ accessory cells (LTi cells analogs for development of lymph nodes). The $CD4^{+}CD3^{-}$ accessory cells present at the T-cell-zone-follicle boundary and within the follicles provide signals through CD134 (also known as OX40) and CD30 at the surface of follicular T cells to direct localization of these cells, maintenance of germinal-centre reactions and the generation of B-cell and $CD4^{+}$ T-cell memory responses.

These follicular T cells strongly express the CD28-related molecule ICOS (inducible T-cell co-stimulator; also known as CD278) and secrete moderate amounts of IL-10, CD278, and IL-10 act as co-stimulatory factors having crucial functions in T helper cell and B-cell responses. After stimulation follicular T cells express CD134 and rapidly upregulate CD40 ligand (CD40L; also known as CD154) from pre-formed stores. This is consistent with their capacity to induce germinal-centre B cells to survive, secrete antibody and express activation-induced (cytidine) deaminase (AID). AID is crucial for immunoglobulin class switching and somatic hypermutation. IL-21 co-stimulation potently induces the expression of both B lymphocyte-induced maturation protein-1 (BLIMP-1) and AID, as well as the production of large amounts of IgG from B cells (Ettinger et al., 2005).

Tfh, Th1, and Th2 cells can all participate in antibody responses. Th1 and Th2 cytokines guide class-switch recombination regulating/determining particular immunoglobulin classes and subclasses, thereby skewing antibody responses depending on the nature of the stimulus. For example, IFN-γ promotes switching to IgG2a during antiviral responses while IL-4 promotes the production of IgE during anti-parasite responses. Both Th1 and Th2 cells are crucial helpers in the extrafollicular pathway of B-cell differentiation and are potent inducers of class switching to T-cell-dependent antibody classes and subclasses outside germinal centers.

$CD4^{+}$ Treg cells suppress Th1/Th2 cells function, the expansion of $CD8^{+}$ cells and B-cell production of antibody. These are $CD4^{+}CD25^{+}CD57^{-}$ T cells that co-express CXCR5 and localize to T cell zone and germinal centers. T cells that express NK1.1 ($CD4^{+}$NKT cells) as well as NK are there.

12.4.4.4. B Cells Differentiation and Humoral Immune Responses

The help that T cells provide to B cells allows for the production of high-affinity memory B cells and long-lived plasma cells specific for foreign antigens. T cell independent immune responses usually occur after direct activation of B cells by bacterial capsular polysaccharides and by microorganism-derived Toll-like-receptor (TLR) ligands. TLR-stimulated T cell independent response is a rapid antibody response to the pathogens through the generation of short-lived, low-affinity extrafollicular plasma cells.

The essential structure for T cell-dependent responses is a germinal center. Germinal centers arise in lymphoid tissues following antigenic stimulation and provide a milieu for proliferation of B cells (known as centroblast). Somatic hypermutation in germinal centers results in increased affinity for antigen in a minority of germinal-center B cells. This process can also generate self-reactive B cells.

Centroblasts exit the cell cycle to become centrocytes and bind antigens that are associated with FDCs using newly expressed cell-surface immunoglobulins. After processing, B cells present antigen in MHC II to T helper cells and elicit help. Interaction with T helper cells has several outcomes: (1) survival and selection of high-affinity centrocytes; (2) class-switch recombination of immunoglobulin and differentiation into long-lived plasma cells and memory B cells; (3) perpetuation of germinal-center reactions by stimulating centrocytes to recycle and become centroblasts. In the absence of antigen-specific T cells, centrocytes are eliminated and germinal centers abort early after induction.

Spontaneous ectopic germinal-center formation has long been recognized to occur in human autoimmune diseases and is a source of somatically mutated high-affinity autoantibodies.

12.4.4.5. NK Cells in the Lymph Node

NK cells are present in secondary lymph nodes, in which they exert important immunoregulatory functions. In humans, NK cells make up as much as 5% of all mononuclear cells in peripheral lymph nodes. These NK cells belong to the $CD56^{bright}$ subgroup, and are mainly cytokine producers in the T-cell area, where they can interact with DCs. They provide IFN-γ for Th1 polarization and act as messengers between innate and adaptive immunity. NK cells are also capable of modulating the function of DCs either by inducing their maturation or by killing them.

12.4.4.6. $CD8^+$ T Cells in the Lymph Node

A small proportion of follicular T cells in human germinal centers are $CD8^+$ (less than 5% in tonsils and 10–15% in lymph nodes). Unlike B cells, which require a significant amount $CD4^+$ T cell help in order to achieve high specificity, magnitude and stability in antibody production (humoral immune responses), $CD8^+$ cell mediated (cytotoxic/cytolitic) cellular immune responses during the first several days after induction are independent from $CD4^+$ help. Only memory formation is dependent on $CD4^+$ T cell help. Mainly it depends upon CD40 ligation and IL-2 production by $CD4^+$ Th1 helper cells and is regulated by co-stimulatory molecules expressed by DCs.

The effector $CD8^+$ T cells possess an arsenal of cellular weapons and are exquisitely refined in their ability to recognize target cells displaying small peptides on a protein scaffold of MHC. Upon recognition of a foreign peptide, $CD8^+$ T cells can use either cytopathic or non-cytopathic mechanisms to rid a cell of an invading pathogen. These mechanisms include deposition of pore-forming molecules and granzyme release, engagement of the apoptosis-inducing ligands (such as Fas ligand), and the release of cytokines such as IFN-γ and TNF-α. All this machinery is programmed in the lymph node during priming and delivered at the site where $CD8^+$ cells perform their function. CTLs will eliminate virally or bacterially infected cells (stromal, macrophages/DCs, or lymphocytes) residing in lymph nodes that would have led to lymphopenia and immune-suppression (HIV/AIDS).

12.4.4.7. Memory Formation and Vaccines

Some polarized antigen-specific $CD4^+$ T cells (Th1, Th2), antigen-primed $CD8^+$ upregulate the expression of receptors for pro-inflammatory chemokines such as CXCR3 (CXCL9, -10, -11, known as α-chemokines IP-10, Mig, I-TAC) and down-regulate CCR7 (lymphoid tissue homing), in order to exit from lymph nodes for immediate participation in peripheral effector responses (von Adrian and Mackey, 2000). In affected tissues, Th1 cells will produce large amounts of IL-2, IL-12 p40, and IFNγ promoting effector function of $CD8^+$ cytotoxic lymphocytes. Th2 cells will produce large amounts of IL-4, IL-5 and IL-13, but not IFN-γ, promoting humoral responses, inflammation and elimination of the pathogen. Most effector cells will die after an antigen is cleared, but a few antigen-experienced memory cells remain for long-term protection. Different subgroups of memory cells stand guard in lymphoid organs and bone marrow in order to patrol peripheral tissues and to be able to mount rapid responses whenever the antigen returns.

The carrier development for T as well B cells is probably programmed during priming, and involves T-T, T-B cell cooperation on the basis of DC co-stimulatory signals. Signaling via CD40, CD27, 4–1BB (CD137), inducible co-stimulatory molecule (ICOS) in the presence of IL-7, IL-15 and IL-21 facilitates the development of memory cells in lymphoid and non-lymphoid tissues. Approximately only 5–10% of effector cells will pass through and generate stem-cell-like capacity for self-renewal, long-term antigen-independent persistence and protective immunity (>20 years). Some of these cells will persist in non-lymphoid tissues after elimination of pathogens (including brain tissue).

Generation of such memory cells is the goal of vaccination against viral and bacterial infections. Today, the best way to achieve this goal is vaccination using attenuated live pathogens.

12.4.4.8. Tolerance

Under normal conditions in normal tissue, liters of interstitial fluid containing lipids, proteins, peptides, apoptotic debris (including brain tissue) and most likely immature dendritic cells loaded with such compounds, pass through lymph nodes. Immature DCs continuously sample and present non-pathologically modified autoantigens to T cells. Due to the low expression levels of class I and class II MHC and co-stimulatory B7-family proteins on immature DC, productive immune responses are not possible. Instead, the cognate T cells undergo limited activation/expansion followed by programmed death, or obtain anergic/tolerant status to the self antigen.

Different types of regulatory cells also actively contribute to the programming and maintenance of peripheral tolerance by production of IL-10 and TGF-β family cytokines.

The stimulatory effects of CD28 ligation are counteracted by the co-inhibitory signaling of cytotoxic T lymphocyte antigen 4 (CTLA-4). CTLA-4 ligation by B7–1 or B7–2 results in inhibition of TCR and CD28 mediated signals (Chen, 2004). Recently discovered B and T lymphocyte attenuator (BTLA) isolated from Th1 cells and expressed mostly by B cells following DCs, Program death-1 (PD1) receptor on lymphocytes and myeloid cells with two counter-receptors B7-H1 (PDL-1) and B7-C (PDL2), respectively, also contribute to tolerance, Figure 12.3.

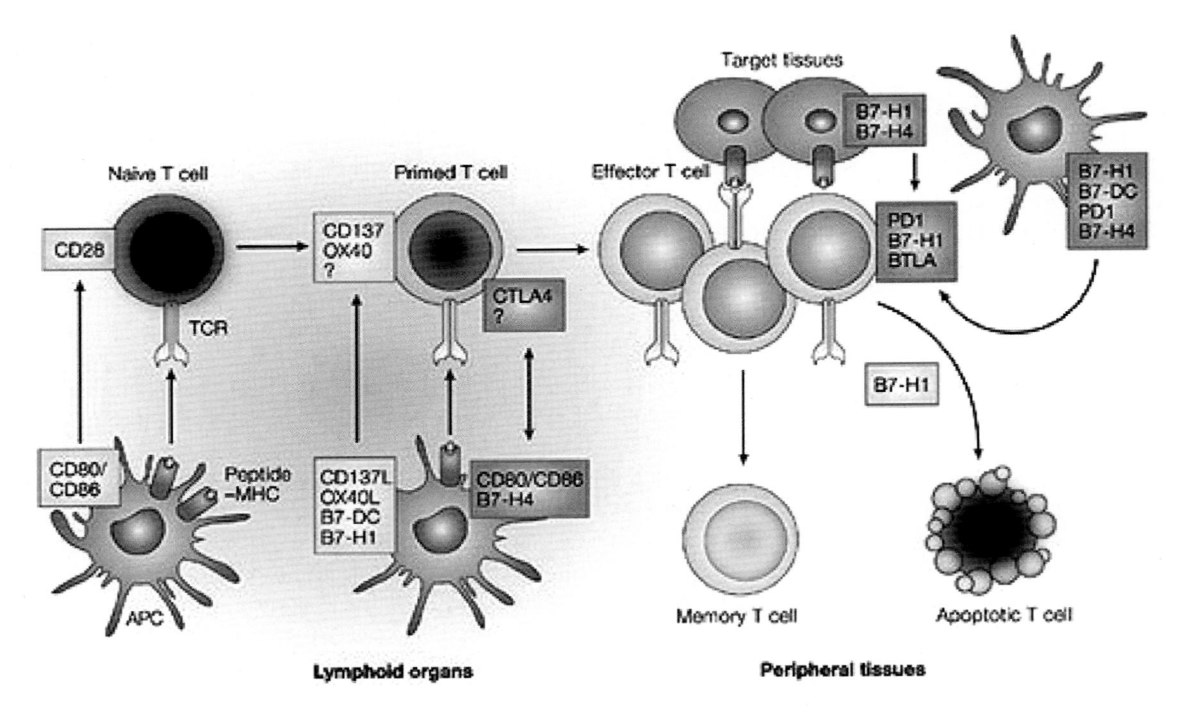

FIGURE 12.3. The co-signal network model. Co-stimulators (light plates) and co-inhibitors (dark plates) have overlapping but preferential functions in controlling one or more stages of T (and probably) B cell activation, including priming, differentiation, maturation and potentially memory responses. From Chen (2004)

12.4.4.9. *Immunologic Memory in Brain*

As in any other tissue, the status of post-capillary vein endothelial cells plays a dominant role in penetration of activated lymphocytes into the brain parenchyma (see Chapter 4). Systemic activation of such post-capillary vein endothelials cells by exo- and endotoxins via increased production of TNF-α and other pro-inflammatory cytokines and up-regulation of adhesion molecules allows lymphocytes to enter perivascular spaces in the meninges and choroids plexus. Neither the meninges nor choroids have an established blood-brain barrier. These two compartments contain potential (macrophages) and actual dendritic cells. Following viral and bacterial infections that might affect APC and induce local immune reactions, memory cells are formed. In healthy individuals at middle age, the CSF contains a very low number of lymphocytes. Over 80% of lymphocytes in CSF are memory T cells ($CD4^+CD45RO^+CD27^+CXCR3^+$). Continuous recirculation of these cells between blood and the CSF is supported by low levels of P-selectin expression on endothelial cells of the deep stromal veins of the choroid plexus and the bridging meningeal veins. Engelhardt and Ransohoff (2005) reviewed lymphocyte migration in the brain in great detail. These memory cells will persist through human life and respond *in situ* to re-appearing antigen (viral, bacterial, etc.). Even so, the memory recall in the brain is delayed due to the immunosuppressive brain environment and activation of pro-apoptotic T-cells-glia interactions via CD95/CD95L (Fas/FasL), CD30/CD153, and other probable TNF-family related factors.

12.5. Neuro-Immune Interaction

12.5.1. Innervation of Lymphoid Tissue

In lymphoid organs, similar to blood vessels, sympathetic/noradrenergic and sympathetic/neuropeptide Y (NPY) innervation predominate parasympathetic (cholinergic) innervation and are actually located along the vessels. In the thymus, at the ultrastructural level, noradrenergic fibers are seen in proximity to thymocytes, mast cells and fibroblasts. Catecholaminergic nerve fibers also run in close contact to TECs. Since TECs form the blood-thymus barrier in the outer thymic cortex, these cells may be targets for circulating epinephrine from the adrenal gland or norepinephrine (NE) released from the perivascular nerves. In the spleen the sympathetic nerve fibers are present among cells in the T-dependent area; macrophages and B cells reside in the marginal zone and the marginal sinus, the site of the lymphocyte entry into the spleen. In lymph nodes, noradrenergic fibers supply paracortical and cortical zones (T cell-rich regions) but are absent from nodular regions and germinal centers that contain B cells. The gut, lung and nasal mucosa have the highest intensity of innervation of associated lymphoid tissues. Nerves predominate in T cell zones of lymphoid aggregates where they contain neuropeptides and the sympathetic neurotransmitter NE. Activation of the sympathetic nervous system and also the resting sympathetic nervous tone are important for controlling innate and adaptive immune responses.

12.5.2. Regulation of Immune Cells Activity by Neurotransmitters and Neuropeptides

As was mentioned above, antigen-presenting cells such as thymic TECs and tissue resident macrophages/DCs play an indispensable role in lymphocyte function ranging from programming central tolerance and immunity to support of their effector function and peripheral tolerance. Cells of macrophage/DC origin express functional receptors for NE ($\alpha1$, $\alpha2$, $\beta1$, $\beta2$), dopamine (D1–5), neuropeptide Y (Y1), opioids (μ, δ, κ), and parasympathetic nicotinic acetylcholine receptors. Complex signaling through these receptors will affect activation status of APCs, expression of co-stimulatory/co-inhibitory molecules and profile of secreted cytokines/chemokines (Elenkov et al., 2000; Tracey, 2002). Signaling through Ach receptors suppresses TNF secretion following the activation of lymphocytes, indicating that the receptors play an anti-inflammatory role.

Macrophages/DCs and lymphocytes express a variety of adrenergic receptors. Increased concentrations of NE or dopamine can modulate T cells polarization toward Th2 profile, stimulate IL-4, IL-5, and IL-13 secretion (Kohm and Sanders, 2001). The same Th2 polarization can be induced by histamine, serotonin, neuropeptides such as substance P, vasoactive intestinal peptide (VIP), pituitary adenylate cyclase-activating polypeptide, calcitonin gene-related peptide, α-melanocyte-stimulating hormone, and leptin (Steinman, 2004).

Summary

After reviewing the development of immune system, it is evident that there is a strong separation between the nervous and immune system based on the absence of professional antigen presenting cells in brain parenchyma and circulating DCs precursors. However, efferent information is actively coming from the brain, but the development of effector reactions is limited by strong central tolerance during education of lymphocyte by TEC expressing brain antigens. The peripheral tolerance is supported by soluble brain antigens filtered through regional lymph nodes and presented by resident DCs in the absence of co-stimulatory signals. The complexity of human adaptive immune responses cannot be reproduced *in vitro*. Each step and hierarchy of interaction of the soluble peptides /proteins with antigen-presenting cells *in situ* and then in draining lymph nodes with different subsets of lymphocytes, as well as, between lymphocytes also can be significantly affected by innervating nervous system. Multiple anatomical and physiological connections exist between the nervous and immune system. Multiple chemical messengers tie these systems together. Thus, these two systems influence each other and interact with each other. Understanding one system may give us insights into the physiology of the other system. The presented review will help to understand brain immunopathology described in the following chapters.

Review Questions/Problems

1. **Briefly summarize the major steps in lymphocyte embryonic development.**
2. **Briefly summarize the major steps in T lymphocyte development.**
3. **Briefly summarize the major steps in B lymphocyte development.**
4. **Briefly summarize the distribution of B cells and their function.**
5. **Briefly summarize the major steps in helper cells polarization.**
6. **Briefly describe the development of central tolerance to myelin.**
7. **Briefly summarize the major factors that limit T cells survival in the brain.**
8. **Briefly summarize the major steps in adaptive immune responses.**
9. **Briefly summarize the major steps in helper cells polarization.**
10. **Briefly summarize the major requirements for B cells differentiation and antibody production.**
11. **Which of the following cell types exist in thymus?**

    a. thymic epithelial cells
    b. dendritic cells
    c. macrophages
    d. CD3-CD4-CD8- cells
    e. CD3+CD4+CD8+ cells
    f. CD3+CD4+ cells
    g. CD3+ CD8+ cells
    h. CD3+CD4+CD25+ cells
    i. all of the above

12. **Which cells interaction leads to development of the central tolerance in thymus?**

    a. thymic epithelial cells with CD3-CD4-CD8- cells
    b. thymic epithelial cells with CD3+CD4-CD8- cells
    c. dendritic cells with CD3+CD4+CD8- cells
    d. dendritic cells with CD3+CD4-CD8+ cells

13. **Which cells interaction leads to development regulatory T cells in thymus?**

    a. thymic epithelial cells with CD3-CD4-CD8- cells
    b. thymic epithelial cells with CD3+CD4-CD8- cells
    c. dendritic cells with CD3+CD4+CD8- cells

d. thymic epithelial cells with CD3+CD4+CD25- cells
e. dendritic cells with CD3+CD4-CD8+ cells

14. Which of the following cell types exit thymus?

a. CD3-CD4-CD8- cells
b. CD3+CD4+CD8+ cells
c. CD3+CD4+CD8- cells
d. CD3+CD4-CD8+ cells
e. CD3+CD4+CD25- cells
f. CD3+CD4+CD25+ cells

15. Which of the following interaction affects the survival of T cells in brain?

a. CD95/CD95L
b. CD200/CD200L
c. CD40/CD154

16. Which of the following is NOT considered to be a site of B cells development?

a. thymus
b. bone marrow
c. spleen
d. liver
e. lymph node

17. Which of the following macrophage-secreted cytokines is NOT considered to be stimulators of NK cells?

a. IL-1
b. IL-10
c. IL-12
d. IL-15
e. IL-18
f. IL-2

18. Which of the following are the major requirements for lymph node development?

a. lymphoid organizer
b. $CD45^{+}$ $CD4^{+}$ $CD3^{-}$ lymphoid tissue inducers
c. LTβR
d. LTα1β2
e. TNF-related activation-induced cytokine (TRANCE)
f. IL-7/IL-7R
g. IL-2

19. Which of the following cell types belong to stromal compartment of lymph nodes?

a. Fibroblastic reticular cells
b. Sinus endothelial cells
c. High endothelial cells
d. T cells
e. B cells
f. DCs

20. Which of the following signaling receptor molecules involved in programming of immunologic memory?

a. CD40
b. CD27
c. CD137
d. ICOS
e. CD28
f. all of the above

21. Which of the following co-signaling molecules involved in programming of peripheral tolerance?

a. CTLA-4
b. BTLA
c. PD-1
d. all of the above

References

Aloisi F, Pujol-Borrell R (2006) Lymphoid neogenesis in chronic inflammatory diseases. Nat Rev Immunol 6:205–217.

Ansel KM, Lee DU, Rao A (2003) An epigenetic view of helper T cell differentiation. Nat immunol 4:616–623.

Blom B, Spits H (2006) Development of human lymphoid cells. Annu Rev Immun 24:287–320.

Bommhardt U, Beyer M, Hunig T, Reichardt HM (2004) Molecular and cellular mechanisms of T cell development. Cell Mol Life Sci 61:263–280.

Bonasio R, Scimone ML, Schaerli P, Grabie N, Lichtman AH, Von Andrian UH (2006) Clonal deletion of thymocytes by circulating dendritic cells homing to the thymus. Nat Immunol 7:1092–1100.

Bruno R, Sabater L, Sospedra M, Ferrer-Francesch X, Escudero D, Martinez-Caceres E, Pujol-Borrell R (2002) Multiple sclerosis candidate autoantigens except myelin oligodendrocyte glycoprotein are transcribed in human thymus. Eur J Immunol 32:2737–2747.

Cardona AE, Pioro EP, Sasse ME, Kostenko V, Cardona SM, Dijkstra IM, Huang D, Kidd G, Dombrowski S, Dutta R, Lee JC, Cook DN, Jung S, Lira SA, Littman DR, Ransohoff RM (2006) Control of microglial neurotoxicity by the fractalkine receptor. Nat Neurosci 9:917–924.

Carlsen HS, Haraldsen G, Brandtzaeg P, Baekkevold ES (2005) Disparate lymphoid chemokine expression in mice and men: No evidence of Ccl21 synthesis by human high endothelial venules. Blood 106:444–446.

Charo IF, Ransohoff RM (2006) The many roles of chemokines and chemokine receptors in inflammation. N Engl J Med 354:610–621.

Chen L (2004) Co-inhibitory molecules of the B7-Cd28 family in the control of T-cell immunity. Nat Rev Immunol 4:336–347.

Choi C, Benveniste EN (2004) Fas ligand/fas system in the brain: Regulator of immune and apoptotic responses. Brain Res Brain Res Rev 44:65–81.

Crivellato E, Vacca A, Ribatti D (2004) Setting the stage: An anatomist's view of the immune system. Trends Immunol 25:210–217.

Cupedo T, Mebius RE (2005) Cellular interactions in lymph node development. J Immunol 174:21–25.

Delgado M, Abad C, Martinez C, Juarranz MG, Arranz A, Gomariz RP, Leceta J (2002) Vasoactive intestinal peptide in the immune system: Potential therapeutic role in inflammatory and autoimmune diseases. J Mol Med 80:16–24.

Derbinski J, Kyewski B (2005) Linking signalling pathways, thymic stroma integrity and autoimmunity. Trends Immunol 26:503–506.

Elenkov IJ, Wilder RL, Chrousos GP, Vizi ES (2000) The sympathetic nerve–an integrative interface between two supersystems: The brain and the immune system. Pharmacol Rev 52:595–638.

Engelhardt B (2006) Molecular mechanisms involved in T cell migration across the blood-brain barrier. J Neural Transm 113:477–485.

Engelhardt B, Ransohoff RM (2005) The Ins and Outs of T-lymphocyte trafficking to the Cns: Anatomical sites and molecular mechanisms. Trends Immunol 26:485–495.

Ettinger R, Sims GP, Fairhurst AM, Robbins R, Da Silva YS, Spolski R, Leonard WJ, Lipsky PE (2005) Il-21 induces differentiation of human naive and memory B cells into antibody-secreting plasma cells. J Immunol 175:7867–7879.

Fogg DK, Sibon C, Miled C, Jung S, Aucouturier P, Littman DR, Cumano A, Geissmann F (2006) A clonogenic bone marrow progenitor specific for macrophages and dendritic cells. Science 311:83–87.

Goldschneider I, Cone RE (2003) A central role for peripheral dendritic cells in the induction of acquired thymic tolerance. Trends Immunol 24:77–81.

Hart DN, Fabre JW (1981) Demonstration and characterization of ia-positive dendritic cells in the interstitial connective tissues of rat heart and other tissues, but not brain. J Exp Med 154:347–361.

Heinemann DE, Peters JH (2005) Follicular dendritic-like cells derived from human monocytes. BMC Immunol 6:23.

Hickey WF (2001) Basic principles of immunological surveillance of the normal central nervous system. Glia 36:118–124.

Kalies K, Blessenohl M, Nietsch J, Westermann J (2006) T cell zones of lymphoid organs constitutively express Th1 cytokine mRNA: Specific changes during the early phase of an immune response. J Immunol 176:741–749.

Kivisakk P, Mahad DJ, Callahan MK, Trebst C, Tucky B, Wei T, Wu L, Baekkevold ES, Lassmann H, Staugaitis SM, Campbell JJ, Ransohoff RM (2003) Human cerebrospinal fluid central memory Cd4+ T cells: Evidence for trafficking through choroid plexus and meninges via P-selectin. Proc Natl Acad Sci USA 100: 8389–8394.

Kivisakk P, Tucky B, Wei T, Campbell JJ, Ransohoff RM (2006) Human cerebrospinal fluid contains Cd4+ Memory T cells expressing gut- or skin-specific trafficking determinants: Relevance for immunotherapy. BMC Immunol 7:14.

Kohm AP, Sanders VM (2001) Norepinephrine and beta 2-adrenergic receptor stimulation regulate Cd4+ T and B lymphocyte function in vitro and in vivo. Pharmacol Rev 53:487–525.

Krumbholz M, Theil D, Cepok S, Hemmer B, Kivisakk P, Ransohoff RM, Hofbauer M, Farina C, Derfuss T, Hartle C, Newcombe J, Hohlfeld R, Meinl E (2006) Chemokines in multiple sclerosis: Cxcl12 and Cxcl13 up-regulation is differentially linked to Cns immune cell recruitment. Brain 129:200–211.

Kyewski B, Klein L (2006) A central role for central tolerance. Annu Rev Immunol 24:571–606.

Massa PT (1993) Specific suppression of major histocompatibility complex class I and class II genes in astrocytes by brain-enriched gangliosides. J Exp Med 178:1357–1363.

Matyszak MK (1998) Inflammation in the CNS: Balance between immunological privilege and immune responses. Prog Neurobiol 56:19–35.

Matyszak MK, Perry VH (1996) The potential role of dendritic cells in immune-mediated inflammatory diseases in the central nervous system. Neuroscience 74:599–608.

Mebius RE (2003) Organogenesis of lymphoid tissues. Nat Rev Immunol 3:292–303.

Niederkorn JY (2006) See no evil, hear no evil, do no evil: The lessons of immune privilege. Nat Immunol 7:354–359.

Ransohoff RM, Kivisakk P, Kidd G (2003) Three or more routes for leukocyte migration into the central nervous system. Nat Rev Immunol 3:569–581.

Singh H, Medina KL, Pongubala JM (2005) Contingent gene regulatory networks and b cell fate specification. Proc Natl Acad Sci USA 102:4949–4953.

Steinman L (2004) Elaborate interactions between the immune and nervous systems. Nat Immunol 5:575–581.

Tracey KJ (2002) The inflammatory reflex. Nature 420:853–859.

Von Andrian UH, Mackay CR (2000) T-cell function and migration. Two sides of the same coin. N Engl J Med 343:1020–1034.

Von Andrian UH, Mempel TR (2003) Homing and cellular traffic in lymph nodes. Nat Rev Immunol 3:867–878.

Watanabe N, Wang YH, Lee HK, Ito T, Wang YH, Cao W, Liu YJ (2005) Hassall's corpuscles instruct dendritic cells to induce Cd4+Cd25+ regulatory T cells in human thymus. Nature 436:1181–1185.

13
Stem Cells

Hui Peng, Nicholas Whitney, Kang Tang, Myron Toews, and Jialin Zheng

Keywords Neurogenesis; Stem cell; Self-renewal; Proliferation; Differentiation; Microglia; Chemokines; Growth factors; Neurotransmitters

13.1. Introduction

It is now accepted that neurogenesis continues throughout life, contrary to the previously held dogma that no new neurons are produced in the adult human brain. Neurogenesis during development and throughout adult life is a result of the proliferation, migration, and differentiation of neural stem cells (NSC). NSC are cells that can differentiate into all the cells of the central nervous system (CNS) and self-renew while retaining their multipotential capabilities. Numerous factors contribute to the survival, proliferation, migration, and differentiation of NSC. These factors include, but are not limited to, chemokines, growth factors, and neurotransmitters.

Even though it is known that new neurons and supporting cells are produced in the adult, it remains unclear how long these new cells survive and how well they integrate into circuitry of the CNS. Much research has focused on replacing damaged neurons by either transplantation of embryonic, fetal, or adult stem cells or by stimulation of endogenous neural stem cells to undergo growth and differentiation. Both concepts have their advantages as well as difficulties that need to be overcome. Future research into the mechanisms by which neurogenesis is stimulated and new neurons are successfully incorporated into the CNS is imperative for the understanding of development and treatment of neurological injury.

13.2. Stem Cells, Neural Stem Cells, and Neural Progenitor Cells

A stem cell is a special kind of cell that has the unique capacity to continually renew itself and also to give rise to additional specialized cell types. It is now known that stem cells, in various forms, can be obtained from the embryo, fetus, and adult (Table 13.1). Stem cell plasticity, the ability to differentiate into more than one kind of cell, may vary depending on where the stem cell originates. Although most of what is known about these cells has been learned by studying rodent stem cells, the emphasis of current research is on the utilization of human stem cells for disease therapy.

13.2.1. Embryonic Stem Cell

Human embryonic stem cells were first collected in 1998 by two different research teams. The cells obtained from the inner cell mass of the blastocyst (4- to 5-day embryo) are embryonic stem cells (ESC); in contrast, cells cultured from the primordial germ cells of 5- to 9-week fetuses are embryonic germ cells (EGC). In the laboratory, ES or EG cells can proliferate indefinitely in an undifferentiated state but can also be manipulated to become specialized or partially specialized cell types, a process known as directed differentiation. Both ES and EG cells are pluripotent, meaning they have the potential to develop into more than 200 different known cell types. This class of human stem cells holds the promise of being able to repair or replace cells or tissues that are damaged or destroyed by many of our most devastating diseases and disabilities.

13.2.2. Adult Stem Cell

An adult stem cell is an undifferentiated cell that is found in an otherwise differentiated (specialized) tissue in the adult. It can renew itself for the lifetime of the organism and yield the specialized cell types of the tissue from which it originated. Adult stem cells usually divide to generate progenitor or precursor cells, which then differentiate and develop into mature cell types that have characteristic shapes and specialized functions. Adult stem cells have been found in tissues that develop from all three embryonic germ layers, but these cells are rare. Often they are difficult to identify, isolate, and purify. Most adult stem cells grown in a culture dish are unable to proliferate in an unspecialized state for long periods

T. Ikezu and H.E. Gendelman (eds.), *Neuroimmune Pharmacology*.
© Springer 2008

TABLE 13.1. Features different stem cell types.

Cell type	Location	Self-renewal capacity	Differential Capacity
Embryonic Stem Cells (ESC)	Inner cell mass of the blastocyst (4- to 5-day embryo)	Infinite	Pluripotent (Can differentiate into every cell of the organism except for the trophoblast)
Embryonic Germ Cells (EGC)	Primordial germ cells of 5- to 9-week fetuses	Infinite	Totipotent (Can differentiate into every cell of the organism)
Adult Stem Cell	Specific niches in adult tissue (Location varies for each tissue/organ)	Infinite	Multipotent (Can differentiate into most cells of the tissue in which they reside. Also, they may have plastisity, which allows them to differentiated into cells of other tissues)
Neural Stem Cell	Embryo: Inner lining of the neuronal tube Fetus: Basal forebrain, cerebral cortex, hippocampus, and spinal cord Adult: Subventricular zone and hippocampal dentate gyrus	Infinite	Multipotent (Can differentiate neurons, astrocytes, and oligodendrocytes)
Neural Progenitor Cell	Fetus: Basal forebrain, cerebral cortex, hippocampus, and spinal cord Adult: Subventricular zone and hippocampal dentate gyrus	Limited (Can self-renew, but only for a limited number of generations)	Multipotent (Can differentiate into at least two cell types of the tissue in which they reside)
Lineage-specific Precursor Cell	Specific niches in adult tissue (Location varies for each tissue or organ)	1. Limited (Can self-renew, but only for a limited number of generations) or 2. Cannot self-renew (It no longer retains the ability to differentiate into more than one cell type), but can amplify in cell number	1. Multipotent (Can differentiate along one specific lineage [e.g., neuronal or astroglial]) or 2. Unipotent (Can only differentiate into one cell type)

of time. The plasticity of adult stem cells appears to be less than that of ES cells; to date there is no isolated population of adult stem cells capable of forming all the kinds of cells of the body.

13.2.3. Neural Stem Cells and Neural Progenitor/ Precursor Cell

Neural stem cells (NSCs) are self-renewing, multipotent cells that generate neurons, astrocytes, and oligodendrocytes (Gage, 2000). During embryonic development, neural stem cells (NSCs) arise from generative zones derived from the inner lining of the neural tube that extend from periventricular regions of the telencephalon to the spinal cord within the mammalian central nervous system (CNS). NSCs can be isolated from the embryonic or fetal CNS, including basal forebrain, cerebral cortex, hippocampus, and spinal cord. In the adult nervous system, NSCs can be isolated from neurogenic zones (the subventricular zone and hippocampal dentate gyrus). In addition, more recent evidence suggests that stem cells can be isolated from non-neurogenic areas, e.g., the spinal cord. NSCs can be grown in culture and retain both their pluripotency and ability to self-renew.

Neural progenitor cells (NPC), in contrast, are multipotent and proliferative cells with only limited self-renewal that can differentiate into at least two different cell lineages ("multipotency," but not "pluripotency") (Gage et al., 1995; Weiss et al., 1996; McKay, 1997). Lineage-specific precursors or progenitors are cells restricted to one distinct lineage (e.g., neuronal, astroglial, and oligodendroglial). The term "precursors is used to encompass both stem and progenitor cells" (Emsley et al., 2005) (Figure 13.1).

13.3. Stem Cells and Neurogenesis During Brain Development

Neurogenesis is primarily a process that involves proliferation, migration, differentiation, and survival of neural stem cells (Palmer et al., 1997; Gage, 2000). Multiple factors have been shown to regulate neurogenesis in the developing and adult nervous system including hormones, neurotransmitters, trophic factors, and chemokines (Cameron et al., 1998; Ferguson and Slack, 2003).

13.3.1. Determination and Formation on Neural Tube

The process of developmental neurogenesis begins during embryonic development as the blastula enters a stage called gastrulation. Gastrulation is identified by a small portion of the blastula that undergoes invagination and involution, where a sheet of cells bends inward and migrates underneath the outer layer of

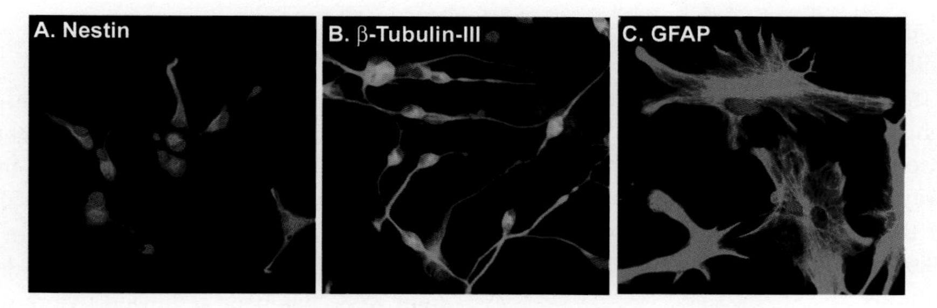

FIGURE 13.1. Characterization of human cortical neural progenitor cells. (A) Cells were isolated from human fetal cortex and culture in NPIM containing EGF, bFGF, and LIF. Cells expressed Nestin, a neural stem/progenitor cell marker. (B-C) After switching to a differentiation medium, NPC were differentiated into neurons (β-tubulin, B) and astrocytes (GFAP, C).

cells. This process forms the endoderm, mesoderm, and ectoderm layers of the embryo. A portion of the mesoderm gives rise to the notochord along the anterior-posterior axis of the embryo. This notochord releases factors that cause primary ectoderm multipotent stem cells to differentiate into more fate-restricted stem cells of the neuroectoderm, forming the neuronal plate. The neuronal plate develops along the dorsal surface with the largest portion, which will form the brain, at the anterior end of the embryo. As the edges of the neuronal plate fold over its center, fuse, and detach from the ectoderm the neuronal plate forms a hollow neural tube with an outer mantle layer and in inner proliferative zone. This neural tube will become the CNS, while a group of cells termed the neural crest on the dorsal side of the neural tube will give rise to the peripheral nervous system.

Upon formation of the neural tube, regionally distinct transcription factors are expressed, designating the configuration of discrete brain regions that will eventually develop into the forebrain, midbrain, cerebellum, hindbrain, and spinal cord (Shimamura et al., 1997; Hatten, 1999). Factors from the microenvironment surrounding neural progenitor cells in the neural tube direct the differentiation of these cells to regionally specific arrays of neurons. These distinct differentiation events include dorsal-ventral polarization in the spinal cord, segmentation in the hindbrain, and lamination in the cerebral cortex (McConnell, 1995).

13.3.2. Neural Stem Cells in CNS Development

The proliferating cells during CNS development reside in the ventricular zone of the neural tube. These cells are a heterogeneous population that exhibit complex patterns of gene expression in both space and time (Temple, 2001). This diversity of gene expression allows differentiation of NSC to astrocytes, oligodendrocytes, and various types of neurons, depending on the signals present in the microenvironment within the developing CNS. As the development of the embryo progresses, multipotent stem cells differentiate into more fate-restricted progenitors, which give rise to macroglial cells and regionally distinct neuron types. These progenitors first give rise to neurons, then to glial cells during embryonic and fetal development, and finally to astrocytes during postnatal growth. Not only is the control of differentiation vital to neurogenesis, but also the regulation of proliferation, migration, and survival of NSC and their differentiated cell progeny. Numerous factors have been identified to regulate these mechanisms, including growth factors, neurotransmitters, and chemokines; these are discussed in detail in later sections.

13.3.3. Stem Cells in Adult Neurogenesis

It is now accepted that neurogenesis continues throughout life, contrary to the previously held dogma that no new neurons are produced in the adult human brain. Two distinct populations of neural progenitor cells (NPC) have been identified: (1) one in the subventricular zone of the lateral ventricles and (2) another in the dentate gyrus of the hippocampus; other, but possibly less significant, populations of NPC may exist throughout the adult CNS. The physiological role of adult neurogenesis and potential to control it for therapeutic benefit are critical areas of current research for the treatment of neurodegenerative disorders and neural damage.

13.4. Stem Cell Signaling Pathways for Migration, Proliferation, and Differentiation

Neurogenesis includes proliferation, neuronal fate specification of neural progenitors (differentiation), migration, maturation, and functional integration of neuronal progeny into neuronal circuits (Ming and Song, 2005). A full understanding of the molecular mechanisms regulating proliferation, migration, and differentiation of these cells is essential if these cells are to be used for therapeutic applications.

13.4.1. Proliferation

Proliferation and survival of stem cells are important for the maintenance of the neural stem cell pool for neurogenesis. The signals involved include members of the fibroblast growth factor (FGF) family, the transforming growth factor-β (TGF-β) superfamily such as the bone morphogenic proteins (BMPs), the growth and differentiation factors (GDFs), and the Hedgehog family. In addition, many other cues are crucial for neural development.

FGFs constitute a large family of structurally related polypeptide growth factors that signal through receptor tyrosine kinases. It has been suggested that cyclin D2 (intimately involved in driving cells through the G1 phase of the cell cycle) is an effector of the FGF-dependent maintenance of the caudal stem zone (Lobjois et al., 2004). Receptor-mediated regulation of proliferation involves calcium signaling and the cAMP and PKA pathway. Activation of the Ras/Raf/Mek/Erk pathway is coupled to the cell cycle machinery via interactions with p53 retinoblastoma tumor suppressor protein (Rb) and E2F families of proteins.

Wnts are a large family of highly conserved secreted signaling proteins related to the Drosophila wingless protein that regulates cell-to-cell interactions during embryogenesis (Nusse, 2003). In vitro as well as in vivo studies have shown that Wnt signaling is required to expand and maintain neural precursor populations in the brain and the spinal cord. As currently understood, Wnt proteins bind to cell surface receptors of the Frizzled family and act on multiple disparate signaling pathways. Three major pathways have been identified: (i) the Wnt/β-catenin pathway, also referred to as the canonical Wnt signaling pathway, (ii) the Frizzled/planar cell polarity (Fz/PCP) pathway, and (iii) the Ca^{++} pathway. The details of these signaling pathways have been reviewed recently (Cayuso and Marti, 2005; Ille and Sommer, 2005). For the canonical Wnt signaling pathway, extracellular Wnt molecules bind to Fz seven transmembrane receptors on the cell surface; these Fz receptors are structurally similar to G-protein coupled receptors (GPCRs). Through several cytoplasmic relay components, the signal is transduced to β-catenin, which then enters the nucleus and forms a complex with TCFs (from T-cell factor) to activate transcription of Wnt target genes (Logan and Nusse, 2004).

Sonic hedgehog (Shh) is one of the members of the hedgehog family of secreted proteins required for multiple aspects of development in a wide range of tissues including the CNS (reviewed in Machold et al., 2003). In addition to its fundamental role played in pattern formation of the ventral CNS, the Shh-Gli pathway has been demonstrated to play a major mitogenenic role in the development of dorsal brain structures, including the cerebellum, neocortex, and tectum (Dahmane and Ruiz i Altaba, 1999; Wallace, 1999; Wechsler-Reya and Scott, 1999; Dahmane et al., 2001; Pons et al., 2001). Shh mediated proliferation is also required for maintenance of neural stem cells in late development and adult. Shh activity is triggered by the binding of the ligand to its receptor Patched (Ptc), an 11-pass transmembrane receptor. Zinc-finger proteins of the Gli family are known to be transcriptional mediators of Shh-signaling. Within the Shh receiving cell, Gli proteins are regulated in the cytoplasm via multiple distinct molecular mechanisms. The cyclic AMP-dependent protein kinase (PKA) acts as a common negative regulator. Gli repressor forms are generated by PKA-mediated phosphorylation and inhibition of PKA activity releases Gli activated forms. Gli proteins then move to the nucleus where they can activate or repress transcription of specific target genes (Jacob and Briscoe, 2003).

13.4.2. Differentiation

Differentiation is the process by which unspecialized cells (such as a stem cell) become specialized into one of the many highly specialized cells that make up the body. There are two directions of differentiation for neural stem cell: neurogenesis and gliogenesis. The regulation of neurogenesis, as well as the neuronal potential and lineage determination by neural stem cells, has been reviewed (Morrison, 2001). Neurogenic factors such as BMPs promote neurogenesis by inducing the expression of proneural basic helix-loop-helix (bHLH) transcription factors to activate the expression of a cascade of neuronal genes. Moreover, the activity of proneural genes inhibits gliogenesis in the CNS; in contrast, mammalian "hairy and enhancer of split homolog" (HES) gene expression promotes a glial fate. It is likely that other genes also influence multipotent stem cells to become glial precursor cells. Olig1 and Olig2 promote the development of oligodendrocytes. Notch activation acts in neural stem cells as a switch that terminates neurogenesis and initiates gliogenesis, even in the continued presence of neurogenic growth factors (Figure 13.2).

It has been shown that mutiple bHLH genes play a critical role in regulation of neural stem cell differentiation (Bertrand et al., 2002; Ross et al., 2003). The activator-type bHLH genes *Mash1, Math*, and *Ngn* are expressed by differentiating neurons. These bHLH factors form a heterodimer with a ubiquitously expressed bHLH factor E47 and activate gene expression by binding to the E box, promoting the neuronal subtype specification. bHLH genes promote neuronal subtype specification later in differentiation as well, as reviewed (Kageyama et al., 2005).

While *Mash1, Math*, and *Ngn* promote neurogenesis, the repressor-type bHLH genes, including *Hes* genes, regulate maintenance of neural stem cells and promote gliogenesis. There are seven members in the *Hes* family; Hes1, Hes3, and Hes5 are highly expressed by neural stem cells. Hes factors have a conserved bHLH domain; they form dimers and bind to DNA. The target genes for Hes factors include the activator-type bHLH genes such as Mash1. Hes1 represses Mash1 expression by directly binding to the promoter. The activator-type bHLH factors form a heterodimer with another bHLH activator E47 and promote neuronal differentiation from neural stem cells. Hes1 forms a nonfunctional heterodimer with E47 and inhibits formation of Mash1-E47 heterodimer. Thus, Hes1 antagonizes Mash1 by two different mechanisms: repressing the expression at the transcriptional level and inhibiting activity at the protein-protein level.

It has been suggested that Notch signaling is important for maintaining multipotency in some neural stem cells but promotes glial differentiation in others or at different times during development. The Notch signaling pathway is best characterized as mediating cell-cell signaling between adjacent cells. The ligands are members of the Delta and Jagged gene families and the receptors are single-pass transmembrane proteins. Upon ligand binding, the intracellular domain of Notch

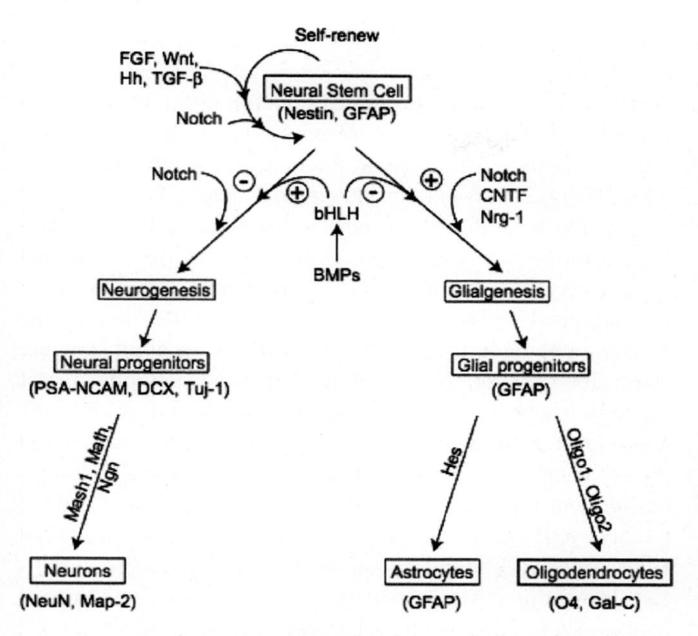

FIGURE 13.2. A proposed mechanism for a select number of factors that regulate self-renew and differentiation of NSC/NPC. Factors that promote self-renewal include but are not limited to fibroblast growth factor (FGF), the transforming growth factor-β (TGF-β), Wnt signaling, and Notch signaling. NSCs are directed toward differentiating along the neuronal lineage by a decrease in Notch signaling and an increase in basic helix-loop-helix (bHLH) transcription factors. As these immature neurons mature they express activator-type bHLH genes *Mash1, Math*, and *Ngn*. NSCs are directed toward the glial lineage by a low level of bHLH and a higher level of factors such as Notch, ciliary neurotrophic factor (CNTF), and Nrg-1. As these glial progenitors mature to astrocytes they express *Hes* and as they differentiate into oligodendrocytes they express *Oligo1* and *Oligo 2*. Specific cell markers can identify each of these cell types. Nestin and GFAP are expressed by NSC, PSA-NCAM, DCX, and Tuj-1are expressed by neural progenitors, NeuN and Map-2 are expressed by maturing neurons, GFAP is expressed by glial progenitors and astrocytes, and O4 and Gal-C are expressed by oligodendrocytes.

(NICD) is released from the plasma membrane and translocates into the nucleus, where it converts the CBF1 repressor complex into an activator complex. The NICD/CBF1 activator complex upregulates targets such as the *Hes* and *Herp* (*Hes*-related protein) genes, which block neurogenesis. This is the central mechanism to the inhibition of neuronal differentiation by Notch (Yoon and Gaiano, 2005). The following figure (Figure 13.2) summarizes a proposed mechanism for a select number of factors that regulate differentiation of NSC/NPC.

13.4.3. Migration Signaling

A remarkable feature of the developing CNS is the extensive migration of cells. As specific classes of cells come to reside in specific layers, migration also reflects the programmed control of neuronal fate. Control of these movements is rather complex, involving multiple regulatory genes and several extracellular molecules. Slit, reelin, and doublecortin may all regulate neuronal cell migration (Nadarajah and Parnavelas, 2002). Slit and its receptor Robos mediate migration of neural progenitor cells crossing the midline in the developing brain (Orgogozo 2004). Doublecortin and reelin regulate cortical neuron migration (Feng and Walsh, 2001; Nadarajah and Parnavelas, 2002). In addition, chemokines, such as SDF-1α, play an important role for neural progenitor cell migration during development, as discussed in Sect. 13.5.1.

13.5. Chemokines and Neurogenesis

Chemokines constitute a family of structurally related low molecular mass (8–11 kDa) proteins with diverse immune and neural functions. In addition to their well-established roles in the immune system, chemokines play a role in the migration, proliferation, differentiation, and survival of neural stem/progenitor cells. (Refer to the Cytokines and Chemokines Chapter 15.5 regarding chemokine and chemokine receptor classification and signaling pathways).

13.5.1. SDF-1 and its Receptor CXCR4

Studies have shown that chemokine receptors are widely expressed in embryonic and adult neural stem/progenitor cells, including CXCR4, CCR2, CCR5, and CX3CR1 (Ji et al., 2004; Tran et al., 2004, Pen et al., 2004; Widera et al., 2004; Ni et al., 2004). CXCR4 is one of the most highly expressed chemokine receptors. Deletion of the gene for the CXCR4 receptor or of its only known ligand, stromal cell-derived factor 1 (SDF-1), results in embryonic lethality. The brains of mouse embryos homozygous for the absence of CXCR4 or SDF-1 have severe abnormalities in the development of the cerebellum (Ma et al., 1998; Zou et al., 1998), hippocampal dentate gyrus (Lu et al., 2001; Bagri et al., 2002), and neocortex (Stumm et al., 2003). In the cerebellum, SDF-1 is highly expressed in the leptomeninx and is the major chemoattractant for external germinal layer cells in the developing cerebellum (Klein et al., 2001; Zhu et al., 2002). In mice lacking SDF-1 or CXCR4, migration of granule cell precursors out of the external germinal layer occurs prematurely, resulting in abnormal development of the cerebellum (Ma et al., 1998). In agreement with a key role for CXCR4, CXCR4 mRNA is expressed at sites of neuronal and progenitor cell migration in the hippocampus at late embryonic and early postnatal ages. The absence of CXCR4 leads to a reduction in the number of dividing cells in the migratory stream and in the dentate gyrus itself; in addition, neurons appear to differentiate prematurely before reaching their target (Lu et al., 2001; Bagri et al., 2002). In the cortex, SDF-1 is highly expressed in the embryonic leptomeninx and is a potent chemoattractant for isolated striatal precursors, while CXCR4 is present in early generated Cajal-Retzius cells of the cortical marginal zone (Stumm et al., 2003). Mice with a null mutation in CXCR4 or SDF-1 show severe disruption of interneuron placement and proliferation, while the submeningeal positioning of Cajal-Retzius cells remains unaffected (Stumm et al., 2003). In conclusion, SDF-1 likely acts both as a chemoattractant and a mitogenic stimulus for neural stem/progenitor cells in the development of the cerebellum, hippocampus, and neocortex.

SDF-1 may also affect spinal cord development. SDF-1/CXCR4 expression is increased in developing spinal cord progenitors. SDF-1 induced chemotaxis in both neural and glial progenitor cell, suggesting that SDF-1/CXCR4 may affect spinal cord development through modulating progenitor cell migration (Luo et al., 2005).

The expression of SDF-1/CXCR4 in the adult CNS suggests that SDF-1/CXCR4 signaling is also important in adult neurogenesis (Bagri et al., 2002; Tran and Miller, 2005). In the adult, neurogenesis continues in the dentate gyrus. Adult neural progenitor cells in the subgranular zone, which produce granule neurons, express CXCR4 and other chemokine receptors, and granule neurons express SDF-1. The expression of SDF-1 and CXCR4 in adults differs from the embryonic patterns but remains consistent with continued functions in granule cell mitogenesis and/or chemotaxis.

Neural stem cell survival is an imperative issue during neurogenesis. Whether SDF-1-CXCR4 interaction is involved in all three steps of neurogenesis is not clear; however, SDF-1-CXCR4 signaling regulates survival of both neural progenitor cells and oligodendrocyte progenitors in vitro (Krathwohl and Kaiser, 2004; Dziembowska et al., 2005).

13.5.2. Other Chemokines and their Receptors

The role of other chemokines and their receptors in neurogenesis has only recently become realized. For example, it has been suggested that CXCR2 plays a vital role in patterning the developing spinal cord; CXCR2 and its ligand, growth-related oncogene alpha (GRO -α, CXCL1) are crucial in arresting embryonic oligodendrocyte precursor migration (Tsai et al., 2002). Spinal cord oligodendrocytes originate in the ventricular zone and subsequently migrate to white matter, stop, proliferate, and differentiate (Tsai et al., 2002). Without CXCR2 signaling, a widespread dispersal of postnatal precursors has been seen (Tsai et al., 2002). This shows that GRO-α-CXCR2 interaction plays an important role in holding a population of presumptive white matter (Tsai et al., 2002), thus, creating an organized and functional spinal cord. Monocyte chemoattractant protein-1 (MCP-1, CCL2) can also activate the migration capacity of rat-derived neural stem cells. With this in mind, it can be assumed that numerous other chemokines and their receptors will be found to play an important role in regulating NPC function, similar to CXCR4.

13.6. Growth Factors and Neurogenesis

Numerous growth factors regulate neurogenesis in the CNS and are valuable tools in the maintenance and differentiation of NPC in culture. Ciliary neurotrophic factor (CNTF) released by astrocytes promotes proliferation of NPC (Emsley and Hagg, 2003). Vascular endothelial growth factor (VEGF) in the subventricular zone (SVZ) promotes survival, differentiation, and release of brain-derived neurotrophic factor (BDNF), which will be discussed in detail in Sect. 13.6.2 (Louissaint et al., 2002; Hagg, 2005). Insulin-like growth factor (IGF) produced endogenously in the brain and from circulating blood stimulates proliferation and differentiation of NPC to a neuronal lineage (Anderson et al., 2002). Transforming growth factor (TGFα) produced by astrocytes promotes proliferation by the activation of epidermal growth factor (EGF) receptor (Enwere et al., 2004; Hagg, 2005). Glial-derived neurotrophic factor (GDNF) has been shown to promote survival of dopaminergic neurons in culture (Nakajima et al., 2001). Platelet-derived growth factor (PDGF) may contribute to survival, differentiation, and migration of NPC (Kwon, 2002). Nerve growth factor (NGF) promotes survival and differentiation (Plendl et al., 1999). Although numerous growth factors are significant in neurogenesis, this section will focus on three growth factors: EGF, basic fibroblast growth factor (FGF2) (bFGF), and BDNF.

13.6.1. Epidermal Growth Factor and Basic Fibroblast Growth Factor

EGF and bFGF regulate survival, proliferation, and differentiation of NPC. Regeneration of neurons after ischemic brain injury in rats was significantly increased by infusion of EGF and bFGF (Nakatomi et al., 2002). NPC culture systems use EGF and bFGF to maintain the cell pool by increasing proliferation and survival and by preventing differentiation. After NPC are stimulated to differentiate into neurons, EGF has neurotrophic properties on the newly forming neurons to promote survival. EGF collaborates with sonic hedgehog (SHH) and non-amyloidogenic amyloid precursor protein (APP) to promote proliferation (Machold et al., 2003; Caille et al., 2004; Hagg, 2005; Palma et al., 2005).

13.6.2. Brain-Derived Neurotrophic Factor

Brain-derived neurotrophic factor (BNDF) promotes survival and differentiation of NPC to a neuronal lineage (Mattson et al., 2004; Hagg, 2005). Infusion of BDNF into the brain of rats has been shown to have antidepressant-like properties, as demonstrated by the learned helplessness test and forced swim test (Siuciak et al., 1997). The antidepressant-like activity may be the result of BDNF-stimulated increase in survival and differentiation of NPC to neurons in the hippocampus. This is supported by the finding that BDNF infusion into the lateral ventricle increases the number of newly generated neurons in various regions of the rat brain (Pencea et al., 2001). Neurodegenerative disorders like bacterial meningitis may cause an increase in BDNF levels, which correlates with an increase in neurogenesis (Tauber et al., 2005). These support the idea that BDNF is a crucial factor in the regulation of neurogenesis, specifically for persuading NPC cultures to differentiate into neurons. BDNF can come from multiple sources, including but not limited to astrocytes, vascular endothelial cells, neurons, and NPC. Other growth factors, hormones, and neurotransmitters may use BDNF as a downstream factor to regulate neurogenesis. BDNF expression in astrocytes can be increased by noradrenaline, serotonin, and glutamate (Zafra et al., 1992; Mattson et al., 2004). Testosterone increases the expression of VEGF, which then stimulates release of BDNF (Palmer et al., 2000; Hagg, 2005). Another regulator of BDNF is nitric oxide production by NPC, which can act in a positive feed back loop with BDNF to shift NPC from proliferation to differentiation (Cheng et al., 2003).

13.7. The Role of Neurotransmitters in the Regulation of Neurogenesis

Neurotransmitters such as acetylcholine (Harrist et al., 2004; Mohapel et al., 2005), dopamine (Baker et al., 2004; Hoglinger et al., 2004; van Kampen and Robertson, 2005), serotonin (Banasr, 2004), norepinephrine (Kulkarni, 2002), GABA (Bolteus and Bordey, 2004; Overstreet Wadiche et al., 2005; Wang et al., 2005), and glutamate (Kitamura et al., 2003; Luk et al., 2003; Brazel et al., 2005), contribute to the regulation of neurogenesis in the dentate gyrus (DG) and the subventricular zone (SVZ) of the lateral ventricles. These neurotransmitters regulate proliferation, differentiation, migration, survival, and synaptic plasticity.

13.7.1. Dopamine in Neurogenesis

Dopamine projections from the midbrain innervate the SVZ and have been shown to promote proliferation and differentiation through the activation of dopamine D2 and D3 receptors (Baker et al., 2004; Hoglinger et al., 2004; Hagg, 2005; van Kampen and Robertson, 2005); however, other evidence suggests that activation of D3 receptors does not affect neurogenesis and D2 receptor activation may inhibit neurogenesis (Baker et al., 2005; Kippin et al., 2005). These apparent contradictions reflect the fact that neurogenesis is complex and is dependent on balance among signals that simulate and inhibit neurogenesis (Hagg, 2005). Specific D3 activation stimulated proliferation and showed preferential differentiation to neuronal phenotype (van Kampen et al., 2004; van Kampen and Robertson, 2005), and dopamine D3 activation has been shown to have neuroprotective effects (Vu et al., 2000; Anderson et al., 2001; Carvey et al., 2001).

13.7.2. Serotonin in Neurogenesis

Serotonin projections innervating the SVZ and DG have been shown to influence proliferation and survival (Simpson et al., 1998; Banasr et al., 2004). Serotonin stimulation of 5-HT1a receptors enhances NPC proliferation and increases levels of BDNF, which promotes survival and stimulates differentiation of NPC to a neuronal lineage (Banasr et al., 2004; Mattson et al., 2004; Hagg, 2005).

13.7.3. Norepinephrine in Neurogenesis

Norepinephrine has been shown to influence proliferation, but it has not been shown to affect survival or differentiation in the DG (Kulkarni et al., 2002). Adrenergic receptors are abundant in the CNS; however, their significance in neurogenesis has not been well established.

13.7.4. GABA in Neurogenesis

Maturing neurons receive GABAergic input for one to two weeks before they form glutamatergic synapses (Overstreet Wadiche et al., 2005). GABA contributes to the stimulation of proliferation, migration, and neurite outgrowth (Owens and Kriegstein, 2002; Bolteus and Bordey, 2004). GABA is essential for the maturation of neurons and formation of functional synapses. GABA's role is not simply the creation of new neurons, but also the formation of functional synapses that ultimately lead to repair or improved neurological function.

13.7.5. Glutamate in Neurogenesis

Glutamate is the primary excitatory neurotransmitter in the mammalian CNS, where it has been shown to mediate synaptic transmission, synaptic plasticity, neuronal toxicity and proliferation, and differentiation of NPC (Dingledine et al., 1999; Arundine and Tymianski, 2004). During many neurodegenerative disorders, extracellular glutamate levels are increased in the CNS. This increased glutamate leads to excitotoxicity of neurons, but some of the negative effects may be counterbalanced by increasing neurogenesis and activation of survival pathways in select cells, including NPC. Glutamate acts on several receptor types, which are classified into "metabotropic" and three "ionotropic" groups: kainate, N-methyl-D-aspartate (NMDA), and α-amino-3hydroxy-5-methyl-4-isoxazolepropionate (AMPA) receptors (Monaghan et al., 1989; Hollmann and Heinemann, 1994; Pin and Duvoisin, 1995). Metabotropic glutamate receptors are GPCRs whose activation leads to release of Ca^{2+} from intracellular stores or inhibition of adenylyl cyclase (Pin and Duvoisin, 1995). Ionotropic receptors are ion channels that open upon binding of glutamate, leading to the influx of sodium and/or calcium and the efflux of potassium, resulting in changes in membrane potential and diverse cellular processes (Arundine and Tymianski, 2004).

NMDA receptor antagonist reduced proliferation of striatal neurons, demonstrating that NMDA receptor activation may induce proliferation (Sadikot et al., 1998; Luk et al., 2003). The metabotropic glutamate receptor subtype mGluR5 is expressed early in development in regions of active proliferation, suggesting that mGluR5 may play a role in proliferation and/or differentiation (Di Giorgi Gerevini et al., 2004).

Numerous studies have shown beneficial effects to neurogenesis by activating and/or potentiating AMPA receptors. By decreasing the rate of desensitization of AMPA receptors, proliferation of NPC increases in rat hippocampus (Bai et al., 2003). Multiple mechanisms have been suggested to explain this AMPA receptor mediated neurogeneration. The increased expression of genes that stimulate proliferation and/or differentiation is one possible method for AMPA-mediated neurogenesis. Activation of AMPA receptors leads to increased expression of immediate early genes, such as: NGFI-A, c-fos, c-jun, and jun-b in oligodendrocyte progenitors (Pende et al., 1994).

Potentiation of AMPA receptors can also increase BDNF expression in the rat hippocampus, especially in the dentate gyrus (Mackowiak et al., 2002). AMPA receptor activation can lead to upregulation of BDNF in a calcium-dependent and calcium-independent manner. Calcium can bind to calcium response elements on the BDNF promoter, resulting in increased BDNF transcription. Also, AMPA receptors can physically associate with the tyrosine kinase Lyn. Lyn activation through AMPA receptors can stimulate mitogen-activated protein kinases (MAPK), leading to increased BDNF expression (O'Neill et al., 2004). As previously described, BDNF promotes survival and differentiation of NPC to a neuronal lineage (Mattson et al., 2004; Hagg, 2005).

13.7.6. Interaction of Neurotransmitters in Neurogenesis

One factor for the creation of the SVZ and DG niches for NPC and neurogenesis could be the overlap of neurotransmitter systems in these two areas (Simpson et al., 1998; Hagg, 2005). Serotoninergic and dopaminergic projections converge almost exclusively over the SVZ. Similarly, serotoninergic and norepinephrinergic projections congregate over the DG (Hagg, 2005). None of these neurotransmitters may be sufficient to stimulate neurogenesis on its own; rather, it is the multitude of neurotransmitters, growth factors, and chemokines that combine to create the niche for NPC and to stimulate successful neurogenesis.

13.8. Brain Inflammation and Neurogenesis

Previously, the brain was thought to be an immune-privileged organ. Now it is clear that the brain does respond to peripheral inflammatory stimuli through both neural and humoral afferent signals. The concept that the microglia act as the brain's immune system has also become widely accepted. The inflammatory responses in the brain have different features from those of nonneuronal tissues. Microglia activation and elevated levels of cytokines and adhesion molecules are hallmarks of CNS inflammation (Matyszak, 1998).

The likely role of immune cells in regulating neurogenesis is of great interest. Their effects may depend on the timing, dynamics, and severity of the inflammation. On the positive side, it has been suggested that the immune system may be protective and even assist regeneration; microglia (Streit, 2002) and macrophages (Rapalino et al., 1998) have been shown to facilitate tissue repair after CNS injury. On the negative side, the immune system may also cause "bystander damage" in the inflamed brain. The limited regeneration capacity and extreme vulnerability of CNS neurons to inflammatory conditions make the inflamed CNS a hostile environment for neurogenesis. Pharmacologic approaches to control neurogenesis and/or neuroinflamation are likely to become important areas of research for treatment of CNS injury and neurodegeneration.

13.8.1. Inflammation can Suppress Neurogenesis

Accumulating evidence suggests that brain inflammation plays an important role in the pathogenesis of chronic neurodegenerative disease such as Alzheimer's disease, Parkinson's disease, and multiple sclerosis (Marchetti and Abbracchio, 2005). Acute brain injury from stroke and status epilepticus is also linked to inflammation. Although it has been reported that these insults trigger increased neurogenesis in the subgranular zone of the dentate gyrus, it has also been shown that newly formed dentate neurons were severely lost, which may be associated with brain inflammation. Arvidsson et al. (2002) showed that more than 80% of new striatal neurons that are generated from the subventricular zone after stroke in rats die within the first week after the insult (Arvidsson et al., 2002). Systemic inflammation triggered by peripheral infection can cause activation of microglia in the brain of Alzheimer's disease patients, leading to aggravation of the cognition decline. Epidemiological studies suggest that patients on long-term anti-inflammatory treatments have a significantly reduced risk of Alzheimer's disease (Zandi and Breitner, 2001).

An invariant feature of damage to the CNS is the migration of microglia to the site of injury and their subsequent activation; however, the specific role of microglia in neurogenesis is still controversial. The activation of microglia can result in either neuroprotective or neurotoxic effects, or both. Previous studies have indicated that microglia are capable of secreting neurotrophic survival factors upon activation (Kim and de Vellis, 2005), and they can direct neural precursor cell migration and differentiation (Kim and de Vellis, 2005). Yet several recent publications (Ekdahl et al., 2003; Monje et al., 2003; Hoehn et al., 2005) report pathogenic roles of microglia on neurogenesis, as well as strengthening the traditional view that immune cells in the CNS have an adverse effect on neurogenesis. In lipopolysaccharide-induced inflammation of the rat CNS, basal hippocampal neurogenesis is strongly impaired and the increased neurogenesis triggered by this brain insult is also attenuated. This impairment is associated with microglial activation. Activated microglia were localized in close proximity to the newly formed cells, and the impairment of neurogenesis depended on the degree of microglia activation. Systemic administration of the microglial activation inhibitor minocycline effectively restored neurogenesis, presumably by suppressing inflammation without affecting neurogenesis (Ekdahl et al., 2003).

Using a coculture system, Monje et al. (2003) found that activated but not resting microglia decreased the differentiation of neural progenitor cells to approximately half of control levels. Furthermore, in vivo models of acute and chronic inflammation also demonstrated that inflammation itself suppresses neurogenesis. Neurogenesis and activated microglia show a striking negative correlation, and decreased microglial activation by inflammatory blockade with the nonsteroidal anti-inflammatory drug indomethacin likely accounts for part of its restoration of neurogenesis (Monje et al., 2003). Hoehn et al. (2005) also showed indomethacin treatment significantly decreased activated microglia in focal cerebral ischemia in rats. This increased the survival of progenitor cells and allowed a higher fraction to differentiate into oligodendrocytes and neurons. The deleterious effect of activated microglia on neurogenesis is most likely mediated through the action of cytokines, such as IL-1β or IL-6, TNFα, IFN-γ, NO, and reactive oxygen species; all of these factors can be released from microglia and are neurotoxic in vitro (Hoehn et al., 2005).

When human neural stem cells were transplanted into the ischemic cortex of rats after distal middle cerebral artery occlusion, transplanted stem cells survived robustly in naive and ischemic brains 4 weeks post-transplant. Survival was influenced by proximity of the graft to the stroke lesion and was negatively correlated with the number of IB4-positive inflammatory cells. The negative correlation of cell survival to lesion size and IB4-positive cells suggests that inflammatory cytokines are detrimental to be the transplanted cells, in accordance with other studies (Arvidsson et al., 2002). Given that the magnitude of the inflammatory response and the types of cytokines present change with time after ischemia, these data imply that the timing of stem cell transplant could significantly influence cell survival, with greater survival predicted if cells are transplanted after inflammation has subsided (Kelly et al., 2004).

13.8.2. Proregenerative Role of Microglia and Brain Immunity

Microglia have unique characteristic of being both supportive glia and immunocompetent defense cells. Many observations strongly support a neuroprotective and proregenerative role of microglia in the injured CNS. In the facial nerve axotomy model of neuron-microglia interaction and neural regeneration (Streit, 2002), microglial cells are found to contact the membrane of neurons that will not undergo apoptosis and will eventually regenerate their axons. Their data suggested that microglia may somehow orchestrate the recovery of these cells by assuming a protective role. Microglia activation after brain injury can also benefit regeneration through the release of neurotrophic molecules and some cytokines (Figure 13.3).

Schwartz and colleagues bring forward a concept of protective autoimmunity (Moalem et al., 1999). Although an uncontrolled immune response impairs neuronal survival and neurogenesis, a local immune response that is properly controlled can support survival and promote recovery. The immune response that causes cell loss under neurodegenerative conditions also blocks neurogenesis in the adult CNS, while the immune response that protects against cell loss also supports neurogenesis (Schwartz, 2003).

The mechanisms by which inflammation regulates neurogenesis remain inconclusive. Recently, investigators have begun to dissect out detrimental aspects of the immune system that may be responsible for unnecessary tissue loss and restrictions on axonal regeneration. Early inhibition of TNF or TGF-β2 significantly decreases scarring and tissue loss, and can lead to improvement in functional outcome after CNS injury (Brewer et al., 1999; Logan et al., 1999). Further investigation of the roles of individual inflammatory factors, growth factors, and neurotransmitters in neurogenesis will help to shed light on this important research field. Figure 13.3 summarizes a proposed mechanism for the simulation of neurogenesis during brain injury and neurodegenerative disorders.

13.9. Stem Cell and Neuronal Repair During Brain Injury and Neurodegenerative Disorders

A variety of insults have been shown to be induce neurogenesis within the adult brain (Parent, 2003), either in regions that have known neurogenic activity, or even in regions where neurogenesis normally does not occur (Yamamoto et al., 2001).

Damage induced by mechanical injury, prolonged seizures, trauma or stroke increases dentate gyrus cell proliferation, where a majority of the newly generated cells differentiate into granule neurons (Parent, 2003). It also has been shown that progenitor cells are capable of proliferation and differentiation into mature myelinating oligodendrocytes in models of acute demyelination (Gensert and Goldman, 1997). In various animal models of seizure, such as kainate and pilocarpine models of temporal lobe epilepsy, chemoconvulsant-induced status epilepticus, or electrical kindling models of epileptogenesis, increased dentate granule cell neurogenesis has been observed after stimulation (Parent, 2003). Roles of stem cells in other specific diseases are elaborated below.

13.9.1. Huntington's Disease and Alzheimer's Disease

Increased neurogenesis has been reported in patients with Huntington's disease and Alzheimer's disease (Curtis et al., 2003; Jin et al., 2004). Compared to control patients, Alzheimer's

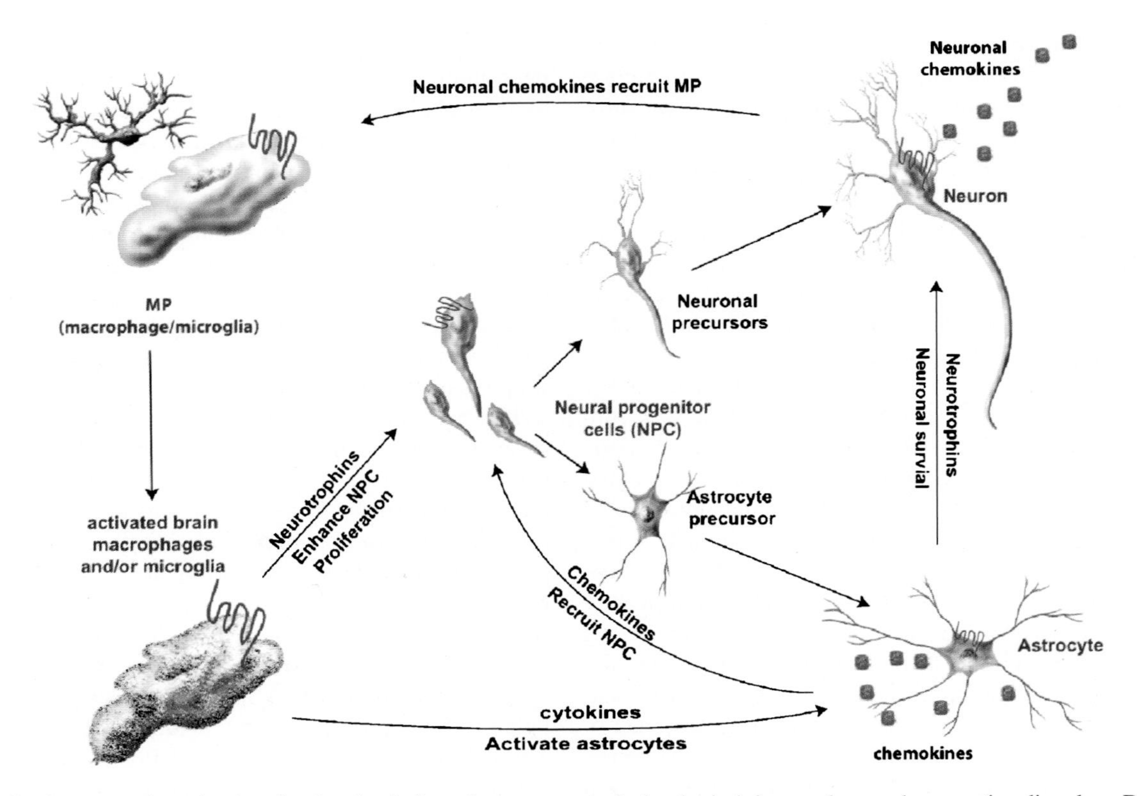

FIGURE 13.3. A proposed mechanism for the simulation of neurogenesis during brain injury and neurodegenerative disorders. During neuronal injury, neurons produce chemokines, which recruit mononuclear phagocytes (MP) into the brain and to the site of injury. As these MP enter an environment of injury or inflammation, they become activated, subsequently releasing factors that might promote neurogenesis. These factors include but are not limited to neurotrophins, which may act directly to enhance the survival and proliferation of NPCs, and cytokines that might activate astrocytes. These activated astrocytes produce chemokines that could promote the migration of NPCs to the site of injury and they can release neurotrophins that can promote neuronal survival. Once the NPCs receive these migratory and neurotrophic signals, the NPCs can migrate, proliferate, and differentiate into neuronal or astrocyte precursors, which then mature into neurons and astrocytes that may integrate into the CNS circuitry.

disease brains showed increased expression of the immature neuronal marker doublecortin and the early neuronal differentiation protein TUC-4. Expression of doublecortin and TUC-4 was associated with neurons in the dentate gyrus, the physiological destination of these neurons, and in the CA1 region, which is the principal site of hippocampal pathology in Alzheimer's disease. A recent paper (Jin et al., 2004) showed enhanced neurogenesis in Alzheimer's disease transgenic mice, in contrast to previous reports (Haughey et al., 2002; Wen et al., 2002). Similarly, there is a significant increase in cell proliferation in human subependymal layer in response to neurodegeneration of the caudate nucleus of Huntingtion's patients compared with control brain. These findings indicate that neuron damage or loss can trigger neurogenesis, and this may represent a mechanism directed toward the replacement of dead or damaged neurons.

13.9.2. Cerebral Ischemia

Recent findings in rodents show that cerebral ischemia is another injury that stimulates neurogenesis in adult brain. Yagita et al. (2001) provide evidence that the proliferating cells found after ischemia are neural progenitor cells (Yagita et al., 2001). Ischemia can lead to the proliferation of newborn cells that migrate into the zone of ischemic injury, differentiate, and express markers of mature striatal neurons (Arvidsson et al., 2002). In general, global cerebral ischemia induces neurogenesis in the dentate gyrus, whereas focal cerebral ischemia additionally increases new neurons in the SVZ (Zhang et al., 2005). Stroke selectively and significantly increases the number of type A and type C cells, which are more actively proliferating neural stem cells. Infusion with the antimitotic agent cytosine-arabinofuranoside to the brain almost completely ablates type A and type C cells in the ischemic subventricular zone, thus indicating that neural stem cells contribute to cerebral ischemia-induced neurogenesis (Chen et al., 2004). In a middle cerebral artery occlusion ischemia model, BrdU labeling was increased in the ipsilateral SVZ and rostral migratory stream, indicating directed migration of neural progenitors toward the infarct area from their sites of origin.

13.9.3. Parkinson's Disease

In contrast to cerebral ischemia, Alzheimer's disease and Huntington's disease, which have clear evidence of increased neurogenesis, the question of whether adult brain generates new dopamine neurons is still controversial. Using confocal laser scanning microscope, Zhao et al. (2003) found dopaminergic neurons with BrdU-positive nuclei in the substantia nigra (Zhao et al., 2003). But a different conclusion was reported in another study (Frielingsdorf et al., 2004). Kay and Blum (2000) also reported the presence of BrdU-positive proliferative cells in the subtantia nigra. Some of cells were identified as microglia, and none of them differentiated into dopamenergic neurons (Kay and Blum, 2000). Another recent study (Yoshimi et al., 2005) revealed lack of neurogenesis of TH-positive neurons from proliferative stem cells in the substantia nigra. These proliferative cells were mainly microglia, but some PSA-positive cells, candidates for newly differentiated young neurons, were present in the substantia nigra and increased in number after dopamine deprivation. Although PSA immunoreactivity is not conclusive evidence of neurogenesis, the results suggests compensatory neuronal differentiation from mitotically silent cells in Parkinson's disease.

13.9.4. Possible Mechanisms of Injury-Induced Neurogenesis

The mechanisms that underlie the injury-induced neurogenesis mostly remain unclear. Growth factors and neurotrophic factors are likely candidates. In the case of forebrain ischemia, it has been shown that expression of bFGF and BDNF increase after injury. Normal neurodevelopment is guided by the spatial and temporal expression of trophic and growth factors, so these factors may represent the brain's attempt to protect injured neurons by activating developmental programs. In addition to protecting neurons, trophic and growth factors have also been shown to stimulate proliferation of adult neural stem cells and to direct their differentiation (Aberg et al., 2000; Pencea et al., 2001).

Activated glial cells are another candidate for directing ischemia-induced neurogenesis. Astrocytes are implicated in the regulation of adult neurogenesis in various settings (Song et al., 2002). Microglia activated during inflammation can be a source of trophic factors and chemokines that influence endogenous stem cell behavior. Endothelial cells are another source of neurotrophic factors, including VEGF, BDNF, bFGF, and chemokines that promote the survival and migration of NSCs. Palmer et al. investigated the relationship of angiogenesis and neurogenesis and referred to the survival coupling of neuroblasts and endothelial cells as the "vascular niche" (Palmer et al., 2000). Treatment with VEGF 24 hours after stroke enhances angiogenesis and neurogenesis, indicating that VEGF plays an important role in brain repair (Zhang et al., 2000).

13.9.5. Functional Significance of Neurogenesis After Injury

While stem cells within the adult brain can be stimulated to replace damaged neurons. Important questions remain: do these newly formed neurons in the adult brain establish appropriate synaptic connections and are they functionally integrated into the existed network? It is still uncertain to what extent neurogenesis contributes to functional recovery. Newly born neurons can establish synaptic contacts and are electrophysiologically functional (Carlen et al., 2002; van Praag et al., 2002). When neurogenesis was blocked, mice exhibited an impaired performance of a hippocampus-dependent memory task, suggesting that the newly born neurons are involved in the formation of certain types of memories (Shors et al., 2001). Using ionizing radiation to decrease neural regeneration after global ischemia, Raber et al. (2004) demonstrated that reduction of neurogenesis reduces functional recovery (Raber et al., 2004). Consistent with these data, Stilley et al. (2004) reported that transplantation of neuronal cells enhanced cognitive function in patients with basal ganglia stroke (Stilley et al., 2004).

An elaborate study by Ramer et al. (2000) finally documented a functional regeneration of sensory axon (Ramer et al., 2000). Not only were specific behavioral tests corresponding to the regenerated fiber phenotypes used, but they also confirmed their findings by reinjury of the regenerated fibers to demonstrate their capability for participation in functional recovery.

13.10. Neural Stem Cells and their Potential Role in Transplant Therapy for Neurodegenerative Disorders

The use of stem cells to generate replacement tissues for treating neurological diseases is a major focus of stem cell research. Spinal cord injury, multiple sclerosis, Parkinson's disease, Huntington's disease, amyotrophic lateral sclerosis and Alzheimer's disease are among those diseases for which the concept of replacing destroyed or dysfunctional cells in the brain or spinal cord is a practical goal. The potential of several promising cell types to serve as an effective transplant has been evaluated in terms of the cell's ability to survive, differentiate, and migrate in an inflammatory milieu of the adult CNS (Magnus and Rao, 2005). Many different cell types, including neural stem cells, precursors, and embryonic stem cells, have been suggested as candidate cells for therapy (Rao and Mayer-Proschel, 2000). Several neurological disorders that show some progress in stem cell therapy are discussed below.

13.10.1. Stem Cells for Transplant Therapy

Embryonic stem cells can proliferate indefinitely and under certain condition they can differentiate to neurons and glia in the laboratory. Even though they hold the promise of being able to repair or replace cells or tissues that are damaged or destroyed in neurodegenerative disorders, transplantation of embryonic stem cells is not feasible with current techniques. The lines of unaltered human embryonic stem cells that exist will not be suitable for direct use in patients. These cells will need to be differentiated or otherwise modified before they can be used clinically. Simply injecting freshly isolated stem cells into the patient would lead to the formation of teratomas, as it occurs when stem cells are injected into mice. Current challenges are to direct the differentiation of embryonic stem cells into specialized cell populations, and also to devise ways to control their development or proliferation once placed in patients.

NSCs isolated from fetal nervous tissue have the potential to differentiate into all types of nervous system cells, including neurons, oligodendrocytes, and astrocytes, so NSCs also have the capacity to replace damaged tissue in both the CNS and PNS. NSCs will restore functional neurons and glia and regenerate injured tissue. It is this characteristic of neural stem cells that makes them a potentially valuable transplantation material in a host of disorders.

Although adult neural stem cells might provide medical solutions that avoid the ethical and legal problems of cloning and fetal stem cell approaches, the limiting factor for the use of adult stem cells in future cell-replacement strategies is that there are insufficient numbers of cells available for transplantation. This is because most adult stem cells that are grown in a culture dish are unable to proliferate in an unspecialized state for long periods of time. In cases where they can be grown under these conditions, they cannot be directed to become specialized as functionally useful cells. New data on the plasticity of adult stem cells may change this situation. A number of studies have suggested that adult stem cells from various organs are plastic, meaning that they can differentiate not only into their original source tissue, but also into cells of unrelated tissue. The limited plasticity of adult stem cells is a major hurdle to overcome before they will be a practical alternative to ES cells, but many believe this feat will be accomplished as study of these cells continues.

13.10.2. Progress in Stem Cell Therapy in Neurological Disorders

13.10.2.1. Parkinson's Disease

Among all the CNS disorders, Parkinson's disease (PD) has the longest history of stem cell transplantation therapy. In Parkinson's disease, the loss of dopaminergic neurons in the substantia nigra is the major pathological change; thus, the major transplant therapeutic strategy has been to restore dopaminergic neurons by fetal tissue throughout the 1990s (Lindvall et al., 1990; Mehta et al., 1998; Dunnett and Bjorklund, 1999). However, fetal tissue grafts showed short graft survival and limited integration of the grafts, which appeared to reduce the usefulness of fetal tissue. The use of stem cells has increased recently, since they appear to be far superior to fetal tissue grafts. Several groups have generated highly enriched populations of dopamine neurons from mesencephalic progenitors. Transplantation of expanded mesencephalic precursors resulted in spontaneous transformation into dopaminergic neurons, and functional recovery has been achieved in Parkinsonian rats (Studer et al., 1998; Sawamoto et al., 2001).

Since only limited numbers of NSCs can be purified from the midbrain, embryonic stem cells may be another candidate for transplantation in PD (Storch et al., 2004; Snyder and Olanow, 2005). McKay's group first generated dopaminergic neurons from mouse embryonic stem cells, with greater enrichment achieved by mimicking the oxygen tension of the developing midbrain. When transplanted into 6-OHDA-lesioned rats, embryonic stem cell-derived dopamine neurons showed functional recovery in this dopamine-depleted animal model (Lee et al., 2000; Studer et al., 2000; Kim et al., 2002). Takagi et al. (2005) generated large numbers of dopaminergic neurons from monkey ES cells in vitro (Takagi et al., 2005). Behavioral studies and functional imaging revealed that the transplanted cells functioned as dopaminergic neurons and attenuated MPTP-induced neurological symptoms. This may be one of the most exciting reports from recent research, for the first time showing that a stem cell-based strategy provides clinical benefits in nonhuman primates.

To reach ultimate functional recovery in Parkinson's disease, transplanted dopaminergic neurons must survive, reinnervate the striatum, and integrate into the host nigrostriatal system. While many reports in recent years are encouraging, to date only limited survival of dopamine neurons derived from stem cells has been observed following transplantation. Even in the best case, more than 99% of implanted dopamine neurons died within 14 weeks. This raises the need to develop optimized protocols of transplantation that provide better graft survival; on the other hand, there is also a need to determine the appropriate number of cells required for transplantation. Even though significant progress has been made in stem cell-based strategy of Parkinson's disease in animal models, it is far from certain that these will directly lead to meaningful treatment for human disease (Snyder and Olanow, 2005).

While the current research focuses on dopaminergic neurons, non-dopaminergic degeneration is also an important part of Parkinson's disease pathology. It remains to be established whether it is favorable to implant a pure population of dopaminergic neurons or whether the graft should also contain a specific mix of other neuron types and glial cells to induce maximum symptomatic relief.

13.10.2.2. Spinal Cord Injury

Spinal cord injury is associated with the loss of both neurons and glial cells; thus, stem cell-based therapy for reconstituting the injured spinal cord could be used to replace multiple cell types.

The extent of spinal cord injury depends on the severity of the initial trauma as well as the level of subsequent injury. The primary impact of contusion injury triggers a cascade of secondary events including hemorrhage, ischemia, excitotoxicity and inflammation, which lead to apoptotic neuronal and oligodendroglial death (Beattie et al., 2002). Transplantation will depend on a thorough understanding of the lesion environment and specific deficits associated with the individual injury. To maximize integration of transplanted cells, a variety of strategies such as reducing inflammation, inhibiting apoptosis of transplanted cells by in vitro preconditioning, and modifying the glial scar are being tested (Ramer et al., 2004).

The potential application of neural stem cells for spinal cord injury has been investigated by numerous studies. It has been reported that neural stem cells induced to neuronal differentiation by neurogenin-2 provided significant functional benefit following transplantation after contusion injury (Hofstetter et al., 2005). Further, undifferentiated cells can achieve the regional appropriate phenotype specification in response to local signals in exclusive niches (Gage, 2000).

Embryonic stem cells also have been tested. Myelination in the injured spinal cord by implanted mouse embryonic stem cells was reported (Brustle et al., 1999). A recent study reported that oligodendrocytes derived form human embryonic stem cells were able to myelinate demyelinated foci in spinal cord contusions (Nistor et al., 2005). The discovery of NG2-expressing progenitors within the spinal cord stirred the hope for using endogenous stem cells in spinal cord injury (Horner and Gage, 2000); however, the current incomplete understanding of the neural stem cell niche makes it impossible to control and manipulate, so this ideal approach remains beyond reach.

13.10.2.3. Multiple Sclerosis

For multiple sclerosis, it has been demonstrated that spontaneous myelin repair may occur as a physiological response to immune-mediated destruction of the myelin sheath. But the spontaneous remyelination is not robust enough and fails over time (Pluchino et al., 2004). Although it remains unclear which type of cell drives axon re-ensheathment in vivo, endogenous oligodendrocyte progenitor cells are believed to play a role in this process (Bruck et al., 2003). Transplanted oligodendrocyte progenitor cells seemed to be more efficient than endogenous oligodendrocyte progenitor cells in repairing the myelin sheath (Blakemore et al., 2000).

The use of neural stem cells in remyelinating therapy has been explored by many researchers. Pluchino et al. (2003) successfully used adult neural stem cells to promote functional recovery in a multiple sclerosis model. Transplanted neural stem cells showed the ability to selectively reach the demyelinating areas, where they differentiated into axon-ensheathing oligodendrocytes and achieved extensive remyelination (Pluchino et al., 2003).

13.10.2.4. Other Diseases

There are relatively fewer reports about the generation of cholinergic neurons, which are the major sites of degeneration in Alzheimer's disease and motor neuron disease. Wu et al. (2002) demonstrate that a "priming" procedure, which involves culturing the fetal human neural stem cells on laminin in the presence of a cocktail of bFGF and heparin in vitro, generates a nearly pure population of neurons after transplantion into adult rat brain. Furthermore, stem cell-derived cholinergic neurons were found following implantation in regions that normally contain many cholinergic neurons, such as the spinal cord and medial septum, but were not seen in regions normally lacking these cells, such as the hippocampus and cortex. In contrast, the human neural stem cells that did not undergo in vitro priming produced primarily glia or undifferentiated cells following implantation (Wu et al., 2002).

Stem cell-based therapy for cerebral ischemia will be more complicated, because the extensive cell death and massive inflammatory response make these brains a more hostile environment for cell grafts. Various sources of cells have been tested for their ability to reconstruct the forebrain and improve function after transplantation in animal models of stroke (Lindvall and Kokaia, 2004). In most cases, only a few grafted cells could survive. Some recent exciting findings in rodents suggest that stroke can induce an increase in neurogenesis; thus, a new therapeutic approach based on self-repair has been brought forth, as discussed in Sect. 13.11.1.

Although stem cell-based strategies have generated some encouraging results, much work is needed to achieve a better understanding of the transplantation efficacy, long-term survival, and functional integration before any future human applications will be feasible.

13.11. Future Directions in Neurogenesis for Neurodegenerative Disorders

Although current research still focuses on cell-based therapies by transplantation of exogenous stem cells or differentiated neurons for neurodegenerative disorders such as Parkinson's disease, Huntington's disease or amyotrophic lateral sclerosis, which all result from a loss of relatively defined phenotypes of cells, accumulating evidence indicates that endogenous neurogenesis may occur as part of an intrinsic brain self-repair process. This raises the possibility of developing therapeutic strategies based on activation of dormant capacities of endogenous neural stem cells.

13.11.1. Endogenous Adult Stem Cells as a Therapeutic Strategy

Recent evidence confirms the widespread occurrence of neural stem cells and the production of new neurons in many parts of adult brain. Although the regeneration of new neurons has been observed only at a very low frequency and can only replace a small fraction of neurons (Fallon et al., 2000; Magavi et al., 2000), these findings encourage work toward a therapeutic strategy based on neurogenesis, especially in neurodegenerative disease where persistent stimuli for neurogenesis exist or where additional stimulation and regulation are possible. For those neural injuries and neurodegenerative disorders that involve relatively broad areas and many types of neurons, this will be an especially attractive direction. With appropriate manipulation, perhaps many types of neurons that have been considered irreplaceable can be regenerated under the right circumstances (Kruger and Morrison, 2002).

Distinct environmental cues may instruct endogenous progenitors to differentiate into specific neuronal subtypes depending on their locations. Adult stem cells can regenerate specific neuronal subtypes appropriate for the sites of damage. For example, targeted degeneration of corticothalamic neurons in the neocortex and dopaminergic neurons innervating the striatum has been shown to induce regeneration of the same cells types.

The strategy of manipulating endogenous stem cell for repairing relies on finding the appropriate factors and signaling molecules that help the patient's own stem cells survive and grow. Previous studies have shown that exogenous growth factors can increase the rate of neurogenesis in the adult brain and a certain combination of exogenous factors and endogenous stimulation may be necessary for boosting the regeneration of neurons at lesion sites (Weiss et al., 1996). By intraventricular infusion of growth factors, CA1 hippocampal pyramidal neurons that were lost to ischemic injury were remarkable regenerated by endogenous neural progenitor cells. These new neurons not only survived for a long period of time but also established synaptic connections and showed functional contributions by ameliorating deficits in hippocampal-dependent spatial cognition in ischemic rat models (Nakatomi et al., 2002).

Although newly born neurons are able to project over relatively long distances to targets, questions remain about whether and how the projections incorporate into local circuits given the complexity of the adult brain network. Nakatomi et al. (2002) noted that the synapses of newly born neurons appeared to remain immature and exhibited altered electrophysiological properties even 3 months after ischemia. Also, not all incorporation of new cells into the brain circuitry is beneficial (Nakatomi et al., 2002). For example, prolonged seizures lead to the generation of ectopic neurons or abnormal projections in the striatum and hippocampus, so abnormal neurogenesis may even worsen seizure damage by its pro-epileptogenic nature (Parent, 2003).

While a growing number of studies report injury-induced enhancement of neurogenesis, it is clear that endogenous stem cells unable to replace most types of neurons lost to injury. There is no evidence connects neurogenesis with improved functions or slowed disease progression in neurodegenerative disease. Thus, it will not be easy to achieve functional neuronal regenesis and brain reconstruction. For example, although hippocampus is a known site of active neurogenesis throughout adulthood, newly generated cells beyond the dentate gyrus appear to be very limited (Rietze et al., 2000). So it is very important to understand how the brain microenvironment changes as a result of injury, and how stem cells respond to the spatial and temporal signals presented. The successful development of therapeutic applications based on endogenous neurogenesis will depend on our ability to manage the proliferation, migration, differentiation, and functional integration of recruited cells. Much work is still required to fulfill such a strategy.

13.11.2. Other Potential Therapeutic Value of NSC

The potential therapeutic value of NSCs is greatly enhanced by the basic biology. NSCs are potent vehicles for gene therapy, so NSCs may be useful to carry copies of genes into tissue that will benefit the disease therapy. NSCs are migratory; thus, transplanted NSCs will migrate from the site of delivery to distant areas of active neurogenesis or to the damaged tissue, where they reconstruct both neural networks and glia support. Because NSCs will migrate to distant, multiple, and extensive regions of the nervous system, they may be useful in "global" degeneration, in which all or most of the nervous system is affected. In addition to regernerative capabilities, stem cells may have an immunoregulatory potential. Mesenchymal stem cells have been shown to inhibit effector T-cell activities, reducing autoimmune inflammation of the CNS (Zappia et al., 2005). This immunoregulatory potential of mesenchymal stem cells suggests that NSC may also possess similar properties, but this is an area of research that still remains to be investigated (Uccelli et al., 2006).

Finally, much of the biology of regeneration of nervous system tissue has yet to be discovered. Many of the factors that contribute to regeneration and/or that prevent it remain unknown. NSCs may provide as yet unidentified factors to host tissue, enabling or enhancing regeneration of damaged tissue.

Summary

Neural stem cells, including early stem cells and late progenitor cells, persist throughout life and provide a source for new neurons, astrocytes, and oligodendrocytes to neural tissue in a process called neurogenesis. These cells are critical in neural development and likely contribute to neuronal repair in disorders such as stroke, multiple sclerosis, Parkinson's disease, HIV-1 associated dementia, and brain inflammation. This chapter reviewed the current understanding of neural stem cells, with a particular focus on the chemokines, growth factors, and neurotransmitters that influence the migration, proliferation, and differentiation of neural stem cells during development and in brain disorders.

Acknowledgments. This work was supported in part by research grants by the National Institutes of Health: R01 NS 41858, P20 RR15635, and P01 NS043985 (JZ). We kindly acknowledge Drs. Howard E. Gendelman and Anuja Ghorpade for valuable suggestion and critics of this work.

Shelley Herek, David Erichsen, Alicia Lopez, Ziyuan Duan, Nathan Erdmann, Dongsheng Xu, Li Wu, Clancy McNalley, and Jianxing Zhao provided technical support for this work. Julie Ditter, Robin Taylor, Myhanh Che, Nell Ingraham, and Emile Scoggins provided outstanding administrative support.

Review Questions/Problems

1. Which cell type is not likely derived from neural stem/ progenitor cells?

a. Neuron
b. Monocyte
c. Astrocyte
d. Oligodendrocyte

2. Which of the following is not a necessary process for successful neurogenesis?

a. Proliferation
b. Migration
c. Differentiation
d. Necrosis of stem cell

3. Neural tissue is derived from which germ layer?

a. Ectoderm
b. Mesoderm
c. Endoderm
d. None of the above

4. What role do basic helix-loop-helix (bHLH) genes Mash1, Math and Ngn have on neurogenesis?

a. Induce differentiation to the glial lineage
b. Induce migration of NPC
c. Maintain the stem cell pool by regulating self-renewal
d. Induce differentiation to the neuronal lineage

5. Which receptor does SDF-1 activate to induce migration of NPC?

a. CCR2
b. CCR4
c. CXCR4
d. CX3CR1

6. What possible effects could selective serotonin reuptake inhibitors (SSOIs) have on neurogenesis?

a. Increase neurogenesis by enhancing proliferation and survival of NPC
b. Increase neurogenesis by inducing migration toward serotonin projections
c. Decrease neurogenesis by down-regulating the expression of brain derived neurotrophic factor (BDNF)
d. Decrease neurogenesis by inhibiting the proliferation of NPC

7. Which cell type releases factor that induce neurogenesis following CNS injury?

a. Astrocytes
b. Microglia
c. Endothelial cells
d. All of the above

8. Which is not a complication for embryonic or fetal stem cell transplant therapy?

a. Transplanted cell are recognized by the immune system as foreign and are rejected
b. Some cultural ethics oppose the use of embryonic and fetal stem cells
c. Transplanted cells preferably differentiate to microglia rather than neurons
d. Transplanted embryonic stem cells may form teratomas

9. Which is not a complication for an autograft (from one to oneself) transplant of adult stem cells?

a. Limited plasticity of adult stem cells hinders the use of non-neuronal stem cells
b. After in vitro culture and transplantation, the immune system recognizes the transplanted cells as foreign and are rejected
c. Adult stem cells grown in vitro are usually unable to proliferate in an unspecialized state
d. In vitro conditions that allow for adequate proliferation of adult stem cells fail to direct the differentiation to functionally useful cells

10. Which is not possible a role (other than neurogenesis) of stem cells in treating CNS disorders?

a. Stem cells can be utilized as vehicles for gene therapy
b. Stem cells can inhibit effector T-cell activities, reducing autoimmune inflammation
c. Stem cells can actively uptake and degrade neurotoxins
d. None of the above

11. What are the two criteria that must be fulfilled in order to identify a cell as a "stem cell"?

12. What is the difference between a stem cell and a progenitor cell?

13. Discuss the positive and negative effects inflammation has on neurogenesis.

14. List some of the roles of stromal cell-derived factor 1 (SDF-1) in CNS development and homeostasis.

15. What are the effects of brain-derived neurotrophic factor (BDNF) on NPC?

16. List the processes that a neural stem cell must undergo to become a functional neuron.

17. Discuss the role of Notch signaling in regulating stem cell self-renewal versus differentiation.

18. Discuss the advantages and shortcomings of embryonic, fetal, and adult stem cells for transplantation therapies.

19. Discuss the benefits and problems associated with therapies using exogenous stem cells as compared to targeting endogenous stem cells to promote neurogenesis.

References

Aberg MA, Aberg ND, Hedbacker H, Oscarsson J, Eriksson PS (2000) Peripheral infusion of IGF-I selectively induces neurogenesis in the adult rat hippocampus. J Neurosci 20:2896–2903.

Anderson DW, Neavin T, Smith JA, Schneider JS (2001) Neuroprotective effects of pramipexole in young and aged MPTP-treated mice. Brain Res 905:44–53.

Anderson MF, Aberg MA, Nilsson M, Eriksson PS (2002) Insulin-like growth factor-I and neurogenesis in the adult mammalian brain. Brain Res Dev Brain Res 134:115–122.

Arundine M, Tymianski M (2004) Molecular mechanisms of glutamate-dependent neurodegeneration in ischemia and traumatic brain injury. Cell Mol Life Sci 61:657–668.

Arvidsson A, Collin T, Kirik D, Kokaia Z, Lindvall O (2002) Neuronal replacement from endogenous precursors in the adult brain after stroke. Nat Med 8:963–970.

Bagri A, Gurney T, He X, Zou YR, Littman DR, Tessier-Lavigne M, Pleasure SJ (2002) The chemokine SDF1 regulates migration of dentate granule cells. Development 129:4249–4260.

Bai F, Bergeron M, Nelson DL (2003) Chronic AMPA receptor potentiator (LY451646) treatment increases cell proliferation in adult rat hippocampus. Neuropharmacology 44:1013–1021.

Baker SA, Baker KA, Hagg T (2004) Dopaminergic nigrostriatal projections regulate neural precursor proliferation in the adult mouse subventricular zone. Eur J Neurosci 20:575–579.

Baker SA, Baker KA, Hagg T (2005) D3 dopamine receptors do not regulate neurogenesis in the subventricular zone of adult mice. Neurobiol Dis 18:523–527.

Banasr M, Hery M, Printemps R, Daszuta A (2004) Serotonin-induced increases in adult cell proliferation and neurogenesis are mediated through different and common 5-HT receptor subtypes in the dentate gyrus and the subventricular zone. Neuropsychopharmacology 29:450–460.

Beattie MS, Hermann GE, Rogers RC, Bresnahan JC (2002) Cell death in models of spinal cord injury. Prog Brain Res 137:37–47.

Bertrand N, Castro DS, Guillemot F (2002) Proneural genes and the specification of neural cell types. Nat Rev Neurosci 3:517–530.

Blakemore WF, Gilson JM, Crang AJ (2000) Transplanted glial cells migrate over a greater distance and remyelinate demyelinated lesions more rapidly than endogenous remyelinating cells. J Neurosci Res 61:288–294.

Bolteus AJ, Bordey A (2004) GABA release and uptake regulate neuronal precursor migration in the postnatal subventricular zone. J Neurosci 24:7623–7631.

Brazel CY, Nunez JL, Yang Z, Levison SW (2005) Glutamate enhances survival and proliferation of neural progenitors derived from the subventricular zone. Neuroscience 131:55–65.

Brewer KL, Bethea JR, Yezierski RP (1999) Neuroprotective effects of interleukin-10 following excitotoxic spinal cord injury. Exp Neurol 159:484–493.

Bruck W, Kuhlmann T, Stadelmann C (2003) Remyelination in multiple sclerosis. J Neurol Sci 206:181–185.

Brustle O, Jones KN, Learish RD, Karram K, Choudhary K, Wiestler OD, Duncan ID, McKay RD (1999) Embryonic stem cell-derived glial precursors: A source of myelinating transplants. Science 285:754–756.

Caille I, Allinquant B, Dupont E, Bouillot C, Langer A, Muller U, Prochiantz A (2004) Soluble form of amyloid precursor protein regulates proliferation of progenitors in the adult subventricular zone. Development 131:2173–2181.

Cameron HA, Hazel TG, McKay RD (1998) Regulation of neurogenesis by growth factors and neurotransmitters. J Neurobiol 36:287–306.

Carlen M, Cassidy RM, Brismar H, Smith GA, Enquist LW, Frisen J (2002) Functional integration of adult-born neurons. Curr Biol 12:606–608.

Carvey PM, McGuire SO, Ling ZD (2001) Neuroprotective effects of D3 dopamine receptor agonists. Parkinsonism Relat Disord 7:213–223.

Cayuso J, Marti E (2005) Morphogens in motion: Growth control of the neural tube. J Neurobiol 64:376–387.

Chen J, Li Y, Zhang R, Katakowski M, Gautam SC, Xu Y, Lu M, Zhang Z, Chopp M (2004) Combination therapy of stroke in rats with a nitric oxide donor and human bone marrow stromal cells enhances angiogenesis and neurogenesis. Brain Res 1005:21–28.

Cheng A, Wang S, Cai J, Rao MS, Mattson MP (2003) Nitric oxide acts in a positive feedback loop with BDNF to regulate neural progenitor cell proliferation and differentiation in the mammalian brain. Dev Biol 258:319–333.

Curtis MA, Penney EB, Pearson AG, van Roon-Mom WM, Butterworth NJ, Dragunow M, Connor B, Faull RL (2003) Increased cell proliferation and neurogenesis in the adult human Huntington's disease brain. Proc Natl Acad Sci USA 100:9023–9027.

Dahmane N, Ruiz i Altaba A (1999) Sonic hedgehog regulates the growth and patterning of the cerebellum. Development 126:3089–3100.

Dahmane N, Sanchez P, Gitton Y, Palma V, Sun T, Beyna M, Weiner H, Ruiz i Altaba A (2001) The sonic hedgehog-gli pathway regulates dorsal brain growth and tumorigenesis. Development 128:5201–5212.

Di Giorgi Gerevini VD, Caruso A, Cappuccio I, Ricci Vitiani L, Romeo S, Della Rocca C, Gradini R, Melchiorri D, Nicoletti F (2004) The mGlu5 metabotropic glutamate receptor is expressed in zones of active neurogenesis of the embryonic and postnatal brain. Brain Res Dev Brain Res 150:17–22.

Dingledine R, Borges K, Bowie D, Traynelis SF (1999) The glutamate receptor ion channels. Pharmacol Rev 51:7–61.

Dunnett SB, Bjorklund A (1999) Prospects for new restorative and neuroprotective treatments in Parkinson's disease. Nature 399: A32–A39.

Dziembowska M, Tham TN, Lau P, Vitry S, Lazarini F, Dubois-Dalcq M (2005) A role for CXCR4 signaling in survival and migration of neural and oligodendrocyte precursors. Glia 50:258–269.

Ekdahl CT, Claasen JH, Bonde S, Kokaia Z, Lindvall O (2003) Inflammation is detrimental for neurogenesis in adult brain. Proc Natl Acad Sci USA 100:13632–13637.

Emsley JG, Hagg T (2003) Endogenous and exogenous ciliary neurotrophic factor enhances forebrain neurogenesis in adult mice. Exp Neurol 183:298–310.

Emsley JG, Mitchell BD, Kempermann G, Macklis JD (2005) Adult neurogenesis and repair of the adult CNS with neural progenitors, precursors, and stem cells. Prog Neurobiol 75:321–341.

Enwere E, Shingo T, Gregg C, Fujikawa H, Ohta S, Weiss S (2004) Aging results in reduced epidermal growth factor receptor signaling, diminished olfactory neurogenesis, and deficits in fine olfactory discrimination. J Neurosci 24:8354–8365.

Fallon J, Reid S, Kinyamu R, Opole I, Opole R, Baratta J, Korc M, Endo TL, Duong A, Nguyen G, Karkehabadhi M, Twardzik D, Patel S, Loughlin S (2000) In vivo induction of massive proliferation, directed migration, and differentiation of neural cells in the adult mammalian brain. Proc Natl Acad Sci USA 97:14686–14691.

Feng Y, Walsh CA (2001) Protein-protein interactions, cytoskeletal regulation and neuronal migration. Nat Rev Neurosci 2:408–416.

Ferguson KL, Slack RS (2003) Growth factors: Can they promote neurogenesis? Trends Neurosci 26:283–285.

Frielingsdorf H, Schwarz K, Brundin P, Mohapel P (2004) No evidence for new dopaminergic neurons in the adult mammalian substantia nigra. Proc Natl Acad Sci USA 101:10177–10182.

Gage FH (2000) Mammalian neural stem cells. Science 287:1433–1438.

Gage FH, Ray J, Fisher LJ (1995) Isolation, characterization, and use of stem cells from the CNS. Annu Rev Neurosci 18:159–192.

Gensert JM, Goldman JE (1997) Endogenous progenitors remyelinate demyelinated axons in the adult CNS. Neuron 19:197–203.

Hagg T (2005) Molecular regulation of adult CNS neurogenesis: An integrated view. Trends Neurosci 28:589–595.

Harrist A, Beech RD, King SL, Zanardi A, Cleary MA, Caldarone BJ, Eisch A, Zoli M, Picciotto MR (2004) Alteration of hippocampal cell proliferation in mice lacking the beta 2 subunit of the neuronal nicotinic acetylcholine receptor. Synapse 54:200–206.

Hatten ME (1999) Central nervous system neuronal migration. Annu Rev Neurosci 22:511–539.

Haughey NJ, Nath A, Chan SL, Borchard AC, Rao MS, Mattson MP (2002) Disruption of neurogenesis by amyloid beta-peptide, and perturbed neural progenitor cell homeostasis, in models of Alzheimer's disease. J Neurochem 83:1509–1524.

Hoehn BD, Palmer TD, Steinberg GK (2005) Neurogenesis in rats after focal cerebral ischemia is enhanced by indomethacin. Stroke 36:2718–2724.

Hofstetter CP, Holmstrom NA, Lilja JA, Schweinhardt P, Hao J, Spenger C, Wiesenfeld-Hallin Z, Kurpad SN, Frisen J, Olson L (2005) Allodynia limits the usefulness of intraspinal neural stem cell grafts; directed differentiation improves outcome. Nat Neurosci 8:346–353.

Hoglinger GU, Rizk P, Muriel MP, Duyckaerts C, Oertel WH, Caille I, Hirsch EC (2004) Dopamine depletion impairs precursor cell proliferation in Parkinson's disease. Nat Neurosci 7:726–735.

Hollmann M, Heinemann S (1994) Cloned glutamate receptors. Annu Rev Neurosci 17:31–108.

Horner PJ, Gage FH (2000) Regenerating the damaged central nervous system. Nature 407:963–970.

Ille F, Sommer L (2005) Wnt signaling: Multiple functions in neural development. Cell Mol Life Sci 62:1100–1108.

Jacob J, Briscoe J (2003) Gli proteins and the control of spinal-cord patterning. EMBO Rep 4:761–765.

Ji JF, He BP, Dheen ST, Tay SS (2004) Expression of chemokine receptors CXCR4, CCR2, CCR5 and CX3CR1 in neural progenitor cells isolated from the subventricular zone of the adult rat brain. Neurosci Lett 355:236–240.

Jin K, Peel AL, Mao XO, Xie L, Cottrell BA, Henshall DC, Greenberg DA (2004) Increased hippocampal neurogenesis in Alzheimer's disease. Proc Natl Acad Sci USA 101:343–347.

Kageyama R, Ohtsuka T, Hatakeyama J, Ohsawa R (2005) Roles of bHLH genes in neural stem cell differentiation. Exp Cell Res 306:343–348.

Kay JN, Blum M (2000) Differential response of ventral midbrain and striatal progenitor cells to lesions of the nigrostriatal dopaminergic projection. Dev Neurosci 22:56–67.

Kelly S, Bliss TM, Shah AK, Sun GH, Ma M, Foo WC, Masel J, Yenari MA, Weissman IL, Uchida N, Palmer T, Steinberg GK (2004) Transplanted human fetal neural stem cells survive, migrate, and differentiate in ischemic rat cerebral cortex. Proc Natl Acad Sci USA 101:11839–11844.

Kim JH, Auerbach JM, Rodriguez-Gomez JA, Velasco I, Gavin D, Lumelsky N, Lee SH, Nguyen J, Sanchez-Pernaute R, Bankiewicz K, McKay R (2002) Dopamine neurons derived from embryonic stem cells function in an animal model of Parkinson's disease. Nature 418:50–56.

Kim SU, de Vellis J (2005) Microglia in health and disease. J Neurosci Res 81:302–313.

Kippin TE, Kapur S, van der Kooy D (2005) Dopamine specifically inhibits forebrain neural stem cell proliferation, suggesting a novel effect of antipsychotic drugs. J Neurosci 25:5815–5823.

Kitamura T, Mishina M, Sugiyama H (2003) Enhancement of neurogenesis by running wheel exercises is suppressed in mice lacking NMDA receptor epsilon 1 subunit. Neurosci Res 47:55–63.

Klein RS, Rubin JB, Gibson HD, DeHaan EN, Alvarez-Hernandez X, Segal RA, Luster AD (2001) SDF-1 alpha induces chemotaxis and enhances sonic hedgehog-induced proliferation of cerebellar granule cells. Development 128:1971–1981.

Krathwohl MD, Kaiser JL (2004) Chemokines promote quiescence and survival of human neural progenitor cells. Stem Cells 22:109–118.

Kruger GM, Morrison SJ (2002) Brain repair by endogenous progenitors. Cell 110:399–402.

Kulkarni VA, Jha S, Vaidya VA (2002) Depletion of norepinephrine decreases the proliferation, but does not influence the survival and differentiation, of granule cell progenitors in the adult rat hippocampus. Eur J Neurosci 16:2008–2012.

Kwon YK (2002) Effect of neurotrophic factors on neuronal stem cell death. J Biochem Mol Biol 35:87–93.

Lee CS, Cenci MA, Schulzer M, Bjorklund A (2000) Embryonic ventral mesencephalic grafts improve levodopa-induced dyskinesia in a rat model of Parkinson's disease. Brain 123 (Pt 7):1365–1379.

Lindvall O, Kokaia Z (2004) Recovery and rehabilitation in stroke: Stem cells. Stroke 35:2691–2694.

Lindvall O, Brundin P, Widner H, Rehncrona S, Gustavii B, Frackowiak R, Leenders KL, Sawle G, Rothwell JC, Marsden CD, et al. (1990) Grafts of fetal dopamine neurons survive and improve motor function in Parkinson's disease. Science 247:574–577.

Lobjois V, Benazeraf B, Bertrand N, Medevielle F, Pituello F (2004) Specific regulation of cyclins D1 and D2 by FGF and Shh signaling coordinates cell cycle progression, patterning, and differentiation during early steps of spinal cord development. Dev Biol 273:195–209.

Logan A, Green J, Hunter A, Jackson R, Berry M (1999) Inhibition of glial scarring in the injured rat brain by a recombinant human monoclonal antibody to transforming growth factor-beta2. Eur J Neurosci 11:2367–2374.

Logan CY, Nusse R (2004) The Wnt signaling pathway in development and disease. Annu Rev Cell Dev Biol 20:781–810.

Louissaint A, Jr., Rao S, Leventhal C, Goldman SA (2002) Coordinated interaction of neurogenesis and angiogenesis in the adult songbird brain. Neuron 34:945–960.

Lu L, Su WJ, Yue W, Ge X, Su F, Pei G, Ma L (2001) Attenuation of morphine dependence and withdrawal in rats by venlafaxine, a serotonin and noradrenaline reuptake inhibitor. Life Sci 69:37–46.

Luk KC, Kennedy TE, Sadikot AF (2003) Glutamate promotes proliferation of striatal neuronal progenitors by an NMDA receptor-mediated mechanism. J Neurosci 23:2239–2250.

Luo Y, Cai J, Xue H, Miura T, Rao MS (2005) Functional SDF1 alpha/CXCR4 signaling in the developing spinal cord. J Neurochem 93:452–462.

Ma Q, Jones D, Borghesani PR, Segal RA, Nagasawa T, Kishimoto T, Bronson RT, Springer TA (1998) Impaired B -lymphopoiesis, myelopoiesis, and derailed cerebellar neuron migration in CXCR4 and SDF 1 deficient mice. Proc Natl Acad Sci USA 95:9448–9453.

Machold R, Hayashi S, Rutlin M, Muzumdar MD, Nery S, Corbin JG, Gritli-Linde A, Dellovade T, Porter JA, Rubin LL, Dudek H, McMahon AP, Fishell G (2003) Sonic hedgehog is required for progenitor cell maintenance in telencephalic stem cell niches. Neuron 39:937–950.

Mackowiak M, O'Neill MJ, Hicks CA, Bleakman D, Skolnick P (2002) An AMPA receptor potentiator modulates hippocampal expression of BDNF: An in vivo study. Neuropharmacology 43:1–10.

Magavi SS, Leavitt BR, Macklis JD (2000) Induction of neurogenesis in the neocortex of adult mice. Nature 405:951–955.

Magnus T, Rao MS (2005) Neural stem cells in inflammatory CNS diseases: Mechanisms and therapy. J Cell Mol Med 9:303–319.

Marchetti B, Abbracchio MP (2005) To be or not to be (inflamed)—is that the question in anti-inflammatory drug therapy of neurodegenerative disorders? Trends Pharmacol Sci 26:517–525.

Mattson MP, Maudsley S, Martin B (2004) BDNF and 5-HT: A dynamic duo in age-related neuronal plasticity and neurodegenerative disorders. Trends Neurosci 27:589–594.

Matyszak MK (1998) Inflammation in the CNS: Balance between immunological privilege and immune responses. Prog Neurobiol 56:19–35.

McConnell SK (1995) Strategies for the generation of neuronal diversity in the developing central nervous system. J Neurosci 15:6987–6998.

McKay R (1997) Stem cells in the central nervous system. Science 276:66–71.

Mehta V, Hong M, Spears J, Mendez I (1998) Enhancement of graft survival and sensorimotor behavioral recovery in rats undergoing transplantation with dopaminergic cells exposed to glial cell line-derived neurotrophic factor. J Neurosurg 88:1088–1095.

Ming GL, Song H (2005) Adult neurogenesis in the mammalian central nervous system. Annu Rev Neurosci 28:223–250.

Moalem G, Leibowitz-Amit R, Yoles E, Mor F, Cohen IR, Schwartz M (1999) Autoimmune T cells protect neurons from secondary degeneration after central nervous system axotomy. Nat Med 5:49–55.

Mohapel P, Leanza G, Kokaia M, Lindvall O (2005) Forebrain acetylcholine regulates adult hippocampal neurogenesis and learning. Neurobiol Aging 26:939–946.

Monaghan D, Bridges R, Cotman C (1989) The excitatory amino acid receptors. Annu Rev Pharmacol Toxicol 29:365–402.

Monje ML, Toda H, Palmer TD (2003) Inflammatory blockade restores adult hippocampal neurogenesis. Science 302:1760–1765.

Morrison SJ (2001) Neuronal potential and lineage determination by neural stem cells. Curr Opin Cell Biol 13:666–672.

Nadarajah B, Parnavelas JG (2002) Modes of neuronal migration in the developing cerebral cortex. Nat Rev Neurosci 3:423–432.

Nakajima K, Hida H, Shimano Y, Fujimoto I, Hashitani T, Kumazaki M, Sakurai T, Nishino H (2001) GDNF is a major component of trophic activity in DA-depleted striatum for survival and neurite extension of DAergic neurons. Brain Res 916:76–84.

Nakatomi H, Kuriu T, Okabe S, Yamamoto S, Hatano O, Kawahara N, Tamura A, Kirino T, Nakafuku M (2002) Regeneration of hippocampal pyramidal neurons after ischemic brain injury by recruitment of endogenous neural progenitors. Cell 110:429–441.

Ni HT, Hu S, Sheng WS, Olson JM, Cheeran MC, Chan AS, Lokensgard JR, Peterson PK (2004) High-level expression of functional chemokine receptor CXCR4 on human neural precursor cells. Brain Res Dev Brain Res 152:159–169.

Nistor GI, Totoiu MO, Haque N, Carpenter MK, Keirstead HS (2005) Human embryonic stem cells differentiate into oligodendrocytes in high purity and myelinate after spinal cord transplantation. Glia 49:385–396.

Nusse R (2003) Wnts and Hedgehogs: Lipid-modified proteins and similarities in signaling mechanisms at the cell surface. Development 130:5297–5305.

O'Neill MJ, Bleakman D, Zimmerman DM, Nisenbaum ES (2004) AMPA receptor potentiators for the treatment of CNS disorders. Curr Drug Targets CNS Neurol Disord 3:181–194.

Orgogozo V, Schweisguth F, Bellaiche Y (2004) Slit-Robo signalling prevents sensory cells from crossing the midline in Drosophila. Mech Dev 121:427–436.

Overstreet Wadiche L, Bromberg DA, Bensen AL, Westbrook GL (2005) GABAergic signaling to newborn neurons in dentate gyrus. J Neurophysiol 94:4528–4532.

Owens DF, Kriegstein AR (2002) Is there more to GABA than synaptic inhibition? Nat Rev Neurosci 3:715–727.

Palma V, Lim DA, Dahmane N, Sanchez P, Brionne TC, Herzberg CD, Gitton Y, Carleton A, Alvarez-Buylla A, Ruiz i Altaba A (2005) Sonic hedgehog controls stem cell behavior in the postnatal and adult brain. Development 132:335–344.

Palmer TD, Takahashi J, Gage FH (1997) The adult rat hippocampus contains primordial neural stem cells. Mol Cell Neurosci 8:389–404.

Palmer TD, Willhoite AR, Gage FH (2000) Vascular niche for adult hippocampal neurogenesis. J Comp Neurol 425:479–494.

Parent JM (2003) Injury-induced neurogenesis in the adult mammalian brain. Neuroscientist 9:261–272.

Pencea V, Bingaman KD, Wiegand SJ, Luskin MB (2001) Infusion of brain-derived neurotrophic factor into the lateral ventricle of the adult rat leads to new neurons in the parenchyma of the striatum, septum, thalamus, and hypothalamus. J Neurosci 21:6706–6717.

Pende M, Holtzclaw LA, Curtis JL, Russell JT, Gallo V (1994) Glutamate regulates intracellular calcium and gene expression in oligodendrocyte progenitors through the activation of DL-alpha-amino-3-hydroxy-5-methyl-4-isoxazolepropionic acid receptors. Proc Natl Acad Sci USA 91:3215–3219.

Peng H, Huang Y, Rose J, Erichsen D, Herek S, Fujii N, Tamamura H, Zheng J (2004) Stromal cell-derived factor 1 mediated CXCR4 signaling in rat and human cortical neural progenitor cells. Journal of Neuroscience Research 76:35–50.

Pin JP, Duvoisin R (1995) The metabotropic glutamate receptors: Structure and functions. Neuropharmacology 34:1–26.

Plendl J, Stierstorfer B, Sinowatz F (1999) Growth factors and their receptors in the olfactory system. Anat Histol Embryol 28:73–79.

Pluchino S, Furlan R, Martino G (2004) Cell-based remyelinating therapies in multiple sclerosis: Evidence from experimental studies. Curr Opin Neurol 17:247–255.

Pluchino S, Quattrini A, Brambilla E, Gritti A, Salani G, Dina G, Galli R, Del Carro U, Amadio S, Bergami A, Furlan R, Comi G,

Vescovi AL, Martino G (2003) Injection of adult neurospheres induces recovery in a chronic model of multiple sclerosis. Nature 422:688–694.

Pons S, Trejo JL, Martinez-Morales JR, Marti E (2001) Vitronectin regulates sonic hedgehog activity during cerebellum development through CREB phosphorylation. Development 128:1481–1492.

Raber J, Fan Y, Matsumori Y, Liu Z, Weinstein PR, Fike JR, Liu J (2004) Irradiation attenuates neurogenesis and exacerbates ischemia-induced deficits. Ann Neurol 55:381–389.

Ramer LM, Au E, Richter MW, Liu J, Tetzlaff W, Roskams AJ (2004) Peripheral olfactory ensheathing cells reduce scar and cavity formation and promote regeneration after spinal cord injury. J Comp Neurol 473:1–15.

Ramer MS, Priestley JV, McMahon SB (2000) Functional regeneration of sensory axons into the adult spinal cord. Nature 403:312–316.

Rao MS, Mayer-Proschel M (2000) Precursor cells for transplantation. Prog Brain Res 128:273–292.

Rapalino O, Lazarov-Spiegler O, Agranov E, Velan GJ, Yoles E, Fraidakis M, Solomon A, Gepstein R, Katz A, Belkin M, Hadani M, Schwartz M (1998) Implantation of stimulated homologous macrophages results in partial recovery of paraplegic rats. Nat Med 4:814–821.

Rietze R, Poulin P, Weiss S (2000) Mitotically active cells that generate neurons and astrocytes are present in multiple regions of the adult mouse hippocampus. J Comp Neurol 424:397–408.

Ross SE, Greenberg ME, Stiles CD (2003) Basic helix-loop-helix factors in cortical development. Neuron 39:13–25

Sadikot AF, Burhan AM, Belanger MC, Sasseville R (1998) NMDA receptor antagonists influence early development of GABAergic interneurons in the mammalian striatum. Brain Res Dev Brain Res 105:35–42.

Sawamoto K, Nakao N, Kakishita K, Ogawa Y, Toyama Y, Yamamoto A, Yamaguchi M, Mori K, Goldman SA, Itakura T, Okano H (2001) Generation of dopaminergic neurons in the adult brain from mesencephalic precursor cells labeled with a nestin-GFP transgene. J Neurosci 21:3895–3903.

Schwartz M (2003) Macrophages and microglia in central nervous system injury: Are they helpful or harmful? J Cereb Blood Flow Metab 23:385–394.

Shimamura K, Martinez S, Puelles L, Rubenstein JL (1997) Patterns of gene expression in the neural plate and neural tube subdivide the embryonic forebrain into transverse and longitudinal domains. Dev Neurosci 19:88–96.

Shors TJ, Miesegaes G, Beylin A, Zhao M, Rydel T, Gould E (2001) Neurogenesis in the adult is involved in the formation of trace memories. Nature 410:372–376.

Simpson KL, Fisher TM, Waterhouse BD, Lin RC (1998) Projection patterns from the raphe nuclear complex to the ependymal wall of the ventricular system in the rat. J Comp Neurol 399:61–72.

Siuciak JA, Lewis DR, Wiegand SJ, Lindsay RM (1997) Antidepressant-like effect of brain-derived neurotrophic factor (BDNF). Pharmacol Biochem Behav 56:131–137.

Snyder BJ, Olanow CW (2005) Stem cell treatment for Parkinson's disease: An update for 2005. Curr Opin Neurol 18:376–385.

Song H, Stevens CF, Gage FH (2002) Astroglia induce neurogenesis from adult neural stem cells. Nature 417:39–44.

Stilley CS, Ryan CM, Kondziolka D, Bender A, DeCesare S, Wechsler L (2004) Changes in cognitive function after neuronal cell transplantation for basal ganglia stroke. Neurology 63:1320–1322.

Storch A, Sabolek M, Milosevic J, Schwarz SC, Schwarz J (2004) Midbrain-derived neural stem cells: From basic science to therapeutic approaches. Cell Tissue Res 318:15–22.

Streit WJ (2002) Microglia as neuroprotective, immunocompetent cells of the CNS. Glia 40:133–139.

Studer L, Tabar V, McKay RD (1998) Transplantation of expanded mesencephalic precursors leads to recovery in parkinsonian rats. Nat Neurosci 1:290–295.

Studer L, Csete M, Lee SH, Kabbani N, Walikonis J, Wold B, McKay R (2000) Enhanced proliferation, survival, and dopaminergic differentiation of CNS precursors in lowered oxygen. J Neurosci 20:7377–7383.

Stumm RK, Zhou C, Ara T, Lazarini F, Dubois-Dalcq M, Nagasawa T, Hollt V, Schulz S (2003) CXCR4 regulates interneuron migration in the developing neocortex. J Neurosci 23:5123–5130.

Takagi Y, Takahashi J, Saiki H, Morizane A, Hayashi T, Kishi Y, Fukuda H, Okamoto Y, Koyanagi M, Ideguchi M, Hayashi H, Imazato T, Kawasaki H, Suemori H, Omachi S, Iida H, Itoh N, Nakatsuji N, Sasai Y, Hashimoto N (2005) Dopaminergic neurons generated from monkey embryonic stem cells function in a Parkinson primate model. J Clin Invest 115:102–109.

Tauber SC, Stadelmann C, Spreer A, Bruck W, Nau R, Gerber J (2005) Increased expression of BDNF and proliferation of dentate granule cells after bacterial meningitis. J Neuropathol Exp Neurol 64:806–815.

Temple S (2001) The development of neural stem cells. Nature 414:112–117.

Tran PB, Ren D, Veldhouse TJ, Miller RJ (2004) Chemokine receptors are expressed widely by embryonic and adult neural progenitor cells. J Neurosci Res 76:20–34.

Tran PB, Miller RJ (2005) HIV-1, chemokines and neurogenesis. Neurotox Res 8:149–158.

Tsai HH, Frost E, To V, Robinson S, Ffrench-Constant C, Geertman R, Ransohoff RM, Miller RH (2002) The chemokine receptor CXCR2 controls positioning of oligodendrocyte precursors in developing spinal cord by arresting their migration. Cell 110:373–383.

Uccelli A, Zappia E, Benvenuto F, Frassoni F, Mancardi G (2006) Stem cells in inflammatory demyelinating disorders: A dual role for immunosuppression and neuroprotection. Expert Opin Biol Ther 6:17–22.

Van Kampen JM, Robertson HA (2005) A possible role for dopamine D3 receptor stimulation in the induction of neurogenesis in the adult rat substantia nigra. Neuroscience 136:381–386.

Van Kampen JM, Hagg T, Robertson HA (2004) Induction of neurogenesis in the adult rat subventricular zone and neostriatum following dopamine D receptor stimulation. Eur J Neurosci 19:2377–2387.

Van Praag H, Schinder AF, Christie BR, Toni N, Palmer TD, Gage FH (2002) Functional neurogenesis in the adult hippocampus. Nature 415:1030–1034.

Vu TQ, Ling ZD, Ma SY, Robie HC, Tong CW, Chen EY, Lipton JW, Carvey PM (2000) Pramipexole attenuates the dopaminergic cell loss induced by intraventricular 6-hydroxydopamine. J Neural Transm 107:159–176.

Wallace VA (1999) Purkinje-cell-derived sonic hedgehog regulates granule neuron precursor cell proliferation in the developing mouse cerebellum. Curr Biol 9:445–448.

Wang LP, Kempermann G, Kettenmann H (2005) A subpopulation of precursor cells in the mouse dentate gyrus receives synaptic GABAergic input. Mol Cell Neurosci 29:181–189.

Wechsler-Reya RJ, Scott MP (1999) Control of neuronal precursor proliferation in the cerebellum by sonic hedgehog. Neuron 22:103–114.

Weiss S, Reynolds BA, Vescovi AL, Morshead C, Craig CG, van der Kooy D (1996) Is there a neural stem cell in the mammalian forebrain? Trends Neurosci 19:387–393.

Wen PH, Shao X, Shao Z, Hof PR, Wisniewski T, Kelley K, Friedrich VL, Jr., Ho L, Pasinetti GM, Shioi J, Robakis NK, Elder GA (2002) Overexpression of wild type but not an FAD mutant presenilin-1 promotes neurogenesis in the hippocampus of adult mice. Neurobiol Dis 10:8–19.

Widera D, Holtkamp W, Entschladen F, Niggemann B, Zanker K, Kaltschmidt B, Kaltschmidt C (2004) MCP-1 induces migration of adult neural stem cells. Eur J Cell Biol 83:381–387.

Wu P, Tarasenko YI, Gu Y, Huang LY, Coggeshall RE, Yu Y (2002) Region-specific generation of cholinergic neurons from fetal human neural stem cells grafted in adult rat. Nat Neurosci 5:1271–1278.

Yagita Y, Kitagawa K, Ohtsuki T, Takasawa K, Miyata T, Okano H, Hori M, Matsumoto M (2001) Neurogenesis by progenitor cells in the ischemic adult rat hippocampus. Stroke 32: 1890–1896.

Yamamoto S, Yamamoto N, Kitamura T, Nakamura K, Nakafuku M (2001) Proliferation of parenchymal neural progenitors in response to injury in the adult rat spinal cord. Exp Neurol 172:115–127.

Yoon K, Gaiano N (2005) Notch signaling in the mammalian central nervous system: Insights from mouse mutants. Nat Neurosci 8:709–715.

Yoshimi K, Ren YR, Seki T, Yamada M, Ooizumi H, Onodera M, Saito Y, Murayama S, Okano H, Mizuno Y, Mochizuki H (2005) Possibility for neurogenesis in substantia nigra of parkinsonian brain. Ann Neurol 58:31–40.

Zafra F, Lindholm D, Castren E, Hartikka J, Thoenen H (1992) Regulation of brain-derived neurotrophic factor and nerve growth factor mRNA in primary cultures of hippocampal neurons and astrocytes. J Neurosci 12:4793–4799.

Zandi PP, Breitner JC (2001) Do NSAIDs prevent Alzheimer's disease? And, if so, why? The epidemiological evidence. Neurobiol Aging 22:811–817.

Zappia E, Casazza S, Pedemonte E, Benvenuto F, Bonanni I, Gerdoni E, Giunti D, Ceravolo A, Cazzanti F, Frassoni F, Mancardi G, Uccelli A (2005) Mesenchymal stem cells ameliorate experimental autoimmune encephalomyelitis inducing T-cell anergy. Blood 106:1755–1761.

Zhang RL, Zhang ZG, Chopp M (2005) Neurogenesis in the adult ischemic brain: Generation, migration, survival, and restorative therapy. Neuroscientist 11:408–416.

Zhang ZG, Zhang L, Jiang Q, Zhang R, Davies K, Powers C, Bruggen N, Chopp M (2000) VEGF enhances angiogenesis and promotes blood-brain barrier leakage in the ischemic brain. J Clin Invest 106:829–838.

Zhao M, Momma S, Delfani K, Carlen M, Cassidy RM, Johansson CB, Brismar H, Shupliakov O, Frisen J, Janson AM (2003) Evidence for neurogenesis in the adult mammalian substantia nigra. Proc Natl Acad Sci USA 100:7925–7930.

Zhu Y, Yu T, Zhang XC, Nagasawa T, Wu JY, Rao Y (2002) Role of the chemokine SDF-1 as the meningeal attractant for embryonic cerebellar neurons. Nat Neurosci 5:719–720.

Zou YR, Kottmann AH, Kuroda M, Taniuchi I, Littman DR (1998) Function of the chemokine receptor CXCR4 in haematopoiesis and in cerebellar development. Nature 393:595–599.

14
Neurobiology and Neural Systems

Tsuneya Ikezu and Howard E. Gendelman

Keywords Active zone; Central nervous system; Dendrites; Dendritic spine; Neuron; Neurotransmitter; Peripheral nervous system; Postsynaptic density; Synapse

14.1. Introduction

Neurobiology is the study and functional organization of the cells that make up the nervous system. The central nervous system (CNS) begins as a simple neural plate that folds to form a groove and then a tube. Then stem cells within the neural tube are directed toward glia and neurons under the influence of various neural developing signaling processes. It is these cells including, but not limited to, neurons, microglia, astrocytes, endothelial cells, their communication and circuitry, one with the other, that lead to our abilities to sense and respond to the environment, think, ambulate, and behave. Neurobiology is at the very interface of biology and neuroscience but is significantly different from each of the fields. Biology is that of all building blocks of cell organization and function regardless of tissue origin. It is broad and without limits. Neuroscience is by its very integration includes works in computation and cognition that are linked to the clinical disciplines of psychiatry and neurology. Each alone or together relate to cell and system analyses and disease. Nonetheless, the disciplines of neuroscience and biology overlap to generate the field of neurobiology. Here there is a central focus on the cell and its functional outcomes. This chapter seeks to describe the discipline of neurobiology starting from the cell and its function to system organization to function. Such processes underlie both nerve cell communication and function as well as the role glial cells can affect the process overall.

14.2. Neuron

Neurons are the core cell elements of the brain, spinal cord and peripheral nerves that process and transmit information. Neurons are composed of a cell body (soma), dendritic tree (arbor, where they receive input) and one axon (where they transmit out) that conduct the nerve signal through electrical impulses. Dendrites are cellular extensions with extensive branches and called, a dendritic tree. The chemical synapse occurs only in one direction from the axon. The complexities of the mammalian nervous system are due, in large measure, to neuronal size, shape and functional heterogeneity. Neurons communicate across synapses (the gap between the axon terminal and the dendrites of the receiving cell) through electrochemical signals to and from the brain at up to 200 mph. The neuron is without a doubt, the basic anatomical building block of the nervous system. Understanding its functions and that of its organelles opens the discovery of nervous system function itself. The German scientist Heinrich Wilhelm Gottfried von Waldeyer-Hartz coined the word "neuron" in 1891. Subsequently Ramón y Cajal established that the two types of neuronal processes, axons and dendrites, do not interconnect in anastomotic continuity and are therefore independent entities. The many neuronal dendrites receive electrical impulses from the axons of other neurons or receptors and conduct this activity to the neuron's own axon for transmission to other cells. This basic tenet is called the "neuron doctrine" (Ramón y Cajal, 1995). The cell soma (so called perikaryon) contains the neuron's nucleus (with DNA and nuclear organelles). The axon, a long extension of the nerve cell, takes information away from the cell body. Bundles of axons are known as nerves or, within the central nervous system (CNS), as nerve tracts or pathways. Myelin coats insulate the axon (except for periodic breaks called nodes of Ranvier), resulting in increased transmission speed along the axon. Myelin is manufactured by Schwann's cells, and consists of 70–80% lipids (fat) and 20–30% protein. Dendrites branch from the cell body and receive messages from for example axons of other neurons across a synapse. A typical neuron has about 1,000–10,000 synapses (that is, it communicates with 1,000–10,000 other neurons, muscle cells, glands, amongst others).

14.2.1. Axon

A neuron usually only has one axon extending from the soma, but it may have branches distal collaterals. The axon

T. Ikezu and H.E. Gendelman (eds.), *Neuroimmune Pharmacology*.
© Springer 2008

emerges from the soma and in rare case from a dendrite, and can extend to distant targets up to a meter or more away from the cell body (e.g. motor neurons and corticospinal projection neurons). Microfilaments and microtubules act together to guide and support the growth and differentiation of axons and dendrites. The exploratory activity of growth cones, as they respond to external guidance cues during the formation of axons and dendrites, are driven by actin filaments; whereas, microtubules stabilize the structure of the newly established axon or dendrite (Mitchison and Kirschner, 1988; Smith, 1988b; Gordon-Weeks, 1991; Avila et al., 1994; Tanaka and Sabry, 1995; Heidemann, 1996; Smith et al., 2000).

14.2.2. Dendrites and Dendritic Arbors

Dendrites are extensions from the soma specialized for receiving and processing synaptic inputs. They make connections relatively close to the soma as compared with axons. Dendrites are rarely longer than 1–2 mm, even in the largest neurons. Larger neurons typically have both a larger soma and more extensive dendritic arbors (Figure 14.1). At one extreme, a dendrite connects a single remote target to the rest of the neuron. This is referred to as sective arborization. At the other extreme, dendritic branches can occupy most of the domain of arborization in a space-filling arborization. An example of this type of arborization is the cerebellar Purkinje cell arbor that forms synapses with at least half of the parallel fiber axons that pass through it. Most dendritic arborizations lie between the selective and space-filling varieties and are referred to as sampling arborizations.

14.2.2.1. Intracellular Components of Dendrites

The contents of large proximal dendrites are similar to those of the soma, including the Golgi apparatus and the rough endoplasmic reticulum (RER). These characteristically organelles diminish in number however, with increasing distance from the soma and decreasing dendrite diameter. The *Smooth endoplasmic reticulum* (SER), which is thought to be involved in the regulation of calcium, unlike the RER is found throughout the dendrites.

Organelles of the early endosomal pathway involved in membrane protein sorting and recycling (*sorting and recycling endosomes*) are common throughout the dendrites. Pits and vesicles representing the initial step in endocytosis are frequently seen in the dendrites. These pits and vesicles have a cytoplasmic coat composed of clathrin (Brodsky, 1988), which gives it a distinctive periodic structure that is also observed on the tips of early endosomes. Clathrin-coated vesicles uncoat less than a minute after their formation from a coated pit (Fine and Ockleford, 1984). For this reason clathrin coated vesicles are only found near their locus of generation within the dendrite.

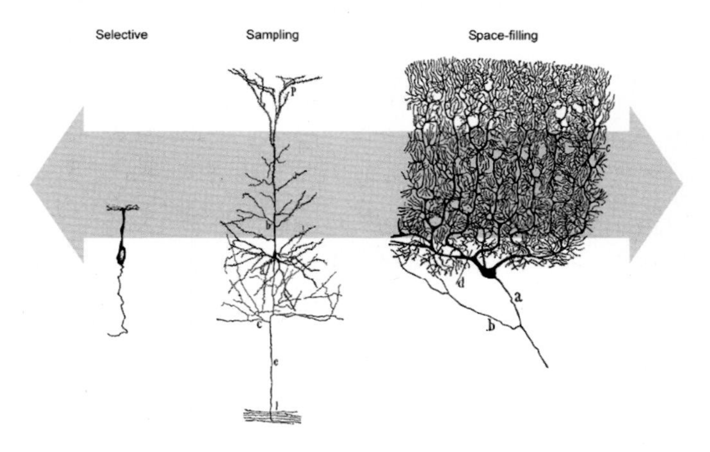

FIGURE 14.1. The densities of dendritic arbors lie on a continuum of values. Differences in arbor densities reflect differences in connectivity. [Drawings of neurons from (Ramón y Cajal, 1995)].

14.2.2.2. Dendritic Spine

The most common synaptic specializations of dendrites are simple spines. Dendritic spines are protrusions from the dendrite of usually no more than 2 µm, often ending in a bulbous head attached to the dendrite by a narrow stalk or neck. The cytoskeleton of dendrites is composed of microtubules, neurofilaments, and actin microfilaments. In an adult brain, the highest concentrations of actin are associated with the dendritic spines which form the postsynaptic component of most excitatory synapses (Matus et al., 1982; Fifkova, 1985; Kaech et al., 1997). The dendritic spine's actin retains a capacity for dynamic activity which can result in rapid changes in the shape of the dendritic spine(Fischer et al., 1998; Dunaevsky et al., 1999; Halpain, 2000; Lendvai et al., 2000; Matus, 2000). By contrast, the highest concentrations of the microtubule components, including tubulin and the microtubule-associated proteins, occur in the shafts of dendrites (Matus et al., 1981; Bernhardt and Matus, 1982; Burgoyne and Cumming, 1984; De Camilli et al., 1984; Huber and Matus, 1984).

14.2.3. Synapse

Neurons communicate with each other at regions of contiguity called synapses. Foster and Sherrington first coined this term in 1897 (Foster and Sherrington, 1897). The synapse is where the axon terminal of one cell links to a dendrite or soma of another. Neurons may have more than 1000 dendrites that connect to even more cells. Synapses function as excitatory or inhibitory in how they affect neuronal activity. When an action potential engages the axonal termini it opens voltage-gated calcium channels. Calcium then enters which causes neurotransmitters within vesicles to be released into the synaptic cleft that then activates postsynaptic neuronal receptors. In mammals, the majority of synapses are chemical rather than electrical. Substances called neurotransmitters and neuromodulators - either a gas or, more commonly, a liquid, rather than an uninterrupted electrical signal - convey information from one neuron to another. The presynaptic part contains vesicles with neurotransmitters. There are at least two common ways to categorize the mammalian synapses: (1) the electrical change in the postsynaptic cell (excitatory or inhibitory), and (2) the neurotransmitter used (for example, cholinergic – acetylcholine and catecholaminergic - dopamine). We the latter classification was used. Synapses are typically surrounded by glial cells, which might perform auxiliary roles such as the secretion of neuromodulatory factors, regulation of extracellular ion concentration, and uptake of neurotransmitters (Haydon, 2001). The specific presynaptic plasma membrane region where synaptic vesicles dock and fuse is called "active zone."

14.2.3.1. Synaptic Adhesion

The synapse is an intercellular junction whose structural organization shares similarities with other types of junctions such as the immunological synapse or the intercellular junctions of epithelial cells (Dustin and Colman, 2002; Jamora and Fuchs, 2002). Functionally, adhesive interactions between presynaptic and postsynaptic components, more specifically the active zone of presynaptic membrane (see 2.3.2) and postsynaptic density (see 2.3.7), will ensure proximity of release and receptor engagement. Several molecules have been identified in relation to synaptic adhesion, including cadherins (Shapiro and Colman, 1999; Yagi and Takeichi, 2000), protocadherins (Frank and Kemler, 2002), neural cell-adhesion molecule (NCAM), L1, fasciclin II (Davis et al., 1997), nectin (Mizoguchi et al., 2002), neurexins and neuroligin (Missler and Sudhof, 1998), integrins (Chavis and Westbrook, 2001), and syndecans (Hsueh and Sheng, 1999).

14.2.3.2. Presynaptic Terminal

A typical presynaptic bouton is a specialized portion of the axon. It is characterized by an active zone, a region where the presynaptic plasma membrane comes into close contact with the postsynaptic plasma membrane and an associated cluster of vesicles (De Camilli et al., 2001). A few synaptic vesicles are adjacent to the active zone and are referred to as docked vesicles. Vesicle exocytosis occurs at the active zone; subsequent endocytic retrieval of vesicular components may occur both at the active zone and in the peri-active zone area (Roos and Kelly, 1999). Vesicle maturation involves acidification of the lumen, loading with the neurotransmitter, association with peripheral membrane proteins needed for exocytosis, and recapture into a vesicle cluster. The vesicle cluster is embedded in an actin-rich area and is generally located next to mitochondria, which provides the energy required for the vesicle cycle and neurotransmitter dynamics and to the endoplasmic reticulum whose function includes the regulation of local cytosolic calcium.

14.2.3.3. Active Zones

The active zone is the specialized region of the cortical cytoplasm of the presynaptic nerve terminal that directly faces the synaptic cleft. It has a dense appearance under the electron microscope reflecting the presence of a protein scaffold—the presynaptic grid—within which docked vesicles are nested (Pfenninger et al., 1969). Functionally, the vesicles that are docked, or a subset of them, comprise a readily releasable pool of vesicles (Murthy et al., 2001) (Schikorski and Stevens, 2001). Scaffolding proteins play a role both in recruiting vesicles to the plasma membrane as well as in the regulation of release. This arrangement ensures the exocytotic release of neurotransmitter in close proximity to their postsynaptic receptors and in setting the release process to the appropriate dynamic range.

14.2.3.4. Exocytosis

It is now well established that *in vivo* efficient membrane fusion requires the interaction of small cytoplasmically exposed membrane proteins called soluble N-ethylmaleimide sensitive factor (NSF) attachment receptors (SNAREs) (Sollner et al., 1993). For synaptic vesicle exocytosis, the relevant SNAREs are synaptobrevin/ vesicle-associated membrane protein (VAMP) 1 and 2, syntaxin 1, and synaptosome-associated protein of 25,000 daltons (SNAP-25). Synaptobrevins/ VAMPs are localized primarily on synaptic vesicles, while syntaxin and SNAP-25 are localized primarily on the plasma membrane. Fusion is driven by the progressive zippering of vesicle and plasma membrane SNAREs forming a four-helix bundle (Sutton et al., 1998). Although many other proteins appear to play critical roles in synaptic vesicle exocytosis, it seems likely that SNAREs are the minimal machinery required for fusion (Weber et al., 1998). Once assembled, SNARE complexes are disassembled by NSF, which functions in conjunction with SNAP proteins.

14.2.3.5. Calcium-Triggered Exocytosis

Synaptic vesicles fuse with the plasma membrane constitutively under resting conditions, but the probability of vesicle fusion is increased dramatically by elevations in cytosolic Ca^{2+}. Synaptotagmin is an integral vesicle membrane protein with two C2 domains that bind Ca^{2+}, SNARE proteins, and phospholipids. A model suggests that Ca^{2+} binding induces the interaction of one or both C2 domains with plasma membrane proteins/lipids to catalyze SNARE-mediated membrane fusion (Fernandez-Chacon and Sudhof, 1999; Chapman, 2002). In addition to synaptotagmin, other proteins with tandem C2 domains are present either on the vesicles [for example, double C2 protein (Doc2) and rabphilin] or at the active zone [for example, Rab3-interacting molecule (RIM) and piccolo] (Rizo and Sudhof, 1998). These together with other calcium-binding proteins within the presynaptic termini, may contribute to Ca^{2+}-sensitivity of exocytosis.

14.2.3.6. Endocytosis: Kiss-and-Run

The concept of synaptic vesicle recycling was conclusively established in the early 1970s (Holtzman et al., 1971; Ceccarelli et al., 1973; Heuser and Reese, 1973), and the role of clathrin-mediated endocytosis in the recycling of synaptic vesicles is now well established. A model of endocytosis referred to as "kiss-and-run" has attracted considerable interest (Ceccarelli et al., 1973; Fesce et al., 1994). In this model, vesicles release neurotransmitter via a transient fusion pore. Newly reformed vesicles may then stay in place, be reloaded, and undergo a new round of exocytosis or may de-dock and allow other vesicles to take their place.

14.2.3.7. Postsynaptic Density

Opposite the active zone on the postsynaptic plasma membrane is an electron dense region referred to as the postsynaptic density (PSD). This membrane structure is thought to be important for both the clustering of postsynaptic receptors and ion channels and for the assembly of the postsynaptic signaling machinery (Hall and Sanes, 1993; Garner and Kindler, 1996). Ultra structural studies have shown that a fine meshwork of filamentous material is assembled at the cytoplasmic face of the active zone and the postsynaptic plasma membrane (Hall and Sanes, 1993; Garner and Kindler, 1996). Electron microscopy also reveals interactions between the PSD and organelles of the dendritic spine. Smooth endoplasmic reticulum cisterns are found in the spine [reviewed by (Fifkova and Morales, 1992)] and in some cases reach the margins of the PSD (Spacek and Harris, 1997). The PSD and spine apparatus appear to be interconnected by actin filaments. However, the actin filaments may be distinct from a PSD core, which contains regulatory components such as *a*CaMKII (Adam and Matus, 1996). The cytoarchitecture of the spine is thought to be in dynamic flux. Actin concentrations are high in the spine, and its polymerization is negatively controlled by intracellular Ca^{2+} fluxes (reviewed by (Fifkova and Morales, 1992)). The PSD-concentrated molecules include, but are not limited to, ionotropic N-methyl-D-acetate (NMDA) and DL-a-amino-3-hydroxy-5-methyl-4-isoxazole propionic acid (AMPA)–type of glutamate receptors, the metabotropic glutamate receptors (Nusser et al., 1994), alpha-actinin-2 (Wyszynski et al., 1997), fodrin (Carlin and Siekevitz, 1983), N-cadherin (Yamagata et al., 1995), calmodulin (Grab et al., 1979), a-calmodulin-dependent protein kinase II (aCamKII) (Kennedy et al., 1983), protein kinase C (PKC) β and γ (Wolf et al., 1986), extracellular signal-regulated kinase 2 (ERK2) (Suzuki et al., 1995), protein kinase A (PKA) (Carr et al., 1992), fyn tyrosine kinase (Suzuki and Okumura-Noji, 1995), TrkB receptor (Wu et al., 1996), and G proteins (Wu et al., 1992). Several PSD-associated molecules are tethered near to or within the PSD via specialized adaptor proteins containing PDZ domains.

14.2.3.8. PDZ Domains

The PDZ domain was originally identified as a common domain found in three structurally similar proteins: PSD-95/SAP90, DLG, and ZO1, which share the Gly-Leu-Gly-Phe sequence motif (Garner and Kindler, 1996; Craven and Bredt, 1998; Hata et al., 1998). Molecular analysis of vertebrate synaptic junctions has revealed that features specific to different types of synapses are achieved through a growing superfamily of proteins containing PDZ domains. The PDZ domain is approximately 90 amino acids in length, and has a high affinity to short peptide sequences located on the C-terminus of the interacting protein. These include, among others, members of the synapse-associated protein (SAP) 90/PSD-95 family [human homologue of the *Drosophila* discs large (hdlg)/SAP97, SAP102 and channel associated protein of synapse-110 (Chapsyn110)/PSD-93], and also CASK, glucocorticoid receptor-interacting protein (GRIP)/ Actin-binding protein (ABP), Synaptic scaffolding molecule (S-SCAM), mammalian homologue of unc-18 (Munc-18) interacting protein 1 (Mint1), and protein interacting with C kinase (PICK1) (Garner and Kindler, 1996; Craven and Bredt, 1998; Hata et al., 1998; Kim and Huganir, 1999). Known interacting molecules of PSD-95 include NMDA receptors, neuroligin, kainate receptor, and guanylate kinase-associated protein (GKAP)/SAP-associated protein (SAPAP) (Kornau et al., 1995; Garcia et al., 1998; Missler et al., 1998; Fujita and Kurachi, 2000). GKAP also interacts with another PDZ-domain containing molecule ProSAP/Shank, which interacts with homer, metabolic glutamate receptors, and cortactin (Boeckers et al., 1999a; Boeckers et al., 1999b; Naisbitt et al., 1999; Tu et al., 1999). Cortactin is an actin-binding molecule, and supports the protein scaffold complex as an architectural anchor on the cytoskeleton. Other known complexes include AMPA receptor subunits GluR2 and GluR3 interaction with PDZ domain of GRIP1 (Dong et al., 1997), GRIP2/ABP (AMPA-receptor binding protein) (Srivastava et al., 1998), and PICK1 (Xia et al., 1999). PDZ-containing molecules also interact with signaling molecules, such as GTPase activating proteins (GAPs) and GDP-GTP exchange factors (GEFs) which regulates the small GTP-binding proteins ras, rac, and rho (Furuyashiki et al., 1999; Alam et al., 1997; Zhang et al., 1999). These signals presumably mediate intracellular dynamics of the PDZ-containing molecular complex, and leads to cytoskeletal movement (such as actin dynamics), complex endocytosis, recycling, and exocytosis.

14.2.4. Axonal Transport

Axon is a polarized extension from cell body, and its plus-end nerve terminal contains presynaptic terminal. The growing axons often contain growth cones at the nerve terminal, mobility of which is critically important for the axonal plasticity. The axon is packed with cargo moving along a unipolar microtubule array either toward the nerve terminal (anterograde transport) or toward the cell body (retrograde transport). This bidirectional transport process, known as axonal transport, is not fundamentally different from the pathways of macromolecular and membrane traffic that occur in all eukaryotic cells. Anterograde motor machinery is fundamental for the transport of presynaptic molecules in the membrane cargo and mitochondria to the nerve terminal. Retrograde machinery is critically important for neurotrophic signaling, which is mediated by retrograde transport of signaling endosomes, containing neurotrophin-receptor complex, into cell soma. Known neurotrophin/receptor complexes include nerve growth factor (NGF)-TrkA, brain-derived neurotrophic factor (BDNF)/neurotrophin (NTF)-4/NTF-5-TrkB, and NTF-3-TrkC. Such retrograde signaling leads to activation of molecules involved in endocytosis, axonal growth, and cell survival, such as phosphatidylisositol-3-kinase (PI3K), ERK1/2/5, Akt, and cAMP-response element binding protein (CREB) (Heerssen and Segal, 2002).

Membranous organelles are the principal cargo of fast axonal transport. Golgi-derived transport vesicles move anterogradely at maximal rates of ~2–5 μm/s, whereas endocytic vesicles, lysosomes, and autophagosomes move retrogradely at maximal rates of ~1–3 μm/s. One exception to the high duty ratio of membranous organelles is mitochondria (Hollenbeck, 1996). These organelles move less rapidly than most Golgi-derived and endocytic vesicles, forming a kinetically distinct component of axonal transport. Membranous and non-membranous cargoes are all transported along axons by the same underlying mechanism but they move at different rates due to differences in their duty ratio.

Anterograde transport of membrane cargo molecules is mediated by kinesins (such as KIF1), whereas retrograde transport is mediated by dynein. Dynein attached to membrane cargo is inactivated during the anterograde transport. Once membrane cargos travel to the nerve terminal at this "turnaround" zone, cytoplasmic dynein and kinesin become activated and repressed respectively. The exact switching mechanism is currently unknown. A zone for switching the direction of organelle movement also may be created when axonal transport is blocked at a site of nerve injury. After a delay, many vesicles accumulating on the anterograde side of the block reverse their direction and travel back to the cell body (Smith, 1988a), where they may instruct the nucleus to respond to the injury. This reversal may be mediated, at least in part, by the recruitment of additional cytoplasmic dynein onto these vesicles (Li et al., 2000).

14.3. Neuronal Classification

There are different types of neurons. Although all neurons carry electro-chemical nerve signals, they can differ in their structure (the number of processes, or axons, emanating from the cell body) and in their location in the body.

14.3.1. Bipolar Neuron

Bipolar neurons have two processes extending from the cell body (Figure 14.2). Most sensory (or afferent) neurons are this type, carrying messages from the body's sense receptors (eyes, ears, etc.) to the CNS. Sensory neurons account for 0.9% of all the neurons. Examples of sensory neurons are retinal cells, olfactory epithelium cells, and the cochlear and vestibular ganglia.

14.3.2. Unipolar Neuron (Formally Called Pseudounipolar Neurons)

Actually, unipolar neurons have two axons rather than an axon and dendrite (Figure 14.3). Mostinterneurons, which form all the neural wiring within the CNS, are of this type. Examples include spinal ganglia, most cranial nerve sensory ganglia, and the trigeminal mesencephalic nucleus. Dorsal root ganglia cells extend one axon centrally toward the spinal cord, and the other axon toward the skin or muscle.

14.3.3. Multipolar Neurons

Multipolar neurons have many processes that extend from the cell body (Figure 14.4). However, each neuron has only one axon. Most motor (or efferent) neurons are this type and account for 9% of all neurons. Multipolar neurons have a diversity of shapes.

Some shapes are so characteristic that those cells are specially named. Examples are:

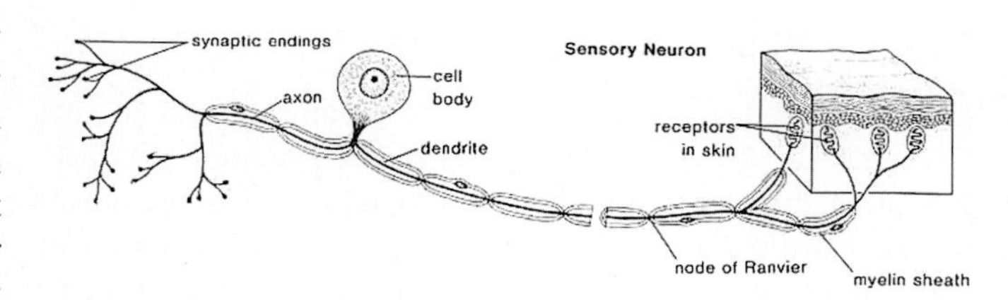

FIGURE 14.2. A sensory neuron leading a signal from a receptor of the skin through the dendrite to the cell body and subsequently further to the axon (figure used with permission from Biology Mad http://www.biologymad.com/NervousSystem/nervoussystemintro.htm).

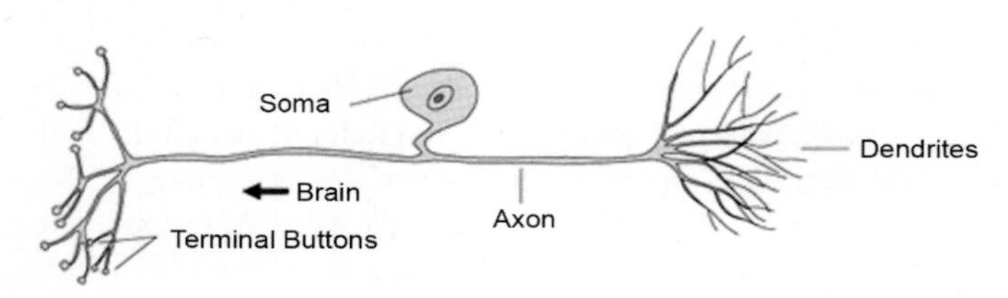

FIGURE 14.3. Drawing of a unipolar neuron (modified with permission from http://homepage.psy.utexas.edu/homepage/class/Psy332/Salinas/Cells/sensoryneuron.gif).

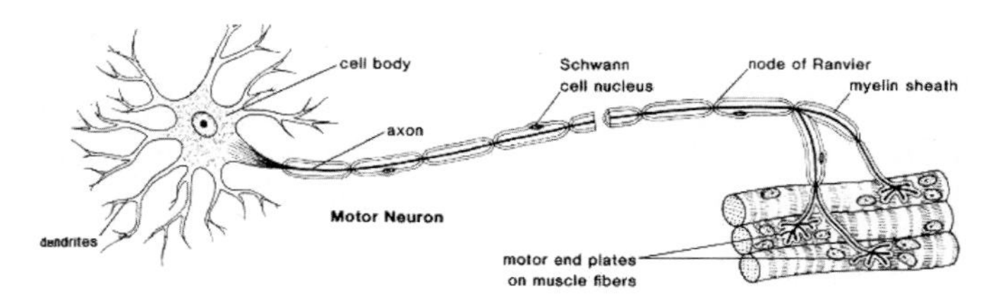

FIGURE 14.4. Drawing of a motor neuron (figure used with permission from Biology Mad http://www.biologymad.com/NervousSystem/nervoussystemintro.htm).

Purkinje cell—A single layer of large cell bodies in the cerebellar cortex, between the molecular (left) and granular (right) layers.

Pyramidal cell—named on the basis of the cell body's shape. It is one of the two major neuronal types found in the cerebral cortex. It has a large apical dendrite, which extends vertically from the top of the pyramid, and basal dendrites, which come off horizontally at the base of the pyramid. The axon also typically extends from the base.

Granule cell of dentate gyrus in the hippocampus (a particular region of the cerebral cortex)—dendrites extend from one end of the cell while the axon comes off from the other.

Cortical interneurons—Interneurons in the cortex constitute a rather diverse group of neurons responsible of establishing mostly local circuits in the cortex.

Motor neuron (motoneuron)—Found in the lamina IX of the spinal cord and certain cranial nerve motor nuclei.

14.4. Diversity in Neuronal Transmission

14.4.1. Glutamatergic Neurons

Glutamate (Glu) is the most abundant amino acid in the CNS. About 30% of the total Glu acts as the major excitatory neurotransmitter in the brain. Glu is synthesized in the nerve terminals from two sources: from glucose via the Krebs cycle and from glutamine by the enzyme glutaminase. The production of the neurotransmitter glu is regulated via the enzyme glutaminase. Glu is stored in vesicles and released by a Ca^{2+} dependent mechanism.

Glutamatergic neurons are widely distributed throughout the entire brain. Prominent glutamatergic pathways are the cortico-cortical projections, the connections between thalamus and cortex, and the projections from cortex to striatum (extrapyramidal pathway) and to brainstem/spinal chord (pyramidal pathway). The hippocampus and the cerebellum also contain many glutamatergic neurons. Glutamatergic neurons and NMDA receptors in the hippocampus are important in the creation of long-term potentiation (LTP), a crucial component in the formation of memory. Cortical neurons use Glu as the major excitatory neurotransmitter. Excess stimulation of glutamatergic receptors, as seen in seizures or stroke, can lead to unregulated Ca^{2+} influx and neuronal damage. Decreased glutamatergic function is thought to be involved in the creation of psychotic symptoms. Phencyclidine and ketamine can induce psychotic symptoms and d-cycloserine or glycine can decrease psychotic symptoms in schizophrenia.

14.4.1.1. Glutamate Receptors

Glu acts at three different types of ionotropic receptors and at a family of G-protein coupled (metabotropic) receptors. Binding of glu to the ionotropic receptors opens an ion channel allowing the influx of Na+ and Ca^{2+} into the cell.

14.4.1.1.1. Ionotropic receptors

- NMDA receptors bind glu and N-methyl-D-aspartate. The receptor is comprised of two different subunits: NR1 (seven variants) and NR2 (four variants). The NMDA receptor is highly regulated at several sites. For example, the receptor is virtually ineffective unless a ligand (such as glycine or D-cycloserine) binds to the glycine site of the receptor which is blocked by the binding of ligands [such as MK-801, ketamine, and Phencyclidine (PCP)] to the PCP site.
- AMPA receptors bind glutamate, AMPA, and quisqualic acid.
- Kainate receptors bind glutamate and kainic acid.

14.4.1.2. The Metabotropic Glutamate Receptor Family

The metabotropic receptor family includes at least seven different types of G-protein coupled receptors (mGluR1–7). They are linked to different second messenger systems and lead to the increase of intracellular Ca^{2+} or the decrease of cAMP. The increase of intracellular Ca^{2+} leads to the phosphorylation of target proteins in the cell.

14.4.2. GABAergic Neurons

Gamma-aminobutyric Acid (GABA) is an amino acid with high concentrations in the brain and the spinal chord. It acts as the major inhibitory neurotransmitter in the CNS. Cortical and thalamic GABAergic neurons are crucial for the inhibition of excitatory neurons. GABA is synthesized via decarboxylation of glutamate by the enzyme glutamate decarboxylase (GAD). Two forms of GAD, GAD65 and GAD67, are found in the brain.

The $GABA_B$ receptor is found postsynaptically (causing decreased excitability) as well as presynaptically (leading to decreased neurotransmitter release). GABA is removed from the synapse by a sodium dependent GABA uptake transporter. GABAergic neurons can be divided into two groups: (1) short-ranging neurons (interneurons, local circuit neurons) in the cortex, thalamus, striatum, hippocampus, cerebellum, and spinal chord, and (2) Medium/Long ranging neurons (projection neurons) in: (1) Basal ganglia (caudate/putamen to globus pallidus to thalamus/substantia nigra), (2) Septum to hippocampus, and (3) Substantia nigra to thalamus, superior colliculus.

Benzodiazepines or barbiturates are helpful in the treatment and prevention of seizures, and modulation of $GABA_A$ receptors is beneficial in the treatment of anxiety disorders, insomnia and agitation—most likely due to a general inhibition of neuronal activity.

14.4.2.1. GABA Receptors

GABA acts at three types of receptors: $GABA_A$, $GABA_B$, and $GABA_C$ receptors. $GABA_A$ and $GABA_C$ receptors are ionotropic, whereas the $GABA_B$ receptor is a G-protein coupled metabotropic receptor. The $GABA_A$ receptor is a fast responding anion channel that is sensitive to bicuculline and picrotoxin blockage. The receptor-channel complex is comprised of five subunits. Activation leads to the opening of the channel, allowing Cl^- to enter the cell, resulting in decreased excitability. Sixteen $GABA_A$ genes encode 5 families of native $GAGA_A$ subtypes. Various compounds can bind to several different sites of the receptor. For example, benzodiazepines with high activity at the a1 are associated with sedation whereas those with higher affinity for $GABA_A$ receptors containing a2 and/or a3 subunits have good anti-anxiety activity.

The $GABA_C$ receptor is insensitive to allosteric modulators, such as benzodiazepines and barbiturates. Native responses of the $GABA_C$ receptor type occur in retinal bipolar or horizontal cells across vertebrate species. Although the term "$GABA_C$ receptors" is still being used frequently they have been reassigned as part of $GABA_A$ receptor family.

The $GABA_B$ receptor is a G-protein coupled receptor with similarity to the metabotropic glutamate receptor that mediates the slow response to GABA. The GABA8 receptor is linked to G_i protein (decreasing cyclic AMP and opening of K+ channels) and G_o protein (closing Ca^{2+} channels). The netto effect is prolonged inhibition of the cell. A well-known agonist is baclofen.

14.4.3. Cholinergic Neurons

Acetylcholine (ACh) is known to be a neurotransmitter since the mid 1920s. ACh is synthesized by the enzyme choline acetyltransferase (CHAT) from the precursors acetylCoA and choline. High-affinity and low-affinity transporters pump choline, the rate-limiting factor in the synthesis of ACh, into the cell. ACh is removed from the synapse through hydrolysis into acetylCoA and choline by the enzyme acetylcholinesterase (AChE). Removing ACh from the synapse can be blocked irreversibly by organophosphorous compounds and in a reversible fashion by drugs such as physostigmine and tacrine.

In the peripheral nervous system, ACh is found as the neurotransmitter in the autonomic ganglia, the parasympathetic postganglionic synapse, and the neuromuscular endplate. Cholinergic neurons in the central nervous system are either wide ranging projection neurons or short ranging interneurons in: (1) the basal forebrain (septum, diagonal band, nucleus basalis of Meynert) projects to the entire cortex, the hippocampus, and the amygdala, (2) the brain stem projects predominantly to the thalamus, and (3) the striatum modulates of the activity of GABAergic striatal neurons. ACh modulates attention, novelty seeking, and memory via the basal forebrain projections to the cortex and limbic structures. Alzheimer's disease and anticholinergic delirium are examples of deficit states. Blocking the metabolism of ACh by AChE with drugs such as tacrine and aricept, strengthens cognitive functioning in AD patients. Brainstem cholinergic neurons are essential for the regulation of sleep-wake cycles through projections to the thalamus. Cholinergic interneurons modulate striatal neurons by opposing the effects of dopamine. Increased cholinergic tone in Parkinson's disease and decreased cholinergic tone in patients treated with neuroleptics are examples of imbalances in these two systems in the striatum.

14.4.3.1. Cholinergic Receptors

ACh acts at two different types of cholinergic receptors: Muscarinic and Nicotinic receptors. Muscarinic receptors (1) bind ACh as well as other agonists (muscarine, pilocarpine, bethanechol) and antagonists (atropine, scopolamine). There are at least 5 different types of muscarinic receptors (M1–M5) and all have slow response times. They are coupled to G-proteins and a variety of second messenger systems. When activated, the final effect can be an opening or closing of channels for K^+, Ca^{2+}, or Cl^-. Presynaptic cholinergic receptors are of the muscarinic or nicotinic type and can modulate the release of several neurotransmitters.

Nicotinic receptors are less abundant then the muscarinic type in the CNS. They bind ACh as well as agonists such as nicotine and antagonists such as d-tubocurarine. The fast acting of the ionotropic nicotinic receptor allows an influx of mainly Na^+, followed by K^+, and Ca^{2+}, into the cell.

14.4.4. Serotonergic Neurons

Serotonin or 5-hydroxytryptamine (5-HT), a monoamine, is widely distributed in many cells of the body and about 1–2% of the entire serotonin body content is found in the CNS. Serotonin is synthesized by the enzyme amino acid decarboxylase from 5-hydroxytryptophan (which is derived from tryptophan via tryptophan hydroxylase). The rate-limiting step is the production of 5-hydroxytryptophan by tryptophan hydroxylase. Serotonin is removed from the synapse by a high-affinity serotonin uptake site that is capable of transporting serotonin in either direction, depending on the concentration.

Serotonergic neurons are restricted to midline structures of the brainstem. Most serotonergic cells overlap with the distribution of the raphe nuclei in the brainstem (but not all raphe neurons are serotonergic). There are three major groups: (1) a rostral group (B6-8 neurons) projects to the thalamus, hypothalamus, amygdala, striatum, and cortex, (2) The remaining two groups (B1-5 neurons) project to other brainstem neurons, the cerebellum and the spinal chord. Serotonin is linked to many brain functions. For example, modulation of serotonergic receptors is beneficial (among others) in the treatment of anxiety, depression, obsessive-compulsive disorder, and schizophrenia. Blockade of 5-HT_3 receptors in the area postrema decreases nausea and emesis, and hallucinogens like LSD modulate serotonergic neurons via serotonergic autoreceptors.

14.4.4.1. Serotonin Receptors

Serotonin acts at two different types of receptors: G-protein coupled receptors and an ion-gated channel. All serotonin receptors, except the 5-HT_3 receptor, are G-protein coupled and can be grouped in four groups. The first group: The 5-HT1 receptors (5-HT_{1A}, 5-HT_{1B}, 5-HT_{1C}, 5-HT_{1D}, 5-HT_{1E}, 5-HT_{1F}) are coupled to G proteins (G_i/G_o), and lead to a decrease of cyclic AMP. The 5-HT_{1A} receptor is also directly coupled to a K^+ channel leading to increased opening of the channel. The 5-HT1 receptors are the predominant serotonergic autoreceptor. The second group: 5-HT_2 receptors (5-HT_{2A}, 5-HT_{2B}, and 5-HT_{2C}) are coupled to phospholipase C, which produce inositol triphosphate (IP_3) and diacylglycerol. IP_3 production leads to a variety of intracellular effects, including IP_3-dependent Ca^{2+} flux. The third group: Three receptors (5-HT_4, 5-HT_6, and 5-HT_7) are coupled to Gs and activate adenylate cyclase. The function of the 5-HT_{5A} and 5-HT_{5B} receptors is poorly understood. The fourth group: The 5-HT_3 receptor is the only ligand-gated Na^+/K^+ ion channel, resulting in a direct plasma membrane depolarization. It is found in the cortex, hippocampus, and area postrema and is typically localized presynaptically and regulates neurotransmitter release. A well-known antagonist is ondansetron.

14.4.5. Noradrenergic Neurons

Norepinephrine (NE), a catecholamine, was first identified as a neurotransmitter in 1946. In the peripheral nervous system, it is found as a neurotransmitter in the sympathetic postganglionic synapse. NE is synthesized by the enzyme dopamine-β-hydroxylase (DbH) from the precursor dopamine (which is derived from tyrosine via DOPA). The rate-limiting step is the production of DOPA by tyrosine hydroxylase, which can be activated through phosphorylation. NE is removed from the synapse by two mechanisms: (1) catechol-O-methyltransferase (COMT), which degrades intrasynaptic NE, and (2) the norepinephrine transporter (NET), the primary way of removing NE from the synapse. Once internalized, NE can be degraded by the intracellular enzyme monoamine oxidase (MAO).

Nonadrenergic neurons in the central nervous system are restricted to the brainstem, especially at the locus coeruleus (LC). They provide the extensive noradrenergic innervation found in the cortex, hippocampus, thalamus, cerebellum, and spinal chord. The remaining neurons are distributed in the tegmental region. They innervate predominantly the hypothalamus, basal forebrain and spinal chord. LC receives afferents from the sensory systems that monitor the internal and external environments. The widespread LC efferents lead to an inhibition of spontaneous discharge in the target neurons. Therefore, the LC is thought to be crucial for fine-tuning the attentional matrix of the cortex. Anxiety disorders may be due to perturbations of this system.

14.4.5.1. Adrenergic Receptors

Norepinephrine acts at two different types of noradrenergic receptors in the CNS: a and b adrenergic receptors. a-adrenoceptors can be subdivided into a1 and a2 receptors. The a1-receptors are coupled to phospholipase and located postsynaptically, while the a2-receptors are coupled to G_i and located primarily presynaptically. Adrenergic b-receptors in the CNS are predominantly of the b1 subtype. The b1-receptors are coupled to Gs and lead to an increase of cAMP. cAMP triggers a variety of events mediated by protein kinases, including phosphorylation of the b-receptor itself and regulation of gene expression via phosphorylation of transcription factors.

14.5. Neuronal Degeneration and Regeneration

14.5.1. Wallerian Degeneration

In either the central or peripheral nervous system (CNS or PNS, respectively) of mammals, if an axon of a neuron is focally destroyed (e.g., cut, crushed, frozen, rendered anoxic) the part of the axon disconnected from the cell body, which would be considered 'distal' relative to the lesion, invariably degenerates. This event is called anterograde or Wallerian degeneration. The part of the axon that remains connected to the cell body may degenerate a short distance (e.g. a millimeter or less), but usually it survives at least in the short term. However, there are often reorganizational changes in the cell body of the damaged neuron, denoted by the term chromatolysis. Collectively, all these changes proximal to the lesion site are referred to as retrograde degeneration; some of them may represent reorganization rather than degeneration changes.

14.5.2. Regeneration

After a focal transection of axons in the mammalian CNS, the proximal cut end of the axon, which is still connected to the cell body, may attempt to regrow by sending out sprouts. But axonal regeneration is limited; in practice functional recovery is usually been meager. However, a similar lesion in the mammalian PNS often results in many of those sprouts from the proximal cut end of the axon regrowing to and reinnervating peripheral targets. The regenerated axons may become myelinated too. Characteristically, the population of regenerated, myelinated axons is smaller in diameter and more numerous than in an undamaged nerve. The difference in regeneration between CNS and PNS is mainly due to the function of Schwann cells in the PNS. Schwann cells provide a "cellular terrain" that is permissive to axonal regrowth. The terrain provided by Schwann cells consists of favorable molecules (e.g., laminin) for growth cone attachment and motility, their intrinsic neurotrophic properties and their myelinating function for regenerated fibers.

14.6. Glia

Glia (microglia, oligodendroglia and astroglia) account for 90% of the total cell numbers in the adult human brain. They have for all but the past half decade been thought to play only scaffolding (surround neurons and hold them in place), nutritional (nutrients and oxygen), and regulatory function (insulate one neuron from another, destroy pathogens, remove debris, and regulate ions and neurotransmitters). Glial cells function to maintain homeostasis, form myelin, and affect neural signal transmission. They are commonly referred to as the "glue" of the nervous system.

A prominent function of glia is to regulate the brain microenvironment and including fluid surrounding neurons and their synapses. With regards to their function glia affect neuron migration both in early ontogeny and development. This is done, in part, through the secretion of growth factors that modify neuronal process growth and differentiation. They are also affect synaptic transmission by regulating clearance of neurotransmitters from the synaptic cleft and by their abilities to modulate presynaptic function. Although glia had been thought to lack chemical synapses or release neurotransmitters this is no longer true. They are no longer considered to be passive bystanders for neural transmission and function. This includes their abilities to prevent toxic accumulations of neurotransmitters such as glutamate as well as release glutamate in response to specific stimulation cues. The most notable differences between neurons and glia rests in their abilities to generate action potentials which neurons have and glia lack. They are also crucial in synaptic plasticity and synaptogenesis.

Astrocytes, the most prominent of glia, play central roles in neuron-to-neuron communication, in neurogenesis and for the developing nervous system. A key advances in neurobiology was the discovery that astroglial can generate neurons not only during development, but also throughout adult life. Few astrocytes maintain their neurogenic potential after development. Although astrocytes can react to brain injury and incite gliosis this rarely affects neurogenesis.

Summary

Neuronal structure has specifically differentiated cellular compartments (axons, dendrites, synapses) for developing interneuronal, neuron-glia, and neuro-immune networks in both the CNS and the PNS. Specific molecules, molecular complexes, motor machinery, and organelles are developed for different neuronal functions, including synaptic transmission, plasticity, degeneration, and regeneration. There is specific anatomical distribution of neurons with significant diversity in neurotransmitters and their projection to other brain subregions is developmentally regulated. Drugs modulating neurotransmitter receptors and ion channels have been clinically applied for the treatment of neurologic and psychiatric symptoms. An important discovery in neurobiology is that astrocytes can generate neurons beyond development to adult life.

Review Questions/Problems

1. Which states that describe the synaptic vesicle are correct?

a. In a neuron synaptic vesicles are also called neurotransmitter vesicles
b. Vesicles store neurotransmitters and are released only during potassium-regulated exocytosis.
c. Vesicles are not required for propagation of nerve impulses.
d. ALL of the above

2. All of the following are involved in axonal transport except

a. Is responsible for movement of mitochondria, lipids, synaptic vesicles, and proteins to and from the neuron.
b. Axons are up to 10,000 times smaller than the length of the cell body.
c. Is responsible for moving molecules destined for degradation to lysosomes
d. Movement toward the synapse is called retrograde transport
e. b and d

3. Unipolar, bipolar, and multipolar neurons are best described by

a. Unipolar neuron has a single neurite with different segments that function as superficial receptors or terminals
b. Multipolar neurons are most common in the nervous system of invertebrates
c. Neurons can never contain a single axon and dendrite at opposite poles of the cell body
d. Unipolar neurons make up a major part of the blood brain barrier
e. All of the above

4. Dysfunction of GABAergic interneurons affects which clinical diseases?

a. Parkinson's disease
b. Alzheimer's disease
c. amyotrophic lateral sclerosis
d. HIV-1 dementia
e. Schizophernia and bi-polar disorders
f. None of the above

5. Which of the following is true to describe Wallerian Degeneration?

a. If a nerve fiber is crushed, the part proximal to the injury (i.e. the part of the axon that is connected from the neuron's nucleus) will degenerate
b. Target tissues of the nerve (typically an innervated muscle) never atrophies
c. Wallerian degeneration is named after Augustus Volney Waller who first described it in 1850, in the Zebra fish
d. None of the above

References

Adam G, Matus A (1996) Role of actin in the organisation of brain postsynaptic densities. Brain Res Mol Brain Res 43:246–250.

Alam MR, Johnson RC, Darlington DN, Hand TA, Mains RE, Eipper BA (1997) Kalirin, a cytosolic protein with spectrin-like and GDP/GTP exchange factor-like domains that interacts with peptidylglycine alpha-amidating monooxygenase, an integral membrane peptide-processing enzyme. J Biol Chem 272:12667–12675.

Avila J, Dominguez J, Diaz-Nido J (1994) Regulation of microtubule dynamics by microtubule-associated protein expression and phosphorylation during neuronal development. Int J Dev Biol 38:13–25.

Bernhardt R, Matus A (1982) Initial phase of dendrite growth: Evidence for the involvement of high molecular weight microtubule-associated proteins (HMWP) before the appearance of tubulin. J Cell Biol 92:589–593.

Boeckers TM, Winter C, Smalla KH, Kreutz MR, Bockmann J, Seidenbecher C, Garner CC, Gundelfinger ED (1999a) Proline-rich synapse-associated proteins ProSAP1 and ProSAP2 interact with synaptic proteins of the SAPAP/GKAP family. Biochem Biophys Res Commun 264:247–252.

Boeckers TM, Kreutz MR, Winter C, Zuschratter W, Smalla KH, Sanmarti-Vila L, Wex H, Langnaese K, Bockmann J, Garner CC, Gundelfinger ED (1999b) Proline-rich synapse-associated protein-1/cortactin binding protein 1 (ProSAP1/CortBP1) is a PDZ-domain protein highly enriched in the postsynaptic density. J Neurosci 19:6506–6518.

Brodsky FM (1988) Living with clathrin: Its role in intracellular membrane traffic. Science 242:1396–1402.

Burgoyne RD, Cumming R (1984) Ontogeny of microtubule-associated protein 2 in rat cerebellum: Differential expression of the doublet polypeptides. Neuroscience 11:156–167.

Carlin RK, Siekevitz P (1983) Plasticity in the central nervous system: Do synapses divide? Proc Natl Acad Sci USA 80:3517–3521.

Carr DW, Stofko-Hahn RE, Fraser ID, Cone RD, Scott JD (1992) Localization of the cAMP-dependent protein kinase to the postsynaptic densities by A-kinase anchoring proteins. Characterization of AKAP 79. J Biol Chem 267:16816–16823.

Ceccarelli B, Hurlbut WP, Mauro A (1973) Turnover of transmitter and synaptic vesicles at the frog neuromuscular junction. J Cell Biol 57:499–524.

Chapman ER (2002) Synaptotagmin: A Ca(2+) sensor that triggers exocytosis? Nat Rev Mol Cell Biol 3:498–508.

Chavis P, Westbrook G (2001) Integrins mediate functional pre- and postsynaptic maturation at a hippocampal synapse. Nature 411:317–321.

Craven SE, Bredt DS (1998) PDZ proteins organize synaptic signaling pathways. Cell 93:495–498.

Davis GW, Schuster CM, Goodman CS (1997) Genetic analysis of the mechanisms controlling target selection: Target-derived Fasciclin II regulates the pattern of synapse formation. Neuron 19:561–573.

De Camilli P, Haucke V, Takei K, Mugnaini E (2001) The Structure of Synapses. Baltimore, MD: Johns Hopkins University Press.

De Camilli P, Miller PE, Navone F, Theurkauf WE, Vallee RB (1984) Distribution of microtubule-associated protein 2 in the nervous system of the rat studied by immunofluorescence. Neuroscience 11:817–846.

Dong H, O'Brien RJ, Fung ET, Lanahan AA, Worley PF, Huganir RL (1997) GRIP: A synaptic PDZ domain-containing protein that interacts with AMPA receptors. Nature 386:279–284.

Dunaevsky A, Tashiro A, Majewska A, Mason C, Yuste R (1999) Developmental regulation of spine motility in the mammalian central nervous system. Proc Natl Acad Sci USA 96:13438–13443.

Dustin ML, Colman DR (2002) Neural and immunological synaptic relations. Science 298:785–789.

Fernandez-Chacon R, Sudhof TC (1999) Genetics of synaptic vesicle function: Toward the complete functional anatomy of an organelle. Annu Rev Physiol 61:753–776.

Fesce R, Grohovaz F, Valtorta F, Meldolesi J (1994) Neurotransmitter release: Fusion or 'kiss-and-run'? Trends Cell Biol 4:1–4.

Fifkova E (1985) Actin in the nervous system. Brain Res 356:187–215.

Fifkova E, Morales M (1992) Actin matrix of dendritic spines, synaptic plasticity, and long-term potentiation. Int Rev Cytol 139:267–307.

Fine RE, Ockleford CD (1984) Supramolecular cytology of coated vesicles. Int Rev Cytol 91:1–43.

Fischer M, Kaech S, Knutti D, Matus A (1998) Rapid actin-based plasticity in dendritic spines. Neuron 20:847–854.

Foster M, Sherrington C (1897) A Textbook of Physiology, (17th edn). New York: The MacMillan Co.

Frank M, Kemler R (2002) Protocadherins. Curr Opin Cell Biol 14:557–562.

Fujita A, Kurachi Y (2000) SAP family proteins. Biochem Biophys Res Commun 269:1–6.

Furuyashiki T, Fujisawa K, Fujita A, Madaule P, Uchino S, Mishina M, Bito H, Narumiya S (1999) Citron, a Rho-target, interacts with PSD-95/SAP-90 at glutamatergic synapses in the thalamus. J Neurosci 19:109–118.

Garcia EP, Mehta S, Blair LA, Wells DG, Shang J, Fukushima T, Fallon JR, Garner CC, Marshall J (1998) SAP90 binds and clusters kainate receptors causing incomplete desensitization. Neuron 21:727–739.

Garner CC, Kindler S (1996) Synaptic proteins and the assembly of synaptic junctions. Trends Cell Biol 6:429–433.

Gordon-Weeks PR (1991) Growth cones: The mechanism of neurite advance. Bioessays 13:235–239.

Grab DJ, Berzins K, Cohen RS, Siekevitz P (1979) Presence of calmodulin in postsynaptic densities isolated from canine cerebral cortex. J Biol Chem 254:8690–8696.

Hall ZW, Sanes JR (1993) Synaptic structure and development: The neuromuscular junction. Cell 72(Suppl):99–121.

Halpain S (2000) Actin and the agile spine: How and why do dendritic spines dance? Trends Neurosci 23:141–146.

Hata Y, Nakanishi H, Takai Y (1998) Synaptic PDZ domain-containing proteins. Neurosci Res 32:1–7.

Haydon PG (2001) GLIA: Listening and talking to the synapse. Nat Rev Neurosci 2:185–193.
Heerssen HM, Segal RA (2002) Location, location, location: A spatial view of neurotrophin signal transduction. Trends Neurosci 25:160–165.
Heidemann SR (1996) Cytoplasmic mechanisms of axonal and dendritic growth in neurons. Int Rev Cytol 165:235–296.
Heuser JE, Reese TS (1973) Evidence for recycling of synaptic vesicle membrane during transmitter release at the frog neuromuscular junction. J Cell Biol 57:315–344.
Hollenbeck PJ (1996) The pattern and mechanism of mitochondrial transport in axons. Front Biosci 1:d91–d102.
Holtzman E, Freeman AR, Kashner LA (1971) Stimulation-dependent alterations in peroxidase uptake at lobster neuromuscular junctions. Science 173:733–736.
Hsueh YP, Sheng M (1999) Regulated expression and subcellular localization of syndecan heparan sulfate proteoglycans and the syndecan-binding protein CASK/LIN-2 during rat brain development. J Neurosci 19:7415–7425.
Huber G, Matus A (1984) Immunocytochemical localization of microtubule-associated protein 1 in rat cerebellum using monoclonal antibodies. J Cell Biol 98:777–781.
Jamora C, Fuchs E (2002) Intercellular adhesion, signalling and the cytoskeleton. Nat Cell Biol 4:E101–E108.
Kaech S, Fischer M, Doll T, Matus A (1997) Isoform specificity in the relationship of actin to dendritic spines. J Neurosci 17:9565–9572.
Kennedy MB, Bennett MK, Erondu NE (1983) Biochemical and immunochemical evidence that the "major postsynaptic density protein" is a subunit of a calmodulin-dependent protein kinase. Proc Natl Acad Sci USA 80:7357–7361.
Kim JH, Huganir RL (1999) Organization and regulation of proteins at synapses. Curr Opin Cell Biol 11:248–254.
Kornau HC, Schenker LT, Kennedy MB, Seeburg PH (1995) Domain interaction between NMDA receptor subunits and the postsynaptic density protein PSD-95. Science 269:1737–1740.
Lendvai B, Stern EA, Chen B, Svoboda K (2000) Experience-dependent plasticity of dendritic spines in the developing rat barrel cortex in vivo. Nature 404:876–881.
Li JY, Pfister KK, Brady ST, Dahlstrom A (2000) Cytoplasmic dynein conversion at a crush injury in rat peripheral axons. J Neurosci Res 61:151–161.
Matus A (2000) Actin-based plasticity in dendritic spines. Science 290:754–758.
Matus A, Bernhardt R, Hugh-Jones T (1981) High molecular weight microtubule-associated proteins are preferentially associated with dendritic microtubules in brain. Proc Natl Acad Sci USA 78:3010–3014.
Matus A, Ackermann M, Pehling G, Byers HR, Fujiwara K (1982) High actin concentrations in brain dendritic spines and postsynaptic densities. Proc Natl Acad Sci USA 79:7590–7594.
Missler M, Sudhof TC (1998) Neurexins: Three genes and 1001 products. Trends Genet 14:20–26.
Missler M, Fernandez-Chacon R, Sudhof TC (1998) The making of neurexins. J Neurochem 71:1339–1347.
Mitchison T, Kirschner M (1988) Cytoskeletal dynamics and nerve growth. Neuron 1:761–772.
Mizoguchi A, Nakanishi H, Kimura K, Matsubara K, Ozaki-Kuroda K, Katata T, Honda T, Kiyohara Y, Heo K, Higashi M, Tsutsumi T, Sonoda S, Ide C, Takai Y (2002) Nectin: An adhesion molecule involved in formation of synapses. J Cell Biol 156:555–565.
Murthy VN, Schikorski T, Stevens CF, Zhu Y (2001) Inactivity produces increases in neurotransmitter release and synapse size. Neuron 32:673–682.
Naisbitt S, Kim E, Tu JC, Xiao B, Sala C, Valtschanoff J, Weinberg RJ, Worley PF, Sheng M (1999) Shank, a novel family of postsynaptic density proteins that binds to the NMDA receptor/PSD-95/GKAP complex and cortactin. Neuron 23:569–582.
Nusser Z, Mulvihill E, Streit P, Somogyi P (1994) Subsynaptic segregation of metabotropic and ionotropic glutamate receptors as revealed by immunogold localization. Neuroscience 61:421–427.
Pfenninger K, Sandri C, Akert K, Eugster CH (1969) Contribution to the problem of structural organization of the presynaptic area. Brain Res 12:10–18.
Ramón y Cajal S (1995) Histology of the Nervous System of Man and Vertebrates. (English translation by N. Swanson and L. W. Swanson) Originally published: Histologie du systeme nerveux de l'homme et des vertebres. (Trans. L. Azoulay), Paris, 1909–1911. New York: Oxford University Press.
Rizo J, Sudhof TC (1998) C2-domains, structure and function of a universal Ca2+-binding domain. J Biol Chem 273:15879–15882.
Roos J, Kelly RB (1999) The endocytic machinery in nerve terminals surrounds sites of exocytosis. Curr Biol 9:1411–1414.
Schikorski T, Stevens CF (2001) Morphological correlates of functionally defined synaptic vesicle populations. Nat Neurosci 4:391–395.
Shapiro L, Colman DR (1999) The diversity of cadherins and implications for a synaptic adhesive code in the CNS. Neuron 23:427–430.
Smith RS (1988a) Studies on the mechanism of the reversal of rapid organelle transport in myelinated axons of Xenopus laevis. Cell Motil Cytoskeleton 10:296–308.
Smith SJ (1988b) Neuronal cytomechanics: The actin-based motility of growth cones. Science 242:708–715.
Smith SL, Somers JM, Broderick N, Halliday K (2000) The role of the plain radiograph and renal tract ultrasound in the management of children with renal tract calculi. Clin Radiol 55:708–710.
Sollner T, Bennett MK, Whiteheart SW, Scheller RH, Rothman JE (1993) A protein assembly-disassembly pathway in vitro that may correspond to sequential steps of synaptic vesicle docking, activation, and fusion. Cell 75:409–418.
Spacek J, Harris KM (1997) Three-dimensional organization of smooth endoplasmic reticulum in hippocampal CA1 dendrites and dendritic spines of the immature and mature rat. J Neurosci 17:190–203.
Srivastava S, Osten P, Vilim FS, Khatri L, Inman G, States B, Daly C, DeSouza S, Abagyan R, Valtschanoff JG, Weinberg RJ, Ziff EB (1998) Novel anchorage of GluR2/3 to the postsynaptic density by the AMPA receptor-binding protein ABP. Neuron 21:581–591.
Sutton RB, Fasshauer D, Jahn R, Brunger AT (1998) Crystal structure of a SNARE complex involved in synaptic exocytosis at 2.4 A resolution. Nature 395:347–353.
Suzuki T, Okumura-Noji K (1995) NMDA receptor subunits epsilon 1 (NR2A) and epsilon 2 (NR2B) are substrates for Fyn in the postsynaptic density fraction isolated from the rat brain. Biochem Biophys Res Commun 216:582–588.
Suzuki T, Okumura-Noji K, Nishida E (1995) ERK2-type mitogen-activated protein kinase (MAPK) and its substrates in

postsynaptic density fractions from the rat brain. Neurosci Res 22:277–285.

Tanaka E, Sabry J (1995) Making the connection: Cytoskeletal rearrangements during growth cone guidance. Cell 83:171–176.

Tu JC, Xiao B, Naisbitt S, Yuan JP, Petralia RS, Brakeman P, Doan A, Aakalu VK, Lanahan AA, Sheng M, Worley PF (1999) Coupling of mGluR/Homer and PSD-95 complexes by the Shank family of postsynaptic density proteins. Neuron 23:583–592.

Weber T, Zemelman BV, McNew JA, Westermann B, Gmachl M, Parlati F, Sollner TH, Rothman JE (1998) SNAREpins: Minimal machinery for membrane fusion. Cell 92:759–772.

Wolf M, Burgess S, Misra UK, Sahyoun N (1986) Postsynaptic densities contain a subtype of protein kinase C. Biochem Biophys Res Commun 140:691–698.

Wu K, Nigam SK, LeDoux M, Huang YY, Aoki C, Siekevitz P (1992) Occurrence of the alpha subunits of G proteins in cerebral cortex synaptic membrane and postsynaptic density fractions: Modulation of ADP-ribosylation by Ca2+/calmodulin. Proc Natl Acad Sci USA 89:8686–8690.

Wu K, Xu JL, Suen PC, Levine E, Huang YY, Mount HT, Lin SY, Black IB (1996) Functional trkB neurotrophin receptors are intrinsic components of the adult brain postsynaptic density. Brain Res Mol Brain Res 43:286–290.

Wyszynski M, Lin J, Rao A, Nigh E, Beggs AH, Craig AM, Sheng M (1997) Competitive binding of alpha-actinin and calmodulin to the NMDA receptor. Nature 385:439–442.

Xia J, Zhang X, Staudinger J, Huganir RL (1999) Clustering of AMPA receptors by the synaptic PDZ domain-containing protein PICK1. Neuron 22:179–187.

Yagi T, Takeichi M (2000) Cadherin superfamily genes: Functions, genomic organization, and neurologic diversity. Genes Dev 14:1169–1180.

Yamagata M, Herman JP, Sanes JR (1995) Lamina-specific expression of adhesion molecules in developing chick optic tectum. J Neurosci 15:4556–4571.

Zhang W, Vazquez L, Apperson M, Kennedy MB (1999) Citron binds to PSD-95 at glutamatergic synapses on inhibitory neurons in the hippocampus. J Neurosci 19:96–108.

15
Cytokines and Chemokines

Yunlong Huang, Nathan Erdmann, Terry D. Hexum, and Jialin Zheng

Keywords Chemokine; Cytokine; Growth factor; HIV-1 associated dementia; Alzeimer's disease; Multiple sclerosis; Inflammation; Chemotaxis

15.1. Introduction

Chemokines, cytokines, and growth factors are signaling molecules that play a crucial role in the coordination of immune responses throughout the body and communication between the immune system and the CNS. Consequently, dysregulation of chemokines, cytokines, and growth factors may facilitate the development of several central nervous system (CNS) diseases. For example, HIV-1 associated dementia (HAD), Alzheimer's disease (AD), and multiple sclerosis (MS) share certain characteristics in their pathogenesis, which include activated macrophage/microglia, production of high levels of proinflammatory cytokines, and impairment of neuronal function. The relationship between disturbances of the immune response and the impairment of neuronal function is the subject of intense investigation. This chapter provides an overview of chemokines, cytokines, and growth factors and their roles in various pathological conditions. Understanding the involvement of chemokines, cytokines, and growth factors in the pathogenesis of CNS diseases may potentially provide the opportunity to identify particular therapeutic targets for the treatment of neurodegenerative disorders.

15.1.1. Classification of Cytokines, Chemokines, and Growth Factors

Cytokines are regulatory proteins secreted primarily by white blood cells but also by a variety of other cells in the body; their functions include numerous effects on cells of the immune system and inflammatory processes (Vilcek, 2003). Most cytokines are single polypeptide chains, although they may form multimers in biological fluids. As an exceptionally large and diverse group of factors, cytokines are very difficult to classify. While many factors share structural similarities or have functional homologues, most generalizations are riddled with exceptions. Similarly, the term "growth factors," used in this chapter, also refers to a broad range of structurally diverse molecular families and individual proteins.

15.1.2. Cytokine Families

Cytokines can be grouped into families based upon their structural homology or structural homology of their receptors. Below, is an effort to group cytokines and their families based on the structural homology of their receptors. This classification helps provide a foundation for thinking about these factors and their role in the brain.

15.1.2.1. Type I Cytokine Family

Despite minimal amino acid sequence homology, type I cytokines and receptors share a similar three-dimensional structure consisting of an extracellular region containing four α helices, a characteristic motif of type I receptors. Members of this family include interleukin-2 (IL-2), IL-3, IL-4, IL-5, IL-6, IL-7, IL-9, IL-12, granulocyte-colony stimulating factor (G-CSF), and granulocyte macrophage-colony stimulating factor (GM-CSF). Among them IL-2, IL-3, IL-4, IL-7 are T cell growth factors; IL-2, IL-6, IL-12 are pro-inflammatory cytokines; and G-CSF and GM-CSF are hematopoietic cytokines important for survival and differentiation of hematopoietic lineages. Interestingly, although belonging to the same family, IL-2 is a Th1 cytokine but IL-4 and IL-5 are Th2 cytokines.

15.1.2.2. Type II Cytokine Family

Type II cytokines include IL-10, IL-19, IL-20, IL-22, and interferons (IFN-α, -β, -ε, -κ, -ω, -δ, -τ and -γ). Functions of this group include induction of cellular antiviral states, modulation of inflammatory responses, inhibition or

T. Ikezu and H.E. Gendelman (eds.), *Neuroimmune Pharmacology*.
© Springer 2008

stimulation of cell growth, and production or inhibition of apoptosis, as well as affecting many immune mechanisms (Renauld, 2003; Pestka et al., 2004). Notably, IL-10 is a potent anti-inflammatory, immunomodulatory cytokine. The IFN system is an important contributor to innate immunity with IFN-γ serving as a potent pro-inflammatory cytokine. The type II cytokine family is also comprised of both Th1 and Th2 cytokines; IFN-γ is a Th1 cytokine while IL-10 is a Th2 cytokine.

15.1.2.3. TNF Family

TNF family is comprised of at least 19 type II transmembrane proteins that have partial homology in their extracellular domains. Members of this superfamily include TNF-α, TNF-β, Fas ligand (FasL), CD40 ligand (CD40L), and TNF-related apoptosis-inducing ligand (TRAIL). Members of the TNF family are known for their apoptosis-inducing ability, though they have additional biological effects. TNF-α is an inflammatory Th1 cytokine released during infection. CD40L is a strong activator of macrophage, while FasL and TRAIL modulate immune responses through the induction of apoptosis. TNF family cytokines also have been suggested to be involved in destructive effects toward tissue in many disease settings.

15.1.2.4. IL-1 Family

The IL-1 family is also called the immunoglobulin superfamily. The IL-1R family includes both transmembrane and soluble proteins with an immunoglobulin-like structure. There are four primary members of the IL-1 family: IL-1α, IL-1β, IL-18, and IL-1Ra. IL-1 ligands (IL-1α and IL-1β, collectively referred to as IL-1) are potent pro-inflammatory cytokines that induce genes associated with inflammation and autoimmune disease. IL-1Ra, on the other hand, is the specific receptor antagonist for IL-1α and IL-1β but not for IL-18.

15.1.2.5. TGF-β Family

Transforming Growth Factor Beta (TGF-β) is the prototypical member of this family, but at least 50 proteins are classified as TGF-β members including activins, inhibins, bone morphogenetic proteins (BMPs), and glial cell line-derived neurotrophic factor (GDNF). The receptors of TGF-β family have characteristic cysteine-rich extracellular domains, kinase domains, GS domains, and a serine/threonine-rich tail (type II receptors). TGF-β has numerous functions including regulation of neuronal survival, orchestration of repair processes, as well as anti-inflammatory properties.

15.1.3. Important Sub-families of Cytokines and Growth Factors

The designation of five cytokine families based upon receptor structural homology is rather vague. Cytokines within each structural family have further diversity in function; hence, smaller groups of cytokines can be designated by the formation of more function-related sub-families. Due to their proliferative and/or differentiative effects, a group of cytokines are collected from different families and referred to as neurotrophic factors, e.g., nerve growth factor (NGF), brain-derived neurotrophic factor (BDNF). Growth factors are known for their ability to enhance cell proliferation and growth. Some growth factors are included in the cytokine family, e.g., platelet-derived growth factor (PDGF). However, other growth factors are classified into different families, e.g., epidermal growth factor (EGF) and fibroblast growth factor (FGF).

15.1.3.1. Neurotrophic Factor Families

The neurotrophin family includes NGF, BDNF, neurotrophin-3 (NT-3), NT-4/5, and NT-6. All neurotrophins are capable of binding to the p75 receptor; each neurotrophin also binds to a specific Trk receptor. Trk is the receptor for NGF; TrkB is the receptor for BDNF and NT-4, while TrkC is the receptor for NT-3. Neurotrophins secreted by cells protect neurons from apoptosis (Korsching, 1993; Lewin and Barde, 1996). Similarly, ciliary neurotrophic factor (CNTF), a structurally related type I cytokine and GDNF, structurally related to TGF-β, each constitute a sub-family of neurotrophic factors.

15.1.3.2. Growth Factor Families

Growth factors are comprised of multiple families, each with unique functions. One family includes vascular endothelial growth factor (VEGF) and PDGF which are potent mitogenic and angiogenic factors with critical roles in embryonic development, wound healing, and the integrity of the blood brain barrier (BBB). Epidermal growth factor (EGF) family members include EGF, neuregulins, amphiregulin, and betacellulin. The EGF family members mediate their growth and proliferative effects on cells of both mesodermal and ectodermal origin. Insulin-like growth factor (IGF)-I and IGF-II belong to the family of insulin-like growth factors that are structurally homologous to proinsulin. IGF-1 has a much higher growth-promoting activity than insulin and is highly expressed in all cell types and tissues. IGF-2 is expressed in embryonic tissues and is related to development. The FGF family is another group of cytokines involved in many aspects of development including cell proliferation, growth, and differentiation. They act on several cell types to regulate angiogenesis, cell growth, pattern formation, embryonic development, metabolic regulation, cell migration, neurotrophic effects, and tissue repair.

15.1.4. Structure and Classification of Chemokines and Their Receptors

First discovered in 1987 (Walz et al., 1987; Yoshimura et al., 1987), chemokines include a group of secreted proteins within the family of cytokines that by definition relate to the induction of migration. These "**chemo**tactic cyto**kines**" are produced by and target a wide variety of cells, but primarily address leukocyte chemoattraction and trafficking of immune cells to locations throughout the body via a gradient. There are two general categories of biological activity for chemokines, the maintenance of homeostasis and the induction of inflammation (Moser and Loetscher, 2001). Homeostatic chemokines are involved in roles such as immune surveillance and the navigation of cells through hematopoesis, and are typically expressed constitutively. Inflammatory chemokines are produced in states of infection or following an inflammatory stimulus, and by targeting cells of the innate and adaptive immune system, facilitate an immune response.

Chemokine activity is mediated by binding a family of seven transmembrane G-protein coupled receptors (GPCR). Chemokine receptors are classified into four families based on the number and position of the N-terminal-conserved cysteine residues within the receptor binding domain. Those receptors with two cysteine residues separated by one amino acid are categorized as α-chemokine receptors (such as CXCR2 and CXCR4); whereas those with two cysteines immediately next to each other are designated as β-chemokine receptors (such as CCR5, CCR4, CCR3 and CCR2). The only γ-chemokine receptor to this point has a single cysteine residue (XCR1) and δ-chemokine receptors have three amino acids separating the two cysteine residues (CX3CR1)(Gabuzda et al., 1998; Hesselgesser and Horuk, 1999; Klein et al., 1999; Miller and Meucci, 1999; van der Meer et al., 2000; Cotter et al., 2002). GPCRs interact with and signal through heterotrimeric guanine-nucleotide-binding regulatory proteins (G-proteins). Upon stimulation by a ligand, GPCRs undergo a conformational change that leads to activation of the G-protein by GDP-GTP exchange, followed by uncoupling of the G-protein from the receptor. Upon activation, G-proteins trigger a cascade of signaling events that regulate various cellular functions (Devi, 2000).

15.1.4.1. CXC Chemokines

CXC chemokines are further separated into two groups based upon the presence or absence of a specific three amino acid sequence found adjacent to the CXC. The Glu-Leu-Arg residues constitute the ELR motif, and if present, the CXC chemokine is considered to be ELR(+). The general function of ELR(+) chemokines revolves around neutrophils, inducing chemotaxis and promoting angiogenesis (Strieter et al., 1995). Chemokines in this group include CXCL1, CXCL2, CXCL3, CXCL5, CXCL6, CXCL7, CXCL8, and CXCL15; unlike ELR(−) members, these factors interact primarily with neutrophils via CXCR1 and CXCR2 receptors. In contrast, ELR(−) chemokines attract lymphocytes and monocytes with little affinity for neutrophils. This subgroup including CXCL4, CXCL9, CXCL10, CXCL11, CXCL12, CXCL13, CXCL14 and CXCL16, has a wider variety of activities, but generally have angiostatic properties and induce chemotaxis in mononuclear cells.

15.1.4.2. CC Chemokines

CC chemokines target primarily mononuclear cells and serve in both homeostatic and inflammatory capacities. The family can be divided into functional groups including allergenic, pro-inflammatory, developmental and homeostatic. The developmental and homeostatic factors are, as expected, constitutively produced, whereas the other groups contain largely inducible signals. Allergenic CC chemokines target eosinophils, basophils and mast cells and are potent attractors and stimulants of histamine release. Inflammatory chemokines are also inducible and can amplify inflammation through the recruitment of effector cells.

15.1.4.3. CX3CL1

CX3CL1 (Fractalkine, FKN) is the lone member of the CX3C subfamily with three intervening residues between the first two cysteines. FKN, the ligand for CX3CR1, is a 373-amino acid, multi-domain molecule found in a wide variety of tissues, including liver, intestine, kidney, and brain. Structural components of FKN include a 76-amino acid chemokine domain (CD) at the N-terminus, which is important in the binding, adhesion, and activation of its target cells (Mizoue et al., 1999; Goda et al., 2000; Haskell et al., 2000; Harrison et al., 2001; Mizoue et al., 2001). FKN also has an 18-amino acid stretch of hydrophobic residues that spans the cell membrane, and an extended carboxyl-terminus that anchors it to the cell surface (Hoover et al., 2000; Cook et al., 2001; Lucas et al., 2001), thus allowing a membrane form as well as a shed soluble form targeting monocytes and T cells (Bazan et al., 1997).

15.1.4.4. XCL1

The final chemokine member is XCL1 (Lymphotactin), the only representative of the C family. This chemokine targets CD4 + and CD8 + lymphocytes, but does not act on monocytes, and binds a unique receptor, XCR1. Although having some homology to ligands CCL3 and CCL8, XCL1 lacks the first and third cysteines characteristic of the CC and CXC chemokines.

All of the above families are imperative for homeostatic functions as well as the orchestration of response to pathogenic insult by the immune system. As investigations carry on, this large family will likely see the emergence of additional cytokines and growth factors, and the family and sub-families of cytokine, growth factor, or chemokine receptors will continue to evolve and expand.

15.2. Cytokines and Growth Factors in the CNS

The view of neuroimmunology has evolved from the dogma that regarded the CNS as an immune privileged site to a view of significant CNS-immune system interactions. Cytokines and growth factors are important to the brain's immune function, serving to traffic leukocytes, maintain immune surveillance, and recruit inflammatory factors. In the normal physiological state immune cells readily cross the blood brain barrier (BBB) and traverse the healthy CNS (Hickey, 1999). As discussed in Chap. 38, CNS inflammation is unique due to the presence of the BBB. A restricted inflammatory process in the CNS is initiated upon encountering foreign antigen, inducing the production of recruitment factors (Irani, 1998). Upon recruitment, monocytes and lymphocytes cross through the BBB to mount immune responses in the CNS. Recruitment is dependent upon the presence of chemotactic factors produced within the CNS that facilitate the crossing of the BBB. This inflammatory state is highly regulated, usually self-limited, and differs from the inflammation in peripheral tissue sites. The threshold to initiate immune responses against antigens in CNS is much higher than that in periphery (Matyszak, 1998; Perry, 1998). Avoidance of uncontrolled activation, release of toxic factors, edema and other effects of robust inflammation are crucial to the maintenance of the vulnerable microenvironment in the CNS. Cytokines and their receptors detectable in brain tissues or neuronal and glial cultures under pathophysiological conditions are summarized in Table 15.1.

CNS-immune system interactions are regulated by cytokines, chemokines, and growth factors. Immune cells, after stimulation by pathogens or abnormal cells, release cytokines, which then interact with immunosensory tissues. Immunosensory structures respond to cytokines, then activate brain neuroimmune circuits responsible for the induction of the defensive response. In the brain, microglia are the resident macrophage and present antigen, orchestrating the innate immune responses. Astrocytes have also occasionally been postulated to have a role in antigen-presentation to $CD4^+$ T cells in the CNS. Other resident cells, such as neurons or oligodendrocytes, are all capable of recruiting immune cells, modulating the immune response through cytokines, chemokines, or growth factors. In addition, neurons may also modulate the antigen-presenting function of microglia. For example, neuronal production of neurotrophins also suppresses IFN-γ induced MHC class II expression (Neumann et al., 1998).

Disease states within the CNS can lead to dysregulation of the inflammatory process. In acute insults such as ischemia or traumatic injuries, evidence suggests cytokines IL-1 and IL-6 are rapidly induced. These pro-inflammatory molecules have been reported to cause neuronal death through different mechanisms. In general, if present for extended periods of time, IL-1 exacerbates acute neurodegeneration through induction of fever, upregulation of neurotoxic molecules such as arachidonic acid, or promoting leukocytes transmigration (Tarkowski

TABLE 15.1. Cytokine ligand and receptor expression in the CNS inflammation.

Cytokines/receptors	Cytokine expression cell types	Receptor expression cell types	Reference
Type I family			
IL-6/IL-6R	Macrophage/microglia, Astrocytes	Macrophage/microglia, Astrocytes	(Perrella et al., 1992; Griffin, 1997; Poluektova et al., 2004)
GM-CSF/ GM-CSFR	Astrocytes	Monocytes/ macrophage	(Perno et al., 1990; Perrella et al., 1992; Lotan and Schwartz, 1994)
IL-4/IL-4R	Macrophage/microglia, Astrocytes	T-cells, Oligodendrocytes	(Wesselingh et al., 1993; Adorini and Sinigaglia, 1997; Hulshof et al., 2002; Kedzierska et al., 2003)
IL-2/IL-2R	Macrophage/microglia	T-cells, Monocyte/macrophage	(Wesselingh et al., 1993; Adorini and Sinigaglia, 1997)
BDNF/TrkB	Astrocytes, Neurons	Neurons, Astrocytes, Oligodendrocytes	(Soontornniyomkij et al., 1998; Boven et al., 1999)
Type II family			
IL-10/IL-10R	Macrophage/microglia, Astrocytes	Macrophage/microglia, T-cells, Oligodendrocytes	(Gallo et al., 1994; Hulshof et al., 2002; Poluektova et al., 2004)
IFN-α/β/IFN-α R	Macrophage/microglia	Macrophage/microglia	(Perrella et al., 2001)
IFN-γ/IFN-γ R	T-cells	T-cells, Macrophage/microglia	(Griffin, 1997)
M-CSF/M-CSF R	Macrophage	Macrophage	(Gallo et al., 1994; Griffin et al., 1995)
TNF superfamily			
TNF-α/TNF-α -R1, -R2	Macrophage/microglia	Astrocytes, Macrophage/microglia, Oligodendrocytes	(Hofman et al., 1989; Matusevicius et al., 1996; Shi et al., 1998; Kast, 2002)
TRAIL/ TRAIL-R1, -R2, -R3, -R4	Monocyte/macrophage, T-cells	Moncyte/macrophage, T-cells, Neurons	(Ryan et al., 2004; Uberti et al., 2004; Aktas et al., 2005)
NGF/TrkA	Neurons, Astrocytes, Macrophage/microglia	Neurons, T-cells	(Soontornniyomkij et al., 1998; Boven et al., 1999; Villoslada et al., 2000)
IL-1 family			
IL-1 α/ IL-1 α R	Macrophage/microglia, Astrocytes	Macrophage/microglia, Astrocytes	(Griffin et al., 1989; Perrella et al., 1992)
IL-1 β/IL-1 β R	Macrophage/microglia, Astrocytes	Macrophage/microglia, Astrocytes	(Wesselingh et al., 1993; Mason et al., 2001; Zhao et al., 2001)
TGF-β family			
TGF-β	Macrophage/microglia, Astrocytes, Neurons	Macrophage/microglia Astrocytes, Neurons	(Logan and Berry, 1993; van der Wal et al., 1993; Perrella et al., 2001)

et al., 1995). Chemokines such as MCP-1 also elevate in acute insults and promote inflammation through recruitment of leukocytes critical to propagate tissue damage (Hausmann et al., 1998). In states of prolonged inflammation, continual activation and recruitment of effector cells may establish a positive feedback loop that perpetuates inflammatory processes and can ultimately lead to neuronal injury and dropout. Similar yet distinct processes occur in multiple CNS disorders such as MS, HAD, and AD. One underlying similarity in each disease state is the cytokine driven inflammatory response (see Figure 15.1). Numerous studies have demonstrated that microglia/macrophage and astrocytes are major sources of inflammatory mediators, including TNF-α and IL-1β, IL-10, and nitric oxide (NO). These mediators initiate or regulate inflammatory processes in the CNS (Aloisi, 2001; Dong and Benveniste, 2001; Hanisch, 2002). In addition to the beneficial effects that glia have in initiating protective immune responses in the CNS, these cells have been implicated in contributing to tissue damage when chronically and/or pathologically activated. For example, many reports demonstrate that activated microglia may exacerbate AD and MS, as described later in this chapter, through secretion of a battery of inflammatory cytokines and cytotoxic agents (Meda et al., 1995; Renno et al., 1995; Benveniste, 1997).

15.3. Molecular Mechanisms of Cytokines, Growth Factors, and Chemokines Activity and Signaling

Multiple signaling pathways have been observed for various cytokines. We will first discuss the molecular signaling of neuronal survival and cell death to better explain the molecular mechanisms involved in cytokine-induced neurodegeneration.

The signaling events leading to apoptosis can be divided into two distinct pathways, involving either the mitochondria (*intrinsic*) or death receptors (*extrinsic*) (for reviews, see Green and Reed, 1998; Green, 2003). The mitochondrial pathway is initiated through various stress signals that impair the mitochondrial membrane integrity. BCL-2 family proteins, including the anti-apoptotic members, i.e., BCL-2 and BCL-XL, and pro-apoptotic members, i.e., Bax, Bak, play a critical role in this pathway (Gross et al., 1999; Wang, 2001; Green, 2003; Danial and Korsmeyer, 2004). The BH3–only BCL-2 family proteins, such as Bid, Bad, Bim, and PUMA, serve as sentinels to these stress signals. Once activated, BH3-only proteins translocate to the outer membrane of mitochondria, where they trigger the oligomerization and activation of both Bax and Bak. In turn, Bax and Bak cause the release of cytochrome c (cyto c) and other apoptogenic factors, including second mitochondrial-derived activator of caspase (SMAC), Htr2, apoptosis inducing factor (AIF), and endonuclease G (EndoG) (Green and Reed, 1998). Cyto c then binds the cytosolic adaptor protein, Apaf-1, mediating the formation of the apoptosis complex "apoptosome." Such complexes lead to the activation of caspase-9, which further processes and activates the effector caspases, pro-caspase-3, 6, or 7. Effector caspases then cleave death substrates and complete apoptosis. The death receptor pathway is initiated through interaction with death receptors and the recruitment of cytoplasmic proteins to specific regions in the intracellular domains of these receptors. The receptor adaptor protein complex then binds pro-caspase-8 and forms a death-inducing signaling complex (DISC) that subsequently releases active caspase-8. Active caspase-8 then activates a caspase cascade and eventually leads to apoptosis of the cell. Many cytokines influence the apoptotic pathways through various receptors and intracellular mediators. Members of TNF superfamily, such as TNF-α, TRAIL, or Fas ligand are able to interact directly with death receptors on neurons initiating neuronal apoptosis through activation of the caspase cascade.

In contrast to the apoptotic pathway, some cytokines and growth factors are able to induce survival and proliferation of neurons through survival pathways. The phosphatidylinositide-3-kinase (PI3K) pathway and Ras- or protein kinase C (PKC)-dependent mitogen-activated protein kinase (MAPK) pathway have been extensively studied as survival pathways. Activated PI3K leads to downstream Akt phosphorylation and activation. Notably, Akt activation converges with death receptor-mediated receptor interacting protein (RIP) activation on the phosphorylation of the IκB kinase (IKK) complex and the resulting IKB degradation. IKB degradation releases active nuclear factor κB (NF-κB) leading to transcription of genes necessary for cell survival. Similarly, activation of MAPK will lead to nuclear translocation of MAPK together with other transcription factors and co-activators and initiate the transcription of a variety of survival genes. Though not limited to those survival pathways, cytokines and growth factors such as NGF, GDNF, and BDNF play a role in protecting, repairing, and/or regenerating neurons. Neurons constantly receive survival and apoptotic signals from the extracellular environment, influencing the lifespan of each individual cell. Various cytokines such as BDNF or NGF increase during diseases and may be a compensatory mechanism to the neural injury of the CNS (Soontornniyomkij et al., 1998; Boven et al., 1999). IL-1β from the IL-1 family or IL-6 from the type I cytokine family can activate the NF-κB pathway, thus potentially promoting survival of neurons. Interestingly, IL-1, TNF and even TRAIL signaling pathways also converge on IKB degradation and NF-κB activation. IL-6 also activates the Ras-dependent MAPK pathway. Regarding IL-1β, the activation of NFκB and MAPK initiates transcription of a variety of genes for either survival or the inflammatory response, dependent upon cell type (Srinivasan et al., 2004).

Although chemokines act through specific GPCRs, chemokines relate to cytokines through many overlapping intracellular signaling pathways. One of the first signaling pathways identified was the inhibition of adenylyl cyclase activity to reduce the intracellular cAMP levels, modulated by the Gα subunit of the Gi/o proteins (Damaj et al., 1996). It was also

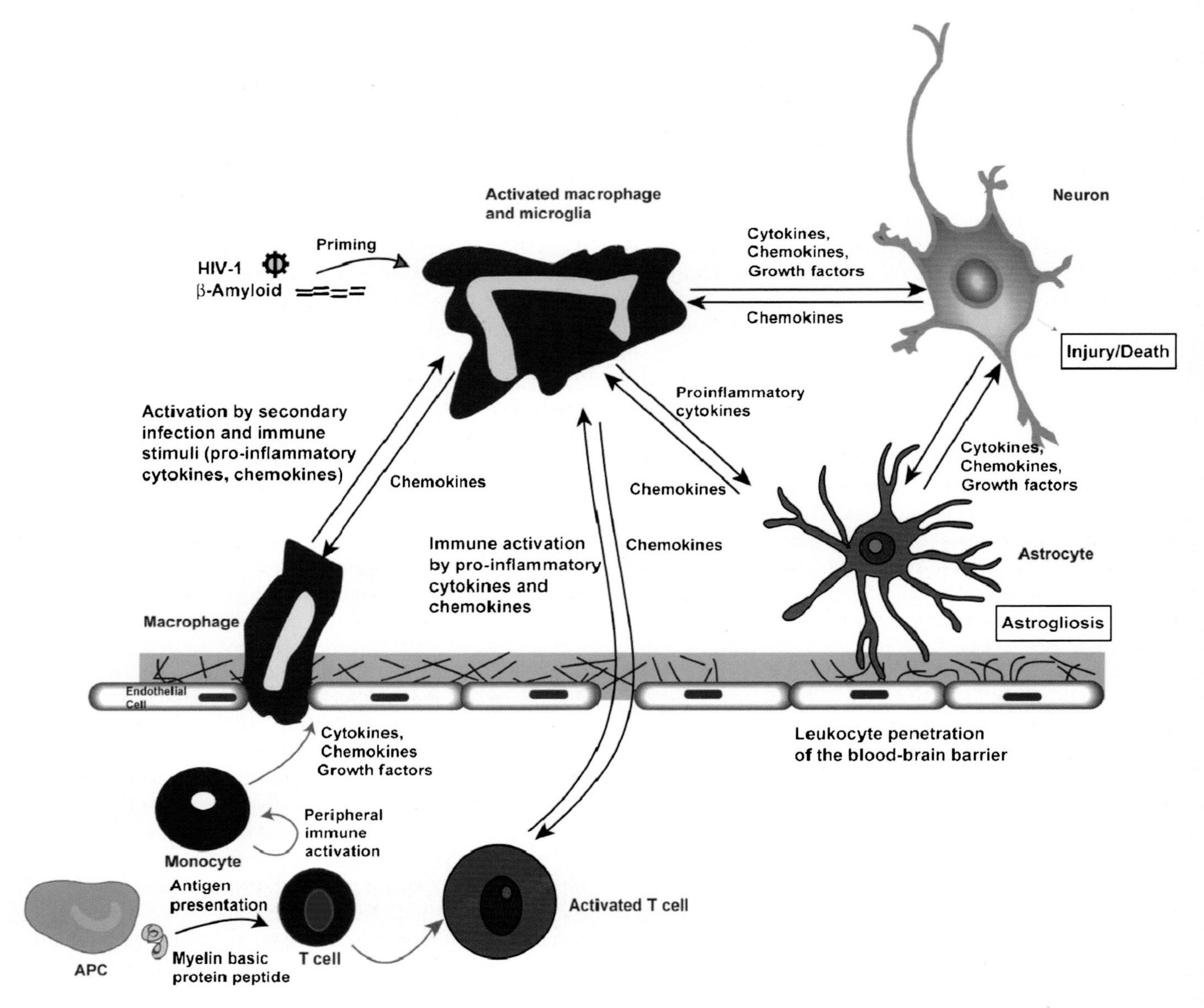

FIGURE 15.1. Cytokine driven inflammatory process in multiple disease states. Multiple disease states have similar cytokine-driven inflammatory process. In HAD, HIV-1 infection of macrophage in the CNS leads to the activation of macrophage and microglia. Activated macrophage and microglia release cytokines, chemokines, and growth factors, and may have overall detrimental effects toward neurons for extended periods of time. Neuronal injury recruits additional macrophages. This injury, recruitment, and activation process forms a positive feedback loop and can aggravate various disease states. In AD and MS, either Aβ priming or T cell-mediated macrophage/microglia activation may utilize similar inflammatory cytokines and chemokines to cause the pathogenic effects on the CNS.

reported that most of the chemokine receptors are able to activate phospholipase C to increase the generation of diacylglycerol and inositol 1,4,5-triphosphate, with a subsequent increase in PKC activity and transient elevations of cytosolic Ca^{2+} concentrations, respectively (Kuang et al., 1996). Several chemokines (SDF-1, for example) have been reported to activate the MAPK cascade, involving ERK1/2 activation. SDF-1 also activates the downstream transcription factor Ets, a substrate for ERK phosphorylation (Ganju et al., 1998). In most cases this activation involves effects mediated by the β subunit of a PTX-sensitive G protein and the activation of the Ca^{2+}-dependent protein kinase Pyk2. PI3-K and subsequent activation of protein kinase B is another transduction pathway activated by multiple chemokines. The JAK/STAT pathway has also been implicated in α and β chemokine receptor signaling (Mellado et al., 2001). Despite the structural similarities among chemokine receptors, they can activate specific transduction pathways leading to diverse responses, an expression of the pleiotropic signal effects carried out by these proteins.

In summary, various families of cytokine, growth factor, and chemokine may greatly influence the apoptotic or survival pathways of neurons and inflammatory state in the CNS. The understanding of signaling pathways activated by cytokines, growth factors, and chemokines continues to increase. However, it is still unclear what specific pathways regulate inflammatory process in neurodegeneration and neuroimmunologic diseases.

15.4. Cytokines in Neurodegenerative and Neuroimmunologic Diseases

Many cytokines have been suggested to participate in neurodegeneration and neurotoxicity. Increases in factors such as TNF-α and IL-1β have been observed before neuronal death (Esser et al., 1996; Matusevicius et al., 1996; Guo et al., 1998; Meda et al., 1999; Little and O'Callagha, 2001). However, the comprehensive effect of pro-inflammatory cytokines such as IL-1β and TNF-α is controversial. Neither of these cytokines causes neuronal death in healthy brain tissue or normal neurons (Gendelman and Folks, 1999). But it is generally believed that in disease states those proinflammatory cytokines contribute to the uncontrolled inflammation in the CNS. The effect of cytokines in each individual disease will be discussed in the following sections.

15.4.1. Cytokines in HAD

HIV-1 infection notoriously attacks the immune system in the periphery, but also can lead to a viral induced dementia. The mechanism has yet to be fully elucidated, but cytokines have been shown to play a significant role in disease progression. HIV-1 usually enters the brain shortly after initial infection, crossing the blood brain barrier through peripherally infected monocytes (Koenig et al., 1986). Brain macrophages and microglia, unlike other cellular residents in the CNS, are able to sustain a productive infection within the brain (Eilbott et al., 1989). Thus, while the rest of the body typically experiences a surge in viral load followed by a gradual reduction, the isolated CNS maintains a low, but sustained level of virus. Although neurons are not infected by HIV-1, a dementia specific to HIV has been described as HAD (Navia et al., 1986; McArthur, 1987).

HAD is the clinical consequence of neuronal damage. The pathologic correlate to HAD, HIV encephalitis (HIVE), is characterized by activated macrophage and microglia, as well as damage of neuronal dendrites and axons and apoptotic neurons. As a result of HIV-1 infection, the immune cells are recruited to the proximate site and produce an array of factors including cytokines and proteases yet are unable to clear the infection. The difficulty in eradicating HIV-1 infection prolongs the immune response leading to a chronic inflammatory state. Chronic inflammation has both detrimental and beneficial effects. On one hand, these responses are essential in limiting viral spread; yet on the other hand, excessive inflammation is detrimental to resident cells such as neurons.

HIV-1 infection leading to neuronal injury and/or loss has been and remains the focus of much investigation. Interest in the toxic effects of cytokines during HAD increased when a study demonstrated HAD involved productive infection of macrophage and microglia (Wiley et al., 1986; Meltzer et al., 1990; Fischer-Smith et al., 2004), and macrophage and microglia become the major cytokine producers in the CNS upon infection (Table 15.1). Two distinct models have been proposed as mechanisms in HAD pathogenesis. The first is the direct neuronal injury model, in which viral proteins (gp120, tat and Vpr) directly interact with neurons causing neuronal injury through various mechanisms. The role of cytokines in this model has been controversial. Cytokines like TNF-α have a synergistic apoptotic effect with viral proteins tat (Shi et al., 1998) and gp120 (Kast, 2002). However, other cytokines inhibit the effects of viral proteins on neurons. For example, TGF-β1 prevents gp120-induced neuronal apoptosis by restoring calcium homeostasis (Scorziello et al., 1997); BDNF and IL-10 can inhibit gp-120 mediated cerebellar granule cell death by preventing gp120 internalization and caspase-3 activation in vitro (Bachis et al., 2001; Bachis et al., 2003). The second model, referred to as the indirect neuronal injury model, proposes that neurons die as bystanders when excessive local concentrations of soluble pro-inflammatory and neurotoxic factors are released by infected MP and astrocytes (Figure 15.1). Studies have supported this notion including the observation that viral protein levels do not correlate with neuronal injury (Petito et al., 1994), while neuronal apoptosis correlates well with microglial activation (Glass et al., 1993; Adle-Biassette et al., 1995). Furthermore, studies have shown that cognitively impaired patients have elevated levels of inflammatory markers and activators in contrast to HIV patients without CNS impairment (Tyor et al., 1992; Sippy et al., 1995).

Although proinflammatory cytokines such as TNF-α and IL-1β are key molecules in the pathogenesis of HAD, other cytokines may play important roles as well. Elevated levels of IL-6, IL-2 and decreased levels of IL-4 have been reported in both CSF and brain sections of HAD patients (Perrella et al., 1992; Wesselingh et al., 1993; Griffin, 1997). These factors directly relate to the inflammatory state of brain. The activation of a Th1 immune response by infected macrophages results in the synthesis of cytokines such as IL-2 and IL-6 that activate macrophage and coordinate the immune response toward HIV-1 infection. IL-2 induces T-cell proliferation and potentiates the release of other cytokines. IL-3 and GM-CSF stimulate the production of new macrophages by acting on hematopoietic stem cells in bone marrow. New macrophages are recruited to the site of infection by TNF-α and other cytokines on the vascular endothelium that signal macrophages to leave the bloodstream and enter the tissue. Moreover, many inflammatory signals have been shown to increase HIV replication. IL-2 and GM-CSF are both potent stimulators of HIV-1 replication in activated CD4 + T cells, and GM-CSF increases HIV-1 replication in macrophage cultures (Perno et al., 1990). The activity of type I cytokines during HIV-1 infection promotes the inflammatory response in an effort to eliminate virus, however in the CNS during HAD, type I cytokine mediated inflammation results in neuronal damage as well as facilitating the replication of HIV.

Type II cytokines are key to the balance of inflammation and regulation of response in HAD. IL-10 downregulates the expression of proinflammatory cytokines and up-regulates the expression of the anti-inflammatory agent IL-1Ra. Increased levels of IL-10 in the CSF have been reported in individuals with HIV-1 encephalitis (Gallo et al., 1994). In a recent study in an HAD SCID mouse model, mRNA levels of IL-10 were increased five-fold as compared to uninfected controls. This change is concurrent with down-regulation of proinflammatory cytokines (IL-1β and IL-6) (Poluektova et al., 2004). Furthermore, pretreatment with IL-10 attenuated the neurobehavioral damage induced by HIV-1 gp120 in an in-vivo animal study (Barak et al., 2002). Increased type II immunomodulatory cytokines are possibly an active attempt to control inflammation and maintain or regain balance in CNS microenvironments.

IFNs have various effects on HAD pathogenesis. IFNs profoundly affect HIV-1 replication in various in-vitro systems. IFN-α and IFN-β are induced by HIV-1 infection and suppress HIV replication at multiple steps of the viral life cycle in macrophage (Gendelman et al., 1992; Gessani et al., 1994). IFN-γ enhances HIV-1 replication in CD4 + T cells in an autocrine manner, while in macrophage culture IFN-γ enhances HIV-1 replication when added prior to infection but inhibits replication when added post-infection. Although levels of IFN-γ have been shown to be elevated in HAD patients (Shapshak et al., 2004), its ultimate role is unclear. IFN-γ shapes the T-cell response and activates MP perhaps limiting viral spread, yet IFN-γ also synergistically enhances the effect of CD40L activation of macrophage. Interferons are known to target infected cells for cell-mediated elimination, but the ultimate role of interferons in HAD pathogenesis is still unclear (Benveniste, 1992).

Among the pro-inflammatory cytokines, TNF-α is the most studied. Levels of TNF-α mRNA are elevated in brain tissue collected from HAD patients (Wesselingh et al., 1993). Studies have shown that TNF-α levels in vulnerable brain regions correlate with neurological disease severity in HIV patients (Gelbard, 1999). During HAD, microglia, macrophages and monocytes show increased expression of TNF-α and TNF receptors (Tyor et al., 1992), an effect promoted by both IFN-γ and IL-1. HAART treated HAD patients have a marked decrease in soluble TNF-α levels in the cerebrospinal fluid; this drop in TNF-α coincides with decreased viral load and improvement in the neurological function of patients (Gendelman et al., 1998). TNF-α may promote neuronal demise through various mechanisms. Increased BBB permeability and recruitment of activated immune cells facilitates viral invasion of the CNS. Synergy of TNF-α, viral proteins and excitotoxic glutamate activates glia to produce neurotoxins or leads to neuronal apoptosis directly (see review Saha and Pahan, 2003). Yet, the effects of TNF-α are complex; differing experimental systems and approaches have shown TNF-α to assume a neuroprotective role. TRAIL, another member of TNF family, is not normally expressed in the CNS. However, TRAIL has been reported to be expressed on the cell membrane of peripheral immune cells and can be cleaved into a soluble, secreted form (Ehrlich et al., 2003). The plasma level of TRAIL is increased to ng/ml ranges in HIV-1-infected patients, particularly those with high viral loads (Herbeuval et al., 2005). Upon HIV-1 infection of CNS, TRAIL is upregulated primarily by infiltrated macrophages that are HIV-1-infected or immune-activated (Ryan et al., 2004). Upregulated TRAIL has been shown to preferentially induce apoptosis in HIV-1-infected macrophages (Huang et al., 2006). Nonetheless, neurons express TRAIL-receptors and TRAIL may then cause neuronal apoptosis through direct interaction with TRAIL receptors on neurons or through macrophage death-mediated release of neurotoxins (see reviews Huang et al., 2005a, b). Thus TRAIL may be an important neurotoxic mediator in the pathogenesis of HAD.

Another proinflammatory cytokine, IL-1 is rapidly produced upon HIV-1 infection within the CNS. The induction of IL-1 has been shown to be associated with HAD (Zhao et al., 2001). IL-1 in this case serves as a very upstream signal for multiple proinflammatory cytokines, notably TNF-α and IL-6, initiating and amplifying inflammation in the brain, which is responsible for the global activation of macrophage and microglia (Chung and Benveniste, 1990; Aloisi et al., 1992; Lee et al., 1993). Direct injections of IL-1 into the CNS result in local inflammatory responses and neural degeneration (Wright and Merchant, 1992). IL-1β directly activates HIV-1 replication in a monocytic cell line by transcriptional and post–transcriptional mechanisms independent of NF-κB. IL-1 also synergistically enhances HIV-1 replication with multiple cytokines including IL-4 and IL-6 (Kedzierska et al., 2003).

TGF-β is expressed in the brain by astrocytes, MP, and oligodendrocytes, and has been shown to exert multiple effects on neurons and glial cells both in vitro and in vivo. Effects of TGF-β include cell cycle control, differentiation, extracellular matrix formation, hematopoesis, and chemotaxis. Importantly in the CNS, TGF-β has key a role in regulating neuron survival and repair processes. TGF-β has also been shown to play a role in several varieties of CNS pathology including ischemia, excitotoxicity and neurodegenerative diseases such as multiple sclerosis. In mild HAD, the cerebral levels of TGF-β has an inverse correlation to IFN-α and HIV RNA; in the severe form of HAD, TGF-β is undetectable (Perrella et al., 2001). In specific culture conditions, TGF-β has neurotrophic effects similar to BDNF and NGF, but a change in conditions can shift the effect of TGF-β toward neurotoxicity (Prehn and Miller, 1996).

In summary, the various cytokines listed here all have certain associations with HAD. This body of evidence shades the inflammatory pattern of cytokine in HAD toward a chronic and detrimental effect that becomes manifest as the disease progresses. However, the cytokines discussed here do not constitute the complete list of players in HAD and other neuroimmune disorder diseases. Chemokines have a unique impact on the immune system; their role will be discussed at length below.

15.4.2. Cytokines in MS

MS is a chronic neurological disease with the hallmark pathological findings of perivascular inflammation and demyelination. Histological sections from MS patients show demyelinating plaques are distributed within the white matter of the CNS, but the most frequently affected sites are the optic nerves, the brainstem, the cerebellum and the spinal cord. Although the classic MS lesion is focal demyelination, axonal damage and neuronal dysfunction are also found in the pathology of MS (Trapp et al., 1999; De Stefano et al., 2001; Bjartmar et al., 2003). The selective loss of axons and myelin sheaths has been postulated to occur via a variety of different mechanisms. Traditionally, MS has been considered an autoimmune disorder consisting of myelin autoreactive T-cells that drive an inflammatory process, leading to secondary macrophage recruitment and subsequent myelin and oligodendrocyte destruction. However, recent detailed studies on a large collection of MS lesions have indicated that structural features of the plaques are extremely variable and the events involved in the immunopathogenesis of MS may be more complicated (Monica et al., 2005).

Recent studies of MS pathology show that the predominant cells in active lesions are lymphocytes, particularly T-cells and macrophages (Lucchinetti et al., 2000; Wingerchuk and Weinshenker, 2000). This pathology is similar to that found in experimental allergic/autoimmune encephalomyelitis (EAE), an animal model of MS, which can be induced in susceptible animals by active immunization with CNS tissue, myelin, myelin proteins, or by the transfer of auto-reactive T-cells. T-cells, upon being activated by antigen-presenting cells, become one of the primary effectors both in MS and EAE (Lucchinetti et al., 2000).

Generally Th1 activated $CD4^+$ cells produce IL-1, IL-2, IFN-γ, and TNF-α to mediate inflammatory pathological processes in immune-mediated tissue damage seen in MS and EAE (Adorini and Sinigaglia, 1997). In contrast, Th2 cells produce IL-4, IL-6, and IL-10 to mediate antibody production and downregulation of Th1 cellular responses (Adorini and Sinigaglia, 1997). The inflammatory reaction in MS and EAE is associated with the upregulation of a variety of Th1 cytokines, including IL-2, IFN-γ, and TNF-α. However, the pathogenesis of MS lesions is more complex, compared to that of a pure Th1-mediated CNS autoimmune disease. Cells other than the classical Th1 T-cells may contribute to the inflammatory process in autoimmune disease (Lucchinetti et al., 2000). For example, CD8 + class I restricted T-cells are also present and clonally expanded in MS lesions (Rohowsky-Kochan et al., 1989; Correale et al., 1997). Axonal destruction in MS lesions correlates better with CD8 + T-cells and macrophages rather than CD4 + T-cells (Kuhlmann et al., 2002). There is also evidence that Th2 cells can participate in pathologic autoimmune processes. Antibodies to both myelin-oligodendrocyte glycoprotein and MBP have been demonstrated in MS lesions and in the serum of MS patients (Wajgt and Gorny, 1983; Bernard and de Rosbo, 1991). Activated macrophage/microglia are a major cell source of excessive production of TNF-α, IL-1β, and NO, thus they are suggested to participate in the pathology associated with MS (Li et al., 1993; Bitsch et al., 2000).

Many inflammatory cytokines, such as Fas ligand and TNF-α can trigger proapoptotic pathways, which are important effectors for neuronal death in EAE. Overexpression of Bcl2, an anti-apoptotic protein that blocks death receptor-initiated signaling events, protects mice primed to develop EAE (Offen et al., 2000). Further, elevated death ligand TRAIL expression level on blood MP of MS patients has been reported (Huang et al., 2000). Oligodendrocytes and neurons have recently been shown to be one of TRAIL's targets (Matysiak et al., 2002; Wosik et al., 2003; Aktas et al., 2005). Although inflammatory mechanisms seem to be an important aspect contributing to tissue injury in MS, the degree to which demyelination and axonal injury are a direct consequence of inflammation is still not clear. Nonetheless, the emphasis on the inflammatory aspect of the MS lesions continues to be a major impetus for therapeutic strategies to date.

Recent observations suggest the potential for T-cells, macrophage, and microglia to produce neurotrophic factors and other neuroprotective factors, which may indicate an important role for inflammation in the repair of MS lesions (Hohlfeld et al., 2000; Stadelmann et al., 2002; Gielen et al., 2003). FGF2 and PDGF have been demonstrated as factors regulating remyelination (Frost et al., 2003). CNTF family factors, such as CNTF, leukemia inhibitory factor, cardiotrophin-1, and oncostatin M have been reported to induce a strong promyelinating effect in a myelination model (Stankoff et al., 2002). NGF enhances neural growth and also has the ability to switch the T-cell phenotype from Th1 to Th2, which reduces CNS tissue damage (Villoslada et al., 2000). Interestingly, IL-1β is crucial to the remyelination of the CNS and this activity is suggested to lead to induction of astrocyte and microglia-macrophage-derived IGF-1 (Mason et al., 2001). IGF-1 may, in addition, have antiapoptotic properties for oligodendrocytes (Mason et al., 2000).

Cytokines themselves may also be used for MS therapy. IFN-β has been in clinical use as an immunomodulatory drug for the treatment of MS for more than 10 years. Administration of IFN-β decreases the relapse rates and new MRI lesions in patients with relapsing/remitting MS (The IFN-β Multiple Sclerosis Study Group, 1993) or delay disease progression in patients with secondary progressive MS (Jacobs et al., 1996;

PRISMS STUDY GROUP, 1998). However, the exact mechanisms of action of IFN-β in MS are unknown. IFN-β is effective in preventing the IFN-γ-induced upregulation of MHC-II on antigen-presenting cells (Yong et al., 1998). In an earlier study, patients treated with IFN-γ exhibited an alarmingly high relapse rate during the initial month of treatment (Panitch et al., 1987). IFN-β can also downregulate the expression of costimulatory molecules and inhibit the activation and proliferation of T cells. Additionally, IFN-β is involved in the alteration of cytokine levels, decreasing the level of Th1 cytokine including IFN-γ, IL-12 and TNF-α (Yong et al., 1998). IFN-β treatment of astrocytes has been reported to stimulate production of NGF in-vitro (Boutros et al., 1997) and may thus promote remyelination in the CNS.

In summary, cytokines and growth factors are very important mediators in the pathogenesis of MS as well as exhibiting therapeutic potential. More effort should be brought to explore the complex interactions among immune cells, cytokines, growth factors and CNS resident cells.

15.4.3. Cytokines in AD

AD is the most common chronic neurodegenerative disorder characterized by the presence of neuritic plaques, neurofibrillary tangles, neurophil threads, and plaque deposits comprised of highly insoluble β-amyloid peptides (Aβ), cellular debris, and inflammatory proteins in the cerebral cortex and hippocampus (Selkoe, 2001; Ringheim and Conant, 2004). Neurons of the hippocampus and cerebral cortex are selectively lost. A number of factors have been thought to contribute redundantly to pathogenesis of AD, including unbalanced calcium homeostasis, cell-cycle protein dysregulation, excitatory amino acids, as well as DNA damage. Deposition of Aβ in the brain parenchyma appears to be the crucial factor for the onset of AD, although mechanisms underlying Aβ effects are far from understood. The precursor of Aβ, amyloid protein precursor (APP), is an acute-phase agent upregulated in neurons, astrocytes, and microglia in response to inflammation, stress, and a multitude of associated cellular stresses. Aβ at high concentrations (generally >10 μM) has been found to act as a potent neurotoxic agent to both neuronal and non-neuronal cells. Interestingly, Aβ induces neurotoxicity through the activation of microglia (Figure 15.1). *In-vitro* studies indicate concentrations of Aβ as low as 1–2 μM can induce neuron cell death but only in the presence of microglia (Qin et al., 2002). Aβ primes microglia and has been shown to directly induce an inflammatory response when injected into the rat brain.

Inflammation has been suggested to contribute to the Aβ deposition and neuropathology associated with AD. For example, many clinical studies have indicated that anti-inflammatory drugs have a protective effect upon the development of AD (Stewart et al., 1997; Broe et al., 2000). AD is associated with the increase of pro-inflammatory cytokines including IL-6, TNF-α, M-CSF, IL-1β, TGF-β, TRAIL, and PDGF (Griffin et al., 1989, 1995; van der Wal et al., 1993; Tarkowski et al., 1999; Akiyama et al., 2000; Uberti et al., 2004). These cytokines are produced by activated glia and some (IL-1β and TNF-α) are capable of perpetuating glial activation, leading to a cycle of overproduction of these potentially neurotoxic molecules, others (TRAIL, for example) have been suggested to induce apoptosis in CNS cells directly. It should be noted, however, that these same cytokines at low concentrations may be neuroprotective. Moreover, a complex relationship seems to exist between Aβ deposition and the Aβ-induced inflammatory response. Because Aβ accumulation appears to trigger inflammation, factors that either promote or inhibit Aβ aggregation would be expected to similarly affect inflammation. Microglia isolated post-mortem from AD patients have been shown to either constitutively express cytokines, including IL-1β, IL-6, and TNF-α, or produce these cytokines in response to Aβ (Lue et al., 2001). Likewise, Aβ and other amyloid-associated proteins are capable of stimulating the production of these same cytokines by rodent glia (Lindberg et al., 2005).

IL-1 regulates the synthesis of the APP from which the pathologic and diagnostic amyloid plaques of AD originate (Goldgaber et al., 1989). The functions of IL-1 suggest that this cytokine plays a key role in the pathogenesis of AD. Brain tissue levels of IL-1 are markedly elevated in AD, and the numbers of IL-1-immunoreactive microglia are increased in tissue sections of Alzheimer brain, mirroring the distribution of microglia within cerebral cortical layers in brains of normal controls (Sheng et al., 1998). This suggests distribution of microglia within the brain in part determines the distribution of Aβ plaques in AD.

In summary, cytokines and growth factors discussed in this section are implicated in the pathophysiological processes of AD of which activated microglia and macrophage perpetuate the inflammatory response that ultimately causes neuronal injury. However, it should be noted that inflammation orchestrated by cytokines, chemokines, and growth factors may only play minor roles, as some researchers regard AD as a neurodegenerative disease rather than neuroinflammatory disease.

15.5. Chemokines in the CNS

While produced most often by immune cells, chemokines are also expressed by cells within the brain including endothelial cells, neurons, astrocytes, microglia and oligodendrocytes where they regulate the migration, recruitment, accumulation, and activation of leukocytes in the brain (Wu et al., 2000). For a summary of chemokines in the CNS, see Table 15.2. Some chemokines such as stromal derived factor-1α (SDF-1α, CXCL12) and fractalkine (FKN, CX3CL1) are constitutively produced in the brain and likely play an important role in central nervous system (CNS) homeostasis and development. Others, such as macrophage inflammatroy protein one alpha and one beta (MIP-1α, CCL3 and MIP-1β, CCL4), monocyte chemotactic

TABLE 15.2. Chemokine ligands and receptor expression in CNS inflammation.

Cell type	Ligands chemokine (systematic name)	Receptors	Reference
Monocytes/Macrophage	MCP-1 (CCL2), MIP-1α/β (CCL3/CCL4)	CCR2	(Bernasconi et al., 1996; Cinque et al., 1998; Kelder et al., 1998; Sorensen et al., 1999; Collman et al., 2000; Gabuzda et al., 2002; Mizuno et al., 2003)
		CCR3, CCR5	
	RANTES (CCL5)	CCR3, CCR5	
	IL-8 (CXCL8)		
	IP-10 (CXCL10)	CXCR3	
		CXCR4, CCR4	
Microglia	MIP-1α/β (CCL3/CCL4)	CCR3, CCR5	(Bernasconi et al., 1996; Ishizuka et al., 1997; Cinque et al., 1998; Kelder et al., 1998)
	RANTES (CCL5)	CCR3, CCR5	
	MCP-1,3 (CCL2,CCL7)	CCR2	
	IL-8 (CXCL8)		
	IP-10 (CXCL10)	CXCR3	
		CCR4, CXCR4, CX3CR1	
Astrocytes	MCP-1 (CCL2)	CCR2	(Conant et al., 1998; Simpson et al., 1998; Zheng et al., 1999b, 2000b)
	MIP-1α/β (CCL3/CCL4)	CCR3, CCR5	
	RANTES (CCL5)	CCR3, CCR5	
	SDF-1α (CXCL12)	CXCR4	
	IL-8 (CXCL8)		
	IP-10 (CXCL10), FKN (CX3CL1)		
		CCR4	
Neurons	FKN (CX3CL1)	CX3CR1	(Xia et al., 1997; Coughlan et al., 2000; Zheng et al., 2000b; Xia and Hyman, 2002)
	MCP-1 (CCL2)	CCR2	
		CCR5, CXCR4	
	GRO-α (CXCL2)	CXCR2	
		CXCR3	
Endothelium	IL-8 (CXCL8)	CXCR4, CCR5	(Gabuzda et al., 2002)
	FKN (CX3CL1)		
	IP-10 (CXCL10)		
	MGSA-α, -β, -γ (CXCL1, 2, 3)		
	MIG (CXCL9)		
	MCP-1,3,4 (CCL2, 7, 13)		

protein-1 (MCP-1, CCL2), and the regulated upon activation normal T-cell expressed and secreted (RANTES, CCL5), are induced by inflammatory stimuli. These chemokines are likely involved in the pathogenesis of a variety of neurodegenerative diseases, where inflammation plays a role in pathogenesis, such as MS, AD, and HAD.

15.6. Chemokines and their Receptors in Neurodegenerative and Neuroimmunologic Diseases

The inflammatory response manifest in the brain during HIV-1 infection of the nervous system leads to the development of a chemo-attractant gradient resulting in the formation of multinucleated giant cells in HIVE (Williams et al., 2002). This enables inflammatory monocyte-derived macrophages (MDM) to enter the brain, become infected and expand the sources of neurotoxic secretory factors that lead to the pathological and clinical aspects of disease (Gendelman et al., 1997). Moreover, chemokine receptors are critical for infection in perivascular macrophages and microglia. Studies have shown chemokines and their receptors play a more direct role in the neuropathogenesis of HIV-1 infection. It is now clear that neurons, glia and neural stem cells express chemokine receptors and the interactions of HIV-1 gp120 with neuronal chemokine receptors leads to apoptosis of neurons (Kaul et al., 2001; Cotter et al., 2002; Ryan et al., 2002; Tran and Miller, 2003; Peng et al., 2004). These effects may be manipulated by chemokines that act on the same receptors. The presence of chemokine receptors on neural cells also supports the notion that chemokines modulate neuronal physiological functions. Thus, chemokine receptors might have a crucial role in the balance between neuronal protection and injury.

15.6.1. Chemokines and Their Receptors in HAD

15.6.1.1. HIV-1, Chemokines and HIV-1 Co-receptors

As previously discussed, chemokines are important players in the development and maintenance of an immune response to foreign insult. In some circumstances however, chemokines play a more central role in the pathogenesis of the disease process. Studies have shown multiple viruses including herpesvirus, poxvirus, retrovirus, and lentivirus take advantage of the chemokine system, posing as analogs, to presumably gain a survival advantage by avoiding or altering immune detection and elimination (Murphy, 2001). Another manipulation of chemokine immune defense first described in 1996, is HIV's use of a chemokine co-receptor in human infection (Feng et al., 1996). Initially, HIV was assumed to rely solely on the CD4 surface protein found on T-cells and macrophages for entry into host cells (CD4 as an HIV receptor is reviewed in Sattentau and Weiss, 1988). However, CD4 alone did not accurately predict the cell interactions of HIV. This eventually led to the breakthrough that chemokine GPCRs mediate viral membrane fusion with human host cells. Each HIV strain has different specificities and interactions with various chemokine receptors, but the two primary coreceptors are CCR5 and CXCR4. Macrophage or M-tropic viral strains utilize CCR5 for infection; T-cell or T tropic viral strains rely upon CXCR4. There is another viral subset, dual tropic or R5X4 strain that employs both coreceptors. Further, additional receptors have been shown to have more limited viral interactions including CCR2, CCR3, CCR8, CX3CR1 and others, but the pathophysiological relevance has yet to be determined (Gabuzda and Wang, 2000).

The co-receptor requirement is a result of receptor ligand interactions between the chemokine GPCRs and the HIV coat protein gp120. Virus-cell interactions characteristically begin with gp120 binding CD4, inducing a conformational change in gp120. This change alters the affinity of gp120 for a coreceptor, either CCR5 or CXCR4 resulting in a trimolecular interaction between gp120, CD4 and the coreceptor (Berson vand Doms, 1998; Dimitrov et al., 1998; Berger et al., 1999). The multi-molecular interaction then permits fusion of HIV viral membrane to the host cell, allowing entry and consequent integration into the host DNA.

Within the past several years the links between chemokines, chemokine receptors, and HIV pathogenesis have been shown to be clear and significant (Cocchi et al., 1995; Deng et al., 1996; Dragic et al., 1996; Feng et al., 1996; Michael, 2002; Rizzardi et al., 2002). Chemokine receptors play a critical role, particularly in the early stages of HIV cell entry both in protective and liable capacities. Chemokine receptors CCR5 and CXCR4 are the major co-receptors for viral entry into CD4+cells (Cocchi et al., 1995; Deng et al., 1996; Dragic et al., 1996; Feng et al., 1996), whereas the presence of chemokines can sometimes help prevent infection. These observations have elicited intense interest in chemokine biology.

Neurons express both chemokines and chemokine receptors, and although not infected by HIV, neurons do express the coreceptors CXCR4 (Zhang et al., 1998) and CCR5 (Rottman et al., 1997). Similar to cells infected by HIV, the neuron coreceptors have affinity for HIV envelope protein gp120, regardless of CD4. Many groups have since shown neuronal toxicity mediated by viral proteins, particularly gp120 (Hesselgesser et al., 1998; Kaul and Lipton, 1999; Ohagen et al., 1999; Zheng et al., 1999b; Chen et al., 2002; Garden et al., 2004). Upon interaction with coreceptors, gp120 induces signaling cascades that may play a role in promoting apoptosis. Blocking of these cascades can block neuronal death in some cases. Interestingly, different viral strains induce varying levels of neuronal toxicity (Gabuzda and Wang, 1999; Zheng et al., 1999a).

15.6.1.2. Neuroprotective and Neurotoxic Effects of Chemokines and Their Receptors in HAD

In contrast to HIV-1 coreceptors, some chemokine ligands have the ability to reduce or ablate neuron toxicity. High levels of chemokines RANTES, MIP-1α, and others have been shown to reduce neuron death (Meucci et al., 1998; Kaul and Lipton, 1999), while SDF-1, at higher concentrations may actually promote neuronal death (Hesselgesser et al., 1998; Kaul and Lipton, 1999; Zheng et al., 1999b). The mechanism is not yet completely understood, but may rely upon simple competitive inhibition, receptor expression changes on the cell surface, or other unknown mechanisms.

Fractalkine (CX3CL1): FKN levels are higher in the CSF of cognitively impaired HIV patients than in infected subjects without cognitive impairment. Moreover, FKN can affect the chemotaxis of primary monocytes across an artificial blood brain barrier and is neuroprotective to cultured neurons (Meucci et al., 2000; Tong et al., 2000). Thus, this neuronal chemokine may serve as a damage signal to recruit macrophages and microglia to the site of injury (Jung et al., 2000; Tong et al., 2000; Zheng et al., 2000a; Zujovic et al., 2000). Subsequent chemokine-MP interactions can initiate inflammatory responses through the production of chemokines/cytokines or protective responses through the production of neurotrophins (Xiao and Link, 1998; Kaul et al., 2001; Cotter et al., 2002).

MCP-1 (CCL2): Despite some association of chemokines and neuroprotection, there are also detrimental effects of chemokine function during HAD pathogenesis. Shown to be expressed in the brains of HAD patients (Conant et al., 1998), MCP-1 (CCL2) is a potent chemoattractant for monocytes and may help fuel the positive feedback loop of inflammation in the HAD brain. MCP-1 recruits monocytic phagocytes to sites of inflammation as was evidenced by a study using a mouse system with elevated levels of MCP-1 resulting in increased phagocytic cells at lesion sites (Fuentes et al., 1995). A high level of MCP-1 in CSF versus plasma was shown to be predictive of dementia development in monkeys (Zink et al., 2001). While clearly possessing chemotactic properties, the ability of MCP-1 to recruit phagocytes through the BBB was further elucidated with a study showing changes in BBB permeability in the presence of MCP-1 (Song and Pachter, 2004). Increased levels of MCP-1 were shown to result in initial protection from infection, however, upon successful HIV-1 infection, increased MCP-1 was shown to lead to increased susceptibility to the development of HAD (Gonzalez et al., 2002).

Interferon-inducible protein 10 (IP10, CXCL10): IP-10 is a CXC chemokine. As indicated by its name, IP-10 is highly induced by interferon as well as other factors, yet is also produced constitutively throughout the body. IP-10 targets multiple subtypes of activated T-cells and macrophages for migration. IP-10 has been found in very high levels in CSF as well as shown to recruit cells into the CNS in the setting of HAD. While clearly a player in recruitment and inflammation, IP-10's role has also been shown to include cytotoxic effects toward neurons (van Marle et al., 2004) and may stimulate HIV-1 replication in macrophages (Lane et al., 2003).

IL-8 (CXCL8): An endogenous ligand for CXCR2, IL-8 is secreted in high levels by HIV-1 infected lymphocytes and macrophages. Although expressed constitutively, immune activation potentiates IL-8 production from infected or uninfected macrophages by agents such as LPS or CD40L (Zheng et al., 2000b). IL-8 levels are increased in the CSF of HAD patients, more so than those lacking cognitive symptoms, supporting the role IL-8 in HAD (Zheng et al., 2001).

SDF-1 (CXCL12): Chemokines have also been shown to have a neuromodulatory capacity, in some cases decreasing excitation and avoiding toxicity. A complicated example is the effect of SDF-1 on glutamate toxicity and uptake, specifically as regulated through astrocytes. SDF-1 is a member of the CXC chemokine subfamily and is the only known physiological ligand for CXCR4 (Rossi and Zlotnik, 2000). CXCR4 is upregulated in HIV and SIV encephalitis, experimental allergic encephalitis (EAE), and brain tumors (Jiang et al., 1998; Sanders et al., 1998; Vallat et al., 1998; Westmoreland et al., 1998). SDF-1 is a potent chemoattractant for resting lymphocytes, monocytes, and CD34-positive hematopoietic progenitor cells (Kim and Broxmeyer, 1999). SDF-1 transcripts are predominantly expressed by oligodendrocytes, astrocytes and neurons in the cortex, hippocampus, and cerebellum (Gleichmann et al., 2000; Stumm et al., 2003). SDF-1 has been shown to be upregulated in the brain of patients with HIVE and in astrocytes by HIV-1-infected and/or immune-activated macrophages *in-vivo* (Langford et al., 2002; Rostasy et al., 2003; Peng et al., 2006). Studies conducted in different settings have shown SDF-1 to promote neuronal survival, reducing glutamate toxicity (Meucci et al., 1998), while other studies have shown SDF-1 to increase neuronal death by interacting with CXCR4 and increasing the release of glutamate and TNF-α from glial cells (Meucci et al., 1998; Kaul and Lipton, 1999; Zheng et al., 1999b; Bezzi et al., 2001; Kaul et al., 2007). This may be due to experimental variation or a concentration dependent effect of SDF-1 and glutamate regulation. Recently, it was suggested that SDF-1 could be cleaved to SDF-1 (5–67) and mediate direct neurotoxicity through CXCR3 (Zhang et al., 2003; Vergote et al., 2006).

15.6.1.3. Therapeutic Avenues Directed Toward Chemokines and Their Receptors

Because HIV requires coreceptors for the induction of productive viral infection and naturally occurring alleles have been shown to effectively limit HIV entry (reviewed in Tang and Kaslow, 2003), potential therapies may rely upon exploitation of CXCR4 and CCR5. However, the ability of HIV to mutate presents a great hurdle in the development of effective therapeutic receptor antagonists. Another layer of complexity is the potential side effects of blocking one or multiple chemokine receptors that have homeostatic and inflammatory roles. Despite the inherent difficulties, multiple approaches have been studied and some are now being tested in clinical trials. Small molecule inhibitors, monoclonal antibodies, and modified chemokine ligands are all related but distinct approaches to limiting HIV entry currently being developed and tested (reviewed in Shaheen and Collman, 2004). Another approach to block initial HIV infection may be to deliver siRNAs targeting the receptors for knockdown (Zhou et al., 2004).

Some chemokine ligands inherently disrupt viral pathogenesis or provide protection against cell death during the disease process. Ligands of HIV coreceptors, such as SDF-1 and RANTES, have been shown to block infection of cells in different systems (Bleul et al., 1996; Lederman et al., 2004). Similarly, ligands to coreceptors have been shown to block HIV envelope protein induced toxicity to neurons (Alkhatib et al., 1996; Meucci and Miller, 1996). The mechanism may be simple blocking or internalization of receptors, or may rely upon the signaling downstream of chemokine interaction.

15.6.2. Chemokines and Their Receptors in MS

Similar to cytokines, the chemokine expression pattern is also a Th1-mediated response in MS. Pro-inflammatory cytokines activate resident macrophages and microglia within the CNS (Figure 15.1). Recruitment and attraction of these cells occurs via integrins and chemokines and is believed to contribute to tissue injury and demyelination. Selective expression of individual chemokines may influence the cellular composition of inflammatory lesions because chemokine receptors are associated with either Th1 or Th2 responses. Th1 proinflammatory cells may express CCR5 (receptor for chemokines RANTES, MIP-1 α and MIP-1β) and CXCR3 (receptor for IP-10 and MIG), whereas Th2 inflammatory cells may shift toward the display of CCR3 (receptor for MCP-3, MCP-4, and RANTES) and CCR8.

In MS autopsy brain sections, CCR5 and CXCR3 are overexpressed in peripheral and lesion associated T-lymphocytes (Simpson et al., 1998; Balashov et al., 1999). The ligands for CCR5 and CXCR3, RANTES and IP-10, also increase in the CSF (Lucchinetti et al., 2003). For the cell source of chemokines, IP-10 was associated with astrocytes and perivascular astrocytic processes within MS lesions (Simpson et al., 1998; Balashov et al., 1999; Sorensen et al., 1999). These results suggested that the interaction between IP-10 and CXCR3 might be the mechanism that traffics T-cells into MS lesions.

Similarly, elevated CCR5 in activated macrophages was found in MS lesions, indicating RANTES might mediate the recruitment and activation of monocytes and macrophages in MS (Sorensen et al., 1999). Further, MCP-1, MCP-2, MCP3, MIP-1α, and MIP-1β immunoreactive astrocytes within MS lesions have been described (McManus et al., 1998; Simpson et al., 1998; van der Voorn et al., 1999; Boven et al., 2000). These results were support by the study in the EAE model, particularly MCP-1 (Ransohoff et al., 1993; Berman et al., 1996), IP-10 (Ransohoff et al., 1993), RANTES (Godiska et al., 1995).

Taken together, these results indicate extensive chemotactic interactions between the glia cells in MS lesion and the infiltrating cells. Whether these interactions ultimately promote destructive inflammation or recruitment of viral and protective regulatory cells that enhance tissue repair is still not clear. This is a crucial question to answer before potential pharmacological intervention based on chemokines and their receptors could be reinforced for the treatment of the MS.

15.6.3. Chemokines and Their Receptors in AD

Immunohistochemical analysis of tissue from human brains with AD have revealed that in AD, there is increased expression of MIP-1β and IP-10 by activated astrocytes, and of the chemokine receptors CCR3 and CCR5 on activated microglia, adjacent to Aβ deposits in AD (Xia et al., 1998, 2000). Dystrophic neuritis has been shown to express the IL-8 receptor CXCR2 within the Aβ plaques of AD (Xia et al., 1997). Aβ is capable of modulating the inflammatory processes involved in AD through chemokine production. MCP-1 is found in activated microglia and within neuritic Aβ plaques, but not in early plaque forms(Ishizuka et al., 1997). Aβ promotes production of MCP-1, MIP-1α, MIP-1β, or IL-8 by monocytes and microglia (Fiala et al., 1998; Meda et al., 1999); Aβ also promotes expression of MCP-1 and RANTES by astrocytes (Johnstone et al., 1999). These observations collectively suggest that chemokines may play a role in the pathogenesis of AD (Xia and Hyman, 1999). It is likely that the production of chemokines by the surrounding cells plays a role in the recruitment and accumulation of astrocytes and microglia in senile plaques (Cartier et al., 2005).

Although the exact role of chemokines in AD is still not yet well defined, neuroprotective effects of chemokines have been noted. Fractalkine/CX3CL1 is suggested to have a protective role in AD. It has been reported that Fractalkine/CX3CL1 inhibits the production of IL-6, TNF-α and NO with Aβ-primed microglia, thereby improving the neuronal survival rate (Zujovic et al., 2000; Mizuno et al., 2003). GRO-α (CXCL2), the ligand for CXCR2 has been suggested to trigger ERK1/2 and PI-3K pathway in a way similar to BDNF, indicating GRO-α, or perhaps other CXCR2 ligands, could have a neutrotrophic effect (Xia and Hyman, 2002). Neuroprotective effects of other chemokines, such as RANTES, SDF-1α, IP-10, and MIP have also been documented (Meucci et al., 1998).

With the current understanding of chemokines and their related regulations, it is unlikely that pharmaceutical intervention will soon be used to treat AD. New therapeutic modalities targeting chemokines and their receptors may become available as a more detailed understanding of chemokines, their receptors, and potential agents in inflammatory changes is garnered.

Summary

The role of cytokines, growth factors, and chemokines in neuroimmune dysregulation is complex, due to the overlapping, synergizing, and antagonizing effects of various factors. Classifying any individual factor or family of factors as beneficial or detrimental oversimplifies the interactions between various cell types and the signaling cascades initiated by them. Instead, cytokines need to be considered as a balanced network, where subtle modifications can shift cells toward different outcomes such as death, proliferation, migration, and induction of inflammation or inhibition of immune responses. Prolonged inflammation has a profound impact on the cytokine network and the cells they target, consequently altering the outcome of cell populations. The CNS is a unique environment that is exquisitely sensitive to cytokines, growth factors, and chemokines; the dysregulation that is rampant during neurodegenerative diseases permanently transforms brain function. The profile of cytokines, growth factors, and chemokines presented in the brain influences the function and ultimately the fate of neurons. Understanding the effects of cytokines, growth factors, and chemokines and how the expression or activity of those factors can be manipulated may provide the key to diagnosing or treating neuroimmunological disease.

Review Questions/Problems

1. **What is a cytokine?**
2. **How does IL-1β contributes to the development of HIV-1 associated dementia, Alzheimer's disease, and multiple sclerosis?**
3. **List five members of TNF family of cytokine and their effects in CNS inflammation.**
4. **List three major families of neurotrophic factors.**
5. **How do different stimuli lead to CNS inflammation in the pathogenesis of HIV-1 associated dementia, Alzheimer's disease, and multiple sclerosis?**
6. **Briefly discuss the approach to classify chemokines and provide an example for each type.**
7. **Describe the two distinct means in which chemokine receptors participate in the pathogenesis of HAD.**
8. **Which chemokine receptors have been identified as coreceptors for HIV?**

9. **Compare two distinct models on how HIV-1 leads to neuronal injury.**
10. **Mitochondria play a pivotal role in mediating signaling events leading to apoptosis. Briefly describe the mitochondrial apoptosis pathway in a sequential order.**
11. **Which of the following statements about chemokines is correct?**
    a. Chemokines and chemokine receptors play a significant role in pathogenesis of inflammatory disorders, such as HIV-1 associated dementia.
    b. Stromal derived factor-1α (SDF-1α, CXCL12) and fractalkine (FKN, CX3CL1) are constitutively produced in the brain.
    c. Macrophage inhibitory proteins one alpha and one beta (MIP-1α, CCL3 and MIP-1β, CCL4), and monocyte chemotactic protein-1 (MCP-1, CCL2) could be induced by inflammatory stimuli in the CNS.
    d. All of the above are correct.
12. **Propose a potential therapeutic approach using a cytokine or chemokine in HIV-1 associated dementia.**
13. **Which of the following statements are FALSE in regards to inflammatory cytokines and chemokines in relationship to neurodegenerative diseases?**
    a. Microglia, brain macrophage and astrocytes are major sources of inflammatory mediators for disease.
    b. Pro-inflammatory cytokines mediate neurodegeneration through the upregulation of neurotrophic factors including NGF, BDNF, and NT-3.
    c. Chemokines such as MCP-1 Promote inflammation through recruitment of monocyte-derived macrophages that propagate tissue damage.
    d. Activated microglia may contribute to tissue damage.
14. **The following belongs to TNF family members EXCEPT?**
    a. CD40L
    b. TRAIL
    c. FasL
    d. TGF
15. **The following cytokines serve as pro-inflammatory factors EXCEPT?**
    a. IL-10
    b. IL-1
    c. IFN-
    d. IL-6
16. **Which of the following statements about type I and type II cytokines are incorrected?**
    a. Type I and type II cytokine families are classified based upon the structural homology of their receptors.
    b. Type I and type II cytokine families are comprised of both Th1 and Th2 cytokines.
    c. Interferons are type II cytokines that promote viral growth while serving as anti-inflammatory and immunomodulatory factors.
    d. Members of the type I cytokine family share a characteristic motif consisting of an extracellular region containing four helices.
17. **Which anti-HIV therapeutics specifically target the infecting agent?**
    a. Co-receptors: CCR5 and CXCR4
    b. Viral receptor: CD4
    c. Viral enzymes required for replication
    d. Adenylate cyclase
18. **The chemokine co-receptors for HIV-infection infection are?**
    a. CXCR7 and CCR1
    b. CCR1 and CCR5
    c. CXCR4 and CCR1
19. **ELR+ chemokines target which cell type?**
    a. Monocytes and lymphocytes
    b. Neutrophils
    c. Progenitor cells
    b. Monocytes
20. **Chemokines are defined by the organization of their N-terminal cysteine residues. Which of the following is not representative of a chemokine classification?**
    a. CC
    b. CXC
    c. CXXC
    d. C
21. **Which of the following is the most potent chemoattractant for monocytes and macrophages?**
    a. CXCL12
    b. CXXXCL3
    c. CCL2
    d. CXCL8

Acknowledgments. This work was supported in part by research grants by the National Institutes of Health: R01 NS 41858, P20 RR15635 and P01 NS043985 (JZ). We kindly acknowledge Dr. Howard E. Gendelman for valuable suggestion and critics of this work. Hui Peng, Robin Cotter, Lisa Ryan, Shelley Herek, David Erichsen, Alicia Lopez, Li Wu, Clancy McNally, Min Cui, Jianxing Zhao, Angelique Walstrom, and Anastasia Persidsky provided technical support for this work. Julie Ditter, Robin Taylor, Myhanh Che, Nell Ingraham and Emilie Scoggins provided outstanding administrative support.

References

Adle-Biassette H, Levy Y, Colombel M, Poron F, Natchev S, Keohane C, Gray F (1995) Neuronal apoptosis in HIV infection in adults. Neuropathol App Neurobiol 21:218–227.

Adorini L, Sinigaglia F (1997) Pathogenesis and immunotherapy of autoimmune diseases. Immunol Today 18:209–211.

Akiyama H, Barger S, Barnum S, Bradt B, Bauer J, Cole GM, Cooper NR, Eikelenboom P, Emmerling M, Fiebich BL, Finch CE, Frautschy S, Griffin WS, Hampel H, Hull M, Landreth G, Lue L, Mrak R, Mackenzie IR, McGeer PL, O'Banion MK, Pachter J, Pasinetti G, Plata-Salaman C, Rogers J, Rydel R, Shen Y, Streit W, Strohmeyer R, Tooyoma I, Van Muiswinkel FL, Veerhuis R, Walker D, Webster S, Wegrzyniak B, Wenk G, Wyss-Coray T (2000) Inflammation and Alzheimer's disease. Neurobiol Aging 21:383–421.

Aktas O, Smorodchenko A, Brocke S, Infante-Duarte C, Topphoff US, Vogt J, Prozorovski T, Meier S, Osmanova V, Pohl E, Bechmann I, Nitsch R, Zipp F (2005) Neuronal damage in autoimmune neuroinflammation mediated by the death ligand TRAIL. Neuron 46:421–432.

Alkhatib G, Combadiere C, Broder CC, Feng Y, Kennedy PE, Murphy PM, Berger EA (1996) CC CKR5: A RANTES, MIP-1alpha, MIP-1beta receptor as a fusion cofactor for macrophage-tropic HIV-1. Science 272:1955–1958.

Aloisi F (2001) Immune function of microglia. Glia 36:165–179.

Aloisi F, Care A, Borsellino G, Gallo P, Rosa S, Bassani A, Cabibbo A, Testa U, Levi G, Peschle C (1992) Production of hemolymphopoietic cytokines (IL-6, IL-8, colony-stimulating factors) by normal human astrocytes in response to IL-1 beta and tumor necrosis factor-alpha. J Immunol 149:2358–2366.

Bachis A, Colangelo AM, Vicini S, Doe PP, De Bernardi MA, Brooker G, Mocchetti I (2001) Interleukin-10 prevents glutamate-mediated cerebellar granule cell death by blocking caspase-3-like activity. J Neurosci 21:3104–3112.

Bachis A, Major EO, Mocchetti I (2003) Brain-derived neurotrophic factor inhibits human immunodeficiency virus-1/gp120-mediated cerebellar granule cell death by preventing gp120 internalization. J Neurosci 23:5715–5722.

Balashov KE, Rottman JB, Weiner HL, Hancock WW (1999) CCR5(+) and CXCR3(+) T cells are increased in multiple sclerosis and their ligands MIP-1alpha and IP-10 are expressed in demyelinating brain lesions. Proc Natl Acad Sci USA 96:6873–6878.

Barak O, Goshen I, Ben-Hur T, Weidenfeld J, Taylor AN, Yirmiya R (2002) Involvement of brain cytokines in the neurobehavioral disturbances induced by HIV-1 glycoprotein120. Brain Res 933:98–108.

Bazan J, Bacon K, Hardiman G, Wang W, Soo K, Rossi D, Greaves D, Zlotnik A, Schall T (1997) A new class of membrane-bound chemokine with a CX3C motif. Nature 385:640–644.

Benveniste EN (1992) Inflammatory cytokines within the central nervous system: Sources, function and mechanisms of action. Am J Physiol 263(1 Pt 1):C1–C16.

Benveniste EN (1997) Role of macrophages/microglia in multiple sclerosis and experimental allergic encephalomyelitis. J Mol Med 75:165–173.

Berger EA, Murphy PM, Farber JM (1999) Chemokine receptors as HIV-1 coreceptors: Roles in viral entry, tropism, and disease. Annu Rev Immunol 17:657–700.

Berman JW, Guida MP, Warren J, Amat J, Brosnan CF (1996) Localization of monocyte chemoattractant peptide-1 expression in the central nervous system in experimental autoimmune encephalomyelitis and trauma in the rat. J Immunol 156:3017–3023.

Bernard CC, de Rosbo NK (1991) Immunopathological recognition of autoantigens in multiple sclerosis. Acta Neurol (Napoli) 13:171–178.

Bernasconi S, Cinque P, Peri G, Sozzani S, Crociati A, Torri W, Vicenzi E, Vago L, Lazzarin A, Poli G, Mantovani A (1996) Selective elevation of monocyte chemotactic protein-1 in the cerebrospinal fluid of AIDS patients with cytomegalovirus encephalitis. J Infect Dis 174:1098–1101.

Berson JF, Doms RW (1998) Structure-function studies of the HIV-1 coreceptors. Semin Immunol 10:237–248.

Bezzi P, Domercq M, Brambilla L, Galli R, Schols D, De Clercq E, Vescovi A, Bagetta G, Kollias G, Meldolesi J, Volterra A (2001) CXCR4-activated astrocyte glutamate release via TNFalpha: Amplification by microglia triggers neurotoxicity. Nat Neurosci 4:702–710.

Bitsch A, Kuhlmann T, Da Costa C, Bunkowski S, Polak T, Bruck W (2000) Tumour necrosis factor alpha mRNA expression in early multiple sclerosis lesions: Correlation with demyelinating activity and oligodendrocyte pathology. Glia 29:366–375.

Bjartmar C, Wujek JR, Trapp BD (2003) Axonal loss in the pathology of MS: Consequences for understanding the progressive phase of the disease. J Neurol Sci 206:165–171.

Bleul CC, Farzan M, Choe H, Parolin C, Clark-Lewis I, Sodroski J, Springer TA (1996) The lymphocyte chemoattractant SDF-1 is a ligand for LESTR/fusin and blocks HIV-1 entry. Nature 382:829–833.

Boutros T, Croze E, Yong VW (1997) Interferon-beta is a potent promoter of nerve growth factor production by astrocytes. J Neurochem 69:939–946.

Boven LA, Middel J, Portegies P, Verhoef J, Jansen GH, Nottet HS (1999) Overexpression of nerve growth factor and basic fibroblast growth factor in AIDS dementia complex. J Neuroimmunol 97:154–162.

Boven LA, Montagne L, Nottet HS, De Groot CJ (2000) Macrophage inflammatory protein-1alpha (MIP-1alpha), MIP-1beta, and RANTES mRNA semiquantification and protein expression in active demyelinating multiple sclerosis (MS) lesions. Clin Exp Immunol 122:257–263.

Broe GA, Grayson DA, Creasey HM, Waite LM, Casey BJ, Bennett HP, Brooks WS, Halliday GM (2000) Anti-inflammatory drugs protect against Alzheimer's disease at low doses. Arch Neurol 57:1586–1591.

Cartier L, Hartley O, Dubois-Dauphin M, Krause KH (2005) Chemokine receptors in the central nervous system: Role in brain inflammation and neurodegenerative diseases. Brain Res Brain Res Rev 48:16–42.

Chen W, Sulcove J, Frank I, Jaffer S, Ozdener H, Kolson DL (2002) Development of a human neuronal cell model for human immunodeficiency virus (HIV)-infected macrophage-induced neurotoxicity: Apoptosis induced by HIV type 1 primary isolates and evidence for involvement of the Bcl-2/Bcl-xL-sensitive intrinsic apoptosis pathway. J Virol 76:9407–9419.

Chung IY, Benveniste EN (1990) Tumor necrosis factor-alpha production by astrocytes. Induction by lipopolysaccharide, IFN-gamma, and IL-1 beta. J Immunol 144:2999–3007.

Cinque P, Vago L, Mengozzi M, Torri V, Ceresa D, Vicenzi E, Transidico P, Vagani A, Sozzani S, Mantovani A, Lazzarin A, Poli G (1998) Elevated cerebrospinal fluid levels of monocyte chemotactic protein-1 correlate with HIV-1 encephalitis and local viral replication. Aids 12:1327–1332.

Cocchi F, DeVico AL, Garzino-Demo A, Arya SK, Gallo RC, Lusso P (1995) Identification of RANTES, MIP-1alpha, and MIP-1beta as the major HIV-suppressive factors produced by CD8 + T cells. Science 270:1811–1815.

Collman RG, Yi Y, Liu QH, Freedman BD (2000) Chemokine signaling and HIV-1 fusion mediated by macrophage CXCR4: Implications for target cell tropism. J Leukoc Biol 68:318–323.

Conant K, Garzino-Demo A, Nath A, McArthur JC, Halliday W, Power C, Gallo RC, Major EO (1998) Induction of monocyte chemoattractant protein-1 in HIV-1 Tat-stimulated astrocytes and elevation in AIDS dementia. Proc Natl Acad Sci USA 95:3117–3121.

Cook DN, Chen SC, Sullivan LM, Manfra DJ, Wiekowski MT, Prosser DM, Vassileva G, Lira SA (2001) Generation and analysis of mice lacking the chemokine fractalkine. Mol Cell Biol 21:3159–3165.

Correale J, Rojany M, Weiner LP (1997) Human CD8 + TCR-alpha beta(+) and TCR-gamma delta(+) cells modulate autologous autoreactive neuroantigen-specific CD4 + T-cells by different mechanisms. J Neuroimmunol 80:47–64.

Cotter R, Williams C, Ryan L, Erichsen D, Lopez A, Peng H, Zheng J (2002) Fractalkine (CX3CL1) and brain inflammation: Implications for HIV-1-associated dementia. J Neurovirol 8:585–598.

Coughlan CM, McManus CM, Sharron M, Gao Z, Murphy D, Jaffer S, Choe W, Chen W, Hesselgesser J, Gaylord H, Kalyuzhny A, Lee VM, Wolf B, Doms RW, Kolson DL (2000) Expression of multiple functional chemokine receptors and monocyte chemoattractant protein-1 in human neurons [In Process Citation]. Neuroscience 97:591–600.

Damaj BB, McColl SR, Mahana W, Crouch MF, Naccache PH (1996) Physical association of Gi2alpha with interleukin-8 receptors. J Biol Chem 271:12783–12789.

Danial NN, Korsmeyer SJ (2004) Cell death: Critical control points. Cell 116:205–219.

De Stefano N, Narayanan S, Francis GS, Arnaoutelis R, Tartaglia MC, Antel JP, Matthews PM, Arnold DL (2001) Evidence of axonal damage in the early stages of multiple sclerosis and its relevance to disability. Arch Neurol 58:65–70.

Deng H, Liu R, Ellmeier W, Choe S, Unutmaz D, Burkhart M, Di Marzio P, Marmon S, Sutton RE, Hill CM, Davis CB, Peiper SC, Schall TJ, Littman DR, Landau NR (1996) Identification of a major co-receptor for primary isolates of HIV-1. Nature 381:661–666.

Devi LA (2000) G-protein-coupled receptor dimers in the lime light. Trends Pharmacol Sci 21:324–326.

Dimitrov DS, Xiao X, Chabot DJ, Broder CC (1998) HIV coreceptors. J Membr Biol 166:75–90.

Dong Y, Benveniste EN (2001) Immune function of astrocytes. Glia 36:180–190.

Dragic T, Litwin V, Allaway GP, Martin SR, Huang Y, Nagashima KA, Cayanan C, Maddon PJ, Koup RA, Moore JP, Paxton WA (1996) HIV-1 entry into CD4 + cells is mediated by the chemokines receptor CC-CKR-5. Nature 381:667–673.

Ehrlich S, Infante-Duarte C, Seeger B, Zipp F (2003) Regulation of soluble and surface-bound TRAIL in human T cells, B cells, and monocytes. Cytokine 24:244–253.

Eilbott DJ, Peress N, Burger H, LaNeve D, Orenstein J, Gendelman HE, Seidman R, Weiser B (1989) Human immunodeficiency virus type 1 in spinal cords of acquired immunodeficiency syndrome patients with myelopathy: Expression and replication in macrophages. Proc Natl Acad Sci USA 86:3337–3341.

Esser R, Glienke W, von Briesen H, Rubsamen-Waigmann H, Andreesen R (1996) Differential regulation of proinflammatory and hematopoietic cytokines in human macrophages after infection with human immunodeficiency virus. Blood 88:3474–3481.

Feng Y, Broder CC, Kennedy PE, Berger EA (1996) HIV-1 entry cofactor: Functional cDNA cloning of a seven-transmembrane, G protein-coupled receptor [see comments]. Science 272:872–877.

Fiala M, Zhang L, Gan X, Sherry B, Taub D, Graves MC, Hama S, Way D, Weinand M, Witte M, Lorton D, Kuo YM, Roher AE (1998) Amyloid-beta induces chemokine secretion and monocyte migration across a human blood–brain barrier model. Mol Med 4:480–489.

Fischer-Smith T, Croul S, Adeniyi A, Rybicka K, Morgello S, Khalili K, Rappaport J (2004) Macrophage/microglial accumulation and proliferating cell nuclear antigen expression in the central nervous system in human immunodeficiency virus encephalopathy. Am J Pathol 164:2089–2099.

Frost EE, Nielsen JA, Le TQ, Armstrong RC (2003) PDGF and FGF2 regulate oligodendrocyte progenitor responses to demyelination. J Neurobiol 54:457–472.

Fuentes M, Durham S, Swerdel M, Letwin A, Barton D, Megill J, Bravo R, Lira L (1995) Controlled recruitment of monocytes and macrophages to specific organs through transgenic expression of monocyte chemoattractant protein-1. J Immunol 155:5769–5776.

Gabuzda D, Wang J (1999) Chemokine receptors and virus entry in the central nervous system. J Neurovirol 5:643–658.

Gabuzda D, Wang J (2000) Chemokine receptors and mechanisms of cell death in HIV neuropathogenesis. J Neurovirol 6(Suppl 1): S24–S32.

Gabuzda D, He J, Ohagen A, Vallat A (1998) Chemokine receptors in HIV-1 infection of the central nervous system. Immunol 10:203–213.

Gabuzda D, Wang J, Gorry P (2002) HIV-1-Associated Dementia. In: Chemokines and the Nervous System (Ransohoff RM, Suzuki K, Proudfoot AEI, Hickey WF, Harrison JK, eds), pp 345–360. Amsterdam: Elservier Science.

Gallo P, Sivieri S, Rinaldi L, Yan XB, Lolli F, De Rossi A, Tavolato B (1994) Intrathecal synthesis of interleukin-10 (IL-10) in viral and inflammatory diseases of the central nervous system. J Neurol Sci 126:49–53.

Ganju RK, Brubaker SA, Meyer J, Dutt P, Yang Y, Qin S, Newman W, Groopman JE (1998) The alpha-chemokine, stromal cell-derived factor-1alpha, binds to the transmembrane G-protein-coupled CXCR-4 receptor and activates multiple signal transduction pathways. J Biol Chem 273:23169–23175.

Garden GA, Guo W, Jayadev S, Tun C, Balcaitis S, Choi J, Montine TJ, Moller T, Morrison RS (2004) HIV associated neurodegeneration requires p53 in neurons and microglia. FASEB J 18:1141–1143.

Gelbard HA (1999) Neuroprotective strategies for HIV-1-associated neurologic disease. Ann N Y Acad Sci 890:312–313.

Gendelman HE, Folks DG (1999) Innate and acquired immunity in neurodegenerative disorders. J Leuk Biol 65:407–409.

Gendelman HE, Baca LM, Kubrak CA, Genis P, Burrous S, Friedman RM, Jacobs D, Meltzer MS (1992) Induction of IFN-alpha in peripheral blood mononuclear cells by HIV-infected monocytes. Restricted antiviral activity of the HIV-induced IFN. J Immunol 148:422–429.

Gendelman HE, Persidsky Y, Ghorpade A, Limoges J, Stins M, Fiala M, Morrisett R (1997) The neuropathogenesis of the AIDS dementia complex. Aids 11:S35–45.

Gendelman HE, Zheng J, Coulter CL, Ghorpade A, Che M, Thylin M, Rubocki R, Persidsky Y, Hahn F, Reinhard J, Jr., Swindells S (1998) Suppression of inflammatory neurotoxins by highly active antiretroviral therapy in human immunodeficiency virus-associated dementia. J Infect Dis 178:1000–1007.

Gessani S, Puddu P, Varano B, Borghi P, Conti L, Fantuzzi L, Gherardi G, Belardelli F (1994) Role of endogenous interferon-beta in the restriction of HIV replication in human monocyte/macrophages. J Leukoc Biol 56:358–361.

Gielen A, Khademi M, Muhallab S, Olsson T, Piehl F (2003) Increased brain-derived neurotrophic factor expression in white blood cells of relapsing-remitting multiple sclerosis patients. Scand J Immunol 57:493–497.

Glass JD, Wesselingh SL, Selnes OA, McArthur JC (1993) Clinical neuropathologic correlation in HIV-associated dementia. Neurology 43:2230–2237.

Gleichmann M, Gillen C, Czardybon M, Bosse F, Greiner-Petter R, Auer J, Muller HW (2000) Cloning and characterization of SDF-1gamma, a novel SDF-1 chemokine transcript with developmentally regulated expression in the nervous system. Eur J Neurosci 12:1857–1866.

Goda S, Imai T, Yoshie O, Yoneda O, Inoue H, Nagano Y, Okazaki T, Imai H, Bloom ET, Domae N, Umehara H (2000) CX3C-chemokine, fractalkine-enhanced adhesion of THP-1 cells to endothelial cells through integrin-dependent and -independent mechanisms. J Immunol 164:4313–4320.

Godiska R, Chantry D, Dietsch GN, Gray PW (1995) Chemokine expression in murine experimental allergic encephalomyelitis. J Neuroimmunol 58:167–176.

Goldgaber D, Harris H, Hla T, Maciag T, Donnelly R, Jacobsen J, Vitek M, Gajdusek D (1989) IL-1 regulates synthesis of amyloid beta-protein precursor mRNA in human endothelial cells. Proc Natl Acad Sci USA 86:7606–7610.

Gonzalez E, Rovin BH, Sen L, Cooke G, Dhanda R, Mummidi S, Kulkarni H, Bamshad MJ, Telles V, Anderson SA, Walter EA, Stephan KT, Deucher M, Mangano A, Bologna R, Ahuja SS, Dolan MJ, Ahuja SK (2002) HIV-1 infection and AIDS dementia are influenced by a mutant MCP-1 allele linked to increased monocyte infiltration of tissues and MCP-1 levels. Proc Natl Acad Sci USA 99:13795–13800.

Green DR (2003) The suicide in the thymus, a twisted trail. Nat Immunol 4:207–208.

Green DR, Reed JC (1998) Mitochondria and apoptosis. Science 281:1309–1311.

Griffin DE (1997) Cytokines in the brain during viral infection: Clues to HIV-associated dementia. J Clin Invest 100:2948–2951.

Griffin W, Stanley L, Ling C, White L, MacLeod V, Perrot L, White C, Araoz C (1989) Brain interleukin 1 and S-100 immunoreactivity are elevated in Down syndrome and Alzheimer's disease. Proc Natl Acad Sci USA 86:7611–7615.

Griffin W, Sheng J, Roberts G, Mrak R (1995) Interleukin-1 expression in different plaque types in Alzheimer's diseases: Significance in plaque evolution. J Neuropathol Exp Neurol 54:276–281.

Gross A, McDonnell JM, Korsmeyer SJ (1999) BCL-2 family members and the mitochondria in apoptosis. Genes Dev 13:1899–1911.

Guo H, Jin YX, Ishikawa M, Huang YM, van der Meide PH, Link H, Xiao BG (1998) Regulation of beta-chemokine mRNA expression in adult rat astrocytes by lipopolysaccharide, proinflammatory and immunoregulatory cytokines. Scand J Immunol 48:502–508.

Hanisch UK (2002) Microglia as a source and target of cytokines. Glia 40:140–155.

Harrison JK, Fong AM, Swain PA, Chen S, Yu YR, Salafranca MN, Greenleaf WB, Imai T, Patel DD (2001) Mutational analysis of the fractalkine chemokine domain: Basic amino acid residues differentially contribute to CX3CR1 binding, signaling, and cell adhesion. J Biol Chem 8:8.

Haskell CA, Cleary MD, Charo IF (2000) Unique role of the chemokine domain of fractalkine in cell capture kinetics of receptor dissociation correlate with cell adhesion. J Biol Chem 275:34183–34189.

Hausmann EH, Berman NE, Wang YY, Meara JB, Wood GW, Klein RM (1998) Selective chemokine mRNA expression following brain injury. Brain Res 788:49–59.

Herbeuval JP, Boasso A, Grivel JC, Hardy AW, Anderson SA, Dolan MJ, Chougnet C, Lifson JD, Shearer GM (2005) TNF-related apoptosis-inducing ligand (TRAIL) in HIV-1-infected patients and its in vitro production by antigen-presenting cells. Blood 105:2458–2464.

Hesselgesser J, Horuk R (1999) Chemokine and chemokine receptor expression in the central nervous system. J Neurovirol 5:13–26.

Hesselgesser J, Taub D, Baskar P, Greenberg M, Hoxie J, Kolson DL, Horuk R (1998) Neuronal apoptosis induced by HIV-1 gp120 and the chemokine SDF-1alpha mediated by the chemokine receptor CXCR4. Curr Biol 8:595–598.

Hickey WF (1999) Leukocyte traffic in the central nervous system: The participants and their roles. Semin Immunol 11:125–137.

Hofman FM, Hinton DR, Johnson K, Merrill JE (1989) Tumor necrosis factor identified in multiple sclerosis brain. J Exp Med 170:607–612.

Hohlfeld R, Kerschensteiner M, Stadelmann C, Lassmann H, Wekerle H (2000) The neuroprotective effect of inflammation: Implications for the therapy of multiple sclerosis. J Neuroimmunol 107:161–166.

Hoover DM, Mizoue LS, Handel TM, Lubkowski J (2000) The crystal structure of the chemokine domain of fractalkine shows a novel quaternary arrangement. J Biol Chem 275:23187–23193.

Huang WX, Huang MP, Gomes MA, Hillert J (2000) Apoptosis mediators fasL and TRAIL are upregulated in peripheral blood mononuclear cells in MS. Neurology 55:928–934.

Huang Y, Erdmann N, Zhao J, Zheng J (2005a) The signaling and apoptotic effects of TNF-related apoptosis-inducing ligand in HIV-1 associated dementia. Neurotox Res 8:135–148.

Huang Y, Erdmann N, Peng H, Zhao Y, Zheng J (2005b) The role of TNF-related apoptosis-inducing ligand in neurodegenerative diseases. Cell Mol Immunol 2:113–122.

Huang Y, Erdmann N, Peng H, Herek S, Davis JS, Luo X, Ikezu T, Zheng J (2006) TRAIL-mediated apoptosis in HIV-1-infected macrophages is dependent on the inhibition of Akt-1 phosphorylation. J Immunol 177:2304–2313.

Hulshof S, Montagne L, De Groot CJ, van der Valk P (2002) Cellular localization and expression patterns of interleukin-10, interleukin-4, and their receptors in multiple sclerosis lesions. Glia 38:24–35.

Irani DN (1998) The susceptibility of mice to immune-mediated neurologic disease correlates with the degree to which their lymphocytes resist the effects of brain-derived gangliosides. J Immunol 161:2746–2752.

Ishizuka K, Igata-Yi R, Kimura T, Hieshima K, Kukita T, Kin Y, Misumi Y, Yamamoto M, Nomiyama H, Miura R, Takamatsu J, Katsuragi S, Miyakawa T (1997) Expression and distribution of CC chemokine macrophage inflammatory protein-1 a/LD78 in the human brain. Neuroreport 8:1215–1218.

Jacobs LD, Cookfair DL, Rudick RA, Herndon RM, Richert JR, Salazar AM, Fischer JS, Goodkin DE, Granger CV, Simon JH, Alam JJ, Bartoszak DM, Bourdette DN, Braiman J, Brownscheidle CM, Coats ME, Cohan SL, Dougherty DS, Kinkel RP, Mass MK, Munschauer FE, 3rd, Priore RL, Pullicino PM, Scherokman

BJ, Whitham RH (1996) Intramuscular interferon beta-1a for disease progression in relapsing multiple sclerosis. The Multiple Sclerosis Collaborative Research Group (MSCRG)s. Ann Neurol 39:285–294.

Jiang Y, Salafranca M, Adhikari S, Xia Y, Feng L, Sonntag M, deFiebre C, Pennel N, Streit W, Harrison J (1998) Chemokine receptor expression in cultured glia and rat experimental allergic encephalomyelitis. J Neuroimmunol 86:1–12.

Johnstone M, Gearing AJ, Miller KM (1999) A central role for astrocytes in the inflammatory response to beta- amyloid; chemokines, cytokines and reactive oxygen species are produced. J Neuroimmunol 93:182–193.

Jung S, Aliberti J, Graemmel P, Sunshine MJ, Kreutzberg GW, Sher A, Littman DR (2000) Analysis of fractalkine receptor CX(3)CR1 function by targeted deletion and green fluorescent protein reporter gene insertion. Mol Cell Biol 20:4106–4114.

Kast RE (2002) Feedback between glial tumor necrosis factor-alpha and gp120 from HIV-infected cells helps maintain infection and destroy neurons. Neuroimmunomodulation 10:85–92.

Kaul M, Lipton SA (1999) Chemokines and activated macrophages in HIV gp120-induced neuronal apoptosis. Proc Natl Acad Sci USA 96:8212–8216.

Kaul M, Garden GA, Lipton SA (2001) Pathways to neuronal injury and apoptosis in HIV-associated dementia. Nature 410:988–994.

Kaul M, Ma Q, Medders KE, Desai MK, Lipton SA (2007) HIV-1 coreceptors CCR5 and CXCR4 both mediate neuronal cell death but CCR5 paradoxically can also contribute to protection. Cell Death Differ 14:296–305.

Kedzierska K, Crowe SM, Turville S, Cunningham AL (2003) The influence of cytokines, chemokines and their receptors on HIV-1 replication in monocytes and macrophages. Rev Med Virol 13:39–56.

Kelder W, McArthur JC, Nance-Sproson T, McClernon D, Griffin DE (1998) b-Chemokines MCP-1 and RANTES are selectively increased in cerebrospinal fluid of patients with human immunodeficiency virus-associated dementia. Ann Neurol 44:831–835.

Kim CH, Broxmeyer HE (1999) Chemokines: Signal lamps for trafficking of T and B cells for development and effector function. J Leukocyte Biol 65:6–14.

Klein R, Williams K, Alvarez-Hernandez X, Westmoreland S, Force T, Lackner A, Luster A (1999) Chemokine receptor expression and signaling in macaque and human fetal neurons and astrocytes: Implications for the neuropathogenesis of AIDS. J Immunol 163:1636–1646.

Koenig S, Gendelman HE, Orenstein JM, Canto MCD, Pezeshkpour GH, Yungbluth M, Janotta F, Aksamit A, Martin MA, Fauci AS (1986) Detection of AIDS virus in macrophages in brain tissue from AIDS patients with encephalopathy. Science 233:1089–1093.

Korsching S (1993) The neurotrophic factor concept: A reexamination. J Neurosci 13:2739–2748.

Kuang Y, Wu Y, Jiang H, Wu D (1996) Selective G protein coupling by C–C chemokine receptors. J Biol Chem 271:3975–3978.

Kuhlmann T, Lingfeld G, Bitsch A, Schuchardt J, Bruck W (2002) Acute axonal damage in multiple sclerosis is most extensive in early disease stages and decreases over time. Brain 125:2202–2212.

Lane BR, King SR, Bock PJ, Strieter RM, Coffey MJ, Markovitz DM (2003) The C-X-C chemokine IP-10 stimulates HIV-1 replication. Virology 307:122–134.

Langford D, Sanders VJ, Mallory M, Kaul M, Masliah E (2002) Expression of stromal cell-derived factor 1alpha protein in HIV encephalitis. J Neuroimmunol 127:115–126.

Lederman MM, Veazey RS, Offord R, Mosier DE, Dufour J, Mefford M, Piatak M, Jr., Lifson JD, Salkowitz JR, Rodriguez B, Blauvelt A, Hartley O (2004) Prevention of vaginal SHIV transmission in rhesus macaques through inhibition of CCR5. Science 306:485–487.

Lee S, Liu W, Dickson D, Brosnan C, Berman J (1993) Cytokine production by human fetal microglia and astrocytes: Differential induction by lipopolysaccharide and IL-1B. J Immunol 150:2659–2667.

Lewin GR, Barde YA (1996) Physiology of the neurotrophins. Annu Rev Neurosci 19:289–317.

Li H, Newcombe J, Groome NP, Cuzner ML (1993) Characterization and distribution of phagocytic macrophages in multiple sclerosis plaques. Neuropathol Appl Neurobiol 19:214–223.

Lindberg C, Selenica ML, Westlind-Danielsson A, Schultzberg M (2005) Beta-amyloid protein structure determines the nature of cytokine release from rat microglia. J Mol Neurosci 27:1–12.

Little AR, O'Callagha JP (2001) Astrogliosis in the adult and developing CNS: Is there a role for proinflammatory cytokines? Neurotoxicology 22:607–618.

Logan A, Berry M (1993) Transforming growth factor-beta 1 and basic fibroblast growth factor in the injured CNS. Trends Pharmacol Sci 14:337–342.

Lotan M, Schwartz M (1994) Cross talk between the immune system and the nervous system in response to injury: Implications for regeneration. FASEB J 8:1026–1033.

Lucas AD, Chadwick N, Warren BF, Jewell DP, Gordon S, Powrie F, Greaves DR (2001) The transmembrane form of the CX3CL1 chemokine fractalkine is expressed predominantly by epithelial cells in vivo. Am J Pathol 158:855–866.

Lucchinetti C, Bruck W, Parisi J, Scheithauer B, Rodriguez M, Lassmann H (2000) Heterogeneity of multiple sclerosis lesions: Implications for the pathogenesis of demyelination. Ann Neurol 47:707–717.

Lucchinetti CF, Brück W, and Lassmann H (2003) Neuroimmunologic Mechanisms in the Etiology of Multiple Sclerosis. In: Neuroinflammation: Mechanisms and Management (Wood PL, ed), pp 359–377. Totowa, N.J.: Humana Press.

Lue LF, Walker DG, Rogers J (2001) Modeling microglial activation in Alzheimer's disease with human postmortem microglial cultures. Neurobiol Aging 22:945–956.

Mason JL, Ye P, Suzuki K, D'Ercole AJ, Matsushima GK (2000) Insulin-like growth factor-1 inhibits mature oligodendrocyte apoptosis during primary demyelination. J Neurosci 20:5703–5708.

Mason JL, Suzuki K, Chaplin DD, Matsushima GK (2001) Interleukin-1beta promotes repair of the CNS. J Neurosci 21:7046–7052.

Matusevicius D, Navikas V, Soderstrom M, Xiao BG, Haglund M, Fredrikson S, Link H (1996) Multiple sclerosis: The proinflammatory cytokines lymphotoxin-alpha and tumour necrosis factor-alpha are upregulated in cerebrospinal fluid mononuclear cells. J Neuroimmunol 66:115–123.

Matysiak M, Jurewicz A, Jaskolski D, Selmaj K (2002) TRAIL induces death of human oligodendrocytes isolated from adult brain. Brain 125:2469–2480.

Matyszak MK (1998) Inflammation in the CNS: Balance between immunological privilege and immune responses. Prog Neurobiol 56:19–35.

McArthur JC (1987) Neurologic manifestations of AIDS. Medicine (Baltimore) 66:407–437.
McManus C, Berman JW, Brett FM, Staunton H, Farrell M, Brosnan CF (1998) MCP-1, MCP-2 and MCP-3 expression in multiple sclerosis lesions: An immunohistochemical and in situ hybridization study. J Neuroimmunol 86:20–29.
Meda L, Cassatella MA, Szendrei GI, Otvos L, Jr., Baron P, Villalba M, Ferrari D, Rossi F (1995) Activation of microglial cells by beta-amyloid protein and interferon- gamma. Nature 374:647–650.
Meda L, Baron P, Prat E, Scarpini E, Scarlato G, Cassatella MA, Rossi F (1999) Proinflammatory profile of cytokine production by human monocytes and murine microglia stimulated with beta-amyloid[25–35]. J Neuroimmunol 93:45–52.
Mellado M, Rodriguez-Frade JM, Manes S, Martinez AC (2001) Chemokine signaling and functional responses: The role of receptor dimerization and TK pathway activation. Annu Rev Immunol 19:397–421.
Meltzer MS, Skillman DR, Gomatos PJ, Kalter DC, Gendelman HE (1990) Role of mononuclear phagocytes in the pathogenesis of human immunodeficiency virus infection. Annu Rev Immunol 8:169–194.
Meucci O, Miller R (1996) gp120-induced neurotoxicity in hippocampal pyramidal neuron cultures: Protective action of TGF-beta1. J Neurosci 16:4080–4088.
Meucci O, Fatatis A, Simen AA, Bushell TJ, Gray PW, Miller RJ (1998) Chemokines regulate hippocampal neuronal signaling and gp120 neurotoxicity. Proc Natl Acad Sci USA 95:14500–14505.
Meucci O, Fatatis A, Simen AA, Miller RJ (2000) Expression of CX3CR1 chemokine receptors on neurons and their role in neuronal survival. Proc Natl Acad Sci USA 97:8075–8080.
Michael NL (2002) Host genetics and HIV—removing the mask. Nat Med 8:783–785.
Miller RJ, Meucci O (1999) AIDS and the brain: Is there a chemokine connection? Trends Neurosci 22:471–479.
Mizoue LS, Bazan JF, Johnson EC, Handel TM (1999) Solution structure and dynamics of the CX3C chemokine domain of fractalkine and its interaction with an N-terminal fragment of CX3CR1. Biochemistry 38:1402–1414.
Mizoue LS, Sullivan SK, King DS, Kledal TN, Schwartz TW, Bacon KB, Handel TM (2001) Molecular determinants of receptor binding and signaling by the CX3C chemokine fractalkine. J Biol Chem 276:33906–33914.
Mizuno T, Kawanokuchi J, Numata K, Suzumura A (2003) Production and neuroprotective functions of fractalkine in the central nervous system. Brain Res 979:65–70.
Monica J, Carson CSA, and Corinne P (2005) Multiple Sclerosis. In: Inflammatory Disorders of the Nervous System: Pathogenesis, Immunology, and Clinical Management (Alireza Minagar JSA, ed), pp 17–40. Totowa, N.J.: Humana Press.
Moser B, Loetscher P (2001) Lymphocyte traffic control by chemokines. Nat Immunol 2:123–128.
Murphy PM (2001) Viral exploitation and subversion of the immune system through chemokine mimicry. Nat Immunol 2:116–122.
Navia BA, Jordan BD, Price RW (1986) The AIDS dementia complex: I. Clinical features I. Ann Neurol 19:517–524.
Neumann H, Misgeld T, Matsumuro K, Wekerle H (1998) Neurotrophins inhibit major histocompatibility class II inducibility of microglia: Involvement of the p75 neurotrophin receptor. Proc Natl Acad Sci USA 95:5779–5784.
Offen D, Kaye JF, Bernard O, Merims D, Coire CI, Panet H, Melamed E, Ben-Nun A (2000) Mice overexpressing Bcl-2 in their neurons are resistant to myelin oligodendrocyte glycoprotein (MOG)-induced experimental autoimmune encephalomyelitis (EAE). J Mol Neurosci 15:167–176.
Ohagen A, Ghosh S, He J, Huang K, Chen Y, Yuan M, Osathanondh R, Gartner S, Shi B, Shaw G, Gabuzda D (1999) Apoptosis induced by infection of primary brain cultures with diverse human immunodeficiency virus type 1 isolates: Evidence for a role of the envelope. J Virol 73:897–906.
Panitch HS, Hirsch RL, Haley AS, Johnson KP (1987) Exacerbations of multiple sclerosis in patients treated with gamma interferon. Lancet 1:893–895.
Peng H, Huang Y, Rose J, Erichsen D, Herek S, Fujii N, Tamamura H, Zheng J (2004) Stromal cell-derived factor 1 mediated CXCR4 signaling in rat and human cortical neural progenitor cells. J Neurosci Res 76:35–50.
Peng H, Erdmann N, Whitney N, Dou H, Gorantla S, Gendelman HE, Ghorpade A, Zheng J (2006) HIV-1-infected and/or immune activated macrophages regulate astrocyte SDF-1 production through IL-1beta. Glia 54:619–629.
Perno CF, Cooney DA, Currens MJ, Rocchi G, Johns DG, Broder S, Yarchoan R (1990) Ability of anti-HIV agents to inhibit HIV replication in monocyte/macrophages or U937 monocytoid cells under conditions of enhancement by GM-CSF or anti-HIV antibody. AIDS Res Hum Retroviruses 6:1051–1055.
Perrella O, Carrievi P, Guarnaccia D, Soscia M (1992) Cerebrospinal fluid cytokines in AIDS dementia. J Neurol 239:387–388.
Perrella O, Carreiri PB, Perrella A, Sbreglia C, Gorga F, Guarnaccia D, Tarantino G (2001) Transforming growth factor beta-1 and interferon-alpha in the AIDS dementia complex (ADC): Possible relationship with cerebral viral load? Eur Cytokine Netw 12:51–55.
Perry VH (1998) A revised view of the central nervous system microenvironment and major histocompatibility complex class II antigen presentation. J Neuroimmunol 90:113–121.
Pestka S, Krause CD, Sarkar D, Walter MR, Shi Y, Fisher PB (2004) Interleukin-10 and related cytokines and receptors. Annu Rev Immunol 22:929–979.
Petito CK, Vecchio D, Chen YT (1994) HIV antigen and DNA in AIDS spinal cords correlate with macrophage infiltration but not with vacuolar myelopathy. J Neuropathol Exp Neurol 53:86–94.
Poluektova L, Gorantla S, Faraci J, Birusingh K, Dou H, Gendelman HE (2004) Neuroregulatory events follow adaptive immune-mediated elimination of HIV-1-infected macrophages: Studies in a murine model of viral encephalitis. J Immunol 172:7610–7617.
Prehn JH, Miller RJ (1996) Opposite effects of TGF-beta 1 on rapidly- and slowly-triggered excitotoxic injury. Neuropharmacology 35:249–256.
PRISMS STUDY GROUP (1998) In.
Qin L, Liu Y, Cooper C, Liu B, Wilson B, Hong JS (2002) Microglia enhance beta-amyloid peptide-induced toxicity in cortical and mesencephalic neurons by producing reactive oxygen species. J Neurochem 83:973–983.
Ransohoff RM, Hamilton TA, Tani M (1993) Astrocyte expression of mRNA encoding cytokines IP-10 and JE/MCP-1 in experimental autoimmune encephalomyelitis. FASEB J 7:592–600.
Renauld JC (2003) Class II cytokine receptors and their ligands: Key antiviral and inflammatory modulators. Nat Rev Immunol 3:667–676.
Renno T, Krakowski M, Piccirillo C, Lin JY, Owens T (1995) TNF-alpha expression by resident microglia and infiltrating leukocytes in the central nervous system of mice with experimental allergic encephalomyelitis. Regulation by Th1 cytokines. J Immunol 154:944–953.

Ringheim GE, Conant K (2004) Neurodegenerative disease and the neuroimmune axis (Alzheimer's and Parkinson's disease, and viral infections). J Neuroimmunol 147:43–49.

Rizzardi GP, Lazzarin A, Pantaleo G (2002) Potential role of immune modulation in the effective long-term control of HIV-1 infection. J Biol Regul Homeost Agents 16:83–90.

Rohowsky-Kochan C, Troiano R, Cook SD (1989) MHC-restricted autoantigen-reactive T cell clones in multiple sclerosis. J Immunogenet 16:437–444.

Rossi D, Zlotnik A (2000) The biology of chemokines and their receptors. Annu Rev Immunol 18:217–242.

Rostasy K, Egles C, Chauhan A, Kneissl M, Bahrani P, Yiannoutsos C, Hunter DD, Nath A, Hedreen JC, Navia BA (2003) SDF-1alpha is expressed in astrocytes and neurons in the AIDS dementia complex: An in vivo and in vitro study. J Neuropathol Exp Neurol 62:617–626.

Rottman JB, Ganley KP, Williams K, Wu L, Mackay CR, Ringler DJ (1997) Cellular localization of the chemokine receptor CCR5. Correlation to cellular targets of HIV-1 infection. Am J Pathol 151:1341–1351.

Ryan LA, Cotter RL, Zink WE, Gendelman HE, Zheng J (2002) Macrophages, chemokines and neuronal injury in HIV-1 associated dementia. Cell Mol Biol 48:125–138.

Ryan LA, Peng H, Erichsen DA, Huang Y, Persidsky Y, Zhou Y, Gendelman HE, Zheng J (2004) TNF-related apoptosis-inducing ligand mediates human neuronal apoptosis: Links to HIV-1 associated dementia. J Neuroimmunol 148:127–139.

Saha RN, Pahan K (2003) Tumor necrosis factor-alpha at the crossroads of neuronal life and death during HIV-associated dementia. J Neurochem 86:1057–1071.

Sanders VJ, Mehta AP, White MG, Achim CL (1998) A murine model of HIV encephalitis: Xenotransplantation of HIV-infected human neuroglia into SCID mouse brain. Neuropathol Appl Neurobiol 24:461–467.

Sattentau QJ, Weiss RA (1988) The CD4 antigen: Physiological ligand and HIV receptor. Cell 52:631–633.

Scorziello A, Florio T, Bajetto A, Thellung S, Schettini G (1997) TGF-beta1 prevents gp120-induced impairment of Ca2 + homeostasis and rescues cortical neurons from apoptotic death. J Neurosci Res 49:600–607.

Selkoe DJ (2001) Alzheimer's disease results from the cerebral accumulation and cytotoxicity of amyloid beta-protein. J Alzheimers Dis 3:75–80.

Shaheen F, Collman RG (2004) Co-receptor antagonists as HIV-1 entry inhibitors. Curr Opin Infect Dis 17:7–16.

Shapshak P, Duncan R, Minagar A, Rodriguez de la Vega P, Stewart RV, Goodkin K (2004) Elevated expression of IFN-gamma in the HIV-1 infected brain. Front Biosci 9:1073–1081.

Sheng J, Griffin W, Royston M, Mrak R (1998) Distribution of interleukin-1-immunoreactive microglia in cerebral cortical layers: Implications for neuritic plaque formation in Alzheimer's disease. Neuropathol Appl Neurobiol 24:278–283.

Shi B, Rainha J, Lorenzo A, Busciglio J, Gabuzda D (1998) Neuronal apoptosis induced by HIV-1 tat protein and TNF-a: Potentiation of neurotoxicity mediated by oxidative stress and implications for HIV-1 dementia. J Neurovirol 4:281–290.

Simpson JE, Newcombe J, Cuzner ML, Woodroofe MN (1998) Expression of monocyte chemoattractant protein-1 and other beta-chemokines by resident glia and inflammatory cells in multiple sclerosis lesions. J Neuroimmunol 84:238–249.

Sippy BD, Hofman FM, Wallach D, Hinton DR (1995) Increased expression of tumor necrosis factor-alpha receptors in the brains of patients with AIDS. J Acquir Immune Defic Syndr Hum Retrovirol 10:511–521.

Song L, Pachter JS (2004) Monocyte chemoattractant protein-1 alters expression of tight junction-associated proteins in brain microvascular endothelial cells. Microvasc Res 67:78–89.

Soontornniyomkij V, Wang G, Pittman CA, Wiley CA, Achim CL (1998) Expression of brain-derived neurotrophic factor protein in activated microglia of human immunodeficiency virus type 1 encephalitis. Neuropathol Appl Neurobiol 24:453–460.

Sorensen TL, Tani M, Jensen J, Pierce V, Lucchinetti C, Folcik VA, Qin S, Rottman J, Sellebjerg F, Strieter RM, Frederiksen JL, Ransohoff RM (1999) Expression of specific chemokines and chemokine receptors in the central nervous system of multiple sclerosis patients. J Clin Invest 103:807–815.

Srinivasan D, Yen JH, Joseph DJ, Friedman W (2004) Cell type-specific interleukin-1beta signaling in the CNS. J Neurosci 24:6482–6488.

Stadelmann C, Kerschensteiner M, Misgeld T, Bruck W, Hohlfeld R, Lassmann H (2002) BDNF and gp145trkB in multiple sclerosis brain lesions: Neuroprotective interactions between immune and neuronal cells? Brain 125:75–85.

Stankoff B, Aigrot MS, Noel F, Wattilliaux A, Zalc B, Lubetzki C (2002) Ciliary neurotrophic factor (CNTF) enhances myelin formation: A novel role for CNTF and CNTF-related molecules. J Neurosci 22:9221–9227.

Stewart W, Kawas C, Corrada M, Metter E (1997) Risk of Alzheimer's disease and duration of NSAID use. Neurology 48:626–632.

Strieter RM, Polverini PJ, Kunkel SL, Arenberg DA, Burdick MD, Kasper J, Dzuiba J, Van Damme J, Walz A, Marriott D, (1995) The functional role of the ELR motif in CXC chemokine-mediated angiogenesis. J Biol Chem 270:27348–27357.

Stumm RK, Zhou C, Ara T, Lazarini F, Dubois-Dalcq M, Nagasawa T, Hollt V, Schulz S (2003) CXCR4 regulates interneuron migration in the developing neocortex. J Neurosci 23:5123–5130.

Tang J, Kaslow RA (2003) The impact of host genetics on HIV infection and disease progression in the era of highly active antiretroviral therapy. Aids 17(Suppl 4):S51–S60.

Tarkowski E, Rosengren L, Blomstrand C, Wikkelso C, Jensen C, Ekholm S, Tarkowski A (1995) Early intrathecal production of interleukin-6 predicts the size of brain lesion in stroke. Stroke 26:1393–1398.

Tarkowski E, Blenow K, Wallin A, Tarkowski A (1999) Intracerebral production of tumor necrosis factor-alpha, a local neuroprotective agent in Alzheimer disease and vascular dementia. Clin Immunol 19:223–230.

Tong N, Perry SW, Zhang Q, James HJ, Guo H, Brooks A, Bal H, Kinnear SA, Fine S, Epstein LG, Dairaghi D, Schall TJ, Gendelman HE, Dewhurst S, Sharer LR, Gelbard HA (2000) Neuronal fractalkine expression in HIV-1 encephalitis: Roles for macrophage recruitment and neuroprotection in the central nervous system. J Immunol 164:1333–1339.

Tran PB, Miller RJ (2003) Chemokine receptors: Signposts to brain development and disease. Nat Rev Neurosci 4:444–455.

Trapp BD, Bo L, Mork S, Chang A (1999) Pathogenesis of tissue injury in MS lesions. J Neuroimmunol 98:49–56.

Tyor WR, Glass JD, Griffin JW, Becker PS, McArthur JC, Bezman L, Griffin DE (1992) Cytokine expression in the brain during acquired immune deficiency syndrome. Ann Neurol 31:349–360.

Uberti D, Cantarella G, Facchetti F, Cafici A, Grasso G, Bernardini R, Memo M (2004) TRAIL is expressed in the brain cells of Alzheimer's disease patients. Neuroreport 15:579–581.

Vallat A-V, Girolami UD, He J, Mhashikar A, Marasco W, Shi B, Gray F, Bell J, Keohane C, Smith TW, Gabuzda D (1998) Localization of HIV-1 co-receptors CCR5 and CXCR4 in the brain of children with AIDS. Am J Path 152:167–178.

Van der Meer P, Ulrich AM, Gonzalez-Scarano F, Lavi E (2000) Immunohistochemical analysis of CCR2, CCR3, CCR5, and CXCR4 in the human brain: Potential mechanisms for HIV dementia. Exp Mol Pathol 69:192–201.

Van der Voorn P, Tekstra J, Beelen RH, Tensen CP, van der Valk P, De Groot CJ (1999) Expression of MCP-1 by reactive astrocytes in demyelinating multiple sclerosis lesions. Am J Pathol 154:45–51.

Van der Wal EA, Gomez-Pinilla F, Cotman CW (1993) Transforming growth factor-beta 1 is in plaques in Alzheimer and Down pathologies. Neuroreport 4:69–72.

Van Marle G, Henry S, Todoruk T, Sullivan A, Silva C, Rourke SB, Holden J, McArthur JC, Gill MJ, Power C (2004) Human immunodeficiency virus type 1 Nef protein mediates neural cell death: A neurotoxic role for IP-10. Virology 329:302–318.

Vergote D, Butler GS, Ooms M, Cox JH, Silva C, Hollenberg MD, Jhamandas JH, Overall CM, Power C (2006) Proteolytic processing of SDF-1{alpha} reveals a change in receptor specificity mediating HIV-associated neurodegeneration. Proc Natl Acad Sci USA 103:19182–19187.

Vilcek J (2003) The cytokines: An overview. In: The Cytokine Handbook, 4th Edition (Angus W Thomson MTL, ed), pp 1–18. San Diego: Academic Press.

Villoslada P, Hauser SL, Bartke I, Unger J, Heald N, Rosenberg D, Cheung SW, Mobley WC, Fisher S, Genain CP (2000) Human nerve growth factor protects common marmosets against autoimmune encephalomyelitis by switching the balance of T helper cell type 1 and 2 cytokines within the central nervous system. J Exp Med 191:1799–1806.

Wajgt A, Gorny M (1983) CSF antibodies to myelin basic protein and to myelin-associated glycoprotein in multiple sclerosis. Evidence of the intrathecal production of antibodies. Acta Neurol Scand 68:337–343.

Walz A, Peveri P, Aschauer H, Baggiolini M (1987) Purification and amino acid sequencing of NAF, a novel neutrophil-activating factor produced by monocytes. Biochem Biophys Res Commun 149:755–761.

Wang X (2001) The expanding role of mitochondria in apoptosis. Genes Dev 15:2922–2933.

Wesselingh SL, Power C, Glass JD (1993) Intracerebral cytokine messenger RNA expression in acquired immunedeficiency syndrome dementia. Ann Neurol 33:576–582.

Westmoreland SV, Rottman JB, Williams KC, Lackner AA, Sasseville VG (1998) Chemokine receptor expression on resident and inflammatory cells in the brain of macaques with simian immunodeficiency virus encephalitis. Am J Pathol 152:659–665.

Wiley CA, Schrier RD, Nelson JA, Lampert PW, Oldstone MBA (1986) Cellular localization of human immunodeficiency virus infection within the brains of acquired immune deficiency syndrome patients. Proc Natl Acad Sci USA 83:7089–7093.

Williams K, Schwartz A, Corey S, Orandle M, Kennedy W, Thompson B, Alvarez X, Brown C, Gartner S, Lackner A (2002) Proliferating cellular nuclear antigen expression as a marker of perivascular macrophages in simian immunodeficiency virus encephalitis. Am J Pathol 161:575–585.

Wingerchuk DM, Weinshenker BG (2000) Multiple sclerosis: Epidemiology, genetics, classification, natural history, and clinical outcome measures. Neuroimaging Clin N Am 10:611–624.

Wosik K, Antel J, Kuhlmann T, Bruck W, Massie B, Nalbantoglu J (2003) Oligodendrocyte injury in multiple sclerosis: A role for p53. J Neurochem 85:635–644.

Wright JL, Merchant RE (1992) Histopathological effects of intracerebral injections of human recombinant tumor necrosis factor-alpha in the rat. Acta Neuropathol (Berl) 85:93–100.

Wu DT, Woodman SE, Weiss JM, McManus CM, D'Aversa TG, Hesselgesser J, Major EO, Nath A, Berman JW (2000) Mechanisms of leukocyte trafficking into the CNS. J Neurovirol 6(Suppl 1): S82–S85.

Xia M, Hyman BT (1999) Chemokines/chemokine receptors in the central nervous system and Alzheimer's disease. J Neurovirol 5:32–41.

Xia M, Hyman BT (2002) GROalpha/KC, a chemokine receptor CXCR2 ligand, can be a potent trigger for neuronal ERK1/2 and PI-3 kinase pathways and for tau hyperphosphorylation-a role in Alzheimer's disease? J Neuroimmunol 122:55–64.

Xia M, Qin S, McNamara M, Mackay C, Hyman B (1997) Interleukin-8 receptor B immunoreactivity in brain and neuritic plaques of Alzheimer's disease. Am J Pathol 150:1267–1274.

Xia M, Qin S, Wu L, Mackay C, Hyman B (1998) Immunohistochemical study of the b-chemokine receptors CCR3 and CCR5 and their ligands in normal and Alzheimer's disease brains. Am J Pathol 153:31–37.

Xia MQ, Bacskai BJ, Knowles RB, Qin SX, Hyman BT (2000) Expression of the chemokine receptor CXCR3 on neurons and the elevated expression of its ligand IP-10 in reactive astrocytes: In vitro ERK1/2 activation and role in Alzheimer's disease. J Neuroimmunol 108:227–235.

Xiao BG, Link H (1998) Immune regulation within the central nervous system. J Neurol Sci 157:1–12.

Yong VW, Chabot S, Stuve O, Williams G (1998) Interferon beta in the treatment of multiple sclerosis: Mechanisms of action. Neurology 51:682–689.

Yoshimura T, Matsushima K, Oppenheim JJ, Leonard EJ (1987) Neutrophil chemotactic factor produced by lipopolysaccharide (LPS)-stimulated human blood mononuclear leukocytes: Partial characterization and separation from interleukin 1 (IL 1). J Immunol 139:788–793.

Zhang K, McQuibban GA, Silva C, Butler GS, Johnston JB, Holden J, Clark-Lewis I, Overall CM, Power C (2003) HIV-induced metalloproteinase processing of the chemokine stromal cell derived factor-1 causes neurodegeneration. Nat Neurosci 6:1064–1071.

Zhang L, He T, Talal A, Wang G, Frankel SS, Ho DD (1998) In vivo distribution of the human immunodeficiency virus/simian immunodeficiency virus coreceptors: CXCR4, CCR3, and CCR5. J Virol 72:5035–5045.

Zhao ML, Kim MO, Morgello S, Lee SC (2001) Expression of inducible nitric oxide synthase, interleukin-1 and caspase-1 in HIV-1 encephalitis. J Neuroimmunol 115:182–191.

Zheng J, Ghorpade A, Niemann D, Cotter RL, Thylin MR, Epstein L, Swartz JM, Shepard RB, Liu X, Nukuna A (1999a) Lymphotropic virions affect chemokine receptor-mediated neural signaling and apoptosis: Implications for human immunodeficiency virus type 1-associated dementia. J Virol 73:8256–8267.

Zheng J, Thylin M, Ghorpade A, Xiong H, Persidsky Y, Cotter R, Niemann D, Che M, Zeng Y, Gelbard H (1999b) Intracellular CXCR4 signaling, neuronal apoptosis and neuropathogenic mechanisms of HIV-1-associated dementia. J Neuroimmunol 98:185–200.

Zheng J, Bauer M, Cotter RL, Ryan LA, Lopez A, Williams C, Ghorpade A, Gendelman HE (2000a) Fractalkine Mediated Macrophage Activation by Neuronal Injury: Relevance for HIV-1 Associated

Dementia. In: 30th Annual Meeting of Society for Neuroscience. New Orleans: Society for Neuroscience.

Zheng J, Niemann D, Bauer M, Leisman GB, Cotter RL, Ryan LA, Lopez A, Williams C, Ghorpade A, Gendelman HE (2000b) a-Chemokines and Their Receptors in the Neuronal Signaling: Relevance for HIV-1-Associated Dementia. In: 7th Conference on Retroviruses and Opportunistic Infections. San Francisco: Foundation for Retrovirology and Human Health.

Zheng J, Niemann D, Bauer M, Williams C, Lopez A, Erichsen D, Ryan LA, Cotter RL, Ghorpade A, Swindells S (2001) HIV-1 Glia Interactions in Interleukin-8 and Growth-Related Oncogene a Secretion, Neuronal Signaling and Demise: Relevance for HIV-1-Associated Dementia. In: 8th Conference on Retroviruses and Opportunistic Infections, p 227. Chicago: Foundation for retrovirology and human health.

Zhou N, Fang J, Mukhtar M, Acheampong E, Pomerantz RJ (2004) Inhibition of HIV-1 fusion with small interfering RNAs targeting the chemokine coreceptor CXCR4. Gene Ther 11:1703–1712.

Zink MC, Coleman GD, Mankowski JL, Adams RJ, Tarwater PM, Fox K, Clements JE (2001) Increased macrophage chemoattractant protein-1 in cerebrospinal fluid precedes and predicts simian immunodeficiency virus encephalitis. J Infect Dis 184:1015–1021.

Zujovic V, Benavides J, Vige X, Carter C, Taupin V (2000) Fractalkine modulates TNF-alpha secretion and neurotoxicity induced by microglial activation [In Process Citation]. Glia 29:305–315.

16
CNS Cell Signaling

Ramendra N. Saha, Keshore R. Bidasee, and Kalipada Pahan

Keywords Signal transduction; CNS homeostasis; Glial activation; JAK-STAT pathway; MAP kinase pathways; Myelination; Neurodegeneration; Neuroinflammation; Neuronal apoptosis; Neuroregeneration

16.1. Introduction

What mysterious forces precede the appearance of these processes... promote their growth and ramification...and finally establish those protoplasmic kisses...which seem to constitute the final ecstasy of an epic love story.

~Santiago Ramón y Cajal [1852–1934]

The science of neurobiology is now almost a century older than times when Spanish neuroanatomist and Nobel laureate Santiago Ramón y Cajal had wondered as above. Yet, these 'mysterious forces' have only been partially illuminated today and the posed question still remains worth pondering upon in contemporary times. What Cajal identified as 'forces' are basically key cellular signals that are transduced preceding growth and ramification. A precipitate of our knowledge today tells us that these 'forces' are mostly generated within and amongst members of the central nervous system (CNS). The present chapter is aimed at appreciating cellular signals and their transduction pathways which underlie the functional output of CNS during normal times, diseased conditions, and regeneration.

16.2. An Introductory Orientation

Signal transduction forms the basis of cellular perception to an external signal. Generally, it refers to defined and regulated cascade of cellular events that identifies a certain signal at cell surface or in intracellular compartments (reception desk) followed by engagement of second messenger pathway(s) that finally enable the cell to respond to the signal.

16.2.1. General Mechanism of Cellular Signal Transduction

Ideally, there are four stages in any signal transduction pathway. The first stage involves binding of receptors by the ligand. These receptors could be intracellular (e.g. nuclear hormone receptors), or may be exhibited on the plasma membrane. The second stage involves activation of receptors in response to ligand binding. Once activated, the receptor recruits several modulators (e.g. G-proteins) as the third step in the cascade. Finally, in the fourth step, second messengers (e.g. cAMP, ceramide) are activated which convey the signal downstream to effecter molecules (e.g. transcription factors, which translocated to the nucleus and induce activation of specific genes). Although most signal transduction pathways are structured around four-stage process, yet variations are also observed.

For a signal to be able to induce an appropriate response to the inducer, it must be specific, fast, and must be amplified along the way of transduction. Indeed, amplification is achieved when one receptor recruits several modulators, which in turn activate several second messengers (Figure 16.1).

16.2.2. Signaling in CNS: A Complex Web of Signaling in Various Cell Types

Previous chapters of this book should have by now impressed the reader with the complexity of cellular types and function in CNS. The main cell types bathing in cerebro-spinal fluid are neurons, astrocytes, microglia, olidendroglia, and Schwann cells. Additionally, there are endothelial cells lining the blood-brain barrier (BBB). These cells, despite having distinguished functions of their own, are remarkably interconnected and demonstrate considerable amount of inter-cellular signaling between similar or dissimilar cells. Such crosstalk between different cell types forms the basis of several physiological outcomes like memory formation and axonal regeneration. Despite sharing several common signaling pathways, yet, sometimes same ligands induce strikingly opposite outcomes

T. Ikezu and H.E. Gendelman (eds.), *Neuroimmune Pharmacology*.
© Springer 2008

in different CNS cells. For example, few inducers of inflammatory response in glia cause degeneration of neurons. This suggests that there are cell-type-specific modulations of certain signaling pathways.

16.3. Signals Maintaining Normal CNS Health and Function

16.3.1. Major Signaling Pathways Maintaining CNS Homeostasis

Regulation and/or maintenance of axonal growth, dendritic pruning, synaptogenesis and synaptic refinement, and neuronal survival/death are essential for the proper functioning of the nervous system. These functions are carried out following the interaction of neurotrophins with their plasma membrane receptors, Trk receptor tyrosine kinases (Trks) and p75 neurotrophin receptor ($p75^{NTR}$) and increase in cytoplasmic Ca^{2+}. In the mammalian brain four neurotrophins have been identified: nerve growth factor (NGF); brain-derived neurotrophic factor (BDNF); neurotrophin 3 (NTF3) and neurotrophin 4 (NTF4, also referred to as NTF4/5) (Zweifel et al., 2005; Lu et al., 2005). Trk family of receptor tyrosine kinases comprises of three different receptors, Trk A, Trk B and Trk C. $p75^{NTR}$ is a member of the tumor necrosis receptor super family (Huang and Reichardt, 2003). In general, activation of Trk receptors stimulates neuronal survival, differentiation, neurite outgrowth, synaptic plasticity, and function. $p75^{NTR}$ acts as a facilitator of Trk-mediated neuronal survival as well as an inhibitor of cell growth and promoter of apoptosis (Nykjaer et al., 2005).

Neurotrophins are synthesized as proneurotrophin precursors of ~27–35 kDa. These precursors of neurotrophins are cleaved either within the cell by the serine protease Furin (trans-Golgi network) and pro-convertase or in the extracellular space by the plasmin and matrix metalloprotineases (MMP3 and MMP7), affording mature neurotrophins of about 13 kDa. While mature NGF preferentially binds to and activates Trk A (k_D ~ 1–10 nM), BDNF and NTF4 (NTF4/5) exhibit high affinity for Trk B. On the other hand, NTF3 binds to and activate Trk C. Mature neurotrophins have slightly lower and similar affinities for $p75^{NTR}$, while proneurotrophins exhibit high affinity for $p75^{NTR}$ (Barker, 2004).

16.3.1.1. Trk Receptor Signaling

The extracellular domain of Trk receptors is made up of three leucine-rich 24 residue motifs flanked on either side by a cysteine cluster (C1 is on the outer side and C2 is in the inner side), followed by two immunoglobulin (Ig)-like domains and a single transmembrane domain. The cytoplasmic domain of Trk receptors contains several tyrosine motifs (Huang and Reichardt, 2003). The major ligand binding site on Trk receptors is located in the region proximal to the Ig-C2 domain. Binding of neurotrophins to Trk receptors triggers receptor dimerization, autophosphorylation of tyrosine residues and activation of several signaling pathways. There are ten conserve tyrosine residues in each Trk receptors. Phosphorylation of Y670, Y672, and Y675 potentiate tyrosine kinase activity by pairing these negatively charged residues with basic residues in their vicinity. Phosphorylation of additional residues creates docking sites for adaptor proteins including Ras-Raf-MEK-Erk-CREB, PI3-kinase-Akt, PLCγ-Ca^{2+}, NF-κB and atypical protein kinase pathways. In Trk A receptor, phosphotyrosine 490 creates a docking site for Shc, fibroblast growth factor receptor substrate 2 (FRS2) which then activates Ras and PI3 kinase. However, phosphorylation of 785 residue recruits PLCγ-1. Activation of these pathways leads to local control of axonal growth, neuronal survival and metabolism.

Neurotrophin-Trk receptor complexes are internalized and retrogradely transported from distal axons to the neuronal cell body where they signal to the soma to mediate target-dependent survival, growth and gene expression. The neurotrophin-Trk receptor complex is internalized by four mechanistically diverse and highly regulated pathways: macropinocytosis; clathrin-mediated endocytosis; caveolae-mediated endocytosis and Pincher-mediated endocytosis. The kinase activity of Trk is probably required for receptor internalization.

16.3.1.2. $p75^{NTR}$ Receptor Signaling

$p75^{NTR}$ is the second class of neurotrophin receptor that is integral for maintaining CNS health and function (Barker, 2004; Lee et al., 2001; Meldolessi et al., 2000). This receptor binds soluble dimeric ligands and often requires (or act as) a co-receptor to facilitate neuronal survival, neuronal death and growth inhibition. Structurally, $p75^{NTR}$ is less complex than Trk receptors. The extracellular domain comprises of four tandemly arranged cysteine-rich motifs that contain the neurotrophin binding site. This is followed by a single transmembrane domain and a cytoplasmic tail. Unlike Trk receptors, the cytoplasmic tail of $p75^{NTR}$ receptor does not possess kinase activity. However, the cytopasmic tail of $p75^{NTR}$ possess three intracellular domains that serve as docking sites for adaptor proteins. They include a domain with homology to the binding site for TNF-receptor mediated factors (TRAFs), a domain homologous but distinct from death domain 1 of typical death receptors and a PSD-binding domain.

Binding of neurotrophins is the primary mechanism by which Trk receptors are activated but the affinity and specificity of neurotrophins for Trk receptors is regulated by $p75^{NTR}$. For example, the association of $p75^{NTR}$ with Trk receptors induces a conformation that has high affinity for NGF. Association of $p75^{NTR}$ also enhances the discrimination of Trk for their preferred neurotrophin ligand (Barker, 2004; Lee et al., 2001; Meldolessi et al., 2000).

The activation of $p75^{NTR}$ plays an important role in neuronal growth. Unliganded $p75^{NTR}$ is an activator of RhoA which mediates the effects of CNS-derived myelin-based growth

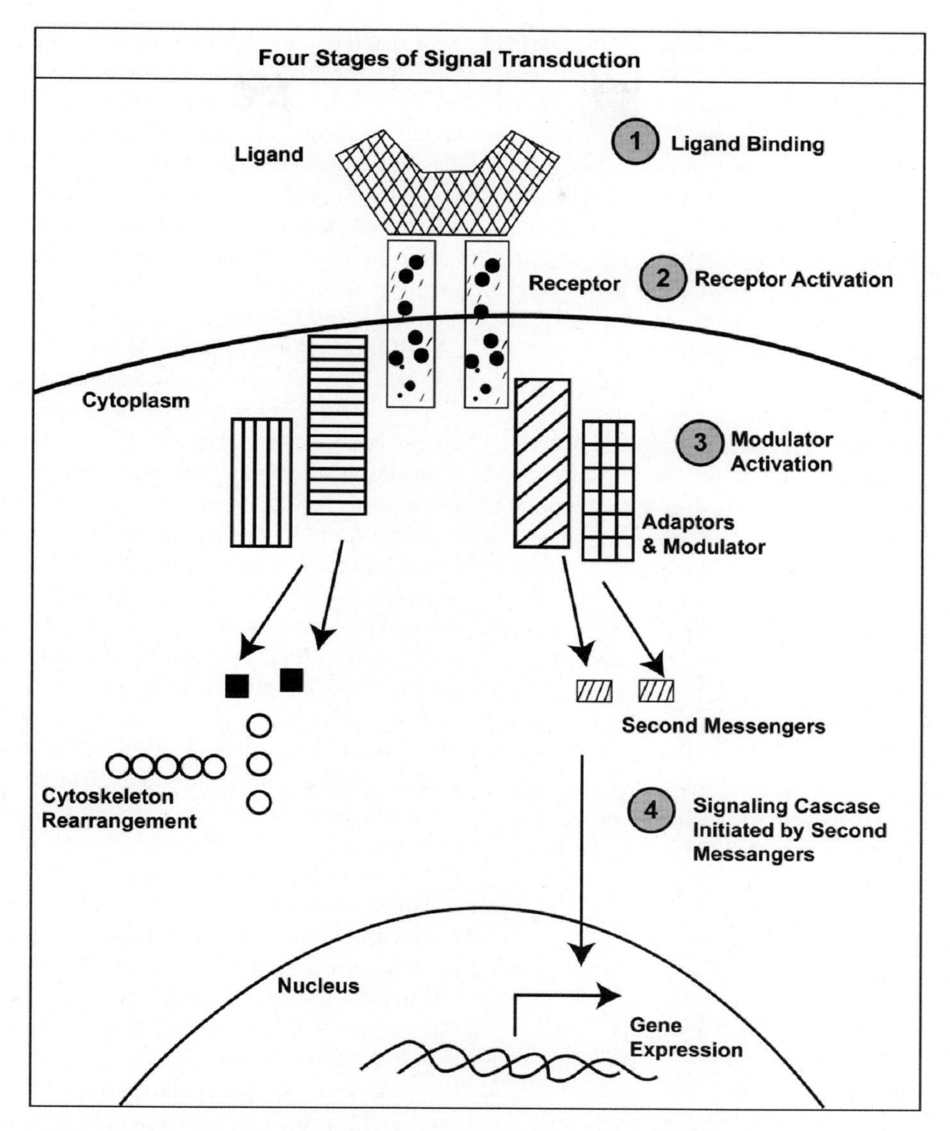

FIGURE 16.1. Basic scheme of signaling. Signal transduction pathways are usually composed of four stages as indicated. After ligand binding, receptor becomes activated and intracellular part of activated receptor recruits adapters and modulators that may produce second messengers. Second messengers may involve activation of enzymes like kinases and phosphatases, and/or transcription factors. Finally, activation of transcription factors results in gene transcription, whereas signals not involving them mostly result in post-translational modification of existing proteins.

inhibitors (MGBIs) that include Nogo, myelin-associated glycoprotein (MAG1) and oligodendrocyte myelin glycoprotein (OMgP). The precise signaling mechanisms by which $p75^{NTR}$-Nogo complex inhibit neuronal growth remains unresolved. However, studies suggest that the binding of MBGIs to $p75^{NTR}$-Nogo complex enhances the association of Rho-GDIα (Rho-GDP dissociation inhibitor α) while NGF abolishes $p75^{NTR}$-Rho-GDIα interaction.

16.3.1.3. Ca^{2+} Signaling

Neurons communicate with each other (as well as with other non-neural cells) either via electrical (action potential) or chemical (neurotrophins and other modulatory ligands) signals. In response to these signals, neurons alter their intracellular free Ca^{2+} levels. This rise in intracellular free Ca^{2+} serves as a ubiquitous second messenger signal to regulate a broad repertoire of neuronal function including axonal and dendritic growth and function, gene transcription, neurotransmitter release, and apoptosis (Berridge, 2005; Ross et al., 2005; DeCoster, 1995). Ca^{2+} can regulate this diverse array of function by virtue of the quantity of release (amplitude), where in the neuron it is release (spatial location) and for how long it was release (time). Under resting conditions, the cytoplasmic free Ca^{2+} in neurons is about 100 nM and upon neuronal activation cytoplasmic free Ca^{2+} increases to about 1–10 μM. The Ca^{2+} that is used for this cytoplasmic increase is mobilized either from external stores (extracellular space) following activation of ligand-gated channel (e.g. NMDA and P2X receptors), voltage-operated Ca^{2+} channels (L-type, T-type and N-type Ca^{2+} channels) or from internal stores (endoplasmic reticulum) via activation of inositol 1,4,5-trisphosphate receptors (InsP3R) and ryanodine receptors (RyR) (Berridge, 2005; Ross et al., 2005; DeCoster, 1995). The rise in cytoplasmic Ca^{2+} maybe sufficient to cause vesicles to fuse to plasma membrane and release their content into the synaptic cleft (neurotransmitter release), overload mitochrondria and induce apoptosis, or activate transcription factors that result in gene expression. After the physiological task is completed, the influxed Ca^{2+} is removed from the cytoplasm by plasma membrane bound Ca^{2+}-ATPases (PMCA) and Na^{+}-Ca^{2+} exchangers. Ca^{2+} that are mobilized from the endoplasmic reticulum are returned to the stores via sarco(endo)plasmic reticulum Ca^{2+}-ATPase (SERCA).

16.3.2. Signaling in Physiological Events of CNS

While there are several interesting aspects of CNS physiology, we will restrain our discussion to signaling in two facets thereof, namely, neuronal plasticity (leading to memory formation) and myelination. While myelination is quite a segregated topic, plasticity and memory formation enjoy viciously overlapping signaling pathways (Matynia et al., 2001), for plasticity is indeed the basis of memory formation/consolidation.

16.3.2.1. Signaling in Neuronal Plasticity and Memory Formation

What is renovating in your brain right now as you are learning about signaling of memory formation, a fraction of which you will perhaps retain in your memory for years to come? The answer is neuronal plasticity, the key prelude to memory formation. It is a consequence of both, qualitative alteration in efficacy of synaptic transmission and quantitative alteration in synapse number due to synaptic growth.

16.3.2.1.1. Signaling in Generation of Short-Term Plasticity and Short-Term Memory

Short-term sensitization of a synapse can occur even in the absence of protein synthesis and is largely dependent on post-translational modification of existing proteins. Assessment of alteration in efficacy of synapses is often performed in laboratories by quantifying long-term potentiation (LTP) or long-term depression (LTD), the artificially induced forms of plasticity. LTP, reflecting short-term plasticity and short-term memory, is often referred to as the early or transient LTP (eLTP) which is induced by weak signals, such as a ringing bell. eLTP involves activation of cAMP and PKA in pre-synaptic neuron and activation of a set of kinases and phosphatases in the post-synaptic neuron that includes their signal dependent transportation to post-synaptic membrane.

In a pioneering effort back in 1976, Eric Kandel, Nobel Laureate for Physiology and Medicine in 2000, had delineated involvement of cAMP in regulation of synaptic transmitter release in giant neurons of *Aplysia* (Kandel, 2001). Subsequently it was elucidated that elevated level of cAMP broadens action potential by limiting certain K^{+} currents while enhancing Ca^{2+} influx into pre-synaptic terminal. In addition to cAMP, protein kinase A (PKA) inhibitors also block pre-synaptic short-term facilitation suggesting a role for this kinase in this process. Activation of PKA, incidentally, is also dependent on elevated cAMP level (Kandel, 2001). Once activated, PKA can regulate release of neurotransmitters and activity of ion channels thereby strengthening synaptic connections for the whole time course of short-term plasticity (Figure 16.2).

In the post-synaptic neuron, binding of the transmitter to its receptor facilitates a brief Ca^{2+} influx which sensitizes the Calcium/Calmodulin-dependent protein kinase II (CaMKII). Since CaMKII remains independently active after transient activation by Ca^{2+} (See Lisman et al., 2002 for more information), this kinase serves well to convert brief synaptic impulse into longer physiological signaling. In addition to its regulation with Ca^{2+}/Calmodulin, activity of CaMKII is enhanced by its binding with cytoplasmic C-terminus tail of NMDA receptor 2B subunit. This binding anchors the enzyme to the membrane, where it phosphorylates GluR1 thereby facilitating conductance of AMPA receptor and their insertion by an indirect mechanism. Physical transport of these receptors in and out of the synaptic membrane contributes to several forms of synaptic plasticity.

Phosphorylation of CaMKII is countered by a specific phosphatase, protein phosphatase 1 (PP1), which remains inactive due to PKA activity. Incidentally, it must be taken into account that PKA activity is not restricted to pre-synaptic cell only, but is also engaged in post-synaptic neuronal signaling. PKA inhibits PP1 activity by phosphorylating the regulatory protein inhibitor-1 (I-1). Protein phosphatase 2b (PP2b) dephosphorylates I-1 to activate PP1, which then attenuates CaMKII activity thereby culminating LTP (Blitzer, 2005). It is thus appropriate to reckon PP1 as the molecule of 'forgetfulness' (Silva and Josselyn, 2002).

16.3.2.1.2. Signaling in Generation of Long-Term Plasticity and Memory

Long-term plasticity and long-term memory essentially is an extended version of short-term plasticity and short-term memory. Both long-term and short-term forms of these processes are dependent on increase in synaptic strength, which in turn is manifestation of enhanced broadening of action potential and release of transmitter in both cases. However, the similarities end there. Subsequently, these processes are different in two aspects. First, long-term changes are contingent on new protein synthesis and secondly, long-term processes involve structural alterations in synapse number and structure (Bailey et al., 2004). In the following lines, we will delineate the signaling responsible for both these operations.

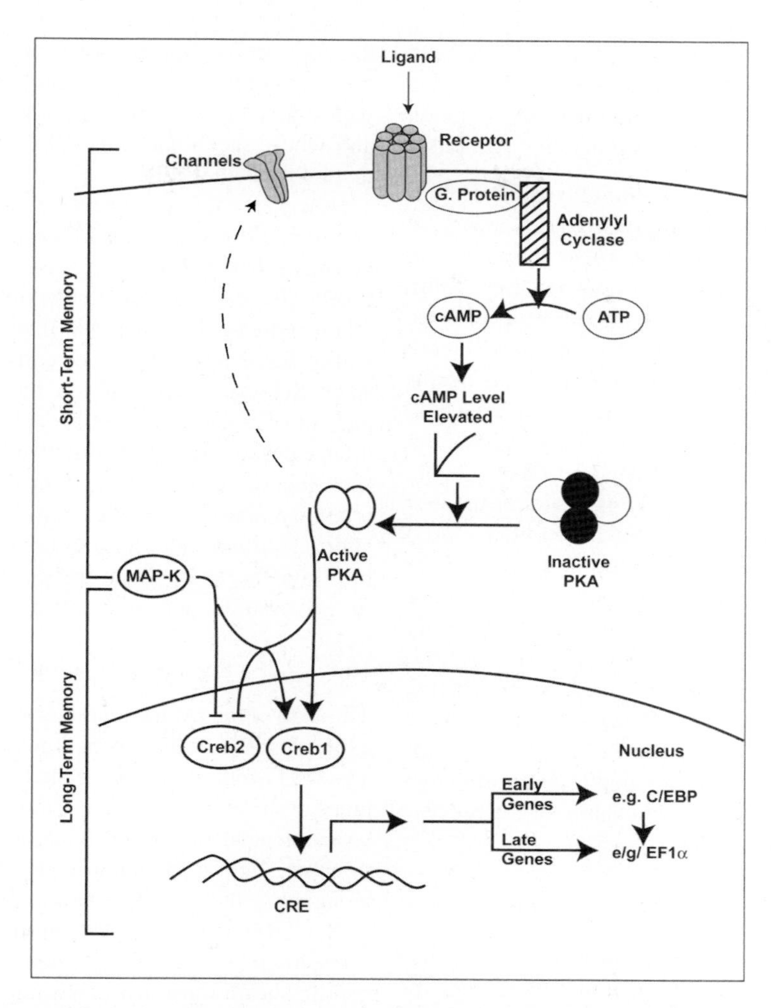

FIGURE 16.2. Signaling for memory formation. Short-term memory formation involves the activation of PKA, but no gene transcription. Activation of G-protein coupled receptors by excitatory stimuli activates adenylyl cyclase. This leads to the elevation of cAMP level, which subsequently activates PKA. Activated PKA undertakes modulation of channels thereby enhancing conductivity. However, prolonged/ repeated activation of this system results in nuclear translocation of PKA, which is the central molecular basis of long-term memory formation. Activated PKA and MAPK activate transcription factor CREB-1 while suppressing the inhibitory CREB-2. Activated CREB-1 binds to CRE region in promoters of early genes like C/EBP. Interestingly, C/EBP itself is a transcription factor that subsequently teams up with CREB to express late memory genes.

Signaling Involved in New Protein Synthesis. We are by now aware of the fact that sensitization of pre-synaptic neurons (by excitatory moieties like 5-HT) elevates cAMP level, which in turn, activates PKA. Interestingly, for long-term memory formation this train of events, responsible for short-term processes, continues further as a conserved central signaling pathway of long-term information processing. PKA activates gene expression by phosphorylating the transcription factor cAMP responsive element binding protein (CREB1), a key 'memory molecule'. Once activated, CREB1 transactivates a set of early genes including two more important transcription factors, CAAT box/ enhancer binding protein (C/EBP) and activation factor (AF) (Kandel, 2001). These transcription factors subsequently regulate expression of important downstream memory genes (Figure 16.2).

However, activation of memory-enhancing genes is only half the tale. The other half involves suppression of memory-suppressing genes by a repressor isoform of CREB1, called CREB2. Interestingly, while one MAPK (ERK) acts to facilitate memory formation, yet another MAPK, called p38, acts a repressor of memory formation. Although exact basis of such inhibition is yet unknown, it has been proposed that p38-MAPK impedes memory formation indirectly by inhibiting activation of ERK (Sharma and Carew, 2004) (Figure 16.2).

In addition to CREB, nuclear factor kappaB (NF-κB) is another architect of importance in facilitating long-term processes. NF-κB p50:p65 is located in synapses of neurons and is engaged in Ca^{2+} responsive pathway. Additionally, it has been proposed that activation of NF-κB p50:p65 is dependent on CaMKII activation (Meffert and Baltimore, 2005). As suggested by impairment of spatial learning by error-prone p65-deficient mice, p65 is involved in long-term processes of information retention. We will hear a lot more about this dimeric transcription factor in neuroinflammation and neurodegeneration sections.

Signaling Involved in Synaptic Remodeling. Synaptic remodeling involved in long-term processes has two facets. Firstly, it involves activation of previously present 'silent' synapses and secondly, it involves engineering brand new synapses. Both ways, the process mainly involves redistribution of synaptic vesicle proteins to reinforce active zone components and more importantly, rearrangement of cytoskeleton. The later forms the structural basis of increment in synaptic contact area and/or formation of new filopodia, many of which are morphological precursors for learning-associated new synapses (Bailey et al., 2004). In dendritic spines of neurons, LTP induction is greatest in spines with greatest F-actin content (Fukazawa et al., 2003) underscoring the importance of F-actin assembly in long-term processes. How is extracellular signal conveyed to achieve actin polymerization? In general, F-actin formation is greatly dependent on signal transduction by small GTPases of Rho family. In *Aplysia*, repeated pulses of 5-HT (capable of inducing long-term processes) selectively activates small GTPase Cdc42, but not Rho or Rac, through the PI3K and PLC pathways. Once activated Cdc42 activates downstream effectors PAK and N-WASP and initiates reorganization of the presynaptic actin network (Udo et al., 2005).

16.3.2.2. Signaling in Axonal Myelination

Myelination, the process of wrapping up byzantine axonal processes with a insulating coat of myelin synthesized by an unique glial population (Oligodendrocytes in CNS and Schwann cells in PNS), involves several receptor signaling pathways in both participating cells, i.e., axons as well as glia. In following lines, we will survey the steadily increasing knowledge base regarding two aspects of myelination; first, signals dictating selection of axons for myelination, and then, signals regulating thickness of myelin sheath. (See Sherman and Brophy, 2005 for additional aspects of myelination.)

16.3.2.2.1. Sorting the Axon toWrap; Signaling in Both Parties

Foremost stages of axonal myelination are contingent on action of neurotrophins. Interestingly, NGF, the prototypical neurotrophin, promotes myelination by Schwann cells, but inhibits myelination by oligodendrocytes (Chan et al., 2004). This is particularly interesting as NGF manipulates myelination by engaging axonal, but not glial, TrkA receptors. How does NGF affect myelination? A previous section of this chapter has illuminated after-effects of ligand dependent engagement of Trk receptors. It is postulated that these signals converge in the nucleus and trigger expression of neuronal genes that modulate glial cells to myelinate. Potential representation of the second category includes small molecules like adenosine that may act on glial purinergic receptors (Stevens et al., 2002) or other moieties such as neuroregulins (Taveggia et al., 2005), which are a family of receptor tyrosine kinases related to EGF and whose receptors (erbB/HER 2–4) are well expressed in myelinating glial cells.

However, NGF responsiveness is not the only event in axon selection for myelination as sensory C-fibers in PNS remain non-myelinated despite expressing TrkA receptor (thus being potentially capable of intercepting up NGF cues). Certain other neurotrophins may be involved. For example, GDNF stimulates the process at an early stage by regulating early stage Schwann cell function by activating PKA and PKC pathway (Iwase et al., 2005). Additionally, considering hindrance posed by NGF in oligodendrocytic myelination, signal transduction in this cell-type must be dependent on some yet unknown non-NGF mechanism. It has been proposed that by altering surface exhibition of certain 'wrap me'/'do not wrap me' signals, axon themselves act as determinants of their myelination (Coman et al., 2005).

16.3.2.2.2. Signals Regulating G-Ratio

The ratio of the axonal diameter divided by the diameter of the axon plus its myelin sheath is referred to as the G-ratio. Usually, the G-ratio is maintained between 0.6 and 0.7. The importance of this constancy lays in the fact that thickness of myelin wraps depend on axonal thickness and demonstrate proportionality. One of the main regulators of the myelin thickness is signal(s) induced by Neuregulin-ErbB system (Michailov et al., 2004). Additionally, neurotrophins like BDNF and neurotrophin p75^{NTR} are also thought to be involved in regulating myelin sheath thickness (Tolwani et al., 2004).

16.4. Signaling During Neuroinflammation

16.4.1. Neuroinflammation

Although inflammation is a self-defense operation, it may attain harmful proportions if not controlled strictly. Evolved as much as we are, human inflammatory system still gets into the overdrive mode in several instances and instead of being efficacious, creates havoc. Neuroinflammation is one such example. Unrestricted inflammatory response in CNS is now considered a root cause of several neurodegenerative diseases, as an overdose of inflammatory moieties tend to injure/ kill neurons.

16.4.1.1. Cells Involved

Considering CNS to be devoid of immune-surveillance has been one of the most coveted myths of biology. We now realize, CNS has its own share of immunological residents (microglia) and is also subjected to infiltration of peripheral immune-cells (macrophages, monocytes and T cells) during diseased states. Additionally, in pathologic brain, astrocytes and endothelial cells of BBB also show an upsurge in expression of immunological moieties. 'Gliosis' is the term commonly used to indicate such inflammatory status of glial cells, which by definition, refers to reactive populating of insulted area in brain or spinal cord by excess growth of glial cells. Reactive gliosis, a neuroinflammatory hallmark, is clinically manifested by enhanced expression of a battery of pro-inflammatory molecules like inducible nitric oxide synthase (iNOS), cyclooxygenase-2 (COX-2), certain adhesion molecules like ICAM and VCAM, and a plethora of pro-inflammatory cytokines like tumor necrosis factor- α (TNF-α) and interleukins (like IL-1, IL-6). In the following lines we shall concentrate on the pathways upregulating these gene products.

16.4.1.2. Inducers

Inducers for neuroinflammation may arise within the CNS (intrinsic) or may be introduced by external agencies (extrinsic). The major intrinsic trigger for neuroinflammation is a necrotic neuron, which tends to dissipate considerable amount of cellular junk around the dying cell, which trigger inflammation in adjacent glial cells. Additionally, pro-inflammatory cytokines, like TNF-α, IL-1β, and IFN-γ, are other major intrinsic inducers of neuroinflammation. Furthermore, contact with certain peripheral inflammatory cells, which sneak into brain through a leaking blood brain barrier during diseases, often triggers inflammatory response in glial cells (Dasgupta et al., 2003).

On the other hand, extrinsic stimuli are often delivered by viruses and bacteria. Bacterial products like lipopolysaccharide (LPS) and DNA with motifs of unmethylated CpG dinucleotides are extremely potent inducers of neuroinflammation. Additionally, viruses themselves, or their products like retroviral coat protein gp41 and gp120, double stranded RNA, and transcription factors like Tat, also induce inflammatory responses in brain cells.

16.4.2. Signaling for Gliosis

16.4.2.1. Positive-Regulatory Signals

16.4.2.1.1. Activation of Pro-inflammatory Transcription Factors (TFs)

Among several TFs involved in mediating neuroinflammation, NF-κB, CCAAT/enhancer-binding protein (C/EBP), activating protein (AP-1), STAT, and interferon regulatory factors (IRF) are the top five TFs that are required for transactivation of almost all pro-inflammatory molecules. Among five members of NF-κB family, dimers of p50:p65 are the most important inflammatory mediators. Activated kinase pathway(s) phosphorylate p50:p65 arresting protein, inhibitory kappaB (IκB) in the cytosol, thereby subjecting it for ubiquitination and subsequent proteosomal degradation. This liberates the p50:p65 heterodimer to enter the nucleus and bind kappaB elements in the target promoter (Li and Verma, 2002). Such targets include almost every gene, whose products are known to be associated with neuroinflammation. C/EBP, a family of six basic leucine zippers, are involved in several cellular responses including, inflammation (Ramji and Foka, 2002). Among them C/EBPβ and C/EBPδ, known to form homo- and heterodimers between themselves, occur most frequently in the neuro-inflammatory radar. Another basic leucine zipper TF is AP-1, which is mainly composed of Jun, Fos, and/or ATF dimers (see Hess et al., 2004 for more details). Leaving out STAT (subsequently discussed elaborately), IRFs are mainly recognized and named after their central role in mediating anti-viral responses. Also, several IRFs (like IRF-1) play an important role in neuroinflammation.

Activation of TFs is the culminating step of several signal transduction pathways, and requires upstream activity of various kinase pathways. Let's roll upwards.

16.4.2.1.2. MAP Kinase Pathway

All three MAP kinase pathways are involved in neuroinflammation in different combination depending on inducing signal and the final product. For example, for iNOS regulation in human astrocytes, both JNK and p38 pathways are recruited in response to IL-1β. However, co-induction by IL-1β along with IFNγ renders the JNK pathway redundant while the p38 pathway is utilized. Once, recruited, MAP kinase cascades ultimately convey inflammatory signal mainly by activating different TFs. Activation of JNK leads to phosphorylation of Jun, which then enters the nucleus to form AP-1 by dimerizing with Fos. Activation of the MEK-ERK cascade mostly activates C/EBP dimers, although activation of other TFs is not ruled out. On the other hand, the hypothesis of NF-κB being downstream of p38 MAP kinase is controversial. In other instances, p38 also regulates TFs like C/EBP, ATF-2, and AP-1 (Figure 16.3B).

In addition to TFs, MAP Kinases may play certain non-canonical roles as well in inducing neuroinflammation. For example, p38 has been shown to phosphorylate Ser10 of Histone3 in promoter region of pro-inflammatory genes (Saccani et al., 2002). Such phosphorylation is postulated to be the epigenetic signature for facilitated docking of pro-inflammatory transcription factors like NF-κB. Additionally, MAP kinases may regulate inflammatory gene expression by regulating certain co-activators.

16.4.2.1.3. JAK-STAT Pathway

While MAP kinases are serine/threonine cascades, the Janus kinase (JAK)-signal transducers and activators of transcription (STAT) pathway epitomizes tyrosine kinase signaling in gliosis. The JAK family is one of ten recognized families of non-receptor tyrosine kinases. Originally identified as the signaling pathway for interferons, JAK-STAT signaling is now known to mediate signals of various cytokines, growth factors and hormones. The basic biology of JAK-mediated signal transduction (Rawlings et al., 2004) is based on ligand-stimulated assembly of receptors into an active complex followed by phosphorylation of the receptor-associated JAKs (JAK1, JAK2, and JAK3) and tyrosine kinase 2 (Tyk2). Subsequently, phosphorylated JAKs phosphorylate inactive cytosolic STATs, which in turn are activated to form homo- or heterodimers. These dimers enter nucleus and bind specific regulatory sequences to activate or repress transcription of target genes (Figure 16.3A). During gliosis, this pathway primarily activates a set of genes including, iNOS, and COX-2.

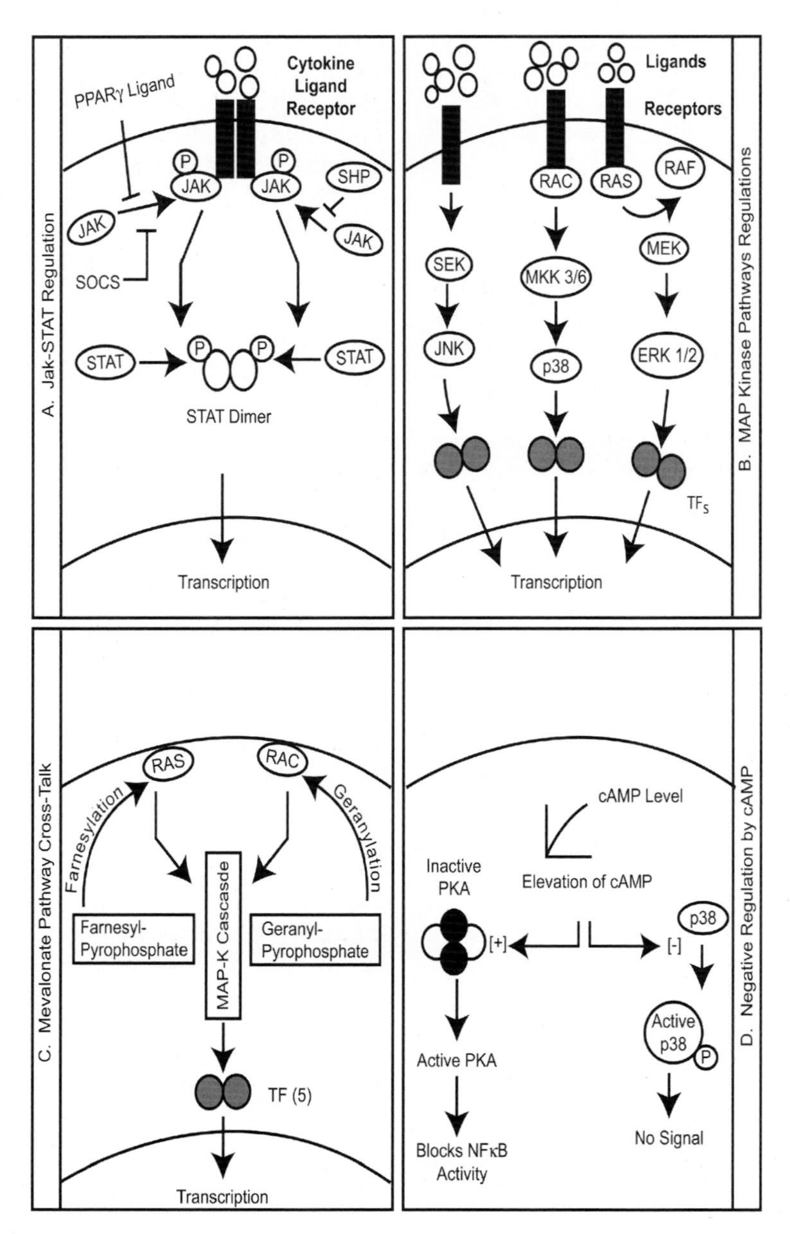

FIGURE 16.3. Various aspects of neuroinflammatory signaling. (A) Ligand-bound receptors auto-activate and then recruit and activate JAK. Subsequently, JAK phosphorylates STAT and facilitates the formation of STAT active dimer that can translocate to the nucleus and participate in gene transcription. Negative regulators of this pathway, like SHP, SOCS and many nuclear receptor ligands, inhibit phosphorylation of JAK and thereby defuse the pathway. (B) All three known MAP kinase pathways, as a common denominator, activate one or more transcription factors, which then mediate gene expression. However, MAP kinases may perform other roles not shown in this diagram. (C) Small G-proteins like Ras and Rac, which acts upstream of MAP-kinase pathways, are at times regulated by cross-talk with other normal biochemical pathways in cell. As shown here, geranylpyrophosphate and farnesylpyrophosphate intermediates of mevalonate cholesterol biosynthesis pathway. These intermediates modify Ras and Rac thereby activating them to induce further signals via MAP kinase pathways. (D) In glial cells, elevation of cAMP leads to the activation of PKA, which in this case, blocks the activation of proinflammatory transcription factors like NF-κB. Elevated cAMP levels also inhibit p38 activity, thus blocking the activation of several transcription factors that function downstream of this kinase.

16.4.2.1.4. Small G-Protein Signaling

We have talked about activation of MAP kinases that subsequently activate TFs. Now, MAP kinases, unlike JAKs, do not get activated by receptor docking. Signals are relayed to MAP kinases from the receptors by an elite group of messengers called small G-proteins (SGP). Based on structural delineations, the SGP super-family is divided into five families; the Ras, Rho/Rac, Rab, Sar1/Arf, and Ran. These 20–30 KDa monomeric GTPases serve as molecular switches of signal transduction by shuttling between two interchangeable forms, the GTP bound active form and the GDP bound inactive form (Takai et al., 2001). During gliosis, the MEK-ERK cascade is sensitive to Ras-Raf activity. On the contrary, activation of MKK3/6, the upstream kinase for p38, is dependent on Rac activity.

How is the signal mediated by SGPs? If the signal originates from G-protein coupled receptors (GPCR), then SGPs may be activated either by signals originating from the classical heterotrimeric G-protein effectors or by activation of receptor tyrosine kinases. The exact mechanisms for these processes are not well understood. However, SGPs also trigger signaling events without involvement of GPCRs. SGPs, like Ras and Rac, are post-translationally modified by metabolites of mevalonate biochemical pathway. Non-saponifiable lipid isoprenoids like farnesyl- and geranylgeranyl-pyrophosphate, are biosynthesized in animals from acetyl-CoA via the mevalonate pathway. These isoprenoids covalently modify and thus modulate the biological activity of SGPs (Maltese, 1990). Upon isoprenylation, these G proteins become membrane-bound and transduce several intracellular signaling pathways leading to activation of MAP kinases (Pahan et al., 1997b) (Figure 16.3C).

16.4.2.1.5. Redox Signaling

Reactive oxygen species (ROS) are multi-potent diffusible molecules capable of carrying out several signal transduction processes in response to several extracellular stimuli. Consistent with their versatile cellular functions, ROS have been also shown to regulate expression of inflammatory products like, iNOS in different brain cells. Antioxidants, like N-acetyl cysteine (NAC), pyrrolidine dithiocarbamate (PDTC) and lycopene are potent inhibitors of inflammatory products in glial cells, thereby proclaiming a role of ROS in mediating gliosis.

Recently, NADPH oxidase has been identified as the ROS-producing molecule in activated glial cells (Pawate et al., 2004). Cytokine stimulation of astrocytes leads to rapid activation of NADPH oxidase and release of ROS followed by expression of pro-inflammatory products like, iNOS. Consistently, attenuated expression of iNOS is observed in primary astrocytes derived from $gp91^{Phox}$-deficient mice (Pawate et al., 2004). ROS are believed to regulate expression of pro-inflammatory gene products via NF-κB. However, the involvement of other transcription factors in ROS-mediated gliosis cannot be ruled out.

16.4.2.1.6. Nitric Oxide Signaling

One of the unavoidable fall-outs of ROS generation is induction of iNOS, which enhances production of nitric oxide (NO) from glial and endothelial cells. NO, the popularly known vasorelaxant, also works as a neurotransmitter when produced in physiological quantities by neurons. However, in excess concentration, NO forms peroxynitrite ($ONOO^-$), a neurotoxic mediator of neuroinflammation. The effects of peroxynitrite in immune regulation are exerted through nitrosylation of cell signaling messengers like cAMP, cGMP, G-protein, JAK/STAT or members of MAP kinase dependent signal transduction pathways (Guix et al., 2005). Nitration of cysteine residues of these proteins may inhibit or activate their functionality. Similar modifications may also manipulate activity of transcription factors like NF-κB, thereby modulating gene expression and encouraging inflammatory outbursts.

16.4.2.1.7. Signaling by Toll-Like Receptors (TLRs)

TLRs are archetypal pattern recognition receptors of innate immune system found in several invertebrates and all vertebrates. Being transmembrane moeties, they recognize a variety of conserved patterns and motifs found in pathogenic entities, and thus serve as sensors of microbial invasion. Once engaged, TLRs generate a complex anti-pathogenic immune response in various cell types, including glial cells in brain. 10 TLRs have been identified in human so far (Konat et al., 2006). All of these posses a structural motif in their cytoplasmic tail [Toll/IL-1 receptor (TIR) domain], which forms the basis of signal transduction by these receptors (Barton and Medzhitov, 2004). Upon ligand binding, adaptor molecules like, MyD88, bind to these receptors and recruit the IL-1 receptor associated kinase (IRAK) leading to their phosphorylation. Phosphorylated IRAK transduces the message downstream and activates MAP kinases and transcription factors like NF-κB. In brain, most of the known TLRs are expressed in brain immune cells, i.e., microglia and astrocytes (Bsibsi et al., 2002; Konat et al., 2006). Furthermore, exposure to specific ligands or certain cytokines induce a rapid upregulation of several TLRs in these cells. TLRs play an important role in mounting immune response in brain abscess as well as other CNS gram-positive infections. However, persistant TLR signaling, which may result from residual microbial products after CNS infection clearance, may potentially damage the brain (Konat et al., 2006).

16.4.2.2. Negative-Regulatory Signals

16.4.2.2.1. Activation of Protein Kinase A

We have already seen a contingency in functioning of cAMP and PKA in previous sections. During glial inflammation, expression of several cytokines and iNOS is dependent on the cAMP-PKA pathway. While expression of cytokines like IL-1β is completely dependent on PKA, expression of pro-inflammatory moieties, like iNOS, are partially dependent on this enzyme (Woo et al., 2004). Despite its partial role, activation of this pathway adequately hinders expression of iNOS (Pahan et al., 1997a). Anti-inflammatory agents like, KL-1037 and N-acetyl-O-methyldopamine, prohibit microglial activation by activating the PKA pathway (Kim et al., 2004, Cho et al., 2001). Taken together, PKA pathway may be considered as a general inhibitory pathway with regard to glial activation (Figure 16.3D). How does PKA block the pro-inflammatory response in glial cells? Recently, it has been shown that cAMP inhibits the activation of p38 MAP kinase in rat primary astrocytes and C6 glial cells (Won et al., 2004). As we have seen in the previous section the p38 MAP kinase plays a pivotal role in glial inflammation. Thus, blocking this kinase will definitely destabilize potential pro-inflammatory signaling intensions (Figure 16.3D).

16.4.2.2.2. Activation of SOCS

In order to counter pro-inflammatory signaling pathways, cells employ a family of proteins called suppressors of cytokine signaling (SOCS). Due to their ability to regulate and subdue a pro-inflammatory signal, these proteins are now considered important regulators of normal immune physiology and immune disease (Leroith and Nissley, 2005). In general, SOCS are present in cells at very low levels. However, they are rapidly transcribed upon exposure of cells to pro-inflammatory stimuli. SOCS can negatively regulate the response of immune cells either by inhibiting the activity of JAK or by competing with signaling molecules for binding to the phosphorylated receptor. Moreover, activators of nuclear hormone receptor PPAR-γ, induce the transcription of SOCS1 and SOCS3 to inhibit the activity of JAK1 and JAK2 in rat primary astrocytes (Park et al., 2003). Both SOCS1 and SOCS3 are capable of binding JAKs to suppress their tyrosine kinase activity. Therefore, PPAR-γ activators reduce the phosphorylation of STAT1 and STAT3 and attenuate pro-inflammatory signals in activated glial cells. These results suggest that up-regulation of SOCS may represent a critical step for suppressing glial inflammation via negative regulation of the JAK-STAT pathway (Figure 16.3A).

16.4.2.2.3. Nuclear Receptor Ligands

Nuclear receptors (NR) are evolutionary conserved lipophilic ligand-regulated transcription factors that control gene expression. NR ligands (NRL) recruit coactivators to the DNA-bound NR thereby transactivating target genes. But we are talking about repression and not transactivation. So how are NRs involved in repression? In the context of neuroinflammation, it is now clear that NRLs repress gene transcription independent of nuclear receptor itself. For example, gemfibrozil, a ligand for peroxisome proliferator-activated receptor-alpha (PPAR-α), inhibits cytokine-induced iNOS expression in human astrocytes independent of PPAR-α (Pahan et al., 2002). Along similar lines, 15-deoxy-12, 14-PGJ2 (15d-PGJ2), a ligand for PPAR-γ, attenuates (LPS + IFN-γ)-induced expression of iNOS in rat primary astrocytes independent of the PPAR-γ itself (Giri et al., 2004). Recently, ligands for other NR such as RAR and RXR have been also shown to suppress the expression of inflammatory products independent of NR (Xu et al., 2005, Royal et al., 2004).

How may NRLs repress iNOS without actually involving the NR? Gemfibrozil, the PPAR-α ligand, strongly inhibited (IL-1β + IFN-γ)-induced activation of NF-κB, AP-1, and C/EBPβ but not that of STAT-GAS in human astroglial cells (Pahan et al., 2002). Furthermore, 15d-PGJ2 inhibits NF-κB pathway at multiple points (Giri et al., 2004). Blocking of NF-κB and other TFs is indeed a handy mode of shutting down stimulus-induced response in a short period of time. Additionally, few mechanisms have been offered to explain the blocking effect of NRLs on pro-inflammatory TFs. PPAR-γ ligands, 15d-PGJ2 and rosiglitazone, reduce phosphorylation of JAK-STAT pathway in activated rat astroglia and microglia thereby leading to the suppression of JAK-STAT-dependent inflammatory responses (Park et al., 2003) (Figure 16.3A). This blockage is not contingent on PPAR-γ and is mediated by rapid transcription of suppressor of cytokine signaling (SOCS) 1 and 3. Additionally, SHP-2 is also involved in the anti-inflammatory action of NRLs. NRL treatment was shown to phosphorylate SHP2 within minutes. As phosphorylated SHP2 dephosphorylates JAK, this creates yet another avenue of blocking the JAK-STAT pathway.

16.4.2.2.4. IL-10 and IL-13 Signaling

IL-10 and IL-13 are anti-inflammatory cytokines. In CNS, systemic inflammation is mediated by expression of these cytokines along with pro-inflammatory ones. The purpose of expressing pro- and anti-inflammatory molecules at the same time is to provide the system with self-antidotes. Similarly, IL-10 knock-out mice demonstrate elevated production of inflammatory gene products in the brain in comparison to their wild-type littermates during encephalitis. IL-13 on the other hand, induces death of activated microglia (Yang et al., 2002), thereby restricting inflammatory output of these cells. Such effect of IL-13 is mimicked by yet another anti-inflammatory cytokine, IL-4. This comes as no surprise since IL-13Rα1-IL-4Rα complex constitutes a receptor for both IL-4 and IL-13 (Hershey, 2003).

How do these anti-inflammatory cytokines work? Essentially, they tend to inhibit biosynthesis of pro-inflammatory cytokines by stimulating biosynthesis of pro-inflammatory cytokine inhibitors like soluble receptors (Burger and Dayer, 1995). Furthermore, they also intercept signals arising from pro-inflammatory receptors-ligand complexes. For example, pro-inflammatory cytokines kick-start the breakdown of plasma-membrane sphingomyelin into ceramide, the second messenger in sphingomyelin pathway, which subsequently acts on downstream JNK pathway. On the other hand, IL-10 and IL-13 inhibit pro-inflammatory cytokine-mediated breakdown of sphingomyelin to ceramide thereby interrupting further pro-inflammatory signaling. This inhibitory signal is mediated via activation of phosphatidylinositol (PI) 3-kinase (Pahan et al., 2000).

16.5. Signaling During Neurodegeneration

Comprehension of the current topic lies in apprehending signaling in neuronal cells. Unlike glial cells, which are threatened with neurodegenerative toxins, neurons tend to face stiff challenges by them. In a neurodegenerative milieu, a particular neuron may undertake signaling to die, or to resist the fulmination, or both. To keep our mind clear, we shall focus only on anti-survival signals leading to death.

16.5.1. Neurodegeneration

16.5.1.1. Apoptosis in Neurons

During development, programmed cell death is a common norm in developing neurons. However, the intrinsic pro-apoptotic pathways are obliterated as neurons mature. Thus, mere withdrawal of trophic factors does not suffice in inducing their death. Additionally, a genuine apoptotic signal is required for onset of self-suicide process in neurons. Specific gene-products undertake the apoptotic task in different types of neurons and different stimuli may induce distinct apoptotic pathways in them (Pettmann and Henderson, 1998).

However, it is important to acknowledge that apoptosis is not the only means for neurons to die. Death of adult neurons in response to pathological challenges also occurs by necrosis, the unregulated cell death mechanism. Necrosis is mediated by increase in intracellular calcium that catalyses activation of Ca^{2+} -dependent cystine proteases like, cathepsins and calpains, which primarily compromise lysosomal integrity. Subsequently, these cystine proteases in the company of released lysosomal enzymes dismantle structural network of neuron. Additionally, the intracellular pH also plays a major role in necrosis (Syntichaki and Tavernarakis, 2003).

Let us get back to apoptosis and start with its inducers.

16.5.1.2. Inducers (Neurotoxins)

Depending on their origin, inducers of neuronal apoptosis can be divided into two categories - extrinsic and intrinsic. Extrinsic inducers are generally of viral or bacterial origin. Viral coat protein gp120, and transcription factors, like, Tat, induce neurodegeneration are often at the root of viral neuropathies like HIV-associated dementia. Similarly, bacterial products like LPS derived from Salmonella has been shown to be neurotoxic (Johansson et al., 2005). Certain other extrinsic conditions leading to neuronal apoptosis include hypoxia, UV radiation, and exposure to steroids.

Among intrinsic inducers, several neurotoxins are generated by inflamed glia, which includes excitotoxins like, kainate and glutamate, peroxynitrite radical, and cytokines. Among cytokines, members of the TNF superfamily are major rogues. FAS ligand (FASL/CD95L) and TNF-related apoptosis inducing ligand (TRAIL) are most noteworthy in this regard where TNF-α itself has a controversial role (Saha and Pahan, 2003). Additionally, several misfolded and/or mutated cellular proteins like protease resistance Prion (PrP^{res}) (Prion disease), Parkin (Parkinson disease), Huntington (Huntington's disease), and amyloid-β (Alzheimer's disease) also cause neurodegeneration. Other intrinsic inducers include intracellular changes like, genotoxic damage, misbalance of intracellular Ca^{2+} and anoikis.

Several of these inducers (like FASL) induce neuronal apoptosis directly while others induce it indirectly (like LPS) via various mechanisms. It is interesting to note that there are several common inducers of neuroinflammation and neurodegeneration. Considering such diverse cellular response from same inducer in different brain cells, the diversity of signaling events in brain cells is quite apparent.

16.5.1.3. Neuronal Receptors

Receptors mediating neuronal apoptosis may be grouped in two types—dependence receptors and non-dependence receptors.

16.5.1.3.1. Dependence Receptors

These are a group of surface receptors that transduce two completely different sets of intracellular signals. In the presence of their respective ligand, these receptors generate a pro-survival signal. But in the absence of ligand, these receptors trigger a pro-apoptotic signal. Thus, survival of the cell is dependent on constant availability of the ligand and hence the receptors have been named as 'dependence' receptors (Mehlen and Bredesen, 2004). These receptors form ligand-dependent complexes that include specific caspases in inactive form. Absence of ligand leads to activation of caspase(s), which then cleaves the receptor itself, releasing a pro-apoptotic peptide fragment from it. $p75^{NTR}$ was described as one of the earliest dependence receptors (Barrett and Bartlett, 1994). Moreover, several receptors, like DCC (Deleted in Colorectal Cancer) and Unc5H2, have been described as dependence receptors playing a major role in neurogenesis.

16.5.1.3.2. Non-Dependence Receptors:

These receptors are straightforward death-receptors without any ambiguity. The best examples of this category applicable to most neuronal types belong to TNF-R family and include TNFR1 and Apoptosis antigen (APO1/Fas/CD95). Ligation of these receptors competently induces death in neurons. In the following lines, we will expand on signaling pathways triggered by these receptors.

16.5.2. Activation of Anti-Survival Pathways

Two pathways lead to apoptosis in mature neurons; the intrinsic mitochondria-dependent pathway and the death receptor-induced extrinsic (mitochondria-independent) pathway. These pathways seldom function exclusively and often converge with each other.

16.5.2.1. Mitochondria-Dependent Pathways

Mitochondrion is one of the most multi-faceted cellular organelle that is involved in energy generation, calcium buffering and regulation of apoptosis. In a happy cell, the mitochondria remain intact and prohibit apoptosis by sequestering a myriad of pro-apoptotic molecules within itself. However, as an apoptotic signal triggers the mitochondria-dependent apoptosis pathway, mitochondria undergo structural changes to become a punctuated and leaky bag of pro-apoptotic molecules.

16.5.2.1.1. Mitochondrial Fission

Mitochondria undergo frequent fission and fusion (Bereiter-Hahn and Voth, 1994), a fact largely understated in cell biology textbooks. A balance between these two processes serves to maintain normal mitochondrial tubular network and thus manifests normal cellular functions. However, during apoptosis onset, mitochondria undergo rapid and frequent rounds of fission thereby generating fragmented punctiform organelles of various sizes. This sets up the stage for apoptosis by leading to mitochondrial DNA loss, respiratory imbalance and ROS generation to alarming proportions (Yaffe, 1999).

16.5.2.1.2. Mitochondrial Leakage

Injurious signal(s) induce translocation of BH3-members of Bcl-2 family (Bim, Bid, Puma, Bad, Noxa, and BMF1) to mitochondrial membrane where they permeabilize it by formation of 'pores' in the outer membrane ('mitochondrial permeability transition'). This leads to cytosolic release of pro-apoptotic molecules cytochrome C, SMAC/DIABLO, apoptosis-inducing factor (AIF), endonuclease G (EndoG), and high temperature requirement serine protease 2 (HTRA2/OMI). Once in the cytoplasm, cytochrome C interacts with cytosolic apoptotic peptidase activating factor-1 (APAF1) resulting in oligomerization of the later. This complex is subsequently called the 'apoptosome' as it binds procaspase-9 and results in its auto-activation to form active caspase-9. The apoptosome complex further mediates downstream caspase activation (Figure 16.4).

Other mitochondrial 'leaked' proteins serve apoptosis from a different angle. SMAC/DIABLO binds and sequesters anti-apoptotic IAP proteins, which otherwise inhibit caspase activity. Similarly, HTRA2 interacts with X-linked IAP (XIAP) thereby interfering with its caspase inhibitory activities. HTRA2 also triggers DNA fragmentation. Also, DNA fragmentation and chromatin condensation is undertaken by the flavoprotein AIF, which interestingly acts without aid of caspases. In addition to HTRA2 and AIF, nuclear DNA also endures direct cleavage activity by the sequence-unspecific DNAase EndoG. (See Lossi and Merighi, 2003 for more details.)

16.5.2.2. Mitochondria-Independent Pathways

Activation of the core apoptotic machinery in mature neurons requires *de novo* transcription and this is not dependent on mitochondria (Figure 16.4). These pathways are discussed below.

16.5.2.2.1. p53 Pathway

p53, a tetrameric transcription factor, has been implicated in neuronal death in several neurodegenerative diseases, including, stroke, AD, PD, and ALS. p53 activity in neurons is upregulated in response to several neurodegenerative stimuli like hypoxic shock, excitotoxicity, DNA damage, and oxidative stress. Such activation is contingent on any of/all three signal-dependent post-translational modifications: phosphorylation (primarily at the N-terminal end), acetylation (primarily at the C-terminal end) and poly-(ADP)-ribosylation.

Once activated, p53 transactivates an array of pro-apoptotic genes products including the death receptors like Fas/CD95, members of mitochondrial apoptotic pathway like, Bax, Puma, Siva, Noxa, Peg3, and Apaf-1 (Culmsee and Mattson, 2005). Furthermore, p53 manipulates several transcription factors, thereby interfering with their normal job. For example, p53 activation blocks NF-κB activity, which in neurons mediates transcription of several pro-survival gene-products. In addition to such nuclear role, p53 can directly trigger synaptic apoptosis by translocating to mitochondria in company of Bax and inducing mitochondrial permeability transition.

16.5.2.2.2. E2F Pathway

Early gene 2 factor (E2F) is a transcription factor that triggers transcription of many genes involved in DNA replication and cell growth control in a dividing cell and acts downstream of several important signaling cascades that regulate cell cycle. E2F remains inactive during early G1 phase of cell-cycle due to sequestering action of hypophosphorylated retinoblastoma (Rb) protein. However, during late G1 phase, hyperphosphorylation of Rb inactivates it, thereby permitting E2F activity whose transcription products drive the cell through cell-cycle routines (Stevaux and Dyson, 2002).

But in post-mitotic neurons, pro-apoptotic stimuli-induced E2F activation triggers core apoptotic machinery (Greene et al., 2004). E2F1 down-regulates the expression of anti-apoptotic factors while upregulating several pro-apoptotic genes like, *apaf1, casp 3, casp 7, and siva.* Furthermore, E2F1 expression leads to the stabilization of p53. But most interestingly, E2F triggers neuronal apoptosis by derepressing cell-cycle regulators (e.g. cyclins and cyclin-dependent kinases) in post-mitotic cells.

16.5.2.2.3. Sphingomyelin-Ceramide Pathway

Ceramide, a lipid second messenger, refers to a family of naturally occurring N-acylated sphingosines. It is generated by the activity of sphingomyelinase (SMase) that breaks sphingomyelin to yield ceramide and phosphocholine. The most common effects of eliciting the sphingomyelin-ceramide pathway are either differentiation or death. Although ceramide at low concentration induces neuronal differentiation, yet non-physiological dose of ceramide generated in a diseased CNS for a prolonged period, induce neuronal apoptosis. Toxic level of ceramide is induced in neurons in response to HIV-1 gp120 and fibrillar Aβ, where neutral SMase, but not the acidic one, plays pivotal role in

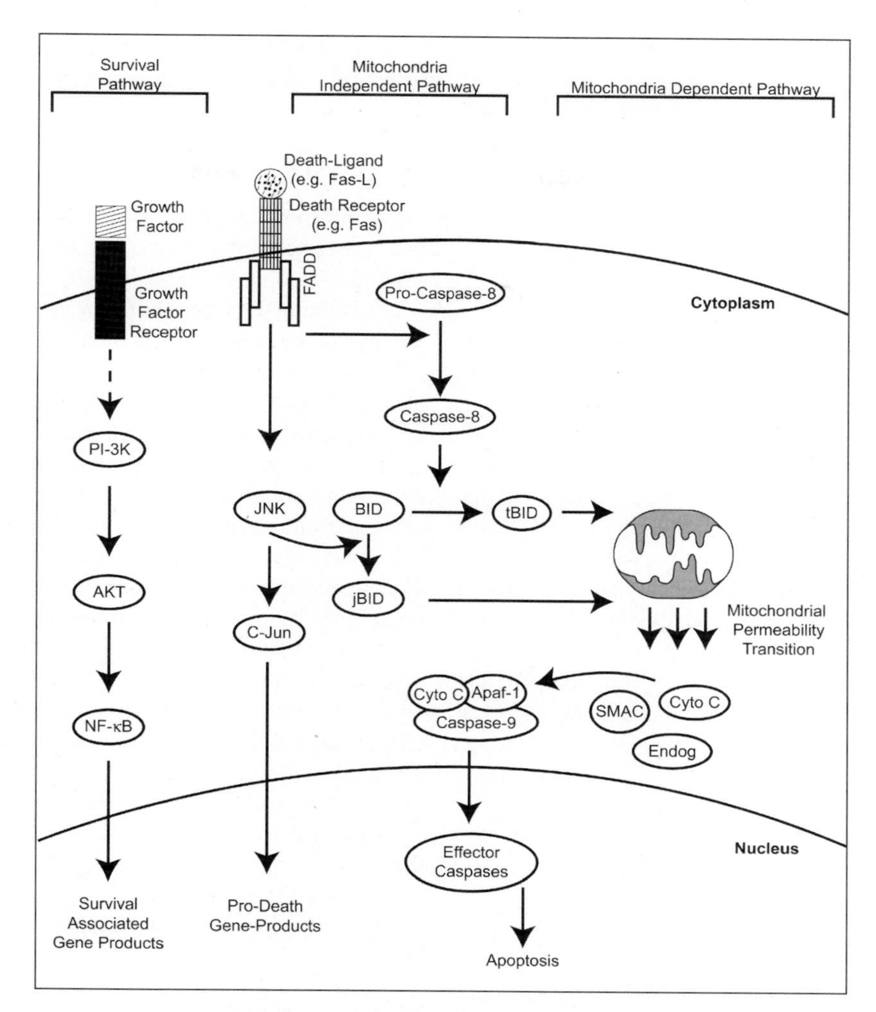

FIGURE 16.4. Pathways of neurodegeneration. Mitochondria-independent pathways usually arise from receptors with death domain (like TNF-R and FAS). Activation of these receptors leads to activation of pro-apoptotic transcription factor like c-Jun, which mediates the expression of a range of pro-apoptotic gene products. On the other hand, the mitochondria-dependent pathway relies on the permeability transition of its membrane, which leads to cytoplasmic release of several pro-apoptotic molecules. Among them, cytochrome C forms the lethal apoptosome complex by binding with cytosolic APF1 and caspase 9. This complex cleaves and activates down-stream caspases thereby ensuring apoptosis. Furthermore, there is ample crosstalk between these two pathways. For example, activation of pro-caspase 8 by ligand-bound death receptors leads to cleavage of Bid to tBid. Along with jBid, formed due to JNK activity, tBid attaches to mitochondrial membrane and form pores to manifest transition in permeability. The survival pathway is shown to appreciate the fact that, neuronal fate due to an insult is often the result of the prevailing pathway amongst ones mediating death and life.

ceramide generation (Jana and Pahan, 2004a,b). Although the mechanisms of ceramide-induced cell death is not fully understood, yet they appear to involve a number of signal transduction pathways, including proline-directed kinases, phosphatases, phospholipases, transcription factors, and caspases (Goswami and Dawson, 2000).

16.5.2.2.4. HAT-HDAC Misbalance

Histone acetyltransferases (HATs) and histone deacetylases (HDACs) represent two enzyme classes that, respectively, catalyze forward and backward reaction kinetics of lysine residue acetylation of nucleosomal histones and various transcription factors. In a normal neuron, enzymatic undertakings of HAT and HDAC remain stoichiometrically balanced that in turn confers stability to the cellular homeostasis by coordinating gene expression and repression on both temporal and spatial basis (Saha and Pahan, 2005). Such equilibrium manifests neuronal homeostasis and is responsible for normal neuronal functions like, long-term potentiation, learning and memory.

However, during neurodegenerative conditions, the neuronal acetylation homeostasis is profoundly impaired. Such impairment is primarily manifested by comprehensive loss of HATs like CBP during various neurodegenerative challenges. However, HDAC protein level does not alter during neurodegeneration. Thus, at the cost of HATs' loss-of-dosage,

HDACs attain facilitated gain-of-function, thereby unsettling the acetylation homeostasis. At this stage, transcription of pro-survival genes is profoundly repressed due to collapsed histone gates at their promoter regions. Furthermore, several transcription factors like CREB, NF-κB, Sp1, and HIF fail to perform their pro-survival transcription duties as all these TFs require to be acetylated for activation.

16.5.2.2.5. A Question to Ponder: Is NF-κB Pro-apoptotic?

NF-κB is critical transcription factor in neurodegeneration. It is a family of five TFs (Li and Verma, 2002) which can dimerize in various combinations amongst themselves leading to regulation of anti-apoptotic well as pro-apoptotic gene products (Barkett and Gilmore, 1999). Additionally, it also upregulates inflammatory gene products as well. With activities spread out at extremes, there is little surprise in the decade old controversy regarding the actual role of NF-κB in neurodegeneration. It is now being accepted that within neurons, NF-κB acts as a pro-survival agent and is responsible for upregulation of survival ensuring gene-products like MnSOD, Bcl-2, Bcl-X_L, and IAPs (Mattson and Camandola, 2001). However, NF-κB becomes the devil's advocate in glia, where it up-regulates neurotoxic and/or pro-inflammatory molecules, like iNOS, interleukins, chemokines (e.g. SDF-1α) and excitatory products. If both statements are true, then is NF-κB neurotoxic or neuroprotective? The debate goes on.

16.6. Signaling During Neuroregeneration

Ramon y Cajól had observed that while PNS tends to repair itself after injury, the CNS does not regenerate. He had concluded from his observations *in vivo* that CNS axons tend not to grow because of certain barriers present within the CNS. He had even suggested that these inhibitors are present in the white matter (myelin). Today, we realize that there are two main obstructions: myelin inhibitors and glial scars. After injury, certain inhibitors displayed on oligodendroglial plasma membrane generate inhibitory signals, which when perceived by an axon blocks its regeneration. Ultimately, scar tissue is formed in the area and regeneration is physically intercepted. However, spontaneous regeneration is manifested in some part of brain and in spinal cord due to amalgamation of several pro-growth signals. In the following lines, we will first talk about inhibitory signals and then about signals that promote regeneration.

16.6.1. Signals Blocking Axonal Regeneration

It is not the lack of neurotrophins, rather the presence of regeneration inhibitors in myelin and glial scars that effectively negate axonal regrowth (McGee and Strittmatter, 2003). Several myelin-derived proteins have been identified as components of CNS myelin, which prevents axonal regeneration in the adult CNS. These inhibitors interact with their neuronal receptors to hamper axonal regeneration.

16.6.1.1. Inhibitor Trio in Myelin: Nogo, Mag, and Omgp

Three different myelin proteins have been identified as of now that are strong inhibitors of axonal growth (Schwab, 2004) (Figure 16.5).

Nogo, named by Martin Schwab of Zurich University, is a member of the endoplasmic reticulum associated protein family, reticulon. Despite being associated with endoplasmic reticulum of oligodendrocytes, Nogo-A is also exhibited on the surface and mediates growth cone collapse (Fournier et al., 2002).

Myelin-associated glycoprotein (Mag), belonging to immunoglobulin superfamily, is a potent inhibitor of post-mitotic neuronal outgrowth. There are two isoforms of this inhibitor that differ only in their cytoplasmic domain. Ability of Mag to bind sialic acid, although not essential for its inhibitory effect, potentiates it nonetheless (De Bellard and Filbin, 1999).

The third and most recent inhibitor is oligodendrocyte myelin glycoprotein (Omgp) (Vourc'h and Andres, 2004). Omgp is linked to outer leaflet of plasma membrane by a GPI-linkage and contains domains of leucine-rich repeats and serine/threonine repeats. Like Nogo and Mag, Omgp induces growth cone collapse and inhibits neurite outgrowth.

In addition to myelin, inhibitory signals also arise from glial scars. Proteoglycans are the main culprit in this regard. Additionally, certain proteins present in the scar, like ephrin-B2 and Sema3, also repel the growth cone (Silver and Miller, 2004).

16.6.1.2. Inhibitory Signaling from Receptor Trio: NgR-Lingo1-p75NTR

It is remarkable that above mentioned all three myelin inhibitory proteins, without any significant domain or sequence similarity, bind and activate a common axonal multi-protein complex. This receptor complex is made up of the ligand binding Nogo-66 receptor (NgR1) and two binding partners responsible for triggering signal transduction, $p75^{NTR}$ and Leucine-rich-repeat and Ig domain containing Nogo-receptor interacting protein-1 (Lingo-1). It is now believed that owing to functional redundancy, presence of any one of the three myelin inhibitory ligands can trigger inhibitory signal(s) through the axonal receptor complex (Filbin, 2003, McGee and Strittmatter, 2003) (Figure 16.5).

After ligand binding of NgR, further neuronal intracellular signal transduction requires either or both $p75^{NTR}$ and Lingo as NgR does not traverse the plasma membrane and is linked to its outer leaflet by GPI-linkage. In both cases, the small GTPase, RhoA is activated, which subsequently mediates the signal via Rho kinase (ROCK) (Filbin, 2003) to cause growth cone collapse by enhancing retrograde F-actin flow (Dickson, 2001) (Figure 16.5). Recently, one more receptor has been described that participates in NgR complex signaling (Shao et al., 2005). Named TAJ/TROY, this orphan receptor belongs to

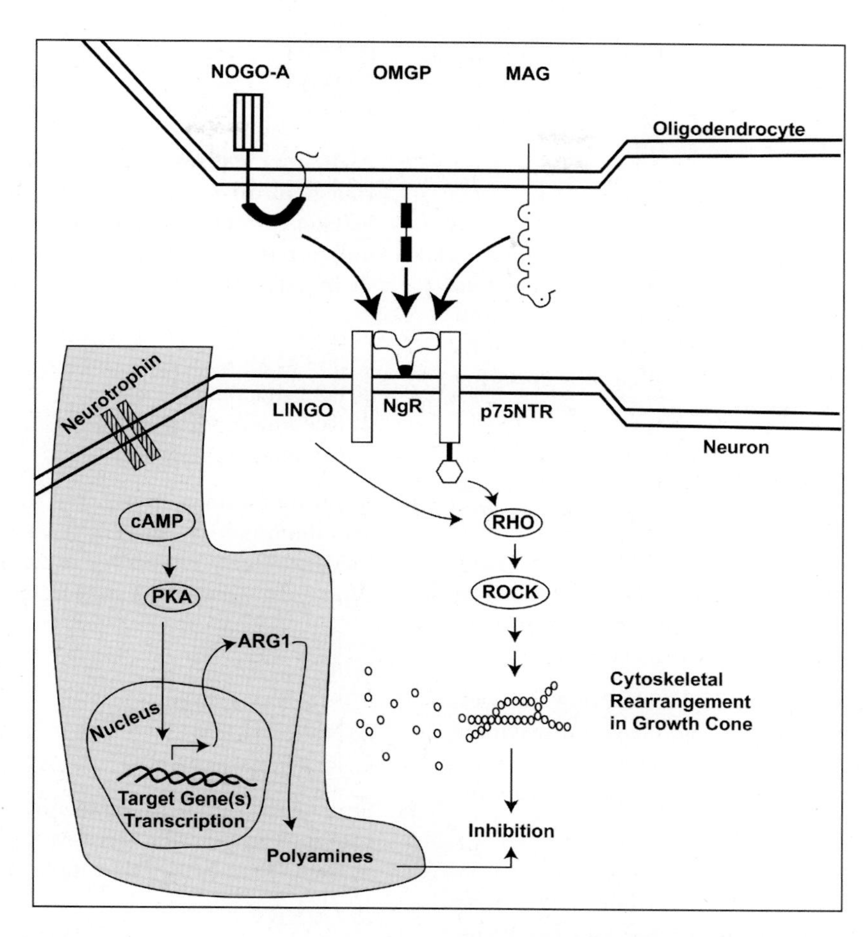

FIGURE 16.5. Pathways of neuroregeneration. Regeneration in the CNS is actively blocked by inhibitors expressed on the oligodendroglial membrane (Nogo, Mag and Omgp). All these inhibitors act on the Nogo66 receptor (NgR), which lacks a cytoplasmic tail. Further signal is conducted from NgR by either or both p75NTR and Lingo. In every case, the Rho GTPase is activated, which alters cytoskeletal arrangement via ROCK and other downstream effectors. However, priming of neurons with neurotrophins results in upregulation of polyamines, which potentially can prohibit the inhibitory effects of NgR complex signaling. Upregulation of polyamines in this case has been shown to be dependent on the cAMP-PKA pathway.

the TNF-superfamily and can functionally replace $p75^{NTR}$ in the NgR complex to activate RhoA in the presence of myelin inhibitors.

16.6.2. Signals Inducing Axonal Regeneration

Neurotrophins can prime neurons for regeneration. They elevate cAMP in neurons in the absence of inhibitory signals, and this elevation sufficiently over-rides any subsequent inhibitory signals. *In vivo,* crushed DRG axons have been shown to regenerate into spinal cord by application of neurotrophins (Ramer et al., 2000). Interestingly, effects of neurotrophins, in most cases, occur through Trk receptor, while inhibitory effects are mediated through $p75^{NTR}$.

Events downstream of cAMP elevation occur in two phases - transcription-independent and transcription-dependent. In the first phase, PKA is activated which is believed to levy a direct effect on the cytoskeleton via Rho GTPase. The second phase begins in a PKA-sensitive manner, but very soon manifests insensitivity to PKA. One of the most important targets of this pathway is the *arg1* gene, whose product (Arginase1) is a key enzyme in biosynthesis of polyamines. Generation of polyamines is the molecular mechanism of this pathway in countering inhibitory signals (Filbin, 2003) (Figure 16.5). Polyamines may exert their effects by influencing chromatin structure and transcription and/or influencing the cytoskeleton directly. They may also influence ion conductance across the axonal membrane.

Summary

Although not every dimension of CNS cellular signaling is covered in entirety due to restriction of publisher's space and reader's patience, yet sincere attempt has been made to introduce every known major signal transduction pathway in CNS. Further reading from referred literature is highly recommended to readers willing to comprehend every known frontier of Cajal's 'epic love story'.

Review Questions/Problems

1. **Neurotrophins are employed for regulation and/or maintenance of axonal growth, dentritic pruning and synaptic refinements. List receptors for neurotrophins and mechanisms by which neurotrophin-receptor complexes become internalized.**
2. **Three Trk receptor tyrosine kinases and four neurotrophins have been identified in the brain. How is affinity and specificity of neurotropins for Trk receptors regulated?**
3. **Ca^{2+} regulates a wide array of neuronal functions. How can this single ion regulate such a diverse functions as neuronal functions, ranging from neurite outgrowth, synaptic plasticity, transmitter release, etc.?**
4. **Ca^{2+} is mobilized from two pools for elevating intracellular levels: the extracellular space and from intracellular endoplasmic reticulum (ER). Give examples ion channels on the plasma and the ER membranes that are used for mobilizing Ca^{2+}.**
5. **After conveying a signal, intracellular Ca^{2+} levels must be restored to resting (basal) levels intracellular levels: Give examples of proteins (pump) on plasma and ER membranes that are used to lower intracellular Ca^{2+} levels.**
6. **Which of the following signaling pathways is expected to antagonize the activation of mitogen-activated protein (MAP) kinase in brain cells?**

   a. JNK signaling
   b. cGMP signaling
   c. cAMP signaling
   d. Akt signaling
   e. JAK signaling

7. **Which of the following signaling pathways will be upregulated in microglia after stimulation with interferon-g (IFN-γ)?**

   a. cAMP-PKA pathway
   b. cGMP-PKG pathway
   c. JAK-STAT pathway
   d. PI-3 kinase-Akt pathway
   e. TLR4 pathway

8. **Multiple sclerosis (MS) is the most co mmon human demyelinating disorder of the CNS. The expression of some genes may increase in the CNS of patients with MS. Identify those genes.**

   a. MOG
   b. MBP
   c. MAG
   d. Nogo
   e. PLP

9. **Alzheimer's disease (AD) is the most common neurodegenerative disease in which a particular set of neurons undergoes apoptotic cell death. Identify a signaling pathway that may attenuate neuronal apoptosis in patients with AD.**

   a. JNK signaling
   b. p38 MAP kinase signaling
   c. bcl2 signaling
   d. Rho kinase signaling
   e. Ceramide signaling

10. **In normal human brain, cells are equipped to counteract inflammatory signaling transduced by proinflammatory cytokines. Which one of the following molecules is expected to counteract such inflammatory signaling?**

    a. Ras signaling
    b. Rac signaling
    c. SOCS signaling
    d. Nitric oxide signaling
    e. Rho kinase signaling

11. **One of the following molecules should play an active role in transcriptional upregulation of memory genes in the CNS. Identify that molecule.**

    a. CREB
    b. bad
    c. bax
    d. HDAC
    e. Lingo

12. **During glial activation, microglia release superoxide that may lead to oxidative stress in the CNS. Which one of the following enzymes should be actively involved in producing superoxide radicals during microglial activation**?

    a. NADPH oxidase
    b. SOD
    c. Catalase
    d. Acyl-CoA oxidase
    e. Glucose oxidase

13. **What is nuclear factor-κB (NF-κB)? Describe the status of NF-κB in normal brain cells. What are the possible signaling mechanisms for the activation of NF-κB? How is the activation of NF-κB related to neuroinflammatory diseases like MS and meningitis?**
14. **What is a mitogen? How do the mitogens generally signal for the abnormal cell growth? How can you possibly inhibit mitogen-induced abnormal cell growth?**
15. **Is mitogen-induced signaling involved in the pathogenesis of brain cancer? If yes, then explain with possible reasons and therapeutic targets.**

Acknowledgements. This study was supported in part by grants from National Institutes of Health (NS39940 and NS48923), National Multiple Sclerosis Society (RG3422A1/1) and Michael J. Fox Foundation to KP and National Institutes of Health (HL85061) and American Diabetes Association (1-06-RA-11) to KRB.

References

Bailey CH, Kandel ER, Si K (2004) The persistence of long-term memory: A molecular approach to self-sustaining changes in learning-induced synaptic growth. Neuron 44:49–57.

Barker PA (2004) p75 is positively promiscuous. Novel partners and new insights. Neuron 42:529–533.

Barkett M, Gilmore TD (1999) Control of apoptosis by Rel/NF-kappaB transcription factors. Oncogene 18:6910–6924.

Barrett GL, Bartlett PF (1994) The p75 nerve growth factor receptor mediates survival or death depending on the stage of sensory neuron development. Proc Natl Acad Sci USA 91: 6501–6505.

Barton GM, Medzhitov R (2004) Toll signaling: RIPping off the TNF pathway. Nat Immunol 5:472–474.

Bereiter-Hahn J, Voth M (1994) Dynamics of mitochondria in living cells: Shape changes, dislocations, fusion, and fission of mitochondria. Microsc Res Tech 27:198–219.

Berridge MJ (2005) Unlocking the secrets of cell signaling. Annu Rev Physiol 67:1–27.

Blitzer RD (2005) Long-term potentiation: Mechanisms of induction and maintenance. Sci STKE (309):tr26.

Bsibsi M, Ravid R, Gveric D, van Noort JM (2002) Broad expression of Toll-like receptors in the human central nervous system. J Neuropathol Exp Neurol 61:1013–1021.

Burger D, Dayer JM (1995) Inhibitory cytokines and cytokine inhibitors. Neurology 45:S39-S43.

Chan JR, Watkins TA, Cosgaya JM, Zhang C, Chen L, Reichardt LF, Shooter EM, Barres BA (2004) NGF controls axonal receptivity to myelination by Schwann cells or oligodendrocytes. Neuron 43:183–191.

Cho S, Kim Y, Cruz MO, Park EM, Chu CK, Song GY, Joh TH (2001) Repression of proinflammatory cytokine and inducible nitric oxide synthase (NOS2) gene expression in activated microglia by N-acetyl-O-methyldopamine: Protein kinase A-dependent mechanism. Glia 33:324–333.

Coman I, Barbin G, Charles P, Zalc B, Lubetzki C (2005) Axonal signals in central nervous system myelination, demyelination and remyelination. J Neurol Sci 233:67–71.

Culmsee C, Mattson MP (2005) p53 in neuronal apoptosis. Biochem Biophys Res Commun 331:761–777.

Dasgupta S, Jana M, Liu X, Pahan K (2003) Role of very-late antigen-4 (VLA-4) in myelin basic protein-primed T cell contact-induced expression of proinflammatory cytokines in microglial cells. J Biol Chem 278:22424–22431.

De Bellard ME, Filbin MT (1999) Myelin-associated glycoprotein, MAG, selectively binds several neuronal proteins. J Neurosci Res 56:213–218.

DeCoster MA (1995) Calcium dynamics in the central nervous system. Adv Neuroimmunol 5(3):233–239.

Dickson BJ (2001) Rho GTPases in growth cone guidance. Curr Opin Neurobiol 11:103–110.

Filbin MT (2003) Myelin-associated inhibitors of axonal regeneration in the adult mammalian CNS. Nat Rev Neurosci 4:703–713.

Fournier AE, GrandPre T, Gould G, Wang X, Strittmatter SM (2002) Nogo and the Nogo-66 receptor. Prog Brain Res 137:361–369.

Fukazawa Y, Saitoh Y, Ozawa F, Ohta Y, Mizuno K, Inokuchi K (2003) Hippocampal LTP is accompanied by enhanced F-actin content within the dendritic spine that is essential for late LTP maintenance in vivo. Neuron 38:447–460.

Giri S, Rattan R, Singh AK, Singh I (2004) The 15-deoxy-delta12,14-prostaglandin J2 inhibits the inflammatory response in primary rat astrocytes via down-regulating multiple steps in phosphatidylinositol 3-kinase-Akt-NF-kappaB-p300 pathway independent of peroxisome proliferator-activated receptor gamma. J Immunol 173:5196–5208.

Goswami R, Dawson G (2000) Does ceramide play a role in neural cell apoptosis? J Neurosci Res 60:141–149.

Greene LA, Biswas SC, Liu DX (2004) Cell cycle molecules and vertebrate neuron death: E2F at the hub. Cell Death Differ 11:49–60.

Guix FX, Uribesalgo I, Coma M, Munoz FJ (2005) The physiology and pathophysiology of nitric oxide in the brain. Prog Neurobiol 76:126–152.

Hershey GK (2003) IL-13 receptors and signaling pathways: An evolving web. J Allergy Clin Immunol 111:677–690.

Hess J, Angel P, Schorpp-Kistner M (2004) AP-1 subunits: Quarrel and harmony among siblings. J Cell Sci 117:5965–5973.

Huang EJ, Reichardt LF (2003) Trk receptors: Role in neuronal signal transduction. Ann Rev Biochem 72:609–642.

Iwase T, Jung CG, Bae H, Zhang M, Soliven B (2005) Glial cell line-derived neurotrophic factor-induced signaling in Schwann cells. J Neurochem 94:1488–1499.

Jana A, Pahan K (2004a) Fibrillar amyloid-beta peptides kill human primary neurons via NADPH oxidase-mediated activation of neutral sphingomyelinase. Implications for Alzheimer's disease. J Biol Chem 279:51451–51459.

Jana A, Pahan K (2004b) Human immunodeficiency virus type 1 gp120 induces apoptosis in human primary neurons through redox-regulated activation of neutral sphingomyelinase. J Neurosci 24:9531–9540.

Johansson S, Bohman S, Radesater AC, Oberg C, Luthman J (2005) Salmonella lipopolysaccharide (LPS) mediated neurodegeneration in hippocampal slice cultures. Neurotox Res 8:207–220.

Kandel ER (2001) The molecular biology of memory storage: A dialogue between genes and synapses. Science 294:1030–1038.

Kim WK, Jang PG, Woo MS, Han IO, Piao HZ, Lee K, Lee H, Joh TH, Kim HS (2004) A new anti-inflammatory agent KL-1037 represses proinflammatory cytokine and inducible nitric oxide synthase (iNOS) gene expression in activated microglia. Neuropharmacology 47:243–252.

Konat GW, Kielian T, Marriott I (2006) The role of Toll-like receptors in CNS response to microbial challenge. J Neurochem 99:1–12.

Lee R, Kermani P, Tang KK, Hempstead BL (2001) Regulation of cell survival by secreted proneurotrophins. Science 294:1945–1948.

Leroith D, Nissley P (2005) Knock your SOCS off! J Clin Invest 115:233–236.

Li Q, Verma IM (2002) NF-kappaB regulation in the immune system. Nat Rev Immunol 2:725–734.

Lisman J, Schulman H, Cline H (2002) The molecular basis of CaMKII function in synaptic and behavioural memory. Nat Rev Neurosci 3:175–190.

Lossi L, Merighi A (2003) In vivo cellular and molecular mechanisms of neuronal apoptosis in the mammalian CNS. Prog Neurobiol 69:287–312.

Lu B, Pang PT, Woo NH (2005) The Ying and Yang of neurotrophin action. Nat Rev 6:603–614.

Maltese WA (1990) Posttranslational modification of proteins by isoprenoids in mammalian cells. FASEB J. 4:3319–3328.

Mattson MP, Camandola S (2001) NF-kappaB in neuronal plasticity and neurodegenerative disorders. J Clin Invest. 107:247–254.

Matynia A, Anagnostaras SG, Silva AJ (2001) Weaving the molecular and cognitive strands of memory. Neuron 32:557–559.

McGee AW, Strittmatter SM (2003) The Nogo-66 receptor: Focusing myelin inhibition of axon regeneration. Trends Neurosci 26:193–198.

Meffert MK, Baltimore D (2005) Physiological functions for brain NF-kappaB. Trends Neurosci 28:37–43.

Mehlen P, Bredesen DE (2004) The dependence receptor hypothesis. Apoptosis 9:37–49.

Meldolessi J, Sciorati C, Clementi E (2000) The p75 receptor: First insights into the trasnduction mechanisms leading to either death or survival. TiPS 21:242–243.

Michailov GV, Sereda MW, Brinkmann BG, Fischer TM, Haug B, Birchmeier C, Role L, Lai C, Schwab MH, Nave KA (2004) Axonal neuregulin-1 regulates myelin sheath thickness. Science 304:700–703.

Nykjaer A, Willnow TE, Petersen CM (2005) p75NTR-live or let die. Curr Opin Neurobiol 15:49–57.

Pahan K, Namboodiri AM, Sheikh FG, Smith BT, Singh I (1997a) Increasing cAMP attenuates induction of inducible nitric-oxide synthase in rat primary astrocytes. J Biol Chem 272: 7786–7791.

Pahan K, Sheikh FG, Namboodiri AM, Singh I (1997b) Lovastatin and phenylacetate inhibit the induction of nitric oxide synthase and cytokines in rat primary astrocytes, microglia, and macrophages. J Clin Invest 100:2671–2679.

Pahan K, Khan M, Singh I (2000) Interleukin-10 and interleukin-13 inhibit proinflammatory cytokine-induced ceramide production through the activation of phosphatidylinositol 3-kinase. J Neurochem 75:576–582.

Pahan K, Jana M, Liu X, Taylor BS, Wood C, Fischer SM (2002) Gemfibrozil, a lipid-lowering drug, inhibits the induction of nitric-oxide synthase in human astrocytes. J Biol Chem 277:45984–45991.

Park EJ, Park SY, Joe EH, Jou I (2003) 15d-PGJ2 and rosiglitazone suppress Janus kinase-STAT inflammatory signaling through induction of suppressor of cytokine signaling 1 (SOCS1) and SOCS3 in glia. J Biol Chem 278:14747–14752.

Pawate S, Shen Q, Fan F, Bhat NR (2004) Redox regulation of glial inflammatory response to lipopolysaccharide and interferon-gamma. J Neurosci Res 77:540–551.

Pettmann B, Henderson CE (1998) Neuronal cell death. Neuron 20:633–647.

Ramer MS, Priestley JV, McMahon SB (2000) Functional regeneration of sensory axons into the adult spinal cord. Nature 403:312–316.

Ramji DP, Foka P (2002) CCAAT/enhancer-binding proteins: Structure, function and regulation. Biochem J 365:561–575.

Rawlings JS, Rosler KM, Harrison DA (2004) The JAK/STAT signaling pathway. J Cell Sci 117:1281–1283.

Ross WN, Nakamura T, Watanabe S, Larkum M, Lasser-Ross N (2005) Synaptically activated Ca2+ release from internal stores in CNS neurons. Cell Mol Neurobiol 25:283–295.

Royal W 3rd, Leander M, Chen YE, Major EO, Bissonnette RP (2004) Nuclear receptor activation and interaction with morphine. J Neuroimmunol 157:61–65.

Saccani S, Pantano S, Natoli G (2002) p38-Dependent marking of inflammatory genes for increased NF-kappa B recruitment. Nat Immunol 3:69–75.

Saha RN, Pahan K (2003) Tumor necrosis factor-alpha at the crossroads of neuronal life and death during HIV-associated dementia. J Neurochem 86:1057–1071.

Saha RN, Pahan K (2005) HATs and HDACs in neurodegeneration: A tale of disconcerted acetylation homeostasis. Cell Death Differ 13(4):539–550, doi: 10.1038/sj.cdd.4401769.

Schwab ME (2004) Nogo and axon regeneration. Curr Opin Neurobiol 14:118–124.

Shao Z, Browning JL, Lee X, Scott ML, Shulga-Morskaya S, Allaire N, Thill G, Levesque M, Sah D, McCoy JM, Murray B, Jung V, Pepinsky RB, Mi S (2005) TAJ/TROY, an orphan TNF receptor family member, binds Nogo-66 receptor 1 and regulates axonal regeneration. Neuron 45:353–359.

Sharma SK, Carew TJ (2004) The roles of MAPK cascades in synaptic plasticity and memory in Aplysia: Facilitatory effects and inhibitory constraints. Learn Mem 11:373–378.

Sherman DL, Brophy PJ (2005) Mechanisms of axon ensheathment and myelin growth. Nat Rev Neurosci 6:683–690.

Silva AJ, Josselyn SA (2002) The molecules of forgetfulness. Nature 418:929–930.

Silver J, Miller JH (2004) Regeneration beyond the glial scar. Nat Rev Neurosci 5:146–156.

Stevaux O, Dyson NJ (2002) A revised picture of the E2F transcriptional network and RB function. Curr Opin Cell Biol 14:684–691.

Stevens B, Porta S, Haak LL, Gallo V, Fields RD (2002) Adenosine: A neuron-glial transmitter promoting myelination in the CNS in response to action potentials. Neuron 36:855–868.

Syntichaki P, Tavernarakis N (2003) The biochemistry of neuronal necrosis: Rogue biology? Nat Rev Neurosci 4:672–684.

Takai Y, Sasaki T, Matozaki T (2001) Small GTP-binding proteins. Physiol Rev 81:153–208.

Taveggia C, Zanazzi G, Petrylak A, Yano H, Rosenbluth J, Einheber S, Xu X, Esper RM, Loeb JA, Shrager P, Chao MV, Falls DL, Role L, Salzer JL (2005) Neuregulin-1 type III determines the ensheathment fate of axons. Neuron 47:681–694.

Tolwani RJ, Cosgaya JM, Varma S, Jacob R, Kuo LE, Shooter EM (2004) BDNF overexpression produces a long-term increase in myelin formation in the peripheral nervous system. J Neurosci Res 77:662–669.

Udo H, Jin I, Kim JH, Li HL, Youn T, Hawkins RD, Kandel ER, Bailey CH (2005) Serotonin-induced regulation of the actin network for learning-related synaptic growth requires Cdc42, N-WASP, and PAK in Aplysia sensory neurons. Neuron 45 :887–901.

Vourc'h P, Andres C (2004) Oligodendrocyte myelin glycoprotein (OMgp): Evolution, structure and function. Brain Res Brain Res Rev 45:115–124.

Won JS, Im YB, Singh AK, Singh I (2004) Dual role of cAMP in iNOS expression in glial cells and macrophages is mediated by differential regulation of p38-MAPK/ATF-2 activation and iNOS stability. Free Radic Biol Med 37:1834–1844.

Woo MS, Jung SH, Hyun JW, Kim HS (2004) Differential regulation of inducible nitric oxide synthase and cytokine gene

expression by forskolin and dibutyryl-cAMP in lipopolysaccharide-stimulated murine BV2 microglial cells. Neurosci Lett 356:187–190.
Xu J, Storer PD, Chavis JA, Racke MK, Drew PD (2005) Agonists for the peroxisome proliferator-activated receptor-alpha and the retinoid X receptor inhibit inflammatory responses of microglia. J Neurosci Res 81:403–411.
Yaffe MP (1999) The machinery of mitochondrial inheritance and behavior. Science 283:1493–1497.
Yang MS, Park EJ, Sohn S, Kwon HJ, Shin WH, Pyo HK, Jin B, Choi KS, Jou I, Joe EH (2002) Interleukin-13 and -4 induce death of activated microglia. Glia 38:273–280.
Zweifel LS, Kuruvilla R, Ginty DD (2005) Functions and mechanisms of retrograde neurotrophin signaling. Nat Rev 6:615–625.

Part 2
Nervous System Diseases and Immunity: Clinical Descriptions

17
Neurodegeneration

Serge Przedborski

Keywords Alzheimer's disease; Amyotrophic lateral sclerosis; Creutzfeldt-Jakob disease; Huntington's disease; Multiple sclerosis; Parkinson's disease; Prion diseases; Striatonigral degeneration; Synucleinopathies; Tauopathies

17.1. Introduction

An examination of the literature indicates that since 1982 the term *neurodegeneration* appears in the title of over 1280 indexed publications. Neurodegeneration is a mentioned topic of virtually all neurological science textbooks. Thus, it is safe to assume that the meaning of the word *neurodegeneration* is universally understood. Most relevant textbooks however will not actually define neurodegeneration but will discuss the issue in bits and pieces as part of the discussion of various diseases of the nervous system rather than as a single chapter. As previously discussed (Przedborski et al., 2003), neurodegeneration is composed of the prefix "neuro-," which denotes relationship to a nerve or the nervous system (http://tropmed.org/dictionary/coverpage9.htm) and of "-degeneration," which here is synonymous of devolution, meaning a process of declining from a higher to a lower level of effective power, vitality or essential quality (http://www.wordreference.com). Thus, neurodegeneration is any pathological condition in which the nervous system or nerve cell (i.e. neuron) loses its function, structure, or both. On a medical point of view however, the term neurodegeneration is used in a more restricted sense. Typically, it represents a large group of heterogeneous disorders in which affected neurons belong to specific subtypes, within specific anatomofunctional territories of the nervous system. Often, but not always, neurodegenerative diseases arise for unknown reasons and progress in a relentless manner. Within the context of this definition diseases of the nervous system can be catalogued into three broad categories: (i) pathologies which are restricted to the nervous system and which are *primary neuronal* diseases (i.e. neurodegenerative diseases *per se*); (ii) pathologies which are restricted to the nervous system but are not *primary neuronal* diseases, such as brain neoplasm or cerebral edema and hemorrhage; and (iii) pathologies provoked by systemic noxious factors which damage the nervous system, such cardiovascular arrest, poison, or infections. Based on this simple categorization, hundreds of disorders of the nervous system including Alzheimer's disease (AD), Parkinson's disease (PD), Huntington's disease (HD), and amyotrophic lateral sclerosis (ALS) clearly fulfill the criteria of neurodegenerative disorder and are unanimously regarded as such. Aside from these unambiguous neurodegenerative diseases, others such as essential tremor, torsion dystonia, Tourette's syndrome, or schizophrenia represent an interesting nosological challenge, as they do not show any distinct neuronal loss. Perhaps they have been traditionally included in this category because they are chronic diseases of the nervous system with an unknown cause. Finally, diseases such as multiple sclerosis (MS) also represent a perplexing situation. Conventionally, MS has been linked to pathology of the myelin that ensheath neuronal axons and not of the neuron *per se*. Several experts, however, have argued that MS is not only a demyelinating disease but also a disease where the neurons die due to a destruction of their axons. Worth noting, long-term disability in MS has a higher correlation with axonal damage than with the degree of demyelination (Bjartmar et al., 2003). For these reasons, several scientific authorities now regard MS as a neurodegenerative disease.

17.2. Frequency, Lifespan, and Co-Morbidity

The two most prevalent neurodegenerative diseases are AD and PD. Epidemiologists, estimate that up to 4.5 million Americans suffer from AD, and 1.2 million from PD. This represents roughly 2.5% of the entire population in the United States. This estimate is likely valid for most, if not all, countries of the world as epidemiological studies have found roughly comparable incidence and prevalence rates of AD and PD around the globe. Nevertheless, some geographic and temporal clusters of neurodegenerative conditions have been described. Two such examples include the marked increase in cases of parkinsonism in connection to the influenza pandemic of 1918 (Casals et al., 1998), and the high incidence of PD-ALS-dementia complex confined to the Chamorros Indians who live on the Western Pacific Island of Guam (Chen and Chase, 1986). These clusters, while quite enlightening, must be regarded as the exceptions rather than the rules. The most consistent risk

T. Ikezu and H.E. Gendelman (eds.), *Neuroimmune Pharmacology*.
© Springer 2008

factor for developing a neurodegenerative disorder, especially in regards to AD and PD, is increasing age (Tanner, 1992) and this fact may have far-reaching implications for the generations to come. As previously noted (Przedborski et al., 2003), over the past century, the growth rate of the population ages 65 and above in industrialized countries has far exceeded that of the population as a whole. Therefore, it can be anticipated that, over the next generations, the proportion of elderly citizens will double. Consequently, unless effective preventive strategies are soon found, the number of persons suffering from a neurodegenerative disorder will rise dramatically and the epidemiological numbers provided above may have to be amended to higher percentages.

It is also important to emphasize the fact that nearly all neurodegenerative disorders shorten the life expectancy of affected patients even if medications which alleviate the symptoms are available. This is well illustrated in the case of PD in which symptoms worsen over time. As discussed elsewhere (Dauer and Przedborski, 2003), before the introduction of the potent symptomatic treatment levodopa, the mortality rate among PD patients was three times that of the normal age-matched subjects. Unexpectedly, population-based surveys suggest that PD patients continue to display a decreased longevity compared to the general population despite the fact that motor problems can be controlled by levodopa (Levy et al., 2002; Morgante et al., 2000; Hely et al., 1989). Yet, only a few neurodegenerative disorders are fatal *per se*. Indeed, only those diseases which affect neurological structures that are implicated in controlling or driving vital physiological functions, such as respiration or circulation, are truly lethal. Amongst these is found ALS, where the losses of lower motor neurons innervating the intercostal and diaphragm muscles lead to fatal respiratory paralysis. In Friedreich ataxia, the association of neurodegeneration with heart disease (Harding, 1981) can also be a cause of death, although, in this case, death is not due to any neurodegenerative event, but instead to cardiodegenerative problems such as congestive heart failure. In most other neurodegenerative disorders, death is attributed neither to the damage of the nervous system nor to any associated extra-nervous system degeneration, but rather to medical problems including fatal falling, aspiration pneumonia, pressure skin ulcers, malnutrition, and dehydration, whose occurrence is favored by immobility, impaired balance, and cognitive decline. The leading causes of death in our industrial societies include cancers and cardiovascular problems. As a result it is worth mentioning that medical co-morbidity such as atrial fibrillation and cancer double the chances of death in nursing homes of demented patients (Van Dijk et al., 1996). Cardiovascular problems, such as high blood pressure, have also been suggested to stimulate the dementing process (Doraiswamy et al., 2002). Thus, medical co-morbidities appear to aggravate the long-term prognosis of patients inflicted with a neurodegenerative disease. However, whether or not patients suffering from a neurodegenerative disease are at greater risk for cancer, stroke, or heart attack, remains to be demonstrated.

17.3. Classification

The estimated number of different neurodegenerative diseases is a few hundred. Among these, many appear to overlap with one another clinically and pathologically. Classification of neurodegenerative diseases, while useful, is quite a complicated task. In neurodegenerative diseases, it is typical that several areas of the brain are affected. Yet, the degrees in which these different brain areas are damaged often vary from one case to another, thus giving rise to different phenotypes. For example, in a disease such as multisystem atrophy where typically parkinsonism is a prominent feature, it may be accompanied with either severe ataxia, autonomic failure, or both, depending on whether, in addition to the basal ganglia, the cerebellum and the intermediolateral column of the thoracic spinal cord also degenerate (Wenning et al., 2004). Even in the case of a defined genetic defect, a similar mutation may also produce varied phenotypes. For instance, in familial parkinsonism linked to mutations in the leucine-rich repeat kinase-2 (LRRK2) gene, a striking pleomorphic pathology has been reported with some members of the affected family, developing a motor neuron disease superimposed to dementia and/or parkinsonism (Zimprich et al., 2004). Despite these difficulties, it remains that the most popular categorization of neurodegenerative disorders is based on either the main clinical feature or the location of the predominant lesion, or often on a combination of the two. This proposed classification has been discussed previously (Przedborski et al., 2003) and will be repeated here for the sake of completeness of this chapter. We have proposed that neurodegenerative disorders may be grouped into diseases of the main anatomical division of the central nervous system: cerebral cortex, basal ganglia, brainstem/cerebellum, and spinal cord. Within each anatomical group, diseases may be sub-grouped based on their main clinical features.

Based on this clinico-anatomical classification, diseases of the cerebral cortex may be divided into cortical diseases associated with dementia (e.g., AD) and without dementia. In theory, circumscribed degeneration of the cerebral cortex can occur in absence of dementia as reported in some cases of progressive primary aphasia (Kirshner et al., 1987). However, because the cerebral cortex is so heavily involved in cognition, neurodegeneration, over time, often spreads beyond the initial locus of pathology, and the majority of these patients end-up with dementia (Le Rhun et al., 2005). It is thus not surprising to find that neurodegenerative diseases of the cerebral cortex are usually equated to dementia. While AD in this group of diseases is by far the most frequent (Sulkava et al., 1983), about 50 other and less prominent dementing cortical diseases can be found (Tomlinson, 1977).

Diseases of the basal ganglia are essentially characterized by abnormal motor activity. Based on the type of motor problem, diseases of the basal ganglia can be classified into hypokinetic or hyperkinetic groups. Hypokinetic basal ganglia disorders include PD, in which the amplitude and velocity of voluntary movements are diminished or, in extreme cases, non-existent.

Aside from PD, parkinsonism—which refers to an association of at least two of the following clinical signs: resting tremor, slowness of movements, stiffness, and postural instability—is also found in a variety of other diseases of the basal ganglia (Dauer and Przedborski, 2003). In some, there is only parkinsonism (e.g., striatonigral degeneration) but in others, often called parkinson-plus syndromes, there is parkinsonism plus signs of cerebellar ataxia (e.g., olivopontocerebellar atrophy), orthostatic hypotension (e.g., Shy-Drager syndrome) or paralysis of vertical eye movements (e.g., progressive supranuclear palsy). Because early on, Parkinsonism may be the only clinical expression of parkinson-plus syndromes, it is difficult to diagnose accurately before the patient reaches a more advanced stage of the disease. This problem is particularly well depicted by the fact that more than 77% of patients with parkinsonism are diagnosed in life as having PD (Stacy and Jankovic, 1992), but as much as a quarter of the diagnosed patients are found at autopsy to have lesions incompatible with PD (Hughes et al., 1992). Of note, it is customary to classify the stiff-person syndrome among the hypokinetic diseases. However, available evidence would argue that the stiff-person syndrome while being a hypokinetic condition is an autoimmune disease involving the nervous system and not a neurodegenerative disease.

Found at the other end of the spectrum are the hyperkinetic basal ganglia disorders, which are represented by HD and essential tremor. In these two conditions, excessive abnormal movements such as chorea or tremor are superimposed onto and interfere with normal voluntary movements. Although hyperkinetic basal ganglia disorders are probably as diverse as hypokinetic basal ganglia disorders, their specific disease markers such as gene mutations, which exist for several of the hyperkinetic syndromes create more accurate, less problematic, classifications.

All neurodegenerative diseases of the cerebellum and its connections are clinically associated with ataxia, meaning that the main criteria of classification will rely less on the clinical presentation than on the loci of pathology. Some diseases of the cerebellum and connections can easily be grouped into three main neuropathological types: cerebellar cortical atrophy (lesion confined to the Purkinje cells and the inferior olives); pontocerebellar atrophy (lesion affecting several cerebellar and brain structures); and Friedreich ataxia (lesion affecting the posterior column of the spinal cord, peripheral nerves, and the heart). However, several other diseases of the cerebellum and its connections fall somewhere in between these three well-delineated categories: dentorubral degeneration, in which the most conspicuous lesions are in the dentate and red nuclei, and Machado-Joseph disease, in which degeneration involves the lower and upper motor neurons, substantia nigra, and the dentate system.

Among the neurodegenerative diseases predominantly affecting the spinal cord are ALS and spinal muscular atrophy, in which the most severe lesions are found in the anterior part of the spinal cord, and the already cited Friedreich ataxia, in which the most severe lesions are found in the posterior part of the spinal cord. Finally, there is one group of neurological diseases which are often, but not always, considered neurodegenerative because of their chronic course and unknown etiopathogenesis, unlike those described above, these show no apparent structural abnormalities. These include torsion dystonia, Tourette syndrome, essential tremor, and schizophrenia. Structural abnormalities are referred to here as a loss of neuronal cell bodies, but it is unknown whether or not structural damage restricted to synaptic connections occurs in torsion dystonia or schizophrenia, which could account for the disease phenotype. Moreover, in all of these singular neurodegenerative disorders, various brain imaging studies and electrophysiological investigations have revealed significant functional abnormalities.

The above discussion was aimed at stressing some of the obtrusive shortcomings of the current classification of neurodegenerative diseases. To circumvent these problems researchers have tried over the recent years to move away from the conventional clinical/neuropathological system of classification and instead invest in a molecular-based classification system. Based on this novel approach, some neurodegenerative diseases, which used to belong to very distinct categories, are now brought together because of a common molecular alteration. For example, HD, spinal-cerebellar atrophy, and myotonic dystrophy now fall under the category of the trinucleotide repeat disorders (Cummings and Zoghbi, 2000); Creutzfeldt-Jakob disease, Gerstmann-Straussler-Scheinker syndrome, and fetal familial insomnia fall under the category of the prion diseases (Prusiner, 1998); PD, progressive supranuclear palsy, and diffuse Lewy body dementia fall under the category of the synucleinopathies (Galvin et al., 2001); and AD, corticobasal degeneration, fronto-temporal dementia with parkinsonism linked to chromosome-17, and Pick disease fall under the category of tauopathies (Goedert and Spillantini, 2001). Although the jury is still out on whether or not this new type of classification will alleviate the problems previously encountered, we strongly believe that a recasting of the conventional neuropathological-based classification to a molecular-based classification holds the promise of a more practical and less ambiguous standing for neurodegeneration from a clinical and therapeutic point-of-view.

17.4. Etiology of Neurodegenerative Diseases

The word *etiology* refers to the initiating factor of a disease. In some cases a given clinical entity is linked to a single etiological factor such as in HD. Interestingly, however, several others can be caused by various and quite distinct factors. PD for example can be caused by overexpression of the small synaptic protein alpha-synuclein (Singleton et al., 2003), loss of parkin E3-ubiquitine ligase activity (Shimura et al., 2000), or by an increase in kinase activity of LRRK2 (West et al., 2005). These observations raise an important issue that plagues the field of neurodegeneration which is whether prominent disorders such as PD or AD are indeed diseases or rather syndromes, i.e. very different diseases are lumped together under a same name because they merely share the same clinical expression.

Regardless of this problem, it should be known that the causes of neurodegenerative diseases or syndromes are in most instances unknown. Even in some rare cases when the etiology has been identified, the mechanism by which the etiological factor leads to neuronal death remains, at best, speculative. For example, while the etiology of HD was identified more than 20 years ago (The Huntington's Disease Collaborative Research Group, 1993), we still do not know with certainty how mutant huntingtin kills striatal neurons. The same circumstances are also true for superoxide dismutase-1 whose mutations are linked to a familial form of ALS (Rosen et al., 1993).

Among the neurodegenerative diseases, only a few arise as a familial condition, supporting a genetic basis. Within the affected members of a family, these genetic diseases can run as an autosomal dominant condition, which is the case for HD and dentatorubral-pallidoluysian atrophy. Less frequently, the disease can run as an autosomal recessive (e.g., familial spastic paraparesis), an X-linked (e.g., spinal and bulbar muscular atrophy), or even a maternally inherited trait (e.g., mitochondrial Leber optic neuropathy). In addition to these genetic-neurodegenerative diseases, others, while primarily sporadic, can also show a small contingent of patients in whom the illness is inherited. This is the case for PD, AD, and ALS, where less than 10% of all cases are generally familial.

For those in whom the disease is truly sporadic, which is the vast majority of patients, it appears that any genetic contribution to the neurodegenerative process would be minimal (Tanner et al., 1999). In these cases, toxic environmental factors may be the prime suspects in initiating neurodegenerative processes. Relevant to this idea is the observation that some neurodegenerative conditions, as mentioned above, might arise in geographic or temporal clusters. For instance, in the case of the PD-ALS-dementia complex of Guam which was introduced previously, it is believed that the pathology is caused by a toxic compound contained in the *Cycas circinalis*, an indigenous plant commonly ingested as food or medicine by the Chamorros Indians (Kurtland, 1988). Intoxication with 1-methyl-4-phenyl-1,2,3,6-tetrahydropyridine, a by-product of the synthesis of a meperidine compound, is also known to produce a severe and irreversible parkinsonian syndrome, which is almost identical to PD (Przedborski and Vila, 2001). Several large-scale epidemiological studies have failed, however, to link in a definitive manner the environmental factors to diseases such as PD (Tanner, 1989). Furthermore, most of the toxic exposures known to cause neurological problems occur within a specific geographic, social, or professional context and in absence of any significant association with an increased incidence of PD, AD, or other prominent neurodegenerative diseases. Collectively, the aforementioned findings argue that sporadic cases are neither clearly genetic nor clearly environmental. This does not exclude the possibility that neurodegenerative disorders can result from a combination of both. Along this line, the demonstration of a non-syndromic familial deafness linked to a mitochondrial point mutation (Prezant et al., 1993) provides compelling support for this view. In this study, family members who harbor the mitochondrial mutation develop a hearing impairment only upon exposure to the antibiotic aminoglycoside, thus illustrating the significant pathogenic interactions between genetics and environment.

17.5. Pathogenesis of Neurodegenerative Diseases

Compared to *etiology*, *pathogenesis* does not refer to the initiation of the disease process, but rather to the cascade of cellular and molecular mechanisms set into motion by the etiological factor(s) which ultimately leads to the demise of the affected neurons. As mentioned above, only very small group of so-called neurodegenerative conditions comprise a set of diseases with no apparent neuropathological changes. In all others, overt neuropathology, mainly in the form of a focal loss of neurons with gliosis, is discernable. Residual neurons may exhibit varying morphologies ranging from an almost normal appearance to a severe distortion with a combination of abnormal features such as process attrition, shape and size alterations of the cell body and nucleus, organelle fragmentation, dispersion of Nissl bodies, cytoplasmic vacuolization, and chromatin condensation. In several neurodegenerative disorders, spared neurons can also present with various types of intracellular proteinacious inclusions, which, in the absence of any definite known pathogenic role, are quite useful in differentiating neurodegenerative disorders. As for the glial response, it is comprised of innate resident immune cells including astrocytes and microglia. To a lesser extent, adaptive immune cells such as T-lymphocytes can also be found in the diseased areas of the brain.

All neurodegenerative disorders associated with cell death have in common the fact that specific subpopulations of neurons degenerate at the level of specific structures of the nervous system. This means that autopsy areas of the brain that are damaged can be surrounded by healthy tissues and that within the affected regions, degenerating and healthy neurons are intermingled. In some neurodegenerative diseases, such as olivopontocerebellar atrophy, multiple brain structures within the nervous system are affected. In these so-called *system neurodegenerative diseases*, the spatial pattern of the lesions often becomes better defined as the disease progresses (Przedborski et al., 2003). In regards to the spatial pattern of the lesions it is essentially unrelated to the distribution of blood vessels. Conversely, as emphasized by Oppenheimer and Esiri (1997) the different lesions appear, at least in some cases, functionally and anatomically interconnected. Such a linked degeneration is observed in ALS, where both the corticospinal track and spinal cord lower motor neurons are affected, or in progressive supranuclear palsy, where both the globus pallidus and the subthalamic nucleus are lesioned. Although such trans-neuronal degeneration is a well-recognized phenomenon (Saper et al., 1987), very little is known about its molecular basis except for the fact that this trans-synaptic demise occurs by programmed cell death (DeGiorgio et al., 1998; Ginsberg and Martin, 2002). Trans-neuronal degeneration is unlikely to account for all of the combinations of lesions that are found in system neurodegenerative diseases. In Friedreich ataxia, there is degeneration of the spinocerebellar tracts and the dentate nuclei, but not of the Purkinje cells, which would have been the link between these two lesions. It is thus more probable that etiological factors are ubiquitous, and that the pathological threshold is attained by specific structures of the nervous system at different times. Perhaps, in some neurodegenerative diseases, some areas of the nervous system may never reach this threshold and will thus remain unaffected throughout the span of the disease. Also, not in all neurodegenerative disorders are large numbers of areas of the nervous system at risk. In some specific neurodegenerative diseases, the lesions appear restricted to one or only a few brain regions. This is particularly well illustrated in spinal muscular atrophy, in which the degenerative process is limited to a loss of lower motor neurons, or in ALS, which is characterized by damage to the upper and lower motor neurons which may represent the sole neuropathological change. Nevertheless, almost all neurodegenerative disorders, which initially appeared *monosystemic,* will eventually become *multisystemic* over the progression of the degenerative process. In ALS for example it is typical to observe overt neuropathological changes mainly at the level of the spinal cord and motor cortex in patients who died after the short course of the disease, for unrelated causes (i.e. car accident or heart attack). However, in ALS patients with a more protracted course, degenerative changes can now also be detected in the substantia nigra (Sasaki et al., 1992) and the oculomotor and trochlear nuclei (Hayashi et al., 1991). Recognition of this temporal spreading of the neurodegenerative process has lead some neuropathologists (Braak et al., 2003) to propose that the neurodegenerative process starts on a specific circumscribed area of the nervous system where it begins a domino-effect: the initial affected area is responsible for the disease occurring in the next region, whereby propagating the degenerative process. Although quite appealing, this model is speculative and would only be true if the different regions affected in the neurodegenerative conditions were anatomic neighbors which is often not the case.

In most, if not all, neurodegenerative disorders the locations of the principal lesions have been well established. However, it often remains difficult to determine the extent of degeneration affecting more than one group of neurons and, consequently, to define the exact neuropathological topography of certain diseases. This problem stems from at least three issues. First, lesions are often missed through incomplete examination of the brain and spinal cord. Second, quantitative morphology in post-mortem samples seldom use the rigorous counting methods necessary to generate reliable neuron numbers (Saper, 1996). The third issue pertains to sick neurons, which will not necessarily die. These compromised neurons often lose their phenotypic markers which are used for identification purposes in order to count them (Clarke and Oppenheim, 1995). Accordingly, the use of phenotypic markers in pathological conditions may lead to erroneous conclusions about the status of a given group of neurons. For these reasons, reported estimations regarding the distribution and magnitude of neuronal loss in neurodegenerative disorders, perhaps with a few exceptions, may have to be taken with caution.

17.6. Onset and Progression of the Disease

As indicated above, prominent neurodegenerative disorders are sporadic and with the absence of pre-symptomatic markers for virtually all of them, patients are seeking medical attention only when the first symptoms of the diseases emerge. Because there is significant cellular redundancy in neuronal pathways, the onset of symptoms does not equate with the onset of the disease. Instead, the beginning of symptoms merely corresponds to a neurodegenerative stage at which the number of residual neurons in a given pathway drops below the number required to maintain its endowed function. It is thus clear that the onset of the disease occurs at some preceding time, which, depending on how fast the neurodegenerative process evolves, can range from a few months to several years. Our ability to determine the actual onset of the disease is at this point unfortunately undermined by the lack of pre-symptomatic markers and the little knowledge about the true kinetics of cell demise.

Occasionally, there is a sudden worsening of a patient's condition. Although a sudden acceleration of the neurodegenerative process cannot be excluded, especially under the effect of intercurrent deleterious factors such as infection, it is more probable that the rate of neuronal death remains about the same throughout the natural course of the disease. However, the relationship between the clinical expression of a disease and the number of residual neurons does not have to be either linear or even constant. For instance, a patient may clinically remain unchanged during a prolonged period of time despite a loss of many cells and then abruptly their condition may deteriorate as the number of neurons drop below a functional threshold. Another important aspect is that virtually all neurodegenerative disorders progress slowly over time, often taking several years to reach end-stage. It must be remembered that neuronal degeneration corresponds to an asynchronous death (Pittman et al., 1999) in that cells within a population die at very different times. The correlate of this fact is that at any given time, only a small number of cells are actually dying, and among these, they are at various stages along the cell death pathway. However, standard clinical, radiological, and biochemical measurements, which are critical to assessing the disease, generate information about the entire population of cells. Therefore, the rate of change in any, of the usual clinical parameters reflects the decay of the entire population of affected cells and provides very little, if any, insight into the pace at which the death of an individual cell occurs. Still, if one looks at the large body of in vitro data, it appears that, once a cell gets sick, the entire process of death proceeds rapidly. Given these facts, it may be possible that the protracted clinical progression is the reflection of only a small number of neurons dying at any given time from a rapid demise.

17.7. Cell-Autonomous of the Degenerative Process

As indicated above, neurodegeneration refers to pathology of neurons. It is thus not surprising to find from an examination of the literature that all theories about neurodegeneration pathogenesis revolves around neurons. This rather *neuronocentrist* view is increasingly at odds, however, with the current concept of neurodegeneration pathogenesis. As stressed above, in most prominent familial neurodegenerative diseases, the mutant proteins are ubiquitously expressed. This is the case, for example, with mutant alpha-synuclein in familial PD and of mutant SOD1 in ALS. While the underlying gene mutations are unquestionably pathogenic, ALS-linked SOD1 mutations expressed selectively in astrocytes (Gong et al., 2000) or in neurons (Pramatarova et al., 2001; Lino et al., 2002) fail to provoke the death of motor neurons in transgenic mice. Conversely, the expression of ALS-linked SOD1 mutations in all cells do cause an adult-onset paralytic phenotype in both transgenic mice and rats (Gurney et al., 1994; Nagai et al., 2001). Further support to the idea that many different cell types collaborate to achieve the demise of neurons in neurodegenerative diseases comes from the work of Clement et al. (2003). In the study, the authors produced chimeric animals made of mixtures of normal and SOD1 mutant-expressing cells. They found that motor neurons that chronically expressed mutant SOD1 did not degenerate if they were surrounded by a sufficient number of normal non-neuronal cells. In addition, normal motor neurons surrounded by mutant-expressing non-neuronal cells were shown to acquire intraneuronal ubiquitinated deposits consistent with the concept that mutant-expressing cells could transfer the diseases phenotype to normal neighboring cells. These data provide a compelling argument in that death of neurons in neurodegenerative diseases such as ALS may not be as cell-autonomous as previously thought. Accordingly, the cellular environment surrounding neurons may play a critical role in determining the fate of specific neurons within these types of diseases of the nervous system. Among the prime candidates that mediate this non-cell autonomous mechanism of neurodegeneration are glial cells. Along this line, neuroinflammation has indeed emerged over the past decade as a key contributor to neurodegeneration in diseases such as AD, PD and ALS (Wyss-Coray and Mucke, 2002; Przedborski and Goldman, 2004; McGeer and McGeer, 2002). Based on both epidemiological and pre-clinical studies in animal models of human diseases (Chen et al., 2003; Wyss-Coray and Mucke, 2002; Przedborski and Goldman, 2004; McGeer and McGeer, 2002), it is now proposed that neuroinflammation could stimulate neurodegeneration.

Summary

Neurodegenerative diseases represent a heterogeneous group of disorders affecting the nervous system. In most instances, they affect adults, their causes are unknown, and progression is relentless. Some are genetic, but most are sporadic. They involve all parts of the nervous system, although the cerebral cortex and the basal ganglia are the most frequent loci of pathology. The historical classification of neurodegenerative diseases, based on clinical and pathological characteristics, is imperfect. New classifications are rather based on molecular determinants. Contrary to common belief, it is now recognized that neurodegenerative disorders are multisystemic, even if specific neuronal pathways are more affected than others. The death of specific types of neurons in neurodegenerative diseases is provoked, not by a single pathogenic factor, but rather by a cascade of multiple deleterious molecular and cellular events.

Acknowledgments. The author wishes to thank Ms. Julia Jeon for assistance in preparing this manuscript. This study is supported by Muscular Dystrophy Association/Wings-over-Wall Street, NIH/NINDS Grants NS38586, NS42269, NS38370, and NS11766–27A2, NIH/AG 021191, the US Department of Defense Grant (DAMD 17–99–1–9474 and DAMD 17–03–1), the Lo'wenstein Foundation, the Smart Foundation, the Parkinson's Disease Foundation.

Review Questions/Problems

1. **Define neurodegeneration.**
2. **What are the three broad categories of diseases of the nervous system?**
3. **Provide examples of diseases from the three categories.**
4. **What is expected to occur with the prevalence of neurodegenerative diseases in the forthcoming generations and why?**
5. **What is the effect of neurodegeneration on the life expectancy and for what reason(s) do patients with neurodegenerative disease typically die?**
6. **What is the common method used in classifying neurodegenerative diseases and what are the difficulties inherent with this type of classification?**

7. Describe the different forms of genetic contributions among the neurodegenerative diseases.

8. What does multisystemic neurodegeneration refer to?

9. Why are neurodegenerative diseases progressive and what does it mean at the level of the whole population of affected cells and a single affected cell?

10. Define the notion of non-cell autonomous neurodegeneration and provide an example.

11. What are the two most prevalent neurodegenerative diseases?

a. Alzheimer's disease and stroke
b. Parkinson's disease and brain tumor
c. Alzheimer's disease and Parkinson's disease
d. Huntington's disease and amyotrophic lateral sclerosis
e. Mitochondrial encephalopathy and multiple sclerosis

12. What are typical neurodegenerative diseases without overt neuropathology?

a. Creutzfeldt-Jakob disease
b. Schizophrenia
c. Essential tremor
d. Torsion dystonia
e. All but a

13. What is the correct statement about the neuropathology of neurodegenerative diseases?

a. Neuropathologic changes in neurodegenerative diseases are always detectable in at least one region of the nervous system.
b. Residual neurons look sick or normal.
c. In addition to the loss of neurons, there is always some gliosis and protein aggregates.
d. If different regions of the brain are affected, the type of neurons degenerating remains identical.
e. The multisystemic nature of a neurodegenerative process is explained by a trans-synaptic phenomenon.

14. What is the correct statement about the etiology of neurodegenerative diseases?

a. The etiology of most neurodegenerative diseases is unknown.
b. Most neurodegenerative diseases are genetic.
c. Neurodegenerative diseases are only linked to nuclear gene defects.
d. Environmental toxic factors have never been implicated in the occurrence of neurodegenerative diseases.
e. Genetic/environmental interactions are unknown causes of neurodegeneration.

15. Which of the following statements is true about the pathogenesis?

a. The death of neurons is caused by a unique pathogenic mechanism triggered by an etiologic factor.
b. Neurodegenerative disorders never become multisystemic upon progression of the disease.
c. It is established that the neurodegenerative process starts in one area of the nervous system and then spreads to others by a domino effect.
d. Before dying, compromised neurons never loose their functions and phenotypic markers.
e. The distribution and magnitude of neuronal loss in neurodegenerative disorders is often difficult to establish with certainty.

16. Which of the following statements are most correct?

a. Onset of symptoms reflects the onset of the disease.
b. The prominent symptoms often reveal the main site of neuropathology.
c. Intercurrent infection can permanently exacerbate the symptoms.
d. Neurodegenerative diseases differ from non-degenerative diseases because the latter never progress in a step-wise manner.
e. b and c

17. Which of the following statements are always correct?

a. Neurodegenerative diseases can involve either the central or the peripheral nervous systems, or both.
b. A familial occurrence of a neurodegenerative disease is the signature of its genetic origin.
c. Cardiovascular problems exacerbate the cognitive decline in patients with dementia and vise-versa.
d. Neurodegenerative diseases are associated with a shortened life-span because they eventually impair respiratory or cardiovascular functions.
e. When available, symptomatic agents normalize life-span in patients suffering from neurodegenerative diseases.

18. The transmissibility of a neurodegenerative phenotype can be?

a. Autosomal dominant
b. Autosomal recessive
c. X-linked
d. Mitochondrial
e. All of the above

19. Regarding the classification of neurodegenerative disorders which statement is not correct?

a. The clinico-anatomic classification is quite popular, but is fraught with many shortcomings.
b. The molecular classification groups together neurodegenerative diseases of very different clinical expression.
c. A parkinsonism can be observed in patients suffering from a wide variety of disorders including Parkinson's disease.
d. Members of a same family carrying an identical mutation always exhibit the same clinical and neuropathological phenotype.
e. A similar clinical phenotype can be caused by distinct etiologic factors.

20. The concept of non-cell autonomy in neurodegeneration implies.

a. A disease phenotype can be transmitted from a mutant cell to a wild-type cell.
b. The whole disease process does not rely solely on pathogenic events taking place in the degenerating cells.
c. The fate of neurons destined to die is at least in part, determined by other cells such as neighboring glia.
d. This concept applies to infectious diseases but not to neurodegenerative disorders.
e. All but d

References

Bjartmar C, Wujek JR, Trapp BD (2003) Axonal loss in the pathology of MS: Consequences for understanding the progressive phase of the disease. J Neurol Sci 206:165–171.

Braak H, Del Tredici K, Rub U, de Vos RA, Jansen Steur EN, Braak E (2003) Staging of brain pathology related to sporadic Parkinson's disease. Neurobiol Aging 24:197–211.

Casals J, Elizan TS, Yahr MD (1998) Postencephalitic parkinsonism—a review. J Neural Transm 105:645–676.

Chen K-M, Chase TN (1986) Parkinsonism-dementia. In: Handbook of Clinical Neurology. Extrapyramidal Disorders (Vinken PJ, Bruyn GW, Klawans HL, eds), pp 167–183. Amsterdam: Elsevier.

Chen H, Zhang SM, Hernan MA, Schwarzschild MA, Willett WC, Colditz GA, Speizer FE, Ascherio A (2003) Nonsteroidal anti-inflammatory drugs and the risk of Parkinson disease. Arch Neurol 60:1059–1064.

Clarke PGH, Oppenheim RW (1995) Neuron death in vertebrate development: In vitro methods. Meth Cell Biol 46:277–321.

Clement AM, Nguyen MD, Roberts EA, Garcia ML, Boillee S, Rule M, McMahon AP, Doucette W, Siwek D, Ferrante RJ, Brown RH, Jr., Julien JP, Goldstein LSB, Cleveland DW (2003) Wild-type non-neuronal cells extend survival of SOD1 mutant motor neurons in ALS Mice. Science 302:113–117.

Cummings CJ, Zoghbi HY (2000) Trinucleotide repeats: Mechanisms and pathophysiology. Annu Rev Genomics Hum Genet 1:281–328.

Dauer W, Przedborski S (2003) Parkinson's disease: Mechanisms and models. Neuron 39:889–909.

DeGiorgio LA, Dibinis C, Milner TA, Saji M, Volpe BT (1998) Histological and temporal characteristics of nigral transneuronal degeneration after striatal injury. Brain Res 795:1–9.

Doraiswamy PM, Leon J, Cummings JL, Marin D, Neumann PJ (2002) Prevalence and impact of medical comorbidity in Alzheimer's disease. J Gerontol A Biol Sci Med Sci 57:M173–M177.

Galvin JE, Lee VM, Trojanowski JQ (2001) Synucleinopathies: Clinical and pathological implications. Arch Neurol 58:186–190.

Ginsberg SD, Martin LJ (2002) Axonal transection in adult rat brain induces transsynaptic apoptosis and persistent atrophy of target neurons. J Neurotrauma 19:99–109.

Goedert M, Spillantini MG (2001) Tau gene mutations and neurodegeneration. Biochem Soc Symp 67:59–71.

Gong YH, Parsadanian AS, Andreeva A, Snider WD, Elliott JL (2000) Restricted expression of G86R Cu/Zn superoxide dismutase in astrocytes results in astrocytosis but does not cause motoneuron degeneration. J Neurosci 20:660–665.

Gurney ME, Pu H, Chiu AY, Dal Canto MC, Polchow CY, Alexander DD, Caliendo J, Hentati A, Kwon YW, Deng H-X, Chen W, Zhai P, Sufit RL, Siddique T (1994) Motor neuron degeneration in mice that express a human Cu, Zn superoxide dismutase mutation. Science 264:1772–1775.

Harding AE (1981) Friedreich's ataxia: A clinical and genetic study of 90 families with an analysis of early diagnostic criteria and intrafamilial clustering of clinical features. Brain 104:589–620.

Hayashi H, Kato S, Kawada A (1991) Amyotrophic lateral sclerosis patients living beyond respiratory failure. J Neurol Sci 105:73–78.

Hely MA, Morris JG, Rail D, Reid WG, O'Sullivan DJ, Williamson PM, Genge S, Broe GA (1989) The Sydney multicentre study of Parkinson's disease: A report on the first 3 years. J Neurol Neurosurg Psychiat 52:324–328.

Hughes AJ, Daniel SE, Kilford L, Lees AJ (1992) Accuracy of clinical diagnosis of idiopathic Parkinson's disease: A clinico-pathological study of 100 cases. J Neurol Neurosurg Psychiat 55:181–184.

Kirshner HS, Tanrida g O, Thurman L, Whetsell WO, Jr. (1987) Progressive aphasia without dementia: Two cases with focal spongiform degeneration. Ann Neurol 22:527–532.

Kurtland LT (1988) Amyotrophic lateral sclerosis and Parkinson's disease complex on Guam linked to an environmental neurotoxin. Trends Neurosci 11:51–54.

Le Rhun E, Richard F, Pasquier F (2005) Natural history of primary progressive aphasia. Neurology 65:887–891.

Levy G, Tang MX, Louis ED, Cote LJ, Alfaro B, Mejia H, Stern Y, Marder K (2002) The association of incident dementia with mortality in PD. Neurology 59:1708–1713.

Lino MM, Schneider C, Caroni P (2002) Accumulation of SOD1 mutants in postnatal motoneurons does not cause motoneuron pathology or motoneuron disease. J Neurosci 22:4825–4832.

McGeer PL, McGeer EG (2002) Inflammatory processes in amyotrophic lateral sclerosis. Muscle Nerve 26:459–470.

Morgante L, Salemi G, Meneghini F, Di Rosa AE, Epifanio A, Grigoletto F, Ragonese P, Patti F, Reggio A, Di Perri R, Savettieri G (2000) Parkinson disease survival: A population-based study. Arch Neurol 57:507–512.

Nagai M, Aoki M, Miyoshi I, Kato M, Pasinelli P, Kasai N, Brown RH, Jr., Itoyama Y (2001) Rats expressing human cytosolic copper-zinc superoxide dismutase transgenes with amyotrophic lateral sclerosis: Associated mutations develop motor neuron disease. J Neurosci 21:9246–9254.

Oppenheimer DR, Esiri MM (1997) Diseases of the basal ganglia, cerebellum and motor neurons. In: Greenfield's Neuropathology (Adams JH, Corsellis JAN, Duchen LW, eds), pp 988–1045. New York: Arnold.

Pittman RN, Messam CA, Mills JC (1999) Asynchronous death as a characteristic feature of apoptosis. In: Cell Death and Diseases of the Nervous System (Koliatsos VE, Ratan RR, eds), pp 29–43. Totowa, New Jersey: Humana Press.

Pramatarova A, Laganiere J, Roussel J, Brisebois K, Rouleau GA (2001) Neuron-specific expression of mutant superoxide dismutase 1 in transgenic mice does not lead to motor impairment. J Neurosci 21:3369–3374.

Prezant TR, Agapian JV, Bohlman MC, Bu X, Oztas S, Qiu W-Q, Armos KS, Cortopassi GA, Jaber L, Rotter JI, Shohat M, Fischel-Ghodsian N (1993) Mitochondrial ribosomal RNA mutation associated with both antibiotic-induced and non-syndromic deafness. Nat Genet 4:289–294.

Prusiner SB (1998) Prions. Proc Natl Acad Sci USA 95:13363–13383.

Przedborski S, Goldman JE (2004) Pathogenic role of glial cells in Parkinson's disease. In: Non-Neuronal Cells of the Nervous System: Function and Dysfunction (Hertz L, ed), pp 967–982. New York: Elsevier.

Przedborski S, Vila M (2001) MPTP: A review of its mechanisms of neurotoxicity. Clin Neurosci Res 1:407–418.

Przedborski S, Vila M, Jackson-Lewis V (2003) Neurodegeneration: What is it and where are we? J Clin Invest 111:3–10.

Rosen DR, Siddique T, Patterson D, Figlewicz DA, Sapp P, Hentati A, Donaldson D, Goto J, O'Regan JP, Deng HX, Rahmani Z, Krizus A, McKenna-Yasek D, Cayabyab A, Gaston SM, Berger R, Tanzi RE, Halperin JJ, Herzfeldt B, Van den Bergh R, Hung W-Y, Bird T, Deng G, Mulder DW, Smyth C, Laing NG, Soriano E, Pericak-Vance MA, Haines J, Rouleau GA, Gusella JS, Horvitz HR, Brown Jr HR (1993) Mutations in Cu/Zn superoxide dismutase gene are associated with familial amyotrophic lateral sclerosis. Nature 362:59–62.

Saper CB (1996) Any way you cut it: A new journal policy for the use of unbiased counting methods. J Comp Neurol 364:5.

Saper CB, Wainer BH, German DC (1987) Axonal and transneuronal transport in the transmission of neurological disease: Potential role in system degenerations, including Alzheimer's disease. Neuroscience 23:389–398.

Sasaki S, Tsutsumi Y, Yamane K, Sakuma H, Maruyama S (1992) Sporadic amyotrophic lateral sclerosis with extensive neurological involvement. Acta Neuropathol (Berl) 84:211–215.

Shimura H, Hattori N, Kubo S, Mizuno Y, Asakawa S, Minoshima S, Shimizu N, Iwai K, Chiba T, Tanaka K, Suzuki T (2000) Familial Parkinson disease gene product, parkin, is a ubiquitin-protein ligase. Nat Genet 25:302–305.

Singleton AB, Farrer M, Johnson J, Singleton A, Hague S, Kachergus J, Hulihan M, Peuralinna T, Dutra A, Nussbaum R, Lincoln S, Crawley A, Hanson M, Maraganore D, Adler C, Cookson MR, Muenter M, Baptista M, Miller D, Blancato J, Hardy J, Gwinn-Hardy K (2003) Alpha-Synuclein locus triplication causes Parkinson's disease. Science 302:841.

Stacy M, Jankovic J (1992) Differentia diagnosis of Parkinson's disease and the parkinsonism plus syndromes. In: Parkinson's Disease (Cedarbaum J, Gancher S, eds), pp 341–359. Philadelphia: W.B. Sauders Company.

Sulkava R, Haltia M, Paetau A, Wikstrom J, Palo J (1983) Accuracy of clinical diagnosis in primary degenerative dementia: Correlation with neuropathological findings. J Neurol Neurosurg Psychiat 46:9–13.

Tanner CM (1989) The role of environmental toxins in the etiology of Parkinson's disease. Trends Neurosci 12:49–54.

Tanner CM (1992) Epidemiology of Parkinson's disease. Neurol Clin 10:317–329.

Tanner CM, Ottman R, Goldman SM, Ellenberg J, Chan P, Mayeux R, Langston JW (1999) Parkinson disease in twins: An etiologic study. JAMA 281:341–346.

The Huntington's Disease Collaborative Research Group (1993) A novel gene containing a trinucleotide repeat that is expanded and unstable on Huntington's disease chromosomes. Cell 72:971–983.

Tomlinson BE (1977) The pathology of dementia. Contemp Neurol Ser 15:113–153.

Van Dijk PT, Dippel DW, Van Der Meulen JH, Habbema JD (1996) Comorbidity and its effect on mortality in nursing home patients with dementia. J Nerv Ment Dis 184:180–187.

Wenning GK, Colosimo C, Geser F, Poewe W (2004) Multiple system atrophy. Lancet Neurol 3:93–103.

West AB, Moore DJ, Biskup S, Bugayenko A, Smith WW, Ross CA, Dawson VL, Dawson TM (2005) Parkinson's disease-associated mutations in leucine-rich repeat kinase 2 augment kinase activity. Proc Natl Acad Sci USA 102:16842–16847.

Wyss-Coray T, Mucke L (2002) Inflammation in neurodegenerative disease-a double-edged sword. Neuron 35:419–432.

Zimprich A, Biskup S, Leitner P, Lichtner P, Farrer M, Lincoln S, Kachergus J, Hulihan M, Uitti RJ, Calne DB, Stoessl AJ, Pfeiffer RF, Patenge N, Carbajal IC, Vieregge P, Asmus F, Muller-Myhsok B, Dickson DW, Meitinger T, Strom TM, Wszolek ZK, Gasser T (2004) Mutations in LRRK2 cause autosomal-dominant parkinsonism with pleomorphic pathology. Neuron 44:601–607.

18
Multiple Sclerosis and Other Demyelinating Diseases

Samantha S. Soldan, Gregory F. Wu, Clyde E. Markowitz, and Dennis L. Kolson

Keywords Multiple sclerosis; experimental allergic encephalomyelitis; central nervous system; virus; cerebrospinal fluid; T-cell receptor; acute disseminated encephalomyelitis; lymphocyte; immune system

18.1. Introduction

Multiple sclerosis (MS) is the most common idiopathic demyelinating disease of the central nervous system (CNS). Although partially effective treatments are now available, MS represents a major target for research into the development of disease-modifying therapies that specifically focus on the neuroimmune pathways of myelin and tissue damage that currently are incompletely understood. Multiple sclerosis is considered to be an example of development of autoimmunity to self-antigens within the CNS through multiple initiating events that include infections and other environmental factors. The direct or indirect induction of immune responses against CNS antigens includes chemotaxis of T cells, B cells, and monocytes, and production of immunoglobulin responses, each of which can act as an effector of myelin damage that occurs in distinct histological patterns. Because a specific cause for MS has not been identified, much MS research has focused on CNS immune responses triggered by unidentified insults that in turn trigger inflammation-mediated cascades of myelin and cellular damage that are likely relevant to other neurodegenerative diseases. This chapter discusses current the epidemiology, etiology, pathophysiology, animal models, virus models and recent advances in the neuroimmunology of MS from the perspective of the potential for development of newer therapies for MS and other inflammatory CNS diseases.

18.2. Clinical Features and Diagnosis of Multiple Sclerosis

Multiple sclerosis (MS) is a disease of unknown etiology that primarily affects the myelin membrane and/or the oligodendroglia (myelin-producing cells) within the central nervous system (CNS). The clinical features are highly variable, but they consistently reflect dysfunction due to inflammation, local edema, and destruction of the "central myelin" of the brain, spinal cord, and/or optic nerves (reviewed in Miller, 1996). Symptoms generally present with gradually increasing severity (days to weeks), followed by gradual resolution (weeks), which may be complete or partial. The frequency of different types of symptoms in MS is difficult to accurately establish, in part due to their highly variable severity and duration (Matthews, 1998). The most common neurological symptoms are focal sensory disturbances including numbness/tingling sensations, dysesthesiae (sensation), paresthesiae (abnormal sensation), L'hermitte's sign (sudden electric-like sensations radiating into the arms or legs while flexing one's neck), and occasionally burning pain. Such sensory disturbances occur in up to 70–80% of individuals during the course of MS (Miller, 1996; Matthews, 1998). Motor manifestations attributable to corticospinal tract dysfunction occur in up to 60% of patients, and often involve the lower extremities. Motor symptoms generally include dyscoordination and weakness that is often described as "heaviness" of an extremity, which may present unilaterally or bilaterally, in the case of spinal cord involvement. Optic nerve involvement, manifested as optic neuritis (inflammation of the optic nerve associated with visual loss) presents as the initial symptom in about 20% of MS patients, but its prevalence during the course of MS is thought to be much higher (Matthews, 1998).

The clinical course of the disease (reviewed in Ebers, 1998) is also variable, with approximately 70–85% of patients starting with a relapsing remitting disease pattern in which remissions are associated with complete or nearly complete recovery (Coyle, 2000). The disease typically gradually progresses (over years) so that remission periods are shorter and neurological recovery is incomplete. In the chronic phase, neurological dysfunction increases without significant improvement. Approximately 15% of patients have primary progressive MS, most commonly expressed as a progressive myelopathy. Other clinical courses have been described including an acute form with rapid neurologic deterioration and sometimes death within a few months, a progressive form

T. Ikezu and H.E. Gendelman (eds.), *Neuroimmune Pharmacology*.
© Springer 2008

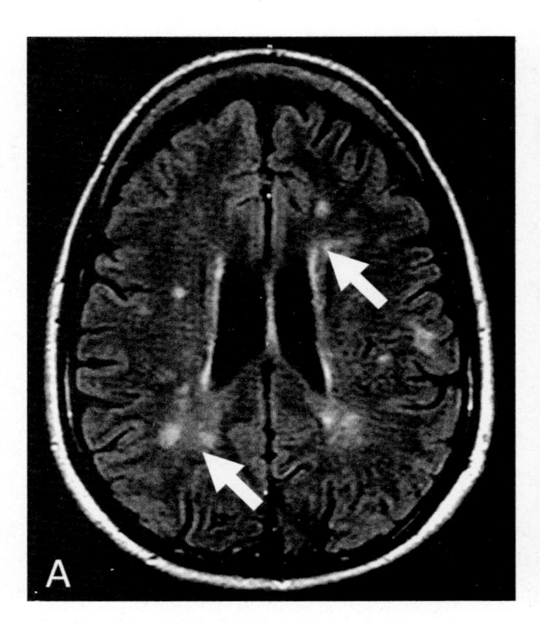

FIGURE 18.1. MRI signal abnormalities in the brain and spinal cord in a 55-year-old woman with multiple sclerosis. (A) FLAIR (Fluid attenuated inversion recovery) image of axial brain image at the level of the lateral ventricles. Arrows point to hyperintense periventricular lesions demonstrating a characteristics MS demyelination pattern; (B) T2W image of sagittal view of the spinal cord of the same individual, demonstrating focal bright signal (arrow) representing demyelination in the cervical spinal cord.

without defined remissions and relapses, and a benign form with a few exacerbations associated with complete recovery. A subclinical form has also been described based upon autopsy findings in asymptomatic individuals. MS can be viewed as a disease with multiple phenotypic presentations that are superimposed over pathological components that range from pure inflammation to gliosis.

Diagnostic criteria intended originally for the purposes of classifying patients for research protocols are now considered standard criteria for clinicians making the diagnosis of MS. The original clinical criteria of Schumacher (Schumacher et al., 1965) (reviewed in Coyle, 2000), were expanded by Poser (Poser et al., 1983) to incorporate paraclinical tests (MRI, cerebrospinal fluid analysis, and evoked potential testing) to increase the certainty of the diagnosis. The MRI typically demonstrates multifocal areas of demyelination surrounding the brain ventricles and within the spinal cord (Figure 18.1). The Poser criteria include the categories of clinically definite, laboratory-supported definite, probable, and possible multiple sclerosis. Previously, practicing clinicians generally attempted only to diagnose cases of definite MS (clinically definite or laboratory-supported definite) because of early recommendations for treatment of definite MS with the first FDA-approved immunomodulating medication, Interferon beta-1b/Betaseron. However, because recent clinical trials in individuals with single, clinically-apparent demyelinating events who did not meet Poser criteria for definite MS demonstrated that Interferon beta-1a (Avonex) greatly reduced the likelihood of development of definite MS over a 2-year period (Kinkel et al., 2006) clinicians have been advised to diagnose and treat patients with such isolated demyelinating events (Frohman et al., 2006b).

These diagnostic criteria, were again modified (Table 18.1) to include specific numbers and locations of MRI-defined lesions that can confirm the diagnosis of MS with a clinically monosymptomatic event or an insidious neurological progression suggestive of MS (McDonald et al., 2001; Polman et al., 2005). Included in Table 18.1 are the Poser criteria (Poser et al., 1983), McDonald criteria (McDonald et al., 2001) and the revised McDonald criteria (Polman et al., 2005). By the Poser criteria (Table 18.1, top two clinical presentations) "definite" multiple sclerosis requires objective evidence of central nervous system (CNS) dysfunction in ≥2 sites of involvement, predominantly in the white matter, relapsing-remitting or chronic progressive (>6 months) in patients between the ages of 10 to 50 years at onset of symptoms. Importantly, there must be no better explanation of the symptoms. These "attacks" may be motor, sensory, visual, or coordination deficits and must last more than 24 h (typically days to weeks). They may be subjective and amnestic (i.e. recalled historically by the patient) or demonstrable by a physician, and separate attacks must be separated in time by at least 30 days of significant improvement to be classified as distinct attacks. In addition, if a physician can demonstrate objective dysfunction in two anatomically separate regions of the CNS, criteria are met for clinically definite multiple sclerosis, assuming no better explanations of the symptoms.

18.3. Epidemiology and Etiology of MS

MS is the most common inflammatory disease of the CNS, affecting an estimated 2,500,000 people worldwide and 350,000 in the United Sates alone (Johnson, 1994) (Kantarci

TABLE 18.1. Diagnostic criteria for multiple sclerosis (Polman et al., 2005).

Clinical presentation	Additional data needed for MS diagnosis
Two or more attacks[a]; objective clinical evidence of two or more lesions	None[b]
Two or more attacks[a]; objective clinical evidence of one lesion	Dissemination in space, demonstrated by: MRI[c] *or* Two or more MRI-detected lesions consistent with MS plus positive CSF[d] *or* Await further clinical attack[a] implicating a different site
One attack[a]; objective clinical evidence of two or more lesions	Dissemination in time, demonstrated by: MRI[e] *or* Second clinical attack[a]
One attack[a]; objective clinical evidence of one lesion (monosymptomatic presentation; clinically isolated syndrome)	Dissemination in space, demonstrated by MRI[c] *or* Two or more MRI-detected lesions consistent with MS plus positive CSF[d] *and* Dissemination in time, demonstrated by: MRI[e] or Second clinical attack[a]
Insidious neurological progression suggestive of MS	One year of disease progression (retrospectively or prospectively determined) *and* Two of the following: a. Positive brain MRI (nine T2 lesions or four or more T2 lesions with positive VEP)[f] b. Positive spinal cord MRI (two focal T2 lesions) c. Positive CSF[d]

Note: If criteria indicated are fulfilled and there is no better explanation for the clinical presentation, the diagnosis is MS; if suspicious, but the criteria are not completely met, the diagnosis is "possible MS"; if another diagnosis arises during the evaluation that better explains the entire clinical presentation, then the diagnosis is "not MS."

[a] An attack is defined as an episode of neurological disturbance for which causative lesions are likely to be inflammatory and demyelinating in nature. There should be subjective report (backed up by objective findings) or objective observation that the event lasts for at least 24 h.

[b] No additional tests are required; however, if tests (MRI, CSF) are undertaken and are negative, extreme caution needs to be taken before making a diagnosis of MS. Alternative diagnoses must be considered. There must be no better explanation for the clinical picture and some objective evidence to support a diagnosis of MS.

[c] MRI demonstration of space dissemination must fulfill the criteria derived from Barkhof and colleagues (Barkhof et al., 2003) and Tintore and coworkers (Tintore et al., 2003).

[d] Positive CSF determined by oligocloncal bands detected by established methods (isoelectric focusing) different from any such bands in serum or by an increased IgG index.

[e] MRI demonstration of time dissemination must fulfill criteria demonstrated in Table 1 in Polman et al. (2005).

[f] Abnormal VEP of the type seen in MS.

and Wingerchuk, 2006). Approximately 1 in 1,000 North Americans will develop MS over their lifetime. While women are 1.5 to 2 times as likely as men to develop MS, males with MS are likely to have a worse prognosis, more rapid disease progression, and later age of disease onset (Coyle, 2000). This evidence of gender dependent differences in disease outcomes suggests that there are sex-specific factors in the phenotypic variability and etiology of MS (Kantarci and Weinshenker, 2005). The etiology of MS is unknown, and the heterogeneous pathology of this disease suggests that several factors might be involved in the spectrum of idiopathic inflammatory demyelinating diseases (IDDs) that are encompassed in the diagnosis of MS (Hafler, 2004). It is believed that genetic, environmental, and immunological factors contribute collectively to the natural history and epidemiology of this disorder (Kurtzke and Wallin, 2000).

18.3.1. Genetic Epidemiology

Epidemiological, familial, and molecular studies of MS have supported a strong influence of genetic background on disease susceptibility (reviewed in Kurtzke and Wallin, 2000). The worldwide distribution of MS is skewed, with areas of high prevalence in North America, Europe, New Zealand, and Australia and areas of lower prevalence in South America, Asia, and Africa. In general, the prevalence and incidence of MS follow a north-south gradient in both hemispheres with individuals of Northern European ancestry more likely to be affected (Compston, 1994; Ebers and Sadovnick, 1994; Sadovnick et al., 1998). Therefore, the north-south gradient observed in the New World may reflect the propensity for individuals from regions of Europe with a high incidence of MS to migrate to the northern regions of the United Sates and Canada, and of individuals from regions of Europe with a lower incidence of MS to migrate to southern regions of the United States and South America (Ebers and Sadovnick, 1994; Sadovnick et al., 1998). In support of this theory, several studies have demonstrated different prevalences of MS among genetically disparate populations residing in the same geographic region. In a 1994 survey of MS in Australia, the prevalence of MS was found to be 37.1/100,000 in New South Wales and 29.4/100,000 in South Australia (McLeod et al.,

1994). Strikingly, no Aborigines or Torres Strait Islanders with MS were identified in this study (McLeod et al., 1994). In addition, the prevalence of MS is higher in Hungarians of Caucasian descent (37/100,000) than in Hungarian gypsies (2/100,000) (Kalman et al., 1991). Similarly, the prevalence of MS among people of Japanese decent living on the Pacific Coast (6.7/100,000) is considerably lower than that of Caucasians living in California 30/100,000). Of interest, it has been demonstrated that people of Japanese ancestry living on the West Coast of the United States have a slightly higher prevalence of MS than those living in Japan (2/100,000) suggesting that environmental factors also have a significant impact on disease susceptibility (Detels et al., 1977).

Family and twin studies clearly support a genetic influence in the development of MS. It has been demonstrated that biological relatives of MS patients have a greater likelihood of developing MS than adoptees and that conversely, family members of adopted individuals with MS do not have an increased risk of developing the disease (Ebers et al., 1995). Among biological relatives, the lifetime risk of developing MS increases with closer biological relationships and approximately 20% of MS patients have an affected family member (Sadovnick, 1993). First-degree relatives, particularly sisters, of an individual with MS are 20–40 times more likely to develop MS and the risk of MS decreases rapidly with second- and third-degree relatives (Sadovnick et al., 1998; Sadovnick and Ebers, 1993; Ebers et al., 1995). In addition, the rate of MS concordance is eight times greater in monozygotic than dizygotic twins. However, the concordance among monozygotic twins is only 25–30%, suggesting that genetic background is not sufficient to cause the disease (Bobowick et al., 1978). Therefore, a combined influence of genetic background and the environment in the development of MS has been proposed.

Considerable effort has been made to assign MS susceptibility to classical models of inheritance. However, the inheritance of MS does not conveniently fit any of these models. The ability to attribute a particular inheritance pattern to MS may be confounded by the difficulty in diagnosing this unpredictable disease. Moreover, because the age of risk ranges from the late teens to the late 50s, an individual cannot be considered unaffected with certitude until they are past the age of high risk. Over the years, several genes associated with immune function have been tentatively associated with an increased risk of MS. A strong association between MS and the major histocompatibility complex (MHC) class II alleles DRB1*1501-DQB1*0602 has been demonstrated (Lincoln et al., 2005). The MHC class I allele A201 has been shown to have protective effects (Sospedra and Martin, 2005). In addition, recent studies have suggested that polymorphisms of the cytokine interferon-γ (IFN-γ) are associated with MS susceptibility in a gender dependent manner (Kantarci et al., 2005).

18.3.2. Environmental Epidemiology

The variation in disease incidence and prevalence according to geography (chiefly by distance from the equator), the influence of migration from low-to-high and high-to-low prevalence areas, and the observation of epidemics and clusters of MS all support an environmental influence on MS etiology (Kurtzke and Wallin, 2000). Migration studies based chiefly on Europeans who immigrated to South Africa, Israel, Australia, and Hawaii have also supported an infectious etiology of MS (Alter et al., 1978; Alter and Okihiro, 1971; Kurtzke et al., 1970; Casetta and Granieri, 2000). In general, individuals who migrate from high risk to low risk areas after the age of 15 maintain their risk of MS. However, individuals who migrate from high risk to low risk areas before the age of 15 acquire a lower risk. These data suggest that an environmental factor must be encountered before the age of 15 in order to influence MS susceptibility. In addition, examples of MS epidemics have been described in the Faroe Islands, the Shetland-Orkney Islands, Iceland, Sardinia, Key West Florida, Mossyrock Washington, South Africa, and Mansfield Massachusetts (Kurtzke, 1995; Pugliatti et al., 2002). Albeit rare, these epidemics support an infectious etiology of MS.

A wide range of environmental factors including vitamin D, smoking, exposure to solar ultraviolet radiation, exposure to organic solvents, household pets, and dietary fatty acids have been associated with the risk of developing MS (Soilu-Hanninen et al., 2005; Schiffer et al., 2001; Neuberger et al., 2004; Hernan et al., 2005; Riise et al., 2002). In addition, infectious agents have been suspected in the etiology of MS for over a century (Johnson, 1994). Dr. Pierre Marie, a pupil of Dr. Jean Martin Charcot who first classified MS and named the disorder *sclerose en plaque*, was the first to suggest an infectious etiology for MS. Marie hypothesized that MS was triggered by an infection that led to changes in blood vessels ultimately resulting in an inflammatory interstitial reaction of glial cells. In addition, Marie believed that many organisms were involved in the pathogenesis of MS based on the anecdotal association of acute infectious diseases (including malaria, typhoid, and childhood exanthem) with the onset of disease.

An infectious etiology of MS is consistent with a number of epidemiological observations as well as the pathological characteristics of this disease (reviewed in Soldan and Jacobson, 2004). It has been speculated that the infectious component in the development of MS might be a virus. Data implicating a virus in the pathogenesis of MS include: (*a*) geographic association of disease susceptibility with evidence of MS clustering (Haahr et al., 1997); (*b*) evidence that migration to and from high risk areas influences the likelihood of developing MS (Alter et al., 1978; Weinshenker, 1996); (*c*) abnormal immune responses to a variety of viruses; (*d*) epidemiological evidence of childhood exposure to infectious agents and an increase in disease exacerbations with viral infection (Johnson, 1994; Weinshenker, 1996); and (*e*) analogy with animal models and human diseases in which viruses can cause diseases with long incubation periods, a relapsing remitting course, and demyelination. It has been hypothesized that infectious diseases could induce MS either through molecular mimicry (the activation of autoreactive cells via cross-reactivity between foreign and self-antigens), bystander activation (activation of autoreactive cells through nonspecific inflammation), or a combination of the two.

Viruses have been implicated in a number of demyelinating diseases of the CNS in humans and other animals. The association of viruses in other demyelinating diseases further suggests a viral influence in the development of MS, by demonstrating that viruses are capable of inducing demyelination, and that they can persist for years in the CNS presenting chronic diseases long after acute infection. Viruses involved in demyelinating diseases of humans and animals include JC virus, measles virus, HTLV-I, canine distemper virus (CDV), murine coronavirus (JHM strain), Theiler's mouse encephalomyelitis virus (TMEV), and Visna virus (reviewed in Soldan and Jacobson, 2004). Over the years, several viruses have been associated with MS, and these associations are based primarily on elevated antibody titers or the isolation of a particular virus from MS material. However, none of these viruses have demonstrated a definitive cause–effect relationship with the disease. Elevated antibody titers to several infectious agents including influenza C, herpes simplex (HSV), human herpes virus-6 (HHV-6), measles, varicella-zoster (VZV), rubeola, vaccinia, Epstein-Barr virus (EBV), simian virus 5 (SV5), multiple sclerosis retrovirus (MSRV), and *Chlamydia pneumoniae* have been reported (reviewed in Soldan and Jacobson, 2004). Although most of these reported agents have been discounted from consideration in the pathogenesis of MS, a few remain viable candidates.

18.3.3. Immunological Influences

In addition to genetic and environmental influences, it is widely accepted that T cell-mediated immune responses are involved in the etiology of MS (reviewed in Frohman et al., 2006a). This is based on the association of MS with genes involved with the immune response, the immunopathology of the disease, the clinical response of MS patients to immunosuppressive and immunomodulatory treatments, and similarities with experimental immune-mediated demyelinating diseases in animals.

A number of immune abnormalities are frequently observed in MS patients and lend support to an immunologic component of the MS disease process. One of the hallmarks against unknown CNS antigens of MS is the intrathecal secretion of oligoclonal antibodies, also known as oligoclonal bands (OCB) as seen by immunofixation electrophoretic separation. OCB are found in the CNS tissue and CSF of greater than 90% of MS patients (when determined by isoelectric focusing electrophoresis) (Fortini et al., 2003) and are helpful in confirming the diagnosis of the disease. OCB are not specific to MS as they are also found in several other chronic inflammatory CNS conditions of either infectious (CNS Lyme disease, bacterial meningitis, human immunodeficiency virus/HIV infection, neurosyphilis) or autoimmune (CNS lupus erythematosus, neurosarcoidosis) origin. Although OCB are not directed against a single antigen, antibody bands specific for viral, bacterial, and self-antigens have been described (Sindic et al., 1994; Sriram et al., 1999). Therefore, it is unclear whether or not the intrathecal synthesis of immunoglobulins observed in MS results from the presence of disease-related lymphocytes or cells that are passively recruited into the CNS after the pathogenically relevant cells crossed the blood brain barrier (BBB).

In addition to the presence of OCB, other immunological markers of disease activity have been described in MS. The overexpression of several proinflammatory cytokines, including tumor necrosis factor-α (TNF-α) and interferon-γ (IFN-γ) have been demonstrated in MS (Ubogu et al., 2006; Frohman et al., 2006b). Treatment of MS patients with IFN-γ resulted in a marked increase in exacerbations which supports the model of MS as an autoimmune disease mediated by T-helper-1 (TH-1) like T-cells. Furthermore, an increase in TNF-α expression has been found to precede relapses and inflammatory activity as measured by MRI, while the mRNA levels of inhibitory cytokines, such as interleukin-10 (IL-10) and transforming growth factor-β (TGF-β), declined (Rieckmann et al., 1995). Recently expression of ADAM-17, a disintegrin and metalloproteinase that is the major proteinase responsible for the cleavage of membrane-bound TNF, was found to be upregulated in MS lesions (Plumb et al., 2006). The overexpression of these cytokines may be involved in disease pathogenesis by causing the upregulation of MHC and adhesion molecule expression on endothelial and glial cells, activation of macrophages and recruitment of TH-1 cells, or by damaging oligodendrocytes and myelin sheaths directly (Selmaj and Raine, 1988). In addition, peripheral blood mononuclear cells (PBMCs) from MS patients have been demonstrated to have increased numbers of chemokine receptor 5 (CCR5) producing cells (Strunk et al., 2000). However, individuals who fail to express a functional CCR5 receptor (due to a common deletion mutation, CCR5 delta 32, carried by 1% of the Caucasian population) are no more likely to develop MS than those with normal CCR5 expression (Bennetts et al., 1997). Finally, soluble adhesion molecules such as intercellular adhesion molecule-1 (ICAM-1) and E-selectin are elevated in MS sera while soluble vascular cell adhesion molecules VCAM-1 and E-selectin are increased in the CSF of MS patients (Dore-Duffy et al., 1995).

Additional support for MS as a disease with an autoimmune component is provided by the clinical improvement obtained with immunomodulatory and anti-inflammatory therapies (Galetta et al., 2002). Although treatment with corticosteroids does not attenuate the long-term course of MS, it is used effectively in the treatment of MS exacerbations. It has been demonstrated that the administration of high-dose steroids immediately stops blood-brain barrier leakage as visualized by gadolinium-enhanced magnetic resonance imaging (MRI) (Simon, 2000). A number of immunosuppressive and chemotherapeutic drugs including cyclophosphamide, and methotrexate have been used in the treatment of MS with variable success. Other immunomodulatory therapies recently used in MS including interferon-β (IFN-β), Copolymer-1 (COP-1) and the humanized monoclonal antibody to the α 4-integrin of the VLA-4 adhesion molecule (Tysabri [Antegren or Natalizumab]) are reviewed in Chap. 11.

18.4. Pathophysiology of MS

18.4.1. Histology and Physiology of MS Lesions

The hallmark of MS is the white matter "plaque," an area of demyelination usually around a blood vessel with perivenular cell infiltration and gliosis (Frohman et al., 2006a). Recent studies have shown that axonal injury is common in chronic plaques and is an important determinant of permanent loss of neurologic function and disability (Trapp et al., 1999). The analysis of biopsy specimens collected for differential diagnosis and autopsy cases from patients with MS with at least one lesion in the active stages of demyelination has demonstrated four histologically-defined patterns of demyelination, Patterns I-IV (Lucchinetti et al., 2000). All four patterns demonstrate macrophage and T-cell inflammation, although they vary in associated pathological findings. Patterns I and II share macrophage and T-cell-associated inflammation and are described as follows: Pattern I ("macrophage mediated," macrophage- and T-cell associated demyelination, perhaps dominated by products of activated macrophages), Pattern II ("antibody mediated," same infiltrates as I, but with prominent IgG and complement/C9neo deposition). These patterns are largely restricted to small veins and venules and are suggestive of T-cell initiated processes in which macrophages and/or antibodies are important effector functions. They are felt to be immunopathologically closest to EAE. Lesions in Patterns III and IV are not typically seen in EAE. Pattern III ("distal oligodendrogliopathy") and Pattern IV ("primary oligodendrocyte damage with secondary demyelination") also demonstrate inflammatory macrophages and T-cells, but deposition of IgG and complement are absent. Pattern III shows degeneration of distal oligodendrocyte processes, loss of oligodendrocytes by apoptosis at the plaque border, and loss of MAG (myelin associated glycoprotein) expression with preservation of PLP (proteolipid protein), MBP (myelin basic protein) and CNPase (cyclic nucleotide phosphodiesterase) staining. This pattern might represent a T-cell-mediated small vessel vasculitis with secondary ischemic damage to the white matter. However, demyelination in Pattern III is distinct from that in I and II in not being centered around veins and venules. Pattern IV also demonstrates prominent oligodendrocyte degeneration without apoptosis, and demyelination may be induced by T-cell and macrophages on the background of metabolically impaired oligodendrocytes. This pattern might be restricted to individuals with the primary progressive form of MS. Remarkably this study found no evidence for intra-individual variability in the plaque morphology: plaques within an individual patient showed similar morphology.

The net physiological consequences of immune-mediated myelin disruption are partial or total conduction block, slowing of conduction, and ephaptic cross-activation due to disruption of salutatory conduction mediated by the nodes of Ranvier in myelinated fibers (reviewed in McDonald, 1998). Demyelinated fibers are incapable of transmitting impulse trains at physiological frequencies, and such a defect likely contributes to the sensory and motor disturbances seen in MS. Such conduction delays are often detected in the optic nerves through visual evoked potential (VEP) testing (McDonald, 1976). An interesting form of intermittent conduction block that is characteristic of multiple sclerosis is that precipitated by changes in temperature. Heat-induced visual acuity loss (Uhthoff's phenomenon), weakness, and fatigue have all been attributed to heat-induced conduction disturbances, and such symptoms often resolve through decreasing body temperature (Guthrie and Nelson, 1995).

18.4.2. Immunopathogenesis of Myelin Damage in MS

Although the etiology of MS is unknown, there is considerable consensus on the role of immune-mediated mechanisms in disease pathogenesis. The following evidence has led to the concept that helper CD4+ T cells coordinate specific autoimmune responses to one or more candidate autoantigens of central myelin: (1) in the white matter lesions, perivenular infiltrates mainly contain lymphomononuclear cells, including CD4+ and CD8+ T cells, macrophages, and B cells; (2) the animal models of MS are primarily mediated by CD4+ T cells, although the humoral immune response also contributes to the pathogenesis; (3) effective treatments seem to preferentially target T-cell responses; and (4) disease susceptibility is influenced by genes that control presentation of antigens to T cells, such as those of the major histocompatibility complex (MHC, HLA in humans).

In addition to the role of CD4+ TH1 cells, recent studies have provided evidence for the involvement of CD8+ T cells and B cells in the pathogenesis of MS (Babbe et al., 2000; Goverman et al., 2005). Clonal expansion of CD8+ T cells, a subset of T cells with cytotoxic functions, has been recently shown to be predominant over CD4+ T cell expansions both in the brain and the spinal fluid of patients with MS (Babbe et al., 2000). Furthermore, the accumulation of B-cell clonotypes in CNS lesions and the spinal fluid of patients suggests that B cells may be chronically stimulated by CNS antigens leading to the synthesis of oligoclonal IgG (Goverman et al., 2005). Myelin-specific antibodies can contribute to myelin damage by inducing complement activation and facilitating macrophage-mediated myelin toxicity. Therefore, demyelination can be mediated through T-cells and macrophages as well as through complement activation. Myelin repair, which is limited *in vivo,* can be promoted *in vitro* with various growth factors (neurotrophin-3, platelet-derived growth factor, glial grow factor-2, others) (reviewed in Lubetzki et al., 2005).

A schematic view on the main processes that lead to autoimmune central demyelination is presented in Figures 18.2 and 18.3. Autoreactive, myelin antigen-specific T cells are part of the normal T-cell repertoire, although at low frequency. These cells are thought to be activated in the peripheral immune system

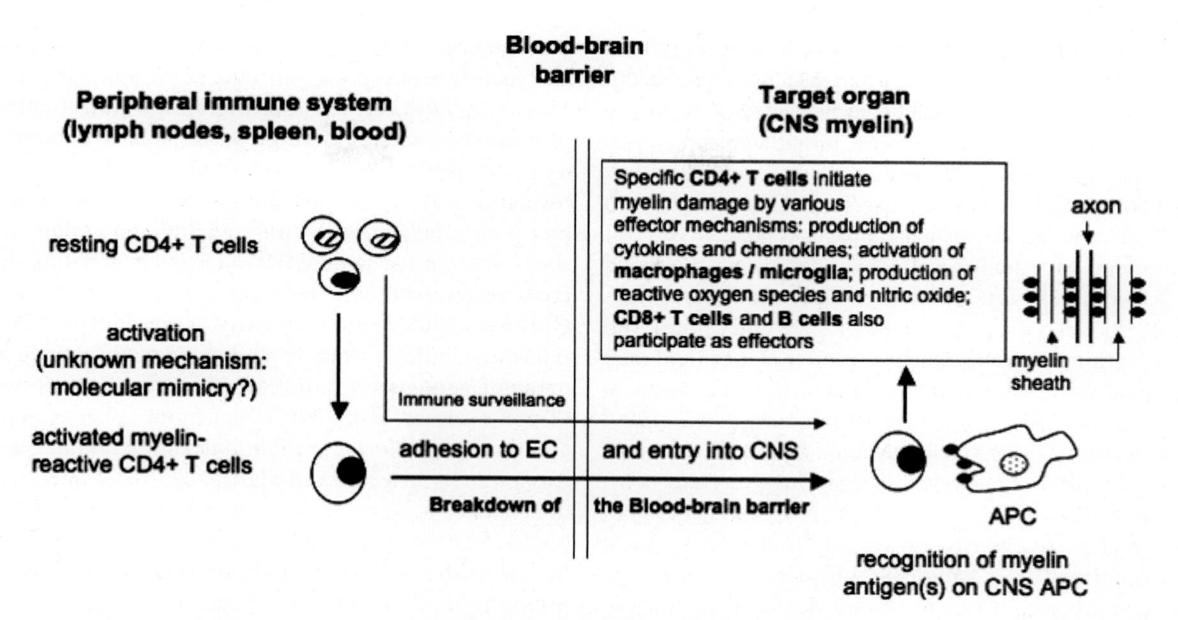

FIGURE 18.2. The immunopathogenesis of MS. A limited number of resting T cells may enter the CNS after crossing the intact blood-brain barrier (BBB). Autoreactive CD4+ T cells may be activated in the periphery by antigens that are structurally similar to myelin antigens (molecular mimicry). When activated T lymphocytes interact with endothelial cells (EC) of CNS venules, cell trafficking across the BBB increases. Reactivation of myelin antigen-specific T cells by local antigen-presenting cells (APC) may initiate myelin damage by several mechanisms: CD4+ T cell production of proinflammatory cytokines and chemokines that further recruit effector cells through an inflamed BBB; CD8+ T inducing cytolysis and chemokine production; activated macrophages producing oxygen radicals and NO (nitric oxide); deposition of myelin antigen-specific antibodies activating complement and facilitating myelin damage by activated macrophages. Resolution of inflammatory damage may be modulated by regulatory cells, that produce anti-inflammatory cytokines, and by elimination of effector cells by apoptosis.

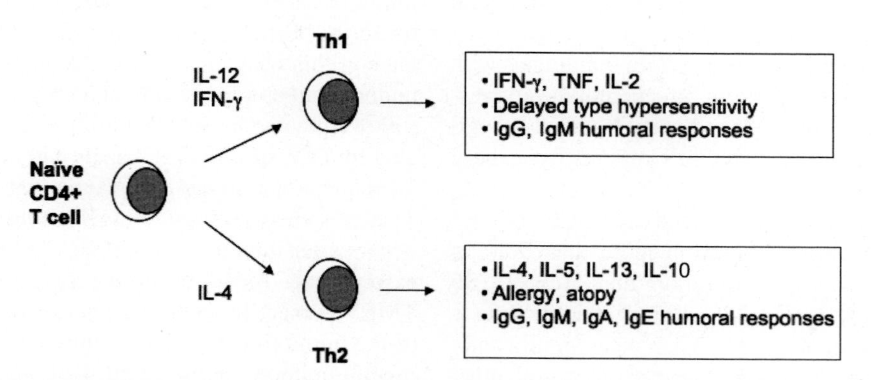

FIGURE 18.3. Comparison of Th1 and Th2 CD4+ T helper cell responses. Th1 and Th2 cells can be discriminated by their ability to produce different patterns of cytokines. Based mainly on the type, dose and route of administration of antigens, the genetic background of the host, and the cytokines present in the microenvironment, naive CD4+ T-cell precursors can develop into T-helper 1 (Th1) or Th2 cells. IL-12 (produced by macrophages) is the dominant factor promoting development of Th1 cells that are involved in cell-mediated immunity to intracellular pathogens and several instances of organ-specific autoimmunity. Th1 development also depends on IFN-γ (mainly produced by T cells). In contrast, early exposure of naive CD4+ T-cell precursors to IL-4 results in the development of Th2 cells, which are involved in atopic and allergic reactions, and in certain types of systemic, antibody-mediated autoimmunity. Regulatory cells that produce the immunomodulatory cytokine TGF-β can modulate the development of both Th1 and Th2 cells and the immunopathology they can induce.

(lymph nodes and spleen) of MS patients by unknown mechanisms. Structural similarity of foreign antigens (such as viral peptides) and myelin proteins may lead to their cross-recognition by myelin-reactive T cells (molecular mimicry). Activated T cells can adhere to endothelial cells in cerebral blood vessels, cross the blood-brain barrier and enter the brain parenchyma. Proteolytic enzymes such as matrix metalloproteases cleave extracellular matrix proteins and allow T-cell entry into the CNS. Upon reactivation by recognition of myelin antigens presented by microglia astrocytes, autoreactive T cells secrete proinflammatory cytokines that may cause direct damage to myelin, such as TNF-α and lymphotoxin, and chemokines that recruit effector cells, such as other T cells and monocytes-macrophages (Figure 18.3).

Recruitment of effector cells by chemokine-regulated chemotaxis of leukocytes and monocyte-macrophages is considered to be critical for effector cell entry into, and migration within, the CNS, each of which is a critical step in the pathogenesis of immune-mediated demyelination in multiple sclerosis (Ransohoff, 1999; Sorensen et al., 1999; Charo and Ransohoff, 2006; Ubogu et al., 2006). Chemokines represent a familty of chemotactic cytokines that are comprised of two major subfamilies: the C-X-C (α chemokine) subfamily that possesses two conserved N-terminal cysteine residues separated by a single amino acid, and the C-C (β chemokine) subfamily that possesses two adjacent cysteine residues. Chemokine-mediated adhesion of cells (tethering), extravasation, and infiltration is a complex process involving multiple signals and receptor-ligand interactions, and it is clear that several major steps are required: chemotaxis, adhesion and transmigration. In individuals with MS as well as in individuals without MS, most lymphocytes within the CSF are CD4+ memory T-cells in proportions that are higher than in the blood (Kivisakk et al., 2004). Such cells consistently express the chemokine receptor CCR7 and most also express CXCR3. Within perivascular MS lesions, however, CXCR3-expressing T-cells are commonly found while CCR7 expression is down-regulated (Sorensen et al., 1999).

Analysis of both CSF and brain tissue from individuals with MS has revealed consistent patterns of chemokine expression that distinguish not only MS patients from neurologically normal controls, but also stable MS patients from those undergoing acute relapses (Ransohoff, 1999; Sorensen et al., 1999). Levels of CXCL10 (CXCR3 ligand) and CCL5 (CCR5 ligand) were undetectable in CSF from more than 90% of control patients, but were elevated in more than 50% of CSF specimens from individuals with MS clinical attacks (Ransohoff, 1999). Furthermore, levels of CCL2 were significantly reduced during acute attacks, while levels of several other chemokines (CCL3, CXCL9, CXCL1, and CXCL8) were not significantly altered. This represented the first demonstration of CNS chemokine alterations related to MS clinical disease activity, and it suggested that TH-1-mediated T-cell signaling clearly plays a role in MS pathogenesis. As a result of such studies in MS patients, clinical trials with selected chemokine receptor antagonists have been undertaken (e.g. a CCR2 antagonist) (Charo and Ransohoff, 2006).

Another critical determinant in the cascade of events leading to MS lesion development is antigen presentation. Antigen presenting cell (APC)/T-cell activation involves at least four major processes: (1) Molecular mimicry, (2) Epitope spreading, (3) Bystander activation, and (4) Cryptic antigen presentation. These mechanisms are not mutually exclusive and each is believed to contribute to the pathogenesis of MS. Molecular mimicry, as described earlier in this chapter, is a phenomenon wherein foreign antigens have enough similarity to the host's endogenous proteins to elicit an autoimmune response. In 1985, it was demonstrated that injection of rabbits with a Hepatitis B virus peptide containing six amino acids in common with MBP induces encephalitis through a cross-reactive immune reaction against MBP (Fujinami and Oldstone, 1985). Since this early observation, the concept of molecular mimicry has been expanded to include involvement of sequence homology, sharing of conserved TCR (T cell receptor) and MHC contact motifs between peptides, additive stimulatory contributions, and structural homology (reviewed in Sospedra and Martin, 2006). Of interest, regions of sequence and structural homology between MBP and two human herpesviruses that are candidate etiologic agents for MS, HHV-6 and EBV have recently been identified (Tejada-Simon et al., 2003; Holmoy et al., 2004).

Epitope spreading occurs when the immune response to a target antigen extends to other epitopes present on the cell that are involved in the primary immune response. These additional epitopes may be within the same molecule (intramolecular) or in different molecules (intermolecular) (Vanderlugt and Miller, 2002). T-cell activation and functional epitope spreading require costimulation via CD28-CD80/86 or CD154-CD40 interactions (Vanderlugt and Miller, 2002). The pathological consequences of epitope spreading have been demonstrated in several mouse models for MS, including Experimental Autoimmune Encephalitis (EAE) and Theiler's Murine Encephalitis Virus (TMEV) induced demyelinating disease (discussed later in this chapter). In EAE, priming with the PLP immunodominant epitope results in an acute clinical episode that is followed by a relapsing remitting disease course that is mediated by PLP specific TH-1 cells and correlates with intramolecular epitope spreading (Vanderlugt and Miller, 2002). In Theiler's virus-induced demyelinating disease, disease onset occurs after infection with TMEV, resulting in a chronic disease course. As the disease progresses, TH-1 responses to TMEV persist. In addition, myelin debris released as a result of the initial tissue damage induces CD4+ TH-1 responses to myelin epitopes starting with PLP and, in late stages of the disease, spreading to MBP and MOG (myelin oligodendrocyte glycoprotein) (Vanderlugt and Miller, 2002).

Bystander activation activation of autoreactive cells through nonspecific inflammation and induction of inflammatory cytokines and chemokines is also of pathological consequence in MS. It has been suggested that bystander activation, induced by persistent virus infection or primed by molecular mimicry may activate autoreactive T-cells specific for the CNS (McCoy et al., 2006). Finally, cryptic antigens may also play a role in immune activation. In other immune-mediated diseases such as Chronic Lymphocytic Thyroiditis and Chagas Heart Disease, exposure of cryptic epitopes leads to the activation of autoimmune cells and further contributes to the pathogenesis (Rose and Burek, 2000; Leon and Engman, 2003). Similarly, it has been suggested that myelin destruction leads to the exposure of cryptic axonal antigens that are normally shielded from immune surveillance by the tightly sealed paranodal loops of myelin. The exposure of these cryptic antigens may activate auto-reactive T-cells and the production of anti-axolemmal antibodies. It has been suggested that these anti-axolemmal antibodies induce neurodegeneration, prevent remyelination and contribute to the axonal loss that correlates with functional loss in MS. (DeVries, 2004).

18.5. Animal Models of MS

18.5.1. Experimental Autoimmune Encephalomyelitis (EAE)

While no single animal model completely recapitulates all features of MS, Experimental autoimmune encephalomyelitis (EAE, Table 18.2) is the most widely used. Rivers' seminal investigation into the iatrogenic human disease, postvaccinal encephalomyelitis was critical in the development of field of EAE (Rivers et al., 1933). Early models of EAE involved the induction of inflammatory demyelinating lesions in experimental animals by immunization with brain or spinal cord tissue (Rivers et al., 1933). Current models rely on active immunization with myelin antigens (myelin basic protein (MBP), myelin associated glycoprotein (MAG), proteolipid protein (PLP) or myelin-oligodendrocyte-glycoprotein (MOG)) in adjuvant or by adoptive transfer of encephalitogenic, myelin antigen specific T cell lines or clones (passive immunization) (Ebers, 1999; Lassmann and Wekerle, 1998; Ercolini and Miller, 2006). Although mouse and rat models are the most commonly used, EAE can be induced in a variety of susceptible animals including guinea pigs, rabbits, pigs, and monkeys.

The clinical signs of EAE reflect the acute inflammatory responses developing in the brain and spinal cord: hindlimb and tail paralysis, quadriparesis, ataxia, abnormal righting responses, and sometimes incontinence (Lublin, 1996). These episodes, whether associated with active or passive immunization, are associated with infiltration of myelin-specific inflammatory TH1 CD4+ T-cells into the CNS (McRae et al., 1992). The disease can be either monophasic (with a single paralytic episode followed by recovery), relapsing-remitting (with repeated cycles of paralysis interrupted with partial or full recovery), or chronic (initial symptoms either stabilize or progress) (Ercolini and Miller, 2006). These later two clinical courses more closely resemble the natural history of MS symptoms. In EAE, TH-1, proinflammatory cytokines contribute to tissue destruction directly, upregulate expression of MHC I and II within the CNS, and lead to macrophage and microglial activation. Paradoxically, treatment with the proinflammatory cytokines IFN-γ and TNF-α is protective in EAE, while similar treatments lead to disease exacerbation in MS (Krakowski and Owens, 1996).

EAE is traditionally regarded as a prototypic TH-1 CD4+ T-cell mediated autoimmune disease of the CNS. However, other cells of the immune system play important roles in the neuropathogenesis of this disorder. Defining the precise role of various cell types in EAE is complicated by the diverse and heterogeneous nature of EAE model systems. Recently, a new class of T-cells, TH-17 cells, has been shown to regulate inflammation in EAE. This unique subset of T helper cells produce IL-17 and develops along a pathway that is distinct from that of TH-1 and TH-2 cell differentiation (Park et al., 2005). It has been demonstrated that neutralization of IL-17 with IL-17-receptor-Fc-protein or an IL-17 monoclonal antibody ameliorates the disease course of EAE (Hofstetter et al., 2005). The production of IL-17 requires upstream activation of IL-23, and neutralizing antibodies specific for

TABLE 18.2. Features of EAE and MS.

Feature	EAE	MS
CNS signs	+++	+++
Relapsing disease	++	+++
Perivascular inflammation	+++	+++
Cellular infiltrate	CD4+ T-cells, MOG-specific CD8+ T-cells	CD8+ T-cells, CD4+ T-cells, B-cells
Demyelination	+, +++, perivenous	+++, diffuse
Remyelination	++	++
Immunogen	MBP, PLP, MOG, others	Unknown
CSF immunology	Antibodies to myelin antigens	Rarely find antibodies to myelin antigens; OCB antigens unknown
Genetic predisposition	++	++
Linked to MHC	++	++
Response to immunomodulation	+++	++

Source: Adapted from Lublin (1996) and Sriram and Steiner (2005).

IL-23 also ameliorate disease progression in EAE (Chen et al., 2006) In addition, EAE is significantly suppressed in IL-17 knockout mice further indicating the importance of IL-17 in the pathogenesis of EAE (Komiyama et al., 2006).

In addition to T-cells, other cell types have been found to play important roles in EAE induced by both active immunization and adoptive transfer. CD8+ cells play a variable role in EAE. In the absence of CD8+ T-cells, more severe disease is observed. However, MOG-specific CD8+ T-cells have been observed and are capable of transferring EAE in SCID mice. In addition, H2k MBP-specific CD8+ T-cell clones recognizing the peptide fragment MBP 79–87 (Huseby et al., 1999). B cells are generally believed to be dispensable in EAE as B-cell deficient (B10.PL × SJL/J) mice develop EAE after immunization (Dittel et al., 2000). In B6 mice, the absence of B-cells exacerbates EAE suggesting a protective role for these cells. However, MOG-induced EAE is aggravated by administration of anti-MOG antibodies, suggesting an important role for the humoral response in some models of EAE.

EAE has been used extensively to gain insight into almost every aspect of the neuroimmunology and neuropathology of MS. In addition, the majority of therapies that are either routinely used or currently under development for MS were initially developed in EAE. However, EAE has received criticism as an incomplete model of MS (Sriram and Steiner, 2005). It has been suggested that EAE is a model of acute CNS inflammation/demyelination such as acute disseminated encephalomyelitis (ADEM) rather than a true counterpart for MS. Critical differences between MS and EAE include differences in lesion composition (CD4+ cells dominate in EAE while in the inflammatory MS lesion macrophages and CD8+ cells are frequently observed), lesion location, affect of immunotherapies (e.g. IFN-γ can ameliorate EAE, but increases exacerbations in MS, IFN-β decreases relapse rate in MS, but can worsen EAE), and the observation that the significant axonal damage observed in MS is absent in most models of EAE (Sriram and Steiner, 2005). Nevertheless, the three dominant, FDA approved treatment modalities for MS (Copolymer-1, IFN-β, and Natalizumab) were all developed in EAE which underscores the great utility of this model in spite of its imperfections (Steinman and Zamvil, 2006).

18.5.2. Viral Models of MS

18.5.2.1. Theiler's Murine Encephalomyelitis Virus

There are two major virus induced, murine models of MS: Theiler's Murine encephalomyelitis viruses (TMEV) and mouse hepatitis virus (MHV). Each is a naturally occurring rodent virus that infects the CNS and induces demyelination in genetically susceptible hosts. Theiler's Murine encephalomyelitis viruses (TMEV) are members of the *Picornaviridae* that naturally cause neurologic and enteric diseases in mice (Rodriguez et al., 1987). Several strains of TMEV are capable of inducing paralysis in susceptible rodents. Interestingly it is the persistent, avirulent strain that causes chronic disease; members of the GDVII subgroup grow to high titers and cause a fatal polioencephalomyelitis, while the Daniels (DA) strain and other members of the Theiler's original (TO) subgroup grow to relatively low titers and induce a chronic inflammatory demyelinaging disease of the spinal cord (Rodriguez and Roos, 1992). In the later case, the disease is bi-phasic with an acute encephalitic phase followed by late chronic demyelinating disease, associated with mononuclear infiltrates and demyelinating lesions (Murray et al., 2000).

Several features of TMEV-induced CNS disease including clinical, genetic, and histopathological similarities with MS make this a relevant experimental model. Chronic and/or relapsing remitting paralytic symptoms are observed in both MS and TMEV encephalomyelitis. In additional, natural infection with the TO strains of TMEV progresses to CNS infection and the development of chronic, progressive neurologic disease in a small percentage of infected mice and intracerebral injection only leads to CNS disease in some mouse strains, suggesting a genetic component to the disease. Both diseases are under multigenic control and susceptibility is associated with genes involved in the immune response including MHC genes (Borrow et al., 1992). In addition, TMEV encephalomyelitis, like MS, is a predominantly TH-1 mediated disorder. A strong, TH-1 response has been demonstrated during the acute phase (Chang et al., 2000) and, in some mouse strains, during the chronic phase (Chang et al., 2000). Recently, matrix metalloproteinases, which degrade extracellular matrix molecules and are involved in demyelination processes, were demonstrated to be upregulated in chronic demyelinating Theiler's Murine encephalomyelitis (Ulrich et al., 2006). Lastly, the cellular makeup of the mononuclear infiltrates in both diseases consists primarily of T lymphocytes, monocytes, and macrophages (the primary host cell of the virus) and demonstrate striking similarities with MS lesions. It has been demonstrated that CD8+ T cells are critical in the prevention of TMEV mediated demyelination and that CD4+ cells are important in the synthesis of neutralizing antibodies. In addition, as TMEV encephalomyelitis is caused by latent and persistent infection of the CNS, it makes this an intriguing model for MS, a disorder for which an infectious etiology is suspected. Recently, recombinant TMEV encoding a mimic peptide for PLP that is naturally expressed by *Haemophilus influenzae* has been described (Olson et al., 2005). Infection with this recombinant virus results in early disease onset. These studies support molecular mimicry as a viable hypothesis in MS.

18.5.2.2. Murine Hepatitis Virus

Murine hepatitis virus (MHV) is a murine coronaviruses that infects the liver. Two strains of MHV, JHM and MHC-A59 are neurotropic variants that are frequently used in studies of MHV infection of the CNS (Matthews et al., 2002). Both viruses readily infect oligodendrocytes, astrocytes and neurons (Matthews et al., 2002). After acute infection, the animals develop a chronic progressive neurologic disease characterized by a single major episode of demyelination and accompanied by hind limb paresis, paralysis, and ataxia. Recovery is mediated by CNS and, sometimes, peripheral nerve remyelination (Matthews et al., 2002). In this model, it is believed that demyelination results from both immune-mediated and direct viral destruction. CD8+ T-cells mediate CNS disease and MHV induces both TH-1 and TH-2 cytokines. Interestingly, partial depletion of CD4+ and CD8+ cells does not eliminate chronic demyelination. In addition, humoral immunity is critical for both clearance of the virus and establishment of persistent infection. Chronic lesions in MHV are found throughout the spinal cord and are similar to chronic lesions in MS. As the disease progresses, lymphocytic infiltration diminishes while demyelination and astrogliosis increases. Unlike TMEV, no autoimmune reaction against brain antigens has been described in the MHV model.

18.6. Recent Advances in the Neuroimmunology of MS

18.6.1. CD4+CD25+ Regulatory T-Cells (T Regs)

While CD4+ T cells are traditionally regarded as pathogenic in MS, CD4+CD25+ regulatory T-cells (T regs) have emerged as major players in inhibiting autoimmune disease in humans and in rodent models. Suppression by T regs is critical in preventing autoreactive T-cells from causing autoimmune disorders and in inducing peripheral tolerance. Accordingly, depletion of T reg cells leads to the onset of systemic autoimmune disorders in mice. Although T reg cells exist in the same frequency in MS patients, the effector function of these cells are significantly impaired in relapsing-remitting MS patients (Viglietta et al., 2004). Interestingly, T reg cell function and expression of the T reg-specific transcription factor, FOXP3, is normal in chronic progressive MS patients (Venken et al., 2006). In EAE, suboptimal T-cell stimulation results in significant pathology and delayed recovery upon depletion of CD4+CD25+FOXP3+ cells, indicating that these cells are highly important in raising the threshold for triggering autoreactive T-cell responses (Stephens et al., 2005). In the future, monitoring the effects of immunomodulatory therapies on T reg cells will help define the role of these cells in autoimmune disorders and lead to the development of novel therapies for MS. Currently, a novel T-cell receptor peptide-based immunotherapy (NeuroVax) that restores normal function of FOXP3+ T regs is entering phase II clinical trials (Darlington, 2005).

18.6.2. B Cell Responses in MS

While the presence of OCB in MS has long been appreciated, recent studies have re-examined the role of B-cells in MS. Antibody binding studies have demonstrated that binding to a neuronal cell line was increased in chronic progressive MS patients compared to relapsing remitting multiple sclerosis patients and other inflammatory CNS disease controls (Lily et al., 2004). This approach could lead to the identification of cell surface autoantigens that may be involved in mediating demyelination or neuronal damage. In another study, proteomics technology was used to characterize autoantibodies directed against candidate antigens, in EAE, MS patients, and controls (Robinson et al., 2003).

In addition, antibodies may play a beneficial role in MS by skewing the immune system away from a TH-1 biased response and toward a TH-2 cytokine pattern or by fostering myelin repair. It has been demonstrated that IgM antibodies to CNS antigens enhance remyelination in MS (Sospedra and Martin, 2005). The beneficial effect of pooled intravenous Ig in MS therapy supports a beneficial role for antibodies in MS. It is believed that this beneficial effect occurs through several mechanisms including inactivation of cytokines, Fc-receptor blockade, blocking of CD4 and MHC, and modulation of apoptosis (Sospedra and Martin, 2005).

18.7. Other Demyelinating Diseases of the CNS

Several other demyelinating disorders such as neuromyelitis optica (NMO/Devic's disease), transverse myelitis, acute necrotizing myelitis and Foix-Alajuanine Syndrome share some features with multiple sclerosis but in some cases represent distinct pathological entities. Neuromyelitis optica that is characterized by a generally acute, severe clinical necrotizing demyelinating syndrome that primarily affects the optic nerves and spinal cord. Unlike classical MS-associated demyelination, the neuromyelitis optica lesions are characterized by necrosis and vascular proliferation associated with a marked elevation of the CSF white blood cell count (>100 wbc/mm^3; often neutrophils) and protein level. OCBs are usually not found in the CSF. Neuromyelitis optica is more often monophasic (35%) than is MS. Pathologically, it is characterized by gray and white matter necrosis, infiltration of macrophages, eosinophils and neutrophils, and deposition of IgM and IgG with perivascular complement activation (Kerr and Calabresi, 2005). Recent discovery of an association between the presence in the serum of an IgG antibody against aquaporin 4, a mercurial-insensitive water channel protein concentrated in astrocytic foot processes at the blood-brain barrier, in patients with neuromyelitis optica, has led to the investigation of an anti-B cell (anti-CD20) humanized monoclonal antibody as treatment for this disease (Pittock et al., 2006; Cree et al., 2005; Lennon et al., 2005).

Demyelinating syndromes such as optic neuritis and transverse myelitis have clearly defined relationships with MS and are felt to represent the typical demyelinating lesions found in the white matter of the brain in MS. Optic neuritis is frequently the initial clinical manifestation of MS, and is typically heralded by a decline in vision associated with eye pain over a 7- to 10-day period (Frohman et al., 2005). Approximately 50–70% of individuals with optic neuritis have periventricular white matter abnormalities consistent with demyelination by MRI assessment, and up to 88% of these individuals develop definite MS within 14 years of presentation of optic neuritis (Brex et al., 2002). Patients respond well to acute corticosteroid treatment as well as prophylactic treatment with IFN-β, similar to individuals with MS (Balcer, 2006). Transverse myelitis refers to the clinical syndrome of partial or complete spinal cord dysfunction resulting from inflammatory demyelination. With partial cord dysfunction, the risk of development of MS is as high as 70%, but it is less than 15% when spinal cord function is completely lost (Coyle, 2000), Transverse myelitis is associated with marked CSF elevations of IL-6, which in model systems can activate inducible nitric oxide synthase (iNOS) in the spinal cord, possibly leading to neuronal injury (Kerr and Calabresi, 2005).

The Foix-Alajuanine Syndrome and acute necrotizing myelitis are pathologically and clinically distinct from demyelinating syndromes associated with MS. Foix-Alajuanine Syndrome is a rare cause of myelopathy (spinal cord dysfunction) that is caused by a dural arteriovenous malformation of the spinal cord (Mishra and Kaw, 2005). It is often confused with transverse myelitis or acute necrotizing myelitis. Rather than immune-mediated cord damage, the syndrome is characterized by subarachnoid hemorrhages that result in severe local arachnoid fibrosis and associated thrombosis of local blood vessels. Patients are usually over 50 years of age, and suffer a slow progression to paraplegia. Acute necrotizing myelitis is distinct from typical demyelinating spinal cord involvement in MS. It is associated with coagulative necrosis of both gray and white matter, infiltration of T-cells and macrophages, a high CSF protein level (>500 mg/dl), and absence of OCB in the CSF, thus distinguishing it pathologically from MS, but suggesting important similarities with neuromyelitis optica (Katz and Ropper, 2000). Clearly, better understanding of the various forms of immune-mediated myelitis (neuromyelitis optica, transverse myelitis, necrotizing myelitis) will likely reveal both common and unique features of each that will guide the development of newer therapies for MS and other demyelinating diseases.

Summary

The pathogenesis of MS is a complex immune-mediated process that can likely be triggered by multiple antigens presented within the CNS, which can elicit T-cell and B-cell responses that promote myelin damage and the resulting neuronal damage. Antigenic responses are thought to be induced and amplified in some cases through molecular mimicry, epitope spreading, bystander activation of antigen-presenting cells, and by cryptic antigens that are normally shielded from presentation. Epidemiological evidence suggests that environmental (infectious) agents can trigger such responses in genetically susceptible hosts, based upon strong associations between geographic residence, occurrence of MS and presence of certain MHC I and II alleles within the affected population. Animal models of induced CNS demyelination (EAE, neurotropic viruses; TMEV, MHV) have largely confirmed that presentation of myelin-associated antigens within the CNS as well as CNS virus infection can elicit strong MS-like T-cell and B-cell responses and MS-like neurological dysfunction in animal hosts. Within the brains of MS patients, pathological studies confirm the infiltration of CD4+ and CD8+ T-cells and macrophages, as well as deposition of IgG and complement to varying degrees in defined subtypes of MS plaques. Along with such infiltrates, robust expression of proinflammatory cytokines and chemokines within the brain and CSF is consistently demonstrated, and such factors can promote the recruitment of reactive T cells and monocytes from the periphery that can amplify inflammatory responses. Clinical responses to different immunomodulating therapies in MS patients further support roles for T-cell- and B-cell-mediated immune dysregulation in MS pathogenesis. Future MS therapies will likely target not only activated lymphocytes (T-cells, B-cells, Tregs) and macrophages, but also the neurons that are rendered vulnerable by associated myelin damage.

Review Questions/Problems

1. **Which of the following statements is true about gender differences in MS?**

   a. Men are more likely than women to develop MS.
   b. Women are likely to have a later age of disease onset.
   c. Women have a more rapid disease progression
   d. Men with MS have a worse prognosis.

2. **The geographic distribution of MS:**

   a. follows an east-west gradient
   b. increases at higher altitudes
   c. increases with distance from the equator
   d. is skewed toward northern latitudes in both hemispheres

3. **An environmental component in the etiology of MS is supported by:**

   a. the geographic distribution of MS
   b. migration studies
   c. relatively low concordance (25–30%) in identical twins
   d. reports of MS epidemics
   e. all of the above

4. **Which of the following immune abnormalities are not associated with MS?**

   a. overexpression of IL-10
   b. increased expression of adhesion molecules
   c. presence of oligoclonal bands in the CSF
   d. increased TNF-α expression preceding clinical relapse

5. Which immunomodulatory therapies are frequently used in MS?

a. IFN-β
b. IFN-γ
c. Copolymer-1
d. corticosteroids
e. all of the above
f. a, b & d

6. Oligoclonal bands represent:

a. Immunoglobulins directed against recently identified myelin epitopes in MS
b. Immunoglobulins directed against unknown CNS epitopes in MS
c. Immunoglobulins that have been shown to deposit around demyelinated plaques in MS
d. Immunoglobulins often detected in the serum of individuals with MS
e. b & e

7. MS plaques have been histologically demonstrated to include:

a. infiltration of CD8+ T-cells
b. infiltration of CD4+ T-cells
c. infiltration of B-cells
d. IgG deposition
e. Complement deposition
f. all of the above
g. a & b
h. a, b & c
i. a, b, c & d

8. Which of the following statements is false?

a. In MS plaques perivenular infiltrates contain mainly mononuclear and lymphocytic cells, including CD8+ T-cells, CD4+ T-cells, macrophages, and B-cells.
b. In the EAE model, the primary infiltrating cell is the CD8+ T-cell.
c. Effective MS treatments generally target T-cell responses
d. In both EAE and MS, disease susceptibility is influenced by genes that control presentation of antigens to T-cells.

9. In the pathogenesis of MS, molecular mimicry implies:

a. Receptors mediating T-cell migration have overlapping function, i.e., they can mimic each other's ligand specificity.
b. Immune modulating chemokines or cytokines can mimic the molecular functions of each other.
c. Structural similarity of foreign antigens and myelin protein components may lead to cross-recognition by myelin-reactive T-cells.
d. Suppression of selected T-cell responses can have a global impact on both CD4+ and CD8+ T-cells.

10. Comparison of EAE with MS shows the following:

a. Both demonstrate monophasic, relapsing, and chronic progressive clinical courses
b. Perivascular inflammation is common in both.
c. Demyelination is diffuse throughout the brain in both.
d. All of the above
e. a & b

11. Which viruses have been associated with an increased risk for development of MS?

a. Herpes simplex (HSV)
b. Human herpes virus-6 (HHV-6)
c. Measles
d. Varicella-zoster
e. Rubeola
f. Epstein-Barr
g. All of the above
h. a, b, c & f

12. Among the viruses listed in question 11, which ones have been demonstrated to be causative agents in MS?

a. a
b. b
c. c
d. d
e. e
f. f
g. all of the above
h. none of the above

13. Which features of Theiler's virus-induced CNS disease make it a useful model for MS?

a. The animals develop chronic and/or relapsing paralytic symptoms.
b. Both diseases are under multigenic control by immune response genes.
c. Both are predominantly TH-1 mediated disorders.
d. B-cell infiltration predominates in Theiler's virus induced disease.
e. a
f. a & b
g. a, b & c
h. a, b, c & d

14. Apoptosis of which cell type is a histological hallmark of at least one pathologic subtype of MS plaque?

a. neuron
b. astrocyte
c. oligodendrocyte
d. macrophage
e. none of the above

15. Optic neuritis is the presenting symptom in what percentage of MS patients?

a. 10%

b. 20%
c. 30%
d. 40%
e. 50%

16. The diagnostic criteria for MS:

a. Depend upon the demonstration of white matter abnormalities in the brain or spinal cord by MRI.
b. Require evidence for the presence of neurological dysfunction for the diagnosis in all cases.
c. Require that there is no better explanation (other than MS) for the clinical presentation.
d. None of the above
e. a, b & c
f. a & c
g. b & c

References

Alter M, Okihiro M (1971) When is multiple sclerosis acquired? Neurology 21:1030–1036.

Alter M, Kahana E, Loewenson R (1978) Migration and risk of multiple sclerosis. Neurology 28:1089–1093.

Babbe H, Roers A, Waisman A, Lassmann H, Goebels N, Hohlfeld R, Friese M, Schroder R, Deckert M, Schmidt S, Ravid R, Rajewsky K (2000) Clonal expansions of CD8(+) T cells dominate the T cell infiltrate in active multiple sclerosis lesions as shown by micromanipulation and single cell polymerase chain reaction. J Exp Med 192:393–404.

Balcer LJ (2006) Clinical practice. Optic neuritis. N Engl J Med 354:1273–1280.

Barkhof F, Rocca M, Francis G, Van Waesberghe JH, Uitdehaag BM, Hommes OR, Hartung HP, Durelli L, Edan G, Fernandez O, Seeldrayers P, Sorensen P, Margrie S, Rovaris M, Comi G, Filippi M (2003) Validation of diagnostic magnetic resonance imaging criteria for multiple sclerosis and response to interferon beta1a. Ann Neurol 53:718–724.

Bennetts BH, Teutsch SM, Buhler MM, Heard RN, Stewart GJ (1997) The CCR5 deletion mutation fails to protect against multiple sclerosis. Hum Immunol 58:52–59.

Bobowick AR, Kurtzke JF, Brody JA, Hrubec Z, Gillespie M (1978) Twin study of multiple sclerosis: An epidemiologic inquiry. Neurology 28:978–987.

Borrow P, Tonks P, Welsh CJ, Nash AA (1992) The role of CD8+T cells in the acute and chronic phases of Theiler's murine encephalomyelitis virus-induced disease in mice. J Gen Virol 73 Part 7:1861–1865.

Brex PA, Ciccarelli O, O'Riordan JI, Sailer M, Thompson AJ, Miller DH (2002) A longitudinal study of abnormalities on MRI and disability from multiple sclerosis. N Engl J Med 346:158–164.

Casetta I, Granieri E (2000) Prognosis of multiple sclerosis: Environmental factors. Neurol Sci 21:S839–S842.

Chang JR, Zaczynska E, Katsetos CD, Platsoucas CD, Oleszak EL (2000) Differential expression of TGF-beta, IL-2, and other cytokines in the CNS of Theiler's murine encephalomyelitis virus-infected susceptible and resistant strains of mice. Virology 278:346–360.

Charo IF, Ransohoff RM (2006) The many roles of chemokines and chemokine receptors in inflammation. N Engl J Med 354:610–621.

Chen Y, Langrish CL, McKenzie B, Joyce-Shaikh B, Stumhofer JS, McClanahan T, Blumenschein W, Churakovsa T, Low J, Presta L, Hunter CA, Kastelein RA, Cua DJ (2006) Anti-IL-23 therapy inhibits multiple inflammatory pathways and ameliorates autoimmune encephalomyelitis. J Clin Invest 116:1317–1326.

Compston A (1994) The epidemiology of multiple sclerosis: Principles, achievements, and recommendations. Ann Neurol 36 Suppl 2:S211–S217.

Coyle PK (2000) Diagnosis and classification of inflammatory demyelinating disorders. In: Multiple Sclerosis: Diagnosis, Medical Management and Rehabilitation (Burks JS, Johnson KP, eds), pp 81–99. New York: Demos Medical Publishing, Inc.

Cree BA, Lamb S, Morgan K, Chen A, Waubant E, Genain C (2005) An open label study of the effects of rituximab in neuromyelitis optica. Neurology 64:1270–1272.

Darlington CL (2005) Technology evaluation: NeuroVax, Immune Response Corp. Curr Opin Mol Ther 7:598–603.

Detels R, Visscher BR, Malmgren RM, Coulson AH, Lucia MV, Dudley JP (1977) Evidence for lower susceptibility to multiple sclerosis in Japanese-Americans. Am J Epidemiol 105:303–310.

DeVries GH (2004) Cryptic axonal antigens and axonal loss in multiple sclerosis. Neurochem Res 29:1999–2006.

Dittel BN, Urbania TH, Janeway Jr. CA (2000) Relapsing and remitting experimental autoimmune encephalomyelitis in B cell deficient mice. J Autoimmun 14:311–318.

Dore-Duffy P, Newman W, Balabanov R, Lisak RP, Mainolfi E, Rothlein R, Peterson M (1995) Circulating, soluble adhesion proteins in cerebrospinal fluid and serum of patients with multiple sclerosis: Correlation with clinical activity. Ann Neurol 37:55–62.

Ebers G (1998) Natural history of multiple sclerosis. In: McAlpines' Multiple Sclerosis, 3rd Edition (Compston A, Ebers G, Lassmann H, McDonald I, Matthews B, Wekerle H, eds), pp 191–221. London: Harcout Brace and Co. Ltd.

Ebers G (1999) Modelling multiple sclerosis. Nat Genet 23:258–259.

Ebers GC, Sadovnick AD (1994) The role of genetic factors in multiple sclerosis susceptibility. J Neuroimmunol 54:1–17.

Ebers GC, Sadovnick AD, Risch NJ (1995) A genetic basis for familial aggregation in multiple sclerosis. Canadian collaborative study group. Nature 377:150–151.

Ercolini AM, Miller SD (2006) Mechanisms of immunopathology in murine models of central nervous system demyelinating disease. J Immunol 176:3293–3298.

Fortini AS, Sanders EL, Weinshenker BG, Katzmann JA (2003) Cerebrospinal fluid oligoclonal bands in the diagnosis of multiple sclerosis. Isoelectric focusing with IgG immunoblotting compared with high-resolution agarose gel electrophoresis and cerebrospinal fluid IgG index. Am J Clin Pathol 120:672–675.

Frohman EM, Frohman TC, Zee DS, McColl R, Galetta S (2005) The neuro-ophthalmology of multiple sclerosis. Lancet Neurol 4:111–121.

Frohman EM, Racke MK, Raine CS (2006a) Multiple sclerosis—the plaque and its pathogenesis. N Engl J Med 354:942–955.

Frohman EM, Havrdova E, Lublin F, Barkhof F, Achiron A, Sharief MK, Stuve O, Racke MK, Steinman L, Weiner H, Olek M, Zivadinov R, Corboy J, Raine C, Cutter G, Richert J, Filippi M (2006b) Most patients with multiple sclerosis or a clinically isolated demyelinating syndrome should be treated at the time of diagnosis. Arch Neurol 63:614–619.

Fujinami RS, Oldstone MB (1985) Amino acid homology between the encephalitogenic site of myelin basic protein and virus: Mechanism for autoimmunity. Science 230:1043–1045.

Galetta SL, Markowitz C, Lee AG (2002) Immunomodulatory agents for the treatment of relapsing multiple sclerosis: A systematic review. Arch Intern Med 162:2161–2169.

Goverman J, Perchellet A, Huseby ES (2005) The role of CD8(+) T cells in multiple sclerosis and its animal models. Curr Drug Targets Inflamm Allergy 4:239–245.

Guthrie TC, Nelson DA (1995) Influence of temperature changes on multiple sclerosis: Critical review of mechanisms and research potential. J Neurol Sci 129:1–8.

Haahr S, Munch M, Christensen T, Moller-Larsen A, Hvas J (1997) Cluster of multiple sclerosis patients from Danish community. Lancet 349:923.

Hafler DA (2004) Multiple sclerosis. J Clin Invest 113:788–794.

Hernan MA, Jick SS, Logroscino G, Olek MJ, Ascherio A, Jick H (2005) Cigarette smoking and the progression of multiple sclerosis. Brain 128:1461–1465.

Hofstetter HH, Ibrahim SM, Koczan D, Kruse N, Weishaupt A, Toyka KV, Gold R (2005) Therapeutic efficacy of IL-17 neutralization in murine experimental autoimmune encephalomyelitis. Cell Immunol 237:123–130.

Holmoy T, Kvale EO, Vartdal F (2004) Cerebrospinal fluid CD4+ T cells from a multiple sclerosis patient cross-recognize Epstein-Barr virus and myelin basic protein. J Neurovirol 10:278–283.

Huseby ES, Ohlen C, Goverman J (1999) Cutting edge: Myelin basic protein-specific cytotoxic T cell tolerance is maintained in vivo by a single dominant epitope in H-2k mice. J Immunol 163:1115–1118.

Johnson RT (1994) The virology of demyelinating diseases. Ann Neurol 36 Suppl:S54–S60.

Kalman B, Takacs K, Gyodi E, Kramer J, Fust G, Tauszik T, Guseo A, Kuntar L, Komoly S, Nagy C, Pálffy, G, Petrányi, GGy (1991) Sclerosis multiplex in gypsies. Acta Neurol Scand 84:181–185.

Kantarci O, Wingerchuk D (2006) Epidemiology and natural history of multiple sclerosis: New insights. Curr Opin Neurol 19:248–254.

Kantarci OH, Weinshenker BG (2005) Natural history of multiple sclerosis. Neurol Clin 23:17–38, v.

Kantarci OH, Goris A, Hebrink DD, Heggarty S, Cunningham S, Alloza I, Atkinson EJ, de Andrade M, McMurray CT, Graham CA, Hawkins SA, Billiau A, Dubois B, Weinshenker BG, Vandenbroeck K (2005) IFNG polymorphisms are associated with gender differences in susceptibility to multiple sclerosis. Genes Immun 6:153–161.

Katz JD, Ropper AH (2000) Progressive necrotic myelopathy: Clinical course in 9 patients. Arch Neurol 57:355–361.

Kerr DA, Calabresi PA (2005) 2004 Pathogenesis of rare neuroimmunologic disorders, Hyatt Regency Inner Harbor, Baltimore, MD, August 19th 2004–August 20th 2004. J Neuroimmunol 159:3–11.

Kinkel RP, Kollman C, O'Connor P, Murray TJ, Simon J, Arnold D, Bakshi R, Weinstock-Gutman B, Brod S, Cooper J, Duquette P, Eggenberger E, Felton W, Fox R, Freedman M, Galetta S, Goodman A, Guarnaccia J, Hashimoto S, Horowitz S, Javerbaum J, Kasper L, Kaufman M, Kerson L, Mass M, Rammohan K, Reiss M, Rolak L, Rose J, Scott T, Selhorst J, Shin R, Smith C, Stuart W, Thurston S, Wall M (2006) IM interferon beta-1a delays definite multiple sclerosis 5 years after a first demyelinating event. Neurology 66:678–684.

Kivisakk P, Mahad DJ, Callahan MK, Sikora K, Trebst C, Tucky B, Wujek J, Ravid R, Staugaitis SM, Lassmann H, Ransohoff RM (2004) Expression of CCR7 in multiple sclerosis: Implications for CNS immunity. Ann Neurol 55:627–638.

Komiyama Y, Nakae S, Matsuki T, Nambu A, Ishigame H, Kakuta S, Sudo K, Iwakura Y (2006) IL-17 plays an important role in the development of experimental autoimmune encephalomyelitis. J Immunol 177:566–573.

Krakowski M, Owens T (1996) Interferon-gamma confers resistance to experimental allergic encephalomyelitis. Eur J Immunol 26:1641–1646.

Kurtzke JF (1995) MS epidemiology worldwide. One view of current status. Acta Neurol Scand Suppl 161:23–33.

Kurtzke JF, Wallin MT (2000) Epidemiology. In: Multiple Sclerosis: Diagnosis, Medical Management and Rehabilitation (Burks JS, Johnson KP, eds), pp 49–71. New York: Demos Medical Publishing, Inc.

Kurtzke JF, Dean G, Botha DP (1970) A method for estimating the age at immigration of white immigrants to South Africa, with an example of its importance. S Afr Med J 44:663–669.

Lassmann H, Wekerle H (1998) Experimental models of multiple sclerosis. In: McAlpines' Multiple Sclerosis (Compston A, Ebers G, Lassmann H, McDonald I, Matthews B, Wekerle H, eds), pp 409–433. London: Harcourt Brace & Co. Ltd.

Lennon VA, Kryzer TJ, Pittock SJ, Verkman AS, Hinson SR (2005) IgG marker of optic-spinal multiple sclerosis binds to the aquaporin-4 water channel. J Exp Med 202:473–477.

Leon JS, Engman DM (2003) The significance of autoimmunity in the pathogenesis of Chagas heart disease. Front Biosci 8:e315–e322.

Lily O, Palace J, Vincent A (2004) Serum autoantibodies to cell surface determinants in multiple sclerosis: A flow cytometric study. Brain 127:269–279.

Lincoln MR, Montpetit A, Cader MZ, Saarela J, Dyment DA, Tiislar M, Ferretti V, Tienari PJ, Sadovnick AD, Peltonen L, Ebers GC, Hudson TJ (2005) A predominant role for the HLA class II region in the association of the MHC region with multiple sclerosis. Nat Genet 37:1108–1112.

Lubetzki C, Williams A, Stankhoff B (2005) Promoting repair in multiple sclerosis: problems and prospects. Curr Opin Neurol 18:237–244.

Lublin F (1996) Experimental models of autoimmune demyelination. In: Handbook of Multiple Sclerosis (Cook SD, ed), pp 119–143. Monticello, NY: Marcel Dekker, Inc.

Lucchinetti C, Bruck W, Parisi J, Scheithauer B, Rodriguez M, Lassmann H (2000) Heterogeneity of multiple sclerosis lesions: Implications for the pathogenesis of demyelination. Ann Neurol 47:707–717.

Matthews B (1998) Differential diagnosis of multiple sclerosis and related disorders. In: McAlpines' Multiple Sclerosis (Compston A, Ebers G, Lassmann H, McDonald I, Matthews B, Wekerle H, eds), pp 223–250. London: Harcout Brace and Co. Ltd.

Matthews AE, Weiss SR, Paterson Y (2002) Murine hepatitis virus—a model for virus-induced CNS demyelination. J Neurovirol 8:76–85.

McCoy L, Tsunoda I, Fujinami RS (2006) Multiple sclerosis and virus induced immune responses: Autoimmunity can be primed by molecular mimicry and augmented by bystander activation. Autoimmunity 39:9–19.

McDonald I (1998) Pathophysiology of multiple sclerosis. In: McAlpines' Multiple Sclerosis (Compston A, Ebers G, Lassmann H, McDonald I, Matthews B, Wekerle H, eds), pp 359–378. London: Harcout Brace and Co. Ltd.
McDonald WI (1976) Pathophysiology of conduction in central nerve fibres. In: Visual Evoked Potentials in Man: New Developments (Desmedt JE, ed), pp 352–354. Oxford: Clarendon Press.
McDonald WI, Compston A, Edan G, Goodkin D, Hartung HP, Lublin FD, McFarland HF, Paty DW, Polman CH, Reingold SC, Sandberg-Wollheim M, Sibley W, Thompson A, van den Noort S, Weinshenker BY, Wolinsky JS (2001) Recommended diagnostic criteria for multiple sclerosis: Guidelines from the International Panel on the diagnosis of multiple sclerosis. Ann Neurol 50:121–127.
McLeod JG, Hammond SR, Hallpike JF (1994) Epidemiology of multiple sclerosis in Australia. With NSW and SA survey results. Med J Aust 160:117–122.
McRae BL, Kennedy MK, Tan LJ, Dal Canto MC, Picha KS, Miller SD (1992) Induction of active and adoptive relapsing experimental autoimmune encephalomyelitis (EAE) using an encephalitogenic epitope of proteolipid protein. J Neuroimmunol 38:229–240.
Miller AE (1996) Clinical features. In: Handbook of Multiple Sclerosis, 2nd Edition (Cook SD, ed), pp 201–221. Monticello, NY: Marcel Dekker, Inc.
Mishra R, Kaw R (2005) Foix-Alajouanine syndrome: An uncommon cause of myelopathy from an anatomic variant circulation. South Med J 98:567–569.
Murray PD, Krivacic K, Chernosky A, Wei T, Ransohoff RM, Rodriguez M (2000) Biphasic and regionally-restricted chemokine expression in the central nervous system in the Theiler's virus model of multiple sclerosis. J Neurovirol 6 Suppl 1:S44–S52.
Neuberger JS, Lynch SG, Sutton ML, Hall SB, Feng C, Schmidt WR (2004) Prevalence of multiple sclerosis in a residential area bordering an oil refinery. Neurology 63:1796–1802.
Olson JK, Ercolini AM, Miller SD (2005) A virus-induced molecular mimicry model of multiple sclerosis. Curr Top Microbiol Immunol 296:39–53.
Park H, Li Z, Yang XO, Chang SH, Nurieva R, Wang YH, Wang Y, Hood L, Zhu Z, Tian Q, Dong C (2005) A distinct lineage of CD4 T cells regulates tissue inflammation by producing interleukin 17. Nat Immunol 6:1133–1141.
Pittock SJ, Weinshenker BG, Lucchinetti CF, Wingerchuk DM, Corboy JR, Lennon VA (2006) Neuromyelitis optica brain lesions localized at sites of high aquaporin 4 expression. Arch Neurol 63:964–968.
Plumb J, McQuaid S, Cross AK, Surr J, Haddock G, Bunning RA, Woodroofe MN (2006) Upregulation of ADAM-17 expression in active lesions in multiple sclerosis. Mult Scler 12:375–385.
Polman CH, Reingold SC, Edan G, Filippi M, Hartung HP, Kappos L, Lublin FD, Metz LM, McFarland HF, O'Connor PW, Sandberg-Wollheim M, Thompson AJ, Weinshenker BG, Wolinsky JS (2005) Diagnostic criteria for multiple sclerosis: 2005 revisions to the "McDonald Criteria". Ann Neurol 58:840–846.
Poser CM, Paty DW, Scheinberg L, McDonald WI, Davis FA, Ebers GC, Johnson KP, Sibley WA, Silberberg DH, Tourtellotte WW (1983) New diagnostic criteria for multiple sclerosis: Guidelines for research protocols. Ann Neurol 13:227–231.
Pugliatti M, Solinas G, Sotgiu S, Castiglia P, Rosati G (2002) Multiple sclerosis distribution in northern Sardinia: Spatial cluster analysis of prevalence. Neurology 58:277–282.
Ransohoff RM (1999) Mechanisms of inflammation in MS tissue: Adhesion molecules and chemokines. J Neuroimmunol 98:57–68.
Rieckmann P, Albrecht M, Kitze B, Weber T, Tumani H, Broocks A, Luer W, Helwig A, Poser S (1995) Tumor necrosis factor-alpha messenger RNA expression in patients with relapsing-remitting multiple sclerosis is associated with disease activity. Ann Neurol 37:82–88.
Riise T, Moen BE, Kyvik KR (2002) Organic solvents and the risk of multiple sclerosis. Epidemiology 13:718–720.
Rivers TM, Sprunt DH, Berry GP (1933) Observations on attempts to produce acute disseminated encephalomyelitis in monkeys. J Exp Med 58:39–53.
Robinson WH, Fontoura P, Lee BJ, de Vegvar HE, Tom J, Pedotti R, DiGennaro CD, Mitchell DJ, Fong D, Ho PP, Ruiz PJ, Maverakis E, Stevens DB, Bernard CC, Martin R, Kuchroo VK, van Noort JM, Genain CP, Amor S, Olsson T, Utz PJ, Garren H, Steinman L (2003) Protein microarrays guide tolerizing DNA vaccine treatment of autoimmune encephalomyelitis. Nat Biotechnol 21:1033–1039.
Rodriguez M, Roos RP (1992) Pathogenesis of early and late disease in mice infected with Theiler's virus, using intratypic recombinant GDVII/DA viruses. J Virol 66:217–225.
Rodriguez M, Oleszak E, Leibowitz J (1987) Theiler's murine encephalomyelitis: A model of demyelination and persistence of virus. Crit Rev Immunol 7:325–365.
Rose NR, Burek CL (2000) Autoantibodies to thyroglobulin in health and disease. Appl Biochem Biotechnol 83:245–251; discussion 251–254, 297–313.
Sadovnick AD (1993) Familial recurrence risks and inheritance of multiple sclerosis. Curr Opin Neurol Neurosurg 6:189–194.
Sadovnick AD, Ebers GC (1993) Epidemiology of multiple sclerosis: A critical overview. Can J Neurol Sci 20:17–29.
Sadovnick AD, Yee IM, Ebers GC, Risch NJ (1998) Effect of age at onset and parental disease status on sibling risks for MS. Neurology 50:719–723.
Schiffer RB, McDermott MP, Copley C (2001) A multiple sclerosis cluster associated with a small, north-central Illinois community. Arch Environ Health 56:389–395.
Schumacher GA, Beebe GW, Kibler RF, Kurland LT, Kurtzke JF, McDowell F, Nagler B, Sibley WA, Tourtelotte WW, Willmon TL (1965) Problems of experimental trials of therapy in multiple sclerosis. Ann NY Acad Sci 122:522–568.
Selmaj KW, Raine CS (1988) Tumor necrosis factor mediates myelin and oligodendrocyte damage in vitro. Ann Neurol 23:339–346.
Simon JH (2000) Magnetic resonance imaging in the diagnosis of multiple sclerosis, elucidation of disease course, and determining prognosis. In: Multiple Sclerosis: Diagnosis, Medical Management and Rehabilitation (Burks JS, Johnson KP, eds), pp 99–126. New York: Demos Medical Publishing, Inc.
Sindic CJ, Monteyne P, Laterre EC (1994) The intrathecal synthesis of virus-specific oligoclonal IgG in multiple sclerosis. J Neuroimmunol 54:75–80.
Soilu-Hanninen M, Airas L, Mononen I, Heikkila A, Viljanen M, Hanninen A (2005) 25-Hydroxyvitamin D levels in serum at the onset of multiple sclerosis. Mult Scler 11:266–271.
Soldan SS, Jacobson S (2004) Infection and multiple sclerosis. In: Infection and Immunity (Shoenfeld Y, Rose NR, eds). Amsterdam, Netherlands: Elsevier Science.
Sorensen TL, Tani M, Jensen J, Pierce V, Lucchinetti C, Folcik VA, Qin S, Rottman J, Sellebjerg F, Strieter RM, Frederiksen JL, Ransohoff RM (1999) Expression of specific chemokines and chemokine receptors in the central nervous system of multiple sclerosis patients. J Clin Invest 103:807–815.

Sospedra M, Martin R (2005) Immunology of multiple sclerosis. Annu Rev Immunol 23:683–747.

Sospedra M, Martin R (2006) Molecular mimicry in multiple sclerosis. Autoimmunity 39:3–8.

Sriram S, Steiner I (2005) Experimental allergic encephalomyelitis: A misleading model of multiple sclerosis. Ann Neurol 58:939–945.

Sriram S, Stratton CW, Yao S, Tharp A, Ding L, Bannan JD, Mitchell WM (1999) Chlamydia pneumoniae infection of the central nervous system in multiple sclerosis. Ann Neurol 46:6–14.

Steinman L, Zamvil SS (2006) How to successfully apply animal studies in experimental allergic encephalomyelitis to research on multiple sclerosis. Ann Neurol 60:12–21.

Stephens LA, Gray D, Anderton SM (2005) CD4+CD25+ regulatory T cells limit the risk of autoimmune disease arising from T cell receptor crossreactivity. Proc Natl Acad Sci USA 102:17418–17423.

Strunk T, Bubel S, Mascher B, Schlenke P, Kirchner H, Wandinger KP (2000) Increased numbers of CCR5+ interferon-gamma- and tumor necrosis factor-alpha-secreting T lymphocytes in multiple sclerosis patients. Ann Neurol 47:269–273.

Tejada-Simon MV, Zang YC, Hong J, Rivera VM, Zhang JZ (2003) Cross-reactivity with myelin basic protein and human herpesvirus-6 in multiple sclerosis. Ann Neurol 53:189–197.

Tintore M, Rovira A, Rio J, Nos C, Grive E, Sastre-Garriga J, Pericot I, Sanchez E, Comabella M, Montalban X (2003) New diagnostic criteria for multiple sclerosis: Application in first demyelinating episode. Neurology 60:27–30.

Trapp BD, Ransohoff R, Rudick R (1999) Axonal pathology in multiple sclerosis: Relationship to neurologic disability. Curr Opin Neurol 12:295–302.

Ubogu EE, Cossoy MB, Ransohoff RM (2006) The expression and function of chemokines involved in CNS inflammation. Trends Pharmacol Sci 27:48–55.

Ulrich R, Baumgartner W, Gerhauser I, Seeliger F, Haist V, Deschl U, Alldinger S (2006) MMP-12, MMP-3, and TIMP-1 are markedly upregulated in chronic demyelinating theiler murine encephalomyelitis. J Neuropathol Exp Neurol 65:783–793.

Vanderlugt CL, Miller SD (2002) Epitope spreading in immune-mediated diseases: Implications for immunotherapy. Nat Rev Immunol 2:85–95.

Venken K, Hellings N, Hensen K, Rummens JL, Medaer R, D'Hooghe MB, Dubois B, Raus J, Stinissen P (2006) Secondary progressive in contrast to relapsing-remitting multiple sclerosis patients show a normal CD4+CD25+ regulatory T-cell function and FOXP3 expression. J Neurosci Res 83:1432–1446.

Viglietta V, Baecher-Allan C, Weiner HL, Hafler DA (2004) Loss of functional suppression by CD4+CD25+ regulatory T cells in patients with multiple sclerosis. J Exp Med 199:971–979.

Weinshenker BG (1996) Epidemiology of multiple sclerosis. Neurol Clin 14:291–308.

19
Guillain-Barré Syndrome, Chronic Inflammatory Demyelinating Polyradiculoneuropathy, and Axonal Degeneration and Regeneration

Ralf Gold and Klaus V. Toyka

Keywords Autoantibodies; Axonal damage; Chronic inflammatory neuropathy; Gangliosides; Guillain-Barré syndrome; Immune neuropathy; Myelin; Nerve conduction; Neuritis; Neurotrophin

19.1. Introduction

Guillain-Barré syndrome (GBS) and chronic inflammatory demyelinating polyradiculoneuropathy (CIDP) are acquired demyelinating diseases of the peripheral nervous system (PNS), characterized by progressive or relapsing proximal and distal muscle weakness with possible sensory loss (Saperstein et al., 2001; Hughes and Cornblath, 2005). Historically GBS was described in 1916 in two French soldiers as an acute, postinfectious paralysis with elevated CSF protein but without cells. At that time also epidemics of poliomyelitis have occurred producing "Landry paralysis", as a direct consequence of enterovirus infection, and in contrast to GBS is usually asymmetrical and associated with CSF pleocytosis. In the following decades, with the availability of electrodiagnosis the clinical hallmarks of GBS have been described more precisely. Estimated annual disease incidence in industrialized countries is up to 2/100,000. In contrast to CIDP, about 50% of GBS cases are linked with preceding infections, typically diarrhea caused by C. jejuni (Enders et al., 1993) or infections of the respiratory tract (Hadden et al., 2001). In GBS acute respiratory failure or cardiac arrhythmias due to dysfunction of the autonomic nervous system cause at least 3% mortality even in specialized centers, especially in elderly patients with rapid disease progression and those suffering from multimorbidity (Hughes and Cornblath, 2005). In addition, both in GBS and CIDP, residual symptoms and chronic fatigue may lead to long-lasting disability in about half (Merkies et al., 1999).

The last decade has seen remarkable progress in understanding cellular and molecular pathways that cause autoimmune damage in these disorders of the PNS (see reviews in Gold et al., 2003, 2005; Kieseier et al., 2004). Based on the observations that a significant proportion of patients with CIDP are responsive to immunotherapy and that an inflammatory response is observed at sites of active disease (Schmidt et al., 1996; Sommer et al., 2005), it is generally accepted that GBS and CIDP are autoimmune disorders with myelin as the likely target for the immune response. There is increasing evidence that GBS and CIDP have heterogeneous pathomechanisms. The dominant subtype is the acute inflammatory demyelinating form (AIDP), a classical demyelinating disease, but an increasing number of axonal variants has been described after the first series linked to preceding infections in China with acute motor axonal neuropathy (AMAN) (McKhann et al., 1991). Whilst first data seemed to associate axonal damage to preceding *C. jejuni* infection, especially with Penner 19 strain (Ho et al., 1995), these regional specificities could not be confirmed in the largest multicenter GBS-study performed to date (Hadden et al., 1998, 2001). Acute motor-sensory axonal neuropathy (AMSAN) is clinically and electrophysiologically similar to AMAN, but has a detectable sensory involvement. Thus, there is probably no single or unique mechanism that leads to axonal damage, but several immunological components will be discussed which may contribute to neuronal dysfunction and degeneration. This heterogeneity of the disease is further underscored by the variant of GBS, the Miller Fisher syndrome (MFS), where cranial nerves and caudal spinal roots are involved, and the disorder is strongly linked with anti-ganglioside GQ1b immunoreactivity.

Chronic inflammatory demyelinating polyneuropathy (chronic polyneuritis, CIDP) characteristically presents as symmetrical weakness, with some impairment of distal sensation. By electrophysiology both motor and sensory nerves are affected with slowing of nerve conduction velocity, especially in proximal nerve segments as reflected by abnormal F-wave examination. Cranial nerves may be involved, but less frequently than in GBS. The disease course may either be relapsing-remitting or stepwise progressive. If the initial attack of CIDP is subacute, disease progression up to reaching a plateau level is longer than 8 weeks. The diagnosis is supported by an elevated CSF protein, and inflammatory demyelination in sural nerve biopsy.

© Springer 2008

19.2. Experimental Models

For many aspects of both diseases, the various models of experimental of allergic/autoimmune neuritis (EAN) have helped us to better understand the immunological mechanisms (see review in Gold et al., 2005). The monophasic EAN of the Lewis rat has its greatest advantage for defining pathogenetic hallmarks and establishing innovative therapeutic principles, similar to the chronic EAN in dark agouti rats. Upon immunization with increasing doses of the neuritogen or when antigen-specific T cell lines are used for adoptive transfer (AT-EAN), many clinical and electrophysiological signs of the human diseases could be reproduced (Heininger et al., 1986; Hartung et al., 1988). In particular, a profound and rapidly evolving nerve dysfunction occurred dose-dependently which correlated with increased axonal damage due to Wallerian degeneration. In the high-dose adoptive transfer model, most likely endoneurial ischemia resulting from severe inflammation and edema was an admixed and critical pathogenic mechanism.

Progress in molecular biology has allowed generation of an increasing number of transgenic mice, which are typically on the genetic background of C57BL/6. For these mouse strains defined peptides of the myelin protein P0 have been described as neuritogenic (Miletic et al., 2005; Visan et al., 2004). Besides the acute and chronic autoimmune models induced by immunization or adoptive transfer of T cells, the immune system has been shown to contribute to disease progression and expression in inherited myelinopathies, which was elucidated by Martini and colleagues (see review in Martini and Toyka, 2004). When the immune system was paralyzed by backcrossing the P0 deficient mice on C57BL/6 background with RAG deficient knockout mice, a remarkable delay of myelin pathology was observed within six months. This was not limited to demyelination, but also axonal damage was reduced which may reflect trophic aspects of the axon-Schwann cell unit. Recent in vitro studies have confirmed that indeed dorsal root ganglia cultures from a rat model for Charcot Marie Tooth disease undergo axonal atrophy over a period of time (Nobbio et al., 2006). This model may be utilized to study the molecular changes underlying demyelination and secondary axonal impairment, in particular in the chronic disorders such as CIDP. As axonal damage may occur after just 3 months and tissue cultures represent a strictly controlled environment, this model may also be ideal for testing neuroprotective therapies.

19.3. Pathogenesis of Axonal Damage in GBS and CIDP

The presence of prominent demyelination, with "onion bulb" formations in chronic disease, and of perivascular inflammatory infiltrates are both hallmarks of CIDP pathology (Bosboom et al., 2001b; Sommer et al., 2005), although they may not always be found on small nerve samples from human biopsies. The "onion bulb" formation is a manifestation of excessive Schwann cell processes and is often produced by segmental repetitive demyelination and remyelination (Krendel et al., 1989; Bosboom et al., 2001a). As the disease progresses, secondary axonal degeneration becomes more established, but it also may occur early in the disease course (Dalakas, 1999). This is very similar to the situation in multiple sclerosis (MS), where the work of Trapp's group (Trapp et al., 1998) quantified progressive axonal damage in late disease stages, complemented by the recent findings of axonal damage in early biopsied MS lesions (Kuhlmann et al., 2002). The precise mechanisms that lead to demyelination and (early) axonal damage in the CNS and PNS are not known, but are thought to be mediated by both cellular and humoral immune factors. In MS, axonal damage has been associated with cytotoxic CD8 T cells (Neumann et al., 2002), nitric oxide (NO) (Smith et al., 2001) and free oxygen radicals. Similar observations relate to the detrimental role of focal nitric oxide in the PNS (Kapoor et al., 2003) although in the whole animal, the opposite effect has also been described (Kahl et al., 2003).

19.3.1. Neurotrophic Factors and Survival in the Inflamed Nervous System

Experimental models have revealed that the immune system is not the only denominator for the extent of damage in the inflamed nervous system. In EAE the neurotrophic cytokine CNTF (ciliary neurotrophic factor) has clear neuroprotective role on survival of oligodendrocyte progenitor cells and mature oligodendrocytes (Linker et al., 2002). This in turn leads to enhanced demyelination in mice where CNTF was lacking, and ultimately was associated with axonal damage. In contrast to other neurotrophins CNTF does not have an effect on immune cells, so that these findings could be attributed only to neuroprotection. A retrospective study in MS patients gave additional evidence that CNTF is also relevant in the human disease MS: those MS patients which have a genetic deficiency of CNTF appeared to exhibit earlier motor symptoms and a more severe disease course (Giess et al., 2002).

Since CNTF has also been described as lesion factor in the PNS (Sendtner et al., 1992), it seemed attractive to substitute neurotrophic factors by exogenous administration. This is a major problem with CNTF, which is mostly absorbed in the liver upon s.c. injection (Dittrich et al., 1994), and did not show protective effects on regeneration of the inflamed nerve in the Lewis rat EAN model (Gold et al., 1996). In contrast LIF (leukemia inhibitory factor), another family member of this IL6-cytokine group turned out to augment survival of oligodendrocytes during EAE (Butzkueven et al., 2002) as confirmed by signaling studies and molecular histology. The administration of LIF had positive effects on the disease course both in a preventive and in a therapeutic setting. Yet LIF also interferes with the immune system and affects T cell priming (Linker, Gold submitted for publication).

Studies with s.c. administration of the neurotrophin BDNF have been performed in Lewis rat EAN (Felts et al., 2002). Treatment of Lewis rats with BDNF (10 mg/kg day) did not significantly affect the neurological deficit, nor significantly improve survival, motor function or motor innervation. The weight of the urinary bladder was significantly increased in control animals with EAN, but remained similar to normal in animals treated with BDNF. These results speak for a very limited effect of exogenous administration of BDNF. For thorough interpretation it would be required that pharmacological tracking studies are performed to verify the effect of BDNF on the target organ.

Thus, it is currently not clear how local factors in the target tissue modulate recovery from an immune attack in the PNS. As yet unknown genetic susceptibility factors may modulate the inflammatory process itself and the response of axons and of myelinating Schwann cells to the inflammatory assault. As an example, axonal degeneration occurring in autoimmune neuropathies clearly affects prognosis (Dalakas, 1999). Indeed, parallel expression of neurotrophic factors and their receptors in CIDP may reflect such survival mechanisms in the PNS (Yamamoto et al., 2002).

19.3.2. Cellular Immune Factors

Inflammatory infiltrates as seen in nerve biopsies, consisting primarily of macrophages (Sommer et al., 2005) and T cells (Schmidt et al., 1996), suggest that a T-cell-mediated delayed hypersensitivity reaction directed toward myelin antigens is a probable cause of inflammatory tissue damage in GBS and CIDP. Inflammatory immune reactions are coordinated by a number of soluble chemical mediators and selective adhesion molecules, including the following: direct differentiation and migration of T cells; translocation of T cells across the vascular endothelium; stimulation of protease release; and recruitment of macrophages and additional T cells to sites of inflammation (see review in Gold et al., 2003). In both CIDP and MS, dysregulation of these chemical mediators (i.e., cytokines, chemokines, and adhesion molecules) is postulated to play a role in the breakdown of the blood-nerve and blood-brain barriers, respectively. Furthermore, dysregulation of chemical mediators may be responsible for aberrant trafficking of T cells into the PNS and perpetuation of inflammatory responses that lead to demyelination. In both the PNS and CNS, inflammatory T cells are eliminated by apoptosis, occurring either during the natural disease course or after treatment with glucocorticosteroids (Zettl et al., 1995).

19.3.3. Humoral Immune Factors

In the last decades, postinfectious molecular mimicry has been postulated as the foremost putative mechanism underlying GBS. This implies that antigenic determinants are shared between an infectious agent and the peripheral nervous system. Thus the initial antigenic stimulation by the infectious agent results in a secondary immune response directed against the nervous system. Several microbial organisms share antigens with the nervous system: C. jejuni serotypes O19 or Lior 11 have lipopolysaccharides (LPS) or lipo-oligosaccharides with ganglioside-like structures (causing anti-GM1 or anti-GQ1b immunoreactivity). In addition, Hemophilus influenza has homologies with GM1 and GT1a, mycoplasma pneumonia with galactocerebroside, and cytomegalovirus has crossreactivity with GM2 ganglioside.

Antibody binding to major glycolipid or myelin protein antigens has been shown in both GBS-CIDP and MS. Antibodies may bind to macrophages via their Fc portion, activating phagocytosis and release of inflammatory mediators toward the myelin sheath. An alternative mechanism is through neuromuscular blocking antibodies. An early study of the functional activity of serum IgG antibodies demonstrated a slowing of nerve conduction in marmoset monkeys upon passive transfer of purified IgG from CIDP patients (Heininger et al., 1984). Recently, IgG antibodies that are capable of blocking neuromuscular transmission were identified in CIDP patient serum (Buchwald, Ahangari and Toyka, unpublished observations). This neuromuscular blockade by IgG antibodies has first been observed in GBS and its variant MillerFisher syndrome (Buchwald et al., 1998, 2002).

Human peripheral nerve myelin contains acidic glycosphingolipids such as sulfated glucuronyl paragloboside (SGPG) and sulfated glucuronyl lactosaminyl paragloboside (SGLPG) (Quarles, 1997; Willison and Yuki, 2002). One study found elevated IgM anti-SGPG antibody titers in six of nine patients (67%) with CIDP (Yuki et al., 1996). In earlier studies, antibodies to a variety of glycolipid antigens were described, including LM1, GM1, GD1a (reviewed in Willison and Yuki, 2002). Another candidate antigen is the HNK-1 carbohydrate epitope, which is common to some glycolipids and other cell adhesion molecules. More evidence for the role of GM1 as target has been given when rabbits were immunized with a ganglioside mixture (Yuki et al., 2001): all of them developed high anti-GM1 IgG antibody titers, flaccid limb weakness of acute onset, and a monophasic illness course. Pathological findings for the peripheral nerves showed predominant Wallerian-like degeneration, with neither lymphocytic infiltration nor demyelination. IgG was deposited on the axons of the anterior roots, and GM1 was proved to be present on the axons of peripheral nerves. Sensitization with purified GM1 also induced axonal neuropathy, indicating that GM1 was the immunogen in the mixture and explaining the association of AMAN with anti-GM1 reactivity.

EAN can be induced by immunization of an animal with myelin proteins such as P2 basic protein, P_0 glycoprotein, and peripheral myelin protein 22 (PMP22) (see review in Gold et al., 2005). Gabriel et al (2000) investigated whether PMP22 may be important in inducing human inflammatory neuropathy. The sera of patients with GBS, CIDP, other neuropathies (ONP), and normal controls were evaluated for IgM and IgG antibodies against PMP22. Antibodies were detected in 52%

of patients with GBS, 35% with CIDP, and 3% with ONP; no antibodies were detected in normal controls (Gabriel et al., 2000). Furthermore, Ritz et al (Ritz et al., 2000) reported PMP22 antibodies in three of six (50%) CIDP patients. In contrast, Kwa et al. (2001) reported the absence of these antibodies in sera from 24 patients with CIDP. The discrepancy among the results of these studies may be due to differences in the PMP22 antigen used to test the sera. When linear peptide epitopes of PMP22 and purified PMP22 protein from overexpression in *E. coli* were used, a higher percentage of patient sera showed reactivity. However, when PMP22 protein was expressed in mammalian cells under native conditions, the sera failed to show any reactivity.

In any chronic autoimmune inflammatory condition, several antigens may be involved via epitope spreading (Lehmann et al., 1992), making it difficult to identify the culprit antigen in an individual patient. Moreover, antibodies display a variety of affinities and avidities and some may activate the complement cascade while others may not (Janeway et al., 2001). Therefore, it is not easy to define the pathogenic role of individual antigen binding specificities. Given the heterogeneity of GBS and CIDP, it is likely that different antibodies are sequentially or even collectively involved in the pathogenesis of this disease.

Recent evidence demonstrates that antibodies against myelin protein zero (P_0), a major structural protein of myelin, may play a role in CIDP (Yan et al., 2001). There is indirect evidence that P_0 may have immunologic relevance. In an experimental study of heterozygous P_0 knockout mice, an animal model of Charcot-Marie-Tooth disease, Schmid et al., 2000 showed that (1) myelin degeneration and impairment in nerve conduction were attenuated when the immune system was impaired, and (2) T cells isolated from these mutant mice demonstrated enhanced reactivity to myelin proteins such as P_0 and P2. P_0, an adhesive cell-surface molecule of the Ig superfamily and the most abundant protein of myelin on the peripheral nerves, has multiple functions in the development and maintenance of myelin (Martini et al., 1995). An increase in macrophages with a subsequent increase in T cells was observed within the nerves of heterozygous P_0 knockout mice. It was hypothesized that a reduction in P_0 could result in an instability of myelin, which then could lead to an attraction of macrophages and a macrophage-mediated attraction of T lymphocytes because these were not tolerized due to lack of myelin protein P0 in thymus (Visan et al., 2004). Activation of autoimmune T cells by antigen-presenting macrophages could lead to cellular and humoral immune reactions, which ultimately could result in demyelination (Schmid et al., 2000).

Yan et al., 2001 studied the sera of 21 CIDP patients for antimyelin activity using immunofluorescence and for binding to myelin proteins using Western blot analysis. Results showed that the sera of six patients (29%) contained anti-P_0 IgG antibodies, and four of these caused conduction block and demyelination when injected intraneurally into experimental animals. These results suggest that P_0 is an autoantigen in some patients with CIDP. More work is needed to ultimately define the precise nature of circulating antibodies and their pathogenic role in CIDP.

Summary

There are similarities in the importance of axonal damage in immune neuropathies and MS. Probably in both cases destructive immune responses are involved, presumably by a direct inflammatory assault. They can be counteracted by endogenous survival factors. As yet only some of them have been identified, mostly belonging to the group of neurotrophins or neurotrophic factors. Their therapeutic application is limited because of their absorption in peripheral tissues such as liver, ultimately preventing them from access to the target organ. With the further progress in molecular and cellular gene therapy we may be able to overcome these obstacles. Currently available immune therapies such as plasmapheresis and intravenous immunoglobulins probably limit the damage by reducing the primary inflammatory assault, or by reducing titers of or neutralizing pathogenic autoantibodies.

Review Questions/Problems

1. **Describe clinical differences between GBS and CIDP.**
2. **Briefly summarize the different experimental neuritis models in relation to the human diseases.**
3. **Describe the key immunological factors contributing to axonal damage in GBS and CIDP.**
4. **Which factors contribute to chronification of the autoimmune inflammation?**
5. **Which potentially pathogenic humoral elements in CIDP patients have been identified?**
6. **Which of the following myelin proteins is not involved in experimental neuritis?**

   a. MAG
   b. P2 Protein
   c. P0 Protein
   d. MOG
   e. PMP22

7. **Which neurotrophic cytokine has been identified as lesion factor in the PNS?**

   a. NGF
   b. LIF
   c. NT-3
   d. CNTF
   e. GMCSF

8. Which is the correct explanation for "onion bulb" formation in CIDP?

a. Macrophages phagocytose myelin debris
b. Abundant T-cell apoptosis
c. Repetitive De- and Remyelination with Schwann cell proliferation
d. Nitric oxide release
e. Genetic myelin deficiency

9. Which of the following infectious agents is linked to axonal damage and GBS?

a. *C.jejuni lior O4*
b. Neisseria meninigidis
c. Listeria monocytogenes
d. Proteus mirabilis
e. *C.jejuni* serotype Penner 19

10. The influence of the immune system on disease progression in genetic myelin disorders.

a. Has been excluded
b. Has been clearly shown in experimental models
c. Has immediate therapeutic implications
d. Is merely speculative
e. Has been confirmed only in the Lewis rat

References

Bosboom WM, Van den Berg LH, Franssen H, Giesbergen PC, Flach HZ, van Putten AM, Veldman H, Wokke JH (2001a) Diagnostic value of sural nerve demyelination in chronic inflammatory demyelinating polyneuropathy. Brain 124:2427–2438.

Bosboom WMJ, Van den Berg LH, Mollee I, Sasker LD, Jansen J, Wokke JHJ, Logtenberg T (2001b) Sural nerve T-cell receptor V beta gene utilization in chronic inflammatory demyelinating polyneuropathy and vasculitic neuropathy. Neurology 56:74–81.

Buchwald B, Toyka KV, Zielasek J, Weishaupt A, Schweiger S, Dudel J (1998) Neuromuscular blockade by IgG antibodies from patients with Guillain-Barre syndrome: A macro-patch-clamp study. Ann Neurol 44:913–922.

Buchwald B, Ahangari R, Weishaupt A, Toyka KV (2002) Intravenous immunoglobulins neutralize blocking antibodies in Guillain-Barre syndrome. Ann Neurol 51:673–680.

Butzkueven H, Zhang JG, Hanninen MS, Hochrein H, Chionh F, Shipham KA, Emery B, Turnley AM, Petratos S, Ernst M, Bartlett PF, Kilpatrick TJ (2002) LIF receptor signaling limits immune-mediated demyelination by enhancing oligodendrocyte survival. Nat Med 8:613–619.

Dalakas MC (1999) Advances in chronic inflammatory demyelinating polyneuropathy: Disease variants and inflammatory response mediators and modifiers. Curr Opin Neurol 12:403–409.

Dittrich F, Thoenen H, Sendtner M (1994) Ciliary neurotrophic factor: Pharmacokinetics and acute-phase response in the rat. Ann Neurol 35:151–163.

Enders U, Karch H, Toyka KV, Michels M, Zielasek J, Pette M, Heesemann J, Hartung HP (1993) The spectrum of immune responses to Campylobacter jejuni and glycoconjugates in Guillain-Barre syndrome and in other neuroimmunological disorders. Ann Neurol 34:136–144.

Felts PA, Smith KJ, Gregson NA, Hughes RAC (2002) Brain-derived neurotrophic factor in experimental autoimmune neuritis. J Neuroimmunol 124:62–69.

Gabriel CM, Gregson NA, Hughes RA (2000) Anti-PMP22 antibodies in patients with inflammatory neuropathy. J Neuroimmunol 104:139–146.

Giess R, Maurer M, Linker R, Gold R, Warmuth-Metz M, Toyka KV, Sendtner M, Rieckmann P (2002) Association of a null mutation in the CNTF gene with early onset of multiple sclerosis. Arch Neurol 59:407–409.

Gold R, Zielasek J, Schroder JM, Sellhaus B, Cedarbaum J, Hartung HP, Sendtner M, Toyka KV (1996) Treatment with ciliary neurotrophic factor does not improve regeneration in experimental autoimmune neuritis of the Lewis rat. Muscle Nerve 19:1177–1180.

Gold R, Dalakas MC, Toyka KV (2003) Immunotherapy in autoimmune neuromuscular disorders. Lancet Neurol 2:22–32.

Gold R, Stoll G, Kieseier BC, Hartung HP, Toyka KV (2005) Experimental autoimmune neuritis. In: Peripheral Neuropathy (Dyck PJ, Thomas PK, eds), pp 609–634. Philadelphia: Elsevier Saunders.

Hadden RD, Cornblath DR, Hughes RA, Zielasek J, Hartung HP, Toyka KV, Swan AV (1998) Electrophysiological classification of Guillain-Barre syndrome: Clinical associations and outcome. Plasma exchange/sandoglobulin Guillain-Barre syndrome trial group. Ann Neurol 44:780–788.

Hadden RDM, Karch H, Hartung HP, Zielasek J, Weissbrich B, Schubert J, Weishaupt A, Cornblath DR, Swan AV, Hughes RAC, Toyka KV (2001) Preceding infections, immune factors, and outcome in Guillain-Barre syndrome. Neurology 56:758–765.

Hartung H-P, Heininger K, Schäfer B, Fierz W, Toyka KV (1988) Immune mechanisms in inflammatory polyneuropathy. Ann N Y Acad Sci 540:122–161.

Heininger K, Liebert UG, Toyka KV, Haneveld FT, Schwendemann G, Kolb-Bachofen V, Ross HG, Cleveland S, Besinger UA, Gibbels E (1984) Chronic inflammatory polyneuropathy. Reduction of nerve conduction velocities in monkeys by systemic passive transfer of immunoglobulin G. J Neurol Sci 66:1–14.

Heininger K, Stoll G, Linington C, Toyka KV, Wekerle H (1986) Conduction failure and nerve conduction slowing in experimental allergic neuritis induced by P2-specific T-cell lines. Ann Neurol 19:44–49.

Ho TW, Mishu B, Li CY, Gao CY, Cornblath DR, Griffin JW, Asbury AK, Blaser MJ, McKhann GM (1995) Guillain-Barre syndrome in northern China. Relationship to Campylobacter jejuni infection and anti-glycolipid antibodies. Brain 118:597–605.

Hughes RAC, Cornblath DR (2005) Guillain-Barre syndrome. Lancet 366:1653–1666.

Janeway CA, Travers P, Walport M, Shlomchik MJ (2001) Immunobiology—the Immune System in Health and Disease. New York: Churchill Livingstone.

Kahl KG, Zielasek J, Uttenthal LO, Rodrigo J, Toyka KV, Schmidt HHHW (2003) Protective role of the cytokine-inducible isoform of nitric oxide synthase induction and nitrosative stress in experimental autoimmune encephalomyelitis of the DA rat. J Neurosci Res 73:198–205.

Kapoor R, Davies M, Blaker PA, Hall SM, Smith KJ (2003) Blockers of sodium and calcium entry protect axons from nitric oxide-mediated degeneration. Ann Neurol 53:174–180.

Kieseier BC, Kiefer R, Gold R, Hemmer B, Willison HJ, Hartung HP (2004) Advances in understanding and treatment of immune-mediated disorders of the peripheral nervous system. Muscle & Nerve 30:131–156.

Krendel DA, Parks HP, Anthony DC, St Clair MB, Graham DG (1989) Sural nerve biopsy in chronic inflammatory demyelinating polyradiculoneuropathy. Muscle Nerve 12:257–264.

Kuhlmann T, Lingfeld G, Bitsch A, Schuchardt J, Bruck W (2002) Acute axonal damage in multiple sclerosis is most extensive in early disease stages and decreases over time. Brain 125:2202–2212.

Kwa MSG, van Schaik IN, Brand A, Baas F, Vermeulen M (2001) Investigation of serum response to PMP22, connexin 32 and P-0 in inflammatory neuropathies. J Neuroimmunol 116:220–225.

Lehmann PV, Forsthuber T, Miller A, Sercarz EE (1992) Spreading of T-cell autoimmunity to cryptic determinants of an autoantigen. Nature 358:155–157.

Linker RA, Maurer M, Gaupp S, Martini R, Holtmann B, Giess R, Rieckmann P, Lassmann H, Toyka KV, Sendtner M, Gold R (2002) CNTF is a major protective factor in demyelinating CNS disease: A neurotrophic cytokine as modulator in neuroinflammation. Nat Med 8:620–624.

Martini R, Toyka KV (2004) Immune-mediated components of hereditary demyelinating neuropathies: Lessons from animal models and patients. Lancet Neurol 3:457–465.

Martini R, Zielasek J, Toyka KV, Giese KP, Schachner M (1995) Protein zero (P0)-deficient mice show myelin degeneration in peripheral nerves characteristic of inherited human neuropathies. Nat Genet 11:281–286.

McKhann GM, Cornblath DR, Ho T, Li CY, Bai AY, Wu HS, Yei QF, Zhang WC, Zhaori Z, Jiang Z, Griffin JW, Asbury AK (1991) Clinical and electrophysiological aspects of acute paralytic disease of children and young adults in northern China. Lancet 338:593–597.

Merkies IS, Schmitz PI, Samijn JP, van der Meche FG, van Doorn PA (1999) Fatigue in immune-mediated polyneuropathies. European Inflammatory Neuropathy Cause and Treatment (INCAT) Group. Neurology 53:1648–1654.

Miletic H, Utermohlen O, Wedekind C, Hermann M, Stenzel W, Lassmann H, Schulter D, Deckert M (2005) P0(106–125) is a neuritogenic epitope of the peripheral myelin protein P0 and induces autoimmune neuritis in C57BL/6 mice. J Neuropathol Exp Neurol 64:66–73.

Neumann H, Medana IM, Bauer J, Lassmann H (2002) Cytotoxic T lymphocytes in autoimmune and degenerative CNS diseases. Trends Neurosci 25:313–319.

Nobbio L, Gherardi G, Vigo T, Passalacqua M, Melloni E, Abbruzzese M, Mancardi G, Nave KA, Schenone A (2006) Axonal damage and demyelination in long-term dorsal root ganglia cultures from a rat model of Charcot-Marie-Tooth type 1A disease. Eur J Neurosci 23:1445–1452.

Quarles RH (1997) Glycoproteins of myelin sheaths. J Mol Neurosci 8:1–12.

Ritz MF, Lechner-Scott J, Scott RJ, Fuhr P, Malik N, Erne B, Taylor V, Suter U, Schaeren-Wiemers N, Steck AJ (2000) Characterisation of autoantibodies to peripheral myelin protein 22 in patients with hereditary and acquired neuropathies. J Neuroimmunol 104:155–163.

Saperstein DS, Katz JS, Amato AA, Barohn RJ (2001) Clinical spectrum of chronic acquired demyelinating polyneuropathies. Muscle & Nerve 24:311–324.

Schmid CD, Stienekemeier M, Oehen S, Bootz F, Zielasek J, Gold R, Toyka KV, Schachner M, Martini R (2000) Immune deficiency in mouse models for inherited peripheral neuropathies leads to improved myelin maintenance. J Neurosci 20:729–735.

Schmidt B, Toyka KV, Kiefer R, Full J, Hartung HP, Pollard J (1996) Inflammatory infiltrates in sural nerve biopsies in Guillain-Barre syndrome and chronic inflammatory demyelinating neuropathy. Muscle Nerve 19:474–487.

Sendtner M, Stöckli KA, Thoenen H (1992) Synthesis and localization of ciliary neurotrophic factor in the sciatic nerve of the adult rat after lesion and during regeneration. J Cell Biol 118:139–148.

Smith KJ, Kapoor R, Hall SM, Davies M (2001) Electrically active axons degenerate when exposed to nitric oxide. Ann Neurol 49:470–476.

Sommer C, Koch S, Lammens M, Gabreels-Festen A, Stoll G, Toyka KV (2005) Macrophage clustering as a diagnostic marker in sural nerve biopsies of patients with CIDP. Neurology 65:1924–1929.

Trapp BD, Peterson J, Ransohoff RM, Rudick R, Mörk S, Bö L (1998) Axonal transection in the lesions of multiple sclerosis. N Engl J Med 338:278–285.

Visan L, Visan IA, Weishaupt A, Hofstetter HH, Toyka KV, Hunig T, Gold R (2004) Tolerance induction by intrathymic expression of P0. J Immunol 172:1364–1370.

Willison HJ, Yuki N (2002) Peripheral neuropathies and anti-glycolipid antibodies. Brain 125:2591–2625.

Yamamoto M, Ito Y, Mitsuma N, Li M, Hattori N, Sobue G (2002) Parallel expression of neurotrophic factors and their receptors in chronic inflammatory demyelinating polyneuropathy. Muscle & Nerve 25:601–604.

Yan WX, Archelos JJ, Hartung HP, Pollard JD (2001) P0 protein is a target antigen in chronic inflammatory demyelinating polyradiculoneuropathy. Ann Neurol 50:286–292.

Yuki N, Tagawa Y, Handa S (1996) Autoantibodies to peripheral nerve glycosphingolipids SPG, SLPG, and SGPG in Guillain-Barre syndrome and chronic inflammatory demyelinating polyneuropathy. J Neuroimmunol 70:1–6.

Yuki N, Yamada M, Koga M, Odaka M, Susuki K, Tagawa Y, Ueda S, Kasama T, Ohnishi A, Hayashi S, Takahashi H, Kamijo M, Hirata K (2001) Animal model of axonal Guillain-Barre syndrome induced by sensitization with GM1 ganglioside. Ann Neurol 49:712–720.

Zettl UK, Gold R, Toyka KV, Hartung HP (1995) Intravenous glucocorticosteroid treatment augments apoptosis of inflammatory T cells in experimental autoimmune neuritis (EAN) of the Lewis rat. J Neuropathol Exp Neurol 54:540–547.

20
Guillain-Barré Syndrome

Helmar C. Lehmann and Kazim A. Sheikh

Keywords Acute flaccid paralysis; Anti-ganglioside antibodies; Campylobacter jejuni; Experimental allergic/autoimmune neuritis (EAN); Gangliosides; Immune neuropathies; Molecular mimicry; T-cells

20.1. Introduction

The term Guillain-Barré syndrome (GBS) is used to denote a group of clinically and pathophysiologically heterogeneous disorders of peripheral nerves that are characterized by acute onset, monophasic course, and potential for substantial recovery, which is expedited by two immunomodulatory therapies. After the near-eradication of polio, GBS is the commonest cause of acute flaccid paralysis worldwide. Current evidence supports the concept that GBS is likely of autoimmune origin. Postinfectious molecular mimicry (autoimmunity) is the currently favored dominant theme in the pathogenesis of GBS. This concept implies antigens shared between the infectious agents and peripheral nerves so that an infection results in an immune response to these crossreactive antigens carried by the organism. The immune response triggered by infection then mediates injury to the peripheral nerves. The results from both clinical and experimental studies on some forms of GBS support this concept. This chapter outlines clinical and pathophysiological features of major variants of GBS and highlights the evidence that supports the hypothesis of postinfectious molecular mimicry in this group of disorders.

20.1.1. Historical Background

The history of GBS is inseparably linked to the seminal paper of Georges Guillain (1876–1961), Jean-Alexandre Barré (1880–1967), and André Strohl (1887–1977) (Guillain et al., 1916). During the weekly meeting of the "*Société médicale des hôpitaux de Paris*" in 1916, they presented the case history of two soldiers of the VI French army, who developed flaccid sensorimotor paralysis. Both patients, without any history of preceding infection, recovered completely after a few weeks. Guillain, Barré, and Strohl were the first to describe elevated spinal fluid protein without cells *"la dissociation cytoalbuminique."*

The presentation of Guillain, Barré, and Strohl was not the first clinical description of the disease. Half a century earlier in 1859, Jean Baptiste Octave Landry (1826–1865) described the case of an 43-year-old man who developed an acute ascending paralysis and died within a few days (Landry, 1859). Landry noted a preceding pulmonary infection in his patient. He brought attention to a disease, which he called *"paralysie ascendante aïgue"* by further reviewing four of his own and five other cases in the contemporary literature. In contrast to the description of Guillain and Barré, two patients of Landry's case series died from the disease. Therefore Landry's acute ascending paralysis had been associated with poor prognosis, whereas Guillain-Barré syndrome was considered to have a mild clinical course with almost complete recovery. Landry noticed preceding infections in two of his cases. Autopsy of his two cases did not show any pathological changes in the central nervous system and this led him to conclude that this disease affects peripheral nerves. In the same year Adolph Kussmaul (1822–1902) (Kussmaul, 1859) reported two cases of a deadly ascending paraplegia. Like Landry, he did not find toxic or anatomical explanation on autopsy. The affection of the peripheral nerve as underlying cause for the disease, was first documented in 1864 when L. Duménil described a case of *"paralysie ascendante aïgue,"* in which he found atrophy of peripheral nerve roots in autopsy material (reviewed in Schott, 1982).

It is noteworthy that Landry's ascending paralysis was a purely clinical diagnosis. It is likely that the early descriptions of Landry and others cover a range of different entities, including atypical forms of polio and infectious neuropathies. Therefore, the application of the lumbar puncture technique by Guillain, Barré, and Strohl was a landmark for classifying and diagnosis of the disease, since it helped to distinguish it from other entities. In 1927 the term Guillain-Barré syndrome was introduced for the first time and become the preferred eponym thereafter (Draganesco and Claudian, 1927). The name of Strohl, who was a medical student in 1916, disappeared unfairly from the original work, although he contributed

T. Ikezu and H.E. Gendelman (eds.), *Neuroimmune Pharmacology*.
© Springer 2008

substantially by his electrophysiological examinations of the two patients.

A fundamental step towards an understanding of the pathogenesis of GBS was made by observations of inflammatory infiltrates and demyelination in peripheral nerves. These pathological studies on autopsy material led to the assumption that GBS was a single pathophysiological entity synonymous with acute inflammatory demyelinating polyradiculoneuropathy (AIDP) (Haymaker and Kernohan, 1949; Asbury et al., 1969). At the same time Waksman and Adams described the induction of an allergic neuritis in rabbits by immunization with peripheral nervous tissue (Waksman and Adams, 1955). Subsequently this model of an experimental allergic/autoimmune neuritis (EAN) was further expanded to other species, and was used as an *in vivo* model for GBS.

The Fisher syndrome (FS) is named after Charles Miller Fisher, a Canadian neurologist. In 1956, Miller Fisher described three patients with acute external ophthalmoplegia, absent tendon reflexes, and ataxia, who recovered spontaneously (Fisher, 1956). Some cases start as FS but subsequently develop weakness. Finally, work over last two decades indicates that some forms of GBS lack features of demyelination and have pathophysiology that is consistent with axonal injury; these cases are termed axonal GBS.

20.1.2. Classification of GBS Variants

Clinical features and/or electrodiagnostic examination provide a framework for classifying GBS into different variants. GBS can be broadly divided into major and minor variants. The major variants typically have muscle weakness or motor neuropathy as the dominant manifestation; they are further divided into demyelinating and axonal variants on the basis of the predominant pathophysiologic process of nerve fiber injury as determined by electrodiagnostic testing, namely, primary demyelination or primary axonal degeneration. Axonal variants are further subclassified according to the fiber type affected. Minor variants are classified based on the constellation of clinical symptoms and not by electrodiagnostic findings. The constellation of symptoms in minor variants is taken to imply regional localization of the pathophysiologic process. On the basis of this schema a simple classification of GBS is proposed in Table 20.1.

TABLE 20.1. Classification of GBS.

Major variants
Demyelinating
AIDP
Axonal
AMAN
AMSAN
AIDP with secondary axonal degeneration
Minor variants
Fisher syndrome
Sensory ataxic variant
Acute idiopathic autonomic neuropathy

The demyelinating form of the disease is termed acute inflammatory demyelinating polyneuropathy (AIDP). Two axonal forms of GBS (Feasby et al., 1986; Yuki et al., 1990; McKhann et al., 1993), however, are now widely recognized on the basis of nerve fiber type affected: acute motor axonal neuropathy (AMAN), the more common form of axonal GBS, is distinguished by nearly pure motor axonal injury; the less common variant is acute motor-sensory axonal neuropathy (AMSAN), characterized by degeneration of both motor and sensory axons. It has been postulated that AMAN and AMSAN represent a pathologic spectrum and that AMSAN actually represents a more severe form of AMAN (Griffin et al., 1996a). Fisher syndrome (FS) is a minor GBS variant characterized by gait disturbance (ataxia), areflexia, and ophthalmoplegia. Other rarer forms without significant motor weakness that may be included under the term GBS include a predominantly sensory variant and acute idiopathic autonomic neuropathy or acute pandysautonomia.

20.1.3. Epidemiology

The incidence of GBS, 1–1.5 per 100,000 (Rees et al., 1998) is surprisingly similar throughout the world despite different infection rates in various geographical regions. Men are slightly more affected than women (Hughes and Cornblath, 2005) and incidence increases with age (Bogliun and Beghi, 2004). Demyelinating forms of the disease are most prevalent in the United States, Europe (Guillain-Barré syndrome variants in Emilia-Romagna, 1998; Hadden et al., 1998a), and most of the developed world, accounting for over 90% of patients. Compared to western world, the incidence of axonal forms of GBS is higher in northern China, Japan, Mexico and other developing countries but are also, less frequently, seen in northern America and Europe (McKhann et al., 1993; Ogawara et al., 2000; Ramos-Alvarez et al., 1969). Notably, in northern China there is a clear seasonal pattern, with peak incidence in summer months, and a predilection of disease in children and villages (McKhann et al., 1991, 1993). The Fisher variant probably represents 5% of cases of GBS and has a similar incidence world-wide.

20.1.4. Clinical Features

The diagnosis of GBS is primarily clinical. In the AIDP variant the majority of cases have some sensory symptoms or paresthesias at the onset of the disease; however, abnormalities on sensory examination are less frequent. Pain, particularly low back, buttock, or thigh pain, is an early symptom in approximately 50% of patients. Subsequently the clinical picture is dominated by weakness often progressing to paralysis. Muscle weakness may begin in the lower limbs and ascend upwards, characteristically involving both proximal and distal muscles. Respiratory muscles can be involved in up to one-third of the hospitalized patients. Complete or partial loss of reflexes is seen in almost all patients. Cranial nerve involvement is seen in two thirds of cases, most commonly causing facial weakness and difficulties of eye closure, ophthalmoplegia, difficulty swallowing or altered taste. Autonomic manifestations include reduced sinus arrhythmia, sinus tachycardia, arrhythmias, labile blood pressure, orthostatic hypotension, abnormal sweating, and pupillary abnormalities. Respiratory and bulbar weakness and autonomic instability are the major cause of morbidity and mortality in GBS.

Clinically, it is difficult to distinguish between axonal and demyelinating forms of GBS. Electrodiagnostic testing is essential to differentiate these variants. AMAN has exclusively motor findings, with weakness typically beginning in the legs, but in some individuals affecting arms or cranial muscles initially (McKhann et al., 1993). Tendon reflexes are preserved until weakness is severe enough to preclude phasic muscle contraction. This probably reflects sparing of muscle afferent fibers. The incidence of dysautonomia has not been systematically examined in axonal cases but it was seen in a small proportion of cases.

Minor variants without significant motor weakness include FS characterized by ataxia, areflexia, and internal and external ophthalmoplegia. Other rarer forms of the disease without significant muscle weakness include a pure sensory variant and acute autonomic neuropathy.

Differential diagnoses include structural lesions, such as myelopathy and infections including HIV, lyme disease, CMV, rarely paralytic rabies (Sheikh et al., 2005), and, in endemic areas, polio. In children botulism should be considered. Toxic and metabolic conditions such as tick bite and porphyria can also mimic GBS. In the intensive care setting, critical illness neuropathy and quadriplegic myopathy may be clinically indistinguishable from GBS.

20.1.4.1. *Investigations*

The main aim of the investigations is to exclude other conditions that can mimic GBS and to confirm the diagnosis. Electrodiagnostic testing is the most critical investigation in the evaluation of patients with GBS; it can potentially provide support for the clinical diagnosis and useful prognostic information. Nerve conduction studies (NCS) are abnormal to some extent in most patients with GBS, but normal studies in the first week do not exclude this diagnosis. Gordon and Wilbourn reported the changes in NCS seen in the first week (Gordon and Wilbourn, 2001). Changes in F wave latencies are probably the most common abnormality early in disease. In AIDP, typical electrophysiological features of demyelination such as slow motor conduction velocities, prolonged distal motor latencies, and partial motor conduction block may not be present until the second or third week. Reduced or absent sensory nerve action potentials (SNAPs) and slowing of sensory conduction velocity are common. In contrast to AIDP, conduction times like distal motor latencies and motor conduction velocities are relatively preserved in the axonal forms, but reduced compound motor action potential (CMAP) amplitudes are characteristic. Inexcitable motor nerves can be seen in more severe cases and in patients

TABLE 20.2. Anti-glycolipid antibodies in different GBS variants.

Variants	Anti-glycolipid antibody
Fisher syndrome	GQ1b/GT1a
AMAN	GD1a, GM1, GM1b, *GalNAc*-GD1a
AIDP	GM1, Asialo-GM1, GD1b, GM2, LM1, GD2, GalC, Forssman antigen
Sensory ataxic variants	GD1b and structurally related gangliosides

with the AMSAN variant. Sensory conductions are usually normal in AMAN, but SNAPs are decreased or absent in the AMSAN variant. Examination of cerebrospinal fluid (CSF) is most useful in excluding other differential diagnoses, particularly infections. Typically, CSF protein is increased in 80–90% of GBS cases without significant pleocytosis. A mild increase in mononuclear cells can be seen in up to 10% of patients. Significant pleocytosis raises the possibility of infection such as Lyme disease or HIV. Serological studies for anti-ganglioside antibodies can be considered for the diagnosis in incomplete forms and unusual variants of GBS, particularly when nerve conductions and CSF are normal. The role of anti-ganglioside serology in routine diagnosis and clinical decision-making remains to be established. Table 20.2 summarizes the common anti-glycolipid antibodies reported in association with various forms of GBS.

20.1.4.2. *Clinical Course*

GBS may progress up to 4 weeks with a nadir being reached within 2–3 weeks in a majority of patients. Recovery usually begins within 2–4 weeks of this nadir, but can be delayed for several months. About one-half of patients become chair- or bed-bound, one-third require intensive care admission, and one-quarter mechanical ventilation (Winer et al., 1988; Rees et al., 1998; Hughes and Cornblath, 2005). Functional recovery is a rule and occurs in a majority of patients over 6–12 months; however 20–30% of patients are left with significant disability and about 10% require assistance with walking. The mortality rate ranges between 3 and 8%, and most deaths are attributed to cardiac arrest due to autonomic disturbance, respiratory failure or infection, or pulmonary embolism.

20.1.4.3. *Prognosis*

The extent and location of axonal injury are the two most important determinants of prognosis after an episode of GBS. Residual disability almost always indicates axonal degeneration. Axon regeneration is required for restoration of function. In GBS, there is characteristically significant pathology in spinal roots and proximal nerves and as peripheral axon regeneration advances at a rate of approximately 1 in./month, recovery is slow and often incomplete. Although peripheral axons have the capacity to regenerate, experimental evidence indicates that: (a) the efficiency of axonal regeneration decreases over time after injury (Fu and Gordon, 1995a); (b) the denervated distal segment of peripheral nerves can optimally support axon regeneration only for a limited time (Fu and Gordon, 1995b);

and (c) the efficiency of reinnervation of original pathways and targets (pathfinding) decreases with advanced age (Le et al., 2001). Poor recovery after axonal GBS is not always the rule; exceptions include children with electrical features of acute denervation, cases with distal axonal degeneration where regeneration is needed over only a short distance, or patients with reversible axonal conduction failure. Poor prognostic factors include advanced age, ventilator dependence, preceding gastrointestinal infection, rapid progression from onset to nadir, severe motor involvement, and electrodiagnostic evidence of extensive axonal injury.

20.1.5. Pathology

Pathology of demyelinating and axonal forms presented below is well established. Pathological changes in minor variants of GBS are not well-characterized.

AIDP: The pathological changes in AIDP have been extensively characterized in postmortem studies. The most prominent feature in AIDP is marked segmental demyelination, which can be found throughout the length of all peripheral nerves including the mixed spinal roots and even the distal terminal nerves (Hall et al., 1992; Massaro et al., 1998). In areas of severe demyelination, signs of secondary axonal degeneration can be observed. Another pathological hallmark is the presence of inflammatory infiltrates, especially in the spinal roots and proximal nerves, which contain T-lymphocytes and macrophages (Asbury et al., 1969; Prineas, 1981). Macrophage-mediated myelin stripping (ingestion/breakdown) is characteristic. Lymphocytic infiltration can be minimal, however, sometimes it may not occur; moreover, localization may differ markedly (Cornblath et al., 1990; Honavar et al., 1991). In a recent postmortem study, inflammatory infiltrates have also been detected in the spinal cord of GBS patients, indicating that there may be subclinical inflammation of CNS structures (Muller et al., 2003). In some cases of AIDP there is breakdown of blood-nerve barrier with deposition of activated complement products on Schwann cells, suggesting a role for antibody-mediated immune injury (Hafer-Macko et al., 1996b). These pathological observations implicate T-cells or autoimmune antibodies in inducing peripheral nerve demyelination in individual cases. Further, T-cell and antibody-mediated immune injury can predominate or act synergistically in an individual case.

Axonal variants: Unlike AIDP, axonal variants of GBS show primary axonal injury without substantial T-cell inflammation or demyelination (Griffin et al., 1995, 1996a). In AMAN, axonal degeneration predominantly involves the motor axons, whereas in AMSAN sensory axons are also involved. It has been postulated that AMAN and AMSAN represent a pathologic spectrum and that AMSAN actually represents a more severe form of AMAN.

In AMAN, deposition of IgG and complement can be detected on the nodal and internodal axolemma (Hafer-Macko et al., 1996a). Macrophages are closely associated with the axons at the nodes and these cells then extend into internodal regions where they surround axons, without disturbing/disrupting the overlying myelin, eventually leading to axon degeneration (Griffin et al., 1995, 1996a). The findings of IgG deposition on axolemma and periaxonal location of macrophages strongly imply that the potential target antigen(s) are expressed on the surface of (motor) axons. Current pathogenetic concepts assume that antibody binding to motor axolemma leads to activation of complement, recruitment of macrophages instead of macrophages, and subsequent degeneration of axons.

Pathologic examination of early cases has shown that at the onset the pathological alterations in AMAN are mainly restricted to the nodes of Ranvier of motor fibers in the ventral roots (Griffin et al., 1996b). In these cases nerve fibers still appear normal except that the nodal gap is lengthened. It is believed that this nodal lengthening is sufficient to cause failure of transmission of nerve impulses and profound clinical weakness. Motor nerve terminals are another site susceptible to injury in AMAN, as indicated by degeneration of intramuscular nerves on muscle biopsy (Ho et al., 1997). These observations suggest that nodes of Ranvier and motor nerve terminals are two sites that are susceptible to injury in AMAN.

In summary, the pathological and immunopathological findings in AMAN suggest that antibody-mediated injury directed against axonal antigens plays a prominent role in the pathogenesis of this disorder.

20.1.6. Treatment

(a) *Supportive care* is the mainstay of medical management in patients with GBS, despite the availability of immunomodulating therapies like plasma exchange and (PE) and intravenous immunoglobulins (IVIg). All patients with GBS should be admitted to a hospital with an ICU experienced in GBS care. The major risks are complications arising from weakness of respiratory and bulbar muscles and autonomic instability. The principles of supportive management include respiratory support in patients with respiratory failure, monitoring and management of dysautonomia, measures to prevent nosocomial infections and complications of immobility, and pain management.

(b) *Immunomodulatory therapies* controlled clinical trials have shown that IVIg and PE are beneficial immunomodulatory treatments in GBS in populations where AIDP is the predominant form of the disease (reviewed in Raphael et al., 2002; Hughes et al., 2004; Hughes et al., 2001). These trials indicate that treatment with PE or IVIg should be considered for all nonambulatory adult patients with AIDP. IVIg is now the first line of treatment for most patients because it can be administered easily and patient acceptance is high. Immunomodulatory treatments should begin as soon as possible after the onset of symptoms to obtain maximal benefit and limit the extent of nerve injury. There are no controlled studies of immunomodulatory therapy in the FS and primary axonal variants of GBS, but anecdotal reports indicate that

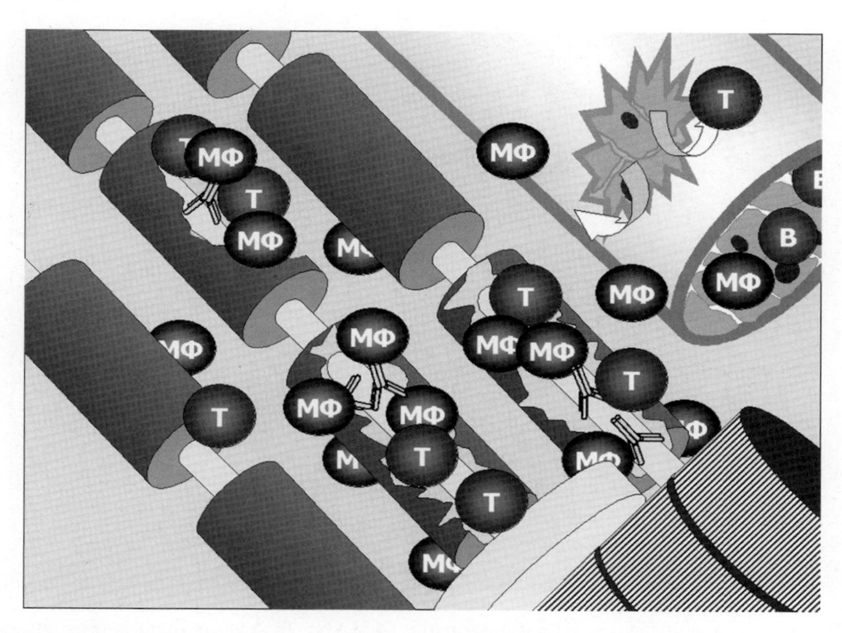

FIGURE 20.1. Hypothetical scheme of the immune response in acute inflammatory demyelinating polyradiculoneuropathy (AIDP): Inflammatory cells migrate from the systemic immune compartment through the damaged blood-nerve barrier into the endoneurium. Inflammatory infiltrates, which contain T-lymphocytes and macrophages cause marked segmental demyelination and secondary axonal degeneration *(B: B-cell; T: T-cell; MΦ: Macrophage).*

both PE and IVIg are beneficial. There is no evidence that corticosteroids are beneficial for treatment of GBS (Hughes and van der Meche, 2000). The current data indicate that the use of multiple immunomodulatory treatment modalities (PE followed by IVIg) is not superior to single therapy (PE or IVIg) for the treatment of GBS.

The precise mechanism(s) of action of PE and IVIg in GBS remain uncertain. PE removes a number of potentially pathogenic circulating factors including autoimmune antibodies, cytokines and complement, and can alter lymphocyte activation (Lehmann et al., 2006). IVIg is proposed to neutralize and inhibit production of autoantibodies, suppress antibody dependent cellular cytotoxicity, decrease natural killer cell function, down-regulate proinflammatory cytokines, and interfere with complement activation (Dalakas, 2002, 2004). It is also proposed that IVIg induces increased catabolism of immunoglobulins including autoantibodies by saturating FcRn transport receptors (Yu and Lennon, 1999).

20.2. Pathogenesis of GBS Variants

This section describes the current concepts and hypotheses about the pathogenesis of different forms of GBS. Because of pathological findings demonstrating the presence of T-cell inflammation in AIDP nerves, most of the research in this disorder has focused on T-cell-orchestrated demyelination. Experimental allergic/autoimmune neuritis (EAN) is a prototypic model of this disease that has provided useful insights into the pathogenesis of AIDP, despite concern about the relevance of this model to human disease because antigen(s) specificity of lymphocyte responses in AIDP is not well defined.

Significant progress has been made in our understanding of the pathogenesis of AMAN and FS variants of GBS. This progress is largely driven by identification of specific anti-ganglioside antibodies in AMAN and FS, which has focused research in this area on antibody-mediated pathophysiologic effects on nerve fibres. Clinical observation that *Campylobacter jejuni* infection commonly precedes GBS and demonstration of ganglioside-like antigens on this infectious organism has provided strong support to the hypothesis of post-infectious molecular mimicry as the predominant pathophysiologic mechanism in AMAN and FS variants of GBS.

The abnormalities of cellular immunity in patients with AIDP and its experimental animal model EAN are discussed to highlight pathogenetic events involved in cell-mediated demyelination. This is contrasted with role of humoral immunity in pathogenesis of AMAN by reviewing pathophysiologic effects of anti-ganglioside antibodies on corresponding antigens (gangliosides) expressed by axons. FS is also included in the discussion because pathophysiologic effects of specific anti-ganglioside antibodies associated with this disorder have also been elucidated. Figures 20.1 and 20.2 summarize the immune mechanisms invoked in AIDP and AMAN variants of GBS, respectively.

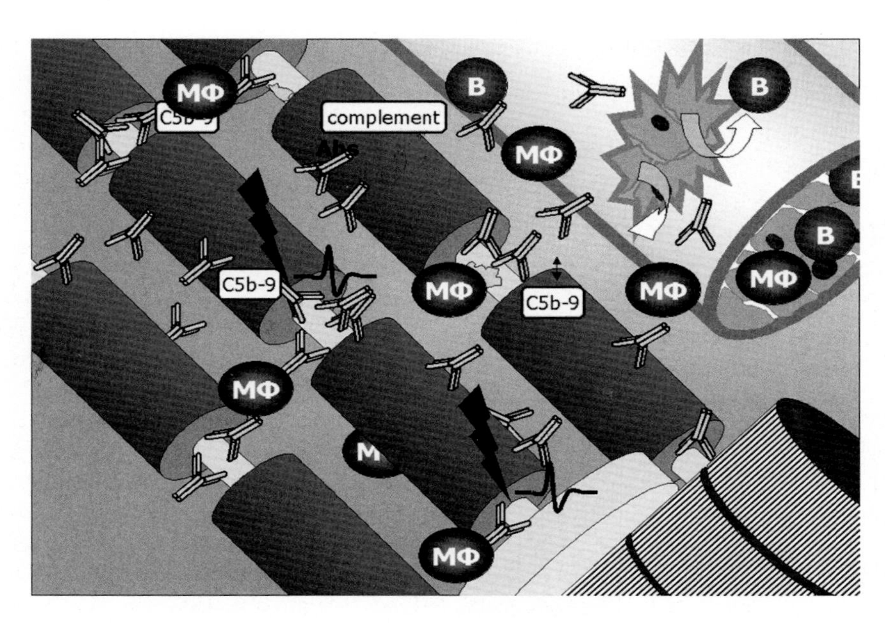

FIGURE 20.2. Proposed pathogenesis in acute motor axonal neuropathy (AMAN): In AMAN there is primary axonal injury without T-cell inflammation and demyelination. Deposition of autoantibodies and complement on the axolemma is followed by structural axonal injury or alteration of axon conduction. Macrophages within the periaxonal space contribute to the axonal damage *(B: B-cell; MΦ: Macrophage, C5b-9: complement factors).*

20.2.1. Acute Inflammatory Demyelinating Polyradiculoneuropathy (AIDP)

Most of our understandings about the cellular mechanisms in the pathogenesis of AIDP have been gained by studies of the experimental allergic/autoimmune neuritis (EAN). In particular, the important role of T-cells in the pathogenesis of EAN has paved the way for the development of hypothetical models for human AIDP. EAN can be induced in different species with whole peripheral nerve tissues or neuritogenic epitopes of peripheral nerve proteins P0, P2, and PMP22 (Hughes et al., 1999). A modified model is the adoptive transfer EAN, where specific T-cell lines against P0 or P2 are passively transferred to the animals induce neuropathy. EAN is a monophasic disease; its severity is associated with cell or antigen dosage (Hartung et al., 1996). Depletion of T-cells by thymectomy or antibodies against T-cells can protect against EAN, indicating that disease induction is dependent on the presence of T-cells (Holmdahl et al., 1985). Further, a normal function of T cells (Holmdahl et al., 1985; Hartung et al., 1987; Jung et al., 1992) and the presence of costimulatory molecules B7.1/CD80 or B7.2/CD86 and CTLA4/CD28 (Kiefer et al., 2000; Zhu et al., 2001a, b, 2003) are required for inducing and supporting EAN, which emphasizes the important role of T-cells as mediators for experimental neuritis.

T cell activation in EAN is accompanied by expression of inflammatory cytokines such as IFN-γ, TNF-α, TNF-β, IL-6 and IL-12 (Zhu et al., 1994, 1997). In sera of GBS, inflammatory cytokines detected include IFN-γ, TNF-α and IL-2, which are increased in patients and partially correlate with disease activity (Taylor and Hughes, 1989; Hartung and Toyka, 1990; Hartung et al., 1991; Exley et al., 1994; Creange et al., 1996). In contrast, levels of TGF-β1, a pleiotrophic cytokine, are lower in GBS patients than in controls (Creange et al., 1998).

Systemic immune activation in AIDP is reflected by an increased number of activated T-cells circulating in the peripheral blood during the early disease course (Taylor and Hughes, 1989; Hartung and Toyka, 1990). Many studies to identify the putative antigen of a specific T cell response in AIDP provided conflicting results. Some studies postulated that peptides from P0 and P2 might be the epitope of specific T-cell lines in GBS, but most of them failed to obtain disease specific T-cell responses against neural antigens. The CD4/CD8 ratios of T-cell populations detected in the peripheral nerves in AIDP closely resemble those in the peripheral blood. Of those, αβT-cells constitutes most of the nerve infiltrates (Cornblath et al., 1990). The usage of Vβ15 T-cell receptor gene suggest the specific T-cell activation by a common antigen or superantigen (Khalili-Shirazi et al., 1997).

In the normal human immune repertoire there are far fewer T-cells that express the γδT-cell receptors than those expressing αβT-cell receptors. Unlike αβT-cells they do not recognise antigenic peptides in the context of MHC molecules and these cells are mainly localised in epithelial barrier tissue. Interestingly, γδT-cells were isolated from sural nerve biopsies of GBS patients (Winer et al., 2002). A specific role of those γδT-cells has been postulated, as they may recognize non-protein antigens. Therefore, they could initiate an immune response to potential carbohydrate and glycolipid antigens (Winer et al., 2002). The proliferation of γδT-cells in response to *C. jejuni* lysates has been shown. Nonetheless, $\alpha\beta$T-cells or stimulatory cytokines such as IL2 are required for expansion of $\gamma\delta$T-cells (Van Rhijn et al., 2003). These studies suggest that T-cell subtypes with specificity for peptide and glycolipid antigens can invade the nerves in AIDP.

A crucial step during the process of inflammation is the migration of activated T-cells from the systemic immune compartment to the sites of inflammation. Like the CNS, the peripheral nerve represents an immunologically privileged environment, protected by an intact blood-nerve barrier. In EAN, the breakdown of the blood-nerve barrier is one of the earliest pathological changes that can be found. The upregulation of specific endothelium-binding molecules on T-cells and their specific ligands on vascular endothelium precedes the migration of T-cells through the blood-nerve barrier. E-Selectins binding sialyl Lewis antigens (Hartung et al., 2002) and VCAM-1/ ICAM-1, are both upregulated in GBS and EAN (Enders et al., 1998; Creange et al., 2001). EAN can be partially inhibited by blocking VCAM-1 or its ligand VLA-4/ α4-β1 integrin (Enders et al., 1998). E-selectins are released into the blood and can be found in elevated levels in GBS patients. Chemokines, which exert chemotactic effects on leucocytes by binding to their specific receptor, contribute to the migration of inflammatory cells (Baggiolini, 1998; Campbell et al., 1998). The differential expression of chemokines and their receptors has been studied in detail in sural nerves of patients with AIDP (Kieseier et al., 2002a). Of those, specific upregulation of CCR-2 and CCR-4 were detected in invading T-cells, whereas endothelial cells expressing the chemokine IP-10 were identified (Kieseier et al., 2002b).

Matrix metalloproteinases (MMPs), a heterogenous group of zinc-dependent endopeptidases, are assumed to contribute to the structural breakdown of the blood-brain barrier, extravasation of leukocytes, and direct demyelination. In EAN, MMP9 and MMP7 are upregulated during early disease course and correlate with disease severity (Kieseier et al., 1998). Further, the administration of MMP inhibitors may attenuate the disease (Redford et al., 1997b). Human CSF samples from patients with GBS show increased gelatinase B activity (Creange et al., 1999). In sural nerve biopsies from GBS patients, augmented proteolytic activity for gelatinase B as well as increased mRNA expression for gelatinase B and matrilysin were detected (Kieseier et al., 1998). MMPs in the inflamed PNS appear not only to promote inflammation, but may also play a role in nerve repair. For example, in EAN, matrilysin remained upregulated throughout the clinical recovery phase, implicating a possible role of this metalloproteinase in restoring the integrity of the PNS (Hughes et al., 1998).

Because macrophages represent the major cell population in infiltrates of affected nerves, they are considered to be key mediators of injury to myelin, Schwann cells, and axons. This hypothesis is reinforced by experimental studies, where depletion of macrophages prevents the development of EAN. Although endoneurial macrophages are present in low frequency in the normal peripheral nerve, most of them migrate through the blood-nerve barrier during the process of inflammation (Hartung et al., 2002). Macrophages may exert their pathological effects by release of inflammatory mediators. Of those, MMPs, TNFα, nitric oxide, and others have been postulated to mediate neurotoxicity (Redford et al., 1997a, b; Hartung et al., 2002). Besides direct harmful effects to nerve fibres, macrophages also perpetuate the inflammatory process by antigen-presenting to T-cells. It is assumed that macrophages also attack Schwann cells in AIDP (Prineas, 1981; Hartung et al., 1996) by antibody-mediated cytotoxicity and activation of complement (Hafer-Macko et al., 1996a, b). Another mechanism for Schwann cell injury/demyelination is via CD8+ T-cells, which are cytolytic T-cells that mediate cytotoxic effects by release of perforin and granzymes. Recently, CD8+ T-cells have been detected in postmortem tissue of GBS patients with prolonged disease course. These results point to an additional direct role of T cell-mediated cytotoxicity in AIDP(Van Rhijn et al., 2003).

There is a growing interest in the role of humoral factor-induced demyelination in AIDP. In demyelinating cases from China, complement activation markers were found on the abaxonal Schwann cell surface. They were associated with vesicular demyelination (Hafer-Macko et al., 1996b) and closely resemble the experimental nerve fiber demyelination induced by anti-galactocerebroside (a glycosphingolipid enriched in myelin) antibody in the presence of complement (Saida et al., 1979a). It is likely that in these cases, the antibody and complement are directly involved in targeting the Schwann cell and myelin, and the role of T-cells may be to open the blood-nerve barrier (Spies et al., 1995a, b). This concept is supported by the observations that disease severity in models of adoptive transfer-EAN is enhanced by transferring antibodies that recognise myelin or Schwann cell epitopes (Hahn et al., 1993; Spies et al., 1995a). Several clinical and experimental observations also support a role for antibody-mediated mechanisms, including the response to plasmapheresis, the presence of anti-myelin and anti-glycoconjugate antibodies (reviewed in (Hughes et al., 1999), and the ability of AIDP sera to induce demyelination after intraneural injection (Saida et al., 1982). Anti-ganglioside antibodies of various specificities have been described in AIDP but their pathogenic role is not accepted because of lack of experimental models demonstrating their demyelinating activity. Antibodies against GM1, GD1b, asialo-GM1, the Gal(β1–3)GalNAc epitope, GM2, LM2, and GT1b (reviewed in (Hughes et al., 1999; Willison and Yuki, 2002) have been described, but antibodies to these gangliosides are not routinely detected.

Although substantial progress has been made in our understanding of the cellular mechanisms that might underlie nerve fibre injury in AIDP, however, definition of target antigens and how the undesirable processes of inflammation, myelin degradation, and subsequent axonal damage are initiated remain unclear.

20.2.2. Acute Motor Axonal Neuropathy (AMAN)

Unlike for AIDP there are almost no clinical or experimental data that have systematically examined the role of cellular immunity in AMAN. Clinical studies over the last 15 years show that patients with AMAN have specific antibodies against two major gangliosides, GM1 and GD1a, and two minor gangliosides, GalNAc-GD1a and GM1b in the peripheral nerves (Figure 20.3A) (Rees et al., 1995a; Jacobs et al., 1996; Hadden et al., 1998b; Ho et al., 1999; Ogawara et al., 2000; Yuki et al., 1993b; Kusunoki et al., 1994, 1996). Most of the data originates from Japan and northern China, where the AMAN form of the disease and preceding *C. jejuni* infections are frequent. Anti-GM1 antibodies have been reported in up to 50% of Japanese patients with AMAN (Ogawara et al., 2000). Anti-GD1a antibodies are present in up to 60% of patients with AMAN compared to 4% in AIDP in northern China (Ho et al., 1999). The frequency of anti-GalNAc-GD1a and -GM1b in motor-predominant GBS is about 10–15% (Kusunoki et al., 1994, 1996; Yuki et al., 1999). Anti-ganglioside antibodies in AMAN are mostly IgG isotype and are commonly IgG1 and IgG3 (Willison and Veitch, 1994; Ogino et al., 1995), which are complement-fixing subtypes in humans. These anti-ganglioside antibodies are oligoclonal or polyclonal and can have a broad range of crossreactivity. In AMAN, the differences of antibody specificity in different populations are not well understood. There are some data to suggest differences in immunogenetic repertoire, and geography may affect the specificity and isotype distribution of anti-ganglioside antibodies in various populations (Ogawara et al., 2000; Ang et al., 2001b).

A brief discussion of peripheral nerve gangliosides is necessary for understanding the pathogenetic effects of anti-ganglioside antibodies. Gangliosides, the target antigens of

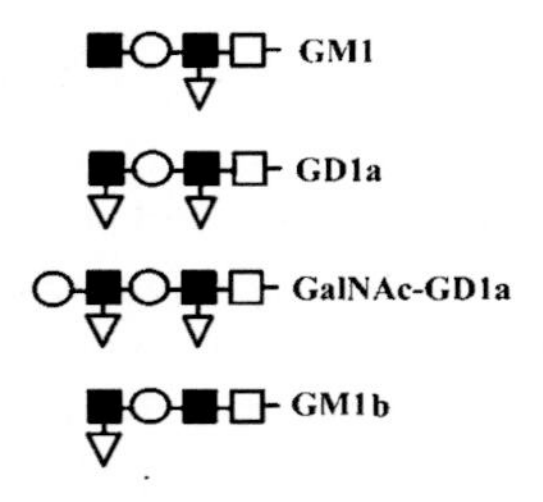

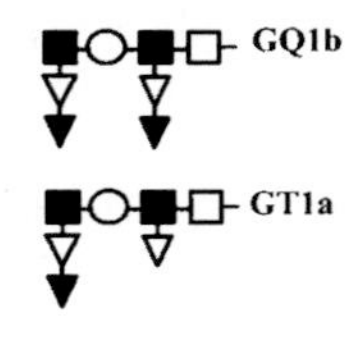

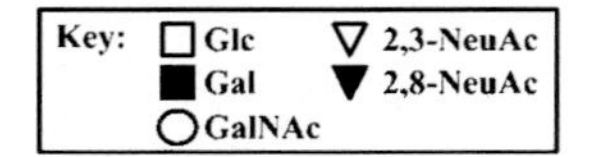

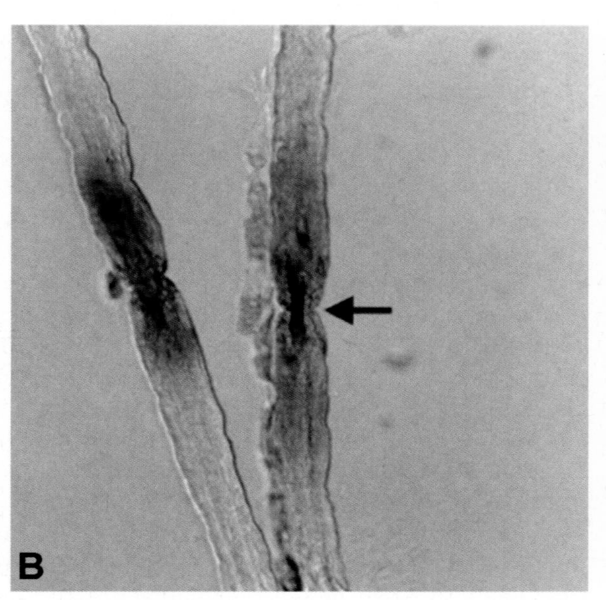

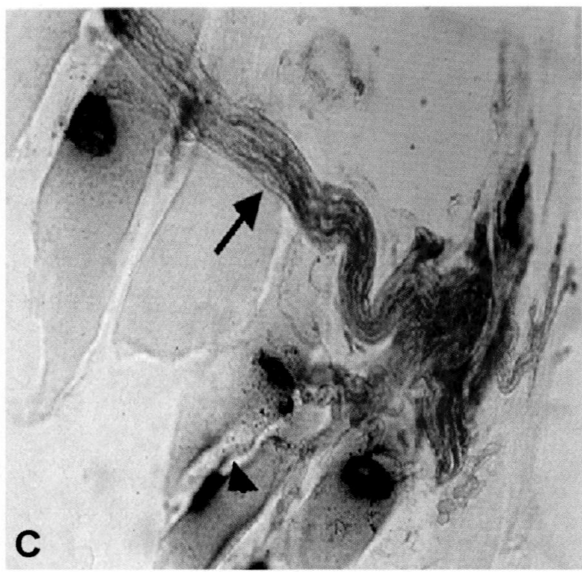

FIGURE 20.3 (A) Schematic diagrams showing the glycan structures of gangliosides implicated as target antigens in AMAN and FS. (B) Teased nerve fibers showing GM1 staining at node of Ranvier (arrow) and paranodal Schwann cells. (C) GM1 is also localized to intramuscular nerve (arrow) and motor nerve terminal (arrowhead); blue staining marks motor end plates (adapted with permission from Sheikh et al., 1999).

anti-ganglioside antibodies, are sialic acid-containing glycolipids enriched in the mammalian nervous system. They contain one or more sialic acids linked to an oligosaccharide chain of variable length and complexity, which is attached to ceramide lipid anchor (Kolter et al., 2002). The ceramide portion of gangliosides anchors them in plasma membranes and glycan moieties are expressed on cellular/axonal surfaces. This organization allows anti-ganglioside antibodies to bind to glycan moieties on cell surfaces. The most abundant gangliosides in the adult mammalian nervous system are GM1, GD1a, GD1b, and GT1b; in peripheral nerves LM1 ganglioside is also enriched, particularly in myelin (Yu and Saito, 1989; Svennerholm et al., 1992, 1994; Ogawa-Goto et al., 1990, 1992). Complex gangliosides are more concentrated in axolemmal fractions; GM1 is enriched in both axons and myelin (Yu and Saito, 1989). Immunolocalization studies indicate that in normal rodents and humans, all complex gangliosides including GM1 and GD1a reside in axons, and GM1 is also found in paranodal Schwann cells, but compact myelin in internodal segments is difficult to stain with anti-ganglioside antibodies or toxins (Ganser et al., 1983; Sheikh et al., 1999; Gong et al., 2002). Minor gangliosides GalNAc-GD1a and GM1b are expressed in peripheral nerves and one study suggests that GalNAc-GD1a may only be expressed by motor neurons and nerve fibers (Ilyas et al., 1988; Yoshino, 1997). In summary, it is important to emphasize that GM1 and GD1a gangliosides are present at the nodes of Ranvier and motor nerve terminals: two sites along the motor nerve fibers reported to be involved in patients with AMAN. Distribution of GM1 gangliosides at nodes of Ranvier and motor nerve terminals is shown as an example in Figure 20.3B-C.

Several anatomical and physiological features make nodes of Ranvier and motor nerve terminals susceptible to antibody-mediated nerve injury: (1) most complex gangliosides are concentrated at the nodes and motor nerve terminals; (2) in myelinated fibers, axonal target antigens are exposed at these sites; (3) sodium and postassium ion channels are clustered at the nodes and disruption of their function can disrupt impulse conduction; (4) structural integrity of the node of Ranvier is critical for nerve fiber conduction; (5) motor nerve terminals are enriched in sodium and calcium channels and disruption of their functions can affect distal impulse conduction or neurotransmitter release at the neuromuscular junctions, respectively. Further, multiple studies indicate that gangliosides are concentrated in microdomains called lipid rafts that are specialized for cell signaling and that modulation of these lipid raft gangliosides can modulate receptor function, including ion channel function.

With this background information, we outline the experimental data that supports the notion that anti-ganglioside antibodies can induce nerve fiber injury in animal models and briefly review the effects of these antibodies on nodes of Ranvier and motor nerve terminals. Two animal models that provide 'proof of concept' that anti-ganglioside antibodies with -GM1 and -GD1a specificities can induce neuropathy. First, Yuki and colleagues used GM1 ganglioside or *C. jejuni* lipopolysaccharide (LPS) with a mixture of keyhole limpet hemocyanin and complete Freund's adjuvant for repeated immunizations in rabbits to induce high titers of anti-GM1 antibodies, clinical paralysis, and electrophysiological and pathological evidence of motor axon injury (Yuki et al., 2001, 2004). The pathological and immunopathological studies in these animals showed features similar to that of human disease, i.e., deposition of IgG on motor axons and the presence of periaxonal macrophages. Further, analysis of animals shortly after the onset of clinical weakness showed only lengthening of the nodes of Ranvier without morphological features of axon degeneration (Susuki et al., 2003). We took an alternative approach to induce neuropathy in mice with anti-GD1a antibodies. This included generation of monoclonal anti-GD1a antibodies (Lunn et al., 2000) and implantation of antibody-secreting hybridoma in mice (Sheikh et al., 2004). Animals implanted with hybridoma develop high titers of anti-GD1a antibodies and axonal neuropathy in peripheral nerves and motor nerve terminals. Our studies showed that hybridoma implantation in transgenic mice lacking complex gangliosides (including GD1a) did not induce neuropathy despite high circulating levels of antibodies, confirming that gangliosides are indeed the target antigens of anti-ganglioside antibodies in this model (Sheikh et al., 2004).

The pathophysiologic effects of anti-ganglioside antibodies on nodes of Ranvier have been examined both *in vitro* and in animal models. Most of the studies have used anti-GM1 antibodies derived from patients with acute and chronic immune neuropathies or produced experimentally. It has been reported that intraneural injections of sera containing anti-GM1 can induce acute conduction block (Santoro et al., 1992; Uncini et al., 1993). We have recently examined the effects of anti-GD1a and GM1 monoclonal antibodies in an intraneural injection model and found that these antibodies induce conduction block and lengthening of the nodes of Ranvier (David et al., 2005). Takigawa and colleagues examined the effects of anti-GM1 sera on ion channel function at the nodes of Ranvier by using a voltage clamp technique on isolated myelinated fibers. They found that in the presence of complement these antibodies induced irreversible decreases in sodium currents and eventual blockade of the channels (Takigawa et al., 1995, 2000). Nonetheless, other investigators have been unable to reproduce these findings (Harvey et al., 1995; Hirota et al., 1997). Possible explanations include differences in experimental paradigms, including sources of antibodies, and affinity and fine specificity of these antibodies. Overall, these studies support the concept that anti-ganglioside antibodies can alter the nodal architecture and directly modulate ion channel function at the nodes of Ranvier.

The effects of anti-ganglioside antibody-mediated injury on motor nerve terminals have been extensively examined by *in vitro* phrenic nerve-diaphragm preparations. Current concepts indicate that anti-ganglioside antibodies can exert immunopharmacological effects on motor nerve terminal physiology and that they can also induce complement-dependent cytotoxic injury to motor nerve terminals. We have recently examined the effects of IgG anti-GM1 and -GD1a antibodies on motor nerve terminals by a perfused macro-patch clamp model with Buchwald's group. Our studies indicate that anti-GM1 and -GD1a antibodies depressed the evoked quantal release (Buchwald et al., 2004). This blockade was reversible or partially reversible after washout of these antibodies and did not require complement. Since calcium channels are critical in evoked quantal release, we examined the effects of these antibodies on depolarization-induced calcium influx by calcium imaging. These studies indicate that anti-GM1 and -GD1a antibodies significantly decrease depolarization-induced calcium influx, suggesting that antibody binding to gangliosides in the presynaptic motor terminals alters calcium channel function (B. Buchwald, K. Sheikh, unpublished observations). Work done by others has characterized the complement-dependent pathophysiologic effects of anti-GD1a antibodies on motor nerve terminals in *ex vivo* hemidiaphragm preparations (Goodfellow et al., 2005). They showed that dense antibody and complement deposits develop over presynaptic motor axons, accompanied by severe ultrastructural damage and electrophysiological blockade of motor nerve terminal function. This pathophysiologic effect, however, required high density expression of GD1a ganglioside (Goodfellow et al., 2005).

In summary, experimental data indicate that active immunization with GM1 ganglioside or *C. jejuni* LPS can reproduce pathophysiological features of AMAN in a rabbit model. Passive transfer with an anti-GD1a monoclonal antibody (by hybridoma implantation) also reproduced axon and motor nerve terminal degeneration in a mouse model. Anti-GM1 and -GD1 antibodies can alter the nodal and motor nerve terminal function and architecture. Tissue and cell culture studies indicate that sodium and calcium ion channel function can be modulated by these antibodies at nodes of Ranvier and presynaptic motor nerve terminals, respectively. These antibodies have both complement-independent immunopharmacologic and complement-dependent cytotoxic effects. The detailed subcellular and molecular effects of these antibodies on motor axons remain to be elucidated. Preferential susceptibility of motor axons to anti-GM1 and -GD1a antibody-mediated injury in AMAN is another fundamental issue that remains unresolved, because biochemical studies indicate that both motor and sensory nerve fibers have similar levels of ganglioside expression (Gong et al., 2002). Despite these and other gaps in our knowledge about the pathogenetic sequence of this disorder, current data support the hypothesis that anti-ganglioside antibodies can induce pathophysiological effects on intact motor nerve fibers.

20.2.3. Fisher Syndrome

Several studies have indicated that anti-GQ1b antibodies are present in more than 80% of the cases with FS (Chiba et al., 1992; Willison et al., 1993; Yuki et al., 1993a). Preceding *C. jejuni* infection is not uncommon in this disease. Anti-GQ1b antibodies in patients with FS can be IgA, IgM, or IgG isotype but the IgG response is most robust and persistent. These IgG antibodies are of complement-fixing isotypes, similar to those in AMAN. Studies indicate that anti-GQ1b antibodies commonly cross-react with a structurally related ganglioside GT1a (Figure 20.3A) (Chiba et al., 1993; Ilyas et al., 1998). Biochemical and immunolocalization studies have mapped the distribution of GQ1b ganglioside in the peripheral nerves: anti-GQ1b antibodies bind to paranodal myelin and nodes of Ranvier, and neuromuscular junctions (NMJs) in extraocular and somatic muscles, and it has been shown that the GQ1b is twice as frequently expressed in the extraocular cranial nerves than in other cranial and somatic peripheral nerves (Chiba et al., 1993, 1997). Pathogenetic studies indicate that anti-GQ1b antibodies bind at the nodes of Ranvier but are unable to induce acute conduction failure (Paparounas et al., 1999). This experimental result has led to the notion that antibody-mediated conduction failure at the levels of the nodes of Ranvier may be less important in the pathogenesis of FS. Because extraocular muscles are paralyzed in FS, anti-GQ1b antibodies do not affect nodal function, and anti-GQ1b antibodies bind to NMJs, experimental approaches have focused on pathophysiologic effects of anti-GQ1b antibodies on NMJs to model ophthalmoplegia seen in FS. In this regard phrenic nerve hemi-diaphragm preparations have been used extensively to study the pathological effects of anti-GQ1b antibodies.

Investigators has examined the effects of FS sera, FS IgG, or human anti-GQ1b antibodies on NMJs and showed that these antibodies bind to NMJs, cause massive quantal release of acetylcholine from nerve terminals and eventually block neuromuscular transmission, primarily through pre-synaptic mechanisms. These pathophysiologic effects were complement-dependent and resembled the effects of paralytic neurotoxin alpha-latrotoxin (Plomp et al., 1999). Willison and colleagues have shown that these antibodies induce degeneration of preterminal motor axons both by direct cytotoxicity and indirectly through damage to perisynaptic Schwann cells at NMJs (O'Hanlon et al., 2001; Halstead et al., 2004, 2005). In contrast, Buchwald et al. (Buchwald et al., 1995, 1998, 2001), reported complement-independent immunopharmacological effects of IgG fractions from both GQ1b-positive and -negative FS on NMJs. They showed that these IgG fractions blocked release of evoked acetylcholine and depressed the amplitude of postsynaptic potentials, implicating both a pre- and postsynaptic blocking effect. This effect was reversible. That the antibody-mediated pathophysiological effects at NMJs may be relevant to clinical disease is supported by an electrophysiological study suggesting that this site is affected in some cases with FS (Uncini and Lugaresi, 1999). There are no existing animal models of FS.

20.3. Molecular Mimicry Hypothesis

In this section we consider microbial agents that commonly cause infections preceding GBS and are reported to carry antigens crossreactive with peripheral nerves. The demonstration of peripheral nerve-crossreactive epitopes in these microbes is the basis for the hypothesis of post-infectious molecular mimicry in GBS. In this regard *Campylobacter jejuni* (*C. jejuni*) is the most well characterized, and stringent data exist that demonstrate the presence of ganglioside-like antigens in these organisms. Recent studies have also implicated *Haemophilus influenzae (H. influenzae), Mycoplasma pneumoniae (M. pneumoniae),* and *Cytomegalovirus* (CMV) as microbes carrying peripheral nerve-crossreactive epitopes. A common theme with these microbes is that they express carbohydrate epitopes that mimic glycolipid antigens in peripheral nerves, and autoimmunity against these epitopes is manifested as anti-carbohydrate antibodies. Most of the discussion highlights molecular mimicry in *C. jejuni*. A brief discussion on *H. influenzae, M. pneumoniae, and* CMV is also included. Glycans mimicking peripheral nerve glycolipids expressed by these microbes are shown in Figure 20.4.

Campylobacter jejuni

GM1-like

GD1a-like

GT1a-like

Haemophilus Influenzae

GM1-crossreactive

GT1a-crossreactive

Mycoplasma pneumonia

cer GalC-crossreactive

GM1-crossreactive

CMV

GM2-crossreactive

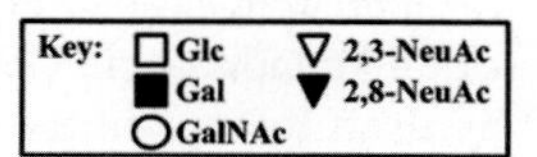

FIGURE 20.4. Glycan structures (mimics) expressed by different microbes that are invoked in molecular mimicry. The term "like" is used when glycan structure has been determined by biochemical/mass spectrometry studies and term "crossreactive" is used when evidence for glycan structure is indirect and based on antibody inhibition or binding studies.

20.3.1. Campylobacter Jejuni

C. jejuni is a gram-negative rod that is one of the most common causes of bacterial gastroenteritis world-wide (Hughes and Rees, 1997; Friedman et al., 2000; Oberhelman and Taylor, 2000). *Campylobacter* infections in US are mostly sporadic and are associated with ingestion of improperly handled or cooked food, particularly poultry products. In northern China contaminated well water was reported as a mode of transmission (McKhann et al., 1993). *C. jejuni* gastroenteritis is reported to be the most frequently recognized event preceding AMAN and other variants of GBS (reviewed in (Hughes and Rees, 1997)). In GBS cases, both stool culture and serologic methods are used to diagnose *Campylobacter* infection, because by the time neurological symptoms develop, the yield of *C. jejuni* from stool culture is relatively low (Nachamkin, 1997). Association of GBS with preceding *C. jejuni* infection was noted in early 1980's (Rhodes and Tattersfield, 1982). Studies indicate that in patients with GBS, the incidence of preceding *C. jejuni* infection varies widely, ranging from 4% in North America to 74% in northern China (Hughes and Rees, 1997; Ho et al., 1999), with an overall prevalence estimated around 30% (reviewed in (Moran et al., 2002)).

Ganglioside-like moieties in *Campylobacter* are contained in its lipopolysaccharide (LPS). *C. jejuni* LPS consists of three components: lipid A, the hydrophobic region inserted into the membrane, an oligosaccharide core divided into inner and outer parts, and capsular polysaccharides also called O-chains. Lipooligosaccharide (LOS) is differentiated from LPS by lack of O-chains. In *C. jejuni* it is the oligosaccharide core region that carries ganglioside-like moieties. Since 1990 a large number of studies have characterized the core regions of LPS/LOS of GBS- and enteritis-associated *C. jejuni* strains. These studies demonstrate that different strains or serotypes of *C. jejuni* LPS/LOS contain several ganglioside-like molecules, including GM1-, GD1a-, GalNAc-GD1a-, GM1b-, GT1a-, GD2-, GD3-, and GM2- like structures (Yuki et al., 1992, 1996; Aspinall et al., 1993, 1994a, b; Sheikh et al., 1998;

Nachamkin et al., 2002). Biochemical/structural studies with mass spectrometry have failed to demonstrate a GQ1b-like structure, target antigen for FS, but antibody binding assays with human or murine monoclonal antibodies did show the presence of GQ1b- and GT1a- cross-reactive moieties in *C. jejuni* LPS/LOS (Jacobs et al., 1997, 1995; Yuki et al., 1994). Figure 20.4 shows the ganglioside-like moieties in *C. jejuni* LOS that are implicated in AMAN and FS. Recent studies show that *Campylobacter* has genetic machinery for LOS biosynthesis with relevant genes clustering in a locus. It has been reported that specific polymorphisms in a sialyltransferase gene are associated with preferential expression of GM1- and GD1a- or GQ1b-like moieties. Based on this observation it has been postulated that these genetic polymorphisms determine the specificity of anti-ganglioside antibody response (anti-GM1 and -GD1a or -GQ1b) and clinical phenotype (AMAN or FS).

Despite the relationship of preceding *Campylobacter* infection, we emphasize that GBS is a very rare complication after *C. jejuni* infection: it is estimated that 1 in 1000 cases of *Campylobacter* infection is complicated by GBS (Allos, 1997). Because GBS follows very rarely after *C. jejuni* infection, investigators have examined *Campylobacter* and host-factors that may lead to this complication. Thus, *Campylobacter* strains isolated from patients with GBS and enteritis have been characterized for ganglioside mimicry. It is known that both GBS- and diarrhea-associated isolates carry ganglioside-like moieties and associated synthetic genes, but GBS-related organisms are more likely to do so (Nachamkin et al., 1999, 2002). Notably, studies indicate that expression of ganglioside-like structures in LOS, even when accounting for increased probability of synthetic machinery for ganglioside-like moieties in GBS-associated isolates, is not sufficient by itself to explain the calculated rate of GBS after *Campylobacter* infection (1 in 1000). Some studies indicate that post-*Campylobacter* GBS cases preferentially associate with certain HLA alleles, the significance of these findings remains unclear because of lack of confirmatory studies (Yuki et al., 1991; Rees et al., 1995b). The host properties that could confer susceptibility to GBS after *C. jejuni* infection are not well established.

Clinical studies showing that GBS sera or purified anti-ganglioside antibodies from GBS sera bind to ganglioside-like moieties in the LPS/LOS (Wirguin et al., 1994; Oomes et al., 1995; Sheikh et al., 1998) promulgated the hypothesis that cross-reactive carbohydrate moieties in LOS incite production of these antibodies. This hypothesis has been examined experimentally by immunization of laboratory animals with *C. jejuni* LPS/LOS to induce anti-ganglioside antibodies (Wirguin et al., 1997; Goodyear et al., 1999; Ang et al., 2001a). There is a high level of tolerance to self-gangliosides, and immunizations in experimental animals have generally induced low affinity non-T cell-dependent antibodies of IgM and IgG3 type (non-complement-fixing isotype in mice), despite the use of adjuvants to recruit T cell help (Wirguin et al., 1997; Goodyear et al., 1999; Bowes et al., 2002). In contrast, serological studies in AMAN and FS indicate class switching to IgG and subclass restriction to IgG1 and IgG3 (complement-fixing isotypes in humans), both usually features of T-cell help. Transgenic mice lacking complex gangliosides are immune naïve to complex gangliosides such as GM1, GD1a, and GQ1b, and immunization with gangliosides or *C. jejuni* LPSs in these animals induces IgG anti-ganglioside antibodies with complement-fixing subclass isotypes, indicating recruitment of T cell help (Lunn et al., 2000; Bowes et al., 2002). These experimental observations suggest that breakdown of tolerance to self-gangliosides might be critical in the pathogenesis of post-*Campylobacter* GBS. Despite these experimental advances the mechanism(s) of anti-ganglioside antibody induction in patients with *Campylobacter* infection remains unclear.

20.3.2. Haemophilus Influenzae

H. influenzae is a gram-negative bacterium that causes human respiratory tract infections. In GBS, preceding infection with *H. influenzae* is rare, ranging between 1% in western Europe to 13% in Japan (Jacobs et al., 1998; Mori et al., 2000; Koga et al., 2005a). In comparison to post-*Campylobacter* GBS, preceding infection with *H. influenzae* is associated with faster recovery, less cranial nerve involvement, and milder disease course (Kuwabara et al., 2001). Based on GBS or FS sera (antibody) binding to LOS/LPS it has been postulated that *H. influenzae* may carry ganglioside-like moieties such as GM1 and GT1a (Koga et al., 2001, 2005a; Mori et al., 1999). Structural/mass spectrometry studies demonstrating ganglioside-like moieties on *H. influenzae* are not yet available. A recent study reports that isolation of *H. influenzae* does not necessarily indicate that this infection is the trigger for induction of anti-ganglioside antibodies because a considerable number (15%) of GBS patients with *H. influenzae* isolated by culture were seronegative for this bacterium, but seropositive for *Campylobacter*. Further, anti-ganglioside antibodies of seronegative patients did not show any crossreactivity with *H. influenzae* LOS (Koga et al., 2005b). Whether immunization with *H. influenzae* triggers induction of anti-ganglioside antibodies in experimental animals remains to be determined.

20.3.3. Mycoplasma Pneumoniae

Antecedent infection with *M. pneumoniae* can be found in 2–12% of patients with GBS (Hao et al., 1998; Jacobs et al., 1998; Ogawara et al., 2000). Studies indicate that some cases of demyelinating form of GBS are associated with *M. pneumoniae* (Hao et al., 1998; Ang et al., 2002). This infection is reported to trigger antibodies against galactocerebroside (GalC), a major glycolipid antigen in myelin, in some patients with GBS. Anti-GalC antibodies have been shown to induce demyelination in experimental animal models (Saida et al., 1979a, b; Hahn et al., 1993). The hypothesis of molecular mimicry is supported by the findings that anti-Gal-C reactivity in GBS sera following

M. pneumoniae infection was specifically inhibited by adding *M. pneumoniae* reagent (Kusunoki et al., 2001). One study has reported the AMAN form of GBS in association with anti-GM1 antibodies and preceding *M. pneumoniae* infections (Susuki et al., 2004). These investigators report that anti-GM1 antibodies bind to lipids extracted from *M. pneumoniae,* suggesting that GM1-like structures are also expressed by this microbe.

20.3.4. Cytomegalovirus (CMV)

In GBS, serological evidence for preceding infection with CMV ranges between 8 and 15% (Dowling and Cook, 1981; Visser et al., 1996; Ogawara et al., 2000; Hadden et al., 2001). Patients with CMV-associated GBS appear to have a different clinical phenotype: significantly younger patients are affected, the disease course is more severe, there is prominent cranial and sensory nerve involvement, and functional recovery is incomplete (Visser et al., 1996). With the availability of modern techniques, PCR is more appropriate to demonstrate infection with CMV. A previous study failed to detect CMV genome in sural nerve biopsies from GBS patients (Hughes et al., 1992) but a recent study demonstrated the presence of CMV DNA in the CSF of about 30% of GBS patients with positive CMV serology (Steininger et al., 2004). Several studies have reported an association of CMV infection with GM2 antibodies in patients with GBS (Irie et al., 1996; Yuki and Tagawa, 1998; Khalili-Shirazi et al., 1999). A recent study showed that anti-GM2 reactivity in GBS sera was abrogated after incubation with fibroblasts infected with a GBS-associated CMV strain, indicating that CMV-infected fibroblasts express GM2-like epitopes recognized by anti-GM2 antibodies (Ang et al., 2000). This intriguing finding raises the possibility that either CMV itself expresses GM2-like epitopes or it induces the expression of this ganglioside in the host cells which renders them immunogenic. The later possibility would suggest that direct molecular mimicry by the infecting agent and target tissue antigens may not be necessary. Besides GM2, antibodies against GalNAc-GD1a and GM1have also been reported in the setting of CMV infection and GBS (Khalili-Shirazi et al., 1999; Kaida et al., 2001).

20.3.5. Gaps in Molecular Mimicry Hypothesis

Despite the accumulation of significant data supporting the hypothesis of post-infectious molecular mimicry in GBS, clearly several fundamental questions remain unresolved: (1) Infectious agents invoked as triggers for anti-glycolipid antibodies in patients with GBS are microbes that induce gastrointestinal and upper respiratory tract infections in a large number of people but only rare individuals develop GBS after these common infections. (2) What are the mechanism(s) of anti-glycolipid antibody induction in patients with post-infectious GBS? The induction of these antibodies in GBS patients is likely to be substantially different from experimental approaches that involve repeated use of infectious organism and adjuvants or immune naïve animals to produce disease-associated anti-ganglioside antibodies. (3) Antibodies against GM1, GM2, and GalC can be seen in other neurological disorders or sometimes in normal controls. What are the antibody-related properties that can distinguish between disease-associated and non-disease-associated antibodies? These and other unresolved issues need further research to increase our understanding of post-infectious autoimmune disorders.

Summary

Autoimmunity is implicated in a small but important group of peripheral nerve diseases. These include the acute inflammatory neuropathies referred to as the Guillain-Barré (GBS) and Fisher syndromes. Early recognition and appropriate management of these neuropathies can prevent significant morbidity and mortality. Substantial evidence exists for an autoimmune pathogenesis in GBS and its subtypes. Although most current evidence supports an antibody driven pathogenesis triggered by infection, T-cell and other cellular immune components are crucial effectors of disease particularly in the demyelinating forms of GBS.

Acknowledgements. KAS is supported by the National Institute of Health (NS42888) and the GBS Foundation. We thank Dr. Pamela Talalay for editorial discussion.

Review Questions/Problems

1. **What is meant by the term molecular mimicry?**
2. **What is the basis of hypothesis of molecular mimicry in the pathogenesis of GBS?**
3. **What are the potential mechanisms of antibody-mediated peripheral nerve dysfunction/injury?**
4. **What are the potential mechanisms of T-cell mediated peripheral nerve dysfunction/injury?**

References

Allos BM (1997) Association between Campylobacter infection and Guillain-Barré syndrome. J Infect Dis 176 Suppl 2:S125–S128.

Ang CW, Jacobs BC, Brandenburg AH, Laman JD, van der Meche FGA, Osterhaus ADME, Van Doorn PA (2000) Cross-reactive antibodies against GM2 and CMV-infected fibroblasts in Guillain-Barré syndrome. Neurology 54:1453–1458.

Ang CW, De Klerk MA, Endtz HP, Jacobs BC, Laman JD, van der Meche FG, Van Doorn PA (2001a) Guillain-Barré syndrome- and Miller Fisher syndrome-associated Campylobacter jejuni lipopolysaccharides induce anti-GM1 and anti-GQ1b antibodies in rabbits. Infect Immun 69:2462–2469.

Ang CW, Koga M, Jacobs BC, Yuki N, van der Meche FG, Van Doorn PA (2001b) Differential immune response to gangliosides in Guillain-Barré syndrome patients from Japan and The Netherlands. J Neuroimmunol 121:83–87.

Ang CW, Tio-Gillen AP, Groen J, Herbrink P, Jacobs BC, van Koningsveld R, Osterhaus ADME, van der Meche FGA, Van Doorn PA (2002) Cross-reactive anti-galactocerebroside antibodies and Mycoplasma pneumoniae infections in Guillain-Barré syndrome. J Neuroimmunol 130:179–183.

Asbury AK, Arnason BG, Adams RD (1969) The inflammatory lesion in idiopathic polyneuritis. Medicine 48:173–215.

Aspinall GO, McDonald AG, Raju TS, Pang H, Moran AP (1993) Chemical structures of the core regions of *Campylobacter jejuni* serotypes O:1, O:4, O:23, and O:36 lipopolysaccharides. Eur J Biochem 213:1017–1027.

Aspinall GO, Fujimoto S, McDonald AG, et al. (1994a) Lipopolysaccharides from *Campylobacter jejuni* associated with Guillain-Barré syndrome patients mimic human gangliosides in structure. Infect Immun 62:2122–2125.

Aspinall GO, McDonald AG, Pang H, Kurjanczyk LA, Penner JL (1994b) Lipopolysaccharides of *Campylobacter jejuni* serotype O:19: Structures of core oligosaccharide regions from the serostrain and two bacterial isolates from patients with the Guillain-Barré syndrome. Biochemistry 33:241–249.

Baggiolini M (1998) Chemokines and leukocyte traffic. Nature 392:565–568.

Bogliun G, Beghi E (2004) Incidence and clinical features of acute inflammatory polyradiculoneuropathy in Lombardy, Italy, 1996. Acta Neurol Scand 110:100–106.

Bowes T, Wagner ER, Boffey J, Nicholl D, Cochrane L, Benboubetra M, Conner J, Furukawa K, Furukawa K, Willison HJ (2002) Tolerance to self gangliosides is the major factor restricting the antibody response to lipopolysaccharide core oligosaccharides in Campylobacter jejuni strains associated with Guillain-Barré syndrome. Infect Immun 70:5008–5018.

Buchwald B, Weishaupt A, Toyka KV, Dudel J (1995) Immunoglobulin G from a patient with Miller-Fisher syndrome rapidly and reversibly depresses evoked quantal release at the neuromuscular junction of mice. Neurosci Lett 201:163–166.

Buchwald B, Weishaupt A, Toyka KV, Dudel J (1998) Pre- and postsynaptic blockade of neuromuscular transmission by Miller-Fisher syndrome IgG at mouse motor nerve terminals. Eur J Neurosci 10:281–290.

Buchwald B, Bufler J, Carpo M, Heidenreich F, Pitz R, Dudel J, Nobile-Orazio E, Toyka KV (2001) Combined pre- and postsynaptic action of IgG antibodies in Miller Fisher syndrome. Neurology 56:67–74.

Buchwald B, Ahangari R, Griffin JW, Sheikh KA, Toyka KV (2004) Different monoclonal anti-ganglioside antibodies show distinct functional effects at the mouse neuromuscular junction: Potential implications for immune neuropathies. J Neurol 251:12.

Campbell JJ, Hedrick J, Zlotnik A, Siani MA, Thompson DA, Butcher EC (1998) Chemokines and the arrest of lymphocytes rolling under flow conditions. Science 279:381–384.

Chiba A, Kusunoki S, Shimizu T, Kanazawa I (1992) Serum IgG antibody to ganglioside GQ1b is a possible marker of Miller Fisher syndrome. Ann Neurol 31:677–679.

Chiba A, Kusunoki S, Obata H, Machinami R, Kanazawa I (1993) Serum anti-GQ1b IgG antibody is associated with ophthalmoplegia in Miller Fisher syndrome and Guillain-Barré syndrome: Clinical and immunohistochemical studies. Neurology 43:1911–1917.

Chiba A, Kusunoki S, Obata H, Machinami R, Kanazawa I (1997) Ganglioside composition of the human cranial nerves, with special reference to pathophysiology of Miller Fisher syndrome. Brain Res 745:32–36.

Cornblath DR, Griffin DE, Welch D, Griffin JW, McArthur JC (1990) Quantitative analysis of endoneurial T-cells in human sural nerve biopsies. J Neuroimmunol 26:113–118.

Creange A, Belec L, Clair B, Raphael JC, Gherardi RK (1996) Circulating tumor necrosis factor (TNF)-alpha and soluble TNF-alpha receptors in patients with Guillain-Barré syndrome. J Neuroimmunol 68:95–99.

Creange A, Belec L, Clair B, Degos JD, Raphael JC, Gherardi RK (1998) Circulating transforming growth factor beta 1 (TGF-beta 1) in Guillain-Barré syndrome: Decreased concentrations in the early course and increase with motor function. J Neurol Neurosurg Psychiatry 64:162–165.

Creange A, Sharshar T, Planchenault T, Christov C, Poron F, Raphael JC, Gherardi RK (1999) Matrix metalloproteinase-9 is increased and correlates with severity in Guillain-Barré syndrome. Neurology 53:1683–1691.

Creange A, Chazaud B, Sharshar T, Plonquet A, Poron F, Eliezer MC, Raphael JC, Gherardi RK (2001) Inhibition of the adhesion step of leukodiapedesis: A critical event in the recovery of Guillain-Barré syndrome associated with accumulation of proteolytically active lymphocytes in blood. J Neuroimmunol 114:188–196.

Dalakas MC (2002) Mechanisms of action of IVIg and therapeutic considerations in the treatment of acute and chronic demyelinating neuropathies. Neurology 59:S13–S21.

Dalakas MC (2004) Intravenous immunoglobulin in autoimmune neuromuscular diseases. J A M A 291:2367–2375.

David MA, Spies JM, Pollard JD, Zhang G, Armati P, Sheikh KA (2005) Intraneural injections of anti-ganglioside antibodies induce reversible conduction failure. J Periphe Nerv Syst 10:18–19.

Dowling PC, Cook SD (1981) Role of infection in Guillain-Barré syndrome: Laboratory confirmation of Herpes viruses in 41 cases. Ann Neurol 9(Suppl):44–55.

Draganesco H, Claudian J (1927) A case of curable radiculo-neuritis (Guillan and Barré's syndrome) appeared during an osteo-myelitis of the arm. Revue Neurol 48:517–521.

Enders U, Lobb R, Pepinsky RB, Hartung HP, Toyka KV, Gold R (1998) The role of the very late antigen-4 and its counterligand vascular cell adhesion molecule-1 in the pathogenesis of experimental autoimmune neuritis of the Lewis rat. Brain 121:1257–1266.

Exley AR, Smith N, Winer JB (1994) Tumor necrosis factor-alpha and other cytokines in Guillain-Barré-syndrome. J Neurol Neurosurg Psychiatry 57:1118–1120.

Feasby TE, Gilbert JJ, Brown WF, et al. (1986) An acute axonal form of Guillain-Barré polyneuropathy. Brain 109:1115–1126.

Fisher M (1956) An unusual variant of acute idiopathic polyneuritis (syndrome of ophthalmoplegia ataxia and areflexia). N Engl J Med 255:57–65.

Friedman CR, Neimann J, Wegener HC, Tauxe RV (2000) Epidemiology of Campylobacter jejuni infections in the United States and other industrialized nations. In: Campylobacter (Nachamkin I, Blaser MJ, eds), pp 121–138. Washington, DC: American Society for Microbiology.

Fu SY, Gordon T (1995a) Contributing factors to poor functional recovery after delayed nerve repair: Prolonged axotomy. J Neurosci 15:3876–85.

Fu SY, Gordon T (1995b) Contributing factors to poor functional recovery after delayed nerve repair: Prolonged denervation. J Neurosci 15:3886–95.

Ganser AL, Kirschner DA, Willinger M (1983) Ganglioside localization on myelinated nerve fibres by cholera toxin binding. J Neurocytol 12:921–938.

Gong Y, Tagawa Y, Lunn MP, Laroy W, Heffer-Lauc M, Li CY, Griffin JW, Schnaar RL, Sheikh KA (2002) Localization of major gangliosides in the PNS: Implications for immune neuropathies. Brain 125:2491–2506.

Goodfellow JA, Bowes T, Sheikh K, Odaka M, Halstead SK, Humphreys PD, Wagner ER, Yuki N, Furukawa K, Furukawa K, Plomp JJ, Willison HJ (2005) Overexpression of GD1a ganglioside sensitizes motor nerve terminals to anti-GD1a antibody-mediated injury in a model of acute motor axonal neuropathy. J Neurosci 25:1620–1628.

Goodyear CS, O'Hanlon GM, Plomp JJ, Wagner ER, Morrison I, Veitch J, Cochrane L, Bullens RW, Molenaar PC, Conner J, Willison HJ (1999) Monoclonal antibodies raised against Guillain-Barré syndrome-associated Campylobacter jejuni lipopolysaccharides react with neuronal gangliosides and paralyze muscle-nerve preparations [published erratum appears in J Clin Invest 1999 Dec;104(12):1771]. J Clin Invest 104:697–708.

Gordon PH, Wilbourn AJ (2001) Early electrodiagostic findings in Guillain-Barré syndrome. Arch Neruol 58:913–917.

Griffin JW, Li CY, Ho TW, Xue P, Macko C, Cornblath DR, Gao CY, Yang C, Tian M, Mishu B, McKhann GM, Asbury AK (1995) Guillain-Barré syndrome in northern China: The spectrum of neuropathologic changes in clinically defined cases. Brain 118:577–595.

Griffin JW, Li CY, Ho TW, Tian M, Gao CY, Xue P, Mishu B, Cornblath DR, Macko C, McKhann GM, Asbury AK (1996a) Pathology of the motor-sensory axonal Guillain-Barré Syndrome. Ann Neurol 39:17–28.

Griffin JW, Li CY, Macko C, Ho TW, Hsieh S-T, Xue P, Wang FA, Cornblath DR, McKhann GM, Asbury AK (1996b) Early nodal changes in the acute motor axonal neuropathy pattern of the Guillain-Barré syndrome. J Neurocytol 25:33–51.

Guillain-Barré syndrome variants in Emilia-Romagna (1998) Italy, 1992–3: Incidence, clinical features, and prognosis. Emilia-romagna study group on clinical and epidemiological problems in neurology. J Neurol Neurosurg Psychiatr 65:218–224.

Guillain G, Barré JA, Strohl A (1916) Sur un syndrome de radiculonéurite avec hyperalbuminose du liquide céphalo-rachidien sans reaction cellulaire. Remarques sur les caractères cliniques et graphiques des réflexes tendineux. Bull Soc Med Hôp Paris 40:1462–70.

Hadden RD, Cornblath DR, Hughes RA, Zielasek J, Hartung HP, Toyka KV, Swan AV (1998a) Electrophysiological classification of Guillain-Barré syndrome: Clinical associations and outcome. Plasma Exchange/Sandoglobulin Guillain-Barré Syndrome Trial Group. Ann Neurol 44:780–788.

Hadden RD, Cornblath DR, Hughes RA, Zielasek J, Hartung HP, Toyka KV, Swan AV (1998b) Electrophysiological classification of Guillain-Barré syndrome: Clinical associations and outcome. Plasma exchange/sandoglobulin Guillain-Barré syndrome trial group. Ann Neurol 44:780–788.

Hadden RD, Karch H, Hartung HP, Zielasek J, Weissbrich B, Schubert J, Weishaupt A, Cornblath DR, Swan AV, Hughes RA, Toyka KV (2001) Preceding infections, immune factors, and outcome in Guillain-Barré syndrome. Neurology 56:758–765.

Hafer-Macko C, Hsieh S-T, Li CY, Ho TW, Sheikh K, Cornblath DR, McKhann GM, Asbury AK, Griffin JW (1996a) Acute motor axonal neuropathy: An antibody-mediated attack on axolemma. Ann Neurol 40:635–644.

Hafer-Macko C, Sheikh KA, Li CY, Ho TW, Cornblath DR, McKhann GM, Asbury AK, Griffin JW (1996b) Immune attack on the Schwann cell surface in acute inflammatory demyelinating polyneuropathy. Ann Neurol 39:625–635.

Hahn AF, Feasby TE, Wilkie L, Lovgren D (1993) Antigalactocerebroside antibody increases demyelination in adoptive transfer experimental allergic neuritis. Muscle Nerve 16:1174–1180.

Hall SM, Hughes RA, Atkinson PF, McColl I, Gale A (1992) Motor nerve biopsy in severe Guillain-Barré syndrome. Ann Neurol 31:441–444.

Halstead SK, O'Hanlon GM, Humphreys PD, Morrison DB, Morgan BP, Todd AJ, Plomp JJ, Willison HJ (2004) Anti-disialoside antibodies kill perisynaptic Schwann cells and damage motor nerve terminals via membrane attack complex in a murine model of neuropathy. Brain 127:2109–2123.

Halstead SK, Humphreys PD, Goodfellow JA, Wagner ER, Smith RA, Willison HJ (2005) Complement inhibition abrogates nerve terminal injury in Miller Fisher syndrome. Ann Neurol 58:203–210.

Hao Q, Saida T, Kuroki S, Nishimura M, Nukina M, Obayashi H, Saida K (1998) Antibodies to gangliosides and galactocerebroside in patients with Guillain-Barré syndrome with preceding Campylobacter jejuni and other identified infections. J Neuroimmunol 81:116–26.

Hartung H-P, Toyka KV (1990) T-cell and macrophage activation in experimental autoimmune neuritis and Guillain-Barré syndrome. Ann Neurol 27(Suppl):S57–S63.

Hartung H-P, Schafer B, Fierz W, Heininger K, Toyka KV (1987) Ciclosporin A prevents P2 T cell line-mediated experimental autoimmune neuritis (AT-EAN) in rat. Neurosci Lett 83:195–200.

Hartung H-P, Reiners K, Schmidt B, Stoll G, Toyka KV (1991) Serum interleukin-2 concentrations in Guillain-Barré syndrome and chronic idiopathic demyelinating polyradiculoneuropathy: Comparison with other neurological diseases of presumed immunopathogenesis. Ann Neurol 30:48–53.

Hartung HP, Willison H, Jung S, Pette M, Toyka KV, Giegerich G (1996) Autoimmune responses in peripheral nerve. Springer Semin Immunopathol 18:97–123.

Hartung HP, Willison HJ, Kieseier BC (2002) Acute immunoinflammatory neuropathy: Update on Guillain-Barré. Curr Opin Neurol 15:571–577.

Harvey GK, Toyka KV, Zielasek J, Kiefer R, Simonis C, Hartung H-P (1995) Failure of anti-GM_1 IgG or IgM to induce conduction block following intraneural transfer. Muscle Nerve 18:388–394.

Haymaker W, Kernohan JW (1949) The Landry-Guillain-Barré syndrome. A clinicopathologic report of fifty fatal cases and a critique of the literature. Medicine 28:59–141.

Hirota N, Kaji R, Bostock H, Shindo K, Kawasaki T, Mizutani K, Oka N, Kohara N, Saida T, Kimura J (1997) The physiological effect of anti-GM1 antibodies on saltatory conduction and transmembrane currents in single motor axons. Brain 120:2159–2169.

Ho TW, Hsieh ST, Nachamkin I, Willison HJ, Sheikh K, Kiehlbauch J, Flanigan K, McArthur JC, Cornblath DR, McKhann GM, Griffin JW (1997) Motor nerve terminal degeneration provides a potential mechanism for rapid recovery in acute motor axonal neuropathy after Campylobacter infection (see comments). Neurology 48:717–724.

Ho TW, Willison HJ, Nachamkin I, Li CY, Veitch J, Ung H, Wang GR, Liu RC, Cornblath DR, Asbury AK, Griffin JW, McKhann GM (1999) Anti-GD1a antibody is associated with axonal but not demyelinating forms of Guillain-Barré syndrome. Ann Neurol 45:168–173.

Holmdahl R, Olsson T, Moran T, Klareskog L (1985) In vivo treatment of rats with monoclonal anti-T-cell antibodies. Scand J Immunol 22:157–169.

Honavar M, Tharakan JKJ, Hughes RAC, Leibowitz S, Winer JB (1991) A clinicopathological study of the Guillain-Barré syndrome. Brain 114:1245–1269.

Hughes RA, Cornblath DR (2005) Guillain-Barré syndrome. Lancet 366:1653–1666.

Hughes RA, Rees JH (1997) Clinical and epidemiologic features of Guillain-Barré syndrome. J Infect Dis 176 Suppl 2:S92–S98.

Hughes RA, van der Meche FG (2000) Corticosteroids for treating Guillain-Barré syndrome. Cochrane Database Syst Rev CD001446.

Hughes R, Atkinson P, Coates P, Hall S, Leibowitz S (1992) Sural nerve biopsies in Guillain-Barré syndrome: Axonal degeneration and macrophage-associated demyelination and absence of cytomegalovirus genome. Muscle Nerve 15:568–575.

Hughes PM, Wells GMA, Clements JM, Gearing AJH, Redford EJ, Davies M, Smith KJ, Hughes RAC, Brown MC, Miller KM (1998) Matrix metalloproteinase expression during experimental autoimmune neuritis. Brain 121:481–494.

Hughes RA, Hadden RD, Gregson NA, Smith KJ (1999) Pathogenesis of Guillain-Barré syndrome. J Neuroimmunol 100:74–97.

Hughes RA, Raphael JC, Swan AV, Van Doorn PA (2001) Intravenous immunoglobulin for Guillain-Barré syndrome. Cochrane Database Syst Rev CD002063.

Hughes RA, Raphael JC, Swan AV, Doorn PA (2004) Intravenous immunoglobulin for Guillain-Barré syndrome. Cochrane Database Syst Rev CD002063.

Ilyas AA, Li SC, Chou DK, Li YT, Jungalwala FB, Dalakas MC, Quarles RH (1988) Gangliosides GM2, IV4GalNAcGM1b, and IV4GalNAcGC1a as antigens for monoclonal immunoglobulin M in neuropathy associated with gammopathy. J Biol Chem 263:4369–4373.

Ilyas AA, Cook SD, Mithen FA, Taki T, Kasama T, Handa S, Hamasaki H, Singhal BS, Li SC, Li YT (1998) Antibodies to GT1a ganglioside in patients with Guillain-Barré syndrome. J Neuroimmunol 82:160–7.

Irie S, Saito T, Nakamura K, Kanazawa N, Ogino M, Nukazawa T, Ito H, Tamai Y, Kowa H (1996) Association of anti-GM2 antibodies in Guillain-Barré syndrome with acute cytomegalovirus infection. J Neuroimmunol 68:19–26.

Jacobs BC, Endtz H, van der Meché FGA, Hazenberg MP, Achtereekte HA, Van Doorn PA (1995) Serum anti-GQ1b IgG antibodies recognize surface epitopes on *Campylobacter jejuni* from patients with Miller Fisher syndrome. Ann Neurol 37:260–264.

Jacobs BC, Van Doorn PA, Schmitz PI, et al. (1996) *Campylobacter jejuni* intections and anti-GM1 antibodies in Guillain-Barré syndrome. Ann Neurol 40:181–187.

Jacobs BC, Hazenberg MP, Van Doorn PA, Endtz HPh, van der Meché FGA (1997) Cross-reactive antibodies against gangliosides and *Campylobacter jejuni* lipopolysaccharides in patients with Guillain-Barré or Miller Fisher syndrome. J Inf Dis 175:729–733.

Jacobs BC, Rothbarth PH, van der Meche FG, Herbrink P, Schmitz PI, De Klerk MA, Van Doorn PA (1998) The spectrum of antecedent infections in Guillain-Barré syndrome: A case-control study. Neurology 51:1110–1115.

Jung S, Kramer S, Schluesener HJ, Hunig T, Toyka KV, Hartung H-P (1992) Prevention and therapy of experimental autoimmune neuritis by an antibody against T cell receptors-alpha/beta. J Immunol 148:3768–3775.

Kaida K, Kusunoki S, Kamakura K, Motoyoshi K, Kanazawa I (2001) Guillain-Barré syndrome with IgM antibody to the ganglioside GalNAc-GDla. J Neuroimmunol 113:260–267.

KhaliliShirazi A, Gregson NA, Hall MA, Hughes RAC, Lanchbury JS (1997) T cell receptor V beta gene usage in Guillain-Barré syndrome. J Neurol Sci 145:169–176.

Khalili-Shirazi A, Gregson N, Gray I, Rees J, Winer J, Hughes R (1999) Antiganglioside antibodies in Guillain-Barré syndrome after a recent cytomegalovirus infection. J Neurol Neurosurg Psychiatry 66:376–379.

Kiefer R, Dangond F, Mueller M, Toyka KV, Hafler DA, Hartung HP (2000) Enhanced B7 costimulatory molecule expression in inflammatory human sural nerve biopsies. J Neurol Neurosurg Psychiatry 69:362–368.

Kieseier BC, Clements JM, Pischel HB, Wells GMA, Miller K, Gearing AJH, Hartung HP (1998) Matrix metalloproteinases MMP-9 and MMP-7 are expressed in experimental autoimmune neuritis and the Guillain-Barré syndrome. Ann Neurol 43:427–434.

Kieseier BC, Dalakas MC, Hartung HP (2002a) Immune mechanisms in chronic inflammatory demyelinating neuropathy. Neurology 59:S7–S12.

Kieseier BC, Tani M, Mahad D, Oka N, Ho T, Woodroofe N, Griffin JW, Toyka KV, Ransohoff RA, Hartung HP (2002b) Chemokines and chemokine receptors in inflammatory demyelinating neuropathies: A central role for IP-10. Brain 125:823–834.

Koga M, Yuki N, Tai T, Hirata K (2001) Miller Fisher syndrome and Haemophilus influenzae infection. Neurology 57:686–691.

Koga M, Gilbert M, Li J, Koike S, Takahashi M, Furukawa K, Hirata K, Yuki N (2005a) Antecedent infections in Fisher syndrome—A common pathogenesis of molecular mimicry. Neurology 64:1605–1611.

Koga M, Koike S, Hirata K, Yuki N (2005b) Ambiguous value of Haemophilus influenzae isolation in Guillain-Barré and Fisher syndromes. J Neurol Neurosurg Psychiatry 76:1736–1738.

Kolter T, Proia RL, Sandhoff K (2002) Combinatorial ganglioside biosynthesis. J Biol Chem 277:25859–25862.

Kussmaul A (1859) Zwei Fälle von Paraplegie mit tödlichem Ausgang ohne anatomisch nachweisbare oder toxische Ursache. Erlangen.

Kusunoki S, Chiba A, Kon K, Ando S, Arisawa K, Tate A, Kanazawa I (1994) N-acetylgalactosaminyl GD1a is a target molecule for serum antibody in Guillain-Barré syndrome. Ann Neurol 35:570–576.

Kusunoki S, Iwamori M, Chiba A, et al. (1996) GM1b is a new member of antigen for serum antibody in Guillain-Barré syndrome. Neurology 47:237–242.

Kusunoki S, Shiina M, Kanazawa I (2001) Anti-Gal-C antibodies in GBS subsequent to mycoplasma infection: Evidence of molecular mimicry. Neurology 57:736–738.

Kuwabara S, Mori M, Ogawara K, Hattori T, Yuki N (2001) Indicators of rapid clinical recovery in Guillain-Barré syndrome. J Neurol Neurosurg Psychiatry 70:560–562.

Landry JBO (1859) Note sur la paralysie ascendante aiguë. Gaz hebdomadaire Méd Chir 6:472–474, 486–488.

Le TB, Aszmann O, Chen YG, Royall RM, Brushart TM (2001) Effects of pathway and neuronal aging on the specificity of motor axon regeneration. Exp Neurol 167:126–132.

Lehmann HC, Hartung HP, Hetzel GR, Stuve O, Kieseier BC (2006) Plasma exchange in neuroimmunological disorders: Part 1: Rationale and treatment of inflammatory central nervous system disorders. Arch Neurol 63:930–935.

Lunn MP, Johnson LA, Fromholt SE, Itonori S, Huang J, Vyas AA, Hildreth JE, Griffin JW, Schnaar RL, Sheikh KA (2000) High-affinity anti-ganglioside IgG antibodies raised in complex ganglioside knockout mice: Reexamination of GD1a immunolocalization. J Neurochem 75:404–412.

Massaro ME, Rodriguez EC, Pociecha J, Arroyo HA, Sacolitti M, Taratuto AL, Fejerman N, Reisin RC (1998) Nerve biopsy in children with severe Guillain-Barré syndrome and inexcitable motor nerves. Neurology 51:394–398.

McKhann GM, Cornblath DR, Ho TW, Li CY, Bai AY, Wu HS, Yei QF, Zhang WC, Zhaori Z, Jiang Z, Griffin JW, Asbury AK (1991) Clinical and electrophysiological aspects of acute paralytic disease of children and young adults in northern China. Lancet 338:593–597.

McKhann GM, Cornblath DR, Griffin JW, Ho TW, Li CY, Jiang Z, Wu HS, Zhaori G, Liu Y, Jou LP, Liu TC, Gao CY, Mao JY, Blaser MJ, Mishu B, Asbury AK (1993) Acute motor axonal neuropathy: A frequent cause of acute flaccid paralysis in China. Ann Neurol 33:333–342.

Moran AP, Prendergast MM, Hogan EL (2002) Sialosyl-galactose: A common denominator of Guillain-Barré and related disorders? J Neurol sci 196:1–7.

Mori M, Kuwabara S, Miyake M, Dezawa M, Adachi-Usami E, Kuroki H, Noda M, Hattori T (1999) Haemophilus influenzae has a GM1 ganglioside-like structure and elicits Guillain-Barré syndrome. Neurology 52:1282–4.

Mori M, Kuwabara S, Miyake M, Noda M, Kuroki H, Kanno H, Ogawara K, Hattori T (2000) Haemophilus influenzae infection and Guillain-Barré syndrome. Brain 2000 Oct; 123 (Pt 10):2171–8 123:2171–2178.

Muller HD, Beckmann A, Schroder JM (2003) Inflammatory infiltrates in the spinal cord of patients with Guillain-Barré syndrome. Acta Neuropathol 106:509–517.

Nachamkin I (1997) Microbiologic approaches for studying Campylobacter species in patients with Guillain-Barré syndrome. J Infect Dis 176:S106–S114.

Nachamkin I, Ung H, Moran AP, Yoo D, Prendergast MM, Nicholson MA, Sheikh K, Ho T, Asbury AK, McKhann GM, Griffin JW (1999) Ganglioside GM1 mimicry in Campylobacter strains from sporadic infections in the United States. J Infect Dis 179:1183–1189.

Nachamkin I, Liu J, Li M, Ung H, Moran AP, Prendergast MM, Sheikh K (2002) Campylobacter jejuni from patients with Guillain-Barré syndrome preferentially expresses a GD(1a)-like epitope. Infect Immun 70:5299–5303.

O'Hanlon GM, Plomp JJ, Chakrabarti M, Morrison I, Wagner ER, Goodyear CS, Yin X, Trapp BD, Conner J, Molenaar PC, Stewart S, Rowan EG, Willison HJ (2001) Anti-GQ1b ganglioside antibodies mediate complement-dependent destruction of the motor nerve terminal. Brain 124:893–906.

Oberhelman RA, Taylor DN (2000) Campylobacter infections in developing countries. In: Campylobacter (Nachamkin I, Blaser MJ, eds), pp 139–153. Washington, D.C.: American Society for Microbiology.

Ogawa-Goto K, Funamoto N, Abe T, Nagashima K (1990) Different ceramide compositions of gangliosides between human motor and sensory nerves. J Neurochem 55:1486–1493.

Ogawa-Goto K, Funamoto N, Ohta Y, Abe T, Nagashima K (1992) Myelin gangliosides of human peripheral nervous system: An enrichment of GM1 in the motor nerve myelin isolated from cauda equina. J Neurochem 59:1844–1849.

Ogawara K, Kuwabara S, Mori M, Hattori T, Koga M, Yuki N (2000) Axonal Guillain-Barré syndrome: Relation to anti-ganglioside antibodies and Campylobacter jejuni infection in Japan. Ann Neurol 48:624–31.

Ogino M, Orazio N, Latov N (1995) IgG anti-GM1 antibodies from patients with acute motor neuropathy are predominantly of the IgG1 and IgG3 subclasses. J Neuroimmunol 58:77–80.

Oomes PG, Jacobs BC, Hazenberg MPH, Banffer JRJ, van der Meché FGA (1995) Anti-GM1 IgG antibodies and *Campylobacter jejuni* bacteria in Guillain-Barré syndrome: Evidence of molecular mimicry. Ann Neurol 38:170–175.

Paparounas K, O'Hanlon GM, O'Leary CP, Rowan EG, Willison HJ (1999) Anti-ganglioside antibodies can bind peripheral nerve nodes of Ranvier and activate the complement cascade without inducing acute conduction block in vitro (see comments). Brain 122:807–816.

Plomp JJ, Molenaar PC, O'Hanlon GM, Jacobs BC, Veitch J, Daha MR, Van Doorn PA, van der Meche FG, Vincent A, Morgan BP, Willison HJ (1999) Miller Fisher anti-GQ1b antibodies: Alpha-latrotoxin-like effects on motor end plates. Ann Neurol 45:189–199.

Prineas JW (1981) Pathology of the Guillain-Barré syndrome. Ann Neurol 9(Suppl):6–19.

Ramos-Alvarez M, Bessudo L, Sabin A (1969) Paralytic syndromes associated with noninflammatory cytoplasmic or nuclear neuronopathy: Acute paralytic disease in Mexican children, neuropathologically distinguishable from Landry-Guillain-Barré syndrome. J A M A 207:1481–1492.

Raphael JC, Chevret S, Hughes RA, Annane D (2002) Plasma exchange for Guillain-Barré syndrome. Cochrane Database Syst Rev CD001798.

Redford EJ, Kapoor R, Smith KJ (1997a) Nitric oxide donors cause a reversible block of conduction in rat central and peripheral demyelinated axons. J Physiology-London 499P:13–14.

Redford EJ, Smith KJ, Gregson NA, Davies M, Hughes P, Gearing AJ, Miller K, Hughes RA (1997b) A combined inhibitor of matrix metalloproteinase activity and tumour necrosis factor-alpha processing attenuates experimental autoimmune neuritis. Brain 120:1895–1905.

Rees JH, Thompson RD, Smeeton NC, Hughes RA (1998) Epidemiological study of Guillain-Barré syndrome in south east England. J Neurol Neurosurg Psychiatry 64:74–77.

Rees JH, Gregson NA, Hughes RAC (1995a) Anti-ganglioside GM1 antibodies in Guillain-Barré syndrome and their relationship to *Campylobacter jejuni* infection. Ann Neurol 38:809–816.

Rees JH, Vaughan RW, Kondeatis E, Hughes RAC (1995b) HLA-class II alleles in Guillain-Barré syndrome and Miller Fisher syndrome and their association with preceding *Campylobacter jejuni* infection. J Neuroimmunol 62:53–57.

Rhodes KM, Tattersfield AE (1982) Guillain-Barré syndrome associated with Campylobacter infection. Br Med J (Clin Res Ed) 285:173–174.

Saida K, Saida T, Brown MJ, Silberberg DH (1979a) In vivo demyelination induced by intraneural injection of antigalactocerebroside serum. Am J Pathol 95:99–116.

Saida T, Saida K, Dorfman SH (1979b) Experimental allergic neuritis induced by sensitization with galactocerebroside. Science 204:1103–6.

Saida T, Saida K, Lisak RP, Brown MJ, Silberberg DH, Asbury AK (1982) In vivo demyelinating activity of sera from patients with Guillain-Barré syndrome. Ann Neurol 11:69–75.

Santoro M, Uncini A, Corbo M, Staugaitis SM, Thomas FP, Hays AP, Latov N (1992) Experimental conduction block induced by serum from a patient with anti-GM1 antibodies. Ann Neurol 31:385–390.

Schott B (1982) History of the Syndrome of Guillain-Barré. Revue Neurol 138:931–938.

Sheikh KA, Nachamkin I, Ho TW, Willison HJ, Veitch J, Ung H, Nicholson M, Li CY, Wu HS, Shen BQ, Cornblath DR, Asbury AK, McKhann GM, Griffin JW (1998) Campylobacter jejuni lipopolysaccharides in Guillain-Barré syndrome: Molecular mimicry and host susceptibility. Neurology 51:371–378.

Sheikh KA, Deerinck TJ, Ellisman MH, Griffin JW (1999) The distribution of ganglioside-like moieties in peripheral nerves. Brain 122:449–460.

Sheikh KA, Zhang G, Gong Y, Schnaar RL, Griffin JW (2004) An anti-ganglioside antibody-secreting hybridoma induces neuropathy in mice. Ann Neurol 56:228–239.

Sheikh KA, Ramos-Alvarez M, Jackson AC, Li CY, Asbury AK, Griffin JW (2005) Overlap of pathology in paralytic rabies and axonal Guillain-Barré syndrome. Ann Neurol 57:768–772.

Spies JM, Pollard JD, Bonner JG, Westland KW, McLeod JG (1995a) Synergy between antibody and P2-reactive T cells in experimental allergic neuritis. J Neuroimmunol 57:77–84.

Spies JM, Westland KW, Bonner JG, Pollard JD (1995b) Intraneural activated T cells cause focal breakdown of the blood-nerve barrier. Brain 118:857–868.

Steininger C, Popow-Kraupp T, Seiser A, Gueler N, Stanek G, Puchhammer E (2004) Presence of cytomegalovirus in cerebrospinal fluid of patients with Guillain-Barré syndrome. J Infect Dis 189:984–989.

Susuki K, Nishimoto Y, Yamada M, Baba M, Ueda S, Hirata K, Yuki N (2003) Acute motor axonal neuropathy rabbit model: Immune attack on nerve root axons. Ann Neurol 54:383–388.

Susuki K, Odaka M, Mori M, Hirata K, Yuki N (2004) Acute motor axonal neuropathy after Mycoplasma infection: Evidence of molecular mimicry. Neurology 62:949–956.

Svennerholm L, Bostrom K, Fredman P, Jungbjer B, Mansson JE, Rynmark BM (1992) Membrane lipids of human peripheral nerve and spinal cord. Biochim Biophys Acta 1128:1–7.

Svennerholm L, Bosträm K, Fredman P, Jungbjer B, Lekman A, Månsson J-E, Rynmark B-M (1994) Gangliosides and allied glycosphingolipids in human peripheral nerve and spinal cord. Biochim Biophys Acta 1214:115–123.

Takigawa T, Yasuda H, Kikkawa R, Shigeta Y, Saida T, Kitasato H (1995) Antibodies against GM1 ganglioside affect K^+ and Na^+ currents in isolated rat myelinated nerve fibers. Ann Neurol 37:436–442.

Takigawa T, Yasuda H, Terada M, Haneda M, Kashiwagi A, Saito T, Saida T, Kitasato H, Kikkawa R (2000) The sera from GM1 ganglioside antibody positive patients with Guillain-Barré syndrome or chronic inflammatory demyelinating polyneuropathy blocks Na+ currents in rat single myelinated nerve fibers. Intern Med 39:123–127.

Taylor WA, Hughes RAC (1989) T lymphocyte activation antigens in Guillain-Barré syndrome and chronic idiopathic demyelinating polyradiculoneuropathy. J Neuroimmunol 24:33–39.

Uncini A, Lugaresi A (1999) Fisher syndrome with tetraparesis and antibody to GQ1b: Evidence for motor nerve terminal block. Muscle Nerve 22:640–644.

Uncini A, Santoro M, Corbo M, Lugaresi A, Latov N (1993) Conduction abnormalities induced by sera of patients with multifocal motor neuropathy and anti-GM1 antibodies. Muscle Nerve 16:610–615.

Van Rhijn I, van den Berg LH, Ang CW, Admiraal J, Logtenberg T (2003) Expansion of human gamma delta T cells after in vitro stimulation with Campylobacter jejuni. Internat Immunol 15:373–382.

Visser LH, van der Meché FGA, Meulstee J, Rothbarth PPh, Jacobs BC, Schmitz PIM, Van Doorn PA, Dutch Guillain-Barré Study Group (1996) Cytomegalovirus infection and Guillain-Barré syndrome: The clinical, electrophysiologic, and prognostic features. Neurology 47:668–673.

Waksman BH, Adams RD (1955) Allergic neuritis: Experimental disease in rabbits induced by the injection of peripheral nervous tissue and adjuvants. J Exp Med 102:213–235.

Willison HJ, Veitch J (1994) Immunoglobulin subclass distribution and binding characteristics of anti-GQ1b antibodies in Miller Fisher syndrome. J Neuroimmunol 50:159–165.

Willison HJ, Yuki N (2002) Peripheral neuropathies and anti-glycolipid antibodies. Brain 125:2591–2625.

Willison HJ, Veitch J, Patterson G, Kennedy PGE (1993) Miller Fisher syndrome is associated with serum antibodies to GQ1b ganglioside. J Neurol Neurosurg Psychiatry 56:204–206.

Winer JB, Hughes RAC, Osmond C (1988) A prospective study of acute idiopathic neuropathy. 1. Clinical features and their prognostic value. J Neurol Neurosurg Psychiatry 51:605–612.

Winer J, Hughes S, Cooper J, Ben Smith A, Savage C (2002) gamma delta T cells infiltrating sensory nerve biopsies from patients with inflammatory neuropathy. J Neurol 249:616–621.

Wirguin I, Suturkova-Milosevic LJ, Della-Latta P, Fisher T, Brown RH, Jr, Latov N (1994) Monoclonal IgM antibodies to GM1 and asialo-GM1 in chronic neuropathies cross-react with *Campylobacter jejuni* lipopolysaccharides. Ann Neurol 35:698–703.

Wirguin I, Briani C, Suturkova-Milosevic L, Fisher T, Della-Latta P, Chalif P, Latov N (1997) Induction of anti-GM1 ganglioside antibodies by Campylobacter jejuni lipopolysaccharides. J Neuroimmunol 78:138–142.

Yoshino H (1997) Distribution of gangliosides in the nervous tissues recognized by axonal form of Guillain-Barré syndrome (in Japanese). Neuroimmunology 5:174–175.

Yu Z, Lennon VA (1999) Mechanism of intravenous immune globulin therapy in antibody-mediated autoimmune diseases. N Engl J Med 340:227–228.

Yu RK, Saito M (1989) Structure and localization of gangliosides. In: Neurobiology of Glycoconjugates (Margolis RU, Margolis RK, eds), pp 1–42. Plenum Publishing Corporation.

Yuki N, Tagawa Y (1998) Acute cytomegalovirus infection and IgM anti-GM2 antibody. J Neurol Sci 154:14–17.

Yuki N, Yoshino H, Sato S, Miyatake T (1990) Acute axonal polyneuropathy associated with anti-GM1 antibodies following *Campylobacter* enteritis. Neurology 40:1900–1902.

Yuki N, Sato S, Itoh T, Miyatake T (1991) HLA-B35 and acute axonal polyneuropathy following *Campylobacter jejuni* infection. Neurology 41:1561–1563.

Yuki N, Handa S, Taki T, Kasama T, Takahashi M, Saito K (1992) Cross-reactive antigen between nervous tissue and a bacterium elicits Guillain-Barré syndrome: Molecular mimicry between gangliocide GM1 and lipopolysaccharide from Penner's serotype 19 of *Campylobacter jejuni*. Biomed Res 13:451–453.

Yuki N, Sato S, Tsuji S, Ohsawa T, Miyatake T (1993a) Frequent presence of anti-GQ1b antibody in Fisher's syndrome. Neurology 43:414–417.

Yuki N, Yamada M, Sato S, Ohama E, Kawase Y, Ikuta F, Miyatake T (1993b) Association of IgG anti-GD1a antibody with severe Guillain-Barré syndrome. Muscle Nerve 16:642–647.

Yuki N, Taki T, Takahashi M, Saito K, Yoshino H, Tai T, Handa S, Miyatake T (1994) Molecular mimicry between GQ_{1b} ganglioside and lipopolysaccharides of *Campylobacter jejuni* isolated from patients with Fisher's syndrome. Ann Neurol 36:791–793.

Yuki N, Taki T, Handa S (1996) Antibody to GalNAc-GD1a and GalNAc-GM1b in Guillain-Barré syndrome subsequent to Campylobacter jejuni enteritis. J Neuroimmunol 71:155–161.

Yuki N, Ho TW, Tagawa Y, Koga M, Li CY, Hirata K, Griffin JW (1999) Autoantibodies to GM1b and GalNAc-GD1a: Relationship to Campylobacter jejuni infection and acute motor axonal neuropathy in China. J Neurol Sci 164:134–138.

Yuki N, Yamada M, Koga M, Odaka M, Susuki K, Tagawa Y, Ueda S, Kasama T, Ohnishi A, Hayashi S, Takahashi H, Kamijo M, Hirata K (2001) Animal model of axonal Guillain-Barré syndrome induced by sensitization with GM1 ganglioside. Ann Neurol 49:712–20.

Yuki N, Susuki K, Koga M, Nishimoto Y, Odaka M, Hirata K, Taguchi K, Miyatake T, Furukawa K, Kobata T, Yamada M (2004) Carbohydrate mimicry between human ganglioside GM1 and Campylobacter jejuni lipooligosaccharide causes Guillain-Barré syndrome. Proc Natl Acad Sci U S A 101:11404–11409.

Zehntner SP, Brisebois M, Tran E, Owens T, Fournier S (2003) Constitutive expression of a costimulatory ligand on antigen-presenting cells in the nervous system drives demyelinating disease. FASEB J 17:1910–1912.

Zhu J, Mix E, Olsson T, Link H (1994) Cellular messenger-Rna expression of interferon-gamma, Il-4 and transforming growth-factor-beta (Tgf-Beta) by rat mononuclear-cells stimulated with peripheral-nerve myelin antigens in experimental allergic neuritis. Clin Exp Immunol 98:306–312.

Zhu J, Bai XF, Mix E, Link H (1997) Experimental allergic neuritis: Cytolysin mRNA expression is upregulated in lymph node cells during convalescence. J Neuroimmunol 78:108–116.

Zhu J, Zou LP, Zhu SW, Mix E, Shi FD, Wang HB, Volkmann I, Winblad B, Schalling M, Ljunggren HG (2001a) Cytotoxic T lymphocyte-associated antigen 4 (CTLA-4) blockade enhances incidence and severity of experimental autoimmune neuritis in resistant mice. J Neuroimmunol 115:111–117.

Zhu Y, Ljunggren HG, Mix E, Li HL, van der Meide P, Elhassan AM, Winblad B, Zhu J (2001b) CD28-B7 costimulation: A critical role for initiation and development of experimental autoimmune neuritis in C57BL/6 mice. J Neuroimmunol 114:114–121.

21
Autoimmune Disease

M. Patricia Leuschen

Keywords Alkylosing spondylitis; Crohn's disease; HLA B27; Keratinocytes; Molecular mimicry; Sjogren's syndrome; Subluxation; Thalidomide

21.1. Introduction

Historically, the concept of autoimmunity centers on the inability to adequately identify some of the body's own proteins. In an autoimmune disease, the immune system inappropriately recognizes those proteins leading to a pathologic humoral and/or cell-mediated immune reaction. Both the "self" protein antigens and their site determine the specific end organ damage. Autoimmune diseases are a diverse group of disorders with relatively poorly understood pathogenesis. Often diagnosed as end-organ specific, autoimmune diseases usually involve multiple sites including the nervous system. In addition, more than one autoimmune disease is frequently diagnosed clinically in the same patient (Brinar et al., 2006).

21.1.1. Autoimmune Nervous System Disease Processes

The classic hypothesis predicts that auto-antibodies are the primary mediators of organ damage in any specific autoimmune disorder. However, this simplistic view does not explain the heterogeneous presentation in patients. In many "autoimmune" disorders evidence supports a role for auto-reactive T cells and genetic control of end organ susceptibility. End organ damage may be due to an interaction between innate and adaptive immunity. While the detection of serum auto-antibodies remains a major marker for clinical diagnosis of disease, auto-antibodies to classic antigens are neither required nor sufficient for end organ damage (Bagavant and Fu, 2005). With multi system involvement and specifically nervous system involvement associated with a wide range of autoimmune disorders, etiology and pathogenesis also become more complex.

Immunosuppressant therapies now make many of the autoimmune disorders treatable but not curable. The role of specific immunotherapies in treatment of specific autoimmune disorders (and not others) will be reviewed. In addition, paradigms relating to the immune system and autoimmune disorders such as the conflicting role of inflammation as both a positive and negative factor and the role of infections in initiating as well as abrogating autoimmunity will be reviewed.

Inflammation is a hallmark of all the autoimmune disorders. The target of either the primary or secondary inflammatory autoimmune response is usually systemic. In Sjogren's syndrome, salivary and lacrimal glands are the primary targets while in rheumatoid arthritis (RA) the primary targets are the synovial joints. Systemic autoimmune responses often go through periods of inflammatory exacerbation and remission. With continuous inflammatory cycles the systemic autoimmune disorder can involve the central nervous system (CNS) demyelination. CNS demyelination can be miss-diagnosed as multiple sclerosis (MS) but more often meets diagnostic criteria for disseminated encephalomyelitis that is the result of vasculitis. CNS demyelination may also be a post-infectious manifestation associated with other autoimmune disorders (Brinar et al., 2006). To understand the transition from a single end organ auto-antibody reaction to multi system involvement that includes the nervous system involves understanding the role of the blood brain barrier and lymphocyte trafficking under inflammatory and pathologic conditions. For example, the increased expression of the matrix metalloproteinase, MMP-9, in the degradation of components of the endothelial extra-cellular matrix is associated with the role of MMPs in vasculitic neuropathy (Gurer et al., 2004).

Vasculitis is a common feature of disseminating autoimmune disorders and refers to inflammation of the blood vessels, both arteries and veins of varying caliber. Neurologically, it is often associated with ischemic injury but may be caused by specific drugs, infectious agents or malignant processes (Younger, 2004). The inflammation associated with vasculitic neuropathy occurs as a hypersensitivity reaction to antigens of perceived non-self proteins. In vasculitic neuropathy, mononuclear cells in the perivascular infiltrate express T cell restricted intracellular antigens,

T. Ikezu and H.E. Gendelman (eds.), *Neuroimmune Pharmacology*.
© Springer 2008

which induce apoptosis of other inflammatory mononuclear cells and may affect recovery (Heuss et al., 2000). Cyclooxygenase-2 mRNA is upregulated in endoneural macrophages resulting in an increased production of prostaglandins. Thus, cyclooxygenase-2 may be a potential target for therapy particularly early in the autoimmune disease process (Hu et al., 2003). Increased cytokine expression in vasculitic neuropathy positively correlates with neuropathic pain (Lindenlaub and Sommer, 2003) and may be influenced by a differential expression of pain related neurotropic factors and their soluble receptors (Yamamoto et al., 2003). Binding of ligand, N^{ε} –(carboxymethyl) lysine, to the receptor for advanced glycation end products results in activation of the pro-inflammatory transcription factor, nuclear factor-[kappa]B (NF[kappa]B) and subsequent expression of NF-[kappa]B-regulated cytokines. This pathway is hypothesized to play a role in both the initiation and the maintenance of inflammation during vasculitic neuropathy (Haselbeck et al., 2004).

21.1.2. Gender and Autoimmune Disorders

Most autoimmune disorders are highly gender specific (Table 21.1). Sjogren's syndrome affects women more frequently than men at a 9:1 ratio while Crohn's disease affects men and women in approximately equal numbers. Alkylosing spondylitis and Reiter's syndrome usually affect more young white men. Often severity of symptoms and progression are more severe in the gender less commonly affected. While MS affects women more frequently than men in a 2:1 or 3:1 ratio, many affected men have a more aggressive and progressive form of MS than women. A correlation also exists between gender specificity of the autoimmune disease and the efficacy of specific immunosuppressant therapies.

An example of the gender dilemma is illustrated in systemic lupus erythematosus (SLE) an autoimmune disease that occurs more frequently in women than men. The gender difference is attributed to differences in the metabolism of sex hormones or regulation by gonadotropin releasing hormone (GnRH). When it occurs in men, SLE tends to be more severe (Yacoub Wasef, 2004). Recent studies also indicate that SLE may be more common in nonwhite women of child bearing age. However, evidence collected over the last quarter century indicates that SLE is increasing in nonwhite South African men (Faller et al., 2005).

The gender bias for some autoimmune disease can have prognostic value. In MS, a CNS demyelinating disease, more women are affected (Paty et al., 1998). However, male gender is predictive of a shorter time to reach a need for assisted walking devices. Men also have a higher rate of cerebellar involvement and a higher risk of primary-progressive disease (Whitacre et al., 1999).

21.1.3. Demographics

Specific autoimmune disorders are often associated with racial or ethnic groups. Genetic haplotypes have been identified in association with some autoimmune disorders. However, genetic factors are not enough to explain the heterogeneity in many of the autoimmune disorders. On the other hand, just as gender often correlates with severity or type of symptoms in most autoimmune disorders, race and ethnicity are associated with specific sub groups within a specific autoimmune disorder.

21.1.4. Proposed Mechanisms and Treatment

The presence of autoantibodies explains some of the symptoms of autoimmune disease but not all. Theories on the role of infection (both viral and bacterial), genetic markers and dys-regulation of T and B cell function are all proposed mechanisms for at least one autoimmune disease. Specific mechanisms will be discussed in conjunction with specific autoimmune disorder. However, one of the paradigms associated with autoimmune disease is the role of infections. In some autoimmune disorders, such as Crohn's disease and alkylosing spondylosis, bacterial infections are linked with initiating the disease in genetically susceptible individuals. In others, such as MS, infections are frequently linked with relapses or exacerbations. Conversely, improved hygiene and living condition worldwide have decreased exposure to foreign antigens and infectious agents but the global incidence of autoimmune disorders has increased (Christen and von Herrath, 2005).

Historically, newly diagnosed patients with autoimmune disorders or those with minor symptoms are treated with symptom

TABLE 21.1. Demographics of Autoimmune Disease.

	Incidence by gender women:men	Demographic considerations
Rheumatoid arthritis	2–3:1	Diagnosis usually 17–35 yrs
Sjogren's syndrome	9:1	Diagnosis > 40 yrs
Systemic lupus erythmatosis	8–9:1	Increased in non-white women
Alkylosing spondylitis	<1:1	Increased in young men
Crohn's syndrome	1:1	
Reiter's Disease (Reactive arthritis)	1:1 after enteric infection	Increased in young white men but increasing in undeveloped world
	1:9 after STD	
Psoriasis	Slight increase in women: men	
Multiple Sclerosis	3:1	Diagnosis usually 20–40yrs

specific therapy such as the use of artificial tearing agents for the dry eyes associated with Sjogren's syndrome. As the understanding of the underlying immune mechanisms has improved, treatment of autoimmune disease has increasingly involved the use of immunologic agents. However, a therapy that is effective in one autoimmune disorder may not work in others. Corticosteroids and other compounds that target the inflammation associated with the disorder are an example of therapies that show some efficacy in most autoimmune disorders. Others such as Interferon (IFN)β for MS and the TNFα anatagonists for Crohn's disease seem relatively disease specific.

Immunomodulator therapy is commonly used for severe autoimmune disease that is refractive to standard therapy. For example, where activated T cells are implicated, cyclophosphamide is often efficacious. It is immunoablative but not myeloablative, permitting the patient's endogenous stem cells to repopulate the hematopoietic/immune systems (Drachman and Brodsky, 2005). A recent review indicates that over 700 patients have received autologous hematopoietic stem cell transplants (AHSCT) as treatment for severe autoimmune disease primarily as part of phase I/II trials following international guidelines (Tyndall and Saccradi, 2005). The results have led to phase III prospective randomized controlled trials of AHSCT in MS, SLE, and RAs in Europe. In the US, similar trials are planned for systemic sclerosis, MS and SLE. A major concern involves which patients would benefit from such invasive therapy. The early trials of AHSCT for MS involved patients with severely progressive disease. While progression was slowed, areas of disability that were present prior to the AHSCT, as might be expected, persisted (Nash et al., 2003;Kozak et al., 2001; Healey et al., 2004). Results suggest the use of AHSCT in MS patients with much lower disability in an attempt to preserve function and to minimize the risk of further degenerative changes (Burt et al., 2003a), However, there are no quantitative markers to identify early who will develop MS severe enough to warrant AHSCT. Similarly, at the one year follow-up patients with severe Crohn's disease treated with AHSCT had a return of low levels of auto-antibodies. At present AHSCT should only be used in clinical trials at experienced centers and long term benefits are not known. Even though each autoimmune disease has a different immune profile, over one third of transplanted patients have sustained remission and those who relapse often respond to immunotherapy that was ineffective prior to transplantation (Tyndall and Saccradi, 2005).

21.2. Autoimmunity as It Applies to the Nervous System

21.2.1. The Blood-Brain Barrier and Lymphocyte Trafficking

For CNS involvement as the primary or secondary target of an autoimmune disease, activated immune cells must be able to enter and leave the CNS. The route for lymphocyte trafficking across the cerebrovascular endothelium into and out of the brain parenchyma is more complex than in systemic tissue. First, the cerebrovascular unit, blood brain barrier (BBB), is designed as a selective barrier consisting of the endothelial cells and associated basement lamina. Unlike microvessels in many areas, the cerebrovascular microvessel basement lamina is well developed; tight junctions between adjacent cerebrovascular endothelial cells prevent molecular and cellular movement between the cells and perivascular astrocytic end plates are positioned between the microvessels and neuronal axons. Second, the cerebrovascular endothelium is designed to facilitate lymphocyte trafficking including attachment, rolling along the plasma membrane and movement through the endothelial cell. Recent reviews discuss the structure and interactions between the cerebrovascular endothelial cells and astrocytes at the BBB (Anderson and Nedergaard, 2003; Nedergaard et al., 2003; Abbott et al., 2006). One role of the BBB and its cellular constituents is to precisely regulate the microenvironment for neuronal signaling. Rather than acting simply as a barrier, the BBB is a dynamic system where permeability and transport are modulated to control the ionic characteristics of the CNS and regulate its volume. The BBB acts to separate pools of neurotransmitters and neuroactive agents that act within the CNS and in the blood stream so that similar agents can be active without 'cross talk" (Cser and Bundgaard, 1984). In vitro studies indicate that brain endothelium is different from endothelium of peripheral microvessels. Tight junctions are more complex and restrict movement of even small ion such as $Na+$ and Cl^- so that the trans-endothelial electrical resistance which is typically 2 to 20 ohm/cm^2 in the peripheral capillaries can be > 1,000 ohms/cm^2 in brain endothelium (Butt et al., 1990). Among the molecules that contribute to the tight junction structure are the transmembrane proteins, occludins and claudins. Occludin is a 60–65 kDa protein with a carboxy - terminal domain that can link with zonula occludens proteins and is hypothesized to regulate tight junctions (Hawkins and Davis, 2005). In the BBB expression of the proteins, claudin 3 (often termed claudin 1/3), claudin 5 and possibly claudin 12 appear to play a role in the high transendothelial electrical resistance (Wolberg and Lippoldt, 2002).

There is strong in vitro evidence that astrocytes up-regulate many of the features related to tight junctions as physical and metabolic barriers. Transport barrier function includes regulating expression and polarized localization of transporters including Pgp (Schinkel, 1999) and GLUT1 (McAllister et al., 2001). Polarization of brain endothelial cells or the presence of orientation and apical-basal properties is a characteristic usually seen in epithelial cells but not endothelial cells. The PAR3-atypical protein kinase C (aPKC-PAR6 complex) may be involved in regulating the cell polarity of brain endothelial cells, (Abbott et al., 2006). Metabolic barrier functions are regulated by still other specialized enzyme systems of astrocytes (Abbott, 2002; Hayashi et al., 1997; Sobue et al., 1999).

Complex mechanisms regulate molecular trafficking as they move into and out of the CNS and pathologic conditions may change this pattern. Since many of the neurologic autoimmune disorders are postulated to involve trafficking of activated T cells across the BBB, the chemokine receptor, CXCR3, has received significant interest. CXCR3 is the receptor for CXCL9/MIG, CXCL101P-10 and CXCL11/I-TAC and is preferentially expressed on activated Th1 T cells. While CXCR3 is hypothesized to play a role in the trafficking of Th1 T cells, studies in animal models such as EAE and transgenic mice using receptor blockade have been problematic (Liu et al., 2005). The conflicting results indicate that the role of CXCR3 and its ligands may be more complex in some types of CNS inflammation than previously thought.

21.2.2. Inflammation: A Double Edged Sword

While inflammation is associated negatively with vasculitis and autoimmune disease, it is also necessary to protect normal cells from viral and bacterial infection. Many of the therapies that act positively as immunosuppressants have an initial inflammatory response. While IFNβ is an effective therapy in the inflammatory phases of MS, the initial response to each injection is an increase in inflammatory cytokines, IL-2 and IFNγ (Elliott et al., 2001). Conversely, natalizumab a therapy that showed great potential in phase I/II trials for aggressive MS by blocking activated lymphocyte trafficking into the CNS was withdrawn because of deaths due to progressive multifocal leukoencephalopathy.

21.3. Autoimmune Diseases with Nervous System Involvement

In reviewing specific autoimmune diseases, the primary involvement is often not the nervous system. However, neurologic symptoms and involvement frequently accompany most autoimmune disorders (Table 21.2).

21.3.1. Rheumatoid Arthritis

21.3.1.1. Epidemiology

Rheumatoid arthritis (RA) is the most common autoimmune diseases and a chronic systemic inflammatory disease affecting both the joints and extra-articular tissue. Often the disease is progressive with the major symptoms including painful, stiffness and swelling of joints culminating in significant morbidity and increased mortality (Khurnana and Berney, 2005). Neurologic complications occur in moderate to severe RA either as a result of disease erosive effects on joints and bones or caused by the disease itself (compressive rheumatoid nodule or rheumatoid vasculitis (Chin and Latov, 2005).

A familial predisposition and genetic markers are linked to an increased risk for developing RA. In the last decade, it became known that there is often an extended period, often years, when specific immunologic markers may be identified before the inflammatory manifestations of RA occur (Masi and Aldag, 2005). Rheumatoid factors and related antibodies occur in conjunction with about half of the presymptomatic manifestations. Significant cigarette smoking is a major risk factor for RA for post-menopausal women and for men. For pre-menopausal onset RA in women, cytokine imbalance and insufficiency of adrenal cortical function may precede inflammatory onset of RA In men, multiple hormonal and cytokine correlations can be found years before onset of RA implicating a long term activation or dysfunction in the neuroendocrine-immune system (Masi and Aldag, 2005).

21.3.1.2. Multi System Involvement with Emphasis on Neurologic Symptoms

Rheumatoid vasculitis affecting the CNS is rare but may present as seizures, dementia, hemiparesis, cranial nerve palsy, blindness, hemispheric dysfunction, cerebellar ataxia or dysphasia (Vollertsen and Conn, 1990; Ando et al., 1995) Expression of

TABLE 21.2. Primary tissue and neurologic involvement in autoimmune diseases.

	Primary tissue involvement	Neurologic Involvement
Rheumatoid Arthritis	Joints & extra-articular tissue	Rare CNS involvement Peripheral neuropathies common
Sjogren's syndrome	Salivary and lacrimal glands	Variable; Sensory ataxic, painful, peripheral and trigeminal neuropathies common
Systemic lupus erythmatosis	Connective tissue	Variable; transient ischemic attacks, seizures, psychosis, cognitive disorders, dementia and delirium reported
Alkylosing spondylitis	Intervertebral joints and sacroiliac joint	Eye inflammations and caudae equinae related neuropathies
Reiter's Disease (reactive arthritis)	Vertebral joints	>15% develop alkylosing spondylitis or severe arthritis with neurologic sequelae
Crohn's syndrome	Intestinal tract	CNS and peripheral involvement seen in > 30%; neurologic features variable
Psoriasis and psoriatic arthritis	Skin Joints and connective tissue	Peripheral neuropathy related to spinal involvement
Multiple Sclerosis	Central nervous system	Central nervous system

matrix metalloproteinases have been linked to the vasculitic neuropathy associated with RA (Gurer et al., 2004).

Peripheral neuropathies are common in RA. The cervical spine is frequently involved in RA and atlantoaxial subluxation is the most common type of instability (Tumialan et al., 2004). Peripheral neuropathy is present in as many as 70% of patients with advanced RA. Those with RA for more than 10 years and onset after age 50 are particularly at risk (Naranjo et al., 2004). Rheumatoid changes to the synovial joints between the dens and the atlas anteriorly and the dens and transverse ligament posteriorly compress the spinal cord. Patients report neck pain and paresthesia with myelopathic findings on hyperflexia, weakness, gait abnormalities, flexor spasms, sphincter disturbances or sensory changes. The degenerative changes in the cervical spine may also compress the vertebral arteries resulting in vertebrobasilar insufficiency manifested by nausea, vertigo, diplopia and dysphasia (Tumialan et al., 2004).

21.3.1.3. Hypotheses for Etiology and Treatment

Inflammation of the synovial joints is the primary symptom in early RA. The dense irregular connective tissue that forms the synovial membrane of the articular capsule thickens. Synovial fluid consists of hyaluronic acid produced by fibroblast-like cells in the membrane and interstitial fluid that may include phagocytic cells filtered from the blood plasma. As the synovial membrane thickens, the synovial fluid accumulates causing pressure and the pain and inflammation characteristic of RA. If the inflammation is untreated, granulation tissue adheres to the articular cartilage and the autoimmune response invades and erodes the cartilage leaving the fibrous tissue to join the ends of the bones. In severe cases, the fibrous tissue ossifies and the bones fuse becoming immobile.

A comprehensive review discusses the therapeutic management of RA (Turesson and Matteson, 2004). Epidemiological studies link extra-articular rheumatoid arthritis manifestations with premature mortality and support aggressive anti rheumatoid therapies for those patients. Cyclophosphamide is favored in patients with systemic rheumatoid vasculitis and methotrexate in those cases with other manifestations of extra-articular rheumatoid arthritis (Turesson and Matteson, 2004). Cyclophosphamide and TNFα inhibitors such as infliximab have some positive success in treatment resistant vasculitis associated with RA (Unger et al., 2003). However, TNFα inhibitors have also been associated with the opposite effect, an induction of extra articular rheumatoid arthritis so their use should be used only in specific cases when close monitoring is in place.

21.3.2. Sjogren's Syndrome

21.3.2.1. Epidemiology

Sjogren's syndrome is a common chronic autoimmune disease that affects the salivary and lacrimal glands, with associated lymphocytic infiltration (Fox, 2005). Primary Sjogren's involves the exocrinopathy alone but Sjogren's can also occur in conjunction with another autoimmune disorder such as RA or SLE. Major symptoms include dryness of the mucous membranes of the mouth, eyes and nose due to the involvement of the lacrimal and salivary glands. Diagnostic criteria require clinical symptoms of dryness, characteristic features on biopsy of a minor salivary gland or auto-antibodies, anti-SS-A or anti-SS-B. Primary Sjogren's syndrome patients also often have antinuclear antibodies.

Numbers vary but it is estimated that between 1 and 4 million people have Sjogren's syndrome (National Institute Arthritis, Musculoskeletal Skin Disease, 2005). The majority of those diagnosed with Sjogren's are women (90%) and the syndrome is usually diagnosed after age 40 but may occur at any age. Sjogren's syndrome affects all races and ethnic backgrounds.

21.3.2.2. Multi System Involvement with Emphasis on Neurologic Symptoms

Multi-system involvement may include the skin, lung, heart, kidneys hematological and lymphoproliferative disorders as well as the nervous system, both peripheral and the CNS. The nature of the neuropathies associated with Sjogren's syndrome varies. In an assessment of 92 patients with primary Sjogren's associated neuropathy, the majority of patients (93%) were diagnosed with Sjogren's syndrome after the neuropathic symptoms appeared (Mori et al., 2005). The study of 76 women and 16 men classified their neuropathies into seven forms: sensory ataxic neuropathy (n = 36), painful sensory neuropathy without sensory ataxia (n = 18), multiple mononeuropathy (n = 11), multiple cranial neuropathy (n = 5), trigeminal neuropathy (n = 15), and autonomic neuropathy (n = 3). Clinicopathological observations suggest sensory ataxic, painful and perhaps trigeminal neuropathy are related to ganglioneuropathic processes. Multiple mononeuropathy and multiple cranial neuropathy is more closely associated with vasculitis. Acute or subacute onset of the neuropathy was seen more frequently in multiple mononeuropathy and multiple cranial neuropathy. Chronic progression predominated in the other forms of neuropathy. Sensory symptoms predominate over motor involvement in sensory ataxia, painful sensory, trigeminal and autonomic neuropathy. Motor weakness and muscle atrophy are observed in multiple mononeuropathy, multiple cranial neuropathy and radiculoneuropathy. Autonomic symptoms are often seen in all forms of neuropathy. Sural nerve biopsy specimens (n = 55) reveal variable axonal loss. Large fiber loss predominates in sensory ataxic neuropathy while small fiber loss occurs in painful sensory neuropathy. Perivascular cell invasion is seen most frequently in multiple mononeuropathy followed by sensory ataxic neuropathy. Multifocal T cell invasion is seen in the dorsal root and sympathetic ganglion, perineural space and vessel walls in the nerve trunks.

21.3.2.3 Hypotheses for Etiology and Treatment

Both environment and genetics are hypothesized to play a role in the etiology of Sjogren's syndrome. A single gene has not been identified. There is evidence that one gene may predispose Caucasian's to the disease since the frequency of the haplotype HLA-DRB1*0301-DRB3*0101-DQA1*0501-DQB1*0201 is significantly increased ($p < 0.001$. Other genes with an increased frequency linked to primary Sjogren's disease in those of Japanese descent include haplotype HLA-DRB1*0405-DRB4*0101-DQA1*0301-DQB1*040 ($p < 0.05$), in those of Chinese descent the haplotype DRB1*0803-DQA1*0103-DQB1*0601 ($p < 0.05$) (Kang et al., 1993). No single class II allele is associated with primary Sjogren's syndrome in the different ethnic groups but a shared amino acid motif in the DQB1 first domain is present in each disease associated haplotype (Kang et al., 1993).

Immunological studies on salivary and lacrimal glands yield conflicting results on the Th1/Th2 balance in Primary Sjogrens's syndrome. A recent study in Finnish Caucasians and healthy controls shows no significant difference in the genes encoding for cytokines involves in the regulation of the Th1/Th2 differentiation (Pertovaara and Hurme, 2006). Their findings support the hypothesis that primary Sjogren's syndrome is primarily a Th1 mediated autoimmune disease.

Sjogren's is also hypothesized to be triggered by a virus, which initiates the inflammatory response in lacrimal and salivary glands. The genetic factors play a role in modulating the immune response to the inflammation. In a normally mounted immune response to inflammation, apoptosis of the activated immune cells reverses the process once the virally infected cells are neutralized. In Sjogren's syndrome the normal apoptotic process does not occur due to a combination of genetic and neuroendocrine factors.

The usual therapy includes topical agents that improve moisture and decrease inflammation. Systemic therapy includes steroids and non-steroidal anti-inflammatory agents; disease modifying agents and cytotoxic agents to address the extra-glandular aspects of the disorder.

21.3.3. Systemic Lupus Erythematosus

21.3.3.1. Epidemiology

Systemic lupus erythematosus (SLE) is a chronic autoimmune, inflammatory disease of connective tissue that is classified as a rheumatoid disease. Tissue damage can occur in any body system and can range from mild to a rapidly fatal disease. As with many autoimmune disorders, SLE usually has periods of exacerbation and remission. Symptoms include painful joints, low grade fever, fatigue, mouth ulcers, weight loss, enlarged lymph nodes and spleen, sensitivity to sunlight, rapid loss of scalp hair and anorexia. A distinguishing feature is the eruption across the bridge of the nose and cheeks often termed a "butterfly" rash. This and other skin ulcerations were originally thought to resemble the bite of a wolf thus the term "lupus" became applied to the disorder. The most serious complications of SLE involve inflammation of kidneys, liver, spleen, lungs, heart, gastrointestinal tact and brain.

The prevalence of SLE is 1 in 2,000 persons with females more likely to be affected by a ratio of 8 or 9 to 1. SLE also occurs more frequently in nonwhite women during their child bearing years. Childhood onset of SLE is relatively rare and includes a diverse array of presenting features sufficient to meet American College of Rheumatology (ACR) criteria for SLE. In a review of childhood onset SLE patients in South Africa, the male to female ratio is 1:2.6 overall with a ratio of 1:1.2 if diagnosed under 10 years of age (Faller et al., 2005).

21.3.3.2. Multi System Involvement with Emphasis on Neurologic Symptoms

SLE and its most serious complications can involve inflammation of the central nervous system as well as kidneys, liver, spleen, lungs, heart, and gastrointestinal tact. Neuropsychiatric SLE became a case definition to describe patients with cerebral involvement in SLE in 1999 (Ad hoc Am College Rheum, 1999). Those CNS syndromes convincingly attributable to SLE in a comprehensive literature review include stroke, transient ischemic attacks, epileptic seizures, psychosis, cognitive disorder and dementia, and delirium (Jennekens and Kater, 2002). The frequency of an association of specific neurologic symptoms with SLE vary. Kwon (Kwon et al., 1999) report stroke accounts for approximately 20% of neurologic events in SLS and is often secondary to hypercoagulable state or cardiogenic embolism. In a study of gender associated neuropsychiatric symptoms, the most frequent neurologic symptoms in women with SLE are psychiatric symptoms and headaches while men with SLE have more seizures and peripheral neuropathy (Yacoub Wasef, 2004). However, a meta-analysis for headache associated with SLE reviewing the incidence of headache of all types, including migraine, in patients with SLE finds the incidence does not differ from control individuals (Mitsikostas et al., 2004).

21.3.3.3. Hypotheses for Etiology and Treatment

Neuropsychiatric SLE symptoms are hypothesized to relate to vascular lesions whereas systemic symptoms may be related to autoantibody-mediated or cytokine mediated impairment of the neuronal function (Katzav et al., 2003). Studies show patients with SLE with neuropsychiatric symptoms have significantly higher antiphospholipid antibody (aPL), particularly anti-cardiolipin antibodies than SLE patients with no neuropsychiatric involvement (Afeltra et al., 2003; Sanna et al., 2003b). The primary target for aPL is beta$_2$ glycoprotein which may be responsible for small vessel thrombosis, vasculopathy or antibody mediated damage (Katzav et al., 2003; DeGroot and Derksen, 2004). Antibodies to microtubules-associated protein 2 are also 17% higher with neuropsychiatric SLE compared to 4% in controls (Williams et al., 2004).

Severe neurological involvement in SLE particularly during acute onset is a particularly devastating diagnosis. Management of CNS SLE is reviewed by Sanna et al., 2003a. Cyclophosphamide is the treatment of choice for severe acute non-thrombotic CNS disease. Treatment with steroids versus immunosuppressant therapy with cyclophosphamide is evaluated in a controlled clinical trial of 32 SLE patients with onset of severe neuropsychiatric involvement for < 15 days. Induction treatment involved IV methylprednisolone (3 g) followed by either IV monthly cyclophosphamide or IV MP every 3 m for 1 year. Overall response rate was 75% for either treatment but those on cyclophosphamide had a significantly better response to treatment ($p < 0.03$) (Barile-Fabris et al., 2005).

Plasmapheresis is effective in patients with severe neuropsychiatric SLE refractory to conventional treatment. Intrathecal methotrexate and dexamethasone is also beneficial to those patients (Dong et al., 2001; Baca et al., 1999). Positive results of a phase I/II trial of autologous hematopoietic stem cell transplantation (AHSCT) at Northwestern University in Chicago, UL has led to a phase III ASCT trial. As with other similar trials involving autoimmune disease, the ASCT trial is designed to include standard of care IV pulse cyclophosphamide (Burt et al., 2003b).

21.3.4. Alkylosing Spondylitis, Reiter's Disease, Psoriasis, Crohn's Syndrome, and Multiple Sclerosis

This broad group of disorders with autoimmune components and the potential for nervous system involvement were chosen as much for their differences than for their similarities to other autoimmune disorders. In alkylosing spondylitis and Reiter's disease the primary end organ target is the vertebral joints but the multi system involvement varies. Both however, affect men more often than women; a characteristic very different than many other autoimmune disease. Psoriasis while usually thought of as a skin disorder is included because the common arthritic involvement is a classic example of multi system involvement in an autoimmune disorder. Crohn's syndrome where the primary target is the gastrointestinal tract affects men and women in equal numbers but has its own pattern of multi system involvement. Finally MS has long been known to have an autoimmune component and a primary target of demyelination in the nervous system but is noteworthy here because of increased evidence of multi system involvement.

21.3.4.1. Alkylosing Spondylitis

21.3.4.1.1. Epidemiology

Alkylosing spondylitis is an inflammatory autoimmune reactive arthritis with a primary end organ target of the intervertebral joints and the sacroiliac joint at the hip. The characteristic features are a bowed spine and inflamed joints. Compared with 8% of Caucasians as a whole, 95% of people with alkylosing spondylitis have the HLA B27 allele. Unlike many autoimmune diseases, alkylosing spondylitis is more common in men and has an early onset between 20 and 40 m years of age.

21.3.4.1.2. Multi System Involvement with Emphasis on Neurologic Symptoms

Alkylosing spondylitis means "stiff vertebra" and is a reactive arthritis that may be triggered by infection. Alkylosing spondylitis can cause complications that involve some aspect of the nervous system. Almost 40% of people with some type of spondylitis will experience inflammation of the eye (iritis or anterior uveitis) that include symptoms of sensitivity to light and skewed vision along with typical inflammatory symptoms of redness and pain. Rarely those with advanced alylosing spondylitis will develop cauda equine syndrome, a scarring of the nerve bundles of the cauda equine at the base of the spine. Depending on the nerves involved, urinary dysfunction, pain and weakness in the lower extremities as well as bowel and sexual dysfunction may result.

21.3.4.1.3. Hypotheses for Etiology and Treatment

B27, the HLA allele proposed to be involved in alkylosing spondylitis, shares an epitope with an enzyme from *Klebsiella pneumoniae*, a common intestinal Gram negative bacterium. Molecular mimicry is thought to result in cross-reactive antibody formation, joint inflammation and spinal damage. Along with many of the immunotherapies described for other similar arthritic autoimmune disorders, alkylosing spondylitis is treated with anti-inflammatory medications and a low carbohydrate diet to reduce intestinal *K. pneumoniae* levels.

21.3.4.2. Reiter's Syndrome (Reactive Arthritis)

21.3.4.2.1. Epidemiology

Reiter's Syndrome is an inflammatory arthritis that produces pain, swelling, redness and heat in the joints. It is one of a family of arthritic disorders, called spondylarthropathies, affecting the vertebral joints The symptoms of Reiter's disease include fever, weight loss, skin rash, inflammation, ulcerations, and pain. Unlike many autoimmune disorders, Reiter's syndrome usually affects young white men between the ages of 20 and 40.

21.3.4.2.2. Multi System Involvement with Emphasis on Neurologic Symptoms

Reiter's syndrome may affect the skin, eyes, bladder, genitalia and mucous membranes throughout the body so that it can initially share characteristics similar to many other autoimmune diseases.

21.3.4.2.3. Hypotheses for Etiology and Treatment

The exact cause is unknown, but there appears to be a genetic link. About 75% of those with the tendency to develop Reiter's syndrome have the same gene marker, HLA-B27, described in the previous description of alkylosing spondylitis. There is also an environmental or infectious component. Reiter's disease can develop in at risk individuals following an infection in the intestines, genital or urinary tract. Bacteria implicated as potential causes of Reiter's syndrome include Chlamydia, a sexually transmitted bacterium and a wide group of bacteria associated with GI infections including Salmonella, Shigella, Yersinia or Campylobacter. Reiter's syndrome usually develops within 2–4 weeks after a bacterial infection in susceptible individuals. While Reiter's syndrome typically has a limited course with symptoms lasting 3 to 12 months, it may recur. For approximately 15–20% of those affected, Reiter's develops into a chronic and often sever arthritis or spondylitis.

Te use of NSAIDS and other anti-inflammatory therapies are similar to those used in other autoimmune arthritic disorders. Corticosteroid injections for severe pain and inflammation at specific joints are standard therapy. For severe forms of the disease immunomodulating anti-rheumatic drugs such as methotrexate and sulfasalazine are effective. As with other similar disorders, the biologic TNF α inhibitors are currently prescribed for severe Reiter's syndrome.

21.3.4.3. Psoriasis

21.3.4.3.1. Epidemiology

Psoriasis is a chronic autoimmune disease with the skin as a primary target and with both genetic and T cell mediated characteristics. Normally, keratinocyte stem cells divide in the deepest layer of the skin's epidermis, the stratum basale, to form immature keratinocytes. These keratinocytes slowly mature and migrate toward the skin surface accumulating more keratin as they go and eventually undergoing apoptotic cell death before being shed to make room for underlying cells. Epidermal growth factor (EGF) plays a major role in regulating the growth process. The cycle, from cell division to apoptosis and shedding usually takes about 30 days. In psoriasis, the keratinocytes move more quickly from the stratum basale to the stratum corneum of the epidermis where they are shed in as little as 7–10 days. Abnormal keratin is associated with these immature keratinocytes that form flaky, scales or red lesions at the skin surface. That can be inflamed and itch, bleed, crack and be painful.

About 2.1% of the U.S. population has psoriasis or more than 4.5 million adults in the U.S. Diagnosis usually is between ages 15 and 35.

21.3.4.3.2. Multi System Involvement with Emphasis on Neurologic Involvement

Psoriasis generally appears at the joints, limbs and scalp but may appear anywhere on the body. At least 10% those with psoriasis develop psoriatic arthritis, a degenerative disease of the joints and connective tissue. Approximately 1 million people in the U.S. have psoriatic arthritis. Psoriatic arthritis usually develops between the ages of 30 and 50 but can develop without the lesions characteristic of psoriasis (Natl Psoriasis Foundation, 2005, Specific forms of psoriasis, psoriasis.org). Of those with psoriatic arthritis, about 20% will develop spinal involvement. The inflammation associated with spinal involvement can lead to vertebral fusion as in alkylosing spondylitis.

21.3.4.3.3. Hypotheses for Etiology and Treatment

Those patients who develop psoriatic arthritis with spinal involvement are most likely to test positive for the HLA-B27 genetic marker also implicated in alkylosing spondylitis and Reiter's syndrome.

The type of lesion and whether the patient also has psoriatic arthritis are important issues in determining therapy. Steroid topical creams, cyclosporine and methotraxate are useful in treatment of psoriasis. Oral tazarotene, a non-biologic retinoid is pending FDA approval for moderate to severe psoriasis. As with most autoimmune disorders, different patients respond differently and newer more targeted therapies are important goals.

Biologically engineered drugs that target proteins associated with the inflammatory response have recently become important in treatment of both psoriasis and psoriatic arthritis. In 2003, Biogen's alefacept (Amevive), a T cell targeted therapy, became the first biologic drug approved for treatment of psoriasis. It was followed by Genetech's efalizumab (Raptiva). Tumor necrosis factor (TNF) α antagonists are taking the forefront with several receiving FDA approval in the last few years. The FDA approved the TNFα inhibitor, etanercept (Embrel) for treatment of chronic, moderate to severe plaque psoriasis in May, 2004 (FDA, 2004) and it shows both clinically significant and patient reported positive outcomes (Krueger et al., 2005). Etanercept was previously approved for psoriatic arthritis (Nestorov et al., 2004). Subcutaneously administered etanercept in patients with psoriasis at either 50 mg once weekly or 25 mg twice weekly indicate similar efficacy and safety profiles (Feldman et al., 2005). Centocor's infliximab (Remicade) and Abbot Laboratories' adalimumab (Humira) are also TNFα inhibitors under investigation for psoriasis and psoriatic arthritis (Weinberg et al., 2005). A major concern of the TNFα antagonist therapies is a possible link to demyelinating disease. Because autoimmune disease often does not occur as a single entity, it is imperative to screen and follow those on the therapy for symptoms of demyelinating disease (Sukal et al., 2006).

21.3.4.4. Crohn's Disease

21.3.4.4.1. Epidemiology

An estimated 500,000 Americans have Crohn's disease, an inflammatory bowel syndrome (IBS) that causes chronic inflammation of the intestinal tract. Unlike many autoimmune diseases, Crohn's disease affects men and women equally. About 20% of people with Crohn's disease have a blood relative with some form of IBS. Crohn's disease and ulcerative colitis are similar and are often mistaken for one another. Both involve inflammation of the lining of the digestive tract, and both can cause severe bouts of watery or bloody diarrhea and abdominal pain. However, Crohn's disease can occur anywhere in the digestive tract, often spreading deep into the layers of affected tissues. Ulcerative colitis, on the other hand, usually affects only the lining of the large intestine (colon) and rectum.

21.3.4.4.2. Multi System Involvement with Emphasis on Neurologic Involvement

CNS and peripheral nervous alterations in patients with Crohns' disease is reported to be as high as 33.2% (Santos et al., 2001) with almost 20% having a direct causal relationship with the Crohn's disease. Other neurologic features associated with Crohn's disease include peripheral neuropathy (axonal, demyelinating and autonomic), myopathies, pseudotumors cerebri, papilliedema, psychiatric disorders (anxiety, phobia and depression) and an association with syndromes such as MS, Cogan's syndrome, Melkersson Rosenthal syndrome, connective tissue disorders and vasculitis. In a retrospective study of peripheral neuropathy associated with inflammatory bowel disease, demyelinating neuropathy is associated with almost 30% of the cases (Gondim et al., 2005).

21.3.4.4.3. Hypotheses for Etiology and Treatment

Mutations in the NOD2/caspase recruitment domain 15 (CARD15) and in the Toll-like receptor 4 (TLR4) gene are associated with an increased susceptibility for Crohn's disease (Leshinsky-Silver et al., 2005).

The neurologic involvement in Crohn's disease can be attributed to several mechanisms including autoimmunity. For the predominant arterial cerebrovascular involvement, the state of hypercoagulation secondary to thrombocytosis and increase in factor V, VIII and fibrinogen are probable mechanisms. However, since anti-phospholipid antibodies and lupus anticoagulant are confirmed in many cases, autoimmune mechanisms are also involved (Santos et al., 2001).

There's no known medical cure for Crohn's disease. However, strong evidence based efficacy of pharmacological and biological therapies has greatly reduced the signs and symptoms of Crohn's disease and even bring about a long-term remission. Sulfasalazine, sulfa-pyridine linked to 5-aminosalicylic acid (5-ASA) via an azo bond, is a mainstay of therapy but is associated with fever, rash, hematolytic anemia, hepatitis, pancreatitis, paradoxical worsening of the colitis and reversible sperm abnormalities (Stein and Hanauer, 2000). Newer 5-ASA agents deliver the active agent, sulfasalazine while minimizing the adverse effects. Olsalazine may cause secretory diarrhea and uncommonly hypersensitivity reactions including a similar worsening of the colitis as well as pancreatitis, pericarditis and nephritis.

Corticosteroids are commonly prescribed for moderate to severe Crohn's disease with short term efficacy probably related to their effect on inflammation. Anti-bacterials are commonly used as primary therapy in Crohn's disease. Common adverse effects of metronidazole are nausea and a metallic taste, however peripheral neuropathy may result from long term use. Ciprofloxacin and similar antibacterials can be beneficial in those patients intolerant to metronidazole.

Crohn's disease patients treated with systemic steroids or azathioprine show significant improvement in patient quality of life scores (Bernklev et al., 2005). However, azathioprine and mercaptopurine are associated with pancreatitis in 3–15% of Crohn's patients. The pancreatitis resolves upon drug cessation. Their effect on bone marrow suppression is dose related. (Stein and Hanauer, 2000).

Biologic agents targeting specific sites in the immunoinflammatory cascade are being used to treat Crohn's disease. TNF α antibody therapies such as infliximab, are a standard therapy for severe Crohn's disease, Thalidomide whose effect is partially mediated by down regulation of TNF α has efficacy in previously refractory Crohn's patients (Ginsberg et al., 2001).

Autologous hematopoietic stem cell transplantation (AHSCT) has undergone phase I/II trials for severe Crohn's disease (Craig et al., 2003). While clinical remission occurred as a result of AHSCT, some evidence of minor laboratory abnormalities and slight inflammation of the colon on colonoscopy were reported at the 1 year post transplant follow up.

21.3.4.5. Multiple Sclerosis (MS)

21.3.4.5.1. Epidemiology of MS

MS is a chronic neurologic disease associated with demyelination and death of neuronal axons. Multiple plaques or scares form within the CNS resulting in a wide range of neurological deficits. Because of the role of demyelination in the disease, it is frequently classified with autoimmune disorders. Information on the neuroimmunologic aspects of the various types of MS are covered in a previous chapter.

The prevalence of MS varies considerably around the world with an increased prevalence historically associated with distance from the equator. Kurtzke classified regions of the world according to prevalence: a low prevalence was considered less than 5 cases per 100,000 persons, an intermediate prevalence was 5 to 30 per 100,000 persons, and a high prevalence was more that 30 per 100,000 persons (Kurtzke, 1991). The prevalence is highest in northern Europe, southern Australia, and the middle part of North America (Wynn et al., 1990; Cottrell et al., 1999; Rosati et al., 1996; Svenningsson et al., 1990; Cook et al., 1985; Midgard et al., 1991). Both genetic and environmental factors are associated with MS. Evidence that genetic factors have a substantial effect on susceptibility to MS is unequivocal. A concordance rate of 31% among monozygotic twins is approximately six times the rate among dizygotic twins (5%) (Sadovnick et al., 1993). The magnitude of the relative risk of MS is associated with the frequency of the HLA-DR2 allele in the general population (Jersild et al., 1973). Populations with a high frequency of the allele have the highest risk of MS. Studies of candidate genes have targeted individual genes with microsatellite markers with the use of association and linkage strategies (Sawcer et al., 1996; Haines et al., 1996; Ebers et al., 1996; Kuokkanen et al., 1997). Rivera (2004) reports that MS is a polygenic disease associated with Class I and Class II HLA antigens. The genes considered most likely associated with MS are found in groups with A3, B7, DR2, and DQW1 haplotypes. These particular haplotypes are prominent in Northern and Central European populations (Caucasians) and rare in East

Asians, Black Africans, and aboriginal Australians. However, while these particular ethnic groups have an increased incidence of MS, identical twin studies show only a 30% correlation in disease occurrence in individuals sharing the same DNA and environmental factors. Epidemiological genetic studies show a relational rather than a causative effect (Barcellos et al., 2002). While regions of interest have been identified, none have been linked to MS with certainty.

21.3.4.5.2. Multi System Involvement with Emphasis on Neurologic Syndromes

The characteristics of MS are covered in previous chapters. In general, MS is a disease of the CNS with other systems involved when neuronal connections are disrupted. As a result, symptoms vary widely. Outcome measures and progression are reviewed in conjunction with chapter 3.1.2 on the therapeutic measures available for treatment of MS.

21.3.4.5.3. Hypotheses for Etiology of MS

Demyelination is a pathological hallmark of MS. Multiple sclerosis lesions stand out as areas of brain white matter completely or partially devoid of myelin. Examined posthumously or by MRI, lesions were originally classified as inactive (little or no evidence of myelin breakdown), subacute (numerous lipid macrophages present at the edges of demyelinated tissue) or acute (numerous lipid macrophages and macrophages containing undigested myelin debris present throughout the demyelinated tissue) (Prineas et al., 1984). Subacute lesions are the most commonly observed. As early as 1958, degenerating myelin was described as irregular swelling of sheaths followed by fragmentation into particles and balls of myelin, later phagocytosed by scavenging microglia (Greenfield, 1958). Lesions have a predilection for the optic nerves, periventricular white matter, brain stem, cerebellum and spinal cord white matter and they often surround one or several medium-sized vessels (Lassmann, 1998). Patients with primary progressive MS often have lesions in the spinal cord, which usually manifest clinically as progressive paraparesis (Bruck et al., 2002). Currently, different patterns of demyelination can be distinguished, including demyelination with relative preservation of oligodendrocytes, myelin destruction with concomitant and complete destruction of oligodendrocytes or primary destruction or disturbance of myelination cells with secondary demyelination (Lucchinetti et al., 1996).

Inflammatory cells associated with MS lesions are typically perivascular in location, but they may diffusely infiltrate the parenchyma of the CNS. The composition of the inflammatory infiltrate varies depending on the stage of demyelinating activity. In general, it is composed of CD4 and CD8 lymphocytes and macrophages; the latter predominate in active lesions. Experimental in vitro and in vivo models of inflammatory demyelination suggest that diverse disease processes, including autoimmunity and viral infection, may induce MS-like inflammatory demyelinated plaques (Hohlfeld et al., 1995; Olsson, 1995). Activated CD4 T cells specific for one or more self antigens are believed to adhere to the luminal surface of endothelial cells in CNS venules and migrate into the CNS at the time of disruption of the blood brain barrier. This process is followed by an amplification of the immune response after the recognition of target antigens on antigen presenting cells. Antibody mediated processes may have an important role in the pathogenesis of MS. Activated macrophages and microglial cells may mediate degradation of myelin and damage oligodendrocytes by production of proinflammatory cytokines. Other factors potentially toxic to oligodendroglial cells include soluble T cell products (such as perforin), the interaction of Fas antigen with Fas ligand, cytotoxicity mediated by the interaction of CD8 T cells with class I major histocompatibility complex antigens on antigen presenting cells, and persistent viral infection (Lucchinetti et al., 1998). Theories on a role for viral infections continue to postulate enhanced sensitization to myelin or CNS autoantigens in childhood and subsequent infection that activates these sub-threshold T cells during adulthood to initiate myelin injury (Raine, 1984; Haegert, 2003). One example is the two hit infectious hypothesis proposed by Haegert (2003) that proposes that a common virus infects the thymus during childhood and is followed by heterogeneous pathogens that fully activate those T cells during adulthood.

Axonal injury may occur not only in the late phases of MS but also after early episodes of inflammatory demyelination (Lucchinetti et al., 1999; Trapp et al., 1998; Bitsch et al., 2000; De Stefano et al., 2001). Gray matter lesions and axonal injury in normal appearing white matter are reported in MS implying that axonal injury could trigger demyelination (Tsunoda and Fujinami, 2002). The pathogenesis of this early axonal injury is still unclear.

Both inflammatory and neurodegenerative components may contribute to the clinical profile of MS. Inflammation appears to be caused by overactive pro-inflammatory T helper 1 cells, initiating the inflammatory cascade. Current IFNβ and glatiramer acetate are most effective in this inflammatory phase of MS. Recent evidence also shows that inflammation may not only be destructive but may also play a part in tissue repair (Grigoriadis et al., 2006).

21.4. Emerging Concepts

Autoimmune disorders are increasing in the general population. With their increase comes an increase in neurologically involvement. While the plethora of new therapies approved in the last 15 years have made many autoimmune diseases treatable, more progress is needed to halt long term progression.

Potential therapeutic benefits of antioxidants to reduce intracellular oxidative stress have not yet been evaluated in vasculitic neuropathy associated with autoimmune disease. Such anti-oxidants as vitamin E, α-lipoic acid or benfotiamine may be effective in interrupting the pro-inflammatory transcription factors and subsequent NF-[kappa]B regulated cytokines that initiate and maintain inflammation associated with vasculitic neuropathy (Haselbeck et al., 2004).

Summary

The general characteristics of autoimmune disease with the implications for neurologic involvement are reviewed. Areas of concentration include the role of inflammation and vasculitis as hallmarks of autoimmune disease in general and the role of a breach of the blood brain barrier with lymphocyte trafficking as specific to involvement of the nervous system. The epidemiology, system involvement including aspects of the nervous system, and current hypotheses for etiology and treatment are discussed for the following autoimmune conditions: Sjogren's Syndrome, Systemic Lupus Erythematosis, Alkylosing Spondylitis, Reiter's Disease, Psoriasis, Crohn's Disease and Multiple Sclerosis.

Review Questions/Problems

1. **This molecule is hypothesized to be involved in degradation of the endothelial extra cellular matrix in vasculitic neuropathy.**

   a. Integrin
   b. Matrix metalloproteinase
   c. TNFα
   d. NF-kappa B

2. **This autoimmune disease affects men more frequently than women.**

   a. Multiple sclerosis
   b. Crohn's Disease
   c. Reiter's Disease
   d. Lupus erythematosus

3. **An autoimmune disease linked directly to an infectious etiology.**

   a. Psoriasis
   b. Rheumatoid arthritis
   c. Alkylosing spondylitis
   d. Systemic Lupus erythematosus

4. **Which of the following is NOT a structural/functional characteristic of the cerebrovascular endothelium?**

   a. Extensive tight junctions
   b. Polarization apical-basal orientation of endothelial cells
   c. Association with astrocytic end plates
   d. Lower trans-endothelial electrical resistance

5. **Inflammation of the synovial joints in rheumatoid arthritis includes this change in the synovial fluid.**

   a. Influx of macrophages and activated T cells
   b. Loss of hyaluronic acid
   c. Thinning of the synovial membrane
   d. Presence of the bacterium *Klebsiella*

6. **HLA B27 allele has been identified in those at greater risk for all except this autoimmune disease.**

   a. Alkylosing spondylitis
   b. Psoriatic arthritis
   c. Reiter's disease
   d. Crohn's disease

7. **The primary type of nervous system involvement in Sjogren's syndrome is**

   a. Spinal cord column involvement
   b. Sensory ataxic neuropathy
   c. Cervical cord involvement at the atlanto-axis junction
   d. Autonomic neuropathy

8. **In systemic lupus erythematosus, this is a common neurologic involvement due to the chronic hyper-coagulable state.**

   a. Autonomic neuropathy
   b. Dementia
   c. Headache
   d. Stroke

9. **Reiter's disease commonly is diagnosed within this age range.**

   a. <20
   b. 20–40
   c. 40–60
   d. >60

10. **Neurologic involvement in Crohn's disease**

    a. Is relatively rare (<5%)
    b. Usually occurs in 20%–30% of Crohn's patients
    c. Occurs in over 90% of Crohn's cases
    d. Is virtually always a peripheral neuropathy

11. **Compare the gender distribution of those diagnosed with Multiple Sclerosis, Crohn's disease and Alkylosing Spondylitis**

12. **Autologous hematopoietic stem cell transplantation has undergone phase I/II trials for several autoimmune disorders. Aside from its invasive nature discuss the pros and cons for its use.**

13. **Discuss the general mechanism of action for biologicals such as infliximab and etanercept.**

14. **Discuss the different hypothesized etiologies for neuropsychiatric SLE versus systemic SLE neuropathies.**

15. **Discuss the postulated role of molecular mimicry in Alkylosing Spondylitis.**

16. **What is the typical short and long-term course of Reiter's disease?**

17. **Thalidomide is receiving new interest for use in some autoimmune conditions. What is the history and concern for its use?**

18. **What is the anatomic basis for psoriasis in the skin?**

19. Discuss how inflammation is both a positive and negative factor in autoimmune disorders.

20. Discuss the causes of vasculitis as it relate to vasculitic neuropathy.

References

Abbott NJ (2002) Astrocyte endothelial interaction and blood brain barrier. Permeability. A Anat 200:629–638.

Abbott NJ, Rohnback L, Hansson E (2006) Astrocyte-endothelial interactions at the blood-brain barrier. Nature Rev Neurosci 7:41–53.

ACR ad hoc committee on neuropsychiatric lupus nomenclature (1999) The American College of Rheumatology nomenclature as case definitions for neuropsychiatric lupus syndrome. Arthritis Rheum 42:599–608.

Afeltra A, Garzia P, Mitterhofer AP, Vadacci M, Galluzzo S, Del-Porte F, Finamore L, Pascucci S, Gaspariiani M, Lagana B, Cacarod D, Ferri GM, Amoro A, Francis A (2003) Neuropsychiatric lupus syndromes: Relationship with autophospholipid antibodies. Neurology 61:108–110.

Anderson CM, Nedergaard M (2003) Astrocyte-mediated control of cerebral microcirculation. Trends Neurosci 26:340–344.

Ando Y, Kai S, Uyama E Iyonaga K, Hashimoto Y, Uchino M, Andro M (1995) Involvement of the central nervous system in rheumatoid arthritis: its clinical manifestations and analysis by magnetic resonance imaging. Intern Med 34:188–191.

Baca V, Lavalie C, Garcia R, Catalan T, Sauceda JM, Sanchez G, Martinez I, Rameriz MI, Marquez LM, Rojas JC (1999) Favorable response to intravenous methylprednisolone and cyclophosphamide in children with severe neuropsychiatric lupus. J Rheumtatol 26:432–439.

Bagavant H, Fu SM (2005) New insights from murine lupus: dissociation of autoimmunity and end organ damage and the role of T cells. Curr Opin Rheumatol 5:523–528.

Barcellos LF, Oksenberg JR, Green AJ, Bucher P, Rimmler JB, Schmidt S, Garcia ME, Lincoln RR, Pericak-Vance MA, Haines JL, Hauser SL, Multiple Sclerosis Genetics Group (2002) Genetic Basis for clinical expression in multiple sclerosis. Brain 125:150–158.

Barile-Fabris L, Ariza-Andraca R, Olguin-Orteg L, Jara LJ, Fraga-Mouret A, Mirand-Limon JM, Fuenta de la Mata J, Clark P, Vargas F, Alocer-Verala J (2005) Controlled clinical trial of IV cyclophosphamide versus IV methylprednisolone in severe neurological manifestations insystemic lupus erythematosus. Ann Rheum Dis 4:620–625.

Bernklev T, Jahnsen J, Schulz T, Sauar J, Lygren I, Henriksen M, Stray N, Kjellevold O, AAdland E, Vatn M, Moun B (2005) Course of disease, drug treatment and health related quality of life in patients with inflammatory bowel disease 5 years after initial diagnosis. Eur J Gastroentrol Hepatol 10:1037–1045.

Bitsch A, Schuchardt J, Bunkowski S, Kuhlmann T, Bruck W (2000) Acute axonal injury in multiple sclerosis. Correlation with demyelination and inflammation. Brain 123:1174–83.

Brinar VV, Petelin Z, Brinar M, Djakoviv V, Zadro I, Vranjes D (2006) CNS demyelination in autoimmune disease. Clin Neurol Neurosurg 108(3):318–326.

Burt RK, Cohen BA, Russell E, Spero K, Joshi A, Oyama Y, Karpus WJ, Luo K, Jovanic B, Traynor A, Karth K, Stefaski D, Burns WH (2003a) Haematopoietic stem cell transplantation for progressive multiple sclerosis: failure of intense immune suppression to prevent disease progression in patients with high disability scores. Blood 102:2373–2378.

Burt RK, Marmont A, Arnold R, Heipi F, Firestein GS, Carrier E, Hahn B, Barr W, Oyama Y, Snowden J, Kalunian K, Traynor A (2003b) Development of a phase III trial of haematopoietic stem cell transplantation for systemic lupus erythmatosis. Bone Marrow Transplant Aug 23 (Suppl 1):S49–S51.

Bruck W, Lucchinetti C, Lassmann H (2002) The pathology of primary progressive multiple sclerosis. Mult Scler 8:93–7.

Butt AM, Jones HC, Abbott NJ (1990) Electrical resistance across the blood brain barrier in unanesthetized rats: A developmental study. J Physiol (London) 429:47–62.

Chin BH, Latov N (2005) Central nervous system manifestations of rheumatologic disease. Curr Opin Rheumatol Jan 17 (1):91–99.

Christen U, von Herrath MG (2005) Infections and autoimmunity – good or bad? J Immunol 174(12):7481–7486.

Cook SD, Cromarty JI, Tapp W, Poskanzer D, Walker JD, Dowling PC (1985) Declining incidence of multiple sclerosis in the Orkney Islands. Neurology 35:545–51.

Cottrell DA, Kremenchutzky M, Rice GP, Hader W, Baskerville J, Ebers GC (1999) The natural history of multiple sclerosis: A geographically based study. Brain 122 (Pt 4):641–647.

Craig RM, Traynor A, Otama Y, Burt RK (2003) Hematopoietic stem cell transplantation for severe Crohn's disease. Bone Marrow Transplant. 32(Suppl 1):S57–59.

Cser HF, Bundgaard M (1984) Blood brain interfaces in vertebrates: a comparative approach. Am J Physiol 3:183–188.

DeGroot PG, Derksen RH (2004) Antiphospholipid antibodies: update on detection, pathophysiology and treatment. Curr Opin Hematol 11:165–169.

De Stefano N, Narayanan S, Francis GS, Armaoutelis R, Tartagia MC, Anteil JP, Matthews PM, Arnold DL (2001) Evidence of axonal damage in the early stages of multiple sclerosis and its relevance to disability. Arch Neurol 58:65–70.

Dong Y, Zhang X, Tang F, Tian X, Zhao Y, Zhang F (2001) Intrathecal injection with methotrexate plus dexamethasone in the treatment of central nervous system involvement in systemic lupus erythematosus. Chin Med J (Engl) 114:764–766.

Drachman DB, Brodsky RA (2005) High-dose therapy for autoimmune neurologic diseases. Curr Opin Oncol 2:83–88.

Ebers GC, Kukay K, Bulman DE, Sadovnick AD, Rice G, Anderson C, Armstrong H, Cousin K, Bell RB, Hader W, Paty DW, Hashimoto S, Oger J, Duquette P, Warren S, Gray T, O'Connor P, Nath A, Auty A, Metz L, Francis G, Paulseth JE, Murray TJ, Pryse-Phillips W, Nelson R, Freedman M, Brunet D, Bouchard JP, Hinds D, Risch N (1996) A full genome search in multiple sclerosis. Nat Genet 3:472–6.

Elliott CL, El-Touny SY, Filipi ML, Healey KM, Leuschen MP (2001) Interferon β1a treatment modulates Th1 expression in γδ + T cells from relapsing-remitting multiple sclerosis patients. J Clin Immun 21(3):200–209.

Faller G, Thomson PD, Kala UK, Hahn D (2005) Demographic and presenting clinical features of childhood systemic lupus erythenatosis. S Afr Med J 6:424–427.

Feldman SR, Kimball AB, Krueger GG, Wooley JM, Lalla D, Jahreis A (2005) Etanercept improves the health-related quality of life of patients with psoriasis: Results of a phase III randomized clinical trial. J Am Acad Dermatol 53:887–889.

Fox RI (2005) Sjogren's syndrome. Lancet 366:321–331.

Ginsberg PM, Dassopoulos T, Ehrenpreis ED (2001) Thalidomide treatment for refractory Crohn's disease: A review of the history, pharmacological mechanisms and clinical literature. Ann Med 33(8):516–525.

Gondim FA, Brannagan TH, Sander HW, Chin RL, Latov N (2005) Peripheral neuropathy in patients with inflammatory bowel disease. Brain 128:867–879.

Greenfield JG (ed) (1958) Neuropathology. London: Edward Arnold, pp. 441–474.

Grigoriadis N, Grigoriadis S, Polyzoidou E, Milonas I, Karussis D (2006) Neuroinflammation in multiple sclerosis: Evidence for autoimmune dysregulation, not simple autoimmune reaction. Clin Neurol Neurosurg 108 (3):241–244.

Gurer G, Erden S, Kocaefe C, Ogzu CM, Tau E (2004) Expression of matrix metalloproteinases in Vasculitic neuropathy. Rheumatol Int 24(5):255–259.

Haegert DG (2003) The initiation of multiple sclerosis: A new infectious hypothesis. Med Hypotheses 60:165–70.

Haines JL, Ter-Minassian M, Bazyk A, Gusells JF, Kim DJ, Terweddow H, Percak-Vance MA, Rimmler JB, Haynes CS, Roses AD, Lee A, Shaner B, Menold M, Sebourn E, Fitoussi RP, Gartioux C, Reyes C, Ribierre F, Gyapay G, Weissenbach J, Hauser SL, Goodkin DE, Lincoln R, Usuku K, Oksenberg JR (1996) A complete genomic screen for multiple sclerosis underscores a role for the major histocompatability complex. The Multiple Sclerosis Genetics Group. Nat Genet 13:469–71.

Haselbeck KM, Bierrhaus A, Trwin S, Kirchner A, Navroth P, Schatzer U, Neudorfer A, Heuss D (2004) Receptor for advanced glycation endproduct (RAGE)-mediated nuclear factor-[kappa]B activation in vasculitic neuropathy. Muscle Nerve 29:853–860.

Hawkins BT, Davis TP (2005) The blood brain barrier/neurovascular unit in health and disease. Pharmacol Rev 57:173–185.

Hayashi Y, Nomura M, Yamagishi S, Haraa S, Yamashita J, Yamamota H (1997) Induction of various blood brain barrier properties in non-neuronal endothelial cells by close apposition to co-cultured astrocytes. Glia 19:13–26.

Healey KM, Pavletic SZ, Al-Omaishi J, Leuschen MP, Pirrucello SJ, Filipi ML, Enke C, Ursick MM, Hahn F, Bowen JD, Nash RA (2004) Discordant functional and inflammatory parameters in multiple sclerosis patients after autologous haematopoietic stem cell transplantation. Mult Scer 10:284–289.

Heuss D, Probst-Cousin S, Kayser C, Neudorfer B (2000) Cell death in vasculitic neuropathy. Muscle Nerve 23:999–1004.

Hohlfeld R, Meinl E, Weber F, Zipp F, Schmidt S, Sotgai S, Goebels N, Voltz R, Spuler A, Iglesius A, Wekerle H (1995) The role of autoimmune T lymphocytes in the pathogenesis of multiple sclerosis. Neurology 45:S33–8.

Hu W, Mathey E, Hartung HP, Kieseier BC (2003) Cyclo-oxygenases and prostaglandins in acute inflammatory demyelination of the peripheral nerve. Neurology. 61:1774–1779.

Jennekens FGI, Kater L (2002) The central nervous system in systemic lupus erythematosus: Part I. Clinical syndromes, a literature investigation. Rheumatol 41:605–618.

Jersild C, Fog T, Hansen GS, Thomsen M, Svejgaard A, Dupont B (1973) Histocompatibility determinants in multiple sclerosis, with special reference to clinical course. Lancet 2:1221–5.

Kang HI, Fei HM, Saito I, Sawada S, Chen SL, Yi D, Chan E, Peeble C, Bugawan TL, Erlich HA, Fox RI (1993) Comparison of HLA class II genes in Caucasian, Chinese and Japanese patients with primary Sjogren's syndrome. J Immunol 150:3615–3623.

Katzav A, Chapman J, Shonfeld Y (2003) CNS dysfunction in the antiphospholipid syndrome. Lupus 12:903–907.

Khurnana R, Berney SM (2005) Clinical aspects of rheumatoid arthritis. Pathophysiology. 3:153–165.

Kuokkanen S, Gschwend M, Rioux JD, Daly MJ, Terwilliger JD, Tienari PJ, Wikstrom J, Palo J, Stein LD, Hudson TJ, Lander ES, Peltonen L (1997) Genome wide scan of multiple sclerosis in Finnish multiplex families. Am J Hum Genet 61:1379–87.

Kozak T, Havrdova E, Pitha J, Gregora E, Oitlik R, Maaloufova J, Kobylka P, Vodrarkora S (2001) Immunoblative therapy with autologous stem cell transplantation in the treatment of poor risk multiple sclerosis. Transplant Proceed 33:2179–2181.

Krueger GG, Langley RG, Finlay AY, Griffiths CE, Wooley JM, Lalla D, Jahreis A (2005) Patient-reported outcomes of psoriasis improvement with etanercept therapy: Results of a randomized phase III trial. Br J Dermatol 153(6):1192–1199.

Kurtzke JF (1991) Multiple sclerosis: Changing times. Neuroepidemiology 10:1–8.

Kwon SU, Koh JY, Kim JS (1999) Vertebrobasilar artery territory infarction as an initial manifestation of systemic lupus erythematosus. Clin Neurol Neurosurg 101(1):62–67.

Lassmann H (1998). Neuropathology in multiple sclerosis: new concepts. Mult Scler 4:93–98.

Leshinsky-Silver E, Karban A, Buzhaker E, Fridlande M, Yakir B, Eliakim R, Reif S, Shaul R, Boaz M, Levi D, Levine A (2005) Is age of onset of Crohn's disease governed by mutation in NOD2/caspase recruitment domain 15 and Toll-like receptor 4? Pediatr Res 58(3):499–504.

Lindenlaub R, Sommer C (2003) Cytokines in sural nerve biopsies from inflammatory and non-inflammatory neuropathies. Acta Neuropathol (Berl) 105:593–602.

Liu L, Callahan MK, Huang D, Ransohoff RM (2005) Chemokine receptor CXCR3: An unexpected enigma. Curr Top Dev Biol 68:149–181.

Lucchinetti C, Bruck W, Rodriguez M, Lassmann H (1996) Distinct patterns of multiple sclerosis pathology indicates heterogeneity in pathogenesis. Brain Pathology 6:259–274.

Lucchinetti CF, Brueck W, Rodriguez M, Lassmann H (1998) Multiple sclerosis: Lessons from neuropathology. Semin Neurol 18:337–49.

Lucchinetti C, Bruck W, Parisi J, Scheithauer B, Rodriguez M, Lassmann H (1999) A quantitative analysis of oligodendrocytes in multiple sclerosis lesions. Brain 122:2279–2295.

National Institute for arthritis, Musculoskeletal and Skin Diseases (2005) Guide to Sjogren's Syndrome. ninds.nih.gov/disorders/sjorgrens/sjorgrens.htm

Masi AT, Aldag JC (2005) Integrated neuroendocrine immune risk factors in relation to rheumatoid arthritis: Should rheumatologists now adopt a model of a multiyear, presymptomatic phase? Scand J Rheumatol 5:342–352.

McAllister MS, Krizanac-Bengez L, Macchia F, Naftalin RJ, Pedley KC, Mayberg MR, Marroni M, Leaman S, Stanness KA, Janigro D (2001) Mechanisms of glucose transport at the blood brain barrier: an in vitro study. Brain Res 904:20–30.

Midgard R, Riise T, Nyland H (1991) Epidemiologic trends in multiple sclerosis in More and Romsdal, Norway: a prevalence/incidence study in a stable population. Neurology 41:887–92.

Mitsikostas DD, Sfikalis PP, Goadby PJ (2004) A meta-analysis for headache in systemic lupus eythematosis: the evidence and the myth. Brain 127:1200–1209.

Mori K, Ijima M, Koike H, Hattori N, Tamaka F, Watanabe H, Katsuno M, Fujita A, Aiba I, Ogata A, Saito T, Asakura K, Yoshida A, Hirayama M, Sobue G (2005) The wide spectrum of clinical manifestations in Sjogren's syndrome associated neuropathy. Brain 128(11):2518–1534.

Naranjo A, Carmona L, Gavrila D, Balsa A, Belmonte MA, Tena X, Rodriguez-Lozano C, Sanmarti R, Gonzalez-Alvaro I; EMECAR Study Group (2004) Prevalence and associated factors of anterior atlantoaxial subluxation in a nation wide sample of rheumatoid arthritis patients. Clin Exp Rheumatol 22:427–432.

Nash RA, Bowen JD, McSweeney PA, Pavletic SZ, Maravilla KR, Park MS, Storek J, Sullivan KM, Al-Omaishi J, Corboy JR, DiPersio J, Georges GE, Gooley TA, Holmberg LA, LeMaistre CF, Ryan K, Openshaw H, Sunderhaus J, Storb R, Zunt J, Kraft GH (2003) High-dose immunosuppressive therapy and autologous peripheral blood stem cell transplantation for severe multiple sclerosis. Blood 102(7):2364–72.

Nedergaard M, Ranson B, Goldman SA (2003) New roles for astrocytes redefining the functional architecture of the brain. Trends Neurosci 26:523–530.

Nestorov I, Zitnik R, Ludden T (2004) Population pharmacokinetic modeling of subcutaneously administered etanercept in patients with psoriasis. J Pharmacokinetic Pharmacodyn 31(6):463–490.

Olsson T (1995) Cytokine-producing cells in experimental autoimmune encephalomyelitis and multiple sclerosis. Neurology 45:S11–S5.

Paty DW, Noseworthy JH, Ebers GC (1998) In Multiple Sclerosis; A Contemporary Series. (Paty DW, Ebers GC eds), pp 48–134. Philadelphia PA: FA Davis Co.

Pertovaara M, Hurme M (2006) Th2 cytokine genotypes are associated with a milder form of primary Sjogren's syndrome. Ann Rheum Dis 65(5):666–670.

Prineas JW, Kwon EE, Cho ES, Sharer LR (1984) Continual breakdown and regeneration of myelin in progressive multiple sclerosis plaques. Ann N Y Acad Sci 436:11–32.

Raine CS (1984) Biology of disease. Analysis of autoimmune demyelination: Its impact upon multiple sclerosis. Lab Invest 50:608–35.

Rivera VM (2005) Clinical characteristics of African-American versus Caucasian Americans with multiple sclerosis. Neurology 64(12);2153.

Rosati G, Aiello I, Pirastru MI, Mannu L, Sanna G, Sau GF, Sotgiu S (1996) Epidemiology of multiple sclerosis in Northwestern Sardinia: Further evidence for higher frequency in Sardinians compared to other Italians. Neuroepidemiology 15:10–9.

Sadovnick AD, Armstrong H, Rice GP, Bulman D, Hashimoto L, Paty DW, Hashimoto SA, Warren S, Hader W, Murray TJ (1993) A population-based study of multiple sclerosis in twins: Update. Ann Neurol 33:281–5.

Sanna G, Bertolaccini ML, Mattheiu A (2003a) Central nervous system lupus: a clinical approach to therapy. Lupus 12:935–942.

Sanna G, Bertolaccini ML, Cuadrado MJ Laing H, Khamashta MA, Mathieu A, Hughes GR (2003b) Neuropsychiatric manifestations in systemic lupus erythematosus: Prevalence and association with antiphospholipid antibodies. J Rheumatol. 30:985–992.

Santos S, Casadevall T, Pascual LF, Tejero C, Larrode P, Iniguez C, Morales F (2001) Neurological alterations related to Crohn's disease. Rev. Neurol. 32(12):1158–1162.

Sawcer S, Jones HB, Feakes R, Gray J, Smaldon N, Chataway J, Robertson N, Clayton D, Goodfellow PN, Compston A (1996) A genome screen in multiple sclerosis reveals susceptibility loci on chromosome 6p21 and 17q22. Nat Genet 13:464–8.

Schinkel AH (1999) P glycoprotein: A gatekeeper in the blood brain barrier. Adv Drug Deliv Rev 36:179–194.

Sobue K, Yamamoto N, Yoneda K, Hodgson ME, Yamashiro K, Tsuruoka N, Tsuda T, Katsuya H, Miura Y, Asai K, Kato T (1999) Induction of blood brain barrier properties in immortalized bovine brain endothelial cells by astrocytic factors. Neurosci Res 35:155–164.

Stein RB, Hanauer SB (2000) Comparative tolerability of treatments for inflammatory bowel disease. Drug Saf. 23(5):429–448.

Sukal SA, Nadiminti L, Granstein RD (2006) Etanercept and demyelinating disease in patients with psoriasis. J Am Acad Dermatol 54(1):160–164.

Svenningsson A, Runmarker B, Lycke J, Andersen O (1990) Incidence of MS during two fifteen-year periods in the Gothenburg region of Sweden. Acta Neurol Scand 82:161–8.

Trapp BD, Peterson J, Ransohoff RM, Rudick R, Mork S, Bo L (1998) Axonal transection in the lesions of multiple sclerosis. N Engl J Med 338:278–85.

Tsunoda I, Fujinami RS (2002) Inside-Out versus Outside-In models for virus induced demyelination: Axonal damage triggering demyelination. Springer Semin Immunopathol 24:105–25.

Tumialan LM, Wippold FJ, Morgan RA (2004) Tortuous vertebral artery injury complicating anterior cervical spinal fusion in a symptomatic rheumatoid cervical spine. Spine 29:E323-E348.

Turesson C, Matteson EL (2004) Management of extra-auricular disease manifestations in rheumatoid arthritis. Curr Opin Rheumatoid 16:206–211.

Tyndall A, Saccradi R (2005) Haematopoietic stem cell transplantation in the treatment of severe autoimmune disease: Results from phase I/II studies, prospective randomized trials and future directions. Clin Exp Immunol 1(1):1–9.

Unger L, Kayser M, Nusselsein HG (2003) Successful treatment of severe rheumatoid vasculitis by infliximab. Ann Rheum Dis 62:587–588.

Vollertsen RS, Conn DL (1990) Vasculitis associated with rheumatoid arthritis. Rheum Dis Clin North Am 16:445–461

Weinberg JM, Bottino CJ, Lindholm J, Buchholz R (2005) Biologic therapy for psoriasis: An update on the tumor necrosis factor inhibitors infliximab, etanercept, and adalimumab and the T cell targeted therapies efalizumab and alefacept. J Drugs Dermatol 4(5):544–555.

Whitacre CC, Reingold SC, O'Looney PA, & The Task Force on Gender (1999) Multiple sclerosis and autoimmunity, a gender gap in autoimmunity. Science 283(5406):1277–1278.

Williams RC Jr, Sugiura K, Tan EM (2004) Antibodies to microtubules associated protein 2 in patients with neuropsychiatric systemic lupus erythematosus. Arthritis Rheum 50:1239–1247.

Wolberg H, Lippoldt A (2002) Tight junctions of the blood brain barrier: development, composition and regulation. Vasc Pharmacol 38:323–337.

Wynn DR, Rodriguez M, O'Fallon WM, Kurland LT (1990) A reappraisal of the epidemiology of multiple sclerosis in Olmsted County, Minnesota. Neurology 40:780–6.

Yacoub Wasef SZ (2004) Gender differences in systemic lupus erythematosus. Gend Med 1:12–17.

Yamamoto M, Ito Y, Mitsuma N, Hattori N, Sobue G (2003) Pain related differential expression of NGF, GDF, IL-6 and their receptors in human vasculitic neuropathy. Intern Med 42:1100–1103.

Younger D (2004) Vasculitis and the nervous system. Curr Opin Neurol 17:317–336.

22
NeuroAIDS

Yuri Persidsky

Keywords Anti-retroviral therapy; HIV-1 associated dementia; Blood–brain barrier; HIV-1 encephalitis; Chemokine macrophage; HIV-1

22.1. Introduction

HIV-1 infection of central nervous system results in significant cognitive abnormalities. While underlying mechanisms of such impairment are still not completely understood, chronic neurotoxin production (pro-inflammatory factors and viral proteins) by virus-infected macrophages are considered to be the cause. A better understanding of HIV-1 neurobiology as it is driven by virus replication and anti-viral immunity will continue to drive the development of new therapeutic strategies to combat cognitive decline.

22.2. Clinical Manifestations, Epidemiology, and NeuroAIDS Disease Evolution in the Antiretroviral Era

22.2.1. Clinical Presentation and Epidemiology

The clinical and neuropsychological disease complex called HIV-1 associated dementia (HAD) affects the frontal-subcortical brain regions. Cognitive, motor and behavioral impairments are characterized by deterioration of fine motor coordination and speed, attention, processing speed, executive function, learning efficiency and working memory (Grant et al., 1995; Heaton et al., 1995). Three forms of disease are manifest, including (1) frank HAD, characterized by severe dysfunction that affects everyday cognitive function; (2) mild neuro-cognitive disorder with impairment in two or more cognitive functions with modest interference with everyday function (equivalent to minor cognitive motor disorder, MCMD); and (3) asymptomatic neuro-cognitive impairment, without apparent effect on everyday function.

One quarter of adults and 50% of children developed significant neuro-cognitive complications as a consequence of infection at a period of profound immunosuppression or the aquired immunodeficiency syndrome (AIDS) before the introduction of anti-retroviral therapy (Navia et al., 1986). While subclinical metabolic and structural abnormalities are detectable in neurologically asymptomatic subjects, overt HAD develops as a late complication of progressive viral infection (Koralnik et al., 1990). Proton magnetic resonance spectroscopy (MRS) shows significant metabolic abnormalities in the frontal white matter, basal ganglia, and parietal cortex of HIV-1 patients with HAD compared to neurologically asymptomatic or seronegative controls (Chang et al., 2004). Such MRS changes are not only for HAD. Indeed, HIV-1-infected individuals with normal cognitive function show higher glial marker myoinositol-to-creatine ratio (MI/Cr) in the white matter than seronegative control subjects. Nonetheless, HAD patients show further increased MI/Cr in the white matter and basal ganglia together with a reduction in the neuronal marker N-acetyl aspartate (NAA)/Cr in the frontal white matter. Glial activation likely occurs during all stages of HIV-1 infection, whereas further inflammatory activity in the basal ganglia and neuronal injury in the white matter is associated with the development of cognitive impairment. Host genetic factors likely play a role (e.g., polymorphisms in the promoters of tumor necrosis factor, TNF-α or monocyte chemo-attractive protein, MCP-1) in disease onset and progression but their contribution to HIV-associated cerebral injury require further study (Gonzalez et al., 2002; Letendre et al., 2004).

22.2.2. Disease Evolution in the Era of Anti-Retroviral Therapy

Introduction of anti-retroviral therapy considerably changed the clinical evolution of HIV-1 infection (Palella et al., 1998). Anti-retroviral therapy has diminished the incidence and prevalence of major opportunistic infections and resulted in

T. Ikezu and H.E. Gendelman (eds.), *Neuroimmune Pharmacology*.
© Springer 2008

improvement in survival rates. 10–15% of patients are now over 50 years of age and with continued advances in treatment, many will reach normal life expectancy (Navia and Rostasy, 2005). As survival rates improved other co-morbid factors emerged in disease evolution including the effects of aging, chronic infection and neurotoxic effects of anti-retroviral medicines themselves. Indeed, nucleoside reverse transcriptase inhibitors (NRTIs) suppress HIV-1 replication, but are also associated with mitochondrial toxicity (Schweinsburg et al., 2005). Reduction in neuronal metabolism was found in individuals taking NRTIs, didanosine and/or stavudine as a result of depleted brain mitochondria and/or alterations in cellular respiration (Schweinsburg et al., 2005).

Anti-retroviral therapy has resulted in decrease in HIV-1 RNA load and in neuroimmune activation, particularly in the cerebrospinal fluid (CSF) (e.g., β2 microglobulin, quinolinate), as well as a decline in the severity and incidence of dementia. Anti-retroviral-treated patients show enhanced attention, executive performance and psychomotor speed when compared those who have not received treatments (Sacktor et al., 2002). Onset of mild cognitive-motor impairment is postponed by anti-retroviral therapy; however, the incidence of mild HAD may increase over time and cognitive impairment can occur at higher CD4 counts (Sacktor et al., 2002). Development of neurovirulent and anti-retroviral-resistant HIV-1 strains was suggested due to the re-emergence of leukoencephalopathy (Langford et al., 2002), and the compartmental differences between peripheral blood and brain isolates found associated with the presence of discordant virus resistance. This suggested autonomous replication of HIV-1 and the independent evolution of drug resistance within the central nervous system (CNS) (Strain et al., 2005). Discordance of resistance was associated with severity of neuro-cognitive deficits. Low CD4+T cell counts were linked to neuro-cognitive impairment and to discordant resistance patterns. Metabolic abnormalities observed by MRS have continued in subjects on anti-retroviral therapy despite significant improvement in CD4+T cell counts (≥300 cells/ml) and suppression of plasma and CSF viral loads (Chang et al., 2003). These studies suggest that HAD has evolved from a subacute, rapidly progressing disease to one more subtle indicating that peripheral suppression of HIV-1 translates into improvement of motor-cognitive function.

22.3. Neuropathology

22.3.1. HIV Encephalitis

It is well accepted that HIV encephalitis (HIVE) is characterized by the presence of HIV-1 infected macrophages in the brain, resulting in the microgliosis and microglia activation (Figure 22.1A), formation of multi-nucleated giant cells (MNGC, Figure 22.1B), and astrogliosis (Figure 22.1C) (Koenig et al., 1986; Budka et al., 1991). Brain mononuclear phagocytes (MP, perivascular macrophages and microglia) are the primary viral targets in the CNS (Figure 22.1D, E); however, restricted viral infection have been detected in other cell populations including astrocytes (Trillo-Pazos et al., 2003). Infection of other cell types (neurons, oligodendrocytes, endothelial cells) reported initially remains controversial. While brain HIV-1 infected MP are most abundant in the frontal cortex, basal ganglia and white matter (Wiley, 1995), less prominent infection is present in the limbic system and brainstem (Masliah et al., 2000) or even white matter of cerebellum seen in severe HIVE. The best correlate to the clinical manifestations of HAD is the number of activated macrophages in the CNS (Glass et al., 1995). Although most patients with HIVE have HAD, not all HAD patients have HIVE (Glass et al., 1995). Even minimal increases in the numbers (Cherner et al., 2002) and perhaps even more importantly, the level of MP immune activation, may affect neurological dysfunction (Persidsky et al., 1999).

Virus likely enters the CNS through infected monocytes and macrophages destined to become brain-resident MP or perivascular macrophages (Davis et al., 1992) or within lymphocytes. It is assumed that HIV-1 enters early after primary infection (at a peak of primary viremia), and HIV-1 infection persists at low levels due to immune privileged status of the CNS. The importance of MP in the neuropathogenesis of HIV-1 brain infection is underscored by the emergence of specific monocyte subsets in the peripheral blood of HAD patients. These cells can, but do not always, express CD14/CD16 and CD163 surface markers (Figure 22.1F, G) and demonstrate enhanced ability to migrate and secrete neurotoxins (Pulliam et al., 1997; Luo et al., 2003). The appearance of this monocyte subset in blood coincides with decreased anti-retroviral peripheral immune responses and precedes development of encephalopathy as demonstrated in pediatric patients (Sanchez-Ramon et al., 2003). Immune activation and failing adaptive immune responses are associated with enhanced monocyte migration through the BBB and the secretion of a variety of immune and viral factors known to affect neuronal function (Gendelman, 2002). Viral load, HIV-1 evolution and sequestration in the CNS suggest that virus replication may be compartmentalized in brain. A number of mechanisms are likely to be involved in the response to therapy in CNS, including among others, the trafficking of cell populations supporting viral replication between blood, CNS and CSF, and the role of the anatomical brain barriers in limiting the access of antiretroviral drugs into the CNS/CSF. A potential risk associated with compartmentalization of HIV infection is of an incomplete suppression of virus replication in the CNS, thus creating the ground for local development of anti-retroviral drug resistance and reseeding blood/peripheral tissues with virus during interruptions of therapy.

There was evidence of HIVE in all patients with MCMD or HAD and 80% of the cases with mild impairment at autopsy (Cherner et al., 2002). Among patients who were neuropsychologically normal at the time of testing, 45% had mild HIVE at death. When all neuropsychologically impaired patients were grouped together, 95% of impaired subjects were found to have HIVE. The correlation between the measure of brain viral burden and the clinical scores of neuro-cognitive deficits showed a significant association. Since most of the correlations between HIV-1 levels and cognitive status were of mod-

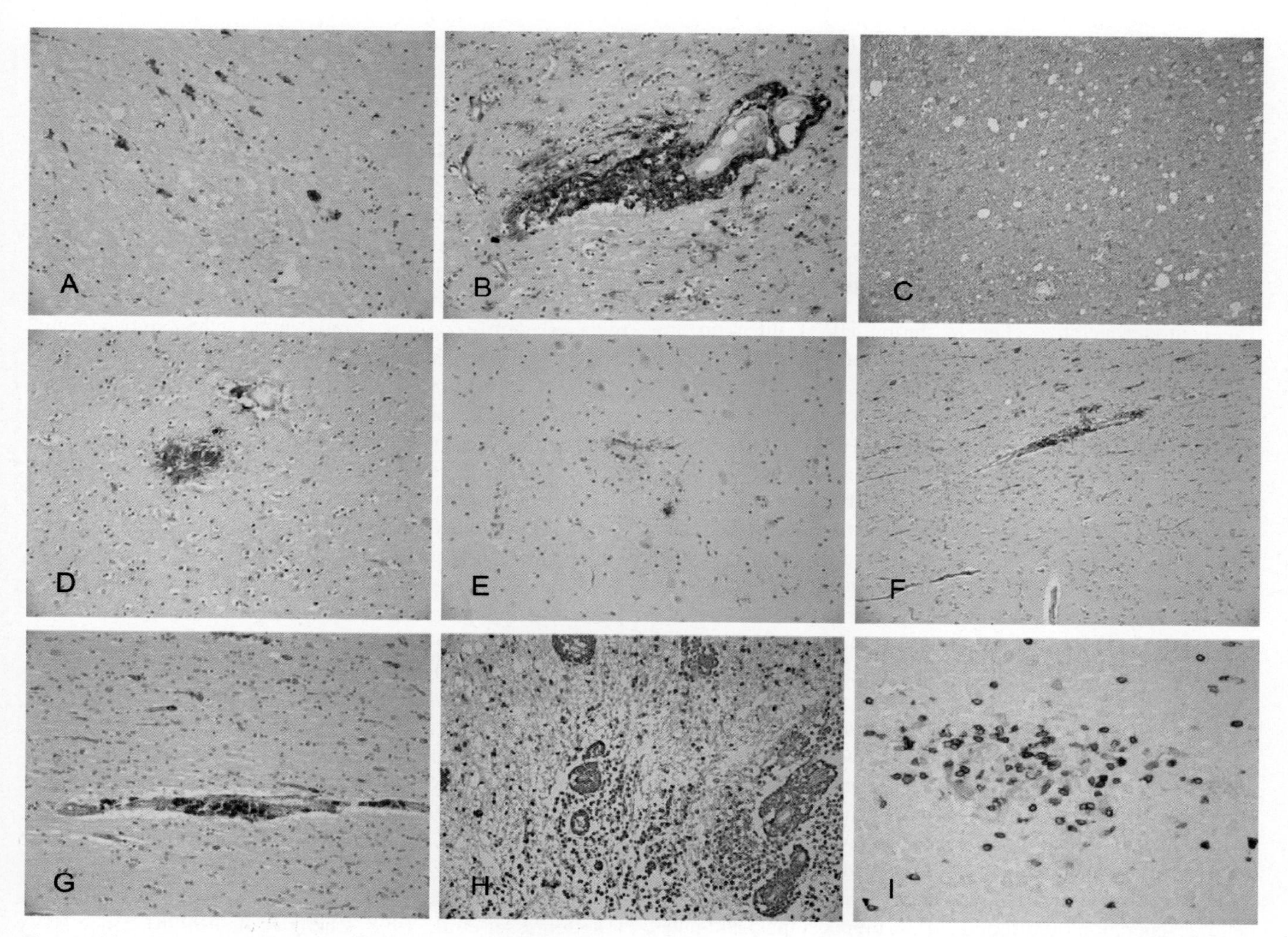

FIGURE 22.1. Neuropathologic manifestations of HIV-1 infection. (A, B) Increased amount of activated brain MP (microglia and macrophages including giant cells) positive for MHC class II (HLA-DR). (C) Wide-spread astrogliosis (GFAP-positive astrocytes) is seen in the areas of the most prominent MP infiltration. (D) Perivascular macrophages and (E) microglia are productively infected HIV-1 p24-positive cells in the brain. (F,G) Infiltration of CD163$^+$ monocytes/macrophages is a hallmark of HIVE (purple) resulting in injury of BBB indicated by decrease of staining for brain endothelial marker, breast cancer resistant protein (brown). (H) Leukoencephalopathy is characterized by massive monocyte/macrophage infiltration and tissue damage. (I) CD8$^+$ T lymphocytes (purple) accumulate around HIV-1 p24-positive brain MP (brown). Original magnification: panels A, B, D, E, G × 200; panels C, F, H × 100; panel I × 400.

erate strength suggests that other factors might also contribute to and mediate the severity of the dementia. Among them, levels of macrophage infiltration and activation in the CNS and the extent of damage to the synapto-dendritic structure of neurons (Glass et al., 1995; Masliah et al., 1997, 2000). Other factors to be taken into account in these clinico-pathological studies are the interval between clinical examination, and antiretroviral status and death.

22.3.2. Neuropathology in the Anti-Retroviral Era

Introduction of anti-retroviral therapy also impacted the patterns in HIV-related neuropathology in the past decade. Anti-retroviral therapy resulted in a decline in opportunistic infections of the CNS and neoplasms, and treated individuals survive longer. This is accompanied by an increase in the frequency of chronic forms of HIVE (Langford et al., 2002; Cysique et al., 2004). Everall et al. (2005) described the emerging variants of HIV neuropathology in ART-treated patients including (1) aggressive forms with severe HIVE and white matter injury, (2) extensive perivascular lymphocytic infiltration, (3) "burnt-out" forms of HIVE, and (4) β-amyloid (Aβ) accumulation with Alzheimer's-like neuropathology. Chronic "burnt-out" forms of CNS infection could be a result of prolonged survival. Such cases are characterized by focal white matter pathology, nonspecific lesions with gliosis and neuronal atrophy (Gray et al., 2003) without overt inflammation and secondary infection. From the onset of the AIDS epidemic, several investigators recognized that HIV-1 involvement of the CNS associates with white matter gliosis and demyelination (Gray et al., 1991). Leukoencephalopathy might be one of the central manifestations of HIVE (Figure 22.1H), and its association with the development of HIV-1 neuro-cognitive disorders is often overlooked due to primary focus on neuronal injury. Increased frequency of HAD (Neuenburg et al., 2002) and in particular of a highly destructive form of HIV-associated leukoencephalopathy (Langford et al., 2002) could be associated with the emergence of anti-retroviral drug-resistant forms of HIV-1. Severe leukoencephalopathy is characterized by significant perivascular infiltration of macrophages and lymphocytes, and it was assumed that this condition is due to an exaggerated response from a reconstituted immune system (Gray et al., 2003).

22.3.3. Patterns of Neuronal Injury

The neurodegenerative changes during HIV-1 infection are characterized by dendritic simplification of pyramidal neurons and selective loss of interneurons (Budka, 1991). The striato-cortical, cortico-cortical, and limbic intrinsic/inhibitory circuitries are mostly affected (Masliah et al., 1996). It is not clear whether they undergo neurodegeneration simultaneously or if there is a temporal progression of the neuronal injury from one site to another. Large pyramidal neurons in the neocortex (Wiley et al., 1991; Masliah et al., 1992), spiny neurons in the putamen (Masliah et al., 1996), medium-sized neurons in the globus pallidus, and interneurons in the hippocampus (Masliah et al., 1995) are among the most severely damaged neuronal populations. There was a 30–50% decrease in large neurons, accompanied by a 20% reduction in neocortical width in the frontal, parietal, and temporal cortices of brains with HIVE (Gray et al., 1991; Asare et al., 1996). Both experimental animal models and studies of human brain tissues suggest that neuronal damage may start in synapses and dendrites and then spread to the rest of the neuron, thereby activating pathways leading to cell death via apoptosis (Garden et al., 2002). Moore and colleagues clearly established the association between markers of regional neurodegeneration (midfrontal cortex, hippocampus, and putamen) at autopsy and levels of neuro-cognitive impairment in HIV-1 infected patients (Moore et al., 2006). The regional combined scores based on the percentage of neuropil occupied by MAP-2 (marker of neuronal cell bodies and dendrites) and synaptophysin (SP, marker of synapses) better correlated with the level of global neuro-cognitive impairment than measures of neuronal markers individually. Regional combined scores for hippocampus and putamen were independent predictors of the degree of neuropsychological impairment before death. Combining information from several markers of neural injury provided the strongest association with degree of neuro-cognitive impairment during life.

22.4. Neuropathogenesis of HIV Infection

22.4.1. Mechanisms of Neuronal Injury During HIV-1 CNS Infection

It is now widely accepted that in the absence of viral infection of neurons or macroglia, indirect mechanisms play a major role in HAD neuronal dysfunction and death. Multiple studies performed over the last decade and a half demonstrated that virus-infected microglia and macrophages secrete, or induce other neural cells to produce, neurotoxic factors that lead to neuronal injury or death during HIVE. These neurotoxins include arachidonic acid and its metabolites (Nottet et al., 1995), platelet activating factor (Gelbard et al., 1994), pro-inflammatory cytokines [tumor necrosis factor (TNF-α) or interleukin-1(IL-1β)], quinolinic acid (Heyes et al., 1991), NTox (Giulian et al., 1996), nitric oxide (Adamson et al., 1996) and glutamate (Jiang et al., 2001). A significant reduction in toxin levels in CSF reversed neuro-cognitive impairment when anti-retroviral therapy was administered to a HAD patient (Gendelman et al., 1998). Neuronal damage and/or dysfunction can be caused by viral proteins such as gp120 (Brenneman et al., 1988), gp41 (Adamson et al., 1996) and Tat (New et al., 1997) secreted by infected brain macrophages. Virions released from macrophages bind to chemokine (chemotactic cytokines) receptors expressed on neurons (Hesselgesser et al., 1998; Zheng et al., 1999a).

In the past few years, significant progress was made in clarifying the biological importance of chemokines in the CNS. Initially viewed as immune modulators, chemokines appear to play much broader biological roles in the CNS including development, neuronal excitability and synaptic transmission, neuro-inflammation, and neurodegeneration. A number of α- and β-chemokine receptors are expressed on neurons, astrocytes, and microglia (Gabuzda et al., 1998; Ghorpade et al., 1998; Luster, 1998; Albright et al., 1999). Two of them, CXCR2 and CXCR4, were found in neuronal cells (Hesselgesser et al., 1998). CXCR4 plays a crucial role in receptor-mediated neuronal apoptosis and cell function (Hesselgesser et al., 1998; Zheng et al., 1999b). Since HIV-1 gp120 can bind to chemokine receptors initiating signal transduction (Weissman et al., 1997), the neuronal chemokine receptor expression and distribution on different types of neurons is significant for understanding of CNS homeostasis and HAD pathogenesis. CXCR4 expression was found on neurons positive for choline acetyltransferase (in the caudate putamen and substantia innominata) and tyrosine hydroxylase (in the substantia nigra pars compacta) (Banisadr et al., 2002). Stimulation of neuronal chemokine receptors, by virions, viral proteins or the CXCR4 natural ligand stromal derived factor-1 [SDF-1α, upregulated in HIVE (Zhang et al., 1998)] may alter intracellular signaling events, and cause neuronal dysfunction and apoptosis (Hesselgesser et al., 1998; Zheng et al., 1999a). Constitutive expression of CXCR4 on neurons suggests that HIV-1 gp120 or SDF-1α could directly interfere with cholinergic and dopaminergic neurotransmission leading to neuronal dysfunction seen in HAD. There is an apparent disagreement between neurotoxic potential of CXCR4 tropic viruses in vitro models and HIV-1 isolates that exist in the brain of patients with HAD. Brain-derived viruses are macrophage tropic and principally use CCR5 for virus entry (Ghorpade et al., 1998). The presence of HIV-1 variants with increased CCR5 affinity and reduced dependence on CCR5 and CD4 in the brains of some HIV-1 infected patients with CNS infection suggested that variants with increased CCR5 affinity may represent a pathogenic viral phenotype contributing to HIV-1 neurodegenerative manifestations (Gorry et al., 2002). This apparent paradox could be associated with the predominance of indirect mechanisms of neurotoxicity (secretory substances of virus-infected and activated MP) in HIV-1 neuropathogenesis.

Laboratory models assaying neuronal function have been developed to mimic neuronal injury during HAD and to link virus infection, MP activation, chemokine production, and neuronal demise. Establishment of connection between neuronal dysfunction with brain MP activation was made possible by placing viral and/or immune products onto neurons and measuring cell signaling events or through ex vivo electrophysiological tests on MP-treated brain slices. Supernatants derived from HIV-1-infected MP were applied to the rat hippocampal brain slices (the site of mammalian learning and memory) and neuronal long-term potentiation (LTP; a neural response linked closely to learning and memory) was assayed. The diminished levels of LTP were detected in brain slices treated with supernatants acquired from virus-infected and activated MP as compared to uninfected or/and non-stimulated cells (Xiong et al., 1999). Establishment of such systems allowed testing of effects of inflammatory stimuli relevant to the neuropathogenesis of HAD and development of potential treatment approaches. IL-8 could represent one of the potential perpetrators in HAD-associated neuronal demise since elevated levels of IL-8 were detected in the brain tissue (Xiong et al., 2003b). When IL-8 effects on LTP were studied in hippocampal slices (Xiong et al., 2003b), this chemokine decreased LTP, and this inhibition was mediated through CXCR2 activation, suggesting a direct effect of IL-8 on synaptic transmission. In vitro studies demonstrated that neurotoxic factors secreted by MP can induce toxicity either directly or indirectly through downstream effects at the N-methyl-D-aspartate-type glutamate receptor leading to apoptosis (Kaul et al., 2001; Xiong et al., 2003a). Quantitative assessment of neuronal injury caused by HIV-1 infected, activated macrophages could be performed in neurotoxicity assays. The determination of neuronal structural proteins [MAP-2 and neurofilaments (NF)] was shown to correlate with neuronal dysfunction and could be used as a read-out system for neuronal survival/injury by computer-based imaging system (Zheng et al., 2004).

22.4.2. Astrocytes

How a relatively small number of infected MP localized in particular areas of the brain can lead to widespread neuronal injury is also unclear. One possibility is that interactions between immune competent MP and astrocytes lead to an amplification of toxic immune responses (Tornatore et al., 1991). Histopathological studies performed with human autopsy tissue showed that reactive astrogliosis, detected by glial fibrillary acidic protein (GFAP) staining, coincided with increased trafficking of activated macrophages and concomitant microglial activation (Persidsky et al., 1999). Neuropathological observations also document that intense astrogliosis and microglial activation is observed in areas of axonal and dendritic damage in HAD (Masliah et al., 1997).

There was consistent evidence that astrocytes become infected with HIV-1 although at low frequency as compared to brain MP. HIV-1 antigens and/or nucleic acid have been identified in astrocytes in brain autopsy tissue from both adult and pediatric AIDS cases (Tornatore et al., 1994; Ranki et al., 1995). In cell cultures, HIV-1 infection of astrocytes results in an initial productive but non-cytopathogenic infection that diminishes to a viral persistence or latent state (Messam and Major, 2000). The major barrier to HIV-1 infection of primary astrocytes appears to be at virus entry and that astrocytes have no intrinsic intracellular restriction to efficient HIV-1 replication (Canki et al., 2001). These cells can play a dual role, amplifying the effects of inflammation and mediating cellular damage as well as protecting the CNS. Cytokine/chemokine communications between MP and astrocytes and viral proteins produced by MP are involved in the balance of protective and destructive actions by these cells. HIV-1 transactivator protein Tat significantly increases astrocytic expression and release of monocyte chemo-attractant protein-1 (MCP-1), β-chemokine (Conant et al., 1998). Interactions between activated MP and astrocytes lead to significant up-regulation of β-chemokines promoting MP infiltration in HIVE (Persidsky, 1999; Persidsky et al., 1999). Secretion of α-chemokines by astrocytes was documented due to pro-inflammatory substances secreted by activated lymphocytes in the setting of HIVE (Poluektova et al., 2001). Pro-inflammatory cytokines (TNF-α, IL-β and IFN-γ) and quinolinic acid present in abundance during HAD are potent inducers of α- and β-chemokines by astrocytes (Guillemin et al., 2003). Astrocytes can suppress HIV-1 replication in monocyte-derived macrophages (MDM) via the production of soluble factors (Hori et al., 1999) and down-regulate secretion of pro-inflammatory factors by HIV-1 infected MP (Nottet et al., 1995). FasL is a molecule promoting apoptosis, and it was shown to be upregulated during HIVE on astrocytes (Ghorpade et al., 2003). Brain extracts from HAD patients had significantly elevated FasL levels compared to HIV-1-seropositive patients and seronegative individuals, and this mechanism can promote neuronal death via Fas–FasL interactions.

Astrocytes directly participate in synaptic integration by releasing glutamate via an exocytosis-like process. This occurs after activation of the receptor CXCR4 by SDF-1 (Bezzi et al., 2001). Autocrine/paracrine TNF-α-dependent signaling leading to prostaglandin formation controls glutamate release and astrocyte communication. Derangement of astrocyte processes is seen when activated microglia enhance release of cytokines in response to CXCR4 stimulation. Changed glial communication has direct neuropathological consequences because agents interfering with CXCR4-dependent astrocyte-microglia signaling prevent neuronal apoptosis induced by HIV-1 gp120. In addition to other functions, astrocytes are a significant component of the BBB ensuring maintenance of unique phenotypic properties of brain microvascular endothelial cells (BMVEC) (Persidsky, 1999) that are altered in HIVE (Dallasta et al., 1999). Clearly, the mechanism of HIV-1 associated brain injury is complex, probably involving multiple neurotoxic factors and pathways.

22.5. HIV-1 Neuroinvasion and Blood–Brain Barrier Compromise

22.5.1. Cellular Neuroinvasion and the BBB

Diffuse myelin pallor is a prominent feature of HIV-induced neuropathogenesis observed both at the early stage of HIV infection and also in HIVE (Gray et al., 1993). It has been shown that white matter changes are caused by alteration of the BBB. Structural and functional abnormalities in the microvasculature during HIV-1 CNS infection have been identified, including serum protein leak, alterations in capillary endothelial cells and basement membranes (Petito and Cash, 1992; Weis et al., 1996). There is a disruption of BBB characterized by decrease in staining for proteins of tight junctions (TJ, ZO-1, occludin and claudin-5) that ensure structural integrity of barrier. Dallasta and colleagues demonstrated significant TJ disruption (as indicated by diminished ZO-1 and occludin staining) in HIVE patients (Dallasta et al., 1999). This correlates with monocyte infiltration into the brain (Persidsky et al., 2006) and is accompanied by up-regulation of adhesion molecules on brain endothelium in HIVE (Nottet et al., 1996).

Originally proposed for visna virus (Peluso et al., 1985), monocyte-associated virus traffic rather than cell-free traffic became the prevailing mode for entry of HIV-1 (Persidsky and Gendelman, 2003). Monocytes are less able to replicate HIV-1 while circulating in the blood, but they can replicate virus upon entering tissues and become macrophages (Gendelman et al., 1988). It is likely that the monocytes being precursors to perivascular macrophages are infected outside the CNS and their traffic result in HIV CNS disease. Indeed, several investigators showed that virus-infected perivascular macrophages are immunophenotypically similar to a subset of blood monocytes (Williams et al., 2001). These blood monocytes expand during HIV infection and preferentially harbor virus, suggesting a strong likelihood of this subset being a vehicle for HIV-1. Phylogenetic analysis of HIV gp160 sequences in a patient with HAD demonstrates a close relationship to the sequences found in deep white matter, blood monocytes, and bone marrow among the tissues examined (Liu et al., 2000). Accumulating evidence suggested that a population of CD14+ monocytes expressing high levels of CD16 is a primary target of HIV-1 in blood (Ellery and Crowe, 2005). It was demonstrated that HIV proviral DNA is preferentially harbored in a minor population expressing CD14/CD16 (Shiramizu et al., 2005), and data obtained in an animal model (rhesus macaques infected with simian immunodeficient virus, SIV) confirmed that SIV DNA is consistently localized in a CD14lowCD16high subset (Williams et al., 2001; Williams et al., 2005). CD16+ monocytes express high levels of CX3CR1 and undergo transendothelial migration in response to chemokine secreted by endothelial cells, fractalkine (Ancuta et al., 2003). Thus, this migratory property possibly makes these cells ideal candidates destined to traffic to the CNS. Importantly, virus-infected monocytes migrate more efficiently in response to relevant β-chemokine, MCP-1 (Persidsky et al., 2006).

Based upon a number of studies, the following scenario of BBB injury during HIVE was suggested: soluble factors and/or interactions with inflammatory cells upregulate adhesion molecules on BMVEC (Nottet et al., 1996; Persidsky et al., 1997), monocytes are activated in peripheral blood (Pulliam et al., 1997; Williams et al., 2001, 2005), attracted to the BBB by increased secretion of β-chemokines within the CNS (Conant et al., 1998; Persidsky, 1999), monocyte–endothelial interactions lead to BBB compromise (Persidsky et al., 2006), and monocyte passage into the CNS. The temporal correlation between BBB disruption and perivascular infiltration of monocytes/macrophages suggests that a part of neuronal injury caused by HIV infection might be initiated at the vasculature. The decrease of BBB compromise in patients treated with anti-retroviral therapy is accompanied by an amelioration of neurocognitive impairment (Avison et al., 2004).

22.5.2. Modeling the BBB

The need for reliable in vitro laboratory systems to study the mechanisms of cell trafficking into the brain is obvious. The difficulties in establishing such a system, however, abound given the physiological and anatomical complexities of the BBB (Abbott, 2005). Modeling systems must take into account the cellular, matrix, immune and physiologic components of this system. The capillaries that form the BBB express a diverse array of functional characteristics that permit the regulation of cell, protein, and macromolecule passage between blood and brain (Hawkins and Davis, 2005). These are all difficult to recapitulate in laboratory cell systems. Nevertheless, diverse BBB models have been developed and include: (1) monolayers of primary BMVEC; (2) co-cultures of astrocytes and endothelial cells derived from organs other than the brain; and (3) co-culture systems of astrocytes and BMVEC expressing TJ, high level of tightness and unique transport systems (Persidsky et al., 1997). These systems allow studies of monocyte migration, effects of viral proteins and inflammatory factors on BBB function (Persidsky et al., 2006; Kanmogne et al., 2007).

22.5.3. Adaptive Immunity

While HIV-1 enters the brain early following viral infection (Davis et al., 1992), detectable productive viral replication and brain macrophage infiltration occur years later and only in some infected people (Sacktor et al., 2002). Explanations are that the virus either re-seeds the brain or endogenous control mechanisms that inhibit HIV-1 breaks down. Such an apparently effective control of viral replication in the brain present over years appears to be mediated by acquired T cell immunity, via cytotoxic CD8+ lymphocytes, (CTLs) (Sethi et al., 1988; Sopper et al., 1998). CTLs provide the principal regulatory elements that control persistent viral production in the periphery as well as the CNS (Dalod et al., 1999; Schmitz et al., 1999). A strong association between HIVE and profound immunodeficiency led to the conclusion that a lack of effective T cell immune responses is associated with ongoing viral production in the brain. Nonetheless, it is likely even more complex, as not all patients with profound immunodeficiency develop HAD and not all HAD patients have significant lymphocyte depletion. In human brain tissue with HIVE, a positive correlation was found between HIV-1pol and IP-10/CD14 (MP marker)/CXCR3 (lymphocyte marker), and between CD14 and CXCR3 mRNA expression reflecting an association between virus replication in MP and MP-mediated attraction of T cells (Poluektova et al., 2001). Neuropathological analyses revealed increased numbers of CD8+ lymphocytes within the neuropil next to virus-infected MP in HIVE brains as compared to brain tissue from seropositive patients without evidence of encephalitis (Figure 22.1I) (Potula et al., 2005). In this setting, the inflammatory cytokines and cytotoxic molecules (like granzyme B or perforin) released by these cells upon activation could contribute to the neurological sequelae of infection. Infiltrating CD8+ T cells may serve as a source of activation stimuli (such as IFN-γ or CD40L) for brain MP.

22.6. Animal Models

22.6.1. Simian Immunodeficiency Virus (SIV)

The SIV model reproduces HIV-1-associated CNS disease including neuropathology, BBB injury, cellular and humoral anti-viral immune responses, and neuro-cognitive abnormalities (Williams et al., 2001; Zink and Clements, 2002). The SIV macaque model allows studies of pathogenesis of lentivirus-induced neurological disease. Disease progression, as seen in SIV infection, demonstrates the same pattern as HIV-1 infection in humans (initial phase of viral replication, period of viral latency, the development of immunosuppression, and death). SIV-infected animals develop neurological symptoms and neuropathologic changes that are similar to those of humans with HIVE (Murray et al., 1992). The SIV model system was instrumental in current understanding of HIV-1 neuropathogenesis including early entrance of virus in the CNS (Chakrabarti et al., 1991), important details on the role of brain MP (microglia, perivascular macrophages) as the vehicle and reservoir for virus replication in the CNS (Kitagawa et al., 1991; Williams et al., 2001), significance of peripheral immune responses in control of virus replication (Williams and Hickey, 2002), inflammatory responses associated with CD8+ lymphocyte infiltration (Marcondes et al., 2001; Burudi et al., 2002; Kim et al., 2004b), inflammatory cytokine profiles similar to ones seen in HIVE (Orandle et al., 2001), and selective replication of neurovirulent stains (Babas et al., 2006). Development of an accelerated model (a combination of neurovirulent molecular clone, immunosuppressive and neurovirulent virus swarm) featured two distinct periods of virus replication in the CNS: (1) parenchymal microglial activation and possibly macrophage infiltration at the period of acute infection with subsequent decline paralleling suppression of viral replication; (2) one month later, a resurgence in virus replication with an increase in macrophage/microglia activation (Zink and Clements, 2002). Similar to human studies (Potula et al., 2005), virus-specific CTLs were found in the brains of SIV-inoculated macaques (Kim et al., 2004b). These cells are capable of eliminating virus-infected macrophages and microglia in vivo and may be causally linked to immune activation. SIV infected macaques were used to link the effects of antiretroviral drugs on CNS viral replication, development of encephalitis and reversible neuronal injury (Williams et al., 2005).

22.6.2. Feline Immunodeficiency Virus

Another model system that offers opportunities for studies of HIV-1 neuropathogenesis is the feline immunodeficiency virus (FIV), a lentivirus inducing an AIDS-like disease in cats that includes severe neurological deficits. Cats infected with FIV initially experience an acute flu-like syndrome including low-grade fever and transient lymphadenopathy, followed by an extended asymptomatic period that culminates in severe immunodeficiency paralleling due to a progressive decline in circulating CD4+ T cells and a sustained increase in activated CD8+ (English et al., 1994). While like HIV-1, FIV infects monocyte-derived cells (Beebe et al., 1994). FIV also infects CD8+ T lymphocytes and B-lymphocytes (English et al., 1993) in vivo, featuring a much broader host cell range than the pattern of cell tropism displayed by HIV-1. Although most primary isolates of FIV may efficiently target both T-lymphocytes and monocytes, productive infection in vivo is largely restricted to FIV-infected T-lymphocytes. Although there are clear differences between the viral genome structures and cellular tropisms of FIV and HIV-1, the systemic and CNS diseases produced in their respective hosts are remarkably similar. Both viruses display a highly conserved tropism for the chemokine receptor, CXCR4 (Willett and Hosie, 1999), and rapidly penetrate the CNS with preferential infection of microglia and macrophages.

Soon after infection, FIV appears in the CNS inducing neurological deficits in some animals. Symptoms seen in these animals include enhanced aggression, increased cortical slow wave activity, abnormal brainstem-evoked potentials, delayed righting and pupillary reflexes, marked changes in sleep architecture, and deficits in cognitive-motor functions (Steigerwald et al., 1999). Similar to other lentiviruses, FIV neurotropism was demonstrated by virus detection in brain tissue and CSF of cats infected by either peripheral or intrathecal inoculation (Johnston and Power, 2002). Neuropathologic examination of FIV-infected cat brains demonstrates microglial nodules with very rare, multinucleated giant cells, gliosis, perivascular lymphocytic infiltrates and myelin pallor. (Hurtrel et al., 1992; Boche et al., 1996). A significant neuronal loss is a feature of FIV infection (Meeker et al., 1997). MRS demonstrated reduced levels in the concentrations of the neuronal marker, N-acetyl-aspartate (NAA) and the NAA/Cr ratio within the brains of FIV-infected cats (Power et al., 1998). FIV can infect a number of different brain cells in vitro including astrocytes (Dow et al., 1990), microglia (Dow et al., 1990; Hein et al., 2003), BMVEC (Steffan et al., 1994), and choroid plexus macrophages (Bragg et al., 2002).

22.6.3. Small Animal Models

The limitations of the primate or cat models of human HIV-1 CNS infection include the costs, special facility requirements, the necessity of using mixtures of viral strains (in order to develop reproducible CNS infection), and the relatively small numbers of animals studied that preclude meaningful statistical analyses. These obstacles have urged researchers to develop small animal models. Several rodent models were established including transgenic mice (expressing viral proteins or relevant inflammatory factors seen in HAD), mice infected primarily with murine retroviruses, or severe combined immunodeficient (SCID) mice inoculated intracerebrally with human HIV-1 infected macrophages and reconstituted with human peripheral blood lymphocytes (PBL).

22.6.3.1. Transgenic Mice

The role of HIV-1 gp120 CNS over expression was studied in transgenic mice in which expression of gp120 derived from CXCR4 tropic virus was placed under regulatory control of the murine GFAP promoter (Toggas et al., 1994). Reactive astrogliosis was prominent in transgenic mice showing high levels of gp120 expression. A 40% reduction in the number of large neurons was found in the neocortex. The loss of this neuronal subpopulation was combined with widespread neuronal dendritic vacuolation. Using the Morris water maze, D'Hooge et al. (1999) showed reduced escape latency and swimming velocity during the acquisition trials in 12-month-old gp120 animals. The spatial memory (retention) was diminished in probe trials in transgenic mice of this age. Cognitive abnormalities correlated with neuronal dysfunction (synaptic plasticity and excitatory postsynaptic potentials measured in CA1 neurons in hippocampal slices) from a gp120 transgenic mice study (Krucker et al., 1998).

HIV-1 Tat protein has neurotoxic affects on in vitro systems and it therefore was proposed to contribute to HIV-1-associated neurological syndromes. Kim and colleagues established and characterized a transgenic mouse model in which Tat expression was regulated by both the astrocyte-specific GFAP promoter and a doxycycline (Dox)-inducible promoter (Kim et al., 2003). Dox-dependent Tat expression occurred exclusively in astrocytes. Tat expression in the brain caused profound abnormalities including failure to thrive, hunched gesture, tremor, ataxia, slow cognition and motor movement, seizures, and premature death. Structural abnormalities of the cerebellum and cortex, brain edema, astrogliosis, degeneration of neuronal dendrites, neuronal apoptosis, and infiltration of activated monocytes and T lymphocytes were shown in these animals. Leukocyte CNS infiltration paralleled overexpression of α- and β-chemokines in these mice (Kim et al., 2004a).

22.6.3.2. Murine Retroviruses

The murine leukemia viruses (MuLV) are retroviruses with significant genetic differences from lentiviruses (like HIV-1) that can induce neurodegenerative disease (spongiform encephalomyelopathy in the hindbrain and spinal cord) in mice (Nagra et al., 1992). Infection results in progressive hind limb paralysis and leukemia that develop within weeks to months after inoculation in susceptible mouse strains. Importantly, intensity and progression of the neurological disease are well correlated with viral replication and accumulation of viral proteins in CNS tissue. Significant differences exist between various viral strains and they are related to the different routes of inoculation, virus and perhaps mouse strain-dependent susceptibility of specific CNS cell types, and strain-dependent speed of disease progression (Nagra et al., 1993). Viral strains resulting in more severe CNS pathology replicate more efficiently and spread in the peripheral organs of mice (Czub et al., 1992). MuLV-infected mice demonstrate neuronal loss, neurotoxin production and behavioral abnormalities, and could be a tool for testing therapeutic agents. However, there are dissimilarities from HIVE and these include the significant genetic differences of MuLV from HIV-1, a lentivirus. In addition, the pathology of MuLV spongiform encephalomyelopathy is generally distinct from that of HIV encephalitis. Spongiform changes are unusual in HIVE and there is a predilection for HIV-1 to infect mononuclear phagocytes in the basal ganglia and white matter of the cerebral hemispheres, whereas there is a relative paucity of MuLV infection in these areas (mostly brainstem and spinal cord). HIV-1 primarily infects microglia and macrophages and to a lesser extent astrocytes; whereas, MuLV, depending on the strain, significantly infects endothelial cells, microglia, oligodendrocytes, astrocytes, and neurons. HIV-1 infection of the CNS results in multinucleated giant cell formation; these types of cells have not been described in MuLV-induced CNS disease. Spongiform changes devoid of significant inflammatory infiltrate are similar to those found in spinal cords of HIV-1 infected patients with vacuolar myelopathy but they are different from the generally non-spongiform pathology of HIVE.

22.6.3.3. Severe Combined Immune Deficient Mouse Model

In order to overcome some of the drawbacks associated with other mouse models, a model of HIVE in SCID mice was developed by Tyor et al. (1993). Human peripheral blood mononuclear cells (PBMC) were injected into the brains of SCID mice as well as cell-free HIV-1 either simultaneously with the PBMC or a day after the PBMC. One to four weeks after injection of human MDM, multinucleated giant cells, some of which were HIV-1 p24-positive, were detected primarily in the area of the needle tract. Mice inoculated with virus only had no HIV detected. HIV strains used included HIV-1IIIB (CXCR4), HIV-1BAL (CCR5), HIV-1MN (CXCR4), and human monocyte derived macrophages (MDM) infected with HIV-1ADA (CCR5) in vitro prior to inoculation. There was evidence that the human macrophages present in HIV-1 infected SCID brains are activated since they were positively stained for MHC class II antigen, TNF-α, and occasionally IL-1 (Tyor et al., 1993). In order to improve the reproducibility of HIVE in SCID mice, key steps for the model development (infection of highly purified human MDM in vitro, placement of equivalent numbers of cells into the same brain regions by stereotactic techniques, mouse brain sampling, quantitation of astrogliosis through use of computer image analysis on serial coronal sections) were standardized. Stereotactic placement of the inoculum in replicate experiments allowed delivery of equal amounts of human HIV-1 infected MDM to the same anatomical areas (basal ganglia and cortex) in each mouse (Persidsky et al., 1996). HIV-1-infected human MDM (placed into the putamen and cortex of SCID mice) remained viable for five weeks. Multinucleated giant cells were located in perivascular spaces. HIV-1 p24 antigen expression detected in 40–60% of inoculated cells in mouse brains was persistent. Human MDM were immune activated (as defined by HLA-DR, IL-1β, IL-6 and TNF-α expression), and neuronal injury (apoptosis) developed after virus-infected MDM placement into mouse brain tissue. Neuro-inflammatory responses (including astrogliosis, microglial reaction and cytokine production) started at three days and peaked two weeks post-inoculation. Similar patterns of cytokine expression were found in human and mouse brains with HIVE. These included the upregulation of both human and mouse TNF-α, VCAM-1 (define), and E-selectin that correlated with the inflammatory reaction in brain tissue and decrease in neuronal dendritic density in MDM injected and contra-lateral hemispheres.

Subsequent experiments demonstrated that SCID mice inoculated with HIV-1-infected MDM in the basal ganglia/cortex showed neuronal dysfunction (detected by impaired LTP, an electrophysiologic response linked closely to learning and memory) and associated cognitive impairment (Zink et al., 2002). Alterations in neuronal physiology paralleled diminished expression of the neuronal synaptic marker (SP) and neuronal axonal marker (NF) assayed in the hippocampus of HIVE mice. By two weeks, HIVE mice expressed 3.8- and 2.6-fold less NF and SP than shams. These findings support the notion that HIV-1-infected and activated brain macrophages can cause neuronal damage at distant anatomic sites. Importantly, the findings confirmed the value of the model in exploring the physiological basis of HAD and development of new therapeutic approaches for its treatment.

Cellular reservoirs and anatomical sanctuary sites, such as the CNS, hamper HIV-1 treatment and allow HIV-1 to persist. To these ends, the SCID mouse model was modified to allow spread of HIV-1 infection between human MDM in mouse brain tissue behind an intact BBB. This model permitted comparative assessment of different anti-retroviral drugs by measurements of viral load and numbers of infected human macrophages (Limoges et al., 2000; Limoges et al., 2001). New drug delivery systems across the BBB and novel neuroprotective therapeutics are under investigation (Persidsky et al., 2001; Dou et al., 2003; Dou et al., 2005).

22.6.3.4. NOD/SCID and Studies of Acquired Immune Responses

What role impaired immune responses play in HIV-1 associated neurodegeneration and HAD pathogenesis is also under investigation (Poluektova et al., 2002). A pattern of human CD8 cells activation and CNS infiltration seen in HIVE was reproduced in response to HIV-1-infected human macrophages injected into the brain of mice reconstituted with autologous peripheral blood lymphocytes (Poluektova et al., 2002). HIV-1-infected MDM were injected into the basal ganglia after syngeneic immune reconstitution by human PBL to generate a human PBL-non-obese diabetic (NOD)/SCID HIVE mouse. Engrafted T lymphocytes produced HIV-1 gag- and HIV-1pol-specific CTLs against virus-infected brain MDM within one week. This was demonstrated by tetramer staining of human PBL in mouse spleens and by IFN-γ ELISPOT. CD8, granzyme B, HLA-DR, and CD45R0-reactive T cells migrated to and were in contact with human MDM in brain areas where infected macrophages were abundant. The numbers of productively infected MDM were markedly reduced (>85%) during two weeks of observation. However, the levels of granzyme-positive cells were significantly reduced and generated CTLs were not able to completely eliminate HIV-1-infected cells.

Elimination of infected macrophages in the lymphocyte reconstituted SCID mouse HIVE model was accompanied by diminished infiltration of CD8+ lymphocytes in the brain. Despite decreased amounts of HIV-1 infected MDM and CD8+ lymphocytes, astrogliosis and microglial reaction persisted. Brain tissue of hu-PBL-NOD/SCID mice showed neuronal injury when compared to SCID mice inoculated with HIV-1 infected MDM only. Expression of neuronal markers was restored after cessation of neuro-inflammation. Mice developed viremia due to infection of circulating CD4+ lymphocytes, and virus-specific antibodies were detected by Western blot in serum. In summary, this model reproduces important features of HIV-1 CNS infection including development of virus-specific CTL, human lymphocyte migration across the BBB, elimination of HIV-1-infected MDM and neuro-inflammation.

Failure to eliminate HIV-1 infected macrophages by CTL in the brain may be due to presentation of HIV-1 antigen by tolerogenic antigen presenting cells (like macrophages) (Persidsky and Gendelman, 2003). A rate limiting enzyme of kynurenine pathway of tryptophan metabolism, indoleamine 2,3-dioxygenase (IDO) is up-regulated in virus-infected macrophages (Grant et al., 2000) and is implicated in the inhibition of antigen-specific T cell proliferation (Mellor et al., 2003). The idea that IDO inhibitor could affect the generation of CTL and clearance of virus-infected macrophages from the brain was tested in hu-PBL-NOD/SCID model. One week after MDM injection in the basal ganglia, animals treated with IDO inhibitor showed increased numbers of CD3+, CD8+ and IFN-γ-producing human T lymphocytes in peripheral blood compared to controls. At week two, mice treated with IDO inhibitor demonstrated a two-fold increase in CD8+ T lymphocytes in the areas of the brain containing HIV-1-infected MDM compared to untreated controls (Potula et al., 2005). By week three, mice exposed to IDO inhibitor showed 89% reduction in HIV-1-infected MDM in the brain as compared to controls. Thus, manipulation of immunosuppressive IDO activity in HIVE may enhance the generation of HIV-1 specific CTL leading to elimination of HIV-1 infected macrophages in brain and attenuation of neuronal injury.

The hu-PBL-NOD/SCID model allowed investigation of potential contributions of co-morbidity factors, like alcohol abuse, in the progression of HIV-1 CNS infection. HIVE hu-PBL-NOD/SCID mice chronically exposed to ethanol demonstrated increased levels of HIV-1 p24, ineffective elimination of HIV-1 infected macrophages by CTLs in the brain, enhanced microglial reaction and prominent BBB damage. These data proved that alcohol could be an exacerbating factor in HIVE (Potula et al., 2006). Understanding of the mechanisms underlying combined toxic effects of HIV-1 brain infection and excessive alcohol use will help to design neuro-protective strategies.

Summary

Studies performed over a period of two decades indicate that HIV-associated cognitive impairment is a chronic neuro-inflammatory condition. Virus productive infection of brain MP is important, however, it alone cannot explain multi-faceted effects of HIV-1 on chronic immune activation within unique brain milieu. Recent advances in understanding of HIV-1 neuropathogenesis identified vulnerable neuronal populations, correlated their loss with cognitive performance and neuro-imaging studies, proved the role of peripheral immune responses in control of HIV-1 replication and chronic immune activation leading to neurodegeneration, underscored the importance of BBB injury, defined specific sub-population of monocytes associated with disease progression and effects of ART. Better understanding of HIV-1 neuropathogenesis guides development of therapeutic approaches that include efficient delivery of anti-retroviral therapy, neuroprotective and anti-inflammatory strategies, and improvement in acquired immunity.

Review Questions/Problems

1. Pathology HIV-1 encephalitis (HIVE) is characterized by

a. accumulation of neutrophils in the neuropil
b. presence of multinucleated giant cells
c. HIV-1 inclusions in the nuclei
d. decrease in number of astrocytes

2. Brain structures most affected by HIV-1 encephalitis are

a. basal ganglia
b. cerebellum
c. frontal cortex
d. brain stem
e. all of the above

3. Causes of neuronal injury in HIV-1 CNS infection are

a. effects of viral proteins (gp41, gp120)
b. cytokines and chemokines
c. glutamate
d. anti-retroviral therapy
e. all of the above
f. a, b, and c only

4. Productively infected cells in HIV-1 CNS infection are

a. neurons
b. endothelial cells
c. oligodendrocytes
d. macrophages/microglia
e. all of the above

5. Virus present in the brain tissues is

a. CCR5-tropic
b. CXCR4-tropic
c. CX3CR-tropic
d. none of the above
e. all of the above

6. HIV-1 CNS infection is best described as

a. acute lytic viral infection of neurons
b. chronic viral infection of oligodendrocytes
c. chronic neuro-inflammatory condition driven by macrophage infection
d. acute lytic infection of astrocytes
e. none of the above

7. HIV-1 brain infection in post-HAART era is characterized by

a. aggressive forms with severe HIVE and white matter injury
b. "burnt-out" forms of HIVE
c. more extensive β-amyloid (Aβ) deposition
d. more often occurrence of opportunistic infections
e. a and d
f. a, b, and c

8. Introduction of HAART resulted in more severe neurocognitive impairment in HIV-1 infected patients

a. true
b. false

9. Peripheral anti-viral immune responses control in measure HIV-1 CNS infection

a. true
b. false

10. Blood–brain barrier is injured in HIV-1 brain infection

a. true
b. false

References

Abbott NJ (2005) Dynamics of CNS barriers: Evolution, differentiation, and modulation. Cell Mol Neurobiol 25:5–23.

Adamson DC, Wildemann B, Sasaki M, Glass JD, McArthur JC, Christov VI, Dawson TM, Dawson VL (1996) Immunologic NO synthase: Elevation in severe AIDS dementia and induction by HIV-1 gp41. Science 274:1917–1921.

Albright AV, Shieh JT, Itoh T, Lee B, Pleasure D, O'Connor MJ, Doms RW, Gonzalez-Scarano F (1999) Microglia express CCR5, CXCR4, and CCR3, but of these, CCR5 is the principal coreceptor for human immunodeficiency virus type 1 dementia isolates. J Virol 73:205–213.

Ancuta P, Rao R, Moses A, Mehle A, Shaw SK, Luscinskas FW, Gabuzda D (2003) Fractalkine preferentially mediates arrest and migration of CD16 + monocytes. J Exp Med 197:1701–1707.

Asare E, Dunn G, Glass J, McArthur J, Luthert P, Lantos P, Everall I (1996) Neuronal pattern correlates with the severity of human immunodeficiency virus-associated dementia complex. Usefulness of spatial pattern analysis in clinicopathological studies. Am J Pathol 148:31–38.

Avison MJ, Nath A, Greene-Avison R, Schmitt FA, Greenberg RN, Berger JR (2004) Neuroimaging correlates of HIV-associated BBB compromise. J Neuroimmunol 157:140–146.

Babas T, Dewitt JB, Mankowski JL, Tarwater PM, Clements JE, Zink MC (2006) Progressive selection for neurovirulent genotypes in the brain of SIV-infected macaques. AIDS 20:197–205.

Banisadr G, Fontanges P, Haour F, Kitabgi P, Rostene W, Melik Parsadaniantz S (2002) Neuroanatomical distribution of CXCR4 in adult rat brain and its localization in cholinergic and dopaminergic neurons. Eur J Neurosci 16:1661–1671.

Beebe AM, Dua N, Faith TG, Moore PF, Pedersen NC, Dandekar S (1994) Primary stage of feline immunodeficiency virus infection: Viral dissemination and cellular targets. J Virol 68:3080–3091.

Bezzi P, Domercq M, Brambilla L, Galli R, Schols D, De Clercq E, Vescovi A, Bagetta G, Kollias G, Meldolesi J, Volterra A (2001) CXCR4-activated astrocyte glutamate release via TNFalpha: Amplification by microglia triggers neurotoxicity. Nat Neurosci 4:702–710.

Boche D, Hurtrel M, Gray F, Claessens-Maire MA, Ganiere JP, Montagnier L, Hurtrel B (1996) Virus load and neuropathology in the FIV model. J Neurovirol 2:377–387.

Bragg DC, Childers TA, Tompkins MB, Tompkins WA, Meeker RB (2002) Infection of the choroid plexus by feline immunodeficiency virus. J Neurovirol 8:211–224.

Brenneman DE, Westbrook GL, Fitzgerald SP, Ennist DL, Elkins KL, Ruff MR, Pert CB (1988) Neuronal cell killing by the envelope protein of HIV and its prevention by vasoactive intestinal peptide. Nature 335:639–642.

Budka H (1991) Neuropathology of human immunodeficiency virus infection. Brain Pathol 1:163–175.

Budka H, Wiley CA, Kleihues P, Artigas J, Asbury AK, Cho ES, Cornblath DR, Dal Canto MC, DeGirolami U, Dickson D, et al. (1991) HIV-associated disease of the nervous system: Review of nomenclature and proposal for neuropathology-based terminology. Brain Pathol 1:143–152.

Burudi EM, Marcondes MC, Watry DD, Zandonatti M, Taffe MA, Fox HS (2002) Regulation of indoleamine 2,3-dioxygenase expression in simian immunodeficiency virus-infected monkey brains. J Virol 76:12233–12241.

Canki M, Thai JN, Chao W, Ghorpade A, Potash MJ, Volsky DJ (2001) Highly productive infection with pseudotyped human immunodeficiency virus type 1 (HIV-1) indicates no intracellular restrictions to HIV-1 replication in primary human astrocytes. J Virol 75:7925–7933.

Chakrabarti L, Hurtrel M, Maire MA, Vazeux R, Dormont D, Montagnier L, Hurtrel B (1991) Early viral replication in the brain of SIV-infected rhesus monkeys. Am J Pathol 139:1273–1280.

Chang L, Ernst T, Witt MD, Ames N, Walot I, Jovicich J, DeSilva M, Trivedi N, Speck O, Miller EN (2003) Persistent brain abnormalities in antiretroviral-naive HIV patients 3 months after HAART. Antivir Ther 8:17–26.

Chang L, Lee PL, Yiannoutsos CT, Ernst T, Marra CM, Richards T, Kolson D, Schifitto G, Jarvik JG, Miller EN, Lenkinski R, Gonzalez G, Navia BA (2004) A multicenter in vivo proton-MRS study of HIV-associated dementia and its relationship to age. Neuroimage 23:1336–1347.

Cherner M, Masliah E, Ellis RJ, Marcotte TD, Moore DJ, Grant I, Heaton RK (2002) Neurocognitive dysfunction predicts postmortem findings of HIV encephalitis. Neurology 59:1563–1567.

Conant K, Garzino-Demo A, Nath A, McArthur J, Halliday W, Power C, Gallo R, Major E (1998) Induction of monocyte chemoattractant protein-1 in HIV-1 Tat-stimulated astrocytes and elevation in AIDS dementia. Proc Natl Acad Sci USA 95:3117–3121.

Cysique LA, Maruff P, Brew BJ (2004) Prevalence and pattern of neuropsychological impairment in human immunodeficiency virus-infected/acquired immunodeficiency syndrome (HIV/AIDS) patients across pre- and post-highly active antiretroviral therapy eras: A combined study of two cohorts. J Neurovirol 10:350–357.

Czub M, McAtee FJ, Portis JL (1992) Murine retrovirus-induced spongiform encephalomyelopathy: Host and viral factors which determine the length of the incubation period. J Virol 66:3298–3305.

D'Hooge R, Franck F, Mucke L, De Deyn PP (1999) Age-related behavioural deficits in transgenic mice expressing the HIV-1 coat protein gp120. Eur J Neurosci 11:4398–4402.

Dallasta LM, Pisarov LA, Esplen JE, Werley JV, Moses AV, Nelson JA, Achim CL (1999) Blood-brain barrier tight junction disruption in human immunodeficiency virus-1 encephalitis. Am J Pathol 155:1915–1927.

Dalod M, Dupuis M, Deschemin JC, Sicard D, Salmon D, Delfraissy JF, Venet A, Sinet M, Guillet JG (1999) Broad, intense anti-human immunodeficiency virus (HIV) ex vivo CD8(+) responses in HIV type 1-infected patients: Comparison with anti-Epstein-Barr virus responses and changes during antiretroviral therapy. J Virol 73:7108–7116.

Davis LE, Hjelle BL, Miller VE, Palmer DL, Llewellyn AL, Merlin TL, Young SA, Mills RG, Wachsman W, Wiley CA (1992) Early viral brain invasion in iatrogenic human immunodeficiency virus infection. Neurology 42:1736–1739.

Dou H, Birusingh K, Faraci J, Gorantla S, Poluektova LY, Maggirwar SB, Dewhurst S, Gelbard HA, Gendelman HE (2003) Neuroprotective activities of sodium valproate in a murine model of human immunodeficiency virus-1 encephalitis. J Neurosci 23:9162–9170.

Dou H, Ellison B, Bradley J, Kasiyanov A, Poluektova LY, Xiong H, Maggirwar S, Dewhurst S, Gelbard HA, Gendelman HE (2005) Neuroprotective mechanisms of lithium in murine human immunodeficiency virus-1 encephalitis. J Neurosci 25:8375–8385.

Dow SW, Poss ML, Hoover EA (1990) Feline immunodeficiency virus: A neurotropic lentivirus. J Acquir Immune Defic Syndr 3:658–668.

Ellery PJ, Crowe SM (2005) Phenotypic characterization of blood monocytes from HIV-infected individuals. Methods Mol Biol 304:343–353.

English RV, Johnson CM, Gebhard DH, Tompkins MB (1993) In vivo lymphocyte tropism of feline immunodeficiency virus. J Virol 67:5175–5186.

English RV, Nelson P, Johnson CM, Nasisse M, Tompkins WA, Tompkins MB (1994) Development of clinical disease in cats experimentally infected with feline immunodeficiency virus. J Infect Dis 170:543–552.

Everall IP, Hansen LA, Masliah E (2005) The shifting patterns of HIV encephalitis neuropathology. Neurotox Res 8:51–61.

Gabuzda D, He J, Ohagen A, Vallat AV (1998) Chemokine receptors in HIV-1 infection of the central nervous system. Semin Immunol 10:203–213.

Garden GA, Budd SL, Tsai E, Hanson L, Kaul M, D'Emilia DM, Friedlander RM, Yuan J, Masliah E, Lipton SA (2002) Caspase cascades in human immunodeficiency virus-associated neurodegeneration. J Neurosci 22:4015–4024.

Gelbard HA, Nottet HS, Swindells S, Jett M, Dzenko KA, Genis P, White R, Wang L, Choi YB, Zhang D, et al. (1994) Platelet-activating factor: A candidate human immunodeficiency virus type 1-induced neurotoxin. J Virol 68:4628–4635.

Gendelman HE (2002) Neural immunity: Friend or foe? J Neurovirol 8:474–479.

Gendelman HE, Orenstein JM, Martin MA, Ferrua C, Mitra R, Phipps T, Wahl LA, Lane HC, Fauci AS, Burke DS, et al. (1988) Efficient isolation and propagation of human immunodeficiency virus on recombinant colony-stimulating factor 1-treated monocytes. J Exp Med 167:1428–1441.

Gendelman HE, Zheng J, Coulter CL, Ghorpade A, Che M, Thylin M, Rubocki R, Persidsky Y, Hahn F, Reinhard J, Jr., Swindells S (1998) Suppression of inflammatory neurotoxins by highly active antiretroviral therapy in human immunodeficiency virus-associated dementia. J Infect Dis 178:1000–1007.

Ghorpade A, Holter S, Borgmann K, Persidsky R, Wu L (2003) HIV-1 and IL-1 beta regulate Fas ligand expression in human astrocytes through the NF-kappa B pathway. J Neuroimmunol 141:141–149.

Ghorpade A, Xia MQ, Hyman BT, Persidsky Y, Nukuna A, Bock P, Che M, Limoges J, Gendelman HE, Mackay CR (1998) Role of the beta-chemokine receptors CCR3 and CCR5 in human immunodeficiency virus type 1 infection of monocytes and microglia. J Virol 72:3351–3361.

Giulian D, Yu J, Li X, Tom D, Li J, Wendt E, Lin SN, Schwarcz R, Noonan C (1996) Study of receptor-mediated neurotoxins released by HIV-1-infected mononuclear phagocytes found in human brain. J Neurosci 16:3139–3153.

Glass JD, Fedor H, Wesselingh SL, McArthur JC (1995) Immunocytochemical quantitation of human immunodeficiency virus in the brain: Correlations with dementia. Ann Neurol 38:755–762.

Gonzalez E, Rovin BH, Sen L, Cooke G, Dhanda R, Mummidi S, Kulkarni H, Bamshad MJ, Telles V, Anderson SA, Walter EA, Stephan KT, Deucher M, Mangano A, Bologna R, Ahuja SS, Dolan MJ, Ahuja SK (2002) HIV-1 infection and AIDS dementia are influenced by a mutant MCP-1 allele linked to increased monocyte infiltration of tissues and MCP-1 levels. Proc Natl Acad Sci USA 99:13795–13800.

Gorry PR, Taylor J, Holm GH, Mehle A, Morgan T, Cayabyab M, Farzan M, Wang H, Bell JE, Kunstman K, Moore JP, Wolinsky SM, Gabuzda D (2002) Increased CCR5 affinity and reduced CCR5/CD4 dependence of a neurovirulent primary human immunodeficiency virus type 1 isolate. J Virol 76:6277–6292.

Grant I, Heaton RK, Atkinson JH (1995) Neurocognitive disorders in HIV-1 infection. HNRC Group. HIV Neurobehavioral Research Center. Curr Top Microbiol Immunol 202:11–32.

Grant RS, Naif H, Thuruthyil SJ, Nasr N, Littlejohn T, Takikawa O, Kapoor V (2000) Induction of indoleamine 2,3-dioxygenase in primary human macrophages by HIV-1. Redox Rep 5:105–107.

Gray F, Hurtrel M, Hurtrel B (1993) Early central nervous system changes in human immunodeficiency virus (HIV)-infection. Neuropathol Appl Neurobiol 19:3–9.

Gray F, Chretien F, Vallat-Decouvelaere AV, Scaravilli F (2003) The changing pattern of HIV neuropathology in the HAART era. J Neuropathol Exp Neurol 62:429–440.

Gray F, Haug H, Chimelli L, Geny C, Gaston A, Scaravilli F, Budka H (1991) Prominent cortical atrophy with neuronal loss as correlate of human immunodeficiency virus encephalopathy. Acta Neuropathol (Berl) 82:229–233.

Guillemin GJ, Croitoru-Lamoury J, Dormont D, Armati PJ, Brew BJ (2003) Quinolinic acid upregulates chemokine production and chemokine receptor expression in astrocytes. Glia 41:371–381.

Hawkins BT, Davis TP (2005) The blood-brain barrier/neurovascular unit in health and disease. Pharmacol Rev 57:173–185.

Heaton RK, Grant I, Butters N, White DA, Kirson D, Atkinson JH, McCutchan JA, Taylor MJ, Kelly MD, Ellis RJ, et al. (1995) The

HNRC 500—neuropsychology of HIV infection at different disease stages. HIV Neurobehavioral Research Center. J Int Neuropsychol Soc 1:231–251.
Hein A, Schuh H, Thiel S, Martin JP, Dorries R (2003) Ramified feline microglia selects for distinct variants of feline immunodeficiency virus during early central nervous system infection. J Neurovirol 9:465–476.
Hesselgesser J, Taub D, Baskar P, Greenberg M, Hoxie J, Kolson DL, Horuk R (1998) Neuronal apoptosis induced by HIV-1 gp120 and the chemokine SDF-1 alpha is mediated by the chemokine receptor CXCR4. Curr Biol 8:595–598.
Heyes M, Brew B, Martin A, Price R, Salazar A, Sidtis J, Yergey J, Mouradian M, Sadler A, Keilp J, et al. (1991) Quinolinic acid in cerebrospinal fluid and serum in HIV-1 infection: Relationship to clinical and neurological status. Ann Neurol 29:202–209.
Hori K, Burd PR, Kutza J, Weih KA, Clouse KA (1999) Human astrocytes inhibit HIV-1 expression in monocyte-derived macrophages by secreted factors. AIDS 13:751–758.
Hurtrel M, Ganiere JP, Guelfi JF, Chakrabarti L, Maire MA, Gray F, Montagnier L, Hurtrel B (1992) Comparison of early and late feline immunodeficiency virus encephalopathies. AIDS 6:399–406.
Jiang ZG, Piggee C, Heyes MP, Murphy C, Quearry B, Bauer M, Zheng J, Gendelman HE, Markey SP (2001) Glutamate is a mediator of neurotoxicity in secretions of activated HIV-1-infected macrophages. J Neuroimmunol 117:97–107.
Johnston JB, Power C (2002) Feline immunodeficiency virus xenoinfection: The role of chemokine receptors and envelope diversity. J Virol 76:3626–3636.
Kanmogne GD, Schall K, Leibhart J, Knipe B, Gendelman HE, Persidsky Y (2007) HIV-1 gp120 compromises blood-brain barrier integrity and enhance monocyte migration across blood-brain barrier: Implication for viral neuropathogenesis. J Cereb Blood Flow Metab 27:123–134.
Kaul M, Garden GA, Lipton SA (2001) Pathways to neuronal injury and apoptosis in HIV-associated dementia. Nature 410:988–994.
Kim BO, Liu Y, Zhou BY, He JJ (2004a) Induction of C chemokine XCL1 (lymphotactin/single C motif-1 alpha/activation-induced, T cell-derived and chemokine-related cytokine) expression by HIV-1 Tat protein. J Immunol 172:1888–1895.
Kim BO, Liu Y, Ruan Y, Xu ZC, Schantz L, He JJ (2003) Neuropathologies in transgenic mice expressing human immunodeficiency virus type 1 tat protein under the regulation of the astrocyte-specific glial fibrillary acidic protein promoter and doxycycline. Am J Pathol 162:1693–1707.
Kim WK, Corey S, Chesney G, Knight H, Klumpp S, Wuthrich C, Letvin N, Koralnik I, Lackner A, Veasey R, Williams K (2004b) Identification of T lymphocytes in simian immunodeficiency virus encephalitis: distribution of CD8 + T cells in association with central nervous system vessels and virus. J Neurovirol 10:315–325.
Kitagawa M, Lackner AA, Martfeld DJ, Gardner MB, Dandekar S (1991) Simian immunodeficiency virus infection of macaque bone marrow macrophages correlates with disease progression in vivo. Am J Pathol 138:921–930.
Koenig S, Gendelman HE, Orenstein JM, Dal Canto MC, Pezeshkpour GH, Yungbluth M, Janotta F, Aksamit A, Martin MA, Fauci AS (1986) Detection of AIDS virus in macrophages in brain tissue from AIDS patients with encephalopathy. Science 233:1089–1093.
Koralnik IJ, Beaumanoir A, Hausler R, Kohler A, Safran AB, Delacoux R, Vibert D, Mayer E, Burkhard P, Nahory A, et al. (1990) A controlled study of early neurologic abnormalities in men with asymptomatic human immunodeficiency virus infection. N Engl J Med 323:864–870.
Krucker T, Toggas SM, Mucke L, Siggins GR (1998) Transgenic mice with cerebral expression of human immunodeficiency virus type-1 coat protein gp120 show divergent changes in short- and long-term potentiation in CA1 hippocampus. Neuroscience 83:691–700.
Langford TD, Letendre SL, Marcotte TD, Ellis RJ, McCutchan JA, Grant I, Mallory ME, Hansen LA, Archibald S, Jernigan T, Masliah E (2002) Severe, demyelinating leukoencephalopathy in AIDS patients on antiretroviral therapy. AIDS 16: 1019–1029.
Letendre S, Marquie-Beck J, Singh KK, de Almeida S, Zimmerman J, Spector SA, Grant I, Ellis R (2004) The monocyte chemotactic protein-1-2578G allele is associated with elevated MCP-1 concentrations in cerebrospinal fluid. J Neuroimmunol 157: 193–196.
Limoges J, Persidsky Y, Poluektova L, Rasmussen J, Ratanasuwan W, Zelivyanskaia M, McClernon D, Lanier E, Gendelman H (2000) Evaluation of antiretroviral drug efficacy for HIV-1 encephalitis in SCID mice. Neurology 54:379–389.
Limoges J, Poluektova L, Ratanasuwan W, Rasmussen J, Zelivyanskaya M, McClernon DR, Lanier ER, Gendelman HE, Persidsky Y (2001) The efficacy of potent anti-retroviral drug combinations tested in a murine model of HIV-1 encephalitis. Virology 281: 21–34.
Liu Y, Tang XP, McArthur JC, Scott J, Gartner S (2000) Analysis of human immunodeficiency virus type 1 gp160 sequences from a patient with HIV dementia: Evidence for monocyte trafficking into brain. J Neurovirol 6(Suppl 1):S70–81.
Luo X, Carlson KA, Wonja V, Mayo R, Biskup T, Stoner J, Anderson J, Gendelman HE, Melendez L (2003) Macrophage proteomic fingerprinting predicts HIV-1 associated cognitive impairment. Neurology 60:1931–1937.
Luste A (1998) Chemokines—chemotactic cytokines that mediate inflammation. N Engl J Med 338:436–445.
Marcondes MC, Burudi EM, Huitron-Resendiz S, Sanchez-Alavez M, Watry D, Zandonatti M, Henriksen SJ, Fox HS (2001) Highly activated CD8(+) T cells in the brain correlate with early central nervous system dysfunction in simian immunodeficiency virus infection. J Immunol 167:5429–5438.
Masliah E, Ge N, Achim CL, Wiley CA (1995) Differential vulnerability of calbindin-immunoreactive neurons in HIV encephalitis. J Neuropathol Exp Neurol 54:350–357.
Masliah E, DeTeresa RM, Mallory ME, Hansen LA (2000) Changes in pathological findings at autopsy in AIDS cases for the last 15 years. Aids 14:69–74.
Masliah E, Ge N, Achim CL, Hansen LA, Wiley CA (1992) Selective neuronal vulnerability in HIV encephalitis. J Neuropathol Exp Neurol 51:585–593.
Masliah E, Sisk A, Malory M, Mucke L, Shenk D, Games D (1996) Comparison of neurodegenerative pathology in transgenic mice overexpressing V717 b-amyloid precursor protein and Alzheimer's disease. J Neurosci 16:5795–5811.
Masliah E, Heaton RK, Marcotte TD, Ellis RJ, Wiley CA, Mallory M, Achim CL, McCutchan JA, Nelson JA, Atkinson JH, Grant I (1997) Dendritic injury is a pathological substrate for human immunodeficiency virus-related cognitive disorders. HNRC Group. The HIV Neurobehavioral Research Center. Ann Neurol 42:963–972.
Meeker RB, Thiede BA, Hall C, English R, Tompkins M (1997) Cortical cell loss in asymptomatic cats experimentally infected

with feline immunodeficiency virus. AIDS Res Hum Retroviruses 13:1131–1140.

Mellor AL, Munn D, Chandler P, Keskin D, Johnson T, Marshall B, Jhaver K, Baban B (2003) Tryptophan catabolism and T cell responses. Adv Exp Med Biol 527:27–35.

Messam CA, Major EO (2000) Stages of restricted HIV-1 infection in astrocyte cultures derived from human fetal brain tissue. J Neurovirol 6(Suppl 1):S90–94.

Moore DJ, Masliah E, Rippeth JD, Gonzalez R, Carey CL, Cherner M, Ellis RJ, Achim CL, Marcotte TD, Heaton RK, Grant I (2006) Cortical and subcortical neurodegeneration is associated with HIV neurocognitive impairment. AIDS 20:879–887.

Murray EA, Rausch DM, Lendvay J, Sharer LR, Eiden LE (1992) Cognitive and motor impairments associated with SIV infection in rhesus monkeys. Science 255:1246–1249.

Nagra RM, Burrola PG, Wiley CA (1992) Development of spongiform encephalopathy in retroviral infected mice. Lab Invest 66:292–302.

Nagra RM, Wong PK, Wiley CA (1993) Expression of major histocompatibility complex antigens and serum neutralizing antibody in murine retroviral encephalitis. J Neuropathol Exp Neurol 52:163–173.

Navia B, Jordan B, Price R (1986) The AIDS dementia complex: I. Clinical features. Ann Neurol 19:517–524.

Navia BA, Rostasy K (2005) The AIDS dementia complex: Clinical and basic neuroscience with implications for novel molecular therapies. Neurotox Res 8:3–24.

Neuenburg JK, Brodt HR, Herndier BG, Bickel M, Bacchetti P, Price RW, Grant RM, Schlote W (2002) HIV-related neuropathology, 1985 to 1999: Rising prevalence of HIV encephalopathy in the era of highly active antiretroviral therapy. J Acquir Immune Defic Syndr 31:171–177.

New DR, Ma M, Epstein LG, Nath A, Gelbard HA (1997) Human immunodeficiency virus type 1 Tat protein induces death by apoptosis in primary human neuron cultures. J Neurovirol 3:168–173.

Nottet HS, Jett M, Flanagan CR, Zhai QH, Persidsky Y, Rizzino A, Bernton EW, Genis P, Baldwin T, Schwartz J, et al. (1995) A regulatory role for astrocytes in HIV-1 encephalitis. An overexpression of eicosanoids, platelet-activating factor, and tumor necrosis factor-alpha by activated HIV-1-infected monocytes is attenuated by primary human astrocytes. J Immunol 154:3567–3581.

Nottet HS, Persidsky Y, Sasseville VG, Nukuna AN, Bock P, Zhai QH, Sharer LR, McComb RD, Swindells S, Soderland C, Gendelman HE (1996) Mechanisms for the transendothelial migration of HIV-1-infected monocytes into brain. J Immunol 156:1284–1295.

Orandle MS, Williams KC, MacLean AG, Westmoreland SV, Lackner AA (2001) Macaques with rapid disease progression and simian immunodeficiency virus encephalitis have a unique cytokine profile in peripheral lymphoid tissues. J Virol 75:4448–4452.

Palella FJ, Jr., Delaney KM, Moorman AC, Loveless MO, Fuhrer J, Satten GA, Aschman DJ, Holmberg SD (1998) Declining morbidity and mortality among patients with advanced human immunodeficiency virus infection. HIV Outpatient Study Investigators. N Engl J Med 338:853–860.

Peluso R, Haase A, Stowring L, Edwards M, Ventura P (1985) A Trojan Horse mechanism for the spread of visna virus in monocytes. Virology 147:231–236.

Persidsky Y (1999) Model systems for studies of leukocyte migration across the blood–brain barrier. J Neurovirol 5:579–590.

Persidsky Y, Gendelman HE (2003) Mononuclear phagocyte immunity and the neuropathogenesis of HIV-1 infection. J Leukoc Biol 74:691–701.

Persidsky Y, Limoges J, Rasmussen J, Zheng J, Gearing A, Gendelman HE (2001) Reduction in glial immunity and neuropathology by a PAF antagonist and an MMP and TNF-alpha inhibitor in SCID mice with HIV-1 encephalitis. J Neuroimmunol 114:57–68.

Persidsky Y, Stins M, Way D, Witte MH, Weinand M, Kim KS, Bock P, Gendelman HE, Fiala M (1997) A model for monocyte migration through the blood-brain barrier during HIV-1 encephalitis. J Immunol 158:3499–3510.

Persidsky Y, Heilman D, Haorah J, Zelivyanskaya M, Persidsky R, Weber GA, Shimokawa H, Kaibuchi K, Ikezu T (2006) Rho-mediated regulation of tight junctions during monocyte migration across blood-brain barrier in HIV-1 encephalitis (HIVE). Blood 107:4720–4730.

Persidsky Y, Ghorpade A, Rasmussen J, Limoges J, Liu XJ, Stins M, Fiala M, Way D, Kim KS, Witte MH, Weinand M, Carhart L, Gendelman HE (1999) Microglial and astrocyte chemokines regulate monocyte migration through the blood-brain barrier in human immunodeficiency virus-1 encephalitis. Am J Pathol 155:1599–1611.

Persidsky Y, Limoges J, McComb R, Bock P, Baldwin T, Tyor W, Patil A, Nottet HS, Epstein L, Gelbard H, Flanagan E, Reinhard J, Pirruccello SJ, Gendelman HE (1996) Human immunodeficiency virus encephalitis in SCID mice [see comments]. Am J Pathol 149:1027–1053.

Petito CK, Cash KS (1992) Blood-brain barrier abnormalities in the acquired immunodeficiency syndrome: Immunohistochemical localization of serum proteins in postmortem brain. Ann Neurol 32:658–666.

Poluektova L, Moran T, Zelivyanskaya M, Swindells S, Gendelman HE, Persidsky Y (2001) The regulation of alpha chemokines during HIV-1 infection and leukocyte activation: Relevance for HIV-1-associated dementia. J Neuroimmunol 120:112–128.

Poluektova LY, Munn DH, Persidsky Y, Gendelman HE (2002) Generation of cytotoxic T cells against virus-infected human brain macrophages in a murine model of HIV-1 encephalitis. J Immunol 168:3941–3949.

Potula R, Poluektova L, Knipe B, Chrastil J, Heilman D, Dou H, Takikawa O, Munn DH, Gendelman HE, Persidsky Y (2005) Inhibition of indoleamine 2,3-dioxygenase (IDO) enhances elimination of virus-infected macrophages in an animal model of HIV-1 encephalitis. Blood 106:2382–2390.

Potula R, Haorah J, Knipe B, Leibhart J, Chrastil J, Heilman D, Dou H, Reddy R, Ghorpade A, Persidsky Y (2006) Alcohol abuse enhances neuroinflammation and impairs immune responses in an animal model of human immunodeficiency virus-1 encephalitis. Am J Pathol 168:1335–1344.

Power C, Buist R, Johnston JB, Del Bigio MR, Ni W, Dawood MR, Peeling J (1998) Neurovirulence in feline immunodeficiency virus-infected neonatal cats is viral strain specific and dependent on systemic immune suppression. J Virol 72: 9109–9115.

Pulliam L, Gascon R, Stubblebine M, McGuire D, McGrath MS (1997) Unique monocyte subset in patients with AIDS dementia. Lancet 349:692–695.

Ranki A, Nyberg M, Ovod V, Haltia M, Elovaara I, Raininko R, Haapasalo H, Krohn K (1995) Abundant expression of HIV Nef and Rev proteins in brain astrocytes in vivo is associated with dementia. AIDS 9:1001–1008.

Sacktor N, McDermott MP, Marder K, Schifitto G, Selnes OA, McArthur JC, Stern Y, Albert S, Palumbo D, Kieburtz K, De Marcaida JA, Cohen B, Epstein L (2002) HIV-associated cognitive impairment before and after the advent of combination therapy. J Neurovirol 8:136–142.

Sanchez-Ramon S, Bellon JM, Resino S, Canto-Nogues C, Gurbindo D, Ramos JT, Munoz-Fernandez MA (2003) Low blood CD8 + T-lymphocytes and high circulating monocytes are predictors of HIV-1-associated progressive encephalopathy in children. Pediatrics 111:E168–175.

Schmitz JE, Kuroda MJ, Santra S, Sasseville VG, Simon MA, Lifton MA, Racz P, Tenner-Racz K, Dalesandro M, Scallon BJ, Ghrayeb J, Forman MA, Montefiori DC, Rieber EP, Letvin NL, Reimann KA (1999) Control of viremia in simian immunodeficiency virus infection by CD8 + lymphocytes. Science 283:857–860.

Schweinsburg BC, Taylor MJ, Alhassoon OM, Gonzalez R, Brown GG, Ellis RJ, Letendre S, Videen JS, McCutchan JA, Patterson TL, Grant I (2005) Brain mitochondrial injury in human immunodeficiency virus-seropositive (HIV+) individuals taking nucleoside reverse transcriptase inhibitors. J Neurovirol 11:356–364.

Sethi KK, Naher H, Stroehmann I (1988) Phenotypic heterogeneity of cerebrospinal fluid-derived HIV-specific and HLA-restricted cytotoxic T-cell clones. Nature 335:178–181.

Shiramizu B, Gartner S, Williams A, Shikuma C, Ratto-Kim S, Watters M, Aguon J, Valcour V (2005) Circulating proviral HIV DNA and HIV-associated dementia. AIDS 19:45–52.

Sopper S, Sauer U, Hemm S, Demuth M, Muller J, Stahl-Hennig C, Hunsmann G, ter Meulen V, Dorries R (1998) Protective role of the virus-specific immune response for development of severe neurologic signs in simian immunodeficiency virus-infected macaques. J Virol 72:9940–9947.

Steffan AM, Lafon ME, Gendrault JL, Koehren F, De Monte M, Royer C, Kirn A, Gut JP (1994) Feline immunodeficiency virus can productively infect cultured endothelial cells from cat brain microvessels. J Gen Virol 75(Pt 12):3647–3653.

Steigerwald ES, Sarter M, March P, Podell M (1999) Effects of feline immunodeficiency virus on cognition and behavioral function in cats. J Acquir Immune Defic Syndr Hum Retrovirol 20:411–419.

Strain MC, Letendre S, Pillai SK, Russell T, Ignacio CC, Gunthard HF, Good B, Smith DM, Wolinsky SM, Furtado M, Marquie-Beck J, Durelle J, Grant I, Richman DD, Marcotte T, McCutchan JA, Ellis RJ, Wong JK (2005) Genetic composition of human immunodeficiency virus type 1 in cerebrospinal fluid and blood without treatment and during failing antiretroviral therapy. J Virol 79:1772–1788.

Toggas SM, Masliah E, Rockenstein EM, Rall GF, Abraham CR, Mucke L (1994) Central nervous system damage produced by expression of the HIV-1 coat protein gp120 in transgenic mice. Nature 367:188–193.

Tornatore C, Nath A, Amemiya K, Major EO (1991) Persistent human immunodeficiency virus type 1 infection in human fetal glial cells reactivated by T-cell factor(s) or by the cytokines tumor necrosis factor alpha and interleukin-1 beta. J Virol 65:6094–6100.

Tornatore C, Chandra R, Berger JR, Major EO (1994) HIV-1 infection of subcortical astrocytes in the pediatric central nervous system. Neurology 44:481–487.

Trillo-Pazos G, Diamanturos A, Rislove L, Menza T, Chao W, Belem P, Sadiq S, Morgello S, Sharer L, Volsky DJ (2003) Detection of HIV-1 DNA in microglia/macrophages, astrocytes and neurons isolated from brain tissue with HIV-1 encephalitis by laser capture microdissection. Brain Pathol 13:144–154.

Tyor WR, Power C, Gendelman HE, Markham RB (1993) A model of human immunodeficiency virus encephalitis in scid mice. Proc Natl Acad Sci USA 90:8658–8662.

Weis S, Haug H, Budka H (1996) Vascular changes in the cerebral cortex in HIV-1 infection: I. A morphometric investigation by light and electron microscopy. Clin Neuropathol 15:361–366.

Weissman D, Rabin RL, Arthos J, Rubbert A, Dybul M, Swofford R, Venkatesan S, Farber JM, Fauci AS (1997) Macrophage-tropic HIV and SIV envelope proteins induce a signal through the CCR5 chemokine receptor. Nature 389:981–985.

Wiley CA (1995) Quantitative neuropathologic assessment of HIV-1 encephalitis. Curr Top Microbiol Immunol 202:55–61.

Wiley CA, Masliah E, Morey M, Lemere C, DeTeresa R, Grafe M, Hansen L, Terry R (1991) Neocortical damage during HIV infection. Ann Neurol 29:651–657.

Willett BJ, Hosie MJ (1999) The role of the chemokine receptor CXCR4 in infection with feline immunodeficiency virus. Mol Membr Biol 16:67–72.

Williams K, Alvarez X, Lackner AA (2001) Central nervous system perivascular cells are immunoregulatory cells that connect the CNS with the peripheral immune system. Glia 36:156–164.

Williams K, Westmoreland S, Greco J, Ratai E, Lentz M, Kim WK, Fuller RA, Kim JP, Autissier P, Sehgal PK, Schinazi RF, Bischofberger N, Piatak M, Lifson JD, Masliah E, Gonzalez RG (2005) Magnetic resonance spectroscopy reveals that activated monocytes contribute to neuronal injury in SIV neuroAIDS. J Clin Invest 115:2534–2545.

Williams KC, Hickey WF (2002) Central nervous system damage, monocytes and macrophages, and neurological disorders in AIDS. Annu Rev Neurosci 25:537–562.

Xiong H, Zeng YC, Zheng J, Thylin M, Gendelman HE (1999) Soluble HIV-1 infected macrophage secretory products mediate blockade of long-term potentiation: A mechanism for cognitive dysfunction in HIV- 1-associated dementia [In Process Citation]. J Neurovirol 5:519–528.

Xiong H, McCabe L, Skifter D, Monaghan DT, Gendelman HE (2003a) Activation of NR1a/NR2B receptors by monocyte-derived macrophage secretory products: Implications for human immunodeficiency virus type one-associated dementia. Neurosci Lett 341:246–250.

Xiong H, Boyle J, Winkelbauer M, Gorantla S, Zheng J, Ghorpade A, Persidsky Y, Carlson KA, Gendelman HE (2003b) Inhibition of long-term potentiation by interleukin-8: Implications for human immunodeficiency virus-1-associated dementia. J Neurosci Res 71:600–607.

Zhang L, He T, Talal A, Wang G, Frankel SS, Ho DD (1998) In vivo distribution of the human immunodeficiency virus/simian immunodeficiency virus coreceptors: CXCR4, CCR3, and CCR5. J Virol 72:5035–5045.

Zheng J, Zhuang W, Yan N, Kou G, Peng H, McNally C, Erichsen D, Cheloha A, Herek S, Shi C (2004) Classification of HIV-1-mediated neuronal dendritic and synaptic damage using multiple criteria linear programming. Neuroinformatics 2:303–326.

Zheng J, Ghorpade A, Niemann D, Cotter RL, Thylin MR, Epstein L, Swartz JM, Shepard RB, Liu X, Nukuna A, Gendelman HE

(1999a) Lymphotropic virions affect chemokine receptor-mediated neural signaling and apoptosis: Implications for human immunodeficiency virus type 1-associated dementia. J Virol 73:8256–8267.

Zheng J, Thylin MR, Ghorpade A, Xiong H, Persidsky Y, Cotter R, Niemann D, Che M, Zeng YC, Gelbard HA, Shepard RB, Swartz JM, Gendelman HE (1999b) Intracellular CXCR4 signaling, neuronal apoptosis and neuropathogenic mechanisms of HIV-1-associated dementia. J Neuroimmunol 98:185–200.

Zink MC, Clements JE (2002) A novel simian immunodeficiency virus model that provides insight into mechanisms of human immunodeficiency virus central nervous system disease. J Neurovirol 8(Suppl 2):42–48.

Zink WE, Anderson E, Boyle J, Hock L, Rodriguez-Sierra J, Xiong H, Gendelman HE, Persidsky Y (2002) Impaired spatial cognition and synaptic potentiation in a murine model of human immunodeficiency virus type 1 encephalitis. J Neurosci 22:2096–2105.

23
HTLV-I/II

Kazunori Fugo, Christian W. Grant, and Steven Jacobson

Keywords CNS, CTL, CSF, HAM/TSP, HLA, HTLV-I, LTR, PBMC

23.1. Introduction

Human T-cell lymphotrophic virus type-I (HTLV-I) is an exogenous human retrovirus that causes adult T cell leukemia (ATL) and a progressive neurological disorder, HTLV-I associated myelopathy/tropical spastic paraparesis (HAM/TSP).

Most HTLV-I infected individuals are asymptomatic carriers. However, 0.25 - 3% infected individuals will develop HAM/TSP. HTLV-I integrated proviral genome is approximately 9 kb including *gag*, *pol* and *env* genes common to all cis-acting retroviruses. A cis-acting enhancer element has been detected within the gag gene of several avian retroviruses, including Rous sarcoma virus, Fujinami sarcoma virus, and the endogenous Rous-associated virus-0. A consensus enhancer core sequence, GTGGTTTG, is present in all of these viral genomes (citation: Arrigo S et al., Mol Cell Biol 1987). The above viral genes are flanked by long terminal repeat (LTR). Human T-cell lymphotrophic virus type-II (HTLV-II) is also an exogenous human retrovirus. Although rare, HTLV-II infection is associated with a chronic encephalomyelopathy similar to that of classic HTLV-I-associated HAM/TSP. In 2005, two new primate retroviruses were identified, Human T-cell lymphotrophic virus type-III (HTLV-III) and Human T-cell lymphotrophic virus type-IV (HTLV-IV), among individuals who hunt, butcher, or keep monkeys or apes as pets in southern Cameroon. However, it is still unknown whether these viruses are pathogenic and can be transmitted among humans.

HAM/TSP is clinically characterized by paraparesis associated with spasticity in the lower extremities, mild peripheral sensory loss, muscle weakness, hyperreflexia, and urinary disturbance associated with preferential damage of the thoracic spinal cord. Most of the patients with HAM/TSP are seropositive for anti-HTLV-I antibodies.

HAM/TSP predominantly affects the thoracic spinal cord. There is degeneration of the lateral corticospinal tract as well as of the spinocerebellar or spino thalamic tract of the lateral column. These lesions are associated with perivascular and parenchymal lymphocytic infiltration with the presence of foamy macrophages, proliferation of astrocytes, and fibrillary gliosis. There is widespread loss of myelin and axons, particularly in the corticospinal tracts. Damage is most severe in the middle to lower thoracic regions of the spinal cord.

Several immunologic abnormalities such as increased spontaneous lymphoproliferation, elevated anti-HTLV-I antibody titers in sera and CSF, increased cytokine production, and cellular immune responses have been observed in patients with HAM/TSP. These abnormalities are more often observed in patients with HAM/TSP compared to HTLV-I infected asymptomatic carriers. Moreover, it has been suggested that an autoimmune mechanism may also be involved in the pathogenesis of HAM/TSP. These evidences suggests that dysregulation in immune system may be associated with pathogenesis of HTLV-I associated neurologic diseases and may be responsible for local parenchymal ("by-stander") damage. This chapter will focus on the abnormality of cytokine expression in peripheral blood and CSF, CD4 + and CD8 + cell response to the HTLV-I, and NK cell activity in patients with HAM/TSP.

23.2. Structure and Gene Regulation of HTLV-I/II

HTLV-I integrated proviral genome is approximately 9 kb including *gag*, *pol* and *env* genes. These genes encode the viral core proteins, reverse transcriptase and envelope proteins, respectively. The above viral genes are flanked by long terminal repeat (LTR). HTLV-I integrated proviral genome contains the unique region at 3' of *env* gene, which is called pX region. pX region encodes the regulatory proteins of HTLV, Tax and Rex (Figure 23.1).

The LTR consists of three regulatory regions, U3, R and U5. The U3 region is the element that regulates transcription of the provirus, mRNA termination and polyadenylation

T. Ikezu and H.E. Gendelman (eds.), *Neuroimmune Pharmacology*.
© Springer 2008

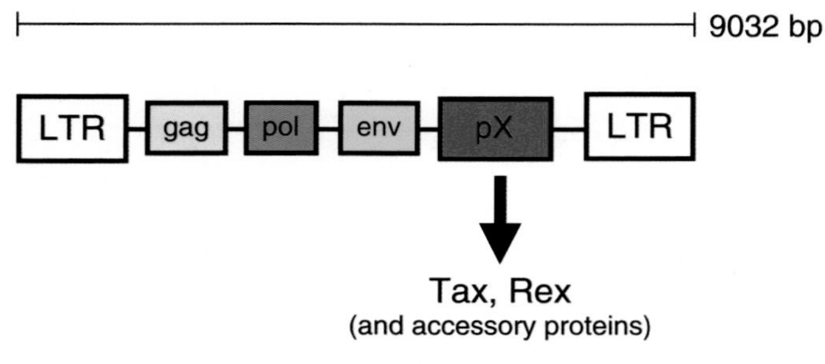

FIGURE 23.1. Scheme of the HTLV-I genome. The HTLV-I proviral genome is 9032 base pairs long. It consists of gag, pol and env genes. The above viral genes are flanked by long terminal repeat (LTR). pX region encodes the regulatory proteins of HTLV, Tax and Rex and accessory proteins.

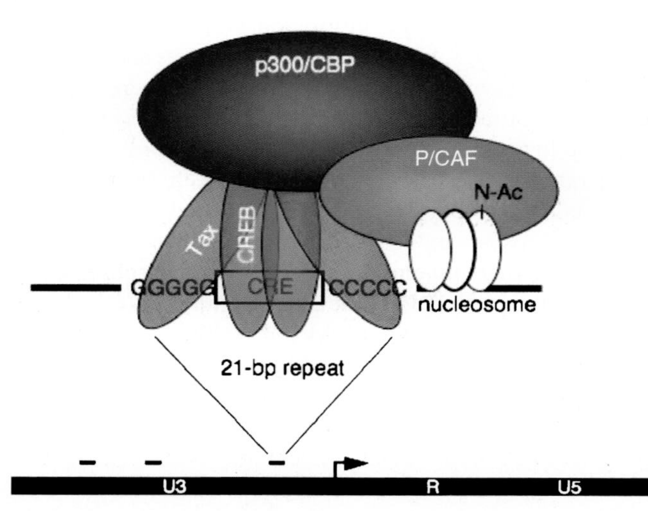

FIGURE 23.2. Scheme of transcriptional activation of the viral promoter by Tax and components of the CREB signaling pathway. U3 element of LTR contains three 21 base pair nucleotide repeats known as Tax-responsive elements (TRE), which are composed of a CRE-like core flanked by GC-rich sequences. Tax interacts with the cellular transcription factor cAMP-responsive element binding protein (CREB) dimer and stimulates its DNA-binding activity to the viral promoter at TRE. Together, the Tax-CREB complexes bound to the viral promoter recruit the co-activator proteins CREB-binding protein (CBP)/p300, and P/CAF, which activates transcription by histone-acetylation/chromatin-remodeling.

signals. These consists of three imperfect 21-base pair nucleotide repeats called Tax responsive elements (TRE) that are necessary for transcriptional activation by the Tax protein.

Tax, a transcriptional activator of the viral genome (Sodroski et al., 1984; Fujisawa et al., 1985), is a 40-kDa phosphoprotein comprised of 353 amino acids. Tax protein is the trans-activator protein, which drives viral gene expression from TRE within U3 region of LTR. Tax also interacts with cellular transcription factors such as cAMP-response element binding protein (CREB) and coactivator p300/CBP, serum-response factor (SRF) and nuclear factor-kappa B (NF-κB) (Johnson et al., 2001) (Figure 23.2). These interactions between Tax and CREB or SRF are responsible for HTLV-I transactivation. It was documented that forkhead box P3 transcription factor p3 (Foxp3) protein, which is specifically expressed in regulatory CD4 + T cells (Tregs) and is required for their development and function, detected in the nucleus of CD4 + CD25 + T cells with immunoregulatory function, is down-regulated in CD4 + CD25 + T cells in patients with HAM/TSP compared to asymptomatic carrier and healthy donor (Yamano et al., 2005). Tax interaction with these factors may lead to immune dysregulations in patients with HAM/TSP.

Rex, a 27-kDa phosphoprotein, is a splicing suppressor of the viral transcripts (Kiyokawa et al., 1985; Seiki et al., 1985). The p27-Rex protein, encoded by the same pX mRNA in an alternate reading frame, accumulates and suppresses splicing of the viral transcripts.

23.3. Clinical Features and Epidemiology of HTLV-Associated Myelopathy and Tropical Spastic Paraparesis (HAM/TSP)

HTLV-I has tropism for T lymphocytes. HTLV-I is the causative agent for adult T-cell leukemia (ATL) and a progressive neurological disease known as HTLV-I associated myelopathy/tropical spastic paraparesis (HAM/TSP) (Gessain et al., 1985; Osame et al., 1986). According to the clinical and laboratory guidelines for the diagnosis of HAM/TSP (Osame, 1990), clinically, HAM/TSP is characterized by muscle weakness, hyperreflexia, spasticity in the lower extremities, and urinary disturbance associated with preferential damage of the thoracic spinal cord. HTLV-I has been also shown to be associated with several other inflammatory diseases, such as alveolitis, polymyositis, and arthritis, uveitis, and Sjogren syndrome. There are also less certain associations with chronic infective dermatitis, Bechet disease, pseudohypoparathyroidism, and systemic lupus erythematosus (Kubota et al., 2000). Another closely related virus, human T-cell lymphotrophic virus type II (HTLV-II), has also been implicated in development of neurological disorders (Lehky et al., 1996; Silva et al., 2002; Orland et al., 2003). In 2005, the new primate retroviruses, Human T-cell lymphotrophic virus type-III (HTLV-III) and Human T-cell lymphotrophic virus type-IV (HTLV-IV) were identified among individuals who hunt, butcher, or keep monkeys or apes as pets in southern Cameroon (Calattini et al., 2005; Wolfe et al., 2005). However, it is still unknown whether these viruses are pathogenic and can be transmitted among humans.

It is estimated that approximately 10–20 million people worldwide are infected with HTLV-I (Poiesz et al., 1980; Edlich et al., 2003). Endemic areas of HTLV-I are in southern Japan, Central and West Africa, the Caribbean, Central and South America, the Middle East, Melanesia (Blattner and Gallo, 1985;

Gessain, 1996), and there are also smaller foci in the aboriginal populations of Australia, Papua New Guinea, and northern Japan. In Europe and North America, the virus is found chiefly in immigrants from these endemic areas and in some communities of intravenous drug users. Within the endemic areas, the seroprevalence varies from 1% to 20%. In contrast to the human immunodeficiency virus (HIV), the majority of infected persons remain asymptomatic and only 5% will develop HAM/TSP or ATL (Gessain et al., 1985; Osame et al., 1986).

HTLV-I is transmitted via three major routes: (i) transmission from mother to child through breast milk; (ii) transmission by sexual contact (mainly from male to female); (iii) transmission by way of infected blood through blood transfusion or contaminated needle usage. Although HTLV-I can infect a wide range of vertebrate cells in vitro (Trejo and Ratner, 2000), it has been thought to preferentially infect CD4 + T cells in vivo (Richardson et al., 1990). Therefore, CD4 + T cells have been considered an important component contributing to the inflammatory process in HAM/TSP. However, several studies have indicated that CD8 + T-cells were also infected with HTLV-I in vivo (Hanon et al., 2000b; Hanon et al., 2000a). The mechanisms through which HTLV-I causes HAM/TSP are not completely understood, although virus-host immune interactions have been suggested to play a role in the pathogenesis of the disorder.

Approximately 5% of HTLV-I infected individuals will develop HAM/TSP or ATL. However, there is no diagnostic tool for early detection or accurate assessment of disease state. Recently, protein profiling in serum from patients to distinguish between these two disease states has been reported by using mass spectrometry (Semmes et al., 2005). This technology may lead new diagnostic tools for HTLV-I infected individuals.

23.4. Pathophysiology in CNS

For HAM/TSP disease development, HTLV-I proviral load is an important factor. Recently, a real time quantitative polymerase chain reaction (PCR) assay (TaqMan) was developed that allows HTLV-I proviral DNA to be measured in peripheral blood mononuclear cells (PBMCs). This assay was used to quantify the respective HTLV-I proviral DNA loads in HAM/TSP patients and asymptomatic carriers where a 16-fold increase in HTLV-I proviral load was observed in HAM/TSP patients compared to asymptomatic carriers (Nagai et al., 1998). Similar results have also been reported in patient cohorts from Jamaica (Manns et al., 1999). Yamano et al. (2002) reported that the HTLV-I tax mRNA load correlated with the HTLV-I proviral DNA load ex vivo, and the HTLV-I tax mRNA load also correlated with the disease severity in HAM/TSP patients (Figure 23.3). Collectively, these observations strongly suggest that a high proviral load plays an important role in the etiology of HAM/TSP. It has been hypothesized that a high HTLV-I proviral load results in higher viral antigen presentation that may drive expansion of antigen-specific T cells and subsequently these T cells are involved in the development of HAM/TSP. Such a model is presented in Figure 23.3. Most individuals infected with HTLV-I have been shown to mount strong CTL responses to viral antigens (Jacobson et al., 1990; Bangham, 2000). It is thought that this strong CTL response functions to protect against disease development by reducing the proviral load (Jeffery et al., 1999). However, this increase in proviral load that results in a marked increase in HTLV-I antigen-specific CTL expansion has also been suggested to contribute to a HTLV-I specific inflammatory process seen in HAM/TSP (Nagai et al., 1998; Jeffery et al., 1999).

Other determinants that are associated with increased risk of developing HAM/TSP are host genetic factors. Nagai et al. (1998) demonstrated higher HTLV-I proviral loads in asymptomatic carriers of families with HAM/TSP patients than those of unrelated asymptomatic carriers. It has been reported that a higher risk of developing HAM/TSP is associated with the human leukocyte antigen (HLA)-A*02-, HLA-Cw*08-, HLA-DR1 +, and tumor necrosis factor (TNF)-863A- (Usuku et al., 1988; Jeffery et al., 1999; Jeffery et al., 2000; Vine et al., 2002).

Histopathological observations have demonstrated that HAM/TSP predominantly affects the thoracic spinal cord

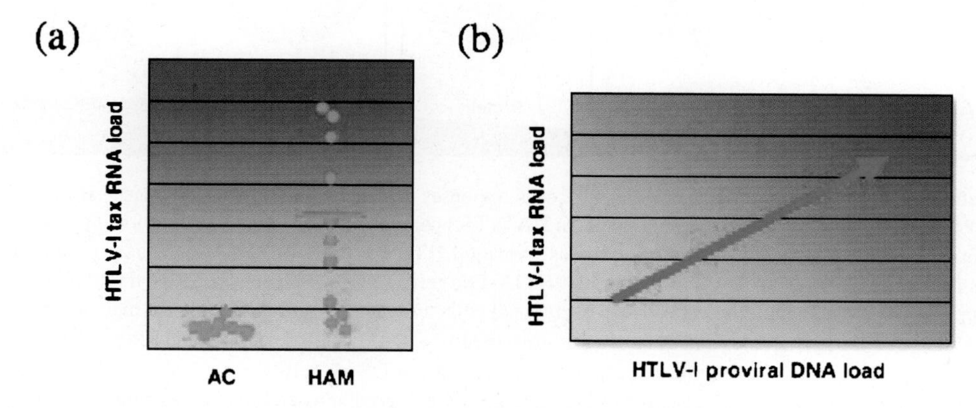

FIGURE 23.3. Schematic representation of the correlation between HTLV-I tax mRNA load in HAM/TSP patients and asymptomatic carriers. (b) Correlation between HTLV-I proviral load and HTLV-I tax mRNA load in HAM/TSP patients.

(Izumo et al., 1989; Levin and Jacobson, 1997). There is degeneration of the lateral corticospinal tract as well as of the spinocerebellar or spino thalamic tract of the lateral column. These lesions are associated with perivascular and parenchymal lymphocytic infiltration with the presence of foamy macrophages, proliferation of astrocytes, and fibrillary gliosis. There is widespread loss of myelin and axons, particularly in the corticospinal tracts of the spinal cord (Osame et al., 1986; Iwasaki, 1990; Umehara et al., 1993). Damage is most severe in the middle to lower thoracic regions of the spinal cord. The proximity to the areas containing inflammation, activation of macrophage/microglia, indicates that axonal damage is closely associated with inflammation in active-chronic lesions. Moreover, perivascular inflammatory infiltration was seen in the brain (deep white matter and in the marginal area of the cortex and white matter) of HAM/TSP patients, and the types of infiltrating cells were similar both in the spinal cords and brains (Aye et al., 2000). A nonrandom distribution of affected regions was suggested by autopsy studies, which showed the regions that are mainly affected are the so-called 'watershed' zones of the spinal cord in HAM/TSP patients (Izumo et al., 1992). This has partly been addressed by a magnetic resonance imaging study that showed increased abnormal-intensity lesions in the white matter of the brain of HAM/TSP patients (Kira et al., 1991). These results suggest that inflammatory changes occurred simultaneously in the spinal cord and in the brain, with the distribution of inflamed vessels closely correlated with the characteristic vascular architecture of the brain and the spinal cord, which led to a slowing of blood flow.

T-cell infiltrations in spinal cord lesions of HAM/TSP patients have been demonstrated by histopathological studies. The proportion of infiltrating cells shifted with the duration of the disease. HAM/TSP patients with short duration of illness, up to 5 years after onset, showed an even distribution of CD4 + T cells, CD8 + T cells, B cells, and foamy macrophages in damaged areas of the spinal cord parenchyma (Akizuki et al., 1989; Umehara et al., 1993). Immunohistochemistry showed that inflammatory cytokines such as TNF-alpha, IL-1beta, and interferon-gamma (IFN-gamma) were expressed on perivascular infiltrating macrophages, astrocytes, and microglia (Umehara et al., 1994a). HLA class I and beta2-microglobulin were expressed on endothelial cells and infiltrating mononuclear cells (Morgan et al., 1989; Wu et al., 1993). Up regulation of HLA class II expression was also found on endothelial cells, microglia, and infiltrating mononuclear cells in the affected lesions. In contrast, in patients with duration of illness from 8 to 10 years, there was a predominance of CD8 + T cells within spinal cord lesions with a concomitant down-regulation of proinflammatory cytokine expression, with the exception of IFN-gamma (Umehara et al., 1993; Umehara et al., 1994a). Such lesions also express increased levels of HLA class I antigens from which HTLV-I specific CD8 + CTLs were isolated (Levin and Jacobson, 1997). Although many hematogenous macrophages could be found in the active-chronic lesions, markers of monocyte/macrophage recruitment and activation were down-regulated as the duration of illness progressed (Abe et al., 1999). These studies demonstrated that immune responses in the spinal cord lesions of HAM/TSP patients gradually change concomitantly with the duration of illness.

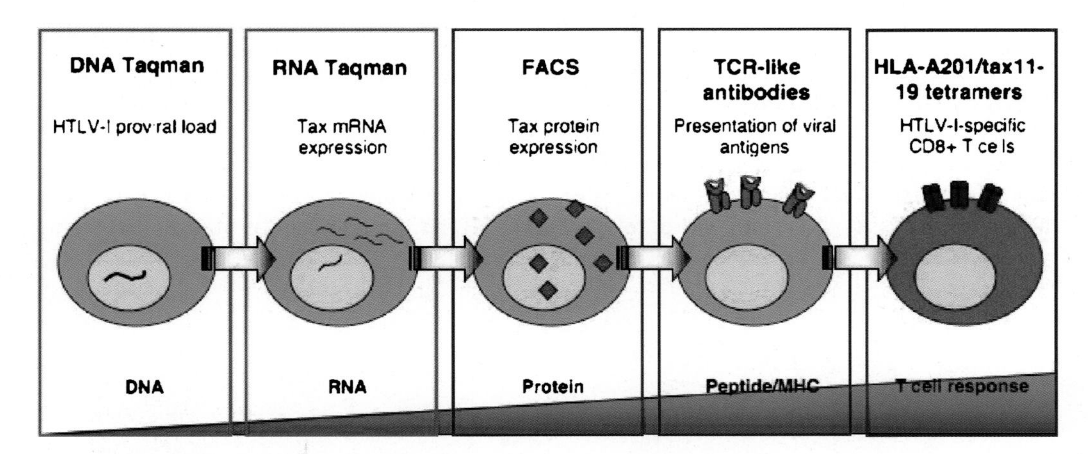

FIGURE 23.4. Schematic of induction of HTLV-I specific CD8+ T cells responses associated with immunopathogenesis of HAM/TSP. Quantitative PCR (DNA Taqman) demonstrates high HTLV-I proviral loads in HAM/TSP patients that are directly proportional to increased HTLV-I mRNA expression. In quantitative RT-PCR (RNA Taqman) analyses, elevated HTLV-I mRNA expression leads to increases of HTLV-I protein expression. This protein can be detected by flow cytometry. HTLV-I proteins (e.g., Tax) can be processed into immunodominant peptides such as Tax11–19 peptide and presented on the cell surface of infected cells in context with MHC class I molecules. HTLV-I Tax11–19 peptide strongly binds to HLA-A*0201 molecules and stimulates a virus-specific CD8+ T cells response. This Tax11–19/HLA-A*0201 complex can be detected with a TCR-like antibody, while the frequency of virus-specific CD8+ T cells can be determined by HLA-A*0201/Tax11–19 tetramers. These antigen-specific responses are expanded in the CSF of HAM/TSP patients and may contribute to disease progression by recognition of HTLV-I-processed antigens in the CNS associated with lysis of HTLV-I-infected inflammatory cells, HTLV-I-infected glial cells, and/or through induction of proinflammatory cytokines and chemokines.

Collectively, these histopathological studies suggested that an inflammatory process in the central nervous system (CNS) is involved in the pathogenesis of HAM/TSP.

CD8 + CTLs in the CNS of HAM/TSP patients was detected by immunohistochemical staining (Anderson et al., 1990; Umehara et al., 1994b). These studies support the hypothesis that HTLV-I infected CD4 + T lymphocytes enter the CNS which may drive local expansion of virus specific CD8 + CTLs. These virus-specific T cells may then either directly lyse virus infected cells and/or release a cascade of cytokines and chemokines that result in pathological changes (Figure 23.5). It is clearly important to define which cells might be targets of CTLs in the CNS.

HTLV-I Tax-specific CD8 + CTLs are hypothesized to play an important role in the development of HAM/TSP. Recently, through the development of tetramer technology, HTLV-I Tax-specific HLA-A*0201-restricted CD8 + T cells could be readily detected in HLA-A*0201 HAM/TSP patients (Altman et al., 1996; Greten et al., 1998). By using such tetramers, HTLV-I Tax 11–19-specific CD8 + cells from the PBMC of HLA-A*0201 HAM/TSP patients were found to represent an extraordinarily high proportion of the total CD8 + population (Greten et al., 1998; Nagai et al., 2001b). Moreover, the frequency of these HTLV-I Tax11–19-specific CD8 + T cells was even higher in the CSF of HAM/TSP patients (Nagai et al., 2001c; Kubota et al., 2002). As we have shown that amount of antigen-specific cells are proportional to the amount of HTLV-I proviral DNA load and the levels of HTLV-I tax mRNA expression from PBMCs (Yamano et al., 2002), the increased frequency of HTLV-I Tax11–19-specific CD8 + T cells in HAM/TSP CSF suggest continuous HTLV-I antigen presentation in vivo (Figure 23.3).

However, it is unclear what cells express HTLV-I antigens in vivo since no HTLV-I protein can be readily detected in PBMCs from HAM/TSP patients ex vivo. This may be reconciled with the observations that virus-specific T cells recognize antigens by engaging the antigen-specific TCR with peptide/HLA complexes displayed on the surface of antigen presenting cells (APCs). While tetramers have been useful for monitoring virus-specific T cells, there has been a shortage of reagents to study and visualize the ligand for the TCR: the peptide/HLA complex (Cohen et al., 2003; Yamano et al., 2004).

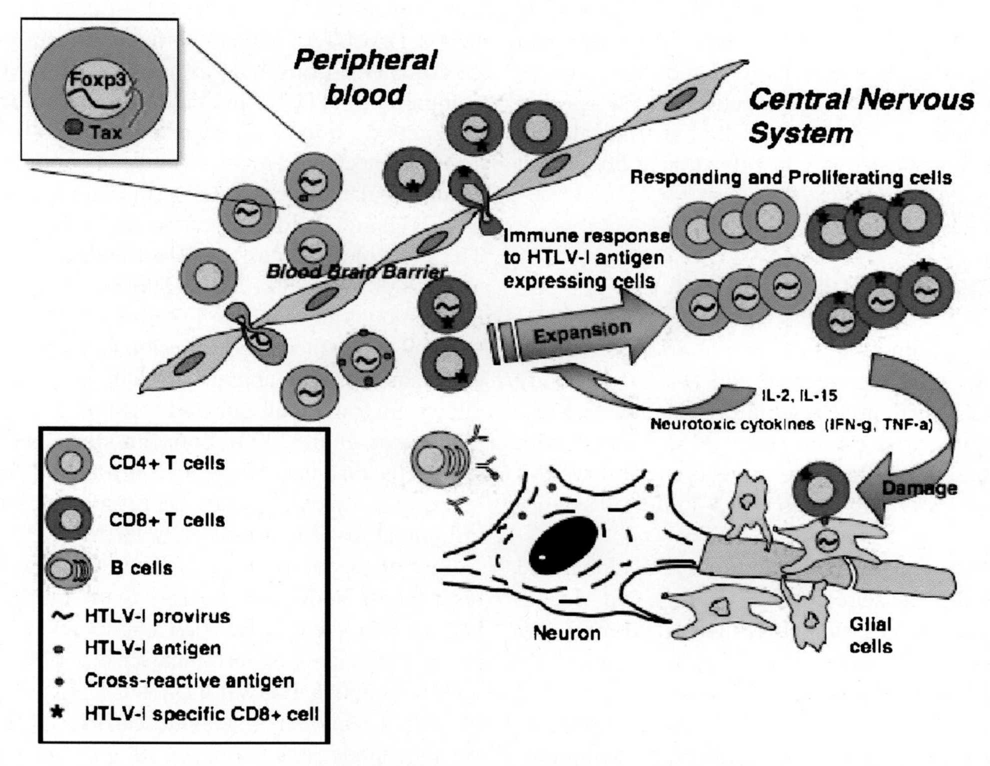

FIGURE 23.5. Model of immunopathogenesis in HAM/TSP. HTLV-I infected CD4+ and CD8+ T cells, as well as antigen-specific T cells migrate across the blood brain barrier from the peripheral blood into the CNS. Recent studies have demonstrated that expression of Tax protein abrogates the regulatory function CD4+ CD25+ T cells by inhibiting expression of Foxp3 at the level of transcription. This event may promote an elevated level of T cell activation in patients with HAM/TSP. As a portion of these HTLV-I infected cells express HTLV-I antigens, antigen-specific T cells recognize antigen-expressing cells in the CNS. By recognition, HTLV-I specific CD8+ T cells can lyse infected target cells or produce proinflammatory chemokines and cytokines. Antibodies produced by B cells that are directed against viral proteins may also cross-react with cellular proteins, in a mechanism known as molecular mimicry. Cytokines such as IL-2 and IL-15 may help bystander T cells expansion. Neurotoxic cytokines such as IFN-gamma and TNF-alpha may contribute to the damage of resident glial cells and neurons in the CNS.

HTLV-I gag, pX, and pol sequences have been localized to the thoracic cord in areas with increased CD4 + cells infiltration using semiquantitative PCR (Akizuki et al., 1989; Yoshioka et al., 1993). The amount of HTLV-I DNA decreased concomitantly with the number of infiltrating CD4 + cells in the spinal cord lesions of HAM/TSP patients having a long duration of illness. HTLV-I DNA has been localized to inflammatory infiltrating UCHL-1 positive cells in affected spinal cord by in situ PCR technique (Matsuoka et al., 1998). HTLV-I tax mRNA was also detected in infiltrating CD4 + T lymphocytes in active lesions in CNS specimens from HAM/TSP patients using in situ hybridization (Moritoyo et al., 1996). In addition, the HTLV-I p19 protein has been localized to spinal cord lesions (Furukawa et al., 1994). Collectively, these findings suggest that the main reservoir of HTLV-I may be infiltrating CD4 + T lymphocytes. Other cells may also harbor the virus. HTLV-I RNA has been shown to localize to astrocytes (Lehky et al., 1995). Recent work using quantitative PCR (TaqMan) indicated that CD8 + T cells were also infected with HTLV-I in vivo (Nagai et al., 2001b). Using a sensitive flow cytometric technique, Hanon et al. also showed that in HTLV-I infection, a significant proportion of CD8 + T cells were infected with HTLV-I (Hanon et al., 2000a). Interestingly, a portion of HTLV-I specific CD8 + CTLs were also infected with HTLV-I and HTLV-I protein expression in infected CD8 + T cells rendered them susceptible to cytolysis mediated by autologous HTLV-I specific CD8 + CTLs (Hanon et al., 2000a). These findings indicate that HTLV-I specific CTLs may serve a dual function as both target and effector cells.

23.5. Immune Response to HTLV-I/II

Several immunologic abnormalities such as increased spontaneous lymphoproliferation, elevated anti-HTLV-I antibody titers in sera and CSF, increased cytokine production, and cellular immune responses have been observed in patients with HAM/TSP. These abnormalities are more often observed in patients with HAM/TSP compared to HTLV-I infected asymptomatic carriers.

The levels of the cytokines IFN-gamma, TNF-alpha, and IL-6 were increased in the sera and CSF (Kuroda and Matsui, 1993). In peripheral blood lymphocytes (PBL) from HAM/TSP patients, mRNA for IL-1beta, IL-2, TNF-alpha, and IFN-gamma are up-regulated (Tendler et al., 1991). Moreover, IL-2 and IFN-gamma in PBMCs isolated from HAM/TSP patients was significantly elevated compared with both asymptomatic carriers and seronegative normal donors (Soldan, 2001). These data suggested that proinflammatory cytokines induced by Tax genes have been suggested to play a role in HAM/TSP pathogenesis.

NK cells are believed to protect against viral invasions at an early stage of an infection before the adaptive immune response is fully activated (Biron, 1997). Therefore, it is conceivable that impaired NK cell activity would result in diminished control of viral infection and increase in viral replication of HTLV-I in HAM/TSP patients. NK cell activity has been reported to be significantly lower in HAM/TSP than in controls (Fujihara et al., 1991). In addition, NK cells from HAM/TSP patients have lower cytotoxic activity and lower antibody-dependent cell-mediated cytotoxicity than those from controls (Yu et al., 1991).

Human NK cell receptors are expressed by NK cells and also some T cells, primarily CD8 + CTLs (Mingari et al., 1998). Inhibitory NK cell receptors (iNKRs) can down-regulate antigen-mediated T-cell effector functions, including cytotoxic activity and cytokine release (Mingari et al., 1998; Biassoni et al., 2001). It is reported that CD8 + T cells that express the NK cell inhibitory receptor were significantly decreased in HAM/TSP patients but not in asymptomatic HTLV-1 carriers (Saito et al., 2003). These receptors are suggested to play a role in regulating CD8 + T cell-mediated antiviral immune responses; therefore, a decrease in their expression may result in higher risk of developing inflammatory diseases such as HAM/TSP.

Although HTLV-I can infect a wide range of vertebrate cells in vitro (Trejo and Ratner, 2000), CD4 + T cells are the main subset of cells infected with HTLV-I in vivo (Richardson et al., 1990). HAM/TSP patients have significantly higher frequencies of HTLV-I Env and Tax antigen specific CD4 + T cells as compared to HTLV-I infected asymptomatic carriers (Goon et al., 2002). It has been reported that CD4 + T cells in HAM/TSP patients have a more Th1-like phenotype as characterized by up-regulated secretion of proinflammatory cytokines such as IFN-gamma and TNF-alpha and down-regulated levels of Th2 cytokines such as IL-4 (Horiuchi et al., 2000; Nakamura et al., 2000; Wu et al., 2000; Hanon et al., 2001). As suggested in the model presented in Figure 23.5, HTLV-I infected CD4 + T cells have increased adhesion activity to endothelial cells and transmigrating activity through basement membrane that allows migration of infected CD4 + T cells into the CNS (Nakamura et al., 2000). Entry into the CNS by HTLV-I CD4 + T cells that have increased production of TNF-alpha, a neurotoxic cytokine, may be responsible for the initiation inflammatory processes seen in HAM/TSP.

One of the most striking features of the cellular immune response in HAM/TSP patients is the highly increased numbers of HTLV-I specific HLA class I restricted CD8 + CTLs in the PBL and CSF (Hinuma et al., 1981; Jacobson et al., 1990). Although HTLV-I CD8 + CTLs have been detected in PBMC of some asymptomatic carriers (Parker et al., 1992), the magnitude and frequency of these responses are higher in patients with neurologic disease (Elovaara et al., 1993). CD8 + CTLs recognize viral and other foreign antigens, usually as small 9-aa peptides, in the context of HLA class I alleles. Although HTLV-I Env, Pol, Rof and Tof (Pique et al., 2000) could be target proteins of HTLV-I specific CTLs, HTLV-I specific CD8 + CTL activity in PBL from HAM/TSP patients is typically restricted to p27x and p40x products of the HTLV-I tax gene (Elovaara et al., 1993). In particular, the

HTLV-I Tax11–9 peptide (LLFGYPVYV) has been defined as an immunodominant epitope presented in the context of HLA-A*0201 and can be recognized by CD8 + CTLs from HAM/TSP patients (Parker et al., 1992; Koenig et al., 1993). Tax11–19 conforms to a known HLA-A*0201 binding motif and has one of the highest affinities known for any peptide-HLA complex (Utz et al., 1992). Recently, HTLV-I Tax peptide loaded HLA-A*0201 dimers and tetramers were developed and used to demonstrate HTLV-I Tax-specific HLA-A*0201 restricted CD8 + T cells (Greten et al., 1998; Bieganowska et al., 1999). HTLV-I Tax11–19-specific CD8 + cells from the PBMC of HLA-A*0201 HAM/TSP patients were found to represent an extraordinarily high proportion of the total CD8 + population (Greten et al., 1998; Nagai et al., 2001c). In patients with both HTLV-1 Tax 11–19 and cytomegalovirus peptide-specific CD8+ T cells, only HTLV-I Tax 11–19 specific CD8+ cells are found to be elevated in the CSF as detected with tetramers (Nagai et al., 2001a; Kubota et al., 2002).

These studies suggest that the HTLV-I specific cells either are specifically expanded in the CSF or are recruited into the CNS from the periphery as depicted in Figure 23.2. Preferential expansion in the CSF may be associated with the recognition of HTLV-I infected cells in this compartment or in the CNS and thus may contribute to the neuropathology associated with HAM/TSP (Nagai et al., 2001a). Recently, it has been reported that HTLV-I tax mRNA expression levels in PBMCs correlated with the amount of HTLV-I Tax11–19-specific CD8+ T cells (Yamano et al., 2002) (Figure 23.4). These data suggest that HTLV-I specific CD8+ T cells may be continuously driven by HTLV-I antigen expression *in vivo*.

HAM/TSP patients have high proviral loads despite vigorous virus-specific CD8+ T cell responses; however, it is unknown whether these T cells are efficient in eliminating the virus in vivo. Sequencing analysis revealed that epitope mutations were remarkably increased in a patient when the frequency and the degeneracy of the HTLV-1 specific CD8+ T cells were at the lowest. It was shown that the frequency and the degeneracy correlated with proviral load in the longitudinal study (Kubota et al., 2003).

The phenomenon of spontaneous lymphoproliferation, defined as the ability of PBMCs to proliferate ex vivo in the absence of exogenous antigens or stimulants such as IL-2, has been well described in PBLs from HAM/TSP patients, from HTLV-I asymptomatic carriers, and from HTLV-II infected persons (Kramer et al., 1989). However, the magnitude of this spontaneous lymphoproliferation is more pronounced in HAM/TSP patients than in asymptomatic HTLV-I carriers. The spontaneous lymphoproliferation of HTLV-I infected PBL is thought to consist of the proliferation of HTLV-I infected CD4+ cells and the expansion of CD8+ cells based on the demonstration of an increase in virus expressing cells concomitant with an increase in the percentage of CD8+ CD28+ lymphocytes. Experimentally, CD8+ T cells, including HTLV-I specific CD8+ T cells, have been shown to be expanded during spontaneous lymphoproliferation (Sakai et al., 2001).

The high frequency of HTLV-I specific CD8+ CTL in HAM/TSP patients correlates with the production of several cytokines. By intracellular cytokine staining coupled with flow cytometry, IFN-gamma, TNF-alpha, and IL-2 were all significantly elevated in the HTLV-I specific CD8+ cells of HAM/TSP patients compared with asymptomatic carriers and HTLV-I seronegative healthy controls (Kubota et al., 1998). In addition, HLA-A*0201 restricted HTLV-I Tax11–19-specific CD8+ CTL lines derived from a HAM/TSP patient released IFN-gamma and IL-2, with higher magnitude upon stimulation with Tax11–19 peptide (Nagai et al., 2001c).

It has been suggested that cytokine expression may be associated with an interaction of the TCR/Ag/HLA trimolecular complex. The molecular characterization of this trimolecular complex has led to major advances in the understanding of how the immune response recognizes antigen and has resulted in technologies that use these MHC-peptide complexes to directly visualize antigen-specific T cells (Greten et al., 1998). HTLV-I Tax11–19-specific CD8+ cells have been shown to possess cytolytic activity directed towards cells expressing HTLV-I Tax peptide via a perforin-dependent mechanism. Increased production of a of MMP-9, chemoattractants (macrophage inflammatory proteins 1alpha and 1beta), and proinflammatory cytokines (TNF-alpha and IFN-gamma) as a result of CD8+ T cell receptor mediated activation by HTLV-I antigens can contribute to damage of CNS tissues (Biddison et al., 1997; Greten et al., 1998; Kubota et al., 1998; Lim et al., 2000).

Moreover, it is documented that an autoimmune mechanism may also be involved in the pathogenesis of HAM/TSP (Levin et al., 1998; Levin et al., 2002b; Levin et al., 2002a; Jernigan et al., 2003). A unique T-cell receptor CDR3 motif, which has been demonstrated in brain lesions of MS and in the animal model experimental autoimmune encephalomyelitis, was also detected in infiltrating lymphocytes in the spinal cord of HAM/TSP patients (Hara et al., 1994). Furthermore, Levin et al. (1998) have provided new evidence in support of molecular mimicry as a possible mechanism in HAM/TSP pathogenesis. Serum immunoglobulin from HAM/TSP patients reacted to neurons in HTLV-I uninfected human CNS but not to cells in the peripheral nervous system or other organs. This reactivity was abrogated by pretreatment with recombinant HTLV-I tax protein. IgG from brain, CSF, and serum of the HAM/TSP patients showed immunoreactivity with heterogeneous nuclear ribonuclear protein-A1 (hnRNP-A1) as the autoantigen. This antibody specifically stained human Betz cells, and infusion of autoantibodies in brain sections inhibited neuronal firing (Levin et al., 2002b; Levin et al., 2002a; Jernigan et al., 2003). These data suggest that molecular mimicry between HTLV-I and autoantigens in CNS might play a role in the pathogenesis of HAM/TSP.

These evidences suggests that dysregulation in immune system may be associated with pathogenesis of HTLV-I associated neurologic diseases and may be responsible for local parenchymal ("by-stander") damage.

23.6. Animal Models of HTLV-I/II Infection

HTLV-I can be consistently infected into rabbits, rats and squirrel monkeys (Lairmore et al., 2005). HTLV-I infectivity for rabbits was first reported by using intravenous inoculations of the MT-2 cell line (Akagi et al., 1985), a T-cell leukemia cell line established from a patient with ATL, and with the Ra-1 cell line (Miyoshi et al., 1985), a rabbit lymphocyte cell line derived from cocultivation of rabbit lymphocytes with MT-2 cells. The rabbit model has provided important information regarding the immune response against HTLV-1 infection (Cockerell et al., 1990). Establishment of a rabbit model of clinical HTLV-1 disease has been more problematic. In the majority of studies, rabbit infection has paralleled the asymptomatic infection of humans. Sporadic clinical lymphoproliferative disorders in HTLV-1-infected rabbits were reported (Simpson et al., 1996; Kindt et al., 2000; Zhao et al., 2002). In each of these cases, clinical disease developed after one year and usually several years after the initial infection.

Rats have been infected with HTLV-1-producing cells and offer a tractable model of HAM/TSP, the neurologic disease associated with the viral infection (Suga et al., 1991; Sun et al., 1999; Hakata et al., 2001). Wistar–King–Aptekman–Hokudai (WKAH) rats emerged as a model of HAM/TSP. HTLV-I-infected WKAH rats develop spastic paraparesis with degenerative thoracic spinal cord and peripheral nerve lesions several months following inoculation (Ishiguro et al., 1992; Kushida et al., 1993). The pathology of rat HAM/TSP differs from that seen in humans. Lesions in humans have a marked T-cell infiltration of affected regions, whereas lymphocytes are not seen in the lesions in rats (Yoshiki, 1995). Subsequent studies defined the time periods over which the pathologic changes occur and indicated that apoptosis of oligodendrocytes and Schwann cells is the primary event leading to demyelination (Yoshiki, 1995; Yoshiki et al., 1997; Ohya et al., 2000). Macrophages are seen in the lesions of rats in response to the demyelination. Production of HTLV-1 pX mRNA, tumor necrosis factor (TNF) alpha mRNA, as well as altered expression of the apoptosis-modifying genes, bcl-2, bax, and p53, have been identified within the lesions (Ohya et al., 2000; Tomaru et al., 2003). In addition, HTLV-1 provirus has been identified in microglial cells and macrophages associated with lesions (Kasai et al., 1999).

Development of rat models for clinical ATL has required the use of immunodeficient rats. Ohashi et al. (1999) demonstrated that an 'ATL-like lymphoproliferative disease' could be established in adult nude (nu/nu) rats following inoculation of some HTLV-1-immortalized cell lines. This led to studies that examined methods of protection against tumor development, including adoptive transfer of T cells (Kannagi et al., 2000) and Tax-specific peptide vaccines (Hanabuchi et al., 2001).

The squirrel monkey has been successfully infected with HTLV-1 and offers an attractive non-human primate model of HTLV-1 for vaccine testing (Kazanji et al., 2000; Kazanji et al., 2001; Mortreux et al., 2001; Sundaram et al., 2004). In this model, peripheral lymphocytes, spleen, and lymph nodes were verified as major reservoirs for HTLV-1 virus during the early phase of infection (Kazanji et al., 1997; Kazanji et al., 2000). It was subsequently established that similar to humans, HTLV-1 infection in squirrel monkeys begins through reverse transcription of the virus genome, which is then followed by clonal expansion of infected cells (Mortreux et al., 2001).

HTLV-I infected mouse models would be useful in providing a small and inexpensive animal for studies of pathogenesis, treatment, and prevention of ATL or other HTLV-I related diseases. However, HTLV-1 does not efficiently infect murine cells. Thus, mouse models must be manipulated to establish HTLV-1 infection.

Transgenic models of HTLV-1 have provided an understanding of the role of Tax and Tax-mediated disruption of lymphocyte function or cytokines in HTLV-1-associated lesions. Arthropathy and other autoimmune diseases developed in Tax transgenic mice and rats controlled through the HTLV-1 promoter (Iwakura et al., 1991; Iwakura et al., 1995; Yamazaki et al., 1997). Transgenic animals expressing Tax manifested mammary adenomas or carcinomas (Yamada et al., 1995; Hall et al., 1998). Overall, these studies showed the strong support that Tax has the oncogenic capabilities, although they did not demonstrate mechanisms of carcinogenesis. Recently, Hasegawa et al. (2006) reported generating HTLV-I tax transgenic mice using the lck proximal promoter. The founder and offspring developed diffuse large cell lymphomas that are histologically similar to ATL cells. Further analysis is expected. Bovine leukemia virus (BLV) infection of sheep offers a reliable model of disease associated with delta retrovirus infections and insight into viral genetic determinants of tumor induction (Willems et al., 2000).

Summary

HTLV-I virus-host immune interactions play a pivotal role in HAM/TSP. HTLV-I has a unique pX region in the genome that might lead to immune dysregulation in patients with the disease. High HTLV-I proviral loads in both CD4+ and CD8+ T cell populations drive increased HTLV-I mRNA levels that result in increased HTLV-I protein expression. Processing of HTLV-I proteins and presentation of HTLV-I specific peptides leads to activation and expansion of antigen-specific T cell responses. The hypothesis that HTLV-I specific CD8+ CTLs play a role in the development of HAM/TSP is supported by localization of these CTL in the CNS. HTLV-I specific CD8+ cells could recognize productively infected cells and respond either by direct lysis of the infected cell or through the release of proinflammatory cytokines and chemokines. These molecules can act to recruit and expand additional inflammatory cells, and have been shown to be toxic to CNS tissue. Several studies suggest that HTLV-I infected lymphocytes may preferentially migrate into the

CSF from peripheral blood or that HTLV-I infected lymphocytes may selectively expand in this compartment. This process is illustrated in Figure 23.5. It is reported that HTLV-I genomic sequences, RNA, and protein have been localized to spinal cord lesions. Therefore, all requirements for CTL recognition, including viral antigen and HLA class I expression, are present in the HAM/TSP lesion, lending support to the argument that CD8+ CTL may be immunopathogenic in this disease. Intensive studies regarding the interaction between HTLV-I specific CD8+ T cells and HTLV-I infected cells will clarify the pathogenesis of HAM/TSP.

Review Questions/Problems

1. Which is most unusual symptom in patients with HAM/TSP?

a. spastic paralysis
b. higher brain dysfunction
c. bladder and rectal disturbance

2. Which antibody is detected in patients with HAM/TSP?

a. anti HTLV-I antibody
b. anti HIV-I antibody
c. anti ganglioside antibody

3. Which inflammatory disease is shown to be associated with HTLV-I infection?

a. gastritis
b. myelitis
c. uvelitis

4. Which level of spinal cord is predominantly affected in patients with HAM/TSP?

a. cervical level
b. thoracic level
c. lumbar level

5. Which cells are infiltrated in the chronic phase of HAM/TSP?

a. CD4 positive T cells
b. CD8 positive T cells
c. B cells

6. Which circulated cytokine is up-regulated in patients with HAM/TSP?

a. IL-2
b. IL-4
c. IL-10

7. Which part is mostly degenerated in spinal cord in patients with HAM/TSP?

a. Anterior column
b. Lateral column
c. Posterior column

8. Which is most endemic area of HTLV-I?

a. North America
b. Caribbean
c. Europe

9. What type of human leukocyte antigen (HLA) is regarded as the higher risk factor associated with HAM/TSP?

a. HLA-A*02(+)
b. HLA-Cw*08(+)
c. HLA-DRB1*0101(+)

10. Which cytokine is mostly detected at degenerated spinal cord in HAM/TSP?

a. TNF-alpha
b. IL-1 beta
c. IFN-gamma

11. Which is characteristically detected in T cells of patients with HAM/TSP?

a. Spontaneous lymphoproliferation
b. Down-regulation of the ratio of CD4/CD8
c. Eruption of the "flower cells"

12. How much is the frequency of the development of HAM/TSP in HTLV-I carrier?

a. Less than 5%
b. 10% to 20%
c. More than 50%

13. Which is inappropriate infection route of the HTLV-I?

a. Transmission through breast milk
b. Sexual contact
c. Droplet infection

14. Which adhesion molecules are mostly detected in the spinal cord of HAM/TSP?

a. ICAM-1
b. VCAM-1
c. ICAM-1 and VCAM-1

15. Which part is histopathologically affected in spinal cord of HAM/TSP?

a. Axon
b. Myelin
c. Axon and Myelin

16. Which inflammatory disease is shown to be associated with HAM/TSP?

a. Sjogren syndrome
b. Atopic dermatitis
c. Crohn's disease

17. Which is predictive finding in regard to CSF of HAM/TSP patients?

a. Leukocyte: Normal range to slightly increasing, Sugar: Normal range

b. Leukocyte: Moderate increasing, Sugar: Normal range
c. Leukocyte: Moderate to severe increasing, Sugar: Decreasing

18. Which sex is much developed in HAM/TSP?

a. Male
b. Female
c. No difference between male and female

19. What are the reservoir cells of the infected HTLV-I?

a. T cells
b. B cells
c. Macrophages

20. Which is strongly associated with the development of HAM/TSP?

a. Leukocyte increasing
b. Hypercalcemia
c. High proviral load

References

Abe M, Umehara F, Kubota R, Moritoyo T, Izumo S, Osame M (1999) Activation of macrophages/microglia with the calcium-binding proteins MRP14 and MRP8 is related to the lesional activities in the spinal cord of HTLV-I associated myelopathy. J Neurol 246:358–364.

Akagi T, Takeda I, Oka T, Ohtsuki Y, Yano S, Miyoshi I (1985) Experimental infection of rabbits with human T-cell leukemia virus type I. Jpn J Cancer Res 76:86–94.

Akizuki S, Yoshida S, Setoguchi M (1989) The neuropathyology of human T cell lymphotropic virus type I associated myelopathy. In: HTLV-I and the Nervous System (Roman GC, Vernant JC, Osame M, eds), pp 253–260. New York: R. Alan Liss.

Altman JD, Moss PA, Goulder PJ, Barouch DH, McHeyzer-Williams MG, Bell JI, McMichael AJ, Davis MM (1996) Phenotypic analysis of antigen-specific T lymphocytes. Science 274:94–96.

Anderson P, Nagler-Anderson C, O'Brien C, Levine H, Watkins S, Slayter HS, Blue ML, Schlossman SF (1990) A monoclonal antibody reactive with a 15-kDa cytoplasmic granule-associated protein defines a subpopulation of CD8+ T lymphocytes. J Immunol 144:574–582.

Aye MM, Matsuoka E, Moritoyo T, Umehara F, Suehara M, Hokezu Y, Yamanaka H, Isashiki Y, Osame M, Izumo S (2000) Histopathological analysis of four autopsy cases of HTLV-I-associated myelopathy/tropical spastic paraparesis: Inflammatory changes occur simultaneously in the entire central nervous system. Acta Neuropathol (Berl) 100:245–252.

Bangham CR (2000) The immune response to HTLV-I. Curr Opin Immunol 12:397–402.

Biassoni R, Cantoni C, Pende D, Sivori S, Parolini S, Vitale M, Bottino C, Moretta A (2001) Human natural killer cell receptors and co-receptors. Immunol Rev 181:203–214.

Biddison WE, Kubota R, Kawanishi T, Taub DD, Cruikshank WW, Center DM, Connor EW, Utz U, Jacobson S (1997) Human T cell leukemia virus type I (HTLV-I)-specific CD8+ CTL clones from patients with HTLV-I-associated neurologic disease secrete proinflammatory cytokines, chemokines, and matrix metalloproteinase. J Immunol 159:2018–2025.

Bieganowska K, Hollsberg P, Buckle GJ, Lim DG, Greten TF, Schneck J, Altman JD, Jacobson S, Ledis SL, Hanchard B, Chin J, Morgan O, Roth PA, Hafler DA (1999) Direct analysis of viral-specific CD8+ T cells with soluble HLA-A2/Tax11–19 tetramer complexes in patients with human T cell lymphotropic virus-associated myelopathy. J Immunol 162:1765–1771.

Biron CA (1997) Activation and function of natural killer cell responses during viral infections. Curr Opin Immunol 9:24–34.

Blattner WA, Gallo RC (1985) Epidemiology of human retroviruses. Leuk Res 9:697–698.

Calattini S, Chevalier SA, Duprez R, Bassot S, Froment A, Mahieux R, Gessain A (2005) Discovery of a new human T-cell lymphotropic virus (HTLV-3) in Central Africa. Retrovirology 2:30.

Cockerell GL, Lairmore M, De B, Rovnak J, Hartley TM, Miyoshi I (1990) Persistent infection of rabbits with HTLV-I: Patterns of anti-viral antibody reactivity and detection of virus by gene amplification. Int J Cancer 45:127–130.

Cohen CJ, Sarig O, Yamano Y, Tomaru U, Jacobson S, Reiter Y (2003) Direct phenotypic analysis of human MHC class I antigen presentation: Visualization, quantitation, and in situ detection of human viral epitopes using peptide-specific, MHC-restricted human recombinant antibodies. J Immunol 170:4349–4361.

Edlich RF, Hill LG, Williams FM (2003) Global epidemic of human T-cell lymphotrophic virus type-I (HTLV-I): An update. J Long Term Eff Med Implants 13:127–140.

Elovaara I, Koenig S, Brewah AY, Woods RM, Lehky T, Jacobson S (1993) High human T cell lymphotropic virus type 1 (HTLV-1)-specific precursor cytotoxic T lymphocyte frequencies in patients with HTLV-1-associated neurological disease. J Exp Med 177:1567–1573.

Fujihara K, Itoyama Y, Yu F, Kubo C, Goto I (1991) Cellular immune surveillance against HTLV-I infected T lymphocytes in HTLV-I associated myelopathy/tropical spastic paraparesis (HAM/TSP). J Neurol Sci 105:99–107.

Fujisawa J, Seiki M, Kiyokawa T, Yoshida M (1985) Functional activation of the long terminal repeat of human T-cell leukemia virus type I by a trans-acting factor. Proc Natl Acad Sci USA 82:2277–2281.

Furukawa K, Mori M, Ohta N, Ikeda H, Shida H, Furukawa K, Shiku H (1994) Clonal expansion of CD8+ cytotoxic T lymphocytes against human T cell lymphotropic virus type I (HTLV-I) genome products in HTLV-I-associated myelopathy/tropical spastic paraparesis patients. J Clin Invest 94:1830–1839.

Gessain A (1996) Epidemiology of HTLV-I and associated diseases. In: Human T-cell Lymphotropic Virus Type I (Hollsgerg P, Hafler D, eds), pp 33–64. Chinchester, UK: John Wiley and Sons.

Gessain A, Barin F, Vernant JC, Gout O, Maurs L, Calender A, de The G (1985) Antibodies to human T-lymphotropic virus type-I in patients with tropical spastic paraparesis. Lancet 2:407–410.

Goon PK, Hanon E, Igakura T, Tanaka Y, Weber JN, Taylor GP, Bangham CR (2002) High frequencies of Th1-type CD4(+) T cells specific to HTLV-1 Env and Tax proteins in patients with HTLV-1-associated myelopathy/tropical spastic paraparesis. Blood 99:3335–3341.

Greten TF, Slansky JE, Kubota R, Soldan SS, Jaffee EM, Leist TP, Pardoll DM, Jacobson S, Schneck JP (1998) Direct visualization of antigen-specific T cells: HTLV-1 Tax11–19- specific CD8(+) T cells are activated in peripheral blood and accumulate in cerebrospinal fluid from HAM/TSP patients. Proc Natl Acad Sci USA 95:7568–7573.

Hakata Y, Yamada M, Shida H (2001) Rat CRM1 is responsible for the poor activity of human T-cell leukemia virus type 1 Rex protein in rat cells. J Virol 75:11515–11525.

Hall AP, Irvine J, Blyth K, Cameron ER, Onions DE, Campbell ME (1998) Tumours derived from HTLV-I tax transgenic mice are characterized by enhanced levels of apoptosis and oncogene expression. J Pathol 186:209–214.

Hanabuchi S, Ohashi T, Koya Y, Kato H, Hasegawa A, Takemura F, Masuda T, Kannagi M (2001) Regression of human T-cell leukemia virus type I (HTLV-I)-associated lymphomas in a rat model: Peptide-induced T-cell immunity. J Natl Cancer Inst 93:1775–1783.

Hanon E, Stinchcombe JC, Saito M, Asquith BE, Taylor GP, Tanaka Y, Weber JN, Griffiths GM, Bangham CR (2000a) Fratricide among CD8(+) T lymphocytes naturally infected with human T cell lymphotropic virus type I. Immunity 13:657–664.

Hanon E, Hall S, Taylor GP, Saito M, Davis R, Tanaka Y, Usuku K, Osame M, Weber JN, Bangham CR (2000b) Abundant tax protein expression in CD4+ T cells infected with human T-cell lymphotropic virus type I (HTLV-I) is prevented by cytotoxic T lymphocytes. Blood 95:1386–1392.

Hanon E, Goon P, Taylor GP, Hasegawa H, Tanaka Y, Weber JN, Bangham CR (2001) High production of interferon gamma but not interleukin-2 by human T-lymphotropic virus type I-infected peripheral blood mononuclear cells. Blood 98:721–726.

Hara H, Morita M, Iwaki T, Hatae T, Itoyama Y, Kitamoto T, Akizuki S, Goto I, Watanabe T (1994) Detection of human T lymphotrophic virus type I (HTLV-I) proviral DNA and analysis of T cell receptor V beta CDR3 sequences in spinal cord lesions of HTLV-I-associated myelopathy/tropical spastic paraparesis. J Exp Med 180:831–839.

Hasegawa H, Sawa H, Lewis MJ, Orba Y, Sheehy N, Yamamoto Y, Ichinohe T, Tsunetsugu-Yokota Y, Katano H, Takahashi H, Matsuda J, Sata T, Kurata T, Nagashima K, Hall WW (2006) Thymus-derived leukemia-lymphoma in mice transgenic for the Tax gene of human T-lymphotropic virus type I. Nat Med 12:466–472.

Hinuma Y, Nagata K, Hanaoka M, Nakai M, Matsumoto T, Kinoshita KI, Shirakawa S, Miyoshi I (1981) Adult T-cell leukemia: Antigen in an ATL cell line and detection of antibodies to the antigen in human sera. Proc Natl Acad Sci USA 78:6476–6480.

Horiuchi I, Kawano Y, Yamasaki K, Minohara M, Furue M, Taniwaki T, Miyazaki T, Kira J (2000) Th1 dominance in HAM/TSP and the optico-spinal form of multiple sclerosis versus Th2 dominance in mite antigen-specific IgE myelitis. J Neurol Sci 172:17–24.

Ishiguro N, Abe M, Seto K, Sakurai H, Ikeda H, Wakisaka A, Togashi T, Tateno M, Yoshiki T (1992) A rat model of human T lymphocyte virus type I (HTLV-I) infection. 1. Humoral antibody response, provirus integration, and HTLV-I-associated myelopathy/tropical spastic paraparesis-like myelopathy in seronegative HTLV-I carrier rats. J Exp Med 176:981–989.

Iwakura Y, Tosu M, Yoshida E, Takiguchi M, Sato K, Kitajima I, Nishioka K, Yamamoto K, Takeda T, Hatanaka M, et al. (1991) Induction of inflammatory arthropathy resembling rheumatoid arthritis in mice transgenic for HTLV-I. Science 253:1026–1028.

Iwakura Y, Saijo S, Kioka Y, Nakayama-Yamada J, Itagaki K, Tosu M, Asano M, Kanai Y, Kakimoto K (1995) Autoimmunity induction by human T cell leukemia virus type 1 in transgenic mice that develop chronic inflammatory arthropathy resembling rheumatoid arthritis in humans. J Immunol 155:1588–1598.

Iwasaki Y (1990) Pathology of chronic myelopathy associated with HTLV-I infection (HAM/TSP). J Neurol Sci 96:103–123.

Izumo S, Usuku K, Osame M, Machigashira K, Johnson M, Hakagawa M (1989) The neuropathology of HTLV-I associated myelopathy in Japan: report of an autopsy case and review of the literature. In: HTLV-I and the Nervous System (Roman GC, Vernant JC, Osame M, eds), pp 261–267. New York: R. Alan Liss.

Izumo S, Ijichi N, Higuchi I, Tashiro A, Takahashi K, Osame M (1992) Neuropathology of HTLV-I-associated myelopathy—a report of two autopsy cases. Acta Paediatr Jpn 34:358–364.

Jacobson S, Shida H, McFarlin DE, Fauci AS, Koenig S (1990) Circulating CD8+ cytotoxic T lymphocytes specific for HTLV-I pX in patients with HTLV-I associated neurological disease. Nature 348:245–248.

Jeffery KJ, Usuku K, Hall SE, Matsumoto W, Taylor GP, Procter J, Bunce M, Ogg GS, Welsh KI, Weber JN, Lloyd AL, Nowak MA, Nagai M, Kodama D, Izumo S, Osame M, Bangham CR (1999) HLA alleles determine human T-lymphotropic virus-I (HTLV-I) proviral load and the risk of HTLV-I-associated myelopathy. Proc Natl Acad Sci USA 96:3848–3853.

Jeffery KJ, Siddiqui AA, Bunce M, Lloyd AL, Vine AM, Witkover AD, Izumo S, Usuku K, Welsh KI, Osame M, Bangham CR (2000) The influence of HLA class I alleles and heterozygosity on the outcome of human T cell lymphotropic virus type I infection. J Immunol 165:7278–7284.

Jernigan M, Morcos Y, Lee SM, Dohan FC, Jr., Raine C, Levin MC (2003) IgG in brain correlates with clinicopathological damage in HTLV-1 associated neurologic disease. Neurology 60:1320–1327.

Johnson JM, Harrod R, Franchini G (2001) Molecular biology and pathogenesis of the human T-cell leukaemia/lymphotropic virus Type-1 (HTLV-1). Int J Exp Pathol 82:135–147.

Kannagi M, Ohashi T, Hanabuchi S, Kato H, Koya Y, Hasegawa A, Masuda T, Yoshiki T (2000) Immunological aspects of rat models of HTLV type 1-infected T lymphoproliferative disease. AIDS Res Hum Retroviruses 16:1737–1740.

Kasai T, Ikeda H, Tomaru U, Yamashita I, Ohya O, Morita K, Wakisaka A, Matsuoka E, Moritoyo T, Hashimoto K, Higuchi I, Izumo S, Osame M, Yoshiki T (1999) A rat model of human T lymphocyte virus type I (HTLV-I) infection: In situ detection of HTLV-I provirus DNA in microglia/macrophages in affected spinal cords of rats with HTLV-I-induced chronic progressive myeloneuropathy. Acta Neuropathol (Berl) 97:107–112.

Kazanji M, Moreau JP, Mahieux R, Bonnemains B, Bomford R, Gessain A, de The G (1997) HTLV-I infection in squirrel monkeys (Saimiri sciureus) using autologous, homologous, or heterologous HTLV-I-transformed cell lines. Virology 231:258–266.

Kazanji M, Ureta-Vidal A, Ozden S, Tangy F, de Thoisy B, Fiette L, Talarmin A, Gessain A, de The G (2000) Lymphoid organs as a major reservoir for human T-cell leukemia virus type 1 in experimentally infected squirrel monkeys (Saimiri sciureus): Provirus expression, persistence, and humoral and cellular immune responses. J Virol 74:4860–4867.

Kazanji M, Tartaglia J, Franchini G, de Thoisy B, Talarmin A, Contamin H, Gessain A, de The G (2001) Immunogenicity and protective efficacy of recombinant human T-cell leukemia/lymphoma virus type 1 NYVAC and naked DNA vaccine candidates in squirrel monkeys (Saimiri sciureus). J Virol 75:5939–5948.

Kindt TJ, Said WA, Bowers FS, Mahana W, Zhao TM, Simpson RM (2000) Passage of human T-cell leukemia virus type-1 during progression to cutaneous T-cell lymphoma results in myelopathic disease in an HTLV-1 infection model. Microbes Infect 2:1139–1146.

Kira J, Fujihara K, Itoyama Y, Goto I, Hasuo K (1991) Leukoencephalopathy in HTLV-I-associated myelopathy/tropical spastic paraparesis: MRI analysis and a two year follow-up study after corticosteroid therapy. J Neurol Sci 106:41–49.

Kiyokawa T, Seiki M, Iwashita S, Imagawa K, Shimizu F, Yoshida M (1985) p27x-III and p21x-III, proteins encoded by the pX sequence of human T-cell leukemia virus type I. Proc Natl Acad Sci U S A 82:8359–8363.

Koenig S, Woods RM, Brewah YA, Newell AJ, Jones GM, Boone E, Adelsberger JW, Baseler MW, Robinson SM, Jacobson S (1993) Characterization of MHC class I restricted cytotoxic T cell responses to tax in HTLV-1 infected patients with neurologic disease. J Immunol 151:3874–3883.

Kramer A, Jacobson S, Reuben JF, Murphy EL, Wiktor SZ, Cranston B, Figueroa JP, Hanchard B, McFarlin D, Blattner WA (1989) Spontaneous lymphocyte proliferation in symptom-free HTLV-I positive Jamaicans. Lancet 2:923–924.

Kubota R, Kawanishi T, Matsubara H, Manns A, Jacobson S (1998) Demonstration of human T lymphotropic virus type I (HTLV-I) tax-specific CD8+ lymphocytes directly in peripheral blood of HTLV-I-associated myelopathy/tropical spastic paraparesis patients by intracellular cytokine detection. J Immunol 161:482–488.

Kubota R, Nagai M, Kawanishi T, Osame M, Jacobson S (2000) Increased HTLV type 1 tas specific CD8+ cells in HTLV type 1-associated myelopathy/tropical spastic paraparesis: Correlation with HTLV type 1 proviral load. AIDS Res Hum Restroviruses 16:1705–1709.

Kubota R, Soldan SS, Martin R, Jacobson S (2002) Selected cytotoxic T lymphocytes with high specificity for HTLV-I in cerebrospinal fluid from a HAM/TSP patient. J Neurovirol 8:53–57.

Kubota R, Furukawa Y, Izumo S, Usuku K, Osame M (2003) Degenerate specificity of HTLV-1-specific CD8+ T cells during viral replication in patients with HTLV-1-associated myelopathy (HAM/TSP). Blood 101:3074–3081.

Kuroda Y, Matsui M (1993) Cerebrospinal fluid interferon-gamma is increased in HTLV-I-associated myelopathy. J Neuroimmunol 42:223–226.

Kushida S, Matsumura M, Tanaka H, Ami Y, Hori M, Kobayashi M, Uchida K, Yagami K, Kameyama T, Yoshizawa T, et al. (1993) HTLV-1-associated myelopathy/tropical spastic paraparesis-like rats by intravenous injection of HTLV-1-producing rabbit or human T-cell line into adult WKA rats. Jpn J Cancer Res 84:831–833.

Lairmore MD, Silverman L, Ratner L (2005) Animal models for human T-lymphotropic virus type 1 (HTLV-1) infection and transformation. Oncogene 24:6005–6015.

Lehky TJ, Fox CH, Koenig S, Levin MC, Flerlage N, Izumo S, Sato E, Raine CS, Osame M, Jacobson S (1995) Detection of human T-lymphotropic virus type I (HTLV-I) tax RNA in the central nervous system of HTLV-I-associated myelopathy/tropical spastic paraparesis patients by in situ hybridization. Ann Neurol 37:167–175.

Lehky TJ, Flerlage N, Katz D, Houff S, Hall WH, Ishii K, Monken C, Dhib-Jalbut S, McFarland HF, Jacobson S (1996) Human T-cell lymphotropic virus type II-associated myelopathy: Clinical and immunologic profiles. Ann Neurol 40:714–723.

Levin MC, Jacobson S (1997) HTLV-I associated myelopathy/tropical spastic paraparesis (HAM/TSP): A chronic progressive neurologic disease associated with immunologically mediated damage to the central nervous system. J Neurovirol 3:126–140.

Levin MC, Krichavsky M, Berk J, Foley S, Rosenfeld M, Dalmau J, Chang G, Posner JB, Jacobson S (1998) Neuronal molecular mimicry in immune-mediated neurologic disease. Ann Neurol 44:87–98.

Levin MC, Lee SM, Morcos Y, Brady J, Stuart J (2002a) Cross-reactivity between immunodominant human T lymphotropic virus type I tax and neurons: Implications for molecular mimicry. J Infect Dis 186:1514–1517.

Levin MC, Lee SM, Kalume F, Morcos Y, Dohan FC, Jr., Hasty KA, Callaway JC, Zunt J, Desiderio D, Stuart JM (2002b) Autoimmunity due to molecular mimicry as a cause of neurological disease. Nat Med 8:509–513.

Lim DG, Bieganowska Bourcier K, Freeman GJ, Hafler DA (2000) Examination of CD8+ T cell function in humans using MHC class I tetramers: Similar cytotoxicity but variable proliferation and cytokine production among different clonal CD8+ T cells specific to a single viral epitope. J Immunol 165:6214–6220.

Manns A, Miley WJ, Wilks RJ, Morgan OS, Hanchard B, Wharfe G, Cranston B, Maloney E, Welles SL, Blattner WA, Waters D (1999) Quantitative proviral DNA and antibody levels in the natural history of HTLV-I infection. J Infect Dis 180:1487–1493.

Matsuoka E, Takenouchi N, Hashimoto K, Kashio N, Moritoyo T, Higuchi I, Isashiki Y, Sato E, Osame M, Izumo S (1998) Perivascular T cells are infected with HTLV-I in the spinal cord lesions with HTLV-I-associated myelopathy/tropical spastic paraparesis: Double staining of immunohistochemistry and polymerase chain reaction in situ hybridization. Acta Neuropathol (Berl) 96:340–346.

Mingari MC, Ponte M, Bertone S, Schiavetti F, Vitale C, Bellomo R, Moretta A, Moretta L (1998) HLA class I-specific inhibitory receptors in human T lymphocytes: Interleukin 15-induced expression of CD94/NKG2A in superantigen- or alloantigen-activated CD8+ T cells. Proc Natl Acad Sci USA 95:1172–1177.

Miyoshi I, Yoshimoto S, Kubonishi I, Fujishita M, Ohtsuki Y, Yamashita M, Yamato K, Hirose S, Taguchi H, Niiya K, et al. (1985) Infectious transmission of human T-cell leukemia virus to rabbits. Int J Cancer 35:81–85.

Morgan OS, Rodgers-Johnson P, Mora C, Char G (1989) HTLV-1 and polymyositis in Jamaica. Lancet 2:1184–1187.

Moritoyo T, Reinhart TA, Moritoyo H, Sato E, Izumo S, Osame M, Haase AT (1996) Human T-lymphotropic virus type I-associated myelopathy and tax gene expression in CD4+ T lymphocytes. Ann Neurol 40:84–90.

Mortreux F, Kazanji M, Gabet AS, de Thoisy B, Wattel E (2001) Two-step nature of human T-cell leukemia virus type 1 replication in experimentally infected squirrel monkeys (Saimiri sciureus). J Virol 75:1083–1089.

Nagai M, Usuku K, Matsumoto W, Kodama D, Takenouchi N, Moritoyo T, Hashiguchi S, Ichinose M, Bangham CR, Izumo S, Osame M (1998) Analysis of HTLV-I proviral load in 202 HAM/TSP patients and 243 asymptomatic HTLV-I carriers: High proviral load strongly predisposes to HAM/TSP. J Neurovirol 4:586–593.

Nagai M, Yamano Y, Brennan MB, Mora CA, Jacobson S (2001a) Increased HTLV-I proviral load and preferential expansion of HTLV-I Tax-specific CD8+ T cells in cerebrospinal fluid from patients with HAM/TSP. Ann Neurol 50:807–812.

Nagai M, Brennan MB, Sakai JA, Mora CA, Jacobson S (2001b) CD8(+) T cells are an in vivo reservoir for human T-cell lymphotropic virus type I. Blood 98:1858–1861.

Nagai M, Kubota R, Greten TF, Schneck JP, Leist TP, Jacobson S (2001c) Increased activated human T cell lymphotropic virus type I (HTLV-I) Tax11–19-specific memory and effector CD8+ cells in patients with HTLV-I-associated myelopathy/tropical spastic paraparesis: Correlation with HTLV-I provirus load. J Infect Dis 183:197–205.

Nakamura T, Furuya T, Nishiura Y, Ichinose K, Shirabe S, Eguchi K (2000) Importance of immune deviation toward Th1 in the early immunopathogenesis of human T-lymphotropic virus type I-associated myelopathy. Med Hypotheses 54:777–782.

Ohashi T, Hanabuchi S, Kato H, Koya Y, Takemura F, Hirokawa K, Yoshiki T, Tanaka Y, Fujii M, Kannagi M (1999) Induction of adult T-cell leukemia-like lymphoproliferative disease and its inhibition by adoptive immunotheraphy in T-cell-deficient nude rats inoculated with syngeneic human T-cell leukemia virus type 1-immortalized cells. J Virol 73:6031–6040.

Ohya O, Ikeda H, Tomaru U, Yamashita I, Kasai T, Morita K, Wakisaka A, Yoshiki T (2000) Human T-lymphocyte virus type I (HTLV-I)-induced myeloneuropathy in rats: Oligodendrocytes undergo apoptosis in the presence of HTLV-I. Apmis 108:459–466.

Orland JR, Engstrom J, Fridey J, Sacher RA, Smith JW, Nass C, Garratty G, Newman B, Smith D, Wang B, Loughlin K, Murphy EL (2003) Prevalence and clinical features of HTLV neurologic disease in the HTLV outcomes study. Neurology 61:1588–1594.

Osame M (1990) Review of WHO kagoshima meeting and diagnostic guidelines for HAM/TSP. In: Human Retrovirology: HTLV (Blattner WA, ed), pp 191–197. New York: Raven Press.

Osame M, Usuku K, Izumo S, Ijichi N, Amitani H, Igata A, Matsumoto M, Tara M (1986) HTLV-I associated myelopathy, a new clinical entity. Lancet 1:1031–1032.

Parker CE, Daenke S, Nightingale S, Bangham CR (1992) Activated, HTLV-1-specific cytotoxic T-lymphocytes are found in healthy seropositives as well as in patients with tropical spastic paraparesis. Virology 188:628–636.

Pique C, Ureta-Vidal A, Gessain A, Chancerel B, Gout O, Tamouza R, Agis F, Dokhelar MC (2000) Evidence for the chronic in vivo production of human T cell leukemia virus type I Rof and Tof proteins from cytotoxic T lymphocytes directed against viral peptides. J Exp Med 191:567–572.

Poiesz BJ, Ruscetti FW, Gazdar AF, Bunn PA, Minna JD, Gallo RC (1980) Detection and isolation of type C retrovirus particles from fresh and cultured lymphocytes of a patient with cutaneous T-cell lymphoma. Proc Natl Acad Sci USA 77:7415–7419.

Richardson JH, Edwards AJ, Cruickshank JK, Rudge P, Dalgleish AG (1990) In vivo cellular tropism of human T-cell leukemia virus type 1. J Virol 64:5682–5687.

Saito M, Braud VM, Goon P, Hanon E, Taylor GP, Saito A, Eiraku N, Tanaka Y, Usuku K, Weber JN, Osame M, Bangham CR (2003) Low frequency of CD94/NKG2A+ T lymphocytes in patients with HTLV-1-associated myelopathy/tropical spastic paraparesis, but not in asymptomatic carriers. Blood 102:577–584.

Sakai JA, Nagai M, Brennan MB, Mora CA, Jacobson S (2001) In vitro spontaneous lymphoproliferation in patients with human T-cell lymphotropic virus type I-associated neurologic disease: Predominant expansion of CD8+ T cells. Blood 98:1506–1511.

Seiki M, Hikikoshi A, Taniguchi T, Yoshida M (1985) Expression of the pX gene of HTLV-I: General splicing mechanism in the HTLV family. Science 228:1532–1534.

Semmes OJ, Cazares LH, Ward MD, Qi L, Moody M, Maloney E, Morris J, Trosset MW, Hisada M, Gygi S, Jacobson S (2005) Discrete serum protein signatures discriminate between human retrovirus-associated hematologic and neurologic disease. Leukemia 19:1229–1238.

Silva EA, Otsuki K, Leite AC, Alamy AH, Sa-Carvalho D, Vicente AC (2002) HTLV-II infection associated with a chronic neurodegenerative disease: Clinical and molecular analysis. J Med Virol 66:253–257.

Simpson RM, Leno M, Hubbard BS, Kindt TJ (1996) Cutaneous manifestations of human T cell leukemia virus type I infection in an experimental model. J Infect Dis 173:722–726.

Sodroski JG, Rosen CA, Haseltine WA (1984) Trans-acting transcriptional activation of the long terminal repeat of human T lymphotropic viruses in infected cells. Science 225:381–385.

Soldan S (2001) Immune response to HTLV-I and HTLV-II. In: Retroviral Immunology: Immune Responses and Restoration (Walker GPB, ed), pp 159–190. Totowa, NJ: Humana Press.

Suga T, Kameyama T, Kinoshita T, Shimotohno K, Matsumura M, Tanaka H, Kushida S, Ami Y, Uchida M, Uchida K, et al. (1991) Infection of rats with HTLV-1: A small-animal model for HTLV-1 carriers. Int J Cancer 49:764–769.

Sun B, Fang J, Yagami K, Kushida S, Tanaka M, Uchida K, Miwa M (1999) Age-dependent paraparesis in WKA rats: Evaluation of MHC k-haplotype and HTLV-1 infection. J Neurol Sci 167:16–21.

Sundaram R, Lynch MP, Rawale SV, Sun Y, Kazanji M, Kaumaya PT (2004) De novo design of peptide immunogens that mimic the coiled coil region of human T-cell leukemia virus type-1 glycoprotein 21 transmembrane subunit for induction of native protein reactive neutralizing antibodies. J Biol Chem 279:24141–24151.

Tendler CL, Greenberg SJ, Burton JD, Danielpour D, Kim SJ, Blattner WA, Manns A, Waldmann TA (1991) Cytokine induction in HTLV-I associated myelopathy and adult T-cell leukemia: Alternate molecular mechanisms underlying retroviral pathogenesis. J Cell Biochem 46:302–311.

Tomaru U, Ikeda H, Jiang X, Ohya O, Yoshiki T (2003) Provirus expansion and deregulation of apoptosis-related genes in the spinal cord of a rat model for human T-lymphocyte virus type I-associated myeloneuropathy. J Neurovirol 9:530–538.

Trejo SR, Ratner L (2000) The HTLV receptor is a widely expressed protein. Virology 268:41–48.

Umehara F, Izumo S, Nakagawa M, Ronquillo AT, Takahashi K, Matsumuro K, Sato E, Osame M (1993) Immunocytochemical analysis of the cellular infiltrate in the spinal cord lesions in HTLV-I-associated myelopathy. J Neuropathol Exp Neurol 52:424–430.

Umehara F, Izumo S, Ronquillo AT, Matsumuro K, Sato E, Osame M (1994a) Cytokine expression in the spinal cord lesions in HTLV-I-associated myelopathy. J Neuropathol Exp Neurol 53:72–77.

Umehara F, Nakamura A, Izumo S, Kubota R, Ijichi S, Kashio N, Hashimoto K, Usuku K, Sato E, Osame M (1994b) Apoptosis of T lymphocytes in the spinal cord lesions in HTLV-I-associated myelopathy: A possible mechanism to control viral infection in the central nervous system. J Neuropathol Exp Neurol 53:617–624.

Usuku K, Sonoda S, Osame M, Yashiki S, Takahashi K, Matsumoto M, Sawada T, Tsuji K, Tara M, Igata A (1988) HLA haplotype-linked high immune responsiveness against HTLV-I in HTLV-I-associated myelopathy: Comparison with adult T-cell leukemia/lymphoma. Ann Neurol 23(Suppl):S143–150.

Utz U, Koenig S, Coligan JE, Biddison WE (1992) Presentation of three different viral peptides, HTLV-1 Tax, HCMV gB, and influenza virus M1, is determined by common structural features of the HLA-A2.1 molecule. J Immunol 149:214–221.

Vine AM, Witkover AD, Lloyd AL, Jeffery KJ, Siddiqui A, Marshall SE, Bunce M, Eiraku N, Izumo S, Usuku K, Osame M, Bangham CR (2002) Polygenic control of human T lymphotropic virus type I (HTLV-I) provirus load and the risk of HTLV-I-associated myelopathy/tropical spastic paraparesis. J Infect Dis 186:932–939.

Willems L, Burny A, Collete D, Dangoisse O, Dequiedt F, Gatot JS, Kerkhofs P, Lefebvre L, Merezak C, Peremans T, Portetelle D, Twizere JC, Kettmann R (2000) Genetic determinants of bovine leukemia virus pathogenesis. AIDS Res Hum Retroviruses 16:1787–1795.

Wolfe ND, Heneine W, Carr JK, Garcia AD, Shanmugam V, Tamoufe U, Torimiro JN, Prosser AT, Lebreton M, Mpoudi-Ngole E, McCutchan FE, Birx DL, Folks TM, Burke DS, Switzer WM (2005) Emergence of unique primate T-lymphotropic viruses among central African bushmeat hunters. Proc Natl Acad Sci USA 102:7994–7999.

Wu E, Dickson DW, Jacobson S, Raine CS (1993) Neuroaxonal dystrophy in HTLV-1-associated myelopathy/tropical spastic paraparesis: Neuropathologic and neuroimmunologic correlations. Acta Neuropathol (Berl) 86:224–235.

Wu XM, Osoegawa M, Yamasaki K, Kawano Y, Ochi H, Horiuchi I, Minohara M, Ohyagi Y, Yamada T, Kira JI (2000) Flow cytometric differentiation of Asian and Western types of multiple sclerosis, HTLV-1-associated myelopathy/tropical spastic paraparesis (HAM/TSP) and hyperIgEaemic myelitis by analyses of memory CD4 positive T cell subsets and NK cell subsets. J Neurol Sci 177:24–31.

Yamada S, Ikeda H, Yamazaki H, Shikishima H, Kikuchi K, Wakisaka A, Kasai N, Shimotohno K, Yoshiki T (1995) Cytokine-producing mammary carcinomas in transgenic rats carrying the pX gene of human T-lymphotropic virus type I. Cancer Res 55:2524–2527.

Yamano Y, Nagai M, Brennan M, Mora CA, Soldan SS, Tomaru U, Takenouchi N, Izumo S, Osame M, Jacobson S (2002) Correlation of human T-cell lymphotropic virus type 1 (HTLV-1) mRNA with proviral DNA load, virus-specific CD8(+) T cells, and disease severity in HTLV-1-associated myelopathy (HAM/TSP). Blood 99:88–94.

Yamano Y, Cohen CJ, Takenouchi N, Yao K, Tomaru U, Li HC, Reiter Y, Jacobson S (2004) Increased expression of human T lymphocyte virus type I (HTLV-I) Tax11–19 peptide-human histocompatibility leukocyte antigen A*201 complexes on CD4+ CD25+ T Cells detected by peptide-specific, major histocompatibility complex-restricted antibodies in patients with HTLV-I-associated neurologic disease. J Exp Med 199: 1367–1377.

Yamano Y, Takenouchi N, Li HC, Tomaru U, Yao K, Grant CW, Maric DA, Jacobson S (2005) Virus-induced dysfunction of CD4+ CD25+ T cells in patients with HTLV-I-associated neuroimmunological disease. J Clin Invest 115:1361–1368.

Yamazaki H, Ikeda H, Ishizu A, Nakamaru Y, Sugaya T, Kikuchi K, Yamada S, Wakisaka A, Kasai N, Koike T, Hatanaka M, Yoshiki T (1997) A wide spectrum of collagen vascular and autoimmune diseases in transgenic rats carrying the env-pX gene of human T lymphocyte virus type I. Int Immunol 9:339–346.

Yoshiki T (1995) Chronic progressive myeloneuropathy in WKAH rats induced by HTLV-I infection as an animal model for HAM/TSP in humans. Intervirology 38:229–237.

Yoshiki T, Ikeda H, Tomaru U, Ohya O, Kasai T, Yamashita I, Morita K, Yamazaki H, Ishizu A, Nakamaru Y, Kikuchi K, Tanaka S, Wakisaka A (1997) Models of HTLV-I-induced diseases. Infectious transmission of HTLV-I in inbred rats and HTVL-I env-pX transgenic rats. Leukemia 11(Suppl 3):245–246.

Yoshioka A, Hirose G, Ueda Y, Nishimura Y, Sakai K (1993) Neuropathological studies of the spinal cord in early stage HTLV-I-associated myelopathy (HAM). J Neurol Neurosurg Psychiatry 56:1004–1007.

Yu F, Itoyama Y, Fujihara K, Goto I (1991) Natural killer (NK) cells in HTLV-I-associated myelopathy/tropical spastic paraparesis-decrease in NK cell subset populations and activity in HTLV-I seropositive individuals. J Neuroimmunol 33:121–128.

Zhao TM, Bryant MA, Kindt TJ, Simpson RM (2002) Monoclonally integrated HTLV type 1 in epithelial cancers from rabbits infected with an HTLV type 1 molecular clone. AIDS Res Hum Retroviruses 18:253–258.

24
Viral Encephalitis

Clinton Jones and Eric M. Scholar

Keywords Encephalitis; Herpes simplex virus type 1; Varicella zoster virus; Cytomegalovirus; West Nile virus; Acyclovir; Valacyclovir; Ganciclovir; Cidofovir; Famciclovir; Foscarnet; Latencyneurological disorders; Lifecycle

24.1. Introduction

Encephalitis is an acute inflammation of the brain that is caused by a viral infection. Several viruses, herpes simplex virus type 1 (HSV-1), Varicella zoster virus (VZV), cytomegalovirus (CMV), West Nile virus, and certain members of the Flavivirus family or Bunyavirus genus for example, can cause encephalitis. Although these viruses have very different biological properties, they have the ability to enter the central nervous system, replicated in neurons, and can all cause encephalitis. In general, patients with encephalitis suffer from fever, headache, seizures, and photophobia. Less commonly, stiffness of the neck can occur with rare cases of patients also suffering from stiffness of the limbs, slowness in movement and clumsiness depending on which part of the brain is involved. Encephalitis is not a typical outcome of viral infections, but it is life threatening and difficult to treat. This chapter discusses the properties of the viruses that cause encephalitis, how they cause clinical disease, and the available therapeutic strategies that are available.

24.2. Herpes Simplex Virus-Mediated Encephalitis (HSE)

24.2.1. Summary of Herpes Simplex Virus Type 1 Productive Infection

Herpes simplex virus type 1 (HSV-1) is a double-stranded DNA virus that has a genome size of 152 Kb, and it encodes at least 84 proteins. Binding and entry of HSV-1 into permissive cells is mediated by viral proteins binding to cellular receptors (Spear, 1993). After uncoating, the viral genome is present in the nucleus and viral gene expression ensues. HSV gene expression is temporally regulated in three distinct phases: immediate early (IE), early (E), or late (L) (Honess and Roizman, 1974). IE gene expression does not require protein synthesis and is stimulated by VP16 (O'Hare, 1993). In general, proteins encoded by IE genes regulate viral gene expression, and as such are important for productive infection. E gene expression is dependent on at least one IE protein, and generally E genes encode nonstructural proteins that play a role in viral DNA synthesis. L gene expression is maximal after viral DNA replication, requires IE protein production, and L proteins comprise the virion particle.

24.2.2. Summary of Latent Infection

Acute infection is typically initiated in the mucosal epithelium, and then HSV-1 establishes latency in sensory neurons located in trigeminal ganglia (TG) or sacral dorsal root ganglia. Despite a vigorous immune response during acute infection, latency is established. As many as 20–30% of sensory neurons are latently infected (reviewed in Jones, 1998, 2003). As a consequence of primary infection, HSV-1 genomic DNA is also present in the central nervous system (CNS) of a significant proportion of the adult human population (Fraser et al., 1981).

The latency associated transcript (LAT) is abundantly transcribed in latently infected neurons (reviewed in Jones, 1998, 2003). Mice, rabbits, or humans latently infected with HSV-1 express LAT. In productively infected cells or latently infected rabbits, an 8.3-Kb transcript is expressed that has the same sense as LAT. Splicing of the 8.3-Kb transcript yields an abundant 2-Kb LAT and an unstable 6.3-Kb LAT. The majority of LAT is not capped, is poly A-, appears to be circular, and is designated as a stable intron. In small animal models, LAT is important but not required for the latency-reactivation cycle (reviewed in Jones, 1998, 2003). The first 1.5 Kb of LAT coding sequences are important for reactivation from latency. It is not clear whether LAT encodes a protein or is a regulatory RNA.

T. Ikezu and H.E. Gendelman (eds.), *Neuroimmune Pharmacology*.
© Springer 2008

LAT interferes with apoptosis in transiently transfected cells and in TG of infected rabbits or mice (Jones, 2003). Inhibiting apoptosis may be the most important function of LAT because two anti-apoptosis genes, the bovine herpesvirus 1 (BHV-1) LAT homologue (Mott et al., 2003; Perng et al., 2002) and the baculovirus IAP gene (Jin et al., 2005), can restore levels of spontaneous reactivation to a LAT null mutant.

24.2.3. Epidemiology of Herpes Simplex Viruses

Approximately 90% of the population is infected with herpes simplex virus type 1 (HSV-1), and at least 10% with HSV-2 (Nahmias and Roizman, 1973; Whitley, 1997). Humans are the only natural reservoir for this infection, and no vectors are involved in transmission of this virus (Stanberry et al., 2004). HSV-1 primary infection occurs mainly in childhood, and HSV-2 infection occurs predominantly in sexually active adolescents and young adults. Aerosols or close contact are the primary mechanisms of viral transmission. Although transmitted by different routes and involving different parts of the body, these two viral subtypes have similar epidemiology and clinical manifestation.

24.2.4. Pathogenesis of HSE

HSV is the most commonly identified cause of acute, sporadic viral encephalitis in the United States, accounting for 10–20% of all cases (Corey, 2005). It is estimated that there are about 2.3 cases per million individuals per year. There are peaks at 5–30 years of age and at more than 50 years of age. Since the 1940s, HSV-1 and HSV-2 have been implicated in the causation of acute necrotizing encephalitis in infants, children, and adults. Encephalitis due to HSV-2 in newborn infants is a widespread disease in the brain and commonly involves a variety of other organs in the body, including the skin, eyes, and lungs (Stanberry et al., 2004).

HSE is characterized by severe destruction of temporal and frontal lobe structures, including limbic mesocortices, amygdala, and hippocampus. Without antiviral therapy, the mortality rate is as high as 70%; but even after antiviral therapy 20% of these patients die. Despite early treatment, chronic progressive tissue damage in magnetic resonance imaging (MRI) can be found up to 6 months following the onset of symptoms. Approximately 2/3 of the HSE cases occur because of reactivation from latency (Yamada et al., 2002), which explains why there is high morbidity and long-term complications despite antiviral treatment (Lahat et al. 1999; McGrath et al., 1997; Skoldenberg, 1991).

In general, HSE is associated with necrotic cell death resulting from virus replication and inflammatory changes secondary to virus-induced immune response (DeBiasi et al., 2002). However, there is not a perfect correlation between virus burden in the brain and the severity of histological changes and neurological symptoms. Furthermore, a small number of HSE patients are negative for HSV-1 DNA early in the course of infection, suggesting that factors other than virus replication are involved in cell death/disease pathogenesis.

24.2.5. Animal Models for Studying HSE

HSE occurs in a certain percentage of mice or rabbits following infection. The frequency of HSE in experimental infections is dependent on the pathogenic potential of the HSV-1 and the mouse strain used for experimental infection. For HSE to occur after ocular infection, the virus must enter the TG, and then spread to the CNS, or the virus directly gains access into the brain via the optic nerve. Models have also been developed in which the virus is directly inoculated into the brain. In this model, transport from the peripheral tissue $\rightarrow$ the peripheral nervous system $\rightarrow$ the CNS is not important. Thus, viral genes necessary for neuronal transport and spread are not crucial for virus infection if the brain is inoculated.

Viral genes necessary for productive infection, inhibiting apoptosis, or inhibiting immune recognition would likely play a significant role in the potential of HSV-1 to initiate encephalitis. Innate immune responses play a significant role in lethal encephalitis because HSV-1 interactions with Toll-like receptor 2 contribute to HSE (Kurt-Jones et al., 2004). A recent study has demonstrated that LAT enhances the frequency of encephalitis in male Balb/C mice (Jones et al., 2005). Thus, viral- and host-specific factors regulate the frequency of HSE.

24.2.6. Clinical Features of HSE

Two recognizable groups of symptoms are seen in most patients. First of all there are nonspecific symptoms that are seen in most forms of encephalitis, including fever, headache, and signs of meningeal irritation such as nausea, vomiting, confusion, generalized seizures, and alteration of consciousness. The second group of changes is referable to focal necrosis of the orbitofrontal and temporal cortexes and the limbic system, and includes anosmia, memory loss, peculiar behavior, speech defects, hallucinations (particularly olfactory and gustatory), and focal seizures. There is rapid progression in some cases, with the appearance of reflex asymmetry, focal paralysis, hemiparesis and coma. Cerebral edema contributes to these symptoms and plays an important role in the outcome (Stanberry et al., 2004).

24.2.7. Therapy

A number of nucleoside analogs are active against primary and recurrent HSV infection. Acyclovir is the major therapy used for HSV encephalitis. It should be given intravenously at a dose of 10 mg/kg every 8 h for 14–21 days. It is usually well tolerated with few side effects. Valacyclovir and famciclovir are alternative drugs used for the treatment of HSV encephalitis (Griffiths, 1995; Stanberry et al., 2002, 2004). For a description of these drugs see §5.

24.3. Varicella Zoster Virus (VZV)-Induced Encephalitis

24.3.1. Summary of Virus Lifecycle and Virus Transmission

Humans are the only known reservoir for VZV, and VZV is a ubiquitous human pathogen. Lke HSV-1, VZV is a large double stranded DNA virus that encodes at least 70 different genes. VZV contains many genes that are similar to HSV-1, suggesting they are functional homologues. Thus, the general steps during productive infection are similar for VZV and HSV-1 and will not be discussed in detail here.

24.3.2. Latency of VZV

VZV appears to be latent only in ganglia, and viral genomes are primarily found in sensory neurons. Analysis of latent VZV is restricted to human ganglia obtained at autopsy. Based on in situ hybridization (ISH) studies combined with sequencing, four transcripts corresponding to VZV genes 21, 29, 62, and 63 have been identified in latently infected human ganglia. A monospecific polyclonal antiserum directed against VZV ORF 63 protein detected this protein in the cytoplasm of neurons (reviewed in Mitchell et al., 2003). These VZV proteins are primarily in the nucleus during productive infection (zoster), suggesting that the cytoplasmic location might maintain VZV in a latent state.

24.3.3. Epidemiology of VZV

A serologic study of 1201 U.S. military trainees indicated that 95.8% had been exposed to virus (Jerant et al., 1998). Chickenpox is common in childhood and affects both genders equally as well as people of all races. VZV is spread by droplet or airborne transmission and is highly contagious. The primary infection is seen as varicella (chickenpox), a contagious and usually benign illness that occurs in epidemics among susceptible children. The VZV virus is highly contagious and can be contacted by aerosol, human contact, or coughing. Primary infection produces varicella (chickenpox), after which VZV becomes latent in neurons of cranial nerve, dorsal root, and autonomic ganglia along the entire neuro-axis. Reactivation from latency can occur decades later, resulting in zoster (shingles), which tends to be more localized than the chickenpox infection.

24.3.4. Clinical Features of VZV

Acute VZV infection leads to chickenpox, which is characterized by malaise, fever, and an extensive vesicular rash (Abendorth and Arvin, 2000). Chickenpox in the immunocompetent child is mostly benign and associated with lassitude and a temperature of 100–103°F for 1–2 days. Other symptoms include malaise, itching, anorexia, weakness, and exhaustion, and these gradually resolve as the illness improves. The hallmark of chickenpox is the skin manifestations that consist of maculopapules, vesicles, and scabs in varying stages. In general immunocompromised children have more numerous lesions. Although chickenpox is usually a mild disease, there are exceptions to this rule. For example, an epidemic of 292 cases of chickenpox occurred in rural India resulted in 3 deaths (Balraj and John, 1994). In the United Kingdom about 25 people die from chickenpox every year, in part because VZV vaccination is not mandatory (Rawson et al., 2001). VZV vaccine effectively prevents varicella; however, breakthrough varicella and virus reactivation can still occur (reviewed by Arvin and Gershon, 1996).

With chickenpox the most frequent organ affected other than the skin is the CNS (Liesegang, 1999). The neurologic abnormalities are often seen as acute cerebellar ataxia or encephalitis. Encephalitis is the most serious complication of chickenpox, and it can be life threatening in adults. It occurs in 0.1–0.2% of patients with chickenpox (Johnson and Milbourn, 1970).

Herpes zoster is characterized by a unilateral vesicular eruption with a dermatome distribution. Postherpetic neuralgia (pain that persists more than 30 days after the onset of rash after cutaneous healing) is the most serious complication in immunocompetent patients. The incidence and duration of postherpetic neuralgia are directly correlated with the age of the patient. Acute retinal necrosis caused by varicella-zoster virus occurs occasionally in immunocompetent patients although more recent reports have focused on ocular disease in HIV-infected patients. In HIV-infected patients the lesions rapidly increase in size and coalesce. These lesions respond poorly to antiviral therapy and almost inevitably cause blindness in the involved eye. The retinitis is less aggressive in immunocompetent patients and can often be treated with antiviral therapy such as acyclovir. Neurologic complications associated with zoster are diverse, including motor neuropathies of the cranial and peripheral nervous system, encephalitis, meningoencephalitis, myelitis, and Guillain-Barré syndrome (Liesegang, 1999). Extracutaneous sites of involvement include the CNS, as shown by meningoencephalitis or encephalitis. The clinical symptoms are similar to those of other viral infections of the brain. This involvement of the CNS with cutaneous herpes zoster is probably more common than recognized clinically. Classically, VZV infection involves dorsal root ganglia. Motor paralysis can occur as a consequence of the involvement of the anterior horn cells, in a manner similar to that encountered with polio. These patients may have severe pain. Herpes zoster in the immunocompromised patient is more severe than in the normal person, but even in these patients disseminated herpes zoster is rarely fatal.

24.3.5. Neurological Disorders Associated with VZV

In general, neurological diseases occur as a result of reactivation from latency. The neurological complications after VZV reactivation are serious and can be life-threatening. VZV is the causal agent in 29% of 3231 cases of encephalitis, meningitis, and myelitis, and VZV is the most common cause of encephalitis in patients over age 65 (Rantalaiho et al., 2001). More than 500,000 Americans develop zoster (severe dermatomal distribution pain and rash) every year. Zoster is frequently followed by postherpetic neuralgia (pain that persists for months and often years after the rash disappears), myelitis, or unifocal or multifocal vasculopathy. Many cases of VZV vasculopathy, myelitis, and polyneuropathy occur in the absence of rash. Since VZV causes a wide spectrum of neurological disorders, testing for VZV DNA and antibody in cerebrospinal fluid (CSF) should be routinely performed. Proper diagnosis is critical because antiviral treatment can be curative, even after weeks to months of chronic VZV infection.

Subsequent reactivation of latent VZV in dorsal root ganglia results in a localized cutaneous eruption termed "herpes zoster" (or shingles). The epidemiology of herpes zoster differs from that of chickenpox. Decreasing virus-specific cell-mediated immune responses, which occur naturally as a result of aging or that result from immunosuppressive illness or medical treatments, increases the risk of shingles (Gnann and Whitley, 2002). Herpes zoster occurs in persons who are seropositive for VZV or more specifically in those who have had chickenpox. Most patients who develop herpes zoster have no history of exposure to other persons with VZV infection at the time of the appearance of lesions. Over 90% of adults in the U.S. have serologic evidence of varicella-zoster infection and are at risk for herpes zoster. The annual incidence is about 1.5–3.0 cases per 1000. Zoster occurs during the lifetime of 10–20% of all persons (Liesegang, 1999). Increasing age is a key risk factor for the development of herpes zoster. The other major risk for herpes zoster is altered cell-mediated immunity. Patients with cancer, those receiving immunosuppressive agents, and organ-transplant recipients are thus at increased risk of shingles. Persons who are positive for human immunodeficiency virus also develop herpes zoster at a higher frequency than those who are negative (Liesegang, 1999).

24.3.6. Potential Models to Study VZV Neuropathogenesis

Unlike HSV-1, VZV does not reactivate from ganglia after experimental infection of primates or rodents. However, after footpad inoculation of rats with VZV, the protein encoded by gene 63 can be detected in lumbar ganglia 1 month after infection. Viral protein is also detected in neurons, both in the nuclei and cytoplasm of infected cells. An independent study using the same rat model detected VZV gene 63 DNA in 5–10% of neurons and VZV RNA in neurons and non-neuronal cells (Kennedy et al., 1999). Simian varicella virus may also be a valuable model to study the pathogenesis of varicella virus–host interactions. Finally, the application of humanized immunodeficient mice (SCID-hu) to VZV has provided new information about viral genes that are necessary for viral growth and pathogenesis (Ku et al., 2005).

24.3.7. Therapy

Therapy for herpes zoster should be aimed at accelerating healing, limiting the severity and duration of acute and chronic pain and reducing any complications associated with the infection. In patients who are immunocompromised, therapy should also be aimed at reducing the risk of viral dissemination. Acyclovir, valacyclovir, and famciclovir are all used in the United States for the treatment of herpes zoster. Acyclovir is approved in the U.S. for the treatment of both chickenpox and herpes zoster in the normal host. Oral acyclovir therapy in normal children, adolescents, and adults shortens the duration of lesion formation by about a day, reduces the total number of new lesions by about 25%, and reduces many of the symptoms in about a third of patients (Gnann and Whitley, 2002; Snoeck et al., 1999).

24.4. Human Cytomegalovirus (CMV)-Induced Encephalitis

24.4.1. Summary of Virus Lifecycle

Human cytomegalovirus (HCMV) is a β-herpesvirus that is an important causative agent of congenital disease and is a significant opportunistic infection (Knipe and Howley, 2001). HCMV is the largest member of the human herpesvirus group. It is estimated to encode at least 165 genes. In contrast to HSV-1, HCMV exhibits a highly restricted host range in cell culture. Primary differentiated fibroblasts show the greatest susceptibility to infection. As with other herpes–viruses, the HCMV-immediate early genes regulate viral transcription and inhibit cellular responses to infection (apoptosis and interferon induction, for example).

24.4.2. Epidemiology of HCMV

CMV is widely distributed among humans—from developed, industrial societies to isolated aboriginal groups (Pass, 1995). As with the other herpesviruses, CMV is readily contracted. The prevalence of CMV increases with age in every group that has been studied. In general, prevalence is greater and acquired earlier in life in developing countries and in the lower socioeconomic sections of developed countries. Infection rates are higher in nonwhites, but these racial differences appear to reflect differences associated with socioeconomic status and living conditions (Pass, 2001).

24.4.3. Pathogenesis and Persistence of HCMV

In healthy individuals, primary infection is usually subclinical, leading to persistence of the virus in a latent state throughout the lifetime of the host. However, primary infection of newborns and reactivation from latent infections in immunocompromised individuals can lead to systemic and chronic disease. Unlike HSV-1 and VZV, HCMV does not typically establish latency in the nervous system. Myeloid-lineage hematopoietic cells (e.g., granulocytes, macrophages, and dendritic cells) are important targets for lifelong latency. HCMV latent gene expression is localized to the immediate early 1 (IE1) and IE2 regions. These transcripts appear to originate from both strands, but it is not clear if IE1 and IE2 proteins are expressed. Since these proteins are known to promote productive infection, high levels of expression do not appear to be consistent with latency. Several open reading frames also exist on the antisense strand of the latency associated transcripts. Antibodies to these ORFs are recognized by serum from healthy seropositive individuals, suggesting they are expressed during natural infection (Kondo et al., 1996).

With respect to HCMV causing disease in immune-suppressed individuals, reactivation from latency appears to be more important than primary infection. It is believed that cytokine-induced differentiation of latently infected cells can lead to a cell type that supports productive infection (Hahn et al., 1998), and viral amplification results from reduced immune surveillance (Fiala et al., 1973). Like many perisistent/latent viral infections, the early events associated with reactivation from latency are poorly understood.

24.4.4. Clinical Features of HCMV

Although HCMV-induced clinical conditionis are usually asymptomatic in patients with intact immune systems, it is a common opportunistic pathogen in immunocompromised patients. In HIV-infected persons CMV disease appears to be due to reactivation of the latent virus in a previously infected host. The clinical manifestations of CMV disease are generally seen when the CD4 T-lymphocyte count falls below 100 cells/ml. Retinitis is perhaps the most well known disease associated with this infection. Other features associated with this infection include gastrointestinal disease, pneumonitis (in the bone marrow transplantation patients), and neurologic disease. HCMV is associated with various neurologic infections in HIV-positive people, particularly inflammation of the ventricle accompanied by encephalitis (ventriculoencephalitis) and ascending polyradiculopathy (disease or injury involving multiple nerve roots). Ventriculoencephalitis usually occurs in advanced HIV infection in persons with a prior CMV diagnosis. Polyradiculopathy is the most common CNS infection caused by CMV and is characterized by urinary retention and progressive bilateral leg weakness. The clinical symptoms generally progress over several weeks to include loss of bowel and bladder control and paralysis. A spastic myelopathy has been reported and sacral paresthesias may occur. The CSF often shows a greater than normal number of cells, a less than normal content of glucose, and elevated protein levels (Cheeseman, 2004; Cheung and Teich, 1999).

24.4.5. HCMV-Induced Encephalitis

Approximately, two out of three newborn children that have symptomatic congenital CMV infection have CNS involvement (Bopanna et al., 1992). The CNS pathology includes microcephaly, elevation of cerobrospinal fluid protein, and neurologic abnormalities (poor feeding, lethargy, or generalized hypotonia, for example). Cranial computed tomography (CT) scans are abnormal in 75% of symptomatic newborns, with the most common abnormality being periventricular calcification (Bopanna et al., 1997). Fortunately, only 5–10% of newborns with congenital HCMV infection are symptomatic.

24.4.6. Treatment

Although antiviral agents currently available for CMV are not ideal, their use has decreased the burden of this disease in transplant patients and has improved the quality of life and survival of patients infected with this virus. Three antiviral agents have been shown to be effective in CMV disease and have been approved for use in the U.S. They are ganciclovir, foscarnet and cidofovir (Cheeseman, 2004; Cheung and Teich, 1999). Antiviral therapy for esophagitis, enterocolitis, endephalitis, peripheral neuropathy, polyradiculoneuropathy, and pneumonitis is usually with intravenous ganciclovir or foscarnet regimens similar to those used for retinitis. Patients with CNS or neurologic disease often respond best to combination therapy (Arribas et al., 1996).

24.5. Therapeutic Agents Available to Treat Herpes Viruses

24.5.1. Acyclovir

Acyclovir (Figure 24.1) is an acyclic guanine nucleoside analog that lacks a 3′-hydroxyl on the side chain. It has good antiviral activity against all herpesviruses and is widely used in the treatment of several different herpesvirus infections. Acyclovir was the first drug clearly demonstrated to be effective against herpes simplex virus infections (Wagstaff et al., 1994).

Acyclovir selectively inhibits viral DNA synthesis. It is preferentially activated in virally infected cells. Cellular uptake and initial phosphorylation are facilitated by the herpes virus thymidine kinase. The affinity of acyclovir for HSV

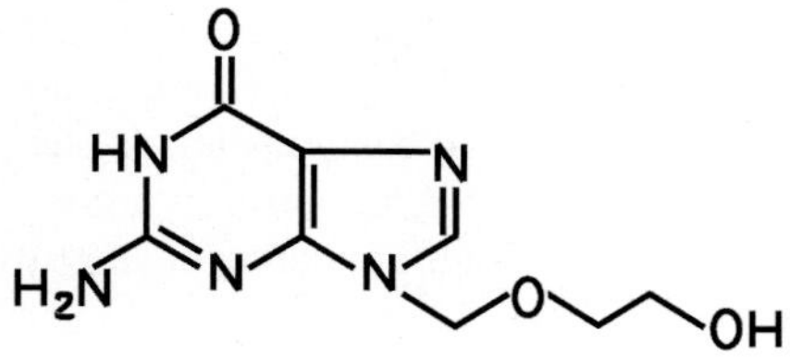

O
HN
N
H_2N
N
N
O
OH

FIGURE 24.1. Acyclovir.

thymidine kinase is about 200-fold greater than for the mammalian enzyme. Cellular enzymes then subsequently convert the monophosphate to acyclovir triphosphate. Acyclovir triphosphate is present in 40- to 100-fold higher concentrations in HSV-infected than in uninfected cells, and competes for endogenous deoxyguanosine triphosphate. The triphosphate competitively inhibits viral DNA polymerases to a much greater extent than cellular DNA polymerases. The triphosphate is also incorporated into viral DNA, where it acts as a chain terminator as a result of the lack of the 3′-hydroxyl group. The terminated DNA template containing acyclovir binds the enzyme and leads to irreversible inactivation of the DNA polymerase (Scholar and Pratt, 2000; Wagstaff and Bryson, 1994).

The oral bioavailability of acyclovir is poor and ranges from 10–30%. The drug is poorly protein bound but is widely distributed throughout body fluids and tissues, including the cerebrovascular fluid. It is primarily excreted unchanged in the urine (Wagstaff et al., 1994).

Acyclovir is useful for treating infections caused by HSV, herpes zoster, and for VZV infections (Whitley and Roizman, 2001). Although HCMV is relatively resistant to acyclovir, some cytomegalovirus infections have responded marginally to large doses of acyclovir, and it seems to be effective for the prophylaxis of cytomegalovirus infections in immunocompromised patients. Epstein-Barr virus is not sensitive to acyclovir, and clinical infections do not respond to the drug.

Oral acyclovir is recommended for the treatment of varicella infection (chickenpox) in patients over 13 years of age who are otherwise healthy, children over 12 months of age with a chronic cutaneous or pulmonary condition or receiving long-term salicylate therapy, and in children receiving short or intermittent courses of aerosolized corticosteroids. Intravenous acyclovir should be used for treatment of varicella infection in immunocompromised children, including those receiving high doses of corticosteroids. Acyclovir has been shown to be the most effective agent for the treatment of infections caused by VZV (Snoeck et al., 1999).

Parenteral acyclovir is the drug of choice for the treatment of initial and recurrent mucosal or cutaneous herpes simplex infections in immunocompromised patients and for the treatment of disseminated, neonatal, encephalitic, and severe first episodes of genital herpes simplex infections in immunocompetent patients (Whitley, 1997). Intravenous acyclovir should also be used for severe diseases such as encephalitis (Brady and Bernstein, 2004).

Acyclovir is generally well tolerated whether administered topically, orally, or by the intravenous route. GI disturbances, headache, and rash may occur. Renal dysfunction due to crystalline nephropathy is more likely with IV administration, rapid infusion, in patients in the dehydrated state and with underlying renal disease, and in large doses. CNS effects are rare but include encephalopathy, tremors, hallucinations, seizures, and coma. Due to an elevated pH, intravenous administration may also cause phlebitis and inflammation at sites of extravasation (Wagstaff et al., 1994).

24.5.2. Valacyclovir

Valacyclovir is an ester prodrug of acyclovir. It provides significantly better oral bioavailability compared to acyclovir. This advantage results in substantially higher serum acyclovir concentrations than is possible with oral acyclovir. In addition, fewer daily doses are required with valacyclovir (Curran and Noble, 2001).

After oral administration and absorption, valacyclovir is converted to acyclovir, which is the active antiviral component of valacyclovir. The antiviral activity and mechanism of action of valacyclovir is identical to that of acyclovir (Perry and Faulds, 1996). The oral bioavailability of valacyclovir is significantly greater than that of acyclovir. Oral administration of valacyclovir results in plasma acyclovir concentrations comparable to those observed with intravenous acyclovir.

Valacyclovir is indicated for the treatment of herpes zoster (shingles), for the treatment of initial and recurrent episodes of genital herpes, and for the suppression of recurrent genital herpes in immunocompetent and HIV-infected patients. It is also indicated for the reduction of transmission of genital herpes in immunocompetent individuals, and for the treatment of cold sores. Valacyclovir appears to be equally effective in treating herpes zoster and recurrent genital herpes in immunocompetent adults. Valacyclovir has shown efficacy in the prophylaxis of cytomegalovirus infections in transplant patients (Perry and Faulds, 1996).

Like acyclovir, valacyclovir is a well-tolerated drug. The most common adverse effects of valacyclovir are headache and nausea. Other adverse events associated with valacyclovir administration include vomiting, weakness, dizziness, and abdominal pain. Thrombotic thrombocytopenic purpura/hemolytic uremic syndrome has also been reported in a few patients after high doses of valacyclovir, as has confusion, hallucinations, and nephrotoxicity (Curran and Noble, 2001; Perry and Faulds, 1996).

24.5.3. Famciclovir

Famciclovir (Figure 24.2) is an oral prodrug of the antiviral agent penciclovir. Famciclovir lacks intrinsic antiviral activity, and this drug owes its activity to formation of penciclovir. Penciclovir is an acyclic guanine nucleoside analog. Penciclovir and its prodrug famciclovir are chemically similar to acyclovir. Penciclovir differs structurally from acyclovir only by the presence of an additional hydroxyl group. Famciclovir is the diacetyl 6-deoxy

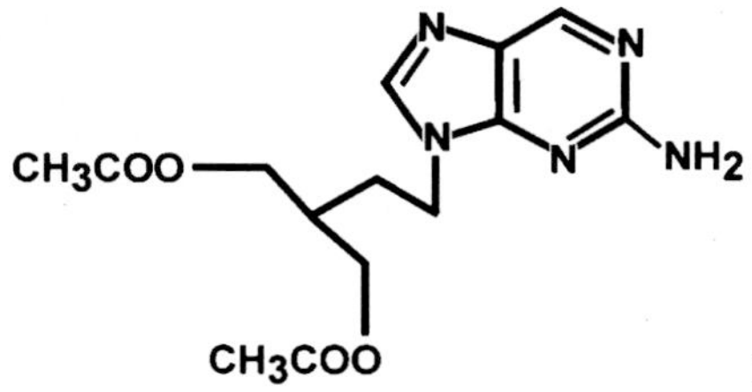

CH_3COO NH_2 CH_3COO

FIGURE 24.2. Famciclovir.

analog of penciclovir. The mechanism of antiviral activity of penciclovir is also similar to that of acyclovir. Both drugs inhibit viral DNA synthesis (Scholar and Pratt, 2000). Penciclovir is rapidly and selectively phosphorylated in virus-infected cells by viral thymidine kinase to the monophosphate, and this is followed by further phosphorylation to the triphosphate, which is the active form of the drug. Over 90% of penciclovir triphosphate in virus cells is the (S)-enantiomer, a competitive inhibitor of DNA polymerases with respect to the natural substrate deoxyguanosine triphosphate (dGTP). Inhibition of the polymerase results in prevention of viral replication by inhibition of viral DNA synthesis. The R enantiomer of penciclovir triphosphate has only minimal activity on viral DNA polymerases (Scholar and Pratt, 2000). In contrast to acyclovir triphosphate, which is an obligate DNA chain terminator, penciclovir triphosphate allows DNA chain extension owing to its free hydroxyl group; however, penciclovir appears at least as effective as acyclovir as an inhibitor of herpes virus DNA synthesis.

Famciclovir is rapidly converted to penciclovir in intestinal and liver tissue after oral administration. More than half of an oral dose of famciclovir is excreted in the urine as unchanged penciclovir. The plasma elimination half-life of penciclovir is about 2 h, similar to that of acyclovir; however, the intracellular half-life of penciclovir in herpes virus-infected cells is considerably longer than that of acyclovir.

Oral famciclovir is used for the treatment of immunocompetent patients with herpes zoster infections and for the treatment and suppression of recurrent genital HSV. It is also indicated for the treatment of recurrent mucocutaneous herpes simplex in both immunocompetent and immunocompromised patients. Famciclovir is fairly well tolerated. Adverse effects include headache, dizziness, nausea, and diarrhea (Scholar and Pratt, 2000).

24.5.4. Ganciclovir

Ganciclovir is an acyclic guanine nucleotide analog with a structure similar to acyclovir but with an additional hydroxymethyl group on the acyclic side chain. It is inhibitory to all herpes viruses but is especially active against cytomegalovirus.

Like acyclovir and penciclovir, ganciclovir inhibits viral DNA synthesis. Also like these other agents, it is first phosphorylated intracellularly by the virus-induced enzyme to the monophosphate form. Further phosphorylation to the di- and triphosphates is catalyzed by cellular enzymes. Intracellular ganciclovir triphosphate concentrations are tenfold higher than those of acyclovir triphosphate and decrease much more slowly with an intracellular half-life of elimination greater than 24 h. These differences probably explain at least in part the greater activity of ganciclovir against cytomegalovirus and provide a rationale for single daily doses in suppressing human cytomegalovirus infections (Markham and Faulds, 1994; Scholar and Pratt, 2000). Ganciclovir triphosphate acts as an inhibitor and substrate for cytomegalovirus DNA polymerase. Ganciclovir triphosphate competitively inhibits the binding of deoxyguanosine triphosphate to DNA polymerase, which results in the inhibition of DNA synthesis and termination of DNA elongation. Ganciclovir is incorporated into both viral and cellular DNA. It appears to limit viral DNA synthesis and the packaging of viral DNA into infectious units.

The bioavailability of orally administered ganciclovir is quite low. In patients with normal renal function the plasma half-life is about 2–4 h. Concentrations of ganciclovir in cerebrospinal fluid are lower than those in serum after intravenous administration (Markham and Faulds, 1994). More than 90% of the drug is eliminated unchanged in the urine. The plasma half-life increases as creatinine clearance declines and may reach 28–40 h in patients with severe renal insufficiency. Ganciclovir is an effective antiviral agent for the treatment of serious life-threatening or sight-threatening cytomegalovirus (CMV) infections in immunocompromised patients. Intravenous ganciclovir, foscarnet, or the combination of both are recommended for the treatment of cytomegalovirus neurological syndromes (Markham and Faulds, 1994).

The dose-limiting and most common adverse effect with intravenous and oral ganciclovir is bone marrow suppression (anemia, leukopenia, neutropenia, and thrombocytopenia). These effects are usually reversible upon withdrawal of the drug. CNS side effects are less common and range in severity from headache to behavioral changes to convulsions and coma. Fever, edema, phlebitis, disorientation, nausea, anorexia, rash, and myalgias have also been reported with ganciclovir therapy (Markham and Faulds, 1994).

24.5.5. Foscarnet

Foscarnet (Figure 24.3) is an inorganic pyrophosphate analog that has antiviral activity against all herpes viruses and the human immunodeficiency virus. Foscarnet is a pyrophosphate analog that acts as a noncompetitive inhibitor of many viral RNA and DNA polymerases as well as HIV reverse transcriptase (Chrisp and Clissold, 1991). It is approximately a 100-fold greater inhibitor of herpes virus DNA polymerase than the cellular DNA polymerase-α; however, some human cell growth suppression has been observed with high concentrations in vitro. Inhibition of DNA polymerase results in inhibition of pyrophosphate exchange, which prevents elongation of the DNA chain (Scholar and Pratt, 2000). Similar to ganciclovir, foscarnet is a virostatic agent. Foscarnet is not a nucleoside and thus not phosphorylated. It reversibly blocks the pyrophosphate-binding site of the viral polymerase in a noncompetitive manner and inhibits cleavage of pyrophosphate from deoxynucleotide triphosphates.

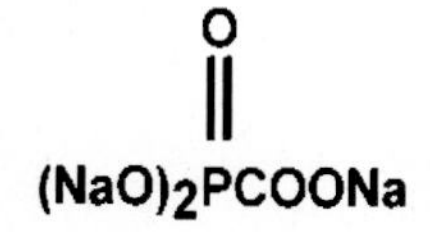

FIGURE 24.3. Foscarnet.

The oral bioavailability of foscarnet is poor, so intravenous therapy is needed to treat viral infections. It is fairly well distributed throughout the body, with CSF levels averaging two-thirds of those in plasma. Foscarnet is taken up slowly by cells, and biotransformation of foscarnet does not occur. The drug is excreted mainly unchanged in the urine (Chrisp and Clissold, 1991).

Foscarnet is used for the treatment of cytomegalovirus (CMV) retinitis and mucocutaneous acyclovir-resistant herpes simplex virus (HSV) infections. It may also be beneficial in other types of CMV or HSV infections (Wagstaff and Bryson, 1994).

In contrast to ganciclovir, foscarnet is not associated with dose-limiting neutropenia, enabling it to be used in combination with zidovudine and other bone marrow suppressant drugs. Nephrotoxicity and hypocalcemia are the major dose-limiting toxicities. Renal toxicity can be minimized with adequate hydration. Changes in serum calcium and phosphate levels may be related to incorporation of the drug into bone. Other adverse effects of the drug include tremor, headache, fatigue, nausea, and vomiting. Some cases of penile and vaginal ulceration have also been reported

24.5.6. Cidofovir

Cidofovir (Figure 24.4) is an antiviral cytidine nucleotide analog with inhibitory activity against HCMV and other herpes viruses. Cidofovir is first converted to an active diphosphate form by cellular enzymes. Antiviral effects of cidofovir are due to inhibition of viral DNA polymerase by the diphosphate metabolite (Neyts and De Clercq, 1994; Plosker and Noble, 1999; Scholar and Pratt, 2000). The diphosphate probably interacts with DNA polymerase either as an alternate substrate (incorporation at the 3′ end or within the interior of the DNA chain) or as a competitive inhibitor (with respect to the normal substrate dCTP). Cidofovir inhibits HCMV DNA synthesis at intracellular concentrations 1000-fold lower than are required to inhibit cellular DNA synthesis (Neyts and De Clercq, 1994). For HSV-1 and HSV-2 corresponding concentrations are at least 50-fold lower.

Cidofovir has poor oral bioavailability (<5%) and is therefore administered intravenously. It has a long intracellular half-life that allows for a prolonged interval between maintenance doses. Cidofovir is excreted extensively by the kidneys and is eliminated almost entirely as unchanged drug in the urine (Plosker and Noble, 1999).

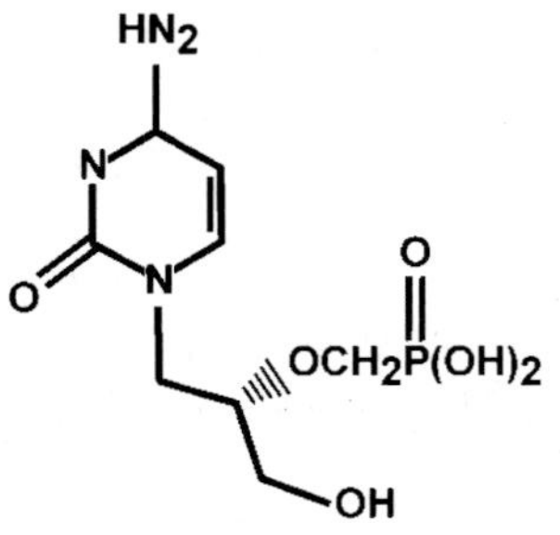

FIGURE 24.4. Cidofovir.

Standard treatment of CMV infections is with ganciclovir or foscarnet. The primary role in therapy of cidofovir is intravenous therapy in AIDS patients with cytomegalovirus retinitis who are unresponsive to or have relapsed on intravenous foscarnet or ganciclovir. It is also used in patients intolerant of these agents. Cidofovir may also be used to treat cytomegalovirus causing gastrointestinal tract infections. Its usefulness in neurological syndromes, pneumonitis, or viremia is unknown (Plosker and Noble, 1999).

Intravenous cidofovir is well tolerated. The major treatment-limiting toxicity of this drug is irreversible nephrotoxicity (Plosker and Noble, 1999). Intravenous pre-hydration with normal saline and administration of oral probenecid must be used with each cidofovir infusion to lessen the effects on the kidney. Serum creatinine and urine protein must be monitored with each infusion and adjusted accordingly. Other adverse effects associated with its use are neutropenia and peripheral neuropathy (Plosker and Noble, 1999).

24.6. West Nile Virus (WNV)-Induced Encephalitis

24.6.1. Summary of Virus Lifecycle and Virus Transmission

West Nile virus (WNV) (Figure 24.5) is a member of the Flaviviridae (Lindenback and Rice, 2001). WNV is a small positive-strand RNA virus that is approximately 11,000 bases long. Genomic RNA serves as a messenger RNA that is translated into a single polyprotein. This polyprotein is then cleaved into at least 10 discrete proteins by cellular proteases and a virally encoded serine protease. Three structural proteins (C, prM, and E) are produced from the polyprotein (Figure 24.5). Seven nonstructural proteins (NS1, NS2A, NS2B, NS3, NS4A, NS4B, and NS5) are generated from the polyprotein. NS5 is the most conserved protein in the Flavirviridae family because this protein is the viral-encoded RNA-dependent RNA polymerase. The other nonstructural proteins are not as well conserved, suggesting that they have virus-specific functions.

WNV has a broad range of antigenic variation and restriction-length polymorphisms, suggesting it is a highly variable virus. For example, strains from Africa, Europe, and the Middle East form a distinct group relative to strains isolated in India and the Far East (Price and O'Leary, 1967). WNV is widely disseminated throughout Africa, Europe, the Middle East, and the Far East. During the summer of 1999, WNV was introduced to the eastern coast of the U.S., and has subsequently moved westward.

WNV can be readily grown in a variety of mammalian cell lines, primary chick embryo fibroblast, primary duck fibroblasts, and Drosophila cells (Lindenback and Rice, 2001). In the wild, WNV is vectored by mosquitoes, and can infect

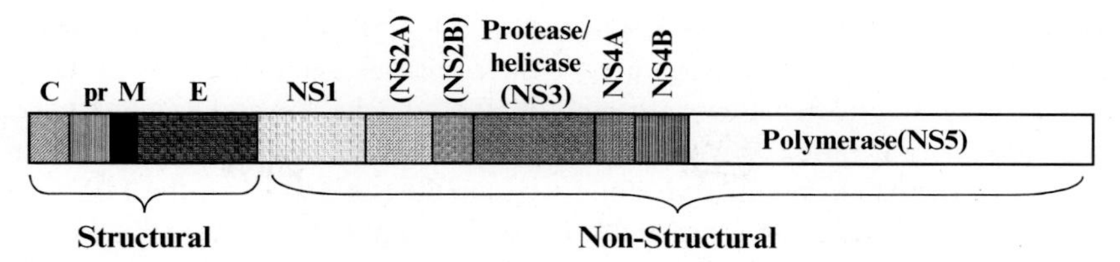

FIGURE 24.5. Schematic of West Nile Virus genome.

a variety of small rodents and birds. In addition, WNV can infect soft and hard ticks under natural and experimental conditions. Almost all species of birds tested (wild species, chickens, and pigeons) develop viremia. In the recent U.S. outbreak, the virus has killed a number of crows, and this has been used to track virus spread (Anderson et al., 1999). Sporadic cases of naturally acquired infections have been reported in horses, and these infections can lead to encephalitis. Bovine species do not develop viremia, but antibodies in cattle are prevalent. Finally, dogs are susceptible to infection, but low viremia levels preclude a significant role in virus transmission. Humans are a dead end host, but as described below, infections can be serious.

24.6.2. Pathogenesis of Encephalitis

The incubation period is 1–6 days, and the typical case is mild. Typical clinical features are fever, headache, backache, generalized myalgia, and anorexia. Rash occurs in about a half of the cases, and the rash is usually roseolar or macuolpapular, and usually involves the chest, back, and upper extremeties. The disease usually runs its course in 3–6 days, and is usually milder in children. Neurologic disease occurs in less than 1% of infected individuals. WNV can also cause severe, potentially fatal neurological disease, including encephalitis, meningitis, paralysis, and anterior myelitis. Although neurons are the primary target of WNV infection, a hallmark of WNV encephalitis is the accumulation of inflammatory infiltrates extending from the meninges in the brain parenchyma. These infiltrates are primarily comprised of lymphocytes and macrophages (Kelley et al., 2003). Most cases of WNV-induced encephalitis generally occur in older individuals, and treatment is supportive.

24.6.3. Animal Models for WNV

Mice have been used extensively to examine virus host interactions, and to examine virus-induced encephalitis. The adaptive immune response plays a crucial role in controlling WNV infections in mice, including encephalitis (Diamond et al., 2003; Shresta and Diamond, 2004; Wang et al., 2003). WNV infection leads to a Toll-like receptor 3-dependent inflammatory response, which promotes brain penetration of the virus, neuronal injury, and enhanced encephalitis (Wang et al., 2004). Since Toll-like receptor 3 recognizes double-stranded RNA and promotes an innate immune response, this finding was somewhat surprising. The innate immune response also protects against WNV-induced encephalitis because when mice lacking the alpha/beta interferon receptor are infected with WNV an increase in the frequency of encephalitis is seen in mice (Samuel and Diamond, 2005). With the availability of a number of knockout mice, utilization of these mouse models to identify cellular factors that regulate WNV-induced encephalitis has a great deal of promise.

24.6.4. Therapy

Currently there is no established treatment for West Nile virus infections. Treatment consists only of supportive and symptomatic care, including support of respiration, intravenous fluids, and prevention of secondary infections.

24.7. Other Viruses that Can Induce Encephalitis

The viruses discussed below sporadically cause encephalitis. In many ways, these viruses can be considered as re-emerging viruses, and thus are important pathogens in certain parts of the world. A brief description of these viruses and how they induce encephalitis is therefore included in this review.

24.7.1. Flavivirus Family Members that Can Cause Encephalitis

24.7.1.1. St. Louis Encephalitis Virus

The St. Louis Encephalitis Virus (SLE) is a member of the Flavivirus family (Burke and Monath, 2001). SLE, like other Flaviviruses, contains a single stranded RNA genome that is approximately 11 kilobases in length. The SLE RNA is an mRNA that is translated into a large precursor protein. The large precursor protein is subsequently processed into the respective functional viral proteins. SLE is carried by a mosquito vector, *Culex* species, which is widely spread throughout the United States (Burke and Monath, 2001). The clinical syndromes associated with infection by SLE include: encephalitis, aseptic meningitis, and febrile headaches. In older patients (>60 years old), the severity of the disease is much worse, and this age group has the highest incidence of encephalitis. Only 2% of younger patients develop encephalitis whereas 22%

of patients who are 60 years or older develop encephalitis. In addition, the frequency of clinically apparent versus non-apparent infections is nearly 10 times higher in patients greater than 60 years old. The incubation period varies between 4–21 days after infection. The onset of disease has numerous non-specific symptoms; general malaise, fever, chills, headache, drowsiness, nausea, sore throat, and/or cough (Quick et al., 1965). Specific diagnosis relies on serologic tests because virus specific antibodies rise sharply after the first week of infection. Virus is not routinely present in CNS fluid making diagnosis difficult.

Between 1964 and 1968, there were 4,478 confirmed cases of SLE, which is considered as a major epidemic (Burke and Monath, 2001, Chamberlin, 1980). Major outbreaks of SLE seem to occur only once in approximately 10 years. During non-epidemic years, less than 50 cases occur in the United States. There is no specific treatment for SLE. Good supportive care is essential, in particular because of high fevers and convulsions caused by the virus.

24.7.1.2. Japanese Encephalitis Virus

The Japanese Encephalitis Virus (JEV) is a typical Flavivirus that is widely distributed in all parts of Asia (Burke and Monath, 2001, Podinger et al., 1996). About 50,000 cases are diagnosed each year in Asia, and of these cases 1000 deaths occur. Within Asia, there is a high degree of sequence variability for JEV suggesting certain sequence variants are more pathogenic than others. Children ranging from 3–15 have a tenfold higher incidence of clinical disease caused by JEV. Older individuals frequently have specific JEV antibodies suggesting they were infected, recovered, and then are protected from infection. Birds and swine are efficient hosts in which the virus can be amplified, and then spread to mosquitoes. Formalin-inactivated JEV is an effective vaccine that prevents encephalitis in humans. Travelers going to Asia are encouraged to get vaccinated.

A JE-like virus, Rocio virus, has been isolated from fatal human encephalitis cases during an outbreak of encephalitis in the 1970s on the south coast of Sao Paulo State, Brazil (Burke and Monath, 2001).

24.7.1.3. Murray Valley Encephalitis Virus

A JEV like virus, Murray Valley Encephalitis Virus (MVE), has been responsible for several sporadic epidemics of encephalitis in southern Australia (Burke and Monath, 2001, Doherty, 1977). Like SLE and JEV, MVE appears to be spread by mosquito vectors. Isolation of MEV from brain tissue or serologic tests is used for diagnosis. There is no specific treatment or vaccine that has been designed for MEV. As with JEV or SLE, treatment is supportive.

24.7.1.4. Tick-Borne Encephalitis Viruses

Tick-borne Encephalitis Viruses (TBE) has been recognized in nearly all countries within continental Europe (Burke and Monath, 2001). Seventeen antigenically related viruses comprise the TBE complex (Holzmann et al., 1993). Thirteen of these viruses can cause human disease, and all are vectored by ixodid ticks. Several vaccines have been developed against certain TBE members. Passive immunization is also effective in 60% of patients if the TBE-specific immunoglobulin is given within 4 days after a tick bite.

24.7.2. Bunyaviridae and Their Role in Encephalitis

24.7.2.1. Summary of Bunyaviridae and Their Replication

The Bunyaviridae encompasses a large group of viruses that have similar morphological, morphogenic, and antigenic properties (Schmaljohn and Hooper, 2001). For example, the Bunyaviridae contains four genera; Bunyavirus, Hantavirus, Nairovirus, and Phlebovirus All members of these genera are arthropod-borne viruses. The Bunyaviridae virions contain three single-stranded RNA genome segments that are replicated and then translated (negative sense strand RNA). A virion associated polymerase replicates the viral RNAs, which subsequently serve as templates for translation.

24.7.2.2. The Bunyavirus Genus and Its Role in Encephalitis

The Bunyavirus genus contains at least 150 different viruses that can be found in all parts of the world except Australia (Nichol, 2001). Nearly all of these viruses are transmitted by mosquitoes and can be transmitted in several vertebrate hosts, including humans. The California encephalitis serogroup viruses cause more cases of human encephalitis than any other Bunyavirus member. Among the California encephalitis serogroup viruses, the LaCrosse virus is the most significant member with respect to causing encephalitis. The primary vector for La Crosse virus is a forest dwelling, tree hole-breeding mosquito (*A. triseriatus*). *A. triseriatus* are primarily found in the northern Midwest and northeastern United States, and thus more than 90% of the cases occur in these areas. Most La Crosse virus induced encephalitis occurs in summer and fall because this is the highest risk for being bitten by an infected female mosquito (Henderson and Coleman, 1971). The vector for Jamestown Canyon virus is *C inonata* mosquitoes and several *Aedes* species, which are broadly distributed across North America. La Crosse virus primarily causes encephalitis in children, whereas the Jamestown Canyon virus causes encephalitis primarily in adults. The California Encephalitis serogroup of viruses is predicted to cause 60–130 cases of encephalitis each year.

In general, the California encephalitis subgroup viruses spread from the site of inoculation to skeletal muscles, which is the major site of replication. The virus then spreads to the lymphatic channels, thus spreading to additional skeletal muscles and cardiac muscles. Ultimately, the virus ends up in the CNS, where viral replication can occur in neurons and glial cells. Due to considerable neuronal necrosis, death can occur 3 or 4 days after infection. Typical lesions in brains of fatal La Crosse encephalitis include cerebral edema, perivascular cuffing, glial nodules, mild leptomeningitis, and areas of focal necrosis (Kalfayan, 1983). There are no specific anti-viral drugs available for the California encephalitis serogroup viruses.

24.7.3. Rabies

The Rabies virus belongs to the Rhabdovirus family (*Lyssavirus* genus), and like all members of the Rhabdovirus family they are enveloped rod-shaped particles. The RNA genome is a non-segmented negative molecule that is approximately 11,000 nucleotides long (de Mattos et al., 2001). The RNA dependent RNA polymerase encoded by this family of viruses is essential for replicating the viral genome (Baltimore et al., 1970). Animal bites by rabid animals, in particular domestic dogs; cause more than 99% of all human rabies cases. The highest mortality rates are due to bites on the face or head because of the close proximity to the brain. Inhalation of aerosols, licks, or transdermal scratches from a rabid animal does not typically cause rabies because the virus cannot enter intact skin. The incubation period is usually 1–2 months, but there can be a great deal of variability for an incubation period. The length of incubation period depends on the bite site and relative proximity to the CNS, amount of virus delivered in the bite, age of the host, and immune status (Fishbein, 1991). Three general phases occur during the development of rabies: (1) the prodromal phase in which the symptoms are nonspecific (fever, chills, headache, nausea, diarrhea, muscle soreness, hydrophobia) that may last up to ten days, (2) acute neurologic phase un which patients exhibit nervous system symptoms (anxiety, agitation, hypersalivation, paralysis, and sporadic episodes of delirium, and (3) coma preceding death (de Mattos et al., 2001). Apart from hydrophobia, rabies induces similar clinical symptoms in animals.

Antibodies confer protective immunity, and thus it has been relatively easy to develop vaccines directed against rabies. The G protein of rabies is the only antigen that induces viral neutralizing antibody, protects against a lethal challenge, and consequently is the basis for a successful vaccine (Cox et al., 1977). Vaccination is crucial because there is no successful treatment directed against rabies. Supportive therapy can prolong life, but does not prevent death. Clinical diagnosis is relatively easy if exposure is documented. Many times, however, there is no history of exposure. Viral neutralizing antibodies, the presence of viral nucleic acids, or viral specific antigens can be readily obtained from patients prior to or following death (Hanlon et al., 1999). As expected, higher levels of viral products can be obtained from brain tissue of infected individuals during the latter stages of infection.

The domestic dog is the major vector for rabies throughout the world (de Mattos et al., 2001). Wild reservoirs for the virus can be found in foxes, raccoons, coyotes, jackals, skunks, or bats. It will be difficult to eradicate the virus from wild animals because of the diverse population of animals that harbor the virus. Current veterinary vaccines are very effective, but like all vaccines they do not work 100% of the time. Domestic dogs are routinely vaccinated, and as such the incidence of rabid dogs has gone down. Wildlife can be effectively vaccinated by providing a potent oral vaccine that is contained within bait. As mentioned above, these vaccines are based on the Rabies virus G protein. Although rabies continues to be a threat to humans, it is not as serious as it was in the twentieth century. For example, from the 1980s to 1990s the annual number of rabies cases decreased from 350 to 114, and mortality rates fell from 1.3 to less than 0.2 per million people.

Summary

Viral induced encephalitis can result from infection by Herpes simplex virus, Varicella zoster virus, human cytomegalovirus, West Nile virus, Bunyavirae, Rabies, or specific family members of the Flavivirus family. Each of these viruses varies in properties; epidemiology, pathogenesis, clinical features and therapeutic approaches. This chapter discusses the clinical features, productive infection, latent infection, epidemiology, and pathogenesis, animal models used to examine encephalitis, and therapy of encephalitis for each of these viruses. The major drugs used to treat encephalitis caused by these viruses are acyclovir, valacyclovir, ganciclovir, cidofovir and foscarnet. For each of these drugs, we discuss the pharmacology, mechanism of action and adverse effects.

Review Questions/Problems

1. The treatment of choice for West Nile virus infection consists mainly of the use of

a. acyclovir
b. steroids
c. ganciclovir
d. supportive and symptomatic care
e. tetracyclines

2. The most well known disease associated with cytomegalovirus infection is

a. genital infections
b. retinitis
c. hepatitis
d. dementia
e. multiple sclerosis

3. Herpes simplex virus-2 infection occurs predominantly in

a. the elderly
b. women

c. homosexual men
d. cancer patients
e. sexually active adolescents

4. Which of the following best describes the mechanism of action of acyclovir?

a. inhibition of reverse transcriptase.
b. stimulation of viral interferon production
c. inhibition of viral DNA polymerase
d. inhibition of viral protein synthesis
e. inhibition of de nove purine synthesis

5. Rabies virus contains

a. a negative stranded RNA viral genome and humans contract it primarily from being bitten by a domestic dog.
b. is only a problem in young children or older adults.
c. can be effectively treated with anti-inflammatory agents.
d. can only be grown in young mice.

6. Intravenous administration of acyclovir is most likely to cause

a. anaphylactic shock
b. renal toxicity
c. liver toxicity
d. bone marrow suppression
e. cardiac arrhythmias

7. The main advantage of valacyclovir compared to acyclovir is

a. a very short half life
b. better oral bioavailability
c. less renal toxicity
d. it is useful for the treatment of HIV infection
e. it has fewer drug interactions

8. Ganciclovir is best used for the treatment of

a. West Nile virus
b. herpes simplex encephalitis
c. bird flu
d. cytomegalovirus
e. herpes zoster

9. Flavivirus family members that can cause encephalitis include

a. herpes simplex virus type 1 and human cytomegalovirus.
b. Rabies and St. Louis encephalitis virus
c. West Nile virus
d. Japanese encephalitis virus and St. Louis encephalitis virus

10. The major adverse effect associated with the use of cidofovir is

a. bone marrow suppression
b. rashes
c. arrhythmias
d. headaches
e. nephrotoxicity

11. Foscarnet is

a. a purine nucleoside analog
b. a pyrimidine nucleoside analog
c. incorporated into DNA resulting in inhibition of chain elongation
d. an inorganic pyrophosphate analog
e. a select inhibitor of RNA polymerase

12. Which of the following viruses causes encephalitis only in newborn children?

a. West Nile virus
b. herpes simplex virus type 1
c. human cytomegalovirus
d. varicella virus zoster
e. all of the above

13. Why is HSV-1 induced encephalitis (HSE) difficult to treat with antiviral therapeutic drugs?

a. Anti-viral drugs do not work because only drug resistant strains cause encephalitis.
b. Many cases of encephalitis are the result of reactivation from latency.
c. Early stages of infection are asymptomatic.
d. Anti-viral drugs do not cross the blood brain barriers.
e. all of the above

14. Effective vaccines have been developed against which virus

a. herpes simplex virus type 1
b. Japanese encephalitis virus
c. rabies
d. SV40

15. Circle all of the viruses that have the potential to induce encephalitis and establish a latent infection in neurons?

a. HSV-1
b. human cytomegalovirus
c. West Nile virus
d. varicella virus zoster
e. SV40

16. The immediate early genes encoded by all herpesviruses regulate

a. viral entry into permissive cells
b. viral DNA replication
c. production of antibodies by B cells
d. viral gene expression
e. all of the above

17. Is there a perfect correlation between the ability of viruses to replicate in the brain and induce encephalitis? Justify your answer.

18. If you were going to study viral induced encephalitis, would you choose HSV-1 or VZV as a model to identify viral genes that regulate the frequency of encephalitis? Give several reasons why you chose the virus that you chose.

Fill in the blanks

19._____ is a member of the Flaviviridae and has been spread by mosquitoes in North America.

20._____ is an alpha-herpesvirus family member that causes encephalitis and expresses LAT in latently infected neurons.

21._____ is an important opportunistic infection that cau es retinitis in AIDS patients.

22._____ is an alpha-herpesvirus family member that causes chicken pox early in childhood and can cause zoster late in life.

23. Describe the clinical features commonly associated with Herpes Simplex encephalitis.

References

Abendroth A, Arvin AM (2000) Host responses to primary infection. In: Varicella-Zoster Virus. (Arvin AM, Gershon AA, eds), pp 142–156. Cambridge: Cambridge University Press.

Anderson JF, Andreadis TG, Vissbrinck CR (1999) Isolation of West Nile virus from mosquitoes, crows, and a Cooper's hawk in Connecticut. Science 286:2331–2333.

Arribas JR, Storch GA, Clifford DB, Tselis AC (1996) Cytomegalovirus encephalitis. Ann Intern Med 125:577–587.

Arvin AM, Gershon AA (1996) Live attenuated varicella vaccine. Annu Rev Microbiol 50:59–100.

Balraj V, John TJ (1994) An epidemic of varicella in rural southern India. J Trop Med Hyg 97:113–116.

Baltimore D, Huang AS, Stampfer M (1970) Ribonucleic acid synthesis of vesicular stomatitis virus, II. An RNA polymerase in the virion. Proc Natl Acad Sci USA 66:572–576.

Bopanna SB, Pass RF, Britt WJ (1992) Symptomatic congenital cytomegalovirus infection: Neonatal morbidity and mortality. Pediatr Infect Dis J 11:93–99.

Bopanna SB, Fowler KB, Vaid Y (1997) Neuroradiographic findings in the newborn period and long-term outcome in children with symptomatic congenital cytomegalovirus infection. Pediatrics 99:409–414.

Brady RC, Bernstein DI (2004) Treatment of herpes simplex virus infections. Antiviral Res 61:73–81.

Burke DS, Monath TP (2001) Flaviviruses. In: Filed Virology. Vol II. pp 1581–1602. Lippincott Williams and Wilkins.

Cheeseman AH (2004) Cytomegalovirus infections. In: Infectious Diseases. (Gorbach SL, Bartlett JG, and Blacklow NR, eds), pp 1543–1551. Philadelphia: Lippincott Williams and Wilkins.

Cheung TW, Teich SA (1999) Cytomegalovirus infection in patients with HIV infection. Mt Sinai J Med 66:113–124.

Chrisp P, Clissold SP (1991) Foscarnet: A review of its antiviral activity, pharmacokinetic properties and therapeutic use in immunocompromised patients with cytomegalovirus retinitis. Drugs 41:104–129.

Corey L (2005) Herpes simplex virus. In: Principles and Practice of Infectious Diseases. (Mandell GL, Bennett JE, and Dolin R, eds). pp 1762–1780. Philadelphia: Elsevier Churchill Livingston.

Cox JH, Dietzschold B, Schneider LG (1977) Rabies virus glycoproteins. II. Biological and serological characterization. Infect Immun 16:754–759.

Curran M, Noble S (2001) Valganciclovir. Drugs 61:1145–1150.

DeBiasi RL, Kleinschmidt-DeMasters BK, Richardson-Burns S, Tyler KL (2002) Central nervous system apoptosis in human herpes simplex virus and cytomegalovirus encephalitis. J Infect Dis 186:1547–1557.

Diamond MS, Shrestha B, Marri A, Engle M (2003) B cells and antibody play critical roles in the immediate defense of disseminated infection by West Nile encephalitis virus. J Virol 77:2578–2586.

Doherty RL (1977) Arthropod-borne viruses in Australia, 1973–1976. Aust J Exp Biol Med Sci, 55:103–130.

Fiala M, Edmonson L, Guze LB (1973) Simplified method for isolation of cytomegalovirus and demonstration of frequent vriemia in renal transplant patients. Proc Soc Exp Biol Med 144:871–875.

Fishbein DB (1991) Rabies in humans. In: The Natural History of Rabies. (Baer GM, ed), pp 519–549. Boca Raton, FL: CRC Press.

Fraser NW, Lawrence WC, Wroblewska Z, Gilden DH, Koprowski H (1981) Herpes simplex virus type 1 DNA in human brain tissue. Proc Natl Acad Sci USA 78:6461–6465.

Gnann JW, Whitley RJ (2002) Herpes zoster. N Engl J Med 347:340–346.

Griffiths PD (1995) Progress in the clinical management of herpes virus infections. Antiviral Chem Chemother 6:191–209.

Hahn G, Jores R, Mocarski ES (1998) Cytomegalovirus remains latent in a common precursor of dendritic and myeloid cells. Proc Natl Acad Sci USA 95:3937–3942.

Hanlon CA, Smith JS, Anderson GR (1999) Recomendations of a national working group on prevention and control of rabies in the United States. Laboratory diagnosis of rabies. The National Working Group on Rabies Prevention and Control. J Am Vet Med Assoc 215:1444–1446.

Henderson BE, Coleman PH (1971) The growing importance of California arboviruses in the etiology of human disease. Prog Med Virol 13:404–461.

Holzmann H, Utter G, Norrby E, Mandl CW, Kunz C, Heinz FX (1993) Assessment of the antigenic structure of tick-borne encephalitis virus by the use of synthetic peptides. J Gen Virol 74:2031–2035.

Honess RW, Roizman B (1974) Regulation of herpes virus macromalecular synthesis: Cascade regulation of three groups of viral proteins. J Virol 14:8–19.

Jerant AF, DeGaetano JS, Epperly TD, Hannapel AC, Miller DR, Lloyd AJ (1998) Varicella susceptibility and vaccination strategies in young adults. J Am Board Fam Pract 11:296–306.

Jin L, Perng G-C, Nesburn AB, Jones C, Wechsler SL (2005) The baculovirus inhibitor of apoptosis gene (cpIAP) can restore reactivation of latency to a herpes simplex virus type 1 that does not express the latency associated transcript (LAT). J Virol 79:12286–12295.

Johnson R, Milbourn PE (1970) Central nervous system manifestations of chickenpox. Can Med Assoc J 102:831–834.

Jones C (1998) Alphaherpesvirus latency: Its role in disease and survival of the virus in nature. Adv Virus Res 51:81–133.

Jones C (2003) Herpes simplex virus type 1 and bovine herpesvirus 1 latency. Clin Microbiol Rev 16:79–95.

Jones C, Inman M, Peng W, Henderson G, Doster A, Perng G-C, Angeletti AK (2005) The herpes simplex virus type 1 (HSV-1) locus that encodes the latency-associated transcript (LAT) enhances the frequency of encephalitis in male Balb/C mice. J Virol 79:14465–14469.

Kalfayan B (1983) Pathology of La Crosse virus infections in humans. Prog Clin Biol Res 123:179–186.

Kelley TW, Prayson RA, Ruiz AI, Isada CM, Gordon SM (2003) The neuropathology of West Nile virus meningoencephalitis: A report of two cases and review of the literature. Am J Clin Pathol 119:749–753.

Kennedy PGE, Grinfield E, Gow JW (1999) Latent varicella-zoster virus in human dorsal root ganglia. Virology 258:451–454.

Knipe DM, Howley PM (ed) (2001) Fields Virology. 4th ed. Philadelphia: Lippincott Williams and Wilkins.

Kondo K, Xu J, Mocarski ES (1996) Human cytomegalovirus latent gene expression in ganulocyte-macrophage progenitors in culture and in sero-positive individuals. Proc Natl Acad Sci USA 93:11137–11142.

Ku C-C, Besser J, Abendroth A, Grose C, Arvin AM (2005) Varicella-zoster virus pathogenesis and immunobiology: New concepts emerging from investigations with the SCIDhu mouse model. J Virol 79:2651–2658.

Kurt-Jones EA, Chan M, Zhou S, Wang J, Reed G, Bronson R, Arnold MA, Knipe DM, Finberg RW (2004) Herpes simplex virus 1 interaction with toll-like receptor 2 contributes to lethal encephalitis. Proc Natl Acad Sci USA 101:1315–1320.

Lahat E, Barr J, Barkai G, Paret G, Brand N, Barzilai A (1999) Long-term neurological outcome of herpes encephalitis. Arch Dis Child 80:69–71.

Liesegang TJ (1999) Varicella zoster viral disease. Mayo Clinic Proc 74:983–998.

Lindenback BD, Rice C (2001) Flaviviridae: The viruses and their replication. In: Fields Virology. (Knipe DM, and Howley PM, eds), 4th ed. Philadelphia: Lippincott Williams and Wilkins.

Markham A, Faulds D (1994) Ganciclovir: An update of its therapeutic use in cytomegalovirus infection. Drugs 48:455–484.

de Mattos CA, de Mattos CC, Rupprecht CE (2001) Rhabdoviruses. In: Filed Virology. Vol II. pp 1245–1277. Lippincott Williams and Wilkins Phil., PA.

McGrath N, Anderson NE, Croxson MC, Powell KF (1997) Herpes simplex encephalitis treated with acyclovir: Diagnosis and long term outcome. J Neurol Neurosurg Psych 63:321–326.

Mitchell BM, Bloom DC, Cohrs RJ, Gilden DH, Kennedy PGE (2003) Herpes simplex virus-1 and varicella-zoster virus latency in ganglia. J Neurovirol 9:194–204.

Mott K, Osorio N, Jin L, Brick D, Naito J, Cooper J, Henderson G, Inman M, Jones C, Wechsler SL, and Perng G-C (2003) The bovine herpesvirus 1 LR ORF2 is crucial for this gene's ability to restore the high reactivation phenotype to a Herpes simplex virus-1 LAT null mutant. J Gen Virol 84:2975–2985.

Nahmias AJ, Roizman B (1973) Infection with herpes-simplex viruses 1 and 2. 3. N Engl J Med 289:781–789.

Neyts J, De Clercq E (1994) Mechanism of action of acyclic nucleoside phosphonates against herpes virus replication. Biochem Pharmacol 47:39–41.

Nichol ST (2001) Bunyaviridae: The viruses and their replication. In: Filed Virology. Vol II. pp 1581–1602. Lippincott Williams and Wilkins Phil., PA.

O'Hare P (1993) The virion transactivator of herpes simplex virus. Semin Virol 4:145–155.

Pass RF (1995) Epidemiology and transmission of cytomegalovirus infection. J Infect Dis 152:243–256.

Pass RF (2001) Cytomegalovirus. In: Fields Virology. (Knipe DM, and Howley PM, eds). 4th ed. pp 2675–2705. Philadelphia: Lippincott Williams and Wilkins.

Perng G-C, Maguen B, Jin L, Mott KR, Osorio N, Slanina SM, Yukht A, Ghiasi H, Nesburn AB, Inman M, Henderson G, Jones C, Wechsler SL (2002) A gene capable of blocking apoptosis can substitute for the herpes simplex virus type 1 latency-associated transcript gene and restore wild-type reactivation levels. J Virol 76:1224–1235.

Perry CM, Faulds D (1996) Valaciclovir: A review of its antiviral activity, pharmacokinetic properties and therapeutic efficacy in herpesvirus infections. Drugs 52:754–772.

Plosker GL, Noble S (1999) Cidofovir: A review of its use in cytomegalovirus retinitis in patients with AIDS. Drugs 58:325–345.

Podinger M, Hall RA, Mackenzie JS (1996) Molecular characterization of the Japanese encephalitis subcomplex of the flavivirus genus. Virology 218:417–421.

Price WH, O'Leary W (1967) Geographical variation in the antigenic character of West Nile virus. Am J Epidemiol 85:84–87.

Quick DT, Thompson JM, Bond JO (1965) The 1962 epidemic of St. Louis encephalitis in Florida. IV. Clinical features in the Tampa Bay area. Am J Epidemiol 81:415–427.

Rantalaiho T, Farkhila M, Vaheri A, Koskiniemi M (2001) Acute encephalitis from 1967 to 1991. J Neurol Sci 184:169–177.

Rawson H, Crampin A, Noah N (2001) Deaths from chickenpox in Engalnd and Wales 1995–1997: Analysis of routine mortalitiy data. Br Med J 323:1091–1093.

Samuel MA, Diamond MS (2005) Alpha/beta interferon protects against lethal West Nile virus infection by restricting cellular tropism and enhancing neuronal survival. J Virol 79:13350–13361.

Schmaljohn CS, Pratt JW (2001) Bunyaviridae: The viruses and their replication. In: Filed Virology. Vol II. pp 1581–1602. Lippincott Williams and Wilkins N.Y., N.Y.

Scholar EM, Pratt WB (2000) Chemotherapy of viral infections, I: drugs used to treat influenza virus infections, herpes virus infections, and drugs with broad-spectrum antiviral activity. In: The Antimicrobial Drugs. pp 491–549. Oxford: Oxford University Press.

Shresta B, Diamond MS (2004) Role of CD8+ T cells in control of West Nile virus infection. J Virol 78:8312–8321.

Skoldenberg B (1991) Herpes simplex encephalitis. Scand J Infect Dis Suppl 80:40–46.

Snoeck R, Andrei G, DeClercq E (1999) Current pharmacological approaches to the therapy of varicella zoster virus infection: A guide to treatment. Drugs 57:187–206.

Spear PG (1993) Entry of alphaherpesviruses into cells. Semin Virol 4:167–180.

Stanberry LR, Oxman MN, Simmons A (2004) Herpes simplex viruses. In: Infectious Dieseases. (Gorbach SL, Bartlett JG, and Blacklow NR, eds), pp 1905–1917. Philadelphia: Lippincott Williams and Wilkins.

Stanberry LR, Spruance SL, Cunningham AL, Bernstein DI, Mindel A, Sacks S, Tyring S, Aoki FY, Slaoui M, Denis M, Vandepapeliere P, Dubin G, GlaxoSmithKline Herpes Vaccine Efficacy Study Group (2002) Glycoprotein-D-adjuvant vaccine to prevent genital herpes. N Engl J Med 347:1652–1661.

Wagstaff AJ, Bryson HM (1994) Foscarnet: A reappraisal of its antiviral activity, pharmacokinetic properties and therapeutic use in immunocompromised patients with viral infections. Drugs 48:153–205.

Wagstaff AJ, Faulds D, Goa KL (1994) Aciclovir: a reappraisal of its antiviral activity, pharmacokinetic properties and therapeutic efficacy. Drugs 47:153–205.

Wang T, Town T, Alexopoulou L, Anderson JF, Fikrig E, Flavell RA (2004) Toll-liker receptor 3 mediates West Nile virus entry into the brain causing lethal encephlaitis. Nat Med 10:1366–1373.

Wang Y, Lobigs M, Lee E, Mullbacher A (2003) CD8+ T cells mediate recovery and immunopathoglogy in West Nile virus encephalitis. J Virol 77:13323–13334.

Whitley R (1997) Herpes Simplex Virus. Philadelphia, New York: Lippincott-Raven Publishers.

Whitley RJ, Roizman B (2001) Herpes simplex virus infections. Lancet 357:1513–1518.

Yamada S, Kameyama T, Nagaya S, Hashizume Y, Yoshida M (2002) Relapsing herpes simplex encephalitis: pathological confirmation of viral reactivation. J Neurol Neurosurg Psych 74:262–264.

25
Alzheimer's Disease

Tsuneya Ikezu

Keywords Amyloid precursor protein; Apolipoprotein E; β-amyloid peptide; Beta-amyloid precursor protein converting enzyme; Insulin degrading enzyme; Neurofibrillary tangle; Oxidative damage; Presenilin-1

25.1. Introduction

Alzheimer's disease (AD) is the most frequent neurodegenerative disease and the most common cause of dementia in humans. Clinically, the classic and most frequent initial symptom is impaired short-term memory that progresses to profound memory failure. AD neuropathology exhibits two hallmark features: (1) senile plaques containing depositions of beta-amyloid protein and (2) neurofibrillary tangles (NFT). Quantitative neuropathological evaluation in early AD shows significant neuronal loss in brain memory regions. Genetic studies demonstrate that aberrant proteolytic processing of the amyloid precursor protein (APP) leads to 1–40 and 1–42 amyloid-β peptide (Aβ) fragments capable of causing AD pathology. However, the mechanisms by which amyloid proteins lead to plaque deposition, NFT, and neuronal cell death is incompletely understood. Thus, the objective of this chapter is to review what is known about disease mechanisms with a focus on the role played by neuroimmune interactions in neurodegenerative processes.

25.2. Clinical Features, Epidemiology, and Neuropathology

25.2.1. Symptoms (Memory Loss to Progressive Dementia)

AD is the leading cause of senile dementia and effective therapy remains in want (Sisodia, 1999). The symptoms of typical AD begin with a loss of short-term memory that slowly progresses to profound memory functions (Corey-Bloom et al., 1994). Initially, impaired short-term memory often manifests itself as losing for forgetting the placement of objects or forgetfulness during conversations. Subsequently, other early cognitive deficits occur that include difficulty in finding words, problem solving and deficits in spatiotemporal awareness. Patients commonly appear depressed and apathetic. During the course of disease agitation and irritability often occur. With further disease progression, memory failure becomes more profound and longer-term memory is affected. Patients with these deficits are frequently disoriented and can easily become lost. Thus supervision of patients may become necessary. About this time, language skills and complex motor skills deteriorate. In still later stages, psychosis and hallucinations can occur along with bradykinesia and rigidity. Patients ultimately become bed-ridden, unable to speak, and succumb from infection or other medical conditions. The average survival time of Alzheimer's patients is 8 years after onset of initial symptoms but the rate of decline is variable.

25.2.2. Diagnosis

Of a large number of potential causes of dementia, AD is the most common. Hence to approach the diagnosis of AD evidence to support this diagnosis need be found to the exclusion of other possibilities. A definitive diagnosis includes: (1) a history obtained from someone who knows the patient well; (2) a physical and neuropsychiatric examination that excludes other neurologic and neurodegenerative disorders; (3) blood tests that exclude metabolic diseases involving the kidneys, liver, or thyroid; (4) a mental status test to determine dementia progression; and (5) brain imaging that exclude focal infection, inflammatory, and cancerous diseases (see Neuroimaging chapter for details). The first cognitive assessment widely used was the Mini-Mental Sate Examination (MMSE). MMSE assesses multiple cognitive tasks including spatiotemporal orientation, immediate/delayed word recall, naming, verbal repetition, reading, writing, and spatial ability (Folstein et al., 1975). There are multiple reports of reduced cognitive performance in preclinical AD, notably impaired function in abstract reasoning, episodic memory, and new learning (Fratiglioni et al., 2001).

T. Ikezu and H.E. Gendelman (eds.), *Neuroimmune Pharmacology.*
© Springer 2008

Episodic memory deficits are dominant among early clinical and preclinical AD. A diagnosis of dementia is usually determined using the Diagnostic and Statistical Manual of Mental Disorders or DSM-IV of the American Psychiatric Association (Association, 1994). A final diagnosis is determined by the National Institute of Neurological Disorders and Stroke and Alzheimer's Disease and Related Disorders Association criteria, NINDS-ADRDA (McKhann et al., 1984). Other diagnosis standards includes the International Classification of Diseases (ICD)-10 guideline (Organization, 1992). Since DSM-IV and ICD-10 criteria closely follow NINDS-ADRDA guidelines, diagnostic procedures for AD are notably consistent worldwide. Given that AD is the primary cause of dementia, AD is often considered to be the likely cause of dementia while clinical tests are used to exclude other causes.

25.2.3. Epidemiology

AD affects more than four million people in the United States and prevalence studies demonstrate an exponential rise of dementia linked to advancing age (Katzman, 2001). The number of cases of dementia in the developed world is projected to rise from 13.5 million in 2000 to 21.2 million in 2025, and to 36.7 million in 2050 (Katzman and Fox, 1999). Currently the number of deaths caused by AD is similar to the number of deaths caused by stroke. AD and stroke together rank as the third most common causes of death (Ewbank, 1999).

In addition to age, other factors are associated with an increased risk of AD. In developed countries, AD appears to be more common in women. Lack of education is a risk factor for senile dementia in China and Europe (Zhang et al., 1990; Schmand et al., 1997). Head trauma is also a risk factor for both sporadic (Mortimer et al., 1991) and familial AD (Guo et al., 2000). Silent myocardial infarcts and coronary stenosis triple the risk for AD (Aronson et al., 1990; Sparks et al., 1990), suggesting the importance of vascular risk factors. Other potential risk factors being studied include diabetes and hypertension (Ott et al., 1999; Peila et al., 2002; Qiu et al., 2005). As discussed below, a large number of genetic mutations are now associated with either early-onset AD or with increased risk of late-onset AD.

Identification of genetic mutations that cause early onset AD has helped to dramatically increase the understanding of the origins of AD. Although cases of familial AD are relatively rare, they represent identified molecular mutations that cause AD. The genes identified to be responsible for early onset familial AD code for the amyloid precursor protein (APP), presenilin-1 and presenilin-2; these are found on chromosomes 21, 14, and 1, respectively. In addition to these dominantly transmitted disease-causing mutations in early onset AD, other studies have identified over 25 genetic risk factors that influence the appearance of the more common sporadic late onset AD. Current research is actively using genetically modified mice to determine how specific protein modifications lead to the pathology of AD.

25.2.4. Pathology

At autopsy, AD brains appear to have widespread neuronal and neuropil loss that thins cortical layers and expands ventricular and sulci spaces. However, there is not an obvious global loss of neurons in the cerebral cortex. Overall neuronal number appears relatively unchanged despite significant loss of brain tissue. Some of this apparent discrepancy is due to preferential loss of large pyramidal neurons in the cortex and relative sparing of smaller, densely packed neurons. Particularly vulnerable regions such as the entorhinal cortex and the CA1 of the hippocampus, show profound neuronal loss early in the disease. Wider involvement of temporal, frontal, and parietal cortices and a preferential loss of large neurons in these areas are seen as the disease progresses.

AD is characterized by an accumulation of senile plaques containing aggregated protein deposits of Aβ fibrils, numerous neurofibrillary tangles (NFTs), reactive astrocytes, and activated microglia in the neocortex and hippocampus (Alzheimer, 1907; Selkoe, 1997). Senile plaques can contain either a diffuse amyloid deposition (presumably early senile plaques) or a dense core of insoluble Aβ with neuritic structures (mature senile plaques). Although Aβ deposition and NFT are commonly observed in non-demented elderly 80 years and older, increased levels of senile plaques are seen in AD brains (Berg et al., 1993; Schmitt et al., 2000). The degree of both Aβ deposition and NFT formation are correlated with cognitive decline (Wilcock and Esiri, 1982; Berg et al., 1993; Braak and Braak, 1995; Naslund et al., 2000).

Although inflammation is a well-recognized phenomenon its precise definition is not yet established (McGeer and McGeer, 2001). Clinically, AD patients exhibit few of the classic signs of inflammation. Histopathologically, the classical acute inflammatory response (neutrophil recruitment), immunoglobulin precipitation, and T-cell accumulation are not seen suggesting that humoral or classical cellular immune-mediated responses are not involved in AD progression (Eikelenboom et al., 1994). However, AD brains show an increase in activated microglial clusters on senile plaques surrounded by astrocytes, indicating focal glial inflammation (Eikelenboom et al., 2002; Wyss-Coray and Mucke, 2002). In particular, volume density of $CD68^+$ brain mononuclear phagocytes (MP; perivascular macrophages and microglia) is correlated with volume density of congophilic compact plaque deposits (Arends et al., 2000). $CD68^+$ cells are present from very early preclinical phases to terminal phases of AD and are involved in disease progression, which will be further described later (Akiyama et al., 2000; Fonseca et al., 2004b). Moreover, deactivation of microglia has implications for therapy since long-term treatment of affected patients with nonsteroidal anti-inflammatory drugs (NSAIDs) has been shown to inactivate microglia and retard the development and progression of AD (in 't Veld et al., 2002), although NSAIDs may also directly affect the APP processing.

25.3. Molecular Pathogenesis

The molecular pathogenesis of AD is centered around Aβ production and its clearance. Aβ is generated by the processing of APP by β and γ-processing enzymes, and cleared from brain by its diffusion, export to vascular system, phagocytosis, or degradation. The following molecules are essential for understanding the molecular mechanism of Aβ processing.

25.3.1. Amyloid Precursor Protein

APP is the first gene to be identified as a causative gene of familial AD (FAD) and has been a key molecule in the study of the molecular mechanism of AD. As shown in Figure 25.1, the APP molecule can be cleaved at different specific cleavage sites by various proteases to generate fragments of different sizes. APP is primarily expressed in neurons and cleaved at the α, β, and γ sites by APP processing enzymes (called α, β, and γ–secretases), producing Aβ (β- and γ-processing product), p3 (α- and γ-processing product), and other APP fragments (Selkoe, 1994). Aβ is composed of 39 to 43 amino acids, mainly 1–40 (Aβ40) and 1–42 (Aβ42). Aβ42 is more amyloidogenic (faster to be aggregated) than Aβ40, and early Aβ deposits are usually detectable with anti-Aβ42 antibodies, but not with anti-Aβ40 antibodies (Selkoe, 1994). FAD mutations on APP gene are all linked to abnormal APP processing (increase in Aβ42/Aβ40 ratio), Aβ aggregation, and Aβ deposition (Scheuner et al., 1996).

25.3.2. PS1 and γ-Processing Enzyme Complex

A combination of the presenilin genes (PS1 and 2), aph-1, pen-2, and nicastrin forms the γ-secretase complex (De Strooper, 2003). The γ-secretase complex cleaves not only APP but also other transmembrane molecules, such as Notch, ErbB-4, and sterol regulatory element binding proteins (SREBPs). PS1 and PS2 have been established as causative genes of early onset FAD (Levy-Lahad et al., 1995; Sherrington et al., 1995). Transgene expression of PS1 FAD mutants enhances the Aβ42/Aβ40 ratio

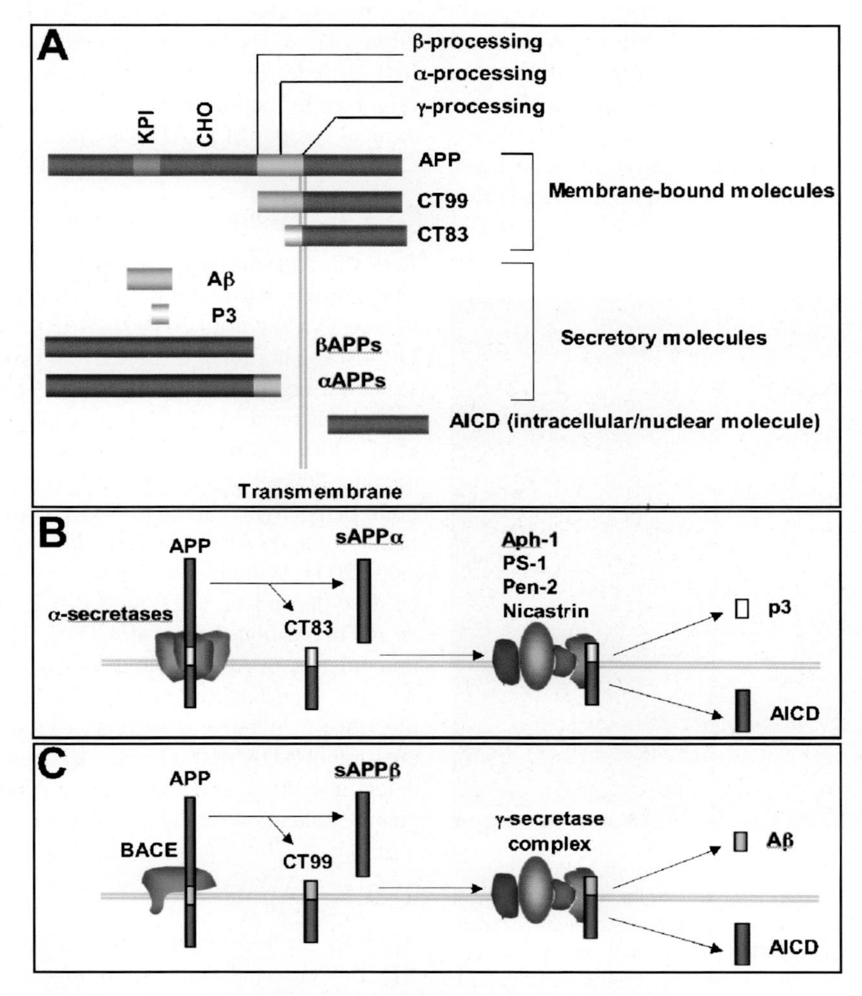

FIGURE 25.1. Scheme of APP processing. A, Cleavage sites and processing enzymes of APP. B, P3 and sAPPα production by α and γ-processing of APP. C, Aβ and sAPPβ production by β and γ-processing of APP.

and accelerates Aβ aggregation and deposition in APP mice, which over-express Swedish FAD APP mutant in brains (Duff et al., 1996; Borchelt et al., 1997; Holcomb et al., 1998). PS2 FAD mutant over-expression converts Aβ40 to Aβ42, whereas PS2 wild type over-expression reduces both Aβ40 and Aβ42 (Mastrangelo et al., 2005). Taken together, these *in vivo* studies demonstrate the effect of PS1/2 FAD mutations on Aβ aggregation through transition of Aβ40 to Aβ42.

25.3.3. APP and PS1 Animal Models

Several established AD mouse models have made significant contributions to the understanding of AD pathogenesis (Price et al., 1998; Spires and Hyman, 2005). Analyses of AD brain tissue obtained from autopsies provide only limited insight into the dynamic nature of the disease process. Currently, no AD mouse model has all of the characteristics of the human disease. However, several lines of transgenic mice expressing mutated APP, such as PDAPP (Games et al., 1995), Tg2576 (Hsiao et al., 1996), APP23 (Sturchler-Pierrat et al., 1997), TgCRND8 (Chishti et al., 2001), APPswe TgC3–3 (Borchelt et al., 1997), PSAPP (Holcomb et al., 1998), and 3xTg-AD mice (Oddo et al., 2003), simulate prominent behavioral and pathological features of AD. Phenotypes displayed by these mice include age-related impairments in learning and memory, electrophysiological abnormalities, neuronal loss, microgliosis, astrogliosis, neuritic changes, amyloid deposition, oxidative stress, and abnormal tau phosphorylation (Price et al., 1998; Smith et al., 1998) (Figure 25.2). Microglia in close proximity to fibrillar Aβ deposits are immunoreactive to interleukin (IL)-1β and TNF-α, whereas activated astrocytes are immunoreactive to IL-1β, transforming growth factor (TGF)-1β, and IL-10 in Tg2576 mice (Frautschy et al., 1998; Benzing et al., 1999; Apelt and Schliebs, 2001).

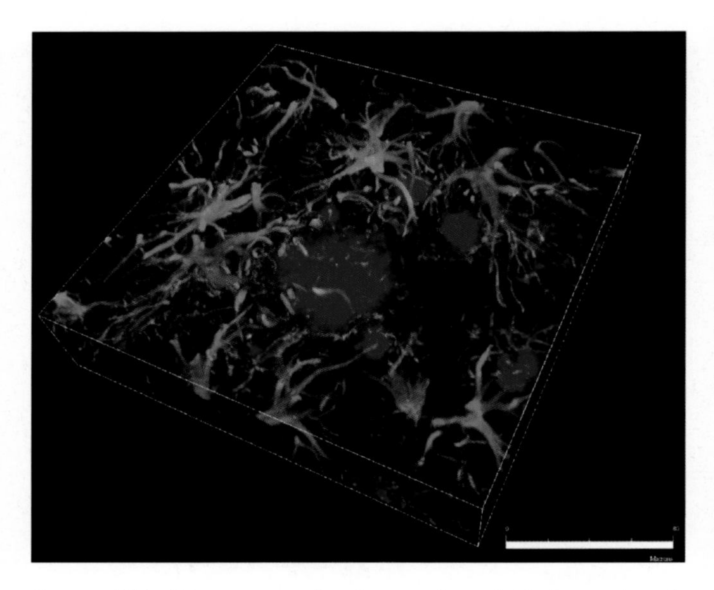

FIGURE 25.2. 3D reconstruction image of amyloid plaque deposition and glial accumulation in APP mouse brain. Cortical section of Tg2576 (14 months of age) was stained by thioflavin-S (compact amyloid plaque, purple), anti-IBA1 rabbit polyclonal (microglia, red), and anti-GFAP mouse monoclonal (astrocytes, green), and imaged using a laser scanning confocal microscopy (Zeiss 510 Meta). Z-stack images were deconvolved and 3D reconstructed using image processing software (AutoQuant).

Tg2576 mice which express Swedish familial AD mutant (K670N/M671L) of APP695 under a hamster prion promoter show memory and learning impairments as determined by Morris water maze (Hsiao et al., 1996). The memory recall is impaired as early as 6 months of age, which persists without significant progression for up to 12 months of age, followed by impairment of both memory acquisition and recall after 12 months of age (Lesne et al., 2006). There is a strong correlation of such age-related memory decline to Aβ accumula tion in the brain (Figure 25.3) (Chen et al., 2000; Janus et al., 2000). A recent study in Tg2576 mice suggests that 9- or 12-mer Aβ oligomers are responsible for such memory loss, which can be isolated by Tg2576 and can induce transient memory loss in injected rodents (Lesne et al., 2006). Contextual fear conditioning tests, which represents hippocampal learning, show memory impairments in PDAPP mice as early as 4 months of age preceding the Aβ deposition (Jacobsen et al., 2006). These studies strongly suggest that onset of memory loss precedes the Aβ deposition, is more correlated with Aβ oligomer formation in the brain, may also be relevant to the clinical symptoms of AD patients.

25.3.4. β- and α-Processing Enzymes

Beta-site APP-cleaving enzyme (BACE1, also called Asp2 or Memapsin2), a β-secretase, is an aspartic transmembrane proteinase (Vassar et al., 1999). BACE2 (also called Asp1 and DRAP) is a homolog of BACE1 and contains the same overall structural organization. BACE1 is essential for Aβ production as demonstrated by BACE1 gene-targeted mouse models (Cai et al., 2001). BACE over-expression enhances total Aβ production, deposition of both diffuse and compact Aβ plaques in brain parenchyma, and reduces cerebrovascular amyloid angiopathy in aged APP/BACE double transgenic mice (Mohajeri et al., 2004; Willem et al., 2004). However, this effect seems to be dose-dependent, since high BACE expression rather inhibits Aβ deposition (Lee et al., 2005). Thus, the effect of BACE upregulation in Aβ pathology *in vivo* is not conclusive. A Disintegrin And Metalloprotease (ADAM) 9, 10, and 17 have been identified as putative α-secretases (Asai et al., 2003). Overexpression of ADAM10 prevents Aβ deposition and hippocampal defect in APP mice (Postina et al., 2004), suggesting its therapeutic application for enhancing α-processing of APP.

25.3.5. ApoE

ApoE is the most well characterized genetic risk factor of AD. Evidence indicates that endogenous apoE in mouse brain enhances Aβ deposition in APP animal models (Bales et al., 1999; Sadowski et al., 2004). Tg2576 mice lacking mouse apoE exhibit diminished compact plaque deposition, while diffuse

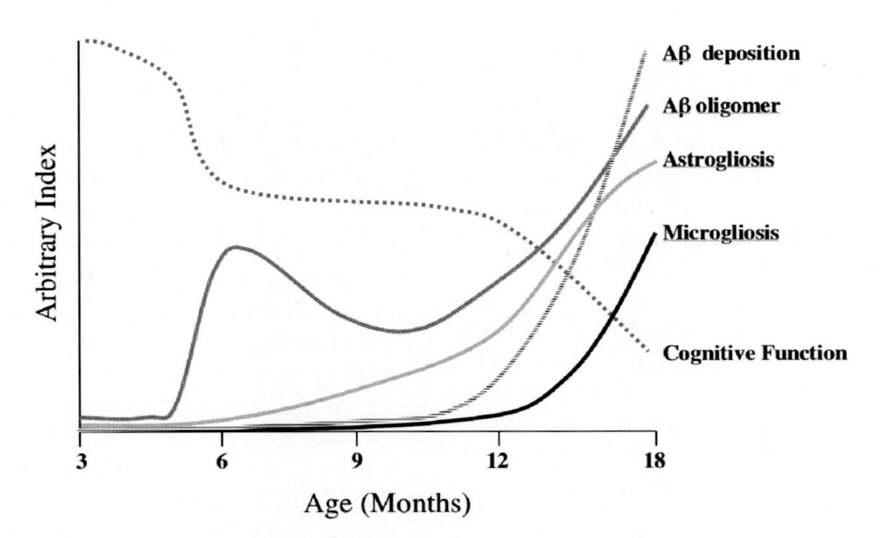

FIGURE 25.3. Schematic representation of the time course of Ab deposition, Ab oligomer accumulation, cognitive dysfunction, astrogliosis, and microgliosis in Tg2586 mice.

plaque load remained unchanged (Bales et al., 1999). Mouse apoE is rather amyloidgenic since apoE disruption significantly reduces the compact plaque formation in APP mice (Bales et al., 1999). In contrast, human apoE3 enhances Aβ clearance in mouse brain, while apoE4 weakly promotes Aβ clearance (Holtzman et al., 1999; Holtzman et al., 2000; Fagan et al., 2002). Although many studies concluded that astrocytes are the main apoE-producing cell in brain (Danik et al., 1999), recent reports suggest that microglial cells are also a source of apoE (Saura et al., 2003; Mori et al., 2004). A recent study of EGFP-ApoE knock-in mice demonstrated that apoE is expressed in 75% of astrocytes, and <10% of microglia after stimulation. Thus, apoE is primarily an astrocyte protein (Xu et al., 2006).

25.3.6. Aβ Degrading Enzymes

The current knowledge of Aβ degradation biology is limited since only a few of the molecules involved have been characterized, namely insulin degrading enzyme (IDE) and neprilysin, which is also known as enkephalinase, and neutral endopeptidase metalloendopeptidase (NEP) (Yamin et al., 1999; Iwata et al., 2000; Selkoe et al., 2001). IDE was originally identified as a microglia-secreted Aβ degrading enzyme (Qiu et al., 1998), and is also found in neurons (Vekrellis et al., 2000). Neprilysin protein level is downregulated during aging, but is locally upregulated in activated astrocytes surrounding Aβ plaques in APP mice (Apelt et al., 2003). Thus the amount of IDE and neprilysine can be altered by chronic inflammatory reaction. Neprilysin, a neutral zinc metallopeptidase, plays a role in degrading insoluble and/or aggregated Aβ42 (Iwata et al., 2000; Selkoe et al., 2001). Transgenic expression of both IDE and neprilysin show inhibition of Aβ accumulation in APP mice (Leissring et al., 2003). In addition, endothelin converting enzyme and matrix metalloprotease-2 and 3 are also new candidates of Aβ degrading enzymes (Choi et al., 2006; White et al., 2006).

25.3.7. Neurofibrillary Tangle Formation

Presence of NFT is one of the original neuropathological hallmarks of AD. NFT are fibrous tangles composed of insoluble, conformationally abnormal, hyperphosphorylated tau protein deposited in neuronal cell bodies. Hyperphosphorylated tau can form a specific insoluble structure known as a paired helical filament (PHF) (Buee et al., 2000; Lee et al., 2001). Six tau protein isoforms are generated by alternative mRNA splicing of the *tau* gene. Based on the repeat of the intermediate microtubule binding domain, the majority of tau is classified as three repeat (3R) or four repeat (4R) tau. The primary function of tau is to bind to and stabilize microtubules, thereby promoting microtubule polymerization. Many tau phosphorylation sites flank microtubule binding domains and are believed to play a significant role in its microtubule binding (Lee et al., 2001; Planel et al., 2002). Hyperphosphorylation of tau appears to cause tau to dissociate from microtubules and form tau protein aggregates, which becomes PHF (Biernat et al., 1993; Bramblett et al., 1993; Alonso et al., 1996). Glycogen synthase kinase 3-β (GSK3β), cyclin dependent protein kinase 5/p25 complex, microtubule affinity regulating kinase, and recently tau-tubulin kinase 1 have been extensively characterized as tau kinases that seem to be involved in the hyperphosphorylation of tau (Ishiguro et al., 1992; Takashima et al., 1993; Drewes et al., 1997; Patrick et al., 1999; Sato et al., 2006).

Tau-related neurodegenerative disorders, or "tauopathies," include AD, argyrophilic grain dementia, corticobasal degeneration, diffuse neurofibrillary tangles with calcification, Down's syndrome, frontotemporal dementia and Parkinsonism linked to chromosome 17 (FTDP-17), Gerstmann-Straussler-Scheinker disease, Niemann-Pick disease, non-Guamanian motor neuron disease with neurofibrillary tangles, Pick's disease, progressive supranuclear palsy, subacute sclerosing panencephalitis, and tangle only dementia (Lee et al., 2001). Genetic studies show that missense mutations in the *tau* gene (*Mtapt*) induce FTDP-17 supporting its importance in tauopathies (Hutton et al., 1998; Lee et al., 2001), although such mutations have not been found in AD. These tau mutations develop similar pathology in transgenic animal models, such as P301L tau mouse (Lewis et al., 2000; Gotz et al., 2001a), R406W tau mouse (Sato et al., 2002), G272V tau mouse (Gotz et al., 2001b), and V337M tau mouse (Tanemura et al., 2002). However, P301L tau mice develop NFT-like tau aggregation not only in CNS but also in peripheral motor neurons, astrocytes and oligodendrocytes, that is correlated with gait disturbances (Lewis et al., 2000).

25.3.8. Neurotoxicity and Synaptic Dysfunction

25.3.8.1. Aβ-Mediated Neurotoxicity

It has been widely accepted that toxicity of Aβ requires aggregation of native Aβ monomers (Hardy and Higgins, 1992; Pike et al., 1993). Aβ (Aβ40 or Aβ42) self-assembles to form low-n oligomers (dimers-hexamers), protofibrils, and fibrils (Roher et al., 1996; Harper et al., 1997; Walsh et al., 1997; Lambert et al., 1998). Since intermediates can further associate into higher-ordered aggregates, it is difficult to determine the oligomerization state of Aβ that causes pathogenicity. Nonetheless, several groups succeeded in isolating spherical non-fibrillar assemblies of synthetic Aβ (Aβ-derived diffusible ligands; ADDLs and amylospheroid; ASPD) (Lambert et al., 1998; Hoshi et al., 2003). These Aβ oligomers show more potent toxicity than fibrillar Aβ at lower concentrations (Lambert et al., 1998) and induce reversible synaptic loss through N-methyl-D-aspartate (NMDA) receptor interaction in organotypic hippocampal culture systems (Shankar et al., 2007). Thus, Aβ oligomers are a likely therapeutic target (De Felice et al., 2004).

Aβ can also potentially mediate neurotoxicity by activating microglia and glia and hereby promoting their secretion of a variety of neurotoxic factors. Aβ and secreted β-amyloid precursor protein (sAPP)-induce microglial activation and neurotoxicity (Giulian and Baker, 1986; Meda et al., 1995; Barger and Harmon, 1997; Ikezu et al., 2003). Neurotoxicity can be mediated through secretion of neurotoxic agents such as FasL, tumor necrosis factor (TNF)-α, reactive oxygen species (ROS), proteases, excitatory amino acids, and nitric oxides. Excitatory amino acids (glutamate, quinolinic acid, D-serine, among others) activate N-methyl-D-aspartate (NMDA) receptors, which are significantly involved in microglial-mediated neurotoxicity (Piani et al., 1992; Giulian et al., 1994). Indeed, both Aβ and sAPP activated microglia increase glutamate secretion (Barger and Basile, 2001; Ikezu et al., 2003). Glutamate and ROS can exacerbate each other's neurotoxic effects due to their relationships with the cystine/glutamate exchange system (Xc^-) antiporter. Aβ/sAPP-induced microglial activation leads to ROS production and glutathione consumption, that in turn promotes cystine import to regenerate glutathione and, as a consequence, increased glutamate is exported in exchange through the Xc antiporter (Ikezu et al., 2003).

25.3.8.2. Aβ-Mediated Synaptic Dysfunction

There are a number of studies documenting the synaptic dysfunction of APP mice, which starts as early as 4 months of age, including enhanced paired-pulse facilitation, disturbed high frequency stimulation (HFS) burst, and rapid decay of LTP in the CA1 field of hippocampal slices (Hsia et al., 1999; Larson et al., 1999). In addition, impaired synaptic transmission and LTP in both the CA1 and DG regions are also evident in Tg2576 mice (Chapman et al., 1999). Aβ oligomers block hippocampal long-term potentiation (LTP) *ex vivo* at nanomolar concentrations. *In vivo* applications of Aβ oligomers potently inhibit rat hippocampal LTP and cognitive function (Walsh et al., 2002; Cleary et al., 2005). These studies strongly suggest that Aβ oligomers induce memory impairment through inhibition of synaptic transmission and long-term potentiation in brain. One of the potential mechanisms of Aβ-induced LTP inhibition is through the activation of a number of inflammation-related molecules, such as c-jun N-terminal kinase, CDK5, p38 mitogen-activated kinase, inducible nitric oxide synthase, and superoxide, (Wang et al., 2004a; Wang et al., 2004b). These data suggest significant involvement of microglial activation in the inhibitory mechanism.

25.3.8.3. Neurotoxicity by Tau Aggregation

Although NFT is a hallmark of AD, its direct involvement in neuronal cell death is still unclear. Presence of NFT does not appear to be highly toxic since clinical studies suggest long-term survival of NFT-bearing neurons (Morsch et al., 1999; Hof et al., 2003). These data are also supported by studies using FTDP-17 tau mutant mice showing NFT formation in the absence of neuronal cell loss (Santacruz et al., 2005) and a lack of cognitive impairment as assessed by a water maze test (Arendash et al., 2004). A new doxycycline-inducible tau transgenic mouse model suggests accumulation of intracellular tau protein rather than NFT formation is responsible for neurodegeneration and cognitive impairment (Ramsden et al., 2005; Santacruz et al., 2005). However, tau is known to be involved in Aβ aggregate-induced neurotoxicity in conjunction with protein kinase activation *in vitro* (Rapoport et al., 2002; Hoshi et al., 2003; Medina et al., 2005). Thus, it is possible that acute neuronal cytoskeletal destabilization by tau phosphorylation and its end-product (globular tau aggregates) are neurotoxic, but its subsequent subcellular compartmentalization as NFT/PHF is a preventive neuroprotective process used to contain neurotoxic aggregates. Further study will be necessary to characterize its effect on neurodegeneration or dementia.

25.4. Immunity and AD

Neuronal loss, edema, and recruited mononuclear phagocytes (moving from blood to brain), are all classical hallmarks of brain inflammation in AD (Arvin et al., 1996). Gliosis, characterized by activation, proliferation, and hypertrophy of astrocytes and microglia, is one important feature of neuroinflammation. Amyloid plaques contain a number of proteins including complement proteins and pro-inflammatory cytokines and chemokines. Many of these proteins are secreted by activated microglia and reactive astrocytes. Activated microglia integrate deeply into neuritic plaques and express major histocompatibility complex type II (MHC-II), a marker for microglial activation. MHC-II expression is significantly upregulated in AD brains (Rogers et al., 1988; Styren et al., 1990).

Neurotoxicity can be mediated through secretion of various factors, ROS, proteases, excitatory amino acids, and nitric oxides. Pro-inflammatory cytokines and chemokines can inhibit fibrillar Aβ phagocytosis by a murine microglial cell line (Koenigsknecht-Talboo and Landreth, 2005). Such T-cell-mediated cytokines also inhibit post-phagocytosis intracellular Aβ degradation in human monocyte-derived macrophages. Aβ-microglia interaction results in two outputs: phagocytosis and signaling for priming and activation. Aβ binding to a complex of cell surface proteins (including CD36, integrin-associated protein (CD47), and $\alpha_6\beta_1$-integrin) leads to src-like tyrosine kinase activation and ROS production (Bamberger et al., 2003). CD36 is necessary for full Aβ-induced mononuclear phagocyte activation and chemotaxis (El Khoury et al., 2003). These findings strongly suggest that Aβ aggregates activates microglia by its binding to specific cell surface molecules, which activates src-like tyrosine kinases and other signaling for Aβ clearance and neuroinflammation in AD brains.

25.4.1. Microglia-Mediated Aβ Clearance

Our current knowledge of Aβ clearance is limited compared to the more detailed understanding of Aβ synthesis and aggregation mechanisms. It is well known that Aβ stimulation enhances microglial phagocytosis, *in vitro* (Kopec and Carroll, 1998). The internalized Aβ in phagosomes initially migrate to acid hydrolase-containing late endosomes and lysosomal compartments, and then moves into perinuclear vesicles, that are morphologically similar to lysosomes, where it stays up to 20 days (Paresce et al., 1997). The long-term accumulation of Aβ has also been observed by using unlabeled Aβ aggregates in bone marrow-derived macrophages (Yamamoto et al., 2007a). In addition to microglia, astrocytes also contribute to the uptake and removal of Aβ in a manner enhanced by a CC-type chemokine, CCL2 (Wyss-Coray et al., 2003).

Microglia constantly take-up Aβ via potential Aβ receptors, including scavenger receptor type A (SR-A), CD36, and receptor for advanced glycation end product (RAGE) (El Khoury et al., 1996; Yan et al., 1996; El Khoury et al., 2003). Although SR-A was characterized as a major receptor for microglial phagocytosis of Aβ (Paresce et al., 1996), APP mice deficient in SR-A show no difference in amyloid plaque formation or synaptic degeneration (Huang et al., 1999). This suggests that other scavenger receptors or Aβ-binding molecules, such as CD36, CD47, and $\alpha_6\beta_1$-integrin, can compensate for the deficiency of SR-A. On the other hand, RAGE seems to enhance Aβ deposition. APP mice deficient in RAGE exhibit reduced Aβ deposition (unpublished observations), which is consistent with enhanced Aβ deposition in mice over-expressing RAGE (Arancio et al., 2004). P-glycoprotein (Pgp), an ATP binding cassette transporter/multidrug resistance-1 α and β gene, is another molecule involved in Aβ clearance from brain parenchyma. Disruption of Pgp exhibits enhanced Aβ deposition, suggesting its role in endothelial efflux of Aβ from the brain (Cirrito et al., 2005).

25.4.2. Glial Inflammation and Innate Immunity

Innate immune responses in AD are mediated by a large number of inflammatory proteins, such as complement factors, acute-phase proteins, pro-inflammatory cytokines, and chemokines (Akiyama et al., 2000; McGeer and McGeer, 2001). Upregulation of the complement system includes activation of both classical and alternative pathways and enhanced secretion of cytokines and chemokines including, but not limited to, IFN-γ, IL-1α/β, IL-6, IL-8, IL-12, TGF-β1/2, TNF-α, CD40L, CCL2 (MCP-1), CCL3 (MIP-1α), CCL4 (MIP-1β), S100-β (for review, see (Akiyama et al., 2000)). The interplay between Aβ aggregates, microglia and astrocytes results in secretion of proinflammatory cytokines that, in turn, may affect local APP expression in nearby neurons. This process results in local upregulation of Aβ and seeding of "new" Aβ deposits, which may therefore be a potential mechanism for the plaque accumulation typically seen in AD brains.

25.4.2.1. Cytokines

25.4.2.1.1. IL-1

IL-1 upregulation seems to occur early in Aβ plaque formation since it is associated with diffuse plaques and also found in young children with Down's syndrome (Griffin et al., 1989; Griffin et al., 1995). IL-1 induces S100β, an astrocyte-derived neurite promoting cytokine (Kligman and Marshak, 1985), and α1-antichymotripsin (Das and Potter, 1995). IL-1 upregulates APP synthesis in primary astrocytes (Rogers et al., 1999), and Aβ production when co-stimulated with IFN-γ *in vitro* (Blasko et al., 2000).

25.4.2.1.2. IFN-γ

IFN-γ is produced by Th1 type T cells and natural killer cells (Young and Hardy, 1995), as well as macrophages (Fultz et al., 1993), astrocytes, and microglia (DeSimone et al., 1998). IFN-γ affects Aβ production, Aβ degradation (Yamamoto et al., 2007b), and neurotoxicity (Meda et al., 1999). IFN-γ, in conjunction with TNF-α, has been shown to enhance APP mRNA transcription and Aβ production (Blasko et al., 1999), and indirectly induce neuronal loss by stimulating release of NO from microglia (Chao et al., 1995). In the Tg2576 mouse model, astrocytes significantly upregulate IFN-γ mRNA *in vivo* (Abbas et al., 2002). We have recently found that transgenic APP mice (Tg2576) lacking IFN-γ receptor type I show marked reduction of Aβ deposition, microgliosis, astrogliosis, and astrocytic BACE expression(Yamamoto et al., 2007b). Thus, IFN-γ plays an important role for AD pathogenesis.

25.4.2.1.3. IL-6

IL-6 is a pleiotropic proinflammatory cytokine, secreted from T cells, macrophages, astro/microglia, and is involved in a number of inflammatory signals, including the acute phase response. IL-6 and its receptor complex, IL-6R and gp130/CD130, are all upregulated in AD brains in a region-specific manner (Hampel et al., 2005). IL-6 immunoreactivity is colocalized with diffuse plaques but not with compact plaques (Hull et al., 1996), suggesting its role in diffuse plaque reaction rather than compact plaque formation. Astrocyte-derived IL-6 induces expression of complement component C3, and may be involved in complement cascade activation in AD (Barnum et al., 1996).

25.4.2.1.4. TGF-β1/2/3

TGF-β1/2/3 are expressed in the CNS and are involved in development, homeostasis, and repair (Finch et al., 1993). TGF-β1 has been detected in plaques (van der Wal et al., 1993) and is elevated in cerebrospinal fluid and serum in AD (Chao et al., 1994b; Chao et al., 1994a). TGF-β2 has also been detected in reactive astrocytes, microglia, and in tangle-bearing neurons in AD brains (Wyss-Coray et al., 2000). TGF-β1 transgenic mice show cerebrovascular amyloid deposition (Wyss-Coray et al., 1997). Over-expression of TGF-β1 in APP mice accelerates vascular amyloid deposition although plaque burden was reduced in brain tissue (Wyss-Coray et al., 1997; Wyss-Coray et al., 2001), suggesting its role in clearing Aβ from brain parenchyma. Since TGF-β1 increases APP expression in astrocytes and microglia (Gray and Patel, 1993; Monning et al., 1994; Harris-White et al., 1998), it may also be in involved in TGF-β1 glial Aβ production. Disruption of TGF-β1 induces neurodegeneration and microgliosis (Brionne et al., 2003). APP mice over-expressing kinase-deficient TGF-β receptor type II, which inhibits endogenous TGF- β signaling, show enhanced Aβ deposition, neurodegeneration, and dendritic loss (Tesseur et al., 2006).

25.4.2.1.5. TNF-α

TNF-α, a pleiotropic proinflammatory cytokine, has been extensively investigated in AD. TNF-α is predominantly expressed by activated microglia, and to a lesser extent, astrocytes and neurons (Meda et al., 1995; Botchkina et al., 1997; Bezzi et al., 2001; Williams et al., 2005). In AD, TNF-α is upregulated in cerebrospinal fluid (Tarkowski et al., 1999), serum (Fillit et al., 1991), and cortex (Tarkowski et al., 1999). The effect of TNF-α on neurons is different dependent on the neuronal preparation and perhaps species; it is neurotrophic to rat cortical, hippocampal, and septal neurons (Cheng et al., 1994; Barger and Harmon, 1997), while directly neurotoxic (or enhances glutamate neurotoxicity) to human cortical neurons (Chao and Hu, 1994; D'Souza et al., 1995; Williams et al., 2005). This discrepancy could be due to the diverse signaling of two distinctly different TNF-α receptors, type I (p55 TNFR-1) and type II (p75 TNFR-2) (Botchkina et al., 1997; Dziewulska and Mossakowski, 2003). TNFR-1, but not TNFR-2, contains an intracellular death domain, common to proteins in the TNF receptor family. TNFR-1 mediates both apoptosis and NF-κB-mediated cell survival (Wallach et al., 1999; Gupta and Gollapudi, 2005). Importantly, primary cultured hippocampal neurons derived from TNFR-1 null mice were resistant to Aβ-mediated neurotoxicity (Li et al., 2004). Thus, TNF-α-mediated neurotoxicity is bidirectional and the outcomes of TNF signaling depend distinctly on cell type and species.

25.4.2.1.6. CD40L

CD40L belongs to the TNF family and one of the most important T-cell mediated cytokines. CD40, the receptor of CD40L, is upregulated by Aβ-stimulation of primary cultured microglia *in vitro* (Tan et al., 1999) and is found to be expressed in reactive microglia in AD patient brain tissue (Togo et al., 2000). Transgenic APP (Tg2576) mice deficient in CD40L show reduced Aβ deposition, astro/microgliosis, and hyperphosphorylation of tau (Tan et al., 2002). In addition, chronic treatment of APP/PS1 biogenic mice with neutralizing anti-CD40L antibody reduces Aβ deposition and causes a modest improvement in cognitive performance (Tan et al., 2002; Todd Roach et al., 2004), suggesting its therapeutic application for AD.

25.4.2.2. Chemokines

25.4.2.2.1. CCL2 (MCP-1)

CCL2 (MCP-1) is a member of the β chemokine subfamily and a candidate molecule for stimulating monocyte chemotaxis into the CNS (Charo et al., 1994). Activated astrocytes and cells of monocytic origin (such as microglia and macrophages) have been shown to express CCL2 in the brain (Calvo et al., 1996; Glabinski et al., 1996). CCL2 has been detected in senile plaques, reactive microglia and microvessels in AD brains (Ishizuka et al., 1997; Xia and Hyman, 1999; Grammas and Ovase, 2001), and is upregulated in cerebrospinal fluid and serum of AD cases and AD animal models (Sun et al., 2003; Galimberti et al., 2005; Janelsins et al., 2005). Over-expression of CCL2 resulted in increased diffuse plaque deposition, microglial accumulation, and apoE expression in APP mice (Yamamoto et al., 2005), accompanied by cognitive impairment (Yamamoto, M et al., unpublished observations). These results show that CCL2/CCR2 signaling is critical in mononuclear phagocyte accumulation and Aβ deposition in CNS.

25.4.2.2.2. CXCR Family

A number of other chemokines and their receptors have been identified in AD. CXCR2 is a member of the CXC chemokines receptor family and is a receptor for multiple ligands, including CXCL1/2/3 (GROα/β/δ) (Zlotnik and Yoshie, 2000). Both CXCR2 and CXCL1 have been detected in AD brains. In particular, CXCR2 expression has been identified in dystrophic neurites of senile plaques (Xia et al., 1997). CXCR3, another CXC receptor family, and its ligand CXCL10 (IP-10) have been found in both cortical and subcortical neurons in normal and AD brains. CXCL10 was significantly upregulated in AD brain (Xia et al., 2000). In addition, expression of CC chomokine receptors, such as CCR3 and CCR5, and their ligands, CCL4 (MIP-1β) and CCL5 (RANTES), is also enhanced on reactive microglia in AD brain (Xia et al., 1998). Although the exact physiological functions of these molecules have yet to be characterized in the context of AD pathogenesis, modulation of such chemokine signaling might be beneficial to correct the detrimental imbalance of metabolic homeostasis and neuroinflammation in brain.

25.4.2.3. Toll-Like Receptors

One of the most important recent findings in the effort to understand innate immunity has been the identification of Toll-like receptors (TLRs). TLRs define a major class of pattern-recognition receptors critical to the initiation and tailoring of both innate and subsequent adaptive immune responses (Beutler, 2004; Iwasaki and Medzhitov, 2004). For example, bacterial cell wall components are recognized by TLR2, and lipopolysaccharide and viral envelope proteins are recognized by TLR4. Double strand RNA, single strand RNA, and unmethylated CpGs are recognized by TLR3, TLR7/8, and TLR9, respectively (for review, see (Akira et al., 2006)). Primary cultured human and mouse microglia express mRNA for TLR1–9, whereas human and mouse astrocytes express high levels of TLR3, and low-level TLR 1, 2, and 4–6 (Olson and Miller, 2004; Jack et al., 2005) (Carpentier et al., 2005; McKimmie and Fazakerley, 2005). A recent study on aged mouse brains demonstrated that TLR1, TLR2, TLR4, TLR5, TLR7 and CD14 expression were upregulated in correlation with age, whereas TLR9 was downregulated (Letiembre et al., 2007). Interestingly, a TLR4 polymorphism was also associated with successful aging (Candore et al., 2006), which further indicates a role of innate immune receptors in aging and potentially AD.

The role of TLRs in neuroinflammation during AD is poorly characterized. However, recent reports indicate that TLR9 ligand (CpG-containing oligonucleotide) induced enhancement of Aβ update by N9 microglia cell line (Iribarren et al., 2005), and Aβ stimulation modulates TLR-specific inflammation (NO and TFN-α release) in primary cultured mouse microglia (Lotz et al., 2005). These results suggest that there is crosstalk between Aβ and TLR signaling that may regulate glial inflammation and Aβ clearance in CNS.

25.4.2.4. Matrix Metalloprotease

Aβ stimulation induces robust expression of matrix metalloprotease (MMP) 1, 3, 10, 10, 12, 19, and disintegrin and metalloprotease (ADAM) 8 in primary cultured human microglia (Walker et al., 2006). Increased expression of MMP 1, 3 and 9 has been documented in AD brains (Backstrom et al., 1996; Leake et al., 2000; Yoshiyama et al., 2000) and may contribute to tissue damage (Cuzner and Opdenakker, 1999). These molecules may be involved in Aβ degradation, cytokine/chemokine processing, and shedding of cell adhesion molecules for enhancing glial motility.

25.4.2.5. Other Inflammatory Molecules: Complement Complex and CD45

Complement C1q, the first complement of the classical complement pathway, is upregulated in AD brains and is found in senile plaques (Afagh et al., 1996). C1q binds to soluble Aβ and modulates its uptake by mononuclear phagocytes (Webster et al., 2000). C1q also directly induces neurotoxicity through its interaction with cell surface calreticulin, a C1q receptor (Luo et al., 2003). However, APP mice lacking C1q exhibited reduced glial and astroglial activation without changes in Aβ deposition (Fonseca et al., 2004a). Thus, C1q may be involved in AD pathogenesis by influencing neurotoxicity and glial inflammation, but its effect on Aβ metabolism is modest.

Complement C3 is upregulated in APP/TGF-β1 biogenic mice and AD brains (Wyss-Coray et al., 2001). Over-expression of soluble complement receptor-related protein y (sCrry), a complement C3 inhibitor, resulted in enhanced Aβ accumulation and hippocampal degeneration (Wyss-Coray et al., 2002). Since C3 is known to promote phagocytosis by binding to specific C3 receptor on specialized cells, C3 is likely to be positively involved in Aβ clearance.

CD45, a transmembrane protein tyrosine phosphatase critically involved in negative regulation of T- and B-cell activation, and an antagonizing partner of the TNF superfamily (Tan et al., 2000a), is also upregulated in AD brains (Masliah et al., 1991). APP mice deficient in CD45 have increased brain levels of TNF-α and NO, suggesting CD45 negative regulation of neuroinflammation in AD (Tan et al., 2000b).

25.4.3. Oxygen Free Radicals

Production of ROS, factors involved in the aging process (Finkel and Holbrook, 2000), is elevated in AD brain and may be an important cause of AD (Martins et al., 1986). Elevated levels of oxidized lipids (lipid peroxidation, malondialdehyde, 4-hydroxynonenal) (Markesbery and Carney, 1999), proteins (advanced glycation end product modifications, tyrosine nitration) (Good et al., 1996; Takeda et al., 1998), and nucleic acids (8-hydroxy-deoxyguanosine) have been documented in AD brains (Lyras et al., 1997). Mitochondria and nicotinamide adenine dinucleotide phosphate (NADPH) oxidase complex dysfunction can generate large amounts of ROS (Beal, 1998), and down regulation of mitochondrial NADPH 15-kD gene have been found in AD cases (Beal, 1998; Manczak et al., 2004). Markers of oxidative stress are elevated not only in the pathologic lesions in AD brain (Good et al., 1996), but also in cerebrospinal fluid of AD patients (Lovell et al., 1997). Aβ peptide can directly activate the NADPH oxidase complex of mononuclear phagocytes (Bianca et al., 1999) and microglia, which are mediated through the interaction of fibrillar Aβ to cell surface molecules (CD36, CD47, and $\alpha6\beta1$-integrin) and tyrosine kinase Vav signaling (Wilkinson et al., 2006). ROS is also known to mediate Aβ-induced neurotoxicity (Behl et al., 1994; Ikezu et al., 2003). In addition to NADPH oxidase, myeloperoxidase (MPO) is expressed by microglia in close proximity to Aβ plaques, and surprisingly, in neurons of AD brains (Reynolds et al., 1999; Green et al., 2004). MPO can catalyze both apoE and Aβ, generating oxidated and fragmentated apoE (Jolivalt et al., 1996) and stable Aβ42 dimers through dityrosine bridge formation (Jolivalt et al., 1996; Galeazzi et al., 1999).

25.4.3.1. Oxidative DNA Damage

DNA damage may also have a role in AD pathology. Recent studies indicate that aging-mediated DNA damage, including alterations in neuronal network and cognitive function-related gene expression profiles, is significantly increased after 40 years of age, (Lu et al., 2004). DNA damage induces cell cycle activation (Kruman et al., 2004), and cell cycle activation has been documented in AD brains (Nagy et al., 1998; Smith et al., 1999). DNA damage also potently induces p53, a tumor suppressor and transcriptional factor involved in neuronal apoptosis (Morrison et al., 2003; Kruman et al., 2004; Culmsee and Mattson, 2005). Neuronal expression of p53 is increased in AD brains (de la Monte et al., 1997) and in APP mouse brains (LaFerla et al., 1996). These data suggest that DNA damage is involved in neurodegeneration in AD brain, or the result of neurodegeneration.

25.5. Immunopharmacology

Epidemiological studies using nonsteroidal anti-inflammatory drugs (NSAIDs) has led to the proposal that NSAIDs reduce the risk of AD (McGeer et al., 1996). These studies have been supported by data showing that treatment of APP mice with NSAIDs (such as ibuprofen and NO-flurbiprofen) *in vivo* results in reduced Aβ deposition and pro-inflammatory cytokine production (Lim et al., 2000; Jantzen et al., 2002; Yan et al., 2003). Microglial activation is suppressed by ibuprofen but rather enhanced by NO-flurbiprofen. Other studies have found that NSAID also affects γ-processing of APP through inhibition of the small GTP-binding protein Rho, which shifts Aβ processing from Aβ42 to Aβ40 both *in vitro* and *in vivo* (Weggen et al., 2001; Eriksen et al., 2003; Zhou et al., 2003). Based on these observations, a new series of drugs derived from NSAIDs having a more specific effect on γ-processing has been developed (Kukar et al., 2005).

Proliferator-activated receptor γ (PPARγ) agonists, such as the thiazolinedione family, are also NSAIDs and can efficiently clear Aβ deposition *in vitro* and *in vivo* (Camacho et al., 2004; Heneka et al., 2005). PPAR is a nuclear transcription factor and its effect is different from ibuprofen and its derivatives. PPARγ transcription factors play important physiological roles in the regulation of lipid metabolism (Mangelsdorf et al., 1995; Lemberger et al., 1996), suppression of inflammation (Heneka et al., 2000), and clinical treatment of diabetes type II (Dormandy et al., 2005). PPARγ agonists enhance Aβ degradation (Camacho et al., 2004), as well as suppress BACE expression and microglial activation (Heneka et al., 2005).

Other anti-inflammatory/antioxidant dietary applications include curcumin, the yellow pigment in turmeric that targets multiple AD pathogenic cascades (Lim et al., 2001; Yang et al., 2005). The dietary omega-3 fatty acid, docosahexaenoic acid (DHA), also reduced amyloid, oxidative damage and synaptic and cognitive deficits in a transgenic mouse model (Calon et al., 2004; Lim et al., 2005). Both DHA and curcumin have favorable safety profiles, epidemiology and efficacy, and may exert general anti-aging benefits (Cole et al., 2005).

Summary

AD is the most common form of elderly dementia and has no effective therapy. Clinical diagnosis is based on the evaluation of cognitive function and lab tests. Pathological diagnosis is based on post-mortem neuropathology, defined by three hallmarks: senile plaque, NFT, and neuronal cell loss. Senile plaque mainly consists of Aβ aggregates and serves as a focal point of astrocyte and microglial activation, while NFT contains hyperphosphorylated tau. Aβ is produced during the processing of APP by BACE and the γ-secretase complex that includes PS1. Both APP and PS1 are causative genes of AD, while apoE4 is a risk factor allele. The Aβ load in CNS is maintained by balancing its production and clearance, which is mediated by passive diffusion to blood stream, degradation by Aβ degrading enzymes (IDE and neprilysin), and microglial phagocytosis. Aβ production, deposition, and neurodegeneration are significantly regulated by astroglial and microglial activation via direct contact as well as secretion of a number of neurotoxicants. Specific pro-inflammatory cytokines, chemokines, and excitotoxins have been characterized as mediators of neuroinflammation and neurotoxicity. Currently their specific roles in mediating the pathogenesis in AD are being explored. Those diverse studies are generating a large number of potential therapeutic targets for treating AD. Beneficial effects have been reported for anti-inflammatory drugs, such as NSAIDs, to treat glial inflammation and disease progression.

Review Questions/Problems

1. Which is NOT the causative gene of familial Alzheimer's disease?

a. APP
b. Presenilin-1
c. Presenilin-2
d. Tau
e. None of the above

2. Presenlin-1 is a member of:

a. α-secretase
b. β-secretase
c. γ-secretase complex
d. Aβ degrading enzyme
e. None of the above

3. Amyloid-beta peptide is a processing product of amyloid precursor protein (APP) through a combination of

a. α-secretase and Beta-site APP cleaving enzyme (BACE1)
b. α-secretase and γ-secretase complex
c. BACE1 and γ-secretase complex
d. BACE1 and insulin-degrading enzyme
e. none of the above

4. Microglia-mediated inflammatory reaction affects AD progression via

a. cytokine production
b. reactive nitrogen/oxygen production
c. excitotoxin production
d. all of the above
e. none of the above

5. Aβ deposition in APP mouse brain can be reduced by vaccination with the following antigens <u>except</u>

a. Aβ
b. apoE4
c. Copaxon-1
d. transgenic potato expressing amyloid-beta peptide
e. none of the above

6. Nonsteroidal anti-inflammatory drugs (NSAIDs) can reduce Aβ deposition in APP mouse brain through

a. suppression of brain inflammation
b. suppression of Aβ42 generation
c. suppression of BACE expression
d. all of the above
e. none of the above

7. Describe the available animal models for studying AD. What is the limitation of the animal models?

8. Describe the potential immunotherapy of AD

References

Abbas N, Bednar I, Mix E, Marie S, Paterson D, Ljungberg A, Morris C, Winblad B, Nordberg A, Zhu J (2002) Up-regulation of the inflammatory cytokines IFN-gamma and IL-12 and down-regulation of IL-4 in cerebral cortex regions of APP(SWE) transgenic mice. J Neuroimmunol 126:50–57.

Afagh A, Cummings BJ, Cribbs DH, Cotman CW, Tenner AJ (1996) Localization and cell association of C1q in Alzheimer's disease brain. Exp Neurol 138:22–32.

Akira S, Uematsu S, Takeuchi O (2006) Pathogen recognition and innate immunity. Cell 124:783–801.

Akiyama H, Barger S, Barnum S, Bradt B, Bauer J, Cole GM, Cooper NR, Eikelenboom P, Emmerling M, Fiebich BL, Finch CE, Frautschy S, Griffin WS, Hampel H, Hull M, Landreth G, Lue L, Mrak R, Mackenzie IR, McGeer PL, O'Banion MK, Pachter J, Pasinetti G, Plata-Salaman C, Rogers J, Rydel R, Shen Y, Streit W, Strohmeyer R, Tooyoma I, Van Muiswinkel FL, Veerhuis R, Walker D, Webster S, Wegrzyniak B, Wenk G, Wyss-Coray T (2000) Inflammation and Alzheimer's disease. Neurobiol Aging 21:383–421.

Alonso AC, Grundke-Iqbal I, Iqbal K (1996) Alzheimer's disease hyperphosphorylated tau sequesters normal tau into tangles of filaments and disassembles microtubules. Nat Med 2:783–787.

Alzheimer A (1907) Ueber eine eigenartige Erkrankung der Hirnrinde. Allg Z Psychiat 64:146–148.

Apelt J, Schliebs R (2001) Beta-amyloid-induced glial expression of both pro- and anti- inflammatory cytokines in cerebral cortex of aged transgenic Tg2576 mice with Alzheimer plaque pathology. Brain Res 894:21–30.

Apelt J, Ach K, Schliebs R (2003) Aging-related down-regulation of neprilysin, a putative beta-amyloid-degrading enzyme, in transgenic Tg2576 Alzheimer-like mouse brain is accompanied by an astroglial upregulation in the vicinity of beta-amyloid plaques. Neurosci Lett 339:183–186.

Arancio O, Zhang HP, Chen X, Lin C, Trinchese F, Puzzo D, Liu S, Hegde A, Yan SF, Stern A, Luddy JS, Lue LF, Walker DG, Roher A, Buttini M, Mucke L, Li W, Schmidt AM, Kindy M, Hyslop PA, Stern DM, Du Yan SS (2004) RAGE potentiates Abeta-induced perturbation of neuronal function in transgenic mice. EMBO J 23:4096–4105.

Arendash GW, Lewis J, Leighty RE, McGowan E, Cracchiolo JR, Hutton M, Garcia MF (2004) Multi-metric behavioral comparison of APPsw and P301L models for Alzheimer's disease: Linkage of poorer cognitive performance to tau pathology in forebrain. Brain Res 1012:29–41.

Arends YM, Duyckaerts C, Rozemuller JM, Eikelenboom P, Hauw JJ (2000) Microglia, amyloid and dementia in alzheimer disease. A correlative study. Neurobiol Aging 21:39–47.

Aronson MK, Ooi WL, Morgenstern H, Hafner A, Masur D, Crystal H, Frishman WH, Fisher D, Katzman R (1990) Women, myocardial infarction, and dementia in the very old. Neurology 40:1102–1106.

Arvin B, Neville LF, Barone FC, Fewerstein GZ (1996) The Role of inflammation and cytokines in brain injury. Neurosci Biobehav Rev 20:445–452.

Asai M, Hattori C, Szabo B, Sasagawa N, Maruyama K, Tanuma S, Ishiura S (2003) Putative function of ADAM9, ADAM10, and ADAM17 as APP alpha-secretase. Biochem Biophys Res Commun 301:231–235.

Association AP (1994) Diagnostic and Statistical Manual of Mental Disorders, 4th edition Edition. Washington, DC: American Psychiatric Association.

Backstrom JR, Lim GP, Cullen MJ, Tokes ZA (1996) Matrix metalloproteinase-9 (MMP-9) is synthesized in neurons of the human hippocampus and is capable of degrading the amyloid-beta peptide (1–40). J Neurosci 16:7910–7919.

Bales KR, Verina T, Cummins DJ, Du Y, Dodel RC, Saura J, Fishman CE, DeLong CA, Piccardo P, Petegnief V, Ghetti B, Paul SM (1999) Apolipoprotein E is essential for amyloid deposition in the APP(V717F) transgenic mouse model of Alzheimer's disease. Proc Natl Acad Sci USA 96:15233–15238.

Bamberger ME, Harris ME, McDonald DR, Husemann J, Landreth GE (2003) A cell surface receptor complex for fibrillar beta-amyloid mediates microglial activation. J Neurosci 23:2665–2674.

Barger S, Harmon A (1997) Microglial activation by Alzheimer amyloid precursor protein and modulation by apolipoprotein E. Nature 388:878–881.

Barger SW, Basile AS (2001) Activation of microglia by secreted amyloid precursor protein evokes release of glutamate by cystine exchange and attenuates synaptic function. J Neurochem 76:846–854.

Barnum SR, Jones JL, Muller-Ladner U, Samimi A, Campbell IL (1996) Chronic complement C3 gene expression in the CNS of transgenic mice with astrocyte-targeted interleukin-6 expression. Glia 18:107–117.

Beal MF (1998) Mitochondrial dysfunction in neurodegenerative diseases. Biochim Biophys Acta 1366:211–223.

Behl C, Davis JB, Lesley R, Schubert D (1994) Hydrogen peroxide mediates amyloid beta protein toxicity. Cell 77:817–827.

Benzing WC, Wujek JR, Ward EK, Shaffer D, Ashe KH, Younkin SG, Brunden KR (1999) Evidence for glial-mediated inflammation in aged APP(SW) transgenic mice. Neurobiol Aging 20:581–589.

Berg L, McKeel DW, Jr., Miller JP, Baty J, Morris JC (1993) Neuropathological indexes of Alzheimer's disease in demented and nondemented persons aged 80 years and older. Arch Neurol 50:349–358.

Beutler B (2004) Inferences, questions and possibilities in Toll-like receptor signaling. Nature 430:257–263.

Bezzi P, Domercq M, Brambilla L, Galli R, Schols D, De Clercq E, Vescovi A, Bagetta G, Kollias G, Meldolesi J, Volterra A (2001) CXCR4-activated astrocyte glutamate release via TNFalpha: Amplification by microglia triggers neurotoxicity. Nat Neurosci 4:702–710.

Bianca VD, Dusi S, Bianchini E, Dal Pra I, Rossi F (1999) beta-amyloid activates the O-2 forming NADPH oxidase in microglia, monocytes, and neutrophils. A possible inflammatory mechanism of neuronal damage in Alzheimer's disease. J Biol Chem 274:15493–15499.

Biernat J, Gustke N, Drewes G, Mandelkow EM, Mandelkow E (1993) Phosphorylation of Ser262 strongly reduces binding of tau to microtubules: Distinction between PHF-like immunoreactivity and microtubule binding. Neuron 11:153–163.

Blasko I, Marx F, Steiner E, Hartmann T, Grubeck-Loebenstein B (1999) TNFalpha plus IFNgamma induce the production of Alzheimer beta-amyloid peptides and decrease the secretion of APPs. FASEB J 13:63–68.

Blasko I, Veerhuis R, Stampfer-Kountchev M, Saurwein-Teissl M, Eikelenboom P, Grubeck-Loebenstein B (2000) Costimulatory effects of interferon-gamma and interleukin-1beta or tumor necrosis factor alpha on the synthesis of Abeta1–40 and Abeta1–42 by human astrocytes. Neurobiol Dis 7:682–689.

Borchelt DR, Ratovitski T, van Lare J, Lee MK, Gonzales V, Jenkins NA, Copeland NG, Price DL, Sisodia SS (1997) Accelerated amyloid deposition in the brains of transgenic mice coexpressing mutant presenilin 1 and amyloid precursor proteins. Neuron 19:939–945.

Botchkina GI, Meistrell ME, 3rd, Botchkina IL, Tracey KJ (1997) Expression of TNF and TNF receptors (p55 and p75) in the rat brain after focal cerebral ischemia. Mol Med 3:765–781.

Braak H, Braak E (1995) Staging of Alzheimer's disease-related neurofibrillary changes. Neurobiol Aging 16:271–278; discussion 278–284.

Bramblett GT, Goedert M, Jakes R, Merrick SE, Trojanowski JQ, Lee VM (1993) Abnormal tau phosphorylation at Ser396 in Alzheimer's disease recapitulates development and contributes to reduced microtubule binding. Neuron 10:1089–1099.

Brionne TC, Tesseur I, Masliah E, Wyss-Coray T (2003) Loss of TGF-beta 1 leads to increased neuronal cell death and microgliosis in mouse brain. Neuron 40:1133–1145.

Buee L, Bussiere T, Buee-Scherrer V, Delacourte A, Hof PR (2000) Tau protein isoforms, phosphorylation and role in neurodegenerative disorders. Brain Res Brain Res Rev 33:95–130.

Cai H, Wang Y, McCarthy D, Wen H, Borchelt DR, Price DL, Wong PC (2001) BACE1 is the major beta-secretase for generation of Abeta peptides by neurons. Nat Neurosci 4:233–234.

Calon F, Lim GP, Yang F, Morihara T, Teter B, Ubeda O, Rostaing P, Triller A, Salem N, Jr., Ashe KH, Frautschy SA, Cole GM (2004) Docosahexaenoic acid protects from dendritic pathology in an Alzheimer's disease mouse model. Neuron 43:633–645.

Calvo CF, Yoshimura T, Gelman M, Mallat M (1996) Production of monocyte chemotactic protein-1 by rat brain macrophages. Eur J Neurosci 8:1725–1734.

Camacho IE, Serneels L, Spittaels K, Merchiers P, Dominguez D, De Strooper B (2004) Peroxisome-proliferator-activated receptor gamma induces a clearance mechanism for the amyloid-beta peptide. J Neurosci 24:10908–10917.

Candore G, Aquino A, Balistreri CR, Bulati M, Di Carlo D, Grimaldi MP, Listi F, Orlando V, Vasto S, Caruso M, Colonna-Romano G, Lio D, Caruso C (2006) Inflammation, longevity, and cardiovascular diseases: Role of polymorphisms of TLR4. Ann N Y Acad Sci 1067:282–287.

Carpentier PA, Begolka WS, Olson JK, Elhofy A, Karpus WJ, Miller SD (2005) Differential activation of astrocytes by innate and adaptive immune stimuli. Glia 49:360–374.

Chao C, Hu S, Erlich L, Peterson P (1995) Interleukin-1 and tumor necrosis factor-alpha synergistically mediate neurotoxicity: Involvement of nitric oxide and of N-methyl-D-aspartate receptors. Brain Behav Immun 9:355–365.

Chao CC, Hu S (1994) Tumor necrosis factor-alpha potentiates glutamate neurotoxicity in human fetal brain cell cultures. Dev Neurosci 16:172–179.

Chao CC, Hu S, Frey WH, 2nd, Ala TA, Tourtellotte WW, Peterson PK (1994a) Transforming growth factor beta in Alzheimer's disease. Clin Diagn Lab Immunol 1:109–110.

Chao CC, Ala TA, Hu S, Crossley KB, Sherman RE, Peterson PK, Frey WH, 2nd (1994b) Serum cytokine levels in patients with Alzheimer's disease. Clin Diagn Lab Immunol 1:433–436.

Chapman PF, White GL, Jones MW, Cooper-Blacketer D, Marshall VJ, Irizarry M, Younkin L, Good MA, Bliss TV, Hyman BT, Younkin SG, Hsiao KK (1999) Impaired synaptic plasticity and learning in aged amyloid precursor protein transgenic mice. Nat Neurosci 2:271–276.

Charo IF, Myers SJ, Herman A, Franci C, Connolly AJ, Coughlin SR (1994) Molecular cloning and functional expression of two monocyte chemoattractant protein 1 receptors reveals alternative splicing of the carboxyl-terminal tails. Proc Natl Acad Sci USA 91:2752–2756.

Chen G, Chen KS, Knox J, Inglis J, Bernard A, Martin SJ, Justice A, McConlogue L, Games D, Freedman SB, Morris RG (2000) A learning deficit related to age and beta-amyloid plaques in a mouse model of Alzheimer's disease. Nature 408:975–979.

Cheng B, Christakos S, Mattson MP (1994) Tumor necrosis factors protect neurons against metabolic-excitotoxic insults and promote maintenance of calcium homeostasis. Neuron 12:139–153.

Chishti MA, Yang DS, Janus C, Phinney AL, Horne P, Pearson J, Strome R, Zuker N, Loukides J, French J, Turner S, Lozza G, Grilli M, Kunicki S, Morissette C, Paquette J, Gervais F, Bergeron C, Fraser PE, Carlson GA, George-Hyslop PS, Westaway D (2001) Early-onset amyloid deposition and cognitive deficits in transgenic mice expressing a double mutant form of amyloid precursor protein 695. J Biol Chem 276:21562–21570.

Choi DS, Wang D, Yu GQ, Zhu G, Kharazia VN, Paredes JP, Chang WS, Deitchman JK, Mucke L, Messing RO (2006) PKC{varepsilon} increases endothelin converting enzyme activity and reduces amyloid plaque pathology in transgenic mice. Proc Natl Acad Sci USA 103(21):8215–8220.

Cirrito JR, Deane R, Fagan AM, Spinner ML, Parsadanian M, Finn MB, Jiang H, Prior JL, Sagare A, Bales KR, Paul SM, Zlokovic BV, Piwnica-Worms D, Holtzman DM (2005) P-glycoprotein deficiency at the blood-brain barrier increases amyloid-beta deposition in an Alzheimer's disease mouse model. J Clin Invest 115:3285–3290.

Cleary JP, Walsh DM, Hofmeister JJ, Shankar GM, Kuskowski MA, Selkoe DJ, Ashe KH (2005) Natural oligomers of the amyloid-beta protein specifically disrupt cognitive function. Nat Neurosci 8:79–84.

Cole GM, Lim GP, Yang F, Teter B, Begum A, Ma Q, Harris-White ME, Frautschy SA (2005) Prevention of Alzheimer's disease: Omega-3 fatty acid and phenolic anti-oxidant interventions. Neurobiol Aging 26:133–136.

Corey-Bloom J, Galasko D, Thal LJ (1994) Clinical Features and Natural History of Alzheimer's Disease. Philadelphia: WB Saunders.

Culmsee C, Mattson MP (2005) p53 in neuronal apoptosis. Biochem Biophys Res Commun 331:761–777.

Cuzner ML, Opdenakker G (1999) Plasminogen activators and matrix metalloproteases, mediators of extracellular proteolysis in inflammatory demyelination of the central nervous system. J Neuroimmunol 94:1–14.

D'Souza S, Alinauskas K, McCrea E, Goodyer C, Antel JP (1995) Differential susceptibility of human CNS-derived cell populations to TNF-dependent and independent immune-mediated injury. J Neurosci 15:7293–7300.

Danik M, Chabot JG, Michel D, Quirion R (1999) Clusterin and Apolipoprotein E Gene Expression in the Adult Brain. Austin: RG Landes Company.

Das S, Potter H (1995) Expression of the Alzheimer amyloid-promoting factor antichymotrypsin is induced in human astrocytes by IL-1. Neuron 14:447–456.

De Felice FG, Vieira MN, Saraiva LM, Figueroa-Villar JD, Garcia-Abreu J, Liu R, Chang L, Klein WL, Ferreira ST (2004) Targeting the neurotoxic species in Alzheimer's disease: Inhibitors of Abeta oligomerization. FASEB J 18:1366–1372.

de la Monte SM, Sohn YK, Wands JR (1997) Correlates of p53- and Fas (CD95)-mediated apoptosis in Alzheimer's disease. J Neurol Sci 152:73–83.

De Strooper B (2003) Aph-1, Pen-2, and Nicastrin with Presenilin Generate an Active gamma-Secretase Complex. Neuron 38:9–12.

DeSimone R, Levi G, Aloisi F (1998) Interferon-gamma gene expresion in rat central nervous system glial cells. Cytokine 10:418–422.

Dormandy JA, Charbonnel B, Eckland DJ, Erdmann E, Massi-Benedetti M, Moules IK, Skene AM, Tan MH, Lefebvre PJ, Murray GD, Standl E, Wilcox RG, Wilhelmsen L, Betteridge J, Birkeland K, Golay A, Heine RJ, Koranyi L, Laakso M, Mokan M, Norkus A, Pirags V, Podar T, Scheen A, Scherbaum W, Schernthaner G, Schmitz O, Skrha J, Smith U, Taton J (2005) Secondary prevention of macrovascular events in patients with type 2 diabetes in the PROactive Study (PROspective pioglitAzone Clinical Trial In macroVascular Events): A randomised controlled trial. Lancet 366:1279–1289.

Drewes G, Ebneth A, Preuss U, Mandelkow EM, Mandelkow E (1997) MARK, a novel family of protein kinases that phosphorylate microtubule-associated proteins and trigger microtubule disruption. Cell 89:297–308.

Duff K, Eckman C, Zehr C, Yu X, Prada CM, Perez-tur J, Hutton M, Buee L, Harigaya Y, Yager D, Morgan D, Gordon MN, Holcomb L, Refolo L, Zenk B, Hardy J, Younkin S (1996) Increased amyloid-beta42(43) in brains of mice expressing mutant presenilin 1. Nature 383:710–713.

Dziewulska D, Mossakowski MJ (2003) Cellular expression of tumor necrosis factor a and its receptors in human ischemic stroke. Clin Neuropathol 22:35–40.

Eikelenboom P, Zhan SS, van Gool WA, Allsop D (1994) Inflammatory mechanisms in Alzheimer's disease. Trends Pharmacol Sci 15:447–450.

Eikelenboom P, Bate C, Van Gool WA, Hoozemans JJ, Rozemuller JM, Veerhuis R, Williams A (2002) Neuroinflammation in Alzheimer's disease and prion disease. Glia 40:232–239.

El Khoury J, Hickman SE, Thomas CA, Cao L, Silverstein SC, Loike JD (1996) Scavenger receptor-mediated adhesion of microglia to beta-amyloid fibrils. Nature 382:716–719.

El Khoury JB, Moore KJ, Means TK, Leung J, Terada K, Toft M, Freeman MW, Luster AD (2003) CD36 mediates the innate host response to beta-amyloid. J Exp Med 197:1657–1666.

Eriksen JL, Sagi SA, Smith TE, Weggen S, Das P, McLendon DC, Ozols VV, Jessing KW, Zavitz KH, Koo EH, Golde TE (2003) NSAIDs and enantiomers of flurbiprofen target gamma-secretase and lower Abeta 42 in vivo. J Clin Invest 112:440–449.

Ewbank DC (1999) Deaths attributable to Alzheimer's disease in the United States. Am J Public Health 89:90–92.

Fagan AM, Watson M, Parsadanian M, Bales KR, Paul SM, Holtzman DM (2002) Human and murine ApoE markedly alters A beta metabolism before and after plaque formation in a mouse model of Alzheimer's disease. Neurobiol Dis 9:305–318.

Fillit H, Ding WH, Buee L, Kalman J, Altstiel L, Lawlor B, Wolf-Klein G (1991) Elevated circulating tumor necrosis factor levels in Alzheimer's disease. Neurosci Lett 129:318–320.

Finch CE, Laping NJ, Morgan TE, Nichols NR, Pasinetti GM (1993) TGF-beta 1 is an organizer of responses to neurodegeneration. J Cell Biochem 53:314–322.

Finkel T, Holbrook NJ (2000) Oxidants, oxidative stress and the biology of ageing. Nature 408:239–247.

Folstein MF, Folstein SE, McHugh PR (1975) "Mini-mental state". A practical method for grading the cognitive state of patients for the clinician. J Psychiatr Res 12:189–198.

Fonseca MI, Zhou J, Botto M, Tenner AJ (2004a) Absence of C1q leads to less neuropathology in transgenic mouse models of Alzheimer's disease. J Neurosci 24:6457–6465.

Fonseca MI, Kawas CH, Troncoso JC, Tenner AJ (2004b) Neuronal localization of C1q in preclinical Alzheimer's disease. Neurobiol Dis 15:40–46.

Fratiglioni L, Small B, Winblad B, Backman L (2001) The transition from normal functioning to dementia in the aging population. In: Alzheimer's Disease: Advances in Etiology, Pathogenesis, and Therapeutics (Iqbal K, Sisodia SS, Winblad B, eds), pp 3–10. West Sussex, UK: John Wiley & Sons Ltd.

Frautschy SA, Yang F, Irrizarry M, Hyman B, Saido TC, Hsaio K, Cole GM (1998) Microglial response to amyloid plaques in APPsw transgenic mice. Am J Pathol 152:307–317.

Fultz M, Barber S, Dieffenbach C, Vogel S (1993) Induction of IFN-g in macrophages by lipopolysaccharide. Int Immunol 5:1383–1392.

Galeazzi L, Ronchi P, Franceschi C, Giunta S (1999) In vitro peroxidase oxidation induces stable dimers of beta-amyloid (1–42) through dityrosine bridge formation. Amyloid 6:7–13.

Galimberti D, Fenoglio C, Lovati C, Venturelli E, Guidi I, Corra B, Scalabrini D, Clerici F, Mariani C, Bresolin N, Scarpini E (2005) Serum MCP-1 levels are increased in mild cognitive impairment and mild Alzheimer's disease. Neurobiol Aging 27(12):1763–1768.

Games D, Adams D, Alessandrini R, Barbour R, Berthelette P, Blackwell C, Carr T, Clemens J, Donaldson T, Gillespie F, et al. (1995) Alzheimer-type neuropathology in transgenic mice overexpressing V717F beta-amyloid precursor protein. Nature 373:523–527.

Giulian D, Baker TJ (1986) Characterization of ameboid microglia isolated from developing mammalian brain. J Neurosci 6:2163–2178.

Giulian D, Li J, Leara B, Keenen C (1994) Phagocytic microglia release cytokines and cytotoxins that regulate the survival of astrocytes and neurons in culture. Neurochem Int 25:227–233.

Glabinski AR, Balasingam V, Tani M, Kunkel SL, Strieter RM, Yong VW, Ransohoff RM (1996) Chemokine monocyte chemoattractant protein-1 is expressed by astrocytes after mechanical injury to the brain. J Immunol 156:4363–4368.

Good PF, Werner P, Hsu A, Olanow CW, Perl DP (1996) Evidence of neuronal oxidative damage in Alzheimer's disease. Am J Pathol 149:21–28.

Gotz J, Chen F, van Dorpe J, Nitsch RM (2001a) Formation of neurofibrillary tangles in P301l tau transgenic mice induced by Abeta 42 fibrils. Science 293:1491–1495.

Gotz J, Tolnay M, Barmettler R, Chen F, Probst A, Nitsch RM (2001b) Oligodendroglial tau filament formation in transgenic mice expressing G272V tau. Eur J Neurosci 13:2131–2140.

Grammas P, Ovase R (2001) Inflammatory factors are elevated in brain microvessels in Alzheimer's disease. Neurobiol Aging 22:837–842.

Gray CW, Patel AJ (1993) Regulation of beta-amyloid precursor protein isoform mRNAs by transforming growth factor-beta 1 and interleukin-1 beta in astrocytes. Brain Res Mol Brain Res 19:251–256.

Green PS, Mendez AJ, Jacob JS, Crowley JR, Growdon W, Hyman BT, Heinecke JW (2004) Neuronal expression of myeloperoxidase is increased in Alzheimer's disease. J Neurochem 90:724–733.

Griffin W, Sheng J, Roberts G, Mrak R (1995) Interleukin-1 expression in different plaque types in Alzheimer's diseases: Significance in plaque evolution. J Neuropathol Exp Neurol 54:276–281.

Griffin WS, Stanley LC, Ling C, White L, MacLeod V, Perrot LJ, White CL, 3rd, Araoz C (1989) Brain interleukin 1 and S-100 immunoreactivity are elevated in Down syndrome and Alzheimer disease. Proc Natl Acad Sci USA 86:7611–7615.

Guo Z, Cupples LA, Kurz A, Auerbach SH, Volicer L, Chui H, Green RC, Sadovnick AD, Duara R, DeCarli C, Johnson K, Go RC, Growdon JH, Haines JL, Kukull WA, Farrer LA (2000) Head injury and the risk of AD in the MIRAGE study. Neurology 54:1316–1323.

Gupta S, Gollapudi S (2005) Molecular mechanisms of TNF-alpha-induced apoptosis in aging human T cell subsets. Int J Biochem Cell Biol 37:1034–1042.

Hampel H, Haslinger A, Scheloske M, Padberg F, Fischer P, Unger J, Teipel SJ, Neumann M, Rosenberg C, Oshida R, Hulette C, Pongratz D, Ewers M, Kretzschmar HA, Moller HJ (2005) Pattern of interleukin-6 receptor complex immunoreactivity between cortical regions of rapid autopsy normal and Alzheimer's disease brain. Eur Arch Psychiatry Clin Neurosci 255:269–278.

Hardy JA, Higgins GA (1992) Alzheimer's disease: The amyloid cascade hypothesis. Science 256:184–185.

Harper JD, Wong SS, Lieber CM, Lansbury PT (1997) Observation of metastable Abeta amyloid protofibrils by atomic force microscopy. Chem Biol 4:119–125.

Harris-White ME, Chu T, Balverde Z, Sigel JJ, Flanders KC, Frautschy SA (1998) Effects of transforming growth factor-beta (isoforms 1–3) on amyloid-beta deposition, inflammation, and cell targeting in organotypic hippocampal slice cultures. J Neurosci 18:10366–10374.

Heneka MT, Klockgether T, Feinstein DL (2000) Peroxisome proliferator-activated receptor-gamma ligands reduce neuronal inducible nitric oxide synthase expression and cell death in vivo. J Neurosci 20:6862–6867.

Heneka MT, Sastre M, Dumitrescu-Ozimek L, Hanke A, Dewachter I, Kuiperi C, O'Banion K, Klockgether T, Van Leuven F, Landreth GE (2005) Acute treatment with the PPARgamma agonist pioglitazone and ibuprofen reduces glial inflammation and Abeta1–42 levels in APPV717I transgenic mice. Brain 128:1442–1453.

Hof PR, Bussiere T, Gold G, Kovari E, Giannakopoulos P, Bouras C, Perl DP, Morrison JH (2003) Stereologic evidence for persistence of viable neurons in layer II of the entorhinal cortex and the CA1 field in Alzheimer disease. J Neuropathol Exp Neurol 62:55–67.

Holcomb L, Gordon MN, McGowan E, Yu X, Benkovic S, Jantzen P, Wright K, Saad I, Mueller R, Morgan D, Sanders S, Zehr C, O'Campo K, Hardy J, Prada CM, Eckman C, Younkin S, Hsiao K, Duff K (1998) Accelerated Alzheimer-type phenotype in transgenic mice carrying both mutant amyloid precursor protein and presenilin 1 transgenes. Nat Med 4:97–100.

Holtzman DM, Bales KR, Wu S, Bhat P, Parsadanian M, Fagan AM, Chang LK, Sun Y, Paul SM (1999) Expression of human apolipoprotein E reduces amyloid-beta deposition in a mouse model of Alzheimer's disease. J Clin Invest 103:R15-R21.

Holtzman DM, Bales KR, Tenkova T, Fagan AM, Parsadanian M, Sartorius LJ, Mackey B, Olney J, McKeel D, Wozniak D, Paul SM (2000) Apolipoprotein E isoform-dependent amyloid deposition and neuritic degeneration in a mouse model of Alzheimer's disease. Proc Natl Acad Sci USA 97:2892–2897.

Hoshi M, Sato M, Matsumoto S, Noguchi A, Yasutake K, Yoshida N, Sato K (2003) Spherical aggregates of beta-amyloid (amylospheroid) show high neurotoxicity and activate tau protein kinase I/glycogen synthase kinase-3beta. Proc Natl Acad Sci USA 100:6370–6375.

Hsia AY, Masliah E, McConlogue L, Yu GQ, Tatsuno G, Hu K, Kholodenko D, Malenka RC, Nicoll RA, Mucke L (1999) Plaque-independent disruption of neural circuits in Alzheimer's disease mouse models. Proc Natl Acad Sci USA 96:3228–3233.

Hsiao K, Chapman P, Nilsen S, Eckman C, Harigaya Y, Younkin S, Yang F, Cole G (1996) Correlative memory deficits, Abeta elevation, and amyloid plaques in transgenic mice. Science 274:99–102.

Huang F, Buttini M, Wyss-Coray T, McConlogue L, Kodama T, Pitas RE, Mucke L (1999) Elimination of the class A scavenger receptor does not affect amyloid plaque formation or neurodegeneration in transgenic mice expressing human amyloid protein precursors. Am J Pathol 155:1741–1747.

Hull M, Berger M, Volk B, Bauer J (1996) Occurrence of interleukin-6 in cortical plaques of Alzheimer's disease patients may precede transformation of diffuse into neuritic plaques. Ann N Y Acad Sci 777:205–212.

Hutton M, Lendon CL, Rizzu P, Baker M, Froelich S, Houlden H, Pickering-Brown S, Chakraverty S, Isaacs A, Grover A, Hackett J, Adamson J, Lincoln S, Dickson D, Davies P, Petersen RC, Stevens M, de Graaff E, Wauters E, van Baren J, Hillebrand M, Joosse M, Kwon JM, Nowotny P, Che LK, Norton J, Morris JC, Reed LA, Trojanowski J, Basun H, Lannfelt L, Neystat M, Fahn S, Dark F, Tannenberg T, Dodd PR, Hayward N, Kwok JB, Schofield PR, Andreadis A, Snowden J, Craufurd D, Neary D, Owen F, Oostra BA, Hardy J, Goate A, van Swieten J, Mann D, Lynch T, Heutink P (1998) Association of missense and 5′-splice-site mutations in tau with the inherited dementia FTDP-17. Nature 393:702–705.

Ikezu T, Luo X, Weber GA, Zhao J, McCabe L, Buescher JL, Ghorpade A, Zheng J, Xiong H (2003) Amyloid precursor protein-processing products affect mononuclear phagocyte activation: Pathways for sAPP- and Abeta-mediated neurotoxicity. J Neurochem 85:925–934.

in 't Veld BA, Launer LJ, Breteler MM, Hofman A, Stricker BH (2002) Pharmacologic agents associated with a preventive effect on Alzheimer's disease: A review of the epidemiologic evidence. Epidemiol Rev 24:248–268.

Iribarren P, Chen K, Hu J, Gong W, Cho EH, Lockett S, Uranchimeg B, Wang JM (2005) CpG-containing oligodeoxynucleotide promotes microglial cell uptake of amyloid beta 1–42 peptide by up-regulating the expression of the G-protein- coupled receptor mFPR2. FASEB J 19:2032–2034.

Ishiguro K, Omori A, Takamatsu M, Sato K, Arioka M, Uchida T, Imahori K (1992) Phosphorylation sites on tau by tau protein kinase I, a bovine derived kinase generating an epitope of paired helical filaments. Neurosci Lett 148:202–206.

Ishizuka K, Kimura T, Igata-yi R, Katsuragi S, Takamatsu J, Miyakawa T (1997) Identification of monocyte chemoattractant protein-1 in senile plaques and reactive microglia of Alzheimer's disease. Psychiatry Clin Neurosci 51:135–138.

Iwasaki A, Medzhitov R (2004) Toll-like receptor control of the adaptive immune responses. Nat Immunol 5:987–995.

Iwata N, Tsubuki S, Takaki Y, Watanabe K, Sekiguchi M, Hosoki E, Kawashima-Morishima M, Lee HJ, Hama E, Sekine-Aizawa Y, Saido TC (2000) Identification of the major Abeta1–42-degrading catabolic pathway in brain parenchyma: Suppression leads to biochemical and pathological deposition. Nat Med 6:143–150.

Jack CS, Arbour N, Manusow J, Montgrain V, Blain M, McCrea E, Shapiro A, Antel JP (2005) TLR signaling tailors innate immune responses in human microglia and astrocytes. J Immunol 175:4320–4330.

Jacobsen JS, Wu CC, Redwine JM, Comery TA, Arias R, Bowlby M, Martone R, Morrison JH, Pangalos MN, Reinhart PH, Bloom FE (2006) Early-onset behavioral and synaptic deficits in a mouse model of Alzheimer's disease. Proc Natl Acad Sci USA 103:5161–5166.

Janelsins MC, Mastrangelo MA, Oddo S, LaFerla FM, Federoff HJ, Bowers WJ (2005) Early correlation of microglial activation with enhanced tumor necrosis factor-alpha and monocyte chemoattractant protein-1 expression specifically within the entorhinal cortex of triple transgenic Alzheimer's disease mice. J Neuroinflammation 2:23.

Jantzen PT, Connor KE, DiCarlo G, Wenk GL, Wallace JL, Rojiani AM, Coppola D, Morgan D, Gordon MN (2002) Microglial activation and beta -amyloid deposit reduction caused by a nitric oxide-releasing nonsteroidal anti-inflammatory drug in amyloid precursor protein plus presenilin-1 transgenic mice. J Neurosci 22:2246–2254.

Janus C, Pearson J, McLaurin J, Mathews PM, Jiang Y, Schmidt SD, Chishti MA, Horne P, Heslin D, French J, Mount HT, Nixon RA, Mercken M, Bergeron C, Fraser PE, St George-Hyslop P, Westaway D (2000) A beta peptide immunization reduces behavioural impairment and plaques in a model of Alzheimer's disease. Nature 408:979–982.

Jolivalt C, Leininger-Muller B, Drozdz R, Naskalski JW, Siest G (1996) Apolipoprotein E is highly susceptible to oxidation by myeloperoxidase, an enzyme present in the brain. Neurosci Lett 210:61–64.

Katzman R (2001) Epidemiology of Alzheimer's disease and dementia: Advances and challenges. In: Alzheimer's Disease: Advances in Etiology, Pathogenesis, and Therapeutics (Iqbal K, Sisodia SS, Winblad B, eds), pp 11–21. West Sussex, UK: John Wiley & Sons Ltd.

Katzman R, Fox P (1999) The world wide impact of dementia in the next fifty years. In: Epidemiology of Alzheimer's Disease: From Gene to Prevention (Mayeux R, Christen Y, eds), pp 1–17. Berlin: Springer.

Kligman D, Marshak DR (1985) Purification and characterization of a neurite extension factor from bovine brain. Proc Natl Acad Sci USA 82:7136–7139.

Koenigsknecht-Talboo J, Landreth GE (2005) Microglial phagocytosis induced by fibrillar beta-amyloid and IgGs are differentially regulated by proinflammatory cytokines. J Neurosci 25:8240–8249.

Kopec KK, Carroll RT (1998) Alzheimer's beta-amyloid peptide 1–42 induces a phagocytic response in murine microglia. J Neurochem 71:2123–2131.

Kruman II, Wersto RP, Cardozo-Pelaez F, Smilenov L, Chan SL, Chrest FJ, Emokpae Jr R, Gorospe M, Mattson MP (2004) Cell cycle activation linked to neuronal cell death initiated by DNA damage. Neuron 41:549–561.

Kukar T, Murphy MP, Eriksen JL, Sagi SA, Weggen S, Smith TE, Ladd T, Khan MA, Kache R, Beard J, Dodson M, Merit S, Ozols VV, Anastasiadis PZ, Das P, Fauq A, Koo EH, Golde TE (2005) Diverse compounds mimic Alzheimer disease-causing mutations by augmenting Abeta42 production. Nat Med 11:545–550.

LaFerla FM, Hall CK, Ngo L, Jay G (1996) Extracellular deposition of beta-amyloid upon p53-dependent neuronal cell death in transgenic mice. J Clin Invest 98:1626–1632.

Lambert MP, Barlow AK, Chromy BA, Edwards C, Freed R, Liosatos M, Morgan TE, Rozovsky I, Trommer B, Viola KL, Wals P, Zhang C, Finch CE, Krafft GA, Klein WL (1998) Diffusible, nonfibrillar ligands derived from Abeta1–42 are potent central nervous system neurotoxins. Proc Natl Acad Sci USA 95:6448–6453.

Larson J, Lynch G, Games D, Seubert P (1999) Alterations in synaptic transmission and long-term potentiation in hippocampal slices from young and aged PDAPP mice. Brain Res 840:23–35.

Leake A, Morris CM, Whateley J (2000) Brain matrix metalloproteinase 1 levels are elevated in Alzheimer's disease. Neurosci Lett 291:201–203.

Lee EB, Zhang B, Liu K, Greenbaum EA, Doms RW, Trojanowski JQ, Lee VM (2005) BACE overexpression alters the subcellular processing of APP and inhibits Abeta deposition in vivo. J Cell Biol 168:291–302.

Lee VM, Goedert M, Trojanowski JQ (2001) Neurodegenerative tauopathies. Annu Rev Neurosci 24:1121–1159.

Leissring MA, Farris W, Chang AY, Walsh DM, Wu X, Sun X, Frosch MP, Selkoe DJ (2003) Enhanced proteolysis of beta-amyloid in APP transgenic mice prevents plaque formation, secondary pathology, and premature death. Neuron 40:1087–1093.

Lemberger T, Desvergne B, Wahli W (1996) Peroxisome proliferator-activated receptors: A nuclear receptor signaling pathway in lipid physiology. Annu Rev Cell Dev Biol 12:335–363.

Lesne S, Koh MT, Kotilinek L, Kayed R, Glabe CG, Yang A, Gallagher M, Ashe KH (2006) A specific amyloid-beta protein assembly in the brain impairs memory. Nature 440:352–357.

Letiembre M, Hao W, Liu Y, Walter S, Mihaljevic I, Rivest S, Hartmann T, Fassbender K (2007) Innate immune receptor expression in normal brain aging. Neuroscience 146(1):248–254.

Levy-Lahad E, Wasco W, Poorkaj P, Romano DM, Oshima J, Pettingell WH, Yu CE, Jondro PD, Schmidt SD, Wang K, et al. (1995) Candidate gene for the chromosome 1 familial Alzheimer's disease locus. Science 269:973–977.

Lewis J, McGowan E, Rockwood J, Melrose H, Nacharaju P, Van Slegtenhorst M, Gwinn-Hardy K, Paul Murphy M, Baker M, Yu X, Duff K, Hardy J, Corral A, Lin WL, Yen SH, Dickson DW, Davies P, Hutton M (2000) Neurofibrillary tangles, amyotrophy and progressive motor disturbance in mice expressing mutant (P301L) tau protein. Nat Genet 25:402–405.

Li R, Yang L, Lindholm K, Konishi Y, Yue X, Hampel H, Zhang D, Shen Y (2004) Tumor necrosis factor death receptor signaling cascade is required for amyloid-beta protein-induced neuron death. J Neurosci 24:1760–1771.

Lim GP, Yang F, Chu T, Chen P, Beech W, Teter B, Tran T, Ubeda O, Ashe KH, Frautschy SA, Cole GM (2000) Ibuprofen suppresses plaque pathology and inflammation in a mouse model for Alzheimer's disease. J Neurosci 20:5709–5714.

Lim GP, Chu T, Yang F, Beech W, Frautschy SA, Cole GM (2001) The curry spice curcumin reduces oxidative damage and amyloid pathology in an Alzheimer transgenic mouse. J Neurosci 21:8370–8377.

Lim GP, Calon F, Morihara T, Yang F, Teter B, Ubeda O, Salem N, Jr., Frautschy SA, Cole GM (2005) A diet enriched with the omega-3 fatty acid docosahexaenoic acid reduces amyloid burden in an aged Alzheimer mouse model. J Neurosci 25:3032–3040.

Lotz M, Ebert S, Esselmann H, Iliev AI, Prinz M, Wiazewicz N, Wiltfang J, Gerber J, Nau R (2005) Amyloid beta peptide 1–40 enhances the action of Toll-like receptor-2 and -4 agonists but antagonizes Toll-like receptor-9-induced inflammation in primary mouse microglial cell cultures. J Neurochem 94:289–298.

Lovell MA, Ehmann WD, Mattson MP, Markesbery WR (1997) Elevated 4-hydroxynonenal in ventricular fluid in Alzheimer's disease. Neurobiol Aging 18:457–461.

Lu T, Pan Y, Kao SY, Li C, Kohane I, Chan J, Yankner BA (2004) Gene regulation and DNA damage in the ageing human brain. Nature 429:883–891.

Luo X, Weber GA, Zheng J, Gendelman HE, Ikezu T (2003) C1q-calreticulin induced oxidative neurotoxicity: Relevance for the neuropathogenesis of Alzheimer's disease. J Neuroimmunol 135:62–71.

Lyras L, Cairns NJ, Jenner A, Jenner P, Halliwell B (1997) An assessment of oxidative damage to proteins, lipids, and DNA in brain from patients with Alzheimer's disease. J Neurochem 68:2061–2069.

Manczak M, Park BS, Jung Y, Reddy PH (2004) Differential expression of oxidative phosphorylation genes in patients with Alzheimer's disease: Implications for early mitochondrial dysfunction and oxidative damage. Neuromolecular Med 5:147–162.

Mangelsdorf DJ, Thummel C, Beato M, Herrlich P, Schutz G, Umesono K, Blumberg B, Kastner P, Mark M, Chambon P, Evans RM (1995) The nuclear receptor superfamily: The second decade. Cell 83:835–839.

Markesbery WR, Carney JM (1999) Oxidative alterations in Alzheimer's disease. Brain Pathol 9:133–146.

Martins RN, Harper CG, Stokes GB, Masters CL (1986) Increased cerebral glucose-6-phosphate dehydrogenase activity in Alzheimer's disease may reflect oxidative stress. J Neurochem 46:1042–1045.

Masliah E, Mallory M, Hansen L, Alford M, Albright T, Terry R, Shapiro P, Sundsmo M, Saitoh T (1991) Immunoreactivity of CD45, a protein phosphotyrosine phosphatase, in Alzheimer's disease. Acta Neuropathol (Berl) 83:12–20.

Mastrangelo P, Mathews PM, Chishti MA, Schmidt SD, Gu Y, Yang J, Mazzella MJ, Coomaraswamy J, Horne P, Strome B, Pelly H, Levesque G, Ebeling C, Jiang Y, Nixon RA, Rozmahel R, Fraser PE, St George-Hyslop P, Carlson GA, Westaway D (2005) Dissociated phenotypes in presenilin transgenic mice define functionally distinct gamma-secretases. Proc Natl Acad Sci USA 102:8972–8977.

McGeer P, Schulzer M, McGeer E (1996) Arthritis and anti-inflammatory agents as possible protective factors for Alzheimer's disease: A review of 17 epidemiologic studies. Neurology 47:425–432.

McGeer PL, McGeer EG (2001) Inflammation, autotoxicity and Alzheimer disease. Neurobiol Aging 22:799–809.

McKhann G, Drachman D, Folstein M, Katzman R, Price D, Stadlan EM (1984) Clinical diagnosis of Alzheimer's disease: Report of the NINCDS-ADRDA Work Group under the auspices of Department of Health and Human Services Task Force on Alzheimer's Disease. Neurology 34:939–944.

McKimmie CS, Fazakerley JK (2005) In response to pathogens, glial cells dynamically and differentially regulate Toll-like receptor gene expression. J Neuroimmunol 169:116–125.

Meda L, Cassatella MA, Szendrei GI, Jr LO, Baron P, Villalba M, Ferrari D, Rossi F (1995) Activation of microglial cells by b-amyloid protein and interferon-g. Nature 374:647–650.

Meda L, Baron P, Prat E, Scarpini E, Scarlato G, Cassatella MA, Rossi F (1999) Proinflammatory profile of cytokine production by human monocytes and murine microglia stimulated with beta-amyloid[25–35]. J Neuroimmunol 93:45–52.

Medina MG, Ledesma MD, Dominguez JE, Medina M, Zafra D, Alameda F, Dotti CG, Navarro P (2005) Tissue plasminogen activator mediates amyloid-induced neurotoxicity via Erk1/2 activation. EMBO J 24:1706–1716.

Mohajeri MH, Saini KD, Nitsch RM (2004) Transgenic BACE expression in mouse neurons accelerates amyloid plaque pathology. J Neural Transm 111:413–425.

Monning U, Sandbrink R, Banati RB, Masters CL, Beyreuther K (1994) Transforming growth factor beta mediates increase of mature transmembrane amyloid precursor protein in microglial cells. FEBS Lett 342:267–272.

Mori K, Yokoyama A, Yang L, Maeda N, Mitsuda N, Tanaka J (2004) L-serine-mediated release of apolipoprotein E and lipids from microglial cells. Exp Neurol 185:220–231.

Morrison RS, Kinoshita Y, Johnson MD, Guo W, Garden GA (2003) p53-dependent cell death signaling in neurons. Neurochem Res 28:15–27.

Morsch R, Simon W, Coleman PD (1999) Neurons may live for decades with neurofibrillary tangles. J Neuropathol Exp Neurol 58:188–197.

Mortimer JA, van Duijn CM, Chandra V, Fratiglioni L, Graves AB, Heyman A, Jorm AF, Kokmen E, Kondo K, Rocca WA, et al. (1991) Head trauma as a risk factor for Alzheimer's disease: A collaborative re-analysis of case-control studies. EURODEM Risk Factors Research Group. Int J Epidemiol 20 Suppl 2:S28–35.

Nagy Z, Esiri MM, Smith AD (1998) The cell division cycle and the pathophysiology of Alzheimer's disease. Neuroscience 87:731–739.

Naslund J, Haroutunian V, Mohs R, Davis KL, Davies P, Greengard P, Buxbaum JD (2000) Correlation between elevated levels of amyloid beta-peptide in the brain and cognitive decline. JAMA 283: 1571–1577.

Oddo S, Caccamo A, Shepherd JD, Murphy MP, Golde TE, Kayed R, Metherate R, Mattson MP, Akbari Y, LaFerla FM (2003) Triple-transgenic model of Alzheimer's disease with plaques and tangles: Intracellular Abeta and synaptic dysfunction. Neuron 39:409–421.

Olson JK, Miller SD (2004) Microglia initiate central nervous system innate and adaptive immune responses through multiple TLRs. J Immunol 173:3916–3924.

Organization WH (1992) International statistical classification of diseases and related health problems. In 10th edition (ICD-10) Edition. Geneva: World Health Organization.

Ott A, Stolk RP, van Harskamp F, Pols HA, Hofman A, Breteler MM (1999) Diabetes mellitus and the risk of dementia: The Rotterdam Study. Neurology 53:1937–1942.

Paresce DM, Ghosh RN, Maxfield FR (1996) Microglial cells internalize aggregates of the Alzheimer's disease amyloid beta-protein via a scavenger receptor. Neuron 17:553–565.

Paresce DM, Chung H, Maxfield FR (1997) Slow degradation of aggregates of the Alzheimer's disease amyloid beta-protein by microglial cells. J Biol Chem 272:29390–29397.

Patrick GN, Zukerberg L, Nikolic M, de la Monte S, Dikkes P, Tsai LH (1999) Conversion of p35 to p25 deregulates Cdk5 activity and promotes neurodegeneration. Nature 402:615–622.

Peila R, Rodriguez BL, Launer LJ (2002) Type 2 diabetes, APOE gene, and the risk for dementia and related pathologies: The Honolulu-Asia Aging Study. Diabetes 51:1256–1262.

Piani D, Spranger M, Frei K, Schaffner A, Fontana A (1992) Macrophage-induced cytotoxicity of N-methyl-D-aspartate receptor positive neurons involves excitatory amino acids rather than reactive oxygen intermediates and cytokines. Eur J Immunol 22:2429–2436.

Pike CJ, Burdick D, Walencewicz AJ, Glabe CG, Cotman CW (1993) Neurodegeneration induced by beta-amyloid peptides in vitro: The role of peptide assembly state. J Neurosci 13:1676–1687.

Pitas RE, Boyles JK, Lee SH, Foss D, Mahley RW (1987) Astrocytes synthesize apolipoprotein E and metabolize apolipoprotein E-containing lipoproteins. Biochim Biophys Acta 917:148–161.

Planel E, Sun X, Takashima A (2002) Role of GSK-3ß in Alzheimer's disease pathology. Drug Dev Res 56:491–510.

Postina R, Schroeder A, Dewachter I, Bohl J, Schmitt U, Kojro E, Prinzen C, Endres K, Hiemke C, Blessing M, Flamez P, Dequenne A, Godaux E, van Leuven F, Fahrenholz F (2004) A disintegrin-metalloproteinase prevents amyloid plaque formation and hippocampal defects in an Alzheimer disease mouse model. J Clin Invest 113:1456–1464.

Price DL, Tanzi RE, Borchelt DR, Sisodia SS (1998) Alzheimer's disease: Genetic studies and transgenic models. Annu Rev Genet 32:461–493.

Qiu C, Winblad B, Fratiglioni L (2005) The age-dependent relation of blood pressure to cognitive function and dementia. Lancet Neurol 4:487–499.

Qiu WQ, Walsh DM, Ye Z, Vekrellis K, Zhang J, Podlisny MB, Rosner MR, Safavi A, Hersh LB, Selkoe DJ (1998) Insulin-degrading enzyme regulates extracellular levels of amyloid beta-protein by degradation. J Biol Chem 273:32730–32738.

Ramsden M, Kotilinek L, Forster C, Paulson J, McGowan E, SantaCruz K, Guimaraes A, Yue M, Lewis J, Carlson G, Hutton M, Ashe KH (2005) Age-dependent neurofibrillary tangle formation, neuron loss, and memory impairment in a mouse model of human tauopathy (P301L). J Neurosci 25:10637–10647.

Rapoport M, Dawson HN, Binder LI, Vitek MP, Ferreira A (2002) Tau is essential to beta -amyloid-induced neurotoxicity. Proc Natl Acad Sci USA 99:6364–6369.

Reynolds WF, Rhees J, Maciejewski D, Paladino T, Sieburg H, Maki RA, Masliah E (1999) Myeloperoxidase polymorphism is associated with gender specific risk for Alzheimer's disease. Exp Neurol 155:31–41.

Rogers J, Luber-Narod J, Styren SD, Civin WH (1988) Expression of immune system-associated antigens by cells of the human central nervous system: Relationship to the pathology of Alzheimer's disease. Neurobiol Aging 9:339–349.

Rogers JT, Leiter LM, McPhee J, Cahill CM, Zhan SS, Potter H, Nilsson LN (1999) Translation of the alzheimer amyloid precursor protein mRNA is up-regulated by interleukin-1 through 5′-untranslated region sequences. J Biol Chem 274:6421–6431.

Roher AE, Chaney MO, Kuo YM, Webster SD, Stine WB, Haverkamp LJ, Woods AS, Cotter RJ, Tuohy JM, Krafft GA, Bonnell BS, Emmerling MR (1996) Morphology and toxicity of Abeta-(1–42) dimer derived from neuritic and vascular amyloid deposits of Alzheimer's disease. J Biol Chem 271:20631–20635.

Sadowski M, Pankiewicz J, Scholtzova H, Ripellino JA, Li Y, Schmidt SD, Mathews PM, Fryer JD, Holtzman DM, Sigurdsson EM, Wisniewski T (2004) A synthetic peptide blocking the apolipoprotein E/beta-amyloid binding mitigates beta-amyloid toxicity and fibril formation in vitro and reduces beta-amyloid plaques in transgenic mice. Am J Pathol 165:937–948.

Santacruz K, Lewis J, Spires T, Paulson J, Kotilinek L, Ingelsson M, Guimaraes A, DeTure M, Ramsden M, McGowan E, Forster C, Yue M, Orne J, Janus C, Mariash A, Kuskowski M, Hyman B, Hutton M, Ashe KH (2005) Tau suppression in a neurodegenerative mouse model improves memory function. Science 309:476–481.

Sato S, Tatebayashi Y, Akagi T, Chui DH, Murayama M, Miyasaka T, Planel E, Tanemura K, Sun X, Hashikawa T, Yoshioka K, Ishiguro K, Takashima A (2002) Aberrant tau phosphorylation by glycogen synthase kinase-3beta and JNK3 induces oligomeric tau fibrils in COS-7 cells. J Biol Chem 277:42060–42065.

Sato S, Cerny RL, Buescher JL, Ikezu T (2006) Tau-tubulin kinase 1 (TTBK1), a neuron-specific tau kinase candidate, is involved in tau phosphorylation and aggregation. J Neurochem 98:1573–1584.

Saura J, Petegnief V, Wu X, Liang Y, Paul SM (2003) Microglial apolipoprotein E and astroglial apolipoprotein J expression in vitro: Opposite effects of lipopolysaccharide. J Neurochem 85:1455–1467.

Scheuner D, Eckman C, Jensen M, Song X, Citron M, Suzuki N, Bird TD, Hardy J, Hutton M, Kukull W, Larson E, Levy-Lahad E, Viitanen M, Peskind E, Poorkaj P, Schellenberg G, Tanzi R, Wasco W, Lannfelt L, Selkoe D, Younkin S (1996) Secreted amyloid beta-protein similar to that in the senile plaques of Alzheimer's disease is increased in vivo by the presenilin 1 and 2 and APP mutations linked to familial Alzheimer's disease. Nat Med 2:864–870.

Schmand B, Smit J, Lindeboom J, Smits C, Hooijer C, Jonker C, Deelman B (1997) Low education is a genuine risk factor for accelerated memory decline and dementia. J Clin Epidemiol 50:1025–1033.

Schmitt FA, Davis DG, Wekstein DR, Smith CD, Ashford JW, Markesbery WR (2000) "Preclinical" AD revisited: Neuropathology of cognitively normal older adults. Neurology 55:370–376.

Selkoe DJ (1994) Cell biology of the amyloid beta-protein precursor and the mechanism of Alzheimer's disease. Annu Rev Cell Biol 10:373–403.

Selkoe DJ (1997) Alzheimer's disease: Genotypes, phenotypes, and treatments. Science 275:630–631.

Selkoe DJ, Xia W, Kimberly WT, Vekrellis K, Walsh D, Esler WP, Wolfe MS (2001) Mechanism of Abeta production and Abeta degradation: Routes to the treatment of Alzheimer's disease. In: Alzheimer's disease: Advances in Ethology, Pathogenesis and Therapeutics (Iqbal K, Sisodia SS, Winblad B, eds), pp 421–432. New York: John Wiley & Sons Ltd.

Shankar GM, Bloodgood BL, Townsend M, Walsh DM, Selkoe DJ, Sabatini BL (2007) Natural Oligomers of the Alzheimer Amyloid-ß Protein Induce Reversible Synapse Loss by Modulating an NMDA-Type Glutamate Receptor-Dependent Signaling Pathway. J Neurosci 27:2866–2875.

Sherrington R, Rogaev EI, Liang Y, Rogaeva EA, Levesque G, Ikeda M, Chi H, Lin C, Li G, Holman K, et al. (1995) Cloning of a gene bearing missense mutations in early-onset familial Alzheimer's disease. Nature 375:754–760.

Sisodia SS (1999) Series introduction: Alzheimer's disease: Perspectives for the new millennium. J Clin Invest 104:1169–1170.

Smith MA, Hirai K, Hsiao K, Pappolla MA, Harris PL, Siedlak SL, Tabaton M, Perry G (1998) Amyloid-beta deposition in Alzheimer transgenic mice is associated with oxidative stress. J Neurochem 70:2212–2215.

Smith MZ, Nagy Z, Esiri MM (1999) Cell cycle-related protein expression in vascular dementia and Alzheimer's disease. Neurosci Lett 271:45–48.

Sparks DL, Hunsaker JC, 3rd, Scheff SW, Kryscio RJ, Henson JL, Markesbery WR (1990) Cortical senile plaques in coronary artery disease, aging and Alzheimer's disease. Neurobiol Aging 11:601–607.

Spires TL, Hyman BT (2005) Transgenic models of Alzheimer's disease: Learning from animals. NeuroRx 2:423–437.

Sturchler-Pierrat C, Abramowski D, Duke M, Wiederhold KH, Mistl C, Rothacher S, Ledermann B, Burki K, Frey P, Paganetti PA, Waridel C, Calhoun ME, Jucker M, Probst A, Staufenbiel M, Sommer B (1997) Two amyloid precursor protein transgenic mouse models with Alzheimer disease-like pathology. Proc Natl Acad Sci USA 94:13287–13292.

Styren SD, Civin WH, Rogers J (1990) Molecular, cellular, and pathologic characterization of HLA-DR immunoreactivity in normal elderly and Alzheimer's disease brain. Exp Neurol 110:93–104.

Sun YX, Minthon L, Wallmark A, Warkentin S, Blennow K, Janciauskiene S (2003) Inflammatory markers in matched plasma and cerebrospinal fluid from patients with Alzheimer's disease. Dement Geriatr Cogn Disord 16:136–144.

Takashima A, Noguchi K, Sato K, Hoshino T, Imahori K (1993) Tau protein kinase I is essential for amyloid beta-protein-induced neurotoxicity. Proc Natl Acad Sci USA 90:7789–7793.

Takeda A, Yasuda T, Miyata T, Goto Y, Wakai M, Watanabe M, Yasuda Y, Horie K, Inagaki T, Doyu M, Maeda K, Sobue G (1998) Advanced glycation end products co-localized with astrocytes and microglial cells in Alzheimer's disease brain. Acta Neuropathol (Berl) 95:555–558.

Tan J, Town T, Paris D, Mori T, Suo Z, Crawford F, Mattson MP, Flavell RA, Mullan M (1999) Microglial activation resulting from CD40-CD40L interaction after b-amyloid stimulation. Science 286:2352–2355.

Tan J, Town T, Mullan M (2000a) CD45 inhibits CD40L-induced microglial activation via negative regulation of the Src/p44/42 MAPK pathway. J Biol Chem 275:37224–37231.

Tan J, Town T, Mori T, Wu Y, Saxe M, Crawford F, Mullan M (2000b) CD45 opposes beta-amyloid peptide-induced microglial activation via inhibition of p44/42 mitogen-activated protein kinase. J Neurosci 20:7587–7594.

Tan J, Town T, Crawford F, Mori T, DelleDonne A, Crescentini R, Obregon D, Flavell RA, Mullan MJ (2002) Role of CD40 ligand in amyloidosis in transgenic Alzheimer's mice. Nat Neurosci 5:1288–1293.

Tanemura K, Murayama M, Akagi T, Hashikawa T, Tominaga T, Ichikawa M, Yamaguchi H, Takashima A (2002) Neurodegeneration with tau accumulation in a transgenic mouse expressing V337M human tau. J Neurosci 22:133–141.

Tarkowski E, Blennow K, Wallin A, Tarkowski A (1999) Intracerebral production of tumor necrosis factor-alpha, a local neuroprotective agent, in Alzheimer disease and vascular dementia. J Clin Immunol 19:223–230.

Tesseur I, Zou K, Esposito L, Bard F, Berber E, Can JV, Lin AH, Crews L, Tremblay P, Mathews P, Mucke L, Masliah E, Wyss-Coray T (2006) Deficiency in neuronal TGF-beta signaling promotes neurodegeneration and Alzheimer's pathology. J Clin Invest 116:3060–3069.

Todd Roach J, Volmar CH, Dwivedi S, Town T, Crescentini R, Crawford F, Tan J, Mullan M (2004) Behavioral effects of CD40-CD40L pathway disruption in aged PSAPP mice. Brain Res 1015:161–168.

Togo T, Akiyama H, Kondo H, Ikeda K, Kato M, Iseki E, Kosaka K (2000) Expression of CD40 in the brain of Alzheimer's disease and other neurological diseases. Brain Res 885:117–121.

van der Wal EA, Gomez-Pinilla F, Cotman CW (1993) Transforming growth factor-beta 1 is in plaques in Alzheimer and Down pathologies. Neuroreport 4:69–72.

Vassar R, Bennett BD, Babu-Khan S, Kahn S, Mendiaz EA, Denis P, Teplow DB, Ross S, Amarante P, Loeloff R, Luo Y, Fisher S, Fuller J, Edenson S, Lile J, Jarosinski MA, Biere AL, Curran E, Burgess T, Louis JC, Collins F, Treanor J, Rogers G, Citron M (1999) Beta-secretase cleavage of Alzheimer's amyloid precursor protein by the transmembrane aspartic protease BACE. Science 286:735–741.

Vekrellis K, Ye Z, Qiu WQ, Walsh D, Hartley D, Chesneau V, Rosner MR, Selkoe DJ (2000) Neurons regulate extracellular levels of amyloid beta-protein via proteolysis by insulin-degrading enzyme. J Neurosci 20:1657–1665.

Walker DG, Link J, Lue LF, Dalsing-Hernandez JE, Boyes BE (2006) Gene expression changes by amyloid {beta} peptide-stimulated human postmortem brain microglia identify activation of multiple inflammatory processes. J Leukoc Biol 79:596–610.

Wallach D, Varfolomeev EE, Malinin NL, Goltsev YV, Kovalenko AV, Boldin MP (1999) Tumor necrosis factor receptor and Fas signaling mechanisms. Annu Rev Immunol 17:331–367.

Walsh DM, Lomakin A, Benedek GB, Condron MM, Teplow DB (1997) Amyloid beta-protein fibrillogenesis. Detection of a protofibrillar intermediate. J Biol Chem 272:22364–22372.

Walsh DM, Klyubin I, Fadeeva JV, Cullen WK, Anwyl R, Wolfe MS, Rowan MJ, Selkoe DJ (2002) Naturally secreted oligomers of amyloid beta protein potently inhibit hippocampal long-term potentiation in vivo. Nature 416:535–539.

Wang Q, Rowan MJ, Anwyl R (2004a) Beta-amyloid-mediated inhibition of NMDA receptor-dependent long-term potentiation induction involves activation of microglia and stimulation of inducible nitric oxide synthase and superoxide. J Neurosci 24:6049–6056.

Wang Q, Walsh DM, Rowan MJ, Selkoe DJ, Anwyl R (2004b) Block of long-term potentiation by naturally secreted and synthetic amyloid beta-peptide in hippocampal slices is mediated via activation of the kinases c-Jun N-terminal kinase, cyclin-dependent kinase 5, and p38 mitogen-activated protein kinase as well as metabotropic glutamate receptor type 5. J Neurosci 24:3370–3378.

Webster SD, Yang AJ, Margol L, Garzon-Rodriguez W, Glabe CG, Tenner AJ (2000) Complement component C1q modulates the phagocytosis of Abeta by microglia. Exp Neurol 161:127–138.

Weggen S, Eriksen JL, Das P, Sagi SA, Wang R, Pietrzik CU, Findlay KA, Smith TE, Murphy MP, Bulter T, Kang DE, Marquez-Sterling N, Golde TE, Koo EH (2001) A subset of NSAIDs lower amyloidogenic Abeta42 independently of cyclooxygenase activity. Nature 414:212–216.

White AR, Du T, Laughton KM, Volitakis I, Sharples RA, Hoke DE, Holsinger RM, Evin G, Cherny RA, Hill AF, Barnham KJ, Li QX, Bush AI, Masters CL (2006) Degradation of the Alzheimer's disease amyloid beta peptide by metal-dependent up-regulation of metalloprotease activity. A quantitative study. J Biol Chem 281(26):17670–17680.

Wilcock GK, Esiri MM (1982) Plaques, tangles and dementia.. J Neurol Sci 56:343–356.

Wilkinson B, Koenigsknecht-Talboo J, Grommes C, Lee CY, Landreth G (2006) Fibrillar beta-amyloid-stimulated intracellular signaling cascades require Vav for induction of respiratory burst and phagocytosis in monocytes and microglia. J Biol Chem 281:20842–20850.

Willem M, Dewachter I, Smyth N, Van Dooren T, Borghgraef P, Haass C, Van Leuven F (2004) beta-site amyloid precursor protein cleaving enzyme 1 increases amyloid deposition in brain parenchyma but reduces cerebrovascular amyloid angiopathy in aging BACE x APP[V717I] double-transgenic mice. Am J Pathol 165:1621–1631.

Williams MA, Turchan J, Lu Y, Nath A, Drachman DB (2005) Protection of human cerebral neurons from neurodegenerative insults by gene delivery of soluble tumor necrosis factor p75 receptor. Exp Brain Res 165:383–391.

Wyss-Coray T, Mucke L (2002) Inflammation in neurodegenerative disease–a double-edged sword. Neuron 35:419–432.

Wyss-Coray T, Masliah E, Mallory M, McConlogue L, Johnson-Wood K, Lin C, Mucke L (1997) Amyloidogenic role of cytokine TGF-beta1 in transgenic mice and in Alzheimer's disease. Nature 389:603–606.

Wyss-Coray T, Lin C, von Euw D, Masliah E, Mucke L, Lacombe P (2000) Alzheimer's disease-like cerebrovascular pathology in transforming growth factor-beta 1 transgenic mice and functional metabolic correlates. Ann N Y Acad Sci 903:317–323.

Wyss-Coray T, Lin C, Yan F, Yu GQ, Rohde M, McConlogue L, Masliah E, Mucke L (2001) TGF-beta1 promotes microglial amyloid-beta clearance and reduces plaque burden in transgenic mice. Nat Med 7:612–618.

Wyss-Coray T, Yan F, Lin AH, Lambris JD, Alexander JJ, Quigg RJ, Masliah E (2002) Prominent neurodegeneration and increased plaque formation in complement-inhibited Alzheimer's mice. Proc Natl Acad Sci USA 99:10837–10842.

Wyss-Coray T, Loike JD, Brionne TC, Lu E, Anankov R, Yan F, Silverstein SC, Husemann J (2003) Adult mouse astrocytes degrade amyloid-beta in vitro and in situ. Nat Med 9:453–457.

Xia M, Qin S, McNamara M, Mackay C, Hyman BT (1997) Interleukin-8 receptor B immunoreactivity in brain and neuritic plaques of Alzheimer's disease. Am J Pathol 150:1267–1274.

Xia MQ, Hyman BT (1999) Chemokines/chemokine receptors in the central nervous system and Alzheimer's disease. J Neurovirol 5:32–41.

Xia MQ, Qin SX, Wu LJ, Mackay CR, Hyman BT (1998) Immunohistochemical study of the beta-chemokine receptors CCR3 and CCR5 and their ligands in normal and Alzheimer's disease brains. Am J Pathol 153:31–37.

Xia MQ, Bacskai BJ, Knowles RB, Qin SX, Hyman BT (2000) Expression of the chemokine receptor CXCR3 on neurons and the elevated expression of its ligand IP-10 in reactive astrocytes: In vitro ERK1/2 activation and role in Alzheimer's disease. J Neuroimmunol 108:227–235.

Xu Q, Bernardo A, Walker D, Kanegawa T, Mahley RW, Huang Y (2006) Profile and regulation of apolipoprotein E (ApoE) expression in the CNS in mice with targeting of green fluorescent protein gene to the ApoE locus. J Neurosci 26:4985–4994.

Yamamoto M, Horiba M, Buescher JL, Huang D, Gendelman HE, Ransohoff RM, Ikezu T (2005) Overexpression of monocyte chemotactic protein-1/CCL2 in beta-amyloid precursor protein transgenic mice show accelerated diffuse beta-amyloid deposition. Am J Pathol 166:1475–1485.

Yamamoto M, Kiyota T, Walsh SM, Ikezu T (2007a) Kinetic analysis of aggregated amyloid-β peptide clearance in adult bone-marrow-derived macrophages from APP and CCL2 transgenic mice. J Neuroimmune Pharm 2:213–227.

Yamamoto M, Kiyota T, Horiba M, Buescher JL, Walsh SM, Gendelman HE, Ikezu T (2007b) Interferon-{gamma} and Tumor Necrosis Factor-{alpha} Regulate Amyloid-{beta} Plaque Deposition and {beta}-Secretase Expression in Swedish Mutant APP Transgenic Mice. Am J Pathol 170:680–692.

Yamin R, Malgeri EG, Sloane JA, McGraw WT, Abraham CR (1999) Metalloendopeptidase EC 3.4.24.15 is necessary for Alzheimer's amyloid-beta peptide degradation. J Biol Chem 274:18777–18784.

Yan Q, Zhang J, Liu H, Babu-Khan S, Vassar R, Biere AL, Citron M, Landreth G (2003) Anti-inflammatory drug therapy alters

beta-amyloid processing and deposition in an animal model of Alzheimer's disease. J Neurosci 23:7504–7509.
Yan SD, Chen X, Fu J, Chen M, Zhu H, Roher A, Slattery T, Zhao L, Nagashima M, Morser J, Migheli A, Nawroth P, Stern D, Schmidt AM (1996) RAGE and amyloid-beta peptide neurotoxicity in Alzheimer's disease. Nature 382:685–691.
Yang F, Lim GP, Begum AN, Ubeda OJ, Simmons MR, Ambegaokar SS, Chen PP, Kayed R, Glabe CG, Frautschy SA, Cole GM (2005) Curcumin inhibits formation of amyloid beta oligomers and fibrils, binds plaques, and reduces amyloid in vivo. J Biol Chem 280:5892–5901.
Yoshiyama Y, Asahina M, Hattori T (2000) Selective distribution of matrix metalloproteinase-3 (MMP-3) in Alzheimer's disease brain. Acta Neuropathol (Berl) 99:91–95.
Young H, Hardy K (1995) Role of interferon-gamma in immune cell regulation. J Leukoc Biol 58:373–381.
Zhang MY, Katzman R, Salmon D, Jin H, Cai GJ, Wang ZY, Qu GY, Grant I, Yu E, Levy P, et al. (1990) The prevalence of dementia and Alzheimer's disease in Shanghai, China: Impact of age, gender, and education. Ann Neurol 27:428–437.
Zhou Y, Su Y, Li B, Liu F, Ryder JW, Wu X, Gonzalez-DeWhitt PA, Gelfanova V, Hale JE, May PC, Paul SM, Ni B (2003) Nonsteroidal anti-inflammatory drugs can lower amyloidogenic Abeta42 by inhibiting Rho. Science 302:1215–1217.
Zlotnik A, Yoshie O (2000) Chemokines: A new classification system and their role in immunity. Immunity 12:121–127.

26
Parkinson's Disease

Serge Przedborski

Keywords Dopamine; Glial-derived neurotrophic factor; Lewy body; Leucine-rich repeat kinase 2; MPTP; Nigrostriatal pathway; Paraquat; Parkin; Rotenone; Alpha-Synuclein

26.1. Introduction

In this chapter the topic of inflammation in Parkinson's disease (PD) and in models of human disease will be reviewed. To set the stage, the biological, clinical and pathological hallmarks of PD and related degenerative conditions will be presented. A more comprehensive review of PD can be found in the following two references (Fahn and Przedborski, 2005; Dauer and Przedborski, 2003).

PD is the second most frequent neurodegenerative disorder of the aging brain after the dementing disorder, Alzheimer's disease (AD). It is estimated that more than one million individuals in the United States are affected with PD and around 50,000 new cases are identified each year. The main clinical features of PD include resting tremor, slowness of movement, rigidity, and postural instability. As many other prevalent neurodegenerative disorders discussed in this book including AD and amyotrophic lateral sclerosis (ALS), PD presents itself essentially as a sporadic condition, i.e. in absence of any genetic linkage. However, as in AD and ALS, only a small fraction of PD patients inherit the disease. In some of these familial cases the genetic defect has been identified and linked to mutations in a variety of genes (Vila and Przedborski, 2004). The clinical manifestations seen in both the sporadic and the familial forms of PD have all been attributed to a profound deficit in brain dopamine (Hornykiewicz and Kish, 1987). Given this fact, it is not surprising that most attention has been geared toward the study of dopaminergic pathways in PD.

Autopsy investigations have revealed that most ascending dopaminergic pathways are affected in PD, though to different degrees (Hornykiewicz and Kish, 1987); the nigrostriatal is consistently more affected than any other (Hornykiewicz and Kish, 1987). The nigrostriatal pathway is made up of dopaminergic neurons whose cell bodies are located in the substantia nigra pars compacta and their projecting axons and terminals in the striatum (Figure 26.1A). Among these dopaminergic neurons, it appears that those with the highest content of brown pigment called neuromelanin are the ones most prone to degeneration (Hirsch et al., 1988). Contrasting with the damage to the ascending pathways, the descending dopaminergic pathways are generally spared (Hornykiewicz and Kish, 1987). If the most salient clinical symptoms of PD are indeed related to the damage of the dopaminergic system in the brain, it cannot be stressed enough that degenerative changes in PD are neither restricted to the nigrostriatal pathway nor other dopaminergic systems (Hornykiewicz and Kish, 1987; Agid et al., 1987). For instance, abnormal histological features can also be found in many non-dopaminergic cell groups including the locus coeruleus, raphe nuclei, and nucleus basalis of Meynert (Braak et al., 1995).

Aside from the loss of neurons, all affected areas of the brain in PD also contain intraneuronal proteinaceous inclusions called Lewy bodies (Galvin et al., 1999). These inclusions can be readily visualized by either classical histological methods, e.g. hematoxylin & eosin, or by immunohistochemistry using antibodies raised against alpha-synuclein or ubiquitin (Figure 26.1B, C). Often, Lewy bodies are large, round, and occupy most of the cytoplasmic area of the few spared neurons. While identification of these proteinaceous inclusions is usually used for neuropathological diagnostic purposes, the actual pathogenic role of Lewy bodies remains controversial. Another, but often overlooked, salient neuropathological feature of PD is the presence of inflammatory change within affected regions (Forno et al., 1992; McGeer et al., 1988). This inflammatory response is the topic of this chapter and will be described in further details below.

T. Ikezu and H.E. Gendelman (eds.), *Neuroimmune Pharmacology*.
© Springer 2008

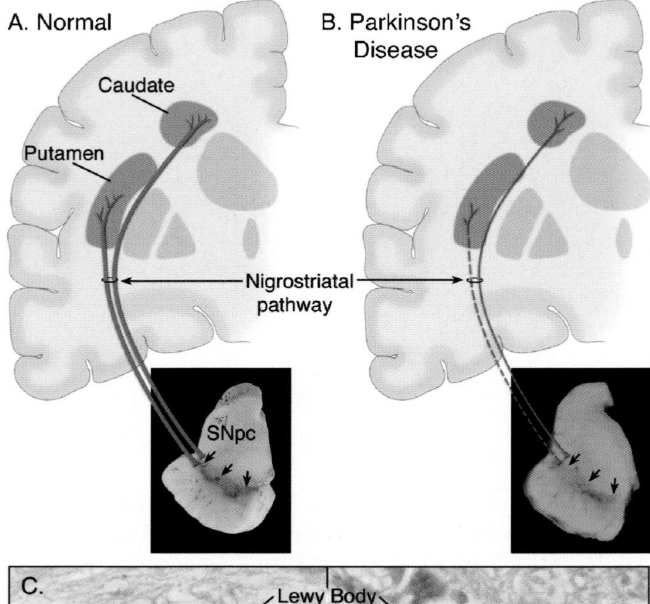

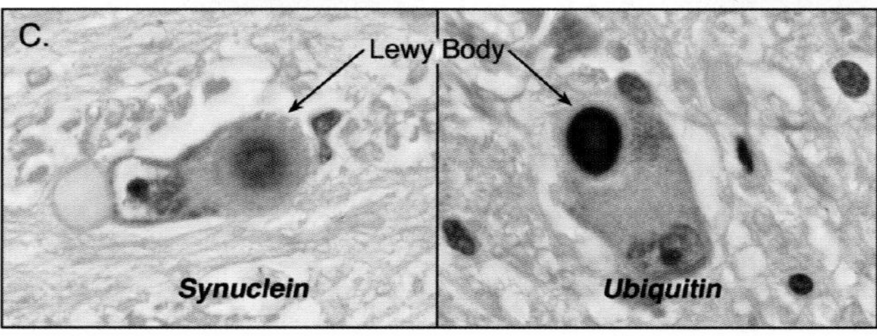

FIGURE 26.1. Neuropathology of Parkinson's disease. (**A**) Schematic representation of the normal nigrostriatal pathway (in red). It is composed of dopaminergic neurons whose cell bodies are located in the substantia nigra pars compacta (SNpc; see arrows). These neurons project (thick solid red lines) to the basal ganglia and synapse in the striatum (i.e. putamen and caudate nucleus). The photograph demonstrates the normal pigmentation of the SNpc, produced by neuromelanin within the dopaminergic neurons. (**B**) Schematic representation of the diseased nigrostriatal pathway (in red). In Parkinson's disease, the nigrostriatal pathway degenerates. There is marked loss of dopaminergic neurons that project to the putamen (dashed line) and a much more modest loss of those that project to the caudate (thin red solid line). The photograph demon strates depigmentation (i.e. loss of dark-brown pigment neuromelanin; arrows) of the SNpc due to the marked loss of dopaminergic neurons. (**C**) immunohistochemical labeling of intra-neuronal inclusions, termed Lewy bodies, in a SNpc dopaminergic neuron. Immunostaining with an antibody against α-synuclein reveals a Lewy body (black arrow) with an intensely immunoreactive central zone surrounded by a faintly immunoreactive peripheral zone (left photograph). Conversely, immunostaining with an antibody against ubiquitin yields more diffuse immunoreactivity within the Lewy body (right photograph). (*From Dauer and Przedborski, Neuron 39:889-909, 2003, with permission.*)

26.2. Inflammatory Response in Parkinson's Disease

Before discussing the issue of inflammation in PD *per se*, it is worth reviewing briefly the question of innate resident immune cell topography in the normal brain. In a healthy adult brain, microglia constitutes roughly 10% of all glial cells and are not evenly distributed (Lawson et al., 1990). In an unstimulated situation, microglia exhibit elongated, bipolar shaped cell bodies with spine-like processes that often branch perpendicularly. Among the areas of the brain susceptible to the PD process, the density of microglia identified by the labeling of the specific plasma membrane glycoprotein F4/80 is higher in the substantia nigra compared to other regions of the brain, at least in adult mice (Lawson et al., 1990). In contrast to microglia, astrocytes in the normal adult brain are homogenously distributed, except in the midbrain where the estimated density of glial fibrillary acidic protein (GFAP)-positive cells varies among the different catecholaminergic groups (Damier et al., 1993). The density of GFAP-positive cells is moderate in the midbrain areas most severely affected in PD, such as in the substantia nigra pars compacta, and high in those least affected, such as the central gray matter (Damier et al., 1993). Furthermore, it has been reported that the density of GFAP-positive cells within the substantia nigra pars compacta is lowest in the calbindin-D_{28K}-poor pockets (Hirsch et al., 1999), which have been previously identified as the zones of the substantia nigra pars compacta where the loss of dopaminergic neurons is the highest (Damier et al., 1999).

Because of the notoriety of the nigrostriatal dopaminergic damage in PD, information regarding inflammation in this common neurodegenerative disorder pertains essentially to the substantia nigra and the striatum. Consistently, it has been found that the loss of dopaminergic neurons in postmortem PD brains is associated with microgliosis and astrocytosis (McGeer et al., 1988; Forno et al., 1992; Banati et al., 1998; Mirza et al., 2000). However, these changes have always been found to be more intense in the substantia nigra pars compacta than in the striatum (McGeer et al., 1988). This situation contrasts with that of the loss of dopaminergic structures, which is more intense in the striatum than in the substantia nigra.

At the cellular level, the majority of the immunostained astrocytes in PD exhibit a resting-like morphology, with thin, like-shaped cell bodies and elongated processes, while only few exhibit a reactive morphology with hypertrophic cell bodies and short processes (Mirza et al., 2000; Forno et al., 1992). Furthermore, there seems to be only a minimal increase in astrocyte numbers when GFAP- or metallothionein I/II-positive cells are quantified (Mirza et al., 2000). Unlike the astrocyte alterations discussed above, the microglial changes in PD are striking (McGeer et al., 1988; Banati et al., 1998; Mirza et al., 2000; Imamura et al., 2003). Morphologically, microglial cells in both the substantia nigra and striata of PD patients, unlike of controls, typically exhibit thick, elongated processes (McGeer et al., 1988; Banati et al., 1998; Mirza et al., 2000). The number of these activated microglia, identified by HLA-DR or ferritin immunostaining, is much higher in PD than in controls (Mirza et al., 2000;

Imamura et al., 2003). Activated microglia are mainly found near free neuromelanin in the neuropil and remaining neurons in the substantia nigra (McGeer et al., 1988). Similar microglial activation is also found in the striatum (Imamura et al., 2003).

26.3. Inflammatory Response in Parkinsonian Syndromes

While PD is the most common of the parkinsonian syndromes, more than 30 different neurological syndromes share the clinical features of PD (Dauer and Przedborski, 2003). Many of these non-PD parkinsonian syndromes are sporadic (Table 26.1) and often exhibit both clinical (e.g. ocular movement) and neuropathological features not typically seen in PD (e.g. striatal or corticospinal track pathology). Of these syndromes most, if not all show at autopsy evidence of nigrostriatal neurodegeneration (not always associated with Lewy bodies) and inflammation (Oppenheimer and Esiri, 1997). Unfortunately, inflammation is casually mentioned in most published reports on non-PD parkinsonian syndromes, and when the authors of these reports do mention inflammation it is often limited to vague, qualitative statements. For instance, in the initial reports of progressive supranuclear palsy (Steel et al., 1964) and striatonigral degeneration (Adams and Salam-Adams, 1986), *gliosis* was recognized as a prominent feature of the pathological changes seen in these syndromes, but is the sole reference to inflammation. However, whether or not gliosis, as mentioned in these studies, refers to alterations in astrocytes, microglia, or to both, remains unknown. Despite this limitation, it is remarkable that gliosis is not only described in these non-PD parkinsonian syndromes at the level of the nigrostriatal pathway, but also at the level of other regions in the brain, which are normally not involved in PD. This indicates a rather widespread neuropathology which is consistent with the usual multisystemic nature of many of these parkinsonian syndromes (Table 26.1).

Aside from sporadic parkinsonian syndromes, several familial forms also exist (Vila and Przedborski, 2004). Like in sporadic PD, histological examinations revealed similar glial alterations in most of the familial forms of parkinsonian syndromes, whether the underlying genetic defect has (Vila and Przedborski, 2004) or has not been identified (Dwork et al., 1993). Among these familial parkinsonian syndromes, the situation of the autosomal dominant form of PD linked to mutations in the gene coding for leucine-rich repeat kinase 2 (LRRK2) is interesting (Zimprich et al., 2004). In the six autopsies from patients carrying a LRRK2 mutation reported by Zimprich and collaborators (2004), all had nigrostriatal dopaminergic neuronal loss and gliosis in the substantia nigra. However, some of these patients had dopaminergic neuronal loss and gliosis without Lewy bodies, whereas others had dopaminergic neuronal loss and gliosis with Lewy bodies (Zimprich et al., 2004). In the latter cases, Lewy bodies were restricted to the brainstem, or widespread in the brainstem and cortex. In one case, there were also *tau*-immunoreactive lesions not only in neurons, but also in glial cells (Zimprich et al., 2004). In the autosomal recessive parkinsonian syndrome linked to *parkin* mutations there is also a loss of dopaminergic neurons associated with gliosis but, as in several LRRK2 mutation cases, not typically with Lewy bodies (Hayashi et al., 2000). These findings support the view that inflammation is a generic phenomenon that arises from neuronal death irrespective of the type of parkinsonian syndrome or neuropathological picture. It also suggests that the presence of Lewy bodies is not a prerequisite for the occurrence of inflammation in PD and related conditions.

TABLE 26.1. Parkinsonian syndromes.

Primary parkinsonism
Parkinson disease (sporadic, familial)
Secondary parkinsonism
Drug-induced: dopamine antagonists and depletors
Hemiatrophy-hemiparkinsonism
Hydrocephalus: Normal pressure hydrocephalus
Hypoxia
Infectious: Post-encephalitic
Metabolic: Parathyroid dysfunction
Toxin: Mn, CO, MPTP, cyanide
Trauma
Tumor
Vascular: Multiinfarct state
Parkinson-plus syndromes
Cortical-basal ganglionic degeneration
Dementia syndromes: Alzheimer disease, diffuse Lewy body disease, frontotemporal dementia
Lytico-Bodig (Guamanian Parkinsonism-dementia-ALS)
Multiple system atrophy syndromes: Striatonigral degeneration, Shy-Drager syndrome, sporadic olivopontocerebellar degeneration (OPCA), motor neuron disease-parkinsonism
Progressive pallidal atrophy
Progressive supranuclear palsy
Familial neurodegenerative diseases
Hallervorden-Spatz disease
Huntington disease
Lubag (X-linked dystonia-parkinsonism)
Mitochondrial cytopathies with striatal necrosis
Neuroacanthocytosis
Wilson disease

Notes: MPTP, 1-methyl-4-phenyl-1,2,3,6-tetrahydropyridine; ALS; amyotrophic lateral sclerosis. (*From Dauer and Przedborski, Neuron 39:889-909, 2003, with permission.*)

26.4. Inflammatory Response in Experimental Models

As discussed above, studies performed in autopsy tissues from patients afflicted with parkinsonian syndromes have led to the conclusion that inflammation is part of the neuropathological picture of these neurodegenerative disorders. While the descriptive data provided by these studies are invaluable, the generated information remains correlative by nature and fails to provide mechanistic insights. This is why researchers rely heavily on experimental models of PD. These models help to define the temporal and topographical relationships between inflammation and dopaminergic neuronal death and, more importantly, to determine the potential pathogenic role of inflammation in the death of dopaminergic neurons.

Experimental models of PD are multiple and can be genetic or toxic (Dauer and Przedborski, 2003). The neuropathological picture created in these models, particularly in the toxic models, is very close to that described in PD. In almost all of the PD models, whether or not there is an overt loss of nigrostriatal dopaminergic neurons, a glial response is found in the substantia nigra. Some data on inflammation are available from rodent models produced by the herbicide paraquat (McCormack et al., 2002) and the mitochondrial poison rotenone (Sherer et al., 2003). Some data can also be found from transgenic mice expressing mutant alpha-synuclein (Gomez-Isla et al., 2003). The limited amount of data on inflammation in these models however is in striking contrast with the wealth of information available for the 6-hydroxydopamine (6-OHDA) and the 1-methyl-4-phenyl-1,2,3,6-tetrahydropyridine (MPTP) toxic models of PD. Perhaps this can be explained by the fact that both the 6-OHDA and MPTP models are not only more popular among PD researchers, but also they have been used for much longer than the other models cited above. Supporting the generic nature of the inflammatory response in neurodegeneration are the observations that both the type and magnitude of the glial alterations in rodents following the administration of 6-OHDA (Nomura et al., 2000; Przedborski et al., 1995; Sheng et al., 1993; Stromberg et al., 1986; Rodrigues et al., 2001; He et al., 1999; Akiyama and McGeer, 1989) are comparable to those seen following the administration of MPTP (see below).

In the early eighties, several drug addicts got intoxicated with MPTP after self-injecting contaminated street-batches of meperidine analog (Langston et al., 1983). Among these individuals, some developed an acute, irreversible parkinsonian syndrome indistinguishable from PD (Ballard et al., 1985). Of the few MPTP-intoxicated individuals who died and underwent an autopsy, post-mortem examination showed a profound loss of nigrostriatal dopaminergic neurons (Davis et al., 1979; Langston et al., 1999). This dramatic loss of large neuronal cells was associated with the obvious abundance of small cells, immunoreactive for either the astrocyte marker GFAP or the microglia marker HLA-DR (Langston et al., 1999). Almost all GFAP- and HLA-DR-positive cells exhibited morphological characteristics of respectively reactive astrocytes and activated microglia (Langston et al., 1999). Images of neuronophagia were also observed (Langston et al., 1999). Similar glial alterations were found in six monkeys, which survived 5–14 years after an acute MPTP intoxication (McGeer et al., 2003). In all of these monkeys, there was also evidence of extracellular neuromelanin and activated microglia in the substantia nigra (McGeer et al., 2003). These monkeys also had an abundance of reactive astrocytes positive for the intracellular adhesion molecule-1 (Miklossy et al., 2005). The presence of intracellular adhesion molecule-1-positive reactive astrocytes, of activated microglia, and of neuronophagia in humans and monkeys subjected to an acute MPTP intoxication many years prior is remarkable as these neuropathological features suggest an active, ongoing inflammatory process. Microglial activation and neuronophagia would be expected to be seen in PD tissues in light of its progressive neurodegenerative nature. However, as stressed by McGeer and collaborators (2003), these neuropathological findings challenge the belief that MPTP, 'produces an acute loss of cells, followed by healing and long-term stabilization of surviving neurons'. Alternatively, these neuropathological data raise the possibility that an acute MPTP intoxication can set in motion a self-sustained cascade of detrimental effects on dopaminergic neurons. Corroborating this view is a positron emission tomography study in which ten individuals exposed acutely to MPTP were scanned twice, seven years apart (Vingerhoets et al., 1994). This work revealed a decrease of [^{18}F]fluorodopa uptake in the striata of these patients between the first and second imaging studies (Vingerhoets et al., 1994). Also remarkable is the fact that three of the ten MPTP-intoxicated participants were asymptomatic at the time of the first scan but became parkinsonian by the time of the second scan (Vingerhoets et al., 1994) supporting a progression of the neurodegenerative process in this toxic parkinsonian syndrome, at least in primates.

In small animals such as mice, time course experiments have been performed to define not only the kinetics of dopaminergic neuronal death but also of the glial response. These studies demonstrated that the appearance of reactive astrocytes paralleled the destruction of dopaminergic structures in both the striatum and the substantia nigra in mice (Czlonkowska et al., 1996; Kohutnicka et al., 1998; Liberatore et al., 1999). Worth noting, GFAP expression remained high even after the main phase of neuronal degeneration (Czlonkowska et al., 1996; Kohutnicka et al., 1998; Liberatore et al., 1999). Also important to note is the fact that upon inhibition of the entry of the active metabolite MPTP, 1-methyl-4-phenylperydinium (MPP^+), into dopaminergic neurons there is not only a protection of substantia nigra dopaminergic neurons, but also a lack of glial response (O'Callaghan et al., 1990). This critical observation indicates that inflammation in the MPTP model is a consequence of the demise of neurons and not the reverse. This interpretation may not be valid for all types of injury used to model PD as we will see for the endotoxin of gram-negative bacteria lipopolysaccharide (LPS). The activation of microglial cells is also quite robust after MPTP administration in mice (Figure 26.2); this has been extensively documented in the MPTP mouse model (Czlonkowska et al., 1996; Kohutnicka et al., 1998; Liberatore et al., 1999; Dehmer et al., 2000). When the astrocytes and microglia responses are compared, however, it appears the microglial activation occurs earlier than that of astrocytes, and peaks just before or coincidentally

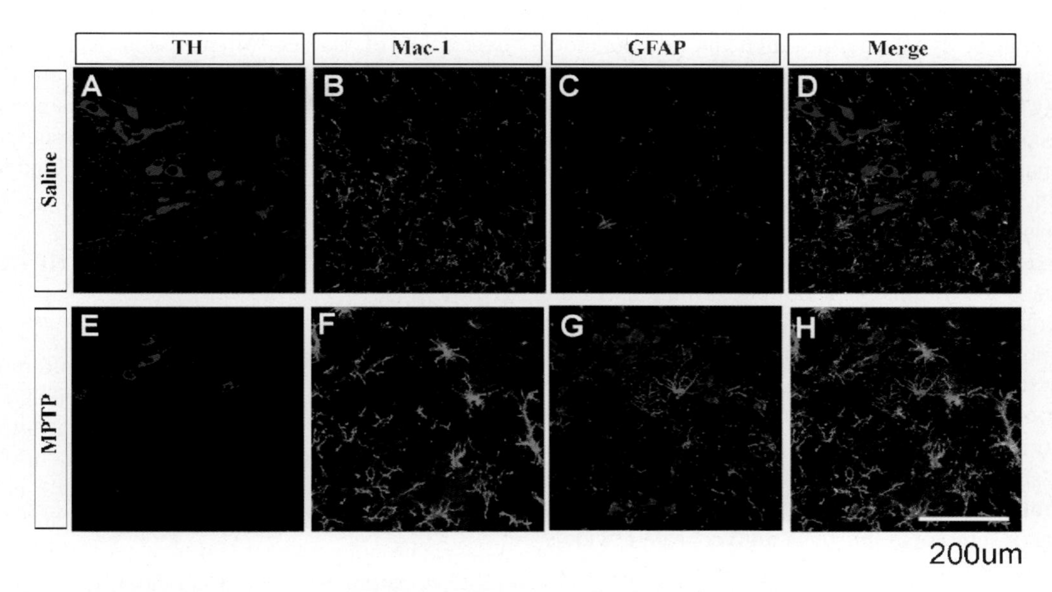

FIGURE 26.2. Effects of MPTP on dopaminergic neurons and glial cells in ventral midbrain of in mice. For this illustration, mice received either 0.3 ml of MPTP (2 mg/ml) or vehicle (saline) in one day. (**A,E**) Two days after the last injection of MPTP, TH immunofluorescence (blue) reveals a marked reduction in dopaminergic neurons in the substantia nigra pars compacta compared to controls (saline). (**B,C,F,G**) The loss of TH labeling coincides with an increase in immunofluorescent labeling of microglial cells (Mac-1; green) and of astrocytes (GFAP; red). (**D,H**) Overlay of the three cellular markers. When MPTP and saline tissue sections are compared, these images show that upon reduction in the dopaminergic neuronal marker, there is an increase in the two glial markers. Ventral midbrain tissue sections (30 μm) were immunostained with a rabbit polyclonal anti-TH (1:1000; Calbiochem, San Diego, CA), a rat anti-MAC-1 (1:1000; Serotec, Raleigh, NC), and a chicken anti-GFAP (1:500, Chemicon, Temecula, CA). *Scale bar = 200 μm.*

to, the climax of dopaminergic neurodegeneration (Liberatore et al., 1999).

In addition to the response of the innate immune system in the MPTP mouse model, there has also been some descriptive data from Członkowska and collaborators (Kurkowska-Jastrzebska et al., 1999a, b) about the response of the adaptive immune system. Although this question will be developed in-depth in the chapter authored by Dr. Lee Mosley, herein it is included as a brief informative aspect. In the series of investigations performed by Członkowska and collaborators cited above, a marked increase of MHC-II antigen expression by microglia, accompanied by a recruitment of T-cells in both the ventral midbrain and striatum of MPTP-injected mice, were found. Contrasting with the infiltration of the diseased areas of the brain with T-cells, there was no B-cell in the tissue samples. The investigators went further into their characterization of the adaptive immune system in this model by showing that the infiltrating T-cells were mainly of the CD8+ type, but some CD4+ were present too, and more than 50% of the observed lymphocytes expressed the CD44 antigen.

26.5. Initiation of the Inflammatory Response in Parkinsonian Brains

If indices of inflammation are well recognized in tissues from virtually all neurodegenerative disorders including PD, how inflammation begins, remains enigmatic. In reviewing the literature, two main hypotheses are emerging. One hypothesis posits that inflammation precedes and triggers the actual loss of dopaminergic neurons. Based on this model, the starting point of the neurodegenerative process would be an insult to the brain by pro-inflammatory factors such as LPS or other infectious-related molecules. Relevant to this possibility is the demonstration that a stereotaxic injection of LPS into a normal substantia nigra pars compacta produces a local inflammatory response associated with a degeneration of dopaminergic neurons in adult rats (Liu et al., 2000). Similarly, LPS-induced inflammation led to the neurodegeneration of dopaminergic MES 23.5 cells or primary ventral midbrain neurons co-cultured with purified microglia (Le et al., 2001). Based on a series of investigations in rodents it seems that even a prenatal exposure to LPS could produce a long-lasting inflammatory response in the brain (Ling et al., 2004). This striking observation led the authors to propose that 'individuals exposed to LPS prenatally, as might occur had their mother had bacterial vaginosis, would be at increased risk for PD'.

As much as the first hypothesis appears tantalizing the most available data are rather consistent with this second hypothesis which posits that inflammation in PD results from the detection of pathological alterations in neighboring neurons by the immune system. According to this model, inflammation in PD and related disorders is thus not a primary event, but instead the consequence of neurodegeneration. However, how dysfunctional or dying neurons elicit the inflammatory response remains unknown. Many glial cells are located in close proximity to neurons. Therefore, it may very well be that inflammation could be initiated by some type of change in the nature or the quality of the neuronal contact with glial cells. In healthy cells including neurons, phosphatidylserine is located at the cytoplasmic side of the plasma membrane, while the outer monolayer is composed of phosphatidylcholine and sphingomyelin. A loss of this phospholipid asymmetry of the plasma membrane, with appearance of phosphatidylserine in the outer membrane leaflet, is a feature of cells undergoing apoptosis (van den Eijnde et al., 1998). It is thus possible that such an alteration of the neuronal membrane could be detected by the neighboring glial cells, leading to their activation.

Cell culture studies also show that inflammation can be triggered by soluble factors, either secreted or leaked by neurons. For instance, chromogranin-A is a glycoprotein widely distributed in the nervous system, which was shown to accumulate in areas of neurodegeneration (Nishimura et al., 1994). Noteworthy is the fact that once released or leaked by neurons, chromogranin-A could activate microglia (Ciesielski-Treska et al., 1998). Furthermore, dopaminergic neurons in both PD and experimental models of PD are the site of a robust induction of cyclooxygenase-2 (Teismann et al., 2003), which is a key prostaglandin-synthesizing enzyme. It is also plausible that once synthesized, prostaglandins exit neurons and activate their cognate receptors expressed by glial cells, whereby triggering inflammatory events. Similarly, neuromelanin upon its release in the extracellular space from dying neurons can strongly activate glial cells (Wilms et al., 2003). Finally, it has also been suggested that misfolded neuronal proteins and protein aggregates may contribute to glial activation in PD. This view has been prompted by the observations that mutations in the genes encoding for parkin and ubiquitin *C*-terminal hydrolase L1—two enzymes of the ubiquitin/proteasome pathway—and for alpha-synuclein—a main component of the intraneuronal proteinaceous inclusions Lewy bodies—are linked to familial PD (Vila and Przedborski, 2004).

Although more work needs to be done to identify the factors that initiate inflammation in PD and related disorders, mounting evidence indicates that this immune response is probably not a standard event of dysfunctional and dying neurons. Currently, most researchers favor the idea that *specific* neuronal-derived molecules bind to *specific* transmembrane receptors, such as toll-like receptors (TLRs) present on glial cells (Bowman et al., 2003). Among the ten different TLRs, each is activated by a distinct ligand (Iwasaki and Medzhitov, 2004). Although activation of TLRs has been studied thus far in the context of infection, it is now believed that inflammation in neurodegenerative disorders arises from the ligation of the particular TLR subtypes by molecules originating from dysfunctional and dying neurons. Relevant to this idea is the demonstration of wild-type alpha-synuclein, which upon overexpression can cause a familial form of PD (Eriksen et al., 2005), and is secreted by neurons through unconventional exocytosis (Lee et al., 2005). Furthermore, it appears that the expression of markers of microglial activation parallels the extent of alpha-synuclein deposits in the substantia nigra of PD patients (Croisier et al., 2005). At this point, it would be quite interesting to know whether alpha-synuclein or any other mutant proteins linked to familial PD can bind to TLRs.

26.6. Role of Inflammation in Parkinson's Disease

Over the past decade, the idea that brain inflammation (sometimes called neuroinflammation) may modulate the neurodegenerative process in disorders like PD has attracted a major interest among neuroscientists. How can this be explained given the fact that inflammation is likely a consequence of neurodegeneration? Neurons in all degenerative diseases die in an asynchronous manner (Pittman et al., 1999). Accordingly, in PD, dopaminergic neurons in the substantia nigra pars compacta do not die simultaneously. Instead, only a small number of dopaminergic neurons are dying at any given time and, among these, many are at various stages along the cell death process. Based on these premises, it may thus be envisioned that the very first neurons to succumb trigger inflammation. Thereafter, the cellular environment in which compromised, living neurons are embedded becomes inflamed. Should this scenario be correct, it is trivial to understand how inflammation could modulate the neurodegenerative process.

While most experts do agree on the idea that glial cells play a role in the neurodegenerative process, the significance and, more importantly, the direction, of the effect remains a matter of fierce debate (Streit, 2002). The dispute is fueled in part by the fact that both *in vitro* and *in vivo* system models of human diseases have revealed both detrimental and beneficial effects of inflammation. For instance, the blockade of the microglial activation by minocycline has been associated with either reduction or augmentation of dopaminergic neurodegeneration after MPTP administration (Du et al., 2001; Wu et al., 2002; Yang et al., 2003; Diguet et al., 2004). These divergent results may stem from the fact that inflammation is capable of exerting both neuroprotective and neurodestructive functions. Thus, depending on local factors, the extent of the degenerative process, and even the etiology of the disease, inflammation can give rise to quite a distinct molecular and cellular phenotype.

26.7. Beneficial Role of Inflammation in Parkinson's Disease

Several *in vivo* studies support the capacity of inflammation in mediating neuroprotective and neurodegenerative properties in the nervous system. One such example is the facial nerve axotomy paradigm in newborn rats and rabbits. In this system model, axotomized motor neurons recover coincidentally with the development of the glial response (Moran and Graeber, 2004). It was also shown that a month after implantation of innate immune resident cells such as microglia into a small mechanically produced cavity in the rat spinal cord, prominent neuritic growth was observed in the microglial grafts (Rabchevsky and Streit, 1997). These results are agreed upon by some experts who believe that under both normal and pathological situations, the function and survival of neurons rely on the presence of glial cells (Streit, 2002).

The basis of glial-derived beneficial role is likely multifactorial involving both cell contact and soluble mediators. For instance, glial cells play a critical role in maintaining ion and pH homeostasis, and extracellular volume. Glial cells can also protect neurons by scavenging and taking-up toxic molecules released by dysfunctional and dying neurons. A striking example of glial assistance in the process of neuronal well-being is glutathione, which is a tripeptide of great importance in the protection of neurons against reactive oxygen species (ROS). Indeed, unlike glial cells, neurons do not possess an uptake system for cysteine (Pow, 2001) which is necessary for the synthesis of glutathione. Instead it has been proposed (Dringen, 2000) that astrocytes assist neurons in their production of glutathione through a complex multistep process. First, the astrocytes would use extracellular substrates as precursors for glutathione. Second, glutathione would be released by astrocytes and then converted by the glial ectoenzyme gamma-glutamyl transpeptidase into the dipeptide CysGly. Third, CysGly would be taken-up by neurons and used as a precursor to neuronal glutathione. It has also been proposed (Dringen, 2000) that glutamine produced by astrocytes would be used by neurons to produce glutamate, which is necessary for the synthesis of glutathione. Finally, astrocytes, can avidly take up extracellular glutamate via the glutamate transporters GLT1 and GLAST. This fact may be quite relevant to PD pathogenesis as not only an excess amount of glutamate in the substantia nigra may emanate from dying neurons but also from the subthalamic glutamatergic input (Benazzouz et al., 2000), which is increased in PD (DeLong, 1990). Thus, due to the ability of glial cells in maintaining the low concentration of extracellular glutamate, they prevent the risk of excitotoxic injury on neurons in the substantia nigra.

Aside from contributing to the protection of neurons against ROS and glutamate, glial and T-cells can produce trophic factors that are essential for the survival of dopaminergic neurons. Among these is the glial-derived neurotrophic factor (GDNF), which seems to be the most potent factor supporting nigrostriatal dopaminergic neurons during their period of natural, developmental death in post-natal ventral midbrain cultures (Burke et al., 1998). GDNF can be released by reactive astrocytes (Schaar et al., 1993) by the activation of microglia following a mechanical lesion of the striatum (Batchelor et al., 2000). After MPTP administration in mice, however, it seems that reactive astrocytes in the substantia nigra do not engage in a significant production of GDNF (Benner et al., 2004). Brain-derived neurotrophic factor (BDNF) is another trophic factor that can also be released by reactive astrocytes (Rubio, 1997; Stadelmann et al., 2002), by activated microglia (Batchelor et al., 1999; Stadelmann et al., 2002), and can support the survival and development of dopaminergic structures in the striatum (Batchelor et al., 1999). Oligodendrocytes whose main function is to provide support to axons and to produce the myelin sheath which insulates axons are also a type of glial cell capable of generating trophic factors (Du and Dreyfus, 2002). For example, striatal oligodendrocytes have been shown to stimulate the survival and expression of phenotypic markers of mesencephalic dopaminergic neurons in culture (Sortwell et al., 2000). Taken together these data support the contention that glial cells, especially astrocytes, could exert neuroprotective effects in PD. Whether any of these, however, do slow the natural course of the disease in parkinsonian patients remains to be demonstrated.

26.8. Detrimental Role of Inflammation in Parkinson's Disease

In addition to the body of literature on the beneficial role of inflammation, there are also a growing number of observations supporting the concept that activation of the innate immune system could worsen neurodegeneration. Three theories have been proposed to explain the detrimental role of glial cells in diseases such as PD. The first theory is called *glial cell senescence* (Streit, 2004) which postulates that glial cells become progressively disabled during normal aging or disease progression. It is assumed with this theory that glial cells loose their functional capacity to exert the type of beneficial effects described above. In this model glial cells do not actively injure neurons, they simply stop supporting them. Although still highly speculative, some investigations would suggest that glial cells may become dysfunctional in PD. For instance, Hishikawa and collaborators (2001) have found argyrophilic, alpha-synuclein-positive inclusions in glial cells from postmortem PD, but not from age-matched control brains. The second theory about the detrimental role of inflammation in PD is called *facilitative neurotoxicity* (Streit, 2002). During this theoretical process, glial cells will eliminate neurons that are compromised beyond viability and functionality by the primary pathological event. Based on this second model, glial cells would assume an active role in the demise of neurons that are destined to die and whose continued presence could hinder neuronal recovery. Finally, the third theory and probably the most popular in regards to the detrimental role of inflammation in PD is called *indiscriminate neurotoxicity*. Based on this third model, glial cells, upon activation, would engage in an array of cytotoxic events, which would stimulate neurodegeneration and promote both the progression and propagation of a disease such as PD. Many studies of experimental models of PD have provided over the past decade strong support for this last concept which is based on the fact that activated glial cells, especially microglia are known to produce a variety of noxious compounds including ROS, reactive nitrogen species (RNS), pro-inflammatory prostaglandins, and cytokines.

Among the range of reactive species produced by glial cells, significant attention has been given to RNS due to the idea that nitric oxide (NO)-mediated nitrating stress could play an important role in the pathogenesis of PD (Przedborski et al., 1996, 2001; Ara et al., 1998; Pennathur et al., 1999; Giasson et al., 2000). Along this line it must be mentioned that the inducible NO synthase (iNOS) is expressed in the substantia nigra pars compacta of both PD patients (Hunot et al., 1996) and MPTP-intoxicated mice (Liberatore et al., 1999; Dehmer et al., 2000) in astrocytes and microglial cells, respectively. Myeloperoxidase and NADPH-oxidase, two pro-inflammatory enzymes which produce strong oxidants, have also been identified in PD and MPTP brain tissues (Choi et al., 2005; Wu et al., 2003). Ablation of these three enzymes have been shown to attenuate MPTP-induced neurodegeneration in mice to varying degrees (Liberatore et al., 1999; Dehmer et al., 2000; Choi et al., 2005; Wu et al., 2003), supporting the involvement of inflammation-mediated oxidative/nitrative stress in the dopaminergic neurodegenerative process.

Prostaglandins and their synthesizing enzymes, such as Cox-2, constitute another group of potential offenders. Over the past few years, Cox-2 has emerged as a key cytotoxic factor associated with inflammation (O'Banion, 1999). As mentioned above, compared to the normal basal ganglia where Cox-2 is minimally expressed, diseased basal ganglia in PD and the MPTP mouse model, Cox-2 is robustly expressed (Teismann et al., 2003). Not surprisingly, the levels of Cox-2 products such as prostaglandin E_2 were also found to be increased in the substantia nigra of PD patients (Teismann et al., 2003; Mattammal et al., 1995). In the MPTP mouse model of PD, Cox-2 induction was shown to be mediated by a c-Jun *N*-terminal kinase (JNK)-dependent mechanism (Teismann et al., 2003; Hunot et al., 2004). Noteworthy, the inhibition of JNK activation, like inhibition of Cox-2, did attenuate MPTP-induced neurodegeneration (Hunot et al., 2004; Teismann et al., 2003). Conversely, while both inhibition and ablation of Cox-2 did attenuate MPTP-induced dopaminergic neurotoxicity, deletion of the constitutively expressed isoenzyme Cox-1 had no effect (Teismann et al., 2003).

Additional glial-derived molecules with cytotoxic potentials include the large group of pro-inflammatory cytokines. The expression level of many of these toxic cytokines assessed by gene profiling appears increased in PD tissues (Mandel et al., 2005). In keeping with this result, the tumor necrosis factor-alpha (TNF-alpha) and interleukine-1beta (IL-1beta) whose levels are increased in both the substantia nigra pars compacta tissues and cerebrospinal fluids of PD patients (Mogi et al., 1994, 1996, 2000). Of note, at this point one cannot exclude that some of these alterations are related to the chronic use of the anti-PD therapy levodopa (Bessler et al., 1999). Nevertheless, in autopsy tissue from PD patients, robust immunostaining for TNF-alpha, IL-1beta, and interferon-gamma (IFN-gamma) is observed in astrocytes at the level of the substantia nigra pars compacta (Hunot et al., 1999). In terms of mechanism, these cytokines may operate in PD on at least two levels. First, once produced by reactive astrocytes, they can stimulate neighboring, quiescent astrocytes and microglia. Consistent with this view is the upregulation of the microglial receptor Fc∞R11/CD23 in response to astrocyte-derived TNF-alpha, IL-1beta, and IFN-gamma (Hunot et al., 1999). In this study, the authors also show that, once upregulated, ligation of Fc∞R11/CD23 stimulates the expression of iNOS and the ensuing production of NO by microglial cells (Hunot et al., 1999). Second, astrocytic and microglial-derived cytokines may also operate directly on dopaminergic neurons by binding to specific cell surface cytokine receptors such as TNF-alpha receptor and FAS. However, the targeting of either TNF-alpha receptor (Rousselet et al., 2002; Ferger et al., 2004) or FAS (Hayley et al., 2004; Landau et al., 2005) has thus far generated conflicting results in the MPTP mouse model. Findings from several studies show that ablation of these receptors decreased MPTP-induced dopaminergic neurotoxicity (Ferger et al., 2004; Hayley et al., 2004) whereas other studies found that it increased MPTP-induced dopaminergic neurotoxicity (Rousselet et al., 2002; Landau et al., 2005). Upon activation, cytokine receptors trigger intracellular death-related signaling pathways, whose molecular correlates include translocation of the transcription nuclear factor-κ-B (NF-κ-B) from the cytoplasm to the nucleus. Relevant to this is the fact that PD patients show 70 times more dopaminergic neurons with nuclear NF-κ-B immunoreactivity compared to control subjects (Hunot et al., 1997). Despite the robust recruitment of NF-κ-B, it is not clear whether this transcriptional factor is instrumental in PD pathogenesis, as mice deficient in one of NF-κ-B main polypeptide, P50, had their nigrostriatal pathway as severely damaged by MPTP as that of their wild-type counterparts (Hunot et al., 2004). As exemplified above, at this point experimental models of PD have not provided much insight into the role of cytokines in neurodegeneration of such condition as PD and related conditions.

Acknowledgements. The author thanks Ms. Julia Jeon for her assistance in preparing this manuscript and is supported by NIH/NINDS Grants RO1 NS38586 and NS42269, P50 NS38370, and P01 NS11766-27A2, NIH/NIA RO1 AG21617-01, NIH/NIEHS R21 ES013177, the US Department of Defense Grant DAMD 17-03-1, the Parkinson Disease Foundation (New York, USA), the Lowenstein Foundation, the MDA/Wings Over Wall Street.

Summary

PD is a common neurodegenerative disorder characterized by motor abnormalities and a loss of primarily but not exclusively dopaminergic neurons. Other neuropathological hallmarks of PD include the proteinaceous inclusion, Lewy body and various inflammatory changes. The inflammation is made up of resident innate and to lesser extent adaptive immune cells. Similar inflammatory changes are found in both sporadic and familial non-PD parkinsonian syndromes and in experimental models of PD. There are some studies that suggest that inflammation initiates PD, but most investigations rather suggest that inflammation in PD is secondary to the demise of neurons. Mounting evidence indicates that inflammation in PD and related conditions can modulate the neurodegenerative process. However, there is still a lack of consensus about whether inflammation in PD plays a beneficial or a detrimental role.

Review Questions/Problems

1. **What are the clinical and neuropathological hallmarks of PD?**
2. **What is the difference between parkinsonian syndrome and Parkinson's disease?**
3. **Provide examples of experimental models of PD.**
4. **What is the topography and composition of the inflammatory response in PD?**
5. **What are the similarities and differences in the inflammatory response among PD, the various parkinsonian syndromes, and the common experimental models of PD?**
6. **What are the arguments in favor of inflammation being a primary or a secondary event in PD?**
7. **How can dying dopaminergic neurons trigger inflammation in PD?**
8. **By which mechanisms could inflammation be beneficial in PD?**
9. **Name the three theories about the detrimental role of inflammation in PD and explain their respective basis.**
10. **Summarize the different cellular and molecular factors of inflammation that can actively mediate the demise of dopaminergic neurons in PD?**
11. **Which statement is incorrect about PD?**
    a. Parkinson's disease is the second most frequent neurodegenerative disorder.
    b. The neurodegenerative process in Parkinson's disease is restricted to the dopaminergic system.
    c. Parkinson's disease is mainly a sporadic condition.
    d. More than 30 other neurodegenerative diseases can clinically look like Parkinson's disease.
    e. Aside from the loss of neurons the neuropathology of Parkinson's disease includes Lewy bodies and gliosis.
12. **What are the main inflammatory cells encountered in the substantia nigra of Parkinson's disease?**
    a. Astrocytes
    b. Microglia
    c. Oligodendrocytes
    d. T-cells
    e. a and b
13. **Which statement is correct concerning inflammation in parkinsonian syndromes?**
    a. It is often noted, but more widespread and less detailed than in Parkinson's disease.
    b. The type of inflammatory response differs between the sporadic and familial parkinsonian syndromes.
    c. The syndrome multisystem atrophy is unique in that inflammation is primarily made up of infiltrating T-cells.
    d. The neuropathological pleomorphism in patients carrying a LRRK2 mutation refers to the fact that neither Lewy bodies nor gliosis is a consistent finding.
    e. None of the above.
14. **Which statement about experimental models of Parkinson's disease is true?**
    a. Both genetic and toxic models exist, but only the former are commonly used.
    b. Inflammation has been described in all popular models of Parkinson's disease.
    c. The MPTP monkey model suggests that an acute intoxication produces an acute neurodegenerative event that is completed in a few days.
    d. The MPTP mouse model suggests that the toxin provokes inflammation, which, in turn, kills dopaminergic neurons.
    e. Neuronophagia which suggests ongoing inflammation has been described in genetics, but not in toxic models of Parkinson's disease.
15. **Which of the following statements is true about inflammation in Parkinson's disease?**
    a. Free neuromelanin fails to activate microglia.
    b. Astrocytosis is as robust as microgliosis.
    c. The propensity of the different dopaminergic structures to degenerate in Parkinson's disease correlates with the basal density of glial cells.
    d. Both prostaglandin and alpha-synuclein count among the factors potentially responsible for triggering inflammation in Parkinson's disease.
    e. It is proven that prenatal infection and subsequent inflammation predispose one to Parkinson's disease.
16. **Which of the following statements is most correct?**
    a. Inflammation can exert both beneficial and detrimental effects.
    b. Most experimental models favor the beneficial role of inflammation
    c. The detrimental role of inflammation in Parkinson's disease is due to the disease-related impairment of glial functions vital to neurons.
    d. Contrary to astrocytes, oligodendrocytes play no role in Parkinson's disease.
    e. Three different theories have been proposed to explain how inflammation may support the survival of dopaminergic neurons.

17. Which of the following glial functions may improve neuronal survival or regeneration?

a. Inhibit phagocytosis.
b. Secrete chemotactic molecules to recruit polynuclear cells.
c. Produce trophic factors.
d. Assist in the synthesis of neuronal superoxide dismutase.
e. Stimulate the formation of myelin to guide new axons.

18. Glial cells can exacerbate neurodegeneration in Parkinson's disease by?

a. Losing their ability to assist neighboring neurons.
b. Accelerating the demise of compromised neurons.
c. A process of indiscriminate toxicity.
d. Decreasing extracellular glutamate levels.
e. a, b and c

19. Regarding the cytotoxicity of inflammation, which statement is not correct?

a. Both oxygen and nitrogen reactive species can participate in the deleterious effects of inflammation.
b. Inactivation of NADPH-oxidase, but not of nitric oxide synthase mitigates MPTP-induced neurodegeneration in mice.
c. The detrimental effects of inflammation on dopaminergic neurons can be mediated by soluble factors.
d. Astrocytes and microglial cells can mutually modulate their degree of activation.
e. Dopaminergic neurons express receptors for various deleterious cytokines.

20. Among the following inflammatory factors, which ones have been identified in the substantia nigra of patients with Parkinson's disease?

a. Tumor necrosis factor
b. Inducible nitric oxide synthase
c. Interferon-gamma
d. Myeloperoxidase
e. All of the above

References

Adams RD, Salam-Adams M (1986) Striatonigral Degeneration. In: HandBook of Clinical Neurology. Extrapyramidal Disorders (Vinken PJ, Bruyn GW, Klawans HL, eds), pp 205–212. New York: Elsevier.

Agid Y, Javoy-Agid F, Ruberg M (1987) Biochemistry of Neurotransmitters in Parkinson's Disease. In: Movement Disorders 2 (C.D. Marsden, S. Fahn, eds), pp 166–230. London: Butterworths.

Akiyama H, McGeer PL (1989) Microglial response to 6-hydroxydopamine-induced substantia nigra lesions. Brain Res 489:247–253.

Ara J, Przedborski S, Naini AB, Jackson-Lewis V, Trifiletti RR, Horwitz J, Ischiropoulos H (1998) Inactivation of tyrosine hydroxylase by nitration following exposure to peroxynitrite and 1-methyl-4-phenyl-1,2,3,6-tetrahydropyridine (MPTP). Proc Natl Acad Sci USA 95:7659–7663.

Ballard P, Tetrud JW, Langston JW (1985) Permanent human parkinsonism due to 1-methyl-4-phenyl-1,2,3,6-tetrahydropyridine (MPTP): 7 cases. Neurology 35:949–956.

Banati RB, Daniel SE, Blunt SB (1998) Glial pathology but absence of apoptotic nigral neurons in long- standing Parkinson's disease. Mov Disord 13:221–227.

Batchelor PE, Liberatore GT, Wong JY, Porritt MJ, Frerichs F, Donnan GA, Howells DW (1999) Activated macrophages and microglia induce dopaminergic sprouting in the injured striatum and express brain-derived neurotrophic factor and glial cell line-derived neurotrophic factor. J Neurosci 19:1708–1716.

Batchelor PE, Liberatore GT, Porritt MJ, Donnan GA, Howells DW (2000) Inhibition of brain-derived neurotrophic factor and glial cell line-derived neurotrophic factor expression reduces dopaminergic sprouting in the injured striatum. Eur J Neurosci 12:3462–3468.

Benazzouz A, Piallat B, Ni ZG, Koudsie A, Pollak P, Benabid AL (2000) Implication of the subthalamic nucleus in the pathophysiology and pathogenesis of Parkinson's disease. Cell Transplant 9:215–221.

Benner EJ, Mosley RL, Destache CJ, Lewis TB, Jackson-Lewis V, Gorantla S, Nemachek C, Green SR, Przedborski S, Gendelman HE (2004) Therapeutic immunization protects dopaminergic neurons in a mouse model of Parkinson's disease. Proc Natl Acad Sci USA 101:9435–9440.

Bessler H, Djaldetti R, Salman H, Bergman M, Djaldetti M (1999) IL-1 beta, IL-2, IL-6 and TNF-alpha production by peripheral blood mononuclear cells from patients with Parkinson's disease. Biomed Pharmacother 53:141–145.

Bowman CC, Rasley A, Tranguch SL, Marriott I (2003) Cultured astrocytes express toll-like receptors for bacterial products. Glia 43(3):281–291.

Braak H, Braak E, Yilmazer D, Schultz C, de Vos RA, Jansen EN (1995) Nigral and extranigral pathology in Parkinson's disease. J Neural Transm Suppl 46:15–31.

Burke RE, Antonelli M, Sulzer D (1998) Glial cell line-derived neurotrophic growth factor inhibits apoptotic death of postnatal substantia nigra dopamine neurons in primary culture. J Neurochem 71:517–525.

Choi DK, Pennathur S, Perier C, Tieu K, Teismann P, Wu DC, Jackson-Lewis V, Vila M, Vonsattel JP, Heinecke JW, Przedborski S (2005) Ablation of the inflammatory enzyme myeloperoxidase mitigates features of Parkinson's disease in mice. J Neurosci 25:6594–6600.

Ciesielski-Treska J, Ulrich G, Taupenot L, Chasserot-Golaz S, Corti A, Aunis D, Bader MF (1998) Chromogranin A induces a neurotoxic phenotype in brain microglial cells. J Biol Chem 273:14339–14346.

Croisier E, Moran LB, Dexter DT, Pearce RK, Graeber MB (2005) Microglial inflammation in the parkinsonian substantia nigra: Relationship to alpha-synuclein deposition. J Neuroinflammation 2:14.

Czlonkowska A, Kohutnicka M, Kurkowska-Jastrzebska I, Czlonkowski A (1996) Microglial reaction in MPTP (1-methyl-4-phenyl-1,2,3,6-tetrahydropyridine) induced Parkinson's disease mice model. Neurodegeneration 5:137–143.

Damier P, Hirsch EC, Zhang P, Agid Y, Javoy-Agid F (1993) Glutathione peroxidase, glial cells and Parkinson's disease. Neuroscience 52:1–6.

Damier P, Hirsch EC, Agid Y, Graybiel AM (1999) The substantia nigra of the human brain. II. Patterns of loss of dopamine-containing neurons in Parkinson's disease. Brain 122:1437–1448.

Dauer W, Przedborski S (2003) Parkinson's disease: Mechanisms and models. Neuron 39:889–909.

Davis GC, Williams AC, Markey SP, Ebert MH, Caine ED, Reichert CM, Kopin IJ (1979) Chronic parkinsonism secondary to intravenous injection of meperidine analogs. Psychiatry Res 1:249–254.

Dehmer T, Lindenau J, Haid S, Dichgans J, Schulz JB (2000) Deficiency of inducible nitric oxide synthase protects against MPTP toxicity in vivo. J Neurochem 74:2213–2216.

DeLong MR (1990) Primate models of movement disorders of basal ganglia origin. Trends Neurosci 13:281–285.

Diguet E, Fernagut PO, Wei X, Du Y, Rouland R, Gross C, Bezard E, Tison F (2004) Deleterious effects of minocycline in animal models of Parkinson's disease and Huntington's disease. Eur J Neurosci 19:3266–3276.

Dringen R (2000) Metabolism and functions of glutathione in brain. Prog Neurobiol 62:649–671.

Du Y, Dreyfus CF (2002) Oligodendrocytes as providers of growth factors. J Neurosci Res 68:647–654.

Du Y, Ma Z, Lin S, Dodel RC, Gao F, Bales KR, Triarhou LC, Chernet E, Perry KW, Nelson DL, Luecke S, Phebus LA, Bymaster FP, Paul SM (2001) Minocycline prevents nigrostriatal dopaminergic neurodegeneration in the MPTP model of Parkinson's disease. Proc Natl Acad Sci USA 98:14669–14674.

Dwork AJ, Balmaceda C, Fazzini EA, MacCollin M, Cote L, Fahn S (1993) Dominantly inherited, early-onset parkinsonism: Neuropathology of a new form. Neurology 43:69–74.

Eriksen JL, Przedborski S, Petrucelli L (2005) Gene dosage and pathogenesis of Parkinson's disease. Trends Mol Med 11:91–96.

Fahn S, Przedborski S (2005) Parkinsonism. In: Merritt's Neurology (L.P. Rowland, ed), pp 828–846. New York: Lippincott Williams & Wilkins.

Ferger B, Leng A, Mura A, Hengerer B, Feldon J (2004) Genetic ablation of tumor necrosis factor-alpha (TNF-alpha) and pharmacological inhibition of TNF-synthesis attenuates MPTP toxicity in mouse striatum. J Neurochem 89:822–833.

Forno LS, DeLanney LE, Irwin I, Di Monte D, Langston JW (1992) Astrocytes and Parkinson's disease. Prog Brain Res 94:429–436.

Galvin JE, Lee VM, Schmidt ML, Tu PH, Iwatsubo T, Trojanowski JQ (1999) Pathobiology of the Lewy body. Adv Neurol 80:313–324.

Giasson BI, Duda JE, Murray IV, Chen Q, Souza JM, Hurtig HI, Ischiropoulos H, Trojanowski JQ, Lee VM (2000) Oxidative damage linked to neurodegeneration by selective alpha-synuclein nitration in synucleinopathy lesions. Science 290:985–989.

Gomez-Isla T, Irizarry MC, Mariash A, Cheung B, Soto O, Schrump S, Sondel J, Kotilinek L, Day J, Schwarzschild MA, Cha JH, Newell K, Miller DW, Ueda K, Young AB, Hyman BT, Ashe KH (2003) Motor dysfunction and gliosis with preserved dopaminergic markers in human alpha-synuclein A30P transgenic mice. Neurobiol Aging 24:245–258.

Hayashi S, Wakabayashi K, Ishikawa A, Nagai H, Saito M, Maruyama M, Takahashi T, Ozawa T, Tsuji S, Takahashi H (2000) An autopsy case of autosomal-recessive juvenile parkinsonism with a homozygous exon 4 deletion in the parkin gene. Mov Disord 15:884–888.

Hayley S, Crocker SJ, Smith PD, Shree T, Jackson-Lewis V, Przedborski S, Mount M, Slack R, Anisman H, Park DS (2004) Regulation of dopaminergic loss by Fas in a 1-methyl-4-phenyl-1,2,3,6-tetrahydropyridine model of Parkinson's disease. J Neurosci 24:2045–2053.

He Y, Lee T, Leong SK (1999) Time course of dopaminergic cell death and changes in iron, ferritin and transferrin levels in the rat substantia nigra after 6- hydroxydopamine (6-OHDA) lesioning. Free Radic Res 31:103–112.

Hirsch E, Graybiel AM, Agid YA (1988) Melanized dopaminergic neurons are differentially susceptible to degeneration in Parkinson's disease. Nature 334:345–348.

Hirsch EC, Hunot S, Damier P, Brugg B, Faucheux BA, Michel PP, Ruberg M, Muriel MP, Mouatt-Prigent A, Agid Y (1999) Glial cell participation in the degeneration of dopaminergic neurons in Parkinson's disease. Adv Neurol 80:9–18.

Hishikawa N, Hashizume Y, Yoshida M, Sobue G (2001) Widespread occurrence of argyrophilic glial inclusions in Parkinson's disease. Neuropathol Appl Neurobiol 27:362–372.

Hornykiewicz O, Kish SJ (1987) Biochemical Pathophysiology of Parkinson's Disease. In: Parkinson's Disease (Yahr M, Bergmann KJ, eds), pp 19–34. New York: Raven Press.

Hunot S, Boissière F, Faucheux B, Brugg B, Mouatt-Prigent A, Agid Y, Hirsch EC (1996) Nitric oxide synthase and neuronal vulnerability in Parkinson's disease. Neuroscience 72:355–363.

Hunot S, Brugg B, Ricard D, Michel PP, Muriel MP, Ruberg M, Faucheux BA, Agid Y, Hirsch EC (1997) Nuclear translocation of NF-kappaB is increased in dopaminergic neurons of patients with Parkinson's disease. Proc Natl Acad Sci USA 94:7531–7536.

Hunot S, Dugas N, Faucheux B, Hartmann A, Tardieu M, Debre P, Agid Y, Dugas B, Hirsch EC (1999) FceRII/CD23 is expressed in Parkinson's disease and induces, in vitro, production of nitric oxide and tumor necrosis factor-alpha in glial cells. J Neurosci 19:3440–3447.

Hunot S, Vila M, Teismann P, Davis RJ, Hirsch EC, Przedborski S, Rakic P, Flavell RA (2004) JNK-mediated induction of cyclooxygenase 2 is required for neurodegeneration in a mouse model of Parkinson's disease. Proc Natl Acad Sci USA 101:665–670.

Imamura K, Hishikawa N, Sawada M, Nagatsu T, Yoshida M, Hashizume Y (2003) Distribution of major histocompatibility complex class II-positive microglia and cytokine profile of Parkinson's disease brains. Acta Neuropathol (Berl) 106:518–526.

Iwasaki A, Medzhitov R (2004) Toll-like receptor control of the adaptive immune responses. Nat Immunol 5:987–995.

Kohutnicka M, Lewandowska E, Kurkowska-Jastrzebska I, Czlonkowski A, Czlonkowska A (1998) Microglial and astrocytic involvement in a murine model of Parkinson's disease induced by 1-methyl-4-phenyl-1,2,3,6-tetrahydropyridine (MPTP). Immunopharmacology 39:167–180.

Kurkowska-Jastrzebska I, Wronska A, Kohutnicka M, Czlonkowski A, Czlonkowska A (1999a) MHC class II positive microglia and lymphocytic infiltration are present in the substantia nigra and striatum in mouse model of Parkinson's disease. Acta Neurobiol Exp (Wars) 59:1–8.

Kurkowska-Jastrzebska I, Wronska A, Kohutnicka M, Czlonkowski A, Czlonkowska A (1999b) The inflammatory reaction following 1-methyl-4-phenyl-1,2,3, 6-tetrahydropyridine intoxication in mouse. Exp Neurol 156:50–61.

Landau AM, Luk KC, Jones ML, Siegrist-Johnstone R, Young YK, Kouassi E, Rymar VV, Dagher A, Sadikot AF, Desbarats J (2005) Defective Fas expression exacerbates neurotoxicity in a model of Parkinson's disease. J Exp Med 202:575–581.

Langston JW, Ballard P, Irwin I (1983) Chronic parkinsonism in humans due to a product of meperidine-analog synthesis. Science 219:979–980.

Langston JW, Forno LS, Tetrud J, Reeves AG, Kaplan JA, Karluk D (1999) Evidence of active nerve cell degeneration in the substantia nigra of humans years after 1-methyl-4-phenyl-1,2,3,6-tetrahydropyridine exposure. Ann Neurol 46:598–605.

Lawson LJ, Perry VH, Dri P, Gordon S (1990) Heterogeneity in the distribution and morphology of microglia in the normal adult mouse brain. Neuroscience 39:151–170.

Le W, Rowe D, Xie W, Ortiz I, He Y, Appel SH (2001) Microglial activation and dopaminergic cell injury: An in vitro model relevant to Parkinson's disease. J Neurosci 21:8447–8455.

Lee HJ, Patel S, Lee SJ (2005) Intravesicular localization and exocytosis of alpha-synuclein and its aggregates. J Neurosci 25:6016–6024.

Liberatore G, Jackson-Lewis V, Vukosavic S, Mandir AS, Vila M, McAuliffe WJ, Dawson VL, Dawson TM, Przedborski S (1999) Inducible nitric oxide synthase stimulates dopaminergic neurodegeneration in the MPTP model of Parkinson disease. Nat Med 5:1403–1409.

Ling ZD, Chang Q, Lipton JW, Tong CW, Landers TM, Carvey PM (2004) Combined toxicity of prenatal bacterial endotoxin exposure and postnatal 6-hydroxydopamine in the adult rat midbrain. Neuroscience 124:619–628.

Liu B, Jiang JW, Wilson BC, Du L, Yang SN, Wang JY, Wu GC, Cao XD, Hong JS (2000) Systemic infusion of naloxone reduces degeneration of rat substantia nigral dopaminergic neurons induced by intranigral injection of lipopolysaccharide. J Pharmacol Exp Ther 295:125–132.

Mandel S, Grunblatt E, Riederer P, Amariglio N, Jacob-Hirsch J, Rechavi G, Youdim MB (2005) Gene expression profiling of sporadic Parkinson's disease substantia nigra pars compacta reveals impairment of ubiquitin-proteasome subunits, SKP1A, aldehyde dehydrogenase, and chaperone HSC-70. Ann N Y Acad Sci 1053:356–375.

Mattammal MB, Strong R, Lakshmi VM, Chung HD, Stephenson AH (1995) Prostaglandin H synthetase-mediated metabolism of dopamine: Implication for Parkinson's disease. J Neurochem 64:1645–1654.

McCormack AL, Thiruchelvam M, Manning-Bog AB, Thiffault C, Langston JW, Cory-Slechta DA, Di Monte DA (2002) Environmental risk factors and Parkinson's disease: Selective degeneration of nigral dopaminergic neurons caused by the herbicide paraquat. Neurobiol Dis 10:119–127.

McGeer PL, Itagaki S, Boyes BE, McGeer EG (1988) Reactive microglia are positive for HLA-DR in the substantia nigra of Parkinson's and Alzheimer's disease brains. Neurology 38:1285–1291.

McGeer PL, Schwab C, Parent A, Doudet D (2003) Presence of reactive microglia in monkey substantia nigra years after 1-methyl-4-phenyl-1,2,3,6-tetrahydropyridine administration. Ann Neurol 54:599–604.

Miklossy J, Doudet DD, Schwab C, Yu S, McGeer EG, McGeer PL (2005) Role of ICAM-1 in persisting inflammation in Parkinson disease and MPTP monkeys. Exp Neurol 192(2):275–283.

Mirza B, Hadberg H, Thomsen P, Moos T (2000) The absence of reactive astrocytosis is indicative of a unique inflammatory process in Parkinson's disease. Neuroscience 95:425–432.

Mogi M, Harada M, Riederer P, Narabayashi H, Fujita K, Nagatsu T (1994) Tumor necrosis factor-alpha (TNF-alpha) increases both in the brain and in the cerebrospinal fluid from parkinsonian patients. Neurosci Lett 165:208–210.

Mogi M, Harada M, Narabayashi H, Inagaki H, Minami M, Nagatsu T (1996) Interleukin (IL)-1 beta, IL-2, IL-4, IL-6 and transforming growth factor-alpha levels are elevated in ventricular cerebrospinal fluid in juvenile parkinsonism and Parkinson's disease. Neurosci Lett 211:13–16.

Mogi M, Togari A, Kondo T, Mizuno Y, Komure O, Kuno S, Ichinose H, Nagatsu T (2000) Caspase activities and tumor necrosis factor receptor R1 (p55) level are elevated in the substantia nigra from parkinsonian brain. J Neural Transm 107:335–341.

Moran LB, Graeber MB (2004) The facial nerve axotomy model. Brain Res Brain Res Rev 44:154–178.

Nishimura M, Tomimoto H, Suenaga T, Nakamura S, Namba Y, Ikeda K, Akiguchi I, Kimura J (1994) Synaptophysin and chromogranin A immunoreactivities of Lewy bodies in Parkinson's disease brains. Brain Res 634:339–344.

Nomura T, Yabe T, Rosenthal ES, Krzan M, Schwartz JP (2000) PSA-NCAM distinguishes reactive astrocytes in 6-OHDA-lesioned substantia nigra from those in the striatal terminal fields. J Neurosci Res 61:588–596.

O'Banion MK (1999) Cyclooxygenase-2: Molecular biology, pharmacology, and neurobiology. Crit Rev Neurobiol 13:45–82.

O'Callaghan JP, Miller DB, Reinhard JF (1990) Characterization of the origins of astrocyte response to injury using the dopaminergic neurotoxicant, 1-methyl-4-phenyl-1,2,3,6- tetrahydropyridine. Brain Res 521:73–80.

Oppenheimer DR, Esiri MM (1997) Diseases of the Basal Ganglia, Cerebellum and Motor Neurons. In: Greenfield's Neuropathology (Adams JH, Corsellis JAN, Duchen LW, eds), pp 988–1045. New York: Arnold.

Pennathur S, Jackson-Lewis V, Przedborski S, Heinecke JW (1999) Mass spectrometric quantification of 3-nitrotyrosine, ortho-tyrosine, and *O,O′*-dityrosine in brain tissue of 1-methyl-4-phenyl-1,2,3, 6-tetrahydropyridine-treated mice, a model of oxidative stress in Parkinson's disease. J Biol Chem 274:34621–34628.

Pittman RN, Messam CA, Mills JC (1999) Asynchronous Death as a Characteristic Feature of Apoptosis. In: Cell Death and Diseases of the Nervous System (Koliatsos VE, Ratan RR, eds), pp 29–43. Totowa, New Jersey: Humana Press.

Pow DV (2001) Visualising the activity of the cystine-glutamate antiporter in glial cells using antibodies to aminoadipic acid, a selectively transported substrate. Glia 34:27–38.

Przedborski S, Levivier M, Jiang H, Ferreira M, Jackson-Lewis V, Donaldson D, Togasaki DM (1995) Dose-dependent lesions of the dopaminergic nigrostriatal pathway induced by intrastriatal injection of 6-hydroxydopamine. Neuroscience 67:631–647.

Przedborski S, Jackson-Lewis V, Yokoyama R, Shibata T, Dawson VL, Dawson TM (1996) Role of neuronal nitric oxide in MPTP (1-methyl-4-phenyl-1,2,3,6-tetrahydropyridine)-induced dopaminergic neurotoxicity. Proc Natl Acad Sci USA 93:4565–4571.

Przedborski S, Chen Q, Vila M, Giasson BI, Djaldatti R, Vukosavic S, Souza JM, Jackson-Lewis V, Lee VM, Ischiropoulos H (2001) Oxidative post-translational modifications of alpha-synuclein in the 1- methyl-4-phenyl-1,2,3,6-tetrahydropyridine (MPTP) mouse model of Parkinson's disease. J Neurochem 76:637–640.

Rabchevsky AG, Streit WJ (1997) Grafting of cultured microglial cells into the lesioned spinal cord of adult rats enhances neurite outgrowth. J Neurosci Res 47:34–48.

Rodrigues RW, Gomide VC, Chadi G (2001) Astroglial and microglial reaction after a partial nigrostriatal degeneration induced by the striatal injection of different doses of 6- hydroxydopamine. Int J Neurosci 109:91–126.

Rousselet E, Callebert J, Parain K, Joubert C, Hunot S, Hartmann A, Jacque C, Perez-Diaz F, Cohen-Salmon C, Launay JM, Hirsch EC (2002) Role of TNF-alpha receptors in mice intoxicated with the parkinsonian toxin MPTP. Exp Neurol 177:183–192.

Rubio N (1997) Mouse astrocytes store and deliver brain-derived neurotrophic factor using the non-catalytic gp95trkB receptor. Eur J Neurosci 9:1847–1853.

Schaar DG, Sieber BA, Dreyfus CF, Black IB (1993) Regional and cell-specific expression of GDNF in rat brain. Exp Neurol 124:368–371.

Sheng JG, Shirabe S, Nishiyama N, Schwartz JP (1993) Alterations in striatal glial fibrillary acidic protein expression in response to 6-hydroxydopamine-induced denervation. Exp Brain Res 95:450–456.

Sherer TB, Betarbet R, Kim JH, Greenamyre JT (2003) Selective microglial activation in the rat rotenone model of Parkinson's disease. Neurosci Lett 341:87–90.

Sortwell CE, Daley BF, Pitzer MR, McGuire SO, Sladek JR, Collier TJ (2000) Oligodendrocyte-type 2 astrocyte-derived trophic factors increase survival of developing dopamine neurons through the inhibition of apoptotic cell death. J Comp Neurol 426:143–153.

Stadelmann C, Kerschensteiner M, Misgeld T, Bruck W, Hohlfeld R, Lassmann H (2002) BDNF and gp145trkB in multiple sclerosis brain lesions: Neuroprotective interactions between immune and neuronal cells? Brain 125:75–85.

Steel JC, Richardson JC, Olszewski J (1964) Progressive supranuclear palsy. Arch Neurol 10:333–358.

Streit WJ (2002) Microglia as neuroprotective, immunocompetent cells of the CNS. Glia 40:133–139.

Streit WJ (2004) Microglia and Alzheimer's disease pathogenesis. J Neurosci Res 77:1–8.

Stromberg I, Bjorklund H, Dahl D, Jonsson G, Sundstrom E, Olson L (1986) Astrocyte responses to dopaminergic denervations by 6-hydroxydopamine and 1-methyl-4-phenyl-1,2,3,6-tetrahydropyridine as evidenced by glial fibrillary acidic protein immunohistochemistry. Brain Res Bull 17:225–236.

Teismann P, Tieu K, Choi DK, Wu DC, Naini A, Hunot S, Vila M, Jackson-Lewis V, Przedborski S (2003) Cyclooxygenase-2 is instrumental in Parkinson's disease neurodegeneration. Proc Natl Acad Sci USA 100:5473–5478.

Van den Eijnde SM, Boshart L, Baehrecke EH, De Zeeuw CI, Reutelingsperger CP, Vermeij-Keers C (1998) Cell surface exposure of phosphatidylserine during apoptosis is phylogenetically conserved. Apoptosis 3:9–16.

Vila M, Przedborski S (2004) Genetic clues to the pathogenesis of Parkinson's disease. Nat Med 10 Suppl:S58–S62.

Vingerhoets FJ, Snow BJ, Tetrud JW, Langston JW, Schulzer M, Calne DB (1994) Positron emission tomographic evidence for progression of human MPTP-induced dopaminergic lesions. Ann Neurol 36:765–770.

Wilms H, Rosenstiel P, Sievers J, Deuschl G, Zecca L, Lucius R (2003) Activation of microglia by human neuromelanin is NF-kappaB dependent and involves p38 mitogen-activated protein kinase: Implications for Parkinson's disease. FASEB J 17:500–502.

Wu DC, Jackson-Lewis V, Vila M, Tieu K, Teismann P, Vadseth C, Choi DK, Ischiropoulos H, Przedborski S (2002) Blockade of microglial activation is neuroprotective in the 1-methyl-4-phenyl-1,2,3,6-tetrahydropyridine mouse model of Parkinson's disease. J Neurosci 22:1763–1771.

Wu DC, Teismann P, Tieu K, Vila M, Jackson-Lewis V, Ischiropoulos H, Przedborski S (2003) NADPH oxidase mediates oxidative stress in the 1-methyl-4-phenyl-1,2,3,6-tetrahydropyridine model of Parkinson's disease. Proc Natl Acad Sci USA 100:6145–6150.

Yang L, Sugama S, Chirichigno JW, Gregorio J, Lorenzl S, Shin DH, Browne SE, Shimizu Y, Joh TH, Beal MF, Albers DS (2003) Minocycline enhances MPTP toxicity to dopaminergic neurons. J Neurosci Res 74:278–285.

Zimprich A, Biskup S, Leitner P, Lichtner P, Farrer M, Lincoln S, Kachergus J, Hulihan M, Uitti RJ, Calne DB, Stoessl AJ, Pfeiffer RF, Patenge N, Carbajal IC, Vieregge P, Asmus F, Muller-Myhsok B, Dickson DW, Meitinger T, Strom TM, Wszolek ZK, Gasser T (2004) Mutations in LRRK2 cause autosomal-dominant parkinsonism with pleomorphic pathology. Neuron 44:601–607.

27
Amyotrophic Lateral Sclerosis

Ericka P. Simpson, Albert A. Yen, and Stanley H. Appel

Keywords Adaptive immune system; ALS; Chemokines; Cytokines; Innate Immune System; Microglia; Neuroinflammation

27.1. Introduction—Clinical Features

Amyotrophic Lateral Sclerosis is a devastating, relentlessly progressive neurodegenerative disease compromising lower and upper motor neurons, and resulting in death, usually from marked weakness of respiratory musculature (Haverkamp et al., 1995). The incidence of ALS is 1–2/100,000 and the prevalence is 4–6/100,000. Ninety percent of ALS cases are sporadic (sALS), while 10% are familial (fALS). The etiology of sALS is unknown, and no known pharmacotherapy has been demonstrated to significantly influence either disease progression or survival. A defect in the gene encoding copper-zinc superoxide dismutase (SOD1) is present in 20% of fALS cases (Rosen et al., 1993). In sALS, the mean age of onset is 57 years and the median survival is approximately four years with younger patients surviving longer than older individuals. Males develop sporadic ALS more frequently than females (ratio 1.7:l). However, in older individuals the frequency of female involvement is increased and the ratio approaches 1:1 over the age of 65, which is similar to the ratio in fALS.

Typical onset is characterized by complaints of weakness or functional impairment, such as difficulty writing, buttoning, or holding onto objects due to involvement of the upper extremities, or frequent stumbling, tripping, and occasional falling indicative of lower extremity involvement. Symptoms are often asymmetric and usually progress from one limb to the contralateral limb, and then proceed rostrally. Clinically, ALS is often misdiagnosed as a painless radiculopathy when initial signs and symptoms are restricted to a single limb or adjacent root and before electromyography (EMG) studies are performed. Therefore, the asymmetry and relentless progression of symptoms are diagnostically important.

In patients with bulbar involvement, symptoms of hoarseness, slurred speech, and drooling most often precede evidence of pharyngeal muscle involvement. These bulbar symptoms, in association with tongue atrophy and fibrillations, may be the presenting feature in 20–25% of patients, for whom the majority are women over the age of 60 (Haverkamp et al., 1995). In general, onset of bulbar signs and symptoms during the disease course carries a poorer prognosis. Cognitive dysfunction due to fronto-temporal lobe involvement is a frequently associated feature in ALS patients and ranges from mild, frontal dysexecutive impairment in the majority to outright fronto-temporal dementia (Ringholz et al., 2005).

The history and physical examination provide the foundation for the diagnosis and should demonstrate the combination of lower motor neuron (LMN) involvement, manifested by weakness and muscle atrophy, and upper motor neuron (UMN) involvement, evidenced by increased tone and hyperreflexia, in at least three areas, including the limbs, tongue, and back muscles. Needle EMG and nerve conduction studies should be performed to confirm the diagnosis in all patients with suspected ALS.

27.1.1. Pathogenesis

Most studies of the pathogenesis of ALS have focused on changes within motor neurons that could reflect the mechanisms of cell injury and cell destruction. Earliest studies suggested a role for glutamate excitotoxicity (Rothstein et al., 1990, 1992; Lin et al., 1998) Other studies emphasized the importance of increased intracellular calcium (Engelhardt et al., 1997; Siklos et al., 1996, 1998) and free-radical mechanisms in mediating cell death in ALS (Bowling et al., 1993; Yim et al., 1996; Simpson et al., 2004). However, none of these explanations are mutually exclusive because altered calcium homeostasis, free radical stress, and glutamate excitotoxicity may all participate in a cell injury cascade leading to motor neuron death. In fact, alterations in one parameter can lead to alterations in other parameters, and each can enhance and propagate the injury cascade. Increased intracellular calcium can enhance free radical production (Dykens, 1994) and glutamate release, (Coyle and Putterfarcken, 1993; Tymianski et al., 1993; Dugan et al.,

© Springer 2008

1995; Carriedo et al., 1998) which in turn can further increase intracellular calcium (Sheen et al., 1992). Increased free radicals can impair the glial uptake of glutamate, increasing the extracellular glutamate available to interact with neuronal AMPA/kainate receptors, which in turn can increase intracellular calcium and/or free radicals. Free radicals can enhance lipid peroxidation leading to increased intracellular calcium (Kakkar et al., 1992). These changes of increased intracellular calcium, increased production of free radicals, and enhanced glutamate excitotoxicity could critically impair motor neuron structures such as mitochondria or neurofilaments, and compromise energy production and axoplasmic flow. Once initiated, the changes could become self-propagating and induce an irreversible cascade resulting in cell death.

Unfortunately, efforts to halt the progression of disease with therapies targeted to these proposed mechanisms of disease have shown extremely modest benefits, at best. It is not that these mechanisms are not involved, but perhaps each individual mechanism may reflect only a limited aspect of the complex pathobiology of disease. Furthermore, most studies focus solely on the neuron and fail to consider the relevance of neuronal-glial interactions. What may be missing is that neuronal injury and cell death do not occur in isolation, but may require the participation of non-neuronal cells such as microglia and astrocytes, as well as other immune cells.

27.1.2. Neuroinflammation

Neuroinflammation is a significant aspect of the neuropathology of ALS, although it is not clear whether the inflammatory cells are the cause or the consequence of motor neuron injury. Inflammatory cells and mediators are present in the vicinity of degenerating motor neurons, and recent studies have suggested that these constituents may, in fact, contribute to motor neuron injury in ALS.

The CNS was long considered to be an immune privileged site. This viewpoint has shifted over the years, as experimental evidence has clearly shown that the CNS is under constant immune surveillance and that the CNS and the peripheral immune system interact extensively. However, inflammatory processes in the CNS differ from those of systemic inflammation. Unlike the rapid, cellular infiltration seen in systemic inflammation, inflammatory responses to a noxious stimulus in the CNS can be less dramatic and may include the activation of the innate immune system, the resident microglial cells, and the presence of the adaptive immune system, lymphocytes and immunoglobulins. There is gradual entry of immune cells including dendritic cells from the peripheral circulation, as well the localized proliferation of dendritic cells within the CNS. The involvement of both innate and adaptive immune systems results in the release of cytotoxic as well as neuroprotective constituents. The initial response to an acute insult (trauma, ischemia, infection) is directed at limiting cellular damage and enhancing repair. However, under chronic insults or a continuing stimulus, a long-lasting inflammatory response may cause additional injury to CNS tissue. In other words, the cells and signaling molecules of neuroinflammation may comprise a "double-edged sword"—sometimes protective and reparative to neurons, and sometimes injurious. Indeed, increasing evidence supports a dual role of immune activation in several CNS diseases, based upon the timing and duration of immune activation, degree and type of injury, intercellular communication, and genetic modifiers (Consilvio et al., 2004; Minagar et al., 2002).

The interrelationships between neurology and immunology have been studied extensively in multiple sclerosis, the prototypical CNS autoimmune disease. More recently, the role of neuroinflammation has been implicated in other neurodegenerative conditions, including Alzheimer's disease and Parkinson disease (Mhatre et al., 2004; McGeer et al., 1991; McGeer and McGeer, 2002a, b, 2003; Hunot and Hirsch 2003).

27.2. Cellular and Biochemical Evidence of Neuroinflammation

27.2.1. Human ALS Studies

Although early studies of ALS tissue did not describe the presence of inflammatory cells in ALS, reports beginning in 1988 suggested that inflammatory cell infiltration might be relatively common (Troost et al., 1988, 1989; Lampson et al., 1988). Further studies (Lampson et al., 1990) demonstrated activated microglia and small numbers of T cells in degenerating white matter of ALS cords. Activated microglia are also prominent in the ventral horn of spinal cords from ALS patients. In contrast to resting microglia, which have long, highly branched processes, activated microglia have hypertrophied soma and thickened, retracted processes, first described by Rio-Hortega (1920). Reactive microglia express surface markers indicative of their activated state. These include the complement component 3 receptor 3 subunit/integrin-αM (CD11b/ITGAM); leukocyte common antigen (LCA); and major histocompatibility complex (MHC) class II glycoproteins, such as human leukocyte antigen DR (HLA-DR). In addition to serving a phagocytic function, activated microglia may also communicate with and affect the function of cells in the local environment, by expressing a broad array of cytokines, chemokines, proteases, and neurotrophic factors (Gonzalez-Scarano and Baltuch 1999).

In our early studies (Engelhardt et al., 1993), we documented the presence of lymphocytes in the spinal cord in 18 of 27 consecutive ALS autopsies. Lymphocytes were predominantly CD4-positive in the vicinity of degenerating cortical spinal tracts, and CD4 and CD8 cells were found in ventral horns. Kawamata et al. (1992) demonstrated the presence of significant numbers of CD8 T cells and, to a lesser extent, CD4 T cells marginating along capillaries in the parenchyma of spinal cord and brains of 13 ALS patients. Lymphocytes and activated microglia expressing MHC Class I and Class II, leukocyte common antigen, FcγRI and β-2 integrins were present in ALS tissue.

The diversity of T cells in ALS spinal cord has also been examined employing the reverse transcriptase polymerase chain reaction (PCR) with variable region sequence specific oligonucleotide primers to amplify Vβ T-cell receptor transcripts (Panzara et al., 1999). A greater expression of $V\beta_2$ transcripts was detected in ALS specimens, independent of the HLA genotype of the individual. As additional confirmation, cells were assayed from the cerebrospinal fluid (CSF) of 22 consecutive ALS patients for the presence of $V\beta_2$ transcripts. This specific T-lymphocyte receptor was demonstrated in 17 of 22 ALS patients' CSF, while only 4 of 19 control patients had similar expression. It is of interest that $V\beta_2$ is a T-cell receptor known to respond to superantigens. Whether superantigen stimulation of lymphocytes with $V\beta_2$ receptors is a primary factor in motor neuron injury or the consequence of motor neuron injury due to some other cause is not known, but the presence of T lymphocyte restriction does suggest the involvement of adaptive immune mechanisms.

Further evidence supporting the importance of immune mechanisms is the presence of dendritic cells in ALS tissues (Henkel et al., 2004). Dendritic cells are potent antigen-presenting cells that initiate and amplify immune responses (Reichmann et al., 2002). They can be derived from the peripheral immune system or from resting CNS microglia. Our laboratory examined mRNA expression of dendritic cell surface markers in individual sporadic ALS (sALS), familial ALS (fALS), and non-neurological disease control (NNDC) spinal cord tissues using semiquantitative and real-time reverse transcription polymerase chain reaction (RT-PCR) (Henkel et al., 2004). Immature (DEC205, CD1a) and activated/mature (CD83, CD40) dendritic cell transcripts were significantly elevated in ALS tissues. The presence of immature and activated/mature dendritic cells (CD1a(+) and CD83(+)) was confirmed immunohistochemically in ALS ventral horn and corticospinal tracts. Monocyte/macrophage/microglia transcripts (CD14, CD18, SR-A, CD68) were increased in ALS spinal cord, and activated CD68(+) cells were demonstrated in close proximity to motor neurons. Increased mRNA expressions of the chemokine monocyte chemoattractant protein-1, MCP-1, which attracts monocytes and myeloid dendritic cells, and of the cytokine macrophage-colony stimulating factor (M-CSF), which stimulates the differentiation of dendritic cells and macrophages, were found in ALS tissues. The MCP-1 protein was expressed in glia in ALS but not in control tissues and was increased in the CSF of ALS patients. Furthermore, those patients who clinically progressed most rapidly expressed significantly more dendritic transcripts than patients who progressed more slowly. These results support the involvement of an adaptive immune/inflammatory response in amplifying motor neuron degeneration in ALS.

Also prevalent in ALS are reactive astrocytes, which stain intensively for the intermediate filament, glial fibrillary acidic protein (GFAP). These reactive astrocytes, with a hypertrophied appearance, are present throughout areas of degeneration. Reactive astrocytes can have multiple roles including a neuroprotective function (Liberto et al., 2004).

Numerous biochemical markers of neuroinflammation have also been found in ALS tissue. The families of compounds include cytokines, chemokines, complement proteins, prostaglandins, interleukins, interferons, integrins, acute phase reactants, apolipoproteins, and (Table 27.1). Many of these biochemical markers have also been found in Alzheimer's disease tissues, suggesting that the inflammatory response in the two conditions may be similar (Eikelenboom and van Gool 2004).

Some of the inflammatory mediators enhance the entry of leukocytes from the periphery into the areas of degeneration. As noted above, MCP-1 is well known to enhance the entry of

TABLE 27.1. Proteins associated with the inflammatory state observed in ALS tissues.

LCA (leukocyte common antigen, CD45)	Membrane-bound protein tyrosine phosphatase; expressed by all hematopoietic cells
LFA-1 (leukocyte function antigen 1, CD11a/CD18 integrin)	Appears on all leukocytes; promotes intercellular adhesion
ICAM-1 (intracellular adhesion molecule 1, CD54)	Cell surface or matrix molecule; promotes binding to LFA-1
GFAP (glial fibrillary acidic protein)	Intermediate filament protein highly expressed by activated astrocytes
Phospholipase A2	Breaks down membrane lipids to form arachidonic acid
COX-1 (cyclooxygenase-1)	Converts arachidonic acid into prostaglandin H2; rate-limiting step in prostaglandin synthesis
COX-2 (cyclooxygenase-2)	Converts arachidonic acid into prostaglandin H2; rate-limiting step in prostaglandin synthesis;
inducible expression	
PGE2 (prostaglandin E2)	Member of prostaglandin family; may have both pro- and anti-apoptotic actions
FcgR-1 (immunoglobulin Fc region receptor 1)	Receptor on phagocytes for IgG bound to an antigen
HLA-DR (human leukocyte antigen DR)	Antigen-presenting surface molecule on immunocompetent cells
IL-6 (interleukin 6)	Inflammatory cytokine released by activated astrocytes and microglia
MCP-1 (monocyte chemoattractant protein-1)	Chemokine that attracts monocytes and myeloid dendritic cells
M-CSF (macrophage-colony stimulating factor)	Growth factor cytokine that acts on monocytes and macrophages to cause them to differentiate
MMP-9 (matrix metalloproteinase-9)	Proteolytic enzyme that can degrade components of the extracellular matrix
C3d (complement C3 fragment d)	Degradation product of activated C3
C4d (complement C4 fragment d)	Degradation product of activated C4
CD-11b (complement receptor 3)	Phagocyte surface receptor recognizing an activated complement fragment from C3
g-Interferon	Inflammatory cytokine released by activated T cells; potent activator of microglia and leukocytes

monocytes and dendritic cells into tissues based on interaction with the cognate receptor CCR2 on invading cells. Leukocytes that are observed in the postcapillary venules also express leukocyte function-associated antigen 1 (LFA-1), a surface molecule, which gives activated leukocytes the ability to adhere to endothelial cells and migrate into CNS tissue. This binding involves a specific interaction between LFA-1 and intercellular adhesion molecule-1 (ICAM-1), an adhesion molecule that is up regulated on endothelial cells in areas of inflammation (McGeer and McGeer, 2003). Another molecule that is up regulated in ALS tissue is matrix metalloproteinase-9 (MMP-9) (Lim et al., 1996). MMP-9 is a proteolytic enzyme that is released by leukocytes and reactive astrocytes to degrade proteins of the extracellular matrix and enhance the entry of leukocytes into areas of inflammation.

Prostaglandins represent inflammatory molecules with both pro-apoptotic and anti-apoptotic actions (Bezzi et al., 1998; Kawamura et al., 1999). In the first step of prostaglandin production, membrane lipids are broken down by phospholipase A_2 to arachidonic acid. Interestingly, phospholipase A_2 has been identified in activated glial cells in areas of neurodegeneration (Stephenson et al., 1999). Next, cyclooxygenase (COX) catalyzes the rate-limiting step of prostaglandin synthesis: the conversion of arachidonic acid to prostaglandin G2 (PGG2). Two distinct COX isoenzymes, COX-1 and COX-2, can mediate this reaction, and they are 65% homologous. COX-1 is constitutively expressed in most tissues, and COX-2 is constitutively expressed in the kidney, stomach, and CNS. However, COX-2 expression is strongly inducible and can be significantly up regulated by several cellular factors, including multiple growth factors, interleukin-1β, tumor necrosis factor-α, lipopolysaccharide (LPS), phorbol ester, and elevated intracellular calcium concentration. A marked increase in COX-2 levels has been seen in ALS spinal cords, consistent with previous reports of increased levels of prostaglandin E_2 (PGE_2), a further product of prostaglandin synthesis (Almer et al., 2002). Both mRNA and protein levels of COX-2 were elevated in ALS spinal cords, and only in pathologically affected areas (Almer et al., 2001; Yasojima et al., 2001).

Additionally, immunoglobulin G (IgG) is present in motor neurons in ALS spinal cords, and also in pyramidal cells of the motor cortex (Engelhardt and Appel, 1990).

27.2.2. Animal Models

Transgenic mice (Gurney et al., 1994; Wong et al., 1995) and rats (Howland et al., 2002) over expressing mutant Cu^{2+}/Zn^{2+} superoxide dismutase 1 (mSOD1) develop a progressive motoneuron disease that resembles the clinical and pathological features of human familial amyotrophic lateral sclerosis (ALS). However, the mechanisms by which mSOD1 causes selective motoneuron death are not clearly defined. Increasing evidence suggests that motoneuron degeneration in ALS is non-cell autonomous, i.e., non-neuronal cells contribute to the pathogenesis of ALS. Although intracellular accumulation of mSOD1 damages motoneurons *in vitro* (Durham et al., 1997) expression of mSOD1 either in astrocytes (Gong et al., 2000) or in neurons (Pramatarova et al., 2001; Lino et al., 2002) does not induce an ALS-like disease in mice. Elegant data from chimeric mice, using the original mSOD1 promoter, in the same genomic location, and with the same number of copies, suggest that mSOD1-overexpressing neurons surrounded by normal glia remained relatively intact, whereas normal neurons surrounded by mSOD1-overexpressing glia showed signs of injury (Clement et al., 2003).

This observation of non-cell autonomy raises the potential importance of non-neuronal cells such as innate immune microglia and adaptive immune lymphocytes in the pathogenesis of motoneuron injury in ALS.

At 40 days of age, a presymptomatic stage, spinal cord from mSOD1 transgenic mice has increased ICAM-1 expression, deposition of immunoglobulin G (IgG) within motor neuron cell bodies, and increased Fc receptor (FcR) expression on microglia (Alexianu et al., 2001). Activated microglia are present as early as 80 days of age, prior to evidence of motor weakness (Hall et al., 1998). In general, the development of these inflammatory changes precedes the onset of motor signs, and their increasing levels parallel disease progression. Multiple studies have confirmed an elevation of pro-inflammatory molecules in the presymptomatic mSOD1 mouse, including such mediators as TNF-α, COX-2, inducible nitric oxide synthase (iNOS), interleukin-1β, IL-1 receptor antagonist (IL1RA), CD86, CD200R, and growth-related oncogene- α, (Gro- α,) (Almer et al., 1999; Hensley et al., 2002; Yoshihara et al., 2002).

Evidence for neuroinflammation has also been found in other animal models of motor neuron degeneration. For example, microglial activation was observed in the Wobbler mouse, a finding that has been confirmed by several groups (Rathke-Hartlieb et al., 1999; Boillee et al., 2001). Elevated TNF- α levels were found in the brain and spinal cord of the *mnd* mouse (Ghezzi et al., 1998). It appears that these other murine models of motor neuron degeneration share some neuroinflammatory features with the mSOD1 model.

27.2.3. In Vitro Studies

To help define the interactions involving non-neuronal cells, such as microglia and astrocytes, we used primary MN cultures to investigate the effects of microglia activated by lipopolysaccharide or IgG immune complexes from patients with amyotrophic lateral sclerosis (Zhao et al., 2004). Following activation, microglia induced MN injury, which was preventable by a microglial inducible nitric oxide synthetase (iNOS) inhibitor as well as by catalase or glutathione. Glutamate was also required to elicit MN injury, as evidenced by the finding that inhibition of the motoneuron AMPA/kainate receptor by CNQX prevented the toxic effects of activated microglia. In fact, peroxynitrite, a toxic product of nitric oxide metabolism, and glutamate were synergistic in producing MN injury. The addition of astrocytes to cocultures of MN and activated microglia prevented MN injury by removing glutamate from the media. Such data suggest that free radicals released from activated microglia may initiate MN injury by increasing the susceptibility of the MN AMPA/kainate receptor to the toxic effects of glutamate. A recently published novel experiment suggests that mSOD1 released from motor neurons derived from transgenic mice could activate microglia, and, in turn, lead to motor neuron injury and cell death (Urushitani et al., 2006).

27.2.4. Conclusions

The cellular evidence for neuroinflammation in ALS consists of activated microglia, reactive astrocytes, macrophages, and a few T cells seen marginating along postcapillary venules. The biochemical evidence consists of the presence of cytokines, chemokines, components of prostaglandin synthesis, and other pro-inflammatory molecules. In vitro data clearly document that immune/inflammatory constituents can mediate cytotoxicity. In vivo data from animal models also provide evidence that glial cells, such as microglia, and infiltrating immune cells can mediate cytotoxicity. The role that these immune responses could play in mediating neuroprotection in ALS is presently being studied in animal model studies. However, it remains for translational research studies of human ALS to document that both neuroprotective and cytotoxic mechanisms are involved in the pathogenesis of ALS, and to harness such processes for therapeutic benefit.

27.3. The Failure of Immunosuppression in ALS

Immunosuppression with steroids, cyclophosphamide, plasmapheresis, and even total lymphoid irradiation have been reported to be ineffective in halting the progression of ALS (Brown et al., 1986; Drachman et al., 1994). Our own study with cyclosporine suggested a possible slowing of the course of ALS in a subpopulation of patients with no effect on the overall population (Appel et al., 1988). The major problem in that study was the toxic side effects of the immunosuppression, and the necessity for lowering medication doses to levels, which failed to suppress the inflammatory responses in the CNS.

The fact that immunosuppressant therapy has not been proven to halt progression has been cited as evidence against an immune-mediated mechanism of motor neuron injury. However, several potential explanations for this lack of efficacy could exist. Immune suppression may be too little and too late. By the time symptoms and signs appear, extensive CNS damage may already have occurred due to increased calcium and free radicals, and it may be difficult to stop the process that the immune reactivity initiated. Further, immunosuppression therapies may not adequately access the CNS and suppress the selective inflammatory reactions there, especially microglial responses and/or dendritic cells. Immunosuppression has similarly failed to ameliorate type-1 diabetes or other immune mediated endocrinopathies without invalidating the important role of immune mechanisms in initiating those disorders. In addition, in chronic progressive multiple sclerosis, characterized by marked CNS immune/inflammatory lesions, immunosuppression is of limited value. Further, what initiates the disease may be different from what propagates the disease (increased calcium, free radicals and glutamate) or administers the final cellular *coup de grace* (e.g. dendritic cells, T cells and/or activated microglia). Finally, if both cytotoxic and neuroprotective mechanisms are being mediated by the immune system, then suppression of such immunity may suppress protective as well as toxic mechanisms.

Clearly, the death of neurons may no longer require the initial immune-mediated cell injury, but may result from alterations in intracellular calcium and free radicals through a number of processes, any of which may initiate an apoptotic cascade. Thus, the failure of immunosuppression does not, per se, invalidate the possible participation of immune/inflammatory mechanisms in the pathogenesis of motor neuron injury and death in ALS. Similarly, the lack of meaningful efficacy of drugs that target glutamate excitotoxicity (e.g. riluzole) does not negate the role of glutamate in the pathogenesis of motor neuron injury in ALS.

27.4. Evidence for a Role in Initiating Disease

The pathology of ALS suggests the activation of both innate and adaptive arms of the immune system, involving antigen capture and presentation, leukocyte migration and trafficking, and microglial activation. Whether these processes are sufficient to initiate motor neuron injury in ALS is unknown, but early studies that searched for an initiating role of an immune response against "self" epitopes (autoimmunity) raised some intriguing possibilities (Appel et al., 1993). For example, an animal model of a primary, immune-mediated motor neuron disease was developed by inoculating guinea pigs with bovine motor neurons. Clinical features of this experimental autoimmune motor neuron disease animal included muscle weakness, electrophysiologic and morphologic evidence of denervation, loss of motoneurons in the spinal cord, and clinical improvement with immunosuppression (Engelhardt et al., 1989; Engelhardt et al., 1990; Garcia et al., 1990; Tajti et al., 1991). These studies documented that a primary immune process is capable of initiating motor neuron disease, but it was still unknown whether autoimmunity was playing a role in ALS. Although the higher incidence of autoimmune diseases and paraproteinemias in ALS patients continues to suggest a link between dysimmunity or paraimmunity and motor neuron disease (Haverkamp et al., 1995; Gordon et al., 1997), the evidence that ALS is a primary, autoimmune disease remains circumstantial.

The possibility that ALS pathogenesis is initiated by a secondary immune response to a primary infectious agent has also been considered, especially in regards to viral exposure. Studies attempting to detect viruses in ALS tissue have been unconvincing even with the application of sensitive detection techniques (Walker et al., 2001; Nix et al., 2004). Following the onset of the HIV/AIDS epidemic, renewed interest arose with reports of several cases of HIV-associated motor neuron disease (Hoffman et al., 1985; Simpson and Tagliati 1994; MacGowan et al., 2001; Moulignier et al., 2001). Compared to classic ALS, HIV-associated ALS occurs in younger patients, progresses more rapidly, and improves with antiretroviral therapy—this response suggests that HIV virus is the etiologic factor in motor neuron dysfunction in these cases. Further studies indicated that the HIV retrovirus infects and replicates within macrophage/microglia cell populations and seems to exert its neurotoxic effects indirectly via Tat, a transcriptional regulator viral protein. The Tat protein affects the functional state of infected microglial cells by stimulating them to secrete neurotoxic molecules, differentially regulating chemokine expression within the brain, and disrupting pathways regulating intracellular calcium and ion channel expression (D'Aversa and Berman 2004). Both *in vitro* models and pathological studies suggest that the HIV-infected/activated microglia and macrophages may selectively damage susceptible neurons, leading to neuropathology (Yi et al., 2004).

Additional support for an association between retroviruses, immune activation, and ALS is provided by a murine neurotropic retrovirus, which can cause motor neuron degeneration in mice (Gardner et al., 1973), and the human T-cell lymphotrophic virus type I (HTLV-1) retrovirus, which predominantly causes myelopathy/tropical spastic paraparesis but has caused an ALS-like syndrome in rare cases (Matsuzaki et al., 2000). Based on this information, it has been proposed that an immune response directed against retroviral proteins induces a chronic neuroinflammatory state that leads to motor neuron degeneration. In accord with this, antibodies against HTLV-1 viral proteins have been reported in some ALS patients, as well as increased reverse transcriptase activity in ALS sera (Ferrante et al., 1995; Andrews et al., 2000). Furthermore, this process may not even require infection and active replication by an exogenous retrovirus. A recent study showed that an immune response can be directed against retroviral proteins produced by activation of endogenous retrovirus sequences that are normally present in the human genome. In that study, the researchers reported that 56% of sporadic ALS patients had IgG antibodies reactive to a protein encoded by the endogenous retroviral sequence, HML-2/Herv-K, and that this immune response was identified in ALS patients more than five times as frequently as in Alzheimer's disease or healthy controls (Hadlock et al., 2004). Together, these findings raise the possibility that the activation of endogenous retroviruses can perpetuate and drive ongoing inflammatory events, at least in a subset of ALS patients.

27.5. Evidence for a Role in Amplifying Disease

Although the evidence that CNS inflammatory events can initiate disease is still questionable, a growing consensus among ALS researchers is that inflammatory cells and mediators play an integral role in affecting motor neuron survival. In addition to the activated microglia, dendritic cells, immunoglobulins, and inflammatory proteins mentioned earlier, we recently documented increased levels of monocyte chemoattractant protein-1 (MCP-1 in ALS spinal cords (Henkel et al., 2004), and that the MCP-1 is expressed mainly by glial cells. MCP-1, a chemokine that attracts myeloid dendritic cells, macrophages, and activated T cells, is also elevated in the CSF and serum of ALS patients, compared to controls (Simpson et al., 2004; Wilms et al., 2003). This elevation may reflect a local inflammatory response that results in an active transmigration of cellular constituents from the periphery into the CNS. Overall, the presence of T cells, immunoglobulins, and dendritic cells suggests that the immune inflammatory response in ALS is involved in CNS tissue injury and likely, in repair.

27.6. Evidence for a Role in Repair and Protection

A traditional viewpoint has been that inflammatory cells and mediators are neurotoxic, and that inhibition of inflammation will delay or prevent neuronal loss. Indeed, therapeutic trials of immunomodulatory agents in the mutant SOD1 mouse has supported this concept (Kriz et al., 2002; Keep et al., 2001; Kirkinezos et al., 2004). However, neuroinflammation is composed of multiple processes, working in series and parallel, which cumulatively can limit cellular injury and promote regeneration, or cause tissue damage and cell loss. An example of this is the dual role of prostaglandins, products of the COX-2 enzyme, in promoting either neuronal survival or apoptotic death. Determining factors may include the target cell and tissue environment, profile of specific prostaglandin(s), and the dose-dependent cellular response to the inflammatory mediators (Consilvio et al., 2004). In the mSOD1 mouse model, increases in COX-2 and PGE2 parallel motor neuron loss, and selective inhibition prolongs survival (Almer et al., 2002; Pompl et al., 2003), implying that COX-2 activation contributes to neuronal degeneration. However, looking at the cellular constituents rather than the biochemical effectors involved may provide greater insight into these processes.

Microglia are able to respond promptly to signals of stress or injury originating within or outside the CNS and direct their responses for purposes of tissue repair and activation of immune responses. This appears to be regulated by neuron-astrocyte-microglia interactions which can either have stimulatory or inhibitory influences in shaping microglial and overall immune response (Aloisi, 2001). Further, microglia produce cytokines and molecules with both inflammatory (TNFα, IL-6, etc.) and anti-inflammatory (TGF-β, IL-10, neurotrophins, etc.) activity, and the balanced interplay between the two is crucial in the propagation or resolution of the inflammatory cascade (Aloisi et al., 2000; Tabakman et al., 2004). Certain factors, which may determine the outcome of this interplay, include the ability of microglia to function as antigen presenting cells, their interaction with antigen-specific T cells, the cytokine milieu, and their ability to sustain T cell growth. Although there is insufficient evidence to confirm that neuroinflammatory responses in ALS are also directed at repair, there is evidence from other CNS inflammatory models, which may provide direction in understanding the relevance of these processes in ALS.

In models of mechanical nerve injury, T lymphocytes have been shown to protect neurons, as evidenced by a greater loss of facial motoneurons after nerve transection in severe combined immunodeficient (*scid*) mice that lack T and B cells (Serpe et al., 1999). This protection is mediated by the T-cell generation of neurotrophic factors, such as brain-derived neurotrophic factor (BDNF) and neurotrophin-3 (NT-3) (Hammarberg et al., 2000). Furthermore, this neuroprotective effect is significantly boosted by myelin-specific T cells (Kipnis and Schwartz, 2002). This suggests that the ability to withstand the consequences of CNS axonal injury is governed by the ability to mount an endogenous Th1 cell-mediated protective response (Schwartz and Kipnis 2002). In animal models, this neuroprotective response can be achieved safely with low-affinity activation of T cells with synthetic altered myelin basic peptides, or copolymer, a synthetic compound of myelin basic protein (Angelov et al., 2003; Kipnis and Schwartz, 2002; Monsonego et al., 2003). Importantly, the beneficial T cell-mediated effect is dependent upon the early activation of microglia and their differentiation into efficient, antigen presenting cells (Shaked et al., 2004). Furthermore, the beneficial reparative effects may be directly mediated by microglial or astroglial release of neurotrophic factors, and indirectly mediated by T cell or dendritic cell signaling to glia.

In ALS, it is unknown whether a significant neuroprotective immune response is present, or if it plays a role in determining populations at risk, disease onset, or rate of progression. Clearly, the potential roles of T cell-, dendritic cell-, and microglia-mediated neuroprotection in neurodegenerative disease, including ALS, may become future targets of therapy. Copolymer 1 has already been shown to delay disease onset, improve life span, and improve motor activity in the mSOD1 mouse model (Angelov et al., 2003), and a phase I study of copolymer 1 has been studied for safety in the ALS population (Gordon et al., 2006). However, a randomized, placebo controlled trial is required to determine therapeutic efficacy.

Summary

There has been a profound paradigm shift regarding the potential importance of neuroinflammation in ALS. Once viewed as an epiphenomenon of motor neuron degeneration, neuroinflammatory changes are now considered to play an integral role in disease pathogenesis. A better understanding of how the cellular and biochemical components of this response affect motor neuron survival will enable us to develop new, immunomodulatory therapies. These novel therapies should be directed at amplifying those pathways involved in neuroprotection, while down-regulating cytotoxic pathways leading to motor neuron injury.

Acknowledgment. We are grateful to the Muscular Dystrophy Association for support of the ALS clinical and research programs at The Methodist Neurological Institute.

Review Questions/Problems

1. Which of the following proposed pathogenic mechanisms in ALS are considered mutually exclusive (present in isolation) of other proposed mechanisms of disease?

a. glutamate excitotoxicity
b. generation of free radicals
c. disrupted axoplasmic flow
d. altered calcium homeostasis
e. all of the above
f. none of the above

2. Which statement best describes the meaning of 'non-cell autonomy' when referring to the pathogenesis of ALS?

a. The risk of developing ALS is independent of cellular factors, and is more dependent on environmental exposures.
b. The progression of disease in ALS is dependent upon therapeutic intervention and early detection
c. Motor neuron injury is independent of interaction with other cell populations within the CNS
d. Motor neuron injury is dependent on interaction with other affected cell populations within the CNS

3. Which line(s) of evidence from human studies support that the immune inflammatory response in ALS has a role in amplifying disease progression?

a. Increased expression of myeloid dendritic cells mRNA transcripts in ALS spinal cord that correlates with a faster rate of progression
b. Increased mRNA expressions of the chemokine MCP-1, and the cytokine macrophage-colony stimulating factor (M-CSF) in ALS spinal cord, which attracts and induces differentiation of monocytes and myeloid dendritic cells, respectively.

c. Both a and b
d. Neither a or b

4. Which line(s) of evidence from animal and *in vitro* studies support that the immune inflammatory response has a role in motoneuron injury?

a. Studies from mSOD transgenic chimeric mice showing that mSOD expressing neurons that are surrounded by wild-type glial cells remain intact
b. Motoneuron injury by activated microglia mediated by glutamate and peroxynitrite
c. Both a and b
d. Neither a or b

5. The failure of immunosuppressive therapy in ALS is sufficient evidence that immune mediated mechanisms are not relevant to disease pathogenesis.

True/False

6. The higher incidence of autoimmune diseases and paraproteinemias in ALS patients confirms that sporadic motor neuron disease is autoimmune in nature.

True/False

7. Association between viral infection and motor neuron disease is based upon cases of motor neuron disease due to retroviral infection (HIV, HTLV-1) and the presence of serum antibodies directed toward endogenous retroviral gene sequences in ALS patients.

True/False

8. In CNS injury, the neuroprotective response of activated T cells is dependent upon early activation and differentiation of microglia into antigen presenting cells which release neurotrophic growth factors.

True/False

9. Although human and animal studies support a role for immune responses in ALS pathogenesis, which statement(s) are reasonable explanations for why immunosuppressive therapy has failed in treating ALS?

1. Immune suppressive therapy is initiated beyond the point when motor neuron recovery and clinical improvement are possible.
2. Inadequate access to or suppression of immune inflammatory cells in the CNS
3. Suppression of both cytotoxic and protective immune mechanisms
4. Immune inflammation does not contribute to ALS disease pathogenesis
   a. 1, 2, 3
   b. 1, 3
   c. 2, 4
   d. 4 only
   e. All of the above

10. Which of the following patient variables are associated with a poor prognosis?

1. Older age of onset
2. Female gender
3. Bulbar onset
4. Sporadic form
   a. 1, 2, 3
   b. 1, 3
   c. 2, 4
   d. 4 only

11. Slurred speech, excessive salivation, and hoarseness may be the presenting feature in what percentage of ALS patients?

a. 15%
b. 25%
c. 35%
d. 45%

12. Cognitive dysfunction is not a typical feature of ALS.

True/False

13. A clinical diagnosis of ALS requires the demonstration of lower motor neuron involvement in which three areas?

a. upper extremity limb, eye muscles, and paraspinous muscles
b. lower extremity limb, eye muscles, and tongue
c. upper extremity limb, tongue, paraspinous muscles
d. lower extremity limb, upper extremity, eye muscles

14. Studies of which disease have provided an increased understanding of the interactions between neurology and immunology?

a. Mollaret's Meningitis
b. Guillan Barre Syndrome
c. poliomyelitis
d. multiple sclerosis

15. The presence of an increased number of dendritic cells in ALS tissues provides evidence for activation of an innate immune response.

True/False

16. The following molecules enhance the entry of leukocytes from the periphery into areas of injury and have been observed in ALS tissues.

a. MCP-1, LFA-1, ICAM-1, AND MMP-9
b. COX-2, GFAP, ICAM-1 MMP-9
c. M-CSF, MCP-1, COX-2, MMP-9
d. COX-2, M-CSF, ICAM-1, MCP-1

17. MCP-1 is a chemokine, which attracts monocytes and myeloid dendritic cells to sites of injury or inflammation.

True/False

18. *In vitro,* motor neuron injury is induced by exposure to activated microglia. Which microglial constituents initiate such injury?

1. kainite and IgG
2. mSOD1 and peroxynitrite
3. glutamate and IgG
4. peroxynitrite and glutamate
   a. 1, 2, 3
   b. 1, 3
   c. 2, 4
   d. 4 only

19. The failure of immunosuppression to ameliorate type-1diabetes or chronic progressive multiple sclerosis, predominantly immune mediated diseases, invalidates the role of immune mechanisms in initiating these disorders.

True/False

20. Prostaglandins have a dual role in either promoting neuronal survival or apoptotic death. What factors may determine which role predominates in CNS injury or degeneration?

1. Target cell and tissue environment, degree of cellular response to the inflammatory mediators, and profile of specific prostaglandin(s)
2. Exposure to prostaglandin and cox-2 inhibitors, target cell and tissue environment, and profile of specific prostaglandins(s)
   a. both 1 and 2
   b. neither 1 or 2
   c. 1
   d. 2

References

Alexianu ME, Kozovska M, Appel SH (2001) Immune reactivity in a mouse model of familial ALS correlates with disease progression. Neurology 57:1282–1289.

Almer G, Vukosavic S, Romero N, Przedborski S (1999) Inducible nitric oxide synthase up-regulation in a transgenic mouse model of familial amyotrophic lateral sclerosis. J Neurochem 72:2415–2425.

Almer G, Guegan C, Teismann P, Naini A, Rosoklija G, Hays AP, Chen C, Przedborski S (2001) Increased expression of the pro-inflammatory enzyme cyclooxygenase-2 in amyotrophic lateral sclerosis. Ann Neurol 49:176–185.

Almer G, Teismann P, Stevic Z, Halaschek-Wiener J, Deecke L, Kostic V, Przedborski S (2002) Increased levels of the pro-inflammatory prostaglandin PGE2 in CSF from ALS patients. Neurology 58:1277–1279.

Aloisi F (2001) Immune function of microglia. Glia 36:165–179.

Aloisi F, Ria F, Adorini L (2000) Regulation of T-cell responses by CNS antigen-presenting cells: Different roles for microglia and astrocytes. Immunol Today 21:141–147. Review.

Andrews WD, Tuke PW, Al-Chalabi A, Gaudin P, Ijaz S, Parton MJ, Garson JA (2000) Detection of reverse transcriptase activity in the serum of patients with motor neurone disease. J Med Virol 61:527–532.

Angelov DN, Waibel S, Guntinas-Lichius O, Lenzen M, Neiss WF, Tomov TL, Yoles E, Kipnis J, Schori H, Reuter A, Ludolph A, Schwartz M (2003) Therapeutic vaccine for acute and chronic motor neuron diseases: Implications for amyotrophic lateral sclerosis. Proc Natl Acad Sci 100:4790–4795.

Appel SH, Stewart SS, Appel V, Harati Y, Mietlowski W, Weiss W, Belendiuk GW (1988) A double-blind study of the effectiveness of cyclosporine in amyotrophic lateral sclerosis. Arch Neurol 45:381–386.

Appel SH, Smith RG, Engelhardt JI, Stefani E (1993) Evidence for autoimmunity in amyotrophic lateral sclerosis. J Neurol Sci 118:169–174. Review.

Bezzi P, Carmignoto G, Pasti L, Vesce S, Rossi D, Rizzini BL, Pozzan T, Volterra A (1998) Prostaglandins stimulate calcium-dependent glutamate release in astrocytes. Nature 391:281–285.

Boillee S, Viala L, Peschanski M, Dreyfus PA (2001) Differential microglial response to progressive neurodegeneration in the murine mutant Wobbler. Glia 33:277–287.

Bowling AC, Schulz JB, Brown RH Jr, Beal MF (1993) Superoxide dismutase activity, oxidative damage, and mitochondrial energy metabolism in familial and sporadic amyotrophic lateral sclerosis. J Neurochem 61:2322–2325.

Brown RH Jr, Hauser SL, Harrington H, Weiner HL (1986) Failure of immunosuppression with a ten- to 14-day course of high-dose intravenous cyclophosphamide to alter the progression of amyotrophic lateral sclerosis. Arch Neurol 43:383–384.

Carriedo SG, Yin HZ, Sensi S, Weiss JH (1998) Rapid Ca2+ entry through Ca2+ -permeable AMPA/Kainate channels triggers marked intracellular Ca2+ rises and consequent oxygen radical production. J Neurosci 18:7727–7738.

Clement AM, Nguyen MD, Roberts EA, Garcia ML, Boillee S, Rule M, McMahon AP, Doucette W, Siwek D, Ferrante RJ, Brown RH Jr, Julien JP, Goldstein LS, Cleveland DW (2003) Wild-type non-neuronal cells extend survival of SOD1 mutant motor neurons in ALS mice. Science 302:113–117.

Consilvio C, Vincent AM, Feldman EL (2004) Neuroinflammation, COX-2, and ALS—a dual role?. Exp Neurol 187:1–10. Review.

Coyle JT, Putterfarcken P (1993) Oxidative stress, glutamate and neurodegenerative disorders. Science 262:689–696.

D'Aversa T, Yu K, Berman J (2004) Expression of chemokines by human fetal microglia after treatment with the human immunodeficiency virus type 1 protein Tat. J Neurovirol 10:86–97.

Drachman DB, Chaudhry V, Cornblath D, Kuncl RW, Pestronk A, Clawson L, Mellits ED, Quaskey S, Quinn T, Calkins A, et al. (1994) Trial of immunosuppression in amyotrophic lateral sclerosis using total lymphoid irradiation. Ann Neurol 35:142–150.

Dugan LL, Sensi S, Canzoniero LMT, Handran SD, Rothman SM, Lin TS, Goldberg MP, Choi DW (1995) Mitochondrial production of reactive oxygen species in cortical neurons following exposure to N-methyl-D-aspartate. J Neurosci 15:6377–6388.

Durham HD, Roy J, Dong L, Figlewicz DA (1997) Aggregation of mutant Cu/Zn superoxide dismutase proteins in a culture model of ALS. J Neuropathol Exp Neurol May;56(5):523–530.

Dykens JA (1994). Isolated cerebral and cerebellar mitochondria produce free radicals when exposed to elevated Ca2+ and Na+: Implications for neurodegeneration. J Neurochem 63:584–591.

Eikelenboom P, van Gool WA (2004) Neuroinflammatory perspectives on the two faces of Alzheimer's disease. J Neural Transm 111:281–294. Review.

Engelhardt JI, Appel SH (1990) IgG reactivity in the spinal cord and motor cortex in amyotrophic lateral sclerosis. Arch Neurol 47:1210–1216.

Engelhardt JI, Appel SH, Killian JM (1989) Experimental autoimmune motoneuron disease. Ann Neurol 26:368–376.

Engelhardt JI, Appel SH, Killian JM (1990) Motor neuron destruction in guinea pigs immunized with bovine spinal cord ventral horn homogenate: Experimental autoimmune gray matter disease. J Neuroimmunol 27:21–31.

Engelhardt JI, Tajti J, Appel SH (1993) Lymphocytic infiltrates in the spinal cord in amyotrophic lateral sclerosis. Arch Neurol 50:30–36.

Engelhardt JI, Siklos L, Appel SH (1997) Altered calcium homeostasis and ultrastructure in motoneurons of mice caused by passively transferred anti-motoneuronal IgG. J Neuropath Exp Neurol 56:21–39.

Ferrante P, Westarp ME, Mancuso R, Puricelli S, Westarp MP, Mini M, Caputo D, Zuffolato MR (1995) HTLV tax-rex DNA and antibodies in idiopathic amyotrophic lateral sclerosis. J Neurol Sci 129 (Suppl):140–144.

Garcia J, Engelhardt JI, Appel SH, Stefani E (1990) Increased MEPP frequency as an early sign of experimental immune-mediated motoneuron disease. Ann Neurol 28:329–334.

Gardner MB, Henderson BE, Officer JE, Rongey RW, Parker JC, Oliver C, Estes JD, Huebner RJ (1973) A spontaneous lower motor neuron disease apparently caused by indigenous type-C RNA virus in wild mice. J Natl Cancer Inst 51:1243–1254.

Ghezzi P, Bernardini R, Giuffrida R, Bellomo M, Manzoni C, Comoletti D, Di Santo E, Benigni F, Mennini T (1998) Tumor necrosis factor is increased in the spinal cord of an animal model of motor neuron degeneration. Eur Cytokine Netw 9:139–144.

Gong YH, Parsadanian AS, Andreeva A, Snider WD, Elliott JL (2000) Restricted expression of G86R Cu/Zn superoxide dismutase in astrocytes results in astrocytosis but does not cause motoneuron degeneration. J Neurosci 20:660–665.

Gonzalez-Scarano F, Baltuch G (1999) Microglia as mediators of inflammatory and degenerative diseases. Annu Rev Neurosci 22:219–240. Review.

Gordon PH, Rowland LP, Younger DS, Sherman WH, Hays AP, Louis ED, Lange DJ, Trojaborg W, Lovelace RE, Murphy PL, Latov N (1997) Lymphoproliferative disorders and motor neuron disease: An update. Neurology 48:1671–1678. Review.

Gordon PH, Doorish C, Montes J, Mosley RL, Diamond B, Macarthur RB, Weimer LH, Kaufmann P, Hays AP, Rowland LP, Gendelman HE, Przedborski S, Mitsumoto H (2006) Randomized controlled phase II trial of glatiramer acetate in ALS. Neurology April 11, 2006; 66(7):1117–1119. Erratum in: Neurology. September 12, 2006; 67(5):920.

Gurney ME, Pu H, Chiu AY, Dal Canto MC, Polchow CY, Alexander DD, Caliendo J, Hentati A, Kwon YW, Deng HX, et al. (1994) Motor neuron degeneration in mice that express a human Cu, Zn superoxide dismutase mutation. Science 264:1772–1775.

Hadlock KG, Miller RG, Jin X, Yu S, Reis J, Mass J, Gelinas DF, McGrath MS (2004) Elevated rates of antibody reactivity to HML-2/Herv-K but not other endogenous retroviruses in ALS. 56th annual meeting of the American Academy of Neurology, San Francisco, CA.

Hall ED, Oostveen JA, Gurney ME (1998) Relationship of microglial and astrocytic activation to disease onset and progression in a transgenic model of familial ALS. Glia 23:249–256.

Hammarberg H, Lidman O, Lundberg C, Eltayeb SY, Gielen AW, Muhallab S, Svenningsson A, Linda H, van Der Meide PH, Cullheim S, Olsson T, Piehl F (2000) Neuroprotection by encephalomyelitis: Rescue of mechanically injured neurons and neurotrophin production by CNS-infiltrating T and natural killer cells. J Neurosci 20:5283–5291.

Haverkamp LJ, Appel V, Appel SH (1995) Natural history of amyotrophic lateral sclerosis in a database population: Validation of a scoring system and a model for survival prediction. Brain 118:707–719.

Henkel JS, Engelhardt JI, Siklos L, Simpson EP, Kim SH, Pan T, Goodman JC, Siddique T, Beers DR, Appel SH (2004) Presence of dendritic cells, MCP-1, and activated microglia/macrophages in amyotrophic lateral sclerosis spinal cord tissue. Ann Neurol 55:221–235.

Hensley K, Floyd RA, Gordon B, Mou S, Pye QN, Stewart C, West M, Williamson K (2002) Temporal patterns of cytokine and apoptosis-related gene expression in spinal cords of the G93A-SOD1 mouse model of amyotrophic lateral sclerosis. J Neurochem 82:365–374.

Hoffman PM, Festoff BW, Giron LT Jr, Hollenbeck LC, Garruto RM, Ruscetti FW (1985) Isolation of LAV/HTLV-III from a patient with amyotrophic lateral sclerosis. N Engl J Med 313:324–325.

Howland DS, Liu J, She Y, Goad B, Maragakis NJ, Kim B, Erickson J, Kulik J, DeVito L, Psaltis G, DeGennaro LJ, Cleveland DW, Rothstein JD (2002) Focal loss of the glutamate transporter EAAT2 in a transgenic rat model of SOD1 mutant-mediated amyotrophic lateral sclerosis (ALS). Proc Natl Acad Sci USA 99:1604–1609.

Hunot S, Hirsch EC (2003) Neuroinflammatory processes in Parkinson's disease. Ann Neurol 53(3):S49–S58; discussion S58–60. Review.

Kakkar P, Mehrotra S, Viaswanathan PN (1992) Interrelation of active oxygen species, membrane damage and altered calcium functions. Mol Cell Biochem 111:11–15.

Kawamata T, Akiyama H, Yamada T, McGeer PL (1992) Immunologic reactions in amyotrophic lateral sclerosis brain and spinal cord tissue. Am J Pathol 140:691–707.

Kawamura T, Horie S, Maruyama T, Akira T, Imagawa T, Nakamura N (1999) Prostaglandin E1 transported into cells blocks the apoptotic signals induced by nerve growth factor deprivation. J Neurochem 72:1907–1914.

Keep M, Elmer E, Fong KS, Csiszar K (2001) Intrathecal cyclosporin prolongs survival of late-stage ALS mice. Brain Res 894:327–331.

Kipnis J, Schwartz M (2002) Dual action of glatiramer acetate (Cop-1) in the treatment of CNS autoimmune and neurodegenerative disorders. Trends Mol Med 8:319–323.

Kirkinezos IG, Hernandez D, Bradley WG, Moraes CT (2004) An ALS mouse model with a permeable blood-brain barrier benefits from systemic cyclosporine A treatment. J Neurochem 88:821–826.

Kriz J, Nguyen MD, Julien JP (2002) Minocycline slows disease progression in a mouse model of amyotrophic lateral sclerosis. Neurobiol Dis 10:268–278.

Lampson LA, Kushner PD, Sobel RA (1988) Strong expression of class II major histocompatibility complex (MHC) antigens in the absence of detectable T-cell infiltration in amyotrophic lateral sclerosis. J Neuropathol Exp Neurol 47:353.

Lampson LA, Kushner PD, Sobel RA (1990) Major histocompatibility complex antigen expression in the affected tissues in amyotrophic lateral sclerosis. Ann Neurol 28:365–372.

Liberto CM, Albrecht PJ, Herx LM, Yong VW, Levison SW (2004) Pro-regenerative properties of cytokine-activated astrocytes. J Neurochem 89:1092–1100. Review.

Lin CL, Bristol LA, Jin L, Dykes-Hoberg M, Crawford T, Clawson L, Rothstein JD (1998) Aberrant RNA processing in a neurodegenerative disease: The cause for absent EAAT2, a glutamate transporter, in amyotrophic lateral sclerosis. Neuron Mar;20(3):589–62.

Lino MM, Schneider C, Caroni P (2002) Accumulation of SOD1 Mutants in Postnatal Motoneurons do not Cause Motoneuron Pathology or Motoneuron Disease. J Neurosci June 15;22(12):4825–4832.

Lim GP, Backstrom JR, Cullen MJ, Miller CA, Atkinson RD, Tokes ZA (1996) Matrix metalloproteinases in the neocortex and spinal cord of amyotrophic lateral sclerosis patients. J Neurochem 67:251–259.

MacGowan DJ, Scelsa SN, Waldron M (2001) An ALS-like syndrome with new HIV infection and complete response to antiretroviral therapy. Neurology 57:1094–1097.

Matsuzaki T, Nakagawa M, Nagai M, Nobuhara Y, Usuku K, Higuchi I, Takahashi K, Moritoyo T, Arimura K, Izumo S, Akiba S, Osame M (2000) HTLV-I-associated myelopathy (HAM)/tropical spastic paraparesis (TSP) with amyotrophic lateral sclerosis-like manifestations. J Neurovirol 6:544–548.

McGeer PL, McGeer EG (2002a) Local neuroinflammation and the progression of Alzheimer's disease. J Neurovirol 8:529–538. Review.

McGeer PL, McGeer EG (2002b) Inflammatory processes in amyotrophic lateral sclerosis. Muscle and Nerve 26:459–470.

McGeer EG, McGeer PL (2003) Inflammatory processes in Alzheimer's disease. Prog Neuropsychopharmacol Biol Psychiatry 27:741–749. Review.

McGeer PL, McGeer EG, Kawamata T, Yamada T, Akiyama H (1991) Reactions of the immune system in chronic degenerative neurological diseases. Can J Neurol Sci 8(3 Suppl):376–379. Review.

Mhatre M, Floyd RA, Hensley K (2004) Oxidative stress and neuroinflammation in Alzheimer's disease and amyotrophic lateral sclerosis: Common links and potential therapeutic targets. J Alzheimers Dis 6:147–157.

Minagar A, Shapshak P, Fujimura R, Ownby R, Heyes M, Eisdorfer C (2002) The role of macrophage/microglia and astrocytes in the pathogenesis of three neurologic disorders: HIV-associated dementia, Alzheimer disease, and multiple sclerosis. J Neurol Sci 202:13–23. Review.

Monsonego A, Beserman ZP, Kipnis J, Yoles E, Weiner HL, Schwartz M (2003) Beneficial effect of orally administered myelin basic protein in EAE-susceptible Lewis rats in a model of acute CNS degeneration. J Autoimmun 21:131–138.

Moulignier A, Moulonguet A, Pialoux G, Rozenbaum W (2001) Reversible ALS-like disorder in HIV infection. Neurology 57:995–1001.

Nix WA, Berger MM, Oberste MS, Brooks BR, McKenna-Yasek DM, Brown RH Jr, Roos RP, Pallansch MA (2004) Failure to detect enterovirus in the spinal cord of ALS patients using a sensitive RT-PCR method. Neurology 62:1372–1377.

Panzara MA, Gussoni E, Begovich AB, Murray RS, Zang YQ, Appel SH, Steinman L, Zhang J (1999) T-cell receptor BV gene rearrangements in the spinal cords and cerebrospinal fluids of patients with amyotrophic lateral sclerosis. Neurobiol Dis 6:392–405.

Pompl PN, Ho L, Bianchi M, McManus T, Qin W, Pasinetti GM (2003) A therapeutic role for cyclooxygenase-2 inhibitors in a transgenic mouse model of amyotrophic lateral sclerosis. FASEB J 17:725–727.

Pramatarova A, Laganiere J, Rousell J, Brisebois K, Rouleau GA (2001) Neuron-specific expression of mutant superoxide dismutase 1 in transgenic mice does not lead to motor impairment. J Neurosci 21:3369–3374.

Rathke-Hartlieb S, Schmidt VC, Jockusch H, Schmitt-John T, Bartsch JW (1999) Spatiotemporal progression of neurodegeneration and glia activation in the wobbler neuropathy of the mouse. Neuroreport 10:3411–3416.

Reichmann G, Schroeter M, Jander S, Fischer HG (2002) Dendritic cells and dendritic-like microglia in focal cortical ischemia of the mouse brain. J Neuroimmunol 129:125–132.

Rio-Hortega P del (1920) El tercer elemento de los centros nerviosos. Poder fagocitario y movilidad de la microglia. Bol Soc Esp Biol 9:154.

Ringholz GM, Appel SH, Bradshaw M, Cooke NA, Mosnik DM, Schulz PE (2005) Prevalence and patterns of cognitive impairment in sporadic ALS. Neurology August 23; 65(4):586–590.

Rosen DR, Siddique T, Patterson D, Figlewicz DA, Sapp P, Hentati A, Donaldson D, Goto J, O'Regan JP, Deng HX, et al. (1993) Mutations in Cu/Zn superoxide dismutase genes are associated with familial amyotrophic lateral sclerosis. Nature 362:59–62.

Rothstein JD, Tsai G, Kuncl RW, Clawson L, Cornblath DR, Drachman DB, Pestronk A, Stauch BL, Coyle JT (1990) Abnormal excitatory amino acid metabolism in amyotrophic lateral sclerosis. Ann Neurol 28:18–25.

Rothstein JD, Martin LJ, Kuncl RW (1992) Decreased glutamate transport by the brain and spinal cord in amyotrophic lateral sclerosis. N Engl J Med 326:1464–1468.

Schwartz M, Kipnis J (2002) Multiple sclerosis as a by-product of the failure to sustain protective autoimmunity: A paradigm shift. Neuroscientist 8:405–413. Review.

Serpe CJ, Kohm AP, Huppenbauer CB, Sanders VM, Jones KJ (1999) Exacerbation of facial motoneuron loss after facial nerve transection in severe combined immunodeficient (scid) mice. J Neurosci 9:RC7.

Shaked I, Porat Z, Gersner R, Kipnis J, Schwartz M (2004) Early activation of microglia as antigen-presenting cells correlates with T cell-mediated protection and repair of the injured central nervous system. J Neuroimmunol 146:84–93.

Sheen VL, Dreyer EB, Macklis JD (1992) Calcium-mediated neuronal degeneration following singlet oxygen production. Neuroreport 3:705–708.

Siklos L, Engelhardt JI, Harati Y, Smith RG, Joo F, Appel SH (1996) Ultrastructural evidence for altered calcium in motor nerve terminals in amyotrophic lateral sclerosis. Ann Neurol 39:203–216.

Siklos L, Engelhardt JI, Alexianu ME, Gurney ME, Siddique T, Appel SH (1998) Intracellular calcium parallels motoneuron degeneration in SOD-1 mutant mice. J Neuropathol Exp Neurol Jun;57:571–587.

Simpson DM, Tagliati M (1994) Neurologic manifestations of HIV infection. Ann Intern Med 121:769–785. Review.

Simpson EP, Henry YK, Henkel JS, Smith RG, Appel SH (2004) Increased lipid peroxidation in sera of ALS patients: A potential biomarker of disease burden. Neurology 62:1758–1765.

Stephenson D, Rash K, Smalstig B, Roberts E, Johnstone E, Sharp J, Panetta J, Little S, Kramer R, Clemens J (1999) Cytosolic phospholipase A2 is induced in reactive glia following different forms of neurodegeneration. Glia 27:110–128.

Tabakman R, Lecht S, Sephanova S, Arien-Zakay H, Lazarovici P (2004) Interactions between the cells of the immune and nervous system: Neurotrophins as neuroprotection mediators in CNS injury. Prog Brain Res 146:387–401. Review.

Tajti J, Stefani E, Appel SH (1991) Cyclophosphamide alters the clinical and pathological expression of experimental autoimmune gray matter disease. J Neuroimmunol 34:143–151.

Troost D, Van den Oord JJ, de Jong JMBV (1988) Analysis of the inflammatory infiltrate in amyotrophic lateral sclerosis. J Neuropathol Appl Neurobiol 14:255–256.

Troost D, Van den Oord JJ, de Jong JMBV, Swaab DF (1989) Lymphocytic infiltration in the spinal cord of patients with amyotrophic lateral sclerosis. Clin Neuropathol 8:289–294.

Tymianski M, Charlton MP, Carlen PL, Tator CH (1993) Source specificity of early calcium neurotoxicity in cultured embryonic spinal neurons. J Neurosci 13:2085–2104.

Urushitani M, Sik A, Sakurai T, Nukina N, Takahashi R, Julien JP (2006) Chromogranin-mediated secretion of mutant superoxide dismutase proteins linked to amyotrophic lateral sclerosis. Nat Neurosci 9:108–118.

Walker MP, Schlaberg R, Hays AP, Bowser R, Lipkin WI (2001) Absence of echovirus sequences in brain and spinal cord of amyotrophic lateral sclerosis patients. Ann Neurol 49:249–253.

Wilms H, Sievers J, Dengler R, Bufler J, Deuschl G, Lucius R (2003) Intrathecal synthesis of monocyte chemoattractant protein-1 (MCP-1) in amyotrophic lateral sclerosis: Further evidence for microglial activation in neurodegeneration. J Neuroimmunol 144:139–142.

Wong PC, Pardo CA, Borchelt DR, Lee MK, Copeland NG, Jenkins NA, Sisodia SS, Cleveland DW, Price DL (1995) An Adverse property of a familial ALS-linked SOD1 mutation causes motor neuron disease characterized by vacuolar degeneration of mitochondria. Neuron Junuary; 14(6):1105–1116.

Yasojima K, Tourtellotte WW, McGeer EG, McGeer PL (2001) Marked increase in cyclooxygenase-2 in ALS spinal cord: Implications for therapy. Neurology. 57:952–956.

Yi Y, Lee C, Liu QH, Freedman BD, Collman RG (2004) Chemokine receptor utilization and macrophage signaling by human immunodeficiency virus type 1 gp120: Implications for neuropathogenesis. J Neurovirol 10 Suppl 1:91–96. Review.

Yim MB, Kang J-H, Kwak H-S, Chock PB, Stadtman ER (1996) A gain-of-function of an amyotrophic lateral sclerosis-associated Cu, Zn superoxide dismutase mutant: An enhancement of free radical formation due to a decreased in Km for hydrogen peroxide. Proc Natl Acad Sci USA 93:5709–5714.

Yoshihara T, Ishigaki S, Yamamoto M, Liang Y, Niwa J, Takeuchi H, Doyu M, Sobue G (2002) Differential expression of inflammation- and apoptosis-related genes in spinal cords of a mutant SOD1 transgenic mouse model of familial amyotrophic lateral sclerosis. J Neurochem 80:158–167.

Zhao W, Xie W, Le W, Beers DR, He Y, Henkel JS, Simpson EP, Yen AA, Xiao Q, Appel SH (2004) Activated microglia initiate motor neuron injury by a nitric oxide and glutamate-mediated mechanism. J Neuropathol Exp Neurol September; 63(9):964–977.

28
Huntington's Disease

Seymour Gendelman, Howard E. Gendelman, and Tsuneya Ikezu

Keywords Huntingtin (htt); CAG repeats; Chorea; Medium spiny neurons (MSNs)

28.1. Introduction

Huntington's disease (HD) is a familial and rare inherited neurological disorder with a prevalence of 5–8 cases per 100,000 worldwide. This makes HD the most common inherited neurodegenerative disorder (Fahn, 2005). HD is passed from parent to child in autosomal dominant fashion. Each child of an HD parent has a 50% chance of inheriting HD. Both sexes are affected equally. It is amongst the first inherited genetic disorders where an accurate test is now available. The disease complex is ascribed to George Huntington who in 1872 described the clinical and neurological manifestations. The neuronal degeneration causes uncontrolled and abnormal body movements called chorea, mental dysfunction, personality changes, and emotional disturbances. Neuropathologically, HD is associated with the death of GABAergic medium sized spiny projection neurons in the caudate nucleus and to neurons in other brain regions. The HD gene, huntingtin (htt), is located on the short arm of chromosome 4, contains N-terminal 18 amino acid-encoding open reading frame, followed by expanded CAG trinucleotide repeat which encodes variable length of poly glutamine tract, and large C-terminal region (>3,100 amino acids). The length of the HD CAG repeats is the primary determinant of the age at which clinical symptoms will appear (Persichetti et al., 1994). Although it is commonly known that the HD polyglutamate tract leads to formation of intracellular inclusions in cytoplasm and/or nucleus in a number of tested species, the correlation of the inclusion with neurotoxicity has been extremely variable. However, the increased number of glutamine is accompanied by the propensity of full-length htt protein to misfold and aggregate; therefore HD belongs to the class of protein misfolding/aggregation disorders. Early HD symptoms are seen in a person's forties, but can occur at any age. Symptoms include mood swings, depression, irritability, difficulties in learning and remembering, and in decision making. As the disease progresses, concentration becomes difficult and daily tasks of maintaining one's self are all but gone; also individuals have difficulty feeding and swallowing. Death in HD typically occurs around 15 years after motor onset due to complications of the disorder, such as aspiration pneumonia. Presymptomic testing is available for individuals who are at risk for carrying the HD gene. There is no known treatment that alters the course of the disease, but symptoms can be managed.

28.2. History

There are at least five reports predating the description of HD by George Huntington's paper published in the Medical and Surgical Reporter on April 12, 1872 (reviewed by Bruyn and Went, 1986). HD was recognized by Charles Waters in 1841 in a letter published by Dunglison in his "Practice of Medicine" and was suggested in the 1846 thesis presentation "On a Form of chorea vulgarly called magrums" by Charles Gorman to the Jefferson Medical College Commentary in 1908 (history reviewed in Harper, 1996). In 1860 Norwegian physician, Dr. Christian Lund wrote another HD report (Folstein, 1989; Harper, 1996). His paper described a hereditary form of St. Vitus Dance.

In 1983 the mutant gene was localized to the short arm of chromosome 4 (Gusella et al., 1983). Subsequent predictive value required the acquisition of both parent's DNA for evaluation (Gusella et al., 1983, 1984; Wexler et al., 1985). A major breakthrough came in 1993 with gene discovery—IT15 (Interesting Transcript 15). It was found to be an expansion of a CAG (Cytosine Adenosine Guanine) trinucleotide repeat, which codes for glutamine, at 4p16.3 (Huntington's Disease Collaborative Research Group, 1993; Ross, 1995; Zoghi and Orr, 2000). A major contribution to the localization and isolation of the gene was from the "Venezuela Project," a comprehensive study of family clusters of HD that has been ongoing for more than four decades (Folstein, 1989; Harper, 1996). Since discovery of the gene, research on the etiology, pathophysiology, biochemistry, polyglutamine toxicity, and potential therapies has greatly intensified.

T. Ikezu and H.E. Gendelman (eds.), *Neuroimmune Pharmacology*.
© Springer 2008

28.3. Epidemiology

HD involves all ethnic groups and is most frequent in patients of Caucasian descent (Harper, 1996). Transmission of the genetic defect that causes HD came to North America was probably through the seafaring colonial powers—England, France, Netherlands, Portugal and Spain—in the seventeenth and eighteenth centuries and then throughout the world (Hayden, 1981; Folstein, 1989). Various migration patterns are certainly a major factor in the appearance of the disease in South Africa and Australia (Folstein, 1989; Goldberg et al., 1996). On the other hand, HD is rare in Japan (approximately 10% of the prevalence in the United States) and in African blacks (in contrast to American blacks and mixed races) (Folstein, 1989; Harper, 1996). Significant foci of families have been noted in Britain, Canada, Sweden, Venezuela, and the United States (Hayden, 1981; Folstein, 1989). Special mention should be made of the Venezuelan occurrence around Lake Maracaibo and the important role of those studies in mapping and isolating the gene for HD (Bruyn and Went, 1986).

28.4. Genetics

The HD gene, *Htt*, was first discovered in 1983 (Gusella et al., 1983), due to and the link between the disease development and the number of CAG repeats (HD Collaborative Research Group, 1993; Ranen et al. 1995; Margolis and Ross, 2001). The gene contains 67 exons and encodes the protein Htt, which is about 350 kDa long, contains 3,144 amino acids, and has no homology with other known proteins (Huntington's Disease Collaborative Group, 1993; Margolis and Ross, 2001). Normal *Htt* contains less than 29 trinucleotide repeats. In most patients with HD, *Htt* gene contains more than 35 repeats of polyglutamine (polyQ) sequence (Ranen et al. 1995; Rubinsztein et al., 1996; Margolis and Ross, 2001; Fahn, 2005) and has 100% penetrance. There is an intermediate group of CAG repeat lengths, 29–35, for which penetrance rates have been difficult to establish due to their infrequent occurrence. HD is a member of the polyglutamine disorders characterized by an unstable CAG trinucleotede repeat including SCA 1,2,3, Machado-Joseph disease, (MJO), 6, 7, 12, 17, dentatorubral-pallidoluysian atrophy (DRPLA), Spinobulbar muscular atrophy (SBMA), and Kennedy's disease (Rubinsztein, 2002) (Figure 28.1).

By 1986, it was noted that patients with a clinical onset prior to age 20 had a preponderant inheritance from affected fathers (Harper, 1996; Fahn, 2005). In those patients with onset prior to 10 years, the incidence approaches 100% (Hayden, 1981). The CAG repeats were found to be unstable in gametes and with high repeat trinucleotide lengths (Riley and Lang, 1991; Duyao et al., 1993; Fahn, 2005). The instability in sperm DNA may explain the increased number of repeats in successive generations of male derived offspring—"anticipation" (Duyao

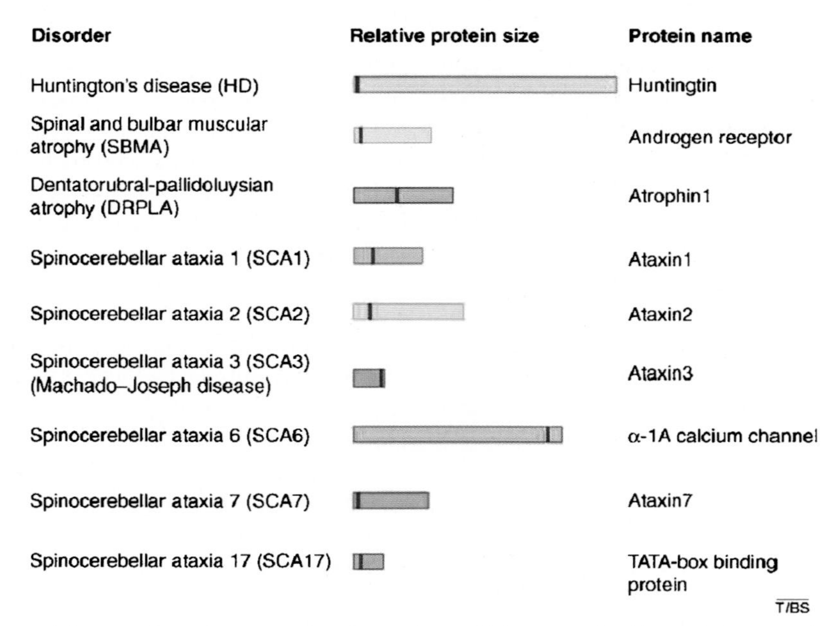

FIGURE 28.1. Protein context of the polyglutamine expansion determines which neuronal cell populations are the most vulnerable. Shown is the location of the polymorphic polyglutamine tract in nine different protein contexts (drawn to scale) that, when expanded, causes the specific loss of neurons from different brain regions and lead to distinct inherited neurodegenerative disorders [*cited from* Gusella and Macdonald (2006). *Huntington's disease: seeing the pathogenic process through a genetic lens. Trends Biochem Sci 31, 533–40* (2006) *with permission from Elsevier*].

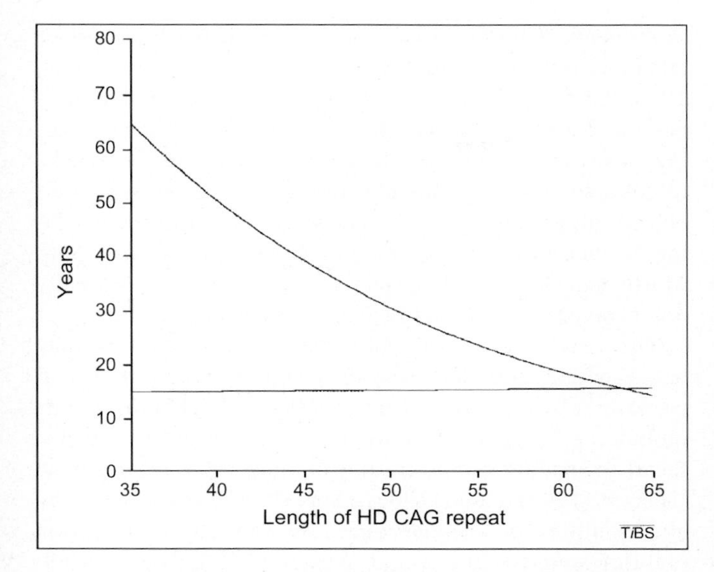

FIGURE 28.2. Correlation of HD CAG-repeat length with age at onset. Best-fit curves for age at neurological onset (red) and duration of disease from onset to death (blue), plotted against CAG-repeat length for the expanded mutant allele from Huntington disease (HD) patients. Age at onset is strongly correlated with the CAG-repeat length (r2 = 0.54; $p < 0.001$), whereas duration of disease shows no correlation with the CAG-repeat length, suggesting that factors independent of the original trigger of pathogenesis predominate after onset of HD to determine rate of progression [*cited from* Gusella and Macdonald (2006). ME. *Huntington's disease: seeing the pathogenic process through a genetic lens. Trends Biochem Sci 31, 533–40 (2006) with permission from Elsevier*].

et al., 1993; Andrew et al., 1993; Ranen et al. 1995; Harper, 1996; Margolis and Ross, 2001; Fahn, 2005). Affected females tend to transmit approximately the same number of repeats (that is, there is an equal probability of increased or decreased repeats); whereas, males tend to transmit a greater (rather than lower) number of repeats (Duyao et al., 1993; Margolis and Ross, 2001; Fahn, 2005). The length of the repeat contributes to the age of onset of symptoms: the greater the number of repeats, the earlier the age of onset (Figure 28.2) (Rubinsztein, 2002). This applies equally to male and female offspring (Harper, 1996; Fahn, 2005). Most adult-onset patients carry 40–50 CAG repeats, whereas greater than 55–60 CAG repeats are more likely to express the juvenile form of HD (Rubinsztein, 2002; Fahn, 2005). The numbers of spontaneous mutations appears to be somewhat greater than initially anticipated, based on increased identification of sporadic cases through DNA analysis (Riley and Lang, 1991; Fahn, 2005).

Homozygotes and heterozygotes have a similar age of onset, though the former, which are very rare, generally experience a much more fulminant course, reflecting a double dose of the abnormal trinucleotide expansion (Harper, 1996; Fahn, 2005). It has not been fully determined that longer repeat patients have a greater rate of progression, though postmortem studies show a greater degree of pathological changes (Kieburtz et al., 1994; Furtado et al., 1996).

28.5. Pathology

The hallmark of HD pathology is the striking regional atrophy and neuronal loss in the striatum (caudate nucleus and putamen) (Roos, 1986; Margolis and Ross, 2001). Macroscopically the brain shows atrophy, predominant in the striatum, with marked ventricular dilatation, and approximately 25–30% decrease in brain weight (Margolis and Ross, 2001). White matter and cortical atrophy are evident in more advanced cases.

Microscopically there is selective loss of small-medium-sized spiny type II neurons with relative sparing of the large neurons of the striatum (Hayden, 1981; Roos, 1986; Margolis and Ross, 2001). There is an associated variable degree of gliosis (Roos, 1986). The earliest and most severe loss is in dorsal and medial regions with progressive severe loss of neurons to the ventral and lateral regions (Harper, 1996; Margolis and Ross, 2001). Other regions of the basal ganglia and areas of the thalamus, cerebellum, brainstem, and spinal cord are affected to a lesser degree (Roos, 1986). The cerebral cortex is also affected, especially the large pyramidal neurons in layers III, V and VI (Roos, 1986; Harper, 1996; Margolis and Ross, 2001).

Striatal atrophy is progressive and may be so severe as to leave the striatum devoid of any cells. In a study comprising 154 HD patients, a classification of pathologic severity was developed based on the degree of striatal cell loss (Vonsattel et al., 1985). Those patients with greater neurodegenerative changes tend to have more severe clinical disease and earlier age of onset as well as a larger number of trinucleotide repeats (Myers et al., 1988; Furtado et al., 1996; Fahn, 2005). Some surviving neurons show evidence of regeneration in dendrites while other neurons are selectively spared (Graveland et al., 1985; Ferrante et al., 1985). The major losses of cortical neurons are those, which project to the thalamus and striatum [superior regions of cortical layer VI; (Margolis and Ross, 2001)]. There is no close correlation between the degree of striatal degeneration and cortical neuronal loss and as such neuropathogenic events may not be wholly secondary to retrograde degeneration (Margolis and Ross, 2001). Analysis of postmortem human HD brain tissue using antibodies directed at the N-terminus of Htt shows intranuclear inclusions in neurons but not glia, most abundant in cortical layers III, V, and VI, and the medium spiny neurons (MSNs) of the striatum. The density of the inclusions correlates with the length of the CAG repeat. However, the inclusions cannot be identified by antibodies directed at internal epitopes of Htt but can be detected (both in full-length wild type and mutated Htt) by immunoflourescence and biochemical fractionation (Margolis and Ross, 2001; Truant, 2003). This suggests that the complete protein may be misfolded in the nucleus as inclusions. Similar inclusions have been detected in transgenic mouse models of HD (Davies et al., 1997; Rubinsztein, 2002). Biochemically, the striatal loss correlates with decreased concentrations of gamma-aminobutyric acid (GABA), glutamic acid decarboxylase (GAD), enkephalin, and substance P (Hayden, 1981; Roos, 1986; Martin and Gusella, 1986). There is relative sparing of cholinergic and somatostain striatal interneurons (Fahn, 2005) but decreased acetylcholine (Ach) and dopamine (DA) receptors in the striatum (Folstein, 1989). This probably results in decreased DA inhibition, increased DA and tyrosine hydroxylase (T-OH) in the substantia nigra, and decreased binding of GABA in the striatum (Roos, 1986; Martin and Gusella, 1986; Folstein, 1989). Neuropeptide Y and somatostatin are increased with relative sparing of their cells of origin in the nucleus accumbens and ventral striatum (Roos, 1986; Martin and Gusella, 1986; Harper, 1996).

N-methyl-D-aspartate receptors (NMDAR) are severely reduced in the striatum and cerebral cortex, suggesting an excitotoxic mechanism for the pathogenesis of HD (Folstein, 1989; DiFiglia, 1990; Fahn, 2005). This is based on the observation that infusion into the brain of glutamate-receptor agonists induces neuronal death and disease manifestations similar to HD (Beal et al., 1986; Vonsattel and DiFiglia, 1998; Bates et al., 2002). Deficits in mitochondrial metabolism were reported in the brains of patients with HD providing another disease hypothesis (Beal, 1998). Caspases, proteases known to be involved in apoptotic cell death are increased in HD, and may participate in neurodegenerative activities (Wellington et al., 1998; Ona et al., 1999;). Htt sequester proteins such as transcription factors, heat-shock proteins, and proteasome subunits may affect the functions of a number of normal neuronal proteins (Davies et al., 1997; Ross et al., 2003).

28.6. Clinical Course

28.6.1. Natural History

Huntington's disease, an autosomal dominant disease affecting males and females equally, is characterized by a triad of major clinical findings: abnormal involuntary movements, personality/behavioral disorder, and cognitive deterioration to dementia (Bruyn and Went, 1986; Folstein, 1989; Harper, 1996; Hayden and Leavitt, 2000; Fahn, 2005). Difficulties in diagnosis arise when there is sporadic disease or an incomplete family history (i.e. adoption or family separation). Though the general age of onset is between 35–45 years (mean age of 40) the range may be from children to the 80s (5% before age 20 and 5% after age 60) (Bruyn and Went, 1986; Folstein, 1989; Riley and Lang, 1991; Hayden and Leavitt, 2000; Margolis and Ross, 2001). Duration of disease usually approximates 15 years (14–20) (Bruyn and Went, 1986; Martin and Gusella, 1986; Harper, 1996). The symptoms may arise almost simultaneously or be spread over years. The behavioral features may be present for years and not considered pathologic prior to the appearance of cognitive decline motor abnormalities. Often the initially noted features are motor—typically of the hyperkinetic involuntary type, choreatic or myoclonic jerking movements, with "cover-up" attempts by the patient to reduce attention to the movement from observers (Bruyn and Went, 1986; Fahn, 2005). The initial onset may be general clumsiness, motor impersistance (the inability to continue an ongoing movement), awkward gait, dropping things, or movements resembling tics. These movements may later progress to overt chorea, myoclonus, dystaxia, and ceaseless writhing, twitching or uncoordinated movements (Bruyn and Went, 1986; Martin and Gusella, 1986; Harper, 1996). Sensation is usually not involved (Hayden and Leavitt, 2000; Fahn, 2005).

The ocular, orofacial, and speech abnormalities may begin with incomplete saccades and difficulty in visual tracking (Bruyn and Went, 1986). Facial movements include appearances of whistling, blowing out of cheeks, facial grimacing, or protruding tongue for a brief period (Harper, 1996; Fahn, 2005). Speech may be altered by abnormalities of oropharyngeal muscles as in repetition of brief sounds (Bruyn and Went, 1986; Folstein, 1989; Hayden and Leavitt, 2000). The more obvious movements may be difficult to separate from other movements as ballism, tardive dyskinesia, or athetosis (Bruyn and Went, 1986). The association with varied stereotyped movements as chorea, myoclonus, general clumsiness, and ataxia help define the situation. Initially the movements are mistaken for tics but the random appearance of chorea is the most striking movement abnormality (Bruyn and Went, 1986; Fahn, 2005) with quick brief movements of proximal, distal, or trunkal muscles. There is gradual development of an awkward, arrhythmic—almost "dancing"—broad-based gait and inability to hold a steady course, lurching to either side with loss of tandem gait (Bruyn and Went, 1986; Hayden and Leavitt, 2000; Fahn, 2005). When attempting to maintain a sustained grip, there are sustained contractions interrupted by sudden relaxations or pauses ("milking grip") (Bruyn and Went, 1986). Ultimately there may be increased tone and even signs of pyramidal dysfunction with clonus and extensor plantar responses, progressing to loss or diminution of choreiform movements and progression to rigidity and dystonia (Folstein, 1989; Hayden and Leavitt, 2000). Terminally patients become incontinent of bowel and bladder and bedridden, unable to perform most aspects of self-care. A disparity may exist in the clinical picture of HD between children and adults (Bruyn and Went, 1986; Hayden and Leavitt, 2000). The juvenile form (Westphal variant, occurring in 5% of patients) is characterized by a fulminate course, rigidity and dystonia progressing to akinesia (Bruyn and Went, 1986). Seizures are common, occurring in more than one-third of younger patients. There is more frequent oculomotor involvement and more rapid decline in cognition (Bruyn and Went, 1986; Hayden and Leavitt, 2000). Chorea is uncommon (Rubinsztein, 2002). Death is usually due to aspiration and respiratory failure. Suicide may be more common in HD at all age groups (Bruyn and Went, 1986; Folstein, 1989; Hayden and Leavitt, 2000; Harper, 1996; Fahn, 2005).

28.6.2. Personality—Behavioral Disorder

Onset of behavioral changes may be quite subtle and difficult to recognize. Time of onset is even more difficult to place. Changes may be interpreted as a reaction to ongoing symptoms but gradually become more overt. In contrast to earlier beliefs that the majority of these behaviors were reactive, more detailed studies have shown these to be distinct neurochemical disorders (Folstein, 1989; Margolis and Ross, 2001). The neurobehavioral disorder has been broadly categorized into abnormalities of arousal, attention, affect, perception, cognition, and personality. Depression, often with progression to frank psychosis, is a more common presentation. There is an initial decrease in interest, lateness at work, taking longer to complete tasks, and withdrawal from family and social interactions (Martin and Gusella, 1986; Folstein, 1989; Goldberg et al., 1996; Hayden and Leavitt, 2000). Irritability, abrupt behaviors and, infrequently, aggressive behaviors are seen (Folstein, 1989; Fahn, 2005). Short temper and tendency to overt obsessive-compulsive behaviors may exist (Bruyn and Went, 1986; Margolis and Ross, 2001). Both apathy and irritability may occur together. Frank psychotic behavior resembling schizophrenia with prominent delusions, hallucinations, and paranoid ideation may occur (Bruyn and Went, 1986; Folstein, 1989; Hayden and Leavitt, 2000; Margolis and Ross, 2001; Fahn, 2005). The psychologic manifestations may respond to specific anti-psychotic, antidepressant or anti-anxiety agents, but there is no treatment to slow the progression of the behavioral changes. Sexual function may be diminished, at times with complaints of impotence. More rarely aggressive sexual behavior occurs (Folstein, 1989). Hospitalization may be necessary because of progressive behavior changes, debilitation, and an inability to provide adequate home care (Bruyn and Went, 1986; Fahn, 2005; Folstein, 1989).

28.6.3. Cognitive Deterioration

Though often intertwined with behavioral aspects, cognitive decline is generally similar to that of the dementias and more difficult to characterize as a distinct diagnostic pattern (Fisher et al. 1983). There is early memory impairment, errors in judgment, poor arithmetic skills, and general mental slowing. Problem solving becomes more difficult, job performance deteriorates, and household responsibilities are unfulfilled. There may be language deficits with or without evidence of aphasia (Folstein, 1989). Despite the cognitive progression, perceptive functions remain generally intact (Bruyn and Went 1986; Martin and Gusella, 1986; Folstein, 1989; Harper, 1996; Margolis and Ross, 2001; Fahn, 2005). Visual memory may be retained until late in the disease. Finally patients may become mute, verbally unresponsive, and progressively apathetic, with a more global rather than subcortical dementia (Bruyn and Went, 1986; Folstein, 1989).

28.6.4. Laboratory

Computerized tomography (CT) or magnetic resonance imaging (MRI) demonstrate enlarged ventricles and usually a striking shrinkage of the caudate nuclei in their ventral paraventricular location, resulting in characteristic ventricular widening in the area of the caudate atrophy. Recent MRI scans of preclinical subjects with HD gene expansions, demonstrate a significant rate of atrophy years prior to their predicted clinical onset (Aylward et al., 2004).

Single photon emission tomography (SPECT) demonstrates reduced blood flow in the area of the caudate and putamen (Harper, 1996). Positron emission tomography (PET) shows reduced glucose metabolism in the striatum (Folstein, 1989; Harper, 1996; Fahn, 2005). Functional MRI (fMRI) shows reduced subcortical (region of caudate and thalamus) participation in a time-discrimination task in preclinical HD subjects (Paulson et al., 2004).

Genetic testing for the HD gene is now widely available to confirm or deny suspected diagnosis. Special care and genetic counseling by knowledgeable and experienced specialists is a crucial component of the testing process. Genetic testing, though relatively straightforward (The World Federation of Neurology Research Group on Huntington's Disease, 1993), carries a number of potential problems (Creighton et al., 2003). When a result of CAG expansion is in the intermediate range, the test should always be repeated. More importantly, there should be careful screening prior to any testing (International Huntington Association and the World Federation of Neurology Research Group on Huntington's Chorea, 1994). How the disease is inherited need to be made very clear. Further, genetic counseling before and after testing must be offered and required prior to further considerations (Folstein, 1989; Riley and Lang, 1991; Fahn, 2005). Psychiatric and/or psychological evaluations should be strongly considered (Folstein, 1989). Legal, ethical and emotional issues are major considerations for predictive testing and should be performed by well-trained and experienced groups. Early knowledge of the disease is also important in regards to treatments as therapeutic trials are being performed in presymptomatic patients as well as those already affected (Hersh, 2003; Taylor, 2004).

28.6.5. Diagnosis

The clinical picture is usually the earliest tool for diagnosis in the adult. A detailed family history, clinical triad, and clear imaging often establish the diagnosis. Gene testing offers the most accurate diagnosis. In the absence of confirmatory gene testing, a differential diagnosis list may need to be considered. HD is a CAG trinucleotide repeat disease in a family of similarly derived diseases including Machado-Joseph disease (SCA-3), a number of spino-cerebellar ataxias (Gusella et al., 1984; Wexler et al., 1985; Bruyn and Went, 1986; Duyao et al., 1993; Harper, 1996; Fahn, 2005). Kennedy's disease, and dentatorubral-pallidoluysian atrophy (DRPLA) (Folstein, 1989; Ravikumar et al., 2003). There are a number of HD-like diseases that ultimately require gene testing for differentiation, including DRPLA in the group manifesting chorea (Harper, 1996; Margolis and Ross, 2001; Fahn, 2005). Neuroacanthocytosis can be difficult to distinguish from HD but atypical retinitis pigmentosa, peripheral neuropathy and the abnormal RBC can be helpful (Fahn, 2005). Adult HD patients may experience seizures early in the disease (Bruyn and Went, 1986). All of the Htt polyglutamine expansion diseases share the characteristics of ataxia and varying degrees of dementia (Harper, 1996; Margolis and Ross, 2001). Other trinucleotide repeats, i.e. CTG repeat in HDL2, may share phenotytic expressions (Fahn, 2005).

Sydenham's chorea occurs at an earlier age than adult onset chorea in HD, is self-limited in its course, and lacks other HD characteristics. Choreiform movements may also occur in systemic lupus erythematosis, tardive dyskinesia, and the dyskinesias due to levodopa toxicity in Parkinson's disease. Brain imaging may be very helpful in these circumstances (Osborn, 1994). Creutzfelt-Jakob disease can show myoclonus and dementia over a relatively rapid period, while multisystem atrophy may present a complicated array of symptoms, though autonomic abnormalities generally appear quite early. The dementias pose a differential problem early on. Alzheimer patients are usually devoid of motor abnormalities, and language difficulties may help differentiate the diagnosis. Subcortical dementias may pose a problem early but should be differentiated with imaging and progression of the disease. The fronto-temporal dementias may manifest with very early, almost isolated, language deficits.

28.6.6. Treatment

Treatment at present is only symptomatic. Appropriate drugs for depression, agitation, irritability, anxiety, or frank psychosis may manage the psychiatric and psychological problems. The motor difficulties may be more complicated and harder to control. Muscle relaxants of central origin and GABAergic medications have not proven to be overly helpful. Neuroleptic agents may partially control the chorea but may cause tardive dyskinesia or enhance cognitive dysfunction in sedated patients.

28.7. Mechanisms of Cell Death

28.7.1. Neurotoxicity

Neurotoxicity in HD has long been thought to reflect a "gain of function" effect of the mutation (i.e. gain of a toxic function by the affected protein, rather than loss of a normal function), leading to excessive neuroexcitation and cell death (Hayden, 1981; Bruyn and Went, 1986; Harper, 1996; Hayden and Leavitt, 2000). Other mechanisms, either related or coincident with neuroexitotoxicity, have been explored since the discovery of the HD gene (Hayden, 1981; Hayden and Leavitt, 2000; Raymond, 2003). Much of this exploration has involved the availability of animal, cell, and neurotoxic models as well as transgenic models (Rubinsztein, 2002; Raymond, 2003; Wellington and Hayden, 2003; Fahn, 2005). Examples of neurotoxic models used to define pathogenesis, mechanisms, and potential therapeutic interventions include quinolinic (Beal et al., 1989) or kainic (McGeer and McGeer, 1976) acid injections into the striatum, or peripheral injections of 3-nitropropionic acid (Beal et al., 1993) in attempts to produce aspects of HD (Margolis and Ross, 2001; Raymond, 2003). The resultant toxicities provided the first evidence suggesting a pathogenic role for excitotoxicity (Margolis and Ross, 2001; Raymond, 2003). Both produced death of medium spiny neurons (MSNs), sparing the large aspiny interneurons in the striatum (Coyle and Schwartz, 1976; McGeer and McGeer, 1976; Margolis and Ross, 2001; Raymond, 2003). In addition, chronic MSN loss induced by 3-nitropropionic acid mitochondrial toxicity is attenuated by NMDAR antagonists (Beal et al., 1993; Raymond, 2003). The results suggest that excessive NMDAR activity with mitochondrial dysfunction may play a major role in striatal MSN death in HD (Margolis and Ross, 2001; Raymond, 2003).

The biochemical loss of GABAergic projection MSNs with preservation of the large aspiny interneurons containing acetylcholinesterase or nitric oxide synthase (NOS), neuropeptide Y and somatostatin is the pattern seen in HD (Ferrante et al., 1985, 1987; Graveland et al., 1985; Margolis and Ross, 2001). These neurotoxic experiments also suggest involvement of metabolic pathways with free radical damage that promote apoptotic progression to cell death triggered by caspase activation and altered GAPDH (glyceraldehyde-3-phosphate dehydrogenase) function (Utz and Anderson, 2000; Wolozin and Behl, 2000; Margolis and Ross, 2001). Free radical tissue damage plays an important role for other CNS diseases, including amyotrophic lateral sclerosis and Parkinson's disease (Facchinetti et al., 1998). Evidence of mitochondrial dysfunction and free radical activation has been found in human HD brain tissue (Browne et al., 1997; Koroshetz et al., 1997).

28.7.2. Biological Function of Htt

Htt protein is composed largely of convective HEAT repeats, which are around 38 amino acid degenerate motifs named for their presence in **H**tt, **E**longation factor 3, protein phosphotase 2A regulatory subunit **A**, and target of rapamycin 1 (**T**OR1) (Andrade and Bork, 1995). Based on the HEAT structure, htt has been suggested to function as a scaffold, organizing members of dynamic complexes for transport and/or transcriptional activity. In addition to the HEAT motifs, many polyQ-containing proteins seem to play a role in protein-protein interactions and regulation of neurogenesis, suggesting that the polyQ regions could be sites for transcriptional alteration (Davies et al., 1997; Margolis et al., 1997; Margolis and Ross, 2001; Sawa et al., 2005). Normal Htt may actually have some anti-apoptotic function, in that reduced levels may contribute to neuronal degeneration (Rigamonti et al. 2000; Leavitt et al., 2001; McMurray, 2001; Wellington et al., 2002). The htt interaction with specific regulators must at least in part occur in the nucleus, and then also plays a role in nuclear entry (Ross, 1997; Margolis and Ross, 2001; Luthi-Carter and Cha, 2003; Truant, 2003). Normally htt is present in cytoplasm with only a small fraction present in the nucleus, in contrast to the disease state (Davies et al., 1997; Di Figlia et al., 1997; Kegel et al., 2002; Luthi-Carter et al., 2003; Truant, 2003). The interaction of transcription-linked proteins and DNA sequences as well as the possible role of mutant htt has been extensively studied (Landles and Bates, 2004). Of the transcription pathways that are affected in HD, the cAMP-responsive element (CRE)- and the specificity protein 1 (SP1)-mediated pathways are the most extensively studied. This owes to their involvement in the expression of gene needed for neuronal survival. Numerous accounts have shown that transcriptional factors such as p53, CREB-binding protein (CBP), SP1, and TATA-binding protein (TBP) can be sequestered into intranuclear aggregates, thus reinforcing the hypothesis of a role of transcriptional dysregulation in HD. Also, the accumulation of chaperones, proteasomes, and ubiquitin in polyQ aggregates suggests that insufficient protein folding and degradation are implicated in the HD pathogenesis (Sakahira et al, 2002; Ciechanover and Brundin, 2003) (Figure 28.3).

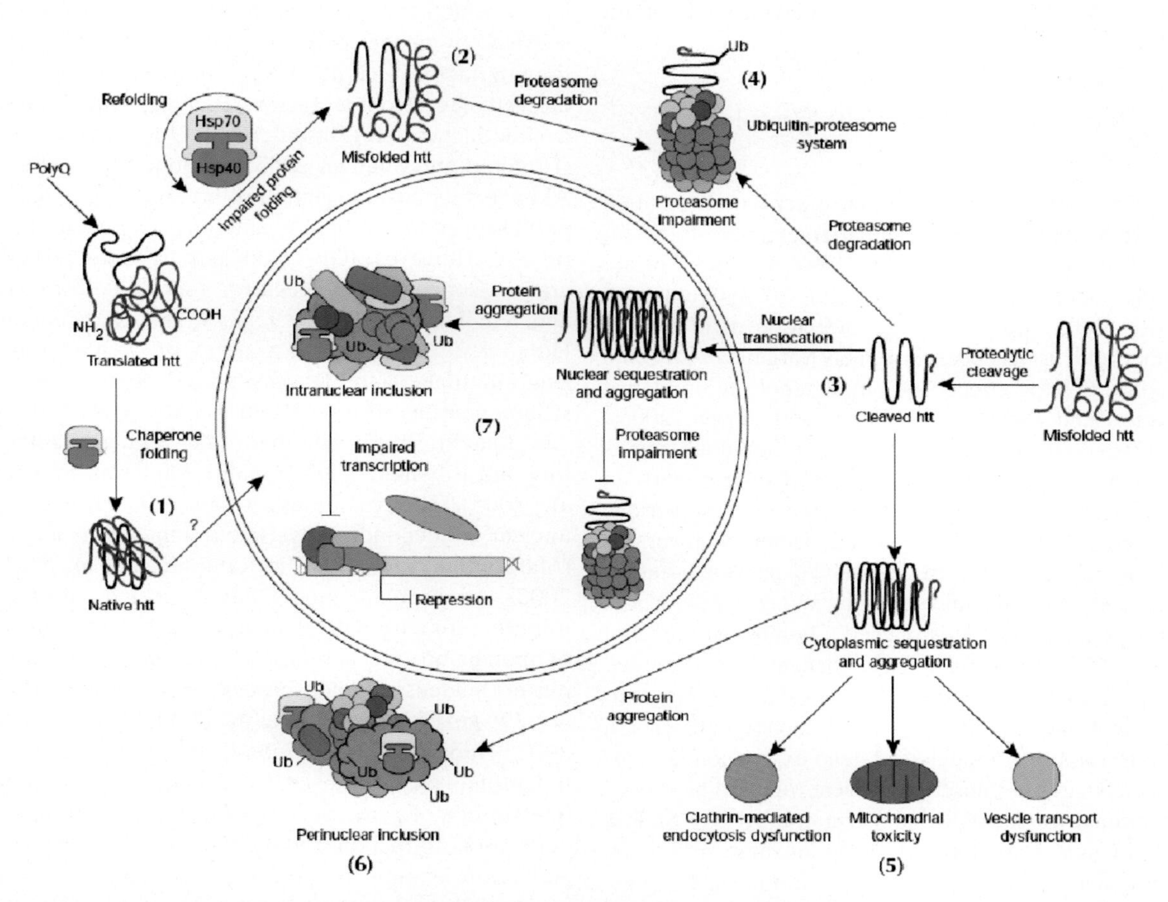

FIGURE 28.3. HD pathogenesis. The molecular chaperones (Hsp70 and Hsp40) promote the folding of newly synthesized htt into a native structure. Wild-type htt is predominantly cytoplasmic and probably functions in vesicle transport, cytoskeletal anchoring, clathrin-mediated endocytosis, neuronal transport or postsynaptic signaling. Htt may be transported into the nucleus and have a role in transcriptional regulation (1). Chaperones can facilitate the recognition of abnormal proteins, promoting either their refolding, or ubiquitination (Ub) and subsequent degradation by the 26S proteasome. The HD mutation induces conformational changes and is likely to cause the abnormal folding of htt, which, if not corrected by chaperones, leads to the accumulation of misfolded htt in the cytoplasm (2). Alternatively, mutant htt might also be proteolytically cleaved, giving rise to amino-terminal fragments that form β-sheet structures (3). Ultimately, toxicity might be elicited by mutant full-length htt or by cleaved N-terminal fragments, which may form soluble monomers, oligomers or large insoluble aggregates. In the cytoplasm, mutant forms of htt may impair the ubiquitin–proteasome system (UPS), leading to the accumulation of more proteins that are misfolded (4). These toxic proteins might also impair normal vesicle transport and clathrin-mediated endocytosis. Also, the presence of mutant htt could activate proapoptotic proteins directly or indirectly by mitochondrial damage, leading to greater cellular toxicity and other deleterious effects (5). In an effort to protect itself, the cell accumulates toxic fragments into ubiquitinated cytoplasmic perinuclear aggregates (6). In addition, mutant htt can be translocated into thenucleus to form nuclear inclusions, which may disrupt transcription and the UPS (7) [*cited from Landles C and Bates GP. Huntingtin and the molecular pathogenesis of Huntington's disease. Fourth in molecular medicine review series. EMBO Rep 5, 958–63 (2004). Reprinted with permission from Nature Publishing Group*].

Nuclear localization of htt fragments probably represents a downstream effect of proteolysis (Wellington and Hayden, 2003; Truant, 2003). Htt is a substrate for a number of proteases including caspases, calpains and aspartyl endopeptidases (Kim et al., 2001; Gafni and Ellerby, 2002; Wellington et al., 2002; Raymond, 2003; Wellington and Hayden, 2003), having 600 potential cleavage sites at the extreme amino terminal (Huntington's Disease Collaborative Research Group, 1993; Wellington and Hayden, 2003). Fragments of normal htt (including polyQ fragments of less than 29 repeats) are probably cleared by the ubiquitin-proteasome system. In contrast, the large expanded polyQ leads ultimately to cytoplasmic and nuclear aggregates (Verhoef et al., 2002; Ravikumar et al., 2003; Wellington et al., 2003).

28.7.3. Polyglutamine Tract/Htt Misfolding and Neurotoxicity

The toxic effect of HD polyglutamine tract has most often been studied within small N-terminal fragments of mutant htt (Beal et al., 1989; Davies et al., 1997; Di Figlia et al., 1997; Ravikumar et al., 2003; Raymond, 2003; Wellington and Hayden, 2003). However, the correlation of inclusions with cellular toxicity has been extremely variable, with some studies reporting toxic effects and others reporting protective effects (Bates and Hockly, 2003; Ross and Poirier, 2005). Both wild type and mutant htt fragments accumulate abnormally in the nucleus in HD brain and model systems early in the disease (Davies et al., 1997; Ross, 1997). Using various mouse model systems, it was shown that decreases in various mRNAs were due to an effect of the polyQ expansions, since it did not occur in transgenic mice carrying a normal polyQ tract (Mangiarini et al., 1996). This opened a number of potential pathways for HD-related pathologic changes and generated even more interest in HD brain and model systems transcription. Using microarray techniques and comparing changes in different transcriptional models and gene expression, at least some of the changes in transcription were found to be related to htt N-terminal fragments (Mangiarini et al., 1996; Steffan et al., 2000; Chan et al., 2002). In some of these strains, N-terminal fragments of mutant htt showed diminished expression of nerve growth factor receptor and deficits in neurite outgrowth (Li et al., 1999; Chan et al., 2002; Luthi-Carter et al., 2003). Changes were also noted in transcription factors in the cell models (Li et al, 1999; Luthi-Carter and Cha, 2003). There are interactions between htt and a number of polyglutamine-rich transcription factors as well as with transcription regulators (Nucifora et al., 2001; Luthi-Carter and Cha, 2003; Bae et al., 2005).

While soluble, mutant protein forms of htt can be degraded by proteosomes, the more stable insoluble aggregated forms tend to be more resistant (Verhoef et al., 2002; Ravikumar et al., 2003). In keeping with proteosomal resistance of aggregated mutant htt, its accumulation may impair proteosomal function ultimately leading to cellular toxicity (Verhoef et al., 2002; Ravikumar et al., 2003) (Figure 28.3). Autophagy (process of bulk degradation of cytoplasmic proteins and damaged organelles) is probably the preferred route for clearance of aggregate-prone protein (Ravikumar et al., 2003). Increased endosomal-lysosomal-like organelles and vesicular bodies are seen in HD patients and transgenic mice (Ravikumar et al., 2003, Davies et al., 1997). The degradation of proteins by the ubiquitin-proteosome/autophagy-lysosome network does not affect nuclear protein since this system remains limited to the cytoplasm (Wellington and Hayden, 2003).

Proteosomal dysfunction leads to endoplasmic reticulum stress, which activates apoptotic signals, which in turn are activated by mutant polyQ expansions (Figure 28.3). Insulin growth factor-1 (IGF-1)/Akt pathway, as well as brain derived neurotrophic factor (BDNF), and ciliary neurotrophic factor (CNTF), block polyQ-htt-induced cell death (Humbert and Saudau, 2003; Saudou et al., 1998). IGF-1/Akt acts by phosphorylation of htt resulting in reduction of polyQ-htt-induced toxicity and thus is a potential therapeutic site. However, IGF-1/Akt activity induces cell proliferation, suggesting the need for a modified form (Humbert and Saudau, 2003). The data suggest that mutant polyQ Htt toxicity results from a series of events, some occurring simultaneously, leading to cell death in the targeted striatal neurons (Harper, 1996; Davies et al., 1997; Hayden and Leavitt 2000; Humbert and Saudau, 2003; Schilling and Borshelt, 2003; Truant, 2003; van Raamsdonk et al., 2005). Though the exact cause of cell death remains unclear, mitochondrial dysfunction must play a crucial role (Margolis and Ross, 2001; Ravikumar et al., 2003; Beal, 2005). A reasonable model was suggested with proteolysis initiating toxicity with production of N-terminal fragments of mutant htt and subsequent cascade of events in cytoplasm, nucleus and cell processes (Harper, 1996; Ross et al., 1998; Hayden and Leavitt, 2000; Margolis and Ross, 2001; Hersh, 2003; Raymond, 2003). These include initial proteolysis, followed by ubiquitinization, autophagy, interruption of gene transcription and regulation, aggregate deposition, mitochondrial dysfunction, and further caspase activation leading to apoptotic cell death, occurring both sequentially and simultaneously (Ross et al., 1998; Margolis and Ross, 2001).

28.7.4. Htt Oligomer as Potential Pathogenic Form of HD Neurotoxicity

As described in the clinical course, neither the progression nor the duration of illness after neurological onset is strongly correlated with CAG length. The fact that the disease process triggered in HD is different from that triggered in other polyQ disorders indicates that some aspect of structure, binding partners, subcellular localization, or activity of htt is crucial to this specificity in neuropathology. As expression of full-length mutant htt occurs throughout life and precedes the detection of N-terminal polyQ fragment in HD patients, it is possible that pathogenesis is triggered via novel property conferred on the htt protein. As an example, htt misfolding and oligomer formation triggered by the aggregation of the expanded polyQ has been proposed as a potential mechanism of htt dysfunction. This is supported by recent studies demonstrating the interaction of htt with a hetero-oligomeric chaperonin TriCcan prevents the aggregation and neurotoxicity of htt (Behrends et al., 2006; Tam et al., 2006). Short G-rich oligonucleotides, capable of adopting a G-quarted conformation, can also inhibit htt aggregation and neurotoxicity *in vitro* (Skogen et al., 2006). Thus, inhibition of htt misfolding may be a potential therapeutic venue for the treatment of HD in the future.

28.8. Glial Inflammation in HD

The neuroimmunological studies on HD are limited. However, PET imaging of HD gene carriers with a specific tracer for microglia, C-(R)-PK11195, specifically detected increased microglial accumulation in striatum in presymptomatic stage as compared to age-matched non-carrier controls (Tai et al., 2007). C-(R)-PK11195 is a radioligand for mitochondria peripheral benzodiazepeine binding sites (PBBS), and its binding in striatum is also strongly correlated with severerity in postsymptomatic HD stages (Pavese et al., 2006). Although this could reflect the microglial accumulation after the neuronal cell death in striatum for the clearance of damaged tissue, these report suggest that microglial accumulation can be used as a presymptomatic diagnosis of the disease for earlier treatment. In terms of immunotherapy on HD, preclinical studies have not been promising. NSAIDs, such as cerecoxib and refecoxib, have shown either no or detrimental effects on the transgenic mouse models of HD (Norflus et al., 2004; Schilling et al., 2004). Minocycline, an anti-inflammatory tetracycline analog extensively tested in multiple neurodegenerative disorder animal models, showed no effect on HD mouse models (Mievis et al., 2007). These studies suggest that glial inflammation can be monitored as one of the earliest sign of neurodegeneration, but may not be a likely target of HD treatment.

28.9. Preclinical Studies

Presently, trials involve an attempt to interrupt the pathways described above—proteolysis, caspase/calpain cleavage, nuclear entry, transcription disruption, growth factor receptor inhibition, mitochondrial dysfunction, ubiquitin and autophagy dysfunction, and aggregate disposition (Marx, 2005). Many of the trials include pre-symptomatic gene carriers, at-risk for the disease, demonstrating a need for reliable markers of disease progression. A full understanding of pathogenesis is not necessarily required to begin therapeutic trials (Feigin and Zgalijardic, 2002; Hersh, 2003).

An understanding of the involved pathways outlined above provides multiple therapeutic targets. Interruption or alteration of these toxic pathways may slow the progression or delay the onset of clinical HD (McMurray, 2001). In transgenic animals various agents have been shown to inhibit caspases, depress aggregate formation, enhance normal cell functions, or combat oxidative stress (Rubinsztein, 2002). Glutamate receptor antagonists, antioxidants, and neurotrophic factors are examples of potential therapies to slow the course of HD. Altering dysfunction in one area may alter the course in other pathways since they are all interdependent to some extent (Marx, 2005).

Transgenic animals, cell systems, and micro array techniques demonstrate a large number of potential agents for human trials. Some of these have already been tested in humans with mixed reuslts—riluzole, coenzyme Q10, remacemide, minocycline, and 10-ethyl-10-deaza-aminopterin (Schilling and Borshelt, 2003; Feigin et al., 2006). Compounds in ongoing trials include: creatine (an endogenous substrate for creatine kinases) (Tabrizi et al., 2005), cystamine (decreases tranglutaminase activity) (Hersh, 2003; Schilling and Borshelt, 2003), amantidine (a noncompetitive NMDAR antagonist) (Hersh, 2003), memantine (an NMDA receptor antagonist) (Beister et al., 2004), clioquinol (a metal-binding compound) (Nguyen et al., 2005), rapamycin (an autophagy inducer) (Berger et al., 2006), and Congo red (protein aggregation inhibitor). Neurotrophic factors (BDNF) and IV infusion of stem cells are among others (Feigin and Zgalijardic, 2002; Bates and Hockly, 2003; Hersh, 2003; Schilling and Borshelt, 2003; Lee et al., 2005). Other candidate compounds include: interleukin-6, adenosine A2A antagonists, dichloroacetate, lipoic acid, lithum and celestrol (Margolis and Ross, 2001; Feigin and Zgalijardic, 2002; Bates and Hockly, 2003; Hersh, 2003; Schilling and Borshelt, 2003).

Summary

Huntington's disease is an autosomal dominant neurological disorder involving the expansion of polyglutamine tract. However, the genetic evidence has indicated that the initial triggering event in the HD pathogenic process occurs at the level of the full-length protein rather than small fragments which appear only later in the disorder. This triggers progressive alterations in cell function that occurs simultaneously, including aggregation of mutant protein in both cytoplasm and nucleus. Following discovery of the mutant gene in 1983, relatively rapid progress has been made with the advent of promising pharmacotherapy to alter disease progression. Though the exact pathogenesis remains unknown, a number of pathologic pathways have been investigated revealing multiple potential therapeutic targets that may alter the course of HD, or delay its onset.

Review Questions/Problems

1. Huntington's disease is a familial and rare inherited neurological disorder affecting up to:

a. 8 people per 100,000
b. 1 per 20,000 people of Western European decent
c. 1 per one million in people of Asian and African descent
d. All of the above

2. What is true about the national history and epidemiology of HD?

a. Men are affected more than woman.
b. It is an inherited genetic disorder where an accurate test is not available.
c. Description of the disease complex was ascribed to Frederick Huntington in 1972.
d. All of the above.
e. None of the above

3. The following is known about the pathogenesis of HD

a. The genetic mutation that causes HD was discovered in 1993.
b. The mechanism by which mutant Htt causes neuronal dysfunction is well known.
c. Excitotoxicity has been proposed as a pathogenic mechanism, on the basis of the observation that infusion of glutamate-receptor agonists into the brain leads to neuronal death and a phenotype similar to HD.
d. Mitochondrial dysfunction is another leading hypothesis for HAD pathogenesis
e. All of the above
f. a, c and d only

4. The genetic defect responsible for HD is

a. A small sequence of DNA on chromosome 4 in which several base pairs are repeated many, many times.
b. The normal gene has three DNA bases, composed of the sequence CAG.
c. In people with HD, the sequence abnormally repeats itself dozens of times.
d. Over time—and with each successive generations—the number of CAG diminishes.
e. All and above
f. a, b, and c only

5. HD involves all ethnic groups and is most frequent in patients of Caucasian descent.

True/False

6. The hallmark of HD pathology is the regional atrophy and neuronal loss in the brain stem.

True/False

7. A triad of major clinical finding characterizes disease: abnormal involuntary movements, personality/behavioral disorders, and cognitive deterioration.
True/False

8. Ocular, orofacial and speech abnormalities are not part of the HD disease complex.
True/False

References

Andrade MA, Bork P (1995) HEAT repeats in the Huntington's disease protein. Nat Genet 11:115–116.

Andrew SE, Goldberg YP, Kremer B, Telenius H, Theilmann J, Adam S, Starr E, Squitieri F, Lin B, Kalchman MA (1993) The relationship between trinucleotide (CAG) repeat length and clinical features of Huntington's disease. Nat Genet 4:398–403.

Aylward EH, Sparks BF, Fields FM, Yallaprogada V, Shprity BD, Rosenblatt A, Brendt J, Gourley LM, Liang K, Zhoni H, Margolis RL, Ross CA (2004) Onset and rate of striatal atrophy in preclinical Huntington's disease. Neurology 63:66–72.

Bae BI, Xu H, Igarashi S, Fujimuro M, Agrawal N, Taya Y, Hayward SD, Moran TH, Montell C, Ross CA, Snyder SH, Sawa A (2005) p53 mediates cellular dysfunction and behavioral abnormalities in Huntington's disease. Neuron 47:1–3.

Bates G, Harper P, Jones L (eds) (2002) Huntington's Disease. New York: Oxford University Press.

Bates GP, Hockly E (2003) Experimental therapeutics in Huntington's disease: Are models useful for therapeutic trials? Curr Opin Neurol 16:465–470.

Beal MF (1998) Mitochondrial dysfunction in neurodegenerative diseases. Biochim Biophys Acta Bio Energetics 1366:211–223.

Beal MF (2005) Mitochondria take center stage in aging and neurodengeration. Ann Neurol 58:495–505.

Beal MF, Kowall NW, Ellison DW, Mazurek MF, Swartz KJ, McEntee WJ (1986) Replication of the neurochemical characteristics of Huntington's disease by quinolinic acid. Nature 321:168–171.

Beal MF, Kowall NW, Ferrante RJ, Ben Cipolloni P (1989) Quinolinic acid striatal lesions in primates as a model of Huntington's disease. Ann Neurol 26:137.

Beal MF, Brouillet E, Jenkins BG, Ferrante RJ, Kowal NW, Miller JM, Storey E, Srivastava R, Rosen BR, Hyman BT (1993) Neurochemical and histologic characterization of striatal excitotoxic lesions produced by the mitochondrial toxin 3-nitropropionic acid. J Neurosci 13:4181–4182.

Behrends C, Langer CA, Boteva R, Bottcher UM, Stemp MJ, Schaffar G, Rao BV, Giese A, Kretzschmar H, Siegers K, Hartl FU (2006) Chaperonin TRiC promotes the assembly of polyQ expansion proteins into nontoxic oligomers. Mol Cell 23:887–897.

Beister A, Kraus P, Kuhn W, Dosse M, Weindl A, Gerlach M (2004) The N-methyl-D-aspartate antagonist memantine retards progression of Huntington's disease. J Neural Transm Suppl 68:117–122.

Berger Z, Ravikumar B, Menzies FM, Oroz L, Underwood BR, Pangalos MN, Schmitt I, Wullner U, Evert BO, O'Kane CJ, Rubinsztein DC (2006) Rapamycin alleviates toxicity of different aggregate-prone proteins. Hum Mol Genet 15:433–442.

Browne SE, Bowling AC, MacGarvey U, Jay Baik M, Berger SC, Muqit MMD, Bird ED, Beal MF (1997) Oxidative damage and metabolic dysfunction in Huntington's disease: Selective vulnerability of the basal ganglia. Ann Neurol 41:646–653.

Bruyn GW, Went LN (1986) Huntington's Chorea. In: Handbook of Clinical Neurology. (Vinken PJ, Bruyn GW, Klawans HL, eds), Vol. 49. pp 267–313. Amsterdam: Elsevier Science Publishers BV.

Chan EYW, Luthi-Carter R, Strand A, Solano SM, Hanson SA, DeJohn MM, Kooperberg C, Olson JM, Cattaneo E (2002)

Increased huntingtin protein length reduces the severity of polyglutamine-induced gene expression changes in mouse models of Huntington Disease. Hum Mol Genet 11:1939–1951.

Ciechanover A, Brundin P (2003) The ubiquitin proteasome system in neurodegenerative diseases: sometimes the chicken, sometimes the egg. Neuron 40:427–446.

Coyle JT, Schwartz R (1976) Lesion of striatal neurons with kainic acid provides a model for Huntington's chorea. Nature 263:244–246.

Creighton S, Almqvist EW, MacGregor D, Fernandez B, Hogg H, Beis J, Welch JP, Riddell C, Lokkesmoe R, Khalifa M, MacKenzie J, Sajoo A, Farrell S, Robert F, Shugar A, Summers A, Meschino W, Allingham-Hawkins D, Chiu T, Hunter A, Allanson J, Hare H, Schween J, Collins L, Sanders S, Greenberg C, Cardwell S, Lemire E, MacLeod P, Hayden MR (2003) Predictive, pre-natal and diagnostic genetic testing for Huntington's disease: The experience in Canada from 1987–2000. Clin Genet 63:462–475.

Davies SW, Turmaine M, Cozens BA, Di Figlia M, Sharp AH, Ross CA, Scherzinger E, Wanker EE, Mangiarini L, Bates GP (1997) Formation of neuronal intranuclear inclusions (NII) underlies the neurological dysfunction in mice transgesnic for the HD mutation. Cell 90:537–548.

DiFiglia M (1990) Excitotoxic injury of the neostriatum: a model for Huntington's disease. Trends Neurosci 13:286–289.

Di Figlia M, Sappe E, Chase KO, Davies SW, Bates GP, Vonsattel JP, Aronin N (1997) Aggregation of huntingtin in neuronal internuclear inclusions and dystrophic neurites in brain. Science 277:1990–1993.

Duyao M, Ambrose C, Myers R, Novelletto A, Parsichetti F, Frontali M, Folstein S (1993) Trinucleotide repeat length instability and age of onset in Huntington's disease. Nat Genet 4:387–392.

Facchinetti F, Dawson VL, Dawson TM (1998) Free radicals as mediator of neuronal injury. Cell Mol Neurobiol 18:667–682.

Fahn S (2005) Huntington's Disease. In: Merritt's Neurology. (Rowland LP, ed) Vol. 11. pp 803–807. Philadelphia, PA: Lippincott Williams and Wilkins.

Feigin A, Zgalijardic D (2002) Recent advances in Huntington's disease: Implications for experimental therapeutics. Curr Opin Neurol 15:483–489.

Feigin A, Ghilardi M-F, Huang C, Ma Y, Carbon M, Guttman M, Paulsen JS, Ghez CP, Eidelberg D (2006) Preclinical Huntington's disease: Compensatory brain responses during learning. Ann Neurol 59:53–59.

Ferrante RJ, Kowall NW, Beal MF, Richardson EP Jr, Bird ED, Martin JB (1985) Selective sparing of a class of striatal neurons in Huntington's disease. Science 230:561–563.

Ferrante RJ, Kowall NW, Beal MF, Martin JB, Bird ED, Richardson EP, Jr (1987) Morphologic and histochemical characteristics of a spared subset of striatal neurons in Huntington's disease. J Neuropathol Exp Neurol 46:12–27.

Fisher JM, Kennedy JL, Caine ED, Shoulson I (1983) Dementia in Huntington's disease: A cross-sectional analysis of intellectual decline. Adv Neurol 38:229–38.

Folstein SE (1989) Huntington's Disease: A disorder of families. Baltimore, MD. Johns Hopkins Univ Press.

Furtado S, Suchowersky O, Rewcastle B, Graham L, Klimek ML, Garber A (1996) Relationship between trinucleotide repeats and neuropathological changes in Huntington's disease. Ann Neurol 39:132–136.

Gafni J, Ellerby LM (2002) Calpain activation in Huntington's disease. J Neurosci 22:4842–4849.

Goldberg YP, Kalchman MA, Metzler M, Nasir J, Zeisler J, Graham R, Koide HB, O'Kusky J, Sharp AH, Ross CA, Jirik F, Hayden MR (1996) Absence of disease phenotype and intergenerational stability of the CAG repeat in transgenic mice expressing the human Huntington's disease transcript. Hum Mol Genet 5(2):177–85.

Graveland GA, Williams RS, Di Figlia M (1985) Evidence of degenerative and regenerative changes in neostriatal spiny neurons in Huntington's diseases. Science 227:770–773.

Gusella JF, Macdonald ME (2006) Huntington's disease: Seeing the pathogenic process through a genetic lens. Trends Biochem Sci 31:533–40.

Gusella JF, Wexler NS, Conneally PM, Naylor SL, Anderson MA, Tanzi RE, Watkins PC, Ottina K, Wallace MR, Sakaguchi AY, Young AB, Shoulson I, Barilla E, Martin JB (1983) A polymorphic DNA marker, genetically linked to Huntington's disease. Nature 306:234–238.

Gusella JF, Tanzi RE, Anderson MA, Hobbs W, Gibbons K, Raschtchian R, Gilliain TC, Wallace MR, Wexler NS, Conneally PM (1984) DNA markers for Nervous System diseases. Science 225:1320–1326.

Harper PS (1996) Huntington's Disease. London: WB Saunders Co. Ltd.

Hayden MR (1981). Huntington's chorea. New York: Springer-Verlag.

Hayden MR, Leavitt BR (2000) Huntington's Disease. In: Kelly's Textbook of Internal Medicine. (Hume HD et al., eds), Vol. 4. pp 2926–2928. Lippincott Williams and Wilkins, Philadelphia, PA.

Hersh SM (2003) Huntington's disease: Prospects for neuroprotective therapy 10 years after the discovery of the causative genetic mutation. Curr Opin Neurol 16:501–506.

Humbert S, Saudau F (2003) Huntingtin phosphorylation and signaling pathways that regulate toxicity in Huntington's disease. Clin Neurosci Res 3:149–155.

Huntington's Disease Collaborative Research Group (1993) A novel gene containing a trinucleotide repeat that is expanded and unstable on Huntington's disease chromosomes. Cell 72:971–983.

International Huntington Association and the World Federation of Neurology Research Group on Huntington's Chorea (1994) Guidelines for the molecular genetics predictive test in Huntington's disease. Neurology 44:1533–1536.

Kegel KB, Meloni AR, Yi Y, Kim YJ, Doyle E, Cuiffo BG, Sopp E, Wang Y, Qin ZH, Chen JD, Nevins JR, Arenin N, Di Figlia M (2002) Huntingtin is present in nucleus, interacts with the transcriptional corepressor C-terminal binding protein and represses transcription. J Biol Chem 277:7466–7476.

Kieburtz K, MacDonald M, Shih C, Feigin A, Steinberg K, Bordwell K, Zimmerman C, Srinidhi J, Sotack J, Gusella J (1994) Trinucleotide repeat length and progression of illness in Huntington's disease. J Med Genet 31:872–874.

Kim YJ, Yi Y, Sapp E, Wang Y, Cuffo B, Kegel KB, Qin ZH, Aronin N, Di Figlia M (2001) Caspase 3-cleaved N-terminal fragments of wild-type and mutant huntingtin are present in normal and Huntington's diseased brains, associate with membranes, and undergo calpain-dependent proteolysis. Proc Natl Acad Sci USA 98:12784–12789.

Koroshetz WJ, Jenkins BG, Rosen BR, Beal MF (1997) Energy metabolism defects in Huntington's disease and effects of coenzyme Q10. Ann Neurol 41:160–165.

Landles C, Bates GP (2004) Huntingtin and the molecular pathogenesis of Huntington's disease. Fourth in molecular medicine review series. EMBO Rep 5:958–963.

Leavitt BR, Guttman JA, Hodgson JG, Kimel GH, Singaraja R, Vohl AW, Hayden MR (2001) Wild-type huntingtin reduces the cellular toxicity of mutant huntingtin in vivo. Am J Hum Genet 68:313–324.

Lee ST, Chu K, Park JE, Lee K, Kang L, Kim SU, Kim M (2005) Intravenous administration of human neural stem cells induces functional recovery in Huntington's disease rat model. Neurosci Res 52:243–249.

Li SH, Cheg AL, Li H, Li XJ (1999) Cellular defects and altered gene expression in PC12 cells stably expressing mutant huntingtin. J. Neurosci 19:5159–5172.

Luthi-Carter R, Cha J-HOJ (2003) Mechanisms of transcriptional dysregulation in Huntington's disease. Clin Neurosci Res 3:265–177.

Mangiarini L, Sathasivan K, Seller M, Cozens B, Harper A, Hetherington C, Lawton M, Trottier Y, Lehrach H, Davies SW, Bates GP (1996) Exon one of the HD gene with an expanded CAG repeat is sufficient to cause a progressive neurological phenotype in transgenic mice. Cell 87:493–506.

Margolis RL, Abraham MA, Gatchell SB, Li SH, Kidwai AS, Breschel TS, Sine OC, Callahan C, McInnis MG, Ross CA (1997) cDNAs with long CAG trinucleotide repeats from human brain. Hum Genet 100:114–122.

Margolis RL, Ross CA (2001) Expansion explosion: new clues to the pathogenesis of repeat expansion neurodegenerative diseases. Trends Mol Med 7:479–482.

Martin JB, Gusella JF (1986) Huntington's disease: Pathogenesis and management. New Eng J Med 315:1267–1276.

Marx J (2005) Neurodegeneration. Huntington's research points to new therapies. Science 310:43–45.

McGeer EG, McGeer PL (1976) Duplication of biochemical changes of Huntington's chorea by intrastriatal injections of glutamic and Kainic acids. Nature 263:517–519.

McMurray C (2001) Huntington's disease: New-Hope for therapeutic trends. Trends Neurosci 24(Suppl):532–538.

Mievis S, Levivier M, Communi D, Vassart G, Brotchi J, Ledent C, Blum D (2007) Lack of Minocycline Efficiency in Genetic Models of Huntington's Disease. Neuromolecular Med 9:47–54.

Myers RH, Vonsattel JP, Stevens TJ, Cupples LA, Richardson EP, Martin JB, Bird ED (1988) Clinical and neuropathologic assessment of severity in Huntington's disease. Neurology 38:341–347.

Nguyen T, Hamby A, Massa SM (2005) Clioqinol down-regulates huntingtin expression in vitro and mitigates pathology in a Huntington's disease mouse model. Proc Natl Acad Sci USA 102:11840–11845.

Norflus F, Nanje A, Gutekunst CA, Shi G, Cohen J, Bejarano M, Fox J, Ferrante RJ, Hersch SM (2004) Anti-inflammatory treatment with acetylsalicylate or rofecoxib is not neuroprotective in Huntington's disease transgenic mice. Neurobiol Dis 17:319–325.

Nucifora FC Jr, Sasaki M, Peters MF, Huang H, Cooper JK, Yamada M, Takakashi H, Tsuji S, Tranceso J, Dawson VL, Dawson TM, Ross CA (2001) Interference by huntingtin and atrophin-1 with CBP-mediated transcription leading to cellular toxicity. Science 291:2423–2428.

Ona VO, Li M, Vonsattel JP, Andrews LJK, Khan SQ, Chung WM, Frey AS, Menon AS, Li XJ, Stief PE, Yuan J, Penney JB, Young AB, Cha JH, Friedlander RM (1999) Inhibition of caspase-1 slows disease progression in a mouse model of Huntington's disease. Nature 399:263–267.

Osborn AG (1994) Diagnostic Neuroradiology. pp 743–744. St Lewis, MO: Mosby Inc.

Paulson JS, Zimbelman JL, Hinten SC, Langbehn DR, Leveroni CL, Benjamin ML, Reynolds NC, Roo SM (2004) fMRI biomarker of early neuronal dysfunction in presymptomatic Huntington's disease. Am J Neuroradiol 10:1715–1721.

Pavese N, Gerhard A, Tai YF, Ho AK, Turkheimer F, Barker RA, Brooks DJ, Piccini P (2006) Microglial activation correlates with severity in Huntington's disease: A clinical and PET study. Neurology 66:1638–1643.

Persichetti F, Srinidhi J, Kanaley L, Ge P, Myers RH, D'Arrigo K, Barnes GT, MacDonald ME, Vonsattel JP, Gusella JF et al. (1994) Huntington's disease CAG trinucleotide repeats in pathologically confirmed post-mortem brains. Neurobiol Dis 1:159–166.

Ranen NG, Stine OC, Abbott MH, Sherr M, Coden AM, Franz ML, Chao NI, Churg AS, Pleasant N, Callhan C (1995) Anticipation and instability of (CAG)n repeats in IT-15 in parent-offspring pairs with Huntington's disease. Am J Hum Genet 57:593–602.

Ravikumar B, Sovan S, Zdenek B, Rubinsztein DC (2003) The roles of the ubiquitin-proteasome and autophagy-lysosome pathways in Huntington's disease and related conditions. Clin Neurosci Res 3:141–148.

Raymond LA (2003) Excitotoxicity in Huntington's disease. Clin Neurosci Res 3:121–128.

Rigamonti D, Bauer JH, De Fraja C, Conti L, Sipione S, Sciiorati C, Clementi E, Hackam A, Hayden MR, Li Y, Cooper JK, Ross CA, Govoni S, Vincenz C, Cattaneo E (2000) Wild-type huntingtin protects from apoptosis upstream of Caspase-3. J Neurosci 20:3705–3713.

Riley BE, Lang AE (1991) Movement Disorders: Huntington's Disease. In: Neurology in Clinical Practice. (Bradley WG, Daroff RB, Fenichel GM, Marsden CD, eds), pp 1585–1586. Stoneham, MA: Butterworth-Heinemann.

Roos RAC (1986) Neuropathology of Huntington's Chorea. In: Handbook of Clinical Neurology. (Vinken PJ, Bruyn GW, Klawans HL, eds), Vol. 49. pp 315–334. Amsterdam: Elsevier Science Publishers BV.

Ross CA (1995) When more is less: pathogenesis of glutamine repeat neurodegenerative diseases. Neuron 65:493–496.

Ross CA (1997) Intranuclear neuronal inclusions: a common pathogenic mechanism for glutamine-repeat neurodegenerative diseases. Neuron 19:1147–1150.

Ross CA, Margolis RL, Beecher MW, Wood JD, Engelender S, Sharp AH (1998) Pathogenesis of Polyglutamine Neurodegenerative Diseases: Towards a Unifying Mechanism. In: Genetic Instabilities in Hereditary Neurological Diseases. (Wells RD, Warren ST, eds), pp 761–776. San Diego, CA: Academic Press.

Ross CA, Poirier MA (2005) Opinion: What is the role of protein aggregation in neurodegeneration? Nat Rev Mol Cell Biol 6:891–898.

Ross C, Poirier MA, Wanker EE, Amzel M (2003) Polyglutamine fibrillogenesis: The pathway unfolds. Proc Natl Acad Sci USA 100:1–3.

Rubinsztein DC (2002) Lessons from animal models of Huntington's disease. Trends Genet 18:202–209.

Rubinsztein DC, Leggo J, Coles R, Almqvist E, Biancalana V, Cossiman JJ, Chotai K, Connarty M, Crauford D, Curtis A, Curtes D, Davidson MJ, Differ AM, Dode C, Dodges A, Frontali M, Ranen NG, Stein OC, Sherr M, Abbott MH, Franz ML, Graham CA, Harper PS, Hedreen JC, Hayden MR (1996) Phenotypic characterization of individuals with CAG repeats and apparently normal elderly individuals with 36–39 repeats. Am J Genet 59:16–22.

Sakahira H, Breuer P, Hayer-Hartl MK, Hartl FU (2002) Molecular chaperones as modulators of polyglutamine protein aggregation and toxicity. Proc Natl Acad Sci USA 99(Suppl 4):16412–16418.

Saudou F, Finkbeiner S, Devys D, Greenberg ME (1998) Huntingtin acts in the nucleus to induce apoptosis but death does not correlate with the formation of intranuclear inclusions. Cell 95:55–66.

Sawa A, Nagata E, Sucliffe S, Dulloor P, Cascio MB, Azeki Y, Roy S, Ross CA, Snyder SH (2005) Huntingtin is cleared by caspases in the cytoplasm and translocated to the nucleus via perinuclear sites in Huntington's disease patient lymphoblasts. Neurobiol Dis 20:267–274.

Schilling G, Borshelt DR (2003) Identifying new therapeutics for Huntington's diseases. Clin Neurosci Res 3:179–186.

Schilling G, Savonenko AV, Coonfield ML, Morton JL, Vorovich E, Gale A, Neslon C, Chan N, Eaton M, Fromholt D, Ross CA, Borchelt DR (2004) Environmental, pharmacological, and genetic modulation of the HD phenotype in transgenic mice. Exp Neurol 187:137–149.

Skogen M, Roth J, Yerkes S, Parekh-Olmedo H, Kmiec E (2006) Short G-rich oligonucleotides as a potential therapeutic for Huntington's Disease. BMC Neurosci 7:65.

Steffan JS, Kazantsev A, Spasic-Baskovic O, Greenwald M, Zhu YZ, Gohler H, Wanker EE, Bates GP, Hausman DE, Thompson LM (2000) The Huntington's disease protein interacts with p53 and CBP and represses transcription. Proc Natl Acad Sci USA 97:6763–6768.

Tabrizi SJ, Blamire AM, Manners DN, Rajagopalan B, Styles P, Schapira AGV, Warner TT (2005) High-dose creatine therapy for Huntington's disease: A 2 year clinical and MRS study. Neurology 64:1655–1656.

Tai YF, Pavese N, Gerhard A, Tabrizi SJ, Barker RA, Brooks DJ, Piccini P (2007) Microglial activation in presymptomatic Huntington's disease gene carriers. Brain 130:1759–1766.

Tam S, Geller R, Spiess C, Frydman J (2006) The chaperonin TRiC controls polyglutamine aggregation and toxicity through subunit-specific interactions. Nat Cell Biol 8:1155–1162.

Taylor SD (2004) Predictive genetic testing decisions for Huntington's disease: Context, appraisal and new moral imperatives. Soc Sci Med 58:137–149.

The World Federation of Neurology Research Group on Huntington's Disease (1993) Presymptomatic testing for Huntington's disease: A world-wide survey. J Med Genet 30:1020–1022.

Truant R (2003) Nucleocytoplasmic transport of huntingtin and Huntington's disease. Clin Neurosci Res 3:157–164.

Utz PJ, Anderson P (2000) Life and death decisions: Regulation of apoptosis by proteolysis of signaling molecules. Cell Death Differ 7:589–602.

Van Raamsdonk JM, Murphy Z, Slow EJ, Leavitt BR, Hayden MR (2005) Selective degeneration and nuclear localization of mutant huntingtin in the YAC 128 mouse model of Huntington's disease. Hum Mol Genet 14:3823–3835.

Verhoef LG, Lindsten K, Masucci MG, Dantuma NP (2002) Aggregate formation inhibits proteosomal degradation of polyglutamine proteins. Hum Mol Genet 11:2689–2700.

Vonsattel JP, Meyers RH, Stevens TJ, Ferrante RJ, Bird ED, Richardson EP, Jr (1985) Neuropathological classification of Huntington's disease. Neurol 44:559–577.

Vonsattel JP, DiFiglia M (1998) Huntington's disease. J Neuropathol Exp Neurol 57:369–384.

Wellington CL, Ellerby LM, Gutekunst CA, Rogers D, Werby S, Graham RK, Loubser O, van Raamsdonk J, Singaraja R, Yang YZ, Gafni J, Bredesen D, Hersh SM, Leavitt BR, Roy S, Nicholson DW, Hayden MR (2002) Caspase cleavage of mutant huntingtin precedes neurodegeneration in Huntington's disease. J Neurosci 22:7862–7872.

Wellington CL, Ellerby LM, Hackam AS, Margolis RL, Trifiro MA, Singaraja R, McCutcheon K, Salvesen GS, Propp SS, Bromm M, Rowland KJ, Zhang T, Rasper D, Roy S, Thornberry N, Pinksy L, Kakizuka A, Ross CA, Nicholson DW, Bredesen DE, Hayden MR (1998) Caspase cleavage of gene products associated with triplet expansion disorders generates truncated fragments containing the polyglutamine tract. J Biol Chem 273:9158–9167.

Wellington CL, Hayden MR (2003) Huntington's disease, editorial. Clin Neurosci Res 3:119–120.

Wexler NS, Conneally PM, Housman D, Gusellа JF (1985) A DNA polymorphism for Huntington's diseases marks the future. Arch Neurol 42:20–24.

Wolozin B, Behl C (2000) Mechanisms of neurodegenerative disorders: Part 2: Control of cell death. Arch Neurol 57:801–804.

Zoghi HY, Orr HT (2000) Glutamine repeats and neurodegeneration. Ann Rev Neurosci 23:217–247.

29
Prion Diseases

Qingzhong Kong and Richard A. Bessen

Keywords Creutzfeldt Jakob disease; Fatal insomnia; Gerstmann-Straussler-Scheinker; Kuru; Neuroinvasion; Prion; Protein-only hypothesis.

29.1. Introduction and History of Prion Diseases

The modern history of the prion diseases is one of novel microbes, anthropological intrigue, and food safety mishaps. The prion diseases, also called the transmissible spongiform encephalopathies, are fatal neurodegenerative diseases that can be sporadic, inherited, or acquired. These multiple origins are unique among human disease. The basis of all prion diseases is the misfolding of the host prion protein into the disease-specific prion protein conformation, called PrP^{Sc}. Transmission of PrP^{Sc} into naïve hosts can lead to additional prion protein misfolding and induction of neurodegenerative disease. Hence, prion diseases are transmissible but are caused by a novel pathogen that lacks a prion-specific nucleic acid genome.

In the 1950s Carleton Gajdusek, an American physician, was attracted to the highlands of New Guinea by reports of a mysterious condition called kuru that was ravaging the Fore stone-age tribes. Kuru primarily afflicts woman and young children and victims succumb to neurological illness that includes shivering and twitching but rapidly progresses to ataxia, paralysis, and eventually death. Gajdusek was perplexed as to the cause of kuru and it was the observation of William Hadlow, a veterinary pathologist, which led to the recognition of the human prion diseases. Hadlow observed the striking similarity in neuropathology between scrapie, a transmissible disease in sheep, and kuru and reasoned that if kuru and scrapie are similar diseases, then kuru should be transmissible to non-human primates (Hadlow, 1959). Gajdusek inoculated kuru into chimpanzees and transmitted neurological disease (Gajdusek et al., 1966). Once the infectious etiology of kuru was established, transmission of kuru was linked to ritualistic endocannibalism among the Fore people. Shortly afterwards, Gajdusek reported transmission of sporadic Creutzfeldt-Jakob disease (CJD), a neurological disease first described in the 1920s, to chimpanzees and they developed neuropathology characteristic of the prion diseases, which includes neuronal loss, spongiform changes, and astrogliosis. These pioneering studies established that kuru and CJD, which were thought to be unrelated diseases of unknown etiology, belong to the same group of human transmissible spongiform encephalopathies. Gajdusek received the Nobel Prize in Physiology or Medicine in 1976 for his work on human prion diseases.

Although the human and animal prion diseases were known to be transmitted by an infectious agent, the unusual resistance of these agents to chemical and physical inactivation was perplexing and led to theories that they were caused by an unconventional slow virus. In 1982, Stanley Prusiner reported that scrapie in sheep was caused by prions, which were defined as infectious proteinaceous particles that are devoid of a nucleic acid (Prusiner, 1982). This hypothesis was met with immense skepticism from the scientific community since it challenged the central principle of molecular biology that genetic information flows from DNA to RNA to protein. However, Prusiner, and others, were able to build upon this theory and his pioneering work was also recognized with a Nobel Prize in Physiology or Medicine in 1997.

Another bizarre chapter in prion diseases began in the mid-1980s with the identification of bovine spongiform encephalopathy (BSE) in cattle. This prion disease was transmitted by industrial cannibalism in which livestock feed was supplemented with meat and bone meal that was unknowingly derived from prion-infected sources, most likely from scrapie-infected sheep. BSE peaked at over 35,000 cases a year in the United Kingdom in the early 1990s but the prevalence has been greatly reduced by the removal of ruminant-derived protein sources in livestock feed. However, the BSE agent is a highly pathogenic strain of prion and is the causative agent for a new human prion disease called variant CJD (Will et al., 1996).

The human prion diseases include CJD, kuru, Gerstmann-Straussler-Scheinker (GSS), and fatal insomnia, which can be classified into three groups based on etiology (Table 29.1).

T. Ikezu and H.E. Gendelman (eds.), *Neuroimmune Pharmacology*.
© Springer 2008

TABLE 29.1. Origin and prevalence of the human prion diseases.

Human prion disease (percent)	Prevalence	Origin (Distribution)
Sporadic (85–90%)		
Creutzfeldt- Jakob Disease	1 in 10^6 per year	Spontaneous (worldwide)
Fatal insomnia	9 cases	Spontaneous
Familial (10–15%)		
Creutzfeldt-Jakob disease E200K-129M	3–5%*	Germ line mutation in *PRNP*
All other families	97 families	Germ line mutations/ insertions in *PRNP*
Fatal familial insomnia	30 families	Germ line mutation in *PRNP*
Gerstmann-Straussler-Scheinker	72 families	Germ line mutations/ insertions in *PRNP*
Acquired (~1%)		
Variant CJD	200 cases	BSE infection (UK, Europe)
Iatrogenic CJD	405 cases	Medical practices (France, Japan, US)
Kuru	~2700 cases	Cannibalism (Papua New Guinea)

* Percent of all human prion diseases based upon epidemiological data from Italy and Japan.

29.2. Prion Protein Gene and Gene Products

29.2.1. Prion Protein Gene and PrP^c

The prion protein is a normal cellular membrane protein that is expressed at high levels in the nervous system, testis, muscle, some hemapoetic cells, and follicular dendritic cells (reviewed by Makrinou et al., 2002). In humans the prion protein is encoded by a single copy gene, which is called *PRNP* and is located on chromosome 20. As illustrated in Figure 29.1A, *PRNP* has two exons and a single intron that span about 15 kb; exon 2 contains the prion protein coding sequence (reviewed by Makrinou et al., 2002). In the *PRNP* coding region there are a large number of pathogenic mutations that are linked to, or have been associated with, the familial prion diseases (Figure 29.1B) (reviewed by Kong et al., 2003). Polymorphisms are also present and the one at codon 129 (methionine or valine) plays an important role in determining susceptibility, clinical phenotype or disease duration in the prion diseases.

The cellular form of the prion protein, called PrP^C, is synthesized as a 253 amino acid precursor protein in humans (Figure 29.1B) (Prusiner, 1991). The N-terminal 22 amino acid residues serve as the signal peptide that allows insertion of the nascent PrP^C peptide into the secretory pathway during biosynthesis. The C-terminal 22 amino acid residues are involved in the covalent addition of the glycosylphosphatidylinositol (GPI) moiety to serine at amino acid 231 (i.e., Ser^{231}). The GPI anchor tethers PrP^C to the extracellular side of the plasma membrane. PrP^C has two N-linked glycosylation sites (Asn^{181} and Asn^{197}) that are the sites of attachment of complex carbohydrates. There are three major PrP^C glycoforms that are defined by the number of N-linked carbohydrate moieties; these are referred to as diglycosylated, monoglycosylated, and nonglycosylated forms (Figure 29.2). The N-terminal region of PrP^C has five octapeptide repeats that bind metal ions (e.g., copper, zinc) but the structure of this domain is undefined, while the C-terminal portion consists of three α-helical and two short β-sheet domains (Figure 29.1B). The cellular function(s) of PrP^C is diverse and it has been implicated in neurogenesis and differentiation (Steele et al., 2006), hematopoetic stem cell self renewal (Zhang et al., 2006), metal ion transportation (Wong et al., 2001), antioxidant activity, signal transduction, apoptosis, and sleep regulation (Tobler et al., 1996; Lasmezas, 2003).

29.2.2. The Protein-Only Hypothesis for the Prion Diseases

Identification of the infectious agent causing prion diseases has been controversial due to its small size and resistance to procedures that inactivate most microorganisms as well as nucleic acids. Extensive efforts to detect a prion-specific small virus or nucleic acid genome have been unsuccessful (Safar et al., 2005). In contrast, procedures that destroy proteins also inactivate prion infectivity. Despite much debate, it is generally agreed that prion diseases are caused by prions, which are defined as proteinaceous infectious particles that lack a nucleic acid genome (Prusiner, 1991). The protein-only hypothesis states that the infectious agent is a misfolded form of PrP^C that is called PrP^{Sc}, the "Sc" refers to scrapie. PrP^{Sc} has distinct biochemical properties from PrP^C that include partial resistance to degradation by proteolytic enzymes, insolubility in detergents, and aggregation into linear fibrils. The basis for these unusual features of PrP^{Sc} is a conformational conversion of α-helical to β-sheet conformation. Replication of the prion agent has been postulated to occur by a protein self-assembly mechanism by which the abnormal PrP^{Sc} fibril can convert the monomeric PrP^C into a subunit of the growing PrP^{Sc} fibril (Jarrett and Lansbury, 1993).

Although the scientific community was initially skeptical of the prion hypothesis, in the past twenty years several key experiments have demonstrated support for this novel concept. PrP null mice, in which the endogenous prion protein gene has been removed, do not demonstrate the ability to produce PrP^{Sc} and prion infectivity nor do they develop prion disease when inoculated with the scrapie agent (Sailer et al., 1994). Additional studies used transgenic mouse models in which the prion protein transgene corresponds to a mutation common in Gerstmann-Straussler-Scheinker disease (GSS), a proline to leucine mutation at codon 102 (i.e., P102L). These transgenic P102L mice spontaneously develop prion disease and the neuropathology is characterized by PrP^{Sc} amyloid plaques, which are similar to plaques found in GSS patients (Hsiao et al., 1990). This classic study demonstrated that mutations in *PRNP* are directly linked to the human prion diseases. Another

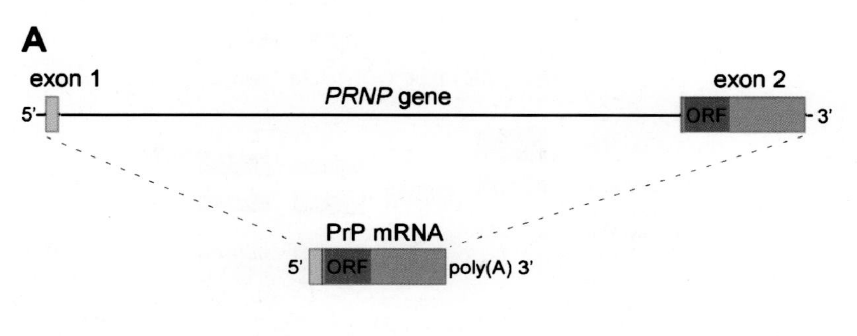

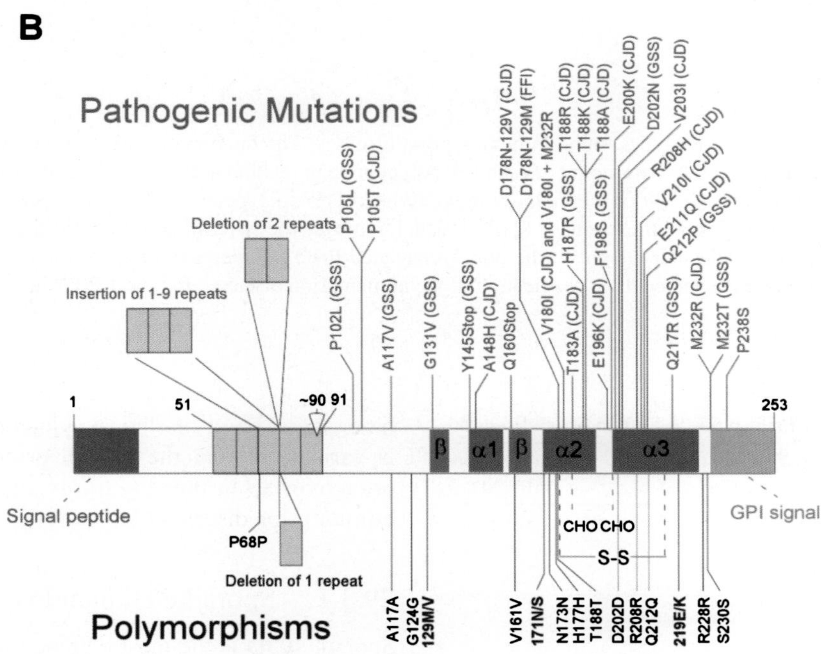

FIGURE 29.1. Diagram of the *PRNP* gene and the PrP protein. (**A**) The prion protein gene and mRNA. The PrP ORF is located within exon 2 and is highlighted in green. (**B**) The PrP protein linear map with posttranslational modifications, polymorphisms, and pathogenic mutations (modified from Figure 29.2 in Kong et al. 2003). The numbers indicate the amino acid residue position in the prion protein and the single letter designation for amino acids is used to denote polymorphisms and mutations. The signal peptides at the amino and carboxyl terminus are indicated in green and orange, respectively. The octapeptide repeats are indicated by grey boxes, while the three α-helices and two short β-sheets are indicated by pale blue and red boxes, respectively. Mutations linked to the human prion diseases are illustrated in red type above the PrP map while normal polymorphisms are indicated below. The arrowhead at amino acid 90 indicates the major cleavage site for proteinase K, which degrades the N-terminal portion of PrP^{Sc}. The two N-linked carbohydrate groups (CHO) and disulfide linkage (-S-S-) are also illustrated.

important proof of the prion hypothesis was the *in vitro* production of prion infectivity following conversion of noninfectious recombinant mouse PrP^{C} into an aggregated fibrillar complex that has a rich β-sheet conformation (Legname et al., 2004).

29.2.3. Molecular Classification of PrP^{Sc} in Human Prion Diseases

Biochemical characterization of PrP^{Sc} reveals heterogeneity in the polypeptide patterns among the human prion diseases and this has been useful for the molecular classification of PrP^{Sc} (reviewed by Gambetti et al., 2003). Limited proteinase K digestion of PrP^{Sc} removes a ~7 kDa N-terminal fragment but leaves intact a protease-resistant PrP^{Sc} core. Typically, there are three protease-resistant polypeptide fragments, referred to as glycoforms, which differ in molecular weight by the addition of zero, one, or two N-linked carbohydrate groups (Figure 29.2). Among the human prion diseases, PrP^{Sc} polypeptide fragments can vary in number (typically 1 to 3), molecular weight, and the ratio of the three glycoforms. Based on these criteria, PrP^{Sc} is classified into three operational groups called type 1, type 2 (a and b), and GSS-type (Table 29.2, Figure 29.2). Type 1 PrP^{Sc} is found in the majority of sporadic CJD (sCJD) cases, some familial CJD (fCJD) cases, and at small

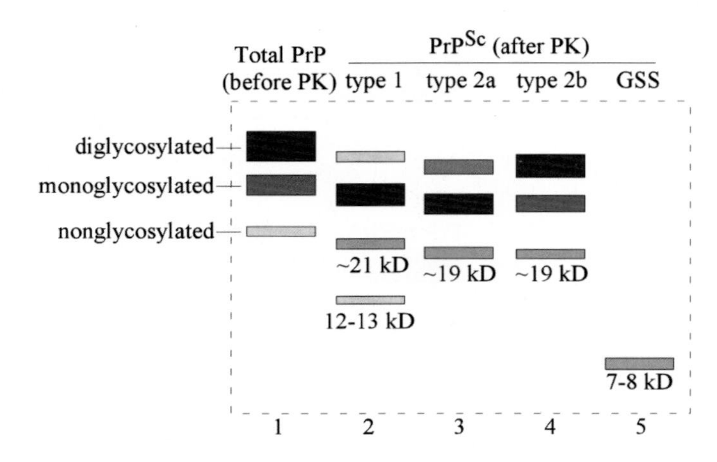

FIGURE 29.2. Schematic diagram of PrP^{Sc} types found in human prion diseases. The three major prion protein polypeptide glycoforms are illustrated on SDS-PAGE (lane 1). The relative amount of each PrP polypeptide in relation to the other glycoforms is indicated by the shading intensity of each band. The molecular weight of the major nonglycosylated PrP^{Sc} polypeptide band is indicated in kilodaltons (kDa). PrP^{Sc} types are defined after limited digestion with proteinase K (PK), which removes the N-terminal portion of the prion protein (lanes 2 to 5). PrP^{Sc} subtypes are defined by the molecular weight of the nonglycosylated PrP^{Sc} polypeptide and the ratio of the three PrP^{Sc} glycoforms. This is one criterion that is used for the molecular classification of the human prion diseases. In type 1 PrP^{Sc} there is an additional 12-13 kDa polypeptide.

TABLE 29.2. Classification of PrP^{Sc} types in human prion diseases.

PrP^{Sc} Type*	N-terminus, amino acid	C-terminus, amino acid	Nonglycosylated PrP^{Sc}, kDa
Type 1	~82	231	~21
Type 2	~97	231	~19
GSS-type	74–90	147–153	7–8

* After limited proteolytic digestion with proteinase K.

amounts in all CJD cases (Yull et al., 2006). The type 2a PrP^{Sc} is associated with a minority of sCJD cases, sporadic fatal insomnia (sFI) and some fCJD cases. Many sCJD patients (20% or more) possess both type 1 and type 2 PrP^{Sc}. In vCJD, PrP^{Sc} polypeptides have a similar molecular weight as found in type 2a PrP^{Sc} but the ratio of the three PrP^{Sc} glycoforms are different. Therefore, vCJD is referred to as type 2b PrP^{Sc} (Figure 29.2). In GSS, a 7–8 kDa PrP^{Sc} polypeptide is referred to as GSS-type PrP^{Sc} and it is only found in GSS cases.

29.3. Epidemiology and Clinical Features of Human Prion Diseases

The human prion diseases are rare but fatal neurodegenerative diseases that have long incubation periods ranging from years to decades. The clinical course is progressive and exhibits a complex neurological pattern that also can last from months to years. Presenting symptoms vary depending on the type of human prion disease, but most will display cognitive, motor, and behavioral deficits during the clinical phase. Some of the prion diseases target to specific brain regions while others have a widespread distribution in the brain. The incubation time, age at onset of clinical symptoms, and disease duration is variable among the human prion diseases and there is often overlap in these features among the major types of human prion diseases (Table 29.1).

29.3.1. Sporadic Human Prion Diseases

Sporadic CJD is the most common human prion disease and accounts for 85–90% of all human cases (reviewed by Gambetti et al., 2003). Sporadic prion diseases are associated with type 1, type 2a (sCJD), or a mixture of type 1 and type 2 PrP^{Sc}. sCJD can be further classified into five subtypes based on a combination of molecular, clinical and pathological features including 1) the genotype at codon 129 of *PRNP* (Methionine or Valine) and 2) the type of protease resistant PrP^{Sc} fragments (1 or 2). These subtypes are designated sCJDMM1/sCJDMV1, sCJDMM2, sCJDVV1, sCJDVV2, and sCJDMV2. These five subtypes of sCJD are associated with distinct clinical and/or pathological characteristics (Parchi et al., 1999; reviewed by Gambetti et al., 2003). In addition to clinical symptoms, periodic sharp wave (PSW) complexes on electroencephalogram (EEG), MRI (hyperintense signal), and the presence of 14–3–3 protein in cerebral spinal fluid (CSF) are useful diagnostic markers for CJD.

The sCJDMM1/sCJDMV1 subtype displays either MM or MV at PrP codon 129 and type 1 PrP^{Sc}. It is the most common subtype and accounts for 60–70% of all sporadic human prion disease. This sCJD subtype has a mean age at clinical onset of 65 years of age and the mean duration of clinical symptoms prior to death is approximately four months. The symptoms at clinical presentation can include cognitive impairment, widened gait or ataxia, behavioral signs (including depression, anxiety, psychosis), and vision defects. Myoclonus and pyramidal signs develop at later stages of clinical disease. PSW complexes on EEG are present in the majority (~80%) of subjects within three months of clinical onset. Protein 14–3–3 in CSF is found in 95% of cases. Therefore, concurrent PSW and 14–3–3 positive CSF are often used as a diagnostic marker for sCJD.

The sCJDVV2 subtype has a VV at PrP codon 129 and type 2 PrP^{Sc}. It is the second most common form of sCJD and accounts for ~16% of all sporadic human prion disease. The mean age at onset is 60 years and the mean duration of the clinical phase is six months. Ataxia is the most common presenting symptom for this subtype. At later stages, dementia almost always develops and is accompanied by myoclonus and pyramidal signs. PSW on EEG is rare and CSF 14–3–3 is positive in ~80% of cases.

The sCJDMV2 subtype has MV at PrP codon 129 and type 2 PrP^{Sc}, and it represents ~9% of the sporadic cases. This subtype has a phenotype very similar to that of sCJDVV2 except for a longer clinical phase before death (mean of 17 months), greater involvement of cognitive and mental symptoms at earlier stages, and frequent aphasia and apraxia at later stages.

The sCJDMM2 subtype has MM at PrP codon 129 and type 2 PrP^{Sc}, and it accounts for 2–8% of the sporadic cases. The mean age at onset is 65 years and the mean clinical duration is 16 months. Cognitive impairment is the universal presenting symptom and is sometimes accompanied by aphasia. At later stages, myoclonus and pyramidal signs, and sometimes Parkinsonism, apraxia, and seizures, develop. There is no PSW on EEG while CSF 14–3–3 is positive in most cases.

The sCJDVV1 subtype has VV at PrP codon 129 and type 1 PrP^{Sc}, and it accounts for 1% of the sporadic cases. The mean age at onset is 39 years and the mean clinical duration is 15 months. This younger age of clinical onset is unusual for a sporadic human prion disease and more closely resembles the age of onset that is characteristic in vCJD. This subtype shows dementia at the early clinical stages and is followed by myoclonus and pyramidal signs. There is no PSW on EEG and the CSF 14–3–3 is positive in all cases tested.

In sporadic fatal insomnia (sFI) there is MM at PrP codon 129 and type 2 PrP^{Sc}. sFI accounts for ~2% of the sporadic cases and the phenotype is similar to that of inheritable FFI, which is characterized by insomnia, but visual signs, cognitive impairment, and motor signs are also common. The mean age at onset is 50 years with a mean clinical duration of 24 months.

29.3.2. Familial Human Prion Diseases

The familial prion diseases are inherited in an autosomal dominant manner and account for 10–15% of all human prion disease. They include familial CJD (fCJD), GSS, and fatal familial insomnia (FFI) (reviewed by Gambetti et al., 2003 and by Kong et al., 2003). The familial prion diseases are caused by pathogenic mutations in the *PRNP* coding sequence. To date, these include 24 mis-sense mutations that result in amino acid changes, 2 nonsense mutations that result in premature stop codons, and about 30 insertion or deletions that lead to changes in the octapeptide repeat number (Figure 29.1B). There are 16 polymorphic sites in *PRNP* (Figure 29.1B). The most important polymorphism is at codon 129, which affects the clinical symptoms in *cis* to the mutant allele and modifies the age of onset and disease duration in *trans* to the mutant allele (reviewed by Kong et al., 2003).

There are eleven mutations in *PRNP* that have been linked to fCJD and these include D187N, V180I, T183A, T188A, E196K, E200K, V203I, H208R, V210I, E211Q, and M232R (Figure 29.1B). In fCJD the disease phenotype is often similar to that found in sCJD, but there can be heterogeneity between individuals carrying different or even the same *PRNP* mutation. This is illustrated in the most common fCJD cases, which is caused by an E200K amino acid substitution in the prion protein. Some individuals with the E200K genotype have a methionine at codon 129 on the mutant allele (i.e., $fCJD^{E200K\text{-}129M}$) and a type 1 PrP^{Sc}, while others with the same E200K mutation have a valine at codon 129 (i.e., $fCJD^{E200K\text{-}129V}$) and a type 2 PrP^{Sc}. In $fCJD^{E200K\text{-}129M}$, the mean age at onset is 58 years and the average clinical duration is 6 months. Symptoms at clinical onset include cognitive and mental impairment that is often accompanied by cerebellar signs and sometimes by visual symptoms and myoclonus. Later in the disease course, seizures and motor and sensory peripheral neuropathy can also be present. These later signs are rare in sCJD. There is PSW on EEG and CSF 14–3–3 is also positive in the majority of affected subjects. In contrast, $fCJD^{E200K\text{-}129V}$ individuals have a phenotype similar to that of sCJDVV2, which presents primarily as ataxia followed by myoclonus with PSW on EEG at late stages. This phenotypic difference between $fCJD^{E200K\text{-}129V}$ and $fCJD^{E200K\text{-}129M}$ illustrates the dramatic influence of the codon 129 polymorphisms on disease phenotype and PrP^{Sc} type. Other common fCJD mutations include D178N-129V and V210I-129M or insertions of one to four octapeptide repeats.

GSS is an inheritable prion disease characterized by PrP^{Sc} deposits in the form of amyloid plaques in the cerebral cortex, degeneration of pyramidal tracts, and a long clinical duration. In GSS there is a highly variable age at onset (between 20 and 73 years) and duration of the clinical phase (mean of 5 years, but range between 5 months and 21 years). The symptoms usually include a slowly progressive cerebellar syndrome, pyramidal signs, and cognitive decline that often develop into dementia. GSS could be mistaken for other common neurodegenerative diseases due to the similarity in clinical signs. These include olivopontocerebellar atrophy, spinocerebellar ataxia, Parkinson disease, amyotrophic lateral sclerosis, Huntington's disease, and Alzheimer disease (Kong et al., 2003). Mutations in *PRNP* that have been linked to GSS include P102L, P105L, A117V, G131V, Y145stop, H187R, F198S, D202N, Q212P, Q217R, and M232T.

$PrP^{P102L\text{-}129M}$ is the most common GSS genotype and is the first reported *PRNP* mutation linked to human prion disease. It has been reported in at least 28 families around the world and it may have occurred independently more than a single time since it is detected in families of different ethnicity. In $GSS^{P102L-129M}$, a protease-resistant PrP^{Sc} polypeptide of 7–8 kDa is the major component of amyloid plaques in the brain. The age at onset is between 30 and 63 years with clinical duration of 1 to 10 years. However, some subjects have a CJD-like phenotype accompanied by a short clinical duration of 5 to 9 months. The initial symptoms include progressive cerebellar signs with ataxia, dysarthria, incoordination of saccadic movements, and occasional pyramidal and pseudobulbar signs. Dementia and akinetic mutism develop at late stages of the disease. Polymorphic variants of the $PrP^{P102L-129M}$ allele include $PrP^{P102L-129V}$ and $PrP^{P102L-129M-219K}$ and these have distinct disease phenotypes.

Fatal familial insomnia is the second most common inherited prion disease and is associated with a D178N-129M mutant allele (reviewed by Kong et al., 2003 and by Montagna et al., 2003). The average age at onset is 49 years but can vary between 20 and 72 years. The duration of the clinical phase also is highly variable with an average of 11 months for 129MM subjects and 23 months for 129MV subjects. Severe insomnia is the most prominent symptom and is likely due to prion targeting to the thalamus, which is reflected in abnormal brain activities in this region on PET and SPECT scans. Other symptoms can include episodes of hallucinations and confusion, myoclonus, spasticity, and seizures. EEG is usually slowed and PSW complexes are often present in patients with long duration. There is a characteristic shortened sleep time and irregular transition between sleep stages on polysomnography. The same D178N mutation is found in fCJD but these individuals have a 129V genotype on the mutant allele and, as a result, the disease phenotype is different and does not include insomnia.

29.3.3. Acquired Human Prion Diseases

The acquired prion diseases account for less than 1% of all human prion disease cases and these include variant CJD, iatrogenic CJD (iCJD), and kuru (reviewed by Will, 2003). In these cases, prion infection is either orally acquired or associated with accidental transmission via medical practices. Many of these latter cases involve contamination with brain tissue, which contains the highest amount of prion infectivity, from the donor host.

Variant CJD was discovered in 1996 and is most likely due to ingestion of BSE-contaminated food products, but direct proof is lacking due to the long interval between exposure and onset of disease. Removal of meat and bone meal supplements from livestock feed in the late 1980s has greatly reduced the prevalence of BSE in the United Kingdom from over 35,000 cases in 1992 to less than 1,000 cases in 2005. As a result, human exposure to BSE has also been greatly reduced. To date, there are 200 cases of vCJD primarily in the United Kingdom. The annual number of vCJD cases reached a peak at 28 in 1999, but this number of cases is relatively low considering that approximately 1,000,000 BSE-infected cattle have entered the human food chain in the UK during the 1980s and 1990s. Due to the long prion incubation period, it is difficult to predict the size of the vCJD epidemic. A retrospective survey of tonsils and appendix specimens (i.e., sites of prion agent replication) removed during routine surgery in the UK revealed only three PrP^{Sc} positive samples in >18,000 cases (Hilton et al., 2004). This suggests that there is no large number of individuals with subclinical vCJD infection as originally estimated. Predictive models currently place the upper number of vCJD cases at 451 by 2010 (Cooper and Bird, 2003).

The clinical and pathological phenotype of vCJD is distinct from most of the other human prion diseases and these unusual features were important in the identification of vCJD (reviewed by Ironside et al., 2005). Unlike the common sporadic forms of human prion disease that primarily afflict individuals in their sixth decade, vCJD is found in younger individuals with mean age of onset at 28 years, ranging between 12 and 74 years of age (reviewed by Will, 2003). The length of the incubation period is uncertain but is estimated to be ~11 years (Cooper and Bird, 2003). The mean clinical duration is ~14 months, which is longer than the 4 months for sCJDMM1 but comparable to that of 16 months for sCJDMM2. All clinical vCJD subjects had PrP^{129MM}. The PrP^{129MM} genotype has a prevalence of 39% in the UK population so the higher distribution in vCJD cases indicates a stronger preference for younger people with a PrP^{129MM} genotype. There could be an age-related exposure or age-dependence for vCJD but none has been demonstrated. The type 2B PrP^{Sc} is characteristic of vCJD (Table 29.2, Figure 29.2) and PrP^{Sc} deposits are present in florid amyloid plaques in the brain. In contrast to the diverse clinical symptoms among sCJD cases, the symptoms of vCJD subjects are more uniform. The presenting symptoms include psychiatric signs and occasionally neurological symptoms (persistent pain, memory impairment). After about 6 months, signs of ataxia, cognitive impairment, and involuntary movements become apparent. Bilateral pulvinar high signal in the thalamus on MRI is prominent in the majority of cases (~78%) (Zeidler et al., 2000). There is general slowing but no PSW complexes on EEG.

Iatrogenic CJD is the second most common acquired human prion disease and these cases are the result of accidental infection due to contact with prion contaminated tissues or instruments during medical procedures (Table 29.1). The mode of prion infection include surgical equipment (e.g., surgical instruments, depth electrodes), transplantation of human tissues (corneal, dura mater), intramuscular injections with growth hormone or gonadotrophin extracted from human pituitary tissues, or blood transfusion (reviewed by Will, 2003). The most likely source of infection is from donors with subclinical sCJD, except for the two transfusion-related cases that have been linked to blood donors who developed vCJD several years later (reviewed by Ironside, 2006). The incubation period in these transfusion related cases was 5 to 6 years, which is shorter than primary vCJD infection in humans.

The most common cases of iCJD are in recipients of human growth hormone therapy that is used to treat children with growth deficits (reviewed by Will, 2003). There have been over 194 cases reported worldwide and these have been linked to growth hormone extracted from cadaveric pituitaries. The clinical and pathological features of growth hormone therapy-related iCJD are distinct from sCJD. The mean incubation period is 12 years (range between 4.5 and 38 years) and the presenting symptoms include progressive cerebellar signs that are sometimes followed by dementia at later stages. The second most common iCJD cases are found in recipients of dura mater homograft (at least 190 cases). Afflicted individuals show typical sCJD symptoms with an incubation period of ~6 years (range between 1.5 and 18 years).

Kuru is an acquired human prion disease of indigenous tribes in Papua New Guinea that is linked to ritualistic endo-cannibalism (Alpers, 1979). More than 2700 cases have been recorded since the 1950s, but this disease has gradually disappeared since the cessation of cannibalism in the late 1950s. The mean incubation period for kuru is about 12 years, but the longest on record is over 50 years. The clinical duration is 6 to 36 months in adults and the main presenting symptom is cerebellar signs, which are often followed by severe dysarthria at late stages in the absence of dementia.

29.4. Neuropathology of Human Prion Diseases

The three principal neuropathological features in human prion diseases are neuronal loss, gliosis, and spongiform changes primarily in grey matter (reviewed by Mikol, 1999). There is considerable diversity among the human prion diseases with respect to the location of pathology in the brain and type of neurons affected. Pathology is not confined to a single brain structure but is widely distributed in the CNS. For each of the human prion diseases, there are typical distinguishing features of neuropathology that are partially defined by the distribution of pathology in the cerebral cortex, subcortical grey matter, and brainstem. Spongiform changes can vary in density and size and range from microvacuolation to the fusion of vacuoles to produce status spongiosis. The degree of spongiform change does not appear to correlate with neuronal loss. Amyloid plaques are also present in the brain in some of the human prion diseases, most notably kuru, vCJD and GSS. These amyloid plaques exhibit birefringence when stained with Congo red and viewed under polarized light.

PrP^{Sc} deposition in the central nervous system is also a hallmark feature of prion disease (Prusiner, 1991). The type of PrP^{Sc} deposit and its location varies among the human prion diseases. PrP^{Sc} is also a component of amyloid plaques. A characteristic of vCJD and kuru is florid amyloid plaques, which is a large cluster of PrP^{Sc} aggregates interspersed with vacuolar pathology. PrP^{Sc} deposition is also present in peripheral tissues of patients with vCJD, primarily in secondary lymphoid tissues. However, no histological changes are apparent in peripheral organs of prion-infected hosts.

In prion diseases, there is a notable absence of encephalitis. The lack of adaptive immune response to PrP^{Sc} is likely due to host tolerance to PrP^{C}. However, there is an increase in the numbers of microglia, activation of astrocytes, upregulation of proinflammatory cytokines, binding of C1q and C3b components to PrP^{Sc}, and complement membrane attack complexes associated with neurons in the brain (Burwinkel et al., 2004; Kovacs et al., 2004; Mabbott, 2004). The precise role of these host responses in neuroprotection and/or neurodegeneration during prion infection is not clear. Interestingly, mice that lack interleukin-10, an anti-inflammatory cytokine, are more susceptible to experimental prion disease (Thackray et al., 2004). In the brains of these mice, there was faster upregulation of pro-inflammatory cytokines (e.g., TNF-α and IL-1β) suggesting that inflammation may be a central event in prion-induced neurodegeneration.

The neuropathology among sCJD cases is not uniform and is variable in nature, severity and location in the brain. In most cases spongiform change is found in the cerebral cortex, cerebellar cortex, and/or the subcortical grey matter (reviewed by Ironside et al., 2005). This is accompanied by reactive gliosis and neuronal loss but there is not a consistent relationship between spongiform change and cell loss. In approximately 10% of sCJD, PrP^{Sc} amyloid plaques are present in the cerebral cortex but most cases have a diffuse synaptic PrP^{Sc} distribution. In contrast to the clinical and pathological diversity present in sCJD, the pathology in vCJD is more uniform. The central pathological features in vCJD are the presence of florid PrP^{Sc} plaques in the cerebral and cerebellar cortex, severe spongiform change in the caudate and putamen, and neuronal loss in the thalamus and midbrain. PrP^{Sc} is also present in affected brain regions. The neuropathology of vCJD is distinct from the other human prion diseases.

In the familial human prion diseases there are also variable clinical and neuropathological features. GSS is characterized by unicentric and multicentric PrP^{Sc} amyloid plaques that are primarily found in the cerebellum, but can also be present in the cerebral cortex (reviewed by Ghetti et al., 1995). Spongiform change and neuronal loss can be focal but some GSS patients have no spongiform changes, while in other cases with a short duration, there is widespread spongiform change. In some GSS families, severe neurofibrillary tangle degeneration is prominent. A feature of FFI is a thalamic degeneration characterized by severe spongiform change, neuronal loss, and astrocytosis in thalamic nuclei (reviewed by Gambetti et al., 1995).

29.5. Peripheral Prion Replication and Neuroinvasion

The sites of prion agent replication and spread are largely dictated by whether the etiology of prion disease is sporadic, familial, or acquired. For example, in sCJD and fCJD, disease initiates in, and is largely confined to, the nervous system. In vCJD, which is acquired via oral ingestion, infection likely begins in secondary lymphoid tissues prior to neuroinvasion and spread to the CNS.

29.5.1. vCJD and Oral Routes of Animal Prion Transmission

Variant CJD is BSE infection in humans and the most probable route of infection is by ingestion of BSE-contaminated food products. The spread of the vCJD agent following oral ingestion has not been directly investigated but evidence suggests it follows a similar pathway to that described for oral ingestion in the animal prion diseases. These pathways have been well described for scrapie in sheep and in rodent models and have three phases: agent entry and replication in secondary lymphoid tissues, spread from secondary lymphoid tissues to the CNS along peripheral nerves, and dissemination and degeneration in the CNS (Figure 29.3). Following oral prion ingestion, the prion agent is initially found in the distal ileum (Heggebo et al., 2000). Entry is likely across the intestinal epithelium via the M cells in the Peyer's patches. Prions then spread to the draining gut-associated lymphoid tissue (GALT) where prion agent is found within months after ingestion (van Keulen et al., 1999). In the GALT, the prion agent is primarily associated with follicular dendritic cells and macrophages in the germinal center of lymphoid follicles. From the GALT the prion agent spreads to other secondary lymphoid tissues including lymph nodes and the spleen over the next several months (van Keulen et al., 2000). The mechanism of prion agent dissemination in secondary lymphoid tissues has not been determined but is thought to be via prion-infected migratory cells that travel in the lymph. A blood borne route is also possible since low levels of prion infectivity have been found in the blood of sheep with scrapie and humans with vCJD (reviewed by Ironside, 2006).

In scrapie-infected sheep, spread of the prion agent from the gut to the CNS occurs via peripheral nerve fibers. From the GALT, the prion agent enters the sympathetic nerves that innervate secondary lymphoid tissues and spreads to the spinal cord via the autonomic nervous system (McBride et al., 2001). The prion agent then ascends the spinal cord along spinal tracts into the brain (Figure 29.3) (van Keulen et al., 2000). In a second pathway, after prion infection of the GALT, there is a delay before agent replication is found in the enteric nervous system. From these sites, the prion agent enters the vagal nerve and spreads to the brainstem where initial agent deposition is found in the dorsal motor nucleus of the vagus (Beekes et al., 1998; van Keulen et al., 2000). Early PrP^{Sc} deposition in this nucleus is indicative of neuroinvasion following oral ingestion. In vCJD, spread from secondary lymphoid tissues to the CNS is likely to take months to years.

Evidence for oral ingestion in vCJD is provided by the peripheral PrP^{Sc} distribution. In humans with vCJD, PrP^{Sc} is present in the ileum and the sympathetic ganglia that provide autonomic innervation to the gut (reviewed by Hilton, 2006). Variant CJD infection is also present in FDCs in secondary lymphoid tissues including the Peyer's patches, tonsil, spleen, lymph nodes, and appendix. This distribution of the prion agent suggests that agent entry via oral ingestion results in the spread of vCJD infection to distal secondary lymphoid tissues (reviewed by Hilton, 2006). This peripheral distribution of the prion agent in lymphoid tissue and autonomic nervous system of the gut is not found in sCJD or the familial human prion diseases suggesting that vCJD is acquired by peripheral exposure.

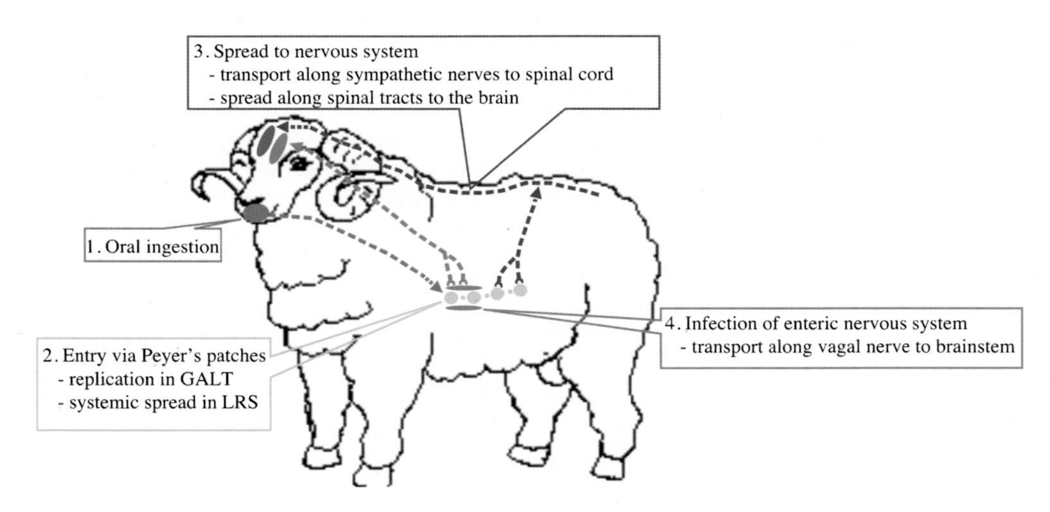

FIGURE 29.3. The stages of prion agent infection and neuroinvasion following oral ingestion of scrapie in sheep.

29.6. CNS Pathogenesis of Sporadic, Familial, and Iatrogenic Prion Diseases

29.6.1. Sporadic CJD

In sCJD the process or events that lead to PrP^{Sc} formation and accumulation are thought to be limited to the CNS. The apparent absence of prion infection in secondary lymphoid tissues in patients with sCJD is cited as evidence against acquisition of prion infection from peripheral exposure. PrP^{Sc} has not been found in the spleen, lymph nodes, tonsil or appendix in patients with sCJD (Head et al., 2004). There is also an absence of infection in the sympathetic ganglia that provide autonomic innervation to the gut, indicating that it is unlikely that there is infection of the enteric nervous system and GALT. However, prion infection has been detected in the peripheral nervous system of patients with sCJD. PrP^{Sc} deposition has been found in retina, trigeminal ganglion, the olfactory nerves and epithelium, dorsal root ganglia and an occasional peripheral nerve (Head et al., 2004). This distribution is consistent with spread of the prion agent away from the CNS along cranial or spinal nerve fibers. PrP^{Sc} has also been found in skeletal muscle of patients with sCJD (Glatzel et al., 2003).

29.6.2. Familial Prion Disease

In familial prion diseases germ line mutations in the prion protein gene leads to PrP^{Sc} deposition in the brain and nervous system. Prion agent has not been found in peripheral tissues in the limited number of familial prion disease cases that have been examined.

29.6.3. Iatrogenic CJD

Iatrogenic CJD can be acquired by either peripheral or CNS exposure to the prion agent from donor tissues or contaminated surgical equipment. The length of time to development of prion disease appears to be related to the site of prion inoculation. Shorter incubation periods are typically found following direct prion exposure to the brain (e.g., contact with sterotactic electrodes or instruments during neurosurgery) while longer incubation periods are seen following peripheral exposure (e.g., intramuscular injection of growth hormone) to the prion agent. In the latter cases, the incubation period can be up to several decades in duration (reviewed by Will, 2003). Due to the low number of iCJD cases, there are few studies examining the peripheral distribution of prion agent. One study reported the presence of PrP^{Sc} in the dorsal root ganglia and a few peripheral nerve fibers of two patients with dura mater related iCJD (Ishida et al., 2005). The pathway of neuroinvasion in iCJD associated with growth hormone therapy is less clear and prion infection was not found in secondary lymphoid tissues of a single patient, but the trigeminal ganglion and dorsal root ganglion were found to be sites of prion infection (Head et al., 2004). This limited amount of data would suggest that the prion agent is restricted to the CNS or PNS in iCJD.

29.7. Prion Agent Interaction with the Immune System

The immune system plays a central role in the pathogenesis of prion infection. In the acquired human prion diseases, it is the initial site of agent replication prior to entry into the nervous system but the immune system does not appear to be essential in the pathogenesis of the familial and sporadic human prion diseases. Following peripheral exposure to the prion agent, the immune system can amplify prion infection but, interestingly, the adaptive immune response does not play a role in the clearance of the prion agent.

29.7.1. Prion Infection of Secondary Lymphoid Tissues

In the 1960s it was first described that the spleen is an early site of prion agent replication following peripheral scrapie infection of mice (Eklund and Hadlow, 1969). Prion replication in the spleen is present shortly after peripheral exposure and reaches a plateau level prior to detection of prion agent in the brain. Mice that receive immunosuppressive therapy or splenectomy prior to peripheral inoculation experience a delay in prion replication in the periphery and spread to the CNS (Kimberlin and Walker, 1989). Mice that are deficient in B cells also are resistant to peripheral prion infection, but B cells do not appear to be a target for prion infection (reviewed by Aguzzi and Heikenwalder, 2005 and by Mabbott and Turner, 2005). One site of PrP^{Sc} deposition in the spleen and lymph nodes is in follicular dendritic cells (FDCs), and PrP^{C} expression on FDCs is required for prion infection of these cells (reviewed by Aguzzi and Heikenwalder, 2005 and by Mabbott and Turner, 2005). FDCs are located in the germinal center of secondary lymphoid tissues and play a role in the presentation of antigen to B cells. B cells are important in FDC maturation and they secrete lymphotoxins (LT) that bind to LT receptors on FDCs. In the absence of B cells or LT, FDCs do not mature and are not susceptible to prion infection.

Several lines of evidence indicate that FDCs are important in prion agent replication in secondary lymphoid tissue (reviewed by Aguzzi and Heikenwalder, 2005 and by Mabbott and Turner, 2005). Temporary de-differentiation of FDCs by injection of antibodies to LT receptors causes a loss of mature FDCs, a disorganization of the cytoarchitecture of the germinal center, and a delay or complete block in peripheral scrapie replication and neuroinvasion. A permanent loss of mature FDCs and disorganization of lymphoid structure is also found in null mice that lack LT or LT receptor genes. When these mice are peripherally challenged with the prion agent, they also demonstrate a greatly reduced capacity for agent replication and, in some cases, a complete block in peripheral prion pathogenesis. These studies indicate that FDCs are important sites of prion agent replication in lymphoid tissues and provide a critical step in the early pathogenesis of prion diseases.

29.7.2. Trafficking of the Prion Agent to Secondary Lymphoid Tissues

The complement system plays a role in the transport of the prion agent from peripheral sites of entry to FDCs in secondary lymphoid tissue (reviewed by Mabbott, 2004). Mice that lack the C3 complement receptor have a delay in prion replication in the spleen and in the onset of neurological disease. Complement component C3 plays a pivotal role in activation of both the classical and alternative complement pathways. Mice that lack the C1qa component or factor B and C2 complement components are deficient in the classical and alternative complement pathways, respectively, and these mice have a delay in prion agent replication in the spleen. As a result, they survive longer than wild type mice following intraperitoneal prion infection. If low doses of the prion agent are inoculated into the periphery of these complement-deficient mice, they do not develop prion disease while wild type mice remain susceptible to prion disease following peripheral infection. Since mice that lack the C1qa component are the most resistant to peripheral prion challenge, this complement factor may directly bind to PrP^{Sc} and activate the classical complement pathway. FDCs have complement receptors on their cell surface and targeting of the prion agent to FDCs may involve binding of PrP^{Sc}-complement complexes to these receptors.

29.7.3. Trafficking of the Prion Agent to the Central Nervous System

The transfer of the prion agent from FDCs to peripheral nerves in the secondary lymphoid tissues is necessary for neuroinvasion. The FDCs are physically separated from the sympathetic nerve fibers but in chemokine receptor CXCR5 null mice there is a reorganization of the germinal center and the sympathetic nerves are now in close proximity to FDCs. Peripheral prion infection in these mice resulted in faster neuroinvasion into the spinal cord indicating that there is accelerated transfer of prion agent from the FDCs to the peripheral nervous system (Prinz et al., 2003). In transgenic mice that overexpress nerve growth factor, there is hyperinnervation of the sympathetic nervous system; these mice have a faster onset of prion agent replication in the spinal cord after peripheral challenge even though the onset of replication in the spleen is similar to wild type mice (Glatzel et al., 2001). Inhibition of sympathetic innervation in newborn mice with antibodies to nerve growth factor or chemical sympathectomy causes a delay in prion agent replication in the CNS (Glatzel et al., 2001). The transfer of the prion agent from FDCs to sympathetic nerves is crucial for prion neuroinvasion into the CNS.

29.7.4. Role of Inflammation in the Peripheral Distribution and Transmission of the Prion Agent

In a healthy host peripheral prion infection primarily involves infection of secondary lymphoid tissues such as the lymph nodes and spleen. In mice with chronic inflammation in a peripheral organ that normally does not contain secondary lymphoid tissue, prion agent replication can also be found. In rodent models of nephritis, pancreatitis and hepatitis in which there is upregulation of lymphotoxins in the kidney, pancreas or liver, respectively, lymphoid tissue containing FDCs can develop in the inflamed organ. Following peripheral inoculation, prion infection can be established in these organs since transient secondary lymphoid tissues are present (Heikenwalder et al., 2005). In one model of prion infection in the kidney of mice with glomerulonephritis, prions could also be found in urine (Seeger et al., 2005). These findings indicate that chronic inflammation can lead to an altered organ tropism of the prion agent and induce excretion of the prion agent from a secretory organ.

In sheep with mastitis, a chronic inflammation of the mammary gland, and concurrent scrapie, prion infection is present in FDCs and macrophages in lymphoid follicles that develop adjacent to mammary ducts (Ligios et al., 2005). Although, there are no accounts of prion shedding from colostrum in scrapie-infected sheep without mastitis, macrophages can be shed into the milk of sheep with mastitis. Secretion of the scrapie-infected macrophages in colostrum of sheep with mastitis could potentially play a role in vertical transmission of scrapie since the route of spread from ewe to lamb has not been determined.

Summary

The human prion diseases are a diverse group of neurodegenerative diseases that are classified by several criteria including disease etiology, *PRNP* polymorphisms at codon 129, PrP^{Sc} types and subtypes, prion incubation period, clinical presentation, neuropathology and disease duration. CJD is unique in that it can have a sporadic, inherited, and acquired etiology, and all three forms can subsequently be transmitted following experimental inoculation of brain tissue. PrP^{Sc} is an unusual pathogen that fails to induce a prion-specific host immune response, but PrP^{Sc}-induced damage to the brain likely causes a proinflammatory response that contributes to neurodegeneration. The human prion diseases are primarily localized to the nervous system, but vCJD also has a peripheral distribution in secondary lymphoid tissues that is consistent with oral exposure to BSE. The prion agent uses the immune system for targeting to lymphoid organs, agent replication, and neuroinvasion via fibers that innervate lymphoid tissues. In chronic inflammatory conditions in which lymphoid organogenesis can occur in non-lymphoid tissues, ectopic prion infection can be established and could play a role in prion transmission.

Review Questions/Problems

1. **What is a prion disease?**
2. **What is the "Protein-only hypothesis"?**
3. **What are the major types of human prion diseases based on clinical phenotypes?**
4. **What are the major PrP^{Sc} types in human prion diseases?**
5. **After peripheral infection, prion neuroinvasion is mediated through what cell type?**
6. **What is the role of B cells in prion neuroinvasion?**

References

Aguzzi A, Heikenwalder M (2005) Prions, cytokines, and chemokines: A meeting in lymphoid organs. Immunity 22:145–154.

Alpers M (1979) Epidemiology and ecology of kuru. In: Slow Transmissible Diseases of the Nervous System. (Prusiner SB, Hadlow WJ, eds), pp 67–92. New York: Academic Press.

Beekes M, McBride PA, Baldauf E (1998) Cerebral targeting indicates vagal spread of infection in hamsters fed with scrapie. J Gen Virol 79:601–607.

Burwinkel M, Riemer C, Schwarz A, Schultz J, Neidhold S, Bamme T, Baier M (2004) Role of cytokines and chemokines in prion infections of the central nervous system. Int J Dev Neurosci 22:497–505.

Cooper JD, Bird SM (2003) Predicting incidence of variant Creutzfeldt-Jakob disease from UK dietary exposure to bovine spongiform encephalopathy for the 1940 to 1969 and post-1969 birth cohorts. Int J Epidemiol 32:784–791.

Eklund CM, Hadlow WJ (1969) Pathogenesis of slow viral diseases. J Am Vet Med Assoc 155:2094–2099.

Gajdusek DC, Gibbs CJJ, Alpers M (1966) Experimental transmission of a kuru-like syndrome in chimpanzees. Nature 209:794–796.

Gambetti P, Parchi P, Petersen RB, Chen SG, Lugaresi P (1995) Fatal familial insomnia and familial Creutzfeldt-Jakob disease: Clinical, pathological and molecular features. Brain Pathol 5:43–51.

Gambetti P, Kong Q, Zou W, Parchi P, Chen SG (2003) Sporadic and familial CJD: Classification and characterization. Br Med Bull 66:213–239.

Ghetti B, Dlouhy SR, Giaccone G, Bugiani O, Frangione B, Farlow MR, Tagliavini F (1995) Gerstmann-Straussler-Scheinker disease and the Indiana kindred. Brain Pathol 5:61–75.

Glatzel M, Heppner FL, Albers KM, Aguzzi A (2001) Sympathetic innervation of lymphoreticular organs is rate limiting for prion neuroinvasion. Neuron 31:25–34.

Glatzel M, Abela E, Maissen M, Aguzzi A (2003) Extraneural pathologic prion protein in sporadic Creutzfeldt-Jakob disease. N Engl J Med 349:1812–1820.

Hadlow WJ (1959) Scrapie and kuru. Lancet 2:289–290.

Head MW, Ritchie D, Smith N, McLoughlin V, Nailon W, Samad S, Masson S, Bishop M, McCardle L, Ironside JW (2004) Peripheral tissue involvement in sporadic, iatrogenic, and variant Creutzfeldt-Jakob disease: An immunohistochemical, quantitative, and biochemical study. Am J Pathol 164:143–153.

Heggebo R, Press CM, Gunnes G, Lie KI, Tranulis MA, Ulvund M, Groschup MH, Landsverk T (2000) Distribution of prion protein in the ileal Peyer's patch of scrapie- free lambs and lambs naturally and experimentally exposed to the scrapie agent. J Gen Virol 81(Pt 9):2327–2337.

Heikenwalder M, Zeller N, Seeger H, Prinz M, Klohn PC, Schwarz P, Ruddle NH, Weissmann C, Aguzzi A (2005) Chronic lymphocytic inflammation specifies the organ tropism of prions. Science 307:1107–1110.

Hilton DA (2006) Pathogenesis and prevalence of variant Creutzfeldt-Jakob disease. J Pathol 208:134–141.

Hilton DA, Ghani AC, Conyers L, Edwards P, McCardle L, Ritchie D, Penney M, Hegazy D, Ironside JW (2004) Prevalence of lymphoreticular prion protein accumulation in UK tissue samples. J Pathol 203:733–739.

Hsiao KK, Scott M, Foster D, Groth DF, DeArmond SJ, Prusiner SB (1990) Spontaneous neurodegeneration in transgenic mice with mutant prion protein. Science 250:1587–1590.

Ironside JW (2006) Variant Creutzfeldt-Jakob disease: Risk of transmission by blood transfusion and blood therapies. Haemophilia 12(Suppl 1):8–15.

Ironside JW, Ritchie DL, Head MW (2005) Phenotypic variability in human prion diseases. Neuropathol Appl Neurobiol 31:565–579.

Ishida C, Okino S, Kitamoto T, Yamada M (2005) Involvement of the peripheral nervous system in human prion diseases including dural graft associated Creutzfeldt-Jakob disease. J Neurol Neurosurg Psychiatry 76:325–329.

Jarrett JT, Lansbury PT, Jr (1993) Seeding "one-dimensional crystallization" of amyloid: A pathogenic mechanism in Alzheimer's disease and scrapie? Cell 73:1055–1058.

Kimberlin RH, Walker CA (1989) The role of the spleen in the neuroinvasion of scrapie in mice. Virus Res 12:201–211.

Kong Q, Surewicz WK, Petersen RB, Zou W, Chen SG, Gambetti P, Parchi S, Capellari L, Goldfarb P, Montagna E, Lugaresi P, Piccardo P, Ghetti B (2003) Inherited prion diseases. In: Prion Biology and Diseases. (Prusiner SB, ed), 2nd edition, pp 673–776. New York: Cold Spring Harbor Laboratory Press.

Kovacs GG, Gasque P, Strobel T, Lindeck-Pozza E, Strohschneider M, Ironside JW, Budka H, Guentchev M (2004) Complement activation in human prion disease. Neurobiol Dis 15:21–28.

Lasmezas CI (2003) Putative functions of PrP(C). Br Med Bull 66:61–70.

Legname G, Baskakov IV, Nguyen HO, Riesner D, Cohen FE, DeArmond SJ, Prusiner SB (2004) Synthetic mammalian prions. Science 305:673–676.

Ligios C, Sigurdson CJ, Santucciu C, Carcassola G, Manco G, Basagni M, Maestrale C, Cancedda MG, Madau L, Aguzzi A (2005) PrPSc in mammary glands of sheep affected by scrapie and mastitis. Nat Med 11:1137–1138.

Mabbott NA (2004) The complement system in prion diseases. Curr Opin Immunol 16:587–593.

Mabbott NA, Turner M (2005) Prions and the blood and immune systems. Haematologica 90:542–548.

Makrinou E, Collinge J, Antoniou M (2002) Genomic characterization of the human prion protein (PrP) gene locus. Mamm Genome 13:696–703.

McBride PA, Schulz-Schaeffer WJ, Donaldson M, Bruce ME, Diringer H, Kretzschmar HA, Beekes M (2001) Early spread of scrapie from the gastrointestinal tract to the central nervous system involves autonomic fibers of the splanchnic and vagus nerves. J Virol 75:9320–9327.

Mikol J (1999) Neuropathology of prion diseases. Biomed Pharmacother 53:19–26.

Montagna P, Gambetti P, Cortelli P, Lugaresi E (2003) Familial and sporadic fatal insomnia. Lancet Neurol 2:167–176.

Parchi P, Giese A, Capellari S, Brown P, Schulz-Schaeffer W, Windl O, Zerr I, Budka H, Kopp N, Piccardo P, Poser S, Rojiani A, Streichemberger N, Julien J, Vital C, Ghetti B, Gambetti P, Kretzschmar H (1999) Classification of sporadic Creutzfeldt-Jakob disease based on molecular and phenotypic analysis of 300 subjects [In Process Citation]. Ann Neurol 46:224–233.

Prinz M, Heikenwalder M, Junt T, Schwarz P, Glatzel M, Heppner FL, Fu YX, Lipp M, Aguzzi A (2003) Positioning of follicular dendritic cells within the spleen controls prion neuroinvasion. Nature 425:957–962.

Prusiner SB (1982) Novel proteinaceous infectious particles cause scrapie. Science 216:136–144.

Prusiner SB (1991) Molecular biology of prion diseases. Science 252:1515–1522.

Safar JG, Kellings K, Serban A, Groth D, Cleaver JE, Prusiner SB, Riesner D (2005) Search for a prion-specific nucleic acid. J Virol 79:10796–10806.

Sailer A, Bueler H, Fischer M, Aguzzi A, Weissmann C (1994) No propagation of prions in mice devoid of PrP. Cell 77:967–968.

Seeger H, Heikenwalder M, Zeller N, Kranich J, Schwarz P, Gaspert A, Seifert B, Miele G, Aguzzi A (2005) Coincident scrapie infection and nephritis lead to urinary prion excretion. Science 310:324–326.

Steele AD, Emsley JG, Ozdinler PH, Lindquist S, Macklis JD (2006) Prion protein (PrPc) positively regulates neural precursor proliferation during developmental and adult mammalian neurogenesis. Proc Natl Acad Sci USA 103:3416–3421.

Thackray AM, McKenzie AN, Klein MA, Lauder A, Bujdoso R (2004) Accelerated prion disease in the absence of interleukin-10. J Virol 78:13697–13707.

Tobler I, Gaus SE, Deboer T, Achermann P, Fischer M, Rulicke T, Moser M, Oesch B, McBride PA, Manson JC (1996) Altered circadian activity rhythms and sleep in mice devoid of prion protein. Nature 380:639–642.

van Keulen LJ, Schreuder BE, Vromans ME, Langeveld JP, Smits MA (1999) Scrapie-associated prion protein in the gastrointestinal tract of sheep with natural scrapie. J Comp Pathol 121:55–63.

van Keulen LJ, Schreuder BE, Vromans ME, Langeveld JP, Smits MA (2000) Pathogenesis of natural scrapie in sheep. Arch Virol Suppl 16:57–71.

Will RG (2003) Acquired prion disease: Iatrogenic CJD, variant CJD, kuru. Br Med Bull 66:255–265.

Will RG, Ironside JW, Zeidler M, Cousens SN, Estibeiro K, Alperovitch A, Poser S, Pocchiari M, Hofman A, Smith PG (1996) A new variant of Creutzfeldt-Jakob disease in the UK. Lancet 347:921–925.

Wong BS, Chen SG, Colucci M, Xie Z, Pan T, Liu T, Li R, Gambetti P, Sy MS, Brown DR (2001) Aberrant metal binding by prion protein in human prion disease. J Neurochem 78:1400–1408.

Yull HM, Ritchie DL, Langeveld JP, van Zijderveld FG, Bruce ME, Ironside JW, Head MW (2006) Detection of type 1 prion protein in variant Creutzfeldt-Jakob disease. Am J Pathol 168:151–157.

Zeidler M, Sellar RJ, Collie DA, Knight R, Stewart G, Macleod MA, Ironside JW, Cousens S, Colchester AC, Hadley DM, Will RG (2000) The pulvinar sign on magnetic resonance imaging in variant Creutzfeldt-Jakob disease. Lancet 355:1412–1418. Zhang CC, Steele AD, Lindquist S, Lodish HF (2006) Prion protein is expressed on long-term repopulating hematopoietic stem cells and is important for their self-renewal. Proc Natl Acad Sci USA 103:2184–2189.

30
Glaucoma

Deven Tuli and Carl B. Camras

Keywords Glaucoma, Intraocular pressure, Optic nerve, Retinal ganglion cell

30.1. Introduction

The word glaucoma came from ancient Greek (Hippocrates, approximately 400 BC), meaning clouded or blue-green hue, most likely describing a patient having a cloudy cornea or advanced cataract precipitated by chronic elevated pressure. Over the last 100 years, further understanding of the concept of glaucoma has accelerated.

Currently, glaucoma is defined as a disturbance of the structural or functional integrity of the optic nerve that causes characteristic changes in the optic nerve, which may also lead to specific visual field defects over time. This disturbance usually can be diminished by adequate lowering of intraocular pressure (IOP). Some subsets of patients can exhibit the characteristic optic nerve damage and visual field defects while having an IOP within the normal range (normal tension glaucoma).

Primary open-angle glaucoma (POAG) is a multifactorial chronic optic neuropathy with a characteristic acquired loss of optic nerve fibers. Cupping and atrophy of the optic disc occur in the absence of other known causes. Such damage develops in the presence of open anterior chamber angles. It produces characteristic visual field abnormalities. IOP is too high for the continued health of the eye.

Individuals who maintain elevated IOP in the absence of nerve damage or visual field loss are considered at risk for developing glaucoma and have been termed ocular hypertensives. POAG is a major worldwide health concern because of its usually asymptomatic but progressive nature, and because it is the leading cause of irreversible blindness in the world. With appropriate screening and treatment, glaucoma usually can be identified and treated to prevent significant visual loss.

30.2. Epidemiology

30.2.1. Prevalence

In the United States (US), multiple population studies [e.g., Framingham (Leibowitz et al., 1980), Baltimore (Tielsch et al., 1991), Barbados (Leske et al., 1997)] have been performed to estimate the prevalence of eye disease, including POAG and ocular hypertension (OHT).

Estimates of the prevalence of glaucoma (Friedman et al., 2004; Table 30.1) and blindness (Congdon et al., 2004; Figure 30.1) in the US demonstrate the following: glaucoma is a leading cause of irreversible blindness, second only to macular degeneration; only one half of the people who have glaucoma may be aware that they have the disease; and more than 2.25 million Americans aged 40 years and older have POAG.

More than 1.6 million have significant visual impairment from glaucoma, with 150,000 bilaterally blind in the US alone (Tielsch et al., 1991). These statistics emphasize the need to identify and closely monitor those at risk for glaucomatous damage.

Approximately 6.7 million people are bilaterally blind from POAG worldwide, and more than 2 million people will develop POAG each year (Quigley, 1996).

30.2.2. Prognosis

Several studies (Gordon et al., 2002) have shown that the incidence of new onset of glaucomatous damage in previously unaffected patients to be approximately 2.6 to 3% in patients with IOPs 21 to 25 mm Hg, 12 to 26% with IOPs 26 to 30 mm Hg, and approximately 42% with IOP higher than 30 mm Hg. The Ocular Hypertension Treatment Study (OHTS) found that approximately 10% of patients with IOPs ranging from 24 to 31 mm Hg but with no clinical signs of glaucoma at baseline develop glaucoma over 5 years, with conversion to glaucoma reduced by 50% if patients are preemptively started on IOP-

T. Ikezu and H.E. Gendelman (eds.), *Neuroimmune Pharmacology*.
© Springer 2008

TABLE 30.1. Prevalence of glaucoma* by age, gender, and race.

	Prevalence/100 Population (95% CI)		
Age, y	White subjects	Black subjects	Hispanic subjects
Women			
40–49	0.83 (0.65–1.06)	16.1 (0.94–2.41)	0.34 (0.15–0.72)
50–54	0.89 (0.78–1.02)	2.24 (1.59–3.14)	0.65 (0.37–1.15)
55–59	1.02 (0.89–1.16)	2.86 (2.16–3.78)	0.98 (0.61–1.58)
60–64	1.23 (1.07–1.41)	3.65 (2.83–4.69)	1.49 (0.97–2.28)
65–69	1.68 (1.37–1.82)	4.64 (3.54–6.05)	2.24 (1.43–3.49)
70–74	2.16 (1.87–2.49)	5.89 (4.28–8.05)	3.36 (2.00–5.60)
75–79	3.12 (2.68–3.63)	7.45 (5.06–10.84)	5.01 (2.68–9.15)
>80	6.94 (5.40–8.88)	9.82 (6.08–16.48)	10.05 (4.35–21.52)
Men			
40–49	0.36 (0.27–0.47)	0.55 (0.31–0.95)	0.39 (0.18–0.85)
50–54	0.61 (0.50–0.74)	1.71 (1.25–2.32)	0.69 (0.39–1.25)
55–59	0.85 (0.72–1.00)	3.06 (2.30–4.04)	1.00 (0.61–1.64)
60–64	1.18 (1.02–1.37)	4.94 (3.69–6.59)	1.44 (0.92–2.24)
65–69	1.64 (1.40–1.91)	7.24 (5.40–9.63)	2.07 (1.32–3.23)
70–74	2.27 (1.90–2.72)	9.62 (7.29–12.59)	2.97 (1.79–4.89)
75–79	3.14 (2.53–3.90)	11.65 (8.81–15.25)	4.23 (2.32–7.60)
>80	5.58 (4.15–7.47)	13.21 (7.85–21.38)	7.91 (3.53–16.77)

Abbreviation: CI, confidence interval.
*Glaucoma indicates primary open-angie glaucoma.
Reproduced with permission from: Friedman DS, Wolfs RC, O'Colmain BJ, Klein BE, Taylor HR, West S, Leske MC, Mitchell P, Congdon N, Kempen J, Eye Diseases Prevalence Research Group (2004) Prevalence of open-angle glaucoma among adults in the United States. Arch Ophthalmol 122:532–538.

lowering therapy (Kass et al., 2002). Significant subsets of higher and lower risk exist when pachymetry (central corneal thickness [CCT]) is taken into account.

Some patients' first sign of morbidity from elevated IOP can be presentation with sudden loss of vision due to a central retinal vein occlusion (CRVO). Elevated IOP is the second most common risk factor for CRVO behind systemic hypertension.

30.2.3. Risk Factors

30.2.3.1. Intraocular Pressure

In eyes suspected of being affected with glaucoma, loss of visual field develops at a rate of approximately 1 to 3 percent per year of observation (Gordon et al., 2002). High IOP is the most important risk factor (Table 30.2). The relative importance of other risk factors also is indicated in Table 30.2. In population-based surveys, between 25 and 50 % of those with glaucomatous damage to the optic nerve have IOP in the normal range. However, the probability of injury increases exponentially with higher IOP (Heijl et al., 2002). Large diurnal fluctuation in IOP is another independent risk factor for glaucoma (Nouri-Mahdavi et al., 2004).

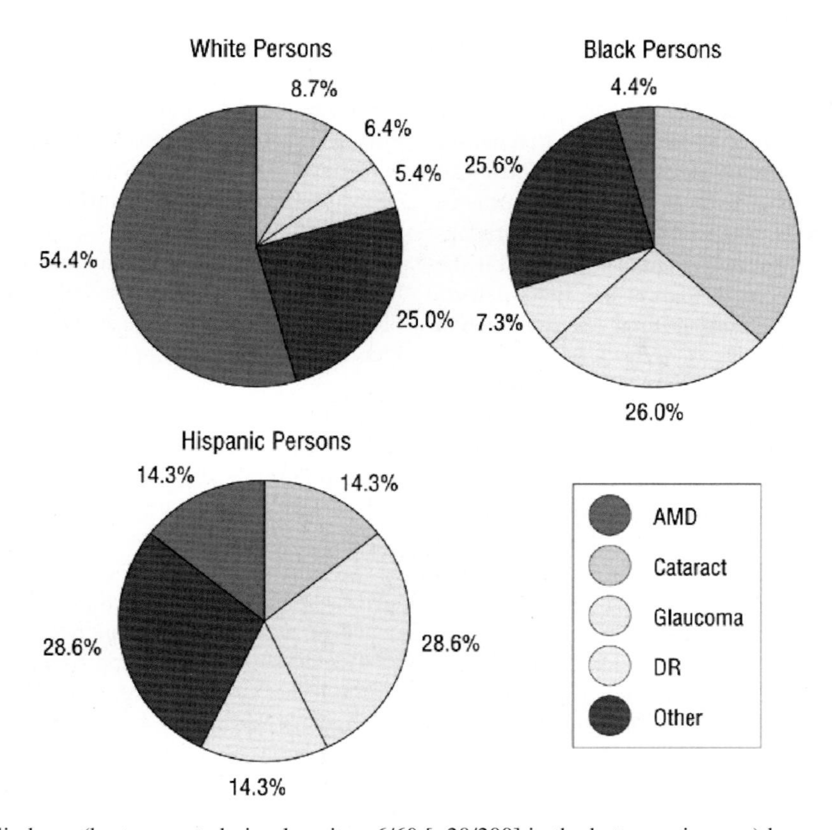

FIGURE 30.1. Causes of blindness (best-corrected visual acuity <6/60 [<20/200] in the better-seeing eye) by race/ethnicity. AMD indicates age-related macular degeneration; DR, diabetic retinopathy. Reproduced with permission from: Congdon N, O'Colmain B, Klaver CC, Klein R, Munoz B, Friedman DS, Kempen J, Taylor HR, Mitchell P, Eye Diseases Prevalence Research Group (2004) Causes and prevalence of visual impairment among adults in the United States. Arch Ophthalmol 122:477–485.

TABLE 30.2. Risk factors for glaucoma.

Established	Highly probable	Possible
Elevated IOP	Race (blacks*)	Presence (or absence) of diabetes
Thin CCT	Positive family history	Hypertension
Increasing age	Low diastolic perfusion pressure	
Large cup to disc ratio	Migraines	
	IOP fluctuations	
	Sleep apnea	
	Immunological disorders	
	Myopia	

IOP = Intraocular Pressure; CCT = Central Corneal Thickness; *In US, West Africa, Caribbean.

30.2.3.2. *Race*

Prevalence of POAG is 3–4 times higher in blacks than in whites in the US (Tielsch et al., 1991; Friedman et al., 2004). Glaucoma is the most common cause of irreversible blindness in the US among people of African or Hispanic descent (Congdon et al., 2004; Figure 30.1). Blacks in the US are more likely to develop glaucoma early in life, and they tend to have a more aggressive form of the disease.

The Barbados Eye Study (Leske et al., 1997) over 4 years showed a 5 times higher incidence of developing glaucoma in a group of black ocular hypertensives as compared with a predominantly white population. Some population studies have found the mean IOP in blacks to be higher than Caucasian controls. Other studies (e.g., Tielsch et al., 1991) found no difference. Consequently, further studies are required to clarify this issue.

The OHTS (Brandt et al., 2001) demonstrated that black patients in the US overall may have reduced CCT, thereby leading to underestimation of true IOP. The reduced CCT was associated with a higher risk of developing glaucoma. Therefore, pachymetry measurement is particularly important in African-American patients who are glaucoma suspects. In blacks, the OHTS also demonstrated a larger cup to disc ratio, which was an independent risk factor for the development of glaucoma (Feuer et al., 2002; Gordon et al., 2002; Table 30.2). These two confounding variables discovered on the multivariate analysis appeared to account for the apparent increased risk in blacks demonstrated in the univariate analysis (Gordon et al., 2002).

30.2.3.3. *Age*

Age older than 40 years is a risk factor for the development of POAG. Up to 15% of African-American men are affected by the ninth decade of life (Friedman et al., 2004; Table 30.1). Consequently, glaucoma is found to be more prevalent in the aging population, even after compensating for the mean rise in IOP with increasing age in the US. However, the disease itself is not limited to only middle-aged and elderly individuals.

30.2.3.4. *Central Corneal Thickness*

CCT is an important risk factor for the development of glaucoma (Gordon et al., 2002; Figure 30.2). CCT likely influences the measurement of IOP with many tonometers, including applanation techniques (Brandt et al., 2001). Increased CCT beyond the mean of 545 μm causes overestimation of IOP; lower CCT translates into underestimation of the IOP. A thin cornea (for example, 480 μm) may occur with glaucomatous visual field loss despite normal applanation IOP because the measurements are fallaciously low. Conversely, a thick cornea (for example, 620 μm) might occur in an eye with high IOP, normal visual fields and a normal optic nerve because it results in false overestimation of true IOP. Ehlers et al. (1975) extrapolated that applanation tonometry is overestimated or underestimated by approximately 5 mm Hg for every 70 μm difference in measured CCT from mean thickness. It is also possible that central corneal thickness may itself constitute an intrinsic risk (or protective) factor for glaucomatous optic nerve damage independent of its ability to affect the IOP measurement.

30.2.3.5. *Family History*

Relatives of those with glaucoma are more likely to have the disease (Drance et al., 1981). However, there is no clear mendelian pattern of inheritance for glaucoma. The Rotterdam Eye Study (Wolfs et al., 2000) concluded that the prevalence of glaucoma was 10.4% in siblings of patients with glaucoma and the relative risk of having POAG increased 9.2-fold for individuals with a relative with POAG. High IOP may be the inherited feature of glaucoma, or inherited risk factors independent of IOP may be involved.

30.2.3.6. *Other Factors*

Some reports have suggested an association between high blood pressure and glaucoma, but this relation is not fully understood and is not uniformly observed. There appears to be a subset of patients with low diastolic perfusion pressures who are at higher risk for POAG (Wilson et al., 1987). Some studies indicate that patients with diabetes have higher IOPs and a higher rate of glaucoma than people without diabetes. However, other studies fail to demonstrate this relationship. In a study in which the presence of diabetes mellitus was self-reported, diabetes protected against the development of glaucoma (Gordon et al., 2002). Other abnormalities in vascular function, including migraine headache (Cursiefen et al., 2000) and vasospasm in the extremities (Broadway and Drance, 1998), appear to be associated with glaucoma as demonstrated in some studies.

Sleep apnea has been shown to be a risk factor for glaucoma in some studies (Pearson, 2000). Myopia may be another risk factor for glaucoma (Daubs and Crick, 1981). The relatively thin eye wall and large globe in severely nearsighted people suggest a possible susceptibility to stretching under the influence of IOP. The association between factors such as concurrent cardiovascular disease (Tielsch, 1991) has not been demonstrated consistently.

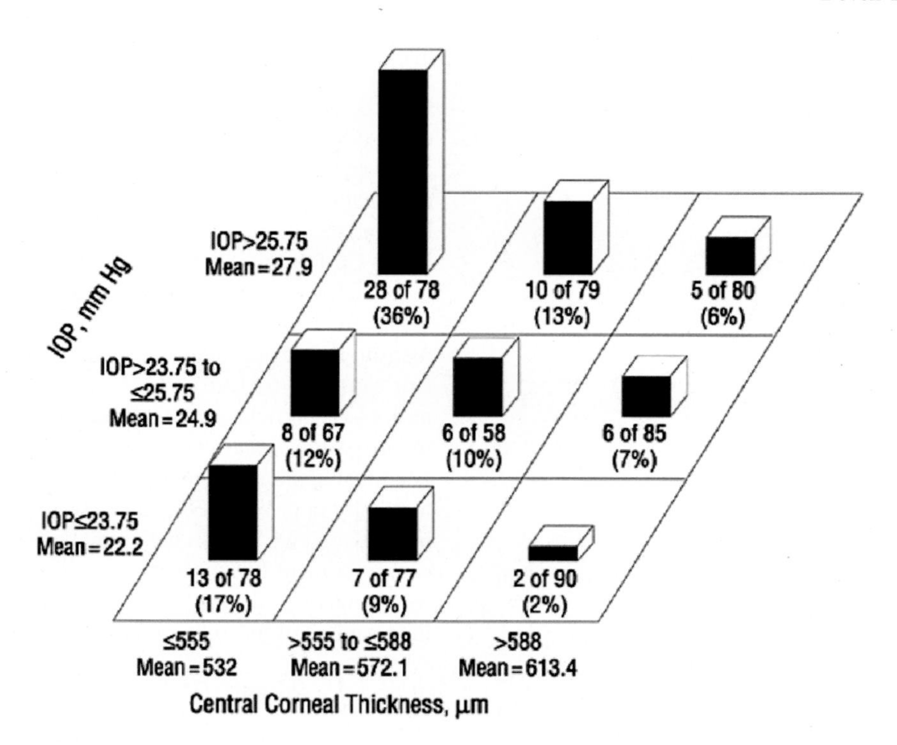

FIGURE 30.2. Risk for developing primary open glaucoma varies with central corneal thickness. The numbers and percentage of participants in the observation group who developed primary open-angle glaucoma (median follow-up, 72 months) are indicated below each bar. Participants are grouped by baseline intraocular pressure of ≤23.75 mm Hg, >23.75 mm Hg to ≤25.75 mm Hg, and >25.75 mm Hg and by central corneal thickness measurements of ≤555 μm, >555 μm to ≤588 μm, and >588 μm. These percentages are not adjusted for length of follow-up. Reproduced with permission from: Gordon MO, Beiser JA, Brandt JD, Heuer DK, Higginbotham EJ, Johnson CA, Keltner JL, Miller JP, Parrish RK 2nd, Wilson MR, Kass MA (2002) The Ocular Hypertension Treatment Study: baseline factors that predict the onset of primary open-angle glaucoma. Arch Ophthalmol 120: 714–20.

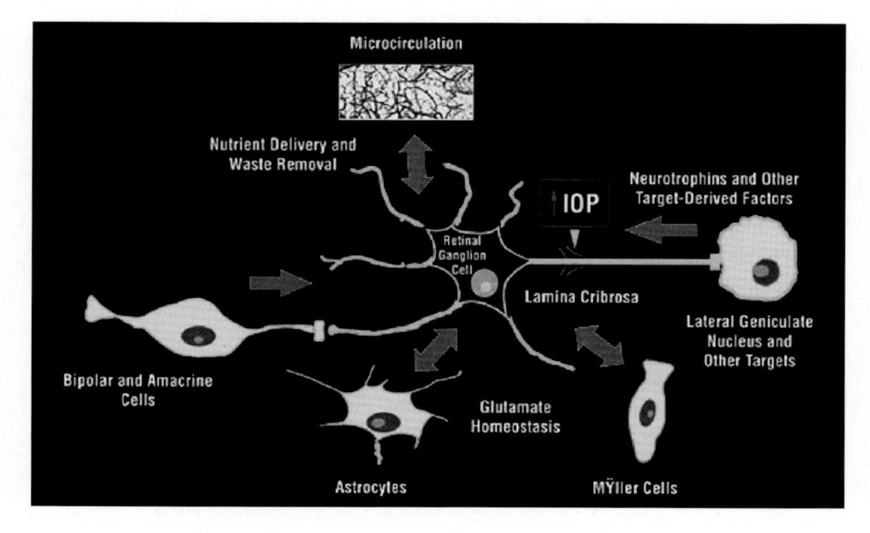

FIGURE 30.3. Several mechanisms have been hypothesized to have a role in causing retinal ganglion cell death in glaucoma. IOP indicate intraocular pressure. Reproduced with permission from: Weinreb and Levin (1999).

30.3. Pathophysiology

In general, the cause of glaucomatous optic neuropathy is unknown. The disease affects the individual axons of the optic nerve, which may die by apoptosis, also known as programmed cell death. Multiple theories exist concerning how IOP can be one of the factors that initiates glaucomatous damage in a patient (Figure 30.3). Two of the fundamental theories include the mechanical and vascular theory.

30.3.1. Mechanical Theory

The mechanical compression theory (Quigley et al., 1981) suggests that elevated IOP causes a backward bowing of the lamina cribrosa, resulting in kinking of the axons as they exit through the lamina pores. This may deprive the axons of neurotrophins or interfere with axoplasmic flow, thereby triggering cell death. Neurotrophins increase RGC survival and may be produced by the RGCs themselves (Perez and Caminos, 1995). Damaged RGC axons are affected by neurotrophin deprivation. With the loss of neurotrophic support of the RGCs, slow death is inevitable. Brain derived neurotrophic factor (BDNF) is one such neurotrophin that can temporarily increase the survival of the RGCs by inhibiting the excitotoxicity-related cell death. Deficiency of these protective factors may contribute to glaucomatous optic neuropathy.

30.3.2. Vascular Theory

The vascular theory (Hayreh et al., 1970) proposes that cell death is triggered by ischemia, whether induced by elevated IOP or as a primary insult. Studies done to evaluate the circulation of the optic nerve using the laser Doppler flowmetry have shown diminished blood flow in the optic nerves of eyes with POAG (Grunwald et al., 1998).

30.3.3. Contributory Mechanisms

Complementary to vascular compromise and mechanically impaired axoplasmic flow, additional pathogenic mechanisms (Figures 30.3 & 30.4) that underlie glaucomatous optic neuropathy include excitotoxic damage from excessive retinal glutamate, peroxynitrite toxicity from increased nitric oxide synthase activity, immune-mediated nerve damage, and oxidative stress (Naskar and Dreyer, 2001).

Beyond the neuronal degeneration that results from the primary insult or risk factors in glaucoma, there is an expanding cascade of events called *"secondary degeneration"*, during which RGCs are gradually damaged from the unfavorable "milieu" of the neighboring degenerating neurons (Schwartz, 2005).

Many pathophysiological mediators common to all neurodegenerative diseases, including glaucoma, have been identified, such as increase in glutamate, conformationally altered self proteins, increase in inflammation-associated factors (cyclooxygenase-2 [COX-2], tumor necrosis factor [TNF-alpha], nitric oxide [NO]), increase in extracellular matrix proteins and growth-associated inhibitors (myelin-related proteins), oxidative stress, and malfunctioning of local immune cells (microglia). These mediators evoke a response for which the tolerance of the neural tissue is minimal.

The important mechanisms of neuronal compromise or death are discussed briefly.

30.3.3.1. Immune Mechanisms

Optic nerve crush injury models have been studied in mice and rats by Schwartz and others (Schwartz, 2004). Experiments conducted with these models have shown that nerve-derived extracellular risk factors induce intracellular death signals, resulting in neuronal death. Neural injury has long been considered self destructive and progressive primarily due to the autoimmunity to self antigen. The anti-self antigen response has been shown in these experiments to have a coincidental beneficial and modulating effect (Schwartz, 2005). The autoimmune T cells limit the activity of the microglial framework and blood-borne macrophages, affecting their ability to deal with a local excitatory situation. Such a controlling effect has a beneficial impact on both survival and regrowth. In the absence of such control, this phenomenon can become destructive, inducing an autoimmune disease. Boosting the beneficial effect by the use of a safe antigen that can cross-react with self-antigens at the site under stress might provide an effective therapeutic approach that influences both rescue and regeneration (Schwartz, 2005). Proper control of autoimmunity that titrates the local microglia response and the infiltrating macrophages may be a therapy that boosts protection and regeneration. Intervention may be applied at the level of the nerve-derived risk factor, at the interface between the secondary mediator and the intracellular signal, or at the level of the intercellular signal itself. Copolymer 1 (Cop-1), an approved drug for the treatment of multiple sclerosis, can be used as a treatment for autoimmune diseases and as a therapeutic vaccine for neurodegenerative diseases (Kipnis and Schwartz, 2002). It has been proposed that the protective effect of Cop-1 vaccination is obtained through a well-controlled inflammatory reaction, and that the activity of Cop-1 in driving this reaction derives from its ability to serve as a "universal antigen" by weakly activating a wide spectrum of self-reactive T cells (Kipnis and Schwartz, 2002). These results have special implications for glaucoma therapy since glaucoma is the most common cause of optic neuropathy.

30.3.3.2. Apoptosis

Apoptosis is classically defined by an orderly pattern of internucleosomal DNA fragmentation, chromosome clumping, cell shrinkage, and membrane blebbing (Wyllie et al., 1980). This is followed by disassembly of the cell into multiple membrane-enclosed vesicles that are engulfed by neighboring cells without inciting inflammation. Apoptosis has often been labeled as *"programmed or physiological cell death."* Necrosis, on the contrary, is characterized by cellular swelling, disruption of organelle and plasma membranes, random DNA fragmentation, and uncontrolled release of cellular contents into the extracellular space, often resulting in inflammation.

Several gene families have been identified that play either positive or negative roles in determining whether a cell (in glaucoma, the RGC) will undergo apoptosis. Caspases are cysteine proteases that play a central role in both propagating apoptotic signals and carrying out disassembly of the cell (Hutchins and Barger, 1998). The prototypical mammalian caspase is interleukin-1 converting enzyme. Caspases are always present in the cytoplasm, ready to be activated as the mediators of apoptosis (Thornberry and Lazebnik, 1998).

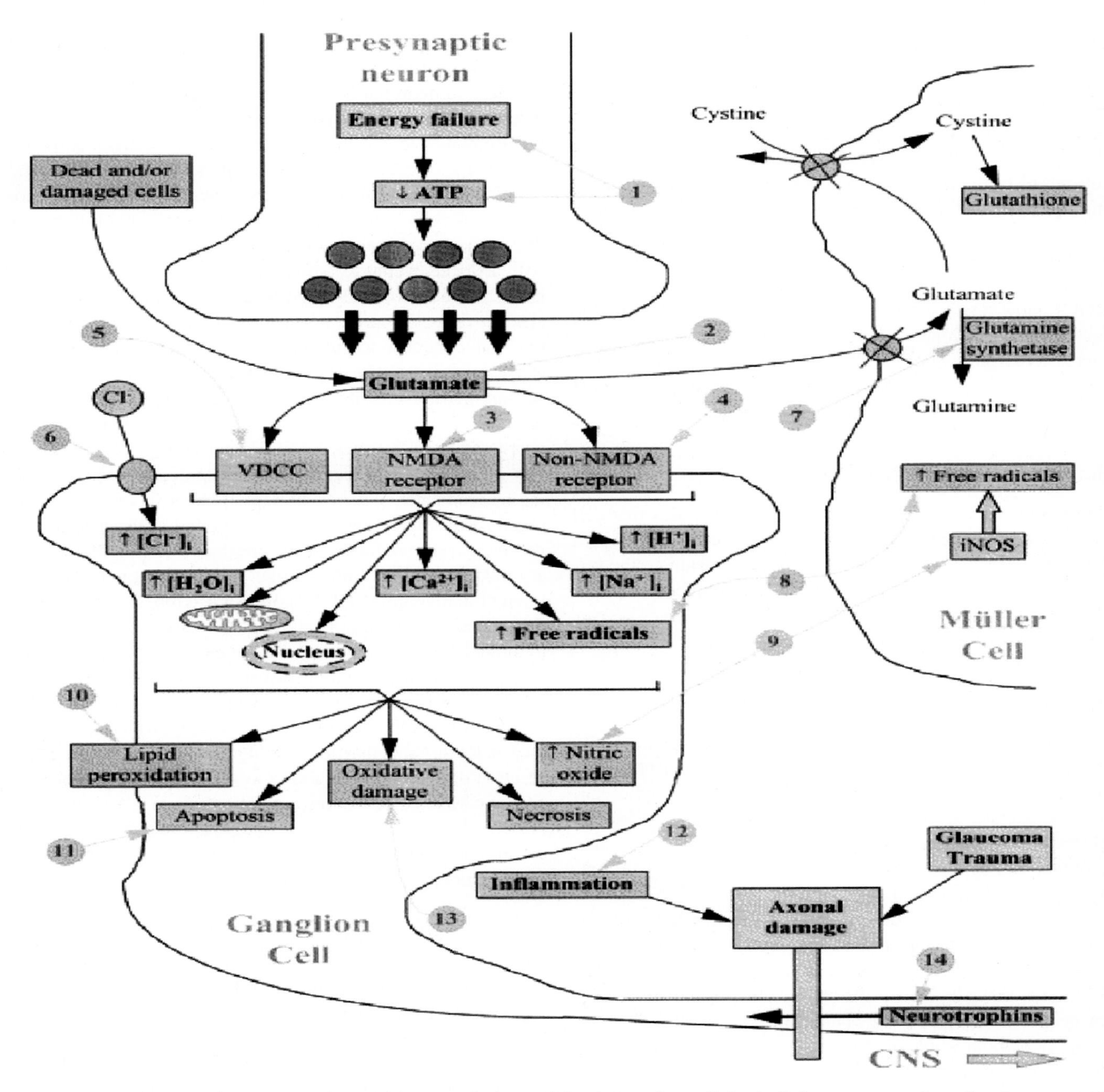

FIGURE 30.4. Diagrammatic summary of potential targets for intervention in ganglion cell death. 1) Prevent energy loss, 2) reduce extracellular glutamate, 3) block ionotropic NMDA receptors, 4) block non-NMDA ionotropic glutamate receptors, 5) block excessive influx of Ca^{2+} through voltage-gated channels, 6) prevent influx of chloride and water, 7) enhance intracellular glutamate catabolism, 8) prevent excessive production of free radicals, 9) prevent detrimental effects of nitric oxide, 10) prevent lipid peroxidation, 11) prevent induction of apoptosis, 12) reduce any inflammation, 13) stop oxidative damage, and 14) maintain neurotrophin support. Reproduced with permission from: Osborne NN, Ugarte M, Chao M, Chidlow G, Bae JH, Wood JP, Nash MS (1999) Neuroprotection in relation to retinal ischemia and relevance to glaucoma. Surv Ophthalmol 43 Suppl 1:S102–S128.

They cleave a wide variety of critical functional and structural cellular proteins, and they also convert other caspases from proenzymes to active forms. Many triggers activate caspases, including increased intracellular calcium, changes in cellular energy balance, and adenosine 3'5' cyclic phosphate.

The main inhibitors of apoptosis are Bcl-2 and related proteins. They have multiple complex functions, such as inhibiting intermediate proteins that activate caspases, gauging intracellular damage and maintaining organelle integrity in ways that are not fully understood. For example, over expression of Bcl-2 inhibits both apoptotic and necrotic cell death by suppressing lipid peroxidation (Adams and Cory, 1998).

30.3.3.3. *Glutamate-Induced Excitotoxicity*

The excitatory amino acid glutamate is known to cause neurotoxicity via binding of glutamate to the *N*-methyl-D-aspartate (NMDA) receptor and the kainate receptors. This can be secondary to many types of cellular damage, including other final common pathways of damage such as oxidative stress or free radical damage (Osborne et al., 1999). Activation of these receptors allows entry of excessive amounts of calcium. Calcium is normally used by cells as a critical intracellular signaling molecule and is involved in many diverse functions, including control of cell-surface ion channels, regulation of enzymes, and even self-regulation of its intracellular level. Abnormally high concentrations of calcium lead to inappropriate activation of complex cascades of proteases, nucleases, and lipases. These directly damage cellular constituents and lead to ganglion cell death.

One neuroprotective strategy, therefore, is to interrupt the resultant excitotoxic cascade by blocking the NMDA receptor. Amongst the several pharmacological drugs tried, MK 801 is one drug that demonstrated neuroprotective properties in pressure-induced and reperfusion–injury animal models of RGC toxicity (Lam et al., 1997). Memantine is a clinically well-tolerated NMDA receptor antagonist (Lipton, 2003). The results of its effectiveness in Alzheimer are encouraging (Reisberg et al., 2003).

30.3.3.4. *Free Radicals*

Free radicals are generated not only through activation of glutamate receptors but also as the inevitable by-product of normal oxidative metabolism. This is especially true in the retina and the RGCs, which have an extremely high metabolic rate. This triggers further changes including calcium release that causes cell damage (as summarized in Sect. 30.3.3.3. above). Endogenous antioxidants such as vitamins E and C, superoxide dismutase (Greenlund et al., 1995), and glutathione normally inactivate free radicals.

30.3.3.5. *Nitric Oxide Synthase*

Nitric Oxide Synthase (NOS) is postulated to have a role in RGC axonal toxicity. Inducible NOS is known to be up regulated in the optic nerve in glaucoma (Liu and Neufeld, 2001). Inhibitors of NOS, such as 3-aminoguanidine, decreased the RGC loss by 70% in rat eyes with raised IOP (Neufeld et al., 1999).

30.3.3.6. *Dopamine Deficiency*

The origin of the pattern electroretinogram (PERG) response is predominantly the RGCs. Parkinson's disease related changes in retinal processing caused by changes in the dopaminergic amacrine, horizontal, and interplexiform cells may alter the receptive field composition of the RGC, thereby altering the PERG response. Delayed PERG latencies in untreated patients with Parkinson's disease were restored with dopaminergic therapy (Jackson and Owsley, 2003). Further support that dopamine deficiency is responsible for changes in the RGC induced PERG is that PERG latencies are delayed in control adults given haloperidol, a dopamine antagonist (Bodis-Wollner and Tzelepi, 1998).

Dopamine has been recognized to have neuroprotective actions in glaucoma. Retinal dopaminergic cells can be detected by the immunohistochemical staining of tyrosine hydroxylase, the rate-limiting enzyme in dopamine synthesis. It has been observed that NMDA induced RGC toxicity dramatically decreased tyrosine hydroxylase immunostaining at the junction between the inner nuclear layer and inner plexiform layer in glaucoma (Kitaoka and Kumai, 2004).

30.3.3.7. *Heat-Shock Proteins*

Heat-shock proteins (Hsp) are known to be an intrinsic protection mechanism associated with an early response against physiologic and environmental stress (Wax and Tezel, 2002). Certain Hsp have special effects upon neurons, including Hsp 72, 27 and 60. Hsp 27 and 60 have been demonstrated to be upregulated in RGCs of glaucoma (Tezel et al., 2000). Recent work has shown that geranylgeranylacetone can induce Hsp 72 in RGCs and protect them from glaucomatous damage in a rat glaucoma model (Ishii et al., 2003).

30.3.3.8. *Activated Microglia*

Yuan and Neufeld (2001) investigated the potential participation of activated microglia, the resident defense cells of the central nervous system in causing optic nerve head (ONH) damage in glaucoma. In normal ONH, microglia do not contain transforming growth factor-beta (TGF-beta), COX-2, or TNF-alpha; however, in glaucomatous ONH, microglia contain abundant TGF-beta, TNF-alpha, and a few microglia are usually positive for COX-2. In glaucomatous ONH, microglia become actively phagocytic and produce cytokines, mediators, and enzymes that can alter the extracellular matrix. Findings from these authors suggest that activated microglia may participate in stabilizing the tissue early in the disease process, but, as the severity of the glaucomatous damage increases, the activated microglia may contribute to glaucomatous optic neuropathy (Yuan and Neufeld, 2001).

30.3.3.9. *Genetic Mutations*

Myocilin (MYOC) is one gene that has been associated with early- and late-onset POAG. The MYOC gene was mapped to the open-angle glaucoma locus (GLC1A) region of chromosome 1q25, and mutations were discovered in patients with adult and juvenile open-angle glaucoma. Mutations in this gene are found in 3 to 5% of adult-onset POAG patients and in 8 to 20% juvenile-onset POAG patients (Wiggs et al., 1998). Different mutations are associated with juvenile and adult glaucoma. In particular, the most common mutation associated with adult glaucoma is a nonsense mutation, whereas all the mutations associated with juvenile glaucoma are missense mutations. Additionally, some of the mutations typically associated with adult glaucoma can cause more severe, earlier onset glaucoma when found in combination with DNA sequence variants in the CYP1B1 gene.

Using single large pedigrees affected by POAG, seven genetic loci have been described (GLC1A-G), and glaucoma-predisposing genes have been identified in three of these loci: GLC1A, myocilin (Wiggs et al., 1998); GLC1E, optineurin (Rezaie et al., 2002); and GLC1G, WDR36 (Monemi et al., 2005). Each of these genes is only responsible for a small fraction of cases of POAG, reflecting the small percentage of POAG that is inherited as a Mendelian trait rather than as a complex trait.

The optineurin protein may function to protect the optic nerve from TNF-alpha–mediated apoptosis, and the loss of function of this protein may decrease the threshold for RGC apoptosis in patients with glaucoma. Surprisingly, mutations in this gene do not appear to contribute significantly to the optic nerve degeneration that is a component of typical high-pressure POAG (Wiggs et al., 2003).

The exact role that IOP plays in combination with these other factors and the latter's significance in the initiation and progression of subsequent glaucomatous neuronal damage and cell death requires further studies.

30.4. Clinical Features

30.4.1. History

The initial patient interview is important in the evaluation for glaucoma. Because of the silent nature of POAG, patients usually will not present with any visual complaints until late in the course of the disease. A history should include the following:

1. Past ocular and medical history including:
   - Eye pain
   - Redness
   - Multicolored halos
   - Headache
   - Previous ocular disease such as cataracts, uveitis, diabetic retinopathy or vascular occlusions
   - Previous ocular laser or surgery
   - Ocular or head trauma
   - Pertinent vasculopathic systemic illnesses, including Raynauds phenomenon or migraine
2. Current medications, including any hypertensive medications (which may affect IOP) or corticosteroids used topically or systemically.
3. Family history of glaucoma.

30.4.2. Physical Exam

Screening the general population for POAG is most effective if targeted at those at high risk, such as African Americans and elderly individuals, especially if the screening includes IOP measurements, assessment of the optic nerve and screening visual fields.

For individuals of age 65 years or above, recommended frequency for a comprehensive adult medical eye evaluation is 6 to 12 months when risk factors for glaucoma are present and 1 to 2 years in the absence of risk factors. For glaucoma suspect individuals under 40 years, the frequency of the eye evaluation should be 2 to 4 years and 5 to 10 years in the presence and absence of risk factors for glaucoma, respectively (American Academy of Ophthalmology: Preferred Practice Patterns, 2005a).

A standard comprehensive eye examination is performed on the initial visit. If any visual field or optic nerve changes consistent with glaucoma are present, additional appropriate testing should be done to establish a diagnosis.

In patients with glaucoma, factors that determine frequency of evaluations include the severity of damage, the stage of the disease, the rate of progression, the extent to which the IOP exceeds the target pressure, and the number and significance of other risk factors for damage to the optic nerve. If the target IOP has been achieved and there is no progression of damage, it is recommended (American Academy of Ophthalmology: Preferred Practice Patterns, 2005b) to follow up with ONH exam and visual field testing at 6 to18-month intervals. On the contrary, if there is progression of damage and the target IOP has not been achieved, more frequent testing is required until stability is achieved.

30.4.2.1. Tonometry

IOP varies from hour-to-hour in any individual. Many studies indicate that the circadian rhythm of IOP peaks in the early hours of the morning. IOP often rises when a patient assumes a supine position. When checking IOP, record measurements for both eyes, the method used (Goldmann applanation is considered the standard), and the time of the measurement.

A difference between contralateral eyes of 3 mm Hg or more indicates greater suspicion of glaucoma. An average of a 10% difference between individual measurements is expected. In most circumstances, the measurements should be repeated on at least 2–3 occasions before deciding on a treatment plan. The measurement should be made at different times of the day to check for diurnal variation. A diurnal variation of more than 5–6 mm Hg indicates an increased risk for glaucoma.

30.4.2.2. Gonioscopy

The angle of the eye is examined by gonioscopy, which requires the use of special lenses. Gonioscopy is performed to rule out angle-closure or secondary causes of IOP elevation, such as angle recession, pigmentary glaucoma, and exfoliation syndrome. The peripheral contour of the iris is examined for plateau iris, and the trabecular meshwork for peripheral anterior synechiae, as well as for neovascular or inflammatory membranes.

Schlemm's canal may be seen if blood refluxes into the canal. This might indicate elevated episcleral venous pressure caused by conditions such as a carotid-cavernous fistula, Graves orbitopathy, or Sturge-Weber syndrome.

Ultrasound biomicroscopy is helpful in assessing the angle, iris, and ciliary body to rule out anatomical pathology and secondary causes of elevated IOP.

30.4.2.3. Optic Disc and Nerve Fiber Layer

The optic disc is examined to estimate the cup-to-disc ratio and to identify excavation typical of glaucoma. Stereoscopic viewing during slit-lamp examination improves the accuracy of disc evaluation, since the disc is a complex, three-dimensional structure (Figure 30.5). Single cutoff values for the cup to disc ratio designed to distinguish between normal subjects and those with glaucoma are imperfect for screening, since the normal ratio varies considerably with optic disc size. When the cup to disc ratio equals or exceeds 0.6, the probability of glaucoma increases dramatically. In screening for glaucoma, however, even combined criteria, including the level of IOP and the cup to disc ratio, identify only two thirds of patients.

The disc should be viewed stereoscopically to assess evidence for glaucomatous damage (Figure 30.5), including the following: cup to disc ratio in horizontal and vertical meridians; color and slope of the cup; appearance of the disc; progressive enlargement of the cup; evidence of nerve fiber layer damage using a red-free filter; notching or thinning of disc rim, particularly at the superior and inferior poles where nerve fibers often can be affected first; pallor; presence of hemorrhage (most common inferotemporally); asymmetry between discs; peripapillary atrophy (possible association with the development of glaucoma); or congenital nerve abnormalities. Stereo fundus photographs are required as a baseline for future comparisons (Figure 30.5).

Other imaging techniques that utilize different physical properties of light for optical analysis can document the status of the optic nerve and the nerve fiber layer. Furthermore, they can be used to detect changes over time (Medeiros et al., 2004). The value of these technologies for diagnosing glaucoma and for determining progression over time is currently under active investigation.

Confocal scanning laser ophthalmoscopy can evaluate the optic disc and peripapillary retina in 3 dimensions and provide quantitative information about the cup, neuroretinal rim, and contour of the nerve fiber layer.

Scanning laser polarimetry measures the change in the polarization state of an incident laser light passing through the naturally birefringent nerve fiber layer to provide indirect estimates of peripapillary nerve fiber layer thickness.

Optical coherence tomography uses reflected light in a manner analogous to the use of sound waves in ultrasonography to create computerized cross-sectional images of the retina and optic disc, and also gives quantitative information about the peripapillary retinal nerve fiber layer thickness.

Fluorescein angiography, ocular blood flow analysis via laser Doppler flowmetry, color vision measurements, contrast sensitivity testing, and electrophysiological tests (e.g., pattern electroretinograms) are used as research tools in the evaluation of POAG patients. Routine clinical use is not advocated at this time.

30.4.2.4. Visual Field Testing

Automated threshold testing (e.g., Humphrey 24–2) is performed to evaluate for glaucomatous visual field defects. If the patient is unable to perform automated testing, Goldmann perimetry may be substituted. New onset glaucomatous defects are found most commonly as an early nasal step, temporal wedge, or paracentral scotoma (more frequent superiorly). Generalized depression also might occur, but is less specific for glaucomatous damage.

Swedish interactive thresholding algorithm (SITA)-based software decreases testing time and boosts reliability, especially in older patients (Schimiti et al., 2002). Short wavelength automated perimetry (SWAP) or blue-yellow perimetry may provide a more sensitive method of detecting visual field deficits, especially in patients with OHT without defects demonstrated by standard testing. If the Humphrey standard automated perimetry results are normal, SWAP should be considered to help detect visual field loss earlier.

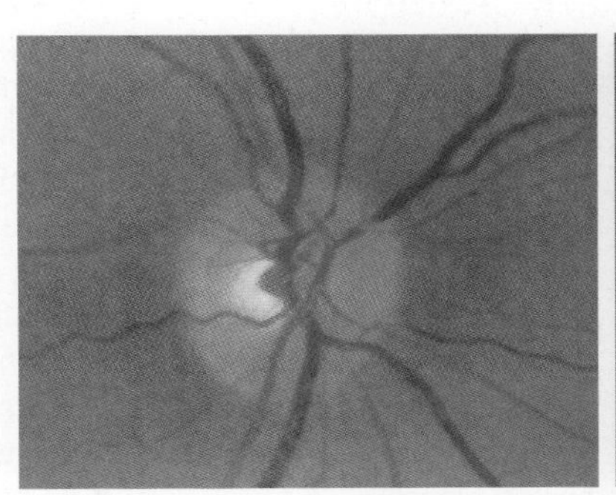
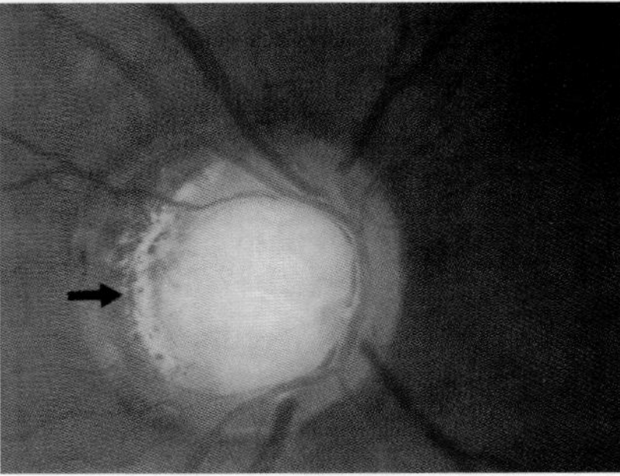

FIGURE 30.5. Stereo photographs representing normal optic nerve head (left) and typical glaucomatous changes (right).

Recent studies suggest SWAP may detect visual loss or progression up to 3 to 5 years earlier than conventional perimetry, as well as in 12 to 42% of patients previously diagnosed with only OHT (Racette and Sample, 2003). Because the testing time may be lengthened, it may be tiring for some patients. However, new SITA-SWAP algorithm software may speed up the testing time and thus improve reliability (Bengtsson and Heijl, 2003).

The pupil size should be documented at each testing session since pupillary constriction can reduce retinal sensitivity and mimic progressive visual field loss.

To establish a reliable baseline, visual fields may have to be repeated several times on successive visits, especially if initial testing shows low reliability indices. Newer glaucoma progression analysis software can help identify reliable perimetric baselines, and probability-based analyses of subsequent fields can assist in distinguishing true progression over time from fluctuation or artifact. In follow-up, if a low risk for onset of glaucomatous damage is present, then repeat testing may be performed once a year. If a high risk of impending glaucomatous damage is present, then testing frequency is adjusted accordingly (as frequent as every few months).

Frequency doubling perimetry (also termed frequency doubling technology or FDT, which is enhanced with new MATRIX software) is a relatively new technology that may detect functional visual loss at an earlier stage in the glaucomatous disease process, thereby detecting more patients who currently are misdiagnosed as having OHT instead of early POAG. Current sensitivities and specificities are continuing to improve, but more baseline data are needed to determine under what circumstances these new techniques will prove to be most useful.

30.5. Differential Diagnosis

Several disorders may mimic glaucomatous ONH and/or visual field changes. Those that resemble structural glaucomatous optic neuropathy or interfere with assessment of glaucomatous cupping include optic disc anomalies, tilted discs, optic nerve drusen, optic nerve pits, optic nerve colobomas, myelinated nerve fibers, ischemic optic neuropathy and optic atrophy.

Factors that mimic glaucomatous visual field loss include: branch retinal artery occlusion; chorioretinal scars; retinal areas treated by photocoagulation or cryotherapy; demyelinating disorders; cerebrovascular accidents, tumors, or other lesions affecting the optic nerve, chiasm, optic tract, optic radiation, and/or the remaining course of nerve fibers to the occipital cortex. Other abnormalities that could account for "pseudo-glaucomatous" visual field defects or vision loss include vitreous hemorrhage, proliferative retinopathy or other retinal disorders.

As discussed in Sect. 30.2.3.4, variation in CCT from normal values may affect IOP measurements by yielding falsely high values when CCT is high.

30.6. Animal Models

Animal models for glaucoma have been used to study the pathophysiological basis of glaucoma. Animal based experiments are more representative than cell culture models particularly in the study of aqueous humor dynamics, the trabecular meshwork-ciliary muscle complex, retinal ganglion cells, ONH axons, ONH framework and vascular flow dynamics in glaucoma subtypes (Weinreb and Lindsey, 2005).

Conducting studies in animal models often are challenging and occasionally require anesthesia, which might affect the parameter being assessed. Conditioned reflex training may help elicit cooperation. Use of remote controlled implants and newer drug delivery options represent advances made on this front.

Animal experiments have provided excellent means to study medical and surgical treatments for glaucoma prior to human trials. Animal models are required to study de novo drugs and their structural and functional influence on ocular tissues.

Genetic modulation of animal models is another area of research. Mice are genetically engineered to suit experimental needs. New models are required and will be developed to advance our understanding of the glaucomatous disease process. Mouse mutants with glaucoma provide models for human glaucoma. From the clinical standpoint, some mouse models share occasional similarities with various aspects of glaucoma occurring in humans. The life span and replication time (one mouse year is equal to 30 human years) allow developmental and invasive studies and make it possible to follow glaucoma progression in a reasonable span of time (Chang et al., 2005). Some of the new mouse glaucoma models used are: Gpnmb with mutation on chromosome 6 to study iris pigment dispersion (Anderson et al., 2002); Trypl with chromosome 4 mutation to study iris stromal atrophy (Anderson et al., 2002); Nm2702 with chromosome 11 mutation to study inner retina loss and optic nerve cupping (Chang et al., 2005); CALB/Rk and si/si model to study early enlargement of optic cup or coloboma (Chang et al., 2005); AKXD-28 for ganglion cell loss and optic nerve cupping (Anderson et al., 2001); and nm 1839- model for Axenfeld's syndrome, anterior segment dysgenesis or buphthalmos (Chang et al., 2005). The effect of glaucoma on the electroretinogram response has been evaluated in mice.

The pig as a model for glaucoma shares many phylogenetic similarities with that of the human and is more readily available compared to the monkey model (Ruiz-Ederra et al., 2005). The porcine retina is more similar to human than other large mammals such as dog, goat or cow. Techniques used for glaucoma diagnosis in humans such as optical coherence tomography or multi-focal electroretinography can be reliably studied in pig eyes. Studies of the pig aqueous outflow system demonstrate it to be an acceptable model for specific glaucomas (McMenamin and Steptoe, 1991).

The monkey model of OHT with its resultant optic neuropathy closely reflects the optic neurodegeneration associated with human glaucoma. Utilization of the glaucoma model in monkeys has helped to further the understanding of aqueous humor

TABLE 30.3. Advantages and limitations of several animal models of glaucoma.

	Monkey	Rabbit	Rat	Mouse
General similarity to human eye	++++	++	++	++
Similarity to human aqueous dynamics	++++	+	+++	++++
Experimental IOP elevation	+++	++	+++	+++
Spontaneous IOP elevation	–	+	–	+++
Cupping, optic axon loss with IOP	++++	–	+++	+++
Genome sequenced	–*	–	–	+++
Many transgenic strains available	–	–	+	++++
Readily available, low cost	–	+	+++	++++
Ease of maintenance and handling	–	+	+++	++++
Ease of testing	+	+++	++	+

*The human and monkey genomes are sufficiently similar that data from the human genome reject often can be applied to experiments using non-human primates.

Reproduced with permission from: Weinreb RN, Lindsey JD (2005), The importance of models in glaucoma research. J Glaucoma 14:302–304.

dynamics, glaucomatous damage in the RGCs and other structures along the visual pathway, and the effects of pharmacological agents on ocular tissues (Rasmussen and Kaufman, 2005). Fundus photographs of glaucomatous monkeys show the morphologic changes consistent with human OHT with the advantage of observing progressive changes over a shorter period of time by withholding treatment despite very high IOPs (Gaasterland and Kupfer, 1974). The ability to correlate histologic changes in glaucomatous monkey eyes with functional and imaging data is a major advantage of the monkey model. Humphrey visual field defects can be performed by trained glaucomatous monkeys (Harwerth et al., 2002). The types of visual field defects are similar to those in glaucomatous humans.

Utilization of an experimental glaucoma model in non-human primates has led to advances in the understanding of glaucomatous changes in the visual pathways from photoreceptors to the visual cortex (Rasmussen and Kaufman, 2005). Newer investigative modalities such as contemporary perimetry techniques have been studied in primate models. Electroretinograms can be performed in monkeys to evaluate changes produced by glaucomatous optic neuropathy.

The choice of animal is governed by the objectives and requirements of the study (Table 30.3). Size of the animal eye, similarity with human eyes, cost and availability are factors that influence the selection of the animal type.

Summary

Glaucoma is defined as a disturbance of the structural or functional integrity of the optic nerve that causes characteristic changes in the optic nerve leading to defects in the visual field. If left untreated, it might eventually lead to blindness. Glaucoma is the leading cause of irreversible blindness throughout the world, and is second only to macular degeneration in the US. Amongst the risk factors, elevated IOP, thin CCT, increasing age and large cup to disc ratio are established. Concerning its pathogenesis, mechanical and vascular theories are fundamental. In addition, many other mechanisms contribute such as immune factors, apoptosis, glutamate-induced excitotoxicity, free radicals, nitric acid synthase and genetic mutations. To assess glaucoma, tonometry, gonioscopy, examination of optic disc, and visual field testing are essential. Animal models for glaucoma have been used to study the pathophysiological basis of glaucoma, aqueous humor dynamics, the trabecular meshwork, the ciliary muscle, and retinal ganglion cells. Animal experiments have provided worthwhile means of evaluating of diagnostic modalities and treatments, both medical and surgical, for glaucoma.

Acknowledgments. Supported in part by an unrestricted grant from Research to Prevent Blindness, New York, NY

Review Questions/Problems

1. The estimated world population that is bilaterally blind from primary open angle glaucoma is:

a. 2–3 million
b. 3–4 million
c. 4–5 million
d. 5–6 million
e. 6–7 million

2. The most important risk factor for developing primary open angle glaucoma is:

a. increasing age
b. thin central corneal thickness
c. African-American race
d. excessive intraocular pressure
e. large cup to disc ratio

3. Mechanisms that might contribute to optic nerve damage in glaucoma include:

a. apoptosis
b. glutamate-induced excitotoxicity
c. nitric oxide synthase
d. free radicals
e. all of the above

4. Mutations in which of the following genes have been shown to be associated with the pathogenesis of glaucoma?

a. GLC1A
b. myocilin
c. optineurin
d. GLC1G
e. all of the above

5. Which of the following testing strategies is NOT considered standard practice in the evaluation of patients with primary open angle glaucoma?

a. tonometry
b. gonioscopy
c. visual field testing
d. optical coherence tomography
e. all of the above

6. All of the following are characteristic changes that might be observed in the optic nerve head in primary open angle glaucoma except:

a. increased cup to disc ratio
b. notching of the neural rim
c. disc hemorrhage
d. new vessels within the disc
e. all of the above

7. Ancillary techniques that might be used to evaluate patients with glaucoma include:

a. nerve fiber layer analyzer
b. confocal scanning laser ophthalmoscopy
c. ultrasound biomicroscopy
d. optical coherence tomography
e. all of the above

8. According to recent estimates, which of the following individuals is at highest risk of developing glaucoma?

a. African-American male, age 80 years
b. Hispanic female, age 60 years
c. Caucasian male, age 80 years
d. African-American female, age 70 years
e. Caucasian female, age 80 years

9. As per the mechanical theory of glaucoma pathogenesis, deprivation of which of the following factors may accelerate retinal ganglion cell loss?

a. heat-shock protein
b. calcium
c. neurotrophins
d. dopamine
e. glutamate

10. All of the following conditions may mimic glaucomatous visual field loss except:

a. branch retinal artery occlusion
b. retinitis pigmentosa
c. cerebrovascular accident
d. chorioretinal scar
e. none of the above

11. Briefly discuss the risk factors for primary open angle glaucoma.

12. Enumerate the important mechanisms involved in pathogenesis of glaucomatous optic neuropathy.

13. Discuss the immune mechanisms contributing to retinal ganglion cell damage in glaucoma.

14. Outline the recent advances in the genetic basis for glaucomatous optic neuropathy.

15. Describe the role of animal models in our understanding of glaucoma.

16. Briefly discuss the prevalence of primary open angle glaucoma in the United States.

17. Outline the importance of central corneal thickness in glaucoma practice.

References

Adams JM, Cory S (1998) The Bcl-2 protein family: arbiters of cell survival. Science 281:1322–1326.

American Academy of Ophthalmology: Preferred Practice Patterns (2005a) Primary Open Angle Glaucoma Suspect.

American Academy of Ophthalmology: Preferred Practice Patterns (2005b) Primary Open Angle Glaucoma.

Anderson MG, Smith RS, Savinova OV, Hawes NL, Chang B, Zabaleta A, Wilpan R, Heckenlively JR, Davisson M, John SW (2001) Genetic modification of glaucoma associated phenotypes between AKXD-28/Ty and DBA/2J mice. BMC Genet 2:1.

Anderson MG, Smith RS, Hawes NL, Zabaleta A, Chang B, Wiggs JL, John SW (2002) Mutations in genes encoding melanosomal proteins cause pigmentary glaucoma in DBA/2J mice. Nat Genet 30:81–85.

Bengtsson B, Heijl A (2003) Normal intersubject threshold variability and normal limits of the SITA SWAP and full threshold SWAP perimetric programs. Invest Ophthalmol Vis Sci 44:5029–5034.

Bodis-Wollner I, Tzelepi A (1998) The push-pull action of dopamine on spatial tuning of the monkey retina: the effects of dopaminergic deficiency and selective D1 and D2 receptor ligands on the pattern electroretinogram. Vision Res 38:1479–1487.

Brandt JD, Beiser JA, Kass MA, Gordon MO (2001) Central corneal thickness in the Ocular Hypertension Treatment Study (OHTS). Ophthalmology 108:1779–1788.

Broadway DC, Drance SM (1998) Glaucoma and vasospasm. Br J Ophthalmol 82:862–870.

Chang B, Hawes NL, Hurd RE, Wang J, Howell D, Davisson MT, Roderick TH, Nusinowitz S, Heckenlively JR (2005) Mouse models of ocular diseases. Vis Neurosci 22:587–593.

Congdon N, O'Colmain B, Klaver CC, Klein R, Munoz B, Friedman DS, Kempen J, Taylor HR, Mitchell P, Eye Diseases Prevalence Research Group (2004) Causes and prevalence of visual impairment among adults in the United States. Arch Ophthalmol 122:477–485.

Cursiefen C, Wisse M, Cursiefen S, Junemann A, Martus P, Korth M (2000) Migraine and tension headache in high-pressure and normal-pressure glaucoma. Am J Ophthalmol 129:102–104.

Daubs JG, Crick RP (1981) Effect of refractive error on the risk of ocular hypertension and open angle glaucoma. Trans Ophthalmol Soc UK 101:121–126.

Drance SM, Schulzer M, Thomas B, Douglas GR (1981) Multivariate analysis in glaucoma: use of discriminant analysis in predicting glaucomatous visual field damage. Arch Ophthalmol 99:1019–1022.

Ehlers N, Bramsen T, Sperling S (1975) Applanation tonometry and central corneal thickness. Acta Ophthalmol (Copenh) 53:1974–1983.

Feuer WJ, Parrish RK 2nd, Schiffman JC, Anderson DR, Budenz DL, Wells MC, Hess DJ, Kass MA, Gordon MO (2002) The Ocular Hypertension Treatment Study: Reproducibility of cup/disk ratio measurements over time at an optic disc reading center. Am J Ophthalmol 133:19–28.

Friedman DS, Wolfs RC, O'Colmain BJ, Klein BE, Taylor HR, West S, Leske MC, Mitchell P, Congdon N, Kempen J, Eye Diseases Prevalence Research Group (2004) Prevalence of open-angle glaucoma among adults in the United States. Arch Ophthalmol 122:532–538.

Gaasterland D, Kupfer C (1974) Experimental glaucoma in the rhesus monkey. Invest Ophthalmol 13:455–457.

Gordon MO, Beiser JA, Brandt JD, Heuer DK, Higginbotham EJ, Johnson CA, Keltner JL, Miller JP, Parrish RK 2nd, Wilson MR, Kass MA (2002) The Ocular Hypertension Treatment Study: baseline factors that predict the onset of primary open-angle glaucoma. Arch Ophthalmol 120:714–720.

Greenlund LJ, Deckwerth TL, Johnson EM, Jr. (1995) Superoxide dismutase delays neuronal apoptosis: a role for reactive oxygen species in programmed neuronal death. Neuron 14:303–315.

Grunwald JE, Piltz J, Hariprasad SM, Dupond J (1998) Optic nerve and choroidal circulation in glaucoma. Invest Ophthalmol Vis Sci 39:2329–2336.

Harwerth RS, Crawford ML, Frishman LJ, Viswanathan S, Smith EL 3rd, Carter-Dawson L (2002) Visual field defects and neural losses from experimental glaucoma. Prog Retin Eye Res 21:91–125.

Hayreh SS, Revie IH, Edwards J (1970) Vasogenic origin of visual field defects and optic nerve changes in glaucoma. Br J Ophthalmol 54:461–472.

Heijl A, Leske MC, Bengtsson B, Hyman L, Bengtsson B, Hussein M, Early Manifest Glaucoma Trial Group (2002) Reduction of intraocular pressure and glaucoma progression: Results from the Early Manifest Glaucoma Trial. Arch Ophthalmol 120:1268–1279.

Hutchins JB, Barger SW (1998) Why neurons die: Cell death in the nervous system. Anat Rec 253:79–90.

Ishii Y, Kwong JMK, Caprioli J (2003) RGC protection with GGA, a heat shock protein inducer, in a rat glaucoma model. Invest Ophthalmol Vis Sci 44:1982–1992.

Jackson GR, Owsley C (2003) Visual dysfunction, neurodegenerative diseases, and aging. Neurol Clin 21:709–728.

Kass MA, Heuer DK, Higginbotham EJ, Johnson CA, Keltner JL, Miller JP, Parrish RK 2nd, Wilson MR, Gordon MO (2002) The Ocular Hypertension Treatment Study: a randomized trial determines that topical ocular hypotensive medication delays or prevents the onset of primary open-angle glaucoma. Arch Ophthalmol 120:701–713.

Kipnis J, Schwartz M (2002) Dual action of glatiramer acetate (Cop-1) in the treatment of CNS autoimmune and neurodegenerative disorders. Trends Mol Med 8:319–323.

Kitaoka Y, Kumai T (2004) Modulation of retinal dopaminergic cells by nitric oxide. A protective effect on NMDA-induced retinal injury. In Vivo 18:311–315.

Lam TT, Siew E, Chu R, Tso MO (1997) Ameliorative effect of MK-801 on retinal ischemia. J Ocul Pharmacol Ther 13:129–137.

Leibowitz HM, Krueger DE, Maunder LR, Milton RC, Kini MM, Kahn HA, Nickerson RJ, Pool J, Colton TL, Ganley JP, Loewenstein JI, Dawber TR (1980) The Framingham Eye Study monograph: an ophthalmological and epidemiological study of cataract, glaucoma, diabetic retinopathy, macular degeneration, and visual acuity in a general population of 2631 adults, 1973–1975. Surv Ophthalmol 24(Suppl):335–610.

Leske MC, Connell AM, Wu SY, Hyman L, Schachat AP (1997) Distribution of intraocular pressure. The Barbados Eye Study. Arch Ophthalmol 115:1051–1057.

Liu B, Neufeld AH (2001) Nitric oxide synthase-2 in human optic nerve head astrocytes induced by elevated pressure in vitro. Arch Ophthalmol 119:240–245.

Lipton SA (2003) Possible role for memantine in protecting retinal ganglion cells from glaucomatous damage. Surv Ophthalmol 48(Suppl 1):S38–46.

McMenamin PG, Steptoe RJ (1991) Normal anatomy of the aqueous humour outflow system in the domestic pig eye. J Anat 178:65–77.

Medeiros FA, Zangwill LM, Bowd C (2004) Comparison of the GDx VCC scanning laser polarimeter, HRT II confocal scanning laser ophthalmoscope, and stratus OCT optical coherence tomograph for the detection of glaucoma. Arch Ophthalmol 122:827–837.

Monemi S, Spaeth G, DaSilva A (2005) Identification of a novel adult-onset primary open-angle glaucoma (POAG) gene on 5q22.1. Hum Mol Genet 14:725–733.

Naskar R, Dreyer EB (2001) New horizons in neuroprotection. Surv Ophthalmol 45(Suppl 3):S250–S255.

Neufeld AH, Sawada A, Becker B (1999) Inhibition of nitric-oxide synthase 2 by aminoguanidine provides neuroprotection of retinal ganglion cells in a rat model of chronic glaucoma. Proc Natl Acad Sci USA 96:9944–9948.

Nouri-Mahdavi K, Hoffman D, Coleman AL, Liu G, Li G, Gaasterland D, Caprioli J, Advanced Glaucoma Intervention Study (2004) Predictive factors for glaucomatous visual field progression in the Advanced Glaucoma Intervention Study. Ophthalmology 111:1627–1635.

Osborne NN, Ugarte M, Chao M, Chidlow G, Bae JH, Wood JP, Nash MS (1999) Neuroprotection in relation to retinal ischemia and relevance to glaucoma. Surv Ophthalmol 43(Suppl 1):S102–128.

Pearson J (2000) Glaucoma in patients with sleep apnea. Ophthalmology 107:816–817.

Perez MT, Caminos E (1995) Expression of brain-derived neurotrophic factor and of its functional receptor in neonatal and adult rat retina. Neurosci Lett 183:96–99.

Quigley HA (1996) Number of people with glaucoma worldwide. Br J Ophthalmol 80:389–393.

Quigley HA, Addicks EM, Green WR, Maumenee AE (1981) Optic nerve damage in human glaucoma. II. The site of injury and susceptibility to damage. Arch Ophthalmol 99:635–649.

Racette L, Sample PA (2003) Short-wavelength automated perimetry. Ophthalmol Clin North Am 16:227–236.

Rasmussen CA, Kaufman PL (2005) Primate glaucoma models. J Glaucoma 14:311–314.

Reisberg B, Doody R, Stoffler A, Schmitt F, Ferris S, Mobius HJ, Memantine Study Group (2003) Memantine in moderate-to-severe Alzheimer's disease. N Engl J Med 348:1333–1341.

Rezaie T, Child A, Hitchings R (2002) Adult-onset primary open-angle glaucoma caused by mutations in optineurin. Science 295:1077–1079.

Ruiz-Ederra J, Garcia M, Hernandez M, Urcola H, Hernandez-Barbachano E, Araiz J, Vecino E (2005) The pig eye as a novel model of glaucoma. Exp Eye Res 81:561–569.

Schimiti RB, Avelino RR, Kara-Jose N, Costa VP (2002) Full-threshold versus Swedish Interactive Threshold Algorithm (SITA) in normal individuals undergoing automated perimetry for the first time. Ophthalmology 109:2084–2092; discussion 2092.

Schwartz M (2004) Optic nerve crush: protection and regeneration. Brain Res Bull 62:467–471.

Schwartz M (2005) Lessons for glaucoma from other neurodegenerative diseases: can one treatment suit them all? J Glaucoma 14:321–323.

Tezel G, Hernandez MR, Wax MB (2000) Immunostaining of heat shock proteins in the retina and optic nerve head of normal and glaucomatous eyes. Arch Ophthalmol 118:511–518.

Thornberry NA, Lazebnik Y (1998) Caspases: enemies within. Science 281:1312–1316.

Tielsch JM (1991) The epidemiology of primary open-angle glaucoma. Ophthalmol Clin North Am 4:649–657.

Tielsch JM, Sommer A, Katz J, Royall RM, Quigley HA, Javitt J (1991) Racial variations in the prevalence of primary open-angle glaucoma: The Baltimore Eye Survey. JAMA 266:369–374.

Wax MB, Tezel G (2002) Neurobiology of glaucomatous optic neuropathy: Diverse cellular events in neurodegeneration and neuroprotection. Mol Neurobiol 26:45–55.

Weinreb RN, Levin LA (1999) Is neuroprotection a viable therapy for glaucoma? Arch Ophthalmol 117:1540–1544.

Weinreb RN, Lindsey JD (2005) The importance of models in glaucoma research. J Glaucoma 14:302–304.

Wiggs JL, Allingham RR, Vollrath D (1998) Prevalence of mutations in TIGR/Myocilin in patients with adult and juvenile primary open-angle glaucoma. Am J Hum Genet 63:1549–1552.

Wiggs JL, Auguste J, Allingham RR (2003) Lack of association of mutations in optineurin with disease in patients with adult-onset primary open-angle glaucoma. Arch Ophthalmol 121:1181–1183.

Wilson MR, Hertzmark E, Walker AM, Childs-Shaw K, Epstein DL (1987) A case-control study of risk factors in open angle glaucoma. Arch Ophthalmol 105:1066–1071.

Wolfs RC, Borger PH, Ramrattan RS, Klaver CC, Hulsman CA, Hofman A, Vingerling JR, Hitchings RA, de Jong PT (2000) Changing views on open-angle glaucoma: definitions and prevalences—The Rotterdam Study. Invest Ophthalmol Vis Sci 41:3309–3321.

Wyllie AH, Kerr JF, Currie AR (1980) Cell death: The significance of apoptosis. Int Rev Cytol 68:251–306.

Yuan L, Neufeld AH (2001) Activated microglia in the human glaucomatous optic nerve head. J Neurosci Res 64:523–532.

31
Stroke and Cerebrovascular Disease

Alison E. Baird

Keywords Atherosclerosis; Complement; Hemorrhagic stroke; Ischemic stroke; Stroke

31.1. Introduction

Strokes result from occlusion or rupture of blood vessels in the brain and are a leading cause of death and disability in the United States (US, Broderick et al., 1998). There are around 700,000 new and recurrent strokes per year in the US, with costs related to stroke estimated to be over $60 billion in 2007 (Rosamond et al., 2007). Around 80–85% of strokes are ischemic and 15–20% of strokes are hemorrhagic. Over the past three decades neuroimaging methods such as computed tomography (CT) and magnetic resonance imaging (MRI) have enormously improved the rapid and accurate diagnosis of ischemic and hemorrhagic stroke (Figure 31.1). More recently, non-invasive vascular imaging methods such as CT angiography (CTA) and MR angiography (MRA) have proven invaluable for the detection of vascular stenoses in the carotid arteries. But the therapeutics of stroke has lagged behind.

Treatment approaches for ischemic stroke have largely focused on reperfusing the ischemic brain tissue—for example with the thrombolytic agent recombinant tissue plasminogen activator (rt-PA, The National Institute of Neurological Disorders and Stroke rt-PA Stroke Study Group, 1995)—and on neuronal protection, mainly from damage caused by excitotoxic amino acids such as glutamate. Over the past 15 years it has been increasingly appreciated that inflammation alone, or in combination with systemic immune system responses, may be an important contributing factor in stroke-related brain injury and stroke outcome, as well as in the development of atherosclerotic cerebrovascular disease. Elements of the inflammatory process may provide promising targets for the treatment of acute ischemic stroke and for vascular disease prevention.

As has been outlined in prior chapters, the immune system is responsible for host responses to infection, and responds in the same way to tissue injury occurring in the body, e.g., as occurs in the brain after stroke. It has been appreciated from post-mortem studies for more than 50 years that leukocyte accumulation—i.e., presumably neutrophil polymorphs and macrophages but the white cell types were not stated in the early studies—in the infarcted brain tissue is a prominent feature of the progression of ischemic brain lesions, and this was thought to be primarily related to the phagocytosis of necrotic brain tissue (Garcia et al., 1994). Inflammation is now considered to actively modify the evolution and repair of ischemic lesions and to probably impact significantly on tissue and clinical outcome after stroke (Garcia, 1992). This stroke-related inflammation is characterized by cellular responses, by activation of immune, endothelial and glial cells, and by the secretion of cytokines and chemokines. Although still somewhat controversial, the current concept that has emerged is of a twofold impact: early on, inflammation appears to exacerbate ischemic and reperfusion injury after stroke, while later on, inflammatory cells may be actively involved in tissue repair and remodeling, and so play a beneficial role (Hallenbeck et al., 2005). Furthermore, exposure to brain antigens may sensitize peripheral leukocytes and it is possible that sensitized memory T cell lymphocytes may either be protective (if sensitized in low doses, resulting in tolerance) or lead to more aggressive immune responses to subsequent stroke (if highly sensitized from exposure to antigens in high doses). Brain-induced systemic inflammation may also increase patients' susceptibility to infections after stroke, most commonly pneumonia and urinary tract infections (Meisel et al., 2005). The role of inflammation in atherosclerotic cerebrovascular disease has also been increasingly recognized to the effect that atherosclerosis is no longer regarded as a disorder of lipid storage within blood vessels but now, primarily, as a disorder of chronic inflammation (Libby et al., 2002).

These changing concepts of the role of the immune system and inflammation in stroke and cerebrovascular disease have led to searches for new treatment and prevention strategies. This chapter will review the current state of knowledge about inflammation and the immune system in stroke and atherosclerosis and discuss potential therapeutic strategies.

T. Ikezu and H.E. Gendelman (eds.), *Neuroimmune Pharmacology*.
© Springer 2008

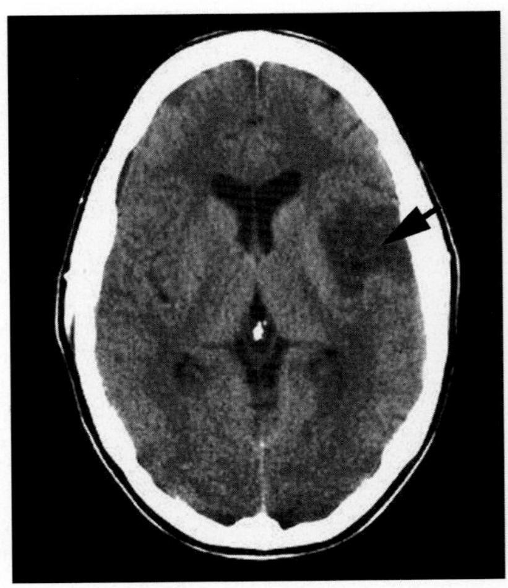
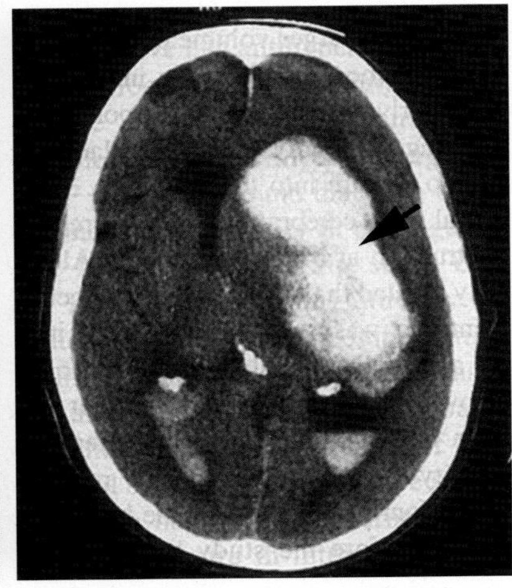

FIGURE 31.1. Ischemic and hemorrhagic stroke. Computed tomography (CT) brain scans of two patients, one with ischemic stroke (left) and one with hemorrhagic stroke (right).

31.2. Immunity and the Central Nervous System

Under physiological conditions there is a small resident population of macrophages—microglia and perivascular macrophages—along with some roaming T cell lymphocytes in the central nervous system. However, most of the immune cells involved in inflammation after stroke are in fact derived from the systemic circulation.

31.2.1. Immune Surveillance and the Central Nervous System

The central nervous system has traditionally been considered to be a severely disadvantaged site for the elicitation of immune responses for the following reasons: (1) the blood brain barrier (BBB) which strictly regulates the passage of molecules and cells into the central nervous system; (2) the poorly developed lymphatic drainage (some cerebrospinal fluid drains into the cervical lymphatics); (3) the low levels of major histocompatibility complex (MHC) molecule expression on microglia, neurons and astrocytes; and (4) the absence of a specific antigen presenting cell type that is particularly important in the initiation of immune responses, the dendritic cell, which is derived from the hematopoietic monocytic cell lineage (Sedgwick, 1995; Karman et al., 2004).

However as stated above, there are some resident immune cells in the brain: these include the microglia, which are derived from the mesoderm and can be replenished from the bone marrow (in contrast to neurons and astrocytes which are ectoderm derived and only replenished in minute numbers). Unlike other tissue macrophages, microglia do not express MHC antigens until they become activated. Another resident immune cell in the brain is the perivascular macrophage that appears to be important in antigen recognition (more so than microglia or neurons). It is also widely accepted that T cell lymphocytes perform surveillance functions in normal brain parenchyma: however, only activated T cells can cross the BBB and they enter the brain in smaller numbers than in other organs of the body. It appears that the entry of T cell lymphocytes is related to a roaming immune surveillance function, rather than being a targeted response to specific brain signaling (Hickey, 2001; Ransohoff et al., 2003). Unlike other organs there are very few resident mast cells in the brain and there are no resident neutrophils. Dendritic cells, although not present in brain parenchyma, are present in the choroid plexus and the meninges; these sites could potentially provide a source of immediate responders after acute brain ischemia, although this has yet to be demonstrated (Karman et al., 2004). From being considered a so-called "immune privileged" site (i.e., severely disadvantaged), the central nervous system is now regarded as an "immunologically specialized" site.

31.2.2. Innate and Adaptive Immunity

After stroke the majority of immune cells in the brain are derived from the circulating blood. Activation of the systemic immune system includes both the innate and adaptive immune systems. The innate immune system is characterized by the immediacy of its response and the non-specificity of its response, and it does not confer long-lasting or protective immunity. Cells of the innate immune system include circulating polymorphonuclear cells (mainly neutrophils), monocyte-macrophages, and natural killer (NK) lymphocytes. The adaptive immune system is specifically characterized by its ability to alter receptor expression and cellular functions when encountering new antigens, whether self or foreign; functions include cell-mediated immunity, humoral immunity, immune response regulation, memory and immunologic tolerance. Cells of the adaptive immune system include T cell lymphocytes—helper T (Th) cells (generally CD4⁺) and cytotoxic T cells (generally CD8⁺)—and B cell lymphocytes. Recent evidence suggests a role of NK cells in memory (Hamerman et al., 2005). After antigenic activation, CD4⁺ T cells can be differentiated into at least three subsets: Th1 (involved in cezll-mediated immunity and which secrete pro-inflammatory cytokines such as gamma-interferon [γ-IFN], interleukin-2 [IL-2] and lymphotoxin), Th2 (which promote humoral immunity and secrete anti-inflammatory cytokines such as IL-4 and IL-10) and regulatory T cells (e.g., CD4⁺ CD25⁺, which secrete transforming growth factor-beta [TGF-β] and IL-10).

31.3. Ischemic Stroke and Inflammation

Ischemic strokes account for around 80–85% of all strokes and are caused by arterial vascular occlusions; rarely occlusion in the cerebral venous system may result in ischemic and/or hemorrhagic stroke. Arterial occlusions resulting from cerebral embolism are the most common causes of ischemic strokes, and by about one week after stroke as many as 70–90% of occlusions will have spontaneously recanalized. Emboli typically originate from atherosclerotic stenoses in the internal carotid artery or from sources in the heart such as clots in the left atrium or the left ventricle. Hypertension-induced vascular disease of the small perforating intracerebral arteries is a common cause of lacunar strokes. A classification of the major stroke subtypes is shown in Table 31.1.

Immediately after the onset of focal brain ischemia, oxygen and glucose deprivation result in loss of adenosine triphosphate (ATP), loss of maintenance of normal ion channels and neuronal depolarization (Wahlgren and Ahmed, 2004) as evidenced by calcium and sodium influx into the cells,

TABLE 31.1. Classification of major stroke subtypes.

	Subtypes	Common causes
Ischemic (80–85%)	Large artery	Carotid artery stenosis
	Cardioembolic	Atrial fibrillation, ventricular clot
	Small vessel disease (i.e., lacune)	Lipohyalinosis of penetrating intracranial arteries
	Other determined	Blood dyscrasias
	Undetermined	
Hemorrhagic (15–20%)	Intracerebral	Hypertension
		Amyloid angiopathy
	Subarachnoid	Aneurysm
		Arteriovenous malformation

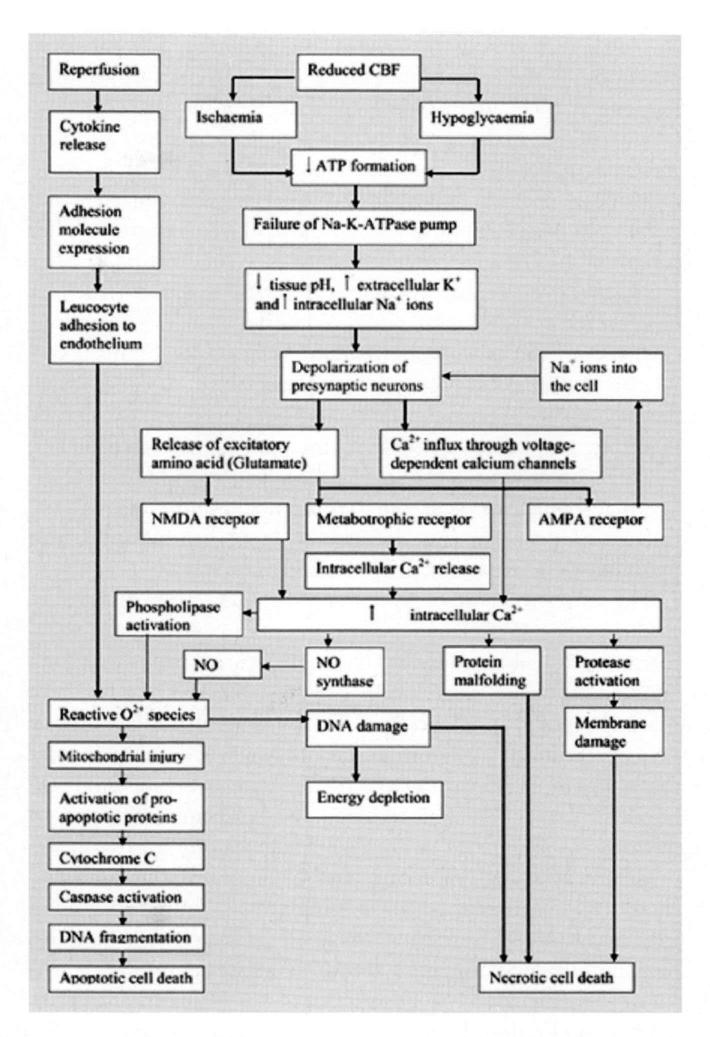

FIGURE 31.2 Cascade of molecular events during focal cerebral ischemia. Cascade of molecular events during ischemia and reperfusion injury. During cerebral ischemia, a series of biochemical events occur that causes depolarization, activation of voltage-operated calcium channels and excessive release of excitatory amino acids, especially glutamate. Activation of glutamate receptors leads to a further increase in intracellular calcium, activation of intracellular enzymes and neuronal death. Successive exploration of this complex pathophysiology has resulted in the development of a great number of candidates for neuroprotective intervention. Many of these candidates have proven to be efficacious in animal models, but so far no trial in stroke patients has confirmed a benefit on primary endpoint measures. CBF = Cerebral blood flow; ATP = adenosine triphosphate; AMPA = *D*-amino-3-hydroxyl-5-ethyl-4-isoxazole propionate; NMDA = N-methyl-*D*-aspartate; NO = nitric oxide. *Reproduced from Wahlgren NG, Ahmed N (2004) Neuroprotection in cerebral ischaemia: facts and fancies—the need for new approaches. Cerebrovasc Dis, 17 Suppl 1:153–166 (permission from S. Karger AG, Medical and Scientific Publishers).*

generation of oxygen free radicals and breakdown of membranes. The cascade of molecular events is shown in Figure 31.2. Brain tissue may be viable for only minutes to 1 h in the core of the resulting infarct, where the blood flow is of the order of 0–10 ml/100 g brain tissue/minute. The ischemic penumbra is the zone of potentially viable tissue surrounding the infarct core; this tissue is nourished in part by collateral blood vessels and may be viable for up to 3–6 h if the perfusion is restored in time. Inflammation is now believed to contribute to brain injury in both the core and the ischemic penumbra, and is believed to be particularly hazardous after reperfusion (after spontaneous arterial recanalization or after rt-PA therapy). Following the return of the blood flow and oxygen the marked increase in oxidative-reduction reactions can lead to the brain's antioxidant systems becoming overwhelmed. Excess oxygen free radicals are generated which damage cerebral tissue directly and through the activation of apoptotic and necrotic cell signaling pathways (Wahlgren and Ahmed, 2004; Ding and Clark, 2006). Neutrophil polymorphs and macrophages exacerbate to reperfusion injury through the release of inflammatory molecules (e.g., pro-inflammatory cytokines) and free radical species.

The current state of knowledge comes from studies in experimental stroke models, from human post-mortem studies and clinical studies. Early magnetic resonance imaging studies showed a correlation between elevated lactate and macrophage brain infiltration. This was identified at autopsy neuropathological examination within a week of the lactate determinations (Petroff et al., 1992). Moreover, elevated lactate also correlated with reduced N-acetylaspartate concentrations my magnetic resonance spectroscopy and with the presence of sheets of foamy macrophages within infarcts. In additional clinical neuroimaging studies employing positron emission tomography (using PK11195 which binds to peripheral benzodiazepine receptors expressed on activated microglia, Figure 31.3, Pappata et al., 2000; Gerhard et al., 2000), single photon emission computed tomography (using indium-111 to detect neutrophil recruitment, Price et al., 2004) and, most recently, magnetic resonance imaging using ultra-small superparamagnetic iron oxide nanoparticles all have demonstrated macrophage infiltration during stroke (Saleh et al., 2004).

31.3.1. Involvement of Immune Cells

The first immune responders after focal cerebral ischemia are the brain microglia. As mentioned above, in the normal brain, microglia are quiescent or inactive, but after an injury, these cells undergo a dramatic change, altering their appearance, and migrating to the site of the damage to help clear away and clean up dead and dying cells and cell debris. Microglia produce cytokines, which trigger astrocytes and cells of the immune system to respond to the injury site (Vilhardt, 2005).

Stroke-related inflammation in the brain appears to largely result from a selective migration and infiltration of systemic (i.e., circulating) leukocytes to the focus of cerebral ischemia. These cells are predominantly those of the innate immune system, at least initially. Activated neutrophil polymorphs begin to infiltrate as early as 4 h after stroke, initially being located at the edges of the infarct, and subsequently in the infarct core. The time course of neutrophil polymorph infiltration is depicted in Figure 31.4: the neutrophil response may be largely completed by 4–7 days in humans (Nilupul Perera et al., 2006). The infiltration of neutrophils overlaps with and is followed by a more prolonged and sustained infiltration of blood-borne monocytes, which become macrophages once they have entered the brain. This starts within the first 24–48 h and may persist for days to weeks to even months. It is not always possible to distinguish between blood-borne macrophages and activated microglia because of similar antigen expression (e.g., CD163, a marker of activated macrophages, is expressed on both). It was noted in early post-mortem studies that the degree of leukocyte infiltration was not always uniform. This cellular inflammation is believed to exacerbate brain injury through the secretion of pro-inflammatory cytokines and through clogging of blood vessels, leading to a no-reflow phenomenon.

Although not commonly recognized, there has been indirect evidence for some years that cells from the adaptive immune system too may play an important role early on in ischemic stroke (Hallenbeck et al., 2005; Schroeter et al., 1994; Becker et al., 2001, 2003). $CD8^+$ lymphocytes have been found in the brain in the first few days in some post-mortem studies and blockade of adaptive immune processes may impact significantly on stroke outcomes. Recently it has been reported in an experimental mouse stroke model (Yilmaz et al., 2006) that $CD4^+$ and $CD8^+$ T lymphocytes and γ-IFN may play important roles in the exacerbation of ischemic reperfusion injury as early as 5 h after stroke. In this study $CD4^+$ and $CD8^+$ T cells and γ-IFN caused the cerebral microvasculature to assume a pro-inflammatory and pro-thrombotic phenotype (as evidenced by increased leukocyte and platelet adhesion to the endothelium). It is possible that lymphocytes may act indirectly, by activating other circulating blood cells and/or extravascular cells such as resident macrophages in the brain. Alternatively lymphocytes may act directly on brain tissue: cytotoxic $CD8^+$ T-cells may induce brain injury through molecules released from their cytotoxic granules. $CD4^+$ Th1 cells, which secrete pro-inflammatory cytokines, including interleukin-2 (IL-2), IL-12, γ-IFN and tumor necrosis factor (TNF), may play a key role in the pathogenesis of stroke, whereas $CD4^+$ Th2 cells may play a protective role through anti-inflammatory cytokines such as IL-4, IL-5, IL-10, and IL-13. Evidence suggests that T cell lymphocytes are likely to be key players in the overall regulation of the immune response.

31.3.2. Molecular Mediators of Inflammation

Inflammatory molecules include adhesion molecules, chemokines and cytokines, many of which are up-regulated in acute cerebral ischemia and contribute to the inflammatory process. The release of pro-inflammatory cytokines is a direct consequence of the ionic imbalances and free calcium accumulation that lead to the release of free fatty acids and other

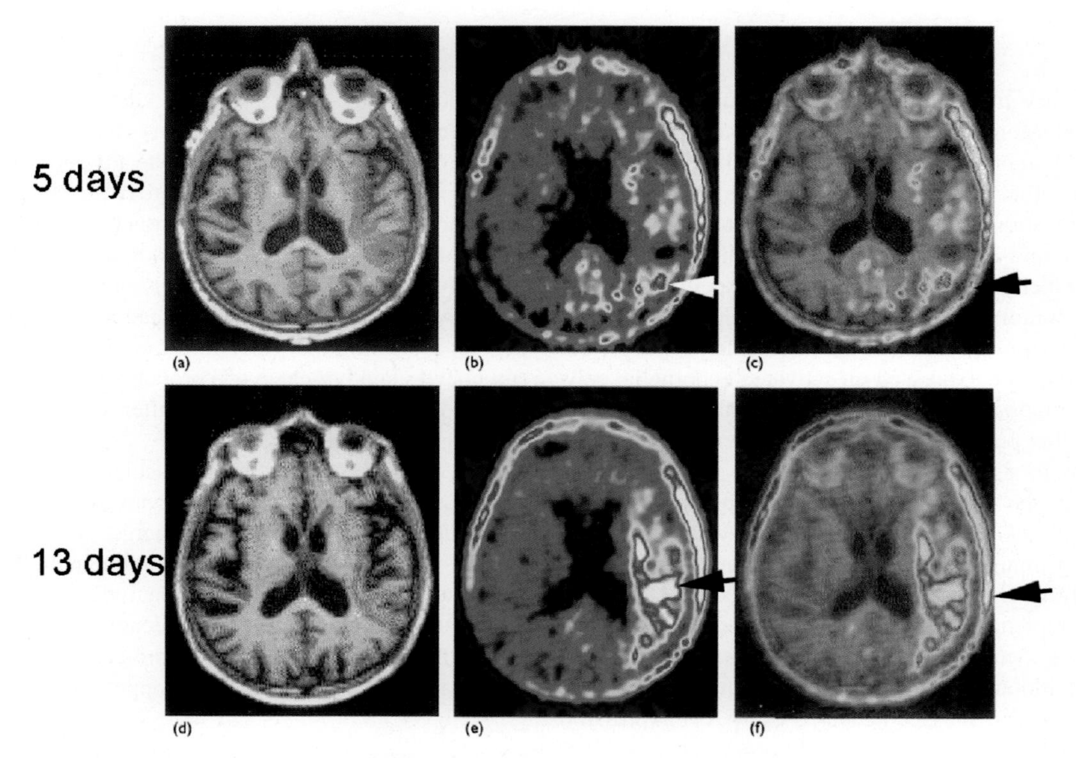

FIGURE 31.3. [11C]PK11195 positron emission tomography (PET) brain imaging in ischemic stroke. [11C]PK11195 is a peripheral benzodiazepine receptor marker expressed on activated microglia/macrophages. [11C]PK11195 positron emission tomography (PET) and T1 weighted NMR of an 87-year-old patient with a left middle cerebral artery territory infarct. The top row shows the images obtained at 5 days and the bottom row the images at 13 days. The T1 weighted MRI is shown in the first column, the PET image in the second column and the co-registered MRI-PET in the third column. Green indicates low receptor binding and red/yellow and white high binding. After 5 days there is only a small degree of PK-binding in the left MCA territory, especially in the temporal lobe, and it is much more pronounced after 13 days. At this time the area of increased binding seems to exceed the area of signal change in the NMR occipitally. Adapted from Gerhard A, Neumaier B, Elitok E, Glatting G, Ries V, Tomczak R, Ludolph AC, Reske SN (2000). In vivo imaging of activated microglia using [11C]PK11195 and positron emission tomography in patients after ischemic stroke. Neuroreport, 11:2957–2960. *Reproduced from Nilupul Perera M, Ma HK, Arakawa S, Howells DW, Markus R, Rowe CC, Donnan GA (2006). Inflammation following stroke. J Clin Neurosci, 13:1–8 (permission from Lippincott, Williams & Wilkins).*

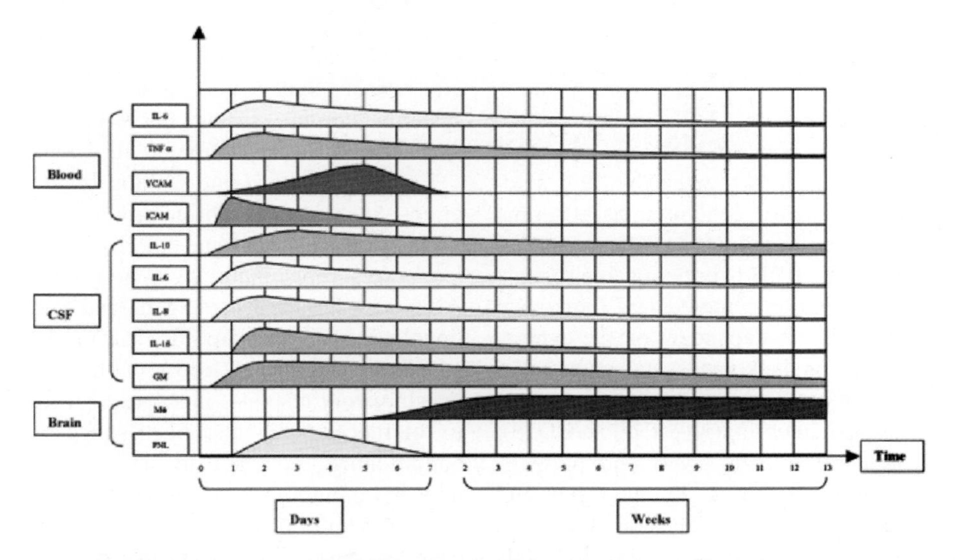

FIGURE 31.4. Time course of cellular and molecular activation folowing stroke. Diagrammatic representation of the time course of cellular activation (PNL, polymorphonuclear leukocytes; Mϕ, macrophages) and inflammatory molecular expression (GM, granulocyte and monocyte colony-stimulating factor; ICAM, soluble intercellular adhesion molecule-1; VCAM, soluble vascular cell adhesion molecule-1; TNF-α, tumor necrosis factor α; IL-1β, interleukin 1β; IL-6, interleukin 6; IL-8, interleukin 8; IL-10, interleukin 10) in human brain, cerebrospinal fluid (CSF) and blood after stroke. *Reproduced from Nilupul Perera M, Ma HK, Arakawa S, Howells DW, Markus R, Rowe CC, Donnan GA (2006). Inflammation following stroke. J Clin Neurosci, 13:1–8 (with permission from Elsevier).*

pro-inflammatory lipid metabolites. These metabolites promote the expression and release of IL-1β and TNF initially from activated microglia; these cytokines then lead to the production of other pro-inflammatory cytokines (e.g., IL-6, IL-8), the activation and infiltration of leukocytes, and the production of anti-inflammatory cytokines (including IL-4 and IL-10), which produces a negative feedback on the cascade. Adhesion molecules include intercellular adhesion molecule (ICAM), vascular cell adhesion molecule (VCAM), and E-selectin, which are expressed on activated endothelial cells during inflammation. Monocyte chemoattractant protein-1 is a chemokine that promotes infiltration of monocytes into the ischemic brain focus.

Other molecular mediators include complement, a host defense system made up of plasma proteins, which identifies pathogens and injured cells, recruits inflammatory cells and induces cell lysis. Early components in the classical, lectin or alternative complement pathways are believed to initiate the cascade leading to microglial activation and the activation and infiltration of blood-borne neutrophils and monocytes. The complement pathway may be activated by ICAM expression on endothelial cells. The matrix metalloproteinases (MMPs) are proteolytic proteins involved in the inflammatory response to stroke, one of the most prominent being MMP-9 which promotes degradation of the extracellular matrix proteins and increases the permeability of the BBB. Myeloperoxidase, an enzyme largely derived from neutrophils but also from some monocytes, catalyzes a number of reactions that generate oxygen free radicals (e.g., superoxide and hydroxyl radicals), particularly after reperfusion. The time course of cytokine changes in the human brain, cerebrospinal fluid and peripheral blood is shown in Figure 31.4 (Nilupul Perera et al., 2006).

31.3.3. Migration of Immune Cells Across the Blood Brain Barrier

The BBB provides the interface between the blood and brain and consists of endothelial cells held together by tight junctions ("zonula occludens"), the matrix containing basal lamina and the astrocyte foot processes that surround 90% or more of the capillary wall. Normally the BBB is impermeable to most cells, and to large and hydrophobic molecules. Leukocyte infiltration after ischemic stroke is dependent on the expression of adhesion molecules on leukocytes and endothelial cells (Barone and Feuerstein 1999). Neutrophils, monocytes and lymphocytes express β2-integrin (i.e., CD11b/CD18), an adhesion molecule that binds to ICAM-1 and ICAM-2 expressed on activated endothelial cells. Lymphocytes and monocytes also express another group of adhesion molecules, ⟨4-integrins, which attach to VCAM-1 on endothelial cells (Figure 31.5). Later on, starting within the first 24 h and going on for several weeks, the BBB breaks down and the passage of immune cells is less impeded.

Blockade of neutrophil infiltration in experimental stroke models, through inhibition of adhesion molecules (ICAM-1, E-selectin [only expressed on activated endothelium], P-selectin) and macrophage-1 antigen (which facilitates leukocyte phagocytosis in addition to leukocyte adhesion and transendothelial migration), has been shown to reduce infarct size (Jiang et al., 1995; Mackay et al., 1996). Furthermore, blockade acutely of ⟨4-integrin (Becker et al., 2001) and CD44, a cell surface glycoprotein involved in cell/cell and cell/matrix interactions (Wang et al., 2002) also reduced infarct size to a similar degree, indicating that early infiltration of mononuclear cells (monocytes and lymphocytes) probably has a more significant impact on tissue outcome early after stroke than may have previously been appreciated.

The concept of the "neurovascular unit"—a functional entity comprising neurons, astrocytes, smooth muscle cells and endothelial cells—has now emerged and it is appreciated that this coordinated unit plays a key role in the hemodynamic response to alterations in brain function and activity. Disruption of this regulatory network occurs after stroke and it is now believed that efforts at neuroprotection should target the entire unit, not just neurons (del Zoppo, 2006).

31.3.4. Effects of Ischemic Stroke on the Systemic Immune System

Stroke induces an acute stress response—i.e., over-activation of the sympathetic nervous system and increased corticosteroid levels (with resultant neutrophilia and lymphocytopenia). This in turn leads to depressed immunity and altered immune responses during the acute phase of stroke and may predispose patients to infections, particularly pneumonia, which is the commonest cause of mortality after the first few days of stroke (Meisel et al., 2005). In the clinical setting, increased total white cell counts and neutrophilia, which correlate with infarct size, are independently associated with worse outcome after stroke. Recently a massive and early activation of the systemic immune system has been shown to occur also in experimental stroke (Offner et al., 2006).

Exposure to local and leaking brain antigens in ischemic brain—such as myelin basic protein—has been shown to sensitize [in high doses] and tolerize [in low doses] peripheral leukocytes (presumably lymphocytes, Becker et al., 2003; Alvord et al., 1974; Youngchaiyud et al., 1974). The functional consequences are not clear but it is possible that such exposure may lead to recurrent strokes or to a more intense inflammatory response in subsequent strokes, or alternatively to tolerance. It has also been suggested that ischemic preconditioning, for which there is evidence in clinical stroke, may be immunologically mediated (i.e., after a recent transient ischemic attack (TIA) it is claimed that patients may have less severe strokes but this is still controversial, Kariko et al., 2004).

Recently, elevated $CD4^+$ $CD28^-$ T cell counts have been shown to be associated with worse outcome after ischemic stroke (Nadareishvili et al., 2004). This pro-inflammatory T cell subset is characterized by longevity, resistance to

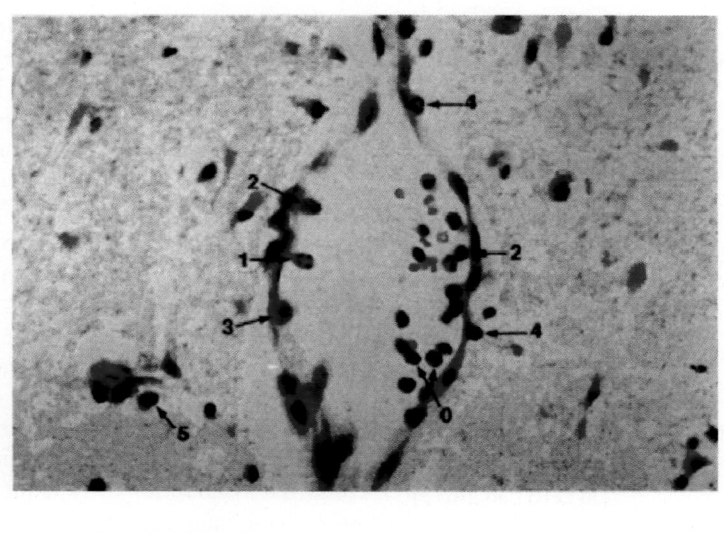

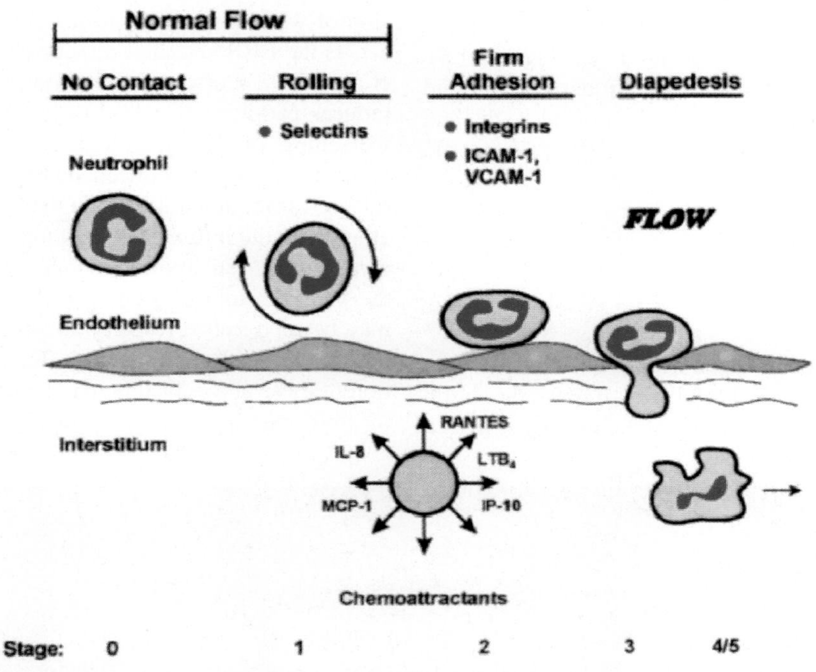

FIGURE 31.5. Leukocyte migration into focally ischemic brain. Histologic and schematic representations of changes in leukocyte behavior in the brain microvessels after focal ischemia. Shortly (within 1–6 h) after experimental stroke, many of the leukocytes, primarily neutrophils, in the ischemic tissue vessels are adherent to the post-capillary venuole and capillary walls. This can modify and exacerbate the decreased blood flow occurring in the already ischemic brain. Then, these neutrophils can find their way outside the vascular walls into the focal ischemic cortex over the next 6–24 h. Macrophages move into the brain later (i.e., over 1–5 days) and significantly accumulate in the infarcted brain. These changes in leukocyte behavior are mediated by increased brain inflammatory cytokine, adhesion molecule(s), and chemokine expression in the ischemic/injured brain. *Reproduced from ref Barone FC, Feuerstein GZ. Inflammatory mediators and stroke: new opportunities for novel therapeutics. J Cereb Blood Flow Metab 1999; 19: 819–834. With permission from Nature.*

apoptosis and secretion of high levels of γ-IFN. Patients with $CD4^+$ $CD28^-$ levels >8% (normally present in <1% of $CD4^+$ cells) during the first 48 h of stroke had a 6-fold increase in dying or developing a recurrent stroke during the following year. A notable finding was that patients with prior stroke had higher levels of these cells, raising the possibility that this population of T cells, already sensitized to brain antigens, contributed to a greater pro-inflammatory reaction to the new stroke. Elevated levels of this pro-inflammatory T cell subset have also been associated with the development of unstable atherosclerotic plaques. With improved typing of lymphocytes it is possible, and even likely, that different functional classes of lymphocytes present prior to or at the time of stroke and which could affect the evolution of ischemic lesions and outcome, will be recognized.

31.3.5. Tissue Repair and Recovery After Ischemic Stroke

The role played by the immune system in the acute phase of ischemic stroke so far appears largely to be unfavorable. A positive role for the immune system may be in tissue remodeling and repair, and the recovery of function days to weeks after stroke. The blood-borne mononuclear cells that have invaded the focally ischemic brain have been shown to up-regulate the expression of neurotrophic and other growth factor genes (e.g., those involved in angiogenesis, Barone and Feuerstein, 1999; Dirnagl et al., 1999). Resident-activated glial cells can also protect cells by facilitating repair and remodeling, and increasing neuronal plasticity, thereby facilitating the recovery of function for the remaining viable neurons and tissue (Figure 31.6). Hematopoietic stem cells may be involved in tissue repair and recovery after stroke, through the generation of microglia and macrophages, and possibly even new neurons (Brazelton et al., 2000; Paczkowska et al., 2005).

31.4. Hemorrhagic Stroke and Inflammation

Hemorrhagic strokes account for 15–20% of all strokes (Figure 31.1). Rupture of a blood vessel into the brain parenchyma results in an intracerebral hemorrhage (ICH); this is commonly caused by hypertension-induced vascular disease of the small perforating intracranial arteries. Chronic hypertension leads to intimal hyperplasia with hyalinosis in the small vessel wall; this predisposes to focal necrosis, causing breaks in the wall of the vessel and "pseudoaneurysms." Another cause of intracerebral hemorrhage is amyloid angiopathy. Rupture of a vessel into the subarachnoid space results in a subarachnoid hemorrhage; this most commonly results from rupture of a brain aneurysm or an arteriovenous malformation (Table 31.1).

After ICH mechanical brain injury results from tissue destruction and increased intracranial pressure from the mass effect of the hematoma occurring within the closed skull. After ICH, as occurs after ischemic stroke there is also an inflammatory reaction, consisting of an initial microglial response followed by an influx of systemic neutrophils and other white blood cells, up-regulation of molecules related to the adhesion of white blood cells to the endothelium, their migration across the BBB and their direction to the site of injury (Gong et al., 2000). It also appears that the hematoma itself actively induces pathways that lead to brain injury, brain edema and worsening of the neurological outcome. These pathways include thrombin produced during clot formation and iron released from hemoglobin via the action of heme oxygenase, along with the influx of neutrophils, microglial activation and complement activation that may exacerbate tissue injury in the peri-hematoma region. It has been hypothesized that this may be a protective response to small hematomas but may be deleterious in large hematomas (Keep et al., 2005): up-regulation of these above factors may be an evolutionary adaptation to limit brain injury during small hematomas (microbleeds),

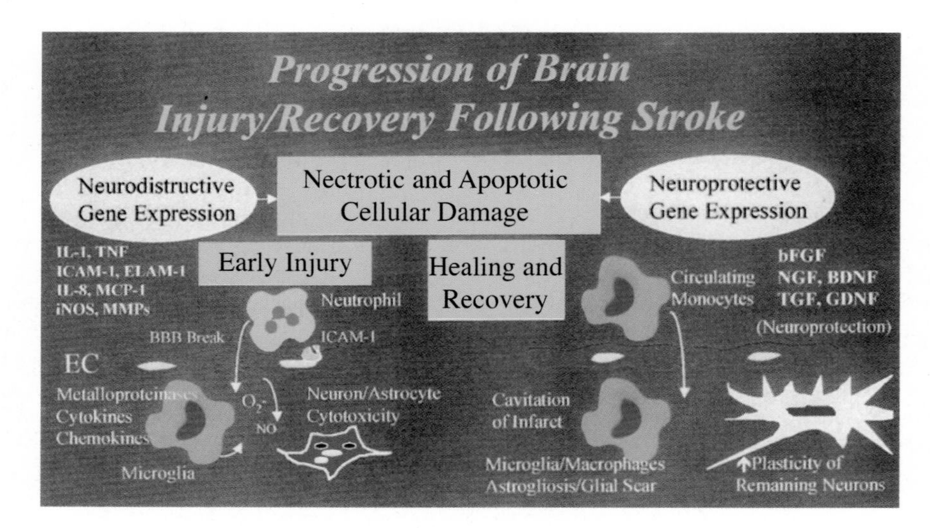

FIGURE 31.6. Brain recovery and repair after stroke. After initiation of brain injury various neurodestructive and neuroprotective genes are expressed. Neurodestructive gene expression (e.g., primarily those inflammatory cytokines, adhesion molecules, chemokines, and inflammatory proteins such as inducible NOS and metalloproteinases) can drive brain inflammation and necrotic/apoptotic cell death. Neuroprotective gene expression includes neurotrophic and growth factors from circulating mononuclear cells that have infiltrated into damaged tissue. In addition, resident-activated glial cells can also protect cells/tissue by facilitating repair and remodeling, and/or increasing neuronal plasticity and thereby facilitating the recovery of function for the remaining viable neurons/tissue. *Reproduced from Barone FC, Feuerstein GZ. Inflammatory mediators and stroke: new opportunities for novel therapeutics. J Cereb Blood Flow Metab 1999; 19: 819–834. With permission from Nature.*

with low levels of these factors being protective but high levels being excessively harmful (as may occur in large hematomas). In experimental models anti-inflammatory drugs such as COX-2 inhibitors, tacrolimus and fucoidan may reduce peri-hematoma inflammation and improve outcome (Nilupul Perera et al., 2006). Administration of the lipid-lowering drug atorvastatin was also associated with improved outcome after experimental ICH through modulation of inflammation in the peri-hematoma region (Jung et al., 2004).

After subarachnoid hemorrhage vasospasm is a common complication that results from irritation of the intracranial vasculature by subarachnoid blood. Vasospasm may lead to delayed cerebral ischemia and ischemic strokes, up to several weeks after the initial hemorrhage. There is preliminary evidence that inflammation plays a role in vasospasm after subarachnoid hemorrhage (Nilupul Perera et al., 2006; Clatterbuck et al., 2003), and anti-inflammatory strategies have been shown to prevent vasospasm in experimental stroke models (Nilupul Perera et al., 2006).

31.5. Atherosclerotic Cerebrovascular Disease and Inflammation

The study of cerebrovascular disease has advanced markedly in recent years with advances in non-invasive imaging methods such as MR angiography and CT angiography as well as an improved understanding of the immune system in the pathogenesis of atherosclerosis. Atherosclerotic cerebrovascular disease is a common cause of strokes and shows a predilection for sites such as the bifurcation of the common carotid artery into the internal and external carotid arteries and the aortic arch and the major intracranial arteries such as the basilar artery and the middle cerebral arteries. Occlusive atherosclerotic vascular disease of these large extracranial arteries is responsible for as many as 20–30% of ischemic strokes and intracranial steno-occlusive disease causes around 5–10% of ischemic strokes.

Atherosclerosis is characterized by the formation of plaques within the innermost layer of artery walls. It has traditionally been regarded as a disorder of lipid storage but is now regarded as a chronic inflammatory disorder. The plaque commences with the deposition of cholesterol-containing lipoproteins in the vessel wall. The first inflammatory cells involved are believed to be circulating monocytes (rather than neutrophil polymorphs). These migrate across the endothelial lining from the blood or from the lymphatics and develop into tissue macrophages or into dendritic cells. The macrophages ("foam cells") accumulate cholesterol in the artery wall. The resulting fatty streaks commence as early as the late teen years or the early 20s and may regress or progress throughout life. Lesions are typically asymptomatic, at least initially. Many fatty streaks develop into atheromatous plaques that contain extracellular debris and cholesterol. A fibrous cap is formed over the lipid-containing plaque by smooth muscle cells from the arterial wall and possibly from blood-derived progenitor cells. The collagen stabilizes the lesion and prevents debris, lipids and pro-coagulant material from reaching the bloodstream (Hallenbeck et al., 2005, Figure 31.7). Pro-inflammatory lymphocytes (i.e., $CD4^+$ Th1 cells) also contribute to and may accelerate atherosclerotic plaque formation (Libby et al., 2002). $CD4^+$ $CD28^-$ cells (mentioned previously) are a rare subset of $CD4^+$ T cells that are long-lived, secrete high levels of γ-IFN and are directly cytotoxic. These cells preferentially infiltrate unstable plaques and contribute to increased endothelial cell lysis in patients with acute coronary syndromes (Nakajima et al., 2002). Naturally arising regulatory T cells ($CD4^+$ $CD25^+$) have recently been shown to be potent inhibitors of the development of atherosclerosis in several mice models and may be a promising target for vascular disease prevention (Ait-Oufella et al., 2006). It has also been suggested that common infections may be associated in the pathogenesis of atherosclerosis. Specific organisms that have been studied include *Chlamydia pneumoniae*, herpes viruses, *human immunodeficiency virus*, *Helicobacter pylori*, and organisms associated with periodontal infections; however definitive proof of causal associations are lacking.

The atheromatous plaque grows silently for years or even decades without producing clinical symptoms. However, trace amounts of pro-inflammatory cytokines released from cells within the plaque, including IL-6, reach the bloodstream and stimulate C-reactive protein (CRP) secretion by the liver. CRP can be detected in the peripheral blood and is an independent marker of risk for myocardial infarction and probably for stroke in healthy individuals. Atherosclerotic plaques precipitate ischemic strokes by rupturing, leading to superimposed thrombus formation and embolism. They may also lead to strokes if the narrowing becomes particularly severe, >50%. The risk of stroke is greater with the greater degrees of narrowing as there is also associated hemodynamic insufficiency. Conventional angiographic techniques and non-invasive modalities (ultrasound, MRA, CTA) are used to detect occlusive arterial disease, and in the case of stroke prevention a narrowing of ≥50% of the internal carotid artery is used in decision-making for referral of patients for carotid endarterectomy.

Certain plaques are particularly prone to rupture. These "vulnerable plaques" contain abundant inflammatory cells and have thin collagen caps. The atheroma's vulnerability to rupture is correlated with higher cellular infiltrates of macrophages and activated T cells. In symptomatic carotid artery plaques elevated ICAM-1 expression (relative to asymptomatic plaques) has been found and the ICAM-1 expression was greatest in the high-grade region of the plaque. Vulnerable plaques are also not necessarily the largest plaques, particularly in patients with coronary artery disease, and current angiographic methods are of limited value in their recognition. However, new MRI methods using novel contrast agents are now being used to better study the features of vulnerable plaques.

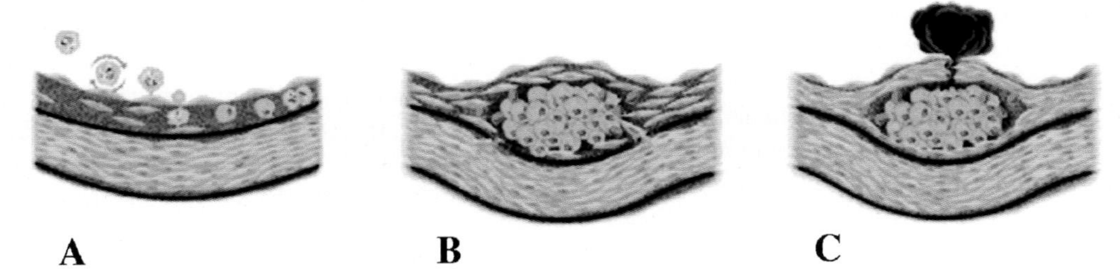

FIGURE 31.7. Inflammation and atherosclerosis. Participation of inflammation in all stages of atherosclerosis. A: Leukocyte (i.e., monocyte) recruitment to the nascent atherosclerotic lesion. Blood leukocytes adhere poorly to the normal endothelium. When the endothelial monolayer becomes inflamed, it expresses adhesion molecules that bind cognate ligands on leukocytes. Selectins mediate a rolling, or saltatory, interaction with the inflamed luminal endothelium. Integrins mediate firmer attachment. Pro-inflammatory cytokines expressed within atheroma provide a chemotactic stimulus to the adherent leukocytes, directing their migration into the intima. Inflammatory mediators such as macrophage colony stimulating factor (M-CSF) can augment expression of macrophage scavenger receptors leading to uptake of modified lipoprotein particles and formation of lipid-laden macrophages. M-CSF and other mediators produced in plaques can promote the replication of macrophages within the intima as well. B: T lymphocytes join macrophages in the intima during lesion evolution. These leukocytes, as well as resident vascular wall cells, secrete cytokines and growth factors that can promote the migration and proliferation of smooth muscle cells (SMCs). Medial SMCs express specialized enzymes that can degrade the elastin and collagen in response to inflammatory stimulation. This degradation of the arterial extracellular matrix permits the penetration of the SMCs through the elastic laminae and collagenous matrix of the growing plaque. C: Ultimately, inflammatory mediators can inhibit collagen synthesis and evoke the expression of collagenases by foam cells within the intimal lesion. These alterations in extracellular matrix metabolism thin the fibrous cap, rendering it weak and susceptible to rupture. Cross-talk between T lymphocytes and macrophages heightens the expression of the potent procoagulant tissue factor. Thus, when the plaque ruptures, as shown here, the tissue factor induced by the inflammatory signaling triggers the thrombus that causes most acute complications of atherosclerosis. (*Reproduced from ref [7]: Libby P, Ridker PM, Maseri A. Inflammation and atherosclerosis. Circulation 2002; 105: 1135–1143. Permission from Lippincott, Williams & Wilkins*).

31.6. Risk Factors for Stroke

As mentioned above, atherosclerotic vascular disease is a common cause of ischemic stroke and also predisposes patients to coronary artery disease, itself a risk factor for stroke. Vascular disease also increases with age and is increased in patients with hypertension, diabetes, a smoking history and hyperlipidemia. Hypertension is the major risk factor for hemorrhagic stroke. Common risk factors for stroke and cerebrovascular disease are listed in Table 31.2. The "metabolic syndrome" is a cluster of risk factors (truncal obesity, hyperglycemia, hypertriglyceridemia, hypertension and low HDL-cholesterol) associated with increased cardiovascular disease risk. Some immune and inflammatory risk factors may include recent infection that increases the risk of stroke to a modest degree and the possible mechanisms for this are being investigated. Periodontal disease, another factor shown to be independently associated with stroke and coronary artery disease, is believed to be associated with chronic inflammation and endothelial dysfunction (Janket et al., 2003). Lipoprotein associated phospholipase A_2 (a pro-inflammatory enzyme associated with low density lipoprotein) is approved by the Food and Drug Administration as a risk marker for stroke and coronary artery disease. Other inflammatory markers in the blood are showing promise as independent risk factors; these include elevated plasma levels of high sensitivity-CRP, IL-6 and the total white blood cell count (Ballantyne et al., 2005).

31.7. Therapeutic Neuroimmune Pharmacological Intervention

31.7.1. Treatment of Ischemic Stroke

As well as other strategies in the treatment and prevention of stroke the improved knowledge of the role of inflammation and the immune system in stroke and cerebrovascular disease has opened the door for research using *immunomodulatory* approaches in the treatment and prevention of stroke. There is only one Food and Drug Administration (FDA) approved treatment for ischemic stroke, the thrombolytic agent rt-PA that enhances brain reperfusion. Neuroprotective trials have been disappointing, with NXY-059—a free radical scavenger—being the latest casualty (Lees et al., 2006; del Zoppo, 2006; http://www.astrazeneca.com, 2007). It is of note that a major limitation of hypothermia therapy for stroke, another form of neuroprotective therapy, has been the excess rate of pneumonic complications that has precluded its use, especially in the elderly.

Immune modulation approaches would seem to be worthwhile pursuing based on the significance of the immune response on tissue and clinical outcomes after stroke. To date, therapeutic strategies for acute ischemic stroke have focused on preventing the recruitment and trafficking of neutrophils into ischemic lesions through inhibition of cellular adhesion molecules, however without success (Enlimomab Acute Stroke Trial Investigators, 2001). Enlimomab, a murine monoclonal antibody against ICAM-1, administered within 6h of

TABLE 31.2. Common risk factors for stroke.

Unmodifiable
Older age
Male sex
Race—African-American
Genetic
Modifiable
Hypertension
Diabetes
Smoking
Atrial fibrillation
Coronary artery disease
Prior stroke
Prior transient ischemic attack
Metabolic syndrome
Illicit drug use
Post-menopausal hormone replacement therapy
Periodontal disease
Inflammatory and immune causes and/or markers
Recent infection
Use of COX-2 inhibitors
Elevated lipoprotein associated phospholipase A_2
Elevated IL-6 (precursor of CRP)
Elevated hsCRP
Elevated total white cell count

onset and continued for 5 days, led to worse stroke outcome: the antibody itself elicited a strong and adverse inflammatory reaction (Furuya et al., 2001). UK-279,276, an antibody that binds selectively to the CD11b/CD18 integrin on neutrophils was not effective in the Acute Stroke Therapy by Inhibition of Neutrophils (ASTIN) study (Krams et al., 2003). It is possible that partial inhibition of neutrophils may be more effective than complete blockade.

Other *immune modulatory* approaches could theoretically include leukocyte-depleting strategies (not appropriate for the clinical population) and general strategies aimed at suppressing immune function (e.g., steroid treatment): these have not proven to be effective in stroke and increase the risk of secondary infection. The antibiotic agent doxycycline appears to reduce leukocyte adhesion and was found to be beneficial in a temporary occlusion experimental stroke model, but has not so far been tested in the clinical setting (Clark et al., 1994). Recombinant human interferon beta-1a (IFN-b1a) is an FDA approved treatment for patients with relapsing remitting multiple sclerosis. IFN-b1a inhibits pro-inflammatory cytokines and prevents BBB disruption and is currently being studied in ischemic stroke in a phase 1 safety study. This agent could lessen reperfusion injury after stroke (and possibly hemorrhagic transformation of infarcts which can be a complication of rt-PA therapy). It too has been noted in a number of observational studies that patients who are taking a statin before a stroke have a better outcome than those who are not (Yoon et al., 2004). It has been proposed that, apart from cholesterol-lowering, statins work through other mechanisms that include *immune modulation* (decreased inflammation), decreased oxidative stress, inhibition of the thrombogenic response and increased atherosclerotic plaque stability. In experimental stroke models statin administration has resulted in reduced size of brain infarcts (Kawashima et al., 2003).

31.7.2. Treatment of Hemorrhagic Stroke

There are few proven treatments for intracerebral hemorrhage, the main management issue being strict control of hypertension. Surgical evacuation is usually reserved for cerebellar hematomas, which show a mass effect (as clinically evidenced by a declining level of consciousness). The coagulation factor Factor VIIa did show some initial promise in phase 2 studies (Mayer et al., 2005) but was recently abandoned (http://www.novonordisk.com, 2007). In terms of *immune modulation*, anti-inflammatory drugs such as COX-2 inhibitors, tacrolimus and fucoidan may reduce peri-hematoma inflammation and improve outcome in experimental stroke models (Nilupul Perera et al., 2006). Administration of the lipid-lowering drug atorvastatin has also been associated with improved outcome after experimental intracerebral hemorrhage, perhaps in part through modulation of inflammation in the peri-hematoma region (Jung et al., 2004).

For subarachnoid hemorrhage surgical clipping of the causative aneurysm or resection of the arteriovenous malformation is the mainstay of treatment. Endovascular coiling of the aneurysm can also be performed. Post-operative infection (either brain or respiratory) is an uncommon complication and not believed to be any more common than after other invasive surgical procedures. Hypervolemic-hemodilution and hypertensive (HHH) therapy is used to prevent spasm. There may be a role for anti-inflammatory measures for the prevention of vasospasm and delayed cerebral ischemia, as shown in a recent pilot study of patients treated with statins (Lynch et al., 2005).

31.7.3. Primary and Secondary Prevention of Stroke

Apart from surgical and interventional therapy of occlusive carotid artery disease, the major approach to preventing vascular disease and subsequent stroke is to pay close attention to the control of modifiable risk factors such as hypertension, smoking, diabetes, and hypercholesterolemia. Coumadin, an anticoagulant, is effective for the primary and secondary prevention of stroke in patients with atrial fibrillation. Aspirin, clopidogrel, and the combination of aspirin and dipyridamole have been proven to be effective for secondary stroke prevention along with the antihypertensive combination of indapamide and perindopril.

In terms of vascular disease prevention, inducing mucosal tolerance to E-selectin may prevent ischemic and hemorrhagic strokes, as shown in an experimental stroke study: low dose exposure to E-selectin is believed to work by diverting the immune response towards a Th2 (anti-inflammatory) response (Takeda et al., 2002). A clinical trial is planned of this agent for secondary stroke prevention. Also, immune modulation using naturally arising regulatory T cells ($CD4^+$ $CD25^+$) may be a promising approach for vascular disease prevention (Ait-Oufella et al., 2006).

Statins have an important role in primary and secondary stroke prevention and possibly work, at least in part as stated previously, by *immune modulation*. The reduction in vascular events with statin use correlates with increased degrees of lowering of the inflammatory marker hsCRP, and is independent of the degree of lipid-lowering. Statins have been shown to modulate gene expression in macrophages *in vitro* (Llaverias et al., 2004; Waehre et al., 2004). Furthermore, the clinical benefit of these drugs is manifested early in the course of lipid-lowering, before plaque regression could occur, and in angiographic studies it has been shown that reversal of arterial narrowing occurs slowly and only to a small extent (Rauch et al., 2000). Statins have been shown to be effective in primary stroke prevention in patients with proven coronary artery disease and in a recent study, high dose statins were proven effective for the secondary prevention of stroke (Amarenco et al., 2006). In observational studies it has also been noted that patients with rheumatoid arthritis who are taking TNF blockers such as infliximab have lower rates of cardiovascular disease (and of pro-inflammatory $CD4^+$ $CD28^-$ T cells); however, the toxicity of these agents precludes their use for vascular disease prevention at the present time.

Summary

The role of inflammation and the immune system in the evolution of ischemic brain lesions, stroke-related reperfusion injury and the development of atherosclerotic vascular disease has been increasingly appreciated since the late 1980s. However, the degree and mechanisms by which inflammation may impact on clinical and tissue outcome after stroke are still being debated. On one hand inflammation is believed to be harmful, by exacerbating ischemic brain injury in the early days after stroke, while, weeks to months after stroke, inflammatory cells may be beneficial through the promotion of tissue repair and remodeling. Brain-induced systemic inflammation may increase patients' susceptibility to infections after stroke. Specific components of the inflammatory process may be potential targets for pharmacological intervention for acute ischemic stroke as well as for the prevention of stroke and cerebrovascular disease. There appears to be an increasingly complex interplay between the immune system and the brain and the vasculature in stroke that could impact on the extent and severity of vascular disease and on stroke outcome and recurrence.

This highlights the rationale for using *newer technologies* for cellular and molecular profiling of blood, brain tissue and atherosclerotic plaques in clinical and translational research studies. This is now possible using multiple combinations of antibodies in flow cytometry studies and with microarray analyses. These studies may permit identification of functional classes of lymphocytes that impact on disease severity and outcome (cf, using flow cytometry, $CD4^+$ $CD28^-CD161^+$ T cells contributing to coronary artery plaque destabilization in acute coronary syndromes, Nakajima et al., 2002), and may permit reclassification of disease (cf, using gene expression profiling, novel subgroups of B cell lymphoma in gene expression studies, Alizadeh et al., 2000). Panels of genes from profiling of peripheral blood mononuclear cells show vascular and inflammatory genes that support current concepts of the stroke-related inflammatory process, and may even eventually provide clues to a possible brain signaling mechanism to the circulating blood after stroke.

Many aspects of the role of inflammation and the immune system in stroke remain controversial (Chamorro and Hallenbeck, 2006). However, more refined characterization of immune responses after stroke may suggest new and different therapeutic targets for acute ischemic stroke, in addition to novel approaches that are ongoing for the prevention of cardiovascular disease.

Review Questions/Problems

1. **Immune cells involved in the evolution of ischemic brain lesions include:**

   a. Microglia
   b. Neutrophils
   c. Monocytes
   d. Lymphocytes
   e. All of the above

2. **Immune cells involved in stroke-related inflammation stroke are predominantly derived from the circulating blood.**

   true/false

3. **Leukocyte infiltration across the blood brain barrier after ischemic stroke is dependent on the expression of adhesion molecules on leukocytes and endothelial cells.**

   true/false

4. **Molecular mediators of inflammation after ischemic stroke include**

   a. cytokines
   b. chemokines
   c. complement
   d. a and b
   e. a, b and c

5. **The first immune cell responders after ischemic stroke are the brain microglia.**

   true/false

6. **The neurovascular unit is comprised of the following:**

   a. astrocytes
   b. neurons
   c. smooth muscle cells
   d. endothelial cells
   e. all of the above

7. The effects of ischemic stroke on the systemic circulation include:

a. increased susceptibility to infection
b. sensitization of circulating lymphocytes to brain antigens
c. both a and b
d. neither a or b

8. Atherosclerosis is now regarded as a chronic inflammatory disorder as opposed to one of lipid storage.

true/false

9. Neutrophils are the major inflammatory cells contained within atherosclerotic plaques.

true/false

10. Features of the vulnerable plaque include:

a. more prone to rupture
b. heavy inflammatory cell infiltrate
c. thin fibrous cap
d. a, b and c
e. a and c

11. Discuss the role of the immune system before and after ischemic and hemorrhagic stroke.

12. Discuss the potential of immunomodulatory approaches for stroke treatment and prevention.

References

Ait-Oufella H, Salomon BL, Potteaux S, Robertson AK, Gourdy P, Zoll J, Merval R, Esposito B, Cohen JL, Fisson S, Flavell RA, Hansson GK, Klatzmann D, Tedgui A, Mallat Z (2006) Natural regulatory T cells control the development of atherosclerosis in mice. Nat Med 12:178–180.

Alizadeh AA, Eisen MB, Davis RE, Ma C, Lossos IS, Rosenwald A, Boldrick JC, Sabet H, Tran T, Yu X, Powell JI, Yang L, Marti GE, Moore T, Hudson J, Jr, Lu L, Lewis DB, Tibshirani R, Sherlock G, Chan WC, Greiner TC, Weisenburger DD, Armitage JO, Warnke R, Levy R, Wilson W, Grever MR, Byrd JC, Botstein D, Brown PO, Staudt LM (2000) Distinct types of diffuse large B-cell lymphoma identified by gene expression profiling. Nature 403:503–511.

Alvord EC, Jr, Hsu PC, Thron R (1974) Leukocyte sensitivity to brain fractions in neurological diseases. Arch Neurol 30:296–299.

Amarenco P, Bogousslavsky J, Callahan A, 3rd, Goldstein LB, Hennerici M, Rudolph AE, Sillesen H, Simunovic L, Szarek M, Welch KM, Zivin JA, Stroke Prevention by Aggressive Reduction in Cholesterol Levels (SPARCL) Investigators (2006) High-dose atorvastatin after stroke or transient ischemic attack. N Engl J Med 355:549–559.

Ballantyne CM, Hoogeveen RC, Bang H, Coresh J, Folsom AR, Chambless LE, Myerson M, Wu KK, Sharrett AR, Boerwinkle E (2005) Lipoprotein-associated phospholipase A2, high-sensitivity C-reactive protein, and risk for incident ischemic stroke in middle-aged men and women in the Atherosclerosis Risk in Communities (ARIC) study. Arch Intern Med 165:2479–2484.

Barone FC, Feuerstein GZ (1999) Inflammatory mediators and stroke: New opportunities for novel therapeutics. J Cereb Blood Flow Metab 19:819–834.

Becker K, Kindrick D, Relton J, Harlan J, Winn R (2001) Antibody to the alpha 4 Integrin decreases infarct size in transient focal cerebral ischemia in rats. Stroke 32:206–211.

Becker K, Kindrick D, McCarron R, Hallenbeck J, Winn R (2003) Adoptive transfer of myelin basic protein-tolerized splenocytes to naive animals reduces infarct size: A role for lymphocytes in ischemic brain injury? Stroke 34:1809–1815.

Brazelton R, Rossi FM, Keshet GI, Blau HM (2000) From marrow to brain: Expression of neuronal phenotypes in adult mice. Science 290:1775–1779.

Broderick J, Brott T, Kothari R, Miller R, Khoury J, Pancioli A, Gebel J, Mills D, Minneci L, Shukla R (1998) The Greater Cincinnati/Northern Kentucky Stroke Study: Preliminary first-ever and total incidence rates of stroke among blacks. Stroke 29:415–421.

Chamorro A, Hallenbeck J (2006) The harms and benefits of inflammatory and immune responses in vascular disease. Stroke 37:291–293.

Clark WM, Calcagno FA, Gabler WL, Smith JR, Coull BM (1994) Reduction of central nervous system reperfusion injury in rabbits using doxycycline treatment. Stroke 25:1411–1415.

Clatterbuck RE, Gailloud P, Ogata L, Gebremariam A, Dietsch GN, Murphy KJ, Tamargo RJ (2003) Prevention of cerebral vasospasm by a humanized anti-CD11/CD18 monoclonal antibody administered after experimental subarachnoid hemorrhage in nonhuman primates. J Neurosurg 99:376–382.

del Zoppo GJ (2006) Stroke and neurovascular protection. N Engl J Med 354:553–555.

Ding Y, Clark JC (2006) Cerebrovascular injury in stroke. Neurol Res 28:3–10.

Dirnagl U, Iadecola C, Moskowitz MA (1999) Pathobiology of ischaemic stroke: An integrated view. TINS 22:391–397.

Enlimomab Acute Stroke Trial Investigators (2001) Use of anti-ICAM-1 therapy in ischemic stroke: Results of the enlimomab acute stroke trial. Neurology 57:1428–1434.

Furuya K, Takeda H, Azhar S, McCarron RM, Chen Y, Ruetzler CA, Wolcott KM, DeGraba TJ, Rothlein R, Hugli TE, del Zoppo GJ, Hallenbeck JM (2001) Examination of several potential mechanisms for the negative outcome in a clinical stroke trial of enlimomab, a murine anti-human intercellular adhesion molecule-1 antibody: A bedside-to-bench study. Stroke 32:2665–2674.

Garcia JH (1992) The evolution of brain infarcts. A review. J Neuropathol Exp Neurol 51:387–393.

Garcia JH, Liu KF, Yoshida Y, Lian J, Chen S, del Zoppo GJ (1994) Influx of leukocytes and platelets in an evolving brain infarct (Wistar rat). Am J Pathol 144:188–199.

Gerhard A, Neumaier B, Elitok E, Glatting G, Ries V, Tomczak R, Ludolph AC, Reske SN (2000) In vivo imaging of activated microglia using [11C]PK11195 and positron emission tomography in patients after ischemic stroke. Neuroreport 11:2957–2960.

Gong C, Hoff JT, Keep RF (2000) Acute inflammatory reaction following experimental intracerebral hemorrhage in rat. Brain Res 871:57–65.

Hallenbeck JM, Hansson GK, Becker KJ (2005) Immunology of ischemic vascular disease: Plaque to attack. Trends Immunol 26:550–556.

Hamerman JA, Ogasawara K, Lanier LL (2005) NK cells in innate immunity. Curr Opin Immunol 17:29–35.

Hickey WF (2001) Basic principles of immunological surveillance of the normal central nervous system. Glia 38:118–124.

http://www.astrazeneca.com/pressrelease/5279.aspx: accessed 6/22/07.

http://www.novonordisk.com/press/sea/sea.asp?NewsTypeGuid = &sShowNewsItemGUID = 0cfc120c-f013-4ad6-9130-55227465e 3e8&sShowLanguageCode = en-GB: accessed 6/22/07.

Janket S-J, Baird AE, Chuang S-K, Jones JA (2003) Meta-analysis of periodontal disease and the risk of coronary heart disease and stroke. Oral Surg, Oral Med, Oral Pathol Oral Radiol Endod 95:559–569.

Jiang N, Moyle M, Soule HR, Rote WE, Chopp M (1995) Neutrophil inhibitory factor is neuroprotective after focal ischemia in rats. Ann Neurol 38:935–942.

Jung KH, Chu K, Jeong SW, Han SY, Lee ST, Kim JY, Kim M, Roh JK (2004) HMG-CoA reductase inhibitor, atorvastatin, promotes sensorimotor recovery, suppressing acute inflammatory reaction after experimental intracerebral hemorrhage. Stroke 35:1744–1749.

Kariko K, Weissman D, Welsh FA (2004) Inhibition of toll-like receptor and cytokine signaling—a unifying theme in ischemic tolerance. J Cereb Blood Flow Metab 24:1288–1304.

Karman J, Ling C, Sandor M, Fabry Z (2004) Dendritic cells in the initiation of immune responses against central nervous system-derived antigens. Immunol 92:107–115.

Kawashima S, Yamashita T, Miwa Y, Ozaki M, Namiki M, Hirase T, Inoue N, Hirata K, Yokoyama M (2003) HMG-CoA reductase inhibitor has protective effects against stroke events in stroke-prone spontaneously hypertensive rats. Stroke 34:157–163.

Keep RF, Xi G, Hua Y, Hoff JT (2005) The deleterious or beneficial effects of different agents in intracerebral hemorrhage: Think big, think small, or is hematoma size important? Stroke 36:1594–1596.

Krams M, Lees KR, Hacke W, Grieve AP, Orgogozo JM, Ford GA, ASTIN Study Investigators (2003) Acute Stroke Therapy by Inhibition of Neutrophils (ASTIN): An adaptive dose-response study of UK-279,276 in acute ischemic stroke. Stroke 34:2543–2548.

Lees KR, Zivin JA, Ashwood T, Davalos A, Davis SM, Diener HC, Grotta J, Lyden P, Shuaib A, Hardemark HG, Wasiewski WW, Stroke-Acute Ischemic NXY Treatment (SAINT I) Trial Investigators (2006) NXY-059 for acute ischemic stroke. N Engl J Med 354:588–600.

Libby P, Ridker PM, Maseri A (2002) Inflammation and atherosclerosis. Circulation 105:1135–1143.

Llaverias G, Noe V, Penuelas S, Vazquez-Carrera M, Sanchez RM, Laguna JC, Ciudad CJ, Alegret M (2004) Atorvastatin reduces CD68, FABP4, and HBP expression in oxLDL-treated human macrophages. Biochem Biophys Res Commun 318:265–274.

Lynch JR, Wang H, McGirt MJ, Floyd J, Friedman AH, Coon AL, Blessing R, Alexander MJ, Graffagnino C, Warner DS, Laskowitz DT (2005) Simvastatin reduces vasospasm after aneurysmal subarachnoid hemorrhage: Results of a pilot randomized clinical trial. Stroke 36:2024–2026.

Mackay KB, Bailey SJ, King PD, Patel S, Hamilton TC, Campbell CA (1996) Neuroprotective effect of recombinant neutrophil inhibitory factor in transient focal cerebral ischaemia in the rat. Neurodegeneration 5:319–323.

Mayer SA, Brun NC, Begtrup K, Broderick J, Davis S, Diringer MN, Skolnick BE, Steiner T, Recombinant Activated Factor VII Intracerebral Hemorrhage Trial Investigators (2005) Recombinant activated factor VII for acute intracerebral hemorrhage. N Engl J Med 352:777–785.

Meisel C, Schwab JM, Prass K, Meisel A, Dirnagl U (2005) Central nervous system injury-induced immune deficiency syndrome. Nat Rev Neurosci 6:775–786.

Nadareishvili ZG, Li H, Wright V, Maric D, Warach S, Hallenbeck JM, Dambrosia J, Barker JL, Baird AE (2004) Elevated pro-inflammatory CD4 + CD28- lymphocytes and stroke recurrence and death. Neurology 63:1446–1451.

Nakajima T, Schulte S, Warrington KJ, Kopecky SL, Frye RL, Goronzy JJ, Weyand CM (2002) T-cell-mediated lysis of endothelial cells in acute coronary syndromes. Circulation 105:570–575.

Nilupul Perera M, Ma HK, Arakawa S, Howells DW, Markus R, Rowe CC, Donnan GA (2006) Inflammation following stroke. J Clin Neurosci 13:1–8.

Offner H, Subramanian S, Parker SM, Afentoulis ME, Vandenbark AA, Hurn PD (2006) Experimental stroke induces massive, rapid activation of the peripheral immune system. J Cereb Blood Flow Metab 26:654–665.

Paczkowska E, Larysz B, Rzeuski R, Karbicka A, Jalowinski R, Kornacewicz-Jach Z, Ratajczak MZ, Machalinski B (2005) Human hematopoietic stem/progenitor-enriched CD34(+) cells are mobilized into peripheral blood during stress related to ischemic stroke or acute myocardial infarction. Eur J Haematol 75:461–467.

Pappata S, Levasseur M, Gunn RN, Myers R, Crouzel C, Syrota A, Jones T, Kreutzberg GW, Banati RB (2000) Thalamic microglial activation in ischemic stroke detected in vivo by PET and [11C]PK1195. Neurology 55:1052–1054.

Petroff OA, Graham GD, Blamire AM, al-Rayess M, Rothman DL, Fayad PB, Brass LM, Shulman RG, Prichard JW (1992) Spectroscopic imaging of stroke in humans: Histopathology correlates of spectral changes. Neurology 42:1349–1354.

Price CJ, Menon DK, Peters AM, Ballinger JR, Barber RW, Balan KK, Lynch A, Xuereb JH, Fryer T, Guadagno JV, Warburton EA (2004) Cerebral neutrophil recruitment, histology, and outcome in acute ischemic stroke: An imaging-based study. Stroke 35:1659–1664.

Ransohoff RM, Kivisakk P, Kidd G (2003) Three or more routes for leukocyte migration into the central nervous system. Nat Rev Immunol 3:569–581.

Rauch U, Osende JI, Chesebro JH, Fuster V, Vorchheimer DA, Harris K, Harris P, Sandler DA, Fallon JT, Jayaraman S, Badimon JJ (2000) Statins and cardiovascular diseases: The multiple effects of lipid-lowering therapy by statins. Atherosclerosis 153:181–189.

Rosamond W, Flegal K, Friday G, Furie K, Go A, Greenlund K, Haase N, Ho M, Howard V, Kissela B, Kittner S, Lloyd-Jones D, McDermott M, Meigs J, Moy C, Nichol G, O'Donnell CJ, Roger V, Rumsfeld J, Sorlie P, Steinberger J, Thom T, Wasserthiel-Smoller S, Hong Y, American Heart Association Statistics Committee and Stroke Statistics Subcommittee (2007) Heart disease and stroke statistics—2007 update: A report from the American Heart Association Statistics Committee and Stroke Statistics Subcommittee. Circulation 115:e69–e171.

Saleh A, Schroeter M, Jonkmanns C, Hartung HP, Modder U, Jander S (2004) In vivo MRI of brain inflammation in human ischaemic stroke. Brain 127:1670–1677.

Schroeter M, Jander S, Witte OW, Stoll G (1994) Local immune responses in the rat cerebral cortex after middle cerebral artery occlusion. J Neuroimmunol 55:195–203.

Sedgwick JD (1995) Immune surveillance and autoantigen recognition in the central nervous system. Aust N Z J Med 25:784–792.

Takeda H, Spatz M, Ruetzler C, McCarron R, Becker K, Hallenbeck J (2002) Induction of mucosal tolerance to E-selectin prevents ischemic and hemorrhagic stroke in spontaneously hypertensive genetically stroke-prone rats. Stroke 33:2156–2163.

The National Institute of Neurological Disorders and Stroke rt-PA Stroke Study Group (1995) Tissue plasminogen activator for acute ischemic stroke. N Engl J Med 333:1581–1587.

Vilhardt F (2005) Microglia: Phagocyte and glia cell. Int J Biochem Cell Biol 37:17–21.

Waehre T, Damas JK, Yndestad A, Tasken K, Pedersen TM, Smith C, Halvorsen B, Froland SS, Solum NO, Aukrust P (2004) Increased expression of interleukin-1 in coronary artery disease with downregulatory effects of HMG-CoA reductase inhibitors. Circulation 109:1966–1972.

Wahlgren NG, Ahmed N (2004) Neuroprotection in cerebral ischaemia: Facts and fancies—the need for new approaches. Cerebrovasc Dis 17 Suppl 1:153–166.

Wang X, Xu L, Wang H, Zhan Y, Pure E, Feuerstein GZ (2002) CD44 deficiency in mice protects brain from cerebral ischemia injury. J Neurochem 83:1172–1179.

Yilmaz G, Arumugam TV, Strokes KY, Granger DN (2006) Role of T-lymphocytes and interferon-γ in ischemic stroke. Circulation 113:2105–2112.

Yoon SS, Dambrosia J, Chalela J, Ezzeddine M, Warach S, Haymore J, Davis L, Baird AE (2004) Rising statin use and effect on ischemic stroke outcome. BMC Med 2:4.

Youngchaiyud U, Coates AS, Whittingham S, Mackay IR (1974) Cellular-immune response to myelin protein: Absence in multiple sclerosis and presence in cerebrovascular accidents. Aust N Z J Med 4:535–538.

32
Neurogenesis and Brain Repair

G. Jean Harry

Keywords Doublecortin; Epidermal growth factor; Fibroblast growth factor; Nestin; Progenitor cells; Stem cells, Radial glia; Rostral migratory stream; Subventricular zone; Subgranular zone

32.1. Introduction

Neurogenesis is a critical process in the formation and development of the neurons that comprise the brain. More recently, the identification of neurogenic regions in the adult brain has suggested that the general process of neurogenesis assumes a role in maintaining the normal brain as well as contributing to the repair of the brain following injury. As identified, adult neurogenesis is the production of new neurons in the adult brain. This is a complex process that is initiated with the division of a precursor cell and, in some cases, neural stem cells that progresses to the generation of functional new neurons. The environments within these neurogenic regions provide intrinsic factors to stimulate and maintain this as a normal process. These include a number of the various growth factors. In addition, there are a number of extrinsic factors that can stimulate this process including life style and environment, as welll as, factors that can inhibit the normal process, such as depression. The potential of these cells to assist in the repair of the brain following injury has fostered the generation of a significant body of research to determine the signals and factors that stimulate the generation and survival of newly generated cells. In addition, efforts to understand the various stimuli that can drive the progenitor or stem cell to a mature neuron or to an oligodendrocyte for the production of myelin have contributed a wealth of information on the nature and plasticity of these cells and this process. The current chapter will address the basic dynamics of neural development with regards to the generation, migration, and maturation of neurons as they provide information to understand the dynamics required for the adult neurogenesis. The identification and details of the neurogenic regions in the adult brain, the growth factors that contribute to maintaining these regions, and the response to injury will be discussed. It is the intent of this chapter to serve as an introduction to adult neurogenesis that can be used as a base for further study and evaluation of the increasing amount of data that is currently being generated.

32.2. Neurogenesis and Classification of Cells

32.2.1. Neural Development

In the normal process of nervous system development, organogenesis occurs during the period from implantation through mid-gestation. Neurogenesis is a complex process involving proliferation, migration, differentiation, and survival. It is characterized by an expansion phase in which stem cells undergo massive symmetric divisions followed by periods where expanded precursor cells give rise to differentiating cells (Gotz and Huttner, 2005). The complex architecture of the brain requires that different cell types develop in a precise spatial relationship to one another. To accomplish this, neural stem cells and their derivative progenitor cells generate neurons as well as astrocytes and oligodendrocytes by asymmetric and symmetric divisions. Stem cells are characterized by their ability for self-renewal with cell division generating at least one identical copy of the mother cell. Symmetric division yields two identical copies and asymmetric division produces one new identical stem cell and one that is determined toward a certain cell lineage. It is these cells that are often referred to as progenitor cells given the reduction in stem cell properties. However, these progenitor cells can dramatically expand in number of new cells. In the development of the brain, such cells can undergo a terminal symmetrical division leading to two differentiating cells (Takahashi et al., 1996).

In some parts of the nervous system, particular kinds of cells are generated from committed progenitor or "blast" cells. These cells proliferate symmetrically and then differentiate. Examples of such progenitors are the sympathoadrenal progenitor and the O2A progenitor from the optic nerve. As initially described, glia O2A progenitor cells first give rise to oligodendrocytes around the time of birth and begin to generate type 2 astrocytes during

T. Ikezu and H.E. Gendelman (eds.), *Neuroimmune Pharmacology.*
© Springer 2008

the second postnatal week (Raff et al., 1983). During this time, platelet-derived growth factor (PDGF) is an important mitogen for the O2A progenitor. The differentiation of type 2 astrocytes is timed by cell-extrinsic factors. For example, ciliary neuronotrophic factor (CNTF) is a diffusible signal that, in association with other signals of the extracellular matrix, is required to induce O2A progenitors to develop into type 2 astrocytes. The current thought is that the NG2 cells of the brain may contain the O2A progenitor cells (Nishiyama et al., 1996; Dawson et al., 2000; see below for additional discussion).

In the formation of the cerebral hemispheres, neurogenesis begins during fetal development; gliogenesis produces astrocytes followed by development of oligodendrocytes. For each cell type, temporal differences in cell production are maintained. For example, the first-generated neurons reach their final position before subsequent generations of neurons. The migration of neuronal precursors plays a role in establishing the identity of some neurons and defining the functional properties and connections of the neuron (Sidman and Rakic, 1973). This translocation is achieved by a combination of the extension of cell process, attachment to the substratum, and subsequent pulling of the entire cell by means of contractile proteins associated with an intracellular network of microfilaments. Directional control occurs as cells move along "guide" cells or according to a concentration gradient of chemotropic molecules. Positional identity of precursor cells is also spatially and temporally regulated by transcription factor patterns.

The radial glial cell processes serve to guide neurons from the zone of neuronal generation to the zones for final settlement and the laminar features of the cortex are generated over time by differential movement of groups of neurons born at different times. These cells also divide asymmetrically to give rise to another radial glial cell while the second cell can differentiate into a neuron. Thus radial glial cells may represent a unique stem cell population (Gaiano et al., 2000; Malatesta et al., 2000). In addition to the radial glia having stem cell potential, astrocytes isolated from the embryonic and early postnatal CNS display stem cell functional features (Laywell et al., 2000). In examining various precursor cells in culture, specific culture conditions can induce some precursors to show a change in commitment such as, the oligodendrocyte precursor cells (Kondo and Raff, 2000a) and transit-activating precursors (Doetsch et al., 2002). These cells can be reprogrammed to multipotent stem cells by the action of cytokines and epidermal growth factor (EGF).

32.2.2. Neurogenic Potential in the Adult

Select areas of the brain contain populations of progenitor cells with various proliferative and migratory potentials (Altman, 1969; Kaplan and Hinds 1977; Altman and Bayer, 1990). In the adult mammalian brain, neurogenesis continues in restricted germinal regions: the subependymal zone of the lateral ventricle subventricular zone (SVZ) and the subgranular zone (SGZ) between the hilus and the granule cell layer of the hippocampal dentate gyrus. Characterization of adult cells found in these proliferative domains demonstrated that the appropriate neurogenic signals are present and continuously support the stem/progenitor cell population. In addition, a third population of stem/progenitor cells may reside within the brain in the form of cells with astrocyte-like properties (Laywell et al., 2000; Magavi and Macklis, 2001).

Mitotically active precursor cells in the adult brain consist of a heterogeneous population of cells including stem cells with the capacity for continual self-renew and undergo multilineage differentiation in that they can differentiate into several distinct cell types (Alvarez-Buylla et al., 2001; Suslov et al., 2002). As compared to the multipotent stem cell, progenitor cells do not have the ability for continual self-renewal and have a limited multipotent capacity. The term "precursor cell" can refer to cells for which there remains limited proliferative ability but the differentiation fate is fairly determined. While a resident stem cell population has been demonstrated within the adult subependyma of the rostral lateral ventricle (Reynolds and Weiss, 1992; Morshead and van der Kooy, 2001; Gritti et al., 1996), such a prominent population has not been clearly demonstrated in the SGZ. Characterization of SGZ cells suggests that while some cells meet the criteria of stem cells, most studies characterizing proliferative SGZ cells suggest that they are primarily progenitor cells (Seaberg and van der Kooy, 2002; Bull and Bartlett, 2005).

The proliferative capacity of cells from these zones can be evaluated with the isolation of cells to form spherical, detached colonies called neurospheres (typically 50–140mm in diameter). The generation of a neurosphere is not conclusive evidence of stem cell presence as both stem cells and progenitor cells are capable of forming such spheres (Reynolds and Rietze, 2005). An additional step for evaluation requires a limiting dilution neurosphere assay in which cells are plated by serial dilutions and analyzed for size of neurospheres and frequency of colony-forming cells. This allows for determination of stem cell and progenitor cell contributions. Under these conditions, the largest colonies (>1.5 mm diameter) originate from stem cells while progenitor cells form small colonies (Reitze and Reynolds, 2006). Cultures of stem cells can be expanded continuously while those of progenitor cells gradually diminish. Other ways to distinguish between the two cell types is in their response to mitogens such as, fibroblast growth factor-2 (FGF-2) and EGF with progenitor cells showing a greater responsiveness to FGF-2. In addition to proliferation, growth factors and other signaling molecules are critical for the differentiation of these newly proliferated cells. Multipotency can be assessed with the removal of EGF and FGF-2 and the addition of other factors to the culture medium. Neurospheres containing stem cells are multipotent and can produce neurons, astrocytes, and oligodendrocytes. If the neurospheres are comprised of progenitor cells, differentiation will generate astrocytes and oligodendrocytes and differentiation to neurons and glia will occur with the addition of brain derived growth factor (BDNF).

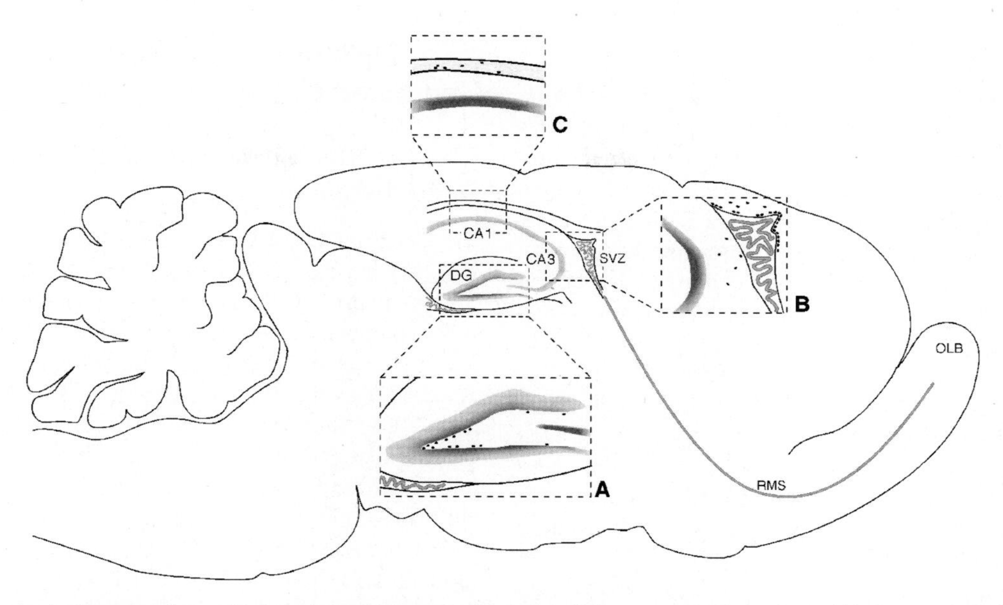

FIGURE 32.1. Schematic diagram of a sagittal section of the mouse brain localizing neurogenic regions. (A) dentate granule cell layer of the hippocampus. Newly proliferating cells in the subgranular zone (SGZ) are represented as dark cells along the inner border of the blades of the dentate. (B) subventricular zone (SVZ). (C) Extension of the ventricular system and the presence of newly proliferating cells within the ventricle noted as dark cell. RMS—rostral migratory stream. OLB—olfactory bulb.

32.2.2.1. Subventricular Zone

Cells generated in the SVZ migrate tangentially through a network of interconnecting pathways distributed throughout the wall of the lateral ventricle. The heterogeneous neuroblasts originating in the adult SVZ migrate in a chain-like manner towards the olfactory bulb (OB) along a defined pathway called the rostral migratory stream (RMS) (Lois et al., 1996). These migrating neuroblasts (type A cells) show a spindle-shaped cell body with one or two cell processes. Their elongated nucleus displays a dispersed chromatin pattern with small aggregates of condensed chromatin masses, 2–3 small nucleoli, and possible nuclear indentations. They have scant and electron-dense cytoplasm with many ribosomes, a small Golgi apparatus, few cisterna of rough endoplasmic reticulum, and many microtubules distributed along the long cell axis. There are no dense bodies, lipid droplets, or microvilli. These cells are immunopositive for nestin, the polysialylated form of the neural cell adhesion molecule (PSA-NCAM), TuJ1 and are immunonegative for glial fibrillary acidic protein (GFAP) and vimentin (Doetsch et al., 1997). (The details of each marker will be discussed later). During their migration, in the absence of radial glia or axonal guidance, the cells maintain contact with each other to reach the OB and differentiate into granule and periglomerular neurons. Cell differentiation is influenced by the expression of specific transcription factors in the SVZ or during migration in the RMS. For example, the transcription factor Pax6 is required for the production of a specific subpopulation of OB neurons (Kohwi et al., 2005). It promotes the generation of neuronal progenitors directing them towards the dopaminergic periglomerular phenotype that is predominantly generated in the RMS (Hack et al., 2005). The transcription factor, Olig2, opposes Pax6 and promotes oligodendrogenesis (Hack et al., 2005).

In the rodent, the primary target site for newborn cells from the SVZ is the OB; however, more recent studies suggest that other brain regions may also receive cells (Arvidsson et al., 2002). This would support a role for stem/progenitor cells characterized in the primate SVZ (Quinones-Hinojosaet al., 2006; Tonchev et al., 2006). Several lines of evidence suggest that the SVZ astrocyte serves as the primary precursor for these new neurons. Astrocytes in the SVZ, type B1 and B2, have distinct morphological characteristics (Doetsch et al., 1997) and express vimentin, nestin, and GFAP but not PSA-NCAM or TuJ1. Transiently amplifying progenitor cells (type C cells) are smooth and spherical, have large nuclei with deep indentations, dispersed chromatin pattern, and large reticulated nucleolus. The Golgi apparatus is large, there are fewer ribosomes than in the type A cells, and they lack intermediate filament bundles. Type C cells stain for nestin, are immunonegative for PSA-NCAM, TuJ1, GFAP, and vimentin. They express the epidermal growth factor (EGF) receptor, express the Dlx2 transcription factor, and are found in small clusters in the SVZ, as well as, isolated cells. NG2 cells in the SVZ display a transit-amplifier type C cell phenotype and may provide a significant proliferative progenitor pool in the brain (Aguirre et al., 2004) Figure 32.1.

32.2.2.2. Subgranular Zone

A relatively small number of proliferating cells derived from the cells of the hippocampal pseudostratified ventricular epithelium during embryonic stages continue to reside in the hippocampal hilus (Nowakowski and Rakic, 1981). In the rodent, this population is established at birth and during the early postnatal period. This intrahilar population produces about 80% of the cells of the dentate granule cell layer (Bayer and Altman, 1975). In humans and monkeys, this is reversed and only about 20% production occurs postnatally (Nowakowski and Rakic, 1981). This zone includes the basal cell band of the granule cell layer and a two-cell-wide layer into the hilus. Several types of neuronal and glial progenitor cells, as well as, astrocytes with radial glial elements, are within this neurogenic zone (Seri et al., 2004; Filippov et al., 2003). Similar to the SVZ, astrocyte-like B cells give rise to the transiently amplifying progenitor C cell to produce the migrating A cell. The putative stem cells morphologically resemble radial glia and have astrocytic properties. They have a large triangular-shaped soma with long apical processes reaching into the granule cell layer. Sparse branching occurs into the outer third of the granule cell band (Filippov et al., 2003). Radial glia-like cells expressing nestin, type-1 cells, consistently express the astrocytic protein, GFAP, and are negative for the astrocytic marker S100β. Type-2 cells are nestin-expressing cells with irregularly shaped nucleus, dense chromatin, and short processes oriented parallel to the SGZ. These cells are either negative (type 2a) or positive (type 2b) for the immature neuronal marker doublecortin (DCX). Type 3 cells are DCX positive but nestin negative, with a rounded nucleus and express PSA-NCAM and Prox-1. The type-3 stage comprises a transition to a postmitotic immature neuron. Adult-generated hippocampal cells migrate primarily to the inner third of the granular cell layer (GL; Kempermann et al., 2004). The exit from the cell cycle into the terminal postmitotic differentiation of granule cells is accompanied by a transient expression of calretinin occurring with expression of Prox-1 and the postmitotic neuronal marker, NeuN. With maturation of the granule cells, calbindin, rather than calretinin, is expressed (Brandt et al., 2003). The immature granule cells have a globular cell body with many fine dendritic processes extending into the molecular layer (Kempermann et al., 2004) and axons into the CA3 pyramidal cell layer (Hastings and Gould, 1999). In the GL these cells assume the nuclear and cytoplasmic morphology of the surrounding neurons, express biochemical markers of immature and mature neurons, and receive functional GABAergic contacts. When the neurons are established in their final position within the GL, they develop spiny dendrites reaching the outer molecular layer and functional glutamatergic afferents increase. With the final stages of maturation, the cells obtain perisomatic GABAergic contacts. When these cells are selectively removed from the brain regions and cultured as neurospheres, they can differentiate into various cells and the neurons can form functional synapses (van Praag et al., 2002; Schmidt-Hieber et al., 2004).

32.3. Methods Used to Detect Newly Generated Cells

32.3.1. Markers Used for Cell Proliferation and Stem/Progenitor Cells

The incorporation of bromodeoxyuridine (BrdU) into replicating DNA during the S-phase of the cell cycle is a marker for DNA synthesis. BrdU incorporation can represent an increase in the number of proliferating cells or simply changes in cell cycle, active DNA repair (Selden et al., 1993), apoptosis (Katchanov et al., 2001), or development of tetraploidy (Yang et al., 2001); thus, in the absence of caution, false conclusions are possible regarding neurogenesis (Rakic, 2002). However, given the usually close link between BrdU incorporation and cell proliferation, it was initially used by Nowakowski et al. (1989) as a tool for studying cell proliferation in the developing nervous system. As an alternative, or complement, Ki-67 is a nuclear protein expressed at different levels in dividing cells for the duration of their mitotic process. There are however, questions with regard to whether the signal is sufficient to detect the early G1 phase of the cell cycle. While Ki-67 immunohistochemistry will provide a snapshot of a specific window in time, BrdU incorporation into cells can be followed over time. In the normal adult brain, the number of cells in the SGL that incorporate BrdU or immunostain for Ki-67 is histologically represented by a sparse number of positive cells along the inner blade of the dentate (Kee et al., 2002). Despite the technical problems with BrdU, results indicating significant amounts of adult-generated cells in the hippocampus (Cameron and McKay, 2001) are supported by labeling of newly generated cells with retroviral vectors and integration into the host genome (Lewis and Emerman, 1994). Co-immunostaining of the cells with neuronal specific markers suggests that these cells can differentiate into neurons. However, most of the newborn cells undergo cell death during the first several weeks after final cell division (Biebl et al., 2000; Dayer et al., 2003).

Additional methods of detection and tracking the fate of newly generated cells have been developed. Fluorescent markers have been combined with retroviruses for birth dating and tracking of cells (van Praag et al., 2002). Continued development of retroviral vectors has resulted in vectors to express transgenes, short-hairpin RNAs, and site-specific recombinase. Transgenic mice have been developed containing a restricted expression of a fluorescent marker in progenitor cells. For example, green fluorescent protein (GFP) under the regulatory control of the nestin gene produces animals with labeled progenitor cells and immature neurons in the dentate gyrus (Yamaguchi et al., 2000; Filippov et al., 2003). Under transcriptional control of proopiomelanocortin (POMC) genomic sequences, cryptic sequences in the transgene provides expression in immature dentate granule cells that are approximately 2 weeks postmitotic (Overstreet et al., 2004). Alternatively, one can examine the progenitor cell population based on their distinct membrane properties. Newborn cells display high input resistance (Ambrogini et al., 2004; Schmidt-Hieber et al., 2004) and a lower threshold for activity-dependent synaptic plasticity such as long-term potentiation is seen in immature granule neurons (van Praag et al., 1999; Schmidt-Hieber et al., 2004).

32.3.2. Neural Cell-Specific Markers

Neuronal Nuclei (NeuN) expression is restricted to postmitotic neurons. The protein is localized in the nucleus with staining occurring in the cytoplasm near the nucleus and in some cases extending into the neurites. NeuN can bind to DNA and may serve as a transcription factor that can be induced with the initiation of terminal differentiation (Sarnat et al., 1998). There is no reported case of NeuN expression in glial cells. NeuN immunoreactivity can be diminished in pathological conditions; however, loss of staining does not necessarily equate to loss of neurons nor does a reemergence of staining indicate neurogenesis (Unal-Cevik et al., 2004).

TuJ1 is the clonal designation for an antibody used against β-III-tubulin, a minor neural isotype of tubulin associated with neurons. Immature neurons in the neurogenic zones express β-III-tubulin; however, there is little data regarding the re-expression in mature neurons and there is some data suggesting a limited overlap with non-neuronal lineages. TUC4 (TOAD-64, Ulip-1, DRP-3, CRMP4) is expressed by postmitotic neurons at the stage of initial differentiation with the highest expression occurring in the growth cone. The duration of the expression after cells have become postmitotic is not known. With additional markers, it may offer information on the development of neurons either in the mature brain or following injury. It does not offer information on the net level of neurogenesis.

In neurogenic regions, PSA-NCAM is a specific marker for the neuronal lineage. The polysialic acid residue reduces the cell adhesion of NCAM; thus, it is found on migratory neuronal cells and identifies type-2b and type-3 cells in the dentate gyrus and C-cells in the SVZ and RMS (Seki, 2002). Expression of PSA-NCAM is thought to occur during post-neuronal differentiation and while, associated with neurogenesis, it is also associated with synaptic plasticity and can be expressed on glial cells in non-neurogenic brain regions (Kiss et al., 1993). In the neurogenic zones, the temporal expression of PSA-NCAM parallels that of doublecortin but offers information of the dendritic arborization of the cells.

Doublecortin (DCX) is a microtubule-associated protein expressed in neuronal cells during migration. It has both cytoplasmic and nuclear localization and is enriched in the leading processes of these cells (Schaar et al., 2004) and in the growth cones of neurites. DCX is transiently expressed during adult neurogenesis and identifies the phase of migration and neurite extension of type-2b and type-3 cells (Brandt et al., 2003; Brown et al., 2003a; Ambrogini et al., 2004). The expression persists into the postmitotic stage of neurons when it overlaps with calretinin expression (Brandt et al., 2003). Throughout the brain, DCX expression overlaps with PSA-NCAM with some evidence of staining of NG2 glial cells.

Nestin is an intermediate filament protein that is localized to a subpopulation of precursor cells in both the SGZ and the SVZ (Filippov et al., 2003). It is expressed at low levels in the neurogenic regions and in blood vessels. In green fluorescent protein (GFP)-nestin transgenic reporter mice, positive cells overlapping with DCX can be found through the entire brain. With injury, both nestin and DCX positive cells are found at the target site and can occur in the absence of neurogenesis. This reflects the residual capability of cells within the non-neurogenic region as induced with changes in the microenvironment or rather, a limited specificity of these markers.

The chondroitin-sulfate proteoglycan, NG2, is a transmembrane protein containing extracellular laminin-like domains. Cells expressing NG2 are a prominent progenitor population in the postnatal brain (Dawson et al., 2000). These cells can be either highly proliferative and migratory or slowly dividing and non-migratory (Belachew et al., 2003; Aguirre et al., 2004). It is thought that NG2 labels primarily precursor cells in the oligodendrocytic lineage and secondarily, a specific type of astrocyte that is closely associated with synapses and axonal structures. The sensitivity of NG2 as a marker for neurogenesis is not clearly defined but may indicate the involvement of radial glia as a precursor cell.

32.4. Regulatory Factors Influencing Adult Neurogenesis

As in many other forms of repair mechanisms for the nervous system, such as neurite outgrowth, reactive synaptogenesis, and remyelination, it is thought that the mechanisms involved recapitulate those that occur in normal development. This assumption has more recently been applied to the mechanisms operating to control stem cell proliferation (Vaccarino et al., 2001). During development of the brain, neural stem cells follow an orderly sequence of events resulting in the coordination of multiple signals to produce specific products at specific critical times. In the adult, this may not be the case, but rather that the system responds to local environmental influences to produce a response appropriate for the situation. Questions that continue to be raised are: What intrinsic properties render a progenitor cell proliferative and migratory? What signaling molecules promote proliferation and migration in quiescent neural progenitors? To gain a better understanding of the plasticity of this process, changes in the number of BrdU+ or viral+ cells within the brain have been examined under various conditions. This process is regulated by a wide range of molecules such as hormones, neurotransmitters, growth factors, and transcription factors. It is also regulated by aging, nutrition, physical exercise, and environmental enrichment. Additional details for some of these modulatory factors will be discussed in the following sections.

32.4.1. Age

While the proliferative activity within the brain germinal zones declines with increasing age, cell proliferation continues throughout life (Altman and Das, 1965; Kuhn et al., 1996; Seki, 2002; Hallbergson et al., 2003; Duman, 2004; Maslov et al., 2004). An age related decrease in hippocampal neurogenesis (Kuhn et al., 1996; Lichtenwalner et al., 2001; Jin et al., 2003; Heine et al., 2004) has been linked to the decline in cognitive function with aging. With a decrease in the generation of new neurons as a function of age, the response to either the environment or injury may override any such changes. For example, an enriched environment initiated in the later half of life significantly increased the level of adult hippocampal neurogenesis in 20-month-old rodents (Kempermann et al., 2002). With transient forebrain ischemia, 2-year-old rats show a greater increase in BrdU incorporation into cells in the SGZ as compared to young adults. However, fewer of these cells survive and proportionally fewer differentiate into mature neurons (Yagita et al., 2001). These age-related changes in neurogenesis may reflect the influences of changes in the host environment with regards to mitogenic and differentiation factors. With increasing age, the ratio of glial cells to neurons increases, often resulting in glial hypertrophy (Cotrina and Nedergaard, 2002). With aging, the increase in glucocorticoids has been associated to synaptic loss and inhibited production of new granule neurons (Nichols et al., 2005). The decreased neurogenesis in the aged brain may be related to a decrease in insulin like growth factor-1 (IGF-1) (Aberg et al., 2000; Lichtenwalner et al., 2001) or increased oxidative stress (Nicolle et al., 2001).

To address such questions of the environment, experimental models of fetal cell transplants into the brain have been employed. Quite often, a decrease in cell survival is seen with increasing age of the host. The survival rate can be improved if the fetal cells are incubated with various signaling factors (e.g., BDNF, NT-3, and caspase inhibitors) prior to transplantation. Using the olfactory bulb as an example, in a developing system, interneurons produced in the SVZ and partly in the RMS (Pencea and Luskin, 2003) populate a newly developing structure; while in the adult, the interneurons are integrated into an existing OB circuit. The origin, final target site, and distribution profiles are age dependent. Newly generated cells are evenly distributed across the anteroposterior axis of the OB and the relative contribution of SVZ versus localized bulbar neurogenesis increases from the neonate to the young adult. Cells derived from the SVZ/RMS in the neonate are targeted to the superficial regions of the GL and in the young adult, they are found primarily in the deeper regions (Lemasson et al., 2005). The fact that newborn cells reach the OB target faster when generated in the young adults as compared to the neonates (Lemasson et al., 2005) suggests the need for maturation of specific processes and anatomical structures to allow for rapid migration of cells.

32.4.2. Life-Style

Conditions normally associated with a decrease in neural activity such as depression or those associated with stress can result in a decrease in an already minimal process (Duman, 2004). In experimental models of stress, cell proliferation within the SVZ and SGZ is decreased and this decrease is often associated with a decrease in memory performance. Pharmacological antidepressant agents such as lithium can enhance hippocampal neurogenesis (Chen et al., 2000). A chronic increase in the levels of corticosteroids has been proposed as one primary mediator of age-related decline in neurogenesis; however, such decline is not necessarily associated with an increase in circulating hormone levels (Heine et al., 2004). Yet, with age, the expression of corticosteroid receptors shifts to a more immature state possibly representing a shift in receptor sensitivity.

Additional life-style modulators have been examined for their stimulatory role in adult neurogenesis. Rodents maintained on dietary restriction show an increase in hippocampal neurogenesis, perform better on learning and memory tasks, show increased resistance to neuronal degeneration, and have increased BDNF levels in the hippocampal and cerebral cortex (Lee et al., 2000, 2002; Duan et al., 2001). Initial studies suggested that neurogenesis might be activity dependent and regulated related to normal behavior (Kemperman et al., 1997). The rate of neurogenesis in animals increases following voluntary exercise, housing in an enriched environment, and with exposure to specific learning and performance paradigms (Gould et al., 1999; Nilsson et al., 1999; van Praag et al., 1999). Environmental enrichment increases neurogenesis in the DG but not the SVZ and is mediated partly by BDNF (Brown et al., 2003b). Such increases can occur with a concurrent improvement on learning tasks dependent upon the hippocampus and inhibition of neurogenesis decreases learning performance. The impact of environmental enrichment is decreased when the basal level of neurogenesis is increased by other factors such as social dominance (Kozorovitskiy and Gould, 2004). These studies suggest that while a number of environmental factors can increase neurogenesis in the normal brain, there are limits to this induction.

32.4.3. Endogenous Factors

Signals in the cell environment regulate the maintenance, proliferation, and neuronal fate commitment of the local stem cell populations (Alvarez-Buylla and Lim, 2004; Gotz and Huttner, 2005). *In vitro* and *in vivo*, numerous factors have been shown to support the production of neural cells from the SVZ and SGZ. Factors derived from cells surrounding the neural stem cells participate in the regulation of neurogenesis such as, growth factors and other small molecules including, FGF2, PACAP, IGF-1, NT3 (Vaccarino et al., 2001, review). As one would expect, the number of factors identified continues to increase with additional work and spans many of the same factors identified as critical for organogenesis and the initial neurogenesis during development.

32.4.3.1. Sex Hormones

The primary support for a hormonal influence in adult neurogenesis comes from the reports of neural progenitor cell proliferation being higher in the female rodent as compared with the male (Tanapat et al., 1999; Abrous et al., 2005). In the female rodent, cell proliferation peaks in proestrous with the increase in estrogen and decreases in phases of low estrogen (Tanapat et al., 1999). This level of proliferation is higher than what is seen in males (Abrous et al., 2005). The effects of estradiol are mixed with the induction of embryonic precursor cell proliferation (Brannvall et al., 2002; Ormerod et al., 2003) yet, a reduction in the mitogenic effects of EGF in rat neural precursor cells. The neuroactive progesterone metabolite, allopregnanolone (3a-hydroxy-5a-pregnan-20-one) induces a significant increase in the proliferation of neuroprogenitor cells derived from the rat hippocampus as well as, cortical derived human neural stem cells (Wang et al., 2005).

32.4.3.2. Growth and Neurotrophic Factors

As extracellular signaling molecules, growth factors have diverse effects on neurogenesis, proliferation, and maintenance of new neurons. Cell surface adhesion and recognition molecules mediate interactions between individual cells and between cells and the extracellular matrix. Additional interactions occur by means of diffusible molecules such as, growth factors and trophic agents. Studies on neural development have identified several trophic factors important for neuronal survival and growth that appear to also contribute to the generation of new cells in the adult brain. Brain-derived neurotrophic factor (BDNF), IGF-1, erythropoietin, epidermal growth factor (EGF), and the basic fibroblast growth factor (FGF2) (bFGF) have been shown to support neural cell production.

EGF is known to be important in the proliferation and maintenance of embryonic and adult neural stem cells (Doetsch et al., 1999, 2002) and EGFR is involved in the radial migration and maturation of neural precursors during embryonic cortical development (Caric et al., 2001). EGF has a strong mitogenic effect on stem cells in culture and upon withdrawal the cells can be induced to differentiate. With the direct infusion of EGF into lateral ventricles proliferation increases within the SVZ (Kuhn et al., 1997) and neurogenesis following ischemia can be augmented (Nakatomi et al., 2002). Both the basic and acidic forms of fibroblast growth factor (FGF) stimulate outgrowth of neurites, reduce the effects of injury, and enhance regeneration. FGF-2 plays a critical role in signaling for hippocampal neurogenesis in normal rodents (Cheng et al., 2002), aged mice (Jin et al., 2003), and following traumatic brain injury (Yoshimura et al., 2001). Following infusion into the ventricle, FGF-2 expands dividing cells of the SVZ and induces net neurogenesis in the OB (Kuhn et al., 1997). FGF-2 protein is up regulated within the hippocampus with seizures induced by kainic acid and after focal cerebral ischemia. With both models of injury, the induction of neurogenesis is significantly decreased in FGF-2 null mice suggesting a regulatory role for this growth factor (Yoshimura et al., 2001).

Nerve growth factor (NGF) is required for survival and neurite outgrowth of cholinergic neurons of the basal forebrain, sympathetic postganglionic neurons, and sensory ganglion cells derived from the neural crest. In the developing and regenerating peripheral nerve, it is produced by both Schwann cells and macrophages. The highest level of NGF occurs in the hippocampus and the cerebral cortex, both of which are targets for cholinergic megnocellular neurons of the basal forebrain. A member of the same family, brain-derived neurotrophic factor (BDNF) is a 12.3 kDa basic protein and, as a growth factor, it has its maximum effect during the time when embryonic neurons contact targets in the CNS. BDNF contributes to brain synaptic plasticity, influences aging, and promotes neurogenesis and cell survival (Mattson et al., 2004). This growth factor assists in maintaining the basal activity in the proliferative zones of the brain (Lee et al., 2002), can directly stimulate neurogenesis (Scharfman et al., 2005), and with voluntary exercise or enriched environment, is upregulated in the dentate gyrus (Farmer et al., 2004). It has also been reported to mediate the effects of antidepressant drugs on hippocampal neurogenesis (Sairanen et al., 2005). However, over expression of BDNF in the hippocampus decreases the neurogenic response to ischemic insult (Larsson et al., 2002) and the blockage of endogenous BDNF increases ischemia induced hippocampal neurogenesis (Gustafsson et al., 2003). The inhibitory effects may be associated with the induction of progenitor cells rather than stem cells in the SGZ and the dysregulatory effects of chronic BDNF exposure as proposed by Larsson et al. (2002). This idea is supported by the work of Cheng et al. (2003) in which BDNF was found to reduce neuroprogenitor cell proliferation and enhance neuronal differentiation. Thus, in an ischemic injury, excess BDNF could serve to regulate neural progenitor cells in the SGZ with a down-regulation of excess proliferation and an induction of differentiation of the cells to the neuronal lineage. With a chronic BDNF administration, the regulatory balance would be disrupted resulting in a decrease in the overall neurogenic response.

The increased number of newly generated cells in the SGZ following life-style modulators, is thought to be due to the increase in cell survival and has been linked to increased protein levels for both BDNF and neurotrophin 3 (NT-3) (Lee et al., 2000, 2002; Duan et al., 2001). Ciliary neurotrophic factor (CNTF) is an acidic protein involved in type-2 astrocyte differentiation. Injection of CNTF into the mouse brain stimulates precursor cell proliferation with the CNTF receptor alpha expressed on GFAP-positive cells of the SVZ (Emsley and Hagg, 2003).

Insulin-like growth factor-1 (IGF-1) mediates BDNF action and, when administered either peripherally or directly into the ventricle, induces both cell proliferation and net neurogenesis (Aberg et al., 2000; Lichtenwalner et al., 2001). It is expressed in neurons with physical activity and many of the protective effects of exercise can be inhibited by blocking the uptake of either IGF-1 (Carro et al., 2001) or vascular endothelial growth factor (Trejo et al., 2001). IGF-1 can also inhibit some of the age-related decline in hippocampal neurogenesis (Lichtenwalner et al., 2001).

32.4.3.3. Cell Adhesion Molecules

Migration is influenced by the adhesion properties of the cells and the direct interactions between a cell and the extracellular matrix. These include cell adhesion molecules (CAMs), intercellular adhesion molecules (I-CAMs), integrins, and cadherins. CAMs are a family of high molecular weight cell surface glycoproteins with regulatory properties during neural development. Family members include neural CAM (N-CAM), neuronal-glial CAM (Ng-CAM-NILE or L1), tenascin, and adhesion molecule on glia (AMOG/beta2 isoform of the membrane Na, K-ATPase pump). N-CAM is widespread early in embryogenesis and in both neurons and glia throughout nervous system development. It mediates Ca2+-independent homophilic binding and aggregation of neuronal cells. N-cadherin is important in Ca2+-dependent cell-cell interactions. Tenascin is a large extracellular matrix glycoprotein implicated in cell proliferation and neural cell attachment. With the decline of tenascin expression begins the expression of L1, a protein involved in heterotypic binding between neuronal and neuroglial cells. Integrins are membrane receptors with ligands consisting of I-CAMs and other matrix components such as collagen, laminin, and fibronectin. Integrin activation can lead to rapid changes in cell adhesion properties in the local environment and can signal intercellular events. These receptors provide the developing neural cells a system for linking adhesion/migration information with other developmental signals controlling proliferation and differentiation. The tangential migration of neuroblasts is influenced by multiple molecular signals. The polysialylated form of the neural cell-adhesion molecule confers a migratory phenotype to neuroblasts (Pencea and Luskin, 2003). Directional migration is mediated by deleted in colorectal carcinoma and integrins (Murase and Horwitz, 2002). The soluble ligands Slit1 and Slit2 serve as guidance signals (Bagri et al., 2002).

32.4.3.4. Other Developmental Signaling Factors

Members of the receptor tyrosine kinase (RTK) family serve as signaling molecules. This large family includes the Eph family and their transmembrane-associated ephrin ligands and members of the Erb RTK family. They serve to influence the proliferation of cells in the SVZ (Conover et al., 2000). The RTK ErbB4 is expressed by neuroblasts located in both the SVZ and within the RMS (Anton et al., 2004). This kinase can be activated by multiple EGF-like domain-containing ligands including the neuregulins (NRGs). The NRGs induce the transcription of genes encoding acetylcholine receptor subunits and affect neuronal migration on radial glial guides in the cerebellum and the cerebral cortex through ErbB4 and ErbB2 receptors (Schmid et al., 2003). It has been proposed that ErbB4 activation helps to regulate the migration of precursors within the RMS and influence placement and differentiation into distinct interneuronal subsets. The Erb RTK family includes the epidermal growth factor receptor (EGFR) Erb10 and its ligand EGF. EGF is known to be important in the proliferation and maintenance of embryonic and adult neural stem cells (Doetsch et al., 1999, 2002; Yarden and Sliwkowski, 2001) and EGFR is involved in the radial migration and maturation of neural precursors during embryonic cortical development (Burrows et al., 1997; Caric et al., 2001; Ciccolini et al., 2005).

Wnt (Wingless) is important for self-renewal in hematopoetic stem cells. In the brain, it is important for induction of neural specification (Munoz-Sanjuan and Brivanlou, 2002). Wnt signaling increased the intracellular concentration of β-catenin which associates with transcription factors TCF/Lef for gene transcription. Adult hippocampal progenitor cells express receptors as well as other components of the Wnt/Beta-catenin signaling pathway (Lie et al., 2005). Wnt3a signaling is required for normal expansion of precursor cells in hippocampal development (Roelink, 2000). Wnt3 stimulates this signaling pathway primarily by astrocytes located in close proximity to neurogenic regions of the brain. The Wnt signaling may serve as a regulatory pathway in adult hippocampal neurogenesis and contribute to the determination of neuronal fate commitment, as well as, cell proliferation.

Sonic hedgehog (Shh) is critical for development and the patterning of the ventral brain (McMahon et al., 2003). It is required for progenitor cell maintenance (Machold et al., 2003). Secretion of the Shh protein increases cell proliferation in the SVZ and SGZ and is required for normal proliferation in the SVZ (Lai et al., 2003). Shh receptors, patched and smoothed, are expressed in the SGZ, as well as, in the hippocampus (Traiffort et al., 1999; Lai et al., 2003).

During development bone morphogenetic proteins (BMPs) activity induced proliferation via activation of the BMP-1A receptor however, the BMP-1B receptor up-regulates $p21^{kip1}$ to inhibit the cell cycle (Panchision et al., 2001) thus, suggesting multiple functions. They can be potent inhibitors of SVZ neurogenesis and are produced by SVZ astrocytes (Lim et al., 2000). BMP2 can promote telencephalic neuroepithelial cells to differentiate as astrocytes. In the presence of leukemia inhibitory factor (LIF), a synergistic action is provided. The antagonist Noggin is secreted by ependymal cells and functions to promote neurogenesis by preventing BMPs from activating their receptors. This would then block the glial promoting effects of BMPs (Lim et al., 2000). The close interactions between the SVZ astrocytes and the ependymal layer maintain the necessary regulatory control or induction of SVZ neurogenesis.

Proneural basic-helix-loop-helix (HLH) transcription factors drive neurogenesis via cell-cycle exit specific protein expression (Guillemot, 1999; Kintner, 2002). The Sox B1 subfamily of the HMG-box transcription factors, (Sox 1–3) is expressed by precursors in the embryonic nervous system. They are expressed by most progenitor cells of the developing CNS. They and are downregulated when the cells exit the cell-cycle and differentiate (Uwanogho et al., 1995; Pevny et al., 1998). It is thought that these factors maintain neural progenitors in an undifferentiated state thus, inhibiting neuronal differentiation. Sox2 is expressed in the germinative zones in the adult rodent brain while Sox3 is expressed transiently by neural progenitors in the SVZ and dentate gyrus (Wang et al., 2006). In human embryonic stem cells, Sox3 is transiently induced with differentiation to neural progenitors, suggesting a role for neural stem cell maintenance. Some bHLH genes are involved in neural determination, others such as *Mash1* and *NeuroD*, are involved in terminal neuronal differentiation. The Proneural protein, Mash1 (mammalian achaete-scute homologue), is essential to the production of neurons in the embryonic ventral telencephalon (Casarosa et al., 1999). With other neurogenin family proteins, Mash1 promotes commitment of multipotent progenitors to neurons and inhibits astrocyte differentiation (Nieto et al., 2001). It is expressed and required for the generation of precursors in the SVZ of the postnatal brain for both oligodendrocytes and olfactory interneurons (Parras et al., 2004). Data is also available to suggest that a similar regulatory pathway involving activity of Mash1/Olig2 and Dlx is involved in the differentiation of both GABAergic interneurons and oligodendrocytes (Fode et al., 2000; Yun et al., 2002 Marshall et al., 2003).

Notch1 activation serves to prevent neuronal differentiation by maintaining cells in a precursor cell state (Chojnacki et al., 2003). Notch receptor signaling restricts the neurogenic potential of precursor cells by activating transcriptional repressors of the *Hes* gene family. These in turn suppress the expression of proneural bHLH proteins that can lead to the development of astrocytes (Tanigaki et al., 2001). A transient over expression of Notch1 in neural crest stem cells switches the cell lineage to glia from neuronal (Morrison et al., 2000).

The transcription factor cAMP response element-binding protein (CREB) is considered to be involved in the regulation of specific phases of adult neurogenesis in the SVZ/olfactory bulb system (Giachino et al., 2005) and in the SGZ for hippocampal neurons (Bender et al., 2001; Nakagawa et al., 2002).

32.5. Contribution of Glial Cells

The stimuli regulating adult neurogenesis also seem to affect gliogenesis. Several studies have demonstrated that cells expressing glial fibrillary acidic protein (GFAP) give rise to neurons in the adult dentate gyrus. Radial glial cells have a bipolar morphology with the cell soma residing in the ventricular zone or SVZ extending long basal processes to the pial surface and a short apical process in contact with the ventricular wall (Levitt and Rakic, 1980; Bentivoglio and Mazzarello, 1999). Notch1 signaling can promote radial glia differentiation (Gaiano et al., 2000). *In vivo* studies suggest that radial glial cells can serve as a source of neuronal precursor cells (Hartfuss et al., 2001; Noctor et al., 2002; Gotz et al., 2002; Ever and Gaiano, 2005). This has lead to the concept that radial glia can serve as a source of new neurons (Alvarez-Buylla et al., 2001; Gregg et al., 2002). However, not all radial glial cells are actively dividing during neurogenesis and not all are neuronal precursors (Schmechel and Rakic, 1979; Gaiano et al., 2000). An additional potential source for radial glia lies in the astrocytes population that has recently been shown to de-differentiate to immature radial glia as a function of cold temperature (Yu et al., 2006). Whether is represents an additional source of new neurons has not been determined.

In vivo, the contribution of GFAP expressing cells to neurogenesis appears to be from a sub-set of cells co-expressing both GFAP and nestin (Filippov et al., 2003), suggesting both a heterogeneity among the GFAP-expressing cells and a contribution to the generation of new neurons. In culture, GFAP-containing cells can show multipotency (Laywell et al., 2000;) while, other studies suggest that any such progenitor cells in the hippocampus require the actual presence of astrocytes to develop into neurons (Song et al., 2002). Kondo and Raff (2000a) showed that oligodendroglia progenitor cells cultured under specific conditions could be reprogrammed to multipotential neural stem cells able to generate both neurons and glia. Their further work proposed a critical role for Mash1 in the timing of oligodendrocyte development (Kondo and Raff 2000b); however, the work of Wang et al. (2001) did not support such a role for Mash1 but rather suggested a critical role for Id2 (inhibitor of DNA binding).

Cell-cell interactions between the newly generated neurons and non-neuronal cells have more recently been explored. Factors derived from local astrocytes participate in the regulation of proliferation and neuronal differentiation (Song et al., 2002). FGF-2 and CNTF are expressed by hypertrophic astrocytes in the denervated outer molecular layer of the hippocampus following lesion (Frautschy et al., 1991; Lee et al., 1997). FGF-2 protein is up-regulated within the hippocampus with seizures induced by kainic acid and after focal cerebral ischemia. With both modes of injury, the induction of neurogenesis was significantly decreased in FGF-2 null mice suggesting a regulation role for this growth factor in adult injury-induced neurogenesis (Yoshimura et al., 2001).

Cells immortalized from early postnatal neural precursors will readily migrate toward glioblastomas (Aboody et al., 2000). In glioblastoma-bearing mice, inoculation of such cells increased survival of the mouse with a greater survival seen if these cells were modified to over express interleukin-4 (Benedetti et al., 2000; Noble, 2000). Increased survival was seen with *in vitro* expanded neural precursor cell lines administered into gliomas (Staflin et al., 2004) and endogenous neural precursor cells show a strong tropism for glioblastomas (Glass et al., 2005). In mice, this attraction of precursors to glioblastomas declines with increasing age (Glass et al., 2005).

In the RMS, neuroblasts migrate tangentially in chains enwrapped by an astrocyte-derived tunnel-like structure termed the glial tube. This migration begins at birth yet, the astrocytes display a relatively homogeneous network and the glial tube does not become obvious until the third postnatal week of life (Peretto et al., 1997, 1999). The formation of the glial tube and the associated release of molecules to promote and accelerate neuroblast migration and the developmentally regulated expression of some integrins (Murase and Horwitz, 2002) may contribute to a more rapid and efficient cell turnover in the adult animal. In the olfactory epithelium, both stimulatory and inhibitory factors influence cell proliferation. *In vitro*, stimulatory factors include the fibroblast growth factors (FGFs), which increase the number of cell divisions that INPs undergo thus generating more neurons. FGFs also support the proliferation and survival of rare progenitors. *In vivo*, FGF8 is considered to be a good candidate for a positive FGF regulator. *In vitro*, other molecules have been reported to stimulate proliferation of OE cells and include: epidermal growth factor (EGF), TGF-β2, TGF-α, and olfactory marker protein (OMP). The anti-neurogenic negative regulators include the bone morphogenetic protein (BMP) family possibly targeting the transcription factor Mash1 (Shou et al., 1999).

32.6. Models of Brain Injury Showing Induction of Neurogenesis

Post-development neurogenesis can be upregulated in response to a variety of stimuli including ischemia, seizures, and head trauma (Kempermann et al., 1997; Magavi and Macklis, 2001; Arvidsson et al., 2002; Nakatomi et al., 2002; Parent and Lowenstein, 2002; Parent et al., 2002; Kokaia and Lindvall, 2003).

32.6.1. Olfactory System Damage

In the olfactory epithelium (OE), undifferentiated progenitor cells generate new neurons throughout life. It has been a source of information regarding cell interactions and the molecular response of progenitor cells (Calof et al., 1996). OE neuronal progenitors lie in close proximity to the olfactory response neurons (ORNs); while, the SVZ as the origin of progenitors for the ongoing production of OB interneurons lies at a distance. Proliferating cells are located throughout the RMS and may respond differently to injury signals depending upon their placement within the migratory process. Physically, one can severe axons of the olfactory receptor neurons, remove the target site with an olfactory bulbectomy, or disrupt sensory input to the olfactory receptor neurons by a naris occlusion to block the nasal opening. Damage can also be induced with inhalation or direct application via nasal irrigation of chemical agents or exposure to corrosive gases.

The extent of recovery of the OB is dependent upon the extent to which ORNs are able to regenerate and re-innervate the tissue. Regeneration is more robust following nasal irrigation with either Triton X-100 or methylbromide as compared to the delayed progression of repair following zinc sulfate. Contact deprivation of olfactory receptor neurons with their target cells by axonal severing or olfactory bulb removal results in rapid apoptosis of mature ORNs followed by degeneration of the olfactory epithelium, and a permanent upregulation in proliferation of cells in the basal compartment of the epithelium for replacement. In the absence of an olfactory bulb with a bulbectomy, these cells turn over with a lifespan of 2 weeks, suggesting that functional connections with the bulb is required for ORN maturation and survival. The high turnover of neurons provides for continued apoptosis in the region and suggests the ORN death may be involved in regulating proliferation of progenitor cells.

Naris occlusion usually involves the closure of one nostril during the early neonatal period and results in decreased sensory input to the ORNs, decreased volume of the ipsilateral olfactory bulb, and decreased proliferation of progenitor cells in the basal OE. When a reversible model is employed, the OE can rapidly recover from these changes with an increase in the number of newly generated neurons in the periglomerular layer, the target of afferent inputs from ORNs (Cummings and Brunjes, 1997).

32.6.2. Seizure-Induced Neurogenesis

Severe and sustained seizures lead to neuronal degeneration in the hippocampus via both a necrotic and apoptotic process. Single or intermittent seizures may or may not lead to neuronal damage. However, the assessment of damage becomes complex given the neurogenic capability of the hippocampus and the current evidence indicating that seizure activity alone will stimulate neurogenesis (Smith et al., 2005). Thus, the number of neurons within the dentate gyrus following insult will depend upon both cell death and cell replacement. It is thought that seizure-induced neurogenesis produces a surplus of granule cell neurons resulting in the formation of abnormal and aberrant neuronal circuits. In rodent models of temporal lobe epilepsy, seizures are associated with an increase rate of neurogenesis in the dentate gyrus within the first week (Bengzon et al., 1997; Gray and Sundstrom, 1998; Parent and Lowenstein, 2002). This induction occurs with a single short seizure period, a prolonged seizure activity, or with repeated kindled seizures induced by direct stimulation of by pilocarpine or kainic acid. Extended seizure activity induced by pilocarpine produces an increase in cell proliferation in the rostral forebrain SVZ with a large proportion of the cells differentiating into neurons. It also produces an induction in the caudal portion of the SVZ but a large proportion of these cells differentiate into glia. Induction in the SGZ also occurs with limbic kindling and intermittent perforant path stimulation. In the rat, both single and intermittent hippocampal kindling stimulations produce a marked increase in apoptosis in neurons along the hilar border of the granule cell layer within 5 h of cessation of stimulation. This is accompanied by an increase in BrdU/NeuN+ cells (Bengzon et al., 1997) and an increase in BDNF protein levels.

32.6.3. Ischemia

Ischemic brain injury results in the death of distinct susceptible populations of neurons in the brain. In addition, the injury concurrently triggers cellular repair mechanisms. SVZ precursors can be recruited following transient middle cerebral artery occlusion (Jin et al., 2001; Zhang et al., 2001). In this stroke model, extensive damage occurs to the striatum and the overlying parietal cortex. The new neurons migrate into the damaged striatum and begin to express striatal specific markers (Arvidsson et al., 2002; Jin et al., 2003; Parent et al., 2002). However, the functional significance of these new neurons remains in question, as the majority die within the first few weeks (Arvidsson et al., 2002). In addition, acute ischemic stroke increases proliferation in the SGL and migration to the hippocampal granule cell layer within 7–10 days post injury (Jin et al., 2001). Treatment with BDNF enhances the repair process and additional neurons are seen to replace pyramidal cells in the damaged CA1 hippocampal region (Nakatomi et al., 2002). While the CA1 region is known to be vulnerable to ischemic injury, cells along the inner blade of the dentate gyrus at the SGL show signs of injury and express active caspase-3, indicative of cell death (Bingham et al., 2005). This could either contribute to the loss of the neurogenic cells or may represent an early signaling event to initiate proliferation. In a rodent model of transient ischemia, reperfusion resulted in a transient decrease in Hes5 mRNA levels with a concurrent increase in Mash1 mRNA levels (Kawai et al., 2005).

32.6.4. Traumatic Brain Injury

Permanent structural changes occur in the brain following a traumatic brain injury (TBI) yet often there is remarkable functional recovery. This is more than likely due to a number of repair mechanisms including altering the connectivity of the remaining neurons and possibly the generation of new neurons. The general cellular characteristics that occur with TBI include a diffuse pattern of cell death. This is accompanied by an intense inflammatory response from both the resident microglia and from infiltrating cells such as monocytes, macrophages, T-cells, and neutrophils, given the damage to the blood brain barrier (Ghirnikar et al., 1998; Raivich et al., 1999). A secondary wave of neuronal death involves what is commonly referred to as delayed neuronal death. Reactive gliosis occurs in the lesion site and the astrocytes form a glial barrier to contain the injury. Thus, TBI results in the upregulation of numerous factors within the lesion site and possibly at more distant non-directly-injured sites, including cytokines, chemokines, and trophic factors. Neurogenesis can be triggered by targeted apoptotic death in cortical neurons (Magavi and Macklis, 2001) and TBI of the cortex can trigger increased proliferation of cells within the SGZ of the DG (Dash et al., 2001; Kernie et al., 2001). At the site of lesion, induction of the transcription factor, Olig2, is a common feature with a significant increase in the number of Olig2-positive cells occurring in multiple nervous system regions (Buffo et al., 2005). During development Olig2 is expressed in neuroepithelial cells, motoneuron precursors, and differentiating oligodendrocytes. It also plays a role in transit-amplifying precursors in the adult subependymal zone. It is possible that upon physical injury or under inflammatory conditions, Olig2 expression may mediate dedifferentiation of glial cells and regulation of neuro/gliogenesis in the brain (Cassiani-Ingoni et al., 2006).

32.6.5. Chemical Injury and Damage to Dentate Granule Neurons in the Hippocampus

Neurogenesis induced in both the SVZ and the SGZ have been examined in animals of various ages following drug or chemical induced injury. Excitotoxicity induced by a direct injection of the glutamate analog, alpha-amino-3-hydroxy-5-methyl-4-isoxazolepropionic acid (AMPA), into the cerebral ventricle of the brain will induce the generation of immature neurons from the SVZ in immature rat pups (Xu et al., 2005). Postnatal excitotoxicity induced by an injection of NMDA into the sensrimotor cortex differentially affected proliferative cells in each of the zones with a decrease of BrdU+ cells in the RMS yet, an increase of labeled cells in the striatum. In the SGZ, this injection and injury did not alter the proliferative nature or level of the cells (Faiz et al., 2005). This finding suggests that different stimuli, location, or level of damage are required to induce proliferation in each of the two regions. Adult hippocampus neurogenesis may be regulated by NMDA receptors present in precursor cells and in newly differentiating granule neurons suggesting an additional property of these receptors in the adult comparable to the critical role during brain development (Nacher and McEwen, 2006; Nacher et al., 2007).

The majority of experimental brain injury models have focused on established models of seizure, ischemia/hypoxia, or traumatic brain injury. However, given that the greatest proportion of cells produced from the SGZ are cells in close proximity, the dentate granule neurons a few studies have been undertaken to examine neurogenesis following a focal injury to this region. Death of dentate granule neurons has often been attributed to apoptotic mechanisms in various animal models. Of these models, adrenalectomy (Nichols et al., 2005), bacterial meningitis (Bogdan et al., 1997), and trimethyltin (Bruccoleri et al., 1998; Geloso et al., 2002; Lefebvre d'Hellencourt and Harry, 2005) show damage localized to the dentate granule neuron and evidence for the induction of proliferation in the SGZ as a mechanism of neural repair. Tauber et al. (2005) reported an increase proliferation and differentiation of neural progenitor cells in the SGZ after bacterial meningitis. This was accompanied by an increased synthesis of BDNF and TrkB, a decrease in NGF mRNA levels, and no changes in GDNF at 30 h following transmittal of the infection. Adrenalectomy has served as a model to examine glucocorticoid regulation in the hippocampus and induces neurogenesis in the dentate gyrus (Fischer et al., 2002; Nichols et al., 2005). Other models of glucocorticoid regulation support a role for this signaling pathway in hippocampal neurogenesis. The trimethyltin model is somewhat unique in the specificity for the dentate granule neurons. This specificity does not appear to be related to glucocorticoid signaling, excitotoxicity, or ischemia but rather, may be related to localized inflammation, oxidative stress, and altered calcium regulation. The timing of neuronal death follows that of "delayed neuronal death" seen with seizure, in that pronounced neuronal death occurs within 24–48 h of a systemic injection of the compound. The initial report of neurogenesis induced by prototypical neurotoxicant trimethyltin (TMT)-induced dentate gyrus degeneration suggested the active period of neural precursor proliferation coincided with the active period of neuronal death and the induction of pro-inflammatory cytokines (Harry et al., 2004). This was followed by two supporting studies in rat (Corvino et al., 2005) and adult mouse (Ogita et al., 2005). In the rat, the primary target site is not the dentate granule neurons but rather the CA 3–4 pyramidal neurons and requires 5–6 days for damage to be clearly evident. In this model, at 14 days, BrdU uptake was seen in NeuN positive cells in the dentate gyrus but co-localization was not yet evident in the CA3–4 region suggestive of undifferentiated cells at this early time point. In the adult mouse, Ogita et al. (2005) reported that the BrdU+ cells that co-expressed NeuN were located not in the SGZ or the granule cell layer but rather in the molecular layer and the hilus. This is in contrast to the data in the weanling mouse (Harry et al., 2004) where co-localization of these two markers was not prominent in these peripheral areas but rather, localization was seen in the dentate granule cell layer. These data suggests that the temporal and spatial progression of neurogenesis following a localized damage to the dentate granule neuron is dependent upon age, level of damage, and localized environmental cues including those associated with a neuroinflammatory response.

32.7. Neuroinflammation

During development, mesodermal cells committed to the macrophage cell lineage infiltrate the CNS to become resident microglia that actively engulf apoptotic neurons (Cuadros and Navascules, 1998). The phagocytosis of apoptotic cells is a process to remove the cellular debris thus, preventing leakage of potentially cytotoxic or antigenic substances. With brain trauma, both the resident microglia and infiltrating monocytes comprise the macrophage population in areas of direct injury and traumatic necrosis of neurons. With transient ischemia, activated microglia are detected as early as 20 min post reperfusion and actively phagocytize the neuronal debris at 24 h. The impact on the neuron of microglia reactivity to an insult and activation to a phagocytic phenotype remains a question. The influence of each of these microglial phenotypes on newly generated neurons is an even greater unknown. In the activated state, microglia can reduce neurogenesis (Monje et al., 2003; Ekdahl et al., 2003). However, microglia activated by interleukin (IL)-4 or interferon gamma can induce the production of neurons or glia from adult progenitor cells (Butovsky et al., 2006). In every injury model the induction of "neurogenesis" is accompanied by a microglial response that may be a source of neurogenic factors. Given their location within the injury, microglia would be in a prime position to provide such signals to the new neurons such as IGFs, EGF, and TGF-β1 (Banati and Graeber, 1994). Microglia also secrete IL-1 that stimulates astroglia hypertrophy and promotes nerve growth factor synthesis by non-neuronal cells (Heese et al., 1998). In addition to monocytes, activated T cells can cross the blood brain barrier and penetrate the CNS (Flugel and Brandl, 2001) and, with appropriate signals, secrete Th2/3 and neurotrophic factors such as, interleukin-10, TGF-β and BDNF (Aharoni et al., 2003). Thus, multiple cells can provide regulatory molecules to the new neurons at a critical time in the injury process.

Summary

The current data demonstrates that the adult brain retains the capacity to generate new neurons. The exact role for these cells, their mechanism of integration into the existing cytoarchitecture, and their ability to restore function following injury remain a basis for extensive ongoing basic research and targeted efforts to identify their therapeutic potential for brain repair (Sohur et al., 2006). The progenitor/stem cell populations in the brain are exposed to multiple signals at the time of origin, during migration, and at the site of differentiation. Distinct signaling pathways that act on a multipotent progenitor cell might inhibit or modulate each other leading to interactive signals and biological processes. The question remains as to how the cells integrate the different extracellular signals present in the injury environment to provide for a successful repopulation of the damaged region.

Review Questions/Problems

1. What is a primary neurogenic region and target site of the adult brain

a. the subventricular zone to the olfactory bulb
b. the subventricular zone to the rostral migratory stream
c. the subgranular zone to the rostral migratory stream
d. the subgranular zone to the pyramidal cell layer of the hippocampus.

2. What types of stimuli have been shown to enhance basal levels of adult neurogenesis?

a. physical exercise
b. enriched environment
c. visual recognition
d. A and B

3. Progenitor cells are distinguished from stem cells by their

a. ability to self-renew
b. multilineage differentiation
c. inability to self-renew
d. B and C

4. In the adult rodent brain, doublecourtin has been used to identify

a. mature neurons
b. oligodendroglia
c. immature neurons
d. stem cells

5. There are both stimulatory and inhibitory factors for proliferation of neural stem/progenitor cells. Which is an inhibitory factor

a. Fibroblast growth factor
b. Transforming growth factor – beta2
c. Bone morphogenetic proteins
d. olfactory marker protein

6. Nestin is a marker for

a. mature glia
b. immature neurons
c. immature neurons and reactive glia
d. olfactory neurons

References

Aberg MA, Aberg ND, Hedbacker H, Oscarsson J, Eriksson PS (2000) Peripheral infusion of IGF-1 selectively induces neurogenesis in the adult rat hippocampus. J Neurosci 20:2896–2903.

Aboody KS, Brown A, Rainov NG, Bower KA, Liu S, Yang W, Small JE, Herrlinger U, Ourednik V, Black PM, Breakefield XO, Snyder EY (2000) Neural stem cells display extensive tropism for pathology in adult brain: Evidence from intracranial gliomas. Proc Natl Acad Sci USA 97:12846–12851.

Abrous DN, Koehl M, LeMoal M (2005) Adult neurogenesis: From precursors to network and physiology. Physiol Rev 85:523–569.

Aguirre AA, Ghittajallu R, Belachew S, Gallo V (2004) NG2-expressing cells in the subventricular zone are type C-like cells and contribute to interneuron generation in the postnatal hippocampus. J Cell Biol 165:575–589.

Aharoni R, Kayhan B, Eilam R, Sela M, Amon R (2003) Glatiramer acetate-specific T cells in the brain express T helper 2/3 cytokines and brain-derived neurotrophic factors in situ. Proc Natl Acad Sci USA 100:14157–14162.

Altman J (1969) Autoradiographic and histological studies of postnatal neurogenesis. IV. Cell proliferation and migration in the anterior forebrain, with special reference to persisting neurogenesis in the olfactory bulb. J Comp Neurol 137:433–458.

Altman J, Bayer SA (1990) Migration and distribution of two populations of hippocampal granule cell precursors during the perinatal and postnatal periods. J Comp Neurol 301:365–381.

Altman J, Das GD (1965) Autoradiographic and histological evidence of postnatal hippocampal neurogenesis in rats. J Com Neurol 124:319–336.

Alvarez-Buylla A, Lim DA (2004) For the long run: Maintaining germinal niches in the adult brain. Neuron 41:683–686

Alvarez-Buylla A, Garcia-Verdugo JM, Tramontin AD (2001) A unified hypothesis on the lineage of neural stem cells. Nat Rev Neurosci 2:287–293.

Ambrogini P, Lattanzi D, Ciuffoli S, Agostini D, Bertini L, Stocchi V, Santi S, Cuppini R (2004) Morpho-functional characterization of neuronal cells at different stages of maturation in granule cell layer of adult rat dentate gyrus. Brain Res 1017:21–31.

Anton ES, Ghashghaei HT, Weber JL, McCann C, Fischer TM, Cheung ID, Gassmann M, Messing A, Klein R, Schwab MH, Lloyd KC, Lai C (2004) Receptor tyrosine kinase ErbB4 modulates neuroblast migration and placement in the adult forebrain. Nat Neurosci 7:1319–1328.

Arvidsson A, Collin T, Kirik D, Kokaia Z, Linvall O (2002) Neuronal replacement from endogenous precursors in the adult brain after stroke. Nature Med 8:963–970.

Bagri A, Marin O, Plump AS, Mak J, Pleasure SJ, Rubenstein JL, Tessier-Lavigne M (2002) Slit proteins prevent midline crossing and determine the dorsoventral position of major axonal pathways in the mammalian forebrain. Neuron 33:233–248.

Banati RB, Graeber MB (1994) Surveillance, intervention and cytotoxicity: Is there a protective role of microglia? Dev Neurosci 16:114–127.

Bayer SA, Altman J (1975) The effects of X-irradiation on the postnatal-forming granule cell populations in the olfactory bulb, hippocampus, and cerebellum of the rat. Exp Neurol 48:167–174.

Belachew S, Chittagallu R, Aguirre AA, Yuan X, Kirby M, Anderson S, Gallo V (2003) Postnatal NG2 proteoglycan-expressing progenitor cells are intrinsically multipotent and generate functional neurons. J Cell Biol 161:169–186.

Bender RA, Lauterborn JC, Gall CM, Cariaga W, Baram TZ (2001) Enhanced CREB phosphorylation in immature dentate gyrus granule cells precedes neurotrophin expression and indicates a specific role of CREB in granule cell differentiation. Eur J Neurosci 13:679–686.

Benedetti S, Pirola B, Pollo B, Magrassi L, Bruzzone MG, Rigamonti D, Galli R, Selleri S, Dimeco F, DeFraja C, Vescovi A, Cattaneo E, Finocchiaro G (2000) Gene therapy of experimental brain tumors using neural progenitor cells. Nat Med 6:447–450.

Bengzon J, Kokaia Z, Elmer E, Nanobashvili A, Kokaia M, Lindvall O (1997) Apoptosis and proliferation of dentate gyrus neurons after single and intermittent limbic seizures. Proc Natl Acad Sci USA 94:10432–10437.

Bentivoglio M, Mazzarello P (1999) The history of radial glia. Brain Res Bull 49:305–315.

Biebl M, Cooper CM, Winkler J, Kuhn HG (2000) Analysis of neurogenesis ad programmed cell death reveals a self-renewing capacity in the adult rat brain. Neurosci Lett 291:17–20.

Bingham B, Liu D, Wood A, Cho S (2005) Ischemia-stimulated neurogenesis is regulated by proliferation, migration, differentiation and caspase activation of hippocampal precursor cells. Brain Res 1058:167–177.

Bogdan I, Leib SL, Bergeron M, Chow L, Tauber MG (1997) Tumor necrosis factor-alpha contributes to apoptosis in hippocampal neurons during experimental group B streptococcal meningitis. J Infect Dis 176:693–697.

Brandt MD, Jessberger S, Steiner B, Kronenberg G, Reuter K, Bick-Sander A, Von deer Behrens W, Kempermann G (2003) Transient calretinin-expression defines early postmitotic step of neuronal differentiation in adult hippocampus neurogenesis of mice. Mol Cell Neurosci 24:603–613.

Brannvall K, Korhonen L, Lindhold D (2002) Estrogen-receptor-dependent regulation of neural stem cell proliferation and differentiation. Mol Cell Neurosci 21:512–520.

Bruccoleri A, Brown H, Harry GJ (1998) Cellular localization and temporal elevation of tumor necrosis factor-alpha, interleukin-1 alpha, and transforming growth factor-beta 1 mRNA in hippocampal injury response induced by trimethyltin. J Neurochem 71:1577–1587.

Brown JP, Couillard-Despres S, Cooper-Kuhn CM, Winkler J, Aigner L, Kuhn HG (2003a) Transient expression of doublecortin during adult neurogenesis. J Comp Neurol 467:1–10.

Brown J, Cooper-Kuhn CM, Kempermann G, Van Praag H, Winkler J, Gage FH, Kuhn HG (2003b) Enriched environment and physical activity stimulate hippocampal but not olfactory bulb neurogenesis. Eur J Neurosci 17:2042–2046.

Bull ND, Bartlett PF (2005) The adult mouse hippocampal progenitor is neurogenic but not a stem cell. J Neurosci 25:10815–10821.

Buffo A, Vosko MR, Erturk D, Hamann GF, Jucker M, Rowitch D, Gotz M (2005) Expression pattern of the transcription factor Olig2 in response to brain injuries: Implications for neuronal repair. Proc Nat Acad Sci 102:18183–18188.

Burrows RC, Wancio D, Levitt P, Lillen L (1997) Response diversity and the timing of progenitor cell maturation are regulated by developmental changes in EGFR expression in the cortex. Neuron 19:251–267.

Butovsky O, Yaniv Z, Schwartz A, Landa G, Talpalar AE, Pluchino S, Martino G, Schwartz M (2006) Microglia activated by IL-4 or IFN-γ differentially induce neurogenesis and oligodendrogenesis from adult stem/progenitor cells. Mol Cell Neurosci 31:149–160.

Calof AL, Hagiwara N, Holcomb JD, Mumm JS, Shou J (1996) Neurogenesis and cell death in olfactory epithelium. J Neurobiol 30:67–81.

Cameron HA, McKay RD (2001) Adult neurogenesis produces a large pool of new granule cells in the dentate gyrus. J Comp Neurol 435:406–417.

Caric D, Raphael H, Viti J, Feathers A, Wancio D, Lillien L (2001) EGFRs mediate chemotactic migration in the developing telecephalon. Development 128:4203–4216.

Carro E, Trejo JL, Busiguina S, Torres-Aleman I (2001) Circulating insulin-like growth factor I mediates the protective effects of physical exercise against brain insults of different etiology and anatomy. J Neurosci 21:5678–5684.

Casarosa S, Fode C, Guillemont F (1999) Mash1 regulates neurogenesis in the ventral telencephalon. Development 126:525–534.

Cassiani-Ingoni R, Coksaygan T, Xue H, Reichert-Scrivner SA, Wiendl H, Rao MS, Magnus T (2006) Cytoplasmic translocation of Olig2 in adult glial progenitors marks the generation of reactive astrocytes following autoimmune inflammation. Exp Neurol 201:349–358.

Chen G, Rajkowska G, Du F, Seraji-Bozorgzad N, Manji HJ (2000) Enhancement of hippocampal neurogenesis by lithium. J Neurochem 75:1729–1734.

Cheng A, Wang S, Cai J, Rao MS, Mattson MP (2003) Nitric oxide acts in a positive feedback loop with BDNF to regulate neural progenitor cell proliferation with differentiation in the mammalian brain. Dev Biol 258:319–333.

Cheng Y, Black IB, DiCicco-Bloom E (2002) Hippocampal granule neuron production and population size are regulated by levels of bFGF. Eur J Neurosci 15:3–12.

Chojnacki A, Shimazaki T, Gregg C, Weinmaster G, Weiss S (2003) Glycoprotein 130 signaling regulates Notch1 expression and activation in the self-renewal of mammalian forebrain neural stem cells. J Neurosci 23:1730–1741.

Ciccolini F, Mandl C, Holzl-Wenig G, Kehlenbach A, Hellwig A (2005). Prospective isolation of late development multipotent precursors whose migration is promoted by EGFR. Dev Biol 284:112–125.

Conover JC, Doetsch F, Carcia-Verdugo JM, Gale NW, Yancopoulos GD, Alvarez-Buylla A (2000) Disruption of Sph/ephrin signaling affects migration and proliferation in the adult subventricular zone. Nat Neurosci 11:1091–1097.

Corvino V, Geloso MC, Cavallo V, Guadagni E, Passalacqua R, Florenzano F, Giannetti S, Mlinari M, Michetti F (2005) Enhanced neurogenesis during trimethyltin-induced neurodegeneration in the hippocampus of the adult rat. Brain Res Bull 65:471–477.

Cotrina ML, Nedergaard M (2002) Astrocytes in the aging brain. J Neurosci Res 67:1–10.

Cuadros MA, Navascules J (1998) The origin and differentiation of microglial cells during development. Prog Neurobiol 56:173–189.

Cummings DM, Brunjes PC (1997) The effects of variable periods of functional deprivation on olfactory bulb development in rats. Exp Neurol 128:124–128.

Dash P, Mach S, Moore A (2001) Enhanced neurogenesis in the rodent hippocampus following traumatic rain injury. J Neurosci Res 63:313–319.

Dawson MR, Levine JM, Reynolds R (2000) NG2-expressing cells in the central nervous system: Are they oligodendroglia progenitors? J Neurosci Res 61:47–479.

Dayer AG, Ford AA, Cleaver KM, Yassaee M, Cameron HA (2003) Short-term and long-term survival of new neurons in the rat dentate gyrus. J Comp Neurol 460:563–572.

Doetsch R, Garcia-Verdugo JM, Alvarex-Buylla A (1997) Cellular composition and three-dimensional organization of the subventricular germinal zone in the adult mammalian brain. J Neurosci 17:5046–5061.

Doetsch F, Caille I, Lim DA, Garcia-Verdugo JM, Alvarez-Buylla A (1999) Subventricular zone astrocytes are neural stem cells in the adult mammalian brain. Cell 97:703–716.

Doetsch F, Petrenu I, Caille I, Garcia-Verdugo JM, Alvarez-Buylla A (2002) EGF converts transit-amplifying neurogenic precursors in the adult brain into multipotent stem cells. Neuron 36:1021–1034.

Duan W, Guo Z, Mattson MP (2001) Dietary restriction stimulates BDNF production in the brain and thereby protects neurons against excitotoxic injury. J Mol Neurosci 16:1–12.

Duman RS (2004) Depression; a case of neuronal life and death? Biol Psychiatry 56:140–145.

Ekdahl CT, Claasen JH, Bonde S, Kokaia Z, Lindvall O (2003) Inflammation is detrimental for neurogenesis in adult brain. Proc Natl Acad Sci USA 100:13632–13617.

Emsley JG, Hagg T (2003) Endogenous and exogenous ciliary neurotrophic factor enhances forebrain neurogenesis in adult mice. Exp Neurol 183:298–310.

Ever L, Gaiano N (2005) Radial "glial" progenitors: Neurogenesis and signaling. Curr Opin Neurobiol 15:29–33.

Faiz M, Acarin L, Castellano B, Gonzalez B (2005) Proliferation dynamics of germinative zone cells in the intact and excitotoxically lesioned postnatal rat brain. BMC Neurosci 6:26.

Farmer J, Zhao X, van Praag H, Wodtke K, Gage FH, Christie BR (2004) Effects of voluntary exercise on synaptic plasticity and gene expression in the dentate gyrus of adult male Sprague-Dawley rats in vivo. Neurosci 124:71–79.

Filippov V, Kronenberg G, Pivneva T, Reuter K, Steiner B, Wang L-P, Yamaguchi M, Kettenmann H, Kempermann G (2003) Subpopulation of nestin-expressing progenitor cells in the adult murine hippocampus shows electrophysiological and morphological characteristics of astrocytes. Mol Cell Neurosci 23:373–382.

Fischer AK, von Rosenstiel P, Fuchs E, Goula D, Almeida OF, Czeh B (2002) The prototypic mineralcorticoid receptor agonist aldosterone influences neurogenesis in the dentate gyrus of the adrenalectomized rat. Brain Res 947:290–293.

Flugel A, Brandl M (2001) New tools to trace populations of inflammatory cells in the CNS. Glia 36:125–136.

Fode C, Ma Q, Casarosa S, Ang SL, Anderson DJ, Guillemot F (2000) A role for neural determination genes in specifying the dorso-ventral identity of telencephalic neurons. Genes Dev 14:67–80.

Frautschy SA, Walicke PA, Baird A (1991) Localization of basic fibroblast growth factor and its mRNA after CNS injury. Brain Res 553:291–299.

Gaiano N, Nye JS, Fishell G (2000) Radial glial identity is promoted by Notch1 signaling in the murine forebrain. Neuron 26:395–404.

Geloso MC, Vercelli A, Corvino V, Repici M, Boca M, Haglid K, Zelano G, Michetti F (2002) Cyclooxygenase-2 and caspase 3 expression in trimethyltin-induced apoptosis in the mouse hippocampus. Exp Neurol 175:152–160.

Ghirnikar RS, Lee YL, Eng LF (1998) Inflammation in traumatic brain injury: Role of cytokines and chemokines. Neurochem Res 23:329–340.

Giachino C, DeMarchis S, Giampietro C, Parlato R, Perroteau I, Schutz G, Fasolo A, Peretto P (2005) cAMP response element-binding protein regulates differentiation and survival of newborn neurons in the olfactory bulb. J Neurosci 25:10105–10118.

Glass R, Synowitz M, Kronenberg G, Qalzlein JH, Markovic DS, Wang LP, Gast D, Kiwit J, Kempermann G, Kettenmann H (2005) Glioblastoma-induced attraction of endogenous neural precursor cells is associated with improved survival. J Neurosci 25:2637–2646.

Gotz M, Huttner WB (2005) The cell biology of neurogenesis. Mol Cell Biol 6:777–788.

Gotz M, Hartfuss E, Malatesta P (2002) Radial glial cells as neuronal precursors: A new perspective on the correlation of morphology and lineage restriction in the developing cerebral cortex of mice. Brain Res Bull 57:777–788.

Gould E, Beylin A, Tanapat P, Reeves A, Shors TJ (1999) Learning enhances adult neurogenesis in the hippocampal formation. Nat Neurosci 2:260–265.

Gray WP, Sundstrom LE (1998) Kainic acid increases the proliferation of granule cell progenitors in the dentate gyrus of the adult rat. Brain Res 790:52–59.

Gregg CT, Chojnacki AK, Weiss S (2002) Radial glial cells as neuronal precursors: The next generation? J Neurosci Res 69:708–713.

Gritti A, Parati EA, Cova L, Frolichsthal P, Galli R, Wanke E, Faravelli L, Morassutti DJ, Roisen F, Nickel DD, Vescovi AL (1996) Multipotential stem cells from the adult mouse brain proliferate and self-renew in response to basic fibroblast growth factor. J Neurosci 16:1091–1100.

Guillemot F (1999) Vertebrate bHLH genes and the determination of neuronal fates. Exp Cell Res 253:357–364.

Gustafsson E, Lindvall O, Kokaia Z (2003) Intraventricular infusion of TrkB-Fc fusion protein promotes ischemia-induced neurogenesis in adult rat dentate gyrus. Stroke 34:2710–2715.

Hack MA, Saghatelyan A, de Chevigny A, Pfeifer A, Ashery-Padan R, Lledo P-M, Gotz M (2005) Neuronal fate determinants of adult olfactory bulb neurogenesis. Nat Neurosci 8:865–872.

Hallbergson AF, Gnatenco C, Peterson DA (2003) Neurogenesis and brain injury: Managing a renewable resource for repair. J Clin Invest 112:1128–1133.

Harry GJ, McPherson CA, Wine RN, Atkinson K, Lefebvre d'Hellencourt C (2004) Trimethyltin-induced neurogenesis in the murine hippocampus. Neurotox Res 5:623–627.

Hartfuss E, Galli R, Heins N, Gotz M (2001) Characterization of CNS precursor subtypes and radial glia. Dev Biol 229:15–30.

Hastings NB, Gould E (1999) Rapid extension of axons into the CA3 region by adult-generated granule cells. J Comp Neurol 413:146–154.

Heine VM, Maslam S, Joels M, Lucassen PJ (2004) Prominent decline of newborn cell proliferation, differentiation, and apoptosis in the aging dentate gyrus, in absence of an age-related hypothalamus-pituitary-adrenal axis activation. Neurobiol Aging 25:361–375.

Heese K, Hock C, Otten U (1998) Inflammatory signals induce neurotrophin expression in human microglial cells. J Neurochem 70:699–707.

Jin K, Minami M, Lan JQ, Mao XO, Batteur S, Simon RP, Greenberg DA (2001) Neurogenesis in dentate subgranular zone and rostral subventricular zone after focal cerebral ischemia in the rat. Proc Natl Acad Sci USA 98:4710–4715.

Jin K, Sun Y, Xie L, Peel A, Mao XO, Batteur S, Greenberg DA (2003) Directed migration of neuronal precursors into the ischemic cerebral cortex and striatum. Mol Cell Neurosci 24:171–189.

Kaplan MS, Hinds JW (1977) Neurogenesis in the adult rat: Electron microscopic analysis of light radioautographs. Science 197:1092–1094.

Katchanov J, Harms C, Gertz K, Hauck L, Waeber C, Hirt L, Priller J, von Harsdorf R, Bruck W, Hortnagl H, Dirnagl U, Bhide PG, Endres M (2001) Mild cerebral ischemia induces loss of cyclin-dependent kinase inhibitors and activation of cell cycle machinery before delayed neuronal cell death. J Neurosci 21:5054–5053.

Kawai T, Takagi N, Nakahara M, Takeo S (2005) Changes in the expression of Hes5 and Mash1 mRNA in the adult rat dentate gyrus after transient forebrain ischemia. Neurosci Lett 380:17–20.

Kee N, Sivalingam S, Boonstra R, Wojtowicz JM (2002) The utility of Ki-67 and BrdU as proliferative markers of adult neurogenesis. J Neurosci Methods 115:97–105.

Kempermann G, Kuhn HG, Gage FH (1997) More hippocampal neurons in adult mice living in an enriched environment. Nature 386:493–495.

Kempermann G, Gast D, Gage FH (2002) Neuroplasticity in old age: Sustained five-fold induction of hippocampal neurogenesis by long-term environmental enrichment. Ann Neurol 52:135–143.

Kempermann G, Jessberger S, Steiner B, Kronenberg G (2004) Milestones of neuronal development in the adult hippocampus. Trends Neurosci 27:447–452.

Kernie SG, Erwin TM, Parada LF (2001) Brain remodeling due to neuronal and astrocytic proliferation after controlled cortical injury in mice. J Neurosci Res 66:317–326.

Kintner C (2002) Neurogenesis in embryos and in adult neural stem cells. J Neurosci 22:639–643.

Kiss JZ, Wang C, Rougon G (1993) Nerve-dependent expression of high polysialic acid neural cell adhesion molecule in neurohypophysial astrocytes of adult rats. Neurosci 53:213–221.

Kohwi M, Osumi N, Rubenstein JLR, Alvarez-Buylla A (2005) Pax6 is required for making specific subpopulations of granule and periglomerular neurons in the olfactory bulb. J Neurosci 25:6997–7003.

Kokaia Z, Lindvall O (2003) Neurogenesis after ischaemic brain insults. Curr Opin Neurobiol 13:127–132.

Kondo T, Raff M (2000a) Oligodendrocyte precursor cells reprogrammed to become multipotential CNS stem cells. Science 289:1754–1757.

Kondo T, Raff M (2000b) Basic helix-loop-helix proteins and the timing of oligodendrocyte differentiation. Development 127:2989–2998.

Kozorovitskiy Y, Gould E (2004) Dominance hierarchy influences adult neurogenesis in the dentate gyrus. J Neurosci 24:6755–6759.

Kuhn HG, Dickinson-Anson H, Gage FH (1996) Neurogenesis in the dentate gyrus of the adult rat: Age-related decrease of neuronal progenitor proliferation. J Neurosci 16:2027–2033.

Kuhn HG, Winkler J, Kempermann G, Thal LJ, Gage FH (1997) Epidermal growth factor and fibroblast growth factor-2 have different effects on neural progenitors in the adult rat brain. J Neurosci 17:5820–5829.

Lai K, Kaspar BK, Gage FH, Schaffer DV (2003) Sonic hedgehog regulates adult neural progenitor proliferation in vitro and in vivo. Nat Neurosci 6:21–27.

Larsson E, Mandel RJ, Klein RL, Muzyczka N, Lindvall O, Kokaia Z (2002) Suppression of insult-induced neurogenesis in adult rat brain by brain-derived neurotrophic factor. Exp Neurol 177:1–8.

Laywell ED, Rakic P, Kukekov VG, Holland EC, Steindler DA (2000) Identification of a multipotent astrocytic stem cell in the immature and adult mouse brain. Proc Natl Acad Sci USA 97:13883–13888.

Lee DA, Gross L, Wittrock DA, Windebank AJ (1997) Differential regulation of ciliary neurotrophic factor (CNTF) and CNTF receptor alpha expression in astrocytes and neurons of the facia dentata after entorhinal cortex lesion. J Neurosci 17:1137–1146.

Lee J, Duan W, Long JM, Ingram DK, Mattson MP (2000) Dietary restriction increases the number of newly generated neural cells and induces BDNF expression in the dentate gyrus of rats. J Mol Neurosci 15:99–108.

Lee J, Seroogy KB, Mattson MP (2002) Dietary restriction enhances neurotrophin expression and neurogenesis in the hippocampus of adult mice. J Neurochem 80:539–547.

Lefebvre d'Hellencourt C, Harry GJ (2005) Molecular profiles of mRNA levels in laser capture microdissected murine hippocampal regions differentially responsive to TMT-induced cell death. J Neurochem 93:206–220.

Lemasson M, Saghatelyan A, Olivo-Marin JC, Lledo PM (2005) Neonatal and adult neurogenesis provide two distinct populations of newborn neurons to the mouse olfactory bulb. J Neurosci 25:6816–6825.

Levitt P, Rakic P (1980) Immunoperoxidase localization of glial fibrillary acidic protein in radial glial cells and astrocytes of the developing rhesus monkey. J Comp Neurol 193:815–840.

Lewis PF, Emerman M (1994) Passage through mitosis is required for oncoretroviruses but not for the human immunodeficiency virus. J Virol 68:510–516.

Lichtenwalner RJ, Forbes ME, Bennett SA, Lynch CD, Sonntag WE, Riddle DR, Lynch CD, Sonntag WE, Riddle DR (2001) Intracerebroventricular infusion of insulin-like growth factor-1 ameliorates the age related decline in hippocampal neurogenesis. Neuroscience 107:603–613.

Lie D-C, Colamarino SA, Song H-J, Desire L, Mira H, Consiglio A, Lein ES, Jessberger S, Lansford H, Dearie AR, Gage FH (2005) Wnt signaling regulates adult hippocampal neurogenesis. Nature 437:1370–1375.

Lim DA, Tramonti AD, Trevejo JM, Herrera DG, Garcia-Verdugo JM, Alvarex-Buylla A (2000) Nogin antagonizes BMP signaling to create a niche for adult neurogenesis. Neuron 28:713–726.

Lois C, Garcia-Verdugo J-M, Alvarex-Buylla A (1996) Chain migration of neuronal precursors. Science 271:978–981.

Machold R, Hayashi S, Rutlin M, Muzumdar MD, Nery S, Corgin JG, Gritli-Linde A, Dellovade T, Porter JA, Rubin LL, dudek H, McMahon AP, Fishell G (2003) Sonic hedgehog is required for progenitor cell maintenance in telencephalic stem cell niches. Neuron 39:937–950.

Magavi SS, Macklis JD (2001) Manipulation of neural precursors in situ: Induction of neurogenesis in the neocortex of adult mice. Neuropsychopharmcology 25:816–835.

Malatesta P, Hartfuss E, Gotz M (2000) Isolation of radial glial cells by fluorescent-activated cell sorting reveals a neuronal lineage. Development 127:5253–5263.

Marshall CA, Suzuki SO, Goldman JE (2003) Gliogenic and neurogenic progenitors of the subventricular zone: Who are they, where did they come from, and where are they going? Glia 43:52–61.

Maslov AY, Barone TA, Plunkett RJ, Pruitt SC (2004) Neural stem cell detection, characterization, and age related changes in the subventricular zone of mice. J Neurosci 24:1726–1733.

Mattson MP, Maudsley S, Martin B (2004) A neural signaling triumvirate that influences ageing and age-related disease: Insulin/IGF-1, BDNF and serotonin. Ageing Res Rev 3:445–464.

McMahon AP, Ingham PW, Tabin CJ (2003) Developmental roles and clinical significance of sonic hedgehog signaling. Curr Topic Dev Biol 53:1–114.

Monje ML, Toda H, Palmer TD (2003) Inflammatory blockade restores adult hippocampal neurogenesis. Science 302:1760–1765.

Morrison SJ, Perez SE, Qiao Z, Verdi JM, Hicks C, Weinmaster G, Anderson DJ (2000) Transient Notch activation initiates an irreversible switch from neurogenesis to gliogenesis by neural crest stem cells. Cell 101:499–510.

Morshead CM, van der Kooy D (2001) A new "spin" on neural stem cells? Curr Opin Neurobiol 11:59–65.

Munoz-Sanjuan I, Brivanlou AH (2002) Neural induction, the default model and embryonic stem cells. Nat Rev Neurosci 3:271–280.

Murase S, Horwitz AF (2002) Deleted in colorectal carcinoma and differentially expressed integrins mediate the directional migration of neural precursors in the rostral migratory stream. J Neurosci 22:3568–3579.

Nacher J, McEwen BS (2006) The role of N-methyl-D-asparate receptors in neurogenesis. Hippocampus 16:267–270.

Nacher J, Varea E, Miguel Blasco-Ibanez J, Gomez-Climent MA, Castillo-Gomez E, Crespo C, Martinez-Guijarro FJ, McEwen BS (2007) N-methyl-d-aspartate receptor expression during adult neurogenesis in the rat dentate gyrus. Neurosci 144:855–864.

Nakagawa S, Kim JE, Lee R, Chen J, Fujioka T, Malberg J, Tsuji S, Duman RS (2002) Localization of phosphorylated cAMP response element-binding protein in immature neurons of adult hippocampus. J Neurosci 22:9868–9876.

Nakatomi H, Kuriu T, Okabe S, Yamamoto S, Hatano O, Kawahara N, Tamura A, Kirino T, Nakafuku M (2002) Regeneration of hippocampal pyramidal neurons after ischemic brain injury by recruitment of endogenous neural progenitors. Cell 110:429–441.

Nichols NR, Agolley D, Zieba M, Bye N (2005) Glucocorticoid regulation of glial responses during hippocampal neurodegeneration and regeneration. Brain Res Brain Res Rev 48:287–301.

Nicolle MM, Gonzalez J, Sugaya K, Kaskerville KA, Bryan D, Lund K, Gallagher M, Mckinney M (2001) Signatures of hippocampal oxidative stress in aged spatial learning-impaired rodents. Neurosci 107:415–431.

Nieto M, Schuurmans C, Britz O, Guillemot F (2001) Neural bHLH genes control the neuronal versus glial fate decision in cortical progenitors. Neuron 29:401–413.

Nilsson M, Perfilieva E, Johanson U, Orwar O, Eriksson PS (1999) Enriched environment increases neurogenesis in the adult rat dentate gyrus and improves spatial memory. J. Neurobiol 39:569–578.

Nishiyama A, Lin XH, Giese N, Heldin CH, Stallcup WB (1996) Co-localization of NG2 proteoglycan and PDGF alpha-receptor on O2A progenitor cells in the developing rat brain. J Neurosci Res 43:299–314.

Noble M (2000) Can neural stem cells be used as therapeutic vehicles in the treatment of brain tumors? Nat Med 6:369–370.

Noctor SC, Flint AC, Weissman TA, Wong WS, Clinton BK, Kriegstein AR (2002) Dividing precursor cells of the embryonic cortical ventricular zone have morphological and molecular characteristics of radial glia. J Neurosci 22:3161–3173.

Nowakowski RS, Rakic P (1981) The site of origin and route and rate of migration of neurons to the hippocampal region of the rhesus monkey. J Comp Neurol 196:129–154.

Nowakowski RS, Lewin SB, Miller MW (1989) Bromodeoxyuridine immunohistochemical determination of the lengths of the cell cycle and the DNA-synthetic phase for a anatomically defined population. J Neurocytol 18:311–318.

Ogita K, Nishiyama N, Sugiyama C, Higuchi K, Yoneyama M, Yoneda Y (2005) Regeneration of granule neurons after lesioning of hippocampal dentate gyrus: Evaluation using adult mice treated with trimethyltin chloride as a model. J Neurosci Res 82:609–621.

Ormerod BK, Lee TT, Galea LA (2003) Estradiol initially enhances but subsequently suppresses (via adrenal steroids) granule cell proliferation in the dentate gyrus of adult female rats. J Neurobiol 55:247–260.

Overstreet LS, Hentges ST, Bumaschny VR, deSouza FSJ, Smart JL, Satangelo AM, Low MJ, Westbrook GL, Rubinstein M (2004) A transgenic marker for newly born granule cells in dentate gyrus. J Neurosci 24:3251–3259.

Panchision DM, Pickel JM, Studer L, Lee SH, Turner PA, Hazel TG, Mckay RD (2001) Sequential actions of BMP receptors control neural precursor cell production and fate. Genes Dev 15:2094–2110.

Parras CM, Galli R, Britz O, Soares S, Galichet C, Battiste J, Johnson JE, Nakafuku M, Vescovi A, and Guillemot F (2004) Mash1 specifies neurons and oligodendrocytes in the postnatal brain. EMBO J 23:4495–4505.

Parent JM, Lowenstein DH (2002) Seizure-induced neurogenesis: Are more neurons good for an adult brain? Prog Brain Res 135:121–131.

Parent JM, Vexler ZS, Gong C, Derugin N, Ferriero DM (2002) Rat forebrain neurogenesis and striatal neuron replacement after focal stroke. Ann Neurol 52:802–813.

Pencea V, Luskin MB (2003) Prenatal development of the rodent rostral migratory stream. J Comp Neurol 463:402–418.

Peretto P, Merighi A, Fasolo A, Bonfanti I (1997) Glial tubes in the rostral migratory stream of the adult rat. Brain Res Bull 42:9–21.

Peretto P, Merighi A, Fasolo A, Bonfanti I (1999) The subependymal layer in rodents: A site of structural plasticity and cell migration in the adult mammalian brain. Brain Res Bull 49:221–243.

Pevny LH, Sockanathan S, Placzek M, Lovel-Badge R (1998) A role for SOX1 in neural determination. Development 125:1967–1978.

Quinones-Hinojosa A, Sanai N, Soriano-Navarro M, Gonzalez-Perex O, Mirzadeh Z, Gil-Perotin S, Romero-Rodriquez R, Berger MS, Garcia-Verdugo JM, Alvarez-Buylla A (2006) Cellular composition and cytoarchitecture of the adult human subventricular zone: A niche of neural stem cells. J Comp Neurol 494:415–434.

Raff MC, Abney ER, Cohen J, Lindsay R, Noble M (1983) Two types of astrocytes in cultures of developing rat white matter: Differences in morphology, surface gangliosides, and growth characteristics. J Neurosci 3:1289–1300.

Raivich G, Bohatschek M, Kloss CU, Werner A, Jones LL, Kreutzberg GW (1999) Neuroglial activation repertoire in the injured brain: Graded response, molecular mechanisms and cues to physiological function. Brain Res Brain Res Rev 30:77–105.

Rakic P (2002) Adult neurogenesis in mammals: An identity crisis. J Neurosci 22:614–618.

Rietze RL, Reynolds BA (2006) Neural stem cell isolation and characterization. Methods Enzymol 419:3–23.

Reynolds BA, Weiss S (1992) Generation of neurons and astrocytes from isolated cells of the adult mammalian central nervous system. Science 255:1707–1710.

Reynolds BA, Rietze RL (2005) Neural stem cells and neurospheres-re-evaluating the relationship. Nat Methods 2:333–336.

Roelink H (2000) Hippocampus formation: An intriguing collaboration. Curr Biol 10:R279–281.

Sairanen M, Lucas G, Emfors P, Castren M, Castren E (2005) Brain-derived neurotrophic factor and antidepressant drugs have different but coordinated effects on neuronal turnover, proliferation, and survival in the adult dentate gyrus. J Neurosci 25:1089–1094.

Sarnat HB, Nochlin D, Born DE (1998) Neuronal nuclear antigen (NeuN): A marker of neuronal maturation in early human fetal nervous system. Brain Dev 20:88–94.

Schaar BT, Kinoshita K, Mcconnell SK (2004) Doublecortin microtubule affinity is regulated by a balance of kinase and phosphatase activity at the leading edge of migrating neurons. Neuron 41:203–213.

Scharfman H, Goodman J, Macleod A, Phani S, Antonelli C, Croll S (2005) Increased neurongenesis and the ectopic granule cells after intrahippocampal BDNF infusion in adult rats. Exp Neurol 192:348–356.

Schmechel DE, Rakic P (1979) Arrested proliferation of radial glial cells during midgestation in rhesus monkey. Nature 277:303–305.

Schmid RS, McGrath B, Berechid BE, Boyles B, Marchionni M, Sestan N, Anton ES (2003) Neuregulin 1-erbB2 signaling is required for the establishment of radial glia and their transformation into astrocytes in cerebral cortex. Proc Natl Acad Sci USA 100:4251–4256.

Schmidt-Hieber C, Jonas P, Bischofberger J (2004) Enhanced synaptic plasticity in newly generated granule cells of the adult hippocampus. Nature 429:184–187.

Seaberg RM, van der Kooy D (2002) Adult rodent neurogenic regions: The ventricular subependyma contains neural stem cells, but the dentate gyrus contains restricted progenitors. J Neurosci 22:1784–1793.

Seki T (2002) Expression patterns of immature neuronal markers PSA-NCAM, CRMP-4 and NeuroD in the hippocampus of young adult and aged rodents. J Neurosci Res 70:327–334.

Selden JR, Dolbeare F, Clair JH, Nichols WW, Miller JE, Kleemeyer KM, Hyland RJ, DeLuca JG (1993) Statistical confirmation that immunofluorescent detection of DNA repair in human fibroblasts by measurement of bromodeoxyuridine incorporation is stoichiometric and sensitive. Cytometry 14:154–167.

Seri B, Garcia-Verdugo JM, Collado-Morente L, McEwen BS, Alvarez-Buylla A (2004) Cell types, lineage, and architecture of the germinal zone in the adult dentate gyrus. J Comp Neurol 478:359.

Shou J, Rim PC, Calof AL (1999) BMPs inhibit neurogenesis by a mechanism involving degradation of a transcription factor. Nat Neurosci 2:339–345.

Sidman RL, Rakic P (1973) Neuronal migration, with special reference to developing human brain; a review. Brain Res 62:1–35.

Smith PD, McLean KJ, Murphy MA, Turnley AM, Cook MJ (2005) Seizures, not hippocampal neuronal death, provoke neurogenesis in a mouse rapid electrical amygdala kindling model of seizures. Neuroscience 136:405–415.

Sohur US, Emsley JG, Mitchell BD, Macklis JD (2006) Adult neurogenesis and cellular brain repair with neural progenitors, precursors and stem cells. Philos Trans R Soc Lond B Biol Sci 361:1477–1497.

Song H, Stevens CG, Gage FH (2002) Astroglia induce neurogenesis from adult neural stem cells. Nature 417:39–44.

Suslov ON, Kukekov VG, Ignatova TN, Steindler DA (2002) Neural stem cell heterogeneity demonstrated by molecular phenotyping of clonal neurospheres. Proc Natl Acad Sci USA 99:14506–14511.

Staflin K, Honeth G, Kalliomaki S, Kjellman C, Edwardsen K, Lindvall M (2004) Neural progenitor cell lines inhibit rat tumor growth in vivo. Cancer Res 64:5347–5354.

Takahashi T, Nowakowski RS, Caviness VS Jr (1996) The leaving or Q fraction of the murine cerebral proliferative epithelium: A general model of neocortical neurogenesis. J Neurosci 16:6183–6196.

Tanapat P, Hastings NB, Reeves AJ, Gould E (1999) Estrogen stimulates a transient increase in the number of new neurons in the dentate gyrus of the adult female rat. J Neurosci 19:5792–5801.

Tanigaki K, Nogaki F, Takahashi J, Tashiro K, Kurooka H, Honjo T (2001) Notch1 and Notch3 instructively restrict bFGF-responsive multipotent neural progenitor cells to an astroglial fate. Neuron 29:45–55.

Tauber SC, Stadelmann C, Sppeer A, Bruck W, Nau R, Gerber J (2005) Increased expression of BDNF and proliferation of dentate granule cells after bacterial meningitis. J Neuropathol Exp Neurol 64:806–815.

Tonchev AB, Yamashima T, Guo J, Chaldakov GN, Takakura N (2006) Expression of angiogenic and neurotrophic factors in the progenitor cell niche of adult monkey subventricular zone. Neurosci 144:1425–1435.

Trejo JL, Carro E, Torres-Aleman I (2001) Circulating insulin-like growth factor 1 mediates exercise-induced increases in the number of new neurons in the adult hippocampus. J Neurosci 21:1628–1634.

Traiffort E, Charytoniuk D, Watroba L, Faure H, Sales N, Ruat M (1999) Discrete localizations of hedgehog signaling components in the developing and adult rat nervous system. Eur J Neurosci 11:3199–3214.

Unal-Cevik I, Kilinc M, Gursoy-Ozdemir Y, Gurer G, Dalkara T (2004) Loss of NeuN immunoreactivity after cerebral ischemia does not indicate neuronal loss: A cautionary note. Brain Res 1015:169–174.

Uwanogho D, Rex M, Cartwright EJ, Perl G, Healy C, Scotting PJ, Sharpe PT (1995) Embryonic expression of the chicken Sox2, Sox3 and Sox11 genes suggests an interactive role in neuronal development. Mech Dev 49:23–36.

Vaccarino FM, Ganat Y, Zhang Y, Zheng W (2001) Stem cells in neurodevelopment and plasticity. Neuropsychopharmacology 25:805–815.

van Praag H, Kempermann G, Gage FH (1999) Running increases cell proliferation and neurogenesis in the adult mouse dentate gyrus. Nat Neurosci 2:266–270.

van Praag H, Schnider AF, Christie BR, Toni N, Palmer TD, Gage FH (2002) Functional neurogenesis in the adult hippocampus. Nature 415:1030–1034.

Wang S, Sdrulla A, Johnson JE, Yokota Y, Barres BA (2001) A role for the helix-loop-helix protein Id2 in the control of oligodendrocyte development. Neuron 29:603–614.

Wang JM, Johnston PB, Ball BG, Brinton RD (2005) The neurosteroid allopregnanolone promotes proliferation of rodent and human neural progenitor cells and regulates cell-cycle gene and protein expression. J Neurosci 25:4706–4618.

Wang TW, Stromberg GP, Whitney JT, Brower NW, Klymkowsky MW, Parent JM (2006) Sox3 expression identified neural progenitors in persistent neonatal and adult mouse forebrain germinative zones. J Comp Neurol 497:88–100.

Yagita Y, Kitagawa K, Ohtsuki T, Takasawa K, Miyata T, Okano H, Hori M, Matsumoto M (2001) Neurogenesis by progenitor cells in the ischemic adult rat hippocampus. Stroke 32:1890–1896.

Yamaguchi M, Saito H, Suzuki M, Mori K (2000) Visualization of neurogenesis I the central nervous system using nestin promoter-GFP transgenic mice. NeuroReport 11:1991–1996.

Yang Y, Geldmacher DS, Herrup K (2001) DNA replication precedes neuronal cell death in Alzheimer's disease. J Neurosci 21:2661–2668.

Yarden Y, Sliwkowski MX (2001) Untangling the ErbB signaling network. Nat Rev Mol Cell Biol 2:127–137.

Yoshimura S, Takagi Y, Harada J, Teramoto T, Thomas SS, Wasber C, Bakowska JC, Breakefield XO, Moskovitz MA (2001) FGF-2 regulation of neurogenesis in adult hippocampus after brain injury. Proc Natl Acad Sci USA 98:5874–5879.

Yu T, Gao G, Feng L (2006) Low temperature induced de-differentiation of astrocytes. J Cell Biochem 99:1096–10107.

Yun K, Fischman S, Johnson J, Hrabe de Angelis M, Weinmaster G, Rubenstein JL (2002) Modulation of the notch signaling by Mash1 and Dix1/2 regulates sequential specification and differentiation of progenitor cell types in the subcortical telencephalon. Development 129:5029–5040.

Xu G, Ong J, Liu YZ, Silverstein FS, Barks JD (2005) Subventricular zone proliferation after alpha-amino-3-hydroxy-5-methyl-4-isoxazolepropionic acid receptor-mediated neonatal brain injury. Dev Neurosci 27:228–234.

Zhang RL, Zhang ZG, Zhang L, Chopp M (2001) Proliferation and differentiation of progenitor cells in the cortex and the subventricular zone in the adult rat after focal cerebral ischemia. Neuroscience 105:33–41.

33
Familial Neurodegenerative Diseases and Single Nucleotide Polymorphisms

Michael Slifer and Jeffery M. Vance

Keywords Complex disease; Genetic heterogeneity; Genomic convergence; Haplogroup; Haplotype; Linkage; Linkage disequilibrium; Mendelian disease; Single nucleotide polymorphism; Susceptibility gene

33.1. Introduction

Neurology is a discipline rich in the study of inherited disorders. Most of these studies have focused on diseases in which changes in a single gene are severe enough to cause a disease by themselves. Examples of such changes are mutations in the dystrophin gene that lead to Duchenne muscular dystrophy, or triplet repeat expansions that lead to Huntington disease. However, as well-known as these examples are, they do not reflect the majority of the diseases that the average physician experiences in his or her practice. Most common disorders have important genetic components, but they are not as obvious. Rather, each contributes a smaller amount of risk, and may require environmental interactions to produce symptoms. The elucidation of these genetic elements has the potential to radically change the way all physicians practice medicine in the future. In this chapter we will begin to explore the nature of these genetic changes (complex genetics) and current knowledge of their contribution to several common neurodegenerative disorders.

33.2. Background

Genetic disorders produced by a single causal gene, such as our example of dystrophin mutations leading to Duchenne muscular dystrophy, are termed "Mendelian" disorders, after the Austrian monk Gregor Mendel (1822–1884), who discovered unitary inheritance in pea plants. There are three primary types of Mendelian inheritance. First, dominant genetic variants or alleles produce a trait or disease if one copy of the allele is sufficient to cause the disease (e.g. Huntington disease). Second are recessive variants, which are only observable if two copies of the disease allele are inherited, generally one from each parent (e.g. Friedrichs ataxia). Third are genes on the X-chromosome producing X-linked traits that may act in either a dominant or recessive fashion. The distinguishing feature of X-linked traits is that a recessive variant will appear to act dominantly in males who each have only one X chromosome and therefore only one allele (e.g. Duchenne muscular dystrophy). In contrast, complex traits are governed by an interaction of multiple genetic variants (polygenic) or multiple genetic variants with or without environmental influences (multifactorial). In a complex disease, the contribution from a single gene is not enough to cause the disease, but does make a person more susceptible. Thus, they are known as susceptibility genes. Any variation in DNA (base pair change, deletion, duplication, or expansion) that occurs at a single location is termed an allele. Allelic mutations so severe that they cause disease by themselves are termed mutations. Variations that are neutral or do not cause disease by themselves are termed polymorphisms. Traditionally, polymorphisms were defined as genetic variations appearing in greater than 1% of the population under investigation. However, with modern techniques, we now know that rarer polymorphisms exist, and so this frequency definition is no longer useful.

The most common currently used polymorphisms are single nucleotide polymorphisms (SNPs), i.e., a variation of a single base pair (usually with only two alleles). They are quite numerous, often every 500–1000 base pairs. Initially, SNPs were difficult to detect easily. In the 1980s only a specialized subset of SNPs, the restriction fragment length polymorphisms (RFLPs), could be practically utilized in studies. RFLPs are defined by short sequences of DNA (four to eight bases) that contain a SNP that is recognized and cut by a specific DNA restriction endonuclease. The SNP variation will either create or eliminate the cutting site. The differing lengths of DNA may then be detected using electrophoresis. When the technique of polymerase chain reaction (PCR) was applied to microsatellites, they became the polymorphism of choice for disease studies (Eisenstein, 1990). Microsatellites contain a

T. Ikezu and H.E. Gendelman (eds.), *Neuroimmune Pharmacology*.
© Springer 2008

variable number of repeating elements (usually di- or tetranucleotide repeats) that are highly polymorphic and widely distributed through the genome, although much less common than SNPs. Using unique sequences surrounding a microsatellite, PCR is able to specifically amplify the DNA region containing the microsatellite. Since the microsatellites vary in size, they can then be separated on electrophoretic gels. Finally, the continued development of technology has brought the field of genotyping full circle, so that we can now easily genotype all SNPs, providing millions of polymorphisms for use in research. The ultimate genotyping will be total human genome sequencing, which in 2006 costs ~10 million dollars per individual. However, technology is progressing to reduce this cost. Current NIH funded projects seek technology that will bring the "$1000" genome to a reality.

The degree or the proportion of variation that is directly attributable to a genetic etiology is termed heritability. It is the ratio of the genetic variance of a trait (e.g. height, blood pressure, body mass index, etc.) to the total phenotypic variance found in a population for that trait. Complicating the evaluation of genetic effects is penetrance. The term penetrance was introduced when molecular and clinical tools were not as sensitive as they are today. Penetrance is an "all or none" phenomena; if something is non-penetrant it was thought that the disease may not manifest in every individual who has the causative genetic variant. With today's improved tools of disease detection, the concept of penetrance is not as useful, as in many cases we can detect some signs or features associated with the disease. Expressivity is generally a more useful concept and refers to variation among affected individuals in severity or manifestation of the disease phenotype. Pentrance and expressivity of causative variants should not be confused with susceptibility in polygenic or multifactorial complex diseases. Finally, the relative risk (λ) is the risk of the disease in first degree offspring of an affected individual relative to the risk of the disease in the general population. Any value of $\lambda > 2$ is considered evidence for a genetic contribution.

The approach used to study a genetic disease depends on available technology and the population under investigation. SNP genotyping is now the least expensive and most reliable technology, so it has become the most common method. However, there are still instances when highly polymorphic microsatellites (often in combination with SNPs) can provide more detailed information in select regions of the genome. Most study designs break down into two types, family-based and case-control. There are several different family-based approaches. Twin studies provide an opportunity to tease apart genetic and familial environmental contributions. The concordance of disease phenotype can be compared between monozygotic and dizygotic twins. Since monozygotic twins share all of their genome while dizygotic twins share on average half of their genomes, greater disease concordance among monozygotic twins is consistent with a genetic component to disease. Limitations to twin studies include difficulty in obtaining reasonable sample sizes, vulnerability to assumptions about shared environments, inability to account for potentially different prenatal environments, and the need to sex and age match the already limited twin groups. Generally, more power and larger samples are needed to localize genes contributing to disease.

A more powerful family-based approach is linkage analysis. Linkage analysis utilizes genetic polymorphisms to determine if a physical region of the genome is inherited along with a trait or disease. Multigenerational families with multiple sampled affected and unaffected individuals are efficient for linkage analysis. However, collection of samples with reliable clinical and phenotypic data may be difficult or impossible in some cases, especially for later onset disorders. Sibling pair designs offer another family-based alternative, and siblings are generally more widely available. These family-based methods rely on identification of polymorphisms that co-occur with disease in families more than we would expect by chance. If a polymorphism travels with disease, then we can conclude that it is likely to be physically near to the gene causing the disease. Sophisticated computer programs are able to account either linkage or association within family-based studies. The difference between linkage and association is often confusing to those beginning the study of complex genetics. Linkage implies causality for a physical region of a chromosome, but not for any particular gene or variation within that region. This is different from association, which implicates a specific allele or variation in that region as statistically coupled to the disease. Association may be due to either (1) true causal association with the disease, or (2) linkage disequilibrium (LD). A marker in LD is so close to the disease locus that recombination or mutation has not yet significantly interrupted its co-inheritance with the disease allele within the population under study. Thus a SNP traveling in a population on the same small piece of DNA as the disease gene (in strong LD) can act as a surrogate for the presence of the actual disease change. LD is measured as one of two terms, the correlation coefficient r^2, or D prime, which are derived independently. The correlation coefficient is usually easier to understand, and has become more commonly used in recent years. The term "linkage disequilibrium" is derived from the fact that physically proximate DNA variations (i.e. tightly linked) do not follow the expectations of Hardy-Weinberg equilibrium (which assumes independent segregation), and thus the variations are in Hardy-Weinberg "disequilibrium." Two polymorphisms in strong LD are inherited together forming a haplotype.

As human genetics moves forward, it relies more heavily on genetic association studies to identify genes contributing to the susceptibility of common disorders and pharmacogenetic interactions (genetic variations that affect any biological interaction with a drug). There are two types of association studies. Case-control studies compare allele frequencies in a set of unrelated affected individuals to those in a set of matched controls. The control populations should be matched with respect to ethnicity as well as other factors such as age. Spurious associations may result from unrecognized population stratification i.e., the existence of multiple population subtypes in what is assumed to be a relatively homogeneous population. Unlike case-control studies, where the unit being studied is the individual, family-based association studies use the individual alleles in a person as the unit to be investigated. Family-based studies control for the possibility of genetic differences between the case and control populations (i.e., stratification) by comparing the frequencies of alleles transmitted to the affected child to the alleles not transmitted. While statistically not as powerful as case-control studies, the ability to greatly reduce this stratification error has led many investigators to prefer family-based designs.

Essentially there are three different research approaches to identifying a disease gene: candidate gene analysis; positional cloning; and the most recent, whole genome association analysis. There are advantages and disadvantages to each. Candidate gene analysis is potentially the fastest, but also the least likely to have success. It assumes that one understands the pathophysiology of the disease, which in most cases one does not. Therefore, it has historically been very inefficient overall. Positional cloning is a "genomic" approach that is unbiased towards the causality of the disease and has been extremely successful. Its strength lies in the use of the fundamental mathematics and the known biology of DNA, i.e., the properties of inheritance. The investigator collects family data and then identifies the location of the disease gene through linkage studies. Then, using a combination of molecular genetics and mathematical genetic techniques, the investigator can identify the genetic change that differentiates the affected from the unaffected individuals. Its disadvantages are that it is more expensive, requires multidisciplinary collaborations, and can take many years. However, its advantages are that it will eventually provide the desired results, it is nonbiased requiring no previous understanding of the disease process, and it will find new, unknown genes. It also handles genetic heterogeneity very well, and thus it is the preferred approach for most investigations. Positional cloning has indeed been the main "engine" that has fostered the genetic revolution in medicine. Whole genome association is very new, and involves simultaneously genotyping a huge number of DNA variations (100,000–500,000 polymorphisms), bypassing linkage analysis, to directly identify the associated genetic variation. It does not require the family-based data that linkage analyses do. While potentially rapid, it is very expensive and its practical value is currently unknown. Genomic convergence (Hauser et al., 2003) describes a rapidly developing evidenced-base approach to designing and evaluating genetic studies. Essentially, candidate genes are prioritized by the number of independent methodologies that support that gene's role in a given disorder. For instance, a whole genome or regional linkage analysis, previous association studies, gene expression studies, as well as evidence from model systems are combined to direct efforts towards those candidates that are most likely to play a significant role in the disease. In addition to identifying those genes that alter susceptibility to disease, we now recognize the importance of modifier genes. A modifier gene effects the phenotypic expression of another gene. The symptoms of many inherited diseases vary widely, even between members of the same family. Severity, age at onset, and symptom profiles may be different for individuals carrying identical genetic mutations. This effect is well known in rodent models, where the homozygous background of different strains can radically alter phenotype.

33.3. Neurodegenerative Disorders

33.3.1. Alzheimer's Disease

Alzheimer's disease (AD) is characterized by an insidious onset and progressive deterioration of memory and at least one other cognitive function (language, praxis, recognition, or executive functioning). It is the leading cause of dementia in the elderly, affecting more than 4.5 million people in the United States (Hebert et al., 2003). Prevalence is strongly dependent upon age.

Several lines of evidence pointed to a genetic diathesis for the development of AD. Initial familial aggregation studies demonstrated clustering of AD within families (Sjogren et al., 1952; Heyman et al., 1983; Nee et al., 1983). Twin studies also demonstrated a genetic component to AD. The concordance rate among monozygotic (identical) twins is significantly higher than the concordance rate among dizygotic twins who on average share only half of their alleles (Breitner et al., 1993; Bergem, 1994; Breitner et al., 1995; Bergem et al., 1997).

33.3.1.1. Mendelian Genes of Early-Onset Alzheimer's Disease (EOAD)

Linkage analysis, followed by positional cloning of candidate genes, was used to identify genetic variants of the three genes that are known to cause early-onset familial AD. The three known AD-causative genes are: amyloid precursor protein (APP) (Goate et al., 1991), presenilin I (PS1), and presenilin II (PS2) (Levy-Lahad et al., 1995; Rogaev et al., 1995; Sherrington et al., 1995; Rogaev et al., 1995; Sherrington et al., 1995). Each of the identified mutations is associated with autosomal dominant inheritance within the affected families.

The APP gene was the initial identified gene, localized to chromosome 21 in 1987 (Goldgaber et al., 1987; Tanzi et al., 1987, 1987). Chromosome 21 was an intriguing location, since Down syndrome patients develop the pathologic signs (neuritic plaques and neurofibrillary tangles) and symptoms of AD at an early age (Lemere et al., 1996). Mutations in APP appear to cause shifts in the proteolytic cleavage of APP toward amyloidogenic pathways. In addition, APP mutations lead to excessive production of the 42 amino acid residue of beta-amyloid. This 42 residue beta-amyloid ($A\beta_{42}$) is less soluble than the alternative 40 residue isoform ($A\beta_{40}$) (Suzuki et al., 1994), and is associated with increased aggregation and neurotoxicity (Hilbich et al., 1991). Worldwide, only about two dozen families have been described carrying mutations of the APP gene.

Subsequent studies in additional AD families with autosomal dominant inheritance demonstrated linkage to chromosome 14 (Schellenberg et al., 1992; St George-Hyslop et al., 1992; Van Broeckhoven et al., 1994; Van Broeckhoven et al., 1994). This led to the identification of the PS1 gene as the cause in these families (Sherrington et al., 1995). Since then, more than 150 different PS1 mutations scattered throughout the gene have been described. Phenotypically, PS1 familial AD is the most aggressive form of AD, and affected individuals generally have disease onset in their fourth or fifth decade of life.

After characterization of APP and PS1, many of the remaining unlinked families originated from the Volga river basin of Russia (Bird et al., 1988), and showed linkage to chromosome 1 (Levy-Lahad et al., 1995). Serendipitously with the cloning of the PS1 gene, a larger 7.5 kb alternative polyadenylation message was identified. The product represented a gene homologous to PS1. The new gene, PS2, was mapped to the known linkage region on chromosome 1, and was found to have the mutation in these families.

Like APP mutations, the PS1 and PS2 mutations can lead to increased levels of $A\beta_{42}$ in the brain. Whether the presenilins interact directly with APP through their putative γ-secretase activity, or act as co-factor for another γ-secretase, remains unclear. Furthermore, the relationship between identified causal mutations and AD phenotype is not necessarily simple. For example, PS1 mutations associated with familial early-onset AD have been identified in individuals with frontotemporal dementia who have no evidence of the $A\beta$ accumulation characteristic of AD.

While variants in these three genes (APP, PS1, and PS2) account for between 30% and 50% of early-onset familial AD, overall they account for less than 2% of all cases of AD (Klaver et al., 1998; Finckh et al., 2000; Liddell et al., 1995; Rosenberg et al., 2000).

33.3.1.2. Late-Onset Alzheimer's Disease (LOAD) and Susceptibility Genes

LOAD patients comprise the majority (90–95%) of individuals afflicted with AD. The prevalence of LOAD increases exponentially with advancing age (Rocca et al., 1991). The relationship between LOAD and the first universally accepted susceptibility gene, Apolipoprotein E (ApoE), was initially discovered using linkage analysis in a subset of AD patients with an age-at-onset greater than 60 (Pericak-Vance et al., 1991). In today's rush to gather larger and larger datasets, it is important to realize that only 33 families were used in this initial analysis, reflecting the strength of ApoE's genetic effect in AD. The ApoE gene lies on the long arm of chromosome 19 (19q13.2) coding for a serum protein involved in the transport, storage, and metabolism of lipids. ApoE in the central nervous system is synthesized by astrocytes (Mahley, 1988). There are three common isoforms for ApoE protein: ApoE2, ApoE3, and ApoE4. These three isoforms result from single amino acid substitutions at two different amino acid residues, 112 and 158 (Weisgraber et al., 1982; Rall et al., 1982). Therefore, the ApoE alleles are actually haplotypes.

The role of ApoE in cardiovascular diseases has been well-documented, and since the initial discovery of its association with AD it has also been shown to be associated with multiple neurodegenerative diseases in a growing number of studies. The ApoE4 allele has consistently been associated with risk for sporadic and familial LOAD, whereas, the ApoE2 allele is inversely associated with risk for LOAD (Corder et al., 1994). The relative risk for LOAD for ApoE4 homozygotes and for ApoE3/4 heterozygotes are ~15 and 3, respectively, compared to ApoE3 homozygotes. A meta-analysis of 40 studies by (Farrer et al., 1997), demonstrates that the effect of ApoE4 allele is weaker in the African-American and Hispanic populations, compared to the Caucasian population, while the effect of ApoE4 allele is greater in the Japanese population, compared to Caucasian individuals. The reasons for this are not clear but may reflect differences in modifier genes.

While ApoE is critically important, it does not account for all of the genetic variation seen in AD. The heritability of AD has been estimated at about 80% (Bergen, 1994), but more than a third of AD cases do not have a single ApoE4 allele. The sibling relative risk (λs) for the ApoE locus is estimated to be about two, and Farrer et al. suggested that ApoE accounts for, at most, 50% of the total genetic effect in AD (Farrer et al., 1997). To date, efforts to identify the remaining AD loci are ongoing.

Variants in more than 200 candidate genes have been tested for association, and variants in 115 genes are reported to be associated with late-onset AD. However, other than ApoE, no candidates have achieved wide support. Several factors account for the difficulties undermining these studies. Most studies have been undersized, the level of genomic detail is small, and locus heterogeneity reduces the statistical impact of any single genetic variant under investigation. Together these elements conspire to make replication or confirmation difficult even if genuine effects are identified. Indeed, recent theoretical work has confirmed what most applied researchers already knew, that validation of association with additional data is more useful than actual separate replication studies (Skol et al., 2006).

Several linkage analyses for LOAD have been published. The most frequently linked chromosomes are 9, 10, and 12. Significant linkage was identified on chromosome 9p using a North American LOAD dataset (Pericak-Vance et al., 2000). Farrer confirmed the linkage to chromosome 9p in a consanguineous Israeli-Arab sample, suggesting a possible autosomal recessive locus (Farrer et al., 2003). In another independent dataset, the linkage signals on 9p and 9q were both confirmed (Kehoe et al., 1999; Myers et al., 2002; Blacker et al., 2003).

However, the actual genetic variants producing the linkage signals remain elusive.

Myers conducted a two-stage genome-wide linkage scan and found linkage to a marker on chromosome 10q (Myers et al., 2000). Subsequently, Ertekin-Taner treated plasma Aβ concentration as a quantitative phenotype and detected linkage to a quantitative trait locus close to the candidate gene α-T catenin on 10q (Ertekin-Taner et al., 2003). Additionally, GSTO1/2 has been identified as an age-at-onset modifier in AD (Li et al., 2003). However, no well replicated candidate gene affecting risk for AD has been identified on chromosome 10.

After the first report identifying linkage evidence on chromosome12p (Pericak-Vance et al., 1997), several follow-up studies have produced mixed results in this region as well as another region on 12q (Pericak-Vance et al., 1997; Pericak-Vance et al., 1998; Scott et al., 2000; Scott et al., 1998). The mixed findings may be attributable to genetic heterogeneity. Scott found linkage evidence suggesting genetic heterogeneity since some families clustered on a smaller region of chromosome 12 and at least one affected individual had Lewy-body dementia but lacked the ApoE4 allele (Scott et al., 2000). Other studies confirmed a risk effect as well as an age-at-onset effect in families that lack the ApoE4 allele (Wu et al., 1998; Mayeux et al., 2002).

Chromosome 12p contains several candidate genes, including the α-2 macroglobulin and DLD receptor related protein-1 genes, which are related to β-amyloid metabolism. However, association studies on these candidate genes have produced conflicting results. Convergent linkage and association evidence has been reported for the glyceraldehyde-3-phosphate dehydrogenase (GAPD) gene on 12p (Li et al., 2004b) but the results could not be replicated in an independent dataset. In short, evidence for susceptibility genes on chromosome 12 remains elusive with conflicting findings, which may in part be due to genetic heterogeneity.

33.3.1.3. Modifier Genes of LOAD

In addition to increasing susceptibility risk for AD, the ApoE gene on chromosome 19 also appears to have modifier effects on the age-at-onset of AD. In one family-based study, the estimated mean age-at-onset for subsets consisting of individuals carrying no ApoE4 allele, one copy of the ApoE4 allele, and two copies of the ApoE4 allele, are 84.3, 75.5, and 68.8 years, respectively (Corder et al., 1993). Additionally, while the ApoE4 allele is a risk factor for LOAD in nearly every population, the magnitude of association between the ApoE4 allele and LOAD may vary by ethnicity.

Although no candidate genes on chromosome 10 have been reported to be significantly associated with risk of LOAD, the Glutathione S-transferase omega-1/2 complex (GSTO1/2) has been associated with variation in age-at-onset of both LOAD and Parkinson disease (Li et al., 2003, 2005). Using a very large pooled AD and PD dataset, Li et al. used linkage analysis to demonstrate overlapping linkage peaks in both disorders. Using gene expression and genomic convergence with the linkage data, Li et al. demonstrated that four genes differentially expressed between AD and control hippocampi were located in the chromosome 10 age-at-onset linkage peak (Li et al., 2003). One of these genes (GSTO1) was highly associated with age-at-onset, as was its homologue, GSTO2, which lies immediately next to it. Subsequent research demonstrates that this association is responsible for the linkage peak but only appears to have its effect in about 25% of AD families, shifting the age-at-onset an average of 8 years in these families (Li et al., 2006). This successful use of genomic convergence is also an excellent demonstration of genetic heterogeneity in both AD and PD. The functions of GSTO1 are not well understood (Board et al., 2000). The Ala140Asp and Thr217Asn variants of GSTO1 display reduced enzyme activity (Tanaka-Kagawa et al., 2003) and therefore might influence the susceptibility to oxidative stress. Recent data suggest that GSTO1 is involved in the post-translational modification of the inflammatory cytokine Interleukin-1β (Laliberte et al., 2003) and therefore contributes to inflammation. This is provocative given reports of the possible role of inflammation in AD and PD (McGeer and McGeer, 2004). Subsequently, four additional case-control studies have been reported for GSTO1 and age-at-onset in AD. One found no age-at-onset effect in Japanese (Nishimura et al., 2004). Lee et al. genotyped only one SNP in GSTO1 and found a weak association for risk of AD in Caribbean Hispanics but not for age-at-onset (Lee et al., 2004). Ozturk reported identical findings in a large Caucasian sample (Ozturk et al., 2005). Kolsch studied GSTO1 in stroke patients and found it was associated with an increase risk for stroke, and also found a mild age-at-onset effect in AD (Kolsch et al., 2004). These present excellent opportunities to discuss some of the difficulties in interpreting association studies. Three of the studies were small, and if only 25% of the AD population is affected (as suggested by Li et al.), then validation of the effect will require large datasets. Li et al. used a family-based design, but the other studies were all case-control. This may be a major point, since the more common background in family-based studies may enable them to be more sensitive at detecting specific gene modifiers. Finally, the difference between a modifier gene and a risk gene may be attributed more to sample selection than anything else. For example, if a sample has primarily earlier age-at-onset cases, then a gene that makes cases fall into that group will be seen as a risk gene.

33.3.1.4. Clinical Implications of Genetic Discoveries for AD

Two broad categories of individuals may present for genetic counseling in AD. Family members are either symptomatic (demonstrating disease symptoms and signs) or are considered presymptomatic and at risk. Application of genetic tests to symptomatic patients seeks to confirm a diagnosis or provide specific mutation information to a family. As its name suggests, presymptomatic testing is for individuals who are currently healthy, but undergo genetic testing to provide information on the likelihood they will get the disease in the future. At present, counseling remains quite controversial in complex disorders since the presence of a single positive test does not provide a clearly defined risk for a single individual. The best example is ApoE and AD. Perhaps more importantly, without useful preventative treatments for such diseases as AD, the justification for testing remains elusive for many clinicians. Thus, we are at an uncomfortable time in medical history, a true "catch 22" since (1) we must know the causes of common diseases to efficiently treat them, (2) this knowledge raises anxiety and suggests risks for asymptomatic individuals, but (3) we have no treatments once we identify these genes. Obviously the hope is that preventative therapies will come quickly once the genetic risk factors are identified.

Currently, genetic counseling for AD families has followed the Huntington disease model recommended by the Canadian Collaborative study (Copley et al., 1995). Generally, high-penetrant mutations, such as PS1, APP, and the Tau gene (for fronto-temporal dementia), appear to serve as better candidates for genetic screening of AD in at-risk families, than low-penetrant mutations, such as PS2 and ApoE (George-Hyslop and Petit, 2005). However, several studies have now been published examining genetic testing for the ApoE4 allele in presymptomatic individuals, with generally very positive results (Drzezga et al., 2005).

33.3.2. Parkinsonian Disorders

33.3.2.1. Parkinson's Disease

Like AD, theories underlying the pathogenesis of Parkinson disease (PD) have evolved since it was first described by James Parkinson in 1817. PD is a neurodegenerative disorder that affects the dopaminergic neurons of the CNS, most notable by the degeneration of cells in the substantia nigra. The primary clinical triad is resting tremor, rigidity and bradykinesia, although many of the non-motor symptoms are also problematic for the patients. Originally believed to be primarily an environmental disease, it is now known to have a strong genetic component. Like AD, the genetic components of PD are split between Mendelian and susceptibility genes. However, several common themes are arising from genetic investigations, notably the importance of mitochondrial dysfunction and inflammation in the pathogenesis of the disorder.

33.3.2.1.1. Mendelian Genes

Most interest in the field has focused on a few Mendelian genes that lead to PD, though it is believed these represent only a small portion (perhaps 10%) of the 1.5 million PD patients in the US. The two most common Mendelian genes are Parkin, on chromosome 6 and Leucine-Rich Repeat kinase 2 (LRRK2) on 12q12. Additionally, less common genes have been identified including DJ1, PINK1, and α-Synuclein.

α-Synuclein: The α-Synuclein gene was the first gene found to have mutations causing PD. However, it remains a rare cause of the disease affecting only a few families. Identification of α-Synuclein's role was relatively straight forward in these families since the large gene triplication followed an autosomal dominant mode of inheritance. It is unclear if more subtle single nucleotide variants contribute to disease in the broader population.

Parkin: Parkin was discovered in juvenile PD (age-at-onset <40 years.) families in Japan using positional cloning techniques. It was quickly realized that this form of PD was not just limited to Japan but was a major contributor to juvenile PD throughout the world. In addition, while the mean age-at-onset of Parkin homozygotes is about 20 years, the age-at-onset distribution is wide, with the oldest age-at-onset reported at 80 years. Thus, Parkin should not be thought of as merely an "early-onset" PD gene but considered in all PD patients with a strong family history.

Parkin is located on the outer mitochondrial membrane, and early research identified it as an ubiquitin E2 ligase, suggesting abnormal protein degradation as its likely mechanism of disease. Recent evidence suggests it is important in the expression of mitochondrial proteins encoded by the mitochondrial genome. Gene expression data also found that mtDNA encoded mitochondrial proteins had increased expression in PD patients versus controls, when compared to their nuclear encoded counterparts (Noureddine et al., 2005). This is also consistent with the van der Walt report of a mtDNA haplogroup association in PD (van der Walt et al., 2003). The initial findings of van der Walt have been confirmed in multiple studies (see below), and raise another possible disease mechanism for Parkin, given the increasing evidence for the importance of mitochondria in PD.

Leucine Rich Repeat Kinase 2 (LRRK2): mutations lead to late-onset PD, in an autosomal dominant pattern, with variable age-at-onset. To date, the S2019G mutation is by far the most common (>50%) and appears to have arisen at different times in the past. However, the biological function of LRRK2 is currently not known.

33.3.2.1.2. Susceptibility Genes

Genomic Linkage in Parkinson Disease: To date, four complete genomic screens have been reported for PD (Scott et al., 2001; Destefano et al., 2001; Pankratz et al., 2002; Hicks et al., 2002). Three linkage screens for age-at-onset genes affecting PD have thus far been performed (Li et al., 2002; Destefano et al., 2002; Pankratz et al., 2004). The most interesting finding is on chromosome 1p, where the age-at-onset linkage peak of Li et al. (Li et al., 2002) is very similar to the risk linkage peak (PARK10) in late-onset Icelandic PD families (Westerman et al., 1999). There is also overlap between the age-at-onset and risk linkage peaks on chromo**some 9q from the GenePD Study** (Destefano et al., 2001; Destefano et al., 2002). It is conceivable that a single gene is influencing both risk and age-at-onset (Apolipoprotein E, for instance, is known to affect both risk and age-at-onset of PD, see below) (Li et al., 2004a), but it is also possible that distinct susceptibility and age-at-onset alleles exist at each one of these loci.

FGF 20: Fibroblast growth factor 20 is a member of the large family of FGF growth factors. It lies within the chromosome 8 linkage peak identified by Scott et al. (Scott et al., 2001) and is notable because it is critical to the culture of dopaminergic neurons derived from embryonic stem cells. In addition, unlike GDNF and BDNF, which are ubiquitous in the CNS, FGF20 is primarily expressed in the substantia nigra, suggesting it has a highly specific relationship with dopaminergic neurons in the brain. van der Walt found strong association of FGF20 alleles and risk for PD (van der Walt et al., 2004). The allelic association was recently validated by Clarimon (Clarimon et al., 2005).

Tau: is a microtubule-associated protein that functions in the assembly and stability of the microtubule. Mutations in Tau are known to cause frontotemporal dementia with Parkinsonism-17 (Hutton et al., 1998). Two different groups initially reported that an intronic microsatellite allele, A_0, was associated with PD (Pastor et al., 2000;Golbe et al., 2001). Oliveira et al. also found association of this allele with PD. Furthermore, it lies under the chromosome 17 peak from a genomic linkage screen, (Scott et al., 2001) supporting its likely importance in PD (Martin et al., 2001). Healy et al. performed a meta-analysis of nine case-control studies plus their own and confirmed that the H1 haplotype (containing the A_0 allele) was highly associated with PD ($p < 10^{-6}$) (Healy et al., 2004). Oliveira demonstrated that the H1 haplotype is extremely large, due to extensive linkage disequilibrium in the region (Oliveira et al., 2004). In fact, the linkage disequilibrium is so large that the H1 haplotype encompassed 3.15 megabases and contains other genes besides Tau, including Corticotropin-releasing Hormone Receptor 1, Presenilin Homolog 2, Saitoin, and KIAA1267. In an attempt to reduce the broad region of linkage disequilibrium, they genotyped a large number of additional markers. They identified a sub-haplotype of H1 (11% of the population) that contains the association to PD ($p = 0.02$) but it still spanned several loci. Thus, while Tau is known to bind to α-Synuclein and is the most likely gene in the associated haplotype, the genetic data at present cannot rule out other genes in the Tau haplotype as the causative factors.

α-Synuclein: Two meta-analyses in the Caucasian and Japanese populations support the association of a complex microsatellite in the α-synuclein promoter (NACP-Rep1) with susceptibility risk for PD (Mizuta et al., 2002;Mellick et al., 2005). However, other SNPs in the a-synuclein promoter have not been associated with PD (Pastor et al., 2001;Holzmann et al., 2003).

Parkin: Association studies in the Parkin gene have diverged widely in their conclusions. No meta-analysis has been published so far, but the largest association study did not find any evidence for association with risk for PD (Oliveira et al., 2003). Investigating late-onset PD has been difficult because many datasets focus on earlier-onset disease phenotypes. However, evidence is mounting that those who are heterozygous for a Parkin mutation within exon 7 of the gene, may be more susceptible to late-onset PD (Oliveira et al., 2003;Foroud et al., 2003). Recent functional protein studies also support this premise. Khan studied 13 unaffected 1st-degree relatives from eight unrelated PD Parkin carriers and found a significant reduction in ^{18}F-dopa uptake in these heterozygotes relative to controls (Khan et al., 2005). Interestingly, similar data exists that supports the same finding for PINK1 and Gaucher carriers.

Mitochondrial Genome (mtDNA): Mitochondria have their own circular genome. The mitochondrial genome is small (16,000 bp) and codes for 26 proteins. Unlike the nuclear genome, it has only one allele, and is not efficient in repairing itself. Mitochondria are inherited primarily from the egg and are quite polymorphic between ethnic groups. Haplotypes in mitochondria are termed haplogroups, and have been very useful in molecular anthropology. van der Walt reported that the J and K haplogroups were associated with decreased risk for PD, particularly in females (609 affected, 340 controls) (van de Walt et al., 2003). Recently, a large study by Pyle of 455 affected, 447 controls supported this finding (Pyle et al., 2005). While van der Walt suggested the causative SNP was the A10398G polymorphism, Pyle et al. felt this was not clear, as the SNP has been found on other haplogroup backgrounds. Another small study looking at A10398G found it trending towards association but was not significant (102 affected, 112 controls) (Otaegui et al., 2004). However, the J haplogroup has been associated with increasing risk for PD in studies in the Irish and Finnish populations (Ross et al., 2001; Autere et al., 2002), but these studies were much smaller in size (90 cases/129 controls and 238 cases/104 controls, respectively).

Inducible NOS (iNOS): Under oxidative stress NOS reacts with superoxide anion to form toxic hydroxyl radicals. Inducible NOS is found in glial cells of the CNS and is induced in response to injury. Levecque and Hague both reported association of iNOS with PD in European and Finnish populations, respectively (Levecque et al., 2003; Hague et al., 2004). Hancock confirmed the association in North Americans, but only in early-onset cases (one member of the family with age-at-onset <40 years) (Hancock et al., 2005). Levecque also found an interaction between smoking and iNOS, which Hancock et al. also recently reported in an even larger dataset. (Hancock et al., 2006).

33.3.2.1.3. Modifier Genes

ApoE has been examined by many groups for association with PD. Several recent meta-analyses concluded that ApoE is associated with risk and age-at-onset for PD, although the effect is not nearly as strong as in AD.

Li demonstrated that GSTO1/2 modifies PD age-at-onset in a manner similar to its effect on AD (Li et al., 2003).

Additionally Oliveira showed that one of the subunits of EIF2B, EIF2B3, located on chromosome 1, has a strong age-at-onset effect in PD (Oliveira et al., 2005). EIF2B is an interesting candidate since mutations in EIF2B are known to cause vanishing white matter disease.

33.3.2.2. Amyotrophic Lateral Sclerosis (ALS)

Amyotrophic lateral sclerosis is an inexorably progressive motor neuron disease, in which both the upper motor neurons and the lower motor neurons degenerate leading to muscle atrophy. Patients eventually experience respiratory failure, usually within three to five years from diagnosis. However, the onset of ALS may be subtle and early symptoms are frequently overlooked. As many as 20,000 Americans have ALS, and an estimated 5,000 people in the United States are diagnosed with the disease each year. Onset is usually in the 5th through 7th decade of life.

33.3.2.2.1. Mendelian Genes

Like AD and PD, the majority of ALS cases have no clear familial pattern and appear multifactorial in nature. However, up to ten percent of all ALS cases are inherited as an autosomal dominant disorder. Twenty percent of these familial cases result from a dominant mutation in the superoxide dismutase 1 gene (SOD1) (Jones et al., 1995; Jackson et al., 1997). Superoxide dismustase normally scavenges free radicals. If not neutralized, free radicals act as promiscuous electron donors causing random damage to DNA and proteins. Mouse models carrying the SOD1 mutation undergo motor neuron degeneration analogous to human ALS. However, mice with deleted SOD1 genes do not get ALS. This suggests the SOD1 mutation is a dominant negative that actively causes disease, as opposed to producing disease through haploinsufficiency, where disease results from a deficiency of functional gene products.

33.3.2.2.2. Susceptibility Genes

ApoE: ApolipoproteinE isoforms are the most thoroughly studied ALS candidates. Like AD and PD, ApoE4 appears to have a role in ALS. However, findings for risk and age-at-onset have been inconsistent. The most promising and well supported role for the ApoE4 isoform has been to accelerate disease progression (Al-Chalabi et al., 1996; Moulard et al., 1996).

VEGF: Initial reports favor a role for vascular endothelial growth factor (VEGF) in ALS (Lambrechts et al., 2003). VEGF is an essential cytokine for angiogenesis and vasculogenesis. Additionally, VEGF receptors (VEGFR-2 and NRP1), signaling effectors, and upstream regulators (HIF1A and HIF2A), all remain intriguing candidate genes.

NGF: Neurotrophic growth factors (NGFs) are a group of neuropeptides that play an important role in regulating the growth, differentiation, and survival of neurons in the peripheral and central nervous systems and are therefore reasonable candidates for a role in ALS. However, there are only few published association studies of NGF-regulating genes in ALS, and one large study failed to support an effect of the ciliary neurotrophic factor (CNTF) locus on ALS risk or clinical phenotype (Al Chalabi et al., 2003). Additionally, mixed results have been reported for the survival motor neuron genes (SMN1 and SMN2) that were originally linked to spinal muscular atrophy.

NF: Abnormal accumulation of intermediate filaments in the perikarya and proximal axons of motor neurons is a common pathological hallmark of ALS. Several studies have examined the neurofilament (NF) subunits and repeating regions, but results are mixed with opposite alleles associated with risk in different studies (Figlewicz et al., 1994; Tomkins et al., 1998; Al Chalabi et al., 1999; Skvortsova et al., 2004).

Glutamate: Gluatamatergic excitotoxicity has long been proposed as a mediator of ALS. The glial glutamate transporter EAAT2 is responsible for glutamate removal from motor neurons and suppressed expression of EAAT2 was observed in 60% of patients with sporadic ALS (Rothstein et al., 1995). Abnormal splice variants of EAAT2 have been detected, however, they may not be due to polymorphisms in the EAAT2 gene (Aoki et al., 1998). Similarly, defective RNA editing of the GluR2 subunit of the glutamate AMPA receptor appears to contribute to motor neuron death in sporadic ALS, but there is no clear role for GluR2 alone.

Free Radicals: Oxidative stress appears to play a role in ALS pathogenesis, and studies of genes in the reactive oxidative pathways and reactive nitrogen pathways are ongoing.

Modifiers: In addition to susceptibility genes, modifier genes may play an important role in ALS pathogenesis and may also direct efforts towards effective interventions. For instance, while variants of ApoE and monoamine oxidase B (MAOB) do not appear to increase risk for ALS, ApoE4 appears to accelerate progression and an MAOB allele may delay age-at-onset (Orru et al., 1999). Ongoing work also explores environmental effects on ALS and the role of candidate gene variants in response to specific environmental stressors (i.e., gene x environment interactions).

33.3.2.3. Progressive Supranuclear Palsy (PSP)

PSP is the second most common cause of Parkinsonism typified by early gait instability and difficulty with vertical eye movement. PSP is characterized by neurofibrillary tangles composed almost entirely of straight filaments of four repeats of Tau protein. Although most cases of PSP appear to be sporadic, genetic diatheses have been implicated. De Yebenes described a pattern of inheritance consistent with a Mendelian autosomal dominant disorder (De Yebenes et al., 1995). Difficulty recognizing the variable phenotypic expression of PSP may be one reason fewer familial cases have been identified than expected (Rojo et al., 1999). The H1 haplotype of the Tau gene has also been found to have association with increased risk for PSP, as it has been for PD (Conrad et al.,

1997). Recently, the H1 and H2 haplotypes have been shown to be directly associated with a large chromosome inversion involving the Tau gene. (Gijselinck et al, 2006).

33.3.2.4. Frontotemporal Dementia with Parkinsonism-17

Frontotemporal dementia (FTD) is characterized by a gradual onset of changes in personality, social behavior, and language. FTD is often associated with parkinsonian symptoms or motor neuron disease. The clinical presentation of FTD is heterogeneous, and FTDs are frequently mistaken for AD or a primary psychiatric disorder. FTD is complex and may present either sporadically or as a familial disorder. In 1994, linkage to chromosome 17 was established in a pedigree with FTD designated disinhibition-dementia-parkinsonism-amyotrophy complex (Lynch et al., 1994). The linkage gave rise to the name frontotemporal dementia with parkinsonism linked to chromosome 17, or FTDP-17. The Tau gene on chromosome 17 was the primary candidate because brains in familial cases of FTD showed neuronal and glial intracytoplasmic Tau inclusions. Since then, more than 25 different mutations, including missense mutations, splice-site mutations, and deletions, have been identified in the Tau gene in more than 50 families with FTD (van Slegtenhorst et al., 2000). Tau mutations can alter Tau splicing, disrupt Tau protein function, or both. However, the frequency of Tau mutations in sporadic cases of FTD, i.e., individuals with no family history of dementia, is rare (Baker et al., 1999). It is important to also realize that a significant number of FTDP-17 patients have not had Tau mutations identified in their families. Very recently these families have been shown to be due to mutations in progranulin, a gene lying near Tau and whose function is not fully understood (Cruts et al., 2006).

33.3.2.5. Multisystem Atrophy (MSA)

MSA is generally regarded as a sporadic, late onset (typically 50–65 years) disease whose clinical presentation overlaps with PD. Unlike PD, however, individuals with MSA experience autonomic failure and early cerebellar signs such as ataxia. MSA is characterized by neuronal loss accompanied by oligodendroglial, and α-Synuclein inclusions. No risk factors, either genetic or environmental, have been identified for MSA and familial cases have not been described. Nevertheless, subsyndromal symptoms of MSA are disproportionately reported by the relatives of patients with MSA compared to controls (Nee et al., 1991). However, a reporting bias among relatives cannot be ruled out. Several studies have looked for polymorphisms in candidate genes, which may predispose an individual towards developing MSA. The apolipoprotein E4 allele is not over-represented in MSA when compared to controls. There are conflicting reports of an association between a cytochrome P-450, CYP2D6 polymorphism, and MSA. Studies have not identified any mutations in the coding regions of the α-Synuclein gene in patients with pathologically confirmed MSA.

33.3.3. Multiple Sclerosis (MS)

Multiple sclerosis is a debilitating autoimmune disorder with a complex genetic component. From the preponderance of evidence, there is compelling support for inherited susceptibility to MS. Consistent with a genetic component to MS, prevalence varies by population, there is familial aggregation of cases, and monozygotic twins (who share all their alleles) consistently have higher concordance rates for MS than dizygotic twins (who share half of their alleles). However, since the concordance rate is less than 100% in monozygotic twins, there appears to be an important environmental component as well, consistent with a multifactorial disease model. The patterns of inheritance have not supported a simple Mendelian model (autosomal dominant, autosomal recessive, or X-linked). However, hidden within the complex inheritance in the majority of cases, Mendelian gene variants could account for a small proportion of disease, as has been described in AD (Pericak-Vance et al., 1995) and PD (Scott et al., 1997). Both candidate gene and linkage approaches have been applied to examine the genetic underpinnings of MS. Candidate genes were chosen from among genes known to have a role in the immune system such as the HLA genes, T-cell receptor genes, and the myelin basic protein gene. One MS susceptibility gene within the major histocompatibility complex (MHC) has been identified and confirmed through linkage and family-based association methods. Allelic association to the HLA-DR2 allele was also confirmed using a family-based association approach (Haines et al., 1998). However, with the exception of the MHC (Bertrams et al., 1972), candidate gene studies have suffered from poor replication of results. Two genomic screens (Sawcer et al., 1996;Haines et al., 1996) have demonstrated suggested linkage to at least three regions in addition to the MHC indicating that further candidate regions should be explored.

Summary

For most of the history of genetic studies, researchers have explored forms of disease that followed the relatively simple rules of Mendelian inheritance. There have been many successes using this strategy, however, the most common disease forms have more complex patterns, and are more challenging to elucidate. Already research has shown us that these common diseases are indeed just clinical phenotypes, in which many different causes lead to a similar clinical presentation. Recognizing this heterogeneity has importance for the development of effective therapies targeted to specific disease etiologies. Additionally, until the genetic heterogeneity is sufficiently characterized, it will be difficult to tease out the environmental contributions to diseases, since different genotypes are likely to response differently to the same environmental stimuli. At this point in time, identifying and modifying risk based on the known genetic underpinnings of many neurodegenerative diseases remains problematic. As a result, genetic testing and counseling are reserved primarily for the most well characterized genetic diseases (e.g. early-onset familial Alzheimers disease). Most neurodegenerative diseases have complex etiologies and the benefits of genetic testing remain controversial. However as we advance our understanding of complex genetic interactions, we will finally bring genetics into mainstream medicine and neurology to significantly transform the way we practice medicine.

Review Questions/Problems

1. The majority of the mutations known to cause neurodegenerative diseases in humans have been discovered for

a. common complex diseases because they affect the most people
b. diseases with a Mendelian inheritance pattern
c. multi-system atrophy
d. non-genetic diseases
e. diseases with no familial component

2. A characteristic of modern genetic research is

a. that very few candidate genes are proposed for any given disorder
b. that positive association studies of candidates have been consistently replicated
c. all of the above
d. none of the above

3. In a genetic study, stratification may refer to

a. unrecognized population sub-types due to recent admixture
b. unrecognized population sub-types due to incorrect matching of cases and controls
c. a method for presumably reducing genetic heterogeneity within a sample by narrowly defining the phenotype
d. all of the above
e. none of the above

4. Heterogeneity within a disease may manifest as

a. allelic heterogeneity, in which many variants of a single gene can cause the same disease
b. locus heterogeneity, in which variants in many different genes may cause the same disease
c. some of the difficulties underlying genetic studies of neurodegenerative diseases
d. all of the above
e. none of the above

5. Penetrance and expressivity refer to

a. the proportion who get the disease and how the disease manifests respectively
b. redundant terms
c. how the disease manifests and the proportion who get the disease respectively
d. all of the above
e. none of the above

6. Microsatellites

a. are relatively highly polymorphic genetic markers
b. can by amplified by polymerase chain reactions
c. contain repeated elements
d. all of the above
e. none of the above

7. An advantage to using single nucleotide polymorphisms in genetic studies is that

a. each marker is highly polymorphic
b. they are common and distributed throughout the genome
c. they are always within exons, and therefore cause a functional change
d. all of the above
e. none of the above

8. Linkage analysis

a. is limited by its exclusive use of restriction fragment length polymorphisms
b. utilizes co-inheritance of a region of the genome with the disease locus to which it is "linked"
c. has the advantage of always identifying causative mutations
d. all of the above
e. none of the above

9. Allelic genetic association analyses may find positive associations because the

a. allele causes the disease
b. allele is in strong linkage disequilibrium with the disease allele
c. result is a false positive
d. all of the above
e. none of the above

10. Genomic convergence provides

a. a method for greatly increasing the resolution of linkage peaks
b. a method for using non-human primates to identify causative alleles
c. an approach to prioritize candidates for further genetic study
d. all of the above
e. none of the above

11. A modifier gene

a. must cause disease.
b. alters the expression of a disease.
c. must be in linkage disequilibrium with the disease allele.
d. all of the above.
e. none of the above.

12. Variants of the GSTO gene are associated with

a. age at onset in AD
b. age at onset in PD
c. modifier effects
d. all of the above
e. none of the above

13. Causative mutations in which of the following genes has been described in early-onset AD?

a. amyloid precursor protein, presenilin-1, presenilin-2
b. presenilin-1, amyloid precursor protein, apolipoprotein C
c. presenilin-1, presenilin-2, apolipoprotein C
d. apolipoprotein C, presenilin-2, amyloid precursor protein
e. huntingtin, presenilin-1, presenilin-2

14. All of the following have been candidates for a role in late-onset AD, but which is the most widely accepted and has the best supporting evidence for increasing susceptibility to AD?

a. UBQLN1
b. APOE
c. GAPD
d. α-T catenin
e. VLDL-R

15. Features of the apolipoprotein E-4 allele include

a. increased susceptibility to late-onset AD.
b. modification of age-at-onset in AD.
c. a dose dependent effect on age-at-onset in AD.
d. an association with other neurodegenerative diseases in addition to AD.
e. all of the above.

16. Characteristic of FTDP-17 is

a. a phenotype that may be mistaken for AD
b. linkage to chromosome 17
c. abnormal Tau accumulation
d. all of the above
e. none of the above

17. Which of the following has been proposed as a candidate for susceptibility to amyotrophic lateral sclerosis?

a. VEGF
b. CNTF
c. glutamate system
d. all of the above
e. none of the above

18. The genetics underlying multiple sclerosis are typical for a neurodegenerative disease in that

a. concordance rates in monozygotic twins is higher than in dizygotic twins.
b. linkage and association approaches have been used to identify likely candidates for a role in the disease
c. there is an important environmental component to the disease
d. all of the above
e. none of the above

19. Huntington disease is an example of a dominantly inherited disease because

a. two copies of the disease allele are needed to express the disease
b. the disease locus is on the X chromosome
c. only one copy of the disease allele is needed to express the disease
d. it exhibits classic anticipation
e. it can only be inherited from an affected mother

20. Huntington disease exhibits all of the following except

a. autosomal dominant inheritance
b. anticipation, where the disease may become more severe with an earlier age-at-onset in subsequent generations
c. it is characterized by expansion of a tri-nucleotide repeat
d. an autosomal recessive form of the disease
e. a causative gene has been identified

References

Al-Chalabi A, Enayat ZE, Bakker MC, Sham PC, Ball DM, Shaw CE, Lloyd CM, et al (1996) Association of apolipoprotein E ε4 allele with bulbar-onset motor neuron disease. Lancet 347:159–160.

Al Chalabi A, Andersen PM, Nilsson P, Chioza B, Andersson JL, Russ C, Shaw CE, Powell JF, Leigh PN (1999) Deletions of the heavy neurofilament subunit tail in amyotrophic lateral sclerosis. Hum Mol Genet 8:157–164.

Al Chalabi A, Scheffler MD, Smith BN, Parton MJ, Cudkowicz ME, Andersen PM, Hayden DL, Hansen VK, Turner MR, Shaw CE, Leigh PN, Brown RH, Jr. (2003) Ciliary neurotrophic factor genotype does not influence clinical phenotype in amyotrophic lateral sclerosis. Ann Neurol 54:130–134.

Aoki M, Lin CL, Rothstein JD, Geller BA, Hosler BA, Munsat TL, Horvitz HR, Brown RH, Jr. (1998) Mutations in the glutamate transporter EAAT2 gene do not cause abnormal EAAT2 transcripts in amyotrophic lateral sclerosis. Ann Neurol 43:645–653.

Autere JM, Hiltunen MJ, Mannermaa AJ, Jakala PA, Hartikainen PH, Majamaa K, Alafuzoff I, Soininen HS (2002) Molecular genetic analysis of the alpha-synuclein and the parkin gene in Parkinson's disease in Finland. Eur J Neurol 9:479–483.

Baker M, Litvan I, Houlden H, Adamson J, Dickson D, Perez-Tur J, Hardy J, Lynch T, Bigio E, Hutton M (1999) Association of an extended haplotype in the tau gene with progressive supranuclear palsy. Hum Mol Genet 8:711–715.

Bergem AL (1994) Heredity in dementia of the Alzheimer type. Clin Genet 46 (1 Spec No):144–149.

Bergem ALM, Engedal K, Kringlen E (1997) The role of heredity in late-onset Alzheimer disease and vascular dementia: A twin study. Arch Gen Psychiatry 54:264–270.

Bertrams J, Kuwert E, Liedtke U (1972) HL-A antigens and multiple sclerosis. Tissue Antigens 2:405–408.

Bird TD, Lampe TH, Nemens EJ, Miner GW, Sumi SM, Schellenberg GD (1988) Familial Alzheimer's disease in American descendants of the Volga Germans: Probable genetic founder effect. Ann Neurol 23:25–31.

Blacker D, Bertram L, Saunders AJ, Moscarillo TJ, Albert MS, Wiener H, Perry RT, Collins JS, Harrell LE, Go RC, Mahoney A, Beaty T, Fallin MD, Avramopoulos D, Chase GA, Folstein MF, McInnis MG, Bassett SS, Doheny KJ, Pugh EW, Tanzi RE (2003) Results of a high-resolution genome screen of 437 Alzheimer's Disease families. Hum Mol Genet 12:23–32.

Board PG, Coggan M, Chelvanayagam G, Easteal S, Jermiin LS, Schulte GK, Danley DE, Hoth LR, Griffor MC, Kamath AV, Rosner MH, Chrunyk BA, Perregaux DE, Gabel CA, Geoghegan KF, Pandit J (2000) Identification, characterization, and crystal structure of the Omega class glutathione transferases. J Biol Chem 275:24798–24806.

Breitner JC, Gatz M, Bergem AL, Christian JC, Mortimer JA, McClearn GE, Heston LL, Welsh KA, Anthony JC, Folstein MF (1993) Use of twin cohorts for research in Alzheimer's disease. Neurology 43:261–267.

Breitner JCS, Welsh-Bohmer KA, Gau BA, McDonald WW, Steffens DC, Saunders AM, Magruder KM, Helms MJ, Plassman BJ, Folstein MF, Brandt J, Robinette CD, Page WF (1995) Alzheimer's disease in the National Academy of Sciences-National Research Council Registry of Aging Twin Veterans III. Detection of cases, longitudinal results and observations on twin concordance. Arch Neurol 52:763–771.

Clarimon J, Xiromerisiou G, Eerola J, Gourbali V, Hellstrom O, Dardiotis E, Peuralinna T, Papadimitriou A, Hadjigeorgiou GM, Tienari PJ, Singleton AB (2005) Lack of evidence for a genetic association between FGF20 and Parkinson's disease in Finnish and Greek patients. BMC Neurol 5:11.

Conrad C, Andreadis A, Trojanowski JQ, Dickson DW, Kang D, Chen X, Wiederholt W, Hansen L, Masliah E, Thlal LJ, Katzman R, Xia Y, Saitoh T (1997) Genetic evidence for the involvement of tau in progressive supranuclear palsy. Ann Neurol 41:277–281.

Copley TT, Wiggins S, Dufrasne S, Bloch M, Adam S, McKellin W, Hayden MR (1995) Are we all of one mind? Clinicians' and patients' opinions regarding the development of a service protocol for predictive testing for Huntington disease. Canadian Collaborative Study for Predictive Testing for Huntington Disease. Am J Med Genet 58:59–69.

Corder EH, Saunders AM, Strittmatter WJ, Schmechel DE, Gaskell PC, Small GW, Roses AD, Haines JL, Pericak-Vance MA (1993) Gene dose of apolipoprotein E type 4 allele and the risk of Alzheimer's disease in late onset families. Science 261:921–923.

Corder EH, Saunders AM, Risch N, Strittmatter WJ, Schmechel DE, Gaskell PC, Rimmler JB, Locke PA, Conneally PM, Schmader KE, Small GW, Roses AD, Haines JL, Pericak-Vance MA (1994) Apolipoprotein E type 2 allele decreases the risk of late onset Alzheimer disease. Nat Genet 7:180–184.

Cruts M, Gijselinck I, van der ZJ, Engelborghs S, Wils H, Pirici D, Rademakers R, Vandenberghe R, Dermaut B, Martin JJ, van Duijn C, Peeters K, Sciot R, Santens P, De Pooter T, Mattheijssens M, Van den BM, Cuijt I, Vennekens K, De Deyn PP, Kumar-Singh S, Van Broeckhoven C (2006) Null mutations in progranulin cause ubiquitin-positive frontotemporal dementia linked to chromosome 17q21. Nature 442:920–924.

De Yebenes J, Sarasa J, Danile S, Lee A (1995) Autosomal dominant progressive supranuclear palsy: Description of a pedigree and review of the literature. Brain 118:1095–1103.

Destefano AL, Golbe LI, Mark MH, Lazzarini AM, Maher NE, Saint-Hilaire M, Feldman RG, Guttman M, Watts RL, Suchowersky O, Lafontaine AL, Labelle N, Lew MF, Waters CH, Growdon JH, Singer C, Currie LJ, Wooten GF, Vieregge P, Pramstaller PP, Klein C, Hubble JP, Stacy M, Montgomery E, MacDonald ME, Gusella JF, Myers RH (2001) Genome-wide scan for Parkinson's disease: The GenePD Study. Neurology 57:1124–1126.

Destefano AL, Lew MF, Golbe LI, Mark MH, Lazzarini AM, Guttman M, Montgomery E, Waters CH, Singer C, Watts RL, Currie LJ, Wooten GF, Maher NE, Wilk JB, Sullivan KM, Slater KM, Saint-Hilaire MH, Feldman RG, Suchowersky O, Lafontaine AL, Labelle N, Growdon JH, Vieregge P, Pramstaller PP, Klein C, Hubble JP, Reider CR, Stacy M, MacDonald ME, Gusella JF, Myers RH (2002) PARK3 influences age at onset in Parkinson disease: A genome scan in the GenePD study. Am J Hum Genet 70:1089–1095.

Drzezga A, Grimmer T, Riemenschneider M, Lautenschlager N, Siebner H, Alexopoulus P, Minoshima S, Schwaiger M, Kurz A (2005) Prediction of individual clinical outcome in MCI by means of genetic assessment and (18)F-FDG PET. J Nucl Med 46:1625–1632.

Eisenstein BI (1990) The polymerase chain reaction: A new method of using molecular genetics for medical diagnosis. N Engl J Med 322:178–183.

Ertekin-Taner N, Ronald J, Asahara H, Younkin L, Hella M, Jain S, Gnida E, Younkin S, Fadale D, Ohyagi Y, Singleton A, Scanlin L, De Andrade M, Petersen R, Graff-Radford N, Hutton M, Younkin S (2003) Fine mapping of the alpha-T catenin gene to a quantitative trait locus on chromosome 10 in late-onset Alzheimer's disease pedigrees. Hum Mol Genet 12:3133–3143.

Farrer LA, Bowirrat A, Friedland RP, Waraska K, Korczyn AD, Baldwin CT (2003) Identification of multiple loci for Alzheimer disease in a consanguineous Israeli-Arab community. Hum Mol Genet 12:415–422.

Farrer LA, Cupples LA, Haines JL, Hyman B, Kukull WA, Mayeux R, Myers RH, Pericak-Vance MA, Risch N, Van Duijn CM (1997) Effects of age, sex, and ethnicity on the association between apolipoprotein E genotype and Alzheimer disease. A meta-analysis. APOE and Alzheimer Disease Meta Analysis Consortium. JAMA 278:1349–1356.

Figlewicz DA, Krizus A, Martinoli MG, Meininger V, Dib M, Rouleau GA, Julien JP (1994) Variants of the heavy neurofilament subunit are associated with the development of amyotrophic lateral sclerosis. Hum Mol Genet 3:1757–1761.

Finckh U, von der KH, Velden J, Michel T, Andresen B, Deng A, Zhang J, Muller-Thomsen T, Zuchowski K, Menzer G, Mann U, Papassotiropoulos A, Heun R, Zurdel J, Holst F, Benussi L, Stoppe G, Reiss J, Miserez AR, Staehelin HB, Rebeck GW, Hyman BT, Binetti G, Hock C, Growdon JH, Nitsch RM (2000) Genetic association of a cystatin C gene polymorphism with late-onset Alzheimer disease. Arch Neurol 57:1579–1583.

Foroud T, Uniacke SK, Liu L, Pankratz N, Rudolph A, Halter C, Shults C, Marder K, Conneally PM, Nichols WC (2003) Heterozygosity for a mutation in the parkin gene leads to later onset Parkinson disease. Neurology 60:796–801.

George-Hyslop PH, Petit A (2005) Molecular biology and genetics of Alzheimer's disease. C R Biol 328:119–130.

Gijselinck I, Bogaerts V, Rademakers R, van der ZJ, Van Broeckhoven C, Cruts M (2006) Visualization of MAPT inversion on stretched chromosomes of tau-negative frontotemporal dementia patients. Hum Mutat 27:1057–1059.

Goate A, Chartier-Harlin MC, Mullan M, Brown J, Crawford F, Fidani L, Giuffra L, Haynes A, Irving N, James L, Mant R, Newton P, Rooke K, Roques P, Talbot C, Pericak-Vance MA, Roses A,

Williamson R, Rossor M, Owen M, Hardy J (1991) Segregation of a missense mutation in the amyloid precursor protein gene with familial Alzheimer's disease. Nature 349:704–706.

Golbe LI, Lazzarini AM, Spychala JR, Johnson WG, Stenroos ES, Mark MH, Sage JI (2001) The tau A0 allele in Parkinson's disease. Mov Disord 16:442–447.

Goldgaber D, Lerman MI, McBride OW, Saffiotti U, Gajdusek DC (1987) Characterization and chromosomal localization of a cDNA encoding brain amyloid of Alzheimer's disease. Science 235:877–880.

Hague S, Peuralinna T, Eerola J, Hellstrom O, Tienari PJ, Singleton AB (2004) Confirmation of the protective effect of iNOS in an independent cohort of Parkinson disease. Neurology 62:635–636.

Haines JL, Ter-Minassian M, Bazyk A, Gusella JF, Kim DJ, Terwedow H, Pericak-Vance MA, Rimmler JB, Haynes CS, Roses AD, Lee A, Shaner B, Menold M, Seboun E, Fitoussi R-P, Gartioux C, Reyes C, Ribierre G, Gyapay G, Weissenbach J, Hauser SL, Goodkin DE, Lincoln R, Usuku K, Garcia-Merino A, Gatto N, Young S, Oksenberg JR (1996) A complete genomic screen for multiple sclerosis underscores a role for the major histocompatability complex. Nat Genet 13:469–471.

Haines JL, Terwedow HA, Burgess K, Pericak-Vance MA, Rimmler JB, Martin ER, Oksenberg JR, Lincoln R, Zhang DY, Banatao DR, Gatto N, Goodkin DE, Hauser SL (1998) Linkage of the MHC to familial multiple sclerosis suggests genetic heterogeneity The Multiple Sclerosis Genetics Group. Hum Mol Genet 7:1229–1234.

Hancock DB, Wheeler BS, Tegnell E, Hauser MA, Martin ER, Vance JM, Scott WK (2005) Association between INOS and early-onset Parkinson disease. American Society of Human Genetics 55th Annual Meeting, Salt Lake City, UT. 10-25-2005.

Hancock DB, Martin ER, Fujiwara K, Stacy MA, Scott BL, Stajich JM, Jewett R, Li YJ, Hauser MA, Vance JM, Scott WK (2006) NOS2A and the modulating effect of cigarette smoking in Parkinson's disease. Annals of Neurology 60(3):366–373.

Hauser MA, Li YJ, Takeuchi S, Walters R, Noureddine M, Maready M, Darden T, Hulette C, Martin E, Hauser E, Xu H, Schmechel D, Stenger JE, Dietrich F, Vance J (2003) Genomic convergence: Identifying candidate genes for Parkinson's disease by combining serial analysis of gene expression and genetic linkage. Hum Mol Genet 12:671–677.

Healy DG, Abou-Sleiman PM, Lees AJ, Casas JP, Quinn N, Bhatia K, Hingorani AD, Wood NW (2004) Tau gene and Parkinson's disease: A case-control study and meta-analysis. J Neurol Neurosurg Psychiatry 75:962–965.

Hebert LE, Scherr PA, Bienias JL, Bennett DA, Evans DA (2003) Alzheimer disease in the US population: Prevalence estimates using the 2000 census. Arch Neurol 60:1119–1122.

Heyman A, Wilkinson WE, Hurwitz BJ, Schmechel D, Sigmon AH, Weinberg T, Helms MJ, Swift M (1983) Alzheimer's disease: Genetic aspects and associated clinical disorders. Ann Neurol 14:507–515.

Hicks AA, Petursson H, Jonsson T, Stefansson H, Johannsdottir HS, Sainz J, Frigge ML, Kong A, Gulcher JR, Stefansson K, Sveinbjornsdottir S (2002) A susceptibility gene for late-onset idiopathic Parkinson's disease. Ann Neurol 52:549–555.

Hilbich C, Kisters-Woike B, Reed J, Masters CL, Beyreuther K (1991) Human and rodent sequence analogs of Alzheimer's amyloid beta A4 share similar properties and can be solubilized in buffers of pH 7.4. Eur J Biochem 201:61–69.

Holzmann C, Kruger R, Saecker AM, Schmitt I, Schols L, Berger K, Riess O (2003) Polymorphisms of the alpha-synuclein promoter: Expression analyses and association studies in Parkinson's disease. J Neural Transm 110:67–76.

Hutton M, Lendon CL, Rizzu P, Baker M, Froelich S, Houlden H, Pickering-Brown S, Chakraverty S, Isaacs A, Grover A, Hackett J, Adamson J, Lincoln S, Dickson D, Davies P, Petersen RC, Stevens M, de Graaff E, Wauters E, van Baren J, Hillebrand M, Joosse M, Kwon JM, Nowotny P, Heutink P (1998) Association of missense and 5'-splice-site mutations in tau with the inherited dementia FTDP-17. Nature 393:702–705.

Jackson M, Al Chalabi A, Enayat ZE, Chioza B, Leigh PN, Morrison KE (1997) Copper/zinc superoxide dismutase 1 and sporadic amyotrophic lateral sclerosis: Analysis of 155 cases and identification of a novel insertion mutation. Ann Neurol 42:803–807.

Jones CT, Swingler RJ, Simpson SA, Brock DJ (1995) Superoxide dismutase mutations in an unselected cohort of Scottish amyotrophic lateral sclerosis patients. J Med Genet 32:290–292.

Kehoe P, Wavrant-De Vrieze F, Crook R, Wu WS, Holmans P, Fenton I, Spurlock G, Norton N, Williams H, Williams N, Lovestone S, Perez-Tur J, Hutton M, Chartier-Harlin MC, Shears S, Roehl K, Booth J, Van Voorst W, Ramic D, Williams J, Goate A, Hardy J, Owen MJ (1999) A full genome scan for late onset Alzheimer's disease. Hum Mol Genet 8:237–245.

Khan NL, Scherfler C, Graham E, Bhatia KP, Quinn N, Lees AJ, Brooks DJ, Wood NW, Piccini P (2005) Dopaminergic dysfunction in unrelated, asymptomatic carriers of a single parkin mutation. Neurology. 64(1):134–136.

Klaver CC, Kliffen M, Van Duijn CM, Hofman A, Cruts M, Grobbee DE, Van Broeckhoven C, de Jong PT (1998) Genetic association of apolipoprotein E with age-related macular degeneration [published erratum appears in Am J Hum Genet 1998 Oct;63(4):1252]. Am J Hum Genet 63:200–206.

Kolsch H, Linnebank M, Lutjohann D, Jessen F, Wullner U, Harbrecht U, Thelen KM, Kreis M, Hentschel F, Schulz A, von Bergmann K, Maier W, Heun R (2004) Polymorphisms in glutathione S-transferase omega-1 and AD, vascular dementia, and stroke. Neurology 63:2255–2260.

Laliberte RE, Perregaux DG, Hoth LR, Rosner PJ, Jordan CK, Peese KM, Eggler JF, Dombroski MA, Geoghegan KF, Gabel CA (2003) Glutathione s-transferase omega 1-1 is a target of cytokine release inhibitory drugs and may be responsible for their effect on interleukin-1beta posttranslational processing. J Biol Chem 278:16567–16578.

Lambrechts D, Storkebaum E, Morimoto M, Del Favero J, Desmet F, Marklund SL, Wyns S, Thijs V, Andersson J, van M, I, Al Chalabi A, Bornes S, Musson R, Hansen V, Beckman L, Adolfsson R, Pall HS, Prats H, Vermeire S, Rutgeerts P, Katayama S, Awata T, Leigh N, Lang-Lazdunski L, Dewerchin M, Shaw C, Moons L, Vlietinck R, Morrison KE, Robberecht W, Van Broeckhoven C, Collen D, Andersen PM, Carmeliet P (2003) VEGF is a modifier of amyotrophic lateral sclerosis in mice and humans and protects motoneurons against ischemic death. Nat Genet 34:383–394.

Lee JH, Mayeux R, Mayo D, Mo J, Santana V, Williamson J, Flaquer A, Ciappa A, Rondon H, Estevez P, Lantigua R, Kawarai T, Toulina A, Medrano M, Torres M, Stern Y, Tycko B, Rogaeva E, George-Hyslop P, Knowles JA (2004) Fine mapping of 10q and 18q for familial Alzheimer's disease in Caribbean Hispanics. Mol Psychiatry 9:1042–1051.

Lemere CA, Blusztajn JK, Yamaguchi H, Wisniewski T, Saido TC, Selkoe DJ (1996) Sequence of deposition of heterogeneous amyloid beta-peptides and APO E in Down syndrome: Implications for initial events in amyloid plaque formation. Neurobiol Dis 3:16–32.

Levecque C, Elbaz A, Clavel J, Richard F, Vidal JS, Amouyel P, Tzourio C, Alperovitch A, Chartier-Harlin MC (2003) Association between Parkinson's disease and polymorphisms in the nNOS and iNOS genes in a community-based case-control study. Hum Mol Genet 12:79–86.

Levy-Lahad E, Wasco W, Poorkaj P, Romano DM, Oshima J, Pettingell WH, Yu CE, Jondro PD, Schmidt SD, Wang K, Crowley AC, Fu YH, Guenette SY, Galas D, Nemens E, Wijsman EM, Bird TD, Schellenberg GD, Tanzi RE (1995) Candidate gene for the chromosome 1 familial Alzheimer's disease locus. Science 269:973–977.

Li Y-J, Scott WK, Hedges DJ, Zhang F, Gaskell PC, Nance MA, Watts RL, Hubble JP, Koller WC, Pahwa R, Stern MB, Hiner BC, Jankovic J, Allen FA, Jr., Goetz CG, Mastaglia F, Stajich JM, Gibson RA, Middleton LT, Saunders AM, Scott BL, Small GW, Nicodemus KK, Reed AD, Schmechel DE, Welsh-Bohmer KA, Conneally PM, Roses AD, Gilbert JR, Vance JM, Haines JL, Pericak-Vance MA (2002) Age at onset in two common neurodegenerative diseases is genetically controlled. Am J Hum Genet 70:985–993.

Li Y-J, Oliveira SA, Xu P, Martin ER, Stenger JE, Scherzer CA, Hulette C, Scott WK, Small GW, Nance MA, Watts RL, Hubble JP, Koller WC, Pahwa R, Stern MB, Hiner BC, Jankovic J, Goetz CG, Mastaglia F, Middleton LT, Roses AD, Saunders AM, Welsh-Bohmer KA, Schmechel DE, Gullans SR, Haines JL, Gilbert JR, Vance JM, Pericak-Vance MA (2003) Glutathione S-transferase omega-1 modifies age-at-onset of Alzheimer disease and Parkinson disease. Hum Mol Genet 12:3259–3267.

Li Y-J, Hauser MA, Scott WK, Martin ER, Booze MW, Qin XJ, Walter JW, Nance MA, Hubble JP, Koller WC, Pahwa R, Stern MB, Hiner CB, Jankovic J, Goetz CG, Small GW, Mastaglia F, Haines JL, Pericak-Vance MA, Vance JA (2004a) Apolipoprotein E controls the risk and age at onset of Parkinson Disease. Neurology 62:2005–2009.

Li Y, Nowotny P, Holmans P, Smemo S, Kauwe JS, Hinrichs AL, Tacey K, Doil L, van Luchene R, Garcia V, Rowland C, Schrodi S, Leong D, Gogic G, Chan J, Cravchik A, Ross D, Lau K, Kwok S, Chang SY, Catanese J, Sninsky J, White TJ, Hardy J, Powell J, Lovestone S, Morris JC, Thal L, Owen M, Williams J, Goate A, Grupe A (2004) Association of late-onset Alzheimer's disease with genetic variation in multiple members of the GAPD gene family. Proceedings of the National Academy of Sciences of the United States of America. 101(44):15688–15693.

Li YJ, Scott WK, Zhang L, Lin PI, Oliveira SA, Skelly T, Doraiswamy MP, Welsh-Bohmer KA, Martin ER, Haines JL, Pericak-Vance MA, Vance JM (2006) Revealing the role of glutathione S-transferase omega in age-at-onset of Alzheimer and Parkinson diseases. Neurobiology of Aging. 27(8):1087–1093.

Liddell MB, Bayer AJ, Owen MJ (1995) No evidence that common allelic variation in the Amyloid Precursor Protein (APP) gene confers susceptibility to Alzheimer's disease. Hum Mol Genet 4:853–858.

Lynch T, Sano M, Marder KS, Bell KL, Foster NL, Defendini RF, Sima AA, Keohane C, Nygaard TG, Fahn S (1994) Clinical characteristics of a family with chromosome 17-linked disinhibition-dementia-parkinsonism-amyotrophy complex. Neurology 44:1878–1884.

Mahley RW (1988) Apolipoprotein E: Cholesterol transport protein with expanding role in cell biology. Science 240:622–630.

Martin ER, Scott WK, Nance MA, Watts RL, Hubble JP, Koller WC, Lyons K, Pahwa R, Stern MB, Colcher A, Hiner BC, Jankovic J, Ondo WG, Allen FH, Jr., Goetz CG, Small GW, Masterman D, Mastaglia F, Laing NG, Stajich JM, Ribble RC, Booze MW, Rogala A, Hauser MA, Zhang F, Gibson RA, Middleton LT, Roses AD, Haines JL, Scott BL, Pericak-Vance MA, Vance JM (2001) Association of single-nucleotide polymorphisms of the tau gene with late-onset Parkinson disease. JAMA 286:2245–2250.

Mayeux R, Lee JH, Romas SN, Mayo D, Santana V, Williamson J, Ciappa A, Rondon HZ, Estevez P, Lantigua R, Medrano M, Torres M, Stern Y, Tycko B, Knowles JA (2002) Chromosome-12 mapping of late-onset Alzheimer disease among Caribbean Hispanics. Am J Hum Genet 70:237–243.

McGeer PL, McGeer EG (2004) Inflammation and the degenerative diseases of aging. Ann N Y Acad Sci 1035:104–116.

Mellick GD, Maraganore DM, Silburn PA (2005) Australian data and meta-analysis lend support for alpha-synuclein (NACP-Rep1) as a risk factor for Parkinson's disease. Neurosci Lett 375:112–116.

Mizuta I, Nishimura M, Mizuta E, Yamasaki S, Ohta M, Kuno S (2002) Meta-analysis of alpha synuclein/ NACP polymorphism in Parkinson's disease in Japan. J Neurol Neurosurg Psychiatry 73:350.

Moulard B, Sefiani A, Laamri A, Malafosse A, Camu W (1996) Apolipoprotein E genotyping in sporadic amyotrophic lateral sclerosis: Evidence for a major influence on the clinical presentation and prognosis. J Neurol Sci 139(Suppl):34–37.

Myers A, Holmans P, Marshall H, Kwon J, Meyer D, Ramic D, Shears S, Booth J, DeVrieze FW, Crook R, Hamshere M, Abraham R, Tunstall N, Rice F, Carty S, Lillystone S, Kehoe P, Rudrasingham V, Jones L, Lovestone S, Perez-Tur J, Williams J, Owen MJ, Hardy J, Goate AM (2000) Susceptibility locus for Alzheimer's disease on chromosome 10. Science 290:2304–2305.

Myers A, Wavrant De-Vrieze F, Holmans P, Hamshere M, Crook R, Compton D, Marshall H, Meyer D, Shears S, Booth J, Ramic D, Knowles H, Morris JC, Williams N, Norton N, Abraham R, Kehoe P, Williams H, Rudrasingham V, Rice F, Giles P, Tunstall N, Jones L, Lovestone S, Williams J, Owen MJ, Hardy J, Goate A (2002) Full genome screen for Alzheimer disease: Stage II analysis. Am J Med Genet 114:235–244.

Nee LE, Polinsky RJ, Eldridge R, Weingartner H, Smallberg S, Ebert M (1983) A family with histologically confirmed Alzheimer's disease. Arch Neurol 40:203–208.

Nee LE, Gomez MR, Dambrosia J, Bale S, Eldridge R, Polinsky RJ (1991) Environmental-occupational risk factors and familial associations in multiple system atrophy: A preliminary investigation. Clin Auton Res 1:9–13.

Nishimura M, Sakamoto T, Kaji R, Kawakami H (2004) Influence of polymorphisms in the genes for cytokines and glutathione S-transferase omega on sporadic Alzheimer's disease. Neurosci Lett 368:140–143.

Noureddine MA, Li YJ, van der Walt JM, Walters R, Jewett RM, Xu H, Wang T, Walter JW, Scott BL, Hulette C, Schmechel DE, Stenger J, Dietrich F, Vance JM, Hauser MA (2005) Genomic convergence to identify candidate genes for Parkinson disease: SAGE analysis of the substantia nigra. Mov Disord 20:1299–309.

Oliveira SA, Li YJ, Noureddine MA, Zuchner S, Qin X, Pericak-Vance MA, Vance JM (2005) Identification of risk and age-at-onset genes on chromosome 1p in Parkinson disease. Am J Hum Genet 77:252–264.

Oliveira SA, Scott WK, Martin ER, Nance MA, Watts RL, Hubble JP, Koller WC, Pahwa R, Stern MB, Hiner BC, Ondo WG, Allen FH, Jr., Scott BL, Goetz CG, Small GW, Mastaglia F, Stajich JM, Zhang F, Booze MW, Winn MP, Middleton LT, Haines JL, Pericak-Vance MA, Vance JM (2003) Parkin mutations and susceptibility alleles in late-onset Parkinson's disease. Ann Neurol 53:624–629.

Oliveira SA, Scott WK, Zhang F, Stajich JM, Fujiwara K, Hauser M, Scott BL, Pericak-Vance MA, Vance JM, Martin ER (2004) Linkage disequilibrium and haplotype tagging polymorphisms in the Tau H1 haplotype. Neurogenetics 5:147–155.

Orru S, Mascia V, Casula M, Giuressi E, Loizedda A, Carcassi C, Giagheddu M, Contu L (1999) Association of monoamine oxidase B alleles with age at onset in amyotrophic lateral sclerosis. Neuromuscul Disord 9:593–597.

Otaegui D, Paisan C, Saenz A, Marti I, Ribate M, Marti-Masso JF, Perez-Tur J, Lopez DM (2004) Mitochondrial polymporphisms in Parkinson's Disease. Neurosci Lett 370:171–174.

Ozturk A, Desai PP, Minster RL, DeKosky ST, Kamboh MI (2005) Three SNPs in the GSTO1, GSTO2 and PRSS11 genes on chromosome 10 are not associated with age-at-onset of Alzheimer's disease. Neurobiol Aging 26:1161–1165.

Pankratz N, Nichols WC, Uniacke SK, Halter C, Rudolph A, Shults C, Conneally PM, Foroud T (2002) Genome screen to identify susceptibility genes for Parkinson disease in a sample without parkin mutations. Am J Hum Genet 71:124–135.

Pankratz N, Uniacke SK, Halter CA, Rudolph A, Shults CW, Conneally PM, Foroud T, Nichols WC (2004) Genes influencing Parkinson disease onset: Replication of PARK3 and identification of novel loci. Neurology 62:1616–1618.

Pastor P, Ezquerra M, Munoz E, Marti MJ, Blesa R, Tolosa E, Oliva R (2000) Significant association between the tau gene A0/A0 genotype and Parkinson's disease. Neurology 47:242–245.

Pastor P, Munoz E, Ezquerra M, Obach V, Marti MJ, Valldeoriola F, Tolosa E, Oliva R (2001) Analysis of the coding and the 5' flanking regions of the alpha-synuclein gene in patients with Parkinson's disease. Mov Disord 16:1115–1119.

Pericak-Vance MA, Bass ML, Yamaoka LH, Gaskell PC, Scott WK, Terwedow HA, Menold MM, Conneally PM, Small GW, Saunders AM, Roses AD, Haines JL (1998) Complete genomic screen in late-onset familial Alzheimer's disease. Neurobiol Aging 19:S39–S42.

Pericak-Vance MA, Bass MP, Yamaoka LH, Gaskell PC, Scott WK, Terwedow HA, Menold MM, Conneally PM, Small GW, Vance JM, Saunders AM, Roses AD, Haines JL (1997) Complete genomic screen in late-onset familial Alzheimer disease. Evidence for a new locus on chromosome 12. JAMA 278:1237–1241.

Pericak-Vance MA, Bebout JL, Gaskell PC, Yamaoka LH, Hung WY, Alberts MJ, Walker AP, Bartlett RJ, Haynes CS, Welsh KA, Earl NL, Heyman A, Clark CM, Roses AD (1991) Linkage studies in familial Alzheimer's disease: Evidence for chromosome 19 linkage. Am J Hum Genet 48:1034–1050.

Pericak-Vance MA, Conneally PM, Gaskell PC, Small GW, Saunders AM, Yamaoka LH, Robinson C, Ter-Minassian M, Locker PA, Pritchard M, Haynes CS, Growden J, Tanzi RE, Gusella JF, Roses AD, Haines JL (1995) The search for additional late-onset Alzheimer disease gene: A complete genomic screen. J Hum Genet 57:A200.

Pericak-Vance MA, Grubber J, Bailey LR, Hedges D, West S, Kemmerer B, Hall JL, Saunders AM, Roses AD, Small GW, Scott WK, Conneally PM, Vance JM, Haines JL (2000) Identification of novel genes in late-onset Alzheimer disease. Exp Gerontol 35:1343–1352.

Pyle A, Foltynie T, Tiangyou W, Lambert C, Keers SM, Allcock LM, Davison J, Lewis SJ, Perry RH, Barker R, Burn DJ, Chinnery PF (2005) Mitochondrial DNA haplogroup cluster UKJT reduces the risk of PD. Ann Neurol 57:564–567.

Rall SC, Jr., Weisgraber KH, Mahley RW (1982) Human apolipoprotein E. The complete amino acid sequence. J Biol Chem 257:4171–4178.

Rocca WA, Hofman A, Brayne C, Breteler MM, Clarke M, Copeland JR, Dartigues JF, Engedal K, Hagnell O, Heeren TJ (1991) Frequency and distribution of Alzheimer's disease in Europe: A collaborative study of 1980–1990 prevalence findings. The EURODEM-Prevalence Research Group. Ann Neurol 30:381–390.

Rogaev EI, Sherrington R, Rogaeva EA, Levesque G, Ikeda M, Liang Y, Chi H, Lin C, Holman K, Tsuda T (1995) Familial Alzheimer's disease in kindreds with missense mutations in a gene on chromosome 1 related to the Alzheimer's disease type 3 gene. Nature. 376(6543):775–778.

Rojo A, Pernaute RS, Fontan A, Ruiz PG, Honnorat J, Lynch T, Chin S, Gonzalo I, Rabano A, Martinez A, Daniel S, Pramstaller P, Morris H, Wood N, Lees A, Tabernero C, Nyggard T, Jackson AC, Hanson A, de Yebenes JG (1999) Clinical genetics of familial progressive supranuclear palsy. Brain 122 (Pt 7):1233–1245.

Rosenberg CK, Pericak-Vance MA, Saunders AM, Gilbert JR, Gaskell PC, Hulette CM (2000) Lewy body and Alzheimer pathology in a family with the amyloid-beta precursor protein APP717 gene mutation. Acta Neuropathol (Berl) 100:145–152.

Ross OA, McCormack R, Curran MD, Duguid RA, Barnett YA, Rea IM, Middleton D (2001) Mitochondrial DNA polymorphism: Its role in longevity of the Irish population. Exp Gerontol 36:1161–1178.

Rothstein JD, Van Kammen M, Levey AI, Martin LJ, Kuncl RW (1995) Selective loss of glial glutamate transporter GLT-1 in amyotrophic lateral sclerosis. Ann Neurol 38:73–84.

Sawcer S, Jones HB, Feakes R, Gray J, Smaldon N, Chataway J, Robertson N, Clayton D, Goodfellow PN, Compston A (1996) A genome screen in multiple sclerosis reveals susceptibility loci on chromosome 6p21 and 17q22. Nat Genet 13:464–476.

Schellenberg GD, Bird TD, Wijsman EM, Orr HT, Anderson L, Nemens E, White JA, Bonnycastle L, Weber JL, Alonso ME, Potter H, Heston LL, Martin GM (1992) Genetic linkage evidence for a familial Alzheimer's disease locus on chromosome 14. Science 258:668–671.

Scott WK, Grubber JM, Abou-donia SM, Church TD, Yamaoka LH, Conneally PM, Small GW, Saunders AM, Roses AD, Haines JL, Pericak-Vance MA (1998) Fine mapping and two-locus maximum lod score analysis in chromosome 12-linked late-onset familial Alzheimer disease (AD). Am J Hum Genet 63(4):A44.

Scott WK, Grubber JM, Conneally PM, Small GW, Hulette CM, Rosenberg CK, Saunders AM, Roses AD, Haines JL, Pericak-Vance MA (2000) Fine mapping of the chromosome 12 late-onset Alzheimer disease locus: Potential genetic and phenotypic heterogeneity. Am J Hum Genet 66:922–932.

Scott WK, Nance MA, Watts RL, Hubble JP, Koller WC, Lyons K, Pahwa R, Stern MB, Colcher A, Hiner BC, Jankovic J, Ondo WG, Allen FH, Jr., Goetz CG, Small GW, Masterman D, Mastaglia F, Laing NG, Stajich JM, Slotterbeck B, Booze MW, Ribble RC, Rampersaud E, West SG, Gibson RA, Middleton LT, Roses AD, Haines JL, Scott BL, Vance JM, Pericak-Vance MA (2001) Complete genomic screen in Parkinson disease: Evidence for multiple genes. JAMA 286:2239–2244.

Scott WK, Staijich JM, Yamaoka LH, Speer MC, Vance JM, Roses AD, Pericak-Vance MA (1997) Genetic complexity and Parkinson's disease Deane Laboratory Parkinson Disease Research Group. Science 277:387–388.

Sherrington R, Rogaev EI, Liang Y, Rogaeva EA, Levesque G, Ikeda M, Chi H, Lin C, Li G, Holman K, Tsuda T, Mar L, Foncin JF, Bruni AC, Montesi MP, Sorbi S, Rainero I, Pinessi L, Nee L, Chumakov

I, Pollen DA, Brookes A, Sanseau P, Polinsky RJ, Wasco W, Da Silva HAR, Haines JL, Pericak-Vance MA, Tanzi RE, Roses AD, Fraser PE, Rommens JM, St George-Hyslop PH (1995) Cloning of a gene bearing missense mutations in early-onset familial Alzheimer's disease. Nature 375:754–760.

Sjogren T, Sjogren H, Lindgren AGH (1952) Morbus Alzheimer and morbus Pick: A genetic, clinical and patho-anatomical study. Acta Psychiatrica Neurol Scandinavica 82:1–66.

Skol AD, Scott LJ, Abecasis GR, Boehnke M (2006) Joint analysis is more efficient than replication-based analysis for two-stage genome-wide association studies. Nat Genet.

Skvortsova V, Shadrina M, Slominsky P, Levitsky G, Kondratieva E, Zherebtsova A, Levitskaya N, Alekhin A, Serdyuk A, Limborska S (2004) Analysis of heavy neurofilament subunit gene polymorphism in Russian patients with sporadic motor neuron disease (MND). Eur J Hum Genet 12:241–244.

St George-Hyslop P, Haines J, Rogaev E, Mortilla M, Vaula G, Pericak-Vance MA, Foncin JF, Montesi M, Bruni A, Sorbi S, Rainero I, Pinessi L, Pollen D, Polinsky R, Nee L, Kennedy J, Macciardi F, Rogaeva E, Liang Y, Alexandrova N, Lukiw W, Schlumpf K, Tanzi R, Tsuda T, Farrer L, Cantu JM, Duara R, Amaducci L, Bergamini L, Gusella J, Roses A, McLachlan DC (1992) Genetic evidence for a novel familial Alzheimer's disease locus on chromosome 14. Nat Genet 2:330–334.

Suzuki N, Cheung T, Cai X, Odeka A, Otvos L, Eckman C, Golde T, Younkin S (1994) An increased percentage of long amyloid beta protein secreted by familial amyloid beta protein precursor (beta APP717) mutants. Science 264:1340.

Tanaka-Kagawa T, Jinno H, Hasegawa T, Makino Y, Seko Y, Hanioka N, Ando M (2003) Functional characterization of two variant human GSTO 1-1s (Ala140Asp and Thr217Asn). Biochem Biophys Res Commun 301:516–520.

Tanzi RE, Bird ED, Latt SA, Neve RL (1987) The amyloid beta protein gene is not duplicated in brains from patients with Alzheimer's disease. Science 238:666–669.

Tomkins J, Usher P, Slade JY, Ince PG, Curtis A, Bushby K, Shaw PJ (1998) Novel insertion in the KSP region of the neurofilament heavy gene in amyotrophic lateral sclerosis (ALS). Neuroreport 9:3967–3970.

Van Broeckhoven C, Backhovens H, Cruts M (1994) APOE genotype does not modulate age of onset in families with chromosome 14 encoded Alzheimer's disease. Neurosci Let 169:179–180.

van der Walt JM, Nicodemus KK, Martin ER, Scott WK, Nance MA, Watts RL, Hubble JP, Haines JL, Koller WC, Lyons K, Pahwa R, Stern MB, Colcher A, Hiner BC, Jankovic J, Ondo WG, Allen FH Jr. Goetz CG, Small GW, Mastaglia F, Stajich JM, McLaurin AC, Middleton LT, Scott BL, Schmechel DE, Pericak-Vance MA, Vance JM, (2003) Mitochondrial polymorphisms significantly reduce the risk of Parkinson disease. American Journal of Human Genetics. 72(4):804–811.

van der Walt JM, Noureddine MA, Kittappa R, Hauser MA, Scott WK, McKay R, Zhang F, Stajich JM, Fujiwara K, Scott BL, Pericak-Vance MA, Vance JM, Martin ER (2004) Fibroblast growth factor 20 polymorphisms and haplotypes strongly influence risk of Parkinson disease. Am J Hum Genet 74:1121–1127.

van Slegtenhorst M, Lewis J, Hutton M (2000) The molecular genetics of the tauopathies. Exp Gerontol 35:461–471.

Weisgraber KH, Innerarity TL, Mahley RW (1982) Abnormal lipoprotein receptor-binding activity of the human E apoprotein due to cysteine-arginine interchange at a single site. J Biol Chem 257:2518–2521.

Westerman AM, Entius MM, Boor PPC, Koole R, de Baar E, Offerhaus GJA, Lubinski J, Lindhout D, Halley DJJ, de Rooij FWM, Wilson JHP (1999) Novel mutations in the LKB1/STK11 gene in Dutch Peutz-Jeghers families. Hum Mutat 13:476–481.

Wu WS, Holmans P, Wavrant-DeVrieze F, Shears S, Kehoe P, Crook R, Booth J, Williams N, Perez-Tur J, Roehl K, Fenton I, Chartier-Harlin MC, Lovestone S, Williams J, Hutton M, Hardy J, Owen MJ, Goate A (1998) Genetic studies on chromosome 12 in late-onset Alzheimer disease. JAMA 280:619–622.

34
The Neuroimmune System in Psychiatric Disorders

L. Charles Murrin and Mark P. Thomas

Keywords Alpha-melanocyte stimulating hormone; Adrenocorticotrophic hormone; Corticotropin releasing hormone; Cytokines; Depression; Hypothalamic-pituitary-adrenal axis; Proenkephalin; Proopiomelanocortin; Schizophrenia; Stress response

34.1. Introduction

The earliest evidence for neuroimmune interaction and regulation is the classical immune conditioning experiment of Ader and Cohen in 1975 (Ader, 1987). Since then, it is well established that the central nervous system (CNS) and the immune system interact and regulate one another's function (Blalock, 1989; Madden and Felten, 1995). This interaction can be particularly important at times of extraordinary stress, whether due to psychological factors or to a physical response to an injury or an infection. The CNS regulates immune system function with classical neurotransmitters acting through the central nervous system and peripherally through the sympathetic nervous system; and with neuropeptides also acting through both the central and peripheral nervous systems. The immune system, in turn, alters CNS function, particularly through the release of cytokines. While this reciprocal regulation between the CNS and the immune system is designed to maintain homeostasis, dysfunction in one can lead to aberrant function in the other, producing a pathological state.

Interactions between the immune and nervous system are complex and attempts to decipher these are still incomplete. One of the first empirical findings in support of an interaction between these systems was the discovery of receptors for neurohormones and neuropeptides on cells of the immune system (Hadden et al., 1970; Wybran et al., 1979; Hazum et al., 1979; Landmann et al., 1981). This implied that immune cells can and do respond to the endogenous ligands for these receptors. Conversely, the demonstration of many cytokine receptors and the cytokines themselves in the CNS suggests not only that these small proteins have a functional role in the CNS, but that cytokines released by the immune system may be able to regulate CNS function (Adler and Rogers, 2005; Cartier et al., 2005). Subsequent studies demonstrated that immune cells also synthesize and release many neurohormones and neuropeptides. This provides a mechanism for local control of the physiological effects of these transmitters, in addition to distant control mediated through the peripheral and central nervous systems. This sharing of ligands and receptors in the nervous and immune systems led to the suggestion that this constituted a complete biochemical information circuit and that this interactive communication system provided the basis for the immune system acting as a sensory system and the brain playing an immunoregulatory role (Blalock, 1994).

In this chapter we will provide an overview of how the hypothalamic-pituitary-adrenal (HPA) axis, the CNS and the immune system are integrated and regulate each other's activity. While there are many neurotransmitters and neurohormones that act directly and indirectly through the CNS on the immune system, such as serotonin, norepinephrine and GABA, we will focus on neuropeptides and cytokines found to regulate the immune response. In particular, we will examine HPA axis regulation of CNS function by release of peptides and glucocorticoids in response to stress and inflammation. We will consider opioid-like peptides as well as CRH and related peptides that are synthesized in the immune system as well as the CNS and how they regulate immune function. We will also discuss cytokines released by the immune system and how they impact CNS function. Finally, we will provide a brief overview of how the neuroimmune system plays a role in psychiatric illnesses such as depression and schizophrenia, a topic explored in more detail in the following two chapters.

34.2. Neurohormones and Their Receptors Associated with Both the Central Nervous System and the Immune System

Most lymphoid tissues are innervated by the sympathetic nervous system, both directly and indirectly. These same tissues express receptors for a variety of hormones, including classical

T. Ikezu and H.E. Gendelman (eds.), *Neuroimmune Pharmacology*.
© Springer 2008

neurotransmitters and hormones released by tissues at a distance. Lymphocytes, for example, express receptors for steroids, enkephalins, endorphins and catecholamines. Expression of these receptors varies with different immune cell types and within the same cell type at different locations or under different physiological or pathological conditions.

There is considerable evidence for the presence of a subset of monoamine receptors on most components of the immune system and for regulation of these immune cells by the respective monoamine neurotransmitters (Sanders et al., 2001; Madden, 2001). Norepinephrine and epinephrine produce their effects via three separate adrenergic receptors, alpha-1, alpha-2 and beta (Bylund et al., 1994; Hein and Kobilka, 1997) and all three appear to be involved in regulating immune function. Data supporting a role for beta-adrenergic receptors is the most detailed but there clearly is a role for alpha1 and alpha2 adrenergic receptors as well. Adrenergic regulation of immune system function, through release of norepinephrine and epinephrine from the sympathetic nervous system or from the adrenal medulla, is an important regulatory factor, particularly in times of stress when the adrenergic system is highly active. Acetylcholine produces its effects through activation of both muscarinic and nicotinic cholinergic receptors. Muscarinic receptors, especially the M1 and M3 subtypes, are found on lymphocytes and they appear to be functional, based on studies with muscarinic agonists such as carbachol and oxotremorine. It is not clear whether nor how much nicotinic cholinergic receptors are involved in immune function. Our understanding of the roles of serotonin and dopamine in regulating lymphocyte function also is less developed. Serotonin, which is released peripherally from platelets and the enteric nervous system, probably acts primarily through 5HT1A and 5HT3 receptors. Radioligand binding data also suggest the D2-family of dopamine receptors predominate over the D1-family in the immune system, although regulatory T cells express only D1 and D5 receptors.

A focus of this chapter is the role of neuropeptides in regulation of the immune system. Neuropeptides are distinct from classical neurotransmitters in that they are usually slower acting and their synthetic process is quite different. Most active neuropeptides are synthesized as part of a larger pro-molecule. These peptidic precursors are then processed to produce the smaller, active neuropeptides, sometimes in a cell specific manner. In addition, neuropeptides are generally not metabolized and hence removed from the system as quickly as the monoamines, and so they may have a longer duration of action. We will discuss peptides derived from proenkephalin and proopiomelanocortin as well as CRH and the related urocortins. Other neuropeptides have also been implicated in regulating immune system function in the periphery and possibly in the CNS, including Substance P and calcitonin-gene related peptide (CGRP). The involvement of peptides in regulating immune and inflammatory reactions has been reviewed (Levite, 2000; McGillis et al., 2001).

34.3. HPA Axis Regulation of CNS Function

The occurrence of physical or psychological events, or "stressors," that threaten the well-being of an organism can lead to activation of a generalized stress response. The stress response consists of a coordinated activation of adaptive processes (collectively referred to as allostasis) that are designed to return the organism to steady-state conditions, that is, homeostasis. This coordinated response is mediated by components of the central nervous system (CNS), the peripheral nervous system (PNS), and the endocrine system. The hypothalamic-pituitary-adrenal (HPA) axis defines the classical core of the neuro-endocrine stress response system. The HPA axis coordinates the component processes of the stress response by serving as the major communication route between the CNS, the PNS and the endocrine system. The HPA axis consists of three major components: the hypothalamic paraventricular nucleus (PVN), corticotrophs in the anterior pituitary gland, and the adrenal cortex. The hypothalamus is situated at the anterior-most end of the brainstem and serves as the final common pathway for regulation of visceromotor and endocrine functions by the CNS. The PVN, situated bilaterally along the midline of the anterior hypothalamus, coordinates endocrine functions via the pituitary gland and visceromotor functions via pathways to the brainstem. The pituitary corticotrophs are one of several distinct cell types in the anterior pituitary, each serving as an interface between the central nervous system and the endocrine system to control some aspect of metabolism or sexual behavior. The adrenal cortex comprises the outer layers of the adrenal gland, which plays a major role in responses to stress, through its innervation by the sympathetic nervous system and via circulating hormones.

The activation of a stress response begins with the appraisal of stressors by the distributed CNS network commonly known as the limbic system. The concept of the limbic system, which was originally defined anatomically, has taken on a functional connotation and now typically refers to CNS systems, which modulate the hypothalamus. These brain areas include the hippocampus, the amygdala, and regions in medial prefrontal cortex. The components of the limbic system all influence hypothalamic output, and constitute a distributed system that regulates affective state and is critically involved in the appraisal and response of the organism to stressors. In general, the hippocampus and the anterior cingulate/prelimbic regions of prefrontal cortex inhibit HPA axis output, while the amygdala and the infralimbic region of prefrontal cortex potentiate HPA axis activity. Thus, different regions of the limbic system modulate the output of the HPA axis in a reciprocal fashion to coordinate the temporal sequence of the stress response, controlling the initiation, propagation, and termination of distinct processes.

34.3.1. Hormones Released by the HPA Axis

Two neuropeptides, corticotrophin-releasing hormone (CRH) and arginine vasopressin (AVP) are released from parvocellular neurons in the hypothalamic PVN to initiate a stress response. The terminal endings of these neurons, located in the median eminence of the hypothalamus, release CRH and AVP into the hypothalamic-hypophysial portal vessel system, where they travel to the anterior pituitary. The two neuropeptides act synergistically on pituitary corticotrophs to activate the synthesis of pro-opiomelanocortin (POMC). This peptide, discussed in detail below, is processed to produce several peptides including adrenocorticotrophic hormone (ACTH), or corticotropin. ACTH released from corticotrophs travels via the bloodstream to act on cells in the zona fasciculata layer of the adrenal cortex, stimulating the synthesis and release of the glucocorticoids, cortisol (in humans) or corticosterone (in rodents).

In addition to regulating the synthesis and release of ACTH from the pituitary, neurons in the PVN also project to areas in the brainstem that control the output of the sympathetic branch of the PNS. Activation of PVN neurons at the onset of a stress response thus leads to increased sympathetic outflow, resulting in release of norepinephrine from sympathetic nerve terminals and epinephrine from the adrenal medulla. The effects of these catecholamines are responsible for the behavioral syndrome known as the "fight-or-flight" response.

The effects of the neuropeptides and glucocorticoids associated with the HPA axis are mediated through distinct receptor subclasses. Two classes of CRH receptors have been identified, type 1 and type 2, with distinct but overlapping distributions in the CNS (Hsu and Hsueh, 2001; Reyes et al., 2001). It is currently thought that CRH acts through type 1 receptors (CRHR1), while novel ligands have been identified for the related type 2 receptors (CRHR2). These ligands, urocortin II (Reyes et al., 2001) and urocortin III (Lewis et al., 2001), play a role in the regulation of appetitive behaviors and are thought to play a role in the adaptive (slow) phase of the stress response (de Kloet et al., 2005) as discussed below. Expression of mRNA for the urocortins has been demonstrated in the hypothalamus, brainstem, and amygdala (Hsu and Hsueh, 2001). Interestingly, the CNS distributions of the CRHR1 and CRHR2 receptors correspond to the CRH and urocortin terminal fields, respectively.

The effects of the glucocorticoids are also mediated through two distinct subclasses of cytosolic steroid receptors. Glucocorticoid receptors (GRs) are found throughout the brain, but GR density is highest in the hypothalamic PVN, ascending aminergic pathways, and in limbic brain regions (Herman et al., 2003). Mineralocorticoid receptors (MR) are named for their primary ligand, the steroid aldosterone, which is produced by the outer zona glomerulosa layer of the adrenal cortex and regulates sodium balance. However, MRs in the brain have a ten-fold higher affinity for glucocorticoids than GRs and thus respond to lower levels of these hormones during the initial phase of a stress response. MRs are also expressed at high levels in limbic brain regions; thus there is considerable overlap between these two receptors in key brain areas that regulate the HPA axis.

34.3.2. The Role of HPA Axis in Response to Stress

Under resting conditions, CRH and AVP are released from the hypothalamus in a pulsatile fashion with a frequency of 2–3 episodes per hour (Engler et al., 1989). The amplitude of the neuropeptide pulses normally increases in the morning, resulting in a circadian fluctuation in ACTH and cortisol levels. This daily rhythm is modulated by feeding and activity schedules, but is particularly perturbed by stressful stimuli originating internally (e.g., anxiety or systemic infection) or externally (e.g., threatening situations). Thus, acute stressors lead to activation of the stress response.

34.3.2.1. Modes of the Acute Stress Response

The stress system is characterized by two modes of operation, a fast phase that serves to mobilize and activate, and a slow phase that promotes recovery and adaptation. Each mode is associated with characteristic behavioral changes that are mediated by distinct cellular and molecular mechanisms.

The initial fast phase is manifested behaviorally by increases in vigilance and arousal, and alterations in attention. This phase is associated with the release of CRH and AVP from the hypothalamic PVN, activation of sympathetic outflow and release of catecholamines from the adrenal medulla, and rising blood levels of cortisol. The effects of CRH in the fast phase are mediated through type 1 receptors (CRHR1), which are responsible for activation of the HPA axis and the sympathetic nervous system. Levels of cortisol rise rapidly due to release of ACTH from the pituitary, and serve in the periphery to mobilize metabolic resources and alter immune system responses (discussed below). The effects of cortisol on brain structures, and thus behavior, in the fast phase are mediated primarily by the higher-affinity MRs, which respond to the rising levels of cortisol at the initiation of a stress response. An important aspect of MR-induced changes involves their role in the appraisal of environmental stimuli in this phase of the response. The alterations in neuronal function mediated by the interaction of cortisol with MRs leads to increases in arousal level, heightened vigilance and alertness, and enhancement of attention.

The slow phase of the stress system is characterized by processes that promote recovery from, and adaptation to, the stressful conditions that prompted the response. At the level of the hypothalamus, this phase is probably mediated by the urocortins acting through CRHR2 receptors (Reul and Holsboer, 2002; de Kloet et al., 2005). In contrast to the fast phase, the slow phase is associated with activation of the parasympathetic nervous system, which promotes the appetitive and metabolic functions, which help to restore homeostasis. As cortisol levels peak and then slowly subside during the slow phase, the lower-affinity GRs are activated and lead to transcriptional changes in target organs that promote recovery. In the brain, GR activation-induced transcriptional changes may be crucial for the consolidation of memory traces involving the association of stressors with sensory stimuli (in the amygdala) (Roozendaal and McGaugh, 1997) and environmental context (in the hippocampus) (Oitzl and de Kloet, 1992).

At all levels of the HPA axis, feedback mechanisms exist that control the output of neuropeptides and circulating hormones. These mechanisms regulate the levels of these factors during normal diurnal cycles, but also act to limit the magnitude of the stress response and terminate its effects. These feedback mechanisms consist of both "short" loops, where each factor acts to inhibit its own secretion, and "long" loops, whereby ACTH acts on PVN neurons to inhibit the secretion of CRH, and cortisol acts on the PVN and the pituitary to inhibit the secretion of CRH and ACTH, respectively (de Kloet et al., 1990; de Kloet, 2003).

34.3.2.2. Chronic Stress Response

It is important to distinguish between the response of the HPA axis to acute stressors, and the response to chronic stress. While the HPA response to acute stressors can clearly be seen as an adaptive response that serves to maintain or re-establish homeostasis, the response to chronic stress can lead to maladaptive conditions, especially in "vulnerable" individuals, that may initiate or facilitate CNS disease processes, as discussed below.

34.3.3. Role of HPA Axis in Inflammation

The activation of the stress systems affects all tissues of the organism, and the peripheral immune system is no exception. These effects are mediated through at least two pathways: via the HPA axis and by virtue of the innervation of lymphatic tissues by autonomic nerve fibers, especially from the sympathetic nervous system. All lymphoid tissues, primary (bone marrow and thymus) as well as secondary (spleen, lymph nodes, and gut-associated lymphoid tissue) are innervated by sympathetic nerve fibers. As discussed above, most lymphoid cells express catecholamine receptors, including B-lymphocytes, CD4- and CD8-positive T cells, dendritic cells, monocytes, and macrophages.

The HPA axis also directly affects peripheral immune functions via receptors for CRH, ACTH and cortisol.

34.3.4. Impact of HPA Axis on Psychiatric Disorders

It is well known that acute and chronic stress can impact certain psychiatric disorders; here we focus on two disorders where the evidence for a role of stress and the activation of the HPA axis is particularly strong. Many lines of evidence suggest that chronic stress is a major risk factor for major depression (MD), and the onset of schizophrenia is often associated with stressful life events.

34.3.4.1. Major Depression

The hyper-reactivity of the HPA axis that occurs with chronic stress is also often observed in depressed patients. (Owens and Nemeroff, 1991)

In animal models, intracerebroventricular injection of CRH induces anxiety and a depression-like phenotype (Heinrichs and Koob, 2004), and CRHR1 antagonists decrease the signs and symptoms of depression. In contrast, the urocortins, thought to act via the CRHR2 system and potentially involved in the adaptive phase of the stress response, seem to have anxiolytic properties (Heinrichs and Koob, 2004). The hypercortisolaemia resulting from chronic stress perturbs monoaminergic systems as observed in depression, and causes dysregulation of anxiety and aggressive behavior. Using neurendocrine tests in longitudinal studies, it has been shown that depressed patients do not respond well to anti-depressant treatment if HPA axis disturbances persist; further, patients showed higher rates of relapse when HPA axis disturbances returned (Zobel et al., 2001). In line with these clinical studies, it is interesting to note that chronic treatment of rats with the anti-depressant amitriptyline resulted in changes in limbic MR and GR that occurred in parallel with HPA axis normalization (Reul et al., 1993). In sum, these observations make a strong case for a major role for the HPA axis in the etiology of depression.

An important aspect of research in depression addresses the role of genetic predispositions and developmental events as causative factors. Based on studies of twins, it is estimated that depression has a heritability of 40% (Sullivan et al., 2000).

34.3.4.2. Schizophrenia

While the etiology of schizophrenia is essentially unknown, it is generally accepted that the disease complex is caused by the abnormal development of limbic system structures that are assembled mainly in the third trimester of human gestation (Weinberger, 1987; Waddington, 1993). Additionally, the hypothesis is gaining support that the disease is precipitated by environmental stressors that occur during the period of final maturation of these limbic structures during adolescence. Given the role of the HPA axis in the stress response, it is thus reasonable to hypothesize that alterations in the HPA axis somehow contribute to the etiology of schizophrenia (reviewed in Corcoran et al., 2003). Several versions of a "two hit" model of schizophrenia hypothesize that the disease is produced in individuals who suffer both developmental insults during in utero or perinatal brain development and abnormal responses to environmental stressors during adolescent maturation (see for example Lieberman et al., 1997; Walker and Diforio, 1997; Waddington et al., 1998).

In fact, differences in HPA axis function have been reported in schizophrenic patients. Diurnal measurements of plasma ACTH revealed elevated levels in drug-naïve schizophrenic patients compared to age and sex matched controls, suggesting basal overactivity of the HPA axis (Ryan et al., 2004). It has been proposed that the loss in hippocampal volume (and the concomitant loss of functional glucocorticoid receptors) demonstrated in schizophrenic patients may contribute to unrestrained and excessive HPA function by decreasing the normal negative feedback control exerted by this structure (Corcoran et al., 2003). In this model, early developmental deficits in hippocampal function would participate in precipitation of schizophrenic symptoms in part by promoting excessive HPA axis responsiveness to stressors. A lack of dexamethasone suppression of cortisol secretion in schizophrenic patients (Sharma et al., 1988; Yeragani, 1990) is consistent with the idea that negative feedback control of the HPA axis is compromised in the disease. However, cause and effect have yet to be determined with confidence. Thus it remains to be determined whether changes in limbic regions that regulate the HPA axis are altered due to chronic stress concomitant with disease onset or whether developmental changes in these structures lead to disregulation of the HPA axis. Abnormal dendritic structure consistent with glucocorticoid-mediated chronic stress has been reported in schizophrenics in the hippocampus (Rosoklija et al., 2000) and prefrontal cortex (Glantz and Lewis, 2001). It is of course possible that early developmental defects in these structures lead to a positive feedback cycle of dendritic remodeling subsequent to HPA axis dysregulation.

The concept of "stress sensitization", whereby early life stressors can induce an enhanced reactivity to later environmental adversities (reviewed in Read et al., 2005), may also be relevant to the etiology of schizophrenia. This phenomenon, which results in abnormal activation of the HPA axis, is thought to involve alterations in the mesolimbic dopamine system (reviewed in Laruelle, 2000) resulting in excessive release of dopamine, a hallmark of psychotic episodes (Laruelle et al., 1996). Taken together, it appears likely that prenatal developmental abnormalities in limbic structures that regulate the HPA axis predispose this system to sensitivity to stressors in early childhood and adolescent maturation. This predisposition may then play a key role in the precipitation of psychotic episodes, as well as the deterioration of brain function that accompanies the progression of the disease.

34.4. Neuropeptides as Modulators of the CNS and Immune Function

There is differential expression of neurotransmitter and neurohormonal receptors among the various cell types of the immune system. This suggests that particular neurohormones produce specific and differential effects on immune cells, and hence on immune function. It also suggests that the pharmacology of immune cells will vary from cell type to cell type and that these pharmacological differences represent potential drug treatment strategies in coping with various pathological states. There is still a great deal of work to be done in this field since there is debate in the literature on receptor expression; that is, which cells consistently express which subset of receptors, and whether and how receptor expression is regulated by the changing milieu.

For example, functional activities of β2-adrenergic, prostaglandin and histamine receptors are increased in T lymphocytes activated by IL-2 (Dailey et al., 1988). This takes place in an apparently coordinated manner even though the prostaglandin receptors are clearly dominant in regulation of signal transduction in resting cells. The change in functional activity is presumably due to up-regulation of receptor expression. These and other studies indicate that receptor expression is a dynamic aspect of immune system function and it that implies alterations in receptor function are important for immune system activities, in this particular case cellular proliferation.

34.4.1. Peptides Derived from Proenkephalin

Two of the most abundant endogenous opioid peptides, methionine-enkephalin (Met-enk; Tyr-Gly-Gly-Phe-Met) and leucine-enkephalin (Leu-enk; Tyr-Gly-Gly-Phe-Leu) are derived from the precursor protein, proenkephalin (PENK). PENK is a 267 amino acid precursor that contains 6 copies of Met-enk, two of which are extended forms (Tyr-Gly-Gly-Phe-Met-Arg-Phe and Tyr-Gly-Gly-Phe-Met-Arg-Gly-Leu) and one copy of Leu-enk. This ratio is essentially the same as the ratio of these peptides found in brain and adrenal chromaffin cells. The fact that there are two distinct peptides closely related in structure and function proved to be a major obstacle that slowed their discovery (Kosterlitz and Hughes, 1977).

The first two opioid peptides were isolated from pig brain and shown to be active in bioassay systems by Hughes and Kosterlitz in 1975 (Hughes et al., 1975). It was some time later, though, that the precursor proteins for these small peptides were discovered. The first of these to be identified was proopiomelanocortin (POMC). It was not until 1982 that PENK was identified and sequenced in multiple species (Udenfriend and Kilpatrick, 1983).

As mentioned above, PENK contains not only both pentapeptide enkephalins, but also a hepta- and octapeptide enkephalin. This structure is maintained across species (Noda et al., 1982; Comb et al., 1982; Gubler et al., 1982). The processing of PENK into smaller peptides, even for an extended period of time, maintains this assortment of four distinct peptides (Fleminger et al., 1983), indicating that all are active products of PENK and that the longer peptides are not simply alternate precursors for Met-enk. Interestingly, while a great deal of attention has been paid to the role of the pentapeptides, Met-enk and Leu-enk, knowledge of the physiological role of the hepta- and octapeptides is still very limited.

There is evidence PENK is not processed in the same way in all tissues and the suggestion has been made that alternative processing may occur in immune system cells containing PENK, compared to processing in neurons (Eriksson et al., 2001). This would open the possibility that differential processing may lead to release of enkephalin-like peptides from the immune system that would have a different spectrum of activities, at least to some extent, than those derived from neural tissues. This, in turn, would provide the opportunity for differential regulation of components of the immune system, depending on the source of the enkephalin-like peptide release.

34.4.1.1. Expression of Enkephalins in the Immune System

PENK is distributed widely throughout the brain (Wamsley et al., 1980; Finley et al., 1981; Harlan et al., 1987), indicating involvement of the enkephalins in a variety of physiological functions. The first suggestion that peptides derived from PENK were involved in immune system function came from the work of Wybran (Wybran et al., 1979), which demonstrated specific binding of both Met-enk and morphine to lymphocytes. Early studies also demonstrated that the enkephalins are concentrated in leukocytes compared to other blood components, supporting the idea that immune cells can synthesize the opioid peptides. Since these initial observations, a large body of evidence has provided more detail supporting the idea that the enkephalin peptides are produced by and have important actions in cells of the immune system.

Determination of which PENK products are formed by different immune cells has been difficult because of the lack of antibodies specific to the various peptides. For instance, many studies use readily available antibodies to Met-enk, but the possibility is strong that these antibodies can also react with larger peptides containing Met-enk. In addition, the variations in results reported in the literature are probably due to differences in cell types as well as the inducing signal. For instance, Th2 cytokines, such as IL-4, IL-6 and TGF-β, produce many-fold increases in PENK mRNA in human thymocytes. On the other hand, Th1 cytokines have variable (IL-1β) or no (IFN-γ) effect (Kavelaars and Heijnen, 2000). Further, Th2 cytokines lead to a 4- to 5-fold increase in Met-enk-like protein in thymocytes, whereas Th1 cytokines produce no such effect. It appears, moreover, that the induced Met-enk itself is not released by thymocytes; rather, larger peptides containing Met-enk are released.

Similar results have been found in T cells. PENK is synthesized in T cells and both synthesized and processed into smaller peptides in macrophages and activated monocytes, but not in unactivated monocytes. Paradoxically, in many cases it has been very difficult to demonstrate release of PENK products. For example, T cells do not appear to secrete detectable levels of PENK-derived peptides whereas monocytes do. On the other hand, under some conditions T cells have been shown to secrete intermediate products of preproenkephalin, the precursor to PENK. These intermediate products appear to inhibit proliferative capacity via unidentified, that is, non-opioid, binding sites. The larger peptides eventually are processed into Met-enk (and Leu-enk) in some cases, whereas in others the larger peptides are probably the active products themselves (Dillen et al., 1993; Hiddinga et al., 1994b).

There are many questions yet to be answered in this area. To compound the problems of determining if the products of PENK are released and are active, the quantity of material provided by the tissues is sometimes small enough to make consistent quantitative data difficult to obtain. This leads to further confusion and contradictions in the literature.

34.4.1.2. Function of the Enkephalins in the Immune System

Met-enk, the most abundant of the enkephalin peptides, produces effects on its target cells that are both concentration-dependent and modified by the presence of other stimuli. The concentration-dependent effects can be bimodal, with low concentrations producing effects opposite those of high concentrations (Oleson and Johnson, 1988). An example is the differing effects of varying concentrations of enkephalins on the production of reactive oxygen species (ROS) by polymorphonuclear leukocytes. ROS production is enhanced by low enkephalin concentrations and suppressed by high concentrations (Roscetti et al., 1988; Marotti et al., 1992). In a similar manner, Met-enk inhibits antibody production by B cells whereas slightly larger peptides also derived from PENK stimulate B cell antibody production (Hiddinga et al., 1994a; Das et al., 1997). One possible explanation of these paradoxical data is that opioid peptides may play a role as modulators of immune system function, fine tuning the current state of activity, rather than being the primary drivers or determinants of function (Eriksson et al., 2001). As such, the effects of these peptides would be highly dependent on the current status of the immune system, whether activated or inactive, and on which other regulatory factors were present and what their overall effects are. In this sense, the endogenous opioids would act more to maintain homeostasis rather than to specifically activate or inhibit the immune system.

The exact role of Met-enk and related peptides derived from PENK remains controversial since there are many studies with seemingly contradictory results (see Eriksson et al., 2001 for review). Many of these studies were carried out in vitro and how well the results translate to in vivo conditions is not clear at present. In addition, relatively few of these studies used experimental conditions that were the same or even similar to those used in other studies, making comparison and resolution of differences difficult.

34.4.2. Peptides Derived from Proopiomelanocortin (POMC)

Another major source of endogenous opioid peptides as well as non-opioid peptides is proopiomelanocortin (POMC). This large peptide contains the sequences of, and is processed into, a number of smaller peptides, including ACTH, β-endorphin, β-lipotropin, α-MSH, β-MSH and γ-MSH. Which products are derived from POMC depends upon the cell in which the processing takes place. In the anterior lobe of the pituitary POMC is processed into ACTH and β-lipotropin. In the neurointermediate lobe the major products are β-endorphin, α-MSH and γ-MSH. POMC was the first of the peptide precursors for which such cell-dependent processing was demonstrated.

34.4.2.1. *Expression of POMC in the Immune System*

The discovery by Blalock and colleagues that POMC was also produced by immune cells was a major event in the developing recognition of interactions between the immune and nervous systems (reviewed in Blalock, 1999). These studies indicated that not only do neurohormones derived from the nervous system alter immune system activity, but also similar or identical hormones can be released from immune cells to regulate the nervous system as well as the immune system itself.

Similar to what was found with the PENK-derived enkephalins, peptides derived from POMC frequently have biphasic dose-response curves; that is, low doses stimulate immune system function whereas high doses usually suppress function. Again, it appears the system is designed to maintain homeostasis in response to a wide range of conditions.

The finding that the immune and neuroendocrine systems both express receptors for opioids and for ACTH and that both systems can synthesize and release peptides active at these receptors, led to the suggestion that the immune system functions as a sensory organ (Blalock, 1984, 1999) and that this forms the basis for the interaction between the two systems. It is well known that the nervous system responds to a variety of stimuli and, when appropriate, releases neurotransmitters and hormones that enable an appropriate reaction to these stimuli or stresses. This can include changes in immune system function. Blalock proposed that the immune system also responds to particular stimuli, in this case environmental changes that would not be readily detected by the nervous system, such as the presence of bacteria or viruses. In response, the immune system releases a variety of compounds, including peptide hormones that will alter both immune and nervous system function.

34.4.2.2. *Beta-Endorphin*

A locally induced inflammation followed by cold-water swimming, a stress that is known to activate intrinsic opioid systems, produces an antinociceptive effect localized to the area of inflammation (Stein et al., 1990a). Through a series of experiments it became apparent that the agent producing this effect was β-endorphin and that it was localized to the area of inflammation; that is, it was not a systemic effect of the peptide. This provided strong evidence that localized release of endogenous opioid peptides in areas of inflammation could play an important role in regulating inflammation and nociception. Subsequent experiments using this same model demonstrated significantly increased levels of both β-endorphin and Met-enk localized to the area of inflammation, although it appeared that β-endorphin was responsible for the antinociceptive effect (Stein et al., 1990b). These studies indicated that localized release of these opioid peptides from immune cells were responsible for the reduction in pain. Further support for the involvement of immune cells in this antinociceptive action came from suppression of the immune system with cyclosporin A, which resulted in the loss of the nociceptive effects of the endogenous opioids. Interestingly, Met-enk did not play an important role under this experimental paradigm and it was not clear whether Met-enk was released from immune cells, even though it was definitely present in the cells. The paradoxical synthesis of two opioid peptides and the release of only one to produce nociception leaves many unanswered questions, such as why is Met-enk synthesized in the first place? This and many other questions remain.

34.4.2.3. *α-MSH*

The tridecapeptide, α-MSH, is also derived from POMC. While originally discovered through its effects on the skin, α-MSH can alter immune system function. Understanding the role of α-MSH in neuroimmune modulation is still in the early stages (Luger et al., 2003). α-MSH acts through melanocortin-1 receptors (MC-1R), which are expressed by immune cells, including macrophages and monocytes. Both β-MSH and γ-MSH, also products of POMC, act through other melanocortin receptors and these receptors have not been found on immune cells to date. This indicates that α-MSH may be the only of these three to be involved in regulating immune function.

Most studies to date suggest that α-MSH modulates or reduces inflammation, down-regulating proinflammatory cytokines and other proinflammatory molecules (Luger et al., 2003). These effects are produced, at least in part, by inhibition of NFκB activity. There is some evidence α-MSH may have biphasic effects in this regard since low concentrations increase antibody production by B cells whereas higher concentrations reduce production. Nevertheless, most studies, including recent studies pointing to regulation of IL-8 (Manna et al., 2006), indicate an antiinflammatory role for α-MSH.

As discussed above, several other peptides may be derived from POMC in a cell-specific manner, dependent upon processing within each cell type. The role of ACTH in nervous system-immune system interactions is discussed elsewhere in this chapter. At present a neuroimmune function for the other POMC peptides has not been demonstrated.

34.4.3. Corticotropin Releasing Hormone

Corticotropin **R**eleasing **H**ormone (CRH; corticotropin releasing factor; CRF) plays a major role in CNS regulation of the immune system. In particular, CRH released from the hypothalamus is the first point in HPA axis regulation of immune function. As a critical player in the response to stress, the HPA axis is a primary point of coordination in the neural-immune interaction. This is discussed in Section 34.3 of this chapter. CRH also plays an important role in many CNS functions, including regulation of the autonomic nervous system and endocrine function. CRH is involved in the central response to anxiety and stress, and hence affects immune system function via that route. There is also convincing evidence that CRH can play a role in depression. The central role of CRH has been reviewed (Dunn and Berridge, 1990; Heim and Nemeroff, 1999). Beyond these well-known roles for CRH, this neuropeptide hormone is also synthesized in immune cells and so appears to play an even more widespread role than originally thought.

Early studies demonstrated a role for CRH in augmenting function of immune cells, such as natural killer cells (Carr et al., 1990). The localization and the synthesis of CRH in the immune system and in immune cells were initially suggested by finding mRNA for CRH in leukocytes (Stephanou et al., 1990) and significant concentrations of CRH in areas of inflammation (Karalis et al., 1991). This suggested CRH was produced locally; that is, in the periphery in areas of inflammation, probably in part by immune cells. Further studies supported these ideas. In Lew/N rats, which are deficient in hypothalamic CRH responses to inflammatory stimuli, there are high levels of CRH in inflamed joints (Crofford et al., 1992). These data suggested that CRH plays the role of an autocrine or paracrine inflammatory factor. The obvious paradox is that CRH plays a powerful antiinflammatory role as the primary CNS activator of the HPA axis, yet in local peripheral sites it has just the opposite effect. Thus, CRH expression and function with respect to inflammation is site specific.

CRH had been shown to have important effects in areas of inflammation, such as augmentation of analgesia by endogenous opioid peptides. It has been further demonstrated that the effects were blocked by local administration of a CRH antagonist, a CRH antibody or by antisense oligonucleotides directed against CRH. This implies that local production and release of this hormone by immune cells do, in fact, occur and are physiologically important (Schafer et al., 1996). It was found that systemic administration of these same various antagonists of CRH action, an approach directed against CRH released from a distant site, was either ineffective or several orders of magnitude less potent than local administration, bolstering this argument.

It proved difficult to definitively demonstrate CRH synthesis from immune cells, although numerous studies provided evidence this does happen (Aird et al., 1993; Ekman et al., 1993). Eventually it became clear that regulation of CRH synthesis and release in immune cells differs from that in hypothalamic neurons. While immune cells may synthesize and release much smaller concentrations of CRH and other neuroimmune peptides, and although their release may require de novo synthesis, an inherently slow process, the fact that immune cells release these hormones locally in the target area compensates for both of these factors to some extent. These data indicating site and tissue specific effects of CRH, sometimes even contradictory effects, point to the complex interrelationship between the nervous, endocrine and immune systems, an interaction that has yet to be deciphered completely.

34.4.4. Nociceptin, Endomorphins, and Urocortins

There are several newly discovered peptides that are related to those already discussed above and that appear to have, in varying degrees, a role in neuroimmune function. **Nociceptin, or Orphanin FQ**, is a 17 amino acid peptide that plays a role in pain sensation as well as other complex CNS functions. The receptor for this peptide, ORL-1, has striking homology with the three classical opioid peptide receptors and so is considered a member of the opioid receptor family, although agonists at ORL-1 are orders of magnitude less potent at the other opioid receptors. The converse is also generally true.

Several studies have shown that ORL-1 are expressed in immune cells, including peripheral blood lymphocytes, T and B cells and polymorphonuclear leukocytes (Halford et al., 1995; Peluso et al., 1998; Hom et al., 1999; Serhan et al., 2001), and these receptors are functionally active (Waits et al., 2004; Halford et al., 1995; Hom et al., 1999; Serhan et al., 2001). Following these findings, nociceptin transcripts were found to be present in peripheral blood lymphocytes (Arjomand et al., 2002). It has been suggested, based on differences in transcripts between immune cells and neurons, that processing of the message and the final form of nociceptin may be cell type specific, similar to other peptides.

The function of nociceptin in the immune system and/or in the interaction between the nervous and immune systems is not clear at present. Studies suggest that it increases T cell activation, suppresses antibody production and is a potent chemoattractant. However, given the presence of both the neuropeptide and its receptor in immune cells, analogous to beta-endorphin and the enkephalins, nociceptin is well positioned to play a role similar to that of the better-known and studied opioid-like peptides.

The **endomorphins** are similar to nociceptin, being endogenous opioid peptides that may play a role in both neural and immune function. Endomorphin-1 and endormorphin-2 are endogenous tetrapeptides that have a very high affinity for mu opioid receptors and are the most selective mu peptide agonists known (Zadina et al., 1997). Originally isolated from brain, the endomorphins have been found in mammalian immune tissues as well. They were first detected in spleen and thymus (Jessop et al., 2000) and subsequently demonstrated in macrophages and lymphocytes (Jessop et al., 2002; Mousa et al., 2002). Expression of both endomorphins is increased in lymph nodes and in areas of inflammation in models of inflammation, but is unchanged in nerve fibers in these models. This, along with a concomitant increase in mu opioid receptor expression and the accumulation of opioid peptide-containing leukocytes in areas of inflammation, suggests the endomorphins play a role in inflammation (Jessop et al., 2002), probably in peripheral control of inflammatory pain (Mousa et al., 2002; Labuz et al., 2006).

The **urocortins** are more recently discovered members of the CRH family (see Reul and Holsboer, 2002; Gysling et al., 2004 for reviews). These three peptides (urocortin, urocortin II, urocortin III) were first isolated from brain. While most of the research involving them has focused on the CNS, it has been shown that the urocortins can alter immune function (Bamberger and Bamberger, 2000; Baigent, 2001; Theoharides et al., 2004). There also have been demonstrations of urocortin expression in immune cells (Bamberger et al., 1998; Bamberger and Bamberger, 2000; Baker et al., 2003), providing a parallel with other neuropeptides discussed above. While the physiological and pathological functions of the urocortins is largely unknown at present, the similarity between the structure, receptors and localization within the immune system for CRH and the urocortins strongly suggests a coordinating or balancing role for this family of peptides in regulation of the immune system.

34.5. Cytokines as Regulators of the CNS and Immune Function

34.5.1. Cytokine Actions in the CNS

Cytokines are peptide factors that are classically associated with humoral communication between immune system elements. However, it has become clear that many members of this large family of molecules also play diverse roles within the nervous system. Cytokine-mediated communication between neural elements has been implicated in all aspects of nervous system development, including cell proliferation, migration, differentiation, and programmed cell death. Further, in the adult brain cytokines play vital roles in the cellular and synaptic plasticity that is requisite for learning and memory processes. During brain injury, cytokines coordinate the response of neuroimmune elements including astrocytes and microglial cells, the brain-resident monocyte-derived phagocytes.

In addition to their role in coordinating immune system functions, cytokines mediate communication between the peripheral immune system and the central and peripheral nervous system (Plata-Salaman, 1991). During an innate immune response, cytokines released by macrophages, vascular endothelial cells, and other cells activate the HPA axis, resulting in the release of adrenal glucocorticoids (Silverman et al., 2005).

In addition to activation of the HPA axis, the pro-inflammatory cytokines also influence behavior; the outcome of this influence is manifested as a syndrome known as "sickness" behavior. This syndrome is characterized by a number of behavioral responses including anhedonia (inability to experience pleasure), decreased general activity and exploratory behavior, decreased feeding and sexual activity, and increased sleep (Dantzer, 2001, 2004; Larson and Dunn, 2001).

34.5.2. Cytokines and the HPA Axis

It has become clear that cytokines are potent activators of the central stress response, serving as the afferent limb of the response system during acute or chronic inflammatory stress. Several pro-inflammatory cytokines, including tumor necrosis factor-α (TNF-α), interleukin-1β (IL-1β), and interleukin-6 (IL-6), can activate the HPA axis (Chrousos, 1995; Tsigos et al., 1997).

34.5.2.1. Interleukin-1β

The cytokine IL-1β serves a central role in the initiation and coordination of inflammatory responses, both in the periphery and in the CNS. Neurons, astrocytes, microglia, and epithelial cells in the brain constitutively express this cytokine, and its central effects in regulating central inflammatory responses are well-documented (Basu et al., 2004). However, IL-1β released by peripheral immune cells can also affect the CNS; in fact this cytokine is believed to play a major part in eliciting the sickness behavioral syndrome mentioned above. Thus, in animal models, peripheral administration of IL-1β causes such diverse responses as decreased exploratory behavior and locomotor activity, decreased operant responding to rewards, anorexia, inhibition of sexual behavior, promotion of sleep, and anxiety reactions. IL-1β also increases fever through actions on the hypothalamus. Thus, human and animal data support the idea that IL-1β is a key regulator of behavioral responses to peripheral immune system activation (Basu et al., 2004).

34.5.3. Impact of Immune System Cytokines in Psychiatric Disorders

The growing awareness of the role of cytokines in regulating CNS function has resulted in a great deal of research in the last decade on their potential role in psychiatric disorders. Initially driven by studies of "sickness" behavior, evidence has accumulated suggesting that levels of pro-inflammatory cytokines are altered during the course of several psychiatric disorders.

34.5.3.1. Role in Major Depression

A large number of studies have demonstrated increases in several pro-inflammatory cytokines in patients diagnosed with MD. It is interesting to note that all of these cytokines are associated with activation of the HPA axis. Thus, elevations in Il-6 have been consistently associated with clinically diagnosed depression (Sluzewska et al., 1995; Maes et al., 1997; Lanquillon et al., 2000; Musselman et al., 2001; Alesci et al., 2005). The circadian pattern of IL-6 secretion is shifted 180 degrees out of phase in depressed patients (Alesci et al., 2005), and IL-6 levels are significantly correlated with severity of depression (Musselman et al., 2001). Prior to administration of antidepressant medication, levels of Il-6 were elevated in patients who did not respond to the treatment, whereas responders showed normal levels before treatment, suggesting that Il-6 levels could be used to predict treatment outcome (Lanquillon et al., 2000). Depression has also been associated with increases in IL-1β (Owen et al., 2001; Thomas et al., 2005) and TNF-α (Hestad et al., 2003; Tuglu et al., 2003). These studies support a role for pro-inflammatory cytokines in MD, and suggest that targeting pro-inflammatory cytokines and their signaling pathways may represent a novel strategy for treatment of the disease (Raison et al., 2006).

34.5.3.2. Role in Schizophrenia

As mentioned above (Section 34.3.4.2) there is growing evidence supporting the hypothesis that schizophrenia is caused by a combination of genetic or epigenetic factors that affect limbic brain structures in the last trimester of development and stress-induced environmental influences during the final maturation of these structures late in human adolescence. This "two hit" idea has led to research directed at two potential targets for cytokine action: the effects of cytokines on the development of the nervous system, and the acute effects of cytokine administration in generating psychotic symptoms.

While the etiology of schizophrenia is essentially unknown, it is generally accepted that the disease complex is caused by the abnormal development of limbic system structures that are assembled mainly in the third trimester of human gestation (Weinberger, 1987; Waddington, 1993). Additionally, the hypothesis is gaining support that the disease is precipitated by environmental stressors around the period of final maturation of these limbic structures. This "two hit" idea has led to research directed at two potential targets for cytokine action: the effects of cytokines on the development of the nervous system, and the acute effects of cytokine administration in generating psychotic symptoms.

A number of environmental factors have been correlated with the development of schizophrenia including birth trauma, maternal viral infection, and season of birth (reviewed in Nawa et al., 2000). Prolonged labor results in abnormally elevated serum Il-6 levels (De Jongh et al., 1997), suggesting a possible link between immune system activation and the perinatal complications associated with the development of schizophrenia. Given the role of proinflammatory cytokines in repair of tissue damage and infection, it is interesting that TGF-alpha and IL1-beta can suppress the normal expression of brain-derived neurotrophic factor (BDNF), a neurotrophin that is a key regulator of neuronal and synaptic development, in the hippocampus (Lapchak et al., 1993). Conversely, IL1-beta causes up-regulation of nerve growth factor (NGF), another factor that regulates growth and development of central neurons, in microglial cells (Heese et al., 1998). These findings compel further research into the interactions between cytokines and the brain-derived trophic factors that may lead to abnormal development of brain regions that ultimately result in the manifestation of schizophrenia in the mature brain.

Several reports of persistent psychosis following the administration of interferon-alpha (Tamam et al., 2003; Thome and Knopf, 2003; Telio et al., 2006) or high doses of IL-2 (Denicoff et al., 1987) demonstrate that cytokines could conceivably mediate some aspects of cognitive impairment observed in schizophrenic patients. These studies are interesting in light of the fact that evidence for abnormal levels of cytokines in schizophrenic patients is accumulating (Malek-Ahmadi, 1996; Prolo and Licinio, 1999). However, further studies are needed to show a causative relationship between cytokine levels in patients and the precipitation of psychotic episodes.

34.6. Overview of the Role of the Neuroimmune System in Depression and Schizophrenia

There has been a long-standing interest in the relationship between psychological disorders and the status of the immune system, based upon the everyday observation that people with significant psychiatric problems also appear to have a high incidence of other physical ailments. Conversely, people suffering from infectious or inflammatory disease appear to have a greater incidence of CNS problems. Two chapters in this book cover this topic extensively. Therefore, only a brief overview is provided here.

34.6.1. Depression and the Immune System

In recent years there has been increased investigation of potential links between immune function and depression. Two large meta-analyses of multiple studies found that patients with major depression show reliable changes in immune function (Herbert and Cohen, 1993; Irwin, 1999), including lowered proliferative response of lymphocytes to mitogens, lower NK cell activity and changes in white blood cell populations. Data suggest that major depressive disorder (MDD) can alter immune function and this is most likely to occur in patients with severe depression (Herbert and Cohen, 1993; Irwin, 1999; Maes, 1999; Frank et al., 2002). While the initial thought was that immune function was depressed in MDD, more recent evidence has suggested that many aspects of the immune system are likely to be activated. Both stress and major depression, two disorders that share many features, can be characterized by activation of some immune capacities and suppression of others (Raison and Miller, 2001).

The mechanisms for these interactions are not well defined, although numerous hypotheses have been advanced. It should be noted at the outset that attempts to determine the relationship between depression and altered immune function are confounded by the multiplicity of factors known to be associated with both and that may alter the interrelationship. Examples include age, gender, sleep status, the likelihood that depression represents a complex of disease states with varying involvement of the immune system, and the frequent presence of other psychiatric or physical disorders which can affect both psychological status and immune function. A detailed review of this topic has been published (Irwin, 2001). Indeed, it has been pointed out by Irwin that current data suggest immune changes in MDD specifically correlated with the disorder are also seen with stress and other psychiatric disorders, suggesting some common characteristic(s) shared by these problems (Irwin, 2001).

In addition, the likelihood that depression represents a complex of disease states with varying involvement of the immune system supports the idea that there is a subgroup of depressed patients who have associated alterations in immune function. Indeed, data suggest that different types of depression display distinct, and even opposite, changes in neuroimmune functions (Antonijevic, 2006). As a result of multiple issues, there are contradictory, or at least inconsistent, findings in the literature. This is characteristic of most types of studies of depression, again probably because depression represents multiple disease states.

There is increasing evidence that aberrations in immune system function induce or at least support the development of

MDD. Activation of some aspects of the immune system is generally associated with MDD, even in patients who are otherwise quite healthy. Many studies support this idea (Raison et al., 2006). Proinflammatory cytokines, such as IL-6 and TNF-α, are increased in plasma and in the CNS. There also are increases in the level of chemokines and adhesion molecules.

Several studies have shown increased plasma levels of IL-6 in patients with MDD. In a more extensive study, Alesci and colleagues found that IL-6 levels were increased in MDD patients throughout the circadian cycle (Alesci et al., 2005). There was a 12-hour shift in the circadian rhythm of IL-6 plasma levels and its complexity was reduced. Even though IL-6 is a known activator of the HPA axis, cortisol levels were not consistently changed in MDD patients compared to controls. Additionally, it was found that IL-6 levels, with their altered rhythm, correlated significantly with mood ratings. IL-6 also induces a "sickness" behavior very similar to depression. These data suggest a direct relationship between IL-6 and depression.

A related example is the effects of interferon-α (IFN-α) on mood and, specifically, its ability to produce significant depression (Trask et al., 2000; Raison et al., 2006). IFN-α is used to treat several serious disorders, including malignant melanoma and hepatitis C. A major side effect of prolonged or high dose therapy with this agent is major depression, even to the point, though rare, of suicidality. Estimates are that this occurs in as many as 30–50% of patients undergoing IFN-α treatment. A history of psychiatric symptoms is considered a relative contraindication to the use of IFN-α, and a psychiatrist should closely monitor treatment of such patients using this drug. Further indications that the syndrome produced by IFN-α is depression are that it is responsive to treatment with antidepressants and that there are indications of changes in serotonin and possibly norepinephrine metabolism centrally. Alterations in both of these neurotransmitters are associated with depression. It is also interesting that IFN-α is a potent inducer of IL-6, which may be associated with the induction of depression.

As indicated by this brief overview there is considerable evidence for a close relationship between changes in the immune system and depression, particularly MDD. Whether there is a clear cause and effect relationship between these is not known in most cases, although there are strong data in the case of IL-6 and IFN-α, as noted. It will be important to determine more clearly the exact relationship between depression and immune system alterations. There is also a need to determine which of the immune alterations that are found are clinically important (Irwin, 2001).

34.6.2. Schizophrenia and the Immune System

The involvement of the immune system in schizophrenia is an even more complicated issue compared to involvement with depression. There have been suggestions for many decades that schizophrenia is associated with immune dysfunction (Vaughan et al., 1949). However, it is only recently that substantial evidence relating to this idea has begun to appear and this is even less definitive than data supporting a link between the immune system and depression. Analogous to depression, there are multiple problems with dissecting a role for the immune system in schizophrenia, including the multiplicity of the disease etiology and the difficulty in controlling for the many factors that are frequently associated with schizophrenia and that also alter immune function themselves. These include age, gender, sleep status, the likelihood that schizophrenia represents a complex of disease states, and the frequent presence of other psychiatric or physical disorders which can affect both psychological status and immune function. The difficulties with such studies have been set forth by Rapaport and Muller (2001).

Nevertheless, despite the difficulties, there are intriguing findings that have strengthened the hypothesis that neuro-immune interactions are altered in schizophrenia patients. One argument for an association between immune function and schizophrenia is that epidemiological data show a linkage between prenatal viral infections and subsequent appearance of schizophrenia. There have also been studies that show an acute exacerbation of schizophrenia can occur in response to immune system dysfunction. These have been reviewed (Rapaport and Muller, 2001). The most extensively examined idea is that schizophrenia is associated with autoimmune diseases or that schizophrenia may, in certain instances, be associated with a lack of normal autoimmune function (Gaughran, 2002; Jones et al., 2005; Kipnis et al., 2006).

Autoimmune diseases are associated with some schizophrenia-like symptoms. There are also associations between the appearance of several autoimmune diseases and schizophrenia, although these are both positive and negative correlations. It has been proposed that there is a genetic linkage between schizophrenia and autoimmune diseases. Again, there are both positive and negative associations. Many of these correlations have been refuted by later data. The extensive literature documenting multiple changes in immune system markers and function in the schizophrenic population has been reviewed extensively (Gaughran, 2002).

An interesting hypothesis advanced recently is that schizophrenia is, at least in some cases, due to loss of a specific aspect of autoimmunity due to prenatal loss of a specific subset of T cells (Kipnis et al., 2006). This idea is based on a series of studies from the Schwartz lab that indicate autoimmune T cells in the CNS play a fundamental physiological role, exerting a protective effect on neurons when they are subjected to stress. The hypothesis is that there is a prenatal loss of the relevant autoimmune clones so that when stress appears later in life, the neuroprotective effects are lost or at least diminished, and ultimately neurodegeneration occurs, leading to schizophrenia. This idea is consistent with a number of characteristics of schizophrenia. The dopamine hypothesis posits abnormally high levels of dopamine in schizophrenia. Dopamine release is increased centrally in response to stress and this normally suppresses regulatory T cells, leading to increased activity in protective autoimmune T cells. In schizophrenia, however, there is a losts of autoimmune protection, consistent with a lack of protective autoimmune T cells. This would result in lack of neuroprotection and greater neurodegeneration. It is generally accepted that schizophrenia is a developmental disorder. There are also considerable data indicating schizophrenia is often associated with maternal inflammation or viral infection. Both of these could lead to loss of immune cells prenatally, perhaps to a specific loss of protective autoimmune clones. There is a great deal of work to be done to prove or disprove this hypothesis, but it does bring together several aspects of schizophrenia and it provides a framework for investigating a potential role of autoimmunity in schizophrenia.

Summary

The studies reviewed in this chapter demonstrate that a strong and reciprocal relationship exists between the central nervous system and the immune system. Indeed, the term "neuroimmune system" is clearly justified and appropriate to emphasize the fact that nervous and lymphoid tissues constitute a unified system that functions in the maintenance of homeostasis. The conventional division between the two systems has blurred, as well as the distinction between neuropeptides on the one hand, and immune cytokines on the other. Two lines of research have altered our perspective on neuroimmune interactions: (a) the identification of conventional neuropeptides and their receptors, especially those related to the HPA axis, in most lymphoid tissues, and (b) the large body of evidence that cytokines, historically associated with immune system communication, play vital roles in nervous function.

The role of the HPA axis in coordinating responses to stressors, and the role of the immune system in mediating the "sickness" response, underscore this reciprocal relationship between the CNS and the immune system. However, the relationship is complex, and it should be clear that we are in the early stages of understanding how immune tissue can contribute to psychiatric disorders. In spite of this nascent state of knowledge, the evidence for hyper-reactivity of the HPA axis and the alterations in cytokine levels observed in both MDD and schizophrenia serve as a foundation for further research establishing a causative relationship between alterations in immune system function and these debilitating CNS disorders.

Review Questions/Problems

1. **Name three of the four peptides for which proenkephalin is the precursor.**

2. **The major physiological role of the enkephalins in regulating immune function appears to be**

   a. stimulation of all immune cells
   b. inhibition of all immune cells
   c. maintenance of homeostasis
   d. increased expression of regulatory T cells
   e. decreasing reactive oxygen species production

3. **Name the three products derived from proopiomelanocortin (POMC) for which evidence is the strongest supporting involvement in regulating immune system function.**

4. **The site of action for α-MSH is**

   a. muopioid receptors
   b. melanocortin-1 receptors
   c. glucocorticoid receptors
   d. melatonin receptors
   e. ACTH receptors

5. **Local release by immune cells of which neuropeptide plays the most important role in nociception produced in areas of inflammation by the immune system?**

   a. methionine-enkephalin.
   b. α-MSH
   c. ACTH
   d. β-endorphin
   e. urocortin-1

6. **Which neuropeptide is antiinflammatory when activating the HPA axis but is proinflammatory when released locally by immune cells?**

   a. ACTH
   b. CRH
   c. β-endorphin
   d. nociceptin
   e. γ-MSH

7. **Endomorphins are endogenous neuropeptides with a very high affinity and selectivity for**

   a. mu opioid receptors
   b. glucocorticoid receptors
   c. melanocortin-1 receptors
   d. ORL-1 receptors
   e. beta adrenergic receptors

8. **Which cytokine has been found to induce major depressive disorder in a high percentage of patients when used to treat malignant melanoma or hepatitis C?**

   a. IL-6
   b. TNF-α
   c. IL-4
   d. interferon-α
   e. IL-10

9. **The neuropeptide synthesized and released from the CNS, particularly the hypothalamus, as well as from immune cells is**

   a. ACTH
   b. urocortin-1
   c. CRH
   d. TNF-α
   e. insulin

10. The proinflammatory cytokine found to have increased levels in plasma and CNS in major depressive illness and whose levels correlate significantly with mood rating is

a. IL-6
b. TNF-α
c. CRH
d. IL-2
e. β-endorphin

References

Ader R (1987) Conditioned immune responses: Adrenocortical influences. Prog Brain Res 72:79–90.

Adler MW, Rogers TJ (2005) Are chemokines the third major system in the brain? J Leukoc Biol 78:1204–1209.

Aird F, Clevenger CV, Prystowsky MB, Redei E (1993) Corticotropin-releasing factor mRNA in rat thymus and spleen. Proc Natl Acad Sci USA 90:7104–7108.

Alesci S, Martinez PE, Kelkar S, Ilias I, Ronsaville DS, Listwak SJ, Ayala AR, Licinio J, Gold HK, Kling MA, Chrousos GP, Gold PW (2005) Major depression is associated with significant diurnal elevations in plasma interleukin-6 levels, a shift of its circadian rhythm, and loss of physiological complexity in its secretion: Clinical implications. J Clin Endocrinol Metab 90:2522–2530.

Antonijevic IA (2006) Depressive disorders—is it time to endorse different pathophysiologies? Psychoneuroendocrinology 31:1–15.

Arjomand J, Cole S, Evans CJ (2002) Novel orphanin FQ/nociceptin transcripts are expressed in human immune cells. J Neuroimmunol 130:100–108.

Baigent SM (2001) Peripheral corticotropin-releasing hormone and urocortin in the control of the immune response. Peptides 22:809–820.

Baker C, Richards LJ, Dayan CM, Jessop DS (2003) Corticotropin-releasing hormone immunoreactivity in human T and B cells and macrophages: Colocalization with arginine vasopressin. J Neuroendocrinol 15:1070–1074.

Bamberger CM, Bamberger AM (2000) The peripheral CRH/urocortin system. Ann N Y Acad Sci 917:290–296.

Bamberger CM, Wald M, Bamberger AM, Ergun S, Beil FU, Schulte HM (1998) Human lymphocytes produce urocortin, but not corticotropin-releasing hormone. J Clin Endocrinol Metab 83:708–711.

Basu A, Krady JK, Levison SW (2004) Interleukin-1: A master regulator of neuroinflammation. J Neurosci Res 78:151–156.

Blalock JE (1984) The immune system as a sensory organ. J Immunol 132:1067–1070.

Blalock JE (1989) A molecular basis for bidirectional communication between the immune and neuroendocrine systems. Physiol Rev 69:1–32.

Blalock JE (1994) The syntax of immune-neuroendocrine communication. Immunol Today 15:504–511.

Blalock JE (1999) Proopiomelanocortin and the immune-neuroendocrine connection. Ann N Y Acad Sci 885:161–172.

Bylund DB, Eikenberg DC, Hieble JP, Langer SZ, Lefkowitz RJ, Minneman KP, Molinoff PB, Ruffolo RR, Jr, Trendelenburg U (1994) International Union of Pharmacology nomenclature of adrenoceptors. Pharmacol Rev 46:121–136.

Carr DJ, DeCosta BR, Jacobson AE, Rice KC, Blalock JE (1990) Corticotropin-releasing hormone augments natural killer cell activity through a naloxone-sensitive pathway. J Neuroimmunol 28:53–61.

Cartier L, Hartley O, Dubois-Dauphin M, Krause KH (2005) Chemokine receptors in the central nervous system: Role in brain inflammation and neurodegenerative diseases. Brain Research Reviews 48:16–42.

Chrousos GP (1995) The hypothalamic-pituitary-adrenal axis and immune-mediated inflammation. N Engl J Med 332:1351–1362.

Comb M, Seeburg PH, Adelman J, Eiden L, Herbert E (1982) Primary structure of the human Met- and Leu-enkephalin precursor and its mRNA. Nature 295:663–666.

Corcoran C, Walker E, Huot R, Mittal V, Tessner K, Kestler L, Malaspina D (2003) The stress cascade and schizophrenia: Etiology and onset. Schizophr Bull 29:671–692.

Crofford LJ, Sano H, Karalis K, Webster EL, Goldmuntz EA, Chrousos GP, Wilder RL (1992) Local secretion of corticotropin-releasing hormone in the joints of Lewis rats with inflammatory arthritis. J Clin Invest 90:2555–2564.

Dailey MO, Schreurs J, Schulman H (1988) Hormone receptors on cloned T lymphocytes. Increased responsiveness to histamine, prostaglandins, and beta-adrenergic agents as a late stage event in T cell activation. J Immunol 140:2931–2936.

Dantzer R (2001) Cytokine-induced sickness behavior: Mechanisms and implications. Ann N Y Acad Sci 933:222–234.

Dantzer R (2004) Innate immunity at the forefront of psychoneuroimmunology. Brain Behav Immun 18:1–6.

Das KP, Hong JS, Sanders VM (1997) Ultralow concentrations of proenkephalin and [met5]-enkephalin differentially affect IgM and IgG production by B cells. J Neuroimmunol 73:37–46.

De Jongh RF, Bosmans EP, Puylaert MJ, Ombelet WU, Vandeput HJ, Berghmans RA (1997) The influence of anaesthetic techniques and type of delivery on peripartum serum interleukin-6 concentrations. Acta Anaesthesiol Scand 41:853–860.

de Kloet ER (2003) Hormones, brain and stress. Endocr Regul 37:51–68.

de Kloet ER, Reul JM, Sutanto W (1990) Corticosteroids and the brain. J Steroid Biochem Mol Biol 37:387–394.

de Kloet ER, Joels M, Holsboer F (2005) Stress and the brain: From adaptation to disease. Nat Rev Neurosci 6:463–475.

Denicoff KD, Rubinow DR, Papa MZ, Simpson C, Seipp CA, Lotze MT, Chang AE, Rosenstein D, Rosenberg SA (1987) The neuropsychiatric effects of treatment with interleukin-2 and lymphokine-activated killer cells. Ann Intern Med 107:293–300.

Dillen L, Miserez B, Claeys M, Aunis D, De PW (1993) Posttranslational processing of proenkephalins and chromogranins/secretogranins. Neurochem Int 22:315–352.

Dunn AJ, Berridge CW (1990) Physiological and behavioral responses to corticotropin-releasing factor administration: Is CRF a mediator of anxiety or stress responses? Brain Res Brain Res Rev 15:71–100.

Ekman R, Servenius B, Castro MG, Lowry PJ, Cederlund AS, Bergman O, Sjogren HO (1993) Biosynthesis of corticotropin-releasing hormone in human T-lymphocytes. J Neuroimmunol 44:7–13.

Engler D, Pham T, Fullerton MJ, Ooi G, Funder JW, Clarke IJ (1989) Studies of the secretion of corticotropin-releasing factor and arginine vasopressin into the hypophysial-portal circulation of the conscious sheep. I. Effect of an audiovisual stimulus and insulin-induced hypoglycemia. Neuroendocrinology 49:367–381.

Eriksson F, Kavelaars A, Heijnen CJ (2001) Preproenkephalin: An Unappreciated Neuroimmune Communicator. In: Psychoneuroimmunology (Ader R, Felten DL, Cohen N, eds), pp 391–403. San Diego: Academic Press.

Finley JCW, Maderdrut JL, Petrusz P (1981) The immunocytochemical localization of enkephalin in the central nervous system of the rat. J Comp Neurol 198:541–565.

Fleminger G, Ezra E, Kilpatrick DL, Udenfriend S (1983) Processing of enkephalin-containing peptides in isolated bovine adrenal chromaffin granules. Proc Natl Acad Sci USA 80:6418–6421.

Frank MG, Wieseler Frank JL, Hendricks SE, Burke WJ, Johnson DR (2002) Age at onset of major depressive disorder predicts reductions in NK cell number and activity. J Affect Disord 71:159–167.

Gaughran F (2002) Immunity and schizophrenia: Autoimmunity, cytokines, and immune responses. Int Rev Neurobiol 52:275–302.

Glantz LA, Lewis DA (2001) Dendritic spine density in schizophrenia and depression. Arch Gen Psychiatry 58:203.

Gubler U, Seeburg P, Hoffman BJ, Gage LP, Udenfriend S (1982) Molecular cloning establishes proenkephalin as precursor of enkephalin-containing peptides. Nature 295:206–208.

Gysling K, Forray MI, Haeger P, Daza C, Rojas R (2004) Corticotropin-releasing hormone and urocortin: Redundant or distinctive functions? Brain Res Brain Res Rev 47:116–125.

Hadden JW, Hadden EM, Middleton EJ (1970) Lymphocyte blast transformation. I. Demonstration of adrenergic receptors in human peripheral lymphocytes. Cell Immunol 1:583–595.

Halford WP, Gebhardt BM, Carr DJ (1995) Functional role and sequence analysis of a lymphocyte orphan opioid receptor. J Neuroimmunol 59:91–101.

Harlan RE, Shivers BD, Romano GJ, Howells RD, Pfaff DW (1987) Localization of preproenkephalin mRNA in the rat brain and spinal cord by in situ hybridization. J Comp Neurol 258:159–184.

Hazum E, Chang KJ, Cuatrecasas P (1979) Specific nonopiate receptors for beta-endorphin. Science 205:1033–1035.

Heese K, Hock C, Otten U (1998) Inflammatory signals induce neurotrophin expression in human microglial cells. J Neurochem 70:699–707.

Heim C, Nemeroff CB (1999) The impact of early adverse experiences on brain systems involved in the pathophysiology of anxiety and affective disorders. Biol Psychiatry 46:1509–1522.

Hein L, Kobilka BK (1997) Adrenergic receptors—from molecular structure to in vivo function. Trends Cardiovasc Med 7:137–145.

Heinrichs SC, Koob GF (2004) Corticotropin-releasing factor in brain: A role in activation, arousal, and affect regulation. J Pharmacol Exp Ther 311:427–440.

Herbert TB, Cohen S (1993) Depression and immunity: A meta-analytic review. Psychol Bull 113:472–486.

Herman JP, Figueiredo H, Mueller NK, Ulrich-Lai Y, Ostrander MM, Choi DC, Cullinan WE (2003) Central mechanisms of stress integration: Hierarchical circuitry controlling hypothalamo-pituitary-adrenocortical responsiveness. Front Neuroendocrinol 24:151–180.

Hestad KA, Tonseth S, Stoen CD, Ueland T, Aukrust P (2003) Raised plasma levels of tumor necrosis factor alpha in patients with depression: Normalization during electroconvulsive therapy. J Ect 19:183–188.

Hiddinga HJ, Isaak DD, Lewis RV (1994b) Enkephalin-containing peptides processed from proenkephalin significantly enhance the antibody-forming cell responses to antigens. J Immunol 152:3748–3759.

Hiddinga HJ, Isaak DD, Lewis RV (1994a) Enkephalin-containing peptides processed from proenkephalin significantly enhance the antibody-forming cell responses to antigens. J Immunol 152:3748–3759.

Hom JS, Goldberg I, Mathis J, Pan YX, Brooks AI, Ryan-Moro J, Scheinberg DA, Pasternak GW (1999) [^{125}I]orphanin FQ/nociceptin binding in Raji cells. Synapse 34:187–191.

Hsu SY, Hsueh AJ (2001) Human stresscopin and stresscopin-related peptide are selective ligands for the type 2 corticotropin-releasing hormone receptor. Nat Med 7:605–611.

Hughes J, Smith TW, Kosterlitz HW, Fothergill LA, Morgan BA, Morris HR (1975) Identification of two related pentapeptides from the brain with potent opiate agonist activity. Nature 258:577–579.

Irwin M (1999) Immune correlates of depression. Adv Exp Med Biol 461:1–24.

Irwin M (2001) Depression and Immunity. In: Psychoneuroimmunology (Ader R, Felten DL, Cohen N, eds), pp 383–398. San Diego: Academic Press.

Jessop DS, Major GN, Coventry TL, Kaye SJ, Fulford AJ, Harbuz MS, De Bree FM (2000) Novel opioid peptides endomorphin-1 and endomorphin-2 are present in mammalian immune tissues. J Neuroimmunol 106:53–59.

Jessop DS, Richards LJ, Harbuz MS (2002) Opioid Peptides Endomorphin-1 and Endomorphin-2 in the Immune System in Humans and in a Rodent Model of Inflammation. Ann N Y Acad Sci 966:456–463.

Jones AL, Mowry BJ, Pender MP, Greer JM (2005) Immune dysregulation and self-reactivity in schizophrenia: Do some cases of schizophrenia have an autoimmune basis? Immunol Cell Biol 83:9–17.

Karalis K, Sano H, Redwine J, Listwak S, Wilder RL, Chrousos GP (1991) Autocrine or paracrine inflammatory actions of corticotropin-releasing hormone in vivo. Science 254:421–423.

Kavelaars A, Heijnen CJ (2000) Expression of preproenkephalin mRNA and production and secretion of enkephalins by human thymocytes. Ann N Y Acad Sci 917:778–783.

Kipnis J, Cardon M, Strous RD, Schwartz M (2006) Loss of autoimmune T cells correlats with brain diseases: Possible implications for schizophrenia. Trends Mol Med 12(3):107–12.

Kosterlitz HW, Hughes J (1977) Opiate receptors and endogenous opioid peptides in tolerance and dependence. Adv Exp Med Biol 85:141–154.

Labuz D, Berger S, Mousa SA, Zollner C, Rittner HL, Shaqura MA, Segovia-Silvestre T, Przewlocka B, Stein C, Machelska H (2006) Peripheral antinociceptive effects of exogenous and immune cell-derived endomorphins in prolonged inflammatory pain. J Neurosci 26:4350–4358.

Landmann RMA, Bittiger H, Buhler FR (1981) High affinity beta-2-adrenergic receptors in mononuclear leucocytes: Similar density in young and old normal subjects. Life Sci 29:1761–1771.

Lanquillon S, Krieg JC, Bening-Abu-Shach U, Vedder H (2000) Cytokine production and treatment response in major depressive disorder. Neuropsychopharmacology 22:370–379.

Lapchak PA, Araujo DM, Hefti F (1993) Systemic interleukin-1 beta decreases brain-derived neurotrophic factor messenger RNA expression in the rat hippocampal formation. Neuroscience 53:297–301.

Larson SJ, Dunn AJ (2001) Behavioral effects of cytokines. Brain Behav Immun 15:371–387.

Laruelle M (2000) The role of endogenous sensitization in the pathophysiology of schizophrenia: Implications from recent brain imaging studies. Brain Res Brain Res Rev 31:371–384.

Laruelle M, Abi-Dargham A, van Dyck CH, Gil R, D'Souza CD, Erdos J, McCance E, Rosenblatt W, Fingado C, Zoghbi SS, Baldwin RM, Seibyl JP, Krystal JH, Charney DS, Innis RB (1996) Single photon emission computerized tomography imaging of amphetamine-induced dopamine release in drug-free schizophrenic subjects. Proc Natl Acad Sci USA 93:9235–9240.

Levite M (2000) Nerve-driven immunity. The direct effects of neurotransmitters on T-cell function. Ann N Y Acad Sci 917:307–321.

Lewis K, Li C, Perrin MH, Blount A, Kunitake K, Donaldson C, Vaughan J, Reyes TM, Gulyas J, Fischer W, Bilezikjian L, Rivier J, Sawchenko PE, Vale WW (2001) Identification of urocortin III, an additional member of the corticotropin-releasing factor (CRF) family with high affinity for the CRF2 receptor. Proc Natl Acad Sci USA 98:7570–7575.

Lieberman JA, Sheitman BB, Kinon BJ (1997) Neurochemical sensitization in the pathophysiology of schizophrenia: Deficits and dysfunction in neuronal regulation and plasticity. Neuropsychopharmacology 17:205–229.

Luger TA, Scholzen TE, Brzoska T, Bohm M (2003) New insights into the functions of alpha-MSH and related peptides in the immune system. Ann N Y Acad Sci 994:133–140.

Madden KS (2001) Catecholamines, sympathetic nerves and immunity. In: Psychoneuroimmunology (Ader R, Felten DL, Cohen N, eds), pp 197–216. San Diego: Academic Press.

Madden KS, Felten DL (1995) Experimental basis for neural-immune interactions. Physiol Rev 75:77–106.

Maes M (1999) Major depression and activation of the inflammatory response system. Adv Exp Med Biol 461:25–46.

Maes M, Bosmans E, De Jongh R, Kenis G, Vandoolaeghe E, Neels H (1997) Increased serum IL-6 and IL-1 receptor antagonist concentrations in major depression and treatment resistant depression. Cytokine 9:853–858.

Malek-Ahmadi P (1996) Neuropsychiatric aspects of cytokines research: An overview. Neurosci Biobehav Rev 20:359–365.

Manna SK, Sarkar A, Sreenivasan Y (2006) Alpha-melanocyte-stimulating hormone down-regulates CXC receptors through activation of neutrophil elastase. Eur J Immunol 36:754–769

Marotti T, Haberstok H, Sverko V, Hrsak I (1992) Met- and Leu-enkephalin modulate superoxide anion release from human polymorphonuclear cells. Ann N Y Acad Sci 650:146–153.

McGillis JP, Fernandez S, Knopf MA (2001) Regulation of immune and inflammatory reactions in local microenvironments by sensory neuropeptides. In: Psychoneuroimmunology (Ader R, Felten DL, Cohen N, eds), pp 217–229. San Diego: Academic Press.

Mousa SA, Machelska H, Schafer M, Stein C (2002) Immunohistochemical localization of endomorphin-1 and endomorphin-2 in immune cells and spinal cord in a model of inflammatory pain. J Neuroimmunol 126:5–15.

Musselman DL, Miller AH, Porter MR, Manatunga A, Gao F, Penna S, Pearce BD, Landry J, Glover S, McDaniel JS, Nemeroff CB (2001) Higher than normal plasma interleukin-6 concentrations in cancer patients with depression: Preliminary findings. Am J Psychiatry 158:1252–1257.

Nawa H, Takahashi M, Patterson PH (2000) Cytokine and growth factor involvement in schizophrenia—support for the developmental model. Mol Psychiatry 5:594–603.

Noda M, Furutani Y, Takahashi H, Toyosato M, Hirose T, Inayama S, Nakanishi S, Numa S (1982) Cloning and sequence analysis of cDNA for bovine adrenal preproenkephalin. Nature 295:202–206.

Oitzl MS, de Kloet ER (1992) Selective corticosteroid antagonists modulate specific aspects of spatial orientation learning. Behav Neurosci 106:62–71.

Oleson DR, Johnson DR (1988) Regulation of human natural cytotoxicity by enkephalins and selective opiate agonists. Brain Behav Immun 2:171–186.

Owen BM, Eccleston D, Ferrier IN, Young AH (2001) Raised levels of plasma interleukin-1beta in major and postviral depression. Acta Psychiatr Scand 103:226–228.

Owens MJ, Nemeroff CB (1991) Physiology and pharmacology of corticotropin-releasing factor. Pharmacol Rev 43:425–473.

Peluso J, LaForge KS, Matthes HW, Kreek MJ, Kieffer BL, Gaveriaux-Ruff C (1998) Distribution of nociceptin/orphanin FQ receptor transcript in human central nervous system and immune cells. J Neuroimmunol 81:184–192.

Plata-Salaman CR (1991) Immunoregulators in the nervous system. Neurosci Biobehav Rev 15:185–215.

Prolo P, Licinio J (1999) Cytokines in affective disorders and schizophrenia: New clinical and genetic findings. Mol Psychiatry 4:396.

Raison CL, Capuron L, Miller AH (2006) Cytokines sing the blues: Inflammation and the pathogenesis of depression. Trends Immunol 27:24–31.

Raison CL, Miller AH (2001) The neuroimmunology of stress and depression. Semin Clin Neuropsychiatry 6:277–294.

Rapaport MH, Muller N (2001) Immunological states associated with schizophrenia. In: Psychoneuroimmunology (Ader R, Felten DL, Cohen N, eds), pp 373–382. San Diego: Academic Press.

Read J, van Os J, Morrison AP, Ross CA (2005) Childhood trauma, psychosis and schizophrenia: A literature review with theoretical and clinical implications. Acta Psychiatr Scand 112:330–350.

Reul JM, Holsboer F (2002) Corticotropin-releasing factor receptors 1 and 2 in anxiety and depression. Curr Opin Pharmacol 2:23–33.

Reul JM, Stec I, Soder M, Holsboer F (1993) Chronic treatment of rats with the antidepressant amitriptyline attenuates the activity of the hypothalamic-pituitary-adrenocortical system. Endocrinology 133:312–320.

Reyes TM, Lewis K, Perrin MH, Kunitake KS, Vaughan J, Arias CA, Hogenesch JB, Gulyas J, Rivier J, Vale WW, Sawchenko PE (2001) Urocortin II: A member of the corticotropin-releasing factor (CRF) neuropeptide family that is selectively bound by type 2 CRF receptors. Proc Natl Acad Sci USA 98:2843–2848.

Roozendaal B, McGaugh JL (1997) Basolateral amygdala lesions block the memory-enhancing effect of glucocorticoid administration in the dorsal hippocampus of rats. Eur J Neurosci 9:76–83.

Roscetti G, Ausiello CM, Palma C, Gulla P, Roda LG (1988) Enkephalin activity on antigen-induced proliferation of human peripheral blood mononucleate cells. Int J Immunopharmacol 10:819–823.

Rosoklija G, Toomayan G, Ellis SP, Keilp J, Mann JJ, Latov N, Hays AP, Dwork AJ (2000) Structural abnormalities of subicular dendrites in subjects with schizophrenia and mood disorders: Preliminary findings. Arch Gen Psychiatry 57:349–356.

Ryan MC, Sharifi N, Condren R, Thakore JH (2004) Evidence of basal pituitary-adrenal overactivity in first episode, drug naive patients with schizophrenia. Psychoneuroendocrinology 29:1065–1070.

Sanders VM, Kasprowicz DJ, Kohm AP, Swanson MA (2001) Neurotransmitter receptors on lymphocytes and other lympohoid cells. In: Psychoneuroimmunology (Ader R, Felten DL, Cohen N, eds), pp 161–196. San Diego: Academic Press.

Schafer M, Mousa SA, Zhang Q, Carter L, Stein C (1996) Expression of corticotropin-releasing factor in inflamed tissue is required for intrinsic peripheral opioid analgesia. Proc Natl Acad Sci USA 93:6096–6100.

Serhan CN, Fierro IM, Chiang N, Pouliot M (2001) Cutting edge: Nociceptin stimulates neutrophil chemotaxis and recruitment: Inhibition by aspirin-triggered-15-epi-lipoxin A4. J Immunol 166:3650–3654.

Sharma RP, Pandey GN, Janicak PG, Peterson J, Comaty JE, Davis JM (1988) The effect of diagnosis and age on the DST: A metaanalytic approach. Biol Psychiatry 24:555–568.

Silverman MN, Pearce BD, Biron CA, Miller AH (2005) Immune modulation of the hypothalamic-pituitary-adrenal (HPA) axis during viral infection. Viral Immunol 18:41–78.

Stein C, Gramsch C, Herz A (1990a) Intrinsic mechanisms of antinociception in inflammation: Local opioid receptors and beta-endorphin. J Neurosci 10:1292–1298.

Stein C, Hassan AH, Przewlocki R, Gramsch C, Peter K, Herz A (1990b) Opioids from immunocytes interact with receptors on sensory nerves to inhibit nociception in inflammation. Proc Natl Acad Sci USA 87:5935–5939.

Stephanou A, Jessop DS, Knight RA, Lightman SL (1990) Corticotrophin-releasing factor-like immunoreactivity and mRNA in human leukocytes. Brain Behav Immun 4:67–73.

Sluzewska A, Rybakowski JK, Laciak M, Mackiewicz A, Sobieska M, Wiktorowicz K (1995) Interleukin-6 serum levels in depressed patients before and after treatment with fluoxetine. Ann N Y Acad Sci 762:474–476.

Sullivan PF, Neale MC, Kendler KS (2000) Genetic epidemiology of major depression: Review and meta-analysis. Am J Psychiatry 157:1552–1562.

Tamam L, Yerdelen D, Ozpoyraz N (2003) Psychosis associated with interferon alfa therapy for chronic hepatitis B. Ann Pharmacother 37:384–387.

Telio D, Sockalingam S, Stergiopoulos V (2006) Persistent psychosis after treatment with interferon alpha: A case report. J Clin Psychopharmacol 26:446–447.

Theoharides TC, Donelan JM, Papadopoulou N, Cao J, Kempuraj D, Conti P (2004) Mast cells as targets of corticotropin-releasing factor and related peptides. Trends Pharmacol Sci 25:563–568.

Thomas AJ, Davis S, Morris C, Jackson E, Harrison R, O'Brien JT (2005) Increase in interleukin-1beta in late-life depression. Am J Psychiatry 162:175–177.

Thome J, Knopf U (2003) Acute psychosis after injection of pegylated interferon alpha-2a. Eur Psychiatry 18:142–143.

Trask PC, Esper P, Riba M, Redman B (2000) Psychiatric side effects of interferon therapy: Prevalence, proposed mechanisms, and future directions. J Clin Oncol 18:2316–2326.

Tsigos C, Papanicolaou DA, Defensor R, Mitsiadis CS, Kyrou I, Chrousos GP (1997) Dose effects of recombinant human interleukin-6 on pituitary hormone secretion and energy expenditure. Neuroendocrinology 66:54–62.

Tuglu C, Kara SH, Caliyurt O, Vardar E, Abay E (2003) Increased serum tumor necrosis factor-alpha levels and treatment response in major depressive disorder. Psychopharmacology (Berl) 170:429–433.

Udenfriend S, Kilpatrick DL (1983) Biochemistry of the enkephalins and enkephalin-containing peptides. Arch Biochem Biophys 221:309–323.

Vaughan WT, Sullivan JC, Elmadjian F (1949) Immunity and schizophrenia. Psychosomatic Medicine 11:327–333.

Waddington JL (1993) Schizophrenia: Developmental neuroscience and pathobiology. Lancet 341:531–536.

Waddington JL, Lane A, Scully PJ, Larkin C, O'Callaghan E (1998) Neurodevelopmental and neuroprogressive processes in schizophrenia. Antithetical or complementary, over a lifetime trajectory of disease a? Psychiatr Clin North Am 21:123–149.

Waits PS, Purcell WM, Fulford AJ, McLeod JD (2004) Nociceptin/orphanin FQ modulates human T cell function in vitro. J Neuroimmunol 149:110–120.

Walker EF, Diforio D (1997) Schizophrenia: A neural diathesis-stress model. Psychol Rev 104:667–685.

Wamsley JK, Young WS, Kuhar MJ (1980) Immunohistochemical localization of enkephalin in rat forebrain. Brain Res 190:153–174.

Weinberger DR (1987) Implications of normal brain development for the pathogenesis of schizophrenia. Arch Gen Psychiatry 44:660–669.

Wybran J, Appelboom T, Famaey JP, Govaerts A (1979) Suggestive evidence for receptors for morphine and methionine-enkephalin on normal human blood T lymphocytes. J Immunol 123:1068–1070.

Yeragani VK (1990) The incidence of abnormal dexamethasone suppression in schizophrenia: A review and a meta-analytic comparison with the incidence in normal controls. Can J Psychiatry 35:128–132.

Zadina JE, Hackler L, Ge LJ, Kastin AJ (1997) A potent and selective endogenous agonist for the mu-opiate receptor. Nature 386:499–502.

Zobel AW, Nickel T, Sonntag A, Uhr M, Holsboer F, Ising M (2001) Cortisol response in the combined dexamethasone/CRH test as predictor of relapse in patients with remitted depression. A prospective study. J Psychiatr Res 35:83–94.

35
Major Depression, Bipolar Syndromes, and Schizophrenia

Frederick Petty, Sriram Ramaswamy, Prasad R. Padala, Jean D. Deupree, and David B. Bylund

Keywords Major depressive disorder; Biopolar disorder; Schizophrenia; Mood disorders; Biogenic amine hypothesis; Learned helplessness; Antidepressant drugs; Mood stabilizers; Dopamine hypothesis; Antipsychotic drugs

35.1. Introduction

Psychiatry is a medical subspecialty dealing with mental and behavioral disorders. Mental illness refers to diseases of the brain, because the mind is a function of the brain, and illness is a synonym for disease. Psychiatric illnesses are generally differentiated from neurological illnesses in that neurological disorders tend to have demonstrable anatomical or physiological abnormalities associated with them. Although the biological basis of psychiatric disorders is well established, the neurochemical and neuroanatomical abnormalities associated with these disorders tend to be subtle. Recent advances in brain imaging and molecular genetics provide optimism regarding our ability to understand psychiatric illnesses on a more biological and molecular basis. However, at this time, psychiatric diagnoses remain clinical and descriptive. That is to say, they are based on observation regarding symptoms, complications, and outcome. The basis of psychiatric diagnosis continues to be the psychiatric interview and mental status examination. At this time, there are no laboratory or imaging tests to diagnose mental illnesses, though laboratory tests may be helpful in diagnosing medical conditions with psychiatric symptoms, such as thyroid disease.

The major or most severe psychiatric disorders include depressive disorders, bipolar disorder (manic-depressive illness), and schizophrenia. Depression and bipolar disorder are classified as mood disorders, because the predominant feature of these conditions is an inappropriate or abnormal emotional state. Schizophrenia, on the other hand, is classified as a thought disorder, because the predominant symptoms involve disturbances in perception and thinking.

Mood disorders and schizophrenia are different from other psychiatric disorders in that they may be associated with psychotic features, including hallucinations and delusions. Also, mood disorders and schizophrenia tend to be more disabling and associated with greater degrees of personal and social impairment.

Other psychiatric disorders, which are not discussed in detail in this review, include anxiety disorders, personality disorders, somatoform disorders, eating disorders, and substance use disorders.

35.2. Major Depressive Disorder

35.2.1. Clinical Diagnosis and Description

Major Depressive Disorder is characterized by a persistent low mood and decreased interest and pleasure. Although feeling depressed as a reaction to unfortunate life events, such as the death of a loved one, is normal, persons with major depressive disorder have a low mood and decreased interest and pleasure which are out of proportion to the environmental stress. Further major depressive disorder persists for a prolonged period of time, long after a "normal" depressive reaction, such as bereavement would have resolved. Additionally, persons suffering depression have a cluster of physical and psychological symptoms, including sleep and appetite disturbance, low energy, and psychomotor retardation (i.e., observably slowed mental and physical processes). Psychological symptoms of depression include feeling worthless, helpless, hopeless and guilty, and difficulty thinking and concentrating. In severe cases, persons with depression may feel that life is not worth living, and develop suicidal thoughts upon which they may act. Further, for a diagnosis of major depressive disorder, these symptoms need to persist for at least two weeks, though usually they are present for much longer before coming to medical attention.

Although, technically, the pathognomonic symptoms of major depressive disorder are low mood and decreased interest or pleasure (anhedonia), from a clinical perspective, the associated symptoms are often those that cause a person

T. Ikezu and H.E. Gendelman (eds.), *Neuroimmune Pharmacology*.
© Springer 2008

to request medical examination. Typically, someone with depression requests help from a primary care provider for complaints of insomnia, fatigue, memory problems, vague aches and pains, and weight loss (Alarcon et al., 1998). The physical and laboratory examination is usually negative, and the patient may resist a psychiatric explanation for these symptoms. Fortunately, there is increased awareness, both in the general public, and in the medical profession, of depressive disorders as treatablemedical illnesses. Nevertheless, epidemiological studies suggest most patients with major depressive disorder do not receive any treatment, let alone adequate treatment (Shapiro et al., 1984).

35.2.2. Diagnostic Criteria

The diagnostic criteria for Major Depressive Disorder (American Psychiatric Association, 2000) are:

1. Sustained low or depressed mood
2. Decreased interest or pleasure
3. Sleep disturbance, decreased or increased
4. Appetite disturbance, weight loss or gain
5. Problems with memory and concentration
6. Feeling worthless or guilty
7. Psychomotor agitation or retardation
8. Fatigue or loss of energy
9. Thoughts of death, suicidal ideation

For the diagnosis of major depressive disorder, a person must have either symptom 1 or 2, plus five out of nine symptoms, and these symptoms must occur most of the day, nearly every day, for at least two weeks. Major depressive disorder is differentiated from other depressive disorders, including minor depression (dysthymia), recurrent brief depression, and adjustment disorder with depressed mood.

Most authorities who have studied major depressive disorder over the last century have concluded that major depressive disorder may be subdivided into two types. These two types have been referred to by variable nomenclature: "psychotic" vs. "neurotic"; "endogenous" vs. "reactive"; "with melancholic features" vs. "without melancholic features"; or "primary" vs. "secondary" (Robins and Guze, 1972). This classification refers in part to severity, and in part to distinct symptoms. There continues to be debate regarding the utility of this classification, though some data suggest that the more severe depressions should receive first line treatment with pharmacological interventions or electroconvulsive shock therapy. The less severe cases may respond as well to psychological treatments, such as cognitive-behavioral psychotherapy.

35.2.3. Epidemiology

The lifetime population prevalence for major depressive disorder in the United States is 5–12% for men, and 10–25% for women (Eaton et al., 1989; Blazer et al., 1998). Part of this discrepancy may result from the methods used in the surveys. Large-scale population surveys using highly structured interviews performed by lay interviewers tend to result in higher prevalence, whereas more focused studies using trained mental health professionals tend to result in lower estimates. Nevertheless, major depressive disorder is one of the most common medical conditions, with a prevalence comparable to diabetes and cardiovascular disease. The health care burden of major depressive disorder is comparable to that of cardiovascular diseases, and is estimated, on a worldwide basis to be among the top three in total expense.

Major depressive disorder has similar prevalence rates in various ethnic, racial, and cultural groups. Some studies have suggested variability of prevalence in some countries, but more transnational and transcultural research is needed.

Epidemiological studies are consistent in finding higher rates of major depression in women than in men, with the usual ratio about 2:1 (Kessler et al., 1993). Also, higher rates of major depressive disorder are found in single or divorced persons, compared to married persons; in urban settings, compared to rural settings, and in young rather than middle aged or elderly persons. The female gender prevalence in major depressive disorder is only found during the years of reproductive potential (Kessler et al., 1993). Prior to menarche, boys have higher rates of depression than girls, and after menopause, men have comparable rates to women. This suggests that hormonal or physiological factors associated with the female menstrual cycle may contribute susceptibility to developing major depressive disorder, though psychosocial factors cannot be altogether excluded.

The age at risk for major depressive disorder is virtually life long. The highest incidence (new cases) is during a person's twenties and thirties. However, the first episode of depression may occur in childhood or adolescence, and in old age. Interestingly, during the last 50–100 years, the age of onset has decreased, with higher rates of major depression seen in young persons, and an increase in childhood and adolescent depression (Klerman and Weissman, 1989). Several factors have been considered in attempting to interpret this historical trend. One factor might be increased sophistication in ascertainment of cases. Another factor is the "period effect" where psychosocial stressors during a particular period might place individuals at greater risk. Also considered is the "cohort effect" in which risk of illness varies between different generations. This effect may be sensitive to genetic influences. Finally, "age effects" are considered, in which risk of illness varies during specific times of the life cycle. Needless to say, these effects probably interact. Nevertheless, it is notable and remarkable that the age of onset of first episode of major depressive disorder was around 40–50 three generations ago, and is now 20–30.

35.2.4. Course of Illness

Major depressive disorder is a recurrent, episodic illness. Untreated episodes of major depressive disorder last, on the average for 6 to 18 months. In 10–20% of cases, however, the major depressive disorder becomes a chronic condition, which may persist for years. About 70% or more of persons who have one episode go on to develop another, and a pattern of recurrent illness, with numerous episodes, is seen in perhaps half of cases (Keller et al., 1992).

Complications of major depression are significant. Persons with major depressive disorder have higher rates of physical illness (Katon and Sullivan, 1990). They are also more likely to develop substance use disorders, particularly alcohol abuse and dependence (Sullivan et al., 2005). Persons with major depressive disorder experience considerable social, vocational, and family impairment, due, in part, to problems with memory, concentration, and poor judgment.

The worst complication of major depressive disorder, of course, is suicide. Persons with major depressive disorder have a lifetime risk of suicide of about 15% (Guze and Robins, 1970). This highlights the fact that major depressive disorder is a potentially lethal illness. Also, this represents one of the major sources of preventable death.

35.2.5. Etiology

Major depressive disorder is familial, with higher prevalence seen in first-degree family members of affected individuals than in the general population (Winokur and Pitts, 1965). This suggests that major depressive disorder may be genetic, and the most compelling evidence that it comes from studies of twins. In these studies, one twin has major depressive disorder, and rates of concordance for the illness are compared between monozygotic twins and dizygotic twins. Monozygotic twins share identical genes, while dizygotic twins are like siblings, sharing half of their genes. Both monozygotic and dizygotic twins are assumed to share similar environmental and familial influences. In major depressive disorder, monozygotic twins are about 40% concordant for the illness, while dizygotic twins are about 10% concordant. For dizygotic twins, the prevalence is comparable to that found in siblings (McGuffin et al., 1991). To date, the precise genetic cause of major depressive disorder remains elusive. It is, like diabetes, a complex genetic disorder, likely involving several genes, and does not follow simple Mendelian genetics.

In addition to genetics, other factors play a role in the etiology of major depressive disorder, or monozygotic twins would be 100% concordant for the illness. Identifying these factors presents a challenge, particularly because many are difficult to measure in a quantitative manner. Childhood abuse and neglect are considered risk factors, as is early parental death or separation (Pine and Cohen, 2002). Stressful life events are also considered to be risk factors (Paykel, 2003). The contribution of life stress to the development of depression is minor, however, because most people who experience even severe stressful life events do not develop major depression. Among the most stressful life events are the death of a spouse or child. Persons who suffer this experience develop a full-blown major depressive disorder (as opposed to grief reaction or bereavement) only about 10–30% of the time. Other stressful life events include catastrophic major medical conditions, such as a myocardial infarction or cancer. Again, a major depression in response to these events is seen in only about 25% of cases. Further, many people develop severe, disabling depressive disorders in the absence of any perceptible life stress.

Of stressful life events, those which are unpleasant or represent loss, and over which the person has no control, are more likely to lead to major depression, according to the learned helplessness model. In this model, inescapable and unpleasant stress leads to the belief that one's response cannot influence outcomes, hence "learned helplessness" (Seligman 1972). This is considered a depressive cognitive attribution, and changing these negative thought patterns is the basis for cognitive psychotherapy.

An interaction between genetic and environmental causes is considered most likely to be causal in many cases. Evidence for this is provided by the recent report that persons who inherit the short allele of the serotonin reuptake site are more likely to develop major depression after several severe stressful life events (Caspi et al., 2003), though this finding has not been widely replicated.

Other psychiatric illnesses, particularly the anxiety disorder, are also risk factors for development of major depressive disorder. Persons with anxiety disorders (panic disorder, obsessive compulsive disorder, social phobia, generalized anxiety disorder, and posttraumatic stress disorder) go on to develop major depressive disorder over the course of 5–20 years in over 50% of cases.

In summary, the etiology of major depressive disorder suggests a genetic component, environmental influences, and other, as yet undetermined factors.

35.2.6. Neurobiology

35.2.6.1. Biogenic Amine Hypothesis

This hypothesis was first formulate around 1965, and encompasses both a catecholamine hypothesis and a serotonin hypothesis. The essential features of the hypothesis involve the idea that decreased levels or activity of norepinephrine and/or serotonin is responsible for the development of the symptoms of depression, and that pharmacological treatments exert their therapeutic effects by restoring biogenic amine function (Schildkraut, 1967). This hypothesis originated from a clinical observation that patients treated with reserpine (the first effective specific antihypertensive agent) often-developed symptoms of major depressive disorder. Reserpine was found to deplete biogenic amines. Subsequently, the first pharmacological treatments for depressive disorders, namely monoamine oxidase inhibitors and tricyclic antidepressants, were found to increase biogenic amine levels in brain.

Most of the evidence supporting the biogenic amine hypothesis is indirect. Specifically, deficits of norepinephrine and/or serotonin have been difficult to demonstrate. Considerable research into urinary levels of methoxy-hydroxy-phenylethanolamine glycol (MHPG), a norepinephrine metabolite, in depressed patients yielded inconsistent results, probably because this metabolite is not a reflection of central noradrenergic activity as once thought (Eisenhofer et al., 2004). Levels of the serotonin metabolite, 5-hydroxyindole acetic acid (5-HIAA) were reported to be low in cerebrospinal fluid of patients with depression (Asberg et al., 1984). However, subsequent research suggests that this reflected impulsive aggressive behavior and suicidality more than the depressive syndrome.

Another test for the biogenic amine hypothesis would involve precursor loading strategies. In other words, compounds which could increase brain levels of norepinephrine and/or serotonin should demonstrate antidepressant efficacy. Results of these studies have been mixed. Several positive findings were reported with treating depression with the serotonin precursors tryptophan and 5-hydroxy tryptophan (5-HT) (Shaw et al., 2001), but these were not consistently replicated. Less positive results were reported with attempts to increase brain norepinephrine levels with precursors.

Possibly the best evidence suggesting involvement of norepinephrine and serotonin in major depressive disorder devolved from depletion studies (Delgado et al., 1990). In these studies, patients who have responded to treatment for depression are given procedures, which deplete brain levels of serotonin or norepinephrine. Serotonin levels are decreased by use of a low monoamine diet, followed by a drink which includes all the amino acids except the serotonin precursor tryptophan. Norepinephrine levels are depleted by administration of alpha-methylparatyrosine. In patients who had responded to treatment with a serotonergic antidepressant, depletion of serotonin caused a prompt and dramatic, but brief reoccurrence of the symptoms of major depression. In patients who had responded to treatment with a noradrenergic antidepressant, depletion of norepinephrine caused a relapse into depression. The converse was not true; in other words, serotonin depletion did not cause relapse in patients who responded to noradrenergic antidepressants, and vice versa.

In addition to the biogenic amines, the amino acid neurotransmitters are also implicated in the neurochemistry of major depressive disorder. Neurotransmitter γ-aminobutyric acid (GABA) levels are low in brain, cerebrospinal fluid, and blood of patients with major depressive disorder (Petty, 1995; Brambilla et al., 2003).

35.2.6.2. Neurotransmitter Receptor Alterations

Changes in adrenergic receptors in the brains of patients diagnosed with major depression have been documented and used to support the biogenic amine hypothesis, although the studies have not been entirely consistent The density of the alpha-2 adrenergic receptor is generally found to be increased, at least when an agonist radioligand is used in the binding assay (Ordway et al., 2003). Alpha-2 adrenergic receptors are also up-regulated in animal models of depression (Flügge et al., 2003).

One of the most consistent findings in preclinical studies of antidepressant treatments (including electroconvulsive shock) is the downregulation (as measured by radioligand binding) of beta-adrenergic receptors. This decrease in receptor density is likely secondary to the increased concentration of norepinephrine in the synaptic cleft. Other alterations in the G protein signaling cascade have been documented, including alterations in G protein and adenylyl cyclase (Donati and Rasenick, 2003; Millan, 2004). Another hypothesis, that decreased neurogenesis might be the cause of depression, is supported by the effects of stress on neurogenesis and the demonstration that neurogenesis seems to be necessary for antidepressant action (Thomas and Peterson, 2003; Henn and Vollmayr, 2004). However, no inclusive hypothesis concerning a mechanism of action of antidepressant therapies has been forth coming.

35.2.6.3. Neuroendocrine Findings

Perhaps the best established and most widely replicated finding in biological psychiatry is that patients with major depressive disorder have elevated levels of the stress hormone cortisol (Carroll et al., 1976). The original finding involved measuring blood levels of cortisol in depressed patients, and these were consistently higher than control over a 24 h day, even allowing for the circadian variation in cortisol levels. Subsequently, the dexamethasone suppression test was used to further test endocrinological abnormalities in patients with major depressive disorder (Carroll, 1985). In this test, the high potency synthetic steroid hormone dexamethasone is administered at 11 pm. In normal controls, levels of cortisol at 8 am are significantly reduced. In about half of patients with major depressive disorder, cortisol levels are not suppressed by dexamethasone. Non-suppression in the dexamethasone suppression test was not found to be specific for depressive disorders, with significant degrees of non-suppression found in other psychiatric syndromes. Further evidence for dysregulation of the hypothalamic-pituitary-adrenal axis in major depressive disorder was provided by studies of corticotrophic releasing hormone. Levels of this hormone, which regulates peripheral cortisol, were elevated in cerebrospinal fluid of patients with major depressive disorder. Whether the hypercortisolemia associated with major depression represents a risk factor or a consequence of illness remains unclear. Certainly, having major depressive disorder, with its attendant physical, psychological, and social stressors, is in and of itself a highly stressful experience. Abnormalities in cortisol function may be a consequence of the illness, rather than a risk factor, because dexamethasone suppression reverts to normal upon recovery from illness. Nevertheless, this line of research has resulted in attempts to study cortisol antagonists as antidepressant agents.

In addition to abnormalities in the cortisol system, abnormal function of growth hormone and of thyroid hormone and of their regulatory mechanisms has been reported. Again, these findings are not specific to depression (Schatzberg et al., 2002).

35.2.6.4. *Neurophysiological Correlates*

Patients with major depressive disorder are reported to have decreased latency to first episode of rapid eye movement (REM) sleep (Kupfer and Foster, 1972). This finding may relate to the hypothesized dysregulation in circadian rhythms found in mood disorders. This interest in circadian rhythms has lead to development of some specific treatments for depression, including sleep deprivation and light therapy.

35.2.6.5. *Brain Imaging*

Numerous studies of brain function in major depression have been undertaken. Studies using positron emission tomography generally show decreased metabolism in the frontal lobes of depressed individuals, more so on the left (Bench et al., 1992). This abnormality is not specific to depression, being observed in other psychiatric conditions such as bipolar disorder and obsessive compulsive disorder. The abnormalities reverse with successful treatment.

35.2.7. Animal Models

Several animal models of depressive disorders have been studied, and space does not permit detailed review of these findings. Animal models of depression have fallen into three general categories: Stress induced, "ethologically relevant," and genetic (Machado-Vieira et al., 2004). Of the stress-induced models, learned helplessness, chronic mild stress, and forced swim (behavioral despair) are best characterized (Willner and Mitchell, 2002).

In the learned helpless model, animals (usually rats) are subjected to a brief (1–2 h) inescapable shock. Subsequently, they are tested in a task in which they can terminate the shock by an operant response. Animals with prior inescapable shock exposure do not perform as well in the test (Maier and Watkins, 2005). Advantages of the learned helplessness model include its use in studies of neurochemical changes, and that it responds to repeated, rather than acute, antidepressant drug administration. Disadvantages of the model include its dependence on acute stress administration, suggesting it may better model posttraumatic stress disorder than major depressive disorder.

The chronic mild stress model involves subjecting rats or mice to unpredictable"mild" stressors over the course of 2–4 weeks. These stressors include soiled bedding, changing light-dark cycle, food deprivation, noise, strange cage mates, and tilted cages. Consumption of a sweet solution is generally decreased in these stressed animals, and this is returned to normal consumption by chronic administration of antidepressant drugs. This decreased response to rewards is thought to model anhedonia or loss of pleasure in depressive disorders. Advantages of the model include its relationship to chronic stress, which may be more relevant to major depressive disorder than acute stress. Disadvantages of the model include its being very labor and time intensive, in addition to difficulties in replication (Willner, 2005).

The forced swim test involves subjecting animals (rats or mice) to a brief (15 min) swim in a restricted space, followed by a test the next day in a similar water tank. Animals will characteristically assume a passive, immobile posture on the second day of the procedure. This is interpreted as representing "behavioral despair," thought to be an analogue of depressive behavior. Administration of antidepressant drugs between the first and second swim leads to decreased immobility in the test (Cryan et al., 2005a). A variant of this model is the tail suspension test in mice (Cryan et al., 2005b). Advantages of the model include its wide spread replicability, ease of use, and ability to predict antidepressant drug efficacy. Disadvantages include the fact that the model responds to acute antidepressant drug treatment, unlike human depression.

The ethologically relevant models include the resident intruder and social hierarchy paradigms (Mitchell and Redfern, 2005). These models are relatively recent, and have the relative advantage of relying on more naturalistic stressor, compared to shock or swim. Also, they respond to chronic, but not acute antidepressant drug administration.

Genetic models hold great promise, particularly in light of the evidence that major depressive disorder has a prominent genetic component. The best studied genetic model is the Flinders Sensitive Line, which was originally bred for increased responsiveness to cholinergic agonists (Overstreet et al., 2005). Several lines of neurochemical and pharmacological evidence support the validity of this model. Another genetic model is congenital learned helplessness, in which rats are bred for susceptibility to shock induced helplessness (King et al., 1993; Henn and Vollmayr, 2005). Interestingly, after several generations, rats no longer require inescapable shock stress to exhibit helpless behavior. More research is needed on the pharmacology of this model.

Other models include the olfactory bulbectomy model (Song and Leonard, 2005). Lesions of the olfactory bulb cause behavioral changes, interpreted to result from disturbed function of the limbic system. These behaviors are reversed by chronic antidepressant administration.

In summary, many animal of models depressive disorders have been developed, each with relative advantages and disadvantages (Nestler et al., 2002). The validity of these models for human depressive disorders continues to be the subject of debate. Probably, this reflects the lack of comprehensive data on the molecular pathophysiology, genetic etiology, and relation to stress in human major depressive disorder.

35.2.8. Treatment

The treatment of major depressive disorder includes both pharmacological and psychological interventions. As to psychological treatments, cognitive and behavioral techniques have demonstrated positive results. Patients who respond to psychological intervention are usually in the range of mild to moderate symptom severity.

Pharmacological treatment for depression involves use of antidepressant drugs. Drugs are available that act by a varity of mechanisms to increase biogenic amines in the nerve endings including: inhibiting monoamine oxidase, an enzyme involved in the metabolism of biogenic amines; and inhibiting the norepinephrine transporter (NET) and/or serotonin transporter (SERT) responsible for the reuptake of biogenic amines into the synapse (Table 35.1). The net effect is to increase levels of the biogenic amines in the synaptic cleft (Table 35.1). In addition to blocking the transporters the action of some drugs is thought to be due to a direct action on biogenic amine receptors. Most of these agents are selective for either the norepinephrine or the serotonin transporters thus resulting in the buildup of norepinephrine and/or serotonin in the synaptic cleft. Some have activity at the dopamine transporter (DAT) (Tables 35.1 and 35.2).

However, these neurochemical effects are acute. That is to say, biogenic amine reuptake is blocked within minutes of drug administration. The therapeutic effects, on the other hand, in ameliorating or reversing the symptoms of major depression, require repeated antidepressant drug administration. Though some improvement in symptoms is often noted during the first week or two of treatment, the full effects of the medication require 4–8 weeks to become manifested. This delayed onset of therapeutic drug action is generally thought to reflect a change in biogenic amine receptor density or sensitivity which requires time to develop, such as down regulation of the beta-adrenergic receptor. However, it remains a clinical challenge and experimental puzzle.

Interestingly, in the learned helplessness animal model of depression, reversal of helpless behavior may be obtained acutely by direct intracerebral drug administration into frontal cortex, though reversal of helpless behavior by systemic drug administration requires several days (Sherman and Petty, 1980). This suggests that the reversal of depressive symptoms by antidepressant drugs may involve pharmacokinetic as well as pharmacodynamic effects (Petty et al., 1982).

The antidepressant drugs are hardly panaceas. Only about 50–60% of patients experience a response to the medications, where "response" is arbitrarily defined as 50% or greater reduction in symptoms (Hirschfeld, 1999). This compares with a 30–40% response to placebo. Rates for remission, with complete recovery and absence of symptoms, are even lower, in the 20–30% range. No one antidepressant is more

TABLE 35.1. Selective antidepressants classified according to their drug class.

Drug class	Drugs	Amine effects
Norepinephrine/Serotonin reuptake inhibitors		
Tertiary amine tricyclics	Amitriptyline	NE, 5-HT
	Clomipramine	NE, 5-HT
	Doxepin	NE, 5-HT
	Imipramine	NE, 5-HT
	Trimipramine	NE, 5-HT
Secondary amine tricyclics	Amoxapine	NE, DA
	Desipramine	NE
	Maprotiline	NE
	Nortriptyline	NE
	Protriptyline	NE
Selective norepinephrine reuptake inhibitor (SNRI)	Reboxetine	NE
Selective serotonin reuptake inhibitors (SSRI)	Citalopram	5-HT
	Fluoxetine	5-HT
	Fluvoxamine	5-HT
	Paroxetine	5-HT
	Sertraline	5-HT
	Venlafaxine	5-HT, NE
Atypical antidepressants	Atomoxetine	NE
	Bupropion	DA, NE?
	Duloxetine	NE, 5-HT
	Mirtazapine	5-HT, NE
	Nefazodone	5-HT
	Trazodone	5-HT
Monamine oxidase (MAO) inhibitors	Phenelzine	NE, 5-HT, DA
	Tranylcypromine	NE, 5-HT, DA
	Selegiline	NE, 5-HT?, DA?
	Moclobemide	NE, 5-HT, DA

Source: Adapted from Table17-1, Baldessarini (2006).

TABLE 35.2. Selectivity of selective antidepressants for NET. The antidepressants are listed according to the selectivity for the NET over SERT or DAT.

Drug	NET[a]	SERT[a]	DAT[a]
NET-selective drugs			
Oxaprotiline	+++	−	−
Maprotiline	++	−	−
Mianserin	++	−	−
Desipramine	+++	+	−
Atomoxetine	+++	++	−
Reboxetine	+++	++	−
Nortriptyline	+++	++	−
Amoxapine	++	++	−
Doxepin	++	++	−
SERT-selective drugs			
Bupropion	−	−	+
Amitriptyline	++	+++	−
Milnacipran	++	+++	−
Imipramine	++	+++	−
Trazodone	−	+	−
Venlafaxine	+	+++	−
Clomipramine	++	++++	−
Fluoxetine	+	++++	−
Paroxetine	++	++++	+
Fluvoxamine	−	+++	−
Sertraline	+	++++	++
S-Citalopram	−	++++	−

[a]The symbols represent the relative affinities (Frazer 1997; Owens et al., 1997; Tatsumi et al., 1997; and Leonard and Richelson, 2000) of the drugs for the NET, SERT, and DAT with (++++) representing drugs with Ki values <1 nM, (+++) Ki from 1 to 10 nM, (++) Ki from 10 to 100 nM, (+) for Ki from 100 to 1000 nM, and (−) for drugs with affinities >1000 nM.

effective than another, and it is not unusual for patients to receive a sequence of several medications, or combinations of medications, before they recover.

In addition to antidepressant drugs, some forms of psychological treatments have been shown effective for treatment of major depressive disorder. These include cognitive behavioral psychotherapy and interpersonal psychotherapy (Weissman, 1979). These therapies differ from traditional psychoanalytically oriented methods in that the therapist takes an active role, the patient is expected to do "homework," and the treatment is time limited, usually for about six months. Little data are available regarding whether the combination of medications and psychotherapy is more effective than either treatment alone, but data are suggestive of an additive effect.

Electroconvulsive therapy is usually reserved for severe or treatment resistant cases of depression, though the response rates to this treatment are higher than for drugs.

35.2.9. Immunological Correlates

This topic is covered in depth in another chapter. However, briefly, therapy with interferon has been shown to induce many symptoms of depression, including fatigue, depressed mood, psychomotor retardation, social withdrawal, and impaired memory and concentration. This may be due to dysregulation of the hypothalamic-pituitary-adrenal axis and stress response systems, and disturbance of serotonin function. Administration of proinflammatory compounds in animal models leads to "sickness" behavior, with correlates to depressive behavior. Long term administration of antidepressant drugs may attenuate this behavior, and inhibit the production of proinflammatory cytokines. The depressive syndrome associated with interferon therapy is responsive to serotonergic antidepressants (Schiepers et al., 2005).

35.3. Bipolar Disorder

35.3.1. Clinical Diagnosis and Description

Persons with bipolar disorder experience episodes of mania or hypomania. These episodes are characterized by hyperactivity, decreased need for sleep, increase in goal directed activity, excessive and rapid speech, starting many new projects, increased engagement in high-risk behavior, and a euphoric or irritablemood. Lack of insight is a hallmark of mania. A person in the manic phase of the illness may impulsively go on spending sprees, gamble, engage in promiscuous high-risk sexual behavior, become violent, and develop bizarre behavior. Although family and friends will recognize these behaviors as maladaptive, if not dangerous, the manic person will be oblivious to the consequences of his or her actions, and often justify them with rationales that may seem plausible. Persons who are experiencing episodes of mania often demonstrate psychotic thoughts; such as feeling they have supernatural powers, or feeling paranoia.

Additionally, persons with bipolar disorder may experience episodes of depression similar to major depressive disorders. Bipolar disorder was previously referred to as "manic-depressive" illness, reflecting the existence of both manic and depressive episodes. Also, some patients with bipolar disorder have mixed episodes, with both manic and depressive symptoms concurrent.

Though most persons who do not have bipolar disorder may consider the manic episodes to be beneficial and a positive experience, given the increased energy and euphoric mood, many patients with bipolar disorder find manic episodes to be very unpleasant, characterized by high levels of irritability and anxiety. Bipolar disorder, like major depressive disorder, is an episodic, recurrent convvdition, and in between episodes, the bipolar person may have function and behavior that appear completely normal.

35.3.2. Diagnostic Criteria

The diagnostic criteria for the depressive episodes in bipolar disorder are similar to those for major depressive disorder. The diagnostic criteria for a manic episode (American Psychiatric Association, 2000) are:

1. A distinct period of elevated, expansive, or irritablemood
2. Grandiosity or increased self-esteem
3. Decreased need for sleep
4. Pressured speech, more talkative than usual
5. Racing thoughts, or flight of ideas (jumping from one thought to another)
6. Distractibility, attention easily drawn to irrelevant stimuli
7. Increased goal directed activity, psychomotor agitation
8. Excessive involvement in pleasurable activities that have high potential for unpleasant consequences (spending sprees, sexual indiscretion)

For a diagnosis of mania, a person must have symptom 1, as well as three or more of symptoms 2–8. If the mood in symptom 1 is only irritable, then four or more of the other symptoms are required. Further, the symptoms must persist for at least a week or lead to hospitalization for the diagnosis of mania to be made. Additionally, the symptoms must cause marked impairment in occupational or social function. Also, the symptoms cannot be due to a medical condition or to a substance use disorder. Manic episodes share many features with amphetamine use disorder.

About half of patients with bipolar disorder experience what are referred to as hypomanic, rather than manic, episodes. Hypomanic episodes are differentiated from manic episodes by being briefer, in that only four days of symptoms are required, and by not having psychotic symptoms, nor marked impairment in functioning, nor requiring hospitalization.

Persons who experience manic episodes are classified as having Bipolar I Disorder, while those who experience only hypomanic episodes are classified as having Bipolar II Disorder.

Technically, having one or more depressive episodes is not required for a diagnosis of bipolar disorder. In other words, a person with bipolar disorder may only experience manic or hypomanic episodes and be classified as bipolar. Practically speaking, virtually all bipolar patients experience depressive episodes at some point during the course of illness.

35.3.3. Epidemiology

The lifetime population prevalence of Bipolar I disorder in the United States is about 1% (Kessler et al., 1994). There is no gender prevalence, with women as likely as men to develop the illness. The prevalence of bipolar I disorder does not appear to be influenced by race, ethnicity, or geography. Similar prevalence rates are reported in most countries, although large scale cross national and cross-cultural studies are needed. The prevalence of bipolar II disorder is less well understood, but is estimated to be about 3–5% (Berk and Dodd, 2005). Needless to say, patients with bipolar II disorder may be less likely to come to medical attention.

The usual age at onset of bipolar disorder is from the late teens into the twenties. The first onset of a manic episode rarely occurs after age 40. In fact, if a person experiences what appears to be the first manic episode after age 50, a medical or substance induced cause is likely (Burke and Wengel, 2003).

35.3.4. Course of Illness

Bipolar disorder is an episodic illness, with an individually variable course. There is no characteristic course that defines most patients. Some individuals may experience few manic episodes, but suffer recurrent depression. Others may have frequent, severe manic episodes, with brief and infrequent depressive episodes. Not infrequently, recurrent depression precedes the development of mania or hypomania. The average length of a manic episode, untreated, is about 3 months. Depressive episodes in bipolar disorder are usually of 3–6 month duration.

Between 10% and 15% of persons with bipolar disorder develop rapid cycling, defined as having four or more episodes per year (Akiskal, 2000). Often patients with rapid cycling will demonstrate marked seasonality with their episodes, such as developing mania in the Spring and Fall, and depression in the Summer and Winter.

There has been some speculation that bipolar disorder worsens with age, with episodes becoming more frequent and severe in middle age. However, there are very few data that support this theory.

Some evidence supports the idea that a manic or hypomanic episode may be induced in vulnerable patients by antidepressant medications (Goldberg and Truman, 2003). This phenomenon, known as the "switch" effect, has been clinically observed in many cases. Some data suggest that treatment with antidepressants, particularly those with dual action at both norepinephrine and serotonin, may precipitate a manic episode in bipolar patients (Stoner et al., 1999; Shulman et al., 2001; Yuksel et al., 2004). Though widely accepted from a clinical perspective, this phenomenon is still somewhat controversial from a research perspective.

35.3.5. Etiology

Bipolar disorder is largely genetic in etiology. Concordance in monozygotic twins is about 70%, which is probably the highest of all psychiatric disorders. Interestingly, the most prevalent psychiatric disorder in first-degree relatives of persons with bipolar disorder is major depressive disorder, suggesting a possible overlapping genetic vulnerability for both mood disorders (Kelsoe, 2000).

The precise mechanism for mode of transmission of bipolar disorder has been the focus of extensive research. To date, the results are not conclusive. Some reports of linkage to chromosome 11 were initially promising, but not replicated. Other reports of linkage to the X chromosome had initial positive findings but difficulty in replication. The strongest data are for linkage to chromosome 18, with several replications of the initial finding (Kelsoe, 2000). A major challenge in genetic studies in bipolar disorder in particular, and of psychiatric illnesses in general, revolves around the heterogeneity of the clinical conditions. These disorders are clinically described, and the diagnosis often involves subjective symptoms and interpretation of behavior. At this time, there are no objective laboratory findings for linkage to genetic studies. "Bipolar disorder" may well include a number of different and distinct biological subtypes. For example, persons who have recurrent brief depression, with occasional mania or hypomania, and a strong family history of major depressive disorder, probably have a different illness than those with recurrent, severe mania and a strong family history of bipolar disorder, and infrequent depression.

Nevertheless, this is an area of intensive research, with the potential for improved diagnosis and treatment. The best estimates are that several genes are involved (5–15). The environmental influence in bipolar disorder is less than that for major depressive disorder. Some risk factors have been identified, including negative stressful life events.

35.3.6. Neurobiology

The neurobiology of bipolar disorder is less well studied than that of major depressive disorder. Early studies of monoamine metabolites found that they were increased in patients with bipolar disorder, but these were subsequently interpreted as due to the increased motor activity seen in these patients.

The "kindling" model of bipolar disorder has attracted considerable interest. This model originally developed after the clinical observation that anticonvulsant drugs were effective for the treatment of bipolar disorder, particularly in the manic phase of the illness. Rodent research on epileptic seizures demonstrated that decreasing electrical stimuli were needed over time to generate seizures, hence the term "kindling." Some clinical evidence suggests, that in some patients with bipolar disorder, the illness progresses, such that in the earlier phases it is more likely to occur in response to stress, while in the later stages, episodes are autonomous and independent of stress. Hence the hypothesis was developed that the reoccurrence of episodes in bipolar disorder might represent a "kindling" phenomenon (Post et al., 1992). The kindling hypothesis has stimulated interest (Ghaemi et al., 1999) but has yet to lead to breakthroughs in understanding the neurobiology of bipolar disorder.

More recent research has focused on examining underlying commonalities in the biochemical actions of the mood stabilizers used to treat bipolar disorder (Zhou et al., 2005), in studies of postmortem brain tissue (Post et al., 2003), and in signal transduction pathways and regulation of gene expression (Bezchlibnyk and Young, 2002). For example, altered levels or function of G-protein alpha subunits and protein kinase A and C have been found in bipolar patients, as well as disruption in second messenger cascades such as the ERK/MAPK pathways.

35.3.7. Animal Models

Developing an animal model of bipolar disorder is challenging, due to the dramatically different clinical presentations of mania and depression. Animal models of depression are described above, and can be considered to model the depressive phase of bipolar disorder. Animal models of mania have included techniques to increase motor activity, such as administration of stimulants, increasing dopamine activity, sleep deprivation, and brain lesions (Machado-Vieira et al., 2004). These models are responsive to medications used to treat bipolar disorder to varying extents.

Recently a hyperactivity model induced by a combination of D-amphetamine and chlordiazepoxide was studied. Lamotrigine, valproate, and carbamazepine, all used to treat bipolar disorder, were all found to decrease this hyperactivity (Arban et al., 2005). Interestingly, while most mania models assume an increase in dopamine, an early model used dopamine depletion with intracerebroventricular injection of 6-hydroxydopamine, which induces hyper-reactivity to environmental stimuli, but not hyperactivity per se (Petty and Sherman, 1981). This model was responsive to chronic lithium and to chronic electroconvulsive shock, and also to acute chlorpromazine, while imipramine worsened the behavior.

Development of an animal model for bipolar disorder probably awaits a better understanding of the genetics of this illness. Development of such a model is further complicated by the fact that clinically some patients respond to lithium, but not valproate, and vice versa.

35.3.8. Treatment

The treatment of bipolar disorder is complex, and depends on the particular phase of illness. The "mood stabilizers" form the foundation of treatment. These include lithium and the anticonvulsant drugs, valproate, and carbamazepine (Bowden, 1998; McElroy and Keck, 2000; Post, 2000). Recently, lamotrigine has been found effective in some patients (Post, 2000). The goal of treatment with mood stabilizers is reduction of frequency and severity of episodes of depression and mania.

When patients present in a manic episode, rapid remission of symptoms is required, particularly if the person is psychotic or experiencing severely disruptive behavior. In these cases, use of an antipsychotic medication is usual. These medications may include use of conventional or "first generation" antipsychotic medications, such as chlorpromazine or haloperidol (Table 35.3). Recently, the "second generation" or "atypical" antipsychotics have also shown efficacy in treatment of acute mania (American Psychiatric Association, 2000). The latter agents are preferred due to their lower likelihood of inducing neuro-

TABLE 35.3. Antipsychotic medications (examples).

Typical	Atypical
Chlorpromazine	Aripiprazole
Haloperidol	Clozapine
Perphenazine	Olanzapine
Thiothixene	Quetiapine
Trifluoperazine	Risperidone
	Ziprasidone

logical side effects. Atypical antipsychotics include clozapine, risperidone, olanzapine, ziprasidone, quetiapine, and aripiprazole, all of which have demonstrated antimanic efficacy.

When bipolar patients present in a depressive episode, initial treatment with a mood stabilizer is recommended (Post, 2000). If a depressed bipolar does not respond to treatment with a mood stabilizer, an antidepressant is prescribed. Most patients with bipolar disorder end up on multiple medications. Electroconvulsive therapy is an effective treatment for bipolar disorder in both the manic and depressed phases of the illness.

35.3.9. Immunological Correlates

Relatively little research has examined the immunological factors in bipolar disorder. Plasma levels of IFN-gamma/TGF-beta 1 and IL-4/TGF-beta 1 were reported to be higher in symptomatic manic patients than controls in one study (Kim et al., 2004). More research is needed to elucidate this relationship.

35.4. Schizophrenia

35.4.1. Clinical Diagnosis and Description

Schizophrenia is a chronic and disabling mental illness, characterized by the presence of psychotic symptoms, including hallucinations and delusions. Hallucinations are sensory misperceptions, in which a person will experience an auditory, visual, or other sensory experience in the absence of an observable stimulus. Thus, persons experiencing a hallucination will, for example, hear voices speaking when there are no persons near by, or see things that other persons do not see. Persons with schizophrenia usually have auditory hallucinations, but may also have hallucinations in other sensory modalities.

Delusions are defined as fixed false beliefs. For example, persons who experience paranoid delusions may believe that they are under surveillance. In some cases this may assume bizarre proportions, such as the belief that aliens from outer space are not only observing, but also controlling the person's actions. Delusions are impervious to a rational explanation. Persons who have delusions will continue in their belief regardless of plausible evidence that their beliefs are wrong.

Persons with schizophrenia also frequently display disorders in the form of their thinking. This may be manifest by difficulty maintaining a systematic train of thought in conversation, and is referred to as "loose associations." Persons with schizophrenia may respond to a question with an answer that is not logical, or change subjects during a conversation in manners that are difficult to follow.

In addition to having hallucinations and/or delusions ("positive symptoms") persons with schizophrenia also have prominent "negative symptoms." These include amotivation, alogia (less speech), and affective flattening (poor emotional response to stimuli).

Generally, schizophrenia leads to severe occupational and social maladjustment. Persons with schizophrenia exhibit "downward drift" in that they have difficulty establishing and maintaining normal social, educational, and occupational activities, considering their background and expectations. Many, if not most, persons suffering from schizophrenia are unable to maintain gainful employment and social interactions.

35.4.2. Diagnostic Criteria

For a diagnosis of schizophrenia (American Psychiatric Association, 2000), a person must have at least two of the following:

1. Delusions
2. Hallucinations
3. Disorganized speech
4. Disorganized behavior
5. Negative symptoms

If the delusions are bizarre, or the hallucinations consist of voices keeping a running commentary on the person's behavior or thought, only one symptom is required. Further, the symptoms have to persist for at least six months, and be accompanied by significant impairment in social or vocational impairment, or impairment in self care. Also, the symptoms must not be due to drug abuse.

Traditionally, schizophrenia has been subclassified into several categories, including paranoid, catatonic, hebephrenic, and undifferentiated. Paranoid schizophrenia is characterized by predominately delusions and hallucinations, with relatively preserved interpersonal abilities. Catatonic schizophrenia is characterized by predominant motor features, while hebephrenic schizophrenia is characterized by a shallow and silly display of emotions. Undifferentiated schizophrenia has predominately negative symptoms. For reasons that are not clear, the prevalence of catatonic and hebephrenic schizophrenia has decreased in the last few decades.

35.4.3. Epidemiology

The lifetime population prevalence of schizophrenia is about 1% (Gottesman, 1989). In a landmark transnational study, similar prevalence was found for schizophrenia in Africa, Asia, Europe, and the Americas (Jablensky and Sartorius, 1988). There is no influence of gender in schizophrenia, with women as likely to develop this condition as men. Males may have an earlier onset of illness, and a more severe course of illness. Some evidence suggests that the prevalence of schizophrenia may be decreasing over time, with fewer new cases of the illness reported. The age at onset of schizophrenia is late teens to early twenties. As in bipolar disorder, the onset of first symptoms after age 40 is rare.

35.4.4. Course of Illness

Schizophrenia is a chronic illness. Though clinical improvement in symptoms may occur, particularly with treatment, remission of the illness is rare. Prior to developing overt symptoms of the illness, such as hallucinations and delusions, many persons with schizophrenia have prodromal symptoms. These include social isolation, idiosyncratic thinking, and feeling suspicious. Once established, schizophrenia is characterized by marked incapacity. Many persons with this condition are unemployable, homeless, and otherwise live on the fringes of society. Not uncommonly, the positive symptoms may ameliorate over the course of time, leaving negative residual symptoms, such as amotivation, to predominate.

Persons who suffer schizophrenia have intellectual impairment, having, on the average, an IQ that is one standard deviation below normal. Complications of schizophrenia include drug and alcohol abuse, depression, and suicide.

35.4.5. Etiology

Schizophrenia is in part a genetic illness. Several studies have consistently shown that monozygotic twins are three to six times more likely to be concordant for the illness than dizygotic twins. Further, there is a 5–10% prevalence of schizophrenia in first degree relatives of schizophrenics. Genetic studies have suggested linkage to chromosomes 22, 6, and 8 (McGuffin et al., 1995; Kendler and Diehl, 1993).

Other causes have also been suggested. There is an increase of schizophrenia in persons born during the winter months (Norquist and Narrow, 2000). Also, increased incidence of schizophrenia is noted in offspring of mothers who suffered influenza during the second trimester of pregnancy (Mednick et al., 1988). A viral hypothesis has been proposed, with some indirect evidence (Torrey and Peterson, 1976; Crow, 1984). Finally, schizophrenia is associated with birth complications, such as transient perinatal hypoxia (Davies et al., 1998).

35.4.6. Neurobiology

The most widely replicated finding in schizophrenia is enlarged ventricles, and reduced brain volume (Weinberger et al., 1979; Pakkenberg, 1987). Functional neuroimaging studies have reported reduced blood perfusion of the frontal lobes during activation with tasks that involve executive function. Some postmortem studies report histological abnormalities, such as disorientation of pyramidal cells in the hippocampus (Falkai and Bogerts, 1986), though these findings have been difficult to replicate.

The predominant neurochemical hypothesis involves the neurotransmitter dopamine. This hypothesis was based on the observation that all drugs that are effective in ameliorating psychotic symptoms have a common mechanism of action in blocking the dopamine D2 receptor. Furthermore, the clinical antipsychotic potencies of these drugs are directly related to their affinities for the dopamine D2 receptor. Thus, a relative excess in dopamine is postulated to account for psychotic symptoms in schizophrenia. Demonstrating excess dopamine in schizophrenia has proved difficult (Owen et al., 1984; Cookson et al., 1989). It is uncertain whether the total density of D2 receptors in schizophrenia is increased. However, recent evidence suggests that the functional state of D2 receptor (i.e., the state of the receptor having high-affinity for dopamine) may be elevated and this could help explain why many (if not most) individuals with schizophrenia are supersensitive to dopamine (Seeman et al., 2005).

A more recent neurochemical hypothesis involves glutamate. This was based on the finding that drugs which block the NMDA glutamate receptor, such as PCP, produce hallucinations, thought disorder, and cognitive deficits, as well as negative symptoms with great similarity to symptoms of schizophrenia (Wilkins, 1989; Meltzer, 1991). Again, demonstrating abnormal glutamate functioning in patients with schizophrenia has proved challenging. Serotonin (5-HT) is also implicated in schizophrenia, due to the mediation of the hallucinogenic effects of drugs such as LSD via 5-HT_{2A} receptors.

As well as hallucinations and delusions, schizophrenia is characterized by deficits in information processing. Schizophrenics have difficulty suppressing irrelevant environmental stimuli. This is referred to as sensory gating or prepulse inhibition, and commonly measured with auditory evoked potentials. When two stimuli, such as sounds, are presented within 30–500 msec, the response to the second stimulus rapidly decreases upon repeated presentations in healthy people, but not in persons with schizophrenia (Braff et al., 2001). This deficit is found in family members of patients with schizophrenia, and has been linked to the alpha-7 nicotinic receptor gene. The deficit is reversed by nicotine on an acute basis, and by some antipsychotic drugs. Alpha-7 nicotinic agonists are under development as treatment for schizophrenia (Martin et al., 2004).

35.4.7. Animal Models

Animal models of schizophrenia are intended to examine specific neurochemical or anatomical theories of the illness, and not to model the cluster of symptoms that characterizes the illness, since these are largely subjective. Models are divided into two types, pharmacological models and lesions models (Marcotte et al., 2001).

The pharmacological models are based on increasing activity of neurotransmitters considered to be involved in schizophrenia, such as dopamine, glutamate, and serotonin. Administration of dopamine agonists, 5-HT_{2A} agonists, and NMDA antagonists disrupts prepulse inhibition in a manner similar to that seen in schizophrenics. This effect is sensitive to some antipsychotic drugs (Geyer et al., 2001).

Lesion models are based on the idea that schizophrenia is, in part, a neurodegenerative or neurodevelopmental disorder. These lesions are generally neurotoxic, and have variable response to antipsychotic drugs (Marcotte et al., 2001). Recently, a novel model was studied in which pregnant mouse dams received a single exposure to a cytokine releasing agent. The offspring demonstrated schizophrenic like behaviors, as well as deficits in prepulse inhibition and enhanced sensitivity to amphetamine (Meyer et al., 2005).

Improved animal models are needed, particularly for the "negative symptoms" of schizophrenia, such as amotivation, and the cognitive symptoms (Geyer and Ellenbroek, 2003).

35.4.8. Treatment

The primary treatment for schizophrenia involves use of antipsychotic medications. These are classified as "typical" or first generation, and "atypical." The atypical antipsychotics differ from the typical in having relatively less extrapyramidal side effects, such as rigidity, dystonia (muscle spasm), akathisia (motor restlessness), and pseudo-Parkinsonian symptoms.

As discussed earlier, these agents are thought to produce their antipsychotic effect by blocking the dopamine D2 receptor in the CNS. These agents are not specific for the dopamine receptors and have various affinities, for example, for the muscarinic acetylcholine receptor, the alpha-1 adrenergic receptor and the H_1 histamine receptor. As a consequence they produce anticholinergic side effects of drug mouth, constipation and cycloplegia, and the adrenergic side effects of postural hypotension, and sedation; and the histaminergic effects of weight gain. The atypical antipsychotics have relatively greater affinity for serotonin receptors, and have demonstrated antidepressant effects in some cases.

These medications can exert a calming and tranquilizing effect with acute administration, and are used for these purposes in treating severely agitated patients. Their antipsychotic effects usually require treatment over the course of a few days or weeks. An exciting aspect of the atypical antipsychotics is their potential therapeutic efficacy for improving negative and cognitive symptoms of schizophrenia.

35.4.9. Immunological Correlates

Phospholipid abnormalities have been reported in schizophrenia, which may be due to altered immune function (du Bois et al., 2005). An intriguing concept is that in some cases schizophrenia my represent an autoimmune illness (Jones et al., 2005). Though the findings are not always consistent, dysregulation of immune system activity is noted in many studies in schizophrenia (Muller et al., 2000; Gaughran, 2002; Leweke et al., 2004).

Summary

Major depression, bipolar syndromes and schizophrenia are common and often severe mental illnesses. All three of these tend to have an onset in late adolescence or young adulthood. Major depression is characterized by persistent low mood and decreased interest and pleasure, as well as physical and psychological symptoms, including sleep disturbance, low energy, and memory problems. Lifetime prevalence for major depressive disorder is about 8% for men and 16% for women. The etiology of major depressive disorder is in part genetic, and in part related to adverse stressful life events. The neurobiology of depressive disorders is linked to alterations in biogenic amine in brain, particularly norepinephrine and serotonin. Additionally, neuroendocrine abnormalities, particularly elevated cortisol, are well documented. Several animal models are well described for depressive disorders, most involving stress. Treatment of major depression can involve psychological intervention, particularly cognitive behavioral therapy, as well as use of antidepressant drugs. Most antidepressant drugs have effects to increase the levels of biogenic amines, but need to be administered on a repeated basis for several weeks for full therapeutic benefit.

Bipolar disorder is characterized by episodes of mania or hypomania, which include hyperactivity, decreased need for sleep, and a euphoric or irritable mood. Additionally, persons with bipolar disorder may have episodes of depression similar to those seen in major depressive disorder. The lifetime prevalence of severe bipolar disorder is about 1% and 3–5% if milder cases are included, afflicting men and women equally. Both bipolar disorder and major depressive disorder tend to be episodic, and in the periods of time between episodes, persons may experience few or no symptoms. The etiology of bipolar disorder is predominately genetic, with a 70% concordance in monozygotic twins. The neurobiology of bipolar disorder is less well understood, and few animal models have been developed. Treatment of bipolar disorder usually involves "mood stabilizer" medications, including lithium, and the anticonvulsants valproate and carbamazepine. At times, antidepressant and antipsychotic medications are also used.

Schizophrenia is a chronic and disabling illness, characterized by psychotic or "positive" symptoms, including hallucinations and delusions. Also, persons with schizophrenia usually have "negative" symptoms, including amotivation, less speech, and poor emotional response to stimuli. Schizophrenia affects about 1% of the population, with an etiology that is, in part, genetic, and possibly related to intrauterine problems. Neurobiological correlates of schizophrenia include enlarged ventricles, and the predominant neurochemical theories involve dopamine and glutamate. As in bipolar disorder, animal models are not well developed. The primary treatment of schizophrenia is with antipsychotic medications, usually the "atypical agents," including clozapine, risperidone, and olanzapine.

Review Questions/Problems

1. The predominate feature of mood disorders is

a. inappropriate or abnormal emotional state
b. disturbance in thinking
c. disturbances in perception
d. hallucinations
e. delusions

2. For a diagnosis of a major depressive disorder the patient must have experienced for at least two weeks of either a sustained low or depressed mood and/or

a. sleep disturbance
b. an appetite disturbance
c. a feeling of worthlessness or guilt
d. loss of energy, fatigue
e. suicidal ideation
f. decreased interest or pleasure

3. The prevalence for major depressive disorder is

a. greater for women than men
b. similar for women and men
c. is greater for Hispanics than Caucasians
d. is higher for Europeans than Asians
e. higher in married than divorced persons

4. Neurobiological studies suggest that primary neurotransmitter(s) involved in major depressive disorders is (are)

a. dopamine
b. serotonin
c. glutamate
d. norepinephrine
e. dopamine and serotonin
f. serotonin and norepinephrine

5. For a diagnosis of mania a person must demonstrate

a. decreased need for sleep
b. racing thoughts
c. hallucinations
d. increased goal directed activity
e. a distinct period of elevated, expansive, or irritablemood, as well as other symptoms
f. grandiosity or increased self-esteem

6. Which one of the following classes of drugs is first line of treatment for bipolar disorders?

a. monoamine oxidase inhibitors
b. selective serotonin reuptake inhibitors
c. cortisol
d. mood stabilizers
e. antibiotics

7. Drugs or drugs of abuse that can produce a manic like state include

a. antiepileptic drugs
b. lithium
c. amphetamine
d. antipsychotics
e. anticholinergic drugs
f. antibiotics

8. Common side-effects of the typical antipsychotics are

a. antidepressant effects
b. insomnia
c. spasticity
d. muscle weakness
e. parkinsonian-like symptoms

9. The lifetime population prevalence for major depressive disorder is

a. 50%
b. 5–25%
c. 2–3%
d. 3–4%
e. unable to determine due to complexity of symptoms

10. The etiology of mood disorders is best conceptualized as involving

a. poor parenting
b. excessive life stress
c. genetic predisposition plus environmental factors
d. maladaptive stress response
e. excessive norepinephrine

11. A commonly replicated finding in major depressive disorder is

a. high cortisol
b. high norepinephrine
c. high dopamine
d. low sodium
e. low cortisol

12. Major depressive disorder is

a. familial and probably genetic
b. due entirely to stress
c. not responsive to treatment
d. related to low cortisol levels
e. shown to be due to excessive biogenic amines

13. A well-established biological correlate for major depressive disorder involves

a. decreased testosterone
b. decreased protein kinase
c. increased testosterone
d. increased sodium
e. none of the above

14. Animal models of depressive disorders include

a. learned helplessness
b. forced swim test
c. chronic mild stress
d. tail suspension test
e. all of the above

15. Drugs used as first line treatment for major depressive disorder include

a. lithium
b. neosporin
c. olanzapine
d. fluoxetine
e. chlordiazepoxide

16. Drugs used to treat bipolar disorder include

a. lithium
b. valproate
c. carbamazepine
d. antidepressants
e. all of the above

17. Persons with bipolar disorder frequently experience

a. decreased need for sleep
b. pressured speech
c. elevated, expansive, or irritable mood
d. increased goal directed activity
e. all of the above

18. The prevalence of bipolar disorder is

a. greater among ethnic minorities
b. greater in european countries
c. less in african-american communities
d. greater in women
e. about 1% in all cultures studied

19. The etiology of bipolar disorder is

a. dependent upon race and social status
b. clearly genetic
c. dependent upon time of birth
d. dependent upon country of origin
e. not due to any genetic factors

20. The diagnosis of schizophrenia requires

a. hallucinations
b. delusions
c. disturbed thought
d. decrease social and vocational abilities
e. all of the above

References

Akiskal HS (2000) Mood disorders: Clinical Features. In: Kaplan and Sadock's Comprehensive Textbook of Psychiatry (Sadock BJ, Sadock VA, eds), pp 1338–1376. Philadelphia: Lippincott, Williams & Wilkins.

Alarcon FJ, Isaacson JH, Franco-Bronson K (1998) Diagnosing and treating depression in primary care patients: Looking beyond physical complaints. Cleve Clin J Med 65:251–260.

American Psychiatric Association (2000) Diagnostic and Statistical Manual of Mental Disorders (4th Ed), DSM-IV-TR (text revision), Washington, DC: American Psychiatric Press.

Arban R, Maraia G, Brackenborough K, Winyard L, Wilson A, Gerrard P, Large C (2005) Evaluation of the effects of lamotrigine, valproate and carbamazepine in a rodent model of mania. Behav Brain Res 158:123–132.

Asberg M, Bertilsson L, Martensson B, Scalia-Tomba GP, Thoren P, Traskman-Bendz L (1984) CSF monoamine metabolites in melancholia. Acta Psychiatr Scand 69:201–219.

Baldessarini RJ (2006) Goodman and Gilmans's The Pharmacological Basis of Therapeutics (Brunton LL, Lazo JS, Parker KL, eds), pp. 432–437. New York: McGraw-Hill.

Bench CJ, Friston KJ, Brown RG, Scott LC, Frackowiak RS, Dolan RJ (1992) The anatomy of melancholia—focal abnormalities of cerebral blood flow in major depression. Psychol Med 22:607–615.

Berk M, Dodd S (2005) Bipolar II disorder: A review. Bipolar Disord 7:11–21.

Bezchlibnyk Y, Young LT (2002) The neurobiology of bipolar disorder: Focus on signal transduction pathways and the regulation of gene expression. Can J Psychiatry 47(135):48.

Blazer DG, Kessler RC, Swartz MS (1998) Epidemiology of recurrent major and minor depression with a seasonal pattern. The National Comorbidity Survey. Br J Psychiatry 172:164–167.

Bowden CL (1998) Treatment of Bipolar Disorder. In: Textbook of Clinical Psychopharmacology (Schatzberg AF, Nemeroff CB, eds), pp 733–745. Washington DC: American Psychiatric Press Inc.

Braff DL, Geyer MA, Swerdlow NR (2001) Human studies of prepulse inhibition of startle: Normal subjects, patient groups, and pharmacological studies. Psychopharmacology 156:234–258.

Brambilla P, Perez J, Barale F, Schettini G, Soares JC (2003) GABAergic dysfunction in mood disorders. Mol Psychiatry 8:721–737.

Burke WJ, Wengel SP (2003) Late-life mood disorders. Clin Geriatr Med 19:777–797.

Carroll BJ (1985) Dexamethasone suppression test: A review of contemporary confusion. J Clin Psychiatry 46:13–24.

Carroll BJ, Curtis GC, Mendels J (1976) Cerebrospinal fluid and plasma free cortisol concentrations in depression. Psychol Med 6:235–244.

Caspi A, Sugden K, Moffitt TE, Taylor A, Craig IW, Harrington H, McClay J, Mill J, Martin J, Braithwaite A, Poulton R (2003) Influence of life stress on depression: Moderation by a polymorphism in the 5-HTT gene. Science 301:386–389.

Cookson JC, Natorf B, Hunt N, Silverstone T, Uppfeldt G (1989) Efficacy, safety and tolerability of raclopride, a specific D2 receptor blocker, in acute schizophrenia: An open trial. Int Clin Psychopharmacol 4:61–70.

Crow TJ (1984) A re-evaluation of the viral hypothesis: Is psychosis the result of retroviral integration at a site close to the cerebral dominance gene? Br J Psychiatry 145:243–253.

Cryan JF, Valentino RJ, Lucki I (2005a) Assessing substrates underlying the behavioral effects of antidepressants using the modified rat forced swimming test. Neurosci Biobehav Rev 29:547–569.

Cryan JF, Mombereau C, Vassout A (2005b) The tail suspension test as a model for assessing antidepressant activity: Review of pharmacological and genetic studies in mice. Neurosci Biobehav Rev 29:571–625.

Davies N, Russell A, Jones P, Murray RM (1998) Which characteristics of schizophrenia predate psychosis? J Psychiatr Res 32:121–131.

Delgado PL, Charney DS, Price LH, Aghajanian GK, Landis H, Heninger GR (1990) Serotonin function and the mechanism of antidepressant action Reversal of antidepressant-induced remission by rapid depletion of plasma tryptophan. Arch Gen Psychiatry 47:411–418.

Donati RJ, Rasenick MM (2003) G protein signaling and the molecular basis of antidepressant action. Life Sci 73:1–17.

du Bois TM, Deng C, Huang XF (2005) Membrane phospholipid composition, alterations in neurotransmitter systems and schizophrenia. Prog Neuropsychopharmacol Biol Psychiatry 29:878–888.

Eaton WW, Kramer M, Anthony JC, Dryman A, Shapiro S, Locke BZ (1989) The incidence of specific DIS/DSM-III mental disorders: Data from the NIMH Epidemiologic Catchment Area Program. Acta Psychiatr Scand 79:163–178.

Eisenhofer G, Kopin IJ, Goldstein DS (2004) Catecholamine metabolism: A contemporary view with implications for physiology and medicine. Pharmacol Rev 56:331–349.

Falkai P, Bogerts B (1986) Cell loss in the hippocampus of schizophrenics. Eur Arch Psychiatry Neurol Sci 236:154–161.

Flügge G, van Kampen M, Meyer H, Fuchs E (2003) Alpha-2A and alpha-2C adrenoceptor regulation in the brain: Alpha-2A changes persist after chronic stress. Eur J Neurosci 17:917–928.

Frazer A (1997) Antidepressants. J Clin Psychiatry 58(supp. 6):9–25.

Gaughran F (2002) Immunity and schizophrenia: Autoimmunity, cytokines, and immune responses. Int Rev Neurobiol 52:275–302.

Geyer MA, Ellenbroek B (2003) Animal behavior models of the mechanisms underlying antipsychotic atypicality. Prog Neuropsychopharmacol Biol Psychiatry 27:1071–1079.

Geyer MA, Krebs-Thomson K, Braff DL, Swerdlow NR (2001) Pharmacological studies of prepulse inhibition of sensorimotor gating deficits in schizophrenia: A decade in review. Psychopharmacology 156:117–154.

Ghaemi SN, Boiman EE, Goodwin FK (1999) Kindling and second messengers: An approach to the neurobiology of recurrence in bipolar disorder. Biol Psychiatry 45:137–144.

Goldberg JF, Truman CJ (2003) Antidepressant-induced mania: An overview of current controversies. Bipolar Disord 5:407–420.

Gottesman IA (1989) Vital statistics, demography, and schizophrenia: Editor's introduction. Schizophr Bull 15:5–7.

Guze SB, Robins E (1970) Suicide and primary affective disorders. Br J Psychiatry 117:437–438.

Henn FA, Vollmayr B (2004) Neurogenesis and depression: Etiology or epiphenomenon? Biol Psychiatry 56:146–150.

Henn FA, Vollmayr B (2005) Stress models of depression: Forming genetically vulnerable strains. Neurosci Biobehav Rev 29:799–804.

Hirschfeld RM (1999) Efficacy of SSRIs and newer antidepressants in severe depression: Comparison with TCAs. J Clin Psychiatry 60:326–335.

Jablensky A, Sartorius N (1988) Is schizophrenia universal? Acta Psychiatr Scand 344(Suppl):65–70.

Jones AL, Mowry BJ, Pender MP, Greer JM (2005) Immune dysregulation and self-reactivity in schizophrenia: Do some cases of schizophrenia have an autoimmune basis? Immunol Cell Biol 83:9–17.

Katon W, Sullivan MD (1990) Depression and chronic medical illness. J Clin Psychiatry 51(Suppl):3–11.

Keller MB, Lavori PW, Mueller TI, Endicott J, Coryell W, Hirschfeld RM, Shea T (1992) Time to recovery, chronicity, and levels of psychopathology in major depression. A 5-year prospective follow-up of 431 subjects. Arch Gen Psychiatry 49:809–816.

Kelsoe JR (2000) Mood Disorders: Genetics. In: Kaplan and Sadock's Comprehensive Textbook of Psychiatry (Sadock BJ, Sadock VA, eds), pp 1308–1318. Philadelphia: Lippincott, Williams & Wilkins.

Kendler KS, Diehl SR (1993) The genetics of schizophrenia: A current genetic-epidemiological perspective. Schizophr Bull 19:261–285.

Kessler RC, McGonagle KA, Swartz M, Blazer DG, Nelson CB (1993) Sex and depression in the National Comorbidity Survey I Lifetime prevalence, chronicity and recurrence. J Affect Disord 29:85–96.

Kessler RC, McGonagle KA, Zhao S, Nelson CB, Hughes M, Eshleman S, Wittchen HU, Kendler KS (1994) Lifetime and 12-month prevalence of DSM-III-R psychiatric disorder in the United States. Arch Gen Psychiatry 51:8–19.

Kim YK, Myint AM, Lee BH, Han CS, Lee SW, Leonard BE, Steinbusch HW (2004) T-helper types 1, 2, and 3 cytokine interactions in symptomatic manic patients. Psychiatry Res 129:267–272.

King JA, Campbell D, Edwards E (1993) Differential development of the stress response in congenital learned helplessness. Int J Dev Neurosci 11:435–442.

Klerman GL, Weissman MM (1989) Increasing rates of depression. JAMA 261:2229–2235.

Kupfer DJ, Foster FG (1972) Interval between onset of sleep and rapid-eye-movement sleep as an indicator of depression. Lancet 2:684–686.

Leonard BE, Richelson E (2000) Synaptic effects of antidepressants. In: Schizophrenia and Mood Disorders: The New Drug Therapies in Clinical Practice (Buckley PF, Waddington JL, eds), pp. 67–84. Boston: Butterworth-Heinemann.

Leweke FM, Gerth CW, Koethe D, Klosterkotter J, Ruslanove I, Krivogosky B, Torrey EF, Yolken RH (2004) Antibodies to infectious agents in individuals with recent onset schizophrenia. Eur Arch Psychiatry Clin Neurosci 254:4–8.

Machado-Vieira R, Kapczinski F, Soares JC (2004) Perspectives for the development of animal models of bipolar disorder. Prog Neuropsychopharmacol Biol Psychiatry 28:209–224.

Maier SF, Watkins LR (2005) Stressor controllability and learned helplessness: The roles of the dorsal raphe nucleus, serotonin, and corticotrophin-releasing factor. Neurosci Biobehav Rev 29:829–841.

Marcotte ER, Pearson DM, Srivastava LK (2001) Animal models of schizophrenia: A critical review. J Psychiatr Neurosci 26:395–410.

Martin LF, Kem WR, Freedman R (2004) Alpha-7 nicotinic receptor agonists: Potential new candidates for the treatment of schizophrenia. Psychopharmacology 174:54–64.

McElroy S, Keck P (2000) Pharmacological agents for the treatment of acute bipolar mania. Biol Psychiatry 48:539–557.

McGuffin P, Katz R, Rutherford J (1991) Nature, nurture and depression: A twin study. Psychol Med 21:329–335.

McGuffin P, Owen MJ, Farmer AE (1995) The genetic basis of schizophrenia. Lancet 346:678–682.

Mednick SA, Machon RA, Huttunen MO, Bonett D (1988) Adult schizophrenia following prenatal exposure to an influenza epidemic. Arch Gen Psychiatry 45:189–192.

Meltzer HY (1991) The mechanism of action of novel antipsychotic drugs. Schizophr Bull 17:263–287.

Meyer U, Feldon J, Schedlowski M, Yee BK (2005) Towards an immuno-precipitated neurodevelopmental animal model of schizophrenia. Neurosci Biobehav Rev 29:913–947.

Millan MJ (2004) The role of monoamines in the actions of established and "novel" antidepressant agents: A critical review. Eur J Pharmacol 500:371–384.

Mitchell PJ, Redfern PH (2005) Animal models of depressive illness: The importance of drug treatment. Curr Pharm Des 11:171–203.

Muller N, Riedel M, Gruber R, Ackenheil M, Schwarz MJ (2000) The immune system and schizophrenia. An integrative view. Ann N Y Acad Sci 917:456–467.

Nestler EJ, Gould E, Manji H, Buncan M, Duman RS, Gershenfeld HK, Hen R, Koester S, Lederhendler I, Meaney M, Robbins T, Winsky L, Zalcman S (2002) Preclinical models: Status of basic research in depression. Biol Psychiatry 52:503–528.

Norquist GS, Narrow WE (2000) Schizophrenia: Epidemiology. In: Kaplan and Sadock's Comprehensive Textbook of Psychiatry (Sadock BJ, Sadock VA, eds), pp 1110–1117. Philadelphia: Lippincott, Williams & Wilkins.

Ordway GA, Schenk J, Stockmeier CA, May W, Klimek V (2003) Elevated agonist binding to alpha-2 adrenoceptors in the locus coeruleus in major depression. Biol Psychiatry 53:315–323.

Overstreet DH, Friedman E, Mathe AA, Yadid G (2005) The Flinders Sensitive Line rat: A selectively bred putative animal model of depression. Neurosci Biobehav Rev 29:739–759.

Owen R, Owen F, Poulter M, Crow TJ (1984) Dopamine D2 receptors in substantia nigra in schizophrenia. Brain Res 299:152–154.

Owens MJ, Morgan WN, Plott SJ, Nemeroff CB (1997) Neurotransmitter receptor and transporter binding profile of antidepressants and their metabolites. J Pharmacol Exp Ther 283:1305–1322.

Pakkenberg B (1987) Post-mortem study of chronic schizophrenic brains. Br J Psychiatry 151:744–752.

Paykel ES (2003) Life events and affective disorders. Acta Psychiatr Scand Suppl 61–66.

Petty F (1995) GABA and mood disorders: A brief review and hypothesis. J Affect Disord 34:275–281.

Petty F, Sherman AD (1981) A pharmacologically pertinent animal model of mania. J Affect Disord 3:381–387.

Petty F, Sacquitne JL, Sherman AD (1982) Tricyclic antidepressant drug action correlates with its tissue levels in anterior neocortex. Neuropharmacology 21:475–477.

Pine DS, Cohen JA (2002) Trauma in children and adolescents: Risk and treatment of psychiatric sequelae. Biol Psychiatry 51:519–531.

Post RM (2000) Mood Disorders: Treatment of Bipolar Disorders. In: Kaplan and Sadock's Comprehensive Textbook of Psychiatry (Sadock BJ, Sadock VA, eds), pp 1385–1430. Philadelphia: Lippincott, Williams & Wilkins.

Post RM, Susan R, Weiss B (1992) Sensitization, kindling, and carbamazepine: An update on their implications for the course of affective illnesses. Pharmacopsychiatry 25:41–43.

Post RM, Speer AM, Hough CJ, Xing G (2003) Neurobiology of bipolar illness: Implications for future study and therapeutics. Ann Clin Psychiatry 15:85–94.

Robins E, Guze SB (1972) Classification of affective disorders: The primary-secondary, the endogenous-reactive, and the neurotic-psychotic concepts. In: Recent Advances in Psychobiology of the Depressive Illnesses, Proceedings of a Workshop Sponsored by the NIMH (Williams TA, Katz MM, Shield JA, eds), Washington, D.C.: U.S. GPO.

Schatzberg AF, Garlow SL, Nemeroff CB (2002) Molecular and cellular mechanisms of depression. In: Neuropsychopharmacology: The Fifth Generation of Progress (Davis KL, Charney D, Coyle JT, Nemeroff C, eds), pp 1039–1050. Philadelphia: Lippincott, Williams & Wilkins.

Schiepers OJ, Wichers MC, Maes M (2005) Cytokines and major depression. Prog Neuropsychopharmacol Biol Psychiatry 29:201–217.

Schildkraut JJ (1967) The catecholamine hypothesis of affective disorders. A review of supporting evidence. Int J Psychiatry 4:203–217.

Seeman P, Weinshenker D, Quirion R, Srivastava LK, Bhardwaj SK, Grandy DK, Premont RT, Sotnikova TD, Boksa P, El-Ghundi M, O'Dowd BF, George SR, Perreault ML, Mannisto PT, Robinson S, Palmiter RD, Tallerico T (2005) Dopamine supersensitivity correlates with D2High states, implying many paths to psychosis. Proc Natl Acad Sci USA 102:3513–3518.

Seligman ME (1972) Learned helplessness. Annu Rev Med 23:407–412.

Shapiro S, Skinner EA, Kessler LG, Von Korff M, German PS, Tischler GL, Leaf PJ, Benham L, Cottler L, Regier DA (1984) Utilization of health and mental health services. Three Epidemiologic Catchment Area sites. Arch Gen Psychiatry 41:971–978.

Shaw K, Turner J, Del Mar C (2001) Tryptophan and 5-hydroxytryptophan for depression. Cochrane Database Syst Rev CD003198.

Sherman AD, Petty F (1980) Neurochemical basis of the action of antidepressants on learned helplessness. Behav Neural Biol 30:119–134.

Shulman RB, Scheftner WA, Nayudu S (2001) Venlafaxine-associated mania. J Clin Psychopharmacol 21:239–241.

Song C, Leonard BE (2005) The olfactory bulbectomized rat as a model of depression. Neurosci Biobehav Rev 29:627–647.

Stoner SC, Williams RJ, Worrel J, Ramlatchman L (1999) Possible venlafaxine-induced mania. J Clin Psychopharmacol 19:184–185.

Sullivan LE, Fiellin DA, O'Connor PG (2005) The prevalence and impact of alcohol problems in major depression: A systematic review. Am J Med 118:330–341.

Tatsumi M, Groshan K, Blakely RD, Richelson E (1997) Pharmacological profile of antidepressants and related compounds at human monoamine transporters. Eur. J Pharmacol 340:249–258.

Thomas RM, Peterson DA (2003) A Neurogenic Theory of Depression Gains Momentum. Mol Interv 3:441–444.

Torrey EF, Peterson MR (1976) The viral hypothesis of schizophrenia. Schizophr Bull 2:136–146.

Weinberger DR, Torrey EF, Neophytides AN, Wyatt RJ (1979) Lateral cerebral ventricular enlargement in chronic schizophrenia. Arch Gen Psychiatry 36:735–739.

Weissman MM (1979) The psychological treatment of depression. Evidence for the efficacy of psychotherapy alone, in comparison with, and in combination with pharmacotherapy. Arch Gen Psychiatry 36:1261–1269.

Wilkins J (1989) Clinical implications of PCP, NMDA and opiate receptors. NIDA Res Monogr 95:275–281.

Willner P (2005) Chronic mild stress (CMS) revisited: Consistency and behavioral-neurobiological concordance in the effects of CMS. Neuropsychobiology 52:90–110.

Willner P, Mitchell PJ (2002) The validity of animal models of predisposition to depression. Behav Pharmacol 13:169–188.

Winokur G, Pitts Jr FN (1965) Affective disorder: VI. A family history study of prevalences, sex differences and possible genetic factors. J Psychiatr Res 3:113–123.

Yuksel FV, Basterzi AD, Goka E (2004) Venlafaxine-induced mania. Can J Psychiatry 49:786–787.

Zhou R, Zarate CA, Manji HK (2005) Identification of molecular mechanisms underlying mood stabilization through genome-wide gene expression profiling. Int J Neuropsychopharmacol 6:1–4.

36
Molecular Pathogenesis for Schizophrenia and Major Depression

Norbert Müller and Markus J. Schwarz

Keywords Schizophrenia; Major depression; Th1/Th2 balance; Pro-inflammatory cytokines; NMDA hypothesis; Glutamate; Serotonin; Tryptophan; Kynurenine; Kynurenic acid; HPA axis

36.1. Introduction

In schizophrenia, dopaminergic hyperfunction in the limbic system and dopaminergic hypofunction in the frontal cortex are discussed to be the main neurotransmitter disturbances. Recent research provides further insight that glutamatergic hypofunction might be the cause for this dopaminergic dysfunction. In major depression, in contrast, glutamatergic hyperfunction seems to be closely related to the lack of serotonergic and noradrenergic neurotransmission and to the core symptoms of major depression. Therefore, glutamatergic dysfunction seems to be a common pathway in the neurobiology of schizophrenia and depression. The function of the glutamatergic system is closely related to the immune system and to the tryptophan-kynurenine metabolism, which both seem to play a key role in the pathophysiology of schizophrenia and major depression (Müller and Schwarz, 2006, 2007).

36.2. Glutamatergic Neurotransmission and NMDA-Receptor Function in Schizophrenia and Major Depression

36.2.1. Schizophrenia

A disturbance in the dopaminergic neurotransmission plays a key-role in the pathogenesis of schizophrenia (Carlsson, 1988). Most drugs ameliorating psychotic symptoms act as dopamine receptor blockers, in particular D2 receptor blockers. In light of only a portion of patients responding to antipsychotic drugs and unsatisfactory long term outcomes, attempts to explain the disease solely in terms of dopaminergic dysfunction leave many aspects of schizophrenia unsolved.

The glutamate hypothesis of schizophrenia postulates an equilibrium between inhibiting dopaminergic and inhibiting glutamatergic neurons; the model of a cortico-striato-thalamo-cortical control loop integrates the glutamate hypothesis with neuroanatomical aspects on the pathophysiology of schizophrenia (Carlsson et al., 2001). Hypofunction of the glutamatergic cortico-striatal pathway is associated with opening of the thalamic filter, which leads to an uncontrolled flow of sensory information to the cortex and to psychotic symptoms. Hypofunction of the glutamatergic neurotransmitter system as a causal mechanism in schizophrenia was first proposed due to the observation of low concentrations of glutamate in the CSF of schizophrenic patients (Kim et al., 1980).

Treatment with NMDA receptor antagonists leads to a marked, dose-dependent increase of amphetamine-induced dopamine release (Miller and Abercrombie, 1996). In schizophrenics, this amphetamine-induced dopamine release is much higher compared to healthy controls (Laruelle et al., 1996). This observation is in accordance with the view that activation of the nigrostriatal dopamine system can take place by opposing activation of inhibitory striatonigral GABAergic projection neurons (Carlsson et al., 2001).

Phencyclidine (PCP), Ketamin, and MK-801 all block the N-methyl-D-aspartate (NMDA) receptor complex and are associated with schizophrenia-like symptoms through hypofunction of the glutamatergic neurotransmission (Krystal et al., 1994; Olney and Farber, 1995). Other NMDA antagonists have psychotogenic properties, too (CPP, CPP-ene, CGS 19755). NMDA receptor hypofunction can explain schizophrenic positive and negative symptoms, cognitive deterioration and structural brain changes (Olney and Farber, 1995).

T. Ikezu and H.E. Gendelman (eds.), *Neuroimmune Pharmacology*.
© Springer 2008

Findings of decreased plasma levels of the NMDA co-agonist glycine in schizophrenics and a correlation of glycine levels with schizophrenic negative symptoms are in line with decreased NMDA-receptor function (Sumiyoshi et al., 2004). Baseline glycine levels predicted the treatment outcome of clozapine on negative symptoms (Sumiyoshi et al., 2005). Clinical investigations targeted the glycine co-agonistic site of the NMDA receptor by administering the amino acids glycine or D-serine, or a glycine pro-drug such as milacemide (Tamminga et al., 1992). Some of these studies have yielded positive results, particularly against the schizophrenic deficit syndrome (Heresco-Levy et al., 1999).

36.2.2. Major Depression

Overwhelming evidence collected over the last 40 years suggests that disturbances in the serotonergic and noradrenergic neurotransmission are the crucial factor in the pathogenesis of major depression (Matussek, 1966; Coppen and Swade, 1988). The common therapeutic mechanism of antidepressant drugs is the increase of serotonergic and/or noradrenergic neurotransmission. Intense research, however, has not yet been able to discover the mechanisms leading to disturbances of the serotonergic/noradrenergic neurotransmission.

Although the glutamatergic system may influence directly or indirectly the serotonergic and noradrenergic neurotransmission, only few data have been published with regard to this interaction. NMDA receptor antagonists increase the serotonin levels in the brain (Yan et al., 1997; Martin et al., 1998). Several studies showed an increased activity of the glutamatergic system in the peripheral blood of depressive patients (Kim et al., 1982; Altamura et al., 1993; Mauri et al., 1998). This result could not be replicated by all authors (Maes et al., 1998). The inconsistency of the findings, however, might be due to methodological problems (Kugaya and Sanacora, 2005).

Support for increased glutamatergic activity in depression comes from magnetic resonance spectroscopy: Elevated glutamate levels were found in the occipital cortex of unmedicated subjects with major depression (Sanacora et al., 2004a). An increased level of a certain neurotransmitter is often associated with a down-regulation of the respective receptor. Accordingly, a reduced glycine binding site of the NMDA receptor was found in the brains of suicide victims and patients with depression (Nowak et al., 1995; Nudmamud-Thanoi and Reynolds, 2004). Moreover, a decrease in the NMDA agonistic MK-801 binding in bipolar patients was observed (Scarr et al., 2003).

Consistent with the view that an increased activity of the glutamatergic system and NMDA receptor agonism is associated with depressed mood, a reduction of the glutamatergic activity through NMDA receptor antagonism might exert antidepressant effects. NMDA antagonists such as MK-801 (Maj et al., 1992; Trullas and Skolnick, 1990), ketamine (Yilmaz et al., 2002), memantine (Ossowska et al., 1997), and others (Kugaya and Sanacora, 2005) exhibited antidepressant effects in different animal models. In humans, D-cycloserine, a partial NMDA receptor agonist, which acts as a NMDA receptor antagonist in high doses, demonstrated antidepressant effects in high doses (Crane, 1959). Slight antidepressive effects have also been observed with amantadine (Huber et al., 1999; Stryjer et al., 2003). Moreover, preliminary data show antidepressant effects of the NMDA receptor antagonist ketamine (Kudoh et al., 2002; Ostroff et al., 2005) and riluzole, an antiglutamatergic agent believed to increase the glutamatergic uptake into the astrocytes, is under intensive investigation for its antidepressant efficacy (Frizzo et al., 2004). A recent series of open-labelled studies and case reports demonstrated the efficacy of riluzole (Coric et al., 2003; Sanacora et al., 2004b; Zarate et al., 2004, 2005).

36.3. Inflammation in Schizophrenia and Depression

36.3.1. Schizophrenia

Infection during pregnancy, in particular in the second trimester, in mothers of off-springs later developing schizophrenia has been repeatedly described (Brown et al., 2004; Buka et al., 2001; Westergaard et al., 1999). As opposed to any single pathogen, the immune response, itself, of the mother may be related to the increased risk for schizophrenia in the offspring (Zuckerman and Weiner, 2005). Indeed, increased IL-8 levels of mothers during the second trimenon were associated with an increased risk for schizophrenia in the offspring (Brown et al., 2004). A fivefold increased risk for developing psychoses later on, however, was also observed after infection of the CNS in early childhood (Gattaz et al., 2004; Koponen et al., 2004).

Signs of inflammation were found in schizophrenic brains (Körschenhausen et al., 1996), and the term 'mild localized chronic encephalitis' to describe a slight but chronic inflammatory process in schizophrenia was proposed (Bechter et al., 2003).

36.3.2. Major Depression

An inflammatory model of major depression (MD) is 'sickness behaviour', the reaction of the organism to infection and inflammation. Sickness behaviour is characterised by weakness, malaise, listlessness, inability to concentrate, lethargy, decreased interest in the surroundings, and reduced food intake—all of which are depression-like symptoms. The sickness-related psychopathological symptomatology during infection and inflammation is mediated by proinflammatory cytokines such as IL-1, IL-6, tumor-necrosis-factor-α (TNF-α), and IFN-γ. The active pathway of these cytokines from the peripheral immune system to the brain is via afferent neurons and through direct targeting of the amygdala and other brain regions after diffusion at the circumventricular organs and choroid plexus (Dantzer, 2001).

Undoubtedly, there is a strong relationship between the cytokine system and the neurotransmitter system, but a more differentiated analysis may be required to understand the specific mechanisms underlying the heterogeneous disease entity of MD.

In humans, the involvement of cytokines in the regulation of the behavioural symptoms of sickness behaviour has been studied by application of the bacterial endotoxin LPS to human volunteers (Reichenberg et al., 2001). LPS, a potent activator of proinflammatory cytokines, was found to induce mild fever, anorexia, anxiety, depressed mood, and cognitive impairment. The levels of anxiety, depression and cognitive impairment were found to be related to the levels of circulating cytokines (Reichenberg et al., 2001; Reichenberg et al., 2002).

Mechanisms that contribute to inflammation and can cause depressive states are:

A direct influence of proinflammatory cytokines on the serotonin and noradrenalin metabolism

1. An imbalance of the type-1 – type-2 immune response leading to an increased tryptophan and serotonin metabolism by activation of IDO in the CNS, which is associated with
2. A decreased availability of tryptophan and serotonin,
3. A disturbance of the kynurenine metabolism with an imbalance in favour of the production of the NMDA-receptor agonist quinolinic acid, and
4. An imbalance in astrocyte- and microglial activation associated with increased production of quinolinic acid.

Effects of antidepressants on the immune function support this view. The mechanisms and the therapeutic implications will be discussed in the following paragraphs.

36.4. Polarized Type-1 and Type-2 Immune Responses

36.4.1. Reduced Type-1 Immune Response in Schizophrenia

A well established finding in schizophrenia is the decreased in vitro production of IL-2 and IFN-γ (Wilke et al., 1996; Müller et al., 2000), reflecting a blunted production of type-1 cytokines. Decreased levels of neopterin, a product of activated monocytes/macrophages, also point to a blunted activation of the type-1 response (Sperner-Unterweger et al., 1999). The decreased response of lymphocytes after stimulation with specific antigens reflects a reduced capacity for a type-1 immune response in schizophrenia, as well (Müller et al., 1991). Decreased levels of soluble (s)ICAM-1, as found in schizophrenia, also represent an under-activation of the type-1 immune system (Schwarz et al., 2000). Decreased levels of the soluble TNF-receptor p55—mostly decreased when TNF-α is decreased—were observed, too (Haack et al., 1999).

A blunted response of the skin to different antigens in schizophrenia was observed before the era of antipsychotics (Molholm, 1942). This finding could be replicated in unmedicated schizophrenic patients using a skin test of the cellular immune response (Riedel et al., 2007).

36.4.2. Increased Type-2 Immune Response in Schizophrenia

Several reports described increased serum IL-6 levels in schizophrenia. IL-6 serum levels might be especially high in patients with an unfavourable course of the disease (Müller et al., 2000). IL-6 is a product of activated monocytes and of the activation of the type-2 immune response. Moreover, several other signs of activation of the type-2 immune response are described in schizophrenia, including the increased production of IgE and an increase of IL-10 serum levels (Cazzullo et al., 1998; Schwarz et al., 2001). In the cerebrospinal fluid (CSF), IL-10 levels were found to be related to the severity of the psychosis (van Kammen et al., 1997).

The key cytokine of the type-2 immune response is IL-4. Increased levels of IL-4 in the CSF of juvenile schizophrenic patients have been reported (Mittleman et al., 1997), which supports that the increased type-2 response in schizophrenia is not only a phenomenon of the peripheral immune response.

36.4.3. Increased Proinflammatory Type-1 Cytokines in Major Depression

Characteristics of the immune activation in MD include increased numbers of circulating lymphocytes and phagocytic cells, upregulated serum levels of indicators of activated immune cells (neopterin, soluble IL-2 receptors), higher serum concentrations of positive acute phase proteins (APP's), coupled with reduced levels of negative APP's, as well as increased release of proinflammatory cytokines, such as IL-1, IL-2, TNF-α, and IL-6 through activated macrophages and IFN-γ through activated T-cells (Müller et al., 1993; Maes et al., 1995a, b; Irwin 1999; Nunes et al., 2002; Müller and Schwarz, 2002; Mikova et al., 2001). Increased numbers of peripheral mononuclear cells in MD have been described by different groups of researchers (Herbert and Cohen, 1993; Seidel et al., 1996b; Rothermundt et al., 2001).

Neopterin is a sensitive marker of the cell-mediated type-1 immunity. The main source of neopterin is monocytes/macrophages. In accordance with the findings of increased monocytes/macrophages, an increased secretion of neopterin has been described by several groups of researchers (Duch et al., 1984; Dunbar et al., 1992; Maes et al., 1994; Bonaccorso et al., 1998).

The increased plasma concentrations of the proinflammatory cytokines IL-1 and IL-6 observed in depressed patients was found to correlate with the severity of depression and with measures of HPA axis hyperactivity (Maes et al., 1993; Schiepers et al., 2005). As genetics plays a role in MD, the genetics of the immune system in relation to MD has also been investigated. Certain cytokine gene polymorphisms, e.g.

in genes coding for IL-1 and TNF-α may confer a greater susceptibility to develop MD (Jun et al., 2003; Fertuzinhos et al., 2004; Rosa et al., 2004).

The production of IL-2 and IFN-γ is the typical marker of a type-1 immune response. In contrary to schizophrenia, IFN-γ is produced in greater amounts by lymphocytes of patients with MD than of healthy controls (Seidel et al., 1996a). Higher plasma levels of IFN-γ in depressed patients, accompanied by lower plasma tryptophan availability were described (Maes et al., 1994). Data on IL-2 in major depression are mainly restricted to the estimation of its soluble receptor sIL-2R in the peripheral blood. Increased sIL-2R levels reflect an increased production of IL-2. The blood levels of sIL-2R were repeatedly found to be increased in MD patients (Sluzewska et al., 1996; Maes et al., 1995a, b).

Since different pathologies may underlie the syndrome of depression, different immunological states might be involved. Indeed, different types of MD were observed to exhibit different immune profiles: The subgroup of melancholic depressed patients showed a decreased type-1 activation—as it was observed in schizophrenic patients (Müller and Schwarz, 2007)—while the non-melancholic depressed patients showed signs of inflammation such as increased monocyte count and increased levels of α_2-macroglobulin (Rothermundt et al., 2001).

Suicidality, observed in a very high amount of depressed patients, seems to be associated with an 'idealtypic' immune activation pattern for depression, since clinical studies have observed higher levels of type-1 cytokines in suicidal patients. In a small study, distinct associations between suicidality and type-1 immune response and a predominance of type-2 immune parameters in non-suicidal patients were observed (Mendlovic et al., 1999). An epidemiological study hypothesized that high IL-2 levels are associated with suicidality (Penttinen, 1995). There have been descriptions of increased levels of serum sIL-2R in medication-free suicide attempters irrespective of the psychiatric diagnosis (Nassberger and Traskman-Bendz, 1993), and treatment with high-dose interleukin-2 has been associated with suicide in a case report (Baron et al., 1993).

These data show that possible different immune states within the 'syndrome' MD have to be more differentiated. The predominant pro-inflammatory, type-1 dominated immune state described in MD maybe an 'idealtypic' state restricted to a majority of patients suffering from MD. Therefore, those and other methodological concerns have to be regarded carefully in future studies.

36.5. Somatic States Associated with Depression and a Proinflammatory Immune Response as Psychoneuroimmunological Model Diseases

36.5.1. Medical Illness Condition

Illness-associated depressive symptoms may not merely be a psychological reaction to the pain, distress and incapacitation that are associated with the physical disease, but may be directly caused by immune activation and the secretion of cytokines. These include acute and chronic infectious conditions (Meijer et al., 1988; Müller et al., 1992; Hall and Smith, 1996) as well as non-infectious conditions associated with chronic inflammation, such as rheumatoid arthritis, cancer and MS (Yirmiya et al., 1999; Pollak et al., 2000). Indeed, several studies have shown that immune dysregulation preceeds the development of depression (Sakic et al., 1996; Yirmiya et al., 1999; Yirmiya, 2000; Pollak et al., 2000).

36.5.2. Therapies Using Type-1 Cytokines

The same mechanism seems to be responsible for depressive states associated with cytokine therapy. Depression is a common side-effect of therapy with IFN-α and IL-2 (Bonaccorso et al., 2002; Schäfer et al., 2004). During these therapies, the enzyme indoleamine 2,3-dioxygenase (IDO) is activated and the tryptophan degradation increases. The psychopathological changes are related to the enhanced tryptophan metabolism and IDO activity: patients developing more severe depressive symptoms during INF-α showed a more pronounced increase in tryptophan metabolism (Bonaccorso et al., 2002; Capuron et al., 2002, 2003). Depression, also a common side-effect during therapy with IFN-ß, often limits the use of IFN-ß in the therapy of multiple sclerosis (Patten et al., 2005); although not a classical type-1 cytokine, IFN-ß, too, leads to an activation of IDO, thereby provoking depressive symptoms (Amirkhani et al., 2005).

36.5.3. Pregnancy and Delivery

Pregnancy is immunologically characterized by a dominance of the type-2 immune response of the mother. In order to protect the fetus from abortion, the maternal organism develops immune tolerance against the non-self organism of the baby (Marzi et al., 1996; Saito et al., 1999). After delivery an activation of the type-1 response of the maternal immune system occurs (Maes et al., 2002; Ostensen et al., 2005). This type-1 dominated maternal immune rebalancing is associated with the postpartum blues (Maes et al., 2002), a state found in 20–75% of mothers (Stein, 1980; Josefsson et al., 2001), or the postpartum depression, a state found in 10–15% of mothers (O'Hara and Swain, 1996). After delivery, an increase of the proinflammatory immune response was described (Maes et al., 2002; Ostensen et al., 2005). Accordingly, postpartum blues is associated with increased activation of the proinflammatory immune response, increased activity of the IDO and a delay in the increase of postpartum tryptophan-levels compared to mothers without 'the blues' (Maes et al., 2002; Kohl et al., 2005). Symptoms of depression and anxiety in the early puerperum are significantly related to the increase of the proinflammatory cytokine IL-8 (Maes et al., 2002). Postpartum depression is associated with the activation of the proinflammatory type-1 immune response and a lack of tryptophan (Gard et al., 1986; Abou-Saleh et al., 1999).

36.6. Therapeutic Mechanisms and the Type-1/Type-2 Imbalance in Schizophrenia and Depression

36.6.1. Schizophrenia: Anti-Psychotic Drugs Rebalance the Type-1/Type-2 Imbalance

In-vitro studies show that the blunted IFN-γ production becomes normalized after therapy with neuroleptics (Wilke et al., 1996). An increase of 'memory cells' (CD4 $^+$ CD45RO $^+$ cells)—one of the main sources of IFN-γ production—during anti-psychotic therapy with neuroleptics was observed by different groups (Müller et al., 1997b). Additionally, an increase of soluble IL-2 receptors (sIL-2R)—the increase reflects an increase of activated, IL-2 bearing T-cells—during anti-psychotic treatment was described (Müller et al., 1997a). The reduced sICAM-1 levels show a significant increase during short term anti-psychotic therapy (Schwarz et al., 2000) and the ICAM-1 ligand leucocyte function antigen-1 (LFA-1) shows a significantly increased expression during anti-psychotic therapy (Müller et al., 1999). The increase of TNF-α and TNF-α receptors during therapy with clozapin was observed repeatedly (Pollmächer et al., 2001). Moreover, the blunted reaction to vaccination with salmonella typhii was not observed in patients medicated with anti-psychotics (Ozek et al., 1971). Recently, an elevation of IL-18 serum levels was described in medicated schizophrenics (Tanaka et al., 2000). Since IL-18 plays a pivotal role in the type-1 immune response, this finding is consistent with other descriptions of type-1 activation during antipsychotic treatment.

Regarding the type-2 response, several studies point out that anti-psychotic therapy is accompanied by a functional decrease of the IL-6 system (Maes et al., 1997; Müller et al., 2000).

36.6.2. Therapeutic Techniques in Depression are associated with Downregulation of the Proinflammatory Immune Response

36.6.2.1. Antidepressant Pharmacotherapy

A modulatory, predominantly inhibitory effect of serotonin reuptake inhibiting drugs on activation of proinflammatory immune parameters was demonstrated in animal experiments (Bengtsson et al., 1992; Song and Leonard, 1994; Zhu et al., 1994).

Several antidepressants seem to be able to induce a shift from a type-1 to a type-2, from a proinflammatory to an anti-inflammatory immune response, since the ability of different antidepressants (sertraline, clomipramine, trazodone) to reduce the IFN-γ/IL-10 ratio significantly was shown in vitro. These drugs reduced the IFN-γ production significantly, while sertraline and clomipramine additionally raised the IL-10 production (Maes et al., 1999). Regarding other in-vitro studies, a significantly reduced production of IFN-γ, IL-2, and sIL-2R was found after antidepressant treatment compared to pre-treatment values (Seidel et al., 1995, b). A down-regulation of the IL-6 production was observed during amitriptyline treatment; in treatment-responders, the TNF-α production decreased to normal (Lanquillon et al., 2000). There are also studies, however, showing no effect of antidepressants to the in-vitro stimulation of cytokines (overview: (Kenis and Maes, 2002), but methodological issues have to be taken into account. There is significant evidence suggesting that antidepressants of different classes show a down-regulation of the type-1 cytokine production in-vitro (Kenis and Maes, 2002).

Regarding serum levels, several researchers observed a reduction of IL-6 during treatment with the serotonin reuptake inhibitor fluoxetine (Sluzewska et al., 1995). A decrease of IL-6 serum levels during therapy with different antidepressants has been observed by other researchers (Frommberger et al., 1997). On the other hand, other groups did not find any effect of certain antidepressants on serum levels of different cytokines (Maes et al., 1995a, 1997).

Since IL-6 stimulates Prostaglandin E_2 (PGE_2) and antidepressants inhibit the IL-6 production, an inhibiting action of antidepressants on PGE_2 would be expected, too (Pollak and Yirmiya, 2002). Over twenty years ago it was suggested that antidepressants inhibit PGE_2 (Mtabaji et al., 1977). A recent in-vitro study showed that both tricyclic antidepressants and selective serotonin inhibitors attenuated cytokine-induced PGE_2 and nitric oxide production by inflammatory cells (Yaron et al., 1999).

36.6.2.2. Non-Pharmacological Therapies: Electro-Convulsive Therapy and Sleep Deprivation

Electroconvulsive therapy (ECT) was found to downregulate increased levels of the proinflammatory cytokine TNF-α in patients with MD (Hestad et al., 2003).

An immune analysis during sleep showed an increase of the type-1 monocyte derived cytokines TNF-α and IL-12 and a decrease of the type-2 IL-10 producing monocytes (Dimitrov et al., 2004). In contrary, continuous wakefulness blocked the increase of type-1 and decrease of type-2 cytokines (T. Lange and S. Dimitrov, personal communication). Thus, sleep deprivation may exert therapeutic effects through a low suppression of type-1 cytokines.

36.7. Divergent Effects of the Role of Type-1/Type-2 Immune Activation are associated with Different Effects to the Kynurenine Metabolism in Schizophrenia and Depression

36.7.1. Schizophrenia

The only known naturally occurring NMDA receptor antagonist in the human CNS is kynurenic acid (KYNA; Stone, 1993). KYNA is one of the three neuroactive intermediate products of the kynurenine pathway. Kynurenine (KYN) is the primary major degradation product of tryptophan (TRP). While the excitatory KYN metabolites 3-hydroxykynurenine (3HK) and quinolinic acid are synthesized from KYN en route to NAD, KYNA is formed in a dead end side arm of the pathway (Schwarcz and Pellicciari, 2002).

KYNA acts both, as a blocker of the glycine coagonistic site of the NMDA receptor (Kessler et al., 1989) and as a non-competitive inhibitor of the $\alpha 7$ nicotinic acetylcholine receptor (Hilmas et al., 2001).

The production of KYNA is regulated by IDO and tryptophan 2,3-dioxygenase (TDO). Both enzymes catalyze the first step in the pathway, the degradation from tryptophan to kynurenine. Type-1 cytokines, such as IFN-γ and IL-2 stimulate the activity of IDO (Grohmann et al., 2003). Figure 36.1 shows the relationship between cytokines, enzymes of the tryptophan metabolism, and formation of neuroactive intermediates.

There is a mutual inhibitory effect of TDO and IDO: a decrease in TDO activity occurs concomitantly with IDO induction, resulting in a coordinate shift in the site (and cell types) of tryptophan degradation (Takikawa et al., 1986). While it has been known for a long time that IDO is expressed in different types of CNS cells, TDO was thought for many years to be restricted to liver tissue. It is known today, however, that TDO is also expressed in CNS cells, probably restricted to astrocytes (Miller et al., 2004).

The type-2 or Th-2 shift in schizophrenia may result in two functional consequences: The expression of IDO, normally increased by type-2 cytokines, in particular INF-γ, is down-regulated, while TDO is up-regulated. The type-1/type-2 imbalance is associated with the IDO/TDO imbalance.

Aditionally, the type-1/type-2 imbalance is associated with the activation of astrocytes and an imbalance in the activation of astrocytes/microglial cells (Aloisi et al., 2000). The functional overweight of astrocytes may lead to a further accumulation of KYNA.

Indeed, a study referring to the expression of IDO and TDO in schizophrenia showed exactly the expected results. An increased expression of TDO compared to IDO was observed in schizophrenic patients and the increased TDO expression was found, as expected, in astrocytes, not in microglial cells (Miller et al., 2004).

36.7.2. Major Depression

The enzyme indoleamine 2,3-dioxygenase (IDO) metabolizes tryptophan to kynurenine, kynurenine is then converted to quinolinic acid via the intermediate 3-HK by the enzyme kynurenine hydroxylase. Both IDO and kynurenine hydroxylase are induced by the type-1 cytokine IFN-γ. The activity of IDO is an important regulatory component in the control of lymphocyte proliferation, the activation of the type-1 immune response and the regulation of the tryptophan metabolism (Mellor and Munn, 1999). It induces a halt in the lymphocyte cell cycle due to the catabolism of tryptophan (Munn et al., 1999). In contrast to the type-1 cytokines, the type-2 cytokines IL-4 and IL-10 inhibit the IFN-γ-induced IDO-mediated tryptophan catabolism (Weiss et al., 1999). IDO is located in several cell types including monocytes and microglial cells (Alberati et al., 1996). An IFN-γ-induced, IDO-mediated decrease of CNS

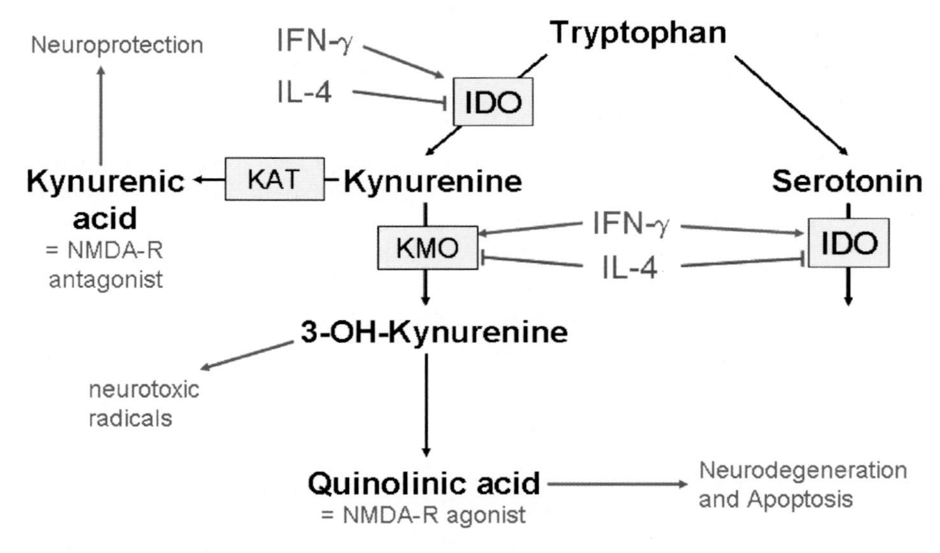

FIGURE 36.1. Neuroimmune interactions of Kynurenine intermediates. Metabolism of tryptophan via the kynurenine pathway leads to several neuroactive intermediates: Kynurenic acid (synthesised by kynurenine aminotransferase, KAT) has neuroprotective properties through antagonism at the NMDA receptor. Quinolinic acid, in contrast, is a NMDA receptor agonist. Both, 3-hydroxykynurenine (3-OH-kynurenine) and quinolinic acid can induce neurodegeneration and apoptosis through induction of excitotoxicity and generation of neurotoxic radicals, respectively. Activity of the key enzyme of the kynurenine pathway, indoleamine 2,3-dioxygenase (IDO), and of the 3-OH-kynurenine forming enzyme kynurenine monoxygenase (KMO) is induced by proinflammatory cytokines like interferon-γ (IFN-γ) and inhibited by anti-inflammatory cytokines like interleukin-4 (IL-4). Serotonin is normally degraded to 5-hydroxyindoleacetic acid (5-HIAA), but the indole ring of serotonin can also be cleaved by IDO. (green arrows = activation; red arrows = inhibition.)

tryptophan availability may lead to a serotonergic deficiency in the CNS, since tryptophan availability is the limiting step in the serotonin synthesis. Other proinflammatory molecules such as PGE_2 or TNF-α, however, induce synergistically with IFN-γ the increase of IDO activity (Braun et al., 2005; Kwidzinski et al., 2005; Robinson et al., 2005). Therefore, not only IFN-γ and type-1 cytokines, but also other proinflammatory molecules induce IDO activity. Since increased levels of PGE_2 and TNF-α were described in MD, other proinflammatory molecules contribute to IDO activation and tryptophan consumption, too (e.g. Linnoila et al., 1983; Mikova et al., 2001).

One of the most consistent findings in biochemical research dealing with mental disorders is that some patients with low 5-hydroxyindoleacetic acid (5-HIAA)—the metabolite of serotonin—in CSF are prone to commit suicide (Lidberg et al., 2000; Mann and Malone, 1997; Nordstrom et al., 1994). This gives additional evidence for a possible link between the type-1 cytokine IFN-γ and the IDO-related reduction of serotonin availability in the CNS of suicidal patients.

A recent study showed that immunotherapy with IFN-α was followed by an increase of depressive symptoms and serum kynurenine concentrations on the one hand, and reduced concentrations of tryptophan and serotonin on the other hand (Bonaccorso et al., 2002). The kynurenine/tryptophan ratio, which reflects the activity of IDO, increased significantly. Changes in depressive symptoms were significantly positively correlated with kynurenine and negatively correlated with serotonin concentrations (Bonaccorso et al., 2002). This study and others (Capuron et al., 2003) clearly show that the IDO activity is increased by IFN, leading to an increased kynurenine production and a depletion of tryptophan and serotonin. The further metabolism of kynurenine, however, seems to play an additional crucial role for the psychopathological states.

In addition to the effects of the proinflammatory immune response on serotonin metabolism, other neurotransmitter systems, in particular the catecholaminergic system, are involved in depression, too. Although the relationship of immune activation and the lack of catecholaminergic neurotransmission has not been well studied, the increase of the monoamino-oxidase (MAO) activity, which leads to decreased noradrenergic neurotransmission, might be an indirect effect of the increased production of kynurenine and quinolinic acid (Schiepers et al., 2005).

36.8. Astrocytes, Microglia, and Type-1/Type-2 Response

The cellular sources for the polarized immune response in the CNS are astrocytes and microglia cells. Microglial cells, deriving from peripheral macrophages, secrete preferably type-1 cytokines such as IL-12, while astrocytes inhibit the production of IL-12 and ICAM-1 and secrete the type-2 cytokine IL-10 (Xiao and Link, 1999; Aloisi et al., 1997). Therefore, the type-1/type-2 imbalance in the CNS seems to be represented by the imbalance in the activation of microglial cells and astrocytes. The view of an over-activation of astrocytes in schizophrenia is supported by the finding of increased levels of S100B—a marker of astrocyte activation—independent of the medication state of the schizophrenic patients (Lara et al., 2001; Schroeter et al., 2003; Schmitt et al., 2005; Rothermundt et al., 2004a, b). Microglia activation, however, was only found in a small percentage of schizophrenics and is speculated to be a medication effect (Bayer et al., 1999). A type-1 immune activation as an effect of neuroleptic treatment has repeatedly been observed.

Since the type-1 activation predominates the response of the peripheral immune system in depression, a dominance of microglial activation compared to astrocyte activation should be observed in depression. Glial reductions were consistently found in brain circuits known to be involved in mood disorders, such as in the limbic and prefrontal cortex (Cotter et al., 2002; Ongur et al., 1998; Rajkowska et al., 1999, 2001; Rajkowska, 2003). Although several authors did not differentiate between microglial and astrocytic loss, this difference is crucial due to the different effects of the type-1/type-2 immune response. Recent studies, however, show that astrocytes are diminished in patients suffering from depression (Johnston-Wilson et al., 2000; Miguel-Hidalgo et al., 2000; Si et al., 2004), although the data are not fully consistent (Davis et al., 2002). A loss of astrocytes was in particular observed in younger depressed patients: the lack of GFAP-immunoreactive astrocytes reflects a lowered activity of responsiveness in those cells (Miguel-Hidalgo et al., 2000). A loss of astrocytes was found in many cortical layers and in different sections of the dorsolateral prefrontal cortex in depression (Rajkowska, 2005).

Moreover, a loss of astrocytes is associated with an impaired reuptake of glutamate from the extracellular space into astrocytes by high affinity glutamate transporters, (Choudary et al., 2005; Gegelashvili et al., 2001). Impaired glutamate reuptake from the synaptic cleft by astroglia prolongs synaptic activation by glutamate (Auger and Attwell, 2000; Danbolt, 2001). Accordingly, increased glutamatergic activity was repeatedly observed in patients with depression (Sanacora et al., 2004a).

36.9. The Glutamate System in Schizophrenia and Depression

36.9.1. The Glutamate System in Schizophrenia: Kynurenic Acid as a Schizophrenogenic Substance

In contrast to microglial cells, which produce quinolinic acid, astrocytes play a key role in the production of kynurenic acid (KYNA) in the CNS. Astrocytes are the main source of KYNA (Heyes et al., 1997). The cellular localization of the kynurenine metabolism is primarily in macrophages and microglial cells, but also in astrocytes (Speciale and Schwarcz, 1993; Kiss et al., 2003). KYN-OHse, however, a critical enzyme in the kynurenine metabolism, is absent in human astrocytes (Guillemin et al., 2001). Accordingly, it has been described that astrocytes cannot produce the product 3HK but they are able to produce large amounts of early kynurenine metabolites, such as KYN and KYNA (Guillemin et al., 2001). This supports the observation that inhibition of KYN-OHse leads to an increase of the KYNA production in the CNS (Chiarugi et al., 1996). The complete metabolism of kynurenine to quinolinic acid is observed only in microglial cells, not in astrocytes. Therefore, due to the lack of KYN-OHse, KYNA accumulates in astrocytes.

A second key-player in the metabolizing of 3-HK are monocytic cells infiltrating the CNS. They help astrocytes in the further metabolism to quinolinic acid (Guillemin et al., 2001). However, the low levels of sICAM-1 (ICAM-1 is the molecule that mainly mediates the penetration of monocytes and lymphocytes into the CNS) in the serum and in the CSF of non-medicated schizophrenic patients (Schwarz et al., 2000) and the increase of adhesion molecules during antipsychotic therapy indicate that the penetration of monocytes may be reduced in non-medicated schizophrenic patients (Müller et al., 1999).

36.9.2. The Glutamate System in Depression: Quinolinic Acid as a Depressiogenic Substance

Apart from certain liver cells, only macrophage derived cells are able to convert tryptophan into quinolonic acid (Saito et al., 1993). Interestingly, in a model of infection, the highest concentrations of quinolinic acid are found in the gray and white matter of the cortex, not in subcortical areas. High levels of quinolinic acid therefore may lead to cortical dysfunction (Heyes et al., 1998).

The strong association between cortical quinolinic acid concentrations and local IDO activity supports the view that the induction of IDO is an important event in initiating the increase of quinolinic acid production (Heyes et al., 1992). In the CNS, invaded macrophages and microglial cells are able to produce quinolinic acid (Saito et al., 1993). Peripheral immune stimulation, however, under certain conditions also leads to increased CNS concentration of quinolinic acid (Saito et al., 1993). During a local inflammatory CNS process, the quinolinic acid production in the CNS increases without changes of the blood-levels of quinolinic acid, While the local quinolinic acid production correlates with the level of ß2 microglobulin, an inflammatory marker. Local CNS concentrations of quinolinic acid are able to far exceed the blood levels (Heyes et al., 1998).

A recent study showed that depressive symptoms are related to an increased ratio of KYN/KYNA in depression (Wichers et al., 2005). The increase of this ratio reflects that in depressed states KYN is preferentially metabolized to quinolinic acid—while the KYNA pathway is neglected. Therefore the preferred metabolism to quinolinic acid—but not the KYNA metabolism—is associated with more pronounced depressive symptoms.

An increase of quinolinic acid is strongly associated with several prominent features of depression: decrease in reaction time (Martin et al., 1992, 1993; Heyes et al., 1991) and cognitive deficits, in particular difficulties in learning (Heyes et al., 1998). In an animal model, an increase of quinolinic acid and 3-hydroxykynurenine was associated with anxiety (Lapin, 2003).

Accordingly, since the activity of the enzyme 3-OHase—directing to the production of quinolinic acid—is inhibited by type-2 cytokines but activated by proinflammatory type-1 cytokines (Alberati et al., 1996; Alberati and Cesura, 1998; Chiarugi et al., 2001), an increased production of quinolinic acid in depressive states would be expected.

The increased levels of quinolinic acid—as NMDA receptor agonist—lead to increased levels of glutamate. Quinolinic acid was shown to cause an overrelease of glutamate in the striatum and in the cortex, presumably by presynaptic mechanisms (Fedele and Foster 1993; Chen et al., 1999). The quinolinic pathway might be the mechanism involved in the increased glutamatergic neurotransmission in MD (Sanacora et al., 2004a).

Moreover, the plasma levels and expression in the brain of ICAM-1 seem to be related to depression: in patients treated with IFN-α, increased sICAM-1 levels were observed to be associated with increased depressive symptoms (Schäfer et al., 2004) and increased expression of ICAM-1 was found in the prefrontal cortex of elder depressive patients (Thomas et al., 2000). ICAM-1 is a type-1 related protein and a cell-adhesion molecule expressed on macrophages and lymphocytes. Increased expression of ICAM-1 is observed in inflammatory processes and promotes the influx of peripheral immune cells through the blood-brain barrier (Rieckmann et al., 1993). By this mechanism, macrophages and co-stimulatory lymphocytes can invade the CNS, further increasing the proinflammatory immune response. Moreover, inflammation is also associated with increased invasion of macrophages/microglia in the CNS (Lane et al., 1996) possibly further switching the balance of type-1/type-2 immune response in favour of type-1 including microglial activation. The alterations of the kynurenine pathway are summarized in Figure 36.2.

36.10. Prostaglandins in Schizophrenia and Depression

The role of PGE_2 in schizophrenia is not well studied; increased levels of PGE_2, however, have been described (Kaiya et al., 1989). PGE_2 induces the production of IL-6, which is consistently described to be increased in schizophrenia. Moreover, an increased COX-2 expression—which is stimulated by PGE_2—was found in schizophrenia (Das and Khan, 1998).

PGE_2 is a molecule of the proinflammatory cascade and stimulates the production of proinflammatory cytokines, e.g. IL-6, the expression of cyclooxygenase-2, and—as a cofactor—the expression of IDO. Therefore an increased secretion of PGE_2 would be expected in depressive disorders,

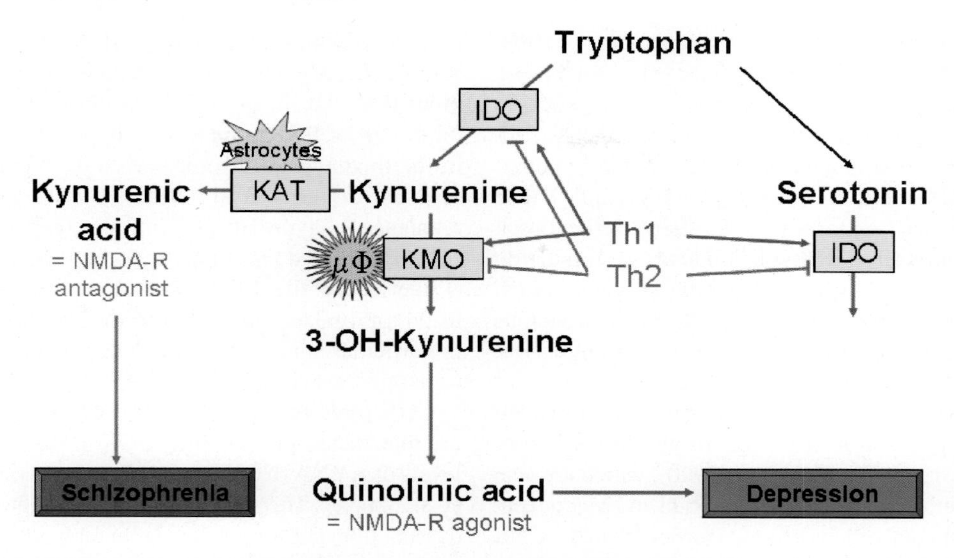

FIGURE 36.2. Alterations of the Kynurenine Metabolism in Schizophrenia and Major Depression Proinflammatory (Th1) cytokines are related to microglia activation, while antiinflammatory (Th2) cytokines are related to activation of astrocytes. Since astrocytes exhibit the kynurenic acid synthesising enzyme kynurenine aminotransferase (KAT), the production of this NMDA receptor antagonist may be induced during Th2 dominated immune response in schizophrenia. Microglia cells, on the other hand, exhibit indoleamine 2,3-dioxygenase (IDO) and kynurenine monoxygenase (KMO), the two key enzymes in the production of the NMDA receptor agonist quinolinic acid. Overproduction of quinolinic acid together with an increased cleavage of the indole ring of serotonin may be involved in the pathophysiology of major depression. (red arrows: mechanisms in schizophrenia; green arrows: mechanisms in depression; ↓ = activation; ⊥ = inhibition.)

too. Accordingly, in-vitro studies show an increased PGE_2 secretion from lymphocytes of depressed patients compared to healthy controls (Song et al., 1998). Increased concentrations of PGE_2 in saliva of depressed patients have been repeatedly described (Ohishi et al., 1988). Moreover, increased levels of PGE_2 have been observed, both in the cerebrospinal fluid and in the serum of depressed patients (Linnoila et al., 1983; Calabrese et al., 1986). As mentioned above, several antidepressants inhibit cytokine-induced PGE_2 production by inflammatory cells (Yaron et al., 1999).

36.11. The Role of the Hormones of the Hypothalamus-Hypophysis-Adrenal-Axis

As a product of activated monocytes and macrophages, IL-6 is one of the most frequently investigated immune parameters in patients suffering from major depression (MD). Most of the publications report a marked increase of in-vitro IL-6 production (Maes et al., 1993) or serum IL-6 levels in depressed patients (Maes et al., 1995a, 1997; Sluzewska et al., 1996; Berk et al., 1997; Frommberger et al., 1997; Song et al., 1998). Contradictory results are very few, indicating reduced (Katila et al., 1994), or not altered serum IL-6 levels (Brambilla and Maggioni, 1998). An age-related increase of IL-6 serum values was reported in patients with major depression (Ershler et al., 1993). The potential influence of possibly interfering variables, however, such as smoking, gender, recent infections and prior medication to IL-6 release and concentration must be considered (Haack et al., 1999).

Although IL-6 is not a type-1 cytokine, it may contribute to IDO activation by the stimulatory effect on PGE_2, which acts as cofactor in the activation of IDO. This fits with a report on the correlation of increased IL-6 production in vitro with decreased tryptophan levels in depressed patients that emphasizes the influence of IL-6 on the serotonin metabolism in depressed patients (Maes et al., 1993).

There is no doubt that IL-6 is involved in the modulation of the HPA axis and increased availability of IL-6 in the hypothalamus is associated with increased HPA activity (Plata-Salaman, 1991). Activation of the HPA axis is one of the best-documented changes in major depression (Roy et al., 1987). Stress acts as a predisposing factor for MD, an increased susceptibility to stress has repeatedly been described in patients with MD, even prior to their first exacerbation of the disorder and psychosocial stressors frequently precede the onset of MD. Additionally, an altered HPA axis physiology and dysfunctions of the extrahypothalamic CRH system have been consistently found in subjects with MD (Hasler et al., 2004). Several studies demonstrate that MD patients exhibit higher baseline cortisol levels or at least much higher cortisol levels during the recovery period after psychological stress (Burke et al., 2005).

The effect of chronic stress on the peripheral immune system and its relevance for major depression has extensively been discussed (O'Brien et al., 2004). Recent in vivo evidence now suggests that stress-induced elevation of glucocorticoids also enhances immune function within the CNS through microglia activation and proliferation. Animal studies show that stress induces an enhanced expression of proinflammatory factors like IL-1β (Pugh et al., 1999; Nguyen et al., 1998), macrophage migration inhibitory factor (MIF) (Bacher et al., 1998; Niino et al., 2000; Suzuki et al., 2000) and cyclooxygenase-2 (COX-2) (Madrigal et al., 2003) in the brain.

On the other hand, pro-inflammatory cytokines such as IL-1 and IL-6 are known to stimulate the HPA-axis via hypothalamic neurons. For example, the release of the CRH and GHRH is stimulated by IL-1 (Besedovsky et al., 1986; Berkenbosch et al., 1987), the central IL-1 upregulation leads to stimulation of CRH, the HPA-axis, and the sympathetic nervous system (Sundar et al., 1990; Weiss et al., 1994). Therefore, a vicious circle may be induced, if the stress response is not limited, as it is discussed in MD (see Figure 36.2).

Elevation of these proinflammatory factors is accompanied with dendritic atrophy and neuronal death within the hippocampus (Sapolsky, 1985; Woolley et al., 1990), which are also found in brains of subjects with MD (Campbell and Macqueen, 2004). These detrimental effects of glucocorticoids in the CNS are mediated by a rise in extracellular glutamate (Moghaddam et al., 1994; Stein-Behrens et al., 1994) and subsequent overstimulation of the NMDA receptor. Such an over-stimulation of the NMDA receptor results in excitotoxic neuronal damage (Takahashi et al., 2002). Nair and Bonneau could demonstrate that restraint-induced psychological stress stimulates proliferation of microglia, which was prevented by blockade of corticosterone synthesis, the glucocorticoid receptor, or the NMDA receptor (Nair and Bonneau, 2006). These data show that stress-induced microglia proliferation is mediated by corticosterone-induced, NMDA receptor-mediated activation within the CNS. Moreover, NMDA receptor activation during stress leads again to increased expression of COX-2 and PGE_2. Both, COX-2 and PGE_2 per se are able to stimulate microglia activation. On the other hand, the functional effects of IL-1 in the CNS—sickness behaviour being one of these effects—were also shown to be antagonized by treatment with a selective COX-2 inhibitor (Cao et al., 1999).

36.12. COX-2 Inhibitors in Schizophrenia and Depression

36.12.1. Cyclooxygenase-2 (COX-2) Inhibitors Inhibit the Production of Kynurenic Acid, Balance the Type-1/Type-2 Immune Response, and Have Therapeutic Effects in Early Stages of Schizophrenia

One class of modern drugs is well known to induce a shift from the type-1 to a type-2 dominated immune response: the selective cyclooxygenase-2 (COX-2) inhibitors. Several studies demonstrated the type-2 inducing effect of PGE_2—the major product of COX-2—while inhibition of COX-2 is accompanied by inhibition of type-2 cytokines and induction of type-1 cytokines (Pyeon et al., 2000; Stolina et al., 2000). Recently, PGE_2 has been shown to enhance the production of type-2 cytokines such as IL-4, IL-5, IL-6, and IL-10; PGE_2 also drastically inhibits the production of the type-1 cytokines IFN-γ, IL-2, and IL-12 (Stolina et al., 2000). Therefore, inhibition of PGE_2 synthesis is hypothesized to be beneficial in the treatment of disorders with dysregulated T-helper cell responses (Harris et al., 2002). COX-2 inhibition seems to rebalance the type-1/type-2 immune response by inhibition of IL-6, PGE_2, and by stimulating the type-1 immune response (Litherland et al., 1999). Therefore COX-2 inhibition seems to be a promising approach in the therapy of schizophrenia, in particular since increased COX-2 expression was found in schizophrenia (Das and Khan, 1998).

COX-inhibition provokes differential effects on the kynurenine metabolism: while COX-1 inhibition increases the levels of KYNA, COX-2 inhibition decreases them (Schwieler et al., 2005). Therefore, psychotic symptoms and cognitive dysfunctions, observed during therapy with COX-1 inhibitors, were assigned to the COX-1 mediated increase of KYNA. The reduction of KYNA levels—by a prostaglandin-mediated mechanism—might be an additional mechanism to the above described immunological mechanism for therapeutic effects of selective COX-2 inhibitors in schizophrenia (Schwieler et al., 2005).

Indeed, in a prospective, randomized, double-blind study of therapy with the COX-2 inhibitor celecoxib add-on to risperidone in acute exacerbation of schizophrenia, a therapeutic effect of celecoxib was observed (Müller and Schwarz, 2002). Immunologically, an increase of the type-1 immune response was found in the celecoxib treatment group (Müller et al., 2004a). The clinical effect of COX-2 inhibition was especially pronounced regarding cognition in schizophrenia (Müller et al., 2005). The finding of a clinical advantage of COX-2 inhibition, however, could not be replicated in a second study. Further analysis of the data revealed that the outcome depends on the duration of the disease (Müller et al., 2004b). The efficacy of therapy with a COX-2 inhibitor seems most pronounced in the first years of the schizophrenic disease process. This observation is in accordance with results from animal studies showing that the effects of COX-2 inhibition on cytokines, hormones, and particularly on behavioural symptoms are dependent on the duration of the preceding changes and the time-point of application of the COX-2 inhibitor (Casolini et al., 2002). A point of no return for therapeutic effects regarding the pathological changes during an inflammatory process has to be postulated.

36.12.2. COX-2 Inhibition as a Possible Antiinflammatory Therapeutic Approach in Depression

Due to the increase of proinflammatory cytokines and PGE2 in depressed patients, antiinflammatory treatment would be expected to show antidepressant effects also in depressed patients. In particular, COX-2 inhibitors seems to show advantageous results: animal studies show that COX-2 inhibition can lower the increase of the pro-inflammatory cytokines IL-1β, TNF-α, and of PGE2, but it can also prevent clinical symptoms such as anxiety and cognitive decline, which are associated with this increase of pro-inflammatory cytokines (Casolini et al., 2002). Moreover, treatment with the COX-2 inhibitor celecoxib—but not with a COX-1 inhibitor—prevented the dysregulation of the HPA-axis, in particular the increase of cortisol, one of the biological key features associated with depression (Casolini et al., 2002; Hu et al., 2005). This effect can be expected because PGE2 stimulates the HPA-axis in the CNS (Song and Leonard, 2000) and PGE2 is inhibited by COX-2 inhibition. Moreover, the functional effects of IL-1 in the CNS—sickness behaviour being one of these effects—were also shown to be antagonized by treatment with a selective COX-2 inhibitor (Cao et al., 1999).

Additionally, COX-2 inhibitors influence—either directly or via CNS-immune mechanisms—the CNS serotonergic system. In a rat model, treatment with rofecoxib was followed by an increase of serotonin in the frontal and the temporo-parietal cortex (Sandrini et al., 2002). Since the lack of serotonin is one of the pinpoints in the pathophysiology of depression, a clinical antidepressant effect of COX-2 inhibitors would be expected due to this effect. A possible mechanism of the antidepressant action of COX-2 inhibitors is the inhibition of the release of IL-1 and IL-6. Moreover, COX-2 inhibitors also protect the CNS from effects of quinolinic acid (Salzberg-Brenhouse et al., 2003).

Accordingly, a clinical antidepressant effect of rofecoxib was found in 2,228 patients with osteoarthritis, 15% of them showing a co-morbid depressive syndrome, which was evaluated by a specific depression self-report. Co-morbid depression was a significant predictor for worse outcome regarding the ostheoarthritis-related pain to rofecoxib therapy. Surprisingly, there was a significant decrease in the rate of substantive depression during therapy with 25 mg rofecoxib from 15% to 3% of the patients (Collantes-Esteves and Fernandez-Perrez, 2003).

Moreover, we were able to demonstrate a significant therapeutic effect of the COX-2 inhibitor on depressive symptoms in a randomized double blind pilot add-on study using the selective COX-2 inhibitor celecoxib in MD (Müller et al., 2006). Although those preliminary data have to be interpreted cautiously and intense research has to be provided in order to evaluate further the therapeutic effects of COX-2 inhibitors in MD, those results are encouraging for further studies dealing with the inflammatory hypothesis of depression with regard to pathogenesis, course and therapy.

Summary

In schizophrenia and depression, opposite patterns of type-1 – type-2 immune activation seem to be associated with differences in the IDO activation and in the tryptophan—kynurenine metabolism resulting in increased production of kynurenic acid in schizophrenia and of increased quinolinic acid in depression. These differences are associated with an imbalance in the glutamatergic neurotransmission, which may contribute to an overweight of NMDA agonism in depression and of NMDA antagonism in schizophrenia. The differential activation of microglia cells and astrocytes may be an additional mechanism contributing to this imbalance. The immunological imbalance results both in schizophrenia and in depression in an increased PGE_2 production and probably also in an increased COX-2 expression. Although there is strong evidence for the view, that the interactions of the immune system, IDO and the serotonergic system, and glutamatergic neurotransmission play a key role in schizophrenia and in depression, several gaps, e.g. the roles of genetics, disease course, sex, different psychopathological states, etc. have to be bridged by intense further research. Moreover, COX-2 inhibition is only one example for possible therapeutic mechanisms acting on these mechanisms. Also the effects of COX-2 inhibition in the CNS as well as the different components of the inflammatory system, the kynurenine-metabolism and the glutamatergic neurotransmission need careful further scientific evaluation.

Acknowledgement. The authors thank Barbara Goldstein-Müller for help in preparation of the manuscript

Review Questions/Problems

1. The NMDA hypothesis of schizophrenia refers to the following neurotransmitter disturbance:

a. mesolimbic hyperactivity of dopamine
b. tuberoinfundibular inhibition of prolactin release by dopamine
c. hypofunction of the glutamatergic cortico-striatal pathway
d. mesolimbic hyperactivity of glutamate
e. all of the above

2. The so far only known endogenous NMDA receptor antagonist is:

a. glutamate
b. glycine
c. ketamine
d. kynurenic acid
e. quinolinic acid

3. The proposed neurotransmitter imbalance in major depression is best characterized by:

a. increased serotonin, increased glutamate, reduced norepinephrine
b. increased serotonin, increased norepinephrine, reduced glutamate
c. increased glutamate, decreased serotonin, decreased norepinephrine
d. none of the above

4. The best replicated immune findings in schizophrenia are:

a. increased in vitro production of IL-2 and decreased serum levels of IL-6
b. decreased in vitro production of IL-2 and IFN-γ and decreased serum levels of IL-6
c. decreased in vitro production of IL-2 and IFN-γ and increased serum levels of IL-6
d. increased serum levels of IL-2 and IFN-γ and decreased in vitro production of IL-6
e. none of the above

5. The following immune response can be frequently detected in depressed patients:

a. an allergy-like Th2 predominance with increased levels of IgE in nearly all depressed patients
b. a decreased Th1 activation in melancholic patients and an increased Th1 activation in non-melancholic depressed patients
c. an increased activation of the adaptive immune system in melancholic patients and an increased activation of the specific immune system in non-melancholic patients
d. an increased Th1 activation in melancholic patients and a decreased Th1 activation in non-melancholic depressed patients

6. The most common psychiatric side effect of IFN-α administration is:

a. paranoid ideation
b. optical and—less frequent—acoustic hallucinations
c. manic episodes
d. depression
e. tics

7. IFN-α administration induces the degradation of the amino acid:

a. choline
b. cysteine
c. serine
d. tryptophan
e. tyrosine

8. The typical immunological effect of antidepressant drugs is:

a. induction of a more specific immune response
b. inhibition of B cell proliferation and antibody production
c. downregulation of Th1 cytokine production
d. downregulation of Th2 cytokine production
e. upregulation of Th1 cytokine production

9. The following enzymes of the kynurenine pathway are induced by pro-inflammatory cytokines such as IFN-g and TNF-a:

a. tryptophan 2,3 dioxygenase, indoleamine 2,3-dioxygenase, kynurenine monoxygenase
b. indoleamine 2,3-dioxygenase, kynurenine hydroxylase (monoxygenase), kynurenine aminotransferase
c. indoleamine 2,3-dioxygenase, kynurenine hydroxylase (monoxygenase)
d. tryptophan 2,3 dioxygenase, kynurenine hydroxylase (monoxygenase)
e. only indoleamine 2,3-dioxygenase

10. The limiting step in serotonin synthesis is in all probability:

a. tryptophan availability
b. serotonin availability in a negative feed back
c. tryptophan hydroxylase activity
d. a and c
e. b and c

11. Glucocorticoids:

a. induce central nervous production of pro-inflammatory mediators
b. inhibit central nervous production of pro-inflammatory mediators
c. induce glutamate signal at the NMDA receptor
d. inhibit glutamate signal at the NMDA receptor
e. a and c
f. b and d
g. b and c

12. Astrocytes are:

a. immunologically ‘silent’ glial cells
b. responsible for a Th1-like immune response within the CNS
c. part of the nerve sheath within the brain
d. responsible for a Th2-like immune response within the CNS
e. not involved in neurotransmitter reuptake and metabolism

13. Proinflammatory cytokines:

a. stimulate the HPA axis
b. have no effect on the HPA axis
c. induce dendritic atrophy
d. none of the above
e. a and c
f. b and c

14. Inhibition of cyclooxygenase isoforms has the following effects:

a. COX-1 inhibition increases the levels of kynurenic acid
b. COX-2 inhibition decreases the levels of kynurenic acid
c. COX-2 inhibition induces a Th1 shift under non-inflammatory conditions
d. all of the above
e. b and c

15. Administration of the COX-2 inhibitor celecoxib was shown to:

a. induce exacerbation of psychotic symptoms in schizophrenia
b. improve psychotic and especially cognitive in schizophrenia
c. be more efficient in chronic schizophrenia than in the early stages
d. enhance antidepressant therapy when added to a conventional antidepressant drug
e. a and c
f. b and d

References

Abou-Saleh MT, Ghubash R, Karim L, Krymski M, Anderson DN (1999) The role of pterins and related factors in the biology of early postpartum depression. Eur Neuropsychopharmacol 9:295–300.

Alberati GD, Cesura AM (1998) Expression of the kynurenine enzymes in macrophages and microglial cells: Regulation by immune modulators. Amino Acids 14:251–255.

Alberati GD, Ricciardi CP, Kohler C, Cesura AM (1996) Regulation of the kynurenine metabolic pathway by interferon-gamma in murine cloned macrophages and microglial cells. J Neurochem 66:996–1004.

Aloisi F, Penna G, Cerase J, Menendez IB, Adorini L (1997) IL-12 production by central nervous system microglia is inhibited by astrocytes. J Immunol 159:1604–1612.

Aloisi F, Ria F, Adorini L (2000) Regulation of T-cell responses by CNS antigen-presenting cells: Different roles for microglia and astrocytes. Immunol Today 21:141–147.

Altamura CA, Mauri MC, Ferrara A, Moro AR, D'Andrea G, Zamberlan F (1993) Plasma and platelet excitatory amino acids in psychiatric disorders. Am J Psychiatry 150:1731–1733.

Amirkhani A, Rajda C, Arvidsson B, Bencsik K, Boda K, Seres E, Markides KE, Vecsei L, Bergquist J (2005) Interferon-beta affects the tryptophan metabolism in multiple sclerosis patients. Eur J Neurol 12:625–631.

Auger C, Attwell D (2000) Fast removal of synaptic glutamate by postsynaptic transporters. Neuron 28:547–558.

Bacher M, Meinhardt A, Lan HY, Dhabhar FS, Mu W, Metz CN, Chesney JA, Gemsa D, Donnelly T, Atkins RC, Bucala R (1998) MIF expression in the rat brain: Implications for neuronal function. Mol Med 4:217–230.

Baron DA, Hardie T, Baron SH (1993) Possible association of interleukin-2 treatment with depression and suicide. J Am Osteopath Assoc 93:799–800.

Bayer TA, Buslei R, Havas L, Falkai P (1999) Evidence for activation of microglia in patients with psychiatric illnesses. Neurosci Lett 271:126–128.

Bechter K, Schreiner V, Herzog S, Breitinger N, Wollinsky KH, Brinkmeier H, Aulkemeyer P, Weber F, Schuttler R (2003) [CSF filtration as experimental therapy in therapyresistant psychoses in Borna disease virus-seropositive patients]. Psychiatr Prax 30:216–220.

Bengtsson BO, Zhu J, Thorell LH, Olsson T, Link H, Walinder J (1992) Effects of zimeldine and its metabolites, clomipramine, imipramine and maprotiline in experimental allergic neuritis in Lewis rats. J Neuroimmunol 39:109–122.

Berk M, Wadee AA, Kuschke RH, O'Neill-Kerr A (1997) Acute phase proteins in major depression. J Psychosom Res 43:529–534.

Berkenbosch F, van Oers J, del Rey A, Tilders F, Besedovsky H (1987) Corticotropin-releasing factor-producing neurons in the rat activated by interleukin-1. Science 238:524–526.

Besedovsky H, del Rey A, Sorkin E, Dinarello CA (1986) Immunoregulatory feedback between interleukin-1 and glucocorticoid hormones. Science 233:652–654.

Bonaccorso S, Lin AH, Verkerk R, van Hunsel F, Libbrecht I, Scharpe S, DeClerck L, Biondi M, Janca A, Maes M (1998) Immune markers in fibromyalgia: Comparison with major depressed patients and normal volunteers. J Affect Disord 48:75–82.

Bonaccorso S, Marino V, Puzella A, Pasquini M, Biondi M, Artini M, Almerighi C, Verkerk R, Meltzer H, Maes M (2002) Increased depressive ratings in patients with hepatitis C receiving interferon-alpha-based immunotherapy are related to interferon-alpha-induced changes in the serotonergic system. J Clin Psychopharmacol 22:86–90.

Brambilla F, Maggioni M (1998) Blood levels of cytokines in elderly patients with major depressive disorder. Acta Psychiatr Scand 97:309–313.

Braun D, Longman RS, Albert ML (2005) A two-step induction of indoleamine 2,3 dioxygenase (IDO) activity during dendritic-cell maturation. Blood 106:2375–2381.

Brown AS, Begg MD, Gravenstein S, Schaefer CA, Wyatt RJ, Bresnahan M, Babulas VP, Susser ES (2004) Serologic evidence of prenatal influenza in the etiology of schizophrenia. Arch Gen Psychiatry 61:774–780.

Buka SL, Tsuang MT, Torrey EF, Klebanoff MA, Bernstein D, Yolken RH (2001) Maternal infections and subsequent psychosis among offspring. Arch Gen Psychiatry 58:1032–1037.

Burke HM, Davis MC, Otte C, Mohr DC (2005) Depression and cortisol responses to psychological stress: A meta-analysis. Psychoneuroendocrinology 30:846–856.

Calabrese JR, Skwerer RG, Barna B, Gulledge AD, Valenzuela R, Butkus A, Subichin S, Krupp NE (1986) Depression, immunocompetence, and prostaglandins of the E series. Psychiatry Res 17:41–47.

Campbell S, Macqueen G (2004) The role of the hippocampus in the pathophysiology of major depression. J Psychiatry Neurosci 29:417–426.

Cao C, Matsumura K, Ozaki M, Watanabe Y (1999) Lipopolysaccharide injected into the cerebral ventricle evokes fever through induction of cyclooxygenase-2 in brain endothelial cells. J Neurosci 19:716–725.

Capuron L, Ravaud A, Neveu PJ, Miller AH, Maes M, Dantzer R (2002) Association between decreased serum tryptophan concentrations and depressive symptoms in cancer patients undergoing cytokine therapy. Mol Psychiatry 7:468–473.

Capuron L, Neurauter G, Musselman DL, Lawson DH, Nemeroff CB, Fuchs D, Miller AH (2003) Interferon-alpha-induced changes in tryptophan metabolism. Relationship to depression and paroxetine treatment. Biol Psychiatry 54:906–914.

Carlsson A (1988) The current status of the dopamine hypothesis of schizophrenia. Neuropsychopharmacology 1:179–186.

Carlsson A, Waters N, Holm-Waters S, Tedroff J, Nilsson M, Carlsson ML (2001) Interactions between monoamines, glutamate, and GABA in schizophrenia: New evidence. Annu Rev Pharmacol Toxicol 41:237–260.

Casolini P, Catalani A, Zuena AR, Angelucci L (2002) Inhibition of COX-2 reduces the age-dependent increase of hippocampal inflammatory markers, corticosterone secretion, and behavioral impairments in the rat. J Neurosci Res 68:337–343.

Cazzullo CL, Scarone S, Grassi B, Vismara C, Trabattoni D, Clerici M, Clerici M (1998) Cytokines production in chronic schizophrenia patients with or without paranoid behaviour. Prog Neuropsychopharmacol Biol Psychiatry 22:947–957.

Chen Q, Surmeier DJ, Reiner A (1999) NMDA and non-NMDA receptor-mediated excitotoxicity are potentiated in cultured striatal neurons by prior chronic depolarization. Exp Neurol 159:283–296.

Chiarugi A, Carpenedo R, Moroni F (1996) Kynurenine disposition in blood and brain of mice: Effects of selective inhibitors of kynurenine hydroxylase and of kynureninase. J Neurochem 67:692–698.

Chiarugi A, Calvani M, Meli E, Traggiai E, Moroni F (2001) Synthesis and release of neurotoxic kynurenine metabolites by human monocyte-derived macrophages. J Neuroimmunol 120:190–198.

Choudary PV, Molnar M, Evans SJ, Tomita H, Li JZ, Vawter MP, Myers RM, Bunney WE, Jr., Akil H, Watson SJ, Jones EG (2005) Altered cortical glutamatergic and GABAergic signal transmission with glial involvement in depression. Proc Natl Acad Sci USA 102:15653–15658.

Collantes-Esteves E, Fernandez-Perrez Ch (2003) Improved self-control of ostheoarthritis pain and self-reported health status in non-responders to celecoxib switched to rofecoxib: Results of PAVIA, an open-label post-marketing survey in spain. Curr Med Res Opin 19:402–410.

Coppen A, Swade C (1988) 5-HT and Depression: The Present Position. In: New Concepts in Depression, (Briley M, Fillion G, eds), pp. 120–136. London: MacMillan Press.

Coric V, Milanovic S, Wasylink S, Patel P, Malison R, Krystal JH (2003) Beneficial effects of the antiglutamatergic agent riluzole in a patient diagnosed with obsessive-compulsive disorder and major depressive disorder. Psychopharmacology (Berl) 167:219–220.

Cotter D, Pariante C, Rajkowska G (2002) Glial Pathology in Major Psychiatric Disorders. In: The Post-Mortem Brain in Psychiatric Research (Agam G, Belmaker RH, Everall I, eds), pp. 291–324. Boston: Kluwer Acad Pub.

Crane GE (1959) Cyloserine as an antidepressant agent. Am J Psychiatry 115:1025–1026.

Danbolt NC (2001) Glutamate uptake. Prog Neurobiol 65:1–105.

Dantzer R (2001) Cytokine-induced sickness behavior: Where do we stand? Brain Behav Immun 15:7–24.

Das I, Khan NS (1998) Increased arachidonic acid induced platelet chemiluminescence indicates cyclooxygenase overactivity in schizophrenic subjects. Prostaglandins Leukot Essent Fatty Acids 58:165–168.

Davis S, Thomas A, Perry R, Oakley A, Kalaria RN, O'Brien JT (2002) Glial fibrillary acidic protein in late life major depressive disorder: An immunocytochemical study. J Neurol Neurosurg Psychiatry 73:556–560.

Dimitrov S, Lange T, Tieken S, Fehm HL, Born J (2004) Sleep associated regulation of T helper 1/T helper 2 cytokine balance in humans. Brain Behav Immun 18:341–348.

Duch DS, Woolf JH, Nichol CA, Davidson JR, Garbutt JC (1984) Urinary excretion of biopterin and neopterin in psychiatric disorders. Psychiatry Res 11:83–89.

Dunbar PR, Hill J, Neale TJ, Mellsop GW (1992) Neopterin measurement provides evidence of altered cell-mediated immunity in patients with depression, but not with schizophrenia. Psychol Med 22:1051–1057.

Ershler WB, Sun WH, Binkley N, Gravenstein S, Volk MJ, Kamoske G, Klopp RG, Roecker EB, Daynes RA, Weindruch R (1993) Interleukin-6 and aging: Blood levels and mononuclear cell production increase with advancing age and in vitro production is modifiable by dietary restriction. Lymphokine Cytokine Res 12:225–230.

Fedele E, Foster AC (1993) An evaluation of the role of extracellular amino acids in the delayed neurodegeneration induced by quinolinic acid in the rat striatum. Neuroscience 52:911–917.

Fertuzinhos SM, Oliveira JR, Nishimura AL, Pontual D, Carvalho DR, Sougey EB, Otto PA, Zatz M (2004) Analysis of IL-1alpha, IL-1beta, and IL-1RA [correction of IL-RA] polymorphisms in dysthymia. J Mol Neurosci 22:251–256.

Frizzo ME, Dall'Onder LP, Dalcin KB, Souza DO (2004) Riluzole enhances glutamate uptake in rat astrocyte cultures. Cell Mol Neurobiol 24:123–128.

Frommberger UH, Bauer J, Haselbauer P, Fraulin A, Riemann D, Berger M (1997) Interleukin-6-(IL-6) plasma levels in depression and schizophrenia: Comparison between the acute state and after remission. Eur Arch Psychiatry Clin Neurosci 247:228–233.

Gard PR, Handley SL, Parsons AD, Waldron G (1986) A multivariate investigation of postpartum mood disturbance. Br J Psychiatry 148:567–575.

Gattaz WF, Abrahao AL, Foccacia R (2004) Childhood meningitis, brain maturation and the risk of psychosis. Eur Arch Psychiatry Clin Neurosci 254:23–26.

Gegelashvili G, Robinson MB, Trotti D, Rauen T (2001) Regulation of glutamate transporters in health and disease. Prog Brain Res 132:267–286.

Grohmann U, Fallarino F, Puccetti P (2003) Tolerance, DCs and tryptophan: Much ado about IDO. Trends Immunol 24:242–248.

Guillemin GJ, Kerr SJ, Smythe GA, Smith DG, Kapoor V, Armati PJ, Croitoru J, Brew BJ (2001) Kynurenine pathway metabolism in human astrocytes: A paradox for neuronal protection. J Neurochem 78:842–853.

Haack M, Hinze-Selch D, Fenzel T, Kraus T, Kuhn M, Schuld A, Pollmacher T (1999) Plasma levels of cytokines and soluble cytokine receptors in psychiatric patients upon hospital admission: Effects of confounding factors and diagnosis. J Psychiatr Res 33:407–418.

Hall S, Smith A (1996) Investigation of the effects and aftereffects of naturally occurring upper respiratory tract illnesses on mood and performance. Physiol Behav 59:569–577.

Harris SG, Padilla J, Koumas L, Ray D, Phipps RP (2002) Prostaglandins as modulators of immunity. Trends Immunol 23:144–150.

Hasler G, Drevets WC, Manji HK, Charney DS (2004) Discovering endophenotypes for major depression. Neuropsychopharmacology 29:1765–1781.

Herbert TB, Cohen S (1993) Depression and immunity: A meta-analytic review. Psychol Bull 113:472–486.

Heresco-Levy U, Javitt DC, Ermilov M, Mordel C, Silipo G, Lichtenstein M (1999) Efficacy of high-dose glycine in the treatment of enduring negative symptoms of schizophrenia. Arch Gen Psychiatry 56:29–36.

Hestad KA, Tonsed S, Stoen CD, Ueland T, Aukrust T (2003) Raised plasma levels of tumor necrosis factor alpha in patients with depression: Normalization during electroconvulsive therapy. J ECT 19:183–188.

Heyes MP, Brew BJ, Martin A, Price RW, Salazar AM, Sidtis JJ, Yergey JA, Mouradian MM, Sadler AE, Keilp J, Rubinow D, Markey SP (1991) Quinolinic acid in cerebrospinal fluid and serum in

HIV-1 infection: Relationship to clinical and neurological status. Ann Neurol 29:202–209.

Heyes MP, Saito K, Crowley JS, Davis LE, Demitrack MA, Der M, Dilling LA, Elia J, Kruesi MJ, Lackner A, Larsen SA, Lee K, Leonard HL, Markey SP, Martin A, Milstein S, Mouradian MM, Pranzatelli MR, Quearry BJ, Salazar A, Smith M, Strauss SE, Sunderland T, Swedo SW, Tourtellotte WW (1992) Quinolinic acid and kynurenine pathway metabolism in inflammatory and non-inflammatory neurological disease. Brain 115:1249–1273.

Heyes MP, Chen CY, Major EO, Saito K (1997) Different kynurenine pathway enzymes limit quinolinic acid formation by various human cell types. Biochem J 326:351–356.

Heyes MP, Saito K, Lackner A, Wiley CA, Achim CL, Markey SP (1998) Sources of the neurotoxin quinolinic acid in the brain of HIV-1-infected patients and retrovirus-infected macaques. FASEB J 12:881–896.

Hilmas C, Pereira EF, Alkondon M, Rassoulpour A, Schwarcz R, Albuquerque EX (2001) The brain metabolite kynurenic acid inhibits alpha7 nicotinic receptor activity and increases non-alpha7 nicotinic receptor expression: Physiopathological implications. J Neurosci 21:7463–7473.

Hu F, Wang X, Pace TW, Wu H, Miller AH (2005) Inhibition of COX-2 by celecoxib enhances glucocorticoid receptor function. Mol Psychiatry 10:426–428.

Huber TJ, Dietrich DE, Emrich HM (1999) Possible use of amantadine in depression. Pharmacopsychiatry 32:47–55.

Irwin M (1999) Immune correlates of depression. Adv Exp Med Biol 461:1–24.

Johnston-Wilson NL, Sims CD, Hofmann J-P, Anderson L, Shore AD, Torrey EF, Yolken RH (2000) Disease-specific alterations in frontal cortex brain proteins in schizophrenia, bipolar disorder, and major depressive disorder. Mol Psychiatry 5:142–149.

Josefsson A, Berg G, Nordin C, Sydsjo G (2001) Prevalence of depressive symptoms in late pregnancy and postpartum. Acta Obstet Gynecol Scand 80:251–255.

Jun TY, Pae CU, Hoon H, Chae JH, Bahk WM, Kim KS, Serretti A (2003) Possible association between -G308A tumour necrosis factor-alpha gene polymorphism and major depressive disorder in the Korean population. Psychiatr Genet 13:179–181.

Kaiya H, Uematsu M, Ofuji M, Nishida A, Takeuchi K, Nozaki M, Idaka E (1989) Elevated plasma prostaglandin E2 levels in schizophrenia. J Neural Transm 77:39–46.

Katila H, Appelberg B, Hurme M, Rimon R (1994) Plasma levels of interleukin-1 beta and interleukin-6 in schizophrenia, other psychoses, and affective disorders. Schizophr Res 12:29–34.

Kenis G, Maes M (2002) Effects of antidepressants on the production of cytokines. Int J Neuropsychopharmacol 5:401–412.

Kessler M, Terramani T, Lynch G, Baudry M (1989) A glycine site associated with N-methyl-D-aspartic acid receptors: Characterization and identification of a new class of antagonists. J Neurochem 52:1319–1328.

Kim JS, Kornhuber HH, Schmid-Burgk W, Holzmuller B (1980) Low cerebrospinal fluid glutamate in schizophrenia patients and a new hypothesis of glutamatergic neuronal dysfunction. Neurosci Lett 20:379–382.

Kim JS, Schmid-Burgk W, Claus D, Kornhuber HH (1982) Increased serum glutamate in depressed patients. Arch Psychiatr Nervenkr 232:299–304.

Kiss C, Ceresoli-Borroni G, Guidetti P, Zielke CL, Zielke HR, Schwarcz R (2003) Kynurenate production by cultured human astrocytes. J Neural Transm 110:1–14.

Kohl C, Walch T, Huber R, Kemmler G, Neurauter G, Fuchs D, Solder E, Schrocksnadel H, Sperner-Unterweger B (2005) Measurement of tryptophan, kynurenine and neopterin in women with and without postpartum blues. J Affect Disord 86:135–142.

Koponen H, Rantakallio P, Veijola J, Jones P, Jokelainen J, Isohanni M (2004) Childhood central nervous system infections and risk for schizophrenia. Eur Arch Psychiatry Clin Neurosci 254:9–13.

Körschenhausen DA, Hampel HJ, Ackenheil M, Penning R, Müller N (1996) Fibrin degradation products in post mortem brain tissue of schizophrenics: A possible marker for underlying inflammatory processes. Schizophr Res 19:103–109.

Krystal JH, Karper LP, Seibyl JP, Freeman GK, Delaney R, Bremner JD, Heninger GR, Bowers MB, Jr, Charney DS (1994) Subanesthetic effects of the noncompetitive NMDA antagonist, ketamine, in humans. Psychotomimetic, perceptual, cognitive, and neuroendocrine responses. Arch Gen Psychiatry 51:199–214.

Kudoh A, Takahira Y, Katagai H, Takazawa T (2002) Small-dose ketamine improves the postoperative state of depressed patients. Anesth Analg 95:114–8.

Kugaya A, Sanacora G (2005) Beyond monoamines: Glutamatergic function in mood disorders. CNS Spectr 10:808–819.

Kwidzinski E, Bunse J, Aktas O, Richter D, Mutlu L, Zipp F, Nitsch R, Bechmann I (2005) Indolamine 2,3-dioxygenase is expressed in the CNS and down-regulates autoimmune inflammation. FASEB J 19:1347–1349.

Lane JH, Sasseville VG, Smith MO, Vogel P, Pauley DR, Heyes MP, Lackner AA (1996) Neuroinvasion by simian immunodeficiency virus coincides with increased numbers of perivascular macrophages/microglia and intrathecal immune activation. J Neurovirol 2:423–432.

Lanquillon S, Krieg JC, Bening-Abu-Shach U, Vedder H (2000) Cytokine production and treatment response in major depressive disorder. Neuropsychopharmacology 22:370–379.

Lapin IP (2003) Neurokynurenines (NEKY) as common neurochemical links of stress and anxiety. Adv Exp Med Biol 527:121–125.

Lara DR, Gama CS, Belmonte-de-Abreu P, Portela LV, Goncalves CA, Fonseca M, Hauck S, Souza DO (2001) Increased serum S100B protein in schizophrenia: A study in medication-free patients. J Psychiatr Res 35:11–14.

Laruelle M, Abi-Dargham A, van Dyck CH, Gil R, D'Souza CD, Erdos J, McCance E, Rosenblatt W, Fingado C, Zoghbi SS, Baldwin RM, Seibyl JP, Krystal JH, Charney DS, Innis RB (1996) Single photon emission computerized tomography imaging of amphetamine-induced dopamine release in drug-free schizophrenic subjects. Proc Natl Acad Sci USA 93:9235–9240.

Lidberg L, Belfrage H, Bertilsson L, Evenden MM, Asberg M (2000) Suicide attempts and impulse control disorder are related to low cerebrospinal fluid 5-HIAA in mentally disordered violent offenders. Acta Psychiatr Scand 101:395–402.

Linnoila M, Whorton AR, Rubinow DR, Cowdry RW, Ninan PT, Waters RN (1983) CSF prostaglandin levels in depressed and schizophrenic patients. Arch Gen Psychiatry 40:405–406.

Litherland SA, Xie XT, Hutson AD, Wasserfall C, Whittaker DS, She JX, Hofig A, Dennis MA, Fuller K, Cook R, Schatz D, Moldawer LL, Clare-Salzler MJ (1999) Aberrant prostaglandin synthase 2 expression defines an antigen-presenting cell defect for insulin-dependent diabetes mellitus. J Clin Invest 104:515–523.

Madrigal JL, Garcia-Bueno B, Moro MA, Lizasoain I, Lorenzo P, Leza JC (2003) Relationship between cyclooxygenase-2 and nitric oxide synthase-2 in rat cortex after stress. Eur J Neurosci 18:1701–1705.

Maes M, Scharpe S, Meltzer HY, Bosmans E, Suy E, Calabrese J, Cosyns P (1993) Relationships between interleukin-6 activity, acute phase proteins, and function of the hypothalamic-pituitary-adrenal axis in severe depression. Psychiatry Res 49:11–27.

Maes M, Scharpe S, Meltzer HY, Okayli G, Bosmans E, D'Hondt P, Vanden Bossche BV, Cosyns P (1994) Increased neopterin and interferon-gamma secretion and lower availability of L-tryptophan in major depression: Further evidence for an immune response. Psychiatry Res 54:143–160.

Maes M, Meltzer HY, Bosmans E, Bergmans R, Vandoolaeghe E, Ranjan R, Desnyder R (1995a) Increased plasma concentrations of interleukin-6, soluble interleukin-6, soluble interleukin-2 and transferrin receptor in major depression. J Affect Disord 34:301–309.

Maes M, Meltzer HY, Buckley P, Bosmans E (1995b) Plasma-soluble interleukin-2 and transferrin receptor in schizophrenia and major depression. Eur Arch Psychiatry Clin Neurosci 244:325–329.

Maes M, Bosmans E, De Jongh R, Kenis G, Vandoolaeghe E, Neels H (1997) Increased serum IL-6 and IL-1 receptor antagonist concentrations in major depression and treatment resistant depression. Cytokine 9:853–858.

Maes M, Verkerk R, Vandoolaeghe E, Lin A, Scharpe S (1998) Serum levels of excitatory amino acids, serine, glycine, histidine, threonine, taurine, alanine and arginine in treatment-resistant depression: Modulation by treatment with antidepressants and prediction of clinical responsivity. Acta Psychiatr Scand 97:302–308.

Maes M, Song C, Lin AH, Bonaccorso S, Kenis G, De Jongh R, Bosmans E, Scharpe S (1999) Negative immunoregulatory effects of antidepressants: Inhibition of interferon-gamma and stimulation of interleukin-10 secretion. Neuropsychopharmacology 20: 370–379.

Maes M, Verkerk R, Bonaccorso S, Ombelet W, Bosmans E, Scharpe S (2002) Depressive and anxiety symptoms in the early puerperium are related to increased degradation of tryptophan into kynurenine, a phenomenon which is related to immune activation. Life Sci 71:1837–1848.

Maj J, Rogoz Z, Skuza G, Sowinska H (1992) Effects of MK-801 and antidepressant drugs in the forced swimming test in rats. Eur Neuropsychopharmacol 2:37–41.

Mann JJ, Malone KM (1997) Cerebrospinal fluid amines and higher-lethality suicide attempts in depressed inpatients. Biol Psychiatry 41:162–171.

Martin A, Heyes MP, Salazar AM, Kampen DL, Williams J, Law WA, Coats ME, Markey SP (1992) Progressive slowing of reaction time and increasing cerebrospinal fluid concentrations of quinolinic acid in HIV-infected individuals. J Neuropsychiatry Clin Neurosci 4:270–279.

Martin A, Heyes MP, Salazar AM, Law WA, Williams J (1993) Impaired motor skill learning, slowed reaction time, and elevated cerebrospinal fluid quinilonic acid in a sub-group of HIV-infected individuals. Neuropsychology 7:147–149.

Martin P, Carlsson ML, Hjorth S (1998) Systemic PCP treatment elevates brain extracellular 5-HT: A microdialysis study in awake rats. Neuroreport 9:2985–2988.

Marzi M, Vigano A, Trabattoni D, Villa ML, Salvaggio A, Clerici E, Clerici M (1996) Characterization of type 1 and type 2 cytokine production profile in physiologic and pathologic human pregnancy. Clin Exp Immunol 106:127–133.

Matussek N (1966) Neurobiologie und Depression. Med Monatsschr 3:109–112.

Mauri MC, Ferrara A, Boscati L, Bravin S, Zamberlan F, Alecci M, Invernizzi G (1998) Plasma and platelet amino acid concentrations in patients affected by major depression and under fluvoxamine treatment. Neuropsychobiology 37:124–129.

Meijer A, Zakay-Rones Z, Morag A (1988) Post-influenzal psychiatric disorder in adolescents. Acta Psychiatr Scand 78:176–181.

Mellor AL, Munn DH (1999) Tryptophan catabolism and T-cell tolerance: Immunosuppression by starvation? Immunol Today 20:469–473.

Mendlovic S, Mozes E, Eilat E, Doron A, Lereya J, Zakuth V, Spirer Z (1999) Immune activation in non-treated suicidal major depression. Immunol Lett 67:105–108.

Miguel-Hidalgo JJ, Baucom C, Dilley G, Overholser JC, Meltzer HY, Stockmeier CA, Rajkowska G (2000) Glial fibrillary acidic protein immunoreactivity in the prefrontal cortex distinguishes younger from older adults in major depressive disorder. Biol Psychiatry 48:861–873.

Mikova O, Yakimova R, Bosmans E, Kenis G, Maes M (2001) Increased serum tumor necrosis factor alpha concentrations in major depression and multiple sclerosis. Eur Neuropsychopharmacol 11:203–208.

Miller CL, Llenos IC, Dulay JR, Barillo MM, Yolken RH, Weis S (2004) Expression of the kynurenine pathway enzyme tryptophan 2,3-dioxygenase is increased in the frontal cortex of individuals with schizophrenia. Neurobiol Dis 15:618–629.

Miller DW, Abercrombie ED (1996) Effects of MK-801 on spontaneous and amphetamine-stimulated dopamine release in striatum measured with in vivo microdialysis in awake rats. Brain Res Bull 40:57–62.

Mittleman BB, Castellanos FX, Jacobsen LK, Rapoport JL, Swedo SE, Shearer GM (1997) Cerebrospinal fluid cytokines in pediatric neuropsychiatric disease. J Immunol 159:2994–2999.

Moghaddam B, Bolinao ML, Stein-Behrens B, Sapolsky R (1994) Glucocorticoids mediate the stress-induced extracellular accumulation of glutamate. Brain Res 655:251–254.

Molholm HB (1942) Hyposensitivity to foreign protein in schizophrenic patients. Psychiatr Quarterly 16:565–571.

Mtabaji JP, Manku MS, Horrobin DF (1977) Actions of the tricyclic antidepressant clomipramine on responses to pressor agents. Interactions with prostaglandin E2. Prostaglandins 14:125–132.

Müller N, Schwarz MJ (2002) Immunology in Anxiety and Depression. In: Handbook of Depression and Anxiety (Kasper S, den Boer JA, Sitsen JMA, eds), pp. 267–288. New York: Marcel Dekker.

Müller N, Schwarz MJ (2006) Inflammation induces serotonergic deficiency and glutamatergic hyperfunction in depression: Towards an integrated view. Mol Psychiatry 12(11):988–1000.

Müller N, Schwarz MJ (2007) The immune-glutamatergic Interaction: Towards an integrated view of schizophrenia. J Neurotransmission Suppl. 72:269–280.

Müller N, Ackenheil M, Hofschuster E, Mempel W, Eckstein R (1991) Cellular immunity in schizophrenic patients before and during neuroleptic treatment. Psychiatry Res 37:147–160.

Müller N, Gizycki-Nienhaus B, Günther W, Meurer M (1992) Depression as a cerebral manifestation of scleroderma: Immunological findings in serum and cerebrospinal fluid. Biol Psychiatry 31:1151–1156.

Müller N, Hofschuster E, Ackenheil M, Mempel W, Eckstein R (1993) Investigations of the cellular immunity during depression and the free interval: Evidence for an immune activation in affective psychosis. Prog Neuropsychopharmacol Biol Psychiatry 17:713–730.

Müller N, Empl M, Riedel M, Schwarz M, Ackenheil M (1997a) Neuroleptic treatment increases soluble IL-2 receptors and decreases soluble IL-6 receptors in schizophrenia. Eur Arch Psychiatry Clin Neurosci 247:308–313.

Müller N, Riedel M, Schwarz MJ, Gruber R, Ackenheil M (1997b) Immunomodulatory Effects of Neuroleptics to the Cytokine System and the Cellular Immune System in Schizophrenia. In: Current Update in Psychoimmunology (Wieselmann G, ed), pp 57–67. Wien, New York: Springer.

Müller N, Riedel M, Hadjamu M, Schwarz MJ, Ackenheil M, Gruber R (1999) Increase in expression of adhesion molecule receptors on T helper cells during antipsychotic treatment and relationship to blood-brain barrier permeability in schizophrenia. Am J Psychiatry 156:634–636.

Müller N, Riedel M, Ackenheil M, Schwarz MJ (2000) Cellular and humoral immune system in schizophrenia: A conceptual re-evaluation. World J Biol Psychiatry 1:173–179.

Müller N, Riedel M, Dehning S, Spellmann I, Müller-Arends A, Cerovecki A, Goldstein-Müller B, Möller H-J, Schwarz MJ (2004a) Is the therapeutic effect of celecoxib in schizophrenia depending from duration of disease? Neuropsychopharmacology 29:176.

Müller N, Ulmschneider M, Scheppach C, Schwarz MJ, Ackenheil M, Möller HJ, Gruber R, Riedel M (2004b) COX-2 inhibition as a treatment approach in schizophrenia: Immunological considerations and clinical effects of celecoxib add-on therapy. Eur Arch Psychiatry Clin Neurosci 254:14–22.

Müller N, Riedel M, Schwarz MJ, Engel RR (2005) Clinical effects of COX-2 inhibitors on cognition in schizophrenia. Eur Arch Psychiatry Clin Neurosci 255:149–151.

Müller N, Schwarz MJ, Dehning S, Douhe A, Cerovecki A, Goldstein-Muller B, Spellmann I, Hetzel G, Maino K, Kleindienst N, Moller HJ, Arolt V, and Riedel M (2006) The cyclooxygenase-2 inhibitor celecoxib has therapeutic effects in major depression: results of a double-blind, randomized, placebo controlled, add-on pilot study to reboxetine. Mol. Psychiatry. 11(7), 680–684.

Munn DH, Shafizadeh E, Attwood JT, Bondarev I, Pashine A, Mellor AL (1999) Inhibition of T cell proliferation by macrophage tryptophan catabolism. J Exp Med 189:1363–1372.

Nair A, Bonneau RH (2006) Stress-induced elevation of glucocorticoids increases microglia proliferation through NMDA receptor activation. J Neuroimmunol 171:72–85.

Nassberger L, Traskman-Bendz L (1993) Increased soluble interleukin-2 receptor concentrations in suicide attempters. Acta Psychiatr Scand 88:48–52.

Nguyen KT, Deak T, Owens SM, Kohno T, Fleshner M, Watkins LR, Maier SF (1998) Exposure to acute stress induces brain interleukin-1beta protein in the rat. J Neurosci 18:2239–2246.

Niino M, Ogata A, Kikuchi S, Tashiro K, Nishihira J (2000) Macrophage migration inhibitory factor in the cerebrospinal fluid of patients with conventional and optic-spinal forms of multiple sclerosis and neuro-Behcet's disease. J Neurol Sci 179:127–131.

Nordstrom P, Samuelsson M, Asberg M, Traskman BL, Aberg WA, Nordin C, Bertilsson L (1994) CSF 5-HIAA predicts suicide risk after attempted suicide. Suicide Life Threat Behav 24:1–9.

Nowak G, Ordway GA, Paul IA (1995) Alterations in the N-methyl-D-aspartate (NMDA) receptor complex in the frontal cortex of suicide victims. Brain Res 675:157–164.

Nudmamud-Thanoi S, Reynolds GP (2004) The NR1 subunit of the glutamate/NMDA receptor in the superior temporal cortex in schizophrenia and affective disorders. Neurosci Lett 372:173–177.

Nunes SOV, Reiche EMV, Morimoto HK, Matsuo T, Itano EN, Xavier ECD, Yamashita CM, Vieira VR, Menoli AV, Silva SS, Costa FB, Reiche FV, Silva FLV, Kaminami MS (2002) Immune and hormonal activity in adults suffering from depression. Braz J Med Biol Res 35:581–587.

O'Brien SM, Scott LV, Dinan TG (2004) Cytokines: Abnormalities in major depression and implications for pharmacological treatment. Hum Psychopharmacol 19:397–403.

O'Hara MW, Swain AM (1996) Rates and risk of post-partum depression—a meta-analysis. In Rev Psychiatry 8:37–54.

Ohishi K, Ueno R, Nishino S, Sakai T, Hayaishi O (1988) Increased level of salivary prostaglandins in patients with major depression. Biol Psychiatry 23:326–334.

Olney JW, Farber NB (1995) Glutamate receptor dysfunction and schizophrenia. Arch Gen Psychiatry 52:998–1007.

Ongur D, Drevets WC, Price JL (1998) Glial reduction in the subgenual prefrontal cortex in mood disorders. Proc Natl Acad Sci U S A 95:13290–13295.

Ossowska G, Klenk-Majewska B, Szymczyk G (1997) The effect of NMDA antagonists on footshock-induced fighting behavior in chronically stressed rats. J Physiol Pharmacol 48:127–135.

Ostensen M, Forger F, Nelson JL, Schuhmacher A, Hebisch G, Villiger PM (2005) Pregnancy in patients with rheumatic disease: Anti-inflammatory cytokines increase in pregnancy and decrease post partum. Ann Rheum Dis 64:839–844.

Ostroff R, Gonzales M, Sanacora G (2005) Antidepressant effect of ketamine during ECT. Am J Psychiatry 162:1385–1386.

Ozek M, Toreci K, Akkok I, Guvener Z (1971) [Influence of therapy on antibody-formation]. Psychopharmacologia 21:401–412.

Patten SB, Francis G, Metz LM, Lopez-Bresnahan M, Chang P, Curtin F (2005) The relationship between depression and interferon beta-1a therapy in patients with multiple sclerosis. Mult Scler 11:175–181.

Penttinen J (1995) Hypothesis: Low serum cholesterol, suicide, and interleukin-2. Am J Epidemiol 141:716–718.

Plata-Salaman CR (1991) Immunoregulators in the nervous system. Neurosci Biobehav Rev 15:185–215.

Pollak Y, Yirmiya R (2002) Cytokine-induced changes in mood and behaviour: Implications for 'depression due to a general medical condition', immunotherapy and antidepressive treatment. Int J Neuropsychopharmacol 5:389–399.

Pollak Y, Ovadia H, Goshen I, Gurevich R, Monsa K, Avitsur R, Yirmiya R (2000) Behavioral aspects of experimental autoimmune encephalomyelitis. J Neuroimmunol 104:31–36.

Pollmächer T, Schuld A, Kraus T, Haack M, Hinze-Selch D (2001) [On the clinical relevance of clozapine-triggered release of cytokines and soluble cytokine-receptors]. Fortschr Neurol Psychiatr 69(Suppl 2):S65–S74.

Pugh CR, Nguyen KT, Gonyea JL, Fleshner M, Wakins LR, Maier SF, Rudy JW (1999) Role of interleukin-1 beta in impairment of contextual fear conditioning caused by social isolation. Behav Brain Res 106:109–118.

Pyeon D, Diaz FJ, Splitter GA (2000) Prostaglandin E(2) increases bovine leukemia virus tax and pol mRNA levels via cyclooxygenase 2: Regulation by interleukin-2, interleukin-10, and bovine leukemia virus. J Virol 74:5740–5745.

Rajkowska G (2003) Depression: What we can learn from postmortem studies. Neuroscientist 9:273–284.

Rajkowska G (2005) Astroglia in the cortex of schizophrenics: Histopathology finding. World J Biol Psychiatry 6:74.

Rajkowska G, Miguel-Hidalgo JJ, Wei J, Dilley G, Pittman SD, Meltzer HY, Overholser JC, Roth BL, Stockmeier CA (1999) Morphometric evidence for neuronal and glial prefrontal cell pathology in major depression. Biol Psychiatry 45:1085–1098.

Rajkowska G, Halaris A, Selemon LD (2001) Reductions in neuronal and glial density characterize the dorsolateral prefrontal cortex in bipolar disorder. Biol Psychiatry 49:741–752.

Reichenberg A, Yirmiya R, Schuld A, Kraus T, Haack M, Morag A, Pollmacher T (2001) Cytokine-associated emotional and cognitive disturbances in humans. Arch Gen Psychiatry 58:445–452.

Reichenberg A, Kraus T, Haack M, Schuld A, Pollmacher T, Yirmiya R (2002) Endotoxin-induced changes in food consumption in healthy volunteers are associated with TNF-alpha and IL-6 secretion. Psychoneuroendocrinology 27:945–956.

Rieckmann P, Nunke K, Burchhardt M, Albrecht M, Wiltfang J, Ulrich M, Felgenhauer K (1993) Soluble intercellular adhesion molecule-1 in cerebrospinal fluid: An indicator for the inflammatory impairment of the blood-cerebrospinal fluid barrier. J Neuroimmunol 47:133–140.

Riedel M, Spellmann I, Schwarz MJ, Strassnig M, Sikorski C, Möller HJ, Müller N (2007) Decreased T cellular immune response in schizophrenic patients. J Psychiatr Res 41:3–7.

Robinson CM, Hale PT, Carlin JM (2005) The role of IFN-gamma and TNF-alpha-responsive regulatory elements in the synergistic induction of indoleamine dioxygenase. J Interferon Cytokine Res 25:20–30.

Rosa A, Peralta V, Papiol S, Cuesta MJ, Serrano F, Martinez-Larrea A, Fananas L (2004) Interleukin-1beta (IL-1beta) gene and increased risk for the depressive symptom-dimension in schizophrenia spectrum disorders. Am J Med Genet B Neuropsychiatr Genet 124:10–14.

Rothermundt M, Arolt V, Fenker J, Gutbrodt H, Peters M, Kirchner H (2001) Different immune patterns in melancholic and non-melancholic major depression. Eur Arch Psychiatry Clin Neurosci 251:90–97.

Rothermundt M, Falkai P, Ponath G, Abel S, Burkle H, Diedrich M, Hetzel G, Peters M, Siegmund A, Pedersen A, Maier W, Schramm J, Suslow T, Ohrmann P, Arolt V (2004a) Glial cell dysfunction in schizophrenia indicated by increased S100B in the CSF. Mol Psychiatry 9:897–899.

Rothermundt M, Ponath G, Arolt V (2004b) S100B in schizophrenic psychosis. Int Rev Neurobiol 59:445–470.

Roy A, Pickar D, Paul S, Doran A, Chrousos GP, Gold PW (1987) CSF corticotropin-releasing hormone in depressed patients and normal control subjects. Am J Psychiatry 144:641–645.

Saito K, Crowley JS, Markey SP, Heyes MP (1993) A mechanism for increased quinolinic acid formation following acute systemic immune stimulation. J Biol Chem 268:15496–15503.

Saito S, Sakai M, Sasaki Y, Tanebe K, Tsuda H, Michimata T (1999) Quantitative analysis of peripheral blood Th0, Th1, Th2 and the Th1:Th2 cell ratio during normal human pregnancy and preeclampsia. Clin Exp Immunol 117:550–555.

Sakic B, Denburg JA, Denburg SD, Szechtman H (1996) Blunted sensitivity to sucrose in autoimmune MRL-lpr mice: A curve-shift study. Brain Res Bull 41:305–311.

Salzberg-Brenhouse HC, Chen EY, Emerich DF, Baldwin S, Hogeland K, Ranelli S, Lafreniere D, Perdomo B, Novak L, Kladis T, Fu K, Basile AS, Kordower JH, Bartus RT (2003) Inhibitors of cyclooxygenase-2, but not cyclooxygenase-1 provide structural and functional protection against quinolinic acid-induced neurodegeneration. J Pharmacol Exp Ther 306:218–228.

Sanacora G, Gueorguieva R, Epperson CN, Wu YT, Appel M, Rothman DL, Krystal JH, Mason GF (2004a) Subtype-specific alterations of gamma-aminobutyric acid and glutamate in patients with major depression. Arch Gen Psychiatry 61:705–713.

Sanacora G, Kendell SF, Fenton L, Coric V, Krystal JH (2004b) Riluzole augmentation for treatment-resistant depression. Am J Psychiatry 161:2132.

Sandrini M, Vitale G, Pini LA (2002) Effect of rofecoxib on nociception and the serotonin system in the rat brain. Inflamm Res 51:154–159.

Sapolsky RM (1985) A mechanism for glucocorticoid toxicity in the hippocampus: Increased neuronal vulnerability to metabolic insults. J Neurosci 5:1228–1232.

Scarr E, Pavey G, Sundram S, MacKinnon A, Dean B (2003) Decreased hippocampal NMDA, but not kainate or AMPA receptors in bipolar disorder. Bipolar Disord 5:257–264.

Schäfer M, Horn M, Schmidt F, Schmid-Wendtner MH, Volkenandt M, Ackenheil M, Müller N, Schwarz MJ (2004) Correlation between sICAM-1 and depressive symptoms during adjuvant treatment of melanoma with interferon-alpha. Brain Behav Immun 18:555–562.

Schiepers OJ, Wichers MC, Maes M (2005) Cytokines and major depression. Prog Neuropsychopharmacol Biol Psychiatry 29: 201–217.

Schmitt A, Bertsch T, Henning U, Tost H, Klimke A, Henn FA, Falkai P (2005) Increased serum S100B in elderly, chronic schizophrenic patients: Negative correlation with deficit symptoms. Schizophr Res 80:305–313.

Schroeter ML, Abdul-Khaliq H, Fruhauf S, Hohne R, Schick G, Diefenbacher A, Blasig IE (2003) Serum S100B is increased during early treatment with antipsychotics and in deficit schizophrenia. Schizophr Res 62:231–236.

Schwarcz R, Pellicciari R (2002) Manipulation of brain kynurenines: Glial targets, neuronal effects, and clinical opportunities. J Pharmacol Exp Ther 303:1–10.

Schwarz MJ, Riedel M, Ackenheil M, Müller N (2000) Decreased levels of soluble intercellular adhesion molecule-1 (sICAM-1) in unmedicated and medicated schizophrenic patients. Biol Psychiatry 47:29–33.

Schwarz MJ, Chiang S, Müller N, Ackenheil M (2001) T-helper-1 and T-helper-2 responses in psychiatric disorders. Brain Behav Immun 15:340–370.

Schwieler L, Erhardt S, Erhardt C, Engberg G (2005) Prostaglandin-mediated control of rat brain kynurenic acid synthesis—opposite actions by COX-1 and COX-2 isoforms. J Neural Transm 112:863–872.

Seidel A, Arolt V, Hunstiger M, Rink L, Behnisch A, Kirchner H (1995) Cytokine production and serum proteins in depression. Scand J Immunol 41:534–538.

Seidel A, Arolt V, Hunstiger M, Rink L, Behnisch A, Kirchner H (1996a) Increased CD56 + natural killer cells and related cytokines in major depression. Clin Immunol Immunopathol 78:83–85.

Seidel A, Arolt V, Hunstiger M, Rink L, Behnisch A, Kirchner H (1996b) Major depressive disorder is associated with elevated monocyte counts. Acta Psychiatr Scand 94:198–204.

Si X, Miguel-Hidalgo JJ, O'Dwyer G, Stockmeier CA, Rajkowska G (2004) Age-dependent reductions in the level of glial fibrillary acidic protein in the prefrontal cortex in major depression. Neuropsychopharmacology 29:2088–2096.

Sluzewska A, Rybakowski JK, Laciak M, Mackiewicz A, Sobieska M, Wiktorowicz K (1995) Interleukin-6 serum levels in depressed patients before and after treatment with fluoxetine. Ann N Y Acad Sci 762:474–476.

Sluzewska A, Rybakowski J, Bosmans E, Sobieska M, Berghmans R, Maes M, Wiktorowicz K (1996) Indicators of immune activation in major depression. Psychiatry Res 64:161–167.

Song C, Leonard BE (1994) An acute phase protein response in the olfactory bulbectomised rat: Effect of sertraline treatment. Med Sci Res 22:313–314.

Song C, Leonard BE (2000) Fundamentals of Psychoneuroimmunology. Chichester, New York: Wiley & Sons.

Song C, Lin A, Bonaccorso S, Heide C, Verkerk R, Kenis G, Bosmans E, Scharpe S, Whelan A, Cosyns P, De Jongh R, Maes M (1998) The inflammatory response system and the availability of plasma tryptophan in patients with primary sleep disorders and major depression. J Affect Disord 49:211–219.

Speciale C, Schwarcz R (1993) On the production and disposition of quinolinic acid in rat brain and liver slices. J Neurochem 60: 212–218.

Sperner-Unterweger B, Miller C, Holzner B, Widner B, Fleischhacker WW, Fuchs D (1999) Measurement of Neopterin, Kynurenine and Tryptophan in Sera of Schizophrenic Patients. In: Psychiatry, Psychoimmunology, and Viruses (Müller N, ed), pp. 115–119. Wien, New York: Springer.

Stein GS (1980) The pattern of mental change and body weight change in the first post-partum week. J Psychosom Res 24: 165–171.

Stein-Behrens BA, Lin WJ, Sapolsky RM (1994) Physiological elevations of glucocorticoids potentiate glutamate accumulation in the hippocampus. J Neurochem 63:596–602.

Stolina M, Sharma S, Lin Y, Dohadwala M, Gardner B, Luo J, Zhu L, Kronenberg M, Miller PW, Portanova J, Lee JC, Dubinett SM (2000) Specific inhibition of cyclooxygenase 2 restores antitumor reactivity by altering the balance of IL-10 and IL-12 synthesis. J Immunol 164:361–370.

Stone TW (1993) Neuropharmacology of quinolinic and kynurenic acids. Pharmacol Rev 45:309–379.

Stryjer R, Strous RD, Shaked G, Bar F, Feldman B, Kotler M, Polak L, Rosenzcwaig S, Weizman A (2003) Amantadine as augmentation therapy in the management of treatment-resistant depression. Int Clin Psychopharmacol 18:93–96.

Sumiyoshi T, Anil AE, Jin D, Jayathilake K, Lee M, Meltzer HY (2004) Plasma glycine and serine levels in schizophrenia compared to normal controls and major depression: Relation to negative symptoms. Int J Neuropsychopharmacol 7:1–8.

Sumiyoshi T, Jin D, Jayathilake K, Lee M, Meltzer HY (2005) Prediction of the ability of clozapine to treat negative symptoms from plasma glycine and serine levels in schizophrenia. Int J Neuropsychopharmacol 8:451–455.

Sundar SK, Cierpial MA, Kilts C, Ritchie JC, Weiss JM (1990) Brain IL-1-induced immunosuppression occurs through activation of both pituitary-adrenal axis and sympathetic nervous system by corticotropin-releasing factor. J Neurosci 10: 3701–3706.

Suzuki T, Ogata A, Tashiro K, Nagashima K, Tamura M, Yasui K, Nishihira J (2000) Japanese encephalitis virus up-regulates expression of macrophage migration inhibitory factor (MIF) mRNA in the mouse brain. Biochim Biophys Acta 1517:100–106.

Takahashi T, Kimoto T, Tanabe N, Hattori TA, Yasumatsu N, Kawato S (2002) Corticosterone acutely prolonged N-methyl-d-aspartate receptor-mediated Ca2 + elevation in cultured rat hippocampal neurons. J Neurochem 83:1441–1451.

Takikawa O, Yoshida R, Kido R, Hayaishi O (1986) Tryptophan degradation in mice initiated by indoleamine 2,3-dioxygenase. J Biol Chem 261:3648–3653.

Tamminga CA, Cascella N, Fakouhl TD, Hertin RL (1992) Enhancement of NMDA-mediated Transmission in Schizophrenia: Effects of Milacemide. In: Novel Antipsychotic Drugs (Meltzer HY, ed), pp. 171–177. New York: Raven Press.

Tanaka KF, Shintani F, Fujii Y, Yagi G, Asai M (2000) Serum interleukin-18 levels are elevated in schizophrenia. Psychiatry Res 96:75–80.

Thomas AJ, Ferrier IN, Kalaria RN, Woodward SA, Ballard C, Oakley A, Perry RH, O'Brien JT (2000) Elevation in late-life depression of intercellular adhesion molecule-1 expression in the dorsolateral prefrontal cortex. Am J Psychiatry 157:1682–1684.

Trullas R, Skolnick P (1990) Functional antagonists at the NMDA receptor complex exhibit antidepressant actions. Eur J Pharmacol 185:1–10.

Van Kammen DP, McAllister-Sistilli CG, Kelley ME (1997) Relationship Between Immune and Behavioral Measures in Schizophrenia. In: Current Update in Psychoimmunology (Wieselmann G, ed), pp. 51–55. Wien, New York: Springer.

Weiss G, Murr C, Zoller H, Haun M, Widner B, Ludescher C, Fuchs D (1999) Modulation of neopterin formation and tryptophan degradation by Th1- and Th2-derived cytokines in human monocytic cells. Clin Exp Immunol 116:435–440.

Weiss JM, Quan N, Sundar SK (1994) Immunological consequences of Interleukin-1 in the brain. Neuropsychopharmacol 10:833.

Westergaard T, Mortensen PB, Pedersen CB, Wohlfahrt J, Melbye M (1999) Exposure to prenatal and childhood infections and the risk of schizophrenia: Suggestions from a study of sibship characteristics and influenza prevalence. Arch Gen Psychiatry 56: 993–998.

Wichers MC, Koek GH, Robaeys G, Verkerk R, Scharpe S, Maes M (2005) IDO and interferon-alpha-induced depressive symptoms: A shift in hypothesis from tryptophan depletion to neurotoxicity. Mol Psychiatry 10:538–544.

Wilke I, Arolt V, Rothermundt M, Weitzsch C, Hornberg M, Kirchner H (1996) Investigations of cytokine production in whole blood cultures of paranoid and residual schizophrenic patients. Eur Arch Psychiatry Clin Neurosci 246:279–284.

Woolley CS, Gould E, McEwen BS (1990) Exposure to excess glucocorticoids alters dendritic morphology of adult hippocampal pyramidal neurons. Brain Res 531:225–231.

Xiao BG, Link H (1999) Is there a balance between microglia and astrocytes in regulating Th1/Th2-cell responses and neuropathologies? Immunol Today 20:477–479.

Yan QS, Reith ME, Jobe PC, Dailey JW (1997) Dizocilpine (MK-801) increases not only dopamine but also serotonin and norepinephrine transmissions in the nucleus accumbens as measured by microdialysis in freely moving rats. Brain Res 765:149–158.

Yaron I, Shirazi I, Judovich R, Levartovsky D, Caspi D, Yaron M (1999) Fluoxetine and amitriptyline inhibit nitric oxide, prostaglandin E2, and hyaluronic acid production in human synovial cells and synovial tissue cultures. Arthritis Rheum 42:2561–2568.

Yilmaz A, Schulz D, Aksoy A, Canbeyli R (2002) Prolonged effect of an anesthetic dose of ketamine on behavioral despair. Pharmacol Biochem Behav 71:341–344.

Yirmiya R (2000) Depression in medical illness: The role of the immune system. West J Med 173:333–336.

Yirmiya R, Weidenfeld J, Pollak Y, Morag M, Morag A, Avitsur R, Barak O, Reichenberg A, Cohen E, Shavit Y, Ovadia H (1999) Cytokines, "depression due to a general medical condition," and antidepressant drugs. Adv Exp Med Biol 461:283–316.

Zarate CA, Jr., Payne JL, Quiroz J, Sporn J, Denicoff KK, Luckenbaugh D, Charney DS, Manji HK (2004) An open-label trial of riluzole in patients with treatment-resistant major depression. Am J Psychiatry 161:171–174.

Zarate CA, Jr., Quiroz JA, Singh JB, Denicoff KD, De Jesus G, Luckenbaugh DA, Charney DS, Manji HK (2005) An open-label trial of the glutamate-modulating agent riluzole in combination with lithium for the treatment of bipolar depression. Biol Psychiatry 57:430–432.

Zhu J, Bengtsson BO, Mix E, Thorell LH, Olsson T, Link H (1994) Effect of monoamine reuptake inhibiting antidepressants on major histocompatibility complex expression on macrophages in normal rats and rats with experimental allergic neuritis (EAN). Immunopharmacology 27:225–244.

Zuckerman L, Weiner I (2005) Maternal immune activation leads to behavioral and pharmacological changes in the adult offspring. J Psychiatr Res 39:311–323.

37
Drugs of Abuse and the Immune System

Toby K. Eisenstein, David E. Kaminsky, Rahil T. Rahim, and Thomas J. Rogers

Keywords Cannabinoids; Chemokines; Cocaine; Cytokines; Immunosuppression; Morphine; Opioids; T-helper 1/T-helper 2 (Th1/Th2); Δ9-tetrahydrocannabinol (Δ^9-THC); Tolerance; Withdrawal

37.1. Introduction

It has been clearly demonstrated that opioids, cannabinoids, and cocaine alter a variety of assays of immune function when added to cells of the immune system in vitro. Receptors and/or mRNA for the receptors for opioids and cannabinoids have been demonstrated in a variety of cells of the immune system. In many cases, especially for the opioids, pharmacological specificity of the action of the drugs has been verified using appropriate antagonists. These observations provide a biological basis for concluding that drugs of abuse have direct affects on immune cells. In addition, all three drugs alter immune responses of rodents, and in some cases, humans, when treated in vivo. In vivo mechanisms of action may be harder to determine as the drugs may activate other physiologic systems in the body that may also impinge on immune responsiveness, such as the HPA and the SNS. Nonetheless, drug abusers suffer the cumulative effects of direct and indirect effects of the drugs on the immune system. Assays of immune function that have been measured include effects on the innate immune system, which encompass effects on macrophages and NK cells, and effects on adaptive immunity, which include the humoral and cellular arms of the immune system. Considerable attention has been focused on cytokine and chemokine mediators of immune function that are modulated by the drugs and on their capacity to modulate activation of the various arms of the immune system by polarization of the cytokine profile. Other commonly used and abused drugs include nicotine and alcohol, both of which have marked effects on the immune system, and are frequently part of the panoply of artificial substances used by addicts. In addition, methamphetamines can alter immune status. Due to space limitations the reader is referred to key papers on these subjects (Sopori, 2002; Gamble et al., 2006; Pacifici et al., 2002)

37.2. Fundamental Concepts

37.2.1. Opioid and Cannabinoid Receptors are on Cells of the Immune System

37.2.1.1. Opioid Receptors

There is now indisputable evidence of natural physiological connections between the neural and immune systems. Much of the evidence supporting this link has come from studies of opioids and cannabinoids and their effects on immune responses. Opioid receptors were first demonstrated biochemically in brain in the early 1970s. Three distinct receptors were discovered in neural tissue and were designated mu, kappa, and delta (Simon, 1991). Their respective natural ligands were found to be the neuropeptides, β-endorphin (predominantly mu), dynorphin (kappa) and methionine-enkephalin (delta) (Akil et al., 1984), although there is considerable cross-reactivity. Alkaloid opioids, including morphine and heroin, bind primarily, but not exclusively, to the mu receptor. Morphine is extracted from opium produced by the poppy plant, and heroin is synthetically diacetylated morphine. In vivo, the major metabolite of heroin is morphine. Most laboratory studies employ morphine because it is less lipophilic than heroin, is used therapeutically, and is easier to dissolve. Since the mu receptor is the major target of both compounds, results obtained with morphine are believed to be directly translatable to heroin abuse, although there is some evidence for unique biologically active heroin metabolites (Rossi et al., 1996).

Credit for recognition of the potential existence of receptors for opioids on cells of the immune system is given to Joseph Wybran in the laboratory of Govaerts. In 1979 he reported modulation of the function of human T cells, purified from normal peripheral blood, by exogenous and

T. Ikezu and H.E. Gendelman (eds.), *Neuroimmune Pharmacology.*
© Springer 2008

endogenous opioids (Wybran et al., 1979). He observed that if morphine was added to human T cells in culture, they lost the ability to rosette with sheep red blood cells (SRBCs), a fortuitous property that was used before flow cytometry to enumerate T cells. In contrast, when the opioid neuropeptide, met-enkephalin, was added to the T cells, rosetting increased. The actions of both opioids were blocked by naloxone, an antagonist for all types of opioid receptors, showing pharmacological specificity of the effects. Wybran proposed that, "normal human blood T lymphocytes bear surface receptor-like structures for morphine and methionine-enkephalin." He concluded that, "Such findings may provide a link between the central nervous system and the immune system." This paper was seminal because it presented strong pharmacologic data to support the existence of opioid receptors on cells of the immune system. However, it has been very difficult to demonstrate opioid receptors routinely by binding studies using cells of the immune system, although scattered reports of positive results are published (Sibinga and Goldstein, 1988; Sharp et al., 1998). The difficulties may be due to low receptor numbers on cells of the immune system, to inability to obtain enough purified subsets of immune cells to do adequate binding studies, or to a need to activate immune cells before the receptors are expressed. With the advent of cloning of the opioid receptors from brain tissue, mRNA transcripts have been detected in primary cells of the immune system or lymphoid or monocytic cell lines for all three classes of opioid receptors (Loh and Smith, 1990; McCarthy et al., 2001; Sharp et al., 1998).

37.2.1.2. Cannabinoid Receptors

Cannabinoid receptors were identified and cloned several years after the opioid receptors. Cannabinoids, like the psychotropic active ingredient extracted from marijuana, Δ^9-tetrahydrocannabinol (Δ^9-THC), were initially believed by some to simply interdigitate into cell membranes. Subsequently, a cannabinoid receptor was cloned from neural tissue (Zhu et al., 1993). When lymphoid tissue was probed for cannabinoid receptor transcripts, a second receptor (designated CB2) was discovered that is primarily expressed in the immune system and has 44% homology to the first receptor (designated CB1) (Munro et al., 1993). mRNA for CB2 is expressed by Natural Killer (NK) cells, B cells, T cells and macrophages (Lee et al., 2001), and radiolabeled cannabinoids have been shown to bind to mouse spleen cells and human T cell lines (Kaminski et al., 1992; Schatz et al., 1997). Endogenous cannabinoid ligands have also been discovered, but they are not peptides like the opioids. Anandamide, or arachidonoylethanolamide, is a polyunsaturated N-acylethanolamine (Devane et al., 1992), as is 2-arachidonoyl-glycerol (Mechoulam et al., 1995; Di Marzo and Deutsch, 1998). It should be appreciated that the cannabinoids may exert their effects by both receptor-dependent and -independent mechanisms.

37.2.2. Mechanisms by Which Drugs of Abuse Can Affect Cells of the Immune System

Testing of whether drugs of abuse can alter immune function has been carried out by in vivo administration of the drug, with assessment of in vivo or ex vivo immune function, or alternatively, by harvesting immune cells from a drug naïve (normal) host and exposing them to the drug in vitro. If a drug is given in vivo, there is the possibility that effects on the immune system are not due to a direct interaction of the drug with cells of the immune system, but that immunomodulatory effects are instead mediated by additional systems in the body. For example, opioids, cannabinoids, and other drugs of abuse, can modulate the Hypothalamic-Pituitary-Adrenal (HPA) axis, which can lead to release of immunosuppressive corticosteroids or to activation of the sympathetic nervous (SNS) system, resulting in downstream alterations in immune function. Effects on immune function could thus be a combination of indirect and direct effects if the drug is given in vivo.

An additional and important consideration is that both the neural system and cells of the immune system produce opioid peptides, which could lead to complicated interacting pathways, especially in vivo (Carr, 1991). In evaluating the studies discussed in this Chapter, it should be recognized that most immune responses were assessed two to three hours after a single injection of a drug. For studies testing longer exposure to morphine, the drug is frequently delivered by subcutaneous implantation of a pellet that slowly releases the drug over a period of a week. Alternatively, mini-pumps dispensing drugs can be implanted. In the case of morphine or other addictive opioids, these methods are used to avoid episodes of withdrawal, which has its own immunomodulatory effects on immune function. In fact, a complication of studies with opioids is that repeated administration can result in tolerance to their analgesic effects, and in many, but not all, cases tolerance to the effects of the drugs on the immune system (Eisenstein et al., 2006). There are a limited number of studies that have addressed the effects of tolerance and of withdrawal from morphine on immune responses (Eisenstein et al., 2006).

37.3. Functional Consequences of Drugs of Abuse on the Immune System

37.3.1. Effects on Innate Immunity and the Inflammatory Response

37.3.1.1. Cells and Molecules of the Innate Immune Response

It is well established that drugs of abuse are involved in a wide range of immunomodulatory activities. Among the areas in immune system function in which these alterations have been observed is innate immunity. This arm of the immune system does not exhibit specificity in recognizing foreign pathogens. It encompasses NK cells and the "professional" phagocytic cells that are the first responders to inflammation or infection. Natural Killer cells are lymphocytes that are involved in recognition and killing of mammalian cells that express foreign antigens, such as tumor antigens or viral antigens. They also produce cytokines (small proteins produced by immune cells that act in an autocrine, paracrine, or endocrine fashion) that can activate the mononuclear phagocytic cells. The phagocytic cells engulf microbes invading the host and include polymorphonuclear (PMN) leukocytes (also called neutrophils), monocytes, and macrophages. Monocytes are formed in the bone marrow and released into the peripheral blood. They are mature, but unactivated cells. When they are stimulated in various ways they can develop into macrophages in vivo or in vitro. Macrophages have several states of activation that are characterized by their production of cytokines and by their levels of microbicidal compounds, including reactive oxygen intermediates (ROIs) and reactive nitrogen intermediates. Macrophages also play an important role in the acquisition and presentation of foreign antigens to lymphocytes, which triggers the adaptive immune response.

37.3.1.2. Effects of Opioids, Cannabinoids, and Cocaine on NK Cells

In the cellular repertoire of the innate immune system, NK cells have the unique ability to kill infected target cells in a manner independent of antigen recognition. Alteration in NK cell function was among the earliest, and has been one of the most consistently observed, effects of drugs on the immune system. Many studies have shown that administration of opioids to rodents results in the suppression of NK cell activity. Leibeskind's laboratory (Shavit et al., 1986b) was the first to show that morphine given to rats subcutaneously for 4 days suppressed NK cell activity. They later concluded that this effect was not directly on the NK cells in the periphery, because N-methylmorphine, which does not pass the blood-brain barrier, did not depress NK function (Shavit et al., 1986a). Subsequently, Weber and Pert (1989) reported that injection of morphine into the periaqueductal gray (PAG) region of the brains of rats resulted in suppression of NK cell activity in the spleens of these animals 3 hr later. Morphine administered into 5 other brain sites was without effect. Moreover, the opioid specific antagonist, naltrexone, blocked the effects in the studies by both groups, implicating the involvement of classic opioid receptors in modulating NK cell activity. Studies have also been carried out in which morphine has been given to normal human subjects under controlled conditions. The results show that peripheral blood NK cell activity is depressed 24 hr after morphine administration (Yeager et al., 1995). The exact mechanism by which engagement of opioid receptors in certain regions of the brain alters peripheral functional capacity of NK cells is not known. However, more recent work implicates a neural pathway involving dopamine in modulating NK activity (Saurer et al., 2006). Nonetheless, the mu opioid receptor (MOR) is certainly involved in the effects on NK cell activity, as mice with a genetic deletion of MOR do not respond to morphine with depressed NK cell activity (Gavériaux-Ruff et al., 1998).

The cannabinoids represent another class of abused drugs affecting immune cell function. Although fewer studies have explored the effects of cannabinoids on NK cells, the literature to date shows that the activity of splenic NK cells is suppressed following in vivo injection of Δ^9-THC into rats and mice (Patel et al., 1985; Klein et al., 1987), and also after in vitro treatment of human peripheral blood NK cells (Specter et al., 1986). Δ^9-THC has also been shown to block proliferation of NK cells in response to the cytokine, IL-2, by down-regulating the IL-2 receptor (Zhu et al., 1993).

There are a limited number of reports of immunomodulation by cocaine. In vivo administration of cocaine i.v. to humans elevated NK cell activity in peripheral blood, but exposure of leukocytes to the drug in vitro had no effect (Van Dyke et al., 1986). It was hypothesized that the in vivo effects occurred via release of epinephrine from the SNS (Van Dyke et al., 1986). Contradictory results have been reported for the effect of cocaine on NK activity in rats, with no effect in vivo (Bayer et al., 1996) or suppression of activity after acute or chronic in vivo administration (Pacifici et al., 2003).

37.3.1.3. Drugs of Abuse Alter Functions of PMN Leukocytes and Monocytes/Macrophages

Opioids given in vivo have been demonstrated to inhibit ex vivo phagocytic cell activity (Tubaro et al., 1983). Opioids selective for all three classes of opioid receptors (mu, kappa, and delta), when added to mouse peritoneal macrophages in vitro, inhibited phagocytosis of the yeast *Candida albicans*, and the effects were blocked by the respective receptor selective antagonists (Szabo et al., 1993). Morphine given in vivo by slow release pellet, or when added to murine bone marrow cells in vitro, blocked the maturation of precursor cells into macrophage colonies (Roy et al., 1991). Morphine pellets also inhibit adhesion of leukocytes to vascular endothelium, thus reducing inflammation (Ni et al., 2000). Studies have also been carried out in which morphine has been added to human peripheral blood mononuclear cells (PBMCs), a semi-purified cell fraction that includes lymphocytes and monocytes. Further work has shown the morphine also inhibits the ability of phagocytic cells to produce superoxide and peroxide (Peterson et al., 1987a). Peterson and colleagues (Chao et al., 1992) have shown that the inhibitory effects of this opioid on phagocytic cell activity are due, at least in part, to the production of Transforming Growth Factor (TGF)-β. Reactive oxygen intermediates, such as superoxide, contribute to the microbicidal activity of phagocytes, so impairment of their production would be consonant with reduced capacity to kill ingested yeast or other microbes. Macrophages taken from mice exposed to morphine can undergo apoptosis (Nair et al., 1997), and evidence supports a role of ROI and nitric oxide (NO) in this process (Singhal et al., 1998; Bhat et al., 2004). Usually depression of NO is associated with decreased macrophage function, so these findings are dichotomous with the other evidence on effects of opioids on these cells.

Δ^9-THC is also reported to decrease the phagocytic activity of human peripheral blood monocytes and mouse peritoneal macrophages (Specter et al., 1991; Lopez-Cepero et al., 1986). Alveolar macrophages of marijuana smokers exhibited decreased microbicidal activity against *Staphylococcus aureus*, which is associated with decreased production of NO (Roth et al., 2004). Δ^9-THC also affects macrophage processing of exogenous protein antigens through the CB2 receptor, as shown pharmacologically and using CB2-deficient mice (McCoy et al., 1999; Buckley et al., 2000).

Cocaine, like Δ^9-THC and morphine, has been reported to decrease the antimicrobial activity of alveolar macrophages obtained from chronic crack cocaine smokers (Roth et al., 2004) and to decrease parameters of mouse macrophage activation (Ou et al., 1989; Pacifici et al., 1993). However, in contrast to the effects of opioids, cocaine has been shown to increase the activation of PMNs in human subjects, as evidenced by increased killing of *Staphylococcus aureus* (Baldwin et al., 1997).

37.3.1.4. Effects of Drugs of Abuse on the Production of Cytokines and Chemokines

One mechanism by which drugs of abuse modulate immune function is through effects on expression of cytokines and chemokines, or on the functional capacity of their receptors. These proteins act as messengers to direct appropriate cellular activation and differentiation (cytokines) and coordinate the migration of cells to particular sites of inflammation or infection (chemokines). Cytokines have a number of functions. They may alter the innate immune response, act as a bridge to influence the type of adaptive immune response, and can also amplify adaptive immune responses. Chemokines can participate in innate immunity, and also influence adaptive immunity by orchestrating the type of cells that migrate to a pathological site. In addition, some of the chemokines can act as co-activators of immune cell function. Proinflammatory and anti-inflammatory cytokines, and also chemokines, which are part of the innate immune response, mediate generalized symptoms of inflammation including fever (IL-1), organ damage (tumor necrosis factor (TNF)-α), and migration of phagocytic cells (IL-8, CCL2), or dampen these responses (IL-10). Cytokines like IL-5 drive B cell maturation and antibody production.

Morphine and other ligands for the mu receptor have been shown to modulate production of both cytokines and chemokines, although the results have been complex. For example, treatment of stimulated PBMCs with morphine decreased IFN-γ production (Peterson et al., 1987b), but increased TGF-β expression (Chao et al., 1992). As TGF-β can be immunosuppressive, the relationship between the levels of the two cytokines may be causal. A biphasic dose response relationship was shown for release of proinflammatory cytokines from stimulated mouse peritoneal macrophages treated with morphine in vitro, with low doses increasing release of IL-6 and TNF-α, and high doses inhibiting them (Roy et al., 1998a). In contrast, when mice were treated with relatively high doses of morphine in vivo and their peritoneal macrophages were stimulated ex vivo, they showed increased production of IL-12 and TNF-α (Peng et al., 2000).

The pattern of proinflammatory cytokines produced has implications for the character of the adaptive immune response, which will occur if the host is exposed to a foreign antigen. A T-helper type 1 (Th1) cytokine profile polarizes the immune response towards a cellular immune response, whereas a T-helper type 2 (Th2) profile favors antibody responses (Mosmann and Coffman, 1987). IFN-γ is a marker of a Th1 response, and IL-4 is indicative of a Th2 response. There is evidence that morphine given in vivo biases the immune response toward a Th2 response (Roy et al., 2004). The consequences of this apparent shift toward Th2-type immunity are not clear, since other studies show that antibody production is suppressed in animals treated similarly with morphine (Bussiere et al., 1993). These results will be discussed in greater detail below.

In regard to chemokines, exposure of PBMCs to DAMGO, a synthetic peptide selective for the mu-opioid receptor, has been shown to increase the expression of pro-inflammatory chemokines CCL2, CCL5, and CXCL10 (Wetzel et al., 2000), and the chemokine receptors CXCR4 and CCR5, the major Human Immunodeficiency Virus (HIV)-1 co-receptors (Steele et al., 2003; Choi et al., 1999), the latter leading to increased replication of HIV. Another major and important finding concerning opioids and chemokines is that there is heterologous desensitization between selective opioid and chemokine receptors (Adler and Rogers, 2005). Both classes of receptors are G protein-coupled. Engagement of one of the receptors leads to failure of the other receptor to signal, probably through transphosphorylation of the second receptor (Steele et al., 2002; Szabo et al., 2002; Szabo et al., 2003; Chen et al., 2004). Prior treatment of rats with CCL5 or CXCL12 into the PAG region of the brain, blocks the antinociceptive effects of morphine similarly administered (Szabo et al., 2002). Heterologous desensitization of chemokine receptors by opioids blocks chemotaxis of human PBMCs to CXCL12 (Szabo et al., 2002; Grimm et al., 1998) and also blocks HIV infection of primary human monocytes in vitro (Szabo et al., 2003). (Figure 37.1)

There are striking parallels between the effects of opioids and cannabinoids on cytokine production. Δ^9-THC has also been shown to suppress proinflammatory cytokines crucial in host defense, such as TNF-α (Fischer-Stenger et al., 1993). Further studies showed a biphasic dose response to cytokine production by Δ^9-THC on human PBMCs, but it was the opposite of that observed with morphine, with low doses inhibiting TNF-α, IL-6 and IL-8, and high doses increasing these cytokines (Berdyshev et al., 1997). In a mouse model of Legionnaires' Disease, Δ^9-THC blocked IFN-γ, IL-12, and IL-12 receptor expression and increased IL-4 levels, reflecting a polarization of the immune response towards a Th2 phenotype (Perez-Castrillon et al., 1992; Klein and Cabral, 2006). Similar

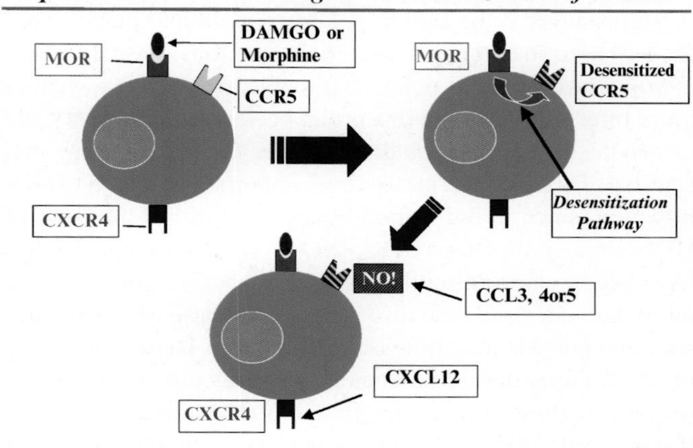

FIGURE 37.1. Heterologous desensitization of chemokine receptors following activation of the MOR. The activation of MOR by either morphine or a MOR-selective agonist such as DAMGO leads to the induction of a cross-desensitization pathway which rapidly leads to the inactivation of some the chemokine receptors, including CCR5. At this point the cell is unresponsive to CCR5-selective chemokines such as CCL3, 4 or 5. However, the MOR-induced cross-desensitization does not effect all chemokine receptors. For example, CXCR4 remains fully responsive to activation by the CXCR4-specific chemokine CXCL12. The cell is also protected from infection with an R5 strain, but not an X4 strain, of HIV-1.

results were obtained using Δ^9-THC on activated human T cells (Yuan et al., 2002). Like morphine, cannabinoids have also been shown to increase the mRNA for the chemokines CCL2, and also for IL-8, in a monocytic cell line transfected with the CB2 receptor (Jbilo et al., 1999).

Cocaine is also reported to modulate the production of cytokines and chemokines. Analysis of *in vitro* cocaine treatment on cytokine and chemokine expression by mouse and human cells has revealed a decreased ability of stimulated immune cells to produce INF-γ, TNF-α, IL-1, and IL-8 (Watzl et al., 1992; Shen et al., 1994). Other studies did not agree, and reported an increase in TNF-α, IL-8, IL-2 and other chemokines (Zhang et al., 1998; Wang et al., 1994). Interestingly, *in vivo* infusion of cocaine showed an increase in IFN-γ and a decrease in the anti-inflammatory cytokine, IL-10 (Gan et al., 1998). The lack of concordance between in vitro and in vivo studies again suggests that drugs may have complex effects in vivo, partly because they can engage mediators from other systems that might affect immune function.

37.3.2. Effects on Adaptive Immunity

37.3.2.1. Humoral Immunity (Antibody Responses)

Morphine given by slow-release pellet has been shown to block the capacity of mouse spleen cells to mount an ex vivo antibody response to the artificial antigen, SRBCs, and a naltrexone releasing pellet blocks the effect of the opioid (Bussiere et al., 1992). Serum antibody responses to tetanus toxoid (Eisenstein et al., 1990) and to trinitrophenylated bovine serum albumin (Weber et al., 1987) were also suppressed in mice by morphine given in vivo. Morphine, U50,488H (kappa agonist) and deltorphin II (delta$_2$ agonist) have all been shown to be immunosuppressive when administered by osmotic mini-pump. All three agonists gave inverted U-shaped dose response curves over a dose range of 0.1 to 10 mg/kg/day, with maximal immunosuppression between 0.5 and 2 mg/kg/day (Rahim et al., 2001). It has been suggested that part of the suppressive effect of morphine pellets on humoral immune responses is due to activation of the HPA-axis (Pruett et al., 1992). However, all of the effects of opioids on the capacity to mount an antibody response cannot be mediated, since spleen cells taken from normal animals that were never exposed to opioids in vivo, have depressed capacity to make antibodies to SRBCs when treated with various opioid agonists in vitro, (Taub et al., 1991; Eisenstein et al., 1995). Also, quaternary naloxone, which does not pass the blood-brain barrier, blocks the immunosuppressive effect of a kappa agonist given in vivo (Radulovic et al., 1995), suggesting that the drug is acting in the periphery. The suppressive effects of morphine and U50,488H in vivo and in vitro appear to result from individual effects on T cells and macrophages (Bussiere et al., 1993; Guan et al., 1994) which are needed to cooperate with B-cells to generate antibody production. Morphine has also been shown to have an effect on the mucosal antibody response. Cultured ileal segments from mice receiving morphine and given cholera toxin orally were suppressed in their ability to produce IgA specific for cholera toxin compared with mice receiving placebo (Peng et al., 2001). The effect of chronic administration of morphine and withdrawal from the drug have been examined for effects on responses to a B-cell mitogen (LPS) (Bryant et al., 1987) and on antibody responses (Rahim et al., 2002). Exposure to morphine from slow-release pellets for 3 to 4 days led to tolerance to the immunosuppressive effects in both assays, and withdrawal led to renewed immunosuppression. Extensive investigation into the mechanisms of withdrawal-induced immunosuppression showed that there was a deficit of macrophage function (Rahim et al., 2003; Rahim et al., 2005). In contrast, other data suggest that withdrawal biases the immune response to a Th2 cytokine phenotype (Kelschenbach et al., 2005). At present it is not known how to reconcile these observations.

Δ^9-THC, like morphine, has been shown to suppress in vitro antibody formation by mouse spleen cells (Kaminski et al., 1992; Kaminski et al., 1994). Interestingly, Δ^9-THC is reported to increase IgG1 responses to *Legionella pneumophila* in drug-treated animals, presumably by polarizing the immune response towards a Th2 phenotype (see below under T cell responses) (Newton et al., 1994).

Cocaine has been investigated for capacity to alter antibody responses and has been found found to be immunosuppressive for some antigens, but not others (Ou et al., 1989; Watson et al., 1983; Havas et al., 1987).

37.3.2.2. T Cell Responses

Morphine given by slow-release pellet also inhibits the ability of splenic T-cells to respond to mitogens (plant products that cause nonspecific proliferation of immune cells). Thus, at 72 hr after pellet implantation splenic murine T cells had a marked inhibition of replication when exposed to Concanavalin A (Con A) (Bryant et al., 1987). Reduction in response to Con A has also been observed using peripheral blood T cells of rats two hours after a single subcutaneous injection of morphine (Bayer et al., 1990). Morphine has also been reported to inhibit thymocyte and lymph node derived T cell proliferation when added to the cells in vitro (Roy et al., 1997; Wang et al., 2001). Other measures of T cell function relate to adaptive immunity in the cellular arm of the immune system. Several studies have shown that chronic morphine exposure inhibits development of delayed-type hypersensitivity (DTH) in rats, mice, and pigs (Pellis et al., 1986; Bryant and Roudebush, 1990; Molitor et al., 1992), and cytotoxic T cells in mice (Carpenter and Carr, 1995).

Δ^9-THC has also been shown to inhibit responses to Con A in a protocol which exposed mouse spleen cells in vitro to the drug (Klein et al., 1985). Cannabinoids have been studied for their ability to polarize T cells toward Th1 or Th2 responses. It has been reported that Δ^9-THC given to mice that are subsequently infected with *Legionella pneumophila* suppressed IL-12Rβ2 expression by a CB1 mediated mechanism and increased GATA3 mRNA via CB2 signaling (Klein et al., 2004). The IL-12 receptor participates in Th1 responses and GATA3 biases towards a Th2 response, leading to the conclusion that cannabinoids can polarize T cell responses towards a Th2 phenotype. Evidence from exposure of activated human peripheral blood T cells to Δ^9-THC also supports a Th2 polarization (Yuan et al., 2002). Δ^9-THC has also been shown to suppress cytotoxic T cell activity against both allogeneic cells (Klein et al., 1991) and virus-infected cells (Fischer-Stenger et al., 1992).

An acute i.v. dose of cocaine has been reported to have an effect similar to that of morphine in suppressing proliferation of T cells in the peripheral blood of rats exposed to Con A two hours after injection of the drug (Bayer et al., 1996). Cocaine has been shown to inhibit production of Th2 cytokines, which would be expected to reduce antibody responses (Wang et al., 1994). Cocaine given orally also suppressed DTH responses in mice (Watson et al., 1983).

37.4. Effects of Drugs of Abuse on Infection

37.4.1. Overview

The majority of publications cited above concluded that the various drugs of abuse were immunosuppressive. A logical conclusion from these observations would be that exposure to the drugs would sensitize to infection. Clinical observations of intravenous drug using (IVDU) populations have noted an increased incidence of bacterial infections (Risdahl et al., 1996). However, it has not been possible to determine whether more infection occurs in this group because of unsanitary use of needles or because the drugs render the person more susceptible. This question has become important in regard to HIV infection, since in the United States one third of all cases of HIV occur in IVDUs (CDC, 2000). Epidemiologic studies have been compromised because drug abusers usually abuse more than one drug, and also smoke and abuse alcohol. Studies have not adequately teased apart these factors, nor controlled for drug dosage or frequency of drug use. Several other reviews explore these issues in depth (Kapadia et al., 2005; Rogers et al., 2005). Laboratory studies using animal models of infections can more definitively clarify whether drugs of abuse can be cofactors in susceptibility to infection. Yet, surprisingly there are only a few studies on drugs of abuse and experimental infections of any type.

37.4.2. Effects of Drugs of Abuse on Infections Other than HIV

Morphine has been shown to sensitize mice to fungal (*Candida albicans*) (Tubaro et al., 1983), parasitic (*Toxoplasma gondii*) (Chao et al., 1990), and bacterial (*Salmonella typhimurium, Streptococcus pneumoniae*) infections (MacFarlane et al., 2000; Wang et al., 2005). Further, morphine and withdrawal from morphine are reported to induce sepsis and enhance sensitivity to bacterial LPS (Hilburger et al., 1997; Roy et al., 1998b), as well as sensitize to *Salmonella* infection (Feng et al., 2005). Δ^9-THC is reported to sensitize to the bacteria, *Legionella pneumophila* and *Listeria monocytogenes* (Newton et al., 1994; Morahan et al., 1979), and to herpes simplex virus (Cabral et al., 1986).

37.4.3. Effects of Drugs of Abuse on HIV or Related Infections

There are several papers showing that opioids up-regulate HIV replication when added to infected human peripheral blood cells or microglia in vitro (Peterson et al., 1990, 1994; Li et al., 2002). As mentioned in Section 37.2.1.4 on cytokines and chemokines, the effects of opioids on HIV infection may be related to drug-induced alterations in cytokine or chemokine production or in expression of chemokine receptors (Peterson et al., 1994; Li et al., 2002; Steele et al., 2003). Proinflammatory cytokines, particularly TNF-α, can activate T cells and lead to increased HIV replication, and chemokine receptors are co-receptors for the virus. It is to be expected that modulation of chemokine and/or chemokine receptor levels can alter HIV infectivity through increasing the pool of infectable cells or by altering availability of the number of co-receptors.

Similar immunodeficiency syndromes may result in monkeys and cats infected with simian immunodeficiency virus

(SIV) and feline immunodeficiency virus (FIV), respectively, but studies with these viruses have not provided a clear indication of the influence of opioids on the development of these immunodeficiency diseases. In one study, monkeys maintained on morphine had a slower progression of SIV infection, although temporary withdrawal augmented viral load (Donahoe et al., 1993), but in another, their progression was more rapid (Chuang et al., 1997). Neither acute nor chronic morphine or withdrawal altered FIV infection in cats, but the number of animals was small and the variance in FIV load large (Barr et al., 2003).

Cocaine has also been investigated in regard to effects on HIV infection and found to potentiate replication in in vitro systems (Peterson et al., 1992; Bagasra and Pomerantz, 1993). Cocaine has also been shown to up-regulate DC-SIGN, a molecule on the surface of human dendritic cells that can act as a receptor for HIV (Nair et al., 2005). In addition, a provocative paper reported that mice with severe combined immunodeficiency, that were reconstituted with human leukocytes and infected with HIV, had enhanced replication of the virus after cocaine treatment (Roth et al., 2002). The fact that opioids and cocaine can alter replication of HIV in cell culture in vitro certainly suggests that the drugs should be able to alter HIV progression in vivo. Isolation of variables in the laboratory permits elimination of confounding factors in epidemiologic studies in humans, such as poly-drug abuse, and also standardizes drug dosage. Thus, there is considerable evidence from laboratory studies supporting a role for drugs of abuse in potentiating HIV infection.

Summary

In general, opioids and cannabinoids have been found to be suppressive in a wide variety of assays of immune responsiveness. The literature on cocaine is smaller and less consistent. An extension of these observations is the assessment of effects of the drugs on resistance to infection, a consequence of alterations in immune function. Both opioids and cannabinoids have been shown to sensitize animals to a variety of experiment infections. Interest in the effects of the opioids on HIV progression is sparked by the epidemiological intersection of intravenous drug abuse and incidence of HIV. In vitro studies give robust results showing augmentation of HIV replication in the presence of opioids. A single study with cocaine, using an immunodeficient mouse repopulated with human lymphoid cells, shows enhanced replication of HIV in the presence of the drug. Together, the literature suggests substantial effects of drugs of abuse on immune competence in general, resulting in immunosuppresion. The implications of the research are that drug abuse increases vulnerability to infection. These studies also reinforce the existence of physiological pathways between the neural and immune systems mediated by opioids and cannabinoids.

Acknowledgments. We thank Dr. Martin Adler, Director of the Center for Substance Abuse Research, Temple University School of Medicine, for critical reading of the manuscript. This work was supported by NIDA grants DA06650, DA11130, DA11134, DA13429, DA14223 and DA14230.

Review Questions/Problems

1. It is observed that morphine given subcutaneously suppresses responses of spleen cells put into tissue culture with the T-cell mitogen Concanavalin A (Con A).

a. Mice with a disruption of the gene coding the mu opioid receptor (Mu Opioid Receptor knock-out mice), would not show suppression to Con A.
b. Morphine acts mainly through the kappa opioid receptor.
c. Morphine is mainly metabolized to heroin when it is injected in vivo.
d. Methyl-morphine could not be used to determine if peripheral receptors or brain receptors are involved in the immunosuppression.
e. None of the above.

2. In regard to cannabinoids:

a. They are proteins.
b. They exert their effects primarily by interdigitating into cell membranes.
c. There is evidence that they polarize the immune response towards a Th1 type phenotype.
d. There is evidence that they polarize the immune response towards a Th2 type phenotype.
e. They have no direct effects on cells of the immune system.

3. In regard to the immune system, morphine has been shown to:

a. Elevate antibody responses to various antigens.
b. Elevate responses to the B-cell mitogen, lipopolysaccharide (LPS).
c. Elevate delayed-type hypersensitivity (DTH) responses.
d. Increase phagocytosis by macrophages.
e. Depress natural killer (NK) cell activity.

4. Cocaine:

a. Has no effect on the immune system.
b. Exacerbates HIV replication in a mouse/human model of viral replication.
c. Binds to the delta opioid receptor.
d. Depresses IL-2 production.
e. All of the above.

5. In regard to infections and drugs of abuse:

a. Morphine sensitizes mice to infection with *Legionella pneumophila* (Legionnaires' Disease).
b. Δ^9-THC sensitizes mice to *Salmonella* infection.

c. Morphine and Δ^9-THC have direct effects on the ability of bacteria to grow in broth cultures.
d. Morphine can affect resistance to infection by altering the activity of phagocytes.
e. Intravenous drug abusers have a rate of infection the same as the general population.

6. Morphine most probably affects the progression of HIV infection by all of the following EXCEPT:

a. Changing the level of chemokine receptors.
b. Changing the level of chemokines.
c. Causing T-cells to undergo uncontrolled cell division.
d. Mediating heterologous desensitization of chemokine receptors.
e. Altering cytokine levels.

7. Which of the following statements is true:

a. Opioids, cannabinoids, and cocaine uniformly suppress the activity of NK cells.
b. The use of N-methylmorphine substantiated the direct effects of opioid-induced NK cell suppression.
c. Mice lacking the μ-opioid receptor do not have suppressed NK cell activity in response to morphine treatment.
d. Δ^9-THC prevents NK cell proliferation by inhibiting the secretion of TGF-β.
e. Morphine inhibits NK cell activity by inducing endogenous cannabinoid levels.

8. Opioids modulate phagocyte activity by:

a. Increasing phagocytic uptake of bacteria.
b. Decreasing apoptosis of phagocytic cells.
c. Enhancing maturation of bone marrow cells into macrophages.
d. Decreasing their ability to make microbicidal compounds in vitro.
e. Decreasing maturation of lymphocytes in the bone marrow.

9. Immune cells treated with morphine:

a. Uniformly have markers of activation.
b. Display altered levels of pro-inflammatory chemokines, but chemokine receptor numbers remained unchanged.
c. Display bi-directional heterologous desensitization between opioid receptors and certain chemokine receptors.
d. Clearly have inhibition of Th2 cytokine responses.
e. None of the above

10. Which of the following statements is true:

a. Δ^9-THC induces its effects through both CXCR4 and CCR5.
b. Both Δ^9-THC and morphine have been shown to induce the expression of pro-inflammatory chemokines, such as MCP-1 and IL-8.
c. Like cannabinoids and morphine, cocaine treatment increases TNF-α and IL-1.
d. Cocaine uniformly suppresses cytokine and chemokine production when given in vivo or applied to cells of the immune system in vitro.
e. Opioids have no effect on antibody formation.

11. Morphine and/or DAMGO treatment results in decreases in which of the following?

a. Apoptosis
b. Pro-inflammatory chemokine production
c. Serum antibody responses
d. TGF-β
e. All of the above

12. Inhibition of T cell responses is observed for which of the following drugs of abuse?

a. Morphine
b. Δ^9-THC
c. Cocaine
d. All of the above
e. None of the above

13. The initial evidence suggesting that opioid receptors are expressed by cells of the immune system was:

a. Results showing impairment of T cells to rosette to sheep red blood cells in the presence of morphine.
b. Studies examining functions of the CB1 and CB2 receptors.
c. Primate studies examining disease incidence in self-administration of heroin
d. Biochemical analyses of lymph nodes in heroin addicts.
e. Studies showing increased HIV infection in heroin abusers.

14. All of the following statements are true EXCEPT?

a. In vitro studies have shown Δ^9-THC biases the immune response towards a Th2 cytokine profile.
b. Cocaine administration stimulated production of the anti-inflammatory cytokine IL-10, while simultaneously suppressing IFNγ.
c. Activated human PMNs show enhanced functional activity in response to cocaine.
d. The effects of opioids, cannabinoids, and cocaine on immune system function are compounded by the involvement of the HPA axis.
e. Chemokine expression is induced by morphine administration.

15. Evaluation of the effects of drugs of abuse can be complicated by a number of factors, including:

a. The effects of the sympathetic nervous system on immune cell function.
b. The effects of the HPA axis on immune cell function.

c. The capacity of cells of the immune system to produce endogenous opioids.
d. The production of endogenous cannabinoids in the periphery.
e. All of the above.

16. The function of G protein-coupled receptors, such as the opioid and cannabinoid receptors, can be regulated by the process of heterologous desensitization. Which of the following statements is true about this regulatory mechanism?

a. This process occurs when the activation of one receptor leads to an increased expression of a second receptor.
b. This process only occurs between receptors expressed on the surfaces of adjacent cells.
c. This process cannot occur between receptors expressed on the same cell.
d. This is a process that appears to involve transphosphorylation of G protein-coupled receptors.
e. All of the above

17. Both opioids and cannabinoids have been shown to sensitize animals to a variety of experimental infections. The impact of drugs of abuse on resistance to infectious agents can be difficult to evaluate because:

a. Drug abuse rarely involves the administration of a single drug, and the effects of poly-drug abuse are poorly understood.
b. Drug abusers are exposed more frequently to pathogenic agents than non-abusers.
c. A number of additional factors, which are hard to control, impact on measurement of the immune competence of drug abusers, including the dose of the drug and the time since it was last taken.
d. The contributions of legal drug use, including nicotine and alcohol, can complicate the effects of illegal drug abuse.
e. All of the above.

18. All of the following influence the effect of opioids on susceptibility to HIV infection EXCEPT:

a. Opioids alter the expression of chemokine receptors that are co-receptors for HIV-1.
b. Opioids increase the expression of some chemokines, which may promote the attraction of additional susceptible target cells for HIV infection.
c. Opioids increase the expression of chemokines, which, for an individual cell, may block viral replication by blockading the chemokine coreceptor.
d. Opioids alter the phagocytic activity of neutrophils, and this would be expected to significantly alter the kinetics of the infection.
e. Opioids would be expected to alter the expression of cytokines that may, in turn, alter the replication rate of the virus in monocytes and T cells.

References

Adler MW, Rogers TJ (2005) Are chemokines the third major system in the brain? J Leukoc Biol 78:1204–1209.

Akil H, Watson SJ, Young E, Lewis ME, Khachaturian H, Walker JM (1984) Endogenous opioids: Biology and function. Ann Rev Neurosci 7:223–255.

Bagasra O, Pomerantz RJ (1993) Human immunodeficiency virus type 1 replication in peripheral blood mononuclear cells in the presence of cocaine. J Infec Dis 168:1157–1164.

Baldwin GC, Buckley DM, Roth MD, Kleerup EC, Tashkin DP (1997) Acute activation of circulating PMNs following in vivo administration of cocaine: A potential etiology for pulmonary injury. Chest 111:698–705.

Barr MC, Huitron-Resendiz S, Sanchez-Alavez M, Hendriksen SJ, Phillips TR (2003) Escalating morphine exposures followed by withdrawal in feline immunodeficincy virus-infected cats: A model for HIV infection in chronic opiate abusers. Drug Alcohol Depend 72:141–149.

Bayer BM, Daussin S, Hernandez M, Irvin L (1990) Morphine inhibition of lymphocyte activity is mediated by an opioid dependent mechanism. Neuropharmacology 29:369–374.

Bayer BM, Hernandez MC, Ding XZ (1996) Tolerance and crosstolerance to the suppressive effects of cocaine and morphine on lymphocyte proliferation. Pharmacol Biochem Behav 53:227–234.

Berdyshev EV, Boichot E, Germain N, Allain N, Anger J-P, Lagente V (1997) Influence of fatty acid ethanolamides and Δ^9-tetrahydrocannabinol on cytokine and arachidonate release by mononuclear cells. Eur J Pharmacol 330:231–240.

Bhat RS, Bhaskaran M, Mongia A, Hitosugi N, Singhal PC (2004) Morphine-induced macrophage apoptosis: Oxidative stress and strategies for modulation. J Leukoc Biol 75:1131–1138.

Bryant HU, Roudebush RE (1990) Suppressive effects of morphine pellet imlants on *in vivo* parameters of immune function. J Pharmacol Exp Ther 255:410–414.

Bryant HU, Bernton EW, Holaday JW (1987) Immunosuppressive effects of chronic morphine treatment in mice. Life Sci 41:1731–1738.

Buckley NE, McCoy KL, Mezey E, Bonner T, Zimmer A, Felder CC, Glass M (2000) Immunomodulation by cannabinoids is absent in mice deficient for the cannabinoid CB(2) receptor. Eur J Pharmacol 396:141–149.

Bussiere JL, Adler MW, Rogers TJ, Eisenstein TK (1992) Differential effects of morphine and naltrexone on the antibody response in various mouse strains. Immunopharmacol Immunotoxicol 14:657–673.

Bussiere JL, Adler MW, Rogers TJ, Eisenstein TK (1993) Cytokine reversal of morphine-induced suppression of the antibody response. J Pharmacol Exp Ther 264:591–597.

Cabral GA, Mishkin EM, Marciano-Cabral F, Coleman P, Harris L, Munson AE (1986) Effect of delta-9-tetrahydrocannabinol on herpes simplex virus type 2 vaginal infections in the guinea pig. Proc Soc Exp Biol Med 182:181–186.

Carpenter GW, Carr DJJ (1995) Pretreatment with β-funaltrexamine blocks morphine-mediated suppression of CTL activity in alloimmunized mice. Immunopharm 29:129–140.

Carr DJJ (1991) The role of endogenous opioids and their receptors in the immune system. Proc Soc Exp Biol Med 198:710–720.

CDC (2000) In: HIV/AIDS surveillance report, 2000. Atlanta, Georgia: US Department of Health and Human Services, Public Health Service.

Chao CC, Sharp BM, Pomeroy C, Filice GA, Peterson PK (1990) Lethality of morphine in mice infected with *Toxoplasma gondii*. J Pharmacol Exp Ther 252:605–609.

Chao CC, Hu S, Molitor TW, Zhou Y, Murtaugh MP, Tsang M, Peterson PK (1992) Morphine potentiates transforming growth factor-β release from human peripheral blood mononuclear cell cultures. J Pharmacol Exp Ther 262:19–24.

Chen C, Li J, Bot G, Szabo I, Rogers TJ, Liu-Chen L-Y (2004) Heterodimerization and cross-desensitization between the μ-opioid receptor and the chemokine CCR5 receptor. Eur J Pharmacol 483:175–186.

Choi Y, Chuang LF, Lam KM, Kung HF, Wang JM, Osburn BI, Chuang RY (1999) Inhibition of chemokine-induced chemotaxis of monkey leukocytes by mu-opioid receptor agonists. in vivo 13:389–396.

Chuang LF, Killam Jr KF, Chuang RY (1997) SIV infection of macaques: A model for studying AIDS and drug abuse. Addict Biol 2:421–430.

Devane WA, Hanus L, Breuer A, Pertwee RG, Stevenson LA, Griffin G, Gibson D, Mandelbaum A, Etinger A, Mechoulam R (1992) Isolation and structure of a brain constituent that binds to the cannabinoid receptor. Science 258:1946–1949.

Di Marzo V, Deutsch DG (1998) Biochemistry of the endogenous ligands of cannabinoid receptors. Neurobiol Dis 5:386–404.

Donahoe RM, Byrd LD, McClure HM, Fultz P, Brantley M, Marsteller F, Ansari AA, Wenzel D, Aceto M (1993) Consequences of opiate-dependency in a monkey model of AIDS. Adv Exp Med Biol 335:21–28.

Eisenstein TK, Meissler Jr JJ, Geller EB, Adler MW (1990) Immunosuppression to tetanus toxoid induced by implanted morphine pellets. Ann N Y Acad Sci 594:377–379.

Eisenstein TK, Meissler Jr JJ, Rogers TJ, Geller EB, Adler MW (1995) Mouse strain differences in immunosuppression by opioids *in vitro*. J Pharmacol Exp Ther 275:1484–1489.

Eisenstein TK, Rahim RT, Feng P, Thingalaya NK, Meissler JJ (2006) Effects of opioid tolerance and withdrawal on the immune system. J Neuroimm Pharmacol 1:237–249

Feng P, Wilson QM, Meissler Jr JJ, Adler MW, Eisenstein TK (2005) Increased sensitivity to *Salmonella enterica* serovar Typhimurium infection in mice undergoing withdrawal from morphine is associated with suppression of interleukin-12. Infect Immun 73:7953–7959.

Fischer-Stenger K, Updegrove AW, Cabral GA (1992) Δ^9-tetrahydrocannabinol decreases cytotoxic T lymphocyte activity to herpes simplex virus type1-infected cells. Proc Soc Exp Biol Med 200:422–430.

Fischer-Stenger K, Dove Pettit DA, Cabral GA (1993) Δ^9-tetrahydrocannabinol inhibition of tumor necrosis factor-α: Suppression of post-translational events. J Pharmacol Exp Ther 267:1558–1565.

Gamble L, Mason CM, Nelson S (2006) The effects of alcohol on immunity and bacterial infection in the lung. Med et Mal Infect 36:72–77.

Gan X, Zhang L, Newton T, Chang SL, Ling W, Kermani V, Berger O, Graves MC, Fiala M (1998) Cocaine infusion increases interferon-γ and decreases interleukin-10 in cocaine-dependent subjects. Clin Immunol 89:181–190.

Gavériaux-Ruff C, Matthes HWD, Peluso J, Kieffer BL (1998) Abolition of morphine-immunosuppression in mice lacking the μ-opioid receptor gene. Proc Natl Acad Sci USA 95:6326–6330.

Grimm MC, Ben-Baruch A, Taub DD, Howard OMZ, Resau JH, Wang JM, Ali H, Richardson R, Snyderman R, Oppenheim JJ (1998) Opiates transdeactivate chemokine receptors: δ and μ opiate receptor-mediated heterologous desensitization. J Exp Med 188:317–325.

Guan L, Townsend R, Eisenstein TK, Adler MW, Rogers TJ (1994) Both T cells and macrophages are targets of κ-opioid-induced immunosuppression. Brain Behav Immun 8:229–240.

Havas HF, Dellaria M, Schiffman G, Geller EB, Adler MW (1987) Effect of cocaine on the immune response and host resistance in BALB/c mice. Int Arch Allergy & Appl Immunol 83:377–383.

Hilburger ME, Adler MW, Truant AL, Meissler Jr JJ, Satishchandran V, Rogers TJ, Eisenstein TK (1997) Morphine induces sepsis in mice. J Infect Dis 176:183–188.

Jbilo O, Derocq J-M, Segui M, Le Fur G, Casellas P (1999) Stimulation of peripheral cannabinoid receptor CB2 induces MCP-1 and IL-8 gene expression in human promyelocytic cell line HL60. FEBS Lett 448:273–277.

Kaminski NE, Abood ME, Kessler FK, Martin BR, Schatz AR (1992) Identification of a functionally relevant cannabinoid receptor on mouse spleen cells that is involved in cannabinoid-mediated immune modulation. Mol Pharmacol 42:736–742.

Kaminski NE, Koh WS, Yang KH, Lee M, Kessler FK (1994) Suppression of the humoral immune response by cannabinoids is partially mediated through inhibition of adenylate cyclase by a pertussis toxin-sensitive G-protein coupled mechanism. Biochem Pharmacol 48:1899–1908.

Kapadia F, Vlahov D, Donahoe RM, Friedland G (2005) The role of substance abuse in HIV disease progression: Reconciling differences from laboratory and epidemiological investigations. Clin Infect Dis 41:1027–1034.

Kelschenbach J, Barke RA, Roy S (2005) Morphine withdrawal contributes to Th cell differentiation by biasing cells toward the Th2 lineage. J Immunol 175:2655–2665.

Klein TW, Cabral GA (2006) Cannabinoid-induced immune suppression and modulation of antigen-presenting cells. J Neuroimm Pharmacol 1:50–64.

Klein TW, Newton CA, Widen R, Friedman H (1985) The effect of delta-9-tetrahydrocannabinol and 11-hydroxy-delta-9-tetrahydrocannabinol on T-lymphocyte and B-lymphocyte mitogen responses. J Immunopharmac 7:451–466.

Klein TW, Newton C, Friedman H (1987) Inhibition of natural killer cell function by marijuana components. J Toxicol Environ Health 20:321–332.

Klein TW, Kawakami Y, Newton C, Friedman H (1991) Marijuana components suppress inducion and cytolytic function of murine cytotoxic T-cells *in vitro* and *in vivo*. J Toxicol Environ Health 32:465–477.

Klein TW, Newton C, Larsen K, Chou J, Perkins I, Lu L, Nong L, Friedman H (2004) Cannabinoid receptors and T helper cells. J Neuroimmunol 147:91–94.

Lee SF, Newton C, Widen R, Friedman H, Klein TW (2001) Differential expression of cannabinoid CB2 receptor mRNA in mouse immune cells subpopulations and following B cell stimulation. Eur J Pharmacol 423:235–241.

Li Y, Wang X, Tian S, Guo C-J, Douglas SD, Ho W-Z (2002) Methadone enhances human immunodeficiency virus infection of human immune cells. J Infect Dis 185:118–122.

Loh HH, Smith AP (1990) Molecular characterization of opioid receptors. Ann Rev Pharmacol Toxicol 30:123–147.

Lopez-Cepero M, Friedman M, Klein T, Friedman H (1986) Tetrahydrocannabinol-induced suppression of macrophage spreading and phagocytic activity in vitro. J Leukoc Biol 39:679–686.

MacFarlane AS, Peng X, Meissler Jr JJ, Rogers TJ, Geller EB, Adler MW, Eisenstein TK (2000) Morphine increases susceptibility to oral *Salmonella typhimurium* infection. J Infect Dis 181:1350–1358.

McCarthy L, Wetzel M, Sliker J, Eisenstein TK, Rogers TJ (2001) Opioids, opioid receptors, and the immune response. Drug Alcohol Depend 62:111–123.

McCoy KL, Matveyeva M, Carlisle SJ, Cabral GA (1999) Cannabinoid inhibition of the processing of intact lysozyme by macrophages: Evidence for CB2 receptor participation. J Pharmacol Exp Ther 289:1620–1625.

Mechoulam R, Ben-Shabat S, Hanus L, Ligumsky M, Kaminski NE, Schatz AR, Gopher A, Almog S, Martin BR, Compton DR, Pertwee RG, Griffin G, Bayewitch M, Barg J, Vogel Z (1995) Identification of an endogenous 2-monoglyceride, present in canine gut, that binds to cannabinoid receptors. Biochem Pharmacol 50:83–90.

Molitor TW, Morilla A, Risdahl JM, Murtaugh MP, Chao CC, Peterson PK (1992) Chronic morphine administration impairs cell-mediated immune responses in swine. J Pharmacol Exp Ther 260:581–586.

Morahan PS, Klykken PC, Smith SH, Harris LS, Munson AE (1979) Effects of cannabinoids of host resistance to Listeria monocytogenes and herpes simples virus. Infect Immun 23:670–674.

Mosmann TR, Coffman RL (1987) Two types of mouse helper T-cell clone: Implications for immune regulation. Immunol Today 8:223–227.

Munro S, Thomas KK, Abu-Shaar M (1993) Molecular characterization of a peripheral receptor for cannabinoids. Nature 365:61–65.

Nair MP, Schwartz SA, Polasani R, Hou J, Sweet A, Chadha KC (1997) Immunoregulatory effects of morphine on human lymphocytes. Clin Diag Lab Immunol 4:127–132.

Nair MPN, Mahajan SD, Schwartz SA, Reynolds J, Whitney R, Bernstein Z, Chawda RP, Sykes D, Hewitt R, Hsiao CB (2005) Cocaine modulates dendritic cell-specific C type intercellular adhesion molecule-3-grabbing nonintegrin expression by dendritic cells in HIV-1 patients. J Immunol 174:6617–6626.

Newton C, Klein TW, Friedman H (1994) Secondary immunity to *Legionella pneumophila* and Th1 activity are suppressed by delta-9-tetrahydrocannabinol injection. Infect Immun 62:4015–4020.

Ni X, Gritman KR, Eisenstein TK, Adler MW, Arfors KE, Tuma RF (2000) Morphine attenuates leukocyte/endothelial interactions. Microvasc Res 60:121–130.

Ou DW, Shen M-L, Luo Y-D (1989) Effects of cocaine on the immune system of Balb/C mice. Clin Immunol Immunopathol 52:305–312.

Pacifici R, di Carlo S, Bacosi A, Zuccaro P (1993) Macrophage functions in drugs of abuse-treated mice. Int J Immunopharmacol 15:711–716.

Pacifici R, Zuccaro P, Farré M, Pichini S, di Carlo S, Roset PN, Palmi I, Ortuño J, Menoyo E, Segura J, De La Torre R (2002) Cell-mediated immune response in MDMA users after repeated dose administration. Ann NY Acad Sci 965:421–433.

Pacifici R, Fiaschi AI, Micheli L, Centini F, Giorgi G, Zuccaro P, Pichini S, di Carlo S, Bacosi A, Cerretani D (2003) Immunosuppression and oxidative stress induced by acute and chronic exposure to cocaine in rat. Int Immunopharm 3:581–592.

Patel V, Borysenko M, Kumar MS, Millard WJ (1985) Effects of acute and subchronic delta 9-tetrahydrocannabinol administration on the plasma catecholamine, beta-endorphin, and corticosterone levels and splenic natural killer cell activity in rats. Proc Soc Exp Biol Med 180:400–404.

Pellis NR, Harper C, Dafny N (1986) Suppression of the induction of delayed hypersensitivity in rats by repetitive morphine treatments. Exp Neurol 93:92–97.

Peng X, Mosser DM, Adler MW, Rogers TJ, Meissler Jr JJ, Eisenstein TK (2000) Morphine enhances interleukin-12 and the production of other pro-inflammatory cytokines in mouse peritoneal macrophages. J Leukoc Biol 68:723–728.

Peng X, Cebra JJ, Adler MW, Meissler Jr JJ, Cowan A, Feng P, Eisenstein TK (2001) Morphine inhibits mucosal antibody responses and TGF-β mRNA in gut-associated lymphoid tissue following oral cholera toxin in mice. J Immunol 167:3677–3681.

Perez-Castrillon J-L, Perez-Arellanos J-L, Carcia-Palomo J-D, Jimeniz-Lopez A, de Castro S (1992) Opioids depress in vitro human monocyte chemotaxis. Immunopharm 23:57–61.

Peterson PK, Sharp B, Gekker G, Brummitt C, Keane WF (1987a) Opioid-mediated suppression of cultured peripheral blood mononuclear cell respiratory burst activity. J Immunol 138:3907–3912.

Peterson PK, Sharp B, Gekker G, Brummitt C, Keane WF (1987b) Opioid-mediated suppression of interferon-γ production by cultured peripheral blood mononuclear cells. J Clin Invest 80:824–831.

Peterson PK, Sharp BM, Gekker G, Portoghese PS, Sannerud K, Balfour Jr HH (1990) Morphine promotes the growth of HIV-1 in human peripheral blood mononuclear cell cocultures. AIDS 4:869–873.

Peterson PK, Gekker G, Chao CC, Schut R, Verhoef J, Edelman CK, Erice A, Balfour Jr HH (1992) Cocaine amplifies HIV-1 replication in cytomegalovirus-stimulated peripheral blood mononuclear cell cocultures. J Immunol 149:676–680.

Peterson PK, Gekker G, Hu S, Anderson WR, Kravitz F, Portoghese PS, Balfour Jr HH, Chao CC (1994) Morphine amplifies HIV-1 expression in chronically infected promonocytes cocultured with human brain cells. J Neuroimmunol 50:167–175.

Pruett SB, Han Y-C, Fuchs BA (1992) Morphine suppresses primary humoral immune responses by a predominantly indirect mechanism. J Pharmacol Exp Ther 262:923–928.

Radulovic J, Miljevic C, Djergovic D, Vujic V, Antic J, von Horsten S, Jankovic BD (1995) Opioid receptor-mediated suppression of humoral immune response in vivo and in vitro: Involvement of κ opioid receptors. J Neuroimmunol 57:55–62.

Rahim RT, Meissler Jr JJ, Cowan A, Rogers TJ, Geller EB, Gaughan J, Adler MW, Eisenstein TK (2001) Administration of mu-, kappa- or delta$_2$-receptor agonists via osmotic minipumps suppresses murine splenic antibody responses. Int Immunopharm 1:2001–2009.

Rahim RT, Adler MW, Meissler Jr JJ, Cowan A, Rogers TJ, Geller EB, Eisenstein TK (2002) Abrupt or precipitated withdrawal from morphine induces immunosuppression. J Neuroimmunol 127:88–95.

Rahim RT, Meissler Jr JJ, Zhang L, Adler MW, Rogers TJ, Eisenstein TK (2003) Withdrawal from morphine in mice suppresses splenic macrophage function, cytokine production, and costimulatory molecules. J Neuroimmunol 144:16–27.

Rahim RT, Meissler Jr JJ, Adler MW, Eisenstein TK (2005) Splenic macrophages and B-cells mediate immunosuppression following abrupt withdrawal from morphine. J Leukoc Biol 78:1185–1191.

Risdahl JM, Peterson PK, Molitor TW (1996) Opiates, infection and immunity. In: Drugs of Abuse, Immunity, and Infections (Friedman H, Klein TW, Specter S, eds), pp. 1–42. Boca Raton: CRC Press, Inc.

Rogers TJ, Bednar F, Kaminsky DE, Davey PC, Meissler Jr J, Eisenstein TK (2005) Laboratory model systems of drug abuse and their relevance to HIV infection and dementia. In: The Neurology of AIDS (Gendelman HE, Grant I, Everall IP, Lipton SA, Swindells S, eds), pp 310–320. Oxford: Oxford University Press.

Rossi GC, Brown GP, Leventhal L, Yang K, Pasternak GW (1996) Novel receptor mechanisms for heroin and morphine-6β-glucoronide analgesia. Neurosci Lett 216:1–4.

Roth MD, Tashkin DP, Choi R, Jamieson BD, Zack JA, Baldwin GC (2002) Cocaine enhances human immunodeficiency virus replication in a model of severe combined immunodeficient mice implanted with human peripheral blood leukocytes. J Infect Dis 185:701–705.

Roth MD, Whittaker K, Salehi K, Tashkin DP, Baldwin GC (2004) Mechanisms for impaired effector function in alveolar macrophages from marijuana and cocaine smokers. J Neuroimmunol 147:82–86.

Roy S, Ramakrishnan S, Loh HH, Lee NM (1991) Chronic morphine treatment selectively suppresses macrophage colony formation in bone marrow. Eur J Pharmacol 195:359–363.

Roy S, Chapin RB, Cain KJ, Charboneau RG, Ramakrishnan S, Barke RA (1997) Morphine inhibits transcriptional activation of IL-2 in mouse thymocytes. Cell Immunol 179:1–9.

Roy S, Cain KJ, Chapin RB, Charboneau RG, Barke RA (1998a) Morphine modulates NFκB activation in macrophages. Biochem Biophys Res Comm 245:392–396.

Roy S, Cain KJ, Charboneau RG, Barke RA (1998b) Morphine accelerates the progression of sepsis in an experimental sepsis model. Adv Exp Med Biol 437:21–31.

Roy S, Wang J, Gupta S, Charboneau R, Loh HH, Barke RA (2004) Chronic morphine treatment differentiates T helper cells to Th2 effector cells by modulating transcription factors GATA 3 and T-bet. J Neuroimmunol 147:78–81.

Saurer TB, Carrigan KA, Ijames SG, Lysle DT (2006) Suppression of natural killer cell activity by morphine is mediated by the nucleus accumbens shell. J Neuroimmunol 173:3–11.

Schatz AR, Lee M, Condie RB, Pulaski JT, Kaminski NE (1997) Cannabinoid receptors CB1 and CB2:a characterization of expression and adenylate cyclase modulation within the immune system. Toxicol Appl Pharmacol 142:278–287.

Sharp BM, Roy S, Bidlack JM (1998) Evidence for opioid receptors on cells involved in host defense and the immune system. J Neuroimmunol 83:45–56.

Shavit Y, Depaulis A, Martin FC, Terman GW, Pechnick RN, Zane CJ, Gale RP, Liebeskind JC (1986a) Involvement of brain opiate receptors in the immune-suppressive effect of morphine. Proc Natl Acad Sci USA 83:7114–7117.

Shavit Y, Terman GW, Lewis JW, Zane CJ, Gale RP, Liebeskind JC (1986b) Effects of footshock stress and morphine on natural killer lymphocytes in rats: Studies of tolerance and cross-tolerance. Brain Res 372:382–385.

Shen HM, Kennedy JL, Ou DW (1994) Inhibition of cytokine release by cocaine. Int J Immunopharmacol 16:295–300.

Sibinga NES, Goldstein A (1988) Opioid peptides and opioid receptors in cells of the immune system. Ann Rev Immunol 6:219–249.

Simon E (1991) Opioid receptors and endogenous opioid peptides. Med Res Rev 11:357–374.

Singhal PC, Sharma P, Kapasi AA, Reddy K, Franki N, Gibbons N (1998) Morphine enhances macrophage apoptosis. J Immunol 160:1886–1893.

Sopori M (2002) Effects of cigarette smoke on the immune system. Nature Rev Immunol 2:372–377.

Specter S, Lancz G, Goodfellow D (1991) Suppression of human macrophage function in vitro by Δ^9-tetrahydrocannabinol. J Leukoc Biol 50:423–426.

Specter SC, Klein TW, Newton C, Mondragon M, Widen R, Friedman H (1986) Marijuana effects on immunity: Suppression of human natural killer cell activity of delta-9-tetrahydrocannabinol. Int J Immunopharmacol 8:741–745.

Steele AD, Szabo I, Bednar F, Rogers TJ (2002) Interactions between opioid and chemokine receptors: Heterologous desensitization. Cytok Growth Factor Rev 13:209–222.

Steele AD, Henderson EE, Rogers TJ (2003) μ-Opioid modulation of HIV-1 coreceptor expression and HIV-1 replication. Virol 309:99–107.

Szabo I, Rojavin M, Bussiere JL, Eisenstein TK, Adler MW, Rogers TJ (1993) Suppression of peritoneal macrophage phagocytosis of *Candida albicans* by opioids. J Pharmacol Exp Ther 267:703–706.

Szabo I, Chen XH, Xin L, Adler MW, Howard OMZ, Oppenheim JJ, Rogers TJ (2002) Heterologous desensitization of opioid receptors by chemokines inhibits chemotaxis and enhances the perception of pain. Proc Natl Acad Sci USA 99:10276–10281.

Szabo I, Wetzel MA, Zhang N, Steele AD, Kaminsky DE, Chen C, Liu-Chen L-Y, Bednar F, Henderson EE, Howard OMZ, Oppenheim JJ, Rogers TJ (2003) Selective inactivation of CCR5 and decreased infectivity of R5 HIV-1 strains mediated by opioid-induced heterologous desensitization. J Leukoc Biol 74:1074–1082.

Taub DD, Eisenstein TK, Geller EB, Adler MW, Rogers TJ (1991) Immunomodulatory activity of μ- and κ-selective opioid agonists. Proc Natl Acad Sci USA 88:360–364.

Tubaro E, Borelli G, Croce C, Cavallo G, Santiangeli C (1983) Effect of morphine on resistance to infection. J Infec Dis 148:656–666.

Van Dyke C, Stesin A, Jones R, Chuntharapai A, Seaman W (1986) Cocaine increases natural killer cell activity. J Clin Invest 77:1387–1390.

Wang Y, Huang DS, Watson RR (1994) In vivo and in vitro cocaine modulation on production of cytokines in C57BL/6 mice. Life Sci 54:401–411.

Wang J, Charboneau R, Balasubramanian S, Sumandeep, Barke RA, Loh HH, Roy S (2001) Morphine modulates lymph node derived T lymphocyte function: Role of caspase-3, -8, and nitric oxide. J Leukoc Biol 70:527–536.

Wang J, Barke RA, Charboneau R, Roy S (2005) Morphine impairs host innate immune response and increases susceptibility to *Streptococcus pneumoniae* lung infection. J Immunol 174:426–434.

Watson ES, Murphy JC, El Sohly HN, El Sohly MA, Turner CE (1983) Effects of the administration of coca alkaloids on the primary immune responses of mice: Interaction with delta 9-tetrahydrocannabinol and ethanol. Toxicol Appl Pharmacol 71:1–13.

Watzl B, Chen G, Scuderi P, Pirozhkov S, Watson RR (1992) Cocaine-induced suppression of interferon-gamma secretion in leukocytes from young and old C57BL/6 mice. Int J Immunopharmacol 14:1125–1131.

Weber RJ, Pert A (1989) The periaqueductal gray matter mediates opiate-induced immunosuppression. Science 245:188–190.

Weber RJ, Ikejiri B, Rice KC, Pert A, Hagan AA (1987) Opiate receptor mediated regulation of the immune response *in vivo*. NIDA Res Mono Ser 48:341–348.

Wetzel MA, Steele AD, Eisenstein TK, Adler MW, Henderson EE, Rogers TJ (2000) μ-opioid induction of monocyte chemoattractant protein-1, RANTES, and IFN-γ-inducible protein-10 expression in human peripheral blood mononuclear cells. J Immunol 165:6519–6524.

Wybran J, Appelboom T, Famaey J-P, Govaerts A (1979) Suggestive evidence for receptors for morphine and methionine-enkephalin on normal human blood T lympohcytes. J Immunol 123:1068–1070.

Yeager MP, Colacchio TA, Yu CT, Hildebrandt L, Howell AL, Weiss J, Guyre PM (1995) Morphine inhibits spontaneous and cytokine-enhanced natural killer cell cytotoxicity in volunteers. Anesthesiology 83:500–508.

Yuan M, Kiertscher SM, Cheng Q, Zoumalan R, Tashkin DP, Roth MD (2002) Δ^9-tetrahydrocannabinol regulates Th1/Th2 cytokine balance in activated human T cells. J Neuroimmunol 133:124–131.

Zhang L, Looney D, Taub D, Chang SL, Way D, Witte MH, Graves MC, Fiala M (1998) Cocaine opens the blood-brain barrier to HIV-1 invasion. J NeuroViro 4:619–626.

Zhu W, Igarashi T, Qi ZT, Newton C, Widen RE, Friedman H, Klein TW (1993) delta-9-tetrahydrocannabinol (THC) decreases the number of high and intermediate affinity IL-2 receptors of the IL-2 dependent cell line NKB61A2. Int J Immunopharmacol 15:401–408.

Part 3
Therapeutics

38
Neuropharmacology

Terry D. Hexum, L. Charles Murrin, and Eric M. Scholar

Keywords Acetylcholine; Autonomic nervous system; Cyclosporine; Epinephrine; General anesthetics; Immunostimulating agents; Immunosuppressive agents; Mycophenolate mofetil; Opioid drugs; Sedative-hypnotic agents

38.1. Introduction

The immune system is subject to modulation by a number of environmental cues including, but not limited to, microbial infections, toxic agents, stress, drugs and trauma. Endogenous immunoregulatory factors include those produced by the nervous and endocrine systems. This chapter discusses how these factors, as well as drugs targeting the immune, autonomic and central nervous systems, affect the immune system.

38.2. The Autonomic Nervous System

This section provides a review of the anatomy and physiology of the autonomic nervous system (ANS) accompanied by a brief description of some of the important drugs which modify its actions. A summary of the interaction between the immune system and the ANS is also included. An extensive discussion of the anatomy and physiology of the ANS can be found in the *Primer on the Autonomic Nervous System* (Robertson, 2004). Detailed information on the pharmacology of the ANS can be found in *Goodman & Gilman's The Pharmacological Basis of Therapeutics* (Brunton et al., 2006).

38.2.1. Anatomy and Physiology

The ANS, a major subdivision of the peripheral nervous system, functions as a regulator of tissues throughout the body. The control of these tissues is the responsibility one or both of the two divisions of the ANS, the parasympathetic nervous system and the sympathetic nervous system including the adrenal medulla, which act either individually or in opposing or complimentary modes. The ANS regulates a variety of activities essential to life including cardiac output, blood flow to various organs, digestion, elimination and various other functions. The other major subdivision of the peripheral nervous system, the somatic nervous system (SNS), controls functions such as locomotion, respiration and posture.

The ANS differs from the SNS in that: (a) the CNS is connected with effector cells on autonomic end organs via two nerve fibers, (b) the synapses (ganglia) linking the CNS with these tissues are located outside the CNS, (c) the ANS uses both norepinephrine (NE) and acetylcholine (ACh) as neurotransmitters rather than only ACh, and (d) the ANS is concerned with the homeostasis of the internal environment of the body rather than locomotion, respiration and posture. The autonomic nerve fibers connecting the CNS to peripheral tissues originate in the spinal and cranial regions (preganglionic) and synapse with the cell bodies of the motor neurons in the autonomic ganglia (postganglionic). This arrangement serves to relay CNS impulses to peripheral effector cells and also to serve as a control loop for autonomic reflex actions. Visceral afferent fibers also provide input to the autonomic nervous system. CNS impulses are influenced by various areas of the brain and integrated primarily by the action of the hypothalamus and the solitary tract nucleus.

The ANS is organized into two divisions, the parasympathetic or craniosacral portion and the sympathetic or thoracolumbar portion. The ratio of pre- to postganglionic fibers is generally 1:1 in the parasympathetic division and not all tissues receive innervation (e.g., spleen, axilla sweat glands, and most blood vessels). In contrast, the ratio of pre- to postganglionic fibers is 1:20 in the sympathetic division with unlimited distribution. This arrangement underscores the more discreet action of the parasympathetic division as compared to the initiation of action in several tissues upon activation of the sympathetic division, the so-called "flight or fight" response. Parasympathetic action is conservative relative to the well-being of the individual whereas sympathetic action is responsive (Table 38.1).

T. Ikezu and H.E. Gendelman (eds.), *Neuroimmune Pharmacology.*
© Springer 2008

TABLE 38.1. Simplified table of ANS activity on some organ systems.

Organ/System	Sympathetic division		Parasympathetic division	
	Action	Receptor[a]	Action	Receptor
BRONCHIOLAR SMOOTH MUSCLE	Relaxes	β_2	Contracts	M
EYE				
Iris	Contracts	α_1	Contracts	M
Radial muscle			Contracts	M
Sphincter muscle				
Ciliary muscle				
GASTROINTESTINAL TRACT				
Smooth muscle	Relaxes	β_2	Contracts	M
Sphincters	Contract	α_1	Relaxes	M
Secretion			Increases	M
GENITOURINARY SMOOTH MUSCLE				
Bladder wall (detrusor)	Relaxes	β_2	Contracts	M
Sphincter & trigone	Contract	α_1	Relax	M
Uterus, pregnant	Relaxes	β_2	Erection	M
Penis, seminal vesicles	Contracts	α_1		
	Ejaculation	α_1		
HEART				
Sinoatrial node	Accelerates	β_1	Decelerates	M
Contractility	Increases	β_1	Decreases (atria)	M
METABOLISM				
Liver	Glycogenolysis	$\alpha_1\beta_2$		
Fat cells	Lipolysis	β_3		
SKIN				
Sweat glands	Increases	M		
VASCULAR SMOOTH MUSCLE				
Skin, splanchnic vessels	Contract	α_1		
Skeletal muscle vessels	Relax or	β_2		
	Contract	α_1		

[a]α—Alpha adrenergic receptor; β—Beta adrenergic receptor; M—Muscarinic receptor.

Neuronal impulses between the CNS and either division of the ANS require both axonal conduction and synaptic or junctional transmission. Synaptic transmission (nerve to nerve) and junctional (nerve to effector organ) transmission are dependent upon the release of neurotransmitter from presynaptic neurons. The majority of clinically significant drugs that alter neuronal activity act at the level of synaptic or junctional transmission. This process is primarily dependent upon the influx of Ca^{+2} through voltage-gated channels to a threshold level that initiates the process of exocytosis. This multi-step process releases significant amounts of neurotransmitter into the synaptic cleft sufficient to open postsynaptic ion channels or activate G protein coupled receptors. The result may be either stimulation or inhibition of the end organ (cardiac muscle, smooth muscle, exocrine glands).

ACh is the transmitter released by: (a) all preganglionic fibers including those innervating the adrenal medulla; (b) postganglionic parasympathetic fibers to smooth muscle, heart and exocrine glands; and (c) postganglionic sympathetic fibers to sweat glands. In contrast, NE is the transmitter released by: (a) postganglionic sympathetic fibers to smooth muscle, heart and glands; and (b) adrenal medulla. The effects of either neurotransmitter are mediated by specific receptors. ACh acts on cholinergic receptors (classified muscarinic or nicotinic) whereas NE acts on adrenergic receptors (classified α or β). Subtypes of muscarinic (M_{1-3}), nicotinic ($N_{N \text{ or } M}$), (α_{1-2}) and (β_{1-3}) receptors exist as well.

38.2.2. Neurotransmission

ACh is synthesized from choline and acetylCoA by the action of choline acetyltransferase in the cytosol of nerve terminals and packaged in synaptic vesicles for release into the synaptic cleft in response to neuronal depolarization. Either muscarinic or nicotinic receptors are responsible for the effects of ACh (e.g., increase in muscle tone or glandular secretion), which are subsequently terminated by the action of acetylcholinesterase (AChE). Choline, as a product of this reaction, can be actively transported into the nerve ending for conversion to ACh once again. ACh synthesis is responsive to the prevailing level of intracellular choline. Muscarinic receptor responses include increases in intracellular Ca^{++} and activation or inhibition of various protein kinases mediated by G protein coupled receptors. Nicotinic receptor activation results in the opening of ligand gated ion channels and the influx of Na^+ and the generation of postsynaptic (ganglia) or endplate (skeletal muscle) potentials.

NE is synthesized by tyrosine hydroxylation (*meta* ring position) followed by decarboxylation and side chain β carbon hydroxylation. The synthesis of this catecholamine is regulated by tyrosine hydroxylase. Tyrosine hydroxylation is also a key step in the synthesis of two other important catecholamines, dopamine and epinephrine. NE is packaged via active transport into synaptic (or chromaffin) vesicles prior to release by neuronal depolarization. The effects of NE are mediated by adrenergic receptors (α or β) which are G protein coupled resulting in either increases or decreases in smooth muscle tone as well as increases in cardiac rate and contractility. These effects arise out of receptor mediated increases in intracellular Ca^{++} and activation or inhibition of various protein kinases. The effects of NE are terminated essentially as a result of its active transport into the presynaptic nerve ending via an energy and Na^{+} dependent process which utilizes the norepinephrine transporter (NET). Ultimately, NE and other catecholamines are metabolized by monoamine oxidase (MAO) and catechol-O-methyltransferase (COMT).

The effects of both ACh and NE can be enhanced by the co-localization of agents such as the neuropeptides vasoactive intestinal peptide (VIP) and neuropeptide Y (NPY). For example, VIP can increase ACh induced salivary secretion (Lundberg and Hokfelt, 1983) and NPY can enhance the vasoconstrictive effects of NE (Wahlestedt et al., 1985). In theory, agents mimicking or blocking the effects of these neuropeptides should expand the opportunities for control of various disease conditions such as Sjögren's disease and hypertension. However, significant advances in such therapy are yet to be realized.

38.2.3. Pharmacology of Autonomic Agents

Neurotransmission at both cholinergic and adrenergic synapses can be influenced by a variety of naturally occurring and synthetic drugs (Table 38.2). For example, ACh secretion is drastically reduced in the presence of botulinum toxin which contains proteases that cleave proteins vital to the process of exocytosis. Nicotine, one of the active ingredients in tobacco can mimic the effects of ACh and increase (or decrease at high doses) autonomic ganglionic transmission as well as that at the skeletal motor endplate. The effects of nicotine on autonomic ganglia can be prevented by prior treatment with agents such as trimethaphan. The widespread effects of this agent due to its effects on all autonomic ganglia have relegated it to use primarily as an experimental agent. Conversely, d-tubocurarine, an active ingredient present in South American arrow poison, is an effective muscle relaxant due its ability to block nicotinic receptors on skeletal muscle and finds use as a preoperative medication. Muscarine, a constituent of certain mushroom species, can mimic the stimulatory effect of ACh on various smooth muscles and exocrine glands. The belladonna alkaloids, which are found in a variety of plants, can antagonize the effects of ACh on muscarinic receptors.

TABLE 38.2. ANS drugs and their receptors.

Cholinergic		Adrenergic	
Muscarinc	Nicotinic	Alpha (α) adrenergic	Beta (β) adrenergic
Acetylcholine	Acetylcholine	Amphetamine	Amphetamine
Atropine	Botulinum toxin	Cocaine	Cocaine
Bethanechol	Neostigmine	Epinephrine	Dobutamine
Botulinum toxin	Nicotine	Norepinephrine	Epinephrine
Ipratropium	Physostigmine	Phenylephrine	Isoproterenol
Muscarine	Rivastigmine	Prazosin	Metoprolol
Physostigmine	Soman	Reserpine	Norepinephrine
Rivastigmine	Trimethaphan	Tyramine	Propranolol
Soman	D-Tubocurarine		Terbutaline
Tropicamide			Tyramine

NE secretion can be effectively decreased by administration of reserpine, an alkaloid isolated from a small woody perennial found in India (Rauwolfia), which has a high affinity for the vesicular monoamine transporter-2 (VMAT-2) and as such prevents NE storage. Alternatively, NE secretion can be increased by administration of tyramine (decarboxylated tyrosine), which is a constituent of a variety of foods including red wine, pickled herring and cheese. Amphetamine has a similar effect, which is most prominently manifested in the CNS. The termination of NE effects can be circumvented by the administration of cocaine, which blocks NE transport into presynaptic nerve endings (NET) an effect, which is also shared by some of the first generation antidepressants, such as imipramine.

The structures of both ACh and NE or alkaloids possessing some of the properties of these agents have been modified to produce a variety of agents that mimic or block the effects of these neurotransmitters. Bethanechol is an ACh derivative with a longer duration of action and selectivity for muscarinic receptors. This agent increases smooth muscle tone and is useful in the treatment of urinary retention and gastroesophageal reflux disease. The structure of the classic AChE inhibitor, physostigmine (eserine, an alkaloid) has been modified (neostigmine) to limit absorption and increase nicotinic receptor stimulation resulting in its use in amelioration of the symptoms of the autoimmune disease, myasthenia gravis. Further modification of the basic structure has yielded agents, e.g., rivastigmine, of some benefit in improving cognition in Alzheimer's disease. AChE is the target for many of the organophosphorus insecticides as well as extremely toxic nerve gases such as soman. Prompt and aggressive treatment with cholinesterase reactivators such as pralidoxime (pyridine-2-aldoxime methyl chloride) reverses the toxic effect of the organophosphorus compounds within moments and restores the response to skeletal motor nerve stimulation.

Perhaps the most widely known agent to alter cholinergic transmission is the belladonna alkaloid, atropine. This agent is relatively selective for muscarinic receptors and as such decreases exocrine gland secretion, smooth muscle tone and the effects of vagal nerve stimulation of the heart. Atropine has several clinical applications including use as a preoperative medication, post-myocardial infarction to increase heart rate and in the treatment of organophosphorus toxicity. Structural modification of this compound has produced shorter acting agents, e.g., tropicamide in eye examinations and agents with limited ability to cross cell membranes, e.g., ipratropium as an airway smooth muscle relaxant. The wide availability of belladonna alkaloid containing plants such as jimson weed in the U.S. has resulted in the misuse of these agents for recreational purposes due to their ability to produce CNS stimulation. As mentioned above, nicotinic receptor antagonists have limited clinical applications.

The naturally occurring catecholamine, epinephrine, has a variety of effects at adrenergic receptors including both smooth muscle stimulation and relaxation, increased cardiac activity and catabolic effects on triglycerides and glycogen. Epinephrine finds use in cardiac stimulation after cardiac arrest, relief of bronchoconstriction, lowering of intraocular pressure in open angle glaucoma and elevation of blood pressure. Modification of the ethylamine side chain of the catecholamine structure results in agents with selective actions on β adrenergic receptors, e.g., isoproterenol, as a smooth muscle relaxant and cardiac stimulant and dobutamine, as a cardiac stimulant. Elimination of one or both of the catechol hydroxyl groups produces compounds with a longer duration of action including additional receptor selectivity dictated by modifications to the ethylamine side chain. For example, phenylephrine is a non-catechol compound containing a hydroxyl group on the side chain β carbon with preference for α receptors. This agent is an effective blood pressure elevating agent that is also used to dilate the pupil prior to an eye examination. Terbutaline is also a non-catechol compound that contains a bulky substituent on the side chain amine resulting in selective action on the β_2 receptor subtype. This agent is similar to isoproterenol and is therefore an effective airway relaxing agent but with reduced cardiac stimulant properties.

Antagonists for α adrenergic receptors are directed primarily toward the α_1 subtype and are therefore useful blood pressure lowering agents, e.g., prazosin. In contrast, agents that block β adrenergic receptor mediated responses are best represented by propranolol, which can block both the β_1 and β_2 adrenergic receptor subtypes. Propranolol is widely used as a cardiac depressant and blood pressure lowering agent. Structural modifications of propranolol produce agents with a longer duration of action and selectivity for β_1 adrenergic receptors. Agents such as metoprolol have a reduced ability to increase airway resistance while still being able to lower blood pressure and depress cardiac activity.

38.2.4. Autonomic Regulation of Immunity

Infection, injury and trauma initiate a series of immune reactions to overcome the effects of various pro-inflammatory mediators resulting in repair of injured tissues and wound healing. The arbiter of this response is the brain, which communicates with the immune system via neurotransmitters, cytokines and hormones (Figure 38.1). Activation of the immune system by the brain occurs through two major pathways: (a) the hormonal response pathway which results in the activation of the hypothalamic-pituitary-adrenal axis releasing glucocorticoids; and (b) the ANS which releases ACh and NE from the parasympathetic and sympathetic divisions, respectively, to modulate the activity of lymphoid tissues. In addition these tissues contain receptors for a variety of neuropeptides (Steinman, 2004) which are co-released with either ACh or NE and are likely to exert prominent effects on the immune system. Classically the sympathetic portion has been viewed as the primary division of the ANS, which influences the immune system arising from the seminal work of Hans Selye (Selye, 1936). However, recent data also demonstrates interplay between the parasympathetic division and the immune system (Czura and Tracey, 2005).

The sympathetic nervous system innervates the major lymphoid organs such as the spleen with nerve fibers reaching both the vasculature and the parenchyma where lymphocytes, primarily T cells (T helper type 1–2, T_H1, T_H2), reside (Friedman and Irwin, 1997). T cells possess receptors for both norepinephrine and neuropeptide Y that are released in response to sympathetic nerve stimulation. The adrenergic receptors are primarily the β_2 subtype, which is consistent with data demonstrating that β_2 agonists can markedly influence the immune system (Kohm and Sanders, 2001). For example, stimulation of T cell receptors results in increased cyclic AMP formation, which can modulate cytokine expression, i.e., decreasing

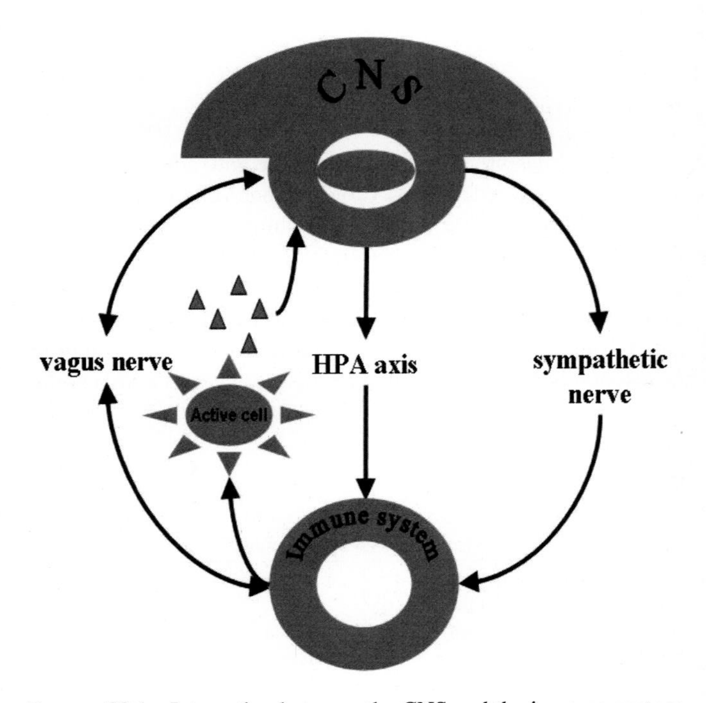

FIGURE 38.1. Interaction between the CNS and the immune system via the ANS. Signals from the CNS influence the immune system through activation of either the parasympathetic (vagus) or sympathetic portions of the ANS. (HPA - hypothalamic-pituitary-adrenal; Δ- cytokine)

TNF-α and increasing IL-8 (Webster et al., 2002). Consistent with this, selective β_2 receptor antagonists such as butoxamine can antagonize the corticotrophin-releasing factor-induced reduction in natural killer (NK) cell activity (Irwin et al., 1990). Sympathetic nerve stimulation also enhances T_H2 cytokine production while inhibiting T_H1 cytokine production. Thus, β_2 receptor agonists can suppress interferon-γ (IFN-γ) production by T_H1 cells, an effect, which can be blocked by propranolol. Interestingly, prion presence in the brain has been linked to an interaction between splenic monocytic cells and the sympathetic nervous system (Steinman, 2004).

In addition to the effects mediated by the sympathetic nervous system, afferent activation of the vagus nerve (parasympathetic nervous system) through the detection of inflammatory stimuli, transmits signals to the brain activating a reflex response (Czura and Tracey, 2005). The vagus nerve is also the efferent arm of this reflex, which results in the release of ACh and binding with macrophage nicotinic receptors. Activation of the $\alpha7$ nicotinic receptor on macrophages leads to cellular deactivation and inhibition of cytokine release. This effect can be blocked by the prior administration of mecamylamine, a nicotinic receptor antagonist with selectivity for neuronal receptors (Saeed et al., 2005). Thus development of agents selective for either adrenergic β_2 receptors or $\alpha7$ nicotinic receptors provides a promising avenue for the treatment of inflammatory diseases.

38.3. CNS Pharmacology

It is well established that the immune system is regulated by the CNS via the HPA axis and the sympathetic nervous system. Regulation of the hypothalamus by other CNS regions is complex but the end result is that stimulation leads to increased release of CRF from the hypothalamus. CRF, in turn, acts on the anterior pituitary to cause release of ACTH into the circulation. ACTH produces release of glucocorticoids from the adrenal medulla. The glucocorticoids, such as cortisol, have powerful regulatory effects on immune system function, suppressing the production and release of pro-inflammatory cytokines and inhibiting other inflammatory processes. This is covered in greater detail in Chapter 34.

More recently it has become apparent that this relationship is reciprocal in nature, i.e., the immune system also can regulate the HPA axis and have significant effects on the CNS (see Carlson, 2003; Eskandari et al., 2003, for reviews). Chemokines, cytokines and other mediators released by the immune system can alter CNS function and abnormally high levels of some cytokines, such as IL-6, are associated with mental disorders, such as depression (see Chapter 34). In addition it is now clear that some neurohormones once thought to be exclusively neuronal are also produced by immune cells and that some immune system mediators are expressed by cells within the CNS. A consequence of this is that drugs whose primary site of action has classically been defined as the central nervous system may play important roles in modulating immune system function as well. This section provides a discussion of the pharmacology of three such classes of drugs. These are the opioid drugs, sedative-hypnotic agents and general anesthetics. The mechanism of action of these drugs will be described insofar as it is known, the traditional clinical uses of these drugs described briefly, and their potential involvement with the immune system will be reviewed.

38.3.1. Central Nervous System

The CNS is one of the most complex organ systems in the body. There are multiple ways in which the CNS can be organized: by anatomic regions based on groupings of cell bodies and neuronal pathways, by neurotransmitter and neurohormonal systems, by cell types (neurons, astrocytes, microglia, blood vessels, etc.), by physiological function and by areas involved in particular disease states. While these organizational systems parallel one another to some extent, there is also considerable intermingling and overlap of one system with another when considering a specific issue. This is made even more complex by the ability of the CNS to compensate for and to use alternate systems in the face of brain areas injured by disease or physical damage. A detailed understanding of these various organizational systems is beyond the scope of this book and can best be obtained from texts dedicated to each.

38.3.2. Opioid Drugs and Opioid Receptors

The opioid drugs are the most commonly used drugs to treat moderate to severe pain throughout the world. Opium, the original substance derived from the opium poppy, *Papaver somniferum*, has been used for millennia as an analgesic agent. The prototype opioid drug, morphine, which is derived from opium, has been used for almost two hundred years and is still the most effective opioid analgesic available. The opioid drugs and their endogenous counterparts, the endorphins, dynorphins and enkephalins, are agonists at opioid receptors. The opioid receptors play a role in numerous physiological and pathological functions. These receptors are important for nociception, emotional behavior (as part of the limbic system) and the reinforcement of behavior (reward system). They are important in regulation of respiration, GI function and immune function. These receptors also play an integral role in dependence upon opioid drugs and appear to play a significant role in dependence produced by other chemically unrelated drugs.

38.3.2.1. Opioid Receptors

There are four major classes of opioid receptors: mu, kappa, delta, and NOP receptors (see http://www.iuphar-db.org/GPCR/ChapterMenuForward?chapterID = 1295). Pharmacological studies in many species and knockout studies in mice indicate that the mu opioid receptors (MOP receptors; MOR; OP_3 receptors) are involved in most of the clinically important effects associated with opioid drugs (Matthes et al., 1996;

Sora et al., 1997). Mu opioid agonists, such as morphine and codeine, are used clinically to treat moderate to severe pain, to control cough and diarrhea, and in the treatment of drug dependence (Gutstein and Akil, 2006). Kappa opioid receptors (KOP receptors; KOR; OP_2 receptors) are also involved in regulation of nociception. Agonists at KOP receptors, such as pentazocine and butorphanol, are used as analgesic agents. High levels of kappa agonists are frequently associated with dysphoria, in contrast to the euphoria that can be produced by agonists acting primarily on MOP receptors. Delta opioid receptors (DOP receptors; DOR; OP_1 receptors) appear to be involved in pain perception, based on their anatomic localization, and may also be involved in drug dependence. There are currently no drugs specifically targeted to DOP receptors in clinical use. While early delta specific agonists were frequently associated with the production of seizures, there is an emerging set of data that suggests seizure liability is not a fundamental characteristic of delta agonists and that agonists at DOP receptors may be useful in treating pain while avoiding the degree of dependence associated with MOP receptor agonists.

The last currently recognized member of the opioid receptor family is the NOP receptor (nociceptin/orphanin FQ receptor; N/OFQ receptor; ORL1 receptor; OP_4 receptor). Most opioid agonists have a much lower affinity for NOP receptors compared to the other opioid receptors. Conversely, the endogenous ligand for the NOP receptors, nociceptin or orphanin FQ (OFQ/N), is structurally similar to dynorphin but has a high affinity only for the NOP receptor (Mogil and Pasternak, 2001). Our understanding of the function of the NOP receptor and its pharmacology is limited and the therapeutic involvement of these receptors, including their relationship with the immune system, is not known.

Within the CNS, opioid receptors are widely distributed and are co-expressed in many of the same regions (Mansour and Watson, 1993; Neal et al., 1999). MOP receptors are found throughout the CNS with dense receptor expression in the basal ganglia, limbic structures, some thalamic nuclei and in the dorsal horn of the spinal cord. They are also found in regions involved in nociception and sensorimotor regulation, including the periaqueductal grey and several thalamic nuclei. KOP receptors have a similar widespread distribution with dense receptor expression in the basal ganglia and limbic structures, some thalamic nuclei and the dorsal horn of the spinal cord. Kappa receptors display striking differences between species, playing a much more prominent role in primates, including humans, than in rodents. Delta receptors are more restricted in distribution but are also found densely expressed in basal ganglia and limbic structures. Finally, NOP receptors are found throughout the CNS, with highest levels in the cortex, hypothalamus, hippocampus, some amygdaloid nuclei, as well as spinal cord (Neal et al., 1999; Florin et al., 2000; Bridge et al., 2003). NOP localization suggests a role in multiple functions, including nociception, the reward system, and stress response. The similar distribution of opioid receptors in many brain regions offers the possibility that they may function as both homomeric and heteromeric receptors (Gomes et al., 2000). A number of studies have shown that heterodimerization of opioid receptors leads to distinct pharmacological characteristics (Gomes et al., 2004; Devi, 2001). It should be noted that there are considerable species differences in the distribution of some opioid receptors (Mansour et al., 1988; Bridge et al., 2003).

38.3.2.2. Endogenous Opioid Neuropeptides

The endogenous ligands for the opioid receptors are peptides known as the endorphins (endogenous morphine) or opiopeptins. These include the pentapeptides methionine-enkephalin and leucine-enkephalin and a heptapeptide and octapeptide version of methionine-enkephalin, all derived from preproenkephalin; β-endorphin derived from proopiomelanocortin; α- and β-dynorphin derived from prodynorphin; endomorphin-1 and -2, whose precursor has not been definitively identified; and orphanin FQ or nociceptin, derived from OFQ/N precursor protein. These peptides are discussed in more detail in Chapter 34.

38.3.2.3. Opioid Drug Effects on the Immune System

The first data indicating that opioid drugs affect immune function came from the work of Wybran, who demonstrated that morphine could modify T cell function (Wybran et al., 1979). Binding of opioid ligands to immune cells has been demonstrated by many laboratories, as described in detail below. The binding sites for opioid drugs include both classical opioid sites, for which binding is blocked by naloxone or another opioid antagonist, and "non-classical" opioid receptors, which are not inhibited by classical antagonists and usually have a low affinity for standard opioid agonists. While the "non-classical" binding sites share many of the properties of G protein coupled receptors (GPCR), such as saturability and sensitivity to cations and GTPγS, their anomalous pharmacology leaves the picture clouded. If these are indeed opioid receptors at which either endogenous or exogenous opioids could be expected to exert an effect, this has yet to be established. Given that the NOP receptor has a distinct pharmacology and low affinity for morphine, and that the NOP receptor appears to be a physiologically important receptor, it is possible that these "non-classical" opioid receptors will prove important in immune function. Similarly, the discovery of splice variants for the MOP receptor that have differing pharmacologies but whose function is currently unknown (Pasternak, 2001; Cadet, 2004) indicates how much is yet to be learned about opioid receptors.

38.3.2.3.1. Opioids and the Immune System (In Vitro Studies)

Studies of opioid effects on immune cell function have relied heavily on the use of specific cell lines in culture. This is due to the expression of opioid receptors on subpopulations of cells, making it difficult to detect receptors or receptor function in a mixed cell population. There have been extensive reviews of the many features of opioid effects on immune cells (Bidlack et al., 2006; McCarthy et al., 2001).

Delta and kappa opioid receptors have been demonstrated on several immune cells including lymphocytes, peripheral blood polymorphonuclear leukocytes and macrophages (Sharp, 2006). Parallel to this, mRNA for MOP, DOP and KOP receptors also have been found in many immune cell types, including T and B cells, lymphocytes, macrophages and peripheral blood mononuclear cells. Both morphine and kappa-selective agonists increase KOP receptor expression in a human lymphocytic cell line (Suzuki et al., 2001). Activation of T cells by anti-CD3-ε leads to increased expression of DOP receptors, priming these cells for greater response to opioids, either endogenous or exogenous (Shahabi et al., 2006; Sharp et al., 2005). Treatment of peripheral blood mononuclear cells with morphine dose-dependently promotes T_H2 cell differentiation, apparently by increasing IL-4 production. This response was lost in cells from MOP receptor knockout mice. These studies point to a suppression of the immune system by morphine. The resulting failure of cell-mediated immunity plays a significant role, in turn, in increased morbidity and mortality following trauma or major surgery (Roy et al., 2001, 2006).

Opioids can regulate intracellular signaling in immune cells. As would be expected, regulation of intracellular signaling depends on the receptor subtype and the cell line being studied. DOP receptors inhibit adenylyl cyclase in human T cells in a pertussis toxin-sensitive manner, indicative of a GPCR linked to Gi/o (Sharp et al., 1996). KOP receptors act in a similar manner in the R1.1 thymoma cell line (Lawrence and Bidlack, 1993). Many other studies have been carried out in the same vein but the results are not always consistent, similar to other studies trying to decipher regulation of immune system function (Sharp, 2006). This supports the idea that regulation of the immune system by opioids, both endogenous and exogenous, is complex and the direction of this regulation is dependent upon many variables. It also suggests that many of the findings reported to date will prove to be highly specific to the experimental paradigm and may not translate well into the *in vivo* situation.

Given the expression of opioid receptors and endogenous opioid peptides by cells in the immune system, the question naturally arises as to whether the opioid drugs have an impact on immune system function. While many studies have addressed this question, there is much to be learned (Bidlack et al., 2006; McCarthy et al., 2001). It appears that KOP receptors play a role in T-cell development since they are highly expressed on immature thymocytes and diminish in expression as these cells mature. KOP receptor-selective agonists inhibit synthesis of IL-1 and TNF-α but not IL-6 in a macrophage cell line (Belkowski et al., 1995). DOP receptors can stimulate T-cell chemotaxis based on the use of highly specific DOP receptor agonists (Heagy et al., 1990).

It has also been shown that morphine, primarily through MOP receptors, can alter viral replication in immune cells. Morphine inhibits the anti-HIV activity of $CD8^+$ T cells *in vitro*, possibly through inhibition of IFN-γ production and activity. This leads to increased HIV production (Wang et al., 2005b). In a similar manner, morphine and, even more so, morphine withdrawal increase Hepatitis C virus replication in human hepatoma cell lines by suppressing interferon-α expression (Wang et al., 2005a). These studies suggest that opioid abuse by people with these and other viral infections can contribute to maintaining and promoting the viral infection. This points to the importance of gaining a thorough understanding of the effects of opioids on the immune system, especially in cases of chronic infection.

38.3.2.3.2. Opioids and the Immune System (in vivo studies)

Are the multiple effects of opioids and opioid receptor stimulation on the cell biology of immune cells reflected clinically? The answer is not clear at present. It is well known that intravenous drug abusers suffer from greater rates of infection than non-abusers. This suggests that chronic drug use reduces function of the immune system. However, it is difficult to make this association clearly since drug abusers typically have many risk factors for developing infections. In addition, attempts to implicate any single drug in potential effects are confounded by the multi-drug use prevalent in this population.

Some animal studies have provided data suggesting opioids do reduce immune function in a clinically relevant manner (see McCarthy et al., 2001, for review). A single dose of morphine in mice or rabbits drastically reduced reticuloendothelial system activity, phagocyte count, phagocytic index, killing properties, and superoxide anion production in polymorphonuclear leukocytes and macrophages (Tubaro et al., 1983). Morphine also increased the susceptibility of animals to fungal infection. The results suggest that there is a relationship between morphine's depression of phagocytic functions and morphine's exacerbation of infections. In a similar study morphine increased the sensitivity of mice to a strain of *Toxoplasma gondii* (Chao et al., 1990). Interestingly in this study, if the mice were made tolerant to morphine, the sensitization effect was lost. A similar protective effect of tolerance was found in pigs infected with swine herpes virus-1 (Risdahl et al., 1993).

In studies examining the human population, results vary. While one study found that survival in HIV-infected intravenous drug abusers was significantly shorter compared to those not using intravenous drugs (Rothenberg et al., 1987), more recent studies have found no difference in disease progression between the two groups (Margolick et al., 1992; Lyles et al., 1997). Based on animal studies it seems that opioid drugs, such as morphine, can reduce immune function in the whole animal, as indicated by both *in vitro* and *in vivo* studies. It is not presently clear how directly these results will translate into the clinical setting or whether they will be relevant to treatment of normal individuals or of those with on-going infections and/or undergoing treatment for drug abuse, such as with methadone or naltrexone.

38.3.3. Sedative-Hypnotic Agents and Drug Abuse

Sedative-hypnotic agents are some of the most widely prescribed and used drugs in the developed world. They are also frequently drugs of abuse. Sedative-hypnotics are used to promote sleep, as their name implies, and to reduce anxiety. Their overall effect is to act as CNS depressants. We will consider the two major classes of drugs in this category, the benzodiazepines and the barbiturates. For a detailed discussion of the pharmacology of these agents, see Charney et al. (2006).

38.3.3.1. Benzodiazepines

The benzodiazepines are a group of chemically related drugs, the best-known members of which are chlordiazepoxide (Librium®) and diazepam (Valium®). There are many others with variations in properties, particularly onset of effects and duration of action. The most recognized effects of the benzodiazepines are centrally mediated and include sedation, hypnosis, decrease in anxiety, muscle relaxation and anticonvulsant effects.

38.3.3.1.1. Benzodiazepine Receptors

The best-characterized molecular target of the benzodiazepines is the GABA-A receptor (Charney et al., 2006; Burt, 2003). Each receptor, which acts as a chloride channel, is composed of five subunits. There are at least seven subunit families composed of at least 19 individual subunits. Some of the subunits undergo alternative splicing, adding to the complexity. Because of the multitude of subunits, there are many thousands of potential subunit combinations. In reality, far fewer are expressed. The most common composition of GABA-A receptors is two alpha, two beta, and one gamma subunit. Which particular subunits compose the GABA-A receptor in different regions of the brain is not well understood. Since the subunit composition determines receptor pharmacology, channel properties and intracellular localization, understanding the composition of the receptors expressed in a particular tissue or cell type is very important. Because of the multitude of possible subunit compositions and the fact that a single cell may express more than one type of GABA-A receptor, understanding the function of GABA-A receptors is a daunting task.

As mentioned above, the GABA-A receptor is a chloride channel. Binding of GABA to the receptor opens the channel to allow an inward flux of chloride ions. Benzodiazepines are modulators of this action. Benzodiazepines are thought to bind at the interface of the alpha and gamma subunits and they do not affect receptor function unless GABA also binds to the receptor. In this case, the benzodiazepines increase the effect of GABA, leading to a greater chloride flux and a greater hyperpolarization of the cell, producing an inhibition of synaptic transmission.

A second receptor for the benzodiazepines is the peripheral benzodiazepine receptor (PBR) which is an enigmatic receptor (Gavish et al., 1999). While there has been a significant amount of work published about it, the PBR remains somewhat obscure and its physiological role is still not clear. The PBR are found primarily on mitochondrial membranes (Zavala, 1997), but are also present on cell surface membranes. In mitochondria the PBR complex has been suggested to make up the mitochondrial permeability transition pore, which plays an important role in regulation of cell survival or death (Casellas et al., 2002). The PBR has been characterized primarily by radioligand binding studies. As the name implies, the PBR was originally discovered as a binding site for benzodiazepines that was pharmacologically distinct from the central binding site and was found outside the CNS. The highest concentration of PBR is found in the adrenal gland but it has also been identified in many other peripheral organs. The pharmacology of the PBR is distinct from that of the central benzodiazepine (GABA-A) receptor in that some of the classical benzodiazepines bind to the PBR with high affinity, while others have only low affinity for the receptor. Some benzodiazepines have a high affinity for both the PBR and GABA-A receptors, while others have high affinity for only one. The clinically useful drugs all have a relatively high affinity for the central receptor.

38.3.3.1.2. Benzodiazepines and the Immune System

Knowledge of the interaction of benzodiazepines and the immune system is much less advanced compared to our understanding of interactions between opioid systems and the immune system. It is clear that the benzodiazepines have strong anti-anxiety effects and that through these actions they reduce the response to stressful situations (Zavala, 1997). This, in turn, will reduce stress-induced activation of the HPA axis and so reduce its regulatory effects on the immune system. These effects are centrally mediated, however, and can be considered indirect effects.

A direct effect of benzodiazepines on immune cells is implied by the presence of GABA-A receptors on some immune cells. Central benzodiazepine receptors form part of and regulate the activity of the GABA-A receptor. Mouse T lymphocytes and human peripheral blood mononuclear cells have been shown to express functional GABA-A receptors (Bergeret et al., 1998; Alam et al., 2006). Treatment of T cells with GABA inhibited proliferation and the effect was identified as being due to GABA-A receptors using pharmacological criteria (Tian et al., 1999). A similar study demonstrated inhibition of several proinflammatory T cell functions by GABA (Tian et al., 2004). More directly, diazepam has been shown to alter neutrophil activity *in vivo*, supporting an effect of benzodiazepines on immune function in the whole animal (da Silva et al., 2003). In this study acute treatment with diazepam increased several measures of activity while chronic treatment increased the PMA-induced neutrophil oxidative burst but decreased neutrophil phagocytosis. There is much more to be learned in this relatively young field.

Numerous studies point to a role for peripheral benzodiazepine receptors (PBR) in regulation of immune function as well (see Covelli et al., 1998; Zavala, 1997). PBR are reported to be expressed on many cells within the immune system, with a rank order of monocytes = polymorphonuclear neutrophils > B cells, NK cells and T cells. It has been shown that benzodiazepines acting on PBR produce a stimulation of chemotaxis, the production of reactive oxygen species and cytokine production. For example, stimulation of PBR induces chemotaxis in human monocytes, an effect not produced by stimulating central benzodiazepine receptors (Ruff et al., 1985). PBR agonists reduce the ability of splenocytes and macrophages to produce an oxidative burst. Conversely, blocking PBR in human neutrophils with antibodies enhances the oxidative burst produced by stimulation with fMLP (Zavala et al., 1990, 1991). In an *in vivo* study, diazepam administration blocked an increase in superoxide anion production in response to a stressful situation, and phagocytosis by peritoneal macrophages was reduced (Salman et al., 2000). Stimulation of mouse spleen macrophages by LPS leads to the production of TNF and this effect can be inhibited by PBR agonists (Zavala et al., 1990). However, studies with human immune cells are not as clear cut and PBR agonists can exhibit bell-shaped dose-response curves, producing just the opposite effects at some concentrations (Taupin et al., 1991). In his review, Zavala concludes that stimulation of peripheral PBR has important consequences on immune cell function, but the data are not always consistent in terms of effects produced (Zavala, 1997).

Unfortunately, for reasons that are not clear, interest in studies of PBR in general, including PBR regulation of immune function has waned in recent years and the lack of progress leaves the entire issue of PBR function in the immune system cloudy. Part of the reason for decreased interest may be the finding that in many cases the benzodiazepines used clinically have been found to have weak or no activity at PBR in some systems, so the relevance to drugs currently used in the clinics is not apparent.

38.3.3.2. Barbiturates

The barbiturates are the other major group of sedative-hypnotic agents. Similar to the benzodiazepines, they are a group of chemically related drugs and there are many variations in properties, particularly onset of effects and duration of action. The most recognized effects of the barbiturates are centrally mediated and include sedation, hypnosis, decreased anxiety, and, at high doses, anesthetic properties. The mechanism of action of the barbiturates is similar to that of the benzodiazepines in that barbiturates increase chloride conductance through the GABA-A receptor. However, there are significant mechanistic differences as well. The exact site of barbiturate action is not certain, although there is some evidence barbiturates act at the interface between the alpha and beta subunits. Gamma subunits do not have to be present in GABA-A receptors for barbiturates to produce their effects, in contrast to the benzodiazepines. In addition, barbiturates potentiate GABA actions by prolonging the periods during which channel openings occur in response to GABA, whereas benzodiazepines increase the frequency of opening bursts.

The impact of barbiturates on immune system function is even more poorly defined than the impact of benzodiazepines. The presence of GABA-A receptors on immune cells offers the possibility that barbiturates can affect immune function. However there has been little examination of this area. Early studies indicated that select barbiturates, such as thiopental, may impair oxygen radical production during phagocytosis of neutrophils (Weiss et al., 1994a; Weiss et al., 1994b), whereas most other barbiturates do not. There appear to be no other detailed studies examining barbiturate impact on immune cells, particularly on specific immune cell types. Based on the little information in the literature, the current conclusion is that most barbiturates, when used at clinically relevant doses, probably do not have a significant impact on immune function.

38.3.4. General Anesthetics

The general anesthetics are a chemically diverse group of drugs used primarily during major surgical procedures. They produce several important effects, including loss of pain perception, loss of consciousness, reduction of autonomic responses, amnesia and respiratory depression. They depress many, if not most, CNS modalities. A detailed account of their pharmacology is presented by Evers and colleagues (Evers et al., 2006).

The mechanism of action for the general anesthetics has been the subject of debate for decades. The unitary theory of anesthetic action held sway for many years and posited that all general anesthetics act by perturbing the physical properties of cell membranes. This all-encompassing and relatively vague explanation has been discarded in recent years. Current understanding is that general anesthetics depress synaptic transmission and some data indicate this is region-specific in the CNS, depending on the agent in use.

In general, there are two broad mechanisms by which the decrease in synaptic transmission is achieved by general anesthetics, either increased inhibitory neurotransmission or decreased excitatory neurotransmission. Many general anesthetics, particularly inhalation anesthetics, appear to act at GABA-A receptors, increasing the sensitivity of the receptors to GABA. This is similar to the effects of the sedative-hypnotic agents, at least at a superficial level, and it results in increased inhibitory neurotransmission. There is evidence that general anesthetics acting through GABA-A receptors bind to a specific site or sites on the receptor, although these have not been defined in most cases. There is limited evidence that other general anesthetics may act by increasing the activation of inhibitory glycine receptors or by reducing the activity of excitatory neuronal nicotinic receptors, both actions that would produce the same end result, decreased activation of postsynaptic neurons. Several other mechanisms of general anesthetic actions are known or have been proposed, including inhibition of excitatory NMDA receptors, and activation of potassium channels leading to a greater cellular hyperpolarization (Evers et al., 2006). While much remains to be learned, the common factor in all of these mechanisms is reduction in postsynaptic neuronal excitability.

38.3.4.1. General Anesthetics and the Immune System

The impact of general anesthetics on immune system function is even less characterized than that of the opioids or the sedative-hypnotic agents. Based on the expression of GABA-A receptors by immune cells, as mentioned above, the molecular target for some of the general anesthetics is in place. The demonstration of NMDA receptors on T cells (Miglio et al., 2005) and of glycine receptors on macrophages and leukocytes (Froh et al., 2002) offers the possibility that other general anesthetics may also affect immune function. Overall, however, the information available is limited. While it is clear that anesthesia and surgery depress immune function, deciphering the role of general anesthetics in this complex situation is difficult. A number of *in vitro* studies have shown that anesthetic agents suppress a number of immune functions, including recruitment and activation of neutrophils, lymphocyte activity and NK cell activity. However, even though there are many *in vitro* studies demonstrating changes in immune cell function, there have been no studies that have shown their importance clinically and it has been suggested that in healthy people these effects are probably negligible (Levy, 2006; Galley and Webster, 2006). On the other hand, it is possible effects of general anesthetics on the immune system may be relevant in patient populations whose overall health is compromised. This question remains to be resolved.

38.4. Immunopharmacologics

This section provides a discussion of the pharmacology of the various agents that directly affect the immune system. It emphasizes their mechanism of action with a brief description of their therapeutic uses. More detailed information on their pharmacology can be found in *Goodman and Gilman's The Pharmacological Basis of Therapeutics* (Brunton et al, 2006). Immunopharmacologics can be divided into two major areas, the immunostimulants and the immunosuppressives.

38.4.1. Immunostimulating Agents

Since many diseases are associated with cellular immunodeficiency there is an important need for immunostimulants. They have potential use as vaccine adjuvants and for the treatment of cancer, human immunodeficiency viral infections and in certain other infections. The immunostimulating agents are probably best known at this time for their use in the treatment of cancer. When used for this purpose these drugs often act through more than one mechanism. For example, in addition to their ability to augment the immune system, they may have direct cytotoxic effects as well as being able to induce differentiation of the cancer cells.

38.4.1.1. Interferons

Interferons were first characterized as antiviral proteins produced by host cells. Later it was found that interferons had antiproliferative activity and also served as modulating agents for macrophages and natural killer cells.

Interferons consist of three families of protein molecules (α, β and γ) whose production can be induced in most cells by several different stimuli. Interferon-α is induced in several types of leukocytes by foreign cells, virus-infected cells, tumor cells, bacterial cells and products and virus envelopes. Interferon-β is produced in fibroblasts, epithelial cells and macrophages by viral and other foreign nucleic acids. Interferon-γ is induced in T lymphocytes by foreign antigens to which the T cells are sensitized (Ballow, 2005; Samuel, 2001). Natural killer cells also produce this interferon under certain conditions. More recent technology allows interferons to be produced by recombinant DNA techniques usually by using a genetically engineered *Escherichia coli* bacterium containing DNA coding for human interferon. Interferons first bind to specific cellular receptors and then initiate a series of intracellular events including induction of certain enzymes, inhibition of cellular proliferation and enhancement of immune activities. The effects of interferon on the immune system include increased phagocytosis by macrophages and augmentation of specific cytotoxicity by T lymphocytes. Interferon may also modulate major histocompatibility molecules, tumor-associated antigens, dendritic cells, Fcγ receptors, and intercellular adhesion molecules (ICAM-1) on tumor cells potentially making them better targets for immune recognition (Wiesemann et al., 2002; Tomkins, 1999; Krensky et al., 2005; Goldstein et al., 2003).

Interferon alpha induces remissions in 90 percent of cases of hairy cell leukemia. It is also effective against Kaposi's sarcoma and certain cases of chronic myelogenous leukemia seem to respond well. A few cases of multiple myeloma, malignant lymphoma and breast cancer have shown complete or partial responses to interferons (Goldstein et al., 2003). Interferons are seldom used as the sole agent, so it is difficult to assign exact values to effectiveness. Indeed, their effectiveness seems to be enhanced when they are used in conjunction with other cancer chemotherapeutic agents, with other cytokines or with surgery or radiation.

38.4.1.2. *Interleukin-2 (IL-2)*

IL-2 is a lymphokine produced by activated helper T-lymphocytes. It induces the proliferation of T cells and promotes the differentiation of lymphocytes into cytotoxic cells. IL-2 also induces interferon gamma production (Lewko and Oldham, 2003). Activation of peripheral leukocytes with the drug produces lymphokine-activated killer cells (LAK) that lyse a variety of tumor cells *in vitro*. IL-2 has been cloned in bacteria through recombinant DNA technology, thus allowing the production of large amounts. IL-2 alone or with LAK cells can cause regression of several established metastatic tumors in animals (Lewko and Oldham, 2003; Dorr, 1993).

IL-2 has been approved for the treatment of renal cell carcinoma and also has shown good activity in malignant melanoma. Both of these tumors are refractory to chemotherapy. IL-2 has also been used investigationally, both alone and in combination with LAK cells, with chemotherapy and with other biological response modifier's (BRM's) to treat a variety of different cancers (e.g., head and neck carcinoma, colorectal cancer, central nervous system cancer etc). In addition to cancer therapy, IL-2 has been used to treat patients infected with human immunodeficiency virus and it has been used *ex vivo* to generate antiviral T cells which were reinfused into patients (Lewko and Oldham, 2003; Dorr, 1993).

38.4.1.3. *Immunoglobulin Intravenous Therapy (IGIV Therapy)*

Immune globulin (IG) is a purified preparation of gamma globulin. It is derived from large pools of human plasma and is comprised of four subclasses of antibodies, approximating the distribution of human serum (Ballow, 2005).

IG is not absorbed to a significant extent following oral administration. Peak levels following IM administration occur in approximately 2 days. When given IV, peak levels are attained immediately but the serum IgG level drops to about 40–50% during the first week after infusion. The half-life of IgG is about 24 days in healthy adults.

Despite its widespread and long history of use its specific mechanism of action is not established. It elicits several biological and immunological responses including anti-infective, anti-inflammatory and immunomodulatory activities. The present evidence seems to be that the multiple effects work in concert with each other. Each mechanism contributes differently depending on the disease (Ballow, 2005; Shah, 2005).

IGIV is used either for antibody replacement or immunomodulation. Some of the indications as replacement therapy include general or specific immunodeficiency states e.g., hepatitis A prophylaxis, chronic lymphocytic leukemia with hypogammaglobulinemia, multiple myeloma with specific antibody deficiency, low birth weight babies at risk for infection and infants/children with HIV. It is also used as an immune modulator in conditions such as idiopathic thrombocytopenic purpura and acquired hemophilia (Krensky et al., 2005; Shah, 2005).

38.4.2. Immunosuppressive Agents

Agents that suppress the immune response have assumed an increasing importance in medicine in recent years. These agents have two important uses. One is for inhibiting organ transplant rejection. The second use of these drugs is in the treatment of autoimmune and inflammatory diseases. Major drugs used as immunosuppressives can be subdivided into several categories.

38.4.2.1. *Corticosteroids*

The adrenal corticosteroids have been extensively used to suppress many different aspects of the immune response. All of the corticosteroids have similar mechanisms of action. They bind to specific corticosteroid binding proteins in the cytoplasm. These complexes are then transported into the nucleus where they bind to discrete portions of the cell's DNA. This binding results in derepression of regulatory genes and the subsequent transcription of new mRNA (Rhen and Cidlowski, 2005). Steroids inhibit the consequences of the immune response. They have several different actions in this regard. One of their most important effects is to transiently alter the number of circulating leukocytes. There is a rapid increase in the number of neutrophils and a concomitant decrease in the number of lymphocytes, monocytes, eosinophils and basophils (Krensky et al, 2005).

Steroids also alter important functional activities of lymphocytes and monocytes. This effect is mainly on T lymphocytes and is at least partly due to a decrease in IL-2 production. Other proinflammatory cytokines such as IL-1 and IL-6 are also down regulated (Barnes, 2001; Krensky et al., 2005). The effect on B lymphocytes is less prominent. Steroids also profoundly impair the function of monocytes/macrophages by significantly decreasing their phagocytic ability. Glucocorticoids also increase I| B expression, thereby curtailing activation of NF-κB, which increases apoptosis of activated cells (Auphan et al., 1995).

Prednisone, prednisolone and other glucocorticoids are used both alone and in combination with other immunosuppressives for treatment of transplant rejection and autoimmune disorders (Barry, 1992; Krensky et al., 2005).

38.4.2.2. *Calcineurin Inhibitors*

38.4.2.2.1. Cyclosporine

With the discovery and development of cyclosporine, a new era in immunopharmacology was born. Cyclosporine was the first agent to affect a specific cell line of the body's immune defenses. It is suppressive mainly to T cells, in contrast to the cytotoxic agents, which affect all cell lines at the same time. Cyclosporine is the forerunner of a group of immunosuppressants that are active against specific components of the immune response.

Cyclosporine predominantly suppresses T-cell dependent immune mechanism such as those underlying transplant rejection and some types of autoimmunity. It has much less effect on humoral immunity (Krensky et al., 2005). It preferentially inhibits antigen-triggered signal transduction in T lymphocytes, inhibiting expression of many lymphokines including IL-2 and the expression of antiapoptotic proteins.

Cyclosporine binds to an intracellular protein, cyclophilin. Cyclophilins and similar binding proteins are now referred to collectively as immunophilins and their enzymatic activities are relevant to the actions of immunosuppressants such as cyclosporine and tacrolimus. This complex inhibits the phosphatase activity of calcineurin, which in turn prevents dephosphorylation and translocation of NFAT. NFAT is required to induce a number of cytokine genes, including that for interleukin-2, which serves as a T-cell growth and differentiation factor (Krensky et al., 2005; Matsuda and Koyasu, 2000). Cyclosporine also increases expression of TGF-β, which is a potent inhibitor of IL-2 stimulated cell proliferation and generation of cytotoxic T lymphocytes (Khanna et al., 1994).

It is important to point out that cyclosporine is not cytotoxic in the ordinary sense and hence does not depress bone marrow. Because it does not encourage (in fact, it suppresses) graft-versus-host immune reactions, it is especially useful in bone marrow transplants.

Cyclosporine is an important drug in preventing rejection after kidney, liver, heart and other organ transplantation (Haberal et al., 2004). Cyclosporine usually is combined with other immunosuppressives especially glucocorticoids and either azathioprine or mycophenolate mofetil and sirolimus (Krensky et al., 2005). In renal allotransplants it has improved graft acceptance in most clinics to 95 percent. In addition to its use in transplantation cyclosporine is used for the treatment of a number of autoimmune diseases. In autoimmune diseases, as might be anticipated, cyclosporine is most effective in those which are T cell mediated. These include several forms of psoriasis, rheumatoid arthritis refractive to all other therapy, uveitis, nephrotic syndrome and type I diabetes mellitus.

38.4.2.2.2. Tacrolimus

Tacrolimus is a macrolide antibiotic produced by the soil fungus *Streptomyces tsukubaensis*. It is one hundred times more potent than cyclosporine *in vitro*, but is equally as toxic if not more so. It suppresses both humoral and cell-mediated immune responses. Although chemically distinct from cyclosporine it elicits similar immunosuppressant effects.

The mechanism of action of tacrolimus is similar to that of cyclosporine at the molecular level (Krensky et al., 2005; Spencer et al., 1997). Tacrolimus exerts potent inhibitory effects on T lymphocyte activation. It binds to an intracellular protein, FKBP-12. A complex is formed which inhibits the phosphatase activity of calcineurin. This in turn prevents dephosphorylation and translocation of nuclear factor of activated T lymphocytes (NFAT) which results in inhibition of T-lymphocyte activation.

Tacrolimus is indicated for prophylaxis of organ rejection in adult patients undergoing transplantation and to treat a variety of autoimmune diseases (Fung, 2004; Krensky et al., 2005; Spencer et al., 1997).

38.4.2.3. Cytotoxic/Antiproliferative Agents

Most anti-proliferative or cytotoxic agents were originally introduced into medicine as anticancer agents. They are able to kill cells that are capable of self-replication, including both normal and neoplastic cells. In the process of responding to an antigen, immunologically competent lymphocytes are transformed from resting cells to actively proliferating cells. Thus they are susceptible to the action of the cytotoxic agents. Several different antiproliferative agents have proved highly active as immunosuppressive agents.

38.4.2.3.1. Cyclophosphamide

The alkylating agent cyclophosphamide is used in cancer chemotherapy. It was one of the first antiproliferative/cytotoxic agents used as an immunosuppressant and is a highly potent agent. It destroys proliferating cells and also appears to alkylate a portion of the DNA in resting cells. Comparatively, cyclophosphamide has a greater effect on humoral antibody responses than on cellular reactions. Its effect on cell mediated immunity is variable with some reactions are inhibited while others are augmented. The inhibition is due to its effects on T suppressor lymphocytes (Tew et al., 2001). Cyclophosphamide has been used to prevent transplant rejection especially in developing countries as a result of its low cost and ready availability. In relatively small doses, cyclophosphamide is also effective in the treatment of some autoimmune disorders such as the childhood nephrotic syndrome, severe systemic lupus erythematosus and Wegener's granulomatosis (Neuhaus et al., 1994; Valeri et al., 1994).

38.4.2.3.2. Azathioprine

Azathioprine is a prodrug in that it is slowly converted in the body to mercaptopurine. This reaction can occur in a nonenzymatic fashion with the aid of glutathione or other sulfhydryl compounds. Perhaps it is this attribute that makes azathioprine superior to mercaptopurine as an immunosuppressant. After conversion to 6-mercaptopurine, its mechanism of action is similar to other purine antimetabolites. It interferes with nucleic acid synthesis through enzyme inhibition and it may be incorporated into DNA after conversion to the nucleotide form (Hande, 2001). Therefore, its action in suppression of the growth of lymphoid cells is not selective. Although both cell-mediated and humoral responses are suppressed it does appear to have a preferential affect to inhibit T cell responses. Azathioprine has been a mainstay drug in all types of transplantation but especially in renal allografts (Hande, 2001). In recent years its use has declined as it is being replaced by mycophenolate mofetil which seems to be safer and more effective. It is indicated as an adjunct for prevention or organ transplant rejection and in severe rheumatoid arthritis (Hong and Kahan, 2000; Krensky et al., 2005).

38.4.2.3.3. Methotrexate

Methotrexate is a folic acid analog. It inhibits folic acid synthetic pathways and DNA synthesis (Bannwarth et al., 1994). Methotrexate is widely used in the treatment of breast and other cancers. Methotrexate has recently gained favor for, and is approved for, the treatment of rheumatoid arthritis in adults who have severe active classical rheumatoid arthritis (O'Dell, 2004). It is also used to treat psoriasis and graft vs host disease.

38.4.2.3.4. Mycophenolate Mofetil

Mycophenolate mofetil is the 2-morpholinoethyl ester of mycophenolic acid (MPA). It is a prodrug that is rapidly hydrolyzed to the active form, mycophenolic acid. Mycophenolic acid is a selective, uncompetitive and reversible inhibitor of inosine monophosphate dehydrogenase (IMPDH). IMPDH is an important enzyme in the *de novo* pathway of purine nucleotide synthesis. This pathway is very important in B and T lymphocytes for proliferation. Other cells can use salvage pathways. Therefore MPA inhibits lymphocyte proliferation and functions. The mofetil ester is first converted to MPA which then is metabolized to an inactive glucuronide (Allison and Eugui, 2000). MPA has a half-life of about 16 hours (Fulton and Markham, 1996).

Mycophenolate mofetil has replaced azathioprine as the major antiproliferative agent used to prevent transplant rejection. It is usually used in combination with glucocorticoids and a calcineurin inhibitor (Mele and Halloran, 2000).

38.4.2.3.5. Sirolimus

Sirolimus is a macrocyclic lactone produced by the bacteria *Streptomyces hygroscopicus*. Like the calcineurin inhibitors cyclosporine and tacrolimus its mechanism of action involves formation of a complex with an immunophilin, in this case, FKBP-12. Unlike cyclosporine and tacrolimus, sirolimus does not affect calcineurin activity but binds to and inhibits the mammalian kinase, target of rapamycin (mTOR.). mTOR is a key enzyme in cell-cycle progression. When inhibited this kinase blocks cell cycle progression at the G1 to S phase transition (Dumont and Su, 1996; Sehgal, 2003).

Sirolimus is used for the prophylaxis of organ transplant rejection in combination with a calcineurin inhibitor and glucocorticoid. In patients at high risk for nephrotoxicity it has been combined with glucocorticoids and mycophenolate to avoid permanent renal damage (Kahan and Camardo, 2001).

38.4.2.3.6. Minocycline

Minocycline is a tetracycline antibiotic. It is also an emerging anti-inflammatory agent. It has been used experimentally and in clinical trials for the treatment of Huntington's disease, Parkinson's disease, multiple sclerosis, AIDS, ALS, and Alzheimer's disease. Minocycline has antiapoptotic effects and is a metalloproteinase inhibitor (Traynor et al., 2006; Yong, 2004; Zink et al., 2005).

38.4.2.4. Antibodies

38.4.2.4.1. Antilymphocyte Antibodies

The production of antisera against a cell type is one of the oldest effective methods of achieving a specific immune action. These antibodies can be divided into two groups: polyclonal antisera that react with multiple antigenic determinants or epitopes and monoclonal antibodies that are directed at only a single epitope.

38.4.2.4.2. Antithymocyte Globulin

Antilymphocyte globulin is a polyclonal antibody that is prepared by immunizing rabbits with human thymocytes (Regan et al., 1999). It contains cytotoxic antibodies that bind to several antigens on the surface of human T lymphocytes. The antibodies deplete circulating lymphocytes by direct cytotoxicity and block lymphocyte function by binding to cell surface molecules involved in the regulation of cell function (Krensky et al., 2005). Clinically this antisera is used primarily to treat acute renal transplant rejections in combination with other immunosuppressants. It is also used for acute rejection of other types of organ transplants and for prophylaxis of rejection (Wall, 1999). This reagent is not specific for T cells and will cross react with other cell types.

38.4.2.4.3. Muromonab-CD3 (Murine Monoclonal Antibody, Anti-CD3)

Anti-CD3 is a monoclonal antibody directed against the ε chain of CD3, a trimeric molecule adjacent to the T-cell receptor on human T lymphocytes (Krensky et al., 2005). It functions as an immunosuppressant, blocking graft rejection during organ transplantation. T cell function is blocked by interfering with CD3, a molecule in the membrane of these cells. CD3 is associated with antigen recognition, proliferation and signal transduction. The number of circulating CD3 positive, CD4 positive and CD8 positive T cells is decreased within minutes of administration of muromonab-CD3. Complement apparently is not involved in this reaction and T cell removal probably results from phagocytic activity of the reticuloendothelial system that follows opsonization by the monoclonal antibody. This is believed to play a major role in the clinical efficacy of this antibody. Recent evidence indicates that other mechanisms also play a role. There is evidence that this antibody induces apoptosis or programmed cell death in certain T cells. A third mechanism is that this antibody may mediate T cell cytolysis by inter-T cell bridging (Hooks et al., 1991; Roitt, 1993; Wilde and Goa, 1996). Monoclonal antibodies like Anti-CD3 have the advantage over polyclonal antibodies of being more selective in their action and of being able to be more precisely administered. In addition they can be more accurately quantitated.

Anti-CD3 is mainly indicated in the treatment of acute allograft rejection in renal transplant patients (Wilde and Goa, 1996; Woodle et al., 1999). Repeated use of muromonab-CD3 can result in the immunization of the patient against the mouse determinants of the antibody rendering it ineffective. Anti-CD3 also produces significant toxicity.

38.4.2.4.4. Daclizumab and Bacluzimab

Daclizumab and bacluzimab are monoclonal antibodies humanized to minimize the occurrence of anti-antibody responses and are mutated to prevent binding to FcRs (Friend et al., 1999). They were developed with the rationale that they could induce selective immunomodulation in the absence of toxicity associated with conventional anti-CD3 monoclonal antibody therapy. Daclizumab and basiliximab are produced by recombinant DNA technology. These antibodies bind with high affinity to the alpha subunit of IL-2 receptor present on the surface of activated, but not resting T lymphocytes and block IL-2 mediated T-cell activation events (Chapman and Keating, 2003; Krensky et al., 2005). Daclizumab has a somewhat lower affinity then does basiliximab (Krensky et al., 2005). These drugs are recommended for prophylaxis of acute organ rejection in adult patients usually as part of combination therapy with glucocorticoids, a calcineurin inhibitor and azathioprine and/or mycophenolate.

38.4.2.4.5. Infliximab

Infliximab is a chimeric anti TNF-α monoclonal antibody containing a human constant region and a murine variable region. It binds with high affinity to TNF-α and prevents the cytokine from binding to its receptors. It is used to treat rheumatoid arthritis and inflammatory bowel disease (Louie et al., 2003).

Summary

Immune system function is regulated or altered by a diverse array of factors. In this chapter we have summarized three seemingly disparate aspects of this array to demonstrate how widespread and varied this regulation may be. The immune system can be affected by the autonomic nervous system, various agents acting on the CNS, including opioid drugs, the sedative-hypnotic agents (the benzodiazepines and barbiturates) and the general anesthetics, and various immunosuppressive and immunostimulating agents. This emphasizes both the complexity of immune system regulation and the sometimes unexpected nature of drug interactions with the immune system. Some of these drugs, which have not been considered as immunotherapeutic agents, offer the possibility of being exploited for their ameliorative effects on diseases affecting the immune system.

Review Questions/Problems

1. Muscarinic receptor activation can be blocked by the presence of

a. neostigmine
b. nicotine
c. trimethaphan
d. botulinum toxin
e. atropine

2. Botulinum toxin acts by

a. blocking Ca^{++} influx into cholinergic nerve endings
b. hydrolyzing the proteins required for docking of the synaptic vesicle with the presynaptic membrane
c. inhibiting choline uptake
d. inhibiting choline acetyltransferase
e. blocking nicotinic receptors

3. The initial action by which cocaine increases heart rate is by

a. blocking the degradation of norepinephrine
b. increasing the release of norepinephrine
c. stimulating autonomic ganglia
d. stimulating α_1 adrenergic receptors
e. preventing the reuptake of norepinephrine

4. Dopamine synthesis is increased by activation of

a. dopamine β-hydroxylase
b. L-aromatic amino acid decarboxylase
c. tyrosine hydroxylase
d. monoamine oxidase
e. choline acetyltransferase

5. The effect of amphetamine on blood pressure will be blunted if an animal has first been treated with

a. reserpine
b. terbutaline
c. trimethaphan
d. physostigmine
e. botulinum toxin

6. The ability of phenylephrine to increase vagal tone would be blunted in the presence of

a. propranolol
b. terbutaline
c. physostigmine
d. cocaine
e. prazosin

7. The vasoconstrictive effect of norepinephrine on the movement of blood through a capillary bed will be enhanced by the concurrent presence of

a. prazosin
b. metoprolol

c. isoproterenol
d. reserpine
e. cocaine

8. Expression of mu and kappa opioid receptors has been found on

a. T cells
b. B cells
c. lymphocytes
d. peripheral blood mononuclear cells
e. all of the above

9. The end result on immune system function of treatment with opioid drugs, such as morphine, is

a. depression of immune function
b. stimulation of immune function
c. increased phagocytic activity
d. decreased viral replication
e. decreased expression of MOP and KOP receptors

10. A binding site (receptor) for both benzodiazepines and barbiturates can be found on the

a. mu opioid receptor
b. NMDA receptor
c. GABA-A receptor
d. peripheral benzodiazepine receptor
e. GABA-B receptor

11. Peripheral benzodiazepine receptors are found primarily

a. at synapses with GABA-A receptors
b. as heterodimers with opioid receptors
c. in the CNS, a paradox
d. on mitochondrial membranes
e. on macrophages

12. A common factor in the proposed mechanisms of action for general anesthetics is

a. a reduction in postsynaptic excitability
b. increased activity at excitatory neurotransmitter synapses
c. perturbation of the physical properties of cellular membranes
d. increased sympathetic nervous system activity
e. blockade of chloride channel function

13. Current research makes it clear that the opioids, sedative-hypnotic agents, and general anesthetics, as used clinically, have a major impact on immune system function

True/False

14. Which of the following statements best describes the mechanism of action of the interferons?

a. selectively stimulate T lymphocytes
b. bind to specific cellular receptors resulting in inhibition of proliferation
c. bind to cyclophilin with resulting inhibition of NFAT translocation to the nucleus
d. inhibit T cell DNA polymerase
e. are antimetabolites and are incorporated into DNA

15. A rapid decrease in the number of lymphocytes, monocytes and eosinophils best describes the mechanism of action of

a. basiliximab
b. mycophenolate mofetil
c. tacrolimus
d. corticosteroids
e. immunoglobulin

16. Which of the following statements best describes the mechanism of action of cyclosporine? It

a. binds to and inhibits m-TOR
b. increases apoptosis of T lymphocytes
c. binds to cyclophilin and inhibits calcineurin
d. binds to steroid nuclear receptors
e. inhibits tyrosine kinase phosphorylation of certain G protein coupled receptors

17. The major clinical use for muromonab-CD3 is

a. acute leukemia
b. rheumatoid arthritis
c. renal cancer
d. HIV infections
e. to prevent transplant rejection

18. Sirolimus, in contrast to tacrolimus

a. is used in cancer chemotherapy
b. is a purine antimetabolite
c. binds to steroid nuclear receptors
d. does not inhibit calcineurin
e. inhibits reverse transcriptase

19. Infliximab is

a. an antibody against the CD3 receptor
b. a humanized anti-TNFα monoclonal antibody
c. a polyclonal antibody
d. a monoclonal antibody to the HER2 oncogene
e. an inhibitor of calcineurin

20. Name three clinical uses for IGIV.

References

Alam S, Laughton DL, Walding A, Wolstenholme AJ (2006) Human peripheral blood mononuclear cells express GABAA receptor subunits. Mol Immunol 43:1432–1442.
Allison AC, Eugui EM (2000) Mycophenolate mofetil and its mechanisms of action. Immunopharmacology 47:85–118.

Auphan N, DiDonato JA, Rosette C, Helmberg A, Karin M (1995) Immunosuppression by glucocorticoids: inhibition of NF-kappa B activity through induction of I kappa B synthesis. Science 270:286–290.

Ballow M (2005) Clinical and investigational considerations for the use of IGIV therapy. Am J Health Syst Pharm 62(Supp 3):S12–S18.

Bannwarth B, Labat L, Moride Y, Schaeverbeke T (1994) Methotrexate in rheumatoid arthritis. An update. Drugs 47:25–50.

Barnes PJ (2001) Molecular mechanisms of corticosteroids in allergic diseases. Allergy 56:928–936.

Barry JM (1992) Immunosuppressive drugs in renal transplantation. A review of the regimens. Drugs 44:554–566.

Belkowski SM, Alicea C, Eisenstein TK, Adler MW, Rogers TJ (1995) Inhibition of interleukin-1 and tumor necrosis factor-alpha synthesis following treatment of macrophages with the kappa opioid agonist U50, 488H. J Pharmacol Exp Ther 273:1491–1496.

Bergeret M, Khrestchatisky M, Tremblay E, Bernard A, Gregoire A, Chany C (1998) GABA modulates cytotoxicity of immunocompetent cells expressing GABAA receptor subunits. Biomed Pharmacother 52:214–219.

Bidlack JM, Khimich M, Parkhill AL, Sumagin S, Sun B, Tipton C (2006) Opioid receptors and signaling on cells from the immune system. J Neuroimmune Pharmacol 1:260–269.

Bridge KE, Wainwright A, Reilly K, Oliver KR (2003) Autoradiographic localization of ^{125}I[Tyr^{14}] nociceptin/orphanin FQ binding sites in macaque primate CNS. Neuroscience 118:513–523.

Brunton LL, Lazo JS, Parker KL (2006) Goodman & Gilman's The Pharmacological Basis of Therapeutics. New York: McGraw Hill.

Burt DR (2003) Reducing GABA receptors. Life Sci 73:1741–1758.

Cadet P (2004) Mu opiate receptor subtypes. Med Sci Monit 10: MS28–MS32.

Carlson SL (2003) Nervous system: immune system interactions. In: Neuroscience in Medicine, 2nd ed. (Conn PM, ed), pp 647–661. Totowa, NJ: Humana Press.

Casellas P, Galiegue S, Basile AS (2002) Peripheral benzodiazepine receptors and mitochondrial function. Neurochem Int 40:475–486.

Chao CC, Sharp BM, Pomeroy C, Filice GA, Peterson PK (1990) Lethality of morphine in mice infected with Toxoplasma gondii. J Pharmacol Exp Ther 252:605–609.

Chapman TM, Keating GM (2003) Basiliximab: a review of its use as induction therapy in renal transplantation. Drugs 63:2803–2835.

Charney DS, Mihic SJ, Harris RA (2006) Hypnotics and sedatives. In: Goodman and Gilman's The Pharmacological Basis of Therapeutics (Brunton LL, Lazo JS, Parker KL, eds), pp 401–427. New York: McGraw-Hill.

Covelli V, Maffione AB, Nacci C, Tato E, Jirillo E (1998) Stress, neuropsychiatric disorders and immunological effects exerted by benzodiazepines. Immunopharmacol Immunotoxicol 20:199–209.

Czura CJ, Tracey KJ (2005) Autonomic neural regulation of immunity. J Intern Med 257:156–166.

da Silva FR, Lazzarini R, de Sa-Rocha LC, Morgulis MS, de Oliveira MC, Palermo-Neto J (2003) Effects of acute and long-term diazepam administrations on neutrophil activity: a flow cytometric study. Eur J Pharmacol 478:97–104.

Devi LA (2001) Heterodimerization of G-protein-coupled receptors: pharmacology, signaling and trafficking. Trends Pharmacol Sci 22:532–537.

Dorr RT (1993) Interferon-alpha in malignant and viral diseases. A review. Drugs 45:177–211.

Dumont FJ, Su Q (1996) Mechanism of action of the immunosuppressant rapamycin. Life Sci 58:373–395.

Eskandari F, Webster JI, Sternberg EM (2003) Neural immune pathways and their connection to inflammatory diseases. Arhtritis Res Ther 5:251–265.

Evers AS, Crowder CM, Balser JR (2006) General anesthetics. In: Goodman and Gilman's The Pharmacological Basis of Therapeutics (Brunton LL, Lazo JS, Parker KL, eds), pp 341–368. New York: McGraw-Hill.

Florin S, Meunier J, Costentin J (2000) Autoradiographic localization of [^{3}H]nociceptin binding sites in the rat brain. Brain Res 880:11–16.

Friedman EM, Irwin MR (1997) Modulation of immune cell function by the autonomic nervous system. Pharmacol Ther 74:27–38.

Friend PJ, Hale G, Chatenoud L, Rebello P, Bradley J, Thiru S, Phillips JM, Waldmann H (1999) Phase I study of an engineered aglycosylated humanized CD3 antibody in renal transplant rejection. Transplantation 68:1632–1637.

Froh M, Thurman RG, Wheeler MD (2002) Molecular evidence for a glycine-gated chloride channel in macrophages and leukocytes. Am J Physiol Gastrointest Liver Physiol 283:G856–G863.

Fulton B, Markham A (1996) Mycophenolate mofetil. A review of its pharmacodynamic and pharmacokinetic properties and clinical efficacy in renal transplantation. Drugs 51:278–298.

Fung JJ (2004) Tacrolimus and transplantation: a decade in review. Transplantation 77:S41–S43.

Galley HF, Webster NR (2006) The immune system. In: Foundations of Anesthesia (Hemmings HC, Hopkins PM, eds), pp 647–665. Philadelphia: Mosby Elsevier.

Gavish M, Bachman I, Shoukrun R, Katz Y, Veenman L, Weisinger G, Weizman A (1999) Enigma of the peripheral benzodiazepine receptor. Pharmacol Rev 51:629–650.

Goldstein D, Jones R, Smalley RV, Borden EC (2003) Inerferons: therapy for cancer. In: Principles of Cancer Biotherapy (Oldham RK, ed), pp 301–327. Boston: Kluwer Academic Publishers.

Gomes I, Jordan BA, Gupta A, Trapaidze N, Nagy V, Devi LA (2000) Heterodimerization of m and d opioid receptors: a role in opiate synergy. J Neurosci 20:RC110.

Gomes I, Gupta A, Filipovska J, Szeto HH, Pintar JE, Devi LA (2004) A role for heterodimerization of mu and delta opiate receptors in enhancing morphine analgesia. Proc Natl Acad Sci USA 101:5135–5139.

Gutstein HB, Akil H (2006) Opioid Analgesics. In: Goodman and Gilman's The Pharmacological Basis of Therapeutics (Brunton LL, Lazo JS, Parker KL, eds), pp 547–590. New York: McGraw-Hill.

Haberal M, Emiroglu R, Dalgic A, Karakayli H, Moray G, Bilgin N (2004) The impact of cyclosporine on the development of immunosuppressive therapy. Transplant Proc 36:143S–147S.

Hande KR (2001) Purine antimetabolites. In: Cancer Chemotherapy and biotherapy. Principles and practice (Chabner BA, Longo DL, eds), pp 295–314. Philadelphia: Lipincott Williams and Wilkins.

Heagy W, Laurance M, Cohen E, Finberg R (1990) Neurohormones regulate T cell function. J Exp Med 171:1625–1633.

Hong JC, Kahan BD (2000) Immunosuppressive agents in organ transplantation: past, present, and future. Semin Nephrol 20:108–125.

Hooks MA, Wade CS, Millikan WJ Jr (1991) Muromonab CD-3: a review of its pharmacology, pharmacokinetics, and clinical use in transplantation. Pharmacotherapy 11:26–37.

Irwin M, Hauger RL, Jones L, Provencio M, Britton KT (1990) Sympathetic nervous system mediates central corticotropin-releasing factor induced suppression of natural killer cytotoxicity. J Pharmacol Exp Ther 255:101–107.

Kahan BD, Camardo JS (2001) Rapamycin: clinical results and future opportunities. Transplantation 72:1181–1193.

Khanna A, Li B, Stenzel KH, Suthanthiran M (1994) Regulation of new DNA synthesis in mammalian cells by cyclosporine. Demonstration of a transforming growth factor beta-dependent mechanism of inhibition of cell growth. Transplantation 57:577–582.

Kohm AP, Sanders VM (2001) Norepinephrine and beta 2-adrenergic receptor stimulation regulate CD4 + T and B lymphocyte function in vitro and in vivo. Pharmacol Rev 53:487–525.

Krensky AM, Vincenti F, Bennett WM (2005) Immunosuppressants, tolerogens and immunostimulants. In: Goodman and Gilman's The Pharmacological basis of therapeutics (Brunton LL, Lazo JS, Parker KL, eds), pp 1405–1431. New York: McGraw Hill.

Lawrence DM, Bidlack JM (1993) The kappa opioid receptor expressed on the mouse R1.1 thymoma cell line is coupled to adenylyl cyclase through a pertussis toxin-sensitive guanine nucleotide-binding regulatory protein. J Pharmacol Exp Ther 266:1678–1683.

Levy JH (2006) The allergic response. In: Clinical Anesthesia (Barash PG, Cullen BF, Stoelting RK, eds), pp 1298–1310. Philadelphia: Lippincott Williams & Wilkins.

Lewko H, Oldham R (2003) Cytokines. In: Principles of Cancer Biotherapy (Oldham RK, ed), pp 183–299. Boston: Kluwer Academic Publishers.

Lundberg JM, Hokfelt T (1983) Coexistence of peptides and classical neurotransmitters. Trends Neurosci 6:325–333.

Lyles CM, Margolick JB, Astemborski J, Graham NM, Anthony JC, Hoover DR, Vlahov D (1997) The influence of drug use patterns on the rate of CD4 + lymphocyte decline among HIV-1-infected injecting drug users. AIDS 11:1255–1262.

Mansour A, Watson SJ (1993) Anatomical distribution of opioid receptors in mammalians: an overview. In: Opioids I (Herz A, ed), pp 79–105. Berlin: Springer-Verlag.

Mansour A, Khachaturian H, Lewis ME, Akil H, Watson SJ (1988) Anatomy of CNS opioid receptors. Trends Neurosci 11:308–314.

Margolick JB, Munoz A, Vlahov D, Solomon L, Astemborski J, Cohn S, Nelson KE (1992) Changes in T-lymphocyte subsets in intravenous drug users with HIV-1 infection. JAMA 267:1631–1636.

Matsuda S, Koyasu S (2000) Mechanisms of action of cyclosporine. Immunopharmacology 47:119–125.

Matthes HWD, Maldonado R, Simonin F, Valverde O, Slowe S, Kitchen I, Befort K, Dierich A, Le Meur M, Doll P, Tzavara E, Hanoune J, Roques BP, Kieffer BL (1996) Loss of morphine-induced analgesia, reward effect and withdrawal symptoms in mice lacking the m-opioid-receptor gene. Nature 383:819–823.

McCarthy L, Wetzel M, Sliker JK, Eisenstein TK, Rogers TJ (2001) Opioids, opioid receptors, and the immune response. Drug Alcohol Depend 62:111–123.

Mele TS, Halloran PF (2000) The use of mycophenolate mofetil in transplant recipients 2. Immunopharmacology 47:215–245.

Miglio G, Varsaldi F, Lombardi G (2005) Human T lymphocytes express N-methyl-D-aspartate receptors functionally active in controlling T cell activation. Biochem Biophys Res Commun 338:1875–1883.

Mogil JS, Pasternak GW (2001) The molecular and behavioral pharmacology of the orphanin FQ/nociceptin peptide and receptor family. Pharmacol Rev 53:381–415.

Neal CR, Jr., Mansour A, Reinscheid R, Nothacker HP, Civelli O, Akil H, Watson SJ, Jr. (1999) Opioid receptor-like (ORL1) receptor distribution in the rat central nervous system: comparison of ORL1 receptor mRNA expression with ^{125}I-[^{14}Tyr]-orphanin FQ binding. J Comp Neurol 412:563–605.

Neuhaus TJ, Fay J, Dillon MJ, Trompeter RS, Barratt TM (1994) Alternative treatment to corticosteroids in steroid sensitive idiopathic nephrotic syndrome. Arch Dis Child 71:522–526.

O'Dell JR (2004) Therapeutic strategies for rheumatoid arthritis. N Engl J Med 350:2591–2602.

Pasternak GW (2001) Insights into mu opioid pharmacology the role of mu opioid receptor subtypes. Life Sci 68:2213–2219.

Regan JF, Campbell K, Van Smith L et al. (1999). Sensitization following Thymoglobulin and Atgam rejection therapy as determined with a rapid enzyme-linked immunosorbent assay. U.S. Thymoglobulin multi-center study group. Transplantation Immunology 7:115–121

Rhen T, Cidlowski JA (2005) Antiinflammatory action of glucocorticoids-new mechanisms for old drugs. N Engl J Med 353:1711–1723.

Risdahl JM, Peterson PK, Chao CC, Pijoan C, Molitor TW (1993) Effects of morphine dependence on the pathogenesis of swine herpesvirus infection. J Infect Dis 167:1281–1287.

Robertson D (2004) Primer on the Autonomic Nervous System. San Diego: Elsevier Academic Press.

Roitt IM (1993) OKT3:Immunology, production, pruification and pharmacokinetics. Clin Transplant 7:367–373.

Rothenberg R, Woelfel M, Stoneburner R, Milberg J, Parker R, Truman B (1987) Survival with the acquired immunodeficiency syndrome. Experience with 5833 cases in New York City. N Engl J Med 317:1297–1302.

Roy S, Balasubramanian S, Sumandeep S, Charboneau R, Wang J, Melnyk D, Beilman GJ, Vatassery R, Barke RA (2001) Morphine directs T cells toward T(H2) differentiation. Surgery 130: 304–309.

Roy S, Wang J, Kelschenbach J, Koodie L, Martin J (2006) Modulation of immune function by morphine: implications for susceptibility to infection. J Neuroimmune Pharmacol 1:77–89.

Ruff MR, Pert CB, Weber RJ, Wahl LM, Wahl SM, Paul SM (1985) Benzodiazepine receptor-mediated chemotaxis of human monocytes. Science 229:1281–1283.

Saeed RW, Varma S, Peng-Nemeroff T, Sherry B, Balakhaneh D, Huston J, Tracey KJ, Al-Abed Y, Metz CN (2005) Cholinergic stimulation blocks endothelial cell activation and leukocyte recruitment during inflammation. J Exp Med 201:1113–1123.

Salman H, Bergman M, Weizman A, Bessler H, Weiss J, Straussberg R, Djaldetti M (2000) Effect of diazepam on the immune response of rats exposed to acute and chronic swim stress. Biomed Pharmacother 54:311–315.

Samuel CE (2001) Antiviral actions of interferons. Clin Microbiol Rev 14:778–809.

Sehgal SN (2003) Sirolimus: its discovery, biological properties, and mechanism of action. Transplant Proc 35:7S–14S.

Selye H (1936) Thymus and adrenals in the response of the organism to injuries and intoxications. Br J Exp Pathol 17:234–248.

Shah S (2005) Pharmacy considerations for the use of IGIV therapy. Am J Health Syst Pharm 62(Supp 3):S5-S11.

Shahabi NA, McAllen K, Sharp BM (2006) d opioid receptors stimulate Akt-dependent phosphorylation of c-jun in T cells. J Pharmacol Exp Ther 316:933–939.

Sharp BM (2006) Multiple opioid receptors on immune cells modulate intracellular signaling. Brain Behav Immun 20:9–14.

Sharp BM, Shahabi NA, Heagy W, McAllen K, Bell M, Huntoon C, McKean DJ (1996) Dual signal transduction through delta opioid receptors in a transfected human T-cell line. Proc Natl Acad Sci USA 93:8294–8299.

Sharp BM, McAllen K, Shahabi NA (2005) Immunofluorescence detection of anti-CD3-e-induced delta opioid receptors by murine splenic T cells. In: Infectious Diseases and Substance Abuse (Friedman H, ed), pp 141–147. New York: Springer.

Sora I, Takahashi N, Funada M, Ujike H, Revay RS, Donovan DM, Miner LL, Uhl GR (1997) Opiate receptor knockout mice define m receptor roles in endogenous nociceptive responses and morphine-induced analgesia. Proc Natl Acad Sci USA 94:1544–1549.

Spencer CM, Goa KL, Gillis JC (1997) Tacrolimus. An update of its pharmacology and clinical efficacy in the management of organ transplantation. Drugs 54:925–975.

Steinman L (2004) Elaborate interactions between the immune and nervous systems. Nat Immunol 5:575–581.

Suzuki S, Chuang LF, Doi RH, Bidlack JM, Chuang RY (2001) Kappa-opioid receptors on lymphocytes of a human lymphocytic cell line: morphine-induced up-regulation as evidenced by competitive RT-PCR and indirect immunofluorescence. Int Immunopharmacol 1:1733–1742.

Taupin V, Herbelin A, Descapms-Latscha B, Zavala F (1991) Endogenous anxiogenic peptide, ODN-diazepam-binding-inhibitor, and benzodiazepines enhance the production of interleukin-1 and tumor necrosis factor by human monocytes. Lumphokine Cytokine Res 10:7–13.

Tew KD, Colvin OM, Chabner BA (2001) Alkylating agents. In: Cancer Chemotherapy and Biotherapy. Principles and Practice (Chabner BA, Longo DL, eds), pp 373–414. Philadelphia: Lipincott Williams and Wilkins.

Tian J, Chau C, Hales TG, Kaufman DL (1999) GABA(A) receptors mediate inhibition of T cell responses. J Neuroimmunol 96:21–28.

Tian J, Lu Y, Zhang H, Chau CH, Dang HN, Kaufman DL (2004) Gamma-aminobutyric acid inhibits T cell autoimmunity and the development of inflammatory responses in a mouse type 1 diabetes model. J Immunol 173:5298–5304.

Tomkins WA (1999) Immunomodulation and therapeutic effects of the oral use of interferon-alpha: mechanism of action. J Interferon Cytokine Res 19:817–828.

Traynor BJ, Bruijun L, Conwit R, Beal F, O'Neill G, Fagan SC, Cudkowicz ME (2006) Neuroprotective agents for clinical trials in ALS: a systematic assessment. Trends Pharmacol Sci 25:609–612.

Tubaro E, Borelli G, Croce C, Cavallo G, Santiangeli C (1983) Effect of morphine on resistance to infection. J Infect Dis 148:656–666.

Valeri A, Radhakrishnan J, Estes D, D'Agati V, Kopelman R, Pernis A, Flis R, Pirani C, Appel GB (1994) Intravenous pulse cyclophosphamide treatment of severe lupus nephritis: a prospective five-year study. Clin Nephrol 42:71–78.

Wahlestedt C, Edvinsson L, Ekblad E, Hakanson R (1985) Neuropeptide Y potentiates noradrenaline-evoked vasoconstriction: mode of action. J Pharmacol Exp Ther 234:735–741.

Wall WJ (1999) Use of antilymphocyte induction therapy in liver transplantation. Liver Transpl Surg 5:S64–S70.

Wang CQ, Li Y, Douglas SD, Wang X, Metzger DS, Zhang T, Ho WZ (2005a) Morphine withdrawal enhances hepatitis C virus replicon expression. Am J Pathol 167:1333–1340.

Wang X, Tan N, Douglas SD, Zhang T, Wang YJ, Ho WZ (2005b) Morphine inhibits CD8 + T cell-mediated, noncytolytic, anti-HIV activity in latently infected immune cells. J Leukoc Biol 78:772–776.

Webster JI, Tonelli L, Sternberg EM (2002) Neuroendocrine regulation of immunity. Annu Rev Immunol 20:125–163.

Weiss M, Buhl R, Birkhahn A, Mirow N, Schneider M, Wernet P (1994a) Do barbiturates and their solutions suppress FMLP-induced neutrophil chemiluminescence? Eur J Anaesthesiol 11:371–379.

Weiss M, Buhl R, Mirow N, Birkhahn A, Schneider M, Wernet P (1994b) Do barbiturates impair zymosan-induced granulocyte function? J Crit Care 9:83–89.

Wiesemann E, Sonmez D, Heidenreich F, Windhagen A (2002) Interferon-beta increases the stimulatory capacity of monocyte-derived dendritic cells to induce IL-13, IL-5 and IL-10 in autologous T-cells. J Neuroimmunol 123:160–169.

Wilde MI, Goa KL (1996) Muromonab CD3: a reappraisal of its pharmacology and use as prophylaxis of solid organ transplant rejection. Drugs 51:865–894.

Woodle ES, Xu D, Zivin RA, Auger J, Charette J, O'Laughlin R, Peace D, Jollife LK, Haverty T, Bluestone JA, Thistlethwaite JR Jr. (1999) Phase I trial of a humanized, Fc receptor nonbinding OKT3 antibody, huOKT3gamma1(Ala-Ala) in the treatment of acute renal allograft rejection. Transplantation 68:608–616.

Wybran J, Appelboom T, Famaey JP, Govaerts A (1979) Suggestive evidence for receptors for morphine and methionine-enkephalin on normal human blood T lymphocytes. J Immunol 123: 1068–1070.

Yong VW (2004) Prospects for neuroprotection in multiple sclerosis. Front Biosci 9:864–872.

Zavala F (1997) Benzodiazepines, anxiety and immunity. Pharmacol Ther 75:199–216.

Zavala F, Taupin V, Descamps-Latscha B (1990) *In vivo* treatment with benzodiazepines inhibits murine phagocyte oxidative metabolism and production of interleukin-1, tumor necrosis factor and interleukin-6. J Pharmacol Exp Ther 255:442–450.

Zavala F, Masson A, Brys L, de Baetselier P, Descamps-Latscha B (1991) A monoclonal antibody against peripheral benzodiazepine receptor activates the human neutrophil NADPH-oxidase. Biochem Biophys Res Comm 176:1577–1583.

Zink MC, Uhrlaub J, De Witt J, Voelker T, Bullock B, Mankowski J, TArwater P, Clements J, Barber S (2005) Neuroprotective and anti-human immunodeficiency virus activity of minocycline. JAMA 293:2003–2011.

39
Neurodegenerative Diseases

Jinsy A. Andrews and Paul H. Gordon

Keywords Alzheimer's disease; Amyotrophic lateral sclerosis; Animal models; Apoptosis; Huntington's disease; Neurodegeneration; Neuroprotection; Parkinson's disease; Randomized controlled trials; Therapeutics

39.1. Introduction

The neurodegenerative disorders, which include Amyotrophic Lateral Sclerosis (ALS), Alzheimer Disease (AD), Parkinson Disease (PD), and Huntington Disease (HD), are clinically heterogeneous. Medications can ameliorate some symptoms, but none reverse the relentless progression of the illnesses. The past several decades have seen a dramatic increase in the understanding of the complex pathophysiology underlying the diseases. Overlapping features at the cellular level provide clues for targets of therapeutic interventions. While the disorders can now be studied using *in vitro* and animal models, they have posed a particular challenge because of their complex biochemistry, pathology and lack of mechanism based treatments. There are still few unequivocal neuroprotective agents, and we await the major breakthrough that will provide truly meaningful clinical improvement. Until then, numerous symptomatic therapies can improve the quality of life of those suffering from neurodegenerative diseases. This chapter outlines the current strategies for the use of symptomatic therapies and the development of neuroprotective treatment for neurodegenerative disorders.

39.1.1. Overview of Mechanism

Progressive and premature neuronal cell death is the common feature among the neurodegenerative disorders. ALS involves progressive degeneration of the motor neurons in the brain and spinal cord, AD shows selective neuron degeneration in the hippocampus and cerebral cortex, PD is caused by degeneration of pigmented dopaminergic neurons in the substantia nigra, and HD has selective neuron loss in the striatum and cortex. It is increasingly recognized that the clinical syndromes may overlap, with features of two or more disorders co-existing. For instance, ALS and dementia, or ALS, parkinsonism and dementia may occur together in the same family or in an individual. Similarly, the disorders have common pathological findings at the cellular level. While the mechanisms that lead to neuronal degeneration are still being sought, in general, it is thought that neurodegenerative diseases are caused by a combination of events including age-associated, genetic and environmental factors in varying proportions (Rowland and Shneider, 2001; Koo and Kopan, 2004). All cases of HD, but only rare cases of ALS and AD are due to single genetic mutations. In the sporadic forms, complex genetic interactions may render an individual susceptible to some, as yet unknown, environmental exposure, with the combination necessary to initiate the disease. Whatever the cause, be it genetic, environmental or some combination of the two, there is set in motion a cascade of events at the cellular level that ultimately results in cell death. Animal models of the genetic forms of the disorders have helped elucidate some of the underlying molecular mechanisms that contribute to neuronal demise. Common processes among the neurodegenerative diseases include oxidative stress, excitotoxic injury, altered metal homeostasis, aberrant protein aggregation, mitochondrial injury, apoptosis, and inflammation (Bossey-Wetzel et al., 2004; Andersen, 2004; Ross and Poirier, 2004). Once the etiologies of individual disorders are identified, more effective treatments will surely follow. Until then, a major goal is to identify therapies that slow the course of the illnesses by targeting cellular changes that occur after the inciting event.

39.1.2. Overview of Animal Models

The inherited forms of neurodegenerative diseases have been utilized in the development of animal models. Rodent models created using transgenic and gene targeting technologies (Deng and Siddique, 2000) now provide the opportunity to better study disease pathology and screen potential therapeutic

T. Ikezu and H.E. Gendelman (eds.), *Neuroimmune Pharmacology*.
© Springer 2008

agents. The advent of the ALS mouse model is one example of the progress that has resulted from this technology. A mutation in the gene encoding the enzyme copper–zinc superoxide dismutase (SOD1) on chromosome 21 was identified as a causative mutation in one form of familial ALS (Rosen et al., 1993). The first mutant transgenic mouse line was created by over expressing high levels of mutant SOD 1 (G93A). This mouse developed a phenotypic and pathologic condition similar to human familial and sporadic ALS (Gurney et al., 1994). The murine model of ALS has since been used to screen a variety of therapeutic agents. Riluzole, the only approved agent for ALS, yielded positive results in the model that correspond very closely to the modest benefit in human clinical trials (Gurney et al., 1996, 1998).

Similarly, mouse models have been created for other neurodegenerative diseases. Transgenic mouse models of AD express mutations in the amyloid precursor protein (APP), presenilin genes, or a combination of the two (German and Eisch, 2004). Insight into the molecular mechanisms of dopaminergic cell death in PD have come from the study of 6-hydroxydopamine (6-OHDA) and 1-methyl 1–4-phenyl-1,2,3,6-tetrahydropyridine (MPTP) animal models (Dawson, 2000). Discovery of genes linked to familial PD have also provided the opportunity to generate animal models that better recapitulate the phenotypic and pathologic features of PD. Similarly, animal models have been created for HD through utilization of the CAG triplet expansion in the HD gene. Homozygous and heterozygous knockout mice did not produce the phenotypic or pathologic features of HD, but transgenic mice models of HD did produce a progressive neurological phenotype that resembles features of HD and has improved the understanding the disease process (Deng and Siddique, 2000).

Many of the past and current therapeutic trials have been aimed at slowing neurodegenerative processes detected in animal models. While it is evident that the models of human diseases have played an essential role in understanding the mechanisms involved in neurodegeneration and in developing therapeutic strategies, the translation of findings in animal models to human therapeutics has been fraught with difficulty. Ten years after the approval of riluzole and following numerous clinical trials based on findings in the model, we still await the second unequivocally positive trial in ALS. Humans are more complex than rodents; the murine models may best be used for establishing the scientific justification for testing a potential agent, with the knowledge that findings in the laboratory, as yet, cannot provide certain hope of success in human trials.

39.2. Disease-Specific Therapy

In certain disorders, such as AD and PD, neurodegeneration leads to deficiencies in specific neurotransmitters. Therapy has been targeted at replacing or increasing levels of the deficient neurotransmitter in patients. Disease specific therapy of this his type does not alter the underlying progression of neurodegeneration, but may improve symptoms, at least temporarily.

39.2.1. Alzheimer's Disease

Changes in the cholinergic system in AD were first studied by Whitehouse et al. (1981), who found atrophy of cholinergic nerve cells in the substantia innominata. It was thought that replacing or increasing acetylcholine in patients with AD could stabilize cholinergic transmission. The cholinergic hypothesis of AD has been the basis of many research efforts and the accumulated evidence has launched the development of a variety of compounds that improve cholinergic neurotransmission (Lanctot et al., 2003; Scarpini et al., 2003). Currently, the cholinergic deficiency is considered to be a secondary event resulting from nerve cell death due to other causes (Albers and Beal, 2000).

39.2.1.1. Cholinesterase Inhibitors

Cholinesterase inhibitors, developed with the goal of stabilizing cholinergic transmission, block the enzyme that degrades acetylcholine in the synaptic cleft and lead to increased levels of acetylcholine in the brain. There are four types of cholinesterase inhibitors currently available in the United States: tacrine, donepezil, rivastigmine and galantamine (Table 39.1). Tacrine, donepezil and galantamine are selective inhibitors of acetylcholinesterase. Galantamine is an allosteric ligand of the nicotinic acetylcholine receptor and is believed to increase the presynaptic release of acetylcholine and prolong postsynaptic neurotransmission (Greenblatt et al., 1999; Scott and Goa, 2000). Donepezil is a piperidine derivative that noncompetitively and reversibly blocks acetylcholinesterase. Rivastigmine is a slow reversible inhibitor of acetylcholinesterase.

Individually, the cholinesterase inhibitors improve memory in patients with AD. There have been no head-to-head comparisons made of the drugs to determine whether one is more effective than the others and no trials of agents in combination. The main known differences are the side effect profiles, titration schedules and the dosing regimens (Table 39.1). The most common side effects of treatment with cholinesterase inhibitors are gastrointestinal and include nausea, vomiting and diarrhea. A meta-analysis of the efficacy and safety of these agents revealed that despite many clinical studies showing

TABLE 39.1. Disease specific and neuroprotective therapy in Alzheimer disease.

Medication	Dosage
Cholinesterase inhibitor	
Tacrine (Cognex)	20–40 mg QID
Donepezil (Aricept)	5–10 mg QD
Rivastigmine (Exelon)	3–6 mg BID
Galantamine (Razadyne)	8–12 mg BID
NMDA antagonist	
Memantine (Namenda)	10 mg BID

short-term benefit, there is insufficient evidence to conclude long term efficacy (Lanctot et al., 2003).

39.2.2. Parkinson's Disease

As in AD, knowledge of the neurotransmitter deficiency underlying PD, in this case dopamine, has been the basis for the development of therapy. Early studies showed that cerebral dopamine was concentrated in the striatum and that levodopa, the precursor to dopamine, could reverse the akinetic effects of the dopamine-depleting agent reserpine in experimental animals (Carlsson et al., 1957, 1958). Eventually, the identification of striatal dopamine depletion as a key neurochemical finding in parkinsonian brains lead to treatment with levodopa in humans and to the subsequent advent of compounds that mimic the effects of dopamine or prolong its action (Table 39.2).

39.2.2.1. Levodopa

Levodopa is the most effective drug for the symptomatic management of PD. Levodopa, which usually produces an excellent initial response in alleviating the motor symptoms of PD, is particularly effective for bradykinesia and tremor (Fahn, 2000). However, side effects such as motor fluctuations and dyskinesias may limit its long-term usefulness for individual patients (Fahn, 2000; Jankovic, 2000). Up to 50% of all patients treated for 5 years or more develop instability in their motor response (Koller et al., 1999; Schrag and Quinn, 2000). As a consequence, levodopa is usually given later in PD, particularly in younger patients who may require many years of anti-Parkinson therapy. In this group, dopa-sparing agents (see below) are usually prescribed until symptoms begin to interfere with independence. Levodopa is often used as the first drug of choice in the elderly, in whom low dose monotherapy is preferable to avoid side effects, and who may not be expected to develop motor fluctuations after years of therapy due to their already advanced age. Immediate and controlled release preparations of levodopa are available.

Carbidopa, a peripheral dopa decarboxylase inhibitor, enhances the therapeutic benefits of levodopa (Opacka-Juffrey and Brooks, 1995). Levodopa is converted to dopamine by dopa decarboxylase, an enzyme that functions both inside and outside the brain. Peripherally produced dopamine directly stimulates the medullary vomiting center, which is not protected by a blood brain barrier, producing anorexia, nausea and vomiting. Carbidopa in doses less than 150 mg per day does not enter the brain and so blocks only the peripheral metabolism of levodopa (Cedarbaum et al., 1986). Co-administration of carbidopa and levodopa therefore has two main effects: It reduces levodopa-related gastrointestinal side effects, and insures that the majority of levodopa is converted to dopamine in the brain where it is needed. Seventy-five mg of carbidopa per day is necessary to completely block peripheral dopa decarboxylase activity. Additional doses can be given to those who are particularly sensitive to levodopa-induced nausea. Other major side effects of levodopa therapy include orthostatic hypotension, chorea, dystonia, myoclonus, akathisia and hallucinations. Studies have not yet shown conclusively whether levodopa alters the underlying rate of neurodegeneration (Fahn and The Parkinson Study Group, 2005).

TABLE 39.2. Disease specific therapy in Parkinson disease.

Medication	Dosage
Dopamine precursor	
Levodopa	300–3,000 mg per day
Dopa decarboxylase inhibitor/dopamine precursor	
Carbidopa/levodopa	25 mg/300 mg – 200 mg/2,000 mg per day
Dopamine agonist	
Bromocriptine	10–30 mg TID
Pergolide	1 mg TID
Ropinirole	3–8 mg TID
Pramipexole	0.5–1.5 mg TID
Cabergoline	2–10 mg per day
COMT inhibitor	
Entacapone	200–400 TID—6×/day
Tolcapone	100–200 TID
NMDA antagonist/anticholinergic	
Amantadine	100–300 per day
Anticholinergic	
Trihexyphenidyl	2 mg TID
Benztropine	1–4 mg QD
MAO inhibitor	
Selegiline	5 mg BID

39.2.2.2. Catechol-O-methyltransferase Inhibitors

Another strategy to enhance dopamine response utilizes the inhibition of catechol-O-methyltransferase (COMT), which metabolizes dopamine once it has been converted from levodopa either in or outside the brain. Agents that block the degradation of dopamine by inhibition of COMT effectively prolong the half-life of dopamine and may increase its peak concentration. The two currently available COMT inhibitors, entacapone (Rinne et al., 1998; Holm and Spencer, 1999) and tolcapone (Kurth and Adler, 1998; Tolcapone Study Group, 1999), reduce motor off time when given with levodopa, but the efficacy of the two drugs has not been directly compared. Tolcapone has a theoretical advantage over entacapone in that it inhibits both peripheral and central COMT, and it has a longer duration of action necessitating administration only three times per day. However, tolcapone can cause explosive diarrhea and abnormal liver function. Fulminant liver failure has occurred, albeit rarely, and therefore frequent monitoring of liver function is necessary in those prescribed tolcapone (Assal et al., 1998). Entacapone usually does not cause diarrhea or abnormal liver function and can be administered with levodopa up to eight times a day.

39.2.2.3. Dopamine Agonists

Dopamine agonists exert their effect by activating dopamine receptors and bypassing the presynaptic synthesis of dopamine. They are often used as part of a dopa-sparing strategy, in which agonists are prescribed early in PD, postponing the need for levodopa and the motor complications associated with exogenous dopamine administration. Activation of the dopamine-2 (D2) receptors is important in mediating the motor effects of dopamine agonists, but to achieve optimal physiologic and behavioral responses, both D1 and D2 receptors must be stimulated (Jenner, 1997; Brooks, 2000).

Bromocriptine was the first dopamine agonist developed followed by pergolide. Pergolide does not require endogenous dopamine for its pharmacological effect, possibly because it activates both D1 and D2 receptors whereas bromocriptine stimulates D2 and inhibits D1 receptors (Perachon et al., 1999). Pergolide is a more potent inhibitor of prolactin secretion and it has a half-life three times longer than bromocriptine. An early study showed that these two medications have similar efficacy in PD (LeWitt et al., 1983), but more recent data suggests that pergolide may be more effective (Pezzoli et al., 1994). Both agents should be increased slowly to prevent orthostatic hypotension and gastrointestinal side effects. Because the compounds are ergot derivatives, they may rarely be associated with retroperitoneal or pericardial fibrosis, including cardiac valvular disease (Van Camp et al., 2004).

In 1997, two additional dopamine agonists, ropinirole and pramipexole, were approved. These agents are nonergolines and therefore may carry a lower risk of ulcerative, vasoconstrictive and fibrotic complications (Shaunak et al., 1999). Pramipexole has been promoted as a D3 agonist (Bennett and Piercey, 1999), but a study that measured the affinities of bromocriptine, pramipexole, pergolide, and ropinirole for human recombinant dopamine D1, D2, and D3 receptors found that all four compounds had high affinity for the D3 receptor (Perachon et al., 1999). The observation that D3 receptors are located primarily in the mesolimbic system may explain the possible antidepressant effect of pramipexole and ropinirole (Corrigan et al., 2000), but may also explain the occurrence of hallucinations noted particularly with higher doses of dopamine agonists.

Pramipexole has been shown to be safe and effective when used as monotherapy early in PD (Parkinson Study Group, 1997, 2000) and in mild to moderate PD (Shannon et al., 1997). Ropinirole has also been shown to be effective in early PD (Rascol et al., 1998; Korczyn et al., 1999). In later stages of PD, dopamine agonists are usually prescribed with levodopa to achieve optimal therapeutic effects and to help moderate the motor fluctuations associated with levodopa (Pinter et al., 1999). Possible side effects of all dopamine agonists include nausea, peripheral edema, somnolence and hallucinations. Pramipexole has also been associated with compulsive behavior (Driver-Dunckley et al., 2003).

39.2.2.4. Anticholinergic Agents

The manipulation of nondopaminergic neurotransmitters can also be used as disease specific therapy in PD. Anticholinergic agents, the first effective therapy for PD, having been described by Charcot in the nineteenth century, are particularly helpful in tremor reduction. The exact mechanism of anticholinergic medications in PD remains unclear. They likely counteract an imbalance between striatal dopamine and acetylcholine that results from degeneration of dopaminergic neurons (Katzenschlager et al., 2003). Medications with anticholinergic properties include amantadine, benztropine and trihexyphenidyl, among others. Amantadine may be the most commonly used anticholinergic medication in PD. In addition to its anti-cholinergic properties, amantadine promotes the release of dopamine and blocks the N-methyl-D-aspartate (NMDA) receptor, which may relieve levodopa-induced dyskinesias. Benztropine has anticholinergic properties and also blocks the reuptake of dopamine.

A meta-analysis of nine double-blind cross-over trials found that anticholinergic therapy is effective as monotherapy or as an adjunct to other anti-Parkinson drugs (Katzenschlager et al., 2003). There have been no studies comparing differences in efficacy and tolerability among anticholinergic medications. The most common side effects include dry mouth, constipation, and urinary hesitancy. These agents may occasionally exacerbate narrow angle glaucoma. Neuropsychiatric and cognitive adverse effects limit their use in the elderly.

39.3. Neuroprotective Therapy

A neuroprotective agent alters the underlying pathophysiology of a disease and therefore delays the onset or slows the progression of neurodegeneration. Much of the current research into the treatment of neurodegenerative diseases focuses on neuroprotective therapy; agents are being tested that impact processes identified at the cellular level.

Glutamate is a major excitatory neurotransmitter in the central nervous system. Theoretically, increased glutamatergic neurotransmission may lead to excitatory neurotoxicity or excitotoxicity. The only two currently approved neuroprotective agents in the United States, riluzole for ALS and memantine for AD, may act to prevent glutamatergic excitotoxicity.

39.3.1. Riluzole

Riluzole, which may block the presynaptic release of glutamate, was initially developed as an anti-epileptic medication. However, after evidence of potential excitotixicty in ALS began to accumulate, riluzole was tested as a neuroprotective agent. It was approved for use in ALS after two double-blind placebo controlled trials showed modest improvement in survival from 3 to 3.25 years (Bensimon et al., 1994; Lacomblez et al., 1996). Post hoc analyses revealed that patients who took riluzole had a slight prolongation in the time to progress from mild to severe disability; however, this effect was not apparent to patients, family or physicians. A meta-analysis of four randomized controlled trials indicated that Riluzole 100 mg per day prolongs survival by approximately 2 months (Miller et al., 2002). Nausea, fatigue, vertigo, and somnolence are the most common side effects of riluzole. Serum transaminase elevations may also occur but rarely to levels that are clinically meaningful.

In a survey of 559 ALS patients from ten centers that was conducted following the approval of riluzole in the U.S., only 49% of patients had taken riluzole even though 90% knew about it (Bryan et al., 1997). The reasons for not taking riluzole included perceived lack of benefit (45%), expense (31%), opinion of the ALS physician (23%), risk of side effects (14%) and the opinion of family and friends (10%). Thirty-four percent of patients discontinued riluzole due to lack of benefit, expense or adverse effects. Among the centers surveyed, the probability of a patient taking riluzole varied from 18–75%. Independent risk factors that significantly influenced the probability of taking riluzole included encouragement or discouragement by the ALS physician and Medicare coverage (no pharmaceutical coverage). This study points out that the personal cost to the patient and the encouragement from the physician are major factors determining who takes riluzole; the decision is influenced by whether or not riluzole is included on drug formularies (Gordon and Mitsumoto, 2006; Ringel and Woolley, 1999). In Europe, riluzole is approved for use in ALS and the health systems cover the cost. Consequently, almost all Europeans with ALS take riluzole (Bradley et al., 2004; Walley, 2004). Riluzole use may increase in the U.S. after institution of the new Medicare medication coverage plan in early 2006.

Once riluzole was shown to be effective for ALS and excitotoxicity was thought to be a component of neurodegeneration in general, studies of riluzole began in PD, HD and AD. The neuroprotective effects of riluzole have been shown in the 6-OHDA and MPTP models of PD (Barneoud et al., 1996; Benazzouz et al., 1995). In a small human trial, riluzole reduced levodopa-induced dyskinesias (Merims et al., 1999). In a separate pilot trial, riluzole extended the "on" state compared to placebo, but the difference was not statistically significant. A placebo-controlled trial in patients with HD showed that riluzole ameliorated chorea intensity (Huntington Study Group, 2003). Riluzole is now being evaluated in phase II clinical trials in AD (Hurko and Walsh, 2000) and in phase III trials in PD (Braz et al., 2004). Large trials with adequate power are still needed to determine whether riluzole is neuroprotective in disorders other than ALS.

39.3.2. Memantine

Memantine, a non-competitive NMDA antagonist, has medium affinity for the phencyclidine binding site of the NMDA receptor. Memantine may block glutamate mediated excitotoxicity, but it leaves the activation of the NMDA receptor during physiological neurotransmission unchanged (Molinuevo et al., 2004). Two randomized placebo controlled clinical trials showed positive effects in later stages of AD (Reisberg et al., 2003; Tariot et al., 2004). These studies led to the approval of memantine for the treatment of moderate and severe AD. Additional clinical data are needed to determine whether memantine can be used as monotherapy or combination therapy in early AD. Common side effects include headache, dizziness, and confusion. Initial pilot trials have already been conducted in HD and PD (Beister et al., 2004; Merello et al., 1999) and are planned in ALS.

39.3.3. Selegiline

Selegiline (Deprenyl) is a monamine oxidase (MAO) inhibitor used in PD. In doses under 10 mg per day, selegiline irreversibly binds MAO-B, preventing the enzyme from degrading dopamine, but does not inhibit MAO-A, thereby avoiding the potential hypertensive crisis, or the "cheese effect," seen with inhibitors of both MAO-A and MAO-B (Yamada and Yasuhara, 2004). The DATATOP Study (Deprenyl and Tocopherol Antioxidative Therapy of Parkinsonism) conducted in 800 patients with early untreated PD, aimed to assess the neuroprotective effects of selegeline 10 mg/day, alpha tocopherol (vitamine E) 2,000 IU/day or a combination of the two (Parkinson Study Group, 1989, 1993, 1998). After a mean follow up of 14 ± 6 months, 154 of 399 selegiline-treated subjects reached the end point, or the time to disability requiring levodopa therapy, compared to 222 of 401 subjects who did not receive selegiline. The projected median length of time to reach the end point was 9 months longer in the selegiline-treated group than in the groups treated with placebo or tocopherol. However, waning effects of selegiline after the first year coupled with slight but significant improvement in motor performance after initiation of selegiline have been used as an argument in favor of a predominantly symptomatic rather than neuroprotective effect. Data from other reports are also conflicting. Several studies suggest that it may exert some sort of protective effect, possibly by a mechanism other than MAO inhibition (Mytilineou et al., 1997, 1998; Maruyama et al., 1997). One study detected no deterioration after washout in *de novo* patients initially treated with selegiline (Palhagen et al., 1998), but most studies have concluded that while selegiline can delay the need for levodopa therapy, this does not translate into meaningful long term therapeutic benefits (Brannan and Yahr, 1995). Furthermore, selegiline does not appear to prevent the development of levodopa-induced motor fluctuations and dyskinesias (Parkinson Study Group, 1996). Further research is needed to fully define the mechanism of action of selegiline in PD, and to clarify its role as a neuroprotective agent.

39.4. Symptomatic Therapy

Even though neurodegenerative diseases still have no cure, specific symptoms may be amenable to medical treatment (Table 39.3). While few randomized clinical trials of symptomatic therapy have been performed in individual neurodegenerative

TABLE 39.3. Symptomatic therapy in neurodegenerative diseases.

Symptom	Management	Dosage
Psychiatric symptoms		
Depression	Tricyclic antidepressants	20–100 mg QD
	Mirtazapine	15–30 mg QHS
	Selective serotonin uptake inhibitors (SSRI)	20–100 mg QHS
	Venlafaxine	37.5–75 mg BID–TID
	Bupropion	100 mg TID
Anxiety	SSRI antidepressants	20–100 mg QHS
	Buspirone	10 mg TID
	Mirtazapine	15–30 mg QHS
	Benzodiazepines	0.5–5 mg BID–TID
	Trazadone	50–100 mg BID–TID
Psychosis/agitation	Neuroleptics (Haloperidol)	0.5–3 mg QD
	Atypical neuroleptics (Risperidone, Olanzepine, Quetiapine, Clozaril)	0.25–200 mg per day
	Benzodiazepines	0.5–30 mg per day
Irritability/aggression	SSRI medications	20–100 mg QD
	Divalproexsodium	250–2,000 mg per day
	Carbamazepine	100–2,600 mg per day
Pseudobulbar affect	SSRI antidepressants	20–100 mg QD
		Tricyclic antidepressants 20–100 mg QHS
		Mirtazapine 15–30 mg QHS
		Venlafaxine 37.5–75 mg BID-TID
		Dextromethorphan/quinidine 30 mg/30 mg BID
		Lithium carbonate 300 mg QD-TID
Sialorrhea	Tricyclic antidepressants	20–100 mg QHS
	Atropine sulfate	0.4 mg Q 4–6 h
	Glycopyrrolate	1–2 mg TID
	Hycosamine sulfate	0.125–0.25 mg Q 4 h
	Diphenhydramine	25–50 mg TID
	Scopolamine transdermal patch	1.5 mg applied behind ear Q 72 h
Motor symptoms		
Spasticity	Baclofen	10–60 mg TID
	Tizanidine	2–8 mg QID
	Dantrolene	25–100 mg TID
	Benzodiazepines	2–10 mg TID
Cramps	Quinine sulfate	260–325 mg QD
	Vitamin E	400 IU TID
	Phenytoin	300 mg QHS
	Diazepam	2–10 mg TID
Tremor	Dopaminergic medication	300–3,000 mg per day
	Amantadine	100–300 mg per day
	Beta blocker	10–30 mg BID-TID
	Primidone	50–250 mg QHS
	Clonazepam	0.5–2 mg QHS
Chorea	Neuroleptics	0.5–100 mg per day
	Atypical neuroleptics	0.5–160 mg per day
	Benzodiazepines	0.5–20 mg per day
	Amantadine	100–300 mg QD
Autonomic symptoms		
Orthostatic hypotension	Fludrocorisone	0.1–1 mg QD
	Midodrine	10 mg TID
Gastrointestinal symptoms		
Thick phlegm	Guaifenesin	200–400 Q 4 h
	Nebulized acetylcysteine	1 NEB Q 6–8 h
	Nebulized saline	1 NEB Q 4–6 h
	Beta blocker	10–30 mg TID
Constipation	Increase fluid intake	N/A
	Increase fiber intake	N/A
	Docusate sodium	100 mg QD-BID
	Milk of magnesia	N/A
	Dulcolax	5–15 mg QD
	Lactulose	15–30 mL QD-BID

(continued)

TABLE 39.3 (continued)

Symptom	Management	Dosage
	Magnesium citrate	120–240 mL PRN
	Polyethylene glycol	3,350 17 g QD
Gentourinary symptoms		
Urinary urgency	Oxybutynin	2.5–5 mg BID
	Tolterodine	1–2 mg BID
	Amitriptyline	25–75 mg QHS
	Oxytrol patch	3.9 mg QD
Erectile dysfunction	Sildenifil citrate	50 mg PRN
Sleep disorders		
REM sleep behavior disorder	Clonazepam	0.5–2 mg QHS
Insomnia	SSRI antidepressants	20–100 mg QD
	Tricyclic antidepressants	20–100 mg QHS
	Mirtazapine	15–30 mg QHS
	Zolpidem	5–10 mg QHS
	Eszopiclone	2–3 mg QHS
	Zaleplon	5–10 mg QHS
	Antihistamines	25–50 mg QHS
	Chloral hydrate	500–1,000 mg QHS
	Benzodiazepine	2–5 mg QHS

diseases, off label use, the prescription of an agent for a symptom or disease other than the one for which it is approved, is often helpful in improving quality of life for patients.

39.4.1. Psychiatric Symptoms

Many patients with neurodegenerative diseases exhibit psychiatric symptoms, including depression and anxiety. Tricyclic antidepressants, such as amitriptyline and nortriptyline, may relieve depression. However, the anticholinergic effects of these medications can lead to confusion and they must be used with caution in patients with cognitive compromise and in the elderly. Tricyclic antidepressants with low anticholinergic properties, such as desipramine or nortriptyline, can also be used. In PD, mirtazapine, which has noradrenergic and serotonergic effects, may relieve tremor in addition to reducing depression (Pact and Giduz, 1999; Gordon et al., 2002). While the selective serotonin reuptake inhibitor (SSRI) medications are helpful in the management of depression in all the disorders, they can exacerbate tremor, and sexual dysfunction is a common side effect.

Agitation, which may occur in patients with dementia, can be treated with anti-psychotics agents, mood stabilizing anticonvulsants, trazadone and anxiolytics (Doody et al., 2001). The atypical anti-psychotic medications are the treatment of choice for psychotic symptoms, such as hallucinations or delusions, particularly in those with Parkinsonism in whom dopamine receptor blockage is contraindicated due to the potential to worsen motor symptoms. In these patients, clozapine, which may reduce tremor in addition to its anti-psychotic effects, is particularly effective. However, rare cases of agranulocystosis necessitate weekly blood counts, and so limit its utility. Quetiapine may be the next agent of choice because it appears to have fewer adverse motor effects than the other medications in the class (Ondo et al., 2005). Irritability and aggression can be seen in HD, and less commonly in PD with dementia, AD and ALS. SSRI medications and anticonvulsants such as valproic acid and carbamazepine can be effective in treating these symptoms.

In ALS, emotional lability or pseudobulbar affect can limit social interactions and quality of life. While there are no currently FDA approved therapies for pseudobulbar affect in ALS, SSRIs, tricylic antidepressants and dopaminergic therapy can be beneficial (Gordon and Mitsumoto, 2006). One randomized controlled trial showed that a combination of dextromethorphan and quinidine reduced emotional lability and improved quality of life in ALS (Brooks et al., 2004).

39.4.2. Sialorrhea

Drooling is an embarrassing symptom that is common in both ALS and PD. Anticholinergic medications are often helpful, but the benefits can be self limited and may require higher doses after initial improvement. Side effects include constipation, fatigue, and impotence. Urinary retention, blurred vision, tachycardia, orthostatic hypotension, dizziness and confusion may also occur. Drug selection depends on the severity and frequency of drooling. Hycosamine, which has a short half-life, is useful for sialorrhea that is associated with mealtimes or a particular time of the day. Transdermal scopolamine and oral glycopyrrolate provide a more continuous effect (Gordon and Mitsumoto, 2006). Botulinum toxin injections have reduced sialorrhea in several small reports of both ALS and PD (Bushara, 1997; Pal et al., 1999; Bhatia et al., 1999), but need to be used with caution in those with dysphagia. Controlled trials are underway to determine the true benefit to risk ratio.

Thick phlegm can also be problematic in ALS. It is exacerbated by inadequate water intake, especially in patients with bulbar weakness, and by use of anti-cholingergic agents, which reduce serous secretions while sparing mucous secretions. Pharmacologic treatments that loosen phlegm include guaifenesin, nebulized acetylcysteine, nebulized saline, or beta adreergic receptor blockers such as propanolol (Gordon and Mitsumoto, 2006). Patients with a weak cough due to respiratory muscle involvement can us a cough-assist device or in-exsufflator to help clear secretions.

39.4.3. Motor Symptoms

In ALS, spasticity can be disabling. Baclofen, a gamma amino butyric acid analog, is the treatment of choice for spasticity, although the benefits are often modest. Patients may develop a sense of looseness or weakness, which can be minimized by slow dose titration. Other side effects include dizziness, fatigue and sedation. Intrathecal baclofen can be tried for those with severe or painful spasticity and inadequate response to oral treatment. Alternative medications to treat spasticity include dantrolene, tizanidine, and benzodiazepines. Dantrolene acts by blocking calcium release at the level of the sarcoplasmic reticulum and has the theoretical benefit of reducing excess neural excitation. At higher doses, liver function abnormalities may occur and require periodic monitoring. Tizanidine, an alpha 2 agonist that has proven benefit for spasticity in multiple sclerosis, has similar side effects to baclofen. Benzodiazepines can be effective for spasticity, painful spasm and cramps. However, these medications can produce sedation and respiratory depression in high doses (Gordon and Mitsumoto, 2006).

In PD, the most common motor symptoms are tremor, bradykinesia and rigidity. Beta blockers, primidone and clonazepam can be used to treat postural tremor. Muscle relaxants are occasionally used for painful rigidity. Most of the typical motor symptoms, including rest tremor, respond to disease-specific therapy, at least early in the disease course. Some levodopa-induced motor complications, including chorea, dystonia, myoclonus and akathisia (Fahn, 2000), can be reduced by prescribing lower more frequent doses of levodopa or by adding an anti-dyskinetic agent such as amantadine (Luquin et al., 1992).

Huntington disease can cause chorea, bradykinesia, rigidity, and postural instability, among other motor symptoms. The chorea is often most noticeable and usually responds to treatment with dopamine receptor blocking or dopamine depleting agents, but only at the expense of worsening bradykinesia. Atypical neuroleptics are also used to treat chorea and some behavioral symptoms. Other medications that are used to treat chorea include benzodiazepines and amantadine. Levodopa is not effective for bradykinesia due to HD because of its potential to worsen chorea and cognitive symptoms.

39.4.4. Autonomic Symptoms

Dysautonomia may produce orthostatic hypotension, sphincter abnormalities and sexual dysfunction. Orthostatic hypotension, particularly problematic in PD and Parkinson-plus syndromes such as multiple system atrophy, can be treated with salt, fludrocortisone, and midodrine (Jankovic et al., 1993). Fludrocortisone is a corticosteroid, but the exact mechanism of its effect on blood pressure is not fully understood. Side effects are similar to other glucorticoids and mineralcorticoids. Midodrine stimulates alpha-one adrenergic receptors. It can cause significant hypertension in the supine position. Other common side effects include parasthesias, piloerection, dysuria, and pruritis.

39.4.5. Gastrointestinal Symptoms

Constipation is a common symptom in the neurodegenerative disorders. Inadequate fluid intake, dysphagia, dysautonomia, and immobility all contribute to constipation. Increasing fluid and fiber intake, along with increasing physical activity can help. Use of docusate sodium, milk of magnesia, dulcolax and senna may also improve symptoms. For severe constipation, lactulose, magnesium citrate and enemas can provide relief.

39.4.6. Genitourinary Symptoms

In neurodegenerative disorders, patients often experience urinary urgency, which is thought to be caused by spasm of the urinary sphincter or detrusor muscle. Oxybutynin (Ditropan) and tolterodine (Detrol) are effective for urinary urgency. For ALS patients and some PD patients, swallowing tablets may be difficult due to dysphagia. Extended release oxybutynin allows for infrequent daily dosing and oxytrol patches avoids the oral route of administration.

Sildenafil citrate (Viagra) has been found to be safe and effective for managing erectile dysfunction in PD, but can unmask orthostatic hypotension in patients with multiple system atrophy (Zesiewicz et al., 2000; Hussain et al., 2001). Sildenafil is also effective for erectile dysfunction in other neurodegenerative disorders, but it should be used cautiously in those with cardiovascular disease.

39.4.7. Sleep Disorders

Sleeplessness in ALS has numerous causes. Respiratory insufficiency, difficulty repositioning in bed, anxiety and depression can all contribute to poor sleep. Treatment of depression with sedating antidepressants such as mirtazapine, tricyclic antidepressants, or trazadone can help promote sleep. Zolpidem, a non benzodiazepine sleep aid, is effective and carries a low risk of respiratory depression. Other medications that can be helpful include anithistamines, chloral hydrate and selective use of benzodiazepines (Gordon and Mitsumoto, 2006). Non-invasive positive pressure ventilation can help relieve orthopnea in those with respiratory muscle weakness, and special equipment, such as a hospital bed, can reduce nighttime discomfort.

The sleep dysfunction in PD may also be multifactorial, but can also be due to an underlying sleep disorder. Parkinson disease is commonly associated with rapid eye movement (REM) sleep disorder (Comella et al., 1998), which can be an initial manifestation of the disease. REM sleep disorder, best diagnosed with an overnight sleep study, can be treated with a nighttime benzodiazepine such as clonazepam.

In AD, non-pharmacological approaches to treating insomnia are undertaken first. Techniques include sleep restriction and keeping patients awake during the day as well as providing a cool comfortable quiet sleep environment at night. If pharmacologic management is necessary, sedating antidepressants or zolpidem may be helpful for promoting sleep in AD (Corey-Bloom and Galaska, 1995). In those with nighttime confusion or sundowning, a low dose of a sedating atypical antipsychotic medication such as quetiapine can be helpful. Anticholinergic hypnotics should be avoided.

39.5. Current and Future Directions for Neuroprotection

39.5.1. Anti-Inflammatory Agents

39.5.1.1. Nonsteroidal Anti-Inflammatory Drugs (NSAIDS)

Because inflammatory cells surround dying neurons in all neurodegenerative disorders, a recent approach to neuroprotection has been the use of anti-inflammatory agents. Nonsteroidal anti-inflammatory drugs (NSAIDs) reduce inflammation in part by inhibiting one or both of the cyclooxygenase (COX) enzyme isoforms. Inhibition of COX reduces prostaglandin synthesis leading to a general decrease in inflammation. While NSAIDS have demonstrated an ability to provide neuroprotection in animal models, human trials have been less rewarding.

Both celecoxib and rofecoxib, selective COX-2 inhibitors, had positive effects in the SOD1 mouse model of ALS (Drachman et al., 2002; Azari et al., 2005). A subsequent randomized controlled trial of celecoxib was negative, however. A statement released by the investigators in 2004 reported that celecoxib did not impart a beneficial effect in 300 ALS patients. There were no differences in adverse effects between treated and control groups and the drug was well tolerated at a dosage of 800 mg/day (www.alsa.org). This trial had a high dropout rate, which may have impaired the ability of the investigators to detect a small impact of the drug if one existed.

Celecoxib and rofecoxib have also been studied in AD. Randomized double blind, placebo controlled trials failed to demonstrate a therapeutic benefit (Sainetti et al., 2000; Aisen et al., 2003). The rofecoxib trial used naproxen as a control; the results were consistent with other studies in which nonselective NSAIDs such as diclofenac, have been ineffective in AD (Scharf et al., 1999). Other NSAIDs including ibuprofen, indomethacin and sulindac sulfide have demonstrated potential efficacy in AD (Rogers et al., 1993; t'Veld et al., 2001), but definitive trials have not yet been conducted.

39.5.2. Immunomodulation

39.5.2.1. Vaccination in ALS

Another approach is to harness the body's own immune system to regulate inflammation. In models of neurodegeneration caused by mechanical or biochemical injury, more neurons survive in the presence of a well-regulated evoked anti-self T cell mediated response (Moalem et al., 1999). Vaccination can be used to induce nonpathogenic T-cell responses, including the activation of anti-inflammatory Th2 cells, which may migrate to lesion sites where they affect innate immune responses through the release of anti-inflammatory cytokines or by inducing the production of neurotrophic factors. Glatiramer acetate (Copaxone, Cop-1), a mix of amino acids in a known molar ratio, was originally designed to mimic myelin basic protein and is now FDA approved for the treatment of multiple sclerosis (MS). Cop-1 can induce a protective T cell mediated response without risk of causing an autoimmune disease. A study of acute and chronic motor neuron disease in animal models demonstrated that twice the number of motor neurons survived in Cop-1 vaccinated mice than control mice (Angelov et al., 2003). This study also showed that Cop-1 vaccination prolonged life span in SOD1 transgenic mice, although pathological findings were not reported. An initial trial in ALS patients revealed the modality to be safe in this disease (Gordon et al., 2006). ALS patients tolerated two different dosing frequencies in conjunction with riluzole, and patients mounted immune responses similar to those seen in MS patients. Larger trials of efficacy are planned.

39.5.2.2. Immunization in AD

In AD, locally generated complement proteins are activated by extracellular amyloid deposits. Microglia aggregate at the site, attacking and destroying neuronal extensions. Research has focused on recruiting the immune system to remove beta amyloid protein deposits. Exogenous antibodies to beta amyloid protein selectively bind to senile plaques and cerebral amyloid *in vivo*, and the amyloid deposition can be ameliorated in beta amyloid precursor protein–transgenic mice by active immunization with the beta amyloid peptide (Walker et al., 1994). Other studies have confirmed that diverse anti-beta amyloid immunization schemes reduce cerebral beta amyloid burden and behavior deficits in AD transgenic mice (Dodel et al., 2003; Lemere et al., 2004).

Beta amyloid immunization was initiated in humans after the successful studies in rodents. Phase I studies were completed without major adverse events and Phase II trials were initiated with AN 1792 (active immunization with amyloid beta 42 and QS-21 adjuvant). The trials were interrupted in 2002 because approximately 6% of those vaccinated developed aseptic meningitis (Orgogozo et al., 2003). Data from the incomplete trial suggest that AD patients who mounted a significant antibody response also showed signs of clinical benefit (Orgogozo et al., 2003; Hock et al., 2003). Autopsy findings from three patients who died approximately 1 year after the first immunization showed evidence of beta amyloid clearance (Nicoll et al., 2003; Ferrer et al., 2004; Masliah et al., 2005). However, two of the three brains also showed changes of encephalitis with infiltration of T lymphocytes, white matter lesions invaded by macrophages, and in one case, severe small vessel disease with multiple cortical hemorrhages (Nicoll et al., 2003; Ferrer et al., 2004). The potential for beta amyloid immunization to be developed further as a disease modifying treatment in AD depends on whether the serious side effects can be effectively controlled. Further studies are needed, especially in aged nonhuman primates, to clearly understand the adverse effects and how to allow for safe application of this therapy to humans (Orgogozo et al., 2003).

39.5.2.3. Immunotherapy in Prion Disease

Antibody based immunotherapy has also been evaluated in the animal models of prion disease. *In vitro* assays have demonstrated that antibodies to the cellular prion protein (PrPc) antagonize the deposition of disease associated prion protein (PrPsc) (Sigurdsson et al., 2002). However, the induction of protective antiprion immune responses has been difficult in wild type animals because of tolerance to endogenous PrPc. Some studies have shown that it might be possible to overcome this tolerance by inducing immune responses to bacterially expressed recombinant PrP (Heppner et al., 2001). However, developing antibodies that are capable of recognizing native cell surface PrPc is more difficult (Heppner and Aguzzi, 2004). One study demonstrated that anti-PrP antibodies cross-linked with PrPc when directly injected into the brains of laboratory animals and subsequently provoked neurotoxicity (Solforosi et al., 2004). However, passive immunization and antibody transgenesis experiments have been successful in animals (White et al., 2003). Further studies are required to determine the mechanism of tolerance needed to safely and successfully implement vaccination as an antiprion regimen.

39.5.2.4. Immunotherapy in Parkinson's Disease

In the MPTP animal model of PD, glatiramer acetate(Cop-1) has also been investigated. In the MPTP model, Cop-1-immune cells provided significantly more neuroprotection than placebo (Benner et al., 2004). Mice received adoptive transfers of splenocytes from Cop-1 or ovalbumin immunized mice because MPTP-immunotoxicity precluded active immunization due to lymphotoxicity (Benner et al., 2004). In animals given splenocytes from Cop-1 immunized mice, T cells entered and accumulated in inflamed areas of the CNS. The T cells secreted interleukin-10, an inhibitory Th2 cytokine, which suppressed microglial activation, and also stimulated local expression of astrocyte-associated glial cell line derived neurotrophic factor. This immunization strategy minimized dopamine loss compared to control animals and resulted in significant protection of nigrostriatal neurons. Human trials are planned in PD, but investigations are needed to define the best routes of administration and dosing frequencies.

Immunophilin binding proteins are being developed as a potential immune based therapy in PD. Immunophilins, intracellular receptor proteins that bind to the immunosuppressive drugs cyclosporine A, FK506, and rapamycin, are 10–50-fold more abundant in the brain than in the immune system (Guo et al., 2001). The immunophilin ligands combine with immunophilin proteins to suppress the immune system by inhibiting the calcium activated phosphate calcineurin, and some promote nerve growth *in vitro* and *in vivo*. GPI 1046 (Guilford Pharmaceutical) and AMG-474-00 or NIL-A (Amgen) are immunophilin ligands that promote neuronal growth without demonstrating immunosuppressive effects (Steiner et al., 1997; Gold et al., 1998). These agents stimulate growth of nigrostriatal dopaminergic neurons spared after MPTP-induced damage to the substantia nigra. Human trials are planned to determine whether these or other immunophilin ligands will be useful in slowing or reversing PD.

39.5.3. Anti-apoptotic Therapies

Neurodegeneration is likely due in part to caspase enzyme driven apoptosis and drugs that inhibit the activation of caspase enzymes are being studied as a potential neuroprotective strategy.

39.5.3.1. Minocycline

Minocycline is a tetracycline antibiotic with anti-inflammatory and anti-apoptotic properties. It has been shown to delay disease onset and prolong survival in transgenic ALS mice (Zhu et al., 2002; Kriz et al., 2002). Two early phase randomized controlled trials indicated that minocycline could be used safely in high doses and in conjunction with riluzole (Gordon et al., 2004). but minocycline caused a worsening of function in a recent phase III trial (Gordon et al., 2007). A separate efficacy trial is just underway in Europe, so that cumulative data may determine the true effect of minocycline in ALS.

Markers of apoptosis are also present in PD (Mogi et al., 2000; Tatton, 2000). Injection of minocycline into the 6-OHDA mouse model of parkinsonism inhibits microglial activation, protects dopaminergic neurons, and reduces markers of apoptosis (He et al., 2001). Minocycline has also been shown to prevent nigrostriatal dopaminergic neurodegeneration in MPTP mice models of PD (Du et al., 2001). Similarly, minocycline delays disease progression in animal models of HD, presumably by inhibiting caspase-1 and caspase-3 mRNA up regulation and decreasing inducible nitric oxide synthase (iNOS) activity (Chen et al., 2000; Tikka et al., 2001). Phase II NINDS-funded clinical trials testing minocycline in PD are underway. The unfortunate untoward effects of a well-known anti-inflammatory agent minocycline, an antibiotic in the tetracycline family, in a randomized phase III clinical trial (Gordon et al., 2007) is cause for pause. Minocycline treatment of ALS patients not only demonstrated no beneficial effect, for some patients it significantly worsened measurable neurologic outcomes. Those who were taking the drug to ALS were asked to disontinue its use. Regrettably, the drug did not affect survival or quality of life measures for people with ALS and raised caution in extrapolating animal model data for human use (Lancet Neurol. 2007 Dec;6(12):1045–53).

39.5.3.2. TCH 346 (Omigapil)

TCH 346, which is structurally related to selegiline, binds to glyceraldehyde 3 phosphate dehydrogenase (GAPDH) in a mechanism that may prevent programmed cell death. It slows the degeneration of motor neurons in *in vitro* models of apoptosis (Carlile et al., 2000), but did not have significant effect on survival in transgenic mice. Phase IIa and IIb trials in ALS showed that the drug did not slow the rate of progression or prolong survival at any of the doses tested (www.als.net/treatments/clintrials/clinics_detailed.asp).

In PD, a nuclear translocation of GAPDH in melanized nigral neurons has been interpreted as a marker for apoptosis (Dastoor and Dreyer, 2001). After MPTP treatment in animal models, GAPDH is up regulated and undergoes nuclear translocation. TCH 346 has been shown to bind to GAPDH and rescue neurons from cell death in this model (Andringa and Cools, 2000). TCH346 also attenuates MPTP-induced toxicity in primates (Andringa et al., 2003). However, further studies are needed to determine whether GAPDH upregulation and nuclear translocation, and whether the effects of TCH346 are meaningful changes in PD. Clinical trials have been initiated.

39.5.3.3. Mitogen Activated Protein (MAP) Kinase Inhibitors

The cascade of apoptosis may be promoted by mitogen activated protein (MAP) kinase activation. CEP1347, a small molecule MAP kinase inhibitor that crosses the blood brain barrier is protective in MPTP-treated mice (Saporito et al., 1999). A clinical trial showed good tolerability, no interference with L-dopa pharmacokinetics, and no symptomatic effects of treatment over a 4-week period (Parkinson Study Group, 2004). A larger phase II trial has commenced.

39.5.4. Antioxidants

Another proposed mechanism of neuronal death is oxidative stress due to the accumulation of free radicals. Normal and abnormal cellular metabolic processes involving molecular oxygen produce free radicals, which can cause oxidative damage to lipids, proteins and nucleic acids (Andersen, 2004).

39.5.4.1. Vitamin E

Alpha tocopherol (vitamin E), a naturally occurring antioxidant, delays disease onset in SOD1 transgenic mice (Gurney et al., 1996). One epidemiological study suggested that regular intake of vitamin E may reduce the risk of contracting ALS (Ascherio et al., 2005). A randomized, placebo controlled trial of vitamin E 500 mg twice daily showed no significant difference in 12-month survival in patients with ALS (Desnuelle et al., 2001). This trial was probably underpowered due to insufficient sample size, and used an unvalidated outcome measure, the Norris Scale. Vitamin E has also been studied in combination with other antioxidants. L-methionine, vitamin E, and selenium were studied in ALS in a double blind placebo controlled trial of the combinations taken three times a day (Stevic et al., 2001). After 12 months there was 50% survival in the control group and 81% in the treatment group, findings that were not statistically significant (Orrell et al., 2005). Vitamin E was ineffective in slowing progression of PD in the DATATOP trial (Parkinson Study Group, 1989).

39.5.4.2. Selegiline

Selegiline may have anti-oxidant and anti-apoptotic properties in addition to inhibiting MAO, and has been reported to increase SOD activity in the basal ganglia of rats (Knoll, 1989). There have been several trials of selegiline 10 mg per day in patients with ALS: a randomized, placebo controlled, double blind trial (Lange et al., 1998), and a placebo-controlled crossover trial (Mitchell et al., 1995). Neither showed improvement in functional or subjective rating scales, but both trials were underpowered due to insufficient sample size and so may be considered inconclusive.

39.5.4.3. N-acetylcysteine

N-acetylcysteine, a precursor of glutathione, which is a natural intracellular antioxidant, has been tested in a randomized, placebo controlled trial in patients with ALS. Thirty-five patients (65%) given N-acetylcysteine 50 mg/kg subcutaneous infusion daily and 30 patients (54%) given placebo were still alive at 12 months, a difference that was not statistically significant (Louwerse et al., 1995).

39.5.4.4. Coenzyme Q10 (CoQ10)

CoQ10 is an essential cofactor of the mitochondrial electron transport chain, and may act as an antioxidant. It has been shown to have neuroprotective effects in models of ALS, PD and HD (Beal, 2002). An open label escalation trial in ALS was completed in 2004; dosages up to 3,000 mg per day were well tolerated (Ferrante et al., 2005). An NINDS-funded randomized controlled phase II trial of CoQ10 is currently underway in patients with ALS.

A phase II trial has also been completed in PD (Shults et al., 2002). Over a period of 16 months, CoQ10 was safe and well tolerated at doses up to 1,200 mg/day. Trends toward less disability occurred in treated patients and the benefits were greater in those receiving the highest dosage. However, larger studies, including an examination of the potential symptomatic effects of CoQ10, over a longer period of time are needed to determine the efficacy of CoQ10 as a neuroprotective agent in PD.

39.5.4.5. AEOL 10150

AEOL 10150 (manganese (III) meso-tetrakis (di-N-ethylimidazole) porphyrin) is a compound that can catalytically decompose biological oxidants such as perioxynitrite via its ability to cycle between MN (III) and MN (IV) states. Several investigators have reported the presence of nitrotyrosine in affected tissue in motor neuron disease; however, no cause and effect relation between protein nitration and neuronal death has been established (Crow et al., 2005). AEOL 10150 prolonged survival in the SOD1 ALS model. A phase I dose ranging clinical trial reported a high degree of safety in this population. Multi-dose studies are underway.

39.5.5. Trophic Factors

39.5.5.1. Insulin-Like Growth Factor

Nerve growth factors promote neuronal sprouting *in vitro*, and reduce nerve cell death *in vivo*. Viral delivery of insulin like growth factor (IGF-1) was shown to prolong survival and delay the onset of disease in the ALS mouse model (Kaspar et al., 2003). Two randomized controlled trials of IGF-1 have been completed in ALS. One showed slowed progression of functional impairment using an unvalidated outcome measure (Lai et al., 1997). The second trial showed no significant change using the same outcome (Borasio et al., 1998). A third trial, using strength testing as the primary outcome, has just completed enrollment. Human trials of viral vector delivery are planned but the safety of the modality must be shown before the FDA will allow trials to proceed.

In PD, animal studies are underway with IGF-1. N-terminal tripeptide of IGF-1 was peripherally administered to 6-OHDA mouse models of PD and results indicated that administration after the onset of nigrostriatal dopamine depletion improved long term parkinsonian motor deficits; however IGF-1 administration did not prevent the loss of tyrosine hydroxylase in the substantia nigra pars compacta or the striatum (Krishnamurthi et al., 2004). Human trials of IGF have not yet been conducted in PD.

39.5.5.2. Vascular Endothelial Growth Factor

Vascular endothelial growth factor (VEGF) is needed for angiogenesis and has been implicated in neurodegeneration (Jin et al., 2000). Researchers have shown that low levels of VEGF in transgenic mice can cause an ALS-like syndrome. Intramuscular delivery of VEGF delays onset and prolongs survival in SOD1 mice (Storkebaum et al., 2005).

VEGF has also been shown to promote neuroprotection in the 6-OHDA mouse model of PD by activating the proliferation of glia and by promoting angiogenesis (Yasuhara et al., 2004). Human trials will likely be initiated in the future, once optimal dosing and route of administration have been established.

39.5.5.3. Brain Derived Growth Factor

Brain derived neurotrophic factor (BDNF) has been shown to prolong survival, protect against glutamate neurotoxicity, and slow disease progression in the wobbler mouse, an early model of ALS. Unfortunately, BDNF did not show benefit in patients with ALS (Anonymous, 1999).

BDNF was also evaluated in the rat model of PD, and was compared to glial cell line derived neurotrophic factor (GDNF); results of the study showed that GDNF was more effective than BDNF in the model, but that there were no significant benefits from either of these neurotrophic factors (Sun et al., 2005).

39.5.5.4. Ciliary Neurotrophic Factor

Ciliary neurotrophic factor (CNTF) was also shown to protect wobbler mice, to be required for preservation of adult motor neurons in knockout mice, and to reduce immunoreactivity in the spinal cord of ALS patients. This drug advanced as far as Phase III clinical trials, however, no significant difference in survival was observed between CNTF and placebo groups (ALS CNTF Treatment Study Group, 1996; Hurko and Walsh, 2000).

CNTF has also been investigated in mouse models of PD. Although CNTF had potent neurotrophic effects for injured adult rat dopaminergic substantia nigra neurons, it did not prevent the disappearance of the transmitter synthesizing enzyme tyrosine hydroxylase (Hagg and Varon, 1993). Human trials were not conducted.

39.5.5.5. Glial Derived Neurotrophic Factor

Glial Derived Neurotrophic Factor (GDNF) has been a particularly promising approach to neuroprotection in ALS and PD (Lapchak, 1998; Gash et al., 1998). GDNF enhances survival of the midbrain dopaminergic neurons *in vitro* and rescues degenerating neurons *in vivo* (Tseng et al., 1997). Intraventricular administration of GDNF in monkeys ameliorated parkinsonism and produced 20% enlargement of nigral neurons accompanied by increased fiber density (Gash et al., 1996). A GDNF-levodopa combination reportedly reduced levodopa side effects in experimental monkeys (Miyoshi et al., 1997). A pilot trial in humans with moderately advanced PD was abandoned, however, because of lack of efficacy and the frequent occurrence of nausea, anorexia, tingling, hallucinations and depression (Nutt et al., 2003). Autopsy of one case showed no evidence of significant trophic effects on nigrostriatal neurons (Kordower et al., 1999). Trials in ALS were similarly abandoned due to high levels of toxicity.

Laboratory studies are now using a lentoviral vector to deliver GDNF to the striatum of monkeys with MPTP-induced Parkinsonism. One study demonstrated reversal of functional deficits, extensive and long term GDNF expression, augmentation of dopaminergic function, and prevention of nigrostriatal degeneration (Kordower et al., 2000). This new delivery system provides a hopeful therapeutic strategy for PD, and has also been beneficial in the SOD1 model of ALS.

39.5.6. Antiglutamatergic Agents

39.5.6.1. Gabapentin

Gabapentin, an antiepileptic agent, may reduce glutamate release and thereby, reduce excitotoxicity. Gabapentin showed a small but significant effect on survival in SOD1 transgenic mice (Gurney et al., 1996). An early phase clinical trial in ALS showed a non-significant slower rate of decline in arm strength due to gabapentin (Miller et al., 1996). A phase III study using a higher dose revealed no change in strength and a more rapid rate of decline of the forced vital capacity (Miller et al., 2001), exemplifying the need to better define dose in early phase trials before preceding to definitive trials.

39.5.6.2. Topiramate

Topiramate reduces glutamate release from neurons and antagonizes activation of the alpha-amino-3-hydroxy-5-methylisoxazole-4-propionic acid (AMPA) receptor, a glutamatergic excitatory amino acid receptor. A double-blind, placebo-controlled trial conducted in 296 ALS patients over 12 months showed a faster rate of decline in arm strength and no effect on survival from topiramate (Cudkowicz et al., 2003). This trial had a large dropout rate and proceeded to a phase III trial without first exploring the effects of topiramate in the SOD model.

Topiramate was recently studied in the marmoset model of PD (Silverdale et al., 2005). Overactive AMPA receptor-mediated transmission may be involved in the pathogenesis of levodopa-induced dyskinesias. Topiramate significantly reduced levodopa-induced dyskinesias without affecting the antiparkinsonian action of levodopa. Further studies are needed to better elucidate the potential benefits of topiramate in PD.

39.5.6.3. Lamotrigine

Lamotrigine is another glutamate release inhibitor that has been studied in neurodegenerative diseases. In ALS, 300 mg of lamotrigine per day had no effect when compared to placebo in a small number of patients (Ryberg et al., 2003). Lamotrigine also failed to show symptomatic benefit in patients with PD (Shinotoh et al., 1997), but neither study had adequate power to truly assess efficacy.

39.5.6.4. Ceftriaxone

Ceftriaxone was selected from the recent NIH-supported high throughput screening initiative in ALS. Approximately 1,050 FDA approved drugs, for which the safety and toxicity are already known, were screened using *in vitro* assay systems (cell survival, cytochrome C release, protein aggregates, etc.). Positive results or "hits" were then tested in the transgenic model. Ceftriaxone was shown to increase glutamate transporter activity and prolong motor neuron survival. Ceftriaxone is a third generation cephalosporin that also protects against SOD toxicity, radiation induced neurodegeneration in *in vitro* models (Tikka et al., 2001), and has the ability to cross the blood brain barrier. Clinical trials funded by the NINDS are expected to begin in 2006, with the goal of determining the pharmacokinetics, safety and efficacy of long-term ceftriaxone treatment in ALS.

39.5.6.5. Talampanel

Talampanel, an AMPA antagonist with antiglutametergic activity, is also being evaluated in neurodegeneration. Talampanel has been shown to be neuroprotective after traumatic brain injury in rats (Belayev et al., 2001), and AMPA receptor antagonists have been shown to prolong survival in the transgenic mouse model of ALS (Van Damme et al., 2005). Phase II trials of talampanel are underway in both PD and ALS.

39.5.6.6. Remacemide

Remacemide, an anticonvulsant and NMDA antagonist, is being studied in PD. It has been shown to enhance the effects of levodopa in rats and monkeys (Greenamyre et al., 1994). A randomized, controlled trial of remacemide in 279 patients with PD and motor fluctuations showed trends towards improvement of "on" time, but no evidence of neuroprotection (Parkinson Study Group, 2001). In a small HD trial, remacemide tended to alleviate chorea but failed to slow functional decline (Huntington Study Group, 2001).

39.5.7. Stem Cell Transplant Therapy

Advances in stem cell therapy have received a great deal of press in recent years. Unfortunately, when public expectations have been raised, patients are likely to search for non-orthodox sources of treatment. There are tens of thousands of internet pages extolling the promise of stem cells, and various forms of stem cell therapy are available on a commercial basis in a number of countries, including China, South America, and Eastern Europe. However, most administer the stem cells in an uncontrolled way and do not have long term follow-up so that rational scientific conclusions cannot be reached. The science provides hope for potential therapeutic interventions in neurodegenerative diseases, but it is in its early stages and much is still needed to be learned about how to control stem cell proliferation, differentiation into specific cells and optimal functional recovery in animal models before human trials. The collection and use of human fetal tissue may also raise ethical concerns.

39.5.7.1. Stem Cell Therapy and Parkinson's Disease

The loss of specific dopaminergic neurons in PD makes the prospect of replacing missing or damaged cells a potential therapy. Recent double blind placebo controlled trials of primary human embryonic dopaminergic tissue used functional neuroimaging and neuropathological investigations to demonstrate integration of transplanted dopaminergic neurons in host striatum (Freed et al., 2001; Olanow et al., 2003). However, only subpopulations of PD patients displayed significant benefits, and some patients developed uncontrollable side effects due to excess dopamine production.

Retinal pigmental epithelia (RPE) are dopaminergic support cells in the neural retina. RPE cells on gelatin beads, also called Spheramine, produce levodopa (Watts et al., 2003), but there are no data yet on the consistency of dopamine production by these cells. An open label clinical trial of transplantation of Spheramine was conducted in six patients with PD (Bakay et al., 2004). Spheramine was transplanted into the striatum and showed clinical effects over 24 months. Spheramine is now undergoing double blind placebo controlled trials in advanced PD.

39.5.7.2. Stem Cell Therapy and Huntington's Disease

In animal models of HD, transplanted striatal cells have survived, grown and established afferent and efferent connections (Freeman et al., 2000). One open label human trial indicated that grafts may have restored function (Hauser et al., 2002). Recipients showed cognitive and motor improvements that were associated with reductions in striatal and cortical hypometabolism. The success of grafting may be sensitive to the age of the donor and to the degree of neuronal loss in the patient. Clinical trials are in early phases and there still needs to be resolution of a number of technical issues (Gardian and Vecsei, 2004).

39.5.7.3. Stem Cell Therapy and Amyotrophic Lateral Sclerosis

Stem cell therapy poses a greater challenge in ALS than in other neurodegenerative diseases because of the length of motor axons and growing evidence that neurodegeneration in ALS may be mediated by neuronal and glial influences. Late stage fetal cortical neurons have been shown to replace apoptotic neurons when grafted into adult mouse neocortex, to receive afferents from host brain, and to project to the contralateral hemisphere (Fricker-Gates et al., 2002). Fetal motor neurons grafted to the adult rat spinal cord lacking motor neurons migrate to the ventral horn and make functional connections with skeletal muscle (Clowry et al., 1991). Whether these neurons are integrated into neuronal circuits is unclear. It is also unknown if similar neuronal replacement could work in the brains of individuals with ALS. Clinical trials have shown the general safety of cerebrospinal fluid injection in small numbers of patients, but there was no means to monitor cell growth, or even to determine if the cells survived (Mazzini et al., 2003). There is still a great deal of research to be done with stem cell therapy before it can be effectively implemented in ALS.

39.5.8. Others

39.5.8.1. Amyotrophic Lateral Sclerosis

39.5.8.1.1. Calcium Channel Blockers

Calcium channel blockers antagonize excitatory amino acid receptor activation. A randomized, double blind, placebo controlled, crossover study of nimodipine was conducted in patients with ALS (Miller et al., 1996b). There was no significant difference in the rate of decline of pulmonary function or limb strength with treatment compared to placebo and the authors concluded that nimodipine was ineffective in slowing the progress of ALS.

The efficacy of verapamil, a calcium channel blocker, was studied in a clinical trial in which the treatment of phase was compared to a natural history phase (Miller et al., 1996c). During months 1 to 3, patients were not given drug and the natural history of their progression was measured. Following the month 3 visit, verapamil was started at 40 mg and titrated to a daily dose of 240 mg per day. The decline in forced vital capacity and limb strength were not significantly different between the treatment phase and natural history phase.

39.5.8.1.2. Creatine

A key pathological feature in the animal model of ALS is abnormality of the mitochondria (Wong et al., 1995). Creatine, which acts as a mitochondrial energy buffer and may also have anti-oxidant properties, provided a dose dependent improvement in survival in the transgenic SOD1-G93A mouse (Klivenyi et al., 1999). A subsequent randomized placebo-controlled trial using 5 g per day showed no significant benefit in survival (Groenveld et al., 2003). This trial broke new ground in several ways. First, the investigators used a sequential analysis plan, so that the trial was stopped once the null hypothesis could no longer be rejected, saving 18 months over a fixed duration trial. Second, the dropout rate was nearly zero, because the investigators made home visits when necessary to obtain data from patients no longer able to travel to the centers. A second double blind placebo controlled trial of 5 g per days also did not show a benefit (Shefner et al., 2004), but this trial likely had inadequate power due to small numbers of patients enrolled. Neither trial may have been dosed adequately.

39.5.8.1.3. Sodium Phenylbutyrate

Transcriptional dysregulation may contribute to the pathogenesis of neurodegeneration (Gonzalez de Aguilar et al., 2000). Sodium phenylbutyrate, a histone deacetylase inhibitor (HDAC inhibitor), regulates transcription and has provided improvement in survival, body weight and motor performance in the SOD mouse model (Ryu et al., 2005). The safety and tolerability of sodium phenylbutyrate is currently being evaluated in a trial funded by the Veterans Administration and the Muscular Dystrophy Association.

39.5.8.1.4. Arimoclomol

Motor neurons may have a high threshold to activation of the heat shock protein pathway, which is involved in protein repair. Arimoclomol, an inducer of heat shock proteins, delays disease progression and improves survival in the murine model (Kieran et al., 2004). Early phase trials with arimoclomol in patients with ALS are currently underway.

39.5.8.1.5. Ribonucleic Acid Interference

Mutant gene expression leading to neurodegeneration might be reduced by gene silencing techniques. RNA interference (RNAi) molecules have been generated that targeted the SOD1 gene and reduce its expression. RNAi has prolonged survival in the SOD1 model (Ralph et al., 2005; Raoul et al., 2005). The technology will only apply to the 2–5% of patients whose ALS is due to mutations in the gene encoding SOD1, and so obtaining adequate numbers of patients to conduct clinical trials may be difficult.

39.5.8.1.6. Combination Therapy

Because of the complexity of the mechanisms underlying neurodegeneration, testing combinations of agents that target different processes could theoretically be more beneficial than individual agents alone. The combinations of minocycline and creatine, and celecoxib and creatine showed additive effects in the SOD1 model (Zhang et al., 2003; Klivenyi et al., 2004). Phase II clinical trials using a novel selection trial design to identify the better performing combination have begun (Gordon et al., 2005).

39.5.8.2. Alzheimer's Disease

39.5.8.2.1. Statins

Statins, used predominantly in the treatment of hypercholesterolemia, act by inhibiting 3-hydroxy-3-methylglutaryl coenzyme A (HMG-CoA) reductase, which regulates the synthesis of cholesterol. Statins are also agonists of peroxisome proliferator activated receptors (PPARs), which are part of the nuclear receptor superfamily and when activated, can suppress transcription of pro-inflammatory genes (Chinetti et al., 2000). *In vitro* and *in vivo* studies have shown that a decrease in serum cholesterol inhibits the production of beta amyloid and plaque (Simons et al., 1998; Fassbender et al., 2001). A randomized, controlled trial evaluating the efficacy of simvastatin on disease progression in patients with AD provided conflicting results. Overall, the plasma levels of beta amyloid protein remained relatively unchanged after 6 months of treatment, but those with better cognitive function demonstrated a greater reduction in beta amyloid protein levels than patients with more severe dementia (Simons et al., 2002). The mechanism of how lowering cholesterol affects beta amyloid production is still unclear; studies have demonstrated that statins do not alter beta amyloid morphology (Hoglund et al., 2004). It may be that the neuroprotective properties of statins are due to antioxidant and anti-inflammatory effects.

39.5.8.2.2. Secretase Inhibitors

Beta amyloid peptide is cleaved from the beta amyloid precursor protein by the beta secretase (or beta amyloid cleaving enzyme, BACE) and gamma secretase enzymes (Cummings, 2004). Experimental studies in animal models have shown that secretase inhibition lowers the amount of beta amyloid protein in the brain and can reduce the formation of beta amyloid protein aggregates (Dewachter and Van Leuven, 2002). While BACE and gamma secretase are viable therapeutic targets, they have thus far been refractory to drug development. Gamma secretase inhibitors diminish beta amyloid protein load in the brain, but also cleave other proteins essential for signaling in cellular proliferation. Gamma secretase is essential for embryogenesis and gamma secretase knockout mice are not viable (Dewachter and Van Leuven, 2002). Until these limitations can be overcome, the potential for gamma secretase inhibitors for AD therapy is restricted.

BACE is another therapeutic target. One study showed that raising BACE expression increased plaque load in beta amyloid precursor protein-transgenic mice (Willem et al., 2004). However, the development of pharmacological inhibitors of BACE has been challenging due to its large catalytic site, which does not bind small molecules effectively.

39.5.8.3. Huntington's Disease

39.5.8.3.1. Sodium Phenylbutyrate

Impaired gene transcription is thought to be a key event in the cascade that leads to neurodegeneration in HD. Mutant huntingtin can bind to histone acetyltransferase, reducing histone acetylation and ultimately suppressing gene transcription. HDAC inhibitors act to relax the DNA conformation and facilitate transcription by keeping DNA acetylated. HDAC inhibitors may also be effective in preventing sequestration of certain transcription factors by mutant huntingtin during the process of aggregation (Ferrante et al., 2003). Sodium phenylbutyrate was tested in the R6/2 mouse model of HD and was shown to significantly increase survival, improve motor function and minimize brain atrophy (Gardian et al., 2005). Sodium phenylbutyrate is currently being tested in human trials.

39.5.8.3.2. Ethyl-EPA

Ethyl-EPA (ethyl-eicopentaenoate) is a semisynthetic, highly purified derivative of the n-3-fatty acid EPA. Ethyl-EPA acts to preserve mitochondrial function by targeting peroxisome proliferator activated receptors. Ethyl-EPA reduces loss of neuronal function by inhibiting caspase activation and apoptosis, and by reducing mitochondrial damage in models of neurodegeneration (Van Raamsdonk et al., 2005). A randomized double blind placebo controlled trial of ethyl-EPA was conducted in patients with HD. This study showed that there was no benefit in the intent to treat cohort of patients with HD. However, exploratory analysis revealed that a significantly higher number of patients in the cohort treated with ethyl-EPA showed stable or improved motor function (Puri et al., 2005). Further studies of ethyl-EPA efficacy are warranted.

39.5.8.4. Parkinson's Disease

39.5.8.4.1. Creatine

Administration of MPTP produces parkinsonism in experimental animals by a mechanism involving impaired energy production. MPTP is converted to 1-methyl-4-phenylperydinium (MPP+), which blocks complex I of the electron transport chain. Supplementation with creatine, a substrate for creatine kinase, may increase phosphocreatine and cyclophosphocreatine levels, buffer against ATP depletion and exert neuroprotective effects. Oral supplementation of creatine produced significant protection against MPTP-induced dopamine depletion in mice (Matthews et al., 1999). Based on these results, a randomized, double-blind, futility clinical trial of creatine was recently conducted (NINDS NET-PD Investigators, 2006). The study found that creatine and separately, minocycline, could not be rejected as futile and that they should be considered for Phase III clinical trials to determine whether they alter the long-term progression of Parkinson disease.

Summary

Until the 1990s, progress in developing therapy for neurodegenerative diseases was slow and there were few clinical trials. However, with the advent of animal models and the recent advances in understanding the basic pathophysiological mechanisms underlying the diseases, potential therapies to prevent, delay the onset or slow the progression of neurodegenerative disease are being identified at an ever increasing rate. High throughput technologies are being used to screen large numbers of potential therapeutic agents and new developments in the realms of immune modulation, RNA interference, viral vector delivery of gene products and stem cell therapy hold great promise for the future. There are still many unanswered questions regarding the mechanisms of disease and why beneficial therapy in animal models has not translated well into human clinical trials. Neurodegenerative diseases are rare, and so identifying enough patients to obtain studies with adequate power has been difficult. Novel phase II designs are now being used to screen greater numbers of agents and to better define correct dosing before proceeding to phase III trials. Multicenter phase III trials are being designed with adequate power, and using meaningful validated outcome measures that reduce the high dropout rates of past trials. Agents are now being tested in combination in order to detect possible additive effects. Focus is also being given to the need to better define the best symptomatic therapies in randomized controlled trials. Further research will prompt more targeted therapies that we hope will soon provide truly meaningful breakthroughs for neurodegenerative disorders.

Review Questions/Problems

1. **Which one of these medications for Alzheimer's disease is NOT a cholinesterase inhibitor?**

   a. galantamine
   b. rivastigmine
   c. tacrine
   d. memantine

2. **What are the side effects that limit the prolonged usefulness of levodopa therapy in Parkinson disease?**

   a. muscle fasciculation
   b. motor fluctuations and dyskinesias
   c. postural instability
   d. bradykinesia

3. **Major side effects of levodopa therapy include which of the following?**

   a. nausea, vomiting, anorexia and orthostatic hypotension
   b. abnormal liver function tests
   c. bradykinesia and tremor
   d. dry eyes and dry mouth

4. **Which class of medications is associated with compulsive behavior?**

   a. COMT (catechol-O-methyltransferase) inhibitors
   b. memantine
   c. dopamine agonists
   d. MAO (monoamine oxidase) inhibitor

5. **Amantadine is a useful medication in the treatment of dyskinesias in Parkinson Disease. What side effect limits its use in the elderly?**

   a. diarrhea
   b. sialorrhea
   c. dyskinesias
   d. neuropsychiatric and cognitive effects

6. **Which is the only FDA approved medication for amyotrophic lateral sclerosis?**

   a. memantine
   b. lamotrigine
   c. riluzole
   d. coenzyme Q10

7. **What are the major factors that influence patients in their decision to take riluzole in the United States?**

   a. cost of the medication and opinion of the physician
   b. medication interaction and dosing regimen
   c. side effects
   d. size of the tablet

8. Memantine and riluzole exert their neuroprotective effects by what proposed mechanism?

a. inhibiting superoxide dismutase
b. inhibiting glutamatergic excitotoxicity
c. inhibiting caspase activity
d. inhibiting mitochondrial injury

9. Memantine is approved for use in which disease?

a. moderate to severe Huntington disease
b. early stages of Huntington disease
c. moderate to severe Alzheimer disease
d. early stages of Alzheimer disease

10. In a patient with Alzheimer disease, what is the ideal class of medication for the treatment of depression?

a. tricyclics
b. selective serotonin reuptake inhibitors
c. anticholinergics
d. neuroleptics

11. Hallucinations in Parkinson disease would ideally be treated with which of the following medications?

a. haloperidol
b. thorazine
c. nortriptyline
d. clozaril

12. Which medication is currently FDA approved for the treatment of pseudobulbar affect in ALS?

a. dextromethorphan/quinidine
b. SSRI antidepressants
c. tricyclic antidepressants
d. none of the above

13. In ALS patients, sialorrhea can be a major symptom as bulbar weakness worsens. Which of these medications is ideal for a patient with swallowing difficulty?

a. amitriptyline
b. scopolamine
c. diphenyhydramine
d. glycopyrrolate

14. Which one of these medications are used to treat chorea in Huntington disease?

a. risperidone
b. levodopa
c. fluoxetine
d. bupropion

15. Sildenafil citrate is effective in treating erectile dysfunction in several neurodegenerative diseases. What side effect can this medication cause in Parkinson disease?

a. bradykinesia
b. tremor
c. orthostatic hypotension
d. chorea

16. Proposed mechanisms in neurodegenerative diseases include all of the following EXCEPT:

a. oxidative stress
b. apoptosis
c. inflammation
d. excess GABA

17. Which of the FDA approved therapy for Multiple Sclerosis is currently being studied in ALS?

a. alpha-interferon
b. beta-interferon
c. glatiramer acetate
d. corticosteroids

18. Phase II clinical trials of beta amyloid immunization in Alzheimer disease were interrupted because:

a. patients vaccinated with the immunization developed aseptic meningitis
b. immunization supplies were not sufficient
c. the trial was not randomized and blinded
d. patients developed reaction at the site they received the immunization

19. Which of these medications investigated in neurodegenerative diseases is/are no longer recommended for treatment?

a. remacemide
b. minocycline
c. insulin growth factor
d. ceftriaxone

20. Which of the following antioxidants have been shown in randomized, placebo controlled trials to be beneficial in any of the neurodegenerative diseases is no longer recommended for treatment?

a. vitamin E
b. n-acetylcycteine
c. selegeline
d. none of the above

References

Aisen P, Schafer K, Grundman M, Pfeiffer E, Sano M, Davis KL, Farlow MR, Jin S, Thomas RG, Thal LJ, Alzheimer's Disease Cooperative Study (2003) Effects of rofecoxib or naproxen vs. placebo on Alzheimer's disease progression: A randomized controlled trial. JAMA 289:2819–2826.

ALS CNTF Treatment Study Group (1996) A double blind placebo controlled clinical trial of subcutaneous recombinant human ciliary neurotrophic factor (rHCNTF) in amyotrophic lateral sclerosis. Neurology 46(5):1244–1249.

Albers DS, Beal MF (2000) Mitochondrial dysfunction and oxidative stress in aging and neurodegenerative disease. J Neural Transm Suppl 59:133–154.

Andersen J (2004) Oxidative stress in neurodegeneration: Cause or consequence? Nat Rev Neurosci 5:S18–S25.

Andringa G, Cools AR (2000) The neuroprotective effects of CGP 3466B in the best in vivo model of Parkinson's disease, the bilaterally MPTP treated rhesus monkey. J Neural Transm 60 Suppl: 215–255.

Andringa G, Eshuis S, Perentes E, Maguire RP, Roth D, Ibrahim M, Leenders KL, Cools AR (2003) TCH346 prevents motor symptoms and loss of striatal FDOPA uptake bilaterally in MPTP treated primates. Neurobiol Dis 14:205–217.

Angelov DN, Waibel S, Guntinas-Lichius O, Lenzen M, Neiss WF, Tomov TL, Yoles E, Kipnis J, Schori H, Reuter A, Ludolph A, Schwartz M (2003) Therapeutic vaccine for acute and chronic motor neuron diseases: Implications for amyotrophic lateral sclerosis. PNAS 100(8):4790–4795.

Anonymous (1999) A controlled trial of recombinant methionyl human BDNF in ALS: The BDNF Study Group (Phase III). Neurology 47:1383–1388.

Ascherio A, Weisskopf MG, O'Reilly EJ, Jacobs EJ, McCullough ML, Calle EE, Cudkowicz M, Thun MJ (2005) Vitamin E intake and risk of amyotrophic lateral sclerosis. Ann Neurol 57(1):104–110.

Assal F, Spahr L, Hadengue A, Rubbia-Brandt L, Burkhard PR (1998) Tolcapone and fulminant hepatitis. Lancet 352:958.

Azari MF, Profyris C, Le Grande MR, Lopes EC, Hirst J, Petratos S, Cheema SS (2005) Effects of intraperitoneal injection of rofecoxib in a mouse model of ALS. Eur J Neurol 12:357–364.

Bakay RA, Raiser CD, Stover NP, Subramanian T, Cornfeldt ML, Schweikert AW, Allen RC, Watts R. Implantation of Spheramine in advanced Parkinson's disease (PD). Front Biosci. 2004;9:592–602.

Barneoud P, Mazadier M, Miquet JM, Parmentier S, Dubedat P, Doble A, Boireau A (1996) Neuroprotective effects of riluzole on a model of Parkinson's disease in the rat. Neuroscience 74:971–983.

Beal MF (2002) Coenzyme Q10 as a possible treatment for neurodegenerative diseases. Free Radic Res 36(4):455–460.

Beister A, Kraus P, Kuhn W, Dose M, Weindl A, Gerlach M (2004) The N-methyl-D-aspartate antagonist memantine retards progression of Huntington's disease. J Neural Transm Suppl 68:117–122.

Belayev L, Alonso O, Liu Y, Chappell AS, Zhao W, Ginsberg MD, Busto R (2001) Talampanel, a novel noncompetitive AMPA agonist, is neuroprotective after traumatic brain injury in rats. J Neurotrauma 18(10):1031–1038.

Benazzouz A, Boraud T, Dubedat P, Boireau A, Stutzmann JM, Gross C (1995) Riluzole prevents MPTP-induced parkinsonism in the rhesus monkey: A pilot study. Eur J Pharmacol 284:299–307.

Benner EJ, Mosley RL, Destache CJ, Lewis TB, Jackson-Lewis V, Gorantla S, Nemachek C, Green SR, Przedborski S, Gendelman HE (2004) Therapeutic immunization protects dopaminergic neurons in mouse models of Parkinson's disease. Proc Natl Acad Sci USA 101(25):9435–9440.

Bennett JP, Piercey MF (1999) Pramipexole-new dopamine agonist for the treatment of Parkinson's disease. J Neurol Sci 163:25–32.

Bensimon G, Lacomblez L, Meninger V (1994) ALS/Riluzole Study Group. N Engl J Med 330:585–591.

Bhatia KP, Munchau A, Brown P (1999) Botulinum toxin is a useful treatment in excessive drooling of saliva. J Neurol Nuerosurg Psych 67:697.

Borasio GD, Robberecht W, Leigh PN, Emile J, Guiloff RJ, Jerusalem F, Silani V, Vos PE, Wokke JH, Dobbins T (1998) A placebo controlled trial of insulin like growth factor-1 in amyotrophic lateral sclerosis. European ALS/IGF-1 Study Group. Neurology 51:583–586.

Bossey-Wetzel E, Schwarzenberger R, Lipton SA (2004) Molecular pathways to neurodegeneration. Nat Med 10:S2–S9.

Bradley WG, Anderson F, Gowda N, ALS CARE Study Group (2004) Changes in the management of ALS since the publication of the AAN ALS practice parameter 1999. Amyotroph Lateral Scler Other Motor Neuron Disord 63:1364–1370.

Brannan T, Yahr MD (1995) Comparative study of selegiline plus L-dopa-carbidopa versus L-dopa-carbidopa alone in the treatment of Parkinson's disease. Ann Neurol 37:95–98.

Braz CA, Borges V, Ferraz HB (2004) Effect of riluzole on dyskinesias and duration of the on state in Parkinson disease patients: A double blind, placebo-controlled pilot study. Neuropharmacology 27(1):25–29.

Brooks DJ (2000) Dopamine agonists; their role in treatment of Parkinson's Disease. J Neurol Neurosurg Psychiatry 68:685–690.

Brooks BR, Thisted RA, Appel SH, Bradley WG, Olney RK, Berg JE, Pope LE, Smith RA; AVP-923 ALS Study Group (2004) Treatment of pseudobulbar affect in ALS with dextromethorphan and quinidine. Neurology 63:1363–1373.

Bryan WW, McIntire D, Camperlengo L et al. (1997). Factors influencing the use of riluzole by ALS patients. 8th International Symposium on ALS/MND. November (abstract presentation).

Bushara KO (1997) Sialorrhea in amyotrophic lateral sclerosis: A hypothesis of a new treatment—botulinum toxin A injections of the parotid glands. Med Hypotheses 48:337–339.

Carlile GW, Chalmers-Redman RM, Tatton NA, Pong A, Borden KE, Tatton WG (2000) Reduced apoptosis after nerve growth factor and serum withdrawal: Conversion of tetrameric glyceraldehyde 3 phosphate dehydrogenase to a dimer. Mol Pharmacol 57:2–12.

Carlsson A, Lindqvist M, Magnusson T (1957) 3,4-dihydrophenylalanine and 5-hydroxytryptophan as reserpine antagonists. Nature 180:1200.

Carlsson A, Lindqvist M, Magnusson T, Waldeck B (1958) On the presence of 3-hydroxytyramine in brain. Science 127:471.

Cedarbaum JM, Kutt H, Dhar AK, Watkins S, McDowell FH (1986) Effect of supplemental carbidopa on bioavailability of L-dopa. Clin Neuropharmacol 9:153–159.

Chen M, Ona VO, Li M, Ferrante RJ, Fink KB, Zhu S, Bian J, Guo L, Farrell LA, Hersch SM, Hobbs W, Vonsattel JP, Cha JH, Friedlander RM (2000) Minocycline inhibits caspase-1 and caspase-3 expression and delays mortality in a transgenic mouse model of Huntington's disease. Nat Med 6:797–801.

Chinetti G, Fruchart JC, Staels B (2000) Peroxisome proliferator activated receptors (PPARs): Nuclear receptors at the crossroads between lipid metabolism and inflammation. Inflamm Res 49:497–505.

Clowry G, Sieradzan K, Vrbova G (1991) Transplants of embryonic motoneurones to adult spinal cord: Survival and innervation abilities. Trends Neurosci 14:355–357.

Comella CL, Nardine TM, Diederich NJ, Stebbins GT (1998) Sleep related violence, injury and REM sleep behavior disorder in Parkinson's disease. Neurology 51:526–529.

Corey-Bloom J, Galaska D (1995) Adjunctive therapy in patients with Alzheimer's disease: A practical approach. Drugs Aging 7(2):79–87.

Corrigan MH, Denehan AQ, Wright CE, Ragual RJ, Evans DL (2000) Comparison of pramipexole, fluoxetine and placebo in patients with major depression. Depress Anxiety 11:58–65.

Crow JP, Calingasan NY, Chen J, Hill JL, Beal MF (2005) Manganese porphyrin given at symptom onset markedly extends survival of ALS mice. Ann Neurol 58:258–265.

Cudkowicz ME, Shefner JM, Schoenfeld DA, Brown Jr RH, Johnson H, Qureshi M, Jacobs M, Rothstein JD, Appel SH, Pascuzzi RM, Heiman-Patterson TD, Donofrio PD, David WS, Russell JA, Tandan R, Pioro EP, Felice KJ, Rosenfeld J, Mandler RN, Sachs GM, Bradley WG, Raynor EM, Baquis GD, Belsh JM, Novella S, Goldstein J, Hulihan J; Northeast ALS Consortium 1 (2003) A randomized placebo controlled trial of topiramate in amyotrophic lateral sclerosis. Neurology 61:456–464.

Cummings JL (2004) Alzheimer's disease. N Engl J Med 351(1):56–67.

Dastoor Z, Dreyer JL (2001) Potential role of nuclear translocation of glyceraldehyde-3-phosphate dehydrogenase in apoptosis and oxidative stress. J Cell Sci 114 Pt 9:1643–1653.

Dawson TM (2000) New animal models for Parkinson's disease. Cell 101(2):115–118.

Deng HX, Siddique T (2000) Transgenic mouse models and human neurodegenerative disorders. Arch Neurol 57:1695–1702.

Desnuelle C, Dib M, Garrel C, Favier A (2001) ALS Riluzole-Tocopherol Study Group. A double blind, placebo controlled randomized clinical trial of alpha tocopherol in the treatment of amyotrophic lateral sclerosis. Amyotroph Lateral Scler Other Motor Neuron Disord 2(1):9–18.

Dewachter I, Van Leuven F (2002) Secretases as targets for the treatment of Alzheimer's disease: The prospects. Lancet Neurol 1(7):409–416.

Doody RS, Stevens JC, Beck C, Dubinsky RM, Kaye JA, Gwyther L, Mohs RC, Thal LJ, Whitehouse PJ, DeKosky ST, Cummings JL (2001) Practice parameter: Management of dementia (an evidence based review). Report of the Quality Standards Subcommittee of the American Academy of Neurology. Neurology 56:1154–1166.

Dodel RC, Hampel H, Du Y (2003) Immunotherapy for Alzheimer's disease. Lancet Neurol 2:215–220.

Drachman DB, Frank K, Dykes-Hoberg M, Teismann P, Almer G, Przedborski S, Rothstein JD (2002) Cyclooxygenase 2 inhibitor protects motor neurons and prolongs survival in transgenic mouse model of ALS. Ann Neurol 52(6):771–778.

Driver-Dunckley E, Samanta J, Stacy M (2003) Pathological gambling associated with dopamine agonist therapy. Neurology 61(3):422–423.

Du Y, Ma Z, Lin S, Dodel RC, Gao F, Bales KR, Triarhou LC, Chernet E, Perry KW, Nelson DL, Luecke S, Phebus LA, Bymaster FP, Paul SM (2001) Minocycline prevents nigrostriatal dopaminergic neurodegeneration in the MPTP model of Parkinson's disease. Proc Natl Acad Sci USA 98(25):14669–14674.

Fahn S (2000) The spectrum of levodopa induced dyskinesias. Ann Neurol 47 Suppl 1:S2–S9.

Fahn S, the Parkinson Study Group (2005) Does levodopa slow or hasten the rate of progression of Parkinson's disease? J Neurol 252 Suppl 4:iv37–iv42.

Fassbender K, Simons M, Bergman C, Stroick M, Lutjohan D, Keller P, Ronz H, Kuhl S, Bertsch T, von Bergman K, Hennerici M, Beyreutner K, Hartman T (2001) Simvastatin strongly reduces Alzheimer's disease beta amyloid 42 and bet amyloid 40 levels in vitro and in vivo. Proc Natl Acad Sci USA 98:5856–5861.

Ferrante RJ, Kubilus JK, Lee J, Ryyu H, Beesen P, Zucker B, Smith K, Kowall NW, Ratan RR, Luthi-Carter R, Hersch SM (2003) Histone deacetylase inhibition by sodium butyrate chemotherapy ameliorates the neurodegenerative phenotype in Huntington's disease mice. J Neurosci 23:9418–9427.

Ferrante KL, Shefner JM, Zhang H, Betensky R, Obrien M, Yu H, Fantasia M, Taft J, Beal MF, Traynor B, Newhall K, Donofrio P, Caress J, Ashburn G, Freberg B, Onell C, Paladenech C, Walker T, Pestronk A, Abrams B, Florence J, Renna R, Schrerbecker J, Malkus B, Cudkowicz M (2005) Tolerance of high dose (3000 mg/day) coenzyme Q 10 in ALS. Neurology 65(11):1834–1836.

Ferrer I, Boada Rovira M, Sanchez Guerra ML, Rey MJ, Costa-Jussa F (2004) Neuropathology and pathogenesis of encephalitis following amyloid beta immunization in Alzheimer's disease. Brain Pathol 14(1):11–20.

Freed CR, Greene PE, Breeze RE, Tsai WY, DuMouchel W, Kao R, Dillon S, Winfield H, Culver S, Trojanowski JQ, Eidelberg D, Fahn S (2001) Transplantation of embryonic dopamine neurons for severe Parkinson's disease. N Engl J Med 344(10):710–719.

Freeman TB, Cicchetti F, Hauser RA, Deacon TW, Li XJ, Hersch SM, Nauert GM, Sanberg PR, Kordower JH, Saporta S, Isacson O (2000) Transplanted fetal striatum in Huntington's disease in Huntington's disease: Phenotypic development and lack of pathology. Proc Natl Acad Sci USA 97:13877–13882.

Fricker-Gates RA, Shin JJ, Tai CC, Catapeno LA, Macklis JD (2002) Late stage immature neocortical neurons reconstruct interhemispheric connections and form synaptic contacts with increased efficiency in adult mouse cortex undergoing targeted neurodegeneration. J Neurosci 22:4045–4056.

Gardian G, Vecsei L (2004) Huntington's disease: Pathomechanism and therapeutic perspectives. J Neural Transm 111:1485–1494.

Gardian G, Browne SE, Choi DK, Klivenyi P, Gregorio J, Kubilus JK, Ryu H, Langley B, Ratan RR, Ferrante RJ, Beal MF (2005) Neuroprotective effects of phenylbutyrate in the N171–82Q transgenic mouse model of Huntington's disease. J Biol Chem 280:556–563.

Gash DM, Zhang Z, Ovadia A, Cass WA, Yi A, Simmerman L, Russell D, Martin D, Lapchak PA, Collins F, Hoffer BJ, Gerhardt GA (1996) functional recovery in parkinsonian monkeys treated with GDNF. Nature 380:252–255.

Gash DM, Zhang Z, Gerhardt G (1998) Neuroprotective and neurorestorative properties of GDNF. Ann Neurol 44 Suppl 1:S121–S125.

German DC, Eisch AJ (2004) Mouse models of Alzheimer's Disease: Insight into treatment. Rev Neurosci 15(5):353–369.

Gold BG, Zeleny-Pooley M, Chaturverdi P, Wang MS (1998) Oral administration of a nonimmunosuppresant FKBP-12 ligand speeds nerve regeneration. Neuroreport 9:553–558.

Gonzalez de Aguilar JL, Gordon J, Rene F, de Tapia M, Lutz-Buchner B, Gaiddon C, Loeffler JP (2000) Alteration of the bcl-x/bax ratio in a transgenic mouse model of amyotrophic lateral sclerosis: Evidence for the implication of the p53 signaling pathway. Neurobiol Dis 7:406–415.

Gordon PH, Moore DH, Miller RG, Florence JM, Verheijde JL, Doorish C, Hilton JF, Spitalny GM, Macarthur RB, Mitsumoto H, Neville HE, Boylan K, Mozaffar T, Belsh JM, Ravits J, Bedlack RS, Graves MC, McCluskey LF, Barohn RJ, Tandan R; for the Western ALS Study Group (2007) Efficacy of minocycline in patients with amyotrophic lateral sclerosis: a phase III randomised trial. Lancet Neurol 6(12):1045–53.

Gordon PH, Mitsumoto H (2006) Symptomatic therapy and palliative aspects of clinical care in ALS. In: Handbook of Clinical Neurology: Motor Neuron and Related Disease (Eisen A and Shaw P, eds). Amsterdam: Elsevier Science Publishers, in press.

Gordon PH, Pullman SL, Louis ED, Frucht SJ, Fahn S (2002) Mirtazapine in Parkinsonian tremor. Parkinsonism Relat Disord 9(2):125–126.

Gordon PH, Moore DH, Gelinas DF, Qualls C, Meister ME, Werner J, Mendoza M, Mass J, Kushner G, Miller RG (2004) Placebo controlled phase I/II studies of minocycline in amyotrophic lateral sclerosis. Neurology 62:1845–1847.

Gordon PH, Cheung YK, Mitsumoto H, Levin B (2005) Novel phase II design for clinical trials of ALS using selection paradigm and group sequential analysis. Amyotroph Lateral Scler 6:14–15.

Gordon PH, Doorish C, Montes J, Mosley RL, Siamond B, MacArthur RB, Wiemer LH, Kaufmann P, Hays AP, Rowland LP, Gendelman HE, Przedborski S, Mitsumoto H (2006) Randomized controlled phase II trial of glatiramer acetate in amyotrophic lateral sclerosis. Neurology, in press.

Greenamyre JT, Eller RV, Zhang Z, Ovadia A, Kurlan R, Gash DM (1994) Antiparkinsonian effects of remacemide hydrochloride, a glutamate antagonist, in rodent and primate models of Parkinson's disease. Ann Neurol 35:655–661.

Greenblatt HM, Kryger G, Lewis T, Silman I, Sussman JL (1999) Structure of acetylcholinesterase complexed with(-)-galantamine at 2.3 A resolution. FEBS Lett 463:321–326.

Groenveld GJ, Veldink JH, Van der Tweel I, Kalmijn S, Beijer C, de Visser M, Wokke JH, Franssen H, van den Berg LH (2003) A randomized sequential trial of creatine in amyotrophic lateral sclerosis. Ann Neurol 53:437–445.

Guo X, Dillman JF, Dawson VL, Dawson TM (2001) Neuroimmunophilins: Novel neuroprotective and neurodegenerative targets. Ann Neurol 50:6–16.

Gurney ME, Pu H, Chiu AY, Dal Canto MC, Polchow CY, Alexander DD, Caliendo J, Hentati A, Kwon YW, Deng HX, Chen W, Zhai P, Sufit R, Siddique T (1994) Motor neuron degeneration in mice that express a human Cu Zn super oxide dismutase mutation. Science 17:1772–1775.

Gurney ME, Cutting FB, Zhai P, Doble A, Taylor CP, Andrus PK, Hall ED (1996) Benefit of vitamin E, riluzole and gabapentin in transgenic models of familial amyotrophic lateral sclerosis. Ann Neurol 39:147–152.

Gurney ME, Fleck T, Himes CS, Hall ED (1998) Riluzole preserves motor function in a transgenic model of familial amyotrophic lateral sclerosis. Neurology 50(1):62–-66.

Hagg T, Varon S (1993) Ciliary neurotrophic factor prevents degeneration of adult rat substantia nigra dopaminergic neurons in vivo. Proc Natl Acad Sci USA 90(13):6315–6319.

Hauser RA, Furtado S, Cimino CR, Delgado H, Eichler S, Schwartz S, Scott D, Nauert GM, Soety E, Sossi V, Holt DA, Sanberg PR, Stoessl AJ, Freeman TB (2002) Bilateral human fetal striatal transplantation in Huntington's disease. Neurology 58:687–695.

He Y, Appel S, Le W (2001) Minocycline inhibits microglial activation and protects nigral cells after 6-hydroxydopamine injection into mouse striatum. Brain Res 909 (1–2):187–193.

Heppner FL, Aguzzi A (2004) Recent developments in prion immunotherapy. Curr Opin Immunother 16(5):594–598.

Heppner FL, Musahl C, Arrighi I, Klein MA, Rulicke T, Oesch B, Zinkernagel RM, Kalinke U, Aguzzi A (2001) Prevention of scrapie pathogenesis by transgenic expression of anti-prion protein. Science 294(5540):178–182.

Hock C, Konietzko U, Streffer JR, Tracy J, Signorell A, Muller-Tillmanns B, Lemke U, Henke K, Moritz E, Garcia E, Wollmer MA, Umbricht D, de Quervain DJ, Hofmann M, Maddalena A, Papassotiropoulos A, Nitsch RM (2003) Antibodies against beta amyloid slow cognitive decline in Alzheimer's disease. Neuron 38(4):547–554.

Hoglund K, Wiklund O, Vanderstichele H, Eikenberg O, Vanmechelen E, Blennow K (2004) Plasma levels of beta amyloid 40 and beta amyloid 42 and total beta amyloid remain unaffected in adult patients with hypercholesterolemia after treatment with statins. Arch Neurol 61:333–337.

Holm KJ, Spencer CM (1999) Entacapone: A review fits us in Parkinson's disease. Drugs 58:159–177.

Huntington Study Group (2001) A randomized placebo controlled trial of coenzyme Q10 and remacemide in Huntington's disease. Neurology 57:397–404.

Huntington Study Group (2003) Dosage effects of riluzole in Huntington's disease: A multicenter placebo controlled study. Neurology 61(11):1551–1556.

Hurko O, Walsh FS. Novel drug development for amyotrophic lateral sclerosis. J Neurol Sci. 2000;180:21–8.

Hussain IF, Brady CM, Swinn MJ, Mathias CJ, Fowler CJ (2001) Treatment of erectile dysfunction with sildenafil citrate (Viagra) in parkinsonism due to Parkinson's disease or multiple system atrophy with observations on orthostatic hypotension. J Neurol Neurosurg Psychiatry 71:371–374.

Jankovic J (2000) Complications and limitations of drug therapy for movement disorders. Neurology 55 Suppl 6:52–56.

Jankovic J, Gilden JL, Hiner BC, Kaufmann H, Brown DC, Coghlan CH, Rubin M, Fouad-Tarazi FM (1993) Neurogenic orthostatic hypotension: A double blind placebo controlled study with midodrine. Am J Med 95:38–48.

Jenner P (1997) Is stimulation of D1 and D2 receptors important for optimal functioning in Parkinson's disease? Eur J Neurol 4 Suppl 3:S3–S11.

Jin KL, Mao XO, Greenberg DA (2000) Vascular endothelial growth factor: Direct neuroprotective effect in vitro ischemia. Proc Natl Acad Sci USA 97:10242–10247.

Kaspar BK, Llado J, Sherkat N, Rothstein JD, Gage FH (2003) Retrograde viral delivery of IGF-1 prolongs survival in a mouse ALS model. Science 301:839–842.

Katzenschlager R, Sampaio C, Costa J (2003) Anticholinergics for symptomatic management of Parkinson's disease. The Cochrane Database of Systematic Reviews, The Cochran Library, Copyright 2005, The Cochrane Collaboration Volume (4).

Kieran D, Kalmar B, Dick JR, Riddoch-Contreras J, Burnstock G, Greensmith L (2004) Treatment with arimoclomol, a coinducer of heat shock proteins, delays disease progression in ALS mice. Nat Med 10:402–405.

Knoll J (1989) The pharmacology of selegiline ((-) deprenyl): New aspects. Acta Neurol Scand Suppl 126:83–91.

Klivenyi P, Ferrante RJ, Matthews RT, Bogdanov MB, Klein AM, Andreassen OA, Mueller G, Wermer M, Kaddurah-Daouk R, Beal MF (1999) Neuroprotective effects of creatine in a transgenic mouse model of ALS. Nat Med 5:347–350.

Klivenyi P, Kiaei M, Gardian G, Calingasan NY, Beal MF (2004) Additive neuroprotective effects of creatine and cyclooxygenase2 inhibitors in a transgenic mouse model of amyotrophic lateral sclerosis. J Neurochem 88:576–582.

Koller WC, Hutton JT, Tolosa E, Capilldeo R (1999) Carbidopa/Levodopa Study Group. Immediate release and controlled release carbidopa/levodopa in PD: A 5 ear randomized study. Neurology 53:1012–1019.

Koo EH, Kopan R (2004) Potential role of presenilin-regulated signaling pathway in sporadic neurodegeneration. Nat Med 10: S26–S33.

Korczyn AD, Brunt ER, Larsen JP, Nagy Z, Poewe WH, Ruggieri S (1999) A 3 year randomized trial of ropinirole and bromocriptine in early Parkinson's disease. Neurology 53:364–370.

Kordower JH, Palfi S, Chen EY, Ma SY, Sendera T, Cochran EJ, Cochran EJ, Mufson EJ, Penn R, Goetz CG, Comella CD (1999) Clinicopathological findings following intraventricular glial derived neurotrophic factor treatment in a patient with Parkinson's disease. Ann Neurol 46:419–424.

Kordower JH, Emborg ME, Bloch J, Ma SY, Chu Y, Leventhal L, McBride J, Chen EY, Palfi S, Roitberg BZ, Brown WD, Holden JE, Pyzalski R, Taylor MD, Carvey P, Ling Z, Trono D, Hantraye P, Deglon N, Aebischer P (2000) Neurodegeneration prevented by lentoviral vector delivery of GDNF in primate models of Parkinson's disease. Science 290:767–773.

Krishnamurthi R, Scott S, Maingay M, Faull RLM, McCarthy D, Gluckman O, Guan J (2004) N-terminal tripeptide of IGF-1 improves functional deficits after 6-OHDA lesion in rats. Neuroreport 15(10):1601–1604.

Kriz J, Nguen MD, Julien JP (2002) Minocycline slows disease progression in a mouse model of amyotrophic lateral sclerosis. Neurobiol Dis 10:268–278.

Kurth MC, Adler CH (1998) COMT inhibition: A new treatment strategy for Parkinson's disease. Neurology 50 Suppl 5:S3–S14.

Lacomblez L, Bensimon G, Leigh PN, Guillet P, Meininger V (1996) Dose ranging study of riluzole in amyotrophic lateral sclerosis. Amyotrophic Lateral Sclerosis/Riluzole Study Group II. Lancet 347:1425–1431.

Lai EC, Felice KJ, Festoff BW, Gawel MJ, Gelinas DF, Kratz R, Murphy MF, Natter HM, Norris FH, Rudnicki SA (1997) Effect of recombinant human insulin like growth factor-1 on progression of ALS. A placebo controlled study. The North America ALS/IGF-1 Study Group. Neurology 49(6):1621–1630.

Lanctot KL, Herrmann N, Yau KK, Khan LR, Liu BA, LouLou MM, Einarson TR (2003) Efficacy and safety of cholinesterase inhibitors in Alzheimer's disease: A meta-analysis. CMAJ 169:557–564.

Lange DJ, Murphy PL, Diamond B, Appel V, Lai EC, Younger DS, Appel SH (1998) Selegiline is ineffective in a collaborative double-blind, placebo-controlled trial for treatment of amyotrophic lateral sclerosis. Arch Neurol 55(1):93–96.

Lapchak PA (1998) A preclinical development strategy designed to optimize the use of glial cell line derived neurotrophic factor in the treatment of Parkinson's disease. Mov Disord 13 Suppl 1:49–54.

Lemere CA, Beierschmitt A, Iglesias M, Spooner ET, Bloom JK, Leverone JF, Zheng JB, Seabrook TJ, Louard D, Li D, Selkoe DJ, Palmour RM, Ervin FR (2004) Alzheimer's disease a beta vaccine reduces central nervous system a beta levels in a non human primate, the Caribbean vervet. Am J Pathol 165(1):283–297.

LeWitt PA, Ward CD, Larsen TA, Raphaelson MI, Newman RP, Foster N, Dambrosia JM, Calne DB (1983) Comparison of pergolide and bromocriptine therapy in parkinsonism. Neurology 33: 1009–1014.

Louwerse ES, Weverling GJ, Bossuyt PM, Meyjes FE, de Jong JM (1995) Randomized double blind controlled trial of acetylcysteine in amyotrphic lateral sclerosis. Arch Neurol 52(6):559–564.

Luquin MR, Scipioni O, Vaamonde J, Gershanik O, Obeso JA (1992) Levodopa induced dyskinesias in Parkinson's disease: Clinical and pharmacological classification. Mov Disord 7:117–134.

Moalem G, Leibowitz-Amit R, Yoles E, Mor F, Cohen IR, Schwartz M (1999) Autoimmune T cells protect neurons from secondary degeneration after central nervous system axotomy. Nat Med 5(1):49–55.

Maruyama W, Naoi M, Kasamatsu T, Hashizume Y, Takahashi T, Kohda K, Dostert P (1997) An endogenous dopaminergic neurotoxin, N-methyl-(R)-salsolinol, induces DNA damage in human dopaminergic neuroblastoma SH-SY5Y cells. J Neurochem 69:322–329.

Masliah E, Hansen L, Adame A, Crews L, Bard F, Lee C, Seubert P, Games D, Kirby L, Schenk D (2005) Abeta vaccination effects on plaque pathology in the absence of encephalitis in Alzheimer disease. Neurology 64(1):129–131.

Matthews RT, Ferrante RJ, Klivenyi P, Yang L, Klein AM, Mueller G, Kaddurah-Douk R, Beal F (1999) Creatine and cyclocreatine attenuate MPTP neurotoxicity. Exp Neurology 157(1):142–149.

Mazzini L, Fagioli F, Boccaletti R, Mareschi K, Oliveri G, Olivieri C, Pastore I, Marasso R, Madon E (2003) Stem cell therapy in amyotrophic lateral sclerosis: A methodologic approach in humans. Amyotroph Lateral Scler Other Motor Neuron Disord 4(3):133–134.

Merello M, Nouzeilles MI, Cammarota A, Leiguarda R (1999) Effect of memantine (NMDA antagonist) on Parkinson's disease: A double blind crossover randomized study. Clin Neuropharmacol 22(5):273–276.

Merims D, Ziv I, Djaldetti R, Melamed E (1999) Riluzole for levodopa induced dyskinesia in advanced Parkinson's disease. Lancet 353:1764–1765.

Miller RG, Moore DH, Young LA, WALS Study Group (1996a) Placebo controlled trial of gabapentin in patients with amyotrophic lateral sclerosis. Neurology 47:1383–1388.

Miller RG, Shepherd R, Dao H, Khramstov A, Mendoza M, Graves J, Smith S (1996b) Controlled trial of nimodipine in amyotrophic lateral sclerosis. Neuromuscul Disord 6(2):101–104.

Miller RG, Smith SA, Murphy JR, Graves J, Mendoza M, Sands ML, Ringel SP (1996c) A clinical trial of verapamil in amyotrophic lateral sclerosis. Muscle Nerve 19:511–515.

Miller RG, Moore DH, Gelinas DF, Dronsky V, Mendoza M, Barohn RJ, Bryan W, Ravits J, Yuen E, Neville H, Ringel S, Bromberg M, Petajan J, Amato AA, Jackson C, Johnson W, Mandler R, Bosch P, Smith B, Graves M, Ross M, Sorenson EJ, Kelkar P, Parry G, Olney R; Western ALS Study Group (2001) Phase III randomized trial of gabapentin in patients with amyotrophic lateral sclerosis. Neurology 56:843–848.

Miller RG, Mitchell JD, Moore DH (2002) Riluzole for amytophic lateral sclerosis/motor neuron disease. Cochrane Database Syst Rev (2):CD001447.

Mitchell JD, Houghton E, Rostron G, Wignall C, Gatt JA, Phillips TM, Kilshaw J, Shaw IC (1995) Serial studies of free radical and antioxidant activity in motor neuron disease ad the effect of selegiline (letter). Neurodegeneration 4(2):233–235.

Miyoshi Y, Zhang Z, Ovadia A, Lapchak PA, Collins F, Hilt D, Lebel C, Kryscio R, Gash DM (1997) Glial cell line derived neurotrophic factor-levodopa interactions and reduction of side effects in parkinsonian monkeys. Ann Neurol 42:208–214.

Mogi M, Togari A, Kondo T, Mizuno Y, Komure O, Kuno S, Ichinose H, Nagatsu T (2000) caspase activities and tumor necrosis factor receptor R1 level are elevated in the substantia nigra from parkinsonian brain. J Neural Transm 107:335–341.

Molinuevo JL, Garcia-Gil V, Vllar A (2004) Memantine: An antiglutamatergic option for dementia. Am J Alzheimers Dis Other Demen 19:10–18.

Mytilineou C, Radcliffe P, Leonardi EK, Werner P, Olanow CW (1997) L-deprenyl protects mesencephalic dopamine neurons from glutamate receptor-mediated toxicity in vitro. J Neurochem 68:33–39.

Mytilineou C, Leonardi EK, Radcliffe P, Heinonen EH, Han SK, Werner P, Cohen G, Olanow CW (1998) Deprenyl and desmethylselegiline protect mesencephalic neurons from toxicity induced by glutathione depletion. J Pharmacol Exp Ther 284:700–706.

NINDS NET-PD Investigators (2006) A randomized, double-blind, futility clinical trial of creatine and minocycline in early Parkinson disease. Neurology 66(5):626–627.

Nicoll JA, Wilkinson D, Holmes C, Steart P, Markham H, Weller RO (2003) Neuropathology of human Alzheimer disease after immunization with amyloid beta peptide: Case report. Nat Med 9(4):448–452.

Nutt JG, Burchiel KJ, Comella CL, Jankovic J, Lang AE, Laws Jr ER, Lozano AM, Penn RD, Simpson Jr RK, Stacy M, Wooten GF; ICV GDNF Study Group (2003) Implanted intracerebroventricular. Glial cell line-derived neurotrophic factor. Randomized, double blind trial of glial derived neurotrophic factor (GDNF) in PD. Neurology 60(1):69–73.

Olanow CW, Goetz CG, Kordower JH, Stoessl AJ, Sossi V, Brin MF, Shannon KM, Nauert GM, Perl DP, Godbold J, Freeman TB (2003) A double-blind controlled trial of bilateral fetal nigral transplantation in Parkinson's disease. Ann Neurol 54(3):403–414.

Ondo WG, Tintner R, Voung KD, Lai D, Ringholz G (2005) Double-blind, placebo-controlled, unforced titration parallel trial of quetiapine for dopaminergic-induced hallucinations in Parkinson's disease. Mov Disord 20(8):958–963.

Opacka-Juffrey J, Brooks DJ (1995) Dihydroxyphenylalanine and its decarboxylase: New ideas on the neuroregulatory roles. Mov Disord 10:241–249.

Orgogozo JM, Gilman S, Dartigues JF, Laurent B, Puel M, Kirby LC, Jouanny P, Dubois B, Eisner L, Flitman S, Michel BF, Boada M, Frank A, Hock C (2003) Subacute meningoencephalitis in a subset of patients with AD after Abeta42 immunization. Neurology 61(1):46–54.

Orrell RW, Lane RJM, Ross M (2005) Antioxidant treatment for amyotrophic lateral sclerosis/motor neuron disease. The Cochrane Database for Systematic Reviews, Volume 4, The Cochrane Collaboration.

Pezzoli G, Martignoni E, Pacchetti C, Angeleri VA, Lamberti P, Muratorio A, Bonuccelli U, De Mari M, Foschi N, Cossutta E, Nicoletti F, Giammona F, Canessi M, Scarlato G, Caraceni T, Moscarelli E (1994) Pergolide compared to bromocriptine in Parkinson's disease: A multicenter, crossover, controlled study. Mov Disord 9:431–436.

Pact V, Giduz T (1999) Mirtazapine treats resting tremor, essential tremor and levodopa induced dyskinesias. Neurology 53:1154.

Pal PK, Calne DB, Calne S, Tsui JK (1999) Botulinum—a toxin in the treatment of sialorrhea in patients with Parkinson's disease. Parkinsonism Relat Disord 5:582.

Palhagen S, Heinonen EH, Hagglund J, Kaugesaar T, Kontants H, Maki-Ikola O, Palm R, Turunen J (1998) Selegiline delays the onset of disability in de novo parkinsonian patients. Neurology 51:520–525.

Parkinson Study Group (1989) Effect of deprenyl on the progression of disability in early Parkinson's disease. N Engl J Med 321:1346–1371.

Parkinson Study Group (1993) Effect of tocopherol and deprenyl on the progression of disability in early Parkinson's disease. N Engl J Med 328:176–183a.

Parkinson Study Group (1996) Impact of deprenyl and tocopherol treatment on Parkinson's disease in DATATOP patients not requiring levodopa. Ann Neurol 39:29–45.

Parkinson Study Group (1997) Safety and efficacy of pramipexole in early Parkinson's disease: A randomized dose ranging study. JAMA 278:125–130.

Parkinson Study Group (2000) Pramipexole vs. levodopa as initial treatment for Parkinson's disease: A randomized controlled trial. JAMA 284:1931–1938.

Parkinson Study Group (2001) A randomized controlled trial of remacemide for motor complication in Parkinson's disease. Neurology 56:455–462.

Parkinson Study Group. Mortality in DATATOP: a multicenter trial in early Parkinson's disease. Parkinson Study Group. Ann Neurol. 1998;43:318–25.

Parkinson Study Group (2004) The safety and tolerability of a mixed lineage kinase inhibitor (CEP 1347) in PD. Neurology 62: 330–332.

Perachon S, Schwartz JC, Sokoloff P (1999) Functional potencies of new anitparkinsonian drugs at recombinant human dopamine D1, D2, and D3 receptors. Eur J Pharm 5:293–300.

Pinter MM, Pogarell O, Oertel WH (1999) Efficacy, safety and tolerability of the non ergoline dopamine agonist pramipexole in the treatment of advanced Parkinson's disease: A double blind, placebo controlled, randomized, multicenter study. J Neurol Neurosurg Psychiatry 66:436–441.

Puri BK, Leavitt BR, Hayden MR, Ross CA, Rosenblatt A, Greenamyre JT, Hersch S, Vaddadi KS, Sword A, Horrobin DF, Manku M, Murck H (2005) Ethyl-EPA in Huntington disease. Neurology 65:286–292.

Ralph GS, Radcliffe PA, Day DM, Carthy JM, Leroux MA, Lee DC, Wong LF, Bilsland LG, Greensmith L, Kingsman SM, Mitrophanous KA, Mazarakis ND, Azzouz M (2005) Silencing mutant SOD1 using RNAi protects against neurodegeneration and extends survival in an ALS model. Nat Med 11:429–433.

Rascol O, Brooks DJ, Brunt ER, Korczyn AD, Poewe WH, Stocchi F (1998) Ropinirole in the treatment of early Parkinson's disease: A 6 month interim report of a 5 year levodopa controlled study. Mov Disord 13:39–45.

Raoul C, Abbas-Terki T, Bensadoun JC, Guillot S, Haase G, Szule J, Henderson CE, Aebischer P (2005) Lentiviral-mediated silencing of SOD1 through RNAinterference retards disease onset and progression in a mouse model of ALS. Nat Med 11:423–428.

Reisberg B, Doody R, Stoffler A, Schmitt F, Ferris S, Mobius HJ; Memantine Study Group (2003) Memantine in moderate to severe Alzheimer's disease. N Engl J Med 348:1333–1341.

Ringel SP, Woolley JM (1999) Economic analysis of neurological services. Eur J Neurol 6(Supp 2):S21–S24.

Rinne UK, Larsen JP, Siden A, Worm-Petersen J (1998) Entacapone enhances the response to levodopa parkinsonian patients with motor fluctuations. Neurology 51:1209–1314.

Rogers J, Kirby L, Hempelman SR, Berry DL, McGeer PL, Kaszniak AW, Zalinski J, Cofield M, Mansukhani L, Willson P, Kogan F (1993) Clinical trial of indomethacin in Alzheimer's disease. Neurology 43:1609–1611.

Rosen DR, Siddique T, Patterson D, Figlewicz DA, Sapp P, Hentati A, Donaldson D, Goto J, O'Regan JP, Deng HX, Rahmani Z, Krizus A, McKenna-Yasek D, Cayabyab A, Gaston S, Berger R, Tanzi R, Halperin J, Herzfeld B, Van der bergh R, Hung WY, Bird T, deng G, Mulder D, Smyth C, Lang N, Soriano E, Pericak-Vance MA, Haines J, Rouleau GA, Gusella JS, Horvitz HR, Brown RH (1993)

Mutations in the Cu, ZN super oxide dismutase gene associated with familial amyotrophic lateral sclerosis. Nature 362:59–62.

Ross C, Poirier MA (2004) Protein aggregation and neurodegenerative disease. Nat Med 10:S10–S17.

Rowland LP, Shneider NA (2001) Amyotrophic lateral sclerosis. N Engl J Med 344(22):1688–1700.

Ryberg H, Askmark H, Persson L (2003) A double blind randomized clinical trial in amyotrophic lateral sclerosis using lamotrigine: Effects on CSF glutamate, aspartate, branched chain amino acid levels and clinical parameters. Acta Neurol Scand 108:1–8.

Ryu H, Smith K, Camelo SI, Carreras I, Lee J, Iglesias AH, Dangond F, Cormier KA, Cudkowicz ME, Brown Jr RH, Ferrante RJ (2005) Sodium phenylbutyrate prolongs survival and regulates expression of antiapoptotic genes in transgenic amyotrophic lateral sclerosis mice. J Neurochem 10:1471–4159.

Sainetti S, Ingrim D, Talwalker S, Geiss GS (2000) Results of a double blind randomized placebo controlled study of celecoxib in the treatment of progression of Alzheimer's disease. In: The Proceedings of the Sixth International Stockholm/Springfield Symposium—Advances in Alzheimer's Therapy, April 5–8, 2000, abstract no. 180.

Saporito MS, Brown EM, Miller MS, Carswell S (1999) CEP 1347/KT 7515, an inhibitor of c-jun N terminal kinase activation, attenuates the 1 methyl 4 phenyl tetrahyopyridine mediated loss of nigrostriatal dopaminergic neurons in vivo. J Pharmocol Exp Ther 288:421–427.

Scarpini E, Scheltens P, Feldman H (2003) Treatment of Alzheimer's disease: Current status and new perspectives. Lancet Neurol 2:539–547.

Scharf S, Mander A, Ugoni A, Vajda F, Christophidis N (1999) A double blind placebo controlled trial of diclofenac/misoprostol in Alzheimer's disease. Neurology 53:197–201.

Schrag A, Quinn N (2000) Dyskinesias and motor fluctuations in Parkinson's disease: A community based study. Brain 123:2297–2305.

Scott LJ, Goa KL (2000) Galantamine: A review of its use in Alzheimer's disease. Drugs 60:1095–1122.

Shannon KM, Bennett JP, Friedman JH (1997) Efficacy of pramipexole, a novel dopamine agonist, as monotherapy in mold to moderate Parkinson's disease. Neurology 49:724–728.

Shaunak S, Wilkins A, Pilling JB, Dick DJ (1999) Pericardial, retroperitoneal and pleural fibrosis induced by pergolide. J Neurol Neurosurg Psychiatry 1163:25–31.

Shefner J, Cudkowicz ME, Shoenfeld D, Conrad T, Taft J, Chilton M, Urbinelli L, Qureshi M, Zhang H, Pestronk A, Caress J, Donofrio P, Sorenson E, Bradley W, Loemen-Hoerth C, Pioro E, Rezania K, Ross M, Pascuzzi R, Heimann-Patterson T, Tandan R, Mitsumoto H, Rothstein J, Smith-Palmer T, MacDonald D, Burke D (2004) A clinical trial of creatine in amyotrophic lateral sclerosis. Neurology 63(9):1656–1661.

Shinotoh H, Vingerhoets FJ, Lee CS, Uitti RJ, Schulzer M, Calne DB, Tsui J (1997) Lamotrigine trial in idiopathic parkinsonism: A double-blind placebo-controlled crossover study. Neurology 48:1282–1285.

Shults CW, Oakes D, Kieburtz K, Beal MF, Haas R, Plumb S, Juncos JL, Nutt J, Shoulson I, Carter J, Kompoliti K, Perlmutter JS, Reich S, Stern M, Watts RL, Kurlan R, Molho E, Harrison M, Lew M; Parkinson Study Group (2002) Effects of Coenzyme Q10 in early Parkinson disease: Evidence of slowing of the functional decline. Arch Neurol 59:1541–1550.

Sigurdsson EM, Brown DR, Daniels M, Kascsak RJ, Kascsak R, Carp R, Meeker HC, Frangione B, Wisniewski T (2002) Immunization delays the onset of prion disease in mice. Am J Pathol 161(1):13–17.

Silverdale MA, Nicholson SL, Crossman AR, Brotchie JM (2005) Topiramate reduces levodopa-induced dyskinesias in the MPTP lesioned marmoset model of Parkinson's disease. Mov Disord 20(4):403–409.

Simons M, Keller P, De Strooper B, Beyreuther K, Dotti CG, Simons K (1998) Cholesterol depletion inhibits the generation of beta amyloid in hippocampal neurons. Proc Natl Acad Sci USA 95:6460–6464.

Simons M, Schwarzler F, Lutjohann D, von Bergmann K, Beyreuther K, Dichgans J, Wormstall H, Hartmann T, Schulz JB (2002) Treatment with simvastatin in normocholesterolemic patients with Alzheimer's disease: A 26 week randomized, placebo controlled, double blind trial. Ann Neurol 52:346–350.

Solforosi L, Criado JR, McGavern DB, Wirz S, Sanchez-Alavez M, Sugama S, DeGiorgio LA, Volpe BT, Wiseman E, Abalos G, Masliah E, Gilden D, Oldstone MB, Conti B, Williamson RA (2004) Cross linking cellular prion protein triggers neuronal apoptosis in vivo. Science 303:1514–1516.

Steiner JP, Hamilton GS, Ross DT, Valentine HL, Guo H, Connolly MA, Liang S, Ramsey C, Li JH, Huang W, Howorth P, Soni R, Fuller M, Sauer H, Nowotnik AC, Suzdak PD (1997) Neurotrophic immunophilin ligands stimulate structural and functional recovery in neurodegenerative animal models. Proc Natl Acad Sci USA 94:2019–2024.

Stevic Z, Nicolic A, Blagojevic D, Saiiz S, Kocev N, Apostolski S, Spasic MB (2001) A controlled trial of combination of methionine and antioxidants in ALS patients. Jugoslovenska Medicinska Biokemija 20:223–238.

Storkebaum E, Lambrechts D, Dewerchin M, Moreno-Murciano MP, Appelmans S, Oh H, Van Damme P, Rutten B, Man WY, De Mol M, Wyns S, Manka D, Vermeulen K, Van Den Bosch L, Mertens N, Schmitz C, Robberecht W, Conway EM, Collen D, Moons L, Carmeliet P (2005) Treatment of motoneuron degeneration by intracerebroventricular delivery of VEGF in a rat model of ALS. Nat Neurosci 8:85–92.

Sun M, Kong L, Wang X, Lu XG, Gao Q, Geller AI (2005) Comparison of the capability of GDNF, BDNF or both to protect nigrostriatal neurons in the rat model of Parkinson's disease. Brain Res 1052(2):119–129.

Tariot PN, Farlow MR, Grossberg GT, Graham SM, McDonald S, Gergel I, Memantine Study Group (2004) Memantine treatment in patients with moderate to severe Alzheimer's disease already receiving donepazil: A randomized, controlled trial. JAMA 291(3):317–324.

Tatton NA (2000) Increased caspase-3 and Bax immunoreactivity accompany nuclear GAPDH translocation and neural apoptosis in Parkinson's disease. Exp Neurol 166:29–43.

Tikka T, Fiebich BL, Goldsteins G, Keinanen R, Koistinaho J (2001) Minocycline, a tetracycline derivative, is neuroprotective against excitotoxicity by inhibiting and proliferation of microglia. J Neurosci 21:2580–2588.

Tseng JL, Baetge EE, Zurn AD, Aebischer P (1997) GDNF reduces drug induced rotational behavior after medial forebrain bundle transaction by a mechanism not involving striatal dopamine. J Neurosci 17:325–333.

Tolcapone Study Group (1999) Efficacy and tolerability of tolcapone compared with bromocriptine in levodopa treated parkinsonian patients. Mov Disord 14:38–44.

t'Veld B, Ruitenberg A, Hofman A, Launer L, van Duijn CM, Stijnen T, Breteler MM, Stricker BH (2001) Nonsteroidal anti-inflammatory drugs and the risk of Alzheimer's disease. N Engl J Med 345:1515–1521.

Van Camp G, Flamez A, Cosyns B, Weytjens C, Muyldermans L, Van Zandijcke M, De Sutter J, Santens P, Decoodt P, Moerman C, Schoors D (2004) Treatment of Parkinson's disease with pergolide and relation to restrictive valvular heart disease. Lancet 363:1179–1183.

Van Damme P, Braeken D, Callewaert G, Robberecht W, Van Den Bosch L (2005) GluR2 deficiency accelerates motor neuron degeneration in a mouse model of amyotrophic lateral sclerosis. J Neuropath Exp Neurol 64(7):605–612.

Van Raamsdonk JM, Pearson J, Rogers DA, Lu G, Barakauskas VE, Barr AM, Honer WG, Hayden MR, Leavitt BR (2005) Ethyl-EPA improves motor dysfunction, but not neurodegeneration in the YAC128 mouse model of Huntington disease. Exp Neurol 196:266–272.

Walley T (2004) Neuropsychotherapeutics in the UK: What has bee the impact of NICE on prescribing? CNS Drugs 18:1–12.

Walker LC, Price ML, Voytko ML, Schenk DB (1994) Labeling of cerebral amyloid in vivo with a monoclonal antibody. J Neuropath Exp Neurol 53(4):377–383.

Watts RL, Raiser CD, Stover NP, Cornfeldt ML, Schweikert AW, Allen RC, Subramanian T, Doudet D, Honey CR, Bakay RA (2003) Stereotaxic intrastriatal implantation of human retinal pigment cells attached to gelatin microcarriers: A potential new cell therapy for Parkinson's disease. J Neural Transm 65:215–217.

White AR, Enever P, Tayebi M, Mushen SR, Linehan J, Brandner S, Anstee D, collinge J, Hawke S (2003) Monoclonal antibodies inhibit prion replication and delay the development of prion disease. Nature 422(6927):80–83.

Whitehouse PJ, Price DL, Clark AW, Coyle JT, DeLong MR (1981) Alzheimer's disease: Evidence for selective loss of cholinergic neurons in the nucleus basalis. Ann Neurol 10:122–126.

Willem M, Dewachter I, Smyth N, Van Dooren T, Borghgraef P, Haass C, Van Leuven F (2004) Beta site amyloid precursor protein cleaving enzyme1 increases amyloid deposition in the brain parenchyma but reduces cerebrovascular amyloid angiopathy in aging BACE x APP double transgenic mice. Am J Path 165(5): 1621–1631.

Wong P, Pardo CA, Borchelt DR, Lee MK, Copeland NG, Jenkins NA, Sisodia SS, Cleveland DW, Price DL (1995) An adverse property of a familial ALS liked SOD1 mutation causes motor neuron disease characterized by vacuolar degeneration of mitochondria. Neuron 14:1105–1116.

Yamada M, Yasuhara H. Clinical pharmacology of MAO inhibitors safety and future. Neurotoxicology 2004; 25:215–221.

Yasuhara T, Shingo T, Kobayashi K, Takeuchi A, Yano A, Muraoka K, Matsui T, Miyoshi Y, Hamada H, Date I (2004) Neuroprotective effects of vascular endothelial growth factor (VEGF) upon dopaminergic neurons in a rat model of Parkinson's disease. Eur J Neurosci 19(6):1494–1504.

Zesiewicz TA, Helal M, Hauser RA (2000) Sildenafil citrate (Viagra) for the treatment of erectile dysfunction in men with Parkinson's disease. Mov Disord 15:305–308.

Zhang W, Narayanan M, Friedlander RM (2003) Additive neuroprotective effects of minocycline with creatine in a mouse model of ALS. Ann Neurol 53:267–270.

Zhu S, Stavrovskaya IG, Drozda M, Kim BY, Ona V, Li M, Sarang S, Liu AS, Hartley DM, Wu du C, Gullans S, Ferrante RJ, Przedborski S, Kristal BS, Friedlander RM (2002) Minocycline inhibits cytochrome c release and delays progression of amyotrophic lateral sclerosis in mice. Nature 417:74–78.

40
Multiple Sclerosis

M. Patricia Leuschen, Kathleen M. Healey, and Mary L. Filipi

Keywords EAE; Gadolinium; Glatiramer acetate; IFNβ-1a; IFNβ-1b; Mitoxantrone; Natalizumab; Neuromyelitis optica; PASAT; Sclerosis

40.1. Introduction

New therapies challenge the very definition of multiple sclerosis (MS). By classic definitions, MS is a chronic neurological disease characterized by plaques or scars in the central nervous system (CNS) as a result of demyelization and atrophy of neuronal axons. However, for most diagnosed with MS, symptoms and progression are much more heterogeneous than that definition would imply. For most with MS, the initial symptoms include a relapsing and remitting inflammatory phase (RR-MS), which responds to immunotherapy. Natural history studies of clinical relapse rates, a primary endpoint used in most phase III treatment trials, report incidence of relapses from 0.5 to 2 per year (Weinshenker and Ebers, 1987). With time, most RR-MS patients transition to what has been characterized as a non-inflammatory secondary progressive form of the disease (SP-MS). Immunotherapies that target the inflammatory phase of MS may not be effective once the transition occurs. In addition, a proportion of those diagnosed with MS exhibit a more aggressive form of MS that does not show significant remission cycles nor respond to the same therapies. This aggressive form of the disease was termed primary progressive (PP-MS) using consensus data from an international survey of clinicians for standard definitions of these common clinical MS courses (Lublin and Reingold, 1996).

40.1.1. Therapeutic Options and Disease Definitions

Luccinetti's group (Lucchinetti et al., 1996; Lucchinetti et al. 2000) changed the way MS disease classifications were viewed when their pathologic studies confirmed that MS plaques differ. Neuroimmunologic evidence now suggests that axonal transaction can precede demyelination rather than occur as a result of it (Trapp et al., 1998; Bitsch et al., 2000; De Stefano et al., 2001; Tsunoda and Fujinami, 2002). Changes in the CNS are associated and most likely precede measured disability by years and possibly decades. Central to these theories about MS is the disconnected temporal relationship between pathological changes in the CNS and disability (Bjartmar and Trapp, 2001). As therapeutic options for MS are reviewed, recent progress in standardizing outcome measures for both evaluating acute inflammatory relapses and long term progression will be discussed. There is no disagreement that the ultimate goal is to develop therapies that prevent disability associated with MS. As the heterogeneity of the disease is better understood, the ability to tailor therapy should also improve.

A definitive diagnosis of MS historically centers on clinical verification of two or more episodes (relapses or exacerbations) separated by time and space and involving separate systems such as optic neuritis and motor involvement of lower limbs (Poser et al., 1983). New guidelines were developed in 2001 by an International Panel on the Diagnosis of Multiple Sclerosis and are known as the "McDonald criteria" (McDonald et al., 2001). For a definitive diagnosis of MS, the McDonald criteria focus on demonstration of lesions in time and space but integrate MRI findings with clinical and diagnostic methods. Worldwide debate over the use of the McDonald criteria led to the convening of a second panel in Amsterdam in 2005 and revisions to the criteria now termed the 2005 Revisions to the McDonald Diagnostic Criteria for MS (Polman et al., 2005). The core of the MS diagnosis remains the clinically objective demonstration of typical disease signs and symptoms in time and space supported by laboratory evidence including MRI. Debate over definitions is important to the development of MS therapies since FDA guidelines insist on clinical outcome measures such as a change in the number of relapses and the time between relapses. Validated qualitative or quantitative outcome measures that specifically define a relapse do not exist.

Initial diagnosis is not the only moving target in defining MS. Debate also continues on when diagnostic criteria support

T. Ikezu and H.E. Gendelman (eds.), *Neuroimmune Pharmacology*.
© Springer 2008

a diagnosis of PP-MS (Thompson et al., 2000) and when or if clinically isolated syndromes such as monosymptomatic optic neuritis convert to MS (Barkhof et al., 1997; Filippi et al., 1994; Jacobs et al., 2001).

40.1.2. Diagnosis and Disease Activity Measures

The level of disability in MS is usually scored using the Kurtzke expanded disability status scale (EDSS), a 10 point scale biased toward motor function (Kurtzke, 1983, 1991). With the addition of 0.5 point steps, the Kurtzke EDSS is often applied with 19 intermediate steps where 0 represents no disability. The lower ratings in the scale are based upon the neurological exam; mid and higher rating scores are based on ambulatory ability and performance of activities of daily living. Data on the average time MS patients remain at a specific EDSS level were published prior to FDA approval of the current therapies (Weinshenker et al., 1991) and illustrate that time at any specific point on the scale is highly variable. Even so, the EDSS is a primary outcome measure in all phase III trials of MS immunotherapies.

The Multiple Sclerosis Functional Composite (MSFC) was developed by the MS Consortium (Whitaker et al., 1995) for evaluating disabilities ranging from motor function to cognitive changes and to replace the EDSS. The three tools included in the MSFC are a timed 25 foot walk, the 9-hole peg test for hand dexterity and the Paced Auditory Serial Addition Test (PASAT). The MSFC is a quantifiable measure of disability but ten years later, it is still infrequently incorporated as a primary outcome measure because of the increased time element necessary to validate the individual tests and the time necessary for their administration.

As a technology, MRI is invaluable in providing information that has greatly advanced knowledge of the disease process of MS. Ian Young is credited with the first image of an MS lesion using MRI and fortuitously predicted that this technique could also measure disease severity and serve as a monitor for effectiveness of treatments (Young et al., 1981). Approximately five years later the contrast medium, gadolinium (Gd) was introduced. This method of injecting a contrast material intravenously is able to demonstrate breech of the blood–brain barrier (BBB) indicating newly acquired active inflammatory lesions of MS (Grossman et al., 1988). It was later verified that extensive sub-clinical activity evidenced by Gd enhancing lesions occurs in the absence of clear relapses or disease worsening at a ratio of 10–15:1 (Isaac et al., 1988; Willoughby et al., 1989; Khoury et al., 1994; Thompson et al., 1992). MRI is now a primary endpoint in phase I and II treatment trials and as a secondary endpoint in phase III trials.

Serial MRI documents that clinical MS exacerbations are associated with focal BBB damage. As early as 1965, experimental allergic/autoimmune encephomyelitis (EAE) animal models revealed lymphocyte adherence to post capillary venules and subsequent migration into CNS parenchyma preceding demyelination (Lampert and Carpenter, 1965). Both serial MRI in MS patients and EAE animal studies support the role of an inflammatory, neuroimmunologic response in the disease. MRI findings during the early stages of MS demonstrate the presence of active inflammatory lesions in the brain or spinal cord (Bjartmar and Trapp, 2003). Axonal injury begins at disease onset and correlates with the degree of inflammation within lesions. This axonal loss remains clinically silent for many years and irreversible neurological disability develops when a threshold of axonal loss is reached and compensatory CNS resources are exhausted. The ratio of silent inflammatory sub-clinical lesions to clinical relapses is reportedly 5–10:1 (Bjartmar et al., 2003; Miller, 2004). This discovery of silent lesion accumulation was instrumental in pushing research for the treatment of MS to the forefront in an attempt to decrease lesion accumulation and subsequent disability progression (Bjartmar and Trapp, 2003). However, the monitoring of disease progression by MRI or quantifying clinical symptoms is not adequate. The presence of MRI lesions correlates poorly with clinical symptoms and does not fully predict or define the level of disability or disease activity. Molecular evidence indicates a continuum of dysfunctional homeostasis and inflammatory changes between areas of lesions and normal appearing white matter. This continuum supports the hypothesis that MS is a generalized process that involves the entire CNS (Lindberg et al., 2004). MS may be active in normal appearing white matter but may not be detectable by current levels of testing.

40.1.3. Emerging Immune Modulators

Prior to the early 1990s, there was no Food and Drug Administration (FDA) approved disease-modifying agents for treatment of any form of MS. In the past ten years, five agents received FDA approval; four are immunomodulators for treatment of RR-MS and one, mitoxantrone, is an immunosuppressant. MS is now a treatable disease. In their Disease Management Consensus Statement (NMMS, 2005), the National Multiple Sclerosis Society (NMMS) recommends that initiation of an immunomodulator should be considered as soon as possible following a definite diagnosis of active MS. The NMMS data also indicate that only 38% of patients with RR-MS are currently on immunomodulatory therapy.

The phase III trials for currently approved MS immunotherapies followed the use of the therapy for 20 months or less. Long term efficacy and compliance can be an important factor since historic data show that people with MS lived on average 35 years following diagnosis even before current therapies (Smith and McDonald, 1999). Usually MS only decreases life expectancy by 2–3 years. However, prior to the advent of current therapies, 50% of those with MS needed assistance (a cane to walk over one half block) 15 years after diagnosis (Weinshenker et al., 1989).

Four of the five current immunomodulatory therapies are only approved for RR-MS and have not shown effectiveness in non-inflammatory SP-MS or more aggressive forms of the disease. MS has been regarded as a single disorder with clinical variants, but there is strong evidence that it may comprise several related disorders with distinct immunological, pathological and genetic features (Compston, 1997). To support this theory, research indicates the demyelination; neuronal atrophy and axonal destruction follow different pathogenetic pathways in subgroups of MS patients (Lassmann, 2001).

The disease, MS, usually does not affect peripheral myelin. There are, however, cases of disease clinically and pathologically indistinguishable from MS in which there is peripheral demyelination in a pattern resembling chronic inflammatory demyelinating polyneuropathy. MS has extremely heterogeneous patterns, with potentially devastating clinical manifestations in some individuals while others have very little progression or disability. Early in the disease, the typical clinical course of MS consists of variable relapsing-remitting symptomatology. While the majority of those with MS at this inflammatory stage respond to immunotherapy with interferon (IFN) β-1a, IFN β-1b or glatiramer acetate, some patients do not.

Natural history studies (Smith and McDonald, 1999), prior to the new therapies show that after 10–15 years the highly inflammatory RR-MS pattern transitions to a more chronic, less inflammatory progressive phase. This chronic phase is characterized by increasing disability with less discreet relapse episodes and a more constant downhill disability course. MS itself has the propensity to dissipate or slow after 20–25 years duration but disability will continue to progress (Paty et al., 1997). Therapies aimed at the inflammatory phase of RR-MS do not halt the less inflammatory progressive phase. This is hypothesized to be due to previously accumulated lesion load within the CNS and the circuitous nature of disability building on disability. With the degree of heterogeneity, a frequent recurrent problem is how to fit the MS treatment to the individual patient's symptoms and progression. The success and approval of immunosuppressant therapy with mitoxantrone for that sub group of MS patients with aggressive inflammatory disease supports the concept that the immune aspects of MS are heterogeneous and that treatment must be tailored to specific disease course.

A similar approach meeting with success is with neuromyelitis optica or Devic's syndrome. While uncommon, neuromyelitis optica preferentially affects the optic nerve and spinal cord. While resembling MS, diagnosis is confirmed by its clinical severity, an MRI pattern with normal brain findings but extensive lesions in the spinal cord, and CSF analysis of polymorphonuclear pteocytosis and no oligoclonal banding. A serum autoantibody marker, NMO-IgG is highly specific for the syndrome (Wingerchuk and Weinshenker, 2005) and the syndrome appears to have a strong humoral component. Open label trials with small numbers of patients indicate that therapy with rituximab, a chemotherapic drug that depletes the B cell population is effective in slowing the relapse rate (Cree et al., 2005).

40.2. Historical Therapies for MS

40.2.1. Hypothetical Benefits Without Improved Clinical Outcomes

40.2.1.1. Interferon γ (IFNγ)

In the 1980s, studies with EAE, an animal model mimicking MS, indicated that interferon γ (IFN γ) was effective in treating that disease and a trial was initiated to evaluate its potential benefit in human MS. Rather than showing efficacy, in 1987, use of IFNγ as a therapy in MS patients caused an increase in clinical exacerbations and forced the clinical trial to terminate early (Panitch et al., 1987). Associated study of IFNγ in 20 MS patients indicates increased concentrations of IFNγ and TNFα precede the observation of clinical defects (Beck et al., 1988). An evaluation of primary RR-MS patient lymphocytes using flow cytometry supports a correlation between EDSS scores and IFNγ secretion (Petereit et al., 2000). Intracellular cytokine immuno-staining of anti CD8+ T cells reveals a correlation with IFNγ and disease phase but not disease activity (Becher et al., 1999). What initially seemed efficacious in the EAE animal model, not only did not decrease MS symptoms but is now felt to be a marker of active inflammatory disease.

40.2.1.2. Histamine

In the 1950s and 1960s, immune therapies including histamine were evaluated and shown to have no significant benefit in decreasing the clinical or laboratory parameters associated with active MS. One variation of a therapy including histamine is still available as a trans-dermal gel patch with a proprietary blend of histamine and caffeine now termed Prokarin. Prokarin was formerly known as "Procarin." The revised spelling was adopted to avoid potential conflicts with another existing trademark. A small double blind pilot study shows Prokarin has a modest effect on fatigue associated with MS but no change in the Multiple Sclerosis Functional Composite (MSFC) (Gillson et al., 2002). Since the trial was only 12 weeks, relapse rate and MRI data were not evaluated. Prokarin use for MS has not been evaluated by the FDA.

40.2.1.3. From Placebos to Bee Stings

No review on therapies for MS would be complete without some mention of the placebo effect and the vast underlying anecdotal stories of agents and treatment with immune effects and hypothetical benefits. Because relapse rates can be relatively low, less than one per year and because fatigue is the single most commonly reported symptom, the placebo effect in many early therapy trials is significant. In addition, reports of MS patients using untested therapies such as daily or weekly exposure to bee stings or removal of all metal teeth fillings continue to appear. Since MS is now treatable with FDA approved therapies, the ability to evaluate such evidence has become problematic.

40.2.2. Beneficial Immune Therapies: Use of Steroids for Acute Episodes RR-MS

Historically, steroids and ACTH have been used in the treatment of MS relapses. While both suppress cell mediated and humoral immune responses, the major effect in MS acute relapse is to suppress inflammation. A meta-analysis of randomized controlled clinical trials (Brusaferri and Candelise, 2000) indicates that any type of steroid or ACTH treatment significantly accelerates short-term recovery from an acute MS relapse. However, there is no evidence that steroids reduce the risk of an MS relapse. A randomized trial of oral versus intravenous (IV) methyl prednisolone for treatment of acute relapses of MS shows no clear advantage of either treatment route (Barnes et al., 1997). Currently methyl prednisolone is considered standard of care for acute MS relapses. Methyl prednisolone (Solu-Medrol) is used with IV treatment for 3–5 days at 500–1000 mg/day. Methyl prednisolone IV therapy may be followed by a oral prednisone taper. The use of low dose oral prednisolone is also effective in conjunction with INFβ therapies when flu-like side effects persist.

40.3. Immunotherapies for RR-MS

40.3.1. FDA Approval of IFNβ-1a, IFNβ-1b, and Glatiramer Acetate for RR-MS

While no cure for MS is known, the FDA approves four immunotherapies for RR-MS. All shorten attacks and lengthen the time between attacks by ~30% (Arnason, 1999a, 1999b). The therapies include variations of IFN β: IFN β-1a, (Avonex and Rebif) and IFN β-1b (Betaseron). The second type of immunotherapy, glatiramer acetate is a copolymer of four amino acids termed Copaxone. All approved therapies involve self injections or injections by a caregiver and are accompanied by side effects.

40.3.1.1. FDA Trials and Their Outcome Measures

The FDA phase III trials leading to approval of the two forms of IFNβ1a, IFNβ-1b and glatiramer acetate for RR-MS used clinical outcome measures such as the number of relapses over a 12 month period as the primary outcome measure for efficacy of the therapies. Along with decreasing the relapse rate, both the IFNβs and glatiramer acetate increase the time between relapses and decrease lesion load on MRI. Since all comparisons between therapies are post-market, the classifications developed by Goodlin (Goodlin et al., 2002) will be used when making comparisons. The longest Class I pivotal trial for an MS drug was up to 36 months (m) for glatiramer acetate (Johnson et al., 1995; Johnson et al., 1998) while the shortest averaged 20 m for intramuscular (IM) IFN β-1a (Avonex) (Jacobs et al., 1996). Formulation, dosing schedules and inclusion criteria differed in the pivotal trials and in the analytical methods used in the trials but the basic outcomes (a decrease in relapses and relapse rate) can be compared according to clinical and MRI findings (Johnson, 2005) (Figure 40.1). All four therapies for RR MS reduce the relapse rate but one, IFNβ-1b did not show a significant reduction in disability (The IFNβ MS Study Group, 1993). MRI outcomes are also positive with a decreased overall burden of disease in three of the four therapies for RR MS. However, IFN β-1a (Avonex) did not show the MRI effect at 2 years (Jacobs et al., 1996). Gd enhancing lesions was not an outcome measure in the initial clinical trial of IFN β-1b but was significant for the other three trials (Johnson et al., 1995; The IFNβ MS Study Group, 1993; PRISMS-Study Group, 1998).

Mitoxantrone (Novantrone) approved by the FDA in 2000, is the only drug approved for treatment of MS once transition

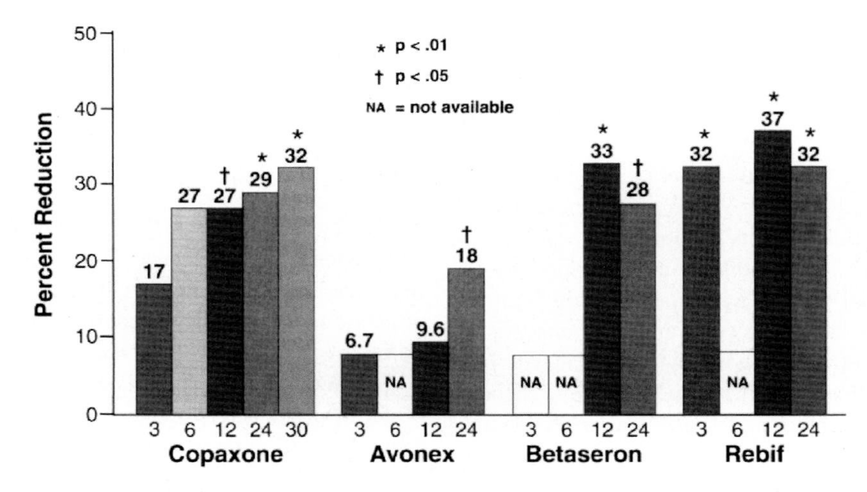

FIGURE 40.1. Relapse Rate Reduction, Phase III Studies (Class I). In the pivotal Class I studies of disease-modifying drugs for multiple sclerosis, gladramer acetate (GA) and higher-frequency (three times weekly) dosing of interferon beta were clearly superior to once-weekly interferon (IM) in reduction of relapse rate. Johnson et al. *Neurology*, 1995.[3] Jacobs et al. *Ann Neurol*, 1996.[5] IFNB Study Group. *Neurology*, 1993.[6] PRISMS Study Group. *Lancet*. 1998.[7] (Adapted from Johnson, 2005).

to the progressive form occurs. Use of mitoxantrone and similar therapy will be discussed in a subsequent section.

40.3.1.2. Side Effects and Compliance

When IFN β therapies were first approved by the FDA for treatment of RR-MS, flu like symptoms were known side effects but consistently described as minimal and transient. However, in the early pivotal trials for IFN β-1b, a significant number of patients withdrew for drug toxicity issues that were often consistent with flu-like symptoms (The IFNβ MS Study Group and The U British Columbia MS/MRI Study Group, 1995). Use of IFN β is now a mainstay of successful long term MS therapy. "Long term" therapy is the key point because, unlike many neurologic diseases, MS is usually diagnosed in young adults and does not significantly decrease life span. Adherence and long term compliance are serious issues particularly for those patients who continue to experience side effects to IFN β injections. It is now known that flu-like side effects are common in RR-MS patients injecting IFN β and are as high as 90% in patients initiating therapy. Use of step wise gradual escalation of dosing when initiating IFN β is standard of care in virtually all MS Centers as a means of decreasing this side effect. In 1999, Rice's group evaluated gradually introducing use of IFN β with or without ibuprofen and reported that the combination did reduce flu-like symptoms (Rice et al., 1999). Ibuprofen was also evaluated against paracetamol and found to be equally effective in managing the flu-like symptoms of IFN β therapy (Reess et al., 2002). In a letter to the editor, Rio (Rio, 2000) supports the gradual introduction of IFN β and use of ibuprofen but also suggests use of oral low dose steroids to minimize side effects. A randomized open label study of naproxen, acetaminophen and ibuprofen for controlling the side effects of IFN β-1a, shows naproxen and ibuprofen are more effective than acetaminophen in reducing the physical side effects. None of the study pain medications were effective in minimizing fatigue or muscle aches and pain in those MS patients on IFN β-1a who had continued to report side effects after >6 m of therapy (Leuschen et al., 2004). In a 1997 study, Munschauer and Kinkel (1997) recommend acetaminophen as the first line adjunct therapy for IFN β associated flu like symptoms. The package insert for IFN-β-1a suggests use of pain medications specifically mentioning acetaminophen but with no specific guidelines. Finally, indomethacin was shown to reduce the side effects of IFN therapy in a study by Mora et al. (1997).

For those RR MS patients whose flu-like symptoms, fatigue or significant muscle and joint pain persist while on IFNβ therapies, use of low dose oral corticosteroids is another option (Rio, 1998; Martinez-Cacenes, 1998).

40.3.1.3. MRI and Bioimaging to Evaluate Clinical Efficacy

Brain atrophy is recognized as a reproducible outcome measure of destructive change in MS. Significant atrophy changes occur within 9–12 months of disease onset in RR-MS in patients treated with the immunomodulators (Hardmeier et al., 2003). This atrophy is detectable even earlier in untreated patients. A direct short-term relationship between inflammation and development of brain atrophy is indistinct, possibly due to the time lag while tissue degradation occurs and lack of testing available that is sensitive enough to pick up minor changes (Hardmeier et al., 2003). It is well documented by MRI that individuals with MS develop varying degrees of brain atrophy, and maintain a 20% greater atrophy than the unaffected population. Previously, brain atrophy was considered a late occurring disease manifestation of MS. Zivadinos and Bakshi (2004) find that brain atrophy can be detected within 3 months of diagnosis. Current thinking is that brain atrophy occurs with normal changes associated with aging, but is considered accelerated if it reaches levels >0.5–1% per year.

Brain atrophy, particularly central atrophy, correlates more closely to increased cognitive impairment than MRI lesion burden. This may be explained in part by atrophy of the thalamus, a deep gray matter structure that mediates cognitive function via cortical and subcortical pathways (Benedict et al., 2004) and by cortical atrophy of the frontal cortex. The frontal cortex atrophy can predict neuropsychological impairments in verbal learning, spatial learning, attention and conceptual reasoning. The regions of the cortex most susceptible to atrophic and cognitive changes in MS are the right and left superior frontal lobes (Benedict et al., 2004). In a related study, MRI brain atrophy was evaluated as a secondary outcome measure in a small group of patients following stem cell transplant for MS (Healey et al., 2004). That study shows ventricular atrophy and thickness of the corpus callosum is a mild but significant correlation to disability. Simon et al. (1999) confirm that substantial neocortical volume loss occurs in MS patients and suggests that neocortical gray matter pathology may occur early in the course of the disease, contributing significantly to neurologic impairment. Although a proportion of this pathology may be secondary to white matter inflammation, the extent of the change suggests that an independent neurodegenerative process is also active.

While many MRI parameters do not correlate well to disability, the presence and degree of atrophy does reflect level of disability. MRI reveals that brain and spinal cord atrophy occur early in the course of MS, far earlier than originally anticipated (Zivadinos and Bakshi, 2004). The degenerative nature of the disease shows increased CNS plaquing with subsequent increases in T1 black holes, T2 total lesion load, and finally increases in brain atrophy. This has important implications for early treatment, as atrophy is thought to reflect destructive, irreversible pathology and sub clinical impairment if not overt disability (Simon, 2001). The assessment of brain volume changes on serial MRI can provide an objective measure of progressive atrophy, reflecting the neurodegenerative component of MS pathology and the long-term efficacy of MS therapies.

40.3.1.4. *Clinical Responses for SP-MS and PP-MS*

The diagnosis of PP-MS remains particularly challenging and its pharmacologic management remains a challenge (Doskoch, 2005). Several clinical trials have focused on these patients and while no treatment efficacy has been demonstrated, information is helping to define this sub group of MS patients. The largest trial (PROMise) began recruitment in 1999 (Wolinsky et al., 2005). Inclusion criteria included progressive neurologic symptoms for at least 6 months with no history of prior exacerbations or a RR pattern. EDSS scores ranged from 3.0 to 6.5 and all patients were screened by MRI for definitive MS. Over 60% of the study group met the classification for definitive PPMS by McDonald criteria. The treatment protocol included glatiramer acetate (20 mg/d) or placebo with a planned 16 m enrollment. The study was terminated in 2002 (Wolinsky et al., 2004). At that point, 37 patients had 41 relapses or 2% per year. Only 14% of the patients with Gd enhanced lesions had a twofold increased risk of relapses. More significant than the low relapse rate was the low rate of progression. With a study prediction rate of 50% among the least effected patients (EDSS 3.0–5.0), the progression rate was actually 16% at 1 year and 28% at 2 years. The unexpectedly slow rate of progression changed the trial's statistical power to detect even a modest therapeutic benefit from use of glatiramer acetate. This slower rate than expected of progression in PP-MS is also reported by E. Waubaunt for a clinical trial of mitoxantrone and one for IFN β-1a discussed at the 19th Annual Meeting of the Consortium of Multiple Sclerosis Centers (Doskoch, 2005). These data support redefining the diagnosis of PP-MS and include the presence of Gd enhanced lesions, positive CSF findings and a high T2 burden of disease as predictors of faster progression (Wolinsky et al., 2004).

Current phase I/II trials in PP-MS are underway and include evaluation of rituximab, a synthetic antibody that binds to and induces lyses of B cells. The trial is scheduled to enroll over 430 patients for 30 m and will monitor clinical outcomes as well as immunologic and MRI measures. Rationale for the trial is based on a efficacy in use with worsening Devic's disease or neuromyelitis optica (Cree et al., 2005).

40.3.2. EAE Animal Models for MS Therapy Development

The animal model for MS, EAE was introduced in 1933 to study CNS reactions to vaccines including measles, small pox and rabies. Although the initial studies were done in monkeys, most studies on MS have used mice. Over time the model has evolved into adoptively transferring myelin basic protein auto reactive T cells into an animal. These programmed T cells cross the BBB producing inflammatory lesions in the CNS as seen in MS. Animals develop functional deficits quickly and in mice scoring of "disability" is done by a 5 point scale. The EAE model continues to be used to investigate new therapies for MS in preclinical testing. However the assumption that this animal model should be central to theory of causation and treatment for MS has been questioned since inception. More recently, EAE models were critically challenged when tissue biopsy of very early MS lesions of seven hour duration exhibited an absence of autoimmune cells within the lesion (Barnett and Prineas, 2004).

While the EAE model has pitfalls, perivascular T cell reactions do mediate the demyelinating process associated with MS when immunologically active cells enter the CNS via activated endothelium (Dorovin-Zis et al., 1992). Myelin is destroyed either as a non-specific reaction by the release of lymphokines, or by antigen specific mechanisms. Several lines of evidence support the primary nature of the inflammatory process in MS. Serial MRI studies reveal that clinical MS attacks are associated with focal BBB damage (Grossman et al, 1988; Kennedy et al., 1988). In the animal model EAE, it is well documented that T cells adhere and subsequently migrate across the cerebrovascular endothelium into the CNS parenchyma where demyelination occurs (Lampert and Carpenter, 1965; Raine, 1994). Activated CD4 + T cells are typically present in MS lesions in areas of active myelin breakdown. In older plaques, gamma delta TCR + T cells are more frequent (Selmaj et al., 1991). Less information is available on when or why there is a shift in the type of T cells present at the demyelination sites as the disease progresses. In addition, recent attention has turned to a role for B cells particularly in aggressive forms of MS.

40.3.3. Failed Clinical Trials

40.3.3.1. *Tysabri (Natalizumab)*

The aftermath of recent clinical trials (Miller et al., 2003), FDA approval (FDA Consum, 2005a) and then withdrawal (FDA Consum, 2005b) after several cases of progressive multifocal leukoencephalopathy were attributed to treatment with Natalizumab (Tysabri) (Langer-Gould et al., 2005; Kleinschmidt-DeMasters and Tyler, 2005) illustrate the promise and the pitfalls of carrying new MS therapies to the marketplace. Tysabri (natalizumab) is a recombinant humanized monoclonal antibody that binds to α4-integrin. The α4 subunits of α4β1 and α4β7 integrins are expressed on the surface of all leukocytes except neutrophils and inhibit the α4 mediated adhesion to their receptors. The α4 family of integrins includes vascular adhesion molecule-1 (VCAM-1), which is expressed on activated vascular endothelium. Rationale for its use centered on data that binding prevents transmigration of leukocytes across the cerebrovascular endothelium into inflammatory brain parenchyma. Evidence also supports a role for Tysabri in inhibition and further recruitment of activated T cells (Rice et al., 2005).

In the phase II trial, all Natalizumab treated MS subjects had fewer relapses and fewer Gd enhancing lesions compared to placebo treated subjects (O'Connor et al., 2005). An open label safety and drug interaction study of the use of Natalizumab

(3.0 or 6.0 mg/kg IV) in combination with IFN β-1a (30 μg IM) found the drug combination generally well tolerated (Vollmer et al., 2004) and had set the stage for larger clinical trials of combination therapies. Even the title of the March-April, 2005 FDA approval report (FDA Consum, 2005a) "New treatment, new hope for those with multiple sclerosis", implied a new approach and drug class with promise for MS and other autoimmune disease. Within two months (FDA Consum, 2005b), the FDA had withdrawn Natalizumab (Tysabri). Reports of progressive multifocal leukoencephalopathy following treatment with Natalizumab (Tysabri) appeared in the summer of 2005 (Langer-Gould et al., 2005) followed by reports of similar findings in the combination Natalizumab and IFN β-1a trial (Kleinschmidt-DeMasters and Tyler, 2005).

40.4. FDA Approved Therapies for MS

All three IFN-β products (Avonex, Rebif, and Betaseron) are approved by the FDA for the treatment of RR-MS. There is variability in IFN-β preparations, administration and their side effects. All produce a similar decrease in relapse rate of ~30% and increase the time between relapses in RR-MS. IFN-β has also been associated with teratogenic effects. Adequate birth control and pregnancy testing are recommended for all women with child bearing potential prior to initiation of any of the IFNβ therapies. Since MS is a disease frequently diagnosed in young adults and with an increased incidence in women, pregnancy testing and reporting is required for all women on the IFNβs.

40.4.1. IFNβ-1a, (Avonex)

40.4.1.1. Nomenclature, Approved Dosing Format, and Dosing Schedule

Avonex was the initial form of IFN β-1a approved by the FDA for use in the United States (Jacobs et al., 1996). The recommended dosing schedule is a once a week IM injection. IFN-β1a. IFN β-1a (Avonex) is supplied as a liquid in single use vials or syringe at the dose of 33 mcg IFN-β1a. The vial also contains 16.5 mg albumin (human), 6.4 mg sodium chloride, 6.3 mg dibasic sodium phosphate and 1.3 mg monobasic sodium phosphate. Of the approved IFNβ therapies, this drug has the lowest dose and longest inter-dosing interval.

40.4.1.2. Adverse Effects of IFN β-1a (Avonex)

The most common side effects of IFN β-1a (Avonex) involve flu-like symptoms including headache, fever, chills, muscle or joint ache, fatigue, pain or injection site inflammation. Initiation of therapy in a step-wise increase in dose and use of pain medications usually alleviate or substantially decrease these side effects. Less common side effects include nausea, stomach upset, diarrhea, a lower red blood cell count, sleep disturbances, dizziness, and infection. As with other IFNβ therapies, which are all classified as Category C, female patients are warned that the therapy has not been evaluated as a human teratogen or during breast feeding.

40.4.1.3. Post Approval Efficacy

A post-approval study evaluated the use of IFNβ-1a following a first demyelinating event when a clinically definitive diagnosis of MS was still not confirmed (Jacobs et al., 2001). Clinically definite MS is defined as two clinical attacks either with clinical evidence of two lesions or clinical evidence of one lesion plus para-clinical evidence (MRI, EEG or CSF changes) of a second lesion. The Controlled High Risk Subjects Avonex Multiple Sclerosis Study (CHAMPS) followed 385 participants in two sub-groups: a placebo group and a group receiving IFNβ-1a. CHAMPS found that initiating therapy with IFNβ-1a at the first demyelinating event is beneficial for patients with evidence of brain lesions on MRI that indicate a high risk of developing clinically definite MS (Jacobs et al., 2001). After 18 months, the median increase of lesion volume in participants receiving IFNβ-1a therapy was 1%, while it was 16% among participants in the placebo group. After three years, the likelihood of participants in the IFNβ-1a group developing clinically definite MS was 35% as compared to 50% in the placebo group (Beck et al., 2002).

The CHAMPS findings provide strong evidence that an IFNβ-1a regimen could be instrumental in delaying the progress of the disease, both by lengthening the time before demyelinating events progress to the level of clinically definite MS and by slowing the effects after a definite diagnosis. CHAMPS also provided data on IFNβ-1a side effects. Within the first six months of the trial, 54% of the IFNβ-1a group reported flu-like symptoms, while 26% of those in the placebo group experienced the same (Jacobs et al., 2001).

40.4.2. IFN β-1b (Betaseron)

40.4.2.1. Nomenclature, Approved Dosing Format, and Dosing Schedule

IFN β-1b (Betaseron) is a purified protein produced by recombinant DNA technology with 165 amino acids and approximate molecular weight of 18,500 daltons. Lyphylized vials contain 0.3 mg IFNβ-1b with mannitol (15 mg) and albumin (15 mg) as stabilizers. Reconstitution with the diluent supplied produces a 0.54% sodium chloride solution. IFN β-1b or Betaseron is also dispensed in a pre-filled syringe at a dose of 22 μg (6MIU) per dose for 66 μg/week. A higher dose of 44 μg (12 MIU) per dose for 132 μg per week is also available. The recommended dosing schedule is three times a week as a subcutaneous injection.

40.4.2.2. Adverse Effects of Betaseron (IFN-β-1b)

The most serious adverse reactions are depression, suicidal ideation and injection site necrosis. A single case of suicide, potentially due to depression, was reported in the initial phase

III trial of IFNβ-1b. The incidence of depression of any severity was 34% in the phase III clinical trial (The IFNβ MS Study Group, 1995). Injection site necrosis was reported in 5% of patients in the early controlled trials. The most commonly reported adverse reactions to IFNβ-1b (Betaseron) are lymphopenia, (lymphocyte cout <1500 mm^3), injection site reaction, asthenia, flu-like symptom complex, headache, and pain. The most frequently reported adverse reactions resulting in some clinical intervention, usually discontinuation of therapy or adjustment in dosage, were depression, flu-like symptom complex, injection site reactions, lymphopenia, increased liver enzymes, asthemia, hypertonia and myasthenia. As with other IFNβ therapy, female patients are warned that the therapy has not been evaluated as a human teratogen or during breast-feeding.

40.4.2.3. Mechanism of Action

The specific mechanism of action of IFNβ-1b in RR-MS is essentially unknown. Immunomodulatory effects of IFNβ-1b include enhancement of suppressor T cell activity, reduction of pro-inflammatory cytokine production, down regulation of antigen presentation and inhibition of lymphocytes trafficking across the BBB.

40.4.3. IFNβ-1a (Rebif)

40.4.3.1. Nomenclature, Approved Dosing Format, and Dosing Schedule

Rebif is the second IFNβ-1a product. This form of IFNβ-1a is a purified 166 amino acid glycoprotein with a molecular weight of ~22,500 daltons. As with Avonex, Rebif is produced by recombinant DNA technology from the genetically engineered human IFNβ gene. Rebif is formulated as a sterile solution in a pre-filled sterile syringe intended for subcutaneous injection, three times each week. Each 0.5 ml of Rebif contains either 22 mcg or 44 mcg IFNβ-1a, 2 or 4 mg human albumin, 10.9 mg mannitol, 0.16 mg sodium actrate and water for injection. While a single subcutaneous injection of the lower dose of this form of IFNβ-1a is below that of the IFNβ-1a IM product, the three times per week dosing schedule raises it above the once weekly dose for the IM preparation.

40.4.3.2. Adverse Effects of Rebif (IFN-β-1a)

The most common side effects involve flu-like symptoms with headache, fever, chills, muscle or joint ache, fatigue, pain or injection site inflammation. Less common side effects include nausea, stomach upset, diarrhea, a lower red blood cell count, sleep disturbances, dizziness, and infection.

40.4.3.3. Efficacy

Class I evidence shows a statistically significant delay in the confirmed progression of disability with IFNβ-1a IM and also shows a statistically significant reduction in relapse rate in a 2 year placebo-controlled clinical trial (PRISMS, 1998). This form of IFNβ-1a was available in Europe and Canada prior to approval by the FDA for use in the United States.

40.4.4. Glatiramer Acetate (Copaxone)

40.4.4.1. Nomenclature, Approved Dosing Format, and Dosing Schedule

Glatiramer acetate was termed copolymer-1 in early clinical trials. Glatiramer acetate is a synthetic compound made up of acetate salts of polypeptides that are found in myelin and contains four naturally occurring amino acids: L-glutamic acid, L-alanine, L-tyrosine, and L-lysine with an average molar fraction of 0.141, 0.427, 0.095, and 0.338, respectively. The average molecular weight of glatiramer acetate is 4,700–11,000 daltons. Copaxone is a sterile, lyophilized powder containing 20 mg of glatiramer acetate and 40 mg of manrntol that is supplied in single-use vials for subcutaneous administration after reconstitution with the diluent supplied (sterile Water for Injection) or in pre-filled syringes.

40.4.4.2. Side Effects of Glatiramer Acetate

Most common adverse effects in controlled trials include injection site reactions, vasodilatation, chest pain, asthenia, infection, pain, nausea, arthralgia, anxiety, and hypertonia. About 10% of patients experience an immediate post-injection reaction (flushing, chest pain, palpitations, anxiety, dyspnea, throat constriction, and urticaria). The symptoms are transient and self-limited, and usually do not require specific treatment. Transient chest pain was noted in 21% of Copaxone patients versus 11% in the placebo group with no long-term sequelae. Unlike therapy with the IFNβs, glatiramer acetate is not associated with flu-like symptoms.

40.4.4.3. Mechanism of Action and Efficacy

Glatiramer acetate is thought to stimulate T cells to change from harmful, pro-inflammatory cytokines to anti-inflammatory cytokines that work to reduce inflammation at lesion sites. In a two-year randomized, double-blind, controlled trial involving 251 ambulatory patients with RR MS, those taking the drug had a 29% reduction in annual relapse rate compared to subjects who were given a placebo (Johnson et al., 1998). Subsequent studies confirm the drug's long-term effectiveness in reducing the number and severity of exacerbations and demonstrate its ability to reduce the number of new, Gd-enhancing brain lesions on MRI (Johnson et al., 2000).

Follow up studies on glatiramer acetate show that the relapse rate continues to be lower after 6 years on therapy (Johnson et al., 2000). The last 3 years of the Johnson study were open label with the initial study a double-blinded protocol.

40.5. Therapies Approved for RR-MS

There are no Class I clinical trials by Goodlin criteria (Goodlin et al., 2002) comparing the approved therapies for RR-MS. The trials discussed in the following section are all post-marking trials of relatively short duration. Even so, valuable information can be found if their shortcomings are remembered.

40.5.1. Glatiramer Acetate Versus IFNβ-1a (IM) and IFNβ-1b

Published head to head trials of the FDA approved therapies for RR MS still involve relatively small numbers and relatively short time periods but they provide important information on efficacy and side effect profiles. The first such study was a retrospective review by Kahn that was followed by a prospective open label 18-month study comparing IFBβ-1a IM, IFNβ-1b and glatiramer acetate (Khan et al., 2001a). The subcutaneous form of IFNβ-1a (Rebif) was not FDA approved at the time. At the end of the 18-month prospective trial, 122 of 156 participants remained on their original therapy (79%). Relapse rate was the primary outcome measure. Participants on Glatiramer acetate had a significant reduction in relapses ($P < 0.0001$), as did those on IFNβ-1b ($p < 0.001$) compared to untreated patients. There was not a significant difference in the relapse rate for the control patients and those on IFβ-1a IM.

40.5.2. IFNβ-1a (IM) and IFNβ-1b

For the INCOMIN Trial, IFNβ-1a (IM Avonex) was compared to IFNβ-1b (Betaseron) in 188 MS patients with minimal baseline disability (EDSS of 1–3.5) (Durelli et al., 2002). The primary endpoint was the number of patients who remained relapse free during the 2 years trial. In this trial, 51% of MS patients who received IFNβ-1b (Betaseron) remained relapse free compared to 36% for those who received IFNβ-1a (IM Avonex) ($p = 0.036$). Disability status was also better for the IFNβ-1b group with only 14% of patients progressing 1 or more EDSS step compared to 20% for those on IFNβ-1a (IM). The INCOMIN Trial was one of the first to provide evidence that IFNβ-1a in the higher dose and dosing schedule was more efficacious than a low dose given once weekly.

40.5.3. IFNβ-1a Subcutaneous Versus IFNβ-1a IM

IFNβ-1a (Rebif) at 44 mcg three times per week was more effective than IFNβ-1a IM (Avonex) in reducing relapses and MRI activity over a 48 week study in a large North American and Candian trial of over 600 patients termed EVIDENCE (Panitch et al. 2002). A significant relapse and MRI activity reductions was reported in patients changing therapy from Avonex to Rebif over an average of 34 weeks. While the higher frequency subcutaneous dosing with Rebif showed a modest positive effect over the IM Avonex form of IFNβ-1a, a higher rate of liver toxicity was reported for the Rebif (18%) versus (9.8%) Avonex ($p < .003$) (Panitch et al., 2002). While verifying that higher dose and more frequent dosing of IFNβ-1a is more effective, this trial brought to the forefront the liver toxicity side effects of the higher dose therapies.

40.5.4. Glatiramer Acetate Versus IFNβ-1a Subcutaneous, IFNβ-1a IM and IFNβ-1b

Almost three hundred patients enrolled in the Haas (2003) prospective head to head study of all four FDA approved therapies for RR-MS. All four therapies show a significant reduction in relapse rate ($p < 0.02$ for the IFNs) with glatiramer acetate showing the most significant reduction ($p < .001$). Discontinuation rates were also higher on the IFNβs; 20–30% by 6 months compared to 8% for those on glatiramer acetate by the end of the 2 year study.

40.6. FDA Approved Therapies for Other Forms of MS

40.6.1. Mitoxantrone (Novantron)

Mitoxantrone (Novantron) received FDA approval for use in progressive MS in December of 2000. The recommended dosage for treatment of multiple sclerosis is 5–12 mg/mL IV every 3 months. Acute side effects of mitoxantrone include nausea and alopecia. Because of cumulative cardiotoxicity, the drug can be used for only two to three years (or for a cumulative dose of 120–140 mg per m^2). A prospective study of 73 MS patients with PP MS who were followed for 10 courses of mitoxantrone (10 mg.mm^2 body surface) combined with methyprednisolone therapy. At mean follow up time of 23.4 months (10–57 month range), no significant changes in end-diastolic diameter, end-systolic diameter, fractional shortening was reported (Zingler et al., 2005). No evidence of signs of congestive heart failure was seen in any of the participants. However ongoing monitoring at higher cumulative doses is warranted. Mitoxantrone is a chemotherapeutic agent that should be prescribed and administered only by experienced health care professionals. A phase-III, randomized, placebo-controlled, multicenter trial found that mitoxantrone, an anthracenedione anti-neoplastic agent, reduces the number of treated MS relapses by 67%. Mitoxantrone slows progression on the EDSS, Ambulation Index, and MRI measures of disease activity (Hartung et al., 2002); it is effective and is recommended for use in patients with worsening forms of MS.

40.7. Promising Areas for a New Generation of MS Therapies

Two general areas hold the most promise in improving treatment of MS. Since all currently approved therapies require injection, the first general area involves the search for oral therapies. The second area builds upon experience with mitoxantrone and involves testing of immunosuppressant cancer chemotherapy for use in MS.

40.7.1. Oral Therapy for MS

Preliminary studies evaluating an oral version of glatiramer acetate did not prove effective in changing the clinical outcome measures related to RR MS. Early trials of ingested IFN-alpha did have some biological effect (Brod et al., 1997) but have not shown the promised efficacy. Currently two other oral therapies have shown positive results in phase II clinical trials (NMMS, 2005).

The first drug, FTY20 (Novartis Pharm Corp), binds to the docking mechanism site (sphingosine-1-phosphate receptor) on immune cells including both T and B cells. The phase II clinical trial was an international double blind placebo controlled study of 281 MS patients with active RR-MS who took one of two doses of FTY20 or placebo daily for 6 months. The primary outcome measure was a change in the number of T2 enhancing lesions on monthly brain MRI. Both treatment groups had significantly fewer enhancing lesions at 6 months compared to the placebo group. The treatment groups also had fewer volume enhancing lesions and new non-enhancing lesions on MRI. Eighty six percent of those in the treatment groups remained relapse free for the 6-month study period compared to 70% for the placebo group. The most frequently reported side effects of FTY20 include mild headache, colds, and gastrointestinal disorders such as diarrhea and nausea.

The promising second oral therapy is Temsirolimus (Wyeth Pharm). Temsirolimus is believed to block the proliferation of immune T cells activated by interleukin, IL-2. The phase II clinical trial of Temsirolimus was also an international double blind placebo controlled trial. The trial involved 296 patients with either RR MS or SP MS with relapses. Participants received one of three doses of oral temsirolimus or placebo daily for 9 months. The primary outcome measure was the number of enhancing lesions after 9 months in study. By 32 weeks into the study, those in the highest treatment dose had 47.8% fewer new enhancing lesions compared to those on placebo. The high dose group also had 51% fewer relapses than the placebo group. Side effects included mouth ulceration or inflammation, menstrual dysfunction, hyperlipidemia and rashes.

40.7.2. Immunosuppressant Chemotherapies and Stem Cell Transplantation for MS

Randomized controlled trials of azathioprine, methotrexate, cladribine, intravenous immunoglobulin and cyclophosphamide have not shown definite modification to the aggressive course of PP MS (Leary and Thompson, 2005). However, these immunosuppressant therapies continue to be evaluated in patients with active aggressive MS.

Evaluation of combination therapies including cyclophosphamide show promise in decreasing the number of Gd+lesions and slowing clinical activity in some patients where IFNβ alone is not successful (Smith et al., 2005; Patti et al., 2004). A small study of 14 rapidly progressing RR MS patients given monthly IV cyclophosphamide indicates improvement in neurologic stability (Khan et al., 2001b). Further evaluation is needed to confirm the efficacy.

Autologous hematopoietic stem cell transplantation (AHSCT) was evaluated in a group of MS patients with aggressive and advanced disease (Nash et al., 2003) with some promise but also with evidence that neurologic loss continued. One important hypothesis from that study was the discordant nature of inflammatory measures and functional deficits (Healey et al., 2004).

Summary

With FDA approval of three forms of IFNβ and glatiramer acetate therapy, RR MS has become a treatable disease. However, disease progression into a non-inflammatory chronic progressive disease is not effected significantly by current therapies. In addition only the immunomodulator, mitoxantrone has been approved for use in the chronic progressive form of MS. The chronic long term nature of the disease and the temporal disconnect between the identification of demyelinating CNS lesions and physical disability continue to pose problems in the development of therapies that successfully halt disability progression in MS.

Review Questions/Problems

Your brother had some "double" vision problems lately and has felt very tired. Your family physician sent him to a neurologist who scheduled several tests. On follow up, the neurologist says that your brother has optic neuritis that may potentially be associated with multiple sclerosis. The neurologist also recommends that your brother consider beginning an interferon β therapy.

1. **What clinical data would support initiating IFNβ therapy even before a definitive diagnosis of MS?**
2. **What clinical information, in association with verification of the clinical symptoms of optic neuritis and fatigue would be sufficient to verify that the diagnosis is definitive MS using the revised MacDonald criteria?**
3. **Your brother has received information on the two FDA approved IFNβ-1a products. What are the differences in dose, dosing schedule, potential for side effects and efficacy for the two products?**
4. **Over the next year, your brother has 2 more episodes that include problems with walking and numbness in his right hand. Are any other criteria necessary for a definitive diagnosis of MS by the revised McDonald criteria?**

5. What would you expect an MRI to show if your brother truly has MS?

6. Your brother has been on IFNβ-1a IM therapy for 9 months and is still complaining of flu like symptoms and muscular aches and pain following each injection. What other therapeutic option(s) does he have?

7. How has the diagnosis of MS changed your brother's life expectancy?

8. Your brother has just had a major relapse that has left him unable to walk. What is the standard therapy for this type of acute relapse and what is the postulated mechanism of action?

9. If the previous scenario had been your sister rather than your brother, would the prognosis have been different? In either case, what is the life expectancy for them if they are in their early 30s at diagnosis?

10. Your brother continues to progress rapidly with no apparent remission even on standard immunotherapy for RR-MS. What therapy is FDA approved for rapidly progressing MS?

11. Which of the FDA approves therapies for MS was originally designed to mimic a portion of the molecular sequence of myelin:

a. Avonex
b. Betaseron
c. Copaxone
d. Rebif

12. Score for an individual with normal motor function on the Kurtzke Expanded Disability Scoring Scale (EDSS)

a. 0
b. 1
c. 5
d. 10

13. The Multiple Sclerosis Functional Composite (MSFC) does not test

a. motor function of the lower extremities
b. motor function of the upper extremities
c. cognitive changes
d. bowel and bladder function

14. The primary outcome measure for phase III FDA trials for the IFNβ therapies

a. number and time between relapses
b. evidence of lesions on MRI
c. a significant change in the MSFC
d. time to complete a 25 foot walk

15. Prokarin is a trans dermal patch containing histamine and this compound

a. glatiramer acetate
b. IFN β
c. caffeine
d. corticosteroida

16. A therapy that has not shown efficacy in the treatment of MS

a. IFNγ
b. IFNβ-1b
c. natalizunab
d. mitoxantrone
e. methyl predisolone

17. The FDA approved therapy for relapsing remitting MS with the smallest % reduction in relapses at 12 months by the Johnson Class I data

a. Avonex
b. Betaseron
c. Copaxone
d. Rebif

18. Brain atrophy is considered significant is it exceeds the these normal changes related to aging

a. <0.5–1%
b. >0.5–1%
c. >1–1.5%
d. >5%

19. This approved therapy for relapsing remitting MS included a warning to monitor depression because of a suicide during the phase III trial.

a. Avonex
b. Betaseron
c. Copaxone
d. Rebif

20. In the Kahn head to head trial which of the following therapies was not evaluated?

a. Avonex
b. Betaseron
c. Copaxone
d. Rebif

References

Arnason BGW (1999a) Treatment of multiple sclerosis with interferon beta. Biomed Pharmacother 53:344–350.

Arnason BGW (1999b) Immunologic therapy of multiple sclerosis. Ann Rev Med 50:291–302.

Barkhof F, Filippi M, Miller DH, Scheltens P, Campi A, Polman CH, Comi G, Ader HJ, Losseff N, Valk J (1997) Comparison of MRI criteria at first presentation to predict conversion to clinically definite multiple sclerosis. Brain 120:2059–2069.

Barnes D, Hughes RA, Morris RW, Wade-Jones O, Brown P, Britton T, Francis DP, Perkin GD, Rudge P, Swash M, Kadifi H, Farmer S, Frankel J (1997) Randomized trial of oral and intravenous methylprednisolone in acute relapses of multiple sclerosis. Lancet 349:902–906.

Barnett MH, Prineas JW (2004) Relapsing and remitting multiple sclerosis: Pathology of the newly forming lesion. Ann Neurol 55(4):458–468.

Becher B, Giacomini PS, Pelletier D, McCrea E, Prat A and Antel JP (1999) Interferon gamma secretion by peripheral blood T cell subsets in MS: Correlation with disease phase and Interferon beta therapy. Ann Neurol 45:247–250.

Beck J, Rondot P, Catinot L, Falcot E, Kirchner H, Wietzerbin J (1988) Increased production of interferon gamma and tumor necrosis factor precedes clinical manifestation in multiple sclerosis: Do cytokines trigger off exacerbations? Acta Neurol Scand 78:318–323.

Beck RW, Chandler DL, Cole SR, Simone JH, Jacobs LD, Kinkel RP, Selhorst JB, Rose JW, Cooper JA, Rice G, Murray TJ, Sandrock AW (2002) Interferon beta 1a for early multiple sclerosis: CHAMPS trial subgroup analysis. Ann Neurol 51(4):481–490.

Benedict RH, Carone DA, Bakshi R (2004) Correlating brain atrophy with cognitive dysfunction, mood disturbances, and personality disorder in multiple sclerosis. J Neuroimaging 14(3 Suppl):36S–45S.

Bitsch A, Schuchardt J, Bunkowski S, Kuhlmann T, Bruck W (2000) Acute axonal injury in multiple sclerosis. Correlation with demyelination and inflammation. Brain 123(Pt 6):1174–1183.

Bjartmar C, Trapp BD (2001) Axonal and neuronal degeneration in multiple sclerosis: Mechanisms and functional consequences. Curr Opin Neurol 14(3):271–278.

Bjartmar C, Trapp BD (2003) Axonal degeneration and progressive neurologic disability in multiple sclerosis. Neurotox Res 5(1–2):157–164.

Bjartmar C, Wujek JR, Trapp BD (2003) Axonal loss in the pathology of MS: Consequences for understanding the progressive phase of the disease. J Neurol Sci 206(2):165–171.

Brod SA, Kernab RH, Nelson LD, Marshall GD Jr, Henninger EM, Khan M, Jin R, Wolinsky JS (1997) Ingested IFN-alpha has biological effects in humans with relapsing remitting multiple sclerosis. Mult Scler Feb 3(1):1–7.

Brusaferri P, Candelise L (2000) Steroids for multiple sclerosis and optic neuritis: A meta-analysis of randomized controlled clinical trial. J Neurol 247:435–442.

Compston A (1997) Genetic epidemiology of multiple sclerosis. J Neurol Neurosurg Psychiatry 62(6):553–561.

Cree BAC, Lamb S, Morgan K, Chen A, Waubant E, Gemain C (2005) An open label study of the effects of rituximab in neuromyelitis optica. Neurology 64:1270–1272.

De Stefano N, Narayanan S, Francis GS, Armaoutelis R, Tartaglia MC, Antel JP, Matthews PM, Arnold DL (2001) Evidence of axonal damage in the early stages of multiple sclerosis and its relevance to disability. Arch Neurol 58:65–70.

Dorovin-Zis K, Bowman PD, Prameya R (1992) Adhesion and migration of human polymorphonuclear leukocytes across cultured bovine brain microvessel endothelial cells. J Neuropathol Exp Neurol. 51(2):194–205.

Doskoch D (2005) Primary progressive MS: The benefits of "negative" trials. MS Exchange 9(3):1–3.

Durelli L, Verdun E, Barbero P, Bergui M, Versino E, Ghezzi A, Montanari E, Zalfaroni M, Independent Comparison of Interferon (INCOMIN) trial Study Group (2002) Every other day interferon beta-1b versus once-weekly interferon beta-1a for multiple sclerosis: Results of a 2 year prospective, randomized multicenter study (INCOMIN). Lancet 35:1453–1460.

FDA Consum (2005a) New treatment, new hope for those with multiple sclerosis. Mar-Apr 39(2):10–17.

FDA Consum (2005b) MS Drug withdrawn from market. May-June 39(3):3.

Filippi M, Horsfiels MA, Morrissey SP, MacManus DG, Rudge P, MacDonald WI, Miller DH (1994) Quantitative brain MRI lesion load predicts the course of clinically isolated syndromes suggestive of multiple sclerosis. Neurol 44:635–641.

Gillson G, Richard TL, Smith RB, Wright JV (2002) A double-blind pilot study of the effect of Prokarin on fatigue in multiple sclerosis. Mult Scler Feb 8(1):20–25.

Goodlin DS, Frohman EM, Garmany GP (2002) Disease modifying therapies in multiple sclerosis: Report of the Therapeurics and Technology Assessment Subcommittee of the American Academy of Neurology and the MS council for Clinical Practice Guidelines. Neurology 58:169–178.

Grossman RI, Braffman BH, Bronson JR, Goldberg HI, Silberberg DH, Gonzalez-Scarano F (1988) Multiple sclerosis: Serial study of gadolinium enhanced MRI imaging. Radiology 169(1):117–122.

Haas J (2003) Onset of clinical benefit of glatiramer acetate (Copaxone) acetate in 255 patients with relapsing remitting multiple sclerosis (RRMS). Neurol 60(suppl):P06–105.

Hardmeier M, Wagenpfeil S, Freitag P, Fisher E, Rudick RA, Kooijmans-Coutinho M, Clanet M, Radue EW, Kappos L, European rIFN beta-1a in relapsing MS dose comparison trial study group (2003) Atrophy is detectable within a 3 month period in untreated patients with active relapsing remitting multiple sclerosis. Arch Neurol 60(12):1736–1739.

Hartung HP, Gonsette R, Konig N, Kwiecinski H, Guseo A, Morrissey SP, Krapf H, Zwinger T, Mitoxantrone in Multiple Sclerosis Study Group (MIMS)l (2002) Mitoxantrone in progressive multiple sclerosis: A placebo-controlled, double-blind, randomized, multicenter trial. Lancet 360:2018–2025.

Healey KM, Pavletic SZ, Al-Omaishi J, Leuschen MP, Pirrucello SJ, Filipi ML, Enke C, Ursick MM, Hahn F, Bowen JD, Nash RA (2004) Discordant functional and inflammatory parameters in multiple sclerosis patients after autologous haematopoietic stem cell transplantation. Mult Scler 10:284–289.

Isaac C, Li DK, Genton M, Jardine C, Grochowski E, Palmer M, Kastrukoff LF, Oger J, Paty D (1988) Multiple sclerosis: A serial study using MRI in relapsing patients. Neurology 38(10):1511–1515.

Jacobs LD, Cookfair DL, Rudick RA, Herndon RM, Richert JR, Salazar AM, Fischer JS, Goodkin DE, Granger CV, Simin JH, Alam JJ, Bartosak DM, Bourdette DN, Braiman J, Brownschneidle CM, Coats ME, Cohan ME, Dougherty DS, Kinkel RP, Mass MK, Munschauer FE, Priori RL, Pullicino PM, Scheokman BJ, Whitman RH (1996) Intramuscular interferon beta 1a for disease progression in relapsing remitting multiple sclerosis. The Multiple Sclerosis Collaborative Research Group (MSCRG). Ann Neurol 39:285–294.

Jacobs LD, Beck RW, Simon JH, Kinkel RP, Brownschneidle CM, Murray TJ, Simonian NA, Slasor PJ, Sandrock AW (2001) Intramuscular interferon beta-1a therapy initiated during a first denyelinating event in multiple sclerosis. CHAMPS Study Group. N Engl J Med 343(13):898–904.

Johnson KP (2005) Practicing responsible evidence-based medicine: Applications for the management of MS patients. Intl J MS Care 6(4):4–8.

Johnson KP, Brooks BR, Cohen JA, Ford CC, Goldstein J, Liska RP, Myers LW, Panitch HS, Rose JW, Schiffer TB, Vollmer T, Weiner LP, Wolinsky JS, Copolymer 1 Multiple Sclerosis Study Group (1995) Copolymer I reduces relapse rate and improves disability in relapsing remitting multiple sclerosis: Results of a phase III multicenter double-blind placebo-controlled trial. Neurol 45:1268–1276.

Johnson KP, Brooks BR, Cohen JA, Ford CC, Goldstein J, Liska RP, Myers LW, Panitch HS, Rose JW, Schiffer TB, Vollmer T, Weiner LP, Wolinsky JS, Copolymer 1 Multiple Sclerosis Study Group (1998) Extended use of glatiramer acetate (Copaxone) is well tolerated and maintains its clinical effect on multiple sclerosis relapse rate and degree of disability. Neurol 50:701–708.

Johnson KP, Brooker BR, Ford CC, Goodman A, Guarnacci J, Lisak RP, Myers LW, Panitch HS, Pruitt A, Rose JW, Kachuck N, Wolinsky JS, COP-1 Study Group (2000) Sustained clinical benefits of glatiramer acetate in relapsing multiple sclerosis patients observed for 6 years. Mult Scler 6:255–266.

Kennedy MK, Dal Canto MC, Trotter JL, Miller SD (1988) Specific immune regulation of chronic-relapsing experimental allergic encephalomyelitis in mice. J Immunol 41(1):2986–2993.

Khan OA, Zvartau-Hind M, Caon C, Din MU, Cochran M, Lisak D, Tselis AC, Kambolz JA, Garbern JY, Lisak RP (2001a) Effect of monthly intravenous cyclophosphamide in rapidly deteriorating multiple scleosis patients resistant to conventional therapy. Mult Scler 3:185–188.

Khan OA, Tslis AC, Kamholz JA, Garbern JY, Lewis RA, Lisak RP (2001b) A prospective open-label treatment trial to compare the effect of IFN beta-1a (Avonex), IFN beta-1b (Betaseron) and glatiramer acetate (Copaxone) on the relapse rate in relapsing-remitting multiple sclerosis: Results after 18 months of therapy. Mult Scler 7:349–353.

Khoury SJ, Guttmann CR, Orav EJ, Hohol MJ, Ahn SS, Hsu L, Kikinis R, Mackin GA, Jolesz FA, Weiner HL (1994) Longitudinal MRI in multiple sclerosis: Correlation between disability and lesion burden. Neurology 44(11):2120–2124.

Kleinschmidt-DeMasters BK, Tyler KL (2005) Progressive multifocal leuko-encephalopathy complicating treatment with natalizumab and interferon beta-1a for multiple sclerosis. N Engl J Med 353:369–374.

Kurtzke JF (1983) Rating neurologic impairment in multiple sclerosis: An expanded disability status scale (EDSS). Neurology 33(11):1444–1452.

Kurtzke JF (1991) Multiple sclerosis: Changing times. Neuroepidemiology 10:1–8.

Lampert P, Carpenter S (1965) Electron microscope studies on the vascular permeability and the mechanisms of demyelination in experimental allergic encephalomyelitis. J Neuropathol Exp Neurol 29:11–24.

Langer-Gould A, Atlas SW, Green AJ, Bollen AW, Pelltier D (2005) Progressive multifocal leukoencephalopathy in a patient treated with natalizumab. NEJM 353:375–381.

Lassmann H (2001) Classification of demyelinating diseases at the interface between etiology and pathogenesis. Curr Opin Neurol 14(3):253–258.

Leary SM, Thompson AJ (2005) Primary progressive multiple sclerosis: Current and future treatment options. CNS Drugs 19(5):369–379.

Leuschen MP, Filipi M, Healey K (2004) A randomized open label study of pain medications (naproxen, acetaminophen, and ibuprofen) for controlling side effects during initiation of IFN β-1a therapy and during its ongoing use for relapsing-remitting multiple sclerosis. Mult Scler 10:636–642.

Lindberg RL, DeGroot CJ, Certa U, Raviol T, Hoffmann F, Kappos L, Leppert D (2004) Multiple sclerosis as a generalized CNS disease—comparative microarray analysis of normal appearing white matter and lesions in secondary progressive multiple sclerosis. J Neuroimmunol 152:154–167.

Lublin FD, Reingold SC (1996) Defining the clinical course of multiple sclerosis: Results of an international survey. National Multiple Sclerosis Society (USA) Advisory Committee on clinical trials of new agents in multiple sclerosis. Neurology 46:907–911.

Lucchinetti CF, Bruck W, Rodriguez M, Lassmann H (1996) Distinct patterns of multiple sclerosis pathology indicates heterogeneity on pathogenesis. Brain 6(3):259–274.

Lucchinetti C, Bruck W, Parisi J, Scheithauer B, Rodriguez M, Lassmann H (2000) Heterogeneity of multiple sclerosis lesions: Implications for the pathogenesis of demyelination. Ann Neurol 47(6):707–717.

Martinez-Cacenes EM (1998) Amelioration of flu-like symptoms at the onset of interferon beta-1b therapy in multiple sclerosis by low dose oral steroids is related to a decrease in interleukin-6 induction. Ann Neurol 44(4):682–685.

McDonald WI, Compston A, Edan G, Goodkin D, Hartung HP, Lublin D, McFarland HF, Paty DW, Polman CH, Reingold SC, Sandberg-Wolheim M, Sibley W, Thompson A, DenNoort S, Weinshenken BV, Wolinsky JS (2001) Recommended diagnostic criteria for multiple sclerosis: Guidelines from the International Panel on the diagnosis of multiple sclerosis. Ann Neurol 50:121–127.

Miller DH (2004) Biomarkers and surrogate outcomes in neurodegenerative disease: Lessons from multiple sclerosis. Neuro Rx Apr (2):284–294.

Miller DH, Khan OA, Sheremata WA, Blumhardt LD, Rice GPA, Libonati MA, Wilmer-Hulme AJ, Dalton CM, Miszkiel KA, O'Connor PW, Intl Natalizumab MS Trial Group (2003) A controlled trial of natalizumab for relapsing remitting multiples sclerosis. N Engl J Med 348(1):15–23.

Mora JS, Kao KP, Munsai TL (1997) Indomethacin reduces the side effects of intrathecal interferon. N Engl J Med 310:126–127.

Munschauer FE, Kinkel RP (1997) Managing side effects of interferon β in patients with relapsing remitting multiple sclerosis. Clin Ther 19(5):883–893.

Nash RA, Bowen JD, McSweeney PA, Pavletic SZ, Maravilla KR, Park MS, Storek J, Sullivan KM, Al-Omaishi J, Corboy JR, DiPersio J, Georges GE, Gooley TA, Holmberg LA, LeMaistre CF, Ryan K, Openshaw H, Sunderhaus J, Storb R, Zunt J, Kraft GH (2003) High dose immunosupreessiove therapy and autologous peripheral blood stem cell transplantation for severe multiple sclerosis. Blood 102:2364–2372.

National Multiple Scleosis Society (2005) Two new oral drugs show positive results in preliminary trials in multiple sclerosis. NMSS Res Clin Updates. June 2, 2005:1–2.

O'Connor P, Miller D, Riester K, Yang M, Panzara M, Dalton C, Mitzkiel K, Khan O, Rice F, Sheremata W, Intl Natalizumab Trial Group (2005) Relapse rates and enhancing lesions in a phase II trial of natalizumab in multiple sclerosis. Mult Scler Oct 11(5):568–572.

Panitch HS, Hirsch RL, Haley AS, Johnson KP (1987) Exacerbation of multiple sclerosis in patients treated with gamma interferon. Lancet 1:893–895.

Panitch H, Goodlin DS, Francis GA Chang P, Cohen PK, O'Connor P, Monaghan E, Li D, Weinshenker B, EVIDENCE Study Group, EVI of Interferon Dose response: European North American Comparative Efficacy; Univ. British Columbia MS/MRI Research Group (2002) Evidence of interferon dose response: European North American comparative efficacy. Neurology 591:1496–1506.

Patti F, Amato MP, Filippi M, Gallo P, Trojano M, Comi GC (2004) A double blind placebo-controlled phase II, add-on study of cyclophosphamide (CTX) for 24 months in patients affected by multiple sclerosis on a background therapy with interferon-beta study denomination: CYCLIN. J Neurol 223(1):69–71.

Paty DW, Noseworthy JH, Ebers GC (1997) Diagnosis of Multiple Sclerosis. In: Multiple Sclerosis Contemporary Neurology Series (Paty DW, Ebers GC, eds), pp 48–134. Philadelphia PA: FA Davis Co.

Petereit HF, Richter N, Pukrop R, Bamborschke S (2000) Interferon gamma production in blood lymphocytes correlates with disability score in multiple sclerosis patients. Mult Scler 6:19–23.

Polman CH, Reingold SC, Edan G, Filippi M, Hartung HP, Kapos L, Lublin FD, Metz LM, McFarland HF, O'Connor PW, Sandberg-Woolheim M, Thompson AJ, Weinshenker BG, Wolinsky JS (2005) Diagnostic criteria for multiple sclerosis: 2005 revisions to the "McDonald Criteria". Ann Neurol 58 (6):840–846.

Poser CM, Paty DW, Schneider L, Scheinberg L, MacDonald WI, Davis FA, Ebers GC, Johnson KP, Sibley WA, Silberberg DH, Tourtellotte WW (1983) New diagnostic criteria for multiple sclerosis: Guidelines for research protocol. Ann Neurol 33:227–231.

PRISMS (Prevention of relapses and Disability by Interferon β 1a subcutaneously in MS Study Group) (1998) Randomized double-blind placebo-controlled study of interferon β1a in relapsing remitting multiple sclerosis. Lancet 352:1498–1504.

Raine CS (1994) The Dale E. McFarlin Memorial Lecture: The immunology of the multiple sclerosis lesion. Ann Neurol 36(Suppl): S61–S72.

Reess J, Haas J, Gabriel K, Fuhlrott A, Fiola M (2002) Both paracetamol and ibuprofen are equally effective in managing flu-like symptoms in relapsing-remitting multiple sclerosis patients during interferon beta-1a (AVONEX) therapy. Mult Scler 8(1):15–18.

Rice GPA, Ebers GC, Lubkin FD, Knobler RL (1999) Ibuprofen treatment versus gradual introduction of interferon β in patients with MS. Neurology 52:1893–1895.

Rice GP, Hartung HP, Calabresi PA (2005) Anti-alpha4 integrin therapy for multiple sclerosis: Mechanisms and rationale. Neurology 26(8):1336–1342.

Rio J (1998) Low-dose steroids reduce flu-like symptoms at the initiation of IFN beta-1b in relapsing remitting MS. Neurology 50(6):1910–1912.

Rio J (2000) Ibuprofen treatment versus gradual introduction of interferon beta-1b in patients with MS. Neurology 54(8):1710.

Selmaj K, Brosman CF, Raine CS (1991) Colocalization of lymphocytes bearing gamma delta T-cell receptor and heat shock protein hsp65+ oligodendrocytes in multiple sclerosis. Proc Natl Acad Sci USA 88(15):6452–6456.

Simon JH (2001) Brain and spinal cord atrophy in multiple sclerosis: Role as a surrogate measure of disease progression. CNS Drugs 15(6):427–436.

Simon JH, Jacobs LD, Campion MK, Rudick R (1999) A longitudinal study of brain atrophy in relapsing multiple sclerosis. Neurology 53:139–148.

Smith KJ, McDonald WI (1999) The pathophysiology of multiple sclerosis: The mechanism underlying the production of symptoms and the natural history of the disease. Philos Trans R Soc Lond B Biol Sci 354:1649–1673.

Smith DR, Weinstock-Guttman B, Cohen JA, Wei X, Gutmann C, Bakshi R, Olek M, Stone L, Greenberg S, Stuart D, Orav J, Stuart W, Weiner H (2005) A randomized blinded trial of combination therapy with cyclophosphamide in patients with active multiple sclerosis on interferon beta. Mult Scler 5:573–582.

The IFNβ Multiple Sclerosis Study Group (1993) Interferon beta 1b is effective in relapsing remitting multiple sclerosis I. Clinical results of a multi-center randomized, double-blinded, placebo-controlled trial. Neurology 43:655–661.

The IFNβ Multiple Sclerosis Study Group and The University of British Columbia MS/MRI Study Group (1995) Interferon β-1b in the treatment of multiple sclerosis: Final outcome of the randomized controlled trial. Neurology 52:1893–1895.

Thompson AJ, Miller D, Youl B, MacManus D, Moore S, Kingsley D, Kendall B, Feinstein A, McDonald WI (1992) Serial gadolinium-enhanced MRI in relapsing/remitting multiple sclerosis of varying disease duration. Neurology 42(1):60–63.

Thompson AJ, Montalban X, Barkhof F, Brochet BL, Filippi M, Miller DH, Polman CH, Stevenson VL, MacDonald WI (2000) Diagnostic criteria for primary progressive multiple sclerosis: A position paper. Ann Neurol 47(6):831–835.

Trapp BD, Peterson J, Ransohoff RM, Rudick R, Mork S, Bo L (1998) Axonal transection in the lesions of multiple sclerosis. N Engl J Med 338:278–285.

Tsunoda I, Fujinami RS (2002) Inside-Out versus Outside-In models for virus induced demyelination: Axonal damage triggering demyelination. Springer Semin Immunopathol 24:105–125.

Vollmer TL, Philips JT, Goodman AD, Agius MA, Libonati MA, Giacchino JL, Grundy JS (2004) An open-label safety and drug interaction study of natalizumab (Antegren) in combination with interferon-beta (Avonex) in patients with multiple sclerosis. Multi Scler Oct 10(5):511–520.

Weinshenker BG, Ebers GC (1987) The natural history of multiple sclerosis. Can J Neurol Sci 14(3):255–261.

Weinshenker BG, Bass B, Rice GP, Noseworthy J, Carriere W, Baskerville J, Ebers GC (1989) The natural history of multiple sclerosis: A geographically based study. 2. Predictive value of the early clinical course. Brain 112(Pt 6):1419–1428.

Weinshenker BG, Rice GP, Noseworthy JH, Carriere W, Baskerville J, Ebers GC (1991) The natural history of multiple sclerosis: A geographic based stuffy 3: Multivariant analysis of predictive factors and models of outcome. Brain 114:1045–1056.

Whitaker JN, McFarland HF, Rudge P, Reingold SC (1995) Outcome assessment in multiple sclerosis clinical trials: A critical analysis. Mult Scler 1:37–47.

Willoughby EW, Grochowski E, Li DK, Oger J, Kastrukoff LF, Paty DW (1989) Serial magnetic resonance scanning in multiple sclerosis: A second prospective study in relapsing patients. Ann Neurol 25(1):43–49.

Wingerchuk DM, Weinshenker BG (2005) Neuromyelitis optica. Curr Treat Options Neurol 7(3):173–182.

Wolinsky JS (2005) Diagnostic criteria for multiple sclerosis: 2005 revisions to the "McDonald Criteria". Ann Neurol 58 (6):840–846.

Young IR, Hall AS, Pallis CA, Legg NJ, Bydder GM, Steiner RE (1981) Nuclear magnetic resonance imaging of the brain in multiple sclerosis. Lancet 2(8255):1063–1066.

Zingler YC, Nabaur M, Jahn K, Gross A, Hohlfeld R, Brandt T, Strupp M (2005) Assessment of potential cardiotoxic side effects of mitoxantrone in patients with multiple sclerosis. Eur Neurol 54(1):28–38.

Zivadinos R, Bakshi R (2004) Central nervous system atrophy and clinical status in multiple sclerosis. J Neuroimaging 14(3 Suppl):27S–35S.

41
HIV-Associated Dementia

Miguel G. Madariaga and Susan Swindells

Keywords HIV infection; dementia; antiretroviral therapy; adjuvant therapy

41.1. Introduction

This chapter reviews the use of antiretroviral therapy (ART) including its impact in altering the natural course of human immunodeficiency virus (HIV) disease; an overview of current regimens in use; factors influencing the choice of initial therapy; complications and adverse effects; and treatment failure, and resistance. The focus is on management of primary HIV-related diseases of the nervous system. While there is no specific therapy for HIV-associated dementia (HAD), both approved and developmental antiretroviral agents are discussed with a view to their effect on central nervous system (CNS) viral infection. Adverse drug effects that impact the nervous system are specifically discussed, as are pharmacological interactions between antiretroviral agents and other medications. The significance of cerebrospinal fluid (CSF) penetration by antiretroviral agents is reviewed. Adjunctive therapies, adherence to ART, treatment of comorbidities and coinfections that may worsen neurologic conditions are reviewed and include psychiatric illnesses and substance abuse. Lastly, non-pharmacological treatment strategies, education of patients and caregivers, graded assistance, behavior modification, and environment modifications are covered.

41.1.1. Overview of HIV-Related Brain Disease

Invasion of HIV into the CNS may occur early after infection, and a picture of aseptic meningitis in HIV primary infection may represent the arrival of infected macrophages. Later on, most chronically infected individuals exhibit some evidence of CNS involvement during progressive disease (Navia and Price, 2005). The target cells for HIV in the CNS are microglia and the macrophages. Once invaded, these cells express both structural and regulatory proteins and produce infectious progeny virions. Astrocytes can also be infected in small numbers, but are only able to produce early gene proteins, and are not capable of producing infected virions (Di Stefano et al., 2005). HIV may invade the central nervous system by one or both of these proposed mechanisms:

The "Trojan Horse" model proposes that infected monocytes (but also macrophages and $CD4^+$ T cells) cross the blood–brain barrier in response to chemokine stimulation. After infiltrating brain parenchyma, these cells also release additional chemokines and further attract additional inflammatory cells (Strelow et al., 2001).

Alternatively, endothelial cells and astrocytes (which compose the blood–brain barrier) are infected directly (Strelow et al., 2001). A variant of this theory is that infectious virions are transported in the cytoplasm of astrocytes and endothelial cells, and eventually transferred to central nervous system cells by transcytosis or macropynocytosis) (Marechal et al., 2001).

41.1.2. Nomenclature

HAD is also referred to as HIV dementia, AIDS Dementia Complex, subacute encephalopathy, HIV encephalopathy, and HIV-1-associated cognitive-motor complex. A staging system of HAD severity was initially developed using clinical manifestations, illustrated in Table 41.1 (Price and Brew, 1988). This classification is also referred to as the Memorial Sloan-Kettering (MSK) staging system. More recently, the World Health Organization and the American Academy of Neurology developed a schema which divides the disorder into two main categories of HIV-associated minor motor cognitive disorder, and HIV-1 associated cognitive-motor complex [American Academy of Neurology (AAN), 1991].

41.1.3. Natural History of HAD

During the 1990s, several investigators identified potential risk factors for the development of HAD. These include increasing age; a history of intravenous drug use, and presence of an AIDS defining illness (Wang et al., 1995). In another cohort study, development of HAD correlated with age, lower hemoglobin,

T. Ikezu and H.E. Gendelman (eds.), *Neuroimmune Pharmacology*.
© Springer 2008

TABLE 41.1. Clinical staging of HIV-associated dementia.

Stage	Clinical description
Stage 0 (normal)	Normal mental and motor function
Stage 0.5 (equivocal/ subclinical)	Absent, minimal, or equivocal symptoms without impairment of work or capacity to perform ADL. Mild signs (snout response, slowed ocular or extremity movements) may be present. Gait and strength are normal
Stage 1 (mild)	Able to perform all but the more demanding aspects of work or ADL but with unequivocal evidence (signs or symptoms that may include performance on neuropsychological testing) of functional intel lectual or motor impairment. Can walk without assistance
Stage 2 (moderate)	Able to perform basic activities of self-care but can not work or maintain the more demanding aspects of daily life. Ambulatory, but may require a single prop
Stage 3 (severe)	Major intellectual incapacity (cannot follow news or personal events, cannot sustain complex conversa tion, considerable slowing of all outputs) or motor disability (cannot walk unassisted, requiring walker or personal support, usually with slowing and clumsiness of arms as well)
Stage 4 (end stage)	Nearly vegetative. Intellectual and social com prehension and output are at a rudimentary level. Nearly or absolutely mute. Paraparetic or paraplegic with urinary and fecal incontinence

Source: Adapted from Price and Brew (1988).

lower body mass index and more constitutional symptoms (McArthur et al., 1993). Psychomotor slowing, a history of HIV related disease, extrapyramidal symptoms, depression and elevated serum beta 2 microglobulin levels have also been shown to be associated with development of dementia (Stern et al., 2001). Other risk factors include $CD4^+$ T cell nadir, the zenith of viral load, and duration of HIV disease (Childs et al., 1999). Individuals with plasma HIV RNA of more than 30,000 copies/mL at baseline had 8.5 times higher chance of developing dementia as compared with those with less than 3,000 copies/mL. Similarly, patients with $CD4^+$ T cell counts below 200 cells/mm^3 had 3.5 more chance of developing dementia as compared with those with CD4 counts of higher than 500 cells/mm^3. Finally, additional risk factors for HIV dementia may incorporate a number of genetic factors such as the presence of ApoE4 genes, MCP-1 mutations, mutations in the CCR2 receptor, and TNF receptor polymorphisms.

Initial estimates on the prevalence of HAD were as high as 66%, but more recently reported at 7% (Navia and Price, 1987; Janssen et al., 1992). With the widespread use of potent, combination ART, the overall incidence of HAD has decreased dramatically. For example, the incidence of newly diagnosed moderate-to-severe dementia fell from 6.6% in 1989, to 1% in 2000 (McArthur et al., 2003). In addition, the mean $CD4^+$ T cell count for patients with HAD has progressively increased with access to improved ART. A study by the National Centre in HIV Epidemiology and Clinical Research at the University of New South Wales demonstrated that the median $CD4^+$ T cell count at diagnosis of HAD increased from 70 cells/mm^3 in 1992–1995, to 120 cells/mm^3 in 1996, and to 170/mm^3 in 1997 ($P = 0.04$) (Dore et al., 1999). These data suggest that ART may have a lesser impact on the incidence of HAD than on other AIDS-defining conditions, and some investigators speculate that poor CNS penetration of many antiretroviral agents is a possible explanation.

Not only is the $CD4^+$ T cell count higher, but also the mean time from diagnosis of HIV dementia to death has extended from 6 months to 44 months, and the clinical presentation has changed with more "cortical" manifestations as compared with previously reported "subcortical" presentations (Brew, 2004). HAD appears to be more stable of late, and does not progress as rapidly (McArthur et al., 2003). Some patients may have residual neurological deficits which are irreversible and do not improve with additional ART (Saksena and Smit, 2005).

41.2. Diagnosis

41.2.1. Clinical Findings

In the early stages of HAD, the most common findings are cognitive difficulties, particularly deficits in attention and concentration. Initial changes may be subtle: losing ones train of thought, requiring lists to remember chores, or inability to perform multistage tasks. In more advanced stages, cognitive impairment is more obvious. Patients are unable to perform complex tasks, lose the ability to read, language skills deteriorate, thought processes are generally slower. Mood changes include increasing irritability and apathy, and later on, frank disorientation may develop.

Motor symptoms are less common, but loss of fine motor movements frequently develops; for example, the dexterity required for playing a musical instrument, or doing and undoing buttons. Tremor and myoclonus are rare. Gait may become hesitant and slow, and patients may lose balance. Deep tendon reflexes in the lower extremities become brisk, and occasionally a Babinski sign or other pyramidal release sign may be observed.

41.2.2. Diagnostic Studies

41.2.2.1. Neuropsychological Testing

Neuropsychological testing in HIV dementia is a useful tool for the differential diagnosis of HAD and to delineate disease extension and progression. The standard Mini-Mental Status is not particularly sensitive for detection of early HAD, however patients may fail in reversing a five-letter word or subtracting from 100 by 7's, and may fail in complex sequential tasks, or remembering three objects. Important information can be obtained by patient self-report and by guided interviews, as well as by behavioral observation. More formal neuropsychological testing is the standard of care, if available. Extended neuropsychological batteries can be used in clinical practice, but are time consuming especially in a busy clinical setting.

TABLE 41.2. Common neuropsychological batteries for HAD.

Organization	Tests included
National Institute of Mental Health (NIMH) Workgroup	WAIS-R Vocabulary
	WMS-R Visual Span
	Paced Auditory Serial Addition Test
	California Verbal Learning Test
	Hamilton Depression Scale
	Speilberger State-Trait Anxiety Scale
	San Diego HIV Neurobehavioral Research Center (HNRC)
	Paced Auditory Serial Addition Test
San Diego HIV Neurobehav ioral Research Center (HNRC)	Boston Naming Test
	Thurstone Written Fluency
	Story Learning and Memory
	Figure Learning and Memory
	Finger Tapping Test
	Grooved Pegboard Test
	WAIS-R Digit Span
Multicenter AIDS Cohort Study (MACS)	Controlled Oral Word Association Test
	Rey Auditory Verbal Learning Test
	Trail Making Test
	Symbol Digit Modalities Test
	Grooved Pegboard Test
	CES Depression Scale

More focused neuropsychological batteries have been developed for this purpose, such as those listed in Table 41.2. The HIV Dementia Scale is also a useful tool and only takes five minutes to administer and score (Power et al., 1995). It is important to remember that interpretation of neuropsychological tests also requires attention to patient demographics such as language skills, presence of comorbid conditions, substance abuse, and concomitant use of psychotropic medications.

41.2.2.2. Neuroimaging and Evoked Potentials

Imaging studies in HAD may include structural brain imaging such as brain magnetic resonance imaging (MRI) and functional or metabolic imaging. Previous cross sectional studies have shown that HIV-infected individuals tend to have generalized brain atrophy, which may be more prominent in individuals with HAD (Aylward et al., 1993, 1995). Particularly, there is gray matter volume reduction in the basal ganglia and posterior cortex, as well as generalized volume reduction of white matter. These findings are consistent with the characterization of HAD as a subcortical dementia. Studies using diffusion tensor imaging (DTI), a magnetic resonance imaging technique suited for the study of subtle white matter abnormalities, have found abnormal fractional anisotropy in the white matter of the frontal lobes and internal capsules of HIV-infected patients (Pomara et al., 2001). This method may be more sensitive than conventional MRI methods for detecting subtle white matter disruptions. Longitudinal studies have shown that with progression of dementia, white matter volume progressively declines and consequently the volume of ventricular CSF increases (Stout et al., 1998). Also, the caudate nucleus sustains accelerated volume loss. All these changes may regress, at least in part, with initiation of potent ART (Filippi et al., 1998).

Several types of functional imaging have been used in HAD including use of positron emission tomography (PET) and single-photon emission computed tomography (SPECT); however the most promising imaging seems to be magnetic resonance spectroscopy. Studies performed using this method showed mild increase in myoinositol and choline-containing compounds in white matter, which correlates with glial damage (McConnell et al., 1994). In more advanced disease states, the increase is marked and also associated with reduction *N*-acetyl aspartate, which is a marker of neuronal loss. These changes may also be reversible with ART. Several studies, but not all, have also found evidence of delayed endogenous event-related potentials in patients with HIV infection (Clifford, 2002b). The delay becomes more prominent in the presence of dementia, and delayed potentials may even precede clinical disease. More recently it has been observed that the classical abnormalities seen on brain imaging in the basal ganglia are now less conspicuous, and mesial temporal lobe abnormalities seem to be more prominent both by PET scanning and on histopathology (Brew, 2004).

41.2.2.3. CSF Studies

While HAD is a diagnosis of exclusion without pathognomic findings in CSF, HIV does elicit an inflammatory response that can be detected. This may manifest as mild lymphocytic pleocytosis, elevation of total protein, increase of total IgG, and/or the presence of oligoclonal bands. In addition, elevation of CSF levels of neopterin, beta 2 microglobulin, quinolinic acid, metalloproteinases and several other cytokines (IL-1, IL-6, TNF-alpha), and chemokines such as monocyte chemotactic protein 1, and IP-10 has been reported (Anderson et al., 2002; Gendelman and Persidsky, 2005). However, correlation of such markers with clinical disease is controversial (Brew, 2004). To complicate matters, ART may suppress markers of inflammation, but the degree of reduction attained by different drugs is variable and the suppression may not be durable (Gendelman et al., 1998). Measurement of virus load in the CSF using polymerase chain reaction (PCR) or other technologies has also produced contradictory findings. In contrast to plasma virus load which correlates well with severity of HIV disease and response to therapy, finding of a positive virus load in the CSF does not necessarily correlate with the diagnosis of HAD and may be present in asymptomatic individuals (Price et al., 2001). Also, viral replication may be inhibited in the periphery but not in the CSF, and neuroinflammation may return despite initial effective therapy (Letendre, 2005).

41.2.2.4. Histopathology

The pathological substratum of HIV dementia is referred to as HIV encephalitis (HIVE). HIVE is a disseminated multifocal process characterized by the presence of multinucleated giant cells of macrophage/microglial origin; microglial nodules; diffuse astrocytosis; myelin pallor; loss of synaptic density; and neuronal cell loss, mainly in the hippocampus, basal ganglia and orbitofrontal cortex, usually sparing of the occipital cortex (Budka, 2005). Necrotic areas may be present. On rare occasions, inflammatory changes may affect the meninges and ventricles causing HIV meningoventriculoencephalitis. Less common subtypes of histopathology include HIV demyelinating leukoencephalopathy, which can be seen in patients failing ART, and is characterized by diffuse damage of white matter, lacking prominent signs of inflammation. HIV vacuolar leukoencephalopathy has histopathology similar to HIV vacuolar myelopathy, and granulomatous angiitis is another less common manifestation.

41.3. Antiretroviral Therapy

The introduction of potent, combination ART and the use of prophylactic medications against opportunistic infections had a significant impact on the natural history of HIV (Palella et al., 1998). Morbidity and mortality have declined dramatically, and HIV has evolved from a fatal disease to a chronic, manageable condition. Immune reconstitution achieved with ART can be potent and sustained enough that primary and secondary prophylaxis against opportunistic infections can be safely discontinued (www.aidsinfo.nih.gov).

Antiretroviral agents currently in use include inhibitors of reverse transcriptase, HIV protease, and fusion inhibitors. Several other compounds are in development including chemokine receptors inhibitors, viral integrase inhibitors, inhibitors of HIV regulatory genes and their byproducts, and inhibitors of viral budding.

41.3.1. Antiretroviral Classes

Currently, there are four classes of antiretroviral agents available in the United States for clinical use:

Nucleoside analogue reverse transcriptase inhibitors (NRTIs),
Nonnucleoside analogue reverse transcriptase inhibitors (NNRTIs),
Protease inhibitors (PIs),
Entry inhibitors.

NRTIs are phosphorylated and converted into triphosphate forms by nucleoside kinases. These activated forms have high levels of affinity for HIV-1 reverse transcriptase and compete with the natural deoxynucleoside triphosphates. Once incorporated into the growing chain of DNA, lack of a 3′-hydroxyl group that can form a phosphodiester bond with the incoming nucleoside causes chain termination. Tenofovir is an exception in this group as it is a nucleotide analogue rather than nucleoside and, as such, requires only two phosphorylation steps instead of three to become the active form. Pharmacological characteristics of FDA approved NRTIs are presented in Table 41.3.

NNRTIs are drugs of diverse chemical structure that act by non-competitive inhibition of HIV-1 reverse transcriptase. They do not require intracellular phosphorylation to be activated. NNRTIs inhibit the enzyme allosterically by binding

TABLE 41.3. Selected characteristics of nucleoside reverse transcriptase inhibitors.

Generic/trade name (abbreviation)	How supplied	Usual adult dose	Major adverse effects	Ratio of CSF level to IC_{50}[a]
Abacavir/Ziagen (ABC)	300 mg tablets	300 mg b.i.d. or 600 mg once daily	Hypersensitivity reaction, which can be fatal	4.0
Didanosine/Videx EC (ddI)	125, 200, 250, or 400 mg enteric-coated capsules	Body weight >60 kg: 400 mg once daily: <60 kg: 250 mg once daily	Peripheral neuropathy, pancreatitis, lactic acidosis with hepatic steatosis	0.11
Emtricitabine/Emtriva (FTC)	200 mg capsules	200 mg q.d.		–
Lamivudine/Epivir (3TC)	150 and 300 mg tablets	150 mg b.i.d. or 300 mg once daily		0.74
Stavudine/Zerit (d4T)	15, 20, 30, or 40 mg capsules	>60 kg: 40 mg b.i.d.; <60 kg: 30 mg b.i.d.	Peripheral neuropathy, pancreatitis, lactic acidosis with hepatic steatosis, neuromuscular weakness (rare)	0.58
Tenofovir/Viread (TDF)	300 mg tablets	300 mg q.d.	GI intolerance, renal insufficiency (rare)	–
Zalcitabine/Hivid (ddC)[b]	0.375 or 0.75 mg tablets	0.75 mg t.i.d.	Peripheral neuropathy, lactic acidosis with hepatic steatosis	0.08
Zidovudine/Retrovir (AZT, ZDV)	100 mg capsules, 300 mg tablets	200 mg t.i.d. or 300 mg b.i.d.	Bone marrow suppression	0.6

[a] The ratio between pharmacokinetic measures, such as the plasma minimum concentration (Cmin), and the 50% inhibitory concentration (IC_{50}) provides a measure of systemic antiviral efficacy. Similarly, the ratio of CSF concentrations to IC_{50} may provide a better estimate of CNS efficacy than CSF concentration alone.
[b] Anticipate discontinuation of distribution in 2006.
Note: Several fixed dose NRTI combination tablets are also available:
- Abacavir 600 mg + lamivudine 300 mg as Epzicom
- Emtricitabine 200 mg + tenofovir 300 mg as Truvada
- Lamivudine 150 mg + zidovudine 300 mg as Combivir
- Abacavir 300 mg + lamivudine 150 mg + zidovudine 300 mg as Trizivir
- Emtricitabine 200 mg + tenofovir 300 mg + efavirenz 600 mg as Atripla

to a hydrophobic pocket in the p66 subunit of reverse transcriptase, which is close to the catalytic site of the enzyme. Several pharmacological properties of NNRTIs are presented in Table 41.4.

PIs act by impeding the action of the HIV-1 protease, an aspartic protease that cleaves the gag and gag-pol precursor molecules into smaller structural proteins and enzymes. By doing this, PIs prevent viral maturation and produce defective viral particles without an electrodense core, and these are unable to infect new cells. Several pharmacological properties of PIs are presented in Table 41.5.

The first entry inhibitor to be licensed in the United States was enfuvirtide. A synthetic peptide, this compound has the ability to inhibit fusion of virions to the host cells by preventing the conformational change of gp41. The peptide is not bioavailable when taken orally and requires subcutaneous administration. The drug is generally well tolerated although can be associated with local reactions at the site of injection. Clinical benefit has mostly been demonstrated in heavily experienced patients with multidrug resistant virus (Trottier et al., 2005). The impact of enfuvirtide on HAD is not known.

TABLE 41.4 Selected characteristics of non-nucleoside reverse transcriptase inhibitors.

Generic/trade name (abbreviation)	How supplied	Usual adult dose	Major adverse effects	CSF/IC_{50}
Delavirdine/Rescriptor (DLV)	100 or 200 mg tablets	400 mg t.i.d.	Rash, increased transaminases	186
Efavirenz/Sustiva (EFV)	50, 100, or 200 mg capsules or 600 mg tablets	600 mg q.h.s.	Neuropsychiatric manifestations, rash, increased transaminases	5.7
Nevirapine/Viramune (NVP)	200 mg tablets	200 mg q.d. for 14 days then 200 mg b.i.d.	Rash, hepatitis	241

41.3.2. Initiating Antiretroviral Therapy: When and What to Start

Recent guidelines for starting of ART recommend that patients with advanced disease, as manifested by the development of an AIDS-defining illness including HAD, or presence of severe symptoms such as unexplained fever, persistent diarrhea, or unexplained weight loss, should be started on ART regardless of their CD4+ T cell count (www.aidsinfo.nih.gov). Asymptomatic patients with CD4+ T cell counts below 200 cells/mm^3 should also be treated. Asymptomatic patients with CD4+ T cell counts between 200–350 cells/mm^3, may be offered treatment at the discretion of the treating clinician and in collaboration with the patient. In patients with CD4+ T cell counts above 350 cells/mm^3, the majority of clinicians would recommend deferring therapy. Some clinicians will treat if patients have a viral load of more than 100,000 copies/mL. Indications for starting of ART are summarized in Table 41.6.

Based on data from clinical trials, either an NNRTI-based or PI-based regimen should be offered to antiretroviral-naïve patients for whom therapy is indicated. The preferred NNRTI-based regimen includes efavirenz and two NNRTIs; either lamivudine or emtricitabine plus zidovudine or tenofovir.

TABLE 41.5. Selected characteristics of protease inhibitors.

Generic/trade name (abbreviation)	How supplied	Usual adult dose	Major adverse effects	CSF/IC_{50}
Atazanavir/Reyataz (ATV)	100, 150, or 200 mg capsules	400 mg q.d. or 300 mg q.d. with ritonavir when used with TDF or EFV	Hyperbilirubinemia, atrio-ventricular block	0.02
Fosamprenavir/Lexiva (f-APV)	700 mg tablets	700 mg b.i.d. or 1400 mg q.d. with ritonavir	GI intolerance, rash	–
Indinavir/Crixivan (IDV)	200, 333, or 400 mg capsules	800 mg q. 8 h or 800 mg b.i.d. with ritonavir	Nephrolithiasis, GI intolerance, hyperbilirubinemia, metabolic complications	108
Nelfinavir/Viracept (NFV)	250 mg tablets	750 mg t.i.d. or 1250 mg b.i.d.	Diarrhea, metabolic complications	4.3
Ritonavir/Norvir (RTV)	100 mg capsules	600 mg q. 12 h (rarely used)	GI intolerance, hepatitis, pancreatitis, metabolic complications	3.3
Saquinavir/Fortovase[a] (SQV-sgc)	200 mg soft gel capsules	1200 mg t.i.d.	GI intolerance, increased transaminases, metabolic complications	4.0
Saquinavir/Invirase (SQV)	200 mg capsules	1,000 mg b.i.d. with 100 mg bid of ritonavir; invirase alone is not recommended	GI intolerance, increased transaminases, metabolic complications	4.0
Tipranavir/Aptivus (TPV)	250 mg capsules	500 mg b.i.d. with ritonavir 200 mg bid	GI intolerance, increased transaminases, metabolic complications	–
Lopinavir 133.3 mg + ritonavir 33.3 mg/Kaletra (LPV/r)	Fixed combination capsules	3 capsules b.i.d.; 4 capsules bid with EFV or NVP	GI intolerance, increased transaminases, metabolic complications	3.3
Darunavir/Prezista (DRV)	300 mg capsules	2 capsules b.i.d. with ritonavir 100 mg b.i.d.	GI intolerance, rash	–

[a] Distribution of Fortovase is anticipated to discontinue in 2006.

TABLE 41.6. Guidelines for the initiation of antiretroviral therapy.

Clinical category	CD4+ T cell count	Plasma HIV RNA level	Recommendation
AIDS-defining illness or severe symptoms	Any value	Any value	Treat
Asymptomatic	CD4+ T cells <200/mm^3	Any value	Treat
Asymptomatic	CD4+ T cells 200–350 mm^3	Any value	Treatment should be offered following full discussion of pros and cons with the patient
Asymptomatic	CD4+ T cells >350/mm^3	>100,000 copies/mL	Most clinicians recommend deferring therapy, but some would treat
Asymptomatic	CD4+ T cells >350/mm^3	<100,000 copies/mL	Defer therapy

Source: Adapted from the Adult and Adolescent Guidelines of the US Department of Health and Human Services, available at www.aidsinfo.nih.gov

However, other NRTIs may be used as a backbone including stavudine, didanosine and abacavir, and nevirapine can be substituted for efavirenz. Important exceptions to NRTI combinations include zidovudine with stavudine (competitive enzyme antagonism), zalcitabine with didanosine (increased risk of mitochondrial damage and peripheral neuropathy), stavudine with didanosine (increased risk of pancreatitis and lactic acidosis, especially in pregnant women), zalcitabine with lamivudine (competitive inhibition of intracellular phosphorylation), and didanosine with tenofovir (high rate of early virological failure). Because of teratogenicity in animals, efavirenz should not be used in pregnant women or those with high risk of pregnancy.

Preferred PI-based regimens are lopinavir/ritonavir plus lamivudine or emtricitabine plus another NRTI, usually zidovudine, stavudine or abacavir. Alternative combinations include other PIs with or without ritonavir, and two NRTIs. The combination of a protease inhibitor with ritonavir provides inhibition of cytochrome p450 enzymes and permits less frequent dosing of amprenavir, indinavir, lopinavir and saquinavir. Use of ritonavir in this setting is also known as "boosting."

Abacavir plus lamivudine plus zidovudine, which is commercially available as a fixed combination in a single tablet, has inferior rates of viral suppression and should only be used when an NNRTI or a PI-based regimen cannot or should not be used as first line of therapy (Gulick et al., 2004).

41.3.3. Antiretroviral Treatment Failure

Despite major advances in treatment of progressive HIV disease, approximately 40% of patients will not achieve sustained viremia below the limits of detection, which is the accepted definition of treatment success (Lucas et al., 1999). Poor adherence is the most important cause of failure of ART, and cognitive impairment is clearly a risk factor for this, such as in patients with HAD. Other reasons include drug intolerance, development of adverse effects, impaired drug absorption and/or metabolism, pharmacokinetic interactions, and pre-existing viral resistance.

The development of viral drug resistance, particularly in patients with poor adherence, is a major concern in clinical practice. A listing of drug resistance mutations in HIV-1 is continuously updated and available from the International AIDS Society-USA at http://www.iasusa.org/resistance_mutations/index.html. If viral resistance is suspected, testing for mutations is helpful in choosing subsequent regimens and has been shown to improve biological outcomes. Both genotyping and phenotyping assays are commercially available, and there are no data to support the utility of one method over the other; although, genotyping is usually less expensive and the results may be available faster.

41.3.4. Importance of CNS Penetration in the Treatment of HAD

With regard to HIV disease, the CNS is both a biological and a pharmacological compartment. It is considered a biological compartment because HIV can replicate independently from virus in the periphery. It is also considered a pharmacological compartment because the blood–brain barrier acts as an obstacle to the penetration of antiretroviral drugs, many of which are highly plasma-protein bound. Both facts may predispose to the development of antiretroviral resistance in CNS, independently from that in plasma. This phenomenon has been observed in patients failing therapy (Cunningham et al., 2000). Hence, the penetration of drugs into the CNS has potential clinical relevance.

Penetration of antiretroviral drugs into the CNS depends on many factors. Protein binding is particularly important because only free drug is able to diffuse across the blood–brain barrier. Other factors that help with penetration into the CNS include small molecular size, a higher degree of acidity; higher lipophilicity; and the presence of inflammation, which may favor the accumulation of drugs in the CNS. The CSF level of an antiretroviral agent is used as a surrogate for drug penetration into the brain parenchyma, although this is not necessarily the optimal approach.

Because it has been assumed that drugs with better CSF penetration would be more effective in treating HAD, particular attention has been paid to antiretroviral agents with higher CSF:plasma concentrations. Among the NRTIs, zidovudine has been widely studied and is known to cross the blood–brain barrier well by passive diffusion. Dideoxyinosine (ddI) and dideoxycytidine (ddC) have poor uptake in the CNS, probably due to their hydrophilic composition. Lamivudine (3TC) has a structural composition similar to ddC and its degree of absorption and efflux are probably similar as well. Stavudine (d4T) may have good penetration into the CSF. Data concerning CSF penetration of licensed antiretroviral agents are included in Tables 41.3, 41.4, and 41.5.

Among the NNRTIs, nevirapine has the highest penetration. Protease inhibitors have poor penetration because they are highly protein bound in plasma. Despite this fact, protease inhibitors have been successful in the treatment of HAD (Gendelman et al., 1998; Saksena and Smit, 2005). Some investigators believe that control of virus replication in the periphery may be sufficient to reverse HIV-related brain disease. Improvement in immune status from ART may also reconstitute the defective blood–brain barrier, which is seen in later stages of HIV infection (Evers et al., 2004). The importance of CNS penetration in the treatment of HAD is therefore controversial, and drugs which do not have adequate CSF levels may still be effective.

Although viral decay in CSF usually parallels that in plasma, occasionally there can be variability and may even be increases of RNA in the CSF despite a decline in plasma levels. These differences in the decay kinetics may be related to the type of cells infected in each compartment: $CD4^{+}$ T cells in plasma, which are rapidly destroyed, versus monocytes and macrophages in the brain, which persist longer (Eggers et al., 2003).

41.3.5. Role of Specific Antiretroviral Agents in the Treatment of HAD

Zidovudine monotherapy was the only available treatment for HIV disease before the 1990s, and was used until 1992. Zidovudine monotherapy has been proven to be efficacious for both the treatment of HAD as well as for HIV-associated minor cognitive/motor disorder (MCMD) (Arendt et al., 1992; Sidtis et al., 1993; Tozzi et al., 1993). Unfortunately, the beneficial effect of zidovudine was transient and an addition of a second NRTI such as dideoxyinosine (ddI), lamivudine (3TC), or dideoxycytidine (ddC), may not further improve psychomotor performance. Stavudine was shown to improve motor performance even after pretreatment with zidovudine (Arendt et al., 2001). A study in 1998, using abacavir versus placebo showed no neurologic deterioration in the abacavir group as compared with the placebo group (Lanier et al., 2001). However, there was no benefit when abacavir was added to a stable ART, despite good proven CNS penetration.

Among the NNRTIs, nevirapine has the best penetration in the CSF with a CSF to plasma ratio up to 40%. Efavirenz is highly protein-bound and has a much lower CSF:plasma ratio, however drug levels in the CSF have been able to successfully inhibit HIV infection in microglia in vitro (Albright et al., 2000). Both NNRTIs have been shown to improve HIV-associated MCMD in both naïve and antiretroviral-experienced patients (Arendt and von Giesen, 2002). There are no data available about the use of delavirdine in HAD.

Protease inhibitors are highly protein-bound. The PIs studied in the CSF include saquinavir, indinavir, and nelfinavir. All of them have very low CSF: plasma ratios. However, several studies have shown some improvement in psychomotor slowing in patients treated with a combination of ART including a PI (Gendelman et al., 1998; Sacktor et al., 1999).

41.4. Adverse Neurologic Effects of Antiretroviral Agents

A detailed reviewed of the toxicity of all antiretrovirals is beyond the scope of this chapter, but information about adverse effects of each drug are available in the package inserts. Commonly occurring neurological toxicities are outlined in the following paragraphs. Management of adverse events from antiretroviral agents typically involves substitution of the offending agent, ideally with a drug from the same class but with a different toxicity profile. Dose modification may be needed to correct for drug interactions when used in certain combinations with NNRTIs and/or PIs.

Distal sensory peripheral neuropathy (DSPN) has been associated with the use of ddC, ddI, and stavudine, alone or in combination. The symptoms of this condition include a burning sensation in feet and hands, numbness and tingling in the feet, cramps in the legs, and absent ankle reflexes. The patient may also exhibit decreased sensation to temperature, pinprick, vibration, and proprioception (Dieterich, 2003). The symptoms are similar to HIV associated axonal neuropathy. This condition may become irreversible, so it is important to diagnose it earlier. The treatment of choice is symptomatic therapy and discontinuation of the toxic agent when possible.

Myopathy mainly affecting the proximal lower limb muscles has been associated with use of zidovudine in up to 18% of cases. The cases have usually occurred after the use of zidovudine for several months. The manifestations of this condition include: myalgias, elevation of creatine kinase, and muscle biopsy revealing ragged root fibers. Zidovudine induced-myopathy may be difficult to distinguish from HIV-associated myopathy. The management may include close monitoring, but in patients who have more severe symptoms, discontinuation of zidovudine may be necessary. A short course of anti-inflammatory drugs or corticosteroids, or even intravenous immunoglobin has been advocated but has not been evaluated in controlled clinical trials. Zidovudine may also cause headaches in up to 50% of cases. The headache induced by zidovudine can be so severe that it may motivate stopping the medication (Enzensberger and von Giesen, 1999).

There have been reports of HIV-associated neuromuscular weakness, probably associated with symptomatic lactic acidosis (Estanislao et al., 2004). Lethal cases have been reported in association with stavudine. Usually the cases have occurred after several months of use of the medication and have been associated with sensory motor neuropathy and elevated serum lactate levels. It is tempting to include this condition as another manifestation of mitochondrial damage induced by NRTIs.

Central nervous toxicity has been seen with efavirenz (Clifford et al., 2005). Manifestations of this process include headache, dizziness, confusion, agitation, amnesia, hallucinations, insomnia, and abnormal vivid dreams. These symptoms occur usually within the first four weeks of treatment, and decline afterwards with time. Dosing the medication at bedtime makes the symptoms more tolerable. Education of the patient about the potential adverse affects of the drug is important to prevent treatment interruptions. Efavirenz may also be associated with depression, anxiety, and suicidal ideation and patients with a history of mental disorders may need to be monitored closely when started on efavirenz. There are no specific therapies for the side effects except discontinuation of the medication if the symptoms are severe enough (Treisman and Kaplin, 2002).

Protease inhibitors are not particularly associated with neurologic side effects but because they tend to cause inhibition of the P450 enzymes, they may alter the metabolism of neurotropic drugs and as a consequence may precipitate neurologic effects (Treisman and Kaplin, 2002).

41.4.1. Pharmacological Interactions Between Antiretrovirals and Other Medications with Activity in the Central Nervous System

The American Psychiatry Association emphasizes that when using psychoactive medications in HIV infected patients, several measures should be considered, such as using lower starting doses and a slow titration, providing simple dosing skills, maintaining awareness of potential drug-drug interactions and of possible adverse effects (http://www.psych.org/psych_pract/treatg/pg/hivaids_revisebook_index.cfm). In particular, PIs and NNRTIs inhibit or induce cytochrome p450, which shares the same metabolic pathway with many psychotropic agents. Among PIs, ritonavir is the most powerful inhibitor, saquinavir being the weakest, and other ones being intermediate. In general, by inhibiting the cytochrome p450 isoenzyme, plasma levels of psychotropic medications are increased. Concomitant use of selective serotonin reuptake inhibitors (SSRIs) and PIs may cause serotonin syndrome (Gillenwater and McDaniel, 2001). Tricyclic antidepressants in higher concentrations may cause delay cardiac conduction, anticholinergic effects and orthostasis. Newer antidepressants such as nefazodone may cause adverse affects if coadministered with ritonavir. The same applies for pimozide, an antipsychotic that may be associated with cardiac arrhythmia; and with benzodiazepines, which can be associated with respiratory depression. Clozapine can also be associated with seizures. In addition, antipsychotics can be associated with extrapyramidal side effects, including severe dystonia, akasthesia, and Parkinsonism.

Sodium valproate is used as a mood stabilizer but has also been shown inhibit an enzyme associated with maintenance of HIV latency (Lehrman et al., 2005). Levels of valproate may be increased by concomitant use of zidovudine. Lithium should be used with caution, especially in patients who have some degree of nephropathy. Carbamazepine is also known to induce activity of the cytochrome p450, and has the potential for bone marrow toxicity, so it is rarely used in HIV-infected patients.

Many recreational drugs such as benzodiazepines, amphetamines, and opioids are also metabolized by the liver. Although information is scant about the clinical significance and interactions between these drugs and antiretroviral agents, unintentional overdoses with methamphetamine and gamma hydroxybutyrate have been reported in patients using PIs, particularly ritonavir. PIs and NRTIs may alter metabolism of methadone and precipitate opioid withdrawal (McCance-Katz et al., 2003).

41.5. Adjuvant Pharmacological Therapy for HAD

Several drugs have been tested as adjunctive agents for the therapy of HAD. This section mainly describes medications that have been subject to clinical trials, with a brief description of other potentially promising compounds.

41.5.1. CPI/1189

CPI/1189 is a lipophylic antioxidant with the chemical structure of a synthetic benzamide, which may have some anti-tumor necrosis alpha activity by scavenging superoxide anion radicals. Tumor necrosis alpha is a critical factor in the pathology of neurologic dysfunction, and has been shown to be associated with HIV encephalitis. CPI/1189 was tested in a clinical trial of 64 patients with HAD. The drug caused improved performance on the Grooved Pegboard Tests, a test for psychomotor speed, but no effect in the composite scores for eight other neuropsychological measures. In addition, although relatively well tolerated, CPI/1189 was associated with some serious adverse effects including: elevation of liver enzymes, possible cataract formation, and decrease in the mean corpuscular volume (Clifford et al., 2002a).

41.5.2. Lexipafant

Neuronal death can be caused by inflammatory mediators including lipid membrane derivatives such as platelet activating factor (PAF). PAF itself may be neurotoxic by increasing the release of glutamate and calcium. Lexipafant is a PAF inhibitor and was studied in a randomized double-blind placebo-controlled trial involving 30 HIV-infected individuals with cognitive impairment (Schifitto et al., 1999). The drug was well tolerated, but there were no significant differences between the placebo and the treatment groups on neuropsychological function.

41.5.3. Memantine

Memantine is a competitive inhibitor of the *N*-methyltaspartate glutamate receptor (NMDA). Clinically, memantine has been effective in the treatment of Alzheimer's dementia and was thought to show promise for the treatment of HAD. A large, double-blind, placebo-controlled trial was designed in the late 1990's (AIDS Clinical Trials Group protocol 301) and showed a trend towards improvement in some neuropsychological testing; but no significant, sustained improvement. (Schifito et al., 2007a).

41.5.4. Minocycline

Minocycline is a long acting tetracycline derivative and has recently been shown to have neuroprotective effects in an animal model of simian immunodeficiency virus (SIV) encephalitis (Zink et al., 2005). The drug crosses the blood barrier and has anti-inflammatory properties by suppressing the p38 mitogen-activated protein kinase. In a small study, comparing minocycline-treated and untreated macaques exposed to SIV, the treated macaques had less severe encephalitis, reduced CNS expression of neuroinflammatory markers (such as major histocompatibility complex class II, macrophage marker CD68, T-cell intracytoplasmic antigen 1, CSF monocyte chemoattractant protein 1), reduced activation of p38 mitogen-activated protein kinase, less axonal degeneration, and lower CNS virus replication. The study also showed that minocycline was able to inhibit SIV and HIV in vitro, probably not by direct antiviral effect but by rendering the intra- or extra-cellular environment hostile for viral proliferation. Another experiment using microglial cultures corroborated that minocycline inhibits HIV production by interrupting nuclear factor NF-kappa B activation and HIV-1 long terminal repeat (LTR)-promoter activity in U38 cells (Si et al., 2004). Clinical trials in humans are in development.

41.5.5. Nimodipine

Glycoprotein 120 (gp120) causes injury in rodent retinal ganglion cells and hippocampal neurons in culture, as a consequence of increased intracellular calcium. The addition of the calcium channel blocker, nimodipine, decreases the levels of intracellular calcium and has been shown to prevent neuronal damage (Dreyer et al., 1990). Similarly, the cytotoxic effect of gp120 was monitored in a population of human neurons. The use of different doses of gp120 was associated with up to 27% loss of viable cells. Nimodipine was able to prevent this cytotoxic effect, presumably by blocking the *N*-methyl-D-aspartate (NMDA) receptor channels, which allow calcium ions to enter the cell (Wu et al., 1996). It has also been suggested that the HIV tat protein may activate a sensitive calcium influx pathway in microglial cells (Contreras et al., 2005).

To date, one clinical trial has evaluated the role of nimodipine in the treatment of HAD. A multicenter phase I/II study by the AIDS Clinical Trial Group enrolled 41 patients with different degrees of HAD (Navia et al., 1998). Patients were assigned to one of three arms: placebo, nimodipine 30 mg orally tid or nimodipine 50 mg five times a day for 16 weeks. Patients in all arms received an NRTI (zidovudine, didanosine or zalcitabine). An intent-to-treat analysis showed no significant difference in the neuropsychological Z score (NPZ-8), a composite for eight neuropsychological tests or in the CSF levels of neopterin or beta-2 microglobulin. There was, however, a nonsignificant improvement in the arm with the highest dose of nimodipine. There was no difference in the toxicity caused by placebo or nimodipine.

41.5.6. OPC-14117

OPC-14117 is a lipophylic compound with a chemical structure similar to vitamin E. The compound is a scavenger of superoxide anion radicals, which may be implicated in the pathogenesis of HAD. In fact, it has been proposed that oxidative stress and excessive activation of glutamate receptors may represent a sequential pathway for cell vulnerability in the brain. Based on this, a double-blind, placebo-controlled, randomized clinical trial was organized by the by the Charles A. Dana Foundation Consortium on the Therapy of HIV Dementia and Related Cognitive Disorders (The Dana Consortium, 1997). Thirty individuals with HIV infection and evidence of cognitive impairment, and who were taking a stable antiretroviral regimen for six weeks prior to randomization, were randomized to receive tablets of OPC-14117 or placebo. The primary outcome measure was safety of the drug including ability of the subjects to complete the study and frequency of adverse effects.

There was no reported differences in tolerability between the treatment groups. The adverse experiences reported to be of severe intensity included metabolic encephalopathy, parotid tumor, diarrhea, fatigue, pneumonia, and nausea and all occurred in the placebo-treated group. In the treated group there was asymptomatic elevation of liver enzymes which required halving the dose of the medication. After 12 weeks of follow up, there was a trend towards improvement in the Rey Auditory Verbal Learning Recall, Delayed Recall Subscales, and in the timed gait test, favoring the OPC group, however the differences were not significant. There were no significant differences between the groups on mean changes in function, mood, CD4 lymphocyte count, and beta-2-microglobulin levels. So, although safe, the medication did not produce significant clinical improvement.

41.5.7. Peptide T

Peptide T or d-ala-peptide-d-amide is an octapeptide (Ala-Ser-Thr-Thr-Thr-Asn-Tyr-Thr), which is so named because of the many threonine aminoacids included in its sequence. This peptide is involved in the attachment sequence of the gp120 molecule of HIV to the CD4 receptor (Ruff et al., 1987). This peptide is a potent competitor of the virus to gp120 binding, and inhibits viral infectivity and blocks the neurotoxic effect of gp120. In addition, peptide T is structurally similar to the neurotransmitter vasoactive intestinal peptide, and may be able to inhibit the production of tumor necrosis factor alpha.

Peptide T was evaluated in a phase I trial of 14 patients who were given intravenous drug for 12 weeks (Bridge et al., 1991). In addition, 6 of the initial 14 received 8 weeks more of intranasal peptide. Results were compared with untreated controls. Findings include increments in cognitive and neuromotor function as well as weight gain of about 2 kg, and improved sense of wellbeing. No significant toxicity was reported. This small study provided the rationale for phase II testing.

A double-blind, placebo-controlled, trial in patients with evidence of cognitive deficit was performed using intranasal peptide (Heseltine et al., 1998). The drug was administered three times a day for six months, and the primary outcome was changes in a score of 23 neuropsychological tests. At the end of the study, there was no significant difference between the treated and placebo groups, however there was some suggestion of cognitive improvement among those patients who had more severe cognitive impairment and those with relatively higher CD4 cell counts. There were significant differences in the development of mood disorders such as depression and irritability, and presence of severe rash in the peptide T group.

41.5.8. Selegiline

Cytokines, such as tumor necrosis alfa and oxygen radicals, may be produced by HIV-infected macrophages or microglia. These substances cause apoptosis in the cerebral cortex and basal ganglia. Selegiline, also called deprenyl, and thioctic acid are monoamine oxidase-B inhibitors that have antioxidative effects and have been evaluated in two clinical trials. Deprenyl has been previously studied in patients with Alzheimer's dementia and Parkinson's disease, showing improvement of memory in these conditions.

A study by the Dana Consortium used deprenyl and thioctic acid in 36 patients with HIV-associated cognitive impairment (The Dana Consortium, 1998). Both medications were well tolerated. The deprenyl group showed some improvement in memory function, with improvement seen on the Rey Auditory Verbal Learning Test. However, other tests of psychomotor speed did not improve significantly. The mechanism of action of deprenyl is not clear, but it may have a trophic effect on injured neurons and may also suppress the formation of free radicals.

A second study used a transdermal system of selegiline administration, which provides constant and higher plasma levels of the drug (Sacktor et al., 2000). Participants were either placed on selegiline or placebo. Selegiline was well tolerated, and the selegiline group did better than the placebo group on some components of the Rey Auditory Verbal Learning Test, however other tests of psychomotor speed, like Cal Cap mean reaction time, Digit Symbol, or Grooved Pegboard, did not improve. Also, there was no significant impact on the functional status of the patients. The study results prompted development of a larger, double-blind, placebo-controlled trial which failed to show evidence of cognitive or functional improvement (Schifito et al., 2007b).

41.6. Additional Management Considerations

41.6.1. Importance of Adherence to Antiretroviral Therapy

ART is the treatment of choice for HAD as this option improves both cognitive function, and overall survival. Adherence to therapy is therefore a crucial issue in this population. Surveys of adherence among HIV-infected patients have showed a percentage of non-adherence ranging from 22% to 59% (Gallant and Block, 1998; Rodriguez-Rosado et al., 1998). Factors that influence adherence to ART include: younger age, African American ethnicity, male sex, lower level of general education, and poorer literacy. In addition, patient beliefs and knowledge about the disease and its treatment may impact adherence. Finally, the physical and psychological condition of patients and the stage of HIV infection also determine non-adherence (Gordillo et al., 1999; Paterson et al., 2000; Castro, 2005). While no studies have specifically targeted adherence levels in patients with HAD, cognitive deficits have been shown to have a negative impact on adherence (Swindells, 2005).

Providers can help to improve treatment adherence by paying attention to issues such as patients' lifestyle/needs, and avoidance of polypharmacy and drugs that may have higher risk of adverse effects. Dietary constraints with certain drugs need to be taken seriously since some patients may be unable to comply with them. Patient education is an important strategy in improving adherence because it improves motivation and intention to adhere to the regimen as prescribed (Williams and Friedland, 1997). Simple reminder devices, such as a portable medication alarm or pill organizers, may be valuable but they may be too sophisticated in certain situations. Use of cues such TV programs or certain daily routines have proven useful (Chesney, 1997). Recently, a programmable prompting device improved adherence to ART in HIV-infected subjects with memory impairment (Andrade et al., 2005). A multidisciplinary team offering a trusting and respectful approach to the patient may be of significant value in improving compliance by implementing strategies for solving potential problems.

41.6.2. Supportive Care

The impact of short-term educational programs in patients and caregivers is usually modest, however, intensive long-term educational programs and support programs for caregivers may delay the time of admission to a nursing home by 12–24 months. Certain non-pharmacological interventions may improve functional performance of patients with dementia. Behavior modification, scheduled toileting, and prompted voiding can reduce urinary incontinence, whereas graded assistance, skills practice, and positive reinforcement can increase functional independence. Patients with HAD may benefit from art therapy, sensory stimulation and other psychosocial interventions, but the impact of such activities is difficult to measure. In addition, education, support and respite care for caregivers improve their emotional well being and quality of life and helps in delaying nursing home placement for patients with dementia.

Summary

Although less common in recent years, HAD remains a challenging condition for patients, families and care providers. Specific diagnostic tests are still lacking. Despite many years of thoughtful investigation and effort, advances in our understanding of the neuropathogenesis of HAD have yet to translate into significant clinical utility. The most noteworthy improvements in treatment of cognitive impairment have come from widespread use of combination ART. Even though many of the most potent compounds have little to no penetration into CSF, control of virus replication in the periphery appears to be sufficient to arrest or reverse the course of HAD. For some patients with chronic neurological disease, changes may irreversible and improvement only modest. In addition, HAD may develop in treated patients, albeit at later stages of disease.

Review Questions/Problems

1. Which of the following drug classes are not currently licensed for the treatment of HIV disease?

a. Nucleoside reverse transcriptase inhibitors
b. Non-nucleoside reverse transcriptase inhibitors
c. Protease inhibitors
d. Entry inhibitors
e. Inhibitors of viral maturation

2.A patient with HIV-associated dementia was started on antiretroviral therapy two weeks ago. He complains of dizziness, insomnia, and vivid dreams. This reaction is most likely related to the use of:

a. Zidovudine
b. Tenofovir
c. Lamivudine
d. Efavirenz
e. Nevirapine

3. Which of the following class of drugs is not recommended for first line therapy in patients with HIV-associated dementia?

a. Nucleoside reverse transcriptase inhibitors
b. Non-nucleoside reverse transcriptase inhibitors
c. Protease inhibitors
d. Entry inhibitors
e. All can be used as first line agents in the treatment of HAD

4. Contemporary treatment of HIV infection requires the use of two nucleoside reverse transcriptase inhibitors (NRTI) combined with either a non-nucleoside reverse transcriptase inhibitors or a protease inhibitor. Which of the following NRTI backbones will you prefer to use in a patient recently diagnosed with HIV-associated dementia:

a. Zidovudine plus stavudine
b. Zidovudine plus lamivudine
c. Zalcitabine plus didanosine
d. Zalcitabine plus lamivudine
e. Didanosine plus tenofovir

5. Which of the following combinations will not be recommended for first line therapy in a patient with HIV associated dementia?

a. Zidovudine plus lamivudine plus efavirenz
b. Tenofovir plus emtricitabine plus lopinavir/ritonavir
c. Tenofovir plus emtricitabine plus atazanavir/ritonavir
d. Lamivudine plus abacavir plus fosamprenavir
e. Zidovudine plus lamivudine plus abacavir

6. Which of the following nucleoside reverse transcriptase inhibitors has the best penetration into CSF?

a. Zidovudine
b. Didanosine
c. Lamivudine
d. Stavudine
e. Abacavir

7. A patient with HIV-associated dementia receiving antiretroviral therapy presents with a burning sensation in feet and hands, numbness and tingling in the feet, cramps in the legs, and absent ankle reflexes. What drug will you be the LEAST likely to blame on his/her new symptoms?

a. Zidovudine
b. Stavudine
c. Didanosine
d. Zalcitabine
e. All of the above can cause the patient's symptoms

8. Which of the following protease inhibitors is the most powerful inhibitor of the P450 enzyme?

a. Atazanavir
b. Fosamprenavir
c. Indinavir
d. Nelfinavir
e. Ritonavir

9. Which of the following drugs has been proven to be effective as adjuvant therapy for HIV associated dementia?

a. Lexipafant
b. Memantine

c. Minocylcine
d. Nimodipine
e. None of the above
f. All of the above

10. In a busy clinical practice setting, the simplest neuropsychological test to diagnose HIV dementia is?

a. Standard Mini-Mental Status test
b. HIV-dementia scale
c. Memorial Sloan Kettering Scale
d. Halstead-Reitan Neuropsychological Battery
e. San Diego HIV Neurobehavioral Research Center

11. HIV-associated dementia is also referred to as:

a. HIV dementia
b. AIDS Dementia Complex
c. Subacute encephalopathy
d. HIV-1-associated cognitive-motor complex
e. All of the above

12. What is correct about the "Trojan Horse" model of HIV invasion into the CNS?

a. Proposes that monocyte-macrophage carry progeny virions from blood to the brain inside cytoplasmic vacuoles and across the blood–brain barrier
b. Requires an inflammatory response in the brain, which elicits a chemokine gradient
c. After infiltrating the brain parenchyma, blood borne monocyte-macrophages also release additional chemokines and further attract additional inflammatory cells
d. Does not require productive viral replication in brain microvascular endothelial cells
e. All of the above

13. The clinical and laboratory features of HIV-associated dementia include all "except" which of the following?

a. Psychomotor slowing
b. History of HIV related disease
c. Extrapyramidal symptoms
d. Depression
e. Reduced viral loads and serum beta 2 microglobulin

14. What is not correct in regards to a diagnosis of HIV-associated dementia?

a. HAD is a diagnosis of exclusion
b. HIV does elicit an inflammatory response. This can be detected by mild lymphocytic pleocytosis, elevation of total protein, increase of total IgG, and/or the presence of oligoclonal bands
c. HAD is always associated with a concurrent diagnosis of progressive multifocal leukoencephalopathy
d. HAD almost always is associated with $CD4^+$ T cell counts of less than 600
e. The clinical manifestations of HAD include behavioral, cognitive, and/or motor abnormalities

15. What aspects of clinical care have impacted the natural history of cognitive impairments in HIV infected individuals?

a. The introduction of potent, combination anti-retroviral therapy
b. The use of prophylactic medications against opportunistic infections
c. Improved biomarkers for disease diagnosis
d. One and two
e. None of the above

16. Antiretroviral agents currently in use include inhibitors of reverse transcriptase, HIV protease, and fusion inhibitors.

True/False

17. Currently there are six classes of antiretroviral agents available in the United States for clinical use:

– Nucleoside analogue reverse transcriptase inhibitors
– Nonnucleoside analogue reverse transcriptase inhibitors
– Protease inhibitors
– Entry inhibitors
– Tat inhibitors
– Nef inhibitors

True/False

18. Nucleoside analogue reverse transcriptase inhibitors are drugs of diverse chemical structure that act by non-competitive inhibition of HIV-1 reverse transcriptase.

True/False

19. Recent guidelines for starting anti-retroviral therapy recommend that patients with advanced disease, as manifested by the development of an AIDS-defining illness including HAD, or presence of severe symptoms such as unexplained fever, persistent diarrhea, or unexplained weight loss, should be started on ART if their $CD4^+$ T cell count is <100.

True/False

20. Abacavir plus lamivudine plus zidovudine, which is commercially available as a fixed combination in a single tablet, has inferior rates of viral suppression and should only be used when an NNRTI or a PI-based regimen cannot or should not be used as first line of therapy.

True/False

References

AIDSinfo [Internet]. Guidelines for the Use of Antiretroviral Agents in HIV-1-Infected Adults and Adolescents. Developed by the Panel on Clinical Practices for Treatment of HIV Infection convened by the Department of Health and Human Services (DHHS) [cited 2005 Oct 6]. Available from: http://www.aidsinfo.nih.gov/

Albright AV, Erickson-Viitanen S, O'Connor M, Frank I, Rayner MM, Gonzalez-Scarano F (2000) Efavirenz is a potent nonnucleoside reverse transcriptase inhibitor of HIV type 1 replication in microglia in vitro. AIDS Res Hum Retroviruses 16:1527–1537.

American Academy of Neurology (AAN) (1991) Nomenclature and research case definitions for neurologic manifestations of human immunodeficiency virus-type 1 infection. Report of a working group of the American academy of neurology AIDS task force. Neurology 41:778–785.

American Psychiatric Association [Internet]. Practice Guideline for the Treatment of Patients with HIV/AIDS. Virginia: American Psychiatric Association; c2000 [cited 2000 Nov]. Available from: http://www.psych.org/psych_pract/treatg/pg/hivaids_revisebook_index.cfm

Anderson E, Zink W, Xiong H, Gendelman HE (2002) HIV-1-associated dementia: A metabolic encephalopathy perpetrated by virus-infected and immune-competent mononuclear phagocytes. J Acquir Immune Defic Syndr 31(Suppl 2):S43–S54.

Andrade AS, McGruder HF, Wu AW, Celano SA, Skolasky RL, Selnes OA, Huang IC, McArthur JC (2005) A programmable prompting device improves adherence to highly active antiretroviral therapy in HIV-infected subjects with memory impairment. Clin Infect Dis 41:875–882.

Arendt G, Hefter H, Buescher L, Hilperath F, Elsing C, Freund HJ (1992) Improvement of motor performance of HIV-positive patients under AZT therapy. Neurology 42:891–896.

Arendt G, von Giesen HJ, Hefter H, Theisen A (2001) Therapeutic effects of nucleoside analogues on psychomotor slowing in HIV infection. AIDS 15:493–500.

Arendt G, von Giesen HJ (2002) Antiretroviral therapy regimens for neuro-AIDS. Current Drug Targets—Infectious Disorders 2:187–192.

Aylward EH, Henderer JD, McArthur JC, Brettschneider PD, Harris GJ, Barta PE, Pearlson GD (1993) Reduced basal ganglia volume in HIV-1-associated dementia: Results from quantitative neuroimaging. Neurology 43:2099–2104.

Aylward EH, Brettschneider PD, McArthur JC, Harris GJ, Schlaepfer TE, Henderer JD, Barta PE, Tien AY, Pearlson GD (1995) Magnetic resonance imaging measurement of gray matter volume reductions in HIV dementia. Am J Psychiatry 152:987–994.

Brew BJ (2004) Evidence for a change in AIDS dementia complex in the era of highly active antiretroviral therapy and the possibility of new forms of AIDS dementia complex. AIDS 18(Suppl 1): S75–S78.

Bridge TP, Heseltine PN, Parker ES, Eaton EM, Ingraham LJ, McGrail ML, Goodwin FK (1991) Results of extended peptide T administration in AIDS and ARC patients. Psychopharmacol Bull 27:237–245.

Budka H (2005) The neuropathology of HIV-associated brain disease. In: The Neurology of AIDS, 2nd ed. (Gendelman HE, Grant I, Everall IP, Lipton SA, Swindells S, eds), pp 376–391. Oxford: Oxford University Press.

Castro A (2005) Adherence to Antiretroviral therapy: Merging the clinical and social course of AIDS. PLoS Med 2:e338.

Chesney MA (1997) New antiretroviral therapies: Adherence challenges and strategies. Presented at the ICAAC 1997 satellite symposium: Evolving HIV treatments: Advances and the challenge of adherence. Available from: http://www.healthcg.com/hiv/treatment/icaac97/adherence/chesney.html

Childs EA, Lyles RH, Selnes OA, Chen B, Miller EN, Cohen BA, Becker JT, Mellors J, McArthur JC (1999) Plasma viral load and CD4 lymphocytes predict HIV-associated dementia and sensory neuropathy. Neurology. 52:607–613.

Clifford DB, McArthur JC, Schifitto G, Kieburtz K, McDermott MP, Letendre S, Cohen BA, Marder K, Ellis RJ, Marra CM; Neurologic AIDS Research Consortium (2002a) A randomized clinical trial of CPI-1189 for HIV-associated cognitive-motor impairment. Neurology 59:1568–1573.

Clifford DB (2002b) AIDS dementia. Med Clin North Am 86:537–550.

Clifford DB, Evans S, Yang Y, Acosta EP, Goodkin K, Tashima K, Simpson D, Dorfman D, Ribaudo H, Gulick RM (2005) Impact of efavirenz on neuropsychological performance and symptoms in HIV-infected individuals. Ann Intern Med 143:714–721.

Contreras X, Bennasser Y, Chazal N, Moreau M, Leclerc C, Tkaczuk J, Bahraoui E (2005) Human immunodeficiency virus type 1 Tat protein induces an intracellular calcium increase in human monocytes that requires DHP receptors: Involvement in TNF-alpha production. Virology 332:316–328.

Cunningham PH, Smith DG, Satchell C, Cooper DA, Brew B (2000) Evidence for independent development of resistance to HIV-1 reverse transcriptase inhibitors in the cerebrospinal fluid. AIDS 14:1949–1954.

Di Stefano M, Sabri F, Chiodi F (2005) HIV-1 structural and regulatory proteins and neurotoxicity. In: The Neurology of AIDS, 2nd ed. (Gendelman HE, Grant I, Everall IP, Lipton SA, Swindells S, eds), pp 49–56. Oxford: Oxford University Press.

Dieterich DT (2003) Long-Term Complications of Nucleoside Reverse Transcriptase Inhibitor Therapy. AIDS Read 13:176–187.

Dore GJ, Correll PK, Li Y, Kaldor JM, Cooper DA, Brew BJ (1999) Changes to AIDS dementia complex in the era of highly active antiretroviral therapy. AIDS 13:1249–1253.

Dreyer EB, Kaiser PK, Offermann JT, Lipton SA (1990) HIV-1 coat protein neurotoxicity prevented by calcium channel antagonists. Science 248:364–367.

Eggers C, Hertogs K, Sturenburg HJ, van Lunzen J, Stellbrink HJ (2003) Delayed central nervous system virus suppression during highly active antiretroviral therapy is associated with HIV encephalopathy, but not with viral drug resistance or poor central nervous system drug penetration. AIDS 17:1897–1906.

Enzensberger W, von Giesen HJ (1999) Antiretroviral therapy (ART) from a neurological point of view. German Neuro-AIDS study group (DNAA). Eur J Med Res 22:456–462.

Estanislao L, Thomas D, Simpson D (2004) HIV neuromuscular disease and mitochondrial function. Mitochondrion 4:131–139.

Evers S, Rahmann A, Schwaag S, Frese A, Reichelt D, Husstedt IW (2004) Prevention of AIDS dementia by HAART does not depend on cerebrospinal fluid drug penetrance. AIDS Res Hum Retroviruses 20:483–491.

Filippi CG, Sze G, Farber SJ, Shahmanesh M, Selwyn PA (1998) Regression of HIV encephalopathy and basal ganglia signal intensity abnormality at MR imaging in patients with AIDS after the initiation of protease inhibitor therapy. Radiology 206:491–498.

Gallant JE, Block DS (1998) Adherence to antiretroviral regimens in HIV-infected patients: Results of a survey among physicians and patients. J Int Assoc Physicians AIDS Care 4:32–35.

Gendelman HE, Zheng J, Coulter CL, Ghorpade A, Che M, Thylin M, Rubocki R, Persidsky Y, Hahn F, Reinhard J Jr, Swindells S (1998) Suppression of inflammatory neurotoxins by highly active

antiretroviral therapy in human immunodeficiency virus-associated dementia. J Infect Dis 178:1000–1007.
Gendelman HE, Persidsky Y (2005) Infections of the nervous system. Lancet Neurol 4:12–13.
Gillenwater DR, McDaniel JS (2001) Rational psychopharmacology for patients with HIV infection and AIDS. Psychiatric Annals 31:28–34.
Gordillo V, del Amo J, Soriano V, Gonzalez-Lahoz J (1999) Sociodemographic and psychological variables influencing adherence to antiretroviral therapy. AIDS 13:1763–1769.
Gulick RM, Ribaudo HJ, Shikuma CM, Lustgarten S, Squires KE, Meyer WA 3rd, Acosta EP, Schackman BR, Pilcher CD, Murphy RL, Maher WE, Witt MD, Reichman RC, Snyder S, Klingman KL, Kuritzkes DR; AIDS Clinical Trials Group Study A5095 Team (2004) Triple-nucleoside regimens versus efavirenz-containing regimens for the initial treatment of HIV-1 infection. N Engl J Med 350:1850–1861.
Heseltine PN, Goodkin K, Atkinson JH, Vitiello B, Rochon J, Heaton RK, Eaton EM, Wilkie FL, Sobel E, Brown SJ, Feaster D, Schneider L, Goldschmidts WL, Stover ES (1998) Randomized double-blind placebo-controlled trial of peptide T for HIV-associated cognitive impairment. Arch Neurol 55:41–51.
IAS-USA HIV Resistance Testing Guidelines Panel [Internet]. California: International AIDS Society-USA [cited Oct/Nov 2005]. Available from: http://www.iasusa.org/ resistance_mutations/index.html
Janssen RS, Nwanyanwu OC, Selik RM, Stehr-Green JK (1992) Epidemiology of human immunodeficiency virus encephalopathy in the United States. Neurology 42:1472–1476.
Lanier ER, Sturge G, McClernon D, Brown S, Halman M, Sacktor N, McArthur J, Atkinson JH, Clifford D, Price RW, Simpson D, Torres G, Catalan J, Marder K, Power C, Hall C, Romero C, Brew B (2001) HIV-1 reverse transcriptase sequence in plasma and cerebrospinal fluid of patients with AIDS dementia complex treated with abacavir. AIDS 15:747–751.
Lehrman G, Hogue IB, Palmer S, Jennings C, Spina CA, Wiegand A, Landay AL, Coombs RW, Richman DD, Mellors JW, Coffin JM, Bosch RJ, Margolis DM (2005) Depletion of latent HIV-1 infection in vivo: A proof-of-concept study. Lancet 366:549–555.
Letendre S (2005) The effects of antiretroviral therapy on viral and nonviral markers in cerebrospinal fluid. In: The Neurology of AIDS, 2nd ed. (Gendelman HE, Grant I, Everall IP, Lipton SA, Swindells S, eds), pp 617–628. Oxford: Oxford University Press.
Lucas GM, Chaisson RE, Moore RD (1999) Highly active antiretroviral therapy in a large urban clinic: Risk factors for virologic failure and adverse drug reactions. Ann Intern Med 131:81–87.
Marechal V, Prevost MC, Petit C, Perret E, Heard JM, Schwartz O (2001) Human immunodeficiency virus type 1 entry into macrophages mediated by macropinocytosis. J Virol 75:11166–11177.
McArthur JC, Hoover DR, Bacellar H, Miller EN, Cohen BA, Becker JT, Graham NM, McArthur JH, Selnes OA, Jacobson LP (1993) Dementia in AIDS patients: Incidence and risk factors. Multicenter AIDS cohort study. Neurology 43:2245–2252.
McArthur JC, Haughey N, Gartner S, Conant K, Pardo C, Nath A, Sacktor N (2003) Human immunodeficiency virus-associated dementia: An evolving disease. J Neurovirol 9:205–221.
McCance-Katz EF, Rainey PM, Friedland G, Jatlow P (2003) The protease inhibitor lopinavir-ritonavir may produce opiate withdrawal in methadone-maintained patients. Clin Infect Dis 37:476–482.
McConnell JR, Swindells S, Ong CS, Gmeiner WH, Chu WK, Brown DK, Gendelman HE (1994) Prospective utility of cerebral proton magnetic resonance spectroscopy in monitoring HIV infection and its associated neurological impairment. AIDS Res Hum Retroviruses 10:977–982.
Navia B, Price RW (2005) An overview of the clinical and biological features of the AIDS dementia complex. In: The Neurology of AIDS, 2nd ed. (Gendelman HE, Grant I, Everall IP, Lipton SA, Swindells S, eds), pp 339–358. Oxford: Oxford University Press.
Navia BA, Price RW (1987) The acquired immunodeficiency syndrome dementia complex as the presenting or sole manifestation of human immunodeficiency virus infection. Arch Neurol 44:65–69.
Navia BA, Dafni U, Simpson D, Tucker T, Singer E, McArthur JC, Yiannoutsos C, Zaborski L, Lipton SA (1998) A phase I/II trial of nimodipine for HIV-related neurologic complications. Neurology 51:221–228.
Palella FJ Jr, Delaney KM, Moorman AC, Loveless MO, Fuhrer J, Satten GA, Aschman DJ, Holmberg SD (1998) Declining morbidity and mortality among patients with advanced human immunodeficiency virus infection. HIV outpatient study investigators. N Engl J Med 338:853–860.
Paterson D, Swindells S, Mohr J, Brester M, Vergis EN, Squier C, Wagener MM, Singh N (2000) Adherence to protease inhibitor therapy and outcomes in patients with HIV infection. Ann Intern Med 133:21–30.
Pomara N, Crandall DT, Choi SJ, Johnson G, Lim KO (2001) White matter abnormalities in HIV-1 infection: A diffusion tensor imaging study. Psychiatry Res 106:15–24.
Power C, Selnes OA, Grim JA, McArthur JC (1995) HIV Dementia Scale: A rapid screening test. J Acquir Immune Defic Syndr Hum Retrovirol 8:273–278.
Price RW, Brew BJ (1988) The AIDS dementia complex. J Infect Dis 158:1079–1083.
Price RW, Paxinos EE, Grant RM, Drews B, Nilsson A, Hoh R, Hellmann NS, Petropoulos CJ, Deeks SG (2001) Cerebrospinal fluid response to structured treatment interruption after virological failure. AIDS 15:1251–1259.
Rodriguez-Rosado R, Jimenez-Nacher I, Soriano V, Anton P, Gonzalez-Lahoz J (1998) Virological failure and adherence to antiretroviral therapy in HIV-infected patients [letter]. AIDS 12:1112–1113.
Ruff MR, Hallberg PL, Hill JM, Pert CB (1987) Peptide T[4-8] is core HIV envelope sequence required for CD4 receptor attachment. Lancet 2:751.
Sacktor N, Schifitto G, McDermott MP, Marder K, McArthur JC, Kieburtz K (2000) Transdermal selegiline in HIV-associated cognitive impairment: Pilot, placebo-controlled study. Neurology 54:233–235.
Sacktor NC, Lyles RH, Skolasky RL, Anderson DE, McArthur JC, McFarlane G, Selnes OA, Becker JT, Cohen B, Wesch J, Miller EN (1999) Combination antiretroviral therapy improves psychomotor speed performance in HIV-seropositive homosexual men. Multicenter AIDS cohort study (MACS). Neurology 52:1640–1647.
Saksena NK, Smit TK (2005) HAART & the molecular biology of AIDS dementia complex. Indian J Med Res 12:256–69.
Schifitto G, Navia BA, Yiannoutsos CT, Marra CM, Chang L, Ernst T, Jarvik JG, Miller EN, Singer EJ, Ellis RJ, Kolson DL, Simpson D, Nath A, Berger J, Shriver SL, Millar LL, Colquhoun D, Lenkinski R, Gonzalez RG, Lipton SA (2007) Adult AIDS Clinical Trial Group (ACTG) 301; 700 Teams; HIV MRS

Consortium. Memantine and HIV-associated cognitive impairment: a neuropsychological and proton magnetic resonance spectroscopy study. AIDS. Sep 12;21(14):1877–1886.

Schifitto G, Sacktor N, Marder K, McDermott MP, McArthur JC, Kieburtz K, Small S, Epstein LG (1999) Randomized trial of the platelet-activating factor antagonist lexipafant in HIV-associated cognitive impairment. Neurological AIDS Research Consortium. Neurology 53:391–396.

Schifitto G, Zhang J, Evans SR, Sacktor N, Simpson D, Millar LL, Hung VL, Miller EN, Smith E, Ellis RJ, Valcour V, Singer E, Marra CM, Kolson D, Weihe J, Remmel R, Katzenstein D, Clifford DB (2007) ACTG A5090 Team. A multicenter trial of selegiline transdermal system for HIV-associated cognitive impairment. Neurology. Sep 25;69(13):1314–1321. Epub 2007 Jul 25.

Si Q, Cosenza M, Kim MO, hao ML, Brownlee M, Goldstein H, Lee S (2004) A novel action of minocycline: Inhibition of human immunodeficiency virus type 1 infection in microglia. J Neurovirol 10:284–292.

Sidtis JJ, Gatsonis C, Price RW, Singer EJ, Collier AC, Richman DD, Hirsch MS, Schaerf FW, Fischl MA, Kieburtz K, Simpson D, Koch MA, Feinberg J, Dafni U, The AIDS Clinical Trials Group (1993) Zidovudine treatment of the AIDS dementia complex: Results of a placebo-controlled trial. Ann Neurol 33:343–349.

Stern Y, McDermott MP, Albert S, Palumbo D, Selnes OA, McArthur J, Sacktor N, Schifitto G, Kieburtz K, Epstein L, Marder KS; Dana Consortium on the Therapy of HIV-Dementia and Related Cognitive Disorders (2001) Factors associated with incident human immunodeficiency virus-dementia. Arch Neurol 58:473–479.

Stout JC, Ellis RJ, Jernigan TL, Archibald SL, Abramson I, Wolfson T, McCutchan JA, Wallace MR, Atkinson JH, Grant I, HIV Neurobehavioral Research Center Group (1998) Progressive cerebral volume loss in human immunodeficiency virus infection: A longitudinal volumetric magnetic resonance imaging study. Arch Neurol 55:161–168.

Strelow LI, Janigro D, Nelson JA (2001) The blood-brain barrier and AIDS. Adv Virus Res 56:355–388.

Swindells S (2005) Current concepts in the treatment of HIV infection with focus on brain disease. In: The Neurology of AIDS, 2nd ed. (Gendelman HE, Grant I, Everall IP, Lipton SA, Swindells S, eds), pp 713–720. Oxford: Oxford University Press.

The Dana Consortium on the Therapy of HIV Dementia and Related Cognitive Disorders (1997) Safety and tolerability of the antioxidant OPC-14117 in HIV-associated cognitive impairment. Neurology 49:142–146.

The Dana Consortium on the Therapy of HIV Dementia and Related Cognitive Disorders (1998) A randomized, double-blind, placebo-controlled trial of deprenyl and thioctic acid in human immunodeficiency virus-associated cognitive impairment. Neurology 50:645–651.

Tozzi V, Narciso P, Galgani S, Sette P, Balestra P, Gerace C, Pau FM, Pigorini F, Volpini V, Camporiondo MP (1993) Effects of zidovudine in 30 patients with mild to end-stage AIDS dementia complex. AIDS 7:683–692.

Treisman GJ, Kaplin AI (2002) Neurologic and psychiatric complications of antiretroviral agents. AIDS 16:1201–1215.

Trottier B, Walmsley S, Reynes J, Piliero P, O'Hearn M, Nelson M, Montaner J, Lazzarin A, Lalezari J, Katlama C, Henry K, Cooper D, Clotet B, Arasteh K, Delfraissy JF, Stellbrink HJ, Lange J, Kuritzkes D, Eron JJ Jr, Cohen C, Kinchelow T, Bertasso A, Labriola-Tompkins E, Shikhman A, Atkins B, Bourdeau L, Natale C, Hughes F, Chung J, Guimaraes D, Drobnes C, Bader-Weder S, Demasi R, Smiley L, Salgo MP (2005) Safety of enfuvirtide in combination with an optimized background of antiretrovirals in treatment-experienced HIV-1-infected adults over 48 weeks. J Acquir Immune Defic Syndr 40:413–421.

Wang F, So Y, Vittinghoff E, Malani H, Reingold A, Lewis E, Giordano J, Janssen R (1995) Incidence proportion of and risk factors for AIDS patients diagnosed with HIV dementia, central nervous system toxoplasmosis, and cryptococcal meningitis. J Acquir Immune Defic Syndr Hum Retrovirol 8:75–82.

Williams A, Friedland G (1997) Adherence, compliance, and HAART. AIDS Clinical Care 9:51–55.

Wu P, Price P, Du B, Hatch WC, Terwilliger EF (1996) Direct cytotoxicity of HIV-1 envelope protein gp120 on human NT neurons. Neuroreport 7:1045–1049.

Zink MC, Uhrlaub J, DeWitt J, Voelker T, Bullock B, Mankowski J, Tarwater P, Clements J, Barber S (2005) Neuroprotective and anti-human immunodeficiency virus activity of minocycline. JAMA 293:2003–2011.

42
Immunotherapy

Jonathan Kipnis

Keywords Adaptive immunity; Copolymer-1 (Copaxone or glatiramer acetate); Immune boost; Immune modulation; Immune suppression; Mucosal tolerance; Nasal tolerance; Oral tolerance; T cell vaccination; Vaccination

42.1. Introduction

The central nervous system (CNS) has been referred as an immune privilege, in which local immune responses are restricted (Hailer et al., 1998; Cohen and Schwartz, 1999; Neumann, 2000; Pachter et al., 2003; Villoslada and Genain, 2004; Hatterer et al., 2005). Unlike most peripheral tissues, the CNS functions through a network of post mitotic cells (neurons) that are incapable of regeneration and hence any immune activity might interfere with cell function. CNS is vulnerable to damage that might be caused by the very means that the immune system uses to defend peripheral tissues from pathogens. Consequently, immune privilege in the CNS has been viewed as an evolutionary adaptation developed to protect the intricate neuronal networks of the CNS from incursion by the immune system (Lotan and Schwartz, 1994; Lotan et al., 1994).

An early definition of CNS immune privilege was based on the assumption that the access of immune system is restricted to the CNS. This assumption was supported by the observed tendency of a very slow rejection of allografts in the CNS (Fuchs and Bullard, 1988; Rao et al., 1989; Broadwell et al., 1990, 1994). Any leukocyte entry into CNS was viewed as evidence of pathology. Several observations have challenged this definition. Firstly, CNS antigens are drained to peripheral (cervical) lymph nodes and induce immune responses in the host (Cserr et al., 1992a, b; Cserr and Knopf, 1992). Second, activated T cells have been found to enter the CNS in the absence of discernible neuropathology (Hickey, 1999; Flugel et al., 2001). Third, leukocyte recruitment into the CNS appears to successfully resolve some CNS viral infections, such as Sindbis virus encephalitis, without the development of any apparent long-term bystander effects (Griffin et al., 1997; Griffin and Hardwick, 1997). Fourth, based on the relatively prolonged survival of xenografts (tissue grafts from different species) in immunosuppressed individuals, it was suggested that the immune system participates in CNS xenograft rejection (Czech et al., 1997). Taken together, these findings indicate that the CNS is accessible to immune cells, and that local immune responses within the CNS are regulated by a number of mechanisms. It seems likely that some of these mechanisms help to limit immune responses, with vital consequences for the functioning of the healthy CNS, however these mechanisms are altered following injury or under other devastating neurodegenerative disorders, thus making injured CNS friendlier and more assessable to immune cells. This includes up-regulation of major histocompatibility complex (MHC) class I and II molecule expression, down-regulation of Fas expression and increased permeability of blood–brain barrier (BBB) (Moalem et al., 1999a).

42.2. Immune Modulation in Neurodegenerative Disorders and Neural Injuries

Acute mechanical or biochemical injury to the mammalian CNS often results in an irreversible functional deficit (Schwab and Bartholdi, 1996; Cottrell et al., 1999; Schwab et al., 2000; Vinters et al., 2000; Gaviria et al., 2002) for several reasons, including the poor ability of injured axons to regrow, and a destructive series of injury-induced events that result in the spread of damage to neurons that escaped the direct injury (Faden, 1993; Faden et al., 1993; Povlishock and Christman, 1995; Povlishock and Jenkins, 1995; Yoles and Schwartz, 1998a; Hauben et al., 2000). This spread of damage is known as secondary degeneration. Attempts to promote CNS recovery have focused on two goals: (1) stimulation of regrowth (Caroni et al., 1988; Caroni and Schwab, 1988; Reier et al., 1992; Cheng et al., 1996; Rapalino et al., 1998; Chong et al., 1999; Neumann and Woolf, 1999), and (2) neuroprotection, or the arrest of self-perpetuating degeneration (Hall et al., 1992; Stolc et al., 1997; Lipton et al., 1998;

T. Ikezu and H.E. Gendelman (eds.), *Neuroimmune Pharmacology*.
© Springer 2008

Yoles and Schwartz, 1998b; Robertson et al., 2000; Roof and Hall, 2000; Schwartz, 2000; Schwartz and Yoles, 2000; Legos et al., 2001; Schwartz, 2001; Schwartz and Kipnis, 2001, 2002; Vajda, 2002; Bialek et al., 2004; Murray et al., 2004; Wu, 2005). The extent of recovery, in the absence of intervention, is a function of the amount of tissue that escaped the initial injury minus the loss associated with a neurodegenerative process of secondary degeneration. Thus the outcome is significantly worse than could have been predicted by the severity of an initial injury, however, could be reduced if initially undamaged or only marginally damaged neurons are rescued from secondary degeneration. Recognition of this fact and comparative ease of achievement as opposed to regeneration, has led to the emphasis on neuroprotection in the approach to CNS therapy. The spread of damage is mediated by numerous factors, among which are glutamate, nitric oxide, deprivation of growth factors, impaired blood supply and thus general metabolic deficit and others. Similar factors operate in different insults and across species, suggesting that the progression of damage might reflect a loss of control over mechanisms that regulate self-components, which, though normally essential, contribute to the damage spread under pathological conditions or at concentrations that exceed physiological. The increased presence of infiltrating or activated immune cells that accompanies degeneration often led scientists and clinicians to conclude that these cells contribute to the pathology and thus immune-suppressive therapies have emerged, without in-depth understanding of the mechanisms underlying their neuroprotective effects. Corticosteroids are the most widely used immune suppressive drugs for neurodegenerative disorders (Faden and Salzman, 1992; Hirschberg et al., 1994; Faden, 1996; Fung and Berger, 2000).

42.2.1. Corticosteroids for Glaucoma and Spinal Cord Injury

In physiology, corticosteroids are a class of steroid hormones that are produced in the adrenal cortex (De Nicola et al., 1998). Corticosteroids are involved in a wide range of physiologic systems such as stress response, immune response and regulation of inflammation, carbohydrate metabolism, protein catabolism, blood electrolyte levels, and behavior (Carlsen, 2005; Chamberlain et al., 2005; Currie et al., 2005; Schumacher and Chen, 2005; Williams, 2005). Corticosteroids have immune-suppressive properties and thus are used to suppress inflammation in the tissues and to suppress immune responses. Corticosteroids are also used to treat various neurodegenerative conditions assuming that suppression of post-injury inflammatory response at the site of injury will benefit tissue survival. Glucocorticoids are used frequently to suppress immunological responses and thus reports of the National Acute Spinal Cord Injury Studies (NASCIS) indicate that the standard method of treatment for patients with SCI currently includes the administration of high intravenous doses of methylprednisolone (MP) (Bracken et al., 1997). Recent reports question the benefit of corticosteroid treatment after CNS trauma; they also point out that the recommended MP dosage is the highest dose of steroids ever used during a 2-day period for any clinical condition, and therefore carries a risk (Hurlbert, 2000; Qian et al., 2000; Short, 2001). Other groups have shown that MP exacerbates axonal loss after optic nerve crush injury in Sprague-Dawley rats (Steinsapir et al., 2000). Corticosteroids are also used in ocular disorders and administered topically for diseases of the outer eye and anterior segment. Topical glucocorticoid therapy frequently increases intraocular pressure in normal eyes and exacerbates intraocular hypertension in patients with glaucoma. The glaucoma is not always reversible on cessation of glucocorticoid therapy (Kanagavalli et al., 2004; Kersey and Broadway, 2005; Kuchtey et al., 2005).

42.2.2. Cyclosporine A and Neurodegenerative Disorders

Cyclosporine A is another immunosuppressant drug. It is used post-alogeneic organ transplant to reduce the activity of the patient's immune system and so the risk of organ rejection. It has been studied in transplants of skin, heart, kidney, lung, pancreas, bone marrow and small intestine. Cyclosporine A (CsA) was also used to treat various neurodegenerative disorders, among which are Amyotrophic Lateral Sclerosis, Alzheimer's and Parkinson's diseases and acute CNS injuries. The micro and macro mechanisms underlying neuroprotection mediated by CsA are not fully understood. CsA has been shown to promote neurite outgrowth in cultured PC12 cells and sensory ganglia. Moreover, application of CsA following spinal cord injury increased GAP-43 expression, a sprouting marker, which was associated with improved functional recovery. CNS injury leads to mitochondrial dysfunction and thus mitochondria isolated from injured CNS show an increased production of reactive oxygen species. Treatment with CsA significantly attenuated mitochondrial dysfunction. In addition, CsA blocks a calcium-induced axonal damage, thus the effect of CsA could be mediated directly via restoration and maintenance of ionic and mitochondrial homeostasis, and not necessarily via suppression of immune response, as was originally thought.

While preclinical data from CsA in neurodegenerative disorders was promising, larger studies in Parkinson's disease did not show any benefit along with side complications. Major neurological complications secondary to cyclosporine are well documented and are known to include confusion, cortical blindness, seizure, spasticity, paresis, ataxia and coma. Most previous reports attribute these to white matter CNS lesions or white/grey matter border lesions. Moreover, several publications appeared recently refuted neuroprotective properties of CsA and similar compounds (immunophilin ligands).

So far, none of the immune suppressive drugs showed a significant and stable neuroprotective effects in any of the neurodegenerative conditions. Moreover, under those conditions when steroids were protective they were suggested to mediate the effect directly rather than via suppressing the immune system.

42.3. Immunomodulatory Therapeutics

42.3.1. T Cell Vaccination

Being not fully satisfied by the immune suppressive therapies and having side complication as a result of these compounds, the idea of modulating rather than suppressing an immune response has been evolved. Very first modulation of immune response was undertaken to treat autoimmune conditions such as multiple sclerosis in animal models of Experimental Autoimmune Encephalitis (EAE). The rationale and practice of T cell vaccination (TCV) against devastating processes underlying pathological autoimmune inflammation was analogous to classical vaccination against infectious disease. However, TCV differs from classical vaccination, as the "pathogen" is the identification marker of patient's own T cell receptor. The vaccine is devised using a component of the immune system itself – creating an autoimmune response against autoimmune T cells. For example, a single inoculation of myelin basic protein (MBP) reactive T cells, previously irradiated, into animals, abrogated an ability to induce autoimmune disease in these recipients. T cells vaccination has been extended to other experimental conditions, like adjuvant arthritis, experimental autoimmune thyroiditis, collagen-induced arthritis, autoimmune uveitis, lupus and diabetes. Therefore, activation of autoimmune T cells against autoimmune T cells causing the disease. It is, however, still a matter of debate whether suppression of autoaggressive T cells, their modulation or the activation of autoimmune T cells due to TCV are the reason for the obtained amelioration of autoimmune destructive symptoms.

42.3.2. Naturally Occurring CD4+ CD25+ Regulatory T Cells

Suppressor T cells are the hot topic of immunology for over last 20 years. Many types of regulatory T cells exist, however, all could be divided into two major populations: naturally occurring regulatory T cells and induced regulatory T cells. While many pathways have been recently elucidated to induce regulatory T cells to treat various autoimmune and neurodegenerative disorders, it is still a matter of debate whether regulatory or suppressor T cells are induced naturally in humans. The sub-population of T cells, termed naturally occurring regulatory T cells (Treg) and bearing a marker of CD25 in their naïve state, exists in both rodents and humans and account for 10% of total CD4+ T cell population. Tregs are also exclusively express Foxp3, transcription factor (Schramm et al., 2004; Viguier et al., 2004; Gavin et al., 2007). Treg cells are antigen specific anergic cells and the mechanism of inhibition requires both soluble factors and cell-to-cell contact. While the biology and the suppressive mechanism of these regulatory T cells are highly complicated and not fully understood, their practical application has been shown in several animal models of autoimmune and neurodegenerative disorders. Injection of Treg cells improved the outcome of EAE in mice. Treg cells accumulated in the site of autoimmune lesion and suppressed autoaggressive T cells. Application of Treg was shown to benefit colitis, autoimmune diabetes and other autoimmune disease. The role of these cells under neurodegenerative conditions is dual – while Treg cells protect neurons directly under certain circumstances, they suppress beneficial inflammatory response under other circumstances and thus exacerbate neuronal survival. Interestingly, injection of these cells into injured mice prone to develop autoimmune disease EAE, increased neuronal survival, whereas in strains that are resistant to EAE, an opposite effect has been obtained. Thus immune modulation using suppressor T cells will most probably benefit patients with autoimmune-mediated neurodegenerative disorders, however, will not be applicable for general population.

Treg cells bear markers of activated cells (e.g. CD25) in their naïve form, therefore, while these cells can be isolated from naïve mice, their isolation from patients with any inflammatory disease is technically impossible, due to the fact that activated effector (non-regulatory) T cells will bear markers identical to those expressed by Treg cells. Therefore, Treg mediated immune modulation does not seem feasible, at least now, until better markers are developed for Treg isolation from a pool of activated CD4+ T cell lymphocytes.

42.3.3. Induction of Regulatory T Cells by Mucosal Immunization

To overcome the need for isolation, activation and re-injection of Treg cells, alternative ways have been developed aimed to induce regulatory T cells or T cells with regulatory phenotype from normal CD4+ effector T cells. The most established approach is induction of mucosal tolerance that gives rise to T cells with anti-inflammatory phenotype, expressing mainly IL-10 and TGF-β cytokines. Evolutionary, mucosal tolerance allows prevention of the immune response toward orally (and nasally) consumed proteins, or in other words, to avoid immune allergy for consumed food. Taking advantage of this natural way to induce anti-inflammatory T cells, self-antigens were given orally in animals suffering from experimental autoimmune diseases. Such for example, mice fed orally with MBP were protective from induction of EAE using conventional vaccination techniques, otherwise inducing autoimmune syndromes. Later, this approach of modulation of immune response by induction of mucosal tolerance was examined on animal models for neurodegenerative disorders. Modulation of autoimmune inflammation, frequently associated with neurodegenerative disorders, was obtained by mucosal (oral or nasal) immunization with proteins abundantly expressed at the site of neurodegeneration, for example, myelin antigen for white matter injuries, retinal proteins for ocular neurodegenerative diseases, etc. Such mucosal vaccination was shown to benefit neuronal survival following acute injuries and chronic neurodegenerative conditions. One concern regarding mucosal vaccination, similar to previously raised concern with regulatory T cell therapy, is that immune response to injury is dependent on many factors and varies among individuals, therefore, mucosal tolerance might not be applicable clinically for use in general population.

42.3.4. Copaxone

Another approach to modulate immune response was initiated in early 70's when artificial compounds (polymers) were synthesized to mimic antigens and thus to induce immune response with partial cross reactivity to self-antigens. Partial recognition of antigen induces either an anti-inflammatory response or leads to anergy. Based on this understanding a synthetic co-polymer (Cop-1) has been synthesized (4.7–11 kDa) from four amino acids: L-alanine, L-lysine, L-glutamic acid and L-tyrosine, in a molar ratio of 4.2:3.4:1.4:1.0, aimed to mimic a dominant epitope of MBP—the most abundant myelin protein. Immunization with Cop-1 induced a T cells response partially cross-reactive with myelin basic protein, however did not induce an autoimmune disease. Moreover, immunization with MBP did not induce EAE if Cop-1 was mixed into the emulsion. Thus, Cop-1 was developed as a drug for multiple sclerosis and was later approved by Food and Drug Administration (FDA). Cop-1 is given daily to MS patients by sub cutaneous injection and has very limited side effects. The mechanism underlying Cop-1 mediated anti-inflammatory effect is based on clonal expansion of T cells with partial cross-reactive with MBP and a strong Th2 phenotype. Thus, Cop-1 reactive T cells were proposed to affect the ongoing inflammation by secretion of anti-inflammatory cytokines, phenomenon named "bystander suppression". Another possible explanation is the ability of Cop-1 to bind to MHC molecules from outside, without being processed, thus displacing dominant myelin proteins and being presented to MBP reactive T cells. Due to non-complete matching between the T cell receptor and the antigen (Cop-1 rather than MBP) the activation of T cells is not complete and such activation induced cellular anergy (no ability to respond or proliferate) or a very strong anti-inflammatory phenotype.

Recently Cop-1 was assessed for its neuroprotective properties, due to its partial cross-recognition with myelin basic protein. Immunization with Cop-1 significantly improved neuronal survival following acute injury of myelinated central and peripheral neurons. Recently Cop-1 was found to be neuroprotective also under various chronic conditions and under severe mental dysfunctions. While the mechanism for its neuroprotective properties is not well understood, it was shown that immunization with Cop-1 induces T cells that can pass through the Blood-Brain Barrier and to accumulate in the injured site. Upon migration to the brain, these cells induce production of growth factors by glial cells as well as produce these factors by themselves. The major growth factor produced and induced by Cop-1 reactive T cells in the brain in the Brain Derived Neurotrophic Factor (BDNF).

42.4. Vaccination for Neurodegenerative Disorders

42.4.1. Harnessing Humoral Immune Responses for Treatments of Spinal Cord Injuries and Alzheimer's Disease

Research over the last few years has demonstrated that myelin-associated inhibitor proteins inhibit sprouting and regeneration in the injured CNS. Antibodies neutralizing myelin inhibitors have been shown to improve spontaneous sprouting and regrowth (Schnell and Schwab, 1990) of injured CNS fibers. Among the factors found to be hostile to regrowth in adult CNS nerves are myelin-associated proteins (Schnell et al., 1999; Schwab et al., 2002; Schwab, 2002; He and Koprivica, 2004; Schwab et al., 2004; Buss et al., 2005) and extracellular matrix proteins such as chondroitin sulfate proteoglycans (CSPG) (Snow et al., 1990; Zuo et al., 1998). Accordingly, for regeneration to occur, these inhibitors must be masked, neutralized, or eliminated. Monoclonal antibodies directed specifically against the myelin-associated inhibitor IN-1 were introduced into partially transected rat spinal cords, with consequent promotion of regeneration (Schnell and Schwab, 1990). More recently it was demonstrated that immunization with myelin-associated proteins, such as NoGo, results in regeneration mediated by antibodies (Ellezam et al., 2003; Skaper, 2005). Interestingly, antibodies to myelin basic protein also were shown to lead to efficient myelin clearance and thus to facilitate recovery by preventive immunization or by passive transfer of specific antibodies (Huang et al., 1999; Ousman and David, 2001; Wong et al., 2003).

Use of antibodies to treat neurological disorders is not restricted to spinal cord injuries and was pioneered in Alzheimer's Disease (AD) research (Imahori and Uchida, 1997; Duyckaerts et al., 1998; Solomon, 2002; Heppner et al., 2004). In AD patients, or in animal models for AD, among various factors causing neurodegeneration, Amyloid-β aggregates are considered to be the major factor (Blass and Gibson, 1991; Richards et al., 1991; Bard et al., 2000). It is agreed that boosting of resident macrophages (microglia) to clear the plaques would benefit disease outcome. The performance of anti-Aβ-antibodies in transgenic mice models of AD prevented formation of new and induced dissolving of existing Aβ plaques (Lambert et al., 2001; Kotilinek et al., 2002; Mohajeri et al., 2002b; Mohajeri et al., 2002a; Nicolau et al., 2002; Rockenstein et al., 2002; Sigurdsson et al., 2002). Moreover, these antibodies protected the mice from learning and age-related memory deficits. Naturally occurring anti-Aβ antibodies have been found in human CSF and in the plasma of healthy individuals, but were significantly lower in AD patients, suggesting that AD patients cannot induce a desired immune reactivity to Aβ plaques and therefore develop a disorder (Monsonego et al., 2003). Active and/or passive immunization against Aβ peptide has been proposed as a method for preventing and/or treating Alzheimer's disease (Das et al., 2003; Frenkel et al., 2003; Furlan et al., 2003; Ghochikyan et al., 2003; Gong et al., 2003; Liu et al., 2004; Manea et al., 2004; Butovsky et al., 2005; Chauhan and Siegel, 2005; Faden et al., 2005). Experimental active immunization with Aβ 1–42 in humans was stopped in phase II clinical trials due to unexpected neuroinflammatory manifestations and death associated with the treatment (Koistinaho et al., 2001; Tariot and Federoff, 2003). Although the trials have been stopped, post-mortem observation of the tissue showed a significant drop in the number of plaques in treated patients. Therefore, there is a hope of developing better antibodies with similar therapeutic potential without such devastating side effects.

Although immune cells are known to play a key role in tissue repair, their activity in the CNS, as outlined above, has mostly been viewed as detrimental. A comparative study of the local inflammatory response in injured peripheral nervous system and CNS axons revealed that in the injured peripheral nervous system (PNS), where regeneration takes place spontaneously, both degeneration and regeneration appear to be brought about and controlled with the prominent participation of macrophages. Macrophages are also a source of cytokines and growth factors that actively participate, both directly and indirectly, in regrowth. As an example, the regeneration of optic nerves in lower vertebrates, as well as of peripheral nerves in mammals, was found to correlate with upregulation of macrophage-derived apolipoproteins, which participate in the recycling of lipids needed for membrane rebuilding (Ignatius et al., 1987; Harel et al., 1989). Similarly, the synthesis of nerve growth factor seems to be regulated by stimulated macrophages. IL-1 and tumor necrosis factor (TNF)-α, both secreted by macrophages, probably induce nerve growth factor transcription in Schwann cells (Lindholm et al., 1987).

In early experiments aimed at boosting of the innate response in the injured CNS to facilitate better phagocytosis of myelin debris, it was found that implantation of homologous macrophages in the completely transected spinal cords of adult rats, it was demonstrated partial recovery of motor function (manifested by locomotor activity scored in an open field) and electrophysiological activity (assessed by motor-evoked potential responses) (Rapalino et al., 1998). These behavioral manifestations were also reflected in the electrophysiological recovery of motor-evoked potential responses in the implanted rats.

42.4.2. Innate and Adaptive Immunity

The first line of defense in any injured tissue is the innate immune response, characterized by activity of the phagocytic immune cells (Paape et al., 2000), however, this response is tightly controlled by adaptive immunity, namely T cells. There are different subpopulations of CD4+ T cells, each responsible for a certain type of immune response. Th1 cells, for example, reinforce cellular immunity, whereas Th2 cells induce humoral (antibody-mediated) immunity. Studies have shown that the autoimmune CD4+ T cells locally boost and control resident microglia and infiltrating blood-borne monocytes, helping them to acquire an activity that allows them to fight off degenerative conditions. It also allows them to buffer toxic compounds without producing excessive amounts of inflammation-associated cytotoxic compounds such TNF-α, nitric oxide (NO), or cyclooxygenase (COX)-2 (Levi et al., 1998; Basu et al., 2002; Janabi, 2002; Schwartz, 2003), and to produce growth factors such as insulin-like growth factor-1 that are critical for neuronal survival and renewal. Thus, the role of CD4+ T cells directed against self-antigens is to activate the innate response in a well-controlled way, enabling it to recognize the threat to the tissue not as a harmful organism that it must kill, but as a toxic substance that it must neutralize or eliminate. In addition, the T cells themselves, if locally activated, can produce protective compounds such as growth factors and neurotrophins (Hammarberg et al., 2000; Moalem et al., 2000; Gielen et al., 2003). All of these tasks can be accomplished by a well-controlled response by Th cells. These cells, in order to do their job, must first home to sites of stress and be locally activated by their specific antigens that reside there. Thus, the homing of T cells to the site at which their local activation can occur is apparently dictated by antigenic specificity. In line with this notion are the findings that T cells having the same antigenic specificity are protective against different types of threatening stimuli occurring at the same site, or against different threatening stimuli at different sites sharing the same immunodominant, constitutively expressed self-proteins (Mizrahi et al., 2002). As a corollary, the same threatening stimulus, if manifested at different sites that do not share common dominant self-antigens, does not benefit from T cells directed against the same antigen (Schori et al., 2001; Mizrahi et al., 2002).

Studies from several laboratories have shown that T cells patrol the healthy CNS but do not accumulate there (Hickey, 1999; Flugel et al., 2001). In the event of an acute injury or chronic neurodegenerative conditions, T cells are recruited by and accumulate in the CNS (Hirschberg et al., 1998; Moalem et al., 1999a), where they might rescue neurons from degeneration if the damage caused by the toxic biochemical environment is not yet irreversible; moreover, the recruited T cells will prevent further deterioration. It is possible that this autoimmune protective mechanism also operates when the threat to the tissue comes from microbial infiltration. In such a case the anti-self response is additional to the anti-microbial response, and might occur without the individual even being aware of it, unless the harnessed autoimmunity gets out of control, in which case its effect is no longer beneficial but destructive, and might result in an autoimmune disease (Schwartz, 2002). This might be the situation in individuals who are predisposed to autoimmune disease development (Kipnis et al., 2001). According to this view, the pathogenic self-proteins that have been implicated in autoimmune diseases are the same proteins against which a well-controlled T-cell response is protective (Moalem et al., 1999b; Kipnis et al., 2002b). It might explain the generally low clinical prevalence of autoimmune diseases and their occurrence mainly in young adults rather than in the elderly population, whereas neurodegenerative diseases and cancer are common and significantly more prevalent in the elderly, in whom the immune system is deteriorating (Linton and Dorshkind, 2004). Boost of autoimmune activity following injury increases the number of surviving neurons, however, it has a risk of development of autoimmune disease. Alternative approach is to boost an immune response with cross-reactivity with self antigens, such as with Cop-1. Immunization with Cop-1 is neuroprotective in various acute and chronic neurodegenerative conditions (Kipnis et al., 2000; Schori et al., 2001; Angelov et al., 2003; Benner et al., 2004; Frenkel et al., 2005). Therefore various ways are searched now to find the best approach under which the boost of autoimmune T cells or of T cells with partial recognition to self-antigens would lead to maximum benefit with minimal destructive consequences of autoimmune disease. These possible therapeutic approaches are discussed in details in the following chapter (Kipnis et al., 2002a).

Summary

From the "immune privilege" status and immunosuppressive treatments considered for almost any neurodegenerative conditions, our understanding of the neuro-immune interactions underwent a significant evolution and current therapies are aimed to modulate and boost rather than suppress and eliminate immune assistance in the brain.

Review Questions/Problems

1. Myelin can cause neuronal growth inhibition.

True/False

2. Injured CNS fibers have the potential for regrowth.

True/False

3. T cells required for neuroprotection are antigen specific.

True/False

4. Naturally occurring regulatory CD4+ CD25+ T cells are derived from the thymus.

True/False

5. Glutamate is a part of both killing machinery and neuronal communication

True/False

6. Immune synapses in the CNS are MHC-II-expressing microglia interacting with T-helper lymphocytes.

True/False

7. Macrophages that can promote regeneration after a spinal cord injury should primarily be

a. cytotoxic
b. antigen presenting
c. phagocytic
d. naïve
e. all of the above

8. Autoimmune T cells required for protection of the injured spinal cord are reactive toward

a. skin proteins
b. heat-shock proteins
c. myelin proteins
d. their antigenic specificity is irrelevant to the type of injury

9. Mucosal tolerance can be viewed as

a. modulation of immune response
b. suppression of immune response
c. boost of the immune response
d. all of the above
e. none of the above

10. Immunization with Copaxone can be viewed as

a. modulation of immune response
b. suppression of immune response
c. boost of the immune response
d. all of the above
e. none of the above

11. Spinal cord injury is associated with

a. axonal regrowth
b. inhibition of axonal regrowth
c. axonal sprouting
d. none of the above

12. Neuroprotection induced by CsA under certain neurodegenerative conditions can be attributed solely to

a. suppression of the immune response
b. activation of the immune response
c. induction of growth factors
d. direct effect on glial cells
e. none of the above

13. The role of immune cells in the CNS – is it destructive or protective? What are benefits of immune boost following injury and what are the possible devastating side effects of such therapy?

14. What is immune modulation and how it could be achieved?

15. Immune suppression is neuroprotective under certain circumstances whereas neurodestructive under others. How this apparent dichotomy could be resolved?

16. What is the evolutionary benefit from having a complex network of autoimmune and regulatory T cells?

17. Does the presence of BBB favors in favor or not of the role of immune system in post-injury recovery?

References

Angelov DN, Waibel S, Guntinas-Lichius O, Lenzen M, Neiss WF, Tomov TL, Yoles E, Kipnis J, Schori H, Reuter A, Ludolph A, Schwartz M (2003) Therapeutic vaccine for acute and chronic motor neuron diseases: Implications for amyotrophic lateral sclerosis. Proc Natl Acad Sci USA 100:4790–4795.

Bard F, Cannon C, Barbour R, Burke RL, Games D, Grajeda H, Guido T, Hu K, Huang J, Johnson-Wood K, Khan K, Kholodenko D, Lee M, Lieberburg I, Motter R, Nguyen M, Soriano F, Vasquez N, Weiss K, Welch B, Seubert P, Schenk D, Yednock T (2000) Peripherally administered antibodies against amyloid beta-peptide enter the central nervous system and reduce pathology in a mouse model of Alzheimer disease. Nat Med 6:916–919.

Basu A, Krady JK, Enterline JR, Levison SW (2002) Transforming growth factor beta1 prevents IL-1beta-induced microglial activation, whereas TNFalpha- and IL-6-stimulated activation are not antagonized. Glia 40:109–120.

Benner EJ, Mosley RL, Destache CJ, Lewis TB, Jackson-Lewis V, Gorantla S, Nemachek C, Green SR, Przedborski S, Gendelman HE (2004) Therapeutic immunization protects dopaminergic neurons in a mouse model of Parkinson's disease. Proc Natl Acad Sci USA 101:9435–9440.

Bialek M, Zaremba P, Borowicz KK, Czuczwar SJ (2004) Neuroprotective role of testosterone in the nervous system. Pol J Pharmacol 56:509–518.

Blass JP, Gibson GE (1991) The role of oxidative abnormalities in the pathophysiology of Alzheimer's disease. Rev Neurol (Paris) 147:513–525.

Bracken MB, Shepard MJ, Holford TR, Leo-Summers L, Aldrich EF, Fazl M, Fehlings M, Herr DL, Hitchon PW, Marshall LF, Nockels RP, Pascale V, Perot PL, Jr., Piepmeier J, Sonntag VK, Wagner F, Wilberger JE, Winn HR, Young W (1997) Administration of methylprednisolone for 24 or 48 hours or tirilazad mesylate for 48 hours in the treatment of acute spinal cord injury. Results of the Third National Acute Spinal Cord Injury Randomized Controlled Trial. National Acute Spinal Cord Injury Study. JAMA 277:1597–1604.

Broadwell RD, Baker BJ, Ebert PS, Hickey WF (1994) Allografts of CNS tissue possess a blood-brain barrier: III. Neuropathological, methodological, and immunological considerations. Microsc Res Tech 27:471–494.

Broadwell RD, Charlton HM, Ebert P, Hickey WF, Villegas JC, Wolf AL (1990) Angiogenesis and the blood-brain barrier in solid and dissociated cell grafts within the CNS. Prog Brain Res 82:95–101.

Buss A, Pech K, Merkler D, Kakulas BA, Martin D, Schoenen J, Noth J, Schwab ME, Brook GA (2005) Sequential loss of myelin proteins during Wallerian degeneration in the human spinal cord. Brain 128:356–364.

Butovsky O, Talpalar AE, Ben-Yaakov K, Schwartz M (2005) Activation of microglia by aggregated beta-amyloid or lipopolysaccharide impairs MHC-II expression and renders them cytotoxic whereas IFN-gamma and IL-4 render them protective. Mol Cell Neurosci 29:381–393.

Carlsen KH (2005) Pharmaceutical treatment of asthma in children. Curr Drug Targets Inflamm Allergy 4:543–549.

Caroni P, Schwab ME (1988) Antibody against myelin-associated inhibitor of neurite growth neutralizes nonpermissive substrate properties of CNS white matter. Neuron 1:85–96.

Caroni P, Savio T, Schwab ME (1988) Central nervous system regeneration: Oligodendrocytes and myelin as non-permissive substrates for neurite growth. Prog Brain Res 78:363–370.

Chamberlain MC, Sloan A, Vrionis F (2005) Systematic review of the diagnosis and management of malignant extradural spine cord compression: The Cancer Care Ontario Practice Guidelines Initiative's Neuro-Oncology Disease Site Group. J Clin Oncol 23:7750–7751; author reply 7751–7752.

Chauhan NB, Siegel GJ (2005) Efficacy of anti-Abeta antibody isotypes used for intracerebroventricular immunization in TgCRND8. Neurosci Lett 375:143–147.

Cheng HL, Sullivan KA, Feldman EL (1996) Immunohistochemical localization of insulin-like growth factor binding protein-5 in the developing rat nervous system. Dev Brain Res 92:211–218.

Chong MS, Woolf CJ, Haque NS, Anderson PN (1999) Axonal regeneration from injured dorsal roots into the spinal cord of adult rats. J Comp Neurol 410:42–54.

Cohen IR, Schwartz M (1999) Autoimmune maintenance and neuroprotection of the central nervous system. J Neuroimmunol 100:111–114.

Cottrell DF, McGorum BC, Pearson GT (1999) The neurology and enterology of equine grass sickness: A review of basic mechanisms. Neurogastroenterol Motil 11:79–92.

Cserr HF, Knopf PM (1992) Cervical lymphatics, the blood-brain barrier and the immunoreactivity of the brain: A new view. Immunol Today 13:507–512.

Cserr HF, Harling-Berg CJ, Knopf PM (1992a) Drainage of brain extracellular fluid into blood and deep cervical lymph and its immunological significance. Brain Pathol 2:269–276.

Cserr HF, DePasquale M, Harling-Berg CJ, Park JT, Knopf PM (1992b) Afferent and efferent arms of the humoral immune response to CSF-administered albumins in a rat model with normal blood-brain barrier permeability. J Neuroimmunol 41:195–202.

Currie GP, Lee DK, Srivastava P (2005) Long-acting bronchodilator or leukotriene modifier as add-on therapy to inhaled corticosteroids in persistent asthma? Chest 128:2954–2962.

Czech KA, Ryan JW, Sagen J, Pappas GD (1997) The influence of xenotransplant immunogenicity and immunosuppression on host MHC expression in the rat CNS. Exp Neurol 147:66–83.

Das P, Howard V, Loosbrock N, Dickson D, Murphy MP, Golde TE (2003) Amyloid-beta immunization effectively reduces amyloid deposition in FcRgamma-/- knock-out mice. J Neurosci 23:8532–8538.

De Nicola AF, Ferrini M, Gonzalez SL, Gonzalez Deniselle MC, Grillo CA, Piroli G, Saravia F, de Kloet ER (1998) Regulation of gene expression by corticoid hormones in the brain and spinal cord. J Steroid Biochem Mol Biol 65:253–272.

Duyckaerts C, Colle MA, Dessi F, Grignon Y, Piette F, Hauw JJ (1998) The progression of the lesions in Alzheimer disease: Insights from a prospective clinicopathological study. J Neural Transm Suppl 53:119–126.

Ellezam B, Bertrand J, Dergham P, McKerracher L (2003) Vaccination stimulates retinal ganglion cell regeneration in the adult optic nerve. Neurobiol Dis 12:1–10.

Faden AI (1993) Comparison of single and combination drug treatment strategies in experimental brain trauma. J Neurotrauma 10:91–100.

Faden AI (1996) Pharmacological treatment of central nervous system trauma. Pharmacol Toxicol 78:12–17.

Faden AI, Salzman S (1992) Pharmacological strategies in CNS trauma. Trends Pharmacol Sci 13:29–35.

Faden AI, Labroo VM, Cohen LA (1993) Imidazole-substituted analogues of TRH limit behavioral deficits after experimental brain trauma. J Neurotrauma 10:101–108.

Faden AI, Movsesyan VA, Knoblach SM, Ahmed F, Cernak I (2005) Neuroprotective effects of novel small peptides in vitro and after brain injury. Neuropharmacology 49:410–424.

Flugel A, Berkowicz T, Ritter T, Labeur M, Jenne DE, Li Z, Ellwart JW, Willem M, Lassmann H, Wekerle H (2001) Migratory activity and functional changes of green fluorescent effector cells before and during experimental autoimmune encephalomyelitis. Immunity 14:547–560.

Frenkel D, Dewachter I, Van Leuven F, Solomon B (2003) Reduction of beta-amyloid plaques in brain of transgenic mouse model of Alzheimer's disease by EFRH-phage immunization. Vaccine 21:1060–1065.

Frenkel D, Maron R, Burt DS, Weiner HL (2005) Nasal vaccination with a proteosome-based adjuvant and glatiramer acetate clears beta-amyloid in a mouse model of Alzheimer disease. J Clin Invest 115:2423–2433.

Fuchs HE, Bullard DE (1988) Immunology of transplantation in the central nervous system. Appl Neurophysiol 51:278–296.

Fung MA, Berger TG (2000) A prospective study of acute-onset steroid acne associated with administration of intravenous corticosteroids. Dermatology 200:43–44.

Furlan R, Brambilla E, Sanvito F, Roccatagliata L, Olivieri S, Bergami A, Pluchino S, Uccelli A, Comi G, Martino G (2003) Vaccination with amyloid-beta peptide induces autoimmune encephalomyelitis in C57/BL6 mice. Brain 126:285–291.

Gavin MA, Rasmussen JP, Fontenot JD, Vasta V, Manganiello VC, Beavo JA, Rudensky AY (2007) Foxp3-dependent programme of regulatory T-cell differentiation. Nature 445:771–775.

Gaviria M, Haton H, Sandillon F, Privat A (2002) A mouse model of acute ischemic spinal cord injury. J Neurotrauma 19:205–221.

Ghochikyan A, Vasilevko V, Petrushina I, Movsesyan N, Babikyan D, Tian W, Sadzikava N, Ross TM, Head E, Cribbs DH, Agadjanyan MG (2003) Generation and characterization of the humoral immune response to DNA immunization with a chimeric beta-amyloid-interleukin-4 minigene. Eur J Immunol 33:3232–3241.

Gielen A, Khademi M, Muhallab S, Olsson T, Piehl F (2003) Increased brain-derived neurotrophic factor expression in white blood cells of relapsing-remitting multiple sclerosis patients. Scand J Immunol 57:493–497.

Gong Y, Chang L, Viola KL, Lacor PN, Lambert MP, Finch CE, Krafft GA, Klein WL (2003) Alzheimer's disease-affected brain: Presence of oligomeric A beta ligands (ADDLs) suggests a molecular basis for reversible memory loss. Proc Natl Acad Sci USA 100:10417–10422.

Griffin D, Levine B, Tyor W, Ubol S, Despres P (1997) The role of antibody in recovery from alphavirus encephalitis. Immunol Rev 159:155–161.

Griffin DE, Hardwick JM (1997) Regulators of apoptosis on the road to persistent alphavirus infection. Annu Rev Microbiol 51:565–592.

Hailer NP, Heppner FL, Haas D, Nitsch R (1998) Astrocytic factors deactivate antigen presenting cells that invade the central nervous system. Brain Pathol 8:459–474.

Hall ED, Braughler JM, McCall JM (1992) Antioxidant effects in brain and spinal cord injury. J Neurotrauma 9(Suppl 1):S165–172.

Hammarberg H, Lidman O, Lundberg C, Eltayeb SY, Gielen AW, Muhallab S, Svenningsson A, Linda H, van Der Meide PH, Cullheim S, Olsson T, Piehl F (2000) Neuroprotection by encephalomyelitis: Rescue of mechanically injured neurons and neurotrophin production by CNS-infiltrating T and natural killer cells. J Neurosci 20:5283–5291.

Harel A, Fainaru M, Shafer Z, Hernandez M, Cohen A, Schwartz M (1989) Optic nerve regeneration in adult fish and apolipoprotein A-I. J Neurochem 52:1218–1228.

Hatterer E, Davoust N, Didier-Bazes M, Vuaillat C, Malcus C, Belin MF, Nataf S (2006) How to drain without lymphatics? Dendritic cells migrate from the cerebrospinal fluid to the B-cell follicles of cervical lymph nodes. Blood 107:806–812.

Hauben E, Butovsky O, Nevo U, Yoles E, Moalem G, Agranov G, Mor F, Leibowitz-Amit R, Pevsner E, Akselrod S, Neeman M, Cohen IR, Schwartz M (2000) Passive or active immunization with myelin basic protein promotes recovery from spinal cord contusion. J Neurosci 20:6421–6430.

He Z, Koprivica V (2004) The Nogo signaling pathway for regeneration block. Annu Rev Neurosci 27:341–368.

Heppner FL, Gandy S, McLaurin J (2004) Current concepts and future prospects for Alzheimer disease vaccines. Alzheimer Dis Assoc Disord 18:38–43.

Hickey WF (1999) Leukocyte traffic in the central nervous system: The participants and their roles. Semin Immunol 11:125–137.

Hirschberg DL, Yoles E, Belkin M, Schwartz M (1994) Inflammation after axonal injury has conflicting consequences for recovery of function: Rescue of spared axons is impaired but regeneration is supported. J Neuroimmunol 50:9–16.

Hirschberg DL, Moalem G, He J, Mor F, Cohen IR, Schwartz M (1998) Accumulation of passively transferred primed T cells independently of their antigen specificity following central nervous system trauma. J Neuroimmunol 89:88–96.

Huang DW, McKerracher L, Braun PE, David S (1999) A therapeutic vaccine approach to stimulate axon regeneration in the adult mammalian spinal cord. Neuron 24:639–647.

Hurlbert RJ (2000) Methylprednisolone for acute spinal cord injury: An inappropriate standard of care. J Neurosurg 93:1–7.

Ignatius MJ, Shooter EM, Pitas RE, Mahley RW (1987) Lipoprotein uptake by neuronal growth cones in vitro. Science 236:959–962.

Imahori K, Uchida T (1997) Physiology and pathology of tau protein kinases in relation to Alzheimer's disease. J Biochem (Tokyo) 121:179–188.

Janabi N (2002) Selective inhibition of cyclooxygenase-2 expression by 15-deoxy-Delta(12,14)(12,14)-prostaglandin J(2) in activated human astrocytes, but not in human brain macrophages. J Immunol 168:4747–4755.

Kanagavalli J, Pandaranayaka E, Krishnadas SR, Krishnaswamy S, Sundaresan P (2004) A review of genetic and structural understanding of the role of myocilin in primary open angle glaucoma. Indian J Ophthalmol 52:271–280.

Kersey JP, Broadway DC (2006) Corticosteroid-induced glaucoma: A review of the literature. Eye 20:407–416.

Kipnis J, Mizrahi T, Yoles E, Ben-Nun A, Schwartz M (2002a) Myelin specific Th1 cells are necessary for post-traumatic protective autoimmunity. J Neuroimmunol 130:78–85.

Kipnis J, Yoles E, Schori H, Hauben E, Shaked I, Schwartz M (2001) Neuronal survival after CNS insult is determined by a genetically encoded autoimmune response. J Neurosci 21:4564–4571.

Kipnis J, Mizrahi T, Hauben E, Shaked I, Shevach E, Schwartz M (2002b) Neuroprotective autoimmunity: Naturally occurring CD4 + CD25 + regulatory T cells suppress the ability to withstand injury to the central nervous system. Proc Natl Acad Sci USA 99:15620–15625.

Kipnis J, Yoles E, Porat Z, Cohen A, Mor F, Sela M, Cohen IR, Schwartz M (2000) T cell immunity to copolymer 1 confers neuroprotection on the damaged optic nerve: Possible therapy for optic neuropathies. Proc Natl Acad Sci USA 97:7446–7451.

Koistinaho M, Ort M, Cimadevilla JM, Vondrous R, Cordell B, Koistinaho J, Bures J, Higgins LS (2001) Specific spatial learning deficits become severe with age in beta -amyloid precursor protein transgenic mice that harbor diffuse beta -amyloid deposits but do not form plaques. Proc Natl Acad Sci USA 98:14675–14680.

Kotilinek LA, Bacskai B, Westerman M, Kawarabayashi T, Younkin L, Hyman BT, Younkin S, Ashe KH (2002) Reversible memory loss in a mouse transgenic model of Alzheimer's disease. J Neurosci 22:6331–6335.

Kuchtey RW, Lowder CY, Smith SD (2005) Glaucoma in patients with ocular inflammatory disease. Ophthalmol Clin North Am 18:421–430, vii.

Lambert MP, Viola KL, Chromy BA, Chang L, Morgan TE, Yu J, Venton DL, Krafft GA, Finch CE, Klein WL (2001) Vaccination with soluble Abeta oligomers generates toxicity-neutralizing antibodies. J Neurochem 79:595–605.

Legos JJ, Lee D, Erhardt JA (2001) Caspase inhibitors as neuroprotective agents. Expert Opin Emerg Drugs 6:81–94.

Levi G, Minghetti L, Aloisi F (1998) Regulation of prostanoid synthesis in microglial cells and effects of prostaglandin E2 on microglial functions. Biochimie 80:899–904.

Lindholm D, Heumann R, Meyer M, Thoenen H (1987) Interleukin-1 regulates synthesis of nerve growth factor in non-neuronal cells of rat sciatic nerve. Nature 330:658–659.

Linton PJ, Dorshkind K (2004) Age-related changes in lymphocyte development and function. Nat Immunol 5:133–139.

Lipton SA, Choi YB, Sucher NJ, Chen HS (1998) Neuroprotective versus neurodestructive effects of NO-related species. Biofactors 8:33–40.

Liu R, Yuan B, Emadi S, Zameer A, Schulz P, McAllister C, Lyubchenko Y, Goud G, Sierks MR (2004) Single chain variable fragments against beta-amyloid (Abeta) can inhibit Abeta aggregation and prevent abeta-induced neurotoxicity. Biochemistry 43:6959–6967.

Lotan M, Schwartz M (1994) Cross talk between the immune system and the nervous system in response to injury: Implications for regeneration. FASEB J 8:1026–1033.

Lotan M, Solomon A, Ben-Bassat S, Schwartz M (1994) Cytokines modulate the inflammatory response and change permissiveness to neuronal adhesion in injured mammalian central nervous system. Exp Neurol 126:284–290.

Manea M, Mezo G, Hudecz F, Przybylski M (2004) Polypeptide conjugates comprising a beta-amyloid plaque-specific epitope as new vaccine structures against Alzheimer's disease. Biopolymers 76:503–511.

Mizrahi T, Hauben E, Schwartz M (2002) The tissue-specific self-pathogen is the protective self-antigen: The case of uveitis. J Immunol 169:5971–5977.

Moalem G, Monsonego A, Shani Y, Cohen IR, Schwartz M (1999a) Differential T cell response in central and peripheral nerve injury: Connection with immune privilege. FASEB J 13:1207–1217.

Moalem G, Leibowitz-Amit R, Yoles E, Mor F, Cohen IR, Schwartz M (1999b) Autoimmune T cells protect neurons from secondary degeneration after central nervous system axotomy. Nat Med 5:49–55.

Moalem G, Gdalyahu A, Shani Y, Otten U, Lazarovici P, Cohen IR, Schwartz M (2000) Production of neurotrophins by activated T cells: Implications for neuroprotective autoimmunity. J Autoimmun 20:6421–6430.

Mohajeri MH, Wollmer MA, Nitsch RM (2002a) Abeta 42-induced increase in neprilysin is associated with prevention of amyloid plaque formation in vivo. J Biol Chem 277:35460–35465.

Mohajeri MH, Saini K, Schultz JG, Wollmer MA, Hock C, Nitsch RM (2002b) Passive immunization against beta-amyloid peptide protects central nervous system (CNS) neurons from increased vulnerability associated with an Alzheimer's disease-causing mutation. J Biol Chem 277:33012–33017.

Monsonego A, Zota V, Karni A, Krieger JI, Bar-Or A, Bitan G, Budson AE, Sperling R, Selkoe DJ, Weiner HL (2003) Increased T cell reactivity to amyloid beta protein in older humans and patients with Alzheimer disease. J Clin Invest 112:415–422.

Murray M, Fischer I, Smeraski C, Tessler A, Giszter S (2004) Towards a definition of recovery of function. J Neurotrauma 21:405–413.

Neumann H (2000) The immunological microenvironment in the CNS: Implications on neuronal cell death and survival. J Neural Transm Suppl 59:59–68.

Neumann S, Woolf CJ (1999) Regeneration of dorsal column fibers into and beyond the lesion site following adult spinal cord injury. Neuron 23:83–91.

Nicolau C, Greferath R, Balaban TS, Lazarte JE, Hopkins RJ (2002) A liposome-based therapeutic vaccine against beta -amyloid plaques on the pancreas of transgenic NORBA mice. Proc Natl Acad Sci USA 99:2332–2337.

Ousman SS, David S (2001) MIP-1alpha, MCP-1, GM-CSF, and TNF-alpha control the immune cell response that mediates rapid phagocytosis of myelin from the adult mouse spinal cord. J Neurosci 21:4649–4656.

Paape MJ, Shafer-Weaver K, Capuco AV, Van Oostveldt K, Burvenich C (2000) Immune surveillance of mammary tissue by phagocytic cells. Adv Exp Med Biol 480:259–277.

Pachter JS, de Vries HE, Fabry Z (2003) The blood-brain barrier and its role in immune privilege in the central nervous system. J Neuropathol Exp Neurol 62:593–604.

Povlishock JT, Christman CW (1995) The pathobiology of traumatically induced axonal injury in animals and humans: A review of current thoughts. J Neurotrauma 12:555–564.

Povlishock JT, Jenkins LW (1995) Are the pathobiological changes evoked by traumatic brain injury immediate and irreversible? Brain Pathol 5:415–426.

Qian T, Campagnolo D, Kirshblum S (2000) High-dose methylprednisolone may do more harm for spinal cord injury. Med Hypotheses 55:452–453.

Rao K, Lund RD, Kunz HW, Gill TJ, 3rd (1989) The role of MHC and non-MHC antigens in the rejection of intracerebral allogeneic neural grafts. Transplantation 48:1018–1021.

Rapalino O, Lazarov-Spiegler O, Agranov E, Velan GJ, Yoles E, Fraidakis M, Solomon A, Gepstein R, Katz A, Belkin M, Hadani M, Schwartz M (1998) Implantation of stimulated homologous macrophages results in partial recovery of paraplegic rats. Nat Med 4:814–821.

Reier PJ, Anderson DK, Thompson FJ, Stokes BT (1992) Neural tissue transplantation and CNS trauma: Anatomical and functional repair of the injured spinal cord. J Neurotrauma 9(Suppl 1):S223–248.

Richards SJ, Waters JJ, Beyreuther K, Masters CL, Wischik CM, Sparkman DR, White CL, III, Abraham CR, Dunnett SB (1991) Transplants of mouse trisomy 16 hippocampus provide a model of Alzheimer's disease neuropathology. EMBO J 10:297–303.

Robertson GS, Crocker SJ, Nicholson DW, Schulz JB (2000) Neuroprotection by the inhibition of apoptosis. Brain Pathol 10:283–292.

Rockenstein E, Mallory M, Mante M, Alford M, Windisch M, Moessler H, Masliah E (2002) Effects of Cerebrolysin on amyloid-beta deposition in a transgenic model of Alzheimer's disease. J Neural Transm Suppl 62:327–336.

Roof RL, Hall ED (2000) Gender differences in acute CNS trauma and stroke: Neuroprotective effects of estrogen and progesterone. J Neurotrauma 17:367–388.

Schnell L, Schwab ME (1990) Axonal regeneration in the rat spinal cord produced by an antibody against myelin-associated neurite growth inhibitors. Nature 343:269–272.

Schnell L, Fearn S, Klassen H, Schwab ME, Perry VH (1999) Acute inflammatory responses to mechanical lesions in the CNS: Differences between brain and spinal cord. Eur J Neurosci 11:3648–3658.

Schori H, Kipnis J, Yoles E, WoldeMussie E, Ruiz G, Wheeler LA, Schwartz M (2001) Vaccination for protection of retinal ganglion cells against death from glutamate cytotoxicity and ocular hypertension: Implications for glaucoma. Proc Natl Acad Sci USA 98:3398–3403.

Schramm C, Huber S, Protschka M, Czochra P, Burg J, Schmitt E, Lohse AW, Galle PR, Blessing M (2004) TGFbeta regulates the CD4 + CD25 + T-cell pool and the expression of Foxp3 in vivo. Int Immunol 16:1241–1249.

Schumacher HR, Chen LX (2005) Injectable corticosteroids in treatment of arthritis of the knee. Am J Med 118:1208–1214.

Schwab JM, Brechtel K, Nguyen TD, Schluesener HJ (2000) Persistent accumulation of cyclooxygenase-1 (COX-1) expressing microglia/macrophages and upregulation by endothelium following spinal cord injury. J Neuroimmunol 111:122–130.

Schwab JM, Beschorner R, Meyermann R, Gozalan F, Schluesener HJ (2002) Persistent accumulation of cyclooxygenase-1-expressing microglial cells and macrophages and transient upregulation by endothelium in human brain injury. J Neurosurg 96:892–899.

Schwab JM, Conrad S, Elbert T, Trautmann K, Meyermann R, Schluesener HJ (2004) Lesional RhoA + cell numbers are suppressed by anti-inflammatory, cyclooxygenase-inhibiting treatment following subacute spinal cord injury. Glia 47:377–386.

Schwab ME (2002) Increasing plasticity and functional recovery of the lesioned spinal cord. Prog Brain Res 137:351–359.

Schwab ME, Bartholdi D (1996) Degeneration and regeneration of axons in the lesioned spinal cord. Physiol Rev 76:319–370.

Schwartz M (2000) Autoimmune involvement in CNS trauma is beneficial if well controlled. Prog Brain Res 128:259–263.

Schwartz M (2001) Neuroprotection as a treatment for glaucoma: Pharmacological and immunological approaches. Eur J Ophthalmol 11(Suppl 2):S7–S11.

Schwartz M (2002) Autoimmunity as the body's defense mechanism against the enemy within: Development of therapeutic vaccines for neurodegenerative disorders. J Neurovirol 8:480–485.

Schwartz M (2003) Macrophages and microglia in central nervous system injury: Are they helpful or harmful? J Cereb Blood Flow Metab 23:385–394.

Schwartz M, Yoles E (2000) Neuroprotection: A new treatment modality for glaucoma. Curr Opin Ophthalmol 11:107–111.

Schwartz M, Kipnis J (2001) Protective autoimmunity: Regulation and prospects for vaccination after brain and spinal cord injuries. Trends Mol Med 7:252–258.

Schwartz M, Kipnis J (2002) Autoimmunity on alert: Naturally occurring regulatory CD4(+)CD25(+) T cells as part of the evolutionary compromise between a 'need' and a 'risk'. Trends Immunol 23:530–534.

Short D (2001) Is the role of steroids in acute spinal cord injury now resolved? Curr Opin Neurol 14:759–763.

Sigurdsson EM, Wisniewski T, Frangione B (2002) A safer vaccine for Alzheimer's disease? Neurobiol Aging 23:1001–1008.

Skaper SD (2005) Neuronal growth-promoting and inhibitory cues in neuroprotection and neuroregeneration. Ann N Y Acad Sci 1053:376–385.

Snow DM, Lemmon V, Carrino DA, Caplan AI, Silver J (1990) Sulfated proteoglycans in astroglial barriers inhibit neurite outgrowth in vitro. Exp Neurol 109:111–130.

Solomon B (2002) Anti-aggregating antibodies, a new approach towards treatment of conformational diseases. Curr Med Chem 9:1737–1749.

Steinsapir KD, Goldberg RA, Sinha S, Hovda DA (2000) Methylprednisolone exacerbates axonal loss following optic nerve trauma in rats. Restor Neurol Neurosci 17:157–163.

Stolc S, Vlkolinsky R, Pavlasek J (1997) Neuroprotection by the pyridoindole stobadine: A minireview. Brain Res Bull 42: 335–340.

Tariot PN, Federoff HJ (2003) Current treatment for Alzheimer disease and future prospects. Alzheimer Dis Assoc Disord 17(Suppl 4):S105–113.

Vajda FJ (2002) Neuroprotection and neurodegenerative disease. J Clin Neurosci 9:4–8.

Viguier M, Lemaitre F, Verola O, Cho MS, Gorochov G, Dubertret L, Bachelez H, Kourilsky P, Ferradini L (2004) Foxp3 expressing CD4 + CD25(high) regulatory T cells are overrepresented in human metastatic melanoma lymph nodes and inhibit the function of infiltrating T cells. J Immunol 173:1444–1453.

Villoslada P, Genain CP (2004) Role of nerve growth factor and other trophic factors in brain inflammation. Prog Brain Res 146:403–414.

Vinters HV, Ellis WG, Zarow C, Zaias BW, Jagust WJ, Mack WJ, Chui HC (2000) Neuropathologic substrates of ischemic vascular dementia. J Neuropathol Exp Neurol 59:931–945.

Williams DM (2005) What does potency actually mean for inhaled corticosteroids? J Asthma 42:409–417.

Wong EV, David S, Jacob MH, Jay DG (2003) Inactivation of myelin-associated glycoprotein enhances optic nerve regeneration. J Neurosci 23:3112–3117.

Wu D (2005) Neuroprotection in experimental stroke with targeted neurotrophins. NeuroRx 2:120–128.

Yoles E, Schwartz M (1998a) Degeneration of spared axons following partial white matter lesion: Implications for optic nerve neuropathies. Exp Neurol 153:1–7.

Yoles E, Schwartz M (1998b) Potential neuroprotective therapy for glaucomatous optic neuropathy. Surv Ophthalmol 42: 367–372.

Zuo J, Neubauer D, Dyess K, Ferguson TA, Muir D (1998) Degradation of chondroitin sulfate proteoglycan enhances the neurite-promoting potential of spinal cord tissue. Exp Neurol 154:654–662.

43
Alzheimer's Disease

Dave Morgan and Marcia N. Gordon

Keywords Active vaccination; Alzheimer's disease; Amyloid; Amyloid precursor protein; Antibody titer; Beta-pleated sheet; Cerebral amyloid angiopathy; Dementia; Immunotherapy; Monoclonal antibody; Passive vaccination; Senility

43.1. Introduction

Historically, the condition of dementia first described by Alois Alzheimer in 1907 was called generically "presenile dementia." This nomenclature distinguished Alzheimer's as a disease, and not the dementia that occurred typically as humans reached advanced years, known commonly as "senility." It was believed that all of us would develop senility after living long enough, and that this was a normal part of the aging process.

One of the defining characteristics of Alzheimer's was its pathology, consisting of senile or neuritic plaques and intracellular inclusions termed neurofibrillary tangles (see below). In the last quarter of the last century it became increasingly clear that many cases of "senile dementia" also had these same plaques and tangles common to the presenile form described by Alzheimer. This led to terminology referring to "senile dementia of the Alzheimer-type" to describe these late-onset cases. Ultimately, it became recognized that the pathologies Alzheimer reported were associated with dementias at any age, and the term "Alzheimer's disease" was used to describe any severe cognitive deterioration that was associated with plaque and tangle pathology. This nomenclature has held for roughly the last 2 decades.

Over this same period, the numbers of individuals diagnosed with Alzheimer's disease increased dramatically. In part, this was due to increased public awareness that severe cognitive deterioration was not normal human aging (senility) but a disease with specific pathological determinants. A second reason was the increasing number of humans living to the age of risk for developing Alzheimer's disease. Although his original patient was quite young (50s), the vast majority of Alzheimer victims are in their 70s and 80s when they begin to show symptoms. The average age of death of an Alzheimer patient is greater than the age of death in the general population. In the past, these were survivors whose physiology protected them from cardiovascular disease and cancers (the leading causes of death for the last 50 years). Today, the remarkable medical success in retarding the age-adjusted incidence of cardiovascular disease has expanded the population reaching these later years, leading to increased numbers of demented patients (currently reaching 4 million in the US, or slightly more than 1 out of every 100 Americans). Medical care costs for Alzheimer patients, who average 10 years survival with the disease, total $100 billion, or 7% of all medical costs in the US. The major reason this disease is so costly is that most are institutionalized for some period preceding their demise. It is estimated that if the onset of dementia could be delayed by as little as 5 years, this would save $50 billion in medical costs (essentially the same as the new Medicare drug benefit). Clearly, effective therapeutic or preventative approaches to this disorder would provide clear benefits not only to the patients and their families, but the nation as a whole.

The pathology of Alzheimer's disease was initially described using silver stains of histological sections. This resulted from the success of the emerging photographic industry in the early part of the last century, where silver chemistry was worked out with great precision. The plaques and tangles described by Alzheimer were by definition, "argyrophilic," meaning they bound silver molecules, which could then be reduced chemically to reveal stained regions.

The neuritic or senile plaques, also now called compact plaques, are comprised of fibrils of a short peptide called the Aß peptide. This 40–42 amino acid peptide aggregates in a beta-pleated sheet structure to form fibrils, which are themselves aggregated into these dense plaques. These fibrils are referred to as "amyloid " because they stain with a dye called Congo red. There are many types of amyloids in the body, all of which share the properties of beta-pleated sheet structures, fibril formation and binding of Congo red dye. One feature of the Congo red dye is that it will shift the plane of polarized light when bound to beta sheet structures, leading to a characteristic red-green birefringence when viewed through

T. Ikezu and H.E. Gendelman (eds.), *Neuroimmune Pharmacology*.
© Springer 2008

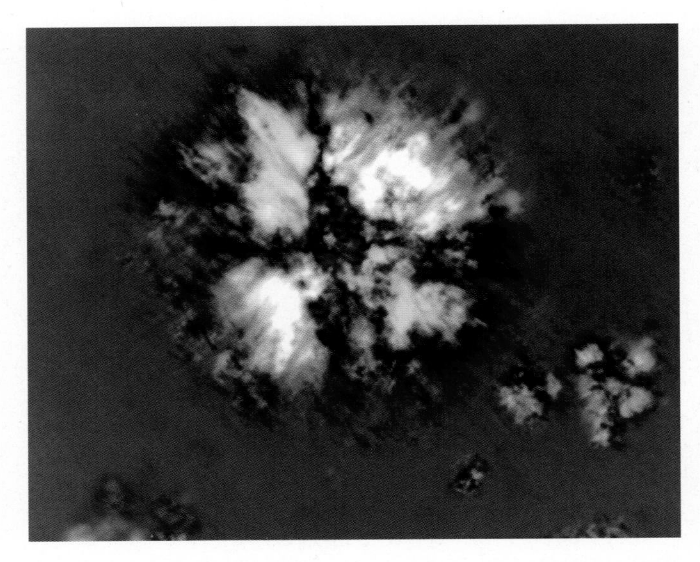

FIGURE 43.1. Compacted parenchymal amyloid plaque in APP transgenic mouse hippocampus. Section was stained with Congo red and viewed with cross-polarized light. Total magnification × 400.

crossed polarizing filters (Figure 43.1). Surrounding these amyloid plaques is a zone of degenerating neuronal processes called neurites. These neurites are swollen, apparently drawn towards the amyloid plaques, and contain proteins or modifications of proteins (such as phosphorylation) not normally found in the larger processes of neurons. In addition to the compacted amyloid plaques, this amyloid material could often be found in association with blood vessels, where it was called cerebral amyloid angiopathy.

In the mid 1980s, George Glenner and his team (Glenner and Wong, 1984) tried to determine the amino acid sequence of the amyloid deposits in Alzheimer's disease. He had already found that other forms of amyloid in the body consisted of small fragments of much larger proteins that aggregated in this beta-pleated sheet secondary structure. When he tried to sequence the amyloid from the compacted plaques, he found variable results because much of the material had been truncated at the N terminal used to initiate the sequencing However, the vascular deposits were more uniform, and he sequenced a 40 amino acid peptide called Aß. This led within a year to the identification of a previously uncharacterized parent molecule containing the Aß peptide sequence called the amyloid precursor protein (APP). This molecule has a single transmembrane spanning domain and the bulk of the 700 amino acid protein is oriented towards the outside of the cell.

Neurofibrillary tangles proved more difficult to identify at the molecular level. They also were fibrillar aggregates of a protein, and they also had a great deal of beta- pleated sheet structure, including staining by Congo red. However, their location within cells precluded them from being referred to as an amyloid. It is now widely accepted that these neurofibrillary tangles are comprised of abnormally and excessively phosphorylated forms of the microtubule associated protein tau. These tau filaments are

TABLE 43.1. The genetics of Alzheimer's disease.

Type of Alzheimer's	Age of onset	Inheritance	Comments
Familial (<2%)	40s–50s	Autosomal dominant	Mutations in amyloid precursor protein, presenilin-1 or presenilin-2 account for most cases
Inherited risk (30–50%)	60s–70s	Not dominant	Apolipoprotein E4 allele increases risk (Earlier age of onset). Pathology more severe.
Sporadic (50–70%)	70s-80s	None identified	Increased risk, Head trauma; decreased risk, Education

largely insoluble by most known solvents, which slowed the identification of their major components.

A detailed account of the genetics of Alzheimer's disease is beyond the scope of this chapter. Others have detailed this quite thoroughly (Hardy, 1997). However, there are at least three genetically identifiable categories of Alzheimer's disease. The first are those rare cases in which Alzheimer's is caused by a dominant genetic mutation (Table 43.1). While accounting for at most 1–2% of all cases, these have been very informative. Three genes are known to carry these mutations. One is the APP itself (Goate et al., 1991). Mutations in APP which modify its cleavage to favor either more Aß or longer forms of Aß (Aß ending at amino acid 42 instead of amino acid 40) can cause an early onset form of Alzheimer's, with symptoms beginning in the 40s to 50s. A second gene that can carry a large number of different mutations and cause Alzheimer's disease is referred to as presenilin-1 (Sherrington et al., 1995). All tested mutations in this gene that cause Alzheimer's disease increase production of the long form of Aß. Presenilin-1 is a component of the complex that cleaves the variable C terminal of Aß. It is also relevant that an extra gene dosage of APP can cause plaque and tangle pathology. A third gene accounting for a small portion of cases is presenilin-2. Most Down's syndrome carriers have an extra copy of the APP gene located on human chromosome 21. By age 40, most will have plaque and tangle pathology upon autopsy, and many will develop a dementia condition in their late 40s and 50s.

A second genetic category includes normally occurring polymorphisms that increase the risk of developing the disease. While many polymorphisms have been mentioned, one that is consistently linked to increased risk is the Apoplipoprotein E4 allele. Individuals carrying this allele have a 3-fold greater risk of developing the disorder (Corder et al., 1993; Poirier et al., 1993). One effect of this allele is to lower the age of onset, so these cases are often found in the 60s. The third category is comprised of those cases for which genetic linkage cannot be identified. It is important to note that the extreme age of onset complicates genetic analyses, as the likelihood of forebears reaching the typical age of onset is quite low. Thus, the size of the idiopathic group is not well defined. As more gene polymorphisms become consistently linked to increased risk, the size of this population may decline.

The recognition that the final common path for all known genetic causes of Alzheimer's disease, including Down's syndrome, is increased amounts of the longer Aß variants led to the development of transgenic mice over-expressing mutated forms of human APP. While many unsuccessful mice were generated in the early 1990s, the first to demonstrate compacted amyloid plaques was the PDAPP mouse (Games et al., 1995). Shortly thereafter, a second APP mouse, the Tg2576, was found to have remarkably similar pathology (Hsiao et al., 1996). In the same month, a presenilin transgenic mouse developed at the University of South Florida was described (Duff et al., 1996). Although the presenilin transgenic mice alone had little pathology, our group found that the presenilin mutation greatly accelerated the accumulation of plaque pathology when crossed with the Tg2576 mouse (Holcomb et al., 1998).

There are now a variety of different APP transgenic mice that develop plaque pathology. The similarities are, in the opinion of these writers, more striking than their differences. All develop compacted plaques in the cerebral cortex and hippocampus. These plaques are associated with dystrophic neurites and the activation of astrocytes and microglial cells. All mice develop a memory dysfunction that is usually correlated with Aß levels in mice of a given age (Morgan, 2003). In none of the mice is there extensive neuron loss, as found in AD. Moreover, none of these mice develop tau pathologies, such as neurofibrillary tangles. Thus, these mice should be characterized as being models of amyloid deposition, not models of Alzheimer's disease.

When Aß is added to tau transgenic mice, the results suggest that amyloid pathology may precipitate tau pathology. Either when APP mice are crossed with tau mice (Lewis et al., 2001) or exogenous Aß is applied to tau transgenic mouse brain (Gotz et al., 2001), there is acceleration of the tau pathology. Recently, a triple transgenic mouse has been generated that harbors APP, presenilin and tau mutations (Oddo et al., 2003), although neuron loss has not been reported to date. These transgenic models of select aspects of Alzheimer pathology have proven useful for preclinical evaluation of experimental therapies, which might be able to slow or halt the progression of Alzheimer's disease, including immunotherapy.

43.2. Early Studies

The first suggestions that anti-Aß antibodies may be useful in ameliorating amyloid pathology came from work by Beka Solomon and colleagues in Israel. They published that anti-Aß antibodies could block the formation of Aß fibrils from monomeric forms in vitro (Solomon et al., 1996). They then proceeded to show that these antibodies could disaggregate preformed amyloid fibrils, and that the antibodies could achieve disaggregation at stoichiometries less than 1:1 (Solomon et al., 1997). This suggested that the antibodies were catalytically dissolving the fibrils, presumably by favoring the formation of non-beta sheet conformations. Inclusion of antibodies could block the in vitro neurotoxicity of Aß. Subsequently, they identified a 4 amino acid sequence in the N terminal domain of Aß they felt was the essential epitope for this mode of action (Frenkel et al., 1998). The catalytic disaggregation hypothesis remains one of the major proposed mechanisms for the action of anti-Aß antibodies (Table 43.2).

The first observation that vaccination against Aß might effectively lower Aß deposition in vivo came from Schenk et al. (1999). These authors demonstrated that immunization of young PDAPP transgenic mice resulted in dramatic reduction in amyloid deposition as the mice aged. Even when started at midlife, the immunization protocol eliminated formation of amyloid deposits. Importantly, they observed the presence of activated microglia in the vicinity of the few remaining deposits. This led them to speculate that opsonization of the amyloid deposits led the microglia to phagocytose this material.

The same group subsequently demonstrated antibody-stimulated phagocytosis could occur in vitro (Bard et al., 2000). They placed microglia on sections of Alzheimer disease brain tissue, with and without anti-Aß antibody present. Only in the cases where the antibody was present did they observe clearance of the amyloid deposits by the microglia. Although cultured primary microglia and cell lines phagocytose fibrillar Aß added to cultures, antibody addition accelerates this process (Webster et al., 2001). Thus, a second mechanism by which anti-Aß antibodies may clear amyloid deposits is by microglial activation and stimulation of phagocytosis (Table 43.2).

A third mechanism by which antibodies might reduce brain Aß levels was suggested by DeMattos et al. (2001). These authors noted that passive administration of a monoclonal antibody generated against Aß resulted in large increases in circulating Aß, analyzed by ELISA (a measurement that may be complicated by circulating antibody under some circumstances). They suggested that the circulating antibodies would sequester Aß in the periphery, thereby increasing the brain to blood concentration gradient leading to a greater net efflux of Aß from the brain to the periphery. It is important to recognize that the deposition of Aß in mouse brain requires considerable over-expression of the APP transgene, always with a mutation that further increases the production of longer forms of Aß. Thus it is conceivable that even minor shifts in the production and clearance of Aß may have profound impacts on whether deposits accumulate or not.

TABLE 43.2. Mechanisms of anti-Aß action.

Catalytic dissolution	Antibody binding converts Aß secondary structure to form incapable of ß-sheet fibril formation	Solomon et al. (1996) Solomon et al. (1997)
Microglial activation	Opsonization of amyloid deposits activates microglia via effector molecules to clear deposits by phagocytosis or other mechanisms	Schenk et al (1999) Bard et al. (2000)
Peripheral sink	Circulating anti-Aß antibodies bind Aß and reduce the free concentration in blood. This leads to increased net efflux from the brain	DeMattos et al. (2001)

For the most part, early studies of Aß vaccination focused upon amyloid deposition. However, some of the APP mice were also being characterized behaviorally, and were found to have deficits in memory function that correlated with the extent of amyloid deposition (reviewed in Morgan, 2003). Thus, shortly after the publication that amyloid vaccination effectively reduced amyloid loads, our group and that of St-George-Hyslop et al began immunizing APP transgenic mice to observe the impact on memory formation. To be honest, our group was concerned that the vaccine might provoke excessive inflammation in the brains of the transgenic mice (by over-activating microglia), and cause premature memory deficits. This was based on a widely-held belief that inflammation associated with amyloid deposits may be a pathogenic mechanism in Alzheimer's disease (Akiyama et al., 2000). However, our testing specifically failed to identify premature memory loss with the immunization, and instead found protection from the development of memory deficits as the mice aged. Similar protection was found by the St-George-Hyslop group in the CRND8 APP transgenic mouse, and we arranged for tandem publication of these results (Morgan et al., 2000; Janus et al., 2000). Subsequent work found that some types of memory deficits could be reversed quite rapidly using passive immunization (Dodart et al., 2002; Kotilinek et al., 2002). This has led to the hypothesis that some pool of Aß other than fibrillar deposits is responsible for some of the memory deficits observed in transgenic mice, and that this pool can be rapidly depleted by some anti-Aß antibodies. Recent studies suggest that it is an oligomeric pool that is responsible for disruption of synaptic plasticity (Walsh et al., 2002; Cleary et al., 2005). Anti-Aß antibodies appear capable of neutralizing the plasticity-disrupting influence of these oligomers (Klyubin et al., 2005).

It is not certain which of the proposed mechanisms of anti-Aß immunotherapy is most correct. It is important to recognize that they are not mutually exclusive, and that all three may be working. In fact, the balance among these three mechanisms may vary for different types of immunotherapy.

43.3. Clinical Trial Experience with Aß Vaccination

Based on the success of the Aß vaccination approach in mouse models of amyloid deposition, Elan and Wyeth teamed to perform phase 1 and phase 2A trials in humans. Phase 1 trials resulted in no overt adverse events in human volunteers. Thus, a phase 2 trial was initiated with 300 patients receiving a vaccine against Aß (AN1792; using QS-21 as an adjuvant) and 60 patients receiving placebo inoculations. The original goal of the trial was to repeatedly vaccinate patients until a predetermined anti-Aß antibody titer was reached. It was noted in the phase 1 trial that only a portion of the patients developed measurable antibody titers against Aß (Schenk, 2002). Even in the mouse studies, repeated inoculation was necessary to achieve high antibody titers (Dickey et al., 2001a). This was further complicated by the advanced age of the patient population, which is known to cause increased variation in the response to vaccines.

Within several months of initiating the trial, the trial was interrupted due to the occurrence of multiple instances of adverse reactions in the patient population. This was characterized as aseptic meningoencephalitis, essentially swelling in the brain that was a form of autoimmune reaction presumably elicited by the vaccine (Orgogozo et al., 2003). Examination of tissue from two patients who ultimately came to autopsy revealed substantial T cell infiltration into the brain (Nicoll et al., 2003; Ferrer et al., 2004). In all, roughly 6% of the patients in the trial developed these symptoms, with the majority recovering after treatment with steroidal anti-inflammatory agents.

Although the inoculations with the Aß vaccines were discontinued, the patients in the trial continued to be monitored both medically and cognitively. One cohort of patients in the trial, those in Zurich, appeared to have benefited from the vaccine. Hock et al. (2003) found that those patients with the highest antibody titers appeared to remain stable cognitively. This stabilization appeared to extend 2 years after the last immunization. Importantly, it was the titers of antibodies that reacted with the amyloid deposits on brain sections that were associated with cognitive benefits (Hock et al., 2002). ELISA assayable titers did not associate with improved behavioral outcomes. When the entire study was analyzed, the linkage between antibody titers and cognitive benefit was less substantial (Gilman et al., 2005). It is unclear whether measurement of brain reactive antibody titers would improve this association as found for the Zurich cohort. Important caveats regarding the implications of this study stem from its truncation. Any results have to be considered within the context that the trial was not completed as planned and the potential benefits or failures must be considered in this context.

The histopathology of the patients in this trial is limited to three published reports. Two from patients that had a meningoencephalitic reaction (Nicoll et al., 2003; Ferrer et al., 2004) and one from a patient lacking any adverse reaction to the vaccine (Masliah et al., 2005). Although control cases were archival and the number of cases is very small, all three reports suggest there was less amyloid deposition than would be expected for an AD patient at each patient's stage of the disease. The reductions did not appear uniform throughout the brain. Moreover, the reductions appeared to be primarily in the neuritic plaque and diffuse Aß deposits, but not in the amyloid deposits associated with the blood vessels.

A curious observation in the trial was the MRI findings of increased hippocampal shrinkage in those patients with the highest antibody titers (Fox et al., 2005). This occurred over the first year of the trial, and was also accompanied by ventricular enlargement. A similar observation was made within the Zurich cohort; however in this case, by the end of the second year, the hippocampal volumes were greater in those patients with the highest brain reactive antibody titers (Nitsch, 2004). Nitsch has speculated that the initial loss of brain volume is secondary to two major changes; elimination of the bulk of the accumulated Aß peptide and reduction of the inflammation and edema associated with the glial reaction to the amyloid deposits. It remains to be seen if the apparent stabilization of brain volume occurs in year 2 for the entire cohort of the clinical trial.

43.4. Other Forms of Active Immunization

Certainly, the AN1792 vaccine developed by Elan was relatively simple. It consisted of full length Aß peptide, incubated in a physiological salt solution to promote fibril formation injected with a QS-21 adjuvant. One of the major issues in developing an effective Aß vaccine is the recognition that Aß is a self-protein to some extent. Aß may be present in normal cells, albeit as a minor and short-lived product of APP processing. Thus vaccines may need to break self-tolerance (Monsonego et al., 2001). A second consideration is that an autoimmune reaction may have undesirable effects, such as that observed in those patients developing meningoencephalitis in the AN1792 trial. Presumably, these adverse reactions were due to development of a T cell response against Aß, a self-antigen. A third consideration is that the relatively short Aß peptide is not by itself a very potent immunogen. Thus, since the original publication by Schenk et al. (1999), there have been a number of alternative vaccine formulations investigated to produce anti-Aß antibodies.

One of the first was an attempt to use mucosal vaccination as an alternative to injections (Weiner et al., 2000; Lemere et al., 2002). This was found to be an effective approach with both production of high titer antibodies and clearance of amyloid deposits in transgenic mice. Another approach was to use a liposome-based therapy with the Aß1–16 peptide rather than full-length molecules (Nicolau et al., 2002). This palmitoylated vaccine construct was effective in breaking down self-tolerance to the vaccine in Aß-overproducing mice. A different rationale was developed by a group at New York University (Sigurdsson et al., 2002). They were concerned that injecting full- length Aß, a known neurotoxin, could produce adverse reactions by forming fibrils. This group has explored use of a truncated and modified Aß peptide, which retains immunogenicity but not the capacity to form fibrils. Importantly, this vaccine can also reverse memory deficits in transgenic mice (Sigurdsson et al., 2004).

Another series of vaccines have used genetic engineering to produce a B-cell epitope towards Aß and a T-cell response towards a non-self antigen. Agadjanyan et al. (2005) used the first 15 amino acids of Aß (where the vast majority of vaccine-generated antibodies bind; (Dickey et al., 2001b; McLaurin et al., 2002) coupled to a synthetic universal T cell epitope (PADRE). This succeeded in producing an immune response with high anti-Aß antibody titers, but with no splenic T cell activation against the Aß peptide. Presumably, a vaccine with these properties would not result in the autoimmune T cell reaction and would avoid the meningoencephalitic reaction found in some patients in the AN1792 trial. Importantly, these same authors found that the adjuvant chosen for the AN1792 trial, QS-21, biased the immune response toward a Th1 type of reaction (Cribbs et al., 2003). Given that Th1 responses are associated with autoimmune reactions, while Th2 responses suppress autoimmune responses, the choice of adjuvant may have contributed to the adverse events that were found with the active vaccine trial.

Another approach has been the development of DNA-based vaccines, where Aß or a component is encoded genetically. Several of these have used viral vectors to deliver the vaccine. Hara et al. (2004) administered adeno-associated virus encoding the Aß peptide orally to mice. This resulted in epithelial cell expression of Aß and prolonged elevation of anti-Aß antibody titers, without eliciting T cell responses against Aß. Kim et al. (2004) used an adenovirus vector encoding Aß plus an adenovirus vector encoding GM-CSF to produce an immune response after intranasal administration. This response was found to have a Th2 bias, and was effective in reducing Aß content in the brains of APP transgenic mice. Lavie et al. (2004) used a filamentous phage vector displaying only 4 amino acids from Aß (EFRH; aa3–6). This vaccine produced a humoral response against Aß, reduced brain amyloid loads and protected APP mice from cognitive deficits. Interestingly, another approach using an HSV amplicon to drive Aß expression with a tetanus toxin fragment produced CNS inflammation, and may serve as a model for the autoimmune reaction apparently found in the AN1792 trial (Bowers et al., 2005).

Thus, there have been a number of clever alternative approaches to developing active immunization protocols for the possible treatment of Alzheimer's disease. Most either enhance the immunogenicity of the vaccination regimen, reduce the possible autoimmunity or both. Nonetheless, there are still potential problems with all of the vaccines. For example, the response will be variable in an aged population, as was clearly found for the AN1792 trial (Bayer et al., 2005). A second consideration is the epitope specificity of the humoral response. Antibodies directed against one epitope may be more effective than others (for example, the apparently greater effectiveness of antibodies reacting with brain amyloid deposits; (Hock et al., 2003). A final consideration is that it is still not certain why the patients developed the meningoencephalitic reaction in the first place. Certainly, the argument that T cell activation against a self-antigen elicited an autoimmune reaction appears logical, but this is not at all proven. There is no good mouse model of the type of reaction found in the two human cases that have come to autopsy (the report by Furlan et al., 2003, has been difficult to replicate). Active immunization reactions can be difficult to control; one cannot unvaccinate someone after the plunger has gone down. Still, if anti-Aß immunotherapy does prove effective in halting or even slowing the progression of Alzheimer's dementia, it is likely that active immunization regimens will become a critical component to therapy for this disease, if only because of their reduced costs compared to passive immunization approaches.

43.5. Passive Immunization

The use of monoclonal antibodies as therapeutics has become commonplace. Antibodies or chimeric proteins against TNF-α are very effective in quelling certain inflammatory diseases. Trastuzumab (against HER2) is effective in reversing the growth of some breast tumors. As of 2005, there were 15 antibody therapeutics approved for use in the US.

Monoclonal antibodies against Aß appear equally effective in reducing brain amyloid deposits as active immunization against Aß. Bard et al. (2000) were the first to demonstrate that systemic administration of anti-Aß antibodies can lower amyloid loads. Importantly, they found that not all antibodies were equally effective in reducing brain Aß content (Bard et al., 2003). In a striking demonstration that amyloid plaques are rapidly reversible structures in transgenic mouse brain, Bacskai et al. (2001) found that amyloid deposits imaged through a craniotomy window with multiphoton microscopy could be removed within 3 days after topical administration of an anti-Aß antibody. Injections of anti-Aß antibodies directly into the brain parenchyma caused a time- dependent clearance of Aß deposits over a period of a week (Wilcock et al., 2003). Similar effects of monoclonal antibodies were found after intraventricular administration (Chauhan and Siegel, 2002; Chauhan and Siegel, 2003). In the triple transgenic mouse model, Oddo et al. (2004) found that antiAß antibody injections into the hippocampus not only cleared amyloid deposits, but also reduced the hyperphosphorylation of tau at some sites. Intriguingly, after 45 days, the amyloid deposits returned as did the tau hyperphosphorylation.

In an attempt to model conditions more proximal to those found in Alzheimer patients, our research group started a passive immunization study in older (19 mo) Tg2576 mice (roughly equivalent to a 65-year-old human with respect to total lifespan). These mice already had considerable plaque accumulation at this age. A time course of the passive immunization therapy (weekly injections) found an initial activation of microglial cells detected by increased expression of Fcγ-receptors and CD45 (Wilcock et al., 2004a). By 2 months, there were reductions in both diffuse and fibrillar forms of Aß, which continued out to three months. By this time the fibrillar amyloid plaques found in the parenchyma were 90% lower in the mice treated with anti-Aß antibodies than in mice given control injections. We also found substantial elevations in circulating Aß levels. By 3 months (but not 1.5 months), these mice also displayed improved cognitive performance. However, in this study and a second study using 5 mo of passive immunotherapy, we observed an increase in vascular amyloid deposits, essentially a murine form of cerebral amyloid angiopathy (Wilcock et al., 2004b). Although this surprised us, an earlier paper found that passive immunization of a different APP mouse, the APP23 mouse, caused an increase in microhemorrhage (Pfeifer et al., 2002). Upon testing in these mice, we confirmed the increase in microhemorrhage as well. However, in spite of this vascular leakage, the learning and memory performance of these old transgenic mice was indistinguishable from nontransgenic mice, and dramatically better than control transgenic animals. A similar observation of increased hemorrhage in mice passively immunized with one antibody, but not another, was reported in parallel by Racke et al. (2005).

It is not entirely clear why some antibodies result in vascular leakage and others do not. It may be that one of the proposed mechanisms of antibody action (microglial activation, for example) predisposes towards this result. Antibodies which do not result in microglial activation may not redistribute amyloid from the parenchyma to the vessels, or result in microhemorrhage. It is also important to recognize that microhemorrhage does not occur in younger mice (Pfeifer et al., 2002). Either the young mice have insufficient amyloid to liberate, or the younger vessels are less sensitive to the weakening caused by cerebral amyloid angiopathy. Nonetheless, it is of some concern that none of the three autopsy studies find reduction in cerebral amyloid angiopathy (in fact it is rated as being high) and one (Ferrer et al., 2004) notes that there is hemorrhage in association with some vascular plaques in this patient.

Within the passive immunization field, there have been discussions of antibody superiority based upon the epitope domain of the antibody (usually divided into N terminal, mid domain or C terminal) or the antibody subtype (IgG1, IgG2a, IgG2b etc), with the conclusion that one type of epitope specificity or subtype is better than another. A third consideration is antibody affinity for Aß. Unfortunately, none of these features has been manipulated independent of the others. Each individual antibody will vary in all of these properties, and, particularly for epitope specificity, it is very difficult to assume equivalence. Thus far, in the opinion of this author, each antibody needs to be considered a separate entity. Each antibody's ability to clear amyloid in vivo, inactivate oligomers, activate microglia, bind Aß, increase circulating Aß and activate effector molecules (such as Fcγ-receptors) must be measured directly. It is plausible that some antibodies may have superior profiles in clinical trials because of one or more of these properties. However, assuming that it is one of these properties, which confer superiority without measuring them, all will likely be misleading.

43.6. Circulating Antibodies

Several groups have found that some individuals have titers of antibodies against the Aß peptide. Unfortunately, consensus regarding a relationship to Alzheimer's disease has not yet emerged as there are reports of increased titers with disease (Nath et al., 2003), decreased titers with disease (Weksler et al., 2002) or no effect of disease (Hyman et al., 2001). It should be noted that the antibody titers observed are extremely low compared to those observed after immunization. Moreover, it is not clear the degree to which endogenous Aß may interfere with the detection of anti-Aß antibodies by ELISA (see Li et al., 2004).

Still, this has led to a short, open label trial of human intravenous immunoglobulin, an FDA approved product that has benefit in several types of disorders, including multiple sclerosis. Dodel et al. (2004) reported that monthly administration of intravenous immunoglobulin to 5 Alzheimer patients reduced Aß in cerebrospinal fluid, increased Aß in serum and improved cognition. Although a definitive conclusion awaits a clinical trial, it is surprising that such a treatment would be effective, even at the large doses administered, because of the vanishingly small amounts of anti-Aß antibodies present. Instead, if confirmed in a larger controlled trial, this observation may suggest a fourth mechanism by which immunotherapy might benefit Alzheimer patients: by modulation of the immune response. One feature of immunoglobulin therapy is that the circulating levels of IgG increase dramatically. This results in feedback regulation of a number of immune processes, as the levels gradually return towards normal. This may also explain the rather curious finding that administration of a proteosome adjuvant with glatiramer acetate alone is adequate to reduce amyloid deposits in transgenic mice (Frenkel et al, 2005). This does not require the presence of antibodies, and appears related to the activation of resident microglial cells near the amyloid deposits. Thus, it is conceivable that immunomodulation may be just as important as a specific immune reaction in the benefits of immunotherapy in APP transgenic mice. Clearly it will be critical to identify if the same holds in human trials.

Summary

Alzheimer's disease is a progressive neurodegenerative disorder in which memory and cognitive dysfunctions are the earliest symptoms. Pathologically, the disorder is characterized by neuron loss in hippocampus and multiple cortical regions, along with the formation of extracellular amyloid plaques and intracellular neurofibrillary tangles. The amyloid in this disorder is comprised largely of a peptide called Aß, which is a degradation product of a protein with unknown function referred to as the amyloid precursor protein. The genetics of dominant inherited forms of Alzheimer's disease implicate increased production of the long form of the Aß peptide as the critical feature leading to unavoidable development of the disease. This recognition has led many to consider anti-amyloid therapeutics as a means of slowing or arresting the disease. One approach that may become the first to test the so-called "amyloid hypothesis" is immunotherapy. Both vaccination and monoclonal antibody therapies have been tested with considerable success in mouse models of amyloid deposition. These approaches can prevent the formation of amyloid deposits, remove already existing deposits, and reverse the memory deficits associated with amyloid deposition in these mice. A human clinical trial was cut short due to an apparent autoimmune reaction in a fraction of the patients, but the results from this truncated trial still suggest that there may be therapeutic benefits of the approach. Current research is focusing on trying to understand the mechanisms by which antibodies directed against Aß aid in clearing the amyloid deposits, how the vaccines might be constructed to overcome self-tolerance without evoking autoimmune reactions, and testing various monoclonal antibody preparations for efficacy in clinical trials. Even partial success in slowing the progression of this disease can have considerable societal and economic impact. Furthermore, verification of the amyloid hypothesis will encourage development of a number of other anti-amyloid therapies that may synergize with the immunotherapeutic approach.

Acknowledgments. Drs. Morgan and Gordon receive support from the following Awards: R01 AG15490, R01 AG 18478, R01 NS48335, R01 AG 25509, P50 AG25701

Review Questions/Problems

1. **What are the two main neuropathological hallmarks of Alzheimer's disease?**
   Compare and contrast their anatomical location and protein composition.
2. **Describe the major types of genetic mutations that are associated with Alzheimer's disease.**
3. **Why is the incidence of Alzheimer's disease increasing?**
4. **What is the "Amyloid Hypothesis" of Alzheimer's disease?**
5. **Review three proposed mechanisms of action for how anti-β–amyloid antibodies might remove amyloid deposits.**
6. **What is the difference between active and passive vaccination?**
7. **List three barriers to the use of active vaccination clinically**
8. **List two advantages of passive immunotherapy over active immunotherapy?**
9. **List two disadvantages of passive immunotherapy over active immunotherapy?**
10. **Which of the following is a risk factor of Alzheimer's disease?**
    a. Head trauma
    b. Cancer
    c. Education
    d. Schizophrenia
    e. All of the above
11. **Which of the following is NOT a proposed mechanism to explain the anti-Aβ action of active immunization?**
    a. Catalytic dissolution of amyloid plaques
    b. Microglial activation and amyloid degradation
    c. Clearance of amyloid through a peripheral sink
    d. Suppression of APP processing
    e. All of the above

12. Amyloid associated with cerebral blood vessels is called:

a. Argyrophilic Alzheimer vasculature
b. Congophilic amyloid angiopathy
c. Senile plaques
d. Neurofibrillary tangles
e. All of the above

References

Agadjanyan MG, Ghochikyan A, Petrushina I, Vasilevko V, Movsesyan N, Mkrtichyan M, Saing T, Cribbs DH (2005) Prototype Alzheimer's disease vaccine using the immunodominant B cell epitope from beta-amyloid and promiscuous T cell epitope pan HLA DR-binding peptide. J Immunol 174:1580–1586.

Akiyama H, Barger S, Barnum S, Bradt B, Bauer J, Cole GM, Cooper NR, Eikelenboom P, Emmerling M, Fiebich BL, Finch CE, Frautschy S, Griffin WS, Hampel H, Hull M, Landreth G, Lue L, Mrak R, Mackenzie IR, McGeer PL, O'Banion MK, Pachter J, Pasinetti G, Plata-Salaman C, Rogers J, Rydel R, Shen Y, Streit W, Strohmeyer R, Tooyoma I, Van Muiswinkel FL, Veerhuis R, Walker D, Webster S, Wegrzyniak B, Wenk G, Wyss-Coray T (2000) Inflammation and Alzheimer's disease. Neurobiol Aging 21:383–421.

Bacskai BJ, Kajdasz ST, Christie RH, Carter C, Games D, Seubert P, Schenk D, Hyman B (2001) Imaging of amyloid-ß deposits in living mice permits direct observation of clearance of plaques with immunotherapy. Nat Med 7:369–372.

Bard F, Cannon C, Barbour R, Burke RL, Games D, Grajeda H, Guido T, Hu K, Huang J, Johnson-Wood K, Khan K, Kholodenko D, Lee M, Lieberburg I, Motter R, Nguyen M, Soriano F, Vasquez N, Weiss K, Welch B, Seubert P, Schenk D, Yednock T (2000) Peripherally administered antibodies against amyloid beta-peptide enter the central nervous system and reduce pathology in a mouse model of Alzheimer disease. Nat Med 6:916–919.

Bard F, Barbour R, Cannon C, Carretto R, Fox M, Games D, Guido T, Hoenow K, Hu K, Johnson-Wood K, Khan K, Kholodenko D, Lee C, Lee M, Motter R, Nguyen M, Reed A, Schenk D, Tang P, Vasquez N, Seubert P, Yednock T (2003) Epitope and isotype specificities of antibodies to beta -amyloid peptide for protection against Alzheimer's disease-like neuropathology. Proc Natl Acad Sci USA 100:2023–2028.

Bayer AJ, Bullock R, Jones RW, Wilkinson D, Paterson KR, Jenkins L, Millais SB, Donoghue S (2005) Evaluation of the safety and immunogenicity of synthetic Abeta42 (AN1792) in patients with AD. Neurology 64:94–101.

Bowers WJ, Mastrangelo MA, Stanley HA, Casey AE, Milo LJ, Jr., Federoff HJ (2005) HSV amplicon-mediated Abeta vaccination in Tg2576 mice: Differential antigen-specific immune responses. Neurobiol Aging 26:393–407.

Chauhan NB, Siegel GJ (2002) Reversal of amyloid beta toxicity in Alzheimer's disease model Tg2576 by intraventricular antiamyloid beta antibody. J Neurosci Res 69:10–23.

Chauhan NB, Siegel GJ (2003) Intracerebroventricular passive immunization with anti-Abeta antibody in Tg2576. J Neurosci Res 74:142–147.

Cleary JP, Walsh DM, Hofmeister JJ, Shankar GM, Kuskowski MA, Selkoe DJ, Ashe KH (2005) Natural oligomers of the amyloid-beta protein specifically disrupt cognitive function. Nat Neurosci 8:79–84.

Corder EH, Saunders AM, Strittmatter WJ, Schmechel DE, Gaskell PC, Small GW, Roses AD, Haines JL, Pericak-Vance MA (1993) Gene dose of apolipoprotein E type 4 allele and the risk of ALzheimer's disease in late onset families. Science 261:921–923.

Cribbs DH, Ghochikyan A, Vasilevko V, Tran M, Petrushina I, Sadzikava N, Babikyan D, Kesslak P, Kieber-Emmons T, Cotman CW, Agadjanyan MG (2003) Adjuvant-dependent modulation of Th1 and Th2 responses to immunization with beta-amyloid. Int Immunol 15:505–514.

DeMattos RB, Bales KR, Cummins DJ, Dodart JC, Paul SM, Holtzman DM (2001) Peripheral anti-A beta antibody alters CNS and plasma A beta clearance and decreases brain A beta burden in a mouse model of Alzheimer's disease. Proc Natl Acad Sci USA 98:8850–8855.

Dickey CA, Morgan DG, Kudchodkar S, Weiner DB, Bai Y, Cao C, Gordon MN, Ugen KE (2001a) Duration and specificity of humoral immune responses in mice vaccinated with the Alzheimer's disease-associated beta-amyloid 1–42 peptide. DNA Cell Biol 20:723–729.

Dickey CA, Morgan DG, Kudchodkar S, Weiner DB, Bai Y, Cao C, Gordon MN, Ugen KE (2001b) Duration and specificity of humoral immune responses in mice vaccinated with the Alzheimer's disease-associated beta-amyloid 1–42 peptide. DNA Cell Biol 20:723–729.

Dodart JC, Bales KR, Gannon KS, Greene SJ, DeMattos RB, Mathis C, DeLong CA, Wu S, Wu X, Holtzman DM, Paul SM (2002) Immunization reverses memory deficits without reducing brain Abeta burden in Alzheimer's disease model. Nat Neurosci 5:452–457.

Dodel RC, Du Y, Depboylu C, Hampel H, Frolich L, Haag A, Hemmeter U, Paulsen S, Teipel SJ, Brettschneider S, Spottke A, Nolker C, Moller HJ, Wei X, Farlow M, Sommer N, Oertel WH (2004) Intravenous immunoglobulins containing antibodies against beta-amyloid for the treatment of Alzheimer's disease. J Neurol Neurosurg Psychiatry 75:1472–1474.

Duff K, Eckman C, Zehr C, Yu X, Prada CM, Perez-tur J, Hutton M, Buee L, Harigaya Y, Yager D, Morgan D, Gordon MN, Holcomb L, Refolo L, Zenk B, Hardy J, Younkin S (1996) Increased amyloid -beta42(43) in brains of mice expressing mutant presenilin 1. Nature 383:710–713.

Ferrer I, Boada RM, Sanchez Guerra ML, Rey MJ, Costa-Jussa F (2004) Neuropathology and pathogenesis of encephalitis following amyloid-beta immunization in Alzheimer's disease. Brain Pathol 14:11–20.

Fox NC, Black RS, Gilman S, Rossor MN, Griffith SG, Jenkins L, Koller M (2005) Effects of Abeta immunization (AN1792) on MRI measures of cerebral volume in Alzheimer disease. Neurology 64:1563–1572.

Frenkel D, Balass M, Solomon B (1998) N-terminal EFRH sequence of Alzheimer's beta-amyloid peptide represents the epitope of its anti-aggregating antibodies. J Neuroimmunol 88:85–90.

Frenkel D, Maron R, Burt DS, Weiner HL (2005) Nasal vaccination with a proteosome-based adjuvant and glatiramer acetate clears beta-amyloid in a mouse model of Alzheimer disease. J Clin Invest 115:2423–2433.

Furlan R, Brambilla E, Sanvito F, Roccatagliata L, Olivieri S, Bergami A, Pluchino S, Uccelli A, Comi G, Martino G (2003) Vaccination with amyloid-beta peptide induces autoimmune encephalomyelitis in C57/BL6 mice. Brain 126:285–291.

Games D, Adams D, Alessandrini R, Barbour R, Berthelette P, Blackwell C, Carr T, Clemens J, Donaldson T, Gillespie F (1995) Alzheimer-type neuropathology in transgenic mice overexpressing V717F beta-amyloid precursor protein. Nature 373:523–527.

Gilman S, Koller M, Black RS, Jenkins L, Griffith SG, Fox NC, Eisner L, Kirby L, Rovira MB, Forette F, Orgogozo JM (2005) Clinical effects of Abeta immunization (AN1792) in patients with AD in an interrupted trial. Neurology 64:1553–1562.

Glenner GG, Wong CW (1984) Alzheimer's disease: Initial report of the purification and characterization of a novel cerebrovascular amyloid protein. Biochem Biophys Res Commun 120:885–890.

Goate A, Chartier-Harlin MC, Mullan M, Brown J, Crawford F, Fidani L, Giuffra L, Haynes A, Irving N, James L (1991) Segregation of a missense mutation in the amyloid precursor protein gene with familial Alzheimer's disease. Nature 349:704–706.

Gotz J, Chen F, Van Dorpe J, Nitsch RM (2001) Formation of neurofibrillary tangles in P301l tau transgenic mice induced by Abeta 42 fibrils. Science 293:1491–1495.

Hara H, Monsonego A, Yuasa K, Adachi K, Xiao X, Takeda S, Takahashi K, Weiner HL, Tabira T (2004) Development of a safe oral Abeta vaccine using recombinant adeno-associated virus vector for Alzheimer's disease. J Alzheimers Dis 6:483–488.

Hardy J (1997) Amyloid, the presenilins and Alzheimer's disease. Trends Neurosci 20:154–160.

Hock C, Konietzko U, Papassotiropoulos A, Wollmer A, Streffer J, Von Rotz RC, Davey G, Moritz E, Nitsch RM (2002) Generation of antibodies specific for beta-amyloid by vaccination of patients with Alzheimer disease. Nat Med 8:1270–1275.

Hock C, Konietzko U, Streffer JR, Tracy J, Signorell A, Muller-Tillmanns B, Lemke U, Henke K, Moritz E, Garcia E, Wollmer MA, Umbricht D, de Quervain DJ, Hofmann M, Maddalena A, Papassotiropoulos A, Nitsch RM (2003) Antibodies against beta-amyloid slow cognitive decline in Alzheimer's disease. Neuron 38:547–554.

Holcomb L, Gordon MN, McGowan E, Yu X, Benkovic S, Jantzen P, Wright K, Saad I, Mueller R, Morgan D, Sanders S, Zehr C, O'Campo K, Hardy J, Prada CM, Eckman C, Younkin S, Hsiao K, Duff K (1998) Accelerated Alzheimer-type phenotype in transgenic mice carrying both mutant amyloid precursor protein and presenilin 1 transgenes. Nat Med 4:97–100.

Hsiao K, Chapman P, Nilsen S, Eckman C, Harigaya Y, Younkin S, Yang F, Cole G (1996) Correlative memory deficits, Abeta elevation, and amyloid plaques in transgenic mice. Science 274:99–102.

Hyman BT, Smith C, Buldyrev I, Whelan C, Brown H, Tang MX, Mayeux R (2001) Autoantibodies to amyloid-beta and Alzheimer's disease. Ann Neurol 49:808–810.

Janus C, Pearson J, McLaurin J, Mathews PM, Jiang Y, Schmidt SD, Chishti MA, Horne P, Heslin D, French J, Mount HT, Nixon RA, Mercken M, Bergeron C, Fraser PE, George-Hyslop P, Westaway D (2000) A beta peptide immunization reduces behavioural impairment and plaques in a model of Alzheimer's disease. Nature 408:979–982.

Kim HD, Kong FK, Cao Y, Lewis TL, Kim H, Tang DC, Fukuchi K (2004) Immunization of Alzheimer model mice with adenovirus vectors encoding amyloid beta-protein and GM-CSF reduces amyloid load in the brain. Neurosci Lett 370:218–223.

Klyubin I, Walsh DM, Lemere CA, Cullen WK, Shankar GM, Betts V, Spooner ET, Jiang L, Anwyl R, Selkoe DJ, Rowan MJ (2005) Amyloid beta protein immunotherapy neutralizes Abeta oligomers that disrupt synaptic plasticity in vivo. Nat Med 11:556–561.

Kotilinek LA, Bacskai B, Westerman M, Kawarabayashi T, Younkin L, Hyman BT, Younkin S, Ashe KH (2002) Reversible memory loss in a mouse transgenic model of Alzheimer's disease. J Neurosci 22:6331–6335.

Lavie V, Becker M, Cohen-Kupiec R, Yacoby I, Koppel R, Wedenig M, Hutter-Paier B, Solomon B (2004) EFRH-phage immunization of Alzheimer's disease animal model improves behavioral performance in Morris water maze trials. J Mol Neurosci 24:105–113.

Lemere CA, Spooner ET, Leverone JF, Mori C, Clements JD (2002) Intranasal immunotherapy for the treatment of Alzheimer's disease: Escherichia coli LT and LT(R192G) as mucosal adjuvants. Neurobiol Aging 23:991–1000.

Lewis J, Dickson DW, Lin WL, Chisholm L, Corral A, Jones G, Yen SH, Sahara N, Skipper L, Yager D, Eckman C, Hardy J, Hutton M, McGowan E (2001) Enhanced neurofibrillary degeneration in transgenic mice expressing mutant tau and APP. Science 293:1487–1491.

Li Q, Cao C, Chackerian B, Schiller J, Gordon M, Ugen KE, Morgan D (2004) Overcoming antigen masking of anti-Abeta antibodies reveals breaking of B cell tolerance by virus-like particles in Abeta immunized amyloid precursor protein transgenic mice. BMC Neurosci 5:21.

Masliah E, Hansen L, Adame A, Crews L, Bard F, Lee C, Seubert P, Games D, Kirby L, Schenk D (2005) Abeta vaccination effects on plaque pathology in the absence of encephalitis in Alzheimer disease. Neurology 64:129–131.

McLaurin J, Cecal R, Kierstead ME, Tian X, Phinney AL, Manea M, French JE, Lambermon MH, Darabie AA, Brown ME, Janus C, Chishti MA, Horne P, Westaway D, Fraser PE, Mount HT, Przybylski M, George-Hyslop P (2002) Therapeutically effective antibodies against amyloid-beta peptide target amyloid-beta residues 4–10 and inhibit cytotoxicity and fibrillogenesis. Nat Med 8:1263–1269.

Monsonego A, Maron R, Zota V, Selkoe DJ, Weiner HL (2001) Immune hyporesponsiveness to amyloid beta-peptide in amyloid precursor protein transgenic mice: Implications for the pathogenesis and treatment of Alzheimer's disease. Proc Natl Acad Sci USA 98:10273–10278.

Morgan D (2003) Learning and memory deficits in APP transgenic mouse models of amyloid deposition. Neurochem Res 28:1029–1034.

Morgan D, Diamond DM, Gottschall PE, Ugen KE, Dickey C, Hardy J, Duff K, Jantzen P, DiCarlo G, Wilcock D, Connor K, Hatcher J, Hope C, Gordon M, Arendash GW (2000) A beta peptide vaccination prevents memory loss in an animal model of Alzheimer's disease. Nature 408:982–985.

Nath A, Hall E, Tuzova M, Dobbs M, Jons M, Anderson C, Woodward J, Guo Z, Fu W, Kryscio R, Wekstein D, Smith C, Markesbery WR, Mattson MP (2003) Autoantibodies to amyloid beta-peptide (Abeta) are increased in Alzheimer's disease patients and Abeta antibodies can enhance Abeta neurotoxicity: Implications for disease pathogenesis and vaccine development. Neuromolecular Med 3:29–39.

Nicolau C, Greferath R, Balaban TS, Lazarte JE, Hopkins RJ (2002) A liposome-based therapeutic vaccine against beta -amyloid plaques on the pancreas of transgenic NORBA mice. Proc Natl Acad Sci USA 99:2332–2337.

Nicoll JA, Wilkinson D, Holmes C, Steart P, Markham H, Weller RO (2003) Neuropathology of human Alzheimer disease after immunization with amyloid-beta peptide: A case report. Nat Med 9:448–452.

Nitsch RM (2004) Immunotherapy of Alzheimer disease. Alzheimer Dis Assoc Disord 18:185–189.

Oddo S, Caccamo A, Shepherd JD, Murphy MP, Golde TE, Kayed R, Metherate R, Mattson MP, Akbari Y, LaFerla FM (2003) Triple-transgenic model of Alzheimer's disease with plaques and tangles: Intracellular Abeta and synaptic dysfunction. Neuron 39:409–421.

Oddo S, Billings L, Kesslak JP, Cribbs DH, LaFerla FM (2004) Abeta immunotherapy leads to clearance of early, but not late, hyperphosphorylated tau aggregates via the proteasome. Neuron 43:321–332.

Orgogozo JM, Gilman S, Dartigues JF, Laurent B, Puel M, Kirby LC, Jouanny P, Dubois B, Eisner L, Flitman S, Michel BF, Boada M, Frank A, Hock C (2003) Subacute meningoencephalitis in a subset of patients with AD after Abeta42 immunization. Neurology 61:46–54.

Pfeifer M, Boncristiano S, Bondolfi L, Stalder A, Deller T, Staufenbiel M, Mathews PM, Jucker M (2002) Cerebral hemorrhage after passive anti-Abeta immunotherapy. Science 298:1379.

Poirier J, Davignon J, Bouthillier D, Kogan S, Bertrand P, Gauthier S (1993) Apolipoprotein E polymorphism and Alzheimer's disease. Lancet 342:697–699.

Racke MM, Boone LI, Hepburn DL, Parsadainian M, Bryan MT, Ness DK, Piroozi KS, Jordan WH, Brown DD, Hoffman WP, Holtzman DM, Bales KR, Gitter BD, May PC, Paul SM, DeMattos RB (2005) Exacerbation of cerebral amyloid angiopathy-associated microhemorrhage in amyloid precursor protein transgenic mice by immunotherapy is dependent on antibody recognition of deposited forms of amyloid beta. J Neurosci 25:629–636.

Schenk D (2002) Amyloid-beta immunotherapy for Alzheimer's disease: The end of the beginning. Nat Rev Neurosci 3:824–828.

Schenk D, Barbour R, Dunn W, Gordon G, Grajeda H, Guido T, Hu K, Huang J, Johnson-Wood K, Khan K, Kholodenko D, Lee M, Liao Z, Lieberburg I, Motter R, Mutter L, Soriano F, Shopp G, Vasquez N, Vandevert C, Walker S, Wogulis M, Yednock T, Games D, Seubert P (1999) Immunization with amyloid-beta attenuates Alzheimer-disease-like pathology in the PDAPP mouse. Nature 400:173–177.

Sherrington R, Rogaev EI, Liang Y, Rogaeva EA, Levesque G, Ikeda M, Chi H, Lin C, Li G, Holman K, Tsuda T, Mar L, Foncin JF, Bruni AC, Montesi MP, Sorbi S, Rainero I, Pinessi L, Nee L, Chumakov I, Pollen D, Brookes A, Sanseau P, Polinsky RJ, Wasco W, Da Silva HAR, Haines JL, Pericak-Vance MA, Tanzi RE, Roses AD, Fraser PE, Rommens JM, George-Hyslop PH (1995) Cloning of a gene bearing missense mutations in early-onset familial Alzheimer's disease. Nature 375:754–760.

Sigurdsson EM, Wisniewski T, Frangione B (2002) A safer vaccine for Alzheimer's disease? Neurobiol Aging 23:1001–1008.

Sigurdsson EM, Knudsen E, Asuni A, Fitzer-Attas C, Sage D, Quartermain D, Goni F, Frangione B, Wisniewski T (2004) An attenuated immune response is sufficient to enhance cognition in an Alzheimer's disease mouse model immunized with amyloid-beta derivatives. J Neurosci 24:6277–6282.

Solomon B, Koppel R, Hanan E, Katzav T (1996) Monoclonal antibodies inhibit in vitro fibrillar aggregation of the Alzheimer beta-amyloid peptide. Proc Natl Acad Sci USA 93:452–455.

Solomon B, Koppel R, Frankel D, Hanan-Aharon E (1997) Disaggregation of Alzheimer beta-amyloid by site-directed mAb. Proc Natl Acad Sci USA 94:4109–4112.

Walsh DM, Klyubin I, Fadeeva JV, Cullen WK, Anwyl R, Wolfe MS, Rowan MJ, Selkoe DJ (2002) Naturally secreted oligomers of amyloid beta protein potently inhibit hippocampal long-term potentiation in vivo. Nature 416:535–539.

Webster SD, Galvan MD, Ferran E, Garzon-Rodriguez W, Glabe CG, Tenner AJ (2001) Antibody-Mediated Phagocytosis of the Amyloid beta-Peptide in Microglia Is Differentially Modulated by C1q. J Immunol 166:7496–7503.

Weiner HL, Lemere CA, Maron R, Spooner ET, Grenfell TJ, Mori C, Issazadeh S, Hancock WW, Selkoe DJ (2000) Nasal administration of amyloid-beta peptide decreases cerebral amyloid burden in a mouse model of Alzheimer's disease. Ann Neurol 48:567–579.

Weksler ME, Relkin N, Turkenich R, LaRusse S, Zhou L, Szabo P (2002) Patients with Alzheimer disease have lower levels of serum anti-amyloid peptide antibodies than healthy elderly individuals. Exp Gerontol 37:943–948.

Wilcock DM, DiCarlo G, Henderson D, Jackson J, Clarke K, Ugen KE, Gordon MN, Morgan D (2003) Intracranially administered anti-Abeta antibodies reduce beta-amyloid deposition by mechanisms both independent of and associated with microglial activation. J Neurosci 23:3745–3751.

Wilcock DM, Rojiani A, Rosenthal A, Levkowitz G, Subbarao S, Alamed J, Wilson D, Wilson N, Freeman MJ, Gordon MN, Morgan D (2004a) Passive amyloid immunotherapy clears amyloid and transiently activates microglia in a transgenic mouse model of amyloid deposition. J Neurosci 24:6144–6151.

Wilcock DM, Rojiani A, Rosenthal A, Subbarao S, Freeman MJ, Gordon MN, Morgan D (2004b) Passive immunotherapy against Abeta in aged APP-transgenic mice reverses cognitive deficits and depletes parenchymal amyloid deposits in spite of increased vascular amyloid and microhemorrhage. J Neuroinflammation 1:24.

44
Parkinson's Disease and Amyotrophic Lateral Sclerosis

R. Lee Mosley, Ashley Reynolds, David K. Stone, and Howard E. Gendelman

Keywords Adaptive immunity; Antibodies; Dopaminergic neurons; Glatiramer acetate; Innate immunity; Motor neurons; Oxidative stress; Superoxide dismutase; α-Synuclein; T cell; T lymphocyte

44.1. Introduction

Parkinson's Disease (PD) and Amyotrophic Lateral Sclerosis (ALS) are common neurodegenerative disorders in the aging population. The primary pathological characteristics of PD are the progressive loss of dopaminergic neurons in the substantia nigra pars compacta (SNpc) and reductions in their termini within the dorsal striatum (see Przedborski, Chapter 26). These lead to profound and irreversible striatal dopamine loss. Cell modeling data indicate that 100–200 SNpc neurons degenerate per day during PD (Orr et al., 2002). ALS, also known as Lou Gehrig's disease, is characterized by gradual degeneration of spinal cord motor neurons eventually leading to progressive weakness, paralysis of muscle, and death (see Simpson et al., Chapter 27). The mechanisms for neurodegeneration in both disorders relate, in part, to the abnormal accumulation, oligomer formation and misfolding of α-synuclein for PD and superoxide dismutase 1 (SOD1) for ALS. These protein aggregates form at synapses and axons leading to signaling abnormalities and neuronal dysfunction. Neuroinflammatory responses, mitochondrial function, glutamate transport toxicity, and free radicals formulation are linked as a consequence of protein toxicities and lead to neuronal destruction in both disorders. It is hypothesized, all together, that changes in the balance between factors promoting aggregation, clearance and synthesis of α-synuclein and SOD1 is a central component of disease pathogenesis. Microglial activation and alterations in lysosomal function are linked to oligomer accumulation. These can accumulate in the membrane and can be recognized by antibodies that promote their clearance (Masliah et al., 2005). Antibodies can lead to decreased accumulation of aggregated α-synuclein in neuronal cell bodies and synapses associated with reduced neurodegeneration. Antibodies can also recognize abnormal α-synuclein associated with the neuronal membrane and promote their degradation through lysosomal-autophagy pathways. Thus, vaccination is effective in reducing the neuronal accumulation of α-synuclein aggregates and further development of this approach might have a potential role in the treatment of PD (Miller and Messer, 2005). Such an anti-amyloidogenic property might also provide a novel strategy for the treatment of other neurodegenerative disorders. Antibodies specific for peptides or conformations of misfolding neurodegenerative disease proteins can be engineered for affinity and stability; and then delivered intracellularly as intrabodies. For SOD1-linked familial ALS, aberrant oligomerization of SOD1 mutant proteins has been implicated. Formation of soluble oligomers suggests a general, unifying picture of SOD1 aggregation linked to neuronal destruction. Thus, vaccination with α-synuclein or SOD1 has been proposed for treatment of disease. In the case of ALS, vaccination with recombinant SOD1 or passive immunization with antibodies against SOD1 species has been tested with success in mouse models of ALS. These parallel similar successes in rodent models of PD. This therapeutic approach is based on reduction of toxic mutant proteins. However, caution is noted in developing such approaches. Work from our own laboratory has shown the importance of the cell-mediated adaptive immune system in neuroprotective treatment strategies for PD and ALS. In particular, T cells have beneficial and harmful affects on neurodegenerative processes. Immunoregulatory treatments [e.g., glatiramer acetate (GA), vasoactive intestinal peptide (VIP), granulocyte macrophage-colony stimulating factor (GM-CSF), and 1,25-dihydroxyvitamin D3] that modulate T cell responses can play an important role in future immunotherapies. In this chapter we review the roles of the innate and adaptive immune system in the pathogenesis of PD and ALS and the recently developed means to harness both for therapeutic benefit (Figure 44.1).

New modalities in genomics, proteomics, and imaging have allowed monitoring of these effects and provide new directives for the future. All together, we provide a balanced review for the role of immunity in the pathogenesis of PD and ALS and the means to modulate it for therapeutic benefit. The strengths and concerns for each of these approaches are discussed.

T. Ikezu and H.E. Gendelman (eds.), *Neuroimmune Pharmacology*.
© Springer 2008

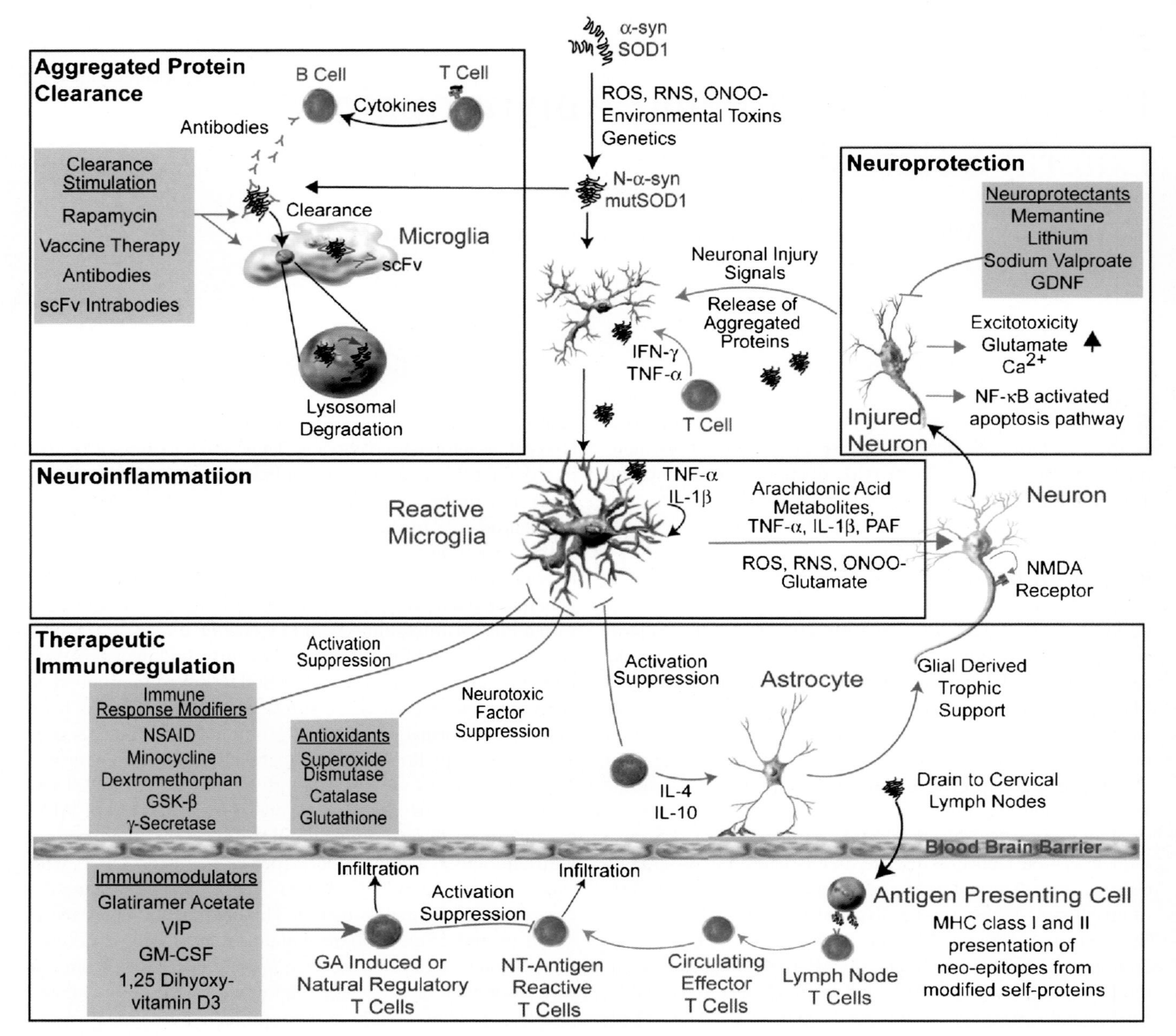

FIGURE 44.1. Immunotherapeutic strategies for PD and ALS. There are several different targeted approaches currently available, or that can be imagined for the treatment of PD or ALS. The first of these is targeting the aggregated or misfolded proteins themselves to prevent oligomer-induced microglial activation. Vaccine-induced antibodies or intracellular-produced single chain antibodies (scFv, intrabodies) directed against misfolded/aggregated proteins, or drugs (e.g., rapamycin) that stimulate microglial/macrophage phagocytosis and lysosomal degradation may prove of therapeutic benefit by clearance of extracellular or intracellular misfolded/aggregated proteins. Another approach may use drugs that directly inhibit neurotoxicity (e.g., by inhibiting excitotoxicity or apoptosis) or promote neuroprotection (e.g., glial cell line-derived neurotrophic factor, GDNF), thus slowing disease progression. Finally, significant efforts are being made towards modulation of the immune response to misfolded/aggregated proteins. This may be accomplished by use of immune response modifiers and antioxidants that attenuate microglial activation. With the realization that disease can either be exacerbated by effector T cells specific for neo-epitopes from modified self-CNS proteins (e.g., NT-modified proteins) or be ameliorated by regulatory T cells (natural or antigen-induced), therapeutic immunoregulation by immunomodulators or adjuvants to induce or upregulate regulatory T cell responses can inhibit exacerbated adaptive and innate immune responses and interdict further neurodegeneration. *α-syn, α-synuclein; GA, glatiramer acetate; GM-CSF, granulocyte macrophage colony stimulating factor; GSK3-β, glycogen synthase kinase 3-β; IFN-γ, interferon-γ; MHC, major histocompatibility complex molecule; mutSOD1, mutant superoxide dismutase-1; N-α-syn, nitrated α-synuclein; NMDA, N-methyl-D-aspartic acid; NSAID, nonsteroidal anti-inflammatory drugs; NT, nitrotyrosine; ONOO⁻, peroxynitrite; PAF, platelet-activating factor; RNS, reactive nitrogen species; ROS, reactive oxygen species; SOD1, superoxide dismutase-1; TNF-α; tumor necrosis factor-α; VIP, vasoactive intestinal peptide.*

44.2. Protein Misfolding and Modifications

44.2.1. α-Synuclein and SOD1 Biology and Biochemistry

α-Synuclein is a major constituent protein in Lewy bodies for which expression of several mutations, duplication or triplication in genes encoding for α-synuclein induce early-onset PD (Polymeropoulos et al., 1997; Kruger et al., 1998; Singleton et al., 2003; Chartier-Harlin et al., 2004; Farrer et al., 2004; Ibanez et al., 2004). Misfolding of this protein and lack of clearance is thought to play a major role in progression of PD. Two major cellular mechanisms of clearance include degradation by the ubiquitin-proteasome and the autophagic-lysosomal pathways. The former is supported by mutations in PD patients for *PARK2* (Waters and Miller, 1994) and *PARK5* (Liu et al., 2002) which respectively encode for proteins with ubiquitin E3 ligase and ubiquitin C-terminal hydrolase activities. The latter mechanism is supported in α-synuclein over-expressing cells by the presence of α-synuclein in organelles with morphological features of autophagic vesicles, increased clearance of α-synuclein with rapamycin, an autophagy stimulator, and increased accumulation of aggregated α-synuclein when this pathway is blocked (Webb et al., 2003; Lee et al., 2004). Moreover, clearance seemingly occurs in an aggregation stage-specific manner where the more toxic oligomeric intermediates are susceptible to clearance, but mature fibrillar inclusion bodies are not. Neutralization of the acidic compartments leads to the accumulation of α-synuclein aggregates and exacerbates α-synuclein-mediated toxicity and cell death (Lee et al., 2004). In ALS, SOD1-positive aggregates are observed in the spinal cords from autopsied familial ALS (FALS) patients, as well as, from mutant SOD1 (mutSOD1) transgenic mice and are thought to play a role in toxicity and neuronal cell death (Shibata et al., 1996; Bruijn et al., 1997; Watanabe et al., 2001; Puttaparthi et al., 2003; Kabashi and Durham, 2006). Clearance of SOD1 aggregates involves proteasome activity, as inhibition of proteasome-mediated proteolysis promotes SOD1 aggregation in tissues from G93A SOD1 mice, whereas restoration of activity reverses aggregate formation (Puttaparthi et al., 2003). However, whether SOD1 impairs proteasomic activity is unclear as analyses from several laboratories yielded mixed results of unchanged, increased, or decreased proteolytic activities measured in mutant SOD1 mice (Kabashi and Durham, 2006). Of interest, proteosomal proteolytic activity in the cytoplasm, but not in the nucleus has been shown to be impaired from motor neurons of SOD1 mice. Additionally, clearance of mutant SOD1 by autophagy is comparable to that of the proteasome pathway as shown under conditions where mutant SOD1 is not toxic, the inhibition of macroautophagy induces mutant SOD1-mediated aggregation and cell death (Kabuta et al., 2006). Thus enhancing proteosomal or lysosomal function may be potential therapeutic strategies to halt disease progression in those afflictions associated with aggregation of misfolded proteins (Figure 44.1).

44.2.2. Protein Nitration

Postmortem analyses of PD patients have consistently demonstrated the increased presence of nitrated residues as biomarkers for oxidative stress. Protein modifications are among the many biomarkers detected in the brains of PD patients. Compared to brains from control donors, elevated levels of nitrated proteins are found in brains and CSF of PD patients (Aoyama et al., 2000). Most notable are 3-nitrotyrosine (NT) modifications of proteins that comprise Lewy bodies (LB), the neuronal inclusions that are considered the hallmarks of PD and consist primarily of α-synuclein, ubiquitin, and lipids (Duda et al., 2000; Giasson et al., 2000) suggesting an increased participation of inflammatory responses and reactive molecular species; however, whether those modifications occur before or after inclusion into LB remain unclear. Also S-nitrosylated forms of parkin (*PARK2*) have been isolated from the temporal cortex from PD patients (Yao et al., 2004). Carbonyl modifications, which are reflective of protein oxidation, are increased in SN, basal ganglia, globus pallidus, substantia innominata, cerebellum and frontal pole compared to controls and patients with incidental LB disease (ILBD), a putatively presymptomatic PD disorder. Whether the involvement of the latter two regions reflects a consequence of L-DOPA treatment or a more global consequence of the inflammatory spread of oxidative stress in PD is unclear. However, increased expression of neural heme oxygenase-1 and glycosylated proteins by nigral neurons provides additional evidence for oxidative damage to proteins in PD.

In ALS, abundant evidence points to the effects of peroxynitrite in affected tissues. Significant increases in levels of NT moieties are detected on CSF proteins from ALS patients (Aoyama et al., 2000; Shaw and Williams, 2000), including Mn superoxide dismutase (Mn-SOD) which is only slightly increased in patients with AD and PD (Chou et al., 1996; Aoyama et al., 2000). Moreover, NT immunoreactivity is associated with motor neurons of the spinal cord and axons (Calingasan et al., 2005), and co-localizes to axonal conglomerates and spheroid neurofilament accumulations of upper and lower motor neurons (Chou et al., 1996) and with Aβ-40 depositions within abnormal neurons (Calingasan et al., 2005).

44.3. Microglial Inflammatory Responses and Therapies

44.3.1. Innate Immunity and Disease

PD is characterized by activation of microglial cells found in and around degenerating neurons. Reactive microglia are commonly seen within the SNpc of PD brains investigated at autopsy (Croisier et al., 2005). A six-fold increase in numbers of reactive microglia has been shown phagocytosing dopaminergic neurons (McGeer et al., 1988b) and correlates with the deposition of α-synuclein (Croisier et al., 2005). These reactive microglia over-express HLA-DR of the human MHC II complex, complement receptor type 3 (CR3, CD11b/CD18, Mac-1, Mo 1), CD68 (EMB11), CD23 (FcεRII, Fc receptor II for IgE), ferritin, CD11a (lymphocyte function-associated antigen-1, LFA-1) and CD54 (ICAM-1). They also secrete a plethora of proinflammatory cytokines such as interferon-γ (IFN-γ) tumor necrosis factor-α (TNF-α), interleukin 1-β (IL-1β), and upregulate enzymes such as inducible nitric oxide synthase (iNOS), and cyclooxygenase (COX) 1 and 2. MHC class II positive microglia also are increased in the putamen, hippocampus, transentorhinal cortex, cingulate cortex and temporal cortex of the PD brain. Reactive microglia serve as *in vivo* indicators of neuroinflammatory responses and contribute significantly to progressive degenerative processes. In early-stage PD imaging, PK11195 ligand binding to peripheral benzodiazepine receptors that are upregulated on reactive midbrain microglia and serve as markers of neuroinflammation, inversely correlates with binding of 2-β-carbomethoxy-3β-(4-fluorophenyl) tropane (CFT) to the dopamine transporter (DAT) in the putamen as a measure of surviving dopaminergic termini and also correlates with the severity of motor impairment (Croisier et al., 2005). Additional evidence for the involvement of neuroinflammation includes epidemiological data that demonstrate daily nonsteroidal anti-inflammatory agents decrease the risk of PD in a large cohort (Chen et al., 2003; Chen et al., 2005). Biochemical and histological evidence in PD brains of increased levels of carbonyl and nitrotyrosine protein modifications, lipid peroxidation, DNA damage, and reduction of glutathione and ferritin, the end results of inflammatory processes, also support the importance of inflammation in PD (Hald and Lotharius, 2005). Postmortem samples of SNpc from sporadic PD patients show elevated levels of the protein $gp91^{phox}$, the main transmembrane component of NADPH-oxidase, which co-localize with microglia and are associated with the production of ROS. Likewise, in 1-methyl-4-phenyl-1,2,3,6-tetrahydropyridine (MPTP)-treated mice (see Przedborski, Chapter 26), large increases in $gp91^{phox}$ immunoreactivity also co-localize in the SNpc with activated (Mac-1 immunopositive) microglia, but not with astrocytes or neurons. Studies of post-mortem brains from three human subjects who, after 3–16 years previously had injected MPTP and developed parkinsonian syndromes, exhibited accumulations of activated microglial cells around dopaminergic neurons (Langston et al., 1999). Thus, an initial acute insult to dopaminergic neurons likely leads to a secondary and perpetuated neuroinflammatory response. This neuroinflammatory reaction serves to alter homeostatic neural mechanisms or exacerbate disease processes by production of proinflammatory factors (Figure 44.1).

In ALS patients an abundance of data indicates the active involvement of the immune system in disease processes. Activated mononuclearphagocytic (MP; macrophages and perivascular microglia) cells are found in affected tissues in ALS patients (Graves et al., 2004; Henkel et al., 2004). Several postmortem studies of tissues from ALS patients exhibiting neuronal loss compared to those of controls demonstrate the increased presence of activated microglia in the ventral horn of the spinal cord, corticospinal tract, motor cortex, and brainstem (Henkel et al., 2004), and are thought to be involved in phagocytosis of degenerating neurons since they are found in close proximity in tissues showing early signs of degeneration. Macrophages and amoeboid microglia within those affected tissues from ALS patients exhibit upregulated HLA-A, -B, -C, and -DR molecules, FcγRI, CR3 (CD11b/CD18), CR4 (CD11c/CD18), macrophage-colony stimulating factor (M-CSF) receptor (CSF-1R), β2 integrins, COX-2, iNOS, and CD40 receptor which strongly suggest those microglia are in a reactive state. Furthermore, *in vivo* PET imaging using PK11195 to visualize upregulated peripheral benzodiazepine receptor by activated microglia show increased levels in the motor cortex, dorsolateral prefrontal cortex, thalamus, and pons of ALS patients. Recent works in mutant SOD1 transgenic mice (see Simpson et al. Chapter 27) suggest a promising therapeutic strategy targeting the innate immune system. The presence of wild type cells in the wild type/SOD1 Tg mouse chimeras enhances survival, whereas mutant SOD1 acting within non-neurons was toxic, no matter whether chimeric construction was with embryonic cells or via bone marrow transplantation to irradiated recipients (Clement et al., 2003; Corti et al., 2004). To assess the role of SOD1 expression in MP cells on survival, mice transgenic for mutSOD1 and flanking LoxP sequences (LoxSOD1) were crossed with mice expressing Cre under control of the CD11b promoter, an integrin exclusively expressed by myeloid cells (Boillee et al., 2006). Those crosses exhibited diminished SOD1 expression in microglia and macrophages, but not other cells, and extended survival compared to LoxSOD1 controls. Although no discernible differences in the activation phenotype of microglia or macrophages could be demonstrated, SOD1 expression was significantly diminished. Finally, reconstitution of G93A-SOD1/PU.$1^{-/-}$ mice with wild-type bone marrow cells, established donor-derived microglia within the CNS, slowed motoneuron loss, prolonged disease duration and survival compared to mice treated with bone marrow cells from SOD1 transgenic animals (Beers et al., 2006). Taken together, these data suggest that targeting innate immunity and expression of SOD1 in the CNS of ALS represent a plausible therapeutic modality for ALS (Figure 44.1).

44.3.2. Oxidative Stress

Once activated, microglia can produce noxious factors including pro-inflammatory cytokines, chemokines, quinolinic acid, arachidonic acid and its metabolites, and excitatory amino acids among others. A major defense mechanism provided by MP is the production of free radicals and reactive molecular species, all potentially toxic to invading organisms. The cellular machinery of myeloid lineage cells that has evolved to produce these toxic products is NADPH oxidase (expressed by microglia, macrophages and neutrophils) and myeloperoxidase (expressed predominately by neutrophils). Among neurons, astrocytes and microglia of the CNS, microglia are, in large measure, responsible for generating a major portion of free radicals (Figure 44.2) (Klegeris and McGeer, 2000).

Reactive species of oxygen, nitrogen, and carbon are thought to play an active role in PD and ALS wherein the very utilization and production of neurotransmitters produces harmful reactive species. In PD, the metabolism of

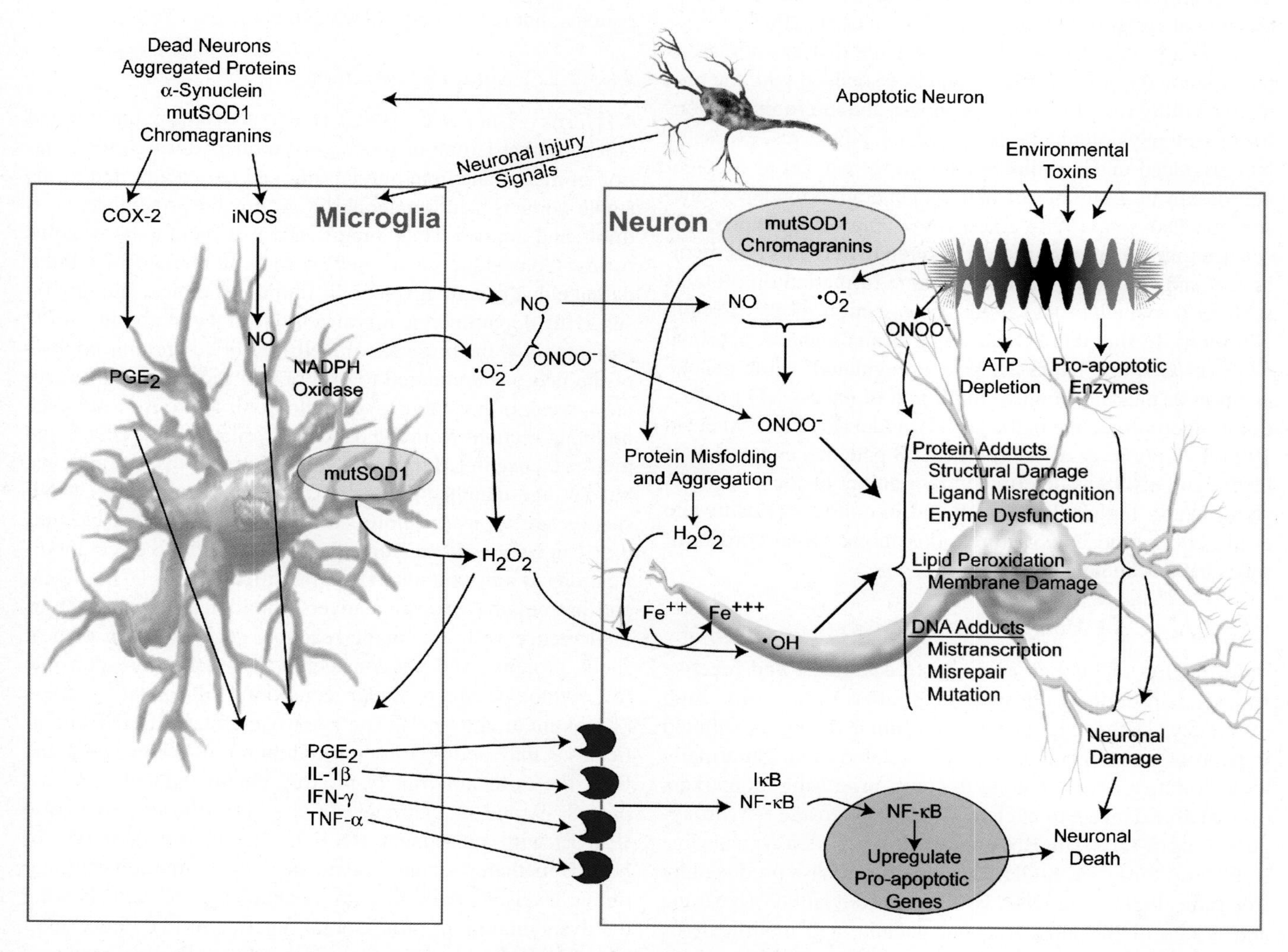

FIGURE 44.2. Neuroinflammatory and oxidative stress pathways in PD and ALS pathogenesis. Free radicals can arise several diverse ways, such as glial cell activation, mitochondrial dysfunction, and protein aggregation. Increased microglia activation is attributable to increased neuronal cell death and cell debris including aggregated proteins. In ALS and related-animal models, mutated SOD1 (mutSOD1) expression increases aggregated proteins, neuronal death and microglial activation. Microglial derived NO and superoxide ($\bullet O_2^-$) species react in extracellular spaces to form peroxynitrite (ONOO-). Peroxynitrite readily crosses cell membranes where it contributes to lipid peroxidation, DNA damage and nitrotyrosine formation in α-synuclein and other cellular proteins. Damaged proteins are targeted to cellular proteosomes for degradation via the ubiquitination pathway. Excess NO produced by activated microglia can lead to S-nitrosylation of cellular proteins, including parkin. Such modifications may diminish E3 ubiquitin ligase activity necessary for efficient protein turnover by proteosomes. Excessive protein damage caused by oxidants and disruptions in the ubiquitin pathways may overload or inhibit protein degradation quality control measures leading to the accumulation of damaged proteins in cells. When reactive species exceed anti-oxidant defenses, oxidative stress is generated; destroying molecular structures, such as proteins, lipids and DNA, causing irreversible and detrimental damage, neuronal cell injury and death. Adapted from Gao et al. (Gao et al., 2003).

dopamine that produces H_2O_2 can exacerbate inflammation and tissue damage by feeding into the ROS cycle and/or by dopamine-quinone modification of protein sulfhydryl groups via nucleophilic additions. Glutamate signaling through the NMDA receptor involves the generation of a hydroxyl radical, the excess of which has been shown to be excitotoxic, while glutamate ligation of the kainite receptor has been shown to induce reactive oxygen and nitrogen species. However, production of the majority of reactive oxygen species is mediated through the superoxide radical produced by NADPH-oxidase; the inhibition of which mitigates neurodegeneration in numerous model systems including MPTP (Wu et al., 2003).

NO is a biological messenger molecule that has numerous physiological roles in the CNS and is associated with the protective killing function of macrophages and microglia. In contrast to the physiological roles of normal NO levels, excessive NO produced under pathological settings can act as a potent neurotoxin in a number of neurodegenerative models (Dawson and Dawson, 1998). Excess NO reacts with superoxide species to form peroxinitrite, which readily crosses cell membranes and contributes to nitrotyrosine formation on proteins such as α-synuclein and SOD1 (see section 44.2.2. Protein Nitration). In sporadic PD and some animal models, neuronal NOS (nNOS) and iNOS are both upregulated, while genetic ablation or pharmacological inhibition of excess NO production is neuroprotective in the MPTP model (Przedborski et al., 1996; Liberatore et al., 1999). In ALS patients, expression of nNOS and eNOS as determined by immunohistochemistry is significantly higher than that found in controls (Kashiwado et al., 2002), and iNOS expression among spinal cord infiltrates have been noted.

44.3.2.1. DNA Modifications

Modification of nucleic acids by free radicals and reactive species can induce chromosomal aberrations with high efficiency, suggesting that chromosomal damage exhibited in neurons of PD patients might be related to an abnormally high oxidative stress. Among the most promising biomarkers of oxidative damage to nucleic acids is nucleoside 8-hydroxyguanosine (8-OHG) for RNA or 8-hydroxy-2′-deoxyguanosine (8-OHdG) for DNA. 8-OHG is an oxidized base produced by free radical attack on DNA by C-8 hydroxylation of guanine and is one of the most frequent nucleic acid modifications observed under conditions of oxidative stress. In PD patients, levels of 8-OHG nucleic acid modifications are commonly increased in the caudate and SN compared to age-matched controls (Zhang et al., 1999). Immunohistochemical characterization of these modifications indicates that the highest levels of 8-OHG modifications are found in neurons of the SN and to a lesser extent in neurons of the nucleus raphe dorsalis and oculomotor nucleus, and occasionally in glial cells. Given that 8-OHG nucleic acid modifications are rarely detected in the nuclear area, mosty restricted to the cytoplasm, and that immunoreactivity is significantly diminished by RNase or DNase and ablated with both enzymes (Zhang et al., 1999), suggests that targets of oxidative attack include both cytoplasmic RNA and mitochondrial DNA. Of particular interest are the findings that concentrations of 8-OHG in CSF of PD patients are higher than in age-matched controls; however, serum concentrations of 8-OHG appear highly variable (Abe et al., 2003).

In ALS, 8-OHdG modified DNA is detected in spinal cord tissues (Calingasan et al., 2005) and CSF (Ihara et al., 2005) from sporadic ALS (SALS) and FALS patients, while levels of nuclear 8-OHdG are increased in motor cortex of SALS patients, but not FALS patients (Shibata et al., 2000).

44.3.2.2. Lipid Peroxidation

4-Hydroxy-2-nonenal (HNE) is a reactive α,β unsaturated aldehyde that is one of the major products during the oxidation of membrane lipid polyunsaturated fatty acids, and forms stable adducts with nucleophilic groups on proteins such as thiols and amines. HNE modification of membrane proteins forms stable adducts that can be used as biomarkers of cellular damage due to oxidative stress. Immunochemical staining on surviving dopaminergic nigral neurons in the midbrains of PD patients show the presence of HNE-modified proteins on 58% of the neurons compared to only 9% of those in control subjects, weak or no staining on oculomotor neurons in the same midbrain sections from PD patients (Yoritaka et al., 1996), and their presence in LB from PD and diffuse LB disease patients, but not age-matched controls (Castellani et al., 2002). HNE species are typically more stable than oxygen species, thus they can easily spread from site of production to effect modifications at a distant site. HNE modifications of DNA, RNA, and proteins have various adverse biological effects such as interference with enzymatic reactions and induction of heat shock proteins, and are considered to be largely responsible for cytotoxic effects under conditions of oxidative stress (Toyokuni et al., 1994). The cytotoxic effects of HNE modifications may be in part due to inhibition of complexes I and II of the mitochondrial respiratory chain; induction of caspase-8, -9, and -3, cleavage of poly(ADP-ribose) polymerase (PARP) with subsequent DNA fragmentation; inhibition of NF-κB mediated signaling pathways; or diminution of glutathione levels. Consistent with an abundance of data showing the dysregulation of proteasomal function in PD, direct binding of HNE to the proteasome also inhibits the processing of ubiquinated proteins. Levels of HNE species that induce no acute change in cell viability *in vitro* initially cause a decrease in the proteasomal catalytic activity to the extent that it those levels induce accumulation of ubiquitinated and nitrated proteins, reductions in glutathione levels and mitochondrial activity, and increased levels of oxidative damage to DNA, RNA, proteins, and lipids (Hyun et al., 2003).

In SALS patients, HNE levels are significantly elevated in the sera, CSF, the ventral motor neurons and surrounding glia (Shibata et al., 2000; Simpson et al., 2004). Levels in serum and CSF are directly correlated with the extent of the disease, but not with the rate of disease progression and are cytotoxic to a hybrid motor neuron cell line. One HNE-modified protein was shown to be the astrocytic transporter EAAT2, (Pedersen et al., 1998) suggesting a role for HNE-mediated impairment of glutamate transport, increased glutamate levels and excitotoxic-induced neurodegeneration in ALS.

44.3.2.3. Glutathione

Production of ROS and NO in neurons is buffered primarily by the glutathione (GSH) system. GSH content in the SNpc of PD patients is decreased by 40–50%, but not in other regions of the brain, nor in age-matched controls or patients with other diseases affecting dopaminergic neurons (Sian et al., 1994b). This diminution continues with progression and severity of disease, suggesting a correlation with concomitant increases in reactive species (Pearce et al., 1997). GSH depletion has been suggested as the first indicator of oxidative stress during PD progression, possibly occurring prior to other hallmarks of PD including the decreased activity of mitochondrial complex I (Andersen, 2004). Also, elevated GSSG/GSH ratios in PD patients (Sian et al., 1994b) argue strongly for a role of oxidative stress in this disease (Dringen, 2000). An increase in glutathione peroxidase immunoreactivity, exclusive to glial cells surrounding surviving dopaminergic neurons, has also been observed in PD brains (Damier et al., 1993). Interestingly, the SN and striatum have lower levels of GSH relative to other regions of the brain, which include, in increasing order: SN, striatum, hippocampus, cerebellum, and cortex (Kang et al., 1999). Although varying in different regions of the brain, all GSH levels diminish by about 30% in the elderly, suggesting a possible link with the age associated risk factor for PD. GSH depletion cannot be explained by increased oxidation of GSH to GSSG as levels of both are diminished in the nigra of PD patients (Chen et al., 1989; Sian et al., 1994b). Diminished GSH levels do not appear to be caused from failure of GSH synthesis as γ-glutamylcysteine synthetase is unaltered, as are glutathione peroxidase and glutathione transferase activities (Sian et al., 1994a). Other possible mechanisms for diminished levels include increased removal of GSH from cells by γ-glutamyltranspeptidase (Sian et al., 1994a) and the formation of adducts of glutamyl and cysteinyl peptides of GSH with dopamine. Nevertheless, depletion of GSH may render cells more sensitive to toxic effects of oxidative stress and potentiate the toxic effects of reactive microglia (Chen et al., 2001).

In contrast, direct evidence for perturbations of the glutathione system in ALS is limited. One report demonstrated in the CSF of SALS patients, increased oxidized NO products, higher GSH levels and lower GSSH levels, thus lower GSSH/GSH ratios (Tohgi et al., 1999); however GSH or GSSH levels in brain or spinal cord tissues remains to be determined. In addition, reports of increased GSH-binding sites in the spinal cord of ALS patients could reflect an upregulation of glutathione receptors (Bains and Shaw, 1997).

44.3.3. Modulation of Innate Immunity as Therapeutic Targets

44.3.3.1. PPAR-γ

Immune suppression through receptor modulation has been another approach attempted to alleviate or reverse PD progression. For example, agonists of peroxisome proliferator-activated receptor-γ (PPAR-γ), a nuclear receptor involved in carbohydrate and lipid metabolism, have been shown to inhibit inflammatory responses in a variety of cell lines, including monocyte/macrophages and microglial cells (Breidert et al., 2002). *In vivo* administration of PPAR-γ agonists modulates inflammatory responses in the brain. Pioglitazone, a PPAR-γ agonist used currently as an anti-diabetic agent, has been shown to have anti-inflammatory effects in animal models of autoimmune disease, attenuate glial activation, and inhibit dopaminergic cell loss in the SN of MPTP treated mice (Breidert et al., 2002). However, pioglitazone treatment had little effect on MPTP-induced changes in the striatum. This result seems to indicate that in the MPTP mouse model of PD, mechanisms regulating glial activation in the dopaminergic terminals compared with the dopaminergic cell bodies are PPAR-γ independent (Breidert et al., 2002).

44.3.3.2. Minocycline and Modulators of Microglial Activation

Minocycline, a semisynthetic second generation tetracycline, easily penetrates the blood-brain barrier and has been shown recently to effectively protect from neurodegeneration in several disease models including ALS, PD, cerebral ischemia, Huntington's disease, multiple sclerosis, and spinal cord injury (Domercq and Matute, 2004), all of which involve microglial activation as the principle effector of secondary neurotoxicity. Tetracyclines prevent cell death in models of neurodegeneration by both attenuation of the innate and adaptive immunity and blockage of apoptotic cascades (Domercq and Matute, 2004). Minocycline specifically inhibits microglial activation and proliferation, the induction of caspase-1and -3, iNOS, and COX-2 (Chen et al., 2000; Domercq and Matute, 2004; Wang et al., 2004). Minocycline also attenuates adaptive immunity by reducing the expression and activity of matrix metalloproteinases, which alter blood-brain barrier permeability (Brundula et al., 2002; Popovich et al., 2003; Zhu et al., 2002; Domercq and Matute, 2004). Oral administration of minocycline was able to attenuate microglial activation, protect dopaminergic neurons in the SNpc, and restore motor function in the MPTP mouse model of PD (Peng et al., 2006). Administration of minocycline was also effective in ALS transgenic mice. Not only was the effect of treatment to delay onset of disease, it resulted in increased motor performance and extended survival by 2–3 weeks compared to non-treated ALS transgenic mice (Zhu et al., 2002; Kriz et al., 2002). Other modalities that have been used successfully to attenuate microglia activation in models of neurodegeneration include non-steriodal anti-inflammatory agents (Dairam, 2006), COX-2 inhibitors such as rofecoxib (Hewett, 2006) and parecoxib (Reksidler, 2007), and dextromethorphan (Zhang, 2006) among others (Figure 44.1).

Free radical production by activated microglia accompanied by dysregulation of antioxidants (e.g., SOD, glutathione and catalase) diminish the protective potential by establishing a pro-oxidative stress environment. In turn, these antioxidants may be used therapeutically to modulate the microglial response. SOD levels have been shown to decline throughout the progression of PD. Direct treatment with SOD or inducers of SOD may be used to inactivate oxygen free radicals produced by activated microglia by converting superoxide to hydrogen peroxide which in turn is cleared by either catalase or glutathione peroxidase upon conversion to water, with GSH as the cosubstrate. Treatments that lead to increased expression of GSH or prevent its degradation may slow disease progression. These antioxidants can be used together to combat the microglial inflammatory response (Figure 44.1).

44.4. Adaptive Immunity

44.4.1. Cell-Mediated Immunity

While naïve T cells are typically precluded from CNS entry, neuroinflammation aggressively recruits activated components of the adaptive immune system to sites of active neurodegeneration by increasing expression of cellular adhesion molecules and inducing chemokine gradients. Moreover, glial cells secrete toxic factors that disrupt blood brain barrier function. Nonetheless, much evidence indicates a far more complex relationship between the CNS and immunological systems than previously thought. Further challenging this view of the "immune privileged" status of the CNS are animal model systems wherein immune deficiencies translate into exacerbated neuronal loss following traumatic injuries. Such injuries are corrected in animals that receive immune reconstitution prior to experimental injury. Rodents and humans that have sustained CNS injuries also have expanded T cell repertoires against myelin-associated antigens, yet do not appear to be at increased risk for the development of CNS autoimmunity. Any functional consequence of such T cell responses against CNS antigens following injury remains to be determined.

Little evidence exists for HLA association with PD. Early reports identified significant association of PD with expression of HLA-B17, -B18, -A2, and –A28, whereas later works failed to detect significant deviations of HLA haplotypes among PD patients compared to unaffected controls. Interestingly, a recent study analyzing HLA class I and II alleles among 45 German PD patients demonstrated a significant increase in the representation of the DQB1*06 allele suggesting an association between idiopathic PD and the immune system (Lampe et al., 2003).

The association of any one HLA type with ALS is controversial. Initial studies found no statistical association with ALS, however increased incidences of ALS patients that express HLA-A3, -Bw40, -Bw35, -B18, -Cw4, as well as decreased incidence of HLA-A9, -B8, -B7, and -DR4 were discovered. Additionally, milder disease progression is associated with HLA-A12 and -Bw40, whereas more aggressive disease is associated with HLA-Bw35.

In PD patients, increased numbers of $CD8^+$ T cells are found in close proximity to activated microglia and degenerating neurons within the SN; however, those numbers are consistently low in frequency (McGeer et al., 1988a). In the MPTP mouse model, numbers of $CD8^+$ and $CD4^+$ T cells are significantly increased as late as day 21 post-intoxication with mean CD4/CD8 ratios of 0.33 ± 0.07 (range 0.19–0.64) (Kurkowska-Jastrzebska et al., 1999). Whether these infiltrating T cells are activated, antigen-specific, or migrating in response to microglial inflammation has yet to be determined, but the presence of major T cell subsets at levels exceeding those typically found in the CNS and in ratios differing from those found in the periphery suggests a role in PD more profound than that associated with surveillance. In that vein, numerous aberrations in peripheral lymphocyte subsets are detectable in PD patients. Compared to age-matched controls, numbers of total lymphocytes in both drug-naïve and -treated PD cohorts have been shown to be diminished by 17%, while $CD19^+$ B cells were diminished by 35% and $CD3^+$ T cells were diminished by 22% (Bas et al., 2001). Among $CD3^+$ T cells, numbers of $CD4^+$ T cells were diminished by 31%; whereas, numbers of $CD8^+$ T cells were not significantly changed. The frequencies of cells within $CD4^+$ T cell subsets are differentially diminished, with a greater loss of naïve helper T cells ($CD45RA^+$) and either unchanged or increased effector/memory helper T cell subset ($CD29^+$ or $CD45RO^+$) (Bas et al., 2001).

That at least some T cell subsets from PD patients are activated is suggested by the increased mutual co-expression of CD4 and CD8 by $CD45RO^+$ T cells, as well as upregulation of CD25 (α-chain of the IL-2 receptor), TNF-α receptors, and significant downregulation of IFN-γ receptors. However, evaluation of these parameters to assess whether activated T cell phenotypes are derived from any one T cell subset or many subsets have yet to be incorporated within one study. Interestingly, a significantly greater number of micronuclei and unrepaired single strand DNA breaks, which have been shown to result from exposure to higher levels of oxidative stress and inflammation (Cerutti, 1985), are detected in lymphocytes and activated T cells from PD patients compared to age-matched controls (Migliore et al., 2002; Petrozzi et al., 2002).

In ALS patients, lymphocytes, as well as myeloid cells are consistently found within or around affected tissues (Graves et al., 2004). Affected tissues with lymphocytic infiltrates include spinal cords, brain, muscles, and associated vessels (Graves et al., 2004). Most reports have identified the infiltrating lymphocytes as T cells, with a minor presence of B cells. Most indicate that both $CD4^+$ and $CD8^+$ T cells are present within affected tissues, but one report indicated the presence of $CD8^+$ T cells, with rare $CD4^+$ cells in anterior and lateral corticospinal tracts and anterior horns (Troost et al., 1990), while another demonstrated the presence of primarily $CD4^+$ T cells with rare $CD8^+$ cells in muscle (Troost et al., 1992). Although not rigorously examined, a consensus suggests that these T cells are in an activated state by the presence of upregulated surface markers such as MHC I and II molecules (Lampson et al., 1990; Troost et al., 1992) and CD40L (CD40 ligand) (Graves et al., 2004). Of particular interest is the finding that PCR amplification of the third complementarity-determining region (CDR3) to examine the T cell receptor (TCR) repertoire of infiltrating T cells in cerebral spinal fluid (CSF), spinal cords and brains of ALS patients showed increased utilization of TCRBV2 (variable region 2 of the T cell receptor β-chain) transcripts which was independent of HLA haplotype (Panzara et al., 1999).

Early studies of ALS patient peripheral blood revealed a general loss of T cells as demonstrated by reduced numbers of E-rosetting T cells (a function mediated by CD2), and later confirmed by flow cytometric analysis showing significant losses of total $CD2^+$ T cells with concomitant increase with increased numbers of surface Ig^+ B cells (Provinciali et al., 1988), whereas others have failed to show significant differences in T or B cell frequency or function (Appel et al., 1986). Analyses of T cell subsets in ALS patients relative to normal controls show variable results, which include diminished or increased frequencies of $CD4^+$ cells, diminished frequencies of $CD8^+$T cells, or no changes in either T cell subset. Of interest, frequencies of T cells that co-express IgM FcR (FcμR), MHC class II, CD38, or IL-13, all phenotypes functionally associated with immunoregulatory functions and/or activated states, are increased in ALS patients (Shi et al., 2006). Increased frequencies of T cells that express MHC class II and $CD4^+IL\text{-}13^+$ T cells are inversely correlated with clinical score, while the latter T cell phenotype directly correlates with the rate of disease progression (Shi et al., 2006). Of interest, in a minor subset of ALS patients with persistent motor conductance blockage, but not those without blockage, the frequencies or $CD3^+$, $CD4^+$, and $CD8^+$ T cells, as well as $CD16^+$ mononuclear cells are increased, suggesting a relationship between peripheral blood abnormalities and the pathogenic processes associated with conduction blockage in ALS patients (Tanaka et al., 1993).

44.4.2. Humoral Immunity

The possibility that humoral immunity may play a role in either the initiation or regulation of PD arose from observations of increased complement components in the SN of PD patients and experimental models wherein dopaminergic degeneration is triggered by adoptive transfer of immunoglobulin from PD patients. In both idiopathic and genetic cases of PD, pigmented dopaminergic neurons immunolabeled with IgG and associated with an increase in activated microglia expressing the high affinity IgG receptor FcγRI (Orr et al., 2005).

Perhaps the most studied of immune-associated aspects in ALS patients are those of humoral immunity. Global alterations of humoral immunity include increased serological complement (C'), γ-globulin and immunoglobulin levels, increased immune complexes, and the presence of immunoglobulin and C' component depositions within kidney, spinal cord and other nervous tissues of ALS patients. These humoral aberrations and depositions of humoral products provoked an increased impetus for an autoimmune approach to ALS etiology. Thus, many studies have assessed the possible antigenic reactivities of those antibodies. Those reactivities from ALS patients' sera and CSF comprise epitopes associated with nervous and non-nervous tissues as well. Nervous system specificities of antibodies in ALS encompass those directed against moieties from spinal cord, motor neurons, neuromuscular junctions, neurofilaments, myelin, and Ca^{2+} channels. The plethora of data showing antibody reactivities to gangliosides especially among IgM antibodies and their localication to motor neurons and spinal grey matter, dorsal and ventral spinal roots, dorsal root ganglion neurons, nodes of Ranvier, neuromuscular junctions and skeletal muscle is thought to reflect a possible pathogenic role in motor neuron diseases; however the mechanism by which those antibodies exact their toll has not been ascertained. Co-culture with neuronal tissues and passive transfer to rodent recipients of sera or immunoglobulin from patients and immunized animals has revealed functional consequences of these reactivities relative to ALS patients. Passive transfer yields increased presence of human IgG in spinal cord motor neurons and neuromuscular junctions with increased miniature end-plate potential (MEPP) frequency (Appel et al., 1991), increases in the rate of spontaneous neurotransmitter release, and axonal degeneration and denervation in most muscles (Uchitel et al., 1992). Within 12–24 hours of IgG transfer, increases are detected in the density of synaptic vesicles, CSF glutamate concentrations, and Ca^{2+} levels in axon terminals of neuromuscular junctions and synaptic boutons on spinal motoneurons (Engelhardt et al., 1995; Engelhardt et al., 1997; La Bella et al., 1997; Pullen et al., 2004). Antibodies are internalized with increased phosphorylation of neurofilament H (Engelhardt et al., 1995) while Golgi system and rough endoplasmic reticulum dilate with concomitant increased Ca^{2+} levels that are precipitously depleted after 24 hours (Engelhardt et al., 1997; Pullen et al., 2004). At 72 hours after transfer, CSF glutamate and aspartate levels increase, without appreciable change in glutamine and glutathione levels (La Bella et al., 1997). By 8 days post-transfer, areas of neuronal cell necrosis are evident (Pullen et al., 2004) and sensitivity to L-type Ca^{2+} channel blockers is retained for up to 4 weeks post transfer (Fratantoni et al., 2000). Thus, passive transfer of immunoglobulin from ALS patients appears to lead to long-lasting effects of motor neurons at the neuromuscular junction and may indicate that such effects may be an early stage event in immune-mediated pathogenesis of ALS.

44.5. Therapeutic Immunoregulation

44.5.1. Cell-Mediated Immunomodulation

Another potential therapeutic avenue for PD may involve T cell mediated immune responses (Figure 44.1). Activation of T cells directed against antigens expressed at the injured areas of the CNS has been shown to be neuroprotective under acute and chronic neurodegenerative conditions (Kipnis et al., 2002). However, immunization with such antigens might lead to development of an autoimmune disease. Immunization with Copolymer-1 (Cop-1, Copaxone, glatiramer acetate) or passive transfer of Cop-1 specific T cells has been shown to be beneficial for protecting neurons from secondary degeneration after injurious conditions (Kipnis et al., 2000). Cop-1 reactive T cells have partial cross-reactivity with myelin basic protein (MBP) and other self-antigens expressed in the brain (Arnon and Sela, 2003). Therefore, immunization with Cop-1 leads to increased accumulation of T lymphocytes in areas of injury within the brain and spinal cord and is neuroprotective without causing any adverse effects; however, the molecular mechanism of this response is not fully understood. T cells reactive to Cop-1 could be a source of brain-derived neurotrophic factor (BDNF) and other neurotrophic factors (Kipnis et al., 2000) or can induce production of neurotrophins by microglial or astroglial cells. Recently, the neuroprotective effect of immunization with Cop-1 was tested in the MPTP model of PD and demonstrated that adoptive transfer of Cop-1-specific T cells, but not ovalbumin-specifc T cells, into MPTP-intoxicated mice attenuates reactive microglial neuroinflammation and inhibits dopaminergic neurodegeneration in both the SNpc and the striatum (Benner et al., 2004). Additionally, adoptive transfer of those T cells protects from the loss of nigral N-acetylaspartate (NAA) levels associated with MPTP-induced neurodegeneration as determined by quantitative proton magnetic resonance spectroscopic imaging (^{1}H-MRSI) (Boska et al., 2005). Suppression of microglial-associated inflammation was associated with T cell accumulation within the SNpc, induction of a T helper type 2 cell (Th2) response with production of anti-inflammatory cytokines (IL-4, IL-10), and increased expression of GDNF by astrocytes, but not by infiltrating T cells or microglia (Benner et al., 2004). Recent studies have shown that $CD4^+$ T cells, rather than $CD8^+$ T cells, possess greater neuroprotective capacity and that passive transfer of anti-Cop-1 antibodies provide no neuroprotection (Laurie et al., 2007). More recently, we demonstrated that adoptive transfer of anti-CD3 activated $CD4^+$ $CD25^+$ regulatory T cells (Tregs), but not activated effector T cells, are capable of ameliorating MPTP-induced neuroinflammation and dopaminergic neurodegeneration (Reynolds et al., 2007). Moreover, Tregs were shown to mediate neuroprotection through suppression of reactive microglia responses to inflammatory stimuli including nitrated α-synuclein, as well as enhancing astrocyte-derived GDNF. In a mouse model of ALS, immunization with Cop-1 has been shown to extend survival by 25% in the G93A mSOD1 transgenic mice (Angelov et al., 2003). These data suggest a putative mechanism for which regulatory T cells, induced by vaccination with cross-reactive epitopes or activated regulatory T cells, extravasate in response to neuroinflammation from neurodegenerative processes; secrete anti-inflammatory cytokines in response to cross-reactive self-epitopes (e.g., myelin basic protein) to attenuate reactive microglia; suppress the inflammatory response; and induce neurotrophic responses by T cells and/or other glia (Figure 44.3).

The recent development of novel immune-mediated therapeutic vaccine strategies for disease intervention and attenuating neuroinflammation are predicated on the ability of regulatory T cells, either induced or naturally occurring, to modulate both innate and adaptive immune responses and suppress inflammation both in autoimmune disease and in models of neurodegeneration. Treatment with VIP is emerging as a therapeutic tool to generate Tregs both *in vitro* and *in vivo*. VIP originates as a neuropeptide that can function as both a neurotransmitter and neuromodulator in many organ systems, including the central and peripheral nervous systems (Said, 1976). VIP-containing neurons are present in the CNS in areas that influence the immune system as well as in lymphoid organs, and are thought to be involved in the recruitment of immune cells, many of which express receptors for VIP (Kaltreider et al., 1997; Reubi et al., 1998; Reubi et al., 2000). VIP is also produced by lymphocytes, preferentially Th2 T cells, in response to different mitogenic or inflammatory stimuli (Gomariz et al., 1990; Gomariz et al., 1992; Leceta et al., 1996; Delgado et al., 2005). Th2-derived VIP also promotes Th2 responses in vivo (Delgado et al., 1999; Delgado et al., 2000; Goetzl et al., 2001; Vassiliou et al., 2001; Delgado et al., 2002; Voice et al., 2003). VIP treatment was recently shown to be efficacious in inducing Tregs in a variety of inflammatory disorders including arthritis, graft versus host disease, and more recently experimental allergic encephalitis (Delgado et al., 2005; Fernandez-Martin et al., 2006). VIP has also been used to expand regulatory T cells *ex vivo* and elicit conversion of $CD4^+$ $CD25^-$ effector T cells to Tr1 regulatory T cells. Thus, VIP can be utilized to generate Tregs specific for self-antigens to promote antigen-specific tolerance and suppress development of autoimmune disorders in a variety of animal model systems (Figure 44.1) (Delgado et al., 2005; Fernandez-Martin et al., 2006; Gonzalez-Rey et al., 2006).

Selective activation of particular subsets of dendritic cells (DCs) with GM-CSF or 1,25-dihydroxyvitamin D3 (Figure 44.1) can not only activate lymphocytes, but also induce T cell tolerance to self-antigens or to specific antigens tandemly administered as immunization or tolerization regimens. DCs also can affect B cell function, antibody synthesis, and isotype switching. Recent studies have shown that DCs stimulated with GM-CSF or 1,25-dihydroxyvitamin D3 may exert their tolerogenic functions through arresting type 1 helper T cells (Th1), skewing the Th1/Th2 balance, and generating regulatory T cells (Gregori et al., 2002; Vasu et al., 2003; Gangi et al., 2005).

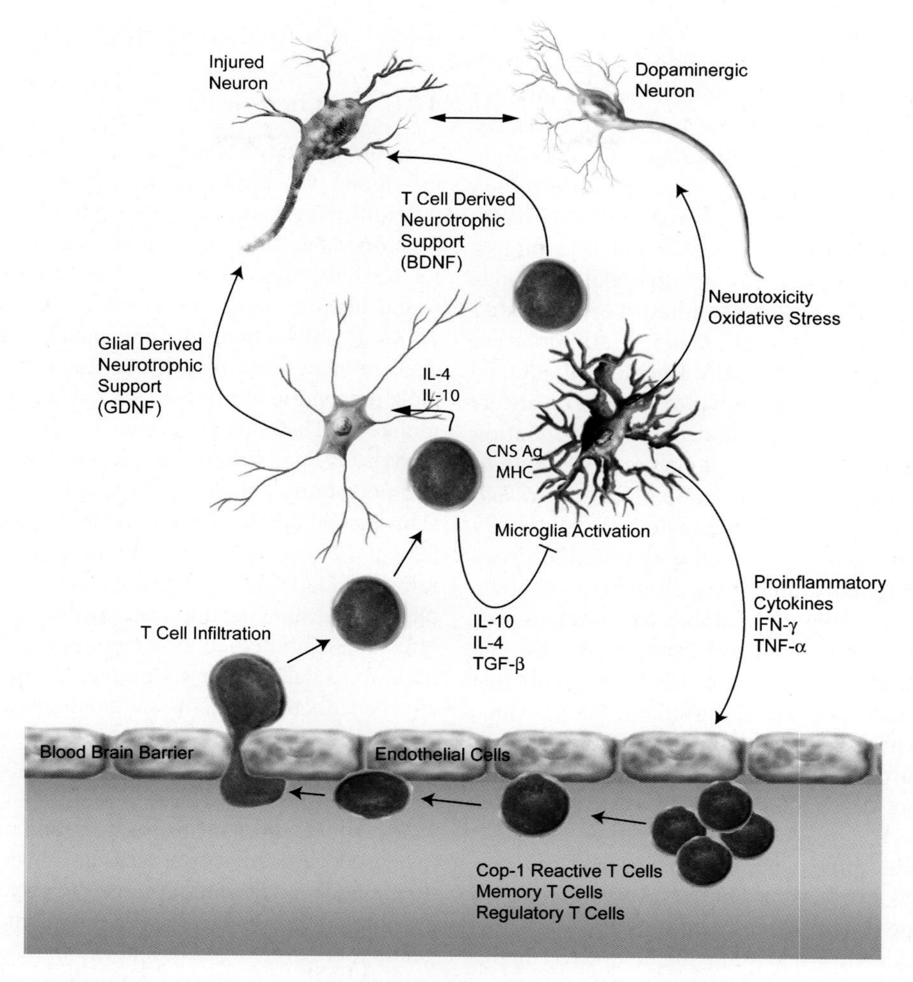

FIGURE 44.3. T cell-mediated neuroprotection in a PD model. In MPTP-intoxicated mice, regulatory T cells infiltrate the inflamed nigrostriatal pathway where they encounter cross-reactive CNS antigens (such as myelin basic protein or reactive species-modified proteins) presented in the context of MHC by resident microglial cells. In response, activated T cells secrete anti-inflammatory cytokines such as IL-4, IL-10, and TGF-β that suppress toxic microglial activities. Neurotrophin expression may occur directly from T cells or T cell-derived IL-4, and IL-10 may induce neurotrophin production in neighboring glia. These activities lead to neuroprotection indirectly by suppression of microglial responses and directly through the local delivery of neurotrophins.

44.5.2. Modulation of Humoral Immunity

Using another vaccine strategy to elicit humoral immune responses directed at cephalic epitopes, immunization of human α-synuclein transgenic mice with mutant human α-synuclein produced high affinity anti-α-synuclein antibodies with concomitant diminution of human α-synuclein inclusions in neurons and synapses, and diminished neurodegeneration (Masliah et al., 2005). Moreover, anti-α-synuclein antibodies recognized abnormal human α-synuclein and supported degradation of α-synuclein aggregates. Similarly, immunization of G37R SOD1 transgenic mice with recombinant mutant SOD1 reduced the amount of mutant SOD1 in spinal cords of immunized mice, increased numbers of surviving motor neurons, and extended their life span by greater than 4 weeks, which significantly correlated with increasing anti-SOD1 antibody titers (Urushitani et al., 2007). In contrast, this strategy failed to protect G93A SOD1 mice, which exhibit a more severe and aggressive phenotype than G37R SOD1 mice, however ventricular infusion of purified anti-human SOD1 antibody alleviated clinical signs and significantly extended life span by 4%. These data provide support for immunotherapeutic strategies that target extracelluar burdens of aggregated or misfolded toxic proteins such as α-synuclein and SOD1 in neurodegenerative disorders (Figure 44.1).

44.5.3. Other Vaccine Strategies

One quasi-immunotherapeutic strategy targets intracellular accumulation of toxic proteins in neuron by using single chain antibodies (scFv) or intrabodies (Chen et al., 1994). This strategy is accomplished by preventing misfolding to the toxic form, interfering with misfolded protein interactions, or enhancing degradation of the toxic form. Clearly keys to this strategy are the production of scFv with affinities that retain epitope specificites in vivo and efficacious delivery to vulnerable or affected neurons. For delivery, genes encoding the scFv must be transfected into targets cells and successfully expressed as a functional intrabodies. To date several scFv to α-synuclein have been prepared (Emadi et al., 2004; Miller et al., 2005; Barkhordarian et al., 2006; Maguire-Zeiss et al., 2006; Emadi et al., 2007). Several specific scFv have been shown to recognize different conformations of α-synuclein and upon co-incubation with monomeric α-synuclein *in vitro,* to decrease the rate of α-synuclein aggregation, inhibit formation of oligomers and protofibrils, and inhibit toxicity. In transfected cells, scFv have been shown to be stably expressed with minimal toxicity for greater than 3 months, bind intracellular α-synuclein, increase the amount of detergent soluble α-synuclein species while decreasing the amount of insoluble species, and counter reduced cell adhesion which characterizes cells that overexpress α-synuclein (Messer and McLear, 2006). Thus, scFv to α-synuclein have the potential to provide a therapeutic modality to control intracellular accumulation of toxic protofibrillar forms of α-synuclein and possibly disease progression (Figure 44.1).

Therapeutic vaccine approaches using neuronal antigens represent a potential efficacious interdictory modality for slowing or halting the progression of neuroinflammation and secondary neurodegeneration and removing neurotoxic aggregates (Figure 44.1). These approaches should be considered in strategies with other anti-inflammatory or anti-oxidant therapies for a combinatorial modality to protect against neuroinflammation and consequent neurodegeneration in PD and ALS. For the latter, clinical trials have recently been completed demonstrating the safety and immunological responsiveness to glatiramer acetate (GA) vaccine therapy in ALS patients (Gordon et al., 2006). Work performed in our own laboratories assessed cell-mediated, cytokine and humoral responses in ALS patients who received GA during the six-month phase II trial. Treated patients showed enhanced lymphocyte proliferation to GA. Plasma samples were evaluated by GA-specific ELISA assays for immunoglobulin (Ig) classes (Ig M, A, and G) and IgG subclasses (IgG1, IgG2, IgG3, and IgG4), and by cytokine bead arrarys (CBA) for Th1 and Th2 cytokine levels (Mosley et al., 2007). Fourteen of 21 GA-immunized patients produced anti-GA Ig responses. All anti-GA IgG subclass concentrations were increased by greater than 4.2-fold in plasma from treated patients, and anti-GA IgG1 comprised the majority of the humoral response. Additionally, changes in plasma Th1 and Th2 cytokine levels were shown to be associated with time of GA treatment. These data show significant humoral responses and cytokine trends following GA immunization in ALS patients.

44.6. Neuroprotective Strategies

44.6.1. Growth Factors

The role of neurotrophins in reducing neurodegeneration and promotion of neuroregenerative processes presents an exciting possibility for therapeutic benefit to PD (Figure 44.1). A study of lentiviral delivery of glial cell line-derived neurotrophic factor (GDNF) showed trophic effects on degenerating nigrostriatal neurons in a primate model of PD (Kordower et al., 2000). Results indicated augmented dopaminergic function in aged monkeys and reversal of functional deficits with complete prevention of nigrostriatal degeneration in MPTP-treated monkeys. These data indicate that GDNF delivery using a lentiviral vector system can prevent nigrostriatal degeneration and potentially induce regeneration in primate models of PD, showing the potential for a viable therapeutic strategy for PD patients. However, recent clinical trials of intraputamenally infused GDNF in PD patients are controversial with one 2-year phase I trial showing improved activity scores and no untoward effects in a limited cohort (Patel et al., 2005), while phase II trials were halted after six months due to lack of efficacy and adverse effects in patients and nonhuman primates.

44.6.2. Neuroprotectants

Many diverse mechanisms, factors, and pathways are involved in neurodegenerative disorders, thus several different therapeutic methods have been developed to target a specific factor or a whole intricate pathway with the intent of ameliorating, preventing, or reversing neuronal cell damage. Inflammation and oxidative stress form a commonality between many neurodegenerative diseases; therefore most therapeutic modalities currently under investigation target MP activation to decrease the magnitude of the inflammatory responses. The targets of these therapies include, but are not restricted to, enhancement of neurotrophic factors such as glia cell-line derived neurotrophic factor (GDNF), up-regulation of anti-inflammatory cytokines (IL-4, IL-10, and TGF-β1), inhibition of enzymatic activities that encourage neurotoxicity (GSK-3β, γ-secretase), Ca^{2+} and glutamate excitotoxicity blockers that inhibit NMDA receptor function, suppression of neuronal cytotoxicity (memantine, lithium, sodium valproate), and attenuation of inflammation by anti-inflammatory drugs (NSAID, minocycline). Anti-inflammatory and/or anti-oxidative therapies could be used in conjunction to form combinational therapies targeting multiple sites of oxidative stress that contribute to inflammatory responses and progressive degenerating disease (Figure 44.1).

44.7. Genetics and Immunity

Recent evidence has shown that genetics may contribute to the onset of neurodegenerative disorders (Li et al., 2002). Linkages to the age at onset (AAO) for PD have been identified on chromosomes 1 and 10. The latter is significantly associated with glutathione s-transferase omega-1 (GSTO1) (Li et al., 2003); a provocative finding since GSTO1 is thought to be involved in the post-translation modification of IL-1, a major component in the regulation of inflammatory responses (Laliberte et al., 2003). One factor associated with the chromosome 1p peak is the embryonic lethal abnormal vision 4 (ELAVL4) gene (Noureddine et al., 2005), a human homologue of the Drosophila ELAV (Good, 1995) and essential for temporal and spatial gene expression during CNS development. Additionally, ELAVL gene products are known to bind to AU-rich response elements (ARE) in the 3′-untranslated region (3′UTR) of inflammation-associated factors (Good, 1995). Interestingly, PD patients homozygotic for allele 1 at position -511 of the IL-1β gene have an earlier onset of the disease than those homozygotic for allele 2, which produces higher amounts of IL-1. Thus, higher production of IL-1β may provide some neuroprotective effect for dopaminergic neurons (Nishimura et al., 2000; Mizuta et al., 2001). Gene expression analyses of post-mortem specimens have shown differentially expressed genes associated with the immune response in ALS, most notably IL-1 receptor accessory protein (IL-1rap), MHC class II, thromboxane synthase, FcεR γ-chain and the cytokine regulated upon activation, normal T-cell expressed and secreted (RANTES) (Malaspina and de Belleroche, 2004; Wang et al., 2006). Of interest, as most of these analyses are performed with tissues from end-stage patients, only approximately 10% of the differentially expressed genes expressed in ALS patients are associated with inflammation/immune function, whereas the majority of genes pertain to stress-activated pathways (Malaspina and de Belleroche, 2004). This is in contrast to mice that express the G93A mutation of the human *SOD1* wherein the majority of early changes are associated with inflammation/immune function genes.

Summary

Although patterns of neuronal degeneration are unique in PD and ALS, both disorders share common pathways and processes that support and possibly initiate neurodegeneration. Most of these processes are associated with induction, propagation, or consequences of neuroinflammation. Increased numbers of microglia that express a reactive phenotype and proximate dying neurons reflect the neuroinflammatory cellular response. Neuroinflammation amplifies oxidative stresses via reactive oxygen, nitrogen, and carbon species that then react with biomolecules and increase molecular modifications of lipids, proteins, and nucleic acids. These reactive modifications eventually become deleterious to biochemical and cellular processes resulting in dysregulation of cellular functions and further neuronal death. Whether neuroinflammatory responses are causal or consequential remains to be determined. Nevertheless, the importance of inflammatory responses to neurodegeneration is underscored in animal models whereby attenuation of neuroinflammation by genetic manipulation or pharmacological agents mitigates neurodegeneration and increases neuronal survival. As such, immunological strategies that target neuroinflammatory processes represent promising candidates for therapeutic intervention in neurodegenerative disorders. These strategies embrace the capacity of regulatory T cells to protect neurons either directly via neurotrophic factors, or indirectly by modulation of microglial function to attenuate neuroinflammatory responses and by induction of astrocyte-derived neurotrophic factors.

Acknowledgments. Support for this research was made by NIH/NINDS (R21 NS049264), the Michael J. Fox Foundation for Parkinson's Research and the Carol Swarts, M.D. Laboratory for Emerging Neuroscience Research.

Review Questions/Problems

1. **Summarize the evidence for innate immune-mediated mechanisms associated with PD.**
2. **Summarize the evidence that suggest a role for adaptive mediated immunity in ALS.**
3. **The cellular target for immunotherapy that would be most beneficial to inhibit secondary neurodegeneration is the**

   a. astrocyte
   b. neuron
   c. microglia
   d. oligodentrocyte

4. **The primary producer of reactive oxygen species by microglia is**

   a. dopamine synthesis
   b. NADPH oxidsase
   c. myeloperoxidase
   d. superoxide dismutase

5. **In PD, postmortem samples show increased**

   a. reactive microglia.
   b. loss of striatal dopaminergic termini
   c. loss of dopaminergic neuronal bodies within the substantia nigra pars compacta
   d. reactive oxygen species modified proteins
   e. all of the above

6. **CD45RO^{+} T cells that mutually express CD4 and CD8 more likely indicate those cells are**

   a. dead
   b. recent thymic emigrants.
   c. anergic
   d. activated

7. An example of free radical modification of RNA is

a. 8-hydroxyguanosine (8-OHG)
b. 4-hydroxy-2-nonenal (HNE)
c. 3-nitrotyrosine (NT)
d. 8-hydroxy-2′-deoxyguanosine (8-OHdG)

8. Free radical modification of DNA is best exemplified by

a. 8-hydroxyguanosine (8-OHG)
b. 4-hydroxy-2-nonenal (HNE)
c. 3-nitrotyrosine (NT)
d. 8-hydroxy-2′-deoxyguanosine (8-OHdG)

9. One molecular marker of nitric oxide modification of proteins is

a. 8-hydroxyguanosine (8-OHG)
b. 4-hydroxy-2-nonenal (HNE)
c. 3-nitrotyrosine (NT)
d. 8-hydroxy-2′-deoxyguanosine (8-OHdG)

10. A protein marker resulting from modifications due to lipid peroxidation is

a. 8-hydroxyguanosine (8-OHG)
b. 4-hydroxy-2-nonenal (HNE)
c. 3-nitrotyrosine (NT)
d. 8-hydroxy-2′-deoxyguanosine (8-OHdG)

11. Immunotherapeutic strategies using scFv directed against α-synuclein targets are designed to

a. prevent misfolding to the toxic form of the protein
b. interfere with misfolded protein interactions
c. enhance degradation of the toxic protein form
d. all of the above

12. Passive transfer of sera or immunoglobulin from ALS patients to rodent recipients results in increases in the following sequelae with the exception of

a. denervation of striatum
b. denervation of muscles
c. human immunoglobulins in spinal cord motor neurons
d. human immunoglobulins in neuromuscular junctions
e. miniature end-plate potential (MEPP)

13. A ligand utilized for in vivo imaging of reactive microglia via upregulated peripheral benzodiazepine receptors is

a. CFT
b. DA
c. GT1b
d. MPTP
e. PK1195

14. The ultimate effect of reactive species on cellular function include

a. ligand misrecognition
b. enzyme dysfunction
c. membrane damage
d. mutation
e. all of above

15. Glatiramer acetate is an immunomodulatory drug that is FDA approved and clinically indicated for

a. amyotrophic lateral sclerosis
b. remitting/relapsing multiple sclerosis
c. Parkinson's disease
d. Alzheimer's disease
e. Huntington's disease

16. Regulatory T cells capable of attenuating microglial responses are more likely to produce and secrete

a. IL-10
b. IL-2
c. IFN-γ
d. TNF-α

17. A first indicator of oxidative stress during progressive disease in PD, occurring prior to other hallmarks including loss of mitochondrial complex I activity, has been suggested to be

a. reactive microglia
b. diminished dopamine
c. increased astrocytosis
d. depletion of glutathione
e. all of the above

18. IL-2/IL-2R interactions are found localized to

a. T cell-T cell interactions
b. striatum
c. frontal cortex
d. cerebellum
e. all of the above

19. The possible mechanism by which glatiramer acetate functions is

a. competition with myelin-basic protein (MBP) for binding to major histocompatibility complex (MHC) molecules
b. competition of GA/MHC with MBP/MHC for binding to the T-cell receptor
c. partial activation and tolerance induction of MBP-specific T cells
d. induction of GA-reactive T-helper 2- (TH2)-like regulatory cells
e. all of the above

20. Levels of glutathione least concentrated in

a. substantia nigra
b. striatum
c. hippocampus
d. cerebellum
e. cortex

References

Abe T, Isobe C, Murata T, Sato C, Tohgi H (2003) Alteration of 8-hydroxyguanosine concentrations in the cerebrospinal fluid and serum from patients with Parkinson's disease. Neurosci Lett 336:105–108.

Andersen JK (2004) Oxidative stress in neurodegeneration: Cause or consequence? Nat Med 10 Suppl:S18–S25.

Angelov DN, Waibel S, Guntinas-Lichius O, Lenzen M, Neiss WF, Tomov TL, Yoles E, Kipnis J, Schori H, Reuter A, Ludolph A, Schwartz M (2003) Therapeutic vaccine for acute and chronic motor neuron diseases: Implications for amyotrophic lateral sclerosis. Proc Natl Acad Sci U S A 100:4790–4795.

Aoyama K, Matsubara K, Fujikawa Y, Nagahiro Y, Shimizu K, Umegae N, Hayase N, Shiono H, Kobayashi S (2000) Nitration of manganese superoxide dismutase in cerebrospinal fluids is a marker for peroxynitrite-mediated oxidative stress in neurodegenerative diseases. Ann Neurol 47:524–527.

Appel SH, Stockton-Appel V, Stewart SS, Kerman RH (1986) Amyotrophic lateral sclerosis. Associated clinical disorders and immunological evaluations. Arch Neurol 43:234–238.

Appel SH, Engelhardt JI, Garcia J, Stefani E (1991) Immunoglobulins from animal models of motor neuron disease and from human amyotrophic lateral sclerosis patients passively transfer physiological abnormalities to the neuromuscular junction. Proc Natl Acad Sci U S A 88:647–651.

Arnon R, Sela M (2003) Immunomodulation by the copolymer glatiramer acetate. J Mol Recognit 16:412–421.

Bains JS, Shaw CA (1997) Neurodegenerative disorders in humans: The role of glutathione in oxidative stress-mediated neuronal death. Brain Res Brain Res Rev 25:335–358.

Barkhordarian H, Emadi S, Schulz P, Sierks MR (2006) Isolating recombinant antibodies against specific protein morphologies using atomic force microscopy and phage display technologies. Protein Eng Des Sel 19:497–502.

Bas J, Calopa M, Mestre M, Mollevi DG, Cutillas B, Ambrosio S, Buendia E (2001) Lymphocyte populations in Parkinson's disease and in rat models of parkinsonism. J Neuroimmunol 113:146–152.

Beers DR, Henkel JS, Xiao Q, Zhao W, Wang J, Yen AA, Siklos L, McKercher SR, Appel SH (2006) Wild-type microglia extend survival in PU.1 knockout mice with familial amyotrophic lateral sclerosis. Proc Natl Acad Sci U S A 103:16021–16026.

Benner EJ, Mosley RL, Destache CJ, Lewis TB, Jackson-Lewis V, Gorantla S, Nemachek C, Green SR, Przedborski S, Gendelman HE (2004) Therapeutic immunization protects dopaminergic neurons in a mouse model of Parkinson's disease. Proc Natl Acad Sci U S A 101:9435–9440.

Boillee S, Yamanaka K, Lobsiger CS, Copeland NG, Jenkins NA, Kassiotis G, Kollias G, Cleveland DW (2006) Onset and progression in inherited ALS determined by motor neurons and microglia. Science 312:1389–1392.

Boska MD, Lewis TB, Destache CJ, Benner EJ, Nelson JA, Uberti M, Mosley RL, Gendelman HE (2005) Quantitative 1H magnetic resonance spectroscopic imaging determines therapeutic immunization efficacy in an animal model of Parkinson's disease. J Neurosci 25:1691–1700.

Breidert T, Callebert J, Heneka MT, Landreth G, Launay JM, Hirsch EC (2002) Protective action of the peroxisome proliferator-activated receptor-gamma agonist pioglitazone in a mouse model of Parkinson's disease. J Neurochem 82:615–624.

Bruijn LI, Becher MW, Lee MK, Anderson KL, Jenkins NA, Copeland NG, Sisodia SS, Rothstein JD, Borchelt DR, Price DL, Cleveland DW (1997) ALS-linked SOD1 mutant G85R mediates damage to astrocytes and promotes rapidly progressive disease with SOD1-containing inclusions. Neuron 18:327–338.

Brundula V, Rewcastle NB, Metz LM, Bernard CC, Yong VW (2002) Targeting leukocyte MMPs and transmigration: minocycline as a potential therapy for multiple sclerosis. Brain 125:1297–1308.

Calingasan NY, Chen J, Kiaei M, Beal MF (2005) Beta-amyloid 42 accumulation in the lumbar spinal cord motor neurons of amyotrophic lateral sclerosis patients. Neurobiol Dis 19:340–347.

Castellani RJ, Perry G, Siedlak SL, Nunomura A, Shimohama S, Zhang J, Montine T, Sayre LM, Smith MA (2002) Hydroxynonenal adducts indicate a role for lipid peroxidation in neocortical and brainstem Lewy bodies in humans. Neurosci Lett 319:25–28.

Cerutti PA (1985) Prooxidant states and tumor promotion. Science 227:375–381.

Chartier-Harlin MC, Kachergus J, Roumier C, Mouroux V, Douay X, Lincoln S, Levecque C, Larvor L, Andrieux J, Hulihan M, Waucquier N, Defebvre L, Amouyel P, Farrer M, Destee A (2004) Alpha-synuclein locus duplication as a cause of familial Parkinson's disease. Lancet 364:1167–1169.

Chen H, Zhang SM, Hernan MA, Schwarzschild MA, Willett WC, Colditz GA, Speizer FE, Ascherio A (2003) Nonsteroidal antiinflammatory drugs and the risk of Parkinson disease. Arch Neurol 60:1059–1064.

Chen H, Jacobs E, Schwarzschild MA, McCullough ML, Calle EE, Thun MJ, Ascherio A (2005) Nonsteroidal antiinflammatory drug use and the risk for Parkinson's disease. Ann Neurol 58:963–967.

Chen M, Ona VO, Li M, Ferrante RJ, Fink KB, Zhu S, Bian J, Guo L, Farrell LA, Hersch SM, Hobbs W, Vonsattel JP, Cha JH, Friedlander RM (2000) Minocycline inhibits caspase-1 and caspase-3 expression and delays mortality in a transgenic mouse model of Huntington disease. Nat Med 6:797–801.

Chen SY, Bagley J, Marasco WA (1994) Intracellular antibodies as a new class of therapeutic molecules for gene therapy. Hum Gene Ther 5:595–601.

Chen TS, Richie JP, Jr., Lang CA (1989) The effect of aging on glutathione and cysteine levels in different regions of the mouse brain. Proc Soc Exp Biol Med 190:399–402.

Chen Y, Vartiainen NE, Ying W, Chan PH, Koistinaho J, Swanson RA (2001) Astrocytes protect neurons from nitric oxide toxicity by a glutathione-dependent mechanism. J Neurochem 77:1601–1610.

Chou SM, Wang HS, Komai K (1996) Colocalization of NOS and SOD1 in neurofilament accumulation within motor neurons of amyotrophic lateral sclerosis: An immunohistochemical study. J Chem Neuroanat 10:249–258.

Clement AM, Nguyen MD, Roberts EA, Garcia ML, Boillee S, Rule M, McMahon AP, Doucette W, Siwek D, Ferrante RJ, Brown RH, Jr, Julien JP, Goldstein LS, Cleveland DW (2003) Wild-type nonneuronal cells extend survival of SOD1 mutant motor neurons in ALS mice. Science 302:113–117.

Corti S, Locatelli F, Donadoni C, Guglieri M, Papadimitriou D, Strazzer S, Del Bo R, Comi GP (2004) Wild-type bone marrow cells ameliorate the phenotype of SOD1-G93A ALS mice and contribute to CNS, heart and skeletal muscle tissues. Brain 127:2518–2532.

Croisier E, Moran LB, Dexter DT, Pearce RK, Graeber MB (2005) Microglial inflammation in the parkinsonian substantia nigra: Relationship to alpha-synuclein deposition. J Neuroinflammation 2:14.

Dairam A, Antunes EM, Saravanan KS, Daya S (2006) Non-steroidal anti-inflammatory agents, tolmetin and sulindac, inhibit liver tryptophan 2,3-dioxygenase activity and alter brain neurotransmitter levels. Life Sci 79:2269–2274.

Damier P, Hirsch EC, Zhang P, Agid Y, Javoy-Agid F (1993) Glutathione peroxidase, glial cells and Parkinson's disease. Neuroscience 52:1–6.

Dawson VL, Dawson TM (1998) Nitric oxide in neurodegeneration. Prog Brain Res 118:215–229.

Delgado M, Leceta J, Gomariz RP, Ganea D (1999) Vasoactive intestinal peptide and pituitary adenylate cyclase-activating polypeptide stimulate the induction of Th2 responses by up-regulating B7.2 expression. J Immunol 163:3629–3635.

Delgado M, Gomariz RP, Martinez C, Abad C, Leceta J (2000) Anti-inflammatory properties of the type 1 and type 2 vasoactive intestinal peptide receptors: Role in lethal endotoxic shock. Eur J Immunol 30:3236–3246.

Delgado M, Leceta J, Ganea D (2002) Vasoactive intestinal peptide and pituitary adenylate cyclase-activating polypeptide promote in vivo generation of memory Th2 cells. Faseb J 16:1844–1846.

Delgado M, Chorny A, Gonzalez-Rey E, Ganea D (2005) Vasoactive intestinal peptide generates CD4+CD25+ regulatory T cells in vivo. J Leukoc Biol 78:1327–1338.

Domercq M, Matute C (2004) Neuroprotection by tetracyclines. Trends Pharmacol Sci 25:609–612.

Dringen R (2000) Glutathione metabolism and oxidative stress in neurodegeneration. Eur J Biochem 267:4903.

Duda JE, Giasson BI, Chen Q, Gur TL, Hurtig HI, Stern MB, Gollomp SM, Ischiropoulos H, Lee VM, Trojanowski JQ (2000) Widespread nitration of pathological inclusions in neurodegenerative synucleinopathies. Am J Pathol 157:1439–1445.

Emadi S, Liu R, Yuan B, Schulz P, McAllister C, Lyubchenko Y, Messer A, Sierks MR (2004) Inhibiting aggregation of alpha-synuclein with human single chain antibody fragments. Biochemistry 43:2871–2878.

Emadi S, Barkhordarian H, Wang MS, Schulz P, Sierks MR (2007) Isolation of a human single chain antibody fragment against oligomeric alpha-synuclein that inhibits aggregation and prevents alpha-synuclein-induced toxicity. J Mol Biol 368:1132–1144.

Engelhardt JI, Siklos L, Komuves L, Smith RG, Appel SH (1995) Antibodies to calcium channels from ALS patients passively transferred to mice selectively increase intracellular calcium and induce ultrastructural changes in motoneurons. Synapse 20:185–199.

Engelhardt JI, Siklos L, Appel SH (1997) Altered calcium homeostasis and ultrastructure in motoneurons of mice caused by passively transferred anti-motoneuronal IgG. J Neuropathol Exp Neurol 56:21–39.

Farrer M, Kachergus J, Forno L, Lincoln S, Wang DS, Hulihan M, Maraganore D, Gwinn-Hardy K, Wszolek Z, Dickson D, Langston JW (2004) Comparison of kindreds with parkinsonism and alpha-synuclein genomic multiplications. Ann Neurol 55:174–179.

Fernandez-Martin A, Gonzalez-Rey E, Chorny A, Ganea D, Delgado M (2006) Vasoactive intestinal peptide induces regulatory T cells during experimental autoimmune encephalomyelitis. Eur J Immunol 36:318–326.

Fratantoni SA, Weisz G, Pardal AM, Reisin RC, Uchitel OD (2000) Amyotrophic lateral sclerosis IgG-treated neuromuscular junctions develop sensitivity to L-type calcium channel blocker. Muscle Nerve 23:543–550.

Gangi E, Vasu C, Cheatem D, Prabhakar BS (2005) IL-10-producing CD4+CD25+ regulatory T cells play a critical role in granulocyte-macrophage colony-stimulating factor-induced suppression of experimental autoimmune thyroiditis. J Immunol 174:7006–7013.

Gao HM, Liu B, Zhang W, Hong JS (2003) Novel anti-inflammatory therapy for Parkinson's disease. Trends Pharmacol Sci 24:395–401.

Giasson BI, Duda JE, Murray IV, Chen Q, Souza JM, Hurtig HI, Ischiropoulos H, Trojanowski JQ, Lee VM (2000) Oxidative damage linked to neurodegeneration by selective alpha-synuclein nitration in synucleinopathy lesions. Science 290:985–989.

Goetzl EJ, Voice JK, Shen S, Dorsam G, Kong Y, West KM, Morrison CF, Harmar AJ (2001) Enhanced delayed-type hypersensitivity and diminished immediate-type hypersensitivity in mice lacking the inducible VPAC(2) receptor for vasoactive intestinal peptide. Proc Natl Acad Sci U S A 98:13854–13859.

Gomariz RP, Lorenzo MJ, Cacicedo L, Vicente A, Zapata AG (1990) Demonstration of immunoreactive vasoactive intestinal peptide (IR-VIP) and somatostatin (IR-SOM) in rat thymus. Brain Behav Immun 4:151–161.

Gomariz RP, De La Fuente M, Hernanz A, Leceta J (1992) Occurrence of vasoactive intestinal peptide (VIP) in lymphoid organs from rat and mouse. Ann N Y Acad Sci 650:13–18.

Gonzalez-Rey E, Fernandez-Martin A, Chorny A, Martin J, Pozo D, Ganea D, Delgado M (2006) Therapeutic effect of vasoactive intestinal peptide on experimental autoimmune encephalomyelitis: Down-regulation of inflammatory and autoimmune responses. Am J Pathol 168:1179–1188.

Good PJ (1995) A conserved family of elav-like genes in vertebrates. Proc Natl Acad Sci U S A 92:4557–4561.

Gordon PH, Doorish C, Montes J, Mosley RL, Diamond B, Macarthur RB, Weimer LH, Kaufmann P, Hays AP, Rowland LP, Gendelman HE, Przedborski S, Mitsumoto H (2006) Randomized controlled phase II trial of glatiramer acetate in ALS. Neurology 66:1117–1119.

Graves MC, Fiala M, Dinglasan LA, Liu NQ, Sayre J, Chiappelli F, van Kooten C, Vinters HV (2004) Inflammation in amyotrophic lateral sclerosis spinal cord and brain is mediated by activated macrophages, mast cells and T cells. Amyotroph Lateral Scler Other Motor Neuron Disord 5:213–219.

Gregori S, Giarratana N, Smiroldo S, Uskokovic M, Adorini L (2002) A 1alpha,25-dihydroxyvitamin D(3) analog enhances regulatory T-cells and arrests autoimmune diabetes in NOD mice. Diabetes 51:1367–1374.

Hald A, Lotharius J (2005) Oxidative stress and inflammation in Parkinson's disease: Is there a causal link? Exp Neurol 193:279–290.

Henkel JS, Engelhardt JI, Siklos L, Simpson EP, Kim SH, Pan T, Goodman JC, Siddique T, Beers DR, Appel SH (2004) Presence of dendritic cells, MCP-1, and activated microglia/macrophages in amyotrophic lateral sclerosis spinal cord tissue. Ann Neurol 55:221–235.

Hewett SJ, Silakova JM, Hewett JA (2006) Oral treatment with rofecoxib reduces hippocampal excitotoxic neurodegeneration. J Pharmacol Exp Ther 319:1219–1224.

Hyun DH, Lee M, Halliwell B, Jenner P (2003) Proteasomal inhibition causes the formation of protein aggregates containing a wide range of proteins, including nitrated proteins. J Neurochem 86:363–373.

Ibanez P, Bonnet AM, Debarges B, Lohmann E, Tison F, Pollak P, Agid Y, Durr A, Brice A (2004) Causal relation between alpha-synuclein gene duplication and familial Parkinson's disease. Lancet 364:1169–1171.

Ihara Y, Nobukuni K, Takata H, Hayabara T (2005) Oxidative stress and metal content in blood and cerebrospinal fluid of amyotrophic lateral sclerosis patients with and without a Cu, Zn-superoxide dismutase mutation. Neurol Res 27:105–108.

Kabashi E, Durham HD (2006) Failure of protein quality control in amyotrophic lateral sclerosis. Biochim Biophys Acta 1762:1038–1050.

Kabuta T, Suzuki Y, Wada K (2006) Degradation of amyotrophic lateral sclerosis-linked mutant Cu,Zn-superoxide dismutase proteins by macroautophagy and the proteasome. J Biol Chem 281:30524–30533.

Kaltreider HB, Ichikawa S, Byrd PK, Ingram DA, Kishiyama JL, Sreedharan SP, Warnock ML, Beck JM, Goetzl EJ (1997) Upregulation of neuropeptides and neuropeptide receptors in a murine model of immune inflammation in lung parenchyma. Am J Respir Cell Mol Biol 16:133–144.

Kang Y, Viswanath V, Jha N, Qiao X, Mo JQ, Andersen JK (1999) Brain gamma-glutamyl cysteine synthetase (GCS) mRNA expression patterns correlate with regional-specific enzyme activities and glutathione levels. J Neurosci Res 58:436–441.

Kashiwado K, Yoshiyama Y, Arai K, Hattori T (2002) Expression of nitric oxide synthases in the anterior horn cells of amyotrophic lateral sclerosis. Prog Neuropsychopharmacol Biol Psychiatry 26:163–167.

Kipnis J, Yoles E, Porat Z, Cohen A, Mor F, Sela M, Cohen IR, Schwartz M (2000) T cell immunity to copolymer 1 confers neuroprotection on the damaged optic nerve: Possible therapy for optic neuropathies. Proc Natl Acad Sci U S A 97:7446–7451.

Kipnis J, Mizrahi T, Hauben E, Shaked I, Shevach E, Schwartz M (2002) Neuroprotective autoimmunity: Naturally occurring CD4+CD25+ regulatory T cells suppress the ability to withstand injury to the central nervous system. Proc Natl Acad Sci U S A 99:15620–15625.

Klegeris A, McGeer PL (2000) Interaction of various intracellular signaling mechanisms involved in mononuclear phagocyte toxicity toward neuronal cells. J Leukoc Biol 67:127–133.

Kordower JH, Emborg ME, Bloch J, Ma SY, Chu Y, Leventhal L, McBride J, Chen EY, Palfi S, Roitberg BZ, Brown WD, Holden JE, Pyzalski R, Taylor MD, Carvey P, Ling Z, Trono D, Hantraye P, Deglon N, Aebischer P (2000) Neurodegeneration prevented by lentiviral vector delivery of GDNF in primate models of Parkinson's disease. Science 290:767–773.

Kriz J, Nguyen MD, Julien JP (2002) Minocycline slows disease progression in a mouse model of amyotrophic lateral sclerosis. Neurobiol Dis 10:268–278.

Kruger R, Kuhn W, Muller T, Woitalla D, Graeber M, Kosel S, Przuntek H, Epplen JT, Schols L, Riess O (1998) Ala30Pro mutation in the gene encoding alpha-synuclein in Parkinson's disease. Nat Genet 18:106–108.

Kurkowska-Jastrzebska I, Wronska A, Kohutnicka M, Czlonkowski A, Czlonkowska A (1999) The inflammatory reaction following 1-methyl-4-phenyl-1,2,3, 6-tetrahydropyridine intoxication in mouse. Exp Neurol 156:50–61.

La Bella V, Goodman JC, Appel SH (1997) Increased CSF glutamate following injection of ALS immunoglobulins. Neurology 48:1270–1272.

Laliberte RE, Perregaux DG, Hoth LR, Rosner PJ, Jordan CK, Peese KM, Eggler JF, Dombroski MA, Geoghegan KF, Gabel CA (2003) Glutathione s-transferase omega 1–1 is a target of cytokine release inhibitory drugs and may be responsible for their effect on interleukin-1beta posttranslational processing. J Biol Chem 278:16567–16578.

Lampe JB, Gossrau G, Herting B, Kempe A, Sommer U, Fussel M, Weber M, Koch R, Reichmann H (2003) HLA typing and Parkinson's disease. Eur Neurol 50:64–68.

Lampson LA, Kushner PD, Sobel RA (1990) Major histocompatibility complex antigen expression in the affected tissues in amyotrophic lateral sclerosis. Ann Neurol 28:365–372.

Langston JW, Forno LS, Tetrud J, Reeves AG, Kaplan JA, Karluk D (1999) Evidence of active nerve cell degeneration in the substantia nigra of humans years after 1-methyl-4-phenyl-1,2,3,6-tetrahydropyridine exposure. Ann Neurol 46:598–605.

Laurie C, Reynolds A, Cuskun O, Bowman E, Gendelman HE, Mosley RL (2007) CD4+ T cells from Copolymer-1 immunized mice protect dopaminergic neurons in the 1-methyl-4-phenyl-1,2,3,6-tetrahydropyridine model of Parkinson's disease. J Neuroimmunol 183:60–68.

Leceta J, Martinez C, Delgado M, Garrido E, Gomariz RP (1996) Expression of vasoactive intestinal peptide in lymphocytes: A possible endogenous role in the regulation of the immune system. Adv Neuroimmunol 6:29–36.

Lee HJ, Khoshaghideh F, Patel S, Lee SJ (2004) Clearance of alpha-synuclein oligomeric intermediates via the lysosomal degradation pathway. J Neurosci 24:1888–1896.

Li YJ, Scott WK, Hedges DJ, Zhang F, Gaskell PC, Nance MA, Watts RL, Hubble JP, Koller WC, Pahwa R, Stern MB, Hiner BC, Jankovic J, Allen FA, Jr., Goetz CG, Mastaglia F, Stajich JM, Gibson RA, Middleton LT, Saunders AM, Scott BL, Small GW, Nicodemus KK, Reed AD, Schmechel DE, Welsh-Bohmer KA, Conneally PM, Roses AD, Gilbert JR, Vance JM, Haines JL, Pericak-Vance MA (2002) Age at onset in two common neurodegenerative diseases is genetically controlled. Am J Hum Genet 70:985–993.

Li YJ, Oliveira SA, Xu P, Martin ER, Stenger JE, Scherzer CR, Hauser MA, Scott WK, Small GW, Nance MA, Watts RL, Hubble JP, Koller WC, Pahwa R, Stern MB, Hiner BC, Jankovic J, Goetz CG, Mastaglia F, Middleton LT, Roses AD, Saunders AM, Schmechel DE, Gullans SR, Haines JL, Gilbert JR, Vance JM, Pericak-Vance MA, Hulette C, Welsh-Bohmer KA (2003) Glutathione S-transferase omega-1 modifies age-at-onset of Alzheimer disease and Parkinson disease. Hum Mol Genet 12:3259–3267.

Liberatore GT, Jackson-Lewis V, Vukosavic S, Mandir AS, Vila M, McAuliffe WG, Dawson VL, Dawson TM, Przedborski S (1999) Inducible nitric oxide synthase stimulates dopaminergic neurodegeneration in the MPTP model of Parkinson disease. Nat Med 5:1403–1409.

Liu Y, Fallon L, Lashuel HA, Liu Z, Lansbury PT, Jr. (2002) The UCH-L1 gene encodes two opposing enzymatic activities that affect alpha-synuclein degradation and Parkinson's disease susceptibility. Cell 111:209–218.

Maguire-Zeiss KA, Wang CI, Yehling E, Sullivan MA, Short DW, Su X, Gouzer G, Henricksen LA, Wuertzer CA, Federoff HJ (2006) Identification of human alpha-synuclein specific single chain antibodies. Biochem Biophys Res Commun 349:1198–1205.

Malaspina A, de Belleroche J (2004) Spinal cord molecular profiling provides a better understanding of amyotrophic lateral sclerosis pathogenesis. Brain Res Brain Res Rev 45:213–229.

Masliah E, Rockenstein E, Adame A, Alford M, Crews L, Hashimoto M, Seubert P, Lee M, Goldstein J, Chilcote T, Games D, Schenk D (2005) Effects of alpha-synuclein immunization in a mouse model of Parkinson's disease. Neuron 46:857–868.

McGeer PL, Itagaki S, Boyes BE, McGeer EG (1988a) Reactive microglia are positive for HLA-DR in the substantia nigra of Parkinson's and Alzheimer's disease brains. Neurology 38:1285–1291.

McGeer PL, Itagaki S, Akiyama H, McGeer EG (1988b) Rate of cell death in parkinsonism indicates active neuropathological process. Ann Neurol 24:574–576.

Messer A, McLear J (2006) The therapeutic potential of intrabodies in neurologic disorders: Focus on Huntington and Parkinson diseases. BioDrugs 20:327–333.

Migliore L, Petrozzi L, Lucetti C, Gambaccini G, Bernardini S, Scarpato R, Trippi F, Barale R, Frenzilli G, Rodilla V, Bonuccelli U (2002) Oxidative damage and cytogenetic analysis in leukocytes of Parkinson's disease patients. Neurology 58:1809–1815.

Miller TW, Messer A (2005) Intrabody applications in neurological disorders: Progress and future prospects. Mol Ther 12:394–401.

Miller TW, Zhou C, Gines S, MacDonald ME, Mazarakis ND, Bates GP, Huston JS, Messer A (2005) A human single-chain Fv intrabody preferentially targets amino-terminal Huntingtin's fragments in striatal models of Huntington's disease. Neurobiol Dis 19:47–56.

Mizuta I, Nishimura M, Mizuta E, Yamasaki S, Ohta M, Kuno S, Ota M (2001) Relation between the high production related allele of the interferon-gamma (IFN-gamma) gene and age at onset of idiopathic Parkinson's disease in Japan. J Neurol Neurosurg Psychiatry 71:818–819.

Mosley RL, Gordon PH, Hasiak CM, Van Wetering FJ, Mitsumoto H, Gendelman HE (2007) Glatiramer acetate immunization induces specific antibody and cytokine responses in ALS patients. Amyotroph Lateral Scler 8:235–242.

Nishimura M, Mizuta I, Mizuta E, Yamasaki S, Ohta M, Kuno S (2000) Influence of interleukin-1beta gene polymorphisms on age-at-onset of sporadic Parkinson's disease. Neurosci Lett 284:73–76.

Noureddine MA, Qin XJ, Oliveira SA, Skelly TJ, van der Walt J, Hauser MA, Pericak-Vance MA, Vance JM, Li YJ (2005) Association between the neuron-specific RNA-binding protein ELAVL4 and Parkinson disease. Hum Genet 117:27–33.

Orr CF, Rowe DB, Halliday GM (2002) An inflammatory review of Parkinson's disease. Prog Neurobiol 68:325–340.

Orr CF, Rowe DB, Mizuno Y, Mori H, Halliday GM (2005) A possible role for humoral immunity in the pathogenesis of Parkinson's disease. Brain 128:2665–2674.

Panzara MA, Gussoni E, Begovich AB, Murray RS, Zang YQ, Appel SH, Steinman L, Zhang J (1999) T cell receptor BV gene rearrangements in the spinal cords and cerebrospinal fluid of patients with amyotrophic lateral sclerosis. Neurobiol Dis 6:392–405.

Patel NK, Bunnage M, Plaha P, Svendsen CN, Heywood P, Gill SS (2005) Intraputamenal infusion of glial cell line-derived neurotrophic factor in PD: A two-year outcome study. Ann Neurol 57:298–302.

Pearce RK, Owen A, Daniel S, Jenner P, Marsden CD (1997) Alterations in the distribution of glutathione in the substantia nigra in Parkinson's disease. J Neural Transm 104:661–677.

Pedersen WA, Fu W, Keller JN, Markesbery WR, Appel S, Smith RG, Kasarskis E, Mattson MP (1998) Protein modification by the lipid peroxidation product 4-hydroxynonenal in the spinal cords of amyotrophic lateral sclerosis patients. Ann Neurol 44:819–824.

Peng J, Xie L, Stevenson FF, Melov S, Di Monte DA, Andersen JK (2006) Nigrostriatal dopaminergic neurodegeneration in the weaver mouse is mediated via neuroinflammation and alleviated by minocycline administration. J Neurosci 26:11644–11651.

Petrozzi L, Lucetti C, Scarpato R, Gambaccini G, Trippi F, Bernardini S, Del Dotto P, Migliore L, Bonuccelli U (2002) Cytogenetic alterations in lymphocytes of Alzheimer's disease and Parkinson's disease patients. Neurol Sci 23 Suppl 2:S97–S98.

Polymeropoulos MH, Lavedan C, Leroy E, Ide SE, Dehejia A, Dutra A, Pike B, Root H, Rubenstein J, Boyer R, Stenroos ES, Chandrasekharappa S, Athanassiadou A, Papapetropoulos T, Johnson WG, Lazzarini AM, Duvoisin RC, Di Iorio G, Golbe LI, Nussbaum RL (1997) Mutation in the alpha-synuclein gene identified in families with Parkinson's disease. Science 276:2045–2047.

Popovich PG, Jones TB (2003) Manipulating neuroinflammatory reactions in the injured spinal cord: back to basics. Trends Pharmacol Sci 24:13–17.

Provinciali L, Laurenzi MA, Vesprini L, Giovagnoli AR, Bartocci C, Montroni M, Bagnarelli P, Clementi M, Varaldo PE (1988) Immunity assessment in the early stages of amyotrophic lateral sclerosis: A study of virus antibodies and lymphocyte subsets. Acta Neurol Scand 78:449–454.

Przedborski S, Jackson-Lewis V, Yokoyama R, Shibata T, Dawson VL, Dawson TM (1996) Role of neuronal nitric oxide in 1-methyl-4-phenyl-1,2,3,6-tetrahydropyridine (MPTP)-induced dopaminergic neurotoxicity. Proc Natl Acad Sci U S A 93:4565–4571.

Pullen AH, Demestre M, Howard RS, Orrell RW (2004) Passive transfer of purified IgG from patients with amyotrophic lateral sclerosis to mice results in degeneration of motor neurons accompanied by Ca2+ enhancement. Acta Neuropathol (Berl) 107:35–46.

Puttaparthi K, Wojcik C, Rajendran B, DeMartino GN, Elliott JL (2003) Aggregate formation in the spinal cord of mutant SOD1 transgenic mice is reversible and mediated by proteasomes. J Neurochem 87:851–860.

Reksidler AB, Lima MM, Zanata SM, Machado HB, da Cunha C, Andreatini R, Tufik S, Vital MA (2007) The COX-2 inhibitor parecoxib produces neuroprotective effects in MPTP-lesioned rats. Eur J Pharmacol 560:163–175.

Reubi JC, Horisberger U, Kappeler A, Laissue JA (1998) Localization of receptors for vasoactive intestinal peptide, somatostatin, and substance P in distinct compartments of human lymphoid organs. Blood 92:191–197.

Reubi JC, Laderach U, Waser B, Gebbers JO, Robberecht P, Laissue JA (2000) Vasoactive intestinal peptide/pituitary adenylate cyclase-activating peptide receptor subtypes in human tumors and their tissues of origin. Cancer Res 60:3105–3112.

Reynolds RD, Banerjee R, Liu J, Gendelman HE, Mosley RL (2007) Neuroprotective activities of CD4+CD25+ regulatory T cells in an animal model of Parkinson's disease. J Leukoc Biol 82(5):1083–1094.

Said SI (1976) Evidence for secretion of vasoactive intestinal peptide by tumours of pancreas, adrenal medulla, thyroid and lung: Support for the unifying APUD concept. Clin Endocrinol (Oxf) 5 Suppl:201S–204S.

Shaw PJ, Williams R (2000) Serum and cerebrospinal fluid biochemical markers of ALS. Amyotroph Lateral Scler Other Motor Neuron Disord 1 Suppl 2:S61–S67.

Shi N, Kawano Y, Tateishi T, Kikuchi H, Osoegawa M, Ohyagi Y, Kira JI (2006) Increased IL-13-producing T cells in ALS: Positive correlations with disease severity and progression rate. J Neuroimmunol 182:232–235.

Shibata N, Hirano A, Kobayashi M, Siddique T, Deng HX, Hung WY, Kato T, Asayama K (1996) Intense superoxide dismutase-1 immunoreactivity in intracytoplasmic hyaline inclusions of familial amyotrophic lateral sclerosis with posterior column involvement. J Neuropathol Exp Neurol 55:481–490.

Shibata N, Nagai R, Miyata S, Jono T, Horiuchi S, Hirano A, Kato S, Sasaki S, Asayama K, Kobayashi M (2000) Nonoxidative protein glycation is implicated in familial amyotrophic lateral sclerosis with superoxide dismutase-1 mutation. Acta Neuropathol (Berl) 100:275–284.

Sian J, Dexter DT, Lees AJ, Daniel S, Jenner P, Marsden CD (1994a) Glutathione-related enzymes in brain in Parkinson's disease. Ann Neurol 36:356–361.

Sian J, Dexter DT, Lees AJ, Daniel S, Agid Y, Javoy-Agid F, Jenner P, Marsden CD (1994b) Alterations in glutathione levels in Parkinson's disease and other neurodegenerative disorders affecting basal ganglia. Ann Neurol 36:348–355.

Simpson EP, Henry YK, Henkel JS, Smith RG, Appel SH (2004) Increased lipid peroxidation in sera of ALS patients: A potential biomarker of disease burden. Neurology 62:1758–1765.

Singleton AB, Farrer M, Johnson J, Singleton A, Hague S, Kachergus J, Hulihan M, Peuralinna T, Dutra A, Nussbaum R, Lincoln S, Crawley A, Hanson M, Maraganore D, Adler C, Cookson MR, Muenter M, Baptista M, Miller D, Blancato J, Hardy J, Gwinn-Hardy K (2003) alpha-Synuclein locus triplication causes Parkinson's disease. Science 302:841.

Tanaka M, Koike R, Kondo H, Tsuji S, Nagai H (1993) Lymphocyte subsets in amyotrophic lateral sclerosis with motor conduction block. Muscle Nerve 16:116–117.

Tohgi H, Abe T, Yamazaki K, Murata T, Ishizaki E, Isobe C (1999) Increase in oxidized NO products and reduction in oxidized glutathione in cerebrospinal fluid from patients with sporadic form of amyotrophic lateral sclerosis. Neurosci Lett 260:204–206.

Toyokuni S, Uchida K, Okamoto K, Hattori-Nakakuki Y, Hiai H, Stadtman ER (1994) Formation of 4-hydroxy-2-nonenal-modified proteins in the renal proximal tubules of rats treated with a renal carcinogen, ferric nitrilotriacetate. Proc Natl Acad Sci U S A 91:2616–2620.

Troost D, Van den Oord JJ, Vianney de Jong JM (1990) Immunohistochemical characterization of the inflammatory infiltrate in amyotrophic lateral sclerosis. Neuropathol Appl Neurobiol 16:401–410.

Troost D, Das PK, van den Oord JJ, Louwerse ES (1992) Immunohistological alterations in muscle of patients with amyotrophic lateral sclerosis: Mononuclear cell phenotypes and expression of MHC products. Clin Neuropathol 11:115–120.

Uchitel OD, Scornik F, Protti DA, Fumberg CG, Alvarez V, Appel SH (1992) Long-term neuromuscular dysfunction produced by passive transfer of amyotrophic lateral sclerosis immunoglobulins. Neurology 42:2175–2180.

Urushitani M, Ezzi SA, Julien JP (2007) Therapeutic effects of immunization with mutant superoxide dismutase in mice models of amyotrophic lateral sclerosis. Proc Natl Acad Sci U S A 104:2495–2500.

Vassiliou E, Jiang X, Delgado M, Ganea D (2001) TH2 lymphocytes secrete functional VIP upon antigen stimulation. Arch Physiol Biochem 109:365–368.

Vasu C, Dogan RN, Holterman MJ, Prabhakar BS (2003) Selective induction of dendritic cells using granulocyte macrophage-colony stimulating factor, but not fms-like tyrosine kinase receptor 3-ligand, activates thyroglobulin-specific CD4+/CD25+ T cells and suppresses experimental autoimmune thyroiditis. J Immunol 170:5511–5522.

Voice JK, Grinninger C, Kong Y, Bangale Y, Paul S, Goetzl EJ (2003) Roles of vasoactive intestinal peptide (VIP) in the expression of different immune phenotypes by wild-type mice and T cell-targeted type II VIP receptor transgenic mice. J Immunol 170:308–314.

Wang J, Wei Q, Wang CY, Hill WD, Hess DC, Dong Z (2004) Minocycline up-regulates Bcl-2 and protects against cell death in mitochondria. J Biol Chem 279:19948–19954.

Wang XS, Simmons Z, Liu W, Boyer PJ, Connor JR (2006) Differential expression of genes in amyotrophic lateral sclerosis revealed by profiling the post mortem cortex. Amyotroph Lateral Scler 7:201–210.

Watanabe M, Dykes-Hoberg M, Culotta VC, Price DL, Wong PC, Rothstein JD (2001) Histological evidence of protein aggregation in mutant SOD1 transgenic mice and in amyotrophic lateral sclerosis neural tissues. Neurobiol Dis 8:933–941.

Waters CH, Miller CA (1994) Autosomal dominant Lewy body parkinsonism in a four-generation family. Ann Neurol 35:59–64.

Webb JL, Ravikumar B, Atkins J, Skepper JN, Rubinsztein DC (2003) Alpha-Synuclein is degraded by both autophagy and the proteasome. J Biol Chem 278:25009–25013.

Wu DC, Teismann P, Tieu K, Vila M, Jackson-Lewis V, Ischiropoulos H, Przedborski S (2003) NADPH oxidase mediates oxidative stress in the 1-methyl-4-phenyl-1,2,3,6-tetrahydropyridine model of Parkinson's disease. Proc Natl Acad Sci U S A 100:6145–6150.

Yao D, Gu Z, Nakamura T, Shi ZQ, Ma Y, Gaston B, Palmer LA, Rockenstein EM, Zhang Z, Masliah E, Uehara T, Lipton SA (2004) Nitrosative stress linked to sporadic Parkinson's disease: S-nitrosylation of parkin regulates its E3 ubiquitin ligase activity. Proc Natl Acad Sci U S A 101:10810–10814.

Yoritaka A, Hattori N, Uchida K, Tanaka M, Stadtman ER, Mizuno Y (1996) Immunohistochemical detection of 4-hydroxynonenal protein adducts in Parkinson disease. Proc Natl Acad Sci U S A 93:2696–2701.

Zhang J, Perry G, Smith MA, Robertson D, Olson SJ, Graham DG, Montine TJ (1999) Parkinson's disease is associated with oxidative damage to cytoplasmic DNA and RNA in substantia nigra neurons. Am J Pathol 154:1423–1429.

Zhang W, Shin EJ, Wang T, Lee PH, Pang H, Wie MB, Kim WK, Kim SJ, Huang WH, Wang Y, Zhang W, Hong JS, Kim HC (2006) 3-Hydroxymorphinan, a metabolite of dextromethorphan, protects nigrostriatal pathway against MPTP-elicited damage both in vivo and in vitro. FASEB J 20:2496–2511.

Zhu S, Stavrovskaya IG, Drozda M, Kim BY, Ona V, Li M, Sarang S, Liu AS, Hartley DM, Wu DC, Gullans S, Ferrante RJ, Przedborski S, Kristal BS, Friedlander RM (2002) Minocycline inhibits cytochrome c release and delays progression of amyotrophic lateral sclerosis in mice. Nature 417:74–78.

45
Protective and Regenerative Autoimmunity in CNS Injury

Jonathan Kipnis and Michal Schwartz

Keywords Adaptive immunity; Autoimmune; CD4CD25; Lymphocytes; Microglia; Neurotrophins; T cells; Spinal cord injury; Vaccination; Wallerian degeneration

45.1. Introduction

Neurodegenerative diseases are generally considered to be non-inflammatory, unlike autoimmune diseases such as multiple sclerosis, which are neurodegenerative diseases that are inflammatory in nature (Trapp et al., 1999a, b; Hohlfeld and Wiendl, 2001; Groom et al., 2003). Nevertheless, most neurodegenerative diseases are accompanied by a local inflammatory response, widely assumed to be unfavorable for CNS recovery (Kurosinski and Gotz, 2002; Jellinger, 2003; Popovich and Jones, 2003). Moreover, the progressive degeneration seen in such diseases is often mediated by compounds and processes that are secondary to the primary risk, e.g., misfolding and aggregation of self-proteins (Shastry, 2003). These primary and secondary risk factors represent a continuous threat to any viable neurons embedded in a chronically diseased tissue; they induce abnormalities in cells in their vicinity, thereby contributing to the chaos rather than helping to resolve it.

Another feature common to most of the chronic neurodegenerative conditions is age dependence. The incidence of Alzheimer's and Parkinson's disease, glaucoma, and many others increases significantly with age (Ossowska, 1993; Mattson, 2003). As will be discussed below the age factor might not be related only to the aging of the brain but also to age-related changes in the immune system. Often the removal of the primary risk factor does not stop disease progression, and the neurodegeneration continues (Weinreb and Levin, 1999; Schwartz and Cohen, 2000). In Parkinson's disease, for example, despite dopamine (L-dopa) treatment as a replacement therapy the dopaminergic neurons continue to die (Hirsch, 1999; Montastruc et al., 1999). In glaucoma, reduction of intraocular pressure often does not stop disease progression (Schwartz et al., 1996). This progressive degeneration, which continues despite removal of the presumed primary risk factor(s), has been linked to what has been recognized as secondary degeneration. Emerging risk factors have been attributed to this phenomenon, including inflammation-associated factors. This has been studied intensively in recent years with a major focus on the balance between the benefit and the risk of uncontrolled immune activity, which exceeds the ability of the CNS to tolerate it.

45.1.1. Inflammation—A Local Response in Acute CNS Insults: Is It Always Bad?

The function of inflammation after any acute or chronic insult to the CNS has long been a matter of debate. Concepts such as the immune-privileged status of the CNS, as well as observations such as the presence of immune cells in the diseased CNS, gave rise to the prevailing belief that immune activity in the CNS is detrimental (Lotan et al., 1994). Many researchers consider inflammation to be an important mediator of secondary damage (Dusart and Schwab, 1994; Carlson et al., 1998; Fitch et al., 1999; Popovich et al., 1999; Mautes et al., 2000). Fitch et al. (1999) demonstrated that inflammatory processes alone can initiate a cascade of secondary tissue damage, progressive cavitation, and glial scarring in the CNS, and they suggested that specific molecules which promote inflammation might play a role in initiating secondary neuropathology (Fitch et al., 1999). This is apparently in line with the finding that removal of macrophages after SCI might improve the functional outcome (Popovich et al., 1999), and with the reported observations that the anti-inflammatory compound methylprednisolone promotes recovery in spinally injured rats (Constantini and Young, 1994). Other studies, however, indicate that local immune activity may have a beneficial effect on the traumatized spinal cord through clearance of cell debris and secretion of neurotrophic factors and cytokines. Macrophages and microglia promote axonal regeneration by clearing the site of injury (David et al., 1990; Perry et al., 1992; Madsen et al., 1998; Rapalino et al., 1998; Fitch et al., 1999; Fischer et al., 2004), and T cells mediate processes of maintenance

T. Ikezu and H.E. Gendelman (eds.), *Neuroimmune Pharmacology*.
© Springer 2008

and repair and promote functional recovery from CNS trauma (Moalem et al., 1999b; Hauben et al., 2000a, b; Schwartz et al., 2001; Ling et al., 2006). Guth and his colleagues suggested a therapeutic combination of anti-inflammatory drugs such as allapenol (to inhibit injury-induced xanthine oxidase), indomethacin (to inhibit constitutive and inducible cyclooxygenase), and bacterial lipopolysaccharide (LPS) to stimulate macrophage activity. The short-term effect of the suggested therapy, however, while including both anti- and pro- inflammatory treatment, appears to be different from its long-run effect. Along the same lines, it was suggested that animals with limited ability to undergo Wallerian degeneration might suffer from limited wound healing and regeneration (Zhang and Guth, 1997). These and other studies led researchers to acknowledge that some aspects of inflammation have negative effects on recovery, whereas others are beneficial and even essential. (Bethea et al., 1999; Mason et al., 2001)As more and more pieces are added to the puzzle of post-traumatic CNS inflammation it becomes increasingly evident that to describe the effect of inflammation on the injured nerve as "good" or "bad" is an oversimplification, as it reflects the common view of inflammation as a single (and deleterious) process rather than as a series of local immune response that is primarily being recruited to cope with threat. Its ultimate benefit, lack of benefit, or even destruction is a reflection of regulation and timing and the ability of the tissue to cope with the harmful effects of the factors produced by the immune response. Conflicting interpretations of inflammation might thus reflect, the experimental injury model employed, the severity of the injury, time elapsed following injury, the markers used to identify locally activated immune cells, the species, and the strain. In addition, the choice immune-based manipulation used to demonstrate adverse effect is also critically affecting the outcome. Immune response being the physiological mechanism by which the body copes with damage, is essential for recovery, but is more constructive when suitably regulated.

45.1.2. Role of Innate Immunity in CNS Repair

Healing of tissue in response to injury involves the synchronized operation of numerous factors and processes, some of them operating in concert and others in sequence. It is essential that the set of processes occurring in the injured axons be synchronized with the set occurring in the cells surrounding the injured axons, such that the axonal environment acquires growth-supportive properties and the axons acquire growth activity. It seems plausible that the two sets of processes operate, but that the proper synchronization is lacking. Acquisition of growth supportive properties by the cellular milieu of the injured axons basically means achievement of a balanced environment for regrowth and cell renewal (Butovsky et al., 2001, 2005a, b, 2006; Ziv et al., 2006b). This synchrony might be achieved by an appropriate postinjury immune response that is compatible with the CNS ability to tolerate.

45.1.3. Lessons from Peripheral Nervous System—Wallerian Degeneration: Is It Needed for Repair?

Central nervous system (CNS) response to injury has long been viewed as though the CNS is a unique tissue whose behavior after injury is governed by different rules than those underlying the response of other types of injured tissues, including the peripheral nervous system (PNS). Accordingly, although failure of the CNS to regenerate has been intensively studied over the years, the usual approach has been to regard the CNS as atypical and therefore to study it as an entity distinct from any other tissue. On the assumption that CNS healing does not necessarily differ in principle from the healing of any other tissue, attempts to uncover the reasons for regeneration failure have begun to focus on comparisons, both phylogenetic and intraspecies, between regenerative and nonregenerative nervous systems. The loss of function following CNS injury has been attributed not only to the failure of regeneration but also to the secondary damage which is a mechanism whereby the spread of damage from directly injured neurons to neurons that escaped the primary lesion (Robertson et al., 2000; Schwartz and Yoles, 2000; Taoka et al., 2000; Schwartz, 2001a; Vajda, 2002; Wu, 2005). In the past two decades it has become clear that failure of CNS regeneration might be partly due to the inability of the cellular elements surrounding the injured axons to create a balanced environment capable of permitting and supporting regrowth (Caroni et al., 1988; Schwab and Bartholdi, 1996; Rapalino et al., 1998; Schwab, 2002). It was shown initially that transected CNS axons, which fail to regenerate in their own degenerative environment, were shown to be capable of growing into transplanted peripheral nerve bridges (Aguayo et al., 1984; So and Aguayo, 1985; Aguayo et al., 1987; Vidal-Sanz et al., 1987). Among the elements that were shown to be hostile to regrowth in adult CNS nerves, and are absent during development, are myelin-associated inhibitors (Tang et al., 2001; Domeniconi et al., 2002; Kim et al., 2004; Schwab et al., 2005). These inhibitors, over the years, have been fully characterized and are known as the NoGo family (Tatagiba et al., 1997; Merkler et al., 2003). NoGo has three major spliced isoforms (termed Nogo-A, -B and -C) that share similar domain structures. Proteins are highly expressed in oligodendrocytes, the longest of these, Nogo-A, has a large N-terminus followed by two putative membrane-spanning domains and a short C-terminal segment. The 66-amino acid segment (known as the Nogo-66 domain) between the two transmembrane domains is extracellular. A neuronal Nogo-66 receptor (NgR) that interacts in trans with the oligodendrocyte Nogo-66 domain has also been identified. Exogeneous NgR expression in neurons that are otherwise not susceptible to Nogo-mediated growth inhibition confers inhibitory susceptibility, indicating that NgR could functionally transduce at least part of the inhibitory signal presented by Nogo.It was suggested that for regeneration to occur, these inhibitors must be either masked or eliminated. In the nervous systems of different species, the levels of such inhibitors and/or the extent of their postinjury removal might correlate with regenerative capacity. For example, in mammalian CNS there is a high level of such inhibitors, whereas in lower vertebrates their level is lower and their removal seems to be more efficient (Lang et al., 1995; Hirsch and Bahr, 1999). Subsequent studies have identified additional myelin-associated proteins as inhibitors of regrowth, such as MAG (Tang et al., 2001; Domeniconi et al., 2002). In addition, it was suggested that the rate of myelin clearance from the CNS is significantly lower than from the PNS (Hofteig et al., 1981; Pellegrino et al., 1986; Stoll et al., 1989). Another set of studies have suggested that astrocytes, thought to be involved in scar formation, are needed for growth support, and their failure to support axonal growth after injury to the mammalian CNS might be related to their cellular properties and the nature of their extracellular milieu, such as production of chondroitin sulphate proteoglycan (Jones et al., 2002; Moon et al., 2002; Jones et al., 2003; Sandvig et al., 2004, Hofteig et al., 1981; Pellegrino et al., 1986; Stoll et al., 1989). Yet, recent studies from our laboratory have suggested that although the overall production of the CSPG might be inhibitory of regrowth, complete inhibition of its production worsened the overall recovery, its early production is needed but should be limited (Rolls et al. unpublished observation). Thus, it appeared that timing and intensity of the glial scar and the local immune response are critically determining the recovery, as is discussed below. These and other related findings suggest that the postinjury behavior of cells surrounding the injured axons might determine the regenerative capacity of the axons. (Dolenc, 1984; Ogawa et al., 1985; Dolenc, 1986; Angaut-Petit and Faille, 1987; Kalichman and Myers, 1987; Gu and Ma, 1991; Berkenbosch, 1992; Sivron et al., 1994; Bregman et al., 1995; Tidball, 1995; diZerega, 1997; Hirsch and Bahr, 1999; Best and Hunter, 2000; Fry, 2001; Kalla et al., 2001; Lakatos and Franklin, 2002; Cui et al., 2004; Fenrich and Gordon, 2004; Ferraro et al., 2004).

45.1.4. Macrophages/Microglia in CNS Repair

Wound healing is a complex, multistep process involving reciprocal interactions between immune cells from the circulation and resident cells of the tissue, with the participation of extracellular matrix proteins and an array of bioactive molecules with multiple actions. In most parts of the body, tissue injury triggers immediate infiltration of circulating immune cells into the damaged area. Twenty four hours after injury, infiltrating monocytes represent the majority of leukocytes at the site of injury. Migration and adhesion of monocytes are controlled by secreted factors, whose autocrine or paracrine activity promotes recruitment of those immune cells (Lazarov-Spiegler et al., 1999; Kalla et al., 2001). On reaching the tissue, the monocytes are locally activated and became 'alternatively' activated macrophages. The macrophages play a central role in wound healing by clearing debris from the injury site (Stoll et al., 1989; Schwab et al., 2001). Macrophages secrete cytokines, growth factors and enzymes into the wound site, and participate in a profusion of autocrine and paracrine reactions with invading immune cells and resident tissue cells. When macrophages are eliminated by local injection of anti-leukocyte serum, or when monocyte production is prevented by injection of glucocorticoids, wound healing proceeds very slowly (Shirafuji et al., 2001, 2003).

In regenerating neural tissues, macrophages appear to be involved in both degeneration and regeneration. In the PNS, there is immediate Wallerian (i.e. anterograde) degeneration of the distal stump, involving the breakdown of axons and the fragmentation of Schwann cell cytoplasm. The invading monocytes play a major role in this process by clearing myelin debris and degenerating fibers, and by facilitating Schwann cell proliferation. In addition, they provide substances that participate in the healing process; for example, macrophage-derived apolipoproteins are expressed at the injury site and probably participate in membrane rebuilding. Nerve growth factor (NGF) synthesis seems to be regulated by the stimulated macrophages. Interleukin 1 and tumor necrosis factor a, secreted by macrophages, probably induce NGF transcription in Schwann cells (Stoll et al., 1989; Frisen et al., 1994; Schwab et al., 2000; Rotshenker, 2003; Stoll et al., 2004). Comparative in vivo and in vitro studies of the mammalian CNS and PNS have pointed to the part played by macrophages in the axonal response to injury, as well as to the link between macrophage activity and the success or failure of regeneration (Griffin et al., 1992; Giulian et al., 1995; Leskovar et al., 2000). In the CNS, in contrast to the PNS, macrophage infiltration following axonal injury is not only delayed but also restricted to the lesion site, rather than being dispersed along the part of the nerve through which the newly growing axons should elongate (Lazarov-Spiegler et al., 1998; Stichel et al., 1999; Wang and Feuerstein, 2000; Dihne et al., 2001; Schwartz, 2001a; Sekiya et al., 2001; Franzen et al., 2004). Furthermore, unlike in regenerative tissues, where invading blood-borne monocytes are the prominent inflammatory cells, in the CNS the resident microglia are considered to be the major mononuclear phagocytic participants at the site of injury. Following injury, activated microglia and infiltrating blood-borne macrophages are immunohistochemically indistinguishable. As the quiescent resident microglia in the intact CNS are thought to be in a down regulated form, it is possible that, following injury, the extent and the nature of their activation, in terms of acquiring healing-supportive activities, is limited compared with that of other tissue-resident macrophages. In vitro studies have shown, for example, that cytokines and growth factors associated with inflammation and wound healing have a significant effect on scar formation and/or dissolution. These factors appear to affect protease production, production of cross-linking enzymes (such as transglutaminase) and production of extracellular matrix proteins, all known significantly to affect the ability of astrocytes to support growth. Other studies have suggested that macrophage-derived factors have a cytotoxic effect on oligodendrocytes (Griot et al., 1989; Griot-Wenk et al., 1991; Zajicek et al., 1992). The above observations, together with other studies point to a link between failure of CNS regeneration and the limitations in rate, activity and distribution of the immune cells in response to CNS injury.

45.1.4.1. The Rationale for Macrophage Therapy and Its Preclinical Characteristics

Experimental results over the last decade suggest that macrophages and brain microglia are multitalented cells that are capable of expressing different functional programs in response to distinct micro-environmental signals. Microbial products and cytokines profoundly affect the differentiation of monocytes towards two phenotypic extremes. Microbial products are associated with the "classical" activation of monocytes/microglia. The "classically" activated macrophages are potent effector cells that kill microorganisms and tumor cells. This "killer instinct" have been long viewed as the only the main function of microglia. In contrast, "alternatively" activated macrophages tune inflammatory response and adaptive immunity, scavenge debris, and promote angiogenesis, tissue remodeling and repair (Summers et al., 1995; Klusman and Schwab, 1997; Mantovani et al., 2002; Bomstein et al., 2003; Gordon et al., 2003; Hauben et al., 2003; Mosser, 2003). Similarly, microglia activated by adaptive immunity are microglia that can present antigens, produce growth factors, buffer glutamate and support cell renewal (Shaked et al., 2004; Butovsky et al., 2005a, b; Shaked et al., 2005; Butovsky et al., 2006).

Initial experiments in animals with complete spinal cord transection demonstrated that local application of macrophages that have been co-incubated with sciatic nerve promoted motor recovery (Lazarov-Spiegler et al., 1996; Rapalino et al., 1998). Subsequently, the experiments were repeated in a model of severe spinal cord contusion. In those experiments blood-borne monocytes were activated by co-incubation with autogeneous skin (Bomstein et al., 2003). The results revealed that this was equally effective for recovery from spinal cord injury. In these and subsequent experiments, the macrophages were characterized phenotypically, and parameters such as site of injection, dosing and therapeutic window were studied. Those macrophages were found to have a dendritic-like phenotype (Bomstein et al., 2003) and expressed features that are reminiscent of 'alternative activation'; production of low levels of proinflammatory cytokines, and production of growth factors and metaloproteases. In subsequent studies aiming at finding the optimal time for intervention with local implantation of macrophages it became clear that as in any other tissue, repair and restoration are not only dependent on location and context, but also timing. Table 45.1 describes time windows representing a different physiological stages following SCI.

Taken together, the results summarized above as well as additional studies, it is suggestive that timely local innate immune cells with "alternative activity," reminiscence of dendritic-like, are required for CNS ability to cope with injurious conditions, and that they are needed at the sub-acute phase following injuries.

45.2. Adaptive Immunity Is Needed to Control Local Innate Response in the CNS

45.2.1. Is Self and Non-Self Discrimination Needed?

"Survival of the fittest" summarizes the essence of Darwinian evolutionary theory. In line with this theory and the pioneering theory of Metchnikoff in the 1890s (Dubos, 1955; Vaughan, 1965), followed by the "clonal expansion" theory of Burnet in the 1950s (Miller, 1994; Silverstein and Rose, 1997; Martini and Burgio, 1999), it was believed that discrimination of self from non-self, thymic education of T cells, and deletion of autoimmune T cells in the thymus are the central features of immunology. Self-tolerance, defined as a state of non-responsiveness to self, was therefore viewed as the optimal condition, and was assumed to enable the fittest to survive (Viret et al., 1999). Studies carried out in rodent models of central nervous system (CNS) insults have suggested that autoimmunity is the body's defense mechanism against any threat to CNS tissue (Schwartz and Cohen, 2000; Schwartz et al., 1999) and that only when the autoimmune response is poorly controlled will an autoimmune disease result. According to these observations and others, it emerged that defining tolerance to self in terms of non-responsiveness is incompatible with the theory of the survival of the fittest. A more appropriate definition of tolerance to self would

TABLE 45.1. Physiological stages following SCI.

Days post spinal cord injury	Description	Reference
3–4 days	A period reflecting the decline of primary infiltration of neutrophils participating in inflammation, and high incidence of apoptotic cells.	(Popovich et al., 1997; Leskovar et al., 2000)
7–10 days	A period of maximum accumulation of activated microglia/macrophages, T cells, and progenitor glial cells.	(Leskovar et al., 2000; McTigue et al., 2001)
14 days	The numbers of ED1 positive cells and T cells are still very high. At the same time different cytokines and chemokines in the injured tissue decrease or disappear.	(Lee et al., 2000; Leskovar et al., 2000; McTigue et al., 2001)
21 days	Many of the injury-induced biochemical and cellular activities in the spinal cord have peaked and begun to return to normal levels.	(Leskovar et al., 2000; McTigue et al., 2001)

be the ability to tolerate an anti-self response without developing an autoimmune disease (Schwartz and Kipnis, 2002). Just as the immune system fights off external pathogens, the autoimmune system fights off threats originating within the body itself (such as cancer, neurodegenerative conditions, tissue injuries), and also serves as a complementary defense mechanism against damage caused by external pathogens. Naturally occurring regulatory T cells ($CD4^+CD25^+$) serve as a physiological safety valve that can be modulated to maintain the fine balance between need and risk (Kipnis et al., 2002; Schwartz and Kipnis, 2002; Kipnis et al., 2004a).

In the early 1990s it became evident that there is little difference, between the T-cell repertoires of healthy individuals and of patients suffering from autoimmune diseases (Lohse et al., 1996). At around the same time it was suggested that the sole function of a group of suppressor T cells newly identified as $CD4^+CD25^+$ was to inhibit the anti-self aggression of any autoimmune T cells that (presumably owing to an evolutionary mistake) had left the thymus and taken up residence in the periphery (Shevach, 2000; Shevach et al., 2001). It is contrary to Darwinian theory, however, to propose that two cell populations exist in the same organism for the sole purpose of inhibiting each other's activity. Survival of the fittest implies that unwanted features, especially if harmful, will disappear, while beneficial features will be transferred to future generations (Paul, 1988; Herman, 1997; Elliott, 2003). Thus, since all humans possess a similar repertoire of autoimmune cells (Finn et al., 1996; Ria et al., 2001), Darwinian theory would presuppose that these cells have a physiological function.

Theoretically, complete elimination of autoimmune T cells would be the best way to prevent autoimmune disease development, whereas uninhibited autoimmunity would be the best way to counteract neurodegenerative disorders and cancer. A Darwinian resolution of these opposing immunological scenarios might have led, as a compromise between risk and benefit, to the concomitant presence of autoimmune T cells and the regulatory T cells that normally suppress them (Schwartz and Kipnis, 2002). Based on the accumulated information describing the role of autoimmunity in the devastating conditions of cancer (Shimizu et al., 1999; Sakaguchi et al., 2001) and neurodegeneration (Kipnis et al., 2002), it seems unlikely that even the most ardent disciple of Burnet would suggest that complete deletion of autoimmunity favors survival of the fittest. Hence a new theory based on solid data suggests that $CD4^+CD25^+$ regulatory T cells do not exist in a permanently suppressive state that keeps autoimmune T cells unresponsive to self-antigens, but are amenable to modulation by physiological signals that weaken or strengthen their suppressive activity according to need.

45.2.2. Autoimmune T Cells Protect Neurons from Degeneration

Autoimmunity has long been viewed as a destructive process. However, a strong body of evidence provides a new view whereby autoimmunity is the body's endogenous response to CNS injury, and that its purpose is beneficial. This notion was based on the observation that in rodents, passive transfer of encephalitogenic (disease-inducing) T cells reactive to myelin basic protein (MBP) reduces postinjury neuronal losses.

During the last two decades it has become increasingly clear that different degenerative diseases of the CNS share a number of primary and secondary features (Evert et al., 2000; Rehman, 2000; Hur et al., 2002; Carri et al., 2003). In many such diseases the local microglial response is often viewed as an unwelcome contributor to the disease pathology (Aschner et al., 2002; Koutsilieri et al., 2002; Liu et al., 2002). Recent data suggest, however, that such a view is an oversimplification, and that a well-controlled glial response is beneficial in protecting the affected tissues (Banati et al., 1994; Aschner et al., 2002), whereas malfunctioning glia contribute to the ongoing neurodegenerative process (Teismann et al., 2003). This spread of neuronal damage is caused, at least in part, by compounds which, though normally essential for the survival and function of neurons, become toxic when their physiological concentrations are exceeded. Among the injury-related mechanisms that might underlie the post-traumatic spread of damage are biochemical and metabolic changes in oxygen and glucose utilization, energy state, lipid-dependent enzymes, free radicals, eicosanoids, tissue ions, biogenic amines, endogenous opioids, and excitatory amino acids. These changes cause alterations in cellular homeostasis, excitotoxicity, local production of agents harmful to nerve cells, and a loss of trophic support from targets, all of which result in secondary neuronal loss.

Immune responses in the CNS are relatively restricted, resulting in the status of the CNS as an immune-privileged site (Streilein, 1995). The unique nature of the communication between the CNS and the immune system can be observed, for example, in the dialog between the CNS and T cells. Under normal conditions activated T cells can cross the blood–brain barrier and enter the CNS parenchyma. However, only T cells capable of reacting with CNS antigens seem to persist there (Hickey et al., 1991). Comparative studies of the T-cell response at sites of axotomy in the CNS and the peripheral nervous system (PNS), using T-cell immunocytochemistry, revealed a significantly greater accumulation of endogenous T cells found in injured PNS axons than in injured CNS axons (Moalem et al., 1999a). Moreover, in cases of inflammation, the CNS showed a marked potential for elimination of T cells via apoptosis, whereas such elimination was less effective in the PNS, and was almost absent in other tissues such as muscle and skin (Gold et al., 1997).

In 1999, it was demonstrated that autoimmune T cells directed against myelin basic protein can protect neurons against degeneration after CNS injury (Moalem et al., 1999b).

To verify that this finding was not merely the result of an experimental manipulation but rather a beneficial physiological response to CNS injury, neuronal degeneration after identical injuries was compared in normal mice and mice devoid of T cells (Kipnis et al., 2001; Yoles et al., 2001). Significantly more degeneration was observed in the nude mice than in the wild type, suggesting that neuroprotection is a physiological, T cell-dependent process (Yoles et al., 2001). To confirm the autoimmune character of this beneficial physiological response, animals were tolerized to myelin antigens at birth. The tolerized animals showed significantly fewer surviving neurons after injury than their matched controls that had been immunized neonatally with an irrelevant (non-myelin) protein (Kipnis et al., 2002).

It had been widely accepted that autoimmune T cells in the periphery are normally kept in a state of tolerance by the suppressive activity of naturally occurring regulatory $CD4^+CD25^+$ T cells (Treg cells) (Thornton and Shevach, 2000), and that elimination or depletion of Treg cells might therefore cause development of an autoimmune disease in susceptible animals (McHugh et al., 2002).

Verification of a new perception of autoimmunity is seen in the experiments in which mice were depleted of Treg cells assuming that such a manipulation would increase the ability to fight off neurodegenerative conditions (Kipnis et al., 2002; Kipnis et al., 2004a). Regardless of the mouse's inherent susceptibility to autoimmune disease, Treg cell depletion resulted in increased neuronal survival (Kipnis et al., 2004b). These and other observations led to propose that the ability to harness a T cell-dependent protective mechanism is controlled by Treg cells, and that the constitutive presence in healthy individuals of both autoimmune T cells and regulatory T cells represents an evolutionary solution to the need for autoimmune T cells for maintenance and repair, with Treg cells acting as a safeguard against the risk of autoimmune disease (Schwartz and Kipnis, 2002). According to these and other results, it was reasonable to assume that weakening of the Treg cell-mediated suppression would benefit both anti-cancer and neuroprotective immunity. It was recently shown that transforming growth factor (TGF)-β sustains the regulatory character of Treg (Fantini et al., 2004; Zheng et al., 2004) cells but no physiological compound has been identified that can weaken Treg. It is plausible that a physiological molecule capable of weakening the activity of Treg cells, at least in cases of CNS injury, must be a brain-derived compound whose half-life is short and whose concentration in the periphery is low.

Among the known neurotransmitters that participate in a stress response, are increased after injury, and are associated with tumors, dopamine seemed to us the most suitable candidate for interaction with Treg cells. Physiological dopamine is increased under stressful conditions (Saha et al., 2001; Vermetten and Bremner, 2002). Its peripheral concentration is low, it is highly unstable in the blood, and it has been shown to participate in neuroimmune dialog (Weihe et al., 1991; Robertson and Jian, 1995; Ilani et al., 2001; Levite et al., 2001; Lemmer et al., 2002). Almost all cells of the immune system, including T cells, bear dopamine receptors (Weihe et al., 1991). Interaction between neurotransmitters and immune cells (e.g. T cells) has been investigated in a whole population of T cells and in a subpopulation of $CD4^+$ T cells, but no distinctions were made in those studies between regulatory and effector (autoimmune) T cells.

Studies of isolated populations of CD4 T cells, revealed that dopamine exerts a direct weakening effect on the suppressive ability of $CD4^+CD25^+$ (Treg) cells (Kipnis et al., 2004b). The receptors that mediate this effect belong to the type-1 family of dopamine receptors (D1R and D5R). These receptors are expressed only weakly, if at all, on effector T cells (Teff cells), but are strongly expressed on Treg cells, allowing preferential action of dopamine on Treg. Short-pulse application of dopamine to Treg significantly decreased the suppressive activity of Treg co-cultured with Teff. The inhibitory effect of dopamine on Treg suppressive activity was manifested in decreased production of interleukin (IL)-10 and TGF-β, which participate in cytokine-mediated suppression. Dopamine was also found to down-regulate CTLA-4, which is expressed constitutively on Treg and is responsible for cell-to-cell contact-mediated suppression. Other catecholamines, such as epinephrine, norepinephrine and serotonin, had no effect on Treg, although they affected Teff and their receptors were found to be expressed by Treg (Kipnis et al., 2004b).

The dopamine-induced weakening of Treg-cell suppressive activity might be considered a first signal in the cascade of events that activates the autoimmune T cells. With the suppression lifted, activation of Teff cells further requires presentation of their specific antigen and of a co-stimulatory molecule by APCs. The availability of antigen is apparently sufficient to maintain homeostasis in the normal healthy CNS, but after an acute injury or under chronic neurodegenerative conditions it appears that antigen availability on its own does not suffice. Thus, in the absence of a mechanism capable of weakening the suppressive effect of Treg cells, the evoked response might not be sufficient to cope with the demands generated by the injury (Schwartz and Kipnis, 2002).

In the case of malignancies, where autoimmunity is required to fight off the cancer, Treg cells interfere with the body's natural propensity to activate a protective mechanism. Depletion of these cells, much as in the case of neurodegenerative conditions, increases the ability to reject tumors (Shimizu et al., 1999). In patients with certain malignancies, dopamine levels in the periphery are increased (Saha et al., 2001). It seems reasonable to suggest that this stress-related compound, possibly in combination with some other compound(s), might be the signal in cancer patients that weakens the Treg-mediated suppressive effect.

Dopamine or its specific D1 agonists have been shown to further block the Treg cell-mediated suppression of Teff cells (Kipnis et al., 2004b), a finding with intriguing implications for therapy. Injection of the D1 agonist SKF-38393, immediately after a traumatic injury, or induction of glutamate neurotoxicity in the mouse CNS, increases neuronal survival. SKF-38393 did not have any effect in mice devoid of T cells, indicating that its beneficial effect in the wild type was not exerted directly on the neurons, but rather via T cells. Since dopamine cannot cross the blood–brain barrier its effect in wild type animals subjected to injury can be assumed to be peripheral. Treatment with dopamine and its relevant agonists might also be considered for cancer rejection. It was shown that regulatory T cells suppress a spontaneous T cell response to cancer and mice depleted of Treg efficiently reject transplanted tumors.

It should be kept in mind that T cells can do only what T cells can do regardless of their antigenic specificity. Production of cytokines and growth factors from T cells is a matter of their activation and taken within the context of the affected/injured tissue. Antigenic specificity of T cells could be viewed only as a means of homing T cells to a tissue/site in need. Once T cells reach the tissue in need and are reactivated, the antigenic specificity becomes irrelevant to the effector phase (Mizrahi et al., 2002). Therefore, in order to achieve any neuroprotective or other effect in the CNS, it would follow that T cells should be directed to CNS antigens. T cells do not have a way of directly communicating with neurons, as neurons do not express MHC class II molecules. Therefore, two routes of interactions are possible; either through cytokines produced by T cells following activation or through an indirect effect. The indirect effect could be mediated via microglia or astrocytes. Both of these cell types are able to express MHC molecules, whose expression is further increased following injury, thus enabling direct interaction of T cells with microglia (Moalem et al., 1999a). The role of microglia and astrocytes is far beyond what was originally thought. Glial cells do not only serve as glue for neuronal tissue but actively participate in excitation and neural firing and generally contribute to brain plasticity. Microglial cells, resident immune cells in the brain, are able to obtain various phenotypes, depending on activators and context of the tissue. Following injury and under chronic neurodegenerative conditions microglia might serve as a primary source for reactive oxygen species, glutamate, nitric oxide and other cytotoxic compounds (Rutkowski et al., 2000; Penkowa et al., 2003; Rostasy, 2005; Stadelmann et al., 2005). On the other hand, microglia, as well as astrocytes, have the potential to buffer glutamate and produce neuronal growth factors. Studies from our laboratory showed that T cells can affect the phenotype of microglia and switch it towards not only a less destructive but also towards a protective phenotype and even a phenotype that support cell renewal. Following activation of microglia with T cells, in-vitro, glutamate-buffering capacity of microglia is significantly increased, along with production of neuronal growth factors, e.g. BDNF, and there is a reduced production of cytotoxic factors, e.g. TNF-alpha (Shaked et al., 2004; Butovsky et al., 2005a, b; Shaked et al., 2005; Butovsky et al., 2006).

Due to the high trafficking ability of T cells and their ability to penetrate tissues and "talk" with the local antigen-presenting cells, they can serve as mobile mini-factories with an ability to produce growth factors and cytokines upon need and thus maintain the homeostasis of the tissue. Under physiological conditions deviations are minimal and glial cells are able to maintain the homeostasis. Under pathological conditions, physiological activation of microglia cannot cope with the extreme deviation and thus T cells are required to facilitate glial activity to regain homeostasis. Therefore a dual effect of T cells in brain maintenance is achieved by prevention of a cytotoxic phenotype of glia and boosting of their protective phenotype (Schadlich et al., 1983; Schwartz, 2004; Schwartz and Kipnis, 2005b).

45.2.3. The Mechanism Underlying Protective Autoimmunity

There are many different subpopulations of $CD4^+$ T cells, each responsible for a certain type of immune response. Th1 cells, for example, reinforce innate immunity and activate $CD8^+$ T cells, whereas Th2 cells recruit and activate B cells. Autoimmune $CD4^+$ T cells (Teff) locally boost and control resident microglia and infiltrating blood-borne monocytes, helping them to acquire the ability to fight off degenerative conditions requiring removal of dead cells and cell debris, as well as buffering of toxic compounds without producing inflammation-associated compounds such tumor-necrosis factor (TNF)-α, NO, or cyclooxygenase (COX)-2 (Levi et al., 1998; Basu et al., 2002; Janabi, 2002; Schwartz, 2003). Thus, according to recent results, the role of $CD4^+$ T cells directed against self-antigens (helper T cells, Th) is to activate the innate response, enabling it to recognize that the threat to the tissue is not a harmful organism that it must kill, but as a toxic substance that it must neutralize or eliminate. In addition, the autoimmune T cells, upon encountering their specific antigens presented by antigen-presenting cells at the lesion site, can produce protective compounds such as growth factors and neurotrophins (Hammarberg et al., 2000; Moalem et al., 2000; Gielen et al., 2003). All of these tasks can be accomplished by a well-controlled response of helper T cells. These Th cells, in order to do their job, require local activation by their specific antigens residing in the site of stress. Thus, antigenic specificity apparently dictates the homing of T cells to the site where their local activation can occur. This is consistent with the findings that T cells having the same antigenic specificity are protective against different types of threatening stimuli occurring at the same site, or against different threatening stimuli at different sites occupied by the same immunodominant self-proteins. As a corollary, the same threatening stimulus, if manifested at different sites that do not share common dominant self-antigens, does not benefit from T cells directed against the same antigens.

Studies from several laboratories have shown that T cells patrol the healthy CNS, but do not accumulate there. The recent data suggest that, in the event of an acute injury or chronic neurodegenerative condition, T cells are recruited by and accumulate in the CNS (Hirschberg et al., 1998; Moalem et al., 1999a), where they might rescue neurons from degeneration if the damage caused by the toxic biochemical environment is not yet irreversible; moreover, the recruited T cells will prevent further deterioration. It is also possible that this autoimmune protective mechanism also operates when the threat to the tissue is from microbial infiltration. In such a case the anti-self response might occur without the individual being cognizant of the response taking place, unless the harnessed autoimmunity gets out of control, in which case its effect is no longer beneficial but destructive, and might result in an autoimmune disease (Dal Canto et al., 2000; Miller et al., 2001). This might be the situation in individuals who are predisposed to autoimmune disease development (Kipnis et al., 2001). According to this theory the pathogenic self-proteins that have been implicated in autoimmune diseases are the very proteins against which a well-controlled T cell response is protective. This might help explain why autoimmune diseases are often attributed to viral infections in the brain. It might also explain the relatively low clinical prevalence of autoimmune diseases and their occurrence mainly in young adults rather than in the elderly population, whereas neurodegenerative diseases and cancer are common and significantly more prevalent in the elderly, in whom the immune system is deteriorating (Linton and Dorshkind, 2004).

45.2.4. The Missing Link—Adaptive Immunity Controls Microglia Phenotype Needed for Survival, Regrowth, and Renewal

It has long been believed that, irrespective of the type or context of an injurious stimulus, microglia show a stereotyped reaction in that they exhibit a predetermined program of executive functions. Experimental evidence supports the notion that some features of the microglial response might indeed originate in a core program of multi-purpose behavior. It should be noted, however, that most of what we know about the diverse activities of microglia emerges from *in-vitro* studies which cannot adequately reflect the complexity of microglial responses *in vivo*. Deprived of their physiological environment and triggered via a single receptor, the *in-vitro* response of isolated microglia is likely to be one-dimensional (Becher et al., 1996; Stalder et al., 1997; Lombardi et al., 1998; Smith et al., 1998; Kloss et al., 2001; Nakajima et al., 2001; He et al., 2002; Saura et al., 2003; Liuzzi et al., 2004; Shin et al., 2004; Vairano et al., 2004; Vegeto et al., 2004). As an example, when activated by bacterial components such as lipopolysaccharide, microglia acquire an inflammatory and cytotoxic phenotype (Lee et al., 1993; Merrill et al., 1993). This emergency scenario, in which the body's fighting force is called upon to attack and kill bacteria, represents only one possible situation involving microglial activation. In contrast to the traditional notion of a stereotyped response, we favor the idea of diversity in microglial behavior. An acquired response that is defined and refined by an ensemble of incoming signals is not a new concept in cell biology. It does, however, represent a departure from the traditional view of microglia.

Danger signals can come from both foreign material (infectious agents) and endogenous sources (damaged cells or tissues, altered molecules, neurotransmitter imbalances). Endogenous toxicity (the 'enemy within' (Schwartz et al., 2003)) might result from membrane breakdown products, the extracellular presence of cytosolic compounds, abnormally processed or aggregated proteins (such as β-amyloid), or abnormal abundance of transmitters (such as glutamate). It appears that microglia fail to distinguish between external and self-derived enemies, and consequently their response to danger signals from endogenous agents resembles their response to invading microbes.

The microglial phenotype can be shaped by adaptive immunity (Butovsky et al., 2005b; Shaked et al., 2005). The chemoattractive message of inducible microglial chemokines can be altered by a single T-cell cytokine: interferon (IFN)-γ, probably serving a feedback mechanism, alters the blend of chemokines, thereby shifting the preference for leukocyte subpopulations and conceivably affecting the composition of further infiltrates. Adding another piece to the puzzle, the nature and intensity of this response can be controlled by interleukin (IL)-4, a cytokine associated with Th2 cells (Butovsky et al., 2005b).

Thus, even standard responses triggered by established stimuli in simplified *in-vitro* settings show substantial variation when another factor is added. Microglia respond differently to the same stimulus if it co-exists with an additional stimulus, suggesting that they should be viewed as cells that acquire different phenotypes rather than behave stereotypically. Two stimuli can generate different effects, depending on the sequence of exposure: "priming" (preconditioning in which the first stimulus prepares the cell for an enhanced response to the second), negative priming (desensitization), or interference (where the second stimulus exercises a veto effect over an ongoing response to the prior stimulus). The two-signal interplay becomes even more complex when the time interval between the two exposures (manifested as 'memory') varies, as discussed below. This suggests that upon arrival of a modulator the executive functions of microglia can change not only in magnitude but also in quality. Variability of microglial activity is therefore not merely a reflection of stimulus strength or persistence, but is largely determined by the nature and context of the stimulus (Butovsky et al., 2005a, 2006).

In our view, microglia function as local sentinels. Under certain circumstances restricted activation of these stand-by immune cells might occur without being detected. If these sentinels fail to correctly read incoming stress signals, however, they will not develop the phenotype needed to fight off the threat, or alternatively, the cost of fighting off the threat is likely to outweigh the benefit (in terms of death of neighboring neurons). In addition, the outcome of a correct response to a particular signal could be detrimental if the signal itself is misleading. Self-compounds such as aggregated β-amyloid, for example, induce microglia to respond to them as if they were invading microorganisms to be killed. The phenotype utilized for that purpose is characterized by the production of cytotoxic molecules in quantities that the brain cannot tolerate. In addition, microglia that encounter, for example, aggregated β-amyloid fail to express class II major histocompatibility complex (MHC-II) molecules, and therefore lack the ability to interact locally with T cells (adaptive immunity) in the way needed for T cell-mediated expression of protective activity against local threats such as oxidative imbalance or cytotoxicity of neurotransmitters. Thus, paradoxically, under such conditions the microglia are precluded from participating in the adaptive immune responses needed to rescue the tissue from the toxicity that they themselves had helped to generate (Shaked et al., 2004, 2005).

Recent studies have shown that an inflammation-associated autotoxic phenotype not only causes neuronal loss but also interferes with neuronal survival, obstructs neurogenesis, and prevents regeneration (Butovsky et al., 2005a, b, 2006). In contrast, microglia that encounter adaptive immunity acquire a phenotype capable of presenting antigens and engaging in dialog with T cells. Such microglia, depending on the nature and amount of T cell-derived cytokines that they encounter, can become activated without producing the potentially cytotoxic cytokine tumor necrosis factor (TNF)-α or can even down-regulate its production. For example, operating via relatively small amounts of IFN-γ, the T-helper (Th)1 cells—classically viewed as pro-inflammatory—can activate microglia to buffer glutamate, a common player in neurodegenerative diseases (Shaked et al., 2005). Likewise, Th2 cells, commonly viewed as anti-inflammatory, by operating via IL-4 can activate microglia to produce insulin-like growth factor (IGF)-I (Butovsky et al., 2005b), known to be associated with cell renewal (Mattson et al., 2004; Varela-Nieto et al., 2004; Shetty et al., 2005; Sonntag et al., 2005; Butovsky et al., 2006). Microglia activated by IFN-γ or IL-4 therefore protect neurons and can support both neurogenesis and oligodendrogenesis (Butovsky et al., 2006).

We maintain that the protective versus destructive dichotomy of microglial effects does not necessarily reflect conflicting or contradictory activities. In normal healthy individuals, these cells stand ready and waiting to perform neural or immune tasks. However, because the CNS has a limited ability of the to tolerate any deviation from homeostasis, even defensive activity on the part of activated microglia can exacerbate a chaotic situation rather than resolve it.

Strategies for preventing overshooting of microglial reactions can be based on pharmacology, by employing selective suppression of undesirable activities while still permitting other executive functions to be performed. An alternative approach of recruiting T cells of a certain phenotype directed against weak agonists of self-antigens, might result in immunomodulation. Examples of such antigens are altered peptide ligands or the synthetic oligopeptide copolymer 1 (Cop-1) (Teitelbaum et al., 1997; Sela, 1999, 2000). By the use of such antigens for vaccination in a context of neurodegenerative conditions, it was possible—irrespective of the primary risk factor—to boost activity in a well-controlled way (Kipnis et al., 2000; Schori et al., 2001; Schwartz, 2001b; Schwartz and Kipnis, 2005a). Paralysis of microglia can be helpful within an experimental setting of a disease model. However, global depletion of microglia for extended periods might impair rather than preserve the structure and function of the CNS.

45.3. Development of Therapeutic Vaccinations

Any immune manipulation, which activates the immune system to induce a well-controlled increase in the likelihood that relevant T cells will home to a site of injury site, can be expected to be beneficial. Three major approaches could be considered: immunization with self-antigen agonists, induction of lymphopenia or functional inactivation of naturally occurring regulatory CD4$^+$CD25$^+$ T cells.

45.3.1. Immune-Based Vaccination for Neurodegenerative Diseases

45.3.1.1. Copolymer-1 (Glatiramer acetate, CA)

Recognizing that T cells are needed for assisting CNS in fighting off neurodegenerative conditions have prompted us to search for safe ways to do so without imposing the risk of developing autoimmune diseases. Data accumulated over the last decade has raised several options, including local transplantation of specially activated macrophages in cases of acute insult. In other situations of either chronic or acute neurodegeneration, the choice depends on the therapeutic window and the condition of the tissue (the critical issue being the bias of the microglia). In searching for active vaccination we considered using agonists of self-antigens. Such agonists can activate a response that weakly cross-reacts with the resident self-antigens. One such antigen is the copolymer glatiramer acetate, known as Cop-1, which is safely used daily for treating multiple sclerosis. Yet, for neuroprotection in cases of noninflammatory neurodegenerative diseases the outcome was critically affected by the dosing, regimen and the choice of the carrier (adjuvant). In the case of animal model of ALS, the use of GA emulsified in complete Freunds adjuvant was beneficial, yet adjuvant-free GA was not found to be effective in any of the tested regimens (Haenggeli et al. 2007) and its daily administration was found to be destructive in female ALS mice (Bukshpan et al., unpublished observations). In a model of glaucoma, weekly or monthly injections of adjuvant-free GA were found to be beneficial, but not daily injections (Bakalash et al. 2005). In an animal model of Alzheimers disease, a weekly injection was found to be beneficial (Butovsky et al. 2006b). Yet translating it into a human therapy requires careful determination of the regimen, which critically determines the T-cell phenotype (Th1, Th2, Treg); and thus, the clinical outcome.

45.3.1.2. Lymphopenia

Induction of lymphopenia significantly increases immunoreactivity towards cancer-specific proteins and efficiently suppresses cancer (Dummer et al., 2002). A sudden drop in the pool of peripheral T lymphocytes stimulates their homeostasis-driven proliferation in order to restore the pool. In response to the stimulus of lymphopenia, naïve peripheral T cells proliferate and acquire a phenotype reminiscent of memory T cells (Ma et al., 2003). The induced proliferation predisposes the individual to development of an autoimmune response, since under lymphopenic conditions T cells can proliferate upon interaction with MHC-II molecules alone, with no need for a co-stimulatory signal (Sara et al., 1999; Gudmundsdottir and Turka, 2001; Elflein et al., 2003). If at the time of lymphopenia induction the body undergoes stress and consequently certain self-antigens are exposed (e.g., antigens related to tissue injury or cancer), an autoimmune response to those antigens will occur, resulting in a high overall incidence of the proliferation of the relevant T lymphocytes (Gelinas and Martinoli, 2002). In rodents suffering from acute or chronic neurodegenerative conditions, induction of lymphopenia significantly benefits post-injury neuronal survival (Kipnis et al., 2004b). Lymphopenia and the subsequent homeostatic proliferation can be induced in a number of ways, the most clinically relevant being low-dose irradiation of the lymphoid organs. As a result of the lymphopenia, T cells proliferate and become activated. They patrol the body, and their patrol route includes the CNS. On reaching the lesion site, and after being activated by the resident cells that present self-antigens in the MHC-II groove, these lymphopenia-derived T cells perform their effector functions, similarly to T cells obtained by immunization with self- or altered self-proteins (Kipnis et al., 2004a).

45.3.1.3. Attenuation of Regulatory T Cell Network

Since the aim is to achieve activation of T cells that cross-react with self-antigens at the site of injury, this can also be done by weakening the naturally occurring Treg cells. In an experimental context, nude mice (devoid of mature T cells) repopulated with a T cell population that did not contain the Treg-cell subpopulation (Kipnis et al., 2002) showed better recovery from a CNS insult than wild-type mice of the same strain. For clinical use, however, what is needed is a reagent that will weaken Treg cells. Dopamine, as mentioned earlier, was found to weaken both the activity and reduce the trafficking of Treg cells (Kipnis et al., 2004b). It is possible that dopamine represents a family of physiological compounds capable of controlling Treg-cell activity, therefore allowing speedy recruitment of the relevant autoimmune T cells. Development of synthetic compounds that can reproduce the dopamine effect is another apparently feasible approach in which a common immune-based therapy could be used to fight off neurodegenerative diseases, irrespective of etiology. Although such compounds might weaken Treg cells nonselectively (i.e., regardless of their antigenic specificity), the subsequently evoked autoimmunity will be restricted to $CD4^+$ cells that encounter their relevant antigens, and will consequently be associated with the site under stress.

Summary

Harnessing the immune system in a well-controlled way might be the therapy of choice for neurodegenerative disorders. As long as the integrity of the immune system is maintained and neurotransmitter imbalance in the brain is within the remediable capacity of the immune system, homeostasis remains intact and the integrity of brain performance is preserved. According to this view, therefore, the widening age-related gap between deteriorating immunity and risk factors for diseases can be narrowed by appropriate activation of the immune system.

Review Questions/Problems

1. **Absence of macrophages increases the rate of Wallerian degeneration**

   YES/NO/UNRESOLVED

2. **Lack of neuroprotection in thymectomized rats points to the role of naturally occurring regulatory $CD4^+CD25^+$ T cells in endogenous neuroprotection**

   YES/NO/UNRESOLVED

3. **Tova cells are not protective following CNS injury but will be protective following PNS injury**

   YES/NO/UNRESOLVED

4. **T cell specific to NoGo protein might induce neuroprotection in injured CNS**

   YES/NO/UNRESOLVED

5. **Absence of macrophages improves neuronal survival following PNS injury**

   YES/NO/UNRESOLVED

6. **Optic-nerve activated macrophages are better phagocytes than sciatic-nerve activated macrophages**

   YES/NO/UNRESOLVED

7. **Lymphatic drainage from the CNS is unique compared to lymphatic drainage from other tissues**

   YES/NO/UNRESOLVED

8. **GAP43 is a good marker for sprouting**

   YES/NO/UNRESOLVED

9. **Neuroprotection and regeneration are similar processes driven by different cell types**

   YES/NO/UNRESOLVED

10. **Perivascular macrophages are important "partners" in T cell-CNS interaction**

    YES/NO/UNRESOLVED

11. Prompt autoimmune response to injury will induce neuroprotection only if this response is tightly controlled.

YES/NO/UNRESOLVED

12. Survival of neurons in the injured CNS is an outcome, at least in part, of protective immune response and a destructive local effect of physiological compounds that exceed their normal concentration.

YES/NO/UNRESOLVED

13. Only EAE-inducing TMBP cells confer neuroprotection following injury

YES/NO/UNRESOLVED

14. T cell specific to NoGo might induce neuroprotection in injured CNS

YES/NO/UNRESOLVED

15. Autoimmune T cell-based therapy can be easily combined with methylprednisolone

YES/NO/UNRESOLVED

16. Which one of the following statements is the most unlikely?

a. Dopamine may affect $CD4^+$ T cells
b. Dopamine may affect Treg cells
c. Treatment with Dopamine might exacerbate cancer
d. Treatment with Dopamine might exacerbate autoimmune disease

17. Wild type Balb/c mice were injected with autologous carcinoma cells. One group was treated with Dopamine and the other with PBS. Mice are euthanized when cancer reaches 2 cm in size. The plot of survival of mice as a function of time is presented bellow. Which of the following statements is correct?

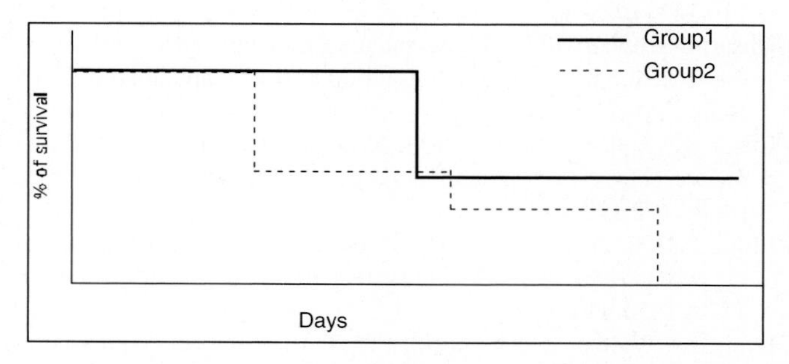

a. Group 1 was injected with PBS and group 2 with Dopamine
b. Group 1 was injected with Dopamine and group 2 with PBS
c. Additional information is required to address this question
d. None of the above

18. Multiple sclerosis patient was treated with a new cellular therapy – activated CD4+ T cells from this patient were isolated (based on the selection for CD25+), anergized in-vitro on immature dendritic cells and returned to the patient's blood stream. This patient has also colon cancer. Which one of the following statements is the most likely?

a. The treatment may exacerbate the ongoing MS disease
b. The treatment may exacerbate the ongoing colon cancer
c. The treatment will not affect any of the conditions
d. The treatment will benefit both diseases

19. Neuroimmunology primarily deals with:

a. Autoimmune inflammations in the brain
b. Autoimmune inflammations in the injured brain
c. Inflammatory pain
d. The effect of mood on activation of T lymphocytes
e. All of the above

20. Patient was treated with Copaxone for multiple sclerosis and has developed breast cancer. Doctors analyzed the blood samples and found an increase in regulatory T cell numbers based on CD25 and Foxp3 markers. The suggestion made by doctors was that Copaxone induces regulatory T cells and this may lead to cancer development. Which of the comments bellow is FALSE.

a. Doctors might be right
b. Doctors did not review the recent literature on Foxp3 in human Treg
c. Only if these cells did not produce IL-2 they could be considered as Treg cells
d. The suppressive activity of these cells in-vitro should be examined before addressing them a suppressor function
e. All of the above

References

Aguayo AJ, Bjorklund A, Stenevi U, Carlstedt T (1984) Fetal mesencephalic neurons survive and extend long axons across peripheral nervous system grafts inserted into the adult rat striatum. Neurosci Lett 45:53–58.

Aguayo AJ, Vidal-Sanz M, Villegas-Perez MP, Bray GM (1987) Growth and connectivity of axotomized retinal neurons in adult rats with optic nerves substituted by PNS grafts linking the eye and the midbrain. Ann N Y Acad Sci 495:1–9.

Al-Ali SY, Al-Zuhair AG, Dawod B (1988) Ultrastructural study of phagocytic activities of young astrocytes in injured neonatal rat brain following intracerebral injection of colloidal carbon. Glia 1:211–218.

Angaut-Petit D, Faille L (1987) Inability of regenerating mouse motor axons to innervate a denervated target. Neurosci Lett 75:163–168.

Aschner M, Sonnewald U, Tan KH (2002) Astrocyte modulation of neurotoxic injury. Brain Pathol 12:475–481.

Bakalash S., Shlomo G.B., Aloni E., Shaked I., Wheeler, L., Ofri, R., Schwartz M. (2005) T-cell-based vaccination for morphological and functional neuroprotection in a rat model of chronically elevated intraocular pressure. J Mol Med 83:904–916.

Banati RB, Schubert P, Rothe G, Gehrmann J, Rudolphi K, Valet G, Kreutzberg GW (1994) Modulation of intracellular formation of reactive oxygen intermediates in peritoneal macrophages and microglia/brain macrophages by propentofylline. J Cereb Blood Flow Metab 14:145–149.

Basu A, Krady JK, Enterline JR, Levison SW (2002) Transforming growth factor beta1 prevents IL-1beta-induced microglial activation, whereas TNFalpha- and IL-6-stimulated activation are not antagonized. Glia 40:109–120.

Becher B, Dodelet V, Fedorowicz V, Antel JP (1996) Soluble tumor necrosis factor receptor inhibits interleukin 12 production by stimulated human adult microglial cells in vitro. J Clin Invest 98:1539–1543.

Berkenbosch F (1992) Macrophages and astroglial interactions in repair to brain injury. Ann N Y Acad Sci 650:186–190.

Best TM, Hunter KD (2000) Muscle injury and repair. Phys Med Rehabil Clin N Am 11:251–266.

Bethea JR, Nagashima H, Acosta MC, Briceno C, Gomez F, Marcillo AE, Loor K, Green J, Dietrich WD (1999) Systemically administered interleukin-10 reduces tumor necrosis factor-alpha production and significantly improves functional recovery following traumatic spinal cord injury in rats. J Neurotrauma 16:851–863.

Bomstein Y, Marder JB, Vitner K, Smirnov I, Lisaey G, Butovsky O, Fulga V, Yoles E (2003) Features of skin-coincubated macrophages that promote recovery from spinal cord injury. J Neuroimmunol 142:10–16.

Bregman BS, Kunkel-Bagden E, Schnell L, Dai HN, Gao D, Schwab ME (1995) Recovery from spinal cord injury mediated by antibodies to neurite growth inhibitors. Nature 378:498–501.

Butovsky O, Hauben E, Schwartz M (2001) Morphological aspects of spinal cord autoimmune neuroprotection: Colocalization of T cells with B7–2 (CD86) and prevention of cyst formation. FASEB J 15:1065–1067.

Butovsky O, Koronyo-Hamaoui M, Kunis G, Ophir E, Landa G, Cohen H, Schwartz M (2006b) Glatiramer acetate fights against Alzheimers disease by inducing dendritic-like microglia expressing insulin-like growth factor 1. Proc Natl Acad Sci 103:11784–11789.

Butovsky O, Talpalar AE, Ben-Yaakov K, Schwartz M (2005a) Activation of microglia by aggregated beta-amyloid or lipopolysaccharide impairs MHC-II expression and renders them cytotoxic whereas IFN-gamma and IL-4 render them protective. Mol Cell Neurosci 29:381–393.

Butovsky O, Ziv Y, Schwartz A, Landa G, Talpalar AE, Pluchino S, Martino G, Schwartz M (2005b) Microglia activated by IL-4 or IFN-gamma differentially induce neurogenesis and oligodendrogenesis from adult stem/progenitor cells. Mol Cell Neurosci 31:149–160.

Butovsky O, Landa G, Kunis G, Ziv Y, Avidan H, Greenberg N, Schwartz A, Smirnov I, Pollack A, Jung S, Schwartz M (2006) Induction and blockage of oligodendrogenesis by differently activated microglia in an animal model of multiple sclerosis. J Clin Invest 116:905–915.

Carlson JA, Ambros R, Malfetano J, Ross J, Grabowski R, Lamb P, Figge H, Mihm MC, Jr (1998) Vulvar lichen sclerosus and squamous cell carcinoma: A cohort, case control, and investigational study with historical perspective; implications for chronic inflammation and sclerosis in the development of neoplasia. Hum Pathol 29:932–948.

Caroni P, Savio T, Schwab ME (1988) Central nervous system regeneration: Oligodendrocytes and myelin as non-permissive substrates for neurite growth. Prog Brain Res 78:363–370.

Carri MT, Ferri A, Cozzolino M, Calabrese L, Rotilio G (2003) Neurodegeneration in amyotrophic lateral sclerosis: The role of oxidative stress and altered homeostasis of metals. Brain Res Bull 61:365–374.

Constantini S, Young W (1994) The effects of methylprednisolone and the ganglioside GM1 on acute spinal cord injury in rats. J Neurosurg 80:97–111.

Cui Q, Cho KS, So KF, Yip HK (2004) Synergistic effect of Nogo-neutralizing antibody IN-1 and ciliary neurotrophic factor on axonal regeneration in adult rodent visual systems. J Neurotrauma 21:617–625.

Dal Canto MC, Calenoff MA, Miller SD, Vanderlugt CL (2000) Lymphocytes from mice chronically infected with Theiler's murine encephalomyelitis virus produce demyelination of organotypic cultures after stimulation with the major encephalitogenic epitope of myelin proteolipid protein. Epitope spreading in TMEV infection has functional activity. J Neuroimmunol 104:79–84.

David S, Bouchard C, Tsatas O, Giftochristos N (1990) Macrophages can modify the nonpermissive nature of the adult mammalian central nervous system. Neuron 5:463–469.

Dihne M, Block F, Korr H, Topper R (2001) Time course of glial proliferation and glial apoptosis following excitotoxic CNS injury. Brain Res 902:178–189.

diZerega GS (1997) Biochemical events in peritoneal tissue repair. Eur J Surg Suppl 577:10–16.

Dolenc VV (1984) Intercostal neurotization of the peripheral nerves in avulsion plexus injuries. Clin Plast Surg 11:143–147.

Dolenc VV (1986) Contemporary treatment of peripheral nerve and brachial plexus lesions. Neurosurg Rev 9:149–156.

Domeniconi M, Cao Z, Spencer T, Sivasankaran R, Wang K, Nikulina E, Kimura N, Cai H, Deng K, Gao Y, He Z, Filbin M (2002) Myelin-associated glycoprotein interacts with the Nogo66 receptor to inhibit neurite outgrowth. Neuron 35:283–290.

Dubos RJ (1955) The micro-environment of inflammation or Metchnikoff revisited. Lancet 269:1–5.

Dummer W, Niethammer AG, Baccala R, Lawson BR, Wagner N, Reisfeld RA, Theofilopoulos AN (2002) T cell homeostatic proliferation elicits effective antitumor autoimmunity. J Clin Invest 110:185–192.

Dusart I, Schwab ME (1994) Secondary cell death and the inflammatory reaction after dorsal hemisection of the rat spinal cord. Eur J Neurosci 6:712–724.

Elflein K, Rodriguez-Palmero M, Kerkau T, Hunig T (2003) Rapid recovery from T lymphopenia by CD28 superagonist therapy. Blood 102:1764–1770.

Elliott P (2003) Erasmus Darwin, Herbert Spencer, and the origins of the evolutionary worldview in British provincial scientific culture, 1770–1850. Isis 94:1–29.

Evert BO, Wullner U, Klockgether T (2000) Cell death in polyglutamine diseases. Cell Tissue Res 301:189–204.

Fantini MC, Becker C, Monteleone G, Pallone F, Galle PR, Neurath MF (2004) Cutting edge: TGF-beta induces a regulatory phenotype in $CD4^+CD25^-$ T cells through Foxp3 induction and down-regulation of Smad7. J Immunol 172:5149–5153.

Fenrich K, Gordon T (2004) Canadian Association of Neuroscience review: Axonal regeneration in the peripheral and central nervous systems—current issues and advances. Can J Neurol Sci 31:142–156.

Ferraro GB, Alabed YZ, Fournier AE (2004) Molecular targets to promote central nervous system regeneration. Curr Neurovasc Res 1:61–75.

Finn OJ, Debruyne LA, Bishop DK (1996) T cell receptor (TCR) repertoire in alloimmune responses. Int Rev Immunol 13:187–207.

Fischer D, Petkova V, Thanos S, Benowitz LI (2004) Switching mature retinal ganglion cells to a robust growth state in vivo: Gene expression and synergy with RhoA inactivation. J Neurosci 24:8726–8740.

Fitch MT, Doller C, Combs CK, Landreth GE, Silver J (1999) Cellular and molecular mechanisms of glial scarring and progressive cavitation: In vivo and in vitro analysis of inflammation-induced secondary injury after CNS trauma. J Neurosci 19:8182–8198.

Franzen R, Bouhy D, Schoenen J (2004) Nervous system injury: Focus on the inflammatory cytokine 'granulocyte-macrophage colony stimulating factor'. Neurosci Lett 361:76–78.

Frisen J, Haegerstrand A, Fried K, Piehl F, Cullheim S, Risling M (1994) Adhesive/repulsive properties in the injured spinal cord: Relation to myelin phagocytosis by invading macrophages. Exp Neurol 129:183–193.

Fry EJ (2001) Central nervous system regeneration: Mission impossible? Clin Exp Pharmacol Physiol 28:253–258.

Gelinas S, Martinoli MG (2002) Neuroprotective effect of estradiol and phytoestrogens on MPP+-induced cytotoxicity in neuronal PC12 cells. J Neurosci Res 70:90–96.

Gielen A, Khademi M, Muhallab S, Olsson T, Piehl F (2003) Increased brain-derived neurotrophic factor expression in white blood cells of relapsing-remitting multiple sclerosis patients. Scand J Immunol 57:493–497.

Giulian D, Li J, Bartel S, Broker J, Li X, Kirkpatrick JB (1995) Cell surface morphology identifies microglia as a distinct class of mononuclear phagocyte. J Neurosci 15:7712–7726.

Gold R, Hartung HP, Lassmann H (1997) T-cell apoptosis in autoimmune diseases: Termination of inflammation in the nervous system and other sites with specialized immune-defense mechanisms. Trends Neurosci 20:399–404.

Gordon T, Sulaiman O, Boyd JG (2003) Experimental strategies to promote functional recovery after peripheral nerve injuries. J Peripher Nerv Syst 8:236–250.

Griffin JW, George R, Lobato C, Tyor WR, Yan LC, Glass JD (1992) Macrophage responses and myelin clearance during Wallerian degeneration: Relevance to immune-mediated demyelination. J Neuroimmunol 40:153–165.

Griot C, Burge T, Vandevelde M, Peterhans E (1989) Antibody-induced generation of reactive oxygen radicals by brain macrophages in canine distemper encephalitis: A mechanism for bystander demyelination. Acta Neuropathol (Berl) 78:396–403.

Griot-Wenk M, Griot C, Pfister H, Vandevelde M (1991) Antibody-dependent cellular cytotoxicity (ADCC) in antimyelin antibody-induced oligodendrocyte damage in vitro. Schweiz Arch Neurol Psychiatr 142:122–123.

Groom AJ, Smith T, Turski L (2003) Multiple sclerosis and glutamate. Ann N Y Acad Sci 993:229–275; discussion 287–228.

Gu YD, Ma MK (1991) Nerve transfer for treatment of root avulsion of the brachial plexus: Experimental studies in a rat model. J Reconstr Microsurg 7:15–22.

Gudmundsdottir H, Turka LA (2001) A closer look at homeostatic proliferation of CD4+ T cells: Costimulatory requirements and role in memory formation. J Immunol 167:3699–3707.

Haenggeli C, Julien JP, Mosely RL, Perez N, Dhar, A, Gendelman HE, Rothstein JD (2007) Therapeutic immunization with a glatiramer acetate derivative does not alter survival in G93A and G37R SOD1 mouse models of familial ALS. Neurobiol Dis 26:146–152.

Hammarberg H, Lidman O, Lundberg C, Eltayeb SY, Gielen AW, Muhallab S, Svenningsson A, Linda H, van Der Meide PH, Cullheim S, Olsson T, Piehl F (2000) Neuroprotection by encephalomyelitis: Rescue of mechanically injured neurons and neurotrophin production by CNS-infiltrating T and natural killer cells. J Neurosci 20:5283–5291.

Hauben E, Nevo U, Yoles E, Moalem G, Agranov E, Mor F, Akselrod S, Neeman M, Cohen IR, Schwartz M (2000a) Autoimmune T cells as potential neuroprotective therapy for spinal cord injury. Lancet 355:286–287.

Hauben E, Butovsky O, Nevo U, Yoles E, Moalem G, Agranov G, Mor F, Leibowitz-Amit R, Pevsner E, Akselrod S, Neeman M, Cohen IR, Schwartz M (2000b) Passive or active immunization with myelin basic protein promotes recovery from spinal cord contusion. J Neurosci 20:6421–6430.

Hauben E, Gothilf A, Cohen A, Butovsky O, Nevo U, Smirnov I, Yoles E, Akselrod S, Schwartz M (2003) Vaccination with dendritic cells pulsed with peptides of myelin basic protein promotes functional recovery from spinal cord injury. J Neurosci 23:8808–8819.

He BP, Wen W, Strong MJ (2002) Activated microglia (BV-2) facilitation of TNF-alpha-mediated motor neuron death in vitro. J Neuroimmunol 128:31–38.

Herman J (1997) Medicine and evolution: Time for a new paradigm? Med Hypotheses 48:403–406.

Hickey WF, Hsu BL, Kimura H (1991) T-lymphocyte entry into the central nervous system. J Neurosci Res 28:254–260.

Hirsch EC (1999) Mechanism and consequences of nerve cell death in Parkinson's disease. J Neural Transm Suppl 56:127–137.

Hirsch S, Bahr M (1999) Growth promoting and inhibitory effects of glial cells in the mammalian nervous system. Adv Exp Med Biol 468:199–205.

Hirschberg DL, Moalem G, He J, Mor F, Cohen IR, Schwartz M (1998) Accumulation of passively transferred primed T cells independently of their antigen specificity following central nervous system trauma. J Neuroimmunol 89:88–96.

Hofteig JH, Vo PN, Yates AJ (1981) Wallerian degeneration of peripheral nerve. Age-dependent loss of nerve lipids. Acta Neuropathol (Berl) 55:151–156.

Hohlfeld R, Wiendl H (2001) The ups and downs of multiple sclerosis therapeutics. Ann Neurol 49:281–284.

Hur K, Kim JI, Choi SI, Choi EK, Carp RI, Kim YS (2002) The pathogenic mechanisms of prion diseases. Mech Ageing Dev 123:1637–1647.

Ilani T, Ben-Shachar D, Strous RD, Mazor M, Sheinkman A, Kotler M, Fuchs S (2001) A peripheral marker for schizophrenia: Increased levels of D3 dopamine receptor mRNA in blood lymphocytes. Proc Natl Acad Sci USA 98:625–628.

Janabi N (2002) Selective inhibition of cyclooxygenase-2 expression by 15-deoxy-Delta(12,14)(12,14)-prostaglandin J(2) in activated human astrocytes, but not in human brain macrophages. J Immunol 168:4747–4755.

Jellinger KA (2003) General aspects of neurodegeneration. J Neural Transm Suppl 65:101–144.

Jones LL, Yamaguchi Y, Stallcup WB, Tuszynski MH (2002) NG2 is a major chondroitin sulfate proteoglycan produced after spinal cord injury and is expressed by macrophages and oligodendrocyte progenitors. J Neurosci 22:2792–2803.

Jones LL, Margolis RU, Tuszynski MH (2003) The chondroitin sulfate proteoglycans neurocan, brevican, phosphacan, and versican are differentially regulated following spinal cord injury. Exp Neurol 182:399–411.

Kalichman MW, Myers RR (1987) Behavioral and electrophysiological recovery following cryogenic nerve injury. Exp Neurol 96:692–702.

Kalla R, Liu Z, Xu S, Koppius A, Imai Y, Kloss CU, Kohsaka S, Gschwendtner A, Moller JC, Werner A, Raivich G (2001) Microglia and the early phase of immune surveillance in the axotomized facial motor nucleus: Impaired microglial activation and lymphocyte recruitment but no effect on neuronal survival or axonal regeneration in macrophage-colony stimulating factor-deficient mice. J Comp Neurol 436:182–201.

Kim JE, Liu BP, Park JH, Strittmatter SM (2004) Nogo-66 receptor prevents raphespinal and rubrospinal axon regeneration and limits functional recovery from spinal cord injury. Neuron 44:439–451.

Kipnis J, Yoles E, Porat Z, Cohen A, Mor F, Sela M, Cohen IR, Schwartz M (2000) T cell immunity to copolymer 1 confers neuroprotection on the damaged optic nerve: Possible therapy for optic neuropathies. Proc Natl Acad Sci USA 97:7446–7451.

Kipnis J, Yoles E, Schori H, Hauben E, Shaked I, Schwartz M (2001) Neuronal survival after CNS insult is determined by a genetically encoded autoimmune response. J Neurosci 21:4564–4571.

Kipnis J, Mizrahi T, Hauben E, Shaked I, Shevach E, Schwartz M (2002) Neuroprotective autoimmunity: Naturally occurring CD4+CD25+ regulatory T cells suppress the ability to withstand injury to the central nervous system. Proc Natl Acad Sci U S A 99:15620–15625.

Kipnis J, Avidan H, Caspi RR, Schwartz M (2004a) Dual effect of CD4+CD25+ regulatory T cells in neurodegeneration: A dialogue with microglia. Proc Natl Acad Sci U S A 101 Suppl 2:14663–14669.

Kipnis J, Cardon M, Avidan H, Lewitus GM, Mordechay S, Rolls A, Shani Y, Schwartz M (2004b) Dopamine, through the extracellular signal-regulated kinase pathway, downregulates CD4+CD25+ regulatory T-cell activity: Implications for neurodegeneration. J Neurosci 24:6133–6143.

Kloss CU, Bohatschek M, Kreutzberg GW, Raivich G (2001) Effect of lipopolysaccharide on the morphology and integrin immunoreactivity of ramified microglia in the mouse brain and in cell culture. Exp Neurol 168:32–46.

Klusman I, Schwab ME (1997) Effects of pro-inflammatory cytokines in experimental spinal cord injury. Brain Res 762:173–184.

Koutsilieri E, Scheller C, Tribl F, Riederer P (2002) Degeneration of neuronal cells due to oxidative stress—microglial contribution. Parkinsonism Relat Disord 8:401–406.

Kurosinski P, Gotz J (2002) Glial cells under physiologic and pathologic conditions. Arch Neurol 59:1524–1528.

Lakatos A, Franklin RJ (2002) Transplant mediated repair of the central nervous system: An imminent solution? Curr Opin Neurol 15:701–705.

Lang DM, Rubin BP, Schwab ME, Stuermer CA (1995) CNS myelin and oligodendrocytes of the Xenopus spinal cord—but not optic nerve—are nonpermissive for axon growth. J Neurosci 15:99–109.

Lazarov-Spiegler O, Solomon AS, Zeev-Brann AB, Hirschberg DL, Lavie V, Schwartz M (1996) Transplantation of activated macrophages overcomes central nervous system regrowth failure. FASEB J 10:1296–1302.

Lazarov-Spiegler O, Rapalino O, Agranov G, Schwartz M (1998) Restricted inflammatory reaction in the CNS: A key impediment to axonal regeneration? Mol Med Today 4:337–342.

Lazarov-Spiegler O, Solomon AS, Schwartz M (1999) Link between optic nerve regrowth failure and macrophage stimulation in mammals. Vision Res 39:169–175.

Lee SC, Dickson DW, Liu W, Brosnan CF (1993) Induction of nitric oxide synthase activity in human astrocytes by interleukin-1 beta and interferon-gamma. J Neuroimmunol 46:19–24.

Lee YL, Shih K, Bao P, Ghirnikar RS, Eng LF (2000) Cytokine chemokine expression in contused rat spinal cord. Neurochem Int 36:417–425.

Lemmer K, Ahnert-Hilger G, Hopfner M, Hoegerle S, Faiss S, Grabowski P, Jockers-Scherubl M, Riecken EO, Zeitz M, Scherubl H (2002) Expression of dopamine receptors and transporter in neuroendocrine gastrointestinal tumor cells. Life Sci 71:667–678.

Leskovar A, Moriarty LJ, Turek JJ, Schoenlein IA, Borgens RB (2000) The macrophage in acute neural injury: Changes in cell numbers over time and levels of cytokine production in mammalian central and peripheral nervous systems. J Exp Biol 203:1783–1795.

Levi G, Minghetti L, Aloisi F (1998) Regulation of prostanoid synthesis in microglial cells and effects of prostaglandin E2 on microglial functions. Biochimie 80:899–904.

Levite M, Chowers Y, Ganor Y, Besser M, Hershkovits R, Cahalon L (2001) Dopamine interacts directly with its D3 and D2 receptors on normal human T cells, and activates beta1 integrin function. Eur J Immunol 31:3504–3512.

Ling C, Sandor M, Suresh M, Fabry Z (2006) Traumatic injury and the presence of antigen differentially contribute to T-cell recruitment in the CNS. J Neurosci 26:731–741.

Linton PJ, Dorshkind K (2004) Age-related changes in lymphocyte development and function. Nat Immunol 5:133–139.

Liu B, Gao HM, Wang JY, Jeohn GH, Cooper CL, Hong JS (2002) Role of nitric oxide in inflammation-mediated neurodegeneration. Ann N Y Acad Sci 962:318–331.

Liuzzi GM, Latronico T, Fasano A, Carlone G, Riccio P (2004) Interferon-beta inhibits the expression of metalloproteinases in rat glial cell cultures: Implications for multiple sclerosis pathogenesis and treatment. Mult Scler 10:290–297.

Lohse AW, Dinkelmann M, Kimmig M, Herkel J, Meyer zum Buschenfelde KH (1996) Estimation of the frequency of self-reactive T cells in health and inflammatory diseases by limiting dilution analysis and single cell cloning. J Autoimmun 9:667–675.

Lombardi VR, Garcia M, Cacabelos R (1998) Microglial activation induced by factor(s) contained in sera from Alzheimer-related ApoE genotypes. J Neurosci Res 54:539–553.

Lotan M, Solomon A, Ben-Bassat S, Schwartz M (1994) Cytokines modulate the inflammatory response and change permissiveness to neuronal adhesion in injured mammalian central nervous system. Exp Neurol 126:284–290.

Ma J, Urba WJ, Si L, Wang Y, Fox BA, Hu HM (2003) Anti-tumor T cell response and protective immunity in mice that received sublethal irradiation and immune reconstitution. Eur J Immunol 33:2123–2132.

Madsen JR, MacDonald P, Irwin N, Goldberg DE, Yao GL, Meiri KF, Rimm IJ, Stieg PE, Benowitz LI (1998) Tacrolimus (FK506) increases neuronal expression of GAP-43 and improves functional recovery after spinal cord injury in rats. Exp Neurol 154:673–683.

Mantovani A, Sozzani S, Locati M, Allavena P, Sica A (2002) Macrophage polarization: Tumor-associated macrophages as a paradigm for polarized M2 mononuclear phagocytes. Trends Immunol 23:549–555.

Martini A, Burgio GR (1999) Tolerance and auto-immunity: 50 years after Burnet. Eur J Pediatr 158:769–775.

Mason JL, Suzuki K, Chaplin DD, Matsushima GK (2001) Interleukin-1beta promotes repair of the CNS. J Neurosci 21:7046–7052.

Mattson MP (2003) Adventures in neural plasticity, aging, and neurodegenerative disorders aboard the CWC beagle. Neurochem Res 28:1631–1637.

Mattson MP, Maudsley S, Martin B (2004) A neural signaling triumvirate that influences ageing and age-related disease: Insulin/IGF-1, BDNF and serotonin. Ageing Res Rev 3:445–464.

Mautes AE, Weinzierl MR, Donovan F, Noble LJ (2000) Vascular events after spinal cord injury: Contribution to secondary pathogenesis. Phys Ther 80:673–687.

McHugh RS, Whitters MJ, Piccirillo CA, Young DA, Shevach EM, Collins M, Byrne MC (2002) CD4(+)CD25(+) immunoregulatory T cells: Gene expression analysis reveals a functional role for the glucocorticoid-induced TNF receptor. Immunity 16:311–323.

McTigue DM, Wei P, Stokes BT (2001) Proliferation of NG2-positive cells and altered oligodendrocyte numbers in the contused rat spinal cord. J Neurosci 21:3392–3400.

Merkler D, Oertle T, Buss A, Pinschewer DD, Schnell L, Bareyre FM, Kerschensteiner M, Buddeberg BS, Schwab ME (2003) Rapid induction of autoantibodies against Nogo-A and MOG in the absence of an encephalitogenic T cell response: Implication for immunotherapeutic approaches in neurological diseases. Faseb J 17:2275–2277.

Merrill JE, Ignarro LJ, Sherman MP, Melinek J, Lane TE (1993) Microglial cell cytotoxicity of oligodendrocytes is mediated through nitric oxide. J Immunol 151:2132–2141.

Miller JF (1994) Burnet oration. The thymus then and now. Immunol Cell Biol 72:361–366.

Miller SD, Katz-Levy Y, Neville KL, Vanderlugt CL (2001) Virus-induced autoimmunity: Epitope spreading to myelin autoepitopes in Theiler's virus infection of the central nervous system. Adv Virus Res 56:199–217.

Mizrahi T, Hauben E, Schwartz M (2002) The tissue-specific self-pathogen is the protective self-antigen: The case of uveitis. J Immunol 169:5971–5977.

Moalem G, Monsonego A, Shani Y, Cohen IR, Schwartz M (1999a) Differential T cell response in central and peripheral nerve injury: Connection with immune privilege. Faseb J 13:1207–1217.

Moalem G, Leibowitz-Amit R, Yoles E, Mor F, Cohen IR, Schwartz M (1999b) Autoimmune T cells protect neurons from secondary degeneration after central nervous system axotomy. Nat Med 5:49–55.

Moalem G, Gdalyahu A, Shani Y, Otten U, Lazarovici P, Cohen IR, Schwartz M (2000) Production of neurotrophins by activated T cells: Implications for neuroprotective autoimmunity. J Autoimmun 20:6421–6430.

Montastruc JL, Rascol O, Senard JM (1999) Treatment of Parkinson's disease should begin with a dopamine agonist. Mov Disord 14:725–730.

Moon LD, Asher RA, Rhodes KE, Fawcett JW (2002) Relationship between sprouting axons, proteoglycans and glial cells following unilateral nigrostriatal axotomy in the adult rat. Neuroscience 109:101–117.

Mosser DM (2003) The many faces of macrophage activation. J Leukoc Biol 73:209–212.

Nakajima K, Honda S, Tohyama Y, Imai Y, Kohsaka S, Kurihara T (2001) Neurotrophin secretion from cultured microglia. J Neurosci Res 65:322–331.

Ogawa K, Suzuki J, Narasaki M, Mori M (1985) Healing of focal injury in the rat liver. Am J Pathol 119:158–167.

Ossowska K (1993) Disturbances in neurotransmission processes in aging and age-related diseases. Pol J Pharmacol 45:109–131.

Paul DB (1988) The selection of the "survival of the fittest." J Hist Biol 21:411–424.

Pellegrino RG, Politis MJ, Ritchie JM, Spencer PS (1986) Events in degenerating cat peripheral nerve: Induction of Schwann cell S phase and its relation to nerve fibre degeneration. J Neurocytol 15:17–28.

Penkowa M, Giralt M, Lago N, Camats J, Carrasco J, Hernandez J, Molinero A, Campbell IL, Hidalgo J (2003) Astrocyte-targeted expression of IL-6 protects the CNS against a focal brain injury. Exp Neurol 181:130–148.

Perry VH, Crocker PR, Gordon S (1992) The blood-brain barrier regulates the expression of a macrophage sialic acid-binding receptor on microglia. J Cell Sci 101 (Pt 1):201–207.

Popovich PG, Jones TB (2003) Manipulating neuroinflammatory reactions in the injured spinal cord: Back to basics. Trends Pharmacol Sci 24:13–17.

Popovich PG, Wei P, Stokes BT (1997) Cellular inflammatory response after spinal cord injury in Sprague-Dawley and Lewis rats. J Comp Neurol 377:443–464.

Popovich PG, Guan Z, Wei P, Huitinga I, van Rooijen N, Stokes BT (1999) Depletion of hematogenous macrophages promotes partial hindlimb recovery and neuroanatomical repair after experimental spinal cord injury. Exp Neurol 158:351–365.

Rapalino O, Lazarov-Spiegler O, Agranov E, Velan GJ, Yoles E, Fraidakis M, Solomon A, Gepstein R, Katz A, Belkin M, Hadani M, Schwartz M (1998) Implantation of stimulated homologous macrophages results in partial recovery of paraplegic rats. Nat Med 4:814–821.

Rehman HU (2000) Progressive supranuclear palsy. Postgrad Med J 76:333–336.

Ria F, van den Elzen P, Madakamutil LT, Miller JE, Maverakis E, Sercarz EE (2001) Molecular characterization of the T cell repertoire using immunoscope analysis and its possible implementation in clinical practice. Curr Mol Med 1:297–304.

Robertson GS, Jian M (1995) D1 and D2 dopamine receptors differentially increase Fos-like immunoreactivity in accumbal projections to the ventral pallidum and midbrain. Neuroscience 64:1019–1034.

Robertson GS, Crocker SJ, Nicholson DW, Schulz JB (2000) Neuroprotection by the inhibition of apoptosis. Brain Pathol 10:283–292.

Rostasy KM (2005) Inflammation and neuroaxonal injury in multiple sclerosis and AIDS dementia complex: Implications for neuroprotective treatment. Neuropediatrics 36:230–239.

Rotshenker S (2003) Microglia and macrophage activation and the regulation of complement-receptor-3 (CR3/MAC-1)-mediated myelin phagocytosis in injury and disease. J Mol Neurosci 21:65–72.

Rutkowski MD, Pahl JL, Sweitzer S, van Rooijen N, DeLeo JA (2000) Limited role of macrophages in generation of nerve injury-induced mechanical allodynia. Physiol Behav 71:225–235.

Saha B, Mondal AC, Basu S, Dasgupta PS (2001) Circulating dopamine level, in lung carcinoma patients, inhibits proliferation and cytotoxicity of CD4+ and CD8+ T cells by D1 dopamine receptors: An in vitro analysis. Int Immunopharmacol 1:1363–1374.

Sakaguchi S, Takahashi T, Yamazaki S, Kuniyasu Y, Itoh M, Sakaguchi N, Shimizu J (2001) Immunologic self tolerance maintained by T-cell-mediated control of self-reactive T cells: Implications for autoimmunity and tumor immunity. Microbes Infect 3:911–918.

Sandvig A, Berry M, Barrett LB, Butt A, Logan A (2004) Myelin-, reactive glia-, and scar-derived CNS axon growth inhibitors: Expression, receptor signaling, and correlation with axon regeneration. Glia 46:225–251.

Sara E, Kotsakis A, Souklakos J, Kourousis C, Kakolyris S, Mavromanolakis E, Vlachonicolis J, Georgoulias V (1999) Post-chemotherapy lymphopoiesis in patients with solid tumors is characterized by CD4+ cell proliferation. Anticancer Res 19:471–476.

Saura J, Petegnief V, Wu X, Liang Y, Paul SM (2003) Microglial apolipoprotein E and astroglial apolipoprotein J expression in vitro: Opposite effects of lipopolysaccharide. J Neurochem 85:1455–1467.

Schadlich HJ, Bliersbach Y, Felgenhauer K, Schifferdecker M (1983) OKT8-binding lymphocytes in diseases of the nervous system. J Neuroimmunol 5:289–294.

Schori H, Kipnis J, Yoles E, WoldeMussie E, Ruiz G, Wheeler LA, Schwartz M (2001) Vaccination for protection of retinal ganglion cells against death from glutamate cytotoxicity and ocular hypertension: Implications for glaucoma. Proc Natl Acad Sci USA 98:3398–3403.

Schwab ME (2002) Increasing plasticity and functional recovery of the lesioned spinal cord. Prog Brain Res 137:351–359.

Schwab ME, Bartholdi D (1996) Degeneration and regeneration of axons in the lesioned spinal cord. Physiol Rev 76:319–370.

Schwab JM, Brechtel K, Nguyen TD, Schluesener HJ (2000) Persistent accumulation of cyclooxygenase-1 (COX-1) expressing microglia/macrophages and upregulation by endothelium following spinal cord injury. J Neuroimmunol 111:122–130.

Schwab JM, Frei E, Klusman I, Schnell L, Schwab ME, Schluesener HJ (2001) AIF-1 expression defines a proliferating and alert microglial/macrophage phenotype following spinal cord injury in rats. J Neuroimmunol 119:214–222.

Schwab JM, Failli V, Chedotal A (2005) Injury-related dynamic myelin/oligodendrocyte axon-outgrowth inhibition in the central nervous system. Lancet 365:2055–2057.

Schwartz M (2001a) Immunological approaches to the treatment of spinal cord injury. BioDrugs 15:585–593.

Schwartz M (2001b) Physiological approaches to neuroprotection. Boosting of protective autoimmunity. Surv Ophthalmol 45 Suppl 3:S256–260; discussion S273–256.

Schwartz M (2003) Macrophages and microglia in central nervous system injury: Are they helpful or harmful? J Cereb Blood Flow Metab 23:385–394.

Schwartz M (2004) Protective autoimmunity and prospects for therapeutic vaccination against self-perpetuating neurodegeneration. Ernst Schering Res Found Workshop 47:133–154.

Schwartz M, Cohen IR (2000) Autoimmunity can benefit self-maintenance. Immunol Today 21:265–268. 00001633 00001633.

Schwartz M, Yoles E (2000) Neuroprotection: A new treatment modality for glaucoma. Current Opinnion in Ophthalmology 11:107–111.

Schwartz M, Kipnis J (2002) Autoimmunity on alert: Naturally occurring regulatory CD4(+)CD25(+) T cells as part of the evolutionary compromise between a 'need' and a 'risk'. Trends Immunol 23:530–534.

Schwartz M, Kipnis J (2005a) Protective autoimmunity and neuroprotection in inflammatory and noninflammatory neurodegenerative diseases. J Neurol Sci 233:163–166.

Schwartz M, Kipnis J (2005b) Therapeutic T Cell-Based Vaccination for Neurodegenerative Disorders: The Role of CD4+CD25+ Regulatory T Cells. Ann N Y Acad Sci 1051:701–708.

Schwartz M, Belkin M, Yoles E, Solomon A (1996) Potential treatment modalities for glaucomatous neuropathy: Neuroprotection and neuroregeneration. J Glaucoma 5:427–432.

Schwartz M, Moalem G, Leibowitz-Amit R, Cohen IR (1999) Innate and adaptive immune responses can be beneficial for CNS repair. Trends Neurosci 22:295–299.

Schwartz M, Shaked I, Fisher J, Mizrahi T, Schori H (2003) Protective autoimmunity against the enemy within: Fighting glutamate toxicity. Trends Neurosci 26:297–302.

Sekiya T, Tanaka M, Shimamura N, Suzuki S (2001) Macrophage invasion into injured cochlear nerve and its modification by methylprednisolone. Brain Res 905:152–160.

Sela M (1999) Specific vaccines against autoimmune diseases. C R Acad Sci III 322:933–938.

Sela M (2000) Structural components responsible for peptide antigenicity. Appl Biochem Biotechnol 83:63–70; discussion 145–153.

Shaked I, Porat Z, Gersner R, Kipnis J, Schwartz M (2004) Early activation of microglia as antigen-presenting cells correlates with T cell-mediated protection and repair of the injured central nervous system. J Neuroimmunol 146:84–93.

Shaked I, Tchoresh D, Gersner R, Meiri G, Mordechai S, Xiao X, Hart RP, Schwartz M (2005) Protective autoimmunity: Interferon-gamma enables microglia to remove glutamate without evoking inflammatory mediators. J Neurochem 92:997–1009.

Shastry BS (2003) Neurodegenerative disorders of protein aggregation. Neurochem Int 43:1–7.

Shetty AK, Hattiangady B, Shetty GA (2005) Stem/progenitor cell proliferation factors FGF-2, IGF-1, and VEGF exhibit early decline during the course of aging in the hippocampus: Role of astrocytes. Glia 51:173–186.

Shevach EM (2000) Regulatory T cells in autoimmmunity. Annu Rev Immunol 18:423–449.

Shevach EM, McHugh RS, Thornton AM, Piccirillo C, Natarajan K, Margulies DH (2001) Control of autoimmunity by regulatory T cells. Adv Exp Med Biol 490:21–32.

Shimizu J, Yamazaki S, Sakaguchi S (1999) Induction of tumor immunity by removing CD25+CD4+ T cells: A common basis between tumor immunity and autoimmunity. J Immunol 163:5211–5218.

Shin WH, Lee DY, Park KW, Kim SU, Yang MS, Joe EH, Jin BK (2004) Microglia expressing interleukin-13 undergo cell death and contribute to neuronal survival in vivo. Glia 46:142–152.

Shirafuji T, Oka T, Sawada T, Tamura K, Kishimoto K, Yamamoto S, Nagayasu T, Takahashi T, Ayabe H (2001) The importance of peripheral blood leukocytes and macrophage infiltration on bronchial wall wound healing in rats treated preoperatively with anticancer agents. Surg Today 31:308–316.

Shirafuji T, Oka T, Sawada T, Tamura K, Nagayasu T, Takeya M, Yoshimura T, Ayabe H (2003) Effects of induction therapy on wound healing at bronchial anastomosis sites in rats. Jpn J Thorac Cardiovasc Surg 51:217–224.

Silverstein AM, Rose NR (1997) On the mystique of the immunological self. Immunol Rev 159:197–206; discussion 207–118.

Sivron T, Schwab ME, Schwartz M (1994) Presence of growth inhibitors in fish optic nerve myelin: Postinjury changes. J Comp Neurol 343:237–246.

Smith ME, van der Maesen K, Somera FP (1998) Macrophage and microglial responses to cytokines in vitro: Phagocytic activity, proteolytic enzyme release, and free radical production. J Neurosci Res 54:68–78.

So KF, Aguayo AJ (1985) Lengthy regrowth of cut axons from ganglion cells after peripheral nerve transplantation into the retina of adult rats. Brain Res 328:349–354.

Sonntag WE, Ramsey M, Carter CS (2005) Growth hormone and insulin-like growth factor-1 (IGF-1) and their influence on cognitive aging. Ageing Res Rev 4:195–212.

Stadelmann C, Ludwin S, Tabira T, Guseo A, Lucchinetti CF, Leel-Ossy L, Ordinario AT, Bruck W, Lassmann H (2005) Tissue preconditioning may explain concentric lesions in Balo's type of multiple sclerosis. Brain 128:979–987.

Stalder AK, Pagenstecher A, Yu NC, Kincaid C, Chiang CS, Hobbs MV, Bloom FE, Campbell IL (1997) Lipopolysaccharide-induced IL-12 expression in the central nervous system and cultured astrocytes and microglia. J Immunol 159:1344–1351.

Stichel CC, Niermann H, D'Urso D, Lausberg F, Hermanns S, Muller HW (1999) Basal membrane-depleted scar in lesioned CNS: Characteristics and relationships with regenerating axons. Neuroscience 93:321–333.

Stoll G, Trapp BD, Griffin JW (1989) Macrophage function during Wallerian degeneration of rat optic nerve: Clearance of degenerating myelin and Ia expression. J Neurosci 9:2327–2335.

Stoll G, Schroeter M, Jander S, Siebert H, Wollrath A, Kleinschnitz C, Bruck W (2004) Lesion-associated expression of transforming growth factor-beta-2 in the rat nervous system: Evidence for down-regulating the phagocytic activity of microglia and macrophages. Brain Pathol 14:51–58.

Streilein JW (1995) Unraveling immune privilege. Science 270:1158–1159.

Summers KL, O'Donnell JL, Hoy MS, Peart M, Dekker J, Rothwell A, Hart DN (1995) Monocyte-macrophage antigen expression on chondrocytes. J Rheumatol 22:1326–1334.

Swartz KR, Liu F, Sewell D, Schochet T, Campbell I, Sandor M, Fabry Z (2001) Interleukin-6 promotes post-traumatic healing in the central nervous system. Brain Res 896:86–95.

Tang S, Qiu J, Nikulina E, Filbin MT (2001) Soluble myelin-associated glycoprotein released from damaged white matter inhibits axonal regeneration. Mol Cell Neurosci 18:259–269.

Taoka Y, Okajima K, Uchiba M, Johno M (2000) Neuroprotection by recombinant thrombomodulin. Thromb Haemost 83:462–468.

Tatagiba M, Brosamle C, Schwab ME (1997) Regeneration of injured axons in the adult mammalian central nervous system. Neurosurgery 40:541–546; discussion 546–547.

Teismann P, Tieu K, Cohen O, Choi DK, Wu du C, Marks D, Vila M, Jackson-Lewis V, Przedborski S (2003) Pathogenic role of glial cells in Parkinson's disease. Mov Disord 18:121–129.

Teitelbaum D, Arnon R, Sela M (1997) Copolymer 1: From basic research to clinical application. Cell Mol Life Sci 53:24–28.

Thornton AM, Shevach EM (2000) Suppressor effector function of CD4+CD25+ immunoregulatory T cells is antigen nonspecific. J Immunol 164:183–190.

Tidball JG (1995) Inflammatory cell response to acute muscle injury. Med Sci Sports Exerc 27:1022–1032.

Trapp BD, Ransohoff R, Rudick R (1999a) Axonal pathology in multiple sclerosis: Relationship to neurologic disability. Curr Opin Neurol 12:295–302.

Trapp BD, Bo L, Mork S, Chang A (1999b) Pathogenesis of tissue injury in MS lesions. J Neuroimmunol 98:49–56.

Vairano M, Graziani G, Tentori L, Tringali G, Navarra P, Dello Russo C (2004) Primary cultures of microglial cells for testing toxicity of anticancer drugs. Toxicol Lett 148:91–94.

Vajda FJ (2002) Neuroprotection and neurodegenerative disease. J Clin Neurosci 9:4–8.

Varela-Nieto I, Morales-Garcia JA, Vigil P, Diaz-Casares A, Gorospe I, Sanchez-Galiano S, Canon S, Camarero G, Contreras J, Cediel R, Leon Y (2004) Trophic effects of insulin-like growth factor-I (IGF-I) in the inner ear. Hear Res 196:19–25.

Vaughan RB (1965) The romantic rationalist: A study of Elie Metchnikoff. Med Hist 19:201–215.

Vegeto E, Ghisletti S, Meda C, Etteri S, Belcredito S, Maggi A (2004) Regulation of the lipopolysaccharide signal transduction pathway by 17beta-estradiol in macrophage cells. J Steroid Biochem Mol Biol 91:59–66.

Vermetten E, Bremner JD (2002) Circuits and systems in stress. I. Preclinical studies. Depress Anxiety 15:126–147.

Vidal-Sanz M, Bray GM, Villegas-Perez MP, Thanos S, Aguayo AJ (1987) Axonal regeneration and synapse formation in the superior colliculus by retinal ganglion cells in the adult rat. J Neurosci 7:2894–2909.

Viret C, Barlow AK, Janeway CA, Jr (1999) On the intrathymic intercellular transfer of self-determinants. Immunol Today 20:8–10.

Wang X, Feuerstein GZ (2000) Role of immune and inflammatory mediators in CNS injury. Drug News Perspect 13:133–140.

Weihe E, Nohr D, Michel S, Muller S, Zentel HJ, Fink T, Krekel J (1991) Molecular anatomy of the neuro-immune connection. Int J Neurosci 59:1–23.

Weinreb RN, Levin LA (1999) Is neuroprotection a viable therapy for glaucoma? Arch Ophthalmol 117:1540–1544.

Wu D (2005) Neuroprotection in experimental stroke with targeted neurotrophins. NeuroRx 2:120–128.

Yoles E, Hauben E, Palgi O, Agranov E, Gothilf A, Cohen A, Kuchroo V, Cohen IR, Weiner H, Schwartz M (2001) Protective autoimmunity is a physiological response to CNS trauma. J Neurosci 21:3740–3748.

Zajicek JP, Wing M, Scolding NJ, Compston DA (1992) Interactions between oligodendrocytes and microglia. A major role for complement and tumour necrosis factor in oligodendrocyte adherence and killing. Brain 115 (Pt 6):1611–1631.

Zhang Z, Guth L (1997) Experimental spinal cord injury: Wallerian degeneration in the dorsal column is followed by revascularization, glial proliferation, and nerve regeneration. Exp Neurol 147:159–171.

Zheng SG, Wang JH, Koss MN, Quismorio F, Jr, Gray JD, Horwitz DA (2004) CD4+ and CD8+ regulatory T cells generated ex vivo with IL-2 and TGF-beta suppress a stimulatory graft-versus-host disease with a lupus-like syndrome. J Immunol 172:1531–1539.

Ziv Y, Ron N, Butovsky O, Landa G, Sudai E, Greenberg N, Cohen H, Kipnis J, Schwartz M (2006) Immune cells contribute to the maintenance of neurogenesis and spatial learning abilities in adulthood. Nat Neurosci 9:268–275.

46
Adjuvants

Sam Sanderson

Keywords Adjuvant; Antigen; C5A; C5a agonist; Complement system; Immunomodulator; Molecular adjuvant; Vehicle

46.1. Introduction

A prominent feature that underscores the current era of vaccinology is the ability to generate a huge variety of antigens (Ag) by various synthetic chemistry and genetic engineering techniques. As a result, Ags now can be produced rapidly and in large quantities with well-defined structures and in highly purified forms—desirable attributes for their use in vaccines intended to generate Ag-specific immune responses in a diverse population. Unfortunately, the routine use of these Ags as integral components of vaccines is encumbered by their inherent lack of immunogenicity. Such vaccines, therefore, require the use of adjuvants in order to potentiate and focus the immune response to the Ag so that optimal immune outcome can be achieved. Thus, the discovery and development of new adjuvants is of growing importance to the design of vaccines capable of meeting the modern threats posed to human and animal populations by new or resurgent infectious and non-infectious diseases.

This chapter will highlight recent trends in the development of novel adjuvants and, to the extent to which it is known, their immunologic mechanisms. A particular emphasis will be place on those adjuvants composed of single molecular entities, the so called molecular adjuvants.

46.2. General Adjuvants: Vehicles and Immunomodulators

An adjuvant is any substance or any formulation of substances that enhances an immune response to an Ag. As a general rule, adjuvants fall into two major categories, vehicles and immunomodulators (Van Regenmortel, 1997).

Vehicles are those adjuvants that help carry Ags to and retain them in proximity to lymphocytes and other auxiliary immune cells, particularly antigen presenting cells (APC), within various lymphoid tissues. Indeed, it is this "depot" effect that is the defining mechanism of vehicle adjuvants. Classic examples of vehicles include liposomes, emulsions, proteosomes, and immunostimulating complexes (ISCOM) (Crouch et al., 2005). Typically, the final vaccine formulation is the Ag contained within the vehicle. Other examples of vaccines that utilize vehicles include formulations in which the Ag is covalently linked to bioadhesive microparticles such as poly(D,L-lactide-coglycolide (PLGA) (Diwan et al., 2004), nanoparticles, or adsorbed on the surface of carriers such as calcium and aluminum salts (phosphate or hydroxide), particularly alum, which is the only adjuvant approved for human use (Olive et al., 2001).

Immunomodulators are adjuvants defined by their ability to activate APCs and/or lymphocytes. Typically, this activation is characterized by the adjuvant-induced release of cytokines form lymphocytes and other auxiliary immune cells. Examples of immunomodulators include muramyl-dipeptide (MDP), monophosphoryl lipid A (MPL) (Ulrich and Myers, 1995), lipopoly-saccharide (LPS) (Johnson et al., 1956), bacterial cell membranes (Muhlradt et al., 1998), certain components of the complement system (Dempsey et al., 1996; Jacquier-Sarlin et al., 1995), cytokines (Afonso et al., 1994; Kurzawa et al., 1998), and oligonucleotide mimics of bacterial DNA. The typical vaccine formulation is one in which the Ag is admixed with the immunomodulator. Variations on this theme include those in which the Ag is covalently linked to the immunomodulator with the rationale that elements of the generalized immune response invoked by the adjuvant might be more directed to the Ag. Some examples of this approach include the incorporation of lipids, lipopolysaccharides, and lipoamino acids into peptide Ags by synthetic methods (Metzger et al., 1991; Defoort et al., 1992; Martinon et al., 1992; Wiesmuller et al., 1992; Olive et al., 2001) and the genetic fusion of chemokines/cytokines to protein Ags (Biragyn et al., 1999).

It should be pointed out that this classification of adjuvants is based more on historical observations than on strict

T. Ikezu and H.E. Gendelman (eds.), *Neuroimmune Pharmacology*.
© Springer 2008

mechanisms of action. In fact, there are several examples of adjuvants belonging to the vehicle family that act as immunomodulators. A noteworthy example of this is the saponins, which are extracted from the plant *Quillaja saponaria* and are used to create emulsions into which Ags are added. Over the years, a variety of chemical modifications of saponins have been used as adjuvants and all appear to stimulate the release of various T helper 1 (Th1) and T helper 2 (Th2) cytokines in addition to their vehicle-like mode of action (Shi et al., 2005).

Finally, there are many examples of vaccine formulations in which individual adjuvants are used in combination with others as a way of optimizing immune efficiency and outcome. Despite the current emphasis placed on evaluating the wide variety of adjuvants in experimental use today and the myriad of possible combinations and formulations, no particular adjuvant or adjuvant system has emerged as an ideal product. Also, given the variation within a data set in which adjuvants are employed, it appears that the effectiveness of an adjuvant or adjuvant system is best evaluated on a case-by-case basis.

46.3. Present-Day Adjuvants in Human and Animal Applications: General Mechanism of Action

As stated above, the principal objective for the use of an adjuvant in a vaccine is to potentiate immune response to an Ag of minimal immunogenicity. How this potentiation is achieved varies from adjuvant to adjuvant and in many cases, the precise mechanism of action is unknown. However, as a rule, immune potentiation is accomplished by the ability of the adjuvant to induce a variety of non-specific activities within the innate arm of the immune system. Once activated, the innate branch of immunity, particularly the complement system, orchestrates the various humoral and cell-mediated responses that operate within and between the innate and acquired arms. The result is a generalized activation and potentiation of the immune system in response to the adjuvant with the hope that this generalized immune priming will allow for a more effective processing and recognition of the Ag contained within the vaccine.

The manner by which an adjuvant induces this immune priming emanates from the exquisite sensitivity of the mammalian immune system to detect the presence of bacteria and bacterial components and respond accordingly in a rapid and vigorous manner. The initial response to these bacterial signals is largely a function of the innate branch of the immune system, which has evolved under continual selective pressure from pathogenic bacteria and other disease-causing microorganisms. It is not surprising, therefore, that the majority of adjuvants in use today are composed of various molecules, components, and structures derived from bacteria. The use of these adjuvants in vaccines, therefore, provides the bacterial signals to which the innate arm of the immune system vigorously reacts, which results in the priming/potentiation of the immune system.

However, this rapid and vigorous innate response to the bacteria-like signals induced by such adjuvants can result in an overly aggressive and misdirected immune response that is accompanied by undesirable side-effects such as anaphylaxis, fever, injection site granulomas or rashes, and local or systemic inflammation. Another drawback is that the adjuvant tends to induce a generalized immune response with little immune specificity directed to the Ag of interest. Also, immunity that might be directed to the Ag can be masked due to the magnitude of the innate response to the adjuvant. In an attempt to overcome these drawbacks, there is a vigorous and concerted research effort worldwide to develop modifications of bacterial-derived adjuvants such that they retain their ability to induce innate responses to bacterial signals, but minimize or eliminate the deleterious inflammatory side-effects that can accompany such signals. Notable examples of such efforts include the many chemical and structural modifications developed and tested on MPL (Ulrich and Myers, 1995) and the saponins (Shi et al., 2005).

46.4. Molecular Adjuvants

46.4.1. Definition

The need for an adjuvant capable of inducing a robust *and* Ag-specific immune response accompanied with minimal inflammatory side-effects has given rise to a new class of adjuvants, which have come to be known as molecular adjuvants (Dempsey et al., 1996). The rationale for a molecular adjuvant is based on the notion that a single, well-defined and well-characterized molecular entity might be better in inducing an immune response more directed to the Ag with fewer non-specific side effects, particularly when the Ag is attached directly to the molecular adjuvant. Thus, a molecular adjuvant can be defined as a single molecular entity that targets an Ag to the cells of the immune system responsible for Ag processing and presentation and/or activates specific pathways of Ag processing and presentation within these cells.

46.4.2. Complement

As described above, "traditional" adjuvants typically are derived from bacterial components in order to provide the bacteria signals necessary for the activation and potentiation of the innate arm of the immune system. One of the first components of innate immunity that becomes activated in response to these bacterial signals is the complement system.

Complement is a plasma system comprised of interrelated proteases arranged in a cascade fashion that becomes activated in response to bacterial signals and Ags associated with bacteria. The role of complement is twofold: 1) to serve as an initial, first line of defense to invading microorganisms and 2) to enhance systemic host defense by activating and orchestrating the various humoral and cell-mediated responses within the acquired and innate arms of the immune system necessary for the effective elimination of the microorganism.

The "first line of defense" is accomplished through the generation of the membrane attack complex (MAC), which is assembled at the end of the complement cascade from the components generated by the proteolytic steps along the cascade. The MAC is directly involved in lysing the membranes of foreign microorganisms.

Secondly, the various humoral and cell-mediated responses operating at the interface of the innate and acquired arms of the immune system are activated and coordinated by pharmacologically active components that are released at certain steps along the complement cascade. These proteolytic byproducts, which are independent of the components that comprise the MAC, induce the various humoral and cell-mediated aspects of the innate and acquired arms of immunity, all of which are necessary for the concerted elimination of the microorganism.

Principally because of this latter reason, these pharmacologically active components of complement are attractive as molecular adjuvants. This is because their use has the potential of enhancing specific pathways of the innate and acquired immune processing and subsequent recognition of an Ag as opposed to the broader and less Ag-focused activation of the innate system in response to bacterial signals that come from of adjuvants derived from bacterial components.

46.4.3. Complement-Derived Molecular Adjuvants

Certain components of complement have been used as molecular adjuvants to enhance Ag-specific immune responses to model Ags. Components of complement that have been used as molecular adjuvants include C3b, C4b (Arvieux et al., 1988, Jacquier-Sarlin et al., 1995), and C3d (Dempsey et al., 1996). These components were chosen primarily for their opsonic properties; i.e., their ability to bind to and coat the surface of a microorganism rendering it more susceptible to immune cell uptake via phagocytosis. The rationale, therefore, was that an Ag covalently attached to these complement fragments similarly would be rendered more susceptible for uptake by APCs and, consequently, would be more effectively processed and presented by the APC. In all cases, the immunogenicity of the model Ag in the Ag-complement vaccine complex was markedly enhanced relative to the Ag alone, suggesting that the complement fragment acted as a molecular adjuvant by enhancing Ag uptake by APCs and, in turn, the ability of the APC to process and present the Ag. Although the precise mechanism by which these molecular adjuvants enhance uptake, processing, and presentation is not fully understood, these studies clearly demonstrate the immunologic potential of using complement components as molecular adjuvants in the design of vaccines.

46.4.4. The Anaphylatoxins

An important group of pharmacologically active byproducts of complement activation are the anaphylatoxins C3a, C4a, and C5a, which are small (74–76 residue) fragments cleaved from the larger, parent complement components C3, C4, and C5, respectively. The principal roles of the anaphylatoxins are to recruit inflammatory cells and lymphocytes to sites of tissue injury and infection and to then activate these cells' various effector responses once recruited (Hugli, 1981). However, the anaphylatoxins play important roles in the activation and regulation of humoral and cell-mediated responses to Ags due to their ability to modulate various humoral and cell-mediated activities between the innate and acquired arms of the immune system (Dempsey *et al.*, 1996; Mastellos *et al.*, 2005; Morgan, 1986). Consequently, the anaphylatoxins are attractive as molecular adjuvants, which may be capable of invoking Ag-specific humoral and/or cell-mediated immune responses.

Unfortunately, the anaphylatoxins are also potent inflammatory mediators and their use as molecular adjuvants would surely be accompanied by local and/or systemic inflammatory side effects. Also, under certain conditions, the anaphylatoxins appear to downregulate immune function. For example, C3a has been regarded as a general suppressor of immune function and has been shown to downregulate certain Th2-mediated responses to Ags (Kawamoto et al., 2004, Morgan, 1986). Thus, while the anaphylatoxins have desirable immune stimulatory activities that make them attractive for use as molecular adjuvants, they carry with them the potential for adverse inflammatory side effects and immune downregulation. In order for an anaphylatoxin to be used as a molecular adjuvant, therefore, its immune stimulatory activities must be enhanced at the expense of its inflammatory activities and any tendency to downregulate immune response.

46.4.5. Immunostimulatory and Inflammatory Properties of C5a

Of the anaphylatoxins, C5a has many immune stimulatory activities that are particularly attractive for its use as a molecular adjuvant. For example, C5a has been shown to enhance Ag-specific and mitogen-induced antibody responses (Goodman et al., 1982; Morgan et al., 1983), Ag-induced T lymphocyte proliferation (Morgan et al., 1983), MHC-restricted T cell proliferation via primary and autologous mixed lymphocyte reactions (Goodman et al., 1982; Morgan et al., 1983; Montz et al., 1990), and the tumoricidal activity of monocytes (Hogan et al., 1989). Also, C5a directly and/or indirectly induces the synthesis and release of a variety of immunoregulatory cytokines from monocytes and macrophages including IL-1 (Okusawa et al., 1987; Goodman et al., 1982; Wetsel, 1995; Buchner et al., 1995), IL-6 (Scholz et al., 1990; Gasque et al., 1995), IL-8 (Ember et al., 1994), IL-12 (Floreani, A., Heires, A., and Sanderson, S.D. personal communication) and TNFα (E.L.

Morgan, personal communication) as well as IL-1β, IL-6, IL-8, IL-12, TNFα, and IFNγ from human dendritic cells (E.L. Morgan, personal communication).

Being an anaphylatoxin, however, C5a also expresses many proinflammatory activities. These include its potent chemotactic activities for the recruitment of inflammatory cells to sites of tissue injury and infection (Shin et al., 1968), its ability to induce smooth muscle contraction (Hugli et al., 1987), increase vascular permeability (Hugli and Muller-Eberhard, 1978; Hugli, 1981, 1990), and induce the release of a variety of secondary inflammatory mediators such as histamine, lysosomal enzymes, and vasoactive eicosanoids from responsive cells such as mast cells, neutrophils, eosinophils, and macrophages (Drapeau et al., 1993; Johnson et al., 1975; Goldstein and Weissmann, 1974; Schorlemmer et al., 1976; Lundberg et al., 1987).

These biologic responses, inflammation and immunomodulation, are induced by the binding of C5a to its specific, high affinity receptor (C5aR, CD88) that is expressed on the surface of the C5a-responsive cell(s). Traditionally, C5aR expression has been viewed as being limited to cells of myeloid origin such as neutrophils, monocytes, mast cells, and eosinophils (Wetsel, 1995). However, it is now known that a variety of human cells of nonmyeloid origin express C5aRs. These include hepatocytes (Buchner et al., 1995), astrocytes (Gasque et al., 1995), bronchial epithelial cells (Floreani et al., 1998), epithelial cells of the gut and kidney (Wetsel, 1995), and osteoblasts (Pobanz et al., 2000). The presence of C5aRs on these cells poses interesting questions about the nature of the C5a-mediated response and its role in the function of these cells under normal and aberrant physiologic conditions and is the focus of an active research program in our and several other laboratories.

Because of its proinflammatory properties, C5a has been implicated in a number of inflammatory disorders such as rheumatoid arthritis (Jose et al., 1990), adult respiratory distress syndrome (Till et al., 1982; Mulligan et al., 1996), gingivitis, (Wingrove et al., 1992), systemic lupus erythematosus (Hopkins et al., 1988) psoriasis (Bergh et al., 1993), reperfusion injury (Amsterdam et al., 1995), and gram-negative bacterial sepsis (Smedegard et al., 1989). This pathogenic involvement has provided the motivation for the development of C5a/C5aR antagonists to inhibit these C5a-mediated inflammatory disorders (Konteatis et al., 1994; Finch et al., 1999).

As the inflammatory properties of C5a have been the motivation behind the development of C5a antagonists, so too have the immunostimulatory properties of C5a been the motivating factor for the development of agonists of C5a that can be used as surrogates of the natural factor for enhancing humoral and cell-mediated immune responses; i.e., as a molecular adjuvant. Using such agonists of C5a as a molecular adjuvant, however, requires developing response-selective agonists that are capable of invoking C5a-like immunostimulatory properties at the expense of C5a-like inflammatory properties.

46.4.6. C5a: Structure-Function Considerations

Human C5a is a 74-residue glycopolypeptide that is comprised of two important structural and functional domains. The first is the well-ordered N-terminal core domain comprised of residue 1–63, which is primarily involved in the recognition and binding of C5aRs (Mollison et al., 1989; Zuiderweg et al., 1989). The second is the C-terminal domain, residues 64–74, which extends from the N-terminal core as a finger-like projection. $C5a_{65-74}$ (ISHKDMQLGR) is a region of considerable backbone flexibility and poorly-defined structure (Zuiderweg et al., 1989), yet the proinflammatory and immunostimulatory activities characteristic of the entire C5a polypeptide reside in this small C-terminal stretch (Ember et al., 1994, 1992; Morgan et al., 1992). Thus, an early objective in developing agonists or antagonists of C5a was to synthetically manipulate this region of C5a with the goal of generating a peptide-based ligand that would either bind the C5aR with high affinity but not transducer a biological signal (a C5a antagonist) or would bind and modulate the C5aR in such a way that only a limited or specific signal transduction event(s) such as immunostimulation would be activated (a response-selective C5a agonist).

Flexibility in the C-terminal, effector region of C5a ($C5a_{65-74}$ or ISHKDMQLGR) is a dominant feature of natural C5a (Mollison et al., 1989; Zuiderweg et al., 1989). It may be that this flexibility allows the C5aR the ability to induce a unique conformation in this region of the C5a ligand that is conducive to the expression of biologic activity characteristic of that particular C5aR-bearing cell. It may be, in fact, that a biologically active conformation in this effector region of C5a responsible for activity in one type of C5aR-bearing cell may not be the ideal conformation for the expression of activity in another. This suggests therefore, that by employing a synthetic strategy in which the flexibility in this effector region of C5a is restricted, one might bias certain conformational features that are important for the expression of certain types of C5aR-mediated activities; i.e., a response-selective agonist. Such a specific and stabilized conformation, when presented to the C5aR, would be more likely to interact with those C5aRs that are capable of accommodating this particular conformation.

46.4.7. Conformationally Restricted Analogues of $C5a_{65-74}$

Over the years, our laboratory has generated a library of analogues of $C5a_{65-74}$ in which backbone flexibility was restricted by certain amino acid substitutions. This was done with the goal of biasing certain conformational features that might be helpful in the search for biologically relevant conformations responsible for the induction of C5a-like immunostimulatory activities versus those responsible for C5a-like inflammation.

Backbone flexibility was restricted by introducing three principal types of residue substitutions within $C5a_{65-74}$: 1) Pro substitutions to restrict ϕ angle flexibility and to narrow the range of sterically allowed backbone conformations in the pre-proline residue (Hruby and Nikiforovich, 1991), 2) Ala substitutions to evaluate the biological importance of the side-chains in the peptide, and 3) D-residue substitutions to assess the contribution of stereoisomeric arrangements. The use of such peptide modifications has an additional advantage in that Pro, Ala, and D-residues occupy well-defined regions of sterically-allowed Ramachandran space (Hruby and Nikiforovich, 1991). Thus, an evaluation of the changes made in the biological activity of these conformationally restricted analogues of $C5a_{65-74}$ can provide information about the specific types of backbone conformation features that are important to specific types of biological activity.

Using this approach of biasing backbone conformation, we showed that one peptide from this library, $C5a_{65-74}$Y65,F67,P69,P71,D-Ala73 or YSFKPMPLaR exhibited about 10% of the potency of natural C5a for its ability to induce the release of spasmogenic eicosanoids from tissue resident macrophages in human umbilical artery, but only about 0.1% of C5a activity in its ability to induce the release of β-glucuronidase from human neutrophils (Finch et al., 1997). This difference in biological potency was reflected by a corresponding difference in binding affinity to the C5aRs expressed on these cells: $IC_{50} = 0.2\,\mu M$ in rat peritoneal macrophages compared to $10\,\mu M$ in human neutrophils (Finch et al., 1997, Vogen et al., 2001). This difference has been expressed as a function of the relative selectivity of YSFKPMPLaR in interacting with C5aRs on macrophages relative to C5aRs on neutrophils: selectivity = antilog[pD_2macrophage activity – pD_2neutrophil activity]. This measure of selectivity showed that YSFKPMPLaR was significantly more potent/selective in inducing eicosanoid release from C5aR-bearing macrophages than enzyme release from C5aR-bearing neutrophils relative to natural C5a and several other analogues less conformationally biased than YSFKPMPLaR (Finch et al., 1997). Another noteworthy property of YSFKPMPLaR, but not natural C5a or more flexible analogues of $C5a_{65-74}$, was the increased stability of the biologically important C-terminal Arg residue to proteolytic cleavage by serum carboxypeptidases (Kawatsu et al., 1996). The presence of this C-terminal Arg residue, particularly in small peptide agonists of C5a, is essential for the expression of full biological activity and its retention *in vivo* is an important consideration in the design of response-selective agonists of C5a for therapeutic use.

Nuclear Magnetic Resonance (NMR) analysis (Vogen et al., 1999b) of YSFKPMPLaR in water indicated the presence of extended, polyproline II (P_{II}) backbone conformation extending through residues 68–70/71 (KPM/P), which was expected from the influence of Pro at position 69 (Hruby and Nikiforovich, 1991). The C-terminal region (PLaR) adopted a distorted type II β-turn or a type II/V β-turn. In the type II/V β-turn, Leu72 exhibited a conformation characteristic of a type II β-turn, whereas D-Ala73 exhibited a conformation characteristic of a type V β-turn. Furthermore, an overlapping inverse γ-turn involving residues LaR was observed within the type II/V β-turn. This combination of conformational features in a peptide of only ten residues is highly unusual and underscores the unique topography that can be introduced in these C-terminal analogues of C5a and their potential for therapeutic use.

Indeed, the increase in potency/selectivity expressed by YSFKPMPLaR for eicosanoid release from macrophages versus enzyme release from neutrophils is attributed to the unique conformational features expressed in YSFKPMPLaR and the fact that they appear to be well accommodated by C5aRs on macrophages, but not by C5aRs on neutrophils. Why this is the case is not understood and it remains to be determined whether this difference might emanate from the ability of YSFKPMPLaR to distinguish between two different subtypes of C5aRs expressed on these two cell types. These studies represent only a beginning in understanding the relationship between structure and function of these C5a agonists and considerable work is required to identify and relate a specific topochemical feature of the agonist to a specific C5aR-mediated biological response(s). Nevertheless, the structure-function information provided by these conformationally biased decapeptide agonists of C5a support the development of highly response-selective agonists of C5a and offers some insight as to the types of backbone conformations that may related to selective activation of the C5aR expressed on different C5aR-bearing cells, particularly C5aR-bearing cells of the immune system (Taylor et al., 2001).

46.4.8. YSFKPMPLaR as a Molecular Adjuvant

YSFKPMPLaR has been used as a molecular adjuvant in *in vivo* studies designed to evaluate its effectiveness as a C5a surrogate capable of inducing Ag-specific humoral and cell-mediated responses to various B and T cell epitopes. This was approached by synthesizing the epitope directly onto the N-terminal end of YSFKPMPLaR thus leaving the C5a agonist moiety of the construct free to interact with C5aRs expressed on APCs. It was hypothesized that the YSFKPMPLaR moiety would carry the epitope to and activate the APC in a manner similar to natural C5a via interaction with surface expressed C5aRs for which YSFKPMPLaR has high affinity and selectivity (Tempero et al., 1997) YSFKPMPLaR, like natural C5a, would activate APCs via the synthesis and/or release of Th1 and Th2 cytokines. Following this APC activation, the C5aR/ligand complex would internalize and carry the attached epitope into intracellular Ag processing pathways where it associates with major histocompatibility complex (MHC) determinants for expression on the APC surface. Of our library of conformationally biased C5a agonists, YSFKPMPLaR was the ideal choice for use as a molecular adjuvant because of its enhanced potency/selectivity for C5aR-bearing APCs such as macrophages versus neutrophils and the stability of its crucial C-terminal Arg reside to cleavage by carboxypeptidases.

46.4.8.1. Generation of Ag-Specific Antibody Responses

In one study (Tempero et al., 1997)) YSFKPMPLaR was used to induce Ag-specific antibody (Ab) responses to a B cell epitope derived from the juxtamembrane region of human mucin type 1 glycoprotein (MUC1). A C5a-active construct was generated by covalently attaching the MUC1 epitope (*YKQGGFLGL*) to the N-terminus of YSFKPMPLaR (*YKQGGFLGL*YSFKPMPLaR). This construct was shown to be a full agonist of natural C5a and expressed the same response-selective properties exhibited by YSFKPMP-LaR alone. A C5a-inactive construct was generated by attaching the MUC1 epitope to the biologically important C-terminal end of YSFKPMPLaR (YSF-KPMPLaR*YKQGGFLGL*). This construct retains the same amino acid composition as the C5a-active construct but is devoid of C5a-like activities and, therefore, serves as an ideal negative control to the C5a-active construct.

C57BL6 and BALB/c mice were immunized with the following constructs: 1) YSFKPMPLaR, 2) *YKQGGFLGL*, 3) YSFKPMPLaR + *YKQGGFLGL*, 4) *YKQ-GGFLGL*YSFKPMPLaR, and 5) YSFKPMPLaR*YKQGGFLGL*. Mice were immunized via intraperitoneal injection of a 100 µl solution containing 100 µg of the peptide. Mice were boosted at 2-week intervals and sera obtained for analysis after 6 weeks. Another set of mice was immunized under identical conditions, but with a more traditional construct of the MUC1 epitope attached to the carrier protein keyhole limpet hemocyanin (KLH). In order to effectively evaluate immune outcome from the peptide-based and KLH-based vaccine constructs, immunizations were performed in the presence of MPL adjuvant so that the KLH construct was given a chance to induce immune response under more typical immunization conditions.

High Ab titers specific for the MUC1 epitope were observed only in mice immunized with the C5a-active construct, *YKQGGFLGL*YSFKPMPLaR and the epitope-KLH construct. These anti-epitope Abs were shown to cross react with rhMUC1 protein expressed on the surface of a transfected pancreatic cell line (Panc-1 M1F.15). This indicates that the anti-*YKQGGFLGL* Abs appear to recognize the *YKQGGFLGL* epitope within intact whole MUC1 protein.

The isotypes of the Abs generated by the molecular adjuvant-containing construct were IgM, IgG2a, and IgG2b. In contrast, Ab isotypes generated by the KLH construct were IgM and IgG1. This suggests that the molecular adjuvant-containing vaccine induced an Ab class switch characteristic of a Th1-like response and, consequently, generated Ab with isotypes distinct from the traditional KLH construct. Thus, the molecular adjuvant appears to induce Ag-specific Abs via an immunologic pathway that differs from the more traditional vaccine design. Finally, mice immunized with the molecular adjuvant-containing construct displayed no outward signs of immediate or delayed inflammatory responses.

Over the years, we have used epitope-YSFKPMPLaR vaccine constructs to immunize rats, hamsters, rabbits, and cattle to a wide variety of B cell epitopes of various lengths (8 to 35 residues) and chemical modifications such as phospho-Tyr/Ser/Thr-containing epitopes. In all cases, immunizations were performed with the epitope-YSFK-PMPLaR construct dissolved merely in unbuffered or phosphate buffered saline (PBS) via simple subcutaneous and/or intraperitoneal injections.

One of the interesting possibilities presented by the molecular adjuvant-containing vaccines is that the molecular adjuvant might enable the processing and presentation of small molecule Ags that are particularly difficult to induce immune response against. In fact, we have generated a molecular adjuvant-containing vaccine to nicotine in which a single nicotine hapten was attached to the N-terminal end of *YKQGGFLGL*YSFKPMPLaR (Nic-*YKQGGFLGL*YS-FKPMPLaR (Sanderson et al., 2003). Rats immunized with this construct dissolved only in saline generated nicotine-specific Ab titers and were shown to be resistant to certain nicotine-induced behavioral effects.

46.4.8.2. Generation of Ag-Specific Cell-Mediated Responses

The effectiveness of YSFKPMPLaR to induce a cytotoxic T lymphocyte (CTL) response to a well defined T cell epitope was evaluated by immunizing mice with C5a-active and C5a-inactive constructs in which a T cell epitope from the hepatitis B surface antigen (HBsAg), *IPQSLDSW-WTSL*, was attached to the N-terminus and C-terminus of YSFKPMPLaR, respectively (Ulrich et al., 2000). BALB/c mice were immunized with the following constructs: (1) YSF-KPMPLaR; (2) *IPQSLDSWWTSL*; (3) *IPQSLDSW-WTSL***RR**; (4) the C5a-active constructs *IPQSLDSWWT-SL*YSFKPMPLaR, *IPQSLDSWWTSL***RR**YSFKPM-PLaR, and *IPQSLDSWWTSL***RVRR**YSFKPMPLaR; and (5) the C5a-inactive constructs YSFKPMPLaR**RR***IPQSLDSW-WTSL*, and *IPQSLDSWWTSL***RR**YSFKPMPLaR**G**. Some of these constructs were designed with a protease-sensitive linker sequence between the CTL epitope and the molecular adjuvant in order to provide a cleavage site for intracellular proteases that would separate the CTL epitope from the molecular adjuvant and facilitate its processing and presentation. Two protease-sensitive linker sequences were used in these designs. One was a double-Arg sequence (**RR**), which is sensitive to proteases of the subtilisin family (Barr, 1991) and other trypsin-like proteases. The other was a sequence specific to the ubiquitous intracellular subtilisin-like protease furin (**RVRR**) (Gordon et al., 1995).

BALB/c mice were immunized subcutaneously with 50–100 µg of the above constructs dissolved only in 100 µl of sterile PBS and were boosted at 21-day intervals. Spleen cells were harvested, incubated with 150 nM of the HBsAg CTL epitope (*IPQSLDSWWTSL*), and used as effector cells against ^{51}Cr-labeled target cells (P815S) that express the HBsAg.

Consistent with the above Ab results, only those mice immunized with the C5a-active constructs generated Ag-specific, CD8^{+} CTL responses. Interestingly, of the C5a-active constructs used in these immunizations, only the two containing the protease-sensitive linker sequences induced a CTL response; i.e., *IPQSLDSWWTSL***RR**YSFKPMPLaR and *IPQSLDSWWTSL***RVRR**YSFKPMPLaR, but not *IPQSLDSWWTSL*YSFKPMPLaR. This observation appears to implicate the importance of an intracellular proteolytic event in separating the epitope from the molecular adjuvant for facilitating Ag/epitope processing and presentation. Finally, mice immunized with either YSFKPMPLaR or the C5a-active epitope-YSFKPMPLaR constructs displayed no outward signs of immediate or delayed inflammatory responses.

In a related study (Pisarev et al., 2005), it was shown that murine dendritic cells pulsed with a C5a-active construct made from a T-cell epitope of the degenerate tandem repeat region of MUC1 (*SAPDNRPAL*) attached to the molecular adjuvant (*SAPDNRPAL***RR**YSFKPMPLaR) induced epitope-specific type 1 T cells. Also, spleen cells from mice transgenic for human MUC1 on the genetic background of C57BL/6 mice were stimulated with irradiated spleen cell pulsed with *SAPDNRPAL***RR**YSFKPMPLaR, *SAPDNRPAL*, or a structurally similar epitope *SAPDTRPAP*. Immunization with *SAPDNRPAL***RR**YSFKPMPLaR induced significant responses to both epitopes, while immunization with *SAPDNRPAL* alone induced responses only to the *SAPDNRPAL* epitope. These results demonstrate the ability of the molecular adjuvant-containing construct, *SAPDNRPAL***RR**YSFKPMPLaR, to induce significant epitope-specific type 1 T-cell responses to cross-reacting epitopes *SAPDNRPAL* and *SAPDTRPAP* derived from different regions of MUC1.

The immunization results obtained from the YSFKPMPLaR-induced humoral and cell-mediated responses suggest that the molecular adjuvant deliver both Ag and stimulatory signals to C5aR-bearing APCs; events that are consistent with the mechanism shown in Figure 46.1. The YSFKPMPLaR moiety of the epitope-YSFKPMPLaR construct interacts with C5aRs expressed on the surface of APCs, particularly dendritic cells, and induces the synthesis/release of cytokines that provide the "help" necessary to engage/activate helper T and/or B cells necessary for the observed CTL and Ab responses, respectively. Following the molecular adjuvant-mediated release of cytokines, the C5aR-ligand complex internalizes and carries the attached epitope into the APC cytoplasm where it is processed by various intracellular Ag processing pathways. The epitope then associates with MHC class I determinants that are subsequently expressed on the APC surface with the bound epitope (Tempero et al., 1997).

Although this mechanism of action for the molecular adjuvant still needs to be effectively characterized and is a major focus of research in our laboratory, it may suggest a novel pathway of exogenous MHC class I Ag presentation such

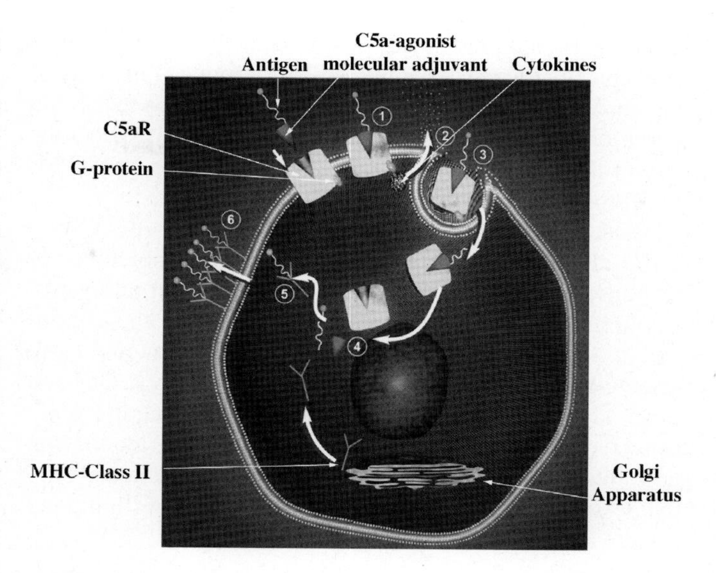

FIGURE 46.1. Proposed Mechanism for C5a Agonist Molecular Adjuvant. (1) The molecular adjuvant with an attached antigen binds to C5aRs on the surface of the APC. (2) The molecular adjuvant induces the release of Th1 cytokines. (3) The molecular adjuvant-vaccine/C5aR complex internalizes. (4) The antigen attached to the molecular adjuvant is processed by intracellular Ag processing pathways. (5) The processed Ag binds to intracellular MHC determinants. (6) The Ag is presented on the surface on the APC.

as that described previously using both *in vitro* and *in vivo* systems (Sigal et al., 1999; Raychaudhuri and Rock, 1998). Since it has been generally assumed that class I-mediated Ag presentation involves the generation of epitopes from endogenously synthesized proteins, the finding that extracellular proteins could be taken up by professional APCs (particularly macrophages and dendritic cells), processed in the cytoplasm or perhaps endosomes to yield antigenic epitopes, which are presented in association with MHC class I determinants, is of considerable significance. The molecular adjuvant moiety of these constructs, therefore, appears to both target the attached epitope to C5aR-bearing APCs *and* enhances the Ag processing/presentation capacity of the APC. Thus, the molecular adjuvant embodies adjuvant properties characteristic of both a vehicle and an immunomodulator.

The expression of an Ag-specific immune response appears to be critically dependent upon the ability of the molecular adjuvant to interact with C5aRs expressed on APCs; however, the extent to which the size of the epitope would sterically interfere with this C5aR/molecular adjuvant interaction it not yet known. All the epitopes we have used to date have been peptides ranging in length from 8 to 35 residues, but it is conceivable that whole proteins could be used as Ags in such molecular adjuvant-containing vaccines. Such a vaccine could be readily generated by attaching a chemically-active or photochemically-active linker to the N-terminal end of YSFKPMPLaR during solid-phase synthesis (Hansen et al., 1996), which then could be used to decorate the surface a protein with multiple copies of YSFKPMPLaR. Such a vaccine construct would have the obvious advantage of containing an Ag having multiple B- and T- cell epitopes that could cover a range of immunogenic responses. Again, whether the size of the protein Ag would disallow the smaller YSFKPMPLaR to interact with C5aRs on APCs is unknown.

These Ab and CTL responses observed with YSFKPMPLaR *in vivo* support the therapeutic potential of response-selective agonists of C5a as molecular adjuvants. Furthermore, such encouraging results justify the continued design of analogues that exhibit more potent and more selective C5a-like immunostimulatory properties. Our and other laboratories are actively engaged in structure-function studies of such analogues with the goal of identifying new topochemical features that could be introduced into the design of C5a agonists that may express an affinity and selectivity for C5aRs on APCs.

46.4.9. Improved Molecular Adjuvant Design: N-Methylation of Backbone Amides

Our structure-function analyses of YSFKPMPLaR implicated the importance of extended backbone conformation as introduced by the two Pro residues at positions 69 and 71 (Finch et al., 1997; Vogen et al., 1999a,b). However, the substitution of a Pro residue to induce such extended conformation also removes the contribution made the side-chain of the substituted residue. Thus, we generated a panel of YSFKPMPLaR analogues in which methyl groups were introduced onto certain backbone amide nitrogen atoms. The presence of an amide methyl group forces an extended backbone conformation of the preceding residue, R_{n-1}, in a manner similar to Pro (Hruby and Nikiforovich, 1991), but retains the biologic contribution made by the side-chain of the residue now bearing the N-methyl group.

From this panel of N-methylated analogues, one (YSFKDMP(MeL)aR) was shown to be considerably more potent in and selective for eicosanoid release from macrophages than enzyme release from neutrophils relative to YSFKPMPLaR (Vogen et al., 2001). Furthermore, binding of YSFKDMP(MeL)aR to C5aRs on human neutrophils could not be detected, but YSFKDMP(MeL)aR binding to C5aRs on macrophages was comparable with YSFKPMPLaR (Vogen et al., 2001). This *in vitro* response-selectiveness was confirmed *in vitro* by a study the vasopressive and neutropenic effects of natural C5a, YSFKPMPLaR, and YSFKDMP(MeL)aR were evaluated in a rat model (Vogen et al., 2001). Rats were anesthetized and administered 2 μg/kg hrC5a and 3 μg/kg YSFKPMPLaR or YSFKDMP(MeL)aR intravenously. YSFKDMP(MeL)aR, identical to YSFKPMPLaR and hrC5a, exhibited a pronounced and rapid decrease in mean blood pressure that recovered to control levels over 90 minutes. However, in marked contrast to YSFKPMP-LaR and hrC5a, YSFKDMP(MeL)aR exhibited no neutropenia or changes in circulating blood volume.

The presence of N-methyl groups on the backbone of the molecular adjuvant YSFKPMPLaR dramatically alters its selectivity for C5aR-mediated activity in macrophages versus neutrophils and may provide a synthetic approach for the development of highly response-selective agonist of C5a and, consequently, molecular adjuvants with little/no inflammatory effects via the engagement of C5aR-bearing neutrophils. Indeed, preliminary results suggest that YSFKDMP(MeL)aR, like YSFKPMPLaR, behaves as an equally effective molecular adjuvant capable of inducing a robust, Ag-specific CTL response to various CTL epitopes (Floreani, A., Heires, A., and Sanderson, S., 2006 unpublished results).

46.5. Advantages of Molecular Adjuvant-Containing Vaccines

Molecular adjuvant-containing vaccines are generated by straightforward, standard, solid phase peptide synthesis, which allows for several advantages over conventional vaccines. Peptide synthesis allows for the elegant control the number of Ags/epitopes attached to the molecular adjuvant. Thus, the exact Ag-to-molecular adjuvant ratio and chemical composition for each lot of vaccine produced can be known, a huge advantage in quality control during manufacture. Peptide synthesis is inexpensive and amenable to scale-up. Thus, hundreds of kilograms of a molecular adjuvant-containing vaccine can be produced in a few days using commercially available equipment. In contrast to conventional vaccines, a molecular adjuvant-containing vaccine is a chemical rather than a biological product. Thus, recombinant methods are not required to generate any vaccine component and, consequently, it can be generated with great purity using standard HPLC methods with no DNA, viral, or bacterial contamination. Vaccine purity and quality can be assured and controlled, therefore, by in-house monitoring. At the end of this purification process, the vaccine is lyophilized to a dry powder, which has a shelf life of years at room temperature. Thus, the vaccine can be readily distributed in this dry powdered form to the clinic where it need only be taken up in saline for the deliverable vaccine formulation when needed. As such, a viable and long shelf-life after distribution can be assured.

Summary

This chapter has highlighted some of the characteristics of traditional and non-traditional adjuvants, the advantages and disadvantages of their use in vaccines, and has introduced the concept of molecular adjuvants as a way to improve immune outcome while minimizing undesirable side-effects. A particular emphasis was placed on the generation of conformationally-restricted, response-selective agonists of the complement component C5a and their potential as molecular adjuvants and at least one synthetic approach by which their molecular adjuvant properties may be improved. Such molecular adjuvant-containing vaccines appear to have tremendous therapeutic, manufacturing, and commercial potential. However, the development and use of such molecular adjuvants and molecular adjuvant-containing vaccines represent only one approach to producing vaccines capable of meeting the modern-day threats posed by antibiotic-resistant strains of bacteria and bioterrorism. It is clear that a well-organized and concerted worldwide effort that utilizes multiple approaches to novel adjuvant and vaccine designs will be required to overcome these growing threats to human and animal populations.

Review Questions/Problems

1. **The majority of adjuvants in experimental use today are composed of various bacterial components. Briefly explain the rationale for the use of such bacterial components in adjuvants.**
2. **Briefly describe the two major biological functions of the complement system.**
3. **The complement-derived anaphylatoxin C5a has many prominent immune stimulatory properties and, consequently, is a good candidate for use as a molecular adjuvant. However, its use as such would be severely limited. Describe the major limitation(s) for the use of natural C5a as a molecular adjuvant.**
4. **Antigen presenting cells (APC) represent an important class of immune cells essential for generating an antigen-specific immune response. Briefly describe the basic cellular mechanism by which APCs contribute to this antigen-specific immune outcome.**
5. **Conventional vaccines in use today typically require the use of added adjuvants to initiate and potentiate an antigen-specific immune response. In contrast, vaccines that incorporate the conformationally-biased agonists of C5a as molecular adjuvants described in this chapter require no added adjuvants – vaccinations are accomplished with just water/saline as the vehicle. Describe why this is possible.**
6. **In addition to the ability to target antigens to C5a receptor-bearing APCs, the conformationally-biased C5a agonists/molecular adjuvants induce C5a-like responses at the C5a receptor on the APC. Describe these C5a-like responses.**
7. **Describe why the conformationally-biased C5a agonists/molecular adjuvants invoke their effects at the C5a receptor expressed on APCs, but not C5a receptors expressed on inflammatory neutrophils.**
8. **Bacteria and various components of bacteria are used as adjuvants in order to provide the bacterial signals that activate principally**
   a) antibody synthesis
   b) the innate arm of immunity
   c) clonal expansion of B cells
   d) oxidative phosphorylation
9. **Adjuvants typically fall into two major categories, which are**
   a) vesicles and nanoparticles
   b) proteins and DNA
   c) live and attenuated
   d) vehicles and immunomodulators
10. **The two principal biological effects of complement component C5a are**
    a) intercalation of DNA and *de novo* protein synthesis
    b) inflammation and immune stimulation
    c) neo-vascularization and spleen development
    d) protease activation and carbohydrate clearance
11. **The C-terminal region of complement-derived C5a (C5a65-74) represents**
    a) the biologically active region of C5a
    b) a post-translational signal sequence
    c) a site for phosphorylation by kinases
    d) a immunogenic site recognized by T lymphocytes
12. **A molecular adjuvant can be defined by all of the following *except*:**
    a) targets antigens to dendritic cells
    b) is a single molecular entity
    c) down regulates costimulatory factors
    d) activates specific pathways of antigen presentation

References

Afonso LCC, Scharton TM, Vieira LQ, Wysocka M, Trinchieri G, Scott P (1994) The adjuvant effect of interleukin-12 in a vaccine against Leishmania major. Science 263:235–237.

Amsterdam EA, Stahl GL, Pan H-L, Rendig SV, Fletcher MP, Longhurst JC (1995) Limitation of reperfusion injury by a monoclonal antibody to C5a during myocardial infarction in pigs. Am J Physiol 268:H448–H457.

Arvieux J, Yssel H, Colomb M-G (1988) Ag-bound C3b and C4b enhance Ag-presenting cell function in activation of human T-cell clones. Immunology 65:229–232.

Barr PJ (1991) Mammalian subtilisins: The long-sought dibasic processing endoproteases. Cell 66:1–3.

Bergh K, Iverson OJ, Lysvand H (1993) Surprisingly high levels of anaphylatoxin C5a des Arg are extractable from psoriatic scales. Arch Dermatol Res 285:131–134.

Biragyn A, Tani K, Grimm MC, Weeks S, Kwak LW (1999) Genetic fusion of chemokines to a self tumor antigen induces protective, T-cell dependent anti-tumor immunity. Nat Biotechnol 17:253–258.

Buchner RR, Hugli TE, Ember JA, Morgan EL (1995) Expression of functional receptors for human C5a anaphylatoxin (CD88) on the human hepatocellular carcinoma cell line HepG2. Stimulation of acute-phase protein-specific mRNA and protein synthesis by human C5a anaphylatoxin. J Immunol 155:308–315.

Crouch CF, Daly J, Henly W, Hannant D, Wilkins J, Francis MJ (2005) The use of a systemic prime/mucosal boost strategy with an equine influenza ISCOM vaccine to induce protective immunity in horses. Vet Immunol Immunopathol 108:345–355.

Defoort J-P, Nardelli B, Huang W, Tam JP (1992) A rational design of synthetic peptide vaccine with a built-in adjuvant. Int J Pept Protein Res 40:214–221.

Dempsey PW, Allison MED, Akkaraju S, Goodnow CC, Fearon DT (1996) C3d of complement as a molecular adjuvant: Bridging innate and acquired immunity. Science 271:348–350.

Diwan M, Elamanchili P, Cao M, Samuel J (2004) Dose sparing of CpG oligodeoxynucleotide vaccine adjuvants by nanoparticle delivery. Curr Drug Deliv 1:405–412.

Drapeau G, Brochu S, Godin D, Levesque L, Rioux F, Marceau F (1993) Synthetic C5a receptor agonists: Pharmacology, metabolism and in vivo cardiovascular and hematologic effects. Biochem Pharmacol 45:1289–99.

Ember JA, Sanderson SD, Taylor SM, Kawahara M, Hugli TE (1992) Biologic activity of synthetic analogues of C5a anaphylatoxin. J Immunol 148:3165–3173.

Ember JA, Sanderson SD, Hugli TE, Morgan EL (1994) Induction of IL-8 release from monocytes by human C5a anaphylatoxin. Am J Pathol 114:393–403.

Finch AM, Vogen SM, Sherman SA, Kirnarsky L, Taylor SM, Sanderson SD (1997) Biologically active conformer of the effector region of human C5a and modulatory effects of N-terminal receptor binding determinants on activity. J Med Chem 40:877–884.

Finch AM, Wong AK, Paczkowski NJ, Wadi SK, Craik DJ, Fairlie DP, Taylor SM (1999) Low-molecular-weight peptidic and cyclic antagonists of the receptor for the complement factor C5a. J Med Chem 42:1965–1974.

Floreani AA, Heires AJ, Welniak LA, Miller-Lindholm A, Clark-Pierce L, Rennard SL, Morgan EL, Sanderson SD (1998) Expression of receptors for C5a anaphylatoxin (CD88) on human bronchial epithelial cells: Enhancement of C5a-mediated release of IL-8 upon exposure to cigarette smoke. J Immunol 160:5073–5081.

Gasque P, Chan P, Fontaine M, Ischenko A, Lamacz M, Gotze O, Morgan BP (1995) Identification and characterization of the complement C5a anaphylatoxin receptor on human astrocytes. J Immunol 155:4882–4889.

Goldstein IM and Weissmann G (1974) Generation of C5a-derived lysosomal enzyme-releasing activity (C5a) by lysates of leukocyte lysosomes. J Immunol 113:1583–1588.

Goodman MG, Chemoweth DE, Weigle WO (1982) Potentiation of the primary humoral immune response in vitro by C5a anaphylatoxin. J Immunol 129:70–75.

Gordon VM, Klimpel KR, Arora N, Henderson A, Leppla SH (1995) Proteolytic activation of bacterial toxins by eukaryotic cells is performed by furin and by additional cellular proteases. Infect Immun 63:82–87.

Hansen PR, Flyge H, Holm A, Lauritzen E, Larsen BD (1996) Photochemical conjugation of peptides to carrier proteins using 1,2,3-thiadiazole-4-carboxylic acid. Int J Pept Protein Res 47:419–426.

Hogan MM, Yancy KB, Vogel SN (1989) Role of C5a in the induction of tumoricidal activity in C3H/HeJ (Lpsd) and C3H/OuJ (Lpsn) macrophages. J Leukoc Biol 46:565–570.

Hopkins P, Belmont HM, Buyon J, Phillips M, Weissmann G, Abramson SB (1988) Increased levels of plasma anaphylatoxins in systemic lupus erythematosus predict flares of the disease and may elicit vascular injury in lupus cerebritis. Arthritis Rheum 31:632–641.

Hruby VJ, Nikiforovich GV (1991) The Ramachandran Plot and Beyond: Conformational and Topographical Considerations in the Design of Peptides and Proteins. In: Molecular Conformation and Biological Interactions (Balaram P and Ramasehan S, eds) pp 429–445. Bangalore: Indian Academy of Sciences.

Hugli TE (1981) The structural basis for anaphylatoxin and chemotactic functions of C3a, C4a, and C5a. Crit Rev Immunol 114:321–366.

Hugli TE (1990) Structure and function of C3a anaphylatoxin. Curr Top Microbiol Immunol 153:181–208.

Hugli TE and Muller-Eberhard HJ (1978) Anaphylatoxins: C3a and C5a. Adv Immunol 26:1–53.

Hugli TE, Marceau F, Lundberg C (1987) Effects of complement fragments on pulmonary and vascular smooth muscle. Am Rev Respir Dis 135:S9–S13.

Jacquier-Sarlin MR, Gabert FM, Villiers M-B, Colomb M-G (1995) Modulation of antigen processing and presentation by covalently linked complement C3b. Immunology 84:164–170.

Johnson AG, Gains S, Landy M (1956) Studies on the O antigen of Salmonella typhosa V Enhancement of antibody responses to proteinantigens by the purified lipopolysaccharide. J Exp Med 103:225–246.

Johnson AR, Hugli TE, Muller-Eberhard HJ (1975) Release of histamine from rat mast cells by the complement peptides C3a and C5a. Immunology 28:1067–1080.

Jose PJ, Moss IK, Maini RN, Williams TJ (1990) Measurement of the chemotactic complement fragment C5a in rheumatoid synovial fluids by radioimmunoassay: Role of C5a in the acute inflammatory phase. Ann Rheum Dis 49:747–752.

Kawamoto S, Yalcindag A, Laouini D, Brodeur S, Bryce P, Lu B, Humbles AA, Oettgen H, Gerard C, Geha RS (2004) The anaphylatoxin C3a downregulates the Th2 response to epicutaneously introduced antigen. J Clin Invest 114:399–407.

Kawatsu R, Sanderson SD, Blanco I, Kendall N, Finch AM, Taylor SM, Colcher DM (1996) Conformationally biased analogs of human C5a mediate changes in vascular permeability. J Pharmacol Exp Ther 278:432–440.

Konteatis ZD, Siciliano J, VanRiper G, Molineaux CJ, Pandya S, Fischer P, Rosen H, Mumford RA, Springer MS (1994) Development of C5a receptor antagonists: Differential loss of functional responses. J Immunol 153:4200–4205.

Kurzawa H, Wysocka M, Aruga E, Chang AE, Trinchieri G, Lee MF (1998) Recombinant interleukin-12 enhances cellular immune responses to vaccination only after a period of suppression. Cancer Res 58:491–499.

Lundberg C, Marceau F, Hugli TE (1987) C5a-induced hemodynamic and hematologic changes in the rabbit: The role of cyclooxygenase products and polymorphonuclear leukocytes. Am J Pathol 128:471–483.

Martinon F, Gras-Masse H, Boutillon C, Chirat F, Deprez B, Guillet J-G, Gomard E, Tartar A, Levy J-P (1992) Immunization of mice with lipopeptides bypasses the prerequisite for adjuvant. J Immunol 149:3416–3422.

Mastellos D, Andronis C, Persidis A, Lambris JD (2005) Novel biological networks modulated by complement. Clin Immunol 115:225–235.

Metzger J, Wiesmuller K-H, Schaude R, Bessler WG, Jung G (1991) Synthesis of novel immunologically active tripalmitoyl-S-glycerylcysteinyl lipopeptides as useful intermediates for immunogen preparations. Int J Pept Protein Res 37:46–57.

Mollison KW, Mandecki W, Zuiderweg ERP, Fayer L, Fey TA, Krause RA, Conway RG, Miller L, Edalji RP, Shallcross MA, Lane B, Fox JL, Greer J, Carter GW (1989) Identification of receptor-binding residues in the inflammatory complement protein C5a by site-directed mutagenesis. Proc Natl Acad Sci (USA) 86:292–296.

Montz H, Fuhrmann A, Schulze M, Gotze O (1990) Regulation of the human autologous T cell proliferation by endogenously generated C5a. Cell Immunol 127:337–351.

Morgan EL (1986) Modulation of the immune response by anaphylatoxins. Complement 3:128–136.

Morgan EL, Thoman ML, Weigle WO, Hugli TE (1983) Anaphylatoxin-mediated regulation of the immune response II. C5a-mediated enhancement of human humoral and T cell-mediated immune responses. J Immunol 130:1257–1261.

Morgan EL, Sanderson SD, Scholz W, Noonan D, Weigle WO, Hugli TE (1992) Identification and characterization of the effector region within human C5a responsible for stimulation of IL-6 synthesis. J Immunol 148:3937–3942.

Muhlradt PF, Kiess M, Meyer H, Sussmuth R, Jung G (1998) Structure and specific activity of macrophage-stimulating lipopeptides from Mycoplasma hyorhinis. Infect Immunity 66:4804–4810.

Mulligan MS, Schmid E, Beck-Schimmer B, Till GO, Friedl HP, Brauer RB Hugli TE, Miyasaka M, Warner RL, Johonson KJ, Ward PA (1996) Requirement and role of C5a in acute lung inflammatory injury in rats. J Clin Invest 98:503–512.

Okusawa S, Dinarello CA, Yancy S, Endres RJ, Lawley TJ, Frank MM, Burke JF, Garland JA (1987) C5a-induction of human interleukin-1: Synergistic effect of endotoxin or interferon-gamma. J Immunol 139:2635–2639.

Olive C, Toth I, Jackson D (2001) Technological advances in antigen delivery and synthetic peptide vaccine development strategies. Mini Rev Med Chem 1:429–438.

Pisarev VM, Kirnarsky L, Caffrey T, Hanisch F-G, Sanderson SD, Hollingsworth MA, Sherman SA (2005) T cells recognize PD(N/T)R motif common n a variable number of tandem repeat and degenerate repeat sequences of MUC1. Int Immunopharmacol 5:315–330.

Pobanz JM Reinhardt RA, Koka S, Sanderson SD (2000) C5a modulation of interleukin-1 beta-induced interleukin-6 production by human osteoblast-like cells. J Periodontal Res 35:137–145.

Raychaudhuri S and Rock KL (1998) Fully mobilizing host defense: Building better vaccines. Nat Biotechnol 16:102510–102531.

Sanderson SD, Cheruku SR, Padmanilayam MP, Vennerstrom JL, Thiele GM, Palmatier MI, Bevins RA (2003) Immunization to nicotine with a peptide-based vaccine composed of a conformationally biased agonist of C5a as a molecular adjuvant. Int Immunopharmacol 3:137–146.

Scholz W, McClurg MR, Cardenas GJ, Smith M, Noonan DJ, Hugli TE, Morgan EL (1990) C5a-mediated release of interleukin-6 by human monocytes. Clin Immunol Immunopathol 57:297–307.

Schorlemmer HU, Davies P, Allison AC (1976) Ability of activated complement components to induce lysosomal enzyme release from macrophages. Nature 261:48–49.

Shi B, Tang P, Hu X, Liu JO, Yu B (2005) OSW saponins: Facile synthesis toward a new type of structures with potent antitumor activities. J Org Chem 70:10354–10367.

Shin HS, Snyderman E, Friedman A, Mellors A, Meyer MM (1968) Chemotactic and anaphylactic fragments cleaved from the fifth component of guinea-pig complement. Science 162:361–363.

Sigal L, Crotty JS, Andino R, Rock KL (1999) Cytotoxic T-cell immunity to virus-infected non-haematopoietic cells requires presentation of exogenous antigen. Nature 398:77–80.

Smedegard G, Cui L, Hugli TE (1989) Endotoxin-induced shock in the rat. A role for C5a. Am J Pathol 135:489–497.

Taylor SM, Sherman SA, Kirnarsky L, Sanderson SD (2001) Development of response-selective agonists of human C5a anaphylatoxin: Conformational, biological, and therapeutic considerations. Curr Med Chem 8:675–684.

Tempero RM, Hollingsworth MA, Burdick MD, Finch AM, Taylor SM, Vogen SM, Morgan EL, Sanderson SD (1997) Molecular adjuvant effects of a conformationally biased agonist of human C5a anaphylatoxin. J Immunol 158:1377–1382.

Till GO, Johnson KJ, Kunkel R, Ward PA (1982) Intravascular activation of complement and acute lung injury. Dependency on neutrophils and toxic oxygen metabolites. J Clin Invest 69:112611–35.

Ulrich JT and Myers KR (1995) Monophosphoryl lipid A as an adjuvant: Past experiences and new directions. Pharm Biotechnol 6:495–524.

Ulrich JT, Cieplak W, Paczkowski NJ, Taylor SM, Sanderson SD (2000) Induction of an antigen-specific CTL response by a conformationally biased agonist of human C5a anaphylatoxin as a molecular adjuvant. J Immunol 164:5492–5498.

Van Regenmortel M (1997) Searching for safer, more potent, better-targeted adjuvants. ASM News 63:136–139.

Vogen SM, Wadi SK, Finch AM, Thatcher JD, Monk PN, Taylor SM, Sanderson SD (1999a) The influence of Lys68 in decepeptide agonists of C5a on C5a receptor binding, activation and selectivity. J Pept Res 53:8–17.

Vogen SM, Prakash O, Kirnarsky L, Sanderson SD, Sherman SA (1999b) Determination of structural elements related to the biological activities of a potent decapeptide agonist of human C5a anaphylatoxin. J Pept Res 54:74–84.

Vogen SM, Paczkowski NJ, Kirnarsky L, Short A, Whitmore JB, Sherman SA, Taylor SM, Sanderson SD (2001) Differential activities of decapeptide agonists of human C5a: The conformational effects of backbone N-methylation. Int Immunopharmacol 1:2151–2162.

Wetsel RA (1995) Structure, function and cellular expression of complement anaphylatoxin receptors. Curr Opin Immunol 7:48–53.

Wiesmuller K-H, Bessler WG, Jung G (1992) Solid phase peptide synthesis of lipopeptide vaccines eliciting epitope-specific B-, T-helper and T-killer cell response. Int J Pept Protein Res 40:255–260.

Wingrove JA, DiScipio RG, Chem Z, Potempa J, Travis J, Hugli TE (1992) Activation of complement components C3 and C5 by a cysteine proteinase (gingipain-1) from Porphyromonas (Bacteroides) gingivalis. J Biol Chem 267:18902–18907.

Zuiderweg ERP, Nettesheim DG, Mollison KW, Carter GW (1989) Tertiary structure of human complement component C5a in solution from nuclear magnetic resonance data. Biochemistry 28:172–185.

47
Polymer Nanomaterials

Alexander V. Kabanov and Elena V. Batrakova

Keywords Block copolymer; Cell-mediated delivery; Liposome; Multidrug resistance; Nanofiber; Nanogel; Nanoparticle; Nanosphere; Nanosuspension; Nanotube

47.1. Introduction

The blood brain barrier (BBB) is one of the most challenging barriers for drug delivery in the body. It significantly restricts the entry of low molecular weight compounds and biomacromolecules to the brain from the periphery. Inefficient delivery of the drugs, DNA and proteins to the brain is a major bottleneck in development of more efficacious and safe modalities for diagnostics and treatment of neurological diseases, especially at early stages of the disease when the BBB remains intact. The low permeability of the BBB is attributed, in large part, to the brain microvessel endothelial cells (BMVEC), which form tight extracellular junctions and have low pinocytic activity (Mayhan, 2001; Pardridge, 2005a). Some relatively lipophilic and low molecular weight substances can transport across the BMVEC by passive diffusion. However, a large number of lipophilic compounds are rapidly effluxed from the brain into the blood by extremely effective drug efflux systems expressed in the BBB (Begley, 1996; Fromm, 2000; Tamai and Tsuji, 2000; Loscher and Potschka, 2005a). These efflux systems include P-glycoprotein (Pgp), Multidrug Resistance Proteins (MRPs), breast cancer resistance protein (BCRP), and the multi-specific organic anion transporter (MOAT). There is also an enzymatic barrier to drug transport in the BMVEC. Activity of many enzymes that participate in the metabolism and inactivation of endogenous compounds, such as γ-glutamyl transpeptidase, alkaline phosphatase, and aromatic acid decarboxylase is elevated in cerebral microvessels (Minn et al., 1991; Abbott and Romero, 1996). These features of the BBB require discovery of new modalities allowing for effective drug delivery to the central nervous system (CNS), which is of great need and importance for treatment of neurodegenerative disorders.

A number of earlier publications and extensive reviews on delivery of small drug molecules and biomacromolecules to the brain are available in the literature (Spector, 2000; Lo et al., 2001; Banks and Lebel, 2002; Cornford and Cornford, 2002; Pardridge, 2002a, b; Begley, 2004a; Kas, 2004; Gaillard et al., 2005; Garcia-Garcia et al., 2005; Liu et al., 2005; Liu and Chen, 2005; Loscher and Potschka, 2005b; Pardridge, 2005a; Roney et al., 2005). The present chapter describes recent findings in the development of a new generation of polymer nanomaterials for drug delivery to the CNS.

47.2. History and Principles of Drug Delivery Using Polymers

The original discovery of a polymer drug delivery system was published in 1964 by Folkman and Long (Folkman and Long, 1964). This work reported that hydrophobic low molecular weight drugs can diffuse through the wall of silicone tubing at a controlled rate. Since then, polymers have occupied a central status in controlled drug release as well as in the fabrication of drug delivery systems (Langer, 2001; Duncan, 2003). During the past decades a large number of drug delivery systems, mostly in the form of microspheres, films, tablets, or implantation devices, have been designed to achieve sustained drug release by taking advantage of the unique properties of polymers.

There are several fundamental properties of polymers that are useful in solving drug delivery problems. First, polymers can be designed to be intrinsically multifunctional and can, for example, be combined either covalently or noncovalently with drugs to overcome multiple problems such as solubility, stability, permeability, etc. Second, polymers can be easily modified with various targeting vectors to direct drugs to specific sites in the body. Third, polymers can be designed to be environmentally responsive materials, allowing for the controlled and sustained release of a drug at its site of action. Finally, polymers themselves can

T. Ikezu and H.E. Gendelman (eds.), *Neuroimmune Pharmacology.*
© Springer 2008

be biologically active, and this property can be exploited in order to modify the activity of various endogenous drug transport systems within the body to improve delivery and, therefore, drug performance.

Numerous polymer-based therapeutics are on the market or undergoing clinical evaluation to treat cancer and other diseases. Many of them are low molecular weight drug molecules or therapeutic proteins that are chemically linked to water-soluble polymers to increase drug solubility, drug stability, or enable site-specific transport of drugs to target tissues affected by the disease.

Recently, as a result of the rapid development of novel nanotechnology-derived materials, a new generation of polymer therapeutics has emerged which uses materials and devices of nanoscale size for the delivery of drugs, genes, and imaging molecules (Kabanov et al., 1995a; Salem et al., 2003; Savic et al., 2003; Kabanov and Batrakova, 2004; Missirlis et al., 2005; Nayak and Lyon, 2005; Trentin et al., 2005). These materials include liposomes, dendrimers, polymer micelles, polymer-DNA complexes ("polyplexes"), and other nanostructured materials for medical use, that are collectively called nanomedicines. Compared to the first generation of polymer therapeutics, the new generation of nanomedicines is more advanced. They entrap small drugs or biopharmaceutical agents, such as therapeutic proteins and DNA, and can be designed to trigger the release of these agents at the target site. Moreover, they can be targeted not only to the particular organ or tissue, but to a particular cell or even an intracellular compartment. The essence of nanotechnology is the ability to work at the molecular level to create large structures with fundamentally new molecular organization. Materials with features on the scale of nanometers often have properties different from their macroscale counterparts. Many nanomedicines are constructed using self-assembly principles, such as spontaneous formation of micelles or interpolyelectrolyte complexes driven by diverse molecular interactions (hydrophobic, electrostatic, etc.). Nanotechnology focuses not only on formulating therapeutic agents in biocompatible nanocomposites but also on exploiting distinct advantages associated with a reduced dimensional scale within 1–100 nm. Some examples of nanoscaled polymeric carriers involve polymer conjugates, polymeric micelles, and polymersomes (Panyam and Labhasetwar, 2003). Because these systems often exhibit similarity in their size and structure to natural carries such as viruses and serum lipoproteins, they offer multifaceted specific properties in drug delivery applications (Lavasanifar et al., 2002). There is still a lack of understanding in terms of why certain arrangements of small molecules cause dramatic changes in their behavior. With this in mind, the goal of this chapter is to explore current research in polymeric nanoparticles, where the specific arrangement of the polymeric matter at the nanoscale imparts new properties that are not attainable from simple polymer solutions, and to summarize their applications for drug delivery systems to the CNS.

47.3. Nanocarriers for Drug Delivery

The need for a protective drug carrier for CNS drugs is highlighted by the fact that many drugs have a low hydrolytic stability and are subject to degradation by blood proteins or by enzymes encountered in the BBB. Furthermore, a drug carrier can be targeted via receptor-mediated transport using a brain-specific vector moiety. A single unit of a given drug carrier can incorporate many drug molecules, resulting in high "payloads" per one targeting moiety. By increasing the payload of the carrier, one might improve the efficacy of the delivery while maintaining a relatively low level of involvement of numbers of targeted moieties and receptors. There are various types of vehicles proposed for transport of neuropharmaceuticals across the BBB: liposomes, degradable nanoparticles, nanospheres, nanosuspensions, polymeric micelles, nanogels, block ionomer complexes, nanofibers and nanotubes (Figure 47.1).

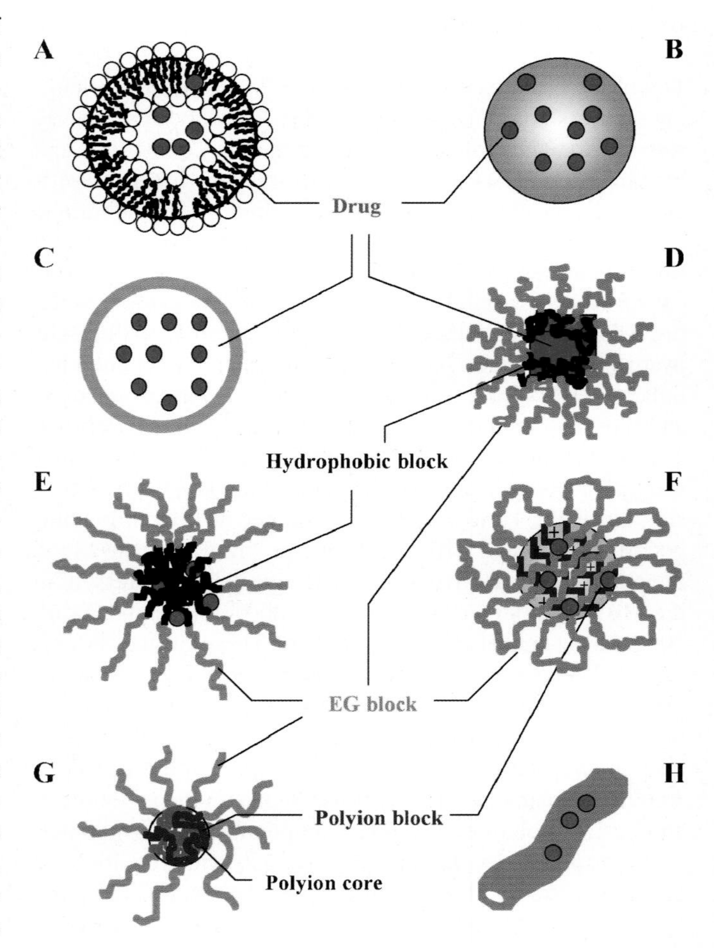

FIGURE 47.1. Types of nanocarriers for drug delivery. A: liposomes; B: nanoparticles; C: nanospheres; D: nanosuspensions; E: polymer micelles; F: nanogel; G: block ionomer complexes; H: nanofibers and nanotubes.

47.3.1. Types of Nanocarriers for Drug Delivery to the Brain

Liposomes are vesicular structures, which are usually composed of unilamellar or multilamellar lipid bilayers (similar to biological membranes) that surround internal aqueous compartments (Figure 47.1A). The size of the vesicles varies from several nanometers to several microns. Relatively large amounts of drug molecules can be incorporated into either the aqueous compartment (water soluble compounds) or the lipid bilayers (lipophilic compounds) providing the possibility to use liposomes as carriers for drug delivery. The use of liposomes for drug delivery across the brain capillaries has been reported extensively (Umezawa and Eto, 1988; Huwyler et al., 1996; Rousseau et al., 1999; Shi et al., 2001; Mora et al., 2002; Thole et al., 2002; Wu et al., 2002; Omori et al., 2003; Schmidt et al., 2003; Zhang et al., 2003b; Aoki et al., 2004; Gosk et al., 2004; Chekhonin et al., 2005; Pardridge, 2005b). In general, encapsulation of a drug into liposomes prolongs the circulation time of the drug in the blood stream, reduces adverse side effects, and in selected cases enhances therapeutic effects of CNS agents. Conventional liposomes containing hydrocortisone were demonstrated to penetrate the BBB in experimental autoimmune encephalomyelitis (Rousseau et al., 1999). In addition, liposomes prepared using lecithin, cholesterol, and *p*-aminophenyl-α-mannoside were efficiently transported across the BBB into mouse via mannose receptor-mediated transcytosis (Umezawa and Eto, 1988). An interesting approach using thermosensitive liposomes loaded with an antineoplastic agent, doxorubicin (Dox), for treatment of malignant gliomas was reported (Aoki et al., 2004). These liposomes released their content when the tumor core was heated to 40°C by a brain heating system. Elevated accumulation of the drug in the brain of the heated animals resulted in a significantly longer overall survival time compared to the non-heated animals.

Conventional liposomes normally are cleared rapidly from the circulation by the reticuloendothelial system. Extended circulation time of these carriers can be accomplished with small-sized liposomes (<100 nm) composed of neutral, saturated phospholipids and cholesterol. Further, it was demonstrated that water-soluble polymers such as polyethylene glycol (PEG) attached to the surface of liposomes reduce adhesion of opsonic plasma proteins, and decrease recognition and rapid removal of liposomes from the circulation by the mononuclear phagocyte system in liver and spleen (Huwyler et al., 1996; Kozubek et al., 2000; Voinea and Simionescu, 2002). Using this approach, PEG-coated ("PEGylated") long-circulating liposomes (or "stealth liposomes") were shown to remain in the circulation with a half-life as long as 50 h in humans (Gabizon et al., 1994). Particularly, commercial PEGylated liposome-encapsulated Dox, Doxil, was already approved for use in the treatment of recurrent ovarian cancer and AIDS-related Kaposi's sarcoma (Gabizon et al., 2003) and was shown to be effective in patients with metastatic breast cancer (Papaldo et al., 2006). Long-circulating PEGylated liposomes were also used for the delivery of high doses of glucocorticosteroids to the CNS to treat multiple sclerosis (Schmidt et al., 2003). The PEGylated liposomes encapsulating prednisolone provided selective targeting to the inflamed CNS in rats (up to 4.5-fold higher accumulation compared to the healthy animals). The mechanism of preferential accumulation of these liposomes in the brain is not fully understood. It appears to be crucial that the liposomes exhibit long-circulating behavior. Their effect is related to either (1) the change in the pharmacokinetics of the encapsulated drug resulting in a greater drug exposure to the BBB or (2) to liposome capture by monocytes/macrophages (in liver, spleen, or blood) followed by cell-mediated transport to the brain.

Attachment of targeting moieties to PEGylated liposomes can direct them to the BBB. Thus, efficient delivery of PEG-liposomes conjugated with transferrin (Tf) to the post-ischemic cerebral endothelium was achieved in rats (Omori et al., 2003). The expression of Tf receptor in the cerebral endothelium was reported to increase with a peak at the first day after induced transient middle cerebral occlusion, which can be employed for drug delivery to the brain after stroke. PEGylated immunoliposomes were also successfully employed to target and transfect brain tissues (Shi et al., 2001). A plasmid DNA was encapsulated within the liposome. Packaging of the DNA in the interior of the liposome prevented degradation of the therapeutic gene by the ubiquitous endonucleases *in vivo*. These liposomes were directed to Tf receptor-rich tissues, such as brain, liver, and spleen by monoclonal antibodies, OX26, that were linked to the free termini of the PEG chains. When a reporter gene was also coupled with a brain-specific glialfibrillary acidic protein promoter, its expression was achieved predominantly in the brain (Shi and Pardridge, 2000).

In addition considerable work was reported on antibodies to transferrin receptor, OX26, that were linked to the surface of PEGylated liposomes via PEG spacers. Such immunoliposome constructs were used to deliver small drugs, Daunomycin (Huwyler et al., 1996), and Digoxin (Huwyler et al., 2002) as well as plasmid DNA (Shi et al., 2001) to the brain. Notably, OX26-conjugated liposomes selectively distributed to BMVEC but avoided choroid plexus epithelium, neurons, and glia (Gosk et al., 2004). In related work OX26 antibodies directly linked to oligonucleotides (Wu et al., 1996) or fibroblast growth factor (Wu et al., 2002) enhanced delivery of these molecules to the brain.

Another example of a brain targeting vector with a remarkable species-specificity is reported by Coloma et al. (Coloma et al., 2000). This vector was genetically engineered from monoclonal antibody to the human insulin receptor, 83-14 MAb, where most of the immunogenic murine sequences were replaced by a human antibody sequence. The intravenously administered antibody showed robust transport to the brain in a rhesus monkey but not in a rat. Furthermore, this vector was also used for targeting liposomes with reporter genes to the brain (Zhang et al., 2003b). As a result, the level of gene expression in the brain was 50-fold higher in the rhesus monkey as compared to the rat.

Monoclonal antibody directed insulin or Tf receptors were also utilized to target PEGylated immunoliposomes for transvascular gene therapy of Parkinson's disease (PD) (Pardridge, 2005b). In the 6-hydroxydopamine rat model of experimental PD, striatal tyrosine hydroxylase (TH) activity was completely normalized after an intravenous administration of TfmAb-targeted liposomes carrying a TH expression plasmid. Treatment of PD using this approach may be possible with dual gene therapy that seeks both to replace striatal TH gene expression with TH gene therapy, and to halt or reverse neurodegeneration of the nigro-striatal tract with neurotrophin gene therapy.

The mechanism by which immunoliposomes penetrate across the BBB is not fully understood. It was hypothesized that the process involves the binding of immunoliposomes to multiple capillary luminal membrane receptors, fusion of the liposomes with several vesicular pits into a large vesicle and transcytosis of this vesicle to the abluminal membrane border (Cornford and Cornford, 2002). Studies by electron microscopy support this hypothesis (Faustmann and Dermietzel, 1985).

In addition to approaches pursuing delivery across the intact BBB, the methods are developed to target sites in the brain under conditions when the integrity of the BBB is compromised by the progressing disease. For example, PEGylated immunoliposomes were coupled with the monoclonal anti-GFAP antibodies (Chekhonin et al., 2005). Experiments with cell cultures demonstrated specific and competitive binding of these immunoliposomes to embryonic rat brain astrocytes. Administered intravenously into rats, the immunoliposomes displayed typical kinetics with elimination half-lives of 8–15 h. Being incapable of penetrating the unimpaired BBB, these immunoliposomes, can be used to deliver drugs to glial brain tumors that express GFAP or to other pathological loci in the brain with a partially disintegrated BBB (Chekhonin et al., 2005).

Another approach involving temporal opening of the BBB using liposomes was proposed by Zhang et al. (Zhang et al., 2004). An endogenetic bioactive peptide, RMP-7 ("cereport") that is known to open tight junctions in the BBB by affecting bradykinin 2 receptors, was linked to conventional liposomes. It was demonstrated that transport of vectorized liposomes loaded with HRP was increased three times in an *in vitro* model of the BBB compared to the non-vecrorized vehicles (Zhang et al., 2004). However, potential toxicological implications of such approach still need to be addressed.

Overall, liposomes have been extensively studied for CNS drug delivery showing increased drug efficacy and reduced drug toxicity. While the examples presented suggest that the liposome technology provides a promising approach for CNS drug delivery, a considerable limitation of liposomes is their relatively low loading capacity for water-insoluble drugs and biomacromolecules.

Nanoparticles are solid species of nanoscale size usually composed of an insoluble and biodegradable polymer or polymer blend (Figure 47.1B). Generally, the methods of preparation of such nanoparticles are simple and easy to scale-up. The drug is captured within the particle upon its preparation and is released upon degradation of the polymer in the body. The use of nanoparticles as carriers for drug and gene delivery has been an area of intensive research and development for over a decade (Moghimi et al., 1990; Gref et al., 1994; Peracchia et al., 1998; Torchilin, 1998; Vinogradov et al., 1999; Lemieux et al., 2000; Alyaudtin et al., 2001; Calvo et al., 2002; Gupta et al., 2003; Moghimi and Szebeni, 2003; Kreuter, 2004; Vinogradov et al., 2004; Cui et al., 2005; Liu et al., 2005; Roney et al., 2005). The surface of such carriers is often modified by a PEG brush to increase the stability of nanoparticles in dispersion and extend circulation time of nanoparticles in the body (Peracchia et al., 1998; Torchilin, 1998; Calvo et al., 2002). To allow for efficient transcytosis across the BMVEC the particle size must be kept small and not exceed ca. 100–200 nm.

For example, poly(butyl)cyanoacrylate nanoparticles were successfully used to deliver a wide range of drugs to the CNS, which were either incorporated into the particle structure or absorbed onto the particle surface (Alyaudtin et al., 2001; Calvo et al., 2001; Calvo et al., 2002; Kreuter et al., 2003; Kreuter, 2004; Steiniger et al., 2004). Prior to administration in the body, the nanoparticles were coated with a PEG-containing surfactant, Tween 80. After the administration, they also absorbed apolipoprotein E available in the blood. The resulting coated nanoparticles were transported across the cerebral endothelial cells by endocytosis via the low density lipoprotein (LDL) receptor. They localized in the ependymal cells of the choroid plexuses, the epithelial cells of via mater and ventricles, and, to a lower extent, in the capillary endothelial cells. The list of the drugs delivered to the CNS in these constructs include, analgesics (dalargin, loperamide), anti-cancer agents (Dox), anticonvulsants (NMDA receptor antagonist, MRZ 2/576), and peptides (dalargin and kytorphin) (Kreuter et al., 2003; Steiniger et al., 2004). Particularly, incorporation of MRZ 2/576, into such nanoparticles prolonged the anticonvulsive activity of the drug almost twofold compared to the drug alone (Friese et al., 2000). Dox bound to the nanoparticles was successfully used for treatment of an aggressive human cancer, glioblastoma, resulting in increased survival times in rats (Steiniger et al., 2004). Enhanced transport of Dox in the nanoparticles to the brain involved bypassing the P-glycoprotein (Pgp) and MRP efflux systems in the BBB that impede the delivery of the drug alone. In another example the nanoparticles were coupled with chelators (desferioxamine or D-penicillamine) and proposed for treatment of Alzheimer disease (AD) and other CNS diseases by reducing oxidative stress in the brain (Liu et al., 2005; Cui et al., 2005). However, a concern regarding the toxicity associated with this delivery approach has somewhat hampered further development for specific therapeutic applications (Olivier et al., 1999).The methods for preparation of nanoparticles commonly employ the use of organic solvents that may result in degradation of immobilized drug agents, especially biomacromolecules.

Nanospheres are hollow nanosized particles (Figure 47.1C) that can be prepared by microemulsion polymerization or covering the surfaces of colloidal templates with thin layers of the desired material followed by selective removal of the templates (Hyuk Im et al., 2005). The carboxylated polystyrene nanospheres (20 nm) were evaluated for drug delivery to CNS (Yang et al., 2004). It was demonstrated that these carriers accumulated in the brain *in vivo* following cerebral ischemia and reperfusion. The intravenously injected nanospheres remained in the vasculature under normal conditions. However, during cerebral ischemia-induced stress that leads to the opening of tight junctions between endothelial cells the extravasation of the nanospheres to the disease site was increased (Kreuter, 2001). Therefore, polystyrene nanospheres may have potential clinical applications for CNS delivery of drugs and imaging agents during ischemia and stroke.

Nanosuspensions of hydrophobic drugs have also attracted attention as drug delivery systems (Rabinow, 2004). These systems represent crystalline particles of a solid drug, which are often stabilized by nonionic PEG-containing surfactants (Figure 47.1D) (Jacobs et al., 2000). They can be manufactured by a variety of techniques such as media milling, high-pressure homogenization or by employing emulsions and microemulsions as templates (Friedrich and Muller-Goymann, 2003; Friedrich et al., 2005). The major advantages of the nanosuspension technology includes its simplicity and general applicability to most drugs including very hydrophobic compounds (Muller et al., 2001; Friedrich et al., 2005). It was suggested that similar to regular nanoparticles (described above), the surface modification of nanosuspensions with Tween 80 may increase the delivery of these particles to the brain (Muller et al., 2001). Currently nanosuspensions of anti-retroviral drugs are evaluated in animal models for CNS delivery and treatment of HIV infection in the brain using cell-mediated delivery approach described below (H. Gendelman, University of Nebraska Medical Center).

Polymeric micelles represent core-shell structures that spontaneously form in aqueous solutions of amphiphilic block copolymers (Figure 47.1E). Block copolymers are polymers containing at least two chains of different chemical nature, for example, a hydrophilic and a hydrophobic chain. In aqueous solutions above a threshold concentration, called a "critical micelle concentration" (CMC) the individual block copolymer molecules termed "unimers" form micelles composed from dozen to a couple of hundred of molecules. The size of the micelles usually varies in the range of from 10 to 100 nm. The hydrophobic chains of the block copolymers segregate forming a hydrophobic core of the micelles while the hydrophilic chains (often PEG) form a hydrophilic protective shell. Polymer micelles can incorporate considerable amounts (up to 20–30% wt.) of hydrophobic compounds. The resulting nanomaterials can serve as carriers for drug delivery ("*micellar nanocontainers*") (Kataoka et al., 2001; Kwon, 2003; Allen and Cullis, 2004; Torchilin, 2004; Moghimi and Agrawal, 2005; Torchilin, 2005; Aliabadi and Lavasanifar, 2006; Tao and Uhrich, 2006). The core-shell architecture of polymeric micelles is essential for their pharmaceutical application. The core is a water-incompatible compartment that is hidden from the aqueous exterior by the hydrophilic chains of the corona, preventing premature drug release and degradation. The hydrophilic shell maintains the micelles in a dispersed state and decreases undesirable drug interactions with cells and proteins through steric-stabilization effects. Upon reaching the target the incorporated drug is released from the micelle via diffusion mechanisms. Polymeric micelles have been extensively evaluated in multiple pharmaceutical applications as drug delivery systems (Kabanov et al., 1989c; Kabanov et al., 2002; Jones et al., 2003; Torchilin et al., 2003; Aliabadi et al., 2005; Bronich et al., 2005; Gaucher et al., 2005; Lee et al., 2005; Miyata et al., 2005; Uchino et al., 2005; Yan and Tsujii, 2005), and carriers for diagnostic imaging agents (Torchilin, 2002). Several clinical trials have been completed or are underway to evaluate polymer micelles for the delivery of anti-cancer drugs for chemotherapy of tumors (Danson et al., 2004; Valle et al., 2004); (Nishiyama et al., 2003). Polymeric micelles formed by Pluronic, PEG-phospholipid conjugates, PEG-b-polyesters, or PEG-b-poly(L-amino acid)s were proposed for drug delivery of poorly water-soluble compounds, such as amphotericin B, propofol, paclitaxel, and photosensitizers (Lavasanifar et al., 2002; Kwon, 2003; Adams and Kwon, 2004; Vakil and Kwon, 2005). It was also emphasized that using polymeric micelles one can significantly increase the drug transport into the brain.

One of the early studies of targeted drug delivery to the brain used micelles of Pluronic block copolymers as carriers for CNS drugs (Kabanov et al., 1989c; Kabanov et al., 1992a). These micelles were conjugated with either polyclonal antibodies against brain α_2-glycoprotein or insulin to target the receptors at the luminal side of BMVEC. Both the antibody-conjugated and insulin-conjugated micelles were shown to effectively deliver a drug or a fluorescent probe incorporated into the micelles to the brain tissue *in vivo* (Kabanov et al., 1992a). Studies that monitored animal behavior as a means to determine the pharmacological activity of dopamenergic compounds, such as mobility and grooming were performed. It was demonstrated that incorporation of a neuroleptic, haloperidol, into the Pluronic micelles vectorized with insulin resulted in 25-fold enhancement of the neuroleptic effects compared with the free drug. Vectorization of the drug-loaded micelles with antibodies lead to an even more pronounced (up to 500-fold) enhancement of the neuroleptic effects. Subsequent studies demonstrated that the insulin vectorized micelles undergo receptor-mediated transcytosis in BMVEC from luminal (blood) to abluminal (brain) side (Batrakova et al., 1998).

A new type of functional polymer micelles with cross-linked ionic cores was recently developed (Bronich et al., 2005). Instead of using amphiphilic block copolymers these micelles are prepared using double hydrophilic block copolymers with nonionic (PEG) and ionic blocks. The micelles are self-assembled by reacting the ionic block with a condensing agent of an opposite charge. The ionic chains incorporated into the core of the micelles were chemically cross-liked and then the condensing agent was removed. The resulting cross-linked micelles contain a hydrophilic ionic core, which is swollen in water, and a hydrophilic PEO shell. The core can incorporate various hydrophilic drugs and imaging agents, which can be then delivered to the target site in the body. Overall, this strategy has potential for the development of novel modalities for delivery of various drugs to the brain, including selected anti-cancer agents to treat metastatic brain tumors as well as HIV protease inhibitors to eradicate HIV virus in the CNS.

Nanogels are a new family of hydrophilic carriers that were recently developed for targeted delivery of biomacromolecules and other drug molecules to the brain (Vinogradov et al., 1999; Vinogradov et al., 2004; Vinogradov et al., 2005a; Vinogradov et al., 2005b). Nanogels represents a nanoscale size polymer network of cross-linked ionic, e.g. polyethyleneimine (PEI), and nonionic PEG chains (Figure 47.1F). It swells in an aqueous solution and collapses upon binding of a drug through ionic interactions. Because of the effect of PEG chains, the collapsed nanogel forms stable dispersions with the particles size of ca. 80 nm. Nanogels can spontaneously absorb biomacromolecules including oligonucleotides (ODNs), plasmid DNA and proteins as well as small charged molecules. A key advantage is that the nanogels display very high loading capacities, 40–60% "payload" by weight, which is not achieved with conventional nanocarrier systems. The transport of ODNs incorporated in the nanogel particles across an *in vitro* model of the BBB was recently reported (Vinogradov et al., 2004). To enhance delivery across the BBB the surface of the nanogels were modified by either Tf or insulin (Vinogradov et al., 2004). Both peptides were shown to increase transcellular permeability of the nanogel and enhance delivery of ODNs across BMVEC monolayers. Overall, the nanogels were shown to be a promising carrier for CNS drug delivery, although only in the early stages of development. *Block ionomer complexes* are formed as a result of the reaction of double hydrophilic block copolymers containing ionic and nonionic blocks with biomacromolecules of opposite charge including oligonucleotides, plasmid DNA and proteins (Figure 47.1G) (Kabanov et al., 1995b; Harada and Kataoka, 1999a, b; Nguyen et al., 2000; Harada and Kataoka, 2003; Zhang et al., 2003a; Jaturanpinyo et al., 2004). For example, block ionomer complexes were prepared by reacting trypsin or lysozyme (that are positively charged under physiological conditions) with an anionic block copolymer, PEG-poly(R,*â*-aspartic acid) (Harada and Kataoka, 1999a; Jaturanpinyo et al., 2004). Such complexes spontaneously assemble into nanosized particles having a core-shell architecture. The core contains polyion complexes of the biomacromolecules and ionic block of the copolymer. The shell is formed by the nonionic block. These nanomaterials were shown to efficiently deliver DNA molecules *in vitro* and *in vivo* into a cell and release them at the site of action (Roy et al., 1999; Nguyen et al., 2000; Harada-Shiba et al., 2002; Junghans et al., 2005). To improve stability of the complexes in the body the biopolymer and block ionomer can be additionally cross-linked with each other (Harada and Kataoka, 2003; Jaturanpinyo et al., 2004) or the polyion chains of the block ionomer within the core can be cross-linked using degradable links. The advantages of such systems in drug delivery applications include simplicity and versatility of the design allowing the incorporation of considerable amounts of different biomacromolecules. Furthermore, block ionomer complexes are environmentally responsive nanomaterials allowing for biomacromolecule release in response to an external stimulus such as change of pH (acidification), concentration and chemical structure of elementary salt, etc.

Nanofibers and nanotubes (Figure 47.1H) have considerable promise in futuristic biomedical applications, for example, for sustained drug release from implants, or as channels for tiny volumes of chemicals in nanofluidic reactor devices, or as the "world's smallest hypodermic needles" for injecting molecules one at a time. These nanomaterials have a prototype in nature. In an organism, microtubules and their assembled structures are critical components for a broad range of cellular functions—from providing tracks for the transport of cargo to forming the spindle structure in cell division. There are various types of synthetic nanofibers and nanotubes manufactured from carbon, silicon, or diverse natural or man-made polymers. They can be vapor-grown (Che et al., 1998), self-assembled from peptide amphiphiles (Bull et al., 2005; Guler et al., 2005) or electrospun manufactured from virtually any polymer material (Dzenis, 2004). Carbon nanotubes and nanofibers have lately attracted great attention in nanomedicine including their potential use as drug carriers, although there are also considerable concerns associated with their safety (Lange et al., 2003; Muller et al., 2005). Self-assembled nanofibers can be designed to incorporate diverse chemical functionalities and thus may have a variety of applications for delivery of small drugs, biomacromolecules or imaging contrast agents (Bull et al., 2005; Guler et al., 2005). Electrospun continuous nanofibers are unique since they represent nanostructures in two dimensions and a macroscopic structures in another dimension (Fong et al., 1999; Dzenis, 2004). They are safer to manufacture than the carbon nanotubes since there is less risk of air pollution.

At present a few studies of nanofibers and nanotubes are focused on CNS drug delivery. One study evaluated electrospun nanofibers of a degradable polymer, PLGA, loaded with anti-inflammatory agent, dexamethasone, for neural prosthetic applications (Abidian and Martin, 2005). A conducting polymer, poly(3,4-ethylenedioxythiophene), was deposited to the nanofiber surface and the coated nanofibers were then mounted on the microfabricated neural microelectrodes, which were implanted into brain. The drug was released by electrical stimulation that induced a local dilation of the coat and increased permeability.

In the future, nanotubes and nanofibers can be administered systemically, if the problem of their toxicity is addressed, for example, by appropriate polymer coating. In this respect, the continuous nanofibers are more likely to be used in implants or tissue engineering applications.

47.3.2. Cell-Mediated Delivery of Nanocarriers to the Brain

A distinct case of the vehicle-mediated CNS drug delivery employs specific cells carriers that can incorporate micro- and nano-containers (such as liposomes) loaded with drugs and act as perfect Trojan horses by migrating across the BBB and carrying drugs to the site of action (Daleke et al., 1990; Fujiwara et al., 1996; Jain et al., 2003; Khan et al., 2005). It is documented that many neurological diseases, such as Parkinson's disease (PD), AD, Prion disease, meningitis, encephalitis and HIV-associated dementia, have in common an inflammatory component (Perry et al., 1995). The process of inflammation is characterized by extensive leukocytes (neutrophils and monocytes) recruitment. These cells have a unique property of migrating toward the site of inflammation via the processes known as diapedesis and chemataxis (Kuby, 1994). Their combat arsenal consists of uptake of the foreign particle, producing toxic compounds, and liberation of substances stored in intracellular vesicles via exocytosis. Therefore, these cells can be used for cell-mediated CNS drug delivery when loaded with a drug and administered into the blood stream.

It has been shown that cells capable of phagocytosis, such as macrophages or monocytes/neutrophils, can endocytose colloidal nanomaterials, for example, liposomes, and subsequently release them into the external media (Daleke et al., 1990; Jain et al., 2003). To accelerate the transport of monocytes/neutrophils to the brain site, the drug-containing liposomes were additionally loaded with magnetic particles (Jain et al., 2003). Magnetic liposomes demonstrated about a tenfold increase in brain levels compared to non-magnetic liposomes when local magnetic field was applied. It is noteworthy that both cell types showed preferential uptake of liposomes containing negatively-charged lipids (such as phosphatidylserine), or liposomes modified with polyanion than liposomes containing only neutral lipids (such as phosphatidylcholine) (Fujiwara et al., 1996). This suggests that by engineering the surface of the nanomaterials one may modulate their uptake into and release from the cell carriers to optimize the therapeutic regimen. The phosphatidylserine-containing negatively charged liposomes were shown to increase therapeutic activity of an encapsulated antifungal agent, chloroquine, against *C. neoformans* infection in the mouse brain (Khan et al., 2005). The chloroquine-loaded liposomes accumulated inside macrophage phagolysosomes and resulted in a remarkable reduction in fungal load in the brain even at low doses compared to the free drug at high doses, thus increasing the antifungal activity of macrophages.

T lymphocytes were also proposed as a potential therapeutic drug carriers for cancer treatment (Steinfeld et al., 2006). The kinetics of loading and release of nanoparticles coated with cytotoxic antibiotic Dox into the cells were examined. It was suggested that the immune cells can accomplish target-specific and sheltered transport to the diseased site.

Ability of host cells to home diseased sites after *ex vivo* manipulation is a fundamental requirement and a major problem for their use as vehicles to target locally acting gene therapy to specific diseased sites. It was demonstrated that in the short term, up to 2h after re-implantation, macrophages accumulate primarily in the lungs and, to a lesser extent, in the liver and spleen rather than in the target diseased tissue. A small proportion of manipulated macrophages (ranging from 0.2 to 28.8%) homes to the diseased site of interest following systemic administration. However, the presence of labeled macrophages in these sites was found to persist for at least 6–7 days in both of these mouse studies (Audran et al., 1995). Therefore, increasing the amount of cell carriers reaching the target site appears to be a crucial point in this type of drug delivery system.

47.3.3. Permeability Enhancers for CNS Drug Delivery

There are a number of efflux mechanisms within the CNS that influence drug concentrations in the brain. Some of them are passive while others are active. Recently much attention has been focused on the so-called multi-drug efflux transporters: Pgp, MRP1, and BCRP, MOAT that belong to the ABC cassette (ATP-binding cassette) family (Pardridge, 1998; Tamai and Tsuji, 2000; Begley, 2004b). Cerebral capillary endothelium expresses a number of efflux transport proteins, which actively remove a broad range of drug molecules before they cross into the brain parenchyma. Pgp is the most thoroughly investigated brain efflux transport protein with broad affinity for dissimilar lipophilic and amphiphilic substrates (Tsuji and Tamai, 1997). As a consequence, the therapeutic value of many promising drugs, such as protease inhibitors for HIV-1 encephalitis (ritonavir, nalfinavir, and indinavir) (Kim et al., 1998), anti-inflammatory drugs (prednesolone, dexamethasone and indomethacin) for treatment of microglial inflammation during idiopathic PD and AD (Tsuji and Tamai, 1997; Perloff et al., 2004), neuroleptic agents (amitriptyline and haloperidol) (Uhr et al., 2000), analgaesic drugs (morphine, beta-endorphin, and asimadoline) (Moriki et al., 2004), as well as anti-fungal agents (itraconazole and ketoconazole) (Miyama et al., 1998), is diminished, and cerebral diseases have proven to be most refractory to therapeutic interventions. Moreover, delivery of many anticancer agents (Dox, vinblastine, taxol, etc.) to the brain for treatment of brain tumors (Tsuji, 1998), and drugs for treatment of epilepsy (carbamazepine, phenobarbital, phenytoin, and lamotrigine) (Potschka et al., 2002) is also restricted by the Pgp drug efflux transporter.

An emerging strategy for enhanced BBB penetration of drugs is co-administration of competitive or noncompetitive inhibitors of the efflux transporter together with the desired CNS drug. First generation low molecular weight Pgp inhibitors (cyclosporine A, verapamil, PSC833, etc.) are substrates of the drug efflux transporter, which compete for the active site with the therapeutic agent (Kemper et al., 2003). Second generation inhibitors (LY335979, XR9576 and GF120918) are non-competitive inhibitors, which allosterically bind to Pgp, inactivating it and increasing drug transport to the brain (Kemper et al., 2004). Despite their high efficiency in cell culture models, the small therapeutic range of these inhibitors, high *in vivo* toxicity, and fast clearance are the main obstacles for their therapeutic application.

Recently, a new class of inhibitors (nonionic polymer surfactants) was identified as a promising component of drug formulations. These compounds are two- or three-block copolymers arranged in a linear AB or ABA structure. The A block is a hydrophilic PEG chain. The B block can be a hydrophobic lipid (BRIJs, MYRJs, Tritons, Tweens, and Chremophor), or a poly(propylene glycol) (PPG) chain (Pluronics) (Figure 47.2).

Studies in multidrug resistant (MDR) cancer cells, polarized intestinal epithelial cells, Caco-2, and polarized BMVEC monolayers provided compelling evidence that selected Pluronic block copolymers can inhibit drug efflux transport systems (Miller et al., 1997; Batrakova et al., 1998; Batrakova et al., 1999a; Batrakova et al., 2001a; Batrakova et al., 2001b; Batrakova et al., 2003b). Specifically, in primary cultured BMVEC monolayers, used as an *in vitro* model of BBB, the inhibition of drug efflux systems, Pgp and MRP was associated with an increased accumulation and permeability of a broad spectrum of drugs in the BBB, including low molecular drugs (Batrakova et al., 1998; Miller et al., 1999) and peptides (Witt et al., 2002). These effects were most apparent at concentrations below the CMC (Miller et al., 1997; Batrakova et al., 1998). It was suggested that the unimers, are responsible for the inhibition of Pgp and MRPs efflux transport system. Incorporation of the probe into the micelles formed at high concentrations of the block copolymer decreases its availability to the cells and reduces the transport of this probe in BMVEC.

Recent findings suggest that effects of Pluronic on drug efflux transport proteins involve interactions of the block copolymers with the cell membranes (Batrakova et al., 2001b; Batrakova et al., 2003b). It was demonstrated that a fine balance is needed between hydrophilic (PEG) and lipophilic (PPG) components in the Pluronic molecule to enable inhibition of the drug efflux systems (Batrakova et al., 2003a). Overall, the most efficacious block copolymers are those with intermediate lengths of PPG block and relatively hydrophobic structure (HLB < 20), such as Pluronic P85 or L61 (Batrakova et al., 1999b). The hydrophobic PPG chains of Pluronic immerse into the membrane hydrophobic areas, resulting in alterations of the membrane structure and

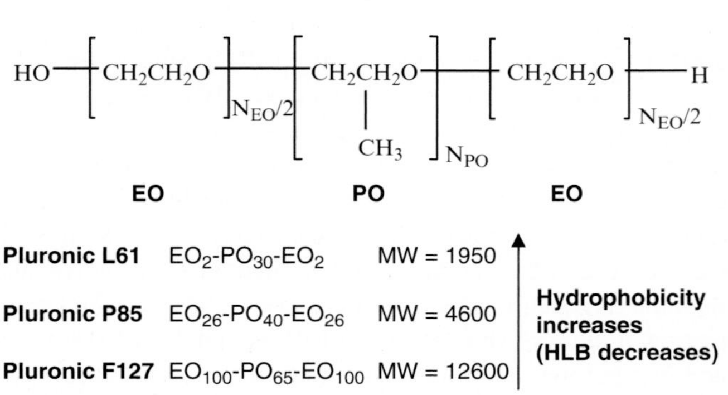

$$HO-\left[CH_2CH_2O\right]_{N_{EO}/2}-\left[CH_2CH(CH_3)O\right]_{N_{PO}}-\left[CH_2CH_2O\right]_{N_{EO}/2}-H$$

EO PO EO

Pluronic L61	EO_2-PO_{30}-EO_2	MW = 1950
Pluronic P85	EO_{26}-PO_{40}-EO_{26}	MW = 4600
Pluronic F127	EO_{100}-PO_{65}-EO_{100}	MW = 12600

Hydrophobicity increases (HLB decreases)

FIGURE 47.2. Pluronic block copolymers with various numbers of hydrophilic EO (*n*) and hydrophobic PO (*m*) units are characterized by distinct hydrophilic-lipophilic balance (HLB). Due to their amphiphilic character these copolymers display surfactant properties including ability to interact with hydrophobic surfaces and biological membranes. In aqueous solutions at concentrations above critical micelle concentration (CMC) these copolymers self-assemble into micelles.

decreases in its microviscosity ("membrane fluidization"). Pluronic at relatively low concentrations (e.g. 0.01%) inhibits the Pgp ATPase activity, possibly due to conformational changes in the transport protein induced by the immersed copolymer chains in the Pgp-expressing membranes (Batrakova et al., 2001b). In particular, Pluronic P85 displayed the effects characteristic of a mixed type enzyme inhibitor—decreasing maximal reaction rate (V_{max}) and increasing Michaelis constant (K_m) for ATP as well as Pgp-specific substrates such as vinblastine (Batrakova et al., 2004a). The magnitude of these effects for vinblasine was as high as over 200-fold V_{max}/K_m change (interestingly, MRP1 ATPase activity was affected less, which could explain somewhat smaller effects of Pluronic on this transporter). In contrast, at high concentrations (e.g. 1%), binding of Pluronic to the membrane actually resulted in restoration of Pgp ATPase activity. This could be due to the segregation of the block copolymer molecules in the 2D clusters in the membrane, which diminishes its interactions with the transport proteins.

Various drug resistance mechanisms, including drug transport and detoxification systems, require consumption of energy to sustain their function in the barrier cells. Because of this fact, mechanistic studies have focused on the effects of Pluronic block copolymers on metabolism and energy conservation in BMVEC (Batrakova et al., 2001a). The basis for such studies was the earlier reports that Pluronic block copolymers can affect mitochondria function and energy conservation in the cells (Kirillova et al., 1993). Recent studies have demonstrated that exposure to Pluronic P85 induced significant decrease in ATP levels in BMVEC monolayers (Batrakova et al., 2001b). The observed energy depletion was due to inhibition of the cellular metabolism rather than a loss of ATP in the environment. The study of Rapoport et al. suggested that Pluronic P85 can be transported into the cells and decrease the activity of electron transport chains in the mitochondria (Rapoport et al., 2000). Remarkably, the ATP depletion induced by Pluronic appears to be tightly linked to the specific cell phenotype, since this effect is observed selectively in the cells that overexpress Pgp (as well as MRPs) (Batrakova et al., 2001b; Kabanov et al., 2003b; Kabanov et al., 2003a; Batrakova et al., 2004a). We suggested that inhibition of ATP production in high energy-consuming cells, such as cells overexpressing Pgp, results in the rapid exhaustion of intracellular ATP, i.e. ATP depletion (Batrakova et al., 2001b). Further, our recent studies indicate that these differences were related to the differences in the major fuel sources used by MDR or sensitive cells (fatty acids vs. glucose respectively) (Rapoport et al., 2006). It was shown that Pluronics affected the stage of the electron entrance in the mitochondrial electron transport chain, presumably due to the interaction of the hydrophobic block of the copolymer with fatty acids, which decreased fuel transport into the mitochondria matrix. In contrast, Pluronics exerted only mild effect on the glucose-based respiration in drug-sensitive cells. Overall, the energy depletion (decreasing ATP pool available for drug transport proteins) and membrane interactions (inhibiting of ATPase activity of drug transport proteins) are critical factors collectively contributing

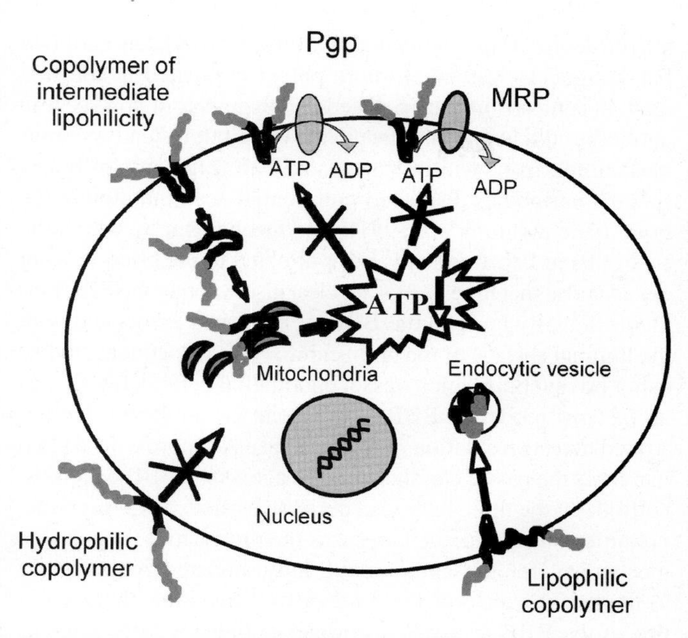

FIGURE 47.3. Schematic illustrating twofold effects of Pluronic block copolymers with intermediate lipophilicity on Pgp and MRPs drug efflux system. These effects include (a) decrease in membrane viscosity ("fluidization") resulting in inhibition of Pgp and MRPs ATPase activity, and (b) ATP depletion in BMVEC. Extremely lipophilic or hydrophilic Pluronic block copolymers do not cross the cellular membranes and do not cause energy depletion in the cells.

to a potent inhibition of the drug efflux systems by Pluronic (Batrakova et al., 2001b) (Figure 47.3).

The effect of Pluronic P85 on drug transport to the brain was evaluated in animal experiments (Batrakova et al., 2001a). Brain delivery of a Pgp substrate, digoxin, administered intravenously in the wild-type mice expressing functional Pgp, was greatly enhanced in the presence of Pluronic P85. It was found that the digoxin brain/plasma ratios in the Pluronic treated animals were practically the same as those in the knockout mice, an animal model that is deficient in both mdr1a and mdr1b isoforms of Pgp. This suggests that co-administration of Pluronic with the drug in mice resulted in inhibition of Pgp in the BBB of the wild-type animals (Batrakova et al., 2001a).

One possible concern in these studies is that by virtue of inhibiting the ATP in BMVEC, the copolymer may display toxic effects on the BBB. However, the ATP depletion was found to be transient; following removal of the block copolymers from BMVEC monolayers the initial ATP levels were restored (Batrakova et al., 2001b). Although there were significant decreases in cellular ATP following Pluronic treatment, even during peak depletion of ATP by Pluronic there was no evidence of loss of barrier functions of BBB as demonstrated using ^{3}H-mannitol as a permeability marker both *in vitro* and *in vivo* (Batrakova et al., 1998; Batrakova et al., 2001a). Moreover, Pluronic does not affect the glucose transporter, GLUT1, and only slightly inhibits the lactate transporter, MCT1, the two transporters playing an important role in the brain metabolism (Batrakova et al., 2004b). Pluronic also does not inhibit the amino acid transporters, LAT1, CAT1, and SAT1, in the BBB (Zhang et al., 2006). A histochemical examination of the tissue sections obtained from animals treated with Pluronic revealed no pathological changes in the BBB. Importantly, no cerebral toxicity of any kind has been observed in the human Phase I and Phase II studies of SP1049C, a Pluronic-based formulation of Dox to treat MDR tumors (Danson et al., 2004; Valle et al., 2004). It is possible, that this formulation, evaluated in human trials, can be adopted for use with CNS drugs to enhance drug delivery to the brain.

47.3.4. Chemical Modification of Polypeptides with Fatty Acids and Amphiphilic Block Copolymers

Delivery of potentially therapeutic polypeptides and proteins to the brain is significantly hampered by the BBB. The hydrophilicity, the lack of stability due to enzymatic or chemical degradation, and the lack of transport carriers capable of shuttling polypeptides across cell membranes all play a part in precluding most polypeptides from transport into the brain (Lee, 1991). Several approaches to modify polypeptides to alter their BBB permeability have been attempted. First, conjugation of proteins with wheatgerm agglutinin (Raub and Audus, 1990; Banks and Broadwell, 1994) or cationic groups ("cationization") (Kumagai et al., 1987; Triguero et al., 1989; Triguero et al., 1991) was shown to enhance delivery of polypeptides to the brain through adsorptive endocytosis. The cationized polypeptides demonstrated enhanced permeability and greater accumulation in the brain *in vitro and in vivo*. However, toxicity and antigenicity of cationized polypeptides could be an issue for their medical use (Bickel et al., 2001). Further, vectors targeting polypeptides to insulin and Tf receptors (monoclonal antibodies) have been considered to enhance passage of these compounds to the brain through receptor-mediated endocytosis (Frank et al., 1986; Duffy and Pardridge, 1987; Friden et al., 1991; Bickel et al., 2001). Thus, potential therapeutic polypeptides, basic fibroblast growth factor and brain-derived neurotrophic factor, were conjugated to OX26, which resulted in increased entry of these polypeptides to the brain and increased drug neuroprotective effects (Zhang and Pardridge, 2001a, 2001b; Song et al., 2002).

Another approach, artificial hydrophobization of polypeptides with a small number of fatty acid residues (e.g. stearate or palmitate) has been shown to enhance cellular uptake (Kabanov et al., 1989a). Specifically, this technique involves point modification of lysine or N-terminal amino groups with one or two fatty acid residues per protein molecule. As a result of such modification, the protein molecule remains water-soluble but also acquires hydrophobic anchors that can target even very hydrophilic proteins to cell surfaces (Slepnev et al., 1995). To obtain low and controlled degrees of modification, a system of reverse micelles of a surfactant, sodium bis-(2-ethylhexyl)sulfosucciate (Aerosol OT) in octane was used as a reaction medium (Kabanov et al., 1987) (Figure 47.4). Over a

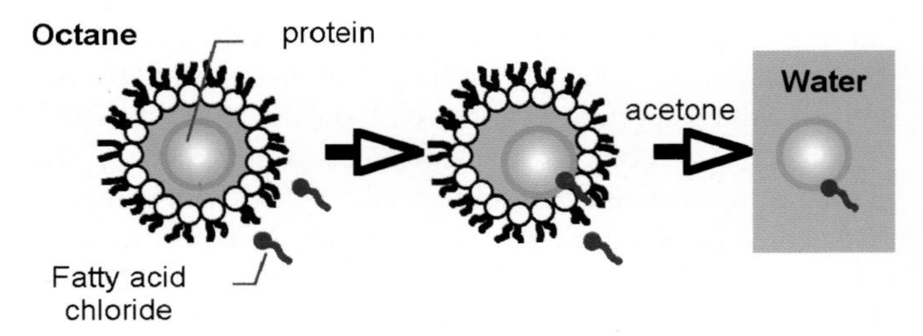

FIGURE 47.4. Chemical modification of the protein with a water-insoluble reagent in the reverse micelles of Aerosol OT in octane. (1) Protein molecule incorporates into the inner water pool of the reverse micelle, acquiring a monolayer cover of the hydrated surfactant molecules. (2) The modifying reagent incorporates into the surfactant layer of the micelle coming into contact with the modified group of the protein. (3) Following the completion of the reaction the modified protein is precipitated and the surfactant and excess of the reagent are removed by adding cold acetone. Proteins modified with fatty acid residues with controlled and low degree of modification are obtained.

dozen water-soluble polypeptides (enzymes, antibodies, toxins, cytokines) have been modified by this method (Kabanov et al., 1987; Kabanov et al., 1989b; Alakhov et al., 1990; Chekhonin et al., 1991; Robert et al., 1993; Robert et al., 1995; Slepnev et al., 1995). Further studies of interactions of the fatty acylated polypeptides with cells were also conducted (Hashimoto et al., 1989; Kabanov et al., 1989b; Alakhov et al., 1990; Colsky and Peacock, 1991; Melik-Nubarov et al., 1993; Ekrami et al., 1995; Slepnev et al., 1995; Chopineau et al., 1998; Kozlova et al., 1999). Modification of water-soluble polypeptides, such as HRP, resulted in enhanced polypeptide binding with the cell membranes and internalization in many cell types (Slepnev et al., 1995). The point modification does not inhibit the specific activity of the polypeptides in the cells. To the contrary, in selected cases, when polypeptides are known to exhibit their effects in cells through binding with a cell surface receptor (e.g. *Staphylococcus aureus* enterotoxin A and recombinant α-interferon) the modification resulted in significant (10 to 100 times) enhancement of these effects (Alakhov et al., 1990; Kabanov et al., 1992b). The increased activity of selected modified polypeptides in cells can possibly be explained by their concentration at the cell membrane, which promotes their binding with receptors (Kabanov et al., 1992b). Furthermore, insulin modified with one palmitic acid residue produced a prolonged hypoglycemic effect compared to the native insulin after intravenous injection and was shown to be less immunoreactive than the native insulin (Hashimoto et al., 1989).

The relevance of this technology to CNS delivery emerged from the studies by Chekhonin et al. (Chekhonin et al., 1991; Chekhonin et al., 1995). Those studies showed that modification of the Fab fragments of antibodies against gliofibrillar acid protein (GFAP) and brain specific alpha 2-glycoprotein (alpha 2GP) with stearate led to an increased accumulation of the modified Fab fragments in the brain in a rat. Furthermore, a neuroleptic drug conjugated with the stearoylated antibody Fab fragments was much more potent compared to the free drug. In comparison, fatty acylated Fab fragments of non-specific antibodies did not accumulate in the brain but instead accumulated in the liver, while stearoylated Fab fragments of brain-specific antibodies displayed preferential accumulation in the brain (Chekhonin et al., 1991). The mechanism by which the stearoylated Fab-fragments were directed to the brain was not elucidated at that time. It was not clear also whether the Fab fragments actually crossed the BBB or remained associated with the luminal surface of the brain capillaries. Subsequent studies using bovine brain microvessel endothelial cells (BBMEC) as an *in vitro* model of BBB (Chopineau et al., 1998) demonstrated that stearoylation of ribonuclease A (approx. 13.6 kDa) increases the passage of this enzyme across the BBB by almost tenfold. Of the three fatty acid derivatives analyzed—myristic, palmitic and stearic, the latter was the most active. A possible mechanism for the entry of the fatty acylated polypeptides to the brain is adsorptive endocytosis. Given the characteristics of the BBB transport for nonessential free fatty acids as reviewed elsewhere (Banks et al., 1997), it is unlikely that the modified polypeptides are able to use those transporters. Thus, use of free fatty acid receptor mediated transport by the modified polypeptides is not likely. However, if such a pathway is feasible, it should be more pronounced when the essential fatty acids, such as linoleic, are used to modify polypeptides (Edmond et al., 1998; Edmond, 2001).

A similar principle was used in the case of a protein modification with Pluronic block copolymers (Figure 47.5); the designed conjugates had different balance between hydrophilic and lipophilic properties (Batrakova et al., 2005). The cell binding and transport of Pluronic P85 and L121 modified HRP were increased up to six and ten times, respectively, compared with the native HRP. A method of conjugation using biodegradable and non-biodegradable links was also varied. As expected, the modification of the protein via a biodegradable link (Figure 47.5A) resulted in the most effective transport of the protein across the brain microvessel endothelial cell monolayers. It is likely, that being highly hydrophilic the HRP molecule requires a highly lipophilic moiety to obtain the optimal hydrophilic-lipophilic balance of the conjugate. Such a lipophilic group has a tendency to remain associated with the cellular membrane as an anchor. Therefore, a biodegradable link allowed the polypeptide molecule to separate from its lipophilic moiety and enter the brain after the conjugate passed membranes of the barrier cells. Data obtained by confocal microscopy visualizes this effect. Modification of the protein with Pluronic P85 via biodegradable link drastically enhanced the transport of HRP into the cells and its accumulation in the cytoplasm, nuclei, and other cellular organelles (Figure 47.5B). It was suggested that the protein conjugate was initially bound to the cellular membrane through the Pluronic moiety incorporated into the lipid bilayer followed by cleavage of the HRP molecule from the Pluronic anchors. *In vivo* studies demonstrated that

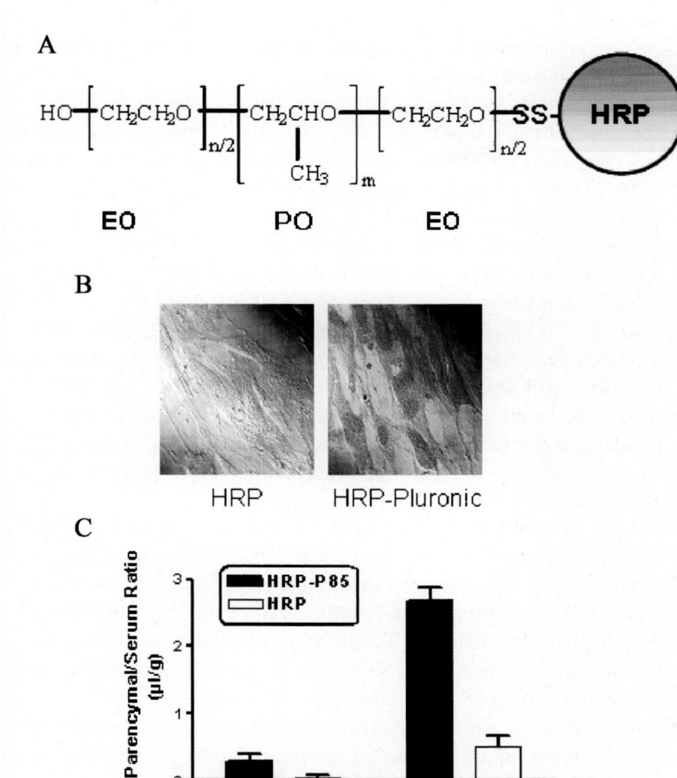

FIGURE 47.5. Effect of HRP modification with Pluronic block copolymer on transport across the BBB in an in vitro and in vivo models. A: HRP conjugated with Pluronic P85 via the biodegradable bond; B: confocal microphotograph of BBMEC monolayers treated with rhodamine-labeled HRP and Pluronic-HRP for 2 h; C: blood-to-brain transport of HRP and Pluronic-HRP in mice.

1) hydrophobization with stearoyl group or 2) amphiphilic modification with a Pluronic block copolymer had increased the rate of HRP penetration across the BBB and increased accumulation of HRP by the brain (Figure 47.5C). Overall modification with Pluronic block copolymers appeared to be more promising for HPR delivery to the brain. In this case the permeability of modified protein in the BBB *in vivo* was increased almost fourfold with no statistically significant effects on the protein peripheral pharmacokinetics or entry into the parenchymal space.

Summary

Tremendous efforts in the last several decades have resulted in numerous inventions of CNS drug delivery systems. Many of these innovative systems have a significant potential for the development of new biomedical applications. The wide variety of strategies reflects the inherent difficulty in transport of therapeutic and imaging agents across the BBB. In fact, the effective combination of several approaches, such as encapsulation of drugs into nanoparticles conjugated with vector moieties or using micelles of Pluronic block copolymers along with Pluronic "unimers" that will inhibit drug efflux transporters in the BMVEC, may give the most promising therapeutic outcomes.

Review Questions/Problems

1. **What are the main features of the BBB that restrict drug transport to the brain?**
2. **Describe the properties of polymers that are useful for drug delivery systems.**
3. **What are the advantages of nanoscaled polymeric carriers?**
4. **List types of nanocarriers for drug delivery. Describe their structure, principal differences, advantages and limitations.**
5. **Describe cell-mediated drug delivery to the CNS.**
6. **Explain how the drug efflux transporters affect drug transport to the brain? List examples of drug efflux transporters expressed in the BBB.**
7. **Describe three generations of inhibitors of drug efflux transporters in the BBB.**
8. **Describe the effects of Pluronic block copolymers on drug efflux transporters in the BBB.**
9. **How is the chemical modification of polypeptides with fatty acids and amphiphilic block copolymers applied to increase delivery of polypeptides to CNS?**
10. **Which of the following molecules can penetrate across the BBB through passive diffusion?**
    a. Lipophilic low molecular weight compounds
    b. Hydrophilic low molecular weight compounds
    c. Lipophilic high molecular weight compounds
    d. Hydrophilic high molecular weight compounds
11. **What size of polymer nanocarriers is the most appropriate for drug delivery?**
    a. 0.1–1 nm
    b. 1–100 nm
    c. 100–1,000 nm
    d. 1–100 µm
12. **Which types of drugs can be incorporated into polymeric micelles?**
    a. lipophilic compounds
    b. hydrophilic compounds
    c. charged compounds
    d. proteins compounds

13. What does CMC mean?

a. colloidal microemulsion complex
b. cell monocarrier
c. critical micelle concentration
d. contrast multi-compound

14. Which polymer is used to stabilize nanoparticles in an aqueous dispersion?

a. polystyrene
b. polymethacrylic acid
c. DNA
d. PEG
e. polycyanoacrylate

15. Which type of cells can be used for cell-mediated delivery?

a. brain microvessel endothelial cells
b. neurons
c. macrophages
d. astrocytes
e. erythrocytes

References

Abbott N, Romero I (1996) Transporting therapeutics across the blood-brain barrier. Mol Med Today 2:106–113.

Abidian M, Martin D (2005) Controlled Release of an Anti-Inflammatory Drug Using Conducting Polymer Nanotubes for Neural Prosthetic Applications. In: MRS Symposium M, p 1. San Francisco.

Adams M, Kwon GS (2004) Spectroscopic investigation of the aggregation state of amphotericin B during loading, freeze-drying, and reconstitution of polymeric micelles. J Pharm Pharm Sci 7:1–6.

Alakhov V, Kabanov A, Batrakova E, Koromyslova I, Levashov A, Severin E (1990) Increasing cytostatic effects of ricin A chain and Staphylococcus aureus enterotoxin A through in vitro hydrophobization with fatty acid residues. Biotechnol Appl Biochem 12:94–98.

Aliabadi HM, Lavasanifar A (2006) Polymeric micelles for drug delivery. Expert Opin Drug Deliv 3:139–162.

Aliabadi HM, Brocks DR, Lavasanifar A (2005) Polymeric micelles for the solubilization and delivery of cyclosporine A: Pharmacokinetics and biodistribution. Biomaterials 26:7251–7259.

Allen TM, Cullis PR (2004) Drug delivery systems: Entering the mainstream. Science 303:1818–1822.

Alyaudtin RN, Reichel A, Lobenberg R, Ramge P, Kreuter J, Begley DJ (2001) Interaction of poly(butylcyanoacrylate) nanoparticles with the blood-brain barrier in vivo and in vitro. J Drug Target 9:209–221.

Aoki H, Kakinuma K, Morita K, Kato M, Uzuka T, Igor G, Takahashi H, Tanaka R (2004) Therapeutic efficacy of targeting chemotherapy using local hyperthermia and thermosensitive liposome: Evaluation of drug distribution in a rat glioma model. Int J Hyperthermia 20:595–605.

Audran R, Collet B, Moisan A, Toujas L (1995) Fate of mouse macrophages radiolabelled with PKH-95 and injected intravenously. Nucl Med Biol 22:817–821.

Banks W, Lebel C (2002) Strategies for the delivery of leptin to the CNS. J Drug Target 10:297–308.

Banks WA, Broadwell RD (1994) Blood to brain and brain to blood passage of native horseradish peroxidase, wheat germ agglutinin, and albumin: Pharmacokinetic and morphological assessments. J Neurochem 62:2404–2419.

Banks WA, Kastin AJ, Rapoport SI (1997) Permeability of the blood-brain barrier to circulating free fatty acids. In: Handbook of Essential, Fatty Acid Biology: Biochemistry, Physiology, and Behavioral Neurobiology. (Yehuda S, Mostofsky DI, eds.), pp 3–14. Totowa, NJ: Humana Press.

Batrakova E, Han H, Miller D, Kabanov A (1998) Effects of pluronic P85 unimers and micelles on drug permeability in polarized BBMEC and Caco-2 cells. Pharm Res 15:1525–1532.

Batrakova E, Li S, Miller D, Kabanov A (1999a) Pluronic P85 increases permeability of a broad spectrum of drugs in polarized BBMEC and Caco-2 cell monolayers. Pharm Res 16:1366–1372.

Batrakova E, Lee S, Li S, Venne A, Alakhov V, Kabanov A (1999b) Fundamental relationships between the composition of pluronic block copolymers and their hypersensitization effect in MDR cancer cells. Pharm Res 16:1373–1379.

Batrakova E, Miller D, Li S, Alakhov V, Kabanov A, Elmquist W (2001a) Pluronic P85 enhances the delivery of digoxin to the brain: In vitro and in vivo studies. J Pharmacol Exp Ther 296:551–557.

Batrakova EV, Li S, Elmquist WF, Miller DW, Alakhov VY, Kabanov AV (2001b) Mechanism of sensitization of MDR cancer cells by Pluronic block copolymers: Selective energy depletion. Br J Cancer 85:1987–1997.

Batrakova E, Li S, Alakhov V, Miller D, Kabanov A (2003a) Optimal structure requirements for pluronic block copolymers in modifying P-glycoprotein drug efflux transporter activity in bovine brain microvessel endothelial cells. J Pharmacol Exp Ther 304:845–854.

Batrakova EV, Li S, Alakhov VY, Elmquist WF, Miller DW, Kabanov AV (2003b) Sensitization of cells overexpressing multidrug-resistant proteins by pluronic P85. Pharm Res 20:1581–1590.

Batrakova EV, Li S, Li Y, Alakhov VY, Kabanov A (2004a) Effect of pluronic P85 on ATPase activity of drug efflux transporters. Pharm Res 21:2226–2233.

Batrakova EV, Zhang Y, Li Y, Li S, Vinogradov SV, Persidsky Y, Alakhov VY, Miller DW, Kabanov AV (2004b) Effects of pluronic P85 on GLUT1 and MCT1 transporters in the blood-brain barrier. Pharm Res 21:1993–2000.

Batrakova EV, Vinogradov SV, Robinson SM, Niehoff ML, Banks WA, Kabanov AV (2005) Polypeptide point modifications with fatty acid and amphiphilic block copolymers for enhanced brain delivery. Bioconjug Chem 16:793–802.

Begley D (1996) The blood-brain barrier: Principles for targeting peptides and drugs to the central nervous system. J Pharm Pharmacol 48:136–146.

Begley DJ (2004a) Delivery of therapeutic agents to the central nervous system: The problems and the possibilities. Pharmacol Ther 104:29–45.

Begley DJ (2004b) ABC transporters and the blood-brain barrier. Curr Pharm Des 10:1295–1312.

Bickel U, Yoshikawa T, Pardridge WM (2001) Delivery of peptides and proteins through the blood-brain barrier. Adv Drug Deliv Rev 46:247–279.

Bronich TK, Keifer PA, Shlyakhtenko LS, Kabanov AV (2005) Polymer micelle with cross-linked ionic core. J Am Chem Soc 127:8236–8237.

Bull SR, Guler MO, Bras RE, Meade TJ, Stupp SI (2005) Self-assembled peptide amphiphile nanofibers conjugated to MRI contrast agents. Nano Lett 5:1–4.

Calvo P, Gouritin B, Chacun H, Desmaele D, D'Angelo J, Noel J, Georgin D, Fattal E, Andreux J, Couvreur P (2001) Long-circulating PEGylated polycyanoacrylate nanoparticles as new drug carrier for brain delivery. Pharm Res 18:1157–1166.

Calvo P, Gouritin B, Villarroya H, Eclancher F, Giannavola C, Klein C, Andreux JP, Couvreur P (2002) Quantification and localization of PEGylated polycyanoacrylate nanoparticles in brain and spinal cord during experimental allergic encephalomyelitis in the rat. Eur J Neurosci 15:1317–1326.

Che G, Lakshmi B, Martin C, Fisher E, Ruoff R (1998) Chemical vapor deposition based synthesis of carbon nanotubes and nanofibers using a template method. Chem Mater 10:260–267.

Chekhonin V, Kabanov A, Zhirkov Y, Morozov G (1991) Fatty acid acylated Fab-fragments of antibodies to neurospecific proteins as carriers for neuroleptic targeted delivery in brain. FEBS Lett 287:149–152.

Chekhonin V, Ryabukhin I, Zhirkov Y, Kashparov I, Dmitriyeva T (1995) Transport of hydrophobized fragments of antibodies through the blood-brain barrier. Neuroreport 7:129–132.

Chekhonin VP, Zhirkov YA, Gurina OI, Ryabukhin IA, Lebedev SV, Kashparov IA, Dmitriyeva TB (2005) PEGylated immunoliposomes directed against brain astrocytes. Drug Deliv 12:1–6.

Chopineau J, Robert S, Fenart L, Cecchelli R, Lagoutte B, Paitier S, Dehouck M, Domurado D (1998) Monoacylation of ribonuclease A enables its transport across an in vitro model of the blood-brain barrier. J Control Release 56:231–237.

Coloma MJ, Lee HJ, Kurihara A, Landaw EM, Boado RJ, Morrison SL, Pardridge WM (2000) Transport across the primate blood-brain barrier of a genetically engineered chimeric monoclonal antibody to the human insulin receptor. Pharm Res 17:266–274.

Colsky A, Peacock J (1991) Palmitate-derivatized antibodies can specifically "arm" macrophage effector cells for ADCC. J Leukoc Biol 49:1–7.

Cornford EM, Cornford ME (2002) New systems for delivery of drugs to the brain in neurological disease. Lancet Neurol 1:306–315.

Cui Z, Lockman PR, Atwood CS, Hsu CH, Gupte A, Allen DD, Mumper RJ (2005) Novel D-penicillamine carrying nanoparticles for metal chelation therapy in Alzheimer's and other CNS diseases. Eur J Pharm Biopharm 59:263–272.

Daleke DL, Hong K, Papahadjopoulos D (1990) Endocytosis of liposomes by macrophages: Binding, acidification and leakage of liposomes monitored by a new fluorescence assay. Biochim Biophys Acta 1024:352–366.

Danson S, Ferry D, Alakhov V, Margison J, Kerr D, Jowle D, Brampton M, Halbert G, Ranson M (2004) Phase I dose escalation and pharmacokinetic study of pluronic polymer-bound doxorubicin (SP1049C) in patients with advanced cancer. Br J Cancer 90:2085–2091.

Duffy KR, Pardridge WM (1987) Blood-brain barrier transcytosis of insulin in developing rabbits. Brain Res 420:32–38.

Duncan R (2003) The dawning era of polymer therapeutics. Nat Rev Drug Discov 2:347–360.

Dzenis Y (2004) Material science. Spinning continuous fibers for nanotechnology. Science 304:1917–1919.

Edmond J (2001) Essential polyunsaturated fatty acids and the barrier to the brain: The components of a model for transport. J Mol Neurosci 16:181–193.

Edmond J, Higa TA, Korsak RA, Bergner EA, Lee WN (1998) Fatty acid transport and utilization for the developing brain. J Neurochem 70:1227–1234.

Ekrami H, Kennedy A, Shen W (1995) Water-soluble fatty acid derivatives as acylating agents for reversible lipidization of polypeptides. FEBS Lett 371:283–286.

Faustmann PM, Dermietzel R (1985) Extravasation of polymorphonuclear leukocytes from the cerebral microvasculature. Inflammatory response induced by alpha-bungarotoxin. Cell Tissue Res 242:399–407.

Folkman J, Long DM (1964) The use of silicone rubber as a carrier for prolonged drug therapy. J Surg Res 71:139–142.

Fong H, Chun I, Reneker D (1999) Beaded nanofibers formed during electrospinning. Polymer 40:4585–4592.

Frank HJ, Pardridge WM, Jankovic-Vokes T, Vinters HV, Morris WL (1986) Insulin binding to the blood-brain barrier in the streptozotocin diabetic rat. J Neurochem 47:405–411.

Friden PM, Walus LR, Musso GF, Taylor MA, Malfroy B, Starzyk RM (1991) Anti-transferrin receptor antibody and antibody-drug conjugates cross the blood-brain barrier. Proc Natl Acad Sci U S A 88:4771–4775.

Friedrich I, Muller-Goymann CC (2003) Characterization of solidified reverse micellar solutions (SRMS) and production development of SRMS-based nanosuspensions. Eur J Pharm Biopharm 56:111–119.

Friedrich I, Reichl S, Muller-Goymann CC (2005) Drug release and permeation studies of nanosuspensions based on solidified reverse micellar solutions (SRMS). Int J Pharm 305:167–175.

Friese A, Seiller E, Quack G, Lorenz B, Kreuter J (2000) Increase of the duration of the anticonvulsive activity of a novel NMDA receptor antagonist using poly(butylcyanoacrylate) nanoparticles as a parenteral controlled release system. Eur J Pharm Biopharm 49:103–109.

Fromm M (2000) P-glycoprotein: A defense mechanism limiting oral bioavailability and CNS accumulation of drugs. Int J Clin Pharmacol Ther 38:69–74.

Fujiwara M, Baldeschwieler JD, Grubbs RH (1996) Receptor-mediated endocytosis of poly(acrylic acid)-conjugated liposomes by macrophages. Biochim Biophys Acta 1278:59–67.

Gabizon A, Catane R, Uziely B, Kaufman B, Safra T, Cohen R, Martin F, Huang A, Barenholz Y (1994) Prolonged circulation time and enhanced accumulation in malignant exudates of doxorubicin encapsulated in polyethylene-glycol coated liposomes. Cancer Res 54:987–992.

Gabizon A, Shmeeda H, Barenholz Y (2003) Pharmacokinetics of pegylated liposomal Doxorubicin: Review of animal and human studies. Clin Pharmacokinet 42:419–436.

Gaillard PJ, Visser CC, de Boer AG (2005) Targeted delivery across the blood-brain barrier. Expert Opin Drug Deliv 2:299–309.

Garcia-Garcia E, Andrieux K, Gil S, Couvreur P (2005) Colloidal carriers and blood-brain barrier (BBB) translocation: A way to deliver drugs to the brain? Int J Pharm 298:274–292.

Gaucher G, Dufresne MH, Sant VP, Kang N, Maysinger D, Leroux JC (2005) Block copolymer micelles: Preparation, characterization and application in drug delivery. J Control Release 109:169–188.

Gosk S, Vermehren C, Storm G, Moos T (2004) Targeting antitransferrin receptor antibody (OX26) and OX26-conjugated liposomes to brain capillary endothelial cells using in situ perfusion. J Cereb Blood Flow Metab 24:1193–1204.

Gref R, Minamitake Y, Peracchia M, Trubetskoy V, Torchilin V, Langer R (1994) Biodegradable long-circulating polymeric nanospheres. Science 263:1600–1603.

Guler MO, Pokorski JK, Appella DH, Stupp SI (2005) Enhanced oligonucleotide binding to self-assembled nanofibers. Bioconjug Chem 16:501–503.

Gupta AK, Berry C, Gupta M, Curtis A (2003) Receptor-mediated targeting of magnetic nanoparticles using insulin as a surface ligand to prevent endocytosis. IEEE Trans Nanobioscience 2:255–261.

Harada A, Kataoka K (1999a) Novel polyion complex micelles entrapping enzyme molecules in the core. 2. Characterization of the micelles prepared at nonstoichiometric mixing ratios. Langmuir 15:4208–4212.

Harada A, Kataoka K (1999b) Chain length recognition: Core-shell supramolecular assembly from oppositely charged block copolymers. Science 283:65–67.

Harada A, Kataoka K (2003) Switching by pulse electric field of the elevated enzymatic reaction in the core of polyion complex micelles. J Am Chem Soc 125:15306–15307.

Harada-Shiba M, Yamauchi K, Harada A, Takamisawa I, Shimokado K, Kataoka K (2002) Polyion complex micelles as vectors in gene therapy–pharmacokinetics and in vivo gene transfer. Gene Ther 9:407–414.

Hashimoto M, Takada K, Kiso Y, Muranishi S (1989) Synthesis of palmitoyl derivatives of insulin and their biological activities. Pharm Res 6:171–176.

Huwyler J, Wu D, Pardridge WM (1996) Brain drug delivery of small molecules using immunoliposomes. Proc Natl Acad Sci U S A 93:14164–14169.

Huwyler J, Cerletti A, Fricker G, Eberle AN, Drewe J (2002) By-passing of P-glycoprotein using immunoliposomes. J Drug Target 10:73–79.

Hyuk Im S, Jeong U, Xia Y (2005) Polymer hollow particles with controllable holes in their surfaces. Nat Mater 4:671–675.

Jacobs C, Kayser O, Muller RH (2000) Nanosuspensions as a new approach for the formulation for the poorly soluble drug tarazepide. Int J Pharm 196:161–164.

Jain S, Mishra V, Singh P, Dubey PK, Saraf DK, Vyas SP (2003) RGD-anchored magnetic liposomes for monocytes/neutrophils-mediated brain targeting. Int J Pharm 261:43–55.

Jaturanpinyo M, Harada A, Yuan X, Kataoka K (2004) Preparation of bionanoreactor based on core-shell structured polyion complex micelles entrapping trypsin in the core cross-linked with glutaraldehyde. Bioconjug Chem 15:344–348.

Jones AT, Gumbleton M, Duncan R (2003) Understanding endocytic pathways and intracellular trafficking: A prerequisite for effective design of advanced drug delivery systems. Adv Drug Deliv Rev 55:1353–1357.

Junghans M, Loitsch SM, Steiniger SC, Kreuter J, Zimmer A (2005) Cationic lipid-protamine-DNA (LPD) complexes for delivery of antisense c-myc oligonucleotides. Eur J Pharm Biopharm 60:287–294.

Kabanov A, Levashov A, Martinek K (1987) Transformation of water-soluble enzymes into membrane active form by chemical modification. Ann N Y Acad Sci 501:63–66.

Kabanov A, Levashov A, Alakhov V (1989a) Lipid modification of proteins and their membrane transport. Protein Eng 3:39–42.

Kabanov AV, Ovcharenko AV, Melik-Hubarov NS, Bannikov AI, Alakhov V, Kiselev VI, Sveshnikov PG, Kiselev OI, Levashov AV, Severin ES (1989b) Fatty acid acylated antibodies against virus suppress its reproduction in cells. FEBS Lett 250:238–240.

Kabanov AV, Chekhonin VP, Alakhov VY, Batrakova EV, Lebedev AS, Melik-Nubarov NS, Arzhakov SA, Levashov AV, Morozov GV, Severin ES, Kabanov VA (1989c) The neuroleptic activity of haloperidol increases after its solubilization in surfactant micelles. Micelles as microcontainers for drug targeting. FEBS Lett 258:343–345.

Kabanov A, Batrakova E, Melik-Nubarov N, Fedoseev N, Dorodnich T, Alakhov V, Chekhonin V, Nazarova I, Kabanov V (1992a) A new class of drug carriers: Micelles of poly(oxyethilene)-poly(oxupropilene) block copolymers as microcontainers for drug targeting from blood in brain. J Control Release 22:141–158.

Kabanov AV, Alakhov VY, Chekhonin VP (1992b) Enhancement of macromolecule penetration into cells and nontraditional drug delivery systems. Sov Sci Rev, Ser D Physicochem Biol 11:1–7.

Kabanov A, Nazarova I, Astafieva I, Batrakova E, Alakhov V, Yaroslavov A, Kabanov V (1995a) Micelle formation and solubilization of fluorescent probes in poly(oxyethylene-b-oxypropilene-b-oxyethylene) solutions. Macromolecules 28:2303–2314.

Kabanov AV, Vinogradov SV, Suzdaltseva YG, Alakhov V (1995b) Water-soluble block polycations as carriers for oligonucleotide delivery. Bioconjug Chem 6:639–643.

Kabanov A, Batrakova E, Alakhov V (2002) Pluronic block copolymers as novel polymer therapeutics for drug and gene delivery. J Control Release 82:189–212.

Kabanov AV, Batrakova EV, Miller DW (2003a) Pluronic block copolymers as modulators of drug efflux transporter activity in the blood-brain barrier. Adv Drug Deliv Rev 55:151–164.

Kabanov AV, Batrakova EV, Alakhov VY (2003b) An essential relationship between ATP depletion and chemosensitizing activity of Pluronic block copolymers. J Control Release 91:75–83.

Kabanov AV, Batrakova EV (2004) New technologies for drug delivery across the blood brain barrier. Curr Pharm Des 10:1355–1363.

Kas HS (2004) Drug delivery to brain by microparticulate systems. Adv Exp Med Biol 553:221–230.

Kataoka K, Harada A, Nagasaki Y (2001) Block copolymer micelles for drug delivery: Design, characterization and biological significance. Adv Drug Deliv Rev 47:113–131.

Kemper EM, van Zandbergen AE, Cleypool C, Mos HA, Boogerd W, Beijnen JH, van Tellingen O (2003) Increased penetration of paclitaxel into the brain by inhibition of P-Glycoprotein. Clin Cancer Res 9:2849–2855.

Kemper EM, Cleypool C, Boogerd W, Beijnen JH, van Tellingen O (2004) The influence of the P-glycoprotein inhibitor zosuquidar trihydrochloride (LY335979) on the brain penetration of paclitaxel in mice. Cancer Chemother Pharmacol 53:173–178.

Khan MA, Jabeen R, Nasti TH, Mohammad O (2005) Enhanced anticryptococcal activity of chloroquine in phosphatidylserine-containing liposomes in a murine model. J Antimicrob Chemother 55:223–228.

Kim RB, Fromm MF, Wandel C, Leake B, Wood AJ, Roden DM, Wilkinson GR (1998) The drug transporter P-glycoprotein limits oral absorption and brain entry of HIV-1 protease inhibitors. J Clin Invest 101:289–294.

Kirillova G, Mokhova E, Dedukhova V, Tarakanova A, Ivanova V, Efremova N, Topchieva I (1993) The influence of pluronics and their conjugates with proteins on the rate of oxygen consumption by liver mitochondria and thymus lymphocytes. Biotechnol Appl Biochem 18(Pt 3):329–339.

Kozlova N, Bruskovskaya I, Melik-Nubarov N, Yaroslavov A, Kabanov V (1999) Catalytic properties and conformation of hydrophobized alpha-chymotrypsin incorporated into a bilayer lipid membrane. FEBS Lett 461:141–144.

Kozubek A, Gubernator J, Przeworska E, Stasiuk M (2000) Liposomal drug delivery, a novel approach: PLARosomes. Acta Biochim Pol 47:639–649.

Kreuter J (2001) Nanoparticulate systems for brain delivery of drugs. Adv Drug Deliv Rev 47:65–81.

Kreuter J (2004) Influence of the surface properties on nanoparticle-mediated transport of drugs to the brain. J Nanosci Nanotechnol 4:484–488.

Kreuter J, Ramge P, Petrov V, Hamm S, Gelperina SE, Engelhardt B, Alyautdin R, von Briesen H, Begley DJ (2003) Direct evidence that polysorbate-80-coated poly(butylcyanoacrylate) nanoparticles deliver drugs to the CNS via specific mechanisms requiring prior binding of drug to the nanoparticles. Pharm Res 20:409–416.

Kuby J (1994) Immunology. New York: Freeman, WH. and Co.

Kumagai AK, Eisenberg JB, Pardridge WM (1987) Absorptive-mediated endocytosis of cationized albumin and a beta-endorphin-cationized albumin chimeric peptide by isolated brain capillaries. Model system of blood-brain barrier transport. J Biol Chem 262:15214–15219.

Kwon GS (2003) Polymeric micelles for delivery of poorly water-soluble compounds. Crit Rev Ther Drug Carrier Syst 20:357–403.

Lange H, Huczko A, Sioda M, Louchev O (2003) Carbon arc plasma as a source of nanotubes: Emission spectroscopy and formation mechanism. J Nanosci Nanotechnol 3:51–62.

Langer R (2001) Drug delivery. Drugs on target. Science 293:58–59.

Lavasanifar A, Samuel J, Kwon GS (2002) Poly(ethylene oxide)-block-poly(L-amino acid) micelles for drug delivery. Adv Drug Deliv Rev 54:169–190.

Lee V (1991) Peptide and protein drug delivery. New York: Dekker.

Lee H, Zeng F, Dunne M, Allen C (2005) Methoxy poly(ethylene glycol)-block-poly(delta-valerolactone) copolymer micelles for formulation of hydrophobic drugs. Biomacromolecules 6:3119–3128.

Lemieux P, Vinogradov SV, Gebhart CL, Guerin N, Paradis G, Nguyen HK, Ochietti B, Suzdaltseva YG, Bartakova EV, Bronich TK, St-Pierre Y, Alakhov VY, Kabanov AV (2000) Block and graft copolymers and NanoGel copolymer networks for DNA delivery into cell. J Drug Target 8:91–105.

Liu G, Garrett MR, Men P, Zhu X, Perry G, Smith MA (2005) Nanoparticle and other metal chelation therapeutics in Alzheimer disease. Biochim Biophys Acta 1741:246–252.

Liu X, Chen C (2005) Strategies to optimize brain penetration in drug discovery. Curr Opin Drug Discov Devel 8:505–512.

Lo E, Singhal A, Torchilin V, Abbott N (2001) Drug delivery to damaged brain. Brain Res Brain Res Rev 38:140–148.

Loscher W, Potschka H (2005a) Drug resistance in brain diseases and the role of drug efflux transporters. Nat Rev Neurosci 6:591–602.

Loscher W, Potschka H (2005b) Role of drug efflux transporters in the brain for drug disposition and treatment of brain diseases. Prog Neurobiol 76:22–76.

Mayhan W (2001) Regulation of blood-brain barrier permeability. Microcirculation 8:89–104.

Melik-Nubarov NS, Suzdaltseva Yu G, Priss EL, Slepnev VI, Kabanov AV, Zhirnov OP, Sveshnikov PG, Severin ES (1993) Interaction of hydrophobized antiviral antibodies with influenza virus infected MDCK cells. Biochem Mol Biol Int 29:939–947.

Miller D, Batrakova E, Waltner T, Alakhov V, Kabanov A (1997) Interactions of Pluronic block copolymers with brain microvessel endothelial cells: Evidence of two potential pathways for drug absorption. Bioconjugate Chem 8:649–657.

Miller D, Batrakova E, Kabanov A (1999) Inhibition of multidrug resistance-associated protein (MRP) functional activity with pluronic block copolymers. Pharm Res 16:396–401.

Minn A, Ghersi-Egea J, Perrin R, Leininger B, Siest G (1991) Drug metabolizing enzymes in the brain and cerebral microvessels. Brain Res Brain Res Rev 16:65–82.

Missirlis D, Tirelli N, Hubbell JA (2005) Amphiphilic hydrogel nanoparticles. Preparation, characterization, and preliminary assessment as new colloidal drug carriers. Langmuir 21:2605–2613.

Miyama T, Takanaga H, Matsuo H, Yamano K, Yamamoto K, Iga T, Naito M, Tsuruo T, Ishizuka H, Kawahara Y, Sawada Y (1998) P-glycoprotein-mediated transport of itraconazole across the blood-brain barrier. Antimicrob Agents Chemother 42:1738–1744.

Miyata K, Kakizawa Y, Nishiyama N, Yamasaki Y, Watanabe T, Kohara M, Kataoka K (2005) Freeze-dried formulations for in vivo gene delivery of PEGylated polyplex micelles with disulfide crosslinked cores to the liver. J Control Release 109:15–23.

Moghimi S, Illum L, Davis S (1990) Physiopathological and physicochemical considerations in targeting of colloids and drug carriers to the bone marrow. Crit Rev Ther Drug Carrier Syst 7:187–209.

Moghimi SM, Szebeni J (2003) Stealth liposomes and long circulating nanoparticles: Critical issues in pharmacokinetics, opsonization and protein-binding properties. Prog Lipid Res 42:463–478.

Moghimi SM, Agrawal A (2005) Lipid-based nanosystems and complexes in experimental and clinical therapeutics. Curr Drug Deliv 2:295.

Mora M, Sagrista ML, Trombetta D, Bonina FP, De Pasquale A, Saija A (2002) Design and characterization of liposomes containing long-chain N-acylPEs for brain delivery: Penetration of liposomes incorporating GM1 into the rat brain. Pharm Res 19:1430–1438.

Moriki Y, Suzuki T, Fukami T, Hanano M, Tomono K, Watanabe J (2004) Involvement of P-glycoprotein in blood-brain barrier transport of pentazocine in rats using brain uptake index method. Biol Pharm Bull 27:932–935.

Muller RH, Jacobs C, Kayser O (2001) Nanosuspensions as particulate drug formulations in therapy. Rationale for development and what we can expect for the future. Adv Drug Deliv Rev 47:3–19.

Muller J, Huaux F, Moreau N, Misson P, Heilier JF, Delos M, Arras M, Fonseca A, Nagy JB, Lison D (2005) Respiratory toxicity of multi-wall carbon nanotubes. Toxicol Appl Pharmacol 207:221–231.

Nayak S, Lyon LA (2005) Soft nanotechnology with soft nanoparticles. Angew Chem Int Ed Engl 44:7686–7708.

Nguyen HK, Lemieux P, Vinogradov SV, Gebhart CL, Guerin N, Paradis G, Bronich TK, Alakhov VY, Kabanov AV (2000) Evaluation of polyether-polyethyleneimine graft copolymers as gene transfer agents. Gene Ther 7:126–138.

Nishiyama N, Okazaki S, Cabral H, Miyamoto M, Kato Y, Sugiyama Y, Nishio K, Matsumura Y, Kataoka K (2003) Novel cisplatin-incorporated polymeric micelles can eradicate solid tumors in mice. Cancer Res 63:8977–8983.

Olivier JC, Fenart L, Chauvet R, Pariat C, Cecchelli R, Couet W (1999) Indirect evidence that drug brain targeting using polysorbate 80-coated polybutylcyanoacrylate nanoparticles is related to toxicity. Pharm Res 16:1836–1842.

Omori N, Maruyama K, Jin G, Li F, Wang SJ, Hamakawa Y, Sato K, Nagano I, Shoji M, Abe K (2003) Targeting of post-ischemic cerebral endothelium in rat by liposomes bearing polyethylene glycol-coupled transferrin. Neurol Res 25:275–279.

Panyam J, Labhasetwar V (2003) Biodegradable nanoparticles for drug and gene delivery to cells and tissue. Adv Drug Deliv Rev 55:329–347.

Papaldo P, Fabi A, Ferretti G, Mottolese M, Cianciulli AM, Di Cocco B, Pino MS, Carlini P, Di Cosimo S, Sacchi I, Sperduti I, Nardoni C, Cognetti F (2006) A phase II study on metastatic breast cancer patients treated with weekly vinorelbine with or without trastuzumab according to HER2 expression: Changing the natural history of HER2-positive disease. Ann Oncol 17:630–636.

Pardridge W (1998) Introduction to the Blood–Brain Barrier. Methodology, Biology and Pathology. p 486. Cambridge University Press.

Pardridge W (2002a) Drug and gene targeting to the brain with molecular Trojan horses. Nat Rev Drug Discov 1:131–139.

Pardridge W (2002b) Drug and gene delivery to the brain: The vascular route. Neuron 36:555–558.

Pardridge W (2005a) The blood-brain barrier: Bottleneck in brain drug development. NeuroRx 2:3–14.

Pardridge WM (2005b) Tyrosine hydroxylase replacement in experimental Parkinson's disease with transvascular gene therapy. NeuroRx 2:129–138.

Peracchia M, Vauthier C, Desmaele D, Gulik A, Dedieu J, Demoy M, D'Angelo J, Couvreur P (1998) Pegylated nanoparticles from a novel methoxypolyethylene glycol cyanoacrylate-hexadecyl cyanoacrylate amphiphilic copolymer. Pharm Res 15:550–556.

Perloff MD, von Moltke LL, Greenblatt DJ (2004) Ritonavir and dexamethasone induce expression of CYP3A and P-glycoprotein in rats. Xenobiotica 34:133–150.

Perry VH, Bell MD, Brown HC, Matyszak MK (1995) Inflammation in the nervous system. Curr Opin Neurobiol 5:636–641.

Potschka H, Fedrowitz M, Loscher W (2002) P-Glycoprotein-mediated efflux of phenobarbital, lamotrigine, and felbamate at the blood-brain barrier: Evidence from microdialysis experiments in rats. Neurosci Lett 327:173–176.

Rabinow BE (2004) Nanosuspensions in drug delivery. Nat Rev Drug Discov 3:785–796.

Rapoport N, Marin A, Timoshin A (2000) Effect of a polymeric surfactant on electron transport in HL-60 cells. Arch Biochem Biophys 384:1000–1008.

Rapoport N, Batrakova E, Timoshin A, Li S, Kamaev P, Alakhov V, Kabanov A (2006) Polymeric Surfactants Selectively Inhibit Respiration in Multidrug Resistant Cancer Cells. Mol Pharm in press.

Raub TJ, Audus KL (1990) Adsorptive endocytosis and membrane recycling by cultured primary bovine brain microvessel endothelial cell monolayers. J Cell Sci 97(Pt 1):127–138.

Robert S, Domurado D, Thomas D, Chopineau J (1993) Fatty acid acylation of RNase A using reversed micelles as microreactors. Biochem Biophys Res Commun 196:447–454.

Robert S, Domurado D, Thomas D, Chopineau J (1995) Optimization of RNase A artificial hydrophobization in AOT reversed micelles. Ann N Y Acad Sci 750:121–124.

Roney C, Kulkarni P, Arora V, Antich P, Bonte F, Wu A, Mallikarjuana NN, Manohar S, Liang HF, Kulkarni AR, Sung HW, Sairam M, Aminabhavi TM (2005) Targeted nanoparticles for drug delivery through the blood-brain barrier for Alzheimer's disease. J Control Release 108:193–214.

Rousseau V, Denizot B, Le Jeune JJ, Jallet P (1999) Early detection of liposome brain localization in rat experimental allergic encephalomyelitis. Exp Brain Res 125:255–264.

Roy S, Zhang K, Roth T, Vinogradov S, Kao RS, Kabanov A (1999) Reduction of fibronectin expression by intravitreal administration of antisense oligonucleotides. Nat Biotechnol 17:476–479.

Salem AK, Searson PC, Leong KW (2003) Multifunctional nanorods for gene delivery. Nat Mater 2:668–671.

Savic R, Luo L, Eisenberg A, Maysinger D (2003) Micellar nanocontainers distribute to defined cytoplasmic organelles. Science 300:615–618.

Schmidt J, Metselaar JM, Wauben MH, Toyka KV, Storm G, Gold R (2003) Drug targeting by long-circulating liposomal glucocorticosteroids increases therapeutic efficacy in a model of multiple sclerosis. Brain 126:1895–1904.

Shi N, Pardridge WM (2000) Noninvasive gene targeting to the brain. Proc Natl Acad Sci U S A 97:7567–7572.

Shi N, Zhang Y, Zhu C, Boado RJ, Pardridge WM (2001) Brain-specific expression of an exogenous gene after i.v. administration. Proc Natl Acad Sci U S A 98:12754–12759.

Slepnev V, Phalente L, Labrousse H, Melik-Nubarov N, Mayau V, Goud B, Buttin G, Kabanov A (1995) Fatty acid acylated peroxidase as a model for the study of interactions of hydrophobically-modified proteins with mammalian cells. Bioconjug Chem 6:608–615.

Song BW, Vinters HV, Wu D, Pardridge WM (2002) Enhanced neuroprotective effects of basic fibroblast growth factor in regional brain ischemia after conjugation to a blood-brain barrier delivery vector. J Pharmacol Exp Ther 301:605–610.

Spector R (2000) Drug transport in the mammalian central nervous system: Multiple complex systems. A critical analysis and commentary. Pharmacology 60:58–73.

Steinfeld U, Pauli C, Kaltz N, Bergemann C, Lee HH (2006) T lymphocytes as potential therapeutic drug carrier for cancer treatment. Int J Pharm 311:229–236.

Steiniger SC, Kreuter J, Khalansky AS, Skidan IN, Bobruskin AI, Smirnova ZS, Severin SE, Uhl R, Kock M, Geiger KD, Gelperina SE (2004) Chemotherapy of glioblastoma in rats using doxorubicin-loaded nanoparticles. Int J Cancer 109:759–767.

Tamai I, Tsuji A (2000) Transporter-mediated permeation of drugs across the blood-brain barrier. J Pharm Sci 89:1371–1388.

Tao L, Uhrich KE (2006) Novel amphiphilic macromolecules and their in vitro characterization as stabilized micellar drug delivery systems. J Colloid Interface Sci 298:102–110.

Thole M, Nobmanna S, Huwyler J, Bartmann A, Fricker G (2002) Uptake of cationized albumin coupled liposomes by cultured porcine brain microvessel endothelial cells and intact brain capillaries. J Drug Target 10:337–344.

Torchilin V (1998) Polymer-coated long-circulating microparticulate pharmaceuticals. J Microencapsul 15:1–19.

Torchilin VP (2002) PEG-based micelles as carriers of contrast agents for different imaging modalities. Adv Drug Deliv Rev 54:235–252.

Torchilin VP (2004) Targeted polymeric micelles for delivery of poorly soluble drugs. Cell Mol Life Sci 61:2549–2559.

Torchilin VP (2005) Lipid-core micelles for targeted drug delivery. Curr Drug Deliv 2:319–327.

Torchilin VP, Lukyanov AN, Gao Z, Papahadjopoulos-Sternberg B (2003) Immunomicelles: Targeted pharmaceutical carriers for poorly soluble drugs. Proc Natl Acad Sci U S A 100:6039–6044.

Trentin D, Hubbell J, Hall H (2005) Non-viral gene delivery for local and controlled DNA release. J Control Release 102:263–275.

Triguero D, Buciak JB, Yang J, Pardridge WM (1989) Blood-brain barrier transport of cationized immunoglobulin G: Enhanced delivery compared to native protein. Proc Natl Acad Sci U S A 86:4761–4765.

Triguero D, Buciak JL, Pardridge WM (1991) Cationization of immunoglobulin G results in enhanced organ uptake of the protein after intravenous administration in rats and primate. J Pharmacol Exp Ther 258:186–192.

Tsuji A (1998) P-glycoprotein-mediated efflux transport of anticancer drugs at the blood-brain barrier. Ther Drug Monit 20:588–590.

Tsuji A, Tamai I (1997) Blood-brain barrier function of P-glycoprotein. Adv Drug Deliv Rev 25:287–298.

Uchino H, Matsumura Y, Negishi T, Koizumi F, Hayashi T, Honda T, Nishiyama N, Kataoka K, Naito S, Kakizoe T (2005) Cisplatin-incorporating polymeric micelles (NC-6004) can reduce nephrotoxicity and neurotoxicity of cisplatin in rats. Br J Cancer 93:678–687.

Uhr M, Steckler T, Yassouridis A, Holsboer F (2000) Penetration of amitriptyline, but not of fluoxetine, into brain is enhanced in mice with blood-brain barrier deficiency due to mdr1a P-glycoprotein gene disruption. Neuropsychopharmacology 22:380–387.

Umezawa F, Eto Y (1988) Liposome targeting to mouse brain: Mannose as a recognition marker. Biochem Biophys Res Commun 153:1038–1044.

Vakil R, Kwon GS (2005) PEG-phospholipid micelles for the delivery of amphotericin B. J Control Release 101:386–389.

Valle JW, Lawrance J, Brewer J, Clayton A, Corrie P, Alakhov V, Ranson M (2004) A phase II, window study of SP1049C as first-line therapy in inoperable metastatic adenocarcinoma of the oesophagus. J Clin Oncol ASCO Annual Meeting Proceedings (Post-Meeting Edition) 22:4195.

Vinogradov S, Batrakova E, Kabanov A (1999) Poly(ethylene glycol)-polyethyleneimine NanoGel particles: Novel drug delivery systems for antisense oligonucleotides. Coll Surf B: Biointerfaces 16:291–304.

Vinogradov SV, Batrakova EV, Kabanov AV (2004) Nanogels for oligonucleotide delivery to the brain. Bioconjug Chem 15:50–60.

Vinogradov SV, Kohli E, Zeman AD (2005a) Cross-linked polymeric nanogel formulations of 5′-triphosphates of nucleoside analogues: Role of the cellular membrane in drug release. Mol Pharm 2:449–461.

Vinogradov SV, Zeman AD, Batrakova EV, Kabanov AV (2005b) Polyplex Nanogel formulations for drug delivery of cytotoxic nucleoside analogs. J Control Release 107:143–157.

Voinea M, Simionescu M (2002) Designing of 'intelligent' liposomes for efficient delivery of drugs. J Cell Mol Med 6:465–474.

Witt KA, Huber JD, Egleton RD, Davis TP (2002) Pluronic p85 block copolymer enhances opioid peptide analgesia. J Pharmacol Exp Ther 303:760–767.

Wu D, Boado R, Pardridge W (1996) Pharmacokinetics and blood-brain barrier transport of [3H]-biotinylated phosphorothioate oligodeoxynucleotide conjugated to a vector-mediated drug delivery system. J Pharmacol Exp Ther 276:206–211.

Wu D, Song BW, Vinters HV, Pardridge WM (2002) Pharmacokinetics and brain uptake of biotinylated basic fibroblast growth factor conjugated to a blood-brain barrier drug delivery system. J Drug Target 10:239–245.

Yan H, Tsujii K (2005) Potential application of poly(N-isopropylacrylamide) gel containing polymeric micelles to drug delivery systems. Colloids Surf B Biointerfaces 46:142–146.

Yang C, Chang C, Tsai P, Chen W, Tseng F, Lo L (2004) Nanoparticle-Based in Vivo Investigation on Blood-Brain Barrier Permeability Following Ischemia and Reperfusion. Anal Chem 76:4465–4471.

Zhang Y, Pardridge WM (2001a) Conjugation of brain-derived neurotrophic factor to a blood-brain barrier drug targeting system enables neuroprotection in regional brain ischemia following intravenous injection of the neurotrophin. Brain Res 889:49–56.

Zhang Y, Pardridge WM (2001b) Neuroprotection in transient focal brain ischemia after delayed intravenous administration of brain-derived neurotrophic factor conjugated to a blood-brain barrier drug targeting system. Stroke 32:1378–1384.

Zhang GD, Harada A, Nishiyama N, Jiang DL, Koyama H, Aida T, Kataoka K (2003a) Polyion complex micelles entrapping cationic dendrimer porphyrin: Effective photosensitizer for photodynamic therapy of cancer. J Control Release 93:141–150.

Zhang Y, Boado R, Pardridge W (2003b) Marked enhancement in gene expression by targeting the human insulin receptor. J Gene Med 5:157–163.

Zhang X, Xie Y, Jin Y, Hou X, Ye L, Lou J (2004) The effect of RMP-7 and its derivative on transporting evens blue liposomes into the brain. Drug Deliv 11:301–309.

Zhang X, Batrakova E, Li S, Yang Z, Li Y, Zhang L, Kabanov A (2006) Effect of Pluronic P85 on Amino Acid Transporters in the Blood Brain Barrier. Pharm Res in press.

48
Gene Therapy and Vaccination

William J. Bowers, Michelle C. Janelsins, and Howard J. Federoff

Keywords Active immunization; Adenovirus; Adeno-associated virus; Alzheimer's disease; Amyloid-beta; Chemokine; Cytokine; Lentivirus; Herpes simplex virus; Isotypes; Parkinson's disease; Passive immunization; Single-chain antibodies; Th1 cells; Th2 cells; Transgene; Vector

48.1. Introduction

Central nervous system (CNS) diseases represent a class of complex disorders for which cures have been largely unmet due to the general lack of knowledge regarding underlying pathogenic mechanisms. Gene-based therapies directed at ameliorating neurodegenerative diseases exhibit great potential due to rapid scientific advances made regarding delivery modalities, neurosurgical methods, neuroimaging, and molecular biological manipulation. Given these breakthroughs, the diseased CNS presents a neuroimmunological challenge as gene delivery many times requires invasive surgical procedures and a majority of the gene delivery platforms incite transient, and sometimes, inflammatory events that possess the potential to exacerbate disease-related processes. In this chapter, we will discuss the most current literature on gene therapy for CNS disorders by detailing the neuroimmunological profiles of presently available gene transfer platforms, approaches that have been made to minimize vector-mediated inflammation, and ways in which the immune system can be harnessed to prevent and/or treat neurodegenerative diseases via gene-based immunotherapy.

48.2. Vector Selection Rationale

To derive an informed decision in the selection of an appropriate gene therapeutic vehicle for the treatment of a specific neurologic disease, the following must be considered: vector capacity, tropism, genome maintenance, vector-mediated transgene expression duration and levels, and safety profile. The size of the therapeutic transcription unit is many times employed as an initial criterion to focus vector choice. This category also includes a given vector's ability to harbor multiple transcription units, thereby potentially affording reconstitution of a complex biochemical pathway (i.e., dopamine biosynthesis for Parkinson's disease). Potential applications for several presently available vector platforms are restricted by insert size limitations and are sometimes excluded if multigene delivery is a prerequisite for therapy.

Cell type specificity is also an important issue when developing a gene-based therapeutic intervention for neurodegenerative disorders. It would be most beneficial if the vector of choice could transduce and express in cell types that comprise only the specific disease-affected pathway. Vector tropism can be regulated through modulation of cellular receptor interactions by one or more of the following approaches: alteration of virus docking proteins, utilization of alternate viral serotypes, pseudotyping, and introduction of tethered ligands for cellular receptors into the viral envelope. Once a vector is optimally targeted to the brain region of interest, therapeutic transgene expression can be restricted to selected cell populations via the utilization of cell type-specific promoters and/or transcriptional elements (Wagner et al., 1992; Zatloukal et al., 1992; Beer et al., 1998; Toyoda et al., 1998; Zabner et al., 1999). Strict spatial control of transgene expression is vital to ensure the correct cells will manufacture the gene product. This control, in turn, reduces the risk that ectopic transgene expression will occur and lead to untoward effects on adjacent neurological pathways.

Most neurodegenerative disorders evolve insidiously over many years thus requiring gene-base modalities to impart therapeutic benefit for several decades of an individual's lifetime. To this end, a vector genome should be stably maintained within the transduced cell for extended periods of time. Vector genome maintenance is, therefore, a critical factor in selection of an appropriate gene therapy vehicle for neurodegenerative diseases. Vector genomes can exist as episomes and/or integrated forms within nuclei of host cells. Mitotically active cells, such as those of the progenitor and glial lineages, eventually exhibit diminished

T. Ikezu and H.E. Gendelman (eds.), *Neuroimmune Pharmacology*.
© Springer 2008

episomal vector-mediated transgene expression. However, the post-mitotic property of CNS neuronal populations does not exclude the utilization of episomal vectors since genomes can be maintained without progressive therapeutic vector loss due to mitosis. Integrating vectors circumvent this issue but their use raises safety concerns including their potential to transactivate proximal proto-oncogenes and to disrupt essential host genes via insertional mutagenesis.

Another similar issue regarding vector selection relates to the desired levels and duration of gene product expression for treatment of neurodegenerative disorders. Depending upon the vector and transcriptional elements chosen, pharmacologic or physiologic levels of transgene expression can be achieved for time periods of short or long duration. As with other selection criteria, the decision of which level/duration of expression is preferred rests heavily on what aspect of the particular disorder is to be targeted and at which time during the disease course the intervention is to be implemented. Early interventions may require maintenance of long-term physiologic levels of transgene expression (i.e., neuroprotective strategies) since the neural networks may be primarily intact at this time. A vector/promoter combination that safely and stably maintains gene expression at nearly physiologic levels in the CNS would serve as a potential candidate for such early treatment approaches. Treatment modalities that are implemented after presentation of clinical symptoms may require long-term pharmacologic levels of transgene product to restore function to a brain region decimated by disease.

Safety is of utmost concern regarding the application of novel gene therapeutic strategies within the brain. Many presently available vectors trigger immunogenic and/or inflammatory responses when introduced into the CNS. These responses are known to arise from the humoral and/or cell-mediated arms of the immune system, and the magnitude differs depending upon which vector type is employed. For example, repeat administration of early generation viral vectors has been shown to lead to lower transgene expression and serious inflammation, likely the result of a primed immune system (Byrnes et al., 1996). Therefore, a vector that is stably maintained and that can express its transgene for extended periods of time would be a more favorable choice as a gene therapeutic vehicle for neurodegenerative conditions. Another aspect that is often overlooked regarding gene therapy safety is the role of transgene products in the elaboration of immune responses and toxicity. Transgene products that are of foreign origin, ectopically expressed, or pharmacologically expressed harbor the potential to induce cytotoxicity and/or immune responses. Research addressing these issues is imperative to elucidate the role of transgene products in the elaboration of these potentially harmful responses, and how such responses can be successfully circumvented. Utilization of regulatable transcriptional or post-transcriptional elements in delivery vectors to provide "fine-tuning" of therapeutic transgene expression levels is a way to minimize harmful clinical outcomes.

48.3. Gene Transfer Platforms

Genetic material can be delivered to cells by two broad classes of vectors: nonviral and viral. Nonviral gene transfer modalities include the use of plasmid DNA directly or in an encapsulated state. Such approaches are deemed the safest due to the lack of inflammatory and/or immunogenic components, but require repeated administration to maintain therapeutic gene expression. Viral vectors exploit the evolutionary achievements of viruses to propagate, package and transfer genetic material from cell to cell and organism to organism. Via genetic modification of mammalian viruses, it is possible to specifically target gene expression in desired cellular populations. Adenovirus, AAV, lentivirus, and HSV-based vectors as they have been utilized for CNS-directed gene therapy will be discussed below. These vector types represent a subset of the more commonly used viral vectors for CNS gene delivery.

48.3.1. Nonviral Gene Transfer

Although viral gene delivery systems have proven more efficient for use in the nervous system, viral vector production requires special expertise and equipment, and remains time and labor intensive. As an alternative to viral vectors, transfection techniques utilizing cationic lipids have shown promise for use with cells of the nervous system and may prove to exact fewer adverse effects. Conventional nonviral systems (reviewed in Anwer et al., 2000), which rely on passive cell targeting through charge interactions, possess several practical advantages over viral delivery of transgenes. These include decreased production time and costs as well as greater safety, since there is no risk of viral vector recombination.

For more than a decade, there have been a variety of reports demonstrating the use of lipid reagents for transfection of cells comprising the nervous system with very modest transfection efficiencies, even after optimization (Holt et al., 1990; Loeffler et al., 1990; Jiao et al., 1992; Matter-Sadzinski et al., 1992; Kaech et al., 1996). However, more recently our laboratory showed dramatically improved transfection efficiencies in the mouse CNS using the cationic lipids Tfx-10 and Tfx-20 (Promega) to deliver DNA with a microprocessor-controlled injector, suggesting further work towards optimization of lipid transfection is warranted for gene transfer applications requiring direct intracranial administration of therapeutic DNA-lipid complexes (Brooks et al., 1998).

Less invasive methods of delivering nonviral vectors to the CNS would be more desirable. Despite the minimal inflammatory profile of intracranial delivery of DNA-lipid complexes, the potential exists for surgery-induced adverse events, while the relatively limited volume of distribution obtained from stereotactic infusion makes the approach rather infeasible when attempting to treat diseases that involve expansive regions of the brain. To that end, methodologies have been developed to enable the injection of modified

DNA-lipid complexes into the bloodstream that are designed to hone to the brain, and sub-regions thereof, to more widely deliver their therapeutic payload. Nonviral gene delivery systems in particular lend themselves to targeting strategies aimed at achieving higher transfer efficiency and tissue specificity, since they can withstand a wider range of chemical and physical conditions used to incorporate targeting moieties. They also allow for more flexibility and ease-of-use when mixing targeting ligands with plasmid DNA and the lipid delivery reagents.

To approach systemic infusion of targeted DNA-encapsulated liposomes, the liposome formulation requires further modification. Whereas in the case of direct stereotactic injections, cationic liposome-DNA complexes can be used, such complexes highly aggregate in physiological saline leading to inefficient and variable dissemination of the transfection complex when delivered via the bloodstream (Mahato et al., 1997; Matsui et al., 1997). The use of neutral (uncharged) liposomal mixtures in combination with inert and biocompatible polymer polyethylene glycol (PEG) produces sterically stabilized liposomal structures (Papahadjopoulos et al., 1991). Inclusion of PEG is also believed to prevent the binding of opsonins circulating within the bloodstream, thereby significantly decreasing phagocytic cell recognition and clearance (Moghimi and Patel, 1992). Physical targeting of these "pegylated liposomes" through the use of ligands and antibodies has been employed to enhance the efficiency of gene transfer to many specific cell types (reviewed in Anwer et al., 2000), including the CNS (Pardridge, 2003).

The blood-brain barrier (BBB) has long-served as a major obstacle to liposome-based gene delivery to the CNS via systemic administration. Zhang and colleagues have employed a pegylated immunoliposome-mediated method for systemic administration of plasmids expressing tyrosine hydroxylase (TH) in a rodent 6-OHDA model of Parkinson's disease [PD; (Zhang et al., 2004)]. A plasmid construct expressing TH, the rate-limiting enzyme in the dopamine biosynthetic pathway, was encapsulated within pegylated immunoliposomes and injected systemically into parkinsonian rats. High-level transgene expression within the striatum and transient correction of apomorphine-induced rotation behavior were observed, indicating that these nonviral gene transfer reagents were capable of traversing the BBB. In addition, a time-course experiment showed that TH enzyme activity within the striatum diminished substantially from Day 3 to 9 post-treatment. In a recent follow-up study, this group demonstrated that transient gene expression profiles derived from immunoliposome-targeted gene delivery appear to be the result of plasmid degradation *in vivo* (Chu et al., 2006). Immune responses generated against pegylated immunoliposomes are thought to be minimal *in vivo*, but if multiple injections are required to subvert disease progression, then the adaptive immune response may play a role in clearing and/or degrading this gene-based therapeutic.

48.3.2. Adenovirus Vectors

Adenoviruses (Ad) are a family of non-enveloped, double-stranded DNA viruses that generally cause mild respiratory infections in mammals. Over 50 serotypes have been identified. Adenovirus-based vectors are attractive candidates for gene delivery to the CNS, as high vector titers can be generated, and these episomally maintained vectors can efficiently infect and express transgenes in a variety of cell types including the non-dividing cells of the CNS (Akli et al., 1993; Bajocchi et al., 1993; Davidson et al., 1993; Le Gal La Salle et al., 1993; Smith et al., 1996a,b,c). First generation vectors consisted of constructs that lacked genes for the potent regulatory proteins E1A, E1B, and E3. Although high titers could be obtained (approximately 10^{13} viable particles per ml) the immunological and physiological complications that resulted from Ad vector transduction in the CNS were serious. This robust response was due to enhanced cytotoxic T lymphocyte (CTL) activity to viral proteins and/or the expressed transgene product (Tripathy et al., 1996; Michou et al., 1997). Vigorous inflammation precludes repeated administration of this vector. As a consequence of these findings, second generation vectors were developed that additionally lacked the E2A gene. The resultant inflammatory response elicited by these vectors was reduced but still fairly substantial. In an effort to completely remove viral genes from the system, helper virus-dependent or so-called "gutless" forms of Ad vectors were developed (Mitani et al., 1995; Parks et al., 1996; Hardy et al., 1997; Morsy et al., 1998). These forms possess a large transgene capacity (up to 28 kb) and do not express any viral proteins. High multiplicities of infection (MOIs) have been used to infect greater than 85% of neurons with little if any cytotoxicity or adverse physiological effects up to 7 days post-infection (Cregan et al., 2000). Gutless Ad vectors direct expression of therapeutic genes *in vivo* with moderately high efficiency among gene therapy vectors. Gutless Ad constructs, but not first-generation Ad vectors, mediate sustained transgene expression in the brain even in the presence of anti-Ad immunity (Thomas et al., 2000, 2001). Another study, conducted by Zou and colleagues, compared transgene expression from first-generation Ad vectors to gutless Ad vectors in brain. Two months following transduction, the helper-dependent vectors exhibited higher transgene expression and elicited lower numbers of brain-infiltrating macrophages and T cells than first generation Ad vectors (Zou et al., 2000), further speaking to the improved safety profile of this Ad vector iteration.

48.3.3. Adeno-Associated Virus Vectors

Adeno-associated virus (AAV) is a non-pathogenic member of the parvovirus family. Its single-stranded DNA viral genome requires co-infection with either adenovirus or HSV for its own replication/propagation. Wild-type AAV encodes two viral gene products, Rep and Cap, which function in replication/integration and structural stability, respectively. Vectors derived from AAV can carry up to a 4.5-kb transgene that is flanked by inverted terminal repeats (ITRs). The relatively small payload capacity can be overcome by the co-injection of multiple vectors, each transducing a different therapeutic gene, and leading to co-expression of two linked biochemical functions (Shen et al., 2000). The ITR sequences promote extrachromosomal replication and Rep-mediated genomic integration of the flanked transgene. Newer methods of packaging have led to titers approaching 10^9 transducing units/10 cm plate and have eliminated the need for helper Ad which assures that negligible levels of contaminating Ad helper virus are present in the preparation (Xiao et al., 1998). Once packaged, the unique Cap protein of AAV allows for high-degree virion purification. Helper virus-free AAV vectors elicit transient and minimal inflammatory reactions when delivered *in vivo*, an observation likely due to the lack of associated viral gene expression and the facility to highly purify vector stocks (Kaplitt and Makimura, 1997; Xiao et al., 1997; Bueler, 1999; Lo et al., 1999).

The existence of numerous characterized AAV serotypes allows for use of alternative capsid types in the event serotype-specific neutralizing antibodies are generated or extant, thus extending the utility of the vector platform in the setting of pre-existing AAV immunity. Moreover, each serotype exhibits distinct patterns of transduction within the major tissues of mammals (reviewed by Wu et al., 2006). In general, AAV1 and 5 exhibit higher transduction efficiencies than vectors packaged with serotype 2 capsid in all regions assessed within the CNS (Alisky et al., 2000; Burger et al., 2004), while AAV4 transduces specific cell types, such as ependyma and astrocytes in the subventricular zone, with higher efficiency (Davidson et al., 2000). The varying serotypes not only dictate transduction efficiency, but they have been shown to modulate transgene expression kinetics. The underlying mechanism(s) is unknown, but may be due to differences in virion uncoating and intracellular trafficking within the transduced cell (Thomas et al., 2004).

AAV vectors when delivered into the CNS are capable of expressing transgene products for periods of time in excess of 4 years. This has been demonstrated in both rodent and non-human primate studies [(Kaplitt et al., 1994; McCown et al., 1996; Guy et al., 1999; Lo et al., 1999; Bjorklund et al., 2000; Chirmule et al., 2000; During et al., 2001) and K. Bankiewicz, personal communication]. The pattern and duration of gene expression in these studies suggests that the AAV vectors once introduced into host cells adopt a transcriptional configuration that supports stable expression. This characteristic could lie in the physical status of the AAV vector genome (integrated vs. episomal) following transduction of a given host cell. The ability of wild-type AAV to specifically integrate into human chromosome 19 is intriguing if this property could be translated to AAV-derived vectors (Kotin et al., 1990). The ability to integrate creates the potential for stable, long-term expression of a transgene for the treatment of neurological disorders. Wu and colleagues presented a method for detecting integrated AAV vector genomes and provided some evidence that integration into CNS cells occurs, however, the specificity of this integration site is yet undetermined (Wu et al., 1998). More recent data generated by the laboratory of Mark Kay suggest that concatenated, non-integrated AAV genomes exist in greater abundance than integrating forms (Chen et al., 2001). To this end, more thorough studies detailing the genome status of AAV vectors *in vivo* must be conducted.

The risk of integration-induced oncogenesis appears to still be a theoretical concern despite a recent report. In long-term rodent studies with rAAV vectors there was a significant increase in the incidence of hepatocellular carcinoma in mice treated with an AAV vector (Daly et al., 2001; Donsante et al., 2001). Systematic investigation was conducted to establish the genesis of the AAV vector/tumor association. These studies employing sensitive quantitative PCR failed to yield evidence that the tumors arose from an insertional mutagenesis event. In a separate study, mice were injected with varying doses of AAV vectors with no evidence of hepatic tumors. These data argue that the hepatic malignancies likely arose by administration of contaminants with the AAV vector stocks and not because of genomic perturbations induced by the vector. However, more recent data compel reexamination of the propensity of rAAV to produce chromosomal alterations. In that study, which was entirely performed in cultured human HeLa cells, AAV vector proviral integrates were found associated with chromosomal deletions and rearrangements most frequently on chromosome 19, nearby but not precisely at the site of wild-type AAV integration (Miller et al., 2002). Multiple explanations for this finding are unexplored and thus its significance is unclear. Assuming safe integration of AAV vectors is demonstrated, a major limitation that will likely persist is the limited transgene size capacity exhibited by these vectors. This shortcoming will prove troublesome when cell-specific promoters and/or transcriptional enhancer elements must accompany the desired therapeutic gene for the purposes of directed expression. One group has attempted to circumvent this issue by developing heterodimeric AAV vectors where transfer of a transcription unit of approximately 9 kb can be achieved (Sun et al., 2000).

48.3.4. Lentivirus Vectors

Replication-defective lentiviral vectors derived from HIV (rHIV) are another promising gene transfer vehicle for CNS applications as they can transduce both dividing and non-dividing cells, do not encode viral proteins, have an approximate 9-kb insert capacity, and are capable of sustaining stable gene expression for greater than a year (Kordower et al., 1999, 2000; Wang et al., 2000; Reilly, 2001). To ensure against the generation of replication-competent HIV virions, the packaging system has been divided into four plasmids and four accessory HIV genes have been deleted from those packaging-assisting plasmids. Titers with this approach yield stocks of 1–5 × 10^9 infectious units (Naldini et al., 1996b; Bensadoun et al., 2000).

HIV-based lentiviral vectors when pseudotyped with the VSV-G protein are principally neurotropic (Desmaris et al., 2001). Entry, proviral derivation and expression appear to occur efficiently in non-dividing neurons without the requirement for disruption of the nuclear envelope (Naldini et al., 1996a,b). Whereas, lentiviral vectors integrate chromosomally in dividing cells, the status of the proviral genome in neurons is less well studied. However, recent evidence garnered from the creation of feline immunodeficiency virus (FIV)-derived vectors devoid of integrase activity suggests that in non-dividing cells, transgene expression levels are equivalent to vectors capable of integrating into the host cell chromosome (Saenz et al., 2004). Efforts to test whether lentiviral vector proviral genomes can be rescued by HIV infection have also been informative. If the vectors retain necessary packaging sequences they are inefficiently rescued (Evans and Garcia, 2000; Follenzi et al., 2000). However, self-inactivating or SIN vectors, which possess major deletions within the U3 and U5 regions of the long terminal repeat (LTR) regions of the lentiviral genome, are unable to be rescued (Yu et al., 1986; Iwakuma et al., 1999).

Lentiviral vectors have been examined with respect to their potential elicitation of immune responses in the brain and within visceral organs. The data available indicate that in the CNS lentiviral vectors are minimally immunogenic, if optimized vector production and purification protocols are followed (Baekelandt et al., 2003). Systemic immunization preceding injection of lentiviral vectors into the CNS determined that pre-existing anti-lentiviral immunity, regardless of the transgene, did not affect transgene expression (Abordo-Adesida et al., 2005). Furthermore, Abordo-Adesida and colleagues showed that the transgene, but not the virion or vector components, is chiefly responsible for providing antigenic epitopes to the activated immune system, following systemic immunization with lentivirus.

Even though they are more limited in number, rodent and non-human primate studies performed to date have shown that rHIV vectors encoding for glial cell line-derived neurotrophic factor (GDNF) provide robust functional and structural neural protection in models of PD and provide an encouraging approach for gene therapy. Kordower and colleagues injected lentiviral vectors expressing GDNF into the striatum and substantia nigra of nonlesioned aged rhesus monkeys and young adult rhesus monkeys treated 1 week prior with the PD symptom-inducing neurotoxicant, 1-methyl-4-phenyl-1,2,3,6-tetrahydropyridine (MPTP) (Kordower et al., 2000). Treatment with these vectors augmented dopaminergic functioning and reversed a set of behavioral deficits. More recently, Brizard et al. were able to utilize a similar GDNF-expressing lentiviral vector to demonstrate functional reinnervation from remaining dopaminergic nerve terminals in the 6-hydroxydopamine (6-OHDA) rat model of PD (Brizard et al., 2006).

Lentivirus vectors have also been used in the development of potential AD therapeutics. Dodart and colleagues recently delivered varying alleles of the apolipoprotein E (apoE) gene, which encodes for a lipid-binding protein with Aβ fibrillogenesis modulation activity, to AD mouse models to determine if expression of different alleles led to differential amyloid deposition (Dodart et al., 2005). Lentiviral delivery of the ε4 allele, one that is believe to enhance the risk of human AD, led to increased Aβ deposition as measured by immunocytochemistry and ELISA. Conversely, lentiviral vector-mediated expression of the ε2 allele resulted in a significant reduction in amyloid burden, which is suggestive of a potent dominant negative effect of apoE2 over mouse apoE on brain Aβ deposition. These findings are exciting but more experimentation is warranted to determine the actual biological function of apoE and how modulation of its activity could alter normal physiology.

48.3.5. Herpes Simplex Virus Vectors

Herpes simplex virus type 1 (HSV-1) is a naturally neurotropic virus capable of establishing latent infection within neurons, but also possesses the ability to infect a wide range of cellular targets. The cellular receptors responsible for virion docking and uptake have been cloned, including the herpesvirus entry mediator A (HveA) and HveC (nectin-1; (Montgomery et al., 1996; Hsu et al., 1997; Kwon et al., 1997; Geraghty et al., 1998; Warner et al., 1998)). Not surprisingly, these receptors (or homologues) are expressed on a variety of cell types. The HSV life cycle involves long periods of latency, and to that end, the virus has evolved a number of elaborate and highly efficient mechanisms to avoid detection and elimination by immune cells (Banks and Rouse, 1992). These properties have led to the development of two forms of HSV-1-based delivery vectors capable of *in vivo* and *in vitro* gene transfer to the nervous system: recombinant and amplicon vectors (Dobson et al., 1990; Breakefield and DeLuca, 1991; Andersen et al., 1992; Breakefield et al., 1992; Fink et al., 1992; Wolfe et al., 1992; Glorioso et al., 1994; Andersen and Breakefield, 1995).

48.3.5.1. HSV Recombinant Vectors

Recombinant HSV vectors comprise a wild-type HSV genome rendered replication defective via disruption/deletion of an indispensable viral gene(s). Typically, the immediate-early gene loci, which encode for potent transactivation proteins that initiate the viral lytic cycle, are targeted for insertion of therapeutic transcription units via homologous recombination (Marconi et al., 1996, 1999). Following construction, recombinant vectors are packaged into infectious virions using an engineered eukaryotic cell line that supplies the absent viral gene product(s) in trans (Dobson et al., 1990; Andersen et al., 1992). The genome of recombinant vectors, at present, can accommodate approximately 30 kb of genetic material. Recombinant vectors are also attractive gene transfer vehicles for CNS disorders because they can be propagated to relatively high titers (10^8–10^9 plaque-forming units/ml). In addition, the threat of insertional mutagenesis is greatly diminished as the vector genome persists episomally within post-mitotic cell nuclei.

Immune responses arising from infusion of recombinant HSV vectors can arise from any of the following sources: viral particle components, co-purified packaging cell debris, low-level *de novo* viral gene product expression, and transgene expression itself. What appears to be a major source of immune response elicitation is related to *de novo* expression of the intact viral transcription units. Recombinant HSV vectors, although replication-defective, harbor virtually intact segments of the HSV genome. These open reading frames (ORFs) are expressed at low levels even in the absence of immediate-early gene products, augmenting the potential for antigen processing and subsequent MHC Class I presentation (Johnson et al., 1992; Krisky et al., 1998). Detailed assessment of immune responses elicited against HSV recombinant vectors within the brain have been lacking. However, insights may be gleaned from studies employing the amplicon vector platform, which have received, immunologically speaking, more investigational attention.

48.3.5.2. HSV Amplicon Vectors

The HSV-1 amplicon is a uniquely designed eukaryotic expression plasmid that harbors two non-protein encoding virus-derived elements: an HSV origin of DNA replication (OriS) and the cleavage/packaging sequence ("a" sequence) (Frenkel, 1981, 1982; Spaete and Frenkel, 1982, 1985; Stow and McMonagle, 1982; Geller and Breakefield, 1988; Federoff et al., 1992; Geschwind et al., 1994). Both *cis* sequences are specifically recognized by HSV proteins to promote the replication and incorporation of the vector genome into viable viral particles, respectively. This highly versatile plasmid can be readily manipulated to contain desired promoters, enhancers, and transgenes of substantial size (~130 kb) (Wade-Martins et al., 2001). Heterologous transcription units either singly or in combination can be cloned into the amplicon plasmid using conventional molecular cloning techniques, and the resultant construct is packaged into enveloped viral particles for subsequent transduction of cells or tissues.

Amplicon plasmids are dependent upon helper virus function to provide the replication machinery and structural proteins necessary for packaging amplicon vector DNA into viral particles. An engineered replication-defective HSV derivative that lacks an essential viral regulatory gene has conventionally provided helper packaging function. These helper viruses are similar to the recombinant HSV vectors discussed above in that they retain a majority of the HSV genome. The final product of helper virus-based packaging contains a mixture of varying ratios of helper and amplicon virions. The titers obtained from helper virus-based amplicon packaging range from 10^8 to 10^9 expressing virus particles/ml. More recently, helper virus-free amplicon packaging methods have been developed by providing a packaging-deficient helper virus genome via a set of five overlapping cosmids or a bacterial artificial chromosome (Fraefel et al., 1996; Saeki et al., 1998, 2001 Stavropoulos and Strathdee, 1998). This packaging strategy requires the co-transfection of eukaryotic cells that are receptive to HSV propagation (i.e., BHK, Vero) with packaging-incompetent HSV genomic DNA, amplicon DNA, and any accessory HSV genes shown to enhance amplicon titers (Bowers et al., 2001) (Figure 48.1). Crude vector lysates are then purified by a series of ultracentrifugation steps and titered by expression or transduction-based methodologies (Bowers et al., 2000). The titers obtained from helper virus-free amplicon packaging typically range from 10^7 to 10^8 expressing virus particles/ml. The lack of contaminating helper virus in these stocks, and thus loss of immunosuppressive proteins like ICP47, has also made the HSV amplicon a powerful delivery platform for infectious disease and cancer

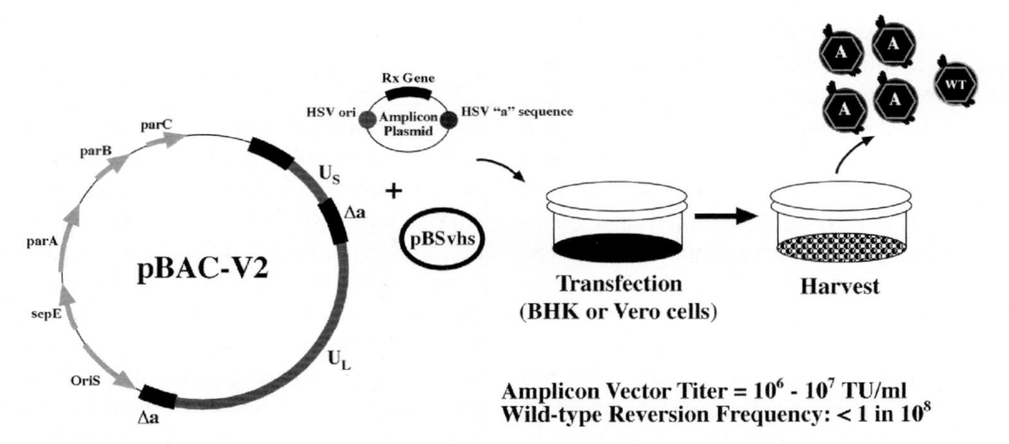

FIGURE 48.1. The helper virus-free method of HSV-1 amplicon packaging involves the use of a bacterial artificial chromosome (BAC) that contains the entire HSV genome minus its cognate cleavage/packaging signals [pBAC-V2; (Stavropoulos and Strathdee, 1998)]. Co-transfection of pBAC-V2, an accessory plasmid encoding HSV virion host shutoff protein (Bowers et al., 2001) and an amplicon plasmid, into a cell line permissible to HSV virion propagation (e.g., Vero, baby hamster kidney cells) results in stocks composed specifically of amplicon-containing virions that exhibit low, if any, cytotoxicity (Olschowka et al., 2003). Titers of stocks produced by this method range from 10^6 to 10^7 transducing units per ml with a very low wild-type reversion frequency (<1 in 10^8).

vaccines (Tolba et al., 2001, 2002; Willis et al., 2001; Hocknell et al., 2002).

Similar to recombinant HSV vectors, immune responses arising from infusion of HSV amplicon stocks can arise from several sources and such responses are dependent upon the packaging system employed. Immune response-eliciting sources include viral particle components, co-purified packaging cell debris, low-level *de novo* viral gene product expression (helper virus-based packaging only), and the expressed transgene. Early generation helper virus-based packaging methods lead to vector stocks that contain substantive levels of contaminating helper virus. Due to the identical physical properties of amplicon and helper virus particles, preferential purification of amplicon particles is not possible. The replication-defective helper virus, comparable to the recombinant HSV vectors described above, expresses viral proteins at low levels within transduced cells. These viral proteins exhibit cytotoxic activity and can potentially undergo antigenic processing and subsequent immune presentation.

Wood and colleagues were the first investigators to examine the immune responses elicited against early iterations of packaged HSV amplicon stocks. For their studies they utilized an amplicon expressing the reporter gene product β-galactosidase and packaged using a recombinant HSV helper virus (*tsK*), which possessed a temperature-sensitive mutation in the ICP4 gene locus (Wood et al., 1994). Although *tsK* is replication-defective at the non-permissive temperature of 39° C, viral gene product-associated cytotoxicity remains observable. Administration of these amplicon stocks induced a vigorous inflammatory response. Elevated MHC Class I expression and microglial activation was evident by 2 days post-infection, which was followed by MHC Class II cell recruitment, T cell activation, and macrophage influx at 4 days following delivery of helper virus-contaminated amplicon stocks.

In a more recent study, our laboratory described the innate responses elicited upon stereotactic delivery of HSV amplicon vectors packaged via two different methods (Olschowka et al., 2003). C57Bl/6 mice were injected with sterile saline, β-galactosidase-expressing amplicon (HSVlac) packaged by a conventional helper virus-based technique, or a helper virus-free HSVlac preparation. Animals were sacrificed at 1 or 5 days post-transduction and analyzed by immunocytochemistry and quantitative RT-PCR for various chemokine, cytokine, and adhesion molecule gene transcripts. All injections induced inflammation with blood-brain barrier opening on Day 1 that was similarly enhanced following all treatments. By Day 5, mRNA levels for the pro-inflammatory cytokines (IL-1β, TNF-α, IFN-γ), chemokines (MCP-1, IP-10) and an adhesion molecule (ICAM-1) had fully resolved in saline-injected mice and to near baseline levels in mice receiving helper virus-free HSVlac. In contrast, mice injected with helper virus-packaged amplicon stocks elicited elevated inflammatory molecule expression and immune cell infiltration even at Day 5. In aggregate, these studies demonstrated helper virus-free amplicon preparations exhibit a safer innate immune response profile when delivered directly to the brain, presumably due to the absence of helper virus gene expression, and provide support for future amplicon-based CNS gene transfer strategies. What remains to be assessed experimentally in greater detail is the role, albeit minor, that virion structural components, packaging cell-derived debris, and transgene product appear to play in the activation of the innate immune response by helper virus-free amplicon stocks. In addition, the role of pre-existing HSV immunity in modulating host responses to brain-delivered HSV vectors remains understudied.

48.4. Neuroimmunotherapy

Awareness of brain "immunocompetence" has increased significantly in recent decades. Immunomodulatory molecules are expressed in the brain by microglia, astrocytes and neurons in the presence of an inducing stimulus, including the infusion of gene transfer vectors. Additionally, macrophages, T cells, and B cells traverse the blood-brain barrier and survey the brain as part of their normal physiologic function (Hickey, 2001). Furthermore, chronic activation of brain-resident inflammatory processes has been shown to underlie the pathophysiology of several neurodegenerative diseases such as Alzheimer's disease (AD) and Parkinson's disease (PD). One of the central features of AD is the excessive accumulation of amyloid beta ($A\beta_{1-42}$), a 42-amino acid peptide, in extracellular senile plaques. Moreover, the AD brain is decorated with complement proteins that bind to Aβ peptides and aid in opsinization for eventual clearance by microglia. In response to accumulating Aβ, microglia in the AD brain express an array of potent cytokines and chemokines that are likely neurotoxic and may contribute to cell loss (Strohmeyer and Rogers, 2001; Hanisch, 2002; McGeer and McGeer, 2003).

Inflammatory processes may also contribute to the degeneration observed in PD brain. PD is characterized by the destruction of dopamine-containing neurons in the substantia nigra (SN), resulting in a loss of dopaminergic afferents to the basal ganglia and eventual motoric dysfunction (Olanow, 2003; Olanow et al., 2003). Pathologically, PD brain exhibits a region-specific accumulation of Lewy bodies, which are intraneuronal inclusions comprised of aggregated α-synuclein and other cellular proteins (Hald and Lotharius, 2005). A variety of pro-inflammatory mediators have been detected in human PD and animal models of the disease (reviewed by McGeer and McGeer, 2004). Moreover, epidemiological evidence has shown that chronic treatment with nonsteroidal anti-inflammatory drugs (NSAIDs) reduces the risk of AD and PD compared to those who take NSAIDs on a non-regular dosing regimen (Schiess, 2003; Standridge, 2004). While it appears that numerous cytokine and chemokine molecules are closely associated with and likely contribute to neurodegeneration in AD and PD, it is uncertain whether these inflammatory mediators impart beneficial effects during pathogenesis.

A promising, yet inherently complex, means of preventing Aβ accumulation in AD or α-synuclein in PD relates to the use of the host immune system to mount specific immune responses against the self-peptides, Aβ and α-synuclein (Klyubin et al., 2005; Masliah et al., 2005). The findings from a variety of immune-based strategies speak to the promise of such approaches, but also reveal the potential for morbid complications, especially in the case of AD vaccination. Therefore, it is imperative that as immunotherapeutics are designed to prevent and/or treat neurodegenerative diseases the underlying inflammatory processes are taken into account.

48.4.1. Immunotherapeutic Approaches to Treating Neurodegenerative Diseases

Two principal immunotherapeutic strategies for diseases afflicting the CNS have been pursued: passive and active vaccine-based approaches. Passive immunization involves the transfer of antiserum, purified antibodies, or an antibody-encoding gene via gene delivery to a recipient to prevent the accumulation of or to promote the removal of a central pathogenic factor. For example, passive immunization achieved by administration of Aβ-specific antibodies has shown promise in preclinical studies. The laboratories of Drs. Holtzman and Ashe have independently reported the benefits of systemic delivery of Aβ-specific antibodies to AD mice (Dodart et al., 2002; Kotilinek et al., 2002). Treated mice exhibited a marked reduction in amyloid accumulation and a significant improvement in memory-oriented behavioral tests. Using a similar passive immunization approach, Bard et al. showed that peripherally administered monoclonal antibodies against the Aβ enter the CNS of PDAPP mice (Bard et al., 2000). Additionally, these passively administered antibodies promoted the clearance of pre-existing amyloid, thus reducing the plaque burden. Prophylaxis against amyloid pathology required multiple injections of high-titer antibody in all studies. Approaches that provide for the stable production of Aβ-specific antibody *in situ*, such as via gene transfer, may represent a more viable therapeutic option.

48.4.2. Single Chain Antibodies as Passive Immunotherapeutics

One gene-based passive vaccination approach that may serve in this capacity involves the use of single-chain antibodies (scFv's). ScFv's are composed of the minimal antibody-binding site formed by non-covalent association of the V_H and V_L variable domains joined by a flexible polypeptide linker. Human scFv-phage libraries are available and allow for high affinity human scFv antibodies to be selected from combinatorial libraries. Thus far, phage display has proven to be a powerful tool for rapidly generating and isolating recombinant antibodies. Selection by phage display entails binding of phage populations expressing antibody molecules on the tip of the phage particle to a specified antigen (i.e. α-Aβ peptide). The antigen is immobilized on a microtiter plate well, incubated with phage and then washed to remove non-specifically bound phage. Phages are eluted from antigen, re-expanded and the entire process is repeated for several more cycles, a process termed "panning." At each cycle, the fraction of specifically bound phage increases while the diversity of antibody sequences decreases until the population consists of only phages that can bind antigen with roughly equivalent panning efficiencies (Malone and Sullivan, 1996). The phage clones identified with strong affinity to the target antigen are then amplified in *E. coli* (Haidaris et al., 2001).

Sequences encoding eukaryotic secretion signals (i.e., kappa light chain leader) can be appended to the scFv genes and subsequently be cloned into gene transfer vectors. Cells transduced by a given gene transfer vector will act as a nexus of scFv expression, leading to a gradient of secreted scFv to act extracellularly on pathogenic factors such as Aβ. Single-chain antibodies are also being investigated for prion disease therapy as a single-chain has been developed that binds to PrP^c and inhibits prion replication (Leclerc et al., 2000; Peretz et al., 2001). Further antibody engineering makes it possible to manipulate the genes encoding these antibodies to allow for expression *within* mammalian cells. Genetically fusing the scFv to intracellular targeting signals allows for specific subcellular expression (Zhu et al., 1999). These intracellular antibodies, termed intrabodies, are capable of modulating target protein function in at least three important ways. Intrabodies can: (1) block or stabilize macromolecular interactions; (2) modulate enzyme function by sequestering substrate, occluding an active site or keeping the enzyme in an active or inactive conformation; and (3) divert proteins to alternative intracellular compartments (for review Richardson and Marasco, 1995). Intrabodies have been utilized for both phenotypic and functional knockouts of target molecules. For example, scFv's directed against the extracellular domain of ErbB-2 fused to an endoplasmic reticulum (ER) signaling domain have been expressed intracellularly and successfully target to the ER lumen. These targeted scFv's were capable of binding newly synthesized ErbB-2 and preventing its transit through the ER to the cell surface resulting in functional inactivation of this protein (Beerli et al., 1994).

Perhaps most relevant to the proposed use of scFv's for a neurological disorder is the recent work from Lecerf et al. (2001). This group identified human scFv intrabodies capable of interacting *in situ* with huntingtin and reducing its aggregation. These scFv's were bound to the N-terminal residues of huntingtin and kept the normally aggregated protein in a soluble complex that then underwent normal protein turnover. Specificity of binding was further determined by fusing the anti-huntingtin scFv with a nuclear localization signal and the subsequent retargeting of soluble huntingtin to cell nuclei. This strategy is currently being used for Parkinson's disease (PD) where single-chains targeted against synuclein are hypothesized to halt the oligomerization and formation of toxic oligomeric species (K. Maguire-Zeiss, personal communication). Given the relatively short half-life of scFv's *in vivo* [<0.7–14 h depending upon the absence or presence of stabilizing agents, respectively; (Chapman, 2002)], the use of integrating viral vectors that exhibit long-term gene expression can provide the means to offer continuous passive immunization for neurodegenerative diseases with underlying abnormal protein accumulation/aggregation.

48.4.3. Active Vaccination

In contrast to passive means of preventing pathogenic protein accumulation in the CNS, active immunization involves using antigen to stimulate the host to produce vigorous immune responses, including the elicitation of antibodies and cytotoxic cells. This more conventional immunization method is many times preferable since long-term immune memory to a specific antigen can be readily maintained.

Recent publications have highlighted the potential of active Aβ peptide-based immunization in the treatment of AD (Schenk et al., 1999; Janus et al., 2000; Morgan et al., 2000). Schenk et al. used a transgenic model that overexpresses a mutant APP (V717F) mini-gene driven by the platelet-derived growth factor promoter (PDAPP; (Schenk et al., 1999)). PDAPP mice immunized and boosted with $A\beta_{1-42}$ peptide with complete and incomplete Freund's adjuvant, respectively, before the onset of AD-like pathology, were protected from development of plaque formation, neuritic dystrophy, and astrogliosis. Treatment of older PDAPP animals (11 months) with existing pathology resulted in reduced plaque burden and slower progression of AD-like neuropathology. Aβ-directed vaccination was assessed by other laboratories in a model of AD that overexpresses mutant human βAPP_{695} (K670N/M671L; Swedish mutation) and PS1 mutations under the control of the hamster prion promoter (PSAPP) (Morgan et al., 2000; Takeuchi et al., 2000). Neuropathologically, these animals demonstrate compact amyloid plaques but no dramatic neuronal loss in either the hippocampus or association cortices. Behaviorally, PSAPP mice exhibit learning and age-related memory deficits as amyloid accumulates. Following vaccination with Aβ, PSAPP mice were protected from learning and age-related deficits as well as a partial reduction in amyloid burden. There were no apparent deleterious effects of the vaccination.

These groundbreaking studies provided impetus for the implementation of Phase I and II clinical trials where human aggregated Aβ and a potent adjuvant (QS-21) was used to immunize individuals with advanced AD. While some of the patients enrolled in the Phase II trial generated anti-Aβ antibody titers (Lee et al., 2005), the trial was halted due to the occurrence of aseptic meningoencephalitis in a subset of responders (based on the antibody titer) and non-responders. Postmortem analyses of two patients that died showed that the encephalitic response was characterized by infiltrating CD4 cells but not CD8s, giant nucleated cells, or macrophages (Ferrer et al., 2004; Gilman et al., 2005). Most intriguing was the observation that these brains harbored low Aβ plaque load strongly suggesting that active Aβ immunization could alter Aβ accumulation patterns (Nicoll et al., 2003). This trial not only demonstrated that immunization against Aβ may be effective but underscored the requirement that AD vaccines require fine-tuning to greatly diminish the likelihood of eliciting brain inflammation.

48.4.4. Immune Shaping: Th1 and Th2 Responses and Antibody Isotypes

A successful immune response against a pathogenic entity as induced by active vaccination, such as in the case of an amyloid-containing plaque in AD, must include at least four major features. These include recognition of the antigens unique to the plaque; induction of the correct effector mechanisms to dissolve the plaque; mobilization of the response in the correct location; and protection of surrounding CNS tissue from excess immune-mediated damage. Different phenotypes of CD4 T cells secreting distinct cytokine patterns have a major role in regulating the type of immune effector functions deployed against pathogens, and protecting host tissues against damage. Th1 cells secrete IL-2, IFNγ, and lymphotoxin and are most useful against intracellular pathogens that are best attacked by a cell-mediated response involving macrophage and granulocyte activation (Sher and Coffman, 1992). Th2 cells secrete IL-4, IL-5, and IL-10, and induce antibody and allergic responses that are useful for combating infections (Finkelman et al., 1997). Although the Th1/Th2 dichotomy is important in a number of mouse and human diseases, several other T cell cytokine secretion phenotypes also exist. Th0 cells secrete both Th1 and Th2 cytokines. Th3 cells (secreting TGFβ; (Chen et al., 1994)), and Treg1 cells [secreting IL-10 but not IL-4; (Groux et al., 1997)] inhibit Th1 responses, and may suppress inflammatory responses that would otherwise cause excessive tissue damage during infections or autoimmunity. CD4 T cells may also remain in an uncommitted state even after activation and proliferation in response to antigen (Sad and Mosmann, 1994; Akai and Mosmann, 1999; Sallusto et al., 1999). These primed precursor cells continue to express IL-2 but not other cytokines, and retain the ability to differentiate into either Th1 or Th2 cells. This expanded pool of antigen-specific cells might then differentiate rapidly in subsequent infections to provide the correct effector functions. In addition to attacking the pathogen, a successful immune response must include regulatory mechanisms that limit damage to host tissues. For example, during Toxoplasma infection in mice, IFNγ is required to defeat the pathogen (Scharton-Kersten et al., 1996), and IL-10 is required to protect the host from an excessive, lethal Th1 response (Gazzinelli et al., 1996). For both mouse and human T cells, IL-4 induces differentiation of naïve CD4 T cells into Th2 cells, and IL-12 induces differentiation of naïve CD4 T into Th1 cells (Seder and Mosmann, 1999). IFNγ also enhances Th1 differentiation, at least partly by enhancing expression of the IL-12 receptor β2 chain. Once differentiated, the Th1 and Th2 effector phenotypes are relatively stable, and so it is difficult to subsequently convert a strongly polarized immune response to a different set of effector functions.

In parallel with the diversity of T cell responses, several antibody responses are possible. The specificity of the antibodies for different antigens, or different epitopes on the same antigen, can markedly affect the usefulness of the antibodies, and the isotype of the antibodies is also important. Under the regulation of T cells and other cytokines, B cells switch to the production of IgA, IgE, or different subsets of IgG, all with unique sets of functions. IgE initiates allergy via triggering of degranulation from mast cells, IgA is selectively secreted at mucosal surfaces, and IgG1 and IgG3 (human) or IgG2a (mouse) are the major isotypes that fix complement and opsonize cells for phagocytosis by macrophages.

48.4.5. Evaluating an Appropriate Active Vaccine Response for Alzheimer's Disease

The ability of APP peptide immunization to reduce plaque formation and improve memory behavior (Schenk et al., 1999; Janus et al., 2000; Morgan et al., 2000) may be explained by solubilization of the protein precipitates by binding of antibody. If this simple model is correct, then an ideal vaccine for Alzheimer's disease should induce a strong antibody response, mainly of isotypes that can cross the blood/brain barrier. Helper T cell function, normally required for a good antibody response, should be limited to Th2-biased responses, because a strong Th1 response carries the risk of inducing a local inflammatory response to the APP antigen, if T cells penetrate the blood/brain barrier for any reason. CTL responses would also be undesirable for similar reasons. In contrast, a concurrent Th3/Treg1 response including TGFβ and IL10 synthesis might be desirable, as these cytokines would help to control any inflammatory response that might develop. Complement activation by antibody/antigen complexes may have either useful or deleterious effects – complement deposition may help to solubilize antigen/antibody complexes that develop in the plaques as a result of antibody binding (Miller and Nussenzweig, 1975), but complement activation may also initiate inflammation. Recruitment of macrophages may also enhance solubilization of precipitates or immune complexes, but activation of the macrophages may lead to local inflammation and tissue damage. It appears as though at least a subset of these negative immune outcomes were at play in the AN-1792 Phase II human clinical trial.

Despite this initial clinical setback, Aβ-directed active vaccination continues to warrant further assessment via careful step-wise refinement and testing of novel immunotherapeutic approaches. It is clearly important to evaluate the antibody isotypes that are induced by candidate vaccine constructs, as well as determining the types of CD4 and CD8 T cell immunity that are elicited. This information will undoubtedly facilitate the design of safer active vaccination strategies to direct the resultant immune response towards more desired effector functions.

48.5. Future Outlook

The future of gene-based immunotherapy for neurological disease remains mired in many unknowns. The most informed choice of vector platform, gene regulation system, and immunotherapeutic strategy will depend on numerous non-mutually exclusive factors: the specific neurodegenerative disease, the disease stage, the underlying inflammatory mechanisms associated with the disease, the interplay between those events and those inherent to the proposed vector systems following CNS delivery, and the therapeutic transgene itself. Another major challenge that faces this field involves the heterogeneity of human immune responses. For example, why did only a subset of individuals enrolled in the AN-1792 Phase II clinical trial present with an encephalitic condition while a majority of others did not? These adverse events may have occurred due to the differing immune profiles of each patient and not specifically a result of disease stage or circulating anti-Aβ antibody titers or isotypes. It is therefore imperative that research continues to be conducted in the area of immune profiling in order to derive highly sensitive measures that are predictive of potential adverse immune-related symptoms. This would facilitate pre-screening of clinical trial participants and may allow for the derivation of custom immunotherapies based on a given patient's immune profile (O'Toole et al., 2005; Rosenberg, 2005). These considerations will be important in the conceptual design and the clinical implementation of gene-based immunotherapeutics for diseases afflicting an organ system formerly regarded, and unfortunately so, as an immunopriviledged region of the human body.

Summary

Clinical implementation of gene-based therapeutics and immunotherapeutics for prevention and/or amelioration of human neurologic diseases eventually will be realized, but this goal inherently presents a series of significant challenges, one of which relates to issues of immune system involvement. Devising gene therapeutics for the diseased central nervous system is particularly challenging from a neuroimmunological perspective given that many targeted neurologic disorders possess an underlying inflammatory component. Presently, direct gene-based delivery to the brain requires invasive surgical procedures and a majority of the gene delivery platforms incite transient, and sometimes, inflammatory events that have the potential to exacerbate disease-related processes. Moreover, peripheral administration of a gene-based immunotherapeutic designed to target pathologic antigens within the brain also carry the untoward potential to amplify extant disease-related inflammation. Studying the immunobiology of viral and nonviral gene delivery platforms in the context of the neurodegenerative condition is therefore imperative to fully appreciate the inflammatory mechanisms at play and provide potential avenues to prevent and/or counter adverse anti-vector immune responses.

Review Questions/Problems

Note: For the following multiple choice questions, there may be more than one correct answer.

You have been recently hired as a consultant at a reputable biotechnology company that develops novel therapies for Sly Syndrome, a lysosomal storage disorder, caused by the absence of an enzyme called β-glucuronidase. Individuals lacking this enzyme exhibit a progressive accumulation of lysosome-harbored undegraded glycosaminoglycans, leading to mental retardation, loss of hearing and vision, joint abnormalities, and enlargement of the liver. This company wants to pursue a gene therapy-based approach to treat the central nervous system-related symptoms of the disease. The stipulations in the company's directive are that the therapy be long lasting (>2 yrs, but preferably life-long), directed to neurons, and be safe and effective. The vector platform the company will eventually pursue needs to effectively deliver a neuron specific enolase (NSE) promoter-driven β-glucuronidase transcription unit (~7 kb in total size). Given your expertise in employing several virus-based gene therapy vector types for *in vivo* applications, you form an opinion on how the company should proceed.

1. **Which of the following vectors would be able to carry the designed transcription unit?**

   a. HSV amplicon vector
   b. Recombinant AAV vector
   c. Lentivirus-based vector
   d. Adenovirus-based vector
   e. All of the above

2. **Which of the following vectors would be most appropriate for this disease application given the description shown above?**

   a. HSV amplicon vector
   b. Recombinant AAV vector
   c. Lentivirus-based vector
   d. Adenovirus-based vector

3. **Select the viral vector(s) that would be able to harbor a 2 kb transcription unit?**

   a. Adenovirus vectors
   b. Adeno-associated virus vectors
   c. Murine oncoretrovirus vectors
   d. Lentivirus vectors
   e. Recombinant HSV vectors
   f. HSV amplicon vectors

4. **Select the viral vector(s) that would be able to harbor a 15 kb transcription unit?**

   a. Adenovirus vectors
   b. Adeno-associated virus vectors
   c. Murine oncoretrovirus vectors
   d. Lentivirus vectors
   e. Recombinant HSV vectors
   f. HSV amplicon vectors

5. **Which of the following considerations takes into account transcriptionally regulating what cell type a particular transgene is expressed within?**

   a. Focal tissue delivery
   b. Promoter selection
   c. Transgene glycosynthesis
   d. Trypan blue
   e. Virus tropism

6. **Which of the following viral vector(s) delivers a genome that is episomally maintained in the transduced host cell nucleus?**

   a. Adenovirus vectors
   b. Murine oncoretrovirus vectors
   c. Lentivirus vectors
   d. Recombinant HSV vectors
   e. HSV amplicon vectors

7. **Which of the following viral vector(s) delivers a genome that integrates in the transduced host cell genome?**

   a. Adenovirus vectors
   b. HSV amplicon vectors
   c. Lentivirus vectors
   d. Replication-restricted HSV vectors
   e. Recombinant HSV vectors

8. **Delivery of viral vectors *in vivo* presents a series of safety and immunological concerns. Name the source(s) of each of the potential dangers that can arise.**
9. **Name the two non-coding sequences derived from HSV that are included in HSV amplicon vectors.**
10. **What is a single-chain antibody?**
11. **What is the difference between active and passive immunization?**
12. **Based upon the scientific data available at present, what would be the ideal immune responses elicited by a vaccine for Alzheimer's disease?**

References

Abordo-Adesida E, Follenzi A, Barcia C, Sciascia S, Castro MG, Naldini L, Lowenstein PR (2005) Stability of lentiviral vector-mediated transgene expression in the brain in the presence of systemic antivector immune responses. Hum Gene Ther 16:741–751.

Akai PS, Mosmann TR (1999) Primed and replicating but uncommitted T helper precursor cells show kinetics of differentiation and commitment similar to those of naive T helper cells. Microbes Infect 1:51–58.

Akli S, Caillaud C, Vigne E, Stratford-Perricaudet LD, Poenaru L, Perricaudet M, Kahn A, Peschanski MR (1993) Transfer of a foreign gene into the brain using adenovirus vectors. Nat Genet 3:224–228.

Alisky JM, Hughes SM, Sauter SL, Jolly D, Dubensky TW, Jr., Staber PD, Chiorini JA, Davidson BL (2000) Transduction of murine cerebellar neurons with recombinant FIV and AAV5 vectors. Neuroreport 11:2669–2673.

Andersen JK, Breakefield XO (1995) Gene delivery to neurons of the adult mammalian nervous system using herpes and adenovirus vectors. In: Somatic Gene Therapy (Wolfe J, ed), pp 135–160. Boca Raton: CRC Press, Inc.

Andersen JK, Garber DA, Meaney CA, Breakefield XO (1992) Gene transfer into mammalian central nervous system using herpes virus vectors: Extended expression of bacterial *lacZ* in neurons using the neuron-specific enolase promoter. Hum Gene Ther 3:487–499.

Anwer K, Kao G, Proctor B, Rolland A, Sullivan S (2000) Optimization of cationic lipid/DNA complexes for systemic gene transfer to tumor lesions. J Drug Target 8:125–135.

Baekelandt V, Eggermont K, Michiels M, Nuttin B, Debyser Z (2003) Optimized lentiviral vector production and purification procedure prevents immune response after transduction of mouse brain. Gene Ther 10:1933–1940.

Bajocchi G, Feldman SH, Crystal RG, Mastrangeli A (1993) Direct *in vivo* gene transfer to ependymal cells in the central nervous system using recombinant adenovirus vectors. Nat Genet 3:229–234.

Banks TA, Rouse BT (1992) Herpesviruses–immune escape artists? Clin Infect Dis 14:933–941.

Bard F, Cannon C, Barbour R, Burke RL, Games D, Grajeda H, Guido T, Hu K, Huang J, Johnson-Wood K, Khan K, Kholodenko D, Lee M, Lieberburg I, Motter R, Nguyen M, Soriano F, Vasquez N, Weiss K, Welch B, Seubert P, Schenk D, Yednock T (2000) Peripherally administered antibodies against amyloid beta-peptide enter the central nervous system and reduce pathology in a mouse model of Alzheimer disease. Nat Med 6:916–919.

Beer SJ, Matthews CB, Stein CS, Ross BD, Hilfinger JM, Davidson BL (1998) Poly (lactic-glycolic) acid copolymer encapsulation of recombinant adenovirus reduces immunogenicity in vivo. Gene Ther 5:740–746.

Beerli RR, Wels W, Hynes NE (1994) Intracellular expression of single chain antibodies reverts ErbB-2 transformation. J Biol Chem 269:23931–23936.

Bensadoun JC, Deglon N, Tseng JL, Ridet JL, Zurn AD, Aebischer P (2000) Lentiviral vectors as a gene delivery system in the mouse midbrain: Cellular and behavioral improvements in a 6-OHDA model of Parkinson's disease using GDNF. Exp Neurol 164:15–24.

Bjorklund A, Kirik D, Rosenblad C, Georgievska B, Lundberg C, Mandel RJ (2000) Towards a neuroprotective gene therapy for Parkinson's disease: Use of adenovirus, AAV and lentivirus vectors for gene transfer of GDNF to the nigrostriatal system in the rat Parkinson model. Brain Res 886:82–98.

Bowers WJ, Howard DF, Federoff HJ (2000) Discordance between expression and genome transfer titering of HSV amplicon vectors: Recommendation for standardized enumeration. Mol Ther 1:294–299.

Bowers WJ, Howard DF, Brooks AI, Halterman MW, Federoff HJ (2001) Expression of vhs and VP16 during HSV-1 helper virus-free amplicon packaging enhances titers. Gene Ther 8:111–120.

Breakefield XO, DeLuca NA (1991) Herpes simplex virus for gene delivery to neurons. New Biol 3:203–218.

Breakefield XO, Huang Q, Andersen JK, Kramer MF, Bebrin WR, Davar G, Vos B, Garber DA, Difiglia M, Coen DM (1992) Gene transfer into the nervous system using recombinant herpes virus vectors. In: Gene Transfer and Therapy in the Nervous System (Gage, FH, Christen, Y, ed), pp 45–48. Heidelberg: Springer-Verlag.

Brizard M, Carcenac C, Bemelmans AP, Feuerstein C, Mallet J, Savasta M (2006) Functional reinnervation from remaining DA terminals induced by GDNF lentivirus in a rat model of early Parkinson's disease. Neurobiol Dis 21:90–101.

Brooks AI, Halterman MW, Chadwick CA, Davidson BL, Haak-Frendscho M, Radel CA, Porter C, Federoff HJ (1998) Reproducible and efficient murine CNS gene delivery using a microprocessor-controlled injector. J Neurosci Meth 80:137–147.

Bueler H (1999) Adeno-associated viral vectors for gene transfer and gene therapy. Biol Chem 380:613–622.

Burger C, Gorbatyuk OS, Velardo MJ, Peden CS, Williams P, Zolotukhin S, Reier PJ, Mandel RJ, Muzyczka N (2004) Recombinant AAV viral vectors pseudotyped with viral capsids from serotypes 1, 2, and 5 display differential efficiency and cell tropism after delivery to different regions of the central nervous system. Mol Ther 10:302–317.

Byrnes AP, MacLaren RE, Charlton HM (1996) Immunological instability of persistent adenovirus vectors in the brain: Peripheral exposure to vector leads to renewed inflammation, reduced gene expression, and demyelination. J Neurosci 16:3045–3055.

Chapman AP (2002) PEGylated antibodies and antibody fragments for improved therapy: A review. Adv Drug Deliv Rev 54: 531–545.

Chen Y, Kuchroo VK, Inobe J, Hafler DA, Weiner HL (1994) Regulatory T cell clones induced by oral tolerance: Suppression of autoimmune encephalomyelitis. Science 265:1237–1240.

Chen ZY, Yant SR, He CY, Meuse L, Shen S, Kay MA (2001) Linear DNAs concatemerize in vivo and result in sustained transgene expression in mouse liver. Mol Ther 3:403–410.

Chirmule N, Xiao W, Truneh A, Schnell MA, Hughes JV, Zoltick P, Wilson JM (2000) Humoral immunity to adeno-associated virus type 2 vectors following administration to murine and nonhuman primate muscle. J Virol 74:2420–2425.

Chu C, Zhang Y, Boado RJ, Pardridge WM (2006) Decline in exogenous gene expression in primate brain following intravenous administration is due to plasmid degradation. Pharm Res 23:1586–1590.

Cregan SP, MacLaurin J, Gendron TF, Callaghan SM, Park DS, Parks RJ, Graham FL, Morley P, Slack RS (2000) Helper-dependent adenovirus vectors: Their use as a gene delivery system to neurons. Gene Ther 7:1200–1209.

Daly TM, Ohlemiller KK, Roberts MS, Vogler CA, Sands MS (2001) Prevention of systemic clinical disease in MPS VII mice following AAV-mediated neonatal gene transfer. Gene Ther 8:1291–1298.

Davidson BL, Allen ED, Kozarsky KF, Wilson JM, Roessler BJ (1993) A model system for *in vivo* gene transfer into the central nervous system using an adenoviral vector. Nat Genet 3:219–223.

Davidson BL, Stein CS, Heth JA, Martins I, Kotin RM, Derksen TA, Zabner J, Ghodsi A, Chiorini JA (2000) Recombinant adeno-associated virus type 2, 4, and 5 vectors: Transduction of variant cell types and regions in the mammalian central nervous system. Proc Natl Acad Sci USA 97:3428–3432.

Desmaris N, Bosch A, Salaun C, Petit C, Prevost MC, Tordo N, Perrin P, Schwartz O, de Rocquigny H, Heard JM (2001) Production and neurotropism of lentivirus vectors pseudotyped with lyssavirus envelope glycoproteins. Mol Ther 4:149–156.

Dobson A, Margolis TP, Sedarati F, Stevens J, Feldman LT (1990) A latent, nonpathogenic HSV-1 derived vector stably expresses β-galactosidase in mouse neurons. Neuron 5:353–360.

Dodart JC, Bales KR, Gannon KS, Greene SJ, DeMattos RB, Mathis C, DeLong CA, Wu S, Wu X, Holtzman DM, Paul SM (2002) Immunization reverses memory deficits without reducing brain Abeta burden in Alzheimer's disease model. Nat Neurosci 5:452–457.

Dodart JC, Marr RA, Koistinaho M, Gregersen BM, Malkani S, Verma IM, Paul SM (2005) Gene delivery of human apolipoprotein E alters brain Abeta burden in a mouse model of Alzheimer's disease. Proc Natl Acad Sci USA 102:1211–1216.

Donsante A, Vogler C, Muzyczka N, Crawford JM, Barker J, Flotte T, Campbell-Thompson M, Daly T, Sands MS (2001) Observed incidence of tumorigenesis in long-term rodent studies of rAAV vectors. Gene Ther 8:1343–1346.

During MJ, Kaplitt MG, Stern MB, Eidelberg D (2001) Subthalamic GAD gene transfer in Parkinson disease patients who are candidates for deep brain stimulation. Hum Gene Ther 12:1589–1591.

Evans JT, Garcia JV (2000) Lentivirus vector mobilization and spread by human immunodeficiency virus. Hum Gene Ther 11:2331–2339.

Federoff HJ, Geschwind MD, Geller AI, Kessler JA (1992) Expression of nerve growth factor *in vivo*, from a defective HSV-1 vector prevents effects of axotomy on sympathetic ganglia. Proc Natl Acad Sci USA 89:1636–1640.

Ferrer I, Boada Rovira M, Sanchez Guerra ML, Rey MJ, Costa-Jussa F (2004) Neuropathology and pathogenesis of encephalitis following amyloid-beta immunization in Alzheimer's disease. Brain Pathol 14:11–20.

Fink DJ, Sternberg LR, Weber PC, Mata M, Goins WF, Glorioso JC (1992) *In-vivo* expression of β-galactosidase in hippocampal neurons by HSV mediated gene transfer. Hum Gene Ther 4:11–19.

Finkelman FD, Shea-Donohue T, Goldhill J, Sullivan CA, Morris SC, Madden KB, Gause WC, Urban JF (1997) Cytokine regulation of host defense against parasitic gastrointestinal nematodes: Lessons from studies with rodent models. Annu Rev Immunol 15:505–533.

Follenzi A, Ailles LE, Bakovic S, Geuna M, Naldini L (2000) Gene transfer by lentiviral vectors is limited by nuclear translocation and rescued by HIV-1 pol sequences. Nat Genet 25:217–222.

Fraefel C, Song S, Lim F, Lang P, Yu L, Wang Y, Wild P, Geller AI (1996) Helper virus-free transfer of herpes simplex virus type 1 plasmid vectors into neural cells. J Virol 70:7190–7197.

Frenkel N (1981) Defective interfering herpesviruses. In: The Human Herpesviruses—an Interdisciplinary Prospective (Nahmias A, Dowdle W, Scchinazy R, eds), pp 91–120. New York: Elsevier-North Holland, Inc.

Frenkel N, Spaete RR, Vlazny DA, Deiss LP, Locker H (1982) The herpes simplex virus amplicon—a novel animal-virus cloning vector. In: Eucaryotic Viral Vectors (Gluzman Y, ed), pp 205–209. New York: Cold Spring Harbor Laboratory.

Gazzinelli RT, Wysocka M, Hieny S, Scharton-Kersten T, Cheever A, Kuhn R, Muller W, Trinchieri G, Sher A (1996) In the absence of endogenous IL-10, mice acutely infected with Toxoplasma gondii succumb to a lethal immune response dependent on CD4+ T cells and accompanied by overproduction of IL-12, IFN-gamma and TNF-alpha. J Immunol 157:798–805.

Geller AI, Breakefield XO (1988) A defective HSV-1 vector expresses *Escherichia coli* β-galactosidase in cultured peripheral neurons. Science 241:1667–1669.

Geraghty RJ, Krummenacher C, Cohen GH, Eisenberg RJ, Spear PG (1998) Entry of alphaherpesviruses mediated by poliovirus receptor-related protein 1 and poliovirus receptor. Science 280:1618–1620.

Geschwind MD, Kessler JA, Geller AI, Federoff HJ (1994) Transfer of the nerve growth factor gene into cell lines and cultured neurons using a defective herpes simplex virus vector. Transfer of the NGF gene into cells by a HSV-1 vector. Mol Brain Res 24:327–335.

Gilman S, Koller M, Black RS, Jenkins L, Griffith SG, Fox NC, Eisner L, Kirby L, Rovira MB, Forette F, Orgogozo JM (2005) Clinical effects of Abeta immunization (AN1792) in patients with AD in an interrupted trial. Neurology 64:1553–1562.

Glorioso JC, Goins WF, Meaney CA, Fink DJ, DeLuca NA (1994) Gene transfer to brain using herpes simplex virus vectors. Ann Neurol Suppl 35:S28-S34.

Groux H, O'Garra A, Bigler M, Rouleau M, Antonenko S, de Vries JE, Roncarolo MG (1997) A CD4+ T-cell subset inhibits antigen-specific T-cell responses and prevents colitis. Nature 389:737–742.

Guy J, Qi X, Muzyczka N, Hauswirth WW (1999) Reporter expression persists 1 year after adeno-associated virus-mediated gene transfer to the optic nerve. Arch Ophthalmol 117:929–937.

Haidaris CG, Malone J, Sherrill LA, Bliss JM, Gaspari AA, Insel RA, Sullivan MA (2001) Recombinant human antibody single chain variable fragments reactive with Candida albicans surface antigens. J Immunol Methods 257:185–202.

Hald A, Lotharius J (2005) Oxidative stress and inflammation in Parkinson's disease: Is there a causal link? Exp Neurol 193: 279–290.

Hanisch UK (2002) Microglia as a source and target of cytokines. Glia 40:140–155.

Hardy S, Kitamura M, Harris-Stansil T, Dai Y, Phipps ML (1997) Construction of adenovirus vectors through Cre-lox recombination. J Virol 71:1842–1849.

Hickey WF (2001) Basic principles of immunological surveillance of the normal central nervous system. Glia 36:118–124.

Hocknell PK, Wiley RD, Wang X, Evans TG, Bowers WJ, Hanke T, Federoff HJ, Dewhurst S (2002) Expression of human immunodeficiency virus type 1 gp120 from herpes simplex virus type 1-derived amplicons results in potent, specific, and durable cellular and humoral immune responses. J Virol 76:5565–5580.

Holt CE, Garlick N, Cornel E (1990) Lipofection of cDNAs in the embryonic vertebrate central nervous system. Neuron 4:203–214.

Hsu H, Solovyev I, Colombero A, Elliott R, Kelley M, Boyle WJ (1997) ATAR, a novel tumor necrosis factor receptor family member, signals through TRAF2 and TRAF5. J Biol Chem 272:13471–13474.

Iwakuma T, Cui Y, Chang LJ (1999) Self-inactivating lentiviral vectors with U3 and U5 modifications. Virology 261:120–132.

Janus C, Pearson J, McLaurin J, Mathews PM, Jiang Y, Schmidt SD, Chishti MA, Horne P, Heslin D, French J, Mount HT, Nixon RA, Mercken M, Bergeron C, Fraser PE, St George-Hyslop P, Westaway D (2000) A beta peptide immunization reduces behavioural impairment and plaques in a model of Alzheimer's disease. Nature 408:979–982.

Jiao S, Acsadi G, Jani A, Felgner PL, Wolff JA (1992) Persistence of plasmid DNA and expression in rat brain cells in vivo. Exp Neurol 115:400–413.

Johnson PA, Miyanohara A, Levine F, Cahill T, Friedmann T (1992) Cytotoxicity of a replication-defective mutant of herpes simplex virus type I. J Virol 66:2952–2965.

Kaech S, Kim JB, Cariola M, Ralston E (1996) Improved lipid-mediated gene transfer into primary cultures of hippocampal neurons. Brain Res Mol Brain Res 35:344–348.

Kaplitt MG, Makimura H (1997) Defective viral vectors as agents for gene transfer in the nervous system. J Neurosci Methods 71:125–132.

Kaplitt MG, Leone P, Samulski RJ, Xiao X, Pfaff DW, O'Malley KL, During MJ (1994) Long-term gene expression and phenotypic correction using adeno-associated virus vectors in the mammalian brain. Nat Genet 8:148–154.

Klyubin I, Walsh DM, Lemere CA, Cullen WK, Shankar GM, Betts V, Spooner ET, Jiang L, Anwyl R, Selkoe DJ, Rowan MJ (2005) Amyloid beta protein immunotherapy neutralizes Abeta oligomers that disrupt synaptic plasticity in vivo. Nat Med 11:556–561.

Kordower JH, Bloch J, Ma SY, Chu Y, Palfi S, Roitberg BZ, Emborg M, Hantraye P, Deglon N, Aebischer P (1999) Lentiviral gene transfer to the nonhuman primate brain. Exp Neurol 160:1–16.

Kordower JH, Emborg ME, Bloch J, Ma SY, Chu Y, Leventhal L, McBride J, Chen EY, Palfi S, Roitberg BZ, Brown WD, Holden JE, Pyzalski R, Taylor MD, Carvey P, Ling Z, Trono D, Hantraye P, Deglon N, Aebischer P (2000) Neurodegeneration prevented by lentiviral vector delivery of GDNF in primate models of Parkinson's disease. Science 290:767–773.

Kotilinek LA, Bacskai B, Westerman M, Kawarabayashi T, Younkin L, Hyman BT, Younkin S, Ashe KH (2002) Reversible memory loss in a mouse transgenic model of Alzheimer's disease. J Neurosci 22:6331–6335.

Kotin RM, Siniscalco M, Samulski RJ, Zhu XD, Hunter L, Laughlin CA, McLaughlin S, Muzyczka N, Rocchi M, Berns KI (1990) Site-specific integration by adeno-associated virus. Proc Natl Acad Sci USA 87:2211–2215.

Krisky DM, Wolfe D, Goins WF, Marconi PC, Ramakrishnan R, Mata M, Rouse RJ, Fink DJ, Glorioso JC (1998) Deletion of multiple immediate-early genes from herpes simplex virus reduces cytotoxicity and permits long-term gene expression in neurons. Gene Ther 5:1593–1603.

Kwon BS, Tan KB, Ni J, Oh KO, Lee ZH, Kim KK, Kim YJ, Wang S, Gentz R, Yu GL, Harrop J, Lyn SD, Silverman C, Porter TG, Truneh A, Young PR (1997) A newly identified member of the tumor necrosis factor receptor superfamily with a wide tissue distribution and involvement in lymphocyte activation. J Biol Chem 272:14272–14276.

Le Gal La Salle G, Robert JJ, Berrard S, Ridoux V, Stratford-Perricaudet LD, Perricaudet M, Mallet J (1993) An adenovirus vector for gene transfer into neurons and glia in the brain. Science 259:988–990.

Lecerf JM, Shirley TL, Zhu Q, Kazantsev A, Amersdorfer P, Housman DE, Messer A, Huston JS (2001) Human single-chain Fv intrabodies counteract in situ huntingtin aggregation in cellular models of Huntington's disease. Proc Natl Acad Sci USA 98:4764–4769.

Leclerc E, Liemann S, Wildegger G, Vetter SW, Nilsson F (2000) Selection and characterization of single chain Fv fragments against murine recombinant prion protein from a synthetic human antibody phage display library. Hum Antibodies 9:207–214.

Lee M, Bard F, Johnson-Wood K, Lee C, Hu K, Griffith SG, Black RS, Schenk D, Seubert P (2005) Abeta42 immunization in Alzheimer's disease generates Abeta N-terminal antibodies. Ann Neurol 58:430–435.

Lo WD, Qu G, Sferra TJ, Clark R, Chen R, Johnson PR (1999) Adeno-associated virus-mediated gene transfer to the brain: Duration and modulation of expression. Hum Gene Ther 10:201–213.

Loeffler JP, Barthel F, Feltz P, Behr JP, Sassone-Corsi P, Feltz A (1990) Lipopolyamine-mediated transfection allows gene expression studies in primary neuronal cells. J Neurochem 54:1812–1815.

Mahato RI, Rolland A, Tomlinson E (1997) Cationic lipid-based gene delivery systems: Pharmaceutical perspectives. Pharm Res 14:853–859.

Malone J, Sullivan MA (1996) Analysis of antibody selection by phage display utilizing anti-phenobarbital antibodies. J Mol Recognit 9:738–745.

Marconi P, Krisky D, Oligino T, Poliani PL, Ramakrishnan R, Goins WF, Fink DJ, Glorioso JC (1996) Replication-defective herpes simplex virus vectors for gene transfer in vivo. Proc Natl Acad Sci USA 93:11319–11320.

Marconi P, Simonato M, Zucchini S, Bregola G, Argnani R, Krisky D, Glorioso JC, Manservigi R (1999) Replication-defective herpes simplex virus vectors for neurotrophic factor gene transfer in vitro and in vivo. Gene Ther 6:904–912.

Masliah E, Rockenstein E, Adame A, Alford M, Crews L, Hashimoto M, Seubert P, Lee M, Goldstein J, Chilcote T, Games D, Schenk D (2005) Effects of alpha-synuclein immunization in a mouse model of Parkinson's disease. Neuron 46:857–868.

Matsui H, Johnson LG, Randell SH, Boucher RC (1997) Loss of binding and entry of liposome-DNA complexes decreases transfection efficiency in differentiated airway epithelial cells. J Biol Chem 272:1117–1126.

Matter-Sadzinski L, Hernandez MC, Roztocil T, Ballivet M, Matter JM (1992) Neuronal specificity of the alpha 7 nicotinic acetylcholine receptor promoter develops during morphogenesis of the central nervous system. EMBO J 11:4529–4538.

McCown TJ, Xiao X, Li J, Breese GR, Samulski RJ (1996) Differential and persistent expression patterns of CNS gene transfer by an adeno-associated virus (AAV) vector. Brain Res 713:99–107.

McGeer EG, McGeer PL (2003) Inflammatory processes in Alzheimer's disease. Prog Neuropsychopharmacol Biol Psychiatry 27:741–749.

McGeer PL, McGeer EG (2004) Inflammation and neurodegeneration in Parkinson's disease. Parkinsonism Relat Disord 1(10 Suppl):S3–S7.

Michou AI, Santoro L, Christ M, Julliard V, Pavirani A, Mehtali M (1997) Adenovirus-mediated gene transfer: Influence of transgene, mouse strain and type of immune response on persistence of transgene expression. Gene Ther 4:473–482.

Miller DG, Rutledge EA, Russell DW (2002) Chromosomal effects of adeno-associated virus vector integration. Nat Genet 30:147–148.

Miller GW, Nussenzweig V (1975) A new complement function: Solubilization of antigen-antibody aggregates. Proc Natl Acad Sci USA 72:418–422.

Mitani K, Graham FL, Caskey CT, Kochanek S (1995) Rescue, propagation, and partial purification of a helper virus-dependent adenovirus vector. Proc Natl Acad Sci USA 92:3854–3858.

Moghimi SM, Patel HM (1992) Opsonophagocytosis of liposomes by peritoneal macrophages and bone marrow reticuloendothelial cells. Biochim Biophys Acta 1135:269–274.

Montgomery RI, Warner MS, Lum BJ, Spear P (1996) Herpes simplex virus-1 entry into cells mediated by a novel member of the TNG/NGF receptor family. Cell 87:427–436.

Morgan D, Diamond DM, Gottschall PE, Ugen KE, Dickey C, Hardy J, Duff K, Jantzen P, DiCarlo G, Wilcock D, Connor K, Hatcher J, Hope C, Gordon M, Arendash GW (2000) A beta peptide vaccination prevents memory loss in an animal model of Alzheimer's disease. Nature 408:982–985.

Morsy MA, Gu M, Motzel S, Zhao J, Lin J, Su Q, Allen H, Franlin L, Parks RJ, Graham FL, Kochanek S, Bett AJ, Caskey CT (1998) An adenoviral vector deleted for all viral coding sequences results in enhanced safety and extended expression of a leptin transgene. Proc Natl Acad Sci USA 95:7866–7871.

Naldini L, Blomer U, Gage FH, Trono D, Verma IM (1996a) Efficient transfer, integration, and sustained long-term expression of the transgene in adult rat brains injected with a lentiviral vector. Proc Natl Acad Sci USA 93:11382–11388.

Naldini L, Blomer U, Gallay P, Ory D, Mulligan R, Gage FH, Verma IM, Trono D (1996b) In vivo gene delivery and stable transduction of nondividing cells by a lentiviral vector. Science 272:263–267.

Nicoll JA, Wilkinson D, Holmes C, Steart P, Markham H, Weller RO (2003) Neuropathology of human Alzheimer disease after immunization with amyloid-beta peptide: A case report. Nat Med 9:448–452.

O'Toole M, Janszen DB, Slonim DK, Reddy PS, Ellis DK, Legault HM, Hill AA, Whitley MZ, Mounts WM, Zuberek K, Immermann FW, Black RS, Dorner AJ (2005) Risk factors associated with beta-amyloid(1–42) immunotherapy in preimmunization gene expression patterns of blood cells. Arch Neurol 62:1531–1536.

Olanow CW (2003) Present and future directions in the management of motor complications in patients with advanced PD. Neurology 61:S24–33.

Olanow CW, Schapira AH, Agid Y (2003) Neuroprotection for Parkinson's disease: Prospects and promises. Ann Neurol 3(53 Suppl):S1–S2.

Olschowka JA, Bowers WJ, Hurley SD, Mastrangelo MA, Federoff HJ (2003) Helper-free HSV-1 amplicons elicit a markedly less robust innate immune response in the CNS. Mol Ther 7:218–227.

Papahadjopoulos D, Allen TM, Gabizon A, Mayhew E, Matthay K, Huang SK, Lee KD, Woodle MC, Lasic DD, Redemann C, et al. (1991) Sterically stabilized liposomes: Improvements in pharmacokinetics and antitumor therapeutic efficacy. Proc Natl Acad Sci USA 88:11460–11464.

Pardridge WM (2003) Gene targeting in vivo with pegylated immunoliposomes. Methods Enzymol 373:507–528.

Parks RJ, Chen L, Anton M, Sankar U, Rudnicki MA, Graham FL (1996) A helper-dependent adenovirus vector system: Removal of helper virus by Cre-mediated excision of the viral packaging signal. Proc Natl Acad Sci USA 93:13565–13570.

Peretz D, Williamson RA, Kaneko K, Vergara J, Leclerc E, Schmitt-Ulms G, Mehlhorn IR, Legname G, Wormald MR, Rudd PM, Dwek RA, Burton DR, Prusiner SB (2001) Antibodies inhibit prion propagation and clear cell cultures of prion infectivity. Nature 412:739–743.

Reilly CE (2001) Glial cell line-derived neurotrophic factor (GDNF) prevents neurodegeneration in models of Parkinson's disease. J Neurol 248:76–78.

Richardson JH, Marasco WA (1995) Intracellular antibodies: Development and therapeutic potential. Trends Biotechnol 13:306–310.

Rosenberg RN (2005) Translational research on the way to effective therapy for Alzheimer disease. Arch Gen Psychiatry 62:1186–1192.

Sad S, Mosmann TR (1994) Single IL-2-secreting precursor CD4 T cell can develop into either Th1 or Th2 cytokine secretion phenotype. J Immunol 153:3514–3522.

Saeki Y, Ichikawa T, Saeki A, Chiocca EA, Tobler K, Ackermann M, Breakefield XO, Fraefel C (1998) Herpes simplex virus type 1 DNA amplified as bacterial artificial chromosome in Escherichia coli: Rescue of replication-competent virus progeny and packaging of amplicon vectors. Hum Gene Ther 9:2787–2794.

Saeki Y, Fraefel C, Ichikawa T, Breakefield XO, Chiocca EA (2001) Improved helper virus-free packaging system for HSV amplicon vectors using an ICP27-deleted, oversized HSV-1 DNA in a bacterial artificial chromosome. Mol Ther 3:591–601.

Saenz DT, Loewen N, Peretz M, Whitwam T, Barraza R, Howell KG, Holmes JM, Good M, Poeschla EM (2004) Unintegrated lentivirus DNA persistence and accessibility to expression in nondividing cells: Analysis with class I integrase mutants. J Virol 78:2906–2920.

Sallusto F, Lenig D, Forster R, Lipp M, Lanzavecchia A (1999) Two subsets of memory T lymphocytes with distinct homing potentials and effector functions. Nature 401:708–712.

Scharton-Kersten TM, Wynn TA, Denkers EY, Bala S, Grunvald E, Hieny S, Gazzinelli RT, Sher A (1996) In the absence of endogenous IFN-gamma, mice develop unimpaired IL-12 responses to Toxoplasma gondii while failing to control acute infection. J Immunol 157:4045–4054.

Schenk D, Barbour R, Dunn W, Gordon G, Grajeda H, Guido T, Hu K, Huang J, Johnson-Wood K, Khan K, Kholodenko D, Lee M, Liao Z, Lieberburg I, Motter R, Mutter L, Soriano F, Shopp G, Vasquez N, Vandevert C, Walker S, Wogulis M, Yednock T, Games D, Seubert P (1999) Immunization with amyloid-beta attenuates Alzheimer-disease-like pathology in the PDAPP mouse. Nature 400:173–177.

Schiess M (2003) Nonsteroidal anti-inflammatory drugs protect against Parkinson neurodegeneration: Can an NSAID a day keep Parkinson disease away? Arch Neurol 60:1043–1044.

Seder RA, Mosmann TR (1999) Differentiation of effector phenotypes of CD4+ and CD8+ T cells. In: Fundamental Immunology (Paul WE, ed), p 879.

Shen Y, Muramatsu S-I, Ikeguchi K, Fujimoto K-I, Fan D-S, Ogawa M, Mizukami H, Urabe M (2000) Triple transduction with adeno-associated virus vectors expressing tyrosine hydoxylae, aromatic-L-amino-acid decarboxylae, and GTP cyclohydolase I for gene therapy of Parkinson's disease. Hum Gene Ther 11:1509–1519.

Sher A, Coffman RL (1992) Regulation of immunity to parasites by T cells and T cell-derived cytokines. Annu Rev Immunol 10: 385–409.

Smith GM, Hale J, Pasnikowski EM, Lindsay RM, Wong V, Rudge JS (1996a) Astrocytes infected with replication-defective adenovirus containing a secreted form of CNTF or NT3 show enhanced support of neuronal populations in vitro. Exp Neurol 139:156–166.

Smith K, Ying B, Ball AO, Beard CW, Spindler KR (1996b) Interaction of mouse adenovirus type 1 early region 1A protein with cellular proteins pRb and p107. Virology 224:184–197.

Smith TA, White BD, Gardner JM, Kaleko M, McClelland A (1996c) Transient immunosuppression permits successful repetitive intravenous administration of an adenovirus vector. Gene Ther 3:496–502.

Spaete RR, Frenkel N (1982) The herpes simplex virus amplicon: A new eucaryotic defective-virus cloning-amplifying vector. Cell 30:305–310.

Spaete RR, Frenkel N (1985) The herpes simplex virus amplicon: Analyses of *cis*-acting replication functions. Proc Natl Acad Sci USA 82:694–698.

Standridge JB (2004) Pharmacotherapeutic approaches to the treatment of Alzheimer's disease. Clin Ther 26:615–630.

Stavropoulos TA, Strathdee CA (1998) An enhanced packaging system for helper-dependent herpes simplex virus vectors. J Virol 72:7137–7143.

Stow ND, McMonagle E (1982) Propagation of foreign DNA sequences linked to a herpes simplex virus origin of replication. In: Eucaryotic Viral Vectors (Gluzman Y, ed), pp 199–204. Cold Spring Harbor: Cold Spring Harbor Laboratory.

Strohmeyer R, Rogers J (2001) Molecular and cellular mediators of Alzheimer's disease inflammation. J Alzheimers Dis 3:131–157.

Sun L, Li J, Xiao X (2000) Overcoming adeno-associated virus vector size limitation through viral DNA heterodimerization. Nat Med 6:599–602.

Takeuchi A, Irizarry MC, Duff K, Saido TC, Hsiao Ashe K, Hasegawa M, Mann DM, Hyman BT, Iwatsubo T (2000) Age-related amyloid beta deposition in transgenic mice overexpressing both Alzheimer mutant presenilin 1 and amyloid beta precursor protein Swedish mutant is not associated with global neuronal loss. Am J Pathol 157:331–339.

Thomas CE, Schiedner G, Kochanek S, Castro MG, Lowenstein PR (2000) Peripheral infection with adenovirus causes unexpected long-term brain inflammation in animals injected intracranially with first-generation, but not with high-capacity, adenovirus vectors: Toward realistic long-term neurological gene therapy for chronic diseases. Proc Natl Acad Sci USA 97:7482–7487.

Thomas CE, Schiedner G, Kochanek S, Castro MG, Lowenstein PR (2001) Preexisting antiadenoviral immunity is not a barrier to efficient and stable transduction of the brain, mediated by novel high-capacity adenovirus vectors. Hum Gene Ther 12:839–846.

Thomas CE, Storm TA, Huang Z, Kay MA (2004) Rapid uncoating of vector genomes is the key to efficient liver transduction with pseudotyped adeno-associated virus vectors. J Virol 78:3110–3122.

Tolba KA, Bowers WJ, Hilchey SP, Halterman MW, Howard DF, Giuliano RE, Federoff HJ, Rosenblatt JD (2001) Development of herpes simplex virus-1 amplicon-based immunotherapy for chronic lymphocytic leukemia. Blood 98:287–295.

Tolba KA, Bowers WJ, Eling DJ, Casey AE, Kipps TJ, Federoff HJ, Rosenblatt JD (2002) HSV amplicon-mediated delivery of LIGHT enhances the antigen-presenting capacity of chronic lymphocytic leukemia. Mol Ther 6:455–463.

Toyoda K, Ooboshi H, Chu Y, Fasbender A, Davidson BL, Welsh MJ, Heistad DD (1998) Cationic polymer and lipids enhance adenovirus-mediated gene transfer to rabbit carotid artery. Stroke 29:2181–2188.

Tripathy SK, Black HB, Goldwasser E, Leiden JM (1996) Immune responses to transgene-encoded proteins limit the stability of gene expression after injection of replication-defective adenovirus vectors. Nat Med 2:545–550.

Wade-Martins R, Smith ER, Tyminski E, Chiocca EA, Saeki Y (2001) An infectious transfer and expression system for genomic DNA loci in human and mouse cells. Nat Biotechnol 19:1067–1070.

Wagner E, Zatloukal K, Cotten M, Kirlappos H, Mechtler K, Curiel DT, Birnstiel ML (1992) Coupling of adenovirus to transferrin-polylysine/DNA complexes greatly enhances receptor-mediated gene delivery and expression of transfected genes. Proc Natl Acad Sci USA 89:6099–6103.

Wang X, Appukuttan B, Ott S, Patel R, Irvine J, Song J, Park JH, Smith R, Stout JT (2000) Efficient and sustained transgene expression in human corneal cells mediated by a lentiviral vector. Gene Ther 7:196–200.

Warner MS, Geraghty RJ, Martinez WM, Montgomery RI, Whitbeck JC, Xu R, Eisenberg RJ, Cohen GH, Spear PG (1998) A cell surface protein with herpesvirus entry activity (HveB) confers susceptibility to infection by mutants of herpes simplex virus type 1, herpes simplex virus type 2, and pseudorabies virus. Virology 246:179–189.

Willis RA, Bowers WJ, Turner MJ, Fisher TL, Abdul-Alim CS, Howard DF, Federoff HJ, Lord EM, Frelinger JG (2001) Dendritic cells transduced with HSV-1 amplicons expressing prostate- specific antigen generate antitumor immunity in mice. Hum Gene Ther 12:1867–1879.

Wolfe JH, Deshmane SL, Fraser NW (1992) Herpesvirus vector gene transfer and expression of β-glucuronidase in the central nervous system of MPS VII mice. Nat Genet 1:379–384.

Wood MJA, Byrnes AP, Pfaff DW, Rabkin SD, Charlton HM (1994) Inflammatory effects of gene transfer into the CNS with defective HSV vectors. Gene Therapy 1:283–291.

Wu P, Phillips MI, Bui J, Terwilliger EF (1998) Adeno-associated virus vector-mediated transgene integration into neurons and other nondividing cell targets. J Virol 72:5919–5926.

Wu Z, Asokan A, Samulski RJ (2006) Adeno-associated Virus Serotypes: Vector Toolkit for Human Gene Therapy. Mol Ther 14:316–327.

Xiao X, Li J, Samulski RJ (1998) Production of high-titer recombinant adeno-associated virus vectors in the absence of helper adenovirus. J Virol 72:2224–2232.

Xiao X, Li J, McCown TJ, Samulski RJ (1997) Gene transfer by adeno-associated virus vectors into the central nervous system. Exp Neurol 144:113–124.

Yu SF, von Ruden T, Kantoff PW, Garber C, Seiberg M, Ruther U, Anderson WF, Wagner EF, Gilboa E (1986) Self-inactivating retroviral vectors designed for transfer of whole genes into mammalian cells. Proc Natl Acad Sci USA 83:3194–3198.

Zabner J, Chillon M, Grunst T, Moninger TO, Davidson BL, Gregory R, Armentano D (1999) A chimeric type 2 adenovirus vector with a type 17 fiber enhances gene transfer to human airway epithelia. J Virol 73:8689–8695.

Zatloukal K, Wagner E, Cotten M, Phillips S, Plank C, Steinlein P, Curiel DT, Birnstiel ML (1992) Transferrinfection: A highly efficient way to express gene constructs in eukaryotic cells. Ann N Y Acad Sci 660:136–153.

Zhang Y, Schlachetzki F, Zhang YF, Boado RJ, Pardridge WM (2004) Normalization of striatal tyrosine hydroxylase and reversal of motor impairment in experimental parkinsonism with intravenous nonviral gene therapy and a brain-specific promoter. Hum Gene Ther 15:339–350.

Zhu Q, Zeng C, Huhalov A, Yao J, Turi TG, Danley D, Hynes T, Cong Y, DiMattia D, Kennedy S, Daumy G, Schaeffer E, Marasco WA, Huston JS (1999) Extended half-life and elevated steady-state level of a single-chain Fv intrabody are critical for specific intracellular retargeting of its antigen, caspase-7. J Immunol Methods 231:207–222.

Zou L, Zhou H, Pastore L, Yang K (2000) Prolonged transgene expression mediated by a helper-dependent adenoviral vector (hdAd) in the central nervous system. Mol Ther 2:105–113.

49
Proteomics and Genomics

Wojciech Rozek and Pawel S. Ciborowski

Keywords Proteomics; Genomics; Neuroproteomics; 2 dimensional electrophoresis; Protein profiling; Protein fingerprinting; Biomarker; Systems biology; Tissue profiling; Bioinformatics

49.1. Introduction

Successful completion of sequencing of the human genome showed considerably smaller number of genes (20,000 to 25,000 protein coding genes) (International Human Genome Sequencing Consortium. Lander ES, 2004) than expected and the number of proteins greatly exceeding the number of genes. This had a big impact on how we envision human and other proteomes and how we will approach proteomic and genomic analyses in the future. Isoforms, products of one gene modified by posttranslational modifications or alternative splicing, is only one part of complexity because function and biological role can also depend on cytoplasmic, subcellular compartments, and/or extracellular localization. Interactions with other proteins and nonproteinacious molecules add yet another layer of complexity.

49.2. Proteomics and Genomics—Technologies for Global Oversight of Complex Biological Systems

Proteomics as a scientific field emerged in mid-1990 along with technological advances in analysis of proteins and peptides. Two-dimensional polyacrylamide gel electrophoresis (2D SDS-PAGE) originally developed in late 70s (Barritault et al., 1976) rapidly advanced to the next level when immobilized pH gradient (IPG strips) gels were introduced (Gelfi and Righetti, 1983) facilitating analysis of gene products containing over 2,000 proteins in a mixture (Gianazza, 1995) with much higher reproducibility. Mass spectrometry of proteins and peptides is developing fast and providing new methods and reagents for quantitative analysis such Isotope Coded Affinity Tags (ICAT). Recently, protein microarrays have emerged, adding yet another powerful tool in global screening of gene products in complex biological systems. Development of user friendly kits for protein sample preparation and extraction enabled nonprotein chemists to isolate proteins, fractionate complex samples, and develop their own protocols in a similar way as development of expression vectors and systems (Feuerstein et al., 2005). Growth of all kind of databases, new and improved algorithms for database searches opened the door for high throughput experiments with substantially increased quality of protein identification and quantitative measurements (Ku and Yona, 2005). Thus it became a reality to identify fingerprints of changes induced by infection, malignant transformation, exposure to toxic agents, or differentiation (Pang et al., 2006). Therefore, substantial progress in prognosis and monitoring of disease progression and discovery of new biomarkers and drug targets is widely expected.

Until now proteomic analysis delivered less than expected new drug targets and biomarkers. Nevertheless, proteomic approach made it possible to accumulate unprecedented amounts of new and original data within a short period of time, which on the other hand, created demand for better and faster tools for data analysis. Another issue is how to manage such large amounts of data and information and how to navigate, for example, through thousands of collected spectra. In response to this demand, Laboratory Information Management Systems (LIMS) became commercially available, yet smaller laboratories in many instances cannot afford expensive softwares and high capacity servers to fully utilize medium to high throughput proteomic analysis.

Advancement in high throughput technology in proteomics and genomics led to studies of systems biology (for definition http://en.wikipedia.org/wiki/Systems_biology), meaning to monitor changes of thousands of individual components within the system at the same time. Thus, top-down approach deals with system modeling leading to hypothesis, prediction, and simulation of behavior of an organism under various physiological conditions. Bottom-up approach will study

T. Ikezu and H.E. Gendelman (eds.), *Neuroimmune Pharmacology.*
© Springer 2008

individual pieces on a discovery basis and combine all the information to understand how the system in question works (Figure 49.1).

The same concept in proteomics studies has technological implications, e.g., which method, sample preparation protocols, and instrumentation will be used. Again, top-down analysis will be based on isolation, analysis, and characterization of an intact protein to reveal its function. Fourier transformed ion cyclotron resonance mass spectrometry (FT-ICR) (Marshall et al., 1998) facilitates such approach in protein identification as a result of random fragmentation of an intact molecule. In contrary, bottom-up approach is based on up-front fragmentation of the protein in question using various proteolytic enzymes with known specificity (Chalmers et al., 2005; Millea et al., 2006). In these experiments, trypsin is most commonly used. An important question that remains is whether more information will be obtained from bottom-up approach, which is protein identification on a cost of protein characterization, or from top-down approach, which is distinguishing differences between two similar proteins/isoforms. Of course, the ideal situation is to obtain both answers from one type of analysis. Handling mixtures of peptides generated from complex a protein mixture (bottom-up approach) will be very different from sample processing at the protein level prior to analysis (top-down approach). Therefore, integration of these two will be a technological challenge in coming years because to make significant contributions, both approaches will require collaboration of scientists with diverse expertise with cross-disciplinary nature.

Regardless of the approach, bottom-up or top-down, cells, their compartments and organelles (nucleus, mitochondria, endosomes etc.) as well as body fluids are complex and dynamic systems comprising of multiple networks of interacting

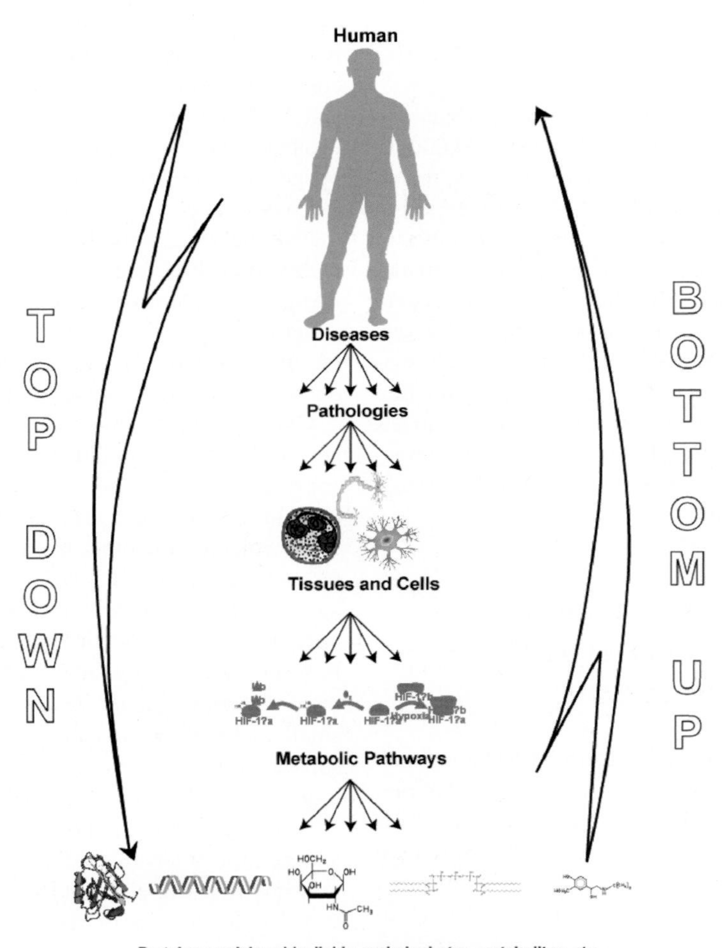

FIGURE 49.1. Top-down and bottom-up approach in studying systems biology. There are three components contributing equally to study biology systems: theoretical, conceptual, and experimental approaches. All of these approaches more and more heavily depend on implementation of effective bioinformatics tools. Hypothesis driven and discovery driven project equally participate in studying systems biology.

molecules of various natures: proteins, glycoproteins, glycolipids, nucleic acids, metabolites etc. An overview, moreover understanding of these components and their mutual interactions and communications, requires specific bioinformatics tools which start emerging. This issue is further amplified by the amount of data/information accumulated in every experiment of a high throughput analyses. Full understanding of interactions between different types of molecules such as DNA, RNA, and proteins in any biological system will require coordination of multiple experimental techniques at the same time which will either require more sample or reduction of sample size for each individual test. Development of application of microfluidics in new analytical instrumentation will definitely facilitate miniaturization efforts in proteomics.

Gene array techniques and gene sequencing which have been used in studying biological systems for many years provided enormous amount of valuable information, just to mention the completion of Human Genome Project (International Human Genome Sequencing Consortium. Lander ES, 2004). Gene arrays are also much more mature technology, with Affymetrix as a leader in this category, than those used in proteomics (Reimers et al., 2005). This technology platform is easy to use in variety of experimental designs and it is the best standardized among other systems. Gene arrays, similarly to proteomic profiling, struggles with reproducibility and number of false positive results (Reimers et al., 2005). Proteomics, on the other hand, offers a number of additional and unique benefits. For example, proteins which are functional products of an expressed gene can be posttranslationally modified, and their subcellular localization can be altered in a different physiological state. This important information cannot be obtained using gene arrays.

Results from studies of biology systems lead to the conclusion that individualized diagnosis, clinical evaluation, and treatment will be inevitable. In many instances, amount of sample available for diagnostic purposes is and will be limited demanding more sensitive techniques and instrumentation on one hand, and development of single cell analysis on the other hand. Automated and simple steps of sample processing combined with highly sensitive and specific measurements will improve what currently contributes to loss of important markers in a limited sample.

49.3. Proteomics Technologies

New and improved technologies are being developed to address important issues in proteomics as a global overview of protein makeup (Rohlff and Southan, 2002; Petricoin and Liotta, 2003). Increasing resolution, sensitivity, speed of analysis (high throughput), and reproducibility are the major areas of concurrent improvements. Summary of technologies most commonly used in proteomics is presented in Table 49.1.

2D SDS-PAGE which combines two modes of separation: isoelectric focusing and polyacrylamide gel elctrophoresis has been on the market for about 30 years (O'Farrell, 1975; Barritault et al., 1976). Initially this technique did not gain much of popularity due to obstacles in part originating from poor

TABLE 49.1. Summary of analytical and preparative methods most commonly used in proteomic studies.

Method	Description	Applications
SDS PAGE	Fractionation based on approximate molecular weight (m.w.).	Preparative/analytical
IEF	Fractionation based on isoelectric point (pI).	Preparative/analytical
2 D electrophoresis (IEF/SDS PAGE)	High resolution separation/fractionation based on pI and m.w., long and complex procedure.	Preparative/analytical
Capillary electrophoresis (CE)	Fractionation based on m.w. or pI, possible direct connection to MS.	Preparative/analytical
HPLC/FPLC	Fractionation based on physico-chemical features of protein (m.w., pI, hydrophobic profile).	Preparative/analytical
SELDI-TOF	Fractionation based on selective binding to activated surfaces followed by MS determination of m.w., automatic fast sample processing combined with mass spectrometry.	Analytical
Protein arrays	Identification/separation techniques based on protein-protein interactions including antigen-antibody interactions.	Analytical
LC-MS/MS	Chromatographic one or two dimensional fractionation followed by mass spectrometric peptide sequencing and protein identification.	Analytical
MALDI-TOF	Requires high resolution mass spectrometry for protein identification based on peptide fingerprinting.	Analytical

reproducibility. It was also not very suitable method for typical routine analyses (Jellum and Thorsrud, 1986) and remained as research tool until early 90s. Development of immobilized pH gradient (IPG) strips made this method much more user friendly, and most importantly, much more reproducible. This allowed creation of computerized protein databases based on two-dimensional separation (Celis et al., 1990). Although the resolution, thus reproducibility, of 2D SDS-PAGE remains a limitation especially in high molecular mass range or for hydrophobic proteins, advancement in protein sequencing by electrospray ionization-mass spectrometry/mass spectrometry (ESI-MS/MS) made possible to identify proteins directly from the protein "spots." Matrix Assisted Laser Desorption Ionization Time-of-Flight (MALDI-TOF) fingerprinting is another approach for protein identification from a mixture of tryptic digested proteins. Further on, development of specialized softwares for analysis of 2D SDS-PAGE gels facilitated automated comparison and quantitation of experiments consisting of multiple gels with large number of proteins separated in a single separation. A disadvantage of 2D SDS-PAGE such as concurrent analysis of high and low molecular weight proteins can be overcome by using a combination of various technologies. This aspect of proteomic analysis is discussed in more detail below in a paragraph describing hyphenated techniques (methods). Difference Gel Electrophoresis (DIGE) from GEHealthcare/Amersham, Inc. is another example of further technological advances. Two samples, one is labeled with 1-(5-carboxypentyl)-1′-methylindodi-carbocyanine halide N-hydroxyl-succinimidyl ester (Cy 5) while the other sample is labeled with 1-(5-carboxypentyl)-1′-propylindo-carbocyanine halide N-hydroxyl-succinimidyl ester (Cy 3) fluorescent dyes. Next, these two samples are mixed together and analyzed in one 2D SDS-PAGE experiment. Common proteins are represented by yellow spots while differentially expressed proteins are represented by different color spots detected with laser based scanner.

Another tool for protein profiling is Surface Enhanced Laser Desorption Ionization Time-of-Flight (SELDI-TOF) (Weinberger et al., 2000; Purohit et al., 2006). This technology utilize MALDI-TOF mass spectrometer as a detection mode and one step chromatography by utilizing proprietary Protein Chips (Ciphergen, Inc.) to reduce complexity of analyzed sample. It is an analytical tool with relatively low resolution and mass accuracy in addition to high variability of intensities. Nevertheless, a big advantage of this method is ease of use for fast screening as compared to, e.g., 2D SDS-PAGE, which requires larger amounts of a sample for analysis. Another advantage of SELDI-TOF is information regarding properties and conditions of binding of proteins of interest to ion-exchange, hydrophobic, IMAC, and/or normal phases which can be utilized with success in translation to preparative methods using corresponding HPLC column (Enose et al., 2005). This technology is also susceptible to relatively high variability in analyses of multiple SELDI-TOF spectra. Jeffries et al. (Jeffries, 2005) developed an algorithm which greatly improved spectra alignment thus improved reproducibility. MALDI-TOF instruments with higher resolution than SELDI-TOF combined with various chromatographic techniques has been also used but did not produce convincingly better results then SELDI-TOF (Coombes et al., 2005). Other mass spectrometers, such as FT-ICR are currently being evaluated for better reproducibility (Johnson et al., 2004).

Other technology platforms used for protein profiling include 2-dimentional HPLC, capillary electrophoresis interfaced with mass spectrometry, tissue profiling and imaging mass spectrometry (Reyzer et al., 2004; Caldwell and Caprioli, 2005), and a combination of IEF (chromatofocusing) in the first dimension followed by reverse-phase high performance liquid chromatography (RP-HPLC) (ProteomeLab PF 2D from Beckman Coulter, Inc.) (Billecke et al., 2006; Soldi et al., 2005). The ProteomeLab PF 2D workstation seems to be a promising alternative to 2D SDS-PAGE because it offers direct connection to an electrospray source of ESI-MS instrument or direct deposition onto MALDI-TOF target (Sheng et al., 2006; Shin et al., 2006). It has been proposed to complement already used profiling methods (Soldi et al., 2005). It may take some time, however, until these methods become more mature and reliable and will gain broader interest in the field of proteomics.

Multidimensional fractionation combined with removal of most abundant proteins is an essential step in detecting low abundant proteins in complex samples, in particular serum, plasma, or cerebrospinal fluid (CSF) (Chromy et al., 2004; Fountoulakis et al., 2004; Ramstrom et al., 2005). Recently published report of Human Proteome Organization (HUPO) Human Plasma Protein Project (HPPP) showed that almost all laboratories participating in this international effort used some mean of sample pre-fractionation. Eventually, 3,020 proteins were included in a database set from 15,677 proteins reported (Omenn et al., 2005). The same approach applies to biomarker discovery based on SELDI-TOF analysis. HPLC pre-fractionation of cell lysates using various columns enhanced sensitivity and specificity of SELDI-TOF determinations. Ciphergen Inc. (Freemont, CA), SELDI-TOF manufacturer recognized the necessity of pre-fractionation and developed spin columns for that purpose. Although such approach seems to be very attractive, it should be used with caution. Fractionation may significantly increase number of samples to be tested substantially increasing costs and time for data analysis.

Recent fast development of nanotechnology not only in drug delivery but also in diagnostics and monitoring combined with the technological boom of miniaturization made it possible to use lower amounts of sample for analysis and sustain at the same level of sensitivity and specificity. Single cell analysis is becoming a reality, although we have to be aware that not all proteins or their modified forms can be effectively detected at this level (Zhang et al., 2001; Diks and Peppelenbosch, 2004). Unlike in genomics, proteomics approach does not have luxury of *in vitro* amplification of proteins. Besides improving the yield of protein of interest during sample preparation, detection limits in laboratory tests lowered by using multi-layered ELISA type of assays in which secondary antibodies labeled with enzyme bind at higher numbers to primary antibodies reacting in 1:1 ratio with specific epitope (Kim et al., 2005). This of course carries a risk of increased background, but for example chemiluminescent detection systems increased sensitivity of western blot analysis by more than order of magnitude comparing to color based detection systems (Sandhu et al., 1991b; Bakkali et al., 1994). There are numerous examples in which proteomics was used to discover new biomarkers while commercial tests are based on assays using different principles, e.g. ELISA (Sandhu et al., 1991a; Hampel et al., 2004).

49.4. Biomarkers

Biomarker in disease can be defined as quantifiable analyte(s) or factor(s) reflecting usually multiple changes in individual's physiological state. These changes, or better said, biological responses are at the molecular, cellular, or whole organism levels. They play critical role in assessing of disease outcome, in particular in patient treatment, and how these treatments will be individualized and/or modified in future clinical trials and how they will affect directions in drug development. Therefore, biomarkers must provide objective measures of normal versus pathogenic processes, responses to treatment whether it leads to a cure, improvement, or only maintenance of patient's health. Thus, biomarkers are absolutely essential for early diagnosis of any disease whether it is slowly progressing pathological process or acutely developing infection.

HIV-1-associated dementia (HAD) is a good example to discuss importance and qualities of good biomarker. HAD is a brain pathology developing in the late stage of HIV-1 infection and it is slowly progressing disease which once started ultimately leads to death. Currently, it is impossible to predict and measure the time of onset and progression of HAD, the two critical parameters determining course of treatment. The main reason of such situation is lack of appropriate biomarkers, which will inform us about mechanism of brain injury caused by mechanisms triggered by HIV-1 in the brain. What will be a good and reliable biomarker of HAD? We expect that whether biomarker is found in CSF or plasma sample of diseased people, it will give us also an insight into cellular mechanisms of an ongoing infection. This part of biomarker characteristics is also important for those who are designing new clinical trials or regimes of existing treatments. Entering clinical trial is based on vague measures, which might be influenced by subjective impact of examiner carries additional risk of inaccurate results. Thus the ideal biomarker should inform us as precisely as possible about the mechanism, status, and prediction of therapy and disease and help in designing of "individualized" therapy for the most effective result. In our example of HAD, an activation of mononuclear phagocytes recruited in high numbers due to ongoing inflammation will lead to conclusions that up- or down regulated proteins originating from these cells due to HIV-1 infection might be good candidates for biomarkers. Such approach, however, should be considered with caution. Our recent findings (Ciborowski et al., 2004) showed that although HIV-1 downregulates MMP-9 expression in monocyte derived macrophages in *in vitro* studies, the net level of this protein in CSF of HAD individuals as well as its net proteolytic activity is increased due to recruitment of high numbers of these cells to the inflamed brain (Ghorpade et al., 2001).

It is important to note that effective and good biomarker can be a specific fragment of protein which is present in diseased state but not under normal conditions. Currently, existing biomarker of this type is $A\beta_{42}$ fragment of α-amyloid protein in Alzheimer's Disease (AD) (Olsson et al., 2003).

49.5. Proteomics in Biomarker Discovery

There are four key components which are necessary for performing proteomic (global) type of experiments: (i) identification of differences generated within the system under various physiological conditions, (ii) fractionation of complex mixtures of proteins and other molecules, (iii) identification differentially expressed proteins and their interacting partners, and (iv) data analysis including statistical analysis, algorithms for database searches, and protein networks. The choice of methods, instrumentation, and experimental approaches, however, should be selected based on the questions which are asked and answers which are sought. A separate field of so called "hyphenated methods" is being developed and deals with combining two or more methods, first one is separation/fractionation of analyzed sample and the second one is in-line connected detection system. Hyphenated techniques were developed and were popular in 70s and 80s—more than 20 years ago and remain vital part of current technological advancement (Murray et al., 1975; Lee et al., 1976). An example of the most commonly used in proteomics is LC-MS (liquid chromatography combined with mass spectrometry) or in other global screening approaches are GC-MS (gas chromatography combined with mass spectrometry) (Kazakevich et al., 2005) and LC-UV-SPE-NMR (liquid chromatography, UV detection, solid phase extraction, and nuclear magnetic resonance) (Exarchou et al., 2006). Sample fractionation prior to detection and/or analysis usually leads to enhanced analytical outcome.

Hyphenated techniques (methods) are meeting a big challenge of proteomic analyses which is an extremely large and dynamic range of protein levels (Julka and Regnier, 2005; Kusnezow et al., 2006). Proteins in serum/plasma may range from 10 to 10^{12} fold difference in abundance making it impossible to observe all proteins in one experiment (Jacobs et al., 2005). Therefore, fractionation of complex mixtures such as whole cell lysates into a reasonable number of samples becomes inevitable and new methods of enriching low abundance proteins are required. Rapidly developing technology of protein arrays discussed in more details below is an attempt to address this issue. From a technical point of view, this technology leads to objective similar to genomic arrays: effective and sensitive protein analysis using automated instrumentation to perform a large number of parallel measurements. This also makes possible to screen functions such as protein-antibody, protein-protein, protein-ligand, protein-drug, and enzyme-substrate interactions, and perform multianalyte diagnostic assays.

A second major challenge for proteomics is transitioning from global analysis and profiling to focused studies on one or several proteins whose function(s) and/or interaction(s) are significant in biological processes and to further use of generated results for more focused screening and developing commercial diagnostic tests. Thus success of proteomic type of analysis is using data and information accumulated from all kind but most importantly from high throughput experiments, to address very specific questions of biological importance. These answers will determine how we will manipulate experimental conditions applied to the systems we study and to these parts of broad proteomics methods, which will be applied as a read-out of changes.

Proteomics cannot exist without interdisciplinary research, grouping experts from a wide variety of specialities, who understand and are able to communicate issues related to protein chemistry and to bring proteomics to a level of functional studies. Currently, research teams combining a relatively narrow group of scientists including protein chemists, mass spectrometrists, and bioinformaticians push the levels of sensitivity and accuracy of analytical instrumentation to unprecedented levels but experiments remain at the stage of cataloging proteins. One example of global proteomics analysis such as the HPPP (Omenn et al., 2005), can be accomplished only through international collaborations of laboratories equipped with various types of mass spectrometers and capable of performing complex analyses. Long term goals of this initiative are to perform comprehensive analysis of plasma and serum protein constituents in people, identify biological sources of variation within individuals over time, across populations and within populations including such factors as age, sex, menstrual cycle, exercise, with validation of biomarkers. Selected diseases, special cohorts, and common medications are also being discussed as parts of HPPP (for more information go to http://www.hupo.org/research/hppp/). Human Brain Proteome Project (HBPP) is even more ambitious in its goals. It includes defining and deciphering normal brain proteome, correlation of the expression pattern of brain proteins and mRNA, identification, validation, and functionally characterization disease-related proteins involved in neurodegenerative diseases and aging by differential protein expression profiling, and establishing a neuroproteomic database accessible to all participating laboratories and the scientific community, just to mentione the few. More details can be found at http://www.hupo.org/research/hbpp/.

Undoubtedly, efforts of this and other currently ongoing international initiatives, will build a foundation for new discoveries and functional studies. Everyone in the broad research community will benefit from this work, particularly a rapidly growing number of smaller proteomics laboratories and programs with research goals focused on specific question(s) of biological importance. The most exciting future lies in functional studies combining proteomic and genomic approaches. However, to link results of genomic and proteomic studies will require innovative approaches because posttranslational modification making protein product biologically active may depend on distant genes and their products but on gene coding for polypeptide of this protein. New tools for profiling focused groups of modifying enzymes such as phosphatases or kinases are emerging and will be effectively used for better understanding of an impact of posttranslational modifications.

49.6. Proteomics of CSF

Total volume of CSF in adult human is about 130–150 ml and production rate is relatively constant under normal physiological conditions, approximately 0.4 ml/ min (more than 500 ml daily). This means that turn over of CSF is at the rate of 4 volumes daily. The significant decrease in CSF secretion and turn over is associated with neurological diseases such as Alzheimer disease and chronic hydrocephalus (Tsunoda et al., 2002; Silverberg et al., 2004) reflecting physiological status of central nervous system (CNS).

CSF surrounds the brain and spinal cord and acts as an intermediate between blood and nervous tissues. Functions of the CSF include buoyancy, acid base buffering and delivery of electrolytes, signaling molecules, transport molecules and micronutrients to the brain parenchyma (Silverberg et al., 2004). CSF is produced predominantly by choroids plexus, specialized vascular tissue located in the brain ventricles (Serot et al., 2003). Choroid plexus tissues are intraventricular structures composed of villi covered by a single layer of ciliated, cuboid epithelium. The plexuses secrete CSF, synthesize numerous molecules, carry nutrients from the blood to CSF, reabsorb brain metabolism by-products and participate in brain immunosurveillance (Serot et al., 2003). Active transport across the choroid plexus is bidirectional so there is also macromolecular transport out of CSF. Blood flow to the choroid plexus is approximately six–times greater than that of the equal volume of brain tissue, suggesting that the choroid plexus is very metabolically active (Silverberg et al., 2004).

Changes in biochemical composition of CSF could serve as a useful tool for investigations of pathological processes in the CNS. CSF is also in contact with the blood plasma through the blood-brain barrier, thus resembling an ultrafiltrate of plasma in its protein constituents. CSF contains sugars, lipids, electrolytes and proteins. Protein concentration in CSF ranges from 0.2 mg/ml to 0.8 mg/ml (0.3 – 1% of serum protein concentration) with more than 70% of the proteins in CSF being isoforms of albumin, transferrin and immunoglobulins (Ogata et al., 2005; Wittke et al., 2005). The proteome of CSF contains proteins that are expressed in, and secreted from circumventricular CNS structures. Therefore, the CSF proteome could provide unique biomarkers for early-stage diagnosis or the staging of a neuronal disease, offer potential insight into the biochemical characterization of affected neuronal population, and clarify the molecular basis of CNS pathologies (Yuan and Desiderio, 2003, 2005).

Previous studies have used proteome comparisons to study such neurological disorders as Alzheimer's disease (AD), Parkinson's disease (PD), frontotemporal dementia, schizophrenia, and Creutzfeldt-Jacob disease. Some examples of utilizing proteomics approaches for discovery of biomarkers for neurological disorders in human CSF are given in Table 49.2.

Investigation of biomarkers in CSF has some important limitations. Availability of larger volumes of CSF samples is limited by the need for invasive lumbar punctures. The large dynamic range of protein concentrations, which can be up to twelve orders of magnitude between the highest and the lowest expressed proteins, makes the analysis difficult (Maccarrone et al., 2004). Various schemes of sample pre-fractionation have been used to circumvent these problems. The widely used albumin removal method is affinity chromatography on Cibacron Blue resins (e.g. Cibacron Blue F3-A agarose); however, this method is not specific only to albumin and it is known to bind numerous other proteins (Gianazza and Arnaud, 1982; Maccarrone et al., 2004). Cibacron blue based depletion could be coupled with utilizing protein G resins to bind immunoglobulins from biological samples (BioRad, Hercules, CA). A newer approach is to use immunoaffinity methods utilizing monoclonal or polyclonal antibodies to deplete not only albumin, but also other highly abundant proteins from biological samples. Resins coupled with antibodies could be packed in HPLC columns for depletion most abundant proteins (e.g. Multiple Affinity Removal System, Agilent Technologies, Inc., Palo Alto, CA) for simultaneous depletion of 6 proteins. Application of that column for CSF preparation before 2D electrophoresis and shotgun mass spectrometry was compared with Cibacron Blue and protein G resins by Maccarrone et al. (2004). Immunoaffinity column showed less nonspecific binding of CSF proteins. Another example of pretreatment CSF samples before proteomic analysis is using the IgY antibody microbeads and spin filters for depletion of most abundant proteins (Seppro Mixed 6, GenWay, San Diego, CA). Comparison of IgY microbeads and immunoaffinity column for pretreatment of CSF samples was performed by Ogata et al (Ogata et al., 2005), the column format removed the major proteins more effectively and approximately 50% more spots were visualized when compared to the 2D gel of CSF without protein depletion. Two other commercially available kit for removing of HSA and HSA/ IgG were used by Ramstrom et al. (2005) for CSF preparation prior to Liquid Chromatography FT-ICR Mass Spectrometry (LC-FTICR MS). Both depletion methods provided a significant reduction of HSA, and the identification of lower abundant components was clearly facilitated.

Every method of removing high abundant proteins from CSF samples before proteomic analysis facilitates identification of discrete proteins, which could be important as a biomarker of physiological stage of CNS. However, the possibility of nonspecific removing of important peptides (through nonspecific binding to resins or forming complexes with target proteins) should be taken into account. Therefore, albumin fraction with co-purified proteins should not be discarded and should be analyzed in separate experiments. One-dimensional electrophoresis followed by mass spectrometric analysis can be applied for such analyses.

49.7. Neuroproteomics

Proteomic studies of the brain pose significant challenges. Brain tissue collection, dissection, preservation, biochemical and molecular integrity are all critical aspects of neuropro-

TABLE 49.2. Summary of protein biomarkers of neurodegenerative disorders discovered using proteomics approach.

Disease	Biomarker (protein)	Applied method	Lit.
Alzheimer's disease	Cystatin C, B-2 microglobulin isoforms, 4.8 kDa VGF polypeptide, unknown 7.7 kDa polypeptide	SELDI-TOF and LC/MS	(Carrette et al., 2003)
	Apolipoprotein A1, apolipoprotein E, apolipoprotein J, B-trace, retinol binding protein, kininogen, alpha-1 antitrypsin, cellcycle progression 8 protein, alpha 1B glycoprotein, alpha - 2 – HS glycoprotein	2D SDS-PAGE and MALDI-TOF	(Puchades et al., 2003)
Frontotemporal dementia	Transthyretin, fragment of VGF S-cysteinylated transthyretin, truncated cystatin C, fragment of chromogranin B	SELDI-TOF and MS/MS	(Ruetschi et al., 2005)
Creutzfeldt-Jacob disease	14-3-3 protein cystatin C	2D SDS-PAGE, SELDI-TOF and LC-MS/MS	(Harrington et al., 1986; Sanchez et al., 2004)
Schizophrenia	Apolipoprotein A-IV	2D SDS-PAGE and LC/MS	(Jiang et al., 2003)
Multiple sclerosis	Cartilage acidic protein, tetranectin, SPARC-like protein, autotaxin t	2D SDS-PAGE and LC -MS/MS	(Hammack et al., 2004)

teomics and neurogenomics. Despite of the fact that many proteins and nucleic acids are stable postmortem (Yates et al., 1990) rapid degradation of other proteins may result in changes in overall protein composition (Abbott, 2003; Choudhary and Grant, 2004). Protein profile can change rapidly within minutes after death making very difficult to sort out changes related to disease from those related to death itself. In many studies done so far, protein extract of entire brain was used without distinguishing specific regions (Wang et al., 2005). To address these issues a variety of techniques of brain dissection have been developed (Sanna et al., 2005). Manual dissection methods are still very useful yielding enough material for microarray analysis with somewhat limited "contamination" of sample of interest with neighboring region which may affect reproducibility of genomic and proteomic analyses. Ultimately, laser dissection technique will be the method of choice for retrieving small, region(s), even single cell of interest, from brain tissue to improve anatomical accuracy (Evans et al., 2003). CSF obtained postmortem might be also valuable source of information, however, it has to be analyzed with caution because of brain necrosis following death (Lescuyer et al., 2004). Nevertheless, Lescuyer et al. found in postmortem collected CSF samples three proteins which have been reported as potential markers of various neurodegenerative disorders. These proteins are prostaglandin D synthase, glial fibrillary acidic protein, and cathepsin D. Other proteins include ubiquitin C-terminal hydrolase L1, peroxiredoxin 5 (Lescuyer et al., 2004).

While some neurological disorders such as AD, PD, HAD, Multiple Sclerosis (MS), or Amyotropic Lateral Sclerosis (ALS) are slowly developing diseases, brain injury due to stroke, trauma, exposure to toxic substances, or hypoxia require immediate intervention. Nevertheless, in all these cases central nervous system is affected and to various extents damaged irreversibly. Thus fast diagnosis and medical intervention are absolutely critical for the outcome of acute diseases such as stroke and bacterial/viral meningitis. Slowly developing diseases such as AD, PD, HAD, MS at this time are diagnosed based on physiological tests, which are not very precise and do not provide precise evaluation of disease progression or treatment efficiency. In both scenarios there is a great need for new, reliable, and accurate biomarkers aiding in objective measurements of disease progression and treatment.

Proteomic and genomic technologies, and in particular when combined as functional genomics are new and promising experimental approaches in creating expression profiles of proteins and their connection with disease specific changes starting with transcription and ending at the level of posttranslational modifications and subcellular localization. Although this area of research is relatively new, it seems to be more convincing that molecular mechanisms underlying many neurodegenerative disorders may have common features. Aggregated and modified proteins (α-synculein in PD and tau in AD), deregulation of mitochondrial functions leading to overproduction of toxic ROS, overproduction of glutamate are factors responsible for neuronal death are observed in many neurodegenerative disorders (Sultana et al., 2006). Nevertheless, triggers of neurotoxic mechanisms are vastly unknown and various proteomic techniques are employed to address these questions. High-throughput shotgun analysis used by Soreghan et al. (2005) to identify targets of protein carbonylation in PS1 + APP transgenic mouse model pointed iNOS-integrin signaling, CRE/CBP transcription regulation, and rab-lyst vesicular trafficking pathways as future research directions on neurodegeneration.

Another neurodegenerative disorder ultimately leading to death is HAD. Despite of dramatic effect of antiretroviral therapy (ART) on slowing disease progression, HAD continues to be major cause of morbidity and death (Gendelman et al., 1998; McArthur et al., 2003) suggesting that ART does not provide complete protection against neurological damage in HIV-infected brains. Laboratory measures of HIV-1 associated inflammation of the brain, HIV Encephalitis (HIVE), or HAD progression developed so far, although valuable, are not diagnostic of HAD (Tornatore et al., 1991; Gendelman et al., 1998; Anderson et al., 2002). This new situation creates a demand for more accurate and reliable markers for monitoring HAD progression. Proteomic analyses of CSF samples obtained from HIV demented individuals is one approach taken by us and others. Another approach is based on the hypothesis that *in vitro* HIV-1 infected macrophage will secrete specific proteins which will be present in CSF samples of HAD patients. Combining both approaches should result in discovering new biomarkers. We also postulate that future diagnosis of HAD progression will be based on a set of multiple markers which levels will be measured based on background levels characteristic for individual being diagnosed and will be combined with psychological and brain imaging tests.

49.8. Functional Genomics

Genomics is the study of genes and their function. It is a broad definition of understanding the structure and molecular mechanisms underlying function of the genome. As the field of genomics is maturing, subfileds emerge. Thus, functional genomics will incorporate protein products into gene characterization, structural genomics will include structural features of genes and chromosomes, comparative genomics will deal with evolutionary aspects of genomic organization and their interspecies relationship, epigenomics is focused on DNA methylation and pharmacogenomics will study targets for new drugs.

Genomics as a scientific field as well as a technology platform is much more advanced than proteomics. Availability of relatively fast sequencing of large sets of genes, in fact entire genomes, is today's reality and the major factor slowing our progress in this area lies in financial resources. Nevertheless, more genomes are sequenced expanding our knowledge which is not only focused on collecting and cataloging of data but mostly on studying how the genome functions.

Sequencing genome primates who are a link to the development of *Homo sapiens* is a challenging and intriguing task. If we can answer a question how our brains evolved and how different they are on the functional level from those of chimpanzees or maccacs, our ability to understand mental disorders would be greatly expanded. There is no doubt that two avenues of genomics, evolutionary and developmental, will be rapidly expanding in the next decade. However, the third avenue, which is our understanding of pathological processes resulting from inflammation and/or trauma (e.g. stroke injury), is much more complex and challenging. Recruitment of mononuclear phagocytes from the periphery and their responses beyond the blood brain barrier add another substantial level of complexity. Do these cells constitute part of the brain or do they switch from bystanders to a role of active players? One of limitations in neurogenomics is the lack of good models to study variety of aspects of brain function. For example, a stimulus leading to change in reaction, e.g. from friendly to aggressive can be quite different between humans and rodents which are very often used in such studies. Another important question is whether these changes occur at the genome level or maybe downstream at the level of activation of certain enzymes without much change in the levels of expression.

49.9. Gene Arrays

In genomic studies microarray technology is used for the rapid quantitative measurement of gene expression in a tissue. mRNA which is directly transcribed from DNA template not only reflects levels of gene expression but also is used directly for hybridization on DNA chip. Levels of complementary hybridization are measured by laser scanning of fluorescently labeled RNA molecules bound to the chips and data are further subjected to computational processing for quality control (sources of errors) and detection of altered gene expression. Multiple experiments can be performed using this experimental approach in which data are combined, normalized, and analyzed for elucidation of distinct pattern characteristic for the studied system. Implementation of chips in microarray assays led to real multiplexing increasing throughput and speed of experimental progress. Miniaturization, improved manufacturing with better quality and affordability make microarray technology widely available. Group effort of technological advancement of more than twenty companies and numerous academic and other laboratories resulted in high density chips becoming available. For example, Affymetrix chips contain 400,000 groups of nucleotides in an area of ~1.6 cm^2. Chips with wide range of genes, focused microarrays narrowed to specific groups of genes as well custom made products make this technology platform suitable for addressing questions of biological importance, for clinical diagnostics and in biomarker discovery. Therefore, microarray technology has broad applications in such areas as expression profiling, functional genomics, pharmacogenomics (Meschia, 2004), and developmental (Jensen et al., 2004) and evolutionary biology (Sikela, 2006).

Despite the fundamental importance of genomic experiments, this technology platform provides only initial and far from completion results. Proteomics, on the other hand, is a follow-up approach, and when results acquired from both technology platforms are combined, more comprehensive picture of functional significance arises. Integration of these two experimental platforms is another challenge because of complexity of systems biology. *First*, a bottom-up or a top-down strategy has to be chosen for experimental strategy. *Second*, which part will be a driving force, genomics or proteomics? *Third*, an integrated bioinformatics platform has to be secured for complementary and coherent transformation of data into information. A possibility to produce custom made DNA microarrays designed and manufactured by commercial entities and/or individual investigators to address specific biological question(s) provide enormous support to such integrated strategies. Extensive literature can be found reviewing technical issues related to gene arrays including hybridization, post-hybridization quality control, normalization and data processing, (Boes and Neuhauser, 2005; Hartmann, 2005; Reimers, 2005). These specific tools allow performing genomic experiments based on results generated in initial proteomic studies. Our experience with such studies, show that experimental design and depth of data mining are critical.

49.10. Genomics in Neuroinflammation

Along with increasing human life expectancy, it is believed that in the next 50 years neurological disorders will exceed as a human health problem such diseases as cancer or heart diseases. Neurological disorders associated with age, e.g., AD PD, ALS, and other diseases such as HIV-1-associated cognitive impairment are correlated with inflammatory response of innate immunity system in the CNS. Response to insult such as bacterial or viral infection, stroke, cancer metastasis, or abnormal expression and accumulation of self proteins is associated not only with activation of resident macrophages, but is also associated with recruitment of large pool of mononuclear phagocytes (MP) from the periphery. These cells migrate through blood brain barrier to clear the insult and protect brain from further damage. In reality, the phenotype of MPs can produce neurotoxic outcome at the site of injury. The traditional approach to study human diseases has been to focus research projects on the study of important pathogenic factors: relevant microorganisms (Qureshi et al., 1999) and other agents linked to pathogenesis such as β-amyloid in AD or α-synuclein in PD with hope for development of effective treatment. This approach limited our progress in understanding of molecular basis of host defense mechanisms. Therefore, understanding what precisely regulates macrophage responses in the brain and how they interact with other types of cells is critical for more effective therapies (Ciborowski and Gendelman, 2006; Foti et al., 2006). New experimental approaches such as gene microarrays now can be used to overcome previous limitations of experimental manipulation and large sets of data, and facilitate genomewide analyses.

Genomics is one approach that can discover new clues leading to better understanding of MPs' function. Under normal conditions, monocytes derived from bone marrow circulate in the blood for approximately 48 h and migrate to tissues where they differentiate into macrophages. Differentiation will be driven by local tissue environment and is not completely reversible when macrophages are removed from various tissues and grown *in vitro* under the same conditions (Enose et al., 2005). Nevertheless, principal functions of these cells – clearance of debris, antigen presentation, phagocytosis, production reactive oxygen and nitrogen species to fight infection will remain the same regardless of tissue localization. Therefore, design of experimental conditions to study MPs' responses will play a critical role in data interpretation and in drawing conclusions. We may expect that genomic studies of, e.g., alveolar macrophages exposed to tobacco smoke (Wu et al., 2005), macrophage from cells generated *in vitro* by treatment of monocyte-derived macrophages (MDM) with oxidized low-density lipoprotein (LDL) (van Duin et al., 2003), or MDM exposed to aggregated α-synuclein will lead to understanding critical similarities and differences underlying inflammatory immune responses. Such comparative genomic studies (Heller et al., 1997) will produce vast amount of data critical for understanding neuroinflammation, which suffers from the lack of a good animal model.

The genomic approach has proven to be indispensable as an initial step in determining any given function. After all genes are coding proteins. On the other hand, the role of gene is finished as soon as transcription is completed. In the following steps, transcripts undergo a complex process leading to functional protein, a product that eventually performs a task. But we have to keep in mind that if an error occurs at transcription level and is reproduced over and over again, a faulty final product is made regardless how good and precise downstream machinery is. Brain is an organ from which biopsies are taken under very specific circumstances, e.g., brain tumors. Such material is further examined for histopathological evaluations. In future, this now well-accepted method will be combined with genomic, proteomic, and metabolomic signatures of tumor cells (Petrik et al., 2006).

Another obstacle in studying brain functions is our limited ability to purify specific populations of cells from either animals or postmortem from human brains. This limitation concerns both proteomics and genomics in a very similar way. Many brain cells such as neurons are terminally differentiated and do not proliferate. On the other hand astrocytes can be grown *in vitro,* therefore, neurons are always "contaminated" with other types of cells (Zheng et al., 1999, 2001). Ability to amplify mRNA in a preparation obtained from cell culture or cell preparation may lead to false conclusions. In situ hybridization can be used for verification; however, this method is laborious and cannot be used in a high throughput analyses.

49.11. Protein Microarrays

Over the last two decades DNA microarrays showed applicability of high throughput studies of gene expression and regulation under various experimental conditions (DeRisi et al., 1996; Khan et al., 1999). Therefore, the idea of developing similar technique for easy analysis of multiple protein samples in parallel was very attractive (Haab et al., 2001). Advancing miniaturization quickly translated to smaller samples required for analysis along with decreasing costs of spotted membranes boosted the demand. Early approach utilized phage displayed of relatively short sequences of amino acids for probing complex mixtures of proteins and screening their binding capacities (Scott and Smith, 1990; Smith and Scott, 1993). Protein microarrays based on antigen-antibody interaction meet such criteria and it obvious that this technology is rapidly filling a gap between genomics and proteomics. The ability of protein quantitation makes this approach even more attractive. Recent years brought a rapid development of various forms of protein microarrays (Templin et al., 2003; Poetz et al., 2005). Limitation of protein microarrays lies in requirement of highly specific capture and detection antibodies or other interacting molecules. Although in DNA microarrays this pre-requisite was accomplished due to high selectivity of hybridization of complementary sequences, they still suffer from some percentage of ambiguous results. In proteomics, posttranslational modifications add another level of molecular diversity and complexity. For example, many proteins, although different, share common modifications such as phosphorylated tyrosines. In these cases it is necessary to identify location of phosphorylated residue on the protein of interest.

Protein microarrays is a powerful tool when proteomic analysis is narrowed to a group of proteins such as growth factors, chemokines, cytokines etc. These classes of proteins already have a great number of pre-required well-characterized and highly specific antibodies reacting with high affinity. Such protein arrays (chips) were immediately developed and are successfully used in monitoring markers of immune responses (Kastenbauer et al., 2005) complementing DNA arrays (Ho et al., 2005). Ho et al. showed that complementation of cDNA and protein arrays in search of biomarkers of Alzheimer's disease can be consistent. Similarly to ELISA system, quality of protein microarrays will depend on the background signal from nonspecific adsorption of labeled protein increasing along with protein concentration. Detection limit of protein of interest will be determined not only by its absolute level in tested sample, but also by the high levels of other proteins with ability to cross-react in a nonspecific manner. Levels of detergents and other components of lysis buffers used prior to analysis have to be taken into account. The latter issue is of critical importance when sample needs to be concentrated and analyzed by MALDI type of mass spectrometer.

Accuracy and reproducibility of protein array data further depends on proper experimental design, normalization procedures, eliminating systematic bias, and appropriate statistical analysis. Eckel-Passow and co-authors made several excellent suggestions how experiments should be designed and analyzed to avoid false results (Eckel-Passow et al., 2005). For example the authors propose to use several different clustering tools instead of single cluster analysis for verification of results in discovering differences in a class of molecules, e.g. cytokines. Another obstacle in working with protein microarrays is stability of protein structure which is critical for proper folding thus interaction with binding partner. Another side of this problem is exposure of interacting surface/epitope while immobilized on the chip. Protein microarrays suffered from skepticism and criticism during early stages of development. Nevertheless, this technology is gaining ground in biomedical research with increasing potential of application in clinical diagnosis. Similarly to DNA microarrays, one can find commercially available protein arrays specific for various microorganisms and diseases, user friendly kits with improved and sophisticated fluorescent tags. For those working in the area of drug developments, protein microarrays are vehicles of accelerated discoveries. For others, protein microarrays are envisioned as tools for exploration and understanding pathological processes, kinetics of changes on molecular level, interactions with molecules different in nature such as lipids, nucleic acids, carbohydrates, drug candidates etc.

49.12. Proteomics and Tissue Profiling

Can tissue be examined for a global protein expression directly and without time consuming sample preparation? Can such profiling be used effectively in a rapid clinical diagnostics to aid pathologists in their work? Can section of tissue be placed on target, covered with matrix and profiled directly in mass spectrometers? If this appealing but challenging idea becomes real, it would be a breakthrough in modern clinical diagnosis.

Early attempts to develop patterns of specific compounds and proteins were based on tissue homogenates (Burns, 1982; Jimenez et al., 1997; Mitchell et al., 1997). Tissue profiling, or more accurately called tissue imaging, based on direct mass spectrometry measurements is a much younger area of research which started being developed in late 90s when MALDI-TOF-MS measurements were performed on intact tissue sections (Chaurand et al., 1999).

MALDI-TOF types of mass spectrometers with high resolution and sensitivity reaching attomolar range is a class of instrumentation of choice for direct tissue analyses. Spectra are generated very fast, a wide range of mass is acquired. Hundreds and even thousands of spots can be analyzed using one tissue section which further processed by computers will create an image. Besides the high throughput of this type of analysis, a major advantage is that collected spectra can be assigned to the localization on the tissue section, thus can localize molecules in question to tissue structures. This, in turn, is important for monitoring such processes as distribution of therapeutics, which can be small molecules as well as peptides and/or proteins. Questions that remain to be addressed in tissue imaging are whether and how important structural analysis (peptide sequence, drug structure) of identified species since downstream processing of spectra is based on the presence, intensity, and signal-to-noise ratio data. Scattered presence of another molecule(s) with close m/z (mass-to-charge ratio) can lead to misleading results. If the analyte's m/z is within 2000, identification based on sequencing can be performed in instruments equipped with post-source decay mode (Stoeckli et al., 2002). Another question is absolute quantitation, which is the weakness of MALDI-TOF comparing to other types of mass spectrometers. Nevertheless, meaningful information can be acquired if the system is properly validated. Validation of proteins can be accomplished using other methods, e.g., fluorescence immunochemistry and co-registartion. It is more challenging to quantitate small molecules (pharmaceuticals) by mass spectrometry without some kind of sample processing. Therefore, proper methods of validation need to be developed. Despite challenges discussed above, application of this technology is advancing resulting in reports such as classification of tumor samples using this technology platform (Chaurand et al., 1999) or single cell analysis (Li et al., 1999).

49.13. Bioinformatics and Information Networks

Expression of proteins and genes on a global scale requires new and efficient tools for database searches and data analysis. Such tools are advanced in genomics because this area of global analysis is much more advanced, while analysis of proteomic data is still in its early development. Algorithms used for proteomic database searching are being constantly improved. Currently, the mostly used algorithms for mass spectrometric peptide sequencing are MASCOT and SEAQUEST. Recently published report from the pilot phase of HPPP experiment compares results from MS/MS spectra analysis using not only MASCOT and SEQUEST but also Spectrum Mill, and X!Tandem concluded that no one of these algorithms outperforms other and much improvement is needed in this area. New approaches were also applied such as processing very large numbers of raw spectra to generate clusters of spectra followed by SEQUEST-like analysis.

An important issue that arouse during early development of proteomic technology is proper comparison of data generated by different centers, laboratories, even individuals. SELDI-TOF is an example of how difficult it was to cross-validate data from laboratory to laboratory which used the same technology platform and instrumentation. Mass spectrometry data can be very susceptible to errors originated from sample processing (Baggerly et al., 2004). The same applies to algorithms used for database searches and accuracy of protein identification based on mass spectrometric sequencing of peptides.

Summary

Proteomics is maturing from its early stages of fascination with ability of high numbers of protein identifications in a single experiment. Quantitative mass spectrometry of proteins is becoming standard method in proteomics laboratories. 2D SDS-PAGE supported by sophisticated software and increasing computer power along with more user friendly supplies providing better reproducibility is another technology advancing global protein outlook. Further technological developments of, i.e. tissue profiling by mass spectrometry and protein arrays, will facilitate a more precise view of changes occurring in proteomes resulting from alterations in biological system. New technology platforms and tools emerging almost every day will overcome current problems and limitations of proteomics such as sample processing, sample availability, reproducibility, quantitation, validation etc. We expect that significant biological advances: biomarker and drug discovery, disease and treatment monitoring, will be made during the coming decade.

This will be very important for understanding mechanisms leading to neurodegeneration such as HAD, AD, PD, ALS, and others. Another big step will be to go beyond slowing down or completely stopping the neurodegenerative process, and pushing immune system to repair the damage and promote regeneration. How much is it possible and how soon it can be accomplished remains an open question. It is without question, however, that understanding functions of complex biological systems such as duality of neurotoxic/neurothrophic functions of MPs in the brain during inflammation will require coordinated monitoring of large number of parameters at the same time and well-designed proteomics, genomics, transcriptomics, metabolomics, as well as neuroimaging experiments.

Review Questions/Problems

1. **What is proteomics?**
2. **What is genomics?**
3. **How does studying the proteome compare to studying the genome?**
4. **Review the challenges in proteomics research?**
5. **What is the difference between biomarker and risk marker?**
6. **Explain in which cases SELDI-TOF technology will be more advantageous than LC-MS/MS and/or MALDI-TOF.**
7. **What are the approaches used in the development of clinical proteomics?**
8. **Advantages and disadvantages in using protein arrays.**
9. **What are biomarkers and how are they identified and validated for use in the clinical setting?**
10. **Describe strategies of sample prefractionation for genomic and proteomic analysis.**
11. **Mathematical models of the molecular pathways in cells will facilitate:**
    a. prediction of previously unknown interactions
    b. combining genomic and proteomic data
    c. discover novel proteins, cellular functions, and pathways
    d. all of the above
12. **Successful translation of results from laboratory to clinic depends on:**
    a. exploratory research studies
    b. validation
    c. prospective screening studies on large-scale population
    d. all of the above
13. **Weakness of 2-dimensional SDS-PAGE is:**
    a. lack of superior reproducibility
    b. lack of protection of proteins from structural destabilization
    c. inability of obtaining protein sequencing data
    d. lack of superior methods of protein labeling
14. **Major difficulty in proteomic studies of certain type of samples such as CSF:**
    a. contamination with drugs penetrating through blood brain barrier
    b. size and availability of the sample
    c. lack of correlation with clinical diagnosis
    d. contamination with red blood cells during spinal tap
15. **Top-down approach in studies of systems biology is preferential over bottom-up approach because:**
    a. it provides more global view
    b. it allows for faster conclusions which will lead to direct development of clinical tests
    c. requires less sophisticated technology and instrumentation
    d. both approaches are equal

16. Major question to be addressed when considering application of genomics and proteomics to screening and testing include selection of:

a. mouse species
b. non-radioactive reagents
c. reputable provider of proteomics and genomics technologies
d. conditions which are true reflections of *in vivo* biological function

17. Commercially available antibody arrays technology has many of the same problems observed with genomics because of problems associated with:

a. data analysis, interpretation, storage, and retrieval
b. high background created by not well developed reagents
c. necessity of using radioactive materials
d. inaccuracy of chips readers

18. The best approach to determine the sensitivity and specificity of these methodologies, and to validate and standardize these technologies?

a. to reduce "biological noise" in the samples and reduce number of replicates
b. to link data obtained from database searches to biological effects via fast computers
c. to develop a standard approach to understanding and communicating variability in experimental data
d. none of the above

19. While planning proteomics experiments, a researcher should:

a. have available all types of mass spectrometers
b. mandatory access to protein microarrays
c. be aware of strength and weaknesses of each technology
d. always be an expert in statistics and computer programming for solving problems associated with bioinformatics

20. Neuroproteomics analysis is most conclusive when

a. experiments are done using entire brain so that all cell types are used in one assay
b. only *in vitro* models are conclusive
c. selected types of cell from a brain are obtained and manipulated
d. two or more cell types are mixed together to facilitate interaction mimicking *in vivo* situation

References

Abbott A (2003) Brain protein project enlists mice in 'dry run'. Nature 425:110.

Anderson E, Zink W, Xiong H, Gendelman HE (2002) HIV-1-associated dementia: A metabolic encephalopathy perpetrated by virus-infected and immune-competent mononuclear phagocytes. J Acquir Immune Defic Syndr 31(Suppl 2):S43–54.

Baggerly KA, Morris JS, Coombes KR (2004) Reproducibility of SELDI-TOF protein patterns in serum: Comparing datasets from different experiments. Bioinformatics 20:777–785.

Bakkali L, Guillou R, Gonzague M, Cruciere C (1994) A rapid and sensitive chemiluminescence dot-immunobinding assay for screening hybridoma supernatants. J Immunol Methods 170:177–184.

Barritault D, Expert-Bezancon A, Milet M, Hayes DH (1976) Inexpensive and easily built small scale 2D electrophoresis equipment. Anal Biochem 70:600–611.

Billecke C, Malik I, Movsisyan A, Sulghani S, Sharif A, Mikkelsen T, Farrell NP, Bogler O (2006) Analysis of glioma cell platinum response by metacomparison of 2-D chromatographic proteome profiles. Mol Cell Proteomics 5(1):35–42.

Boes T, Neuhauser M (2005) Normalization for Affymetrix GeneChips. Methods Inf Med 44:414–417.

Burns MS (1982) Applications of secondary ion mass spectrometry (SIMS) in biological research: A review. J Microsc 127 (Pt 3): 237–258.

Caldwell RL, Caprioli RM (2005) Tissue profiling by mass spectrometry: A review of methodology and applications. Mol Cell Proteomics 4:394–401.

Carrette O, Demalte I, Scherl A, Yalkinoglu O, Corthals G, Burkhard P, Hochstrasser DF, Sanchez JC (2003) A panel of cerebrospinal fluid potential biomarkers for the diagnosis of Alzheimer's disease. Proteomics 3:1486–1494.

Celis JE, Honore B, Bauw G, Vandekerckhove J (1990) Comprehensive computerized 2D gel protein databases offer a global approach to the study of the mammalian cell. Bioessays 12:93–97.

Chalmers MJ, Mackay CL, Hendrickson CL, Wittke S, Walden M, Mischak H, Fliser D, Just I, Marshall AG (2005) Combined top-down and bottom-up mass spectrometric approach to characterization of biomarkers for renal disease. Anal Chem 77:7163–7171.

Chaurand P, Stoeckli M, Caprioli RM (1999) Direct profiling of proteins in biological tissue sections by MALDI mass spectrometry. Anal Chem 71:5263–5270.

Choudhary J, Grant SG (2004) Proteomics in postgenomic neuroscience: The end of the beginning. Nat Neurosci 7:440–445.

Chromy BA, Gonzales AD, Perkins J, Choi MW, Corzett MH, Chang BC, Corzett CH, McCutchen-Maloney SL (2004) Proteomic analysis of human serum by two-dimensional differential gel electrophoresis after depletion of high-abundant proteins. J Proteome Res 3:1120–1127.

Ciborowski P, Gendelman HE (2006) Human immunodeficiency virus-mononuclear phagocyte interactions: Emerging avenues of biomarker discovery, modes of viral persistence and disease pathogenesis. Curr HIV Res 4:279–291.

Ciborowski P, Enose Y, Mack A, Fladseth M, Gendelman HE (2004) Diminished matrix metalloproteinase 9 secretion in human immunodeficiency virus-infected mononuclear phagocytes: Modulation of innate immunity and implications for neurological disease. J Neuroimmunol 157:11–16.

Coombes KR, Morris JS, Hu J, Edmonson SR, Baggerly KA (2005) Serum proteomics profiling—a young technology begins to mature. Nat Biotechnol 23:291–292.

DeRisi J, Penland L, Brown PO, Bittner ML, Meltzer PS, Ray M, Chen Y, Su YA, Trent JM (1996) Use of a cDNA microarray to analyse gene expression patterns in human cancer. Nat Genet 14:457–460.

Diks SH, Peppelenbosch MP (2004) Single cell proteomics for personalised medicine. Trends Mol Med 10:574–577.

Eckel-Passow JE, Hoering A, Therneau TM, Ghobrial I (2005) Experimental design and analysis of antibody microarrays: Applying methods from cDNA arrays. Cancer Res 65:2985–2989.

Enose Y, Destache CJ, Mack AL, Anderson JR, Ullrich F, Ciborowski PS, Gendelman HE (2005) Proteomic fingerprints distinguish microglia, bone marrow, and spleen macrophage populations. Glia 51:161–172.

Evans SJ, Choudary PV, Vawter MP, Li J, Meador-Woodruff JH, Lopez JF, Burke SM, Thompson RC, Myers RM, Jones EG, Bunney WE, Watson SJ, Akil H (2003) DNA microarray analysis of functionally discrete human brain regions reveals divergent transcriptional profiles. Neurobiol Dis 14:240–250.

Exarchou V, Fiamegos YC, van Beek TA, Nanos C, Vervoort J (2006) Hyphenated chromatographic techniques for the rapid screening and identification of antioxidants in methanolic extracts of pharmaceutically used plants. J Chromatogr A 1112(1–2):293–302.

Feuerstein I, Rainer M, Bernardo K, Stecher G, Huck CW, Kofler K, Pelzer A, Horninger W, Klocker H, Bartsch G, Bonn GK (2005) Derivatized cellulose combined with MALDI-TOF MS: A new tool for serum protein profiling. J Proteome Res 4:2320–2326.

Foti M, Granucci F, Pelizzola M, Beretta O, Ricciardi-Castagnoli P (2006) Dendritic cells in pathogen recognition and induction of immune responses: A functional genomics approach. J Leukoc Biol 79:913–916.

Fountoulakis M, Juranville JF, Jiang L, Avila D, Roder D, Jakob P, Berndt P, Evers S, Langen H (2004) Depletion of the high-abundance plasma proteins. Amino Acids 27:249–259.

Gelfi C, Righetti PG (1983) Preparative isoelectric focusing in immobilized pH gradients. II. A case report. J Biochem Biophys Methods 8:157–172.

Gendelman HE, Zheng J, Coulter CL, Ghorpade A, Che M, Thylin M, Rubocki R, Persidsky Y, Hahn F, Reinhard J, Jr., Swindells S (1998) Suppression of inflammatory neurotoxins by highly active antiretroviral therapy in human immunodeficiency virus-associated dementia. J Infect Dis 178:1000–1007.

Ghorpade A, Persidskaia R, Suryadevara R, Che M, Liu XJ, Persidsky Y, Gendelman HE (2001) Mononuclear phagocyte differentiation, activation, and viral infection regulate matrix metalloproteinase expression: Implications for human immunodeficiency virus type 1-associated dementia. J Virol 75:6572–6583.

Gianazza E (1995) Isoelectric focusing as a tool for the investigation of post-translational processing and chemical modifications of proteins. J Chromatogr A 705:67–87.

Gianazza E, Arnaud P (1982) Chromatography of plasma proteins on immobilized Cibacron Blue F3-GA. Mechanism of the molecular interaction. Biochem J 203:637–641.

Haab BB, Dunham MJ, Brown PO (2001) Protein microarrays for highly parallel detection and quantitation of specific proteins and antibodies in complex solutions. Genome Biol 2: RESEARCH0004.

Hammack BN, Fung KY, Hunsucker SW, Duncan MW, Burgoon MP, Owens GP, Gilden DH (2004) Proteomic analysis of multiple sclerosis cerebrospinal fluid. Mult Scler 10:245–260.

Hampel H, Buerger K, Zinkowski R, Teipel SJ, Goernitz A, Andreasen N, Sjoegren M, DeBernardis J, Kerkman D, Ishiguro K, Ohno H, Vanmechelen E, Vanderstichele H, McCulloch C, Moller HJ, Davies P, Blennow K (2004) Measurement of phosphorylated tau epitopes in the differential diagnosis of Alzheimer disease: A comparative cerebrospinal fluid study. Arch Gen Psychiatry 61:95–102.

Harrington MG, Merril CR, Asher DM, Gajdusek DC (1986) Abnormal proteins in the cerebrospinal fluid of patients with Creutzfeldt-Jakob disease. N Engl J Med 315:279–283.

Hartmann O (2005) Quality control for microarray experiments. Methods Inf Med 44:408–413.

Heller RA, Schena M, Chai A, Shalon D, Bedilion T, Gilmore J, Woolley DE, Davis RW (1997) Discovery and analysis of inflammatory disease-related genes using cDNA microarrays. Proc Natl Acad Sci USA 94:2150–2155.

Ho L, Sharma N, Blackman L, Festa E, Reddy G, Pasinetti GM (2005) From proteomics to biomarker discovery in Alzheimer's disease. Brain Res Brain Res Rev 48:360–369.

International Human Genome Sequencing Consortium. Lander ES, Birren B, Nusbaum C, Zody MC, Baldwin J, Devon K, Dewar K, Doyle M, FitzHugh W, Funke R, Gage D, Harris K, Heaford A, Howland J, Kann L, Lehoczky J, LeVine R, McEwan P, McKernan K, Meldrim J, Mesirov JP, Miranda C, Morris W, Naylor J, Raymond C, Rosetti M, Santos R, Sheridan A, Sougnez C, Stange-Thomann N, Stojanovic N, Subramanian A, Wyman D, Rogers J, Sulston J, Ainscough R, Beck S, Bentley D, Burton J, Clee C, Carter N, Coulson A, Deadman R, Deloukas P, Dunham A, Dunham I, Durbin R, French L, Grafham D, Gregory S, Hubbard T, Humphray S, Hunt A, Jones M, Lloyd C, McMurray A, Matthews L, Mercer S, Milne S, Mullikin JC, Mungall A, Plumb R, Ross M, Shownkeen R, Sims S, Waterston RH, Wilson RK, Hillier LW, McPherson JD, Marra MA, Mardis ER, Fulton LA, Chinwalla AT, Pepin KH, Gish WR, Chissoe SL, Wendl MC, Delehaunty KD, Miner TL, Delehaunty A, Kramer JB, Cook LL, Fulton RS, Johnson DL, Minx PJ, Clifton SW, Hawkins T, Branscomb E, Predki P, Richardson P, Wenning S, Slezak T, Doggett N, Cheng JF, Olsen A, Lucas S, Elkin C, Uberbacher E, Frazier M, Gibbs RA, Muzny DM, Scherer SE, Bouck JB, Sodergren EJ, Worley KC, Rives CM, Gorrell JH, Metzker ML, Naylor SL, Kucherlapati RS, Nelson DL, Weinstock GM, Sakaki Y, Fujiyama A, Hattori M, Yada T, Toyoda A, Itoh T, Kawagoe C, Watanabe H, Totoki Y, Taylor T, Weissenbach J, Heilig R, Saurin W, Artiguenave F, Brottier P, Bruls T, Pelletier E, Robert C, Wincker P, Smith DR, Doucette-Stamm L, Rubenfield M, Weinstock K, Lee HM, Dubois J, Rosenthal A, Platzer M, Nyakatura G, Taudien S, Rump A, Yang H, Yu J, Wang J, Huang G, Gu J, Hood L, Rowen L, Madan A, Qin S, Davis RW, Federspiel NA, Abola AP, Proctor MJ, Myers RM, Schmutz J, Dickson M, Grimwood J, Cox DR, Olson MV, Kaul R, Raymond C, Shimizu N, Kawasaki K, Minoshima S, Evans GA, Athanasiou M, Schultz R, Roe BA, Chen F, Pan H, Ramser J, Lehrach H, Reinhardt R, McCombie WR, de la Bastide M, Dedhia N, Blocker H, Hornischer K, Nordsiek G, Agarwala R, Aravind L, Bailey JA, Bateman A, Batzoglou S, Birney E, Bork P, Brown DG, Burge CB, Cerutti L, Chen HC, Church D, Clamp M, Copley RR, Doerks T, Eddy SR, Eichler EE, Furey TS, Galagan J, Gilbert JG, Harmon C, Hayashizaki Y, Haussler D, Hermjakob H, Hokamp K, Jang W, Johnson LS, Jones TA, Kasif S, Kaspryzk A, Kennedy S, Kent WJ, Kitts P, Koonin EV, Korf I, Kulp D, Lancet D, Lowe TM, McLysaght A, Mikkelsen T, Moran JV, Mulder N, Pollara VJ, Ponting CP, Schuler G, Schultz J, Slater G, Smit AF, Stupka E, Szustakowski J, Thierry-Mieg D, Thierry-Mieg J, Wagner L, Wallis J, Wheeler R, Williams A, Wolf YI, Wolfe KH, Yang SP, Yeh RF, Collins F, Guyer MS, Peterson J, Felsenfeld A, Wetterstrand KA, Patrinos A, Morgan MJ, de Jong P, Catanese JJ, Osoegawa

K, Shizuya H, Choi S, Chen YJ (2004) Finishing the euchromatic sequence of the human genome. Nature 431:931–945.

Jacobs JM, Adkins JN, Qian WJ, Liu T, Shen Y, Camp DG, 2nd, Smith RD (2005) Utilizing human blood plasma for proteomic biomarker discovery. J Proteome Res 4:1073–1085.

Jeffries N (2005) Algorithms for alignment of mass spectrometry proteomic data. Bioinformatics 21(4):3066–3073.

Jellum E, Thorsrud AK (1986) High resolution two-dimensional protein electrophoresis in clinical chemistry. Scand J Clin Lab Invest Suppl 184:71–76.

Jensen P, Magdaleno S, Lehman KM, Rice DS, Lavallie ER, Collins-Racie L, McCoy JM, Curran T (2004) A neurogenomics approach to gene expression analysis in the developing brain. Brain Res Mol Brain Res 132:116–127.

Jiang L, Lindpaintner K, Li HF, Gu NF, Langen H, He L, Fountoulakis M (2003) Proteomic analysis of the cerebrospinal fluid of patients with schizophrenia. Amino Acids 25:49–57.

Jimenez CR, Li KW, Dreisewerd K, Mansvelder HD, Brussaard AB, Reinhold BB, van der Schors RC, Karas M, Hillenkamp F, Burbach JP, Costello CE, Geraerts WP (1997) Pattern changes of pituitary peptides in rat after salt-loading as detected by means of direct, semiquantitative mass spectrometric profiling. Proc Natl Acad Sci USA 94:9481–9486.

Johnson KL, Mason CJ, Muddiman DC, Eckel JE (2004) Analysis of the low molecular weight fraction of serum by LC-dual ESI-FT-ICR mass spectrometry: Precision of retention time, mass, and ion abundance. Anal Chem 76:5097–5103.

Julka S, Regnier FE (2005) Recent advancements in differential proteomics based on stable isotope coding. Brief Funct Genomic Proteomic 4:158–177.

Kastenbauer S, Angele B, Sporer B, Pfister HW, Koedel U (2005) Patterns of protein expression in infectious meningitis: A cerebrospinal fluid protein array analysis. J Neuroimmunol 164(1–2):134–139.

Kazakevich YV, LoBrutto R, Vivilecchia R (2005) Reversed-phase high-performance liquid chromatography behavior of chaotropic counteranions. J Chromatogr A 1064:9–18.

Khan J, Saal LH, Bittner ML, Chen Y, Trent JM, Meltzer PS (1999) Expression profiling in cancer using cDNA microarrays. Electrophoresis 20:223–229.

Kim MG, Shin YB, Jung JM, Ro HS, Chung BH (2005) Enhanced sensitivity of surface plasmon resonance (SPR) immunoassays using a peroxidase-catalyzed precipitation reaction and its application to a protein microarray. J Immunol Methods 297:125–132.

Ku CJ, Yona G (2005) The distance-profile representation and its application to detection of distantly related protein families. BMC Bioinformatics 6:282.

Kusnezow W, Syagailo YV, Goychuk I, Hoheisel JD, Wild DG (2006) Antibody microarrays: The crucial impact of mass transport on assay kinetics and sensitivity. Expert Rev Mol Diagn 6:111–124.

Lee ML, Novotny M, Bartle KD (1976) Gas chromatography/mass spectrometric and nuclear magnetic resonance determination of polynuclear aromatic hydrocarbons in airborne particulates. Anal Chem 48:1566–1572.

Lescuyer P, Allard L, Zimmermann-Ivol CG, Burgess JA, Hughes-Frutiger S, Burkhard PR, Sanchez JC, Hochstrasser DF (2004) Identification of post-mortem cerebrospinal fluid proteins as potential biomarkers of ischemia and neurodegeneration. Proteomics 4:2234–2241.

Li L, Garden RW, Romanova EV, Sweedler JV (1999) In situ sequencing of peptides from biological tissues and single cells using MALDI-PSD/CID analysis. Anal Chem 71:5451–5458.

Maccarrone G, Milfay D, Birg I, Rosenhagen M, Holsboer F, Grimm R, Bailey J, Zolotarjova N, Turck CW (2004) Mining the human cerebrospinal fluid proteome by immunodepletion and shotgun mass spectrometry. Electrophoresis 25:2402–2412.

Marshall AG, Hendrickson CL, Jackson GS (1998) Fourier transform ion cyclotron resonance mass spectrometry: A primer. Mass Spectrom Rev 17:1–35.

McArthur JC, Haughey N, Gartner S, Conant K, Pardo C, Nath A, Sacktor N (2003) Human immunodeficiency virus-associated dementia: An evolving disease. J Neurovirol 9:205–221.

Meschia JF (2004) Clinically translated ischemic stroke genomics. Stroke 35:2735–2739.

Millea KM, Krull IS, Cohen SA, Gebler JC, Berger SJ (2006) Integration of multidimensional chromatographic protein separations with a combined "top-down" and "bottom-up" proteomic strategy. J Proteome Res 5:135–146.

Mitchell AE, Morin D, Lakritz J, Jones AD (1997) Quantitative profiling of tissue- and gender-related expression of glutathione S-transferase isoenzymes in the mouse. Biochem J 325 (Pt 1): 207–216.

Murray JF, Jr., Gordon GR, Gulledge CC, Peters JH (1975) Chromatographic-fluorometric analysis of antileprotic sulfones. J Chromatogr 107:67–72.

O'Farrell PH (1975) High resolution two-dimensional electrophoresis of proteins. J Biol Chem 250:4007–4021.

Ogata Y, Charlesworth MC, Muddiman DC (2005) Evaluation of protein depletion methods for the analysis of total-, phospho- and glycoproteins in lumbar cerebrospinal fluid. J Proteome Res 4:837–845.

Olsson A, Hoglund K, Sjogren M, Andreasen N, Minthon L, Lannfelt L, Buerger K, Moller HJ, Hampel H, Davidsson P, Blennow K (2003) Measurement of alpha- and beta-secretase cleaved amyloid precursor protein in cerebrospinal fluid from Alzheimer patients. Exp Neurol 183:74–80.

Omenn GS, States DJ, Adamski M, Blackwell TW, Menon R, Hermjakob H, Apweiler R, Haab BB, Simpson RJ, Eddes JS, Kapp EA, Moritz RL, Chan DW, Rai AJ, Admon A, Aebersold R, Eng J, Hancock WS, Hefta SA, Meyer H, Paik YK, Yoo JS, Ping P, Pounds J, Adkins J, Qian X, Wang R, Wasinger V, Wu CY, Zhao X, Zeng R, Archakov A, Tsugita A, Beer I, Pandey A, Pisano M, Andrews P, Tammen H, Speicher DW, Hanash SM (2005) Overview of the HUPO plasma proteome project: Results from the pilot phase with 35 collaborating laboratories and multiple analytical groups, generating a core dataset of 3020 proteins and a publicly-available database. Proteomics 5:3226–3245.

Pang RT, Poon TC, Chan KC, Lee NL, Chiu RW, Tong YK, Wong RM, Chim SS, Ngai SM, Sung JJ, Lo YM (2006) Serum Proteomic Fingerprints of Adult Patients with Severe Acute Respiratory Syndrome. Clin Chem 52(3):421–429.

Petricoin EF, Liotta LA (2003) Clinical applications of proteomics. [Review] [83 refs]. J Nutr 133:2476S–2484S.

Petrik V, Loosemore A, Howe FA, Bell BA, Papadopoulos MC (2006) OMICS and brain tumour biomarkers. Br J Neurosurg 20:275–280.

Poetz O, Schwenk JM, Kramer S, Stoll D, Templin MF, Joos TO (2005) Protein microarrays: Catching the proteome. Mech Ageing Dev 126:161–170.

Puchades M, Hansson SF, Nilsson CL, Andreasen N, Blennow K, Davidsson P (2003) Proteomic studies of potential cerebrospinal fluid protein markers for Alzheimer's disease. Brain Res Mol Brain Res 118:140–146.

Purohit S, Podolsky R, Schatz D, Muir A, Hopkins D, Huang YH, She JX (2006) Assessing the utility of SELDI-TOF and model averaging for serum proteomic biomarker discovery. Proteomics 6(24):6405–6415.

Qureshi ST, Skamene E, Malo D (1999) Comparative genomics and host resistance against infectious diseases. Emerg Infect Dis 5:36–47.

Ramstrom M, Hagman C, Mitchell JK, Derrick PJ, Hakansson P, Bergquist J (2005) Depletion of high-abundant proteins in body fluids prior to liquid chromatography fourier transform ion cyclotron resonance mass spectrometry. J Proteome Res 4:410–416.

Reimers M (2005) Statistical analysis of microarray data. Addict Biol 10:23–35.

Reimers M, Heilig M, Sommer WH (2005) Gene discovery in neuropharmacological and behavioral studies using Affymetrix microarray data. Methods 37:219–228.

Reyzer ML, Caldwell RL, Dugger TC, Forbes JT, Ritter CA, Guix M, Arteaga CL, Caprioli RM (2004) Early changes in protein expression detected by mass spectrometry predict tumor response to molecular therapeutics. Cancer Res 64:9093–9100.

Rohlff C, Southan C (2002) Proteomic approaches to central nervous system disorders. Curr Opin Mol Ther 4:251–258.

Ruetschi U, Zetterberg H, Podust VN, Gottfries J, Li S, Hviid Simonsen A, McGuire J, Karlsson M, Rymo L, Davies H, Minthon L, Blennow K (2005) Identification of CSF biomarkers for frontotemporal dementia using SELDI-TOF. Exp Neurol 196:273–281.

Sanchez JC, Guillaume E, Lescuyer P, Allard L, Carrette O, Scherl A, Burgess J, Corthals GL, Burkhard PR, Hochstrasser DF (2004) Cystatin C as a potential cerebrospinal fluid marker for the diagnosis of Creutzfeldt-Jakob disease. Proteomics 4:2229–2233.

Sandhu FA, Salim M, Zain SB (1991a) Expression of the human beta-amyloid protein of Alzheimer's disease specifically in the brains of transgenic mice. J Biol Chem 266:21331–21334.

Sandhu GS, Eckloff BW, Kline BC (1991b) Chemiluminescent substrates increase sensitivity of antigen detection in western blots. Biotechniques 11:14–16.

Sanna PP, King AR, van der Stap LD, Repunte-Canonigo V (2005) Gene profiling of laser-microdissected brain regions and subregions. Brain Res Brain Res Protoc 15:66–74.

Scott JK, Smith GP (1990) Searching for peptide ligands with an epitope library. Science 249:386–390.

Serot JM, Bene MC, Faure GC (2003) Choroid plexus, aging of the brain, and Alzheimer's disease. Front Biosci 8:s515–521.

Sheng S, Chen D, van Eyk JE (2006) Multidimensional liquid chromatography separation of intact proteins by chromatographic focusing and reversed phase of the human serum proteome: Optimization and protein database. Mol Cell Proteomics 5:26–34.

Shin YK, Lee HJ, Lee JS, Paik YK (2006) Proteomic analysis of mammalian basic proteins by liquid-based two-dimensional column chromatography. Proteomics 6:1143–1150.

Sikela JM (2006) The jewels of our genome: The search for the genomic changes underlying the evolutionarily unique capacities of the human brain. PLoS Genet 2:e80.

Silverberg GD, Mayo M, Saul T, Carvalho J, McGuire D (2004) Novel ventriculo-peritoneal shunt in Alzheimer's disease cerebrospinal fluid biomarkers. Expert Rev Neurother 4:97–107.

Smith GP, Scott JK (1993) Libraries of peptides and proteins displayed on filamentous phage. Methods Enzymol 217:228–257.

Soldi M, Sarto C, Valsecchi C, Magni F, Proserpio V, Ticozzi D, Mocarelli P (2005) Proteome profile of human urine with two-dimensional liquid phase fractionation. Proteomics 5(10): 2641–2647.

Soreghan BA, Lu BW, Thomas SN, Duff K, Rakhmatulin EA, Nikolskaya T, Chen T, Yang AJ (2005) Using proteomics and network analysis to elucidate the consequences of synaptic protein oxidation in a PS1 + AbetaPP mouse model of Alzheimer's disease. J Alzheimers Dis 8:227–241.

Stoeckli M, Staab D, Staufenbiel M, Wiederhold KH, Signor L (2002) Molecular imaging of amyloid beta peptides in mouse brain sections using mass spectrometry. Anal Biochem 311:33–39.

Sultana R, Poon HF, Cai J, Pierce WM, Merchant M, Klein JB, Markesbery WR, Butterfield DA (2006) Identification of nitrated proteins in Alzheimer's disease brain using a redox proteomics approach. Neurobiol Dis 22(1):76–87.

Templin MF, Stoll D, Schwenk JM, Potz O, Kramer S, Joos TO (2003) Protein microarrays: Promising tools for proteomic research. Proteomics 3:2155–2166.

Tornatore C, Nath A, Amemiya K, Major EO (1991) Persistent human immunodeficiency virus type 1 infection in human fetal glial cells reactivated by T-cell factor(s) or by the cytokines tumor necrosis factor alpha and interleukin-1 beta. J Virol 65:6094–6100.

Tsunoda A, Mitsuoka H, Bandai H, Endo T, Arai H, Sato K (2002) Intracranial cerebrospinal fluid measurement studies in suspected idiopathic normal pressure hydrocephalus, secondary normal pressure hydrocephalus, and brain atrophy. J Neurol Neurosurg Psychiatry 73:552–555.

Van Duin M, Woolson H, Mallinson D, Black D (2003) Genomics in target and drug discovery. Biochem Soc Trans 31:429–432.

Wang H, Qian WJ, Mottaz HM, Clauss TR, Anderson DJ, Moore RJ, Camp DG, 2nd, Khan AH, Sforza DM, Pallavicini M, Smith DJ, Smith RD (2005) Development and evaluation of a micro- and nanoscale proteomic sample preparation method. J Proteome Res 4:2397–2403.

Weinberger SR, Morris TS, Pawlak M (2000) Recent trends in protein biochip technology. Pharmacogenomics 1:395–416.

Wittke S, Mischak H, Walden M, Kolch W, Radler T, Wiedemann K (2005) Discovery of biomarkers in human urine and cerebrospinal fluid by capillary electrophoresis coupled to mass spectrometry: Towards new diagnostic and therapeutic approaches. Electrophoresis 26:1476–1487.

Wu HM, Jin M, Marsh CB (2005) Toward functional proteomics of alveolar macrophages. Am J Physiol Lung Cell Mol Physiol 288: L585–L595.

Yates CM, Butterworth J, Tennant MC, Gordon A (1990) Enzyme activities in relation to pH and lactate in postmortem brain in Alzheimer-type and other dementias. J Neurochem 55: 1624–1630.

Yuan X, Desiderio DM (2003) Proteomics analysis of phosphotyrosyl-proteins in human lumbar cerebrospinal fluid. J Proteome Res 2:476–487.

Yuan X, Desiderio DM (2005) Proteomics analysis of human cerebrospinal fluid. J Chromatogr B Analyt Technol Biomed Life Sci 815:179–189.
Zhang HT, Kacharmina JE, Miyashiro K, Greene MI, Eberwine J (2001) Protein quantification from complex protein mixtures using a proteomics methodology with single-cell resolution. Proc Natl Acad Sci USA 98:5497–5502.
Zheng J, Thylin MR, Persidsky Y, Williams CE, Cotter RL, Zink W, Ryan L, Ghorpade A, Lewis K, Gendelman HE (2001) HIV-1 infected immune competent mononuclear phagocytes influence the pathways to neuronal demise. Neurotox Res 3:461–484.
Zheng J, Thylin MR, Ghorpade A, Xiong H, Persidsky Y, Cotter R, Niemann D, Che M, Zeng YC, Gelbard HA, Shepard RB, Swartz JM, Gendelman HE (1999) Intracellular CXCR4 signaling, neuronal apoptosis and neuropathogenic mechanisms of HIV-1-associated dementia. J Neuroimmunol 98:185–200.

50
Neuroimaging

Michael D. Boska and Matthew L. White

Keywords Magnetic resonance imaging; Magnetic resonance spectroscopy; Neurodegeneration; Positron emission tomography; Single photon emission computed tomography

50.1. Introduction

A wide range of neuroimaging techniques have been developed which aid in clinical diagnosis of neurological diseases and are valuable tools for research into basic physiological mechanisms and effects of new therapies to combat neurological disease. This is a rapidly evolving field as new imaging techniques with enhanced sensitivity and specificity are constantly being developed and made available to the clinician and research community. This chapter will provide an introduction to the basic mechanisms, capabilities and applications of neuroimaging methods, allowing the neuroscience student to gain an appreciation of the range of available imaging techniques and the potential for use as serial non-invasive monitors of the progression and treatment of neurodegenerative disease in animals and humans.

Neuroimaging techniques include nuclear medicine studies such as positron emission tomography (PET) and single photon emission computed tomography (SPECT). Morphological imaging methods include magnetic resonance imaging (MRI) and computed tomography (CT). Studies of brain physiology and biochemistry include PET, SPECT, CT perfusion, MRI perfusion, magnetization transfer MRI (to assess myelin loss), ^{1}H and ^{31}P magnetic resonance spectroscopy (MRS) and magnetic resonance spectroscopic imaging (MRSI). These methods provide a broad array of techniques for serial non-invasive measurement of alterations in brain physiology and biochemistry, which are early and sensitive indicators of many neurological diseases. Although structural imaging techniques are not always sensitive indicators of neurological disease, functional and molecular imaging methods, including SPECT, PET and physiological MR techniques, have great value as they can reveal abnormalities before structural atrophy or focal CNS lesions are visible.

Neuroimaging techniques can be broadly classified by the information content of the images. Morphological information can be obtained from MRI and CT, sensitivity and specificity of MRI for detecting brain lesions is improved over CT (Brugieres et al., 1995) due to improved soft tissue contrast of MRI (Ell et al., 1987; Kim et al., 1996). In addition, MRI or CT can be combined with nuclear medicine techniques to provide anatomical detail allowing clear identification of the location of radioactive probes. Physiological information, including tissue perfusion mapping, mapping of receptor densities, and mapping of tissue biochemistry can be accomplished using MRI, MRSI, SPECT, and PET. Functional information, mapping of neuronal activity, can be obtained using functional MRI and PET, as well as other imaging modalities not covered in this chapter including electroencephalography (EEG), which records the electrical activity of the brain and magnetoencephalography (MEG) which provides spatially resolved maps of the electrical activity of the brain. This chapter will cover the basics of signal generation, signal acquisition and some examples of the information obtained in MRI, MRSI, SPECT and PET as an introduction to this broad and fascinating field.

50.2. Principles of Imaging

50.2.1. Basic Principles of MRI

MRI methodology is a complex and intriguing field of study. In this chapter, the basic mechanisms of MRI signal production and acquisition are introduced to allow the student of the neurosciences to develop a base understanding. Excellent textbooks are available for more advanced study of the details (Slichter, 1996; Haacke et al., 1999; Liang and Lauterbur, 2000; Gillard et al., 2005).

50.2.1.1. Signal Source

The signal in magnetic resonance imaging (MRI) and spectroscopy (MRS) rely on the magnetic properties of nuclei (Baier-Bitterlich et al., 1998). Nuclei with a non-zero *spin quantum number* ($I \neq 0$) will posses a *magnetic moment vector* μ. The magnetic moments are a result of the paramagnetic fields produced by the interaction of charged particles within the nucleus. The spin quantum number of the nucleus will be zero ($I = 0$) if both the mass number and the atomic number

T. Ikezu and H.E. Gendelman (eds.), *Neuroimmune Pharmacology.*
© Springer 2008

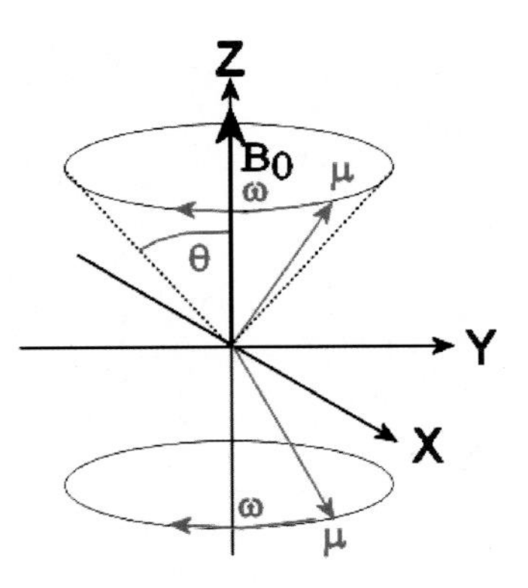

FIGURE 50.1. The precession of the nuclear magnetic moment vectors μ about the static magnetic field vector B_0. The angle of the precessional cone is given by θ. The precessional frequency is given by the Larmor frequency ω.

of the nucleus are even. These nuclei will not have a nuclear magnetic moment and hence are not observable using magnetic resonance techniques. Any nucleus with an odd atomic number or and odd mass number will have a non-zero spin quantum number ($I \neq 0$). These nuclei will have a nuclear magnetic moment and thus can produce a signal.

For nuclei with $I = 1/2$, including ^1H observed in MRI, the magnetic moment vector μ of the nuclei will either be parallel or anti-parallel to the externally applied magnetic field (Figure 50.1). The magnetic moment of a nucleus can be related to its angular moment by the magnetogyric ratio of that nucleus defined by (Harris, 1983; Krane, 1987):

$$\gamma = \frac{\mu}{I} \tag{1}$$

where μ is the nuclear magnetic moment vector of a nuclear isotope, and I is the angular momentum (spin) quantum number of a nucleus. The magnetogyric ratio is constant for each nuclear isotope. The sum of all magnetic moments is the macroscopic magnetic moment, named *the Net Magnetization Vector **M***, which is defined as:

$$M = \sum(\mu_+ + \mu_-) \tag{2}$$

where μ_+ and μ_- represent magnetic moments in the parallel and anti-parallel state for a two quantum states ($m_1 = \pm 1/2$) of a spin 1/2 magnetic nucleus ($I = 1/2$) such as the proton.

50.2.1.2. *Precession and the Larmor Equation*

The two states of different energy are not exactly parallel and anti-parallel to the external magnetic field. The magnetic moments are at an angle **θ** relative to the magnetic field due to the quantized energy states of the subatomic particles. The force of the magnetic field $\mathbf{B_0}$ on the magnetic moments **μ** forces these nuclear magnetic moments into a precessional motion (Figure 50.1). The frequency of precession, also called *Larmor* frequency, is proportional to the magnetic field according to the Larmor equation:

$$\omega = \gamma \text{ x } B_0 \tag{3}$$

where ω is the rotational frequency of the net magnetization produced by excitation of the nuclear magnetic moments (rad/s), and γ is the gyromagnetic ratio, constant for a given nucleus. $\mathbf{B_0}$ is the magnetic field strength at the nucleus measured in Tesla (T) or Gauss (G). 1 T = 10,000 G. As an example, the resonance frequency of protons is 64 MHz at 1.5 T.

The net magnetization vector **M** is the source of MR signal. The time evolution of the net magnetization vector **M** is the signal source in magnetic resonance. Motion of **M** will create an oscillating magnetic field. This oscillating magnetic field will create an induced voltage in a receiving coil. This induced voltage in the receiving coil over time is the signal.

50.2.1.3. *Resonance*

When a nucleus is exposed to a secondary magnetic field ($\mathbf{B_1}$) that has an oscillation identical in frequency and direction to its own Larmor precession, the nuclei absorb energy from the external source. The secondary magnetization is applied at the Larmor frequency, which is typically in the radiofrequency (RF) range, thus the brief application of the B_1 field is referred to as an RF pulse. Absorption of energy from the RF pulse results in a phase coherence in individual nuclei during the process of changing from the low energy to the high energy state. The net effect of $\mathbf{B_1}$ is a tipping of $\mathbf{M_0}$ (net equilibrium magnetization) away from the Z (longitudinal) axis (parallel to $\mathbf{B_0}$) towards the X,Y (transverse) plane. The force of $\mathbf{B_0}$ on the transverse component of **M** now rotates $\mathbf{M_{xy}}$ at the Larmor frequency. The rotating magnetization causes a voltage in a coil "tuned" to the Larmor frequency to pick up the signal by magnetic induction (Figure 50.2). The resulting angle between the Z axis and **M** (typically 90°) is called *flip* angle and depends upon the amplitude and duration of the RF pulse.

50.2.1.4. *Signal Detection and Time Evolution*

Faraday's law of induction states that if a receiver coil is magnetically coupled to an oscillating magnetic field, a voltage is induced in the receiver coil. Therefore, as **M** precesses at the Larmor frequency in the transverse plane, a voltage is induced in the coil. This voltage constitutes the *MR signal*. The magnitude of the signal depends on the magnitude of **M** in the transverse plane. The amount of magnetization in the longitudinal plane gradually increases (T_1 relaxation). At the same time, but independently, the net magnetization in the transverse plane gradually decreases due to loss of phase coherence (T_2 relaxation). As the magnitude of transverse magnetization decreases, so does the magnitude of the voltage induced in the receiver coil (Figure 50.2). The evolution and decay of the net magnetization, **M** is picked up by the receiver coil resulting in the magnetic resonance signal, the *free induction decay (FID)*.

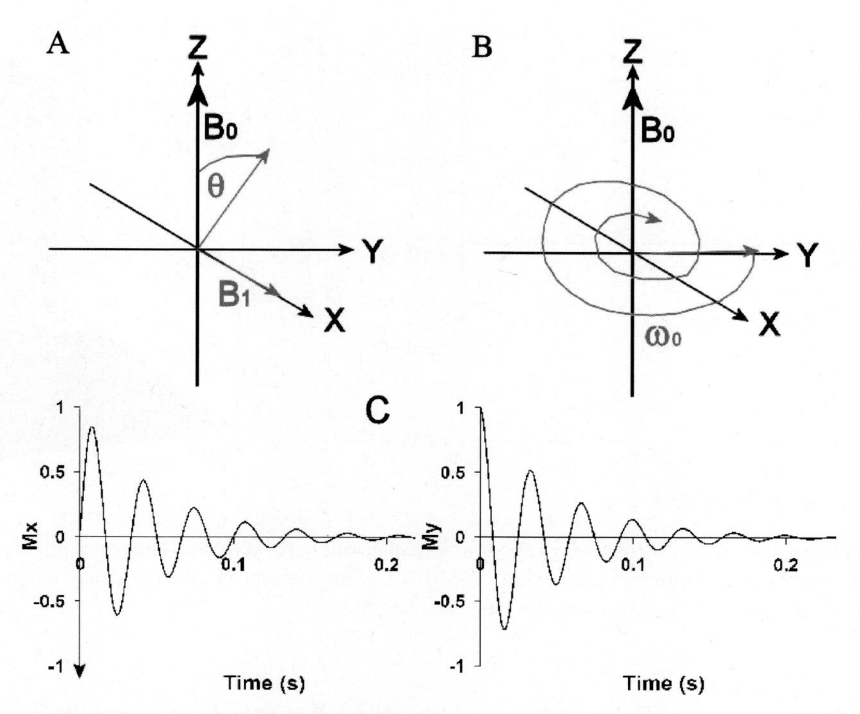

FIGURE 50.2. Application of a secondary magnetic field (B_1) rotating at the Larmor frequency has the effect of tipping the net magnetization (M) away from the equilibrium position (M_0) along the longitudinal axis and towards the transverse plane. After the brief application of B_1, rotation of M in the transverse (X-Y) plane at the Larmor frequency (B) induces a Free Induction Decay (FID) signal in a pickup coil (C).

When the RF pulse is switched off, **M** realigns with the Z axis at a rate ($1/T_1$) slower than the rate of signal loss ($1/T_2$). In order to do so, **M** must transfer the energy gained to the surrounding molecules. This process is referred to as T_1 *relaxation* (spin-lattice relaxation). As T_1 relaxation occurs, **M** returns to *equilibrium* (**M_0**).

50.2.1.5. *Magnetic Field Gradient Used to Generate MRI*

The magnetic field dependence of the signal's frequency is used as a basis for spatial encoding of the magnetic resonance signal. This is done using magnetic field gradients. Magnetic field gradients cause a linear shift in magnetic field with position. This allows one to use a single chemical species, such as water, as a signal source for imaging (MRI). Sophisticated spatial encoding schemes have been developed to allow multi dimensional spatial encoding.

The Larmor equation demonstrates the basis of these encoding schemes. As the magnetic field changes, so does the frequency. In one dimension, application of a magnetic field gradient of 10 gauss/cm [1 Tesla (T) = 10,000 gauss (g)] produces a field shift in the water frequency of 64 MHz (1.5 T) of 42667 Hz/cm. The basis of the use of the magnetic field gradient for spatial encoding is displayed in Figure 50.3.

50.2.1.6. *Fourier Transform: Frequency Analysis of the Time Domain MR Signal*

The time domain MR signal is difficult to visually interpret; hence, the signal is transformed into a spectrum, which is intensity on the vertical axis and frequency on the horizontal axis. This is performed mathematically by the Fourier transform. The Fourier transform is performed according to the equation:

$$f(\omega) = \int_{-\infty}^{\infty} f(t)e^{-i\omega t}\,\delta t \qquad (4)$$

where ω = frequency (radians/s), t = time (s), $i = \sqrt{-1}$, $f(\omega)$ is the frequency domain signal (spectrum) and $f(t)$ is the time domain signal (MR signal, free induction decay or spin echo).

50.2.1.7. *Spin Echo*

Signal loss in the free induction decay (FID) can be partially refocused using a second application of B_1 with a duration and amplitude that produces an 180° flip angle of **M** (180° pulse). This has the effect of placing individual magnetic moments, which are no longer in phase with the central frequency, in a position to realign with spins at the central frequency.

Pulsed magnetic resonance experiments are represented by parallel time lines of the applied pulses and the signal (Figure 50.4A). We will introduce this formalism here to explain the spin-echo and, in the next section, the spin-echo MRI pulse sequence. The first line is the time line of the applied B_1 pulses, also referred to as radiofrequency, or RF pulses because they are applied at the Larmor frequency, which are typically in the radiofrequency range.

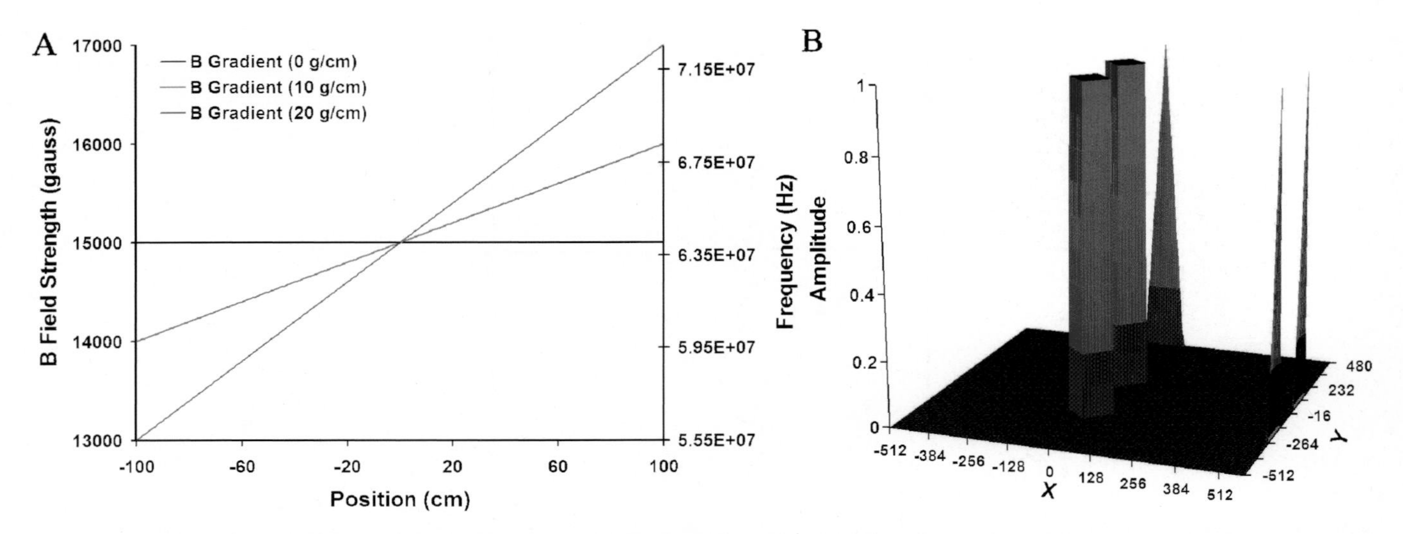

FIGURE 50.3. Magnetic field gradients used for frequency encoding spatial position. A: Magnetic field strength as a function of position in the magnet for three gradient strengths. B: Arbitrary object within the MRI system (center) and the frequency readout (signal) from a gradient applied during signal acquisition along the x-axis (rear) and the y-axis (right).

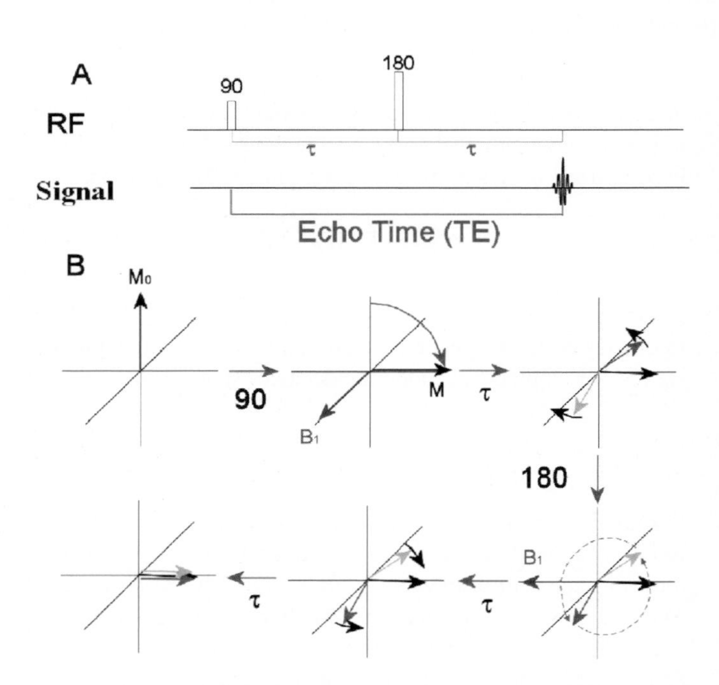

FIGURE 50.4. Time evolution of the spin-echo signal. A: Pulse sequence diagram showing (top) the timing of the applied 90 and 180° RF pulses (B_1) and (bottom) the time evolution of the spin-echo signal. B: Vector diagram of the time evolution of the spin-echo signal. Clockwise from top-left: Equilibrium magnetization is rotated by the 90° RF pulse, signal evolves and individual spins lose phase coherence, reducing signal intensity, a 180° RF pulse is applied which places out of phase magnetic moments in position to rephase at a time (τ) equal to the time of dephasing before the 180° pulse, allowing the refocusing of the signal (spin-echo).

50.2.1.8. Spin-Echo Magnetic Resonance Imaging (MRI)

Three dimensions of spatial encoding are required for MRI. These three dimensions are independently applied by (1) signal production from a thin slice by using a frequency selective excitation in the presence of a magnetic field gradient (slice selection), (2) acquiring the spin-echo signal in the presence of a magnetic field gradient (frequency or read-out gradient) and acquiring the spin-echo signal after producing a spatially dependent phase shift which is incremented in the second dimension (phase encoding gradient). The pulse sequence diagram showing the application of the frequency selective RF pulses, the slice, phase and frequency encoding gradients and the spin-echo signal acquired are shown in Figure 50.5. The spin-echo acquisition is repeated for each value of phase encoding gradients to fill in the second dimension of the single slice MRI, requiring repetition of the sequence 256 times for 256 × 256 image resolution. The raw MRI is acquired in the time domain, which requires Fourier transform of the signal into the frequency domain for display (Figure 50.6).

50.2.2. Modification of Signal Intensity

The power and versatility of magnetic resonance techniques lie in the ability to modify signal intensity based on the biophysical properties of the spins of interest (primarily tissue water). Signals can be "weighted" (signal amplitude altered) based on the magnetic relaxation properties (T_1 and T_2), water diffusion (random molecular motion), or the interaction of the hydration layers of membranes and proteins with free water (magnetization transfer). Tissue perfusion can be mapped by the introduction of contrast agents or inversion of signal from water in arterial blood supply of the tissue. In addition, signal can be suppressed from tissue based on frequency (fat or water suppression), position (suppressing motion artifacts from unwanted tissue), or preferentially viewed based on moving vs static water (angiography or suppression of signal from flowing blood).

50.2.2.1. Magnetic Relaxation, T1 and T2

50.2.2.1.1. Spin-Lattice Relaxation (T_1)

T_1 relaxation is the process of releasing the energy absorbed by the magnetic nuclei during the application of the RF pulse. This process occurs by releasing the energy to the surrounding molecules (lattice) returning the spin system to the equilibrium condition (M_0). The rate at which this process occurs in tissue depends on the molecular environment and varies with tissue type, providing a means of generating contrast between tissues with otherwise similar signal intensities (water content). T_1 relaxation occurs exponentially after an RF pulse with a time constant of T_1 according to the equation:

$$Mz\,(t) = M_0(1-e^{-t/T_1}) \quad (5)$$

where: $M_z(t)$ is the time dependent amplitude of the Z component of the magnetization (along B_0), which approaches the equilibrium magnetization (M_0) over time (t) with the time constant T_1. Since MRI acquisition typically requires multiple excitations, adjusting the repetition time (TR) between successive excitations reduces the net magnetization according to TR/T_1 and alters the T_1 based contrast. As a result, signal intensity within each volume element in the MR image (voxel) has intensity modified by the average T_1 of tissue within the voxel according to the repetition time of the acquired image. This occurs as illustrated in Figure 50.7A. Figure 50.7A shows the Z-magnetization recovery curves for two different tissues, one with a T_1 of 400 ms and one with a T_1 of 900 ms. In order to recover full magnetization, a delay (TR) of $5xT_1$ is required. Using a short TR (500 ms, dashed lines) the tissue with the 400 ms T_1 has recovered to 71% of M_0 while the tissue with the 900 ms T_1 has only recovered to 43% of M_0. The signal intensity will be proportional to the recovery of the net magnetization after the second 90° RF pulse, providing a signal that is 65% more intense for the shorter T_1 tissue even if the original signal intensity (M_0, proportional to water content) is identical, providing improved tissue contrast and a mechanism for detecting altered cell content (pathology) within tissue.

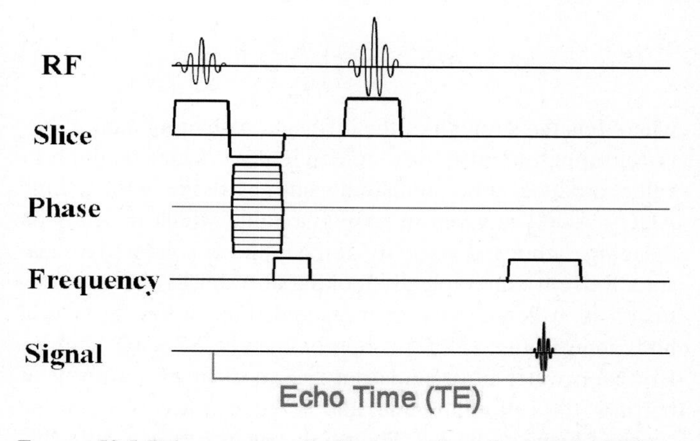

FIGURE 50.5. Pulse sequence for the generation of spin-echo MRI

50.2.2.1.2. Spin-Spin Relaxation (T_2)

The spin echo (Section 50.2.1.6) demonstrates that the 180° RF refocusing pulse reverses off-resonance effects (spins at a slightly different frequency than the Larmor frequency) after the 90° RF pulse allowing a signal to be generated at a time after the original signal loss. However, the natural process of energy exchange between adjacent molecules causes a loss of signal coherence by exchange of spin states with loss of phase coherence. This loss of signal cannot be refocused. The exchange of spin states with no net loss of energy in the spin system is referred to as spin-spin relaxation (T_2). Signal loss as a function of echo time (or TE) is given by:

$$M = M_0 e^{-TE/T_2} \quad (6)$$

where TE = echo time and T_2 is the average T_2 of the tissue. As a result, signal intensity within each volume element in the MR image (voxel) has intensity reduced by the average T_2 of tissue within the voxel according to the echo time of the acquired image. T_2 effects on the signal intensity in two different

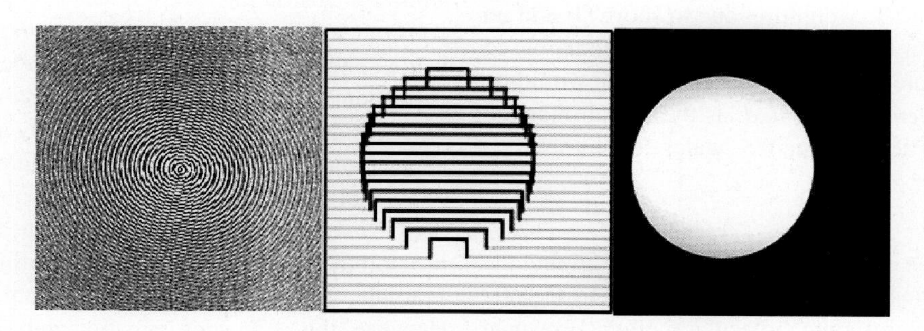

FIGURE 50.6. MRI signal, two dimensional Fourier transform and grey scale representation of the Fourier transformed data. Left: grey scale representation of the single slice MRI of a spherical phantom filled with water. Center: spectral representation of the Fourier transformed data. Right: image represented in gray scale, the standard viewing method.

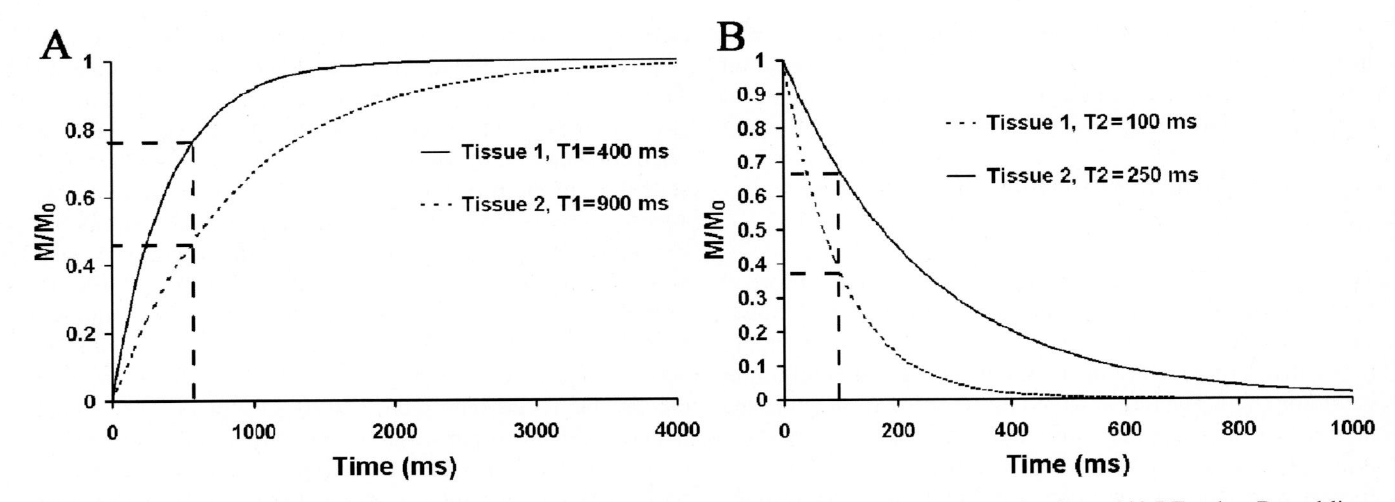

FIGURE 50.7. T_1 and T_2 relaxation curves. A: Recovery of the net magnetization along the Z (B_0) axis after a 90° RF pulse. Dotted lines demonstrate the amount of Z magnetization after a 500 ms delay (TR), proportional to the signal intensity after the second 90° RF pulse. This demonstrates the contrast available in tissue of similar water content (M_0) with different T_1 by reducing TR. B: decay of signal in the transverse plane (X-Y) with echo time (TE). Dotted lines demonstrate the amount of longitudinal (X-Y) magnetization (signal) with a 100 ms spin echo (TE), proportional to the signal intensity during the acquisition of a 100 ms echo. This demonstrates the contrast available in tissue of similar water content (M_0) with different T_2 by increasing TE.

tissues with the same water content (M_0) at an echo time of 100 ms is shown by the dashed lines in Figure 50.7B. Tissue with shorter T_2 (100 ms) refocuses 37% while the longer T_2 tissue (250 ms) refocuses 67% of the original signal, demonstrating significant contrast between tissues of different T_2 using a long echo time acquisition. This demonstrates the contrast available in tissue of similar water content (M_0) with different T_2 by increasing TE. It should also be noted that the signal intensity will be modified at the same time by TR and T_1 through the choice of TR, thus a long TR is typically chosen for purely T_2 weighted images. An example of T_2 weighted imaging and calculation of a T_2 map from multiple TE images from mouse brain can be seen in Figure 50.8.

Multiple echoes may be refocused from one excitation by application of multiple 180°- RF pulses, allowing the acquisition of a series of echoes within one TR. For example, if one echo is acquired at a very early echo time and a second echo is acquired at 100 ms, two images can be acquired. If a long TR is used so that the second echo is purely T_2 weighted, then the first echo with almost no T_2 weighting due to short TE and no T_1 weighting because of the long TR will have contrast based purely on the tissue water content (M_0). This image would be referred to as *proton density weighted*, as the signal intensity in each voxel is proportional to the free water content.

50.2.2.2. Diffusion

Symmetrical application of magnetic field gradients around the refocusing pulse in a spin echo will refocus the signal from static molecules. However, microscopic molecular motion (Brownian motion) will cause dephasing of individual magnetic moments to a degree which is dependent on the freedom of the molecular motion of the water in cells. Freedom of the molecular motion is affected by the size and shape of the cell and is modified in different cell types. Signal intensity is modified by:

$$ADC_w = \log\left(\frac{S_0}{S}\right) \times b \qquad (7)$$

where b is the strength of the diffusion weighting, and ADC_ω is the apparent diffusion coefficient of water. A high b-value produces low signal intensity in tissue with a high ADC_ω. ADC_ω is given in mm²/s, a value which depends on the temperature and viscosity of the liquid or solid under measure. Since we are typically looking at water in MRI, the diffusion of water is what is measured. Free water, at typical body temperature (38°C) is approximately 2.2×10^{-3} mm²/s. However, water in tissue is not free to diffuse as a result of barriers, the cell membrane, and is reduced according to the size and shape of the cell. Shorter distances between cell walls result in a lower diffusion rate. Areas of free water in tissue, such as cerebral spinal fluid or large arteries and veins, show similar diffusion rates to free water. Tissue diffusion values in cells range from 0.6 to 1.0×10^{-3} mm²/s. In addition, if the cell is not spherical, which is typically the case in brain, ADC_ω is dependent on the direction of diffusion weighting.

Diffusion can be measured separately in various directions by application of diffusion gradients in different directions during individual measurements. In order to obtain an approximation of the average diffusion of water in an asymmetrical cell, measurement of diffusion in three orthogonal directions (for example, along X, Y, and Z directions) are obtained and averaged. However, this is an approximation and a true value can only be obtained by determination of the ***diffusion tensor***. This requires a minimum of six individual measurements. This can be understood once the student is familiar with the definition of a tensor.

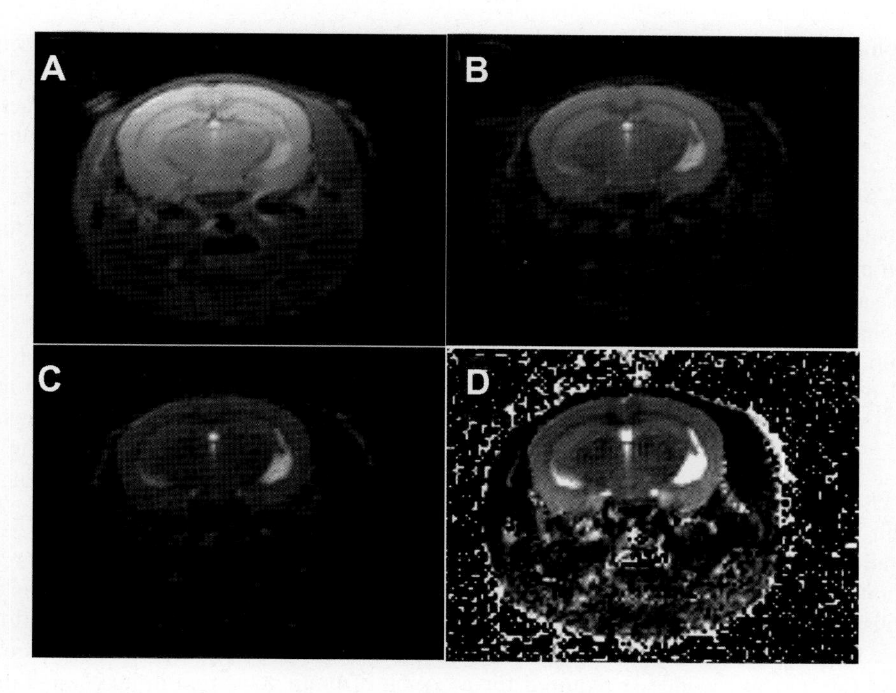

FIGURE 50.8. T_2 mapping data set, 3/8 T_2 weighted images with (A) TE = 15 ms, (B) TE = 60 ms, (C) TE = 120 ms and (D) T_2 map created by fitting eight echo imaging sequence to Eq. 6.

A tensor is a mathematical representation of a three dimensional shape using a three by three matrix (Figure 50.9, Eq. 8). The matrix represents the total magnitude of the object in each of nine directions, three of which are equal due to symmetry (e.g. Dxy = Dyx) of the other three cross components, reducing the problem to six dimensions. Hence, six directions of diffusion measurement are the minimum needed to define the diffusion tensor. More measures can be used to improve the accuracy of the measurement through cross correlation of multiple measures.

$$\vec{D} = \begin{pmatrix} D_{xx} & D_{xy} & D_{xz} \\ D_{yx} & D_{yy} & D_{yz} \\ D_{zx} & D_{zy} & D_{zz} \end{pmatrix} \quad (8)$$

The diffusion tensor can than be *diagonalized* to generate a diffusion tensor of the form:

$$\vec{D} = \begin{bmatrix} \lambda_1 & 0 & 0 \\ 0 & \lambda_2 & 0 \\ 0 & 0 & \lambda_3 \end{bmatrix} \quad (9)$$

to determine the primary *eigenvalues* (λ_1, λ_2, λ_3) as designated in Figure 50.9. These data can are used to generate pixel by pixel maps of mean diffusivity [$D_{av} = 1/3 \times (\lambda_1 + \lambda_2 + \lambda_3)$] and fractional anisotropy (FA), which is given by the equation:

$$FA = \frac{1}{\sqrt{2}} \sqrt{\frac{(\lambda_1 - \lambda_2)^2 + (\lambda_2 - \lambda_3)^2 + (\lambda_1 - \lambda_3)^2}{\lambda_1^2 + \lambda_2^2 + \lambda_3^2}} \quad (10)$$

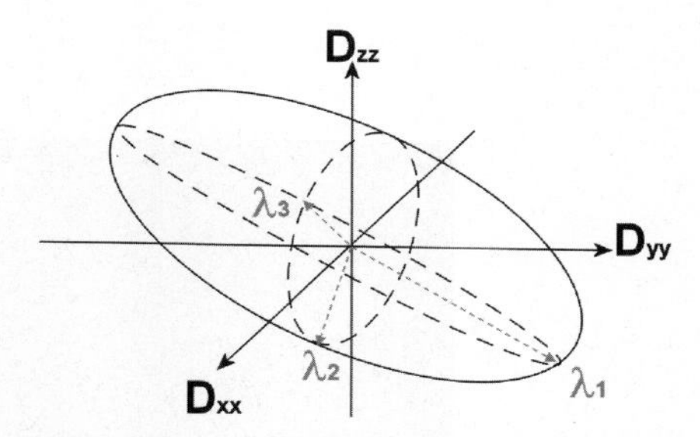

FIGURE 50.9. Example of a non-spherical cell with directional dependence of diffusion. Primary laboratory frame of reference is given by x, y and z axes, while the primary axes of diffusion are given as λ_{11} (major), λ_{22} (intermediate) and λ_{33} (minor). Determination of the values of elements of the diffusion tensor allows determination of the magnitude and direction of the primary axes of diffusion (eigenvalues)

FA ranges from zero for a spherical cell to one for an infinitely long cell with zero cross sectional area.

Determination of the principle eigenvalues (λ_1, λ_2, λ_3) also allows determination of the principle direction of the cell orientation, which is the orientation of λ_1 in the laboratory (X, Y, Z) frame. Principle direction of the cell can be used to determine fiber orientation in white matter tracts and provide additional delineation of anatomical structure. A map of λ_1 orientation is produced using a red-green-blue encoding in each pixel to represent the direction of the primary eigenvalue, with red representing L-R orientation, blue representing inferior-superior (foot to head) direction and green representing the anterior-posterior direction. Intermediate angles are represented by linear combinations of the three primary colors. An example of DTI from a mouse brain showing the D_{av}, FA, and RGB encoded λ_1 direction maps are displayed in Figure 50.10.

50.2.2.3. *Magnetization Transfer*

T_2 of protons on molecules are dependent on molecular motion. Large molecules such as membranes and proteins with slow molecular motion have a very short T_2 (<1 ms). As a result, the linewidth of these molecules is very broad compared to free water. The relationship between linewidth (full width at half maximum) and the T_2 of molecules is given by:

$$\Delta \nu = \frac{1}{\pi \times T_2} \quad (11)$$

where $\Delta\nu$ = linewidth (full width at half maximum). As a result, selective saturation of the protons on large molecules and membranes without affecting the magnetization of free water can be accomplished by a narrow bandwidth RF pulse 4–20 kHz off resonance from the free water signal. Application of the saturation pulse for several seconds is required for a steady state to develop between the saturated macromolecules and exchange of the free water with the hydration layer of the macromolecules, developing a partial saturation (signal loss) in the free water. Signal loss by this mechanism is proportional to the exchange rate of water in the hydration layer of the membranes and macromolecules (Figure 50.11). Determination of the degree of magnetic saturation is done using two images, one without the saturation pulse and one with the saturation pulse. The two images are then divided to determine the spatial distribution of the magnetization transfer ratio (MTR) according to the equation:

$$MTR = \frac{M_0 - M_{ss}}{M_0} \quad (12)$$

where M_{ss} = signal intensity in the presence of steady state saturation and M_0 = signal intensity in the absence of the off-resonance macromolecule saturation. A more complete description of the magnetization transfer effect can be found in a review article by Henkelman (Henkelman et al., 2001).

50.2.2.4. *Perfusion*

Cerebral perfusion is a marker for metabolic activity of the brain with normal hemodynamic function. Cerebral perfusion has been used extensively for detection of disease and mapping of brain activation in health and disease. MRI methods for mapping cerebral perfusion include (1) qualitative bolus tracking after injection of magnetic contrast agents and (2) quantitative arterial spin-tagging. Each method is described along with the relative strengths and weaknesses, below.

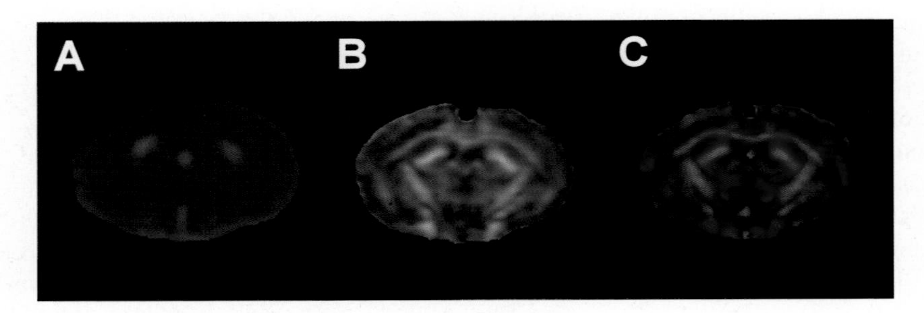

FIGURE 50.10. Diffusion tensor imaging results from a mouse brain imaging study at 7 T. (A) mean diffusisivity (D_{av}) (B) Fractional Anisotropy (FA), (C) Color coded FA orientation map.

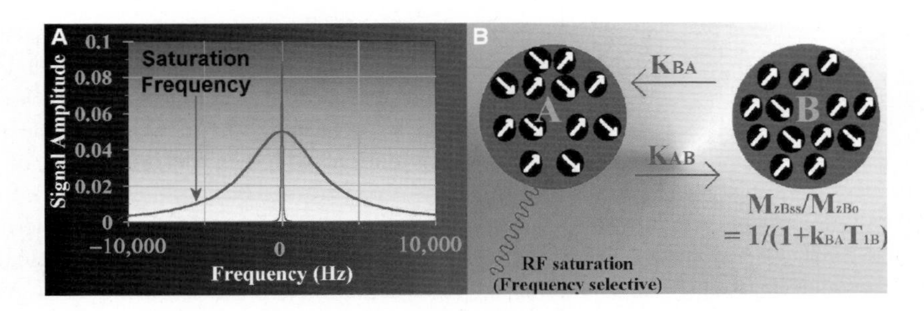

FIGURE 50.11. Magnetization transfer MRI. A: Off-resonance application of a saturation pulse will saturate the signal from macromolecules with short T2 (broad line) while leaving the narrow water resonance (center line) uneffected. B: Saturation of macromolecules (spin pool A) causes reduction of the free water signal (spin pool B) measured as reduction in intensity of MRI (M_{zBss}/M_{zB0}) inversely proportional to the exchange rate of free water with the water in the macromolecular hydration layer (k_{BA}) times the T_1 of free water (T_{1B}).

50.2.2.4.1. Bolus Tracking Perfusion MRI

Injection of a bolus (typically in 5s or less) of magnetic contrast agent causes signal intensity changes in tissue as the bolus passes through the tissue. Magnetic contrast agents used are typically chelated gadolinium (e.g. Gd-DTPA) complexes, which reduce T_1 and T_2* of blood and tissue, increasing signal in T_1 and reducing signal in T_2* weighted images. Advanced rapid magnetic resonance imaging techniques, such as echo planar imaging (EPI) are capable of acquiring multiple slices covering the human brain in 1–3s, allowing the dynamic tracking of the spatial distribution of the injected contrast agent. Signal intensity versus time is then analyzed for each voxel (volume element) in each acquired image to determine (1) time to peak contrast (mean transit time, MTT) and (2) area under the bolus passage signal, proportional to cerebral blood volume (CBV). By mass analysis, cerebral blood flow (CBF) can then be determined by the relation:

$$MTT = \frac{CBF}{CBV} \tag{13}$$

Examples of the signal change over time after bolus injection is shown in Figure 50.12 for a representative normal flow area (MTT_1) and an area with compromised cerebral perfusion (MTT_2). While this method is robust, it suffers from a number of issues related to absolute quantitation of the CBF including, but not limited to partial volume effects, nonlinear relationship between signal intensity change and contrast agent concentration, and determination of the input function of the bolus required for accurate quantitation. In clinical practice, MTT maps are the preferred method for visualizing areas of reduced CBF.

50.2.2.4.2. Arterial Spin-Tagged Perfusion MRI

Quantitation of the CBF can be an important factor in studies of therapeutic intervention. The method of arterial spin-labeled (ASL) perfusion MRI allows quantitative determination of the perfusion rate in units of ml/100 g tissue/min. However, the technique requires exceptional instrumental stability, as the result depends on accurate determination of the difference between two images on the order of a few percent. With the appropriate acquisition parameters to achieve sufficient signal to noise ratio (>100) and current instrument stability, this technique is feasible, especially in high field clinical systems. Currently, arterial spin-tagged perfusion measures are not widely used in clinical settings.

Arterial spin-tagged perfusion MRI is performed by inverting the water protons in the blood of the arteries supplying blood to the tissue of interest. Typically, in brain, the carotid arteries are used as the supply. The original description of the method (Detre et al., 1992; Williams et al., 1992) uses a continuous narrow band RF inversion of the protons in the arterial blood, outside of the imaged slice, before and after excitation to cause a steady-state loss of signal in cerebral tissue. The signal loss is related to tissue perfusion(Detre et al., 1992) by:

$$f = \frac{\lambda}{T_{1app}}\left(1 - \frac{M_{ss}}{M_0}\right) \tag{14}$$

where:

$$\frac{1}{T_{1app}} = \frac{1}{T_1} + \frac{f}{\lambda} \tag{15}$$

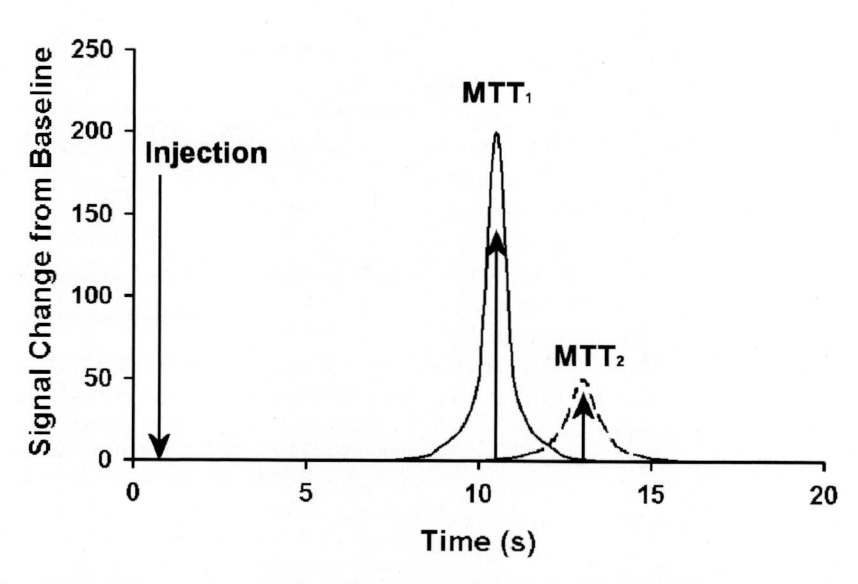

FIGURE 50.12. Simulated examples of the time course of signal changes after Gd-bolus administration. Voxelwise analysis of the time series of images is used to create maps of mean transit time (MTT), cerebral blood volume (CBV), corresponding to the area under the curve, and cerebral blood flow (CBF) according to Eq. 13. The delayed curve (MTT_2) corresponds to an area of compromised CBF.

And: f = flow rate, λ = water extraction fraction of tissue from blood (0.8–1.0), M is the MRI signal intensity with (M_{ss} = steady state) and without (M_0) inversion of the protons in the water of the inflowing blood. One potential complication with this method is the effect of magnetization transfer, which occurs with the continuous inversion pulse. This can be balanced by reversing the gradient so that the spatial location of the inversion is above the head when acquiring the M_0 image. This effect is reduced, but not eliminated using more recent pulsed arterial spin labeling techniques(Kim, 1995; Kwong et al., 1995).

50.2.2.5. Relaxometry, Quantitative DWI and Quantitative MT

Perfusion is not the only quantitative method available using magnetic resonance techniques. T_1, T_2, apparent diffusion coefficient of water (ADC_ω), and MT effects can be quantitatively mapped using the appropriate series of measurements. For example, acquisition of MRI data collected in a series of echo times can be processed to create an image, which is a quantitative map of T_2 (Figure 50.8). Each effect requires its own set of measures, but in principle all methods for modifying signal intensity in MR can be quantified. With the exception of DWI, quantitative mapping is not typically done in a clinical setting due to the time required for additional measures and additional processing required for these data. Advances in MRI are increasing the speed of acquisition, which may allow quantitative techniques to precisely define alterations in neuronal cell, physiology and metabolism as research into multidimensional quantitative MRI defines the sensitivity and specificity of combining these methods to diagnose neurological disease.

50.2.3. Magnetic Resonance Spectroscopy and Spectroscopic Imaging

Prior to the development of MRI in the 1970s magnetic resonance was a well-developed technique used for chemical analysis by spectroscopic analysis of chemicals in sample preparations. This capability has been extended to image guided localized magnetic resonance spectroscopy of tissue in-vivo. Localized spectroscopy uses frequency selective pulses in the presence of a gradient for volume selective excitation and spectroscopic imaging uses phase encoding gradients for spatial encoding of the spectroscopic signals. Many sophisticated pulse sequences have been developed for in-vivo localized spectroscopy and are in use in research and clinical MRI systems worldwide. The difficulty of in-vivo spectroscopy is primarily low spatial resolution due to the low signal intensity. As a result long acquisition times are required to overcome the low signal intensity by signal averaging. Despite this difficulty, in-vivo spectroscopy is unique in the ability to perform non-invasive analysis of chemical composition of tissue without biopsy and allows the serial non-invasive determination of tissue chemistry with treatment. While any nucleus with a magnetic moment can be studied using MRS and/or MRSI, here we will focus on the most frequently used nucleus for study of brain, the proton (^{1}H).

Frequency shifts of nucleii on different molecules (spectroscopy) are caused by the electron cloud of a molecule. Electron cloud density, and hence field shifts due to the electron cloud, are specific to each molecule. According to the Larmor equation, this results in a frequency shift of the ^{1}H signal for each proton on each molecule, developing a specific "frequency fingerprint" for each molecule. The areas of the peaks are proportional to the concentration of each molecule in the region sampled. These areas are analyzed by fitting the peak areas with one of a variety of spectroscopic analysis software packages. An example of the spectrum from a region of mouse brain acquired at 7 Tesla is shown in Figure 50.13.

^{1}H MRS applies frequency selective pulses in order to generate signal from a single volume of interest defined using MRI. The signal is acquired in the absence of a magnetic field gradient in order to detect the minor frequency shifts due to

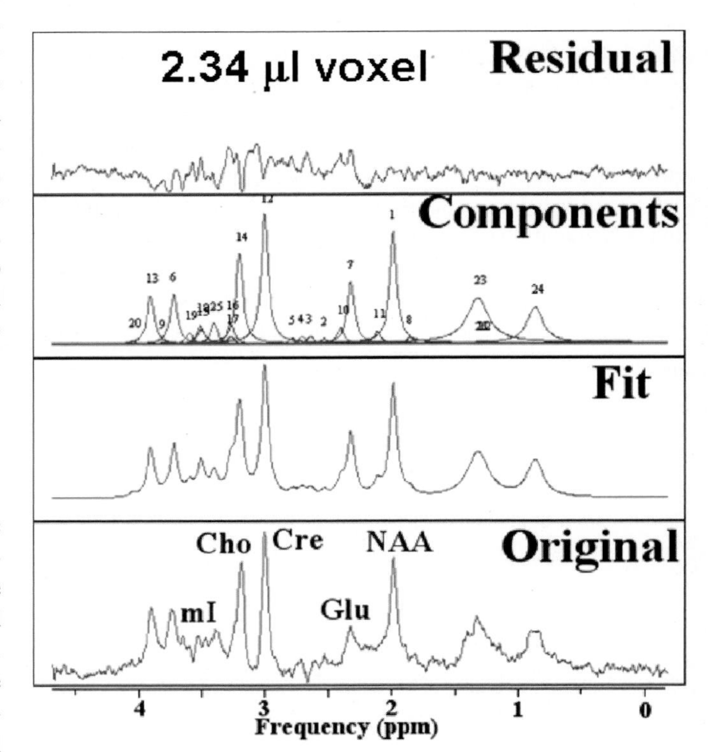

FIGURE 50.13. Example of a 1H MR spectrum from a 1H MRSI data set with a nominal spatial resolution of 2.34 µl (1.5 × 1.25 × 1.25 mm) in mouse brain. Original: Spectrum obtained from the region. Fit: Overall fit using models of known spectra of metabolites including, but not limited to N-acetyl aspartate (NAA), glutamate (Glu), glutamine, creatine (Cre), choline (Cho) and myoinostitol (mI). Components: Individual peaks contributing to the overall spectra of metabolites and Residual: Subtraction of the fit from the original spectrum. The frequency scale is given in ppm [(frequency/Larmor frequency) × 10^6] to create a frequency scale insensitive to the magnetic field strength.

the electron clouds around atoms (chemical shift).^{1}H MRSI uses either a slice selective excitation or volume selective excitation with spatial encoding over the region using phase encoding gradients. This produces an array of volume selective spectra over the region selected. This technique provides improved spatial resolution and the ability to reformat data post-acquisition to allow precise localization of spectra within the slice or volume selected. These advantages come at the expense of acquisition time (typically 15–30 min in human clinical systems).

Several technical challenges need to be overcome for successful application of ^{1}H MRS/MRSI. The range of metabolite frequencies are relatively narrow, requiring excellent B_0 homogeneity over the region from which data are obtained. This requirement is significantly more difficult to achieve in MRSI (collecting multiple spatially resolved spectra simultaneously over large brain regions) than localized MRS (single spectrum from a small region in the brain). While MRSI is in general used in research laboratories, localized ^{1}H MRS is most frequently used in clinical applications due to speed of acquisition and reduced technical challenges.^{1}H MRS/MRSI requires the suppression of water, as the metabolite concentrations are in the millimolar (mM) concentration range while pure water is 55 M, each water molecule having two equivalent protons, produces a signal intensity of 110 M, five orders of magnitude greater than the signals from the metabolites. Water suppression requires excellent system stability and the elimination of "stimulated echoes" from residual water after application of the suppression technique(s).

Most MRS/MRSI studies of brain use creatine as an internal concentration standard, as creatine has a long T_2 and has relatively constant concentration in most cell types. Advanced methods exist to allow quantitation of the metabolites measured using MRS/MRSI. These methods include the referencing of signal to the internal water signal and referencing in-vivo signal amplitude to signal amplitudes from chemical samples of known concentration. Quantitation requires the correction of signal amplitude for T_1 and T_2 of the metabolites for each acquisition technique. Ideal acquisition parameters for quantitative measurements employ minimum echo time (TE) and a long repetition time (TR) to minimize magnetic relaxation corrections and reduce the chance that changes in relaxation times with pathology could effect the determination of metabolite concentrations.

50.2.4. Computed Tomography Methods, X-Ray Computed Tomography, Single Photon Emission Computed Tomography and Positron Emission Tomography

Computed tomography (CT) techniques share a common mechanism for spatial encoding. A two dimensional projection of the object (similar to a standard x-ray or a single read out in MRI without phase encoding) at multiple angles can be reconstructed into two or three dimensional images. Data are acquired by rotation of a set of detectors either opposite a radiation source as in CT, or around the subject in the case of internal radiation sources as in Single Photon Emission Computed Tomograph (SPECT) and Positron Emission Tomograph (PET). This method can also be used with MRI by combining gradients to provide a series of readouts at multiple angles (typically from 64 to 256), however modern MRI acquisition techniques are more efficient, therefore this technique is rarely used in MRI.

Reconstruction of images from tomographic methods are performed using the reverse Radon transform (Herman, 1980) which uses the series of angular projections to reconstruct images. The resulting data set can be displayed as a rotating three-dimensional movie or resliced in any direction to display a series of two-dimensional slices.

X-ray CT uses a high intensity X-ray beam and a series of detectors that rotate around the subject. Modern CT scanners use multiple sets of X-ray sources and detectors to simultaneously acquire multiple slices (currently, up to 64) to speed acquisition. This allows full brain coverage at high spatial resolution in a matter of seconds. High doses of X-ray (about 5 rads) are typical exposures for this type of exam.

SPECT detects emitted gamma rays from radioactively labeled molecules or cells. The position of the emitted gamma ray is determined by an array of crystal detectors covered by a collimator, which limits the angles from which gamma rays can enter the detector. Collimators consist of either a lead shield with a pinhole for three-dimensional tomographic acquisitions or a series of slots for two dimensional, planar (2D) acquisitions. Sensitivity of SPECT is limited by the area of the collimators opening. This concept can be understood by thinking of a pinhole collimator placed at a distance from the center of the subject being scanned. For example, if a 1 mm diameter pinhole is placed 10 cm from a mouse, the area covered on the surface of the sphere prescribed by the opening of the collimator (radius = 10 cm) is 1257 cm^2 ($4\pi r^2$). The gamma ray capture area of a 1 mm pinhole (0.5 mm = 0.05 cm radius) is 0.008 cm^2 (πr^2). Emission of gamma rays from the radioactive tracers is isotropic (equal probability of travel in any direction) the fraction of total gamma rays captured by the pinhole will be $0.008/1257 = 6 \times 10^{-6}$. Sensitivity can be increased by increasing the size of the pinhole, at the expense of spatial resolution or increasing the number of detectors in order to increase the counts for each angular projection. Recent technical developments have introduced multiple pinhole collimators (9–12) per detector array and multiple detectors for a significant increase (factor of 40) in detection sensitivity. The relatively low sensitivity of SPECT is typically compensated by using high doses of radioactive label, typically 50–500 microcuries to allow detection of labeled molecules or cells down to the nanomolar (10^{-9} M) concentration range.

PET uses positron emission to increase sensitivity of detection over SPECT. Once a positron is emitted from the radioactive atom on the labeled molecule, it travels a short distance (2–5 mm, depending on the energy of the emitted positron) and undergoes an annihilation reaction with an electron. Annihilation, occurs because the positron is antimatter to the electron. The reaction creates two 0.511 MeV gamma rays which then travel 180° away from each other. PET scanners employ a ring of detectors (typically 1024) and coincidence detection (simultaneous detection of two gamma rays) to improve detection sensitivity and filter out background radiation. As a result of the high capture cross section of the ring of detectors and the virtual elimination of background radiation, sensitivity is enhanced approximately four orders of magnitude compared to SPECT. This allows detection of lower concentrations of tagged molecules and cells down to the picomolar (10^{-12} M) concentration range with lower doses of radioactivity injected into the subject (typically around 10 microcuries).

Positron emitting nuclides have very short half lives, on the order of minutes to two hours. This makes operation of a cyclotron and a radiochemistry laboratory essential to the use of PET scanners. ^{18}F is the longest radionuclide with a half-life of 1.87 h, making a central production facility within a city feasible for radiopharmaceuticals employing this nuclide. Most clinical PET facilities have on-site cyclotrons and radiopharmaceutical laboratories to allow the use of short-lived isotopes in clinical studies.

One difficulty with some SPECT and PET images is the potential lack of anatomical detail, especially in studies of receptor distribution where widespread distribution of the label in tissue may not occur. Two methods are used to overcome this problem, coregistration to anatomical images and simultaneous acquisition of anatomical images. Coregistration is typically performed for images lacking anatomical detail using fiducial markers visible in both scan sets to allow spatial registration of the SPECT or PET images to anatomical images such as CT or MRI. A more robust and time efficient method is the use of SPECT/CT or PET/CT scanners in which anatomical and tracer studies are acquired simultaneously providing anatomical images acquired simultaneously with SPECT or PET image acquisition.

50.3. Information Content

Effective use of imaging methods requires the understanding of the type of information available from each scanning technique. As neuroscientists, it is necessary to consider pathophysiological correlates of disease when designing and interpreting imaging studies. Some current research is focusing on correlation of histopathology and neuroimaging in animal models of neuronal disease (Boska et al., 2005; Nelson et al., 2005) which will serve to validate and refine the interpretation of imaging results. Consideration of expected pathology is required for experimental design and application of the most appropriate imaging methods to specifically address a particular physiological question. In the previous section, the basic principles of imaging methods were presented to allow the neuroscience student to understand the signal source and methods used to obtain imaging information. We will build on this to obtain a broad view of the basic capabilities of each imaging method and examples of the information available by use of the method.

50.3.1. Magnetic Resonance Techniques

The wide variety of morphological, physiological, and biochemical information available from the multitude of MR techniques, which allow sophisticated combinations of methods to be applied to individual neurological studies. This is one reason that MRI is often the best choice for neuroimaging studies, as the combination of multiple methods in one study can provide a wide array of coregistered information on the effects of disease and experimental therapies on brain function and biochemistry.

50.3.1.1. T_1, T_2, and Proton Density Weighted MRI

The use of T_1, T_2 and proton density weighted images comprise the bulk of clinical MRI scans due to the ability to detect cellular alterations in tissue and determine morpholgical changes with exquisite soft tissue contrast. As a result, brain lesions, especially white matter lesions, tumors, and damaged regions from vascular disease are visible typically as dark areas on T_1 and bright areas on T_2 weighted images.

While these imaging modalities are sensitive to cell type, the combination of T_1, T_2 and proton density are not specific to cell type or the type of alteration in normal cell physiology. Improved specificity can be obtained by combining these MRI scans with other types of information including clinical manifestations of disease, other neuroimaging scans and advanced analytical methods to improve diagnostic accuracy.

Use of T_1 and T_2 weighted images can be used to determine grey matter, white matter, and CSF content in neuroimages and determine the volume of multiple brain structures. Reduced volumes of gray matter, basal ganglia, and subthalamic nuclei have been found in various neurological diseases, the details of which are beyond the scope of this chapter. Good reviews are available for findings from HIV dementia (Boska et al., 2004), Alzheimer's disease (Kantarci and Jack, 2004; Chong and Sahadevan, 2005; Masdeu et al., 2005), Parkinson's disease (Krabbe et al., 2005; Seppi and Schocke, 2005) and the hippocampus in a wide range of neurological diseases (Geuze et al., 2005). Generally, decreased volumes of affected brain structures, such as the substantia nigra in Parkinson's disease and atrophy of the grey matter in dementia such as Alzheimer's disease have been found in these studies. Scientific investigation continues into this area and as imaging and analysis software improves and the definition of changes due to disease are better defined, the use of these data in clinical diagnosis and patient management will be put into practice.

Recent studies have also shown that neuronal depolarization will cause increases in T_1 and T_2. T_1 was shown to increase by 13% while T_2 increased by 88% at 1.5 T in a rat model of spreading depression (Stanisz et al., 2002). While this effect has not been studied in great detail, it is possible that this effect could be a source of detecting abnormalities in neuronal polarization with careful quantitation of T_1 and T_2 in neuronal diseases.

It will be apparent to the neurosciences student that this field, even in the case of one of the best characterized and most studied imaging modalities, standard MRI, that new information is acquired rapidly, advanced through the acquisition of data in biologically characterized animal studies and translated to well characterized human disease states. This process is cyclical, as advances in imaging methods will improve the classification of human neurological disease, leading to improved characterization and more refined human study. Advances in data processing methods through computer science advances, developments of imaging data correlated with biological data, and advanced statistical analyses will be primary areas of imaging research for the advancement of disease characterization and detection sensitivity and specificity of imaging methods for many years to come.

50.3.1.2. Diffusion Weighted Imaging (DWI)

The primary use of DWI in clinical practice is detection of acute stroke lesions. Within minutes after a stroke, a low diffusion area of brain is visible in the majority of stoke cases. In a diffusion weighted image, this appears as a bright area, as tissues with low diffusion lose less signal during the diffusion weighting. Signal abnormalities usually do not appear before 16h on T_1 and 8h on T_2 weighted images (Yuh et al., 1991). Diffusion weighting requires a long echo acquisition, so the appearance of bright areas on DWI can be either from low diffusion or increased T_2 signal which occurs in late acute to chronic infarcts. In order to eliminate confusion from this effect, two images are acquired, one with no diffusion weighting and one (or more) with high diffusion weighting. This allows the calculation of a parametric image of the apparent diffusion coefficient of water (ADC_ω). Acute stroke lesions typically appear as bright on DWI and dark on ADC_ω maps. As stroke lesions age, the cells in the lesion begin to break down after cell death and leave an area of edema. As the stroke lesion progresses towards cellular breakdown and edema, both diffusion and T_2 increase. After days to weeks, the diffusion "pseudonormalizes" to a value equivalent to normal brain tissue, progressing to high diffusion. A combination of ADC_ω, T_1 and T_2 maps can be used to characterize the age of the stroke lesion (Nagesh et al., 1998; Welch et al., 2001).

Diffusion is altered in other types of brain lesions and is not specific for stroke (Wang et al., 1998). The range and types of diffusion alterations are an area of active research. Combination of DWI, T_1, T_2, MT and perfusion may serve to more precisely characterize lesions from neurological diseases (Dijkhuizen and Nicolay, 2003).

50.3.1.3. Diffusion Tensor Imaging

Multiple metrics can be calculated from diffusion tensor imaging (DTI) data, typically mean diffusivity (D_{av}) and fractional anisotropy (FA). Analysis of brain pathology using DTI metrics can be performed by lesion analysis, whole brain histograms of FA and D_{av}, tractography and analysis of otherwise normal appearing white matter. Fractional anisotropy has been studied in numerous diseases and detects abnormalities in otherwise normal appearing white matter by conventional MRI. This demonstrates the increased sensitivity to disease within the brain that DTI metrics can resolve. Abnormal FA in normal appearing white matter has been found to occur in multiple sclerosis, radiation injury, aging, schizhophrenia, and amyotropic lateral sclerosis (Foong et al., 2000; Filippi et al., 2001; Cosottini et al., 2005; Kitahara et al., 2005; Salat et al., 2005). Fractional anisotropy maps or DTI in general for acute stroke analysis has not been found to add significant clinical information in the setting of acute stroke therapy, although, active investigation in this area continues.

Brain lesion analysis is also an active area of research. DTI metrics have been found to help differentiate between tumor types and promises to provide a sensitive method for abnormal tissue characterization. Fractional anisotropy has already been found to correlate with the cell density of brain tumors (Beppu et al., 2003; Beppu et al., 2005). The ability of FA to correlate with cell density likely accounts for the sensitivity of FA to tissue typing in brain lesions. Fractional anisotropy has been found to differentiate between low grade and anaplastic astrocytomas, to help differentiate between brain metastasis and high grade gliomas, and there is evidence that FA may help to detect tumor infiltration not visible by conventional MRI (Holmes et al., 2004; Tsuchiya et al., 2005; Goebell et al., 2006). Other DTI metrics have been utilized in brain lesion analysis and the early findings point to the breadth of future possibilities.

The diffusion tensor data can also be utilized to perform white matter tractography (Jellison et al., 2004). Tractography is the virtual dissection of white matter tracts within the brain and provides a novel method to analyze the white matter tracts in disease states. Tractography in combination with other DTI metrics and other MRI techniques such as perfusion and spectroscopy opens the door for currently uncharted areas of research. Previously there was nearly no indication of the intricate connections present in the white matter on conventional MRI. DTI tractography can be used to map out numerous fiber tracts including but not limited to the corticospinal tracts, superior occipitofrontal fasciculus, inferior occipitofrontal fasciculus, uncinate fasciculus, superior longitudinal (arcuate) fasciculus, inferior longitudinal (occipitotemporal) fasciculus, (Jellison et al., 2004). With advancements and refinements in aquisition of DTI, performing tractography provides further advances in the "virtual dissection" of the white matter tracts.

A very exciting and innovative application of DTI and tractography that will have profound clinical application is analysis of brain lesions. There has already been early work defining tumor growth and metastases within major white matter tracts (Witwer et al., 2002; Yu et al., 2005; Hlatky et al.,

2006). This application will provide measures of white matter tract displacement and tumor invasion that will be defined with great specificity. Previously estimates of white matter tract invasion could only be made by gross estimates of where the white matter tracts were expected to lie based on cross-sectional MRI or CT studies. However, current mothods of DTI tractography can not reproducibly demonstrate the the true size of fiber bundles (Kinoshita et al., 2005).

50.3.1.4. *Magnetization Transfer (MRI)*

Magnetization Transfer (MT) MRI is sensitive to areas of demyelination, as the MT effect from myelin is particularly strong. The use of MT MRI for the detection and characterization of lesions due to multiple sclerosis has been well studied (Atalay et al., 2005). Quantitative analysis of the magnetization transfer ratio (MTR) of gray matter in the earliest stages of the disease demonstrate statistically significant reduction in the basal ganglia of these patients (Audoin et al., 2004). In general multiple advanced MR techniques, including MTR maps, ^{1}H MRSI, quantitative T_1 and T_2 mapping, diffusion tensor imaging, and studies of blood flow have lead to a more sophisticated understanding of the disease from the early view of simply an immune mediated demyelination process to the understanding of disease effects on neurons, axons, and oligodendrocytes (Minagar et al., 2005).

MT MRI has been shown to be sensitive to other minor structural abnormalities caused by neurological disease. MTR maps were analyzed with a statistical package demonstrated that areas of epileptic activity which were not detected visually had statistically reduced MTR in the epileptic focus (Rugg-Gunn et al., 2003). When MT MRI was combined with DWI, reductions in both MTR and ADC_{ω} were found in the epileptic focus of patients (Ferini-Strambi et al., 2000).

More advanced quantitation methods for MT MRI have been shown to be sensitive to white matter tracts and axonal density (Yarnykh and Yuan, 2004) as well as being sensitive to the earliest events in stroke lesions (Jiang et al., 2001) including early detection of ruptures in the blood-brain barrier which may be susceptible to hemorrhagic transformation (Knight et al., 2005).

50.3.1.5. *Perfusion MRI*

Perfusion in cerebral tissue with normal vasculature is controlled by metabolic demand, allowing use of regional quantitative perfusion measures to detect neurodegeneration in a variety of diseases. Perfusion measures have been a cornerstone of nuclear medicine (SPECT and PET) studies for characterizing brain regions affected in neurological disease.

MRI perfusion studies have been used primarily to study perfusion deficits in stroke lesions. The mismatch between DWI and perfusion deficit may represent tissue at risk for damage (Neumann-Haefelin et al., 1999) which may be salvaged using thrombolytic therapy. However this interpretation of mismatch between DWI and perfusion MRI is controversial and study of this effect is an area of active research (Rowley, 2005).

Brain tumor imaging is a growing area of clinical application of MRI perfusion studies. MRI perfusion studies have been found to provide valuable information when evaluating multiple facets of the imaging of gliomas, metastatic disease, abscesses, radiation necrosis, and meningiomas. The grading of gliomas pre-operatively has been found to be enhanced by utilizing MRI perfusion (Law et al., 2003; Law et al., 2004). Also, the determination if a lesion is a high-grade glioma or a metastatic lesion, the differentiation of which is markedly important clinically, can be facilitated utilizing MRI perfusion as well as MRS (Law et al., 2002). The differentiation of an abscess from a ring enhancing glioma is also potentially facilitated by MRI perfusion (Holmes et al., 2004). A significant problem that can occur with MRI perfusion is leakage of the contrast through the blood brain barrier on the first pass so that the appearance of the perfusion curve is distorted and a true evaluation of the tumor perfusion is not obtained. This is unfortunately a potentially frequent occurrence since blood brain barrier disruption is a hallmark of high-grade gliomas.

Studies of perfusion deficits in other neurological diseases have been hindered in MRI by the difficulties in quantitation of cerebral perfusion in bolus tracking perfusion measures. Arterial spin-labeling techniques have been successfully employed in recent studies of human disease including Alzheimer's disease (Alsop et al., 2000; Johnson et al., 2005). Arterial spin-labeling perfusion measures are greatly enhanced at high field due to increased signal to noise and increased T_1, improving reliability of the technique (Wang et al., 2002). Increased availability of high field (3 T and above) clinical MRI systems will allow greater use of quantitative perfusion measures to assess neurodegenerative disease in clinical practice.

50.3.1.6. *^{1}H MRSI*

Proton MRS detectable metabolites are primarily N-acetyl aspartate (NAA), choline (Cho), creatine (Cre), glutamate (Glu), glutamine (Gln), and myoinostitol (mI). NAA is found predominately in the neurons and the typical physiological intraneuroanal concentrations is 9–12 mM (Urenjak et al., 1992), and as such is widely used as a marker of neuronal density. However, biochemical studies suggest the possibility of multiple roles for the metabolite NAA including an energy metabolism reserve (Heales et al., 1995; Bates et al., 1996), a supply of acetyl-Coenzyme-A for the tricarboxylic acid cycle in mitochondria (Mehta and Namboodiri, 1995; Miller et al., 1996), a source of N-acetyl-aspartylglutamate (a neuropeptide that is involved in neurotransmission), a source of glutamate (a neuroexcitatory molecule which itself is a source for gamma-amino butyric acid) (Miller et al., 1996) and a role as an osmotic regulator (Taylor et al., 1995). NAA is synthesized exclusively in the mitochondria of neurons and decreases in NAA may be due, at least in part, to mitochondrial dysfunction (Goldstein, 1969; Heales et al., 1995; Mehta and Namboodiri, 1995). Neuronal cell death induces an irreversible loss of NAA due to localization of NAA in the neuron (Urenjak et al., 1992), while reversible loss of NAA is likely associated with reversal of mitochondrial dysfunction or replacement of damaged mitochondria within the neurons. The choline peak (Cho) is due to total levels of mobile choline containing compounds. Cho mainly consists of glycerophosphocholine and phosphocholine, compounds involved in phospholipid metabolism in brain tissue. Choline concentrations are known to be elevated in multiple sclerosis (MS) during periods of active demyelination (Simone et al., 2001). Total creatine (Cre) concentration [Cre] is the amount of creatine plus phosphocreatine in brain tissue. Cre is part of the creatine kinase energy metabolism buffer system used to maintain ATP levels in times of acute mismatch between oxidative adenosine triphosphate (ATP) supply and ATP demand. The [Cre] reflects the health of systematic energy use and storage (Miller, 1991). Total [Cre] tends to remain relatively constant in most cell types, although the [Cre] in glial cells is four times that in neuronal cells (Petroff et al., 1995) and is known to be elevated in multiple sclerosis and epilepsy where gliosis is significant. mI may also act as a marker of glial cell numbers (Chang et al., 1999) and as an osmoregulator or intracellular messenger (Ross, 1991).

Lactate [Lac] is seen as a result of numerous etiologies such as secondary to anaerobic glycolysis in tumors. Hypoxic insult can also cause an elevation of Lac. Lipid occurs in the same spectral region as Lac and can obscure Lac. It is often difficult to separate these two peaks from each other but lactate has a classic doublet appearance. The presence of lactate can be confirmed by utilizing a TE of 136 ms causing the lactate doublet to invert (appear upside down in the spectrum) due to j-coupling (Yablonskiy et al., 1998). Lipids indicate the presence of necrosis or disruption of myelin sheaths. A higher level of lipid in a brain tumor generally indicates a high-grade malignancy. Lipids are best seen on spectra obtained with short TEs. However, contamination of the spectrum with scalp fat or bone marrow in the skull can result in a lipid peak, so care must be taken to interpret spectra with an understanding of the technical details of the acquisition.

^{1}H MRS and MRSI have been demonstrated to be sensitive methods for early detection of neuronal dysfunction or loss, reflected by loss of NAA. The application of MRS adds specificity to the diagnosis of a lesion relative to MR imaging alone (Ross and Michaelis, 1994; Howe and Opstad, 2003; Nelson, 2003).

MR spectroscopy is used extensively in clinical practice to help differentiate tumor from non-tumor lesions in the brain. The Cho/Cr and NAA/Cho ratios can be utilized to help accomplish this task (Poptani et al., 1995; Moller-Hartmann et al., 2002). Increased Cho/Cr and decreased NAA/Cho are positive inidications of a brain tumor (Poptani et al., 1995; Moller-Hartmann et al., 2002). The lipid and lactate peaks are more variable in tumor and non-tumor lesions but can also aid in tumor diagnosis. There are a number of disease processes, such as multiple sclerosis plaques, that cannot always be differentiated by spectroscopy from a brain tumor.

Tumor grading can be challenging with MR spectroscopy and there are overlaps between the appearance of the spectra from different grades of astrocytomas and oligodendrogliomas. The Cho peak and the Lip/Lac peaks tend to demonstrate increases as an astrocytoma progresses from low-grade to glioblastoma multiforme (GBM) (Nelson et al., 2002; Howe et al., 2003). Due to the complexity and necrosis in GBMs the peak values are somewhat variable (Howe et al., 2003). The Cr and the NAA peaks are decreased in low-grade GBM (Nelson et al., 2002). Myo-inositol is increased in low grade astrocytomas and is decreased in high grade astrocytomas(Castillo et al., 2000). The Lip and Lac peaks can be used to separate low-grade from high-grade (anaplastic) oligodendrogliomas in which they have been found to be significantly elevated (Rijpkema et al., 2003). In low grade oligodendrogliomas, glutamate and glutamine have been found to be elevated and this may help to differentiate oligodendrogliomas from other brain tumors such as astrocytomas (Rijpkema et al., 2003).

Gliomatosis cerebri can demonstrates subtle changes on MR spectroscopy. The subtle changes likely result in part from the infiltrative nature of the tumor and that a component of the spectrum represents normal brain. It has been described that the Cho/Cr levels may be normal and that only an elevated MI level will be detected in gliomatosis cerebri (Mohana-Borges et al., 2004). Also, there may be normal Cho/Cr levels, associated decreased NAA and Lip/Lac present (Pyhtinen, 2000).

It can be critical to obtain a MRSI study over the entire complex heterogeneous areas of a brain lesion in order to institute appropriate therapy, plan surgeries and accurately diagnose the brain lesion (Nelson, 2003). MRSI helps to define the spatial heterogeneity that is present in brain lesions and also contributes to defining the spatial extent of brain lesions. Two-dimensional MRSI performed at multiple levels or a single three-dimensional MRSI sequence can be performed to accomplish this task. MRSI can be used to help differentiate infiltrating primary brain tumors from metastatic disease. With metastatic tumors in the brain the spectral abnormality will not extend beyond the area of enhancement but with primary infiltrating brain tumors the spectral abnormality will persist outside of the area of enhancement.

Diagnosing radiation necrosis versus tumor recurrence is also a useful and accurate application of intracranial MR spectroscopy (Schlemmer et al., 2001; Nelson, 2003). Radiation necrosis will have large Lip and Lac peaks without elevated Cho whereas tumor would have elevated Cho. MRSI is particularly useful for distinguishing recurrent tumor versus radiation necrosis given its ability to spatially sample multiple areas of a lesion (Chernov et al., 2005).

50.3.2. Computed Tomography

Computed tomography (CT) of the brain revolutionized the care of neurological patients and was instrumental in helping to elucidate many disease processes after its introduction.

However, the advantages of utilizing CT to image the brain have been greatly reduced by MRI's markedly better CNS contrast, MRI's physiologic imaging abilities and the metabolic information obtainable from PET or SPECT imaging. CT still plays an important role in the setting of trauma due to its proven ability to detect hemorrhages that require surgical intervention. CT also is advantageous in the trauma setting since it is often too difficult to adequately screen a patient for an emergent MRI and to monitor the acute trauma patient in the MRI scanner. Hemorrhages appear hyperdense acutely on CT and CT can accurately demonstrate subdural, epidural, parenchymal, intraventricular, and subarachnoid hemorrhages. CT exquisitely demonstrates the skull and skull base and is unsurpassed in detecting fractures.

CT has limited ability to differentiate the type of insult occurring in the brain and usually a brain lesion results in a lower density than the surrounding tissues. Lesions with calcification, hemorrhage, or high cellularity (such as lymphoma, primitive neural ectodermal tumors, medulloblastomas or small cell lung cancer metastasese) will result in increased density. Iodinated contrast agents help evaluate lesions in the brain for blood brain barrier breakdown and contrast helps improve lesion conspicuity. However, a gadolinium contrast enhanced MRI is more sensitive to detecting evidence of blood brain barrier breakdown than a contrast enhanced CT (Mihara et al., 1989).

CT has been shown to be able to detect early changes in the development of Alzheimer's disease (de Leon et al., 1993), although it has been found that other technologies, such as MRI and PET imaging, show changes better than CT. Also, in the analysis of seizure patients CT can demonstrate numerous parenchymal pathologies but a MRI examination is often a better evaluation (Heinz et al., 1989). CT is particularly useful when evaluating seizure patients for detecting small calcified lesions such as in neurocysticersosis, periventricular calcifications of tuberous sclerosis, or the cortical calcifications present in Sturge-Weber syndrome.

A non-contrasted CT is often performed as the initial evaluation of the brain in the stroke patient secondary to CTs availability and CTs ability to exclude hemorrhage. The non-contrasted brain CT is also the neuroimaging study that was utilized in major stroke treatment studies (Furlan et al., 1999). The non-contrasted brain CT offers limited information compared to MRI but recently with the development of CT perfusion of the brain additional information can now be obtained. CT perfusion moves CT imaging from static morphological imaging to providing physiological data that has already been shown to be useful for stroke imaging (Lev et al., 2001; Wintermark et al., 2006).

Multiple parameters can be calculated with CT perfusion imaging of the brain and include cerebral blood flow, cerebral blood volume, time-to-peak, and mean transit time (Wintermark et al., 2006).

The information obtainable with CT perfusion is very powerful especially when combined with computed tomographic angiography (CTA). These data can be used as a substitute for MRI when evaluating stroke patients (Schramm et al., 2004).

CTA is a method for relatively non-invasively imaging the arteries and veins of the brain. The ability to create a CTA examination of the circle-of-Willis became possible with the advent of spiral CT, however, the quality of the examinations and the robustness of the techniques were greatly improved with the advent of multislice CT. The speed of acquisition continues to advance at a remarkable rate. With a 64-slice CT scanner the whole brain can be imaged in seconds with submillimeter slice thickness. The ability to scan fast when performing CTA of the brain cannot be overstated given that a main application is evaluating patients with subarachnoid hemorrhage for the presence of cerebral aneurysms. Patients with subarachnoid hemorrhage often have mental status changes that make it difficult or impossible for them to follow directions and lie still for an extended period of time. CTA is now often used to screen patients with subarachnoid hemorrhage for cerebral aneurysms in order to determine if an aneurysm is present and what type of therapy (surgical clipping or intravascular coiling) is appropriate.

The evaluation of acute stroke patients with CTA is being performed to evaluate for the underlying vascular etiology (Smith et al., 2006). CTA can nicely demonstrate areas of stenosis or thrombosis in the arteries. The presence of a large vessel intracranial occlusion in an acute stroke patient has been found to even be an independent predictor of poor outcome (Smith et al., 2006). Also, computed tomographic venography has been found to be beneficial for evaluating patients for the presence of venous thrombosis.

50.3.3. Single Photon Emission Computed Tomography

Single Photon Emission Computed Tomography (SPECT) in general has a greater clinical availability compared to Positron Emission Tomography (PET). Secondary to SPECT's greater availability, proven capabilities and its lower cost it is an often used clinical and research imaging modality. SPECT is very useful for evaluating brain perfusion and has demonstrated utility for evaluating the brain perfusion of many disease processes including stroke, Alzheimer's disease and other dementias, Parkinson's disease, seizure patients, and brain tumors. Perfusion is just one parameter that can be measured with SPECT and there are many different radiotracers that can be used with SPECT for neurotransmitter system imaging (Kegeles and Mann, 1997).

Thallium-201 SPECT and fluoro-deoxyglucose (FDG) PET are both used to help differentiate residual brain tumor and brain tumor recurrence from radiation necrosis. Thallium-201 SPECT has been found to be just as good or even better for tumor detection than FDG PET (Kahn et al., 1994; Maria et al., 1998; Stokkel et al., 1999). Thallium-201 SPECT can also be utilized to help differentiate lymphoma from toxoplasmosis which is an important question in immunocompromised patients particularly patients with human immunodeficiency virus (Young et al., 2005). Difficulties have been found when evaluating immunocompromised patients in the clinical setting and this may in part be due to the difficulty of evaluating small lesions (Licho et al., 2002).

There are numerous psychiatric applications of SPECT and these include evaluation of major depressive disorder, obsessive compulsive disorder, and schizophrenia (Novak et al., 2005; Topcuoglu et al., 2005; Alves et al., 2006). Studies have been performed evaluating these diseases and other psychiatric disorders at their initial presentations, during disease progression, and what the associated imaging changes are with medical intervention. There are many continued opportunities for investigation utilizing SPECT in psychiatry particularly whether it is utilizing a newly formulated radiotracer, utilizing advanced analytical techniques or combining SPECT data with MRI data.

SPECT has been found to be quite useful when evaluating patients with seizures (McNally et al., 2005; Tae et al., 2005). SPECT imaging is commonly used for the ictal study of seizure patients in order to define areas that are the seizure focus. It has in general been much easier to perform ictal SPECT than ictal PET. Advanced analytical techniques of SPECT imaging have helped to improve the data analysis and utilization of the SPECT data in the evaluation of seizure patients but further improvements and validation of the techniques are required (Knowlton et al., 2004).

50.3.4. Positron Emission Tomography

The information obtained from a Positron Emission Tomography (PET) scan depends on the radiotracer used and there is a wide array of radiotracers that can be used by the neuroscientist when analyzing the brain. PET using fluoro-deoxyglucose (PET-FDG) has wide clinical and research availability and has many applications in the neurosciences. PET-FDG can be used to study numerous pathologies including tumors, strokes, dementia, addiction disorders, schizophrenia, depression, seizures, and developmental abnormalities. There are a greater number of tracers that have been developed for PET than SPECT that allow in-vivo mapping of numerous neurotransmitter systems in humans (Kegeles and Mann, 1997).

PET-FDG has been used extensively to investigate brain tumors including astrocytomas, oligodendrogliomas, meningiomas, lymphomas, metastatic disease, and the differentiation of tumor necrosis from tumor recurrence. In the evaluation of brain tumors PET-FDG has been found to help determine the grade of malignancy and can provide predictive prognostic information (Kaschten et al., 1998; De Witte et al., 2000; Borgwardt et al., 2005). Carbon-11-methionine has also been found useful, if not better, than PET-FDG for predicting histological grade and the prognosis of gliomas (Kaschten et al., 1998). Carbon-11-methionine is felt to be superior to FDG secondary to its sensitivity and clearer delineation of tumors (Pirotte et al., 2004; Van Laere et al., 2005), although, it is not as readily available for clinical use and investigation as FDG. Thallium-201 SPECT as mentioned in the previously section also has been shown to perform as good if not better than PET-FDG in brain tumor analysis (Kahn et al., 1994; Maria et al., 1998; Stokkel et al., 1999).

Evaluating dementias with PET imaging is another area that has generated a lot of attention as well as clinical and research applications. The ability of PET scanning to help differentiate whether a patient has Alzheimer's disease has generated great interest due to the large number of individuals affected. PET-FDG is the radiotracer that is routinely utilized in the evaluation of dementias. PET-FDG can help to determine if a patient has Alzheimer's disease, mild cognitive impairment, and if a patient with mild cognitive impairment is likely to convert to Alzheimer's disease (Mosconi et al., 2004; Anchisi et al., 2005). Frontal-temporal dementia, vascular dementia and lewy body dementia diagnosis can also be facilitated with PET-FDG (Mirzaei et al., 2003; Higdon et al., 2004; Kerrouche et al., 2006).

The utility of imaging patients who have or potentially have Parkinson's disease can be valuable in multiple ways including early confirmation of the diagnosis, predicting clinical course, and to evaluate the results of therapies. When evaluating Parkinson's disease both PET-FDG and 6-[(18)F]fluoro-L-dopa (FDOPA), a common presynaptic dopaminergic radiotracer, can be effectively utilized (Kaasinen et al., 2006). FDG and FDOPA PET has been shown to be an indicator of the severity of Parkinson's disease and at patient's initial evaluations can be used to predict the clinical prognosis (Kaasinen et al., 2006). The differentiation by PET-FDG of Parkinson's disease, progressive supranuclear palsy and multiple system atrophy may also be facilitated given differences of metabolism that have been found (Juh et al., 2004). Therapeutic induced changes of brain metabolism can be evaluated for both surgical and medical interventions (Hilker et al., 2002; Hershey et al., 2003; Zhao et al., 2004). These type of inquiries provide knowledge that can be used to judge an intervention's success, how an intervention's success may have been produced, and potentially to predict early on in a treatment paradigm what long term success may be obtained.

There are several possible applications of PET imaging for seizure evaluation. PET-FDG again has been the predominant radiotracer that has been utilized for seizure evaluation, although, numerous radiotracers have been studied. Preoperative evaluation of temporal lobe epilepsy is a frequently performed application of PET-FDG and PET-FDG is able to help predict the success of surgical intervention and can help guide the type of surgical resection (Theodore et al., 1997; Kim et al., 2003). PET as already mentioned usually studies the seizure patient in the interictal state since it is difficult to perform ictal PET studies. PET-FDG can also be a predictor of verbal decline after anterior temporal lobectomy – a surgery performed to help control complex partial seizures (Griffith et al., 2000). It has also been shown that PET-FDG is beneficial for evaluating for seizure foci in a general epilepsy population (Swartz et al., 2002).

Summary

This chapter has presented an overview of the mechanisms and applications of neuroimaging for the student of neuroscience with a view towards providing an understanding of the potential of neuroimaging techniques in neuroscience research. As the field progresses, methods for whole brain analyses, development of new detection methods and new radiotracers combined and coregistered will provide whole brain non-invasive histology for diagnosis of disease and tracking the effects of new therapies in animal models of neurological disease and, eventually, human subjects. Research techniques being developed will combine multiple coregistered techniques combined with molecular and histological analyses to provide sensitive and specific characterization of the gamut of combined neuroimaging methods for diagnosis and monitoring of neurological diseases.

Review Questions/Problems

1. Diffusion weighted images become positive in acute stroke in:

a. 16 h
b. minutes
c. 8 h
d. 1 day

2. Advantages of diffusion tensor imaging include:

a. demonstrates abnormalities when conventional MRI is normal.
b. provides functional maps of brain activity.
c. it is indispensable for planning stroke therapy.
d. can be used to calculate perfusion metrics.

3. Perfusion imaging data can be used to create maps that:

a. allow for "virtual dissection" of white matter tracts.
b. are used for creating MR angiograms.
c. can be utilized to determine tissue at risk for damage in patients with stroke.
d. show peaks of lactate in tissue damaged by stroke.

4. Magnetic resonance spectroscopic imaging (MRSI) does *not*:

a. help define the spatial heterogeneity of brain lesions.
b. provide less useful information than single voxel spectroscopy for distinguishing radiation necrosis versus recurrent tumor.
c. allow for accurate diagnosis of brain lesions.
d. demonstrate the same metabolic peaks as single voxel spectroscopy.

5. Computed tomographic brain imaging:

a. continues to have great utility when evaluating trauma patients.
b. demonstrates acute hemorrhages as hypodense.
c. has replaced MRI for many imaging applications.
d. provides better metabolic information than PET or SPECT.

6. Computed tomographic perfusion imaging:

a. has already been utilized in major stroke treatment studies.
b. can be utilized to evaluate brain diffusion.
c. has limited utility for evaluating stroke patients.
d. allows multiple parameters such as cerebral blood volume and mean transit time to be calculated.

7. When evaluating seizure patients CT is particularly useful for:

a. demonstrating the periventricular calcifications of tuberous sclerosis.
b. demonstrating small soft tissue cortical lesions not detected by MRI.
c. demonstrating the cortical tubers in patients with tuberous sclerosis.
d. demonstrating the brainstem calcifications of Sturge-Weber syndrome.

8. SPECT imaging of the brain:

a. is more expensive than PET imaging.
b. has been nearly replaced with PET imaging.
c. availability is a major problem in the United States.
d. is very useful for evaluating brain perfusion.

9. SPECT imaging does not have:

a. many radiotracers that can be utilized.
b. many brain perfusion applications.
c. the ability to analyze brain tumors with Thallium-201.
d. undisputed ability to evaluate for lymphoma in patients with human immunodeficiency virus.

10. PET imaging:

a. has fewer radiotracers that can be utilized than SPECT.
b. has few imaging applications for which FDG is used.
c. has the ability to differentiate tumor necrosis from tumor recurrence.
d. lacks clear ability to differentiate Alzheimer's disease from mild cognitive impairment.

11. Which of these statements is true:

a. T2-weighted images become positive in a stroke at around 24 h.
b. Diffusion images allow for an estimation of the mean transit time.
c. MR spectroscopy utilizes radiotracers.
d. CTA is used to screen for cerebral aneurysms.

12. MR spectroscopy does *not*:

a. demonstrate increased myo-inositol in high grade astrocytomas.
b. potentially only demonstrates subtle findings for gliomatosis cerebri.

c. demonstrate increased myo-inositol in low grade astrocytomas.
d. show glutamate and glutamine elevations in low-grade oligodendrogliomas.

References

(1995) Tissue plasminogen activator for acute ischemic stroke. The National Institute of Neurological Disorders and Stroke rt-PA Stroke Study Group. N Engl J Med 333:1581–1587.

Alsop DC, Detre JA, Grossman M (2000) Assessment of cerebral blood flow in Alzheimer's disease by spin-labeled magnetic resonance imaging. Ann Neurol 47:93–100.

Alves TC, Rays J, Fraguas Jr R, Wajngarten M, Telles RM, Luis DESDF, Meneghetti JC, Chow Robilotta C, Prando S, Campi DECC, Buchpiguel CA, Busatto GF (2006) Association between major depressive symptoms in heart failure and impaired regional cerebral blood flow in the medial temporal region: A study using 99m Tc-HMPAO single photon emission computerized tomography (SPECT). Psychol Med 36(5):597–608.

Anchisi D, Borroni B, Franceschi M, Kerrouche N, Kalbe E, Beuthien-Beumann B, Cappa S, Lenz O, Ludecke S, Marcone A, Mielke R, Ortelli P, Padovani A, Pelati O, Pupi A, Scarpini E, Weisenbach S, Herholz K, Salmon E, Holthoff V, Sorbi S, Fazio F, Perani D (2005) Heterogeneity of brain glucose metabolism in mild cognitive impairment and clinical progression to Alzheimer disease. Arch Neurol 62:1728–1733.

Atalay K, Diren HB, Gelmez S, Incesu L, Terzi M (2005) The effectiveness of magnetization transfer technique in the evaluation of acute plaques in the central nervous system of multiple sclerosis patients and its correlation with the clinical findings. Diagn Interv Radiol 11:137–141.

Audoin B, Ranjeva JP, Au Duong MV, Ibarrola D, Malikova I, Confort-Gouny S, Soulier E, Viout P, Ali-Cherif A, Pelletier J, Cozzone PJ (2004) Voxel-based analysis of MTR images: A method to locate gray matter abnormalities in patients at the earliest stage of multiple sclerosis. J Magn Reson Imaging 20:765–771.

Baier-Bitterlich G, Tretiakova A, Richardson MW, Khalili K, Jameson B, Rappaport J (1998) Structure and function of HIV-1 and SIV Tat proteins based on carboxy-terminal truncations, chimeric Tat constructs, and NMR modeling. Biomed Pharmacother 52:421–430.

Bates TE, Strangward M, Keelan J, Davey GP, Munro PM, Clark JB (1996) Inhibition of N-acetylaspartate production: Implications for 1H MRS studies in vivo. Neuroreport 7:1397–1400.

Beppu T, Inoue T, Shibata Y, Kurose A, Arai H, Ogasawara K, Ogawa A, Nakamura S, Kabasawa H (2003) Measurement of fractional anisotropy using diffusion tensor MRI in supratentorial astrocytic tumors. J Neurooncol 63:109–116.

Beppu T, Inoue T, Shibata Y, Yamada N, Kurose A, Ogasawara K, Ogawa A, Kabasawa H (2005) Fractional anisotropy value by diffusion tensor magnetic resonance imaging as a predictor of cell density and proliferation activity of glioblastomas. Surg Neurol 63:56–61; discussion 61.

Borgwardt L, Hojgaard L, Carstensen H, Laursen H, Nowak M, Thomsen C, Schmiegelow K (2005) Increased fluorine-18 2-fluoro-2-deoxy-D-glucose (FDG) uptake in childhood CNS tumors is correlated with malignancy grade: A study with FDG positron emission tomography/magnetic resonance imaging coregistration and image fusion. J Clin Oncol 23:3030–3037.

Boska MD, Mosley RL, Nawab M, Nelson JA, Zelivyanskaya M, Poluektova L, Uberti M, Dou H, Lewis TB, Gendelman HE (2004) Advances in neuroimaging for HIV-1 associated neurological dysfunction: Clues to the diagnosis, pathogenesis and therapeutic monitoring. Curr HIV Res 2:61–78.

Boska MD, Lewis TB, Destache CJ, Benner EJ, Nelson JA, Uberti M, Mosley RL, Gendelman HE (2005) Quantitative 1H magnetic resonance spectroscopic imaging determines therapeutic immunization efficacy in an animal model of Parkinson's disease. J Neurosci 25:1691–1700.

Brugieres P, Combes C, Ricolfi F, Houhou S, Sadik JC, Thomas P, Gray F, Gaston A (1995) [Imagery of human immunodeficiency virus (HIV) encephalitis]. J Neuroradiol 22:163–168.

Castillo M, Smith JK, Kwock L (2000) Correlation of myo-inositol levels and grading of cerebral astrocytomas. AJNR Am J Neuroradiol 21:1645–1649.

Chang L, Ernst T, Leonido-Yee M, Walot I, Singer E (1999) Cerebral metabolite abnormalities correlate with clinical severity of HIV-1 cognitive motor complex. Neurology 52:100–108.

Chernov M, Hayashi M, Izawa M, Ochiai T, Usukura M, Abe K, Ono Y, Muragaki Y, Kubo O, Hori T, Takakura K (2005) Differentiation of the radiation-induced necrosis and tumor recurrence after gamma knife radiosurgery for brain metastases: Importance of multi-voxel proton MRS. Minim Invasive Neurosurg 48:228–234.

Chong MS, Sahadevan S (2005) Preclinical Alzheimer's disease: Diagnosis and prediction of progression. Lancet Neurol 4:576–579.

Cosottini M, Giannelli M, Siciliano G, Lazzarotti G, Michelassi MC, Del Corona A, Bartolozzi C, Murri L (2005) Diffusion-tensor MR imaging of corticospinal tract in amyotrophic lateral sclerosis and progressive muscular atrophy. Radiology 237:258–264.

de Leon MJ, Golomb J, George AE, Convit A, Tarshish CY, McRae T, De Santi S, Smith G, Ferris SH, Noz M, et al. (1993) The radiologic prediction of Alzheimer disease: The atrophic hippocampal formation. AJNR Am J Neuroradiol 14:897–906.

De Witte O, Lefranc F, Levivier M, Salmon I, Brotchi J, Goldman S (2000) FDG-PET as a prognostic factor in high-grade astrocytoma. J Neurooncol 49:157–163.

Detre JA, Leigh JS, Williams DS, Koretsky AP (1992) Perfusion imaging. Magn Reson Med 23:37–45.

Dijkhuizen RM, Nicolay K (2003) Magnetic resonance imaging in experimental models of brain disorders. J Cereb Blood Flow Metab 23:1383–1402.

Ell PJ, Costa DC, Harrison M (1987) Imaging cerebral damage in HIV infection. Lancet 2:569–570.

Ferini-Strambi L, Bozzali M, Cercignani M, Oldani A, Zucconi M, Filippi M (2000) Magnetization transfer and diffusion-weighted imaging in nocturnal frontal lobe epilepsy. Neurology 54:2331–2333.

Filippi M, Cercignani M, Inglese M, Horsfield MA, Comi G (2001) Diffusion tensor magnetic resonance imaging in multiple sclerosis. Neurology 56:304–311.

Foong J, Maier M, Clark CA, Barker GJ, Miller DH, Ron MA (2000) Neuropathological abnormalities of the corpus callosum in schizophrenia: A diffusion tensor imaging study. J Neurol Neurosurg Psychiatry 68:242–244.

Furlan A, Higashida R, Wechsler L, Gent M, Rowley H, Kase C, Pessin M, Ahuja A, Callahan F, Clark WM, Silver F, Rivera F (1999) Intra-arterial prourokinase for acute ischemic stroke. The PROACT II study: A randomized controlled trial. Prolyse in Acute Cerebral Thromboembolism. JAMA 282:2003–2011.

Geuze E, Vermetten E, Bremner JD (2005) MR-based in vivo hippocampal volumetrics: 2 Findings in neuropsychiatric disorders. Mol Psychiatry 10:160–184.

Gillard JH, Waldman AD, Barker PJ (2005) Clinical MR Neuroimaging. Cambridge Press.

Goebell E, Paustenbach S, Vaeterlein O, Ding XQ, Heese O, Fiehler J, Kucinski T, Hagel C, Westphal M, Zeumer H (2006) Low-Grade and Anaplastic Gliomas: Differences in Architecture Evaluated with Diffusion-Tensor MR Imaging. Radiology 239(1):217–222.

Goldstein FB (1969) The enzymatic synthesis of N-acetyl-L-aspartic acid by subcellular preparations of rat brain. J Biol Chem 244: 4257–4260.

Griffith HR, Perlman SB, Woodard AR, Rutecki PA, Jones JC, Ramirez LF, DeLaPena R, Seidenberg M, Hermann BP (2000) Preoperative FDG-PET temporal lobe hypometabolism and verbal memory after temporal lobectomy. Neurology 54:1161–1165.

Haacke EM, Brown RW, Thompson MR, Venkatesan R (1999) Magnetic Resonance Imaging, Physical Principles and Sequence Design. New York, NY: John Wiley and Sons.

Harris R (1983) Nuclear Magnetic Resonance Spectroscopy. Massachussets: Pitnam Publishers.

Heales SJ, Davies SE, Bates TE, Clark JB (1995) Depletion of brain glutathione is accompanied by impaired mitochondrial function and decreased N-acetyl aspartate concentration. Neurochem Res 20:31–38.

Heinz ER, Heinz TR, Radtke R, Darwin R, Drayer BP, Fram E, Djang WT (1989) Efficacy of MR vs CT in epilepsy. AJR Am J Roentgenol 152:347–352.

Henkelman RM, Stanisz GJ, Graham SJ (2001) Magnetization transfer in MRI: A review. NMR Biomed 14:57–64.

Herman GT (1980) Image Reconstruction from Projection: The Fundamentals of Computerized Tomography. New York, NY: Academic Press.

Hershey T, Black KJ, Carl JL, McGee-Minnich L, Snyder AZ, Perlmutter JS (2003) Long term treatment and disease severity change brain responses to levodopa in Parkinson's disease. J Neurol Neurosurg Psychiatry 74:844–851.

Higdon R, Foster NL, Koeppe RA, DeCarli CS, Jagust WJ, Clark CM, Barbas NR, Arnold SE, Turner RS, Heidebrink JL, Minoshima S (2004) A comparison of classification methods for differentiating fronto-temporal dementia from Alzheimer's disease using FDG-PET imaging. Stat Med 23:315–326.

Hilker R, Voges J, Thiel A, Ghaemi M, Herholz K, Sturm V, Heiss WD (2002) Deep brain stimulation of the subthalamic nucleus versus levodopa challenge in Parkinson's disease: Measuring the on- and off-conditions with FDG-PET. J Neural Transm 109:1257–1264.

Hlatky R, Jackson EF, Weinberg JS, McCutcheon IE (2006) Intraoperative Neuronavigation Using Diffusion Tensor MR Tractography for the Resection of a Deep Tumor Adjacent to the Corticospinal Tract. Stereotact Funct Neurosurg 83:228–232.

Holmes TM, Petrella JR, Provenzale JM (2004) Distinction between cerebral abscesses and high-grade neoplasms by dynamic susceptibility contrast perfusion MRI. AJR Am J Roentgenol 183:1247–1252.

Howe FA, Opstad KS (2003) 1H MR spectroscopy of brain tumours and masses. NMR Biomed 16:123–131.

Howe FA, Barton SJ, Cudlip SA, Stubbs M, Saunders DE, Murphy M, Wilkins P, Opstad KS, Doyle VL, McLean MA, Bell BA, Griffiths JR (2003) Metabolic profiles of human brain tumors using quantitative in vivo 1H magnetic resonance spectroscopy. Magn Reson Med 49:223–232.

Jellison BJ, Field AS, Medow J, Lazar M, Salamat MS, Alexander AL (2004) Diffusion tensor imaging of cerebral white matter: A pictorial review of physics, fiber tract anatomy, and tumor imaging patterns. AJNR Am J Neuroradiol 25:356–369.

Jiang Q, Ewing JR, Zhang ZG, Zhang RL, Hu J, Divine GW, Arniego P, Li QJ, Chopp M (2001) Magnetization transfer MRI: Application to treatment of middle cerebral artery occlusion in rat. J Magn Reson Imaging 13:178–184.

Johnson NA, Jahng GH, Weiner MW, Miller BL, Chui HC, Jagust WJ, Gorno-Tempini ML, Schuff N (2005) Pattern of cerebral hypoperfusion in Alzheimer disease and mild cognitive impairment measured with arterial spin-labeling MR imaging: Initial experience. Radiology 234:851–859.

Juh R, Kim J, Moon D, Choe B, Suh T (2004) Different metabolic patterns analysis of Parkinsonism on the 18F-FDG PET. Eur J Radiol 51:223–233.

Kaasinen V, Maguire RP, Hundemer HP, Leenders KL (2006) Corticostriatal covariance patterns of 6-[(18)F]fluoro-L-dopa and [(18)F]fluor odeoxyglucose PET in Parkinson's disease. J Neurol 253:340–348.

Kahn D, Follett KA, Bushnell DL, Nathan MA, Piper JG, Madsen M, Kirchner PT (1994) Diagnosis of recurrent brain tumor: Value of 201Tl SPECT vs 18F-fluorodeoxyglucose PET. AJR Am J Roentgenol 163:1459–1465.

Kantarci K, Jack CR, Jr. (2004) Quantitative magnetic resonance techniques as surrogate markers of Alzheimer's disease. NeuroRx 1:196–205.

Kaschten B, Stevenaert A, Sadzot B, Deprez M, Degueldre C, Del Fiore G, Luxen A, Reznik M (1998) Preoperative evaluation of 54 gliomas by PET with fluorine-18-fluorodeoxyglucose and/or carbon-11-methionine. J Nucl Med 39:778–785.

Kegeles LS, Mann JJ (1997) In vivo imaging of neurotransmitter systems using radiolabeled receptor ligands. Neuropsychopharmacology 17:293–307.

Kerrouche N, Herholz K, Mielke R, Holthoff V, Baron JC (2006) (18)FDG PET in vascular dementia: Differentiation from Alzheimer's disease using voxel-based multivariate analysis. J Cereb Blood Flow Metab 26(9):1213–1221.

Kim DM, Tien R, Byrum C, Krishnan KR (1996) Imaging in acquired immune deficiency syndrome dementia complex (AIDS dementia complex): A review. Prog Neuropsychopharmacol Biol Psychiatry 20:349–370.

Kim SG (1995) Quantification of relative cerebral blood flow change by flow-sensitive alternating inversion recovery (FAIR) technique: Application to functional mapping. Magn Reson Med 34:293–301.

Kim YK, Lee DS, Lee SK, Kim SK, Chung CK, Chang KH, Choi KY, Chung JK, Lee MC (2003) Differential features of metabolic abnormalities between medial and lateral temporal lobe epilepsy: Quantitative analysis of (18)F-FDG PET using SPM. J Nucl Med 44:1006–1012.

Kinoshita M, Yamada K, Hashimoto N, Kato A, Izumoto S, Baba T, Maruno M, Nishimura T, Yoshimine T (2005) Fiber-tracking does not accurately estimate size of fiber bundle in pathological condition: Initial neurosurgical experience using neuronavigation and subcortical white matter stimulation. Neuroimage 25:424–429.

Kitahara S, Nakasu S, Murata K, Sho K, Ito R (2005) Evaluation of treatment-induced cerebral white matter injury by using diffusion-tensor MR imaging: Initial experience. AJNR Am J Neuroradiol 26:2200–2206.

Knight RA, Nagesh V, Nagaraja TN, Ewing JR, Whitton PA, Bershad E, Fagan SC, Fenstermacher JD (2005) Acute blood-brain barrier opening in experimentally induced focal cerebral ischemia is preferentially identified by quantitative magnetization transfer imaging. Magn Reson Med 54:822–832.

Knowlton RC, Lawn ND, Mountz JM, Kuzniecky RI (2004) Ictal SPECT analysis in epilepsy: Subtraction and statistical parametric mapping techniques. Neurology 63:10–15.

Krabbe K, Karlsborg M, Hansen A, Werdelin L, Mehlsen J, Larsson HB, Paulson OB (2005) Increased intracranial volume in Parkinson's disease. J Neurol Sci 239:45–52.

Krane K (1987) Introductory Nuclear Physics. New York: John Wiley and Sons.

Kwong KK, Chesler DA, Weisskoff RM, Donahue KM, Davis TL, Ostergaard L, Campbell TA, Rosen BR (1995) MR perfusion studies with T1-weighted echo planar imaging. Magn Reson Med 34:878–887.

Law M, Cha S, Knopp EA, Johnson G, Arnett J, Litt AW (2002) High-grade gliomas and solitary metastases: Differentiation by using perfusion and proton spectroscopic MR imaging. Radiology 222:715–721.

Law M, Yang S, Wang H, Babb JS, Johnson G, Cha S, Knopp EA, Zagzag D (2003) Glioma grading: Sensitivity, specificity, and predictive values of perfusion MR imaging and proton MR spectroscopic imaging compared with conventional MR imaging. AJNR Am J Neuroradiol 24:1989–1998.

Law M, Yang S, Babb JS, Knopp EA, Golfinos JG, Zagzag D, Johnson G (2004) Comparison of cerebral blood volume and vascular permeability from dynamic susceptibility contrast-enhanced perfusion MR imaging with glioma grade. AJNR Am J Neuroradiol 25:746–755.

Lev MH, Segal AZ, Farkas J, Hossain ST, Putman C, Hunter GJ, Budzik R, Harris GJ, Buonanno FS, Ezzeddine MA, Chang Y, Koroshetz WJ, Gonzalez RG, Schwamm LH (2001) Utility of perfusion-weighted CT imaging in acute middle cerebral artery stroke treated with intra-arterial thrombolysis: Prediction of final infarct volume and clinical outcome. Stroke 32:2021–2028.

Liang Z-P, Lauterbur PC (2000) Principles of Magnetic Resonance Imaging: A Signal Processing Perspective. New York, NY: IEEE Press Marketing.

Licho R, Litofsky NS, Senitko M, George M (2002) Inaccuracy of Tl-201 brain SPECT in distinguishing cerebral infections from lymphoma in patients with AIDS. Clin Nucl Med 27:81–86.

Maria BL, Drane WE, Mastin ST, Jimenez LA (1998) Comparative value of thallium and glucose SPECT imaging in childhood brain tumors. Pediatr Neurol 19:351–357.

Masdeu JC, Zubieta JL, Arbizu J (2005) Neuroimaging as a marker of the onset and progression of Alzheimer's disease. J Neurol Sci 236:55–64.

McNally KA, Paige AL, Varghese G, Zhang H, Novotny EJ, Jr., Spencer SS, Zubal IG, Blumenfeld H (2005) Localizing value of ictal-interictal SPECT analyzed by SPM (ISAS). Epilepsia 46:1450–1464.

Mehta V, Namboodiri MA (1995) N-acetylaspartate as an acetyl source in the nervous system. Brain Res Mol Brain Res 31:151–157.

Mihara F, Hirakata R, Hasuo K, Yasumori K, Yoshida K, Kuroiwa T, Masuda K, Fukui M (1989) Gd-DTPA administered MR imaging of intracranial mass lesions: A comparison with CT and precontrast MR. Radiat Med 7:227–235.

Miller BL (1991) A review of chemical issues in 1H NMR spectroscopy: N-acetyl-L-aspartate, creatine and choline. NMR Biomed 4:47–52.

Miller SL, Daikhin Y, Yudkoff M (1996) Metabolism of N-acetyl-L-aspartate in rat brain. Neurochem Res 21:615–618.

Minagar A, Gonzalez-Toledo E, Pinkston J, Jaffe SL (2005) Neuroimaging in multiple sclerosis. Int Rev Neurobiol 67:165–201.

Mirzaei S, Rodrigues M, Koehn H, Knoll P, Bruecke T (2003) Metabolic impairment of brain metabolism in patients with Lewy body dementia. Eur J Neurol 10:573–575.

Mohana-Borges AV, Imbesi SG, Dietrich R, Alksne J, Amjadi DK (2004) Role of proton magnetic resonance spectroscopy in the diagnosis of gliomatosis cerebri: A unique pattern of normal choline but elevated Myo-inositol metabolite levels. J Comput Assist Tomogr 28:103–105.

Moller-Hartmann W, Herminghaus S, Krings T, Marquardt G, Lanfermann H, Pilatus U, Zanella FE (2002) Clinical application of proton magnetic resonance spectroscopy in the diagnosis of intracranial mass lesions. Neuroradiology 44:371–381.

Mosconi L, Perani D, Sorbi S, Herholz K, Nacmias B, Holthoff V, Salmon E, Baron JC, De Cristofaro MT, Padovani A, Borroni B, Franceschi M, Bracco L, Pupi A (2004) MCI conversion to dementia and the APOE genotype: A prediction study with FDG-PET. Neurology 63:2332–2340.

Nagesh V, Welch KM, Windham JP, Patel S, Levine SR, Hearshen D, Peck D, Robbins K, D'Olhaberriague L, Soltanian-Zadeh H, Boska MD (1998) Time course of ADCw changes in ischemic stroke: Beyond the human eye! Stroke 29:1778–1782.

Nelson JA, Dou H, Ellison B, Uberti M, Xiong H, Anderson E, Mellon M, Gelbard HA, Boska M, Gendelman HE (2005) Coregistration of quantitative proton magnetic resonance spectroscopic imaging with neuropathological and neurophysiological analyses defines the extent of neuronal impairments in murine human immunodeficiency virus type-1 encephalitis. J Neurosci Res 80:562–575.

Nelson SJ (2003) Multivoxel magnetic resonance spectroscopy of brain tumors. Mol Cancer Ther 2:497–507.

Nelson SJ, McKnight TR, Henry RG (2002) Characterization of untreated gliomas by magnetic resonance spectroscopic imaging. Neuroimaging Clin N Am 12:599–613.

Neumann-Haefelin T, Wittsack HJ, Wenserski F, Siebler M, Seitz RJ, Modder U, Freund HJ (1999) Diffusion- and perfusion-weighted MRI. The DWI/PWI mismatch region in acute stroke. Stroke 30:1591–1597.

Novak B, Milcinski M, Grmek M, Kocmur M (2005) Early effects of treatment on regional cerebral blood flow in first episode schizophrenia patients evaluated with 99Tc-ECD-SPECT. Neuro Endocrinol Lett 26:685–689.

Petroff OA, Pleban LA, Spencer DD (1995) Symbiosis between in vivo and in vitro NMR spectroscopy: The creatine, N-acetylaspartate, glutamate, and GABA content of the epileptic human brain. Magn Reson Imaging 13:1197–1211.

Pirotte B, Goldman S, Massager N, David P, Wikler D, Vandesteene A, Salmon I, Brotchi J, Levivier M (2004) Comparison of 18F-FDG and 11C-methionine for PET-guided stereotactic brain biopsy of gliomas. J Nucl Med 45:1293–1298.

Poptani H, Gupta RK, Roy R, Pandey R, Jain VK, Chhabra DK (1995) Characterization of intracranial mass lesions with in vivo proton MR spectroscopy. AJNR Am J Neuroradiol 16:1593–1603.

Pyhtinen J (2000) Proton MR spectroscopy in gliomatosis cerebri. Neuroradiology 42:612–615.

Rijpkema M, Schuuring J, van der Meulen Y, van der Graaf M, Bernsen H, Boerman R, van der Kogel A, Heerschap A (2003) Characterization of oligodendrogliomas using short echo time 1H MR spectroscopic imaging. NMR Biomed 16:12–18.

Ross BD (1991) Biochemical considerations in 1H spectroscopy. Glutamate and glutamine; myo-inositol and related metabolites. NMR Biomed 4:59–63.

Ross B, Michaelis T (1994) Clinical applications of magnetic resonance spectroscopy. Magn Reson Q 10:191–247.

Rowley HA (2005) Extending the time window for thrombolysis: Evidence from acute stroke trials. Neuroimaging Clin N Am 15:575–587.

Rugg-Gunn FJ, Eriksson SH, Boulby PA, Symms MR, Barker GJ, Duncan JS (2003) Magnetization transfer imaging in focal epilepsy. Neurology 60:1638–1645.

Salat DH, Tuch DS, Hevelone ND, Fischl B, Corkin S, Rosas HD, Dale AM (2005) Age-related changes in prefrontal white matter measured by diffusion tensor imaging. Ann N Y Acad Sci 1064:37–49.

Schlemmer HP, Bachert P, Herfarth KK, Zuna I, Debus J, van Kaick G (2001) Proton MR spectroscopic evaluation of suspicious brain lesions after stereotactic radiotherapy. AJNR Am J Neuroradiol 22:1316–1324.

Schramm P, Schellinger PD, Klotz E, Kallenberg K, Fiebach JB, Kulkens S, Heiland S, Knauth M, Sartor K (2004) Comparison of perfusion computed tomography and computed tomography angiography source images with perfusion-weighted imaging and diffusion-weighted imaging in patients with acute stroke of less than 6 hours' duration. Stroke 35:1652–1658.

Seppi K, Schocke MF (2005) An update on conventional and advanced magnetic resonance imaging techniques in the differential diagnosis of neurodegenerative parkinsonism. Curr Opin Neurol 18:370–375.

Simone IL, Tortorella C, Federico F, Liguori M, Lucivero V, Giannini P, Carrara D, Bellacosa A, Livrea P (2001) Axonal damage in multiple sclerosis plaques: A combined magnetic resonance imaging and 1H-magnetic resonance spectroscopy study. J Neurol Sci 182:143–150.

Slichter C (1996) Principles of Magnetic Resonance, 3 Edition. Heidelberg, Germany: Springer-Verlag.

Smith WS, Tsao JW, Billings ME, Johnston SC, Hemphill JC, 3rd, Bonovich DC, Dillon WP (2006) Prognostic significance of angiographically confirmed large vessel intracranial occlusion in patients presenting with acute brain ischemia. Neurocrit Care 4:14–17.

Stanisz GJ, Yoon RS, Joy ML, Henkelman RM (2002) Why does MTR change with neuronal depolarization? Magn Reson Med 47:472–475.

Stokkel M, Stevens H, Taphoorn M, Van Rijk P (1999) Differentiation between recurrent brain tumour and post-radiation necrosis: The value of 201Tl SPET versus 18F-FDG PET using a dual-headed coincidence camera—a pilot study. Nucl Med Commun 20:411–417.

Swartz BE, Brown C, Mandelkern MA, Khonsari A, Patell A, Thomas K, Torgersen D, Delgado-Escueta AV, Walsh GO (2002) The use of 2-deoxy-2-[18F]fluoro-D-glucose (FDG-PET) positron emission tomography in the routine diagnosis of epilepsy. Mol Imaging Biol 4:245–252.

Tae WS, Joo EY, Kim JH, Han SJ, Suh YL, Kim BT, Hong SC, Hong SB (2005) Cerebral perfusion changes in mesial temporal lobe epilepsy: SPM analysis of ictal and interictal SPECT. Neuroimage 24:101–110.

Taylor DL, Davies SE, Obrenovitch TP, Doheny MH, Patsalos PN, Clark JB, Symon L (1995) Investigation into the role of N-acetylaspartate in cerebral osmoregulation. J Neurochem 65:275–281.

Theodore WH, Sato S, Kufta CV, Gaillard WD, Kelley K (1997) FDG-positron emission tomography and invasive EEG: Seizure focus detection and surgical outcome. Epilepsia 38:81–86.

Topcuoglu V, Comert B, Karabekiroglu A, Dede F, Erdil TY, Turoglu HT (2005) Right basal ganglion hypoperfusion in obsessive compulsive disorder patients demonstrated by Tc-99m-HMPAO brain perfusion spect: A controlled study. Int J Neurosci 115:1643–1655.

Tsuchiya K, Fujikawa A, Nakajima M, Honya K (2005) Differentiation between solitary brain metastasis and high-grade glioma by diffusion tensor imaging. Br J Radiol 78:533–537.

Urenjak J, Williams SR, Gadian DG, Noble M (1992) Specific expression of N-acetylaspartate in neurons, oligodendrocyte-type-2 astrocyte progenitors, and immature oligodendrocytes in vitro. J Neurochem 59:55–61.

Van Laere K, Ceyssens S, Van Calenbergh F, de Groot T, Menten J, Flamen P, Bormans G, Mortelmans L (2005) Direct comparison of 18F-FDG and 11C-methionine PET in suspected recurrence of glioma: Sensitivity, inter-observer variability and prognostic value. Eur J Nucl Med Mol Imaging 32:39–51.

Wang A, Shetty A, Woo H, Roo SK, Manzione J, Moore J (1998) Diffusion Weighted MR Imaging in Evaluation of CNS Disease. Rivista di Neuroradiologia 11:109–112.

Wang J, Alsop DC, Li L, Listerud J, Gonzalez-At JB, Schnall MD, Detre JA (2002) Comparison of quantitative perfusion imaging using arterial spin labeling at 1.5 and 4.0 Tesla. Magn Reson Med 48:242–254.

Welch KM, Nagesh V, D'Olhaberriague LD, Zhang ZG, Boska MD, Patel S, Windham JP (2001) Automated three-dimensional signature model for assessing brain injury in emergent stroke. Cerebrovasc Dis 1(11 Suppl):9–14.

Williams DS, Detre JA, Leigh JS, Koretsky AP (1992) Magnetic resonance imaging of perfusion using spin inversion of arterial water. Proc Natl Acad Sci USA 89:212–216.

Wintermark M, Flanders AE, Velthuis B, Meuli R, van Leeuwen M, Goldsher D, Pineda C, Serena J, van der Schaaf I, Waaijer A, Anderson J, Nesbit G, Gabriely I, Medina V, Quiles A, Pohlman S, Quist M, Schnyder P, Bogousslavsky J, Dillon WP, Pedraza S (2006) Perfusion-CT assessment of infarct core and penumbra. Receiver operating characteristic curve analysis in 130 patients suspected of acute hemispheric stroke. Stroke 37:979–985.

Witwer BP, Moftakhar R, Hasan KM, Deshmukh P, Haughton V, Field A, Arfanakis K, Noyes J, Moritz CH, Meyerand ME, Rowley HA, Alexander AL, Badie B (2002) Diffusion-tensor imaging of white matter tracts in patients with cerebral neoplasm. J Neurosurg 97:568–575.

Yablonskiy DA, Neil JJ, Raichle ME, Ackerman JJ (1998) Homonuclear J coupling effects in volume localized NMR spectroscopy: Pitfalls and solutions. Magn Reson Med 39:169–178.

Yarnykh VL, Yuan C (2004) Cross-relaxation imaging reveals detailed anatomy of white matter fiber tracts in the human brain. Neuroimage 23:409–424.

Young RJ, Ghesani MV, Kagetsu NJ, Derogatis AJ (2005) Lesion size determines accuracy of thallium-201 brain single-photon emission tomography in differentiating between intracranial malignancy and infection in AIDS patients. AJNR Am J Neuroradiol 26:1973–1979.

Yu CS, Li KC, Xuan Y, Ji XM, Qin W (2005) Diffusion tensor tractography in patients with cerebral tumors: A helpful technique for neurosurgical planning and postoperative assessment. Eur J Radiol 56:197–204.

Yuh WT, Crain MR, Loes DJ, Greene GM, Ryals TJ, Sato Y (1991) MR imaging of cerebral ischemia: Findings in the first 24 hours. AJNR Am J Neuroradiol 12:621–629.

Zhao YB, Sun BM, Li DY, Wang QS (2004) Effects of bilateral subthalamic nucleus stimulation on resting-state cerebral glucose metabolism in advanced Parkinson's disease. Chin Med J (Engl) 117:1304–1308.

Glossary

Δ⁹THC: tetrahydrocannabinol is the major psychoactive ingredient in the Cannabisplant. **Δ⁹THC** is responsible for both the psychiatric and therapeutic effects obtained from marijuana. Its receptor, the cannabinoid receptor, is located mainly tat the presynaptic gap. The areas of the brain most affected are the basal ganglia, cerebellum, cerebral cortex, and the hippocampus. The acute effects consist of degradation in short term memory, changes in sensory perception, reduced concentration, disturbances in motor abilities, hypothermia, increased blood pressure and heart rate, and reduced pain perception.

6-Hydroxydopamine: a neurotransmitter analogue that depletes noradrenergic stores in nerve endings and reduces dopamine levels in brain. Its mechanism of action is linked to cytolytic free-radical production.

A

ACAID: Anterior chamber-associated immune deviation-a unique form of immune tolerance. ACAID is a selective, systemic immune deficiency where Antigen-specific delayed hypersensitivity and complement-fixing antibody responses are impaired by lymphocyte responses while other immune effectors are left preserved.

Ach: Acetylcholine is the chemical responsible for transmitting impulses between neurons in the central and peripheral (parasympathetic and somatic) nervous systems. It is the chemical that allows nerve cells to communicate with one another.

acetylcholinesterase: the enzyme responsible for the destruction of acetylcholine at the synaptic cleft (region between nerve cells) after its neuronal release during neurotransmission.

action potential: Is a spike of electrical discharge that is propagated alone the cell membrane. It is essential for information transmission amongst and from nerve cells to tissues throughout the body.

active transport: movement of materials across cell membranes and epithelial layers against an electrochemical gradient, requiring the expenditure of metabolic energy. The process involves the use of cellular energy and commonly ATP to actively pump substances into or out of the cell.

active vaccination: the process of injecting a protein (an antigen) into a host, usually together with an adjuvant (a molecule to enhance immunity) for the purpose of preventing disease. In response, the host immune system generates an immune response that includes specific cells and antibodies against the antigen. Although most commonly used to prevent microbial infections, active vaccination has been used to ameliorate disease.

active zone: specialized region of the cortical cytoplasm of the presynaptic nerve terminal that faces the synaptic cleft. It is the place for synaptic vesicle docking and neurotransmitter release.

ADEM: acute disseminated encephalomyelitis is an immune mediated disease of brain. It is brief but significant and results in direct myelin damage. It usually occurs following a viral infection or vaccination (commonly for measles, mumps or rubella), but it may also appear spontaneously.

acute flaccid paralysis: loss of voluntary motion, due to temporary or permanent dysfunction of lower motor neurons or associated nerve fibers, neuromuscular junction, or muscle fibers. It is the most common sign of acute polio but is also associated with West Nile Virus infection. In chronic cases of permanent flaccid paralysis muscle tone is decreased and muscle atrophy may occur.

adaptive immune system: the adaptive immune system evolved in early vertebrates and allows for a specific and targeted response based upon immunological memory of antigens or related antigens previously encountered by the host innate immune systems. This requires the recognition of specific "non-self" antigens. The effectors of the adaptive immune response include T cells and B cells.

adenoviruses: these are a family of non-enveloped medium-sized (90–100 nm), icosahedral double-stranded DNA viruses that infect various mammalian species, including humans. These viruses commonly cause mild respiratory infections. Adenovirus-based vectors have been used for gene delivery into the nervous system, as high vector titers can be generated, and these episomally maintained vectors efficiently infect and express transgenes in a variety of cell types including the non-dividing cells.

AAV: adeno-associated virus is a non-pathogenic member of the parvovirus family that harbors a single-stranded DNA viral genome. Vectors derived from AAV can carry up to a 4.5-kb transgene and stably express a transgene *in vivo* months to several years.

adjunctive: something joined or added to another thing but not essentially a part of it, used in this context to refer to additional therapies for HIV-related neurological conditions.

adjuvant: a formulation of substances that is used to enhance/potentiate immune response to an antigen. Complete adjuvants are water in oil emulsions composed of inactivated or killed mycobacteria-usually *Mycobacteriumm tuberculosis*. Whereas incomplete adjuvants are the same adjuvants, but without the mycobacterial components.

(adsorptive) endocytosis: cellular uptake of extracellular materials within membrane-limited vacuoles or microvesicles. ENDOSOMES play a central role in endocytosis.

ACTH: adrenocorticotrophic hormone is a peptide hormone released by anterior pituitary cells in response to stressful stimuli that causes the synthesis and release of cortisol (corticosteroid) from the adrenal cortex. It is an important component of the hypothalamic-pituitary-adrenal (HPA) axis. ACTH is released from pro-opiomelanocortin and secreted from corticotropes in response to corticotropin-releasing hormone (CRH) released by the hypothalamus.

(Adsorptive) transcytosis: a mechanism for transcellular transport in which a cell encloses extracellular material in an invagination of the cell membrane to form a vesicle, then moves the vesicle across the cell to eject the material through the opposite cell membrane by the reverse process.

adult stem cell: a cell found in adult somatic tissue that can differentiate into the cells of the particular tissue in which it resides and can self-renew while maintaining its differential capability.

AIDS: Acquired Immunodeficiency Syndrome (AIDS) is an acquired 1defect of cellular immunity associated with infection by the human immunodeficiency virus (HIV), a CD4-positive T-lymphocyte count under 200 cells/μL or less than 14% of total lymphocytes, and increased susceptibility to opportunistic infections and malignant neoplasms. Clinical manifestations also include emaciation (wasting) and dementia. These elements reflect criteria for AIDS as defined by the CDC in 1993.

Akt: serine/threonine protein kinase that is a cellular homologue of the viral oncogene v-Akt. There are three genes in the Akt family: Akt1, Akt2, and Akt3. Akt1 is involved in cell survival pathways by inhibiting apoptosis. Akt2 is an important signaling molecule in the Insulin signaling pathway, and is required to induce insulin transport. The role of Akt3 is unknown, but it is predominately expressed in the brain.

alkylosing spondylitis: literally "stiff vertebra" is an autoimmune inflammatory disease that primarily affects the intervertebral joints (joints of the spine) and the sacro-iliac joint.

allele: the name given to a specific polymorphic variation.

allele association: association with a specific allele but not necessarily a genotype.

allograft: tissue transplanted from the same species but of a different genotype.

alpha adrenergic receptor: one of two general classes of G protein-coupled receptors that can be further sub-classified and responds to the neurotransmitter norepinephrine.

α-MSH: alpha melanocyte stimulating hormone is a tridecapeptide derived from proopiomelanocortin that stimulates production of melanin and acts on melanocortin-1 receptors found on immune cells.

alpha synuclein: an abundant presynaptic neuronal protein found in the brain of uncertain function, but thought to act as a molecular chaperone involved in vesicular transport. Mutation and/or oxidation of alpha-synuclein increase its propensity to aggregate and is a major component of Lewy bodies that plays a role in neurodegeneration and neuroprotection.

AD: Alzheimer's disease is the most common degenerative brain disease of unknown cause resulting in progressive mental deterioration with disorientation, memory disturbance and confusion. AD leads to progressive dementia, often accompanied by dysphasia and/or dyspraxia. The condition may also give rise ultimately to spastic weakness and paralysis of the limbs, epilepsy and other variable neurological signs. Alzheimer disease is marked histologically by the degeneration of brain neurons especially in the cerebral cortex and by the presence of neurofibrillary tangles and plaques comprised of highly insoluble β-amyloid peptides in the CNS.

amacrine cell: lateral inhibitory interneuron in the inner retina located at the inner plexiform layer where bipolar cells and ganglion cells synapse. Amacrine cells deliver the majority of the ganglion cells input, and also regulate the output of the cone bipolar cells. These cells are thus responsible for the complex processing of the retinal image by adjusting brightness as well as detecting motion.

amphetamine: a central nervous stimulant that increases neuronal activity by releasing norepinephrine, dopamine and/or serotonin from nerve endings.

amygdala: an almond-shaped mass of gray matter, one in each hemisphere of the brain, associated with feelings of fear and aggression and important for visual learning and memory.

Aβ: Amyloid-beta is a peptide of 39–43 amino acids that is the main constituent of amyloid plaques in the brains of Alzheimer's disease patients. Similar plaques appear in some variants of Lewy body dementia and in inclusion body myositis, a muscle disease. Aβ also forms aggregates coating cerebral blood vessels in cerebral amyloid angiopathy.

APP: amyloid precursor protein is a single transmembrane molecule, a causative gene of FAD, and is acted upon by proteases to form the amyloid-β peptide.

ALS: Amyotrophic lateral sclerosis is a rare fatal progressive degenerative disease of unknown cause that affects motor neurons, usually begins in middle age, and is characterized by muscular weakness; called also Lou Gehrig's disease.

amyloid: an abnormal aggregation of proteins or protein fragments with a beta-pleated sheet secondary structure.

animal models: normal animals modified mechanically, genetically or chemically to create laboratory paradigms with infrahuman species (usually rodents) which study behavior, neurochemistry, or physiology to demonstrate all or part of the characteristics of a disease.

antibodies repertoire: antibody genes are pieced together from widely scattered bits of DNA. The possible combinations are nearly endless. As this gene forms, it assembles segments that will determine the variable-V, diversity-D, joining-J, and constant-C segments of this antibody molecule. For example, variable-region and joining-region rearrangements of the immunoglobulin light chain called V_LJ_L rearrangements.

antibody titer: the concentration of a specific antibody in blood.

antidepressant drugs: pharmacological agents used for ameliorating symptoms of depressive disorders, including monoamine oxidase inhibitors, selective serotonin reuptake inhibitors, serotonin/norepinephrine reuptake inhibitors, and other agents.

anti-ganglioside antibodies: circulating autoantibodies directed against gangliosides. Anti-ganglioside antibodies are associated with axonal forms of GBS and can be induced experimentally in laboratory animals.

antigen presenting cells (APC): a cell that displays foreign antigens complexed with MHC on its surface. T-cells may recognize this complex using their T-cell receptor (TCR).

antigen: the molecule against which an immune response is raised.

anti-inflammatory cytokines: a series of immunoregulatory molecules that suppress proinflammatory responses.

antipsychotic drugs: pharmacological agents used to treat psychotic disorders, including schizophrenia, mania, and psychotic depression. These drugs are usually classified as "typical" or "atypical," depending on their affinity for the dopamine D_2 receptor, lack of extrapyramidal side effects, and action upon serotonin receptors.

antiretroviral therapy: use of anti-HIV agents to treat disease by lowering viral titer and reducing viral replication, usually in combination of at least three drugs.

antithymocyte globulin: purified gamma globulin from the serum of rabbits immunized with human thymocytes. It is used for induction of immunosuppression in the treatment of acute renal transplant rejection.

apoE: apolipoprotein E is a component of lipoprotein complex both in plasma and cerebrospinal fluid, responsible for cholesterol transport in the blood stream. One copy of the ApoE4 allele is an established risk factor of AD.

apoptosis: a genetically determined process resulting in cell self-destruction that is marked by the fragmentation of nuclear DNA, and is a normal physiological process eliminating DNA-damaged, superfluous, or unwanted cells (as immune cells targeted against the self in the development of self-tolerance). Apoptosis may occur prematurely in neurodegenerative conditions; also called programmed cell death.

arbor: dendritic arbor, a single dendrite, the arbor density is called arborization.

association: describes a relationship between a disease and a specific allele or haplotype of a polymorphism.

astrocytes: major glial cells in the CNS (star-shaped cell as of the neuroglia). Astrocytes are characterized by histopathologically by glial fibrillary associated protein (GFAP). Astrocytes perform many functions including formation of the blood brain barrier, provide neurotrophic factors and metabolic support to the CNS, and have a role both in the repair and scarring process in the brain.

atropine: an alkaloid isolated from the belladonna plant that blocks the action of acetylcholine on muscarinic cholinergic receptors.

autonomic nervous system: the portion of the peripheral nervous system which acts autonomously through either of two divisions (parasympathetic and sympathetic) and regulates the action of smooth muscle, exocrine glands and the heart.

axon: the long extension of a neuron that carries nerve impulses away from the cell body.

axonal transport: bidirectional transport process, consisting of anterograte (transport of presynaptic molecules) and retrograte transport systems (neurotrophic signaling).

axon terminals: the hair-like ends of the axon.

atherosclerosis: characterized by the formation of plaques within the innermost layer of artery walls and narrowing of the vessel lumen. Vascular occlusion in the brain leads to stroke and in the coronary arteries leads to heart attack.

autoantibodies: immunoglobulins directed against self-antigens, representing an autoimmune process.

autoimmune disease: a condition in which a person's immune cells recognize self-produced molecules (self-antigens), leading to the inappropriate destruction of host tissue, e.g. rheumatoid arthritis, multiple sclerosis, and possibly Addison's disease.

autoimmune T cells: T cells that recognize self-antigens and are responsible for cell-mediated destruction of host tissue in an autoimmune disease.

autoimmunity: a condition in which an individual's immune system reacts against his or her own tissues, causing diseases such as lupus erythemathosus.

axonal damage: via inflammation not only the myelin sheath, but also the axon itself is destroyed.

B

Bad: a pro-apoptotic member of the Bcl-2 gene family involved in initiating apoptosis. BAD-Bcl-2-associated death promoter protein.

β-amyloid peptide (Aβ): a principal component of senile plaque, a pathological hallmark of AD, which is generated by β and γ-processing of APP.

barbiturates: a group of sedative-hypnotic agents that also have anti-anxiety effects. Barbiturates act on GABA-A receptors.

basiliximab: humanized monoclonal antibody directed against the IL-2 receptor on T cells. It is used for the prophylaxis of acute organ rejection in adult patients.

Bax: a protein of the Bcl-2 gene family that promotes apoptosis by competing with Bcl-2.

Bcl2: the prototype for a family of mammalian proteins located in the membranes of the endoplasmic reticulum (ER), nuclear envelope, and in the outer membranes of the mitochondria. They can be either pro-apoptotic (Bax, BAD, Bak and Bok among others) or anti-apoptotic.

BCR: B cell receptor is an antigen-specific receptor is a sample of the antibody it is prepared to manufacture; it recognizes antigen in its natural state.

Bcl11a: C2H2 type zinc-finger protein. The corresponding mouse gene is a common site of retroviral integration in myeloid leukemia, and may function as a leukemia disease gene, in part, through its interaction with BCL6. During hematopoietic cell differentiation, this gene is down-regulated. It is possibly involved in lymphoma pathogenesis since translocations associated with B-cell malignancies also deregulate its expression.

BDNF: brain derived neurotrophic factor is a member of the nerve growth factor family of trophic factors. In the brain, BDNF has a trophic action on retinal, cholinergic, and dopaminergic neurons, and in the peripheral nervous system it acts on both motor and sensory neurons. (From Kendrew, The Encyclopedia of Molecular Biology, 1994).

belladonna alkaloid: any of several compounds that block the effect of acetylcholine on muscarinic cholinergic receptors.

benzodiazepines: a group of anti-anxiety agents that can also produce sedation, hypnosis and that have anti-convulsant effects. In the CNS, benzodiazepines act on GABA-A receptors. They also act on peripheral benzodiazepine receptors in the immune system.

beta-amyloid precursor protein converting enzyme (BACE): a single transmembrane aspartic proteinase, which processes APP at β-site; the β-secretase.

beta adrenergic receptor: one of two general classes of G protein-coupled receptors that can be further sub-classified and responds to the neurotransmitter norepinephrine.

beta-pleated sheet: a molecular structure where proteins spontaneously fold into a layered shape, analogous to the accordion folds of a fan.

bethanechol: a derivative of acetylcholine with stimulant effects primarily on the smooth muscle of the gastrointestinal tract and the urinary bladder.

bipolar cell: retinal interneuron that transfers visual information from photoreceptors to ganglion cells.

biogenic amine hypothesis: the hypothesis that major depressive disorder and bipolar disorder are functionally related to abnormal levels of catecholamines and/or serotonin.

Bioinformatics: field of science in which biology and computer science merge into a single discipline to create, manage and interpret massive sets of complex biological data.

Biomarker: distinctive biological indicator of biological state, process or condition.

bipolar disorder: a mental disorder characterized by cyclic episodes of mania or hypomania, usually accompanied by depressive episodes.

block copolymer: a polymer in which all of one type of monomer are grouped together, and all of the other type are grouped together. A block copolymer can be thought of as two homopolymers joined together at the ends.

block ionomer complex: a complex between two oppositely charged block copolymers.

blood-brain barrier: specialized non-fenestrated tightly-joined ENDOTHELIAL CELLS with TIGHT JUNCTIONS that form a transport barrier for certain substances between the cerebral capillaries and the BRAIN tissue.

BLIMP1: a protein that acts as a repressor of beta-interferon gene expression. The protein binds specifically to the PRDI (positive regulatory domain I element) of the β-IFN gene promoter. Transcription of this gene increases upon virus induction.

botulinum toxin: a group of zinc proteases that can degrade one or more of the key proteins required for the process of exocytosis thus preventing the neuronal release of acetylcholine.

brain: the part of the central nervous system contained within the cranium, comprising the prosencephalon, mesencephalon, and rhombencephalon. It is derived from the anterior part of the embryonic neural tube.

brain endothelial cell: highly specialized cells that form the blood-brain barrier.

butoxamine: an antagonist of norepinephrine action selective for the subtype two beta adrenergic receptor.

C

C5a: the 74-residue anaphylatoxin derived from C5 during complement activation that possesses inflammatory and immune stimulatory properties causing the accumulation of white blood cells at the site of complement activation.

C5a agonist: a peptide mimetic of natural C5a that exhibits certain C5a-like activities.

Ca^{2+} signaling: signal transduction mechanism in which Ca^{2+} is recruited from extracellular or intracellular stores to the cytoplasm in response to external stimuli. Elevation of intracellular Ca^{2+} activates a series of enzymes.

CAG repeats: primary determinant of the onset of HD and other neurodegenerative disorders, including SBMA, DRPLA, SCA 1–3, 6, 7 and 17. CAG codes for the amino acid Glutamine, and extra repeats is thought to lead to a "toxic gain of function," and increased propensity of the translated protein to form aggregates. Genetic testing of the HD CAG repeat is a useful tool for clinical diagnosis.

calcium excitability: refers to the dynamics of calcium homeostasis.

cAMP signaling: signal transduction cascade initiated by the action of external stimuli on binding to a G-protein couple receptor (Gs) resulting in activation of adenylyl cyclase, production of cyclic AMP and activation of cyclic AMP-dependent protein kinase.

Campylobacter jejuni: Gram-negative bacterium, which is one of the most frequent causes of bacterial gastroenteritis world-wide. *Campylobacter jejuni* gastroenteritis is the most frequently recognized preceding infection in GBS.

cannabinoids: substances that bind to cannabinoid receptors.

case-control: study describes an association study that compares allele frequencies using unrelated affected cases and matched controls.

caspase: known as proteases, which play essential roles in apoptosis (cell death) and inflammation.

CD1d: the sole group-2 member of the CD1 family of major histocompatibility (MHC)-like glycoproteins.

CD3: complex of delta, epsilon, gamma, zeta and eta chains of integral membrane glycoproteins that associates with T cell antigen receptor (TCR), and is required for TCR cell surface expression and signal transduction. CD3 is a universally expressed by T cells. Member of immunoglobulin superfamily.

CD4: nonpolymorphous glycoproteins belonging to immunoglobulin superfamily. Expressed on surface of T helper cells, accessory cells, macrophages and serves as co-receptor in MHC class II-restricted antigen induced T cell activation. Major receptor for HIV-1.

CD5: belongs to ancient scavenger receptor superfamily. CD5+ B cells, which may arise from B-1 cells (subset of B cells) produce "generalist antibodies"—polyreactive low affinity "natural" antibodies to exogenous antigens (tetanus toxoid, lipopolysaccharide) as well as autoreactive antibodies. CD5 may serve as a dual receptor, giving either stimulatory or inhibitory signals depending both on the cell type and the development stage. Key regulator of immune tolerance; abnormalities may produce autoimmunity.

CD8: cell surface glycoprotein, member of immunoglobulin superfamily. Heterodimer of an alpha and a beta chain linked by two disulfide bonds; heterodimer on thymocytes and homodimer on peripheral blood T cells MHC class I restricted receptor; binds to nonpolymorphic region of class I molecules; may increase avidity of interactions between cytotoxic T cell and target cell during antigen-specific activation. Can kill target cells by recognizing peptide-MHC complexes on them or by secreting cytokines capable of signaling through death receptors on target cell surface. CD8 alpha cells promote survival and differentiation of activated T cells into memory CD8+ T cells, which may become clonal (but not malignant) in the elderly.

CD11b: member of integrin receptor family; also called integrin alpha M, Mac-1. Mediates adhesion to substrates by opsonization with iC3b and subsequent phagocytosis, neutrophil aggregation, and chemotaxis. Ligands include fibrinogen,

Factor X, ICAM1, iC3b, some bacteria. CD11b is a marker for both macrophages and microglia, and expression increases with cell-activation.

CD11c: member of b2 of integrin receptor family; also called integrin alpha X, CR4, LeuM5. Clears opsonized particles and immune complexes; also binds to fibrinogen and is involved in adhesion of monocytes and neutrophils to endothelium. Myeloid cell marker.

CD19: response regulator that plays a dominant role in establishing signaling thresholds for antigen receptors and other surface receptors on B cells; also regulates B cell development, activation and differentiation. Assembles with the antigen receptor of B lymphocytes in order to decrease the threshold for antigen receptor-dependent stimulation. Co-receptor with CD21. Earliest B cell antigen in fetal tissue. B cell marker.

CD21: complement component receptor-2 (CR2) that binds to C3d.

CD25: IL-2 receptor alpha chain; exists in at least three forms. Limited expression may safeguard against catastrophic T-cell proliferation by immunogenic stimulus. Expressed, typically at high levels, on PHA-stimulated T cells, and is commonly used as a marker for activated T cells. Expressed on B cells stimulated with anti-IgM antibody. Expressed on monocytes/macrophages stimulated with LPS. Full signaling of IL-2 requires trimer with CD122 (IL-2R beta) and CD132 (IL-2R gamma) chains.

CD27: marker of T cell activation. CD27/CD70 interactions also regulate B cell proliferation and T cell differentiation. The protein is a member of the TNF-receptor superfamily. This receptor is required for generation and long-term maintenance of T cell immunity. It binds to ligand CD70, and plays a key role in regulating B-cell activation and immunoglobulin synthesis. This receptor transduces signals that lead to the activation of NF-kappaB and MAPK8/JNK. Adaptor proteins TRAF2 and TRAF5 have been shown to mediate the signaling process of this receptor. CD27-binding protein (SIVA), a pro-apoptotic protein, can bind to this receptor and is thought to play an important role in the apoptosis induced by this receptor.

CD28: cell adhesion molecule; co-receptor for B cell-T cell cooperation. Promotes T cell activation. Receptor for CD80 (B7.1) & CD86 (B7.2), found on activated B cells. Constitutive, high abundance, low affinity receptor; opposite signals are mediated by CTLA4 (CD152). Result of T cell antigen stimulation depends on sum of effects of T cell receptor, CD28 and its ligand, CTLA4 and its ligand. CD8+, CD28+ T cells mediate antigen specific cytotoxic T cells (class I restricted) (90% of CD8+ T cells). CD8+, CD28− T cells mediate suppressor T cells (10% of CD8+ T cells). CD28 costimulation is essential for CD4-positive T-cell proliferation, survival, interleukin-2 production, and T-helper type-2 (Th2) development.

CD30: tumor necrosis factor receptor superfamily, member 8, is a member of the TNF-receptor superfamily. This receptor is expressed by activated, but not by resting, T and B cells. TRAF2 and TRAF5 can interact with this receptor, and mediate the signal transduction that leads to the activation of NF-kappaB. This receptor is a positive regulator of apoptosis, and also has been shown to limit the proliferative potential of autoreactive CD8 effector T cells and protect the body against autoimmunity. Two alternatively spliced transcript variants of this gene encoding distinct isoforms have been reported.

CD34: cell-cell adhesion molecule and cell surface glycoprotein. May mediate attachment of stem cells to bone marrow extracellular matrix or directly to stromal cells. Constitutively expressed on endothelial cells.

CD38: a multifunctional ectoenzyme widely expressed in cells and tissues especially in leukocytes and endothelial cells. CD38 also functions in cell adhesion, signal transduction and calcium signaling. Positive and negative regulator of cell activation and proliferation, depending on the cellular environment.

CD40: member of the TNF-receptor superfamily. This receptor has been found to be essential in mediating a broad variety of immune and inflammatory responses including T cell-dependent immunoglobulin class switching, memory B cell development, and germinal center formation. AT-hook transcription factor AKNA is reported to coordinately regulate the expression of this receptor and its ligand, which may be important for homotypic cell interactions. Adaptor protein TNFR2 interacts with this receptor and serves as a mediator of the signal transduction. The interaction of this receptor and its ligand is found to be necessary for amyloid-beta-induced microglial activation, and thus is thought to be an early event in Alzheimer disease pathogenesis. Two alternatively spliced transcript variants of this gene encoding distinct isoforms have been reported.

CD44: family of cell surface glycoproteins with isoforms generated by alternate splicing of mRNA. CD44 is important in epithelial cell adhesion (cell-cell and cell-matrix) to hyaluronate in basement membranes and in maintaining polar orientation of cells; also binds laminin, collagen, osteopontin, matrix metalloproteinases (MMPs) and fibronectin. Involved in leukocyte attachment and rolling on endothelial cells, homing to peripheral lymphoid organs and sites of inflammation and leukocyte aggregation, hematopoiesis, and tumor metastasis.

CD45: leukocyte common antigen (LCA); a protein tyrosine phosphatase. Critical requirement for T and B cell antigen receptor-mediated activation. Expressed on all hematopoietic cells except ethrocytes and platelets at different levels. Some isoforms of CD45 are CD45RO, CD45RA, and CD45RB. Each CD45 isoform is distinguished from one another depending on the type of exon the CD45 has or the exons the CD45 does not have. The CD45RA isoform contains the A exon only and the CD45RB has the B exon only whereas the CD45RO has neither the none of the exons: A, B, or C.

CD45RA: CD45RA—expressed by naive/resting T cells, medullary thymocytes

CD45RO: CD45RO—expressed by memory/activated T cells, cortical thymocytes

CD56: N-CAM (neuronal adhesion molecule). Regulates homophilic (like-like) interactions between neurons and between neurons and muscle. Also associates with fibroblast growth factor receptor and stimulates tyrosine kinase activity of receptor to induce neurite outgrowth. When neural crest cells stop making N-CAM and N-cadherin and start displaying integrin receptors, cells separate and migrate. Contributes to cell-cell or cell-matrix adhesion during development. Lymphocyte activated killer phenomenon mediated by IL-2 activated CD56+, CD3-, NK cells. Prototypic marker of NK cells, also present on subsets of CD4+ and CD8+ T cells.

CD62L: L selectin; LECAM-1. Mediates lymphocyte homing to high endothelial venules of peripheral lymphoid tissue, leukocyte rolling on activated endothelium at inflammatory sites.

CD64: high affinity receptor binds to Fc region of IgG. CD64 is important in phagocytosis via receptor-mediated endocytosis of IgG-antigen complexes. Mediates antigen capture for presentation to T cells, antibody-dependent cellular cytotoxicity, release of cytokines and reactive oxygen intermediates.

CD68: a 110-kDa transmembrane glycoprotein that is highly expressed by human monocytes and tissue macrophages. It is a type I integral membrane protein with a heavily glycosylated extracellular domain. May have a role in macrophage phagocytic activities. Specific to lysosomes, not cell lineage. Expressed by macrophage/monocytes, basophils, dendritic cells, mast cells, myeloid cells, CD34+ progenitor cells, neutrophils, osteoclasts, activated platelets, B and T cells.

CD117 (c-Kit): the tyrosine kinase type receptor. The c-Kit gene encodes a 145-kDa transmembrane protein that is a member of the receptor tyrosine kinase subclass III family that includes platelet-derived growth factor receptor (PDGF-R), macrophage colony-stimulating factor receptor (M-CSF-R or c-fms), and fms-like tyrosine kinase-3/fetal liver kinase-2 (Flt-3 /Flt-2). The ligand for c-Kit, Steel factor (SF) or kit ligand, stem cell factor, and mast cell growth factor, exists both as a secreted and membrane-bound form where the latter appears to be most important for biologic activity *in vivo*. Important for development and survival of mast cells, hemopoietic stem cells, melanocytes, germ cells, and interstitial cells of Cajal.

CD122: IL-2 receptor beta chain.

CD123: interleukin 3 receptor, alpha (low affinity). The receptor is comprised of a ligand specific alpha subunit and a signal transducing beta subunit shared by the receptors for interleukin 3 (IL-3), colony stimulating factor 2 (CSF2/GM-CSF), and interleukin 5 (IL-5). The binding of this protein to IL-3 depends on the beta subunit. The beta subunit is activated by the ligand binding, and is required for the biological activities of IL-3.

CD134 (OX40): tumor necrosis factor receptor superfamily, member 4. This receptor has been shown to activate NF-kappaB through its interaction with adaptor proteins TRAF2 and TRAF5. Knockout studies in mice suggest that this receptor promotes the expression of apoptosis inhibitors BCL2 and BCL2lL1/BCL2-XL, and thus suppresses apoptosis. The knockout studies also suggest the roles of this receptor in $CD4^+$ T cell response, as well as in T cell-dependent B cell proliferation and differentiation.

CD154: CD40 ligand (TNF superfamily, member 5, hyper-IgM syndrome) is expressed on the surface of T cells. It regulates B cell function by engaging CD40 on the B cell surface. A defect in this gene results in an inability to undergo immunoglobulin class switch and is associated with hyper-IgM syndrome.

CD278: inducible T-cell co-stimulator, belongs to the CD28 and CTLA-4 cell-surface receptor family. It forms homodimers and plays an important role in cell-cell signaling, immune responses, and regulation of cell proliferation.

cell body: the body of the neuron containing the nucleus (also called the soma or perikaryon).

cell-mediated delivery: delivery of therapeutic agents to the site of action inside carrier cells.

central memory t cells: Memory T cells that express L-selectin and CC-chemokine receptor 7 (CCR7) and have the capacity to circulate from the blood to the secondary lymphoid organs. They have a nonpolarized differentiation state in that they secrete IL-2 but not IFN-γ or IL-4; however, on restimulation, they rapidly differentiate into cytokine-producing effector cells.

central nervous system (CNS): The main information-processing organs of the nervous system, consisting of the brain, spinal cord, and meninges that controls and coordinates most function of the body and mind.

centroblast: a proliferating germinal-center B cell, which undergoes somatic hypermutation and immunoglobulin class switching.

centrocyte: the non-dividing progeny of a centroblast. These cells need to be selected on the basis of affinity for antigen, following interaction with immune complexes that are associated with follicular dendritic cells, and ability to elicit help from follicular B helper T (T_{FH}) cells.

ceramide: a sphingosine-based lipid molecule that regulates a wide spectrum of biological processes such as cellular differentiation, proliferation, apoptosis and senescence.

CSF: cerebrospinal fluids are fluid in the ventricles of the brain, between the arachnoid and pia mater, and surrounding the spinal cord that absorbs shocks and maintains uniform pressure.

cerebral amyloid angiopathy: the deposition of β-amyloid in brain blood vessels.

chemoattractant-receptor homologous molecule expressed by t_h2 cells (crth2): A cell-surface marker for human T helper 2 cells.

chemokines: a group of secreted proteins within the family of cytokines that by definition relate to the induction of migration. These "**chemo**tactic cyto**kines**" are produced by and target a wide variety of cells, but primarily address leukocyte chemoattraction and trafficking of immune cells to locations throughout the body via a concentration-dependent gradient. Chemokines are categorized on the basis of the protein structure according to the types of cysteine motifs (e.g., C–C, C–X–C)

chemotaxis: attraction and trafficking of cells to locations via a gradient.

choline acetyltransferase: the enzyme present in nerve endings that is responsible for the esterification of choline yielding the neurotransmitter acetylcholine.

cholinergic receptors: one of two general classes of receptors (either G protein-coupled or ligand-gated) that can be further sub-classified and responds to the neurotransmitter acetylcholine.

chorea: an abnormal voluntary movement disorder, and subcategory of dyskinesias, which are caused by overactivity of the neurotransmitter dopamine in the areas of the brain that control movement. Chorea is a primary feature of HD and important diagnostic symptom.

chromogranin A: a type of chromogranin which was first isolated from chromaffin cells of the adrenal medulla but is also found in other tissues and in many species including human, bovine, rat, mouse, and others. It is an acidic protein with 431 to 445 amino acid residues. It contains fragments that inhibit vasoconstriction or release of hormones and neurotransmitter, while other fragments exert antimicrobial actions.

chronic inflammatory neuropathy: a chronic autoimmune neuritis, which involves demyleination of peripheral nerves and spinal roots. This disease is thought to be mediated by autoantibody production to peripheral myelin. The effector cell types include macrophages, B cells, and CD4+ T cells.

choroid: a layer of the eye adjacent to the RPE containing a dense bed of fenestrated capillaries (also called the choriocapillaris).

clade: a group of biological taxa (as species) that includes all descendants of one common ancestor.

class-switch recombination: the process by which proliferating B cells rearrange their DNA to switch from expressing IgM (or another class of immunoglobulin) to expressing a different immunoglobulin heavy-chain constant region, thereby producing antibody with different effector functions.

clinical trials: an experiment to test quality, value, or usefulness of a therapeutic agent in human subjects.

CNPase: refers to 2′,3′-cyclic nucleotide 3′-phosphodihydrolase that is present in myelin.

CNS homeostasis: level of maintenance needed for normal central nervous system health and function.

combination therapy: the use of two or more therapies, especially drugs, to treat a disease or condition.

cocaine: a compound present in the leaves of coca trees that is a central nervous stimulant, which increases neuronal activity by lengthening the duration of action of norepinephrine, dopamine and/or serotonin in the neuronal synapse.

complement system: a host defense mechanism made up of a group of serum proteins involved in the control of inflammation, the activation of humoral and acquired immune responses, and the lytic attack on cell membranes.

complex disease: is a disease due to the interaction of multiple susceptibility genes and/or environmental risk factors. Most common medical problems fall within this category

cone: photo-sensitive retinal cell mediating color vision and responsive to higher light levels than rods.

connexins (Cx): connexins, or gap junction proteins, are a family of structurally-related transmembrane proteins that assemble to form vertebrate gap junctions.

corticosteroids: a group of natural and synthetic hormones that are immunosuppressive agents and are used for treatment of transplant rejection and autoimmune disorders. Natural corticosteroids are synthesized from cholesterol in the adrenal cortex and are involved in the stress response and regulation of inflammation, carbohydrate metabolism, protein catabolism, blood electrolyte levels, and behavior. Corticosteroids are further classified as glucocorticoids and mineralocorticoids.

CRH: corticotropin releasing hormone is a peptide hormone produced by cells in the hypothalamus in response to stressful stimuli that causes the synthesis and release of ACTH from anterior pituitary cells.

Creutzfeldt-Jakob disease (CJD): the most common prion disease of humans and can have a sporadic, familial or acquired etiology.

Crohn's disease: is a chronic autoimmune disease affecting the gastrointestinal tract, and can affect any region from mouth to anus.

CXCR4 receptor: the receptor for the chemokine stromal cell-derived factor (SDF-1).

cyclic nucleotide 3′ phosphohydrolase (CNPase): a component of myelin that demonstrates increased activity in active stages of MS:

cyclooxygenase-2 (COX-2): an inducibly-expressed subtype of prostaglandin-endoperoxide synthase. It plays an important role in many cellular processes and inflammation. It is the target of cox2 inhibitors.

cyclophosphamide: a nitrogen mustard alkylating agent that is used as an anticancer agent and immunosuppressive agent mainly in the treatment of autoimmune diseases.

cyclosporine: a cyclic polypeptide that acts as a calcineurin inhibitor. It is an immunosuppressive agent that is used in organ transplantation and autoimmune diseases.

cytokines: regulatory soluble glycoproteins (8–30 kDa), such as interleukins and lymphokines, secreted by inflammatory leukocytes and some non-leukocytic cells that act as intercellular mediators. They differ from classical hormones in that they are produced by a number of tissue or cell types rather than by specialized glands. They generally act locally in a paracrine or autocrine rather than endocrine manner, and facilitate communication among immune system cells and between immune system cells and the rest of the body.

cytotoxic T lymphocytes (CTL): T cells that can kill other cells. Most cytotoxic T cells are MHC class I-restricted CD8+ T cells. Cytotoxic T cells are important in host defense against cytosolic pathogens.

D

daclizumab: humanized monoclonal antibody directed against the IL-2 receptor on T cells. It is used for the prophylaxis of acute organ rejection in adult patients.

delusions: fixed false beliefs

dementia: a term describing memory loss and other cognitive impairments that result from a number of causes, including Alzheimer's disease; marked by the development of multiple cognitive deficits (including memory impairment, aphasia, and inability to plan and initiate complex behavior). Deteriorated mentality is often accompanied by emotional apathy.

demyelination: the act of demyelinating, or the loss of the myelin sheath insulating the nerves, and is the hallmark of some neurodegenerative autoimmune diseases, including: multiple sclerosis, transverse myelitis, chronic inflammatory demyelinating polyneuropathy, Guillain-Barre Syndrome, and adrenoleukodystrophy. When myelin degrades, conduction of signals along the nerve can be impaired or lost, and the nerve eventually withers.

dendrites: the branching structures projecting from the cell body of a neuron that receive messages from other neural cells and work to transmit these messages via electrical impulses to the cell body from which they project.

dendritic spine: a small protrusion from the dendrite comprised of a head that forms one half of a synapse and a thinner neck that connects the head to the dendrite. Responsible for the controlled diffusion of molecules from the synapse to the dendrite. They contain the highest concentrations of actin, which retains dynamic activity and can drive rapid changes in its shape and therefore its function.

dentate granule cell layer: tiny neurons (10 μm in diameter) located in the dentate gyrus of the hippocampus, which contain glutamatergic projection axons, and are one of only two major types of adult neuronal populations to undergo neurogenesis.

dentate gyrus: a subfield in hippocampus.

depression: mood disorder (clinically, major depressive disorder or a milder form referred to as dysthymia) characterized by persistent sadness, hopelessness, cognitive impairment and other symptoms that interfere with work, sleep, etc. and in severe cases can lead to suicide.

dexamethasone suppression test: measures cortisol levels in response to dexamethasone (synthetic cortisol) administration. Specifically, it tests the response of adrenal glands to ACTH (adrenocorticotropic hormone which is secreted from the anterior pituitary gland of the brain). Cortisol levels should decrease following dexamethasone administration. Abnormally high cortisol levels are seen following dexamethasone administration in up to 50% of patients with a major depressive disorder.

differentiation: the ability of a pluripotent cell to become a specialized cell type by expressing distinct lineage specific genes and undergoing morphological changes.

dobutamine: a norepinephrine derivative that mimics norepinephrine action primarily at the subtype one beta adrenergic receptor.

dopamine: one of the catecholamine neurotransmitters in the brain. It is derived from tyrosine and is the precursor to norepinephrine and epinephrine. Dopamine is a major transmitter in the extrapyramidal system of the brain, and important in regulating movement. A family of receptors (receptors, dopamine) mediate its action.

dopamine hypothesis: the hypothesis that schizophrenia is functionally related to excessive dopamine activity. Evidence for this is the observation that many antipsychotics have DA-antagonistic effects.

dopaminergic neurons: neurons that utilize dopamine as a neurotransmitter.

doublecortin: binds to the microtubule cytoskeleton, stabilizing them and causing bundling of microtubules.

drug delivery: the transport of therapeutic agents to the site of action.

d-tubocurarine: an alkaloid that blocks the action of acetylcholine on skeletal muscles causing paralysis

dynorphins: endogenous opioid peptides, 17 amino acids in length, derived from prodynorphin.

dysthymia: minor depression, with fewer and less intense symptoms as major depressive disorder.

E

E2A: immunoglobulin enhancer binding factors, also E12/E47. Binds the immunologlobulin κ chain enhancer and is involved in promoter regulation.

EAE: experimental allergic/autoimmune encephalitis is an animal model for demyelinating diseases, such as multiple sclerosis. It is induced by injection of myelin basic protein or whole CNS tissue together with adjuvants.

EBF: early B-cell factor.

effector memory t cells: cells that have an L-selectin$^-$ CCR7 phenotype. They have immediate effector functions, including rapid production of cytokines (IFN-γ or IL-4), and they migrate to sites of inflammation, such as the skin and the gut. They recognize foreign invaders with a faster and stronger immune response than the initial immune response.

embryonic germ cell (EG): a totipotent cell that can differentiate into all the cells of the organism, self-renew while maintaining its differential capability, the precursor for adult germ cells, and can be isolated from primordial germ cells of 5–9-week human fetuses.

embryonic stem cell (ES): a multipotent stem cell that can differentiate into all the cells of the organism except the trophoblast, self-renew while maintaining its differential capability, and is isolated from the inner cell mass of the blastocyst stage embryo.

encephalopathy: a disease of the brain; especially one involving alterations of brain function and/or structure.

endocytosis: uptake of molecules that cannot readily pass through the cell membrane through engulfment by the membrane resulting in the formation of a vesicle. This process is employed for a variety of functions within the cell such as reloading of synaptic vesicles with vesicular components at the active site through receptor-mediated endocytosis.

endorphins: endogenous opioid peptides, 31 amino acids in length, derived from proopiomelanocortin.

endothelins: 21-Amino-acid peptides produced by vascular endothelial cells and functioning as potent vasoconstrictors. The endothelin family consists of three members, ENDOTHELIN-1; ENDOTHELIN-2; and ENDOTHELIN-3. All three peptides contain 21 amino acids, but vary in amino acid composition. The three peptides produce vasoconstrictor and pressor responses in various parts of the body. However, the quantitative profiles of the pharmacological activities are considerably different among the three isopeptides.

enkephalins: short endogenous opioid peptides derived from proenkephalin and proopiomelanocortin.

epidemiology: science of the incidence, distribution, and control of disease in a population.

epidermal growth factor (EGF): a 6-kDa polypeptide growth factor initially discovered in mouse submaxillary glands. Human epidermal growth factor was originally isolated from urine based on its ability to inhibit gastric secretion and called urogastrone. EGF exerts a wide variety of biological effects including the promotion of proliferation and differentiation of mesenchymal and epithelial cells.

epinephrine: an analog of norepinephrine that is formed in the adrenal medulla and acts primarily as a hormone on either alpha or beta adrenergic receptors. Responsible for the "fight or flight" response to danger.

EPSC: excitatory postsynaptic currents are the flow of ions resulting in EPSP

EPSP: excitatory postsynaptic potentials are temporary increases in postsynaptic membrane potentials caused by the flow of positively charged ions into the postsynaptic cell. The influx of positively charged ions leads to an excitatory response therefore simplifying the neuron's generation of an action potential.

Epstein Barr virus (EBV): a gamma human herpesvirus.

exocytosis: vesicles carrying molecules to be secreted from the cell merge with the cell membrane resulting n the vesicular contents being released to the exterior of the cell. Thi is the method by which neurotransmitters are released at the active site.

experimental allergic encephalomyelitis (EAE): an animal model of brain inflammation, used mostly with rodents, and serves as a model of the human disease multiple sclerosis.

experimental allergic neuritis (EAN): animal model of acute inflammatory demyelinating polyradiculoneuropathy that shares many features with human GBS. EAN can be induced in rodents, rabbits, and monkeys by immunization with peripheral nerve homogenate, myelin, and myelin proteins such as P0.

F

facilitated diffusion: A process of passive transport where molecules move across membranes with the assistance of transport proteins

family-based association study: is an association study that uses families as the source, where the parents are the controls and affected are their offspring.

Fas: Fas/APO-1/CD95 (36 kDa) is a member of the tumor necrosis factor (TNF) receptor superfamily. Fas has been shown to be an important mediator of apoptotic cell death, as well as being involved in inflammation.

fatal insomnia: rare prion disease of humans that is initially characterized by sleeplessness, later developing into hallucinations, followed by extreme weight loss, dementia and finally death

ferritin: iron-containing proteins that are widely distributed in animals, plants, and microorganisms. Their major function is to store iron in a nontoxic bioavailable form. Each ferritin molecule consists of ferric iron in a hollow protein shell (apoferritins) made of 24 subunits of various sequences depending on the species and tissue types.

fibroblast growth factor (FGF)-2: a single-chain polypeptide growth factor that plays a significant role in the process of wound healing and is a potent inducer of physiologic angiogenesis. Several different forms of the human protein exist ranging from 18–24 kDa in size due to the use of alternative start sites within the fgf-2 gene. It has a 55% amino acid residue identity to fibroblast growth factor 1 and has potent heparin-binding activity. The growth factor is an extremely potent inducer of DNA synthesis in a variety of cell types from mesoderm and neuroectoderm lineages. It was originally named basic fibroblast growth factor based upon its chemical properties and to distinguish it from acidic fibroblast growth factor (Fibroblast Growth Factor 1).

fetal liver kinase (Flk)-2: a member of the tyrosine kinase receptor family, this 135–150-kDa molecule is expressed by multipotential progenitor cells including primitive B cell and myelomonocytic progenitors in fetal liver and adult bone marrow. Also called as Fms-like tyrosine kinase 3 (Flt3) and CD135 in human.

follicular dendritic cells (FDCs): cells found in lymph follicles that present immune complexes to B cells, aiding in B cell maturation.

fovea: the high acuity region of the retina upon which the center of the visual image is focused. The fovea contains only cones.

G

gadolinium: a radio-opaque dye injected in conjunction with magnetic resonance imaging (MRI) to evaluate the brain parenchyma for breaches in the cerebrovascular system.

ganglia: a collection of cell bodies of the postganglionic motor neurons which join synapse with nerves from the central nervous system; they act as relay stations for CNS impulses to peripheral effector cells and as control loops for reflex actions.

gangliosides: sialic acid-containing glycosphingolipids which are widely distributed in mammalian tissues but highly enriched in the nervous system, due to their incorporation into the myelin sheath. Gangliosides consist of a tetraose core to which variable numbers of sialic acids are attached; the ceramide or lipid portion of the molecule is attached to the internal glucose. GM1, GD1b, GD1a, and GT1b are the most abundant complex gangliosides in the nervous system.

gap junctions: gap junction or nexus is a junction between certain animal cell-types that allows cell-to-cell passage of ions, hormones, and neurotransmitters. One gap junction is composed of two connexons (or hemichannels), which connect across the intercellular space, therefore connecting the cytoplasm of the cells. They are analogous to the plasmodesmata that join plant cells.

GATA3: binding protein 3. GATA-3 expression inhibits the differentiation of Th1 cells *in vivo*, induces Th2 cell differentiation, and increases functional capacity to secrete Th2 cytokines and enhance surface expression of markers for antigen-experienced Th2-committed cells. GATA-3 expression must be sustained to maintain the Th2 phenotype.

GDNF: glial cell line-derived neurotrophic factor is the founding member of the glial cell line-derived neurotrophic factor family. It was originally characterized as a nerve growth factor promoting the survival of midbrain dopaminergic neurons, and it has been studied as a potential treatment for Parkinson disease.

general anesthetics: a group of diverse agents used in surgery to produce loss of pain perception and loss of consciousness. They depress most CNS modalities, including respiration.

gene arrays: high throughput technique used to analyze different expression profiles using a collection of gene-specific nucleic acids.

gene modifiers: are genes that affect the expression of the phenotype of a trait.

genetic heterogeneity: describes when a disorder has more than one genetic cause that causes the same clinical presentation.

genome: one haploid set of chromosomes with the genes they contain; *broadly*: the genetic material of an organism.

genomic convergence: a process in which multiple independent genomic research techniques are used to identify possible causal genes, with the overlap between the results in each technique converged to produce the genes with the highest likelihood of being involved in the disorder.

Genomics: field of science that combines genetics and molecular biology techniques to create genetic maps and DNA sequences; analyze and organize genetic information into databases.

genotype association: association with a specific genotype but not necessarily an allele.

germinal center: a lymphoid structure that arises within follicles after immunization with, or exposure to, a T-cell-dependent antigen. It is specialized for facilitating the development of high-affinity, long-lived plasma cells and memory B cells.

Gerstmann-Straussler-Scheinker (GSS): syndrome is a prion disease linked to germ line mutations or insertions in the human prion protein gene resulting in a neurodegenerative brain disorder.

GFAP: glial fibrillary acidic proteins form the main intermediate filament in astrocytes and are therefore used as the marker protein of astrocytes.

glatiramer acetate: an FDA approved therapy for relapsing remitting multiple sclerosis that

consists of acetate salts of polypeptides that are found in myelin and contains four naturally

occurring amino acids randomly synthesized: L-glutamic acid, L-alanine, L-tyrosine, and L-lysine; available as Copaxone, 20 mg of glatiramer acetate and 40 mg of mannitol for subcutaneous injection daily.

glia: non-neural cellular element that has diverse functions in the nervous system including participation in innate immunity, myelin formation, signal transmission, and homeostasis maintenance.

glial activation: refers to activation of glial cells (astrocytes and microglia) in response to proinflammatory stimuli triggered by injury or disease. This can usually be detected by the presence of glial fibrillary acidic protein (GFAP) in astrocytes.

glial scar: scar formed due to insults to the CNS.

glucocorticoids: steroid hormones (esp. cortisol in humans) produced by the adrenal cortex in response to stressful stimuli.

glutamate: the most important excitatory neurotransmitter in the CNS. Glutamate binds to either metabotropic, or ionotropic receptors. Ligation of glutamate at the NMDA receptor—one of the ionotropic receptors—induces calcium influx into the neuron. Astrocytes are crucially involved in glutamate metabolism, storage and re-uptake.

glutamine: an amino acid that is formed from glutamate. Glutamine in excess is toxic for the brain.

growth factor: growth factors stimulate cell proliferation and/or differentiation during embryonic development, tissue growth and wound healing. The term "growth factors," used in this chapter, also refers to a broad range of structurally diverse molecular families and individual proteins.

guanine nucleotide binding proteins (G-proteins): membrane-associated, heterotrimeric proteins composed of three subunits: alpha, beta and gamma subunits. G proteins and their receptors (GPCRs) form one of the most prevalent signaling systems in mammalian cells, regulating systems as diverse as sensory perception, cell growth and hormonal regulation Binding of ligands such as hormones and neurotransmitters to a G-protein coupled receptors causes a conformational change, which in turn activates the bound G protein on the intracellular-side of the membrane. The activated receptor promotes the exchange of bound GDP for GTP on the G protein alpha subunit.

Guillain Barré syndrome: an acute monophasic neuropathy characterized by an autoimmune response to peripheral myelin, and is the most common acquired demyelinating peripheral neuropathy. Pathogenesis involves mononuclear cell infiltration in the peripheral nerves, along with both B cells and CD4+ helper T cells. Symptoms begin as tingling sensations and weakness in the legs and in extreme cases can lead to the development of total paralysis. Current treatment include high-dose intravenous Ig and plasma exchange.

H

hallucinations: sensory misperceptions in which a person will experience an auditory, visual, or other sensory experience in the absence of an observable stimuli

haplogroup: a haplotype of the mitochondrial genome.

Haplotype: with two polymorphisms in linkage disequilibrium the alleles of those two polymorphisms travel together as a single combined allele or haplotype.

herpes simplex virus (HSV): A naturally neurotropic virus capable of establishing latent infection within neurons, but also possesses the ability to infect a wide range of cellular targets. Recombinant HSV vectors comprise a wild-type HSV genome rendered replication defective via disruption/deletion of an indispensable viral gene(s). Typically, the immediate-early gene loci, which encode for potent transactivation proteins that initiate the viral lytic cycle, are targeted for insertion of therapeutic transcription units via homologous recombination. The HSV-1 amplicon, another type of HSV vector, is a uniquely designed eukaryotic expression plasmid that harbors two non-protein encoding virus-derived elements: an HSV origin of DNA replication (OriS) and the cleavage/packaging sequence ("a" sequence). Both *cis* sequences are specifically recognized by HSV proteins to promote the replication and incorporation of the vector genome into viable viral particles, respectively. This highly versatile plasmid can be readily manipulated to contain desired promoters, enhancers, and transgenes of substantial size (~130 kb).

hemorrhagic stroke: accounts for 15–20% of all strokes. Rupture of a blood vessel into the brain parenchyma results in an intracerebral hemorrhage (ICH); this is commonly caused by hypertension-induced vascular disease of the small perforating intracranial arteries. Rupture of a vessel into the subarachnoid space results in a subarachnoid hemorrhage; this most commonly results from rupture of a brain aneurysm or an arteriovenous malformation.

high endothelial venules (hevs): specialized venules found in lymphoid tissues, which are the site of entry for lymphocytes from the bloodstream into lymph nodes and Peyer's patches.

hippocampus: part of the brain located inside the temporal lobe. It forms a part of the limbic system and plays a part in memory and spatial navigation.

histone acetyl transferase: enzymes that acetylate conserved lysine residues on histone proteins by transferring an acetyl group from acetyl CoA to form -N-acetyl lysine.

histone deacetylase: a class of enzymes that remove acetyl groups from an -N-acetyl lysine on histones.

HIV-1 associated dementia: HIV-1 associated dementia is a cognitive disorder specific to HIV, HAD is the clinical consequence of HIV-1 infection in the CNS and chronic inflammatory response that ultimately causes neuronal damage.

horizontal cell: lateral inhibitory interneuron in the outer retina.

HTLV-I-associated myelopathy/tropical spastic paraparesis (HAM/TSP): a chronic progressive inflammatory disorder of the CNS caused by infection with HTLV. This disorder affects 0.25–3% of HTLV-I-infected individuals.

human herpes virus 6 (HHV-6): a beta human herpesvirus, which is a common opportunistic viral infection for the immunosupressed.

hypothalamo-pituitary-adrenocortical (HPA) axis: the major endocrine route in stress response. Activation of the HPA axis results in the secretion of corticotropin-releasing factor (CRF) from the hypothalamic paraventricular nucleus (PVN), adrenocorticotropin (ACTH) from the pituitary, and glucocorticosteroids (cortisol) from the adrenal cortex. Cytokines such as IL-1, IL-6, TNF-α are known to activate the HPA axis.

human leukocyte antigen (HLA): the genetic designation for the human major histocompatibility complex (MHC). This group of genes resides on chromosome 6, and encodes cell-surface antigen-presenting proteins and many other genes. Individual loci are designated by upper-case letters, as in HAL-A, and alleles are designated by numbers, as in HLA-A0201. HLA-B27 is a genetic marker present in patients alkylosing spondylitis and Reiter's syndrome

human T cell lymphotropic virus type I (HTLV-I): an exogenous human retrovirus that infects 10–20 million people worldwide. Endemic areas of HTLV-I infection include southern Japan, certain regions of Africa and South American, and the Caribbean. HTLV-I is the etiologic agent of both an aggressive T cell leukemia and an inflammatory disorder of the CNS.

Huntingtin (htt): a causative gene of Huntington's Disease (HD), with its N-terminal CAG trinucleotide repeat expansion correlated with the onset of the disease. Predicted to function as a scaffolding molecule for multiple transporting/transcriptional molecules in cytoplasm and nucleus. Inclusion of htt and/or its N-terminal polyglutamine tract have been proposed as a potential mechanism of neurodegeneration in HD.

Huntington Disease: a progressive neurodegenerative disorder that is inherited as an autosomal dominant trait, that usually begins in middle age, that is characterized by choreiform movements (involuntary, rapid, jerky movements) and mental deterioration leading to dementia. Histologically, Huntington disease shows atrophy of the caudate nucleus and the loss of brain cells with a decrease in the level of several neurotransmitters—called also Huntington's—abbreviation HD:

hydrophilicity: a physico-chemical characteristic of a compound attributed to its ability to dissolve in water solutions.

hyphenated techniques:combination of individual techniques, such as liquid chromatography (LC) and mass spectrometry (MS), to create one encompassing method (LC-MS) capable of new analytical possibilities.

hypomania: a less severe and less persistent form of mania. Seen in such disorders as bipolar II and cyclothemia.

hypothalamic-pituitary-adrenal (HPA) axis: neuroendocrine pathway that mediates the stress response by regulating systemic cortisol levels and brain CRH levels.

I

ID2: a helix-loop-helix protein that can inhibit transcription factors with a basic helix-loop-helix motif. Besides its role in lymphoid organogenesis, ID2 is required for the generation of natural killer cells.

IFNβ-1a: an FDA approved therapy for relapsing remitting multiple sclerosis. IFNβ-a is available as Avonex (Biogen-Idec) 33 mcg for intramuscular injection (IM) once a week and Rbif (Serono) with doses of 22 mcg and 44 mcg for subcuntaneous (SQ) injection 3×/week.

IFNβ-1b: an FDA approved therapy for relapsing remitting multiple sclerosis, available as Betaseron for subcutaneous injection 22 μg (6 MIU) per dose for 66 μg/week and 44 μg (12 MIU) per dose for 132 μg/week.

IKAROS: this gene encodes a family of zinc-finger transcription factors that regulate transcription required for the development of all lymphoid lineages, as well as lymph nodes and Peyer's patches.

IL-7R: a receptor for interleukine 7 (IL-7). The function of this receptor requires the interleukin 2 receptor, gamma chain (IL2Rγ), which is a common gamma chain shared by the receptors of various cytokines, including interleukines 2, 4, 7, 9, 15, 21. This protein has been shown to play a critical role in the V(D)J recombination during lymphocyte development. This protein is also found to control the accessibility of the TCR gamma locus by STAT5 and histone acetylation. Knockout studies in mice suggested that blocking apoptosis is an essential function of this protein during differentiation and activation of T lymphocytes. The functional defects in this protein may be associated with the pathogenesis of the severe combined immunodeficiency (SCID).

imipramine: an agent effective in the treatment of depression that acts to increase neuronal activity by lengthening the duration of action of norepinephrine and/or serotonin in the neuronal synapse.

immune deviation: modification of cells having or producing antibodies or lymphocytes capable of an immune response to an antigen after prior exposure to that antigen.

immune neuropathies: heterogeneous group of neuropathies with an assumed autoimmune pathogenesis. This includes acute forms such as Guillain-Barré Syndrome (GBS) and chronic conditions such as chronic inflammatory demyelinating polyneuropathy (CIDP) as well as polyradiculoneuropathies associated with monoclonal paraprotein.

immune neuropathy: disease of the peripheral nerves that is mediated by autoimmune processes, either cellular or humoral autoimmunity

immune privileged: organs or tissues where entry of effector lymphocytes is blocked by anatomic barriers and is not recognized by immune cells due to absence or low levels of expression of MHC class I and II antigens.

immunoglobulins: natural immunomodulatory compounds that provide passive immunization and are used in a variety of diseases.

immunomodulator: a type of adjuvant that directly activates cells of the immune system, usually via cytokine release

immunosuppression: a decrease in any functional assay of immune competence from its baseline level in normal animals.

immunotherapy: the treatment of a disease by vaccination or by administering antibodies.

infliximab: chimeric anti-TNF-α monoclonal antibody. It is used in the treatment of rheumatoid arthritis and other autoimmune diseases.

innate immune system: the dominant system of host defense in most organisms. Innate immune defenses are non-specific, and do not confer long-lasting immunity against a pathogen.

inflammation: a local response to cellular injury that is marked by capillary dilatation, leukocytic infiltration, and often loss of function. Inflammation serves as a mechanism initiating the elimination of noxious agents and of damaged tissue:

insulin degrading enzyme (IDE): responsible for the degradation of the β–chain of insulin, expressed abundantly in microglia and also other neuronal cells in brain.

insulin like growth factor: resembles insulin but is contained in part of a complex used to communicate with their physiologic environment, contributing to cell such processes as proliferation and inhibition of cell death. The complex consists of two cell-surface receptors, two ligands, and six high affinity binding proteins.

intercellular adhesion molecule 1 (ICAM-1): molecules that promote adhesion between cells.

interferon-gamma: the major interferon produced by mitogenically or antigenically stimulated lymphocytes. It is structurally different from type I interferon and its major activity is immunoregulation. It has been implicated in the expression of class II histocompatibility antigens in cells that do not normally produce them, leading to autoimmune disease.

interferons (IFNs): natural and recombinant compounds with important immunomodulatory activities, such as inflammation, antigen presentation, and antiviral activity. IFN type I binds to IFN-α receptor complex, and IFN type II binds to IFN-γ receptor complex.

interleukin: a group of cytokines (secreted signaling molecules) that mediate autocrine, paracrine and endocrine communication between cells of the immune system and other systems of the body. They enhance cell proliferation and differentiation, proteins DNA produced by the cells of the immune system of most vertebrates in response to challenges by foreign agents such as viruses, bacteria, parasites and tumor cells. They were first seen to be expressed by white blood cells (leukocytes, hence the -leukin) as a means of communication (inter-). The name is something of a relic though; it has since been found that interleukins are produced by a wide variety of bodily cells. The function of the immune system depends in a large part on interleukins. They are designated by IL and a number. IL-1 causes fever. IL-12 biases towards a Th1 response. IL-10 is anti-inflammatory. IL-8 is chemotactic. The interleukins are comprised of some of the cytokines and some of the chemokines.

interleukin-2-recombinant form of IL-2: an immunostimulant used to treat adults with metastatic renal carcinoma and melanoma.

intraocular pressure: a measurement of the fluid pressure inside the eye. This fluid, or aqueous humor, nourishes the cornea, iris, and lens, and it helps the eye maintain its globular shape. Normal eye pressure, as measured by an eye doctor, usually ranges between 10 and 21 mm of mercury, with an average of 16.

ion channel: protein that mediates the transport of ions across membranes through the formation of a pore in the cell membrane, therefore allowing for the establishment of a voltage gradient across the cell membrane.

ipratropium: a derivative of atropine that blocks the action of acetylcholine on muscarinic cholinergic receptors.

IRF4 and IRF8: IFN regulatory factors (IRFs) are a family of transcription factors involved in the regulation of both innate and adaptive immunity. By gene knockout studies, IRF1, IRF4, and IRF8 have been found to be essential for the proper differentiation and functions of immune cells. Lymphocyte development requires IRF1 and IRF4, whereas IRF7 and IRF8 are needed for monocyte-to-macrophage differentiation, and together with IRF1, IRF8 orchestrates the development of Th1 immune responses by regulating the gene expression of IL-12, the major Th1 immune response-inducing cytokine. Recently, IRF4 and IRF8 have also been shown to participate in DC differentiation and functions. An interesting feature of IRF4 and IRF8 is their ability to function as transcriptional activators and repressors, depending on the promoter context.

ischemic stroke: account for around 80–85% of all strokes and are nearly always caused by arterial vascular occlusions; rarely occlusion in the cerebral venous system may result in ischemic and/or hemorrhagic stroke.

isoproterenol: an analog of norepinephrine that mimics norepinephrine action primarily at the beta adrenergic receptors. It is used to treat asthma, chronic bronchitis, and emphysema; by relaxing the airways to allow for increased airflow.

isotypes: classification of various immunoglobulins by the type of chain they are composed of. This classification varies between species.

J

JAK–STAT pathway: a signaling pathway utilized by cells to respond to cytokines and growth factor stimuli. This pathway transduces the signals carried by cytokines to the cell nucleus, where activated STAT proteins modify gene expression. JAKs—Janus Kinases; STATs—Signal Transducers and Activators of Transcription.

K

keratinocytes: epithelial cells dividing in the stratum basale of the epidermis, they accumulate keratin, undergo apoptosis and are sluffed off at the skin surface. In psoriasis, the cycle is accelerated from 30 days to 7–10 days.

KIRs: killer immunoglobulin like receptors are MHC-binding receptors on NK cells that triggers NK cell-mediated responses. KIRs diversify human natural killer cell populations and T cell subpopulations. Ly49, the analogous system in rodents.

Kuru: a prion disease of the Fore speaking tribes in the New Guinea highlands in which disease was transmitted via endocannibalism.

kynurenic acid: the only known endogenous NMDA receptor antagonist. Kynurenic acid is the end product of a side arm of the kynurenine pathway. Besides its neuroprotective NMDA binding properties, kynurenic acid also blocks the α7 nicotinic acetylcholine receptor.

kynurenine pathway: more than 95% of tryptophan in the organism are metabolized through the kynurenine pathway, which is entitled by the first stable intermediate, kynurenine. The activity of one of the key enzymes, indoleamine 2,3-dioxygenase, is regulated by cytokines: IFN-γ and TNF-α are inducers, while IL-4 and IL-10 are inhibitors of IDO. Neuroactive kynurenine pathway intermediates include the free-radical generators 3-hydroxykynurenine, 3-hydroxyanthranilic acid, and quinolinic acid. The latter is also an excitotoxic NMDA receptor agonist. The major end product of the kynurenine pathway is nicotinamide adenine dinucleotide phosphate ($NADP^+$/NADPH), a major source of electrons e.g. for reductive biosynthesis.

L

learned helplessness: a condition induced by inescapable, unavoidable, unpleasant stress, in which impaired coping and maladaptive behavior results.

LEF1: lymphoid enhancer-binding factor 1 is a 48-kDa nuclear protein expressed in pre-B and T cells. LEF1 binds to a functionally important site in the T-cell receptor-alpha enhancer and confers maximal enhancer activity. LEF1 belongs to a family of regulatory proteins that share homology with high mobility group protein-1.

lentivirus: a member of the retrovirus family of enveloped RNA viruses associated with causing serious immunosuppressive diseases (i.e., AIDS). Vectors derived from lentiviruses can harbor a transgene up to 9 kb in size and express the transgene *in vivo* months to years.

leucine-rich repeat kinase-2 (LRRK2): a complex protein with multiple domains, including a leucine-rich repeat (LRR), a ROC-COR GTPase, a mitogen-activated protein kinase kinase kinase, and WD40 domains. Mutations have been found in all domains in LRRK2 and have been identified as the cause of the late-onset, autosomal-dominant Park8 type of Parkinson's disease.

Lewy body: Intracytoplasmic neuronal inclusions which are intensely eosinophilic under routine haematoxylineosin stain. Immuocytochemically, Lewy bodies share epitopes with phosphorylated and non-phosphorylated α-synclein, neurofilament subunits, tubulin, microtubule-associated protein 1 and 2, and positively immunostain with ubiqiutin. Lewy bodies are classically associated with Parkinson's disease, but also are present in a distinct form of dementia termed diffuse Lewy body disease, described recently in the elderly.

limbic system: an interconnected system of brain nuclei associated with basic needs and emotion, for example hunger, pain, pleasure, satisfaction and instinctive motivation.

lin: lineage marker negative cells are those lacking distinctive antigen associated with lymphoid, myeloid, erythroid or natural killer cells.

linkage: describes the relationship between two measurable traits that travel together through a pedigree more than expected by chance, suggesting the two lie physically near each other.

linkage disequilibrium: two genetic variations that lie physically close to each other (linkage) and travel together through multiple generations in a population unchanged. This means the allele of one variation will be on the same piece of DNA as a specific allele of another variation, and thus they are correlated. This correlation can be measured and is referred to as linkage disequilibrium. The actual term comes from the fact that these variations do not follow the expectations of Hardy-Weinberg equilibrium and thus are in disequilibrium do to linkage.

lipophilicity: a physico-chemical characteristic of a nonpolar compound attributed to its ability to dissolve in organic solvents.

lipopolysaccharide (LPS): large molecule consisting of a lipid and a polysaccharide (carbohydrate) joined by a covalent bond. LPS is a major component of the outer membrane of Gram-negative bacteria, contributing greatly to the structural integrity of the bacteria, and protecting the membrane from certain kinds of chemical attack. LPS is an endotoxin, and induces a strong response from normal animal immune systems. LPS acts as the prototypical endotoxin, because it binds the CD14/TLR4/MD2 receptor complex, which promotes the secretion of proinflammatory cytokines in many cell types, particularly in macrophages.

liposome: a spherical vesicle with a membrane composed of a phospholipid and cholesterol bilayer.

long terminal repeat: a region of the retroviral genome principally involved in mediating insertion of the viral genome into the host chromosome.

long-term potentiation: a long-lasting enhancement of synaptic efficacy. Regarded as a primary mechanism in memory formation.

Ly49: killer cell lectin-like receptor subfamily A, member 1. Analogous for human KIRs. A natural killer cell receptor.

lymphoid chemokines (CCL-13, -21): these chemotactic cytokines are constitutively expressed in lymphoid tissues and mediate the formation and maintenance of micro-domains in lymphoid organs. (See chemokines)

lymphotoxin (lt): protein belongs to the tumor-necrosis factor family and can be produced as a secreted homotrimer, LTβ, or as a membrane-bound heterotrimer, $LT\alpha_1\beta_2$. The heterotrimer $LT\alpha_1\beta_2$ binds to the lymphotoxin-receptor (LTR). LTβ is an inducer of the inflammatory response system and involved in normal development of lymphoid tissue.

M

macrophage: a phagocytic cell of the reticuloendothelial system or mononuclear phagocyte system that may be fixed or freely motile. Myeloid-derived cell of the innate immune system matures from a monocyte, and functions in the protection of the body against infection and noxious substances.

magnetic resonance imaging (MRI): the use of a nuclear magnetic resonance spectrometer to non-invasively produce electronic images of specific atoms and molecular structures in solids, especially human cells, tissues, and organs.

magnocellular ganglion cells: large retinal ganglion cells with low resolution but high sensitivity to intensity changes and motion.

major depressive disorder: a mental illness characterized by a persistent low mood and decreased interest and pleasure, as well as symptoms such as loss of energy and sleep and appetite disturbances

major histocompatibility complex (MHC): a large genomic region with the primary immunological function to bind and "present" antigenic peptides on the surfaces of cells for recognition (binding) by the antigen-specific T cell receptors (TCRs) of lymphocytes. Differential structural properties of MHC class I and class II molecules account for their respective roles in activating different populations of T lymphocytes. Class I molecules play a central role in the immune system by presenting peptides derived from the endoplasmic reticulum lumen. They are expressed in nearly all cells. The class II molecule plays a central role in the immune system by presenting peptides derived from extracellular proteins. Class II molecules are expressed in antigen presenting cells (APC: B lymphocytes, dendritic cells, macrophages).

mania: a phase of bipolar disorder characterized by elevated or irritable mood, as well as symptoms such as increased high risk behavior and decreased need for sleep.

mantle zone: the area of a secondary follicle that surrounds the germinal center and contains IgD^+ naive, resting B cells.

mass spectrometry: technique used to measure the mass-to-charge ratio of ions to identify the components of a sample.

MAP kinase pathways: signaling pathway utilized by cells to respond to extracellular stimuli. Activation of this pathway regulates varied cellular activities including gene expression, mitosis, differentiation, and cell survival/apoptosis.

MDR inhibitor: a compound terminating multidrug resistance in cells.

mecamylamine: a synthetic compound which blocks the effect of acetylcholine at autonomic ganglia.

medium spiny neurons (MSNs): the most numerous and principal projection neurons from the striatum. The death of GABAergic MSNs in the striatum is associated with Huntington's Disease.

melanocortin-1 receptors: G-protein-coupled receptors that are stimulated by α-MSH (melanocyte stimulation hormone). These receptors are found in pigmented cells as well as immune cells.

melanopsin: the photopigment found in intrinsically light-sensitive ganglion cells involved in setting circadian rhythms.

mendelian disease: a disease caused by a single gene mutation such as sickle-cell anemia.

meningitis: inflammation of the meninges and especially of the pia mater and the arachnoid.

metallothionein(MT)-I/II: a family of Cys-rich, low molecular weight (MW ranging from 3500 to 14000 Da) proteins. MTs have the capacity to bind both physiological (such as Zn, Cu, Se) and xenobiotic (such as Cd, Hg, Ag) heavy metals through the thiol group of its cysteine residues, which represents nearly the 30% of its amino acidic residues. Important to uptake, transport and retention of metals.

mGluR6: type III metabotropic glutamate receptor that mediates responses of ON type bipolar cells in the vertebrate retina. mGluR6 couples to the G protein, G_o, stimulating a signaling cascade that causes the closing of cation channels.

micelle: an aggregate of surfactant molecules dispersed in a liquid colloid. A typical micelle in aqueous solution forms an aggregate with the hydrophilic "head" regions in contact with surrounding solvent, sequestering the hydrophobic tail regions in the micelle center.

microglia: cells derived from myeloid progenitor cells (as are macrophages and dendritic cells) which come from the bone marrow and are key cellular mediators of neuroinflammatory processes in the CNS.

mitogen: A substance that induces cell division in lymphocytes in an antigen nonspecific way. Concanvalin A (Con A) is extracted from the Jack Bean plant and triggers T cell proliferation. Lipopolysaccharide (LPS) is part of the outer membrane of Gra-mnegative bacteria and triggers B cell proliferation.

mitoxantrone: FDA approval for use in progressive MS; available as (Novantron) with recommended dosage of 5–12 mg/mL IV every 3 months. Because of cumulative cardiotoxicity, the drug can be used for only 2–3 years for a cumulative dose of 120–140 mg/m^2.

molecular adjuvant: an adjuvant comprised of a single molecular entity that targets antigens to and/or activates specific pathways of antigen processing and presentation.

molecular mimicry: immunological mechanism, whereby epitopes incidentally shared by microbial antigens and target tissues elicit an autoreactive T- or B-cell response in the wake of an infective illness. This autoimmune response then injures the "self" target tissues.

monoclonal antibody: a laboratory-produced antibody clone of one isotype that reacts to a single epitope.

mononuclear phagocytes (MP): phagocytic cells that typically initiate the host immune response. Mononuclear phagocytes include monocytes, mature macrophage, microglia and dendritic cells. These cells are widely distributed throughout the body in blood, tissue, and immune tissue identifying pathogens and steering the immune response, as well as clearing debri.

mood disorders: mental disorders characterized by inappropriate or abnormal emotional states, such as depression and bipolar disorder.

mood stabilizers: pharmacological agents used to treat bipolar disorder, including lithium, valproate, carbamazepine, and lamotrigine.

morphine: an alkaloid present in opium, an extract of the poppy plant, which binds principally to the mu opioid receptor and relieves pain.

MPP$^+$: a product of MPTP metabolism, capable of inducing a Parkinson like state through inhibition of complex I of the electron transport chain. (see MPTP)

MPTP: 1-methyl-4-phenyl-1,2,3,6-tetrahydropyridine, a lipophilic pro-neurotoxin that is metabolized by monoamine oxidase B to the active dopaminergic toxin, 1-methyl-4-phenylpyridinium (MPP+).

muromonab-CD3: a monoclonal antibody directed at the epsilon chain of CD3 adjacent to the T cell receptor on human T lymphocytes, preventing T-cell activation. It is indicated for the treatment of acute organ transplant rejection.

muscarine: an alkaloid which blocks the effects of acetylcholine at G protein-coupled cholinergic receptors in the peripheral (parasympathetic) and central nervous systems.

Muller cell: A radial glial cell that is the predominant glial cell type in retina.

multidrug resistance: the ability of pathologic cells to withstand chemicals that are designed to aid in the eradication of such cells.

Multiple Sclerosis: (abbreviated MS, also known as disseminated sclerosis or encephalomyelitis disseminata) a chronic, inflammatory, demyelinating disease that affects the central nervous system (CNS). MS can cause a variety of symptoms, including changes in sensation, visual problems, muscle weakness, depression, difficulties with coordination and speech, severe fatigue, cognitive impairment, problems with balance, overheating, and pain.

mutation: is a genetic change so severe that the disruption in the processing or gene function causes a disease by itself.

mycophenolate mofetil: a prodrug that is hydrolyzed to the active agent mycophenolate. Mycophenolate inhibits *de novo* purine synthesis and is indicated for prophylaxis of transplant rejection.

myelin: ensheathes the axons of the nervous system

myelin associated glycoprotein (MAG): a myelin protein that is a candidate autoantigen in multiple sclerosis.

myelin basic protein (MBP): an abundant component of myelin that is a candidate autoantigen in MS.

myelin inhibitors: adult mammalian central nervous system (CNS) lack the ability to regenerate after injury. This is due in part to inhibitors within myelin, These inhibitors are myelin–Nogo, myelin-associated glycoprotein (Mag) and oligodendrocyte myelin glycoprotein (Omgp).

MOG: myelin oligodendrocyte glycoprotein is a glycoprotein believed to be important in the process of myelination of nerves in the central nervous system. The gene for MOG, found on chromosome 6, was first sequenced in 1995. It is a transmembrane protein expressed on the surface of oligodendrocyte cell and on the outermost surface of myelin sheaths. Interest in MOG has centered on its role in demyelinating diseases, particularly MS. Several studies have shown a role for antibodies against MOG in the pathogenesis of MS.

myelin sheath: the fatty substance that surrounds and protects some nerve fibers.

myelination: the formation of the myelin sheath around a nerve fiber.

N

NADPH oxidase: nicotinamide adenine dinucleotide phosphate oxidase, a primary producer of reactive oxygen species.

nanofiber: a microscopic fiber whose diameter is measured in nanometers.

nanogel: a nanoscale size polymer network of cross-linked chains.

nanoparticle: a microscopic particle whose size is measured in nanometers.

nanosphere: a microscopic hollow particle whose size is measured in nanometers.

nanosuspension: a colloidal dispersion (mixture) in which a finely-divided species of a nanoscale size does not rapidly settle out.

nanotube: a microscopic tube whose diameter is measured in nanometers.

natalizumab: Tysabri is a recombinant humanized monoclonal antibody that binds to the $\alpha4$ subunits of $\alpha4\beta1$ and $\alpha4\beta7$ integrins that are expressed on the surface of all leukocytes except neutrophils and inhibit the <4 mediated adhesion to their receptors. Effective in phase I/II FDA trials for MS it was withdrawn due to deaths from progressive multifocal Leukoencephalopathy

natural killer cell: a lymphocyte (white blood cell) that can kill target cells that express nonself proteins, such as viral antigens or tumor antigens.

neostigmine: an inhibitor of acetylcholinesterase which prolongs the actions of acetylcholine.

neprilysin: a neutral zinc metallopeptidases and important $A\beta$ degrading enzyme, which is involved in the clearance **of $A\beta$ in the brain.**

nerve conduction: salutatory propagation of nerve impulses along the axon

nestin: a large (200 kDa) intermediate filament protein found in developing rat brain. Often used as a specific marker of neuronal progenitor cells.

neural progenitor cells (NPC): the term progenitor cell is used to refer to immature or undifferentiated cells, typically found in post-natal animals. While progenitor cells share many common features with stem cells, the term is far less restrictive.

neuritis: inflammation of the peripheral nerve

neurodegeneration: a progressive loss of brain tissue structure or function due to a disease process; in the case of Alzheimer's and Parkinson's disease, due to accumulation of aggregated cellular products.

neurodegenerative diseases: the subset of neurological disorders that include neuron pathologies, but exclude diseases of the nervous system due to cancer, edema, hemorrhage, trauma, poisoning, hypoxia, etc.

neurofibrillary tangle (NFT): intraneuronal protein aggregates and a pathological hallmark of AD, composed of filamentous hyperphosphorylated tau (PHF-tau) and other ubiquitinated proteins.

neurogenesis: the formation of nervous tissue through a process that involves proliferation, migration, differentiation, and survival of neural stem cells.

neuroimmune process: the biochemical and electrophysiological interactions between the NERVOUS SYSTEM and IMMUNE SYSTEM that result in regulation of the immune system by the nervous system.

Neuroinflammation: chronic, CNS-specific, inflammation that is mediated predominantly by microglia that may engender neurodegenerative events.

Neuroinvasion: the process by which prions enter the nervous system.

neuromelanin: A modified form of melanin pigment normally found in certain neurons of the nervous system, especially in the substantia nigra and locus ceruleus.

neuromyelitis optica: also known as Devic's disease, is a severe variation of a demyelinating disease affecting the optic nerve and spinal cord.

neuron: nerve cell that transmits nerve signals to and from the brain, consisting of a cell body, axon(s) and dendrite(s).

neuronal apoptosis: death of neurons via programmed cell death.

neuronal plasticity: refers to the changes that occur in the organization and/or function of the brain as a result of experience.

neuropeptides: endogenous compounds made of amino acids that act upon receptors within the nervous system.

neuroprotection: mechanisms that protect the brain from apoptosis (programmed cell death) or degeneration, following a brain injury or in the course of a neurodegenerative disorder.

neuroproteomics: analysis of proteins and protein complexes of the nervous system (see Proteomics).

neuropsychology: the science of integrating psychological observations on behavior and the mind with neurological observations of the brain and nervous system.

neuroregeneration: the regrowth or repair of the nervous system or its components: tissues, cells or cell products.

neurosphere: a spherical group of neural progenitor cells growing in a suspension culture.

neurotoxin: a toxic complex typically of protein that acts preferentially on the nervous system.

neurotransmitter: chemical that mediates information transfer between neurons.

neurotrophin: a small secretory molecule needed for growth, maintenance, or repair in the nervous system

neurovascular unit: a composition of endothelial cells, astrocytes, neurons, and a contractile apparatus of either smooth muscle cells or pericytes that form a functionally integrated network.

NF-κB: transcription factor found in all cell types and is involved in cellular responses to stimuli such as stress, cytokines, free radicals, ultraviolet irradiation, and bacterial or viral antigens.

NG2: expressed by a variety of immature glia in the CNS including oligodendrocyte progenitor cells, paranodal astrocytes and perisynaptic glia. The protein has a large extracellular domain with two LNS/Lam G domains at the N-terminus and a short intracellular tail with a PDZ-recognition domain at the C-terminus.

nicotine: an alkaloid that blocks the effects of acetylcholine at ligand-gated ion channels in the peripheral (parasympathetic and somatic) and central nervous systems.

nigrostriatal pathway: the neural pathway connecting the substantia nigra to the striatum.

nitric oxide: a free radical gas produced endogenously by a variety of mammalian cells, synthesized from arginine by nitric oxide synthase. Nitric oxide is one of the endothelium-dependent relaxing factors released by the vascular endothelium and mediates vasodilation. It also inhibits platelet aggregation, induces disaggregation of aggregated platelets, and inhibits platelet adhesion to the vascular endothelium. Nitric oxide activates cytosolic guanylate cyclase and thus elevates intracellular levels of cyclic gmp.

nitric oxide signaling: nitric oxide is a gaseous molecule released from the vascular endothelium in response to a variety of chemical and physical stimuli. It triggers smooth muscle cells in the vessel wall to relax by activating soluble guanylate cyclases (sGC), increasing the cyclic guanosine monophosphate (cGMP) concentration and activating the protein kinase G.

NMDA hypothesis: based on the observation that NMDA receptor antagonists like phencyclidine frequently induce schizophrenia-like psychotic symptoms, the NMDA hypothesis of schizophrenia proposed an endogenous NMDA receptor antagonist that may be expressed in higher amounts in individuals suffering from schizophrenia. Because of the close functional relationship between glutamatergic nerurotransmission via NMDA receptors and dopaminergic neurotransmission, the NMDA hypothesis includes the generally accepted disturbance of dopamine signaling in schizophrenia.

NMDA receptor: n-methyl-d-aspartate receptor, named for the chemical agonist NMDA. The receptor is a heterodimer, forming an ionic channel that opens upon the binding of agonists glutamate and glycine. NMDA receptors are permeable to a number of ions, but primarily sodium and calcium. NMDA function is fundamental to proper CNS function, but overstimulation is harmful resulting in excitotoxicity.

nociception: pain perception.

nociception/orphanin FQ: an endogenous opioid peptide that acts on the nociceptin/Orphanin FQ receptor.

node of ranvier: one of the many gaps in the myelin sheath—this is where the action potential occurs during saltatory conduction along the axon.

norepinephrine: the chemical responsible for transmitting impulses between neurons in the central and peripheral (sympathetic) nervous systems and acts on either alpha or beta adrenergic receptors.

notch: family of evolutionarily conserved proteins that regulates a broad spectrum of cell-fate decisions and differentiation processes during fetal and post-natal development. The best characterized role of Notch signaling during mammalian hematopoiesis and lymphopoiesis is the essential function of the Notch1 receptor in T-cell lineage commitment. More recent studies have addressed the roles of other Notch receptors and ligands, as well as their downstream targets, revealing additional novel functions of Notch signaling in intra-thymic T-cell development, B-cell development and peripheral T-cell function.

nuclear hormone receptor: receptor in which hormone binding is required to control the activity of the nuclear receptor.

nucleus: the organelle in the cell body of the neuron that contains the genetic material of the cell.

O

ocular immunology: a branch of immunology studying immune processes in the eye

ocular inflammation: an inflammatory process affecting the eye:

olfactory bulb: ovoid body resting on the cribriform plate of the ethmoid bone where the olfactory nerve terminates. The olfactory bulb contains several types of nerve cells including the mitral cells, on whose dendrites the olfactory nerve synapses, forming the olfactory glomeruli. The accessory olfactory bulb, which receives the projection from the vomeronasal organ via the vomeronasal nerve, is also included here.

oligoclonal bands (OCB): bands of immunoglobulins that are seen when a patient's blood plasma or cerebrospinal fluid (CSF) is analyzed by protein electrophoresis:

oligodendrocytes: are myelin-synthesizing cells of the CNS.

opiate peptides: endogenous peptides with opiate-like activity. The three major classes currently recognized are the ENKEPHALINS, the DYNORPHINS, and the ENDORPHINS. Each of these families derives from different precursors, proenkephalin, prodynorphin, and PRO-OPIOMELANOCORTIN, respectively. There are also at least three classes of OPIOID RECEPTORS, but the peptide families do not map to the receptors in a simple way.

opioids: substances typically used to relieve pain through the binding of opioid receptors.

opioid receptors: G-protein-coupled receptors that are the targets of the endogenous opioid peptides and of the opioid drugs, such as morphine.

opiopeptins: a general name for the whole class of endogenous peptides acting upon opioid receptors.

opsin: G-protein coupled receptor activated by light in photoreceptor outer segments. Cones have cone opsins and rods possess rhodopsin.

ORL-1 receptor: also known as the nociceptin/orphanin FQ receptor—a G-protein-coupled receptor for which the endogenous ligand is nociceptin.

oxidative damage: the oxidation of proteins, lipids, and DNA, which results in neuronal dysfunction. An important component of neuronal damage and aging, where neuroinflammation is significantly involved.

oxidative stress: an imbalance between formation and neutralization of reactive oxygen species.

outer segment: photoreceptor cell structure that contains visual pigments and the enzymatic machinery for phototransduction.

P

p53: protein 53 (TP53) is a transcription factor critical factor to the prevention of mutation that regulates the cell cycle and hence functions as a tumor suppressor.

p75^{NTR} signaling: a member of the tumor necrosis factor receptor superfamily that facilitates apoptosis during development and after injury to the CNS. It activates Jun kinase (JNK), caspases 9, 6, and 3 and accumulates cytochrome c within the cytosol.

paraquat: trade name for the herbicide N,N′-dimethyl-4,4′-bipyridinium dichloride. Dangerously poisonous to humans as the compound is easily reduced resulting in superoxide generation.

parasympathetic: the division of the autonomic nervous system that regulates various body functions and is active primarily during periods of rest.

Parkin: an E3 ligase in the ubiquitin-proteasome system. Parkin mutations have been associated with famial Parkinson's disease.

Parkinson's Disease (PD): a chronic progressive neurodegenerative disease chiefly of later life that is linked to decreased dopamine production due to loss of dopaminergic neurons in the substantia nigra. The clinical features are tremor of resting muscles, rigidity, slowness of movement, impaired balance, and a shuffling gait—called also paralysis agitans. Current therapies include L-DOPA and deep brain stimulation.

Pax-5: paired box gene 5 (B-cell lineage specific activator). The PAX proteins are important regulators in early development, and alterations in the expression of their genes are thought to contribute to neoplastic transformation. The PAX5 gene encodes the B-cell lineage specific activator protein (BSAP) that is expressed at early, but not late stages of B-cell differentiation. Its expression has also been detected in developing CNS and testis. Therefore, PAX5 gene product may not only play an important role in B-cell differentiation, but also in neural development and spermatogenesis.

parvocellular ganglion cells: small retinal ganglion cells that possess the high resolution and color sensitivity required for fine feature analysis and color vision.

passive immunization: administering antibodies for temporary immune protection against an antigen. Immunity exists as long as the antibodies remain in the body.

PASAT: paced auditory sequential addition test is a segment of the multiple sclerosis functional composite that tests one component of cognitive function.

pathogen-associated molecular patterns (PAMPs): small molecular motifs consistently found on pathogens; recognized by toll-like receptors and other pattern recognition receptors (PRRs) in plants and animals.

peripheral benzodiazepine receptor: a G-protein-coupled receptor for benzodiazepines found primarily in the periphery, particularly in the immune system, such as microglia in brain.

peripheral blood mononuclear cells (PBMC): lymphocytes and monocytes isolated from the peripheral blood, usually by Ficoll Hypaque density centrifugation.

peripheral nervous system: part of the nervous system consisting of the nerves and neurons outside the central nervous system.

phenylephrine: a norepinephrine analog that mimics norepinephrine action primarily at the alpha adrenergic receptors.

photoreceptor: principle photo-sensitive cell in the retina

physostigmine: an inhibitor of acetylcholinesterase which acts to prolong the actions of acetylcholine.

PI-3 kinase Phosphoinositide 3-kinases: a family of enzymes that are capable of phosphorylating the 3 position hydroxyl group of the inositol ring of phosphatidylinositol.

PK11195: 1-(2-chlorophenyl-N-methylpropyl)-3-isoquinoline-carboxamide, a ligand for peripheral benzodiazepine receptors.

PLP: refers to proteolipid protein that is present in myelin.

polymorphism: a genetic variation in a base pair, repeat, deletion or duplication that is neutral or acts as a susceptibility allele

postsynaptic density: an electrondense region located opposite the active zone on the postsynaptic plasma membrane.

pralidoxime: a reactivator of acetylcholinesterase that has been inhibited by organophosphorus compounds.

prazosin: synthetic compound that blocks the action of norepinephrine on alpha adrenergic receptors.

preproenkephalin: a precursor peptide for the enkephalins. It is processed into proenkephalin before being converted to multiple enkephalin molecules.

presenilin-1 (PS1): a principal component of γ-secretase complex, processes APP at the γ-site, and a causative gene of the early onset of FAD. The γ-secretase complex is also critically involved in Notch processing and its signaling.

prion: infectious protein agent that induces a fatal neurodegenerative disease in animals and humans that is characterized by the accumulation of the abnormal isoform of the prion protein.

progenitor/precursor cell: a cell that can differentiate into multiple cell types and has limited self-renewal capacities, so that the cell can proliferate while maintaining its differential capability for a short period of time but not for the entire life of the organism.

proinflammatory cytokines: cytokines that induce an inflammatory response such as IFN-γ, IL-1(β), IL-18, and TNF-α. some authors also include IL-6, though this cytokine has an extremely broad range of actions.

proliferation: the ability to multiply cell population numbers by undergoing self-renewal or division.

proopiomelanocortin: a large precursor peptide that is the source of many active peptides, including some endorphins, enkephalins, alpha-MSH and others.

propranolol: a synthetic compound that blocks the action of norepinephrine on beta adrenergic receptors.

protein fingerprinting: method of classifying new protein families based on conserved regions within alignments of related proteins.

protein microarrays: high throughput technology to detect differentially expressed proteins using protein-protein interactions.

protein-only hypothesis: proposed by Dr. Stanley Prusiner to describe the novel properties of prions; prions are proteinaceous infectious particles that lack an agent-specific nucleic acid genome.

protein profiling: identification of proteins in a sample.

proteolipid protein (PLP): the most abundant myelin protein; a candidate autoantigen in MS:

proteomics: field of science using high throughput profiling techniques to analyze proteins expressed under a certain set of conditions within an individual cell or organism.

PU.1: spleen focus forming virus (SFFV) proviral integration oncogene spi1. The PU.1 plays indispensable and distinct roles in hematopoietic development through supporting HSC self-renewal as well as commitment and maturation of myeloid and lymphoid lineages.

R

radial glia: neuronal progenitor cells that arise from neuroepithelial cells after the onset of neurogenesis with more restricted differentiation abilities. In the developing nervous system, radial glia function both as neuronal progenitors and as a scaffold upon which newborn neurons migrate. In the mature brain, the cerebellum and retina retain characteristic radial glial cells. In the cerebellum, these are Bergmann glia, which regulate synaptic plasticity. In the retina, the radial Müller cell is the principal glial cell, and participates in a bidirectional communication with neurons.

randomized controlled trials: an experiment in which investigators randomly allocate eligible people into (e.g. treatment and control) groups to receive or not to receive one or more interventions that are being compared. The results are assessed by comparing outcomes in the treatment and control groups. This form of trial is often used in performing scientific research due to its reliability in avoiding false causality.

reactive oxygen species (ROS): generally very small molecules that are highly reactive due to the presence of unpaired valence shell electrons. They include oxygen ions, free radicals and peroxides both inorganic and organic. ROS are formed as a natural byproduct of the normal metabolism of oxygen and have important beneficial roles in cell signaling and innate immunity. However, during times of environmental stress ROS levels can increase dramatically, potentially resulting in significant damage to cell structures. This culminates into what is known as oxidative stress and can lead to such results apoptosis or necrosis.

redox signaling pathways: signal transduction pathways based on electron-transfer processes that play important messaging roles in biological systems. These pathways are activated by free radicals, reactive oxygen species (ROS), and other electronically-activated species and are turned off by reducing species.

regulatory T cells: also called suppressor T cells, regulatory T cells are a small population of $CD4^+$ T cells that have regulatory (that is, suppressor) activities towards other cells of the immune system and serve to limit immune responses, particularly those which may be against host tissue. These cells express the transcription factor forkhead box P3 (FOXP3) which is used as a specific marker for these cells. An absence of T_{Reg} cells or their dysfunction is associated with severe autoimmunity.

relative risk: a statistical tool used to measure the ratio of two conditional probabilities. For example, in genetic disease inheritance, relative risk can measure the degree of genetic contribution to a trait, defined as the risk of the disease in first degree offspring (sibs, parents, offspring) of an affected individual relative to the risk of the disease in the general population.

reserpine: an alkaloid that prevents the accumulation of catecholamines such as norepinephrine and dopamine in the synapses. Reserpine performs this action by irreversibly binding to these neurotransmitters and inhibiting their uptake into synaptic vesicles. The compound has been used as an antihypertensive as well as a sedative but is rarely used any longer due to its risk of depression and the availability of better drug therapies.

resting membrane potential: membrane potential prevailing in a resting unstimulated cell. The value is often negative, indicating an excess of negative potential inside the cell in comparison with the external environment. The difference in potential is based on the ion concentrations within and surrounding the cell membrane, as well as the ion transport proteins embedded in the cell membrane.

retina: a thin sheet of neural cells at the back of the eye that is continuous with the optic nerve. The retina is responsible for phototransduction and the initial stages of vision. The retina's photoreceptor cells, called rods and cones, produce neural signals in response to light that then undergo complex processing by ganglion cells of the retina. The resulting information is sent to the brain via the optic nerve to be interpreted into a visual image.

retinal ganglion cell: the output cell of the retina whose axons form the optic nerve and project to higher visual center. These cells process the signals produced by the photoreceptor cells of the retina and then send the resulting information to the brain via the optic nerve for interpretation.

retinal pigment epithelium: the pigmented epithelial cells forming the outermost layer of the retina. This layer of cells is attached to the choroid vascular tissue and provides nourishment to the cells of the retina.

retinoid-related orphan receptor (ROR): an orphan nuclear hormone receptor that regulates the survival of $CD4^+CD8^+$ doublepositive (DP) thymocytes and is essential for the development of the lymph nodes and Peyer's patches.

retrovirus: any of a family (Retroviridae) of enveloped RNA viruses (such as HIV and numerous tumorigenic viruses) that replicate in a host using a DNA intermediate produced using their RNA template and reverse transcriptase. Through the process, their viral genome is incorporated into the genome of infected cells. There are three common genes found in most retroviruses, the gag, pol, and env genes, which encode for the viral proteins, enzymes, and envelope of the viruses respectively.

ribbon: structure in synaptic terminals of photoreceptors, retinal bipolar cells, and hair cells that is specialized for sustained neurotransmitter release.

rivastigmine: an inhibitor of acetylcholinesterase which prolongs the actions of acetylcholine with prominent effects in the central nervous system. The compound has been shown effective in treating the symptoms of dementia associated with Alzheimer's disease.

RNA interference: a posttranscriptional genetic mechanism, which suppresses gene expression. In this mechanism, double-stranded RNA cleaved into small fragments initiates the degradation of a complementary messenger RNA. This pathway is thought to have developed as part of the innate immune system to protect against viruses, as in controlling development and genome preservation. This mechanism is used as a technique (as the introduction of double-stranded RNA into an organism) that artificially induces mRNA inhibition and is used for studying or regulating gene expression.

rod: a type of Photo-sensitive retinal cell mediating scotopic vision near visual threshold. These cylindrical-shaped photoreceptor cells are located primarily on the outer edges of the retina and are therefore used also for most peripheral vision.

rostral migratory stream: the route for neuronal precursors (periglomerular and granule cells) to migrate to the olfactory bulb where they differentiate into interneurons throughout the life of rodents. Overlying the rostral migratory stream is the anterior or olfactory limb of the anterior commissure (or commissura anterior pars bulbaris), which carries centrifugal fibers, including some from the medial forebrain bundle, to the region.

rotenone: a botanical insecticide that is an inhibitor of mitochondrial electron transport. Inhibition is achieved through the hindrance of the electron transfer chain preventing NADH from converting to ATP thus limiting usable cellular energy. The compound is considered moderately hazardous and mildly toxic to mammals, including humans.

S

Sca-1: a mouse member of the interferon-inducible Ly-6 family of genes that can conveniently be used as a marker for stem/early progenitor cells. Sca-1 is used to isolate many stem cells in the mouse. The density of this antigen declines with differentiation. Sca-1 belongs to a family of proteins bearing a UPAR (urokinase plasminogen activator receptor) domain. The UPAR protein plays a role in cellular adhesion and migration by modulating integrin function and degradation of the extracellular matrix.

schizophrenia: a mental illness characterized by cognitive dysfunction, hallucinations, delusions, thought disorder and poor social and vocational functioning. The term can also be used to describe a group of mental disorders characterized by the same or similar indicators.

Schwann's cells: cells that produce myelin in the peripheral nervous system. These cells are located between the axon and axon terminal of neurons and create the myelin sheath which aides in insulating axons and in increasing impulse speeds as they are propagated through the neuron (allowing for salutatory conduction).

sclerosis: a hardening of tissue as a result of scaring or plaques such as in the central nervous system when formed by inflammatory perivascular lesions.

senility: a term once used to describe dementia.

sickness behavior: one of the immunological models of major depression. The cytokines IL-1, IL-6, and TNF-α have been identified to induce depression-like mood disturbance upon administration or during disease states that are accompanied by increased or over-production of these cytokines.

self-renewal: the ability of a cell to divide into two daughter cells with identical properties and retaining equal differential capabilities as the parent cell.

SCID: severe combined immunodeficiency, a genetic disorder in which both "arms" (B cells and T cells) of the adaptive immune system are crippled. This disease is fatal if untreated due to extreme susceptibility to infectious diseases.

single-chain antibodies: engineered antibodies composed of the minimal antibody-binding site formed by non-covalent association of the V_H and V_L variable domains joined by a flexible polypeptide linker. Sequences encoding eukaryotic secretion signals can be appended to the scFv genes and subsequently be cloned into gene transfer vectors for prolonged *in vivo* expression. These antibodies are significantly smaller than natural antibodies but retain the same affinity as natural antibodies and can be produced in large quantities at relatively low costs using biotechnology.

signal serotonin: serotonin (5-hydroxytryptamine, 5-HT) is a biogenic indoleamine derived from the dietary amino acid tryptophan. Serotonergic neurons are located almost exclusively in the raphe nuclei of the brainstem, but their projections innervate virtually all other parts of the central nervous system. A reduced serotonergic neurotransmission is supposed to play a crucial role in the pathophysiology of major depression among other disorders.

signal transduction: mechanism by which external stimuli are interpreted by cells. This transformation is often done through biochemical reactions involving secondary and cascade pathways within and between cells.

simian virus 5 (SV-5): a paramyxovirus belonging to the family Paramyxoviridae. The genomes of these viruses are composed of negative-sense single-stranded RNA and this family of viruses is responsible for a number of human and animal diseases.

single nucleotide polymorphism: the most common polymorphism in the human genome, a nucleotide polymorphism at a single base. The result of these differences creates alleles.

sirolimus: a macrolide antibiotic that inhibits T lymphocyte activation and proliferation. It is an immunosuppressive agent that is used mainly to prevent rejection during organ transplantation. The compound works to inhibit a response to interleukin-2 which is necessary to activate T cells.

Sjogren's syndrome: an autoimmune disorder affecting the salivary and lacrimal glands. In this disorder, immune cells attack host exocrine glands that are responsible for the production of tears and saliva.

Smac/Diablo: a mitochondrial protein that promotes some forms of apoptosis by neutralizing one or more members of the IAP family of apoptosis inhibitory proteins.

substantia nigra pars compacta (SNpc): the portion of the substantia nigra within the midbrain that encompasses the dopaminergic neurons which innervate the caudate nucleus and putamen of the basal ganglia. This area of the brain plays an important role in the pleasure/reward system and in addiction. Additionally, degeneration of the neurons in this area of the brain is an indicator of Parkinson's disease.

suppressor of cytokine signaling (SOCS) proteins: a family of proteins that inhibit the JAK-STAT signaling pathway by competitively binding to phosphotyrosine binding sites on cytokine receptors.

superoxide dismutase-1, copper-zinc superoxide dismutase (SOD1): a cytoplasmic enzyme that works as an antioxidant converting superoxide to oxygen and hydrogen peroxide. Mutations in the enzyme have been shown to be a factor in the motor neuron disease amyotrophic lateral sclerosis (ALS).

soman: an extremely toxic long acting organophosphorus compound with prominent inhibitory effects on acetylcholinesterase causing interference with nerve functioning

somatic hypermutation: point mutations that occur in cycling centroblasts and are targeted to the immunoglobulin variable-region gene segments. Some mutations might generate a binding site with increased affinity for the specific antigen, but others can lead to loss of antigen recognition by the B-cell receptor and generation of a self-reactive B-cell receptor. Each mutation occurs only within individual cells and the mutations are not passed on to offspring.

Sox4: a member of the SOX (SRY-related HMG-box) family of transcription factors involved in the regulation of embryonic development and in the determination of the cell fate. The encoded protein may act as a transcriptional regulator after forming a protein complex with other proteins, such as syndecan binding protein (syntenin). The protein may also function in the apoptosis pathway leading to cell death as well as to tumorigenesis and may mediate downstream effects of parathyroid hormone (PTH) and PTH-related protein (PTHrP) in bone development.

spatial memory: a type of memory responsible for recording information about one's environment and its spatial orientation. For example, a person's spatial memory is required in order to navigate around a familiar city, just as a rat's spatial memory is needed to learn the location of food at the end of a maze.

STAT4: a member of the STAT family of transcription factors. In response to cytokines and growth factors, STAT family members are phosphorylated by the receptor associated kinases, and then form homo- or heterodimers that translocate to the cell nucleus where they act as transcription activators. This protein is essential for mediating responses to IL-12 in lymphocytes, and regulating the differentiation of T helper cells.

STAT6: protein that plays a central role in exerting IL-4 mediated biological responses. It is found to induce the expression of BCL2L1/BCL-X(L), which is responsible for the anti-apoptotic activity of IL-4. Knockout studies in mice suggested the roles of this gene in the differentiation of T helper 2 (Th2) cells, expression of cell surface markers, and class switching of immunoglobulins.

stem cell: a single cell that can differentiate along multiple lineages and can self-renew throughout the life of an organism while maintaining its differential capability. The three primary categories of stem cells are embryonic stem cells (originating from blastocysts), adult stem cells (derived from adult tissues), and cord blood stem cells (found in the umbilical cord).

stem cell therapy: Treatments and related research involving the use of undifferentiated cells which have self-renewal capacity and the unique potential to produce any kind of cell in the body.

stress: any physical or psychological stimulus that disrupts homeostasis.

striatum: a continuous part of the basal ganglia comprised of the caudate nucleus and putamen. This portion of the brain plays roles in modulating movement pathways as well as taking part in a variety of cognitive functions.

stroke: brain injury caused by occlusion or rupture of blood vessels into or around the brain causing a subsequent loss of oxygen and glucose to the affected brain area. Stroke is commonly termed a "cerebrovascular accident" or a "brain attack" and can cause severe neurological damage or death.

stromal cell-derived factor 1 (SDF-1): the ligand for CXCR4 receptor, this CXC chemokine (also designated CXCL12) is involved in chemotaxis, brain development, and immune cell recruitment.

subgranular zone: a brain region deep within the hippocampal parenchyma, at the interface between the granule cell layer and the hilus of the dentate gyrus, where adult neurogenesis occurs. It is one of the two major known adult neurogenesis sites of the brain, along with the subventricular zone.

subluxation: partial dislocation of a joint or organ.

substantia nigra: A layer of large pigmented nerve cells in the midbrain that produce dopamine and whose destruction is associated with Parkinson's disease.

subventricular zone: a paired brain structure situated throughout the lateral walls of the lateral ventricles. Along with the subgranular zone of dentate gyrus, this zone serves as a source of neural stem cells in the process of adult neurogenesis. It harbors the largest population of proliferating cells in the adult brain of rodents, monkeys and humans. Neurons generated in this zone travel to the olfactory bulb via the rostral migratory stream.

susceptibility gene: a gene with multiple polymorphic forms, of which one or more forms causes differences in the processing or function of the gene that makes an individual more susceptible to a specific disease, but only in the presence of other interacting or contributing susceptibility genes or environmental factors.

sympathetic: the division of the autonomic nervous system that regulates various body functions and is always active on a basal level, but becomes more active during periods of stress and/or exertion. It causes the "fight-or-flight" response during times of stress by promoting the release of adrenaline and noradrenaline which bind to adrenergic receptors in peripheral tissues and cause such reactions and increased heart rate and blood pressure and pupil dilation.

symptomatic therapy: treatment or remedy for physical disturbance observed by the patient due to an underlying disease. Also called palliative care, this form of treatment does not address the basic cause or provide a cure for the underlying disease but rather focuses on preventing or easing suffering associated with the disease.

synapse: from the Greek meaning "to grasp," a junction between two neural cells where the cell body (or dendrite) of one nerve is positioned across a small gap from the axon of another nerve. Signal transmission occurs across these gaps via neurotransmitters, allowing the neurons to communicate with each other through the conversion of electrical impulses into chemical signals.

systems biology: a field of science that studies the interactions between the components of biological systems and how these interactions give rise to the function and behavior of that system.

T

tacrolimus: a macrolide antibiotic that acts as a calcineurin inhibitor to hinder T cell signal transduction and IL-2 transcription. It is an immunosuppressive agent that is used mainly to prevent rejection during allogeneic organ transplantation.

T-bet: T-box genes encode transcription factors involved in the regulation of developmental processes. This gene is the human ortholog of mouse Tbx21/Tbet gene. Studies in mouse show that Tbx21 protein is a Th1 cell-specific transcription factor that controls the expression of the hallmark Th1 cytokine, IFNg. Expression of the human ortholog also correlates with IFNg expression in Th1 and natural killer cells, suggesting a role for this gene in initiating Th1 lineage development from naive Th precursor cells.

T-cell receptor (TCR): a heterodimeric molecule found on the surface of T lymphocytes (T-cells) that is responsible for recognizing antigens bound to major histocompatibility complex (MHC) molecules. Upon antigen-receptor binding with MHC, a series of biochemical reactions occur which activate the associated T-cell.

T-cells: T (thymus derived) cell. These white blood cells, known as T lymphocytes, mature in the thymus and have regulatory immune functions. As part of the adaptive (cell-mediated) immune system, these cells can recognize antigens processed by antigen-presenting cells in the context of specific receptors. The cell surface protein CD3 is a marker for these cells.

terbutaline: a norepinephrine derivative that mimics norepinephrine action primarily at the subtype two beta adrenergic receptor. This compound is used as a bronchodilator and tocolytic.

temporal lobe: either of two lobes of the brain, located on either side of the cerebral hemisphere and part of the cerebrum. This portion of the brain is involved in auditory processing, semantics, and also contains the hippocampus which is responsible for learning and memory.

Th1 cells: also called inflammatory CD4 T cells, a subset of T cells that secrete the cytokines IL-2, IFNg, and lymphotoxin and are most useful against intracellular pathogens that are best attacked by a cell-mediated response involving macrophage and granulocyte activation.

Th2 cells: also called helper CD4 T cells, a subset of T cells that secrete the cytokines IL-4, IL-5, and IL-10, and induce antibody and allergic responses that are useful for combating infections.

Th1/Th2 balance: according to the Th1/Th2 paradigm, naïve CD4+ T helper cells can be activated to differentiate into two types of effector cells: either Th1 or Th2 cells. In general, macrophages and dendritic cells tend to stimulate the Th1 differentiation, whereas extracellular antigens tend to stimulate Th2 cell differentiation. Th1 cells in turn stimulate macrophages and induce IgG production by B cells, while Th2 cells stimulate the humoral immune response including IgA, IgE, and certain IgG subtypes. An imbalance between Th1 and Th2 cells is observed in several immunological disease conditions; typical examples are Th1 predominance in rheumatoid arthritis and Th2 predominance in IgE-mediated allergy.

Th1/Th2: subclasses of T helper cells that are distinguished by their cytokine profiles. Th1 cells are characterized by production of IFN-g. Th1 cells are characterized by production of IL-4. Th2 cells polarize the immune response towards cellular immunity and Th2 cells polarize towards antibody formation.

thalidomide: a drug originally used to treat morning sickness during pregnancy that caused severe limb deformities. This compound downregulates TNFa and is under investigation for treatment of severe Crohn's syndrome and other autoimmune disorders.

therapeutics: a branch of medical science dealing with the application of remedies to diseases.

tissue profiling: a tool for rapid detection of molecules, usually proteins and peptides, obtained directly from intact tissues to identify pathological changes.

T lymphocyte or T cell: Please see "T cell."

tolerance: refractoriness to physiological effects of a stimulus which develops with prolonged exposure, such as with drug tolerance. Alternatively, the failure of the immune system to react to an antigen, such as in the case of tolerance to self-antigens

toll-like receptors (TLR): a class of single membrane-spanning non-catalytic receptors that recognize structurally conserved molecules derived from microbes and activate immune cell responses. TLRs are believed to play a key role in the innate immune system. They are a type of pattern recognition receptors (PRRs) and recognize molecules that are broadly shared by pathogens but distinguishable from host molecules, collectively referred to as pathogen-associated molecular patterns (PAMPs). These receptors are present in both vertebrates and invertebrates and have similarities with receptors found in bacteria and plants. Due to this, these receptors are thought to be one of the oldest and most conserved components of the immune system.

transcription nuclear factor-κ-B: a protein complex found in all cell types that is involved in cellular responses to stimuli such as stress, cytokines, and foreign antigens. This transcription factor plays a regulatory role in immune response to infection and has also been suggested to be involved in processes of synaptic plasticity and memory. When improperly regulated, this transcription factor has also been implicated in causing cancer, autoimmune diseases, and other immune system malfunctions.

transducin: G protein expressed in retinal photoreceptor cells that is activated upon absorption of a photon by opsin. Activation of this protein stimulates cGMP-specific phosphodiesterase which leads to the vertebrate phototransduction cascade.

transgene: a segment of DNA (generally coding DNA) which is taken from one cell or organism and is introduced into different cells or organisms to modify a phenotype.

transmembrane protein transport (diffusion): the process of moving proteins from one cellular compartment (including extracellular) to another by various sorting and transport mechanisms such as gated transport, protein translocation, and vesicular transport. The tendency of a gas, solute or protein is to pass from a point of higher pressure or concentration to a point of lower pressure or concentration and to distribute itself throughout the available space:

transport: movement or transference of biochemical substances that occurs in biological systems.

trimethaphan: a synthetic compound which blocks the effect of acetylcholine at autonomic ganglia thereby blocking the sympathetic and parasympathetic nervous system. This compound has been used to reduce bleeding during neurosurgery, as well as to treat such emergencies as pulmonary edema and hypertensive crisis.

tropicamide: an atropine derivative which blocks the action of acetylcholine on muscarinic cholinergic receptors. This compound is often used during eye examinations to dilate the pupil and may also be used before or after eye surgery.

Trk receptor: cell surface proteins that are involved in transducing the actions of neurotrophins to promote neuronal survival, proliferation, migration, axonal and dendritic outgrowth and patterning, synapse strength and plasticity, injury protection, as well as controlling the activity of ion channels and neurotransmitter receptors.

tryptophan: L-tryptophan (TRP) is an essential amino acid, present in nearly all natural proteins. It is the precursor of the neurotransmitters serotonin and melatonin and of kynurenine, which is quantitatively the most important TRP metabolite.

tumor necrosis factor: serum glycoprotein produced by activated macrophages and other mammalian mononuclear leukocytes that acts as a cytokine to help stimulate inflammation and the acute phase reaction of the immune system. It has necrotizing activity against tumor cell lines and increases ability to reject tumor transplants. Also known as TNF-alpha, it is only 30% homologous to TNF-beta (lymphotoxin), but they share TNF receptors. Disregulation of this cytokine has been implicated in causing cancer, among other human diseases.

tyramine: the decarboxylated form of the amino acid tyrosine which is present in food (e.g., cheese, beer, wine) and promotes the release of norepinephrine from sympathetic nerve endings. Significant displacement of norepinephrine by this compound when ingested in large quantities or while taking monoamine oxidase inhibitor medication is thought to cause vasoconstriction and increased heart and blood pressure.

U

urocortins: most recently discovered peptides belonging to the corticotropin-releasing factor family whose functions are generally unknown, but may play roles in immune function in addition to having anorexigenic and hypotensive effects, among other functions.

V

vaccination: the introduction of antigenic stimulants to the immune systems of humans or animals for the purpose of inducing the artificial development or regulation of immunity. Multiple kinds of vaccines exist to serve this purpose including inactivated, live attenuated, subunit, and DNA vaccines.

varicella zoster virus (VZV): an alpha human herpesvirus that causes the diseases chickenpox and, if later reactivated, shingles. A vaccine is currently available against the virus.

vector: a "vehicle," such as a modified virus or DNA molecule, used to deliver genetic material into the body for gene therapy. Alternatively, an organism that does not cause disease itself but can transmit infection through transferring pathogens among different hosts (such as a mosquito).

vehicle: a type of adjuvant that possesses no medicinal action of its own, but delivers or targets an antigen to cells of the immune system.

vimentin: a member of the intermediate filament family of proteins that is an important component of eukaryotic cell cytoskeletons. These proteins are responsible for cell flexibility and resilience under mechanical stress as well as the transport of low density lipoprotein intended to undergo esterification.

visual-evoked potential-(VEP) viruses: minute infectious agents whose genomes are composed of DNA or RNA, but not both. They are characterized by a lack of independent metabolism and the inability to replicate outside living host cells.

W

withdrawal: signs and symptoms that occur as a result of a discontinuation of certain activities or drugs (such as opioids) and includes perturbations in various physiological systems that are indications of physical dependence.

Definitions provided by contributing authors or the National Library of Medicine unless otherwise noted.

Index

H

M

Q

R